12th International Conference on Numerical Methods in Industrial Forming Processes (NUMIFORM 2016)

MATEC Web of Conferences Volume 80 (2016)

Troyes, France
4 - 7 July 2016

Editor:

Khemais Saanouni

ISBN: 978-1-5108-3167-4

Printed from e-media with permission by:

Curran Associates, Inc.
57 Morehouse Lane
Red Hook, NY 12571

CURRAN ASSOCIATES INC. proceedings.com

Some format issues inherent in the e-media version may also appear in this print version.

Printed by Curran Associates, Inc. (2016)

For additional information, please contact EDP Sciences – Web of Conferences at the address below.

EDP Sciences – Web of Conferences
17, Avenue du Hoggar
Parc d'Activité de Courtabœuf
BP 112
F-91944 Les Ulis Cedex A
France

Phone: +33 (0) 1 69 18 75 75
Fax: +33 (0) 1 69 28 84 91

contact-edps@webofconferences.org

Additional copies of this publication are available from:

Curran Associates, Inc.
57 Morehouse Lane
Red Hook, NY 12571 USA
Phone: 845-758-0400
Fax: 845-758-2633
Email: curran@proceedings.com
Web: www.proceedings.com

TABLE OF CONTENTS

MATEC Web of Conferences
Open access conference proceedings
in Materials science, Engineering
and Chemistry
A flexible, fast and efficient
publication service,
for an open access recording
of your event
and a wide dissemination of
your proceedings
www.matec-conferences.org
edp open Open Access Publishing
from EDP Sciences

Aims & scope

MATEC Web of Conferences is an open-access journal devoted to the publishing of conference proceedings dealing with Materials science, Engineering and Chemistry.

The journal aims to make conference proceedings visible, citable and widely usable by the scientific community. It offers outstanding services for the publication and dissemination of contents in order to ensure long-term archiving and maximum exposure of scientific events.

Coverage includes all engineering disciplines: civil, naval, mechanical, chemical, and electrical engineering as well as nanotechnology and metrology. The journal also deals with the physical-chemical characterization of materials, their implementation, their resistance in their environment, etc. Others subdisciplines of chemistry, such as analytical chemistry, petrochemistry, organic chemistry, etc., and even pharmacology, are also welcome.

Why publish with MATEC Web of Conferences?

- **Open access:** conference papers available worldwide.
- **Fast publication:** less than 3 months.
- **Total flexibility:** no page restrictions, free unlimited use of colour in the online version, any kind of scientific document welcomed (traditional articles, posters, abstracts and slideshows).
- **High visibility:** all documents indexed in main databases and easily found online.
- **Fair author rights:** CC-BY Creative Commons license.
- **Range of additional publishing options:** printed proceedings, CD-ROMs, copy-editing, etc.

Contact: Isaline Augusto
matec@webofconferences.org

ISSN (electronic edition): 2261-236X

www.matec-conferences.org

edp open Open Access Publishing from EDP Sciences

Committees:

- Chairmen:
  - Khemais Saanouni, Professor at UTT (Chairman)
  - Jean-Loup Chenot, Scientific Director at Transvalor SA (Co-Chairman)
  - Jean-Loup Duval, Scientific Director at ESI Group (Co-Chairman)

- Members of the Organizing Committee
  - T. Adnot (LASMIS/UTT)
  - P.A. Adragna (LASMIS/UTT)
  - H. Badreddine (LASMIS/UTT)
  - P. Belloy (LASMIS/UTT)
  - L. Caillot (LASMIS/UTT)
  - J.L. Chenot (Transvalor/CEMEF)
  - P. Conraux (ESI-Group)
  - L. Daniel (LASMIS/UTT)
  - S. De Wreede (LASMIS/UTT)
  - R. Ducloux (Transvalor)
  - I. Dugourd (ESI-Group)
  - J.L. Duval (ESI-Group)
  - A. Ebel (LASMIS/UTT)
  - M. François (LASMIS/UTT)
  - S. Fleury (LASMIS/UTT)
  - C. Gallerne (ESI-Group)
  - D. Gangarossa (LASMIS/UTT)
  - C. Garnier (LASMIS/UTT)
  - B. Guelorget (LASMIS/UTT)
  - C. Labergère (LASMIS/UTT)
  - P. Lafon (LASMIS/UTT)
  - G. Lecomte (ESI-Group)
  - L. Moreau (GAMMA3/UTT)
  - B. Panicaud (LASMIS/UTT)
  - N. Plet (LASMIS/UTT)
  - O. Reimbold (Transvalor)
  - D. Retraint (LASMIS/UTT)
  - L. Rigaud (LASMIS/UTT)
  - E. Rouhaud (LASMIS/UTT)
  - K. Saanouni (LASMIS/UTT)
  - L. Stiot (LASMIS/UTT)
  - C. Yendjadj (LASMIS/UTT)

- International Scientific Committee:
  - Agassant J.F., (CEMEF, France)
  - Altan T. (Ohio State University, USA)
  - Angelet de Saracibar C. (Univ.of Barcelona, Spain)
  - Asserin O. (CEA, Saclay, France)
  - Ayad R. (URCA, France)
  - Bachelet E. (SAFRAN, France)
  - Badreddine H. (UTT, France)
  - Balan T. (ENSAM/Metz, France)
  - Banabic D. (Technical Univ. Cluj-Napoca, Romania)
  - Bariani P. F. (University of Padova, Italy)
  - Barlat F. (Postech, Korea)
  - Batalha G F. (Sao Paulo University, Brazil)
  - Bay F. (CEMEF, France)

- Bay N. (Technical University of Denmark)
- BelhadjSalah H. (ENIM, Tunisia)
- Bellet M. (CEMEF, France)
- Bergheau J.M. (ENISE, France)
- Binetruy C. (ECNNantes, France)
- Boisse P., (INSA Lyon, France)
- Cao J. (Northwestern University, , USA)
- Cardoso R.P.R. (Univ West England, Bristol, UK)
- Celentano D. (Pontificia Cato. Univ. Santiago, Chile)
- Ceretti E. (University of Brescia, Italy)
- Cesar de Sa J., (University of Porto, Portugal)
- Chastel Y. (Renault, France)
- Chatti S. (IUL, Dortmund, Germany)
- Chen F.K. (Taipei Univ, Taiwan)
- Chenot JL (Transvalor, France)
- Chinesta F., (ECN, France)
- Choquart F. (CERA-Treves, Reims, France)
- Chung K. (Seoul National Univ, Korea)
- Coupez T. (EC Nantes, France)
- Cueto E. (University of Zaragoza, Spain)
- David L. (PSA, France)
- Delannay L. (Univ Catolique Louvain, Belgium)
- Hu P. (Dalian University of Technology, China)
- Huerta A. (Barcelona, Spain)
- Huétink H. (University of Twente, The Netherlands)
- Jiang Z-Y (Wollogong Univ., Australia)
- Kaftanoglu B. (Altim Univ, Ankara, Turky)
- Kawalla R. (IFM, TU Freiberg, Germany)
- Kuroda M. (Univ Yamagata, Japan)
- Kusiak J. (A.G.H., Krakow, Poland)
- Kuwabara T. (Tokyo Univ, Japan)
- Labergère C. (UTT, France)
- Lafon P. (UTT, France)
- Larsson R. (Chalmers Univ. of Technology, Sweden)
- Li D. Y. (Shanghai Jiaotong Univ, China)
- Li D. Z. (IMR, China)
- Liewald M. (IFU, University of Stuttgart, Germany)
- Lin J.G. (Imperial College, UK)
- Loge R. (EPFL, Lausanne, Switzerland)
- Lorong P. (PIMM ENSAM/Paris, France)
- Lubineau G. (KAUST, Saudi Arabia)
- Macret J.L. (Airbus-Defense and Space, France)
- Maier C. (Dunarea de Jos University of Galati, Romania)
- Manach P.Y., (UBS, France)
- Maniatty A. (Rensselaer Polytechnic Institute, USA)
- Dixit P.M. (Indian Institute of Tech. Kanpur, India)
- Dogui A. (ENIM, Tunisia)
- Dubar L. (Univ. Valencienne, France)
- Ducloux R. (Transvalor, France)
- Duroux P. (Arcelor Mittal, France)
- Duval J.L. (ESI, France)
- Flores F. (Univ National de Cordoba, Argentina)
- Fourment L. (CEMEF, France)
- Fratini L. (University of Palermo, Italy)
- Geers M.G.D (T. U. Eindhoven, the Neederlands)
- Geijselaers B. (Univ of Twente, the Neederlands)
- Gelin J.C., (ENSMM, France)
- Gerard P. (Arkema, France)
- Ghosh S., (Ohio State University, USA)
- Gilles P. (AREVA, France)
- Guo Y.Q. (URCA, France)
- Habraken A.M. (University of Liège, Belgium)
- Hartley P. (Birmingham University, UK)
- Hirt, G. (IBF, RWTH-Aachen)
- Hogdson P. (Deakin Univ, Australia)
- Hora P. (ETH, Zurich, Switzerland)
- Martins P.A.F. (University of Lisbon, Portugal)
- Massoni E., (CEMEF, France)
- Mattiasson K. (Volvo, Sweden)
- Menezes L.F. (University of Coimbra, Portugal)
- Micari F. (University of Palermo, Italy)
- Montmitonnet P. (CEMEF, France)
- Moon Y.H. (Pusan Univ, Korea)
- Mori KI, (Toyohashi University, Japan)
- Music O. (Altim Univ, Istambul, Turky)
- Nielsen K.B. (Aalborg University, Denmark)
- Oldenburg M. (Lulea Univ. of Technology, Sweden)
- Onate E. (University of Barcelona, Spain)
- Owen D.R.J (Swansea University, UK)
- Perez L. (U.T.F, Santa Maria, Chile)
- Picart P. (ESMM, Besançon, France)
- Pietrzyk M. (A.G.H., Krakow, Poland)
- Ponthot P. (University of Liège, Belgium)
- Potier-Ferry M. (Univ. Metz, France)
- Qin Y. (Glasgow, UK)
- Rahmoun M. (ENSA, Morocco)
- Rassili A. (University of Liège, Belgium)
- Rassineux A. (UTC, France)
- Reese S. (Aachen, Germany)
- Regnier G. (PIMM ENSAM/Paris, France)
- Rouhaud E. (UTT, France)
- Saanouni K. (UTT, France)

List of Partners and Exhibitors:

edp sciences

MATEC Web of Conferences

Statement of Peer review

In submitting conference proceedings to MATEC Web of Conferences, I certify to the Publisher that I adhere to the **Policy on Publishing Integrity** of the journal in order to safeguard good scientific practice in publishing.

1. All articles have been subjected to peer review administered by the proceedings editors.
2. Reviews have been conducted by expert referees, who have been requested to provide unbiased and constructive comments aimed, whenever possible, at improving the work.
3. Proceedings editors have taken all reasonable steps to ensure the quality of the materials they publish and their decision to accept or reject a paper for publication has been based only on the merits of the work and the relevance to the journal.

Title, date and place of the conference

- NUMIFORM 2016 (The 12th International Conference on Numerical Methods in Industrial Forming Processes)
- July 4th - 7th 2016
- University of Technology of Troyes, France

Proceedings editor(s):

- Prof. Khemais SAANOUNI

Date and editor's signature 25/08/2016

17, avenue du Hoggar - PA de Courtabœuf – BP 112 - 91944 Les Ulis Cedex A (France)
Tél. : 33 (0)1 69 18 75 75 - Fax : 33(0)1 69 07 45 17 – www.edpsciences.org

Preface

Since the sixties, numerical methods for solving complex multiphysic problems have undergone continuous improvements thanks to the availability of more and more powerful computers and the development of interactive robust software based on realistic material behavior and sophisticated algorithms. Like all other sectors of engineering sciences, the field of material forming simulation or "virtual forming" takes full benefit from these scientific and technological developments.

The NUMIFORM triennial international conferences series, introduced by the late Prof. O.C.Zienkiewicz in 1982, aim to provide a forum where recent advances and future directions in the numerical simulations of forming and manufacturing processes are discussed by scientists and engineers from academia and industry worldwide. The topics covered are of great interest not only for academics from theoretical and computational mechanics but also for professionals (engineers and researchers) from industry involved in traditional and novel forming and manufacturing technologies for conventional and emerging materials.

This twelfth edition of the NUMIFORM international conferences series (NUMIFORM2016) was held at University of Technology of Troyes (UTT) in France from July 4th to 7th 2016, co-organized with two renowned French companies in the field of virtual forming technologies: ESI Group and Transvalor SA. As the eleven previous editions of NUMIFORM international conferences series, this edition provides a forum for scientists and practicing engineers alike to present the latest findings in their research endeavor and at the same time to explore future research directions in the field.

In this proceeding, the reader can find 105 peer reviewed papers selected among the 199 oral presentations given during NUMIFORM2016. These papers are organized by topics following the sixteen topics of the conference.

We address our warm thanks to all the authors and participants to this twelfth edition of NUMIFORM international conferences series. In addition, our thanks are due to public and private sponsors who, through their financial support or their participation to the industrial exhibition, helped significantly the organization of this event: UTT, ESI Group, Transvalor SA, Région Champagne-Ardenne, Conseil Général de l'Aube, Grand Troyes, l'Association Française de Mécanique (AFM), MECAMAT, MK2i, Composite Integrity, QForm. Special thanks are due to the colleagues who organized and fully managed the papers submitted to their MS/IMS, as well as many colleagues from the international Scientific Committee, the Steering Committee and the Advisory committee who kindly performed the reviewing of some full papers. Many thanks are due to the colleagues from UTT/LASMIS, ESI Group and Transvalor SA involved in the practical organization of NUMIFORM2016.

The Editors

K. Saanouni (UTT), Chairman

J.L. Chenot (Transvalor SA), Co-Chairman

J.L. Duval (ESI Group), Co-Chairman

MATEC Web of Conferences **80**, 01001 (2016) DOI: 10.1051/matecconf/20168001001
NUMIFORM 2016

Advanced numerical simulation based on a non-local micromorphic model for metal forming processes

Evangelia Diamantopoulou[1,a], Carl Labergere[1], Houssem Badreddine[1] and Khemais Saanouni[1]

[1]*University of Technology of Troyes, ICD/LASMIS UMR-CNRS 6281, 12 rue Marie Curie CS 42060 Troyes Cedex, France*

Abstract. An advanced numerical methodology is developed for metal forming simulation based on thermodynamically-consistent nonlocal constitutive equations accounting for various fully coupled mechanical phenomena under finite strain in the framework of micromorphic continua. The numerical implementation into ABAQUS/Explicit is made for 2D quadrangular elements thanks to the VUEL users' subroutine. Simple examples with presence of a damaged area are made in order to show the ability of the proposed methodology to describe the independence of the solution from the space discretization.

1 Introduction

Nowadays, the fully local constitutive equations have been well established to model the induced material softening behaviour due to thermal, damage and other microstructure-dependent phenomena. However, the solutions of the fully local constitutive equations are highly sensitive to the space and time discretization. The mechanics of generalized continua makes possible the straightforward introduction of the characteristic lengths into the constitutive equations of materials with microstructure in order to overcome this drawback [7]. Among others, the most recent and comprehensive review of these generalized continua and their use to solve various problems in mechanics of solids and fluids is found in the recent books and publications ([22,23], [24], [2], [7]). The micromorphic theory was proposed simultaneously by (Eringen and Suhubi, 1964[25]; Mindlin, 1964[26]). It consists of introducing a general non-compatible full field of micro-deformation as an extra degree of freedom, in addition to the classical displacement field. The micromorphic approach can, in fact, be applied to any macroscopic quantity in order to introduce a character length scale in the original classical continuum model in a systematic way, as presented by ([7], [8]). From the comparison between nonlocal and micromorphic theories, Forest and Aifantis [27] concluded that when the micromorphic variable remains as close as possible to the plastic strain, the micromorphic model reduces to the strain gradient theory.

The main goal of this paper is to show the efficiency of the formulated fully coupled nonlocal micromorphic constitutive equations in getting really mesh independent solutions even with high values of the ductile damage. The concept of internal lengths, that have to be experimentally determined from accurate measurements by using advanced methods, is introduced.

The introduction of the new balance equations in the nonlocal model with the effect of the ductile damage demands the construction of an element based on the assumed strain method combined with the Hu-Washizu variational form. The standard B matrix obtained from the displacement field analytically and including the hourglass control part, is projected to eliminate volumetric and shear locking phenomena. An air bending test of a DP1000 dual phase steel sheet is performed to validate and to show the mesh independent solution predicted by the micromorphic nonlocal model. This kind of nonlocal elastoplastic constitutive equations with damage can be later used for the numerical simulation of various sheet and bulk metal forming processes.

2 Theoretical aspects

2.1 The generalized principle of virtual power

The virtual power of internal forces of the classical local continuum is extended to the micromorphic continuum by using the virtual micromorphic damage ([2, 7]):

$$P_{\text{int}}\left(\vec{u}^*, \breve{d}^*\right) = -\int_V \left(\underline{\sigma} : \left(\vec{\nabla}\dot{\vec{u}}\right)^* + \breve{Y}\dot{\breve{d}}^* + \vec{\breve{Y}} \cdot \nabla\dot{\breve{d}}^*\right) dV \qquad (1)$$

where, $\underline{\sigma}$ is the Cauchy stress tensor, the scale $\breve{Y}$ and vector $\vec{\breve{Y}}$ are the stress-like variables with respect to micromorphic damage $\breve{d}$ and its first gradient $\vec{\nabla}\breve{d}$. Similarly, the virtual power of external forces can be also extended with the simple and generalized body forces $\vec{f}^u$, f^d and $\vec{f}^{gd}$, and contact forces $\vec{F}^u$ and F^d:

[a] evangelia.diamantopoulou@utt.fr

$$P_{ext}\left(\vec{u}^*,\breve{d}^*\right)=\rho\int_V\left(\vec{f}^u\cdot\dot{\vec{u}}^*+f^d\dot{\breve{d}}^*+\vec{f}^{gd}\cdot\vec{\nabla}\dot{\breve{d}}^*\right)dV+\int_\Gamma\left(\vec{F}^u\cdot\dot{\vec{u}}^*+F^d\dot{\breve{d}}^*\right)dS \quad (2)$$

The virtual power of the inertia forces are enhanced by the micromorphic phenomena:

$$P_a\left(\vec{u}^*,\breve{d}^*\right)=\rho\int_V\left(\ddot{\vec{u}}\cdot\dot{\vec{u}}^*+\zeta_{\breve{d}}\ddot{\breve{d}}\dot{\breve{d}}^*\right)dV \quad (3)$$

where, $\zeta_{\breve{d}}$ is a scale factor which maps the local density to the micromorphic level.

The generalized virtual power enhanced with the micromorphic damage takes the following form for any given kinematically admissible fields:

$$P_{int}+P_{ext}=P_a \qquad \forall\vec{u}^*,\breve{d}^*\ K.A. \quad (4)$$

Substitute Eq. (1) to Eq. (3) into the generalized form of virtual power Eq. (4), and by using the divergence theorem to transform the volume integrals, leads to the two momentum equations:

$$\begin{cases}\vec{\nabla}\cdot\underline{\sigma}+\rho\vec{f}^u=\rho\ddot{\vec{u}} & in\ \Omega\\ \underline{\sigma}\cdot\vec{n}=\vec{F}^u & on\ \Gamma\end{cases} \quad (5a)$$

$$\begin{cases}\left(\vec{\nabla}\cdot\vec{\breve{Y}}+\breve{Y}\right)+\rho\left(f^d-\vec{\nabla}\cdot\vec{f}^{gd}\right)=\rho\zeta_{\breve{d}}\ddot{\breve{d}} & in\ \Omega\\ \left(\vec{\breve{Y}}-\rho\vec{f}^{gd}\right)\cdot\vec{n}=F^d & on\ \Gamma\end{cases} \quad (5b)$$

The two partial differential equations are the strong forms with respect to displacement and micromorphic damage with Initial Boundary Value Problem (IBVP). And the stress-like variables $\underline{\sigma}$, $\breve{Y}$ and $\vec{\breve{Y}}$ will be defined by the state relations developed under the framework of thermodynamic irreversible processes.

2.2 State relations

In this section, a specific state potential as well as yield function and dissipation potential are constructed for an isothermal isotropic elastoplasticity fully coupled with isotropic local damage, isotropic and kinematic hardening and isotropic micromorphic damage. The damaged state variables are applied to the local state variables according to the assumption of the equivalence of total energy developed in (see [12-15], [16], [17], [18], [19], [7], [21, 22]). And for simplification, we also postulate that only the local damage is coupled with the micromorphic damage. With these assumptions, a quadratic the Helmholtz free energy can be expressed as:

$$\rho\psi=\frac{1}{2}(1-d)\left(\underline{\varepsilon}^e:\underline{\underline{\Lambda}}:\underline{\varepsilon}^e+\frac{2}{3}C\underline{\alpha}:\underline{\alpha}\right)+\frac{1}{2}\left(\left(1-d^\gamma\right)Qr^2+\breve{H}\left(d-\breve{d}\right)^2+\breve{H}^g\vec{\nabla}\breve{d}\cdot\vec{\nabla}\breve{d}\right) \quad (6)$$

where $\underline{\underline{\Lambda}}=\lambda_e\underline{1}\otimes\underline{1}+2\mu_e\underline{\underline{1}}$ is the fourth-rank elastic tensor expressed with the classical Lame's parameters. C and Q are the macro modules of kinematic and isotropic hardening respectively. $\breve{H}$ is the micromorphic module with respect to local damage and micromorphic damage. $\breve{H}^g$ is the micromorphic module related to the gradient of micromorphic damage.

The local and nonlocal stress-like variables are derived in the following forms:

$$\underline{\sigma}=\rho\frac{\partial\psi}{\partial\underline{\varepsilon}^e}=(1-d)\underline{\underline{\Lambda}}:\underline{\varepsilon}^e=(1-d)\left(\lambda_e tr\left(\underline{\varepsilon}^e\right)\underline{1}+2\mu_e\underline{\varepsilon}^e\right) \quad (7)$$

$$\underline{X}=\rho\frac{\partial\psi}{\partial\underline{\alpha}}=\frac{2}{3}(1-d)C\underline{\alpha} \quad (8)$$

$$R=\rho\frac{\partial\psi}{\partial r}=\left(1-d^\gamma\right)Qr \quad (9)$$

$$Y=-\rho\frac{\partial\psi}{\partial d}=\frac{1}{2}\left(\underline{\varepsilon}^e:\underline{\underline{\Lambda}}:\underline{\varepsilon}^e+f_h\frac{2}{3}C\underline{\alpha}:\underline{\alpha}+f_h(\gamma/2)d^{\gamma-1}Qr^2\right)-f_h\breve{H}\left(d-\breve{d}\right)=Y^e+f_h(Y^\alpha+Y^r+\breve{Y}) \quad (10)$$

$$\breve{Y}=\rho\frac{\partial\psi}{\partial\breve{d}}=-\breve{H}\left(d-\breve{d}\right) \quad (11)$$

$$\vec{\breve{Y}}=\rho\frac{\partial\psi}{\partial\vec{\nabla}\breve{d}}=\breve{H}^g\vec{\nabla}\breve{d} \quad (12)$$

In Eq. (10) γ is the parameter governing the effect of the damage on the isotropic hardening and f_h a simple function equal to 1 if the hydrostatic part of the stress is positive, else it is equal to 0.

The energy density release rate given by state relation with respect to damage in Eq. (10) can be decomposed into local part and coupled nonlocal part:

$$\begin{cases}Y=Y_{loc}+Y_{nloc}\\ Y_{loc}=\frac{1}{2}\left(\underline{\varepsilon}^e:\underline{\underline{\Lambda}}:\underline{\varepsilon}^e+\frac{2}{3}C\underline{\alpha}:\underline{\alpha}+\gamma d^{\gamma-1}Qr^2\right)\\ Y_{nloc}=\breve{Y}=-\breve{H}\left(d-\breve{d}\right)\end{cases} \quad (13)$$

2.3 Evolution equations

Since there's no dissipation from micromorphic damage, to account for the damage effects, we limit ourselves to the isotropic plasticity for simplicity. The von Mises yield function and dissipation potentials used in ([7], [12-14], [16], [17], [18], [19]) are used:

$$f\left(\underline{\sigma},\underline{X},R;d\right)=\frac{\|\underline{\sigma}-\underline{X}\|}{\sqrt{1-d}}-\frac{R}{\sqrt{1-d^\gamma}}-\sigma_y\leq 0 \quad (14)$$

$$F\left(\underline{\sigma},\underline{X},R,d\right)=f+\frac{3}{4}\frac{a\underline{X}:\underline{X}}{C(1-d)}+\frac{1}{2}\frac{bR^2}{Q\left(1-d^\gamma\right)}+\frac{S}{(s+1)(1-d)^\beta}\left\langle\frac{Y-Y_0}{S}\right\rangle^{(s+1)} \quad (15)$$

where, $\|\underline{\sigma}-\underline{X}\| = \sqrt{(3/2)(\underline{\sigma}^{dev}-\underline{X}):(\underline{\sigma}^{dev}-\underline{X})}$ defines the well-known von Mises equivalent stress and $\underline{\sigma}^{dev}$ is its deviatoric part of the stress tensor. The parameters a and b reflect the nonlinear property of the kinematic and isotropic hardening respectively and S, s and β characterize the nonlinear evolution of the ductile damage.

Clearly, the yield function and plastic potential are identical to the local theory. However, the indirect contribution of the micromorphic state variables is introduced from the state relations. Applying the generalized normality rule to the above defined local yield function and dissipation potential lead to the following evolution equations:

$$\underline{\dot{D}}^p = \dot{\lambda}\frac{\partial F}{\partial \underline{\sigma}} = \dot{\lambda}\underline{\tilde{n}} = \dot{\lambda}\frac{\underline{n}}{\sqrt{1-d}} \tag{16}$$

$$\underline{\dot{\alpha}} = -\dot{\lambda}\frac{\partial F}{\partial \underline{X}} = \dot{\lambda}\left(\frac{\underline{n}}{\sqrt{1-d}} - a\underline{\alpha}\right) \tag{17}$$

$$\dot{r} = -\dot{\lambda}\frac{\partial F}{\partial R} = \dot{\lambda}\left(\frac{1}{\sqrt{1-d^{\gamma}}} - br\right) \tag{18}$$

$$\dot{d} = \dot{\lambda}\frac{\partial F}{\partial Y} = \frac{\dot{\lambda}}{(1-d)^{\beta}}\langle\frac{Y-Y_0}{S}\rangle^{s} \tag{19}$$

where $\underline{n} = 3(\underline{\sigma}^{dev}-\underline{X})/2\,\|\underline{\sigma}-\underline{X}\|$ is the outside normal to the yield surface in the stress space and $\dot{\lambda}$ is the plastic multiplier.

2.4 Transformation of the micromorphic balance equations

Considering the generalized balance equation of micromorphic momentum Eq. (5b), it can be easily expressed in strain-like variables space by using micromorphic state relations Eq. (11) and Eq. (12):

$$\begin{cases} \left(\breve{H}^{g} Lap\left(\breve{d}\right) - \breve{H}\left(d-\breve{d}\right)\right) + \rho\left(f^{\breve{d}} - \vec{\nabla}\cdot\vec{f}^{g\breve{d}}\right) = \rho\zeta_{D}\ddot{\breve{d}} & in\ \Omega \\ \left(\breve{H}^{g}\left(\vec{\nabla}\breve{d}\right) - \rho\vec{f}^{g\breve{d}}\right)\cdot\vec{n} = F^{\breve{d}} & on\ \Gamma \end{cases} \tag{20}$$

By neglecting the micromorphic body forces , i.e. $f^{\breve{d}}=0$, $\vec{f}^{g\breve{d}}=0$ and $F^{\breve{d}}=0$, the above equation can take the following form:

$$\begin{cases} l_{\breve{d}}^2 Lap\left(\breve{d}\right) - \left(d-\breve{d}\right) = \rho\frac{\zeta_{\breve{d}}}{\breve{H}}\ddot{\breve{d}} & in\ \Omega \\ \left(\breve{H}^{g}\left(\vec{\nabla}\breve{d}\right)\right)\cdot\vec{n} = 0 & on\ \Gamma \end{cases} \tag{21}$$

where, $l_{\breve{d}}$ is the intrinsic internal length scale parameter relative to the micromorphic damage defined as the ratio of the micromorphic moduli:

$$l_{\breve{d}}^2 = \frac{\breve{H}^g}{\breve{H}} \tag{22}$$

The Eq. (21) is used as a strong form from which the associated weak form will be deduced in order to solve the fully coupled IBVP.

3 Internal lengths

The development of an 'advanced' modelling of multiphysic thermomechanical coupling in the framework of the generalized continuum mechanics (micromorphic theory) is proposed in order to introduce the concept of internal lengths that are representative of the materials microstructures while accounting for the various initial and induced anisotropies under large deformations. These internal lengths have to be experimentally determined from accurate measurements of highly localized displacement/strain (or velocity/strain rate) fields by using advanced methods to measure the kinematic fields at the relevant scales. For that, the electronic speckle pattern interferometry (ESPI) method will be used.

We seek to locally measure the displacement and velocity fields in order to access to the local strain/strain rate fields inside the localized zones. An example of a localization pattern obtained with the ESPI technique on a 460 steel sheet is shown in Figure 1 [6]. The tensile direction is horizontal and the interference fringes due to the displacement field are shown in the left where each grey level corresponds to 2.8 nm and the total relative displacement is about 3 µm. The corresponding strain rate field (tensile component) extracted with the geometrical model is shown in the right. The widths B1 and B2 of the localization bands are obtained directly from the experimental measure. These experimental measurements will provide inputs for the advanced micromorphic constitutive equations and mainly to determine the internal lengths related to the targeted micromorphic phenomena [5].

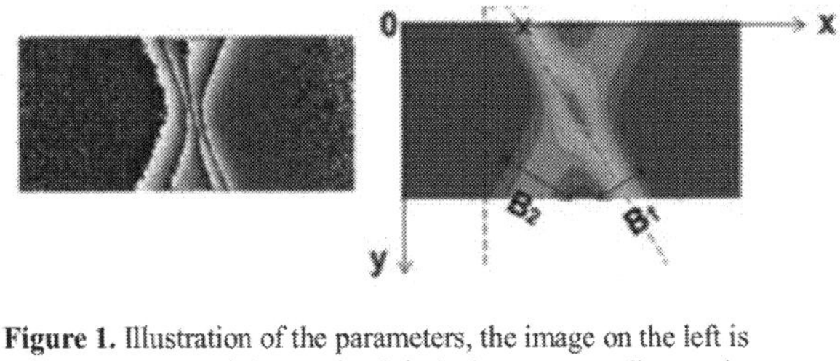

Figure 1. Illustration of the parameters, the image on the left is a fringe pattern and that on the right is the corresponding strain rate map obtained [6].

A purely geometric (or kinematic) model used to describe the strain rate pattern and extract global parameters. It is independent of the constitutive behaviour of the material. The strain rate distribution in each band is depicted by a linear combination of a Gaussian and a Lorentzian function:

$$\dot{\varepsilon}_{xx}(x,y)=\dot{\varepsilon}_1^{\max}(\eta_1 G_1+(1-\eta_1)L_1)+\dot{\varepsilon}_2^{\max}(\eta_2 G_2+(1-\eta_2)L_2) \quad (23)$$

where $\dot{\varepsilon}_i^{\max}$ is the maximum strain rate, G_i is a Gaussian function, L_i is a Lorentzian function, η_i and $1-\eta_i$ are the weights of the Gaussian and the Lorentzian functions respectively ($i \in \{1, 2\}$).

The Gaussian and Lorentzian functions are defined as following:

$$G(a,b,B,x,y)\exp(-\pi(\frac{x-ay-b}{B})^2)$$
$$L(a,b,B,x,y)=\frac{1}{1+(\pi\frac{x-ay-b}{B})^2} \quad (24)$$

with:

- a, the inclination of the band with respect to the transverse direction.
- b, the location of the band.
- B, the integral width of the band.

In all cases of the present study, the specimens failed in the same direction, along the band that was oriented almost perpendicular to the rolling direction or along band 1 as illustrated in **Fig.1**. Thereby to simplify the explications, this band will be called dominant band and the other one disappearing band, and the superscript "dom" and "dis" will be used to indicate them.

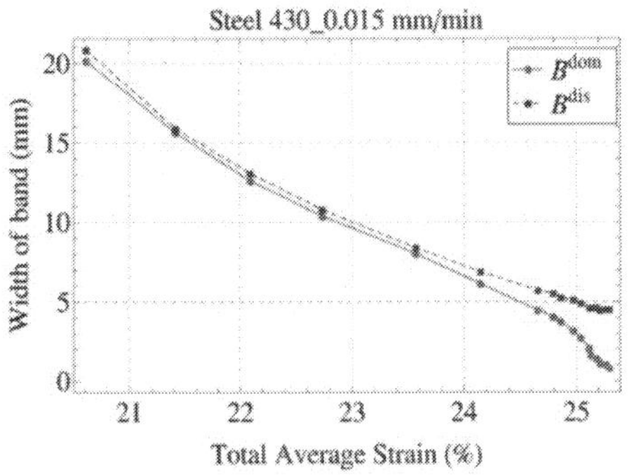

Figure 2. Evolutions of the localization band characters vs total average strain (or time) with different tensile speed. Maximum strain rate of the bands vs total average strain, bandwidth vs total average strain [6].

For the identification of the internal lengths, an effort is under progress. According to the analytical-experimental study described above, an experimental-numerical study is evolving in order to determine the values of $\breve{H}$ and $\breve{H}^g$ and verify the characteristic length for each material and mesh size in total agreement with the experiments and the information extracted by **Fig.2** (bandwidths, total average strain, estimated error between numerical and experimental strain rate etc).

Let us note here, that the above study concerns a 460 (X8Cr17) steel specimen.

For this paper and the air bending application presented in section 5, a DP1000 dual phase sheet is used. The internal length for this type of material is determined empirically and fixed at $\sim 0.2mm$ by following the rule that **Mesh size (Ms)** $< l_d$.

4 Numerical aspects

From the numerical point of view, an advanced numerical procedure is presented. The finite element code that is used to implement the micromorphic model is developed in ABAQUS®. The numerical integration in mixed one/four quadrature(s) (or Gauss) point for all the coupled constitutive equations is necessary in order to compute the stress tensor and the overall stress-like internal variables which appear in the weak forms associated to the balance equations. For the time discretization of the equations of our behaviour model, the fully implicit (iterative) Euler scheme in combination with an asymptotic one applied to the hardening equations are used. The overall constitutive equations are then reduced to two nonlinear and strongly coupled algebraic equations (for the isotropic plasticity) with two unknowns: $\Delta\lambda$ and d_{n+1}. The elastic prediction-plastic correction method is then used together with the Newton-Raphson resolution scheme to solve the two equations. The behaviour model defined in the framework of the micromorphic continua is implemented into ABAAQUS/Explicit® thanks to the VUMAT user's subroutine.

The introduction of the new micromorphic balance equations in the behaviour model with nonlocal damage demands the construction of a special element based on a mixed variational form and having $\breve{d}$ as additional d.o.f. Fish and Belytschko (see [4]) proposed an extension of the Hu-Washizu weak form of the variational principle in nonlinear mechanics of solids. We consider two variational forms and introduce balance equation with the contribution of the micromorphic damage discretized on a domain Ω.

The weak form of the balance equations can be obtained thanks to the weighted residual method [4] in order to get:

$$\begin{cases}\int_\Omega \underline{\sigma}_a : \delta\underline{\varepsilon}_a d\Omega + \delta\int_\Omega \underline{\sigma}_a : \left(\vec{\nabla}^{sym}(\vec{u})-\underline{\varepsilon}_a\right)d\Omega - \delta W_{ext}=0 \\ \int_\Omega \left(\breve{d}\delta\breve{d}+\ell_d^2(\vec{\nabla}\breve{d}).(\vec{\nabla}\delta\breve{d})\right)d\Omega - \int_\Omega \left(d\delta\breve{d}\right)d\Omega = 0\end{cases} \quad (25)$$

where δW_{ext} is the external forces virtual work, δ represents the classical variation operator, $\vec{u}$ is the displacement vector, $\underline{\sigma}$ the Cauchy stress tensor, $\underline{\varepsilon}_a$ the tensor of the assumed strains, $\underline{\sigma}_a$ the assumed stress tensor and $\underline{\nabla}^s(\vec{u})$ is the symmetric part of the displacement gradient. Simo and Hughes [9] suggested the construction of the projection of the velocity field's discretized gradient $\left[B^e_{n+\frac{1}{2}}\right]$. This new operator is chosen in order to evaluate the assumed strain tensor $\{\varepsilon^a\}_{n+1} = \left[B^e_{n+\frac{1}{2}}\right]\{\dot{u}^e\}_{n+\frac{1}{2}}$ and to compute the associated assumed stress tensor $\{\sigma^a_{n+1}\}$ which is orthogonal to the difference between the symmetric part of the velocity gradient and the assumed strain tensor. This choice allows us to simplify the Hu-Washizu weak form as follows:

$$\begin{cases} \int_\Omega \underline{\sigma}_a : \delta\underline{\varepsilon}_a d\Omega - \delta W_{ext} = 0 \\ \int_\Omega \left(\breve{d}\delta\breve{d} + \ell_d^2(\vec{\nabla}\breve{d}).(\vec{\nabla}\delta\breve{d})\right)d\Omega - \int_\Omega \left(d\delta\breve{d}\right)d\Omega = 0 \end{cases} \quad (26)$$

The matrix $\left[B^e_{n+\frac{1}{2}}\right]$ is decomposed in two parts: $\left[B^0_{n+\frac{1}{2}}\right]$ for the tensile and shear modes that gives the stress tensor components in the centre of the element, and $\left[B^h_{n+\frac{1}{2}}\right](\xi,\eta) = \xi\left[B^\xi_{n+\frac{1}{2}}\right] + \eta\left[B^\eta_{n+\frac{1}{2}}\right]$ for the control of the hourglass modes for all the elements. Finally, the expression of the total strain tensor, according to the assumptions above, is written as follows:

$$\{\varepsilon^a\}_{n+1} = \left[B^e_{n+\frac{1}{2}}\right]\{\hat{u}^e\}_{n+\frac{1}{2}} = \underbrace{\left[B^0_{n+\frac{1}{2}}\right]\{u^e\}_{n+\frac{1}{2}}}_{\{D^0\}_{n+1}} + \underbrace{\left[B^h_{n+\frac{1}{2}}\right]\{u^e\}_{n+\frac{1}{2}}}_{\{D^h\}_{n+1}}$$
$$= \{\varepsilon^0\}_{n+1} + \{\varepsilon^h\}_{n+1} \quad (27)$$

Wang et al [1] proposed the construction of a quadrangular element to the field of the assumed strains in a 2D plane strain in order to control the shear and volumetric locking. The isoparametric shape functions for the 4-node quadrilateral can be written in terms of a set of orthogonal base vectors as:

$$N_i(\xi,\eta) = \frac{1}{4}\mathbf{s} + \frac{1}{4}\boldsymbol{\xi}\xi + \frac{1}{4}\boldsymbol{\eta}\eta + \frac{1}{4}\mathbf{h}\xi\eta \quad (28)$$

where (ξ,η) are the coordinates in the reference space with $\boldsymbol{\xi}^T = [-1,1,1,-1]$, $\boldsymbol{\eta}^T = [-1,-1,1,-1]$, $\mathbf{s}$ is the translation vector with $\mathbf{s} = [1,1,1,1]$ and $\mathbf{h}$ is the hourglass vector with $\mathbf{h} = [1,-1,1,-1]$.

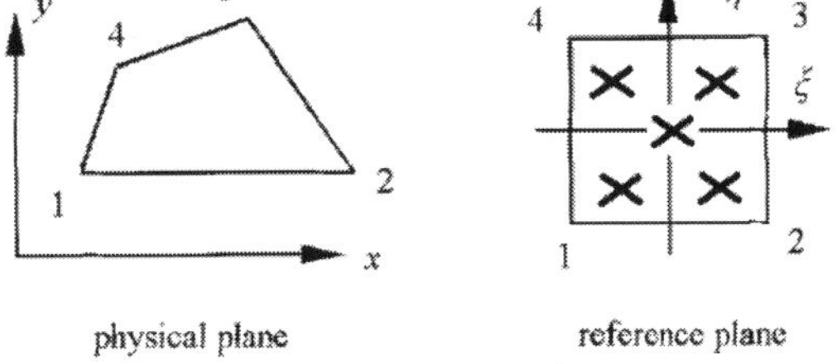

Figure 3. A 4-node quadrilateral element in physical and natural coordinate systems.

This element is used in our formulations by adding a new degree of freedom associated to the nonlocal damage. In order to ensure the objectivity of the model, corotational local system is built in the centre of the element using the shape functions defined in the reference space (ξ,η). The orientation of this corotational triad is governed by rotation tensor $\underline{\underline{Q}}$ which is expressed under the following form:

$$[Q] = \begin{bmatrix} e_1^1 & e_1^2 \\ e_2^1 & e_2^2 \end{bmatrix}, \; \vec{g}_1 = \frac{\partial x}{\partial \xi}\hat{x} + \frac{\partial x}{\partial \eta}\hat{y}, \; \vec{e}_1 = \frac{\vec{g}_1}{\|\vec{g}_1\|}, \; \vec{e}_2.\vec{e}_1 = 0, \; \|\vec{e}_2\| = 1 \quad (26)$$

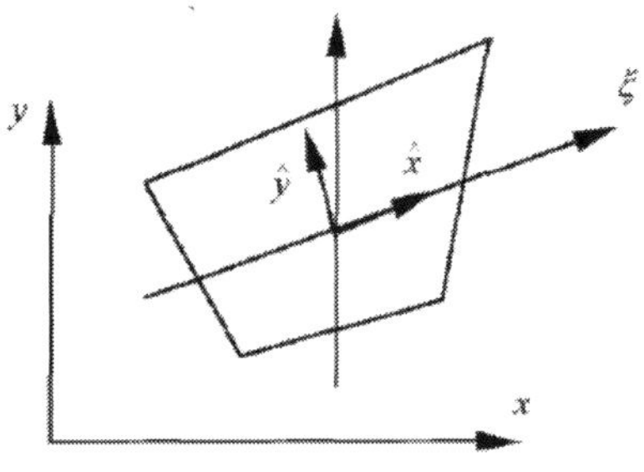

Figure 4. Corotational system definition of the Q4-URI element [1].

5 Numerical simulation of the air bending test

The air bending test was modelled in a 2D plane strain configuration. The sheet is discretized with the Q4 quadrangular element implemented in a VUEL ABAQUS/Explicit® subroutine.

The DP1000 steel material parameters have been determined by an inverse approach [11] based on the experimental results: E=*220.0 GPa*, ν=*0.3*, σ_y=*810.0 MPa*, Q=*3800.0 MPa*, b=*15.0*, C=*69500.0 MPa*, a=*240.0*, S=*12.0 MPa*, s=*1.2*, β=*1.0*, Y_0=*0.0*, γ=*4.0*.
3D measurement of strain conducted via Aramis-GOM system make to measure the strain distribution over the specimen and fracture strain, respectively.

5.1 Behaviour model with local damage

Initially, $\breve{H} = \breve{H}^g = 0$ is chosen. Figure 5 shows the isovalues of the internal damage 'd' for the three different mesh sizes. It is clear that the size of the localization zone depends on the mesh size.

For this case the results of damage mapped the at final fracture stage are obtained for different mesh sizes. This figure shows clearly that the mesh size affects not only localization zone but also the moment of the final fracture (i.e. global displacement when the damage of an element reaches the value '$d=1$', namely U_{rupt}). The dependence of the localization band to the mesh size leads to different values of fracture displacement in such a manner that smallest mesh size allows the smallest fracture displacement. This ascertainment is confirmed through global force/displacement curves given in Figure 6.

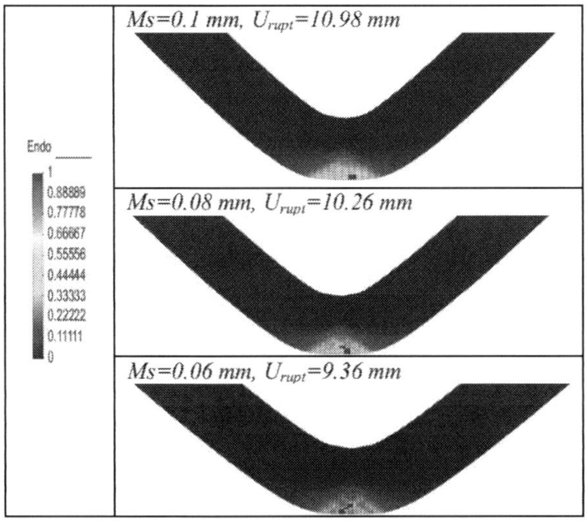

Figure 5. Damage isovalues of a ruptured DP1000 specimen in the framework of the local continua for different Ms.

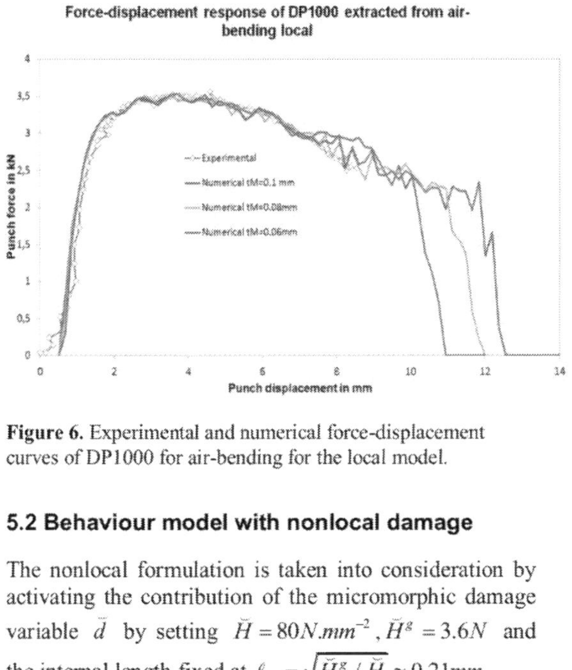

Figure 6. Experimental and numerical force-displacement curves of DP1000 for air-bending for the local model.

5.2 Behaviour model with nonlocal damage

The nonlocal formulation is taken into consideration by activating the contribution of the micromorphic damage variable $\breve{d}$ by setting $\breve{H} = 80N.mm^{-2}$, $\breve{H}^{g} = 3.6N$ and the internal length fixed at $\ell_{d} = \sqrt{\breve{H}^{g} / \breve{H}} \approx 0.21mm$.

For this case, the results considering the damage are also obtained after testing on three different mesh sizes (Figure 7). We can see that the damage localization bandwidth becomes smoother than the local damage case (i.e. the damage gradient varies less than in the local case). The length of damage gradient appears virtually identical for the three mesh cases but the crack always propagates in final stage along a single row of elements. The regularization length makes it possible to calibrate the amount of damage dissipation that crosses a band and its influence is significant for the model's response. Unlike the local model, the force/displacement curve (see Figure 8) gives the same fracture displacement value at $U_{rupt} \approx 10.6mm$.

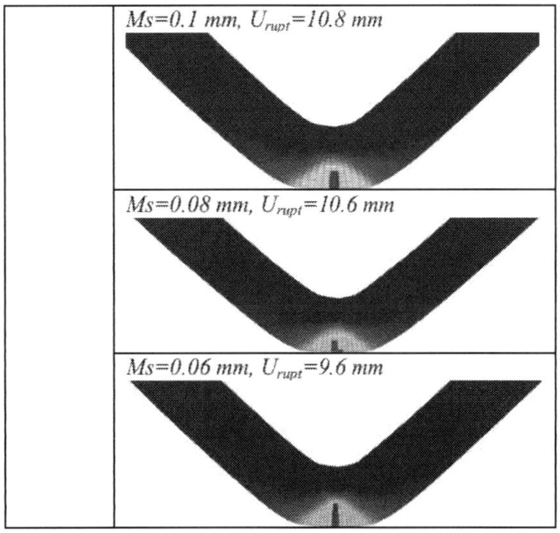

Figure 7. Damage isovalues of a ruptured DP1000 specimen in the framework of the nonlocal continua for different Ms

However, we observe a slight difference of the three results in the softening phase. This difference can be explained by the need to introduce a micromorphic variable associated with isotropic hardening in order to regularize this stage.

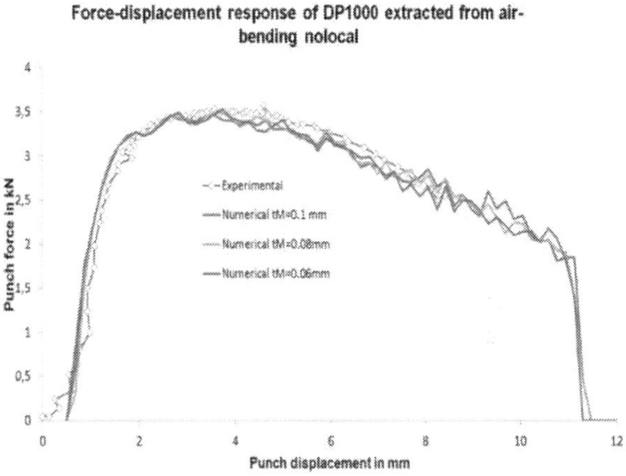

Figure 8. Experimental and numerical force-displacement curves of DP1000 for air-bending for the non-local model.

Acknowledgements

The authors acknowledge the financial support of the National Research Agency (ANR) as well as the contribution of their partners from LRM/UTC and CdM/Ecole des Mines in the project "MICROMORFING". Also Prof. M. François (UTT/LASMIS) and Ph.D C.Bao (UTT/LASMIS) are gratefully acknowledged for the experimental work.

References

1. J. Wang, J. Chen, M. Li, ACTA MECHANICA SNICA, Vol.20, No.6, December 2004.
2. K. Saanouni, ISTE, London, UK and John Wiley & Sons, Inc., Hoboken, NJ, USA, 2012.
3. M. Hamed, Ph.D. thesis, Université de Technologie de Troyes, 2012.
4. T. Belytscho, W.K. Liu, B. Moran, John Wiley & Sons Ltd, The Atrium, Southern Gate, Chichester, West Sussex, England , 2000.
5. C. Labergere, B. Guelorget, M. François, INTERNATIONAL JOURNAL OF SOLIDS AND STRUCTURES 51(S 23–24), pp. 3944–3961, 2014.
6. C. Bao, M. François, L. Le Joncour, Applied Mechanics and Materials Vol. 784, pp. 514-519, 2015.
7. K. Saanouni, M. Hamed, INTERNATIONAL JOURNAL OF SOLIDS AND STRUCTURES 50(S 14–15), pp. 2289–2309, 2013.
8. S. Forest, ASCE journal of engineering Mechanics, vol 135, pp. 177-131, 2009.
9. Simo J.C., Hughes T.J.R, Springer Verlag, New York, 1998.
10. C. Soyarslan, M. Malekipour Gharbi, A.E. Tekkaya, International Journal of Solids and Structures 49 (2012), pp. 1608–49 (13):1608-1626, 2012.
11. M. Gharbi, C. Labergere, H. Badreddine, C. Soyarslan, A. Weinrich, M. Hermes., S. Chatti, H. Sulaiman, K. Saanouni , A.E. Tekkaya., 10th International conference on technology of plasticity, pp. 25-30, September 2011.
12. Saanouni K, Forster C, Hatira FB, International Journal of Damage Mechanics 3 (2):140-169, 1994.
13. Saanouni K, Nesnas K, Hammi Y, International Journal of Damage Mechanics 9 (3):196-240, 2000.
14. Saanouni K, Cherouat A, Hammi Y, Revue européenne des éléments finis 2-3-4:327-351, 2001.
15. Saanouni K, Badreddine H, Ajmal M, Journal of engineering materials and technology 130 (2):02102201-02102211, 2008.
16. Saanouni K, Hammi Y, Elsevier, New York, 2000
17. Saanouni K, Chaboche J, Numerical and Computational Methods in Comprehensive Structural Integrity 3, 2003
18. Labergère C, Rassineux AA, Saanouni KK, International journal of material forming 4 (3):317-328, 2011
19. Issa M, Labergère C, Saanouni K, Rassineux, CIRP Journal of Manufacturing Science and Technology 5 (3):175-195, 2012
20. Badreddine H, Saanouni K, Dogui A, International Journal of Plasticity 26 (11):1541-1575, 2010.
21. Badreddine H, Saanouni K, Nguyen TD, International Journal of Solids and Structures 63:11-31, 2015.
22. Eringen AC, *Microcontinuum field theories: I,* Foundations and Solids. Springer Verlag, New York, 1999.
23. Eringen AC, *Nonlocal continuum field theories,* Springer Science & Business Media, 2002.
24. Forest S, *Milieux continus généralisés et matériaux hétérogènes,* Presses des MINES, 2006.
25. Eringen AC, Suhubi ES, International Journal of Engineering Science 2 (2):189-203, 1964.
26. 7. Mindlin RD, Arch Rational Mech Anal 16 (1):51-78, 1964.
27. Forest S, Aifantis EC, International Journal of Solids and Structures 47 (25–26):3367-3376, 2010.

Nonlocal constitutive equations of elasto-visco-plasticity coupled with damage and temperature

Weijie Liu[1,2], Khemais Saanouni[2,a], Carl Labergère[2], Houssem Badreddine[2] Samuel Forest[3] and Ping Hu[1]

[1]*Faculty of Vehicle Engineering and Mechanics, Dalian University of Technology, 2 Rue Linggong, 116024 Dalian, China*
[2]*ICD/LASMIS, UMR-CNRS 6281, University of Technology of Troyes, 12 Rue Marie Curie BP2060, 10000 Troyes Cedex, France*
[3]*Centre des matériaux, UMR-CNRS 7633, MINES ParisTech, BP 87, 91003 Evry Cedex, France*

Abstract. In this paper, the nonlocal anisothermal elasto-visco-plastic constitutive equations strongly coupled with ductile isotropic damage, nonlinear isotropic hardening and kinematic hardening are developed to model the material behaviour under finite strain. The new micromorphic variable of damage is introduced into the principle of virtual power and new additional balance equations are obtained. Thermodynamically-consistent nonlocal constitutive equations are then deduced. The evolution equations are deduced from the generalized normality rule for the Norton-Hoff visco-plastic potential. This model is used to simulate various material responses under different velocities at high temperature. The micromorphic parameters of damage: micromorphic density and H moduli are studied to examine the effects of micromorphic damage. Biaxial tension is performed to make a comparison between the local damage model and the micromorphic damage model.

1 Introduction

Nowadays, economic and environmental demands are forcing the industrialists to design lightweight mechanical components. The large-scale complex metallic components are manufactured using hot or cold forming processes under relatively extreme conditions, such as complex multi-axial loading, high strain rate, and high temperature. Even for the experienced engineers, it is hardly possible to estimate the finite deformations.

The fully local Cauchy constitutive equations have been well established to model the induced material softening behaviour due to thermal, damage and other microstructure-dependent phenomena. However, the solutions of the fully local constitutive equations are highly sensitive to the space and time discretization. The natural way to overcome this drawback is to account for an appropriate neighbourhood effect of each material point by introducing some characteristic lengths, representative of the materials' microstructures, into constitutive equations[1].

The generalized continuum mechanics makes possible the straightforward introduction of characteristic lengths into the constitutive equations of materials with microstructure. During the 1960s, motivated by the original works of the Cosserat[2] brothers, many theoretical works have been devoted to the mechanics of generalized continua to solve various problems in mechanics of solids and fluids left without satisfactory solution in the framework of classical local continuum mechanics as found in Mindlin[3-5]; Eringen and Suhubi[6]; Eringen[7].

As summarised by Forest[8, 9], all these generalized continuum theories, which are still based on the assumption of local action (Truesdell and Noll[10]), can be classified into two classes: (i) the higher grade continua and (ii) the higher order continua. Higher grade continua are those based on higher order spatial derivatives of the displacement field. While, higher order continua are based on the introduction of additional degrees of freedom[11]. A third class of generalized continuum theories is the so called strictly nonlocal continuum field theories, well summarized by Eringen[12] where a unified foundation of the basic field equations are presented and are not based on the principle of the local action.

The micromorphic theory was proposed by (Eringen and Suhubi[6]; Mindlin[3]). It envisions a material body as a continuous collection of deformable particles; each possesses finite size and inner structure and requires supplying additional degrees of freedom (dof) to a material point[13]. On the other hand, classical continuum mechanics considers a material body as a continuous collection of material points, each with infinitesimal size and no inner structure and usually needs displacements as dofs. The purpose to go beyond the classical continuum mechanics is to take into account the microstructure of the material in question while still keeping the advantages of continuum theory. The micromorphic approach can be applied to any macroscopic quantity in order to introduce a character length scale in the original classical continuum model in a systematic way, as presented by (Forest[9]). From the comparison between nonlocal and micromorphic theories, Forest and Aifantis[14] concluded

[a] Corresponding author: khemais.saanouni@utt.fr

that when the micromorphic variable remains as close as possible to the plastic strain, the micromorphic model reduces to the strain gradient theory.

The main goal of this paper is to develop and examine the generalized anisothermal elasto-visco-plastic constitutive equations strongly coupled with ductile damage, nonlinear isotropic and kinematic hardening. The effects on the thermal viscoplasticity from the micromorphic damage are examined from the parametrical study and biaxial tension test.

2 Theoretical aspects

In this section, a set of fully coupled constitutive equations will be derived from the state and dissipation potentials in the form of elasto-visco-plasticity and anisothermal formulations, coupled with the local isotropic and kinematic hardening, local isotropic damage, micromorphic isotropic and kinematic hardening, and micromorphic damage, under the framework of thermodynamic irreversible processes with state variables.

2.1 The Extended principle of virtual power

The generalized virtual power of internal forces is extended by the micromorphic damage, isotropic and kinematic hardenings, and their first gradients:

$$P_{\text{int}}\left(\vec{u}^*, \breve{d}^*\right) = -\int_V \left(\underline{\sigma} : \vec{\nabla}\dot{\vec{u}}^* + \breve{Y}\dot{\breve{d}}^* + \breve{\vec{Y}} \cdot \vec{\nabla}\dot{\breve{d}}^* \right) dV \qquad (1)$$

where, $\underline{\sigma}$ is the Cauchy stress tensor. $\breve{Y}$ and $\breve{\vec{Y}}$ are the stress-like variables with respect to micromorphic damage $\breve{d}$ and its first gradient respectively.

The extended virtual power of external forces can be written in the additive form: classical local contributions and additional micromorphic contributions:

$$\begin{aligned} P_{ext} &= \rho\int_V \left(\vec{f}^u \cdot \dot{\vec{u}}^*\right) dV + \int_\Gamma \left(\vec{F}^u \cdot \dot{\vec{u}}^*\right) dS \\ &+ \rho\int_V \left(f^d \dot{\breve{d}}^* + \vec{f}^{gd} \cdot \vec{\nabla}\dot{\breve{d}}^* \right) dV + \int_\Gamma \left(F^d \dot{\breve{d}}^* \right) dS \end{aligned} \qquad (2)$$

where, the small f indicates the generalized body force associated with the displacement, micromorphic degree of freedoms and their first gradients. The capital F is the generalized contact surface forces of displacement and micromorphic dofs respectively.

In the same way, the generalized virtual power of inertia forces is given below:

$$P_a = \rho\int_V \left(\ddot{\vec{u}} \cdot \dot{\vec{u}}^* + \zeta_d \ddot{\breve{d}}\dot{\breve{d}}^* \right) dV \qquad (3)$$

where, ζ_d is a scale factor which maps the local density to the micromorphic level[15].

Applying the generalized virtual power for any given kinematical admissible fields, we may obtain the classical equilibrium equation and three additional balance equations:

$$\begin{cases} \vec{\nabla} \cdot \underline{\sigma} + \rho\vec{f}^u = \rho\ddot{\vec{u}} & in\ \Omega \\ \underline{\sigma} \cdot \vec{n} = \vec{F}^u & on\ \Gamma \end{cases} \qquad (4)$$

$$\begin{cases} \left(\vec{\nabla} \cdot \breve{\vec{Y}} + \breve{Y}\right) + \rho\left(f^d - \vec{\nabla} \cdot \vec{f}^{gd}\right) = \rho\zeta_D \ddot{\breve{d}} & in\ \Omega \\ \left(\breve{\vec{Y}} - \rho\vec{f}^{gd}\right) \cdot \vec{n} = F^d & on\ \Gamma \end{cases} \qquad (5)$$

2.2 Thermodynamic irreversible processes

We assume that the energy conservation holds for the micromorphic continuum of the first order. Using the kinetic energy theorem, the local form of the rate of specific internal energy (1st principle of thermodynamics) is enhanced with the additional micromorphic dofs:

$$\rho\dot{e} = \underline{\sigma} : \dot{\underline{\varepsilon}} + \left(\rho\varsigma - \vec{\nabla} \cdot \vec{q}\right) + \left(\breve{Y}\dot{\breve{d}} + \breve{\vec{Y}} \cdot \vec{\nabla}\dot{\breve{d}}\right) \qquad (6)$$

where, ς is the volume heat generation. $\vec{q}$ is the heat flux vector across the boundary surface.

According to the well-known Clausius-Duhem inequality and the relations between the specific internal energy and specific Helmholtz free energy, we may have the following inequality:

$$\underline{\sigma} : \dot{\underline{\varepsilon}} - \rho\left(s\dot{T} + \dot{\psi}\right) - \frac{\vec{q}}{T} \cdot \vec{\nabla}T + \left(\breve{Y}\dot{\breve{d}} + \breve{\vec{Y}} \cdot \vec{\nabla}\dot{\breve{d}}\right) \geq 0 \qquad (7)$$

where, s is the specific entropy. T represents the temperature.

As in local continuum mechanics, the state potential is assumed to be a closed concave function of temperature and convex function of all the classical local state variables elastic strain, isotropic and kinematic hardening, and local damage as well as the micromorphic damage, micromorphic isotropic and kinematic hardening, and their first gradients. With the assumption of additive decomposition of total strain rate, the Clausius-Duhem inequality Eq.(7) can be expressed as:

$$\begin{aligned} &\left(\underline{\sigma} - \rho\frac{\partial\psi}{\partial\underline{\varepsilon}^e}\right) : \dot{\underline{\varepsilon}}^e + \rho\left(s + \frac{\partial\psi}{\partial T}\right)\dot{T} + \left(\breve{Y} - \rho\frac{\partial\psi}{\partial\breve{d}}\right)\dot{\breve{d}} \\ &+ \left(\breve{\vec{Y}} - \rho\frac{\partial\psi}{\partial\vec{\nabla}\breve{d}}\right) \cdot \vec{\nabla}\dot{\breve{d}} + \underline{\sigma} : \dot{\underline{\varepsilon}}^{vp} \\ &- \left(\rho\frac{\partial\psi}{\partial d}\dot{d} + \rho\frac{\partial\psi}{\partial r}\dot{r} + \rho\frac{\partial\psi}{\partial\underline{\alpha}} : \dot{\underline{\alpha}}\right) - \frac{\vec{q}}{T} \cdot \vec{\nabla}T \geq 0 \end{aligned} \qquad (8)$$

Generally, if we assume that overall the formulas $\left(\underline{\sigma} - \rho\frac{\partial\psi}{\partial\underline{\varepsilon}^e}\right)$, $\left(s + \frac{\partial\psi}{\partial T}\right)$, $\left(\breve{Y} - \rho\frac{\partial\psi}{\partial\breve{d}}\right)$ and $\left(\breve{\vec{Y}} - \rho\frac{\partial\psi}{\partial\vec{\nabla}\breve{d}}\right)$ do not depend on their rates respectively and the micromorphic variables do not dissipate, the following state relations and pure local residual dissipation are obtained:

- The classical local state relations:

$$\begin{aligned} &\underline{\sigma} = \rho\frac{\partial\psi}{\partial\underline{\varepsilon}^e} \quad s = -\frac{\partial\psi}{\partial T} \\ &\underline{X} = \rho\frac{\partial\psi}{\partial\underline{\alpha}} \quad R = \rho\frac{\partial\psi}{\partial r} \quad Y = -\rho\frac{\partial\psi}{\partial d} \end{aligned} \qquad (9)$$

- The micromorphic (nonlocal) state relations:

$$\breve{Y} = \rho\frac{\partial\psi}{\partial\breve{d}} \quad \breve{\vec{Y}} = \rho\frac{\partial\psi}{\partial\vec{\nabla}\breve{d}} \qquad (10)$$

- The residual dissipation:

$$\phi_{in} = \underline{\sigma} : \underline{\dot{\varepsilon}}^{vp} - \underline{X} : \underline{\dot{\alpha}} - R\dot{r} + Y\dot{d} - \frac{\vec{q}}{T} \cdot \vec{\nabla} T \geq 0 \quad (11)$$

2.3 The heat equation

Using the local form of 1st principle of thermodynamic Eq.(6), the state relations Eqs.(9)-(10) and the relations of internal energy and Helmholtz free energy, we may have the following enhanced heat equation:

$$\underline{\sigma} : \underline{\dot{\varepsilon}}^{vp} - \underline{X} : \underline{\dot{\alpha}} - R\dot{r} + Y\dot{d} + \left(\rho\varsigma + \vec{\nabla} \cdot \vec{q}\right) - \rho\dot{s}T = 0 \quad (12)$$

Substituting the state relation of entropy into the above equation:

$$\underline{\sigma} : \underline{\dot{\varepsilon}}^{IR} + Y\dot{d} - R\dot{r} - \underline{X} : \underline{\dot{\alpha}} + \left(\rho\varsigma - \vec{\nabla} \cdot \vec{q}\right) - \rho T \frac{\partial(s)}{\partial T} \dot{T}$$
$$+T \begin{pmatrix} \frac{\partial \underline{\sigma}}{\partial T} : \underline{\dot{\varepsilon}}^{e} - \frac{\partial Y}{\partial T} \dot{d} + \frac{\partial R}{\partial T} \dot{r} + \frac{\partial \underline{X}}{\partial T} : \underline{\dot{\alpha}} \\ + \frac{\partial \breve{Y}}{\partial T} \dot{\breve{d}} + \frac{\partial \vec{\breve{Y}}}{\partial T} \cdot \vec{\nabla} \dot{\breve{d}} \end{pmatrix} = 0 \quad (13)$$

If we assign:

$$R_{pl} = \begin{pmatrix} \frac{\partial \underline{\sigma}}{\partial T} : \underline{\dot{\varepsilon}}^{e} - \frac{\partial Y}{\partial T} \dot{d} + \frac{\partial R}{\partial T} \dot{r} + \frac{\partial \underline{X}}{\partial T} : \underline{\dot{\alpha}} \\ + \frac{\partial \breve{Y}}{\partial T} \dot{\breve{d}} + \frac{\partial \vec{\breve{Y}}}{\partial T} \cdot \vec{\nabla} \dot{\breve{d}} \end{pmatrix} \quad (14)$$

The final heat equation can be rewritten as:

$$\rho C_v \dot{T} - \rho\varsigma + \vec{\nabla} \cdot \vec{q} - \phi_{in} - R_{pl} T = 0 \quad (15)$$

where, C_v is the specific heat capacity for constant volume.

Clearly, R_{pl} comprises the additional contributions to the heat equation from the micromorphic variables.

3 An example of micromorphic constitutive and evolution equations

In this section, a specific state potential is constructed for the anisothermal isotropic elasto-visco-plasticity model fully coupled with isotropic local damage, isotropic and kinematic hardening and isotropic micromorphic damage. The damaged state variables are applied to the local state variables according to the assumption of the equivalence of total energy developed in references[1, 11, 16-18].

3.1 Micromorphic state potential and associated state relations

For simplification, we also postulate that only the local damage is coupled with the micromorphic damage. With these assumptions, a quadratic the Helmholtz free energy can be expressed as:

$$\begin{aligned} \rho\psi &= \frac{1}{2}\left(\underline{\tilde{\varepsilon}}^{e} : \underline{\underline{\Lambda}} : \underline{\tilde{\varepsilon}}^{e} + \frac{2}{3} C \underline{\tilde{\alpha}} : \underline{\tilde{\alpha}} + Q\tilde{r}^{2}\right) \\ &+ \frac{1}{2}\breve{H}\left(d - \breve{d}\right)^{2} + \frac{1}{2}\breve{H}^{g} \vec{\nabla}\breve{d} \cdot \vec{\nabla}\breve{d} \\ &- \rho C_v \frac{\left(T - T_0\right)^{2}}{2T_0} - \left(T - T_0\right) P \underline{1} : \underline{\tilde{\varepsilon}}^{e} \end{aligned} \quad (16)$$

where, $\underline{\underline{\Lambda}} = \lambda_e \underline{1} \otimes \underline{1} + 2\mu_e \underline{\underline{1}}$ is the fourth order elastic tensor expressed with Lamé's parameters. C and Q are the macro moduli of local kinematic and isotropic hardening respectively. $\breve{H}$ and $\breve{H}^{g}$ are the moduli of coupled local and micromorphic damage and its first gradient respectively. P is a scalar modulus for isotropic thermal expansion.

Using the state relations Eqs.(9)-(10), we may have the following expressed local stress-like variables:

$$\begin{aligned} \underline{\sigma} &= \underline{\underline{\tilde{\Lambda}}} : \underline{\varepsilon}^{e} - \left(T - T_0\right) \tilde{P} \underline{1} \\ s &= \frac{C_v}{T_0}\left(T - T_0\right) + \frac{1}{\rho} \tilde{P} \underline{1} : \underline{\varepsilon}^{e} \\ \underline{X} &= \frac{2}{3} \tilde{C} \underline{\alpha} \qquad R = \tilde{Q} r \\ Y &= \frac{1}{2}\left(\underline{\varepsilon}^{e} : \underline{\underline{\Lambda}} : \underline{\varepsilon}^{e} + \frac{2}{3} C \underline{\alpha} : \underline{\alpha} + Q r^{2}\right) \\ &- \breve{H}\left(d - \breve{d}\right) - \frac{1}{2\sqrt{1-d}}\left(T - T_0\right) P \underline{1} : \underline{\varepsilon}^{e} \end{aligned} \quad (17)$$

And nonlocal stress-like variables:

$$\breve{Y} = -\breve{H}\left(d - \breve{d}\right) \qquad \vec{\breve{Y}} = \breve{H}^{g} \vec{\nabla}\breve{d} \quad (18)$$

It's clear that if the micromorphic effects are neglected, $\breve{H} = \breve{H}^{g} = 0$, then all the micromorphic state variables given by Eq.(18) vanish leading to zero contribution of the nonlocal terms. In that case the classical fully coupled local state relations originally given in Saanouni et al[1, 16]. are recovered.

3.2 Dissipation analysis and evolution equations

For the case of the single potential, fully isotropic damageable thermo-elasto-visco-plasticity with ductile damage, we postulate that the overall viscoplastic potential $\varphi^{*}\left(\underline{\tilde{\sigma}}, \underline{\tilde{X}}, \tilde{R}; T\right)$ is additively decomposed[1] into two contributions representing respectively the Norton-Hoff viscoplastic flow $\varphi_{vp}^{*}\left(\underline{\tilde{\sigma}}, \underline{\tilde{X}}, \tilde{R}; T\right)$ and the damage potential $\varphi_{d}^{*}\left(d; T\right)$:

$$\begin{aligned} \varphi_{vp}^{*} &= \frac{K^{v}(T)}{m_v(T)+1} \left\langle \frac{F_p\left(\underline{\tilde{\sigma}}, \underline{\tilde{X}}, \tilde{R}; T\right)}{K^{v}(T)} \right\rangle^{m_v(T)+1} \\ \varphi_{d}^{*} &= \frac{S_d(T)}{\left(s_d(T)+1\right)\left(1-d\right)^{\beta_d(T)}} \left\langle \frac{Y - Y_0(T)}{S_d(T)} \right\rangle^{\left(s_d(T)+1\right)} \dot{p} \end{aligned} \quad (19)$$

where, K^{v} and m_v are two constants which are characteristic of material viscosity and $F_p\left(\underline{\tilde{\sigma}}, \underline{\tilde{X}}, \tilde{R}; T\right)$ is expressed in term of fully isotropic case:

$$F_p = f_p + \frac{3}{4}\frac{a\underline{X}:\underline{X}}{C(1-d)} + \frac{1}{2}\frac{bR^2}{Q(1-d)} - \frac{1}{3}aC\underline{\tilde{\alpha}}:\underline{\tilde{\alpha}} - \frac{1}{2}bQ\tilde{r}^2 \quad (20)$$

with $f_p = \|\underline{\tilde{\sigma}} - \underline{\tilde{X}}\| - \tilde{R} - \sigma_y$ is the classical yield function.

$\dot{p}$ is the accumulated viscoplastic strain rate.

The evolution equations deduced from these potentials, using the generalized normality rule are:

- Viscoplastic strain rate:

$$\underline{D}^{vp} = \left\langle \frac{f_p}{K^v} \right\rangle^{m_v} \frac{\partial f_p}{\partial \underline{\sigma}} = \dot{\lambda}_{vp}\underline{\tilde{n}}_f \quad (21)$$

where, $\underline{\tilde{n}}_f = \frac{\partial f_p}{\partial \underline{\sigma}} = \frac{3}{2}\frac{1}{\sqrt{1-d}}\frac{(\underline{\sigma}^{dev} - \underline{X})}{\|\underline{\sigma}^{dev} - \underline{X}\|}$ and the multiplier $\dot{\lambda}_{vp}$, depending on the form of the potential, is given as follow:

$$\dot{\lambda}_{vp} = \left\langle \frac{f_p}{K^v} \right\rangle^{m_v} \quad (22)$$

- Kinematic hardening strain rate:

$$\underline{\dot{\alpha}} = -\left\langle \frac{f_p}{K^v} \right\rangle^{m_v} \frac{\partial F_p}{\partial \underline{X}} = \dot{\lambda}_{vp}\left(\underline{\tilde{n}}_f - a\underline{\alpha}\right) \quad (23)$$

- Isotropic hardening strain rate:

$$\dot{r} = -\left\langle \frac{f_p}{K^v} \right\rangle^{m_v} \frac{\partial F_p}{\partial R} = \frac{\dot{\lambda}_{vp}}{\sqrt{1-d^\gamma}}(1 - b\tilde{r}) \quad (24)$$

- Isotropic ductile damage rate:

$$\dot{d} = \frac{\dot{p}}{(1-d)^{\beta_d}}\left\langle \frac{Y - Y_0}{S_d} \right\rangle^{s_d} = \dot{p}\hat{Y} \quad (25)$$

From Eq.(22), we define the equation for the actual viscoplasticity yield surface in the form:

$$f_{vp} = f_p - K^v\left(\dot{\lambda}_{vp}\right)^{\frac{1}{m_v}} = f_p - \sigma^v = 0 \quad (26)$$

As expected, viscous stress is very dependent on the form of the viscoplastic potential and is defined as follow:

$$\sigma^v = K^v\left(\dot{\lambda}_{vp}\right)^{\frac{1}{m_v}} \quad (27)$$

4 Numerical implementation

In this section, we perform the parametrical study of the proposed model of anisothermal damaged isotropic elasto-visco-plasticity fully coupled isotropic, kinematic hardening and micromorphic isotropic damage. The material property under the reference temperature (20°C) is given in the below table.

Table 1. Material parameters under reference temperature.

Name	Symbol	Value
Young's modulus	$E(T)$	205000 MPa
Poisson ratio	v	0.3
Yield stress	$\sigma_y(T)$	792 MPa
Isotropic hardening modulus	$Q(T)$	320 MPa
	b	0.6
Kinematic hardening modulus	$C(T)$	5000 MPa
	a	20
Damage factors	$S_d(T)$	20
	s_d	2
	β_d	1
	Y_0	0
	γ	1.8
Visco-plasticity factors	K^v	793.8MPa
	m_v	2.7

As usual the material moduli are temperature dependent, the following relations are assumed:

$$P(T) = P_0\left[1 - G^\alpha(T)\right] \quad (28)$$

where, $G(T) = \frac{T - T_0}{T_f - T_0} \geq 0$ and T_0 is the reference temperature. T_f is the fusion temperature for the specific material. α is a coefficient. For Young's modulus $\alpha = 4$; for capital S of damage factor $\alpha = 1.08$ and for other modules it equals to 1.03.

4.1 Anisothermal viscoplasticity with local damage for a material point

In this section, the parametrical studies of temperature, load velocity, micromorphic density and its moduli are performed in a material point to examine the response of the proposed anisothermal elasto-visco-plastic model coupled with local and micromorphic isotropic damage, local isotropic hardening and kinematic hardening, .

Figure 1 illustrates the thermal effects on the evolutions of equivalent stress based on the proposed constitutive equations coupled with local damage and local mixed hardening. The accumulated plastic strain (max: 0.652; min: 0.592) increases and the tensile stress (max: 1108.72MPa; min: 416.91MPa) decreases steadily as temperature increases.

Figure 2 shows the four types of evolutions of local damage under different strain rates (velocity) (from 1.0E-04s^{-1} to 0.1s^{-1}) under the initial temperature 1000°C. Clearly, the evolution of local damage is accelerated by the increasing strain rate. The fracture plastic strains are 0.652, 0.648, 0.636 and 0.605 respectively.

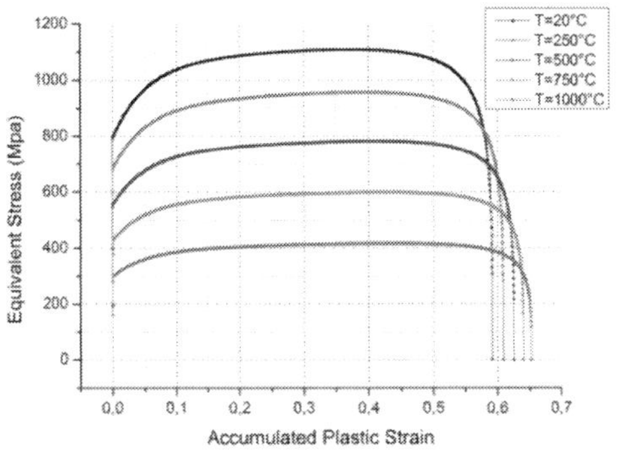

Figure 1. Stress-strain curves for different temperatures with the same strain rate 1.0E-04s^{-1}.

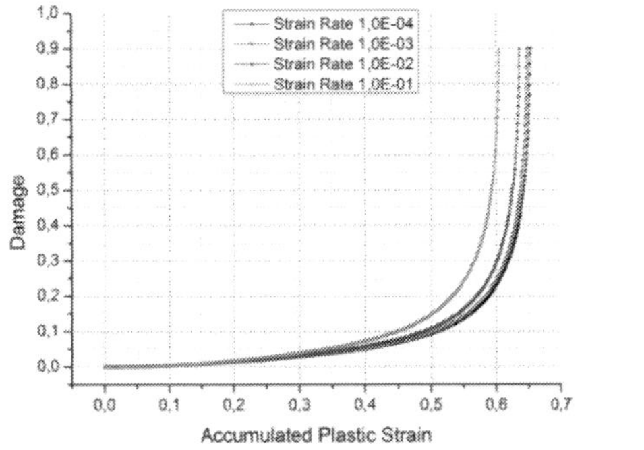

Figure 2. Evolution of damage for different loading velocity with the same temperature 1000°C.

4.2 Biaxial tension for local damage with FE

The geometric size and boundary conditions of the sample are shown in Figure 3. For simplifications, three different initial temperatures (20°C, 500°C and 1000°C), and four velocities(2.0E-04mm/s, 2.0E-05mm/s, 2.0E-06mm/s and 2.0E-07mm/s) are performed on a single finite element in this example. The approximate strain rates are 1.0E-01s^{-1}, 1.0E-02s^{-1}, 1.0E-03s^{-1} and 1.0E-04s^{-1} respectively.

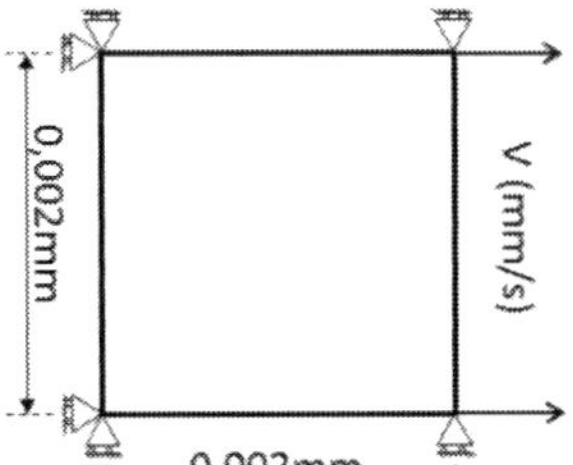

Figure 3. Scheme of tension test.

Figure 4 summarizes the material responses of the local damage model, in terms of equivalent stress – accumulated plastic strain curves, for 7 cases including three temperatures and three velocities. From this figure the viscoeffects are very clear: when the temperature increases the ductile plasticity is extended and the final fracture plastic strain is also enlarged (0.591 in 20°C, 0.626 in 500°C and 0.653 in 1000°C with velocity 2.0E-07mm/s). However, both the initial yield stress (792.00MPa, 547.09MPa and 281.12MPa) and hardening effects are decreased. The opposite tendency of the fracture plastic strain is observed for the raise of loading velocity at each temperature, while the equivalent stress after yield point shows significant differences (31.24MPa for 500°C and 66.85MPa for 1000°C). In Figure 5, it illustrates the different evolutions of local damage for the same seven cases. It's clear that the evolution of damage is delayed by the raise of the temperature and the decelerated load velocity.

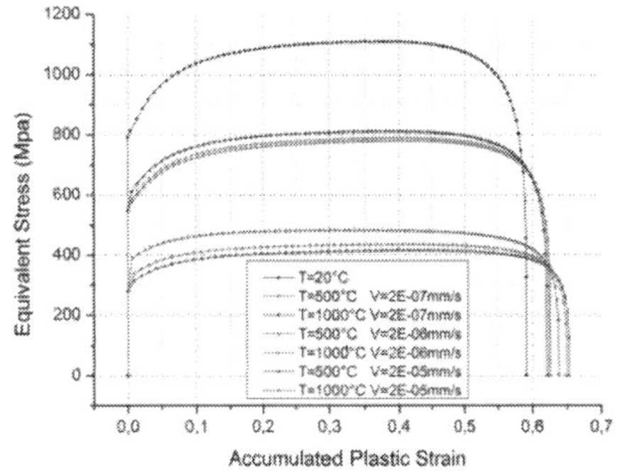

Figure 4. Stress-strain curves for different temperature and load velocity coupled with local damage.

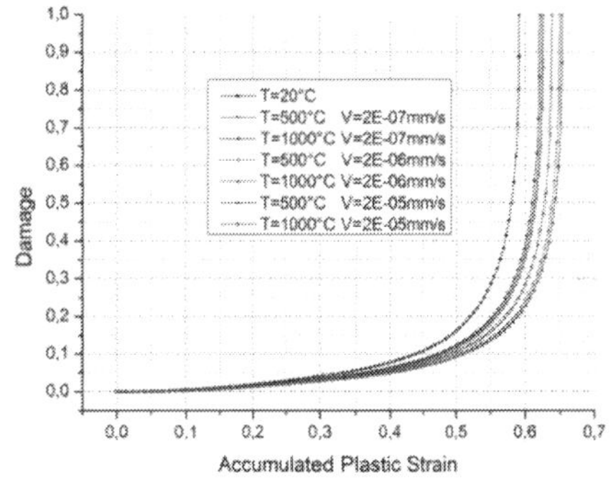

Figure 5. Evolution of local damage for different temperature and load velocity.

4.3 Choices of micromorphic parameters with FE

Let us consider the choices of micromorphic parameters including micromorphic density, the H moduli for the coupled local damage and micromorphic damage and the gradient of micromorphic damage. In this section, all the simulations are performed under temperature 500°C and load velocity 2.0E-05mm/s.

In Figure 6, it illustrates the stress-strain curves and the evolutions of damage for different micromorphic density (0.1~100 times of local density) with given H modules equal to 1000MPa. From the theoretical formula Eq.(5), we can see that the micromorphic density will be used to calculate the accelerate of micromorphic damage at each time step, which determines the incremental micromorphic damage and affects the local damage. In this figure, it clearly represents that the larger micromorphic density evidently slows the evolutions of local damage and enlarged the ductile plasticity of the material.

Figure 7 represents the effects of different choices of H modules (1.0~1000MPa) for the stress-strain curves and the evolution of local damage. From the state relations of stress-like variable with respect to local damage Eq.(17), we know that the evolution of the local damage is affected by the micromorphic damage through the H modules. In this figure, it describes that the larger H modules the slower evolution of local damage. We can also find this character from Eq.(17), the term of coupled local and micromorphic damage has a negative sign, which induces the reduction of the stress-like variable Y of local damage.

Compared with the curves for local damage, both of the figures imply that when the micromorphic density or the H modules approach to zero, the material response becomes closer and closer to the one from local damage.

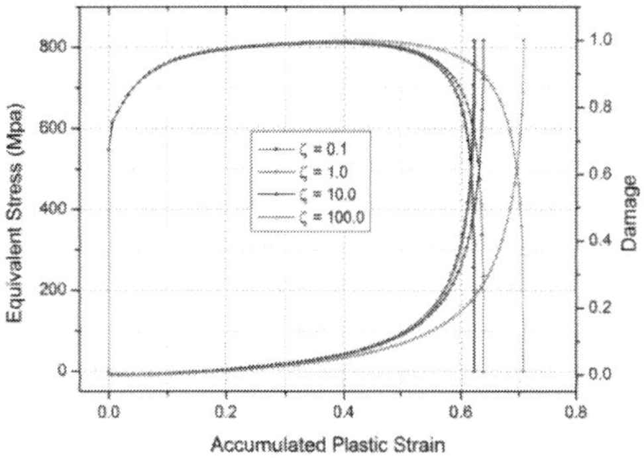

Figure 6. Stress-strain curves and evolution of damage for some values of micromorphic density.

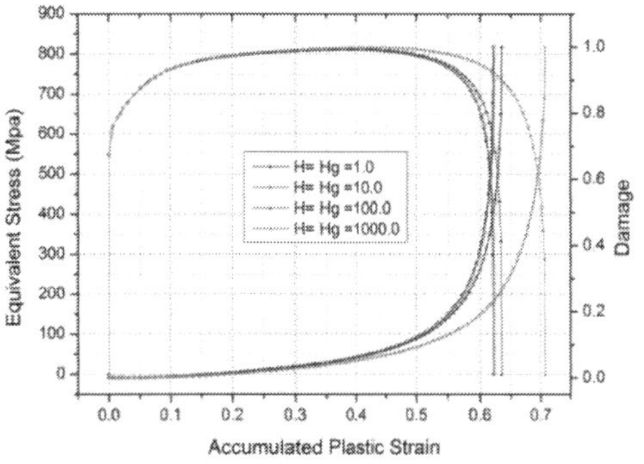

Figure 7. Stress-strain curves and evolution of damage for some values of micromorphic H modules.

4.4 Biaxial tension for micromorphic damage with FE

After the analysis of parameters, we have a conclusion that the smaller micromorphic density and H modules gives the more approximate result to the ones from the local damage formulas. In absence of any experimental indication, the micromorphic parameters: density 100 times, H modules 1000MPa, will be considered to perform the biaxial tension test in this section.

In Figure 8 and Figure 9, it gives the stress-strain curves and the evolutions of local damage of 9 cases (three temperatures and four velocities). It's clear that, in this choice of parameters, the fracture plastic strain is significantly improved and the evolution of local damage is apparently delayed. Compared with the curves from local damage formulas Figure 4 and 5, it's apparently that the maximum tensile stresses have very little differences for the four velocities on each temperature. However, the final fracture accumulated plastic strain grows larger significantly (maximum roughly 1.556 for temperature 1000°C and load velocity 2.0E-04mm/s).

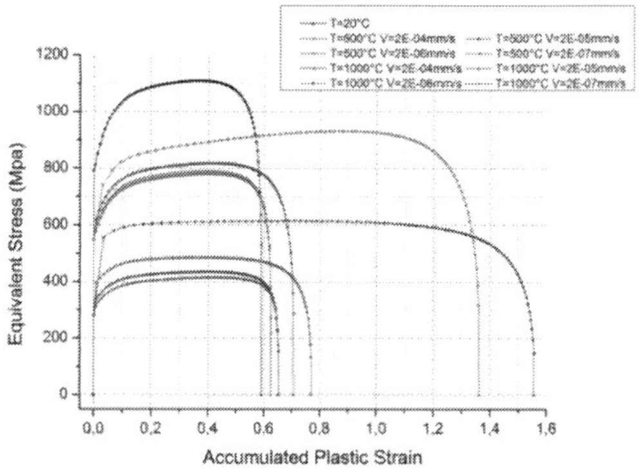

Figure 8. Stress-strain curves for different temperature and load velocity with micromorphic damage.

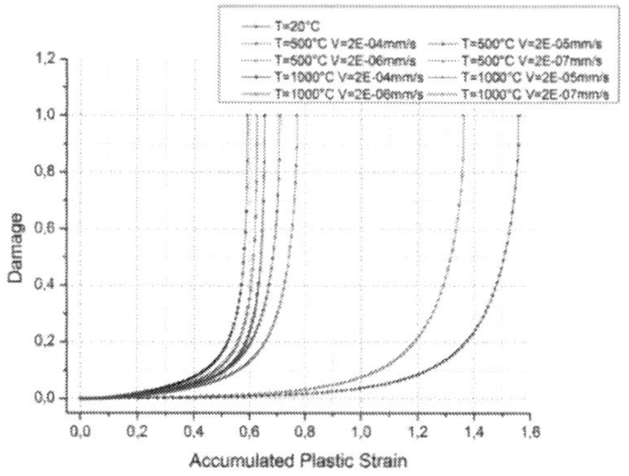

Figure 9. Evolution of damages for different temperature and load velocity with micromorphic damage.

5 Conclusions

The nonlocal anisothermal elasto-viscoplastic constitutive equations strongly coupled with ductile damage, nonlinear isotropic hardening and kinematic hardening are developed under the framework of the irreversible thermodynamic processes.

The new micromorphic damage is introduced into the principle of virtual power and a new balance equation is obtained. From the comparisons of the biaxial tension between local and micromorphic damage, we can make a conclusion that when micromorphic density and H modules approach to zero, the micromorphic effect is negligible and it degenerates to the local damage case; otherwise, the micromorphic damage significantly delays the evolution of local damage especially under cases of high temperature or high strain rate.

Acknowledgement

The financial support of ANR through the Program Micromorfing with contract ANR-14-CE07-0035-01 is fully acknowledged. The authors gratefully acknowledge financial support from China Scholarship Council.

References

1. K. Saanouni, Damage mechanics in metal forming: advanced modeling and numerical simulation. (John Wiley & Sons, 2013).
2. E. Cosserat and F. Cosserat, Reprinted by Cornell University Library. English translation by DH Delphenich (2009).
3. R. D. Mindlin, Arch. Rational Mech. Anal. 16 (1), 51-78 (1964).
4. R. D. Mindlin, International Journal of Solids and Structures 1 (4), 417-438 (1965).
5. R. D. Mindlin and N. N. Eshel, International Journal of Solids and Structures 4 (1), 109-124 (1968).
6. A. C. Eringen and E. S. Suhubi, International Journal of Engineering Science 2 (2), 189-203 (1964).
7. A. C. Eringen, Microcontinuum field theories: I. Foundations and Solids. (Springer Verlag, New York, 1999).
8. S. Forest and R. Sievert, International Journal of Solids and Structures 43 (24), 7224-7245 (2006).
9. S. Forest, Journal of Engineering Mechanics 135 (3), 117-131 (2009).
10. C. Truesdell and W. Noll, The non-linear field theories of mechanics. (Springer, 2004).
11. K. Saanouni and M. Hamed, International Journal of Solids and Structures 50 (14–15), 2289-2309 (2013).
12. A. C. Eringen, Nonlocal continuum field theories. (Springer Science & Business Media, 2002).
13. J. D. Lee and X. Wang, International Journal of Engineering Science 49 (12), 1378-1387 (2011).
14. S. Forest and E. C. Aifantis, International Journal of Solids and Structures 47 (25–26), 3367-3376 (2010).
15. D. P. Flanagan and T. Belytschko, International Journal for Numerical Methods in Engineering 17 (5), 679-706 (1981).
16. K. Saanouni, C. Forster and F. B. Hatira, International Journal of Damage Mechanics 3 (2), 140-169 (1994).
17. K. Saanouni and J. L. Chaboche, in Comprehensive Structural Integrity, edited by I. M. O. R. Karihaloo (Pergamon, Oxford, 2003), pp. 321-376.
18. H. Badreddine, K. Saanouni and A. Dogui, International Journal of Plasticity 26 (11), 1541-1575 (2010).

Modelling of flow behaviour and dynamic recrystallization during hot deformation of MS-W 1200 using the phase field framework

Jan Hiebeler[1,a], Kirill Khlopkov[1], Oleg Shchyglo[2], Thomas Pretorius[1], Ingo Steinbach[2]

[1]*ThyssenKrupp Steel Europe, Department of modelling and simulation, 47166 Duisburg, Germany*
[2]*Interdisciplinary Centre for Advanced Materials Simulation (ICAMS), Ruhr-Universität Bochum, Universitätsstr. 150, Germany*

Abstract. A new simulation environment is developed to simulate the evolution of microstructure and the corresponding flow stress during rolling. An orientation dependent crystal plasticity hardening model is coupled to grain evolution-, recovery- and recrystallization kinetics within a phase field framework. Hardening and softening kinetics are treated consecutive to differentiate between individual effects. Simulation results are compared to hot compression tests at 1373 K with a strain rate of 1 s^{-1}.

1 Introduction

In order to predict material properties after hot forming of steel, it is necessary to understand the microstructural processes during forming. The demand for microstructure simulations is driven by ecological and economic aspects. The first existing simulations were focused on empirically predicting experimental measurable parameters which characterize the material [1–3]. Following the transition towards more physical based approaches leads to an understanding of the importance of a defined microstructure. With the ability to model complex structures today the phase field method is a well-established way to approach the field of thermomechanical microstructure evolution [4–9]. However to successfully setup a simulation environment it is necessary to differentiate between single effects to weigh the influence.

In this work four major modules were coupled to model the hot forming process: I) Crystal plasticity, II) Recovery, III) Recrystallization, IV) Grain evolution.

2 Experimental section

In order to determine crystal plasticity input parameters hot compression test of MS-W 1200 steel at T = 1373 K and deformation rate $\dot{\varphi}$ = 1 s^{-1} were performed on a deformation dilatometer Bähr DIL 805 D. The decrease of the flow stress shown in Fig. 1 after the true strain reaches approx. 0.4, indicate the presence of dynamic recrystallization. Initial austenite grain size was measured by microscopic analysis after quenching and an annealing treatment at T_{anneal} = 823 K for 900 s. Resulting microstructure analysis shown in Fig. 2 lead to an average grain diameter of $\bar{d}$ = 35 µm [10].

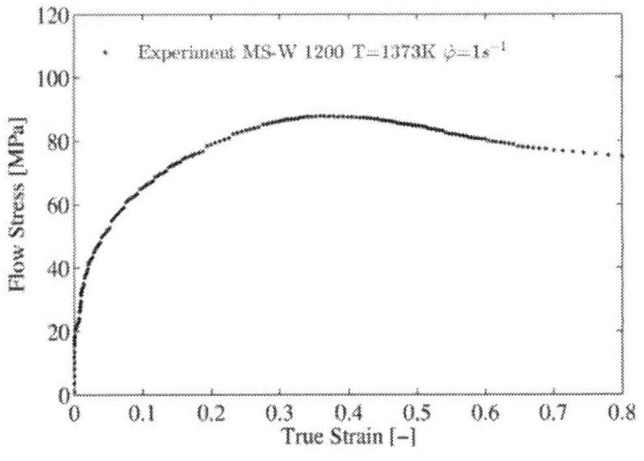

Figure 1. Results of hot compression experiment at T = 1373 K and deformation rate $\dot{\varphi}$ = 1 s^{-1}.

Figure 2. Initial grain structure with average grain size $\bar{d}$ = 35 µm [10].

[a] Corresponding author: Jan.Hiebeler@ruhr-uni-bochum.de

3 Modelling section

3.1 Crystal plasticity

The crystal plasticity kinetics is based on a model from Hutchinson [11–13] and were incorporated into the phase field framework by Borukhovich and Engels [14]. Starting with the approach by Borukhovich and Engels in this study we implemented the physically based recovery model by Verdier et al. [15] in order to differentiate between single effects. The different phenomena in question are illustrated in Fig. 3.

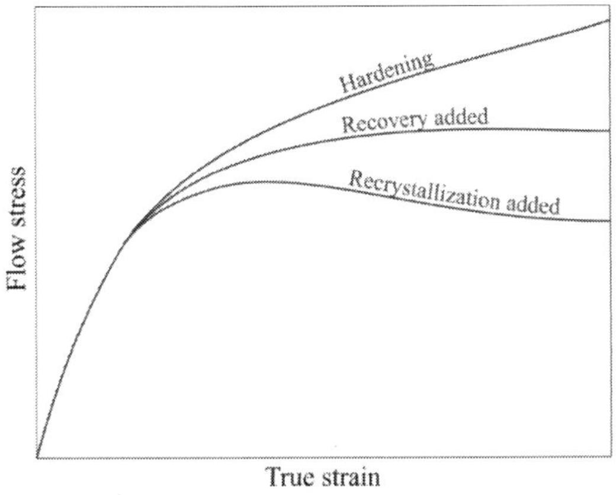

Figure 3. Differentiation of microstructure evolution effects.

In this study we use the stress dependent shear rate which is given by:

$$\dot{\gamma} = \dot{\gamma}_0 \cdot \left(\frac{\tau}{\tau_c}\right)^{Fpow}, \quad (1)$$

with the initial glide velocity $\dot{\gamma}_0$, shear stress τ, critical shear stress τ_c and hardening exponent $Fpow$. The corresponding hardening rate is given by:

$$\dot{\tau}_c = H_0 \cdot |\dot{\gamma}|, \quad (2)$$

with initial hardening parameter H_0 and shear rate $|\dot{\gamma}|$. The corresponding dislocation density is given by the Taylor relation [16]:

$$\rho = \left(\frac{\tau}{\alpha \cdot \mu \cdot b}\right)^2, \quad (3)$$

where α is the Taylor factor, μ is the shear modulus and $\boldsymbol{b}$ is the Burgers vector.

3.2 Dynamic recovery

The dynamic recovery model is based on the idea of treating the stress relaxation as an effect of thermally activated annihilation and reorganization of dislocations. The model is derived by expressing the strain rate due to the Orowan relation together with the temperature driven activation criterion for dislocation movement [15]:

$$\dot{\tau} = \frac{-E\rho b^2 v_D}{M} exp\left(\frac{U_0}{k_B T}\right) sinh\left(\frac{V_{act}}{k_B T}\right), \quad (4)$$

with the bulk modulus E, Debye frequency v_D, Taylor factor M, activation energy U_0, activation volume V_{act}, Boltzmann constant k_B and temperature T.

3.3 Dynamic recrystallization

The dynamic recrystallization model is divided into two parts: I) nucleation and II) growth.
While the classical nucleation theories consider energy as key factor for the onset of dynamic recrystallization [17–20] a different approach from [21] introduces a minimum amount of work necessary. By identifying the stored deformation energy as work the authors of [22–24] also incorporate the idea of a relaxation as dissipation part and relate nucleation to the rate of hardening. When an external applied work $\sigma_c\dot{\varepsilon}$ exceeds the internal entropy production, P_d^c, another process within the microstructure has to take place which supposed to be the nucleation of new grains with the condition:

$$\sigma_c\dot{\varepsilon} > P_d^c \quad (5)$$

To catch the exact moment of nucleation one has to analyze the hardening rate over the flow stress where the inflection point indicates the nucleation moment as shown in Fig 4. The advantage lies in the connection to the crystal plasticity kinetic which leads to an implicit orientation dependent recrystallization criterion that is not based on a fixed value but on the individual hardening of each grain within the microstructure.

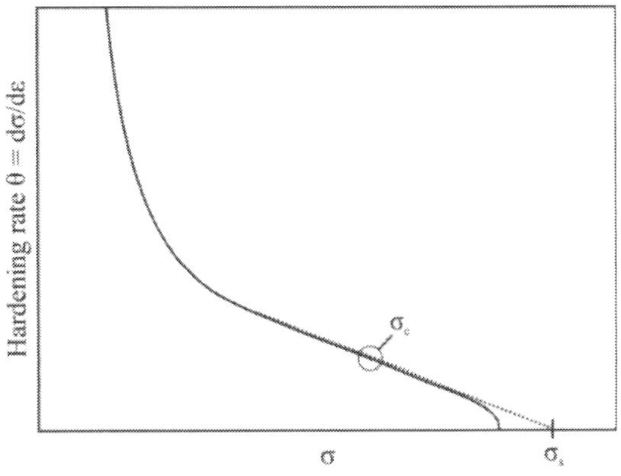

Figure 4. Schematic illustration of the nucleation criterion for dynamic recrystallization from Poliak [22].

While a new nucleus is created, it exhibits a crystal orientation. In order to take the line of least resistance the most energetically favoured orientation, related to the local stress field and elastic anisotropy, is chosen. The energy distribution for a uniaxial local stress field compared to the crystal orientation is illustrated in Fig. 5. In our test case the minimal elastic energy is established for an alignment of the crystal orientation towards [1 0 0] direction.

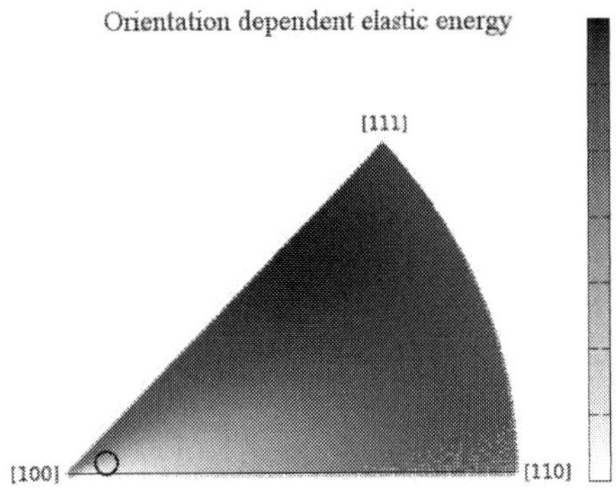

Figure 5. Different crystal orientations and the relating local energy for a uniaxial stress fields. Favoured orientation is indicated by the encircled area.

Subsequently to model grain structure evolution an approach from Beck was implemented which states that grains grow faster towards higher deformation areas [25]. This effect is schematically displayed in the following Figure.

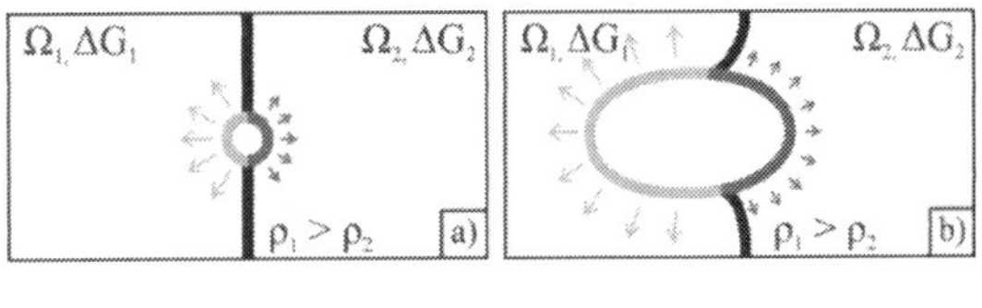

Figure 6. Schematic illustration of the stored energy based boundary migration by [25].

The grain growth velocity is determined by driving the force ΔG and mobility m:

$$v_{gb} = \Delta G \cdot m \tag{6}$$

while the driving force results from the difference of stored energies at the grain boundary:

$$\Delta G = \gamma_\alpha - \gamma_\beta, \tag{7}$$

with the grain energies γ_α and γ_β being calculated from stored deformation energy γ_d, shear modulus μ, burgers vector $\boldsymbol{b}$, local volume element V and dislocation density ρ:

$$\gamma_d = \frac{1}{2}\mu b^2 \frac{1}{V}\int_V \Delta\rho dx \tag{8}$$

4 Simulation and Results

The simulation was optimized for reasonable simulation time and resolution. The initial grain structure is generated by the OpenPhase code with normal grain size distribution and consists of fourteen grains with different orientations and average grain diameter of $\bar{d}$=35 µm [26, 27]. The simulation geometry input parameters are given in Tab. 1. The input parameters for the mechanical solver are given in Tab. 2. These parameters cover elasticity and crystal plasticity at the experimental temperature T = 1373 K and strain rate $\dot{\varphi}$ = 1 s^{-1}. Elastic constants were taken from a reference data collection of Ledbetter and Reed [28]. Plasticity related input was derived from hot compression experiments and the dislocation glide velocity was estimated by a work of Güvenc [9]. Recovery kinetics is characterized by the input parameters shown in Tab. 3. Microstructure evolution and recrystallization kinetics input is given in Tab. 4. The simulation procedure is illustrated in the flowchart in the appendix B. Within each time increment of the simulation the four key modules: I) crystal plasticity, II) recovery, III) recrystallization and IV) grain evolution are called. It is important to note that in our study there are no distinct generations of new grains but rather a continuous nucleation and growth of new grains as long as the nucleation and growth conditions are met. Whether a grain grows or vanishes, forms a specific shape or size is not predetermined, but a result of the systems energy minimization.

4.1 Microstructure evolution

In Fig. 8 the microstructure evolution during the deformation simulation is shown. The color scale indicates the hardening and the shape of the grain boundaries is illustrated by the gray lines. At first for a deformation of φ = 0.12 the initial grain structure can clearly be identified. Small inhomogeneity of hardening between different grains is observed with an increased hardening at the grain boundary areas. The deformation process continues for φ = 0.25 with increasing hardening inhomogeneity. Overall hardening is the dominant factor in the earlier state of deformation while grains get compressed in loading direction and stretched in orthogonal directions which can be clearly seen by comparing the simulation results for φ = 0.12 and φ = 0.40. For the latter it is notable that although no recrystallization can be observed, there are already several nuclei present in the microstructure, which do influence the evolution, but are not yet resolved in the figure. With further deformation and growth of the nuclei, the recrystallization and corresponding softening process can be seen at φ = 0.52. Several grains have evolved from the substructure and contribute to the grain growth process. The softening effect starts to level out during ongoing deformation due to the dynamic recrystallization and growth of already recrystallized but further deformed grains at φ = 0.65 and reaches its plateau at φ = 0.80.

In order to compare the simulation and experimental results the evolution of the flow stress is derived by a homogenization of the critical resolved shear stress over the simulation volume to model the global evolution. The comparison of flow stress evolution between the hot compression experiment and the simulation of MS-W 1200 steel at T = 1373 K and strain rate $\dot{\varphi}$ = 1 s-1 is shown in Fig. 7. It is important to note that the obtained flow stress evolution is the result of complex interaction between microstructure evolution and elasto-plastic deformation.

5 Conclusions

A consistent full-field model for hot deformation of the steel MS-W 1200 was developed to capture the effects of crystal plasticity, dynamic recovery, dynamic recrystallization and grain evolution. Incorporated into the phase field framework it is possible to keep track of the microstructure evolution during forming. Hot compression experiments were used to identify initial crystal plasticity parameters. Microscopic analysis was used to parametrize the initial grain structure. The simulated flow stress evolution results show good agreement with the experiment.

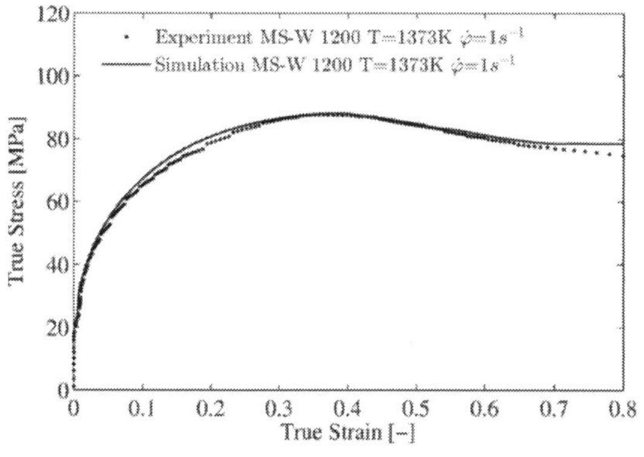

Figure 7. Comparison of flow stress evolution between experiment and simulation of MS-W 1200 steel at T = 1373 K and strain rate $\dot{\varphi}$ = 1 s^{-1}.

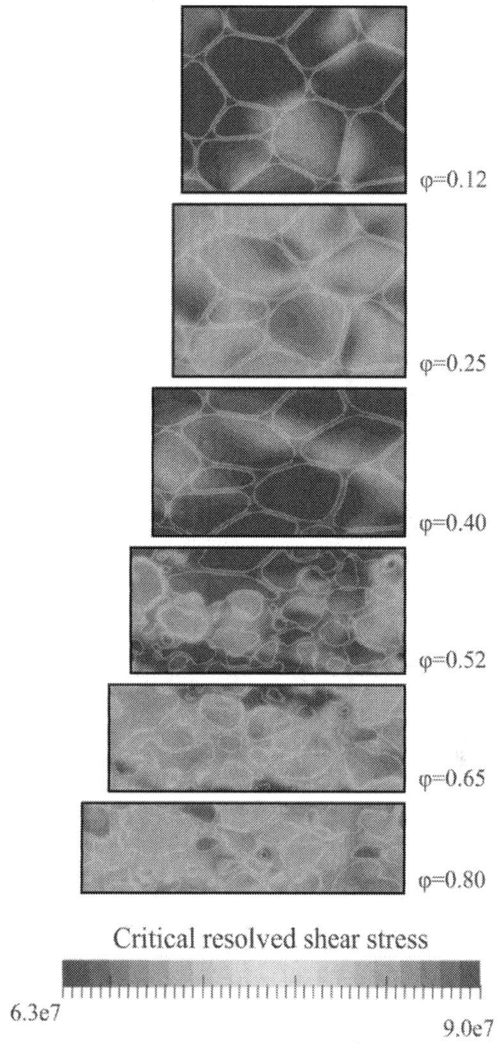

Figure 8. Simulation results of microstructure evolution of MS-W 1200 steel at T = 1373 K and strain rate $\dot{\varphi}$ = 1 s^{-1}.

References

[1] M. Avrami, "Kinetics of phase change. III: Granulation, Phase Change and Microstructure," *Journal of Chemical Physics*, pp. 177–184, 1941.

[2] A. N. Kolmogorov, "Statistical theory of crystallization of metals," *Izvestia Akademia Nauk SSSR Ser. Mathematica*, pp. 355–359, 1937.

[3] W. A. Mehl and R. F. Johnson, "Reaction Kinetics in processes of nucleation and growth," *Transactions of American Institute of Minin and Metallurgical Engineers*, pp. 416–458, 1939.

[4] I. Steinbach and O. Shchyglo, "Phase-field modelling of microstructure evolution in solids: Perspectives and challenges," *Current Opinion in Solid State and Materials Science*, pp. 87–92, 2011.

[5] I. Steinbach, "Phase-Field Model for Microstructure Evolution at the Mesoscopic Scale," 01 Jan, pp. 89–107, 2013.

[6] L. Q. Chen, "Phase-field models for microstructure evolution," *Annu. Rev. Mater. Res,* 01 Jan, 2002.

[7] M. Muramatsu, Y. Aoyagi, Y. Tadano, and K. Shizawa, "Phase-field simulation of static recrystallization considering nucleation from subgrains and nucleus growth with incubation period," *Computational Materials Science*, 01 Jan, pp. 112–122, 2014.

[8] N. Moelans, B. Blanpain, and P. Wollants, "An introduction to phase-field modeling of microstructure evolution Comput. Coupling Phase Diagr," *Thermochem*, 01 Jan, 2008.

[9] O. Güvenc, M. Bambach, and G. Hirt, "Coupling of Crystal Plasticity Finite Element and Phase Field Methods for the Prediction of SRX Kinetics after Hot Working," *steel research international*, 01 Jan, pp. 999–1009, 2014.

[10] T. J. Baron, K. Khlopkov, T. Pretorius, D. Balzani, D. Brands, and J. Schröder, "Modeling of Microstructure Evolution with Dynamic Recrystallization in Finite Element Simulations of Martensitic Steel," *Steel research*, 01 Jan, 2015.

[11] D. Peirce, A. Asaro, and R. J. Needleman, "An analysis of non-uniform and localized deformation in ductile single crystals," *Acta Metallurgica*, 01 Jan, pp. 1087–1119, 1982.

[12] J. W. Hutchinson, "Elastic-Plastic Behaviour of Polycrystalline Metals and Composites," *Proc. R. Soc. Lond. A,* 01 Jan, pp. 247–272, 1970.

[13] J. Pan and R. J. R, "Rate Sensitivity of Plastic Flow and Implications for Yield Surface Vertices," *International Journal of Solids and Structures*, 01 Jan, pp. 973–987, 1983.

[14] E. Borukhovich, P. S. Engels, T. Böhlke, and O. Shchyglo, "Large strain elasto-plasticity for diffuse interface models," *Modelling and Simulation in Materials Science and Engineering*, 01 Jan, 2014.

[15] M. Verdier, Y. Brechet, and P. Guyot, "Recovery of AlMg Alloys: Flow Stress and Strain-Hardening Properties," *Acta Materialia*, 01 Jan, 1999.

[16] G. I. Taylor, "The Mechanism of Plastic Deformation of Crystals. Part I. Theoretical," *Proceedings of the Royal Society of London*, 01 Jan, pp. 362–387, 1934.

[17] C. Zener and J. H. Hollomon, "Effect of Strain Rate Upon Plastic Flow of Steel," *J. Appl. Phys,* 01 Jan, 1944.

[18] R. D. Doherty, D. A. Hughes, F. J. Humphreys, J. J. Jonas, J. J. D, M. E. Kassner, W. E. King, T. R. McNelley, H. J. McQueen, and A. D. Rollett, "Current issues in recrystalliszation: a review," *Material Science and Engineering*, 01 Jan, pp. 219–274, 1997.

[19] F. J. Humphreys and M. Hatherly, *Recrystallization and related Annealing Phenomena*: Pergamon, 1995.

[20] C. M. Sellars and J. A. Whiteman, "Recrystallization and Grain Growth in Hot Rolling," *Metall Science*, 01 Jan, pp. 187–194, 1997.

[21] P. J. Wray, "On set of recrystallization during the tensile deformation of austenitic iron at intermediate strain rates," *Metall. Trans,* 01 Jan, pp. 1197–1203, 1975.

[22] E. I. Poliak and J. J. Jonas, "A one-parameter approach to determining the critical conditions for the initiation of dynamic recrystallization," *Acta Materialia*, 01 Jan, 1996.

[23] J. Kestin, *Article in Patterns, Defects and Microstructures in Non-Equilibrium Systems*: Martinus Nijhoff, 1987.

[24] M. G. Beghi, C. E. Bottani, and G. Caglioti, "Thermoelastic-Plastic Transition and Dislocation Dynamics in Metals by Deformation Calorimetry," *NATO ASI Series*, pp. 187–201.

[25] P. A. Beck and P. R. Sperry, "Strain induced boundary migration in high purity aluminum," *Journal of Applied Physics*, 01 Jan, pp. 150–152, 1950.

[26] R. D. Kamachali, *Grain boundary motion in polycrystalline materials*. Bochum, 2012.

[27] I. Steinbach, "Phase-field models in materials science," *Modelling Simul. Mater. Sci. Eng,* 01 Jan, pp. 1–31, 2009.

[28] H. M. Ledbetter and R. P. Reed, "Elastic Properties of Metals and Alloys," *J. Phys. Chem. Ref. Data*, 01 Jan, pp. 531–617, 1973.

DOI: 10.1051/matecconf/20168001003

Appendix A

Table 1. Environmental input parameters for simulation.

Simulation edge length	-	100	µm
Average gran diameter	$\bar{d}$	35	µm
Simulation resolution	-	100^2	data points
Simulation increments	N	200	#
Number of grains	-	14	#
Boundary conditions	-	periodic	

Table 2. Input parameters for the mechanical solver for temperature T = 1373 K and strain rate $\dot{\varphi} = 1\ s^{-1}$.

Anisotropic elastic constants	C11	$160 \cdot 10^9$	Pa
	C12	$120 \cdot 10^9$	Pa
	C44	$100 \cdot 10^9$	Pa
Initial critical resolved shear stress	$T_{c,0}$	$20 \cdot 10^6$	Pa
Strain rate	$\dot{\varphi}$	1	s^{-1}
Exponent for flow evolution	F_{pow}	2.02	-
Hardening parameter	H_0	$1.1 \cdot 10^{13}$	-
Burgers vector	b	$2.56 \cdot 10^{-10}$	M
Shear modulus	µ	$20 \cdot 10^9$	Pa
Initial dislocation glide velocity	v_o	10^{-3}	ms^{-1}

Table 3. Input parameters for the recovery kinetics.

Debye frequency	v_D	$5.4 \cdot 10^{12}$	s^{-1}
Taylor factor	a	0.3	-
Activation energy	U_0	327300	$Jmol^{-1}$
Activation volume	V_{act}	$28b^3$	m^3
Temperature	T	1373	K

Table 4. Input parameters for the recrystallization and micro-structure evolution kinetics.

Mobility of high angle grain boundaries	m	$1.2 \cdot 10^{-11}$	$m^4J^{-1}s^{-1}$
Energy of high angle grain boundary	γ_{gb}	0.6	Jm^{-2}
Average energy per unit length of dislocation	U_c	$1.31 \cdot 10^{-9}$	Jm^{-1}
Cut-off radius for nucleation	R_{cutoff}	4	µm
Mobility factor for nuclei	C_m	18.6	-
Burgers vector	b	$2.56 \cdot 10^{-10}$	m

DOI: 10.1051/matecconf/20168001003

Appendix B

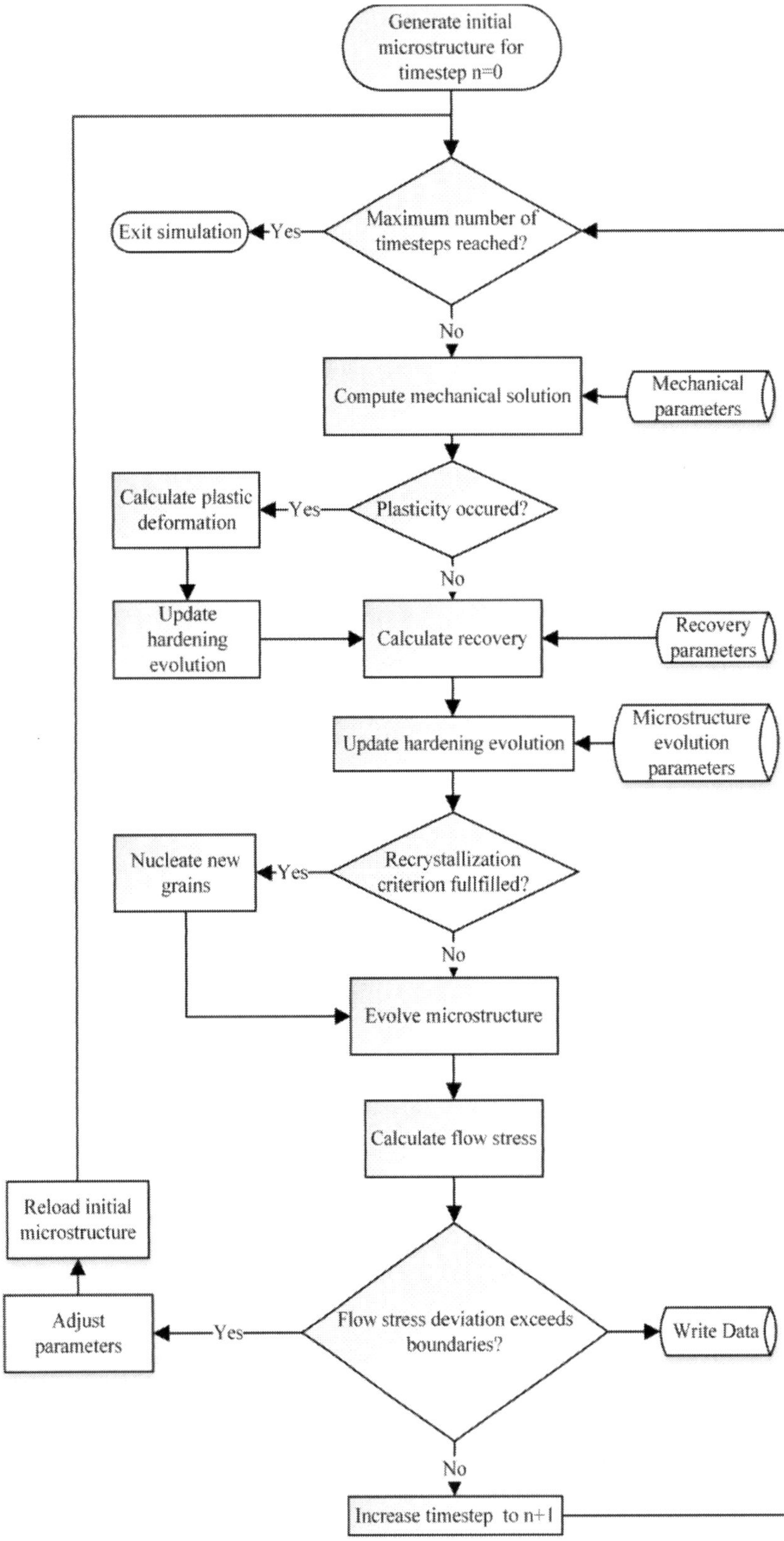

Application of a second-gradient model of ductile fracture on a Dissimilar Metal Weld

Jun Yang[1,2], Rémi Lacroix[2,a], Jean-Michel Bergheau[3], Jean-Baptiste Leblond[4] and Gilles Perrin[5]

[1] *ENSTA, 828 boulevard des Maréchaux, 91762 Palaiseau Cedex, France*
[2] *ESI France, Le Recamier, 70 rue Robert, 69458 Lyon Cedex 06, France*
[3] *Université de Lyon, ENISE, LTDS, UMR 5513 CNRS, 58, rue Jean Parot, 42023 Saint-Etienne Cedex 2, France*
[4] *∂'Alembert, boîte 162, 4 place Jussieu 75252 Paris Cedex 05, France*
[5] *AREVA NP, Tour Areva, 1 place Jean Millier, 92400 Courbevoie, France*

Abstract. A "micromorphic", second-gradient model applicable to ductile porous materials has been proposed, as an improvement from the fundamental work of Gurson that take into account the physical mechanisms responsible for ductile damage. The model has been applied to the study of fracture of the decarburized layer of a Dissimilar Metal Weld. The model successfully reproduces the crack path experimentally observed in a notched tensile sample extracted from this weld, different from the one predicted by the first gradient model.

1 Introduction

Ductile failure is a major mode of fracture of metals under monotonic loading, at medium to high temperature. It can occur during an accidental condition provoking crack propagations associated with large plastic deformations, eventually leading to the catastrophic failure of industrial components. However the numerical reproduction of crack propagation still requires improvements of the theory of Fracture Mechanics in order to be truly predictive. The developments of this theory is therefore crucial to the improvement of industrial safety regulations.

The elastoplastic constitutive models developed from the fundamental works of Gurson [1] take into account the physical mechanisms responsible for ductile damage: nucleation (usually through debonding of the metallic matrix around inclusions), growth and coalescence of cavities. These local approaches of ductile fracture were significantly improved over the last 10 years by taking into account the effects of work hardening, viscosity, void shape and contact of the voids' boundaries with the enclosed inclusions (works of Gologanu, Siruguet, Flandi).

Additionally, a "micromorphic", second-gradient model applicable to ductile porous materials has been proposed [2]. The goal of this model is to solve the problem of unlimited localization of strain and damage due to softening, inevitable in conventional softening models. As its implementation in a finite element code involves a large number of degrees of freedom, a specific method of numerical resolution has been used to permit the convergence of the model [3][4]. It has been successfully applied to study the failure behaviour of standard fracture specimens such as tensile axisymmetric (TA) pre-cracked specimens of A508 steel, and Compact Tension (CT) specimens of 316L stainless steel.

This paper presents the application of this new model to the more complex case of crack propagation in a Dissimilar Metal Weld. It organizes as follows:

- Section 2 is dedicated to a brief introduction of the second gradient model for porous metals.
- Section 3 presents the experimental facture test of a Dissimilar Metal Weld.
- Section 4 describes the numerical model representing the experiment.
- Section 5 presents the results of both the first and second gradient model on the case studied
- Section 6 discusses their performances to reproduce the experimental observations. Their relevance is discussed in terms of failure mode and in terms of physical interpretation of the internal parameters of the model.

2 First and Second gradient model for porous metals

The problem of unlimited strain localization is well known in local approaches of failure. It is inherently related to the softening nature of such models, which softens the most loaded regions, which in turn will be more severely loaded and softened. This phenomenon results in the accumulation of very large strains in the smallest elementary volume: one finite element. Ultimately, the prediction of softening is dependent on the mesh size, which is not acceptable.

[a] Corresponding author: remi.lacroix@esi-group.com

This issue is most commonly solved by introducing a heuristic inner-length to the ductile model. An averaging of the local damage or load state is then performed within this inner-length, thus spreading the softening over this length. This can efficiently make the prediction of softening independent of the mesh size, provided that the mesh size is small enough compared to the inner-length.

This is for example the case of Gurson's original model for porous plasticity, modified by Tvergaard and Needleman [6][7]. This model has been extensively used to predict the ductile failure behaviour of fracture specimens and pipes [10][11]. It considers metals as an incompressible matrix in which an initial porosity can grow and coalesce. Its formulation classically involves the first gradient of the nodal field, with the following yield surface:

$$F(\sigma,\sigma_y,f^*) = \frac{{\sigma_{eq}}^2}{{\sigma_y}^2} + 2qf^*ch\left(\frac{3}{2}\frac{\sigma_m}{\sigma_y}\right) - 1 - q^2f^{*2}$$

σ_{eq} is the Von Mises equivalent stress and σ_m is the mean stress. Coalescence is approached by $f^* = f$ if $f < f_c$ and $f^* = f_c + \delta.(f - f_c)$ if $f > f_c$, with f_c and δ heuristic parameters representing the void coalescence over a critical porosity f_c [6]. The porosity f evolves from an initial value f_o with $\dot{f}$ given by the mean plastic strain rate $\dot{\varepsilon}^p_{kk}$ and the porosity by the relation $\dot{f} = (1-f)\dot{\varepsilon}^p_{kk}$.

q is a the Tvergaard parameter [7]. The value of $q = 4/e$ is used following the works of Perrin and Leblond [8]. It represents an equivalent porous matrix surrounding the porosity. Matrix hardening is introduced by $\sigma_y(\varepsilon^p_{eq})$, being the stress associated to the plastic strain ε^p_{eq} on the uniaxial tensile curve.

In the Gurson model above, independence on the mesh size is given by a heuristic averaging of the porosity rate $\dot{f}$ on a volume given by an inner-length [9]. This model is labelled in this paper as the first gradient Gurson model.

The second approach used on this study to solve the problem of dependence on the mesh size is the one proposed by Gologanu et al [2]. This approach lays on micromorphic basis, as it extends the original homogenization of a hollow sphere of Gurson to take into account the local variation of the strain field at the scale of the porosity. This extended Gurson criterion then writes as follow:

$$F(\sigma,m,\sigma_y,f^*) = \frac{1}{{\sigma_y}^2}\left({\sigma_{eq}}^2 + \frac{Q^2}{b^2}\right) + 2qf^*ch\left(\frac{3}{2}\frac{\sigma_m}{\sigma_y}\right) - 1 - q^2f^{*2}$$

Additionally to the variables defined above, m is a third-rank tensor representing the local moment, and Q^2 is a quadratic form of the components of m. This model has been has been improved by the works of Enakusta et al [3], and further implemented by Bergheau et al [4]. The originality of the latter implementation is the fact that it does not introduce additional degrees of freedom. The reader is invited to refer to this latter work for further details on the theoretical basis of this model.

The inner-length b is therefore inherently accounted for in this extended yield criterion. Independence on the mesh size has been proven, as well as the ability to successfully reproduce the ductile failure of TA and CT steel samples [4]. This model is referred in this paper as the second gradient Gurson model.

3 Crack propagation in a Dissimilar Metal Weld on a Notched Tensile Test

Dissimilar Metal Welds (DMW) are of major interest in Nuclear Power Plants. They are principally used in the primary circuit as connections between large ferritic components such as the pressure vessel, and the austenitic cooling lines. The fracture toughness of the assembly highly depends on the fracture properties of the DMWs, and these are therefore of high interest. **Figure 1** shows a cross section of such a DMW.

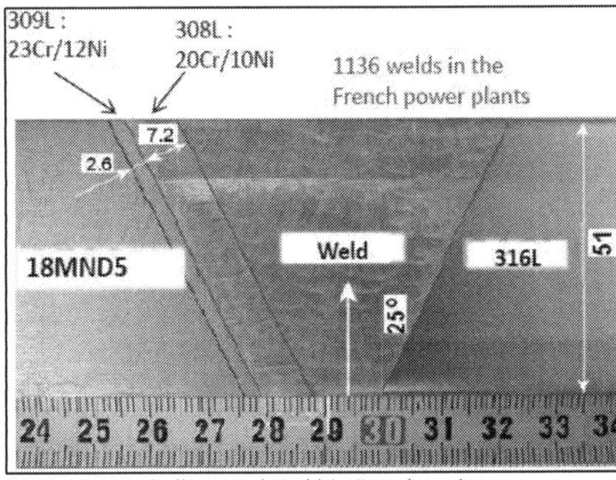

Figure 1 - Dissimilar Metal Weld in French nuclear power plants [5].

The experimental investigation presented below has been carried out in the framework of the doctorate of Fanny Mas [5]. It focuses on the interface between the ferritic steel and the 309L buttering, and more particularly on the failure properties of this interface.

For the sake of simplicity, the case of the internal cladding of ferritic components has be chosen, exhibiting chemical and microstructural gradients across the interface similar to those found on other DMWs joining ferritic and austenitic pipes. The cladding is made of a first layer of 309L and a second layer of 308L, joined by Submerged Arc Welding to the ferritic internal surface.

After the welding process, a Post-Welding Heat Treatment (PWHT) is performed in order to reduce the residual stresses in the DMW. The component is heated up at 610°C at a slow heating rate, hold at this temperature for 8 hours, and cooled down at a slow cooling rate.

Solute diffusion, precipitation or dissolution can occur extensively during this treatment, given the large duration and large volume of material involved. On the opposite, the welding process involves significantly shorter durations for a given volume of material, insufficient for these phenomena to become significant.

DOI: 10.1051/matecconf/20168001004

3.1 Metallurgical observation of a Dissimilar Weld Metal after PWHT

Micrographic views of the interface between the 18MND5 and the 309L strip before and after PWHT are shown in **Figure 2**.

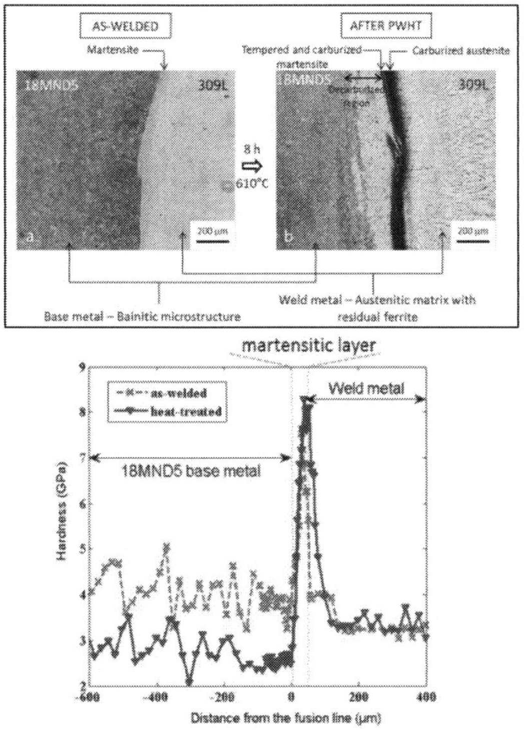

Figure 2 – (Top) Micrographic views of the interface between the 18MND5 and the 309L after etching. (Left) Before PWHT. (Right) After PWHT. (Bottom) Nano-hardness measurements across the fusion line, before and after the heat-treatment [5].

Observations conducted in the work quoted above concluded in the identification of the following zones before PWHT [5]:

- The base metal 18MND5 with a bainitic microstructure (ferritic lath and Fe_3C carbides).
- A Heat Affected zone in the 18MND5. Its width can stretch over 10 mm.
- A thin martensitic layer is found at the interface. Its width is irregular and varies from 5 to 200 microns.
- A fully austenitic zone is seen next to the martensitic layer on the side of the 309L. Its width is about 70 microns.
- The 309L weld metal exhibiting $(\delta + \gamma)$ microstructure, with residual δ-ferrite embedded in the γ-austenitic matrix.

During heat treatment, dissolution of the carbides in the 18MND5 and diffusion of the carbon from the 18MND5 to the 309L are observed. This results in the following modifications of the microstructure [5]:

Formation of large ferritic grains of 18MND5 (over 200 microns) next to the fusion line. They show a very low carbon contents and no or little carbides. Further in the 18MND5, the nominal content of carbon is found only at 500 microns from the fusion line.

Carbon enrichment of the martensitic layer and of the purely austenitic layer in the weld side. Precipitation in carbides takes place over 70 microns from the interface.

Nano-indentations across the fusion line before and after the heat-treatment has been performed [5], and confirms these observations. Figure 2 shows these measurements, exhibiting a decrease in hardness in the 18MND5, and an increase in the weld metal. This latter increase takes place in a thin zone next to the fusion line, in the carburized martensite and austenite.

3.2 Notched Tensile Test

Decarburization of the 18MND5 such as observed above might impact the overall toughness of the DMW. To address this question, a specific notched tensile test has been designed to assess the failure behaviour of this zone particularly. **Figure 3** shows the geometry of one test sample. The test sample is flat, 0.81 mm thick, normal to the fusion surface. It includes both a 18MND5 zone and a 309L zone, with the fusion line.

A notch reduces the cross section of the specimen right on the decarburized zone. The dimensions of the specimen are detailed in the **Table 1**.

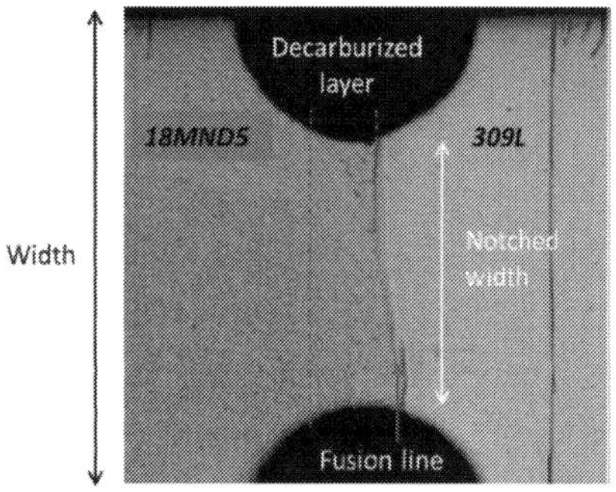

Figure 3 - Geometry of Notched Tensile Test [5].

Table 1 - Dimension of the notched tensile specimen [5].

Thickness	0,81 mm
Width	1,995 mm
Notched width	1,004 mm
Effective length	8,025 mm
Notched cross-section	0,846 mm²
Size of decarburized zone	0,2 mm
Mean thickness of martensite layer	5 µm

The tensile load is applied at the specimen ends by clamping. The notch opening distance is measured on the optical recordings carried out during the test. The average between the opening distance of the top and the bottom notch is considered. The result of the tensile test is shown in **Figure 6.** Failure occurs in the decarburized zone.

DOI: 10.1051/matecconf/20168001004

4 Numerical model of the specimen

Appropriate assessment of the failure behaviour of the decarburized zone is proposed through the local approach to fracture. This section presents the numerical set up of the model. The SYSTUS™ software [12] has been used to carry out the simulations, and Visual Mesh™ [13] for the model geometry.

4.1 Specimen geometry and behaviour zones

One fourth of the specimen is modelled by mean of symmetry, including only one half of its thickness and one half of its width. The entire length is modelled because of the dissymmetric nature of a Dissimilar Metal Weld. The model is discretized with linear 8-nodes hexahedral elements, with selective reduced integration. 16550 elements and 20559 nodes are used for one fourth of the specimen. **Figure 4** illustrates the geometry of the numerical model.

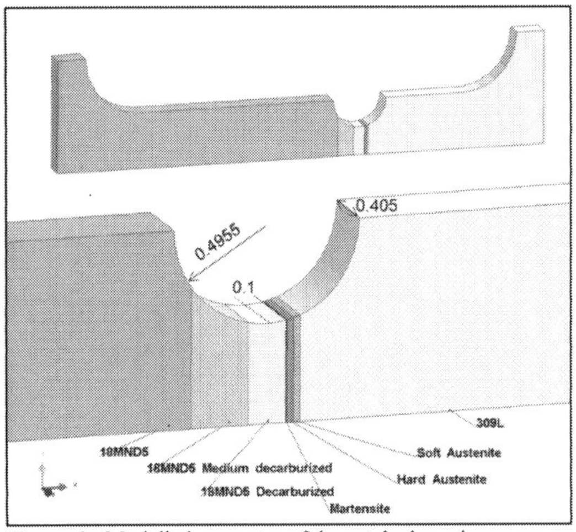

Figure 4 - Modelled geometry of the notched specimen.

The gradients of mechanical properties in the specimen has been modelled by various homogeneous zones representing their average expected behaviour. For instance, the progressive decarburization observed on the 18MND5 when increasing the distance to the fusion line is approached by only 2 behaviour zones (labelled 'Decarburized' and 'Medium decarburized'). Identically, the progressive carburization of the austenite when decreasing the distance to the fusion line is approached by only 2 behaviour zones (labelled 'Hard and 'Soft').

The 18MND5 found in the specimen with the nominal carbon content (over 500µm from the fusion line) is considered homogeneous, as well as the 309L. In total, seven different behaviour zones have been defined, as shown in Figure 4. The

Table 2 lists their dimensions and mesh sizes.

Table 2 - Dimensions and mesh sizes of the various behaviour zones.

Behaviour Zone	Thickness (µm)	Mesh size (µm)
18MND5	on the left end	60
18MND5 Medium decarburized	300	17
18MND5 Decarburized	200	11
Martensite	5	5
Hard austenite	35	9
Soft austenite	35	9
309L	on the right end	60

On the basis of experiments carried out in the works quoted above [5], the monotonic stress-strain curves of each of the behaviour zone have been identified. The Table 3 details their elastoplastic properties.

Table 3 - Elastoplastic properties of the various behaviour zones.

Behaviour Zone	Young Modulus (GPa)	Yield Stress (MPa)	Strain hard-ening	Ductile failure
18MND5	237	530	Isot.	-
18MND5 Medium decarburized	237	347	Isot.	-
18MND5 Decarburized	230	297.5	Isot.	Yes
Martensite	230	1814	-	-
Hard austenite	234	2275	-	-
Soft austenite	229	1261	-	-
309L	196	349.3	Isot.	-

The strain hardening of the softer materials are shown in Figure 5, and represented by an isotropic model. The harder materials, i.e. the martensite and the carburized austenite, have shown little or no strain hardening, and are modelled with nearly perfect plasticity.

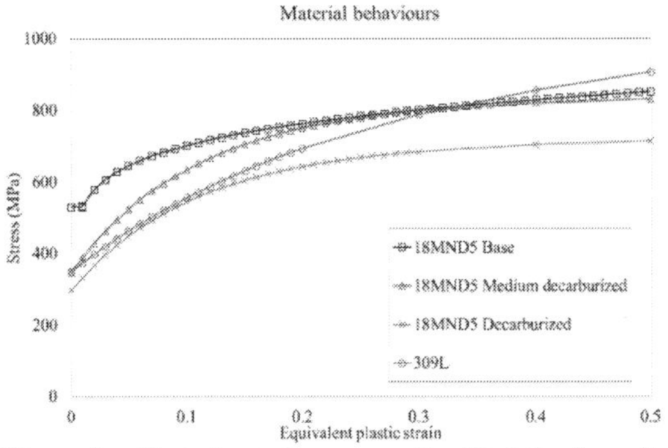

Figure 5 - Strain hardening curves identified for the softer behaviour zones.

4.2 First and second gradient models of ductile damage for the decarburized zone

The study of the ductile failure of the decarburized zone has been carried out with the local approach to failure. The failure model is applied only on the behaviour zone labelled '18MND5 Decarburized'. The two variants of the original Gurson model presented in section 2 are compared: the first gradient and second gradient variants. The first gradient model has been used in this behaviour zone with selective reduced integration. The second gradient model has been used with one modification of the integration scheme.

Selective reduced integration is appropriate with linear elements submitted to incompressible plasticity. It is however irrelevant for porous plasticity. The integration schema has therefore been adapted to dynamically switch on the selective reduced integration only for the elements showing compressive stresses and porosity inferior or equal the initial porosity.

Both models share the same definition of parameters left to the user to reproduce the experiments. The Table 4 presents the values of these parameters. The initial porosity is chosen as a reasonable density of inclusions present in the decarburized layer: 10-4. The inner-length is approximately deduced from the fractured surfaces as an average distance between the voids: 50 µm. The coalescence parameters are left for fitting the experimental load versus notch opening distance.

After fitting the coalescence parameters, very different values of the porosity at coalescence were found between the two variant models. The one found for the first gradient model being extremely close to the initial porosity.

Table 4 - Parameters set for the two variants of the Gurson model to represent the experimental curve.

Gurson variant	f_0	f_c	δ	b (µm)
First gradient	10^{-4}	$10.5 \cdot 10^{-4}$	6.0	50
Second gradient	10^{-4}	$5.0 \cdot 10^{-2}$	2.0	50

5 Results

The comparison of the performances of the two variants of the Gurson model (first gradient and second gradient) are detailed in regards of two experimental outputs. The first one is the predicted load versus notch opening curve, and the second one is the failure mode of the decarburized layer. The comparison of the load versus notch opening is not as interesting as the comparison of the failure mode, as the coalescence parameters have been specially set for a satisfactory representation of the measured curve.

The simulation with the first gradient model has completed 30 time steps in 1 hour an 30 minutes with 6 processors, while the simulation with the second gradient model has completed 28 time steps in about 32 hours with 6 processors.

5.1 Load versus notch opening curve

Satisfactory agreement have been reached between the load versus notch opening curves as measured and as predicted by the two variants of the Gurson model, as shown in Figure 6. The measurement of Notch opening has been slightly offset by a value of -8µm for better agreement with calculations. This is assumed acceptable.

The two variants of the Gurson model can successfully reproduce the measured curve up to instability by adjusting the coalescence parameters. The second gradient model encountered a convergence issue during the final collapse of the notched section, stopping the simulation as shown in Figure 6. The first gradient model did not encountered such issues. Apart from this, no significant differences can be noted in the output of the two models.

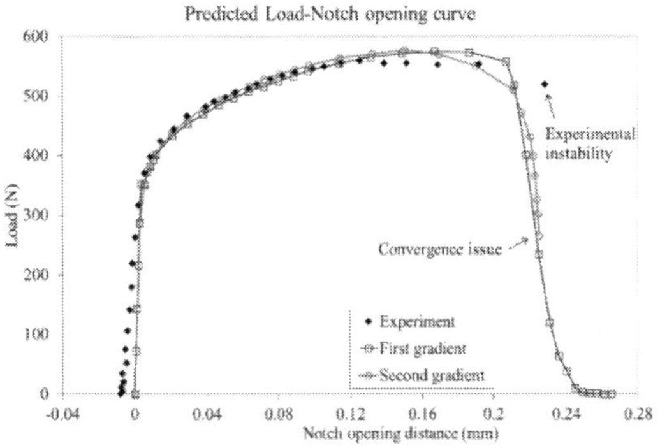

Figure 6 - Load versus notch opening curves as measured and as predicted by the two variant of the Gurson model.

5.2 Failure mode of the decarburized layer

The location of failure in the decarburized layer is analysed in this section. The Figure 7 and Figure 8 illustrate the failure location as predicted by the two variants of the Gurson model at similar level of damage. For that purpose, locations where the porosity is greater than 0.03 are displayed in black, while other locations are displayed in white. Figure 7 illustrates the porosity field seen from the outside of the specimen, while Figure 8 illustrates it from the inside, in the bulk material located on the planes of symmetry.

In the case of the first gradient model, the porosity starts to grow in the centre of the notched section, at about 150 µm from the fusion line in the decarburized zone. It then develops to reach the free surfaces and comes closer to the fusion line, at about 40 µm from the fusion line on the external corner. With the second gradient model, the porosity grows first right next to the fusion line, around the external corner. It then develops to cover most of the notched section next to the fusion line, and later tends to distance itself from the fusion line when coming closer to the centre of the notched section. The predictions of failure mode by the two variants of Gurson model are thus almost opposite.

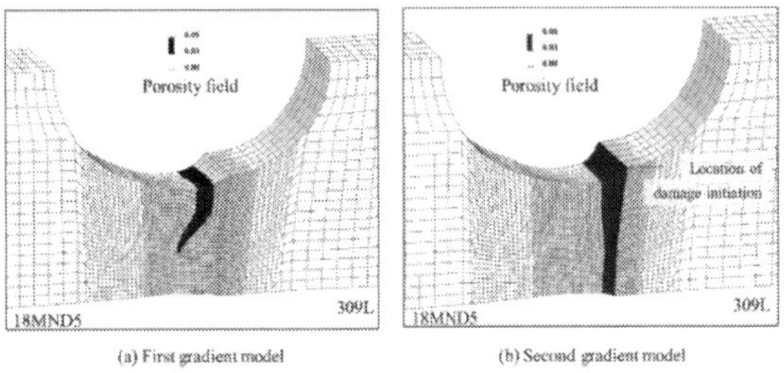

Figure 7 - Comparison of the failure location in the decarburized layer as seen from the outside. Left: First gradient model, notch opening=0,236 mm. Right: Second gradient model, notch opening=0,225 mm.

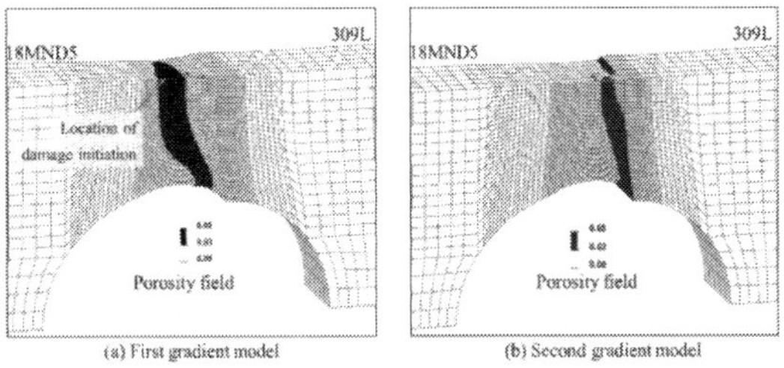

Figure 8 - Comparison of the failure location in the decarburized layer as seen from the mid specimen symmetry planes. Left: First gradient model, notch opening=0,236 mm. Right: Second gradient model, notch opening=0,225 mm.

The presence of harder behaviour zones can be noticed on the side of the 309L, either with the first gradient or second gradient model. They show minimal deformation compared to the soft decarburized zone, acting as a shield on the side of the 309L.

6 Discussions

This paper have shown the performances of two variants of the Gurson original model of ductile fracture applied to a 3D simulation of the failure of a Dissimilar Metal Weld. Both models are non-local, i.e. solve the problem of unlimited strain localization. The first gradient model solves it on a purely heuristic way, while the second gradient model solves it on a micromorphic basis. Both models are able to reproduce the global measurements of load and notch opening measured experimentally.

The following remarks can be drawn from this study:

- To the knowledge of the authors, this study is the first investigation of the ductile failure of a Dissimilar Metal Weld with a local approach solving the issue of unlimited strain localization on a micromorphic basis.
- The performances of the second gradient model in terms of CPU time and convergence is satisfactory in regards to the level of complexity of this model. It justifies further efforts to improve its performances.
- As previously noted [4], appropriate values of porosity at coalescence required for an experimental match for the second gradient model are very different from those required for the first gradient model.
  Indeed, the first gradient model knows only the local mean stress as driving force for void growth. The second gradient model however knows an additional driving force for void growth, i.e. the local 'moment tensor'. For a given loading situation, void growth will therefore be stronger with the second gradient model than with the first gradient model.
  The value of porosity at coalescence of the second gradient model are much more in accordance with micromechanical data either approached by measurements or by simulations. This observation clearly argues in favour of the second gradient model.
- The development of damage in the decarburized zone such as presented by the works of F. Mas [5] is a clear discriminant between the first gradient model and the second gradient model. They predict an almost opposite failure mode. One can explain this difference by the different driving forces taken into account for void growth.
- Mean stress is first found the centre of the notched section, sufficiently away from the hard zones that shield their immediate surrounding. The first gradient model hence predicts the first void growth in this zone. The 'moment tensor' is strong along the fusion line due to the strong gradient of mechanical properties. It therefore strongly modifies the prediction of the failure mode of the second gradient model.
- Clear experimental observations are missing to give a final appreciation of the differences between the two variants of the model. They should state whether the damage starts at the centre of the notched section away from the fusion line, or along the fusion line on the external surface.

Acknowledgments

The authors are grateful to Fanny Mas and her directors Yves BRECHET and Catherine TASSIN from the SIMaP laboratory in Grenoble, France, for the collaboration during her PhD project.

References

[1]. A. Gurson. Continuous Theory of Ductile Rupture by Void Nucleation and Growth: Part I – Yield Criteria and Flow Rules for Porous Ductile Media. ASME J. of Eng. Material. and Technology, 99 (1997), 2-15.
[2]. M. Gologanu, J.B. Leblond, G. Perrin, J. Devaux, Recent extensions of Gurson's model for porous ductile metals. In: Suquet, P. (Ed.), Continuum Micromechanics. In: CISM Courses and Lectures, vol. 377. Springer, pp. 61–130, 1997.
[3]. K. Enakoutsa, J.B. Leblond, European Journal of Mechanics A/Solids, 28, 445-460 (2009).
[4]. J.M. Bergheau, J.B. Leblond, G. Perrin, Computer Methods in Applied Mechanics and Engineering, vol. 268, pp. 105-125 (2014).
[5]. F. Mas, Solidification and phase transformations in a dissimilar Steel weld 18MND5/309L/308L: Evolution of microstructure and mechanical properties, PhD thesis, Université de Grenoble.
[6]. V. Tvergaard, A. Needleman, Analysis of cup-cone fracture in a round tensile bar, Acta Metall. 32 (1984) 157–169.

DOI: 10.1051/matecconf/20168001004

[7]. V. Tvergaard, Influence of voids on shear band instabilities under plane strain conditions, Int. J. Fract. 17 (1981) 389–407.

[8]. G. Perrin, J.B. Leblond, Analytical study of a hollow sphere made of plastic porous material and subjected to hydrostatic tension – Application to some problems in ductile fracture of metals, International Journal of Plasticity,6 , 677-699, 1990.

[9]. J.B. Leblond, G. Perrin, J. Devaux, Bifurcation effects in ductile metals with nonlocal damage. ASME J. Appl. Mech. 61, 236–242, 1994.

[10]. P. Gilles, V. Robin, J. Devaux, Ductile Tearing Behavior near a Bimetallic interface, MECAMAT, 9th European Mechanics of Materials Conference, 2006.

[11]. Assessment of Dissimilar Weld Integrity (ADIMEW): Final Report of the NESC-III Project, Joint Research Center, European Commision, EUR 22510 EN, 2006.

[12]. SYSTUS v17.0 Reference Manual, ESI Group, 2015.

[13]. Visual Mesh v11.0, ESI GROUP, 2015.

MATEC Web of Conferences **80**, 01005 (2016) DOI: 10.1051/matecconf/20168001005
NUMIFORM 2016

Finite Similitude in Metal Forming

Keith Davey[1a], Rooholamin Darvizeh[1] and Anees Al-Tamimi[1]

[1]*School of MACE, The University of Manchester, Manchester, UK*

Abstract. Scaled experimentation founded on dimensional analysis has a long history but has to-date achieved little success in complex metal-forming applications. This is particularly true for thermo-mechanical processes, where scaled experimentation is recognised to provide little insight. It is shown in this paper that current similarity approaches in continuum physics can be overly prescriptive. The concept of finite similitude is introduced, which is shown to be better suited to finite-element analysis for the design of scaled experiments.

1 Introduction

Scaling and in particular dimensional analysis plays an important role in experimental continuum physics. The most well-known and important theorem in modern dimensional analysis is the Buckingham Pi theorem [1], which quantifies the number of dimensionless quantities. If it transpires that the number and magnitude of the dimensionless quantities of the full-scale model and the scaled model coincide, then the two models are denoted similar and accordingly produce corresponding physics. But in reality, metal forming processes are complex and it is not generally possible to find an exact similar model. Although it is evident that dimensional analysis has an important part to play in scaled experiments the inability to match dimensionless quantities imposes severe limitations on its use. These limitations are particularly pronounced for metal forming processes, where a review of the literature reveals a dearth of serious significant technical works on the subject. Although numerous text books on dimensional analysis exist, few cover solid mechanics [1, 2] and fewer still cover forming (none in fact found). The number of articles involving both dimensional analysis and forming [3, 4] is very limited. Similarity solutions simply do not exist in practical metal forming and hence a fundamental difficulty exists when designing scaled experiments. A fundamental concern of any scaling process is whether the results obtained can be guaranteed to be representative of the associated real physics. *Does the physics involved scale?* The short answer is no, but despite this limitation, physical models can still have a key role to play.

The presented work introduces a novel scaling methodology which is based on scaling of the integral (transport) form of the conservation laws [5]. Unlike differential representations of the conservation laws, which are point-based, the transport equations are applied to a finite space and this feature facilitates a direct map between two control volumes in the full and scaled models. In other words, using transport equations permits the direct scaling of the full-scale model. This direct scaling can be achieved by the use of Nanson's identities and time scaling to represent the physics of the full scale process at the smaller scale or vice versa.

2 Preliminary concepts

The presented scaling theory is based on the generalised integral form of the conservation equations [5], which can be represented mathematically as

$$\frac{D^*}{D^*t}\int_{\Omega_{ps}} \rho_{ps}\psi_{ps}dV_{ps} + \int_{\Gamma_{ps}} \rho_{ps}\psi_{ps}\left(\underline{v}_{ps}-\underline{v}^*_{ps}\right)\cdot \underline{n}_{ps}d\Gamma_{ps}$$
$$= -\int_{\Gamma_{ps}} \underline{J}^{\psi}_{ps}\cdot \underline{n}_{ps}d\Gamma_{ps} + \int_{\Omega_{ps}} \rho_{ps}b^{\psi}_{ps}dV_{ps} \quad (1)$$

where ρ, ψ, $\underline{v}$, $\underline{J}$, b, $\underline{n}$ are respectively density, physical field, velocity field, flux , source, unit normal to the boundary Γ of the control volume Ω. The subscript *ps* in equation (1) denotes the variable associated with the physical space (unscaled model) and the star derivative $D^*/D^*t\left(=\partial/\partial t|_{X^*}\right)$ is temporal derivative corresponding to the map $\mathbf{x}\left(\mathbf{X}^*,t\right):\Omega^*_{PS}\mapsto\Omega_{ps}$ between the reference space Ω^*_{PS} and the moving control volume Ω_{ps} in the physical space [5-6]. This star-time derivative plays a similar rule as the well-known material derivative and gives rise to the definition of the control volume velocity $\underline{v}^*_{ps}=D^*\mathbf{x}/D^*t$.

[a] Corresponding author: keith.davey@manchester.ac.uk

A similar diffeomorphism map can be considered for the trial-scaled space, i.e. $\mathbf{s}(\mathbf{S}^*,\tau):\Omega_{TS}^* \mapsto \Omega_{ts}$ which gives the velocity of Ω_{ts} as $\underline{v}_{ts}^* = D^*\mathbf{s}/D^*\tau$, where $D^*/D^*\tau \equiv \partial/\partial\tau|_{\mathbf{S}^*}$. The symbol used for time in this map (τ) is purposely different from the physical-space map symbol (t), which indicates that time in the trial space can run at a different rate from that in the physical space. Based on this newly defined diffeomorphism the transport equation for the trial space is

$$\frac{D^*}{D^*\tau}\int_{\Omega_{ts}} \rho_{ts}\psi_{ts}dV_{ts} + \int_{\Gamma_{ts}} \rho_{ts}\psi_{ts}\left(\underline{v}_{ts} - \underline{v}_{ts}^*\right)\cdot\underline{n}_{ts}d\Gamma_{ts} = -\int_{\Gamma_{ts}} \underline{J}_{ts}^{\psi}\cdot\underline{n}_{ts}d\Gamma_{ts} + \int_{\Omega_{ts}} \rho_{ts}b_{ts}^{\psi}dV_{ts} \quad (2)$$

Although presently there is no obvious relationship between equations (1) and (2), the transport phenomena in the physical and trial spaces can be related by assuming the existence of a smooth map $\mathbf{x}(\mathbf{s},\tau)$ such that

$$dx_i = \frac{\partial x_i}{\partial s_j}ds_j + \left.\frac{\partial x_i}{\partial \tau}\right|_{\mathbf{s}} d\tau \quad (3)$$

In addition the time variables in the physical and trial spaces can be related by

$$dt = h(\tau)d\tau \quad (4)$$

where h is a bijection map.

It is convenient here to assume $\mathbf{x}$ to be a function of $\mathbf{s}$ only; then equation (3) can be reduced to $\mathbf{dx} = F_s \cdot \mathbf{ds}$, where $F_s = \partial\mathbf{x}/\partial\mathbf{s}$. This deformation tensor gives rise to Nanson's geometric identities $dV_{ps} = |F_s|dV_{ts}$ and $\mathbf{d\Gamma}_{ps} = |F_s|\mathbf{d\Gamma}_{ts}\cdot F_s^{-1}$, where $\mathbf{d\Gamma}_{ps} = \underline{n}_{ps}\mathrm{d}\Gamma_{ps}$ and $\mathbf{d\Gamma}_{ts} = \underline{n}_{ts}\mathrm{d}\Gamma_{ts}$. By substituting these relationships into equation (1), it can rewritten as

$$\frac{1}{h(\tau)}\frac{D^*}{D^*\tau}\int_{\Omega_{ts}^*} \rho_{ps}\psi_{ps}|F_s|dV_{ts} + \int_{\Gamma_{ts}^*} \rho_{ps}\psi_{ps}|F_s|F_s^{-1}\cdot\left(\underline{v}_{ps} - \underline{v}_{ps}^*\right)\cdot\underline{n}_{ts}d\Gamma_{ts} = -\int_{\Gamma_{ts}^*} |F_s|F_s^{-1}\cdot\underline{J}_{ps}^{\psi}\cdot\underline{n}_{ts}d\Gamma_{ts} + \int_{\Omega_{ts}^*} \rho_{ps}b_{ps}^{\psi}|F_s|dV_{ts} \quad (5)$$

Equations (2) and (5) are identical if

$$\rho_{ts}\psi_{ts} = \alpha^{\psi}\rho_{ps}\psi_{ps}|F_s| \quad (6a)$$

$$\underline{v}_{ts} - \underline{v}_{ts}^* = hF_s^{-1}\cdot\left(\underline{v}_{ps} - \underline{v}_{ps}^*\right) \quad (6b)$$

$$\underline{J}_{ts}^{\psi} = \alpha^{\psi}h|F_s|F_s^{-1}\cdot\underline{J}_{ps}^{\psi} \quad (6c)$$

$$\rho_{ts}^{\psi}b_{ts} = \alpha^{\psi}h\rho_{ps}b_{ps}^{\psi}|F_s| \quad (6d)$$

where the scalar α^{ψ} is involved for slightly greater generality. Decoupling of control volume and material velocities is achieved on setting $\underline{v}_{ts}^* = hF_s^{-1}\cdot\underline{v}_{ps}^*$ and $\underline{v}_{ts} = hF_s^{-1}\cdot\underline{v}_{ps}$, which is sufficient for Equation (6b).

3 Isotropic scaling

The theoretical and practical aspects of isotropic scaling are presented in this section. In isotropic similarity the physical parameters are scaled with the same ratio in all principal directions. This condition simplifies the second-order tensor F_s to βI where β is scalar scaling factor and I is identity tensor. Substitution of $|F_s| = \beta^3$ into equations (6a-6d) results

$$\rho_{ts}\psi_{ts} = \alpha^{\psi}\beta^3\rho_{ps}\psi_{ps} \quad (7a)$$

$$\underline{v}_{ts}^* = h\beta^{-1}\underline{v}_{ps}^* \quad (7b)$$

$$\underline{v}_{ts} = h\beta^{-1}\underline{v}_{ps} \quad (7c)$$

$$\underline{J}_{ts}^{\psi} = \alpha^{\psi}h\beta^2\underline{J}_{ps}^{\psi} \quad (7d)$$

$$\rho_{ts}b_{ts}^{\psi} = \alpha^{\psi}h\beta^3\rho_{ps}b_{ps}^{\psi} \quad (7e)$$

It is evident that there exists a set of physical variables ρ_{ts}, ψ_{ts}, $\underline{v}_{ts}^*$, $\underline{v}_{ts}$, $\underline{J}_{ts}^{\psi}$ and b_{ts}^{ψ} along with scaling parameters α^{ψ} and h which enables equations (1) and (2) to match. But the challenge is how to practically match the physical and trial models transport equations in the case of metal forming processes.

3.1. Practical aspects

The transport conservation relations involved in thermo-mechanical metal forming are continuity, momentum, energy and entropy which for the physical space (i.e. full-scale model) can be formulated as

$$\frac{D^*}{D^*t}\int_{\Omega_{ps}} \rho_{ps}dV + \int_{\Gamma_{ps}} \rho_{ps}\left(\underline{v}_{ps} - \underline{v}_{ps}^*\right)\cdot\underline{n}_{ps}d\Gamma_{ps} = 0 \quad (8)$$

$$\frac{D^*}{D^*t}\int_{\Omega_{ps}} \rho_{ps}\underline{v}_{ps}dV + \int_{\Gamma_{ps}} \rho_{ps}\underline{v}_{ps}\left(\underline{v}_{ps} - \underline{v}_{ps}^*\right)\cdot\underline{n}_{ps}d\Gamma = \int_{\Gamma_{ps}} \underline{\underline{\sigma}}_{ps}\cdot\underline{n}_{ps}d\Gamma_{ps} + \int_{\Omega} \rho_{ps}\underline{b}_{ps}dV_{ps} \quad (9)$$

$$\frac{D^*}{D^*t}\int_{\Omega_{ps}}\rho_{ps}e_{ps}dV_{ps}+\int_{\Gamma_{ps}}\rho_{ps}e_{ps}\left(\underline{v}_{ps}-\underline{v}^*_{ps}\right)\cdot\underline{n}_{ps}d\Gamma_{ps}$$
$$=\int_{\Gamma_{ps}}\underline{v}_{ps}\cdot\underline{\underline{\sigma}}_{ps}\cdot\underline{n}_{ps}d\Gamma_{ps}-\int_{\Gamma_{ps}}\underline{q}_{ps}\cdot\underline{n}_{ps}d\Gamma_{ps}+ \quad (10)$$
$$\int_{\Omega_{ps}}\rho_{ps}Q_{ps}dV+\int_{\Omega_{ps}}\rho_{ps}\underline{v}_{ps}\cdot\underline{b}_{ps}dV_{ps}$$

$$\frac{D^*}{D^*t}\int_{\Omega_{ps}}\rho_{ps}s_{ps}dV_{ps}+\int_{\Gamma_{ps}}\rho_{ps}s_{ps}\left(\underline{v}_{ps}-\underline{v}^*_{ps}\right)\cdot\underline{n}_{ps}d\Gamma_{ps}$$
$$=-\int_{\Gamma_{ps}}T^{-1}_{ps}\underline{q}_{ps}\cdot\underline{n}_{ps}d\Gamma_{ps}+\int_{\Omega_{ps}}\rho_{ps}T^{-1}_{ps}Q_{ps}dV_{ps}+\dot{S}_i \quad (11)$$

where $e=u+\frac{1}{2}\underline{v}\cdot\underline{v}$, u is stationary internal energy, and s is specific entropy, $\underline{q}\cdot\underline{n}$ is heat flux, Q represents a heat source, $\underline{\underline{\sigma}}$ is the Cauchy stress tensor, $\underline{b}$ is a body force and $\dot{S}_i\geq 0$ is associated with irreversibility.

3.1.1 Scaling of Mass

For the conservation of mass or continuity transport equation $\psi=1$, $\underline{J}=\underline{0}$ and $b=0$ where for similarity system of equations (7a-e) give $\rho_{ts}=\alpha^{\rho}\beta^3\rho_{ps}$, $\underline{v}^*_{ts}=h\beta^{-1}\underline{v}^*_{ps}$, $\underline{v}_{ts}=h\beta^{-1}\underline{v}_{ps}$, $\underline{J}^{\psi}_{ts}=\underline{0}$ and $\rho_{ts}b^{\psi}_{ts}=0$. The relationship of principal interest here is

$$\rho_{ts}=\alpha^{\rho}\beta^3\rho_{ps} \quad (12)$$

For example for steel-aluminium case where the physical model material is steel and the trial model material is aluminium, the density scaling factor α^{ρ} is $\left(1/\beta^3\right)\left(\rho_{\text{aluminium}}/\rho_{\text{steel}}\right)$.

3.1.2 Scaling of Momentum

As equation (9) indicates the physical field and the flux terms in the conservation of momentum are respectively $\underline{v}$ and $\underline{\underline{\sigma}}$, which on replacing them in equations (7a-e) provides $\rho_{ts}\underline{v}_{ts}=\alpha^{\underline{v}}\beta^3\rho_{ps}\underline{v}_{ps}$, $\underline{v}^*_{ts}=h\beta^{-1}\underline{v}^*_{ps}$, $\underline{v}_{ts}=h\beta^{-1}\underline{v}_{ps}$, $\underline{\underline{\sigma}}_{ts}=\alpha^{\underline{v}}h\beta^2\underline{\underline{\sigma}}_{ps}$ and $\rho_{ts}\underline{b}_{ts}=\alpha^{\underline{v}}h\beta^3\rho_{ps}\underline{b}_{ps}$.

Combination of relation $\rho_{ts}\underline{v}_{ts}=\alpha^{\underline{v}}\beta^3\rho_{ps}\underline{v}_{ps}$ with equation (12) and the constraint $\underline{v}_{ts}=h\beta^{-1}\underline{v}_{ps}$ results

$$\alpha^{\underline{v}}=h\beta^{-1}\alpha^{\rho} \quad (13)$$

Also the relation $\underline{\underline{\sigma}}_{ts}=\alpha^{\underline{v}}h\beta^2\underline{\underline{\sigma}}_{ps}$ can be used in any constitutive model. For example in the case of von-Mises plasticity model, $\underline{\underline{\sigma}}'_{ts}:\underline{\underline{\sigma}}'_{ts}=\frac{2}{3}\bar{\sigma}^2_{ts}$ and $\underline{\underline{\sigma}}'_{ps}:\underline{\underline{\sigma}}'_{ps}=\frac{2}{3}\bar{\sigma}^2_{ps}$ where $\bar{\sigma}$ and $\underline{\underline{\sigma}}'$ are the effective stress and the deviatoric stress tensor such that $\bar{\sigma}_{ts}=\alpha^{\underline{v}}h\beta^2\bar{\sigma}_{ps}$ and $\underline{\underline{\sigma}}'_{ts}=\alpha^{\underline{v}}h\beta^2\underline{\underline{\sigma}}'_{ps}$. Using equation (13) and considering $\bar{\sigma}_{ts}=Y_{ts}$ and $\bar{\sigma}_{ps}=Y_{ps}$ results in

$$h=\sqrt{\frac{1}{\alpha^{\rho}\beta}\frac{Y_{ts}}{Y_{ps}}} \quad (14)$$

where Y_{ts} and Y_{ps} are the trial and physical material models yield stresses.

3.1.3 Scaling of Energy

In this case the physical field ψ, the flux $\underline{J}$ and the body force b are respectively e, $\underline{v}\cdot\underline{\underline{\sigma}}$ and $\underline{v}\cdot\underline{b}$. Adopting the same methodology by using equations (7a,d,e) provides $\rho_{ts}e_{ts}=\alpha^{e}\beta^3\rho_{ps}e_{ps}$, $\underline{v}_{ts}\cdot\underline{\underline{\sigma}}_{ts}=\alpha^{e}h\beta^2\underline{v}_{ps}\cdot\underline{\underline{\sigma}}_{ps}$, $\rho_{ts}\underline{b}_{ts}=\alpha^{\underline{v}}h\beta^3\rho_{ps}\underline{b}_{ps}$. By using equation (12) the first relationship can be expanded as $\alpha^{\rho}\left(u_{ts}+\frac{1}{2}v^2_{ts}\right)=\alpha^{e}\left(u_{ps}+\frac{1}{2}v^2_{ps}\right)$ and consequently be separated into two parts as

$$\alpha^{\rho}u_{ts}=\alpha^{e}u_{ps} \quad (15)$$

and

$$\alpha^{\rho}\underline{v}_{ts}\cdot\underline{v}_{ts}=\alpha^{e}\underline{v}_{ps}\cdot\underline{v}_{ps} \quad (16)$$

where the last relationship together with $\alpha^{\rho}\underline{v}_{ts}=\alpha^{\underline{v}}\underline{v}_{ps}$, obtained from the scaling of momentum, gives

$$\alpha^{\rho}\alpha^{e}=\alpha^{\underline{v}}\alpha^{\underline{v}} \quad (17)$$

Also applying the similarity relations to the heat flux and heat-source terms give

$$\underline{q}_{ts}=\alpha^{e}h\beta^2\underline{q}_{ps} \quad (18)$$

$$\rho_{ts}Q_{ts}=\alpha^{e}h\beta^3\rho_{ps}Q_{ps} \quad (19)$$

Based on equation (17), the other two energy similarity relations, $\underline{v}_{ts}\cdot\underline{\underline{\sigma}}_{ts}=\alpha^{e}h\beta^2\underline{v}_{ps}\cdot\underline{\underline{\sigma}}_{ps}$ and

$\rho_{ts}\underline{\text{v}}_{ts}\cdot\underline{b}_{ts}=\alpha^{e}h\beta^{3}\rho_{ps}\underline{\text{v}}_{ps}\cdot\underline{b}_{ps}$, do not provide any extra information and result in the same relationships as determined by the momentum scaling, i.e. $\underline{\underline{\sigma}}_{ts}=\alpha^{\text{v}}h\beta^{2}\underline{\underline{\sigma}}_{ps}$ and $\rho_{ts}\underline{b}_{ts}=\alpha^{\text{v}}h\beta^{3}\rho_{ps}\underline{b}_{ps}$.

3.1.4 Scaling of Entropy

Applying similarity equations (7a-e) to the entropy equation results in 6 relationships. Only two of them are not duplications of the relationships determined on scaling of the energy and momentum equations. These relationships are $\rho_{ts}s_{ts}=\alpha^{s}\beta^{3}\rho_{ps}s_{ps}$ or equivalently $\alpha^{\rho}s_{ts}=\alpha^{s}s_{ps}$ and the entropy-transfer similarity relationship provides

$$\frac{\underline{q}_{ts}}{T_{ts}}=\alpha^{s}h\beta^{2}\frac{\underline{q}_{ps}}{T_{ps}} \tag{20}$$

but note that equation (18) is $\underline{q}_{ts}=\alpha^{e}h\beta^{2}\underline{q}_{ps}$, which on substitution reveals

$$\frac{\alpha^{e}}{T_{ts}}=\frac{\alpha^{s}}{T_{ps}} \tag{21}$$

which provides a proportional relationship between temperatures.

3.2 Scaling of constitutive laws

The scaling details presented in the previous section are only for the conservation laws but in practice experimentalists need to consider the scaling of any associated material-constitutive relationships as well. In the presented methodology, the scaling of the constitutive relations is determined by the scaling applied to the conservation equations. For example, in metal forming, stress is typically a function of strain, strain rate and temperature. For this case a Johnson-Cook material model is a common choice, which for the physical-material model can be written as

$$\bar{\sigma}_{ps}=\left(Y_{ps}+B_{ps}\bar{\varepsilon}_{ps}^{n}\right)\left(1+C\ln\left(\frac{\dot{\bar{\varepsilon}}_{ps}}{\dot{\bar{\varepsilon}}_{ps}^{0}}\right)\right)\left(1-\left(\frac{T_{ps}-T_{ps}^{0}}{T_{ps}^{melt}-T_{ps}^{0}}\right)^{m}\right) \tag{22}$$

where Y_{ps} is the initial yield stress, $\dot{\bar{\varepsilon}}_{ps}^{0}$ is a reference strain rate, T_{ps}^{0} is a reference temperature, T_{ps}^{melt} is the solidus temperature and; *B, n, C and m* are coefficients determined by experimental means.

In view of the identity $\bar{\sigma}_{ts}=\alpha^{\rho}h^{2}\beta\bar{\sigma}_{ps}$, established in section 3.1.2 and the proportional relationships for temperature and strain rate with strain being dimensionless, equation (22) transforms into

$$\bar{\sigma}_{ts}=\left(Y_{ts}+B_{ts}\bar{\varepsilon}_{ts}^{n}\right)\left(1+C\ln\left(\frac{\dot{\bar{\varepsilon}}_{ts}}{\dot{\bar{\varepsilon}}_{ts}^{0}}\right)\right)\left(1-\left(\frac{T_{ts}-T_{ts}^{0}}{T_{ts}^{melt}-T_{ts}^{0}}\right)^{m}\right) \tag{23}$$

where $Y_{ts}=\alpha^{\rho}h^{2}\beta Y_{ps}$, $B_{ts}=\alpha^{\rho}h^{2}\beta B_{ps}$ and $\bar{\varepsilon}_{ts}=\bar{\varepsilon}_{ps}$.

Another material property of interest to thermo-mechanical analysis is thermal conductivity and with application of a classical Fourier law $\underline{q}_{ps}=-k_{ps}\nabla_{ts}T_{ps}$ but for similarity $\underline{q}_{ts}=\alpha^{e}h\beta^{2}\underline{q}_{ps}$ and $\nabla_{ts}=\beta\nabla_{ps}$, and since $\alpha^{s}T_{ts}=\alpha^{e}T_{ps}$, it follows that

$$k_{ts}=\alpha^{s}h\beta k_{ps} \tag{24}$$

which may or may not match the behaviour of a real material.

3.3 Practical constraints

According to the results of section 3.1, there are 6 scaling parameters which are β, h, α^{ρ}, α^{v}, α^{e} and α^{s}. But based on equations (13) and (17), α^{v} and α^{e} can be expressed in terms of other scaling factors and therefore there are only 4 independent factors. The factor β is usually given, which means an experimentalist has 3 degrees of freedom to be quantified, i.e. h, α^{ρ} and α^{s}.

In specific cases where the number of model-material properties and constitutive coefficients are less than or equal to 3, then there can be a perfect match for the trial material model. An example might be pure-mechanical analysis of a rigid, perfectly-plastic material, where the unknowns are only ρ_{ts} and Y_{ts}. But in general metal forming problems there are typically more than 3 unknowns for the trial-material model and therefore the material properties determined for the scaled model cannot be matched with a real material model. In other words, the calculated material model for the scaled experiment does not exist in reality. However, the experimentalist can attempt to select real-material models which approximate the unscaled model (see the case studies present in section 4).

As mentioned above there are 3 free scaling factors which their values can be chosen by an experimentalist. One approach therefore is to identify parameters of the scaled-material model that can be fixed. Deciding which material parameters should be chosen depends on the nature of the process. For example in the case studies presented in the next section ρ_{ts}, Y_{ts} and T_{ts} are chosen to be fixed, i.e. their values set equal to the density, yield stress for aluminium 7039 and room temperature. Fixing these 3 values results in a material model which does not exist in reality (i.e. its other material parameters do not match aluminium 7039 values) despite its density and yield stress being equal to those of aluminium 7039. Therefore direct scaling does not necessarily result in a real material model for the scaled experiment. The steps of the direct scaling can be summarised as

Step 1. Selection of a trial material model for the scaled experiment (e.g. aluminium 7039).
Step 2. Fix 3 scaled model physical parameters depending on their importance for the process (e.g. yield stress, density and temperature).
Step 3. Calculation of scaling parameters, i.e. h , α^{ρ} , $\alpha^{\underline{v}}$, α^{e} and α^{s} .
Step 4. Calculation of the material properties and constitutive relation coefficient for the scaled model (e.g. coefficients of the Johnson-Cook plasticity model).

Applying equations (7a-e) to evaluate *ps* variables using *ts* values is called reverse scaling. It is evident that applying the reverse scaling to the scaled material model calculated in step 4 gives exactly the same material model as the full-scale physical model. On the other hand, applying the reverse scaling to the real trial material model (e.g. aluminium 7039) does not match the unscaled material model but it provides an approximation for the full-scale model behaviour.

4 Case studies

In this section hot forging of a cylinder is considered as the case study. The dimensions of the unscaled physical model are radius (r_{ps})=*15 mm* and length (l_{ps})=*30 mm* and it is made of steel 4340. The dimensional scaling factor β is set to be 2 and therefore the dimensions of the scaled trial model are r_{ts}=*7.5 mm* and l_{ts}=*15 mm*. The cylinder is placed between two rigid plates and deformed in the axial direction by half of its length (see Figure 1). The finite element analyses are performed using ABAQUS software [7]. The consistent system of units used are *kg, mm, ms, K, kN, GPa* and *J* respectively for the mass, length, time, temperature, force, stress and energy.

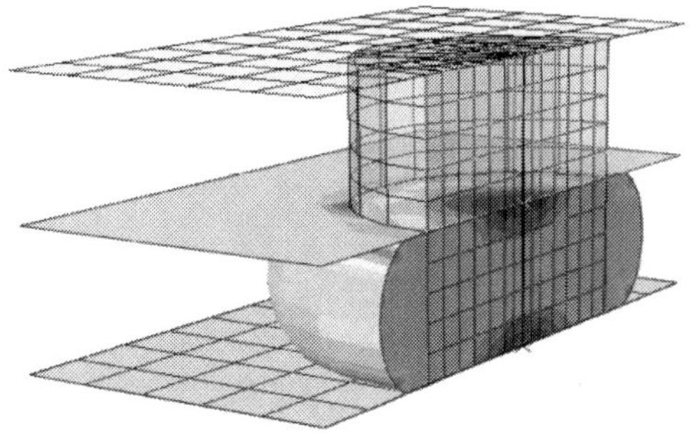

Figure 1. Un-deformed/deformed full-scale (half) steel model

The process is thermo-mechanical and consists of two steps:

Step 1. Ensure the temperature of the full-scale model reaches *700 K* (scaled model to run at the room temperature *293 K*).
Step 2. Applying axial displacement by amount of $l_{ps}/2$ for unscaled model and $l_{ts}/2$ for the scaled model.

The natural heat convection between the surface of the cylinder and the surrounding area (dry air) is modelled during step 2. The time process of the unscaled model for step 1 is *0.01 ms* and for step 2 is *1 ms*, i.e. an high-loading rate typical to impact forming is considered. And the time processes of the scaled model depend on the selected trial material, i.e. time scaling factor *h*. Two possibilities are considered for the trial materials in the next subsections. In the first trial the behaviour of aluminium is adopted as the trial material model and its action under reverse scaling is studied. A tungsten alloy is considered as the second trial material in section 4.2.

4.1 Direct/Reverse scaling I: steel-aluminium

For the first trial, aluminium 7039 is used as the scaled experiment material. The thermo-mechanical material properties and Johnson-Cook relation coefficients for the full-scale model (steel 4340) and the scaled model (Al 7039) are presented in Table 1. The symbols c, α_{exp}, h_{conv} and μ represent respectively the specific heat, thermal expansion, convection film coefficient and friction coefficients. The values of the variables are given in consistent system of units (*kg, mm, ms, K*).

Table 1. Material constants for scaled and unscaled models [8].

Material model	ρ	E	ν	Y	B	ε^{0}
PS: Steel 4340	7.83×10^{-6}	200	0.3	0.792	0.51	0.1
TS: Al 7039	2.77×10^{-6}	70	0.33	0.337	0.34	0.21
TS: Scaled material	2.77×10^{-6}	85.1	0.33	0.337	0.21	0.21

C	n	m	T^{melt}	T^0	c	K	α_{exp}	h_{conv}	μ
0.014	0.26	1.03	1793	293	475	4.45×10^{-3}	1.23×10^{-5}	5×10^{-6}	0.6
0.01	0.41	1	877	293	880	1.40×10^{-2}	2.34×10^{-5}	5.57×10^{-6}	0.6
0.01	0.41	1	750	123	1360	2.48×10^{-3}	2.94×10^{-5}	5.57×10^{-6}	0.6

As it is highlighted in Table 1, the density and yield stress of the scaled material is set to be equal to the trial material model (Al 7039). This is due to the two degrees of freedom provided by free scaling factors h and α^ρ. Since there are in total 3 free scaling factors (h, α^ρ and α^s), the scaling factor α^s is still free and it allows to fix another physical parameter of the scaled experiments. In this case study the applied temperature to the scaled cylinder is fixed to be the room temperature *(293 K)* while the applied temperature to the full-scale model is *700 K*. The other two scaling factors α^{v} and α^e can be calculated using the values of the free scaling factors h, α^ρ and α^s through equations (13) and (17). The scaling parameters values for steel-aluminium combination are presented in Table 2.

Table 2. Scaling factors for steel-aluminium combination.

Factor	β	h	α^ρ	α^{v}	α^e	α^s
Value	2	2.19	4.42 $\times10^{-2}$	4.85 $\times10^{-2}$	5.32 $\times10^{-2}$	1.27 $\times10^{-1}$

The process time of the unscaled model for the step 2 is *1 ms* and for the scaled model is *0.455 ms*, i.e. *1/h*. It is clear from Table 1 data that the scaled material model does not exist in reality but the selected trial material (Al 7039) provides a first attempt at the scaling experiment. The finite element results for the axial force-displacement of the scaled and trial material models are depicted in Figure 2. The erratic behaviour at the initial stages of the results is due to the thermal step (i.e. step 1).

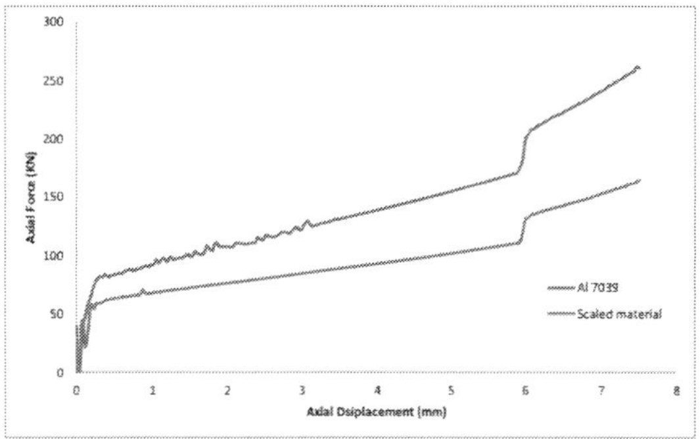

Figure 2. Force-displacement plots for scaled and trial material models

Applying the reverse scaling (see section 3.3) to the scaled material presented in Table 1 gives exactly the same material properties as the unscaled material model (steel 4340) while the reverse map of the Al 7039 material model results in a different material model with the following constants: $\rho=7.83\times10^{-6}$, $E=164.5$, $\nu=0.33$, $Y=0.792$, $B=0.80$, $\dot{\varepsilon}^0=0.1$, $C=0.014$, $n=0.41$, $m=1$, $T^{melt}=2095.2$, $T^0=700$, $c=306.24$, $K=2.51\times10^{-2}$, $\alpha_{exp}=9.79\times10^{-6}$, $h_{conv}=5\times10^{-6}$, and $\mu=0.6$. These values are worked out in the consistent system of units (*kg, mm, ms, K*). The axial force-displacement behaviour of the physical model (steel 4340), reversed scaled material and reversed Al 7039 models under the original (full-scale) loading conditions are compared in Figure 3.

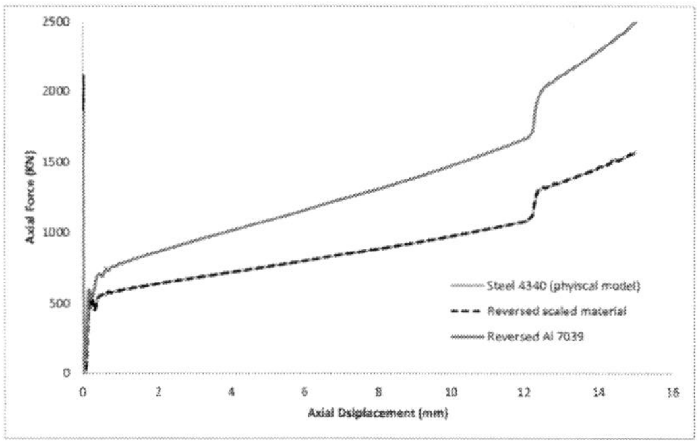

Figure 3. Force-displacement plots for physical and reversed material models.

The results show that the selection of Al 7039 as the trial material model gives rises to significantly higher loads than what is required in the design of the scaling experiment. In next subsection, another trial material (tungsten alloy) is used to provide a closer approximation for the thermo-mechanical behaviour of the full-scale physical model.

4.2 Direct/Reverse scaling II: steel-tungsten

The geometry, dimensions and thermo-mechanical loadings of this case study are exactly similar to the previous case study. The material of the full scale physical model is the same (i.e. steel 4340) but the trial material model is selected to be tungsten alloy (0.07Ni, 0.03Fe). The material properties and Johnson-Cook coefficients for the selected tungsten alloy are: $\rho=1.70\times10^{-5}$, $E=450$, $v=0.28$, $Y=1.506$, $B=0.177$, $\dot{\varepsilon}^0=0.1$, $C=0.016$, $n=0.12$, $m=1$, $T^{melt}=1723$, $T^0=293$, $c=134$, $K=1\times10^{-2}$, $\alpha_{exp}=5\times10^{-6}$, $h_{conv}=5\times10^{-6}$, and $\mu=0.6$ [8]. The calculated scaling factors values for the steel-tungsten scaling experiment are presented in Table 3.

Table 3. Scaling factors for steel-tungsten combination.

Factor	β	h	α^ρ	α^v	α^e	α^s
Value	2	1.87	2.71×10^{-1}	2.54×10^{-1}	2.38×10^{-1}	5.68×10^{-1}

By adopting the same approach used for the previous case study the scaled material constants can be worked out as follow: $\rho=1.70\times10^{-5}$, $E=380$, $v=0.3$, $Y=1.506$, $B=0.97$, $\dot{\varepsilon}^0=0.18$, $C=0.014$, $n=0.26$, $m=1.03$, $T^{melt}=750$, $T^0=123$, $c=994$, $K=9.46\times10^{-3}$, $\alpha_{exp}=2.94\times10^{-5}$, $h_{conv}=2.13\times10^{-5}$, and $\mu=0.6$. The process time of the unscaled model for the step 2 is *1 ms* and for the scaled model is *0.534 ms*, i.e. *1/h*. The axial force-displacement behaviour of the scaled models (i.e. trial and scaled materials) is depicted in Figure 4.

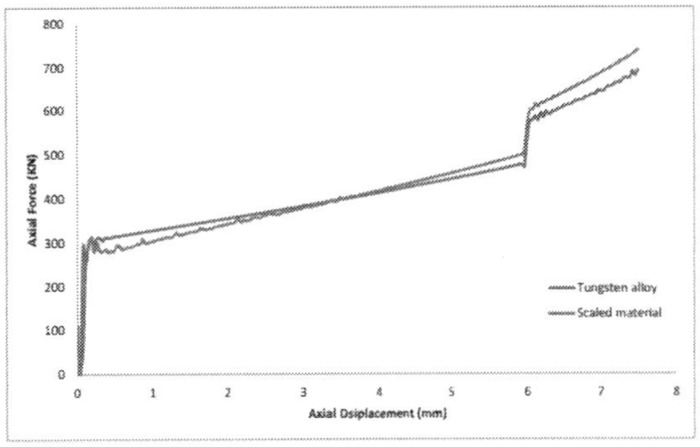

Figure 4. Force-displacement plots for scaled and trial materials models

Applying the reverse scaling (see section 3.3) to the scaled material gives exactly the same material properties as the full-scale model material (steel 4340) while the reverse map of the trial material (tungsten alloy) results in a different material model with the following constants: $\rho=7.83\times10^{-6}$, $E=236.65$, $v=0.28$, $Y=0.792$, $B=0.093$, $\dot{\varepsilon}^0=0.1$, $C=0.016$, $n=0.12$, $m=1$, $T^{melt}=4116.3$, $T^0=700$, $c=64.04$, $K=4.70\times10^{-3}$, $\alpha_{exp}=2.09\times10^{-6}$, $h_{conv}=5\times10^{-6}$, and $\mu=0.6$. The axial force-displacement behaviour of the physical model (steel 4340), reversed scaled material and reversed tungsten alloy models are shown in Figure 5.

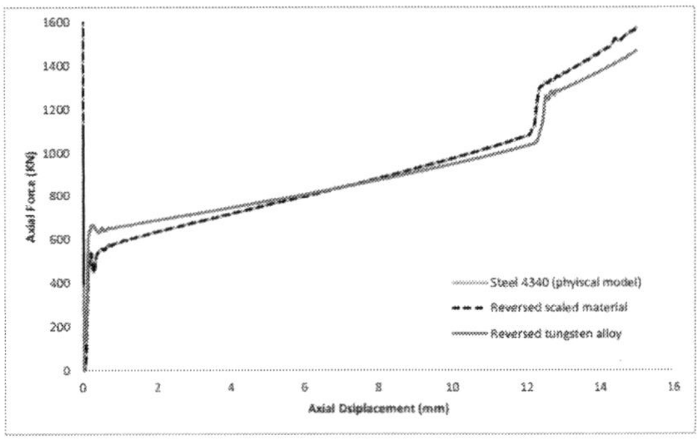

Figure 5. Force-displacement plots for physical and reversed material models.

These results indicate that tungsten alloy (0.07Ni, 0.03Fe) as the trial material model provides a close match for the scaling of this forming process. The results provide confidence for adopting a tungsten-alloy scaled model to represent the physics of the steel billet.

5 Conclusions

The paper is concerned with a novel methodology in isotropic scaling of thermo-mechanical processes and the following conclusions can be drawn from the presented work:

- The generalised integral form of the conservation laws (i.e. the transport approach) can be scaled and used to represent continuum physics on a scaled space.
- The presented theory provides a general framework for both isotropic and anisotropic scaling although its application in this paper is limited to isotropic scaling.
- The presented theory and numerical results confirm the applicability of the presented scaling methodology to thermo-mechanical processes.

References

1. L.I. Sedov, *Similarity and dimensional methods in mechanics* (CRC Press, 10th Ed. 1993)
2. T. Qing-Ming, *Dimensional analysis: with case studies in Mechanics* (Springer, 2011)
3. J.S. Ajiboye, K.H. Jung, Y.T. Im, J. Mech. Sci. Technol. **24** 115-118 (2010)
4. A.A. Ezra, F.A. Penning, Exp. Mech. **2** 234-239 (1962)
5. R. Darvizeh, K. Davey, Int. J. Impact Eng. **82** 59-73 (2015)
6. R. Darvizeh, K. Davey, Comput. Struct. **164** 145-160 (2016)
7. ABAQUS 6.14, Dassault Systèmes (Providence, RI, USA)
8. M.A. Meyers, *Dynamic behavior of materials* (Wiley 1994)

Simulation on microscopic deformation behavior of engine gasket

Ke CAO[1], Yixi ZHAO[1,a] and Sun JIN[1]

[1] *Shanghai Key Laboratory of Digital Manufacture for Thinwalled Structures, Shanghai Jiao Tong University, China*

Abstract. Engine gasket is the thin steel plate with coatings. Its compressive deformation affects the height of engine assembly, which changes combustion chamber's volume and then influences engine performance. In this paper, the assembly simulation of head, block and gasket was studied. Then a further simulation takes the thin coatings on steel sheet into consideration. In the second simulation, gasket's partial micro-model in micron-scale meshes was established and deformation of coatings was analysed. The combination of deformations explores the cause of fluctuations of gasket's tightened thickness and obtains more precise numerical results of tightened thickness at different positions. Moreover, a lead pellet test was carried out to verify the results of tightened thickness calculation.

1 Introduction

As the modern automobile engines keep on pursuing higher power, economy and emission performance, the requirements of manufacture precision and reliability for each part are increasing. The cylinder gasket is the most important component in sealing system which directly affects the performance.

There are already some researches on the relationship between mechanical structures of gasket and its sealing ability which are the focus of engine sealing study.

Hettich et al. studied the characteristic of gasket under dynamic shear stress, and established corresponding constitutive relations [1]. Micheely and Kämpkes studied the overall performance of the connection in consideration of asbestos rubber gasket sealing performance, and discussed the effect of the gasket's microstructure to leakage characteristics [2]. Since 1990, nonlinear finite element simulation started to be used to research the stress distribution and deformation of the gaskets.

Ghasemi studied the deformation of the multilayer metal gasket by finite element simulation and discussed all aspects of IC engine sealing analysis [3]. S.-S. CHO carried on the parameter study of the beads on the gasket to analyze the effect of structure features to sealing ability by FEA methods [4]. H. Zhang studied the sealing pressure of the gasket's features by simulation on ANSYS. His colleague, M. Xi, studied the relationship between the parameters of beads and the sealing pressure. Although the thickness of coatings was not considered in their simulations, the experiments of Fuji pressure-sensitive film which can show different colors under different pressure confirmed that the contact stress results of simulation are precise and acceptable[5].

These researches paid more attention to the macrostructure design of steel sheets of gasket and analyzing the relationship between structures and the sealing ability. But during the assembly of engine, besides the deformation of steel sheets of gasket, there will be the hyperelastic deformation of coatings as well.

Although the thickness of coatings is small (0.01mm-0.015mm for each layer), the hyperelastic deformation will result in the fluctuation of tightened thickness of gasket at different positions and then affect compression ratio. Till now, little study has been done about the calculation of these microscopic deformations and the relationship between this deformation and fluctuation of tightened thickness.

In this research, the assembling simulation of head, block and gasket of a certain engine is studied. Based on the stress results of assembling simulation, a further microscopic simulation was studied to analyze the deformation behavior of coatings and sheets of the gasket.

The influence of both deformation of metal gasket and the hyperelastic deformation of micron-scale coatings to the tightened thickness was investigated. A lead pellet experiment was designed to verify the simulation results. Further, several sensitive components for tightened thickness are found which could be helpful for the quality control in the engine assembly.

2 Effect of gasket's thickness deviation to compression ratio

As the liaison between cylinder head and block, gasket is also the component of the combustion chamber. In compression ratio calculation its thickness and diameter should be concerned. The fluctuation of thickness will

[a] Corresponding author: E-mail: yxzhao@sjtu.edu.cn

affect combustion chamber's volume and further affect compression ratio.

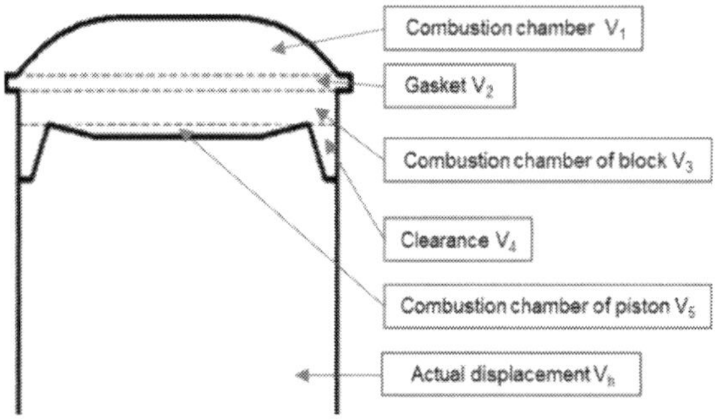

Figure 1. Volumetric components of combustion chamber

The volume of combustion chamber can be divided into several parts which are shown in Figure 1. $V_1 \sim V_5$ present the volume of combustion chamber of cylinder head, combustion chamber of gasket, combustion chamber of block, clearance and combustion chamber of piston respectively. V_h means the volume of cylinder displacement.

Compression Ratio of engine is a value that represents the ratio of the volume of its combustion chamber from its largest capacity to its smallest capacity. It could be presented as:

$$\varepsilon = 1 + \frac{v_h}{v_1+v_2+v_3+v_4+v_5} \quad (1)$$

The volume component of gasket is:

$$V_2 = \pi \frac{d_n^2 * h_n}{4} \quad (2)$$

d_n: Diameter of the cylinder

h_n: tightened thickness

Now supposing that the other components (exclude the gasket height parameters) are constants, compression ratio can be expressed as:

$$\varepsilon = f(h_n) \quad (3)$$

So the variation of compression ratio can be expressed as equation (4):

$$\Delta\varepsilon = f'(h_n)\Delta h_n \quad (4)$$

If the tolerance of tightened thickness h_n is $\pm$ 0.05mm, it will result in 1.051% variation of compression ratio. So controlling the deviation of tightened thickness Δh_n will be significant for the stability of compression ratio.

3 Research strategy

3.1 Gasket Geometric characteristics

Gaskets with structure of multilayer steel plate have good mechanical properties on capacity of compression and resilience.

In this research, the gasket consists of two main plates. The upper plate (No.1 in Figure 2) contains the full seal ribs (bead) around cylinders and half seal ribs (bead) beside the edges of gaskets. The lower plate (No.2 in Figure 2) contains flangings (stopper) around cylinders. The structure of gasket is shown in Figure 2. The bead plate is about twice as thick as the stopper plate.

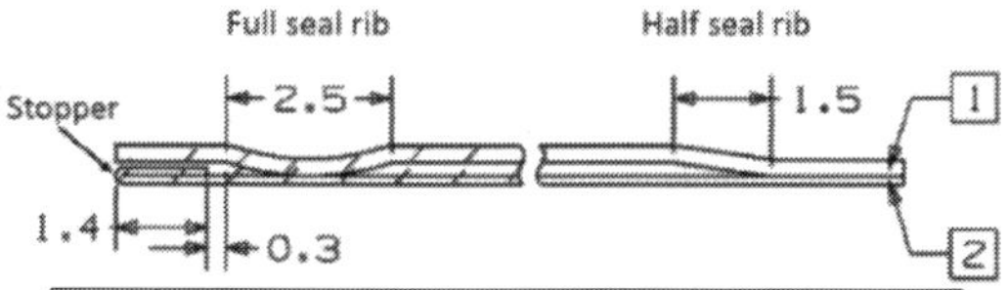

No.	Plate	Material	Thickness (mm)
1	Upper (Bead plate)	SUS301H-TH	0.25
2	Lower (Stopper plate)	SUS304BA	0.12

Figure 2. Structure of gasket

To achieve a good seal, RFC101D coating layers are attached on the surfaces of sheets. As shown in Figure 3, there are both unvulcanized and vulcanized layers on the outer surfaces of bead plate and stopper plate while it just needs one vulcanized layer between two plates. Coatings make it capable to resist high temperature and corrosion. Coatings' thickness increases with the roughness of machined surfaces of head and block. The coatings are very thin compared to steel sheets. But the hyperelastic deformation will occur in the engine assembly process, and which results the fluctuation of tightened thickness on the micron scale. So, coatings' deformation can never be ignored in tightened thickness analysis.

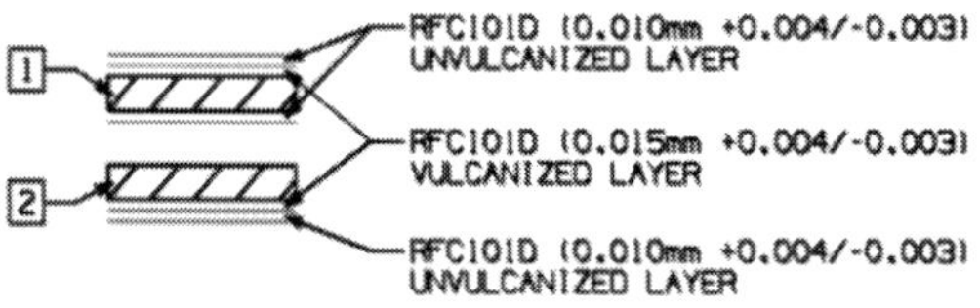

Figure 3. Coatings on the steel sheets

3.2 Research strategy analysis

The aim of the research is to calculate the tightened thickness by FEA method and then analyse the fluctuation of thickness. The most direct method is to create all the layers in the model of assembly (including the steel sheets, coatings and other parts of engine) and use this integrated model to calculate tightened thickness. But the problem is that thickness of coatings is too thin to create in the same size model with steel sheets and other parts of engine. Actually, if the tiny meshes of coatings are created, there will be too many meshes in the simulation along with the complex contact and friction conditions which will result in non-convergence and long computing time.

The two-step solution is proposed in this research. First is the macroscopic simulation of engine assembly without a consideration of the coatings. In this situation, result of thickness is none of significance (because coatings are not contained in model, the thickness results are untruthfulness) while the results of contact stress is reasonable (because the contact area which transmits specific bolt loads does not change with or without coatings).

The second is to establish a microscopic simulation for a partial model which is very small and considers coatings.

If the contact stress can be extracted in the first simulation and used as the input to the microscopic one, the tightened thickness can be calculated.

So, the strategy of this research consists of these two steps:

1. Use the macroscopic simulation of engine assembly to calculate the contact stress distribution on the gasket (without a consideration of the coatings).
2. Use the varying contact stress as the input, create a partial tiny model (including coatings) to calculate the tightened thickness and analyze the fluctuation of thickness.

4 Simulation on deformation of engine assembly

In this section, the macroscopic simulation on deformation of macroscopic geometrical features is carried out which aims to obtain the contact stress on the gasket.

4.1 Assumptions of simulation

To realize the deformation behaviour of gasket when the assembly is tightened, a holistic analysis should be carried on.

In the simulation, three assumptions are made.

1. Quarter of the model is used to replace the whole part.
2. Cylinder block and the stainless stopper plate of gasket are treated as the rigid body.
3. As the initial simulation, coating thickness is only 25μm. Compared to the engine height 70mm, this value can be ignored.

4.2 Model simplification and the assembly

The simplification of the models retains main geometric features, including the combustion chamber, inlet, outlet and water channels of cylinder head, beads and stoppers of gasket. Figure 4 presents models before and after simplification and the assembly of simulation.

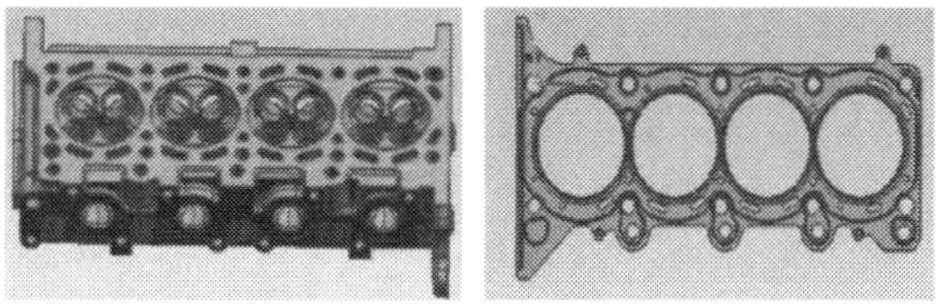

(a) Original models of cylinder head and gasket

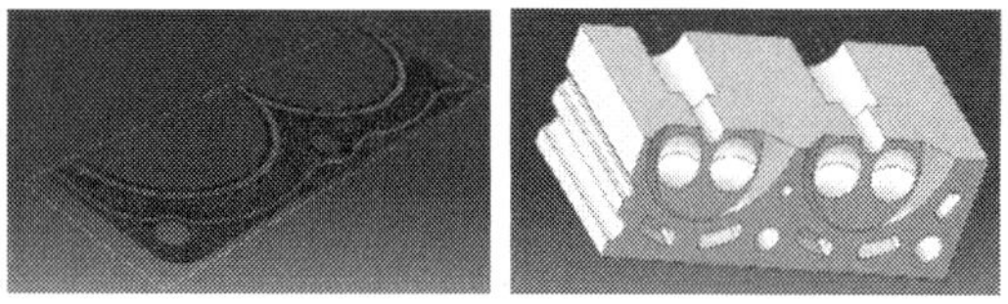

(b) Simplified quarter models of cylinder head and gasket

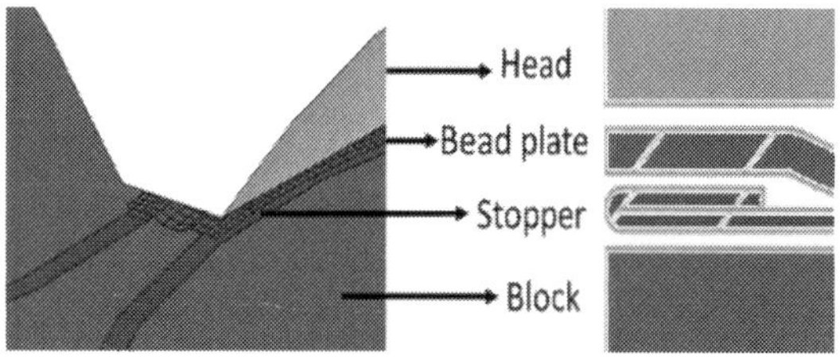

(c) The assembly of model and its section view

Figure 4. Parts before and after simplification

The simplified cylinder head has a height 70mm, length 170mm and width 60mm. Meanwhile, the thickness of bead plate is 0.25mm.

4.3 Loads

The engine in the research uses torque-angle tightening process with parameters 20±2Nm and 80±2° .

During the assemblage of cylinder head, the torque data obtained by torque wrench are about 60Nm. By the equation (6), the torque can be turned into the down force from bolts to cylinder head.

$$F = \frac{T}{K \times D} \quad (6)$$

F: bolt tightening force
T: Torque of bolt
K: torque coefficient
D: nominal diameter of bolts

The equivalent down force calculated is 30KN per bolts. This loads were applied onto three bolt holes of the quarter model by distribution force loads of Abaqus.

4.4 Material Properties

Material properties of each parts are shown in Table 1. Besides, other parts in this simulation are viewed as rigid bodies.

The steel plates of gasket in this paper is made of 0Cr18Ni9 stainless steel by stamping while the cylinder head is made of AC4B-T6 aluminium alloy by casting. The mechanical properties of these two materials can be found from Table 1.

Table 1. Material properties of parts

Part	Gasket	Head
Material	0Cr18Ni9 stainless steel	AC4B-T6 aluminium alloy
Young modulus	210GPa	74GPa
Poisson's ratio	0.3	0.33
Density	7.8g /mm^3	2.68g /mm^3

DOI: 10.1051/matecconf/20168002001

4.5 Element Meshing

Considering the large difference of the dimensions between length and thickness, the shell elements are used to present gasket in the simulation. C3D8R elements are used to present the structure of cylinder head. Because the cylinder block are simplified as a rigid, it doesn't need to apply meshes on it.

Approximate sizes and quantity of the models are shown in Table 2. In this simulation, the global size of gaskets (quarter of the model) is 1mm and quantity is more than 13000. The mesh density like this aims to guarantee that every bead which will deform during the simulation could consist of the meshes small enough. If the mesh density is smaller, the bead of the model will be consist of only one or two line of meshes which is not able to present the deformation. Considering the cylinder head will not have a large deformation and efficiency of this simulation, the globe size of meshes for cylinder head is chosen as 3mm.

Table 2. Meshes of parts

Part	Gasket	Head
Type of meshes	Shell S4	Solid C3D8R
Approximate global size (mm)	1	3
Quantity	13701	14582

4.6 Boundary Conditions

The boundary condition of contact between gasket and other parts is the most complex process in simulation. For the contact pairs, friction formulation is penalty in the tangential direction. Meanwhile, the normal behavior is Hard Contact. The sliding formulation is small sliding [5]. Because stopper hardly moves in the tangential direction, it can be considered to bear only normal force and be tied to block [6]. With the meshes extremely small, the surfaces of gasket are chosen as the slave face in contact pair [7].

Table 3. Definition of contact

Contact pairs	**Type of contact**	**Friction coefficient**
Head to bead plate	Surface to surface standard	0.15
Bead plate to stopper	Surface to surface standard	0.15
Bead plate to block	Surface to surface standard	0.15
Stopper to Block	Tie	—

4.7 Analysis Steps

If the cylinder head is directly put onto gasket and applied bolt force, the complex surface topography of gasket will result to no convergence. To solve this problem, three analysis steps are created.

In the initial step, cylinder head and gasket are separated to a distance of 0.05mm.

In the first step, contacts are defined. Cylinder head moves down 0.1mm slowly and makes parts into contact smooth and stable.

In the second step, down forces of bolts are applied onto the bolt hole.

4.8 Results of simulation

In Figure 5, the area between the full bead and half bead shows the largest displacement of the normal direction, which is about 0.2mm. Nevertheless, in the distribution of the contact stress, the force on this area is zero. The result presents that although the bead plate deforms visibly, it does not completely touch the upper surface of the cylinder block. Moreover, Figure 6 shows that the largest contact stress appears on inner cycles close to cylinder (To be exact, the max appears on Arc3-4). According to these results and the structure of the assembly, stoppers and full seal beads bear the majority of the down force and transmit it to cylinder block.

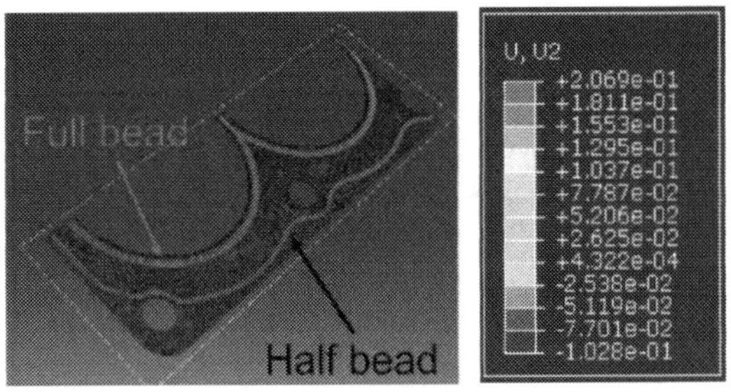

Figure 5. Result of displacement on normal direction

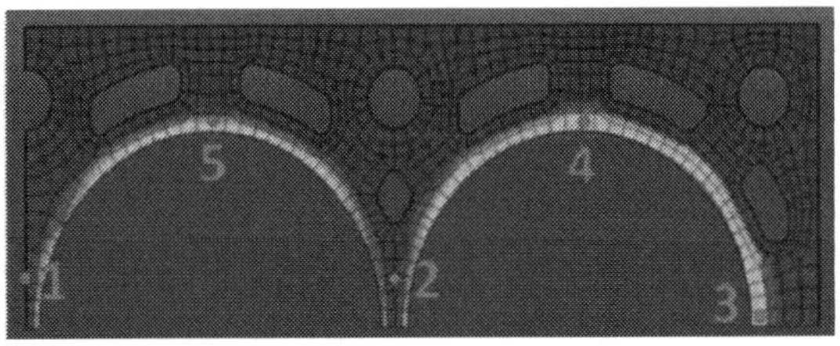

Figure 6. Result of contact stress on the lower surface of cylinder head

In my model, 5 meshes on the cylinder head are chosen to introduce the surface pressure of these meshes as the input of the simulation microscopic. The 5 points in Figure 6 present the position of the chosen meshes.

By reading the pressure data of 4 nodes of each mesh, the average pressure of the strip close to the cylinder can be calculated according the simulation result.

Table 4. Pressure results at different positions

Position	Description	Average Pressure (MPa)
1	Between cylinders	64.21
2	Between cylinders	62.16
3	On the stopper	158.12
4	On the stopper	159.78
5	On the stopper	104.54

The pressure data of the 5 positions are shown in Table 4. Three steps can be found in Table 4. The pressure at position 1 and 2 are about 60MPa while position 4 and 5 have the pressure of 160MPa. And pressure of position 5 is about 105MPa.

These data are used as the down force input in the microscopic Simulation.

4.9 Shortage of the simulation on deformation of engine assembly

The simulation of this section doesn't consider thin coating layers. So although the data of pressure at 5 positions are quite different, there is not an obvious difference about the tightened thickness in the macroscopic simulation (compression of stainless sheets under 60MPa is 0.14μm while the compression under 160MPa is 0.4μm).

So there is no significance to use the simulation in this section to analyse the change of thickness. That's why it needs to carry out microscopic simulation of the model with coatings to calculate the tightened thickness.

5 Simulation on coatings' microscopic deformation

The macroscopic simulation in section 3 does not contain coatings on the sheet for their extremely tiny scales. But actually, because the material of coatings, fluororubber, is very soft compared to the metal sheet, coatings undertake the vast majority of deformation. So it is necessary to analyse the microscopic deformation behaviour of coatings in a partial model.

5.1 Model of microscopic simulation

Figure 7 shows the section view of the gasket itself and the assembly including gasket, cylinder head and block (Another clearer schematic in Figure 4(c) can help to present the structure of this tiny region). As shown in this figure, three coatings are created on steel plates. The pressure is applied on the cylinder head and then transmitted to gasket. Block in this simulation is considered as a rigid which is fixed as a support.

Because of the rotational symmetry of the cylinder, the simplified axisymmetric model in Figure 7 can represent almost all the features of the assembly. The model takes coatings into consideration and therefore it can be used to calculate the tightened thickness.

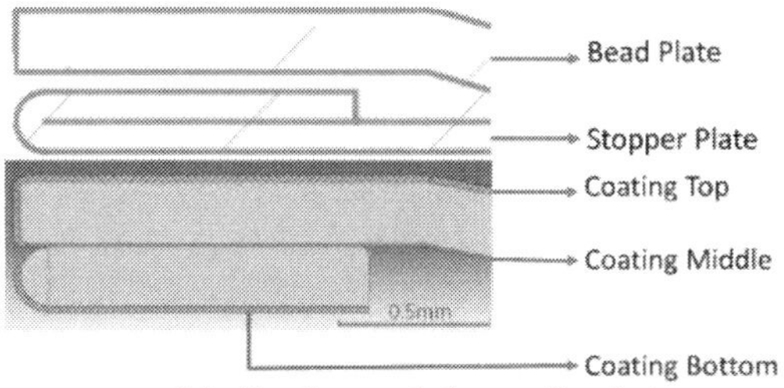

(a) Coatings and sheets of gasket

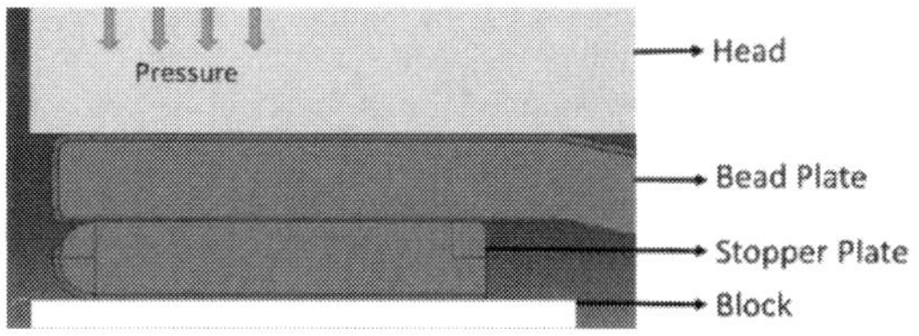

(b) The assembly of model in simulation

Figure 7. Section view of gasket and axisymmetric model

Actually, this structure (including head, bead plate, stopper and block) is same at each position along the inner edge of the cylinder. Nevertheless, because of the specific positions of the tightening bolts, data of pressure along the edge are quite different (as shown in Table 4). If the different pressures are used as the input to the model in Figure 7, the fluctuation of tightened thickness can be expressed numerically.

5.2 Materials

The coatings are made from RFC101D fluororubber. In the simulation, the property of the fluororubber can be expressed by Mooney-Rivlin model. The strain energy density function for an incompressible Mooney–Rivlin material is as:

$$U = C_{10}(\bar{I}_1 - 3) + C_{01}(\bar{I}_2 - 3) + \frac{1}{D}(J^{el} - 1)^2$$

In the equation, where C_{10} and C_{01} are empirically determined material constants, and I_1 and I_2 are the first and the second invariant of the unimodular component of the left Cauchy–Green deformation tensor. And D is a material constant related to the volumetric response. In the simulation, hardness of fluororubber is 70. C_{10} and C_{01} are assigned as 0.4276 and 0.1007. D is about 0.03.In many research of simulation of rubber, D can be considered as 0 while the rubber is considered as the completely incompressible material. But in the ABAQUS/explicit product, a completely incompressible material is not allowed. So the D is assigned as a tiny value.

Materials of gaskets and cylinder head are same as the macroscopic simulation.

5.3 Meshes

Meshes used in this section have several properties. In the abaqus/explicit, hyperelastic material can be only expressed by hybrid type elements. Moreover, to use the ALE adaptive method, the reduced integrated type

DOI: 10.1051/matecconf/20168002001

elements should be used. So, CAX4R elements (4-node bilinear axisymmetric quadrilateral, reduced integration, hourglass control) are used for the fluororubber material.

Arbitrary Lagrangian Eulerian (ALE) adaptive meshing used in the simualtion can maintain a high-quality mesh throughout an analysis, even when large deformations.

5.4 Loads

Table 4 in chapter 3.8 shows the input of the simulation in this section. These inputs are applied as the pressure on the simplified cylinder head which will transmit the pressure to the gasket.

5.5 Results of simulation under different pressures

Figure 8 presents the extrusion of the coatings. The coatings can hardly be compressed because of the high Poisson ratio of material. But the material will deform toward the gap and because of the extrusion of material, thickness of coatings will reduce. The phenomenon is shown in Figure 8.

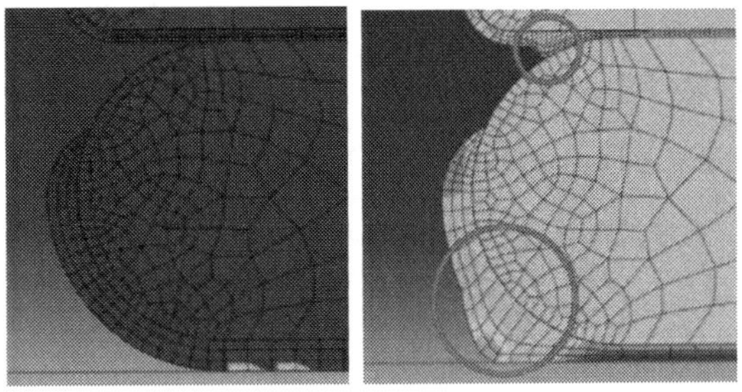

Figure 8. Examples for deformation of coatings

The thickness results of simulation are obtained by measure the distance between the lower surface of head and upper surface of block.

After measuring the displacement of the nodes on lower surface of cylinder head, groups of data under a specific pressure can be obtained. Table 6 shows the average thickness of each group. These results of thickness will be compared to experiments in next chapter.

Table 5. Average thickness under different pressures

Position	Pressure (MPa)	Average thickness (mm)
1	64.21	0.5332
2	62.16	0.5338
3	158.12	0.5148
4	159.78	0.5146
5	104.54	0.5223

6 Experiment comparison

6.1 Material and Boundary conditions

This paper uses the lead pellet test which is commonly used industrially to measure the true value of tightened thickness.

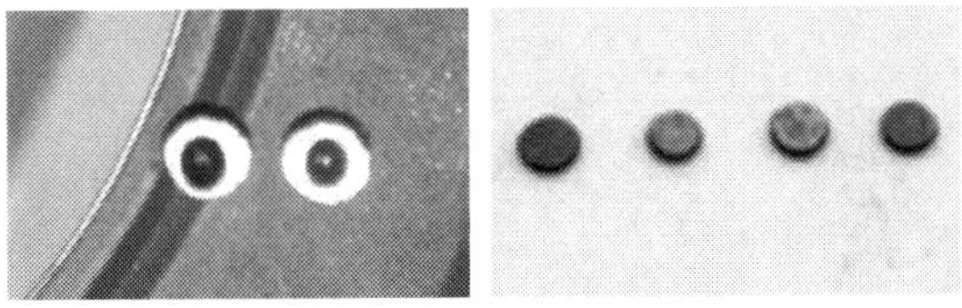

Figure 9. Lead pellets in holes on gasket and deformed pellets

After clean up the surfaces of head and block, lead pellets with a diameter of about 1.5mm are placed in holes which are 4mm in diameter (as shown in Figure 9 (left)). The holes are processed on the gasket by wire-electrode cutting.

The cylinder head is installed and bolts are tightened with the standard torque and angle. After standing for 5 minutes, the cylinder head is removed and the deformed pellets are collected to measure the thickness by micrometer caliper.

Figure 9 (right) shows the deformed pellets, the pellets are turned into discs with smooth surfaces. Lead is one of the softer metals which could easily deformer and will not spring back after removing the loads. So a lead pellet in the slit will express the tightened thickness at its position.

6.2 Positions measured

The experiment chose 2 groups of positions for detection which is shown in Figure 10.

Group A is placed at the edge of cylinder. Group B is placed at the corners between two cylinders.

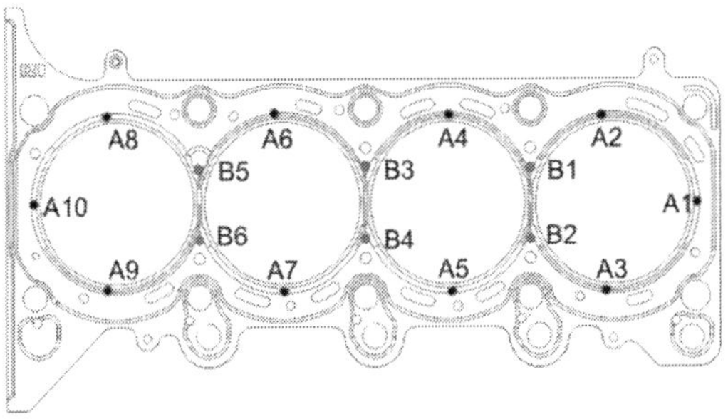

Figure 10. Positions to place the lead pellets

6.3 Comparison of results from simulation and experiment

With 6 gaskets of same type, 6 experiments are carried out. The 6 groups of data are averaged and the tightened thickness of each position can be obtained.

Moreover, the simulation obtains 5 data of thickness. According to the symmetry of cylinder head and gasket. The quarter model in Figure 4 can be generalized so that these 5 points can describe the thickness distribution simplistically.

The results of simulation and experiment have good agreement in figure 11 and Table 7 which verify the validity of the simulation method in this paper.

Table 6. Comparison of the thickness measured and simulated

Position Group A	Thickness (measured) (mm)	Thickness (simulated) (mm)	Relative error (%)
A1	0.5128	0.5148	0.3835
A2	0.5138	0.5146	0.1492
A3	0.5133	0.5146	0.2468
A4	0.5192	0.5223	0.6035
A5	0.5183	0.5223	0.7653
A6	0.5188	0.5223	0.6682
A7	0.5185	0.5223	0.7329
A8	0.5178	0.5146	0.6244
A9	0.5137	0.5146	0.1817
A10	0.5110	0.5148	0.7436

Position Group B	Thickness (measured) (mm)	Thickness (simulated) (mm)	Relative error (%)
B1	0.5298	0.5338	0.7487
B2	0.5308	0.5338	0.5589
B3	0.5303	0.5332	0.5405
B4	0.5328	0.5332	0.0688
B5	0.5328	0.5338	0.1814
B6	0.5330	0.5338	0.1501

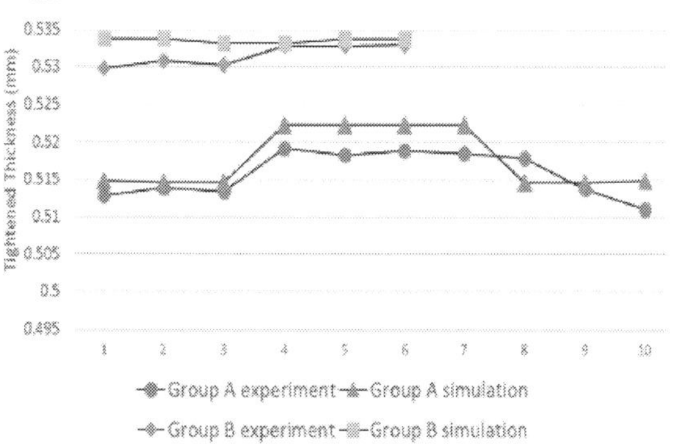

Figure 11. Comparison between simulation and experiment

The data from the lead pellets test and simulation can provide two main results:

1. From Table 7, an obvious difference between group A and group B can be found. Tightened thickness data of group B are all above 0.530mm. Meanwhile, data of group A are all smaller than 0.520mm.
2. For group A, positions on the middle region (A4~A7) of gasket have a larger tightened thickness than the positions of two sides (A1~A3, A8~A10). The reason is the tiny deformation of cylinder head.

Both the results of experiments and simulation show the tightened thickness distribution when the cylinder head is installed. Under the high load of bolts, the head deforms to an arch. The middle region has a larger thickness than side region.

7 Conclusion

1. In this research, a method to study tiny deformation of gasket is established. Deformation of gasket compressed by cylinder head and block was divided into two steps.

   First step focuses on the deformation of macroscopic geometrical structure including beads and stoppers of gasket. But because the simulation does not consider coatings which occupy large share of compressive deformation, the data of tightened thickness are not reasonable. So, the microscopic simulation of a tiny piece model near the edge of cylinder is carried out which focuses on the pure compression of the sheet and soft coatings.

   These two simulations with quite different scales are linked by the contact stress. The contact stress results are taken from the first simulation and used as the input of the second microscopic simulation.

   The combination of these two steps of simulation embodies almost all the deformation behaviours near the contact surface during the assembly of engine completely from which a conclusion is obtained that the fluctuation of tightened thickness mainly comes from the deformation of coatings under different contact pressure.

   The lead pellet test compares the result of simulations and verifies the feasibility of the method.
2. The aim of the research is to analyse the fluctuation of tightened thickness. The results of simulations and experiments present the diversity of tightened thickness at different positions. The two side cylinders are compressed tighter than two middle cylinders. The diversity mainly comes from the structure of cylinder head and the bolt load. To avoid the difference of thickness, an effective method is to reduce the torque of bolts located at four corners. But it will result in the leakage. If it is possible to reduce the fluctuation of tightened thickness by this adjustment will be included in the next step of research.

Acknowledgment

This work is supported by the National Natural Science Foundation of China under Grant No. 51535007 and National Science and Technology Major Project of the Ministry of Science No.2014ZX04002011

References

1. V. Hettich, M. Jaeckle, G. Lechner, Tribologie und Schmierungstechnik, **42(2)**: 80-85 (1995)
2. C. Bramsiepe, L. Pansegrau, G. Schembecker, Chemical Engineering Journal, **159(1)**: 11-16 (2010)
3. A. Ghasemi, SAE Technical Paper, (2012)
4. Cho S S, Han B K, Lee J H, et al, International journal of automotive technology, **7(1)**: 109-115 (2006)
5. C. Lee, K. Chiang, W. Chen, et al, Finite Elements in Analysis and Design, **41(11)**: 1160-1174 (2005)
6. H. Zhang, J. Ni, Q. Du, et al, Chinese Internal Combustion Engine Engineering, **5**: 017 (2011)
7. P. Yu, X. Shi, G. Li, et al, Lubrication Engineering, **5**: 035 (2006)

MATEC Web of Conferences **80**, 02002 (2016) DOI: 10.1051/matecconf/20168002002
NUMIFORM 2016

A new micro scale FE model of crystalline materials in micro forming process

Liang Luo[1], Haibo Xie[1], Dongbin Wei[1, 2], Xiaogang Wang[3], Cunlong Zhou[3] and Zhengyi Jiang[1, a]

[1]*School of Mechanical, Materials & Mechatronic Engineering, University of Wollongong, Wollongong NSW 2500, Australia*
[2]*School of Electrical, Mechanical and Mechatronic Systems, University of Technology, Sydney, NSW 2007, Australia*
[3]*School of Material Science and Engineering, Taiyuan University of Science and Technology, Taiyuan, Shanxi 030024, China*

Abstract. Micro forming of metals has drawn global attention due to the increasing requirement of micro metal products. However, the size effects become significant in micro forming processes and affect the application of finite element (FE) simulation of micro forming processes. Dividing samples into small areas according to their microstructures and assigning individual properties to each small area are a possible access to micro forming simulation considering material size effects. In this study, a new model that includes both grains and their boundaries was developed based on the observed microstructures of samples. The divided subareas in the model have exact shapes and sizes with real crystals on the sample, and each grain and grain boundaries have their own properties. Moreover, two modelling methods using different information from the microstructural images were introduced in detail. The two modelling methods largely increase the availability of various microstructural images. The new model provides accurate results which present the size effects well.

1 Introduction

Micro manufacturing has drawn global attention due to the increasing requirement on the micro products. Once scaled down to a micro scale, significant size effects on the micro fabrication processes and on the micro products become considerable [1-3]. Individual grain impacts on deformation behaviour substantially and causes great data scatter in experiments. Therefore, the finite element (FE) simulation, which generally cannot reflect these phenomena, needs to consider the size effects. The modified Swift and Hall-Petch (H-P) models were introduced to present strain-stress relationships on the micro scale [4, 5]. Thus, simulation using these accurate micro scale strain-stress relationships can provide acceptable results. Additionally, the surface layer model and the modified surface layer model were adopted to explain the decrease of flow stress with decreased sample sizes [6, 7]. However, all the aforementioned models cannot explain the increasing data scatter from experiments and the localised deformation behaviour caused by the individual grain. To further reveal the size effects, the Voronoi models were developed [8-10]. The sample was divided into small subareas according to the Voronoi structures and the small subareas were assigned different mechanical properties. Consequently, the size effects, such as the decrease of the flow stress and the locally special deformation, can be expressed by the distribution of mechanical properties on the subareas. Furthermore, the centroidal Voronoi structure with similar Voronoi cells is comparable to the annealed metals with equiaxed crystals. Thus the Voronoi models can present the sample on a grain level. Nevertheless, not every metal has the equiaxed crystalline microstructure and the equiaxed crystals are not exactly the same as the centroidal Voronoi cells. Although the distribution and sizes of the Voronoi cells can be adjusted by a density function, the cells distribution can hardly match the real microstructures of the samples. Due to a large quantity of grains on the sample, the grains are generally classified into several groups, and each group has one kind of mechanical properties. As there are no clear classification criteria, all the Voronoi cells are organised randomly. Grain boundaries, that play an important role in the micro forming process, are generally ignored in the Voronoi simulation models. Thus, the Voronoi models are not accurate enough compared to the real micro samples. To precisely account for the size effects, a new micro scale model is highly required. The subareas on the new model should have exact shapes and sizes to the crystals on the sample. The grains' arrangement is based on reasonable classification rules, such as the actual grains sizes or grains geometrical orientation. The grains and grain boundaries should be considered separately.

In this study, a real microstructure simulation model considering grains and their boundaries separately was developed by two different modelling methods. The two modelling methods were explained and the arrangement of grains was discussed. Next, the developed models were utilised in the micro deep drawing simulation.

[a] Corresponding author: jiang@uow.edu.au

Finally, simulation results were discussed and compared with experimental results.

2 Modelling Method I

The first modelling method uses the microstructural images with clear grain boundaries from various instruments, such as metallurgic microscope and electron backscatter diffraction (EBSD). The grains were detected and then all the grain boundaries were identified through an image processing program in MATLAB. Fig. 1(a) shows a typical microstructure image from the EBSD. The microstructural image was firstly read into MATLAB and then transferred into a binary image where there are only black grain boundaries and white grains. During the transformation, noise and small holes were eliminated. Therefore, non-grain information on the image was omitted. Subsequently, the binary image was treated with the Moore-Neighbour tracing algorithm modified by Jacob's stopping criteria. All the grains, the individual closed subareas, were then detected and sorted, as shown in Fig. 1(b), and their information was stored in MATLAB. During the detection, each grain's area, geometrical centre and geometrical orientation can be calculated. The blue ports on Fig. 1(b) are the grains geometrical centres.

Simultaneously, a normal finite element model was generated where all the elements were in a single part with the same mechanical properties. The elemental and nodal information was then imported into MATLAB. The centre of each element was used to present the elemental positon. Then, the elements were mapped onto the grains detected by MATLAB. By comparing all the grains and the elements positions, the elements inside one grain were assigned the corresponding grain's ID number. Accordingly, the elements in one grain were set as one individual part with separate mechanical properties. By contrast, the elements that belong to none of the gains were defined as the grain boundaries. Finally, the updated relating elemental information, part and material properties indexes, were written into the pre-built normal finite element model. Fig. 1(c) displays the final finite element model based on the microstructure as shown in Fig. 1(a).

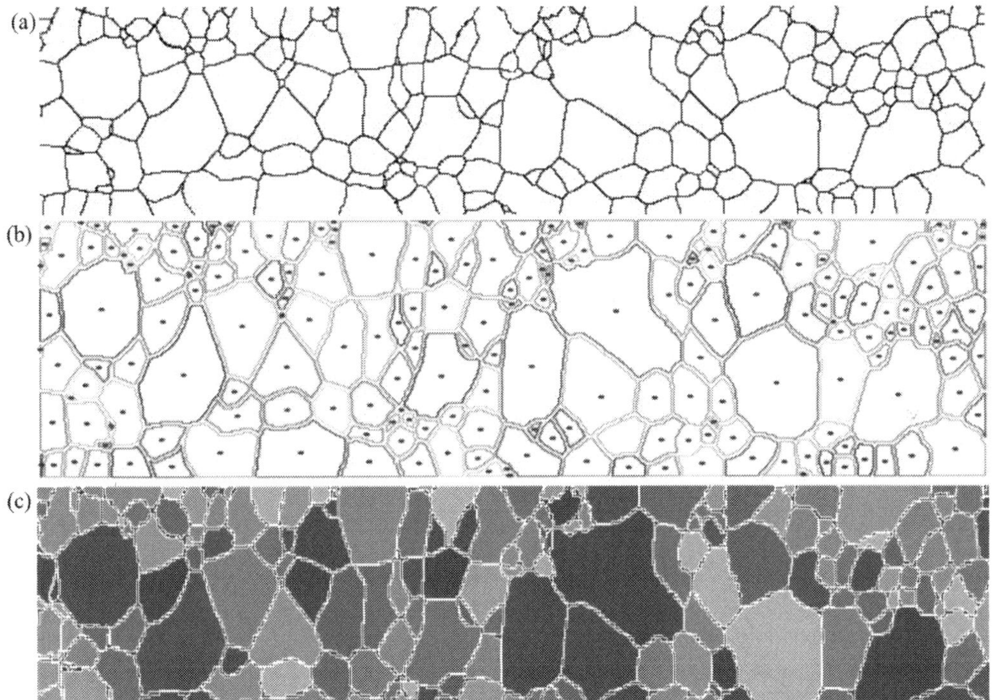

Figure 1. (a) Microstructure of a sample from EBSD, (b) its corresponding geometry detected by MATLAB and (c) corresponding simulation model.

As the simulation model is on the micro scale, there are a great quantity of grains and their accordant mechanical properties in the simulation model. Thus, the simulation model can be very large and over the executable size of the commercial simulation software. Therefore, all the grains should be categorised as several groups, and each group has one type of mechanical properties. The overall simulation model size can then be greatly reduced due to the limited information about material properties. In the aforementioned Voronoi models, the grains were classified randomly or based on the grains' statistic information. By contrast, the grains areas and their geometrical orientations can be utilised for reasonable arrangement in the new model. Generally, samples with relatively large grains have relatively low flow resistance and low tensile strength. Accordingly, large grains were set low mechanical properties, while small grains were assigned high mechanical properties.

Furthermore, the grain boundaries were allocated mean mechanical properties of all grain groups. Fig. 2(a) shows the grain area distribution of the sample illustrated in Fig. 1(a), and Fig. 2(c) displays the corresponding finite element simulation model. Five colours indicated five area levels of grains and five grain groups. Regarding grain orientation, although the orientation is calculated based on grain's geometry, this geometry indicates the former processing of the material and thus the preferred deformation direction of the grain. The geometrical orientation is the angle (in degrees ranging from 0 to 180 degrees) between the x-axis (horizontal axis) and the major axis of the ellipse that has the same second-moments as the detected grain. Thus, all the grains orientations were analysed and all the grains were classified into several groups accordingly. In addition, the anisotropic material model can be utilised and relating parameters indicating anisotropy were set according to the orientation angle. For example, the in-plane anisotropic value, the R value defined as Eq. (1), is generally used for presentation of material anisotropy. At any angle $\emptyset$, $R_\emptyset$ can be calculated from Eq. (2). According to the 3-parameter Barlat yielding criteria, Eq. (3), anisotropic properties of the grains with different orientation angles can be considered. Fig. 2(b) presents the orientation distribution of the grains and Fig. 2(d) exhibits the corresponding simulation model.

$$R = \frac{\frac{dW}{d\varepsilon}/W}{\frac{dT}{d\varepsilon}/T} \tag{1}$$

$$R_\emptyset = \frac{2m\sigma_Y^m}{(\frac{\partial\Phi}{\partial\sigma_x}+\frac{\partial\Phi}{\partial\sigma_y})\sigma_\emptyset} - 1 \tag{2}$$

$$2\sigma_Y^m = a|K_1 + K_2|^m + a|K_1 - K_2|^m + c|2K_2|^m \tag{3}$$

$$K_1 = \frac{\sigma_x + h\sigma_y}{2} \tag{4}$$

$$K_2 = \sqrt{\left(\frac{\sigma_x - h\sigma_y}{2}\right)^2 + p^2\tau_{xy}^2} \tag{5}$$

where R is the in-plane anisotropy value, W and T are the width and thickness respectively, $\sigma_\emptyset$ is the uniaxial tension in the $\emptyset$ direction, σ_Y is the yield stress, m is 8 for face cantered cubic materials, σ_x and σ_y are the stress along x and y directions respectively, τ_{xy} is the shear stress, K_1 and K_2 are given by Eqs. (4) and (5) respectively, a, c, h and p are the orientation related material parameters and for isotropic material a=c=h=p=1.

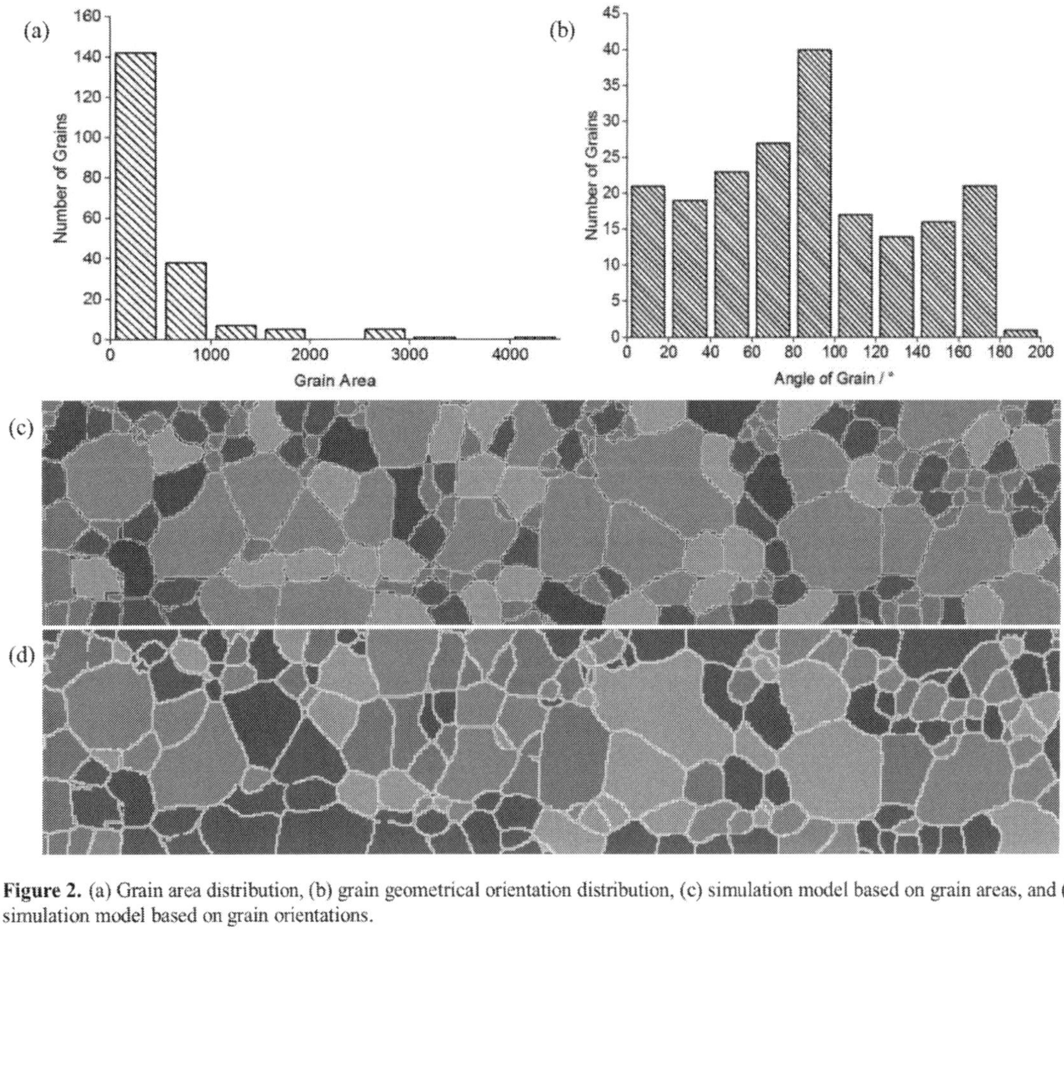

Figure 2. (a) Grain area distribution, (b) grain geometrical orientation distribution, (c) simulation model based on grain areas, and (d) simulation model based on grain orientations.

If the microstructural information is difficult to obtain for some tiny samples, such as in-plane microstructures of ultra-thin sheets used for micro sheet forming, the Voronoi structures can also be used via the first modelling method. The grains and grain boundaries can also be considered separately. The grains areas and orientations can be computed and used for classification of the Voronoi cells. Fig. 3(a) illustrates a centroid Voronoi structure, Fig. 3(b) the detected grains from the Voronoi structure and Fig. 3(c) its corresponding simulation model considering both grains and grain boundaries.

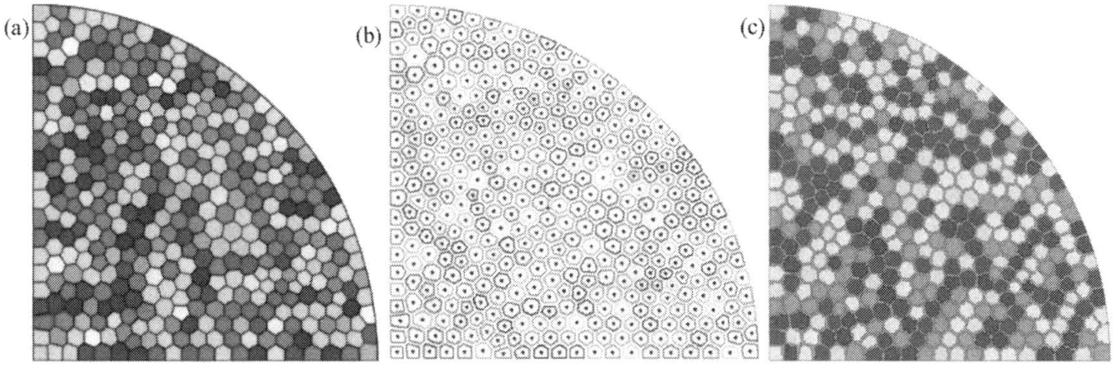

Figure 3. (a) Centroidal Voronoi structure, (b) detected grains and (c) corresponding simulation model from the first modelling method.

3 Modelling method II

There is another method to generate the real microstructure model from colourful or grey level images. If the images are coloured by the orientation or the Eulerian angles, the second modelling method can be used to build the real microstructural simulation model. Fig. 4(a) exhibits the orientation information obtained from EBSD where the colour indicates different orientations of the grains. Unlike the first modelling method, the colourful image was read into MATLAB without binarisation process. The colour information was discretised into the red, green and blue (RGB) system and kept as a 3D matrix composited by three 2D matrixes. Moreover, every pixel on the image has the red, green and blue values ranging from zero to 255 from the three 2D matrixes respectively. The position of a data point in the 2D matrix is the pixel positon on the image. Therefore, the pre-built mesh was mapped onto the colourful image, and each element was projected to a pixel according to their central coordinates. Similarly, all the grains were arranged as several groups with individual mechanical properties based on the RGB colour values. As the grain boundaries on the image were of black, the elements being of black were defined as the grain boundaries. Fig. 4(b) shows the developed simulation model based on the colourful image including the orientation information. Therefore, the real microstructural simulation model considering grains and their boundaries respectively can be generated. If grey level images were used, the RGB values were replaced by the grey level value ranging from zero to 255 too. Similarly, all the elements were arranged according to the grey level values. As the colour information indicates orientation, grains with close colour were naturally defined as one part with a type of mechanical properties. The grain boundaries were also set the average mechanical properties of all grain groups.

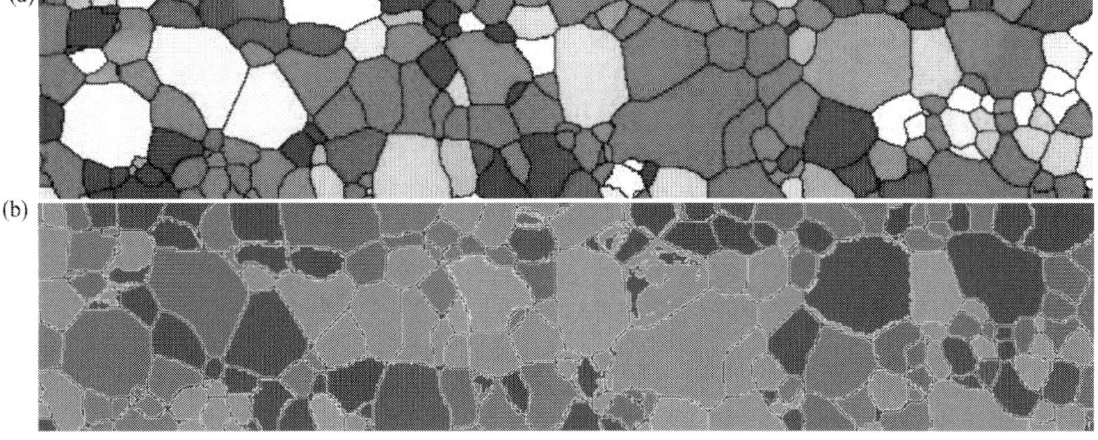

Figure 4. (a) Orientation image from EBSD and (b) its corresponding simulation model.

If the microstructural information is impossible, the Voronoi structures can also be utilised in the second modelling method. Fig. 5(a) shows the centroidal Voronoi structure representing the annealed metal with equiaxed crystals, and Fig. 5(b) displays its FE simulation model. All the grains were classified based on the Voronoi cell colours and grain boundaries were identified by black.

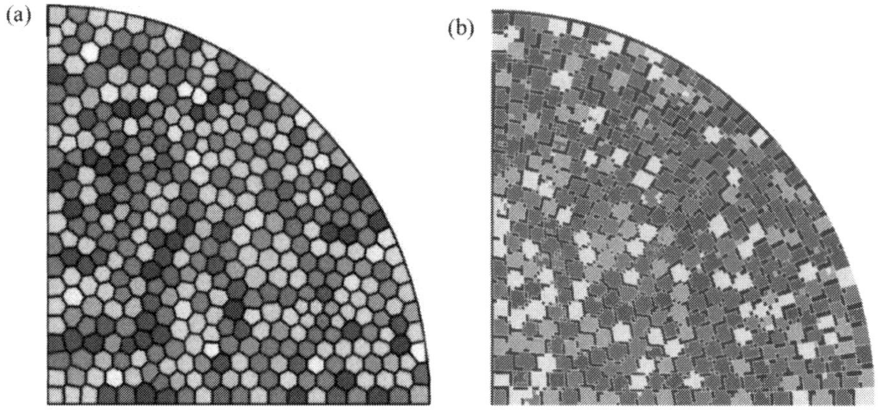

Figure 5. (a) Centroidal Voronoi and (b) its FE simulation model from the second modelling method.

4 Discussion

The mesh of the new model is independent on the images or the Voronoi structures for both two modelling methods. Thus, the total simulation model size can be small and executable for the commercial software while the mesh quality can be kept at a high level. However, it is better to generate the mesh based on the grains and their boundary sizes. With relatively large elemental sizes compared to the grain sizes, the grains cannot be presented smoothly and the grain boundaries might be lost.

Regarding the first modelling method, all the grains are detected as the closed subareas. Therefore, grain geometrical information, such as the area and geometrical orientation, can be obtained and used for further sorting of the grains. However, for the ultra-small grains on a low-resolution image, the detected grains may be of lines or curves but not closed geometries due to the boundary searching algorithm applied in the program. Although this information can be used, it is generally deleted as noise.

The second modelling method links the element to its corresponding RGB or grey level values. Therefore, the meaning of the colour values is included in the model naturally. Moreover, the grains are grouped automatically and grain boundaries can be found as they have fixed colour value (normally black). Due to the colour transition and limited number of grain groups, it is possible to arrange the elements belonging to one grain in different grain groups.

To validate this new model, micro deep drawing models were built and run. The Voronoi structure was utilised and both grains and their boundaries were considered in the new model. As the Voronoi structure was used, the new models built by the two modelling methods were almost the same, except the grain group distribution. Fig. 6(a) and (b) display the blank models from the first and the second modelling methods respectively, Fig. 6(c) and (d) show the effective stress of the drawn cups using the first and second blank models with an average grain size of 40 μm respectively. Localised deformation and material inhomogeneity can be reflected by the new models. Therefore, the stress is uneven at the same height level on the drawn cup, as shown in Fig. 6(c) and (d).

A Voronoi model without consideration of grain boundaries and a normal deep drawing model were also developed for comparison. In addition, the blanks with thickness of 50 μm were annealed at 1100 ℃ and drawn into micro cups. The microstructure of the blank was of equiaxed crystals and the average grain size was 40 μm. Fig. 7(a) displays the mouth of the drawn cup and Fig. 7(b-d) exhibit thickness distribution of the drawn cups mouths from the developed new model, a Voronoi model without consideration of grain boundaries and a normal model without account of material inhomogeneity, respectively. The maximum wall thickness at the cup mouth area was selected for comparison between the simulation and the experimental results. Without consideration of the material microscopic heterogeneity, localised deformation cannot be presented and therefore the maximum thickness was averaged resulting in the lowest peak thickness values among all the simulation. By contrast, the new model and the Voronoi model considered microscopic heterogeneity. Higher maximum thickness than that in the normal model was obtained from these two models. Furthermore, without buffer of the grain boundaries, the Voronoi model presented the largest thickness among all simulations. Fig. 8 compares the maximum thickness between both simulation and average experimental results. The developed model presented closer results to the experimental results than the other simulation models.

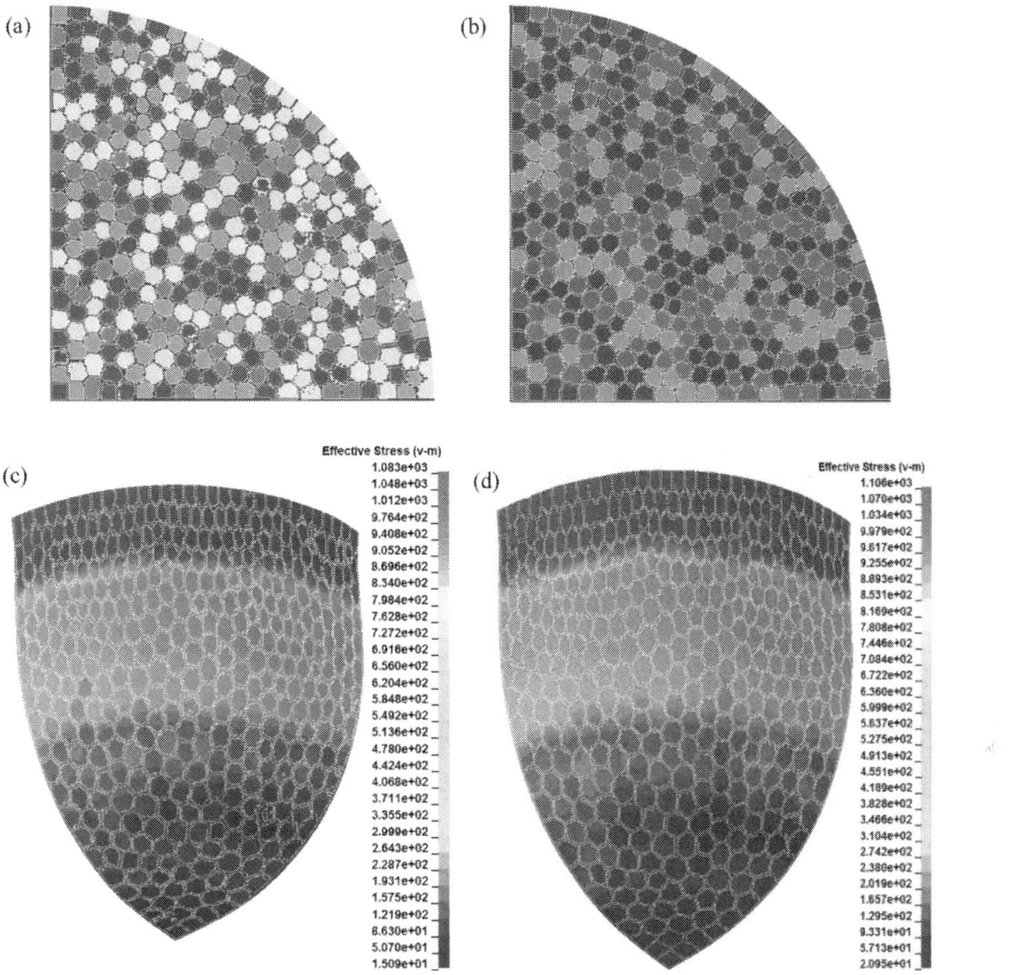

Figure 6. (a), (b) Blank models from the first and second modelling methods respectively and (c), (d) effective stress of the drawn cup from the first and the second models.

5 Conclusions

The micro scale simulation model considering both grains and their boundaries can be developed successfully through the two modelling methods. Moreover, the two modelling methods can include the primary microstructural information.

The first method utilises various microstructural information with clear boundaries, such as the images from the metallurgic microscope and EBSD, to detect the grains and then the grain boundaries. The grain sizes and geometrical orientations can be calculated and used for classification of grains. Therefore, the grains can be reasonably categorised based on these geometrical information. Furthermore, the Voronoi structures can also be applied via this modelling method, and therefore grain boundaries can be considered. The second modelling method utilised the colour information of the image to detect grain boundaries and classify grains naturally. Similarly, the Voronoi structures can be applied via the second method, if the real microstructure is difficult to obtain.

The developed new model via either of the two modelling methods can present the samples on the grain level considering grains and grain boundaries individually. Therefore, the size effects, especially the localised deformation and influences of the individual grain, can be well presented. The developed new model provides closer results to the experimental results, regarding the maximum thickness, than the Voronoi model and the normal model. This indicates the importance of consideration of grain boundaries in micro FE simulation.

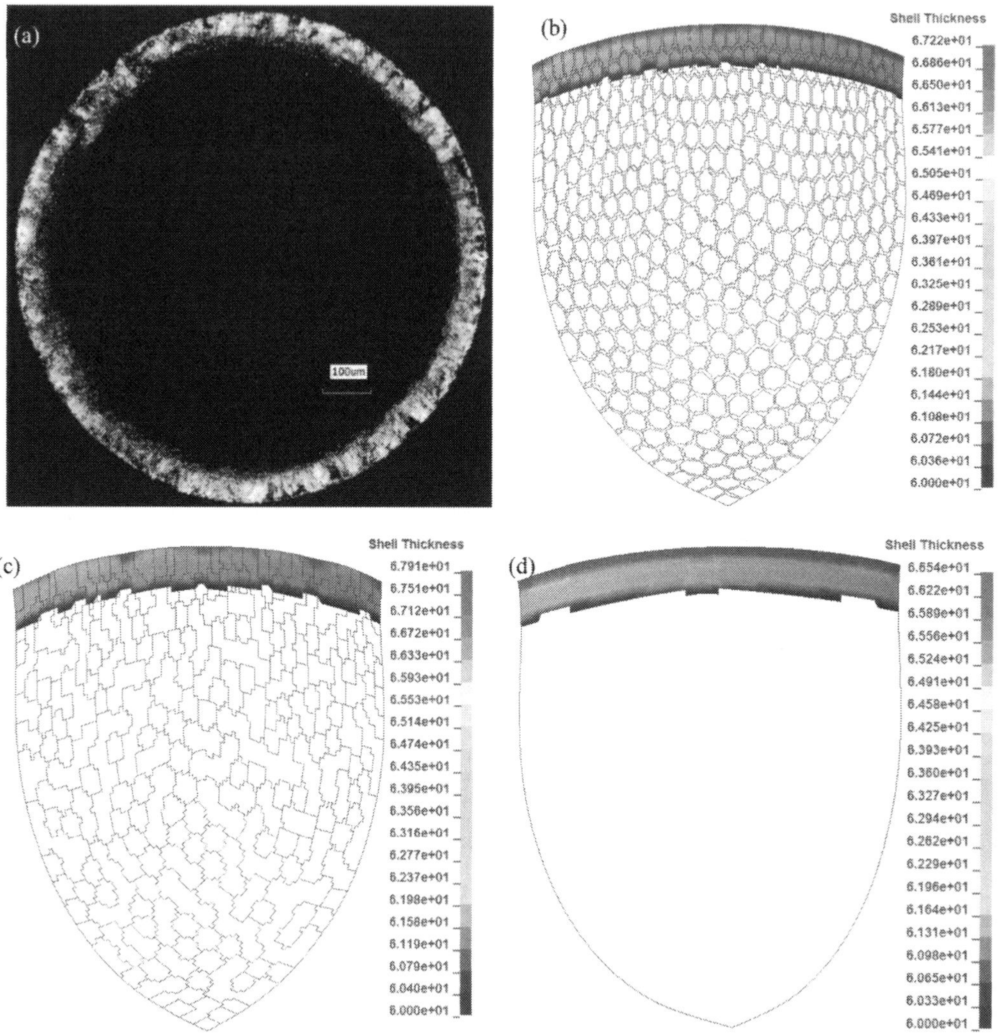

Figure 7. (a) Drawn cup with 1100 ℃ annealed blank and maximum thickness distribution from the simulation with (b) the developed new model, (c) the Voronoi model and (d) the normal model.

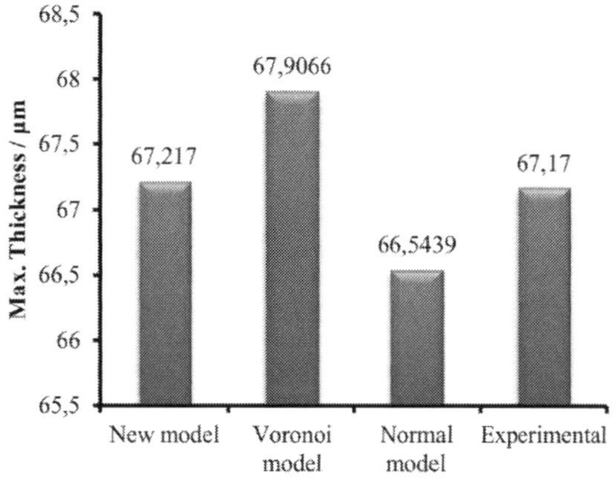

Figure 8. Comparison of the maximum thickness among different simulation models and average experimental results.

Acknowledgements

The first author greatly thanks for the finical support from the Chinese Scholar Council (CSC 201206160011) and the International Postgraduate Tuition Award (IPTA) offered by the University of Wollongong. This work is also supported by Australian Research Council (ARC) for Future Fellowship, and Open Research Fund from the Sate Key Laboratory of Rolling and Automation, Northeastern University, China (Contract No.: 2014003). The authors also thank that Mr. Guoqing Zu provided the experimental images of Figs. 1(a) and 4(a).

References

1. Z. Xu, L. Peng, X. Lai, M. Fu, Mater. Sci. and Eng.: A. **611** 345-353 (2014)
2. A. Molotnikov, R. Lapovok, C.F. Gu, C.H.J. Davies, Y. Estrin, Mater. Sci. and Eng.: A. **550** 312-319

(2012)
3. F. Vollertsen, Key Eng. Mater. **473** 3-12 (2011)
4. G.C. Wang, W. Zheng, T. Wu, H. Jiang, G.Q. Zhao, D.B. Wei, Z.Y. Jiang, J. Mater. Process Technol. **212** 678-684 (2012)
5. F.H. Yeh, C.L. Li, Y.H. Lu, J. Mater. Process Technol. **201** 237-241 (2008)
6. N. Ma, Huazhong Univ. of Sci. and Technol. (2005)
7. N. Ma, X.H. Dong, J. Shanghai Jiaotong Univ. **12.3** 323-327 (2007)
8. L. Luo, Z.Y. Jiang, H.N. Lu, D.B. Wei, K.Z. Linghu, X.M. Zhao, D. Wu, Procedia Eng. **81** 2409-2414 (2014)
9. H.N. Lu, D.B. Wei, Z.Y. Jiang, X.H. Liu, K. Manabe, Comp. Mater. Sci. **77** 44-52 (2013)
10. C.A. Sweeney, P.E. McHugh, J.P. McGarry, S.B. Leen, Int. J. Fatigue **44** 202-216 (2012)

Introduction to the level-set full field modeling of laths spheroidization phenomenon in α/β titanium alloys

D. Polychronopoulou[1,a], N. Bozzolo[1], D. Pino Muñoz[1], J. Bruchon[2], M.Shakoor[1], Y. Millet[3], C. Dumont[4], I. Freiherr von Thüngen[5], R. Besnard[6] and M. Bernacki[1]

[1]*MINES ParisTech, PSL Research University, CEMEF - Centre de mise en forme des matériaux, CNRS UMR 7635, CS 10207, rue Claude Daunesse, 06904 Sophia Antipolis Cedex, France*
[2] *École Nationale Supérieure des Mines de Saint-Étienne. Centre Sciences des Matériaux et des Structures. Département Mécanique et Procédés d'Elaboration. 158 cours Fauriel, 42023 Saint-Etienne Cedex 02, France*
[3]*Timet Savoie, 62 avenue Paul Girod, 73400 Ugine, France*
[4]*Aubert & Duval, rue des Villas, 63770 Les Ancizes-Comps, France*
[5]*Safran, SafranTech - Pôle Matériaux et procédés, Rue des Jeunes Bois, Châteaufort, CS 80112, 78772 Magny-Les-Hameaux, France*
[6]*CEA Valduc, 21120 Is-Sur-Tille, France*

Abstract. Fragmentation of α lamellae and subsequent spheroidization of α laths in α/β titanium alloys occurring during and after deformation are well known phenomena. We will illustrate the development of a new finite element methodology to model them. This new methodology is based on a level set framework to model the deformation and the *ad hoc* simultaneous and/or subsequent interfaces kinetics. We will focus, at yet, on the modeling of the surface diffusion at the α/β phase interfaces and the motion by mean curvature at the α/α grain interfaces.

1 Introduction

Two-phase α/β titanium alloys are materials with numerous applications in different industrial domains, mostly due to their attractive mechanical properties. They exhibit different microstructures depending on the applied thermomechanical path. Starting from an initial stable microstructure of α lamellae inside β grains, the phenomenon of spheroidization can be observed during and after the microstructure deformation. More precisely, during the deformation, the fragmentation of α lamellae and the subsequent spheroidization of α laths occurs.
Spheroidization has received considerable attention due to its importance in microstructural control. The new spheroidized microstructure shows enhanced strength and ductility, so evidently, this phenomenon raises high interest for the industrial applications [1].

In this paper, we will illustrate a new finite element (FE) methodology in order to model these microstructural evolutions. The interest is focused on the first mechanisms occurring during spheroidization at the lamellae interfaces without considering the microstructure deformation modeling. The α/α grain interfaces are introduced arbitrarily leading to surface diffusion at the α/β phase interfaces and the motion by mean curvature at the α/α grain interfaces. In order to model efficiently this interfacial kinetics, a level set framework was introduced.

Some basic cases of surface diffusion will be detailed in order to illustrate the first steps and the numerical choices of the methodology.

2 Describing the physical problem

According to Semiatin [2], the first basic mechanisms occurring during hot deformation, in order to get the spheroidized microstructures, are the surface diffusion at the α/β interfaces and the motion by the mean curvature at the α/α grain interfaces. The motivation for the formation of grooves in α lamellae and the consequent splitting of them, is the tendency of the different interfaces to evolve in order to minimize the total interfacial energy. Grooving is usually initiated by atomic scale processes near the region of α/α grain boundary and α/β interface

[a] Corresponding author: danai.polychronopoulou@mines-paristech.fr

intersection. Next section illustrates the governing equations of this phenomenon.

2.1 Motion by surface diffusion

According to Mullins [3], in order to describe the atoms flow at the α/β interface we can consider a surface flux $\vec{j}$:

$$\vec{j} = \nu\vec{\upsilon}, \quad (1)$$

where ν denotes the number of drifting atoms per unit area and $\vec{\upsilon}$ denotes the average velocity of these drifting atoms. Assuming local equilibrium we can express $\vec{\upsilon}$ with the Nerst-Einstein formula as following:

$$\vec{\upsilon} = \frac{D_{\alpha\beta}}{\dot{k}T}\nabla_s \mu, \quad (2)$$

where $D_{\alpha\beta}$ denotes the surface diffusivity of the α/β interface, μ the chemical potential, $\dot{k}$ the Boltzmann constant and T the absolute temperature. The ∇_s operator corresponds to the surface gradient operator defined as the tangential component of the gradient:

$$\nabla_s\mu = \nabla\mu - (\nabla\mu \cdot n)n = P\nabla\mu, \quad (3)$$

with $\boldsymbol{n}$ the outward-pointing unit vector normal to the surface and $P = I - \boldsymbol{n}\otimes\boldsymbol{n}$.

By considering Eq.(1) and Eq.(2), the following equation is obtained:

$$\vec{j} = \frac{\nu D_{\alpha\beta}}{\dot{k}T}\nabla_s \mu. \quad (4)$$

Assuming that there is mass conservation, the surface motion can then be described by:

$$v_n = -\dot{\Omega}(\nabla_s \vec{j}), \quad (5)$$

where v_n denotes the normal velocity of the surface and $\dot{\Omega}$ the atomic volume. By combining Eq. (4) and Eq. (5), we obtain:

$$v_n = -\frac{\nu\dot{\Omega}D_{\alpha\beta}}{\dot{k}T}\Delta_s\mu, \quad (6)$$

with $\Delta_s = \nabla_s \cdot \nabla_s$ the surface Laplacian operator (or Laplace-Beltrami operator).

From Eq. (6), it is notable that the normal velocity is associated with the chemical potential of the atoms.

Considering κ as the mean curvature (sum of the principal curvatures in 3D) and $\gamma_{\alpha\beta}$ the α/β interface energy and by ignoring the possible effects of anisotropy, the following relationship is obtained:

$$\mu = -\gamma_{\alpha\beta}\dot{\Omega}\kappa. \quad (7)$$

Thanks to Eq.(6) and Eq.(7), we obtain:

$$v_n = \frac{\gamma_{\alpha\beta}\nu\dot{\Omega}^2 D_{\alpha\beta}}{\dot{k}T}\Delta_s\kappa = B\Delta_s\kappa, \quad (8)$$

with $B = \frac{\gamma_{\alpha\beta}\nu\dot{\Omega}^2 D_{\alpha\beta}}{\dot{k}T}$, the kinetic coefficient. Eq. (8) describes the relation between the motion by surface diffusion and the surface Laplacian of the mean curvature [4] [5] [6] [7].

2.2 Motion by mean curvature

In order to describe precisely the surface evolution of an α lamella, the influence of the mean curvature should be also considered [5] [7]. The $\gamma_{\alpha\alpha}$ grain boundary energy is indeed very important for the lamellae splitting.

The grain boundary energy is given by the well-known Gibbs-Thompson relationship where the normal velocity v_n of the grain boundary is described proportionally to the mean curvature κ:

$$v_n = -\kappa\frac{\gamma_{\alpha\alpha} b\check{\nu}\dot{\Omega}}{\dot{k}T} e^{-\frac{\Delta E}{\dot{k}T}} = -A\kappa, \quad (9)$$

with $A = \frac{\gamma_{\alpha\alpha} b\check{\nu}\dot{\Omega}}{\dot{k}T} e^{-\frac{\Delta E}{\dot{k}T}}$, where $\gamma_{\alpha\alpha}$ denotes the grain boundary energy, b is the burgers vector norm associated with the hoping event, $\check{\nu}$ is the Debye frequency and ΔE is the difference in free energy between the two grains.

2.3 Motion of the interfaces

Surface diffusion and mean curvature motions are taking part simultaneously during the phenomenon of spheroidization. A global velocity combining both of these motions can then be summarized as:

$$v_n = B(\Delta_s\kappa) - A\kappa, \quad (10)$$

where B is defined in α/β phase interfaces and A is defined in α/α grain interfaces.

3 Level set formulation

A level-set model was formulated in order to deal with the topological changes at the α/β interface. The level set method was chosen due to its capability to immerse/describe/capture easily in a FE context the interfaces [8] [9] [10] [11] and also due to the fact that geometrical quantities as the mean curvature κ and the outside normal n can be obtained as:

$$n = \frac{\nabla\varphi}{\|\nabla\varphi\|}, \quad (11)$$

and

$$\kappa = div(\boldsymbol{n}) = \boldsymbol{\nabla}\frac{\nabla\varphi}{\|\nabla\varphi\|}, \quad (12)$$

with

$$\varphi(x,t) = \pm d(x,\Gamma(t)). \quad (13)$$

φ is then defined over the domain Ω as the signed distance function to the interface Γ of the subdomain of interest that we will denote Π.

The sign convention of Eq. (11) corresponds to a distance function negative inside Π and positive outside.

Thus the interface velocity can be rewritten in a level set form as:

$$\vec{v} = v_n\frac{\nabla\varphi}{\|\nabla\varphi\|} = v_n n = (B(\Delta_s\kappa) - A\kappa)n, \quad (14)$$

with:

$$B = \begin{cases} \dfrac{\gamma_{\alpha\beta} v \grave{\Omega}^2 D_{\alpha\beta}}{\grave{k}T}, & \text{at the } \alpha/\beta \text{ interface} \\ 0, & \text{otherwise} \end{cases} \quad (15)$$

$$A = \begin{cases} \dfrac{\gamma_{\alpha\alpha} b\check{v}\grave{\Omega}}{\grave{k}T} e^{-\frac{\Delta E}{kT}}, & at\ the\ \alpha/\alpha\ interface \\ 0, & otherwise \end{cases} \quad (16)$$

It can also be proved that in the considered level-set formulation [4] [12], v_n can be re-written as:

$$v_n = B(\Delta_s\kappa) - A\kappa$$
$$= \frac{B}{\|\nabla\varphi\|}\nabla\cdot(\|\nabla\varphi\|P\nabla\kappa) - A\kappa \quad (17)$$

The velocity is then defined in the entire domain and corresponds in the vicinity of the zero level-set function of φ, i.e. Γ, to the interface velocity [4] [5].

4 Generation of the microstructure

As already mentioned, the α lamellae and their shape evolution are going to be represented with the use of a level set function. A new Representative Volume Element class was built in our fully parallel multi-component C++ Library called CIMLib [13] and used for these numerical developments in order to describe α colonies in β grains by using the level set approach.

The basic steps of this method are:

a) consideration of a given domain as a β grain.

b) division of the domain in colonies (by Voronoï tessellation for example [14]).

c) In each colony, Voronoï seeds belonging to a prescribed common axis are introduced.

d) Partition of each colony into regions, which are based to the minimum distance of the points of the domain from the Voronoï seeds of the corresponding colony. Each region consists of all the integration points closer to the corresponding seed that any other, called as Voronoï cells [14].

e) A variable ε is introduced for shifting the Voronoï cells in order to manage the description of the two different phases. This variable can be defined from a distribution or from the desired volume fraction of the α phase.

f) All these operations can be described in a exact analytical way in the considered LS framework.

Figure 1. Representation of an initial microstructure of a β grain with three different α colonies thanks to a LS approach.

Figure 1 illustrates a 2D case in a 1mm x 1mm domain. The red regions describe the α lamellae and the blue regions the β media. Furthermore, in order to emphasize the efficiency of the developed method to describe statistically the microstructure of interest, a corresponding illustrative back-scattered electrons (BSE) image obtained thanks to a scanning electron microscope (SEM) from an initial microstructure of TA-6V alloy is displayed in Figure 2.

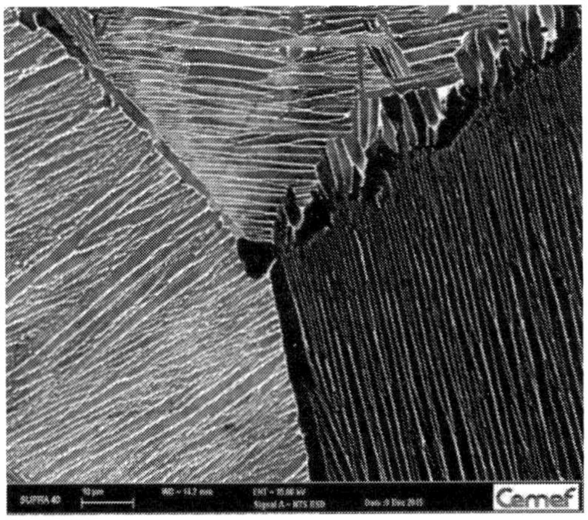

Figure 2. An initial microstructure of TA6V, with 3 different colonies.

5 A surface diffusion methodology

For the modelling of the induced flow from the surface diffusion mechanism at the α/β interface, a finite element solver that follows a level set method is used. At any time t, the transport velocity $\vec{v}$ is defined by:

$$\vec{v} = \mathrm{B}(\Delta_s \kappa)\boldsymbol{n} = \frac{B}{\|\nabla\varphi\|}\nabla\cdot(\|\nabla\varphi\|P\nabla\kappa)\boldsymbol{n}, \quad (18)$$

with $\mathrm{B} = \frac{\gamma_{\alpha\beta}\nu\Omega^2 D_{\alpha\beta}}{kT}$. The B coefficient is defined as a constant and it is chosen to neglect any anisotropy concerning the interface energy and the diffusivity. Additionally, isothermal conditions are assumed. The time evolution of $\Gamma(t)$ is then obtained by solving the following convective system:

$$\begin{cases} \frac{\partial\varphi}{\partial t} + \vec{v}\cdot\overrightarrow{\nabla\varphi} = 0 \\ \varphi(x,0) = -d\big(x,\Gamma(0)\big)\chi_\Omega + d\big(x,\Gamma(0)\big)\chi_{\bar{\Omega}} \end{cases} \quad (19)$$

The interface can then be obtained at each time step as the 0-isovalue of the distance function and the velocity is updated by following Eq.(18) before the following time step. At the following subsections, more extensive details are given for the resolution algorithm.

5.1. Surface diffusion velocity identification and transport resolution

The methodology used to obtain the surface diffusion velocity is based on the finite element based strategy introduced by Bruchon et al. in [4] [15].

Indeed, as P1 description of the LS is considered in the proposed methodology, one of the basic issues in the problem of surface diffusion is that the velocity is defined by the Laplacian of the curvature, which means that the velocity is a function of the fourth order spatial derivative of φ. The numerical strategy proposed by Bruchon et al. consists to solve this problem by considering a "regularized" formulation. More precisely Eq.(18) is solved in a weak form by using a FE formulation. Further informations can be found in [4] [15].

5.2. Convection-Reinitialization methodology

By assuming the appropriate calculation of the surface velocity, the traditional strategy of convection and subsequent reinitialization steps is used. The main idea is to solve the advection equation and to rebuild the metric properties of the level-set function in order to keep a distance function ($\|\nabla\varphi(x,t)\| = 1$) at least near the interface $\Gamma(t)$. Classical approaches consist in solving, separately, the convective part and the reinitialization part thanks to the resolution of a classical Hamilton-Jacobi system [11] or to adopt an unified advection and renormalization methodology by solving one single equation based on a smooth description of the level-set [15]. Here, a new approach is proposed.

This new approach is based on a separate resolution of the transport and of the reinitialization part. Convective equation is firstly solved thanks to a stabilized P1 solver (SUPG or RFB method). Then, a parallel and direct reinitialization algorithm detailed in [16], which has been proven to be extremely fast and accurate is used. In this algorithm, the $\Gamma(t)$ interface is firstly discretized into a collection of segments (respectively triangles in 3D) and the nodal values of the level-set function are then updated by finding the nearest element of the collection and calculating the distance between the considered node and this nearest element. This method takes advantage of a space-partitioning strategy using *k-d* tree and an efficient bounding box strategy enabling to maximize the numerical efficiency for parallel computations.

Moreover, this methodology presents two other interest:

-it enables to avoid the validation/calibration of unphysical parameters necessary to reinitialize [11] [15],

-it enables to obtain directly an exact P1 description of $\boldsymbol{n}$ [17] before to solve Eq. (18) rather than following the classical less precise methodology where the normal is computed by performing a P1 interpolation of the first derivative of the level-set function.

6 First academic case

In this section we examine an ellipsoid shape under surface diffusion. We want to test the efficiency of the proposed formulation for a simple case with an analytical solution. Considered computational domain is a *1mm x 1mm* square centered *in (0,0)*. An initial ellipse (*a=0.3mm* and *b=0.2mm)* of equation $x^2/a^2 + y^2/b^2 = 1$ is considered. Of course, this shape is going to evolve towards a circle shape while conserving its area. Thus limit radius, i.e. limit value of a and b is given by the value $\sqrt{\pi ab} \approx 0.43416\, mm$.

Initially, in order to test the efficiency of the approach without dealing with the effect of the meshing and remeshing in the results, a fixed mesh is considered during simulations. An initial isotropic mesh adaptation is considered in a ring centered in *(0,0)* and defined as $0.19mm \leq r \leq 0.31mm$ in order to keep a very fine mesh, defined as h in Table 1, in all the zone crossed by the zero isovalue of the level-set function during the simulation.

The figure 3 shows the FE mesh used (a), the initial distance function field (b), a zoom on the FE mesh (c), the curvature field near the interface (d) and the normal velocity field near the interface obtained initially thanks to the FE resolution of Eq. (18) (e). Red or white line corresponds to the initial ellipse interface (*0* isovalue of the level-set function).

In all simulations B is assumed to be homogeneous and equal to $1\, mm^4/s$. Exact velocity of the point *(a(t),0)* is known [15] and given by:

$$\vec{v_a}(t) = B\frac{3\,(b^2-a^2)a}{b^6}\vec{i} = v_a(t)\vec{i}. \quad (20)$$

Hence *a(t+dt)* can be easily evaluated thanks to a forward Euler method:

$$\mathrm{a(t + dt) = a(t) + dt\, v_a(t)}, \quad (21)$$

and *b(t+dt)* can be easily obtained by verifying the area conservation at anytime:

$$\mathrm{b(t + dt) = a(0)b(0)/a(t + dt)}. \quad (22)$$

This method is then used with a time step of 1*ms* to evaluate the "exact" evolution of a and b values during the shape evolution. Concerning FE simulations, at each time step, the positions of *(a(t),0)* and *(0,b(t))* are determined on the zero-isovalue of the distance function and then compare to the "exact" solution. Final time of all simulations is fixed to 1*s*.

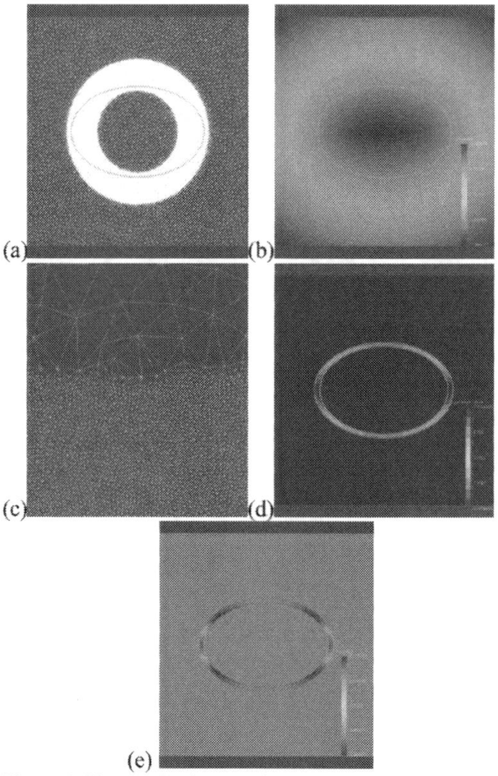

Figure 3. First academic case: the FE mesh used (a), the initial distance function field (b), zoom on the FE mesh at the interface between coarse and fine mesh (c), $\boldsymbol{\kappa}$ near the interface (d) and $\boldsymbol{v_n}$ near the interface (e). Red or white line corresponds to the initial ellipse interface (zero-isovalue of the level-set function).

Table 1 summarizes all the parameters (time step, mesh size in the fine mesh zone, number of elements of the used mesh, method used), the CPU time of the simulations and the corresponding precision of the results obtained concerning the positions of *(a(t),0)* for $t \in [0,1s]$ by using the unified convective-renormalized approach described in [15] and the new one proposed here with different time step and h values. These cases are representatives of an important number of other performed simulations.

Errors are defined as:

$$e_1 = \frac{\|a_{sim}(t)-a_{exact}(t)\|_{L_1}}{\|a_{exact}(t)\|_{L_1}} = \frac{\sum_i \left|a_{sim_i}-a_{exact_i}\right|}{\sum_i \left|a_{exact_i}\right|} \text{ and,} \quad (23)$$

$$e_2 = \frac{\|a_{sim}(t)-a_{exact}(t)\|_{L_2}}{\|a_{exact}(t)\|_{L_2}} = \frac{\sqrt{\sum_i \left(a_{sim_i}-a_{exact_i}\right)^2}}{\sqrt{\sum_i a_{exact_i}^2}}, \quad (24)$$

where i denotes the discretization in time.

Table 1. Summarized data and results of some tested configurations

Data	**Case1**	**Case2**	**Case3**	**Case4**
h in fine zone in μm	1	1	2	1
#Elt	2,74e5	2,74e5	7,1e4	2,74e5
Time step in ms	1	5	10	0.1
Conv + exact Reinit	X	X	X	
Unified approach				X
Calculation time on 12CPU	1h	17mins	1min	6h28mins
e1 in %	2.5	2.6	2.1	2.2
e2 in %	3	3.1	2.9	2.8

From different simulations (Cases 1 to 4 of Table 1 are representatives), we can summarize the following comments:

-Considering the precision of *a(t)* predictions (see e_1 and e_2 errors on Table 1), both approaches are relevant to model surface diffusion. Figure 4 illustrates at *t*=1*s*, the difference of the exact interface and the results obtained with the Case1.

-The time step of the Case4 ($dt = 0.1\ ms$) is the maximal value usable for $h = 1\mu m$ and the unified formulation to avoid numerical instabilities. Such instabilities were not identified for the new proposed approach even for important time step and mesh size (Case 3 for example).

-As illustrated in Table 1, calculation time of the new proposed approach is then clearly very attractive comparatively to the unified approach.

-Even for the new proposed approach, decreasing the dt and h values below, respectively, $10\ ms$ and $2\ \mu m$ seems not improve the results quality ($e_2 \approx 3\%$). It can be explained by the fact that the residual error is due to the FE resolution of Eq. (18).

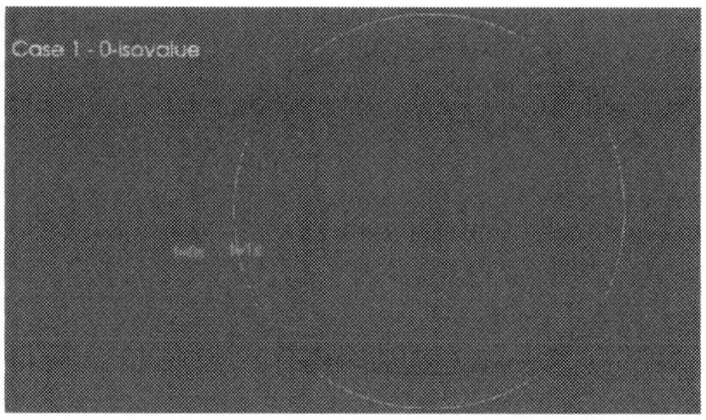

Figure 4. Comparison at *t=0s* and *t=1s* of the exact solution (red lines) and the Case 1 0-isovalue (white lines).

7 Second academic case: volume loss and mesh adaptation

Next, we consider a more realistic shape of a long ellipse with *a*=0.5mm and *b*=0.1mm and mesh adaptation. Indeed, in order to propose acceptable calculation time to model surface diffusion of complex microstructure such as one of the Figure 1, fixed meshing strategy of the previous section is not an option. Thus a meshing and remeshing strategy must be used.

To begin, two based-metric meshing strategies associated with the MTC topological mesher were tested. MTC is a P1 automatic remesher based on elements topology improvement that was developed for Lagrangian simulations under large strains. This tool was extended to anisotropic mesh adaptation, for which it was extensively used in context of FE microstructure description [11] [10] [14] [15] [4] [17].

The first metric considered is the metric describes in [14], this metric enables to impose isotropic or anisotropic (in the normal direction of the interface) fine mesh in the vicinity of the interface without considering its curvature. Here, isotropic adaptation was considered (Method 1 in Table 2).

The second used metric field is based on an a priori error estimator linking the interpolation error on the LS function to its gradient vector and hessian matrix. As already described, these variables represent respectively the normal vector to the interface and its main curvatures. Using this data, a metric field can be built in order to have a very fine mesh size in the normal direction to the interface, and to control the mesh size in the other directions depending on local curvature [18] (Method 2 in Table 2).

Simulation and data results are reported in Table 2 for both meshing/remeshing methods. The "Conv + exact Reinit" strategy was used. Final time of both simulations is fixed to 1*s*. As the mesh size near the interface is of the same order than the mesh size used in Case1 and Case2, same precision could be expected. However, remeshing is also synonymous of diffusion concerning the FE fields carrying the interface at each remeshing operation necessary to follow interface motion. Then, volume conservation was tracked for these cases.

Table 2. Summarized data and results with mesh adaptation.

Data	Case5	Case6
h_min in the fine zone in μm	1	1
#Elt	<10000	<10000
Time step in *ms*	1	1
Conv + exact Reinit	X	X
Remeshing period	5dt	5dt
Remeshing method	Method 1	Method 2
Calc. Time on 12CPU	7 mins	4 mins
Volume loss in %	7.5	1

Results described in Table 2 illustrate that the mesh adaptation technique come with good precision and faster calculation time. Meshing adaptation based on the curvature enables to obtain a very good conservation of the volume. This aspect will be of course very important for real configurations where the initial thin shape of the α lamellae implies very high ratio between the minimal and the maximal values of the interface curvature. So, this meshing strategy in terms of metric seems particularly indicated.

8 Discussions

If the results described previously in terms of calculation time and precision could be sufficient to study the surface diffusion of one thin ellipse, it seems clear that to further improve our numerical framework, it is important to deal with real 2D or 3D configurations.

In order to face this problem, we adopt a new topological mesher, Fitz, developed by Shakoor et al. [18]. With Fitz, a body fitted meshing and remeshing is possible while using complex metric as previously described. It was then proved in [19] that this new mesher associated with a volume conservation constraint which is compatible both with implicit and body-fitted interfaces, the modification of the interface due to remeshing can be delayed enough so that a Lagrangian Level-Set method becomes more than interesting compared to an Eulerian Level-Set method, even when large deformations or displacements occur.

First tests are very promising, allowing in the Case6 configuration to obtain a precision of 2% concerning the volume conservation with a calculation time of 30s on 12 CPU. Figure 5 illustrates a result obtained with this numerical framework for a "a=0.5mm and b=0.05mm" configuration.

Moreover we anticipate with this mixed implicit/explicit description of the interfaces an easier coupling between surface diffusion at the α/β interfaces and motion by mean curvature at the α/α grain interfaces.

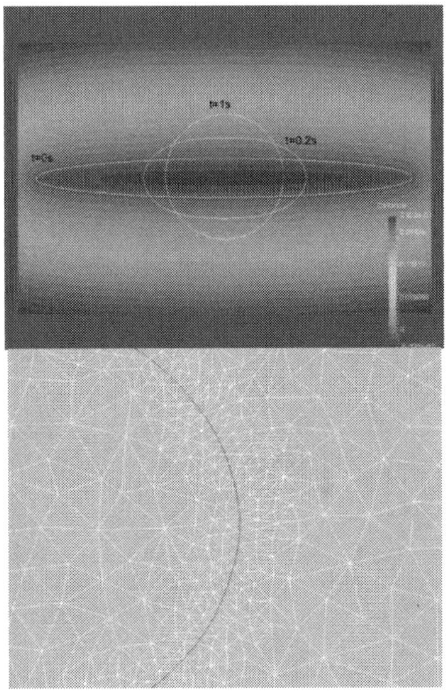

Figure 5. A "a=0.5mm and b=0.05mm" configuration with a mixed implicit/explicit description of the interfaces: (top) initial distance function field and the obtained interfaces (white lines) at t=0s, 0.2 s and 1s. (Bottom) Zoom on the conforming mesh at t=0.2s, the interface is defined by the red line.

9 Conclusions

The first steps of a new FE numerical framework dedicated to the modelling of the mechanisms of spheroidization in α/β titanium alloys have been detailed. This numerical framework has been illustrated on the surface diffusion of α laths. Several cases were studied in order to check the efficiency of the proposed approach. Optimal algorithm of resolution and meshing strategies were proposed in terms of precision and calculation time. A new way was also opened by considering a mixed implicit/explicit description of the interfaces. All this numerical tools will now be exploited to deal with coupling between surface diffusion at the α/β interfaces and motion by mean curvature at the α/α grain interfaces. An existing crystal plasticity finite element approach developed in a level-set context [20] [17] will also be used to model the deformation of the α lamellae.

Acknowledgements

The authors would like to gratefully thank Aubert & Duval, CEA, Timet and Safran for funding this research through the SPATIALES consortium.

References

1. G. Lütjering, J. Williams, *Titanium* (Springer: Heidelberg, 2007)
2. S. Semiatin, D. Furrer, Metals Branch (Metals, Ceramics, and NDE Division, 2008)
3. W. Mullins, Acta Metallurgica 6, 414-427 (1958)
4. J. Bruchon, D. Pino Muñoz, F. Valdivieso, S. Drapier, G. Pacquaut, European Journal of Computational Mechanics 19, 281-292 (2010)
5. V. Derkach, M.Sc. thesis, Department of Mathematics, Technion–Israel Institute of Technology (2010)
6. V. Derkach, A. Novick-Cohen, A. Vilenkin, E. Rabkin, Acta Materialia 65, 194-206 (2014)
7. P. Smereka, Journal of Scientific Computing 19, 439-456 (2003)
8. S. Osher, F. Fedkiw, Journal of Computational Physics 169, 463-502 (2001)

[9] J. Sethian, Level Sets Methods and Fast Marching Methods (Cambridge Monograph on Applied and Computational Mathematics, 1999)

10. M. Bernacki, R. Logé, T. Coupez, Scripta Materialia 64, 525–528 (2011)
11. M. Bernacki, Y. Chastel, T. Coupez, R. Logé, Scripta Materialia 58, 1129-1132 (2008)
12. M. Burger, F. Hausser, C. Stocker, A. Voigt, Journal of Computational Physics **225**, 183-205 (2007)
13. H. Digonnet, L. Silva, T. Coupez, , in Proceedings of the 9th International Conference on Numerical Methods in Industrial Forming Processes (2007)
14. M. Bernacki, H. Resk, T. Coupez, R. Logé, Modelling and Simulation in Materials Science and Engineering **17**, 064006 (2009)
15. J. Bruchon, S. Drapier, F. Valdivieso, International Journal for Numerical Methods in Engineering **86**, 845-861 (2011)
16. M. Shakoor, B. Scholtes, P.-O. Bouchard, M. Bernacki, Applied Mathematical Modelling **39**, 7291-7302 (2015)
17. B. Scholtes, R. Boulais-Sinou, A. Settefrati, D. Pino Muñoz, I. Poitrault, A. Montouchet, N. Bozzolo, M. Bernacki, Computational Materials Science **122**, 57–71 (2016)
18. M. Shakoor, M. Bernacki, P.-O. Bouchard, Engineering Fracture Mechanics **147**, 398-417 (2015)
19. M. Shakoor, P.-O. Bouchard, M. Bernacki, International Journal for Numerical Methods in Engineering (In press)
20. H. Resk, L. Delannay, M. Bernacki, T. Coupez, R. Logé, Modelling and Simulation in Materials Science and Engineering **17**, 075012 (2009)

A new algorithm for dense ellipse packing and polygonal structures generation in context of FEM or DEM

Dmitrii N. Ilin[a], Marc Bernacki

MINES ParisTech, PSL-Research University, CEMEF Centre de mise en forme des matériaux, CNRS UMR 7635, CS 10207 rue Claude Daunesse, 06904 Sophia Antipolis, France

Abstract. A new constructive ellipse packing algorithm is presented. It allows to respect the imposed area, shape and spatial orientation distribution (i.e. the inertia tensor) and achieve high packing densities. The packing density decreases with increasing particles aspect ratio what is in agreement with results reported in the literature. The generated packings with complex imposed area, shape and spatial orientation distributions with densities in the range of 0.74 and 0.8 are presented. The efficiency of the algorithm is demonstrated by comparison with the Optimized Dropping and Rolling method for disk packing. Moreover, the proposed packing strategy enables to generate very easily non-equiaxed polygonal structures by using Laguerre-Voronoï tessellation of the generated disk-based ellipse packing.

1 Introduction

The problem of packing of objects with different geometrical shape attracted considerable amount of attention during the last few decades due to its application in modelling of granular materials, porous media, foams, polycrystalline structures and so on. Most studies were carried out for the packing of uniform and arbitrary sized spheres [1], considering the problem of optimal packing [2-5], (and its dual, the sphere cutting problem [6, 7]) or space-filling [8-14].

In the context of space-filling problem, the sphere packing can be obtained by the dynamic methods, in which particles usually change their position and/or their size during the filling process to improve the local and global densities, or by the constructive methods, generating particles sequentially and calculating their positions thanks to geometrical rules.

The dynamic methods were widely used for the ellipsoid packing and high densities were reported in published works [15-18]. For the methods in which ellipsoidal particles are generated by choosing randomly initial position and orientation and letting them fall in an open box under the gravity, an anisotropic behaviour of the packing system was reported [18, 19]. In the case when the spatial orientation of ellipses is fixed, it was found that the packing by dynamic methods can result in low densities [20].

The constructive method of axisymmetric elliptical particles approximated by four connected arcs was proposed in [21] and it was extended to three dimensional (3D) axisymmetric ellipsoidal forms in [22]. In [23] the authors developed an advancing front packing algorithm. Recently, an overlapping detection algorithm for the sequential packing of elliptical particles was proposed and generalized in [24]. Relatively high packing densities were reported, but the generation of packing with an imposed spatial orientation distribution was not considered. It should be noted that due to numerous rejections of randomly generated positions of particles during the filling process in the sequential inhibition method, the computation time for achieving high and even moderate packing densities is typically long.

A constructive ellipse packing method is proposed in the present work. Each elliptical particle is approximated by a set of circles. Recently, this method was examined for applying in discrete element simulations [25]. In the present study, we focus on the packing of particles which can be applied not only to consider modelling of granular materials, but also for the generation of polygonal structures (which can be used, for example, to describe polycrystalline materials statistically). In this case, higher packing densities are necessary to provide good accuracy in respecting the desired polygon area, shape and spatial orientation distributions. In the proposed packing method, the list of accommodated in domain ellipses is built in which they are sorted according to the position in the domain (height). To reach high packing densities, each new particle is placed at the lowest possible position by testing pairs of ellipses from the "Height list".

The paper is organized as follows. At first, the discretization of ellipses on multi-circles is described. Then, the algorithm of packing is presented with introducing the Height list used to optimize packing density. Some numerical tests are reported for imposed area, shape and spatial orientation distributions. Finally,

[a] Corresponding author: dmitrii.ilin@mines-paristech.fr

densities and computation time provided by the proposed algorithm are compared with the Optimized Dropping and Rolling Method (ODRM) [9] in particular case of disk packing.

2 Ellipse packing algorithm

In this section, the approach used to approximate an ellipse by a set of circles is described in the first part. In the second part, the algorithm of ellipse packing is detailed.

2.1 Multi-circles

Assuming that the principal axes coincide with the axes of local Cartesian coordinates *Oxy*, the corresponding ellipse equation is:

$$\left(\frac{x}{a}\right)^2 + \left(\frac{y}{b}\right)^2 = 1\,, \qquad (1)$$

where a and b are the major and minor half-lengths of principal axes.

A random ellipse can be modelled by a set of circles as follows [25]:

- All the centers of circles are on the major axis and, hence, the circle equation is

$$(x - x_i)^2 + y^2 = R_i^{\,2}\,, \qquad (2)$$

where x_i is the abscissa of the center and R_i is the radius of the circle with $i = 0$ for the central circle and $i > 0$ for the right part of the ellipse.

- All the circles are tangent to the ellipse. If (x_{it},y_{it}) is the point of tangency for the circle i with center $(x_i,0)$, after calculating the derivative with respect to x for the left and right parts of Eq.(1) and Eq.(2) and substituting $x = x_{it}$, we obtain:

$$\frac{x_{it}}{a^2} + \frac{y_{it}}{b^2}\frac{dy}{dx}(x_{it}) = 0\,, \qquad (3)$$

$$(x_{it} - x_i) + y_{it}\frac{dy}{dx}(x_{it}) = 0\,, \qquad (4)$$

which gives:

$$x_{it} = \frac{a^2 x_i}{a^2-b^2} = \frac{\lambda^2 x_i}{\lambda^2-1}\,, \qquad (5)$$

with $\lambda = a / b$ the ellipse shape ratio. Consequently, the radius of the circle R_i can be found from Eq.(1) and Eq.(2) by eliminating y^2 and substituting $x = x_{it}$ where x_{it} is given by Eq.(5):

$$R_i^{\,2} = b^2\left(1 - \frac{x_i^{\,2}}{a^2-b^2}\right) \qquad (6)$$

- The center of the largest circle $(0,0)$ is in the center of the ellipse and its radius is equal to the semi-minor axis b.

- The center of the circle which is the nearest neighbour of the largest (central) circle, $(x_1,0)$, is imposed by the user (the closer it is to the central circle, the better the final approximation of the ellipse is obtained).

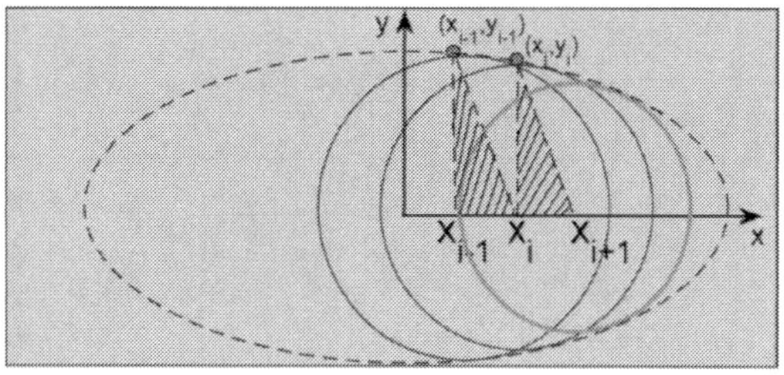

Figure 1. Approximation of the ellipse by a set of circles.

- If positions of the circles $C_{i-1}((x_{i-1},0),R_{i-1})$ and $C_i((x_i,0),R_i)$ are known, the center of the circle $C_{i+1}((x_{i+1},0),R_{i+1})$ is calculated subsequently from the condition of similarity of the triangle with vertices $(x_i,0)$, $(x_{i+1},0)$ and (x_i,y_i) and triangle formed by vertices $(x_{i-1},0)$, $(x_i,0)$ and (x_{i-1},y_{i-1}) thanks to Eq.(7) (see also Fig.1):

$$\frac{y_i}{y_{i-1}} = \frac{x_{i+1} - x_i}{x_i - x_{i-1}} \qquad (7)$$

- The point of tangency for the smallest circle of center $(x_{N_l},0)$ is imposed as $(a,0)$ and, hence, from (5) and (6) we obtain:

$$R_N = b^2/a\,, \qquad (8)$$

$$x_N = a - R_N = a\left(1 - \frac{b^2}{a^2}\right) = a\left(1 - \frac{1}{\lambda^2}\right). \qquad (9)$$

If Eq.(7) leads to $x_{i+1} > x_N$, we impose $x_{i+1} = x_N$ and the iteration is terminated. After reflecting the circles $\{C_i((x_i,0),R_i),\ i=1..N\}$ over the y-axis, the description of the ellipse by a set of $2N+1$ circles is obtained. The total number of circles approximating an ellipse depends on the ellipse shape ratio, λ, and the precision imposed by the user, x_1.

2.2 Packing algorithm

The proposed ellipse packing method, called Advancing Layer algorithm (ALA), is described in this subsection. The flow-chart of the algorithm for placing the first row (ellipses are in contact with the bottom of the domain) is summarized in Fig.2. When the first row is completed, the bulk of the domain is filled as described in the flow-chart detailed in Fig.3.

Each new ellipse is generated according to the imposed area, A, shape (aspect ratio), λ, distribution. Then, the ellipse is decomposed on circles as it is described in the previous subsection with the precision imposed by the user and rotated by the angle α (see Fig.4) generated according to the imposed spatial orientation distribution.

To place the ellipse E_i in contact with the bottom edge, the lowest circle should be determined and y-coordinate of its centre is imposed to be equal to its radius. Then, y-coordinates of other circles of the particle are computed. When the particle is placed in the left

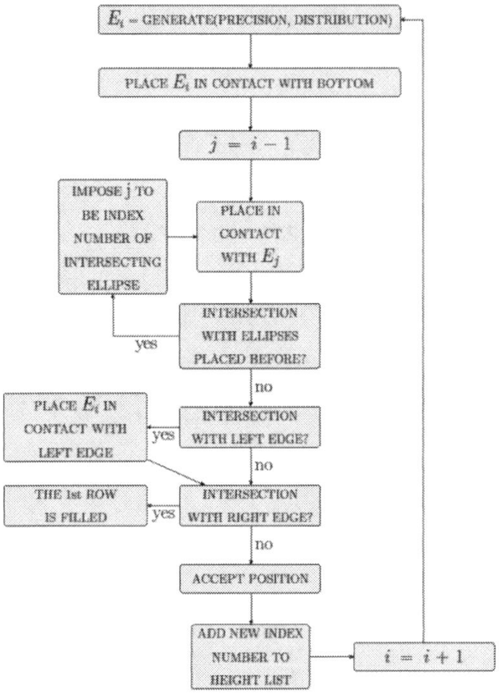

Figure 2. Flow-chart of the filling algorithm for the first row.

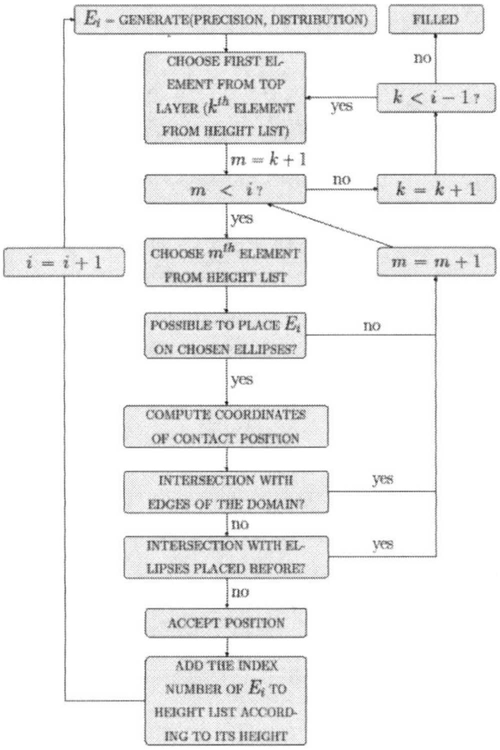

Figure 3. Flow-chart of the algorithm for filling the bulk.

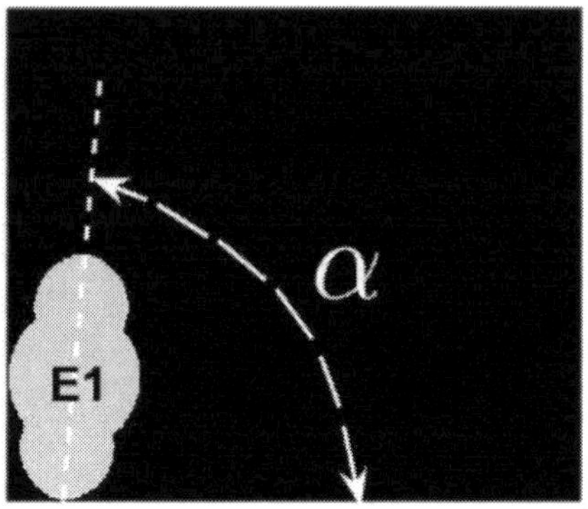

Figure 4. The first ellipse in the bottom corner of the domain. The spatial orientation angle, α, is shown.

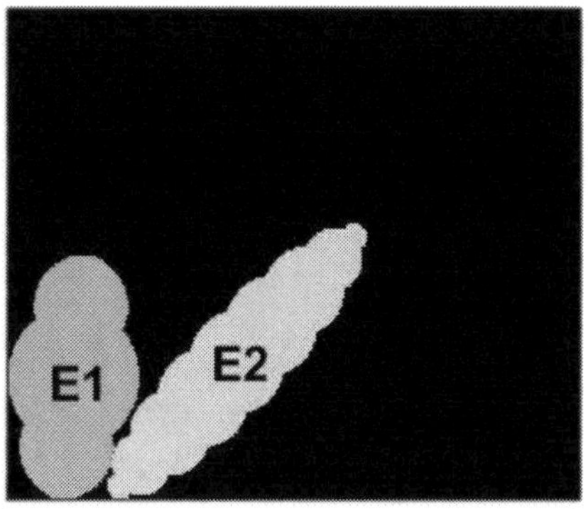

Figure 5. Placement of the second ellipse in contact with the first one (the first row).

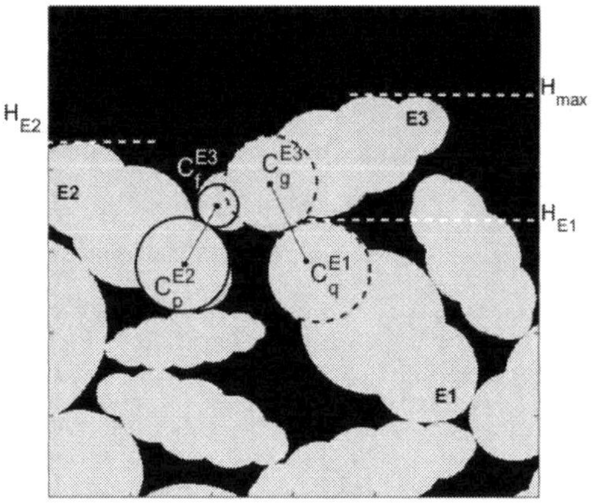

Figure 6. Placement of the ellipse E3 on E1 and E2. The corresponding pairs of circles C_q^{E1} - C_g^{E3}, C_p^{E2} - C_f^{E3} in contact and ellipses heights (H_{E1}, H_{E2} and H_{max}) are shown.

bottom corner of the domain, x-coordinates are defined analogically.

After accommodating the first elliptical particle in the corner, other ellipses are generated and placed as described in the flow-chart for filling the first row (Fig.2). Fig.5 illustrates an example of two particles placed in contact with bottom: after placing E_1 in the corner, the position of E_2 is calculated by searching one of its circles to be in contact with the circles of E_1 with no overlapping of other circles.

When the first row of ellipses is filled, the index number of each particle is recorded in the Height list at the position which corresponds to the height of the particle (its maximal y-coordinate). The ellipses in the Height list are sorted in ascending order, i.e. from the lowest ellipse (the first position in the list) to the highest one (the last element in the list).

For completing the bulk of the domain, the flow-chart described in Fig.3 is used for any tested ellipse. After its generation (following imposed distributions), its decomposition and rotation, a new ellipse should be placed in contact with two other particles. Pairs of previously placed ellipses are tested starting from the first element to the last one recorded in the Height list. Fig. 6 illustrates the placement of the ellipse E_3 on the pair of E_1 (with the height H_{E1}) and E_2 (with the height H_{E2}). The contact position is determined from two conditions of the contact for two pairs of circles (C_q^{E1} with C_g^{E3}, C_p^{E2} with C_f^{E3} in Fig. 6) without any intersection with other circles. Two equalities of the sum of radii to the distance between centres of the corresponding circles from the chosen pairs give two equations with two unknown variables. The system can have two, one or zero solutions. Two solutions correspond to two possible contact positions for the new ellipse – "beneath" and "above". When the new ellipse is placed in contact with the two others, the intersection with other previously placed particles should be checked. In the case of no intersection, the position is accepted and the index number of the new ellipse is recorded in the Height list according to its computed height. In Fig.6 the ellipse E_3 is placed above E_1 and E_2 (since the position beneath E_1 and E_2 cannot be accepted due to the intersection with other particles). When the overlapping with other particles is detected, another pair of elements is chosen from the Height list. If after testing all possible pairs of ellipses for placing the new particle the contact position was not accepted, the packing is stopped.

In the presented form, the packing algorithm is quite time consuming and generates packings with inhomogeneous densities due to the fact that the pairs of particles to place the new ellipse are tested from the bottom to the top. This strategy is synonymous of higher packing density in the bottom part and in longer computation time since a lot of unsuccessful tests (with overlapping) should be performed before to reach the good position for a given new ellipse. To avoid it, the definition of the Top layer was introduced. It is defined as follows. Let H_{max} be the height (maximal y-coordinate) of the highest ellipse placed in the domain (the last element in the Height list) and H_{Ei} be the height of the ellipse E_i (see Fig.6). The Top layer of the packing is

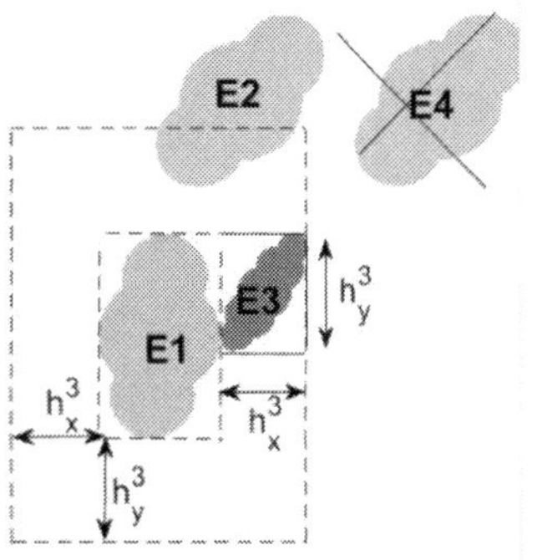

Figure 7. Ellipse cells for E_1 and E_3 and candidate domain (the biggest rectangle, see the text for details).

composed of the particles E_i for which

$$H_{max} - H_{E1} < h^{TL} a_{max} , \quad (10)$$

where a_{max} is the maximal semi-major axis of the ellipses and h^{TL} is a coefficient imposed by the user. The Top layer contains all the ellipses recorded in the Height list at last positions. The number of particles in the Top layer depends on the value of h^{TL}. In the present work, the values from 1.1 to 2 were tested to optimize the computational time and the global packing density.

In Fig.7 the new ellipse E_3 is placed on two previously accommodated ellipses: the first ellipse E_1 is chosen from the Top layer and the second ellipse is chosen from the rectangular domain (the biggest rectangle in Fig.7, called further candidate domain) defined by the rectangular cells of E_1 and E_3 (two small rectangles in Fig.7). Fig.7 shows the case when the ellipse E_4 is not tested for placing E_3 since it does not intersect the candidate domain, the ellipse E_2 overlaps with the candidate domain and, hence, can be considered for placing E_3.

In addition, when the tests of intersection with the previously placed ellipses are performed, first the overlapping of the rectangular cells of ellipses is checked. If the cells overlap, the intersection test for the ellipses is then performed.

In the described form, the ALA was applied for dimensionless ellipse and disk packings and the results are reported in the next section.

3 Results

3.1 Ellipse packing

Most of the works dealing with ellipse packing consider the case when ellipse orientation (ellipsoids in 3D case) is not conserved during the filling process. According to

[20], when particles are allowed to rotate, the ellipse

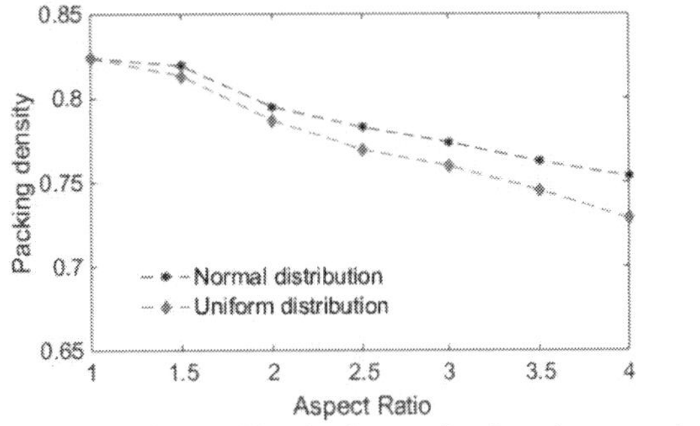

Figure 8. Ellipse packing density as a function of aspect ratio with imposed uniform A distribution law and uniform (shown in red) and normal (shown in blue) α distributions.

packing density increases when the ellipticity increases until reaching a maximum value and then drops below the density of disk packing. Higher densities compared to the packing of disks can then be obtained but without respecting data concerning α. When no rotation is allowed, it was also shown that the packing density decreases as the ellipticity increases.

In this 2D work, we focus on the problem of ellipse packing with spatial orientations conserved during the filling. A uniform ellipse area distribution law with normal and uniform α distributions on $]-\pi/2, \pi/2]$ range for the angle of orientation of the semi-major axis packing of ellipses have been used to generate packing by using the ALA. Depending on the aspect ratio, 4600 to 5100 ellipses were generated (a high number of particles is used for better respecting the imposed distributions). Evolution of the packing density as a function of aspect ratio is described in Fig.8. These results are in agreement with [20]: density decreases with increasing aspect ratio. The obtained values are lower in the case of the uniform α distribution than the ones with normal α distribution. One may conclude that for a material with a preferential direction of ellipses elongation, packing with higher densities can be generated.

Two examples of ellipse packing with imposed uniform A and α distributions are shown in Fig.9 and Fig.10 for the aspect ratios $\lambda = 1.5$ and $\lambda = 4$, respectively. One can see that there is no preferential direction in the orientation of the ellipses. More voids are observed in the case of higher ratio which is clearly indicated by the obtained packing densities: 0.7284 in the case of $\lambda = 4$ and 0.8133 for $\lambda = 1.5$.

Finally, to demonstrate the performance of the ALA, it was tested for a more complex case when a bimodal distribution is imposed for all three parameters, A, λ and α. The size of the considered domain is 30×30. A varies from 0.0001 up to 0.181 with two modes of 0.0386 and 0.1271. λ is in the range of 1.5 to 4.4375 with two modes of 2.125 and 3.5625. Finally, α varies from $-\pi/2$ to $\pi/2$ with two modes of -0.287π and 0.2π. The generated packing of 10696 ellipses with a density of 0.7979 is shown in Fig.11.

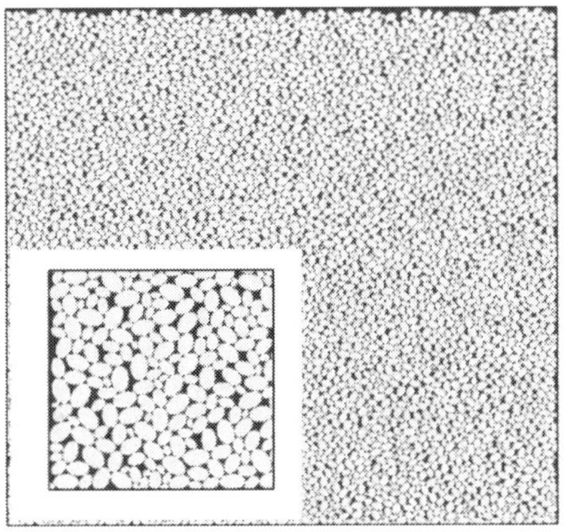

Figure 9. 5081 ellipses with uniform A and α distributions and $\lambda = 1.5$, the packing density is 0.8133.

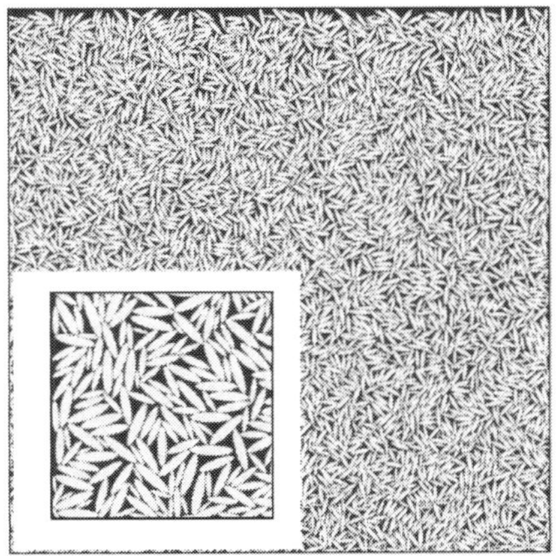

Figure 10. 4596 ellipses with uniform A and α distributions and $\lambda = 4$. The packing density is 0.7284.

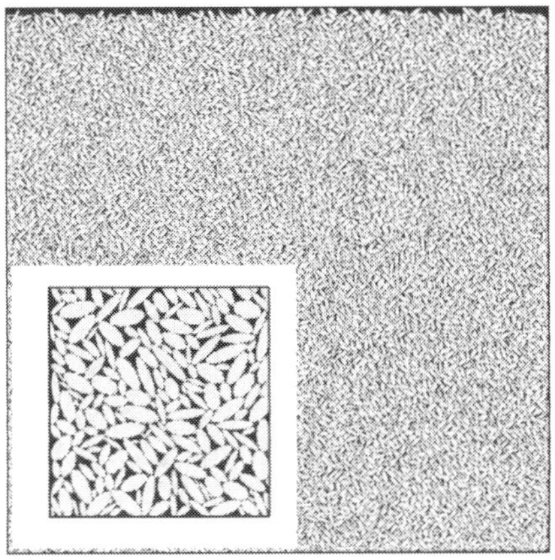

Figure 11. 10696 ellipses with imposed bimodal A, α and λ distributions. The density is 0.7979.

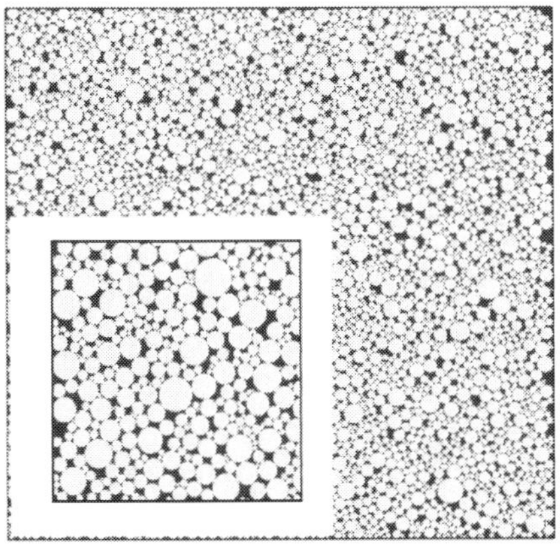

Figure 12. Packing of disks with log-normal type size distribution by ODRM (5139 disks, density is 0.8083, CPU time is 1 s).

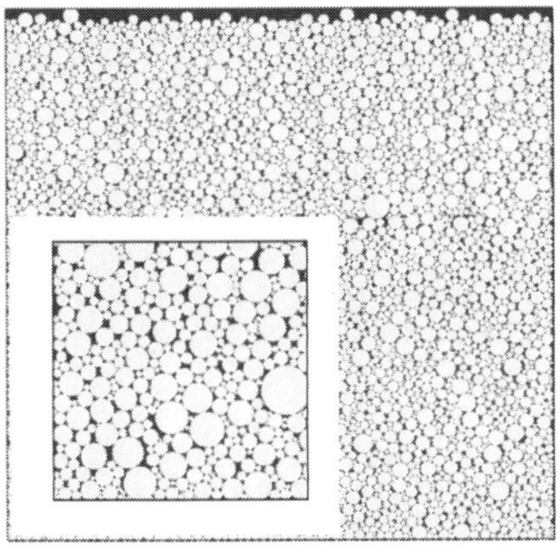

Figure 13. Packing of disks with log-normal type size distribution by ALA (5220 disks, density is 0.8493, CPU time is 1.9 s).

3.2 Comparison with Optimized Dropping and Rolling method

To demonstrate the efficiency of the ALA for disk packing it was compared with the ODRM [9]. The packings of disks by ODR and ALA were generated with an imposed log-normal radius distribution. The distribution width was fixed to $R_{max}/R_{min} = 48$, where R_{max} and R_{min} are, respectively, the maximal and the minimal disk radius. To define the top layer in the ALA, the parameter h^{TL} was chosen to be equal to 1.1. The computation results are presented in Fig.12 and Fig.13. One can see that the ALA allows to place disks in the domain more homogeneously with higher densities compared to ODR. Packing densities obtained by ODR and ALA are 0.8083 and 0.8493, respectively. The CPU time needed for the ALA computation (1.9 s for the packing of 5220 disks) is about two times longer than the

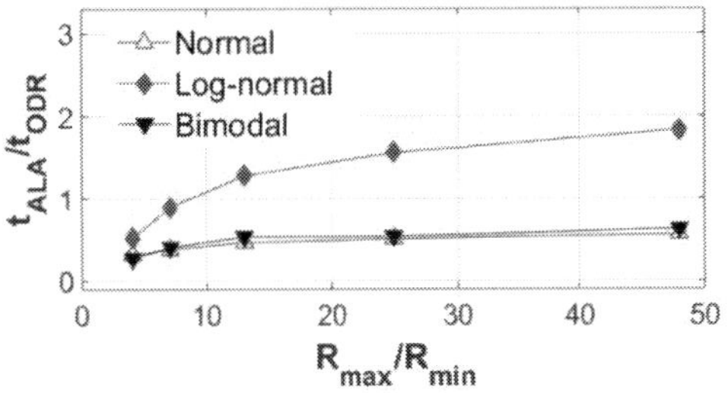

Figure 14. Comparison between the ALA and the ODR CPU times for different disk size distribution width.

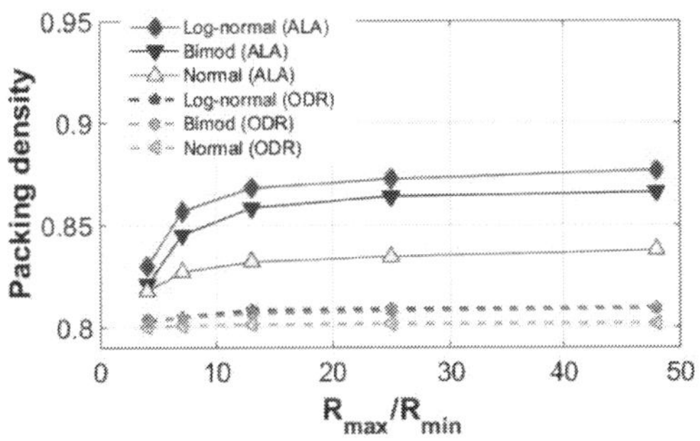

Figure 15. Dependence of the packing density provided by ALA on the disk size distribution width for packing about 50000 disks with normal, log-normal and bimodal size distributions.

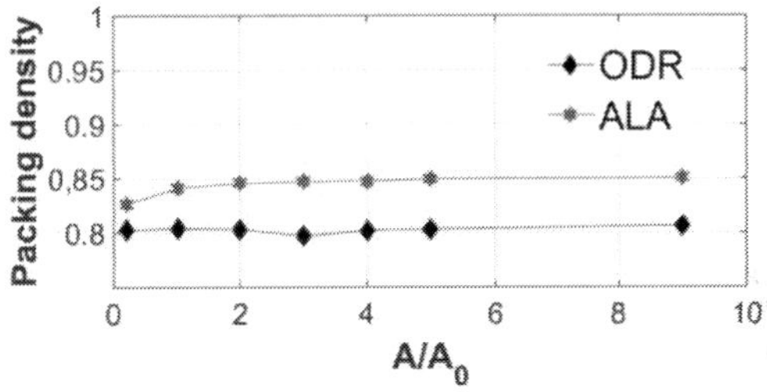

Figure 16. Dependence of packing density obtained by ODR and ALA on the domain size for the imposed log-normal type distribution.

time required for disk packing by ODR method (about 1 s for the packing of 5139 disks). The computation time for the ODR (t_{ODR}) and ALA (t_{ALA}) was compared for filling domains of different sizes (from 2000 to 110000 disks) with the same (log-normal type) size distribution. In all the cases, the time required for the AL method is about two times longer. However, the analysis shows that the ratio t_{ALA}/t_{ODR} decreases when the size polydispersity R_{max}/R_{min} decreases. This dependence is illustrated in Fig.14 (red curve). The ALA is about two times faster than ODR algorithm when the size polydispersity is equal to 4 and about two times slower when $R_{max}/R_{min} = 50$.

This property is explained by the fact that, when the variability in disk size decreases, the number of particles tested in the top layer for placing a new one decreases

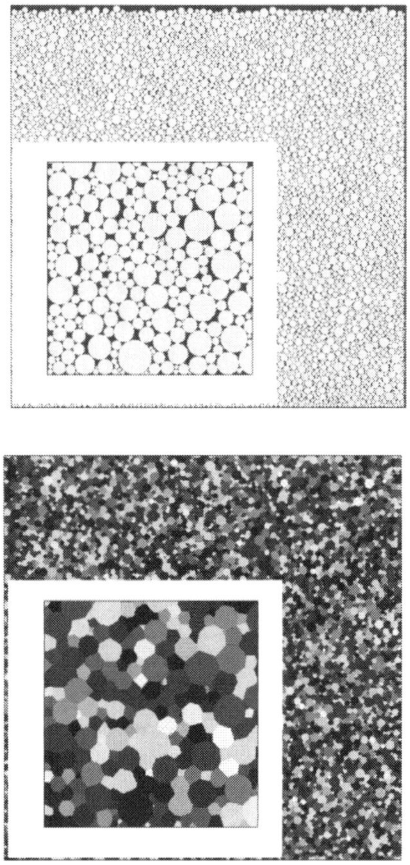

Figure 17. Packing of 10221 disks (density: 0.8597) and corresponding cell structure (Laguerre-Voronoï tessellation) generated with imposed log-normal type size distribution.

also (since less small particles are generated). The same analysis was performed for other types of distributions. The results are shown in Fig.14 as well. In contrast to the case of log-normal type A distribution, the ALA is faster than the ODR method for disk packing with imposed bimodal and normal A distribution laws.

The results shown in Fig.14 were obtained for packings of about 50000 disks (the corresponding domain size is different for all the distributions and polydispersities), and a similar dependence of the CPU time on the polydispersity was found for packings of about 10000 disks for all three distributions.

Packing densities obtained with the ALA for all the distributions (see Fig.15) and polydispersities were found to be higher than ones provided by the ODR algorithm. Fig.15 illustrates that in the case of the AL method the densities increase with increasing the polydispersity and can achieve very high values (0.88) whereas they are typically in the range of 0.8 and 0.81 for packings generated by the ODR method. In addition, the study of the packing density as a function of the domain size was conducted. Fig.16 illustrates the evolution of density as a function of the domain size for the imposed log-normal size distribution law.

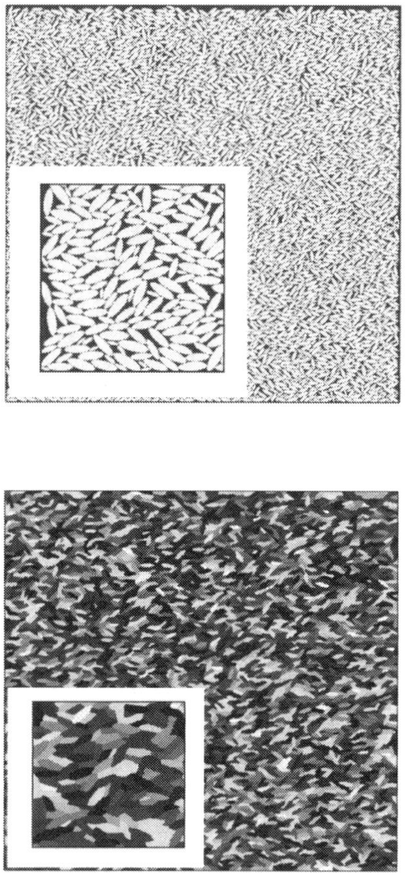

Figure 18. Packing of 4892 ellipses (density: 0.7737) and corresponding cell structure (Laguerre-Voronoï tessellation) generated with imposed aspect ratio λ = 3, uniform A and normal α distributions.

Therefore, the proposed ALA is more effective and provides higher packing densities than the classical or Optimized Dropping and Rolling methods in the considered 2D context.

4 Conclusions

The results reported in this paper demonstrate that the Advancing Layer Algorithm for packing of disks and ellipses provide high packing densities which are achieved thanks to the constructed Height list. The comparison with recently developed Optimized Dropping and Rolling method shows that the proposed packing algorithm has lower computational cost in 2D case for low and moderate variabilities in disks' size. In addition, the generated packing has higher densities than the ones obtained by the ODR method.

The dense packing of ellipses represented by multi-circles was constructed for different imposed A, λ and α distributions. The analysis exhibits that ellipse packing densities are lower than the ones obtained for disks and decrease as the ellipticity increases. Nevertheless, density of 0.8 for ellipse packing can be achieved and provides good agreement with imposed area, shape and orientation distribution.

The packing method developed in the present work is based on the approximation of ellipses by a set of circles. Firstly, this facilitates the calculation of contact position and overlapping detection. Secondly, since the obtained packing is in fact a disk packing, the Laguerre-Voronoï Tessellation Method can be easily applied for generating polygonal structures [9]. Indeed, the high packing densities achieved thanks to the Height list used in the ALA provide good accuracy in respecting the desired inertia matrix distributions imposed for generating statistically equivalent cell structures (see Fig. 17-18).

The extension of ALA to 3D packing and polygonal structure generation is also currently developed. In this case, triplets of ellipsoids (instead of pairs of ellipses in 2D) are tested for placing each new particle. Some additional optimization techniques need to be implemented to improve the efficiency of the algorithm and to enable parallel processing.

References

1. Castillo, F. J. Kampas, and J. D. Pintér. Solving circle packing problems by global optimization: Numerical results and industrial applications. Eur. J. Oper. Res., **191**:786-802, (2008).
2. W. Q. Huang, Y. Li, B. Jurkowiak, Ch. M. Li, and R. Ch. Xu. A two-level search strategy for packing unequal circles into a circle container. *In Principles and Practice of Constraint Programming*, (2003).
3. W. Q. Huang, Y. Li, H. Akeb, and C. M. Li. Greedy algorithms for packing unequal circles into a rectangular container. J. Oper. Res. Soc., **56**:539-548, (2005).
4. Yu. G. Stoyan and G. Yas'kov. A mathematical model and a solution method for the problem of placing various-sized circles into a strip. Eur. J. Oper. Res., **156**:590-600, (2004).
5. J. Kallrath. Cutting circles and polygons from area-minimizing rectangles. J. Glob. Optim., **43**:299-328, (2009).
6. L. Madej, K. Pasternak, J. Szyndler and W. Wajda. Development of the modified cellular automata sphere growth model for creation of the digital material representations. Key Eng. Materials, 611-612:489-496, (2014).
7. J. Kallrath and S. Rebennack. Cutting ellipses from area-minimizing rectangles. J. Glob. Optim., **59**:405-437, (2014).
8. K. Hitti. Direct numerical simulation of complex Representative Volume Elements (RVEs): Generation, resolution and homogenization. PhD thesis, Ecole Nationale Supérieure des Mines de Paris, (2011).
9. K. Hitti and M. Bernacki. Optimized dropping and rolling (ODR) method for packing of poly-disperse spheres. Appl. Math. Model., **37**:5715-5722, (2013).
10. Y. T. Feng, K. Han, and D. R. J. Owen. Filling domains with disks: an advancing front approach. Int. J. Numer. Meth. Engng, **56**:699-713, (2003).
11. Y. Shi and Y. Zhang. Simulation of random packing of spherical particles with different size distributions. Appl. Phys. A, **92**:621-626, (2008).
12. A. Benabbou, H. Borouchaki, P. Laug, and J. Lu. Sphere packing and applications to granular structure modeling. *In RaoV. Garimella, Editor, Proceedings of the 17th International Meshing Roundtable*, pages 1-18. Springer Berlin Heidelberg, (2008).
13. A. Benabbou, H. Borouchaki, P. Laug, and J. Lu. Numerical modelling of nanostructured materials. Finite Elem. Anal. Des., **46**(1-2):165-180, (2010).
14. K. Bagi. An algorithm to generate random dense arrangements for discrete element simulations of granular assemblies. Granul. Matter., **7**:31-43, (2005).
15. A. Donev, S. Torquato, and F. H. Stillinger. Neighbor list collision-driven molecular dynamics simulation for nonspherical hard particles. I. Algorithmic details. J. Comput. Phys., **202**:737-764, (2005).
16. A. Donev, S. Torquato, and F. H. Stillinger. Neighbor list collision-driven molecular dynamics simulation for nonspherical hard particles. II. Applications to ellipses and ellipsoids. J. Comput. Phys., **202**:765-793, (2005).
17. A. Donev, R. Connelly, F. H. Stillinger, and S. Torquato. Underconstrained jammed packings of nonspherical hard particles: Ellipses and ellipsoids. Phys. Rev. E, **75** 051304, (2007).
18. R. M. Baram and P. G. Lind. Deposition of general ellipsoidal particles. Phys. Rev. E, **85** 041301, (2012).
19. J. Buchalter and R. M. Bradley. Orientational order in amorphous packings of ellipsoids. Europhys. Lett, **26**(3):159-164, (1994).
20. G. Delaney, D. Weaire, S. Hutzler, and S. Murphy. Random packing of elliptical disks. Phil. Mag. Lett., **85** (2):89-96, (2005).
21. C.-Y. Wang and V.-C. Liang. A packing generation scheme for the granular assemblies with planar elliptical particles. Int. J. Numer. Anal. Meth. Geomech., **21**:347-358, (1997).
22. C.-Y. Wang, C.-F. Wang, and J. Sheng. A packing generation scheme for the granular assemblies with 3D ellipsoidal particles. Int. J. Numer. Anal. Meth. Geomech., **23**:815-828, (1999).
23. Y. T. Feng, K. Han, and D. R. J. Owen. An advancing front packing of polygons, ellipses and spheres. *In R. P. Jensen B. K. Cook, editor, Discrete Element Methods. ASCE: New York*, (2002).
24. W. X. Xu, H. S. Chen, and Z. Lv. An overlapping detection algorithm for random sequential packing of elliptical particles. Physica A, **390**:2452-2467, (2011).
25. D. Markauskas, R. Kacianauskas, A. Dziugys, and R. Navakas. Investigation of adequacy of multi-sphere approximation of elliptical particles for dem simulations. Granular Matter, **12**:107-123, (2010).

Full field modeling of dynamic recrystallization in a global level set framework, application to 304L stainless steel

Romain Boulais-Sinou[1,a], Benjamin Scholtes[1], Daniel Pino Muñoz[1], Charbel Moussa[1], Isabelle Poitrault[2], Isabelle Bobin[3], Aurore Montouchet[3], Marc Bernacki[1]

[1]*MINES ParisTech, PSL Research University, CEMEF - Centre de mise en forme des matériaux, CNRS UMR 7635, CS 10207, rue Claude Daunesse 06904 Sophia Antipolis Cedex, France*
[2]*Industeel, ArcelorMittal, CRMC 71200 Le Creusot, France*
[3]*Areva Creusot Forge, 6 allée J. Perrin 71200 Le Creusot, France*

Abstract. A new full field numerical approach for the simulation of dynamic and post-dynamic recrystallization will be detailed. A level Set framework is employed to link a crystal plasticity finite element method with the modeling of recrystallization. Plasticity is calculated through the activation of slip systems and provides predictions for both SSDs and GNDs densities. These predictions control the activation and kinetics of recrystallization. All the developments are applied on 304L stainless steel.

Keywords: Recrystallization, CPFEM, Level Set.

1 Introduction

The optimization of industrial forming processes relies on the understanding of the material behaviour under the deformation conditions. The mechanical and thermal properties of metallic materials are inherited from their microstructure properties. Metals are formed of microscopic-length crystals, also called grains, which evolve when they undergo large plastic strain and high temperature. It is essential to control the microstructural evolutions with regard to the final in-use properties.

During hot working, the energy is mainly released through heat dissipation. However, a small part of the energy remains stored in the material under the form of crystallographic defects, principally dislocations. Many mechanisms can then be activated to minimize this energetic potential, such as recovery, grain growth (GG) or recrystallization (ReX). The rearrangement of the microstructure during ReX is a complex phenomenon that is highly investigated in the literature [1]. During ReX, dislocations-free grains, called nuclei, emerge from atomistic rearrangement and growth over the old deformed grains. The nucleation of new grains can occur during the deformation (dynamic recrystallization, DRX) and continue during an annealing treatment (post-dynamic recrystallization, PDRX), or it can be fully activated after the dynamic step during this post-treatment (static recrystallization, SRX).

An increasing number of industrial companies show a strong interest in modeling the microstructure evolutions at small scale. In that context modeling means to understand/predict some physical phenomena at work in ReX mechanisms, such as the evolution of the stored energy, the strain hardening effect of plastic deformation or the prediction of the grain size and orientation distributions. This understanding can be obtained either through the development of simplified models (mean field approaches, MFA) or through the development of deterministic finite element (FE) methods where the microstructure topology is described (full field approaches, FFA). Both approaches need to be linked with experimental data to ensure that the models fit well with the real physical phenomena. The FFA can be divided into two main categories (probabilistic or deterministic). The probabilistic methods, such as the Monte-Carlo [2] or the Cellular Automata [3] approaches which are based on a stochastic evolution of the state variables on a voxel grid. The deterministic methods rely on solving systems of partial differential equations (PDEs). The later are more precise but also more time consuming. Due to this limitation, the PDE systems are solved on a Representative Volume Element (RVE). RVE's must be as small as possible while being big enough to contain all the heterogeneities of the microstructure. The Vertex method [4] is based on a front tracking description of the microstructure. This explicit description can encounter difficulties to handle topological events such as the appearance or disappearance of grains mainly in 3D. Thus front capturing approaches, like the Phase field method (PF) [5][6] and the level set method (LS) [6-9] seem more adapted to the simulation of ReX. The use of a FE

[a] Corresponding author: romain.boulais-sinou@mines-paristech.fr

framework and the *ad hoc* meshing/remeshing capabilities allows also to reach large deformations of RVEs, which is not straightforward when dealing with regular grids such as in the Fast Fourier Transform methods (FFT) [10]. The PF formulation requires additional numerical parameter, like the interface width, whereas the LS formulation only relies on measurable physical and measurable quantities, which makes it more direct to use for solving the PDE systems of Crystal Plasticity Finite Element Method (CPFEM) and DRX.
This paper presents a full field approach in a LS framework and the first steps of its application to the simulation of DRX in 304L stainless steel.

2 Modelling of dynamic recrystallization

In this section, the solving of crystal plasticity problem using the FE method is presented. First we introduce the governing equations and the crystal plasticity PDE systems. Then, we describe the LS formulation used to represent the polycrystal RVE and the grain boundary motion due to the stored energy.

2.1 Crystal Plasticity Finite Element Method

The behaviour of the polycrystal during the deformation is described by a CPFEM model, based on a classical elasto-viscoplastic formulation in a Lagrangian formalism. We use the classic decomposition of the deformation gradient into an elastic and a plastic part.

$$F = F_e\, F_p. \qquad (1)$$

In crystalline matter, the accommodation of plastic deformation F_p is done through the activation of slip systems at the atomistic level. When the local stress on an atomic plan reaches a critical value, the atoms start to glide to accommodate the macroscopic deformation and create atomistic defects, mainly dislocations. In FCC material, there are twelve crystallographic slip systems {111} <110> that are likely to be activated with the plastic deformation. For each slip system α, we need to determine the average velocity of a dislocation gliding. Here, the slip rate $\dot{\gamma}_\alpha$ is calculated from the following viscoplastic exponential law:

$$\dot{\gamma}_\alpha = \dot{\gamma}_0 \left|\frac{\tau^\alpha}{\tau_c^\alpha}\right|^{1/m} sign(\tau^\alpha), \qquad (2)$$

where $\dot{\gamma}_0$ is the reference slip rate, m is the sensitivity exponent, τ^α is the local shear stress on slip system and τ_c^α is the critical resolved shear stress (CRSS) required to activate the gliding. τ^α is related to the macroscopic stress $\bar{\sigma}$ through the Schmidt tensor $\bar{\bar{T}}^\alpha$, which is calculated for each slip system α,

$$\tau^\alpha = \bar{\bar{T}}^\alpha : \bar{\sigma}. \qquad (3)$$

For a FFC material, τ_c^α is supposed to be the same for all the slip systems, noted τ_c, and it is obtained solving the following dislocation-based hardening law:

$$\tau_c = \tau_0 + \psi\mu b\sqrt{\rho}, \qquad (4)$$

with τ_0 the initial CRSS, b the Burger's vector norm, μ the shear modulus and ψ a hardening coefficient to be identified with experimental data and chosen in the following as 1/2. The prediction of the dislocation density ρ is then of prime importance for the integration of the behavior law. In this work, two type of dislocation densities are considered and the total dislocation density is the sum of those two dislocation populations. The statistically stored dislocation (SSD) is related to the velocity gradient V_p with the Yoshie-Laasraoui-Jonas formulation [11]:

$$\begin{cases} \dot{\rho}_{SSD} = \left(\frac{K_1}{M} - \frac{K_2}{M}\rho_{total}\right) V_p, \\ V_p = \sum_\alpha |\dot{\gamma}_\alpha|, \\ \rho_{total} = \rho_{SSD} + \rho_{GND}, \end{cases} \qquad (5)$$

where K_1 and K_2 are material parameters related to the storage and annihilation of dislocations, respectively, and M is the Taylor factor. The geometrically necessary dislocations (GND) accumulate in regions of plastic gradient such as the grain boundaries in order to satisfy the geometric continuity. Hence they are predicted from the gradient of plastic deformation $\mathbf{F}_p$ presented in [12]:

$$\begin{cases} \dot{\boldsymbol{\rho}}_{GND_\alpha} = \frac{1}{b}\nabla\times(\dot{\gamma}_\alpha \boldsymbol{n}_\alpha \mathbf{F}_p), \\ \dot{\rho}_{GND} = \sum_\alpha \left\|\dot{\boldsymbol{\rho}}_{GND_\alpha}\right\|_{L_1}, \end{cases} \qquad (6)$$

with $\boldsymbol{n}_\alpha$ the slip direction of the α system. In the end, the PDE system can be summarized as shown in figure 1.

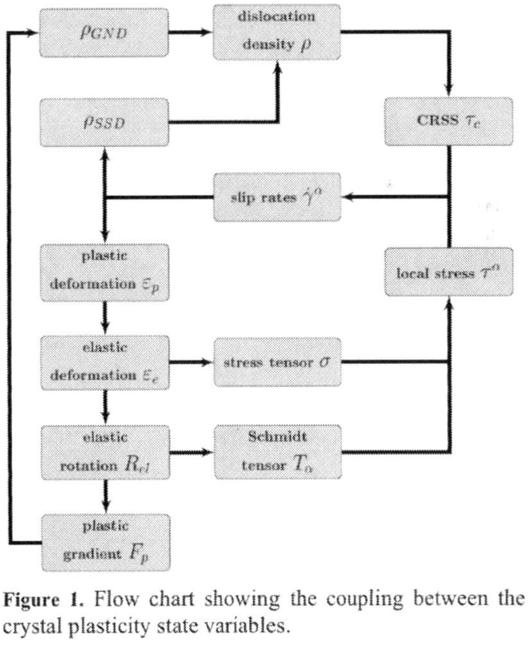

Figure 1. Flow chart showing the coupling between the crystal plasticity state variables.

2.2 Level-Set formulation

The LS method is used to describe the interfaces, here the grain boundaries. At each interpolation point of the finite element mesh (P1 interpolation), the LS function ψ

corresponds to the signed distance to the interface Γ of a sub-domain Ω. The position of an interface is known though the isovalue $\psi = 0$ of its corresponding LS function. The sign convention used here is $\psi \geq 0$ inside Ω and $\psi \leq 0$ outside Ω.

$$\begin{cases} \psi(x,t) = \pm d(x, \Gamma(t)), x \in \Omega, \\ \Gamma(t) = \{\psi(x,t) = 0, x \in \Omega\}. \end{cases} \quad (7)$$

In the case of ReX, the displacement of the LS interface is done in an Eulerian way by solving the following transport equation according to a given velocity field $\vec{v}$.

$$\begin{cases} \frac{\partial\psi(x,t)}{\partial t} + \vec{v} \cdot \nabla\psi(x,t), \\ \psi(x, t=0) = \psi_0(x). \end{cases} \quad (8)$$

For a grain boundary, the velocity $\vec{v}$ is a combination of the stored energy potential between the neighboring grains and the capillarity force at the interface,

$$\vec{v} = \vec{v}_{\Delta Energy} + \vec{v}_{capillarity}. \quad (9)$$

The energy gradient between two neighboring grains $\Delta Energy = E_{grain_i} - E_{grain_j}$, is calculated from the average dislocation densities of the grain i and j.

$$E_i = \int_{x \in grai\square\, i} \left[\frac{1}{2}\mu b^2 \rho(x)\right] dx. \quad (10)$$

When working with large polycrystal, the number of LS functions required to describe all the grains can be very high. In this paper, we work with the Global Level Set (GLS) framework developed by Scholtes et al. and described in [13] for GG and [14] for SRX. The GLS methodology enables us to describe more than one grain with one LS function. The full fields and mean fields inherent to each grain are conserved individually and the problematics due to topological events, like grain coalescence, are handled by swapping the grains from one GLS to another. In the following, GGReX will denote the algorithm developed in [7][8] and improved in [13][14] to model grain boundary motion due the capillarity force and the stored energy.

2.3 Grain Growth and Recrystallization model

If we consider the ReX phenomenon modeling, we observe that the overall stored energy of the system must decrease. Physically, when a grain boundary is moving, it leaves a dislocation-free area behind. To reproduce this phenomenon in our FFA model, we chose to reset the dislocation densities SSD and GND behind the recrystallization front. Thanks to the LS functions describing the grain boundaries, it is easy to identify which area has been recrystallized and reset the energy field at the corresponding mesh elements. Figure 2. illustrates how elements are identified behind the front (red hatched elements). At the atomistic level, the rearrangement of atoms behind the recrystallization front changes the local lattice orientation. It is commonly accepted that the new orientation in this area will be similar to the orientation of the grain it has grown from. To deal with this local orientation change, we chose to extend the orientation field of the growing grain behind the recrystallized front as shown in Figure 3. Capillarity effects (GG) are also taken into account on the proposed formalism as described in [8][13][14].

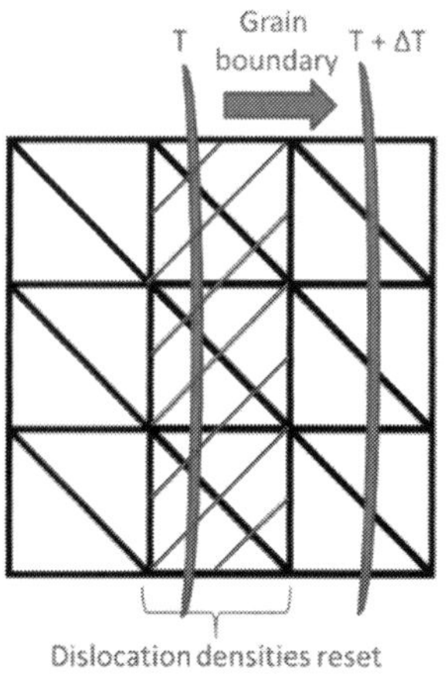

Figure 2. Identification of the recrystallized element and resetting of the dislocation density.

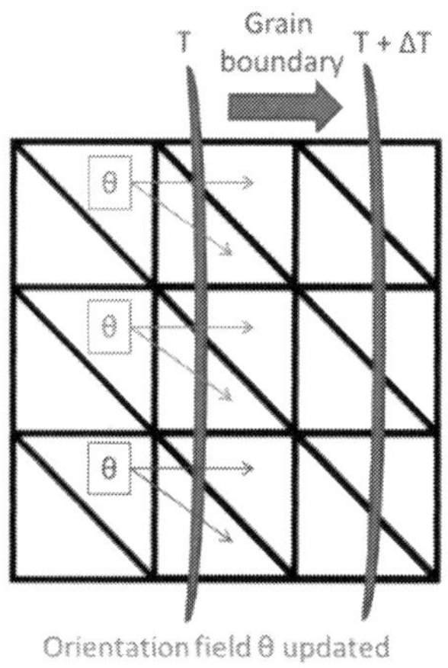

Figure 3. Extension of the orientation field from the growing grain onto the recrystallized elements.

2.4 Nucleation model

The nucleation events occur when the dislocation density reaches a critical value, meaning that the material has stored more energy than it can handle. In previous work at CEMEF [15][16], criteria for both SRX and DRX nucleation were proposed. For the DRX model, the critical dislocation density ρ_{cr} is obtained as a limit from a sequential equation:

$$\begin{cases} \rho_{cr_0} = \left(\frac{20K_1\gamma\dot{\varepsilon}}{3M\tau^2}\right)^{\frac{1}{3}}, \\ \rho_{cr_{n+1}} = \left[\frac{-2\gamma\dot{\varepsilon}\frac{K_2}{M\tau^2}}{ln\left(1-\frac{K_1}{\theta K_2}\rho_{cr_n}\right)}\right]^2, \end{cases} \quad (11)$$

with $\dot{\varepsilon}$ the macroscopic strain rate, γ the interfacial energy, M the mobility of the interface, θ a constant whose value is close to 0.3 [15] and τ the dislocation line energy. When an element of the domain reaches the critical value ρ_{cr}, a spherical new grain is introduced. In order to ensure that the nucleus do not shrink as soon as it appears, its initial radius R_{ini} is chosen greater than the local critical radius R_{cr} defined as follow:

$$R_{ini} > R_{cr} = \frac{2\gamma}{\rho_{cr}\tau}. \quad (12)$$

All nuclei formed are dislocation-free new crystals. The initial dislocation density of a nucleus is then set equal to an annealing configuration, $\rho_{ini} = 10^{11}m^{-2}$. At their creation, we assume that their orientation θ is the same as the region of the parent grain they grow from. To model this aspect, the initial orientation field inside is set to the value θ_0 from the center element of the nucleus, i.e. the element chosen to introduce the nucleus. Figure 4 illustrates how a new grain is introduced on the FE mesh.

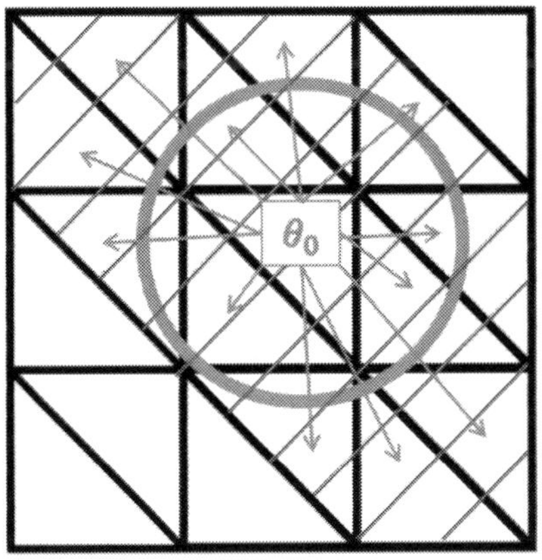

Figure 4. Introduction of a new germ (blue LS). Dislocation density and orientation field are reset on the red hatched element.

3 Case Studied

In order to test our formulation we built a classical Channel-Die test where a 3D polycrystal RVE is deformed. The material investigated here is 304L stainless steel. All experimental data were obtained from hot torsion tests under constant strain rate $\dot{\varepsilon} = 10^{-2}\,s^{-1}$ and constant temperature $T = 1000°C$ [14-16]. However, for the first simulations described in the following, we fixed a lower strain rate value equal to $\dot{\varepsilon} = 2.78 \cdot 10^{-6}\,s^{-1}$ that corresponds to 1% strain over one hour. The objective is to accurately observe the evolution of grain boundaries, orientation field and dislocation densities while the deformation on the RVE does not exceed 3%. The initial RVE is a 1×1×1 mm^3 cubic polycrystal that contains 200 grains for a mean grain size $\bar{D} = 212\,\mu m$. To generate the microstructure, a Voronoï tessellation and a LS immersion are used [17]. Boundary conditions are imposed on all RVE external faces and respect the constant strain rate $\dot{\varepsilon}$ described previously. The mesh used for this simulation contained roughly 100 000 elements. Also, an adaptive mesh refinement [18] is performed every 5% strain level. At this validation stage, the impact of the chosen mesh size was not yet investigated. In order to represent an initial annealed state for the material, the initial GND density is null $\rho_{GND_0} = 0\,m^{-2}$ and the SSD density is set at an estimated value of $\rho_{SSD_0} = 10^{11}\,m^{-2}$. The CPFEM and GGReX models are coupled following the global algorithm detailed in the flow chart in Figure 5. The global time step is fixed and adapted during the simulation as the minimum of the time step values required by the CPFEM and GGReX models.

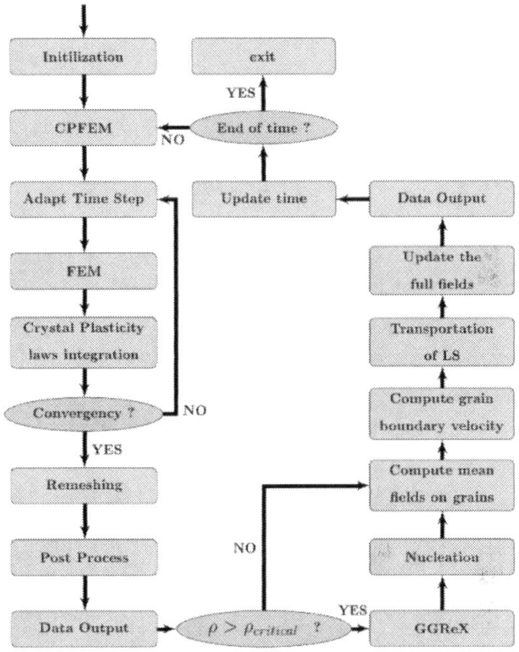

Figure 5. Coupling between CPFEM and GGReX.

4 Primary results

In the early stage of the simulation, only CPFEM is at work and the grains are accumulating energy. When the energy gradient between grains becomes significant it activates the GG/ReX mechanisms. Under the constant strain rate and temperature condition imposed from experimental data, we observe that the energy reduction through the grain boundary migration is too important to reach the experimental critical value calculated at $\rho_{cr} = 1.8 \cdot 10^{13}m^{-2}$ from Eq. (11). Hence there is no nucleation step in these results and a Strain Induced Boundary Migration (SIBM) configuration is observed. Fig. 6 illustrates the decrease of the dislocation density due to the effect of the grain boundary migration (SIBM). Fig. 7 and Fig. 8 illustrate, respectively, the evolution of the first component of the orientation field and the evolution of the dislocation density field (in m^{-2}).

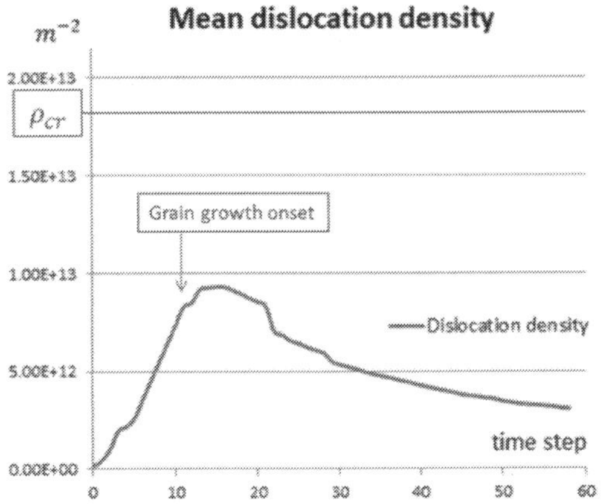

Figure 6. Evolution of the mean dislocation density in the RVE.

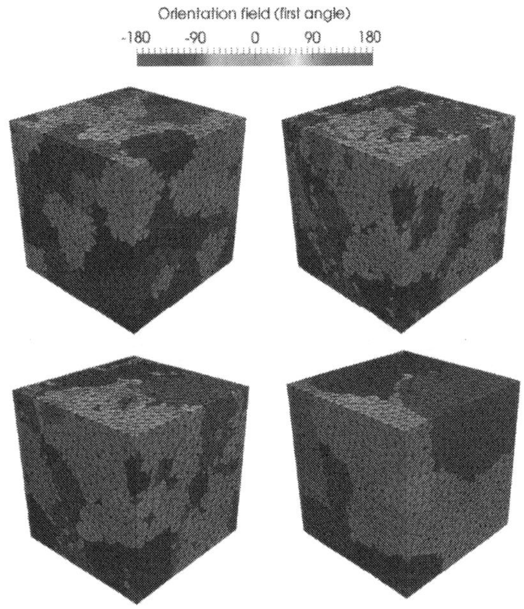

Figure 7. First component of the orientation field ∈ $[-180°, 180°]$. From left to right and top to bottom: initial field and fields after 150, 300 and 600 time steps.

5 Perspectives

This coarse case is a first step toward the simulation of DRX. Further testing of the described coupling between the crystal plasticity model and the GGReX model for finer FE mesh must be carried out. First, we will perform simulations using the same thermomechanical conditions as the experimental torsion tests. Our numerical DRX model will then be validated thanks to experimental data concerning 304L steel material.

Then, it will be interesting to observe the RVE behavior with the introduction of the dislocation-free grains. Once the nucleation happened, the softening effect on the strain stress response will be investigated. For instance, we could observe the effects of higher strain rate conditions or lower temperatures. The orientation dependence of the nuclei could also be an interesting topic research. Complex industrial pass leading to DRX phenomena will be investigated. Finally, these developments will be implemented in the DIGIMU® software package [19].

Figure 8. Evolution of the dislocation density field ρ (m^{-2}). From left to right and top to bottom: initial field and fields after 150, 300 and 600 time steps.

Acknowledgements

The authors would like to gratefully thank ArcelorMittal Industeel, Areva, Ascometal, Aubert & Duval, CEA and Safran for funding this research through the MICROPRO2 consortium.

References

1. A. D. Rollett, Overview of modeling and simulation of recrystallization, *Progress in Materials Science*, **42**, 79 – 99 (1997).
2. K. Okuda and A.D. Rollett, Monte Carlo simulation of elongated recrystallized grains in steels, *Computational Materials Science*, **34**, 264 – 273 (2005).
3. G. Kugler and R. Turk, Modeling the dynamic recrystallization under multi-stage hot deformation, *Acta Materialia*, **52**, 4659 – 4668 (2004).
4. K. Piekos, J. Tarasiuk, K. Wierzbanowski, B. Bacroix, Generalized vertex model of recrystallization – Application to polycrystalline copper, *Computational Materials Science* **42**, 584–594 (2008).
5. R. Darvishi Kamachali, S. Kim, I. Steinbach, Texture evolution in deformed AZ31 magnesium sheets: Experiments and phase-field study, *Computational Materials Science* **104**, 193–199 (2015).

6. Y. Jin, N. Bozzolo, A.D. Rollett, and M. Bernacki, 2D finite element modeling of anisotropic grain growth in polycrystalline materials: level set versus multi-phase-field method, *Computational Materials Science*, **104**, 108–123 (2015).
7. M. Bernacki, H. Resk, T. Coupez, and R. Logé, Finite element model of primary recrystallization in polycrystalline aggregates using a level set framework, *Modelling and Simul. Mater. Sci. Eng.* **17**, 064006 (2009).
8. M. Bernacki, R. Logé, and T. Coupez, Level set framework for the finite-element modelling of recrystallization and grain growth in polycrystalline materials, *Scripta Materialia*, **64**, 525–528 (2011).
9. M. Shakoor, B. Scholtes, P.-O. Bouchard and M. Bernacki, An efficient and parallel level set reinitialization method – Application to micromechanics and microstructural evolutions, *Applied Mathematical Modelling*, **39**, 7291 – 7302 (2015).
10. B. Liu, D. Raabe, F. Roters, P. Eisenlohr and R.A. Lebensohn, Comparison of finite element and fast fourier transform crystal plasticity solvers for texture prediction, *Modelling and Simul. Mater. Sci. Eng.* **18**, 085005 (2010).
11. A. Laasraoui and J.J. Jonas, Prediction of steel flow stresses at high temperatures and strain rates, *Metallurgical Transactions A*, **22**, 1545-1558 (1991).
12. E.P. Busso, F.T. Meissonnier and N.P. O'Dowd, Gradient-dependent deformation of two-phase single crystals, *Mechanics and Physics of Solids*, **48**, 2333-2361 (2000).
13. B. Scholtes, M. Shakoor, A. Settefrati, P.-O. Bouchard, N. Bozzolo and M. Bernacki, New finite element developments for the full field modeling of microstructural evolutions using the level-set method, *Computational Materials Science*, **109**, 388–398 (2015).
14. B. Scholtes, R. Boulais-Sinou, A. Settefrati, D. Pino Muñoz, I. Poitrault, A. Montouchet, N. Bozzolo, and M. Bernacki, 3D level set modeling of static recrystallization considering stored energy fields, *Computational Materials Science*, **122**, 57-71 (2016).
15. K. Huang, Towards the modelling of recrystallization phenomena in multi-pass conditions: application to 304L steel, PhD thesis, MINES ParisTech (2011).
16. A. L. C. Fabiano, Modelling of crystal plasticity and grain boundary motion of 304L steel at the mesoscopic scale, PhD thesis, MINES ParisTech (2012).
17. K. Hitti, P. Laure, T. Coupez, L. Silva and M. Bernacki, Precise generation of complex statistical Representative Volume Elements (RVEs) in a finite element context, *Computational Materials Science*, **61**, 224 – 238 (2012).
18. H. Resk, L. Delannay, M. Bernacki, T. Coupez and R. Logé, Adaptive mesh refinement and automatic remeshing in crystal plasticity finite element simulations, *Modelling and Simul. Mater. Sci. Eng.* **17**, 075012 (2009).
19. B. Scholtes, M. Shakoor, N. Bozzolo, P.-O. Bouchard, A. Settefrati, and M. Bernacki, Advances in level-set modeling of recrystallization at the polycrystal scale - development of the digi-μ software, *Key Engineering Materials*, **651–653**, 617–623 (2015).

Simulation of the combined process "helical rolling-pressing" in the software package Simufact. Forming

Naizabekov Abdrakhman, Lezhnev Sergey*, Panin Evgeniy, Arbuz Alexandr

Rudny industrial institute, Rudny, Republic of Kazakhstan

Abstract. This work is devoted to the simulation of new energy-saving combined process "helical rolling-pressing" in program Simufact Forming. During the simulation were studied the main parameters of stress-strain state and microstructure. The variation of the main technological parameters showed that the influence of the heating temperature of the workpiece plays a significant role in the degree of grain grinding. While the speed of deformation, allowed by the mill design limits, has no significant effect on the microstructure.

1 Introduction

The combination of helical rolling and subsequent pressing in equal-channel angular matrix allows to implement the scheme of stress-strain state with the intense shift of metal layers at helical rolling in the first deformation zone and the unification of intense shear deformation angular pressing with torsion in the second deformation zone (Fig. 1).

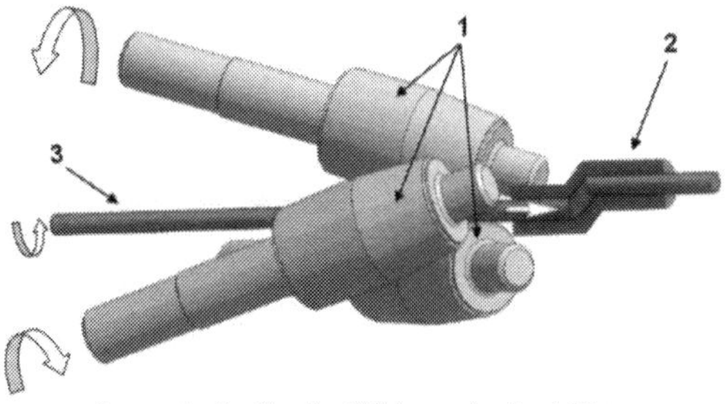

1 – conical rolls; 2 – ECA-matrix; 3 – billet
Figure 1. Combined process "helical rolling-pressing"

Thus non-deformed central part of the workpiece in the first deformation zone will receive intensive deformation in another diagram in the second zone – in ECA-matrix. In addition, it is interested to study the parameters of angular pressing with torsion. In fact, this is the process of continuous extrusion with the elements of the torsion of the workpiece inside the matrix. It will increase the level of accumulated plastic deformation in the volume of the workpiece and contribute to the elaboration of the central zone. This creates good conditions for the formation of ultrafine-grained structure throughout the volume of a round workpiece.

In the study of any process forming a key position ahead of laboratory or industrial tests is to study the stress-strain state (SSS). This will allow to identify the distribution of stresses and strains in the present process, and also to determine their critical values, which, in turn, will allow you to check the working tool for strength.

To determine the values of stress and strain need to find the values of the components of the corresponding tensors, for which a three-dimensional flow of metal is very difficult to visualize. So usually when considering the parameters of the SSS use simple indicators of the intensity of strain and stress, or so-called equivalent strain and equivalent stress, which include components of deformations and stresses in the following form:

$$\varepsilon_{EQ} = \frac{\sqrt{2}}{3}\sqrt{(\varepsilon_1-\varepsilon_2)^2+(\varepsilon_2-\varepsilon_3)^2+(\varepsilon_3-\varepsilon_1)^2}, \quad (1)$$

$$\sigma_{EQ} = \frac{1}{\sqrt{2}}\sqrt{(\sigma_1-\sigma_2)^2+(\sigma_2-\sigma_3)^2+(\sigma_3-\sigma_1)^2}, \quad (2)$$

where ε_1, ε_2, ε_3 - main strains,
σ_1, σ_2, σ_3 - main stresses.

Also in the study of SSS is very useful is the study of the temperature conditions of the process, since the temperature change of deformable metal significantly affects the energy-power parameters of deformation.

2 Initial model and temperature distribution

In result of computer simulation of the combined process "helical rolling – pressing" in the software package Simufact.Forming was obtained a successful model of this process with the following optimum values:

- coefficient of friction in the rolls – 0,8;
- coefficient of friction in the matrix - 0,1;

* Corresponding author: sergey_legnev@mail.ru

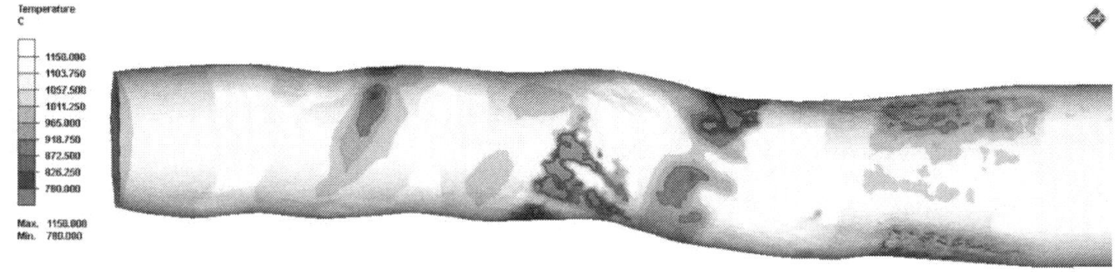

Figure 2. Temperature distribution

- the angle of intersection of channels in the matrix – 150°;

- the distance between matrix and deformation zone of the rolls - 15 mm.

The initial workpiece was a rod with a diameter of 25 mm and a length of 150 mm. Steel 15 was selected as material. Roll dimensions and their structural location consistent with the design of the laboratory rolling mill 10-30. The roll rotation speed was 100 rpm. Equal channel step matrix had channel diameter of 21 mm. Temperature of the billet was accepted by 1000° C, all tools in the model had a temperature of 20° C.

In figure 2 a,b showed the temperature distribution of the workpiece. In the contact zone of the workpiece with the rolls the temperature is reduced to 950-970° C. In the zone of junction of the channels of the matrix temperature is lowered to 850° C. In further temperature drop across the section is leveled due to heat transfer from the central layer to surface.

3 Stress state

When considering the equivalent stress it should understand that this option does not show, what stress operates at a specific point - tensile or compressive. As a radical expression, its value is always positive. It shows the intensity of the action of the stress, i.e., is there at this point stress or not. Its value characterizes the average value of all stresses acting at a given point.

At the first stage, during helical rolling (figure 3 a) stress covers the entire deformation zone of the rolls. In zones of contact of metal with rolls its value reaches 150 MPa, in zones free of contact, the value of this parameter reaches 120 MPa.

In the second stage, during pressing (figure 3 b) stress covers a vast area, connecting two of the deformation. In longitudinal rolling with the support of such a merger does not occur, because there is no twisting of the blank, causing the growth of shear stresses. Due to the action of backpressure, stress within the deformation zone of the rolls is increased to 190 MPa in the contact zones of the metal with the rolls. Free from contact zones, the value of this parameter is in the limit of 140 MPa. In the zone of junction of the channels of the matrix the magnitude of equivalent stresses up to 160 MPa.

There is also clearly tracks the forces of backpressure. Behind the zones of contact of metal with rolls arise extent of physical deformation area that is a consequence of two concurrent factors – backpressure from the side of the matrix and twisting of the workpiece.

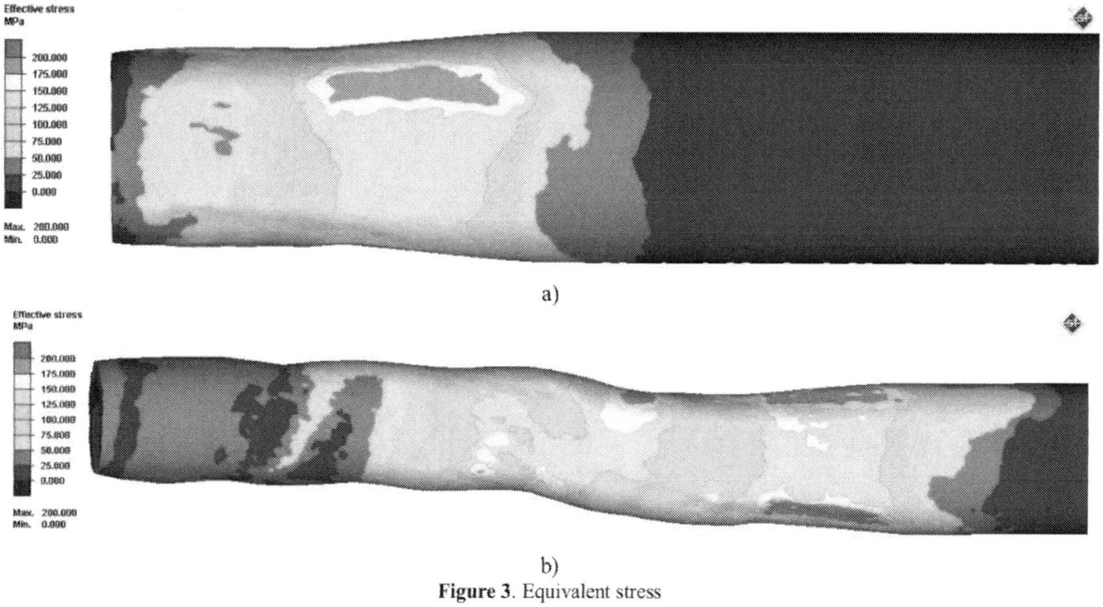

b)

Figure 3. Equivalent stress

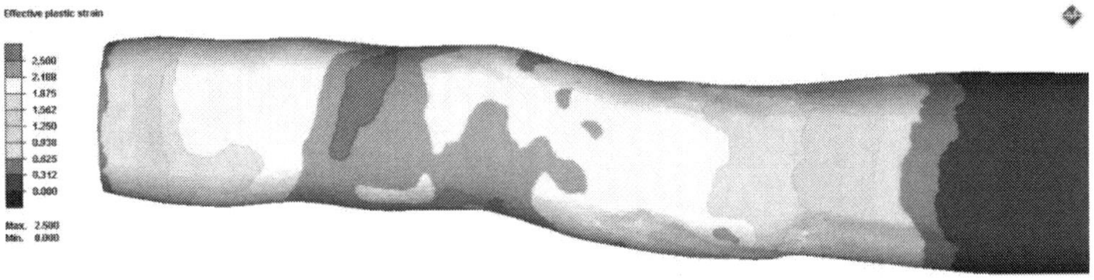

a)

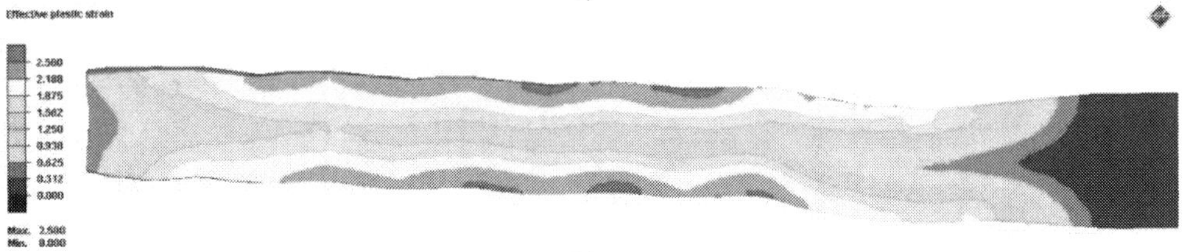

b)

Figure 4. Equivalent strain

4 Strain state

As noted above, to study the strain state use the indicator of intensity of deformation – equivalent strain, which includes major components of deformation. This option allows to monitor the degree of accumulated strain, which is cumulative.

When studying the strain state is necessary not only to ensure a high level of equivalent strain, required for the formation of subalternity patterns, but a uniform distribution of this parameter over the cross section of the workpiece.

During helical rolling (figure 4) deformation develops in the deformation zone of the rolls in the following form: its value reaches 1.2 in the surface layers; in central layers this parameter value is at 0.7 (figure 4 b), i.e. the difference of the values of the equivalent strain reaches 72%.

During pressing there is a significant increase of this parameter due to the implementation of shear deformation during the movement of billet through the channels of the matrix. The value of equivalent strain is 2.3 in the surface layers, in central layers the value of this parameter is at the level of 1.4, i.e. the difference of the values of the equivalent strain reaches 64%. Thus, the difference of values of this parameter after passing through the channels of the matrix decreased by 8%.

Comparison of the values of equivalent strain in the surface and central layers of the billet showed that the distribution of this parameter over the cross section is rather uneven. It is therefore important to achieve a more uniform distribution of the equivalent strain over the cross section of the rod.

It can be concluded that the difference in equivalent strain in the central zone to the entrance to the matrix and after the matrix, equal to 0.7, the difference is the portion of deformation corresponding to simple shear in the matrix. That is, the center of the workpiece when passing the channel matrix gains equivalent strain of 0.7 in the direction perpendicular to the direction of deformation of the previous stage and also perpendicular to the orientation of the structure of the central part. Several cycles of such deformation should contribute to the transformation oriented streaky structures in a more equiaxed, thus reducing the anisotropy of properties on the section of the rod.

5 Modelling of microstructure evolution

To study the evolution of microstructure was used the specialized database of the microstructure of the Matilda program, which uses the data of stress-strain state, strain rate and temperature of ready designed models to SIMUFACT.forming, complements them with the data of physic-chemical properties and their behavior for a given material and its structure from the database, then using the algorithm Yada [1], calculates the process parameters of static and dynamic recrystallization that may cause changes in grain size. Grain size is also calculated for each node of the finite element model and displayed at the end of the simulation at any convenient time for rendering the form.

During deformation there are two kinds of recrystallisation (static and dynamic), which have influence on initial grain size. Calculation of dynamic recrystallisation is the part of model, which refers to the processes in the forming zone. It starts when the effective log. strain ε exceeds the critical strain ε_k. Main data in this model is Zener-Hollomon parameter:

$$Z = \dot{\varepsilon} \exp\left(\frac{Q}{RT}\right) \tag{3}$$

where $\dot{\varepsilon}$ - strain rate, s^{-1};
Q - activation energy for forming, J/mol;
R - gas constant = 8,3144 J/(mol * K);
T - forming temperature, K.

Critical strain for beginning of dynamic recrystallisation:

$$\varepsilon_k = [a1 \cdot D_0^{a2} \cdot Z^{a3} + a4] \cdot a5 \tag{4}$$

where: a1÷ a5 - model coefficients; D_0 - initial gain size, μm; Z - Zener-Hollomon parameter.

Log. strain to 50 % of dyn. recrystallisation:

$$\varepsilon_{50} = [c1 \cdot D0^{c2} \cdot \exp(c3 / T) \cdot \dot{\varepsilon}^{c4}] + c5 \quad (5)$$

where: c1÷ c5 - model coefficients.

Dynamic recrystallised part:

$$X_{Dyn} = 1 - \exp[f1 \cdot ((\varepsilon - \varepsilon_k)/ \varepsilon_{50})^{f2}) \quad (6)$$

where: f1, f2 - model coefficients.

Dynamic recrystallised grain size, μm;

$$D_{Dyn} = d1 \cdot Z^{d2} \quad (7)$$

where: d1, d2 - model coefficients.

Calculation of static recrystallisation is the part of model, when the workpiece segment leave the forming zone.This is valid until the recrystallisation process is interrupted to a new forming or a phase transition. It starts when the effective log. strain ε exceeds the critical strain ε_k and it ends when the temperature is below the lower application range boundary.

Mean log. strain after starting of dynamic recrystallisation:

$$\varepsilon_m = p2 \cdot(\varepsilon_k + \varepsilon_{50} \cdot \{1 - \exp[-(\varepsilon - \varepsilon_k)/ \varepsilon_{50}]\}) \quad (8)$$

where: p2 - coefficent for model adaptation.

Time to 50% of static recrystallisation, s:

$$t_{50} = g1 \cdot \varepsilon^{g2} \cdot D_0^{g3} \cdot \exp(\frac{g4}{T}) \cdot \dot{\varepsilon}^{g5} \quad (9)$$

where: g1÷ g5 - model coefficients.

Time for begin and end of static recrystallisation, s:

$$\begin{cases} t_{05} = b_1 \times t_{50} \\ t_{95} = b_2 \times t_{50} \end{cases} \quad (10)$$

where: b1, b2 - model coefficients.

Temperature corrected pause time, s:

$$t_{P\_kor} = \sum \left\{ d_{Ti} \cdot \exp\left[-\left(\frac{Q_{ST}}{1000 \cdot R \cdot T} \right) \right] \right\} \quad (11)$$

where: d_{Ti} - length of time increment i, s;

Q_{ST} - activation energy for static recrystallisation, J/mol.

Static recrystallised part:

$$X_{Stat} = 1 - \exp[m2 \cdot (t_{P\_kor}/t_{50})^{m1}] \quad (12)$$

where: m1, m2 - model coefficients.

Static recrystallised grain size, μm:

$$D_{Stat} = [s1 \cdot \varepsilon^{s2} \cdot D_0^{s3} \cdot Z^{s4}] + s5 \quad (13)$$

where: s1 ÷ s5 - model coefficients.

Mean grain size at the end of recrystallisation:

$$D = X_{Dyn} D_{Dyn} + (1 - X_{Dyn}) X_{Stat} D_{Stat} + \left[1 - X_{Dyn} - \left(1 - X_{Dyn} \right) X_{Stat} \right] D_0 \quad (14)$$

In large forming pauses the process of grain growth starts directly after the end of dynamic and static recrystallisation When $(X_{Dyn} + X_{Stat}) > 95$ % and $(t_{P\_kor} - t_{95}) > 0$ the new mean grain size mKG in μm is calculated according to:

$$mKG^{w1} - D^{w1} = w2 \cdot (t_{P\_kor} - t_{95}) \cdot \exp[-Q/(RT)] \quad (15)$$

where: w1, w2 - model coefficients.

Thus, in the initial model were imported property data of AISI1015 steel corresponding to steel 15. In the simulation of the microstructure in the program uses the assumption that the workpiece before the deformation has a uniform structure with the same grain size at any point. As the source size adopted, the average grain diameter of 30 μm, which corresponds to the 7th score [2].

After calculation were obtained the results of the microstructure evolution, which were considered separately on the surface and in the longitudinal section of the billet. During helical rolling the surface layers of the workpiece are subjected to intense deformation, leading to considerable grain refinement from 30 μm to 20 μm, which corresponds to the 8th score (figure 5). After passing through the channels of the matrix through the implementation of shear deformation there is a further grinding of the grain to 14 μm, which corresponds to 9th score.

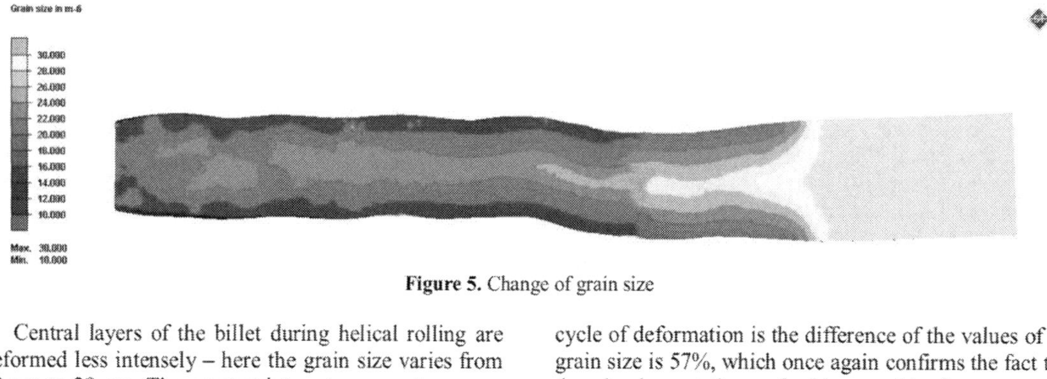

Figure 5. Change of grain size

Central layers of the billet during helical rolling are deformed less intensely – here the grain size varies from 30 μm to 28 μm. The greatest interest represents an area of the workpiece after passing through the channels of the matrix. Here there is a grinding of grain along the entire cross section of the workpiece in the central zone, the grain size changes from 28 μm to 22 μm. Thus, after one cycle of deformation is the difference of the values of the grain size is 57%, which once again confirms the fact that the implementation of this combined process of elaboration of the workpiece in the cross section is uneven and levelling properties over the cross section it is necessary to conduct several cycles of deformation. It should also be noted that the simulation results of the microstructure quite well correlate with the results of a study of equivalent strain, the distribution of which in the cross section is largely similar to the picture of the microstructure evolution.

After study of basic model was given the task to identify how to affect the grain size change of the main technological parameters, namely, temperature, heating of the workpiece and the speed of deformation. These parameters were chosen from the considerations that they are the most flexible in terms of changing values. Temperature is easy enough to vary during the billet heating process in the furnace. The rate of deformation in this combined process is determined by the rolling speed on the mill screw, the value of which is also easy to change when the frequency converter.

5.1 Microstructure evolution in models with varying values of heating temperature of the workpiece

For studying the effect of temperature on the evolution of microstructure were built models with the heating temperatures of the workpiece 1150 °C and 850 °C. As shown the study of these models, the effect of temperature on the change of grain size is very significant. Lowering the temperature causes the intensification of grinding of the original grain, while the increasing temperature, the process slows down. This is well illustrated by the histogram shown in figure 6. This is because with increasing temperature of heating of metal in it significantly accelerates the process of static recrystallization, which is an undesirable character because it leads to the emergence and growth of new grains. At lower temperatures, this type of recrystallization is slowed down significantly, and the grain refinement is under the influence of dynamic recrystallization, which takes place directly in the deformation process.

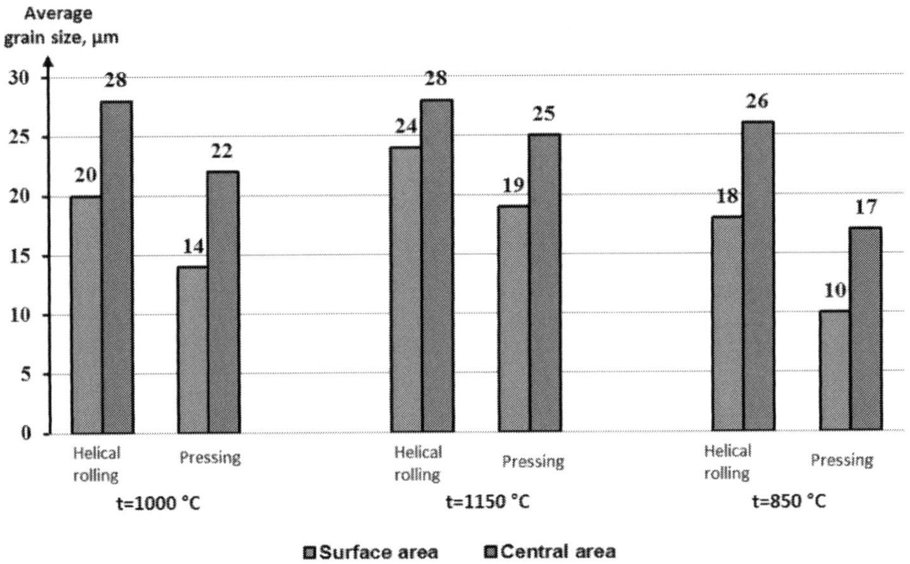

Figure 6. Change of grain size with variation of the heating temperature of the workpiece

In the analysis of the dispersion of values, it was found that in a model with a heating temperature of the workpiece 1150 °C during helical rolling the difference of grain size between the central and surface areas is 17%; after pressing in the matrix, this difference is 32%. The increased scatter is due to the fact that, despite the increased value of a heating temperature of the workpiece at its contact with the working tool surface layers cool more rapidly than central layers. Therefore, when deformation in the matrix, where contact with the tool has a greater length than helical rolling, there is an intensive grain refinement in the surface area.

In the model with a heating temperature of the workpiece 850 °C, during helical rolling the difference of grain size between central and surface areas is 44 %; after pressing in the matrix, this difference reaches 70 %. It is necessary to note not only the increase of dispersion between central and surface areas, but also the decline in overall grain size compared to the base model. Both of these factors are a consequence of lowering the heating temperature of the workpiece, which leads to a decrease of static recrystallization.

It should be noted that despite the increased level of variance model with low temperature heating is optimal in terms of more intensive grinding grain. This is particularly evident when comparing values of grain size in central part, which worked out much worse surface area. As has been shown in studies of the deformed state, to align the values of accumulated strain needed to perform multiple passes of deformation. In this case, it will also reduce the variation in grain size. Therefore, for the subsequent multi-pass deformation it was decided to use a heating temperature of the workpiece equal to 850 °C.

5.2 Microstructure evolution in models with variable speed of deformation

To study the influence of speed of deformation on the microstructure evolution it was necessary to construct additional models with different rolling velocities and compare them with the base model, where the roll rotation speed is 100 rpm.

Because the helical rolling mill is equipped with system regulation of roll rotation speed using a frequency converter, it was decided to study the change in grain size under extreme values of rolling speed. In particular, the minimum technologically possible roll rotation speed is 10 rpm; the maximum possible roll speed is 300 rpm. This choice was made due to the fact that in many works [3-7] in the study of ECAP has been proven that changing the speed of deformation in a relatively small range (3-5 times) had no effect on change of grain size.

The simulation results of the microstructure shown in figure 7, confirmed the results of the above studies. By varying the rate of deformation of from 10 to 300 rpm, the microstructure of the billet is almost constant. A small change in the mean diameter of the grains (1-2 μm) occurs only in the surface area at low speed rolling. This is directly related to the fact that when reducing the speed of deformation of the workpiece has a more prolonged contact with the tool, resulting in stronger surface layers cool down.

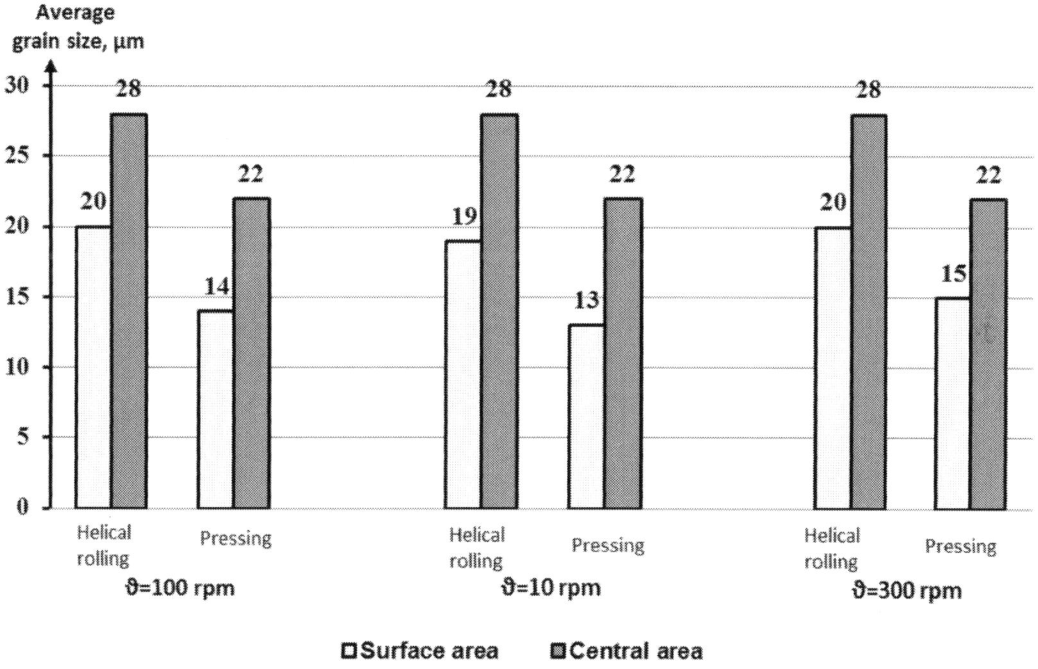

Figure 7. Change of grain size with variation of speed of deformation

It should be noted here that the variation of speed of deformation will be impractical decision. Increasing the speed of rotation of the rolls leads to increased dynamic loads on the frame of a rolling mill, which is a negative factor. An excessive decrease in strain rate is also undesirable because it leads to a substantial cooling of the workpiece and possible growth of grain in central parts of large workpieces due to static recrystallization.

Despite the fact that in the study of the ECAP process, it was found that the decrease in the deformation temperature is a favorable condition for the formation of fine patterns, in the implementation of this combined process may also occur such situation when the workpiece has cooled so that the forces generated by the mill will not be enough to push the workpiece through the channels of the matrix. Therefore, for the subsequent multi-pass deformation it was decided to use the rate of deformation of the basic model, equal to 100 rpm.

5.3 Microstructure evolution after several cycles of deformation

Initial model for multipass deformation was the model with low heating temperature of the workpiece - 850 °C, which was identified as the most optimal. The deformation in the 2nd and 3rd passage was effected under the following conditions:

- second pass: the workpiece with diameter of 20 mm was rolled to a diameter of 17 mm and passed through a matrix with a channel diameter of 18 mm;
- third pass: the workpiece with a diameter of 17 mm was rolled to a diameter of 14 mm and passed through a matrix with a channel diameter of 15 mm.

As a result the following data were obtained (Fig. 8). In the first pass after the helical rolling the difference in average grain diameter between central and surface area was 8 μm; after pressing in the matrix this difference reached 7 μm.

During the second pass the grinding of grain during helical rolling is not as intense as in the first pass. This is because the compression in this passage was 3 mm, which is less than in the first passage (5 mm). Here, the average grain size in the surface area was 8 μm, in the central zone was 13 μm, i.e. the difference was 5 μm. During subsequent pressing in the matrix, the average diameter of the grain was reduced to 6 microns in surface area and 10 μm in the central zone. In this case, the difference value is decreased to 4 μm.

During the third pass after the helical rolling average diameter of the grain in the surface area was 5 μm, in the central zone was 9 μm, i.e., the difference amounted to 4 microns. When pressing in the matrix average grain size decreased to 3 μm in the surface area and 6 μm in the central zone, i.e. the difference of the values was only 3 μm.

Thus, after study of model of multi-pass deformation it was found that with increase in the number of passes there is not only a general decrease of average diameter of the grain, and the gradual equalization of this parameter between the central and surface area.

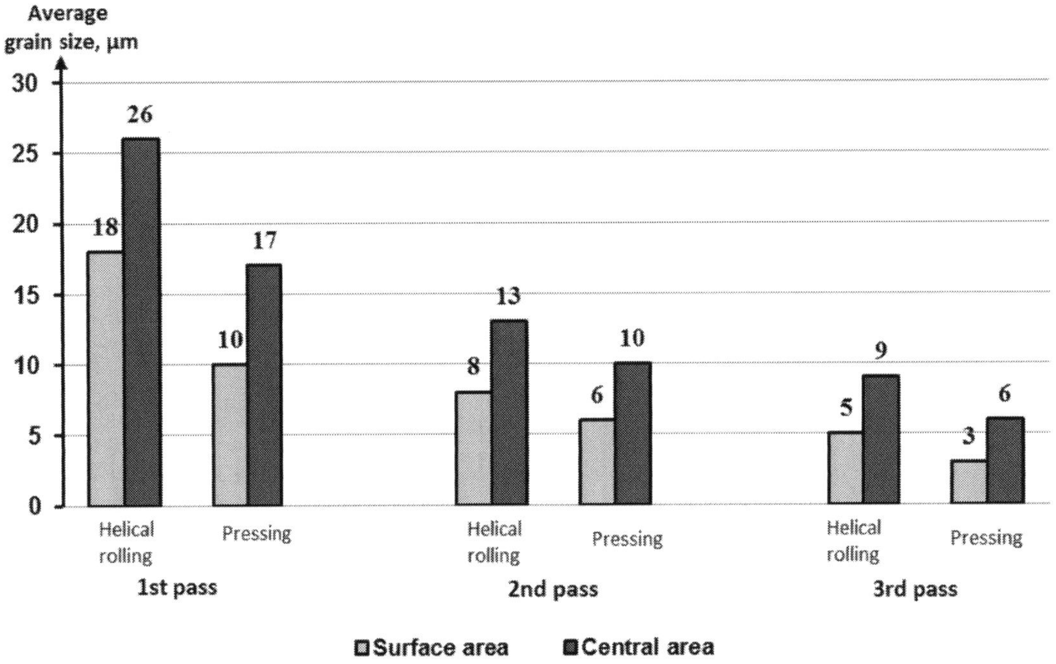

Figure 8. Change of grain size during multipass deformation

6 Conclusion

A study of the microstructure evolution in the combined process "helical rolling-pressing", it was found that this combined method of deformation can significantly grind the original grain size. The variation of the main technological parameters showed that the influence of the heating temperature of the workpiece plays a significant role in the degree of grinding grain. While the speed of deformation, within the permissible installation designs, has no significant effect on the microstructure.

Due to the helical rolling the grain milling occurs mainly in the surface layers of the workpiece. The subsequent pressing in the matrix leads to a change in grain size over the entire cross section of the workpiece. But despite this, the difference values of the mean diameter of the grain between the central and surface area remains. To reduce the observed dispersion of values and the alignment of the grain size in the cross section it is necessary to expose the workpiece to several cycles of deformation. So, after the third pass, the difference of the average values of grain size between the central and surface area is only 3 μm, which is significantly lesser than that after the first passage (7 μm). This allows to speak about quite uniform in this study on the cross-section of the workpiece.

It should be noted that the obtaining of such results by ECAP method requires 4-6 cycles of pressing, while the presented results were obtained after three passes, with time of each pass for rods with 250 mm length is 5-8 seconds. Performance, in this case, (according to the claimed performance of mill-bases a combined installation) will be up to 100 kg/h of high quality round steel bar without length restrictions with sub-ultra-fine-grained structure. This is significantly higher than the main competing processes and provides an opportunity to talk about the greater efficiency of the proposed process.

Acknowledgement

Work is performed under the state budget-funded theme "Innovative energy - saving technology to obtain sub-ultra-fine-grained structural metals and alloys new combined process "helical rolling-pressing" with the program "Grant financing of scientific researches on 2015-2017" (Customer - the Ministry of education and science of the Republic of Kazakhstan).

References

1. H. Yada, N. Matsuzu, K. Nakajima, K. Watanabe and H. Tokita. Trans. ISIJ 23.100–109 (1983)
2. GOST 5639-82. Steel and alloys. Methods of detection and determination of grain size
3. R. Valiev, I. Alexandrov. Bulk nanostructured metal materials (Moscow, Akademkniga, 2007)
4. P. Berbon, N. Tsenev, R. Valiev, M. Furukawa, Z. Horita, M. Nemoto, T. Langdon. Metall Mater Trans A 29, 9, 2237-2243 (1998)
5. Z. Horita, D.J. Smith, M. Furukawa, M. Nemoto, R. Valiev, T. Langdon. J Mater Res 11, 8, 1880-1890 (1996)
6. R. Valiev, R. Islamgaliev, I. Alexandrov. Prog Mater Sci 45, 2, 103-189 (2000)
7. R. Valiev, T. Langdon. Prog Mater Sci 51, 881–981 (2006)

Ductility prediction of substrate-supported metal layers based on rate-independent crystal plasticity theory

Holanyo K. Akpama[1], Mohamed Ben Bettaieb[1,2, a] and Farid Abed-Meraim[1,2]

[1]*LEM3, UMR CNRS 7239 - Arts et Métiers ParisTech, 4 rue Augustin Fresnel, 57078 Metz Cedex 3, France*
[2]*DAMAS, Laboratory of Excellence on Design of Alloy Metals for low-mAss Structures, Université de Lorraine, France*

Abstract. In this paper, both the bifurcation theory and the initial imperfection approach are used to predict localized necking in substrate-supported metal layers. The self-consistent scale-transition scheme is used to derive the mechanical behavior of a representative volume element of the metal layer from the behavior of its microscopic constituents (the single crystals). The mechanical behavior of the elastomer substrate follows the neo-Hookean hyperelastic model. The adherence between the two layers is assumed to be perfect. Through numerical results, it is shown that the limit strains predicted by the initial imperfection approach tend towards the bifurcation predictions when the size of the geometric imperfection in the metal layer vanishes. Also, it is shown that the addition of an elastomer layer to a metal layer enhances ductility.

1 Introduction

The ductility of a material is characterized by its ability to deform homogeneously under some imposed loading. For sheet metals undergoing in-plane biaxial loading, at a certain limit strain, the deformation starts concentrating in narrow bands. The occurrence of such localization bands marks the onset of localized necking in the sheet. Predicting the different limit strains that lead to localized necking is crucial for designing functional or structural components used in industrial devices. To this end, several numerical models have been developed to predict localized necking, which is represented in the form of forming limit diagram (FLD). This FLD concept was initially introduced by Keeler and Backofen [1], for representing the limit strains in the range of positive strain paths, and has been extended by Goodwin [2] to the range of negative strain paths. Despite the wide range of FLD prediction models available in the literature, very few of them have been devoted to metal/elastomer bilayers. However, the latter have proven better stretchability than traditional freestanding metal layers, and are being increasingly used in the industry. For instance, in the design of electronic devices that require high levels of stretchability, substrate-supported metal layers are often used. This is the case of stretchable conductors used in biomedical applications, and interconnects that are used in large-scale integrated circuits [3,4].

The current paper proposes an efficient tool for the prediction of localized necking in substrate-supported metal layers. The mechanical behavior of a representative volume element of the metal layer is determined from the mechanical behavior of its microscopic constituents by using the self-consistent scale-transition scheme. Such a micromechanical approach allows an accurate description for the mechanical behavior of the metal layer. Indeed, the self-consistent model takes into account essential microstructure-related features that are relevant at the microscale. These microstructural aspects include key physical mechanisms, such as initial and induced crystallographic textures, morphological anisotropy, and interactions between grains and their surrounding medium.

It is noteworthy that, in the literature, other alternatives to the self-consistent scheme are used as well. These alternatives are generally based on the finite element method (CPFEM), where spatial discretization by finite elements is generally required at both the polycrystalline aggregate and the single crystal scales. The mechanical behavior at the single crystal scale is described by a finite strain rate-independent constitutive framework, where the Schmid law is used to model the plastic flow. This rate-independent formulation is more suitable for the modeling and the simulation of cold forming processes, where viscous effects are limited. The developed model is applied in this paper to metal layers with FCC crystallographic structure.

On the other hand, the mechanical behavior of the elastomer layer is assumed to obey the hyperelastic neo-Hookean constitutive law. The adherence between the two layers is assumed to be perfect. In order to predict localized necking in the bilayer, the overall mechanical behavior is coupled with two different localization criteria: the bifurcation theory, initially developed by Rice [5], and the imperfection approach initiated by Marciniak and Kuczynski [6]. The use of the Schmid law at the single crystal scale allows predicting limit strains at

[a] Corresponding author: Mohamed.BenBettaieb@ensam.eu

realistic levels when the bifurcation theory is used as localization criterion. One of the main conclusions of this paper is that the addition of an elastomer layer can significantly retard the occurrence of localized necking in the whole bilayer. It is also demonstrated that the FLDs of the bilayer predicted by the Marciniak–Kuczynski (M–K) analysis tend towards those predicted by the bifurcation theory in the limit of a vanishing size for the geometric imperfection of the metal layer.

The current paper may be viewed as an extension of the investigations carried out in [7], which were restricted to a phenomenological description for the mechanical behavior of the metal layer.

The remainder of the paper is organized as follows:

- The constitutive equations describing the mechanical behavior of the metal and elastomer layers will be outlined in the second section.
- In the third section, the theoretical framework for the two localization criteria will be presented.
- Various numerical results obtained by the application of the developed tool will be presented and discussed in the fourth section.

Notations

The following notations and abbreviations are adopted in this paper:

- Mechanical fields corresponding to the polycrystal (resp. single crystal) scale are denoted by capital (resp. small) letters. To be consistent with the notations adopted for the metal layer, the different mechanical fields corresponding to the elastomer layer are also denoted by capital letters.

- $\bullet^{PS}$: the in-plane part of the field • (in relation with the plane-stress assumption).

- $\bullet_{\blacklozenge}$: quantity • associated with layer ♦ (metal or elastomer layer).

- •(B): quantity • associated with the band (M–K analysis).

- •(S): quantity • associated with the safe zone (M–K analysis).

- SC and FC will be used, as needed, to specify the self-consistent scheme and the full-constraint Taylor model, respectively.

- FM (resp. BL) refers in figures of Section 4 to the freestanding metal layer (resp. metal/elastomer bilayer).

These notations may be combined. For instance, tensor $\mathbf{X}$ in the elastomer layer of the band is denoted $\mathbf{X}_E(\mathrm{B})$.

2 Mechanical behavior of the bilayer

2.1 Metal layer

2.1.1 Constitutive equations at the polycrystalline scale

Let us consider a polycrystalline aggregate, which is assumed to be statistically representative of the metal layer. The modeling of the mechanical behavior of this aggregate is thus sufficient to accurately describe the mechanical behavior of the whole metal layer. To derive the mechanical behavior of the polycrystalline aggregate from the behavior of its microscopic constituents, the self-consistent model is used. Only the main lines of this scheme are presented in the current section. Further details on this model are provided in [8]. Compared to the full-constraint Taylor model, which is much more commonly used due to its simplicity, the self-consistent scheme presents a number of advantages. Indeed, through the formulation of this scale-transition scheme, the equilibrium condition at the single crystal level is satisfied. Also, the grain morphology and the interactions between the grain and its surrounding medium are accounted for.

Because the behavior of the polycrystalline aggregate is modeled within the framework of finite strains, the nominal stress rate $\dot{\mathbf{N}}$ and the velocity gradient $\mathbf{G}$ are used as appropriate stress and strain measures, respectively. The macroscopic tangent modulus $\mathbf{L}$ linking $\dot{\mathbf{N}}$ to $\mathbf{G}$ is then obtained by using the self-consistent model

$$\dot{\mathbf{N}} = \mathbf{L} : \mathbf{G} . \qquad (1)$$

The macroscopic velocity gradient $\mathbf{G}$ and the nominal stress rate $\dot{\mathbf{N}}$ can be derived from their microscopic counterparts $\mathbf{g}$ and $\dot{\mathbf{n}}$ by using the averaging Hill theorem [9]

$$\mathbf{G} = \overline{\mathbf{g}(\mathbf{x})} \quad ; \quad \dot{\mathbf{N}} = \overline{\dot{\mathbf{n}}(\mathbf{x})} , \qquad (2)$$

where $\mathbf{x}$ is a material point in the polycrystalline aggregate, and $\bar{\mathbf{a}}$ the average of field $\mathbf{a}$ over the volume V of the polycrystalline aggregate

$$\bar{\mathbf{a}} = \frac{1}{V}\int_V \mathbf{a}(\mathbf{x})\,\mathrm{d}\mathbf{x} . \qquad (3)$$

Conversely, the microscopic velocity gradient and nominal stress rate are linked to their macroscopic counterparts by the following relations:

$$\mathbf{g}(\mathbf{x}) = \mathbf{A}(\mathbf{x}) : \mathbf{G} \quad ; \quad \dot{\mathbf{n}}(\mathbf{x}) = \mathbf{B}(\mathbf{x}) : \dot{\mathbf{N}} , \qquad (4)$$

where $\mathbf{A}(\mathbf{x})$ and $\mathbf{B}(\mathbf{x})$ are fourth-order concentration tensors for the velocity gradient and the nominal stress rate, respectively.

At the microscopic level, a behavior law similar to Eq. (1) can be derived by combining the constitutive relations of the single crystal

$$\dot{\mathbf{n}}(\mathbf{x}) = \mathbf{l}(\mathbf{x}) : \mathbf{g}(\mathbf{x}) , \qquad (5)$$

Where $\mathbf{l}$ is the microscopic tangent modulus. Furthermore, it is assumed that all mechanical variables are homogeneous within the individual single crystals

$$\mathbf{g}(\mathbf{x}) = \sum_{I=1}^{N_g} \mathbf{g}^I \theta^I(\mathbf{x}) \quad ; \quad \mathbf{l}(\mathbf{x}) = \sum_{I=1}^{N_g} \mathbf{l}^I \theta^I(\mathbf{x}) , \qquad (6)$$

where N_g is the number of single crystals that compose the polycrystalline aggregate, and θ^I is an indicator function defined as

$$\theta^I(\mathbf{x}) = 1 \text{ if } \mathbf{x} \in V^I \quad ; \quad \theta^I(\mathbf{x}) = 0 \text{ if } \mathbf{x} \notin V^I . \quad (7)$$

Thus, for any material point $\mathbf{x}$ within single crystal I, relation (5) becomes

$$\dot{\mathbf{n}}^I = \mathbf{l}^I : \mathbf{g}^I . \quad (8)$$

The relation between $\mathbf{L}$ and $\mathbf{l}$ can be obtained by combining Eqs. (1)-(5)

$$\mathbf{L} = \overline{\mathbf{l}(\mathbf{x}) : \mathbf{A}(\mathbf{x})} \quad (9)$$

By using Green's tensor, the analytical expression of the concentration tensor $\mathbf{A}^I$ is obtained after some developments

$$\mathbf{A}^I = (\mathbf{I} - \mathbf{T}^{II} : (\mathbf{l}^I - \mathbf{L}))^{-1} : \overline{(\mathbf{I} - \mathbf{T}^{II} : (\mathbf{l}^I - \mathbf{L}))^{-1}}^{-1} ,(10)$$

where $\mathbf{T}^{II}$ is a fourth-order tensor, which is function of $\mathbf{L}$, that describes the interaction between grain I and its surrounding medium. The macroscopic tangent modulus derived by the 1-site self-consistent version of the incremental scheme of Hill [9] can be finally obtained as follows:

$$\mathbf{L} = \sum_{I=1}^{N_g} f^I \mathbf{l}^I : \mathbf{A}^I , \quad (11)$$

where f^I is the volume fraction of single crystal I.

The above equations (10) and (11) represent a non-linear problem, which is incrementally solved by using the iterative fixed point method. At each time increment, the main unknown of this iterative problem is the macroscopic tangent modulus $\mathbf{L}$. It must be noted that the full-constraint Taylor model can be easily deduced from the self-consistent approach, by considering that the concentration tensors $\mathbf{A}^I$ of the different grains are equal to the fourth-order identity tensor. In this particular case, no iteration is required. To calculate $\mathbf{L}$ by Eq. (11), the microscopic tangent modulus $\mathbf{l}^I$ of all individual grains should be computed. To this end, the following section is dedicated to the derivation of the analytical expression of the microscopic tangent modulus.

2.1.2 Constitutive equations at the single crystal scale

The mechanical behavior of the single crystals that make up the polycrystalline aggregate is described by a rate-independent constitutive framework. Because the single crystals undergo finite strains, the effect of lattice rotation is accounted for.

The microscopic velocity gradient is additively split into its symmetric and skew-symmetric parts, denoted as $\mathbf{d}$ and $\mathbf{w}$, respectively

$$\mathbf{d} = (1/2)(\mathbf{g} + \mathbf{g}^T) \quad ; \quad \mathbf{w} = (1/2)(\mathbf{g} - \mathbf{g}^T) . (12)$$

The strain tensor $\mathbf{d}$ and the spin tensor $\mathbf{w}$ are split into their elastic and plastic parts

$$\mathbf{d} = \mathbf{d}^e + \mathbf{d}^p \quad ; \quad \mathbf{w} = \mathbf{w}^e + \mathbf{w}^p . \quad (13)$$

The rotation of the single crystal lattice frame $\mathbf{r}$ is related to the elastic part of the spin tensor $\mathbf{w}^e$ by the following relation:

$$\dot{\mathbf{r}}.\mathbf{r}^T = \mathbf{w}^e . \quad (14)$$

In order to satisfy the objectivity principle, the co-rotational rate $\boldsymbol{\sigma}^\nabla$ of the Cauchy stress tensor $\boldsymbol{\sigma}$, with respect to the lattice rotation, is related to the elastic strain rate $\mathbf{d}^e$ by the following hypoelastic law:

$$\boldsymbol{\sigma}^\nabla = \dot{\boldsymbol{\sigma}} - \mathbf{w}^e.\boldsymbol{\sigma} + \boldsymbol{\sigma}.\mathbf{w}^e = \mathbf{C}^e : \mathbf{d}^e , \quad (15)$$

where $\mathbf{C}^e$ is the fourth-order elasticity tensor. The elastic behavior of the metal layer is assumed to be isotropic and therefore $\mathbf{C}^e$ is defined by the Young modulus E and the Poisson ratio ν.

The inelastic deformation is only due to the slip on the crystallographic planes. Thus, $\mathbf{d}^p$ and $\mathbf{w}^p$ can be defined by the following relations:

$$\mathbf{d}^p = \sum_{\beta=1}^{N_s} \dot{\gamma}^\beta \operatorname{sgn}(\tau^\beta) \mathbf{R}^\beta \; ; \; \mathbf{w}^p = \sum_{\beta=1}^{N_s} \dot{\gamma}^\beta \operatorname{sgn}(\tau^\beta) \mathbf{S}^\beta ,(16)$$

where:

- N_s is the total number of slip systems (equal to 12 for FCC crystallographic structure).
- $\dot{\gamma}^\beta$ is the slip rate of the slip system β.
- $\mathbf{R}^\beta$ (resp. $\mathbf{S}^\beta$) is the symmetric part (resp. skew-symmetric part) of the Schmid orientation tensor.
- τ^β is the resolved shear stress of the slip system β, which is equal to $\boldsymbol{\sigma} : \mathbf{R}^\beta$.

The Schmid law is used to model the plastic flow of the single crystal, as follows:

$$\forall \beta = 1,...,N_s : \quad \begin{cases} |\tau^\beta| < \tau_c^\beta \Rightarrow \dot{\gamma}^\beta = 0 \\ |\tau^\beta| = \tau_c^\beta \Rightarrow \dot{\gamma}^\beta \geq 0 \end{cases} , \quad (17)$$

where τ_c^β is the critical shear stress of the slip system β.

The Cauchy stress tensor $\boldsymbol{\sigma}$ is related to the nominal stress tensor $\mathbf{n}$ by the following relation:

$$\mathbf{n} = j\,\mathbf{f}^{-1}.\boldsymbol{\sigma} , \quad (18)$$

where j is the Jacobian of the microscopic deformation gradient $\mathbf{f}$. The nominal stress rate $\dot{\mathbf{n}}$ required in Eq. (5) can be easily obtained from Eq. (18)

$$\dot{\mathbf{n}} = j\,\mathbf{f}^{-1}.\left(\dot{\boldsymbol{\sigma}} + \boldsymbol{\sigma}\,\mathrm{Tr}(\mathbf{d}) - \mathbf{g}.\boldsymbol{\sigma}\right). \tag{19}$$

In the current paper, an updated Lagrangian approach is adopted. Thus, in Eq. (19), $\mathbf{f}$ is set to the second-order identity tensor $\mathbf{I}_2$ and j is set to 1, which reduces to

$$\dot{\mathbf{n}} = \dot{\boldsymbol{\sigma}} + \boldsymbol{\sigma}\,\mathrm{Tr}(\mathbf{d}) - \mathbf{g}.\boldsymbol{\sigma}. \tag{20}$$

By combining Eqs. (12), (13), (15), (16) and (20), $\dot{\mathbf{n}}$ can be expressed as a function of the slip rates $\dot{\gamma}^\beta$

$$\begin{aligned}\dot{\mathbf{n}} = &\left(\mathbf{C}^e + \boldsymbol{\sigma}\otimes\mathbf{I}_2\right) : \mathbf{d} - \boldsymbol{\sigma}.\mathbf{w} - \mathbf{d}.\boldsymbol{\sigma}\\ &- \sum_{\beta=1}^{N_s} \mathrm{sgn}(\tau^\beta)\left(\mathbf{C}^e : \mathbf{R}^\beta + \mathbf{S}^\beta.\boldsymbol{\sigma} - \boldsymbol{\sigma}.\mathbf{S}^\beta\right)\dot{\gamma}^\beta\end{aligned}. \tag{21}$$

Let us now introduce the set of active slip systems $\mathcal{A}$, defined as the slip systems for which the slip rates are strictly positive. The set of slip systems β used in Eq. (21) is then reduced to the set $\mathcal{A}$

$$\begin{aligned}\dot{\mathbf{n}} = &\left(\mathbf{C}^e + \boldsymbol{\sigma}\otimes\mathbf{I}_2\right) : \mathbf{d} - \boldsymbol{\sigma}.\mathbf{w} - \mathbf{d}.\boldsymbol{\sigma}\\ &- \sum_{\beta\in\mathcal{A}} \mathrm{sgn}(\tau^\beta)\left(\mathbf{C}^e : \mathbf{R}^\beta + \mathbf{S}^\beta.\boldsymbol{\sigma} - \boldsymbol{\sigma}.\mathbf{S}^\beta\right)\dot{\gamma}^\beta\end{aligned}. \tag{22}$$

In order to obtain the expression of the microscopic tangent modulus $\mathbf{l}$ from relation (22), the slip rates of the active slip systems $\dot{\gamma}^\beta$ should be expressed as functions of $\mathbf{g}$. To this end, the consistency condition, restricted to the active slip systems, is used

$$\forall\beta\in\mathcal{A}: \quad \dot{\chi}^\beta = \mathrm{sgn}(\tau^\beta)\,\dot{\tau}^\beta - \dot{\tau}_c^\beta = 0 \quad ; \quad \dot{\gamma}^\beta > 0. \tag{23}$$

By using the definition of the resolved shear stress as well as Eqs. $(13)_1$ and (15), the resolved shear stress rate $\dot{\tau}^\beta$ can be expressed as

$$\forall\beta\in\mathcal{A}: \quad \dot{\tau}^\beta = \mathbf{R}^\beta : \boldsymbol{\sigma}^\nabla = \mathbf{R}^\beta : \mathbf{C}^e : (\mathbf{d} - \mathbf{d}^p). \tag{24}$$

As to the time derivative of τ_c^β, it is given by the following isotropic hardening law:

$$\begin{aligned}&\forall\beta\in\mathcal{A}: \quad \dot{\tau}_c^\beta = h\sum_{\alpha\in\mathcal{A}}\dot{\gamma}^\alpha\\ &h = h_0\left(1 + \frac{h_0\,\Gamma}{\tau_0\,n}\right)^{n-1} \quad ; \quad \Gamma = \sum_{\alpha=1}^{N_s}\gamma^\alpha\end{aligned}, \tag{25}$$

where h_0 is the initial hardening rate and n the power-law hardening exponent. τ_0 is the initial critical shear stress, which is assumed to be the same for the different slip systems.

The expression of the slip rates for the active slip systems is finally obtained by inserting Eqs. $(16)_1$, (24) and (25) into the consistency condition (23)

$$\begin{aligned}\forall\beta\in\mathcal{A}: \quad \dot{\gamma}^\beta &= \sum_{\alpha\in\mathcal{A}} M^{\beta\alpha}\mathrm{sgn}(\tau^\alpha)\,\mathbf{R}^\alpha : \mathbf{C}^e : \mathbf{d}\\ &= \mathbf{y}^\beta : \mathbf{d}\end{aligned}, \tag{26}$$

where $\mathbf{M}$ is the inverse of matrix $\mathbf{P}$ of rank $\mathrm{Card}(\mathcal{A})$, which is defined by the following index form:

$$\forall\alpha,\beta\in\mathcal{A}: \quad P^{\alpha\beta} = \left(h + \mathrm{sgn}(\tau^\alpha)\,\mathrm{sgn}(\tau^\beta)\,\mathbf{R}^\alpha : \mathbf{C}^e : \mathbf{R}^\beta\right). \tag{27}$$

Combining Eqs. (5), (21) and (26), one can obtain the analytical expression of the microscopic tangent modulus $\mathbf{l}$

$$\begin{aligned}\mathbf{l} = &\mathbf{C}^e + \boldsymbol{\sigma}\otimes\mathbf{I}_2 - {}^1\boldsymbol{\Lambda}(\boldsymbol{\sigma}) - {}^2\boldsymbol{\Lambda}(\boldsymbol{\sigma})\\ &- \sum_{\alpha\in\mathcal{A}} \mathrm{sgn}(\tau^\alpha)\left(\mathbf{C}^e : \mathbf{R}^\alpha + \mathbf{S}^\alpha.\boldsymbol{\sigma} - \boldsymbol{\sigma}.\mathbf{S}^\alpha\right)\otimes\mathbf{y}^\alpha\end{aligned}, \tag{28}$$

where ${}^1\boldsymbol{\Lambda}$ and ${}^2\boldsymbol{\Lambda}$ are fourth-order tensors that reflect the contribution of Cauchy stress convective terms

$$\begin{aligned}{}^1\Lambda_{ijkl}(\boldsymbol{\sigma}) &= \frac{1}{2}(\delta_{lj}\sigma_{ik} - \delta_{kj}\sigma_{il})\\ {}^2\Lambda_{ijkl}(\boldsymbol{\sigma}) &= \frac{1}{2}(\delta_{ik}\sigma_{lj} + \delta_{il}\sigma_{kj})\end{aligned}. \tag{29}$$

In order to update the different microscopic variables, the above constitutive equations should be integrated incrementally over each time increment. Analyzing these constitutive equations, one can observe that the main unknowns are the set of active slip systems and the corresponding slip rates. An implicit ultimate algorithm, very similar to the one developed in [10], is used to integrate these equations. The microscopic tangent modulus is determined from the updated microscopic variables at the end of the time increment.

2.2 Elastomer layer

In contrast to the metal layer, the mechanical behavior of the elastomer layer is assumed to be homogeneous and is modeled by a hyperelastic neo-Hookean law [11]. The formulation of this model allows linking the Cauchy stress tensor $\boldsymbol{\Sigma}$ to the left Cauchy-Green tensor $\mathbf{V}$

$$\boldsymbol{\Sigma} = q\,\mathbf{I}_2 + \mu\,\mathbf{V}^2, \tag{30}$$

where μ is the shear modulus, q an unknown hydrostatic pressure to be determined by applying the incompressibility constraint, and $\mathbf{V}$ is defined by the following relation:

$$\mathbf{V}^2 = \mathbf{F}.\mathbf{F}^T, \tag{31}$$

with $\mathbf{F}$ being the deformation gradient tensor of the elastomer layer.

3 Strain localization criteria

Let us consider a bilayer comprised of a metal layer M and an elastomer layer E. The two layers are assumed to be perfectly adhered and sufficiently thin. The small thickness assumption allows formulating the two above-mentioned localization criteria, namely the bifurcation theory and the initial imperfection approach, under the plane-stress conditions.

3.1 Bifurcation theory

The bilayer is submitted to uniform strain, where the in-plane strain rates are equal to $\dot{E}_{11} = 1$, $\dot{E}_{22} = \rho$ and $\dot{E}_{12} = 0$. ρ is the strain-path ratio ranging from −1/2 (uniaxial tensile state) to 1 (equibiaxial tensile state). Considering the plane-stress conditions, this specific loading implies that the macroscopic velocity gradient and the macroscopic nominal stress rate tensor have the following generic forms:

$$\mathbf{G} = \begin{pmatrix} 1 & 0 & 0 \\ 0 & \rho & 0 \\ 0 & 0 & ? \end{pmatrix} \quad ; \quad \dot{\mathbf{N}} = \begin{pmatrix} ? & ? & 0 \\ ? & ? & 0 \\ 0 & 0 & 0 \end{pmatrix}, \qquad (32)$$

where symbol ? designates the unknown components in both tensors. These generic forms are valid for both layers.

The bifurcation criterion states that strain localization occurs when the acoustic tensor becomes singular. Hence, this criterion is expressed in the following form:

$$\det\,(\vec{\mathcal{N}}^{PS} \, . \mathbf{L}^{PS} . \vec{\mathcal{N}}^{PS}) = 0\,, \qquad (33)$$

where:

- $\vec{\mathcal{N}}^{PS}$ is the unit vector normal to the localization band. Here, it is taken equal to $(\cos\theta, \sin\theta)$.
- $\mathbf{L}^{PS}$ is the averaged plane-stress tangent modulus of the bilayer. This modulus is defined by the following relation:

$$\mathbf{L}^{PS} = \frac{h_M\,\mathbf{L}_M^{PS} + h_E\,\mathbf{L}_E^{PS}}{h_M + h_E}, \qquad (34)$$

where h_M (resp. h_E) is the initial thickness of the metal (resp. elastomer) layer and $\mathbf{L}_M^{PS}$ (resp. $\mathbf{L}_E^{PS}$) is the plane-stress tangent modulus of the metal (resp. elastomer) layer.

$\mathbf{L}_M^{PS}$ is derived from the 3D expression of the metal layer tangent modulus $\mathbf{L}_M$ (see Section 2.1.1) by using the following condensation relation:

$$\forall \alpha, \beta, \gamma, \delta = 1,2: \quad L^{PS}_{M\alpha\beta\gamma\delta} = L_{M\alpha\beta\gamma\delta} - \frac{L_{M\alpha\beta 33}\, L_{M33\gamma\delta}}{L_{M3333}}. \qquad (35)$$

$\mathbf{L}_E^{PS}$ is defined by the following relation [7]:

$$\mathbf{L}_E^{PS} = \boldsymbol{\mathcal{L}} + \boldsymbol{\Sigma} \otimes \mathbf{I}_2 - {}^1\boldsymbol{\Lambda}(\boldsymbol{\Sigma}) - {}^2\boldsymbol{\Lambda}(\boldsymbol{\Sigma})\,, \qquad (36)$$

It is worth noting that the tensors ${}^1\boldsymbol{\Lambda}$ and ${}^2\boldsymbol{\Lambda}$ used in relation (36) are given in Eq. (29). The non-zero components of the fourth-order tensor $\boldsymbol{\mathcal{L}}$ are defined by the following relations:

$$\begin{cases} \mathcal{L}_{1111} = 2\mu[e^{2E_{11}} + e^{-2(E_{11}+E_{22})}] \\ \mathcal{L}_{2222} = 2\mu[e^{2E_{22}} + e^{-2(E_{11}+E_{22})}] \\ \mathcal{L}_{1122} = 2\mu\, e^{-2(E_{11}+E_{22})} \\ \mathcal{L}_{1212} = \dfrac{\mu}{2}[e^{2E_{11}} + e^{2E_{22}}] \end{cases} . \qquad (37)$$

For each strain-path ratio ρ, and at each time increment, the bifurcation criterion (33) is checked for all possible band orientations ($\theta \in [0,\ \pi/2]$). When the acoustic tensor becomes singular for a given band orientation, the computation is stopped. The overall major strain E_{11} that is thus calculated corresponds to the localization limit strain and the band orientation, which zeroes the determinant of the acoustic tensor, corresponds to the necking band orientation.

3.2 Initial imperfection approach

In order to apply the initial imperfection approach (called hereafter M–K analysis for the sake of brevity) to the bilayer, the assumption of the preexistence of a groove in the form of a band in the metal layer (Figure 1) is made. Thus, in accordance with various experimental observations, the localization initiates always in the metal layer.

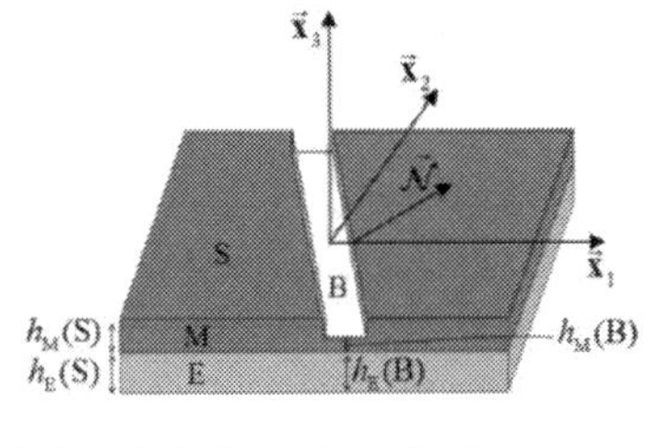

Figure 1. M–K analysis for a bilayer (initial geometry and band orientation).

In order to present the equations of the M–K analysis related to the metal/elastomer bilayer, the following notations are adopted:

- $h_M(B)$: initial thickness of the metal layer inside the band.
- $h_M(S)$: initial thickness of the metal layer outside the band.
- $h_E(B)$: initial thickness of the elastomer layer inside the band.
- $h_E(S)$: initial thickness of the elastomer layer outside the band (equal to $h_E(B)$).

On the basis of these notations, the initial geometric imperfection ratio ξ_0 (corresponding to the metal layer only), can be specified as

$$\xi_0 = 1 - \frac{h_M(B)}{h_M(S)} . \tag{38}$$

The M–K analysis is based on four main equations:

- As a consequence of the perfect adherence between the metal and the elastomer layer, the following equalities between the in-plane velocity gradients in the metal layer and their counterparts in the elastomer layer are satisfied:

$$\begin{aligned} \mathbf{G}_M^{PS}(B) = \mathbf{G}_E^{PS}(B) = \mathbf{G}^{PS}(B) \\ \mathbf{G}_M^{PS}(S) = \mathbf{G}_E^{PS}(S) = \mathbf{G}^{PS}(S) \end{aligned} \tag{39}$$

When the adherence between the two layers is not perfect, the above equalities between the in-plane velocity gradients in the metal and elastomer layers are not satisfied. In this case, the beneficial effect of necking retardation due to the elastomer layer is reduced to some extent, which should be proportional to the extent of surface debonding.

- The kinematic compatibility condition between the band and the uniform zone (i.e., outside the band): this condition requires that the displacement increments should be continuous across the band and it is mathematically expressed as

$$\mathbf{G}^{PS}(B) = \mathbf{G}^{PS}(S) + \dot{\mathbf{C}}^{PS} \otimes \vec{\mathcal{N}}^{PS} . \tag{40}$$

- The equilibrium balance across the interface between the band and the homogeneous zone:

$$\begin{aligned} \vec{\mathcal{N}}^{PS}.(h_M(B)\dot{\mathbf{N}}_M^{PS}(B) + h_E(B)\ \dot{\mathbf{N}}_E^{PS}(B)) \\ = \vec{\mathcal{N}}^{PS}.(h_M(S)\ \dot{\mathbf{N}}_M^{PS}(S) + h_E(S)\ \dot{\mathbf{N}}_E^{PS}(S)) \end{aligned} . \tag{41}$$

- The behavior law of both the metal and the elastomer layer, restricted to the plane dimension inside and outside the band, respectively: these constitutive equations are expressed in the following generic form:

$$\begin{aligned} \dot{\mathbf{N}}_M^{PS}(B) = \mathbf{L}_M^{PS}(B) : \mathbf{G}^{PS}(B) \quad ; \quad \dot{\mathbf{N}}_E^{PS}(B) = \mathbf{L}_E^{PS}(B) : \mathbf{G}^{PS}(B) \\ \dot{\mathbf{N}}_M^{PS}(S) = \mathbf{L}_M^{PS}(S) : \mathbf{G}^{PS}(S) \quad ; \quad \dot{\mathbf{N}}_E^{PS}(S) = \mathbf{L}_E^{PS}(S) : \mathbf{G}^{PS}(S) \end{aligned} . \tag{42}$$

By inserting the constitutive relations (42) into the equilibrium equation (41), this latter becomes

$$\begin{aligned} \vec{\mathcal{N}}^{PS}.(h_M(B)\mathbf{L}_M^{PS}(B) + h_E(B)\ \mathbf{L}_E^{PS}(B)) : \mathbf{G}^{PS}(B) \\ = \vec{\mathcal{N}}^{PS}.(h_M(S)\ \mathbf{L}_M^{PS}(S) + h_E(S)\ \mathbf{L}_E^{PS}(S)) : \mathbf{G}^{PS}(S) \end{aligned} . \tag{43}$$

For each strain-path ratio ρ and each initial band orientation θ_0, the equations that govern the M–K approach are integrated incrementally over each time increment. Indeed, by analyzing Eqs. (40) and (43), it can be seen that the main incremental unknowns of the M–K approach are the two components of the jump vector $\dot{\mathbf{C}}^{PS}$. For each loading (determined by a strain path and a band orientation), the computations are stopped when the norm of the jump vector $\dot{\mathbf{C}}^{PS}$ increases rapidly, thus marking the localization of the deformation in the band zone. Further numerical and algorithmic details regarding the M–K approach can be found in [7].

4 Prediction results

4.1 Material and geometric data

The polycrystalline aggregate studied in this paper is made of 2000 grains. Its initial crystallographic texture is generated randomly (see Figure 2) in such a way that it is orthotropic with respect to the rolling and transverse directions. It is widely recognized that the initial crystallographic texture strongly affects both the shape and the overall level of the predicted FLDs [12]. The influence of texture on localized necking in substrate-supported metal layers will be the matter of future investigations. Initially, all of the grains are assumed to be spherical with the same volume fraction. We assume here that the adopted polycrystalline aggregate is representative of the studied metal sheet. Several literature studies have suggested that at least 1000 grains are required for the polycrystalline aggregate to be representative of the studied material. From a number of numerical simulations (that have been conducted, but not included in this paper for conciseness), it is observed that beyond 2000 grains, the response of the polycrystalline aggregate representing the metal layer remains almost unchanged. As plastic strain localization occurs at relatively large strains, the values of limit strains are almost unaffected by the elastic behavior. This justifies the consideration of simple isotropic elastic behavior in the current study.

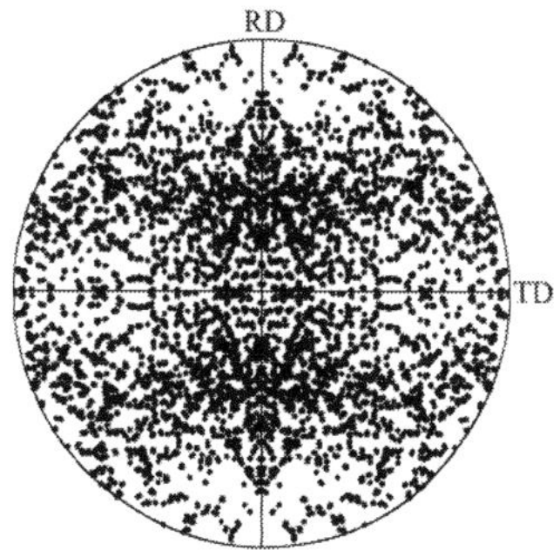

Figure 2. Initial crystallographic texture of the studied polycrystalline aggregate: {111} pole figure.

The material parameters of the single crystals are given in Table 1. The elastic parameters are representative of steel materials. As to hardening parameters, they are the same as those used in [12].

Table 1. Material parameters of the single crystals used in the metal layer

Elasticity		Hardening		
E [GPa]	ν	τ_0 [MPa]	h_0 [MPa]	n
210	0.3	40	390	0.35

The shear modulus of the elastomer layer is set to 22 MPa. This choice is based on data for polyurea [13]. The ratio between the initial thickness of the elastomer layer and the initial thickness of the metal layer is fixed to 0.5.

4.2 Bifurcation theory predictions

The evolution of the minimum of the cubic root of the determinant of the acoustic tensor $\vec{\mathcal{N}}^{PS}.\mathbf{L}^{PS}.\vec{\mathcal{N}}^{PS}$, over all possible band orientations, as a function of the major strain E_{11} is displayed in Figure 3. The onset of strain localization occurs when this minimum reaches 0 as defined by the bifurcation criterion (33). The SC approach is used as scale-transition scheme. Four representative strain paths are considered in this figure: $\rho=-0.5$, $\rho=0$, $\rho=0.5$, and $\rho=1$. By comparing Figure 3 (a) and Figure 3 (b), one can easily observe that the presence of the elastomer layer substantially retards the occurrence of strain localization. This result is the consequence of the fact that under biaxial loading, the tangent modulus of the elastomer remains unchanged, or increases, while the tangent modulus of the metal layer steadily decreases.

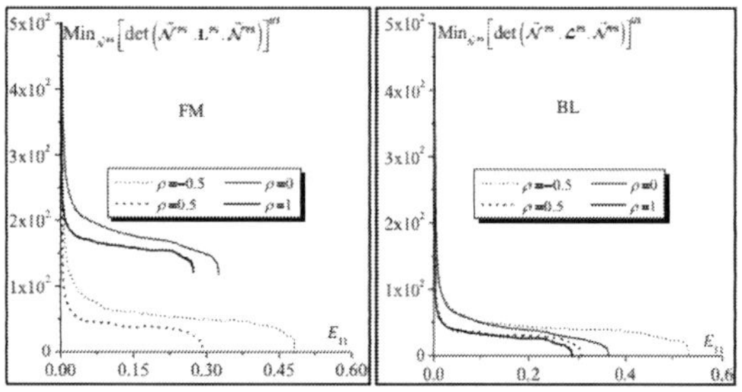

(a) (b)

Figure 3. Evolution of the minimum of the cubic root of the determinant of the acoustic tensor as a function of E_{11} for four different strain paths ($\rho=-0.5$, $\rho=0$, $\rho=0.5$, and $\rho=1$): (a) Freestanding metal layer; (b) substrate-supported metal layer.

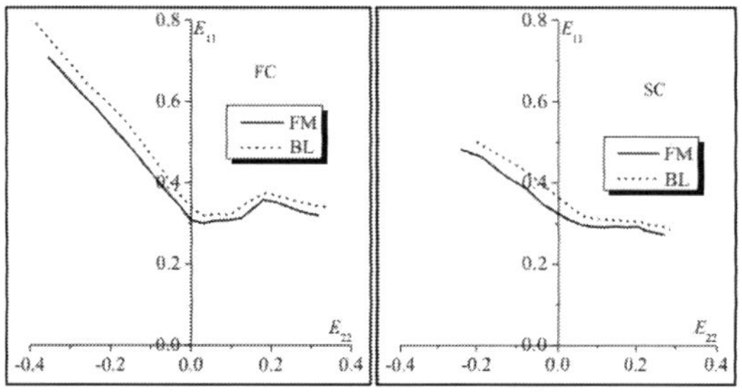

(a) (b)

Figure 4. Effect of the elastomer layer on the improvement of the FLD of the bilayer: (a) SC model; (b) FC model.

The effect of the elastomer layer on necking retardation for all strain paths $\rho\in[-1/2,1]$ is investigated in Figure 4. This figure, confirms the preliminary result obtained in Figure 3: the elastomer layer permits to shift the FLD monotonically upwards, and thus enhances the ductility of the bilayer.

4.3 M–K analysis predictions

In order to illustrate the onset of strain localization, the evolution of the in-plane components of the jump vector $\dot{\mathbf{C}}$ and the determinant of the acoustic tensor in the band are plotted in Figure 5 as function of the major strain in the safe zone E_{11}. In this simulation, the strain path ρ, the initial imperfection ratio ξ_0 and the initial band orientation θ_0 are fixed to 0, 10^{-3} and 0°, respectively. The SC model is used to ensure the transition between the microscopic and macroscopic scales of the metal layer. It is clear from Figure 5 (a) that the jump vector $\dot{\mathbf{C}}$ remains very close to $\mathbf{0}$ before strain localization. This jump vector, and especially its first component, increases very abruptly when the strain in the safe zone E_{11} is about 0.28, thus leading to the localization of deformation within the band. The evolution of the determinant of the acoustic tensor in the band is given in Figure 5 (b). The limit strain (0.28) is reached when this determinant becomes equal to 0. This result is expectable considering the similarity in the mathematical formulation of the bifurcation theory and the M–K approach. The evolution of this determinant can be used as a reliable alternative indicator of strain localization for the different strain paths.

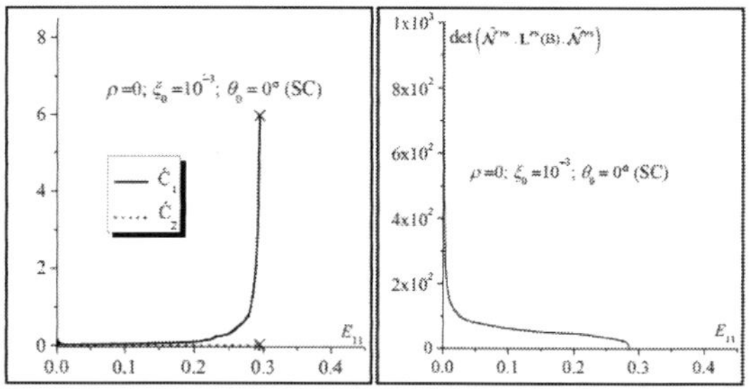

(a) (b)

Figure 5. Illustration of the onset of strain localization (bilayer; $\rho=0$; $\xi_0=10^{-3}$; $\theta_0=0°$; SC): (a) Evolution of the in-plane components of the jump vector $\dot{\mathbf{C}}$ as a function of the major strain in the safe zone E_{11}; (b) Evolution of the determinant of the acoustic tensor in the band as a function of the major strain in the safe zone E_{11}.

The comparison between the FLDs predicted by bifurcation theory and those determined by M–K analysis is shown in Figure 6. Three different values of the initial imperfection ratio ξ_0 are considered: 10^{-4}, 10^{-3} and 10^{-2}. It is clear that the shape and the level of the predicted FLDs are significantly influenced by the amount of initial geometric imperfection. It is also found that for all strain paths, the limit strains predicted by bifurcation theory set an upper bound to those yielded by the M–K approach.

DOI: 10.1051/matecconf/20168002007

Moreover, this result is valid for both scale-transition schemes, namely the FC and SC models.

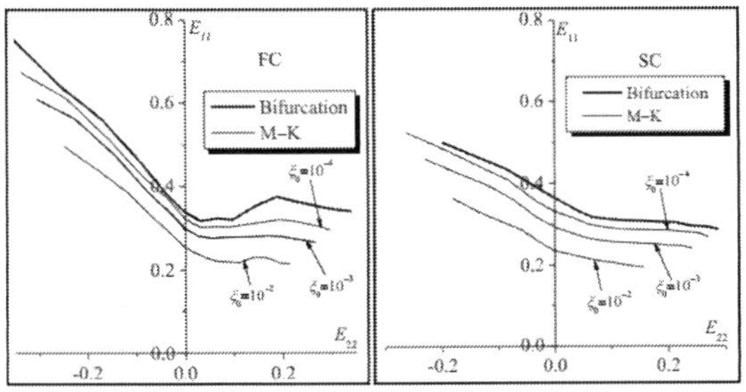

(a) (b)

Figure 6. Effect of the initial geometric imperfection on the shape and the level of the FLDs of the bilayer: (a) SC model; (b) FC model.

5 Concluding remarks

A powerful tool to predict the onset of localized necking in substrate-supported metal layers has been developed in this paper. In this tool, the mechanical behavior of the metal (resp. elastomer) layer is modeled by the self-consistent micromechanical model (resp. neo-Hookean hyperelastic model). The mechanical modeling of the bilayer is coupled with two strain localization criteria to predict the limit strains: the bifurcation theory and the initial imperfection approach. From the numerical predictions obtained by applying this tool, three main conclusions can be drawn:

- The presence of an elastomer layer increases substantially the level of the limit strains for the bilayer.
- The shape and the level of the predicted FLDs are significantly influenced by the amount of initial geometric imperfection, which is assumed to initiate in the metal layer.
- The limit strains of the whole bilayer predicted by bifurcation theory set an upper bound to those yielded by the M–K approach.

References

1. S.P. Keeler, W.A. Backofen, Trans. ASM **56**, 25 (1963)
2. G.M. Goodwin, Metallurgia Italiana **60**, 767 (1968)
3. S.L. Chiu, J. Leu, P.S. Ho, J. Appl. Phys. **76**, 5136 (1994)
4. M. Hommel, O. Kraft, Acta Mater. **49**, 3935 (2001)
5. J.R. Rice, *14th International Congress of Theoretical and Applied Mechanics*, 207 (1976)
6. Z. Marciniak, K. Kuczynski, Int. J. Mech. Sci. **9**, 609 (1967)
7. M. Ben Bettaieb, F. Abed-Meraim, Int. J. Plast. **65**, 168 (2015)
8. P. Lipinski, M. Berveiller, E. Reubrez, J. Morreale, Arch. Appl. Mech. **65**, 291 (1995)
9. R. Hill, Proc. Roy. Soc. London, **326**, 131 (1972)
10. H.K. Akpama, M. Ben Bettaieb, F. Abed-Meraim, Int. J. Num. Meth. Eng., doi: 10.1002/nme.5215 (2016)
11. S.C. Hunter, Int. J. Mech. Sci., **21**, 203 (1979)
12. K. Yoshida, M. Kuroda, Int. J. Solids Struct. **49**, 2073 (2012).
13. A.V. Amirkhizi, J. Isaacs, J. McGee, S. Nemat-Nasser, Philos. Mag. **86**, 5847 (2006).

Direct micro-to-macro modelling of the cold rolling of pearlitic steel

Laurent Delannay[1,a], Jeroen Tacq[2], Didier Bardel[1] and Marc Seefeldt[2]

[1]*iMMC, Université catholique de Louvain, 1348 Louvain la Neuve, Belgium*
[2]*MTM, Katholieke Universiteit Leuven, 3001 Heverlee, Belgium*

Abstract. A crystal plasticity model is adapted in order to predict strength, internal stresses and texture development in pearlite. The model is used as the user-defined material law in macroscopic finite element simulation of rolling. It is shown that the material response depends strongly on the reorientation of cementite lamellae. The model produces good estimates of the lattice strains measured by neutron diffraction as well as the uniaxial tensile curve.

1 Introduction

Metal forming processes such as rolling typically involve large plastic strains and significant changes of both the microstructure and the crystallographic texture. Such microstructural changes, which are a consequence of dislocation slip at the micron scale, may strongly influence the macroscopic plastic flow (e.g. the anisotropic hardening). In the present study, finite element (FE) modelling serves to predict strain heterogeneity across the thickness of a rolled workpiece, while the microscopic material response is computed based on an advanced crystal plasticity model.

The material under investigation is pearlitic steel [1], which has a characteristic dual-phase microstructure consisting of a thin stacking of ferrite and cementite (Fig. 1). Cementite forms a-few-nanometers-thick, extremely strong lamellae, whereas the strength of the softer ferrite depends on the inter-lamellar spacing. The composite response results from the contribution of both phases and it is expected to be anisotropic due to crystallographic texture and to the preferential alignment of cementite lamellae in the course of rolling.

The present study demonstrates that it is feasible to conduct direct micro-to-macro modelling of the microstructure (and strength) evolution during multi-pass rolling of pearlite. It is shown that texture and internal stresses are strongly influenced by the model assumptions about microscopic strain heterogeneity between the two constitutive phases as well as between adjacent grains (or adjacent pearlite "colonies").

2 Model description

Cold rolling simulations were performed following the procedure proposed by Bardel et al. [2] in order to study through-thickness texture gradients in rolled aluminium.

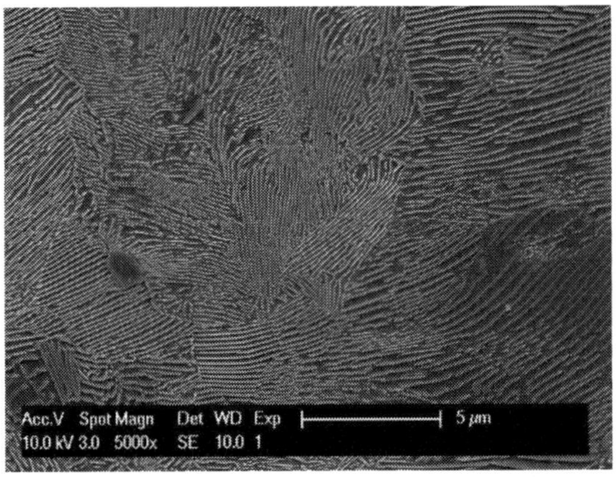

Figure 1. Microstructure of "patented" pearlite samples (before rolling). The cementite lamellae appear lighter. From [1].

Simulations, using the standard (implicit version of abaqus), assumed plane strain deformation, rigid rolls and finite relative displacement along the contact line between roll and workpiece. The half-thickness was discretized into 11 linear elements with reduced integration (C2D4R). Each rolling pass was simulated separately of the other passes, starting from square elements and using the state variables at the exit of a previous pass as input for the current pass (Fig. 2).

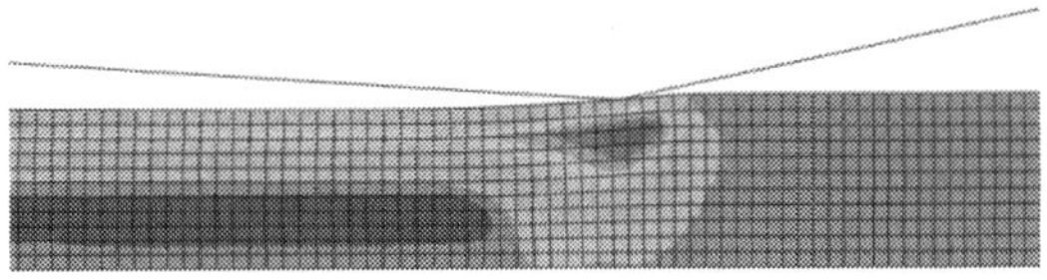

Figure 2. Illustration of the heterogeneity of shear strains across the workpiece thickness when steady state is detected at the exit of the rolling gap.

[a] Corresponding author: laurent.delannay@uclouvain.be

Inside one layer of elements across the thickness of the workpiece, the material response was computed using crystal plasticity theory. The model polycrystal consisted of 2000 grains and the interaction of adjacent grains was computed using the ALAMEL approximation [3-4]. This means that regions on either sides of a planar grain boundary underwent opposite shear superimposed to the macroscopic strain. The amount of the local strain heterogeneity was computed by minimizing the plastic work [2,4]. Within each grain crystal plasticity was applied to a stacking of ferrite and cementite lamellae (Fig. 3) while enforcing geometrical compatibility and stress equilibrium across the planar interface.

The initial lattice orientations were random in ferrite. The cementite lamellae were introduced by randomly selecting a variant of the Isaichev [1] orientation relationship within each grain. Hence, the interface normal was (110) in cementite and (112) in ferrite, whereas (010) in cementite was parallel to (111) in ferrite.

Dislocation glide occurred along {110}<111> and {112}<111> systems in ferrite, whereas (010)[001] and (110)[$1\bar{1}1$] type slip systems were considered in cementite [5] which has orthorhombic symmetry (a = 5.0816 Å, b = 6.7446 Å, c = 4.5206 Å). An isotropic Voce representation of hardening was used in each phase:

$$\tau_c = \tau_{sat} - (\tau_{sat} - \tau_{c0})\ exp(\Gamma/\Gamma_0) \quad (1)$$

where τ_{c0} and τ_{sat} were the initial and the saturating value of the critical resolved shear stress (CRSS) τ_c, whereas Γ was the accumulated slip on all slip systems and Γ_0 determined the initial hardening rate.

Time integration of the crystal plasticity law was fully implicit and a consistent tangent operator was provided to the FE solver [2,4].

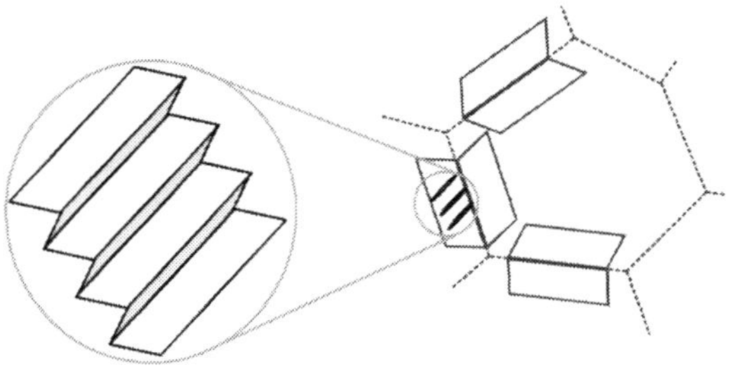

Figure 3. Representation of the strain heterogeneity considered between adjacent grains (right) and within each grain due to the stacking of soft ferrite and hard cementite (left).

3 Results

The experimental data used to assess the model was obtained based on uniaxial tensile tests on dog-bone-shaped samples of C75 fully pearlitic steel sheet with a cementite volume fraction of about 12%. Tests were done after patenting at 500°C, giving a mean inter-lamellar spacing of 140 nm, and after cold rolling to an equivalent strain of 1.1 with subsequent stress-relief treatment at 470°C for 15 min. The tensile axis was aligned with the rolling direction [2].

3.1. Determination of the hardening parameters

The elastic constants and the hardening parameters of the two phases are listed in Table 1. The elastic constants of (embedded) cementite have been assumed as equal to the ones of ferrite; the hardening parameters have been determined by fitting the experimental tensile curves obtained on the pearlite samples before and after rolling (Fig. 4). The same set of parameters was thus used for the simulation of the two tensile tests as well as the intermediary rolling.

Table 1. Material parameters.

	Ferrite	**Cementite**
C_{11}, C_{12}, C_{44} [GPa]	231, 135, 116	231, 135, 116
τ_{c0} [MPa]	250	1600
τ_{sat} [MPa]	380	2000
Γ_0 [-]	65	10

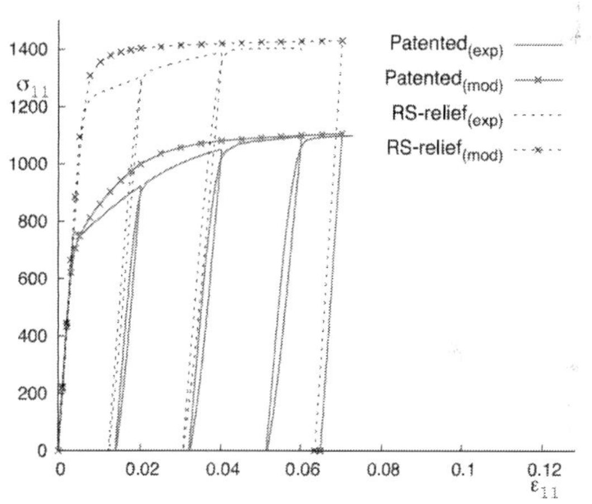

Figure 4. Uniaxial tensile stresses [MPa] obtained on pearlite samples before ("patented") and after cold rolling to a strain of 1.1 (and partial relaxation of internal stresses "RS-relief").

The initial CRSS value for ferrite is in the range that one would expect for an as-patented material and the present mean inter-lamellar spacing: Gil Sevillano [6] proposed, based on the assumptions of intralamellar dislocation glide in the ferrite channels and predominant activation of intralamellar multiplicative dislocation sources, a lower and an upper bound of the critical shear stress of

$$\tau_b \cong 1.2 \frac{Gb}{2\pi S} \ln \frac{S}{b} \quad (2)$$

and

$$\tau_m \cong 1.2 \frac{Gb}{\pi S} \ln \frac{S}{2b}, \quad (3)$$

DOI: 10.1051/matecconf/20168002008

respectively. For the present mean interlamellar spacing S this gives a critical shear stress range from about 175 to about 315 MPa. Note that these estimates do neither include Peierls stress, nor solution strengthening, nor dislocation storage strengthening, compare, for instance, [7]. With appropriate update of the interlamellar spacing, however, they can form the basis of a physics- and microstructure-based hardening law [8]. The initial and saturation values for cementite correspond to estimated cementite strengths of about 5-8 GPa reported in literature [6,9].

3.2 Internal stresses

Internal stresses developing during the two uniaxial tensile tests have been measured *in situ* using neutron diffraction (in transmission) on the strain diffractometer beamline HK9 at the Nuclear Physics Institute of ASCR, Řež near Prague, Czech Republic. Tensile rig and sample were positioned such that the tension and rolling direction was aligned with the diffraction vector (Young setup) and the entire width of the sample was immersed in the neutron beam. In the plastic regime, each neutron diffraction measurement was done at constant extensometer strain, implying some stress relaxation during the measurement. The ferrite lattice strains in tensile and rolling direction were monitored for the <200>, <211>, <220> and <310> orientation fibres. Three unloading-reloading cycles were carried out in order to study possible reverse yielding of ferrite [1].

Fig. 5 demonstrates that the crystal plasticity model produced fair predictions of the longitudinal lattice strains in different crystallographic fibres. The model predictions presented in these plots correspond to the average elastic strain developed in the tensile direction in the sampling of grains oriented in such a way that a chosen crystal direction is tilted by less than 7° from the tensile axis.

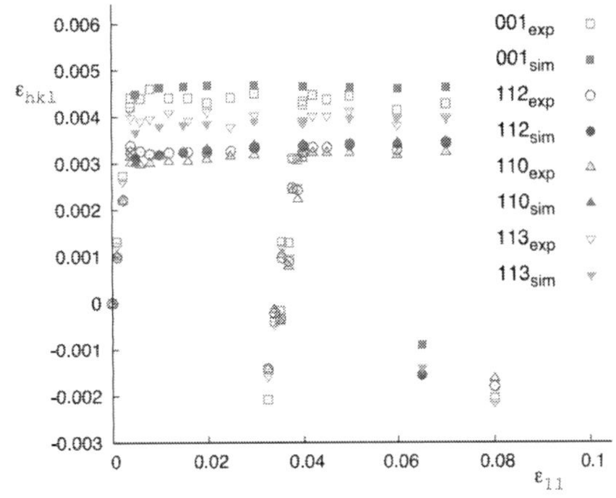

(a)

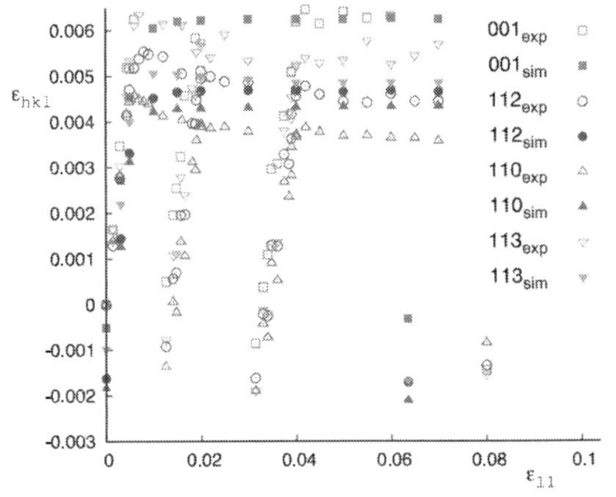

(b)

Figure 5. Assessment of the model predictions of the elastic lattice strains during the uniaxial tensile test on (a) the patented sample and (b) the cold rolled sample (after a heating treatment relaxing the internal stresses due to rolling).

3.3 Cold rolling texture

The measured and predicted textures of ferrite after cold rolling are presented in Fig. 6. The predicted texture development was somewhat too rapid, leading to larger intensities of the Orientation Distribution Function (ODF) as compare to experiment. Yet the model predicted the correct stable texture components.

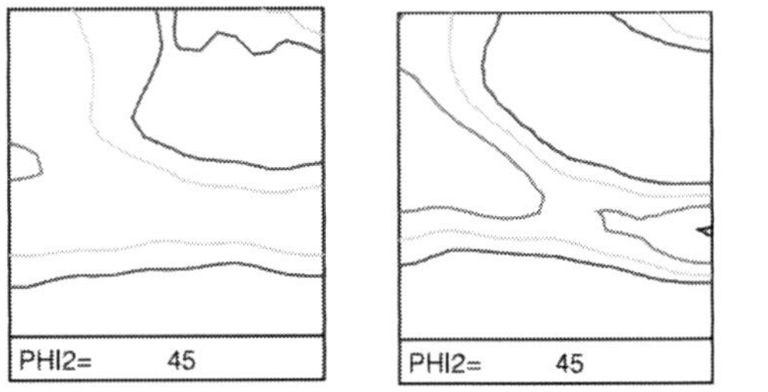

Figure 6. Section φ_2=45° of the orientation distribution function (ODF) of the ferrite phase after cold rolling. The experimental texture is on the left and the predicted one is on the right. The contour lines intensities are: 1 – 2 – 4 – 8 [10].

4 Discussion

In Fig. 4, the macroscopic tensile curves are reproduced only approximately when using a Voce law to represent the strain hardening of dislocation slip systems. The model overestimates the yield stress but it predicts a valid estimate of the strength after a small elongation. The larger overestimation of the yield stress of the cold rolled and stress-relieved sample may be expected because the stress relief treatment will have resulted in some recovery due to dislocation annihilation and rearrangement, which is not incorporated in the Voce law as it was used for both cold rolling and subsequent tensile straining. Moreover, both the cold rolled and the stress-relieved samples show an extended micro yield regime, which

 DOI: 10.1051/matecconf/20168002008

may be due to microstructural heterogeneity, as revealed by SEM. The fact that the deviation is limited and fading out with increasing strain thus indicates that the recoverable part of work-hardening linked to dislocation storage remains small compared to the non-recoverable one linked to lamellae narrowing and alignment.

The ferrite lattice strains measured during the tensile tests by in situ neutron diffraction are globally well predicted (Fig. 5). In particular, the ordering of the reflections is properly reproduced within the range of experimental uncertainty for the loading part of both curves. This ordering can be understood as a consequence of elastic anisotropy: the <200> grain family is the most compliant one and thus develops the highest lattice strains; the <110> grain family is the stiffest one and thus develops the lowest lattice strains. With increasing plastic deformation one would expect gradual load transfer from the softer <200> and <310> orientation fibres towards the stronger <110> and <211> orientation fibres. However, within the limited plastic strain range of the test, this is observed neither in the experiment, nor in the simulations. This might be an indication that for plastic anisotropy of ferrite the morphology of cementite is the dominating factor over the lattice orientation of ferrite.

The model does not cover the observed slight relaxations after yielding of the cold rolled and stress-relieved sample. In this sample, the ferrite and cementite lamellae are largely aligned with the rolling plane. The ferrite cold rolling texture is almost fully developed. Ferrite softening might then be due to (micro) shear banding and/or due to rupture of some of the cementite lamellae, both not incorporated in the model. The model does capture the average amplitude of the lattice strains that reflect the residual stress state after unloading (bottom-right of the plots), but here the ordering of the diffraction peaks is incorrect.

The cold rolling texture of ferrite (at the sheet mid-thickness) is correctly simulated but its intensity is larger than the intensity of the experimental texture (Fig. 6). It is possible that such discrepancy is due to the localization of plastic deformation in microscopic shear bands (MSB). The influence of MSB on the evolution of microstructure and texture of pearlite is currently under investigation, using MSB model suggested in [11].

An important feature of the model considered here is that it accounts for the progressive reorientation of the cementite lamellae during rolling. Whereas cementite lamellae are randomly oriented in the initial model polycrystal, they then tend to align parallel to the rolling plane (Fig. 7). As shown in Fig. 8, this contributes significantly to the macroscopic strength of the rolled sheet. Indeed, cementite lamellae are loaded less when they are randomly oriented than when they are aligned parallel to the tensile direction.

The model predictions rely of the simplified ALAMEL assumption about the interaction of adjacent grains. Crystal plasticity based finite element modelling (CPFEM) is likely to provide a more accurate reproduction of the grain-to-grain strain heterogeneity [4]. This has been checked by modeling the pearlite polycrystal with a finite element mesh in which each grain was represented by cubic element with eight integration points. Fig. 9 compares the phase stress obtained using either the ALAMEL model or CPFEM, demonstrating that the difference is rather small.

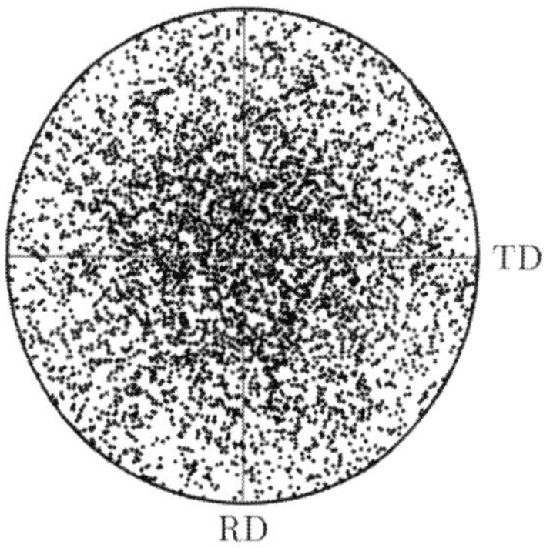

(a)

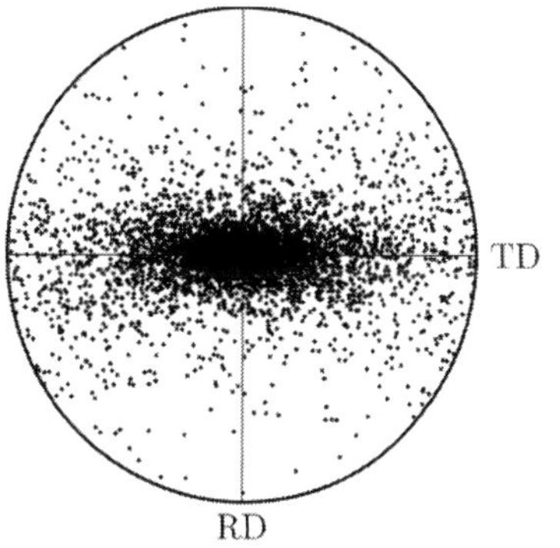

(b)

Figure 7. Pole figures showing the distribution of orientations of the cementite lamellae relative to the rolling direction (RD), the transverse direction (TD) and the sheet normal direction (ND) before (a) and after (b) rolling.

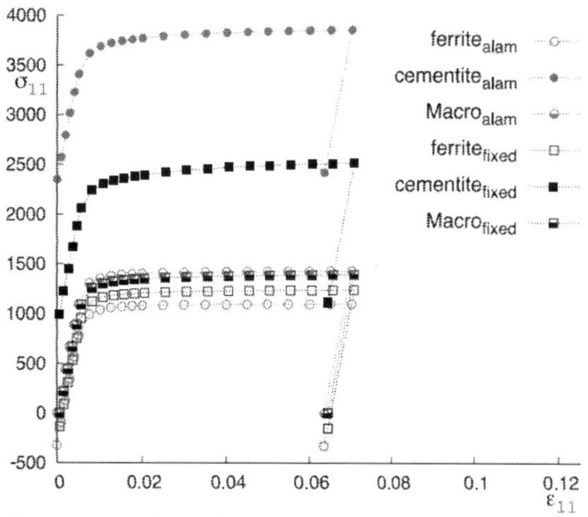

Figure 8. Model prediction of the phase and macroscopic stresses [MPa] when cementite lamellae are either re-oriented during rolling or not (resp. "alam" or "fixed").

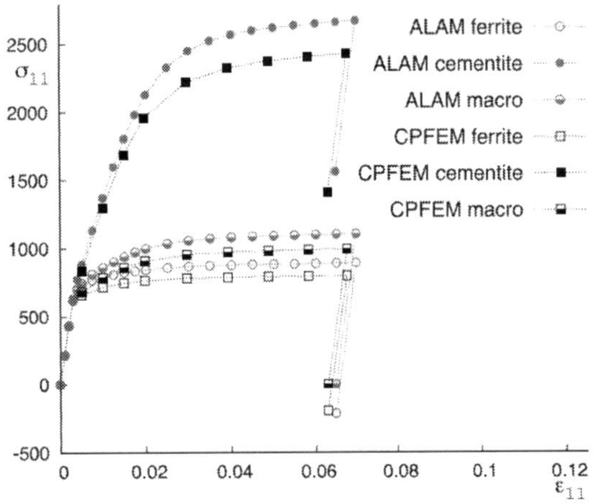

Figure 9. Model prediction of the phase and macroscopic stresses [MPa] when using either the ALAMEL model or CPFEM.

The microscopic CPFEM prediction can be improved by using more than a single finite element to represent each grain. However, an explicit representation of pearlite lamellae would imply refining the mesh so much that the computation time would become excessive.

5 Conclusion

A crystal plasticity model has been adapted in order to account for the cooperative deformation of ferrite and cementite phases in pearlite. The model has been implemented as a user-defined material law in a finite element code allowing macro scale simulation of rolling. Comparison of the predictions to experimental measurements of internal stresses and texture indicate that the mechanical response is strongly influenced by the progressive rotation of cementite lamellae, which preferentially align parallel to the rolling plane. Accounting for such microstructural evolution during metal forming is a strong motivation for the development of micro-to-macro modelling schemes.

Acknowledgements

Funding was obtained through the IAP project 7/21 "Multiscale Mechanics of Interface Dominated Materials" of the Belgian Science Policy BELSPO. LD is mandated by the FSR-FNRS, JT held a fellowship of The Research Foundation - Flanders (FWO). The neutron diffraction experiments were conducted at the Czech Center of Accelerators and Nuclear Analytical Methods (CANAM, LM2011019) with funding from the Ministry of Education, Youth and Sports of the Czech Republic. CANAM is gratefully acknowledged for granting beam time and support, in particular through the beam time scientist Dr. Jan Pilch. The European Union's Seventh Framework Program for research, technological development and demonstration supported the neutron experiments under the NMI3-II Grant number 283883.

References

1. J. Tacq, *Residual and Internal Stress Development Resulting from Plastic Deformation of Multi-Phase Alloys: the Case of Pearlite. Ph.D. Thesis*, KU Leuven, Belgium (2015).
2. D. Bardel, L. Delannay, O.V. Mishin, Int. J. Plast. (in preparation).
3. P. Van Houtte, S. Li, M. Seefeldt, L. Delannay, Int. J. Plast., **21** (2005), 589-624.
4. L. Delannay, M.A. Melchior, J.W. Signorelli, J-F. Remacle, T. Kuwabara, Comp. Mater. Sci., **45** (2009), 739-743.
5. J. Gil Sevillano, J. Mater. Sci. Eng., **21** (1975), 221.
6. J. Gil Sevillano, J. Phys. III, **1** (1991), 967-988.
7. X. Zhang, A. Godfrey, X. Huang, N. Hansen, Q.Liu, Acta mater., **59** (2011), 3422-3430.
8. P. Watté, P. Van Houtte, E. Aernoudt, J. Gil Sevillano, W. Van Raemdonck, I. Lefever, Mater. Sci. Forum, **157-162** (1994), 1689-1694.
9. Y. Tomota, P. Lukáš, D. Neov, S. Harjo, Y.R. Abe, Acta mater., 51 (2003), 805-817.
10. P. Van Houtte, MTM-FHM software, Department MTM, KU Leuven, Belgium (2015).
11. F. Lin, L. Delannay, J. Mech. Phys. Sol. (submitted).

Study on micro hydro-mechanical deep drawing using finite element method

Xiaoguang Ma[1], Jingwei Zhao[1], Wei Du[2], Xin Zhang[2], Laizhu Jiang[2] and Zhengyi Jiang[1,a]

[1]*School of Mechanical Materials and Mechatronic Engineering, University of Wollonggong, NSW 2522, Australia*
[2]*Stainles Steel Technical Centre, Baosteel Research Institude (R&D Centre), Baoshan Iron & Co., Ltd, Shanghai 200431, China*

Abstract. A numerical model was established to investigate the micro hydro-mechanical deep drawing process of austenitic stainless steel 304 foil (0.05 mm thickness). Due to the miniaturisation of the specimen size, the effect of grain size, gap distance and radial pressure during drawing process could be prominent. The results indicate that the appropriate radial pressure and gap distance could improve the limit drawing ratio (LDR) of manufactured cylindrical cups by reducing the friction resistance. The maximum LDR obtained in the present work reaches 3.2, which is much higher than that obtained by conventional deep drawing process.

1 Introduction

In recent decade, micro forming has been an attractive research topic due to its wide applications in fabricating micro parts for micro electro mechanical system, microelectronic technology and precision machinery [1] With high productivity, high utilisation of raw material and good mechanical properties, micro forming process is considered to be significant for development of manufacturing engineering. Micro deep drawing, which is a key branch of micro forming process, is widespread used in a variety of fields including automobile body parts, painless needles and micro electronic component. Thus, the significance of micro deep drawing investigation should not be ignored.

Furushima et al. [2] studied the effects of oxidation and surface roughening on limiting drawing ratio (LDR) in dieless drawing process of austenitic stainless steel 304. It was found that the LDR in dieless drawing strongly depends on heating temperature. The effect of die/blank holder and punch radius on LDR in angular deep drawing dies was studied by Ozek et al. [3]. It was concluded that both the LDR and punch force increased with increasing the die/blank holder and punch radius. The microdeep drawability of very thin sheet steels was studied by Saotome et al. [4] to clarify drawing characteristics. For sheet steels of below 0.2 mm thickness, the LDR was found to decrease with the increase of relative punch diameter (D_p/t) from 10 to 100. It is worth mentioned that for cases with D_p/t=10, the bending effect is dominant since blank holder pressure has little effect on drawability. A new deep drawing method using incremental electromagnetic assisted stamping with radial magnetic pressure was proposed by Cui et al. [5]. Compared to conventional deep drawing, the new method could significantly increase the value of material flow, decrease the tensile stress and thickness reduction at flange area, thereby improving LDR. Irthiea et al. [6] proposed a technique through adopting an initial gap between the adjustment ring and blank holder using flexible tools in which the blank holder was moveable and supported by a particular spring that provided the required holding force Both numerical and experimental results showed that relatively bigger initial gaps and smaller friction coefficient increased the formability whereas the friction coefficient between the blank and flexible die did not make a big difference.

Significant work has been done by researchers on deep drawing process to obtain high LDR, which is defined to be the ratio of the maximum blank diameter that can be safely drawn into a cup without flange to the punch diameter. Sato et al. [7] designed and manufactured new apparatus for micro-complex-shape components. It was found that an appropriate counter pressure during hydroforming process could eliminate the generation of wrinkles and reduces friction force, thereby improving drawing performance. The cylindrical cups with LDR of 2.1 have been successfully produced using designed tools. The diamond like carbon film coated blank holders and dies were designed by Gong et al. [8] to reduce the friction effect. The average friction coefficient of 0.169 was reached during micro deep drawing process and the cups were well formed with LDR of 2.1. Abdelmaguid et al. [9] proposed a dynamic programming method to reduce the number of drawing stages through heat treatment. The shell diameters at different stages could be optimised with a novel combination of dynamic programming, thereby improving LDR. Mostafapur et al. [10] investigated the pulsating blank holder effect on drawing to improve the

[a] Corresponding author: jiang@uow.edu.au

formability of aluminum 1050 alloy by means of numerical and experimental methods, and found that the pulsating blank holder increased the depth of cylindrical parts without partial thinning and tearing compared with the static blank holder. The key parameters influencing deep drawing performance were investigated and optimised by Zein et al. [11] using finite element modelling. The recommended value of die shoulder radius, punch nose radius, fluid friction coefficient and radial clearance were given to increase LDR. Besides, too much blank holding force was found to induce significant thinning phenomenon. Vollertsen and Hu [15] studied the effect of punch velocity on friction coefficient and they also investigated the LDR with different friction conditions. Gau et al. [14] designed singles dies for a whole manufacture process in order to obtain products with high LDR with one stage drawing and two stages ironing at high temperatures. The influence of die/blank holder and punch radius on LDR was studied by Ozek and Bal [16], and it was found that the LDR could be effectively improved by increasing the die/blank holder angle.

Based on existing publications, even though a relative high LDR could be obtained, it is still no more than 2.1 for conventional one stage drawing.

In this study, austenitic stainless steel 304 blank was drawn through micro deep drawing with radial pressure by using FEM. The purpose of this work is to obtain high LDR relative to previous work through optimisation of key processing parameters. The effects of the average grain size, radial pressure and gap distance were discussed to indicate the role of key parameters in the micro deep drawing process.

2 FE modelling

2.1 Solution method

Given that significant deformation could be observed during micro deep drawing process, the whole forming process was considered to be nonlinear to describe the elastic-plastic behaviour of austenitic stainless steel 304. The Newton–Raphson iterations were subsequently performed within an increment step to acquire the equilibrium condition between the externally applied load and the internal structural resistance [21]. The scheme is given by

$$x_{n+1} = x_n - \frac{f(x_n)}{\frac{df(x_n)}{dx}} \qquad (1)$$

The iterative procedure would be terminated when following condition applies.

$$|x_{r+1} - x_r| < \varepsilon \qquad (2)$$

where ε is a value that depends on the accuracy set in advance and on machine precision.

2.2 Flow rule

In metal plasticity, it is typically assumed that the plastic strain increment and the normal to the pressure-dependent yield surface have the same direction. Therefore, when the material begins its plastic deformation, the plastic strain increment could be defined as:

$$d\varepsilon_p = d\lambda \frac{\partial f}{\partial \sigma} \qquad (3)$$

where $d\lambda > 0$ is a hardening parameter, the function f is the plastic potential function that describes the relationship between the stress and plastic strain.

2.3 Damage model

The ductile damage evolution was also defined for investigation. Besides, Holloman hardening law was considered to describe the work hardening effect of material. The Hollomon equation is given below

$$\overline{\sigma_Y} = K(\varepsilon_0 + \overline{\varepsilon})^n \qquad (4)$$

where $\overline{\sigma_Y}$ is the effective stress, K is the strength constant, ε_0 and $\overline{\varepsilon}$ are the pre-strain and effective strain respectively, and n is the strain hardening exponent. The damage would be generated when strain rate reaches 0.8.

2.4 Material model

The material used in this research is an austenitic stainless steel (304) ultrathin foil with thickness of 0.05 mm. To eliminate the effect of divergent grain properties on deformation mode, Voronoi diagram is utilised in numerical model.

The Voronoi diagram is the tuple of cells (R_k), which is the set of all points in whose distance to site P_k is not greater than their distance to the other sites P_j $(j \neq k)$. The equation could be written as:

$$R_k = \{x \in X | d(x, P_k) \leq d(x, P_j)\} \; (j \neq k) \qquad (5)$$

Generally, there are two ways to generate Voronoi tessellation in commercial software ABAQUS /CAE. One way is to directly run the program in MATLAB and write PY files, then import file into ABAQUS and mesh the model. This method is convenient and fast, however, a great number of elements would be generated with this method, leading to great time consumption in further investigation. Moreover, the quality of elements is quite poor, making it hard to ensure the high accuracy of final result. Another common method is to mesh the model in ABAQUS first, then run the program in MATLAB. Based on the obtained results, the INP file could be reedited and imported to ABAQUS program. With this way, the amount of elements could be determined by users and a relatively high accuracy could be achieved despite much more time would be spent in pre-processing. In this study, the latter method was chosen to set up the numerical model of blank. To investigate the effect of grain size on micro deep drawing process, three average grain sizes were applied in this study. Figure 1 shows the Voronoi diagram of workpiece utilised in this study.

Figure 1. Voronoi diagram of blank with 49555 grains

Since the geometry feature of grains was obtained in Voronoi diagram, all grains should be assigned with certain elastic-plastic properties to show the deformability of individual grains. Hardness describes the resistance to plastic deformation, therefore, the hardness of grains from micro hardness test was considered to be effective to evaluate the grain heterogeneity. From Figure 2, it can be observed that 7 groups of Vickers hardness values are obtained to represent 7 grain properties in austenitic stainless steel 304, the inhomogeneity coefficient ξ represents the ratio of the hardness at each category divided to the average hardness. Based on results acquired from micro hardness tests, all grains are considered to be described with bilinear isotropic hardness model, with the same Young's modulus and dissimilar tangent modulus. As shown in Figure 3, seven classes of heterogeneity coefficients are utilised to describe the plastic performance of material with Young's modulus of 192 GPa, and initial yield stress σ_s of 205 MPa.

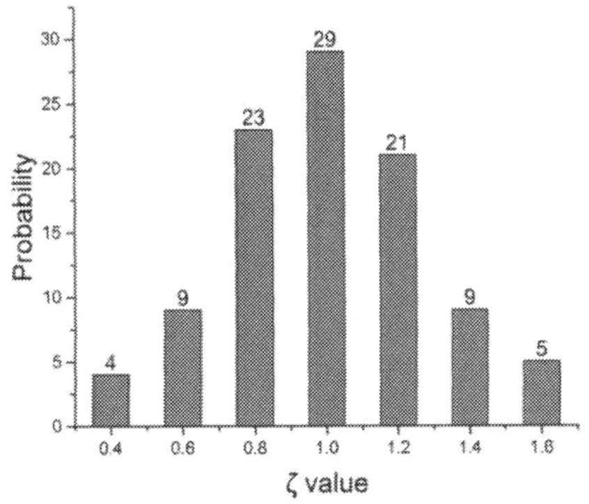

Figure 2. Proportion of grains with inhomogeneity coefficients

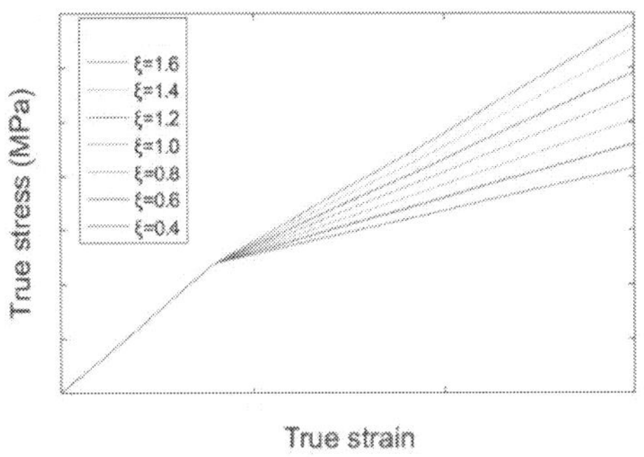

Figure 3. The flow stress curves with different inhomogeneity coefficients

Other parameters used in this study are shown in Table 1. Density, Young's modulus and Possion's ratio are obtained from tensile tests while the F and n values followed Saotome [19].

Table 1. Material properties used in simulation

Density ($g \cdot \mu m^{-1}$)	Young's modulus (GPa)	Possion's ratio	F-value (GPa)	*n*-value
$7.93e^{-12}$	192	0.25	1.55	0.44

2.5 Simulation condition

The austenitic stainless steel foil and tools were meshed with 8-node linear brick, reduced integration, hourglass control elements (C3D8R). The properties of tools are shown in Table 2 and the dimension of microtools used for micro deep drawing is listed in Table 3. The dimension of tools was optimised with numerical simulation and these parameters were found to be beneficial for drawability of workpiece in this study. Besides, full constrains were assigned to both two parts and a velocity of 0.4 mm/s is defined for punch during drawing process. The whole drawing process is carried out at room temperature. Given that the drawing process is completed in several seconds, the variance of temperature during forming process is negligible in this study.

Table 2. Properties of tools

Density (kg/m³)	Young's modulus (GPa)
8000	177

Table 3. Dimension of experimental tools

Blank diameter (mm)	Punch diameter (mm)	Punch corner radius (mm)	Die diameter (mm)	Die corner radius (mm)
2.64	0.8	0.1	0.9	0.3

The FE model of micro deep drawing process is illustrated in Figure 4.

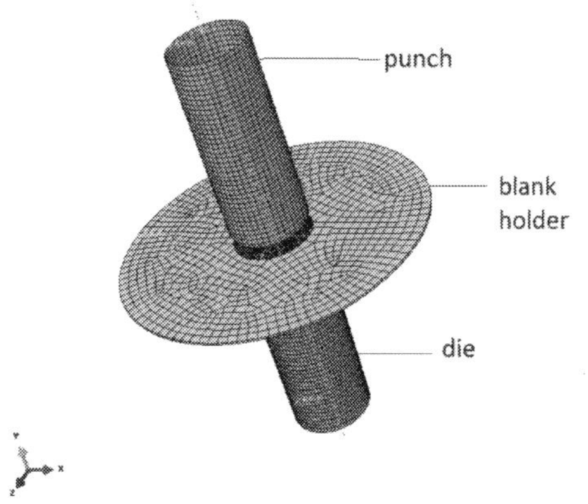

Figure 4. The FE of micro deep drawing process

3 Results and discussion

3.1 Effect of average grain size

For micro forming process, the effect of average grain size is quite significant due to size effect. When the average grain size increases, the number of grains at thickness direction would be reduced, causing that the deformation of individual grain become dominant during forming process. Instead, with smaller average grain size and more grains at thickness direction, the increasing grain boundaries would weaken the influence of dislocation and slip systems, thereby reducing the effect of crystal plasticity.

Figure 5 presents the equivalent plastic strain distribution of micro deep drawing process with different grain sizes. The forming limit is found to increase with grain size ranging from 0.015 to 0.025 mm. However, when the average grain size below 0.015 mm, that is, there are only one or two grains at thickness direction, and the LDR would be much worse due to the different behaviours of individual grains and different performance of crystal plasticity. This finding is also supported by experiments conducted by Xu et al. [23].

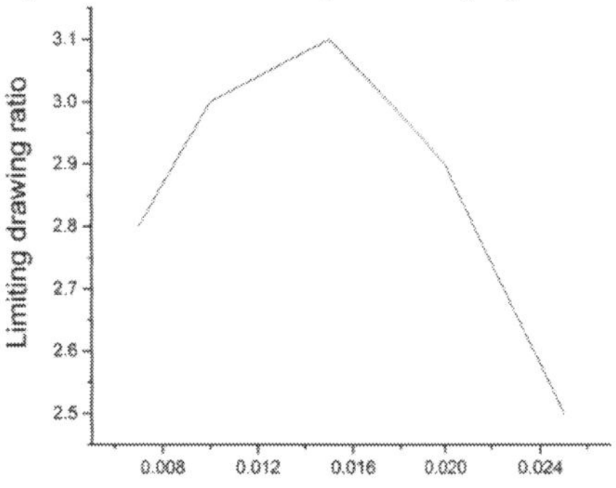

Figure 5. The equivalent plastic strain distribution of micro deep drawing process with different grain sizes

3.2 Effect of hydro-pressure

In micro deep drawing process, one of the key reasons leading to failure is the friction between die and blank at flange area, to reduce the effect of friction, radial pressure is applied in this study to improve the formability of workpiece. It is concluded by Wang et al. [24] that a radial pressure could reduce the drawing force and increase LDR. However, it should be noted that too much radial pressure would also give a rise to the occurrence of wrinkles. Moreover, Cui et al. [25] advocated that radial pressure could significantly decrease the tensile stress and thickness reduction at flange area, thereby improving the formability of material.

Figure 6 shows the dependence of LDR on radial pressure. The maximum LDR obtained is 3.2, which is much higher than that of conventional deep drawing (2.1). It can also be observed that with the radial pressure increases from 0 to 240 MPa, the LDR rises from 1.8 to 3.2. The enhancement of LDR with increasing the radial pressure is mainly due to the fact that the radial pressure is effective to reduce the stress at flange area and punch force. By applying appropriate radial pressure, the friction effect could be minimised, reducing the cracks at flange area. However, with the range from 240 to 400 MPa, the LDR decreases with increasing radial pressure due to cracks and wrinkles generated from excess radial pressure. Therefore, it can be concluded that an optimal value of radial pressure could promote the LDR, which matches the conclusions from Wang et al. [24] and Cui et al[25].

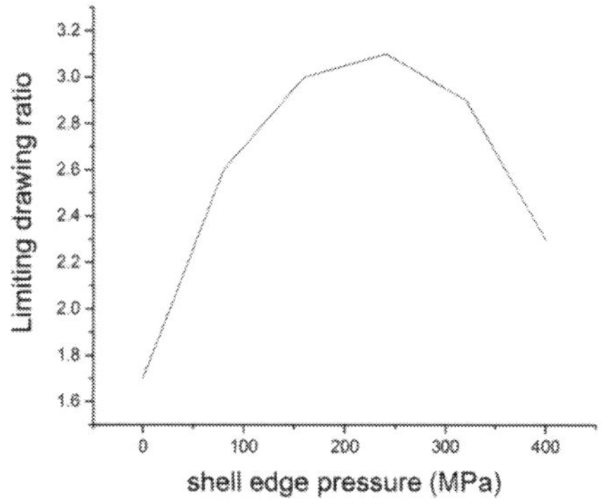

Figure 6. The dependence of LDR on radial pressure

3.3 Effect of gap distance

In micro deep drawing process, the gap distance between blank holder and blank is one of most crucial parameters in formability of materials. Irthieea et al. [26] concluded that the reduction of initial gaps increased the thickness reduction especially at the side wall of formed product. Besides, it was further announced that a higher initial gaps could improve the LDR of obtained cups. However, too much gap would also increase the occurrence of wrinkles. Therefore, a balance is required to select an optimal gap distance, which critically affects the formability of cylindrical cups. This point is supported by Gavas and Izciler [27] who declared that excess blank holder gap would cause excessive wrinkling and buckling.

Figure 7 shows the dependence of LDR on initial gap. The LDR is observed to increase from 2.8 to 3.2 with the increase of initial gap from 0 to 0.005 mm, which indicates that a larger gap distance would promote the formability of foil, however, when gap distance increases from 0.005 to 0.01 mm, the occurrence of wrinkles would prevent drawing process and reduce the LDR. Thus, it is recommended to use a higher initial gap below limitation in the case of generation of wrinkles.

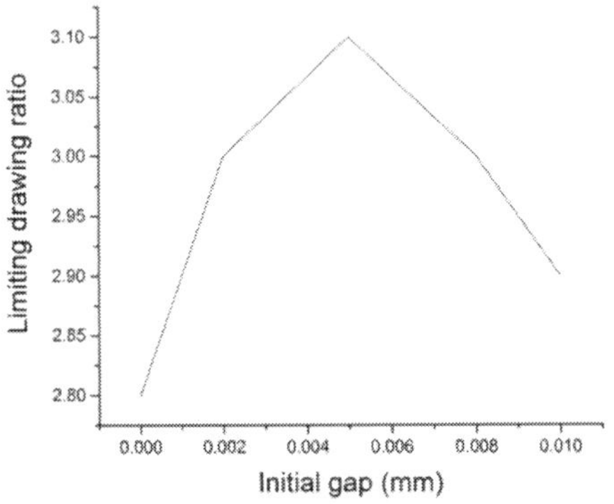

Figure 7. The dependence of LDR on initial gap

4 Conclusions

In this paper, a numerical model was established to investigate the formability of austenitic stainless steel 304 cups with consideration of the effects of radial pressure, grain size and initial gap distance. Based on numerical results, several conclusions are drawn as follows:

(1) The LDR for austenitic stainless steel foil (0.05 thick) with suitable radial pressure attains 3.2, which is 1.6 times of the conventional deep drawing.

(2) Suitable relative punch diameter in micro deep drawing should be applied to minimise the bending resistance and to avoid fractures generated during forming process.

(3) The increase of initial gaps gives a rise to LDR. Use of a higher initial gap is recommended to increase formability. However, the gap distance should not exceed the limitation beyond which wrinkles will be caused.

Acknowledgements

The first author would like to thank the financial support by IPTA scholarship from University of Wollongong for his PhD study. The study is also supported by Baosteel Australia Joint Centre (BAJC).

References

1. F. Gong, Z. Yang, Q. Chen, Z. Xie, J. Yang, Precision Eng. **42**, 224 (2015)
2. T. Furushima, Y. Imagawa, K. Manabe, T. Sakai, J. Mater. Process. Technol. **223**, 186 (2015)
3. C. Ozek, M. Bal, Int. J. Adv. Manuf. Technol. **40**, 1077 (2009)
4. Y. Saotome, K. Yasuda, H. Kaga, Japan J. Mater. Process. Technol. **113**, 641 (2001)
5. X.H. Cui, J.H. Mo, J.J. Li, Procedia Eng. **81**, 813 (2014)
6. I. Irthiea, G. Green, S. Hashim, A. Kriama, J. Mach. Tools Manuf. **76**, 21 (2014)
7. H. Sato, K. Manabe, K. Ito, D. Wei, Z. Jiang, J. Mater. Process. Technol. **224**, 233 (2015)
8. F. Gong, B. Guo, C. Wang, D. Shan, Diamond Relat. Mater. **20**, 196 (2011)
9. D. Min, B. Jeon, H.Kim, N. Kim, J. Mater. Process. Technol. **54**, 230 (1995)
10. A. Molotnikov, R. Lapovok, C.F. Gu, C.H.J. Davies, Y. Estrin, Mater. Sci. Eng. A **550**, 312 (2012)
11. Z.T. Xu, L.F. Peng, X.M. Lai, M. W. Fu, Mater. Sci. Eng. A **611**, 345 (2014)
12. A. Mostafapur, S. Ahangar, R. Dadkhah, Int. J. Adv. Manuf. Technol. **69**, 1113 (2013)
13. H. Zein, M. El-Sherbiny, M. Abd-Rabou, M. El Shazly, American J. Mech. Eng. **1**, 20 (2013)
14. JT. Gau, S. Teegala, KM. Huang, TJ. Hsiao, BT. Lin, J. Manuf. Process. **15**, 298 (2013)
15. F. Vollertsen, Z.Y. Hu, Prod. Eng. Res. Dev. **4**,553 (2010)
16. C. Özek, M. Bal, Int. J. Adv. Manuf. Technol. **40**,1077 (2009)
17. L. Peng, F. Liu, J. Ni, X. Lai, Mater. Des. **28**,1731 (2007)
18. C.H. Chen, H.T. Gau, R.S. Lee, Mater. Manuf. Process. **24**,1256 (2009)
19. Y. Saotome, K. Yasuda, H. Kaga, J. Mater. Process. Technol. 641 (2001)
20. G. Faraji, M.M. Mashhadi, R Hashemi, J. Manuf. Sci. Technol. **3**,262 (2010)
21. F. Qu, Z. Jiang, H. Lu, Int. J. Mech. Sci. **105**, 182 (2016)
22. Z. Jiang, H. Lu, D. Wei, K.Z. Linghu, X. Zhao, X. Zhang, D. Wu, Procedia Eng. **81**, 2463 (2014)
23. Z.T. Xu, L.F. Peng, X.M. Lai, M.W. Fu, Mater. Sci. Eng. A **611**, 345 (2014)
24. H. Wang, L. Gao, M. Chen, J. Mech. Sci. **53**,793 (2011)
25. X. Cui, J. Mo., J. Fang, J. Li, J. Procedia Eng. **81**, 813 (2014)
26. I. Irthiea, G. Green, S. Hashim, A. Kriama. Int. J. Mach. Tools Manuf. **76**, 21 (2014)
27. M. Gavas, M. Izciler, J. Mater. Des. **28**, 1641 (2007)

Modeling grain orientation of DP600 steel by Nd:YAG laser

Shibo Liu[1,2,a], Afia Kouadri-Henni[1,2] and Adinel Gavrus[1,2]

[1]*Université Européenne de Bretagne, France*
[2]*INSA-LGCGM-EA 3913, 20 avenue des Buttes de Coësmes, 35708, Rennes, France*

Abstract. A FE-CA (Finite Element-Cellular Automaton) model is built to simulate the microstructure evaluation during laser welding of DP600 high strength steel. A moving conical heat source is performed in ABAQUS to achieve temperature field during welding. A linear interpolation method is used to transfer node temperature of ABAQUS to cell temperature of Cellular Automaton. With the consideration of temperature and solute concentration, the solidification of DP600 after weld is simulated. The established FE-CA model can simulate 2D equiaxed and column dendrite grain growth. The competitive growth between different dendrite arms is presented. The weld microstructure simulation result shows good agreement with experiment results.

1 Introduction

To relieve energy crisis and environment problems, high strength material is applied in automotive industries in order to reduce the car weight [1]. As a type of AHSS (advanced high-strength steel), DP steel is a low carbon steel which contains 3-30% hard martensite phase in a soft ferrite matrix [2]. The hard phase martensite contributes to the high strength while soft ferrite helps to improve ductility [3, 4], which make the steel a good combination of adequate formability and high strength. Compared with traditional high-strength low-alloy (HSLA) steels, DP steel has no yield point elongation and a higher strain-hardening rate [5]. The high strain-hardening rate enhance the absorption of energy when deformation is happened, which helps decrease of damage in car collision [6].

Laser welding process efficiently integrates DP600 in the automobile industry. By employing appropriate welding parameters, the technique produces a strong weld joint altering the crystallographic texture. As the texture is the cause of anisotropy in rolled plates [7-8], the anisotropy may be changed by laser welding process. Arminjon [9] relates macroscopic anisotropy parameters using texture coefficients in one of mathematical models, where an appropriate grain orientation is an important parameter to calculate anisotropic property. Usually, to know experimentally the grain orientation of a material requires high price, either with advanced equipment or with tedious work. The aim to obtain the grain orientation by simulation, the Finite Element Method combined with Cellular Automaton Model (FE-CA) can be a successful way to simulate a 2D weld microstructure during solidifications.

A welding heat source plays an important role for FEM simulation of temperature distribution history during solidification. Rosenthal [10] developed Point and Line Heat Source Models for moving heat resource to analyze weld thermal field history which has been popular in 1930s. Later, a Disc Model Heat Source Model with a Gaussian distribution of flux deposited on the surface of the plate was proposed by Pavelic et al. [11] and combined with FEM by Krutz et al. [12] and Friedman [13]. The model achieves more realistic temperature distribution than those calculated by Rosenthal model. Since the Disc Model is not suitable for high energy density welding such as laser welding, a Hemispherical Power Density Distribution volume source model was proposed by Paley et al. [14] and finally developed into Double Ellipsoid Three-Dimensional Heat Source Model [15]. After the popular double ellipsoid model, a Three-Dimensional Conical Heat Source seems to be more suitable for deep penetration welding [16]. In this paper, a Conical Heat Source is applied and a interpolation method is used to transfer temperature field from FEM to Cellular Automaton model.

Cellular Automaton (CA) was invented in late 1940s by Stanislaw Ulam and John von Neumann. After dozens of years' development, the CA method began to model grain structure formation in solidification by M.Rappaz and Ch.A. Gandin [17]. Six years later, Nastac [18] proposed a comprehensive CA model for simulating dendritic crystals growth during the solidification of binary alloys, in which work mathematical description, solution methodology and program algorithm are depicted in detail. In 2003, Wang et al. [19] successfully simulates equiaxed and columnar dendritic grains in 3D. In 2007, Y.H. Wei et al. [20] first attempt CA model on 2D simulation of welding solidification. Contributed to by many works [21–24], CA method has become a powerful tool in simulating microstructure evolution of welding solidification process.

[a] Corresponding author: shibo.liu@insa-rennes.fr

2 Simulation

The grain orientation analysis is sequentially coupled as a thermal-microstructure analysis. The simulation model consists in two parts:

1. Thermal analysis part:

• Heat source, heat transfer and boundary condition of radiation, convection and conduction are considered to analysis the thermal data of the model.

• The calculated thermal data is stored in a data file which is prepared for CA analysis.

2. Microstructure analysis part:

• An interpolation method for welding is proposed to import the thermal data of ABAQUS analysis model to the CA model.

• With the consideration of cell temperature, undercooling degree, nuclei density, solute concentration at the interfaces and between cells, dendrite tip curvature, growth velocity, and time increment, the condition of grain growth is simulated.

2.1 ABAQUS temperature model

2.1.1 Dimensions

The dimension of thermal analysis model of a lap-joint weld is presented in the Figure 1. The 0.1mm gap is used to improve the weld quality. The zoomed zone of A-A part is shown together with the width of weld line equal to 1.2 mm.

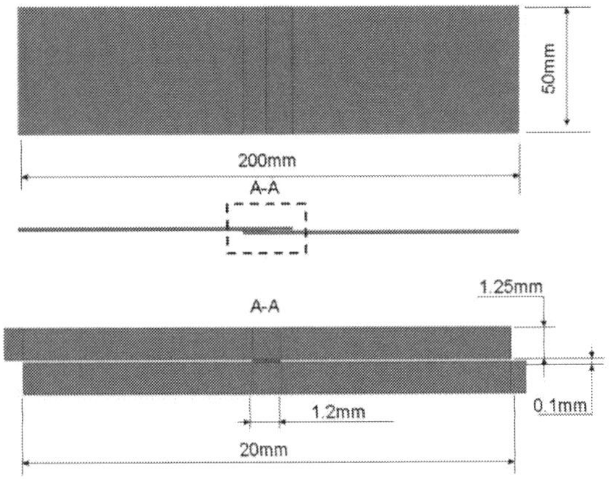

Figure 1. The dimension of welding simulation model

2.1.2 Conical heat source

In the case of deep penetration welding simulation, conical heat source with a Gaussian distribution is used as a suitable heat source model [16]:

$$Q_v(r,z)=\frac{9\eta Pe^3}{\pi(e^3-1)}\times\frac{e^{-\frac{3r^2}{r_c^2}}}{(z_e-z_i)(r_e^2+r_e r_i+r_i^2)} \quad (1)$$

where r_c is

$$r_c = r_i + \frac{(r_e-r_i)(z-z_i)}{z_e-z_i} \quad (2)$$

Table 1. Parameters of heat source.

Symbol	Meaning	Magnitude
$Q_v(r,z)$	total volumetric heat flux	$W\cdot m^{-3}$
r_e	superior radius	0.8 mm
r_i	inferior radius	0.35 mm
r_c	variational radius	$r_i \leq r_c \leq r_e$
Z_e	top surface position	0.26 mm
Z_i	bottom surface position	0 mm
P	laser power	4 KW
η	laser absorption efficieny	42 %
v	welding speed	3 m/min

2.1.3 Heat transfer and boundary conditions

The equation of heat transfer is governed by energy equilibrium balance:

$$\rho\frac{dH}{dt}-div(\lambda gradT)-Q=0 \quad (3)$$

The convection limits condition on the surrounding surface can be written in the following form:

$$\lambda(T)gradT(x_i,t)_{surface}+h(T(x_i,t)-T_0)=0 \quad (4)$$

The radiation limits condition on the weld pool surface is written by

$$\lambda(T)gradT(x_i,t)_{surface}+\varepsilon\sigma(T(x_i,t)^4-T_\infty^4)=0 \quad (5)$$

In Equation 3, 4 and 5, ρ is the density, H the enthalpy, h the convection coefficient, λ the thermal conductivity, t the time, T the temperature, Q the internal heat, ε the emissivity with a value of 0.25, σ the Stefan Boltzman with value 5.6704 × 10-8W/ (m·K4). The surrounding temperature is supposed to be 25℃.

2.1.4 Material parameters

Table 2 shows that the conductivity of DP600 material is temperature dependent.

Table 3 is presented the temperature dependent density of DP600.

In Table 4, the specific heat of dual phase material is defined to change with temperature.

Table 2. Conductivity of model.

Conductivity (W/(m·℃))	Temperature (℃)
34.8	25
35	200
35	254
21.4	1100
25.5	1200
60	1300

Table 3. Density of model

Mass density ρ (Kg/m^3)	Temperature (℃)
7594	0
7300	1000
7210	1200
7200	1500

Table 4. Specific heat of model

Specific heat c (J/(Kg·℃))	Temperature (℃)
520	0
550	200
600	400
748	600
690	800
520	1100
610	1200
620	1500

2.2 Cellular Automaton microstructure model

2.2.1 Computation procedure

The program procedure of weld solidification simulation is proposed with a CA computation method in a 2D mathematical model. The concentration of all cells are supposed to be the same at beginning of the procedure.

1. Temperature data is read from a 3D FEM ABAQUS Simulation and imported into a scanned 2D CA model.
2. The state of all cells, according to each cell temperature, are determined to be liquid, growing or solid.
3. If all cell temperatures is above melting temperature, time is increased, and the procedure returns to step 1. If not, nucleation and growth of liquid cells is computed.
4. For the growing cell, solid fraction, dendrite tip curvature, mesh anisotropy and concentration is computed.
5. Solute diffusion between cells is computed.
6. Crowing velocity (of solid/liquid interface moving) of growing cell is computed.
7. Time increment is computed.
8. Directions of cells are displayed using different colors.
9. If all the cell is solid, data is saved and computation stop. If not, the procedure return to step 1.

2.2.2 Interpolation of Cell Temperature

As is seen in Figure 2, the temperature history data, T_a, T_b, T_c and T_d, are obtained from ABAQUS heat model and transferred to 2D Cellular Automaton by procedure interpolation on indirect method. The ABCD zone is meshed by m×n cells, and each cell is displayed by a pixel.

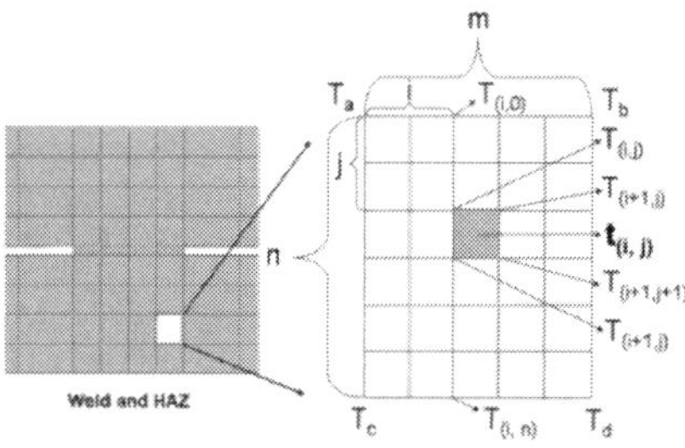

Figure 2. The schema of procedure linear interpolation method

First, the node temperature at the edge $T_{(i,0)}$ and $T_{(i,n)}$ are interpolated. Second, node temperature $T_{(i,j)}$ is calculated. Finally, the cell temperature $t_{(i,j)}$ is estimated as a mean value of the four node temperature values:

$$T_{(i,0)} = \frac{i}{\mathrm{m}}(T_b - T_a) + T_a = \frac{i}{\mathrm{m}}T_b + \frac{(\mathrm{m}-i)}{\mathrm{m}}T_a \quad (6)$$

$$T_{(i,n)} = \frac{i}{\mathrm{m}}(T_d - T_c) + T_c = \frac{i}{\mathrm{m}}T_d + \frac{(\mathrm{m}-i)}{\mathrm{m}}T_c \quad (7)$$

$$\begin{aligned} T_{(i,j)} &= \frac{j}{\mathrm{n}}(T_{(i,\mathrm{n})} - T_{(i,0)}) + T_{(i,0)} \\ &= \frac{1}{\mathrm{m}\cdot\mathrm{n}}[(\mathrm{m}-i)(\mathrm{n}-j)\cdot T_a + i\cdot(\mathrm{n}-j)\cdot T_b \\ &+ (\mathrm{m}-i)\cdot j\cdot T_c + i\cdot j\cdot T_d] \end{aligned} \quad (8)$$

$$t_{(i,j)} = \frac{1}{4}[T_{(i,j)} + T_{(i+1,j)} + T_{(i,j+1)} + T_{(i+1,j+1)}] \quad (9)$$

2.2.3 Nucleation

According to the kinematic model described by Thevoz et al. [25], the nucleation density increases as undercooling increases. Because the undercooling of edge is bigger than that of bulk, the nucleation density of the edge is different from that of the bulk:

The edge of the weld pool ("e" stands for edge) [17] follows the kinematic equation:

$$\delta n_e = \int_{\Delta T}^{\Delta T+\delta(\Delta T)} \frac{n_{e,max}}{\sqrt{2\pi}\Delta T_{e,\sigma}}(1-f_{e,s})e^{-\frac{1}{2}(\frac{x-\Delta T_{e,max}}{\Delta T_\sigma})^2} dx \quad (10)$$

The bulk of weld pool ("b" stands for bulk) is governed by:

$$\delta n_b = \int_{\Delta T}^{\Delta T+\delta(\Delta T)} \frac{n_{b,max}}{\sqrt{2\pi}\Delta T_{b,\sigma}}(1-f_{b,s})e^{-\frac{1}{2}(\frac{x-\Delta T_{b,max}}{\Delta T_\sigma})^2} dx \quad (11)$$

In the above formula, ΔT is the whole undercooling of the bulk or edge i.e. $\Delta T=T_{liq}-T_{cell}$. T_{liq} is the liquidus temperature according to phase diagram. n_e is the nucleation density of the edge of weld pool. n_b is the nucleation density of the bulk of weld pool. $n_{e,max}$ and $n_{b,max}$ are the maximum nucleation density for edge and bulk respectively. ΔT_σ is the standard deviation of the degree of undercooling. $\Delta T_{e,max}$ and $\Delta T_{b,max}$ is the maximum degree of undercooling for edge and bulk respectively. $f_{e,s}$ and $f_{b,s}$ are fractions of solid already formed at the edge and the bulk. The increase of nucleation density, δn_e or δn_b, multiplied by the edge cell number (Ne) or the bulk number (Nb), results to the number of cells nucleated at step n.

The nucleation position is chosen randomly among all remaining liquid cells. The nucleation probabilities [17] of the edge and bulk (dp_e and dp_b) during each time step are given by

$$dp_e = \frac{\delta N_e}{N_{edge}} \quad (12)$$

And

$$dp_b = \frac{\delta N_b}{N_{bulk}} \quad (13)$$

where N_{edge} and N_{bulk} are the cell number of the edge and the bulk of weld pool, respectively. During nucleation, each liquid cell is scanned and a random number, Rand, is assigned ($0 \le Rand \le 1$). The nucleation of a "liquid" cell will occur only if $Rand \le dp_e$ or $Rand \le dp_b$.

2.2.4 Solid fraction

The solid fraction of growing cells can be determined by [18]

$$\delta fs = \frac{\delta t}{a}\left(V_x + V_y - V_x V_y \frac{\delta t}{a}\right) \quad (14)$$

Here "a" is the mesh size, V_x is the maximum growth velocity of growing cell and δt is the increment of time. If $fs \ge 1$, the cell state is changed from growing to solid and states of neighbor cells are changed from liquid to growing according to the limited angles method [21] to keep the growth angle. The limited angles method are presented in Figure 3. With many cells to stand for one cell, the growth direction can be fixed. The growth direction will always be kept in any circumstances.

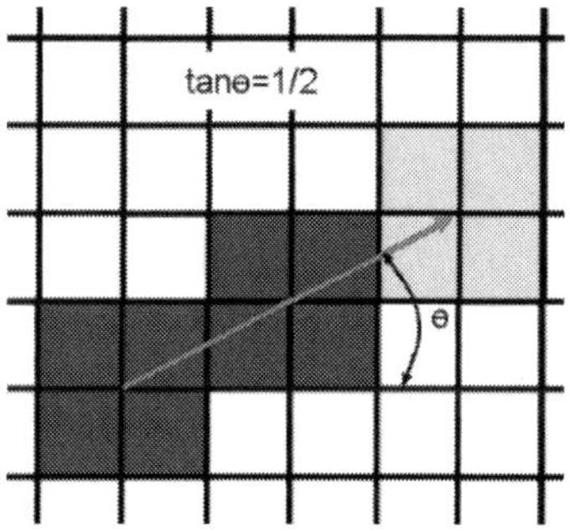

Figure 3. The growth direction with limit angle method

2.2.5 Interface Curvature

The solid/liquid interface curvature K for a cell is given by [18]:

$$K = \frac{1}{a}\left(1 - 2\frac{fs + \sum_{k=1}^{N} fs(k)}{N+1}\right) \quad (15)$$

In the equation (15), a is the mesh size, fs is the solid fraction, N stands for the number of neighbor cells. The result of the curvature calculation range from -1/a to 0 for concave interface and from 0 to 1/a for convex interface. With increment of m and n in Figure 2, the value of a decrease and can be very small.

2.2.6 Interface Concentration

For growing cell, the interface temperature (T^*) is given by [26]:

$$T^* = T_L^{EQ} + (C_L^* - C_0)m_L - \Gamma K f(\varphi,\theta) - \frac{v}{\mu_k} \quad (16)$$

where T_L^{EQ} is the equilibrium liquidus temperature and mL is the liquidus slope of phase diagram, Γ is the Gibbs-Thomson coefficient and f (φ,θ) is a coefficient used to account the growth anisotropy. In laser welding process, the high velocity of grain growth induced by high temperature gradient is considered in terms of kinetic undercooling, where the interface temperature is affected by v/μ_k. Here v is the growth velocity of solid-liquid interface and μ_k is the dynamic coefficient of grain growth.

For growing cell, the interface solute concentration C_L^* can be calculated from Equation 16:

$$C_L^* = C_0 + \frac{T^* - T_L^{EQ} + \Gamma \bar{k} f(\varphi,\theta) + \frac{v}{\mu_k}}{m_L} \quad (17)$$

The solute concentration gradient G_C in growing cell can be estimated:

$$G_C = \frac{C_L^* - C_L}{(1-f_S)\Delta x/2} \tag{18}$$

where Δx is the mesh size in the x direction. Solute concentration in liquid cell is calculated from Equation 18:

$$C_L = C_L^* - \frac{1-f_S}{2}\Delta x G_C \tag{19}$$

2.2.7 Surface tension anisotropy

The anisotropy of surface tension is estimated [18]:

$$f(\varphi,\theta) = 1 + \delta cos[4(\varphi-\theta)] \tag{20}$$

Here, δ = 0.04, θ is the angle between growth direction and +x direction, and φ is computed from the formula:

$$\varphi = arcos\left(\frac{V_x}{\left[V_x^2 + V_y^2\right]^{1/2}}\right) \tag{21}$$

2.2.8 Diffusion Equation

Starting from 3D cells, Figure 4 illustrates the diffusion of 2D cells. For the left cell, the upper part is viewed as solid and the lower as liquid. For the right cell, the front part is viewed as liquid and the back as solid.

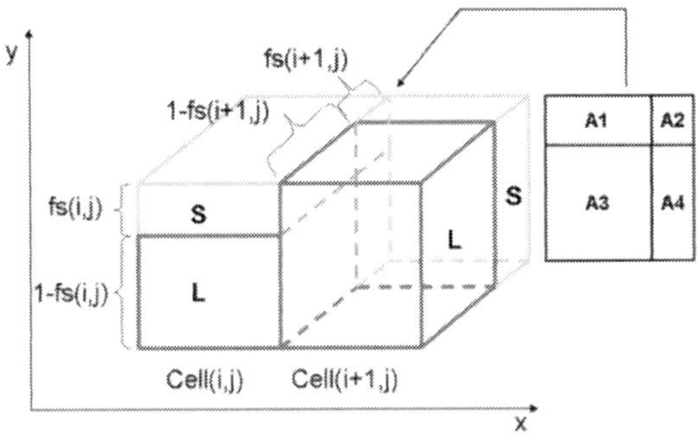

Figure 4. Schematic diagram of diffusion between CA cell

Using Fick's second law along the x direction, it is obtained:

$$\frac{\partial C(x,t)}{\partial t} = -\frac{\partial J}{\partial x} \tag{22}$$

where C is the concentration, t the time, and J the solute flux. The total contact area is A. A1 and A4 are contact region between liquid and solid. A2 is solid contact region. A3 is liquid contact region. The analyses have following expressions:

$$A1 = fs(i,j)\cdot[1-fs(i+1,j)]\cdot \mathrm{A} \tag{23}$$

$$A2 = fs(i,j)\cdot fs(i+1,j)\cdot \mathrm{A} \tag{24}$$

$$A3 = [1-fs(i,j)]\cdot[1-fs(i+1,j)]\cdot \mathrm{A} \tag{25}$$

$$A4 = fs(i+1,j)\cdot[1-fs(i,j)]\cdot \mathrm{A} \tag{26}$$

Every contact part is added together and finally the solute concentration has the following kinematic equation:

$$\begin{aligned}\frac{\partial C(x,t)}{\partial t} &= \mathrm{d_s}\cdot\frac{C_S(i+1,j)-C_S(i,j)}{a^2}\cdot fs(i,j)\cdot fs(i+1,j)\\ &+\sqrt{\mathrm{d_l}\cdot\mathrm{d_s}}\cdot\frac{C_S(i+1,j)-C_L(i,j)}{a^2}\cdot[1-fs(i,j)]\cdot fs(i+1,j)\\ &+\sqrt{\mathrm{d_l}\cdot\mathrm{d_s}}\frac{C_L(i+1,j)-C_S(i,j)}{a^2}\cdot fs(i,j)\cdot[1-fs(i+1,j)]\\ &+\mathrm{d_l}\frac{C_L(i+1,j)-C_L(i,j)}{a^2}\cdot[1-fs(i,j)]\cdot[1-fs(i+1,j)]\end{aligned} \tag{27}$$

d_l is diffusion coefficient of liquid and d_s is diffusion coefficient of solid. C_L and C_S are solute concentration of liquid and solid respectively. (i, j) refers to the position of the i column and j row. f_s are solid fraction of cell.

2.2.9 Growth Velocity of the Interface

According to growth velocity expression by Nastic [18] in x-direction(y-direction is similar):

$$\begin{aligned}V_x(i,j) &= \frac{\mathrm{d_l}}{\mathrm{a(1\text{-}k)}}[(1-\frac{C_L(i-1,j)}{C_L^*(i,j)})f_L(i-1,j)\\ &+(1-\frac{C_L(i+1,j)}{C_L^*(i,j)})f_L(i+1,j)]\\ &+\frac{\mathrm{kd_s}}{\mathrm{a(1\text{-}k)}}[(1-\frac{C_S(i-1,j)}{kC_L^*(i,j)})f_S(i-1,j)\\ &+(1-\frac{C_S(i+1,j)}{kC_L^*(i,j)})f_S(i+1,j)]\end{aligned} \tag{28}$$

k is partition coefficient.

2.2.10 Time step

The time step is computed as [18]:

$$\delta t = \frac{1}{5}min\left(\frac{\mathrm{a}}{V_{max}}, \frac{\mathrm{a}^2}{\mathrm{d_l}}, \frac{\mathrm{a}^2}{\mathrm{d_s}}\right) \tag{29}$$

Here, a is the mesh size, Vmax is the maximum growth velocity of growing cell.

2.2.11 Parameters of the proposed 2D CA model

Table 5 presents the parameters of the proposed CA model.

Table 5. Parameters of CA model.

Symbol	Meaning	Magnitude
k	Partition coefficient	0.3
d_l	Liquid diffusion coefficient	$3.0\times10^{-9}\ m^2s^{-1}$
d_s	Solid diffusion coefficient	$3.0\times10^{-12}\ m^2s^{-1}$
a	Cell size	10^{-7} m
T_L^{EQ}	Liquidus temperature	1530 °C
Γ	Gibbs Thomson Coefficient	1.9×10^{-7}Km
m_L	Liquidus slope	-80 °C/%wt
$\Delta T_{e,\sigma}$	Edge undercooling deviation	20°C
$\Delta T_{b,\sigma}$	Bulk undercooling deviation	50°C
$\Delta T_{e,max}$	Edge mean undercooling	20°C
$\Delta T_{b,max}$	Bulk mean undercooling	50°C

3 Simulation results and discussion

3.1. Temperature evolution simulation

According to the ABAQUS FEM, the temperature history are drawn in Figure 5. T_A, T_B, T_C and T_D are temperatures of four node A, B, C and D. The four nodes all experience a sudden increment and decrement of temperature. The maximum temperature of node A and C is higher than the liquidus temperature while node B and D not.

3.2 Single nucleation supposing on isotropic temperature distribution

As is shown in Figure 6, the simulation domain is made up of 300×313 cells with cell length of 1μm. Under an supposing isotropic temperature environment and starting from a nucleation along the direction of θ= arctan(1/2), an equiaxed dendrite growth is simulated. All cells follow the temperature of the node A. A small nuclei gradually grows to a big equiaxed dendrite which takes up the whole simulation zone. The interaction of secondary and tertiary arms can be clearly seen in Figure 6(c). The dendrite grows until meeting the entire simulation boundaries.

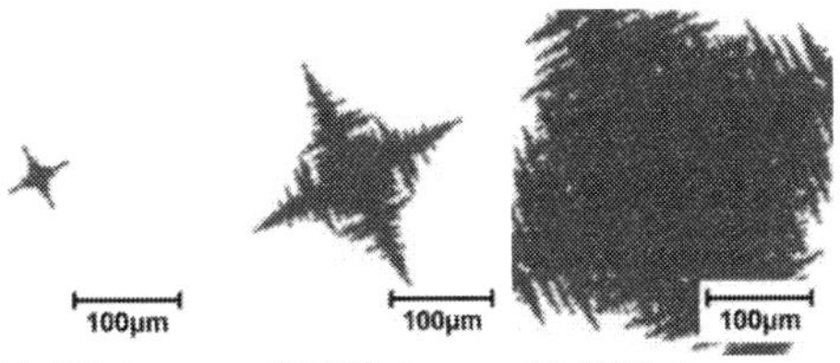

(a) 576 steps (b) 879 steps (c) 1399 steps

Figure 6. The growth of 2D equiaxed dendrite for three different time steps

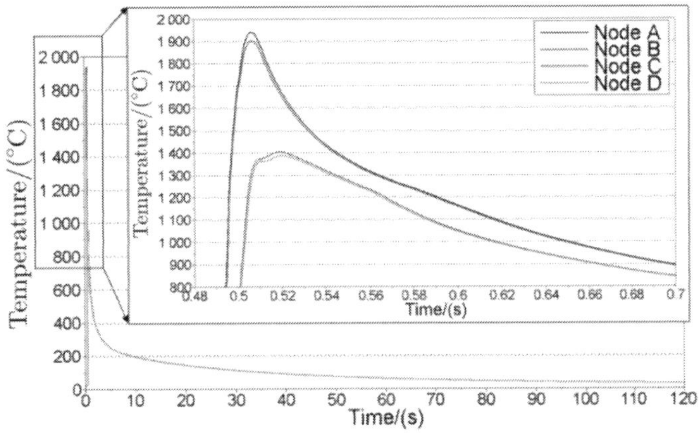

Figure 5. The temperature histories of four nodes corresponding to 2D CA domain.

3.3 Edge dendrites growth supposing on isotropic temperature distribution

As can be seen in the figure 7, the simulation domain is made up of 300×313 cells with cell length of 1μm. Under an isotropic temperature environment and several small dendrite at edge with the same growth direction, the growth of columnar dendritic grains is simulated. Starting from small nucleus, 11 dendrites grow up to columnar dendrites at the same velocity towards the same direction. The columnar dendrites grow until they take up the whole simulation zone. The interaction of dendrite arms between different grains are presented in Figure 7(b) and 7(c).

 DOI: 10.1051/matecconf/20168002010

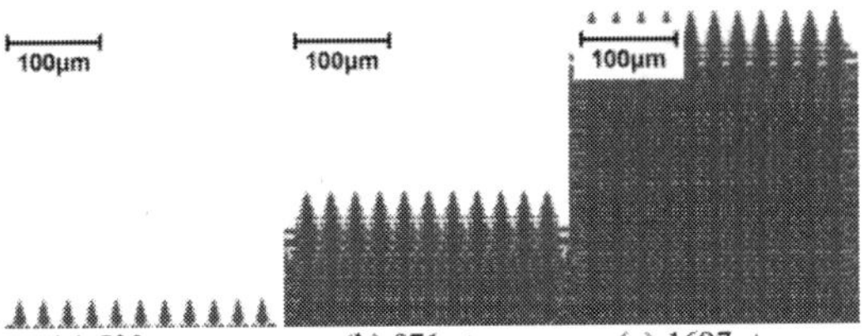

(a) 533 steps (b) 971 steps (c) 1697 steps

Figure 7. The growth of 2D columnar dendrites for three different time steps

3.4 Microstructure simulation from the welding temperature distribution

Unlike the isotropic temperature distribution assumptions used in section 3.2 and section 3.3, the temperature distribution used here are weld temperatures from FEM results in Figure 5. The simulation domain is made up of 900×939 cells with cell length of 0.33μm. The interpolation of temperature is described in the previous section 2.2.2. The results are presented in Figure 8.

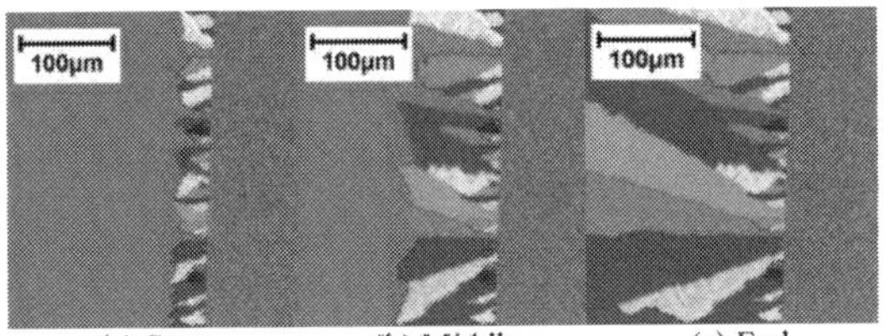

(a) Start (b) Middle (c) End

Figure 8. Weld Microstructure Simulation for three different steps

So in figure 8(a), the beginning of the dendrite growth is presented. The left part with red colour stands for melting alloy, the middle part with different colours represent dendrites with different growth directions and the right part with random distributed colours are the never melt alloy. Nucleuses with random directions are randomly set at the edge. Figure 8(b) shows the competition of dendrite growth. Dendrites with the directions near to the preferential direction grow faster than those with directions far from the preferential. Growths of some dendrites are depressed by dendrites with faster growth directions. In figure 8(c), the final simulated microstructure of the 2D CA model is presented. Only several dendrites with preferential growth directions grow up. The grains that go through melting and solidifying present an orientation towards the centre weld.

3.5 Equiaxed dendrite

Simulation results of 2D dendrite with different cell length are compared with the SEM observation of microstructure of DP600 weld in Figure 9(d) [27].

In microscope simulation, the equiaxed dendrite morphologies with different cell length are presented in Figure 9 (a), (b) and (c). All the simulation results provide similar morphology with dendrite in experiment observations. As the cell length of 2D CA model decrease, the resolution of simulation increase and more detailed morphologic information can be achieved. If proper parameters are chosen, the cell length can be increased and the computation efficiency improved only with less precision for the CA model.

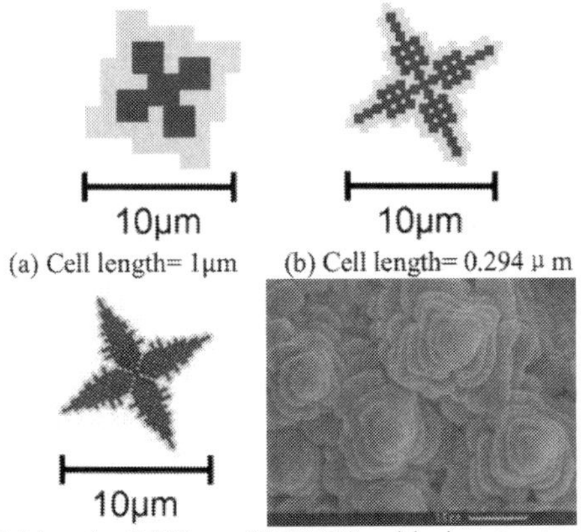

(a) Cell length= 1μm (b) Cell length= 0.294 μ m

(c) Cell length= 0.067μm (d) DP600 weld microstructure [27]

Figure 9. Dendrite morphology of 2D simulations and experimental SEM observation

3.6 Microstructure of laser welding

In Figure 10, a laser welding experiment is done according to the parameters described in Section 2.1. The corresponding simulation results are also presented in Figure 10.

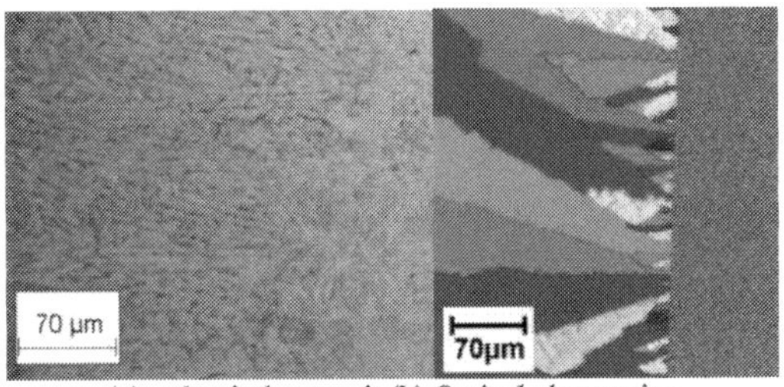

(a) simulation result (b) Optical observation

Figure 10. Microstructure of FZ and HAZ of DP600 weld

Using the temperature interpolation technique, a part of weld zone is simulated and compared with the optic observation. In a cross-section of weld simulation, orientations of dendrites are mainly from the edge to centre of the weld according well with experiment. At the beginning edge of molten pool, dendrites have random growth orientations. Under weld temperature distribution and with dendrites interaction, only grains with preferential orientations succeed to grow up. The final simulation result is similar to experiment observation except that the width of simulation dendrite still need to be improved.

4 Conclusions

A 2D Finite Element Cellular Automaton model for simulating microstructure evolution of DP600 by laser welding is developed. The chosen temperature interpolation method can successfully combine the simulation model of FEM with the microstructure model of Cellular Automaton. With the interpolation method, local microstructure of weld can be simulated based on the FEM temperature simulation results.

The CA model is capable to simulate equiaxed dendrite growth with one nucleation in the centre of simulation zone using the isotropic temperature distribution. The interaction of secondary and tertiary arms of equiaxed dendrites is also presented by the model. Simulation results are compared with experimental observation in the weld, and similar morphology is found.

The CA model is also able to simulate columnar dendrite growth with several nucleations at the edge of simulation zone with the isotropic temperature distribution. With interactions of dendritic arms between different dendrites, the simulated dendrites grow with the same velocity toward the same direction until the whole simulation zone is taken up.

The above FE-CA model coupling can predict the weld microstructure with random nucleations and weld temperature distribution. The results of simulation are compared with experimental observation and a good accordance is found. The FE-CA model can also successfully simulation local microstructure of weld based on the FEM temperature output data, which is important to further computation of grain orientation value. This work of 2D FE-CA coupling is just a beginning of a general model applied for laser welding simulation.

Acknowledgement

The work is financially supported by the China Scholarship Council and the laboratory LGCGM of INSA de Rennes.

References

1. G. Davies, Materials for automobile bodies (Elsevier,Butterworth-Heinemann, 2012)
2. M. Calcagnotto, Y. Adachi, D. Ponge, D. Raabe, Acta Materialia, 59, 658 (2011)
3. D. Chen, Z. Wang, X. Jiang, S. Ai, C. Shih, Mater. Sci. Eng., A, 108, 141 (1989)
4. S. Kuang, Y.l. Kang, H. Yu, R.d. Liu, Int J. Miner. Metall. Mater. 16, 159 (2009)
5. J. Fekete, A. Stibich, IBEC, 01, 3101 (2001)
6. O. Bouaziz, H. Zurob, M. Huang, Steel Res. Int. 84, 937 (2013)
7. B. Duggan, C.S. Lee, R. Smallman, Mater. Sci. Forum, 157, 1759 (1994)
8. B. Duggan, Y. Tse, H. Ning, M. Quadir, Mater. Sci. Forum, 408, 1151 (2002)
9. M. Arminjon, B. Bacroix, Acta Mechanica, 88, 219 (1991)
10. D. Rosenthal, ASME, (1946)
11. V. Pavelic, R. Tanbakuchi, O. Uyehara, P. Myers, Weld. J. 48, 295 (1969)
12. P. Tekriwal, J. Mazumder, Weld. J. 67, 150 (1988)
13. E. Friedman, J. Pres. Ves. Tech. 97, 206 (1975)
14. Z. Paley, P. Hibbert, Weld. J. 54, 385s (1975)
15. J. Goldak, A. Chakravarti, M. Bibby, Metall. Trans. B, 15, 299 (1984)
16. C. Wu, H. Wang, Y. Zhang, Weld. J. New. 85, 284 (2006)
17. M. Rappaz, C.A. Gandin, Acta Metall. Mater. 41, 345 (1993)
18. L. Nastac, Acta Mater. 47, 4253 (1999)
19. W. Wang, P.D. Lee, M. McLean, Acta Mater. 51, 2971 (2003)
20. Y.H. Wei, X.H. Zhan, Z.B. Dong, L. Yu, Sci. Tech.Weld. Join. 12, 138 (2007)
21. X. Zhan, Y.Wei, Z. Dong, J. Mater. Proce. Tech. 208, 1 (2008)
22. Y. Liu, Q. Xu, B. Liu, Tsinghua Sci. Tech. 11, 495 (2006)
23. M.F. Zhu, D.M. Stefanescu, Acta Mater. 55, 1741 (2007)
24. L. Beltran-Sanchez, D. Stefanescu, Metall. Mater. Trans. A, 34, 367 (2003)
25. P. Thevoz, J. Desbiolles, M. Rappaz, Metall. Trans. A 20, 311 (1989)
26. X.h. Zhan, Z.b. Dong, Y.h. Wei, Y. Wang, Chinese J. Nonferr. Metals. 19, 1431 (2009)
27. M. Marya, X.Q. Gayden, Welding Journal, 84, 172 (2005)

Compaction simulation of nano-crystalline metals with molecular dynamics analysis

A.R. Khoei [a], A. Rezaei Sameti, H. Mofatteh and M. Babaei

Center of Excellence in Structures and Earthquake Engineering, Department of Civil Engineering, Sharif University of Technology, P.O. Box. 11365-9313, Tehran, Iran

Abstract. The molecular-dynamics analysis is presented for 3D compaction simulation of nano-crystalline metals under uniaxial compaction process. The nano-crystalline metals consist of nickel and aluminum nano-particles, which are mixed with specified proportions. The EAM pair-potential is employed to model the formation of nano-particles at different temperatures, number of nano-particles, and mixing ratio of Ni and Al nano-particles to form the component into the shape of a die. The die-walls are modeled using the Lennard-Jones inter-atomic potential between the atoms of nano-particles and die-walls. The forming process is model in uniaxial compression, which is simulated until the full-dense condition is attained at constant temperature. Numerical simulations are performed by presenting the densification of nano-particles at different deformations and distribution of dislocations. Finally, the evolutions of relative density with the pressure as well as the stress-strain curves are depicted during the compaction process.

1 Introduction

Powder metallurgy is assumed as a method of producing the near-to-net shape industrial components from the loose powders by executing the pressure with or without applying the heat. In this process, the densification of powder is in accordance with the structural features, e.g. the hardness, plastic behavior, and surface characteristics, and the geometrical features of powder, e.g. the particle size, shape, and distribution **[1]**. A remarkable privilege of this method is the facility in mixing various types of powder materials to obtain the required characteristics **[2]**. This capability of powder metallurgy enables one to fabricate the products with variant mechanical features.

In compaction simulation of metal powders, the size of particles dramatically affects the densification process and varies from the nano- to micro-scales. Basically, the discrete (micro) and continuum (macro) techniques are developed to model the powder compaction process **[3–8]**. In discrete (micro) techniques, there have been taken several simplifying assumptions into the computation to alleviate the problem complexity and computational cost **[3–5]**. However in the continuum (macro) models, the use of appropriate plasticity models have been a great interest in modeling the micro-powder forming **[6–8]**. In nano-scale particles, the ratio of surface atoms to volume atoms is significantly considerable in comparison to the micro or macro-particles **[9]**. Thus, it can be expected that the behavior of nano-scale particles in compaction process can differ noticeably from the behavior of micro-scale particles that highlight the necessity of simulating the nano-particles compaction individually.

There are several numerical simulations reported in literature for modeling of the nano-powder compaction. Boltachev et al. **[10]** employed the granular dynamics method to simulate the quasi-static uniaxial compaction of nano-powders. They investigated the efficiency of particles size in the quasi-static compaction of nano-powders using the granular dynamic method **[11]**. Suryanarayanan and Sastry **[12]** developed a theoretical model on the basis of a micro-mechanism model to simulate the densification of nano-particles under uniaxial and hydrostatic pressing. Balakrishnan et al. **[13]** simulated the consolidation of nano-particles by applying the discrete element method (DEM).

One of the most popular techniques proposed by researches to study the behavior of nano-particles is based on the molecular dynamics (MD) method **[14–19]**. In this regard, Henz et al. **[14]** performed the MD analysis to verify the energetic reaction of the nickel and aluminum nano-particles. In their study, two distinct models were considered for Ni and Al nano-particles, including the separated nano-particles and coated nano-particles. Zhang et al. **[16]** investigated the behavior of copper nano-particles in the die-compaction process by applying the MD method. They studied the compaction behavior for various particle sizes, temperatures, and packing arrangements, and deduced that these parameters can affect the mechanical properties of the compacted components. Stone et al. **[17]** determined the coefficients of inter-particle friction under the uniaxial compression by applying the MD technique. Kiselev **[18, 19]** simulated the exertion of external force on the spherical piston comprised of copper and copper-molybdenum mixture

[a] Corresponding author: arkhoei@sharif.edu

nano-particles with MD analysis. They indicated that because of the surface traction, the copper nano-particles fill the voids and the crystalline structures change into amorphous forms under a small movement of the piston.

Considering the extremely nonlinear behavior of nano-particles through the process of compaction, it is required to implement a powerful technique in compaction simulation of nano-crystalline metals. In this study, the molecular dynamics technique is employed to study the forming process of nano-particles under die-pressing. The nano-crystalline metals consist of nickel and aluminum nano-particles. The simulation of forming process is accomplished by imposing the uniaxial compression on a set of nano-particles at the constant temperature. In order to study the densification behavior of combined Ni/Al nano-particles, numerical simulations are performed for various mixing ratios of Ni and Al nano-particles. Moreover, the effect of temperature and number of nano-particles are investigated through the compaction process. The plan of the paper is as follows; in Section 2, the methodology and desired numerical simulations are described. The geometrical characteristics of specimens together with the inter-atomic potential and the procedure of modeling are described in this section. In Section 3, the results of computational modeling are presented by illustrating the evolutions of stress versus strain and the relative density versus pressure for various parameters of forming process.

2 Methodology

The compaction simulation of metal nano-particles is performed using three-dimensional uniaxial compression on a set of metallic powders consisting of nickel and aluminum nano-particles with various mixing ratios. In order to investigate the influence of the number of nano-particles, three types of nano-particles arrangement, including 4, 16 and 64 nano-particles, are considered. To simulate the formation of nano-particles, the die-walls are positioned as the rectangular prism with the interior dimensions of $10.7\times10.7\times5.5\ nm^3$, $21.1\times21.1\times5.5\ nm^3$ and $41.9\times41.9\times5.5\ nm^3$ corresponding to 4, 16 and 64 nano-particles, respectively. In view of crystallography, the metallic atoms are arranged initially as the F.C.C structure with the lattice constants of 0.352 nm for nickel and 0.405 nm for aluminum nano-particles, respectively. Each nano-particle is considered as a spherical shape with the diameter of 5 nm that contains about 4000 atoms of aluminum and 6000 atoms of nickel. In order to investigate the performance of various mixing ratios in die-forming process, the nano-particles are combined with five distinct ratios of Ni and Al as follows; (1) pure Ni nano-particles, (2) pure Al nano-particles, (3) mixed 25%Ni–75%Al, (4) mixed 50%Ni–50%Al, (5) mixed 75%Ni–25%Al, in which for all mixed Ni–Al, the positions of nano-particle are chosen randomly. The initial configurations of components with 4, 16 and 64 nano-particles for the mixed 50%Ni–50%Al are shown in Figure 1. The nano-particles are set at the distance of 0.2 nm from each other, and 0.25 nm from the die-walls.

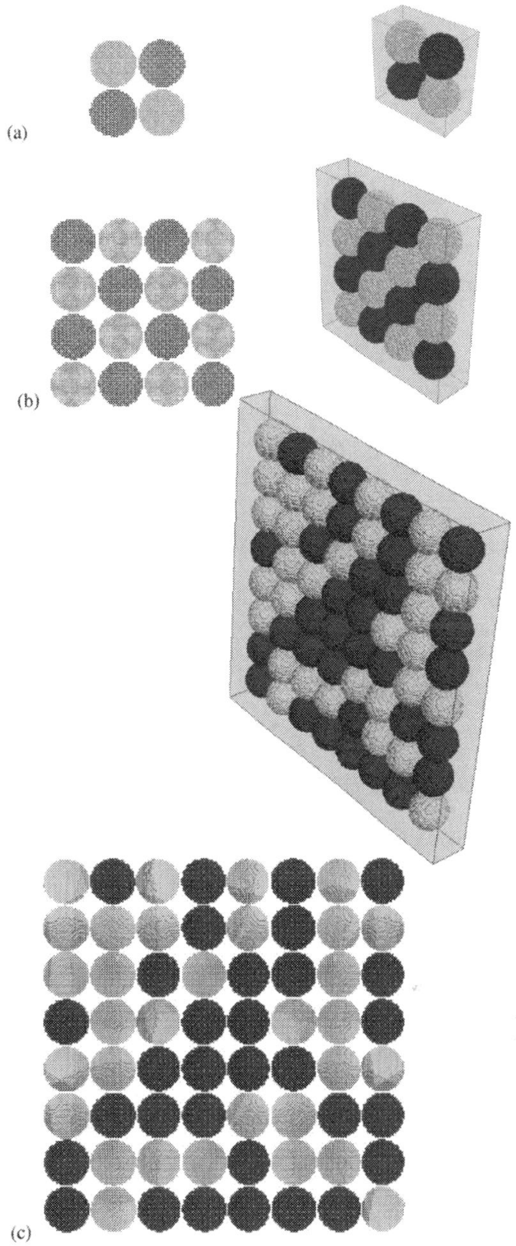

Figure 1. The initial configurations of Al and Ni nano-particles for the mixed 50%Al–50%Ni with (a) 4, (b) 16 and, (c) 64 nano-particles (the blue atoms corresponds to Al and green atoms corresponds to Ni).

It has been shown that the inter-atomic potential of the embedded-atom method (EAM) can be applied efficiently in molecular dynamics analysis of Ni, Al and mixed Ni–Al atoms with sufficient accuracy **[15, 20, 21]**. Due to versatility and accuracy of the EAM inter-atomic potential in modeling of large atomic systems, Henz et al.

[15] studied the sintering of Ni and Al nano-particles using the EAM potential. Olmsted et al. [20] employed the EAM potential to investigate the velocity of edge and screw dislocations in the Al and Ni materials. Zhang et al. [21] simulated the Ni–Al interface diffusion using the EAM potential. In the current study, the inter-atomic interaction is defined based on the EAM potential, which is basically expressed as [22, 23]

$$E_{\text{tot}} = \sum_{i} F_i(\rho_i) + \frac{1}{2}\sum_{i \neq j} \Phi_{ij}(R_{ij}) \qquad (1)$$

where E_{tot} is the total energy of the atomic system, ρ_i is the total host electron on the atom i, R_{ij} is the distance between atom i and atom j, Φ_{ij} is the pair inter-atomic potential, and F_i is the embedding energy of the atom i. The constants of the EAM potential were obtained for Ni and Al by Pun and Mishin [24]. The interaction between the die-walls and the atoms of nano-crystalline particles are characterized by the Lennard-Jones inter-atomic potential. In order to alleviate the effect of compaction velocity, the top-punch movement is assumed with the speed of 0.4 Å/picosec in the vertical direction. The compaction process is followed until all voids between nano-particles are vanished and the full-dense condition is achieved. The compaction simulation is performed under the canonical ensemble in which the temperature of system is preserved nearly constant. To manifest the influence of system temperature on the forming process, all numerical simulations are performed in two distinct values of temperature, i.e. 300 and 600 K. To accomplish the numerical simulations, the open-source LAMMPS software is utilized [25].

In order to trace the distribution of dislocation in nano-particles, the parameter of centro-symmetry is computed over the component at different stages of the compaction process. This parameter is defined as [26]

$$CS_i = \sum_{j=1}^{N/2} \left| \mathbf{R}_{ij} + \mathbf{R}_{i(j+N/2)} \right|^2 \qquad (2)$$

in which CS_i is the centro-symmetry parameter for the atom i, N is the number of ith atom neighbors, and $\mathbf{R}_{ij}$ is the vector of atom i to atom j. This parameter is approximately zero for atoms in the perfect F.C.C structure and has a positive value for the rest of atoms.

3 Numerical simulation results

In the process of powder forming, the nano-particles present the severe nonlinear deformations; so the molecular dynamics analysis is employed to model this nonlinear behavior. Despite the use of parallel processing technique, the maximum number of atoms is limited to 384000 atoms at each specimen due computational costs, which are placed in 64 nano-particles with the diameter of $5\,\text{nm}$. As mentioned earlier, during the compaction process of nano-particles, several parameters, such as the temperature, number of nano-particles, the type of nano-powder, size of nano-particles, arrangement of different nano-particles and etc can have considerable influence in the mechanical properties of compacted component. On the other hand, in order to investigate the effect of each parameter given above, several numerical simulations should be performed. In this study, numerical simulations are performed employing two metallic nano-powders, i.e. the nickel and aluminum nano-particles, at different temperatures, various mixing ratios, and different number of nano-particles.

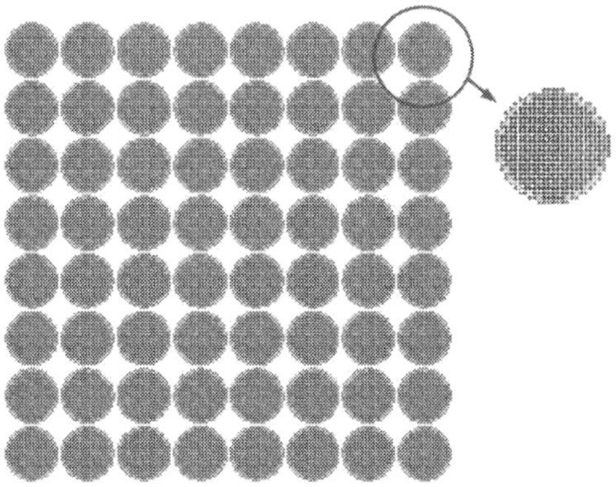
(a)

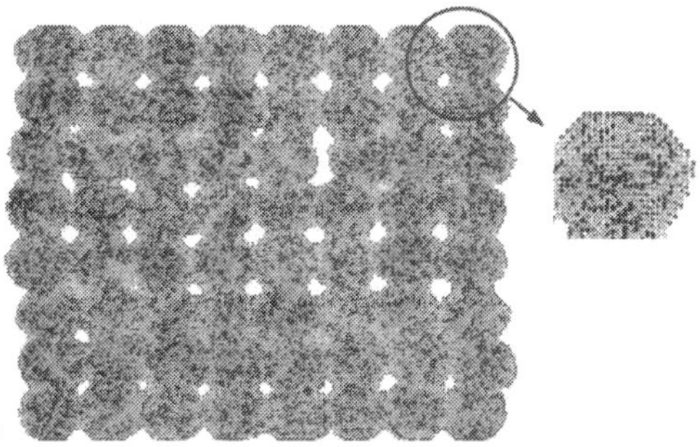
(b)

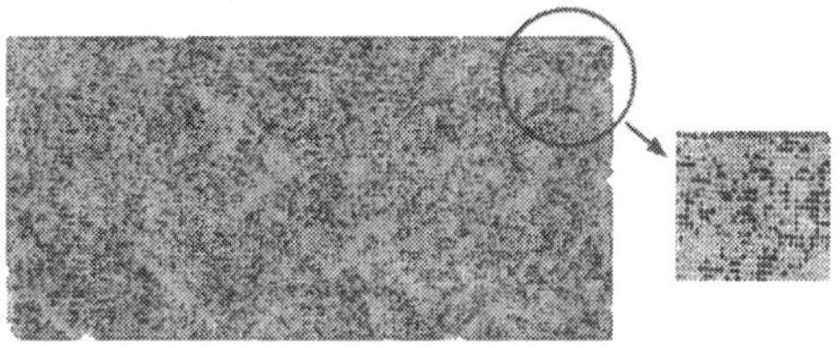
(c)

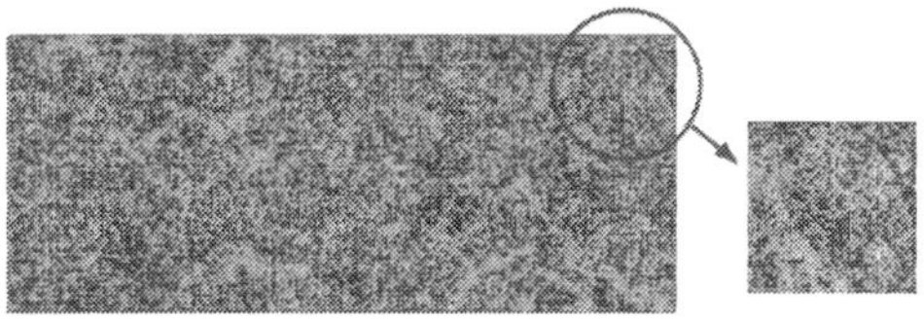
(d)

Figure 2. The snapshots of 64 aluminum nano-particles at different stages of compaction process in the temperature of 300K; (a) initial configuration, (b) 47%, (c) 68% and, (d) 83% of relative density. The colors of atoms are displayed according to the normalized centro-symmetry parameter, in which the bold blue is atoms of the perfect F.C.C structure, the pale blue is atoms of partial dislocations in F.C.C structure, the green is atoms of stacking fault, and the red color is atoms of the surface

In order to track the densification behavior of nano-particles during the compaction process, the snapshots of deformed configurations are illustrated in Figure 2 at different levels of powder die-pressing for aluminum nano-particles. Obviously, by increasing the top punch movement, the nano-particles stick together due to their surface tension, and deform in a way that the voids between nano-particles reduce gradually until the nano-particles form into the shape of the die. In this figure, the colors of atoms are presented according to the normalized centro-symmetry parameter computed with respect to the square of lattice constant for Al nano-particles at different stages of compaction process. In view of the aluminum crystallography, the normalized centro-symmetry value is zero for atoms of the perfect F.C.C structure displayed by the bold blue, however for atoms of partial dislocations in the F.C.C structure, it is almost between 0.03 and 0.18 (pale blue), for atoms of the stacking faults it is between 0.18 and 0.96 (green color), and for atoms of the surface it is more than 0.96 (red color) [27]. Obviously, the distribution of dislocations can be seen in this figure at different levels of die-forming process. The presence of dislocations clearly demonstrates the plastic behavior of nano-particles through the densification process.

In order to quantitatively demonstrate the nonlinear behavior of nano-particles during the forming process, the evolutions of the compaction pressure, relative density, the mean-stress of atoms in y–direction, and the strain of nano-particles are computed through the compaction process for each top punch movement of 0.2 nm, and the results are plotted in the form of the relative density – pressure and stress–strain graphs. In Figures 3 and 4, the evolutions of relative density versus pressure and stress versus strain are plotted for a set of nine specimens, including the 4, 16, and 64 nano-particles of the pure Al, pure Ni, and mixed Al/Ni with mixing ratio of 50%Al–50%Ni at the temperature of 300K. It can be observed that for the relative density less than 75 percents, the graphs of relative density with 4, 16 and 64 nano-particles are almost identical. In fact, up to 75 percents of the relative density, the difference in the number of nano-particles have no effect on the values of density. The reason of this phenomenon can be revealed from the corresponding stress-strain curves, as shown in Figure 4. Obviously, up to 75 percents of the relative density, the values of mean-stress in y–direction are approximately zero. It indicates that in this phase of forming process, the nano-particles do not experience a considerable deformation, and the compaction process is performed mainly by rearranging the nano-particles and decreasing the voids in the component. Evidently after this phase of forming process, the values of mean-stress of component develop rapidly and the densification curves corresponding to compaction of 4, 16 and 64 nano-particles differ from each other. Obviously, in the last phase of forming process when the material becomes harden, the compaction pressure and the mean-stress value of compacted component increase as the number of nano-particles increases. Clearly, increasing the number of nano-particles results in a considerable discrepancy in the densification behavior corresponding to compaction of 4 and 16 nano-particles, while the graphs obtained from the compaction of 16 and 64 nano-particles represent a slight difference. Thus, it can be concluded that an appropriate number of 16 and/or 64 nano-particles results in a desired solution for the evolutions of relative density versus pressure and stress versus strain. In this study, the 16 nano-particles are used efficiently to capture the densification behavior of the pure Al, pure Ni, and mixed Al/Ni with various mixing ratios at the temperature of 300K. Moreover, as it can be expected from Figures 3 and 4, the compaction pressure and the mean-stress value of the nickel nano-powders are clearly greater than those values of the aluminum nano-powders. In addition, the graphs of mixed Al/Ni with mixing ratio of 50%Al–50%Ni are obtained in the region between the graphs of pure Al and pure Ni nano-particles.

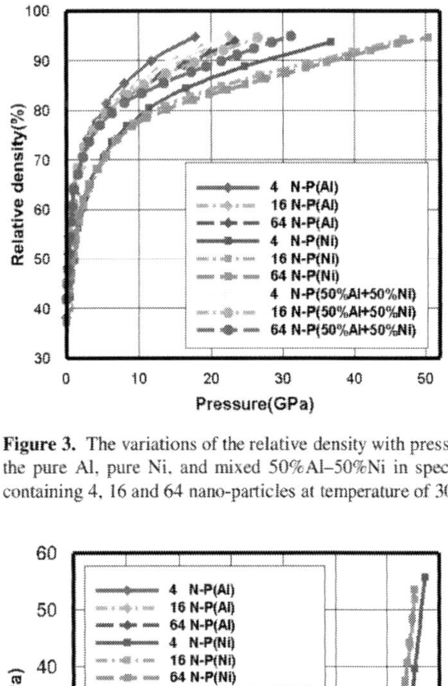

Figure 3. The variations of the relative density with pressure of the pure Al, pure Ni, and mixed 50%Al–50%Ni in specimens containing 4, 16 and 64 nano-particles at temperature of 300K

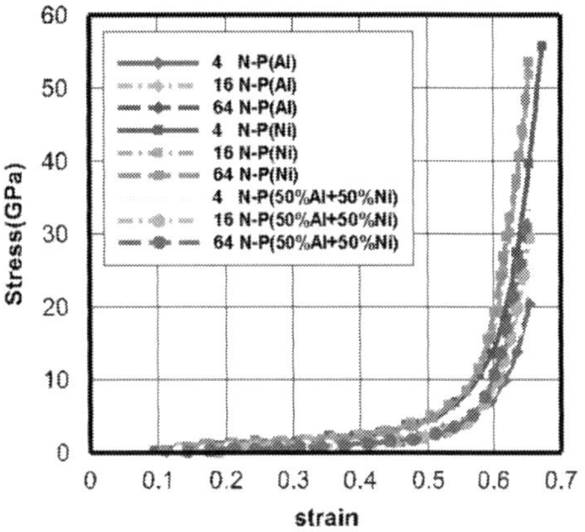

Figure 4. The stress – strain curves of the pure Al, pure Ni, and mixed 50%Al–50%Ni in specimens containing 4, 16 and 64 nano-particles at temperature of 300K

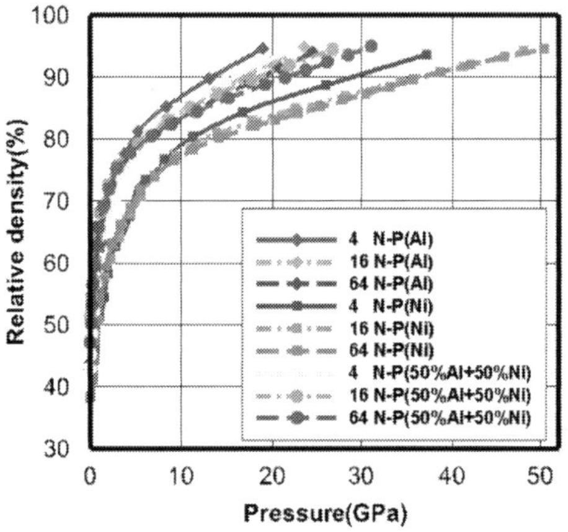

Figure 5. The variations of the relative density with pressure of the pure Al, pure Ni, and mixed 50%Al–50%Ni in specimens containing 4, 16 and 64 nano-particles at temperature of 600K

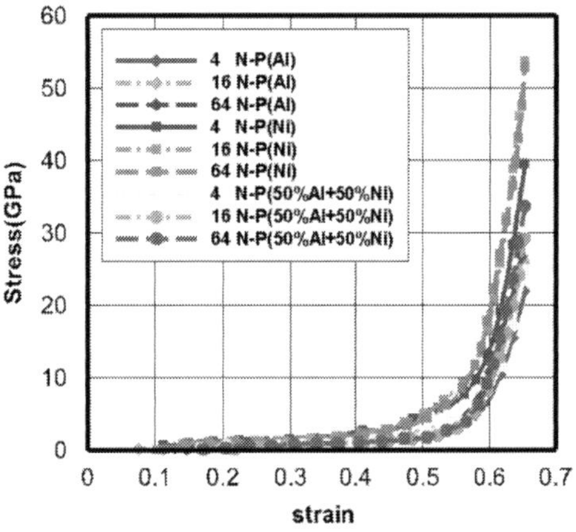

Figure 6. The stress – strain curves of the pure Al, pure Ni, and mixed 50%Al–50%Ni in specimens containing 4, 16 and 64 nano-particles at temperature of 600K

In Figures 5 and 6, the variations of relative density with pressure as well as the evolutions of stress versus strain are plotted at the temperature of 600K for different cases of pure Al, pure Ni, and mixed Al/Ni with mixing ratio of 50%Al–50%Ni. A comparison between the graphs plotted in these figures for the temperature of 600K and those reported in Figures 3 and 4 for the temperature of 300K reveals that the thermal compaction process affects the densification behavior of nano-powders. Basically, the trends of stress–strain curves as well as the relative density – pressure graphs are similar for the two temperatures of 300 and 600 K, however, the temperature of the specimen moderately affects the values of compaction pressure and the mean-stress of the component. Obviously, by increasing the temperature, the velocity and the thermal vibration of the atoms increase in the component, which result in an increase of the inter-atomic forces and the atomistic stresses.

Accordingly, it can be deduced from the graphs of relative density versus pressure and stress versus strain obtained from molecular dynamics analyses that the compaction pressure and the mean-stress of the atoms increase during the forming process of nano-powders by increasing the temperature of the system. Since there are two different metallic nano-powders, i.e. the nickel and aluminum nano-particles, proposed in the numerical simulations, the effect of temperature is not identical on the densification behavior of various specimens. As can be observed from Figures 3–6, the specimens containing the aluminum nano-powders are more sensitive to the temperature in comparison with the specimens containing nickel nano-powders. On the other hand, the nickel nano-particles display more resistant against the increasing of temperature in comparison with the aluminum nano-particles during the compaction process.

In order to investigate the densification behavior of various combinations of Ni and Al nano-particles during the powder forming process, numerical simulations are performed for five mixing ratios, i.e. the pure Al, pure Ni, mixed 75%Al–25%Ni, mixed 50%Al–50%Ni, and mixed 25%Al–75%Ni nano-particles. In Figures 7 and 8, the evolutions of relative density versus compaction pressure and the normal stress versus normal strain are plotted during the compaction simulations of 64 nano-particles at the temperature of 300K for various mixing ratios. As can be seen from these figures, by increasing the percentage of nickel nano-particles, the value of compaction pressure as well as the mean-stress value of component increase, for which the maximum values of pressure and stress are obtained for the pure Ni nano-particles. It is obvious from Figure 7 that the mixed Al/Ni with higher percentage of aluminum has a tendency to the pure Al curve, in a way that the difference between the curve of pure Al and that of mixed 75%Al–25%Ni nano-particles are negligible. In contrast, there is a remarkable gap between the graph obtained from the mixed 25%Al–75%Ni and that of the pure Ni nano-particles. This trend can be justified using the material properties of nickel and aluminum.

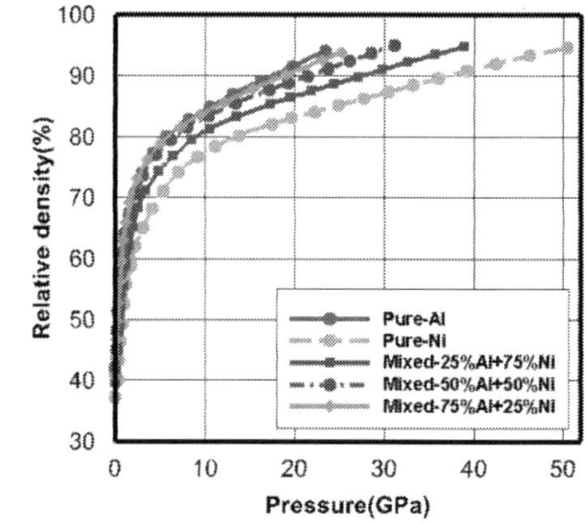

Figure 7. The variations of relative density with pressure of the pure Al, pure Ni, 75%Al–25%Ni, 50%Al–50%Ni, and 25%Al–75%Ni in specimen of 64 nano-particles at temperature of 300K

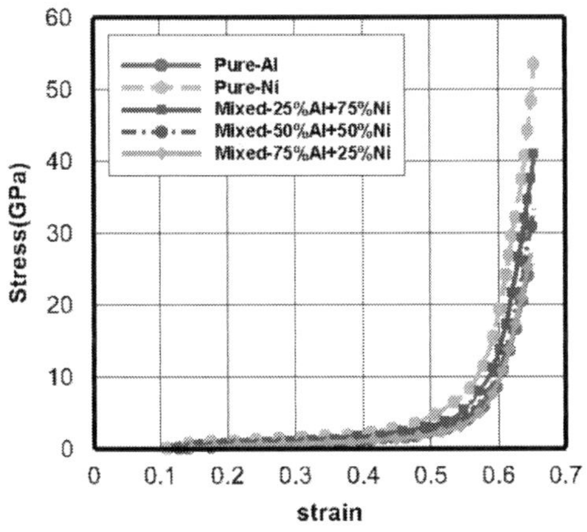

Figure 8. The stress – strain curves of the pure Al, pure Ni, mixed 75%Al–25%Ni, 50%Al–50%Ni, and 25%Al–75%Ni in specimen of 64 nano-particles at temperature of 300K

In view of the mechanical behavior, nickel has a greater stiffness in comparison with aluminum, and has a more resistance against the deformation. Hence, in the compaction of Ni/Al nano-particles, the deformation of aluminum nano-particles is dominant and as a result, the corresponding graph has a tendency to the pure Al curve.

In order to investigate the temperature effect during the compaction process of mixed Ni/Al nano-particles, numerical simulations are performed at the temperature of 600K. In Figures 9 and 10, the variations of relative density with compaction pressure are plotted together with the evolutions of normal stress versus normal strain for various mixing ratios of the pure Al and pure Ni, mixed 75%Al–25%Ni, mixed 50%Al–50%Ni, and mixed 25%Al–75%Ni nano-particles. It can be observed from various combinations of Ni/Al nano-particles that by increasing the temperature of the system, the values of stress and pressure increase during the compaction process. Note that although the temperature has different effects on the behavior of nickel and aluminum, it can be seen from Figures 9 and 10 that the influence of temperature is more important in the cases of pure Al and mixed 75%Al–25%Ni, and the increase of temperature does not make a considerable effect on the stress–strain and relative density – pressure graphs of the remaining mixed Ni/Al combinations. In Figure 11, the contours of normal stress are presented for various mixing ratios of 64 nano-particles at the final stage of compaction process in the temperature of 300K. Obviously, the value of normal stress increases as the percentage of nickel nano-particles increases in the component for which the maximum stress value can be observed for the pure Ni nano-particles. Finally, the distributions of dislocations are shown in Figure 12 for various mixing ratios of 64 nano-particles at the relative density of 85%. It can be seen that the number of dislocations increases as the percentage of nickel nano-particles increases for which the maximum number of dislocations is obtained for the pure Ni nano-particles.

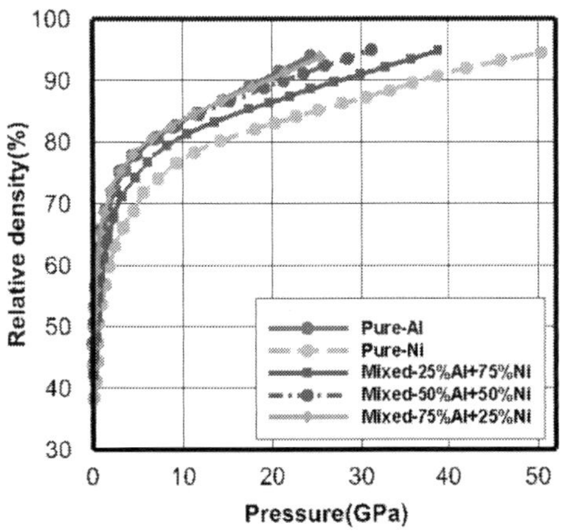

Figure 9. The variations of the relative density with pressure of the pure Al, pure Ni, mixed 75%Al–25%Ni, mixed 50%Al–50%Ni, and mixed 25%Al–75%Ni in the specimen of 64 nano-particles at the temperature of 600K

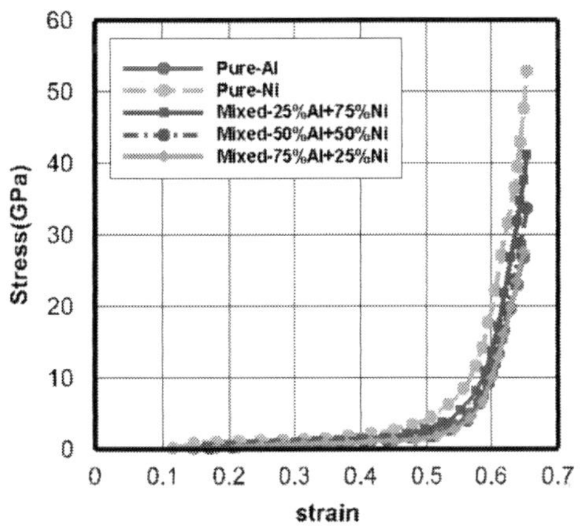

Figure 10. The stress – strain curves of the pure Al, pure Ni, mixed 75%Al–25%Ni, mixed 50%Al–50%Ni, and mixed 25%Al–75%Ni in the specimen of 64 nano-particles at the temperature of 600K

4 Conclusion

In the present paper, the three-dimensional compaction simulation of nano-crystalline metal powders was presented based on the molecular dynamics method. The EAM pair-potential was used to model the forming process of nano-particles at different temperatures, various number of nano-particles, and different mixing ratios of nickel and aluminum. The Lennard-Jones inter-atomic potential was utilized to model the interaction between the atoms of nano-particles and die-walls. The forming process was model in uniaxial compression, and the compaction process was followed until the full-dense condition was attained at the constant temperature.

DOI: 10.1051/matecconf/20168002011

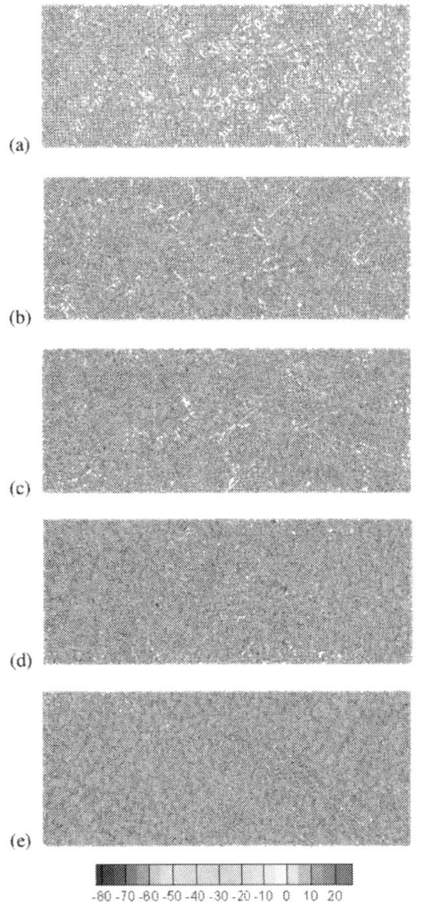

Figure 11. The contours of normal stress (GPa) for 64 nano-particles in the temperature of 300K at the final stage of compaction process; (a) pure Al, (b) mixed 75%Al–25%Ni, (c) mixed 50%Al–50%Ni, (d) mixed 25%Al–75%Ni, (e) pure Ni

The numerical simulations were performed on a set of specimens with different number of particles containing the pure Ni, pure Al, and mixed Ni/Al nano-powders under the uniaxial compression. In order to study the influence of various parameters in the compaction process of nano-powders, the numerical simulations were carried out with different number of nano-particles, various temperatures, and different mixing ratios of Ni/Al. The distribution of dislocations was traced by presenting the contours of centro-symmetry parameter at different levels of the compaction process; it was shown that the voids between the nano-particles reduce, the material becomes harden, and the plasticity behavior of nano-particles plays an important role in the forming process. It was shown that increasing the number of nano-particles results in a significant discrepancy in the densification behavior corresponding to compaction of 4 and 16 nano-particles, however, the results have a slight difference in the compaction of 16 and 64 nano-particles.

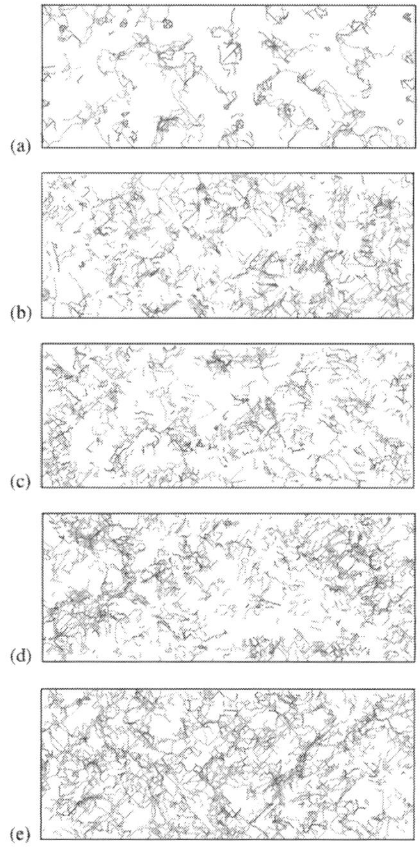

Figure 12. The distributions of dislocations for 64 nano-particles in the temperature of 300K at the powder compaction of 85%; (a) pure Al, (b) mixed 75%Al–25%Ni, (c) mixed 50%Al–50%Ni, (d) mixed 25%Al–75%Ni, (e) pure Ni

It has been concluded that an appropriate number of 16 nano-particles can be used to obtain a desired solution for the evolutions of relative density versus pressure and stress versus strain. Furthermore, it was shown from molecular dynamics analyses that by increasing the temperature of the system the compaction pressure and the mean-stress of the atoms increase during the forming process of nano-powders. It was observed that the nickel nano-particles present more resistant against the increase of temperature comparing to the aluminum nano-particles during the compaction process. In addition, it was shown from the evolutions of compaction pressure and the mean-stress of compacted component that the mixed Al/Ni nano-particles with higher percentage of aluminum has a tendency to the pure Al behavior, in which the difference between the curve of pure Al and that of mixed 75%Al–25%Ni are negligible. In fact, it was observed that in the compaction of mixed Ni/Al nano-particles, the deformation of aluminum nano-particles is dominant and as a result, the corresponding result has a tendency to the pure Al behavior. The numerical simulations clearly demonstrated that the molecular dynamics method is capable to model the compaction process of nano-crystalline metal powders.

References

1. A.R. Khoei, *Computational Plasticity in Powder Forming Process*, (Elsevier 2005)
2. J.H. Cho, K.T. Kim, Int. J. Mech. Sciences **43**, 921 (2001)
3. R.S. Ransing, D.T. Gethin, A.R. Khoei, P. Mosbah, R.W. Lewis, Mater. Design **21**, 263 (2000)
4. P. Redanz, N.A. Fleck, Acta Mater. **49**, 4325 (2001)
5. J.F. Jerier, B. Hathong, V. Richefeu, B. Chareyre, D. Imbault, F.V. Donze, P. Doremus, Powder Tech. **208**, 537 (2011)
6. A.R. Khoei, S.O.R. Biabanaki, S.M. Parvaneh, Appl. Math. Model. **37**, 443 (2013)
7. A.R. Khoei, A.R. Azami, Int. J. Mech. Sciences **47**, 94 (2005)
8. A.R. Khoei, H. DorMohammadi, Comp. Mater. Sci. **41**, 1 (2007)
9. T. Pu, Annu. Rep. Prog. Chem. Sect. C: Phys. Chem. **104**, 142 (2008)
10. G.S. Boltachev, K.E. Lukyashin, V.A. Shitov, N.B. Volkov, Phys. Rev. E **88**, 012209 (2013)
11. G.S. Boltachev, N.B. Volkov, Technic. Phys. Let. **36**, 823 (2010)
12. R. Suryanarayanan, S.M.L. Sastry, Acta Mater. **47**, 3079 (1999)
13. A. Balakrishnan, P. Pizette, C.L. Martin, AIP Conf. Proc. **1145**, 105 (2009)
14. B.J. Henz, T. Hawa, M. Zachariah, J. Appl. Phys. **105**, 124310 (2009)
15. B.J. Henz, T. Hawa, M. Zachariah, Molecul. Simul. **35**, 804 (2009)
16. Y.W. Zhang, P. Liu, C. Lu, Acta Mater. **52**, 5105 (2004)
17. T.W. Stone, M.F. Horstemeyer, Y. Hammi, P.M. Gullett, Acta Mater. **56**, 3577 (2008)
18. S.P. Kiselev, J. Appl. Mech. Technic. Phys. **48**, 412 (2007)
19. S.P. Kiselev, J. Appl. Mech. Technic. Phys. **49**, 712 (2008)
20. D.L. Olmsted, L.G. Hector, W.A. Curtin, R.J. Clifton, Model. Simul. Mat. Sci. Eng. **13**, 371 (2005)
21. C. Zhang, H. Wang, Y. Qiu, Eng. **3**, 227 (2011)
22. M.S. Daw, M.I. Baskes, Phys. Rev. Let. **50**, 1285 (1983)
23. M.S. Daw, M.I. Baskes, Phys. Rev. B **29**, 6443 (1984)
24. G.P. Pun, Y. Mishin, Philosoph. Mag. **89**, 3245 (2009)
25. LAMMPS Molecular Dynamics Simulator Code. http://lammps.sandia.gov
26. C.L. Kelchner, S.J. Plimpton, J.C. Hamilton, Phys. Rev. B **58**, 11085 (1998)
27. B. Tang, R. Yang, Chin. J. Phys. **53**, 120802 (2015)

Prediction of Mean Grain Size Evolution during the Forging of IN718 Turbine Disc

Zhang Haiyan[1, a], Zhang Shihong[2] and Cheng Ming[2]

[1] *School of Mechanical Engineering, Ningbo University of Technology, Zhejiang, Ningbo, 315016, China*
[2] *Institute of Metal Research, Chinese Academy of Sciences, Shenyang, Liaoning, 110016, China*

Abstract. The microstructure evolution during the cogging, upsetting and hot die forging in the manufacture of IN718 turbine disc has been predicted through coupling the microstructure evolution prediction system developed with a commercial finite element software. The prediction results were in good agreement with the actual ones. Furthermore, the effects of process parameters on the microstructure development during the hot die forging of IN718 turbine disc have been investigated. To design the hot die forging, a uniform function of microstructure has been proposed as the goal function.

1 Introduction

Superalloy IN718 is an important material used for aero-engine turbine disks because of its excellent mechanical properties in the temperature range from -253 to 650°C [1, 2]. To ensure the required mechanical properties, it is crucial to obtain a uniform and fine microstructure. Generally, the sizes of turbine disks are large, and the shapes are complex. In addition, the microstructure of IN718 is sensitive to the hot deformation parameters. Therefore, the defects of coarse grain and duplex grain always appear in the forgings. Figure 1 shows the manufacturing of IN718 turbine disc [3]. At first, the as-cast structure has been changed to as-forged structure during the cogging, and a big forged bar with the diameter lager than 200 mm has been gained. Then, though the subsequent forging processes which mainly included upsetting and closed die forging, the turbine disc forging has been obtained. At finally, the required turbine disc has been obtained by the heat treatment and the machining. As the high alloying, the thermal deformation area of alloy IN718 is narrow, and the hot deformation ability is low. In addition, the structure of alloy IN718 has transmissibility. In order to obtain the turbine disc with a uniform and fine structure, the structure in each process should been controlled in the manufacturing. Thus, it is important to understand the microstructure evolution in each process.

In the present work, the microstructure development during the cogging, upsetting and hot die forging of the turbine disc have been investigated through coupling the microstructure evolution prediction system developed with a commercial FE Software, MSC.Superform.

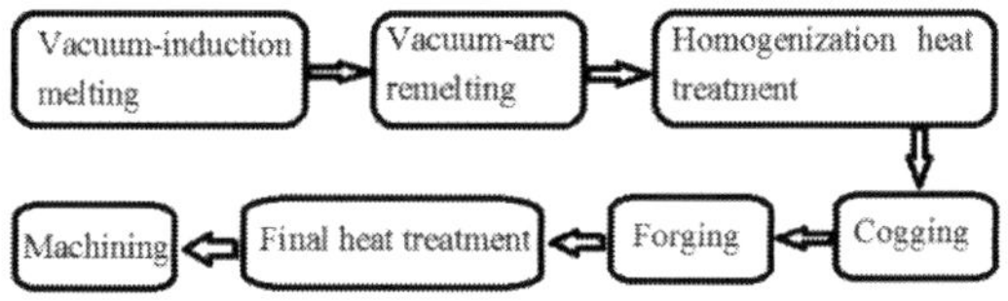

Figure 1. Manufacture of Superalloy turbine disc

2 Microstructure simulation of the cogging

The mechanism of microstructure evolution during the cogging is complex, which includes dynamic recrystallization (DRX), metadynamic recrystallization (MDRX), static recrystallization (SRX) and grain growth up. The microstructure evolution system during the cogging has been developed based on the molds in Ref. [4]. Figure 2 is the calculation flowchart of the microstructure evolution system during the cogging. When the recrystallized volume fraction is higher than 95%, the recrystallization occurs completely, and then the grain grows up in the following deformation. For the partial recrystallization, the mean grain size is the average grain size.

Moreover, the cogging is a complex deformation process, and there are many parameters influencing the deformation. Figure 3 shows the finite element model for the cogging process. The deformation and microstructure development of section A in the middle of the billet have been analyzed. The deformation temperature of billet was 1100°C, and the initial grain size was 2mm. The die temperature was 400°C, and the reduction of the top die was 30mm each pass.

[a] Corresponding author: zhang-hai-yan@hotmail.com

DOI: 10.1051/matecconf/20168002012

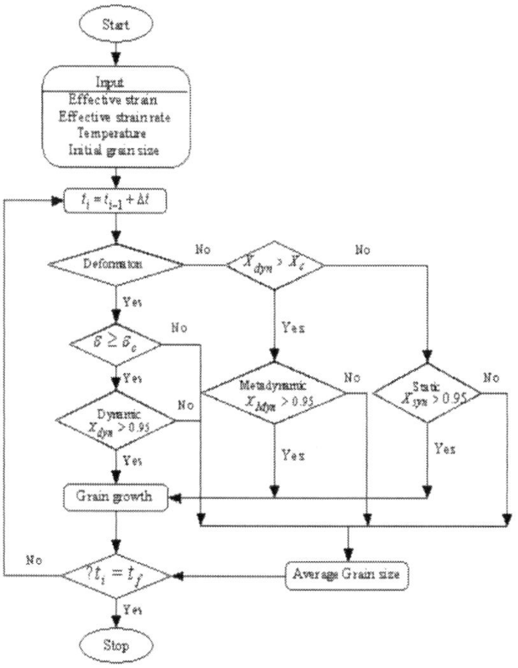

Figure 2. Calculation flow chart of the microstructure evolution system

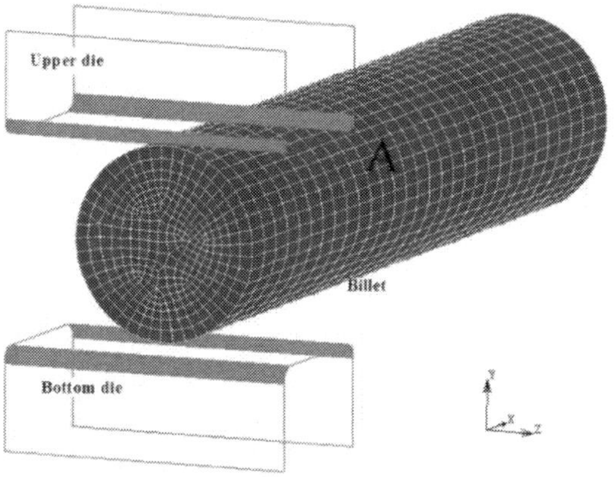

Figure 3. Finite element model for cogging process

Figure 4a-f shows the distribution of temperature, effective strain, DRX fraction, MDRX fraction, SRX fraction and mean grain size in section A at the end of the fourth pass, respectively.

In Figure 4, it showed that the grain in the half radius area in the radial sections was finer due to the deformation condition in favor of the occurrence of recrystallization, and the MDRX was important for the grain refining. Furthermore, as the SRX fraction in the radial sections was higher than the DRX fraction or the MDRX fraction, the SRX was the main microstructure evolution mechanism during the cogging process of IN718 alloy.

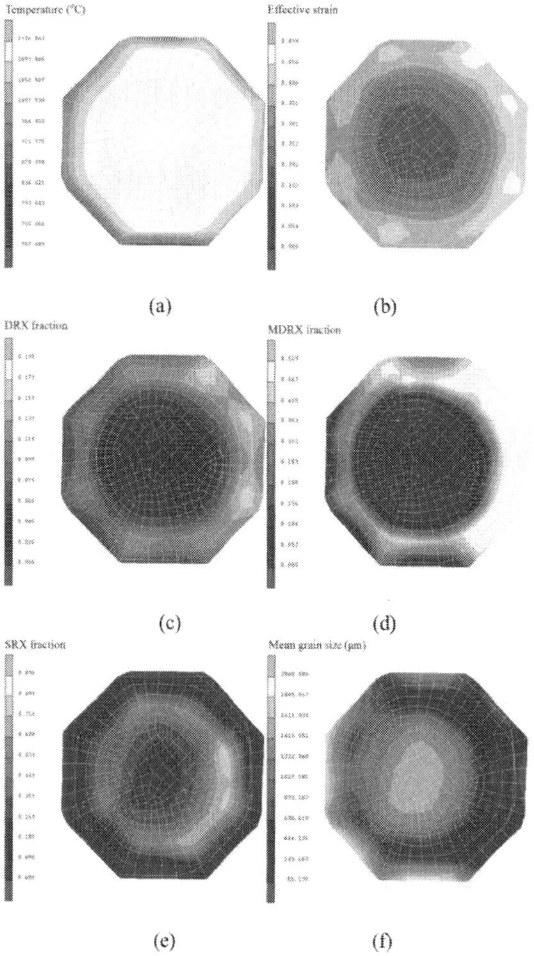

Figure 4. Distribution of temperature (a), effective strain (b), DRX fraction (c), MDRX fraction (d), SRX fraction (e) and mean grain size (f) in section A at the end of fourth pass

3 Microstructure simulation of the upsetting

The original structure of the alloy during the forging is different with that during the cogging. So, the deformation constitutive equation and the microstructure models in terms of critical strain, DRX and grain growth for forged IN718 have been established based on the isothermal compression tests and heat treatment tests. Figure 5 is the calculation flowchart of DRX and grain growth. When the recrystallized volume fraction X_{dyn} is higher than 95%, DRX occurs completely, and the mean grain size is the recrystallized grain size, and then the grain grows up in the following deformation. For the partial dynamic recrystallization, the mean grain size is the average grain size davg, calculated as: $d_{avg}=(1-X_{dyn})d_0+X_{dyn}d_{dyn}$, where d_{dyn} and d_0 are the dynamic recrystallized grain size and initial grain size, respectively.

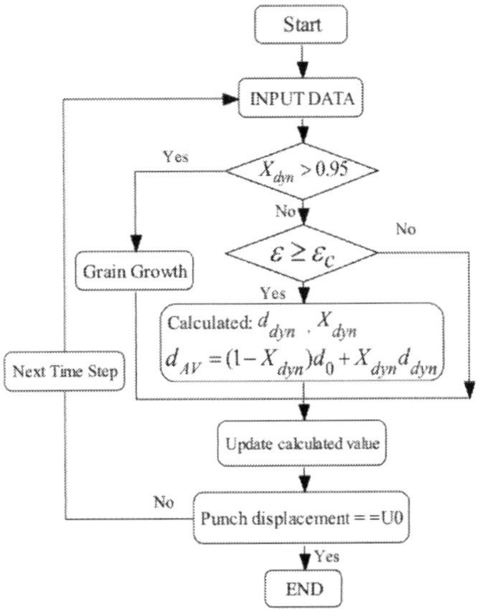

Figure 5. Calculation flowchart of DRX and grain growth

Non-isothermal upsetting experiments have been carried out to refine and validate the microstructure evolution models. Cylindrical specimens with a diameter of 53 mm and a height of 100 mm were used for the upsetting tests. Figure 6 shows the prediction of mean grain size during non-isothermal forging by FEM. The deformation temperature of workpiece was 1030°C, and the initial grain size was 32μm. The die temperature was 300°C, and the workpiece was deformed to 70% reduction in height. It can be clearly identified three deformation zones in Figure 6: dead zone A, free deformation zone B and easy deformation zone C, and. Figure 7 shows the microstructures observed in the corresponding zones of the workpiece. The predicted mean grain size is in good agreement with the observations.

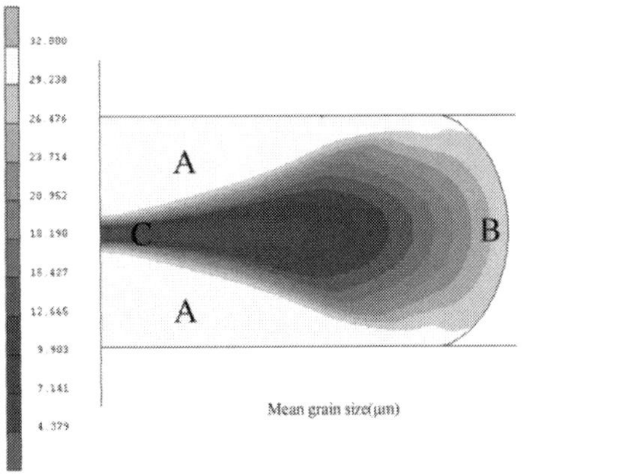

Figure 6. Prediction of mean grain size during Superalloy IN718 non-isothermal forging

Due to the effect of heat exchange and friction with dies, the top and bottom of workpiece are the dead zones during deformation. For the material located in these zones, the DRX takes place hardly (Figure 7a). The mean grain size in zone A after deformed was 45μm, which was larger than the initial grain size of 32μm. The reason is that the workpiece was held for 30 minutes after heated to deformation temperature to make the temperature uniform in the experiment and the original grain grew up during the holding period. As the DRX occurs completely, a uniform and fine microstructure is obtained in the easy deformation zone C (Figure 7b). In the free deformation zone B, there is a duplex grain microstructure for the uncompleted DRX (Figure 7c).

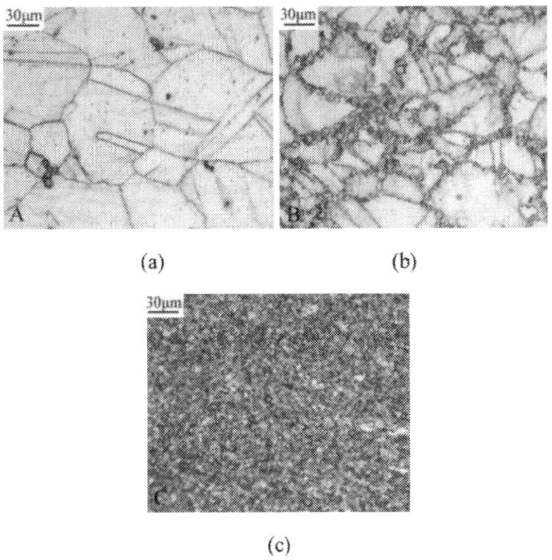

Figure 7. Microstructure in different zones of workpiece: dead zone A (a), free deformation zone B (b) and easy deformation zone C (c)

Figure 8 shows the mean grain size distribution of the billet deformed under the actual upsetting condition. The size of billet is 250mm in diameter and 470mm in height. The initial temperature of the billet was 1030°C, and the billet was deformed to 70% reduction in height. Due to the grain growth up during the holding period after heated to the deformation temperature, the initial grain size was 45μm in the FEM.

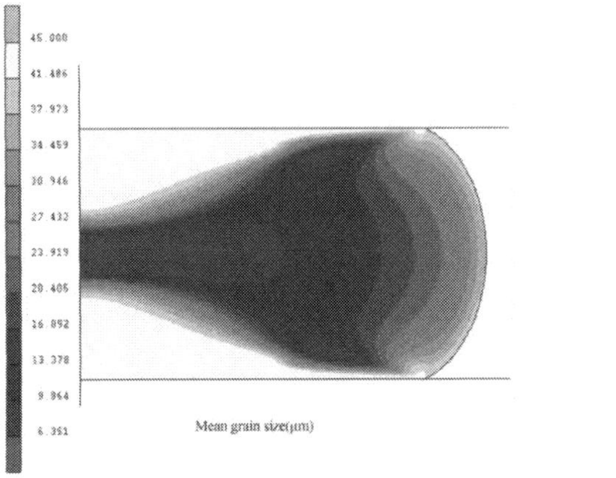

Figure 8. Mean grain size distribution of the billet deformed under the actual upsetting condition

4 Microstructure simulation of the hot die forging

Figure 9 shows the mean grain size distribution in the turbine disk at the end of forging in the actual process. Five zones of coarse grain (A, B, C, D and E) can be distinctly seen in Figure 9. For the zone A, C and E, the temperature and strain are low during forging, so it is difficult for the occurrence of DRX. The material in the zone B and D flows difficultly due to the chilling of billet surface for the heat exchange with die, and it is the deformation dead zone. In the zone F, for the temperature and strain are higher, the XRD occurs abundantly and the mean grain size is smaller. Figure 10 shows the microstructures observed in the zone A and F of the turbine disk. It can be seen that the predicted mean grain size is in good agreement with the observations.

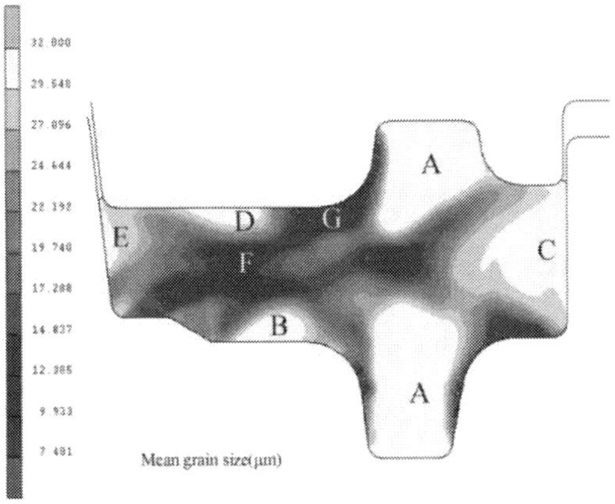

Figure 9. Mean grain size distribution in the turbine disk at the end of forging in the actual process

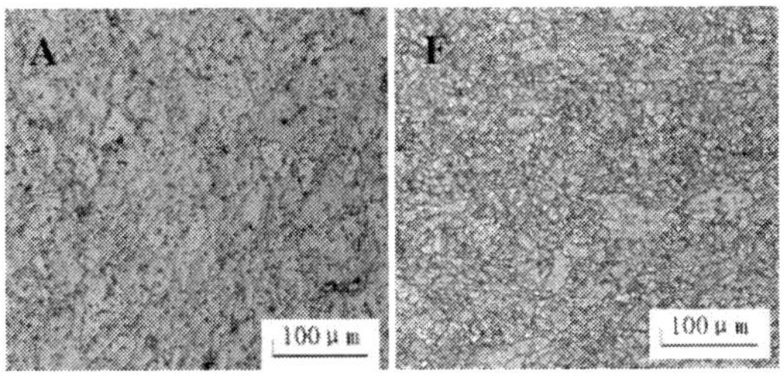

(a) (b)

Figure 10. Microstructures observed in different zones of turbine disk: zone A (a) and zone F (b)

During the hot die forging of IN718 turbine disks, billet size, die temperature, billet temperature and deformation speed would affect the grain size. It is important to obtain a turbine disk with uniform grain size. A uniform function of microstructure (W) was put forward as the objective function to design the forging of IN718 turbine disk, which is expressed as Eq. (1).

$$W = \sum (d_i - d_{avg})^2 / n_1 \quad (1)$$

where d_i is the mean grain size of each element, d_{avg} is the average grain size, and n_1 is the sum of elements. The smaller the value of W is, the more uniform the grain size in turbine disk is. Figure 11 is effect of parameters on the uniformity of the deformation distribution function W. As the height and the die temperature increasing the value of W increased; as the deformation speed increasing, the value of W decreased. The influence of the inner diameter and die temperature on W was smaller. In order to prevent the grain grow up in the high-temperature heating and heat insulation process, the optimization parameters of the turbine disk forging process are as follows: billet temperature 1000°C, die temperature 900°C.

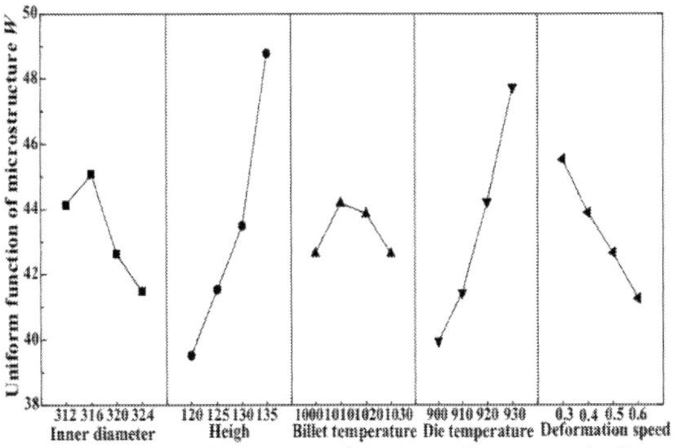

Figure 11. Effect of process parameters on the uniform function of microstructure W

Figure 12 shows the mean grain size distribution in the turbine disk at the end of forging under the optimum hot die forging condition. Compare to the actual forging process (Figure 9), the mean grain size distribution in the zone F is more uniform under the optimization parameters (Figure 12). The mean grain size in the zone A and C are finer than that in the actual forging process. But there is no improvement for the coarse grain in the zone B, D and E (Figure 12). The value of W under the actual forging process and the optimum forging process is 50.49 and 32.29, respectively. Therefore, the optimum process is helpful to obtain the turbine disk with a uniform grain size.

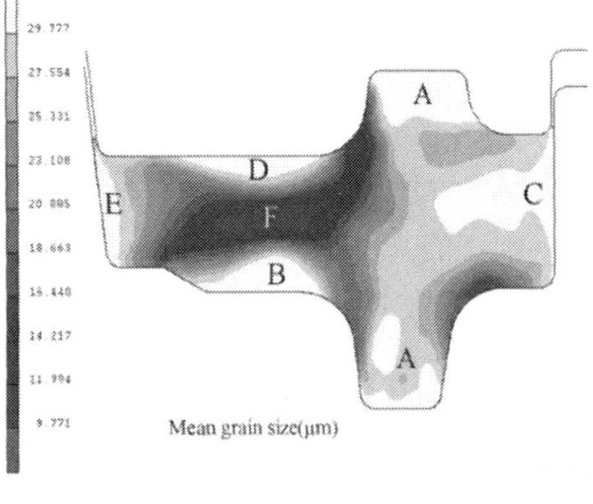

Figure 12. Mean grain size distribution in the turbine disk at the end of forging under the optimum forging condition

5 Conclusions

(1) During the cogging of IN718, the grain in the half radius area in the radial sections was finer as the deformation condition in favor of the occurrence of recrystallization. MDRX was important for the grain refining, and SRX was the main microstructure evolution mechanism during the cogging process.
(2) During the upsetting, there were three deformation zones in the billet. In the dead zone, the DRX took place hardly. In the free deformation zone, there was a duplex grain microstructure for the uncompleted DRX. In the easy deformation zone, a uniform and fine microstructure was obtained as the DRX occurred completely.
(3) The effect of billet size, die temperature and deformation speed on the microstructure evolution during the hot die forging of IN718 turbine disk were investigated by FEM. A uniform function of microstructure was put forward as the objective function to optimize the hot die forging process.

Acknowledgments

This work has been supported by the National Natural Science Foundation of China with the Grant Number: 51505240, the Natural Science Foundation of Zhejiang with the Grant Number: LY15E050008, and the Natural Science Foundation of Ningbo with the Grant Number: 2015A610068. In addition, the author gratefully acknowledges the support of K.C.Wong Education, Hong Kong.

References

1. R.E. Schafrik, D.D. Ward, J.R. Groh, *Superalloys 718, 625, 706 and Various Derivatives*, 1 (TMS, Warrendale, 2001)
2. D.F. Paulonis, J.J. Schirra, *Superalloys 718, 625, 706 and Various Derivatives*, 13 (TMS, Warrendale, 2001)
3. S. Tin, P.D. Lee, A. Kermanpur, Metal. Mater. Trans. A, **36**, 2493 (2005)
4. J.T. Yeom, C.S. Lee, J.H. Kim, N.K. Park, Mater. Sci. Engng. A, **449-451**, 722 (2007)

DOI: 10.1051/matecconf/20168002013

Research on hot stamping for a typical part of B1500HS boron steel using experiment and numerical simulation methods

Huiping Li[1,a], Bingtao Tang[2,a], Lianfang He[1] and Cheng Wang[1]

[1] *School of Materials Science and Engineering, Shandong University of Science and Technology, Qingdao 266590, PR China*
[2] *School of Materials Science and Engineering, Shandong Jianzhu University, Jinan, PR China*

Abstract. In the paper, a typical part of B1500HS boron steel was formed using the hot stamping tools, and the effect of austenitization temperature on the microstructure and mechanical properties of B1500HS steel was studied by the experiment and finite element methods. The results show that, the temperature of steel plate has a significant effect on the temperature of hot stamping tools, and the temperature of punch rises at a faster speed than that of die in the hot stamping process. The austenitization temperature and time both have significant effects on the size of martensite, but have not obvious effects on the hardness. The cooling rate of steel plate has a significant effect on the tensile strength when the austenitization temperature is 870 ° C. The fracture of sample austenitized at 870 ° C or 900 ° C is the dimple, the fracture of sample austenitized at 930 ° C or 960 ° C is the mixture of quasicleavage and dimple.
Key words: Hot stamping, Austenitization temperature, Microstructure, Mechanical properties.

1 Introduction

In order to reduce vehicle weight and tail-gas emission without compromising crash performance, most of automotive manufacturers focus upon application of new materials besides optimizing framework of vehicle. Body-in-white or BIW, in which a car body's sheet metal components have been welded together, is the largest structure of vehicle, and main contributor to the vehicle weight, so it is one of the best objectives of weight reduction. Mechanical performance of BIW can be improved with the weight reduction by using advanced high-strength steel (AHSS) or ultra high-strength steel (UHSS). But formability of AHSS or UHSS reduces remarkably with the enhancement of strength, and some defects, such as poor forming quality, cracking and spring-back, are easy to appear in the cold stamping.

At present, lots of research about the hot stamping of quenchable boron steel has been done in the world, such as the investigation of high-strength steel forming technology, design of hot stamping tools with cooling system, constitutive relationships of boron steel at high temperature, formability of quenchable boron steel in the hot stamping and so on. Kolleck et al. [1] studied the induction heating of boron steel in the hot stamping. The grade of austenitization as well as the mechanical properties of the quenched component had been taken into account in the research. This heating method can reduce the heating time, the investment costs and the floor space of heating device. Min et al. [2] presented a prediction model for hot forming limits of steel 22MnB5 based on Storen and Rice's Vertex theory and Logan-Hosford yield criterion. By comparison, the calculated FLD based on the Vertex theory and four-order Logan-Hosford yield criterion was in good accordance with the measured FLD. Nikravesh et al. [3] described the effect of hot plastic deformation and cooling rate on the final properties of 22MnB5 steel in hot deformation process by means of deformation dilatometer. Li et al. [4] described the hot deformation behavior of B1500HS steel with the microstructure of austenite, ferrite + pearlite, bainite or martensite using the modified Arrhenius and Johnson - Cook models. Mori [5], Yanagimoto [6], Khan [7] and et al. studied the bending spring-back of UHSS by experiment. Lim [8] and Xu [9] studied the cooling system design in the hot stamping die.

At present, many researchers have focused their emphasis on the study of numerical simulation of hot stamping process. Tekkaya et al. [10] presented the thermo-mechanical coupled simulation of hot stamping process by means of experimentally calculated material data and described a number of procedures for the simulation of hot stamping, aiming at a notable decrease in computation time. So et al. [11] studied the warm blanking and cold blanking of 22MnB5 steel, and compared the results of experiment and finite element simulation. Some recommendations for the manufacturer of 22MnB5 steel products were presented. Xing et al. [12] built a material model for the hot stamping of quenchable boron steel according to the experimental data of mechanics and thermal physical properties. The numerical simulation of hot stamping was done by ABAQUS software. The simulation results were basically in agreement with the experimental results. Lei et al. [13]

[a] Corresponding author: lihuiping99@163.com; tangbingtao@gmail.com

simulated the cooling process of hot-stamping dies by CFX software, and studied the effect of processing parameters, such as pressure holding time and cooling water velocity, on the temperature of hot-stamping dies. The numerical simulation results had a good agreement with the experimental results. Bao et al. [14] studied the differences of effective quenching and forming precision between heat free bending and thermal contact bending by ABAQUS software. Liu et al. [15,16] researched the hot stamping of 22MnB5 using the simulation and experiment methods. Bok et al. [17] studied the forming process of B pillar, and the data from simulation and experiment could provide some important information about the temperature, martensite distribution and technological parameters for the design of hot stamping. Cai et al. [18] studied the distributions of metallography, hardness and yield stress of 22MnB5 in the hot stamping process by using Ansys/Ls-Dyna and Kirkaldy-Venugopalan austenitic decomposition model.

In this paper, the effect of austenitization temperature on the microstructure and mechanical properties of B1500HS boron steel is studied by the experiment and finite element methods.

2 Experiments

2.1 Materials

The material used in the paper is B1500HS boron steel sheet with 1.6 mm thickness and non-coating. Microstructure of boron steel is about 25 % pearlite and 75 % ferrite, the yield strength and tensile strength are about 340 MPa and 500 MPa. The chemical compositions of B1500HS are listed in Table 1. The size of steel blank used in the experiment is about 115 × 115 mm. The samples which are used to test the microstructure and mechanical performance are cut from the region I, II and III of stamped parts by the wire electric discharge machine (WEDM), as shown in Figure 1.

Table 1. Chemical composition of B1500HS (wt %)

C	Si	Mn	S	P	Cr	Mo	Ti	B
0.23	0.25	1.35	0.006	0.015	0.19	0.04	0.03	0.003

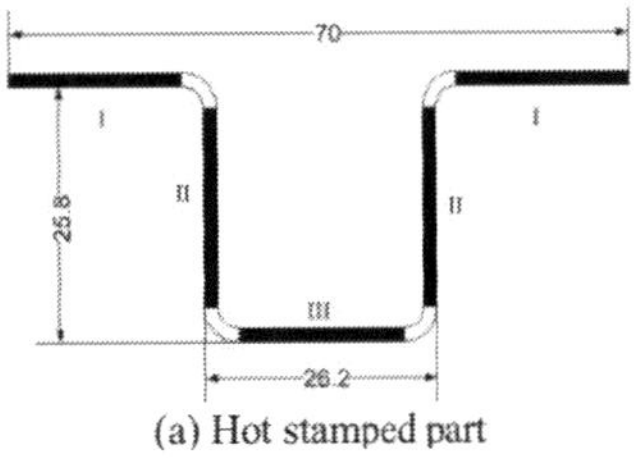

(a) Hot stamped part

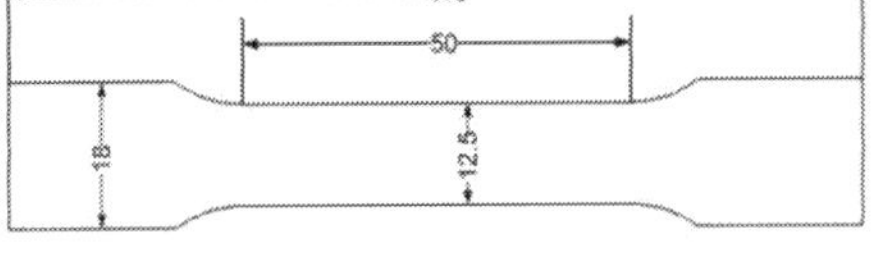

(b) Tensile sample

Figure 1. Schematic diagrams of part and sample (unit: mm)

2.2 Quenching experiments

Hot stamping tools used in the experiment are shown in Figure 2. The thermocouple positions are marked as symbol ①, ②, ③and ④. The material of tools is H13 steel. NiCr-NiSi thermocouples with the diameter of 2 mm and length of 100 mm are used to measure the temperature of tools. Data acquisition system TC-08 is used to convert the signal from thermocouples. Testing machine CMT5000 and other equipments are used to measure the hardness, tensile strength and microstructure of samples.

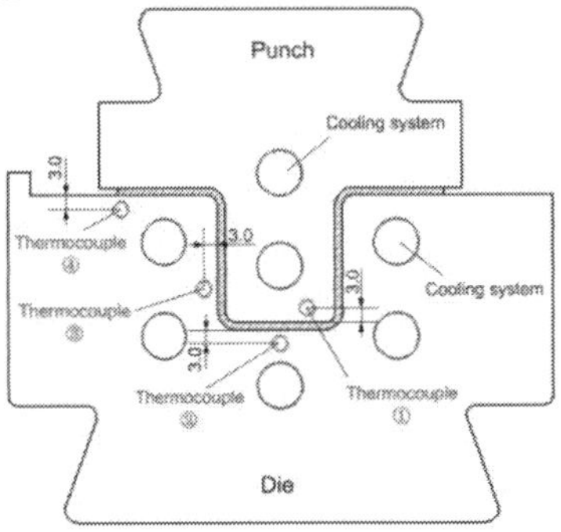

Figure 2. Schematic diagram of hot stamping tools

The temperature of resistance furnace is set to 870°C. When the temperature arrives at about 870°C, five steel plates are put into the furnace for austenitization. After heated for about 5 min, the first steel plate is quickly transferred into the hot stamping tools, formed and cooled in the tools, and then moved out of the tools. Other four steel plates also undergo these steps in order. So the heating times of five steel plates are respectively about 300 s, 420 s, 540 s, 660 s and 780 s. For the hot stamping of steel plates at the austenitization temperature of 900 °C, 930 °C and 960 °C, the operation steps are same as that at 870 °C.

3 EXPERIMENT RESULTS AND DISCUSSIONS

3.1 Temperature of completely austenitization

The expansion curve attained by the quenching and deformation dilatometer DIL 805A/D shows that, the temperature (Acl) at which pearlite transforms into austenite is about 725°C, and the temperature (Ac3) at which ferrite transforms into austenite is about 810°C, when the heating rate is about 200 /h.

The heating rate of steel plates in the heating furnace SX2-4-10G is much higher than 200 °C/h, and the heating rate has a significant effect on the temperature Acl and Ac3 [19]. In order to know the temperature curve of steel plate in the heating and cooling, a thermocouple is welded on the side middle of steel plate. The position of thermocouple is shown in Figure 3.

The temperature curve can be divided into six stages: 1 - heating in the furnace, 2 - transporting (free cooling), 3 - contacting with the punch, 4 - forming in the tools, 5 -

cooling in the tools, 6 - out of die and cooling in the air. According to the variation tendency of temperature curve, the temperature Ac1 and Ac3 are about 776.8 °C and 824.4 °C when B1500HS steel is heated in the furnace.

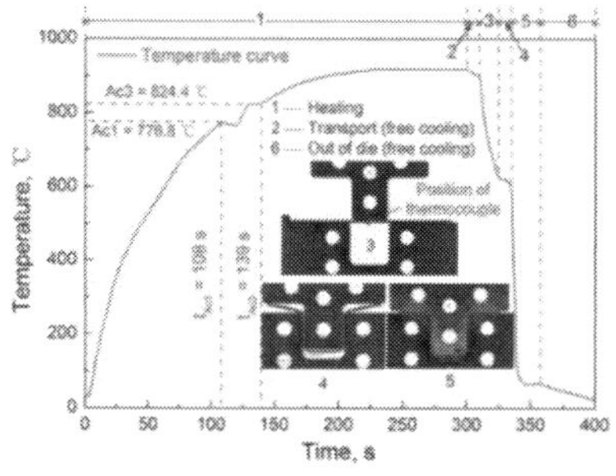

Figure 3. Temperature curve in the heating and cooling process

3.2 Temperature of tools

In the hot stamping, temperatures of four positions marked with symbol ①, ②, ③and ④shown in Figure 2 are measured by thermocouples. The temperature curves are shown in Figure 4.

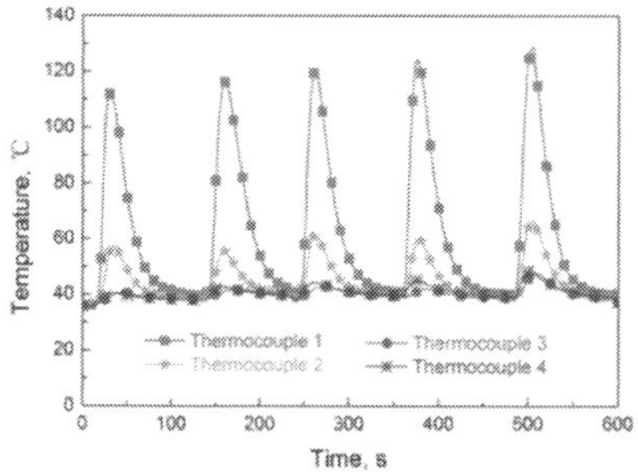

(a) 870°C

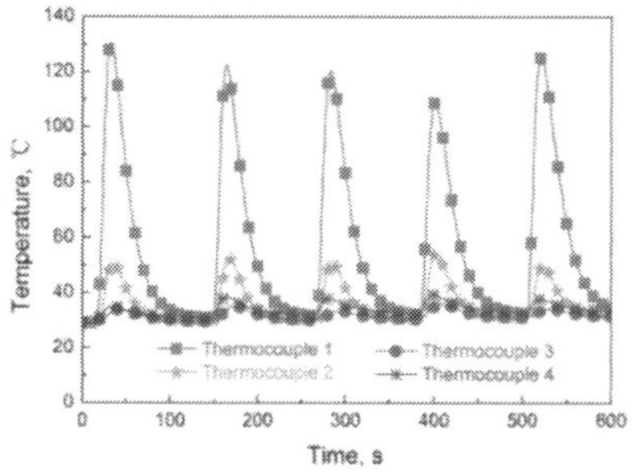

(b) 900°C

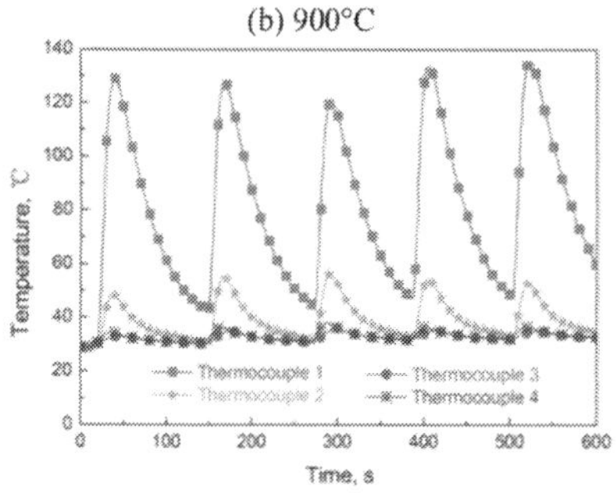

(c) 930°C

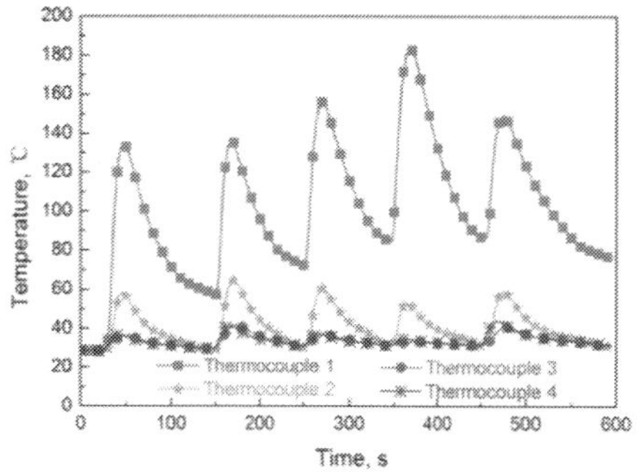

(d) 960°C

Figure 4. Time-temperature curves measured by thermocouples

Before the punch contacts with the heated boron steel plate, the steel plate only contacts with the upper surface of die. Only a little heat energy transfers from the steel plate to the die, as the boundary heat transfer coefficient is smaller due to no pressure between them [20]. When the die contacts with the plate, the plate instantaneously separates from the upper surface of die due to the bending of steel plate, and begins deforming and sliding into the die with the punch, as shown in Figure 5. So the contact time is very short, and the temperature variation at the positions ④is small, as shown in Figure 4.

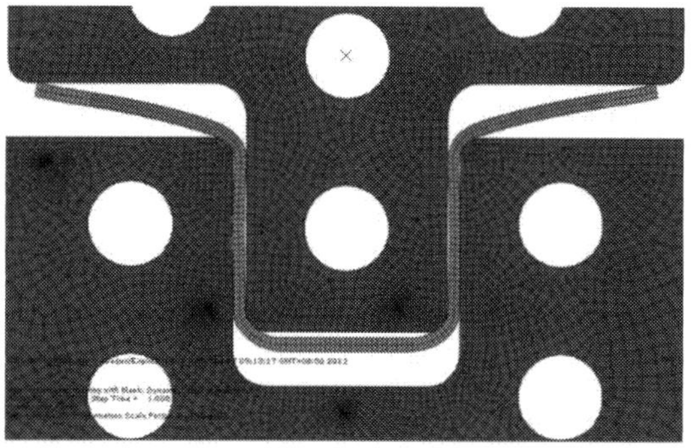

Figure 5. Hot stamping of steel plate simulated by FEM

When the steel plate completely contacts with the surface of punch and die, as shown in Figure 6, the pressure among them rapidly reaches to over 30MPa. Much heat energy transfers from the steel plate to the die and punch, as the boundary heat transfer coefficient is relative to the pressure [21].

The temperature variation at the position ① is significant, but the temperature variation at the position ②is smaller than that at the position ①, as shown in Figure 4. The main reason is that, the thickness of punch is small (about 24 mm). Moreover, the volume of die is much larger than that of punch, so the temperature variation of the die is slower than that of the punch when the same heat energy is absorbed.

 DOI: 10.1051/matecconf/20168002013

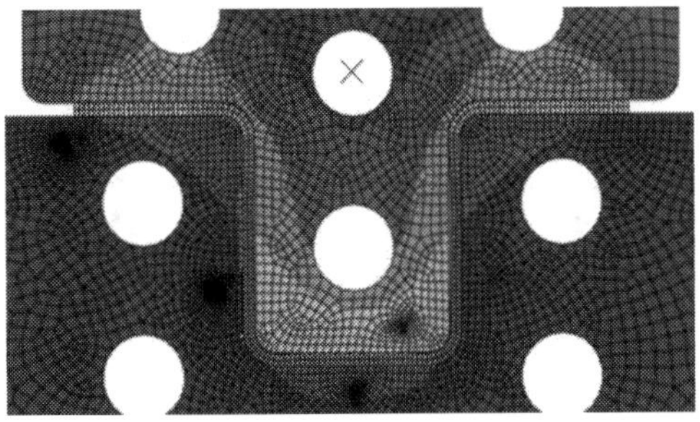

Figure 6. Diagram of stamped part completely contacting with punch and die

For the region II shown in Figure 1, it is in the gap of punch and die. Because the gap is a little larger than the thickness of plate, the clearance may appear among the punch, die and stamped part. The heat exchange between the forming part, punch and die is similar to the heat exchange without pressure. So the temperature variation at the position ③is smaller, as shown in Figure 4.
The temperature-time curves show that, with the increase of austenitization temperature, the temperature peak at the region ①of punch increases, but there is a little effect on other regions of tools. After the die temperature increases to the peak, the heat energy taken away from the die and punch by the cooling water is more than that from the hot stamped part to the die and punch, so the temperatures of punch and die decrease with time.

The temperature variations of punch and die in Figure 4 show that, the temperatures of punch and die instantaneously rise and rapidly decrease under the action of the hot stamped part and tool cooling system, and the punch is easier to get higher temperature. The temperature of punch and die is one of the key factors to affect the microstructure and mechanical performance of forming part. If no cooling system works in the tools, the temperature of tools would become higher and higher. The microstructure and mechanical performance of hot stamped part would be affected. In addition, the service life of tools would be affected.

3.3 Microstructure

The samples are cut from the regions I, II and III of part shown in Figure 1 by WEDM. The samples are polished, and then etched using 4% nitric acid alcohol solution. The microstructure observed by the optical microscope in the region I, II, and III is the lath martensite. The numerical simulation result of phase - transformation by finite element method shows the volume fraction of martensite is more than 98 %, as shown in Figure 7.

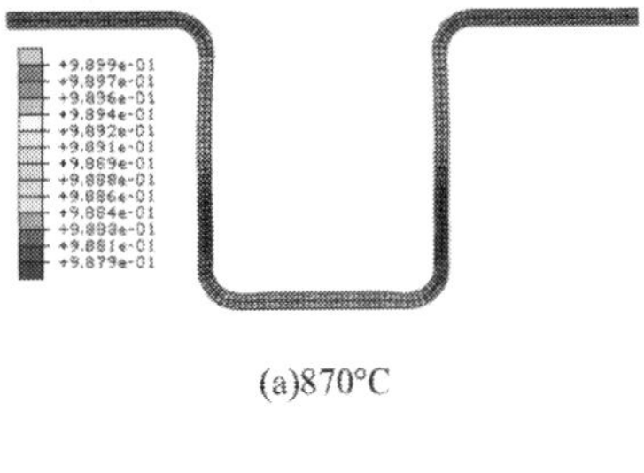

(a)870°C

(b)900°C

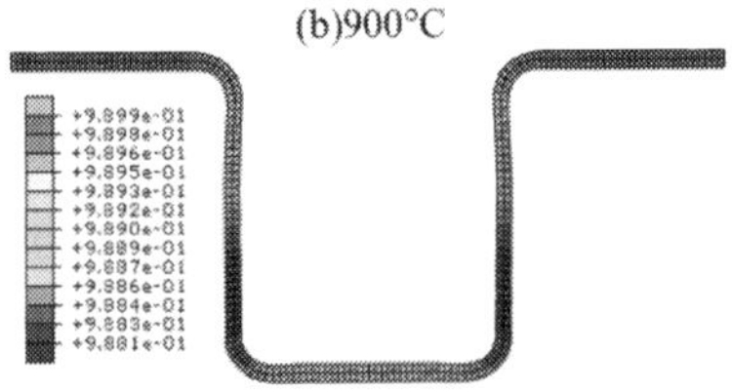

(c)930°C

(d)960°C

Figure 7. Distribution and volume fraction of martensite

For the same hot stamped part, the microstructure in the region I, II, and III is very similar with each other, when the austenitization temperature is in the range of 870°C - 960 °C. In the heating process, five steel plates are simultaneously put into the furnace for austenitization, their austenitization time are about 300 s, 420 s, 540s, 660 s and 780 s. The results of microstructure show that, the austenitization temperature and time both have a significant effect on the size of martensite, the size of lath martensite becomes bigger and bigger with the increase of austenitization temperature and time, as the grain size of austenite is coarser and coarser with the increase of austenitization temperature and time. The microstructures relative to different austenitization temperature and time are shown in Fig. 8.

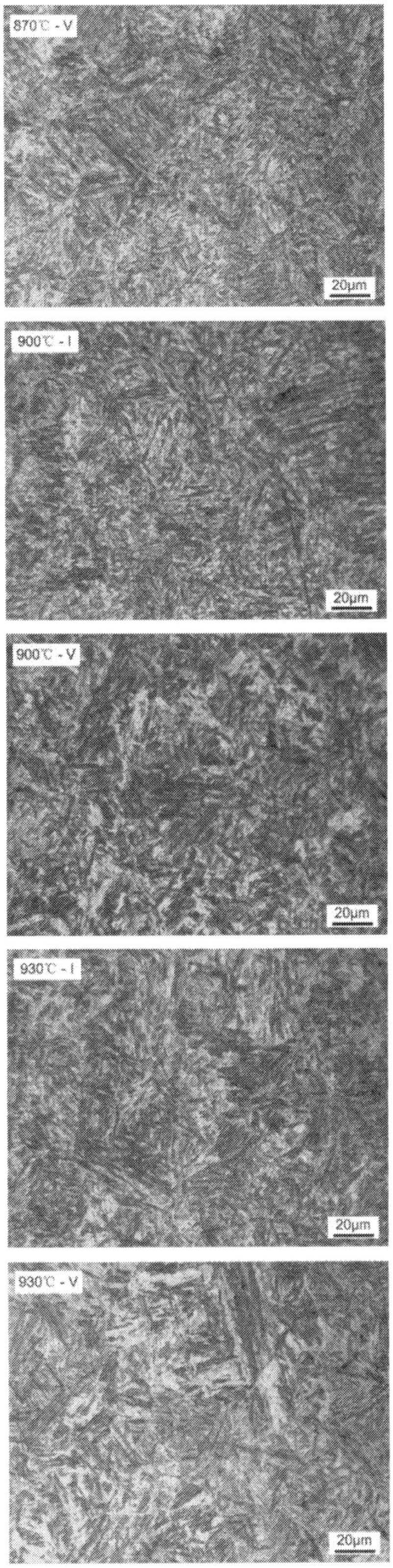

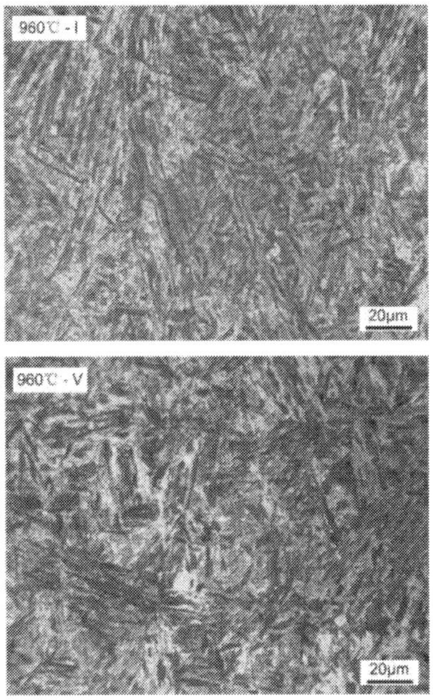

Figure 8. Microstructures relative to different austenitization temperature and time

3.4 Hardness

For the hot stamped parts at the different austenitization temperature, the averages of hardness tested by Rockwell hardness tester at the region I, II and III are shown in Table 2.

Table 2. Hardness of specimen in marked regions (HRC)

	Region I	Region II	Region III
870° C	43.8	42.7	44.1
900° C	45.5	42.8	43.5
930° C	45.7	44.4	45.4
960° C	45.3	43.7	44.7

The testing result of hardness shows that, the austenitization temperature and time have a little effect on the hardness of samples, the values of hardness relative to the different regions and austenitization temperatures are very close. For the same part, the hardness at the region I and III is slightly higher than that at the region II due to the difference of cooling rate. In the hot stamping, the boundary heat transfer coefficient (BHTC) at the region I and III is larger than that at the region II due the higher pressure at the region I and III [22]. The heat flux from the part to the tools in the hot stamping can be described with Newton's law of cooling,

$$q = H(T_w - T_c) \tag{1}$$

Where, q represents the heat flux, H represents the BHTC, T_w represents the surface temperature of part, and T_c represents the temperature of tool.

According to Eq.(1), the heat flux q will increase for fixed T_w and T_c, if the BHTC rises. So the cooling rate at the region I and III is less than that at the region II.

Hardness of steel is the function of the cooling rate at 700°C and chemical composition [23].

The hardness calculated by finite element method is shown in Table 2 and Figure 9. The calculation results show that, the hardness at the region I and III is slightly higher than that at the region II, which is basically consistent with the experiment results.

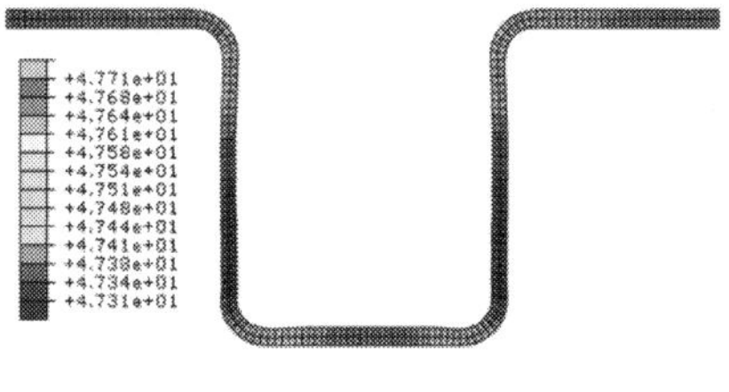

(a) 870°C

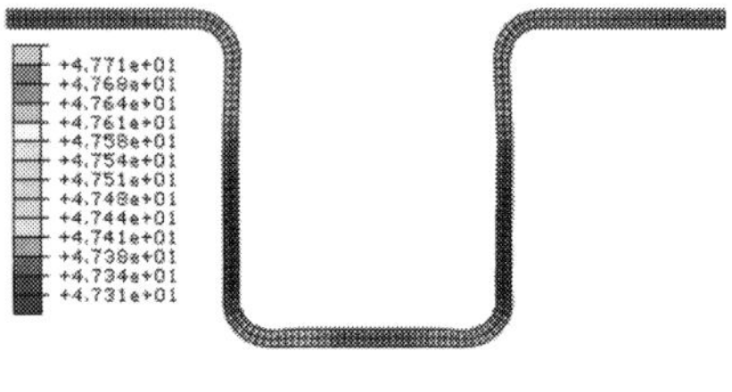

(b) 900°C

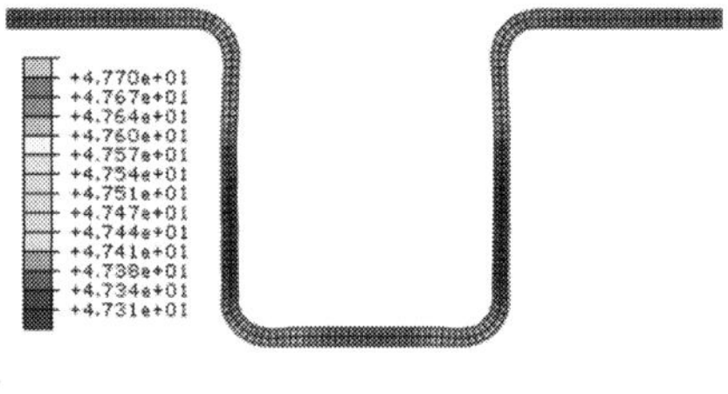

(c) 930°C

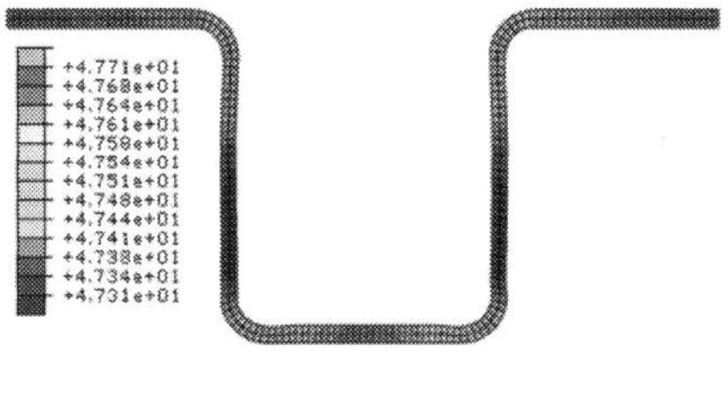

(d) 960°C

Figure 9. Hardness distribution simulated by FE method

3.5 Stress and strain

The tensile samples are cut from the hot stamped part by WEDM, as shown in Figure 1. All the samples are done the static tensile test by CMT5000 tensile testing machine.

According to the definition, the true stress and true strain can be expressed as

$$\sigma_t = \frac{F}{A_1} = \frac{F}{A_0 l_0} l_1 \tag{2}$$

$$\varepsilon_t = \ln \frac{l_0 + \Delta l_1}{l_0} \tag{3}$$

where, F is the instantaneous load, ε_t is the true stress, σ_t is the true strain, A_1 is the instantaneous section area of sample, A_0 is the original section area of sample, l_1 is the instantaneous length of sample, l_0 is the original gauge length of sample, and Δl_1 is the instantaneous variation of length in the tension testing.

The true stress and strain curves can be attained using the Eq.(1), Eq.(2) and the data measured by CMT5000, as the curves shown in Figure 10. The tensile strengths of at the region I, II and III are listed in table 3.

Table 3. Tensile strengths at the region I, II and III (Mpa)

	Region I	Region II	Region III
870 ° C	1806.3	1679.2	1758.6
900 ° C	1768	1651.8	1720.6
930 ° C	1771.3	1709.2	1724.6
960 ° C	1753.9	1714.1	1716.9

The curves in Figure 10 and the data in Table 3 show that, the elongation and tensile strength of the sample at austenitization temperature 870°C is a little higher than that at other austenitization temperatures, as the grain size of austenite and size of lath martensite are finer, as shown in Figure 8.

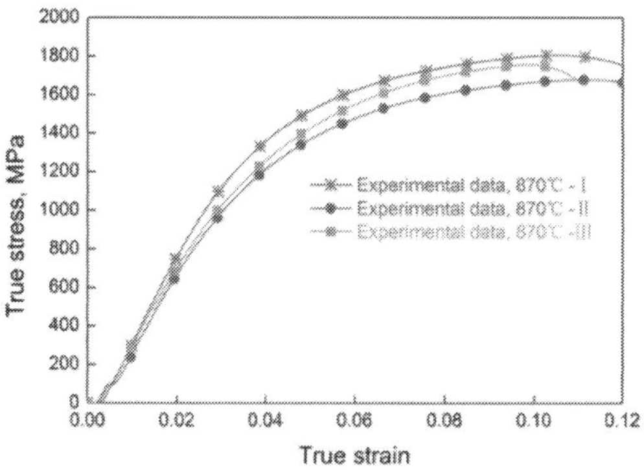

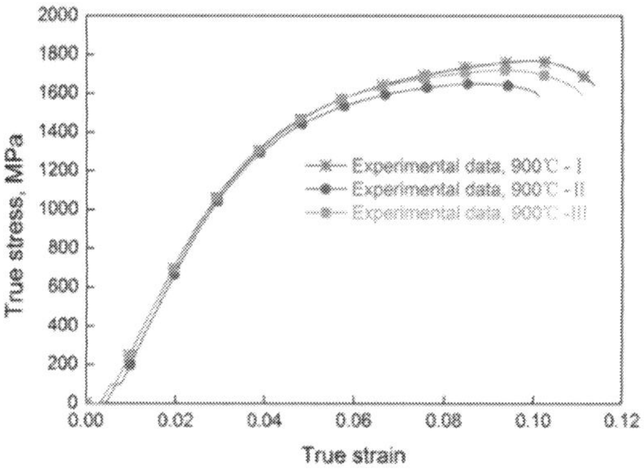

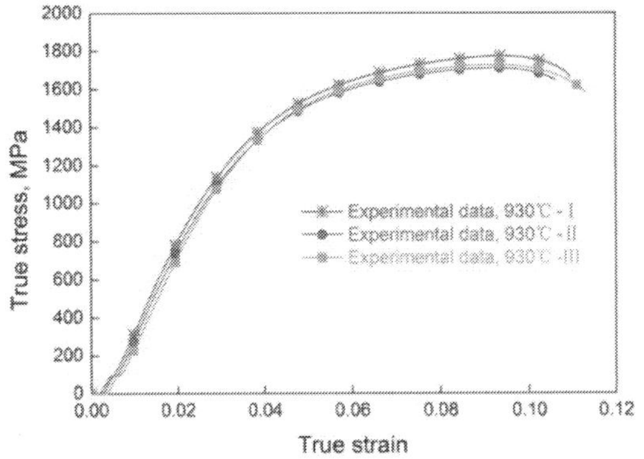

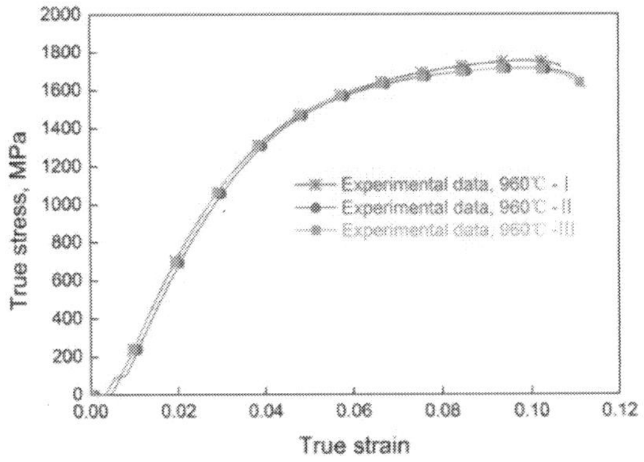

Figure 10. True strain-stress curves of sample at the region I,II and III

The tensile strength of the sample at the region II is a little less than that at the region I and III, as the cooling rate at region II is lower than that at other regions, and this is similar to the hardness distributions of hot stamped part. When the austenitization temperature of steel plate is 870°C or 900°C, the difference of tensile strengths of sample at the region I, II and III is about 130 MPa. When the austenitization temperature of steel plate is 930°C or 960°C, the tensile strengths at the region I, II and III are approximately same.

Except the part with the high and uniform strength, the hot stamping can also be used to manufacture the structural components with distributed properties. For instance, the crash performance of components like the B-pillar of car is improved when softer phases (e.g. ferrite or bainite) with lower strength but higher ductility are able to absorb a greater amount of energy during the crash [25].

The curves in Figure 10 and the data in Table 3 show that, the lower austenitization temperature is helpful for manufacturing the structural components with distributed properties by introducing a thin air gap between the tools and steel plate [26], or using the tool materials with the lower thermal conductivity [27], as the austenite attained at the lower austenitization temperature is particularly susceptible to the cooling rate [28].

3.6 Fracture morphology

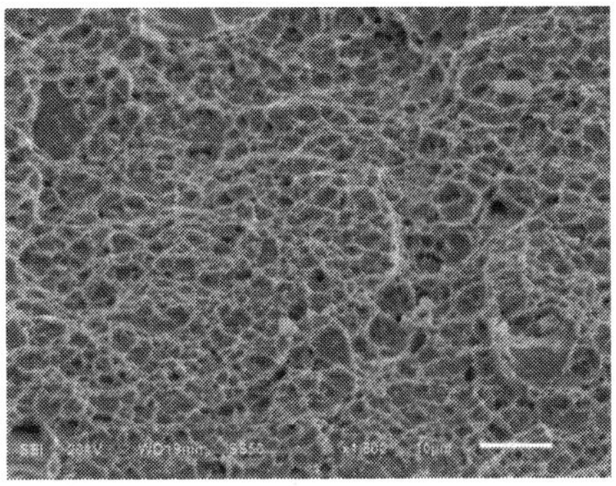

(a) 870°C

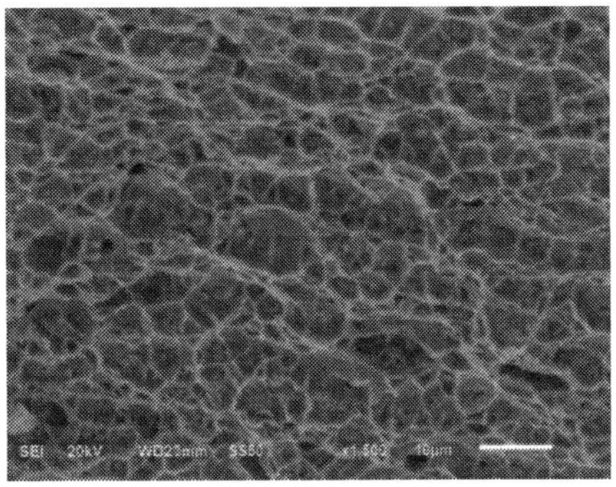

(b) 900°C

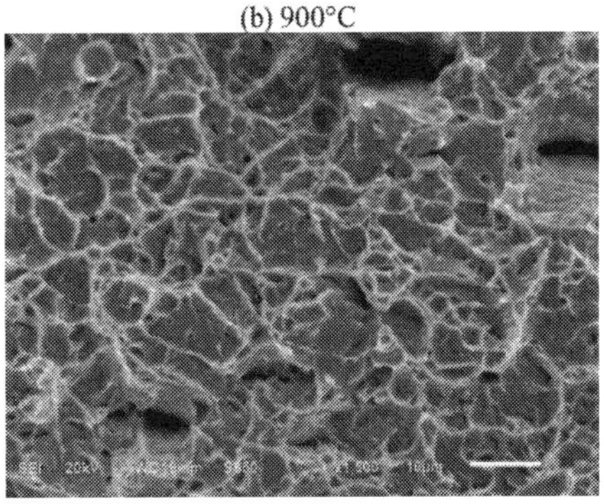

(c) 930°C

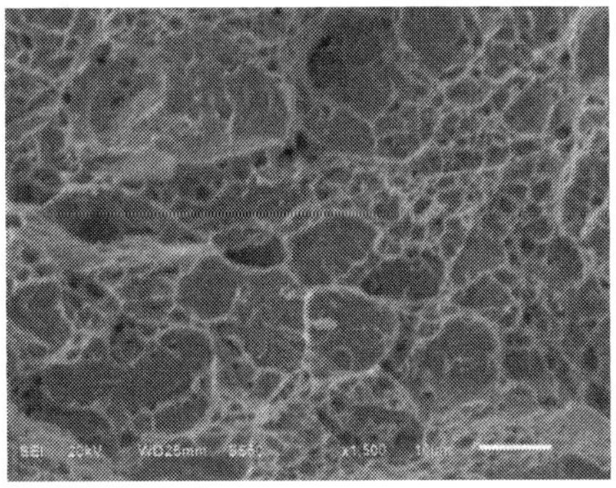

(d) 960°C

Figure 11. Fracture morphologies at different austenitization temperature

The fracture morphology of sample austenitized at 870 °C, 900 °C, 930 °C or 960 °C is shown in Figure 11.

All the fracture are the ductile fracture, the fracture of sample austenitized at 870 °C or 900 °C is the dimple, the fracture of sample austenitized at 930 °C or 960 °C is the mixture of quasicleavage and dimple, and the size of dimple rises with the increase of austenitization temperature.

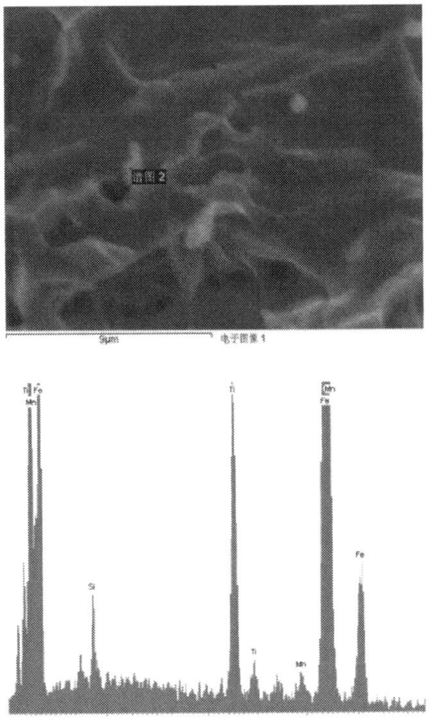

Figure 12. SEM / EDS analysis of fracture

According the temperature curve shown in Figure 3, pearlite and ferrite can fully transform into austenite when the austenitization temperature is over 824.4 °C. In the holding temperature process, the segregation of impurity elements (such as P, S, Si, Ti, etc) for B1500HS can occur on the grain boundary of austenite due to higher austenitization temperature, as shown in Figure 12. Some defect such as voids is easy to appear on the grain boundary due to segregation of impurity elements, and it can grow with the increase of austenitization temperature.

When the tension is put on the sample with the void defect, the voids can grow during deformation until they link together. Some continuous fracture paths may come into being, and some dimples can appear on the fracture of sample. The smaller dimple due to the fine grain size of austenite is helpful for the increase of tensile strength and elongation of sample. With the increase of austenitization temperature, the grain size of austenite and size of lath martensite become larger and larger, more and more quasicleavages occur on the fracture. All these are harmful to the tensile strength, can cause the tensile strength of the sample austenitized at 900 °C, 930 °C or 960 °C is less than that at 870 °C, as shown in Figure 8, Figure 10 and Figure 11.

4 CONCLUSIONS

The effect of austenitization temperature on the microstructure and mechanical properties of B1500HS boron steel is studied by the experiment and finite element methods. The conclusions are as follows:
(1) The temperatures of punch and die rise due to the heated steel plate, and the punch is easier to get higher temperature than the die. Cooling system in the punch and die is very necessary for improving the microstructure and mechanical performance of hot stamped part.
(2) The austenitization temperature and time both have significant effects on the size of martensite, the size of lath martensite becomes bigger and bigger with the increase of austenitization temperature and time
(3) The austenitization temperature and time have not a distinct effect on the hardness of hot stamped part.
(4) The cooling rate has a significant effect on the tensile strength when the austenitization temperature is 870 °C. The lower austenitization temperature is helpful for manufacturing the structural components with distributed properties.
(5) The fracture of hot stamped part austenitized at 870 °C or 900 °C is the dimple, the fracture of part austenitized at 930 °C or 960 °C is the mixture of quasicleavage and dimple, and the size of dimple rises with the increase of austenitization temperature.

ACKNOWLEDGEMENTS

This work was financially supported by the National Natural Science Foundation of China (51175302), the Program for New Century Excellent Talents in University (NCET-12-0342), The Science and Technology Development Program of Shandong and Huangdao (2014GGX103024, 20140132) and Scientific Research Foundation of Shandong University of Science and Technology for Recruited Talents (2013RCJJ002, 2013RCJJ005).

References

1. R. Kolleck, R. Veit, M. Merklein, et al. Cirp Ann.-Manuf. Techn., **58**, 275 (2009)
2. J.Y. Min, J.P. Lin, J.Y. Li, W.H. Bao. Comp. Mater. Sci., **49**, 326 (2010)
3. M. Nikravesh, M. Naderi, G.H. Akbari. Mat. Sci. Eng. A, **540**, 24 (2012)
4. H.P. Li, L.F. He, G.Q. Zhao, L. Zhang. Mat. Sci. Eng. A, **580**, 330 (2013)
5. K. Mori, K. Akita, Y. Abe. Mach. Tool. Manu., **47**, 321 (2007)
6. J. Yanagimoto, K. Oyamada. Cirp Ann.-Manuf. Techn., **56**, 265 (2007)
7. S.H. Khan, M. Saeed Ahmed, F. Ali, et al. Eng. Fail. Anal., **16**, 128 (2009)
8. Woo-Seung Lim, Hong-Seok Choi, Seok-young Ahn, et al. Int. J. Adv. Manuf. Tech., **70**, 1189 (2014)

9. Y. Xu, Z.D. Shan. Int. J. Adv. Manuf. Tech., **70**, 257 (2014)
10. A.E. Tekkaya, H. Karbasian, W. Homberg, M. Kleiner. Prod. Eng. Res. Devel., **1**, 85 (2007)
11. H. So, D. Fabann, H. Hoffmann, et al. J Mater. Process. Tech., **212**, 437 (2012)
12. Z.W. Xing, J. Bao, Y.Y. Yang. Mat. Sci. Eng. A, **499**, 28 (2009)
13. C.X. Lei, J.J. Cui, Z.W. Xing, et al. Physics Procedia, **25**, 118 (2012)
14. J. Bao, Z.W. Xing, Y.Y. Yang, et al. Mater. Sci. Tech., **17**, 326 (2009)
15. H.S.Liu, Z.W. Xing, J. Bao, et al. J. Mater. Eng. Perform., **19**, 325 (2010)
16. H.S. Liu, W. Liu, J. Bao, et al. J. Mater. Eng. Perform., **20**, 1 (2011)
17. H.H. Bok, M.G. Lee, H.D. Kim, et al. Metal. Mater. Int., **16**, 185 (2010)
18. Y. Cai, W. Li, G. Li. Mater. Res. Innov., **18**, 509 (2014)
19. L.F. He, H.P. Li, K. Gai, et al. Trans.Mater. Heat Treat., **36**, 199 (2015)
20. B. Abdulhay, B. Bourouga, C. Dessain, et al. Heat Transfer Eng., **32**, 497 (2011)
21. P. Hu, L. Ying, Y. Li, Z.W Liao. J Mater. Process. Tech., **213**, 1475 (2013)
22. A. Azushima, K. Uda, H. Matsuda. J Mater. Process. Tech., **214**, 3031 (2014)
23. P. Maynier, J. Dollet, P. Bastien. *Hardenability concepts with applications to steel* (AIME, New York, 1978)
24. H.P. Li, G.Q. Zhao, S.T. Niu, C.Z. Huang. Mat. Sci. Eng. A, **452-453**, 705 (2007)
25. E.J.F.R. Caron, K.J. Daun, M.A. Wells. Metall. Mater. Trans. B, **44**, 332 (2013)
26. H. Karbasian, A.E. Tekkaya. J Mater. Process. Tech., **210**, 2103 (2010)
27. M. Maikranz-Valentin, U. Weidig, U. Schoof, et al. Steel Res. Int., **79**, 92 (2008)
28. Q.Z. Gao, Y.C. Liu, X.J. Di, et al. Nuc. Eng. Des., **256**, 148 (2013)

Full-field simulation of solidification and forming of polycrystals

Efim Borukhovich[1,a], Martin Boeff[1], Alexander Monas[1], Marvin Tegeler[1], Se-Jong Kim[2], Chang-Seok Oh[2], Ingo Steinbach[1]
[1]*Interdisciplinary Centre for Advanced Materials Simulation (ICAMS), Ruhr-Universität Bochum, Universitätsstr. 150, Germany*
[2]*Korea Institute of Material Science (KIMS), Changwon, South Korea*

Abstract. The phase-field method has emerged as the method of choice for simulation of microstructure evolution and phase-transformations in material science. It has wide applications in solidification and solid state transformations in general. Recently, the method has been generalized to treat large deformation and damage in solids. A through process full-field simulation will be presented starting from solidification and ending with the evolution of damage during large deformation. Aspects of numerical discretization, efficient numerical integration and massive parallelization will be discussed.

1 Introduction

Due to their low density, magnesium alloys are of high importance for modern lightweight structures and are widely used for various applications in automotive and aerospace industries as well as in the consumer electronics. The most applied alloying element in Mg alloys is aluminum which is used for AZ91, AM50 (cast) or AZ31 (wrought) technical alloys. The mechanical properties of Mg-Al alloys are primarily determined by their microstructure which consists of two main phases: the Mg-rich hexagonal close-packed (HCP) α-phase and a near stoichiometric $Mg_{17}Al_{12}$ β-phase, and further minor secondary phases due to third alloying elements.

An exemplary SEM-micrograph of the later simulated Mg-5%Al is reproduced in fig. 1. The addition of Al in small concentrations has a positive effect on the corrosion properties of the Mg-Al alloys, mainly due to the formation of a percolating β-phase network [3] which is less electrochemically active compared to the α-phase. Experiments, as in [4] revealed that the formation of a closed shell of β-phase, preventing α-phase dendrites from building networks, is a good strategy to significantly increase the corrosion resistance of cast Mg-Al alloys. Furthermore, it severely hinders the galvanic corrosion if Mg-alloys are brought into contact with more noble metals. Thus, tailoring the Mg-Al alloy microstructure is of great importance for controlling its corrosion resistance [5]. In this work, we apply the phase-field method [6, 7, 8] to simulate the solidification microstructure of Mg-Al alloys. We outline the strategy for the full solidification cycle simulation using the phase-field method followed by a damage simulation.

The information about the as cast morphology and the distribution of the secondary β-phase in particular shall point towards improved materials and process design,

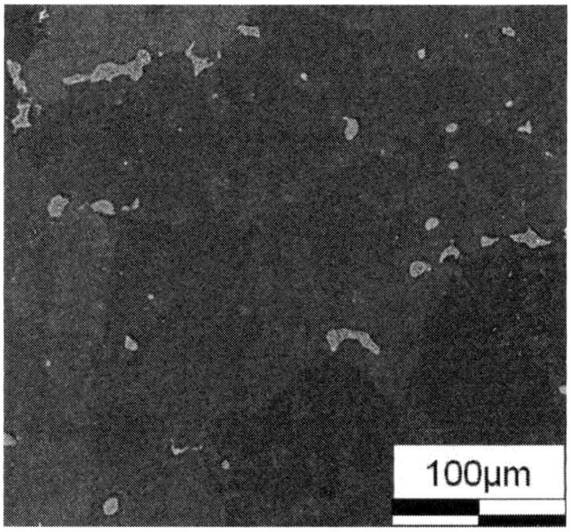

Figure 1. SEM-image of as-cast Mg-5%Al microstructure. dark: α-phase, bright: β-phase

also in terms of corrosion. The proposed approach enables metallurgical process control by correlation of metallurgical parameters to metallic phase fractions and mechanical properties. This will significantly contribute to the development of Integrated-Computational-Materials-Engineering (ICME) by adding service-life aspects to computational materials design.

2 Phase-Field Model

In this study, we apply the multi-phase-field method on a meso-scale with locally linearized phase diagram information [9]. In this method each grain of a distinct crystallographic phase is attributed by its own phase-field variable $\phi_p(\boldsymbol{x}, t) \in [0,1]$.

[a] Corresponding author: Efim.Borukhovich@rub.de

Three distinct crystallographic phases are considered: liquid melt, Mg-rich α-phase and a near stoichiometric $Mg_{17}Al_{12}$ β-phase, where each solid grain is attributed by an identifier of the crystallographic phase and its orientation in space.

The governing equations for the evolution of phase fields and the concentration are given below (for details see [7, 9]).

$$\dot{\phi}_k = \sum_{l=1}^{N} \frac{\mu_{kl}}{N} \left\{ \sum_{m=1}^{N} [\sigma_{lm} - \sigma_{km}] I_m + \frac{\pi^2}{8\eta} h' \Delta g_{kl} \right\} \quad (1)$$

$$\text{with } I_m = \nabla^2 \phi_m + \frac{\pi^2}{\eta^2} \phi_m \quad (2)$$

$$\dot{c} = \nabla \left\{ \sum_{k=1}^{N} \phi_k [D_k \nabla c_k] + \sum_{k,l=1}^{N} \mathrm{j}_{kl} \right\} \quad (3)$$

Here ϕ is the phase field parameter; k, l and m are the phase field indexes; μ_{kl} is the interface mobility between phases k and l; N is the number of non-vanishing phase fields/grains in a given point; I is the generalized curvature term; η is the interface thickness; h is the shape function related to the phase field contour; Δg_{kl} is the thermodynamic driving force; D_k is the diffusion coefficient of Al in phase k; $\mathbf{j}_{kl}$ is the antitrapping current.

The domain size of our simulations is set to 300μm in every direction with a grid spacing of 1μm resulting in 300^3 cells. The interface thickness η is set to the length of 5 grid-cells.

In order to consider the effect of the external cooling rate on nucleation and growth of solid phases we solve the heat balance in average over the whole domain with n_{xyz} discrete elements:

$$\rho c_p \dot{T} = \sum_{n_{xyz}} \sum_{k \neq l} \dot{\psi}_{kl} L_{kl} + \dot{Q}_{ext} \,. \quad (4)$$

Here, T is the temperature of the system, $\dot{Q}_{ext}$ is the heat extraction rate, ρ and c_p are the specific heat and the mass density which are taken as constant throughout the simulation.

$\dot{\psi}_{kl}$ are the local transformation rates between pairs of phases which sum up to the total transformation rate of individual phases $\dot{\phi}_k = \sum_{l \neq k}^{N} \dot{\psi}_{kl}$ in the interface or a junction point with N phases (for details see [11]). L_{kl} is the latent heat associated with the individual transformations between phases k and l. In the present study, we treat all possible transformations between liquid and α, liquid and β and α and β phases.

3 Nucleation model

In order to capture the dependence of the nucleation density on the cooling rate, we employ the so-called "free growth model" by Greer and coworkers [12]. This model is based on a given distribution of seed particles of different size. Assuming perfect wetting condition between the primary α-phase and the seed particles, the hemispheric cap of the α-phase on the seed particles defines the critical undercooling for free growth of the nucleating phase which is inversely proportional to the size of the seed particles [13].

In our simulations, we start from a random distribution of particles with a given density and size distribution over the entire simulation domain initially filled with liquid melt.

The utilized normal distribution is given by:

$$N(d) = \frac{K}{b\sqrt{2\pi}} exp\left[-\frac{(d-a)^2}{2b^2}\right] \quad (5)$$

Each particle is assigned an effective size d which determines the critical undercooling for its activation.

The resulting number of particles is typically higher than the actual number of the activated particles (active nucleation sites).

Activation happens if a grid cell, which contains a particle, reaches its critical undercooling, i.e., if the difference between actual temperature and melting temperature dependent on the local alloy concentration exceeds the local critical undercooling. A similar model has also been applied in [9, 10]. For further details and used parameters see [14].

4 Mechanical framework

Using standard notation in continuum mechanics, the deformation mapping maps a material point ξ (position vector) belonging to the reference configuration to its position in the actual deformed configuration $\mathbf{x}(\xi)$. Locally, this mapping can be described by the deformation gradient:

$$\mathbf{F} = \frac{\partial \mathbf{x}}{\partial \boldsymbol{\xi}} = \mathbf{1} - \frac{\partial \boldsymbol{u}}{\partial \boldsymbol{\xi}}, \quad (6)$$

with $\mathbf{u}$ being the displacement vector. The deformation gradient has the polar decomposition

$$\mathbf{F} = \mathbf{RU}, \quad (7)$$

with the rotation tensor $\mathbf{R}$ and the right stretch tensor $\mathbf{U}$. For small deformation, where $\frac{\partial \mathbf{x}}{\partial \xi} \ll \mathbf{1}$ can be assumed, and for defining a predictor strain increment in the iterative solution procedure as outlined below, we neglect the nonlinear terms, approximating $\frac{\partial}{\partial \xi} \sim \frac{\partial}{\partial \mathbf{x}}$. Hence, this assumption results in the definition of the engineering strain $\Delta\boldsymbol{\varepsilon}$ and the rotation increments $\Delta\mathbf{R}$:

$$\Delta\boldsymbol{\varepsilon} = \frac{1}{2}(\mathbf{F} + \mathbf{F}^T) - \mathbf{1}, \quad (8)$$

$$\Delta\mathbf{R} = \frac{1}{2}(\mathbf{F} - \mathbf{F}^T). \quad (9)$$

Although the given description of deformation does not rely on tracing back the current configuration to the original configuration, it is helpful to integrate the strain (either total, elastic or plastic) by a proper strain measure. For the sake of additivity, we use the logarithmic strain:

$$\boldsymbol{\varepsilon} = \ln(\mathbf{U}). \quad (10)$$

The basic idea is to formulate an iterative scheme that, first, by iterative solution of the small strain mechanical problem finds the mechanical equilibrium :

$$\boldsymbol{\nabla}\boldsymbol{\sigma} = \mathbf{0}. \quad (11)$$

And, second, at the same time applies the geometry change in an Eulerian setting. The geometry update is given by the previously defined rotation increment (eq. 4) and by the displacement $\mathbf{u}$ itself that can be applied to any conserved scalar variable ρ by the compressible transport equation:

$$\Delta\rho + \boldsymbol{\nabla}(\mathbf{u}\rho) = 0. \quad (12)$$

Rotation increments $\Delta\mathbf{R}$ are explicitly applied to any tensorial variables $\boldsymbol{\tau}$

$$\boldsymbol{\tau} \to \Delta\mathbf{R}\,\boldsymbol{\tau}\,(\Delta\mathbf{R})^T. \quad (13)$$

In detail, the scheme is structured as follows. Assuming that at time t the system starts in mechanical equilibrium, the reference state is identified with the current state:

$$\boldsymbol{\xi}^0 = \mathbf{x}^0. \quad (14)$$

The system will be driven out of mechanical equilibrium by:

1) changing the boundary conditions, e.g., by applying external load,
2) changing the material's constants, e.g., by diffusion of species or temperature and by the impact of damage,
3) plastic relaxation $\boldsymbol{\varepsilon}_p$ or
4) phase transformation or grain boundary motion associated with volume change and the eigenstrain $\boldsymbol{\varepsilon}^*$.

In the following, we will consider in detail the effect of damage on the elasticity tensor C, plastic relaxation, and phase transformation summing up to stress-free strain increment. The present mechanical state is no longer divergence free:

$$\boldsymbol{\nabla}\boldsymbol{\sigma}_n^0 - \boldsymbol{\nabla}(\mathbf{C}^0\Delta\tilde{\boldsymbol{\varepsilon}}^0) \neq 0. \quad (15)$$

Here the superscript $()^0$ denotes the zeroth iteration. An elastic corrector strain $\boldsymbol{\varepsilon}^1$ for the first iteration is calculated defining a corrector displacement $\mathbf{u}^1$:

$$\boldsymbol{\nabla}\boldsymbol{\sigma}_n^0 - \boldsymbol{\nabla}(\mathbf{C}^0(\Delta\tilde{\boldsymbol{\varepsilon}}^0 - \boldsymbol{\varepsilon}^1)) = 0. \quad (16)$$

In the next stage, the intermediate displacement $\mathbf{u}^1$ is applied by a geometry update according to eq. (12) and the intermediate rotation $\Delta\mathbf{R}^1$ is applied to update the Cauchy stress

$$\boldsymbol{\sigma}^1 = \Delta\mathbf{R}^1\boldsymbol{\sigma}^0(\Delta\mathbf{R}^1)^T + \Delta\boldsymbol{\sigma}^1, \quad (17)$$

with the intermediate stress increment

$$\Delta\boldsymbol{\sigma}^1 = \mathbf{C}^0(\Delta\tilde{\boldsymbol{\varepsilon}}^0 - \boldsymbol{\varepsilon}^1). \quad (18)$$

With the new stress increment, the stress-free strain $\Delta\tilde{\boldsymbol{\varepsilon}}^0$ has to be updated according to the plastic relaxation and the stiffness tensor $\mathbf{C}^0$ according to the damage evolution as described in the damage section. Eq. (12) describes an update of the actual stress state of a material's point that is accumulated from the incremental stresses and rotations based on a so-called co-rotational objective time derivative in line with [17]. With the new stress predictor $\Delta\boldsymbol{\sigma}^1$ at hand, the iteration procedure is repeated. Convergence is reached after n iterations when the corrector strain vanishes $\boldsymbol{\varepsilon}^n \to 0$ as well as the rotation increment $\Delta\mathbf{R}^n \to \mathbf{1}$. Then the stress tensor $\boldsymbol{\sigma}^n$ becomes divergence-free.

The presented approach can be coupled with any plasticity framework in a straightforward manner. Here, we apply a simple damage independent viscoplastic model in order to demonstrate the effect of damage on elastic softening during plastic yielding. The influence of damage instead will be considered by differentiation of nominal and effective stress space (see [18]) where the effective stress is the driving force for creep. With a strain rate constant A and a stress exponent k we define the plastic strain rate as

$$\dot{\boldsymbol{\varepsilon}}_p = A\boldsymbol{\sigma}^k. \quad (19)$$

Thus, accumulation of large plastic strain is straight forward, simplifying its coupling with plastic strain dependent models, like the damage model.

In order to model failure of the modeled

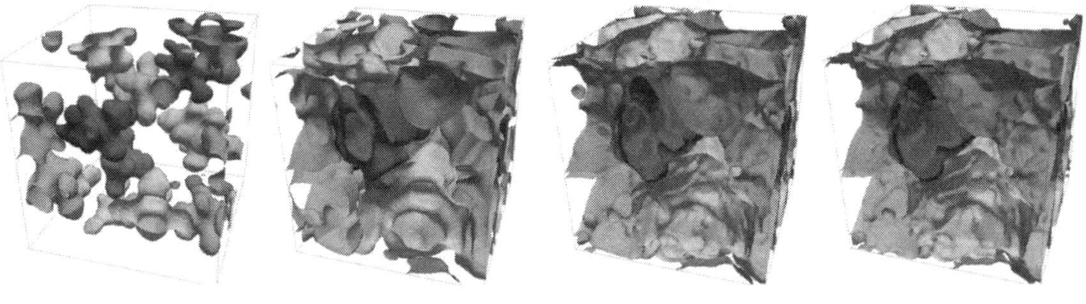

Figure 2. Microstructure evolution during solidification of Mg-5wt.% Al alloy.

microstructure, a damage model published in [18] is used. It assumes the stiffness tensors being dependent on the equivalent plastic strain:

$$\mathbf{C} = \mathbf{C}_0(1 - D), \tag{20}$$

$$D = \begin{cases} 0 \; if \; p < p_{min}, \\ \dfrac{p - p_{min}}{p_{max} - p_{max}} \; if \; p_{min} < p < p_{max}, \\ 0.95 \; if \; p > p_{max}. \end{cases} \tag{21}$$

Where the equivalent plastic strain p is calculated via integration of the equivalent plastic strain rate:

$$\dot{p} = \sqrt{\frac{2}{3}\dot{\boldsymbol{\varepsilon}}_p : \dot{\boldsymbol{\varepsilon}}_p}. \tag{22}$$

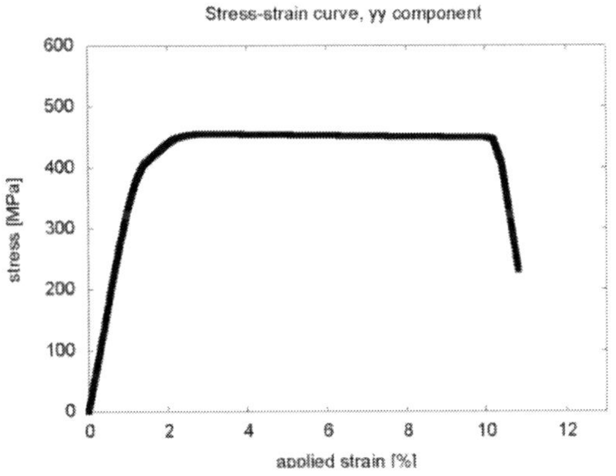

Figure 3. Stress-strain curve obtained during the simulation of damage in the Mg-5wt.% Al system.

5 Result

The governing equations (eq. 1-3) are solved with an explicit Euler integration scheme. Our simulations start with a cubical fully liquid domain with an edge length of 300µm size at 900K (above the melting temperature). From this stage, a fixed heat extraction is applied at each time step. After a few simulated time steps, the system is undercooled and nucleation occurs. The nucleated grains of primary α-phase grow in a dendritic fashion and fill up most of the volume. The nucleation density is highly dependent on the heat extraction rate applied (see [15]).

After the growth slows down, the heat extraction rate becomes again increasingly more dominant than the latent heat production and the system cools down further. When the domain gets cooled below the eutectic temperature, nucleation and growth of the secondary β-phase occurs. The resulting microstructure is shown in Fig. 2. The eutectic solidification happens on a different scale and is usually treated in a multi-scale approach (see [14]). In this work we can assume the rest of the melt at this point in time to decompose into secondary β-phase and tertiary α-phase and form a fine-grained eutectic structure. Because of the scale difference of our primary grain structure and the interdendritic eutectic, we treat the eutectic as one phase with continuous properties rather than a mix of two phases.

Having the solidified microstructure, a tensile test like calculation is performed. In general, all through the mechanical simulation, the phase-field evolution can be considered to ensure the proper interface profile. It has however reduced dynamics due to the low temperatures of mechanical testing. The system is strained along the y-axis until failure is observed. By failure we mean the formation of a large region of maximal damage $D = 0.95$ and a stress drop as seen in fig. 3. In the current simulation, this happens after the applied strain of approximately 10%.

The stiffness coefficients tensor used here is taken from [16] and can be seen in tab.1. The plasticity model is parametrized by k = 5 and A = 1e-13 Pa^{-5}. The damage parameters are set as $p_{min} = 0.08$ and $p_{max} = 0.10$.

Table 1: Stiffness coefficients [16]

C_{11}	C_{33}	C_{44}	C_{55}	C_{12}	C_{13}
59.40	61.60	16.40	16.90	25.61	21.44

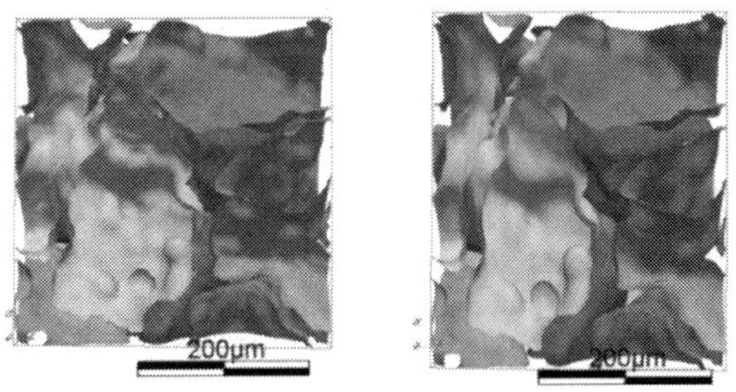

Figure 4. Microstructure before and after 10% straining deformation.

The 10% deformed system is shown in fig. 4. Each of the grains is modelled to be individually oriented. Due to the stiffness anisotropy, the different grain orientations result in a stress field that starts plastic strain evolution and finally creates damage and softens the material. Since there is only a slight variation in the stiffness coefficients, the resulting damage does not localise in the grain boundaries. Instead, after there is a maximally damaged material, the damaged region developes in an almost spherical manner. However, the stress-strain curve presented in fig. 3 reproduces qualitatively realistic behaviour (compare Song et al. [19]).

6 Parallelism

Large scale phase-field simulations, such as the simulation of Mg-Al alloy solidification mentioned above, require significant amounts of memory and computation time, thus exceeding the capabilities of single node computers. Distributed-memory parallelism using MPI allows us to overcome these limitation and makes simulations of large computational domains feasible. A simple domain decomposition that divides the domain into equally sized sub-domains can yield an

DOI: 10.1051/matecconf/20168002014

insufficient performance due to load-imbalances caused by an inhomogeneous distribution of the computational load in the domain.

The cause for these inhomogeneous load distributions is the active parameter tracking as proposed by Vedantam et al. [1].

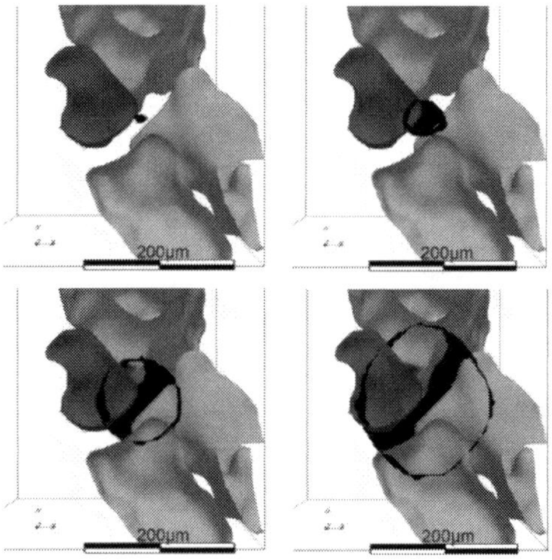

Figure 5. Damage evolution (black surface) during the last calculation steps in zoomed view.

Here, only the interactions between locally existing phase fields in each grid point are calculated, which significantly reduces the total computational complexity when dealing with a large number of different phase fields. This practice, however, can lead to inhomogeneous load distributions as the computational complexity in a grid point is dependent on the number of locally active phase fields in this grid point.

In order to fully utilize the computational resources of modern parallel computer architectures, the load-balancing techniques presented by Tegeler et al. [2] are applied. This involves an over-decomposition of the computational domain, i.e. more sub-domains than MPI-processes as seen in fig. 6, so that a greedy graph-partitioning algorithm can be used to assign sub-domains to processes with the goal of minimizing the load-imbalance and communication cost.

Even if the computational load is balanced between processes, waiting times can occur at synchronization points, i.e., halo exchange between the sub-domains if the computational load is not balanced between all synchronization points.

Therefore, multiple synchronization points within one time step are avoided by increasing the size of the halo to allow multiple stencil operations per time step.

In order to improve the worst case load-imbalance guaranteed by the graph-partitioning algorithm, we use bisection of computationally costly sub-domains into two smaller sub-domains of reduced computational load along a cutting plane, which splits the computational load into two approximately equal parts.

As the bisection of a sub-domain introduces additional computations on the halo as well as communication, an optimal sub-domain load is determined at runtime taking into account the hardware properties and the specific load distribution and only sub-domains of higher load are split into smaller domains.

For the simulation of Mg-Al alloy solidification as seen in fig. 2 a total of 6 compute nodes with 16 cores and 128GB of RAM each were used.

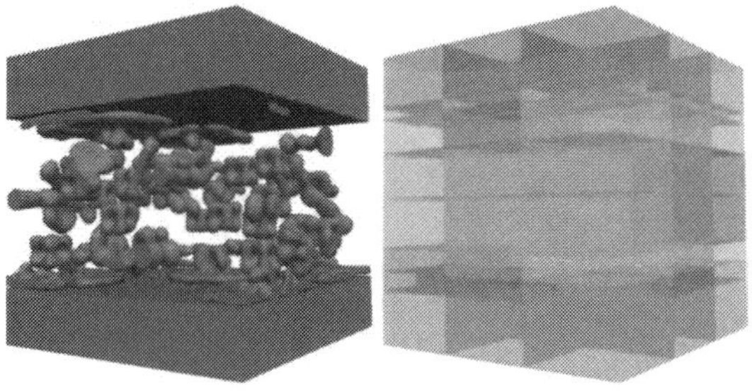

Figure 6. Interdendritic Mg-Al alloy solidification with the domain decomposition shown on the right-hand side. Different colours indicate different process ranks.

In order to benchmark the scaling properties of the parallelization, a set performance tests using the simulation of Mg-Al alloy solidification with the multiscale approach mentioned above have been conducted. In a channel between two primary α-phase dendrites a secondary β-phase nucleates and grows much faster than the primary α-phase. This is depicted in fig. 6 together with the domain decomposition generated by the presented load balancing techniques.

It shows that the sub-domains near the border of the channel have been split into smaller sub-domains as the computational load is high in this region, due to interfaces between α- and β-phase and the liquid.

The results of the scaling benchmark are shown in fig. 7 and it can be seen that if the computational domain per core is kept constant some fluctuation in the weak scaling performance are observable, but no significant decline in performance as the largest time per step on the largest domain is even slightly smaller compared to smallest domain in the set of benchmark runs.

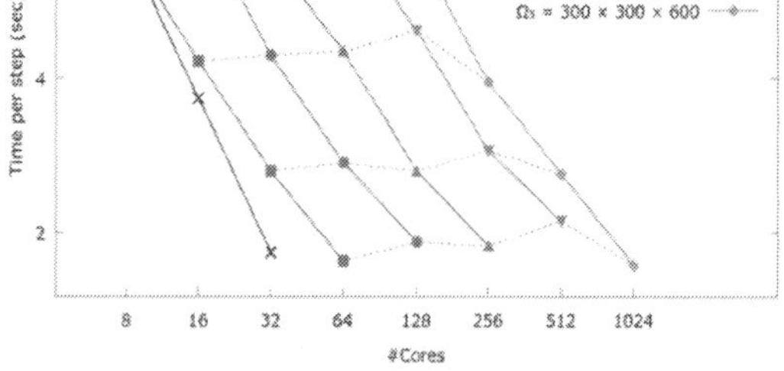

Figure 7. Combined weak (dashed lines) and strong (solid lines) scaling performance for the interdendritic Mg-Al alloy solidification as seen in fig. 6.

7 Conclusion

The presented work incorporates modelling of two physical processes occurring in a magnesium system: the solidification process that results in a typical Mg microstructure and a subsequent mechanical testing process quantifying the quality of the resulting alloy microstructure. The used approach follows the spirit of computational execution of experimental work and defines a possible workflow corresponding to modelling of a material processing chain. Both parts are executed within the phase-field framework profiting on its morphology describing potential accompanied by mathematical simplicity. Due to specification of realistic model parameters the observed results are comparable to experimental observations. With this prove of concept for the used work procedure at hands, the authors plan to use precise mechanical data for a specified system and a crystal structure informed plasticity to improve the mechanical model gaining a one to one relation between the experimental findings and the simulation predictions.

Acknowledgements

This work was supported by Fundamental Research Program of the Korea Institute of Materials Science (KIMS).

References

1. S. Vedantam and B. S. V. Patnaik, Phys. Rev. E 73, 016703 (2006)
2. M. Tegeler, A. Monas, and G. Sutmann, in Proceedings of Fourth International Conference on Parallel, Distributed, Grid and Cloud Computing for Engineering, 5 (2015)
3. M. C. Zhao, M. Liu, G. Song and A. Atrens: Corros. Sci., 2008, 50, pp. 1939-1953.
4. M. Jönsson, D. Thierry and N. LeBozec: Corros. Sci., 2006, 48, pp. 1193-1208.
5. D. Höche and J. Isakovic: Chem. Ing. Tech., 2013, Volume 85, Issue 12, pp. 1951-1956.
6. [6] L-Q Chen: Annu. Rev. Mater. Res., 2002, 32, pp. 113-140.
7. I. Steinbach: Modell. Simul. Mater. Sci. Eng., 2009, 17, pp. 073001-31.
8. I. Steinbach,: Annu. Rev. Mater. Res., 2013, 43, pp. 89-107.
9. B. Böttger, J. Eiken and I. Steinbach: Acta Mater. 2006, 54, pp. 2697-2704.
10. J. Eiken: Int. J. Mat. Res. (formerly Z. Metallkd.) 101 (2010) 4
11. I. Steinbach and A. Pezzolla: Phys. D (Amsterdam, Neth.), 1999, 134, pp. 385-393.
12. A. L. Greer, A. M. Bumm, A. Tronche, P. V. Evans and D. J. Bristow: Acta Mater., 2000, 48, pp. 2823-2835.
13. I. Maxwell and A. Hellawell: Acta Metall., 1975, 23, pp. 229-237.
14. A. Monas, O. Shchyglo, S. Kim, C. Yim, D. Höche, I. Steinbach: JOM, 67, 1805-1811.
15. A. Monas, O. Shchyglo, D. Höche, M. Tegeler, I. Steinbach: IOP Conference Series: Materials Science and Engineering, 84, 012069.
16. L. J. Slutsky and C.W. Garland, Phys. Rev. 107 (1957) 972-976
17. T. J. R. Hughes J. Winget, International Journal of Numerical Methods in Engineering, 15 (1980) 1862-1867.
18. M. Boeff, F. Gutknecht P.S. Engels A. Ma A. Hartmaier, Engineering Fracture Mechanics, 147 (2015) 373-387.
19. Jie Song, Shou-Mei Xiong, Mei Li, John Allison Journal of Alloys and Compounds 477 (2009) 863–869

Crystal plasticity extend FEM implementation of thermal-tensile aluminum alloy

Yang Liu[1], Yiguo Zhu[1], Liang Ying[1] and Ping Hu[1,a]

[1]*State Key Laboratory of Structural Analysis for Industrial Equipment, Faculty of Vehicle Engineering and Mechanics, Dalian University of Technology, 2 Linggong Road, Dalian 116023, China*

Abstract. Multi-level approach has been used to simulate the thermal deformation of aluminium alloy at different temperature and strain rate. The crystal plasticity model is extended in the finite element method and the thermal behaviour is integrated in the constitutive equations. Moreover, the damage evolution is also reflected in the simulation using continuum damage mechanics model. Thus, the void evolution and thermal effect could both be shown in the simulation. A new shear strain rate model is constructed with the thermal activated mechanism to describe the rate dependent behaviours during tensile test. The thermal parameters are determined in a fitting test of representative volume element to compare with the experimental data. The results prove that the mechanical tensile behaviour of 5052 aluminium alloy could be well described at different temperatures. The damage evolution process is expressed by the stress concentration and strain concentration in the finite element simulation, which are also confirmed by the experiments.

1 Introduction

Aluminium alloy has become one of most popular metal in the vehicle industry. The steel is replacing by the aluminium alloy for reducing the weight and protecting the environment. The high specific strength and good resistance allow aluminium alloy to form car components, such as trunk, bumper, and engine hood with good properties. To produce the components using aluminium alloy, the thermal mechanical behaviours should be simulated and controlled in the appropriate range, since the formability of aluminium alloy is small in the room temperature. The traditional phenomenological models lack inner illustration of material mechanism and metallographic essence of microstructure. In the microscopic view, anisotropic plastic behaviour is a primary phenomenon in the crystallography and the dislocation dynamics could be its most significant cause. In addition to the deformation, the crystallographic anisotropy also leads to the different kinds of orientation dependent behaviours, such as strain hardening and void evolution.

In this sense, to better understand the relationship between material property and microstructure of aluminium alloy, various multi-level polycrystalline schemes are built and developed to describe this orientation property such as Taylor and Saches models. These early attempts have already simulated the texture and lattice reorientation well, but the grain interaction and complicated boundary condition could not be explicitly revealed in these homogenization methods. After the 1980s, a crystal plasticity extended finite element method (CPFEM) occurred in 2D or 3D framework [1-3]. Nowadays, it has become a widely-used method to simulate different kinds of materials and complicated outer boundary conditions. The stress heterogeneity and material behaviours near grain boundaries are betters understood using CPFEM. Each grain scale level of CPFEM is hierarchically constructed by associated constitutive laws accounting for microstructure properties. Green's function fast Fourier transform (FFT) and spectral interpolation is also combined within CPFEM to efficiently and accurately simulate the orientation-dependent mechanism at the grain interface [4, 5].

Considering more complex outer and inner effects, such as the temperature, strain rate, recrystallization, and dislocation mobility, the mechanical response could be influenced and deviated enormously from the ordinary condition. Therefore, more complicated model are needed to describe these effects, and the constitutive laws are generally revised to satisfy more complicated cases. As for thermal effects, a visco-plastic self-consistent (VPSC) model is generally used to homogenize the polycrystal and calculate the mechanical behaviour as a function of strain rate and temperature [6]. This advanced mean-field scheme is able to simulate the temperature field coupled with mechanical field, but still lack of capturing co-deformation within grains. In view of spatial stress distribution, 3D CPFEM framework is used and the thermodynamically-consistent framework is integrated to characterize the heat conduction and the heat boundary

[a] Corresponding author: pinghu@dlut.edu.cn

condition. Another thermomechanical coupling could calculate the heat generation of deformation recovery, plastic work dissipation and stored cold work motivated by the rearrangement of defects [7, 8].

Here, integrated in the CPFEM simulation, the damage evolution of continuum damage mechanics model and thermal activated mechanism are involved in describing the thermomechanical behaviours and void evolution process of aluminium alloy at elevated temperature. The thermal-associated parameters should be obtained by the fitting test, which could be done by introducing a representative volume element (RVE). Along with the jump test, all the parameters could be confirmed in the RVE simulation and later used in the CPFEM simulation of actual thermal tensile test. The practical specimen of aluminium alloy is simulated and the necking behaviour is also compared to a reduced domain, and localized deformation and void evolution are shown in the 3D large plastic strain simulation. This article is constructed as follows: In chapter 2, the multi-scale framework is introduced; in chapter 3, the thermal constitutive equations and the damage functions are briefed; then, the thermal parameters are obtained in the RVE simulation and the jump test is also introduced; in chapter 4, the simulation of actual thermal tensile test is briefed; the conclusion and discussion are brought in chapter 5.

2 Framework of Crystal Plasticity Extended Finite Element Method

This hierarchical model is based on the framework of crystal plasticity and implemented in the finite element method. In the finite element method, the displacement field is calculated by the discretized virtual work principle and it will be checked at each increment. Thus, the stress equilibrium is satisfied at the boundary of grains, moreover, the strain compatibility is also fulfilled among grains in a weak numerical sense.

$$F = F^e F^p F^\theta \quad (1)$$

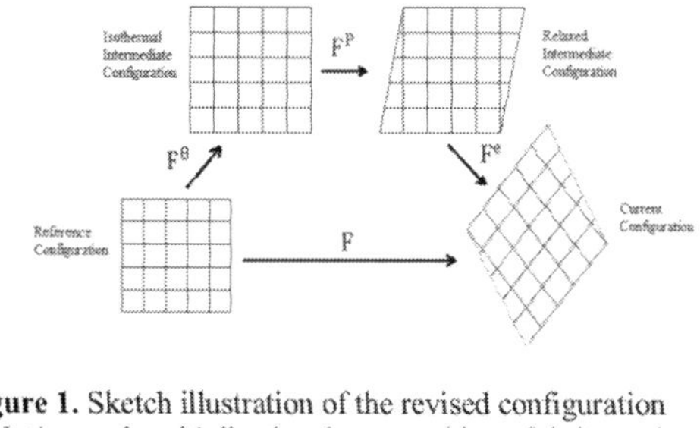

Figure 1. Sketch illustration of the revised configuration evolution and multiplicative decomposition of deformation gradient.

The crystal plasticity model is coupled with thermal effect and damage evolution, which is inserted in the subroutine Use-defined Material Mechanical Behaviour (UMAT) of commercial software ABAQUS. The constitutive model of crystal plasticity method is revised. As for the kinematic relation, the multiplicative decomposition of deformation gradient is modified due to thermal influence.

The original elastic and plastic deformation gradients are $\mathbf{F}^e$ and $\mathbf{F}^p$, respectively. It should be noticed that $\det \mathbf{F}^e > 1$ and $\det \mathbf{F}^p = 1$, which are due to the lattice stretching and the incompressible plastic volume. $\mathbf{F}^\theta$ accounts for the isothermal deformation gradient, which is defined as,

$$F^\theta = [1 + \mu(\theta - \theta_0)]I \quad (2)$$

where, θ and θ_0 are the current and initial absolute temperature, and $\mathbf{I}$ is the unit tensor.

The deformation gradient is related to the subroutine UMAT at the start of each step in ABAQUS. And the basic constitutive equations are used.

$$E = \frac{1}{2}({F^e}^T F^e - I) \quad (3)$$

$$T = {F^e}^{-1}\{(\det F^e)\sigma\}{F^e}^{-T} \quad (4)$$

Other traditional kinematic equations are defined as follows to relate the microscopic crystal structure to the macroscopic mechanical deformation field.

$$\dot{F}^p = L^p F^p \quad (5)$$

$$L^p = \sum_\alpha \dot{\gamma}^\alpha \bar{P}_0^\alpha \quad (6)$$

$$F_{t+\Delta t}^p \doteq (I + \Delta t L_{t+\Delta t}^p) F_t^p \quad (7)$$

where $\bar{P}_0^\alpha$ is the initial Schmidt tensor, $\mathbf{L}^p$ the plastic velocity gradient, $\dot{\gamma}^\alpha$ the shear strain rate in the αth slip system. Equation (7) is the time incremental representation of updating $\mathbf{F}^p$.

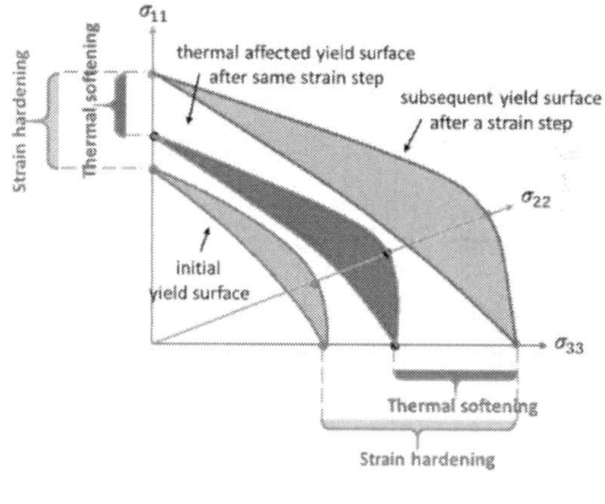

Figure2. The yield surface changing with strain hardening and thermal softening in the stress space.

According to the thermal activated mechanism, the evolution of shear strain rate is revised instead of the power-law equation.

$$\dot{\gamma}^\alpha = \frac{\rho_m v_{id} b^2}{2} exp\left(\frac{-\Delta F}{k\theta}\right) exp\left(\frac{(\tau^\alpha - s^\alpha)\Delta V}{k\theta}\right) \quad (8)$$

where, ρ_m is the mobile dislocation density, v_{id} is the vibration frequency of this segment of dislocation, b is the Burger's vector, v_g the average gliding dislocation velocity, k the Boltzmann constant, ΔF and ΔV the thermal activated energy and volume.

In addition to the shear strain rate function, the strain hardening rate is also revised.

$$\dot{s}^\alpha = \sum_\beta h^{\alpha\beta}\left|\dot{\gamma}^\beta\right|,\ h^{\alpha\beta} = q^{\alpha\beta}h^{(\beta)}, \quad (9)$$

$$h^{(\beta)} = h_0\left(\frac{s^s-s^\beta}{s^s-s_0}\right)^a,\ s^s = s_0^s\left(\frac{\dot{\gamma}}{\dot{\gamma}_s}\right)^{\hat{m}} \quad (10)$$

where, $h^{\alpha\beta}$ is the strain hardening rate in the slip system α shearing on the slip system β, $q^{\alpha\beta}$ is the hardening matrix. $h^{(\beta)}$ is defined by initial CRSS s_0, saturation strain rate value $\dot{\gamma}_s$ and other material-independent parameter, reference CRSS value s_0^s, the reference hardening rate h_0, exponential value $\hat{m}$. The parameters of the strain rate function and the strain hardening equation could be obtained in the fitting test of representative volume element (RVE).

As for the damage evolution, a simple continuum damage equation is used to couple with the finite element method [3].

$$D = \begin{cases} 0 & (\gamma_m \le \gamma_{m,ini}) \\ D_{max}\left(\frac{\gamma_m-\gamma_{m,ini}}{\gamma_{m,max}-\gamma_{m,ini}}\right)^M & (\gamma_{m,ini} < \gamma_m < \gamma_{m,max}) \\ D_{max} & (\gamma_m \ge \gamma_{m,max}) \end{cases} \quad (11)$$

where, $\gamma_{m,ini}$ is the initial shear strain, D_{max} the maximum value of damage factor, $\gamma_{m,max}$ is the maximum shear strain of the 12 slip systems in FCC crystal, M is the exponent that controls the shape of these curves. The damage initiates when γ_m is higher than $\gamma_{m,ini}$ in each integration point of finite elements. As γ_m reaches the $\gamma_{m,max}$ in one integration point, the damage reaches its maximum value.

3 Figures and tables

The conventional tensile test is modified into thermal tensile equipment. The copper electrode is used to attain the thermocouple onto aluminum alloy to control the current temperature and the tensile behavior is controlled by the computer to precisely measure the stress-strain curves. Other strain rate and temperature jump tests are also used to obtain the thermal activated energy and volume, ΔF and ΔV [9].

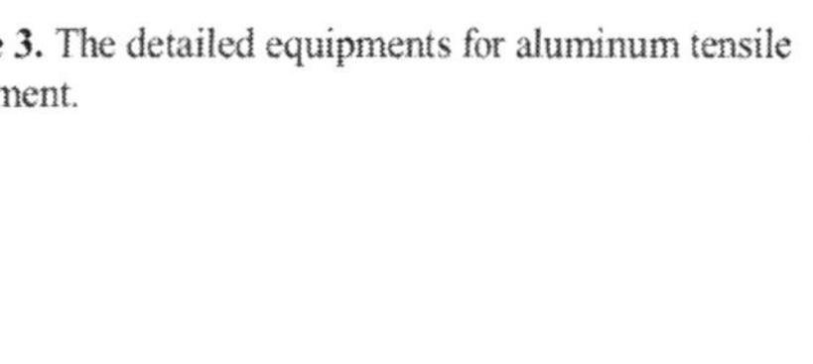

Figure 3. The detailed equipments for aluminum tensile experiment.

In this section, the RVE simulation test is used to obtain the 12 material parameters ρ_m, v_{id}, ΔF, ΔV, q, h_0, a, s_0^s, $\hat{m}(\theta = 300K)$, $\hat{m}(\theta = 373K)$, $\hat{m}(\theta = 473K)$, $\hat{m}(\theta = 573K)$ and to verify the experimental parameters acquired from the jump tests. The cube size is about $0.3mm \times 0.3mm \times 0.3mm$, which contains 50 crystallographic orientations, representing an aluminium aggregate. The Voronoi tessellation is used to divide the cube, and the morphology of each single crystal is similar to the real grain morphology. 5659 tetrahedron elements are conducted to mesh the polycrystalline to better suit the element shape and obtain a better grain mesh. The mechanical fitting test is revealed in the stress-strain curves and a good agreement is acquired to simulate the material parameters, which can be further used in the practical simulation of realistic tensile behavior in Section 4.

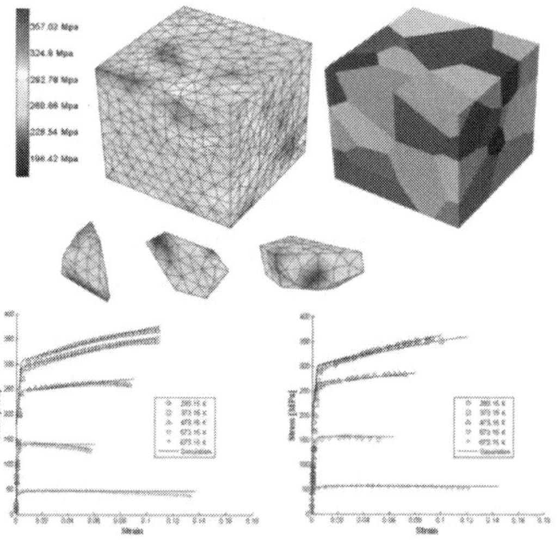

Figure 4. Schematic illustration of parameters determination of RVE finite element simulation.

4 Thermal tensile simulation

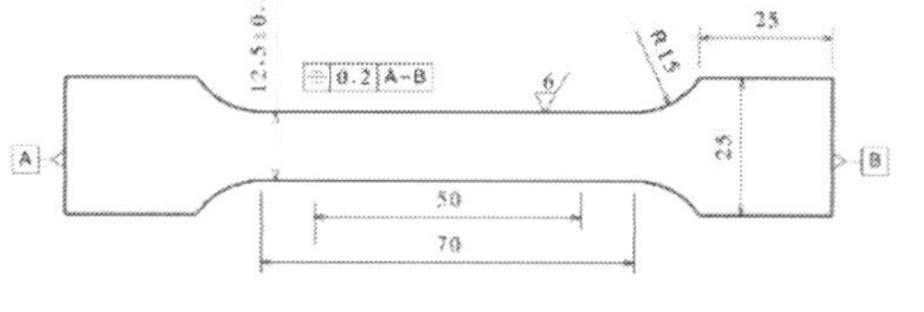

Figure 5. The dimension parameter of specimen for tensile test.

To verify this multiscale framework of CPFEM method, an actual tensile test is compared with the simulation results on 5052 aluminum alloy.

Moreover, a Taylor-based homogenization model is also implemented with the crystal plasticity model to compare the mechanical results. The material parameters are obtained using RVE simulation. Here, as shown in Fig. 5, the CPFEM simulation results agrees well with the experimental data and also the Taylor model at all the three elevated temperatures. In both the simulation and experiment, as the temperature gets higher, the strain hardening phenomenon is numerically weakened by the thermal affected strain hardening equation. The damage point of each temperature varies because of the damage model.

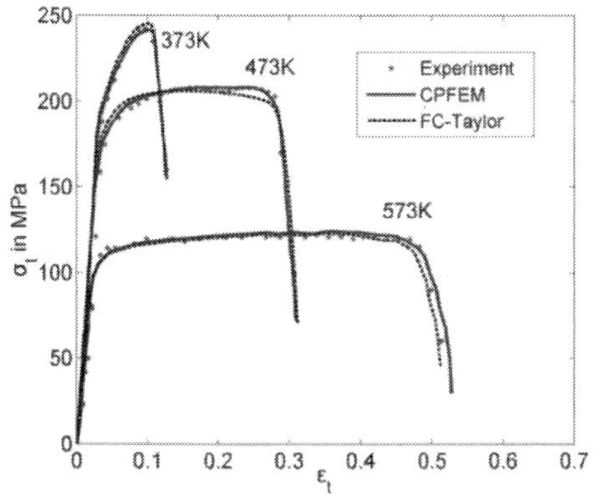

Figure 6. Comparison between experimental data and crystal plasticity methods at various temperatures.

The stress concentration is shown in Fig. 6 according to the orientation mismatch among adjacent grains, which is originally caused by the single crystal anisotropy. Generally isotropic J2 plasticity cannot reveal this kind of concentration. The damage evolution process is simulated in the ABAQUS, and the subroutine UMAT is implemented.

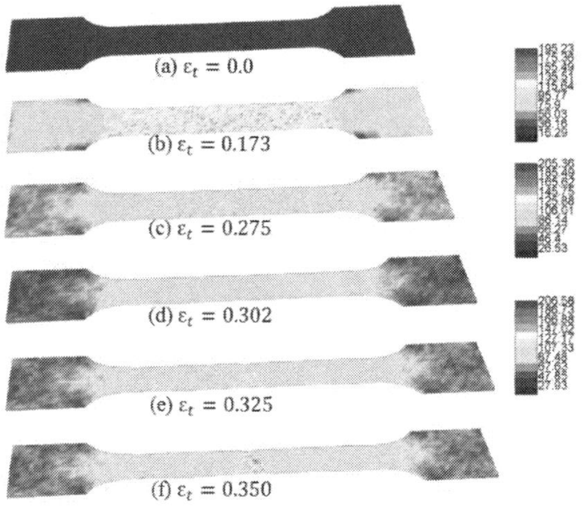

Figure 7. Von Mises stress distribution (maximum shear strain damage model: $\theta = 473K, \gamma_{m,ini} = 0.223, \gamma_{m,max} = 0.249$, $D_{max} = 0.8, M = 3.0$).

As shown in Fig. 7, the damage evolution is revealed as the logarithmic strain exceeds a certain value. The most significant phenomenon occurs as the $\varepsilon_t = 0.325$, when all the elements of the middle section has entered into the damage level. The damage elements increases as the strain is enhanced and it is similar to the void evolution in the material behavior of aluminum alloy.

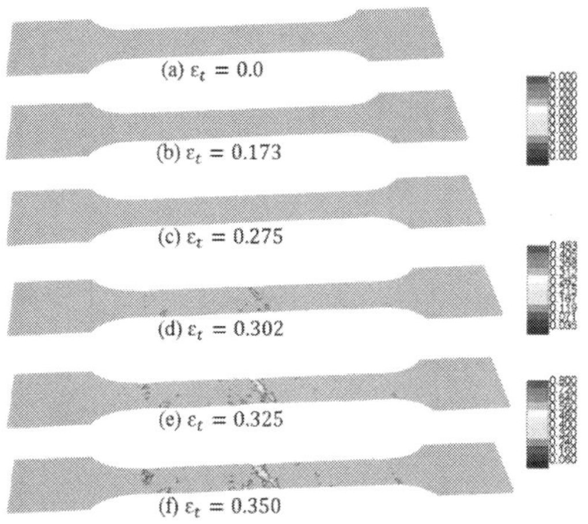

Figure 8. Damage factor distribution (maximum shear strain damage model: $\theta = 473K, \gamma_{m,ini} = 0.223, \gamma_{m,max} = 0.249$, $D_{max} = 0.8, M = 3.0$).

5 Conclusion and discussion

A crystal plasticity extended finite element method is implemented in a multiscale framework, and it could be used to consider the thermal effects and the damage evolution. The local damage evolution is shown in the simulation of commercial software ABAQUS, and the mechanical response is calculated using the subroutine UMAT. The crystal plasticity model is revised by the thermal activated mechanism and a RVE simulation is able to obtain all the material parameters. Therefore, any FCC-structure material behavior could be calculated according to the fitting test of RVE simulation. The mechanical response agrees well with the experimental data.

The stress concentration is shown in Fig. 6 according to the orientation mismatch among adjacent grains, which is originally caused by the single crystal anisotropy. Generally isotropic J2 plasticity cannot reveal this kind of concentration. The damage evolution process is simulated in the ABAQUS, and the subroutine UMAT is implemented.

In our future study, different crystal structures are considered to develop the subroutine and other polycrystalline behaviors are the goals of our research, such as the grain boundary property, the recrystallization behavior. The temperature field is also a significant problem to be solved coupled with the mechanical field in the finite element analysis.

Acknowledgements

This work was funded by the Key Project of the National Natural Science Foundation of China (No.11272075), China Postdoctoral Science Foundation (2014M561223) and the Fundamental Research Funds for the Central Universities (DUT14RC(3)032).

DOI: 10.1051/matecconf/20168002015

References

1. B. Peeters, S.R. Kalidindi, P. Van Houtte and E. Aernoudt, Acta Mater. 48, 2123-2133 (2000)
2. C.C., Tasan, J.P.M. Hoefnagels, M. Diehl, D. Yan, F. Roters and D. Raabe, Int. J. Plasticity 63, 198-210 (2014)
3. J. Kim and J.W. Yoon, Int. J. Plasticity 73, 3-23 (2015)
4. L. Chen, J. Chen, R.A. Lebensohn, Y.Z. Ji, T.W. Heo, S. Bhattacharyya, K. Chang, S. Mathaudhu, Z.K. Liu and L.Q. Chen, Comput. Method Appl. M. 285, 829-848 (2015)
5. D. Raabe and F. Roters, Int. J. Plasticity 20, 339-361 (2004)
6. J.R. Mayeur, I.J. Beyerlein, C.A. Bronkhorst and H.M. Mourad, Int. J. Plasticity 65, 206-225 (2015)
7. S. Keshavarz and S. Ghosh, Acta Mater 61, 6549-6561 (2013)
8. G.B. Sarma and B. Radhakrishnan, Mater. Sci. Eng., A 385, 91-104 (2004)
9. D. Caillard and J.L. Martin, *Thermal activated mechanisms in crystal plasticity* (ELSEVIER Ltd, Oxford, 2003)

Biaxial experiments and numerical analyses on damage prediction in metal forming processes

Michael Brünig[1,a], Steffen Gerke[1], and Marco Schmidt[1]

[1] *Institut für Mechanik und Statik, Universität der Bundeswehr München, 85577 Neubiberg, Germany*

Abstract. The paper discusses an anisotropic continuum damage and failure model for ductile metals. The phenomenological approach is based on kinematic definition of damage tensors and takes into account the effect of stress state on damage conditions and damage strain evolution laws. Different branches of these criteria based on experimental studies and numerical simulations are considered corresponding to various microscopic damage and fracture mechanisms depending on stress intensity, stress triaxiality and the Lode parameter. Experiments with biaxially loaded specimens and corresponding numerical simulations have been performed showing that they cover a wide range of stress states in combined tension, shear and compression regimes. Digital image correlation technique has been used to analyze current strain states in critical regions of the specimens. Damage behavior for shear-compression loading conditions appearing in metal forming processes is discussed in detail.

1 Introduction

Numerical calculations for optimization of forming processes currently receive remarkable attention because products of modern metal forming processes have to fulfill economic, environmental and material strength requirements caused by the increasing demands of the customers. Therefore, the use of high quality metals like high strength steels, advanced high strength steels and aluminum alloys has been increased during the last decades. There are, for example, requirements on lightweight design leading to improved energy consumption or cost efficiency and, at the same time, to enforce the safety requirements. These aspects demand enhancement of material properties to avoid early localization of inelastic strains as well as damage and fracture of structural components under complex loading conditions. Hence, numerical simulation of various metal forming processes becomes a indispensable practical tool in modern technologies. This requires highly predictive and practically applicable constitutive theories as well as corresponding efficient numerical methods. Therefore, accurate modeling and numerical simulation of large inelastic deformations as well as of damage and failure behavior in materials and structures are of keen interest for many metal forming processes. In various experiments and forming operations with ductile metals large and often localized inelastic deformations occur which may be accompanied by different damage and fracture mechanisms on the micro- and meso-scales. Formation of damage on the micro-level may then lead to macro-cracks and to final failure of structural elements. Since industrial processes like rolling and forging often show remarkable negative hydrostatic stress states damage and fracture mechanisms at negative stress triaxialities are here of special interest.

[a]Corresponding author: michael.bruenig@unibw.de

Characteristics of damage and failure mechanisms depend on stress state acting in a material point. For example during tension loading with high positive stress triaxialities damage in ductile metals is mainly caused by nucleation, growth and coalescence of micro-voids whereas during shear and compression loading with nearly zero or negative stress triaxialities the main damage mechanisms on the micro-scale are formation, growth and coalescence of micro-shear-cracks. Combination of these basic microscopic damage mechanisms occurs for moderate positive stress triaxialities whereas no damage has been observed in ductile metals for high negative stress triaxialities. Thus, ductile damage behavior and crack formation strongly depend on the stress triaxiality. Hence, development of accurate and realistic phenomenological continuum models must be based on detailed experimental and numerical analyses on both the macro- and the micro-level to investigate and to understand these complex stress-state-dependent damage and fracture processes. Corresponding constitutive equations have to be evaluated and quantification of damage amount at various stress states needs to be performed to be able to predict and to judge quality of metal forming processes.

Combination of investigations on both the micro- and the macro-level has widely enriched the understanding of damage and fracture processes in metals. In particular, from microscopic point of view ductile fracture in metals is the integral process of nucleation, growth and coalescence of the different micro-defects discussed above. For example, detailed information on stress state dependence of these damage and fracture processes on the micro-level can be obtained by numerical simulations considering representative volume elements under various loading conditions [1–5]. These microscopic numerical calculations allow detailed analysis of individual behavior of micro-defects and it was possible to detect a number of damage and failure mechanisms which have not been revealed by experiments alone. Interpretation of numerical results of these calculations on the micro-scale allow proposition of evolution equations for chosen damage parameters and of equations for damage and fracture criteria as well as of evolution laws for damage strain rate tensors depending on the stress state [4, 5]. Since the proposed criteria and evolution equations are only based on numerical analyses experimental validation is still necessary.

Experiments with carefully designed metal specimens have been proposed and discussed in the literature. In particular, uniaxial tension tests with un-notched and notched specimens and corresponding numerical calculations have been performed to study the dependence of damage and failure on different positive stress triaxialities [6–9]. In addition, new geometries of uniaxially loaded specimens have been developed to investigate the behavior under nearly zero stress triaxialities where shear mechanisms occur in the critical parts of the specimens [6, 7, 9, 10]. Furthermore, to examine the behavior under wider regions of stress triaxialities butterfly specimens with complex geometries have been developed which can be loaded in uniaxial tension tests in different directions using special experimental equipment [8, 11]. Alternatively, two-dimensional experiments and different geometries of cruciform specimens have been investigated [12]. In addition, biaxial experiments with new shear-tension-specimens in combination with numerical simulations have been proposed by [13, 14] to study stress-state-dependent damage and failure processes. Scanning electron microscope analyses of fracture surfaces elucidate different mechanisms of ductile fracture on the micro-level.

From theoretical point of view, anisotropic damage models taking into account tensorial damage variables are more suitable to simulate stress-state-dependent behavior but their practical applicability is often limited by large number of material parameters and difficulties in their identification. Therefore, in the present paper a phenomenological continuum damage approach based on simplified damage mode functions will be discussed. Experiments with bi-axially loaded specimens will be presented where evolution of strain fields are analyzed by digital image correlation in critical regions of the specimens where damage and failure occur. Numerical simulations based on the proposed continuum model have been performed and numerical results will be used to explain stress-state-dependent damage and failure processes especially in the regime of negative stress triaxialities.

2 Continuum damage model

The anisotropic continuum damage model proposed by [15] is used to predict the inelastic deformation behavior as well as the evolution of ductile damage and failure in ductile metals. This phenomenological approach applicable for different loading conditions is based on introduction of micro-structurally based damage variables that not only represent the volume fraction of the micro-defects but also take into account their current shape and orientation. In this context, in the proposed continuum damage model the damage process is governed by the evolution of damage strains on the macro-level caused by various stress-state-dependent microscopic mechanisms of the micro-defects. Thus, the thermodynamically consistent continuum framework is based on kinematic description of damage leading to definition of damage strain rate tensors. The corresponding kinematic approach takes into account additive decomposition of the strain rate tensor into elastic, $\dot{\mathbf{H}}^{el}$, effective plastic, $\dot{\bar{\mathbf{H}}}^{pl}$, and damage parts, $\dot{\mathbf{H}}^{da}$. Furthermore, free energy functions with respect to undamaged and damaged configurations are introduced leading to elastic laws which in the damaged configuration are influenced by increasing damage to model the deteriorating effect of damage on elastic material properties, see [15] for further details. Considering the undamaged configurations plastic behavior is governed by a yield condition and a flow rule and, in a similar way, considering the damaged configurations damage behavior is characterized by a damage condition and a damage rule both depending on the stress triaxiality and the Lode parameter.

Equations characterizing elastic-plastic behavior of the undamaged matrix material are based on consideration of the effective undamaged configurations. The elastic behavior is described by an isotropic hyper-elastic law leading to the effective Kirchhoff stress tensor

$$\bar{\mathbf{T}} = 2G\,\mathbf{A}^{el} + \left(K - \frac{2}{3}G\right)\,\mathrm{tr}\mathbf{A}^{el}\,\mathbf{1} \qquad (1)$$

where G and K are the constant shear and bulk modulus of the undamaged matrix material, respectively, and $\mathbf{A}^{el}$ represents the elastic part of the strain tensor. In addition, plastic behavior is governed by the yield condition

$$f^{pl}\left(\bar{I}_1, \bar{J}_2, c\right) = \sqrt{\bar{J}_2} - c\left(1 - \frac{a}{c}\bar{I}_1\right) = 0\,, \qquad (2)$$

with the first and second deviatoric invariant $\bar{I}_1 = \mathrm{tr}\bar{\mathbf{T}}$ and $\bar{J}_2 = \frac{1}{2}\,\mathrm{dev}\bar{\mathbf{T}}\cdot\mathrm{dev}\bar{\mathbf{T}}$ of the effective stress tensor (1), the strength coefficient c of the matrix material and the hydrostatic stress coefficient a. In the present paper, the hydrostatic-stress-dependent isotropic yield function (2) is taken into account to adequately model the plastic behavior of metals. It has been shown in carefully performed experiments by [16] that even plastic yielding and hardening of metals shows small dependence on the hydrostatic stress. Numerical analyses discussed in [17] have shown that variation of the parameter a in (2) remarkably affects the strain localization and failure behavior of specimens which is an important aspect in metal forming processes. On the other hand, metal sheets usually show anisotropic plastic behavior caused by the manufacturing process. However, in many metals the plastic anisotropy is small and, thus, the isotropic yield condition (2) seems to be an adequate approximation. For larger anisotropy the use of an anisotropic yield criterion as discussed, for example, in [18] is recommended.

Furthermore, plastic volumetric strains are marginal in metals [16]. Thus, the plastic potential function

$$g^{pl}(\bar{\mathbf{T}}) = \sqrt{\bar{J}_2} \tag{3}$$

is introduced leading to the isochoric effective plastic strain rate

$$\dot{\bar{\mathbf{H}}}^{pl} = \dot{\lambda}\,\frac{\partial g^{pl}}{\partial \bar{\mathbf{T}}} = \dot{\lambda}\,\frac{1}{2\sqrt{\bar{J}_2}}\,\mathrm{dev}\bar{\mathbf{T}} = \dot{\gamma}\,\bar{\mathbf{N}}\,. \tag{4}$$

phenomenologically modeling the evolution of plastic deformations in ductile metals in an accurate manner. In Eq. (4) $\dot{\lambda}$ is a non-negative scalar-valued factor, $\bar{\mathbf{N}} = \frac{1}{\sqrt{2\bar{J}_2}}\,\mathrm{dev}\bar{\mathbf{T}}$ represents the normalized deviatoric stress tensor and $\dot{\gamma} = \bar{\mathbf{N}}\cdot\dot{\bar{\mathbf{H}}}^{pl} = \frac{1}{\sqrt{2}}\,\dot{\lambda}$ denotes the equivalent plastic strain rate measure used in the present constitutive model to characterize the amount of plastic deformation.

Moreover, the damaged configurations are considered to model the deformation behavior of anisotropically damaged ductile metals. It has been observed in many experiments that damage has remarkable influence on the elastic behavior of solids leading to material deterioration. Thus, the elastic constitutive law of the damaged material element takes into account both the elastic and the damage strain tensors, $\mathbf{A}^{el}$ and $\mathbf{A}^{da}$. Then, the Kirchhoff stress tensor with respect to the damaged configurations is given by

$$\begin{aligned}\mathbf{T} = \; & 2\,\left(G + \eta_2\,\mathrm{tr}\mathbf{A}^{da}\right)\,\mathbf{A}^{el}\\ & + \left(K - \frac{2}{3}G + 2\eta_1\,\mathrm{tr}\mathbf{A}^{da}\right)\mathrm{tr}\mathbf{A}^{el}\,\mathbf{1}\\ & + \eta_3\,\left(\mathbf{A}^{da}\cdot\mathbf{A}^{el}\right)\,\mathbf{1}\\ & + \eta_3\,\mathrm{tr}\mathbf{A}^{el}\mathbf{A}^{da}\\ & + \eta_4\,\left(\mathbf{A}^{el}\mathbf{A}^{da} + \mathbf{A}^{da}\mathbf{A}^{el}\right)\end{aligned} \tag{5}$$

where the additional constitutive parameters η_1, ..., η_4 are related to deteriorating effects of the macroscopic elastic material properties caused by different damage and failure mechanisms on the micro-level.

Determination of onset and continuation of damage is based on the concept of damage surface formulated in stress space [15, 19]. Thus, the damage condition

$$f^{da} = \alpha I_1 + \beta\sqrt{J_2} - \sigma = 0 \tag{6}$$

is expressed in terms of the stress invariants I_1 and J_2 of the Kirchhoff stress tensor and the damage threshold σ represents the material toughness to micro-defect propagation. In Eq. (6) the variables α and β represent damage mode parameters corresponding to the different damage mechanisms acting on the micro-level: shear modes for negative stress triaxialities, void-growth-dominated modes for large positive triaxialities and mixed modes (simultaneous growth of voids and evolution of micro-shear-cracks) for lower positive stress triaxialities. In addition, the influence of the Lode parameter is taken into account because it has been shown that its effect on the evolution of the microstructure is remarkable especially in low stress triaxiality regions. Therefore, the damage mode parameters α and β in Eq. (6) depend on the stress intensity $\sigma_{eq} = \sqrt{3J_2}$ (von Mises equivalent stress), the stress triaxiality

$$\eta = \frac{\sigma_m}{\sigma_{eq}} = \frac{I_1}{3\sqrt{3J_2}} \tag{7}$$

defined as the ratio of the mean stress σ_m and the von Mises equivalent stress σ_{eq} as well as on the Lode parameter

$$\omega = \frac{2T_2 - T_1 - T_3}{T_1 - T_3} \quad \text{with } T_1 \geq T_2 \geq T_3 \tag{8}$$

expressed in terms of the principal stress components T_1, T_2 and T_3.

For the tested aluminum alloy the dependence of α and β on stress state has been examined in detail performing numerical simulations on the micro-scale by [4]. For practical applications simplified functions have been proposed by [14] still allowing accurate modeling of failure behavior observed in different experiments. It is worthy to note that the parameters α and β correspond to different damage and fracture mechanisms acting on the micro-level. Therefore, they can be seen to be typical parameters for a wide class of metals and not only for the aluminum alloy under investigation. Thus, only the damage threshold σ in the damage criterion (6) is a material parameter whereas α and β are not.

In particular, the parameter α is taken to be

$$\alpha(\eta) = \begin{cases} -0.15 & \text{for } \eta \leq 0 \\ 0.33 & \text{for } \eta > 0 \end{cases} \tag{9}$$

and the parameter β is given by the non-negative function

$$\beta(\eta, \omega) = \beta_0(\eta, \omega = 0) + \beta_\omega(\omega) \geq 0\,, \tag{10}$$

with

$$\beta_0(\eta) = -1.28\,\eta + 0.85 \tag{11}$$

and

$$\beta_\omega(\omega) = -0.017\,\omega^3 - 0.065\,\omega^2 - 0.078\,\omega\,. \tag{12}$$

Furthermore, the damage strain rate tensor can be determined by the damage rule

$$\dot{\mathbf{H}}^{da} = \dot{\mu}\left(\bar{\alpha}\,\frac{1}{\sqrt{3}}\,\mathbf{1} + \bar{\beta}\,\mathbf{N}\right) \tag{13}$$

where $\dot{\mu}$ is a non-negative scalar-valued factor. In Eq. (13) the normalized stress related deviatoric tensor $\mathbf{N} = \frac{1}{2\sqrt{J_2}}\,\mathrm{dev}\tilde{\mathbf{T}}$ has been used and $\dot{\mu}$ represents in the proposed continuum damage approach the equivalent damage strain rate measure characterizing the amount of increase in irreversible damage. The parameters $\bar{\alpha}$ and $\bar{\beta}$ are kinematic variables describing the portion of volumetric and isochoric damage-based deformations. These parameters also correspond to different damage and fracture mechanisms on the micro-scale. Again, they are based on numerical analyses with void-containing unit cells under various three-dimensional loading conditions [4] as well as on comparison of test results and numerical simulations of different experiments with uniaxially and biaxially loaded specimens [14].

In particular, the parameter $\bar{\alpha}$ characterizing the amount of volumetric damage strain rates caused by isotropic volume change of micro-defects is given by

$$\bar{\alpha}(\eta) = \begin{cases} 0 \text{ for } \eta \leq 0 \\ 0.5714\,\eta \text{ for } 0 < \eta \leq 1.75 \\ 1 \text{ for } \eta > 1.75 \end{cases} . \quad (14)$$

The parameter $\bar{\beta}$ corresponding to the amount of anisotropic isochoric damage strain rates caused by evolution of micro-shear-cracks is taken to be

$$\bar{\beta}(\eta, \omega) = \bar{\beta}_0(\eta) + f_\beta(\eta)\, \bar{\beta}_\omega(\omega) \quad (15)$$

with

$$\bar{\beta}_0(\eta) = \begin{cases} 0.87 \text{ for } \eta \leq \frac{1}{3} \\ 0.979 - 0.326\,\eta \text{ for } \frac{1}{3} < \eta \leq 3 \\ 0 \text{ for } \eta > 3 \end{cases} , \quad (16)$$

$$f_\beta(\eta) = -0.0252 + 0.0378\,\eta \quad (17)$$

and

$$\bar{\beta}_\omega(\omega) = \begin{cases} 1 - \omega^2 \text{ for } \eta \leq \frac{2}{3} \\ 0 \text{ for } \eta > \frac{2}{3} \end{cases} . \quad (18)$$

It can be summarized that the macroscopic damage rule (13) takes into account volumetric parts (first term in Eq. (13)) corresponding to isotropic growth of voids on the micro-scale as well as deviatoric parts (second term in Eq. (13)) corresponding to anisotropic evolution of micro-shear-cracks, respectively. Therefore, both basic damage mechanisms discussed above (growth of isotropic voids and evolution of micro-shear-cracks) acting on the micro-level are involved in the macroscopic damage rule (13) of the phenomenological continuum model. In addition, it has been shown in [20] that the shear deformation component activated damage mechanisms in the small negative stress triaxiality regime. In numerical simulations this behavior can be taken into account by the deviatoric parts in Eq. (13) in a phenomenological manner.

3 Experiments and numerical simulations

The investigated material is an aluminum alloy of series 2017. The phenomenological continuum model discussed above takes into account different material parameters which have to be identified because they are needed as input data in numerical calculations. Determination of constitutive parameters is based on fitting of experimental curves of uniaxial tests and has been discussed in detail by [14]. Briefly, elastic behavior is modeled with Young's modulus $E = 65,000$ MPa and Poisson's ratio $\nu = 0.3$. Plastic behavior is governed by the hardening law

$$c = c_0 \left(\frac{H \gamma}{n\, c_0} + 1 \right)^n \quad (19)$$

with the initial yield strength $c_0 = 174$ MPa, the hardening modulus $H = 1800$ MPa and the hardening exponent $n = 0.23$. Onset of damage is characterized by the damage threshold $\sigma = 295$ MPa whereas the effect of damage on elastic behavior is simulated by the parameters $\eta_1 = -100,000$ MPa, $\eta_2 = -50,000$ MPa, $\eta_3 = -12,500$ MPa, and $\eta_4 = -2500$ MPa.

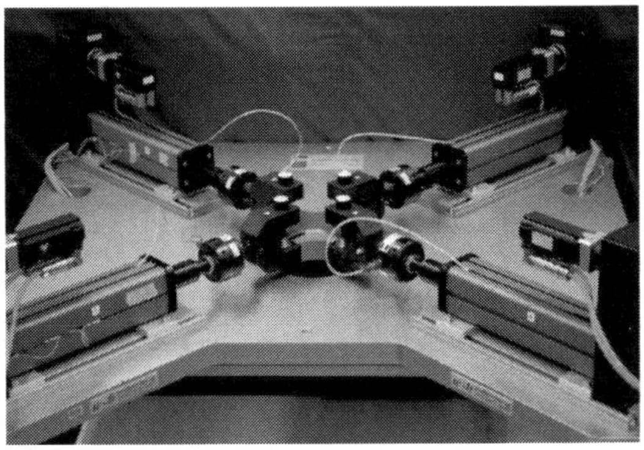

Figure 1. Biaxial testing machine

Many metal forming processes are carried out under compressive loading conditions and, therefore, there is remarkable demand for experiments to examine in detail inelastic deformation behavior as well as damage and failure in the regime of negative stress triaxialities. A new experimental program has been developed to propose new tests revealing the effect of stress state on inelastic behavior, damage and fracture in ductile metals undergoing combined tensile, shear and compressive loading conditions [13]. The experiments are performed using the biaxial testing machine type LFM-BIAX 20 kN (produced by Walter & Bai, Switzerland) shown in Fig. 1. It contains four electro-mechanically, individually driven cylinders with load maxima and minima of ± 20 kN (tension and compression loading are possible). The respective specimens are fixed in the four heads of the cylinders where clamped or hinged boundary conditions are possible.

Specimens with special geometry are biaxially loaded leading to shear-tension and shear-compression mechanisms in their central parts where localized inelastic deformations and final failure occur. The experimental equipment discussed above and this geometry of specimens are used here to study the deformation and failure behavior in the range of negative stress triaxialities. In addition, corresponding numerical calculations deliver detailed information on distributions and amounts of stress and strain measures especially in critical regions of the specimens. The numerical analyses are carried out using the finite element program ANSYS enhanced by a user-defined ma-

DOI: 10.1051/matecconf/20168003001

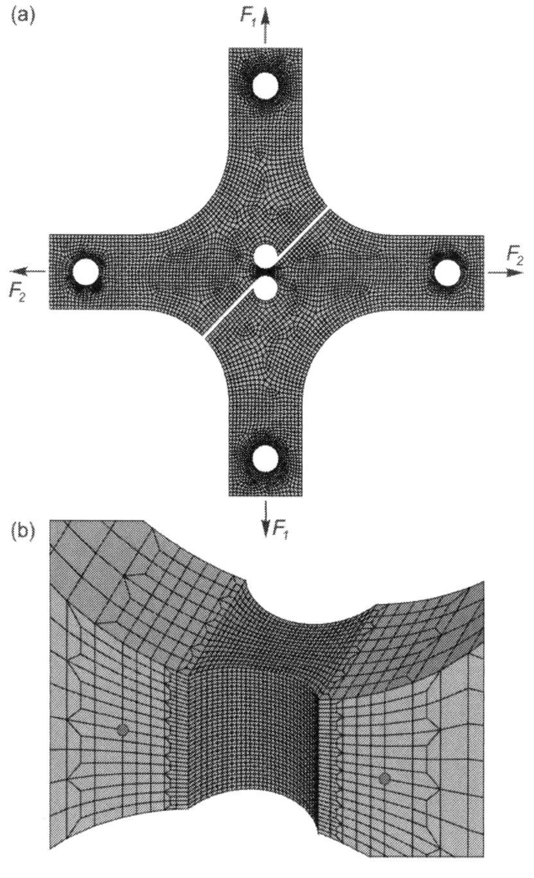

Figure 2. (a) Specimen, loading conditions and finite element mesh, (b) Detail of the notched central part

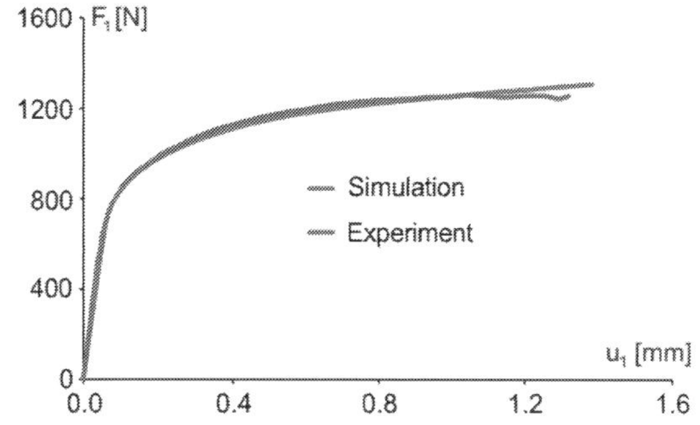

Figure 3. Load-displacement curves $F_1 - u_1$

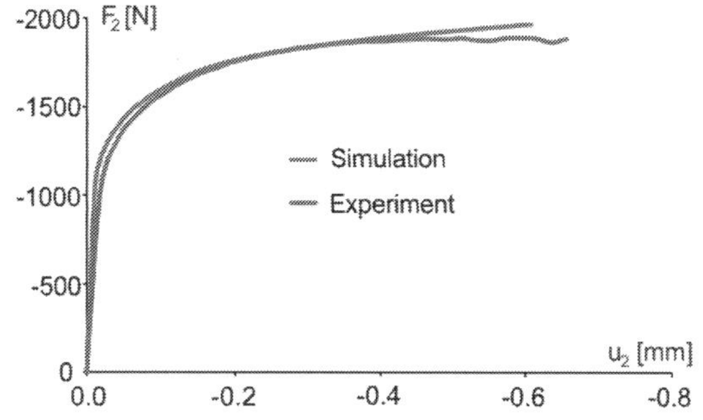

Figure 4. Load-displacement curves $F_2 - u_2$

terial subroutine based on the continuum model discussed above.

Geometry, finite element mesh and loading conditions of the flat specimen are shown in Fig. 2(a). The numerical calculations are based on finite element meshes with 42,248 eight-node-elements of type Solid185. During the tests the specimens are simultaneously loaded in vertical and horizontal direction by the forces F_1 and F_2. The vertical load F_1 leads to shear mechanisms in the center of the specimen whereas the load F_2 leads in this part to superimposed tension or compression modes. In the present paper, results of experiments and of corresponding numerical simulations for the load ratio $F_1 : F_2 = 1 : -1.5$ will be discussed in detail which are seen to be relevant for metal forming processes. In the center of the specimen additional notches in thickness direction have been milled, see Fig. 2(b), leading to localization of inelastic deformations, damage and final fracture in this part. Here, refinement of the finite element mesh has been used to accurately cover high gradients of stress and strain quantities. This part of the specimen is 3mm high with thickness of 2mm and the radius of the notch is 2mm. Displacement measures analyzed and discussed in the present paper correspond to movements of the red points shown in Fig. 2(b). For example, the vertical displacement u_1 is the vertical relative movement of these points whereas the displacement u_2 is the horizontal one. During the experiments, these displacement are recorded by digital image correlation discussed below and in the numerical simulations displacements of corresponding finite element nodes are computed.

Results of experiments and of corresponding numerical simulations are analyzed and compared to elucidate localized deformation and failure behavior. In particular, load-displacement curves $F_1 - u_1$ are shown in Fig. 3. After elastic loading a large region with inelastic behavior including hardening occurs. The load maximum is about $F_1 = 1300N$. Final fracture of the specimen occurred at $u_1 = 1.3mm$ but no remarkable softening behavior can be observed before this point has been reached. The numerically predicted curve based on the continuum model discussed above shows very good agreement with the experimental one. In addition, the load-displacement curves for the other direction, $F_2 - u_2$, are shown in Fig. 4 with load minimum of $F_2 = -1900N$ and the final displacement $u_2 = -0.65mm$. Again, very good agreement of experimental and numerically predicted curves can be observed.

In the load-displacement curves (Figs. 3 and 4) small waves can be observed especially shortly before final fracture occurred. This behavior is typical for shear-compression loading conditions and has also been observed for other negative stress ratios, see for example

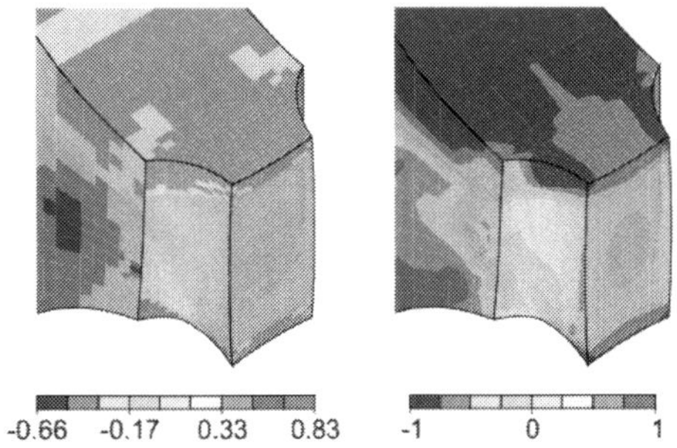

Figure 5. Stress triaxiality η and Lode parameter ω

[14]. One reason for this behavior may be evolution of localized inelastic shear deformations caused by formation of shear-cracks. Due to superimposed compression loading continuation of these shear deformations stops because crack areas come into contact and are pressed together. It is worthy to note that this behavior does not mean healing of the locally failed material because the shear-cracks still exist and will be again active when associated shear forces increase during further loading of the specimen.

The load-displacement curves discussed above only demonstrate the global behavior of the specimen for the chosen loading condition. Therefore, further analyses are recommended to study the local damage and failure behavior especially in critical regions where inelastic deformations are localized and final fracture will occur. For example, numerically predicted distributions of the stress triaxiality η (Eq. 7) and the Lode parameter ω (Eq. 8) at the end of the loading process shortly before final fracture occurs will be discussed. These stress parameters in a vertical cut in the central region of the specimen are shown in Fig. 5. In particular, for the considered load ratio $F_1 : F_2 = 1 : -1.5$ the stress triaxiality is here nearly constant in the central area with $\eta = -0.2$ and only in the upper and lower boundaries positive stress triaxialities of about $\eta = 0.33$ occur caused by the tensile deformation of these parts caused by the shear loading. In addition, similar stress parameter distribution can be seen for the Lode parameter. It is about $\omega = 0.0$ in the center and reaches $\omega = -0.7$ at the boundaries. For these values in the central area of the specimen damage and failure will be caused by formation and growth of micro-shear cracks whereas the effect of nucleation and growth of micro-voids will be marginal which is typical for shear-compression loading conditions. Combination of voids and shear-cracks on the micro-level will only occur in the boundaries of the notched region.

The amount of damage strains can be quantified by the equivalent damage strain measure μ, see Eq. (13). Figure 6 shows its numerically predicted distribution in the notched part of the specimen at the end of the numerical simulation, shortly before fracture occurred in the experiments. The equivalent damage strain μ shows its maxima of about $\mu = 4\%$ at the boundaries of the notch but also about $\mu = 2\%$ in the band between these points. Since

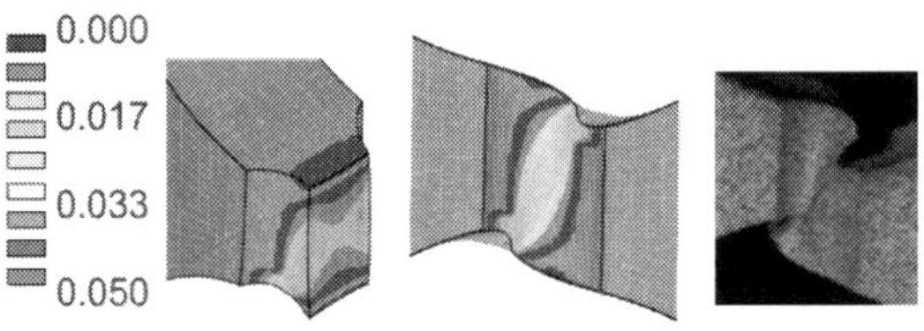

Figure 6. Equivalent damage strain μ and experimental fracture line

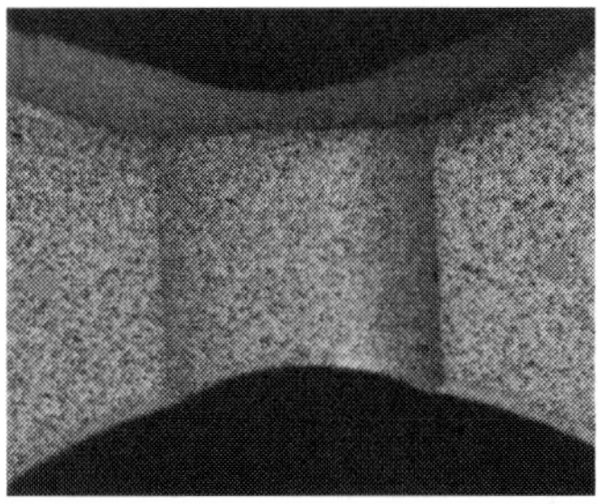

Figure 7. Speckle pattern of the specimen's center incl. emphasis of points (red) for displacement measurement

these values of the equivalent damage strain are predicted to only occur in this band, the onset of macro-cracking will be in this part and the crack then runs along the band of moderate equivalent damage strains which are the weakest points of the damaged material. This numerically predicted band nicely corresponds to the fracture line of the tested specimen also shown in Fig. 6. In addition, the numerically predicted geometry of the inelastically deformed central part of the specimen nicely corresponds to the geometry of the tested specimen shown in the photo.

Digital image correlation (DIC) technique has been applied in the present experimental program to analyze deformation behavior in the critical zones of the specimens. During the shear-compression test the three-dimensional displacement field of the specimen surface was measured with a Q-400 digital 3D image correlation system provided by Dantec/Limess. The stereo setting consists of two 2.0Mpx cameras equipped with Ricoh FL-CC7528-2M 75mm F/2.8 lenses and extension tubes of 10mm. The evaluation was realized with the corresponding Istra 4D software provided with the system. Specimen preparation was realized shortly before the experiment started. One side of the specimen was firstly sprayed with a white acrylic lacquer and then the speckle pattern shown in Fig. 7 was sprayed on with a graphite-based coating. It is worthy to note that no ex-foliation of the coating was observed during the experiment. The displacements of the selected points marked in red in Fig. 7 as well as the forces in axes directions were transferred and stored with the data sets of the image correlation system at a frequency of 0.25Hz while the velocity of the machine was 0.04mm/min.

Strain fields of the tested specimens analyzed by DIC in the load steps shortly before final fracture occurs are

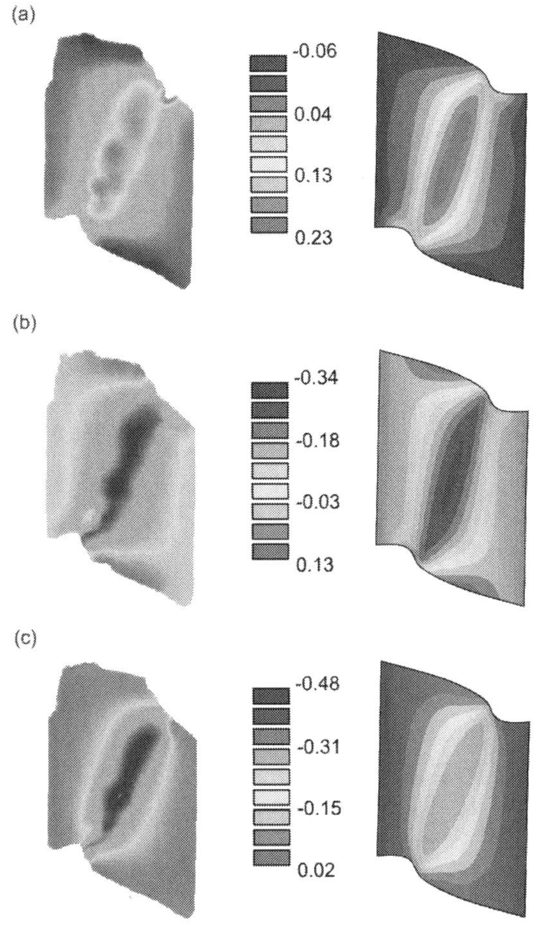

Figure 8. Strain fields taken from DIC (left) and numerical simulation (right): (a) Normal strain in direction 1, (b) Normal strain in direction 2, (c) Shear strain

compared with corresponding fields predicted by numerical simulations based on the continuum model discussed above. Since the theoretical approach takes into account elastic, plastic and damage strain rate tensors the comparison of measured and numerically predicted strain fields can also be seen as validation of the flow rule (4) and the damage rule (13). To visualize numerically obtained strain fields at the boundaries of the specimen's center thin shell elements with marginal stiffness have been overlayed and numerically predicted strain measures are taken from these elements. The experimentally measured strains shown in Fig. 8 are represented with respect to a coordinate system with origin at the center of the notch where the z-axis is defined perpendicular to the specimen surface and a further axis-direction is defined in tensile direction (Fig. 2) of the current step while the predicted numerical results are represented with respect to the global coordinate system. Thus, both coordinate systems coincide quite well, which is not easy to achieve on the experimental side.

Figure 8(a) shows the distribution of the normal strain component in direction 1 (see Fig. 2) whereas the normal strain component in direction 2 is visualized in Fig. 8(b) and the shear strain in Fig. 8(c). In particular, the normal strain component in direction 1 analyzed by DIC (left picture in Fig. 8(a)) is localized in a small band with maximum values of 23%. This distribution and these values are also obtained from the corresponding numerical simulation (right picture in Fig. 8(a)). In addition, the distributions of the normal strain component in direction 2 analyzed by DIC and that one obtained by the numerical simulation are visualized in Fig. 8(b). A similar localized band of this normal strain component can be seen with minimum values of -34% caused by the superimposed compression loading. Again, for these experimental and numerically predicted normal strains, distributions and values agree very well. Furthermore, distributions of the shear strains is shown in Fig. 8(c). Experimental shear strains analyzed by DIC (left picture in Fig. 8(c)) are also localized in a small band with values up to 48%. In comparison, numerically predicted shear strains (right picture in Fig. 8(c)) show similar distribution but the highest values are only about 30%. One reason for this quantitative difference is the possible occurrence of larger shear-cracks in this band. Their effect has also been observed in the load-displacement curves as mentioned above (small waves shortly before final fracture occurs, see Figs. 3 and 4). These shear-cracks lead to further softening of the damaged material and, thus, to larger shear strains in this localized band but due to remarkable superimposed compression and associated contact of the crack areas this will not lead to subsequent final fracture. However, this formation of larger shear-cracks cannot be modeled by the proposed continuum model discussed above because such discontinuities cannot be taken into account. It is worthy to note that this discrepancy only in quantities of experimental and numerically predicted shear strains has also be observed for other load ratios in the shear-compression range [14] and will be analyzed in more detail by the authors in near future.

4 Conclusions

The paper has discussed an anisotropic continuum damage and failure model for ductile metals. The phenomenological approach is based on kinematic definition of damage tensors and takes into account the effect of stress state on damage conditions and damage strain evolution laws. Different branches of these criteria based on experimental studies and numerical simulations are considered corresponding to various microscopic damage and fracture mechanisms depending on stress intensity, stress triaxiality and the Lode parameter. Damage mode functions for a wide class of metals have been proposed which are able to simulate all relevant effects observed in different experiments and will allow practical applicability.

Experiments with biaxially loaded specimens undergoing shear-compression loading appearing in metal forming processes have been performed. Digital image correlation technique has been used to analyze current strain fields in critical regions of the specimens.

Numerical simulations corresponding to the experiments have been performed. Detailed information on load-deformation behavior as well as on stress and strain states have been analyzed and compared with available experimental data especially in the critical regions of the specimens where large inelastic deformations and ductile failure occurred. The results of the presented experimental-numerical procedure allow validation of the proposed stress-state-dependent constitutive equations for moderate negative stress triaxialities. It seems that under shear-compression loading conditions cracks occur in the center of the specimen but their areas come into contact and are pressed together. Thus, the activity of these cracks will be interrupted for a while, but they are still existent. This failure may play an important role for later use of formed products and might lead to remarkable reduction in their safety and lifetime. This effect will be analyzed in more detail by the authors in near future.

In summary, the proposed constitutive model can be used in numerical calculations to optimize sheet and bulk metal forming processes. In addition, it has been shown that the final macro-cracks are initiated in points of moderate equivalent damage strains and following fracture lines of the respective failed specimens occur in the band of maximum equivalent damage strains which might give reference to formulation of alternative engineering fracture criteria based on damage strain measures.

References

[1] M. Kuna, D. Sun, Int. J. Fract. **81**, 235-258 (1996)
[2] X. Gao, J. Kim, Int. J. Sol. Struct. **43**, 6277-6293 (2006)
[3] I. Barsoum, J. Faleskog, Int. J. Sol. Struct. **48**, 925-938 (2011)
[4] M. Brünig, S. Gerke, V. Hagenbrock, Int. J. Plast. **50**, 49-65 (2013)
[5] M. Brünig, S. Gerke, V. Hagenbrock, Int. J. Plast. **63**, 49-63 (2014)
[6] Y. Bai, T. Wierzbicki, Int. J. Mech. Sci. **46**, 81-98 (2004)
[7] X. Gao, G. Zhang, C. Roe, Int. J. Damage Mech. **19**, 75-94 (2010)
[8] M. Dunand, D. Mohr, J. Mech. Phys. Sol. **59**, 1374-1394 (2011)
[9] M. Brünig, O. Chyra, D. Albrecht, L. Driemeier, M. Alves, Int. J. Plast. **24**, 1731-1755 (2008)
[10] L. Driemeier, M. Brünig, G. Micheli, M. Alves, Mech. Mat. **42**, 207-217 (2010)
[11] Y. Bai, T. Wierzbicki, Int. J. Plast. **24**, 1071-1096 (2008)
[12] T. Kuwabara, Int. J. Plast. **23**, 385-419 (2007)
[13] M. Brünig, D. Brenner, S. Gerke, Eng. Fract. Mech. **141**, 152-169 (2015)
[14] M. Brünig, S. Gerke, M. Schmidt, Int. J. Fract. (accepted for publication) (2016)
[15] M. Brünig, Int. J. Plast. **19**, 1679-1713 (2003)
[16] W.A. Spitzig, R.J. Sober, O. Richmond, Metall. Trans. **7A**, 1703-1710 (1976)
[17] M. Brünig, Int. J. Plast. **15**, 1237-1264 (1999)
[18] H. Badreddine, K. Saanouni, T.D. Nguyen, Int. J. Sol. Struct. **63**, 11-31 (2015)
[19] C. Chow, J. Wang, Eng. Fract. Mech. **27**, 547-558 (1987)
[20] S. Kweon, Eur. J. Mech. A/Solids **31**, 203-212 (2012)

Ductile failure analysis of high strength steel in hot forming based on micromechanical damage model

Liang Ying [1], Wenquan Liu[2], Dantong Wang[1] and Ping Hu[1,a]

[1] *School of Automotive Engineering, Dalian University of Technology, 116024 Dalian Liaoning, China*

[2] *Department of Engineering Mechanics, Dalian University of Technology, 116024 Dalian Liaoning, China*

Abstract: The damage evolution of high strength steel at elevated temperature is investigated by using the Gurson-Tvergaard-Needleman (GTN) model. A hybrid method integrated thermal tensile test and numerical technique is employed to identify the damage parameters. The analysis results show that the damage parameters are different at different temperature as the variation of tested material microstructure. Furthermore, the calibrated damage parameters are implemented to simulate a bugling forming at elevated temperature. The experimental results show the availability of GTN damage model in analyzing sheet formability in hot forming.

1 Introduction

High strength steel (HHS) components have been widely used for the purpose to reduce the weight of automobile and guarantee crash safety as well. Nevertheless, a major drawback of the HHS is the low formability and severe spring-back at ambient temperature. Hot forming of high strength steel, especially 22MnB5, is an innovative technology which integrates the hot forging and cold forming, so that it can solve these defects quite well[1]. However, the elastic modulus and tensile strength decrease markedly and the frictional coefficient increases significantly with the increase of temperature. Therefore, forming fracture is also a major problem to study during hot forming process.

The efficient method for predicting material formability is based on the damage mechanics theory. The microscopic damage analysis proposed by McClintock[2], developed by Gurson[3] and lots of later authors, has been widely used to investigate the fracture behavior of metal sheet. GTN model is one of the most widely used damage models for sheet forming. However, there are few reports on the damage investigation of high strength steels based on GTN model for hot forming process. Klocker et al[4] have used an axisymmetric unit cell model to study the void growth and coalescence in metals deformed at elevated temperature. And the critical void volume fraction f_c shows strong dependency upon temperature. Moreover, the similar results were discovered by Tinet et al[5] experimentally, but the failure void fraction f_F nearly kept nearly in constant. Moreover, the microstructure of boron steel 22MnB5 will vary with the variation of temperature. Thus, the investigation of damage evolution for hot forming process must fully consider the influence of temperature.

In this work, the ductile damage of 22MnB5 at elevated temperature has been studied. The damage parameters of GTN model are optimized by a hybrid method integrated thermal tensile test and numerical technique for different tested temperature. Eventually, the damage model with the calibrated parameters are implemented to simulate a high temperature bugling process by means of the finite element code ABAQUS. The corresponding experiment is also conducted and the comparison of simulation data and experiment results has been carried out.

2 Material damage model

2.1 GTN damage model

The damage model proposed by Gurson and modified by Tvergaard and Needleman[6] known as GTN model, which has the expression:

[a] Corresponding author: pinghu@dlut.edu.cn.cn

DOI: 10.1051/matecconf/20168003002

$$\varphi(\sigma_{eq},\sigma_h,f^*)=(\frac{\sigma_{eq}}{\bar{\sigma}_y})+2f^*q_1\cosh(-\frac{3q_2\sigma_h}{2\bar{\sigma}_y})-(1+q_3f^{*2})=0 \quad (1)$$

Where σ_{eq} is the macro equivalent stress of Cauchy stress tensor, σ_h is the macro hydrostatic stress, $\bar{\sigma}_y$ is the flow stress of matrix material, the parameters q_1, q_2, q_3 are introduced by Tvergaard[7] to improve the prediction accuracy and the suggestive value for metal have been adopted, $q_1 = 1.5$, $q_2 = 1.0$, $q_3 = q_1^2$. Here, f^* is the function of damage volume fraction f:

$$f^* = \begin{cases} f & f \le f_c \\ f_c + (f - f_c)\frac{f_u - f_c}{f_F - f_c} & f_c < f \le f_F \\ 1/q_1 & f \ge f_F \end{cases} \quad (2)$$

Where f_c is the critical void volume fraction at which void coalescence starts, $f_u = 1/q_1$ is the ultimate value of f^* at material failure, f_F is the final void volume fraction at material fracture. It can be seen when $f_c < f \le f_F$, there is an acceleration of f^*. It was introduced by Tvergaard and Needleman to characterize the damage accelerated evolution process caused by void coalescence.

In general, the damage evolution of metal materials during plastic deformation includes two parts, the growth and nucleation of voids as shown in Eq. (3), the detailed descriptions can be found in paper[8]:

$$\dot{f} = \dot{f}_{growth} + \dot{f}_{nucleation} \quad (3)$$

The equivalent plastic strain $d\bar{\varepsilon}^{pl}$ of matrix material can be determined by the equivalent plastic work rule:

$$(1-f)\bar{\sigma}_y d\bar{\varepsilon}^{pl} = \boldsymbol{\sigma} d\boldsymbol{\varepsilon}^{pl} \quad (4)$$

Where $\boldsymbol{\sigma}$ and $\boldsymbol{\varepsilon}^{pl}$ are the macro plastic strain tensor and macro stress tensor, respectively.

For hot forming process, the microstructure and mechanical property of high strength steel are different with the variation of temperature. Therefore, to investigate the damage evolution behavior under hot stamping process, the damage parameters under various temperatures should be determined.

2.2 The hardening law of matrix material

The flow behavior of matrix material 22MnB5 at elevated temperature is characterized by the modified Norton-Hoff hardening law:

$$\sigma = K \cdot \exp\left(\frac{\beta}{T}\right) \cdot (\varepsilon_0 + \varepsilon)^{n(T)} \cdot \dot{\varepsilon}^{m(T)} \quad (5)$$

with

$$n(T) = n_0 \exp(-c_n(T_i - T_0)) \quad (6)$$

$$m(T) = m_0 \exp(c_m(T_i - T_0)) \quad (7)$$

where $K, \beta, \varepsilon_0, n_0, m_0, c_n, c_m$ are material constants and T_0 is the reference temperature. For tested material at different temperature, the material constants are determined from the thermal tensile test in next section. The determined constants are listed in Table1. It can be seen that the effect of temperature on hardening exponent is vanished for matrix-material without damage. The method to obtain material mechanical property without damage can be found in literature[9].

TABLE1. Norton-Hoff constants for 22MnB5 at elevated temperature

K	β	ε_0	n_0	m_0	c_n	c_m
26.647	2500	0.007389	0.2653	0.0101	0.0	0.000828

3. Hot tensile test and parameter calibration

3.1 Hot tensile test

According to actual hot forming process, the temperature-time procedure of hot tensile test with a strain rate $0.1\,s^{-1}$ is executed as shown in Fig 1. The experimental material is hot rolled boron steel 22MnB5 with a thickness of 2mm. The dog bone shaped samples with gauge length 50mm and width 20mm are used in the test.

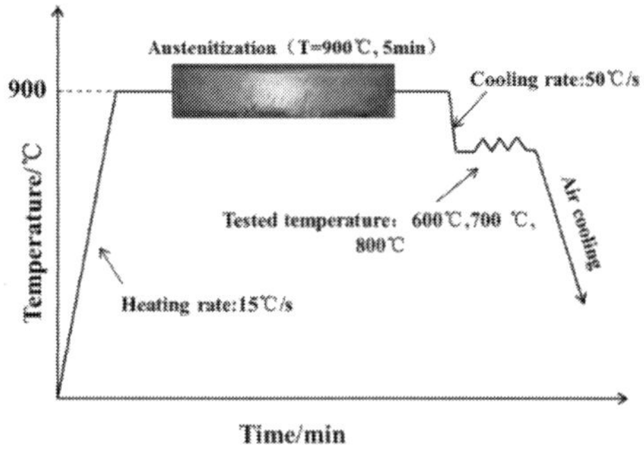

Figure 1. Procedure of hot tensile test

The symbols in Fig 2(a) represent the true stress-strain relation of boron steel 22MnB5 at 600℃, 700℃ and 800℃. It can be seen that deformation temperature has a strong influence on tensile force. The peak stress decreases with the increase of temperature markedly. For each test temperature, the flow stress firstly increases due to work hardening. After reaching peak stress, the flow curve begins to drop until the occurrence of final fracture as the domination of voids damage. The stress-plastic strain curves without damage for different tested temperature predicted by Norton-Hoff model are shown in Fig 2(b) which are

DOI: 10.1051/matecconf/20168003002

consistent with the experimental results.

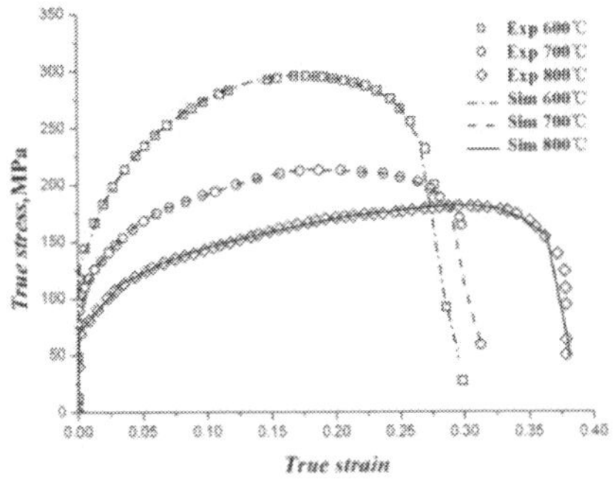

(a) comparsion of experimental (symbol) and predicted (solid) stress-strain curves of whole process

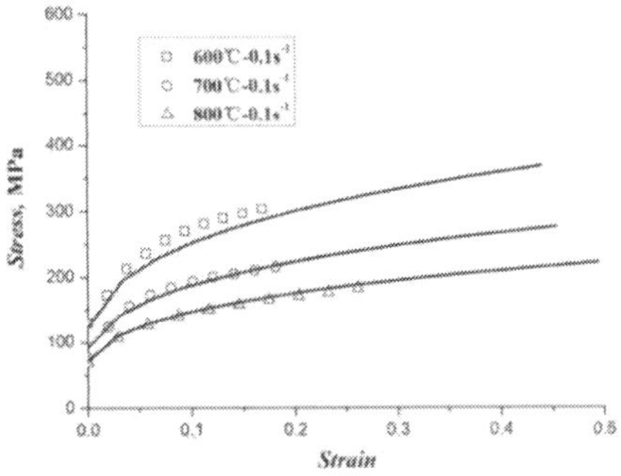

(b) comparsion of experimental (symbol) and fitted (solid) stress-strain relationships of hardening phase

Figure 2. Comparsion of experimental and predicted curves

3.2 Calibration of damage parameters

In general, the GTN model has 9 parameters as mentioned above. To minimize the calibration procedure for these parameters, $q_0, q_1, q_3, \varepsilon_N, S$ have been kept constants as identified in section2.

Hence, f_0, f_N, f_c, f_F are the final damage parameters needed to be determined for boron steel 22MnB5 at different temperature.

To calibrate the damage parameter, a hybrid method integrated experiment and numerical technique is employed. The thermal tensile simulations at various temperatures are conducted as the hot tensile test mentioned above. Naturally, the simulation curve and experimental one can match each other very well if the peak point and fracture point from these two ways match each other, respectively.

The procedure of this hybrid method to identity these four damage parameters can be summarized as follows:

a. Obtaining the experimental stress-strain data at different temperatures.

b. Simulation design for tensile test by central composite experimental design

c. Extracting response variables, peak point and fracture point, for each simulation.

d. Establishing response surface models for peak stress, peak strain, fracture stress and fracture strain.

e. Calculating damage parameters by Genetic Algorithm method. The GA program is realized through MATLAB software.

f. Using the optimum value of damage parameters to simulate tensile test and compare with experimental results.

The optimal value of damage parameters are listed in Table 2 and the predicted stress-strain data (solid curves) using these damage parameters are shown in Fig. 2(a), which show a good accordance with experimental results. From Table 2, we can know that the damage parameters vary with the variation of tested temperature which can be explained by the microstructure transformation of boron steel 22MnB5 at different temperatures.

Table2. Damage parameters for 22MnB5 at different temperatures

Temperature	f_0	f_N	f_c	f_F
600℃	0.0005	0.0110	0.0401	0.1345
700℃	0.0007	0.0141	0.0495	0.1410
800℃	0.0008	0.0165	0.0592	0.1550

4. Numerical simulation of bulging experiment

To verify the calibrated damage parameters, a bugling process at elevated temperature is simulated and the corresponding experiment is also carried out.

4.1 Bugling apparatus and experiment process

To investigate the formability of metal sheet at elevated temperature, a bugling apparatus is designed and developed as shown in Fig 4. The device consists a hemispherical punch, a blank holder and a die with draw-bead. This device can be heated up to 900℃ and the temperature of tested blank is kept during forming process.

Figure 3. Experiment apparatus for high temperature bugling

In this work, the bugling experiment at 700℃ has been carried out to identify the effectiveness of above damage parameters. The same procedure as hot tensile test was adopted: firstly, the metal sheet is heated to 900℃ in furnace and this temperature is preserved 5min for soak austenitization. Then the sheet metal blank is rapidly transferred into the bulging apparatus and rapidly cooled to the tested temperature with an average cooling rate 50℃/s and deformed at this temperature until failure. Meanwhile, the tools are also heated up to tested temperature and remained all the process. The punch velocity is set to 500 mm/min and a blank holder force of 100 kN is provided by the upper die. The mixture of graphite and molybdenum disulfide is laid on the punch surface before bugling to reduce friction between the sheet metal specimen and the punch.

4.2 Numerical simulation for bugling test

Consistent with the experiment apparatus, the finite element model is established as shown in Fig 4. Both the metal blank and tools are meshed using quadrilateral shell element but the tools are set as coupled rigid bodies to avoid deformation. Because of the constant temperature condition during whole bugling process, the thermal coefficient is not needed. The hardening relation predicted by modified Norton-Hoff and material damage parameters at 700℃ obtained above are implemented in the simulation model. The friction coefficient is set to 0.22 which is less than the common value 0.4 in hot forming as the addition of lubricant. And the sample is a 180*180mm square blank with a thickness of 2mm.

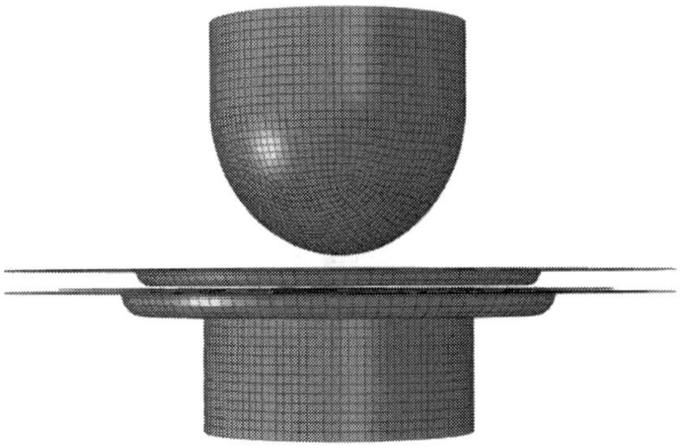

Figure 4. Simulation model of bugling forming

Fig 5 shows the comparison between the experimental and numerical punch force-stroke curves. The solid curve is the experimental result; the symbol point is the predicted result with material damage while the dotted line is the predicted result without damage. It can be seen that the curve predicted with damage can match the experimental curve very well which proves the availability of GTN model and the calibrated damage parameter. One more thing should be noted is that the initial simulation curve has some fluctuation, the reason for this probably causing by the instability of contact algorithm. So some work to smooth the curves has been done but the effectiveness is still maintained.

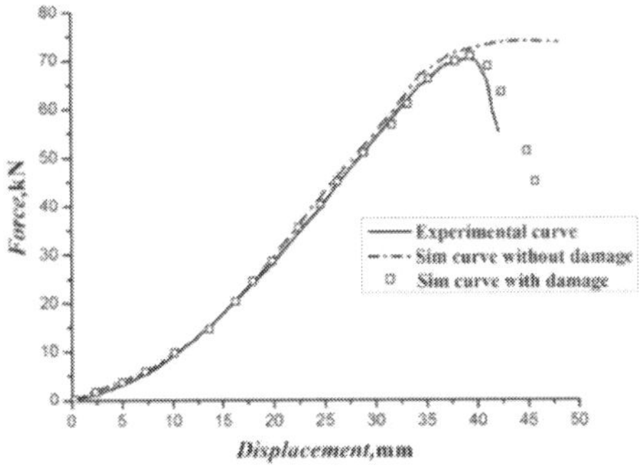

Figure 5. Comparison between the experimental and numerical punch force-stroke curves

Fig 6 reports the void damage distribution of tested sample at different deformation phases. It shows that damage mainly distributes at the location where the blank contacts the punch and draw bead firstly. Along with the downward of punch, void damage begins to gather near the dome of the blank which is accordance with the experimental result as shown in Fig 7.

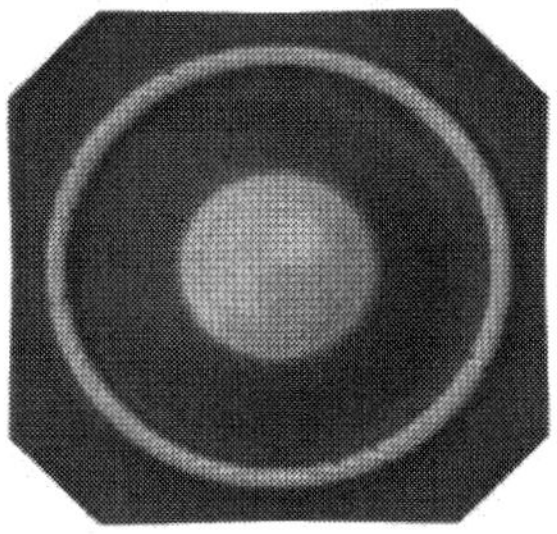

(a) Initial period of bugling

(b) Deformation phase before fracture

Figure 6. Numerical void damage distribution at different bugling phases

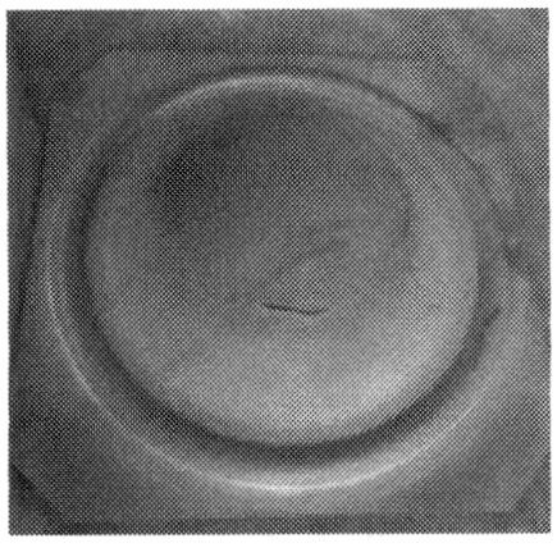

Figure 7. Sample after bugling forming at 700℃

DOI: 10.1051/matecconf/20168003002

To study the damage evolution in detail, the effective strain and void volume fraction of 5 elements nearby the fracture location are extracted as shown in Fig 8. From Fig 9, we can know that the void distribution of these elemets relatively homogeneous before 0.04s, but after that, the void fraction of these elemets begin to separate while the growth rate of element 1 is larger than others. Then, the void volume fraction nearly keeps constant after necking happened. The same rule can be summaried from the evolution of these elements' strain.

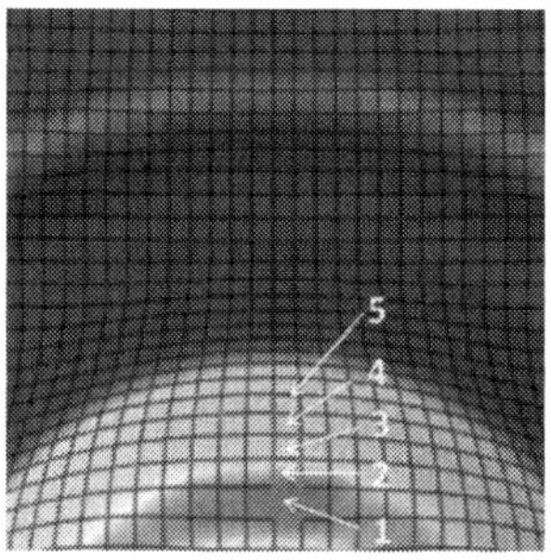

Figure 8. Extraction of analytical elements

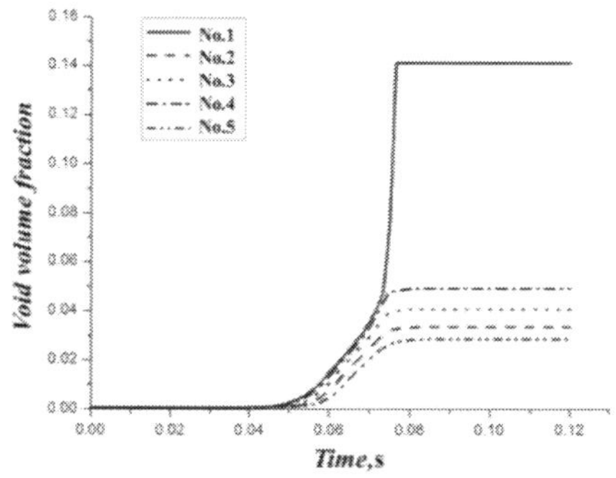

(a) Void damage evolution of 5 elements

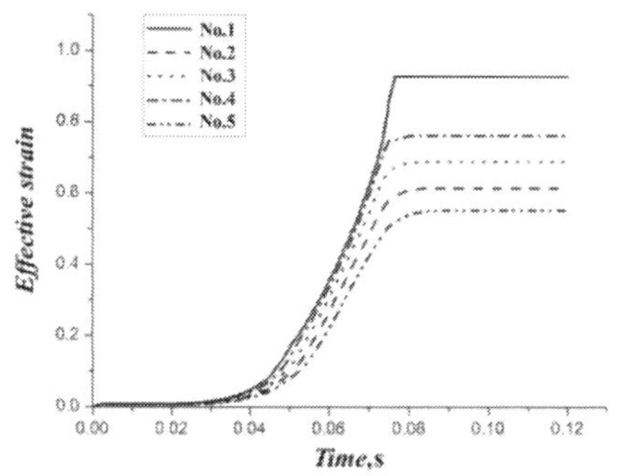

(b) Effective strain evolution of 5 elements

Figure 9. Comparison of void fraction and effective strain for different elements

5. Conclusions

The GTN damage model is used to research the damage evolution of high stength steel 22MnB5 at elevated temperature. A hybrid method integrated experiment and numerical technique is employed to identify the damage parameters. The results show that the damage parameters are different with the variation of temperature. This may be explained by the variation of 22MnB5 microstructure and further investigation like SEM analysis should be done to provide more detail state. The bugling forming at a deformation temperature of 700℃ has been conducted as well as the corresponding simulation. The compraison between experimental and numerical results show the availability of GTN damage model in analyzing sheet formability at elevated temperature condition.

Acknowledgement

This work was funded by the Key Project of the National Natural Science Foundation of China (No.11272075) and China Postdoctoral Science Foundation (2014M561223) and the Fundamental Research Funds for the Central Universities (DUT15TD31)

References

1. A. Mecke, I. Lee, J.R. Baker jr., M.M. Banaszak Holl, B.G. Orr, Eur. Phys. J. E 14, 7 (2004)
2. 1. H. Karbasian, A.E. Tekkaya, J MATER PROCESS TECH, 210, 15 (2010) 2103-2118.
3. 2. F.A. McClintock, Journal of applied mechanics, 35, 2 (1968) 363-371.
4. 3. A. L. Gurson, Journal of Engineering Materials and Technology-Transactions of the ASME, 99 (1977) 2-15.
5. 4. H. Klöcker, V. Tvergaard, INT J FRACTURE, 106, 3 (2000) 259-276.
6. 5. H. Tinet, H. Klöcker, J. Le Coze, ACTA MATER, 52, 13 (2004) 3825-3842.
7. 6. V. Tvergaard, A. Needleman, Acta Metallurgica, 32, 1 (1984) 157-169.
8. 7. V. Tvergaard, INT J FRACTURE, 18, 4 (1982) 237-252.
9. 8. C.C. Chu, A. Needleman, Journal of Engineering Materials and Technology, 102, 3 (1980) 249-256.
10. 9. I.I. CUESTA, J.M. ALEGRE, R. LACALLE, FATIGUE FRACT ENG M, 33, 11 (2010) 703-713.

Prediction of Forming Limit Diagrams under combined Bending-Stretching loadings

Mohamed Ben Bettaieb[1,2, a], Honoré Lagaza[1], Farid Abed-Meraim[1,2], and Xavier Lemoine[1]

[1]*LEM3, UMR CNRS 7239 - Arts et Métiers ParisTech, 4 rue Augustin Fresnel, 57078 Metz Cedex 3, France*
[2]*DAMAS, Laboratory of Excellence on Design of Alloy Metals for low-mAss Structures, Université de Lorraine, France*

Abstract. A number of parts and components involved in the automotive industry are made of thin bent sheets. Unfortunately, the classical predictions based on traditional Forming Limit Diagrams are not relevant when the strain distribution is heterogonous over the thickness, as is the case for bent sheet metals. The aim of the present contribution is to propose an extension of the well-known Marciniak–Kuczynski approach to account for the effect of bending on formability prediction. The new developed tool allows predicting the limit strains for the whole range of strain paths. The mechanical behavior of the studied sheets follows the rigid–plastic flow theory. Through numerical results, it is shown that bending tends to decrease the formability limit of the sheet metal.

1 Introduction

The study of ductility of metallic components and parts is an ambitious challenge in academic and industrial applications. This ductility is often characterized by a Forming Limit Diagram (FLD). The FLD concept has been initially introduced by Keeler and Backofen [1], for representing the limit strains in the range of positive strain paths, and has been extended by Goodwin [2] to the whole range of strain paths. Due to the complexity of the experimental characterization of FLDs and its relatively high cost, a number of theoretical and / or numerical models have been developed in the literature. These models generally require the use of an instability criterion along with a constitutive law to describe the evolution of the mechanical state of the studied sheet. In the literature, a large amount of models have been developed to numerically determine FLDs.

From these theoretical /numerical approaches, we can quote the work of Marciniak and Kuczynski [4], who coupled the initial imperfection approach with a rigid–plastic flow theory in order to predict the limit strains in the range of positive strain paths. Later, Hutchinson and Neale [5] extended the initial imperfection approach, initially developed in [4], to the range of negative strain paths. In the latter work, both the rigid–plastic flow theory and the deformation theory of plasticity were used to model the mechanical behavior of the studied sheet. In most of the approaches proposed in the past, the effect of bending deformation on formability is not taken into account, since homogeneous deformation through the sheet thickness is assumed. However, it is not uncommon that metallic sheets used in industrial applications (automotive, aeronautic...) undergo combination of bending and stretching deformations. Such modes of deformation may have a substantial impact on the formability of the studied parts. Several experimental studies revealed that bending tends to increase the formability of metal sheets. Nowadays, there are many attempts to develop FLD prediction models that take into account bending effects. In this field, Shi and Gerdeen [6] integrated the effect of strain gradient and curvature in the prediction of FLDs for anisotropic materials. Sriram et al. [7] developed an empirical model to characterize fracture behavior during forming of advanced high strength steels under bending dominated conditions. He et al. [8] extended the initial imperfection approach developed in [4] to the case of combined stretching-bending loading.

Nonetheless, their work was restricted to the right-hand side of the FLD, where bending is along the major stretch direction. Two different constitutive laws were used in [8] to model the mechanical behavior of the sheet: the flow theory and the deformation theory of plasticity. The obtained numerical results show that the bending process decreases sheet metal formability when the flow theory is used, whereas the opposite trend is observed with the deformation theory. Furthermore, the model developed in [8] was combined with finite element analysis in order to predict the forming limits of a sheet metal undergoing continuous bending-under-tension loading [8]. More recently, Safdarian [10] developed a new model for predicting FLDs of tailor welded blanks. In this model, the M–K approach developed in [5] has been enriched by taking into account the effect of

[a] Corresponding author: Mohamed.BenBettaieb@ensam.eu

 DOI: 10.1051/matecconf/20168003003

bending strain on sheet metal formability. The flow theory of plasticity was used to describe the mechanical behavior of the studied sheet. Contrary to the approach developed in [8], the modified M–K model developed in [10] allows predicting the limit strains for the whole range of strain paths. However, this model presents some drawbacks and limitations. Indeed, in this model, the strain paths, both within the band and outside it, are assumed to remain linear during deformation. Consequently, the increment of strain components is replaced by total strain components in the constructive equations governing the associated modified M–K model. This assumption is not relevant in the band zone, where it is known that the strain path is constantly changing along deformation. Besides, the shear stresses have been withdrawn in the formulation of the equilibrium equations. This point represents the second limitation of the model. Indeed, the shear stresses cannot be neglected, especially when the band is not perpendicular to the major strain direction. In the present paper, a new modified M–K model is developed to address the above-mentioned issues. The main result of this paper is that the addition of bending effects tends to decrease sheet metal formability for the whole range of strain paths. This result confirms the numerical predictions reported in [8].

The remainder of the paper is organized as follows:
- The equations governing the new modified M–K model will be detailed in the second section.
- In the third section, the numerical and algorithmic aspects related to the model will be presented.
- Various numerical results obtained by application of the developed model will be presented and discussed in the fourth section.

Notations

The following notations and abbreviations are adopted in this paper:
- $\bullet^{T}$: transpose of tensor •.
- $\bullet^{B}$: quantity • associated with the band.
- $\bullet^{H}$: quantity • associated with the zone located outside the band.
- $\bullet_{ST}$: the stretching part of quantity •.
- $\bullet_{BE}$: the bending part of quantity •.
- •(t): value of quantity • at time t.
- $\bullet_{i}$: quantity • associated with the integration point i.

These notations can be combined. For instance, the stretching part of the strain tensor in the band is denoted by $\boldsymbol{\varepsilon}^{B}_{ST}$.

2 Constitutive equations

2.1 Mechanical behavior

Elasticity is neglected in the subsequent constitutive equations. This assumption is justified, because strain localization occurs at relatively large strains. Moreover, the plastic flow is assumed to be isotropic and incompressible. The rigid–plastic flow theory is then used to model the mechanical behavior of the studied sheet metal. Hence, the strain rate tensor $\dot{\boldsymbol{\varepsilon}}$ is derived by using the normality flow rule

$$\dot{\boldsymbol{\varepsilon}} = \dot{\lambda}\frac{\partial \sigma_{eq}}{\partial \boldsymbol{\sigma}}, \qquad (1)$$

where $\dot{\lambda}$ is the plastic multiplier (equal here to the von Mises equivalent strain rate $\dot{\varepsilon}_{eq}$), $\boldsymbol{\sigma}$ is the Cauchy stress tensor, and σ_{eq} is the von Mises equivalent stress. By using the definition of σ_{eq}, the rigid–plastic constitutive law can be derived from Eq. (1)

$$\dot{\boldsymbol{\varepsilon}} = \frac{3\dot{\varepsilon}_{eq}}{2\sigma_{eq}}\mathbf{S} \Leftrightarrow \mathbf{S} = \frac{2\sigma_{eq}}{3\dot{\varepsilon}_{eq}}\dot{\boldsymbol{\varepsilon}}, \qquad (2)$$

where $\mathbf{S}$ is the deviatoric part of $\boldsymbol{\sigma}$. The two latter tensors are linked by the following relation:

$$\boldsymbol{\sigma} = \mathbf{S} + p\,\mathbf{I}_2, \qquad (3)$$

where p is the hydrostatic pressure equal to $1/3\,\mathrm{tr}(\boldsymbol{\sigma})$ and $\mathbf{I}_2$ is the second-order identity tensor.

The equivalent stress σ_{eq} is related to the equivalent strain ε_{eq} by the Hollomon isotropic hardening law

$$\sigma_{eq} = K\,\varepsilon_{eq}^{n}, \qquad (4)$$

where K et n are two material hardening parameters.

By combining Eqs. (2) and (4) one obtains

$$\mathbf{S} = \frac{2K\,\varepsilon_{eq}^{n}}{3\dot{\varepsilon}_{eq}}\dot{\boldsymbol{\varepsilon}}. \qquad (5)$$

2.2 Modified M–K approach

2.2.1 Assumptions related to the Bending-Stretching process

The sheet is assumed to be initially flat, with X and Y the two in-plane principal directions and Z the through-thickness direction, as illustrated in Figure 1. It should be understood that the X-Y-Z coordinate system represents a material coordinate frame, which will rotate as material deforms. The following assumptions and choices are made in the development of this modified M–K approach:
- The sheet is first subjected to a bending moment M applied in Y-direction, resulting in the sheet being curved along X-direction, with a uniform curvature R. Hence, there is a single curvature along X-direction, while the sheet remains flat along the Y-direction (see, Figure 1).
- The neutral axis is assumed to remain at the middle layer of the sheet metal during deformation.

- The studied sheet is assumed to be wide enough (along the *Y*-direction) relative to its thickness. Consequently, the strain component ε_{yy} can be neglected during the bending process.
- After the bending stage, an axial major strain along X and a minor strain along Y are then applied to the sheet under a constant strain path ρ, while the bending radius is kept constant equal to R.
- The studied sheet is assumed to be thin. Therefore, the plane-stress assumption is adopted, in accordance with several literature works [7]. This assumption can be expressed in the following index form:

$$\sigma_{xz} = \sigma_{yz} = \sigma_{zz} = 0 . \qquad (6)$$

This plane-stress condition allows us to determine the hydrostatic pressure p introduced in Eq. (3)

$$p = S_{xx} + S_{yy} . \qquad (7)$$

Combining the plane-stress and the incompressibility conditions, one can easily obtain

$$\varepsilon_{xz} = \varepsilon_{yz} = 0 \quad ; \quad \varepsilon_{zz} = -(\varepsilon_{xx} + \varepsilon_{yy}) . \qquad (8)$$

Equations (6)-(8) are valid for any material point of the sheet.

2.2.2 Modified M–K equations

The M–K approach is based on the assumption of the preexistence of an initial geometric imperfection in the form of a band across the thickness of the sheet, as illustrated in Figure 1. This band is defined by its normal unit vector $\vec{n}$. The angle between vector $\vec{n}$ and the major strain direction X is denoted θ. The initial thickness of the band (resp. zone outside the band) is equal to h_0^{B} (resp. h_0^{H}).

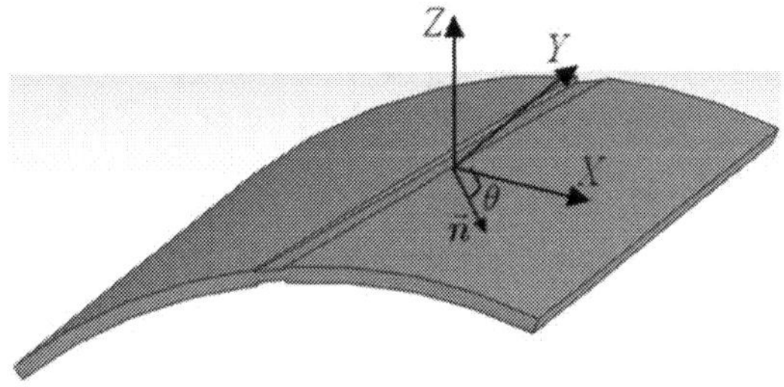

Figure 1. Initial imperfection approach for the sheet metal

As a consequence of the assumptions made in Section 2.2.1, it can be shown that the different mechanical variables (strain, stress…) are independent of the *x* and *y* coordinates, both inside and outside the band. They depend only on the z coordinate. Under the incompressibility and plane-stress conditions, it is more convenient to rewrite in what follows the governing equations in the form of an in-plane formulation (in the plane (*X Y*)). The out-of-plane components can be easily derived from Eqs. (6)-(8).

The total strain $\boldsymbol{\varepsilon}$ is additively decomposed into a bending part $\boldsymbol{\varepsilon}_{BE}$ and a stretching part $\boldsymbol{\varepsilon}_{ST}$

$$\boldsymbol{\varepsilon} = \boldsymbol{\varepsilon}_{BE} + \boldsymbol{\varepsilon}_{ST} . \qquad (9)$$

The above decomposition is valid both inside and outside the band.

However, the bending part $\boldsymbol{\varepsilon}_{BE}$ of the deformation tensor has the same expression in both zones

$$\boldsymbol{\varepsilon}_{BE} = \begin{pmatrix} \mathrm{Ln}(1 + z / R) & 0 \\ 0 & 0 \end{pmatrix}, \qquad (10)$$

where z is the third coordinate of the material point considered.

The stretching part $\dot{\boldsymbol{\varepsilon}}_{ST}^{H}$ is assumed to be uniform over the thickness of the sheet and is given by the following expression:

$$\dot{\boldsymbol{\varepsilon}}_{ST}^{H} = \begin{pmatrix} \dot{\varepsilon}_{ST\,xx}^{H} & 0 \\ 0 & \rho\,\dot{\varepsilon}_{ST\,xx}^{H} \end{pmatrix}, \qquad (11)$$

where the strain-path ratio ρ is varied in the range $-1/2 \le \rho \le 1$ to span the complete FLD.

The stretching part of the strain rate in the band $\dot{\boldsymbol{\varepsilon}}_{ST}^{B}$ is related to that outside the band $\dot{\boldsymbol{\varepsilon}}_{ST}^{H}$ by the following kinematic compatibility condition:

$$\dot{\boldsymbol{\varepsilon}}_{ST}^{B} = \dot{\boldsymbol{\varepsilon}}_{ST}^{H} + (1/2)((\dot{\mathbf{c}} \otimes \vec{n}) + (\dot{\mathbf{c}} \otimes \vec{n})^{T}) , \qquad (12)$$

where $\dot{\mathbf{c}}$ is the jump vector. Vectors $\dot{\mathbf{c}}$ and $\vec{n}$ are assumed to be uniform over the thickness of the sheet. The evolution of the band orientation θ is given by the following relation:

$$\theta = \arctan\left[\tan(\theta_0) \exp(1 - \rho)\, \varepsilon_{ST\,xx}^{H} \right], \qquad (13)$$

where θ_0 is the initial value of the band orientation.

The global equilibrium equation across the band can be expressed as follows:

$$\left(\int_{-h^{H}/2}^{h^{H}/2} \boldsymbol{\sigma}^{H}\, dz \right) . \vec{n} = \left(\int_{-h^{B}/2}^{h^{B}/2} \boldsymbol{\sigma}^{B}\, dz \right) . \vec{n} , \qquad (14)$$

where h^{H} and h^{B} are the current thicknesses of the sheet outside and in the band zone, respectively. They are expressed as functions of the initial thicknesses h_0^{H} and h_0^{B} and of the components of $\dot{\boldsymbol{\varepsilon}}_{ST}^{H}$ and $\dot{\boldsymbol{\varepsilon}}_{ST}^{B}$

$$h^{H} = h_0^{H}\, e^{-(\varepsilon_{ST\,xx}^{H} + \varepsilon_{ST\,yy}^{H})} \quad ; \quad h^{B} = h_0^{B}\, e^{-(\varepsilon_{ST\,xx}^{B} + \varepsilon_{ST\,yy}^{B})} . \qquad (15)$$

It is difficult to determine an exact value for the two integrals introduced in Eq. (14). Therefore, they are rather numerically approximated by the trapezoidal method, after a geometric discretization of the sheet thickness is performed

$$\int_{-h^{H}/2}^{h^{H}/2} \boldsymbol{\sigma}^{H}\,dz = \sum_{i=1}^{i=N_p-1}\left(\frac{(z_{i+1}^{H}-z_{i}^{H})(\boldsymbol{\sigma}_{i+1}^{H}-\boldsymbol{\sigma}_{i}^{H})}{2}\right),\quad \int_{-h^{B}/2}^{h^{B}/2} \boldsymbol{\sigma}^{B}\,dz = \sum_{i=1}^{i=N_p-1}\left(\frac{(z_{i+1}^{B}-z_{i}^{B})(\boldsymbol{\sigma}_{i+1}^{B}-\boldsymbol{\sigma}_{i}^{B})}{2}\right) \quad (16)$$

where N_p is the number of integration points used in the numerical quadrature, which is assumed, for convenience, to be the same for both zones. z_i^H (resp. z_i^B) is the current position of the ith integration point over the thickness in the zone outside the band (resp. in the band). Hence, z_1^H, $z_{N_p}^H$, z_1^B, and $z_{N_p}^B$ are equal to $-h^H/2$, $h^H/2$, $-h^B/2$, and $h^B/2$, respectively.

Furthermore, z_i^H (resp. z_i^B) is related to its initial value z_{i0}^H (resp. z_{i0}^B) by

$$z_i^H = z_{i0}^H\, e^{-(\varepsilon_{ST\,xx}^H+\varepsilon_{ST\,yy}^H)} \quad ; \quad z_i^B = z_{i0}^B\, e^{-(\varepsilon_{ST\,xx}^B+\varepsilon_{ST\,yy}^B)}. \quad (17)$$

3 Algorithmic aspects

3.1 Bending process

Before the stretching loading is applied, the sheet metal is first bent by applying a bending moment M. During this bending stage, the curvature radius decreases from $+\infty$ (which corresponds to a flat sheet) to a finite value R. This bending is moderate compared to stretching (typically the ratio h_0^H / R does not exceed 0.2). Therefore, it is legitimate to consider the distance z of any integration point to the neutral axis as a constant value, which is taken equal to its initial value z_0. At the end of this bending process, only the equivalent strains corresponding to the different points located inside and outside the band are computed and stored

$$\forall i = 1,\ldots,N_p : \begin{cases} \varepsilon_{eq\,i}^H = \dfrac{2}{\sqrt{3}}\,\mathrm{Ln}(1 + z_{i0}^H / R) \\ \varepsilon_{eq\,i}^B = \dfrac{2}{\sqrt{3}}\,\mathrm{Ln}(1 + z_{i0}^B / R) \end{cases}. \quad (18)$$

3.2 Algorithm for the FLD prediction

The general algorithm used to predict the FLD is based on the following three nested loops:

- For each strain path ρ ranging from $-1/2$ to 1 (with increments $\Delta\rho$ of 0.1).
  - For each initial band orientation θ_0, spanning the admissible range of inclination angles (i.e., between $0°$ and $90°$), at user-defined intervals (here, we take intervals of $1°$).
    - For each time interval $[t_0, t_0+\Delta t]$, apply the implicit incremental algorithm described in Section 3.3 to integrate the governing equations detailed in Section 2. The application of this incremental integration scheme is stopped when the following criterion is satisfied:

$$\Delta\varepsilon_{ST\,zz}^B \,/\, \Delta\varepsilon_{ST\,zz}^H \geq 10\,. \quad (19)$$

The strain component $\varepsilon_{ST\,xx}^H$, thus obtained once the criterion (19) is satisfied, is considered as being the critical strain ε_{xx}^* corresponding to the initial band inclination θ_0 and strain path ρ. The factor 10 in Eq. (19) is rather arbitrary and any other relatively large positive value can be used without loss of accuracy. Indeed, the effect of this value on the predicted critical strain is minimal, since the absolute value of the band thickness strain rate $\dot{\varepsilon}_{ST\,zz}^B$ increases very rapidly towards infinity at incipient necking.

The lowest critical strain ε_{xx}^*, over all initial angles θ_0, and the corresponding current angle, define the necking limit strain ε_{xx}^L and the necking band orientation, respectively, for the current strain path ρ.

3.3 Incremental integration of the governing equations

The main purpose of this incremental algorithm is to integrate the equations governing the modified M–K approach over a typical time increment $[t_0, t_0+\Delta t]$. In this aim, we assume that, at each integration point located in both zones, outside and inside the band, the following quantities are known at time t_0:

- The coordinate through the thickness direction z.
- The equivalent strain ε_{eq}.

In order to simplify notations, the argument $t_0+\Delta t$ will be omitted hereinafter, with the implied understanding that the corresponding quantity is evaluated at $t_0+\Delta t$, unless otherwise indicated.

The increment $\Delta\varepsilon_{ST\,zz}^B$ is chosen as loading parameter over $[t_0, t_0+\Delta t]$. It is denoted α (and is typically fixed to 0.001).

The increments $\Delta\boldsymbol{\varepsilon}_{ST}^H$ and $\Delta\boldsymbol{\varepsilon}_{ST}^B$ are derived from the integration of Eqs. (11) and (12), respectively

$$\Delta\boldsymbol{\varepsilon}_{ST}^H = \begin{pmatrix} \Delta\varepsilon_{ST\,xx}^H & 0 \\ 0 & \rho\,\Delta\varepsilon_{ST\,xx}^H \end{pmatrix}. \quad (20)$$

$$\Delta\boldsymbol{\varepsilon}_{ST}^B = \begin{pmatrix} \Delta\varepsilon_{ST\,xx}^H + \Delta c_1\, n_1 & (1/2)(\Delta c_1\, n_2 + \Delta c_2\, n_1) \\ (1/2)(\Delta c_1\, n_2 + \Delta c_2\, n_1) & \rho\,\Delta\varepsilon_{ST\,xx}^H + \Delta c_2\, n_2 \end{pmatrix} \quad (21)$$

It is noteworthy that Eq. (21) has been derived from Eq. (12) by implicitly assuming that, all along the time increment, the components of vector $\vec{n}$ keep their initial values at t_0.

Furthermore, Eq. (21) in conjunction with the incompressibility condition leads to the following relation:

$$(1+\rho)\,\Delta\varepsilon^{H}_{ST\,xx} + \Delta c_1\, n_1 + \Delta c_2\, n_2 = -\Delta\varepsilon^{B}_{ST\,zz}\,. \quad (22)$$

In the band, the position of the different integration points at $t_0+\Delta t$ can be obviously expressed in terms of $\Delta\varepsilon^{B}_{ST\,zz}$

$$\forall i = 1,\ldots,N_p:\; z_i^{B} = z_i^{B}(t_0)\,e^{\Delta\varepsilon^{B}_{ST\,zz}}\,. \quad (23)$$

By using Eqs. (10) and (23), the increment of $\boldsymbol{\varepsilon}^{B}_{BE}$ over $[t_0, t_0+\Delta t]$ for the different integration points within the band is derived as

$$\forall i = 1,\ldots,N_p:\; \Delta\boldsymbol{\varepsilon}^{B}_{BE\,i} = \begin{pmatrix} \mathrm{Ln}\left(\dfrac{R+z_i^{B}}{R+z_i^{B}(t_0)}\right) & 0 \\ 0 & 0 \end{pmatrix}. \quad (24)$$

The position at $t_0+\Delta t$ of the integration points located outside the band can be determined by the following equation:

$$\forall i = 1,\ldots,N_p:\; z_i^{H} = z_i^{H}(t_0)\,e^{\Delta\varepsilon^{H}_{ST\,zz}}\,, \quad (25)$$

where $\Delta\varepsilon^{H}_{ST\,zz}$ is determined by applying the incompressibility condition

$$\Delta\varepsilon^{H}_{ST\,zz} = -(1+\rho)\,\Delta\varepsilon^{H}_{ST\,xx}\,. \quad (26)$$

The increment of $\boldsymbol{\varepsilon}^{H}_{BE}$ over $[t_0, t_0+\Delta t]$ for the different integration points located outside the band is determined by a relation very similar to Eq. (24)

$$\forall i = 1,\ldots,N_p:\; \Delta\boldsymbol{\varepsilon}^{H}_{BE\,i} = \begin{pmatrix} \mathrm{Ln}\left(\dfrac{R+z_i^{H}}{R+z_i^{H}(t_0)}\right) & 0 \\ 0 & 0 \end{pmatrix}. \quad (27)$$

The increment over $[t_0, t_0+\Delta t]$ of the strain tensors $\boldsymbol{\varepsilon}^{H}$ and $\boldsymbol{\varepsilon}^{B}$ at the different integration points located outside and in the band zone, respectively, is finally determined as follows:

$$\forall i = 1,\ldots,N_p:\; \begin{cases} \Delta\boldsymbol{\varepsilon}_i^{H} = \Delta\boldsymbol{\varepsilon}^{H}_{ST} + \Delta\boldsymbol{\varepsilon}^{H}_{BE\,i} \\ \Delta\boldsymbol{\varepsilon}_i^{B} = \Delta\boldsymbol{\varepsilon}^{B}_{ST} + \Delta\boldsymbol{\varepsilon}^{B}_{BE\,i} \end{cases}, \quad (28)$$

where the expressions of $\Delta\boldsymbol{\varepsilon}^{H}_{ST}$, $\Delta\boldsymbol{\varepsilon}^{H}_{BE\,i}$, $\Delta\boldsymbol{\varepsilon}^{B}_{ST}$, and $\Delta\boldsymbol{\varepsilon}^{B}_{BE\,i}$ are given in Eqs. (20), (27), (21), and (24), respectively. Accordingly, for a given strain path ρ, the strain increments $\Delta\boldsymbol{\varepsilon}_i^{H}$ and $\Delta\boldsymbol{\varepsilon}_i^{B}$ are functions of three scalar unknowns: $\Delta\varepsilon^{H}_{ST\,xx}$, Δc_1 and Δc_2.

The expression of the deviatoric stress $\mathbf{S}$ at the different integration points inside and outside the band is derived from the incremental form of Eq. (5)

$$\mathbf{S} = \frac{2K\,\varepsilon_{eq}^{n}}{3\,\Delta\varepsilon_{eq}}\,\Delta\boldsymbol{\varepsilon}\,. \quad (29)$$

Combining Eqs. (3) and (7), one can easily derive the expression of the Cauchy stress tensor $\boldsymbol{\sigma}$ at the different integration points

$$\boldsymbol{\sigma} = \mathbf{S} + (S_{xx} + S_{yy})\,\mathbf{I}_2\,. \quad (30)$$

The above expression for $\boldsymbol{\sigma}$ is then inserted in the approximation (16), which in turn is inserted in the equilibrium equation (14). Analyzing the previous developments, Eq. (14) may be regarded as a system of two equations with three scalar unknowns: $\Delta\varepsilon^{H}_{ST\,xx}$, Δc_1 and Δc_2. By considering Eq. (14) along with Eq. (22), a system of three scalar equations with the three above unknowns is obtained. This system should be solved iteratively by using the Newton–Raphson method in order to determine the above-mentioned scalar unknowns. Ultimately, the determination of these unknowns allows computing the different quantities at $t_0+\Delta t$: the distance of the different integration points to the neutral axis of the sheet, the distribution of the equivalent strain through the thickness in both zones, and the band orientation. In particular, the values of the strain components $\Delta\varepsilon^{H}_{ST\,zz}$ and $\varepsilon^{H}_{ST\,xx}$ (the latter being equal to $\varepsilon^{H}_{ST\,xx}(t_0) + \Delta\varepsilon^{H}_{ST\,xx}$) are of special interest; the former being required for the application of the necking criterion (19), while the latter is needed in the algorithm of Section 3.2.

4 Prediction results

A DP600 sheet steel with the hardening parameters K = 945 MPa and n = 0.16 is used in the following simulations. The initial thicknesses h_0^{H} and h_0^{B} are fixed to 1 mm and 0.99 mm, respectively. The number of integration points N_p is set to 11. To simplify the notations used in the following results, the stretching strains $\varepsilon^{H}_{ST\,xx}$, $\varepsilon^{H}_{ST\,yy}$, $\varepsilon^{H}_{ST\,zz}$, and $\varepsilon^{B}_{ST\,zz}$ are simply denoted $\varepsilon^{H}_{11}, \varepsilon^{H}_{22}$, ε^{H}_{33}, and ε^{B}_{33}, respectively. In fact, the latter simplified notations are more common for the presentation of results corresponding to FLD predictions.

Figure 2 shows the onset of localized necking in the case of plane-strain state (ρ =0). In this figure, three bending radii are considered: R =3 mm; R =10 mm; R =100 mm. The results corresponding to a classical flat sheet are also included in this figure. The occurrence of strain localization is predicted when the ratio $\Delta\varepsilon^{B}_{33} / \Delta\varepsilon^{H}_{33}$ exceeds 10, as stated by criterion (19). It is clear from this figure that the limit strain increases with the bending

 DOI: 10.1051/matecconf/20168003003

radius R. This means that bending has as effect to accelerate the occurrence of localized necking, especially when the bending radius is small. For large values of the bending radius R (e.g., more than 100 mm), the bending effect is very limited. In such conditions, the prediction with high bending radius is almost the same as that corresponding to a flat sheet, as observed in Figure 2.

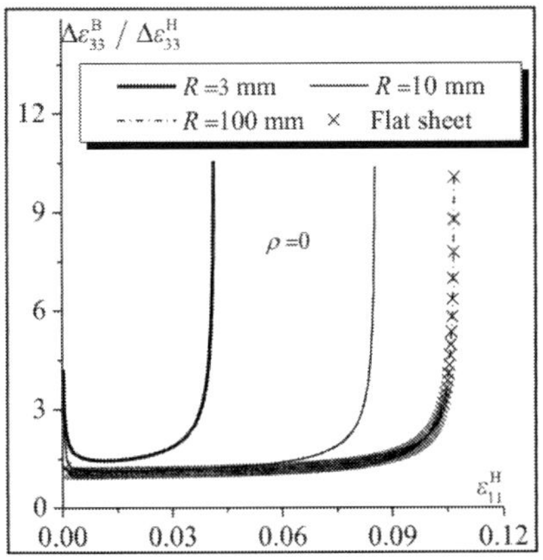

Figure 2. Illustration of the onset of localized necking in the case of plane-strain state (ρ =0)

The evolution of the critical strain ε_{11}^{*} as a function of the curvature $1 / R$ is plotted in Figure 3 for four representative strain paths (ρ =–0.5; ρ =0; ρ =0.5; and ρ =1). The initial band orientation is taken equal to 30° for the first strain path (which is different from the necking band orientation), and 0° for the other strain paths. For the different strain paths, the effect of curvature on necking acceleration is obvious, especially for the uniaxial tensile state (ρ =–0.5). The same trend was observed in [8]. Small vertical lines indicate the results for the bending radii of 100 mm, 10 mm, and 3 mm.

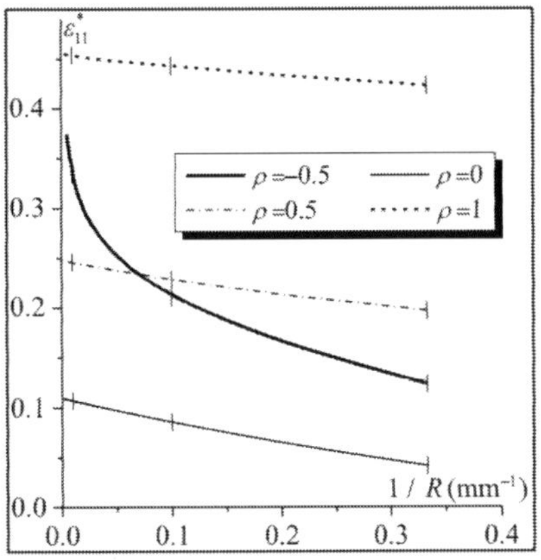

Figure 3. Evolution of the critical strain ε_{11}^{*} as a function of the curvature $1 / R$

The effect of bending radius on the evolution of the critical strain ε_{11}^{*}, as a function of the initial band orientation θ_0, is investigated in Figure 4 for the case of uniaxial tensile strain path. Note that the necking band orientation is not perpendicular to the major strain direction, whatever the value of the bending radius R. Indeed, the critical strain ε_{11}^{*} decreases with the initial band orientation θ_0 until reaching its lowest value, which corresponds to the limit strain ε_{11}^{L}, and increases afterwards. This result is common to all negative strain paths, as demonstrated in Figure 5. However, for positive strain paths, the necking band is perpendicular to the major strain direction. These observations are classical for flat sheets when bending effects are not considered. The effect of bending radius on the band orientation θ is not very significant, as shown in Figure 5.

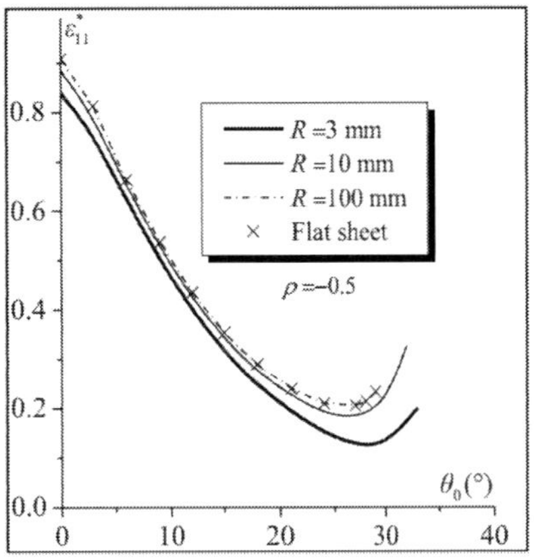

Figure 4. Effect of the bending radius on the evolution of the critical strain ε_{11}^{*} as a function of the initial band orientation θ_0 for uniaxial tensile strain path

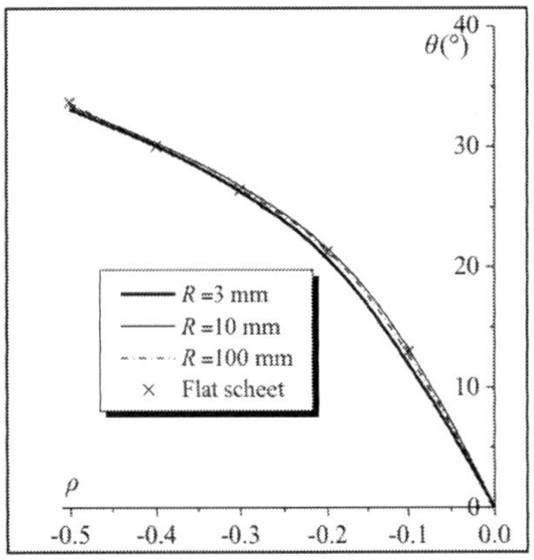

Figure 5. Effect of the bending radius on the evolution of the necking band orientation θ as a function of the strain path ρ

The results presented in Figure 4 are extended in Figure 6 to the whole range of strain paths, where the effect of bending radius on the location and shape of FLDs is analyzed. It appears from this figure that the bending effect is more important in the left-hand side of the FLD than in its right-hand side. The predictions reported in this figure confirm once again that the classical FLD results, corresponding to a flat sheet, are naturally recovered in the limit of very large curvature radius.

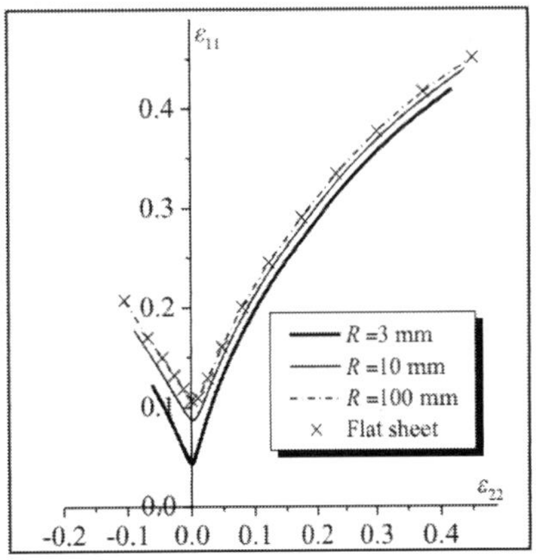

Figure 6. Effect of the bending radius on the location and the shape of FLDs

5 Concluding remarks

A numerical tool has been developed in this paper for the prediction of localized necking in sheet metals undergoing combined bending-stretching loadings. This model may be useful for understanding how the bending effects involved in forming processes would affect sheet metal formability. Contrary to previous models, specifically devoted to this type of combined loadings, the current model is able to predict the occurrence of localized necking for the whole range of strain paths. In the current approach, the mechanical behavior of the sheet metal is described by the rigid–plastic flow theory, while localized necking prediction is based on the imperfection approach. From the numerical predictions obtained by applying this tool, it is concluded that the addition of bending loading to traditional stretching tends to accelerate the occurrence of localized necking.

References

1. S.P. Keeler, W.A. Backofen, Trans. ASM **56**, 25 (1963)
2. G.M. Goodwin, Metallurgia Italiana **60**, 767 (1968)
3. S.L. Chiu, J. Leu, P.S. Ho, J. Appl. Phys. **76**, 5136 (1994)
4. Z. Marciniak, K. Kuczynski, Int. J. Mech. Sci. **9**, 609 (1967)
5. J.W. Hutchinson, K.W. Neale, *Koistinen, D.P., Wang, N.M. (Eds.), Mechanics of Sheet Metal Forming. Plenum*, 127 (1978)
6. M. Shi, J. Gerdeen, J. Mater. Shaping Technol. **9**, 253 (1991)
7. S. Sriram, H. Yao, N. Ramisetti, J. Manuf. Sci. Eng. **134**, 031003 (2012)
8. J. He, Z.C. Xia, S. Li, D. Zeng, J. Manuf. Sci. Eng. **135**, 227 (2013)
9. J. He, Z.C. Xia, D. Zeng, S. Li, J. Eng. Mat. Tech. **135**, 031009 (2013)
10. R. Safdarian, Mech. Res. Commun. **67**, 47 (2015)

MATEC Web of Conferences **80**, 03004 (2016) DOI: 10.1051/matecconf/20168003004
NUMIFORM 2016

Evaluation of ductile failure models in Sheet Metal Forming

Rui Amaral[1,a], Pedro Teixeira[2], Erfan Azinpour[2], Abel D. Santos[2], J. Cesar de Sa[2]

[1]*INEGI, Institute of Science and Innovation in Mechanical and Industrial Engineering, R. Dr. Roberto Frias 400, 4200-465 Porto, Portugal*
[2]*FEUP, Faculty of Engineering, University of Porto, R. Dr. Roberto Frias, 4200-465 Porto, Portugal*

Abstract. Traditionally, combination of equivalent plastic strain and stress triaxiality parameters are taken into account when performing characterization of material ductility. Some well-established models like Lemaitre model, GTN based models and many others perform relatively well at high-triaxiality stress states but fail to give adequate answers to low-triaxiality states. In this work, three damage models are presented, applied and assessed to a cross-shaped component. Concerning material, AA5182-O, corresponding damage parameters are characterized by an inverse analysis procedure for each damage model.

1 Introduction

When manufacturing automotive body structures, sheet metal forming processes are commonly used, which include drawing, bending and stamping. During this processing the sheet metal can be subjected to large localized deformations with significant through-thickness necking in which 3D stress states develop and dictate the fracture event of the metal blank.

The use of numerical methods such as the finite-element method to handle large plastic deformations has created the possibility to analyse, with a relative success, a forming process during its development stage, including damage and fracture [1].

New generation materials used in automobile industry, as a need to face environmental restrictions (low exhaust emissions) without compromising and even improving safety specifications, are hampered by their lower ductility leading to premature fracture and by the difficulties in predicting such behaviour, thus challenging numerical simulations.

In this paper three damage models are used: Lemaitre's ductile damage model, GTN Gurson-Tvergaard-Needleman (GTN) model, which is an extended version of the model proposed by Gurson [2] and Johnson-Cook model. In the framework of Continuous Damage Mechanics, a local damage model is implemented in Abaqus/Explicit code [3] and corresponding constitutive equations for the coupled model were developed and implemented. It is based on the Lemaitre's ductile damage evolution law, fully coupled with Hill's orthotropic plasticity criterion. Lemaitre model is implemented in a user material subroutine, which was developed to work in the FE package Abaqus. The GTN and Johnson-Cook models were already implemented in FE package, but corresponding damage parameters were identified by current research.

The selected sheet material is an aluminium alloy, AA5182-O. Corresponding characterization for damage parameters is performed by an inverse analysis procedure, using reference experimental tests.

Numerical simulations are presented and the corresponding results are compared with an experimental failure obtained for a cross-shaped component, in order to test and validate the different damage models and their ability for prediction of damage growth and fracture initiation.

2 Damage models

2.1 Lemaitre Damage Model

Induced by large plastic flow, the internal deterioration of the material, which may lead to macroscopic collapse, accompanies the deformation process and therefore, although different in nature, the two dissipative processes influence each other and should, therefore, be coupled at the constitutive level. The model formulated by Lemaitre [4] defines the damage variable as the neat area of a unit surface cut by a given plane corrected for the presence of existing cracks and cavities. By assuming homogenous distribution of micro-voids and the hypothesis of strain equivalence, the effective stress tensor can be represented as:

$$\tilde{\boldsymbol{\sigma}} = \frac{\boldsymbol{\sigma}}{1-D} \qquad (1)$$

where is $\boldsymbol{\sigma}$ the Cauchy stress tensor. The damage variable D value lies between 0 (undamaged state) and 1 (rupture). The evolution law for the internal variables are derived

[a] Corresponding author: ramaral @inegi.up.pt

from a potential of dissipation and considering the damage value, it can be written as:

$$\dot{D} = \frac{\dot{\gamma}}{1-D}\left(\frac{-Y}{S}\right)^{s} \quad (2)$$

where $\dot{\gamma}$ is the plastic consistency parameter and Y is damage energy release rate given by:

$$Y = -\frac{1}{2E(1-D)^2}[(1-\upsilon)\boldsymbol{\sigma}:\boldsymbol{\sigma} - \nu(tr\,\boldsymbol{\sigma})^2] \quad (3)$$

where E and υ are the Young's modulus and Poisson's ratio, respectively.

2.2 GTN Damage Model

Based on a micromechanical approach, Gurson [2] has proposed an internal variable to describe material degradation, which represents the volume fraction of micro-cavities that nucleate, grow and coalesce during loading. The original model was further extended by Tvergaard [5] and Needleman [6] replacing the original internal variable f by a modified parameter f^* in order to take into account the accelerated void coalescence process after a critical void volume fraction f_c is reached. The yield function of the GTN damage model is given by:

$$\phi = \left(\frac{\bar{\sigma}}{\sigma_Y}\right)^2 + 2q_1 f^* cosh\left(\frac{3}{2}q_2\frac{\sigma_H}{\bar{\sigma}}\right) - \left(1 + q_3 f^{*2}\right) \quad (4)$$

where q_1, q_2 and q_3 are the model parameters, σ_Y the flow stress and σ_H and $\bar{\sigma}$ the hydrostatic stress and equivalent von Mises stress, respectively. The modified porosity parameter f^* is calculated by:

$$f^* = \begin{cases} f & f \leq f_c \\ f_c + \dfrac{1/q_1 - f_c}{f_F - f_c}(f - f_c) & f > f_c \end{cases} \quad (5)$$

The evolution law of the void volume fraction is given by:

$$\dot{f} = \dot{f}_G + \dot{f}_N \quad (6)$$

accounting for void growth $\dot{f}_G$ and nucleation $\dot{f}_N$, both variables expressed by:

$$\dot{f}_G = (1-f)tr\,\dot{\boldsymbol{\varepsilon}}_p \quad (7)$$

$$\dot{f}_N = \frac{f_N}{S_N\sqrt{2\pi}} exp\left[-\frac{1}{2}\left(\frac{\bar{\varepsilon}_p - \varepsilon_N}{S_N}\right)^2\right]\dot{\bar{\varepsilon}}_p \quad (8)$$

where f_N, S_N and ε_N are material parameters and $\dot{\boldsymbol{\varepsilon}}_p$ is the plastic strain rate tensor.

2.3 Johnson-Cook model

Damage indicators that can express the material's degradation under plastic straining are widely used in industry due to their relative simplicity. These models are often calculated using a fully uncoupled approach, meaning that the progressive degradation does not affect the plastic behaviour of the material. The most used model is the one proposed by Johnson and Cook [7], which, besides the postulation of a phenomenological strain rate and temperature dependent hardening law, have also defined the fracture strain ε_f as a function of stress triaxiality η, plastic strain rate $\dot{\varepsilon}_p$ and temperature θ as:

$$\varepsilon_f = (d_1 + d_2 e^{d_3\eta})\left[1 + d_4 ln\left(\frac{\dot{\varepsilon}_p}{\dot{\varepsilon}_0}\right)\right](1 + d_5\theta) \quad (9)$$

where d_1, d_2, d_3, d_4 and d_5 are material parameters. The corresponding damage indicator is calculated by the expression:

$$D = \sum \frac{\Delta\bar{\varepsilon}_p}{\varepsilon_f} \quad (10)$$

where $\Delta\bar{\varepsilon}_p$ is the equivalent plastic strain increment.

3 Material and damage parameters

3.1 Material characterization

In this article, a mechanical characterization of the aluminium alloy AA5182-O, with 1 mm thickness, was performed [8]. Tensile tests of AA5182-O specimens were conducted for different angles according to the rolling direction. Also monotonic simple shear tests were performed for different angles. The corresponding flow curves are shown in figure 1 and 2.

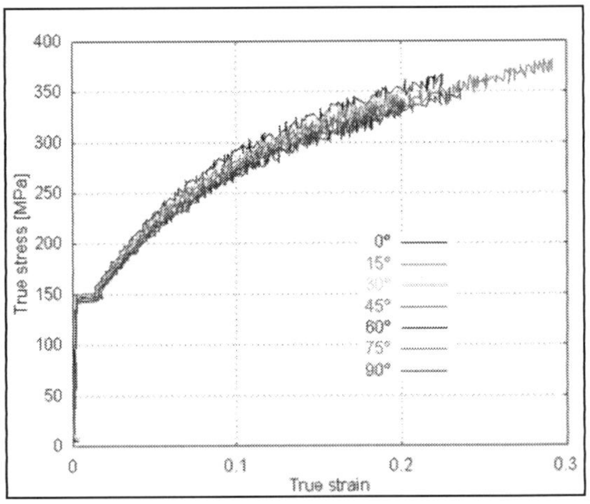

Figure 1. Flow stress-strain curve obtained from uniaxial tensile test for AA5182-O, according to different directions relative to the rolling direction [8].

The fundamental properties of the material obtained from stress-strain curves, such as Young modulus (E), Poisson coefficient (v), yield stress (Rp), ultimate tensile strength (R_m) and total elongation (e_t), are presented in table 1, as well as the anisotropy coefficients (*r-value*) for different directions relative to rolling direction.

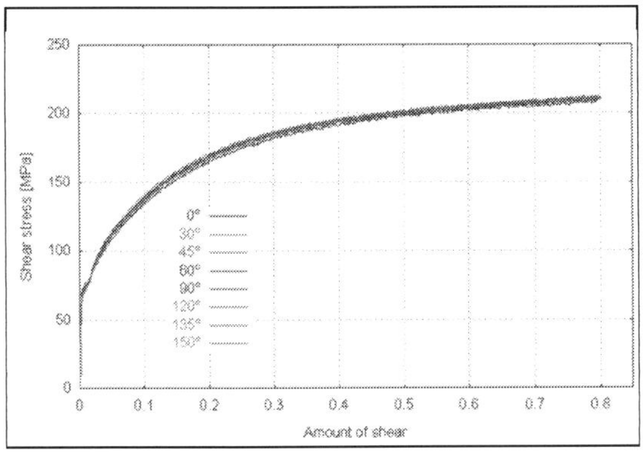

Figure 2. Flow curve obtained from monotonic simple shear test for AA5182-O, according to different directions relative to the rolling direction [8].

Table 1. Mechanical properties for AA5182-O.

Property	Value
Young modulus [GPa]	69
Poisson coefficient	0.3
Yield stress [MPa]	149.5
Ultimate tensile strength [MPa]	283.8
Total elongation [%]	24.4
r_0, r_{45}, r_{90}	0.79, 0.85, 0.7

The behaviour of the aluminium alloy AA5182-O can be well described by using an isotropic hardening given by the Voce Law. Table 2 gives the corresponding parameters obtained for this constitutive model.

Table 2. Voce Law parameters for AA5182-O.

Voce Law $Y = Y_0 + R_{sat}\left[1-\exp\left(-C_r\cdot\bar{\varepsilon}^p\right)\right]$	
Parameter	**Value**
Y0 [MPa]	149
Rsat [MPa]	208.7
Cr	12.1

3.2 Damage parameter identification

Parameters for presented damage models (section 2), were determined by an inverse numerical analysis based on optimization algorithms. Due to the fact that an inhomogeneous state is prevalent during material softening, experimental and numerical force-elongation curves are considered [9]. The minimization of the mean square error (MSE) between experimental and numerical data is calculated using an optimization algorithm implemented on MATLAB. As a result of the iterative optimization procedure, the best fit values for the presented damage models are obtained. The used methodology is shown in figure 3.

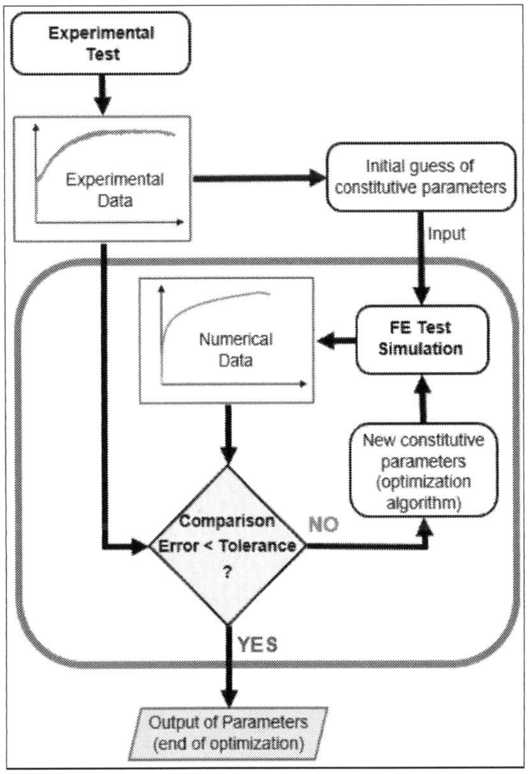

Figure 3. Methodology used for the identification of the damage parameters.

Lemaitre damage model parameters have been identified based on the results obtained from the tensile test [10]. The parameters for Lemaitre damage model are presented in table 3.

Table 3. Lemaitre damage model parameters for AA5182-O.

Parameter	Value
Damage Exponent - *s*	1.0
Damage Denominator – *S* [MPa]	1.25

The GTN damage model has 7 parameters that characterize the behaviour of the material, but there is a mutual dependency that must be considered [11]. According to literature, the parameters q_1, q_2 and S_N are kept constant [12, 13] and the initial void volume fraction f_0 can be obtained through an analytic expression based on chemical composition [14], which means that finally only 4 parameters will be determined. In table 4 the parameters of GTN damage model are presented, obtained from the identification procedure and using the experimental uniaxial tensile and biaxial bulge test.

Table 4. GTN damage model parameters for AA5182-O.

Parameter	q_1	q_2	q_3	f_0	f_C	f_F	ε_N	S_N	f_N
Value	1.5	1	2.25	0.01	0.021	0.04	0.3	0.1	0.001

The Johnson-Cook damage model is employed without strain rate and temperature effects, which means that parameters d_4 and d_5 are not considered. Since this model accumulates failure, in which the failure strain is more dependent to stress triaxiality than the others variables, it is more important the determination of parameters d_1, d_2 and d_3 [15]. In table 5 the parameters of Johnson-Cook damage model are presented, obtained from the identification procedure and using uniaxial tensile and simple shear test. The obtained parameters are similar to those identified by Sun *et al* [16], also for AA5182-O.

Table 5. Johnson-Cook damage model parameters for AA5182-O.

Parameter	**Value**
d_1	0.024
d_2	0.38
d_3	1.5
d_4, d_5	0

4. Results and discussion

4.1 Cross Tool geometry

The applicability of the described damage models is illustrated by means of an experimental deep drawing failure case of a cross-shaped (figure 4) sheet metal component [17].

Figure 4. Tool geometry and active parts. The punch cross geometry has 487 mm length.

Using an aluminium alloy AA5182-O sheet with a thickness of 1 mm, failure occurs as shown in figure 5.

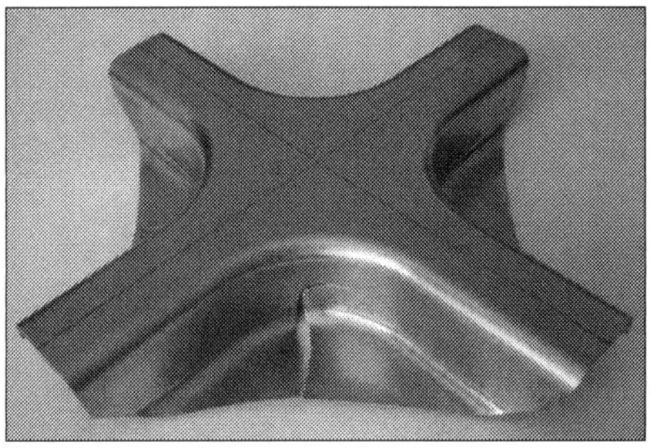

Figure 5. Cross-shaped component with breakage.

4.2 Numerical model

A 3D FEM explicit analysis (Abaqus/Explicit) model is considered with one quarter of the real setup, due to symmetry. The tools are modeled as fully rigid surfaces with discretization performed by three noded rigid elements (R3D3 type from Abaqus Library). The blank discretization uses a double layer of deformable eight noded solid elements with reduced integration (C3D8R type from Abaqus Library), making a total of 1186 elements. Regarding the material behaviour, the blank was modeled as an elasto-plastic material with isotropic hardening described by Voce Law. The blank material properties used, were already presented in section 3. The GTN and Johnson-Cook models are implemented in FE package Abaqus, but in the case of Lemaitre model a user material subroutine was developed and implemented in the same FE program.

4.3 Results

Figures 6, 7 and 8 present the contours of the equivalent plastic strain and the damage value for the three considered damage models, respectively, Lemaitre, GTN and Johnson-Cook.

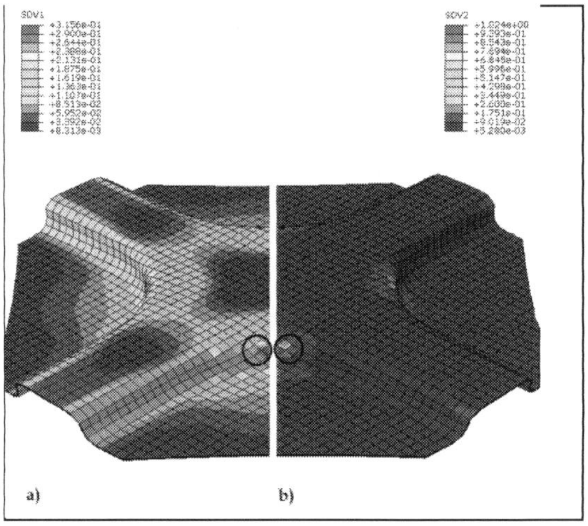

Figure 6. Obtained results using the identified parameters of Lemaitre damage model. a) Equivalent plastic strain; b) Damage contour prediction.

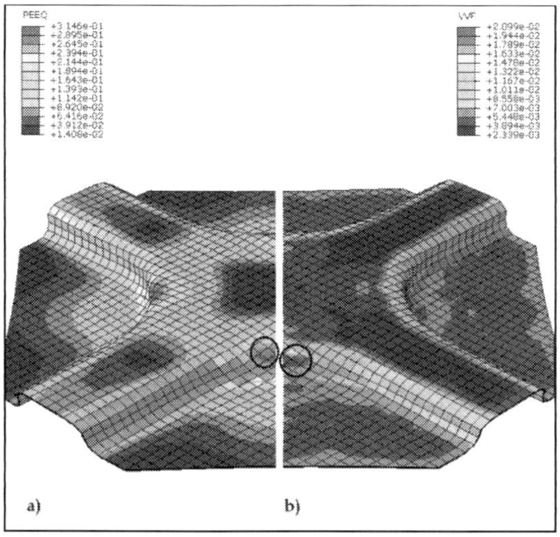

Figure 7. Obtained results using the identified parameters of GTN damage model. a) Equivalent plastic strain; b) Damage contour prediction.

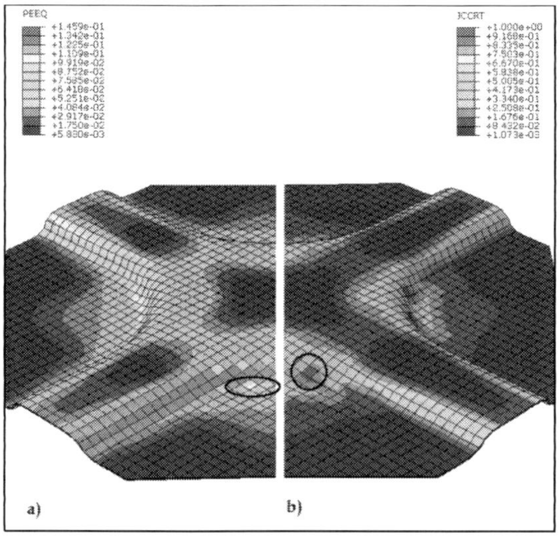

Figure 8. Obtained results using the identified parameters of Johnson-Cook damage model. a) Equivalent plastic strain; b) Damage contour prediction.

The figure 9 shows the levels of triaxiality in the cross-shaped component. Figures 6 to 8 shows the fracture initiation for the damage models presented in this study. For current proposed experimental component, both maximum strain and damage variable have very close locations (at the vertical wall of cross-center geometry). Usually, when analyzing fracture location, maximum damage is attained in regions where triaxiality is higher. All damage models in this study show similar trend, which is in accordance with experimental evidence. However the punch displacement to failure and maximum strain are different.

Although Lemaitre and GTN models predict fracture initiation at the same punch displacement, Johnson-Cook predicts failure at lower punch displacement and lower plastic strain. This difference can be attributed to inexistence (not considered) of plastic softening modeling in Johnson-Cook model, while the other models include such softening.

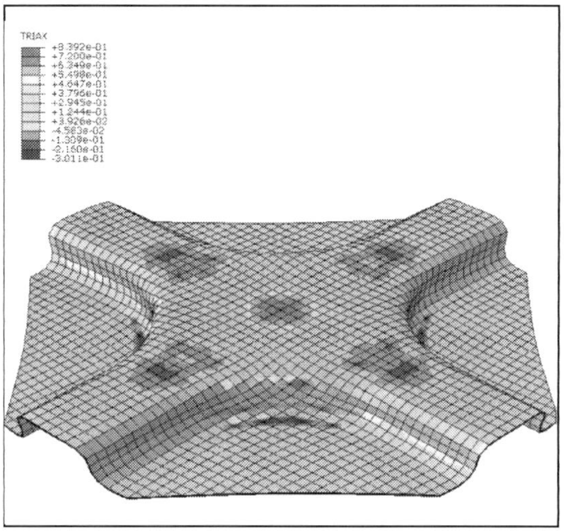

Figure 9. Triaxiality levels present in cross-shaped component.

Another reason can be due to the parameter identification basis for each model, which in turn is related with triaxiality values. The tests used for Johnson-Cook parameter identification are related with low triaxiality tests (shear+tensile), although in case of GTN and Lemaitre, the tests used have higher triaxiality values (tensile+bulge). Since experimental component shows critical deformation localization for higher triaxialities, this means that GTN and Lemaitre and corresponding damage parameters are more adjusted to predict failure in this component, which is in accordance to obtained numerical results and experimental evidence.

5 Conclusions

In this paper, three different ductile damage models, GTN, Johnson-Cook and Lemaitre were described and compared for prediction of damage in sheet metal forming processes. A finite element analysis of a cross-shaped component, using the parameters obtained for the aluminum alloy AA5182-O, has been taken as a basis of comparison between the damage models, in order to assess the accuracy of the results. For this particularly case, the simulations of GTN and Lemaitre show a good agreement with the damage location observed in the experimental component. The parameters obtained for the Johnson-Cook damage model, from tensile and simple shear tests, overestimate the development of damage at high triaxialities.

It is known that to cover the range of the relation between stress triaxiality and strain at failure, namely on low stress triaxiality states, other failure models are needed [18, 19] and therefore they will be tested in future studies.

Acknowledgements

The authors gratefully acknowledge the financial support of the Portuguese Foundation for Science and Technology (FCT) under project PTDC/EMS-TEC/6400/2014.

References

1. P. Teixeira, A.D. Santos, J.M.A. César de Sá, F.M. Andrade Pires, A. Barata da Rocha. Int. J. Mat. Form., 2, 463-466, (2009)
2. L. Gurson. J. Eng. Mat. Tech., 99, 2–15, (1977)
3. P. Teixeira, A.D. Santos, F.M. Andrade Pires, J. César de Sá. J. Mat. Proc. Tech., **177**, 278-281, (2006)
4. J. Lemaitre. J. Eng. Mat. Tech, 107, 83-89, (1985)
5. V. Tvergaard V. Int. J. Fract, 18, 237-252, (1982)
6. Needleman, V. Tvergaard. Acta metall., **32**, 157-169, (1984)
7. G. R. Johnson, W. H. Cook. Eng. Fract. Mech., **21**, 31-48, (1985)
8. 3DS report, 2001. Selection and identification of elastoplastic models for the materials used in the benchmarks. 18-Months Progress Report, Inter-regional IMS contract ''Digital Die Design Systems (3DS)'', IMS 1999 000051
9. Tsoupis, M. Merklein. 13th LS-Dyna Conf., (2013)
10. Lemaitre, R. Desmorat, *Engineering Damage Mechanics*, (2005)
11. D.A. Wang, W.Y. Chien, K.C. Liao, S.C. Tang, Ch. J. Mech Series A **19**, 161-168, (2003)
12. Z.L. Zang, Fatig. Fract. Eng. Mat. Struct 19, 561-570, (1996)
13. D.Z. Sun, A. Hönig, W. Böhme, W. Schmitt. Frac. Mec. **25**, 343-357, (1995)
14. M. Doing, K. Roll. AIP Conf., 1541-1546, (2011)
15. Marzbanrad, S. Noruzi, A. Ahmado, S.K. ISME2013, (2013).
16. H.T. Sun, J. Wang, G.Z. Shen, P. Hu. Int. J. Auto. Tech. 14, 605-610, (2013)
17. Pinto, A.D. Santos, P. Teixeira, P.J. Bolt. J. Nat. Proc. Tech., **202**, 47-53, (2008)
18. L. Malcher, F.M. Andrade Pires , J.M.A. César de Sá, Int. J. Plast, **30–31**, 81–115, (2012)
19. L. Malcher , F.M. Andrade Pires, J.M.A. César de Sá, Int. J. Plast., **54**, 193–228, (2014)

MATEC Web of Conferences **80**, 03005 (2016) DOI: 10.1051/matecconf/20168003005
NUMIFORM 2016

Prediction of wrinkling and springback in sheet metal forming

D.M. Neto[1,a], M.C. Oliveira[1], J.L. Alves[2], A.D. Santos[3] and L.F. Menezes[1]

[1]*CEMUC, Department of Mechanical Engineering, University of Coimbra, Rua Luís Reis Santos, 3030-788 Coimbra, Portugal*
[2]*CMEMS, Center for MicroElectroMechanics Systems, University of Minho, Campus de Azurém, 4800-058 Guimarães, Portugal*
[3]*FEUP, Faculty of Engineering, University of Porto, Rua Dr. Roberto Frias, 4200-465 Porto, Portugal*

Abstract. The finite element simulation is currently a powerful tool to optimize forming processes in order to produce defect-free products. Wrinkling and springback are main geometrical defects arising in sheet metal forming. Nevertheless, the prediction of such defects requires accurate numerical models. This study presents the experimental and numerical analysis of a rail with high tendency to develop both wrinkling (top surface of geometry) and springback (flange). The punch force evolution and the final geometry of the rail, evaluated in four different cross-sections, are the main variables analysed. Globally, the numerical results are in good agreement with the experimental measurements. However, the shape of the wrinkle is significantly influenced by the symmetry conditions considered in the model (1/4 of the blank). In fact, considering the full model of the blank, the numerical results are in better agreement with the experimental ones. On the other hand, the computational cost of the numerical simulation considering the full blank is approximately 12 times higher than using 1/4 of the blank.

1 Introduction

Sheet metal forming processes are commonly used in the automotive industry to produce several metal parts with complex geometry [1]. Currently, the major concerns of the automotive industry are the environmental protection (low fuel consumption and, consequently, low exhaust emissions) and the safety specifications. Accordingly, in order to meet these requirements, new materials have been introduced in car body manufacturing, such as high-strength steels as well as aluminium alloys [2–4]. Nevertheless, these material are more prone to develop geometrical defects, namely springback [5]. Another important geometrical defect arising in sheet metal forming is the wrinkling behaviour, which results from the instability under compressive stresses [6].

The integration of numerical simulation in the design and development of sheet metal forming processes was a key factor for the fulfilment of the increasing requirements for time and cost efficiency [7, 8]. In fact, the typical experimental trial-and-error die design has been progressively replaced by the finite element simulation of the process. This demands accurate numerical models, namely the introduction of new constitutive models [9], new types of finite elements [10] and the accurate treatment of the frictional contact conditions [11]. Despite the many advances in the numerical simulation tools development, the accurate prediction of geometrical defects still represents a challenge for the simulation [12]. For instance, it is known that wrinkling prediction is strongly affected by the finite element discretization [6]. Also, springback prediction is influenced by the model selected to describe the material kinematic hardening behaviour [13].

This study presents the experimental and numerical analysis of a rail with a high tendency to develop both wrinkling (top surface of geometry) and springback (at the flange), thus representing an additional challenge to simulation, due to a possible interaction of effects. The experimental procedure is described in Section 2, while the adopted finite element code is presented in Section 3. The proposed numerical models of the forming process are defined in Section 4 and the comparison between numerical and experimental results is presented in Section 5. The main conclusions are outlined in Section 6.

2 Experimental procedure

This study deals with the sheet metal forming of a rail, which is prone to 2D springback behaviour and wrinkling on the top surface. In order to obtain a representative set of reference experimental results, required for validation the different strategies that may be adopted in the numerical simulation, special care was devoted to the standardization of the experimental procedure [14]. The experimental setup involves a punch, a die and a blank-holder, as shown in Figure 1. The forming process comprises three phases. In the first one, the die moves downward, clamping the blank between the die and the blank-holder with an initial 90 kN prescribed force. This blank-holder force is defined by six nitrogen gas springs (see Figure 1), being all connected, in order to assure the same pressure in each of them. In the second phase, the die and blank-holder move

[a] Corresponding author: diogo.neto@dem.uc.pt

downward together, forming the part and defining a punch penetration of 60 mm. At this stage the gas springs are being compressed, thus increasing the blank-holder force from 90 to 130 kN. The third phase is the part removal from tools, in which the springback or elastic recovery takes place (see Figure 2).

Figure 1. Experimental tool used to perform the forming process [14].

Figure 2. Experimental geometry of the rail after being removed from the forming tools. Two holes are trimmed on the top of the rail for fixing the part during its measurement.

The blank sheet (mild steel DC06) is a square with 300.0 mm length and 1.0 mm thickness. The experimental tests are carried out with the rolling direction of the blank sheet aligned with the longitudinal direction of the rail. In order to try to guarantee a correct amount of lubricant and its uniform distribution in the blank, the lubrication method suggested by Santos et al. [15] was adopted. The amount of lubricant (Quaker 6130) is defined as 1.4 g/m^2/face.

The evolution of both the punch and the blank-holder force was recorded by load cells placed under each gas spring and under the punch. The punch displacement is experimentally evaluated through a displacement transducer. At least five tests were performed in order to check the reproducibility of the experiments. Accordingly, a representative result is selected, for each measurement under analysis. At the end of the second phase of the forming process, two holes are punched on the top of the rail, as seen in Figure 2. These holes are used to obtain a reference position in the jig, which was defined for experimental measurements. The definition of the coordinate system used in the evaluation of the cross-section profiles of the rail is coincident with one of the holes, as shown in Figure 3. For this geometry, four cross-section profiles are defined, which are marked in Figure 3. Five sets of measurements are performed for each cross-section at every 3 mm [14].

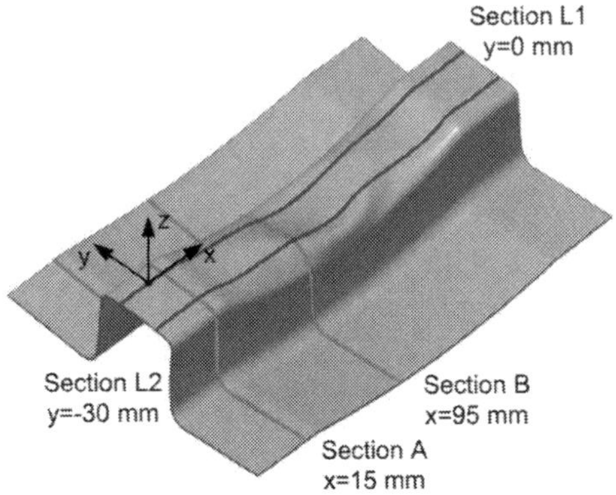

Figure 3. Definition of the four cross-sections used to evaluate experimentally and numerically the final geometry of the rail.

3 DD3IMP – static implicit FE code

The in-house static implicit finite element code DD3IMP [16], which has been specifically developed to simulate sheet metal forming processes, is adopted in the present study to carry out the numerical simulations. The mechanical model takes into account large elastoplastic strains and rotations, while the evolution of the deformation process is described by an updated Lagrangian formulation.

3.1 Variational principle

The rate form of the equilibrium equations and boundary conditions can be expressed by the principle of virtual velocities [17] in the form:

$$\int_V \{(\dot{\boldsymbol{\sigma}}^J - 2\boldsymbol{\sigma}\mathbf{D}) : \delta\mathbf{D} + (\mathbf{L}\boldsymbol{\sigma}) : \delta\mathbf{L}\} dV = \int_S \overline{\mathbf{t}} \cdot \delta\mathbf{v} dS, \quad (1)$$

where V denotes the domain occupied by the body and S represents the surface on which the external forces $\overline{\mathbf{t}}$ are prescribed. $\delta\mathbf{v}$ denotes the virtual velocity field, $\mathbf{L}$ is the velocity gradient tensor and $\mathbf{D}$ is the strain rate tensor, which is the symmetric part of $\mathbf{L}$. The Jaumann derivative of the Cauchy stress tensor $\boldsymbol{\sigma}$ is given by:

$$\dot{\boldsymbol{\sigma}}^J = \dot{\boldsymbol{\sigma}} + \boldsymbol{\sigma}\mathbf{W} - \mathbf{W}\boldsymbol{\sigma}, \quad (2)$$

where $\dot{\boldsymbol{\sigma}}$ stands for the time derivative of the Cauchy stress tensor and $\mathbf{W}$ is the total spin tensor (antisymmetric part of $\mathbf{L}$).

3.2 Constitutive material models

The elastoplastic constitutive model adopted considers isotropic elastic behaviour and anisotropic plastic behaviour. The differential form of the constitutive equation is given by:

$$\overset{\nabla}{\boldsymbol{\sigma}}^{\mathrm{J}} = \mathbf{C}^{\mathrm{ep}} : \mathbf{D}, \tag{3}$$

where $\mathbf{C}^{\mathrm{ep}}$ is the fourth-order tensor that defines the elastoplastic modulus in the current configuration, according to the adopted constitutive model. The expression for this tensor depends of the algorithms adopted in the integration of the constitutive model and the type of relation considered between the states at the beginning and at the end of the loading increment. Thus, it is possible to consider the tangent elastoplastic modulus or the consistent elastoplastic modulus.

Several advanced yield criteria and work hardening laws (constitutive models) have been implemented in DD3IMP to allow a better description of the different material mechanical behaviour [13,18]. In each load step, an explicit approach is used to calculate the trial solution, which is adjusted using a generalization of the $r_{\min}$ strategy [19]. In order to guarantee the equilibrium of the deformable body, this trial configuration is successively corrected using an implicit method. The Newton–Raphson method is used to solve the nonlinear system of equations arising from the spatial and temporal discretization of the weak form. Some high performance computing techniques have been incorporated to take advantage of multi-core processors, namely OpenMP directives in the most time consuming branches of the code [20].

3.3 Frictional contact conditions

The modelling of the blank with solid elements allows the accurate evaluation of the contact forces and the through-thickness gradients (stress and strain) [21]. On the other hand, the forming tools are considered rigid and their outer surfaces are modelled by Nagata patches [22]. The friction between the blank and the forming tools is described by the classical Coulomb's law. The frictional contact problem is regularized through the augmented Lagrangian method [23], leading to a mixed system of equations involving both displacements and contact forces as unknowns [24]. The Newton–Raphson scheme is used to solve, in a single iterative loop, the non-linearities associated with both the contact and the elastoplastic behaviour of the deformable body [16].

4 Numerical model

In order to analyse the influence of the symmetry conditions on the wrinkling phenomena, two distinct finite element models are studied. The first one considers only 1/4 of the blank due to geometric and material symmetry conditions of the forming process. On the other hand, the entire forming process is simulated in the second model. Further, in the full model, both the sheet and the rolling direction are rotated 1° with the Ox direction, which induces non-symmetrical conditions on the forming process.

4.1. Material behaviour

The mechanical behaviour of the mild steel DC06 is assumed elastoplastic. The elastic behaviour is considered isotropic and constant, which is described by Hooke's law with Young's modulus of 210 GPa and Poisson ratio of 0.30. Regarding the plastic behaviour, the isotropic hardening is described by the Swift law, which is combined with kinematic hardening of Armstrong–Frederick type. The yield stress defined by the Swift law is given by:

$$Y = K(\varepsilon_0 + \bar{\varepsilon}^{\mathrm{p}})^n \quad \text{with} \quad \varepsilon_0 = \left(\frac{Y_0}{K}\right)^{1/n}, \tag{4}$$

where K, ε_0 and n are the material parameters, while $\bar{\varepsilon}^{\mathrm{p}}$ denotes the equivalent plastic strain. The kinematic part of the work hardening [25], i.e. the evolution of the back-stress tensor $\mathbf{X}$, is described by:

$$\dot{\mathbf{X}} = C_{\mathrm{X}}\left[\frac{X_{\mathrm{sat}}}{\bar{\sigma}}(\boldsymbol{\sigma}' - \mathbf{X}) - \mathbf{X}\right]\dot{\bar{\varepsilon}}^{\mathrm{p}} \quad \text{with} \quad \mathbf{X}(0) = \mathbf{0}, \tag{5}$$

where X_{sat} characterizes the saturation value of $\mathbf{X}$, while the material parameter C_{X} characterizes the rate of approaching the saturation. The equivalent effective stress is denoted by $\bar{\sigma}$, while the deviator of the Cauchy stress tensor is represented by $\boldsymbol{\sigma}'$.

The material parameters of the Swift law combined with kinematic hardening were identified using experimental results of a uniaxial tensile test as well as monotonic and Bauschinger simple shear tests, all of them performed with specimens cut along the rolling direction [26]. The five material parameters of the isotropic–kinematic hardening law are listed in Table 1. The adopted model provides a satisfactory fit of the experimental data (DC06), as highlighted in the stress-strain curves presented in Figure 4.

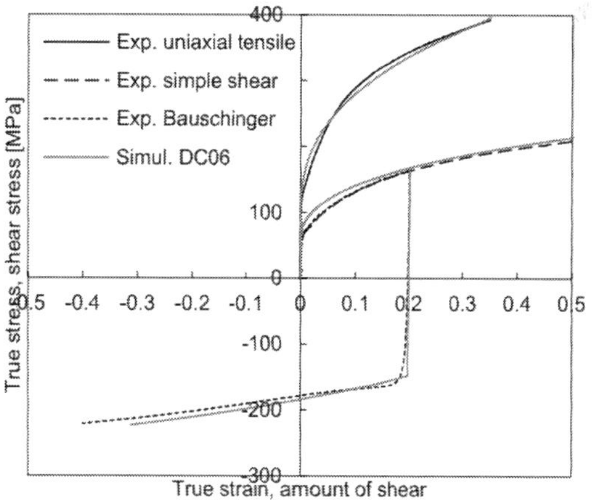

Figure 4. Comparison of the stress-strain curves predicted by the material model with the experimental ones for uniaxial tensile test, simple shear and Bauschinger simple shear after 20% forward shearing [26].

DOI: 10.1051/matecconf/20168003005

Table 1. Material parameters for the isotropic–kinematic hardening described by Swift law (DC06).

Y_0 [MPa]	K [MPa]	n	C_X	X_{sat} [MPa]
122.2	435.0	0.219	1.45	116.7

In order to deal with the anisotropic behaviour of the metal sheet, the Hill'48 yield criterion and the associated flow rule are adopted. Accordingly, the equivalent effective stress is defined by the quadratic yield function:

$$\bar{\sigma}^2 = F(\hat{s}_{22}-\hat{s}_{33})^2 + G(\hat{s}_{33}-\hat{s}_{11})^2 + H(\hat{s}_{11}-\hat{s}_{22})^2 + +2L\hat{s}_{23}^2 + 2M\hat{s}_{13}^2 + 2N\hat{s}_{12}^2, \quad (6)$$

where $\hat{s}_{ij}$ stands for the components of the effective stress tensor in the axes of orthotropy. The six material constants F, G, H, L, M and N are calculated from the experimental r-values obtained by uniaxial tensile tests carried out at three different orientations with respect to the rolling direction. The obtained values are presented in Table 2.

Table 2. Anisotropy parameters for the Hill'48 yield criterion (DC06).

F	G	H	L	M	N
0.264	0.283	0.717	1.500	1.500	1.279

4.2 Sheet and tools discretization

The blank is discretized with 8-node hexahedral finite elements associated with a selective reduced integration scheme [27]. Since the prediction of wrinkling phenomena in sheet metal forming simulation requires a fine mesh [6], a regular mesh is adopted in the Oxy plane (element size of 1.0 mm in the central region), as shown in Figure 5. The discretization of the blank comprises two layers of finite elements through the thickness, which enable an accurate prediction of the forming forces and the springback [13]. Hence, the modelling of the full blank comprises 130,000 finite elements, while 1/4 of the blank requires 32,500 finite elements (see Figure 5).

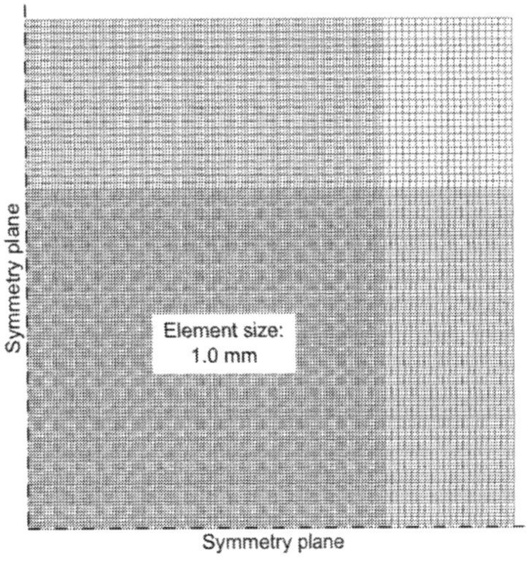

Figure 5. Discretization of the blank (1/4) using 32,500 hexahedral finite elements (2 layers through the thickness).

The geometry of the forming tools is modelled by Nagata patches [22], where the nodal normal vectors required for the contact surface smoothing approach are evaluated through the IGES file [28]. The discretization of the three tools involved in the numerical simulation is presented in Figure 6. The surface of the punch is described by 660 patches, the blank-holder comprises 300 patches and the surface of the die is discretized by 805 patches.

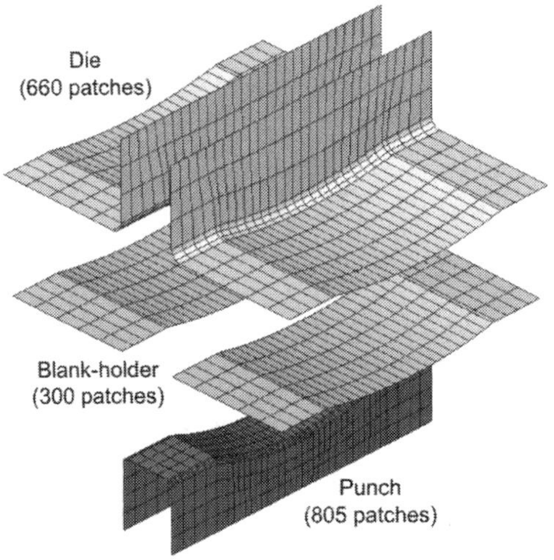

Figure 6. Discretization of the forming tools using Nagata patches.

4.3 Friction model

The friction coefficient is influenced by a wide set of parameters, such as: (i) micro and macro geometry of contacting surfaces; (ii) sliding velocity; (iii) contact pressure and (iv) temperature. Thus, an advanced friction models is adopted in the present study, where the friction coefficient μ is function of the normal contact pressure [29]. The evolutional law for Coulomb's friction coefficient is defined by a Hockett-Sherby law:

$$\mu = B - (B - A)\exp(-mP^n), \quad (7)$$

where A, B, m and n are numerical fitting parameters, while P is the normal contact pressure.

The four parameters involved in the selected friction model (see Table 3) were determined using the flat die test [30]. The steel sample is maintained between two small flat dies, while the holding force is accurately controlled by load cells. The sliding speed is constant and equal to 1.5 m/min. The normal pressure is kept constant during the test. The flat dies were produced using the same steel selected for the forming tools and the lubrication conditions adopted are also the same (Quaker 6130 using 1.4 g/m^2/face) [30]. Five pressure levels are considered during the tests, providing five different values of friction coefficient, as shown in Figure 7. The pressure strongly influences the friction coefficient. Indeed, the friction coefficient is greater at low pressure than at higher pressure. The fitting of the measured values by the advanced friction law is presented in Figure 7, showing a good agreement between them.

Table 3. Parameters of the adopted friction law for DC06.

A	*B*	*m*	*n*
0.31	0.16	0.61	0.71

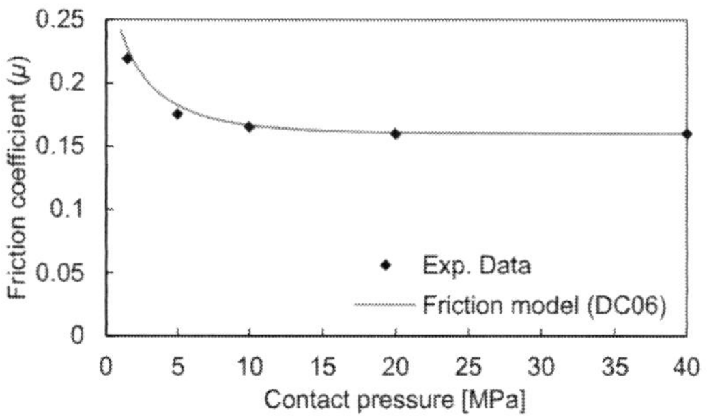

Figure 7. Evolution of the friction coefficient with the normal contact pressure, expressed by the adopted friction law. Comparison with experimental data from [30].

5 Results and discussion

The comparison between numerical and experimental results is presented in this section, namely the punch force evolution and the final geometry of the rail. The springback of the flange and the wrinkles on the top of the rail are the defects evaluated in the present study. Besides, the two finite element models are compared both in terms of accuracy and computational cost.

5.1. Forming forces

Both the punch and the blank-holder force evolutions are presented in Figure 8, comparing numerical and experimental results. The numerical blank-holder force increases linearly with the punch displacement, which is imposed in the numerical model, according with the experimental evolution. The predicted punch force evolution is identical for both finite element models presented. Nevertheless, the experimental value of the force is slightly overestimated by the numerical simulation, as shown in Figure 8. In fact, the initial slope predicted by the numerical simulation is higher than the one experimentally measured. The occurrence of wrinkling on the top surface of the rail induces a drop in the numerical force evolution at approximately 20 mm of punch displacement, which is more pronounced in the model considering the full blank.

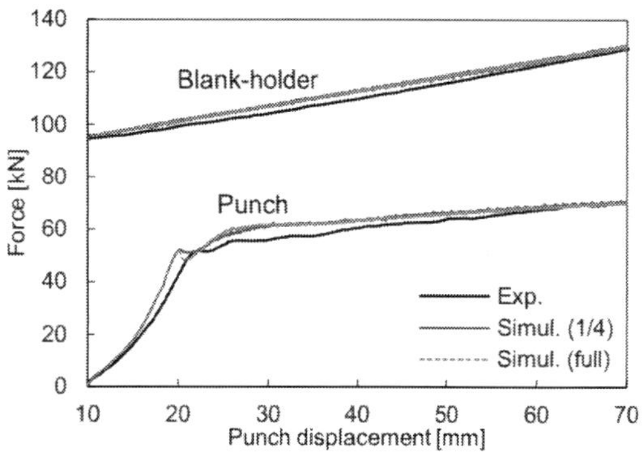

Figure 8. Comparison between experimental and numerical punch force evolution in the forming of the rail (DC06).

Considering the modelling of the entire blank, the total blank-holder force is equally divided by each flange of the rail. Thus, the vertical displacement of each flange of the blank-holder is adjusted independently in each increment, in order to guarantee a linear increase with the punch displacement. This procedure allow to obtain identical draw-in for each flange and, consequently, symmetry in the springback angle.

5.2 Cross-section profiles

The final geometry of the rail is evaluated through the four cross-section profiles defined in Figure 3, allowing to evaluate simultaneously the springback of the flange and the wrinkle on the top surface. The geometry of the rail after springback, predicted by the finite element simulation is shown in Figure 9. The numerical simulation of the entire blank provides an unsymmetrical wrinkle, which is accordance with the experimental observation (see Figure 2).

In order to assess the accuracy of each finite element model, the cross-section profiles obtained by numerical simulation are compared with the experimental ones. The comparison between experimental and numerical profiles in the cross-section A (x=15 mm) and B (x=95 mm) is presented in Figure 10 (a) and (b), respectively. Regarding the cross-section A, the numerical profile is in very good agreement with the experimental one. Besides, both numerical models (1/4 and full blank) provide identical results. Concerning the cross-section B, the numerical profile is in good agreement with the experimental one, excluding the top surface of the rail, where the two numerical models predict distinct geometries (see Figure 10 (b)). In fact, the wrinkle predicted by numerical simulation is symmetric using 1/4 of blank (see Figure 9 (a)), while it is unsymmetrical using the full model of the blank, which in accordance with the experimental result. Nevertheless, the geometry of the wrinkle is not perfectly defined, as shown in Figure 10 (b). Since the material adopted presents a clear anisotropic distribution of the *r*-values (see Table 2), a slight rotation the rolling direction with the symmetric full blank provides an identical wrinkle shape.

DOI: 10.1051/matecconf/20168003005

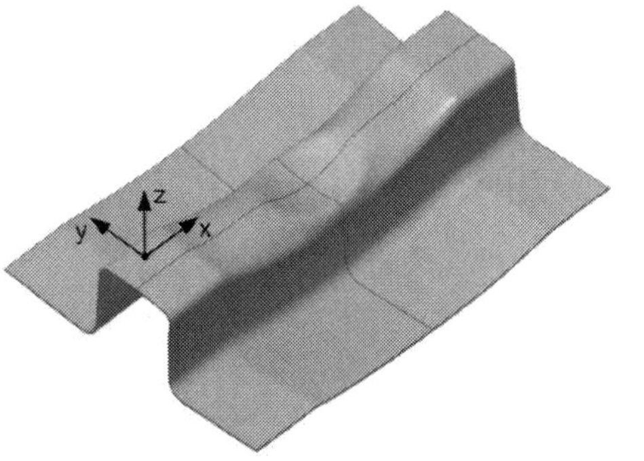

(a)

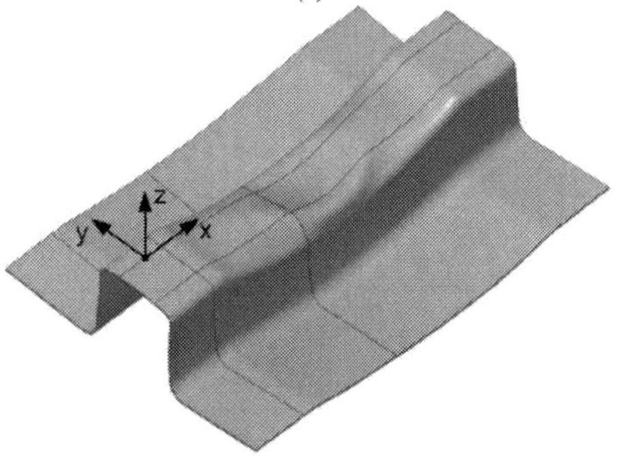

(b)

Figure 9. Final geometry of the rail after springback predicted by finite element simulation: (a) 1/4 of blank; (b) full blank including the four cross-sections used to evaluate the geometrical defects.

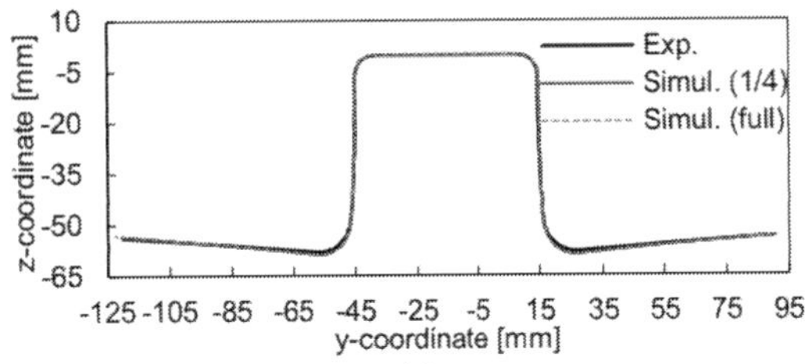

(a)

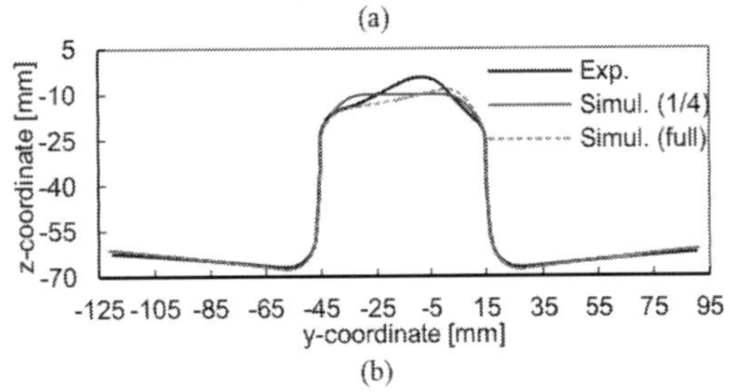

(b)

Figure 10. Comparison between experimental and numerical profile of the cross-section: (a) A (x=15 mm); (b) B (x=95 mm).

The springback of the rail is quantified in this study by the angle between the flange and the horizontal plane. The comparison between experimental and numerical values is shown in Table 4, for both cross-sections previously studied. Since the rail after springback is rather symmetric (see Figure 10), the presented values for numerical model with the full blank are the average of the two flanges. The flange springback angle is slightly overestimated by the numerical simulation, as shown in Table 4. The difference is about 0.5° in the cross-section A and 1.1° in the cross-section B. Nevertheless, both numerical models provide identical results, as shown in Table 4. Thus, the prediction of the flange springback behaviour is not considerably influenced by the shape of the wrinkle arising on the top surface.

Table 4. Comparison between experimental and numerical (full blank) springback angle of the flange.

Cross-section	Exp. [°]	1/4 model [°]	Full model [°]
A	3.9	4.39	4.40
B	4.3	5.41	5.37

The profile of the rail measured in the cross-sections L1 and L2 (longitudinal direction) is directly dictated by the wrinkle arising on the top surface. The comparison between experimental and numerical profiles in the cross-section L1 (y=0 mm) and L2 (y=-30 mm) is presented in Figure 11 (a) and (b), respectively. Globally, the numerical predictions are in good agreement with the experimental measurements. On the other hand, the numerical shape of the rail is significantly influenced by the symmetry conditions adopted in the finite element model (1/4 model). Indeed, considering the full model of the blank, the numerical results are in better agreement with the experimental measurements, as shown in Figure 11.

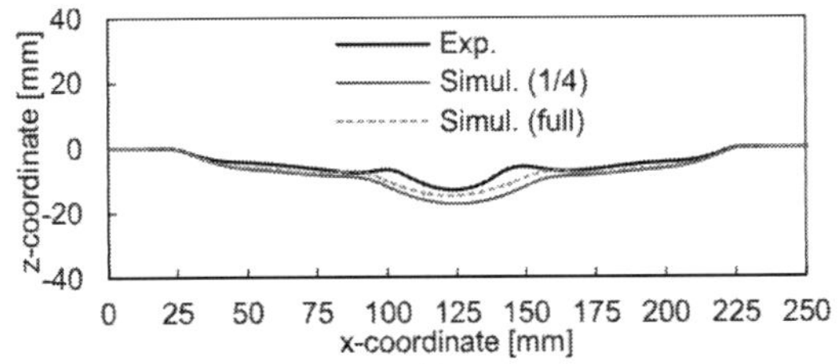

(a)

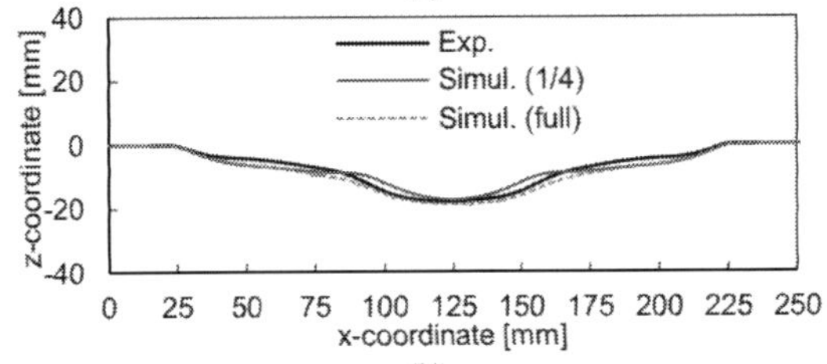

(b)

Figure 11. Comparison between experimental and numerical profile of the cross-section: (a) L1 (y=0 mm); (b) L2 (y=-30 mm).

5.3 Computational performance

The computational performance of the proposed finite element models is presented in Table 5. The adoption of the numerical model that takes into account the full blank leads to a significant increase of both the computational time and the number of increments. In fact the computational cost of the numerical simulation considering the full blank is approximately 12 times higher than using 1/4 of the blank. The same behaviour is observed for the number of increments, as shown in Table 5. This increase of the computational cost is a consequence of the higher number of finite elements adopted and the numerical complexity required to control the blank-holder force independently in each flange.

Table 5. Computational performance of both finite element models.

	1/4 model	Full model
Nº increments	1823	4839
Computational time [h]	30.4	384.7

6 Conclusions

The application of the numerical simulation in sheet metal forming processes has been a key factor for the fulfilment of the increasing requirements for time and cost efficiency, as well as quality improvement. This study presents the experimental and numerical analysis of a rail prone to 2D springback and wrinkling defects. The finite element results are compared with the experimental ones in order to assess the accuracy on the proposed numerical models. The punch force evolution and the final geometry of the rail, evaluated in several cross-sections, are the main variables under analysis.

Two distinct finite element models are compared in this work, both in terms of accuracy and computational cost. Due to geometric and material symmetry conditions, only one quarter of the blank is considered in the first model. On the other hand, the second model takes into account the entire blank, which is rotated 1° relative to the forming tools. Both numerical models provide identical punch force evolutions, as well as flange springback angles. However, these two variable are slightly overestimated by the numerical simulation. Regarding the wrinkle arising on the top surface of the rail, its shape is predicted differently by each finite element model. In fact, considering the full model of the blank, the numerical results are in better agreement with the experimental measurements, where the wrinkle tend to be unsymmetrical. Nevertheless, the computational cost of the numerical simulation considering the full blank is approximately 12 times higher than using 1/4 of the blank.

Acknowledgements

The authors gratefully acknowledge the financial support of the Portuguese Foundation for Science and Technology (FCT) under projects with reference UID/EMS/00285/2013, PTDC/EMS-TEC/0702/2014 (POCI-01-0145-FEDER-016779) and PTDC/EMS-TEC/6400/2014 (POCI-01-0145-FEDER-016876) by UE/FEDER through the program COMPETE2020. The first author is also grateful to the FCT for the Postdoctoral grant SFRH/BPD/101334/2014.

References

1. A. Makinouchi, J. Mater. Process. Technol. **60**, 19–26 (1996)
2. R. Kuziak, R. Kawalla, S. Waengler, Arch. Civ. Mech. Eng. **8**, 103–117 (2008)
3. W.S. Miller, L. Zhuang, J. Bottema, A.J. Wittebrood, P. De Smet, A. Haszler, A. Vieregge, Mater. Sci. Eng. A **280**, 37–49 (2000)
4. M.K. Kulekci, Int. J. Adv. Manuf. Technol. **39**, 851–865 (2007)
5. H. Laurent, J. Coër, P.Y. Manach, M.C. Oliveira, L.F. Menezes, Int. J. Mech. Sci. **93**, 59–72 (2015)
6. D.M. Neto, M.C. Oliveira, R.E. Dick, P.D. Barros, J.L. Alves, L.F. Menezes, Int. J. Mater. Form. (to be published)
7. A.E. Tekkaya, J. Mater. Process. Technol. **103**, 14–22 (2000)
8. G. Gantar, T. Pepelnjak, K. Kuzman, J. Mater. Process. Technol. **130-131**, 54–59 (2002)
9. B. Revil-Baudard, O. Cazacu, P. Flater, N. Chandola, J.L. Alves, J. Mech. Phys. Solids **88**, 100–122 (2016)
10. R. Alves de Sousa, J. Yoon, R. Cardoso, R. Valente, J. Gracio, Int. J. Plast. **23**, 490–515 (2007)
11. D.M. Neto, M.C. Oliveira, L.F. Menezes, Arch. Comput. Methods Eng. (to be published)
12. M. Banu, M. Takamura, T. Hama, O. Naidim, C. Teodosiu, A. Makinouchi, J. Mater. Process. Technol. **173**, 178–184 (2006)
13. M.C. Oliveira, J.L. Alves, B. Chaparro, L.F. Menezes, Int. J. Plast. **23**, 516–543 (2007)
14. A.D. Santos, J. Ferreira Duarte, A. Reis, A. Barata da Rocha, L. Menezes, M. Oliveira, A. Col, T. Ono, J. Mater. Process. Technol. **125-126**, 798–805 (2002)
15. A.D. Santos, A. Reis, J.F. Duarte, P. Teixeira, A.B. Rocha, M.C. Oliveira, J.L. Alves, L. Menezes, J. Mater. Process. Technol. **155-156**, 1980–1985 (2004)
16. L.F. Menezes, C. Teodosiu, J. Mater. Process. Technol. **97**, 100–106 (2000)
17. R.M. McMeeking, J.R. Rice, Int. J. Solids Struct. **11**, 601–616 (1975)
18. D.M. Neto, M.C. Oliveira, J.L. Alves, L.F. Menezes, Mater. Des. **60**, 368–379 (2014)
19. Y. Yamada, N. Yoshimura, T. Sakurai, Int. J. Mech. Sci. **10**, 343–354 (1968)
20. L.F. Menezes, D.M. Neto, M.C. Oliveira, J.L. Alves, AIP Conf. Proc. **1353**, 1220–1225 (2011)
21. C. Teodosiu, D. Daniel, H.-L. Cao, J.-L. Duval, J. Mater. Process. Technol. **50**, 133–143 (1995)
22. D.M. Neto, M.C. Oliveira, L.F. Menezes, J.L. Alves, Comput. Methods Appl. Mech. Eng. **271**, 296–320 (2014)
23. P. Alart, A. Curnier, Comput. Methods Appl. Mech. Eng. **92**, 353–375 (1991)
24. A. Heege, P. Alart, Int. J. Numer. Methods Eng. **39**, 165–184 (1996)
25. J. Lemaitre, J.-L. Chaboche, *Mechanics of Solid Materials*, Cambridge University Press (1990)
26. H. Haddadi, S. Bouvier, M. Banu, C. Maier, C. Teodosiu, Int. J. Plast. **22**, 2226–2271 (2006)
27. T.J.R. Hughes, Int. J. Numer. Methods Eng. **15**, 1413–1418 (1980)

28. D.M. Neto, M.C. Oliveira, L.F. Menezes, J.L. Alves, Finite Elem. Anal. Des. **72**, 35–46 (2013)
29. M.C. Oliveira, J.L. Alves, L.F. Menezes, Arch. Comput. Methods Eng. **15**, 113–162 (2008)
30. P. Wouters, D. Daniel, C. Magny, Selection and Identification of Friction Models for the 3DS Materials-Identification of the Frictional Behaviour, Digital Die Design Systems (3DS) IMS 1999 000051, 2002.

Numerical prediction of ductile failure in the blanking process by means of uncoupled and coupled phenomenological damage models

Cristian Canales[1,a] and Jean-Philippe Ponthot[1]

[1]*Department of Aerospace and Mechanical Engineering, University of Liège, 13A Allée de la découverte, 4000 Liège, Belgium*

Abstract. Fracture prediction in blanking has gained great attention due to increasing requirements of high-quality products. In this work, the predictive capabilities of different uncoupled and coupled damage models, recently implemented in a fully implicit homemade FE code, for blanking process are compared. Some advanced models considered here include damage sensitivity to both triaxiality and Lode angle in their formulation. The material characterization is based on the history of internal variables during loading, where different mechanical tests are considered. Finally, numerical and experimental results are compared for a wide range of different geometries of blanking.

1 Introduction

The blanking process has now become an important cutting technique in mass production industries. The quality of the cut edge is strongly influenced by the geometrical parameters of the process and the complex material behavior of the plate. In general, a numerical approach must be able to deal with several issues involved in the blanking operation. The most critical challenge in the simulation of this manufacturing process is to accurately predict the onset of fracture and further propagation of the cracks in the workpiece. This has recently gained great attention due to the increasing requirements of the market to manufacture high-quality products in a small amount of time.

In general, the prediction of failure during metalforming processes has been focus of extensive research efforts and several models have been proposed over the years. The phenomenological damage models currently available in the literature can be grouped in two main families: uncoupled and coupled models. On the one hand, the uncoupled approach defines a fracture locus (similar to yielding) without any effect in the plasticity description of the material i.e., a failure surface is constructed that works as a damage indicator. This type of models have been extensively applied on industrial applications due to the simplicity of their material characterization and numerical implementation in simulation software. On the other hand, coupled damage models introduce a characteristic variable counting for the damage accumulation into the constitutive formulation, which models the softening effects of damage into the material.

Classically, the sensitivity of the damage accumulation to the stress states has been solely introduced by means of the stress triaxiality ratio (related to the second invariant of the stress tensor). The first insight of the role of this variable in ductile fracture was presented by the early studies of McClintock [1], Rice and Tracey [2] and Hancock [3], starting from the theoretical analysis of voids grow under different conditions. Later on, Gurson [4] proposed a fracture model based on the impact of the hydrostatic pressure in the behavior of porous media, followed by several enhancements of the model in order to take into account different mechanisms affecting ductile failure (see e.g. the work of Tvergaard and Needleman [5], Leblond et al. [6], Pardoen and Hutchinson [7] and Xue [8]). Thanks to the experimental observations recently performed by Bao and Wierzbicki [9,10] and Barsoum and Faleskog [11] the influence of the Lode angle (related to the third invariant of the stress tensor) on damage was putted in evidence, where non-smooth fracture locus were exhibited for low levels of triaxiality (shear dominated fracture). This clearly contrast with the formulation followed by former damage models.

In this work, the onset of fracture during blanking is studied by means of different uncoupled and coupled damage models. First, a short description of the selected models is done. Then, the identification procedure of material constants is performed considering different mechanical tests in a hybrid experimental-numerical framework. The optimal set of material parameters is then used to predict fracture initiation in blanking for a wide range of different geometries. The computed results are compared with experimental ones in order to analyze the predictive capabilities of the ductile fracture models.

[a] Corresponding author: ccanales@ulg.ac.be

2 Damage models

As mentioned in the previous section, there are two important quantities that take a major role in the accumulation of damage during any loading process and which depend on the stress state. The first quantity is the stress triaxiality ratio η, which is defined as:

$$\eta = \frac{p}{\bar{\sigma}} \quad (1)$$

where p is the hydrostatic pressure and $\bar{\sigma}$ is the von Mises equivalent stress.

The second quantity corresponds to the Lode angle parameter $\bar{\theta}$,

$$\bar{\theta} = 1 - \frac{6}{\pi}\theta = 1 - \frac{2}{\pi} arccos\left(\frac{det(s)}{\bar{\sigma}}\right) \quad (2)$$

This parameter is a normalized version of the Lode angle θ, which values are always between -1 and 1. The definition of both η and $\bar{\theta}$ have defined a new set of parameters able to characterize the stress states encountered in several mechanical test, which has recently helped to define a whole new family of damage models.

Shear dominated failure is the main fracture mode that promotes crack propagation in the blanking process. Consequently, the selection of damage models in this work was based on their potential capabilities to predict ductile fracture in different metal forming applications, specially for the ones involving low triaxialities.

All the models considered in the present work have been implemented in a fully implicit homemade Finite Element code called Metafor [12, 13]. This code relays in a hypoelastic-based formulation to integrate the constitutive equations along with the classical return mapping algorithm. Furthermore, an updated Lagrangian approach is here adopted.

2.1 Uncoupled phenomenological models

For the case of uncoupled models, a cumulative framework is adopted. Here, the onset of fracture is governed by a scalar valued function D defined in the space of equivalent plastic strain $\bar{\varepsilon}^p$, stress triaxiality η and Lode parameter $\bar{\theta}$, as follows:

$$D = \int_0^{\bar{\varepsilon}^p} \frac{1}{\bar{\varepsilon}_f^p(\eta, \bar{\theta})} d\bar{\varepsilon}^p \quad (3)$$

where $\bar{\varepsilon}_f^p$ is the equivalent plastic strain to fracture expressed in terms of η and $\bar{\theta}$. Failure is then assumed to take place when the damage variable reaches the value D=1. It is worth mentioning that the influence of loading paths in the development of damage is intrinsically taken into account in this cumulative approach, where each increment of damage during the loading process is integrated. This is of great importance for non-proportional loading paths or non-linear evolution of plastic strain leading to fracture.

In this work, three models recently developed in the literature are selected to study the onset of ductile fracture during blanking.

2.1.1 Bai and Wierzbicki, BW

Bai and Wierzbicki [14, 15] proposed a fracture model based on the experimental observations of non-smooth fracture locus previously done by Bao and Wierzbicki [9, 10]. In this model, the influence of stress triaxiality on damage is represented by a series of exponential functions, leading to higher fracture strains for lower triaxialities. This type of exponential dependency has been widely used in early studies [1–3]. Furthermore, the effect of the Lode parameter is included by means of a quadratic function. A symmetric form of this fracture locus with respect to $\bar{\theta}$ = 0 is here considered, which is defined as:

$$\bar{\varepsilon}_f^p = (c_1 e^{-c_2\eta} - c_3 e^{-c_4\eta})\bar{\theta}^2 + c_3 e^{-c_4\eta} \quad (4)$$

where c_1, c_2 c_3 and c_4 are material parameters to be identified. In addition, a constant cut-off value for stress triaxiality $\eta = -1/3$ is adopted in this model. This means that there is no damage accumulation when η is below this value, as proposed by Bao and Wierzbicki [16].

2.1.2 Modified Mohr-Coulomb, MMC

Recently, the classical Mohr-Coulomb fracture model [17] has been rewritten in terms of η and $\bar{\theta}$ in order to consider their possible effects on damage into its formulation [15, 18]. When the classical Swift law is used to describe the strain hardening of the material in a J2 plasticity framework, the Modified Mohr-Coulomb takes the form:

$$\bar{\varepsilon}_f^p = \left(\frac{1}{K}\left\{c_2\left[\sqrt{\frac{1+c_1^2}{3}}\cos\left(\frac{\bar{\theta}\pi}{6}\right) + c_1\left(\eta + \frac{1}{3}\sin\left(\frac{\bar{\theta}\pi}{6}\right)\right)\right]^{-1} - \sigma_0\right\}\right)^{1/n} \quad (5)$$

where c_1 and c_2 are material constants related to the damage model. In Eq. 5, the parameters K and n are related to the Swift law.

2.1.3 Lou, Yoon and Huh, LYH

Lou et al. [19] has been recently developed a model based on the underlying mechanisms leading to fracture in shear dominated processes. This model also includes a variable cut-off value for η, which depends on the value of the Lode angle. This fracture model is defined as,

$$\bar{\varepsilon}_f^p = c_3\left(\frac{2}{\sqrt{L^2+3}}\right)^{-c_1}\left(\left\langle\frac{1}{1+C}\left[\eta + \frac{3-L}{3\sqrt{L^2+3}} + C\right]\right\rangle\right)^{-c_2} \quad (6)$$

with,

$$L = \frac{3\tan(\theta) - \sqrt{3}}{3\tan(\theta) + \sqrt{3}} \quad (7)$$

where c_1, c_2 and c_3 are material parameters. The constant C is fixed to 1/3 in this study, as proposed by Lou et al. In Eq. 7, L corresponds to an alternative definition of the Lode angle and the $\langle\cdot\rangle$ symbol denotes the MacAuley brackets. It can be noticed that the Lode dependent cut-off value for damage accumulation is below -1/3 for any set of η and L.

2.2 Coupled phenomenological model

For the case of coupled approaches, the Lemaitre model [20, 21] is selected. In this model, the softening behavior of the material is governed by an internal damage variable D (scalar for isotropic damage), which affects the constitutive behavior as follows:

$$\tilde{\sigma} = \frac{\sigma}{1 - D} \quad (8)$$

where $\tilde{\sigma}$ is the effective stress tensor and the σ is the Cauchy stress tensor. The evolution of damage in this model is derived from the dissipation potential of a damaged material, which finally takes the form:

$$\dot{D} = \left\{ \frac{\tilde{\sigma}\left[\frac{2}{3}(1 + \nu) + 3(1 - 2\nu)\eta^2\right]}{2ES(1 - D)^2} \right\}^s \dot{\bar{\varepsilon}}^p \quad (9)$$

where ν is the Poisson ratio, E is the Young's modulus, $\dot{\bar{\varepsilon}}^p$ is the equivalent plastic strain rate, s and S are material constants. In this model, two conditions are considered to prevent the development of damage: a constant cut-off value of η=-1/3 and an equivalent strain threshold ε_0. Failure is assumed to occur when the damage variable reaches a certain critical value D_c.

Table 1. Material properties of the X30Cr13 stainless steel.

Young's modulus, E	1.87e5 (MPa)
Poisson's ratio, ν	0.28
Density, ρ	7.0e3 (kg/m^3)
Swift hardening law	$\sigma_Y = \sigma_0 + K[\bar{\varepsilon}^p]^n$ $= 366 + 736[\bar{\varepsilon}^p]^{0.41}$ (MPa)

3 Material characterization

3.1 Plasticity

The material under study corresponds to a ferritic stainless steel (X30Cr13), where the Hooke's law is selected to replicate the material behavior in the elastic regime. Potential effects of damage in the elastic behavior of the material are not considered in this work.

A J2 elasto-plastic model with a non-linear isotropic hardening law described by the Swift model is assumed to represent the material behavior in the plastic regime. Experimental yielding points obtained by performing several tensile tests are extracted from the work of Goijaerts et al. [22], where rolling operations were used to deform the specimens at different levels before performing the tests. These experimental results are then used to characterize the hardening law thanks to a least square procedure. The elastic properties and resulting material parameters are presented in Table 1.

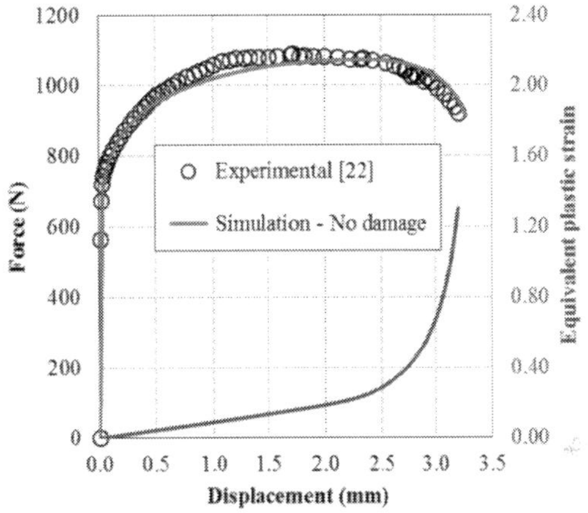

Figure 1. Evolution of force and equivalent plastic strain with respect to displacement for the tensile test at 0 (MPa) superposed pressure used for material characterization.

Table 2. Geometrical dimensions of the blanking test.

Dimension	Value (mm)
R_d	5.0
r_d	0.14
R_s	15.0
T	1.0
C	0.15
r_p	0.088

3.2 Uncoupled damage models

The uncoupled models were characterized considering 4 different mechanical tests under different stress states. The experiments were performed by Goijaerts et al. [22] and they consisted in three tensile test under different superposed hydrostatic pressures i.e., 0, 250 and 500 MPa, and a blanking test with a relatively large clearance (C) of

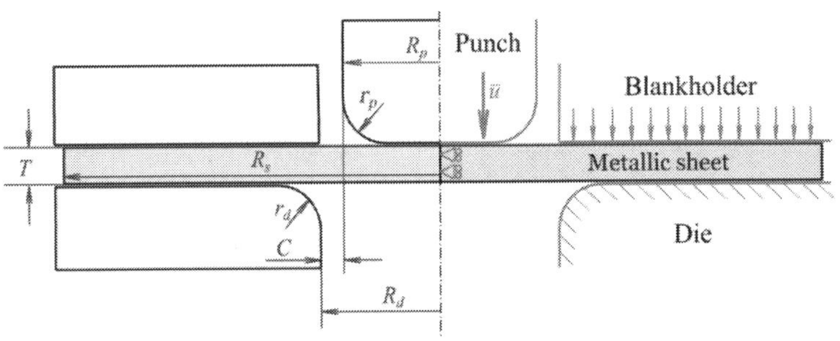

Figure 2. Geometry (left) and boundary conditions (right) considered for blanking.

15% of the sheet thickness. All the tests were simulated using Metafor, where the history of the relevant variables (i.e., η, $\bar{\theta}$ and $\bar{\varepsilon}^p$) was recorded up to failure at the zone where fracture was assumed to start with the nearest integration point.

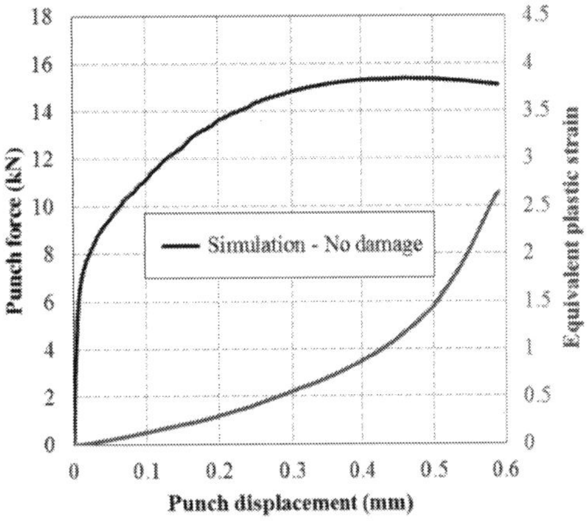

Figure 3. Evolution of punch force and equivalent plastic strain with respect to punch displacement for blanking considering C = 0.15 (mm) used for material characterization.

The simulation of the tensile test required a full 3D FE model due to necking, see [22] for geometrical details. Classical linear hexahedral elements with constant pressure and 8 quadrature points per element are considered here. Due to symmetry, only an eight of the specimen was modeled along with the corresponding boundary conditions related to this geometrical choice. One end of the specimen was fixed while in the other end displacement was prescribed. Experimentally, the force-displacement curves did not exhibited any influence of the superposed pressures, which reinforce the choice of using J2-based plasticity model to describe the material behavior. This also means that only the simulation of the test under 0 MPa superposed pressure is needed to record the history of internal variables, which can be used a posteriori to compute the evolution of η, $\bar{\theta}$ and $\bar{\varepsilon}^p$ for the other tensile tests. The simulation was done until the minimum thickness on the necking area (center of the specimen) after fracture was reached by the numerical model. The characteristic force-displacement curve obtained from experiments and simulation are compared in Figure 1 and good agreement is found between both correlations. The strain hardening and the necking point are thus well captured by the elastoplastic model (no damage).

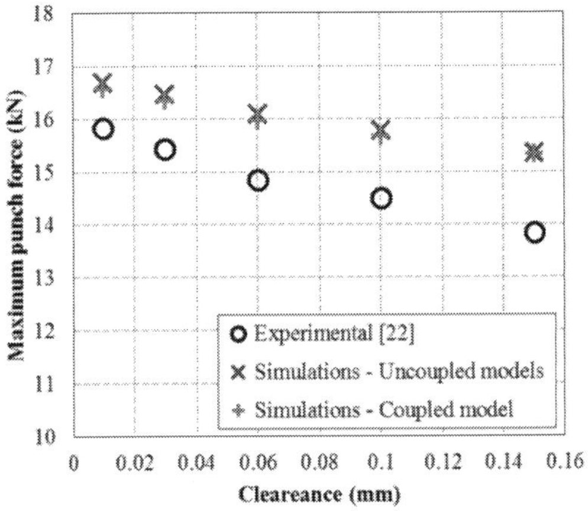

Figure 4. Maximum punch forces during blanking for different clearances obtained from experiments and simulations.

For the case of the blanking test, the numerical framework described by Canales et al. [23] for quasi-static blanking operations was adopted. Here, a 2D axisymmetric FE model was considered and global remeshing steps were required to avoid high distortion of elements during the simulation. Classical Q4 axisymmetric elements with constant pressure and 4 quadrature points per element are used to avoid locking due to quasi-incompressibility. The initial geometry along with the boundary conditions are shown in Figure 2; the dimensions therein are detailed in Table 2. The tools are taken as perfectly rigid and represented by analytical curves. The die is completely fix and a pressure of 5.5 MPa is applied trough the blankholder during the simulation. Vertical displacement is prescribed on the punch in order to trigger the cut of the sheet. The classical Coulomb law with a friction coefficient of $\mu = 0.1$ was used to model the contact interactions between the workpiece and the tools. The simulation was performed until the punch displacement at fracture registered in the experiment was reached. The computed force-displacement curve of the punch is shown in Figure 3. It can be noticed that fracture occurs after reaching the maximum level of punch force. It is worth to mention that as no material softening is considered in the uncoupled damage models, the maximum punch force

Table 3. Comparison of measured and predicted strain to fracture for all damage models.

Test	Equivalent plastic strain to fracture (error %)				
	Measured	BW	MMC	LYH	Lemaitre
Tensile, p = 0 (MPa)	1.302	1.487 (14.1%)	1.364 (4.7%)	1.399 (2.8%)	1.234 (-5.1%)
Tensile, p = 250 (MPa)	1.655	1.887 (14.0%)	1.900 (14.7%)	1.779 (7.4%)	-
Tensile, p = 500 (MPa)	1.934	1.977 (2.2%)	2.051 (6.0%)	2.935 (51.7%)	-
Blanking, C = 0.15 (mm)	2.906	2.733 (3.2%)	2.654 (0.25%)	2.722 (2.8%)	2.622 (-9.7%)

exhibited during blanking is not expected to change when considering any of these models. The maximum punch forces during blanking simulations considering both uncoupled and coupled damage models are shown in Figure 4. For each clearance, almost no difference between the computed results is detected.

The evolution of the equivalent plastic strain with respect to the stress triaxiality and the Lode parameter (loading paths) for all tests are shown in Figure 5. For the tensile test under different superposed pressures, the development of plastic strain is highly nonlinear; the lowest and the highest levels of η and $\bar{\theta}$ are reached, respectively, at the beginning of each test. It can be noticed that the plots of the blanking experiment (BC0.15) are almost entirely developed in the vicinity of $\eta = 0$ and $\bar{\theta} = 0$, reveling shear as the dominated mode leading to fracture in this test. In all cases, the evolution of stress states is far from being constant and the use of average values of η and $\bar{\theta}$ in the identification of material constants is not suitable. Moreover, the use of a cut-off value for damage accumulation does not seem to be relevant in these experiments because plastic strain is completely developed above the limiting level of stress triaxiality considered in the models. Nevertheless, this limit could prevent the models to incorrectly predict fracture initiation in locations under high negative triaxialities.

Finally, the curves in Figure 5 were used in a full inverse analysis to characterize the damage models. Here, the difference between the full damaged condition (D = 1) and the computed damage variable (Eq. 3) during the loading process for each test was minimized. The optimization procedure was performed by means of an interior-point algorithm for nonlinear optimization. The measured and computed equivalent plastic strains to fracture for all test are compared in Table 3. The obtained material parameters are summarized in Table 4.

3.3 Coupled damage model

The calibration of the Lemaitre model was based on the softening behavior of the material in the tensile test under 0 MPa superposed pressure, following the identification procedure described by Lemaitre and Desmorat [21]. The material parameters in Eq. 8 are computed to be s = 2.98 and S = 6.1, with a strain threshold of $\varepsilon_0 = 0.22$. Finally, the critical value of the damage variable for which fracture is assumed to occur is $D_c = 0.37$. In all simulations (tensile test and blanking), the smaller mesh size is fixed to 0.02 mm in order to avoid mesh dependency of the results.

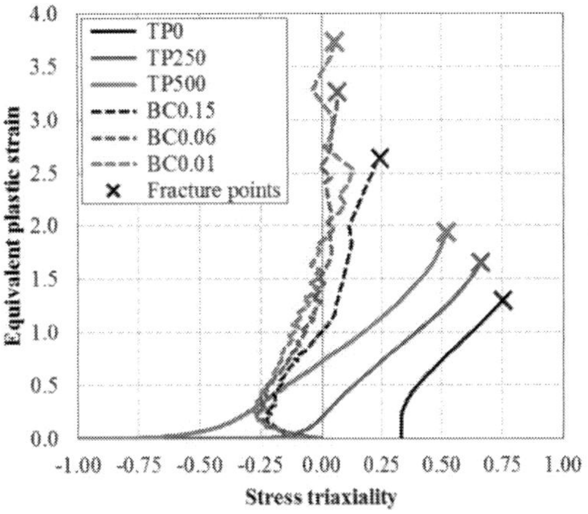

(a)

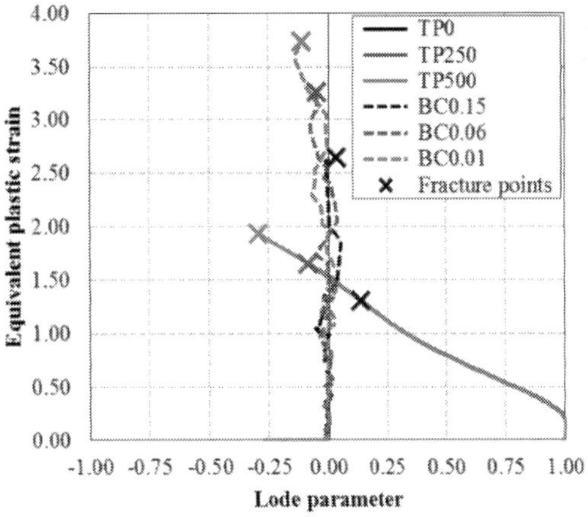

(b)

Figure 5. Evolution of equivalent plastic strain with respect to stress triaxiality (a) and Lode parameter (b) considering no damage. TP0, TP250 and TP500 correspond to tensile tests under 0 MPa, 0 MPa and 0 MPa, respectively. BC0.01, BC0.06 and BC0.15 correspond to blanking test with clearances of 0.01 mm, 0.06 mm and 0.15 mm, respectively.

 DOI: 10.1051/matecconf/20168003006

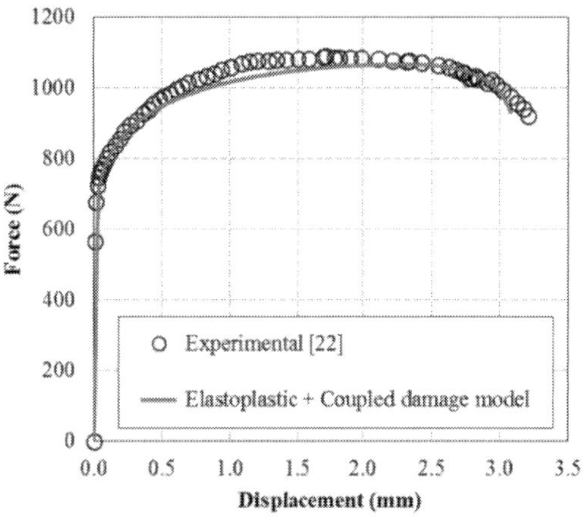

Figure 6. Evolution of force with respect to displacement for the tensile test at 0 (MPa) superposed pressure considering the Lemaitre model.

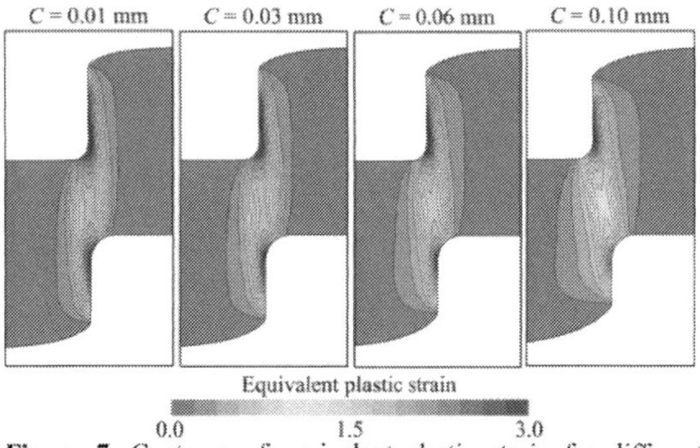

Figure 7. Contours of equivalent plastic strain for different clearances at 0.6 mm of punch displacement considering no damage.

The resulting force-displacement curve for the tensile test employing the Lemaitre model is shown in Figure 6. The measured and computed equivalent plastic strains to fracture for the tensile and the blanking tests are compared in Table 3. The measured fracture strains are underestimated in both cases for the Lemaitre model.

4 Numerical predictions in blanking

In this section, different geometries of the axisymmetric blanking tests described in the previous section are used to validate the characterized damage models. The pairs of clearances and punch radius used in this section are detailed in Table 5. The contours of equivalent plastic strain in the shearing zone (area delimited by R_p and R_d) for the different clearances are shown in Figure 7. As expected, plastic deformation is almost entirely developed inside this critical zone and a more concentrated distribution is exhibited for smaller clearances. The change in the geometry induces a different evolution of strain with respect to η and $\bar{\theta}$, as can be seen in Figure 5. Here, the curves are obtained numerically by performing the simulations until the punch displacement to fracture obtained from experiments was reached in each case. Moreover, larger fracture strains are obtained for smaller clearances.

In Figure 4, the maximum forces during blanking obtained from experiments and simulations are compared. It can be noticed that the model (considering no damage) correctly predicts the shift in the magnitude of the maximum force for the entire range of clearances considered in this work.

Table 4. Optimized material parameters for the uncoupled damage models.

Model	Material parameter			
	c_1	c_2	c_3	c_4
BW	3.79	1.30	2.72	1.87
MMC	0.48	1.04e3 (MPa)	-	-
LYH	6.04	1.67	2.79	-

In Figure 8, the computed punch displacements at which fracture is initiated into the sheet considering all the damage models are compared with the experimental results. It can be seen that the uncoupled models overestimate the experimental values for all clearances, excepting for the case of C = 0.01 mm where all models are inside the experimental error. This overestimation coincides with the differences made in the characterization procedure (see Table 3). Furthermore, the shift of the punch displacement at fracture for the range of clearances is well captured and small differences are encountered between the computed points for each clearance.

Table 5. Different clearances and corresponding punch radius considered in the validation.

C (mm)	r_p (mm)
0.01	0.0824
0.03	0.0880
0.06	0.0859
0.10	0.0975

For the case of the Lemaitre model, the computed values clearly underestimate the experimental findings for all clearances, despite the good performance of this model in the parameter identification step. The maximum difference (36%) is encountered for the smallest clearance, which decreases for larger clearances. Additionally, the shift of the values for the range of clearances studied here is not well predicted. In this case, the use of a higher critical damage value could prevent the underestimations exhibited by the model, but will not change the deviation of the numerical predictions from the experimental results for values of C lower than 0.06 mm.

Finally, the modified Mohr-Coulomb damage model presents the best overall performance in comparison with the rest of the models. In addition, the computed values by this model for both C = 0.01 mm and C = 0.03 mm are within the experimental error. These findings are surely relevant, taking into consideration that the MMC model has only two material parameters to be identified. This also highlights the importance of correctly expressing the sensitivity of damage to the set of relevant variables i.e., stress triaxiality and the Lode angle, above the use of a large number of fitting parameters.

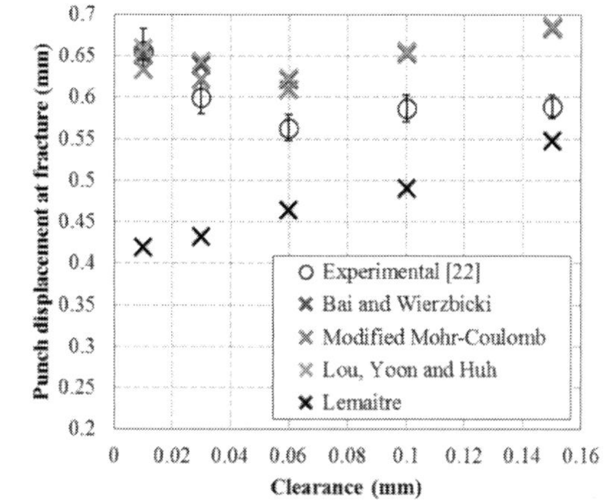

Figure 8. Punch displacement at fracture obtained from experiments and simulations.

The distribution of the damage variable prior to fracture for the different damage models and considering a clearance C = 0.1 mm are shown in Figure 9. The uncoupled models present a small difference in the distribution of D right before fracture, while is highly concentrated for the Lemaitre model. Nevertheless, for all the damage models the onset of fracture is predicted to be at the same location i.e., near the minor radius of the punch where a clear concentration of strains is computed (see Figure 7).

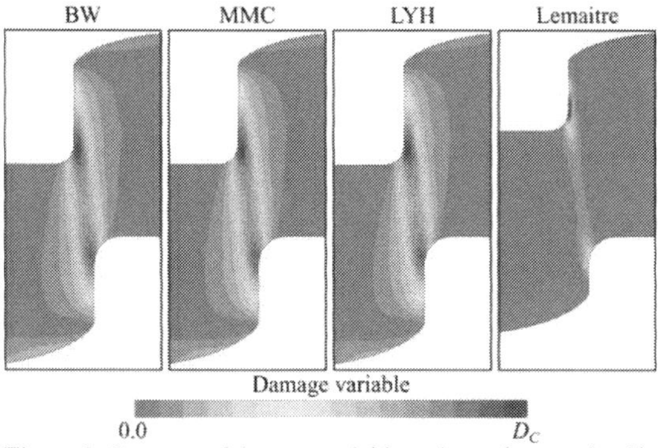

Figure 9. Contours of damage variables prior to fracture for all damage models.

In Figure 10, the evolution of the damage variables for the different models considering C=0.1 mm are compared. It can be noticed that the results of all the uncoupled models present almost the same behavior during the entire process. For the Lemaitre model, the evolution of D is also highly nonlinear, with two different stages of damage accumulation that could be explain by the shift of the stress triaxiality magnitude with respect to plastic strain accumulation (see Figure 5a). It is worth mentioning that the coupled damage model considered in this work only takes into account the value of stress triaxiality to characterize the stress state, despite of the clear difference of stress conditions between the tests used for the material characterization (tensile test) and the final application (blanking). This indicates that the use of the Lode parameter is of great importance in the development of advanced damage models, specially in the case of shear-dominated failure as suggested in recent works [9, 17, 19, 24].

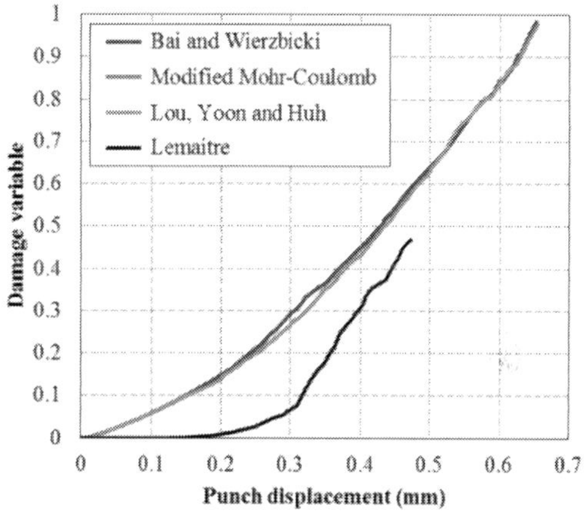

Figure 10. Evolution of damage variables with respect to equivalent plastic strain for all damage models.

5 Conclusions

In this work, the onset of failure in the blanking process was predicted by means of uncoupled (BW, MMC and LYH) and coupled (Lemaitre) damage models. This complex application presents a shear dominated mode of failure, with a highly nonlinear development of plastic strain. The identification of material parameters was based on a hybrid experimental-numerical framework, where different mechanical tests were considered. Then, different geometries of an axisymmetric blanking test was analyzed in order to validate the failure models.

The uncoupled damage models presented better results for the entire range of clearances in this study, with a small overestimation of the punch displacement at fracture. In particular, the modified Mohr-Coulomb damage model presented the best overall performance for the tests used in the material calibration and validation steps. By the contrary, the Lemaitre model exhibited an important underestimation of the values in the experimental validation, despite of the good results exhibited by this model in the material parameter identification. These findings suggest that the MMC model is a powerful option to predict failure occurrence when the blanking process is modeled, taking into consideration that only 2 parameters

 DOI: 10.1051/matecconf/20168003006

are needed to be identified in this case. This also indicates that the right description of the damage dependency to the stress state (i.e. stress triaxiality and Lode angle) is more important that the number of material parameters included into the model.

A comparative analysis of these models to predict the final shape of the sheared edge will be the subject of a future contribution.

Acknowledgments

This research has been partially funded by the Interuniversity Attraction Poles Program (IAP 7/21) initiated by the Belgian Science Policy Office. C. Canales would also like to thank the FRIA-FNRS fund for its financial support.

References

1. F.A. McClintock, S.M. Kaplan, C.A. Berg, Int. J. Fract. **2**, 614 (1966)
2. J.R. Rice, D.A. Tracey, J. Mech. Phys. Solids. **17**, 201 (1969)
3. J.W. Hancock, A.C. Mackenzie, J. Mech. Phys. Solids. **24**, 147 (1976)
4. A.L. Gurson, J. Eng. Mater. Technol. **99**, 2 (1977)
5. V. Tvergaard, A. Needleman, Acta Metall. **32**, 157 (1984)
6. J.B. Leblond, G. Perrin, J. Devaux, Eur. J. Mech. A-Solids **14**, 499 (1995)
7. T. Pardoen, J.W. Hutchinson, J. Mech. Phys. Solids. **48**, 2467 (2000)
8. L. Xue, Eng. Fract. Mech. **75**, 3343 (2008)
9. Y.B. Bao, T. Wierzbicki, Int. J. Mech. Sci. **46**, 81 (2004)
10. Y.B. Bao, T.Wierzbicki, J. Eng. Mater. Technol. **126**, 314 (2004)
11. I. Barsoum, J. Faleskog, Int. J. Solids Struct. **44**, 1768 (2007)
12. J.P. Ponthot, Int. J. Plas. **18**, 91 (2002)
13. P.P. Jeunechamps, J.P. Ponthot, Int. J. Numer. Meth. Engng. **94**, 920 (2013)
14. Y. Bai, T. Wierzbicki, Int. J. Plas. **24**, 1071 (2008)
15. Y. Bai, T. Wierzbicki, Int. J. Fract. **161**, 1 (2010)
16. Y.B. Bao, T. Wierzbicki, Eng. Fract. Mech. **72**, 1049 (2005)
17. O. Mohr, Zeitschrift des Vereins Deutscher Ingenieure. **44**, 1524 (1900)
18. D. Bigoni, A. Piccolroaz, Int. J. Solids Struct. **41**, 2855 (2004)
19. Y. Lou, J.W. Yoon, H. Huh, Int. J. Plas. **54**, 56 (2014)
20. J. Lemaitre, Eng. Fract. Mech. **25**, 523 (1986)
21. J. Lemaitre, R. Desmorat, Engineering damage mechanics: ductile, creep, fatigue and brittle failures (Springer, Berlin, 2005) 380
22. A.M. Goijaerts, L.E. Govaert, F.P.T, Baaijens, J. Manuf. Sci. Eng. **122**, 476 (2002)
23. C. Canales, P. Bussetta, J.P. Ponthot, Int. J. Mater. Form. (2015) DOI: 10.1007/s12289-015-1270-7
24. T.S. Cao, J.M. Gachet, P. Montmitonnet, P.O. Bouchard, Eng. Fract. Mech. **124-125**, 80 (2014)

Determination of instability of a DP 980 steel sheet under different stress states based on experiment and theoretical models

Hong-Wu Song[1, a], Dong-Zhi Sun[2], Florence Andrieux[2] and Shi-Hong Zhang[1]

[1]*Institute of Metal Research, Chinese Academy of Sciences, 110016 Shenyang, China*
[2]*Fraunhofer Institute for Mechanics of Materials IWM, 79108 Freiburg, Germany*

Abstract. It is generally accepted that the formability of sheet metal is limited by the onset of instability which can be described by forming limit diagram (FLD) in ε_1-ε_2 space. Common tests for FLD data usually cover strain ratio range from uniaxial to equal biaxial stretching, $-1/2 \le \varepsilon_2/\varepsilon_1 \le 1$. In this paper a new design of cruciform type specimen for uniaxial stretching is proposed to test formability with strain ratios in the range from shear to tension, $-1 \le \varepsilon_2/\varepsilon_1 \le -1/2$ which can be controlled by the geometry parameters. The forming limit of a DP 980 steel sheet was determined using both conventional method and proposed specimen with strain ratios from $-1 \le \varepsilon_2/\varepsilon_1 \le 1$. Then the experimental results were drawn in both ε_1- ε_2 and ε_{eq}-σ_m/σ_{eq} spaces, and the predicted results based on Hill, Swift, M-K, and Hora instability models were compared with the experimental ones. The formability features of the studied steel in whole strain ratio range and differences among the investigated theoretical models were finally discussed. The results indicate that the M-K instability model shows better prediction of the studied steel compared to the other models investigated in this research.

1 Introduction of theoretical models and experimental methods for sheet metal formability

It is generally accepted that the formability of sheet metal is limited by the onset of instability which can be described by forming limit diagram (FLD) which concept was published by Keeler [1] and then developed by Goodwin [2] almost 50 years ago. During the last decades different assessment methods for determination of metal sheet formability have been developed.

Many theoretical models are proposed and are presented in detail (formulation of the model, solving methods, numerical aspects, advantages and limitations) in [3]. Some of the well-established theoretical models which will be used in this paper are briefly introduced in the following.

The first models proposed by Swift [4] and Hill [5] assuming homogeneous sheet metals were so-called models of diffuse necking and localized necking, respectively. Hill suggested the onset of localized necking when the increase of stress by strain hardening is balanced by geometrical softening due to thickness reduction. Swift assumed the onset of diffuse necking occurs when the load reaches maximum in both principal directions. According to Hill approach the localized neck is expected along a line of zero extension and a strain increment has to be negative. The Swift criterion is considered to underestimate the formability for negative minor strain. So both criteria are commonly combined together with the Hill criterion for the left side of the FLD and Swift model for the right one.

Later on, the Swift model has been further developed by Hora and Tong named Modified Maximum Force Criterion (MMFC) [6] for theoretical prediction of the localized necking considering the additional hardening effect to postpone localization by the stress transformation, and it was referred as Hora model in this paper.

Marciniak and Kuczynski [7] proposed a model taking into account the geometrical nonhomogenity of the sheet metal, known as M-K method. The M-K model proposed a physical approach by introducing the idea of flow localization in presence of an initial imperfection.

The mentioned theoretical models have their own advantages and limitations which are described in detail in [3]. However, there is still lack of systemically study on their applicability for different kind of materials. In this investigation, comparative research will be carried out to investigate the difference for forming limit prediction given by different theoretical models for a DP 980 steel.

Over the years, many researchers also studied on experimental determination methods for FLD. Some of these tests are summarised in Fig. 1 [8]. The most established testing procedure to determine FLD was proposed by Nakajima et al. [9]. A hemi-spherical punch and several specimen geometries are used for testing material formability under different strain paths.

[a] Corresponding author: hwsong@imr.ac.cn

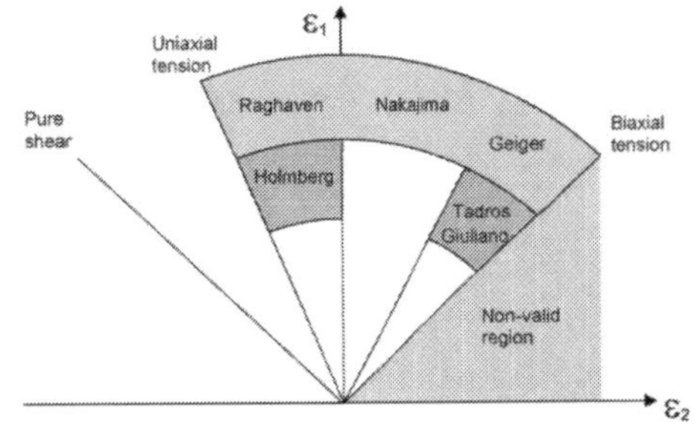

Figure 1. Summary of standard sheet material testing procedures with tested path range

However, as is shown in Figure 1, the standard methods for testing FLD are mainly used to characterise formability under certain in-plane load conditions from uniaxial tension, through plane-strain to biaxial stretching. But in industrial forming processes not only in-plane loads, but a mixture of in-plane, pure-shear and combined shear-and tension loads can also be observed [8, 10]. Today there is still a lack in testing equipment for those combined load conditions because with conventional testing set-ups no superposed shear rates can be realized.

Even for uniaxial tension it is not easy to achieve a constant loading path from onset of instability to rupture due to the necking effects observed in most steels. Smooth tensile specimen with a hole in the specimen center is very often used to approximate the stress state of uniaxial tension [11]. However, the disadvantage of the specimen is that the strain field in the specimen is very inhomogeneous and only a small region near the hole shows the stress state close to the uniaxial tension.

Although biaxial loading with cruciform specimens can be used to obtain shear-tension strain paths [12], the set-up is so complex and not easy to handle.

Recently, Shouler et al. [8] and Held et al. [10] proposed similar type of samples for characterising material formability under shear-tension loads with relatively simple set-up. However, the obtained strain paths are almost pure shear, which not easily occur during forming process. In addition, for advanced high strength steels such as DP steels fracture happens normally without instability under pure shear loading.

Considering the possible strain paths during forming process with mainly tension superimposed with some shear components, new sample geometry based on cruciform specimen is proposed to test formability under shear-tension strain paths which are closer to uniaxial tension. The set-up is only uniaxial tension loading, so the main advantage of the proposed testing is easy to handle and the measurements can also be easy to repeat.

2 FLD obtained by standard Nakajima testing and comparison with different theoretical models

For a DP 980 steel sheet with a thickness 1.5 mm, the standard Nakajima testing-DIN EN ISO 12004-2 was applied to measure the FLD and the failure curve [13, 14]. Instability limits under different strain paths were achieved by changing the notched width of the specimen as shown in Figure 2. Different theoretical models including Swift, Hill, Hora and M-K models which were briefly described in the introduction part were applied to predict the formability. A Von Mises isotropic yield function and a Swift power law hardening equation were used. For M-K model, the initial imperfection parameter f_0 is fixed to 0.995 calculated by surface roughness of the delivered sheets using methods proposed in [15].

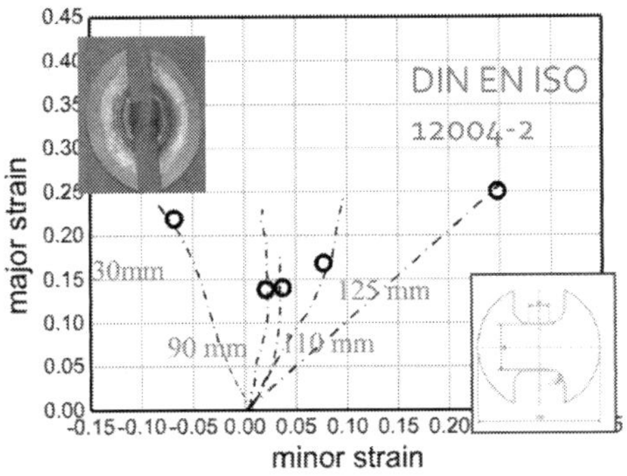

Figure 2. Measured FLD data by standard Nakajima testing [13]

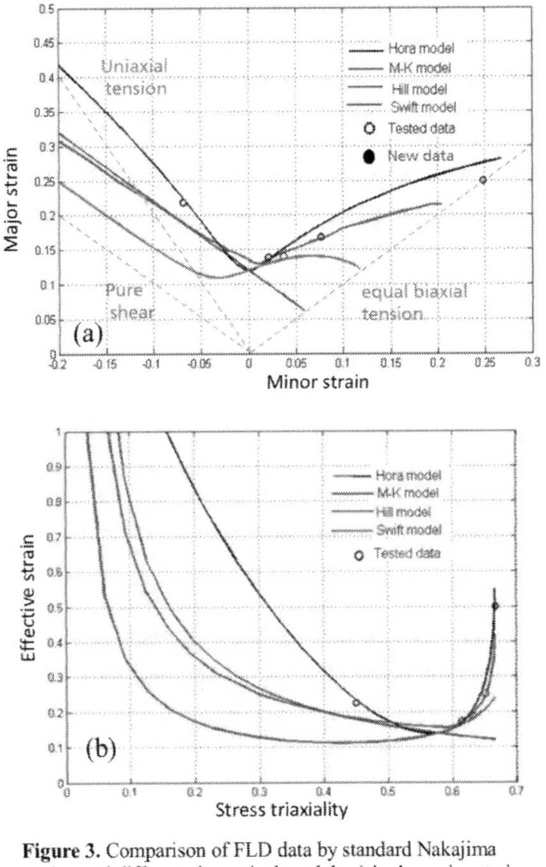

Figure 3. Comparison of FLD data by standard Nakajima testing and different theoretical models a) in the major strain-minor strain space, b) in the effective strain-stress triaxiality space

Comparison of tested data and prediction results in both major strain-minor strain space and effective strain-stress triaxiality space is given in Figure 3. The triaxiality is defined as the ratio of the mean stress σ_m to the von Mises stress σ_{eq}.

It can be seen that in the right side of FLD, the Hora and M-K models predict similar results and have good agreement with tested data. The Swift model obviously underestimates the formability and the Hill model is not relevant on the right side. In the left side, remarkable difference occurs among the theoretical models. However, as there is only one tested data in the left side of FLD, it is difficult to say which model is more applicable for formability prediction of the studied steel. As suggested in [15] the effective strain related FLD (shown in Figure 3 (b)) is less strain path dependent than the traditional FLD in major strain-minor strain space (shown in Figure 3 (a)), the effective related FLD is more suitable for practical forming process. As there is obvious difference of the theoretical FLD models on the left side, more testing data are needed to make better comparison and evaluation of the different theoretical models.

3 Proposed instability tests for shear-tension strain paths

3.1 Design of a new specimen for instability test under shear-tension strain path

Normal uniaxial tension loading can be used to get shear-tension strain path with the proposed specimen with a geometry given in Figure 4. It is a modified cruciform specimen which only needs uniaxial loading. The strain paths in the specimen can be changed by adjusting the three dimensional parameters *W*, *R*, *t* shown in Figure 4.

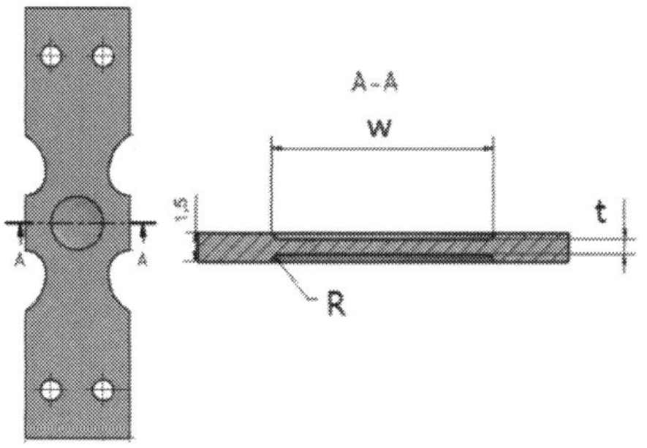

Figure 4. The geometrical design of the proposed specimen

The test on the new specimen is modelled with LS-Dyna using solid elements with edge length of 0.5 mm. The same yield function and hardening law as for theoretical prediction of FLD are used. The distribution and evolution of thickness strain were examined during deformation and the most thinned region in the specimen can be considered where to be the location of the onset of instability. The simulated results of strain paths in the most thinning region for different specimens are shown in Figure 5. These two regions are located in the center and at the notch. It can be seen that, the strain paths in the region where instability can possibly occur lie under shear-tension (near the specimen center) or near uniaxial tension (near the notch) dependent on where instability occurs. The position for the onset of instability can be controlled by adjusting the dimension parameters. All the predicted strain paths are linear in the major strain-minor strain space and almost linear in the effective strain-stress triaxiality space. These verified the applicability of the propose specimen test.

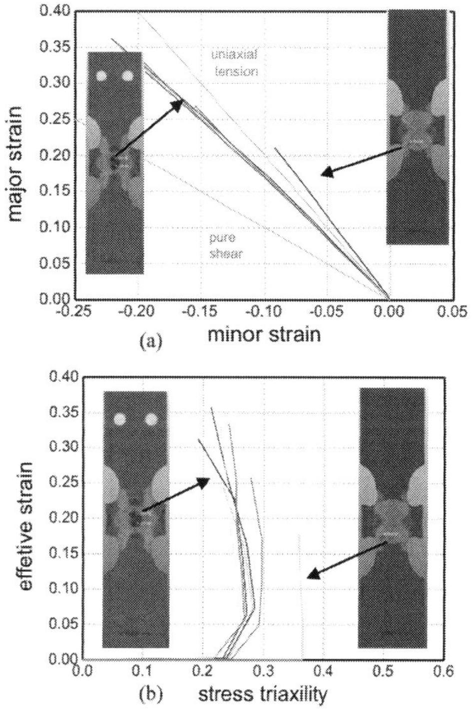

Figure 5. Simulated results for strain paths in the most thinned region in the specimen during uniaxial tension loading a) in the major strain-minor strain space, b) in the effective strain-stress triaxiality space

3.2 Experimental evaluation of the proposed tests

Several specimens with different geometries were extracted are prepared with different dimension parameters by sparkling cutting. Static uniaxial tension tests are carried out and the DIC (Digital image correlation) technique based on using ARAMIS system is applied to measure strain evolution during deformation.

Figure 6 shows the measured load-displacement curves for different specimen geometries and the typical broken specimens, and the dashed line is the simulation results of specimen 1 which indicates good agreement with experiment data

The thickness reductions were evaluated and the instability occurrence is defined as follows:

First, according to the measured distribution of thickness reduction as shown in Figure 7 (a) and Figure 7 (c), the most thinning regions with higher thinning can be captured.

DOI: 10.1051/matecconf/20168003007

Then points are selected across the thinned regions to evaluate the evolution of thickness reductions as shown correspondingly in Figure 7 (b) and Figure 7 (d). Finally when thickness reduction of at one point starts to increase faster than at the nearly ones from both side, the instability onset is assumed. .

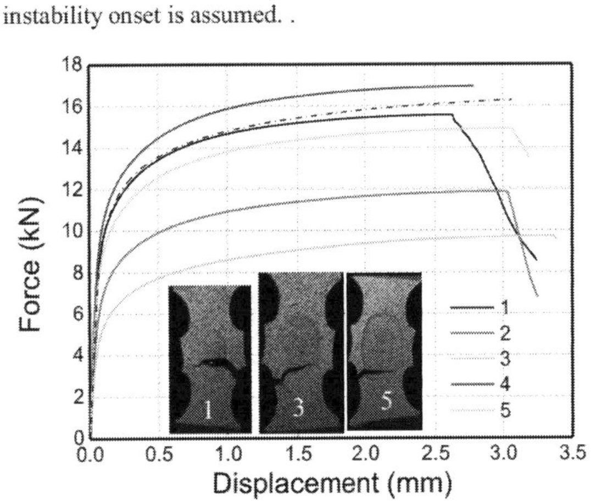

Figure 6. Measured load-displacement curves for different specimen geometries and the typical broken specimens.

The thickness reductions were evaluated and the instability occurrence is defined as follows:
First, according to the measured distribution of thickness reduction as shown in Figure 7 (a) and Figure 7 (c), the most thinning regions with higher thinning can be captured.

Then points are selected across the thinned regions to evaluate the evolution of thickness reductions as shown correspondingly in Figure 7 (b) and Figure 7 (d).

And finally when thickness reduction of at one point starts to increase faster than at the nearly ones from both side, the instability onset is assumed.

The defined instability occurrence for two specimens are given as examples indicated by dot line in Figure 7 (b) and Figure 7 (d) for different positions where instability occurs at points 2, 3 and point 0 respectively.

Using this definition, five specimens with different dimension parameters are tested, and the measured strain paths and strains for instability in both traditional major stain-minor strain space and effective strain-stress triaxiality space are given in Figure 8. It can be clearly seen that as FE predicted, all the tested strain paths where instability occurs are almost linear and lie in shear-tension or near uniaxial tension dependent on the position where instability occur controlled by specimen dimensional parameters. It can also be found that shear-tension strain paths were achieved when fracture failure initiates near the centre of the specimens and near uniaxial tension strain paths were achieved when fracture failure initiates at the edge notches.

The failure can be controlled by adjusting dimensional parameters of tested specimens.

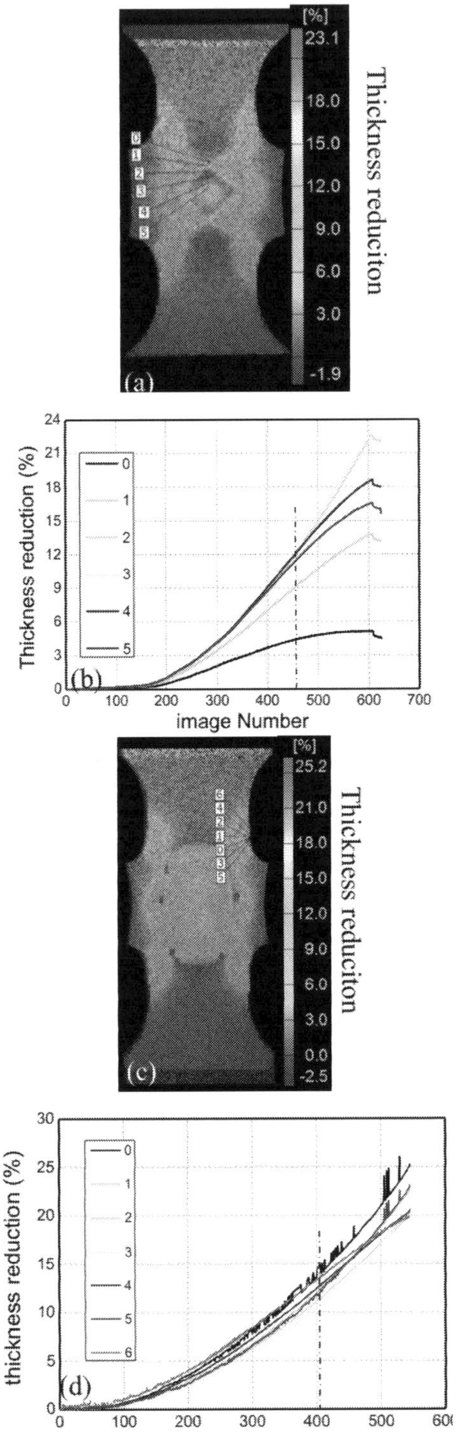

Figure 7. Definition of instability occurrence in the tested specimens a) distribution of thickness reduction of specimen 1, b) evolution of thickness reduction of selected point in specimen 1, c) distribution of thickness reduction of specimen 2and d) evolution of thickness reduction of selected point in specimen 2

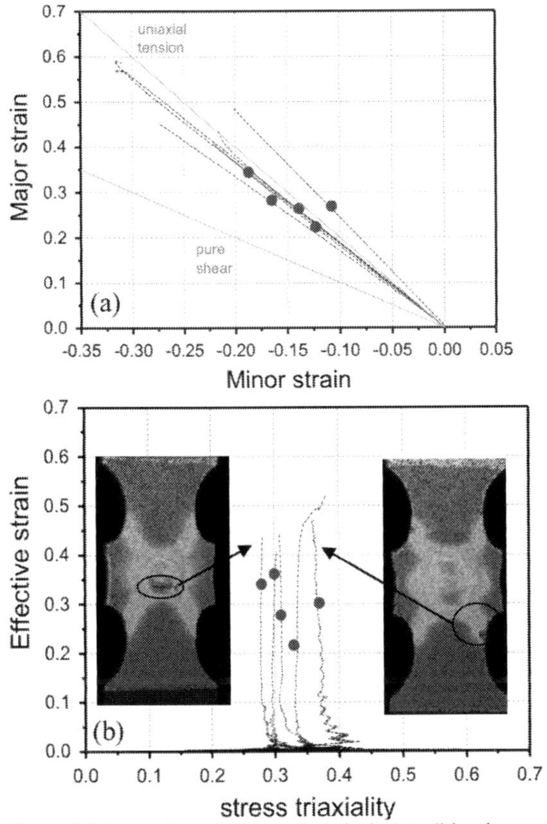

Figure 8. Measured instability results in both a) traditional FLD and b) effective related space

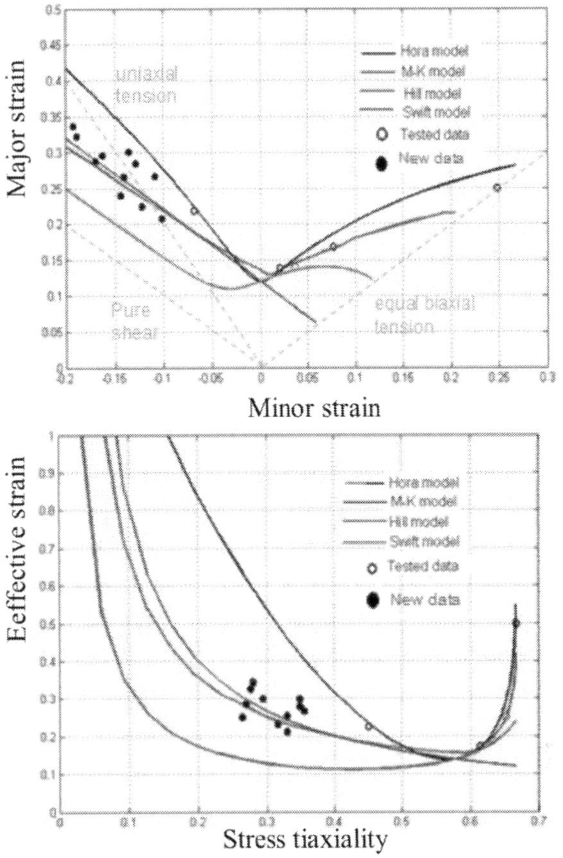

Figure 9. The comparison of measured instability results by proposed testing with different theoretical models in both traditional FLD and effective related space

4 Discussions

Each of the five specimens used to get results in Figure.8 are tested twice for repeatable evaluation. And the measured instability results from the ten tests (named new data) and from the standard Nakajima tests (named tested data) were drawn together in both traditional FLD and effective strain-stress triaxiality space, which are drawn in Figure 9. The comparison of the predictions by the investigated theoretical models are also presented in the figure.

With the help of new testing data, an evaluation of the theoretical models in the left side of the FLD can be done. The M-K and Hill models seem to give better formability prediction for the studied DP 980 steel in region between shear and uniaxial tension. Hora and Swift model respectively overestimates or underestimates the formability.

In order to confirm this result other variations of the proposed specimen geometry have to be determined in order to cover the whole range between shear and uniaxial tension with experimental results.

5 Summaries

A new testing method was proposed for formability measurement under shear-tension strain path using a simple experimental set up (uniaxial tension). Both FE simulation and experimental evaluations validated the applicability of the proposed testing. The comparison of tested formability with different theoretical prediction for the studied DP 980 steel shows that the M-K model gives better results especially under shear-tension region in the less path dependent effective strain-stress triaxiality space.

Acknowledgement

The support of CSC (Chinese scholarship council) is greatly appreciated.

References

1. S. P. Keeler, W. A. Backofen, Trans. ASM **56**, 25 (1963)
2. G. M. Goodwin, Society of Automotive Engineers **No. 680093**, 380 (1968)

3. D. Banabic, *Sheet Metal Forming Processes*, (Springer, Heidelberg, 2010)
4. H. W. Swift, J. Mech. Phys. Sol. **1**, 1 (1952)
5. R. Hill, J. Mech. Phys. Sol. **1**, 19 (1952)
6. P. Hora, L. Tong, *Proc. of the IDDRG Congress.* (Lisbon, 1994)
7. Z. Marciniak, K. Kuckzynski, Int. J. Mech. Sci. **9**, 609 (1967)
8. D. R. Shouler, J. M. Allwood, J. Mater. Proc. Tech. **210**, 1304 (2010)
9. K. Nakajima, T. Kikuma, K. Hasuka, *Proc. of the* 7th *Biennial IDDRG Congress.* (Amsterdam, 1972)
10. C. Held, R. Schleich, M. Sindel, M. Liewald, Int. J. Mater. Form. **2**, 467 (2009)
11. A. Trondl, S. Klitschke, W. Böhme, D.-Z. Sun, Fraunhofer IWM 2015 Schlussbericht, IWM-Berichts-Nr.: 1235/2015, Stiftungs-Nr. S24/10195/12
12. A. Hannon, P. Tiernan, J. Mater. Proce. Tech. **198**, 1 (2008)
13. S. Klitschke, W. Böhme, J. Schüler, J. Skoeries, D. Neumann, AVIF-Projekt (Stiftungs-Nr. S24/10195/12), Zwischenbericht PbA-Sitzung am 18.03.2015
14. S. Klitschke, W. Böhme, Tagungsband Werkstoffprüfung **S. 213-218**, (2014)
15. M. Nurcheshmeh, D. E. Green, Int. J. Mater. Form. **7**, 1 (2014)

FEM-DEM coupling simulations of the tool wear characteristics in prestressed machining superalloy

Peng Ruitao[1a], Tang Heng[1,2], Tang Xinzi[1] and Zhou Zhuan[3]

[1] *Xiangtan University, Xiangtan, Hunan 411105, China*
[2] *South China University of Technology, Guangzhou, Guangdong 510640, China*
[3] *Hunan Nanfang Aviation Industry Co., Ltd, Zhuzhou, Hunan 412002, China*

Abstract. Due to the complicated contact loading at the tool-chip interface, ceramic tool wear in prestressed machining superalloy is rare difficult to evaluate only by experimental approaches. This study aims to develop a methodology to predict the tool wear evolution by using combined FEM and DEM numerical simulations. Firstly, a finite element model for prestressed cutting is established, subsequently a discrete element model to describe the tool-chip behaviour is established based on the obtained boundary conditions by FEM simulations, finally, simulated results are experimentally validated. The predicted tool wear results show nice agreement with experiments, the simulation indicates that, within a certain range, higher cutting speed effectively results in slighter wear of Sialon ceramic tools, and deeper depth of cut leads to more serious tool wear.

1 Introduction

Nickel-based superalloys are widely utilized in industrial application due to their excellent performance, as a hard-to-machine material there's still a challenge to cut the alloy efficiently, prestressed cutting method has been validated as an efficacious way to control surface residual stress [1, 2], however characteristics of superalloy induce severe thermo-mechanical loads at the tool-chip interface which still result in significant wear of the tool.

Tool wear phenomena are complicate and experimental, analytical and numerical investigations are important keys to reveal the tool wear process and mechanism. In the last decade, experimental studies are performed to investigate the tool wear in machining superalloy. Rahman [3] explored the influence of cutting condition on the machinability of Inconel 718 by experiments. Kitagawa [4] investigated the temperature and wear of ceramic tools in high-speed machining of Inconel 718, boundary wear has been observed as the main form of tool wear. Costes [5] investigated the tool-life and wear mechanisms of CBN tools in machining of Inconel 718 and reported that the adhesion wear and diffusion wear were the main wear styles in the cutting process. These studies show that the influence of tool wear on machining process and more investigation are required.

More recently, with the rapid development of multi-core CPU technique, numerical methods are widely used to study the cutting process beside the tool wear, such as finite element method (FEM), molecular dynamics (MD), discrete element method (DEM), and so on.

The FEM based on continuum mechanics, has already been widely used to solve tool abrasion problem. Yao [6] built a FEM model for studying the ceramic tool abrasion, effect of the physical properties such as hardness elastic modulus and fracture toughness to tool abrasion performance were researched, it's concluded that the higher of the hardness and elastic modulus, the higher of abrasion rate; and the higher of the fracture toughness, the lower of abrasion rate. Lorentzon [7] predicted the cutting force, chip thickness, cutting temperature, the contact stress and relative rate between chip and tool in cutting process via the FEM simulation, and rake face crater wear and flank wear were also analyzed. Monaghan [8] revealed the effect of rake angle on the tool wear via FEM simulations, and it's proved that negative rake angle may aggravate tool adhesion wear.

Discrete element method (DEM), evolved from molecular dynamics theory, which has widely been implemented in the study of geotechnics, concretes, ceramics and other brittle materials [9]. It shows great potential in simulating granular material and analyzing cracks. Tan [10] analyzed the mechanisms of the formation and propagation of cracks in machining SiC and Al2O3 via DEM simulations; it indicates good agreement with the experimental validation. Fillot [11] provided a fundamental approach to study the wear process by considering the detachment of particles, the particles were judged whether they're in the contact zone or not, it's validated that discrete element method is feasible to simulate wear behavior. More recently, scientists have successfully utilized the DEM to study material's fracture, damage,

Corresponding Peng Ruitao: pengruitao@xtu.edu.cn

wear behaviors and contact mechanics between rough surfaces [12-17].

The FEM-based method is capable to solve non-linear contact problem with reasonable accuracy and computational efficiency, however, it is lack of capability or qualification to reproduce the discontinuous behavior of material during cutting process, such as crack path and third body particles. The DEM has a practical potential to predict the evolution of crack initiation and propagation as well as the behavior of third body particles with high fidelity. However, the DEM is a time consuming algorithm in nature. Therefore, it is merely adopted to simulate assembly systems constructed by relatively small amount of elements. Considering the strengths and weakness of FEM and DEM for solving tool wear problem [18], a FEM-DEM coupled method is introduced here to research the ceramic tool abrasion behavior.

The aim of our current work is to develop a fundamental method to numerically analyze the wear problem taking into account the discontinuous behavior of ceramic tools. This paper is composed of five sections; In Section 2 an experimental setup was designed for prestressed machining superalloy. In Section 3, stress analysis of prestressed loading was performed, subsequently a finite element model for prestressed cutting superalloy was built and experimentally validated, and the contact stress between chip and tool, and the temperature distribution of tool as well as the speed of chip flow were extracted for the DEM models. In Section 4, a discrete element modeling technique was proposed and the microscopic parameters required in DEM simulations are determined and calibrated, effects of cutting factors and prestress on crack initiation and propagation were studied. And the DEM simulated results are validated by experimental studies. Finally, the paper was concluded in Section 5.

2 Experimental work

In our research, an experimental setup was carried out to verify the accuracy and feasibility of the FEM-DEM model. The tool-workpiece configuration, machining device and experimental setup used for measurements are illustrated in Figure. 1.

In order to load prestress to the workpiece, a special prestressed loading device is designed, as shown in Fig.1(a). The middle of fixture body and the inner circle of taper-lock are both conical shape and tapered consistently. To twist the nut on the right, the workpiece will be expanded. Moreover, a tailor-made workpiece is designed for orthogonal cutting to validate the FEM models.

The cutting tool is a rhombus uncoated insert, of a Sialon ceramic material, designated by CNGN120408, and is fixed on tool holder of the standard ISO designation DCLNR2525M12. KISTLER Model 9257B dynamometer is used to measure the three components of the cutting forces (main cutting force "tangential" Fc, feed force "axial" Ff, and depth of cut force "radial" Fp) during cutting tests. The basic geometry characteristics of Sialon ceramic cutting tools are listed in Table 1. The workpiece is a ring of the superalloy GH4169, with inner diameter of 40 mm, and outside diameter of 60 mm

Use A4 paper size (210 x 297 mm) and adjust the margins to those shown in Table 1. The final printed area will be 172 x 252 mm.

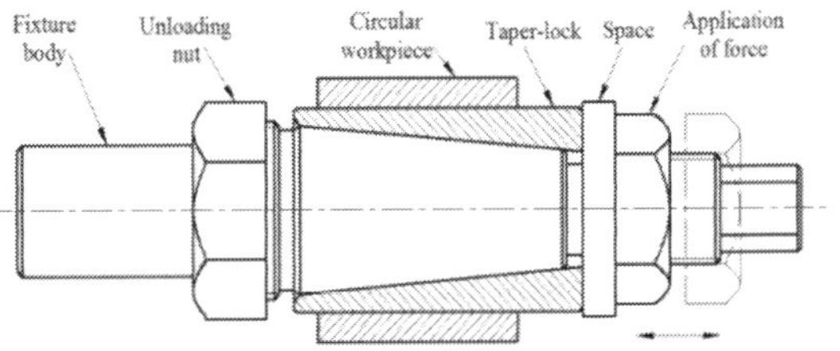

(a) Scheme of prestressed loading device

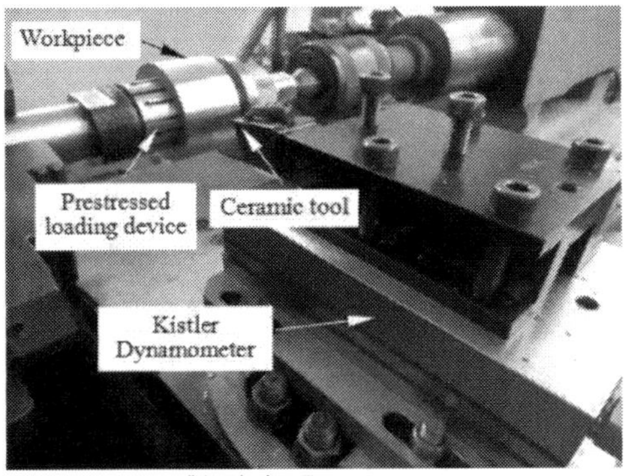

(b) Kistler Dynamometer

Figure 1. Experimental setup for machining

Table 1. Geometry characteristics of Sialon ceramic tools

κ_r/°	λ_s/°	γ_0/°	α_0/°	κ_r'/°	Cutting nose radius r_ε/mm
95	-6	-6	6	5	0.8

Aiming to reveal the influence of parameters including prestress, cutting speed, cutting depth on tool wear in prestressed cutting, single factor scheme, shown in Table 2, is designed for FEM simulations, DEM simulations and experimental validation.

Table 2. Single factor scheme

Case	Prestress	Cutting speed	Cutting depth	Feed rate
1	0	90	0.2	0.2
2	150	90	0.2	0.2
3	300	90	0.2	0.2
4	400	90	0.2	0.2
5	400	30	0.2	0.2
6	400	60	0.2	0.2
7	400	130	0.2	0.2
8	400	90	0.1	0.2
9	400	90	0.3	0.2
10	400	90	0.4	0.2

3 Finite element simulations

3.1 Constitutive model

In order to accurately simulate the process of cutting superalloy using FE method, the material constitutive model must be calibrated to account for the material response under high strain rate impact. To analyze the machining process, a Johnson-Cook model [19] has been used to represent the thermo-viscoplastic behavior of GH4169, where the flow stress is given as follows:

$$\sigma_p = (A + B\varepsilon_p^n)[1 + C\ln(\frac{\dot{\varepsilon}_p}{\dot{\varepsilon}_{p0}})][1 - (\frac{T - T_r}{T_m - T_r})^m] \quad (1)$$

Where A and B are the material constants, C is the strain rate sensitivity, n is the strain-hardening, m is the thermal softening exponent, εp is the equivalent plastic strain, έp is the plastic strain rate (s-1), έp0 is the reference of plastic strain rate (s-1), T is the temperature, Tm is the material melting temperature and Tr is the reference temperature. The superalloy GH4169 parameters for Johnson-Cook model are given in Table 3.

Table 3. Johnson-Cook parameters of GH4169[20-21]

A (MPa)	B (MPa)	n	C	m	$\dot{\varepsilon}_{p0}$ (s^{-1})	T_r (^0C)	T_m (^0C)
1241	622	0.6522	0.0134	1.3	1	20	1297

3.2 FE modelling and boundary conditions

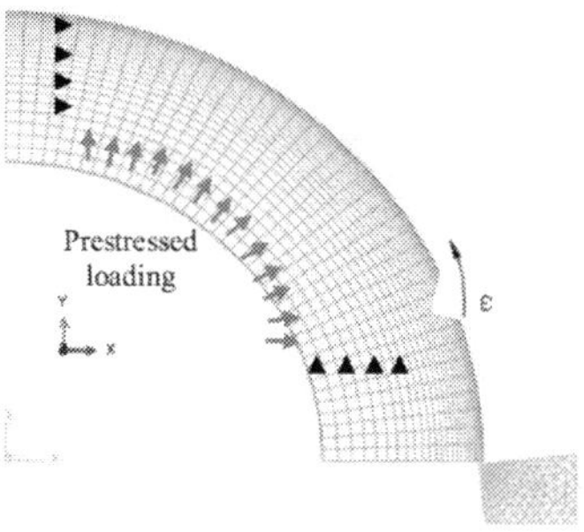

Figure 2. The FEM mesh model of prestressed cutting

The FE model of prestressed cutting is shown in Fig.2. Base on the method of prestress loading proposed by the author's previous literatures [1-2], the internal expansion method is adopt to load the prestress on the workpiece. Because of the stresses and displacements of the circular workpiece is axial symmetry, a quarter of the workpiece is considered to establish the model. The fixed constraint is applied to the horizontal and vertical boundary layers to prevent the rigid body movement. The cutting tool and workpiece are meshed with a plane strain coupled linear temperature-displacement 4 nodes element. Moreover, the outer ring of workpiece and cutting tool near of tool/workpiece contact zone are refined with a minimum element size. The cutting conditions are given in Table 2. The thermal boundary condition is the room temperature which is 27°C.

3.3 Model validation

In order to validate the FE model for prestressed cutting, an orthogonal cutting experiment was designed base on the mentioned experimental setup in Section 2. In orthogonal prestressed cutting, measured cutting forces are compared with those obtained in FEM simulations. Influence of prestress, cutting speed and cutting depth on cutting force Fc and thrust force Fp are shown in Fig.3(a-c). It is observed that all the values of cutting forces predicted by FEM simulation are little lower than that in the experiment, and the overall trend of the simulated results shows a reasonably good agreement with the experimental results.

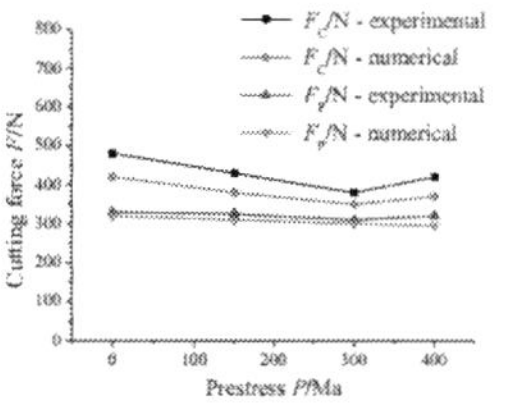

(a) Effect of prestress

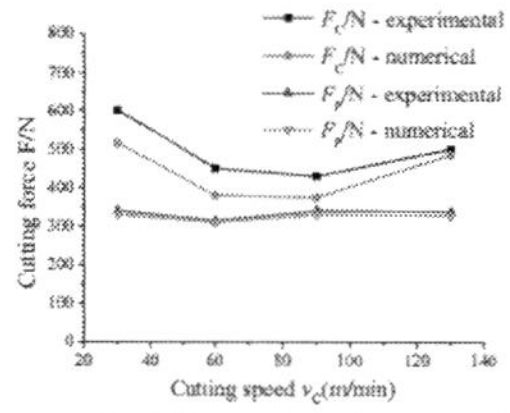

(b) Effect of cutting speed

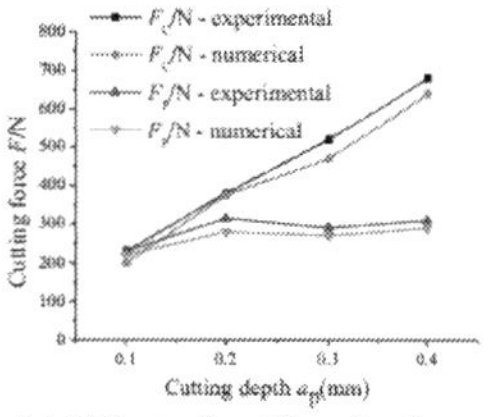

(c) Effect of cutting depth

Figure 3. Effect of cutting parameters on cutting force

As mentioned above, the simulation results of the cutting force have been found in good agreement with experimental results, considering the possible errors, the established FE models are deemed to be efficient and acceptable.

3.4 Results

The contact stress and contact length between tool and chip, temperature distribution of tool as well as speed of chip flow are obtained based on FEM simulations. The contact length between chip and tool is defined as the distance between the first nodal which commences to contact and the last nodal starts to detach.

The speed of the chip flow means the average velocity of the nodes of chip on the contact area while in steady cutting state. The velocities of the chip flow along the tool-chip contact area are shown in Table 4.

The results indicate that the higher the cutting speed, the higher the chip velocity, and the values generally consistent with the theoretical calculation [22]. The theoretical formula is shown as follows:

$$v_{cf} = \frac{v_c \sin\phi}{\cos(\phi - \gamma_0)} \quad (2)$$

where v_{cf} is the velocity of chip flow along the tool, v_c means the cutting velocity, represents the shear angle, γ_0 is the rake angle of tool.

Table 4. Chip flow velocity based on FEM simulation

Case	1	2	3	4	5
Chip flow velocity v_{cf}/m·min^{-1}	1.04	1.12	1.12	1.12	0.38
Case	6	7	8	9	10
Chip flow velocity v_{cf}/m·min^{-1}	0.75	1.61	1.11	1.13	1.14

The contact stresses between chip and tool extracted from the nodes on the tool's rake face are illustrated in Fig.4(a-k). From this figure, it is found that the maximum contact stress generally occurs approximately 0.02mm from the tool tip. The contact stresses of the FEM model on the cut boundary are calculated by interpolating the FE simulated results. In DEM models, the nodal force will be specified as boundary condition for corresponding discrete element particle.

Fig.5 shows the temperature distribution of the tool. There is a decreasing temperature distribution trend along the thickness direction of the tool. The red folder line in the enlarged view is the temperature extraction path. The point-tracking temperature extracted from FEM simulations are shown in Table 5. It indicates that the highest nodal temperature and distribute tendency are varying with cutting conditions.

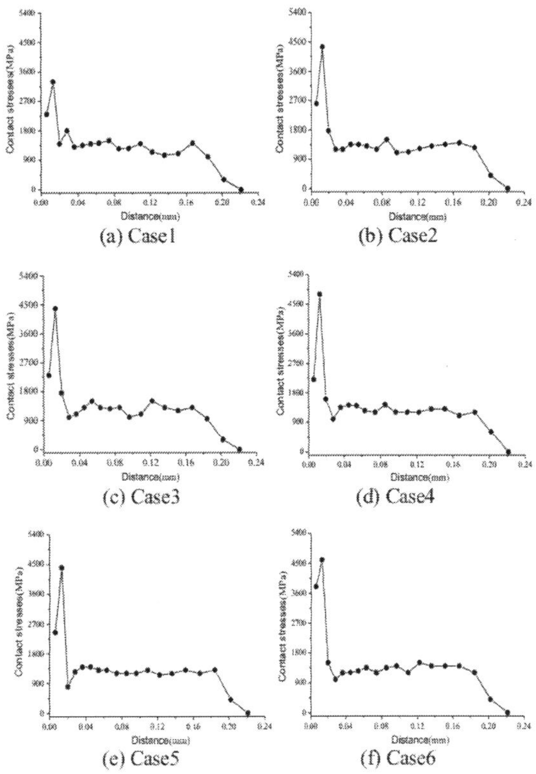

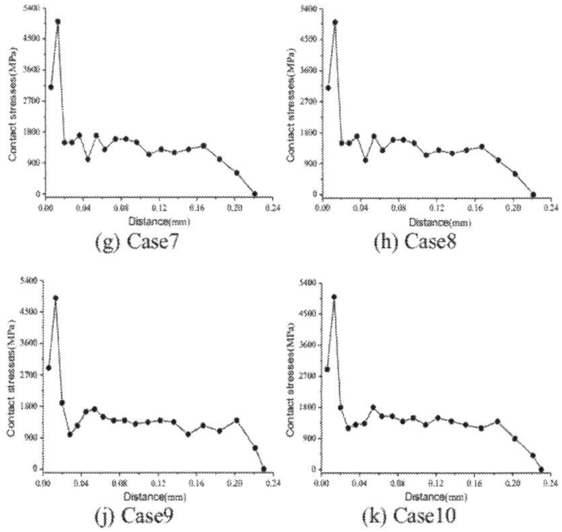

Figure 4. The contact stresses of nodes in tool's rake face

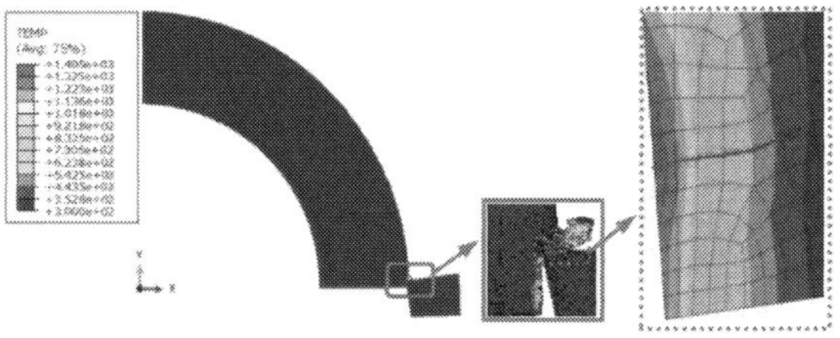

Figure 5. Temperature field of tool and extraction path (/°C)

Table 5. Point-tracking temperature from FEM simulation

Case	TP$_1$ (°C)	TP$_2$ (°C)	TP$_3$ (°C)	TP$_4$ (°C)	TP$_5$ (°C)
1	962	852	756	667	584
2	963	853	758	667	584
3	964	855	758	668	586
4	964	856	759	669	587
5	880	829	783	737	692
6	936	891	843	795	747
7	1062	962	863	768	679
8	894	842	791	740	692
9	1054	953	857	767	683
10	1073	958	855	762	676

4 Discrete element simulations

In PFC2D software, the deformational behavior between units such as relative displacement generally can be realized by the parallel-bond to connect the joint. For continuous medium material, the parallel-bond glue is of a finite size that acts over either a circular or rectangular cross-section lying between the particles; it can transmit both forces and moments between particles. If either of the maximum stresses exceeds its corresponding bond strength, the parallel bond breaks and the particles break away from the contact area. Therefore, the damage of materials can be simulated by the connect bond's breakage

DOI: 10.1051/matecconf/20168004001

between units. DEM method can dispose the procedure of materials' crack formation and propagation intuitively.

4.1 Macro-parameters' Correction

In order to match the established DEM model's mechanical property with the actual materials, it is necessary to adopt different mechanical model to adjust the DEM simulation macro-parameters (particle radius, particle stiffness, bond strength) [23]. The micro-parameter setting in the model is regarded reasonable when simulated results matched well with the practical material mechanics test value. Mechanical model such as uniaxial compression test, three-point bending and single edge shearing and single edged notch beam were used to calibrate the macro-parameters of the ceramic material[24]. The calibrated results of Sialon ceramic material and GH4169 material are shown in Table 6 and Table 7 respectively [25].

Table 6. Simulated material properties of Sialon ceramic compared with referenced ones

Mechanical Property	Referenced Result	Simulated Result	Relative Error
Elastic Modulus/GPa	300	305	1.6%
Poisson ratio	0.218	0.231	5.7%
Uniaxial compression strength/MPa	3600	3669	1.9%
Bend strength/MPa	750~1200	1152	—
Brittle strength KIC/MPa·m$^{1/2}$	5.5~6.0	5.265	4.3%

Table 7. Simulated material properties of GH4169 compared with tested ones

Mechanical Property	Tested Result	Simulated Result	Relative Error
Elastic Modulus/GPa	206	192	6.8%
Poisson ratio	0.3	0.274	8.7%
Bend strength/MPa	1110	1208	—
Brittle strength KIC/MPa·m$^{1/2}$	0.233	0.217	6.8%

4.2 Temperature field

Cutting temperature is an important factor which directly affects the cutting tool wear, workpiece surface integrity and machining precision in machining process. Kitagawa and Maekawak [4] utilized the embedded thermocouple method to determine the temperatures of Si3N4 ceramic tool rake face in machining of Inconel 718. The authors reported that the temperatures of tool rake face will reach 1100°C when the cutting speed over 100m/min, and the higher temperature can aggravate the wear of tool. The influence of temperature is not considered in previous discrete element simulation of machining. It's known that the temperature of contact area in cutting nickel-based superalloy is much high and it has greater effects on the machinability of cutting tool and the workpiece, so the temperature field should be considered in the discrete element simulation. In this paper, the coupled thermo-mechanical models are established with the consideration of thermos influence on tool wear.

4.3 The friction and wear model between tool and chip

As mentioned before, the FEM simulation provides the boundary conditions for the DE model, including force boundary, temperature boundary and chip flow velocity.

The friction and wear model of tool and chip was established based on the simulated microcosmic parameters, the temperature distribution of the discrete element model was set according to the finite element simulation, and finally a coupled thermo-mechanical analysis was realized, as shown in Fig.6. In order to predict the friction and wear behavior between tool and chip in a limited size model, the chip's model was established by adopting the periodic boundary, namely, when a particle moves out of the system from the up side in y direction, consequently it will move into the system from the below side at the same time. The contact stresses extracted from FEM simulations were loaded on the right boundary of tool, the right boundary was modeled by vertically arranged assembly of discrete elements with unified radius. Contact pressure was imposed to the right wall (blue grain) of tool model (yellow grain), and the interpolated nodal stresses obtained from FEM simulations were applied to the corresponding discrete elements, note that a certain contact stresses between chip and tool should always be maintained. The chip (red grain) was given a constant upward flow velocity, which moved into the field from the below side of the periodic boundary, then flows along the tool surface, as soon as the particle arrived at the up side of the periodic boundary, it then break away from abrasion zone, and these particles then approach the below side of the periodic boundary repeatedly, the continuous movement from the below side to the up side of chip particles was achieved.

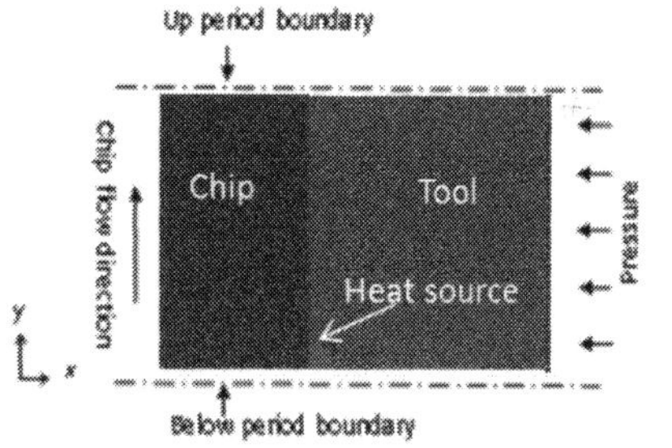

Figure 6. Tool-chip thermo-mechanical model

4.4 Results and Analysis

The DEM simulation results are shown in Figure.7 and Figure.8, the white curves refers to the detached particles, which are named cracks in the cutting process, it is convenient to analyze the concealed cracks of the tool, the tool's crack produced in the rake face can be observed distinctly which has a certain distance to the tool tip, and it gradually extends to the inner layer until to the rear face, consequently lamellar exfoliation is occurred on the tool tip. The tip of tool prone to wear when the force loaded on tool tip exceeded the limited value.

The simulated results under different cutting speed are presented in Figure.7. From these pictures, it is found that breakup and exfoliation of tool tip increase with cutting speed to a maximum first achieved at the speed of 90 m/min and then decreasing. In other words, tool wear is the most serious at the speed of 90 m/min. Figure.8 shows the simulation results under different cutting depth. The results represent that the crack length and the peeling area produced in the tool tend to be increased with the increase of cutting depth. Figure.10 illustrates the simulated results of tool wear under different prestress. It can be seen that tool wear shows more serious without prestress than that with prestress. In addition, within a certain range, the tool wear will be ameliorated with the increase of prestress.

The contact surface between tool and chip exists micro-cracks and deciduous particles (gray particle), which are shown in Figure.9. Shedding particle are mingled with the two contact surface, the wear form between chip and tool turns into 3-body wear state [26] and this layer of particles intensifies the wear of tool.

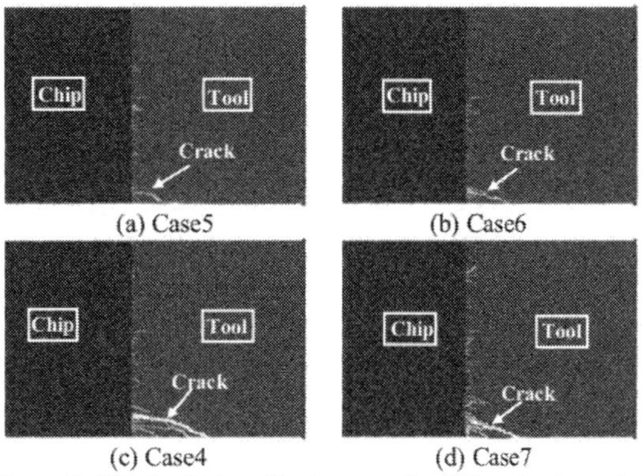

Figure 7. DE simulation of tool wear under different cut speeds (P=400MPa,a_p=0.2mm,f=0.2mm/r)

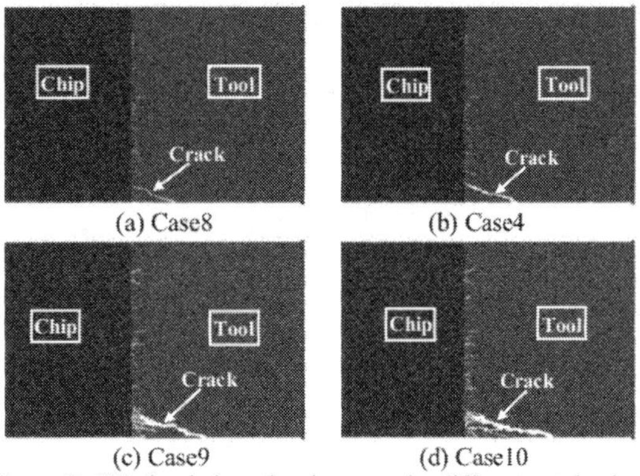

Figure 8. DE simulation of tool wear under different cut depths (P=400MPa,v_c=90m/min,f=0.2mm/r)

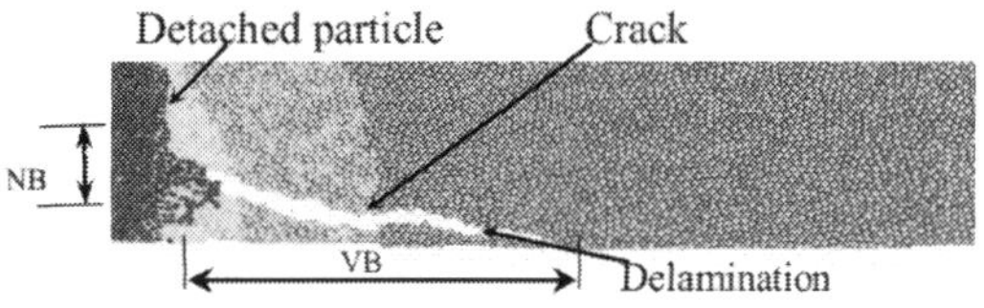

Figure 9. Partial enlargement of cracks in Fig.8. (c)

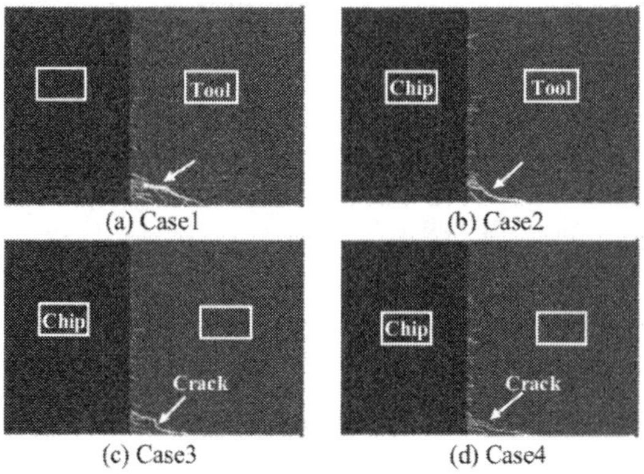

Figure 10. DE simulation of tool wear under different prestress (a_p=0.2mm,v_c=90m/min,f=0.2mm/r)

Figure.11 clearly shows the number of peeled particle from tool and the number of micro-crack, which are calculated automatically by the software, vary along with the cutting speed at the cutting depth of 0.2mm. It suggests that with the increase of cutting speed, cutting tool micro-crack number and particle peeling number were increased first and then decreased. The main reason is that as cutting speed increased to a certain range, the increase of temperature in workpiece will result in the mollification of workpiece [27-28], and the cutting force is reduced with the decreased strength and stiffness of workpiece. Meanwhile, the increased cutting temperature improves the friction deformation between chip and tool, thereby leads to the decreased of the additional deformation and chip outflow resistance in the area.

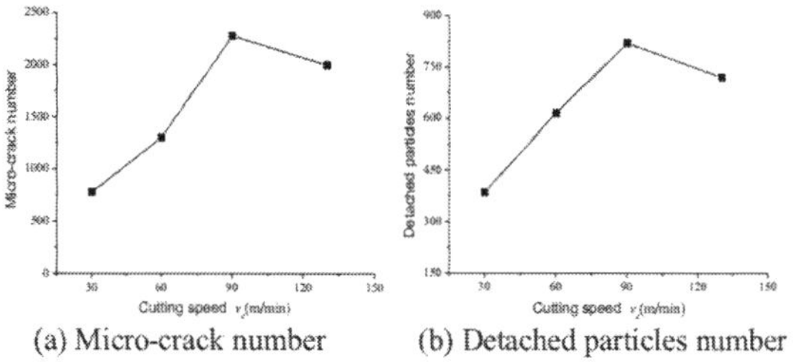

Figure 11. Effect of cutting speed on micro-crack number, detached particles number and wear rate

The number of shedding particles and micro crack of tool vary with the cutting depth are shown in Figure.12. It's noted that the number of shedding particles and micro cracks increase with the increase of cutting depth. It further explains that increased cutting depth can aggravate the damage of tool.

The number of shedding particles and micro crack of tool vary with the prestress are presented in Fig.13. It is shown from Fig.13 that, as the prestress increases, the number of shedding particles and micro cracks initially decrease and then begin to stabilize. From these results it is clearly seen that, within a certain range, the prestress conditions present effective wear resistance of flank wear.

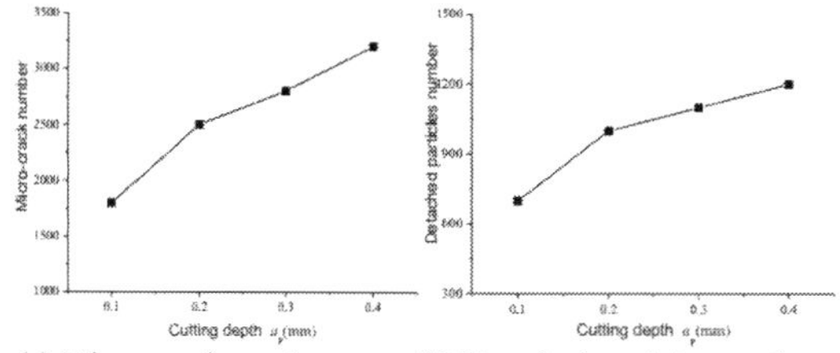

(a) Micro-crack number (b) Detached particles number

Figure 12. Effect of cut depth on micro-crack number, detached particles number and wear rate

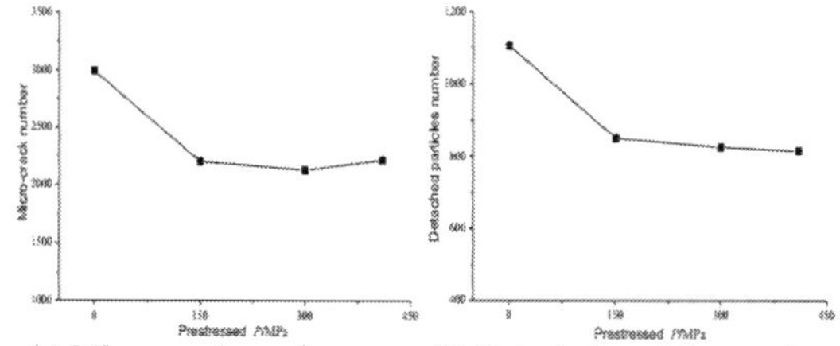

(a) Micro-crack number (b) Detached particles number

Figure 13. Effect of prestress on micro-crack number, detached particles number and wear rate

4.5 Model validation

According to the experiment and DEM simulation, the influence of cutting depth, cutting speed and prestress on tool flank wear (VB) are revealed. The measured results and predicted values are compared in Fig.14.

It can be observed that the cutting depth has greatest influence on the tool flank wear. And the value of VB increases with the increases of cutting depth. Meanwhile, it can also be found that the tool flank wear increase with cutting speed to a maximum first achieved at the speed of 90 m/min and then decreasing. Both FEM results and experimental results indicate that prestress conditions are beneficial to mitigate the tool flank wear. As shown in Fig.14, the overall trend of the simulated results shows a reasonably good agreement with the experimental results, the established DE models are deemed to be acceptable as an efficient tool to predict tool wear in prestressed machining superalloy.

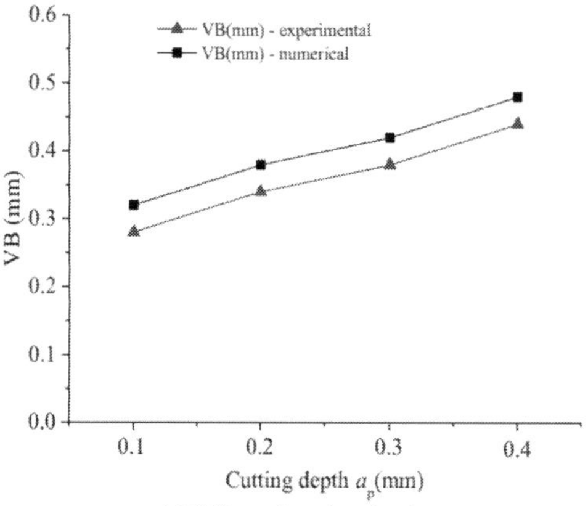

(a) Effect of cutting depth

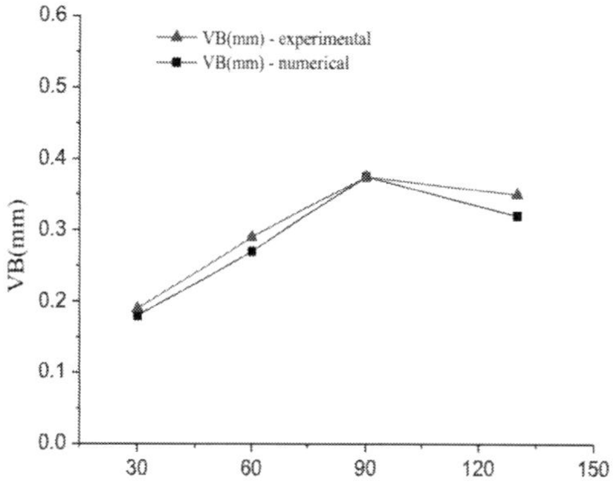

(b) Effect of cutting speed

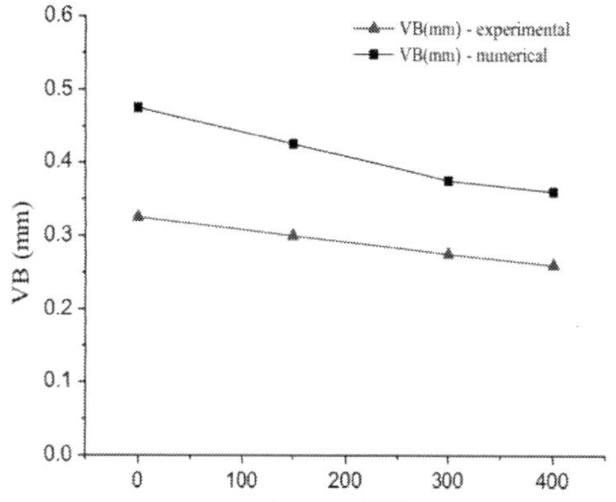

(c) Effect of cutting prestress

Figure 14. Effect of the cutting parameters on tool wear

5 Conclusions

Through combined FEM-DEM simulation and experimental validation, this paper discusses the ceramic tool wear behavior in prestressed machining superalloy GH4169. According to results presented in this paper, the following conclusions can be addressed:

(1) A finite element model of the prestressed cutting process is established on the basis of the prestress loading mechanism and validated by orthogonal cutting experiments, the contact stress and contact length between tool and chip, temperature distribution of tool as well as speed of chip flow in steady cutting state are obtained based on FEM simulations, which provide necessary boundaries for DEM models.

(2) The inter-element contact constitutive model as well as corresponding microscopic parameters are determined and calibrated to describe continuous and discontinuous behaviors of material in DEM simulation. The discrete element model of tool-chip wear was established, effect of prestress, cutting depth and cutting speed on tool wear were simulated and also validated by experiments under different cutting conditions, and the simulated results were consistent with the experiment results. The both results show that the combined FEM-DEM methodology is an efficient tool to qualitatively and quantitatively analyze the

tool wear behavior, especially in simulating the generation of the micro-crack and delamination.

(3) Within a certain range, prestress conditions are beneficial to mitigate the tool flank wear, higher cutting speed effectively results in slighter wear of ceramic tool, and deeper depth of cut leads to more serious tool wear.

Acknowledgments

This study was supported by the Joint Fund of Provincial Natural Science Foundation between Hunan Province and Zhuzhou City (13JJ8007), and also funded by the National Natural Science Foundation of China (No. 51475404, No.51375418, No. 51305377), and the Foundation of Hunan Educational Committee (No. 14B166).

References

1. R.T. Peng, X.Z. Tang, Y.Q. Tan, X.W. Liu. Finite Proc. Ame. Inst. Phys, **1532**, 962-968 (2013).
2. R.T. Peng, M. Liao, X.Z. Tang, Y.Q. Tan, X.W. Liu. J. Mech. Engi, **19**, 186-191 (2012).
3. M. Rahman, W.K.H. Seah, T.T. Teo. J. Mate. Proc. Tech, **63**, 199-204 (1997).
4. T. Kitagawa, A. Kubo, K. Maekawa. Wear, **202**, 142-148 (1997).
5. J.P. Costes, Y. Guillet, G. Poulachon, et al. Int. J. Mach. Tool. Manu, **47**, 1081-1087 (2007).
6. S.Q. Yao, S.M. Xing, J.X. Deng. Tribol. Int, **26**, 566-569 (2006).
7. J. Lorentzon, N. Jarvstrat. J. Mach. Tool. Manu, **48**, 1072-1080 (2008).
8. J. Monaghan, T. Macginley. Comp. Mater. Sci, **16**, 275-284 (1999).
9. S.Q. Jiang, Y.Q. Tan, S.J. Nie, et, al. J. Inorg. Mater, **25**, 1286-1290 (2010).
10. Y.Q. Tan, D.M. Yang, Y. Sheng. J. Eur. Ceram. Soc, **29**, 1029-1037 (2009).
11. N. Fillot, I. Iordanoff, Y. Berthier. Tribol. Int, **40**, 973-981 (2007).
12. M. Nouari, I. Iordanoff. Tribol. Int, **40**, 1351-1359 (2007).
13. Y.Q. Tan, D.M. Yang, Y. Sheng. J. Eur. Ceram. Soc, **29**, 1029-1037 (2009).
14. J.F. Jerier, J.F. Molinari. Tribol. Int, **47**, 1-8 (2012).
15. A. Delaplace. Geomech. Geoengin, **4**, 79-89 (2009).
16. I. Iordanoff, A. Battentie, J. Neauport, et al. Tribol. Int, **41**, 957-64 (2008).
17. D. Andre, I. Lordanoff, J.L. Charles, et al. Comp. Methods Appl. Mech. Engrg, **213**, 113-25 (2012).
18. B.D. Leonard, P. Patil, T.S. Slack, et al. J. Tribol-the ASME, **133**, 1-12 (2011).
19. G. Johnson, W. Cook. Engin. Fracture. Mech, **2**, 31-48 (1985).
20. P.J. Arrazola, A. Kortabarria, A. Madariaga, et al. Pract. Th, **41**, 87-103 (2014).
21. J. Lorentzon, N. Järvstråt, B.L. Josefson. J. Mater. Process. Tech, **209**, 4645-4653 (2009).
22. M.E. Merchant. J. Appl. Phys, **16**, 267-275 (1945).
23. D.O. Potyondy, P.A. Cundall. Int. J. Rock. Mech. Min. Sci, **41**, 1329-1364 (2004).
24. S.Q. Jiang, Y.Q. Tan, D.M. Yang, et al. J. Chin. Ceram. Soc, **5**, 918-923 (2010).
25. R.T. Peng, F. Lu, X.Z. Tang, Y.Q. Tan. Mater. Sci. Forum, **800-801**, 385-389 (2014).
26. M. Godet. Wear, **100**, 437-452 (1984).
27. J. Zhao, J. Deng, J. Zhang, et al. Wear, **208**, 220-225 (1997).
28. E. Oñate, J. Rojek. Comput. Methods. Appl. Mech. Engrg, **193**, 3087-3128(2004).

Finite element method analysis of surface roughness transfer in micro flexible rolling

Feijun Qu[1], Haibo Xie[1] and Zhengyi Jiang[1,a]

[1]*School of Mechanical, Materials and Mechatronic Engineering, University of Wollongong, Northfields Avenue, Wollongong 2522, Australia*

Abstract. Micro flexible rolling aims to fabricate submillimeter thick strips with varying thickness profile, where the surface quality of products is mainly determined by initial workpiece surface roughness and subsequent surface asperity flattening process, which is affected by process parameters during rolling. This paper shows a 3D finite element model for flexible rolling of a 250 μm thick workpiece with reduction of 20 to 50%, and rolling phase with thinner thickness indicates a better ability to decrease the surface roughness. Four types of initial workpiece surface roughness are studied in the simulation, and the influences of process parameters, such as friction coefficient, rolling speed and roll gap adjusting speed, on surface asperity flattening of workpieces with different initial surface roughness have been numerically investigated and analysed.

1 Introduction

Flexible rolling has been rapidly developed and gradually used on a large industrial scale in recent years to meet the rising demand for weight-saving and cost-efficient products as the roll gap may be adjusted to achieve a variable thickness along the metal sheet's length during the rolling process [1]. In the meantime, numerous research activities have been conducted to obtain a better understanding and make further development of this forming technology. Engler et al. [2] presented the process optimisation and control of material properties in the flexible rolling of aluminium alloy sheet based on lab-scale experiments, which elucidated the range of properties that could be realised through flexible rolling while maintaining reproducible constant material properties. Huang et al. [3] proposed a flexible rolling simulation model on the basis of Deform-3D and verified the feasibility of flexible rolling by comparison between simulation curve and target profile curve. They later discussed the influence of different process parameters on the quality of flexibly rolled products and optimised these parameters to obtain better results via orthogonal trial [4]. Kopp et al. [5] reviewed the flexible rolling process launched at the Institute of Metal Forming and the company Mubea. With a controlled online adjustment of the roll gap, any thickness transitions could be selected and adapted exactly to the demanded load of the product in use. Zhang et al. [6] derived a mathematical formula to control the vertical motion of the roll in flexible rolling, which proved effective by simulation outcomes, and they also found that the vertical velocity of the roll was significantly impacted by the length of the transition zone of flexibly rolled blank and the diameter and rotation speed of the roll. Liu et al. [7-9] set up formulas for calculating parameters, such as roll force, forward slip and horizontal velocity in flexible rolling, and performed finite element simulations and experiments to further verify the theoretical calculations. This technology had proved feasible and would become a new growth point in rolling area.

Additionally, exploration of flexible rolling at the microscale has currently begun to follow the technological development towards miniaturisation of engineering systems and devices. For instance, Qu et al. [10, 11] numerically modelled the flexible rolling process with initial workpiece thickness decreased to the micrometer range. They concentrated on looking into the springback of the workpiece along the thickness direction with consideration of the microstructure of the material. Nonetheless, few papers report the tribological feature as well as surface roughness transfer in this novel rolling technology. Most discussion about these issues has taken place in the field of conventional hot and cold rolling. Jiang et al. [12] established a finite element model to probe the effects of the flow stress of the oxide scale, the friction coefficients at the roll-scale and the scale-steel interfaces, the rolling speed and the temperature on the final surface roughness in hot strip rolling. The simulation results exhibited good agreement with the calculated ones, showing that the developed model was applicable in hot strip mills. Tang et al. [13] also

[a] Corresponding author: jiang@uow.edu.au

simulated the evolution of surface roughness during hot strip rolling with oxide scale. They revealed that both the scale layer thickness and the surface roughness of work roll would affect the final surface roughness significantly, while the impact of asperity wavelength on the final roughness was not significant. Le and Sutcliffe [14] identified the way, in which different wavelengths of roughness behave by analysis of the spectrum of the surface roughness, utilising the fast Fourier transform method and quantified the area of the micro-pits with roll marks subtracted off via a newly developed image analysis technique in cold rolling of aluminium alloy. Some researchers have also studied the surface roughness transfer in the specific case of skin pass rolling. Kijima [15] experimentally and analytically investigated the influence of roll radius on surface roughness transfer in skin pass rolling of steel strip, taking into account the elastic deformation and surface roughness of the roll. The results indicated that the skin pass rolling related to roughened rolls was appropriately simulated with large radius rolls, and the peak pressure from vertical compression was the most significant parameter for roughness transfer under this condition. Afterwards, Kijima [16] conducted experiments to examine the effect of lubrication on surface roughness transfer in skin pass rolling with roughened rolls. It was pointed out that the rolled surface roughness decreased to a larger degree at a larger rolling force with lubrication than that with the dry condition, which was not clearly observed in the case of the small roll due to a short contact length in the roll bite. Kainz et al. [17] investigated and analysed the surface roughness transfer characteristics for both wet and dry skin pass rolling of hot-dip galvanised steel strips by utilising the finite element method. The simulation outcomes clarified that the basic transfer mechanisms of the 3D elasto-plastic forming processes were a combination of penetration processes and reverse extrusion phenomena. Moreover, the asperity flattening, another major mechanisms of contact interactions in metal forming processes, has been widely studied. Sadowski and Stupkiewicz [18] constituted a micromechanical framework to enquire into the impact of friction on asperity flattening in the presence of bulk deformation. Their findings showed that the friction promoted asperity flattening at high flattening rates, while an anomalous regime existed in which the friction slowed down asperity flattening when the flattening rate was low compared to the bulk deformation rate. Lu et al. [19] did experimental and theoretical work on the asperity flattening process under large bulk strain. They used specimens with two types of model asperities and the results disclosed that the pressure, bulk strain and asperity angles had effects on the asperity flattening and the ridge-shaped asperity exhibited a higher contact ratio than the pyramid-shaped asperity. Sutcliffe [20] analysed the flattening of random rough surfaces on a workpiece undergoing macroscopic deformation with a model of the surface composed of long and short wavelength components, which differed by a factor of 10. The asperities were found to be flattened at a rate depending on the ratio of the initial root mean square amplitudes of both wavelength components, and the flattening behaviour of the long wavelength asperities became dominant when the amplitude of the long wavelength asperities was much bigger than that of the short wavelength asperities. Consequently, it can be seen from the literatures that the investigation of the surface asperity flattening as well as the evolutionary changes of the surface in the particular case of flexible rolling is still a blank field, particularly when the workpiece is less than a millimeter in thickness.

In this study, a 3D micro flexible rolling finite element model is set up to analyse the surface asperity deformation and the surface roughness transfer of a 250 μm thick workpiece with reductions of 20 to 50%. Four kinds of initial workpiece surface roughness are separately applied to the workpiece, and how the friction coefficient, rolling speed and roll gap adjusting speed influence the final surface roughness are evaluated for each sort of initial surface roughness.

2 Analysis of surface roughness and deformation of surface asperity

The actual rough surface of the workpiece is approximated as an ideal rough surface in this investigation, which comprises a series of convex and concave surfaces of identical radii of curvature along rolling direction, as shown in Figure 1.

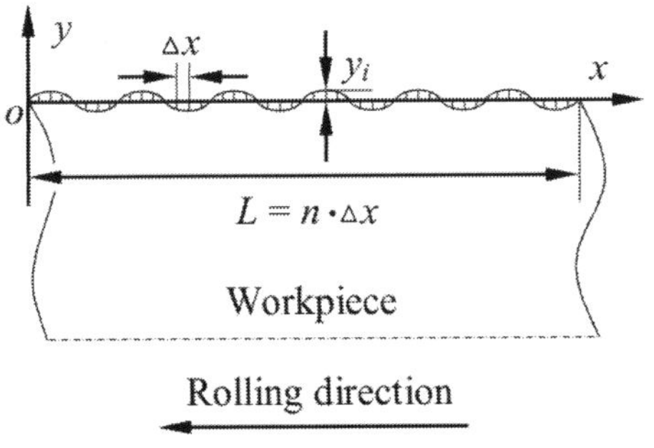

Figure 1. Idealised surface roughness model for the workpiece

Consider the workpiece segment of length L, which is divided into n subsegments by (n-1) points evenly spaced on this workpiece segment. The central line average roughness R_a may be calculated as [21]:

$$R_a = \frac{1}{n}\sum_{i=1}^{n}|y_i| \qquad (1)$$

where y_i is the distance between the curve and the x-axis which represents the central line. The R_a value can be computed with sufficient accuracy if n is large enough.

Figure 2 illustrates the surface asperity deformation mechanism during micro flexible rolling process. In the interaction between the roll and the workpiece, the roll surface is assumed to be smooth and rigid, and the workpiece asperities are flattened on account of the combined action of normal and tangential contact forces [22], namely both convex and concave surfaces get flatter as their radii of curvature become larger, as depicted in Figure 2. The final surface roughness profile depends on all deformations of the asperities in mixed characteristic, i.e. the asperities may be stretched or compressed in disparate directions, and the surface roughness transfer behaviour can be quantitatively estimated by the difference in R_a value before and after flexible rolling.

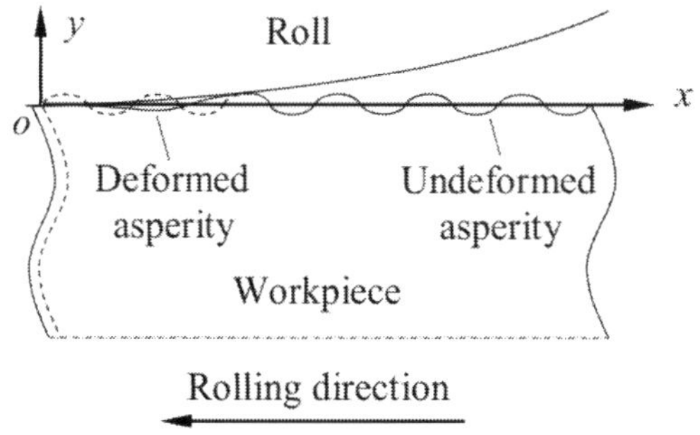

Figure 2. Surface asperity deformation in micro flexible rolling

3 FEM Simulation conditions

3.1 Geometric modelling and mesh generation

A 3D half model symmetric about the z-x plane has been constructed in ABAQUS/CAE, which consists of the smooth-surfaced upper roll of diameter 4.2 mm and length 6 mm, and the top half of the rough-surfaced workpiece with dimensions 6.5×1.5×0.25 mm in the rolling (x), transverse (z) and thickness (y) directions respectively, as displayed in Figure 3.

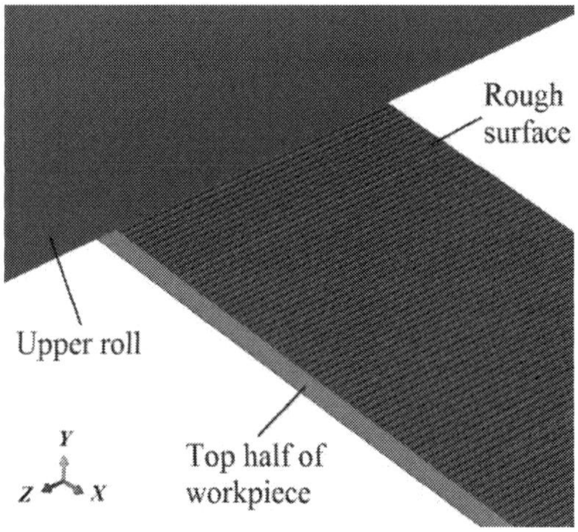

Figure 3. Geometric model of micro flexible rolling considering workpiece surface roughness

The roll was regarded as a rigid body, adopting 4-node 3D bilinear rigid quadrilateral elements (R3D4) with uniform mesh size of 0.05 mm, in order to eliminate the effect of its elastic deformation on the deformation pattern of the workpiece asperities. Whereas the workpiece was set as an elastoplastic body and meshed with 10-node modified quadratic tetrahedron elements (C3D10M), as these elements provide a better approximation of the waveform on the workpiece surface, as exposed in Figure 4. The element size of the workpiece matrix was selected to be 0.02 mm, which moderately decreased to a much finer order of 0.005 mm for the workpiece surface layer, so that penetration can be avoided in the zone of contact between the roll and the workpiece.

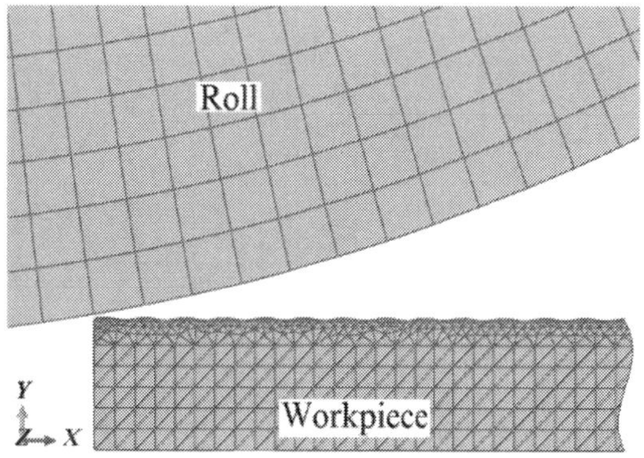

Figure 4. Mesh distributions of the micro flexible rolling model

3.2 Steps and boundary conditions

The simulation of micro flexible rolling process employed the implicit method and involved four general, static analysis steps, which are as follows: initial contact, 20% reduction, thickness transition and 50% reduction.

At the first stage, the workpiece was made to move in the minus x direction to come into contact with the roll. After that the rolling phase with 20% reduction began and the workpiece continued forward movement with the aid of friction force between the two objects. When entering the stage of thickness transition, the roll was bestowed with a speed in the minus y direction to obtain the designed profile. This speed was suppressed immediately after reduction of 50% was reached, which also proclaimed the commencement of the rolling phase with 50% reduction.

The boundary conditions for each step are listed in Table 1, where A_u is the geometric centre of the upper roll, α is the middle surface of the workpiece to the thickness direction, v_x, v_y and v_z are the velocity components in directions x, y and z, respectively, ω_x, ω_y and ω_z are the angular velocity components of the roll rotating around the axes x, y and z, separately, Δh is the reduction amount, D is the diameter of the roll and l is the length of the zone of contact between the roll and the workpiece.

3.3 Mechanical properties and constitutive model for material

Table 1. Boundary conditions for different analysis steps

Step	Position	Boundary conditions
Initial contact	A_u	$v_x = v_y = v_z = \omega_x = \omega_y = 0$ $\omega_z \neq 0$
	a	$v_y = 0$ $v_x \neq 0$
20% reduction	A_u	$v_x = v_y = v_z = \omega_x = \omega_y = 0$ $\omega_z \neq 0$
	a	$v_y = 0$
Thickness transition	A_u	$v_x = v_z = \omega_x = \omega_y = 0$ $v_y = -\frac{\Delta h \omega_z D}{2l}$ $\omega_z \neq 0$
	a	$v_y = 0$
50% reduction	A_u	$v_x = v_y = v_z = \omega_x = \omega_y = 0$ $\omega_z \neq 0$
	a	$v_y = 0$

The workpiece was assigned with mechanical properties of 304 stainless steel including density (ρ), elastic modulus (E), Poisson's ratio (v), initial yield stress (σ_{s0}) and tangent modulus (E_{TAN}) at differing strain rates, whose values were originated from Ref. [23] and are given in Table 2.

Table 2. Workpiece mechanical properties: (a) 0.0001/s strain rate; (b) 0.01/s strain rate; (c) 0.1/s strain rate

(a)

Material	ρ (kg/m^3)	E (GPa)	v	σ_{s0} (MPa)	E_{TAN} (MPa)
304 stainless steel	8000	193	0.29	190	1549

(b)

Material	ρ (kg/m^3)	E (GPa)	v	σ_{s0} (MPa)	E_{TAN} (MPa)
304 stainless steel	8000	193	0.29	205	1630

(c)

Material	ρ (kg/m^3)	E (GPa)	v	σ_{s0} (MPa)	E_{TAN} (MPa)
304 stainless steel	8000	193	0.29	215	1685

The bilinear true stress-strain constitutive model with isotropic hardening has been utilised to simulate the material behaviour of the workpiece, which is shown in Figure 5 [10]. In regard to isotropic hardening, the yield surface of the von Mises yield condition in stress space grows in size while retaining its shape and position during plastic flow [24].

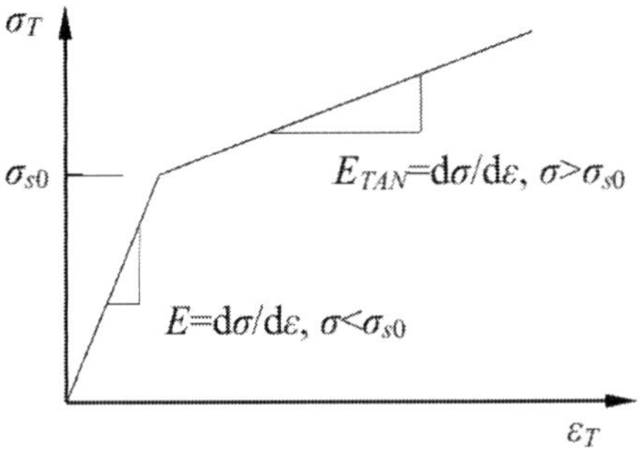

Figure 5. Bilinear isotropic hardening constitutive model for 304 stainless steel

4 Results and discussion

Initial surface roughness of 1, 2, 5 and 10 μm, in conjunction with dissimilar rolling parameters, was successively adopted in the simulation to examine their interactive effects on the flattening process of surface asperities. Figure 6 presents the changes of workpiece surface morphology after flexible rolling, where initial surface roughness is 1 μm, friction coefficient 0.1, rolling speed 20 mm/s and roll gap adjusting speed 2.5 mm/s, respectively. It is observed that the surface asperities are largely flattened and the final surface profile approaches a smooth line.

In addition, a variety of mesh sizes were individually selected for the surface layer of the workpiece in order to optimise the mesh size, which yields satisfactory computational accuracy with least running time. Table 3 exhibits the influence of mesh size on the computational time and accuracy with other simulation parameters identical to those for Figure 6. As can be seen in Table 3, for either reduction the final surface roughness decreases sharply as the mesh size descends to 5 μm, and then makes a very slight decrease as the mesh size keeps continually dropping to 2.5 μm. Although the mesh size of 2.5 μm seems to be able to obtain more accurate results than other mesh sizes, the computational time soars to 226 h which is nearly 7 times longer than that for the mesh size of 5 μm. Based on this evidence, the mesh size of 5 μm has been used in the surface layer of the workpiece in the simulation.

Table 3. Influence of mesh size on computational time and accuracy

Mesh size (μm)	20	10	5	2.5
Number of meshes	8,483	12,986	30,383	140,285
Computational time (h)	14	21	34.5	226
Final surface roughness for 20% reduction (μm)	0.684	0.639	0.603	0.601
Final surface roughness for 50% reduction (μm)	0.597	0.562	0.548	0.547

4.1 Effect of friction coefficient

As the friction coefficient usually ranges from 0.1 to 0.3 in cold rolling without lubrication, typical values of 0.1, 0.2 and 0.3 have been chosen for micro flexible rolling [25]. Figure 7 demonstrates the relationship between the final surface roughness and the friction coefficient for different initial surface roughness with regard to the reductions of 20 and 50%.

It can be seen from Figure 7 that for both reductions, the final surface roughness of the workpiece decreases significantly with an increase of the friction coefficient for each level of initial surface roughness. This is mainly because the friction formulation between the roll and the workpiece employed the penalty method to place constraint to the workpiece [26]. The constraint grows stronger as the friction coefficient turns out to be greater, which leads to a more powerful prevention of workpiece deformation but a quicker occurrence of surface asperity

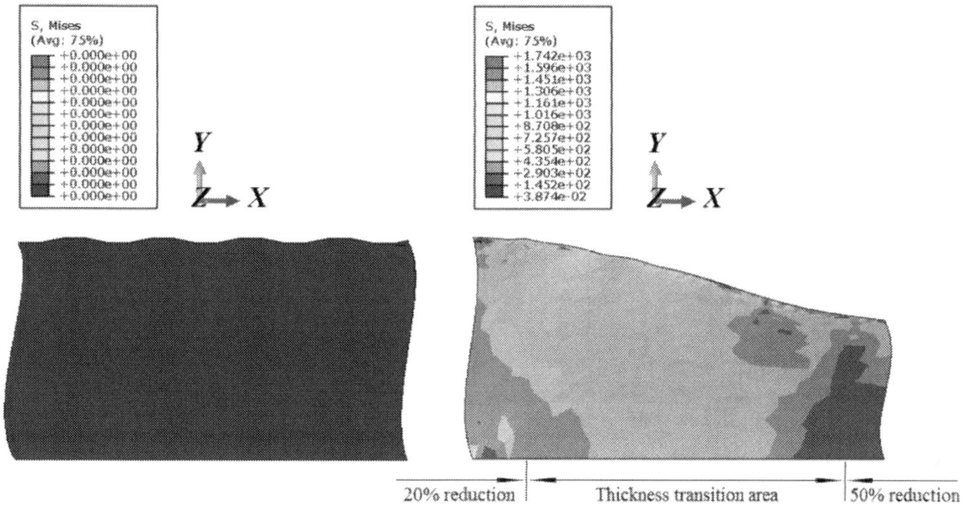

Figure 6. Comparison of workpiece surface morphology before and after flexible rolling

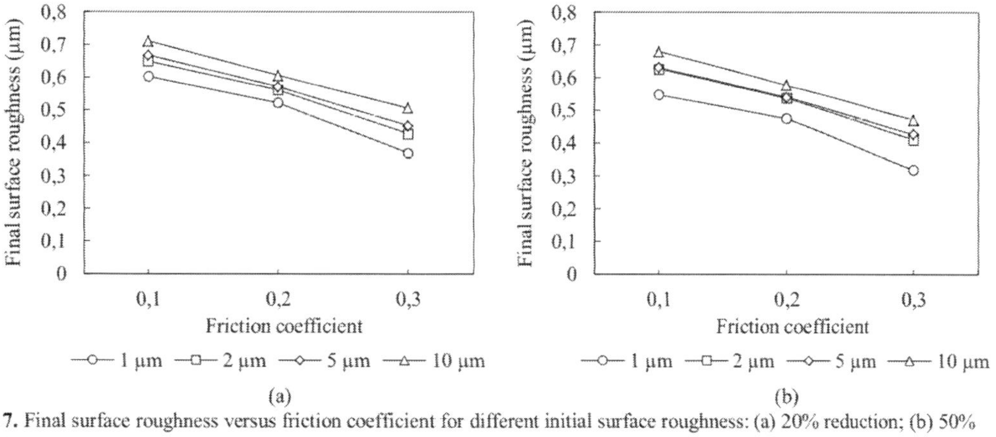

Figure 7. Final surface roughness versus friction coefficient for different initial surface roughness: (a) 20% reduction; (b) 50% reduction

flattening. Therefore an increasing friction coefficient has an accelerating effect on the surface asperity flattening process.

Table 4 shows the comparison between the simulation and experimental results on the effect of friction coefficient on final surface roughness. The experimental data are approximate values which were originated from Ref. [27], where the initial surface roughness was very close to the value of 0.75 μm, hence simulations with initial surface roughness of 0.75 μm have been performed and the results are utilised in Table 4. It is seen that the simulation results are in agreement with the experimental ones, namely the final surface roughness can be further reduced by increasing the friction coefficient. Another point to be noticed is that the increase in reduction gives a rise to the decrease of the final surface roughness, which is rather obviously reflected in the experimental data.

Table 4. Comparison between simulation and experiment: (a) 20% reduction; (b) 50% reduction

(a)

Friction coefficient	Final surface roughness (μm)	
	Simulation	Experimental
0.1	0.352	0.245
0.3	0.273	0.238
Difference (%)	22.44	2.86

(b)

Friction coefficient	Final surface roughness (μm)	
	Simulation	Experimental
0.1	0.226	0.101
0.3	0.164	0.075
Difference (%)	27.43	25.74

The positive effect of friction coefficient on the flattening of surface asperities can also get convinced by Bay et al.'s work [28, 29], where they argued that the real area of contact increased as the surface asperity flattening process was enhanced by the large friction factor. The exemplary simulation results with initial surface roughness of 10 μm are displayed in Table 5, together with the experimental data appearing in Ref. [29]. As is seen from Table 5 that the final surface roughness may be decreased with increasing friction coefficient/factor, which results in a bigger real area of contact irrespective of the reduction or normal pressure.

Table 5. Effect of friction coefficient/factor on surface asperity flattening: (a) simulation; (b) experiment

(a)

Friction coefficient	Final surface roughness (μm)	
	20% reduction	50% reduction
0.1	0.711	0.680
0.3	0.507	0.471

(b)

Friction factor	Real area of contact	
	1 unit of normal pressure	2 units of normal pressure
0	0.405	0.750
0.3~0.4	0.495	0.850

4.2 Influence of rolling speed

The influence of rolling speed on the final surface roughness is shown in Figure 8, where the friction coefficient is 0.1 and roll gap adjusting speed is 2.5 mm/s for all cases.

As is noticed in Figure 8, the increase in rolling speed can promote the process of surface asperity flattening and acquire a rolled surface with lower surface roughness, no matter what the original surface roughness of the workpiece and the reduction are. This being the case a higher rolling speed means a higher strain rate, which speeds up the progressive shearing process especially in the surface zone, and the applied stress may rise up to a higher level in the surface area under this condition, which eventually results in a larger amount of surface deformation of the workpiece [30].

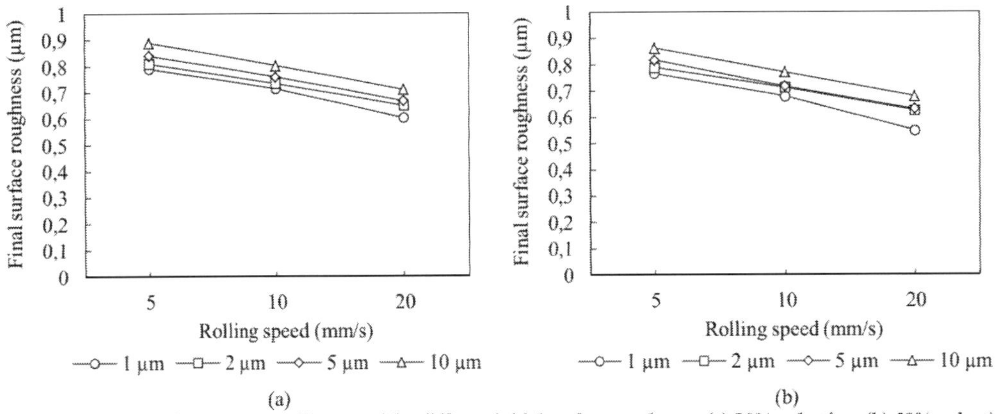

Figure 8. Final surface roughness versus rolling speed for different initial surface roughness: (a) 20% reduction; (b) 50% reduction

 DOI: 10.1051/matecconf/20168004002

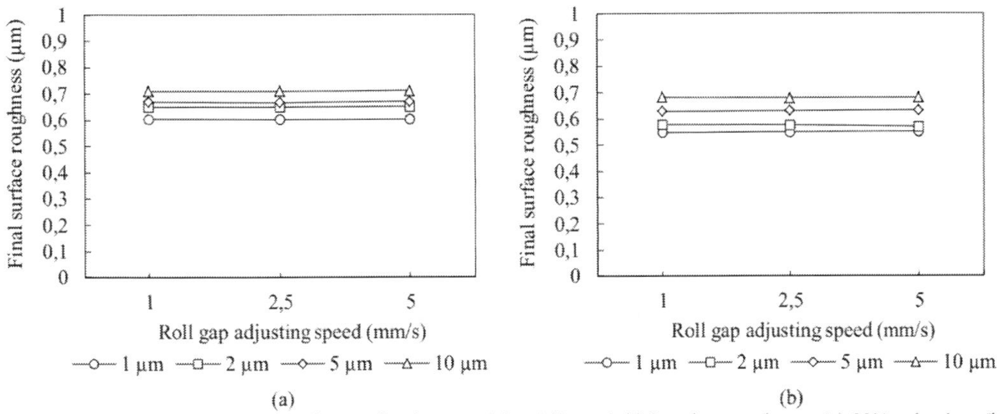

Figure 9. Final surface roughness versus roll gap adjusting speed for different initial surface roughness: (a) 20% reduction; (b) 50% reduction

4.3 Effect of roll gap adjusting speed

Figure 9 shows the relationship between the final surface roughness and the roll gap adjusting speed with respect to the same friction coefficient and rolling speed for all instances. As can be noted in Figure 9, the final surface roughness changes little with increasing roll gap adjusting speed, inasmuch as the variation of roll gap adjusting speed alters only the strain rate in the thickness transition area. In other words, the deformation history within the zones of 20 and 50% reductions hardly changes even though the roll gap adjusting speed varies in different cases. Consequently the roll gap adjusting speed plays an insignificant role in the surface asperity flattening process in the rolling phase with an invariable thickness.

4.4 Effect of reduction and initial surface roughness

It can be found out from Figures 7-9 that the final surface roughness decreases with reduction and also reduces considerably as the value of the initial surface roughness falls down. The reason for this phenomenon is that the bigger reduction causes the asperity to be more depressed, thus the contact area between the roll and workpiece surface increases, making the whole surface layer deform more evenly and the rolled surface smoother. According to the same reasoning, the asperities with smaller initial roughness are easier to be flattened and attain a larger contact area during the rolling process, which brings about better surface finish of the rolled workpiece.

5 Conclusions

This research analyses the surface roughness transformation in flexible rolling of a 250 μm thick workpiece with reduction increasing from 20 to 50% by finite element modelling. The simulation outcomes can generate the following conclusions:

1. Lower surface roughness can be achieved in the rolling phase with thinner thickness or smaller surface asperity in respect of a constant reduction.
2. The increase of friction coefficient accompanies an obvious acceleration of the surface asperity flattening process.
3. The final surface roughness in the rolling phase with a uniform thickness decreases as the rolling speed increases, but shows little change as the roll gap adjusting speed increases, while a higher roll gap adjusting speed leads to a lower level of final surface roughness in the thickness transition area.
4. The developed finite element model is applicable for investigating the workpiece surface roughness transfer in the longitudinal direction during micro flexible rolling process.

Acknowledgements

The authors are greatly obliged to University of Wollongong and Australian Research Council (ARC, Grant no. FT120100432) for their support to this study.

References

1. M. Grüber, R. Kopp, G. Hirt, *60 excellent inventions in metal forming*, 213-218 (2015)
2. O. Engler, C. Schäfer, H.-J. Brinkman, J. Brecht, P. Beiter, K. Nijhof, J. Mater. Process. Technol, **229**, 139-148 (2016)
3. C. Huang, Y. Gan, J.T. Du, C.Z. Chen, Q.J. Chen, Appl. Mech. Mater, **101-102**, 897-900 (2012)
4. J.T. Du, C. Huang, C.Z. Chen, Q.J. Chen, Appl. Mech. Mater, **80-81**, 546-550 (2011)
5. R. Kopp, International Sheet Metal Review, **4**, 20-24 (2005)
6. Y. Zhang, J. Tan, J. Iron Steel Res. Int, **22(8)**, 703-708 (2015)
7. X.H. Liu, J. Iron Steel Res. Int, **18(1)**, 01-07 (2011)
8. G.J. Zhang, X.H. Liu, X.L. Hu, Y. Zhi, J. Iron Steel Res. Int, **20(10)**, 10-16 (2013)
9. G.J. Zhang, X.H. Liu, Adv. Mat. Res, **418-420**, 1232-1236 (2012)

10. F.J. Qu, Z.Y. Jiang, H.N. Lu, J. Appl. Math, **2015** (2015)
11. F.J. Qu, Z.Y. Jiang, H.N. Lu, Int. J. Mech. Sci, **105**, 182-190 (2016)
12. Z.Y. Jiang, J. Tang, W. Sun, A.K. Tieu, D. Wei, Tribo. Int, **43**, 1339-1345 (2010)
13. J.N. Tang, A.K. Tieu, Z.Y. Jiang, Key Eng. Mat, **274-276**, 499-504 (2004)
14. H.R. Le, M.P.F. Sutcliffe, Wear, **244**, 71-78 (2000)
15. H. Kijima, J. Mater. Process. Tech, **214**, 1111-1119 (2014)
16. H. Kijima, J. Mater. Process. Tech, **225**, 1-8 (2015)
17. A. Kainz, D. Paesold, G. Riha, K. Krimpelstätter, K. Zeman, *ABAQUS Austria Users' Coference*, (2005)
18. P. Sadowski, S. Stupkiewicz, Tribol. Int, **43**, 1735-1741 (2010)
19. C. Lu, D. Wei, Z. Jiang, K. Tieu, P. I. Mech. Eng. J-J. Eng, **222**, 271-278 (2008)
20. M. P. F. Sutcliffe, J. Tribol.-T. ASME, **121**, 433-440 (1999)
21. L. Chen, A. Ahadi, J.M. Zhou, J.-E. Ståhl, Modeling and Numerical Simulation of Material Science, **4**, 153-162 (2014)
22. S. Stupkiewicz, Z. Mróz, *Contact mechanics*, 179-186 (2002)
23. B. Antoun, J. Korellis, B. Song, *Proceedings of the 2007 SEM annual conference on experimental & applied mechanics*, (2005)
24. T.D. Dudderar, J. Duffy, J. Appl. Mech.-T. ASME, **34(1)**, 200-206 (1967)
25. H.Y. Zhao, PhD. Thesis, (2008)
26. Ö. Yeniay, Mathematical and Computational Applications, **10(1)**, 45-56 (2005)
27. H.J. Li, PhD. Thesis, (2012)
28. T. Wanheim, N. Bay, A.S. Petersen, Wear, **28**, 251-258 (1974)
29. N. Bay, T. Wanheim, Wear, **38**, 201-209 (1976)
30. H.J. Li, Z.Y. Jiang, D.B. Wei, Wear, **301**, 11-18 (2013)

Advanced Wear Simulation for Bulk Metal Forming Processes

Bernd-Arno Behrens [1], Anas Bouguecha [1], Milan Vucetic [1], Alexander Chugreev [1,a]

[1] *Institute of Forming Technology and Machines (IFUM), Leibniz Universität Hannover, 30823 Garbsen, Germany*

Abstract. In the recent decades the finite element method has become an essential tool for the cost-efficient virtual process design in the metal forming sector in order to counter the constantly increasing quality standards, particularly from the automotive industry as well as intensified international competition in the forging industry. An optimized process design taking precise tool wear prediction into account is a way to increase the cost-efficiency of the bulk metal forming processes. The main objective of the work presented in this paper is a modelling algorithm, which allows predicting die wear with respect to a geometry update during the forming simulation. Changes in the contact area caused by geometry update lead to the different die wear distribution. It primarily concerns the die areas, which undergo high thermal and mechanical loads.

1 Introduction

In bulk metal forming industry the tooling and setup costs amount primarily from 5 to 15 % of the total production costs depending on the process type [1]. In the case of precision forging they can even arise up to 35 % [2]. In this context virtual tool life estimation is gaining importance in the recent years. In addition, new challenges in the bulk metal forming industry are high-strength materials, trends for the near net shape production and increasing performance of the forming machines. High strength of the formed material and short cycle times are tightly related to the prevailing contact pressures and thus to the die load and its failure.

Numerous factors including plastic deformation, thermal and mechanical fatigue affect the die service life during sheet [3] and bulk [4] metal forming. Extensive investigations on the tool life in hot forging processes have shown that more than 70% of tool failure is caused by the die wear [5]. In this context precise evaluation of die wear can help to avoid an improper tool design already at early development stages. On the other hand, it can also significantly contribute to improve the current tool geometry and thus the cost-efficiency of the forming process.

2 Theory

2.1 Conventional modelling of die wear

Wear is described as a progressive loss of material from the surface caused by a mechanical action of the opposite contact body [6]. To allow a realistic prediction of the tool wear during metal forming processes, research works have been carried out by several authors to estimate the die wear experimentally as well as numerically.

The most widely used models for die wear prediction are based on the Eq. 1 provided by Archard [7]. Archard has defined the wear volume (W) as a function of load on the contact area (P), relative sliding distance (L), local surface hardness of the wearing material (H) and wear coefficient (k):

$$W = k\frac{PL}{H} \quad (1)$$

Relative movements between the forming tool and the formed material during hot forging occur in combination with thermal loads and high contact pressures acting at the die surface simultaneously. These factors are being variously superimposed during the forming process and lead to a pronounced die wear, which differs by a location.

Doege et al. [8] have used Archard's equation in a commercial finite element code for the analysis of die wear in a precision hot forging process, taking pressure dependent heat transfer into account. Due to complexity of the occurring microstructural and tribological effects, not all process-specific factors have been taken into account. Behrens [9] has improved the Eq. 1 by extending it with a temperature and time dependence of the tool surface hardness (Eq. 2).

$$\dot{w} = k\frac{\sigma_N}{H(\vartheta,t)}v_{rel} \quad \text{on} \quad \partial\Omega^{tool} \quad (2)$$

In this equation the normal velocity of tool wear $\dot{w}$ is defined as a product of normal contact stress σ_N, relative sliding velocity v_{rel} and is inverse proportional to the current die surface hardness $H(\vartheta,t)$. Through

[a] Corresponding author: chugreev@ifum.uni-hannover.de

DOI: 10.1051/matecconf/20168004003

adjusting the wear coefficient k in the equation (2) the results have shown a good qualitative agreement with experimental data for different forming processes, e. g. for hot extrusion [10] and hot forging [9]. In an earlier work Behrens et al. [11] have extended the equation (Eq. 1), in order to take tangential friction stress component into account. However, determination of the wear coefficient is an extremely time- and cost-consuming procedure. In addition, the coefficient obtained is mostly bound to a particular hot forging process. Thus the classical wear simulation can mainly be used for a qualitative analysis to detect some critical areas or to compare different tool designs [12].

2.1 Advanced Algorithm for Wear Prediction

According to Hoffmann [12], the wear rate does not undergo a linear propagation. In general, wear rate is changing through three different stages: early run-in period with instable wear rate, middle stage with stable low wear rate and the third stage which is characterized by accelerated wear (Fig. 1).

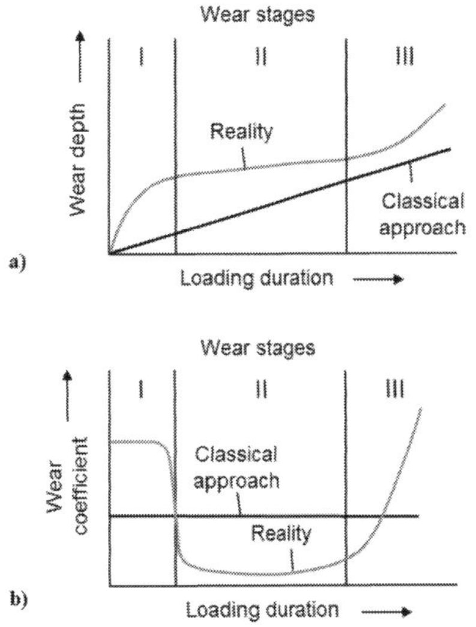

Figure 1. Three stages of wear progression: illustrated on wear coefficient (top) and on wear depth (bottom) (Based on [12]).

In addition, the tool geometry varies over the production time due to wear and plastic deformation, changing the contact area and thus the local stress state in the active contact zone. The effect of plastic deformation decreases with increasing amount of forging cycles [5]. Hence, the effect of tool wear is crucial for the tool design especially in high-volume serial production [13]. For this purpose several researches have been carried out to observe the effect of updating the geometry due to the tool wear [14, 15]. Hoffmann et al. [16] have shown that the geometry update scheme significantly affects the die wear prediction in cold sheet metal forming, which then leads to more realistic results. Considering these effects, a more realistic quantitative estimation of the tool wear can be performed.

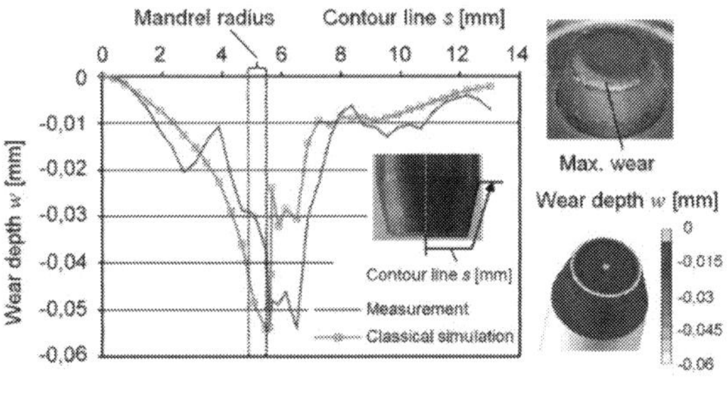

Figure 2. Comparison of measured and simulated wear profiles at the die mandrel after 2,000 forging cycles: with classical simulation algorithm [9].

As shown in Fig. 2, the die wear calculated with the classical approach is concentrated locally at the mandrel radius. However, the real tool material at the sharp die radius gets worn and thus, the contact pressures and the resulting wear gets resolved in its vicinity, making the area with the highest wear at the sharp die radius more extended contrary to the computed one. Besides, worn tool geometry affects the final shape of the formed material. To maintain tight tolerances over the whole tool service life it is also important to have a reliable simulation tool for quantitative die wear prediction. In order to take changes of the worn tool geometry into account an iterative algorithm as shown in Fig. 3 was developed.

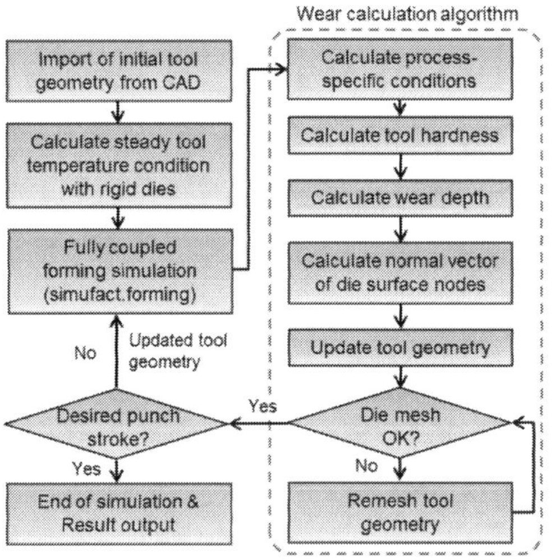

Figure 3. Shortened flowchart of the simulation algorithm involving tool wear estimation with geometry update.

In order to implement the geometry update algorithm, a fully coupled forming simulation intending the calculation of the steady state of tool temperature distribution is to be carried out at first. After that, absolute incremental values of the local wear depth are computed accordingly to Eq. 2 for each tool node which comes in contact with the workpiece. Finally, the affected tool nodes are moved in the normal direction out of the contact body (workpiece) at a distance equal to the calculated tool wear depth increment.

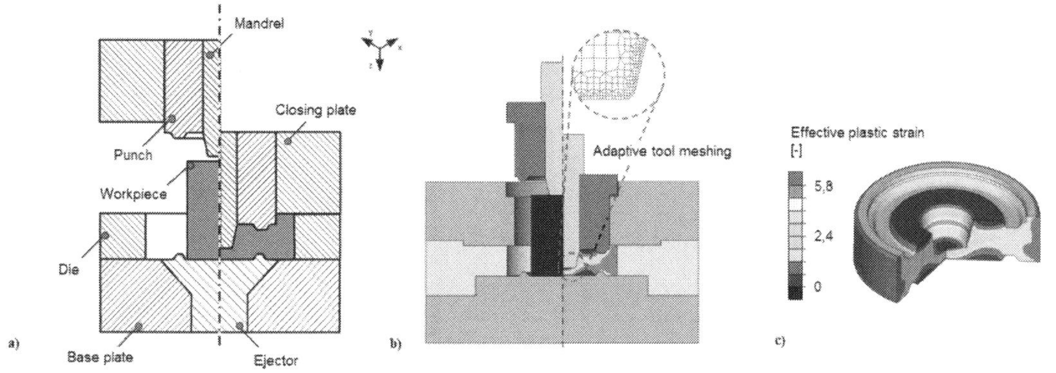

Figure 4. Schematics of the considered precision forging process: (a) tooling system, (b) scheme of the used FE-Model and (c) plastic strain contour in the forged part.

It must be noted, that for the wear prediction with the geometry update scheme it is still vital to employ an experimentally validated wear coefficient *k*, i.e. using the experimental results shown in Fig. 2, in order to calculate the absolute values of the wear depth for the subsequent node displacement algorithm. Due to possible node displacements during the wear calculation algorithm die elements can be distorted or turn inside out, thus an additional remeshing step for tool geometry has been included in the simulation algorithm.

3 Application

In order to verify whether the geometry update algorithm has any significant deviation from the conventional wear prediction scheme, both of these algorithms were employed in the FE simulation of a precision hot forging process (Fig. 4).

The investigated process was analysed with a nonlinear implicit FE-solver of Simufact.forming. Workpiece was modelled with quadrilateral elements having elastic-viscoplastic material behaviour of typical medium-carbon steel (DIN 1.0503). The forming temperature of the raw part is 1200 °C. Non-uniform initial temperature of the forming tools as considered in [5] was assumed.

The forming tools are modelled as heat conducting deformable bodies which can undergo elastic deformation during the forming process. Surface hardness of the dies is computed based on the nodal temperature and amount of forging cycles, as discussed in [9]. Particular attention needs to be paid to modelling of friction [17] as well as heat transfer [18] behaviour during the hot forging processes. The combined Coulomb-Tresca model with a friction coefficient of 0.1 and the friction shear factor of 0.4 based on the work done by Jeong et al. [19] has been used in the simulation. Since, the authors have already broadly investigated this process in the previous work, the other simulation parameters such as heat transfer coefficients, film coefficients to environment etc. as well as experimental data were extracted from [5]. In order to decrease computation time, the full process was modelled in accordance with the rotational symmetry of the workpiece.

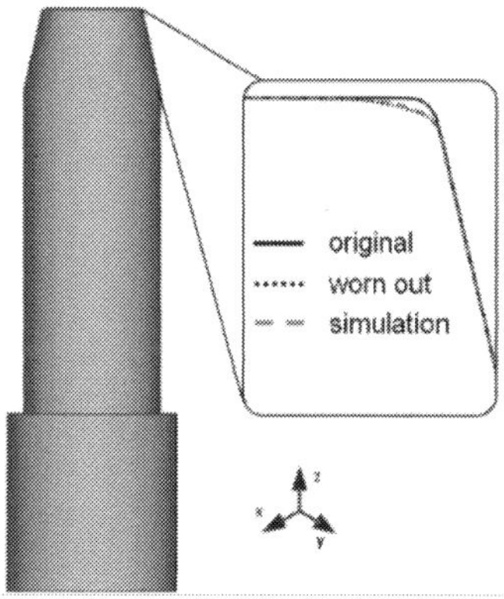

Figure 5. Facing of original, worn out and simulated geometries of the mandrel after 2000 forging cycles: advanced simulation algorithm (based on [5]).

As shown in the Fig. 5, significant deviation in the tool geometry can occur depending on the process-specific conditions as well as on the amount of forging cycles of the respective processes.

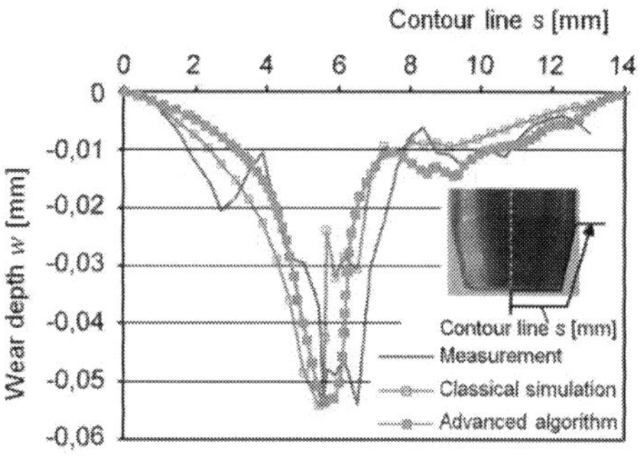

Figure 6. Comparison of measured and simulated wear profiles at the die mandrel after 2,000 forging cycles: with advanced simulation algorithm.

Furthermore, changes in the tool geometry influence the effective contact area and thus the occurring contact stresses. This effect is especially evident on the high loaded die areas, such as the die radius. Therefore, deviations of the tool geometry as well as changes in the contact pressure distribution result in a different tool wear progress compared to the conventional simulation scheme (Fig. 6). The most significant contrast arises at the die areas which undergo high thermal and mechanical loads such as the die radius in the investigated process.

4 Conclusions

Optimized process design taking tool wear into account is an effective way to extend the life of forming dies and thus the economics of the forming process. For this purpose, an advanced simulation algorithm taking tool geometry deviations into account has been proposed in this paper. In addition to the temperature-dependent tool hardness and elastic die deformation, tool geometry update caused by die wear has been integrated in the wear prediction algorithm. Due to the arising geometry changes, tool mesh can become distorted. Thus a special remeshing step for tool mesh has been also incorporated. Furthermore, tool wear of the discussed hot forging process was investigated numerically with the means of the introduced algorithm.

The conducted investigations have shown a strong contrast in comparison to the results computed using the conventional simulation scheme, especially at the high loaded die areas. Particularly, for the case of precision forging it is vital to provide reliable tool wear estimation for high loaded die areas. In the further works authors would like to extend the presented computational model by considering complex stress states, more complicated tool geometries and to calibrate the model based on the experimental results from the other forming processes. Due to the fact that plastic tool deformation can occur at such high forging temperatures, but the use of plastically deformable dies will crucially increase the computation time and make the simulation inefficient for industrial process design, the used method for tool wear prediction should be further developed in order to consider the plastic deformation of tools. Furthermore, progressive nature of the wear coefficient, die fatigue due to the crack formation as well as microstructure phenomena at the tool surface will be the subjects of future works.

Acknowledgments

The authors wish to express sincere thanks to the German Research Foundation (DFG) for supporting the project T12 “Beschichtungen als Verschleißschutz für Werkzeuge der Warmmassivumformung“ within the scope of the collaborative research center 489 (SFB 489). Moreover, the authors are very grateful to the Simufact Engineering GmbH for their kind support and conducive discussions.

References

1. M. Knörr, *Auslegung von Massivumformwerkzeugen gegen Versagen durch Ermüdung,* Springer Verlag (1996)
2. R. Geigert, R. Balbach, Neue Einsatzbereiche durch erhöhte Präzision, *Fertigungstechnisches Kolloquium 85*, pp. 92-103 (1985)
3. E. Doege, Maschinen fur die Warmumformung und Blechformung. *9. Umformtechnisches Kolloquium, pp. 155-167 (1977).*
4. M. Shirgaokar, Ph. D. Thesis (2008)
5. B.-A. Behrens, F. Schaefer, J. Mat. Proc. Tech., **167**, pp. 309-315 (2005)
6. E. Rabinowicz, *Friction and Wear of Materials*, John Wiley and Sons (1995)
7. J. F. Archard, Wear Theory and Mechanism, *Wear Control Handbook,* ASME, New York, pp. 35-80 (1980)
8. E. Doege, H. Naegele, U. Schliephake, *Proceedings of the Institution of Mechanical Engineers*, **280**, pp. 111-119 (1993)
9. B.-A. Behrens, CIRP Annals - **57**, pp. 305-308 (2008)
10. A. Groseclose, C. Choi, J. L. Gonzalez-Mendez, T. Altan, Estimation of Die Stresses and Wear in Warm Forging of Steel Pinion Shafts (2009)
11. B.-A. Behrens, A. Bouguecha, M. Vucetic, A. Chugreev, D. Rosenbusch, ESAFORM 2016 Proceedings (2016, accepted)
12. K. Ersoy, M. Labermeyer, Verschleiß sicher vorhersagen, Blech InForm, 3 (2007)
13. B.-A. Behrens, A. Bouguecha, T. Hadifi, A. Klassen, Key Eng. Mat., **504-506**, pp. 163-168 (2012)
14. M. Oqvist, Wear, **249**, pp. 6-1 (2001)
15. C. Wang, J. Chen, Int. J. Adv. Man. Tech., **69**, pp. 797-803 (2013)
16. H. Hoffmann, C. Hwang, K. Ersoy, CIRP Annals – Man. Tech., **54**, pp. 217-220 (2005)
17. B.-A. Behrens, A. Bouguecha, T. Hadifi, J. Mielke, Prod. Eng. Res. Devel., **5**, pp. 621-627 (2011)
18. B.-A. Behrens, A. Bouguecha, I. Lüken, J. Mielke, M. Bistron, Tribology in Hot Forging, Comprehensive Materials Processing, **5**, pp. 211-234 (2010)
19. D. J. Jeong, D. J. Kim, J. H. Kim, B. M. Kim, T. A. Dean, J. Mat. Proc. Tech., **113**, pp. 544-550, (2001).

Comparative analysis of bonding mechanism in solid state metal working processes

G. Buffa[1,a], S. Pellegrino[1], E. Lo Valvo[1], L. Fratini[1]

[1]*University of Palermo, Department of chemical engineering, management, computer, mechanical-DICGIM, Viale delle Scienze 128, 90128 Palermo, Italy, http://www.dicgim.unipa.it*

Abstract. The Piwnik and Plata pressure-time bonding criterion was applied to Friction Stir Welding, Linear Friction Welding, Porthole Extrusion and Roll Bonding. A neural network was set up, trained and used to predict the bonding occurrence starting from the main field variable distributions calculated through specific numerical models developed for each process. The analysis of the results permitted to predict the occurrence of solid bonding and to highlight differences and analogies between the processes in order to obtain sound solid welds.

1 Introduction

Solid state bonding processes as Friction Stir Welding (FSW), Linear Friction Welding (LFW), Porthole Die Extrusion (PDE), and Accumulative Roll Bonding (ARB) are currently widely used in industry [1] due to their ability of not requiring the melting of the material, making them particularly suitable for all those materials difficult to weld with traditional techniques, such as aluminum, magnesium and titanium alloys.

However, often the process design is based on the knowledge of the operator rather than upon scientific knowledge. In fact, solid bonding occurrence can hardly be predicted by FEM models of a given manufacturing processes due to the different physical phenomena taking place at the same time and affecting the onset of solid state welding. The extrusion of hollow or semi-hollow profiles (PDE) is a common industrial process which is often made by using the "porthole die" [2]. During the process the material separates into two or more seams. The two material flows eventually join in the welding chamber, under proper conditions of temperature and pressure. If the appropriate conditions of temperature and pressure are achieved, the welding process takes place in a solid state condition. Welding lines in the longitudinal extruded profiles are determined and the quality of the extruded joints is closely related to the efficiency of the mechanical longitudinal welds [3]. Accumulative Roll Bonding (ARB) is a process in which two sheets metal are overlapped and rolled with thickness reduction of 50 % [4]. The two metal sheets, by passing through the rollers, are welded forming the final laminate, with reduced thickness compared to the original sheets, due to the rolling ratio imposed. Also in this process, solid bonding occurs under proper conditions of temperature and pressure.

Friction Stir Welding (FSW) is a solid state welding process patented by The Welding Institute in Cambridge (UK) in 1991 [5]. Solid welding of the sheet metals is produced by the heat generated by the friction between the tool and the sheets [6]. This process is becoming increasingly important due to the exceptional results obtained for the mechanical junction of materials, such as aluminum alloys, referred to as "not weldable" by conventional welding technologies. A key factor to obtain sound joints is the proper choice of the process parameters, which in turn affects the bonding conditions determining the mechanical resistance of the obtained joints. During the process, the tool rotation speed (R) and feed rate (V), determining the specific thermal contribution conferred to the joint, are combined in a way that an asymmetric metal flow is obtained. In particular, an advancing side and a retreating side are observed: the former being characterized by the "positive" combination of the tool feed rate and of the peripheral tool velocity, the latter having velocity vectors of feed and rotation opposite each other [7].

Proper values of the main field variables, as temperature, strain and strain rate are needed in order to get the final effective bonding [8]. The Linear Friction Welding process (LFW) is a solid-state joining process [9] in which the two pieces are placed in contact by means of axial force and, at the same, a reciprocating linear motion is activated [10]. The formation of a uniform weld bead is obtained due to the heat generated by the friction forces work. The weld bead consists of the plasticized material, part of which is ejected from the interface as flash. When a reciprocating motion occurs under an assigned pressure, a significant amount of heat is produced. Due to the concurrent effect of oscillation and pressure, the material at the interface is forced to flow out of the joint. Part of this plasticized material,

[a] Corresponding author: gianluca.buffa@unipa.it

called flash, is removed during welding. In this way, any pre oxidized surface and other impurities are removed through the flash, allowing intimate contact between the materials and, hence, the creation of the joint.

A few research groups worked on the development of a solid bonding criterion for accumulative roll bonding (ARB) and porthole die extrusion (PDE). The first criterion in literature was proposed by Akeret [11]. In his study, the author proposed a simple formulation taking into account just the maximum value of the contact pressure. Due to its extreme simplicity, the criterion was widely used in the past years. More recently, Azushima [12] successfully used this approach in the prediction of solid bonding in ARB of different metals. Unfortunately, due to its extreme simplicity, this criterion is not suitable for processes characterized by a more complex mechanics, namely PDE, FSW and LFW. Piwnik and Plata [13] proposed an enhanced formulation of the welding criterion based on the integral on time of the ratio between the contact pressure and the material flow stress. Due to the variation of the flow stress during the time in which material self-contact occurs, it can be stated that this criterion takes into account the local and instant values of temperature, strain, strain rate and pressure.

A few studies can be found in literature as examples of successful application of the Piwnik and Plata criterion. Ceretti [14] identified the threshold value, as a function of temperature, for aluminum alloy AA6061 during ARB experiments. Then, the authors validated the formulation obtained though PDE tests. The aim of the paper is the development of an integrated numerical tool able to predict the occurrence of solid bonding and the quality of a weld as a function of the technological parameters of the processes. In particular, the FEM models allowed to correlate the process parameters to the main field variables distributions. The latter were used by the NN in order to predict the occurrence of solid bonding. Different process input parameters have been used in order to obtain different qualities of the welds, including sound and not welded joints. A numerical campaign was developed for each process to calculate the main field variables, i.e. temperature, strain, strain rate and stress. A Neural Network (NN) for the four different processes PDE, ARB and FSW and LFW was developed and integrated into the FEM tool. The NN is able to provide, for each point identified of the transverse sections of the joints, a qualitative output indicating occurrence of solid state welding as well as a quantitative output based on the Piwnik and Plata criterion indicating the level of "soundness" of the weld.

2 Experimental data

2.1 Friction stir welding

The effects of the operating parameters have been highlighted on some of the fundamental aspects of the FSW process in order to obtain sound joints and defective joints [15]. The sheets thickness was 2.5 mm. Butt joints were obtained out of aluminum alloys AA6061-T6, 100 mm x 200 mm in dimensions. The sets of geometric and technological parameters were chosen in order to obtain intentionally different quality of welded joints (Table 1). As regards to the technological parameters, fixed tilt angle of the tool of 2 ° was selected.

Table 1 FSW parameters used for the numerical campaign

Parameter	*Value*
Feed rate V [mm/min]	100, 200, 400
Rotational speed R [rpm]	500, 1000
Tilt angle	2°
Tool plunge[mm]	2.2

2.2 Accumulative roll bonding

As far as the Roll Bonding process is regarded, experimental data were taken from the paper by D'Urso et al [16]. In the paper, two sheets were rolled together to a final welded sheet thickness of 10 mm. The tests were carried out at the varying of the initial sheets thickness, in order to obtain different rolling ratio values - ranging from 50 % to 83.3 % - and of sheets temperature - ranging from 300 °C to 530 °C (table 2).

Table 2 Roll Bonding parameters used

thickness [mm]	*Threshold [°C]*	*Temperature [°C]*
12	520	530, 510, 490, 470.
13	490	530, 510, 490, 470, 450, 430.
14	420	470, 450, 430, 410, 390, 360.
15	380	430, 410, 390, 360, 330, 300.
16	340	410, 390, 360, 330, 300.
18	320	390, 360, 330, 300.
20	300	360, 330, 300.

2.3 Porthole die extrusion

Experimental data for the PDE process were taken from the paper by Ceretti et al. [17]. In particular, PDE tests were conducted with varying Welding Chamber Height (WCH) and Rib Thickness (RT) on AA6061 aluminum alloy. Other technological and geometrical parameters were kept constant (table 3). The weld limit was experimentally obtained and numerical simulations were run for each case studying in order to highlight the process conditions resulting in effective bonding or in defective parts.

Table 3 PDE parameters used for the numerical campaign

Parameter	*Value*
Welding chamber width [mm]	140
Hole Width [mm]	30
Welding chamber height [mm]	30, 50
Rib Thickness [mm]	30, 60
Billet temperature [°C]	480
Dies Temperature [°C]	450
Punch Speed [mm/s]	8

2.4 Linear friction welding

LFW joints were obtained out of AA6061-T6 aluminum alloy specimens. In order to develop the experimental campaign, a previously in-house designed and built LFW machine was equipped with measuring sensors [18]. The contact surface of the utilized specimens was 10 x 7 mm^2 [19]. The tests were carried out using varying oscillation and contact pressure. In particular, three values of oscillation frequency and five values of contact pressure were selected. Process time and oscillation amplitude were kept constant for all the tests. In this way, a total of 15 different welding configurations were investigated (Table 4).

Table 4 LFW parameters used for the numerical campaign

Parameter	*Value*
Frequency f [Hz]	36, 45, 58
Pressure p [MPa]	20, 30, 40, 50, 60
Amplitude A [mm]	2
Time t [sec]	1.25

3 Numerical Models for the Considered Processes

The commercial FEA software DEFORM-3D™, Lagrangian implicit code designed for metal forming processes, is used to model the four processes. A rigid visco-plastic material model was utilized to simulate the four processes. In particular, a strain, strain rate and temperature dependent flow stress was utilized [20]. For the thermal characteristics of the AA6061 aluminum alloys taken into account, the following values were utilized: thermal conductivity k1 = 166 [N/(s °C)] and thermal capacity c = 2.3 [N/ (mm^2 °C)] taken from literature [15]. No variation of k1 and c with temperature was taken into account; this assumption linearizes the thermal equation and results in better convergence. Additionally, thermal exchange with environment was considered, with coefficient equal to 0.02 [N/(mm-s-°C)].

3.1 Friction stir welding model

A fully 3D FEM model for the FSW process is proposed [21], that is thermo-mechanically coupled and with rigid-viscoplastic material behavior.

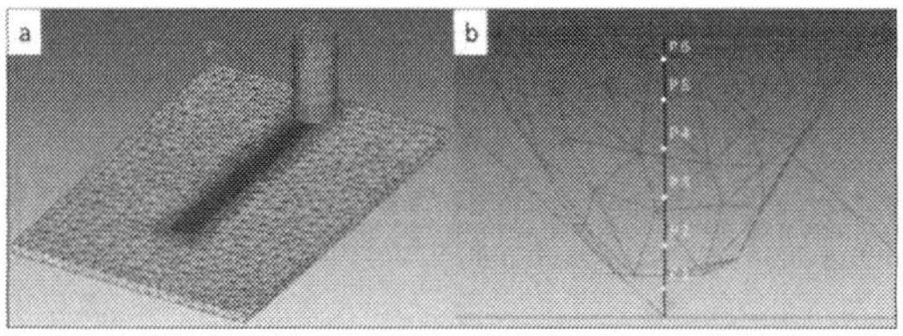

Figure 1 a) FSW model; b) Initial position of the identified points

A unique feature of this model is the representation of sheet seam (abutting edges) as a continuum. This continuum assumption avoids the numerical instabilities resulting from the discontinuities present at the edge of the two sheets. Predicted results are compared with the experimental data to validate this model. The workpiece is modeled as a rigid visco-plastic material, and the welding tool is assumed rigid. This assumption is reasonable as the yield strength of the sheet (conventionally aluminum alloy) is significantly lower than the yield strength of the tool (tool steel or carbide).

This model was first calibrated by comparing calculated force and temperature distribution with experimental results [21]. Then it was used to investigate the distribution of the main field variables in the heat affected zone and the weld nugget. The tool is modeled as a rigid object and meshed, for the thermal analysis only, with about 4,000 tetrahedral elements. The blank is meshed with about 10,000 tetrahedral elements with finer mesh along the weld seam (Fig. 1a). The clamping fixture and the backplate were taken into account by proper boundary conditions on the top and bottom surface of the blank. An adaptive remeshing algorithm is used to overcome convergence problems arising from the severe deformation the material undergoes during the process. A constant time increment of 0.001 s was used. A constant shear friction factor of 0.46 was used for the tool-sheet interface on the basis of a previous experimental thermal characterization and of a numerical sensitivity analysis for the shear friction factor m [22]. At the end of the simulation, the material flow was investigated through the analysis of the nodes movement and the main field variables history that they experience. The "node tracking" option of the software was utilized, highlighting, for a set of nodes initially placed along the sheets separation line in a transverse section, their final position after deformation. The identified points were monitored throughout the process: six points equally spaced along the joint thickness were identified as shown in Fig. 1b. The reference transverse section was taken after 40 mm of weld length, when the process has already entered a steady state and the obtained data are free from transient effects.

3.2 Accumulative roll bonding model

The 2D numerical model for the Roll Bonding process was developed starting from the experimental campaign developed by D'Urso et al. [23].

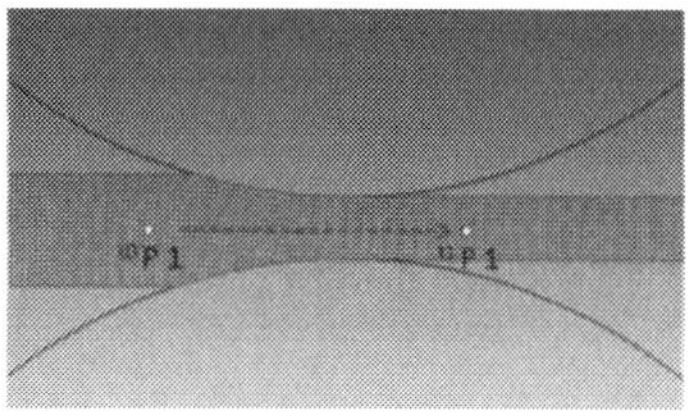

Figure 2 Observation point on the sheets separation line for the roll bonding process

A plane strain approach was selected and, similarly to what was done for the previous model, the two sheets were modeled as a "single block". Two refinement mesh windows were used close to the rolls in order to get an average element size of 0.2 mm in that area. A constant time increment of 0.001 s was used. A shear friction model was adopted with value equal to 0.8. The heat exchange coefficient with environment was set equal to zero because of the isothermal conditions of the experimental tests. An initial temperature corresponding to the considered case study condition was given to the sheet. Finally, due to the stationary conditions of the rolling process, just one point for each case study was taken into account for the analysis of the field variables histories leading to the bonding. The point P1 was placed along the horizontal symmetry line, i.e. on the separation surface between the two sheets, as illustrated in Figure 2 together with the process model.

3.3 Porthole die extrusion model

The numerical model for PDE was developed starting from the results found in the paper [17] by Ceretti et al. as indicated in Table 3. As far as the modeling of the process is regarded, a 2D model was developed taking advantage of the existing symmetry. A refining mesh window was placed close to the rib with minimum element size of about 0.2 mm (Fig. 3 a and b). It is worth noticing that a simplified configuration of the process was used, based on the assumptions adopted in [17], in which the bridges are absent. However, the features associated with welding in PDE are present in the model, making it a relevant yet simple case study for the scope of the paper. A constant time increment of 0.001 s was used. A shear friction model was selected with friction factor equal to 0.4.

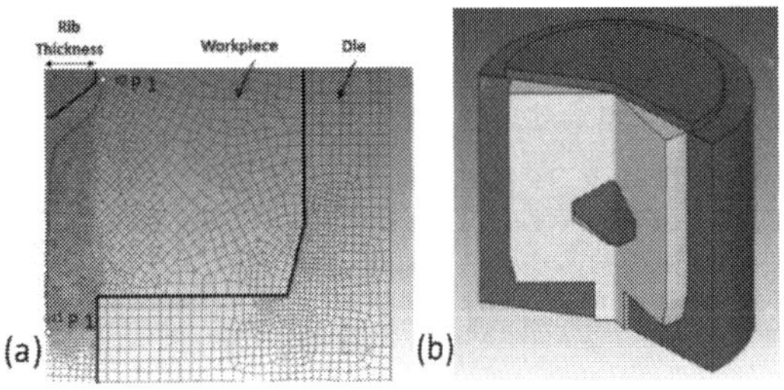

Figure 3 (a)Sketch of the observation point flow for the PDE process and (b) 3D view of the model

Five tracking points have been considered for the monitoring of the field variables evolution. Finally, a slightly different approach was followed for the PDE process. The five considered tracking points were placed along the bonding line, i.e. that vertical symmetry axis of the process, and "tracked back" to their starting position. This approach was followed in order to be sure that the observed points are involved in the bonding phenomenon at the contact interface between the two adjoining metal flows.

3.4 Linear friction welding model

A 3D model was developed for the LFW process. The workpiece, reproducing the top specimen (Fig. 4a,) was modeled as a rigid visco-plastic object. A longitudinal symmetry plane, parallel to the oscillation direction (red transparent plane in figure 4a) was used in order to model just half of the top specimen thus saving CPU time. Further details on model setup can be found in [24]. The actual dimensions of the top specimen were reproduced. The bottom specimen was modeled as a rigid plate 60 mm in width, 2.5 mm in thickness and 90 mm in length. With this assumption, significant saving in CPU time can be achieved with acceptable accuracy in the field variables prediction [24]. This choice was made in order to simplify the contact condition avoiding numerical instabilities due to the deformable-deformable object contact.

The software allows only discretization with tetrahedral elements, (linear shape functions). It was necessary to introduce a denser discretization close to the welding surface where high gradient of the main field variables exist. Additionally, a re-meshing referring volume was identified all along the specimen movement. Furthermore, it was necessary to make preliminary tests, with increasing mesh density, to determine the optimal dimension of the finite elements in these critical areas. Consequently, two windows with different discretization were created. The model has approximately 32,000 elements with the smallest elements characterized by edge of about 0.15 mm. The specimen is constrained against x and y translation for 7/8 of its height for increased stability, while an axial load was imposed on the top surface of the workpiece along the z axis.

A friction window was used to simulate the real physical contact conditions, i.e. the top specimen contact surface varies with time according to the oscillation frequency. Outside the friction window, the constant shear friction factor was set equal to zero. A constant time increment of 0.0001 s was used for all the simulations. No heat exchange was considered between the top and bottom objects because of the symmetry of the process. The oscillation was assigned to the bottom specimen while the pressure was applied to the upper surface of the top specimen.

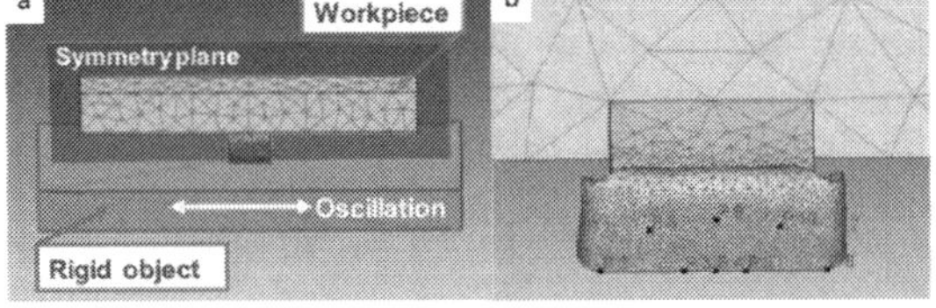

Figure 4 a) LFW model; b) Final position of the identified points

At the end of the simulation the material flow was investigated through the analysis of the nodes position and the main field variables history that they experience. The "point tracking" option of the software was utilized, highlighting, for a set of points initially placed at the contact interface, their final position after deformation.

Ten points were identified and monitored throughout the process as shown in Figure 4b(please notice that Fig 4b refers to a bottom view rotated of about 45° with respect to an horizontal axis)

4 Numerical results

In order to build the training data set for the neural network, the material flow has been studied for each of the numerical simulations implemented, for each processes. In this study the pressure-time (W) welding criterion was utilized:

$$W = \int_{t0}^{t1} \frac{p}{\bar{\sigma}} dt \qquad (1)$$

For the aluminum alloy AA6061 taken into account in to this study the W limit curve was identified and a regression was performed obtaining the following expression as a function of temperature [13]:

$$W^{LIM} = 13425.437T^{-1.41} \qquad (2)$$

Regarding the FSW process, in order to calculate, for each of the observation points highlighted in Fig. 1b, the value of the welding criterion, the material flow occurring during the FSW process must be properly predicted. Buffa et al. [15] proved the effectiveness of the developed model for the prediction of the material flow by comparing the calculated results with experimental measurements of the zig-zag line due to the oxides particles dispersed in the transverse section. Following the approach proposed in the above cited paper, the reference points were tracked during the process. It is worth noticing that the observation time interval begins as the points experience non-zero pressure values, i.e. before the tool reaches the reference transverse section.

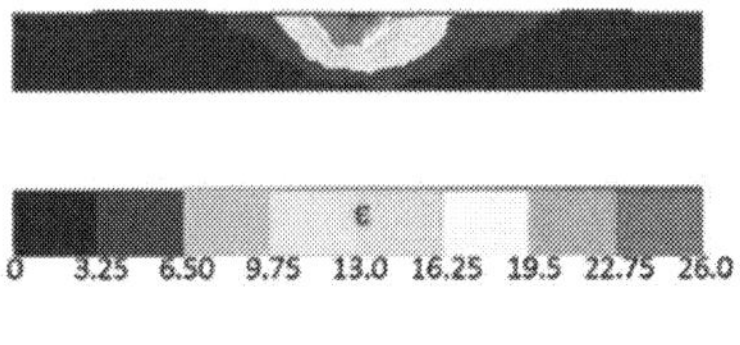

Figure 5 Strain distribution in the specimen

The field variable distribution highlights some of the characteristics of the FSW process: strain is not symmetric with respect to the welding line (Fig. 5). The strain peaks are shifted towards the advancing side. As far as the roll bonding process is regarded, a similar approach used for the FSW process was followed. In particular, the selected point P1 lies on the initial separation line between the sheets (Figure 2, initial time t0) and, during the process, moves horizontally as indicated by the red dotted line in Figure 2. The final position (t1) corresponds to the time increment at which the bonding process is completed. For the roll bonding process the pressure that generates the welding can be easily calculated considering the stress in the vertical direction.

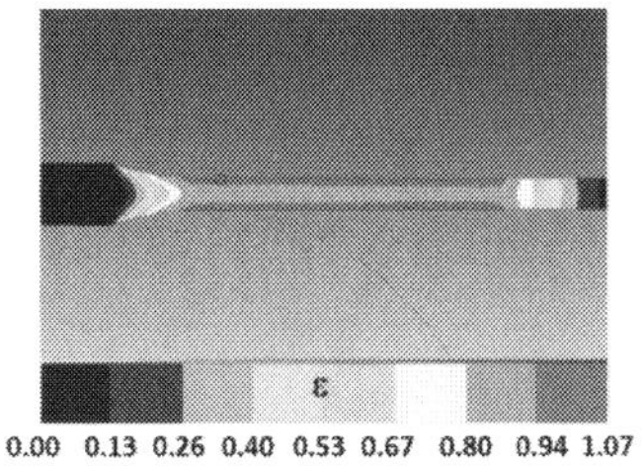

Figure 6 Strain distribution in the specimen

Smaller strain values are observed, the similar trends of the observed field variables with respect to FSW show that the two techniques, characterized by totally different process mechanics, induce the same solicitations in the material in order to obtain the solid bonding. A slightly different approach was followed for the PDE process. The five considered tracking points were placed along the bonding line, i.e. that vertical symmetry axis of the process, and "tracked back" to their starting position. This approach was followed in order to be sure that the observed points are involved in the bonding phenomenon at the contact interface between the two adjoining metal flows. Fig. 7a shows the strain distribution during the process. It is worth noticing that a peak of strain is found in the die corner. As it is shown in Fig. 7b, this is due to a flow vortex taking place in this area due to flat geometry of the bottom die surface. Additionally, it should be observed that some slave penetration in the master surface occurs in this area. This numerical issue contributes to slightly decrease the actual pressure in the welding chamber.

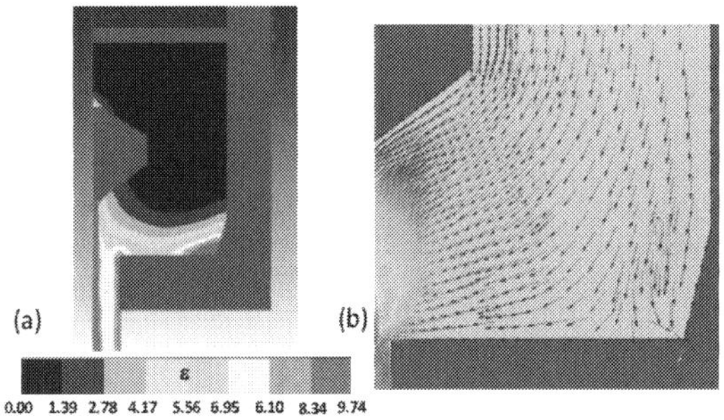

Figure 7 (a) strain distribution in the specimen and (b) velocity vectors

Finally, regarding the LFW process, a time step of 0.01 s was selected to extract the data for each of the identified points (Fig. 4b) and calculate the sum indicated in Eq. 1. In order to build the training data set for the neural network, the value of the welding criterion was calculated for each of the observation points and the final position of the observation points was highlighted (Fig. 4b).

In this way, 150 values were obtained. As already mentioned, due to the relatively simple material flow, ten tracking point for each process condition were utilized. In the linear friction welding process the pressure p generating the solid bonding (Eq. 1) can be easily calculated considering the actual contact pressure at interface.

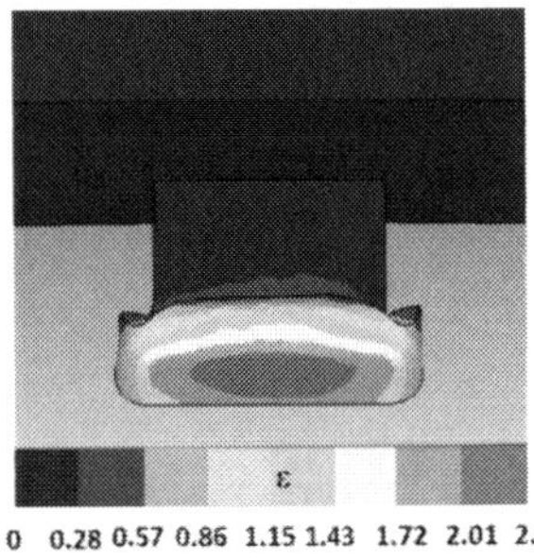

Figure 8 Strain distribution in the specimen

Figure 8 shows the strain distribution in the specimen, at the end of the process. Maximum strain is reached at the center of the contact inteterface. The field variable distribution highlight some of the characteristics of the linear friction welding process: the strain is higher at the welding surface. Two main zones can be identified regarding the effective strain: the zone still in in contact with the bottom specimen experience very high strain values; the zones of the material belonging to the flash, which have been in contact with the bottom specimen for a smaller time, show lower values.

5 Neural Network set up

Based on these observations, the development of the neural network has been evaluated as the most effective for the prediction of the bonding phenomenon. The neural network was designed with the aim to get two separated outputs: a qualitative output, indicating if the bonding process for the considered observation point occurred; a quantitative output, indicating the level of "soundness" of the bonding. In order to build the latter indicator, a parameter Q was defined starting from the Piwnik and Plata parameter W for each observed point i:

$$Q_i = \frac{W_i}{W_i^{LIM}} \qquad (3)$$

Where

$$W = \int_{t0}^{t1} \frac{p}{\vartheta} dt \cong \sum_j \frac{p_j}{\vartheta_j} \Delta t_j \qquad (4)$$

W^{LIM} is the critical value of W that must be reached in order to obtain the solid bonding (eq. 2). It is worthy of notice that a summation was used instead of the integral due to the discrete time increments j used for the simulations. The training data set was build calculating an average value of the considered field variable for a specific tracking point and a given process, according to the following:

$$T_i = \frac{\int_{t0}^{t1} T dt}{t1-t0} \cong \sum_j \frac{T_j \Delta t_j}{t1-t0} \qquad (5)$$

$$p_i = \frac{\int_{t0}^{t1} p dt}{t1-t0} \cong \sum_j \frac{p_j \Delta t_j}{t1-t0} \qquad (6)$$

$$\varepsilon_i = \frac{\int_{t0}^{t1} \varepsilon dt}{t1-t0} \cong \sum_j \frac{\varepsilon_j \Delta t_j}{t1-t0} \qquad (7)$$

$$\dot{\varepsilon}_i = \frac{\int_{t0}^{t1} \dot{\varepsilon} dt}{t1-t0} \cong \sum_j \frac{\dot{\varepsilon}_j \Delta t_j}{t1-t0} \qquad (8)$$

In this way, 36 data were derived from FSW (6 points, 6 process conditions), 34 from ARB (one point, 34 process conditions), 20 from PDE (5 points, 4 process conditions) and 150 from LFW, corresponding to 15 process conditions and 10 points for each process. The network architecture consists of 5 hidden layers: the input layer is characterized by 4 neurons corresponding to the average values of temperature, strain, strain rate and pressure as calculated in equations 5-8. Three hidden layers were used, with 5, 4 and 3 neurons, respectively. Finally, an output layer, with 2 neurons corresponding to the qualitative "welded-not welded" output and the quantitative Q parameter is found. All the data were normalized to assume values between -1 and 1. Each layer was fully connected to the next and, according to the back-propagation rule, the weights of the connections linking a neuron belonging to a certain layer to a neuron belonging to the next were adjusted in the learning stage with the aim to minimize the error between the desired output and the calculated one. The topology of the utilized network, reported in figure 9, was determined on the basis of an optimization procedure aimed to improve the network performances.

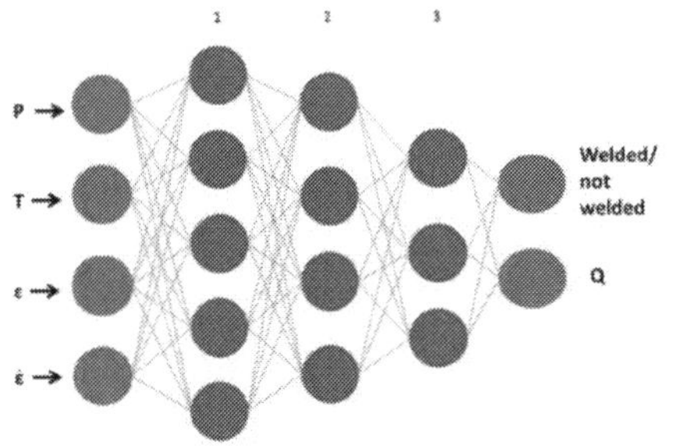

Figure 9 Architecture of the Utilized Neural Network

As far as the training is regarded, it was developed providing to the network the input data coming from the simulations, as previously described; target data came from the experimental observations (welded/not-welded qualitative output) and combined experimental and numerical results (Q parameter). About 7 % of the available input and target data, i.e. 16 values, were used for the test of the network. In particular, 4 observation points for each process were selected, including both sound and not effective bonding.

6 Neural Network Results

Table 5 shows the 16 data used for the test of the NN. For each observation point, the correspondent process parameters are given together with the target output. As far as FSW is regarded, points P1 and P2, namely the closest points to the bottom of the joint, have been chosen for different process conditions. As a matter of fact, when insufficient heat is given to a joint, the so called tunnel defect is found at the advancing side close to the bottom tip of the tool pin.

Table 5 Observation points selected for the test of the developed Neural Network

Id	Point	Process	Process Parameters	Welded	Q
1	P 1	PDE	WCH=30mm RT=30mm	N	0.27
2	P 2	PDE	WCH=30mm RT=60mm	N	0.30
3	P 2	PDE	WCH=50mm RT=60mm	Y	2.73
4	P 4	PDE	WCH=50mm RT=60mm	Y	2.68
5	P 1	RB	Si =13mm T=450°C	N	1.04
6	P 1	RB	Si =14mm T=390°C	N	1.04
7	P 1	RB	Si = 15mm T=300°C	N	0.87
8	P 1	RB	Si = 20mm T=330°C	Y	1.62
9	P 1	FSW	R=500rpm v=200mm/min	N	0.77
10	P 1	FSW	R=500rpm v=400mm/min	N	0.29
11	P 2	FSW	R=1000rpm v=100mm/min	Y	3.95
12	P 1	FSW	R=1000rpm v=200mm/min	Y	1.36
13	P7	LFW	36 Hz-20 MPa	N	0.26
14	P 4	LFW	45 Hz-30 MPa	Y	1.98
15	P 8	LFW	45 Hz-30 MPa	Y	1.96
16	P2	LFW	58 Hz-30 MPa	Y	3.27

The two outputs of the developed Neural Network are showed in figures 10 and 11 and compared with experimental results. When the Boolean network output is equal to -1, no solid bonding is obtained. On the contrary, an output equal to 1 indicates that solid bonding was obtained.

As it can be observed, a perfect prediction of the occurrence of the bonding is found for all the case studies. Looking at the Q parameters, i.e. the discrete one, the network shows a satisfactory prediction capability (Fig. 11).

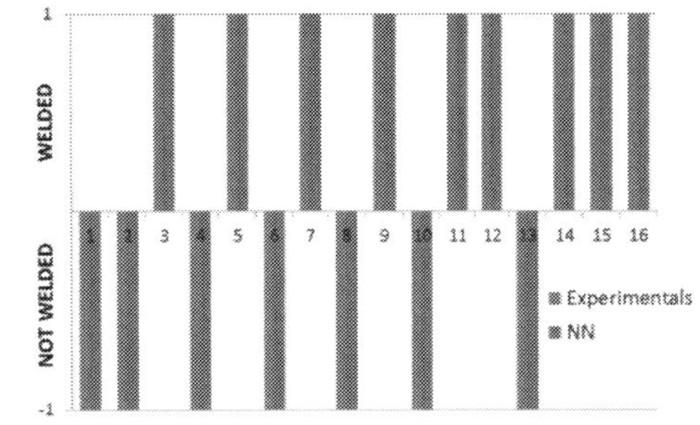

Figure 10 Neural Network response: solid bonding occurrence

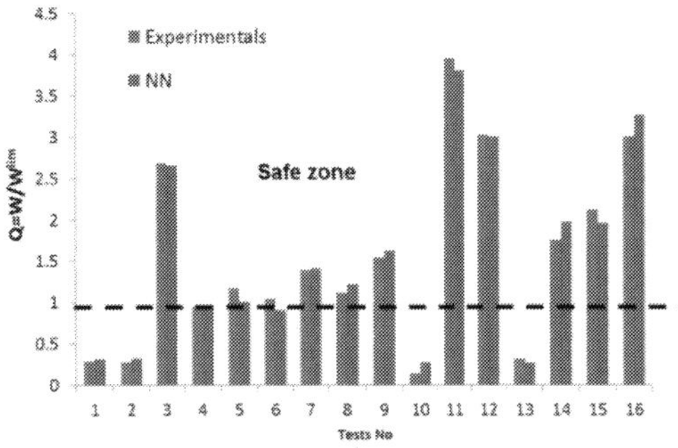

Figure 11 Neural Network response: Q parameter

The maximum difference between the calculated target and the one predicted by the NN is found for ID 5 and 6. It is noted that these points are experimentally not bonded, as correctly predicted by the qualitative network output. The prediction of the network proves to be correct in almost all of the points analyzed, with deviations from the actual values that, for most case studies, do not exceed 5 %. The insufficient heat area can be identified considering all the points for which Q<1 is obtained.

7 Conclusions

A numerical tool aimed to predict the occurrence and the quality of solid bonding in four different manufacturing processes is presented. The results from the FE models were used to set up, train and test a dedicated neural network. The obtained output shows that the developed NN allows the correct identification of the occurrence of solid bonding for all the considered case studies. Additionally, the parameter Q, index of the soundness of the weld, is predicted with maximum error lower than 5%. Regardless of the peculiar process mechanics of a given process, the developed NN can be used as a process design tool in order to select proper process variables leading to the production of sound parts.

References

1. Nicholas, E.D., and Thomas, W.M., International Journal of Materials and Product Technology, 1998, 13, (1-2), pp. 45-55

2. Valberg, H., and Malvik, T., International Journal of Materials and Product Technology, 1994, 9, (4-6), pp. 428-463
3. Liu, G., Zhou, J., and Duszczyk, J., Journal of Materials Processing Technology, 2008, 200, (1-3), pp. 185-198
4. Valiev, R.Z., Materials Science and Engineering A, 1997, 234-236, pp. 59-66
5. Thomas, M.W., Nicholas, E.D., Needham, J.C., Murch, M.G., Templesmith, P., Dawes, C.J.December 1991
6. London, B., Mahoney, M., Bingel, W., Calabrese, M., Bossi, R.H., and Waldron, D., in Editor (Ed.)^(Eds.): 'Book Material flow in friction stir welding monitored with Al-SiC and Al-W composite markers' (2003, edn.), pp. 3-12
7. Liu, G., Murr, L.E., Niou, C.S., McClure, J.C., and Vega, F.R., Scripta Materialia, 1997, 37, (3), pp. 355-361
8. Mishra, R.S., and Ma, Z.Y., Materials Science and Engineering R: Reports, 2005, 50, (1-2), pp. 1-78
9. Crossland, B., Contemp Phys, 1971, 12, (6), pp. 559-574
10. Richter, W., Patent No. DE477084, 1929
11. Akeret, R., Proceedings of the Fifth International Aluminum Extrusion Technology Seminar, 1992, 1, pp. 319-336
12. Azushima, A., Kopp, R., Korhonen, A., Yang, D.Y., Micari, F., Lahoti, G.D., Groche, P., Yanagimoto, J., Tsuji, N., Rosochowski, A., and Yanagida, A., CIRP Annals - Manufacturing Technology, 2008, 57, (2), pp. 716-735
13. Plata, M., and Piwnik, J., 7th International Aluminum Extrusion Technology, 2000, 1, pp. 205-211
14. Ceretti, E., Fratini, L., Gagliardi, F., and Giardini, C., CIRP Annals - Manufacturing Technology, 2009, 58, (1), pp. 259-262
15. Buffa, G., Pellegrino, S., and Fratini, L., Journal of Materials Processing Technology, 2014, 214, (10), pp. 2102-2111
16. D'Urso, G., Longo, M., Giardini, C., and Ceretti, E., Proc. Technical Paper - Society of Manufacturing Engineers2012 pp. Pages
17. Ceretti, E., Mazzoni, L., and Giardini, C., International Journal of Material Forming, 2009, 2, (SUPPL. 1), pp. 101-104
18. Fratini, L., Buffa, G., Campanella, D., and La Spisa, D., Materials and Design, 2012, 40, pp. 285-291
19. Fratini, L., Buffa, G., Cammalleri, M., and Campanella, D., CIRP Annals - Manufacturing Technology, 2013, 62, (1), pp. 295-298
20. Fratini, L., and La Spisa, D., in Editor (Ed.)^(Eds.): 'Book Numerical simulation of linear fiction welding (LFW) processes' (2011, edn.), pp. 1284-1289
21. Buffa, G., Hua, J., Shivpuri, R., and Fratini, L., Materials Science and Engineering A, 2006, 419, (1-2), pp. 389-396
22. Fratini, L., Beccari, S., and Buffa, G., in Editor (Ed.)^(Eds.): 'Book Friction stir welding fem model improvement through inverse thermal characterization' (2005, edn.), pp. 259-266
23. D'Urso, G., Longo, M., Ceretti, E., and Giardini, C., in Editor (Ed.)^(Eds.): 'Book Coupled simulative-experimental procedure for studying the solid state bonding phenomena' (2012, edn.), pp. 181-188
24. Buffa, G., Cammalleri, M., Campanella, D., and Fratini, L., Materials and Design, 2015, 82, pp. 238-246

Heat transfer modeling in asymmetrical sheet rolling of aluminium alloys with ultra high shear strain

Alexander Pesin [a], Denis Pustovoytov

Nosov Magnitogorsk State Technical University, Department of Material processing, 455000, Magnitogorsk, Lenin prospect, 38, Russia

Abstract. Asymmetrical sheet rolling is a method of severe plastic deformation (SPD) for production of aluminium alloys with UFG structure. Prediction of sheet temperature during SPD is important. The temperature of sheet is changed due to the conversion of mechanical work into heat through sliding on contact surfaces and high shear strain. Paper presents the results of FEM simulation of the effect of contact friction, rolling speed and rolls speed ratio on the heating of aluminium sheets during asymmetrical rolling.

1 Introduction

The aluminium alloys have been used extensively in aerospace and automobile industries because of their low density and high specific strength [1]. Conventional strengthening mechanisms applicable to aluminium alloys like solid solution strengthening, work hardening and precipitation hardening have their own limitations [2]. Although these mechanisms cause considerable strengthening, they cannot improve the properties to match steel. Hence, the need arises for novel processing methods and new techniques to improve properties of aluminium alloys. A possible way to further increase strength of aluminium alloys is to form an ultrafine grain (UFG) structure using severe plastic deformation (SPD) methods [3]. Nowadays many variants of SPD techniques are available [4]. Asymmetric rolling process is one of them [5-8]. Contrary to other SPD methods such as high pressure torsion and equal channel angular pressing, the asymmetric rolling allows to provide the possibility for overcoming the limitation of producing ultrafine grained materials with large dimensions due to its continuous feature. Among the asymmetric rolling processes, the differential speed rolling in which the speeds of the upper and lower rolls are different, is considered to be the most effective for achieving ultrafine grained structure of aluminium alloys [9-10].

It is well known that shear strain plays a critical role in the grain refinement of aluminium alloys processed by asymmetric rolling [11-15]. The strength of the material continuously increases with increasing the applied shear strain. However the resulting plastic work is converted into heat energy. Therefore it is necessary to estimate the actual sheet temperature during the asymmetric rolling for a precise control of recovery/recrystallization, and hence of the grain size.

Many works studied the shear strain during asymmetric rolling by using finite element method (FEM). Ji and Park [16] analysed various asymmetric rolling processes by the rigid-viscoplastic FEM. The results of the numerical simulation demonstrated that shear strain is more severe in the material layers, where the circumferential speed or friction coefficient is greater. Ji et al. [17] investigated the deformation mechanics of differential speed rolling with a high-speed ratio between the rolls by rigid-plastic FEM. It was found out that the shear strain takes a great portion of the total effective strain in average. Kim et al. [18] analysed the effect of speed ratio on the development of shear strain and texture during differential speed rolling. FEM simulation results showed that the effective strain accumulated during asymmetric rolling increased with high-speed ratio. It was demonstrated that high-ratio differential speed rolling, where a large shear strain is induced during asymmetric rolling at high speed ratios of ≥ 2, is quite effective for grain refinement. FE analysis of SPD in differential speed rolling was studied in [19]. Effects of the speed ratio, the friction coefficient and the length of deformation zone were investigated by the rigid-plastic finite-element analysis. As a result, mechanisms in development of plastic deformation and an upper limit of the speed ratio were found. Double-pass differential speed rolling was introduced to achieve the symmetric distribution of the effective strain through thickness of the sheet. It was shown that an excessive speed ratio unnecessarily wastes a portion of the power through friction at the interface between the sheet and rolls. This also exerts a tensile stress in the sheet, which could result in fracture.

Influence of process parameters on distribution of shear strain through sheet thickness during asymmetric rolling was investigated by FEM in [20]. It was found that shear strain strongly depends on opposite contact

[a] Corresponding author: pesin@bk.ru

friction forces operated in the deformation zone. Ultra high shear strain through the sheet thickness can be obtained with the following parameters: 1) high thickness reduction per pass (≥50%); 2) rolls speed ratio (≥2) should be equal to thickness reduction per pass; 3) friction coefficient must be very high, so the asymmetric rolling must be carried out without lubrication. The problem with dry cold asymmetric rolling is the high heat generated in the roll gap by contact friction and the plastic deformation of the aluminium alloys. The heat generation is usually around 90-95% of the mechanical work. The temperature rise during the asymmetric rolling can be as large as to increase the workpiece temperature above the recovery temperature or the grain growth temperature. Therefore by creating the UFG structure in aluminium alloys it is important to know the value of deformation heat because of recovering processes which are activated at particular temperature level.

The temperature rise during cold asymmetric rolling depends on the thickness reduction per pass, the friction coefficient, rolling speed and rolls speed ratio. Therefore the asymmetric rolling with high-speed ratio and high contact friction is a process that may be limited by temperature rise of the rolled aluminium sheets. There are many publications on the deformation behaviour during asymmetric rolling by using FEM. However, only a few activities on the analysis of temperature and heat transfer have been reported [21-22]. In the present study, finite element simulation was conducted to analyse the heat transfer and temperature field of the sheet during cold asymmetric rolling of aluminium alloys. In order to better understand both temperature distribution and strain characteristics during the processing of aluminium alloys, FEM simulations were carried out for asymmetric rolling at room temperature under different rolling parameters. The heat transfer analysis presented in this study can be used to optimize the asymmetric rolling process as SPD method to improve microstructure and mechanical properties of aluminium sheets.

2 Research method

A coupled deformation and heat transfer simulation of the asymmetric rolling was carried out using the commercial FEM code DEFORM 3D. FEM simulation leads to reduction of costs in comparison with physical modeling. The geometry model and FE meshing of asymmetric rolling process is shown in Fig. 1. Process simulation was performed at room temperature with taking into account the increment of the metal's temperature due to the conversion of mechanical work into heat.

The governing equation for heat transfer is expressed as [23, 24]:

$$k\nabla^2 T + \dot{q} = \rho C \frac{\partial T}{\partial t} \quad (1)$$

where ρ is the density, C is the specific heat capacity, T is the temperature, t is time, k is the thermal conductivity, and $\dot{q}$ is a heat generation term.

Heat generation in asymmetrical rolling is due to work of plastic deformation and friction. Heat generation due to plastic deformation is given by [23, 24]:

$$\dot{q}_{pw} = \xi \int \bar{\sigma}\dot{\bar{\varepsilon}} dV \quad (2)$$

where ξ is a deformation efficiency term, representing the fraction of the work of deformation converted to heat and is assumed as 0.9, $\bar{\sigma}$ is the flow stress, $\dot{\bar{\varepsilon}}$ is strain rate.

The boundary condition for the roll-sheet contact surface includes friction heating and heat exchange via temperature difference of two objects [23, 24]:

$$\dot{q}_1 = \int_{S_1} \eta f_{S_1} \left| u_{S_1} \right| dS_1 + \int_{S_1} H \Delta T dS_1 \quad (3)$$

where u_{S_1} is the sliding velocity, f_{S_1} is the friction stress, η is the fraction of friction energy absorbed by the object and is assumed as 0.9, H is the lubricant heat transfer coefficient, and ΔT is the temperature difference between two objects.

The boundary condition of the free surface includes convection heat and radiation heat from/to the environment [23, 24]:

$$\dot{q}_2 = \int_{S_2} h_c (T - T_\infty) dS_2 + \int_{S_2} \sigma\varepsilon \left(T^4 - T_\infty^4\right) dS_2 \quad (4)$$

where h_c is the convection heat transfer coefficient, T_∞ is the environment temperature, σ is the Stefan-Boltzmann radiation constant, and ε is the emissivity of the surface.

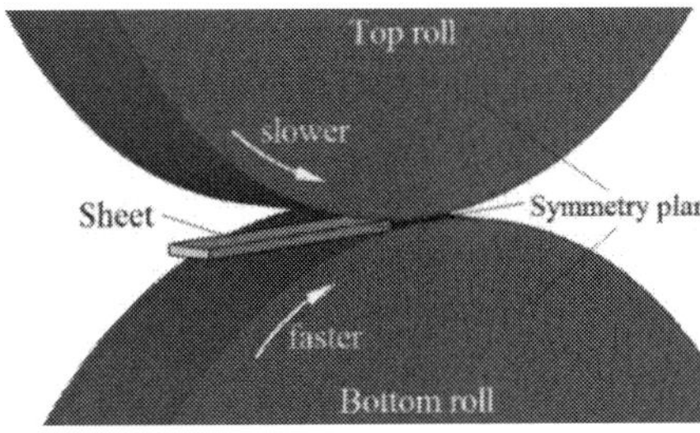

Figure 1. Geometry model and FE meshing of asymmetric rolling process

Pure aluminium (20-250°C, 1-100 s^{-1}), aluminium alloys Al 5083 (20-440°C, 1-100 s^{-1}) and Al 7075 (20-440°C, 1-100 s^{-1}) from DEFORM 3D material library were chosen as a hardened rigid-plastic materials for the sheet. As an example, the stress-strain curves of the materials at 20°C and strain rate 1.0 s^{-1} are shown in Fig. 2.

The thermal constants of pure Al and its alloys Al 5083 and Al 7075 used in FEM simulation are shown in Table 1. The diameters of the rolls were 200 mm, and the rolls were considered as rigid. High-carbon chromium alloy steel AISI D2 from DEFORM 3D material library was chosen as a material for rolls. The simulations were carried out at three different heat transfer coefficients H between rolls and sheet (Table 1). Thermal properties were taken as constants to improve computational efficiency.

DOI: 10.1051/matecconf/20168004005

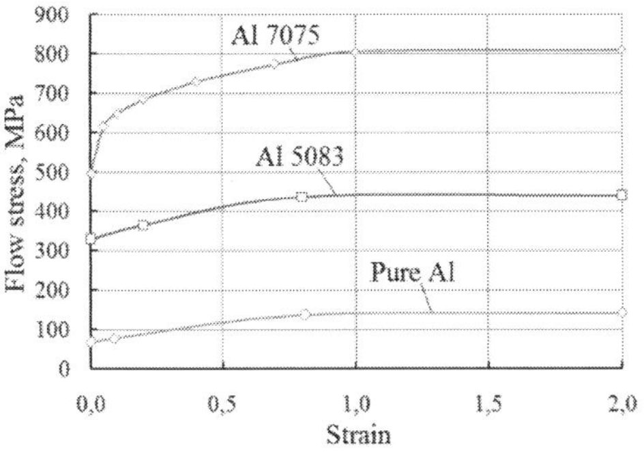

Figure 2. Stress-strain curves of Al alloys used in FEM simulations

Table 1. Properties of materials used in FEM simulation

Properties	Pure Al, Al 5083, Al 7075	AISI D2
Thermal conductivity k, W/(m×K)	180.2	50.71
Emissivity ε	0.05	0.7
Heat capacity C, J/(kg×K)	901.1	485.4
Density ρ, kg/m^3	2700	7850
Heat transfer coefficient H between rolls and sheet, W/(m^2×K)	2000; 11000; 20000	
Heat transfer coefficient h_c between rolls/sheet and air, W/(m^2×K)	20	
Stefan-Boltzmann radiation constant, W/(m^2×K^4)	5.6703×10^{-8}	
Environment temperature T_∞, °C	20	

The simulations were carried out at three different angular speeds of the bottom roll: 1; 5; 10 rad/s. It corresponded to 0.1 m/s, 0.5 m/s and 1.0 m/s of circumferential speed. Speed of the top roll was reduced by 10; 30; 50% during asymmetric rolling. Sheet with initial dimensions of 2.0×20.0×50.0 mm was used for all simulations. The number of initial brick elements was 14220 for the sheet. Rolls were meshed with 123035 tetrahedral elements. An automatic remeshing was used in all the simulations. The boundary conditions were set in order to constrain the symmetry plane of the rolls and sheet along the rolling direction. A Coulomb friction model was used between rolls and sheet:

$$f_{S_1} = \mu p \qquad (5)$$

where f_{S_1} is the friction stress, μ is the friction coefficient, p is the interface pressure between two bodies.

The influence of the friction coefficient (0.1…0.4), rolling speed (0.1…1.0 m/s) and rolls speed ratio (up to 50%) on the temperature field of asymmetrically rolled aluminium sheets were analysed.

3 Simulation results and discussion

The aluminium alloy sheets are heated by friction and plastic deformation work during symmetric and asymmetric cold rolling processes. Friction heat is generated on the interfaces and completely distributed between the sheet and rolls. Plastic work of deformation occurs with heat generation in the interior of the sheet. It is expected that the sheets could be heated differently during symmetric and asymmetric cold rolling processes. So the temperature change during symmetric and asymmetric rolling was analysed and compared. Fig. 2 shows the temperature change of the Al 7075 sheet during symmetric rolling where thickness reduction is 50%, friction coefficient is 0.4 and heat transfer coefficient H=11000 W/(m^2×K). With increasing the rolling speed from 0.1 m/s to 1.0 m/s, the maximum temperature of rolled sheets is increased from 159 °C (Fig. 2, a) to 231 °C (Fig. 2, b).

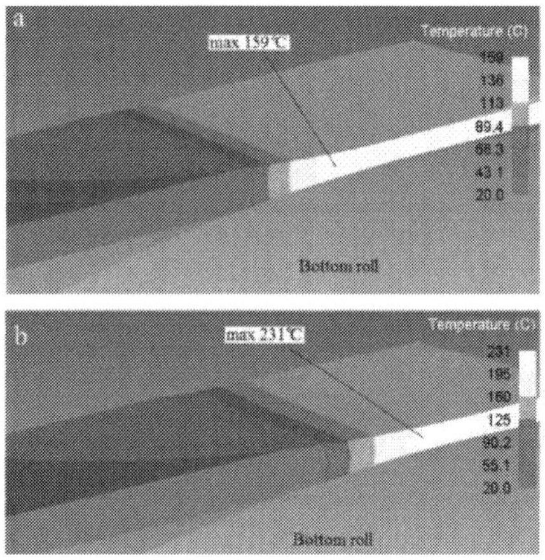

Figure 3. Temperature field in the deformation zone during symmetric rolling: a – rolling speed is 0.1 m/s; b – 1.0 m/s

Fig. 3 shows the temperature change of the Al 7075 sheet during asymmetric rolling where heat transfer coefficient H=11000 W/(m^2×K), thickness reduction is 50%, rolls speed ratio is 50% and friction coefficient is 0.4. It should be noted that the rolling speed is the most important factor that influences the temperature rise of Al sheets during asymmetric rolling. With increasing the rolling speed from 0.1 m/sec to 1.0 m/s, the maximum temperature in rolled Al 7075 sheets increases from 217 °C (Fig. 3, a) to 328 °C (Fig. 3, b). So the temperature rise can be controlled through adjustment of the rolling speed.

Homogeneous temperature distribution through sheet thickness was obtained at the exit of deformation zone in all cases of asymmetric rolling. But temperature distribution along the width of the sheet was inhomogeneous (Fig. 3).

High shear strain through thickness of the sheet can be obtained only with a high value of friction coefficient. But it leads to temperature rise of the rolled aluminium sheets. Decreasing of the contact friction coefficient is beneficial to reduce the temperature in deformation zone

(Fig. 4), but it also leads to significantly decreasing of the shear angle and shear strain during asymmetric rolling.

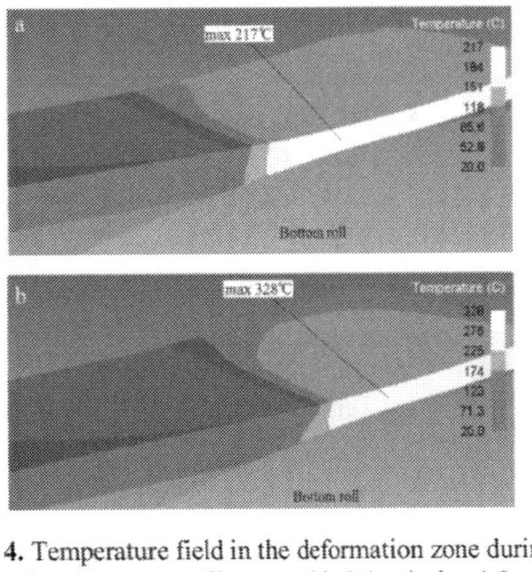

Figure 4. Temperature field in the deformation zone during asymmetric rolling: a – rolling speed is 0.1 m/s; b – 1.0 m/s

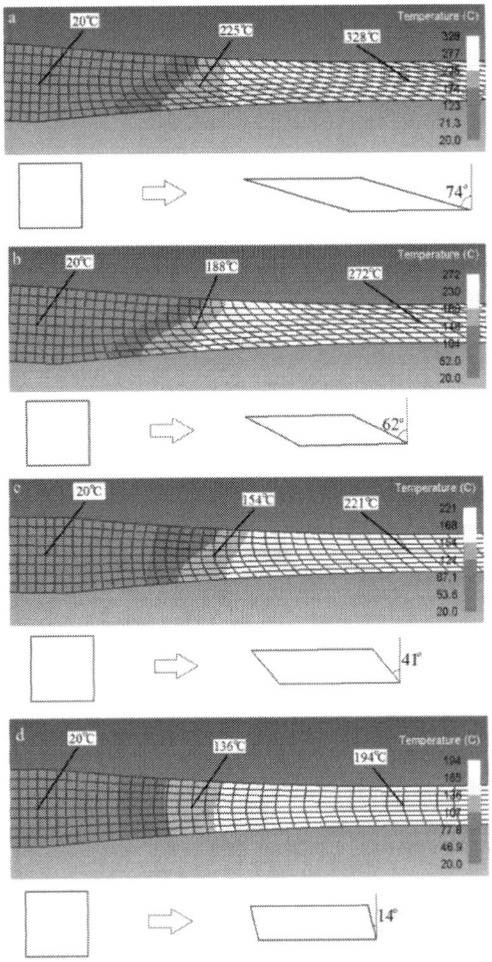

Figure 5. Temperature field and shear angle during asymmetric rolling with different friction coefficient μ =0.4 (a); μ =0.3 (b); μ =0.2 (c); μ =0.1 (d) (Al 7075, thickness reduction is 50%, rolls speed ratio is 50%; rolling speed is 1.0 m/s; heat transfer coefficient is 11000 W/(m^2×K) in all calculation variants).

Contact friction is one of the main factors that affect the shear strain during asymmetric rolling. High friction coefficient increases the opposite contact friction forces which operates in the deformation zone and it creates an additional shear strain. Decreasing of the friction coefficient leads to serious decrease of the effective strain (Fig. 5).

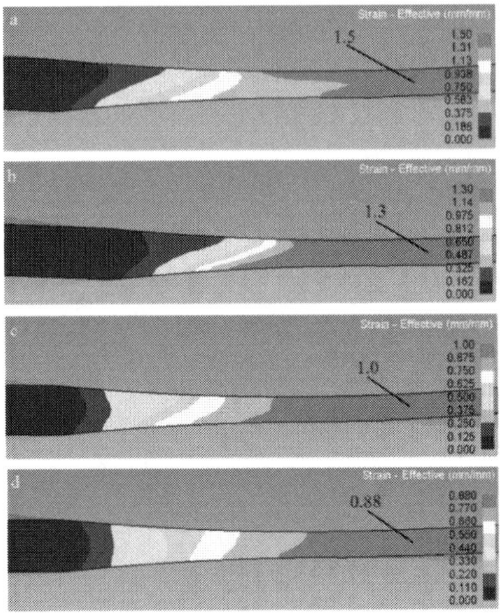

Figure 6. Effective strain during asymmetric rolling with different friction coefficient μ =0.4 (a); μ =0.3 (b); μ =0.2 (c); μ =0.1 (d) (Al 7075, thickness reduction is 50%, rolls speed ratio is 50%; rolling speed is 1.0 m/s; heat transfer coefficient is 11000 W/(m^2×K) in all calculation variants)

Increasing the rolling speed leads to serious increase of the temperature in the deformation zone (Fig. 6). The friction heat is generated on the interfaces and it is distributed between the aluminium sheet and rolls. Increasing the rolling speed leads to decreasing of the contact time. Therefore heat is less intensive transferred to the rolls and to environment.

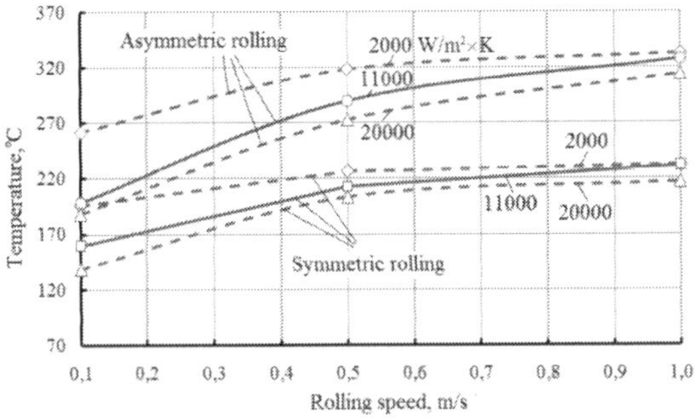

Figure 7. Interrelation of the temperature in the deformation zone and rolling speed during symmetric and asymmetric rolling of Al 7075 with different heat transfer coefficient H between rolls and sheet (thickness reduction is 50%, friction coefficient is 0.4, rolls speed ratio is 50% for asymmetric rolling)

Fig. 7 shows the interrelation of the temperature in the deformation zone and rolls speed ratio during rolling. With increasing the rolls speed ratio to 50%, the temperature of rolled Al 7075 sheets is increased from 230 °C to 328 °C (for H=11000 W/m^2×K). It can be explained by more serious sliding velocity on the contact surfaces. As a result the friction heat is generated.

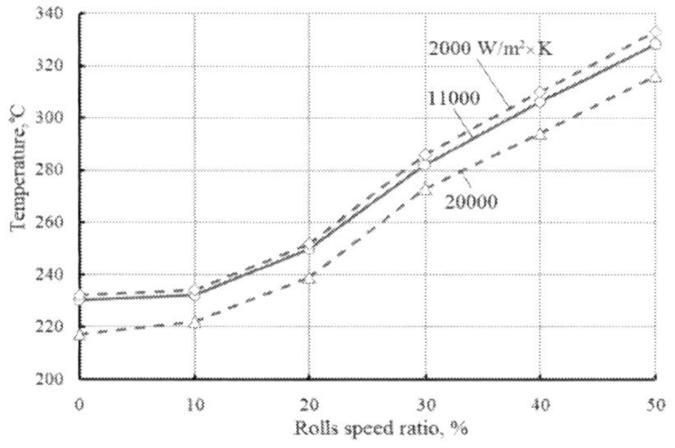

Figure 8. Interrelation of the temperature in the deformation zone and rolls speed ratio (Al 7075, thickness reduction is 50%, friction coefficient is 0.4, rolling speed is 1.0 m/s)

Fig. 8 shows the interrelation of the sliding velocity and friction coefficient during symmetric rolling. Negative sliding velocity corresponds to backward slip zone. Positive sliding velocity corresponds to forward slip zone. Sliding velocity on the contact surfaces has an influence on temperature through frictional heat. On the one hand increasing contact friction coefficient leads to decrease of sliding in the backward slip zone. On the other hand increasing friction coefficient leads to increase of sliding in the forward slip zone. Although sliding in the backward slip zone is more serious.

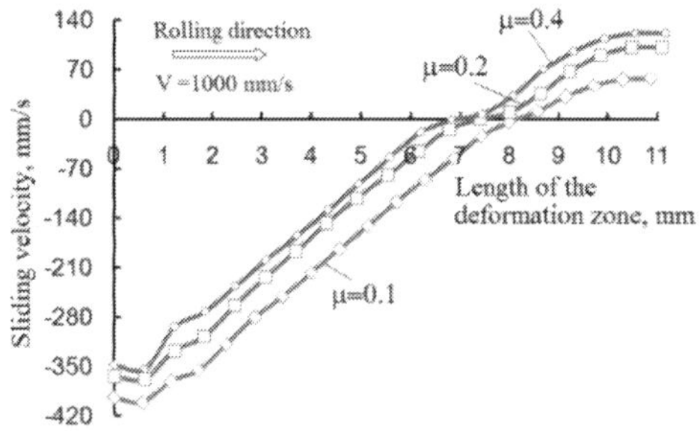

Figure 9. Sliding velocity on the contact surfaces during symmetric rolling with different friction coefficients (Al 7075, thickness reduction is 50%, rolling speed is 1.0 m/s, heat transfer coefficient is 11000 W/(m^2×K)

Sliding velocity on the contact of the top roll during asymmetric rolling with different friction coefficients is shown in Fig. 9. High friction coefficient increases the sliding velocity in the forward slip zone.

Forward slip zone is absent on the contact with the bottom (fast) roll (Fig. 10). It corresponds to the cases when rolls speed ratio is equal to thickness reduction per pass. Increasing of the friction coefficient leads to decrease of the sliding velocity. As a result the friction heat is more serious on the contact with the top (slow) roll.

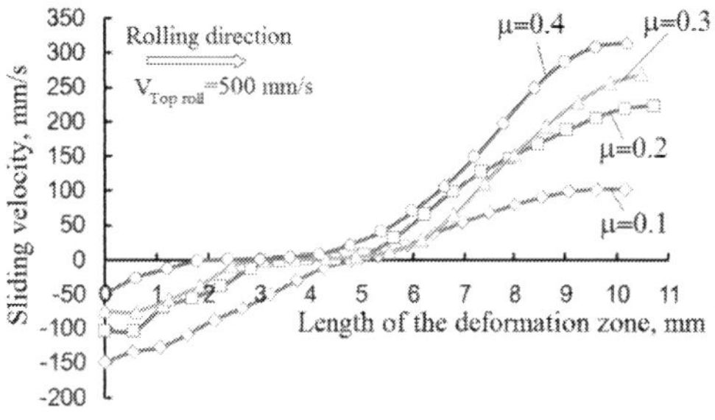

Figure 10. Sliding velocity on the contact of the top roll during asymmetric rolling with different friction coefficients (Al 7075, thickness reduction is 50%, rolls speed ratio is 50%, heat transfer coefficient is 11000 W/(m^2×K)

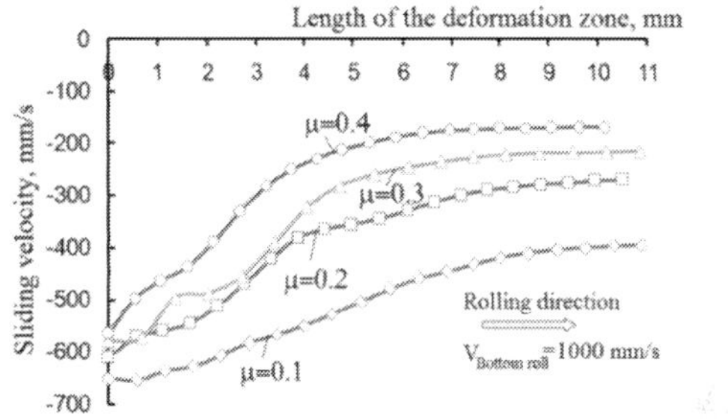

Figure 11. Sliding velocity on the contact of the bottom roll during asymmetric rolling with different friction coefficients (Al 7075, thickness reduction is 50%, rolls speed ratio is 50%, heat transfer coefficient is 11000 W/(m^2×K)

Fig. 11 shows the interrelation of the temperature in the deformation zone and friction coefficient during asymmetric rolling. Rise of temperature in deformation zone is significantly different during asymmetric rolling of pure aluminium and its alloys 7075, 5083. With increasing the friction coefficient from 0.1 to 0.4, the temperature of rolled sheets is increased from 50 °C to 78 °C for pure aluminium, from 127 °C to 236 °C for Al 5083 and from 194 °C to 328 °C for Al 7075 (for H=11000 W/m^2×K). It can be explained by more serious work of plastic deformation during asymmetric rolling of much stronger material.

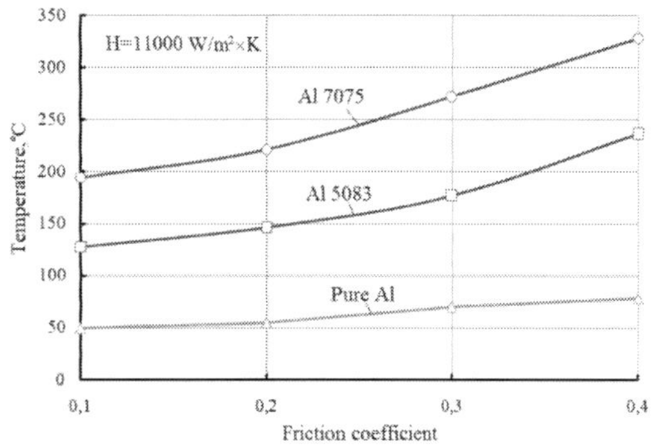

Figure 12. Interrelation of the temperature in the deformation zone and friction coefficient during asymmetric rolling of pure

Al and alloys Al 7075, Al 5083 (thickness reduction is 50%, rolling speed is 1.0 m/s, rolls speed ratio is 50%)

The microstructure of Al alloys is affected by the thermal field during asymmetric rolling. The temperature level of dynamic recovery and grain growth should be further studied in future.

4 CONCLUSIONS

The aluminium alloy sheets are heated by friction and plastic deformation work during asymmetric cold rolling. This paper presents the results of FEM simulation of the effect of contact friction, rolling speed and rolls speed ratio on the heating of aluminium sheets. Asymmetric rolling process leads to serious increase of the temperature in the deformation zone in comparison with conventional rolling. The temperature rise can be controlled through adjustment of the rolling speed.

The heat transfer analysis presented in this study can be used to optimize the asymmetric rolling process as SPD method to improve microstructure and mechanical properties of aluminium sheets. The temperature level of dynamic recovery and grain growth should be further studied in future.

ACKNOWLEDGEMENTS

The study was supported by a grant of the Russian Science Foundation (project No15-19-10030)

REFERENCES

1. I. Sabirov, M.Yu. Murashkin, R.Z.Valiev, Materials Science & Engineering A. 560 (2013) 1-24
2. V.I. Elagin, Metal Science and Heat Treatment. 9 (2007) 3-11
3. R.K. Islamgaliev, et al., Materials Science and Engineering: A. 319-321 (2001) 877-881
4. Y. Estrin, A. Vinogradov, Acta Materialia. 61 (2013) 782-817
5. Q. Cui, K. Ohori, Materials Science and Technology. 16 (2000) 1095-1101
6. J. Jiang, Y. Ding, F. Zuo, A. Shan, Scripta Materialia. 60 (2009) 905-908
7. Loorentz, Young Gun Ko, Journal of Alloys and Compounds. 536S (2012) S122-S125
8. Loorentz, Y.G. Ko, Journal of Alloys and Compounds. 586 (2014) S205-S209
9. J.K. Lee, D.N. Lee, International Journal of Mechanical Sciences. 50 (2008) 869-887
10. K. Bobor, Z. Hegedus, J. Gubicza, I. Barkai, P. Pekker, G. Krallics, Periodica Polytechnica Mechanical Engineering. 56 (2012) 111-115
11. F. Zuo, J. Jiang, A. Shan, Transactions of Nonferrous Metals Society of China, 18 (2008) 774-777
12. J.K. Lee, D.N. Lee, Key Engineering Materials. 340-341 (2007) 619-626
13. A. Pesin, M. Chukin, A. Korchunov, D. Pustovoytov, Key Engineering Materials. 622-623 (2014) 912-918
14. A. Pesin, A. Korchunov, D. Pustovoytov, Key Engineering Materials. 685 (2016) 162-166
15. M. Sverdlik, A. Pesin, D. Pustovoytov, A. Perekhozhikh, Advanced Materials Research. 742 (2013) 476-481
16. Y.H. Ji, J.J. Park, Materials Science and Engineering: A. 499 (2009) 14-17
17. Y.H. Ji, J.J. Park, W.J. Kim, Materials Science and Engineering: A. 454-455 (2007) 570-574
18. W.J. Kim, B.G. Hwang, M.J. Lee, Y.B. Park, Journal of Alloys and Compounds. 509 (2011) 8510-8517
19. J.J. Park, Computational Materials Science. 100 (2015) 61-66
20. A. Pesin, D. Pustovoytov, Key Engineering Materials. 622-623 (2014) 929-935
21. O. Yevtushenko, Acta Mechanica et Automatica, 2 (2008) 99-102
22. W.R.D. Wilson, C.T. Chang, C.Y. Sa, Journal of Materials Shaping Technology, 6 (1989) 229-240
23. Modern Manufacturing Engineering, Edited by J. Paulo Davim, Springer International Publishing (2015)
24. Handbook of Workability and Process Design, Edited by George E. Dieter, Howard A. Kuhn, S. Lee Semiatin, ASM International (2003)

Modeling of the roll wear and material damage during high-ratio differential speed rolling of aluminium alloy 7075

Alexander Pesin[a], Denis Pustovoytov, Natalya Lokotunina

Nosov Magnitogorsk State Technical University, Department of Material processing, 455000, Magnitogorsk, Lenin prospect, 38, Russia

Abstract. The technology of high-ratio differential speed rolling (HRDSR) with high contact friction can be used for production of aluminium alloys with UFG structure. But a high friction coefficient is not desirable because of the increasing of the material damage. The paper presents a numerical analysis of the rolls wear and material damage during HRDSR of aluminium alloy 7075. The FEM was used to calculate the rolls wear by Archard's model and material damage by normalized Cockroft-Latham model.

1 Introduction

Aluminium and its alloys play a key role in modern engineering since they are the most used non-ferrous materials [1]. They have been widely used as structural materials in aerospace, automotive and other industries due to low density and high specific strength. Conventional approaches to increasing strength properties of aluminum alloys based on modification of their chemistry by alloying and on development of heat treatment and thermomechanical treatment methods are almost exhausted. A more efficient way to further increase strength and other operational and technological properties of aluminium alloys is to form an ultrafine grain (UFG) structure using severe plastic deformation (SPD) methods.This can be achieved through any of a number of different processes such as equal channel angular pressing, multi-directional forging, accumulated roll bonding, high-pressure torsion, twist extrusion and high-ratio differential speed rolling (HRDSR) in which the speeds of the upper and lower rolls are different. Since the HRDSR is a continuous process, it has great potential for industrial production of ultrafine grained aluminium alloys [2-7].

It is well known that the mechanism of SPD during HRDSR comes from its large equivalent strain, which is composed of compressive strain and additional shear strain. For a given thickness reduction, the equivalent strain is proportional to shear strain because the normal strain is the same. Shear strain during HRDSR can be presented as tangent of shear angle. During HRDSR shear angle strongly depends on opposite contact friction forces operated in the deformation zone. High shear strain through thickness of the sheet can be obtained only with a high value of friction coefficient [8-11]. So HRDSR should be carried out without lubrication. But a high friction coefficient is not desirable, not only because of the energy wasted but also because of the wear and damage increasing of the sliding system [12].

It is well accepted that adhesive, abrasive, fatigue, and corrosive wear may each play a role in cold rolling wear. The wear patterns may exist separately or in a mixed regime.

Experimental investigations of the roll wear during HRDSR have been the subject of several publications.In [13] an investigation of the wear mechanisms in the cold rolling of aluminium strip was reported. The effects of the pass reduction, the viscosity of the lubricant and the additives in the lubricant on wear were researched by measuring the relative mass wear rate and by observation of the rolled surface by means of optical microscopy and scanning electron microscopy.

HRDSR with high contact friction is a process that may be limited by damage and edge cracks of the rolled Al sheet.Therefore an important concern in the HRDSR of high-strength Al alloys is whether the desired strain can be performed without any fracture of the material. Fracturing by ductile damage occurs in HRDSR process due to the development of microcracks associated with high shear strain.In order to investigate this phenomenon, simulation programs based on the Cockroft-Latham fracture model is usedin engineering analysis [14].The Cockcroft-Latham fracture criterion assumes that the maximum principal stress is the most relevant in the initiation of fracture. The criterion is defined in terms of traction plastic work associated with the principal stress along the path of the equivalent plastic strain [14].

Many researchers have used the finite element method to investigate the influence of cold rolling process parameters (e.g., thickness reduction, friction conditions) on the damage and edge cracks of the rolled Al sheets. A computational and experimental study of damage evolution during conventional cold rolling of aluminium alloys was reported in [15]. Although many studies have

[a]Corresponding author: pesin@bk.ru

been done, there are no numerical data on the effect of the HRDSR process parameters on the roll wear and damage of the rolled Al sheets. The goal of this research is the numerical simulation and comparison of the rolls wear and material damage during conventional cold rolling and HRDSR. The results of investigation can be used to optimize the high-ratio differential speed rolling process to improve the quality of UFG aluminium sheets.

2 Research method

Software DEFORM 3D, based on finite element method (FEM), was used to analyze the three-dimensional metal flow during conventional cold rolling and HRDSR. DEFORM 3D is an effective tool that allows to predict the character of the metal flow, roll wear and material damage during cold rolling.

The roll wear phenomenon in cold rolling is very complex. Many wear mechanisms exist described as adhesive wear, abrasive wear, diffusive wear, fatigue wear, etc. DEFORM 3D offers two kinds of tool wear model: Archard model and Usui model. Archard abrasive wear model can be viewed as a primary wear mechanism where the wear of the sliding system greatly differ in hardness. Usui model is applicable to the processes where diffusion is a major contributor to wear. Therefore, we choose Archard as the mathematical model of roll wear simulation analysis. Refer to Archard's wear model (1) it is clear that the interface pressure and sliding velocity are the main factors to affect the wear of work rolls.

$$w = \int K \frac{P^a V^b}{H^c} dt, \quad (1)$$

where w – wear depth, mm; P – interface pressure, MPa; V – sliding velocity, m/s; H – Rockwell C hardness of roll material; a=1, b=1, c=2 – calibration coefficients (standard values for common tool steels); K – wear factor ($K=5\times10^{-7}$ is a typical value for cold rolling processes).

Damage in the cold rolling process is a three dimensional phenomenon that requires a 3D model for accurate predictions. In this study, the normalized Cockroft-Latham damage model was applied to predict the material fracture during HRDSR. The Cockroft-Latham model defines a damage parameter D as shown in following equation:

$$D = \int^{\bar{\varepsilon}} \frac{\sigma_{max}}{\bar{\sigma}} d\bar{\varepsilon} \le C, \quad (2)$$

where σ_{max} – the maximum positive principal stress; $\bar{\varepsilon}$ – the equivalent plastic strain; $\bar{\sigma}$ – the effective stress; C – the critical damage value at fracture.

This model determines the likelihood of fracture from tensile plastic deformation, by consideration of the plastic work to failure. When the damage value reaches the constant C, the material is fractured. While it is necessary to determine the constant C by experimentation, a general rule is that materials with very good workability will fracture at a damage value between 0.8 and 1.0 and materials with poor workability will fracture at values between 0.3 and 0.4 [15].

The modeling of the process was performed at room temperature taking into account the increment of the metal's temperature due to the thermal effect of deformation. Heat transfer coefficient at the interface in all calculation variants was set constant and equal to 11000 $W/(m^2\times K)$. The work rolls were assumed as rigid. High-carbon chromium alloy steel AISI D2 was chosen as a material for rolls. In the HRDSR process the temperature of the rolls does not exceed 50°C. So the hardness (62 HRC) of the AISI D2 rolls was a constant. Work roll radii were set constant and equal to 100 mm. A Coulomb friction model was used between rolls and sheet, which assumes that no relative motion occurs if the equivalent frictional stress is less than a critical value. The friction coefficient varied within the following range: 0.1...0.4. The thickness reduction per pass was 50% in all calculation variants. Aluminium alloy 7075 from DEFORM 3D material library was chosen as a hardened rigid-plastic material for sheet.

Differential speed rolling process was carried out at three different angular speeds of the bottom roll: 1; 5; 10 rad/s. It corresponded to 0.1 m/s, 0.5 m/s and 1.0 m/s of circumferential speed. Speed of the top roll was reduced by 10; 30; 50% during asymmetric rolling. Sheet with initial dimensions of 2.0×20.0×50.0 mm was meshed with 14220 brick elements. Rolls were meshed with 117000 tetrahedral elements. The boundary conditions were set in order to constrain the mid-width plane (symmetry plane) of the sheet along the rolling direction.

In the course of simulation of conventional cold rolling and HRDSR the influence of the process parameters on roll wear and material damage was studied. In all calculation variants roll wear depth corresponds to the one revolution of the roll.

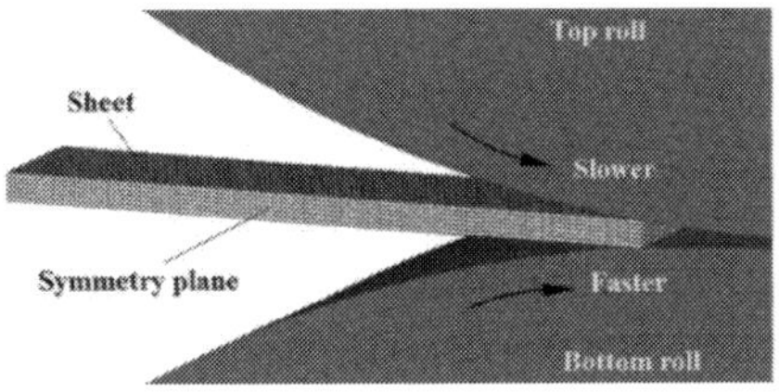

Figure. 1. FEM model of the process

3 Simulation results and discussion

Fig. 2 shows the interrelation of roll wear and friction coefficient during conventional (symmetric) cold rolling and HRDSR. The larger the contact friction is, the more serious is roll wear for conventional rolling. It can be explained by increasing of the interface pressure. The wear of both rolls is the same for conventional cold rolling. However, the roll wear is significantly differs for HRDSR. Increasing the contact friction coefficient from 0.1 to 0.4 leads to increasing of the roll wear up to 1.9 times for the top roll. But it is decreased up to 2.0 times for the bottom roll (Fig. 2). Decreasing of the friction coefficient less than 0.3 is not desirable because of significantly decreasing of the shear strain during HRDSR.

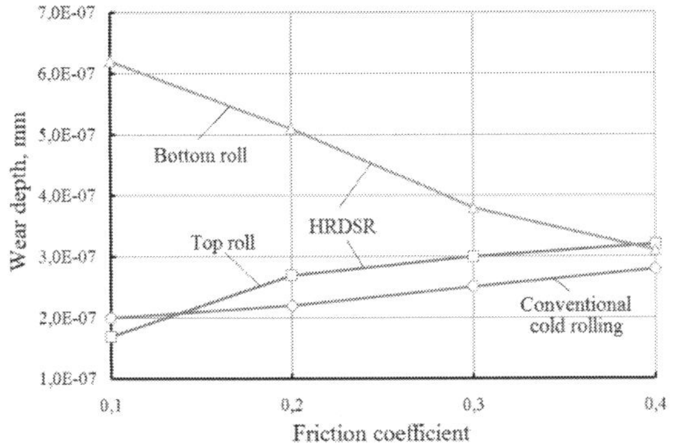

Figure. 2. Interrelation of rolls wear and friction coefficient during conventional cold rolling and HRDSR (thickness reduction is 50%; rolls speed ratio is 50%; rolling speed is 1.0 m/s)

Fig. 3 shows the fields of roll wear depth (mm) during conventional cold rolling and HRDSR. By analyzing the results of simulation, it can be inferred that at a low friction coefficient the wear of bottom (fast) roll is much higher during HRDSR. It can be explained by different influence of the pressure and sliding velocity on the contact surfaces.

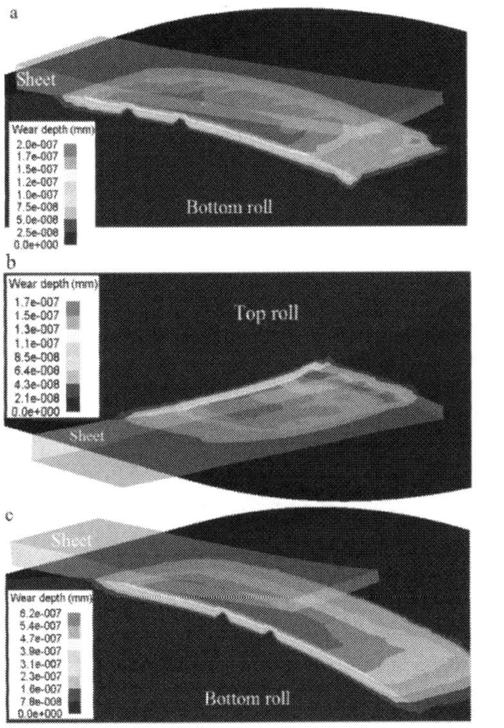

Figure 3. Rolls wear during conventional cold rolling (a) and HRDSR (b, c) (thickness reduction is 50%; rolling speed is 1.0 m/s; friction coefficient is 0.1; rolls speed ratio is 50% for HRDSR)

Increasing the friction coefficient leads to decreasing of the sliding velocity on the contact with both rolls during conventional cold rolling (Fig. 4a, b). At the same time the interface pressure is increased significantly (Fig. 6a, b). As a result, roll wear depth is increased during conventional cold rolling.

Another situation is during HRDSR. On the one hand, increasing friction coefficient leads to increasing of the sliding velocity on the contact with the top (slow) roll (Fig. 4c, d). On the other hand it leads to decreasing of the sliding velocity on the contact with the bottom (fast) roll (Fig. 5). Increasing the friction coefficient leads to insignificant increasing of the interface pressure during HRDSR (Fig. 6c, d). As a result, roll wear depth is increased for the top roll and it is decreased for the bottom roll.

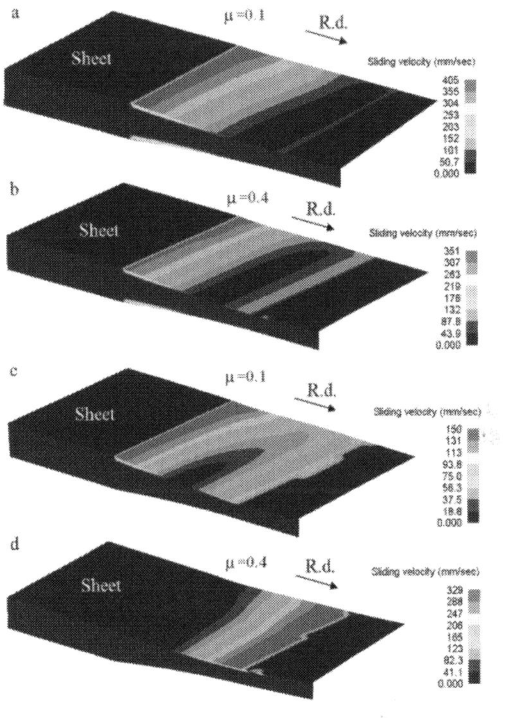

Figure 4. Sliding velocity on the contact with the top roll during conventional cold rolling (a, b) and HRDSR (c, d) (thickness reduction is 50%; rolling speed is 1.0 m/s; rolls speed ratio is 50% for HRDSR; R.d. – rolling direction)

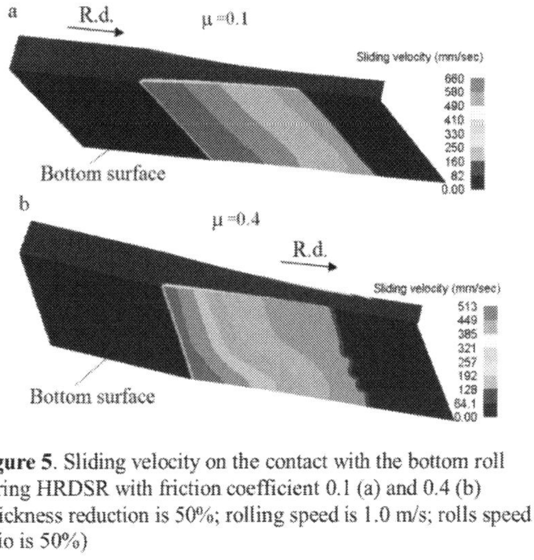

Figure 5. Sliding velocity on the contact with the bottom roll during HRDSR with friction coefficient 0.1 (a) and 0.4 (b) (thickness reduction is 50%; rolling speed is 1.0 m/s; rolls speed ratio is 50%)

DOI: 10.1051/mateconf/20168004006

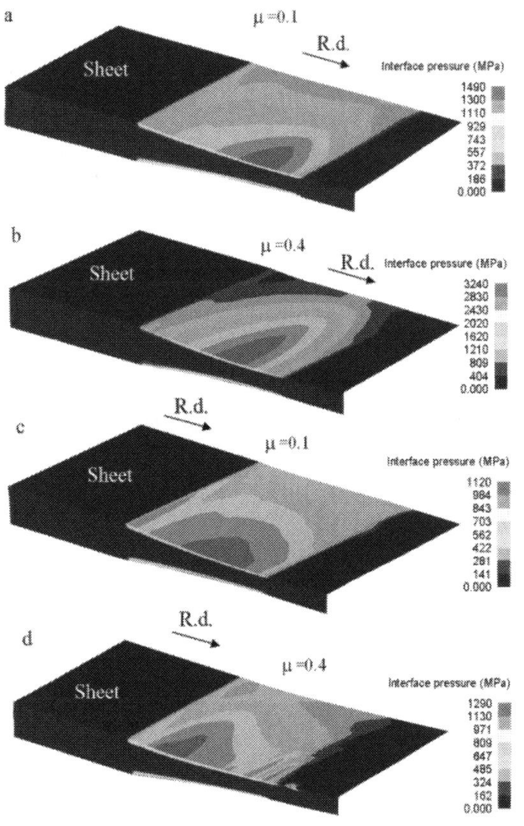

Figure 6. Interface pressure during conventional cold rolling (a, b) and HRDSR (c, d) (thickness reduction is 50%; rolling speed is 1.0 m/s; rolls speed ratio is 50% for HRDSR)

Decreasing the rolling speed from 1.0 m/s to 0.1 m/s leads to decreasing of the roll wear (Fig. 7).

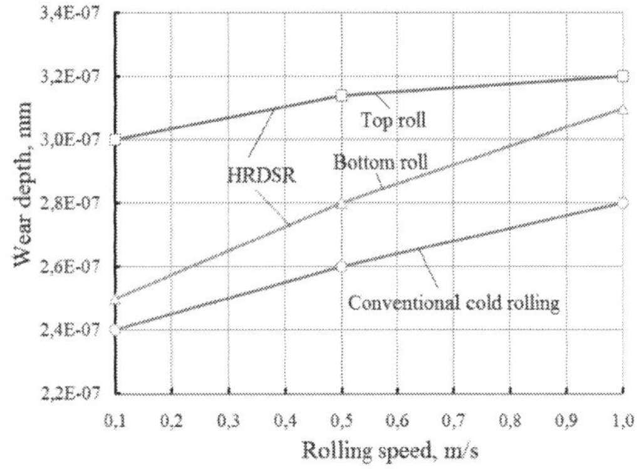

Figure 7. Interrelation of roll wear and rolling speed during conventional cold rolling and HRDSR (thickness reduction is 50%; friction coefficient is 0.4; rolls speed ratio is 50%)

Higher is the speed ratio, more serious is roll wear depth (Fig. 8). Increasing the rolls speed ratio till 50% leads to increasing of the wear up to 1.14 times for the top roll and up to 1.10 times for the bottom roll when friction coefficient is 0.4. Increasing the rolls speed ratio more than thickness reduction per pass (50%) leads to significant increasing of the wear of the bottom (fast) roll up to 2.2 times.

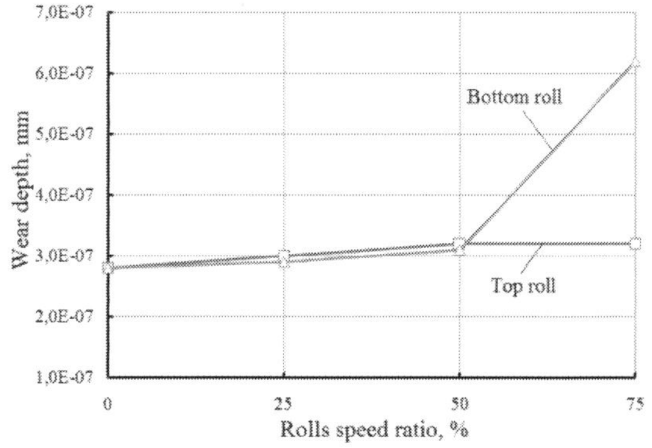

Figure 8. Interrelation of roll wear and rolls speed ratio during HRDSR (thickness reduction is 50%; friction coefficient is 0.4; rolling speed is 1.0 m/s)

The friction coefficient and rolls speed ratio are the most two important factors that influence on the damage and fracture of Al sheets during HRDSR in comparison with conventional cold rolling.

Fig. 9 shows the contours of the damage parameter D according to the Eq. (2) with the normalized Cockroft-Latham model for conventional cold rolling. It can be seen that the damage concentration occurs on the free edge after exiting the rolls, with the maximum at the middle of the sheet thickness. It corresponds to regions where the tensile principal stresses occur. Very high tensile stresses also occur in the rear noncontact deformation zone (highlighted by arrow in Fig. 9b).

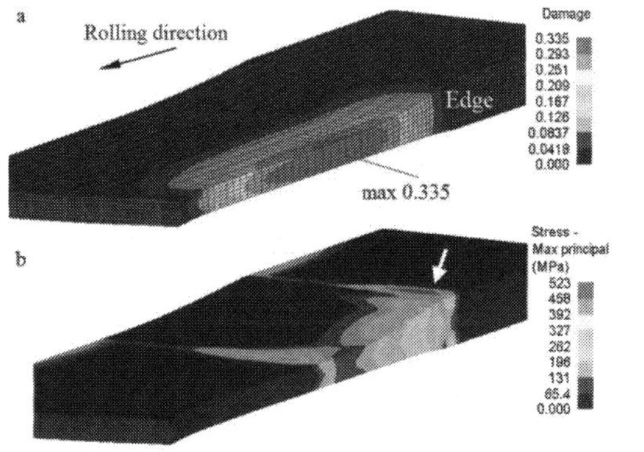

Figure 9. Field of the material damage (a) and tensile stresses (b) during conventional cold rolling (friction coefficient is 0.1; thickness reduction is 50%; rolling speed is 1.0 m/s)

HRDSR leads to increasing of the material damage (Fig. 10) in comparison with conventional cold rolling because of additional shear strain. It should be noted that damage parameter D is higher for the bottom surface of the sheet. Higher tensile stresses are found directly under the bottom roll in the longitudinal direction at the edge where stress concentration is observed. Asymmetry of HRDSR process is the cause of the sheet curvature. It also leads to increasing of the tensile stresses in the bottom surface of the sheet. Very high tensile stresses occur in the front and rear noncontact deformation zones (highlighted by arrows in Fig. 10b). FEM simulations give damage parameter values very close to that needed for fracture.

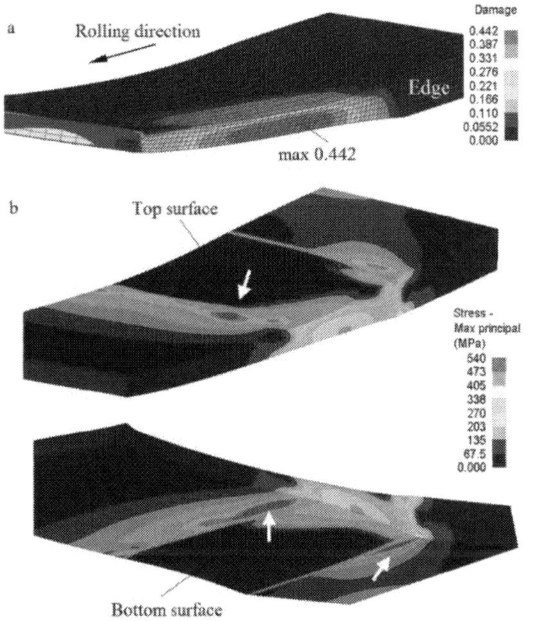

Figure 10. Field of the material damage (a) and tensile stresses (b) during HRDSR (friction coefficient is 0.4; thickness reduction is 50%; rolling speed is 1.0 m/s; rolls speed ratio is 50%)

Contact friction is one of the main factors that affect the material damage during cold rolling. Increasing the friction coefficient leads to increasing of the material damage from 0.335 to 0.364 for conventional rolling and from 0.347 to 0.442 for HRDSR (Fig. 11). The damage is also increased when the rolls speed ratio is increased. Decreasing of the friction coefficient is beneficial to reduce the edge damage, but it also leads to significantly decreasing of the shear strain during HRDSR.

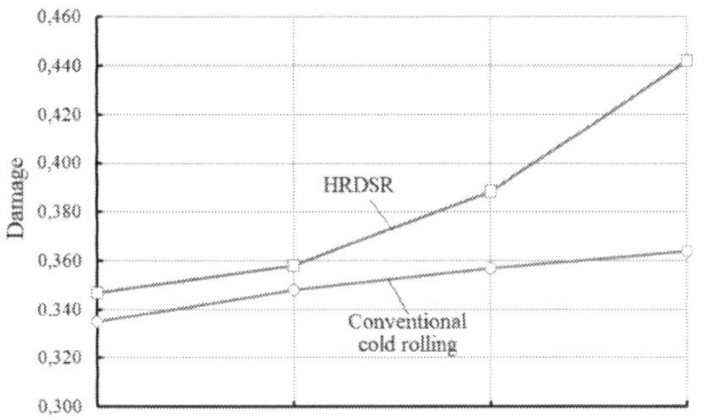

Figure 11. Interrelation of material damage and friction coefficient during conventional rolling and HRDSR (thickness reduction is 50%; rolls speed ratio is 50%; rolling speed is 1.0 m/s)

4 CONCLUSIONS

The numerical simulation allowed to carry out a comparison between conventional cold rolling and HRDSR in order to highlight differences in the roll wear and material damage imparted by each processes.

Increasing the friction coefficient leads to decreasing of the sliding velocity and to increasing of the pressure on the contact with both rolls during conventional cold rolling. As a result, roll wear depth is increased during conventional cold rolling.

Another situation is during HRDSR. On the one hand, increasing friction coefficient leads to increasing of the sliding velocity on the contact with the top (slow) roll. On the other hand it leads to decreasing of the sliding velocity on the contact with the bottom (fast) roll. Increasing the friction coefficient leads to insignificant increasing of the interface pressure during HRDSR. As a result, roll wear depth is increased for the top (slow) roll and it is decreased for the bottom (fast) roll.

Increasing the friction coefficient leads to increasing of the material damage during HRDSR. The damage is also increased when the rolls speed ratio is increased. Damage is higher for the bottom surface of the sheet where higher tensile stresses were found. FEM simulations give damage parameter values very close to that needed for fracture of Al 7075.

The results of simulation can be used to optimize the HRDSR process to improve the quality of UFG aluminium sheets. Further experimental investigation of the roll wear and material damage during HRDSR is required.

ACKNOWLEDGEMENTS

The study was supported by a grant of the Russian Science Foundation (project No15-19-10030)

REFERENCES

1. I. Sabirov, M.Yu. Murashkin, R.Z.Valiev, Materials Science & Engineering A. 560 (2013) 1-24
2. Y.H. Ji, J.J. Park, W.J. Kim, Materials Science and Engineering: A. 454-455 (2007) 570-574
3. Y.H. Ji, J.J. Park, Materials Science and Engineering: A. 499 (2009) 14-17
4. J. Jiang, Y. Ding, F. Zuo, A. Shan, Scripta Materialia. 60 (2009) 905-908
5. F. Zuo, J. Jiang, A. Shan, Transactions of Nonferrous Metals Society of China, 18 (2008) 774-777
6. Q. Cui, K. Ohori, Materials Science and Technology. 16 (2000) 1095-1101
7. W.J. Kim, B.G. Hwang, M.J. Lee, Y.B. Park, Journal of Alloys and Compounds. 509 (2011) 8510-8517
8. A. Pesin, D. Pustovoytov, Key Engineering Materials. 622-623 (2014) 929-935
9. A. Pesin, M. Chukin, A. Korchunov, D. Pustovoytov, Key Engineering Materials. 622-623 (2014) 912-918
10. A. Pesin, A. Korchunov, D. Pustovoytov, Key Engineering Materials. 685 (2016) 162-166

11. M. Sverdlik, A. Pesin, D. Pustovoytov, A. Perekhozhikh, Advanced Materials Research. 742 (2013) 476-481
12. A. Pesin, D. Pustovoytov, T. Shveyova, M. Sverdlik, MATEC Web of Conferences. 26 (2015) 1-4
13. Zhang Hui, Journal of Materials Processing Technology. 31 (1992) 235-243
14. H. Yu, et al., Scientific reports. 5 (2015) 1-9
15. S. Ghosh, M. Li, D. Gardiner, Journal of Manufacturing Science and Engineering. 126 (2004) 74-82

MATEC Web of Conferences **80**, 05001 (2016) DOI: 10.1051/matecconf/20168005001
NUMIFORM 2016

On twinning and anisotropy in rolled Mg alloy AZ31 under uniaxial tension

Xiaoqian Guo[1], Adrien Chapuis[2], Xianbiao Mao[1], Qing Liu[2] and Peidong Wu[3,a]

[1]*State Key Laboratory for Geomechanics and Deep Underground Engineering, China University of Mining and Technology, Xuzhou, Jiangsu 221116, China*
[2]*College of Materials Science and Engineering, Chongqing University, Chongqing 400044, China*
[3]*Department of Mechanical Engineering, McMaster University, Hamilton, Ontario, Canada*

Abstract. The Elastic Visco-Plastic Self-Consistent (EVPSC) model, with the recently developed Twinning and De-Twinning (TDT) description, is applied to study the mechanical behaviour of hot-rolled Mg alloy AZ31 under uniaxial tension. Numerical results are compared to the experimental uniaxial tensile tests reported earlier by Chapuis et al. [29] for the out-of-plane directions of a thick plate along angles of α=0°, 30°, 45°, 60° and 90° between the normal direction and longitudinal specimen axis. It is shown that accounting for the initial texture and calibrating the EVPSC-TDT model by using uniaxial tension tests along the rolling direction and normal direction permits prediction of the strength anisotropy and strain hardening behavior along all five tensile directions, i.e. for cases in which the contribution of twinning is dominating, negligible or intermediate.

1. Introduction

Modeling mechanical behaviour of textured magnesium (Mg) alloys has been a hot research topic (see e.g. [1-23]).However, it is noted that most of the research has been restricted to uniaxial straining along directions of orthotropic symmetry, where the straining is either essentially twinning dominated or essentially devoid of twinning. The works of Oppedal et al. [24] and Guo et al. [25] are exceptional. Oppedal et al. [24] performed uniaxial compression testing along off-axis directions within an extruded AM30 bar. They employed these data to illustrate the need for a new model, which specifically accounts for the dislocation density evolution, as well as the "transmutation" of that dislocation density due to twinning. Guo et al. [25] studied, both experimentally and numerically, the anisotropy in a hot-rolled AZ31 plate under uniaxial compression. The experimental tests covered a wide range of tensile twinning activity: from twinning dominated, through intermediate twinning activity to negligible twinning. The numerical simulations were based on the Elastic Visco-Plastic Self-Consistent (EVPSC) model of Wang et al. [26], with the Twinning and De-Twinning (TDT) description later introduced by Wang et al. [27,28]. It has been demonstrated that accounting for the initial texture and calibrating the EVPSC-TDT model using in-plane uniaxial tension and compression along the rolling direction (RD) permits prediction of the strength anisotropy and monotonic strain hardening behavior along any tilt angle with respect to the normal direction (ND). It is very interesting to note that this capability is achieved without requiring a complex, dislocation density-based latent hardening approach.

In the present paper, the mechanical anisotropy of hot-rolled Mg alloy AZ31 under uniaxial tension is investigated numerically, based on the EVPSC-TDT model. Numerical results are compared with the experimental results reported by Chapuis et al. [29]. In the experimental work, five uniaxial tension tests along angles of α=0°, 30°, 45°, 60° and 90° between the ND and longitudinal specimen axis were performed to experimentally study the mechanical anisotropy. The paper is organized as follows. Section 2 introduces the EVPSC-TDT model. In Section 3 we show both the experimental and predicted results with emphasizing on relationships between twinning and anisotropy in the sheet under uniaxial tension. Conclusions are drawn in Section 4.

2. Constitutive Model

In this section, we briefly recapitulate the EVPSC-TDT model. For details, we refer to Wang et al. [26,28]. The plastic deformation of a crystal is assumed to be due to crystallographic slip and twinning on crystallographic system $(\boldsymbol{s}^{\alpha},\boldsymbol{n}^{\alpha})$, with $\boldsymbol{s}^{\alpha}$ and $\boldsymbol{n}^{\alpha}$ being the slip/twinning direction and the normal of the slip/twinning plane for system α , respectively. For Mg alloys, Basal <*a*>, Prismatic <*a*>, and Pyramidal <*c*+*a*>slip systems, and $\{10\bar{1}2\}<\bar{1}011>$ extension twin system are usually considered.

[a] Corresponding author: peidong@mcmaster.ca

Regardless of slip or twinning, the driving force for shear rate, $\dot{\gamma}^{\alpha}$, is the resolved shear stress, τ^{α}. For slip,

$$\dot{\gamma}^{\alpha} = \dot{\gamma}_0 \left|\tau^{\alpha}/\tau_{cr}^{\alpha}\right|^{\frac{1}{m}} \operatorname{sgn}(\tau^{\alpha}) \qquad (1)$$

For twinning,

$$\dot{\gamma}^{\alpha} = \begin{cases} \dot{\gamma}_0 \left|\tau^{\alpha}/\tau_{cr}^{\alpha}\right|^{\frac{1}{m}} & \tau^{\alpha} > 0 \\ 0 & \tau^{\alpha} \le 0 \end{cases} \qquad (2)$$

where $\dot{\gamma}_0$ is a reference shear rate, τ_{cr}^{α} is the critical resolved shear stress (CRSS), and m is the strain rate sensitivity. The TDT model assumes that a grain has four potential operations associated with twinning and detwinning. Operation A is twin nucleation and initiates a twin band or 'child'. Operation B is a propagation of the child into the parent grain. Operations A and B increase the twin volume fraction and thus correspond to twinning. Operation C is a propagation of the parent into the child. Operation D splits the twin band and decreases the twin volume fraction through re-twinning. Operations C and D decrease the twin volume fraction and thus correspond to detwinning. Furthermore, the TDT model treats new twin band (child) as a new grain, and all twin variants are possibly present.

For both slip and twinning, the evolution of CRSS, τ_{cr}^{α}, is given by:

$$\dot{\tau}_{cr}^{\alpha} = \frac{d\hat{\tau}^{\alpha}}{d\Gamma} \sum_{\chi} h^{\alpha\chi} \left|\dot{\gamma}^{\chi}\right| \qquad (3)$$

where $\Gamma = \sum_{\beta} \int \left|\dot{\gamma}^{\alpha}\right| dt$ is the accumulated shear strain in the grain, and $h^{\alpha\chi}$ are the latent hardening coupling coefficients, which empirically account for the obstacles on system α associated with system χ. $\hat{\tau}^{\alpha}$ is the threshold stress and is characterized by:

$$\hat{\tau}^{\alpha} = \tau_0^{\alpha} + (\tau_1^{\alpha} + h_1^{\alpha}\Gamma)(1 - \exp(-\frac{h_0^{\alpha}}{\tau_1^{\alpha}}\Gamma)) \qquad (4)$$

Here, τ_0, h_0, h_1, and $\tau_0 + \tau_1$ are the initial CRSS, the initial hardening rate, the asymptotic hardening rate, and the back-extrapolated CRSS, respectively.

Because it is rare that a grain can be fully twinned, a threshold twin volume fraction is defined in the model to terminate twinning. Consequently, the TDT model introduces two statistical variables: accumulated twin fraction, V^{acc}, and effective twinned fraction, V^{eff}. More specifically, V^{acc} and V^{eff} are the weighted volume fraction of the twinned region and volume fraction of twin terminated grains, respectively. The threshold volume fraction, V^{th}, is defined as $V^{th} = \min\left(1.0,\ A_1 + A_2 \frac{V^{eff}}{V^{acc}}\right)$, where A_1 and A_2 are two material constants.

The response of a polycrystal comprised of many grains is obtained by applying the self-consistent approach. Various self-consistent schemes have been proposed. Wang et al. [26] have demonstrated that the predicted response of a polycrystal is very sensitive to the self-consistent scheme employed. Wang et al. [30] have evaluated various self-consistent schemes used in self-consistent modeling by applying them to the large strain behavior of magnesium alloy AZ31B plate under different deformation processes. It was found that the Affine self-consistent scheme gave the best overall performance among the self-consistent approaches examined. Therefore, the Affine self-consistent scheme is employed in the present study.

3. Results and discussion

The material reported by Chapuis et al.[29] and considered in the present paper is a 50 mm thick rolled magnesium alloy AZ31, annealed for homogenization at 400^{0}C for 1h. The rolled and annealed plate has an average grain size of about 40μm. In the present paper the RD, TD and ND stand for the rolling, transverse and normal directions, respectively. The initial texture of the material was measured using an EBSD map with a step size of 2μm. It is shown in Fig. 1 in terms of the {0001}, {11-20} and {10-10} pole figures. Clearly, the material exhibits a strong basal texture.

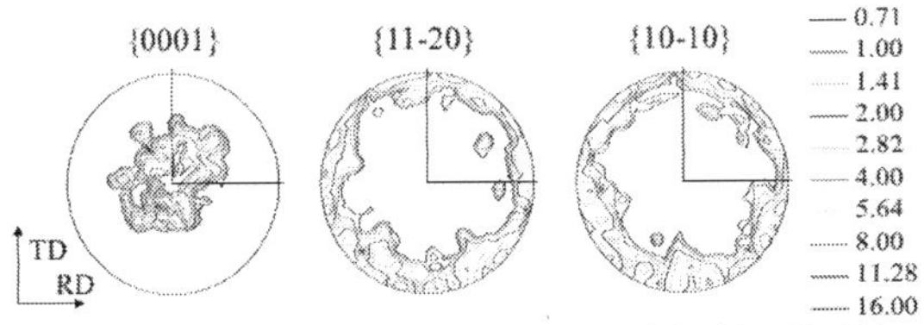

Figure 1. Initial texture presented in terms of the {0001}, {11-20} and {10-10} pole figures.

Uniaxial tension tests along different directions with respect to the ND have been performed by Chapuis et al. [29]. More specifically, as shown in Fig. 2, five different specimen orientations with tilt angles of α =0°, 30°, 45°, 60° and 90° between the ND and loading direction (LD) were used to study the mechanical anisotropy under uniaxial tension. These tests are very similar to those carried out by Wang and Choo [31] on samples with different mean grain sizes in the range of 26 ~127 μm.

In all the simulations reported in the present paper, the reference slip/twinning rate $\dot{\gamma}_0$ and the rate sensitivity m are prescribed to be the same for all slip/twinning systems: $\dot{\gamma}_0 = 0.001s^{-1}$ and $m = 0.05$, respectively. The room temperature elastic constants of the magnesium single crystal are taken as $C_{11} = 58.0$, $C_{12} = 25.0$, $C_{13} = 20.8$, $C_{33} = 61.2$ and $C_{44} = 16.6$ (units of GPa). Values of the hardening parameters are estimated by curve-fitting numerical simulations of monotonic uniaxial tension along the ND and uniaxial tension along the RD to the corresponding experimental data.

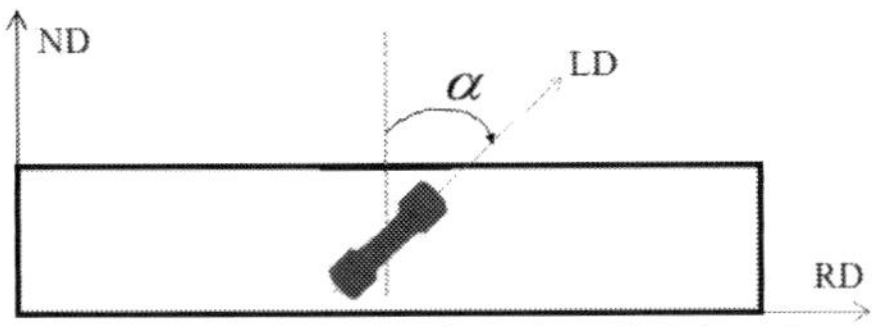

Figure 2. Schematic representation of a uniaxial tension sample with an angle α rotating around the TD from the ND to the loading direction (LD).

Fig. 3 presents the measured and simulated true stress and true strain curves under uniaxial tension along $\alpha=0^\circ$ (ND) and $\alpha=90^\circ$ (RD). The importance of extension twinning under uniaxial tension along the ND ($\alpha=0^\circ$) is clearly revealed by the characteristic *S*-shape of the flow curve, while uniaxial tension along the RD ($\alpha=90^\circ$) exhibits a typical slip dominated plastic deformation. It is found that the EVPSC-TDT model can fit the experimental curves very well. The determined values of the material constants are listed in Table 1.

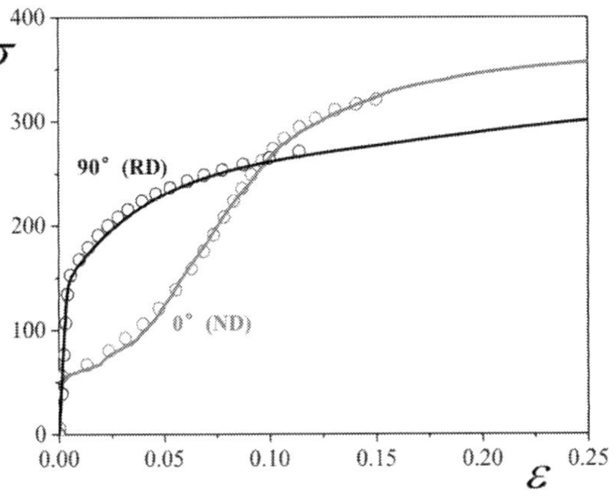

Figure 3. Measured (symbols) and simulated (solid lines) true stress and true strain curves under uniaxial tension along $\alpha=0^\circ$ (ND) and $\alpha=90^\circ$ (RD). The experimental data are taken from Chapuis et al. [29].

Fig. 4 presents the relative activities of slip/twinning under uniaxial tension along $\alpha=0^\circ$ and $\alpha=90^\circ$. The relative activities are defined as the ratios of the plastic shear rates of each deformation mode to the total plastic shear rate due to all the deformation modes at the applied strain. Under uniaxial tension along $\alpha=0^\circ$ and at strains $\varepsilon \le 0.04$, tensile twinning is very active, and the remaining plasticity is accommodated by Basal slip. For strains $\varepsilon \ge 0.05$, the tensile twining activity is significantly reduced, and the EVPSC-TDT model predicts increasing Prismatic slip activity and little Pyramidal slip (Fig. 4a). As expected, there is negligible twinning activity under uniaxial tension along $\alpha=90^\circ$. The model predicts that, under uniaxial tension along $\alpha=90^\circ$, mostly Basal and Prismatic slip accommodate the plastic deformation.

Table 1. List of values of the hardening parameters involved in the EVPSC-TDT model.

Mode	τ_0 /MPa	τ_1 /MPa	h_0 /MPa	h_1 /MPa	h^{st}	A_1	A_2
Basal	12	1	10	10	1		
Prismatic	78	55	650	15	1.4		
Pyramidal	100	140	2000	40	1		
Extension twin	35	10	200	10	1	0.2	0.8

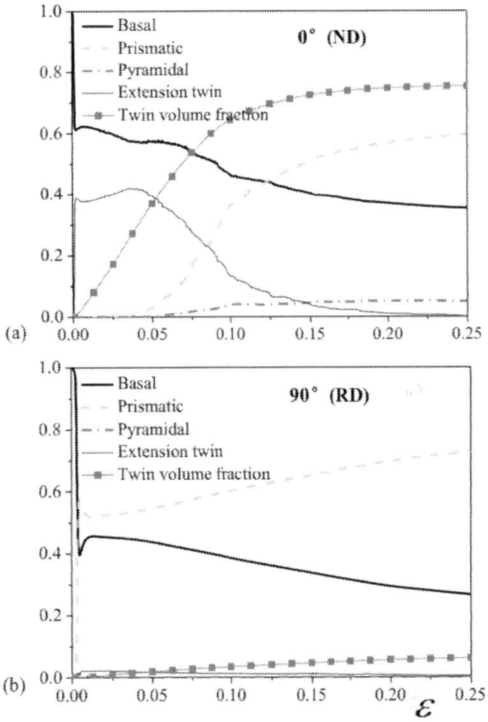

Figure 4. Predicted relative slip and twinning activities and twin volume fraction under uniaxial tension along $\alpha=0^\circ$ (a), and $\alpha=90^\circ$ (b).

It is noted that Pyramidal slip is almost inactive in uniaxial tension along $\alpha=90^\circ$ and little (<6%) in uniaxial tension along $\alpha=0^\circ$. It is found that Pyramidal slip activity is always below 6% in all the uniaxial tension tests. This implies that the hardening parameters associated with the Pyramidal slip cannot be confidently determined from the uniaxial tension cases performed in the present study. However, it is also noted that during in-plane uniaxial compression of a rolled plate with strong basal texture, Pyramidal slip is very active at relatively large strains [4, 19]. Thus, uniaxial compression allows us to confidently fit the material parameters associated with the Pyramidal slip system [25]. However, in-plane uniaxial compression was not carried out in the work of Chapuis et al.[29]. Nevertheless, based on the previous simulations performed on similar rolled AZ31 plates [19, 25], the values of the material parameters associated with the

Pyramidal slip system reported in Table 1 are found to be in a reasonably realistic range.

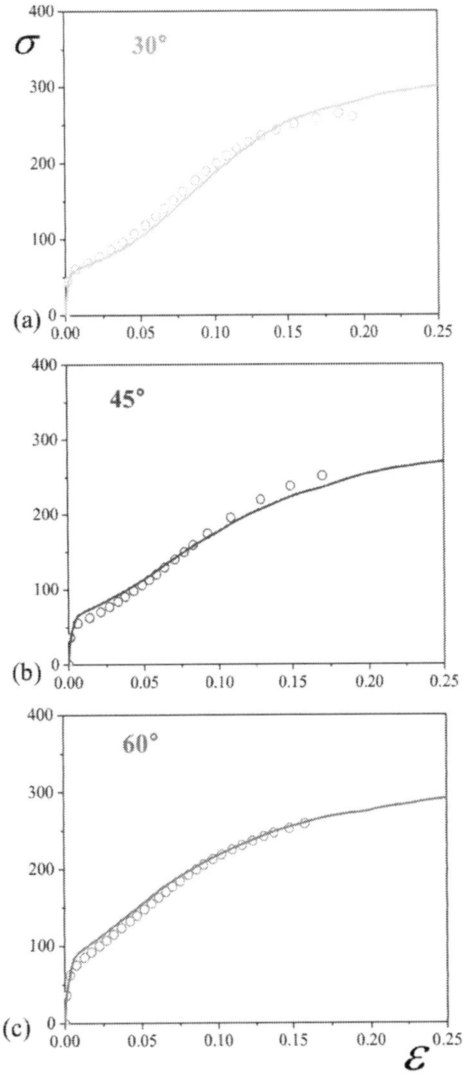

Figure 5. Measured (symbols) and simulated (solid lines) true stress and true strain curves under uniaxial tension along α=30° (a), α=45° (b), and α=60° (c).The experimental data are taken from Chapuis et al. [29].

We proceed by studying uniaxial tension along α=30°, α=45° and α=60°. Fig. 5 shows the measured and predicted stress and strain curves. In the case of α=30° (Fig. 5a), both the measured and predicted stress and strain curves are similar to the case of α=0° (i.e. along the ND). An *S*-shaped stress and strain curve indicates that tensile twinning is an important plastic deformation mechanism. The EVPSC-TDT model reproduces very well the measured flow stress. For α=60° (Fig. 5c), a "concave-down" stress and strain curve is observed both experimentally and numerically. This implies that the deformation under uniaxial tension along α=60° is slip dominated. The agreement between the measured and predicted stress and strain curves is again very good. The uniaxial tension behavior of the α=45° case (Fig. 5b) is between those of the α=30° and α=60° cases. The EVPSC-TDT model slightly over-estimates the flow stress at small strains and slightly under-estimates the flow stress at large strain levels.

Although not shown in the present paper, it was found that the EVPSC-TDT model predicts very well the measured texture evolution reported in Chapuis et al. [29].

4. Conclusions

We have numerically studied the twinning and anisotropy in a hot-rolled AZ31 plate under uniaxial tension. It has been demonstrated that accounting for the initial texture and calibrating the EVPSC-TDT model by using uniaxial tension along the RD and ND permits prediction of the strength anisotropy and strain hardening behavior along other directions, for cases for which the contribution of twinning is large, small and intermediate. Notably, it is not required to parameterize any complex dislocation density-based hardening model.

Acknowledgements

X.Q. Guo and X.B. Mao acknowledge the support of the State Key Development Program for Basic Research of China (Grant No. 2013CB227900), the Joint Funds of the National Natural Science Foundation of China (Grant No.U1261201). A. Chapuis is co-supported by the National Basic Research Program of China ("973" Project) (Grantno.2013CB632204) and the National Natural Science Foundation of China (51101175, 51371203 and 51350110332). P.D. Wu is supported by the Natural Sciences and Engineering Research Council of Canada (NSERC).

References

1. H. Abdolvand, M.R. Daymond. Acta Mater. **60**, 2240(2012).
2. S. Graff, W. Brocks, D. Steglich. Int. J. Plasticity **23**, 1957(2007).
3. C.F. Gu, L.S. Toth, M. Hoffman. Acta Mater. **62**, 212(2014).
4. X.Q. Guo, W. Wu, P.D. Wu, H. Qiao, K. An, P.K. Liaw. Scripta Mater. **69**, 319(2013).
5. T. Hama, Y. Kuchinomachi, N. Hosokawa, H. Fujimoto, H. Takuda. Mater. Sci. Eng. **A551**, 209(2012).
6. J. Jain, P. Cizek, W.J. Poole, M.R. Barnett. Acta Mater. **61**, 4091(2013).
7. F. Kabirian, A.S. Khan, T. Gnäupel-Herlod. Int. J. Plasticity **68**, 1(2015).
8. J.H. Kim, D. Kim, Y.-S. Lee, M.-G. Lee, K. Chung, H.-Y. Kim, R.H. Wagoner. Int. J. Plasticity **50**, 66(2013).

9. M. Knezevic, A. Levinson, R. Harris, R.K. Mishra, R.D. Doherty, S.R. Kalidindi. Acta Mater. **58**, 6230(2010).
10. S.Y. Lee, H. Wang, M.A. Gharghouri, G. Nayyeri, W. Woo, E. Shin, P.D. Wu, W.J. Poole, W. Wu, K. An. Acta Mater.**73**, 139(2014).
11. J. Levesque, K. Inal, K.W. Neale, R.K. Mishra. Int. J. Plasticity **26**, 65(2010).
12. O. Muránsky, M.R. Barnett, D.G. Carr, S.C. Vogel, E.C. Oliver. Acta Mater. **58**, 1503(2010).
13. H. Qiao, S.R. Agnew, P.D. Wu. Int. J. Plasticity **65**, 61(2015).
14. A. Staroselsky, L. Anand. Int. J. Plasticity **19**, 1843(2003).
15. H. Wang, P.D. Wu, M.A. Gharghouri. Mater. Sci. Eng. **A527**, 3588 (2010).
16. H. Wang, P.D. Wu, K.P. Boyle, K.W. Neale. Int. J. Solids Structures **48**, 1000(2011).
17. H. Wang, P.D. Wu, C.N. Tomé, J. Wang. Int. J. Solids Structures **49**, 2155(2012).
18. H. Wang, P.D. Wu, J. Wang. Int. J. Plasticity **47**, 49(2013).
19. W. Wu, H. Qiao, K. An, X.Q. Guo, P.D. Wu, P.K. Liaw. Int. J. Plasticity **62**, 105(2014).
20. P.D. Wu, X.Q. Guo, H. Qiao, D.J. Lloyd. Mater. Sci. Eng. **A625**, 140(2015).
21. G.W. Zhou, M.K. Jain, P.D. Wu, Y.C. Shao, D.Y. Li, Y.H. Peng. Int. J. Plasticity **79**, 19(2016).
22. J.J. Bhattacharyya, F. Wang, P.D. Wu, W. Whittington, H. El Kadiri, S.R. Agnew. Int. J. Plasticity **81**, 123(2016).
23. H. Wang, S.Y. Lee, M.A. Gharghouri, P.D. Wu, S.G. Yoon. Acta Mater. **107**, 414(2016).
24. A.L. Oppedal, H.El Kadiri, C.N. Tomé, S.C. Vogel, M. Horstemeyer.Philos. Mag. **93**, 4311(2013).
25. X.Q. Guo, A. Chapuis, P.D. Wu, S.R. Agnew. Int. J. Solids Structures **64-65**, 42(2015).
26. H. Wang, P.D. Wu, C.N. Tomé, Y. Huang. J. Mech. Phys. Solids **58**, 594(2010).
27. H. Wang, P.D. Wu, C.N. Tomé, J. Wang. Mater. Sci. Eng. **A555**, 93(2012).
28. H. Wang, P.D. Wu, J. Wang, C.N. Tomé. Int. J. Plasticity **49**, 36(2013).
29. A. Chapuis, P. Lei, Q. Liu. Mater. Sci. Eng. **A561**, 167(2013).
30. H. Wang, B. Raeisinia, P.D. Wu, S.R. Agnew, C.N. Tomé. Int. J. Solids Structures **47**, 2905(2010).
31. Y. Wang, H. Choo. Acta Mater. **81**, 83(2014).

Tension-compression asymmetry modelling: strategies for anisotropy parameters identification

Pedro Barros[1,a], José Luís Alves[2], Marta Oliveira[1] and Luís Filipe Menezes[1]

[1] *CEMUC, Department of Mechanical Engineering, University of Coimbra, Pólo II, Rua Luís Reis Santos, Pinhal de Marrocos, 3030-788 Coimbra, Portugal*
[2] *CMEMS, Microelectromechanical Systems Research Unit, University of Minho, Campus de Azurém, 4800-058 Guimarães, Portugal*

Abstract. This work presents details concerning the strategies and algorithms adopted in the fully implicit FE solver DD3IMP to model the orthotropic behavior of metallic sheets and the procedure for anisotropy parameters identification. The work is focused on the yield criterion developed by Cazacu, Plunkett and Barlat, 2006 [1], which accounts for both tension–compression asymmetry and orthotropic plastic behavior. The anisotropy parameters for a 2090-T3 aluminum alloy are identified accounting, or not, for the tension-compression asymmetry. The numerical simulation of a cup drawing is performed for this material, highlighting the importance of considering tension-compression asymmetry in the prediction of the earing profile, for materials with cubic structure, even if this phenomenon is relatively small.

1 Introduction

Plastic forming of metals with hexagonal closed packed (HCP) structure is still challenging due to their low ductility at room temperature and their unusual deformation characteristics, related with very pronounced anisotropy with strong asymmetry between tensile and compressive behavior. This tension-compression asymmetry is present because, unlike cubic metals (both face centered (FCC) and body centered (BCC)), hexagonal metals deform due to the activation of mechanical twinning or non-Schimd type slip at single crystal level. In contrast to slip, twinning is a directional shear mechanism: in general, shear in one direction can produce twinning while shear in the opposite direction cannot. Although less pronounced, the strength differential (SD) effect is also present in materials with cubic structure. However, this effect has been typically neglected, with the yield surface assumed to show a point-symmetry, with respect to the center, such that a stress state and its reverse state have the same flow stress absolute value [2–5].

For the correct modelling of metallic sheets mechanical behavior, it is necessary to perform experimental characterization tests. The most usual is the uniaxial tensile test, allowing to have the yield stresses and anisotropy coefficients in several directions of the sheets plane (generally, 0, 15, 30, 45, 60, 75 and 90 degrees from the rolling direction). In order to improve the description of the yield surface, it is also recommended to determine experimentally the biaxial yield stress and the biaxial anisotropy coefficient [6–8]. However, in order to characterize the metallic sheets mechanical behavior for compression stress states, it is necessary to avoid buckling effects. This requires the use of smaller specimens and, consequently, leading to supplementary difficulties in the acquisition and analysis of the experimental results, particularly for high strain values (see e.g.[6,9–11]).

This work is focused on evaluating the influence of taking into account tension-compression asymmetry in the numerical simulation of a cup drawing. Some details concerning the non-quadratic yield criterion proposed by Cazacu, Plunkett and Barlat (2006) are presented in the first section. Following, a description of the anisotropy parameters identification is shown for the description of the mechanical behavior of a 2090-T3 aluminum alloy. The simulation of a cup drawing is performed in order to evaluate the influence of taking into account, or not, tension-compression asymmetry in the prediction of the earing profile.

2 Constitutive model

The description of both anisotropic behavior and tension-compression asymmetry is possible using the model proposed by Cazacu, Plunkett and Barlat (2006) [1]. The equivalent stress $\bar{\sigma}$ associated with the orthotropic form of the CPB06 yield criterion is defined as

$$\bar{\sigma} = B\left[\left(|s_1| - k\,s_1\right)^a + \left(|s_2| - k\,s_2\right)^a + \left(|s_3| - k\,s_3\right)^a\right]^{\frac{1}{a}}, \quad (1)$$

where a and k are material parameters and s_1, s_2 and s_3 are the principal values of $\mathbf{s} = \mathbf{C}\boldsymbol{\sigma}'$, determined following Barlat *et al.* (1997) [12]. $\boldsymbol{\sigma}'$ is the deviatoric stress tensor

[a] Corresponding author: pedro.barros@dem.uc.pt

and $\mathbf{C}$ is a constant 4th-order tensor which, for 3-D stress conditions involves 9 independent anisotropy coefficients, given by

$$\mathbf{C}=\begin{bmatrix} C_{11} & C_{12} & C_{13} & 0 & 0 & 0 \\ C_{12} & C_{22} & C_{23} & 0 & 0 & 0 \\ C_{13} & C_{23} & C_{33} & 0 & 0 & 0 \\ 0 & 0 & 0 & C_{44} & 0 & 0 \\ 0 & 0 & 0 & 0 & C_{55} & 0 \\ 0 & 0 & 0 & 0 & 0 & C_{66} \end{bmatrix}. \quad (2)$$

Considering isotropic conditions ($\mathbf{C}$ equal to the 4th-order identity tensor), the coefficient k alone allows for the description of the SD effect, giving a direct measure of the ratio between tensile and compressive yield stresses, as

$$k=\frac{1-h}{1+h}, \text{ with } h=\left[\frac{2^a-2\left(\dfrac{\sigma^{\mathrm{T}}}{\sigma^{\mathrm{C}}}\right)^a}{\left(2\dfrac{\sigma^{\mathrm{T}}}{\sigma^{\mathrm{C}}}\right)^a-2}\right]^{1/a}. \quad (3)$$

This means that the $\sigma^{\mathrm{T}}/\sigma^{\mathrm{C}}$ ratio, i.e. the ratio between the yield stress in uniaxial tension and the yield stress in uniaxial compression, for the three principal axes is the same. However, for an anisotropic material it is not possible to define a single $\sigma^{\mathrm{T}}/\sigma^{\mathrm{C}}$ ratio and, consequently, the physical meaning of the k parameter is lost [13]. B is a constant (see Eq. (1)) defined such that $\bar{\sigma}$ reduces to the tensile yield stress in the rolling direction and is given by

$$B=\left[\frac{1}{\left(|\phi_1|-k\,\phi_1\right)^a+\left(|\phi_2|-k\,\phi_2\right)^a+\left(|\phi_3|-k\,\phi_3\right)^a}\right]^{\frac{1}{a}}, \quad (4)$$

with

$$\begin{Bmatrix}\phi_1\\ \phi_2\\ \phi_3\end{Bmatrix}=\begin{Bmatrix}(2/3)C_{11}-(1/3)C_{12}-(1/3)C_{13}\\ (2/3)C_{21}-(1/3)C_{22}-(1/3)C_{23}\\ (2/3)C_{31}-(1/3)C_{32}-(1/3)C_{33}\end{Bmatrix}. \quad (5)$$

For $a=2$, $k=0$ and $\mathbf{C}$ equal to the 4th-order identity tensor, the von Mises yield criterion is recovered. Convexity is guaranteed for any integer $a\geq 2$ and $k\in[-1,1]$ [1]. In this work the analysis performed considers $a=2$.

In order to highlight the influence of taking, or not, into account tension-compression asymmetry, the yield criterion proposed by Barlat *et al.* (1991), labeled YLD91, is also used in this work. The equivalent stress $\bar{\sigma}$ associated with the orthotropic form of the YLD91 yield criterion is given as

$$\bar{\sigma}=\left\{\frac{1}{2}\left[|s_1-s_2|^m+|s_2-s_3|^m+|s_1-s_3|^m\right]\right\}^{\frac{1}{m}}, \quad (6)$$

where s_1, s_2, s_3 are the principal values of $\mathbf{s}=\mathbf{L}\boldsymbol{\sigma}'$ and m is the exponent that controls the shape of the isotropic yield surface, usually taken as 6 or 8 for body centered (BCC) and face centered (FCC) materials, respectively [14]. $\boldsymbol{\sigma}'$ is the deviatoric stress tensor and $\mathbf{L}$ is a constant 4th-order tensor given by

$$\mathbf{L}=\begin{bmatrix} (c_2+c_3)/3 & -c_3/3 & -c_2/3 & 0 & 0 & 0 \\ -c_3/3 & (c_3+c_1)/3 & -c_1/3 & 0 & 0 & 0 \\ -c_2/3 & -c_1/3 & (c_1+c_2)/3 & 0 & 0 & 0 \\ 0 & 0 & 0 & c_4 & 0 & 0 \\ 0 & 0 & 0 & 0 & c_5 & 0 \\ 0 & 0 & 0 & 0 & 0 & c_6 \end{bmatrix}, \quad (7)$$

which, for a 3-D stress state, involves only 6 independent anisotropy coefficients. Therefore, although $\mathbf{C}$ and $\mathbf{L}$ are defined using the same formalism of the linear transformation, proposed by Barlat *et al.* (1991) [12], they lead to different sets of anisotropic parameters.

The yield condition is defined as

$$F(\bar{\sigma},Y)=\bar{\sigma}-Y=0, \quad (8)$$

where Y is the flow stress, which depends of the hardening law selected, such that the yield stress is Y_0. The adopted constitutive model considers an associated flow rule, meaning that the yield function serves as the plastic potential for determining the plastic strain increment. The plastic strain rate tensor, $\mathbf{D}^{\mathrm{p}}$, is determined according to

$$\mathbf{D}^{\mathrm{p}}=\dot{\lambda}\frac{\partial F(\bar{\sigma},Y)}{\partial\boldsymbol{\sigma}}, \quad (9)$$

The equivalent stress is given by Eq. (1) or (6), depending on the yield criterion adopted, and $\dot{\lambda}$ is a scalar value designated by plastic multiplier that can be demonstrated to be equal to the equivalent plastic strain rate, $\dot{\bar{\varepsilon}}^{\mathrm{p}}$. The equivalent plastic strain is defined as

$$\bar{\varepsilon}^{\mathrm{p}}=\int_0^t\dot{\bar{\varepsilon}}^{\mathrm{p}}\mathrm{d}t=\int_0^t\frac{\boldsymbol{\sigma}'}{\bar{\sigma}}:\mathbf{D}^{\mathrm{p}}\mathrm{d}t. \quad (10)$$

3 Material parameters identification

The anisotropy coefficients should be determined such that the yield criterion reproduces the material's mechanical behavior as close as possible. The procedure adopted is based on the minimization of an error function, evaluating the difference between the estimated values and the experimental ones, as follows

$$F(\mathbf{A}) = \sum_{\theta=0}^{90} w_{\sigma_\theta^T}\left(\sigma_\theta^{Y_T}(\mathbf{A})/\sigma_\theta^{Y_T}-1\right)^2 + \sum_{\theta=0}^{90} w_{\sigma_\theta^C}\left(\sigma_\theta^{Y_C}(\mathbf{A})/\sigma_\theta^{Y_C}-1\right)^2 + \sum_{\theta=0}^{90} w_{r_\theta}\left(r_\theta(\mathbf{A})/r_\theta-1\right)^2 + w_{\sigma_b}\left(\sigma_b(\mathbf{A})/\sigma_b-1\right)^2 + w_{r_b}\left(r_b(\mathbf{A})/r_b-1\right)^2 \quad (11)$$

where $\mathbf{A}$ represents the set of yield parameters. $\sigma_\theta^{Y_T}$, $\sigma_\theta^{Y_C}$ and r_θ are the experimental yield stresses in tension, compression and r-values, respectively, obtained from the uniaxial tensile and compression tests for a specific orientation (θ) to RD. σ_b is the experimental yield stress obtained from the equibiaxial test and r_b can be obtained from the disc compression test. $\sigma_\theta^{Y_T}(\mathbf{A})$, $\sigma_\theta^{Y_C}(\mathbf{A})$, $r_\theta(\mathbf{A})$, $\sigma_b(\mathbf{A})$ and $r_b(\mathbf{A})$ are the correspondent values predicted from the adopted yield criterion. The weighting factors, $w_{\sigma_\theta^T}$, $w_{\sigma_\theta^C}$, w_{r_θ}, w_{σ_b} and w_{r_b} are used to balance the influence of the experimental data. The anisotropy coefficients C_{44} and C_{55}, of CPB06, and c_4 and c_5, for the YLD91, are associated with out-of-plane properties, difficult to evaluate for metallic sheets. Therefore, they are considered equal to the isotropic value of 1.0. In brief, the number of parameters to be identified for the CPB06 yield criterion is eight: seven anisotropy parameters and k. For the YLD91 yield criterion this number reduces to four, i.e. the anisotropy parameters, since $m = 8$.

Both yield criteria were implemented in the in-house code DD3MAT, enabling the use of all the experimental data defined in Eq. (11) in the parameters identification procedure. The optimization algorithm implemented in DD3MAT is based on a downhill simplex method, which is a derivative free method. For the CPB06 yield criterion, the identification procedure assumes a range of $[-1,1]$ for k and a range of $[-5,5]$ for all the other parameters. For the YLD91, the range considered for the anisotropy parameters is $[0,5]$.

3.1 Application to a 2090-T3 aluminum alloy.

The experimental data for the 2090-T3 aluminum alloy is presented in Table 1, and considers results from both uniaxial tension and compression tests, as well as for biaxial tension and disk compression [15].

The anisotropy parameters were identified for both the CPB06 and the YLD91, in order to compare their ability to accurately describe the materials' mechanical behavior. Figure 1 presents the comparison between numerical and experimental r-values as well the yield stress in tension (labelled "T") and compression (labelled "C"), for both yield criteria.

Table 1. Experimental data obtained from uniaxial tension, compression, biaxial tension and disk compression tests, for the 2090-T3 aluminum alloy [15].

Test direction [°]	r-value	$\sigma_\theta^{Y_T}$ [MPa]	$\sigma_\theta^{Y_C}$ [MPa]
0	0.21	279.62	248.02
15	0.33	269.72	260.75
30	0.69	255.00	255.00
45	1.58	226.77	237.75
60	1.05	227.50	245.75
75	0.55	247.20	263.75
90	0.69	254.45	266.48
σ_b		289.40	
r_b	0.670		

Regarding the r-values, both yield criteria seem to describe well the materials mechanical behavior, except for the test performed at 45°. The CPB06 yield criterion presents a better prediction, mainly between 60° and 90°. Also, even though no experimental data is available, the numerically predicted compression r-values are also represented in this figure, highlighting the difference in behavior of this material in tension and compression. Note that the r-values in compression are higher than the ones in tension for the rolling direction until about 45°, at which they become lower until the transverse direction.

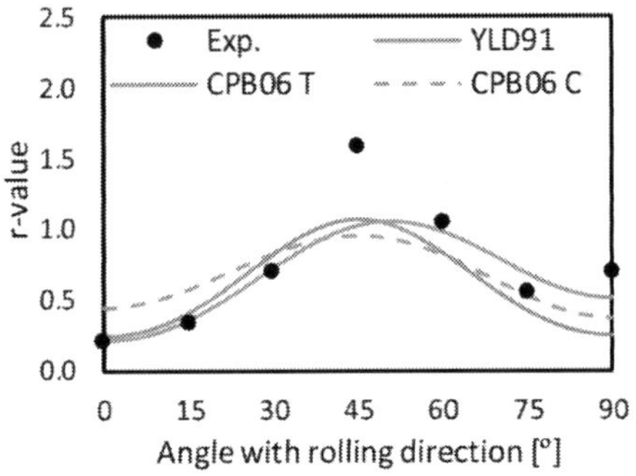

(a)

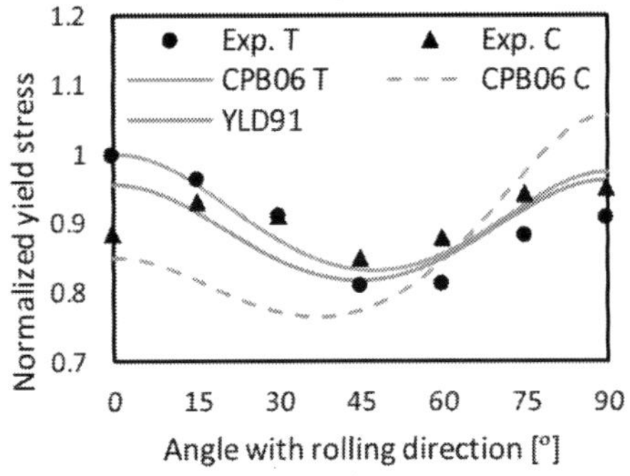

(b)

Figure 1. Experimental and predicted (a) r-values and (b) normalized yield stresses for the 2090-T3 aluminum alloy.

As for the yield stresses in tension both yield criteria show a similarly good agreement with the experimental results, with the CPB06 being more accurate mainly for angles closer to the rolling direction. Considering the yield stresses in compression, note that the YLD91 prediction is the same as for the tension case. The CPB06 seems to predict the main shape of the yield stresses evolution, even though it under predicts the values from 0° to 45° and over predicts from 45° to 90°.

Figure 2 presents the normalized yield surfaces ($Y_0 = 279.62$ MPa) predicted, for both yield criteria, for the 2090-T3 aluminum alloy, in the plane σ_1, σ_2, with $\sigma_3 = 0$. The YLD91 predicted surface presents a point symmetry, implying that the behavior of the material in tension is the same as in compression. The CPB06 predicted yield surface is not symmetric, thus better representing the material behavior for different stress states. Note that the condition $Y_0 = \sigma_0^{Y_T}$ is imposed by the CPB06 yield criterion [13], while for the YLD91 this condition is not imposed in the anisotropy parameters identification procedure.

When compared with the experimental results, the CPB06 closely predicts the yield stresses both in tension and compression, thus predicting the different ratios obtained for the two principal directions shown. Table 2 shows the ratios obtained for these axes and also the one predicted along the thickness direction, corresponding to the biaxial one (with $\sigma_3^T/\sigma_3^C = \sigma_b^C/\sigma_b^T$). In fact, the numerically obtained ratios for the σ_1 and σ_2 axis are very close to the ones obtained experimentally. Note that the experimental ratio for the σ_3 is not available, thus only the numerically predicted one is shown. Regarding the biaxial values, the experimental and predicted σ_b and r_b are presented in Table 3.

As for the σ_b value, none of the yield criteria show a good agreement with the experimental value, with the YLD91 yield criterion predicting a slightly closer value. As for the r_b value, both yield criteria over predict its value when comparing with the experimental one. Finally, Table 4 shows the anisotropy parameters obtained for both the CPB06 and YLD91 yield criteria.

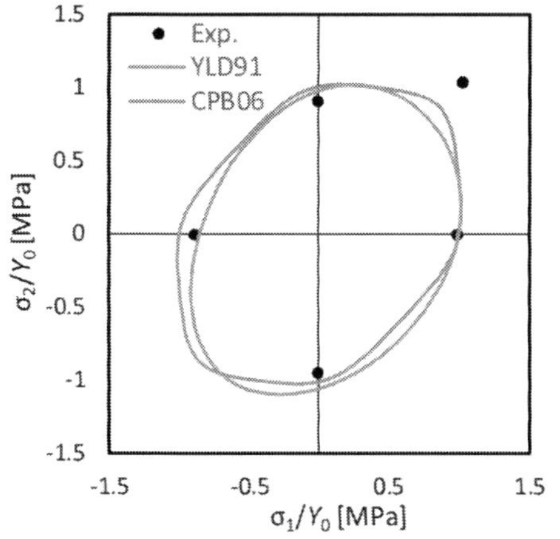

Figure 2. Predicted yield surfaces, in the σ_1, σ_2 plane, for the YLD91 and CPB06 yield criteria.

Table 2. Ratios obtained for the three principal axis for the 2090-T3 aluminum alloy.

	(σ_1^T/σ_1^C)	(σ_2^T/σ_2^C)	(σ_3^T/σ_3^C)
Experimental	1.1274	0.9549	-
CPB06	1.1756	0.9217	1.0361

Table 3. Experimental and numerically predicted biaxial tensile values for the 2090-T3 aluminum alloy.

	σ_b	r_b
Experimental	289.40	0.670
YLD91	230.83	0.971
CPB06	219.42	0.968

Table 4. Anisotropy parameters identified for both the YLD91 and CPB06 yield criteria, for the 2090-T3 aluminum alloy.

CPB06		YLD91	
C_{11}	0.9353	c_1	1.0692
C_{22}	0.4171	c_2	1.3002
C_{33}	0.7278	c_3	0.8560
C_{66}	0.7758	c_6	1.2132
C_{23}	-0.1194	m	8
C_{13}	0.0971		
C_{12}	0.1591		
k	0.0922		

4 Cup drawing of a circular blank

The numerical simulation of a cup drawing of a circular cup was performed in order to analyze the influence of the SD effect in the earing profile. This test is based on the work of Yoon et al. (2000) [10].

4.1 Problem description

The schematic of the cup drawing process, together with the main dimensions, is shown in Figure 3. Due to geometrical and material symmetries, only a quarter of the global structure is modelled. The contact with friction conditions is described by Coulomb's law, using a constant friction coefficient, μ, of 0.1. The blank sheet is circular in shape with a radius of 158.76 mm and thickness of 1.6 mm. The blank-holder force has a value of 22.2 kN, which correspond to the minimum value predicted to avoid wrinkles [10].

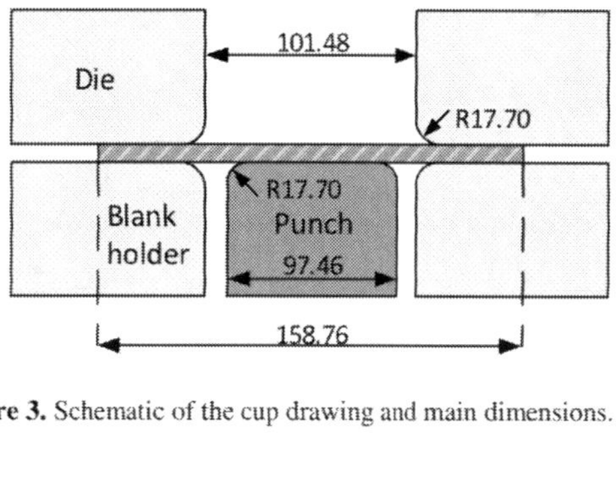

Figure 3. Schematic of the cup drawing and main dimensions.

The sheet is discretized with 8-node hexahedral finite elements, combined with a selective reduced integration technique [16]. Figure 4 shows the in-plane sheet discretization used. Two elements were used in the thickness direction, leading to a total of 19648 elements.

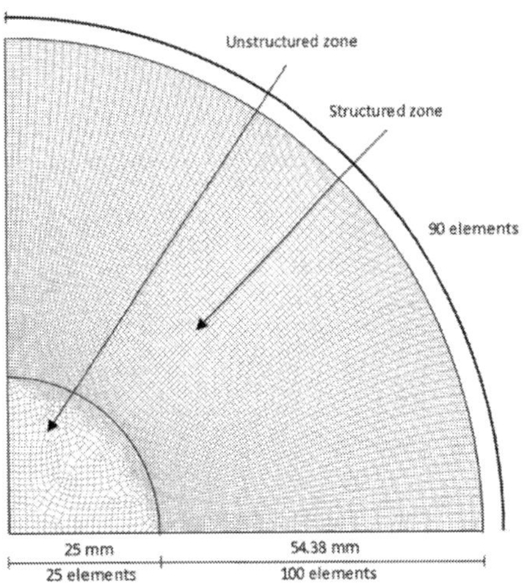

Figure 4. In-plane blank sheet discretization.

The material's mechanical behavior is assumed to be isotropic in the elastic regime, being described by the Young's modulus, E, and the Poisson ratio, υ. The plastic behavior is described using an isotropic work hardening Swift type law, given by

$$Y\left(\bar{\varepsilon}^{\mathrm{p}}\right)=K\left(\varepsilon_0+\bar{\varepsilon}^{\mathrm{p}}\right)^n, \qquad (12)$$

where Y is the flow stress and $\bar{\varepsilon}^{\mathrm{p}}$ is the equivalent plastic strain. The elastic properties and material parameters of the hardening law used for the simulation are presented in Table 5. The YLD91 and CPB06 anisotropy parameters used in the simulation are the ones presented in Table 4.

Table 5. Elastic properties and material parameters of the work hardening law.

Elastic properties	Isotropic hardening (Swift law)
$E = 74$ [GPa]	$K = 646$ [MPa]
$\upsilon = 0.34$	$\varepsilon_0 = 0.025$
	$n = 0.227$

4.2 Results and discussion

The numerically predicted punch force and blank holder displacement with punch displacement are presented in Figure 5. Regarding the blank holder displacement, the results indicate a higher gap with the die for the CPB06, which corresponds to a thicker flange. In order to avoid excessive ironing of the flange by the blank holder, when the displacement inverts, a limit of 0.2 mm is imposed.

The punch force evolution is very similar for both yield criteria throughout the process. At around 35 mm a pronounced decrease in the force is noticed. For the CPB06, this stepped decrease occurs sooner, indicating that the blank loses contact with the blank holder sooner in the process. This indicates that the height of the formed cup is lower, for the CPB06.

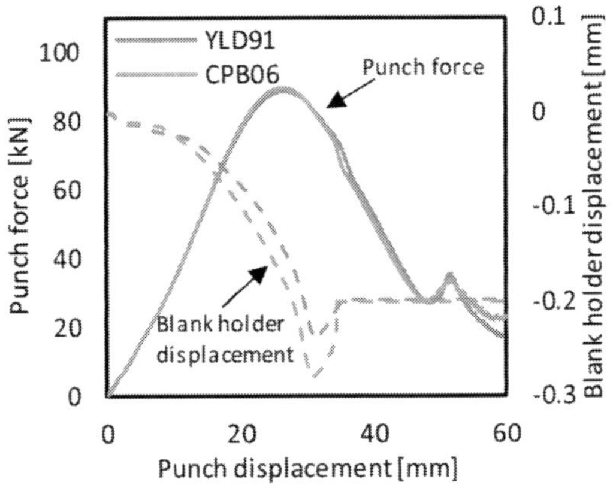

Figure 5. Numerically predicted punch force and blank holder displacement with punch displacement.

The comparison between experimental and numerically predicted cup height *vs.* angle from rolling direction, after drawing, is presented in Figure 6. The YLD91 predicts a height that is higher than the one predicted with the CPB06, between 0° and 75°. Note that the shape of the ear for both yield criteria is nearly the same. Negligible differences can be seen between 60° and 90°. Also, the CPB06 predicted location for the maximum height seems to correlate better with the experimental one.

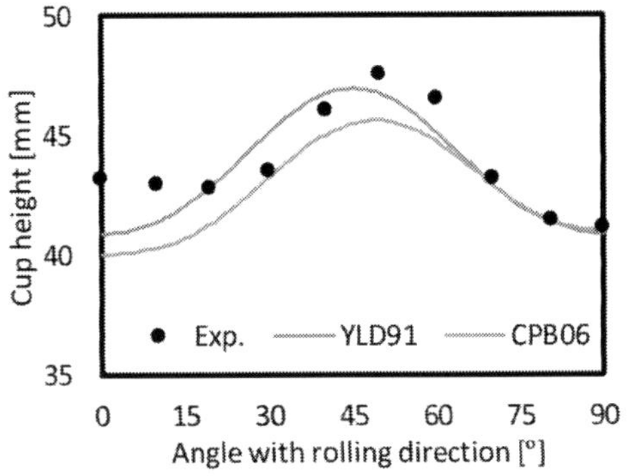

Figure 6. Comparison between experimental and numerically predicted cup height vs. angle from rolling direction.

In order to understand the differences in the predicted cup height, the stress and strain states evolutions during the cup forming process were analyzed for points located on the outermost flange, at every 15° from the rolling direction. This analysis was performed assuming that, as long as the stress component in the thickness direction is small, the outer flange will be submitted to a compression stress state, as shown in Figure 7. Therefore, the material behavior will be dictated by the stress and r-values predicted for this stress state, which for the YLD91 are the same as for a tensile stress state.

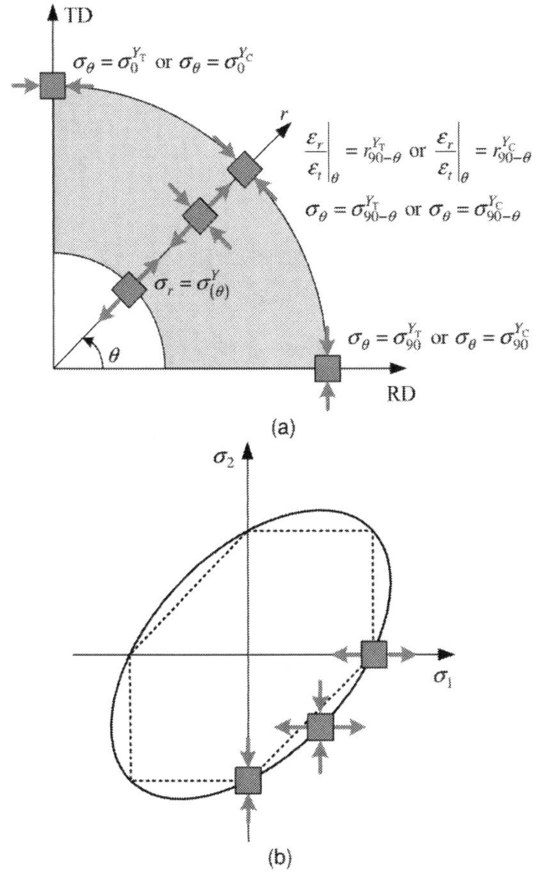

Figure 7. Deformation of an element on the flange: (a) stress states on the flange and (b) stress states on the yield surface (adapted from Yoon et al. 2011 [17]).

Figure 8 presents the evolution of the ratio between the plastic strain along the radial and the thickness direction, estimated from the numerical results, and the r-values estimated, for both yield criteria. The results are presented to a maximum punch stroke of 50 mm, corresponding approximately to the instant at which the cup completely enters the die cavity. Considering the YLD91 yield criterion, the $\varepsilon_r / \varepsilon_t$ and estimated r-values show a good agreement, meaning that the stress state component in the thickness direction is small, thus not altering the compression stress state in the circumferential direction, for the material points located in the outer flange. Also, it is possible to observe that the yielding of the points occurs at approximately the same instant (8 mm). Regarding the CPB06 yield criterion, the agreement between the $\varepsilon_r / \varepsilon_t$ and estimated r-values (compression stress state, as shown in Figure 1 (a)) is also very good. Note that, in this case, the material points located in the outermost flange do not yield at the same instant. The point corresponding to 0° yields last, for a displacement of about 11.5 mm, followed of the point at 15°, for a displacement of about 8.5 mm. This behavior is coherent with the fact that the cup height predicted by the CPB06 yield criterion is lower for angles closer to the rolling direction, i.e. since it yields later, the final cup height is lower.

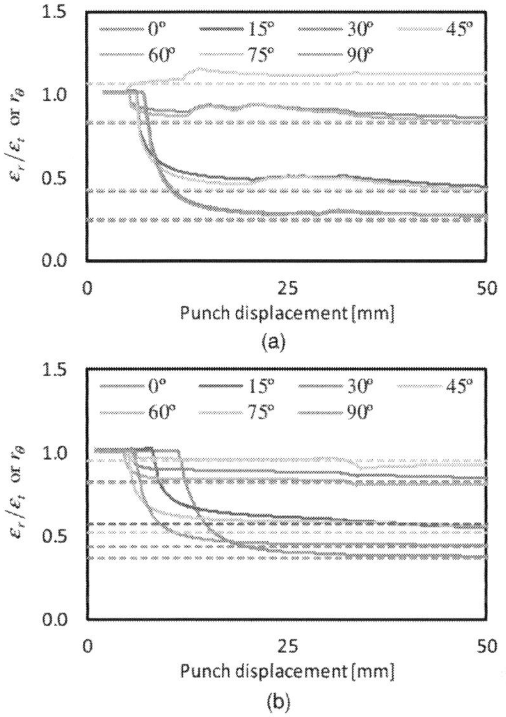

Figure 8. Evolution, with the punch displacement, of $\varepsilon_r / \varepsilon_t$ (solid lines) and r-values (dashed lines) estimated with (a) YLD91 and (b) CPB06 yield criteria.

Figure 9 presents the evolution of the ratio between the circumferential stress, estimated from the numerical results, and the ratio between the yield stress, estimated with both the YLD91 and CPB06 yield criteria, and Y_0. For the CPB06 yield criterion, the yield stresses considered are the ones predicted for compression (Figure 1 (b)). For both yield criteria there is a good correlation between the circumferential stress normalized by the flow stress, σ_θ / Y, and the normalized yield stress values estimated with both the YLD91 and CPB06 yield criteria, $\sigma_{90-\theta} / Y_0$. These results are in accordance with the fact that the compression yield stresses in the rolling direction will have a direct impact on the material behavior at the transverse direction, as shown in Figure 7. In fact, the behavior of the rim in the direction defined by θ (with the rolling direction) is controlled by the material compression properties in the direction defined by $90-\theta$ [17][18]. The fact that the CPB06 compression yield stress at 90° is over predicted, has as consequence the pronounced decrease in height of the earing profile, mainly between 0° and 60°. Also, the contact with friction, in this problem, is not expected to alter the geometry of the earing profile, since the blank holder force is almost uniformly distributed in the flange [18].

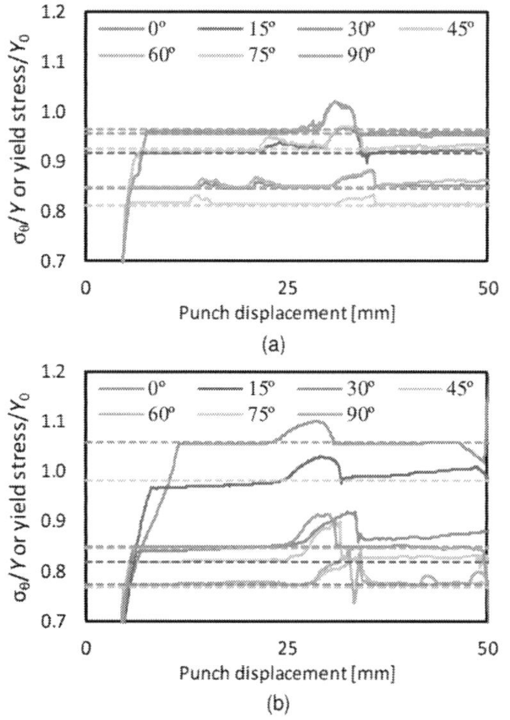

Figure 9. Evolution, with punch displacement of σ_θ / Y (solid lines) and the ratio between yield stress (dashed lines) estimated with (a) YLD91 and (b) CPB06 yield criteria and Y_0 .

From Figure 9 it is possible to note that, regarding the YLD91 yield criterion, all the points of the flange yield at approximately the same time, thus being coherent with the findings from Figure 8. For the CPB06, for angles of 0° and 15° from the rolling direction, the material yields for a punch displacement higher than for the other directions. Again, this is coherent with the results from Figure 8, but also from Figure 1 (b). The fact that the normalized yield stress values estimated for a compression test performed at 75° and 90° to the rolling direction are higher than 1.0, on the contrary with the others, delays the yielding for material points located between the rolling direction and 15°. This also allows saying that the fact that the cups height predicted with the CPB06 is lower for points closer with the rolling direction is directly related with the later yielding of this material.

Figure 10 presents the thickness strains evolution measured experimentally and the ones predicted, for both the YLD91 and CPB06 yield criteria, for the rolling and transverse direction. Regarding the YLD91 yield criterion, the evolution in the rolling and transverse directions are the same, with no observable differences. On the other hand, the predictions for the CPB06 yield criterion show that the thickness strain in the transverse direction is always higher than the one in the rolling direction. This result is, in fact, coherent with the experimental results even thought, for both yield criteria, the strains are globally over predicted. The fact that the thickness strain in the rolling direction is lower, for the CPB06, also correlates with the lower cup height in this direction.

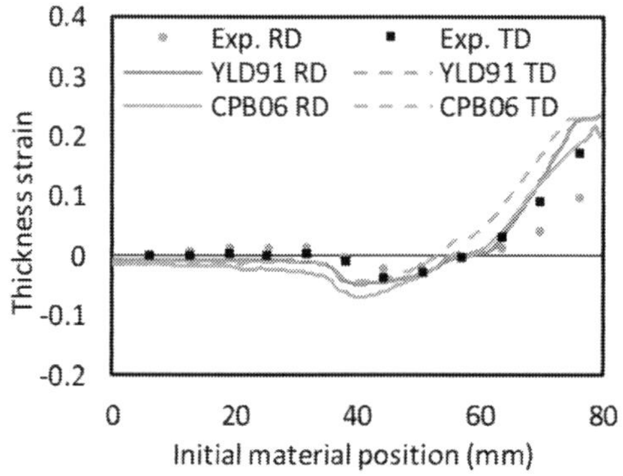

Figure 10. Evolution of the predicted and measured strain, for the YLD91 and CPB06 yield criteria, regarding the rolling and transverse directions.

5 Conclusions

In this work, the anisotropy parameters for the CPB06 yield criterion, which accounts for tension-compression asymmetry, as well as for the YLD91, were determined. The results show that both yield criteria can describe the yield stresses and r -values in tension, with the CPB06 also describing the materials' mechanical behavior for compression stress states.

The results for the numerical simulation of the cup drawing show that there is a direct influence of the yield criteria prediction of the compression yield stresses on the final shape of the cup, since this is the main stress state in the cups' flange during drawing.

Thus, a good prediction of the material directionalities, including the compression yield stresses, is important for the numerical simulation of drawing processes, even for materials exhibiting a small tension-compression asymmetry since it controls the overall accuracy of the earing profile.

Acknowledgements

The authors gratefully acknowledge the financial support of the Portuguese Foundation for Science and Technology (FCT) under projects with reference PTDC/EMS-TEC/0702/2014 and PTDC/EMS-TEC/6400/2014. The first author is also grateful to the FCT for the PhD grant SFRH/BD/98545/2013.

References

1. O. Cazacu, B. Plunkett, F. Barlat, Int. J. Plast. 22, 1171–1194 (2006).
2. O. Cazacu, F. Barlat, Math. Mech. Solids 6, 613–630 (2001).
3. F. Barlat, H. Aretz, J.W. Yoon, M.E. Karabin, J.C. Brem, R.E. Dick, Int. J. Plast. 21, 1009–1039 (2005).
4. D. Banabic, Sheet Metal Forming Processes, Springer-

Verlag Berlin Heidelberg, 2010.
5. J. Yoon, F. Barlat, R. Dick, M. Karabin, Int. J. Plast. 22, 174–193 (2006).
6. D.J. Lege, F. Barlat, J.C. Brem, Int. J. Mech. Sci. 31, 549–563 (1989).
7. K. Pöhlandt, D. Banabic, K. Lange, in:, ESAFORM 2002, Krakow, Poland., 2002, pp. 723–727.
8. F. Barlat, J.C. Brem, J.W. Yoon, K. Chung, R.E. Dick, D.J. Lege, F. Pourboghrat, S.H. Choi, E. Chu, Int. J. Plast. 19, 1297–1319 (2003).
9. F. Barlat, D.J. Lege, J.C. Brem, C.J. Warren, in:, T.. Lowe, A.. Rollett, P.. Follansbee, G.. Daehn (Eds.), TMS Annu. Meet., TMS, New Orleans, LA, 1991, pp. 189–203.
10. J.W. Yoon, F. Barlat, K. Chung, F. Pourboghrat, D.Y. Yang, Int. J. Plast. 16, 1075–1104 (2000).
11. M. Tritschler, a. Butz, D. Helm, G. Falkinger, J. Kiese, Int. J. Mater. Form. 7, 259–273 (2014).
12. F. Barlat, Y. Maeda, K. Chung, M. Yanagawa, J.C. Brem, Y. Hayashida, D.J. Lege, K. Matsui, S.J. Murtha, S. Hattori, R.C. Becker, S. Makosey, J. Mech. Phys. Solids 45, 1727–1763 (1997).
13. P.D. Barros, M.C. Oliveira, J.L. Alves, A. Andrade-Campos, L.F. Menezes, in:, P. Hora (Ed.), Institute of Virtual Manufacturing, ETH Zurich, Zurich, 2015, pp. 149–154.
14. F. Barlat, D.J. Lege, J.C. Brem, Int. J. Plast. 7, 693–712 (1991).
15. F. Moayyedian, M. Kadkhodayan, (2014) 2804–2826.
16. T.J.R. Hughes, Int. J. Numer. Methods Eng. 15, 1413–1418 (1980).
17. J.W. Yoon, R.E. Dick, F. Barlat, Int. J. Plast. 27. 1165–1184 (2011).
18. P.D. Barros, D.M. Neto, J.L. Alves, M.C. Oliveira, L.F. Menezes, Rom. J. Tech. Sci. - Appl. Mech. 60, 105–136 (2015).

Fracture prediction with a material model based on stress-rate dependency related with non-associated flow rule

Tetsuo Oya[1,a], Jun Yanagimoto[2], Koichi Ito[3], Gen Uemura[3] and Naomichi Mori[3]

[1]*Keio University, Graduate School of Science and Technology, 2238522 Hiyoshi 3-14-1, Kohoku-ku, Yokohama, Japan*
[2]*The University of Tokyo, Institute of Industrial Science, 1538505 Komaba 4-6-1, Meguro-ku, Tokyo, Japan*
[3]*M & M Research, 1500001 Jingumae 4-15-3, Shibuya-ku, Tokyo, Japan*

Abstract. In this paper, a novel material model for anisotropic metals is proposed. Fracture prediction using the proposed model is based on stress-rate dependency related with non-associated flow rule in which an arbitrary order function for both the yield and plastic potential functions are used. In addition, an explicit formulation of the equivalent plastic strain increment that is plastic-work-conjugated with the equivalent stress is also presented. Then, fracture prediction analyses are conducted by using the 3D local bifurcation theory.

1 Introduction

Modern demand for material models is various and high level; therefore, more precise and physically valid material model has been required. Fracture prediction for sheet and bulk metal forming is one of such demands from industry and this paper focuses on it. To conduct reliable fracture prediction during metal forming, a material model that has high physical resolution should be constructed and used together with objective fracture criterion. For this purpose, a novel material model for anisotropic metals has been developed based on the non-associated flow rule by the authors. This paper shows its outline and application on the fracture prediction problem, which has been carried out using the 3D local bifurcation theory.

Most of the phenomenological plastic constitutive models are based on the framework called the associated flow rule, which is a type of normality rule having a yield function as its plastic potential function. In metal forming analysis, the use of the associated flow rule is not necessarily a best choice for anisotropic metals such as cold-rolled steels and aluminum alloys. In metal forming research, many material models have been studied to overcome such problems. The origin of the anisotropic material model is considered to be Hill's model [1]. This model was developed on the basis of von Mises's isotropic material model and is still frequently used because of its acceptable prediction results for most general metals.

Many phenomenological yield functions have been proposed. For example, Karafillis and Boyce [2], and Barlat and Lian [3] presented yield functions for anisotropic metals. Most of their expressions are too complex to explicitly define the equivalent plastic strain increment, which is work-conjugated with the equivalent stress. This is because the yield function is used to express both flow stress anisotropy and deformation anisotropy. A few researchers have proposed material models based on the non-associated flow rule. Stoughton and Yoon [4] proposed a model based on the non-associated flow rule that adopted Hill's quadratic yield function. Recently, Safaei et al. [5] proposed a model based on the non-associated flow rule that adopted Barlat's yield function [6]. These models include disadvantages of the original yield functions they made the basis; therefore, there remains difficulties toward the practical use of non-associated flow rule models.

In this paper, a novel material model for anisotropic metals is proposed. The model is based on the non-associated flow rule with arbitrary order function for both the yield and plastic potential functions. In addition, an explicit formulation of the equivalent plastic strain increment that is plastic-work-conjugated with the equivalent stress is also presented. Then, fracture prediction analyses are conducted by using the 3D local bifurcation theory.

[a] Corresponding author: oya@sd.keio.ac.jp

2 Material model

In this section, the orthodox anisotropic plasticity model proposed by Hill, which is called Hill48 model in this paper, is briefly introduced first because its formulation is essential to our model. Then, the proposed model is introduced, which is a natural extension of Hill48 for arbitrary order functions and based on the non-associated flow rule. The definitions of the yield function, plastic potential function, and the plastic strain increment are presented.

2.1 Hill48 model

To obtain a concise expression for the yield function, the following deviatoric stress vector $\boldsymbol{s}$ and flow stress anisotropy matrix $\boldsymbol{A}$ are introduced:

$$\begin{aligned}\boldsymbol{s} &= \left\{s_1, s_2, s_3, s_4, s_5, s_6\right\}^T \\ &= \left\{s_y - s_z, s_z - s_x, s_x - s_y, t_{yz}, t_{zx}, t_{xy}\right\}^T ,\end{aligned} \quad (1)$$

$$\boldsymbol{A} = \begin{bmatrix} F & 0 & 0 & 0 & 0 & 0 \\ 0 & G & 0 & 0 & 0 & 0 \\ 0 & 0 & H & 0 & 0 & 0 \\ 0 & 0 & 0 & 2L & 0 & 0 \\ 0 & 0 & 0 & 0 & 2M & 0 \\ 0 & 0 & 0 & 0 & 0 & 2N \end{bmatrix} , \quad (2)$$

where F, G, H, L, M, and N are Hill's anisotropic parameters. Then, the yield function $f(\boldsymbol{\sigma})$ is defined as follows:

$$f(\boldsymbol{\sigma}) = \bar{\sigma} = \sqrt{\frac{3}{2(F+G+H)}\left(\boldsymbol{s}^T \cdot \boldsymbol{A} \cdot \boldsymbol{s}\right)} , \quad (3)$$

where $\bar{\sigma}$ is the equivalent stress. Using the plastic work conjugacy, the definition of the equivalent plastic strain increment can be explicitly obtained as

$$d\varepsilon^P = \sqrt{\frac{2(F+G+H)}{3}\left(\boldsymbol{B} \cdot d\boldsymbol{e}^P\right)^T \cdot \left(\boldsymbol{A} \cdot \boldsymbol{B} \cdot d\boldsymbol{e}^P\right)} , \quad (4)$$

where $d\boldsymbol{e}^P$ is the strain increment vector. Here, the matrix $\boldsymbol{B}$ is an inverse matrix of a compliance matrix $\boldsymbol{C}$ that is obtained from the following flow rule relationship with a plastic multiplier $d\Lambda$:

$$d\boldsymbol{e}^P = \frac{\partial g(\boldsymbol{\sigma})}{\partial \boldsymbol{\sigma}} d\Lambda = \boldsymbol{C} \cdot \boldsymbol{s} d\Lambda . \quad (5)$$

2.2 Proposed model

After the Hill48 model, many useful material models have been developed based on the associated flow rule and provided good results in some cases; however, the intrinsic problem was not solved. Namely, in an anisotropic sheet metal, the directions of yield stress evolution and plastic strain increment do not generally coincide; in other words, the shape of the yield surface and the plastic potential, which prescribes the direction of the plastic strain increment, do not match. If a yield function is devised to exhibit good agreement with yield stresses obtained from experiments, the representation of stress anisotropy will be achieved but that of deformation anisotropy will be lost. Therefore, the authors consider it is natural to adopt the non-associated flow rule to represent two different anisotropy independently [7].

Following the formulation of the above-explained Hill's anisotropic model, the proposed model is developed. A vector $\boldsymbol{s}_m$, which is a deviatoric stress vector with its components having powers of m, is defined.

$$\begin{aligned}\boldsymbol{s}_m &= \left\{s_1^m, s_2^m, s_3^m, s_4^m, s_5^m, s_6^m\right\}^T \\ &= \left\{(\sigma_y - \sigma_z)^m, (\sigma_z - \sigma_x)^m, (\sigma_x - \sigma_y)^m, \tau_{yz}^m, \tau_{zx}^m, \tau_{xy}^m\right\}^T\end{aligned} \quad (6)$$

Using the same anisotropic matrix $\boldsymbol{A}$ as that in the Hill48 model, the quadratic form $\boldsymbol{s}_m \cdot \boldsymbol{A} \cdot \boldsymbol{s}_m$ is obtained. This takes the form of Hill's second-order yield function with its components having powers of $2m$. In the proposed model, we define the yield function $f(\boldsymbol{\sigma})$ as being equal to the equivalent stress $\bar{\sigma}$; namely, the expression is

$$f(\boldsymbol{\sigma}) = \bar{\sigma} = \sqrt[2m]{\frac{3}{2(F+G+H)}\left(\boldsymbol{s}_m^T \cdot \boldsymbol{A} \cdot \boldsymbol{s}_m\right)} . \quad (7)$$

This higher-order function preserves the form of Hill's quadratic yield function, that is, it contains the same anisotropic parameters F, G, H, L, M, and N. This feature is important because it is possible to construct a higher-order yield function by changing the power m without increasing the number of undetermined variables. Therefore, the burden of determining anisotropic parameters through numerous experiments can be avoided.

In our non-associated flow rule, a function different from the yield function is adopted as a plastic potential function, which provides the direction of the plastic strain increment of the subsequent state of current stress. In this paper, the previously introduced function $f(\boldsymbol{\sigma})$ is used as the yield function, and another function $g(\boldsymbol{\sigma})$ that takes the same form as $f(\boldsymbol{\sigma})$ but has different anisotropic parameters F^*, G^*, H^*, L^*, M^*, and N^* is adopted in the plastic potential function. In the expression, asterisks are used to distinguish $f(\boldsymbol{\sigma})$ from $g(\boldsymbol{\sigma})$. For example, the anisotropy matrix $\boldsymbol{A}$ is changed to $\boldsymbol{A}^*$, in which the original parameters F, G, H, L, M, and N are also changed to F^*, G^*, H^*, L^*, M^*, and N^*, respectively. To express another order of the function, the power variable n is used instead of m. Note that the parameters m and n are positive integer in this study. Thus, the plastic potential function of the proposed model takes the form of

$$g(\boldsymbol{\sigma}) = \bar{\sigma}^* = \sqrt[2n]{\frac{3}{2(F^*+G^*+H^*)}\left(\boldsymbol{s}_n^T \cdot \boldsymbol{A}^* \cdot \boldsymbol{s}_n\right)} . \quad (8)$$

In the next subsection, how to determine the both set of anisotropic parameters is described.

From the definition of the plastic work, an explicit expression for our equivalent plastic strain increment is obtained as

$$d\varepsilon^P = \frac{n\bar{\sigma}^{*n}}{\bar{\sigma}}\sqrt{\frac{2(F^*+G^*+H^*)}{3}\left(\boldsymbol{B}_n^* \cdot d\boldsymbol{e}^P\right)^T \cdot \left(\boldsymbol{A}^* \cdot \boldsymbol{B}_n^* \cdot d\boldsymbol{e}^P\right)} . \quad (9)$$

2.3 How to determine anisotropic parameters

The main disadvantage of the non-associated flow rule models would be the increase in the number of unknown variables. Usually, these variables can be specified from experiments such as tensile tests; therefore, the use of the non-associated flow rule model leads to an increased burden of experiments and measurements. In addition, if a higher-order function is required, the burden increases, making this approach impractical. Thus, to enjoy the benefits of the non-associated flow rule model, an increase in the numbers of unknown variables should be avoided, and this is achieved with the proposed model.

The plastic anisotropy characteristics of materials are classified into two categories; namely, plastic flow stress anisotropy and plastic deformation anisotropy. The former and latter should be incorporated in the yield function and the plastic potential function, respectively. Consider cold-rolled metal sheets under a plane stress state. Under the plane stress condition, the number of variables is halved. The anisotropic variables that must be determined are F, G, H, and N and F^*, G^*, H^*, and N^*. The former set expresses the yield stress anisotropy and the latter set expresses the deformation anisotropy.

The parameters about stress anisotropy, F, G, and H, are determined by the yield stresses that are obtained by tensile test in rolling direction, transverse direction, and equibiaxial test. The parameters about deformation anisotropy, F^*, G^*, H^*, and N^*, are determined by the r-values in the direction of rolling, diagonal, and transverse. The remaining parameter N is determined by optimization using the tensile test data in the diagonal direction. Note that the order of the functions, m and n, should be determined before these anisotropic parameters are determined because these values specify the function type, and have own physical meaning different from the anisotropic parameters.

3 Fracture prediction theory

In this section, three key theories adopted in this study will be briefly explained. These theories have been proposed by Ito et al. [8]. The proposed material model explained in the previous section is combined with these theories.

3.1. 3D local bifurcation theory

Based on Hill's general bifurcation theory, bifurcation occurs at when the following condition is satisfied.

$$I\left[\Delta v\right] = \int \Delta L : \Delta \dot{S}\, dV = 0 \ , \tag{10}$$

where $\boldsymbol{v}$ is velocity field, $\boldsymbol{L}$ and $\dot{S}$ are velocity gradient tensor and 1st Piola-Kichhoff stress tensor rate, respectively. $\dot{S}$ can be represented by the Cauchy stress tensor $\boldsymbol{\sigma}$ as

$$\dot{S} = D : \dot{\varepsilon} + \sigma \mathrm{tr}(L) + \omega \cdot \sigma - \sigma \cdot \omega - L \cdot \sigma = A : L \ , \tag{11}$$

where $\dot{\boldsymbol{\varepsilon}}$, $\boldsymbol{\omega}$, $\boldsymbol{D}$ are strain rate tensor, spin tensor, and tangent stiffness tensor, respectively. And $\boldsymbol{A}$ is a fourth rank tensor that relates the nominal stress rate $\dot{S}$ and velocity gradient tensor $\boldsymbol{L}$. To characterize the bifurcation mode, the velocity gradient tensor is allowed to be discontinuous when the velocity gradient tensor crosses the bifurcation border Γ. $\boldsymbol{L}$ can be represented by the normal vector $\boldsymbol{n}$ on the bifurcation border and the local deformation mode vector $\boldsymbol{m}$ that is normal to $\boldsymbol{n}$

$$L = m \otimes n \ . \tag{12}$$

Moreover, the mode vector $\boldsymbol{m}$ can be decomposed into three different vectors. Note that the vectors $\boldsymbol{m}$ and $\boldsymbol{n}$ are different from the parameters in the yield and plastic potential functions. Substituting Eq. (11) and (12) into (10), we have

$$I\left[m, n; \sigma\right] = hH\left[m, n; s\right] - \sigma\, \Sigma\left[m, n; s\right] \ , \tag{13}$$

where h is a hardening rate and the first and second term of this formula are described as

$$H\left[m, n; s\right] = m \cdot n \cdot \bar{D}(s) \cdot n \cdot m \ , \tag{14}$$

$$\Sigma\left[m, n; s\right] = \frac{1}{2}\left[m \cdot s \cdot m - n \cdot s \cdot nI\right] \ . \tag{15}$$

In an elasto-plastic material subjected to large strain, ignoring elastic deformation, the tangent stiffness tensor can be assumed to be proportional to the hardening rate h. Under this assumption, the tangent stiffness tensor can be represented with the polar decomposition form.

Then, the current stress is represented by

$$\boldsymbol{\sigma} = \sigma s, \quad \sigma = \sqrt{\boldsymbol{\sigma} : \boldsymbol{\sigma}}, \quad s = \boldsymbol{\sigma} / \sigma \ , \tag{16}$$

where σ and $\boldsymbol{s}$ are the norm of the current stress tensor and normalization tensor which gives stress ratios for each stress component, respectively.

Based on these relations, we finally have the following local bifurcation criterion.

$$\left(\frac{\sigma}{h}\right)_{cr} = \min\left(\frac{H\left[m, n; s\right]}{\Sigma\left[m, n; s\right]}\right) \ . \tag{17}$$

3.2 Ito-Goya's plastic constitutive model

Local bifurcation brings abrupt change on the current strain rate direction. Since the classical J_2 theory does not allow the rotation of the strain rate direction caused by the subsequent stress rate direction, it is not appropriate to the bifurcation problems. Therefore, in this study, Ito-Goya's plastic constitutive equation is applied, because this model can take the dependency of the strain rate direction on the stress rate direction into account. In this study, the revised Ito-Goya's plastic constitutive equation represented as follows was used.

 DOI: 10.1051/matecconf/20168005003

$$d\boldsymbol{\varepsilon}^p = \Lambda(\boldsymbol{n}_F : \boldsymbol{l}_p)\left|d\boldsymbol{\sigma}'\right|\left[K_C d\boldsymbol{l} + (1-K_C)\boldsymbol{n}_N\right], \quad (18)$$

where $\boldsymbol{n}_N$ is an unit tensor called natural direction. This tensor indicates the direction of the deviatoric stress rate $d\boldsymbol{\sigma}'$ that is identical to that of the plastic strain rate. The other terms introduced in this equation can be defined as:

$$\boldsymbol{n}_F = \frac{\nabla\bar{\sigma}(\boldsymbol{\sigma})}{\left|\nabla\bar{\sigma}(\boldsymbol{\sigma})\right|},\quad \boldsymbol{n}_N = \frac{\nabla\bar{\sigma}^*(\boldsymbol{\sigma})}{\left|\nabla\bar{\sigma}^*(\boldsymbol{\sigma})\right|},\quad \boldsymbol{l} = \frac{d\boldsymbol{\sigma}'}{\left|d\boldsymbol{\sigma}'\right|},\ \boldsymbol{l}_p = \frac{\boldsymbol{\sigma}'}{\left|\boldsymbol{\sigma}'\right|}. \quad (19)$$

In Eq. (18), the parameter K_C, which takes a value between 0 and 1, shows the dependency of the direction of the strain rate on stress rate. When K_C is equal to 1, the Ito-Goya's constitutive equation becomes the J_2 flow theory.

3.3 Evaluation of sheet formability

The bifurcation criterion represented by Eq.(17) indicates that the local bifurcation, which is specified by the mode vectors $\boldsymbol{m}$ and $\boldsymbol{n}$ that are based on the current stress ratio tensor $\boldsymbol{s}$, should be identified by the ratio σ / h that means the ratio of the stress level to the work-hardening. In another words, the formability represented in the σ / h plane is completely free from the strain path dependency that is usually observed in a typical FLD (Forming Limit Diagram) represented in strain space. Mechanically, stress σ means the intensity of fracture initiation, and the hardening coefficient h means the material's resistance against fracture. Therefore, exhibiting FLDs on the σ / h plane is mechanically reasonable.

The fracture limit lines in this FLD show both the SR (Stören-Rice) limit and 3D local bifurcation limit. The typical bifurcation that leads to final fracture would be lie between the lower bound represented by the SR limit and the upper bound represented by the 3D local bifurcation limit.

4 Fracture prediction analyses

In this section, the effect of the orders of the yield function and the plastic potential function in the proposed anisotropic plastic constitutive equation on the fracture prediction is numerically investigated. Using the constitutive relation in Eq.(18), assuming proportional loading, integrations on the stress increment is carried out from zero value with variable stress ratio $\alpha = \sigma_2 / \sigma_1$. The stress ratio α takes values between 0 and 1, which means the stress state changes from uniaxial to equi-biaxial via plane strain state. For each stress state, the functional $\boldsymbol{I}$ can be calculated, whose sign is available to judge the 3D local bifurcation criterion.

A virtual material is used for this analysis. Its material properties are as follows;

$$\bar{\sigma} = 500(\varepsilon^P)^{0.2},\quad K_C = 0.2\ . \quad (20)$$

The calculation results are shown in Figures 1 and 2. In these figures, the SR limits and 3D limits, calculated for each stress ratio, are plotted as upper lines and lower lines, respectively. Note that the calculated values indicate necking initiation based on each theory and we regard these values as potential onset of fracture. Since the purpose of this analysis is to investigate the effect of the order of the proposed yield and plastic potential functions on fracture prediction, an isotropic material is assumed in these calculations. When anisotropy is assumed, the shape of the yield and plastic potential function differs, which would affect the calculation and bring more complicated results.

First, the order of the plastic potential function n is fixed to 2, and that of the yield function m is varied as 1, 2, 3, and 4. The result is shown in Fig.1. It can be observed that the increase in the order of the yield function did not affect each fracture initiation predicted by both the SR theory and 3D theory; in other words, all the fracture limit curves coincided. We considered this result as follows: the order of the yield function affected both the stress σ and hardening ratio h; therefore, the effect of the order was offset and the fracture limit curves became the same.

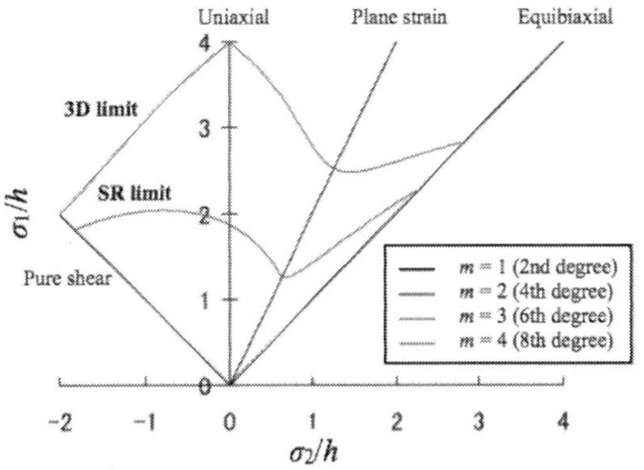

Figure 1. Predicted fracture limits for varied yield function's order; n=2, m = 1, 2, 3, and 4.

Second, the order of the yield function m is fixed to 2, and that of the plastic potential function n is varied as 1, 2, 3, and 4. The result is shown in Fig.2. In contrast to the first case, the predicted fracture limits in uniaxial, plane strain, and equi-biaxial directions are identical for any n; however, deteriorations are observed at in-between regions. The reason is assumed that the natural direction in Eq.(18) becomes identical in these directions for each n value, which resulted in these fracture limit curves. Especially in plane strain direction, the higher-order plastic potential function brings relatively flat shape in this direction, which could be the reason for flat shape in the fracture prediction curves.

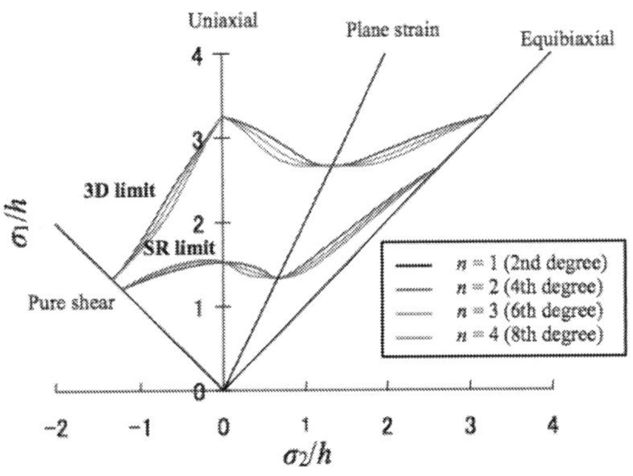

Figure 2. Predicted fracture limits for varied plastic potential function's order; m=2, n = 1, 2, 3, and 4.

Some mechanical characteristics are revealed through these simulations; however, there are many issues to be verified and considered because the effect of separation of the yield and plastic potential functions on the fracture limit prediction have never been sufficiently investigated before this research. It seems apparent that this approach will bring deep insight into the material behaviour in large deformation and near the fracture mechanism due to its considerable physical resolution; therefore, more effort should be paid to numerical and experimental investigations for this model.

5 Conclusions

A novel material model for anisotropic metals is proposed. The proposed model is based on the non-associated flow rule with arbitrary order function for both the yield and plastic potential functions. In addition, an explicit formulation of the equivalent plastic strain increment that is plastic-work-conjugated with the equivalent stress is also presented. Then, fracture prediction analyses are conducted by using the 3D local bifurcation theory. Numerical simulations of the proposed model are conducted with virtual material data. Physically reasonable results were obtained by these analyses.

References

1. R. Hill, Proc. Roy. Soc. London, **197**, 281-297 (1948).
2. A. P. Karafillis and M. C. Boyce, J. Mech. Phys. Solids, **41**, 12, 1859-1886 (1993).
3. F. Barlat and J. Lian, Int. J. Plasticity, **5**, 51-66 (1989).
4. T. B. Stoughton and J. W. Yoon, Int. J. Plasticity, **25**, 1777-1817 (2009).
5. M. Safaei, S. L. Zang, M. G. Lee, and W. D. Waele, Comput. Mater. Sci., **81**, 15-29 (2014).
6. F. Barlat, J. C. Brem, J. W. Yoon, K. Chung, R. E. Dick, D. J. Lege, F. Pourboghrat, S. H. Choi, and E. Chu, Int. J. Plasticity, **19**, 1297-1319 (2003).
7. T. Oya, J. Yanagimoto, K. Ito, G. Uemura, and N. Mori, *Procedia. Engineering*, **81**, 1210-1215(2014).
8. K. Ito, K. Sato, M. Goya, and T. Yoshida, Int. J. Mech. Sci., **42**, 2233-2248 (2000).

Creep age forming of Al-Cu-Li alloy: Application to thick sheet forming of double curvature aircraft panel

Wael Younes [1,a], Eliane Giraud[2], Zouari Ahmed[2], Philippe Dal Santo[2] and Sjoerd van der Veen[3]

[1]*IRT Jules Verne, Chemin du Chaffault, 44340 Bouguenais, France*
[2]*LAMPA, ENSAM, 2 boulevard du Ronceray, 49035 Angers Cedex 01, France*
[3]*Airbus Structures Research, Materials & Processes, 18 rue Marius Terce, 31300 Toulouse, France*

Abstract. Creep-age-forming of a thick Al-Cu-Li sheet is studied. An industrial stamping press is used to form a double curvature panel at a reduced scale. This forming, which includes several relaxation steps, is modelled using ABAQUS. A material model describing an elasto-viscoplastic behaviour with anisotropy effect has been identified and implemented in ABAQUS using Fortran subroutine. The numerical model is validated by comparing experiments and numerical results in terms of deformed shapes and an improved forming cycle is suggested.

1 Introduction

In recent years, lightening aeronautical structures is one of the main axes of improvement. An interesting perspective is to add lithium to conventional aluminum alloys, which improves their mechanical characteristics and reduces their density [1]. The aluminum-copper-lithium alloy (Al-Cu-Li) is one of these materials recently developed. However, its formability by hot or cold sheet forming processes is not well known. Numerical simulations are generally used to predict the final shape of the formed parts. But, the quality of the prediction is directly linked to the material laws implemented in the codes which must describe as precisely as possible the alloy behavior under the thermomechanical process conditions.

Generally, the Creep-Age-Forming (CAF) process [2] includes four main stages: (i) the heating of the sheet until an ageing temperature (here, 150°C); (ii) the deformation of the sheet by applying a pressure on its upper surface using a punch; (iii) the ageing phase in which the sheet, always clamped between the die and the punch, is held at high temperature and stress relaxation occurs and (iv) the cooling of the sheet. Thus, during the forming cycle, two metallurgical phenomena occur: stress relaxation and age-hardening. It is therefore possible to underline the main advantages of this type of process: low springback (due to stress relaxation) and forming procedure simplification (due to the change of microstructure generated during the forming cycle).

In this study, the Al-Cu-Li alloy has been characterized under thermomechanical conditions representative of the forming cycle and modeled using classical laws (in first approach). A numerical model has then been developed using the ABAQUS software and used to simulate the experimental forming of a double curvature panel performed on an industrial press. After the validation of this model by comparing experimental and numerical results, an improved forming cycle is suggested.

2 Characterization of the Al-Cu-Li alloy

2.1 Experimental procedure

Two types of experiments have been performed: tensile experiments and relaxation experiments. The testing conditions are given in Table 1.

Tensile tests have been performed on plate specimens at different strain rates. They allow identifying the yield locus criterion parameters and studying the strain rate dependence of the material behavior. Tensile specimens were taken from the quarter of the sheet thickness and in three directions α (0°, 45° and 90°).

Relaxation tests were carried on round specimens at different level of strains. The specimens are maintained during 8h at a given level of strain and the stress evolution as a function of time is registered. The holding time for the relaxation corresponds to the duration of the age-hardening treatment for the studied alloy.

For each type of experiments, a Gleeble 3500 was used. The specimen was heated by Joule effect at 10 K/s until the fixed temperature (150°C). The temperature was measured by a K-type thermocouple welded in the central part of the sample. The experiment was started after a holding time of 20s at the test temperature to ensure a homogeneous temperature in the specimen.

[a] Corresponding author: wael.younes@irt-jules-verne.fr

Table 1. Testing conditions

Tensile experiments									
Temperature (°C)	150								
Strain rate (s-1)	10-1			10-2			5.10-3		
Direction α (°)	0	45	90	0	45	90	0	45	90
Relaxation experiments									
Temperature (°C)	150								
Strain rate (s-1)	10-2								
Strain level	4%		8%			12%		16%	

2.2 Experimental results and discussion

Experimental results are normalized by means of a reference state corresponding to the yield stress σ0 measured in the rolling direction.

2.2.1 Anisotropy and viscoplasticity

Normalized stress-strain curves obtained after tensile tests performed at 150°C with different strain rates are shown in Figure 1. The alloy exhibits strain hardening: stress increases continuously until the fracture of the specimen. The orientation α has not great influence on the material behavior. It seems however that the orientation of 45° leads to the lowest stress and highest strain at failure. It is also possible to notice that the alloy does not show a strong viscoplastic behavior: the same level of stress is reached whatever the considered strain rate.

2.2.2 Relaxation

Normalized stress-time curves obtained at 150°C after relaxation experiments at different strain levels are shown in Figure 2. The stress evolution shows two zones. The first zone is illustrated by a fast decrease in stress until a stable value 45% lower than the initial one. In this second zone, the stress keeps a constant value. The decrease rate of the stress in the first zone is the same for all tests: a period between 1500 and 1600 seconds is necessary to reach the second zone.

This evolution in two zones of stress as a function of time has already been observed in aluminum alloys [3]. As explained by Chen et al. [3], the stress drop is linked to a decrease of the dislocations speed, and the stabilization of the stress is due to the cessation of the dislocations movements since precipitates are formed in advance.

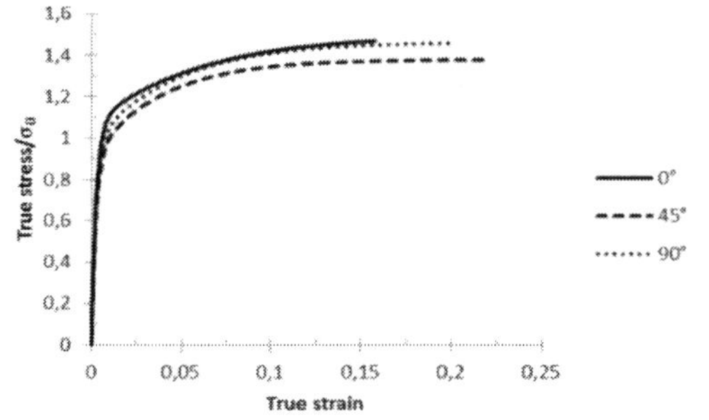

(a)

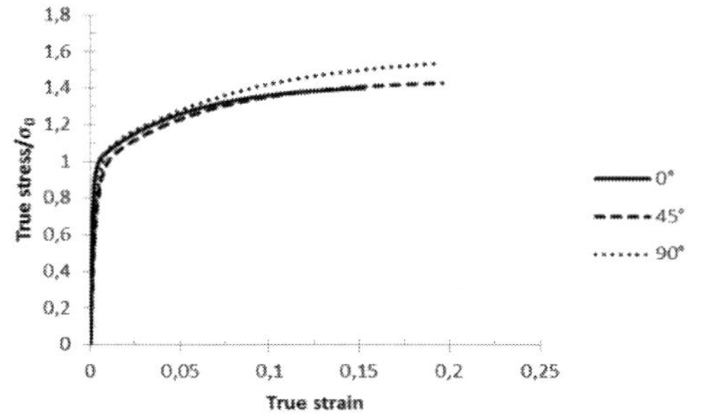

(b)

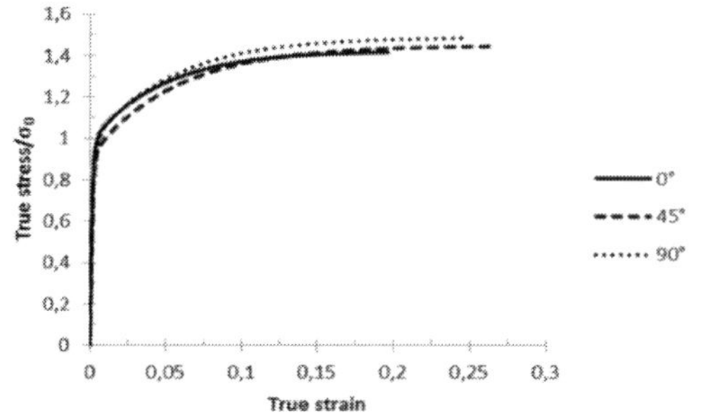

(c)

Figure 1. Normalized stress-strain curves obtained at 150°C for three orientations and with a strain-rate of (a) 10-1, (b) 10-2 and (c) 5x10-3 s-1

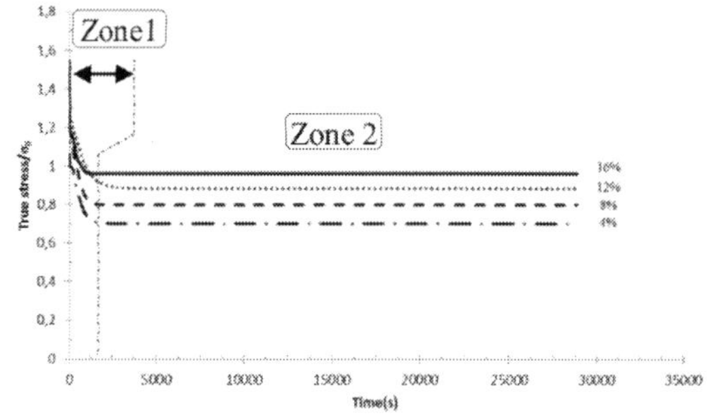

Figure 2. Stress relaxation as a function of time with different strain rate

3 Forming of a double curvature panel in Al-Cu-Li

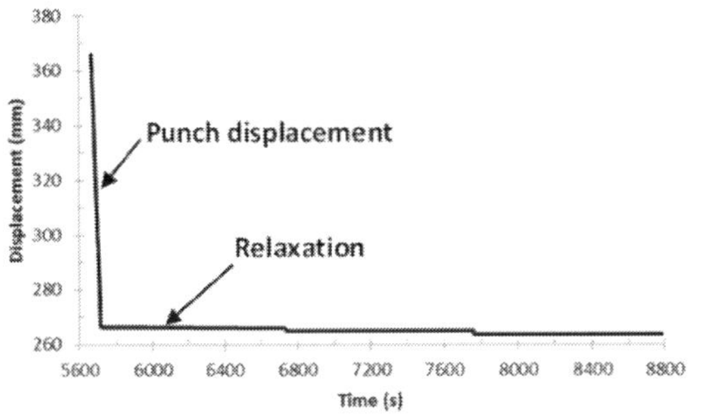

(a)

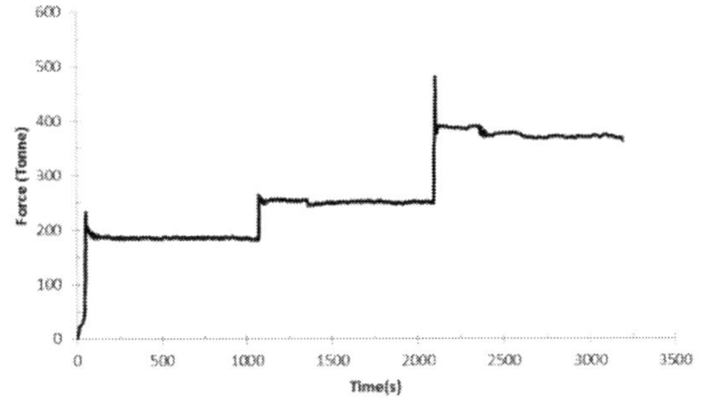

(b)

Figure 3. (a) Experimental punch displacement as a function of time, (b) Experimental punch force as a function of time

The hydraulic press of IRT "Jules VERNE" has been used for stamping the double curvature panel. The sheet dimension is about: 1000x550x40mm. The forming process is performed in three sequences composed of a forming step followed by a relaxation phase (Figure 3.a). During the two first sequences, the displacement of the punch is monitored without restriction on the force. In the final stage, a limit force of 500T is fixed (which corresponds to the maximum performance of the press). A Coulomb friction coefficient of about 0.2 is ensured between the sheet and the die and between the sheet and the punch by applying a boron nitride powder on both sides of the sheet.

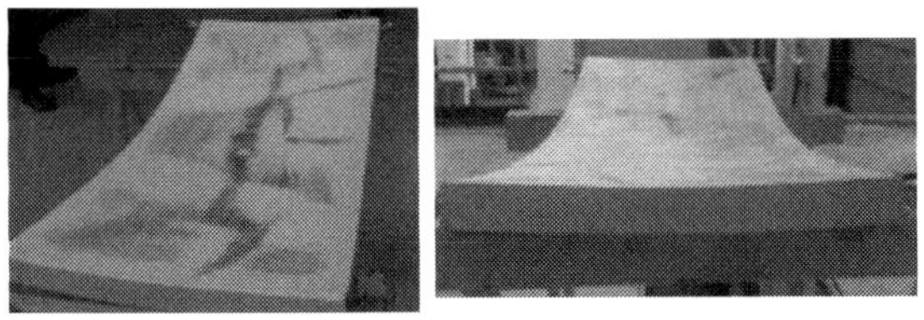

Figure 4. Experimental formed part

The measured punch force evolution during the forming process is shown in Figure 3.b and is in agreement with the three sequences describes above. The maximum force reached during the last sequence is of 480T (and not 500T as asked) due to the precision of the press control system. It is also possible to notice that the threshold stress during relaxation phases is generally reached after 200s.

A picture of the formed part is shown in Figure 4. It is possible to note that the forming cycle applied on the sheet does not allow to fill completely the die and that the double curvature is not clearly visible: a strong springback occurs.

4 Numerical simulation of the forming process

4.1 Material models

Two different models have been used: one for the forming phase and one for the relaxation phase of each sequence.

For the forming phase of each sequence, an elasto-plastic model with anisotropy criterion is used. The elastic behavior is described by a Young modulus previously identified as a function of temperature: a decrease of 7% of its value is found when the temperature increases from 25°C to 150°C. The plastic behavior is described using a tabulated law from the stress-strain curve obtained experimentally. The anisotropy criterion of Hill48 [4] has also been identified using the experimental data previously obtained.

For the relaxation phase of each sequence, a classical power law implemented in ABAQUS thru the Creep user subroutine has been used:

$$\dot{\varepsilon} = A\sigma^n \qquad (1)$$

The module A and the exponent n were identified from the experimental tests.

The material model has been validated by simulating tensile experiments followed by a relaxation step at different level of strains.

4.2 Numerical model

A 3D model of the forming process (Figure 5) of the double curvature panel has been realized in ABAQUS®. A half of the part is considered with a symmetry boundary condition along the Y-axis. The punch and the die have been modeled as discrete rigid surfaces. A displacement along the Z-axis has been imposed on the punch positioned above the piece. The die has been fixed below the sheet. After stamping, a displacement along the Z-axis has been applied in order to release the punch action on the sheet which allows the springback.

A C3D8H element type with full integration is used. The size of each element is about $13x13x7mm^3$. The numerical forming steps and the Z-axis displacement conditions are controlled by specific amplitude curves that describe exactly the different forming stages programmed on the machine during the experimental test.

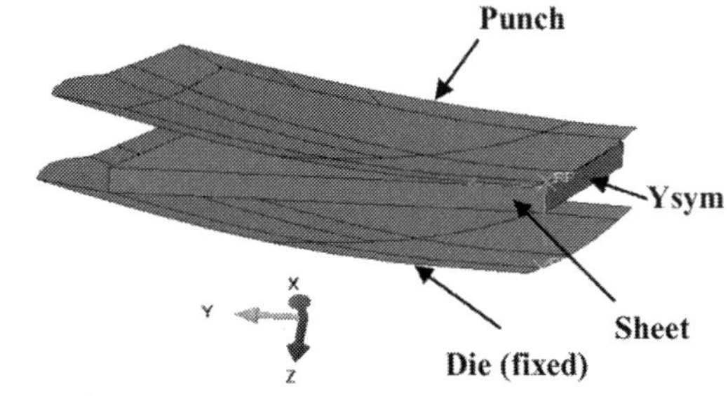

Figure 5. 3D model of the forming process

4.3 Numerical results

The simulated punch force evolution as a function of time is shown in Figure 6. A fill rate of only 99.67% has been obtained by following the experimental strategy of stamping. The fill rate is defined as the ratio of displacement allowing 100% of filling to the actual displacement.

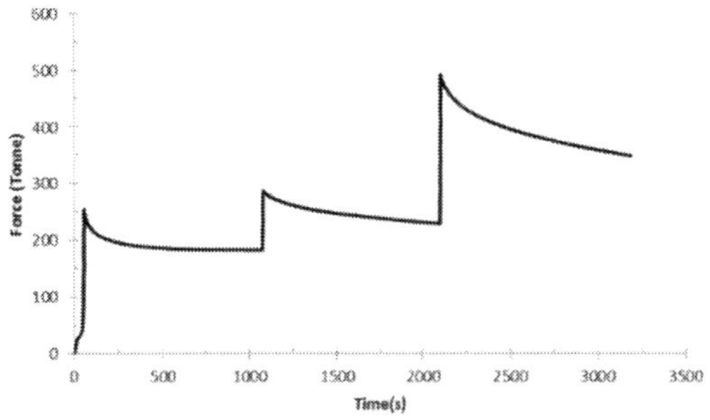

Figure 6. Numerical force as a function of time

The final shape has been reached after three increasing forming phases, each followed by three relaxation steps as seen in Figure 6. A numerical springback value of 13.5mm has been evaluated by measuring the distance between the lower surface of the deformed sheet before and after the springback (Figure 7) while the measure of experimental shapes gives a springback value of 21 mm. This difference can be explained by the experimental measurement technique which must be improved and the actual material model that doesn't take into account the kinematic hardening.

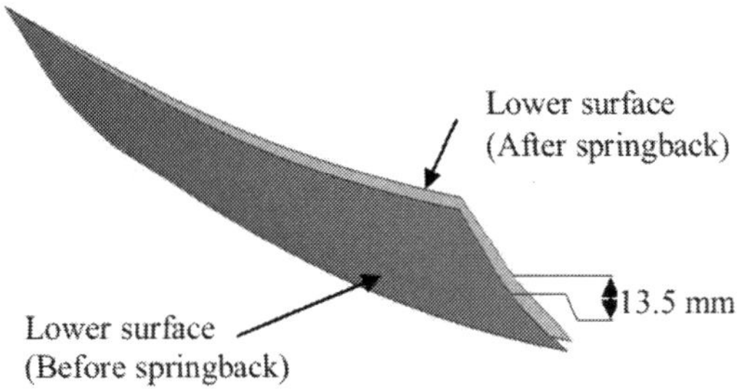

Figure 7. Springback after numerical forming and relaxation

5 Comparison between experimental and numerical results

5.1 Force evolution as a function of time

A comparison between experimental and numerical force evolution is given in Figure 8. A good prediction of the effort is obtained with a maximum error of 5% in the final forming step.

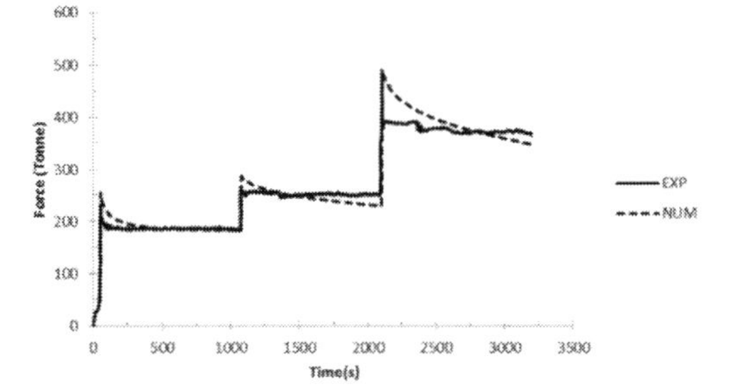

Figure 8. Comparison between experimental and numerical results

5.2 Contact zones on the upper surface

Figure 9 shows the contact play between the punch and the sheet for each sequence of the process. As summarized in Table 2, the contact play evolves continuously during the forming in different zones of the sheet (defined in Figure 10). During sequence 1, contact is present only in the central zone. During sequence 2, contact is present on the majority of the sheet except in the central zone and on the edges of the sheet. During sequence 3, some loss of contact occurs and the edges of the sheet are always not in contact. During sequence 4, the edges of the sheet are in contact with the punch.

Table 2. Contact play evolution

	Zone 1	Zone 2	Zone 3	Zone 4	Zone 5
Seq. 1	No contact	No contact	No contact	Contact	No contact
Seq.2	Contact	Contact	Contact	No contact	No contact
Seq. 3	No Contact	Contact	Contact	No contact	No contact
Seq. 4	No contact	No contact	No contact	No Contact	Contact

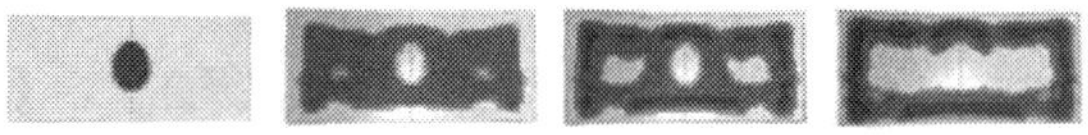

(a) (b) (c) (d)

Figure 9. (a) Seq.1, (b) Seq.2, (c) Seq.3, (d) Seq.4

These numerical results are compared to the experimental results. As visible in Figure 10, the boron nitride coatings is modified on some zones of the sheet, this is due to the contact with the punch. The confrontation between numerical and experimental results, in terms of contact zones, shows a good prediction of these zones by the simulation.

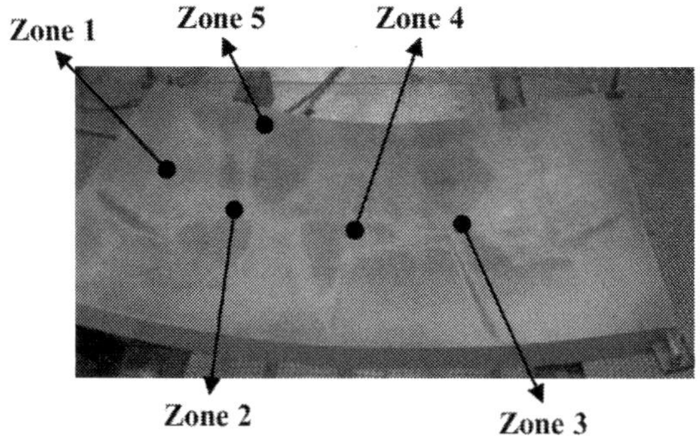

Figure 10. Experimental contact zones of the upper sheet

5.3 Contact zones on the lower surface

The observation of the lower surface of the sheet (i.e. the zone theoretically in contact with the die) does not shown a modification of the boron nitride coating except on the edges of the sheet (Figure 11.a). These contact zones are in agreement with the numerical results (Figure 11.b) and confirms that the filling is not complete.

(a) (b)

Figure 11. Contact zones on the lower part of the sheet, a- experimental result, b- numerical result

6 Proposition of an improved forming cycle

Based on the previous observations, two conclusions can be made:

(i) The forming cycle applied on the thick sheet has not allowed a complete forming of the double curvature panel. Thus, a modification of the forming cycle is necessary;

(ii) The numerical tool developed above is in good agreement with the experimental results. Thus, this model can be used to find the most appropriate forming cycle;

The principle of the new forming strategy is to relax the sheet during 1000s every time the maximum force of the press is reached (500T) until obtaining a filling rate of 100%.

An ABAQUS user subroutine UAMP has been developed to drive the loading force. The subroutine uses a numerical force sensor that has been defined at the punch reference point. When the force reaches the maximum set value, the punch velocity is set to zero allowing the material relaxation.

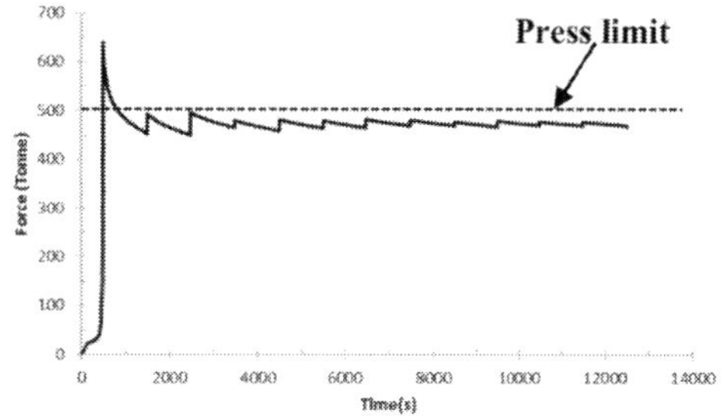

Figure 12. Numerical force evolution with the new strategy

The result of the simulation is provided in Figure 12 which shows punch force evolution as a function of time. It is observed that the force in the first step exceed 500T. This due to the large time increment that is automatically imposed by ABAQUS® at the beginning of the analysis. But, it can be underlined that twelve forming/relaxation sequences are necessary to obtain the appropriate shape (with a limited springback).

Conclusion and perspectives

A double curvature panel has been realized by Creep-Age-Forming at 150°C on a 40mm thick sheet in Al-Cu-Li. A numerical model with ABAQUS software has been developed to predict the forming of this part following the forming cycle applied (i.e. several forming/relaxation sequences). A material model describing an elasto-viscoplastic behavior with anisotropy effect has been identified and implemented using Fortran subroutine. This model has been validated since numerical results are in good agreement with experimental results in terms of force punch and contact zones (punch/piece and die/piece). A new forming strategy has been finally proposed in order to obtain the best shape part.

Future works will consist in: (i) modeling kinematic hardening for a good prediction of the springback and (ii) developing a unified law to take into account the forming and relaxation phases in the process cycle control.

Acknowledgements

This paper is part of the METAFOR project managed by IRT Jules Verne. The authors wish to associate the industrial and academic partners of this project; respectively ACB, STELIA, AIRBUS, CONSTELLIUM, ENSAM.

References

1. R.J.H. Wanhill, Int. J. Fatigue **16**, 3-20 (1994).
2. L. Zhan, J. Lin and T.A. Dean, Int. J. Machine Tools & Manufacture **51**, 1-17 (2011)
3. J.F. Chen, J.T. Jiang, L. Zhen, W.Z. Shao, J. of Mat. Proc. Tech. **214**, 775-783 (2014).
4. R. Hill, Proc. Roy. Soc. Lond. Series A, Math. Phys. Sci. **193**, 281-297 (1948).

Multi Scale Models for Flexure Deformation in Sheet Metal Forming

Edmondo Di Pasquale[1,2,3,a]

[1]*SIMTECH, 2 rue AlbertSamain, F-75017 Paris, France*
[2]*Univ. Lille Nord de France, F-59000 Lille, France*
[3]*ENSIAME, Mont Houy, F-59313 Valenciennes, France*

Abstract. This paper presents the application of multi scale techniques to the simulation of sheet metal forming using the one-step method. When a blank flows over the die radius, it undergoes a complex cycle of bending and unbending. First, we describe an original model for the prediction of residual plastic deformation and stresses in the blank section. This model, working on a scale about one hundred times smaller than the element size, has been implemented in SIMEX, one-step sheet metal forming simulation code. The utilisation of this multi-scale modeling technique improves greatly the accuracy of the solution. Finally, we discuss the implications of this analysis on the prediction of springback in metal forming.

1 Introduction and state of the art

After the first industrial applications [1]-[2], sheet metal forming simulation is nowadays part of industrial design processes. Most of the time, we use so called incremental codes for this task, where there is a Finite Element model of physical forming tools and a step-by-step simulation of their contact with the blank.

An alternative technique, one-step or inverse simulation[3], uses only the final geometry of the stamped part to predict metal deformation. This feature makes it easier to integrate this kind of simulation in the design cycle. For example, we can use the same Finite Element model for crash and metal forming simulations, making it much easier to couple the two processes and leading to much better accuracy in the prediction of performances.

This paper presents an original analytic approach to the study of the cycles of bending and unbending. This approach was developed in 2003, but it is receiving attention in this past months. Flexure bending and unbending is a process that takes place over a distance comparable to the thickness of the blank. Capturing the gradient of deformation would thus require a discretization of the order of the tenth of a millimeter, whereas typical mesh size for this kind of simulation is one hundred times larger. We have used multi scale techniques to integrate this analysis into our simulation.

Since the beginning of scientific study of metal forming [4]. Cycles of bending and unbending induced by tool geometrical features such as die radii or drawbeads to generate both a restraining force and a permanent plastic deformation.

Cycles of bending and unbending also create residual stresses, which, in turn generate springback deformations when the tools are released [5]-[8]).

In this paper, we describe the formulations of the code SIMEX as an example of one-step sheet metal forming simulations. The first version of SIMEX was released in 1994, and since 2004 it is integrated in the solution package of ESI Group. Other commercial one-step simulation codes are marketed by ALTAIR, AUTOFORM, LSTC, etc ...

This paper describes the theoretical foundation and the implementation in SIMEX of an equivalent model for the cycles of bending and unbending undergone by the blank over the die radius. This development greatly improved the accuracy of SIMEX simulation.

2 One step metal forming simulation

2.1 Formulation

The starting point of the inverse simulation is the finite element model of the stamped part. The algorithm of inverse simulation enables the user to find the position of the stamped part's nodes on the blank's initial geometry (Figure 1). This research is carried out without taking into account the history of the deformation, hence the naming of "one-step". The calculated displacements field ensures the equilibrium of the stamped part, through the calculations of the deformation (thinning, elongation) and of the constraints.

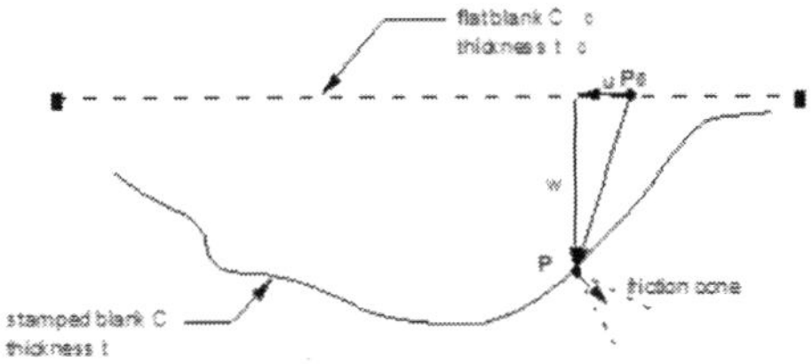

Figure 1: One step sheet metal forming simulation

The main feature of the inverse problem is that the FE model is built on the final shape, whereas the unknowns are distributed between the flat sheet metal (result of the calculation) and the stamped part (starting point of the calculation).

The material model implemented in SIMEX follows the plasticity laws of Henki-Mises, which are based on the following hypotheses:

The deformation's elastic component can be ignored with respect to the plastic component.

The deformation paths are radial (Figure 1), i.e. at each time $t > 0$

[a] Corresponding author: edmondo@simtech.fr

$$\frac{\varepsilon_1(t)}{\varepsilon_2(t)} = \alpha(constant) \tag{1}$$

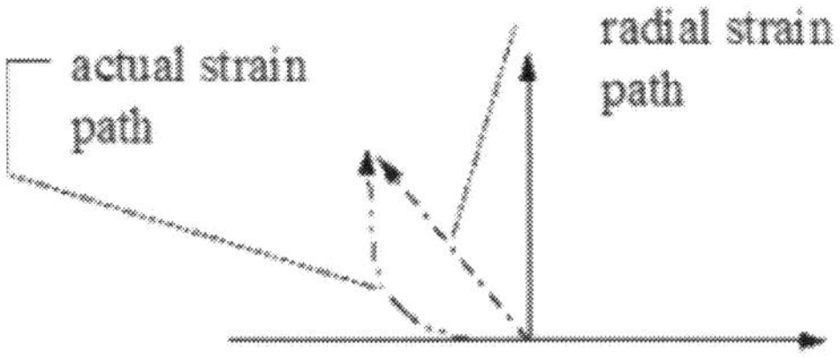

Figure 2: Radial strain path vs. actual strain path

On the basis of these hypotheses, the equations of the material flow (associative plasticity of Von Mises and Hill, [9]) can be integrated to give rise to and explicit relation between the eigenvalues of the stress and the strain tensors:

$$\frac{\sigma_1}{\sigma_2} = E_s P^{-1} \frac{\varepsilon_1}{\varepsilon_2} \tag{2}$$

Where Es is the secant modulus, i.e. the ration between □ and □p from the equation:

$$\sigma = k\left(\varepsilon_0 + \varepsilon_p\right)^n \tag{3.1}$$

And P is a matrix, function of the chosen plasticity criterion. For the Henki-Mises criterion, we have the following P matrix, where R is the Lankford anisotropy coefficient:

$$P^{-1} = \begin{bmatrix} \frac{(1+R)^2}{(1+R)(1+2R)} & \frac{R(1+R)}{1+2R} \\ \frac{R(1+R)}{1+2R} & \frac{(1+R)^2}{(1+R)(1+2R)} \end{bmatrix} \tag{3.2}$$

As for the friction, we use Coulomb's law. Two friction-contact models are established:

Unilateral contact between the metal sheet and the punch.

Bilateral contact between metal sheet, die and blankholder.

2.2 Modeling issues

The most important simplification of the inverse method is therefore the search of the equilibrium in the final configuration of the process. There are two important factors which are not taken into account:

The history of deformation: The history of the contact between the sheet metal and the tool. Achieving accuracy with one-step simulation implies the elaboration and development of equivalent models, which reconstruct the history of contact and/or deformation.

In order to understand how equivalent models work in one-step sheet metal forming simulation, let us consider a simple, U-shaped forming operation. The model represents the final shape of the part, it contains 1905 nodes and 1600 quadrilateral elements. In order to define equivalent models, it must be partitioned according to the role of the different areas in the forming operations (Figure 3, which shows also the profile sections on which we shall take profiles of thinning strains). We should point out that the Finite Element model presents sharp edges for the punch and die line, whereas in reality there will be fillets. This is consistent with the modeling procedures for stress analysis and crash simulation.

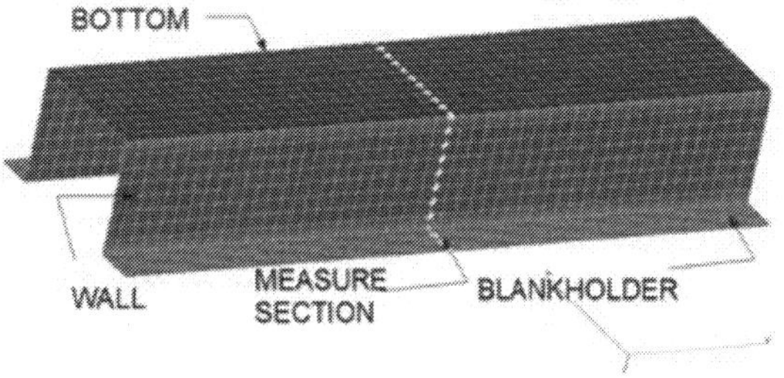

Figure 3: U-shape finite element model

In an actual punch die forming process, the advancing punch would adhere to the surface of the blank and prevent this portion of material from deformation. As one-step simulation does not take into account the history of contact, this effect is not simulated and the blank at the bottom of the U shape shows significant deformation (see the "standard bottom" thinning profile in Figure 4). Assigning a stick feature to the bottom of the punch triggers an approximate simulation of the history of contact. The deformation under the punch is negligible and the results are much more realistic.

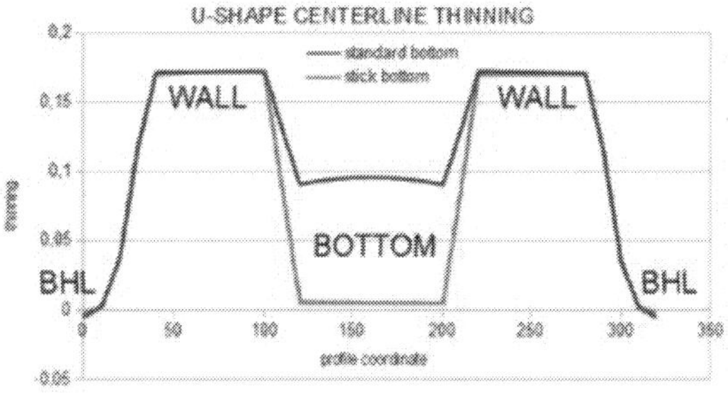

Figure 4 : Approximate simulation of history of contact using the stick feature

Taking into account the history of contact is important but not enough to assure the accuracy of the simulation. Figure 5 shows the thinning distribution (relative to the case where the bottom is modeled with stick feature. We can point out the following problems:

The wall deformation (about 17% thinning) is caused essentially by the blankholder force through the friction coefficient. The force corresponding to this deformation is 200 tons, corresponding to 667 Mpa of pressure for a friction coefficient of 0.2. This is an extremely high value for the blankholder pressure with respect to industrial standards.

The thinning distribution on the wall shows significant reduction on the extremities. In reality we would have small gradient of thinning but by no means comparable with what we see in Figure 5.

In the physical process, the process of bending/unbending over the die radius contributes greatly to the wall deformation. Among other things, it would induce a state of deformation close to a plane strain, greatly reducing the gradient of deformation along the y axis of our model.

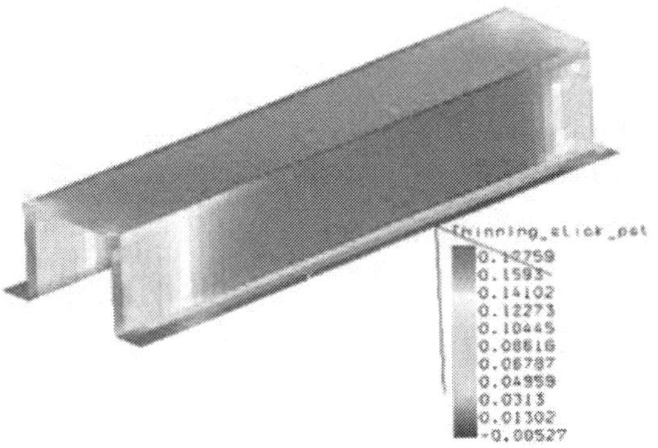

Figure 5 : Thinning distribution on U-shape part

In the following, we describe the theoretical foundation and the implementation of an equivalent model for the cycles of bending and unbending undergone by the blank over the die radius. We shall see that this improve the qualitative distribution of thinning on the U-shape part and allows for an excellent comparison between experimental and SIMEX results.

3 Flexural deformation in metal forming

3.1 Description of the model

In order to understand the physical processes involving flexure in sheet metal forming, let us consider the very simple U-shape of Figure 5. This simple shape can be obtained from two different processes: bending and forming with punch and die tooling.

In bending, the initially flat blank is bent on the four corners. Several sequences of bending are possible depending on the tooling setup. Generally speaking, the deformations of the different corners are independent on each other and the portion of materials between the bend are not affected by the deformation process. When we release the tools, each corner will undergo its own springback and recover part or the imposed deformation.

In punch and die tooling, the initially flat blank is set over a die. Prior to actual forming, a blankholder presses the blank against the die. The forming of the U-shape is obtained when a male tool (punch) pushes the blank into the die cavity. The material flowing into the die cavity undergoes a cycle of bending/unbending when it flows around the die entry radius. As a result, although the shape of the part wall may look at first sight identical to that obtained from a bending process, this material locks in both a plastic deformation and significant residual stresses. The shape of such a part after springback shows a typical curved wall.

For the case of bending and punch and die forming alike, the final shape presents four bends, i.e. areas of significant curvature. However, the deformation processes of the blank over these bends are very different.

In the case of a bending process, the deformation over the bends will be purely flexural, and can be written, as long as the radius of curvature is large with respect to the material thickness as:

$$\varepsilon_x = \frac{y}{R_D + h/2} \tag{4}$$

The x and y axis belonging to the curvilinear reference frame defined in Figure 6.

In the case of a form and die punch, this distribution of deformation will essentially be correct for the two bends on the punch nose or punch line. However, over the die radius, the distribution of deformation will vary greatly along the curvilinear coordinate, as it will be shown in the following.

For the sake of simplicity, we assume that the effect of friction is negligible. We also assume that the blank is in the elastic phase before entering the die radius. The blank, goes through the following process (Figure 6):

A: Unloaded elastic state. A relatively small, uniform stress and strain across the thickness can be present. The resultant force per unit width T will be present, for equilibrium considerations, over all the sections of the die radius.

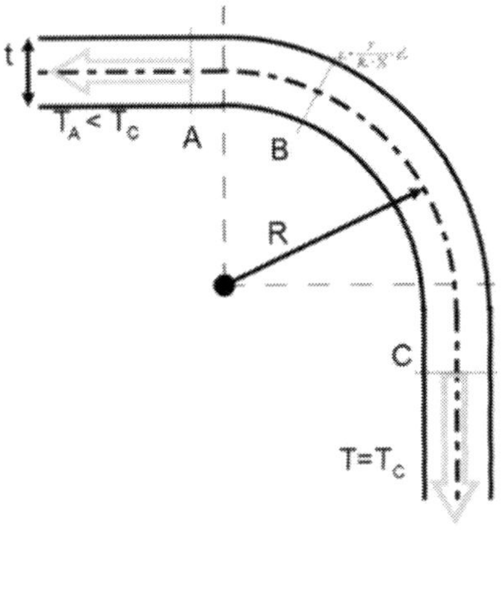

Figure 6: History of deformation over the die radius

B: After entering the die radius (classical theory of continuum mechanics suggests a distance approximately equal to the blank thickness) the section will have a curvature radius equal to r_D+h/2. The energy required for this deformation is provided by a traction force T applied at the end of the die radius by the punch via the formed material. This force must be balanced, so that the deformation state of the blank will be a combination of the flexural deformation (1) and of a membrane deformation $\varepsilon_M{}^B$.

$$\varepsilon_x = \frac{y}{R_D + h/2} + \varepsilon_M^B \quad (5)$$

C: At the exit of the die radius, the blank will be flat again, or in any case the residual radius of curvature will be much higher than the die radius r_D. The deformation will be again constant across the thickness, equal to a new value ε_M^C.

3.2 Analytical model of deformation

We study the deformation of a material segment (in the sense of [10]) of blank flowing over the die radius. This treatment gives us values for the permanent plastic deformation ε_M^C and for the traction T = TC at the end of the die bend.

We should point out that the traction force T may be below yield level (i.e., the ration T/h may be less than the yield stress □y). However, plastic deformation is usually present across the whole thickness of the blank. Further, depending on the strain hardening characteristics, there is usually a gradient of stress across the thickness, leading to a springback deformation when the residual stresses are released.

Before implementation into SIMEX, we present the results of a semi-analytical analysis. In this connection, we make a few simplifying assumptions.

The first assumption is that plain sections remain plane, i.e. we can define a flexural component of the deformation equal to:

$$\varepsilon_f = \frac{y}{R_D + h/2} \quad (6)$$

Maximum flexural deformation will take place on the upper (sup) and lower (inf) skin of the blank:

$$\varepsilon_f^{sup} = \frac{h}{2\left(R_D + h/2\right)}, \quad \varepsilon_f^{inf} = -\frac{h}{2\left(R_D + h/2\right)} \quad (8)$$

The second assumption is that bending and unbending cause the same amount of membrane deformation across the thickness of the blank:

$$\varepsilon_M^C = 2\,\varepsilon_M^B = \varepsilon_M \quad (9)$$

The third assumption is that the material hardening behavior is isotropic, i.e. the yield surface expands without shifting when the Von Mises stress increases according to the Krupkowsky-Swift law (3.1). Analysis of the effect of different hardening behaviour will be presented in a future publication.

Finally, we shall neglect the elastic component of the deformation.

Even with the above simplifying assumptions, the deformation pattern is relatively complicated.

The first results presented are relative to a typical low grade steel blank of thickness h=0.67, drawn over a die radius R=5 mm. The hardening curve is given by:

$$\sigma = 491\left(0.00725 + \varepsilon_p\right)^{0.182} \quad (10)$$

No restraining force T_A is present at this time. Figure 7 shows the deformation profiles across the thickness for section B (bending loading) and C (bending unloading).

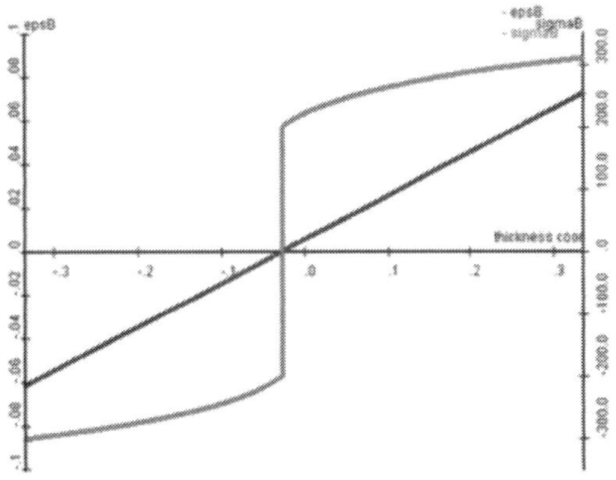

(a) Loading

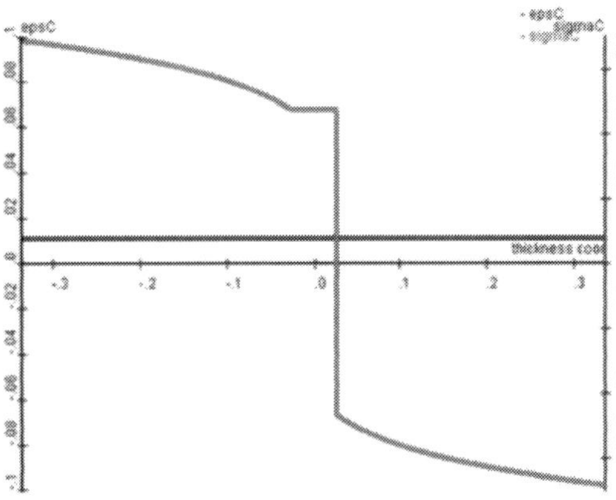

(b) Unloading

Figure 7 : Deformation and stress after bending

After the bending loading (section B) the fibers of the material are in traction above the neutral fiber y_0, in compression below the neutral fiber. The value of y_0 is given by:

$$y_0 = \varepsilon_M^B R_h = \frac{\varepsilon_M^B}{2} R_h \quad (11)$$

where

$$R_h = R + \frac{h}{2}$$

Fiber stress will be above σ_y in the extended fibers and below $-\sigma_y$ in the compressed fibers.

During the unbending process, the deformation will be brought to a constant value ε_M, but the different fibers will follow different paths (Figure 8)

For $y < -y_0$ (lower fibers) the strain will increase to ε_M while the stress will revert above σ_y.

For $y > y_0$ (upper fibers) the strain will decrease to ε_M while the stress will revert below $-\sigma_y$.

Finally, for $-y_0 < y < y_0$, the strain will increase to ε_M while the stress keeps the constant value corresponding to this plastic deformation.

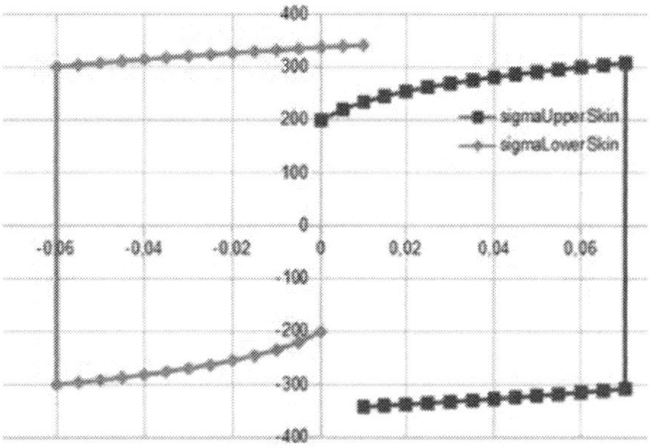

Figure 8: Stress-strain path of upper and lower skin

Finally, we can point out that the stress distribution at the end of the die radius (point C) gives rise to a moment M_C, defined as:

$$M_C = \int_{-h/2}^{h/2} y \, \sigma_x^C \, dy \qquad (12)$$

Our model predicts the restraining (traction) force T and the residual moment M. We present the results for this two quantities, normalized with respect to the values corresponding to a totally plastified section:

$$T_{norm} = \frac{T}{\sigma_y h} \qquad (13)$$

$$M_{norm} = \frac{M}{\sigma_y h^2/4} \qquad (14)$$

We can now analyze the behavior of the metal sheet going through the die radius. It will be determined by the die geometry, modeled via the flexural strain □f, defined as:

$$\varepsilon_f = \frac{h}{2R_h} \qquad (15)$$

We analyze here the effect of the material properties, by comparing the behavior of a ductile material (low grade steel) with a perfectly plastic material.

Permanent deformation after die radius has a very similar pattern for different materials as a function of flexural strain (Figure 9).

Traction force, on the other hand, is significantly affected by material hardening. In both cases, value of traction force is very small for small flexural strains, but it increases much more rapidly in the presence of significant hardening, to a value of about 40% of T_{norm} (Figure 10).

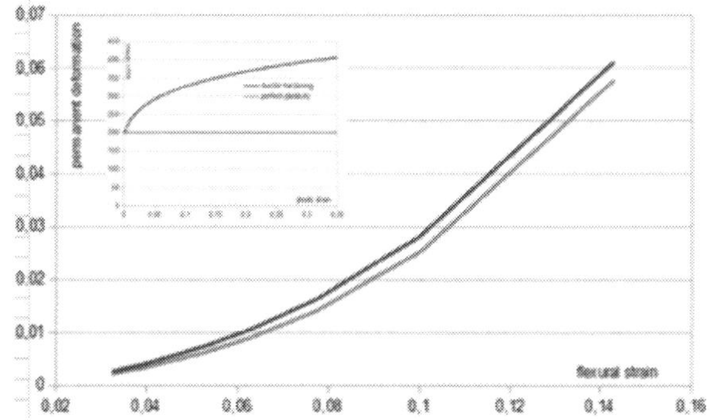

Figure 9 : permanent deformation vs. flexural strain

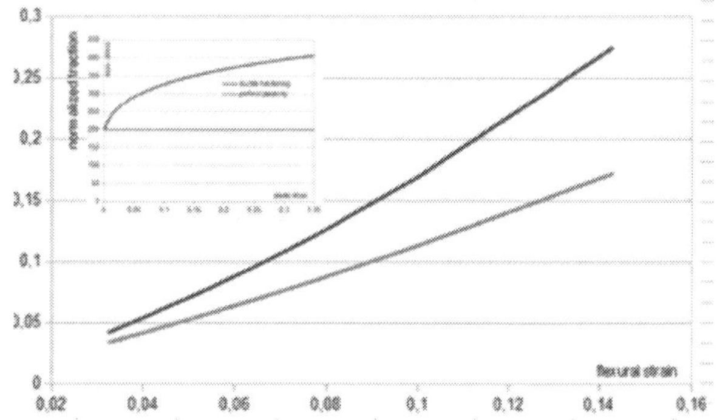

Figure 10 : traction force vs. flexural strain

Material hardening affects greatly also the resultant moment of residual stresses (Figure 11).

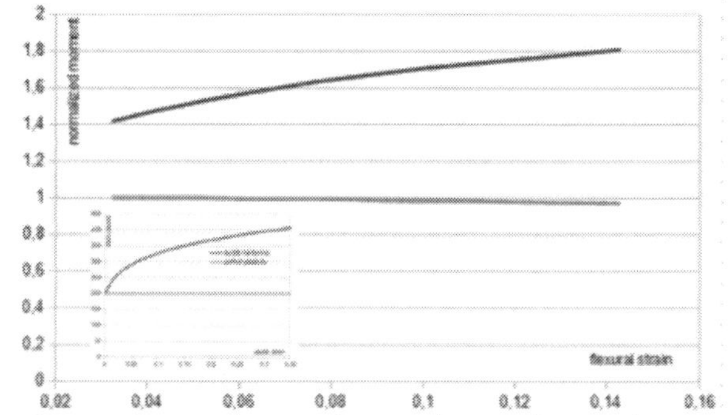

Figure 11 : residual moment vs. flexural strain

4 Application to industrial part

The above described methodology has been imbedded in the "bending" boundary condition of SIMEX code. Even though the mesh size is roughly the order of magnitude of the die radius, this boundary condition models the bending/unbending cycle which takes place on a much lower scale.

The French RNTL MACARENA project addressed the general issue of coupling crash computation with metal forming simulation. SimTech was a partner in

MACARENA alongside, among others, PCA Peugeot Citroen and RENAULT.

One of the problems addressed within MACARENA was the accuracy of one-step sheet metal forming simulations. In order to validate the new features of SIMEX, we referred to the component shown in Figure 12. RENAULT had designed, built, instrumented and measured this component to reproduce deformation patterns of actual industrial components, even though it not correspond to any actual vehicle component.

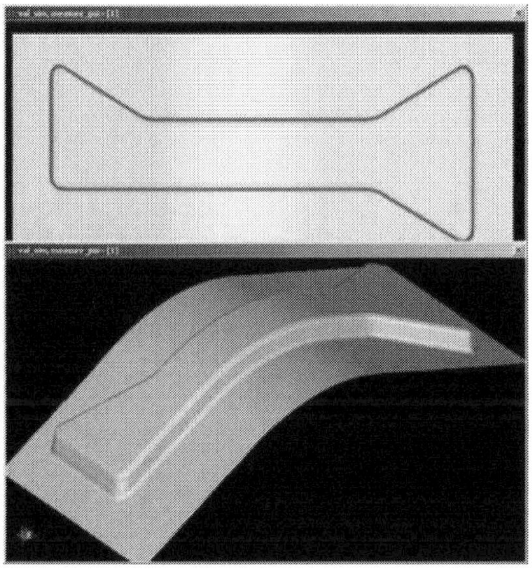

Figure 12 : RENAULT MACARENA rail

The formed material was a steel with the following characteristics:

- thickness = 1.97 mm
- yield strength = 315 Mpa
- ultimate strength = 425 Mpa
- necking elongation = 18%
- Lankford coefficient = 1.43

The relevant process data were:

- Blankholder force = 32 tons
- Friction coeff. = 0.15
- die entry radius = 3 mm

After forming, measurements of residual thickness have been carried out on 55 points on the part surface (5 replications). 17 points were situated along a section at part center, whereas the other 38 points were spread in significant positions over the part surface (Figure 13).

The mesh chosen for simulation (Figure 14) is representative of coupling with crash, fatigue or other engineering analyses. It consists of a total of 5321 elements (4034 trias, 1287 quads).

The mesh is divided in three parts, approximatively modeling the forming process:

- A bottom part, sticking to the punch
- A wall part, which flows over the die radius during the forming process
- A blankholder part, subjected to the 32 ton holding force

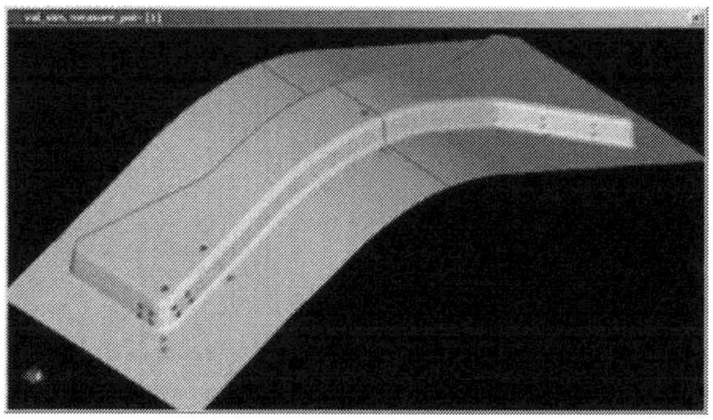

Figure 13: MACARENA rail with measurement points

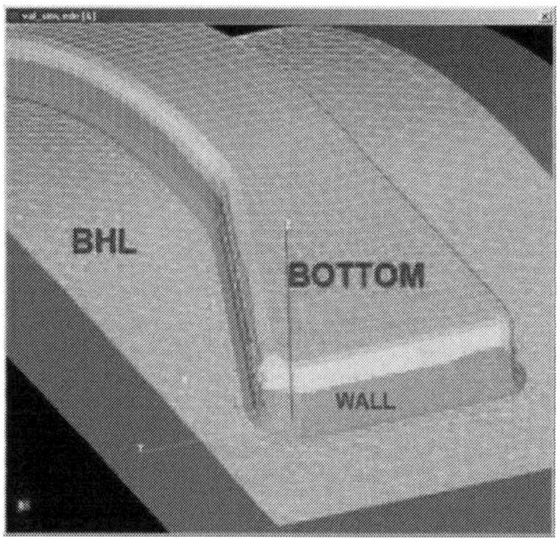

Figure 14: SIMEX mesh for MACARENA rail

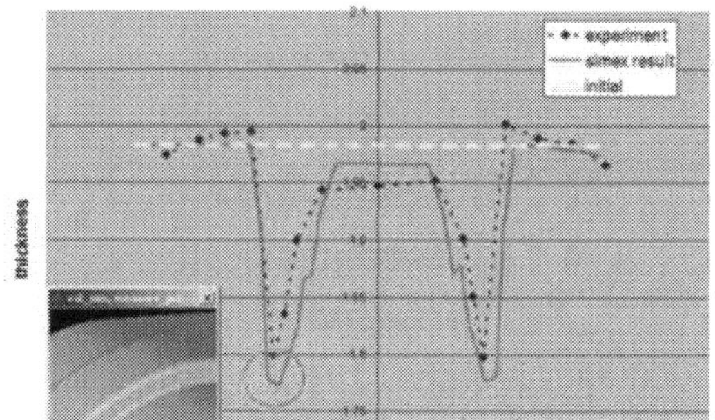

Figure 15: SIMEX vs. measured thickness of central section

Flexure conditions are generated along the die entry radius to simulate the flexure cycle.

Figure 15 presents the comparison of the residual thickness along the central line. SIMEX results are shown with respect to experimental measurements. The qualitative comparison is excellent, The peaks of deformation are correctly reproduced as well as the region of low deformation. As for quantitative results,

DOI: 10.1051/matecconf/20168006002

maximum difference between experimental measurement and simulation is 30 μm.

5 Conclusions

Material flow over a die radius and the associated cycles of bending and unbending happen on a relatively small scale, with respect to the average element size in one step simulation. We have developed a semi-analytical model for the prediction of the residual strains and stresses in the sections after the die radius and we have implemented it into SIMEX code.

The application of this model has greatly improved the prediction capabilities of the code. On the industrial parts studied, the agreement between numerical results and SIMEX prediction is excellent.

Acknowledgments

Part of the work presented has been carried out within the framework of the RNTL MACARENA project. The support of the French Ministry of Industry is dutifully acknowledged.The author would like to thank the RENAULT partcipants to MACARENA project (in particular Mr. Pierre LORY) for their cooperation

References

[1] Zhou, D., Wagoner, R.H., J. Materials Processing Technology, 50 (1995), 1-16.

[2] Makinouchi, A., J. Materials Processing Technology, 60 (1996), 19-26.

[3] El Mouatassim, M. , Thomas, B., Jameux, J-P. DiPasquale, E., Simulation of Materials Processing : Theory, Methods and Applications, Shen & Dawson (eds), Balkema, Rotterdam (1995).

[4] Nine, H. D., "New drawbead concepts for sheet metal forming", Journal of Applied Metal Working, Vol. 2, No. 3, July 1982.

[5] Firat, M., Kaftanouglu, B., Eser, O., J. Materials Processing Technology, 196 (2008) 135-148.

[6] Sanchez, L.R., Int. J. of Mechanical Sciences, 53 (2010) 429-439.

[7] Nanu, N., Brabie, G., Int. J. of Mechanical Sciences, 64 (2012) 11-21

[8] Ghaei, A.,Green, D.E., Aryanpour, A., Materials and Design, 88 (2015) 461-470.

[9] Lemaitre, J., Chaboche, J.-L., « Mechanics of solid materials », Cambridge Univ. Press, 1994

[10] Malvern, L.E., "Introduction to the mechanics of a continuous medium", Prentice-Hall 1969.

DOI: 10.1051/matecconf/20168006003

Residual stresses of a magnesium alloy (AZ31) welded by the friction stir welding processes

A. Kouadri-Henni[1a], L. Barrallier[2], Riad Badji[3]

[1] *Université Européenne de Bretagne, France, INSA, LGCGM, EA 3913, F-35708 Rennes, FRANCE*
[2] *Arts et Métiers Paris Tech, Mécasurf, JE 2504, F-13617 Aix-en-Provence, FRANCE*
[3] *Research Center in Industrial Technologies CRTI, P.O Box 64, Cheraga, Algeria*

Abstract. The objective of this study was to evaluate the residual stresses of FSW welding magnesium alloys (AZ31). The results show that the FSW processes lead to the formation of several distinct zones with differing mechanical properties. The residual stresses evolution have been explained by the heterogeneous modifications of the microstructure particularly a marked decrease in the grain size, a high modification of the crystallographic texture and the different anisotropic properties resulting from plasticity induced by the FSW process.

1 Introduction

Currently, industrial welding processes are rapidly evolving, this being driven by the need for process optimisation, cost effectiveness and ecological impact. It is for these reasons that friction stir welding (FSW) has started to appear in production workshops, notably in the automobile, aeronautical and mechanical industries. Furthermore these processes are being studied for welding materials that are reputed to be difficult to weld, such as aluminium or magnesium alloys. Magnesium alloys are potentially useful because of their low density, which is 35% less than that of aluminium, and their good mechanical characteristics [1, 2]. Experimental work carried out in this area has shown that welding magnesium by FSW results in a better quality weld than that obtained by conventional methods (arc welding); however both methods require a precise evaluation of the parameters involved in the processes [3,4]. These studies have enabled a better understanding of the relationship between the welding processes, the weld quality, the microstructure and the residual stresses of the weld [3, 4]. Several studies have shown that the plasticity is a fundamental factor that governs the mechanical properties of the welded joints [5, 6]. Most studies have shown the complexity of FSW and associated heterogeneities. From a mechanical point of view, considering the initial nature of the base metal (AZ31, AZ91, WE45, and AZ61), results has shown anisotropy in mechanical properties. These variation of mechanical properties have been linked principally to the initial fabrication process of the base metal and then to the welding process [6]. Welding processes equally produce residual stresses which are linked to the large deformation in FSW. For FSW peaks in residual tensile stress are observed in the TMAZ in the longitudinal direction and compressive residual stress peaks occur in a transversal direction in the fusion zone or nugget [6]. In the case of this study, a rolled magnesium alloy AZ31 was used for the welding tests to understand the evolution of microstructure and mechanical properties in particular the calculus of residual stresses.

2 Principle of FSW

The principle of FSW is based on the conversion of mechanical energy into thermal energy by friction generated by tool, which is generally threaded, on the material to be welded. This consists of a metal cylinder with a profiled and. threaded nib that is rotated at a high speed (Figure 1).

[a] Corresponding author: *afia.kouadri-henni@insa-rennes.fr*

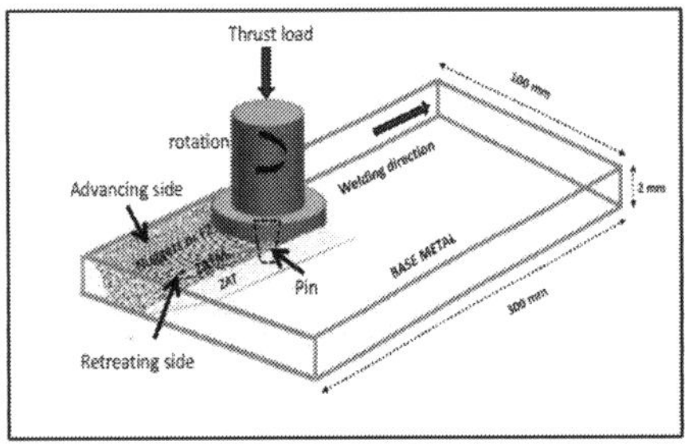

Figure 1. Principe of FWS process

The tool pin penetrates the plane of the joint at the interface between the two sheets to be welded and thoroughly mixes the metal of the two pieces in an operation which is a combination of forging and extrusion. The heat locally generated transforms the metal into a plastic state which allows the tool to be plunged into the material and then permits displacement along the line of the joint to form, after cooling, the weld. The melting point of the material is not reached in the process. The material undergoes a plasticizing process during the passage of the tool and a dynamic recrystallization during cooling [7, 8, 9]. At the end of this process one can distinguish several zones in the region of the weld in which the evolution of the microstructure is controlled by a complex interaction between plastic strain, recovery and recrystallization. These interactions will also include homogeneous and heterogeneous precipitation phenomena depending on the nature of the alloy.

3 Experimental methods

3.1 Sample presentation

The alloy used in the study FSW welding was a ternary magnesium - aluminium - zinc alloy with the designation AZ31 according ASTM standards. AZ31 magnesium alloy was chosen due to its good rolling capabilities. 2 mm thick plates were hot rolled by several passes and then levelled and 300°C annealed. Table 1 shows the chemical specification of the AZ31 magnesium alloy supplied.

Table 1: Principal chemical composition of AZ31 alloy in base metal, BM, (EDS analysis) [10]

BM	Al	Zn	Mn	Si	Cu
Element (weight %)	2.5-3.5	0.6-3.4	0.2-0.6	< 0.05	< 0.008

AZ31 magnesium alloys used were rolled to produce 2 mm thin plates from 100 mm ingots which were hot rolled with several passes using a 4 rollers Quarto rolling mill. Strain rates between 5-20% (thickness reduction per pass) and rolling speeds between 0.5-2 m/s were used. A strong basal texture resulted from this process. These plates were used as raw material for welding and as samples for base material analysis. For the experimental study performed within the framework of our research, samples were machined into parallelepidic shapes with dimensions of 300 x 100 x 2 mm^3.

3.2 Parameters of the welding process

Two plates were welded together side by side using a FSW process. To validate the optimisation strategy, a plan of experiments has been carried out following a simplified plan of Tagushi.

FSW processing was carried out with machines located at the Welding Institute (IW-G-ESAB). The processing parameters influencing the weld quality were studied. The visual aspect of the FSWelds was improved with increasing the load applied (F), welding speed (V) and tool rotation rate (W) with a smaller flash, less surface corrosion and less tunneling phenomena occurring. Experiments performed showed that increasing welding speed (V) at a constant tool rotation rate (W) resulted in internal voids and lack of bonding caused by the insufficient material flow. For a constant welding speed, low tool rotation rates lead to the formation of internal voids because the frictional heat was not sufficient to promote material flow. These defects disappeared with increasing W, but with a further increase, internal voids, lack of bonding and surface cracking occurred due to excess expelled material. These results are consistent with the literature [5, 6, 7, 8]. The tool used had a 5 mm pin diameter and 10 mm shoulder. The process parameters that were varied were the welding speed (V) and tool rotation rate (W). Various clamping systems and welding pressures were tested. The optimized domains are summarised in table 2. After welding, the welded samples had dimensions of 300 x 200 x 2 mm^3.

Table 2: Welding FSW parameters choice

Welding pression (kN)	Welding speed V (mm/min)	Tool rotation rate W (RPM)
1.5	90	1200

3.3 Location of the analysed zones

Figure 2 shows the location of the zones that were analysed in the work pieces welded by FSW.

Measurements were carried out in each of the zones modified by the welding process; the measurements were uniquely at the surface and ranged from the centre of the weld to the base metal. In our case, with the FSW process, two measures were carried out on either side of the fusion zone which consisted of two zones: the central stir friction zone (SWZ, zone 4); the thermo mechanically affected zone (TMAZ) on the left of the weld, advancing side, (TMAZ, zone 3); and the TMAZ to the right, the retreating side, referred to as TMAZ, zone 5. The HAZ to the left and right of the weld are respectively referred to as HAZ, zones 2 and 6 and finally the base metal BM, zones 1 and 7.

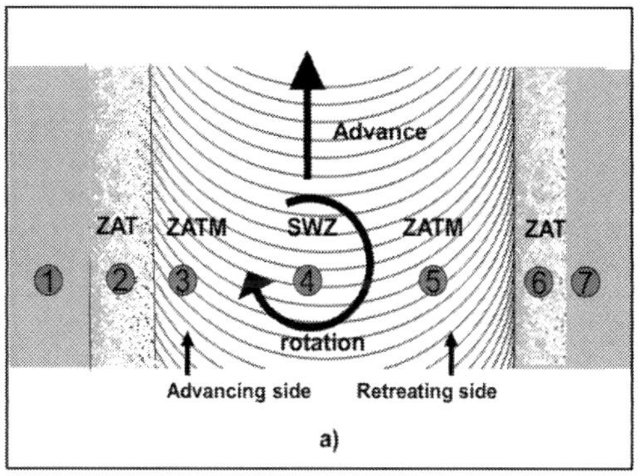

Figure 2: Location of the analysed zones

3.4 Metallurgical analysis

Before welding, each sample was machined to remove the oxide layer and cleaned with acetone. After mechanical polishing to a mirror finish, the welds were examined by optical microscope. The microstructure of the magnesium alloy was exposed by Nital Etchant (4 % and 20% HNO3, 96 % C_2H_5OH) and by the reaction of Keller Etchant (2.5 ml of HNO3, 1.5 ml of HCl, 1.0 ml of HF and 95 ml of water). Observations were also made using a scanning electron microscope (SEM); point chemical analysis was carried out on the samples using an electron microprobe (EDS).

3.5 Mechanical analysis

3.5.1 Hardness measurement

The micro hardness was measured by a Vickers indenter under a load of 100 g on cross sections applied for 15 sec. A program on a conventional optical microscopy measured the diagonal lines of the mark. A series of measurements were carried out close to the surface, at every location, from the SWZ, FZ and to the BM.

3.5.2 Tensile tests

The evaluation of the mechanical behaviour of the weld was carried out using transverse tension test. Dumbbell specimen preparation was performed using water jet cutting. The welded plates were cut transversely across the weld. The lateral section was then polished to avoid crack initiation during tensile test. An Instron machine with mechanical grips and a 50 kN load cell was used and tensile tests were performed at 5 mm/min strain rate.

3.5.3 Calculation of residual stresses

The method used to calculate the residual stresses by X-ray diffraction uses the distance between crystallographic planes as an internal strain gauge. The diffractometer used for these studies was equipped with a 4 circle goniometer - (Seifert MZ VI - TS) and a PSD detector. The X-ray source was a 1 mm collimator and the wave length used was Cr-Kα. The intensity and voltage of the source were 30 mA and 40 kV respectively. A masking system was used to limit the irradiated zone to 0.8 mm. Under these conditions the depth of penetration into the AZ31 was estimated to be 7 µm. The calculation of residual stress required several steps:

a) Identification of crystalline phases

Identification of the phases was carried out by indexing diffraction peaks. The position of the peaks was defined by optimisation of parameters of the pseudo-Voigt function (Kα_1 and Kα_2) of the diffraction intensities. The information thus obtained was used to determine which family of crystalline planes would be used as the internal gauge and consequently the settings to adopt for the measurement of strain $\varepsilon_{\phi\psi}$. In this study, only the α-phase was taken into consideration.

b) Calculation of Orientation Distribution Function (ODF)

The AZ31 base metal was in hot rolled condition. Using this deformation process, highly textured resulting material was expected. FSW induced high deformation gradients which may cause changes in crystalline orientations. The study of the texture evolution is then required to understand the anisotropic characteristic of the welds and its influence on mechanical properties. The texture of the base metal and the texture evolution in FSWelds was studied using XRD techniques. Analysis of the structure was only carried out for the predominant phase: the α-Mg matrix. The textures were measured by the choice of 5 pole figures corresponding to the diffraction planes of the principal α-Mg phase: the prismatic plane $\{10\bar{1}0\}$, basal $\{0002\}$, pyramidal $\{10\bar{1}1\}$, planes of the type $\{10\bar{1}2\}$ and $\{11\bar{2}0\}$. The diffraction intensities, expressed as the number of counts per second, were measured by the variation of two angles: the polar angle ψ, varying from 0 to 70° starting in the axial direction of the sample and a step size of 5°; the azimuth angle ϕ, varying from 0 to 360° with a step size of 10°. The ODF was then calculation from the 5 pole figures [10, 11, 12]. We extracted the information's such as the volume fraction of the principal orientation from the ODF (example is given in figure 3 for the zones welded FSW process).

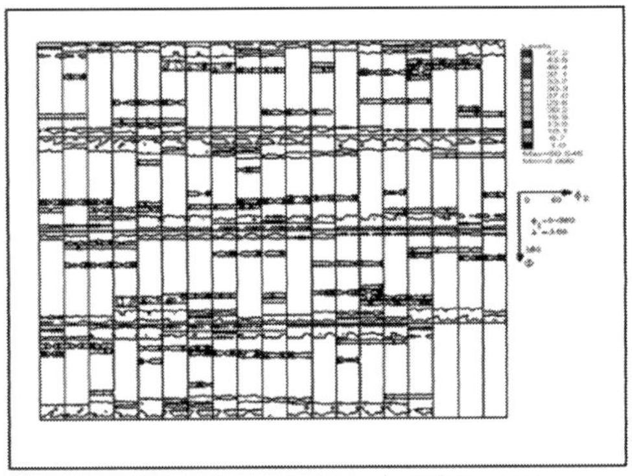

Figure 3: ODF of the welded zone by FSW process

Every orientation and fraction volume in every zone has been extracted from the ODF. We introduced the data in the calculation of residual stresses in the case of a presence of a texture.

c) Strain Measurement $\varepsilon_{\phi\psi}$

The strains $\varepsilon_{\phi\psi}$ were deduced from the angular shift of the diffraction peaks. For the measurement of strain the diffraction plane chosen was the plane $\{10\bar{1}3\}$ of the magnesium phase. The measurements were carried out by the simultaneous variation of 11 polar angles Ψ varying from ± 65° with a $\sin^2\Psi$ distribution, and 6 azimuth angles varying from 0 to 360°.

d) Calculation of residual stresses

When a material is not textured, therefore isotropic, the residual stresses were calculated by the $\sin^2\Psi$ law [10].

$$\varepsilon_{\phi\psi} = \frac{1}{2} S_2 \sigma_\phi \sin^2\psi + S_1(\sigma_{11}^S + \sigma_{22}^S) \quad (1)$$

When the material is textured, the residual stresses were calculated taking into account the grain orientation: the orientation distribution function (ODF).

$$\varepsilon_{\phi\psi}(\mathrm{hkl}) = \mathrm{F}_{\phi\psi}^{\mathrm{hkl}}(\phi, \Psi, \Omega, \mathrm{s}(\Omega), \mathrm{S}, \mathrm{ODF}) : \sigma_{\mathrm{ij}} \quad (2)$$

The components of the compliance tensor S_1 and S_2 were calculated with the aid of the elastic constants of a mono-crystal of magnesium taken from published literature (equation 1). The values used in the calculation of residual stress were as follows: S_1 (- 5, 58 10^{-6} MPa^{-1}) and S_2 (26.04. 10^{-6} MPa^{-1}) [13, 14, 15]. In the case of textured zones, the residual stresses were calculated taking into account the anisotropic polycrystalline constants. In this case the radio-crystallographic elastic constants $\mathrm{F}_{\phi\psi}^{\mathrm{hkl}}$ (equation 2) were determined from the ODF. In both cases a homogeneous Kröner-Eshelby self-consistent model was used [14, 15]. With the X-ray measurement technique used only the superficial layer of the metal (several microns) interact with the X-ray beam. The stress state was taken as plane stress ($\sigma_3 = 0$).

4 Results and discussion

The FSW welding processes result in different microstructures and mechanical properties.

4.1 General properties of the base metal

The AZ31 base metal revealed a fine equiaxed grain structure, with very few deformation twins. Two grain size populations could be observed (5μm and 15μm) and the average grain size is 10 μm (figure 4).

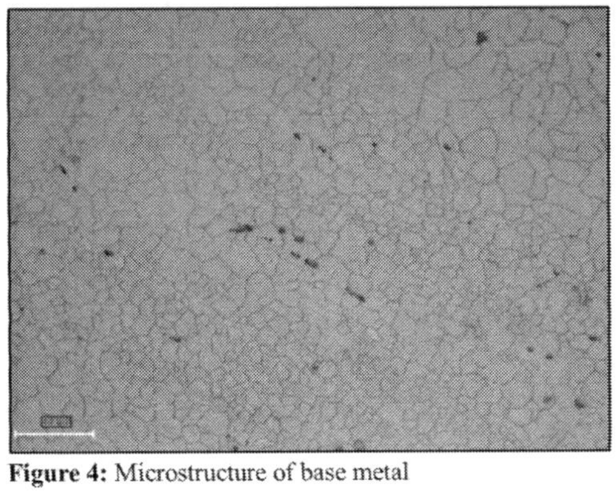

Figure 4: Microstructure of base metal

AZ31 optical micrograph study showed that the microstructure consisted in principally of a primary α-Mg phase structure.

4.2 Evolution of the microstructure in the welded zones

In the case of the FSW process, the base metal which is remote from the weld site is not subjected to any significant deformation or heating; that is to say it conserves its original properties. Figure 5 shows the different zones resulting from the FSW process.

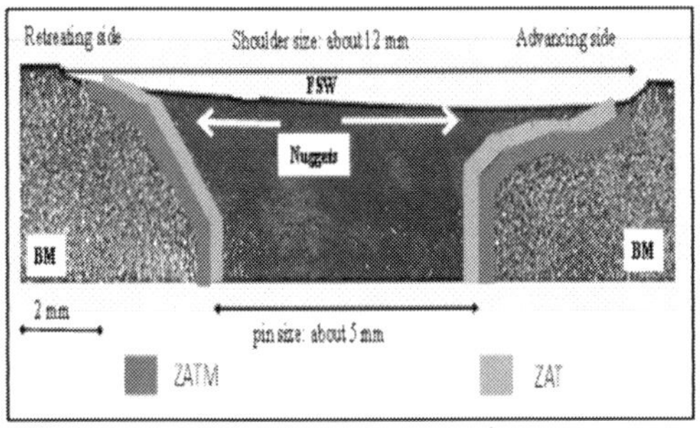

Figure 5: Microstructure of FSWeld section

The figure 6 shows the microstructure evolution:

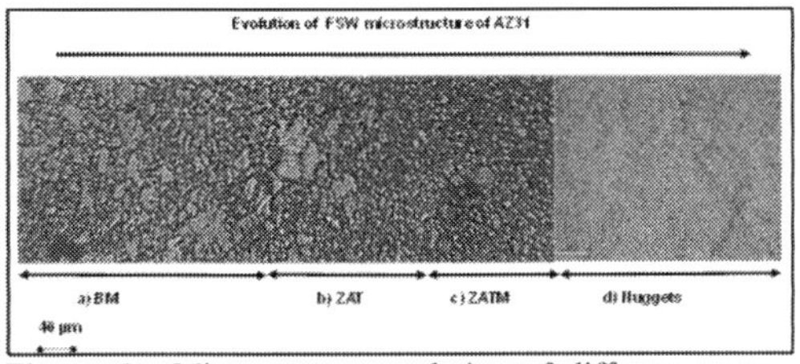

Figure 6: Microstructure evolution of different zones resulting from the FSW process.

The FSW process leads to a welded zone with two distinct zones: The HAZ and the friction zone. The latter zone is comprised of two zones: the thermo-mechanically altered zone TMAZ (figure 6c) and the stir welded zone, SWZ (6d), equally referred to as the nugget. In the FSW process, the HAZ (6b) is subjected to heating without notable deformation.

A large microstructure gradient is obtained by FSW process. Dynamic recrystallization occurs in the FSW nugget, leading to a fine grain size structure (about 2 μm). In the TMAZ, microstructure consists in not fully recrystallized grains. The TMAZ grain size ranges between 2 to 15 μm. The zone, situated at the boundary between the base metal and the highly deformed zone (TMAZ) can be characterised by the similarity of the grains to those of the base metal and the fine grains, interposed between the coarse grains, resulting in a dynamic recrystallization.

4.3 Results Orientations Overlap Volume Fraction

Superficial texture analysis was performed using laboratory XRD. Measurements were made in the nugget, TMAZ, HAZ and BM. The pole figures were examined with respect to the surface and central regions on both sides of the centerline (i.e., on the advancing and retreating sides) BM and HAZ. The volume fraction was extracted from ODF.

4.3.1 BM and HAZ

The base metal pole figures showed a strong texture, with the {0002} basal plane normal, parallel to the normal direction (ND), which is the typical hot rolling texture of magnesium alloys [4]. Concerning the HAZ, the results show a texture near of that of the BM. The fraction volume is slowly low compared of BM. These results (table 3) are reasonable because the HAZ retains the same grain structure as the parent material. Indeed, this zone does not undergo any plastic deformation but experiences thermal cycling. This temperature leads to some microstructural change but is not sufficient to modify the crystallographic texture of the materials. ODF calculation indicates that more than about 90 pct of the grains are oriented.

Table 3: Orientations volume fraction in the BM and HAZ obtained by FSW process

Volume	**Delta**	**Delta**	**Delta**	
Fraction	**Phi1**	**Phi**	**Phi2**	**Orientation**
89.35	**205.00**	**55.00**	**12.50**	**{0 0 1} < 1 0 0 >**
1.35	**10.00**	**10.00**	**10.00**	**{0 0 1} < 1 -2 0 >**
9.33	**The Rest Volume Fraction**			
90.70	**Orientations Overlap Volume Fraction**			

4.3.2 Welded Zone

With FSW process, the ODF calculation indicates a preferential orientation of the grains in the nuggets (table 4a) with the $\{002\}<210>$ orientation (about 90 %). which correspond to the three Euler angles ($\varphi_1 = 205°$, $\Phi = 55°$, $\varphi_2 = 12.5°$).

Tables 4a: Orientations volume fraction in the nuggets obtained by FSW process

Volume	**Delta**	**Delta**	**Delta**	
Fraction	Phi1	Phi	Phi2	Orientation
94	205.00	55.00	12.50	{0 0 2} < 2 1 0 >
4.63	45.00	45.00	45.00	{0 0 1} < 7 1 -1 >
2.37	The Rest Volume Fraction			
97.63	**Orientations Overlap Volume Fraction**			

In the nugget zone, the orientation with the {002} basal plane normal parallel to the normal direction (ND) is the typical friction texture of magnesium alloys. By comparing the ODF calculation (Tables 4b and 4c) along both sides of the central region, i.e., in the fracture region (TMAZ), the results show that most of the TMAZ has orientation components with a strong tendency for the (001) basal plane to align with the surface of the hard pin of the welding tool. This zone presents only the principal orientation of the (0002) basal plane. These texture components were rotated around the 'normal direction', the direction of the axis of the pin. The tilt is different along the side. In the literature, these local orientation evolutions are explained as being mainly due to shear plastic flow along the pin column surface [6, 7]. These zones were deformed and influenced by heat and high plastic deformation. Our results show that these zones are characterized by the modification of a crystallographic texture. The pole figures show that a texture developed along the top surfaces of the welded zones, with (0002) parallel to the plate surface.

Tables 4b: Orientations volume fraction in TMAZ (advancing side) zone obtained by FSW process

Volume Fraction	Delta Phi1	Delta Phi	Delta Phi2	Orientation
11.14	45.00	42.50	40.00	{0 0 1 }< 14 15 0>
83.97	45.00	45.00	40.00	{0 0 1 }< 1 7 0>
4.89	The Rest Volume Fraction			
95.11	**Orientations Overlap Volume Fraction**			

Tables 4c: Orientations volume fraction in TMAZ (retreating side) zone obtained by FSW process

Volume Fraction	Delta Phi1	Delta Phi	Delta Phi2	Orientation
76.60	45.00	42.50	40.00	{0 0 1 }< 14 15 0>
19.37	45.00	45.00	40.00	{0 0 1 }< 1 7 0>
4.03	The Rest Volume Fraction			
95.97	**Orientations Overlap Volume Fraction**			

The results show that most of these zones had a texture with a strong tendency for the (0002) basal plane to align with the surface of the hard pin of the welding tool. The appearance of other texture components in the stirred zone in addition to that of the basal plane indicates that friction stir welding leads to several mechanisms in the stirred zone: a shear deformation process and dynamic recrystallization.

4.3 Evolution of micro-hardness

Average values of micro hardness measures in the different zones are shown in figure 7.

In the case of the FSW process the average micro hardness in the fusion zone is 85 ± 2 HV which is slightly less than that of the base metal: 90 ± 5 HV. The micro hardness decreases slightly in the fusion zone and in the TMAZ (82 ± 2 HV); the slight decrease in the HAZ, where the grain sizes are identical to those in the BM, can be explained by the fact the material has been submitted to a thermal annealing cycle [6, 8].

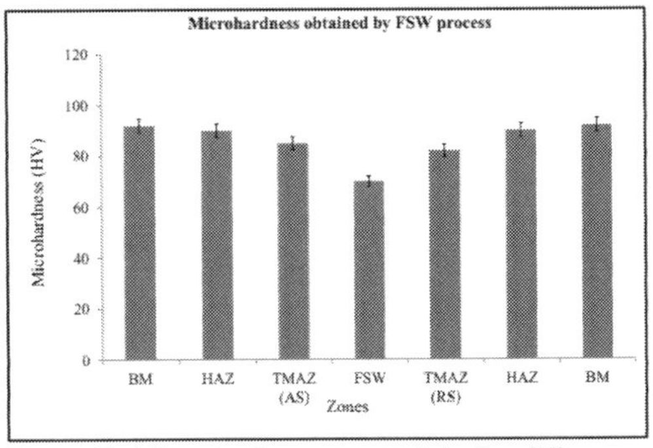

Figure 7: Micro hardness measured in the difference zones

In the welded zone there is a slight reduction in micro hardness despite the fact that the grain size is greatly reduced; the values obtained do not obey the Hall and Petch Law [10]. Within the welded zone the variation of micro hardness can be explained by the fact that the high degree plastic deformation and heating are conducive to the phenomenon of dynamic recrystallization resulting from dynamic recovery. This effect predominates over the reduction in grain size and leads to a reduction in the density of dislocation resulting from the restoration phenomenon. This phenomenon occurs in the case of non-age hardened alloys that have been heavily strain hardened as in the case of AZ31. Close the TMAZ / nugget the hardening due to work hardening resulting from the action of heavy deformation that gives rise to a partial recrystallization and splitting up of the intermetallic which could explain why the micro hardness remains relatively constant compared with the large decrease in the grain size.

4.4 Evolution of residual stresses

Figure 8 shows the average calculated values of the longitudinal and transversal residual stresses in the various zones welded by FSW.

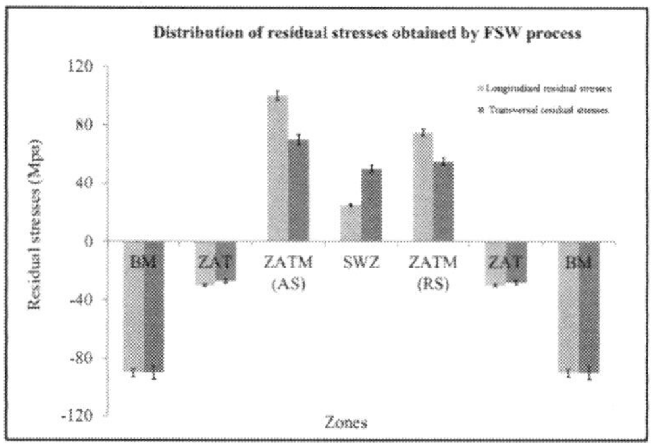

Figure 8: Evolution of residual stresses

The base metal has a compressive residual stress of the order of 100 MPa. In the zones welded the FSW process highly modifies the distribution of residual stress in the various zones. The HAZ has weak compressive residual stresses and is of similar magnitude in the longitudinal and transversal directions to the weld seam (30 MPa). In both the TMAZ and FZ the residual stresses are tensile. The residual stress profile in both of these zones shows two pronounced peaks in the TMAZ whereas the centre of the FSW has very low tensile residual stresses. Furthermore the profile is partially non-symmetrical transversal to the weld. It appears that the TMAZ does not have the same behaviour on the advancing side (AS) and the retreating side (RS) of the weld. The advancing side of the weld has a higher level of residual stress in comparison with the retreating side. This variation in stress corresponds with the asymmetrical plastic deformation imposed on the material which is inherent in this process: the force causing plastic deformation is higher on the AS side of the weld [11, 26]. The heterogeneity of stresses is in turn linked to the FSW method which induces local and different plastic deformations in the weld zone. These heterogeneities can equally be attributed to the formation of a crystallographic texture with differing orientations depending on the zone analysed [12] and to the presence of shear bands.

5. Conclusion

The objective of this study was to compare the local properties: microstructure, micro hardness, and residual stresses) of the same metal: thin sheets of magnesium alloy AZ31.

The FSW process does not lead to chemical modifications and significant phase changes. On the other hand it does lead to the phenomenon of partial recrystallization in the TMAZ and total recrystallization in the SWZ. Furthermore the welded zone could be characterised by tensile residual stresses and a heterogeneity depending on which zone was analysed. These variations could be attributed to the presence of a textured crystallography and shear bands which cause local plastic strain within these zones. The dissymmetry of stresses between the advancing and retreating sides of the joint can be attributed to the higher plastic force on one side of the weld and is inherent in the FSW process.

References

1. C.E.D. Rowe and W.M. Thomas, Advances in tooling materials for friction stir welding, Technical report, TWI Ltd Cambridge, UK.
2. R.S. Mishra, Z.Y. Ma, Mater. Sc. and Eng. R 50, 1–78 (2005)
3. H. Zhang, S. B. Lin, L. Wu and J. C. Feng. Act. metta .sin.17, 747-753(2004)
4. M.M. Avedesian and H. Baker, ASM Specialty Handbook, (1999)
5. R.S. Mishraa and Z.Y. Ma. Mater. Sc. and Eng. Reports 50, 1-78(2005).
6. T. Nagasawa, M. Otsuka, T. Yokota, and T. Ueki. Chapter Magnesium Technology 2000. The Miner., Metal., Mater. Society, 383–387 (2000).
7. A. Scialpi, L.A.C. De Filippis, P. Cavaliere, Mater. and Des.28, 1124–1129(2007).
8. K. Elangovan, V. Balasubramanian, Mater.ials Sc. and Eng. 459A, 7-18(2007).
9. Mustafa Boz, Adem Kurt, Mater. and Des. 25, 343-347(2004).
10. A. Kouadri, L. Barrallier, Mater. Sci. Eng. 429A, 11- 17(2006).
11. W. Woo, H. Choo, D.W. Brown, P.K. Liaw , Z. Feng, Scr. Mater. 54, 1859-1864(2006).
12. A.P. Reynolds, Elizabeth Hood, Wei Tang, 21S, Scr. Mater.52, 491-494(2005)., p. 491–494.
13. F. Caleyo, T. Baudin, M.H Mathon and R. Penelle, Eur. Phys. J. AP, 15, 85-96(2001).
14. H.R. Wenk, K. Pawlik, J. Pospiech and J.S Kallend, Textures and microstructures 22, 233-260(1994).
15. K. Pawlik, J. Pospiech, K. Lücke K., Proceeding of ICOTOM-8, Santa Fe, USA, 531(1987).

Modeling Metadynamic Recrystallization of a Die Steel during Ingot Breakdown Process

Kanwal Chadha , Davood Shahriari and Mohammad Jahazi

Department of Mechanical Engineering, École de Technologie Supérieure, Montréal, Canada

Abstract. Ingot forging processes often consist of several successive deformation steps with high interpass times, during which metadynamic recrystallization (MDRX) occurs. Two-stage isothermal compression tests were carried out at 1150°C and 1200°C with strain rates of 0.25-2s^{-1} and interpass times of 5-25s. Based on the experimental results, a material model for MDRX is proposed. The constitutive model was implemented in Forge NxT 1.1® software to simulate the multistage compression. Results from the material model are consistent with the numerical analysis and experimental results.

1 Introduction

Forging of ingot after casting process starts with ingot breakdown process which is generally performed at very high temperatures (0.75 of melting point "M_P") so as to breakdown the chemical (macrosegregation) and microstructural structure (as-cast dendritic) inhomogeneity [1]. The forging process usually consists of various deformation passes which includes interpass periods between deformations. During interpass periods, dynamic recovery [2], static recrystallization [3] and metadynamic recrystallization [4-7] may occur. Hot workability is defined as the amount of deformation that a material can undergo without cracking and reach desirable mechanical properties and microstructure at a given temperature and strain rate. It is generally evaluated by various parameters like strain rate, strain, temperature and interpass time. It is therefore a challenging task to determine the optimum thermomechanical route to apply in order to achieve the desired mechanical property keeping in view the microstructural changes and softening mechanisms involved. Kinetic equations are often used to describe the plastic flow properties of metals and alloys. Many research groups have attempted to develop kinetic equations using Avrami equation to describe the flow behavior of various alloys using experimental data [5, 8-10]. Despite large efforts being made on the development of kinetic equations for wrought ferrous alloys, little information is available on the flow behavior of as-cast structures at very high temperature (~0.7M_p) and prolonged interpass times. This corresponds to the ingot breakdown process during which the porosities inherited from solidification are close, macrosegregation levels are reduced and the as-cast structure is converted into a much finer and more isotropic wrought structure. The objective of this study is to investigate the influence of strain rate, temperature and interpass time on compressive deformation characteristics of as-cast 42CrMo using double hit hot compressive tests. Avrami equation describing the relationship of flow stress, strain rate, strain, interpass time and temperature was investigated and used to simulate real time analysis of the ingot breakdown process using FEM analysis.

2 Experimental

The composition of the as-cast 42CrMo high strength steel is shown in Table 1. The specimens were provided by Finkl Steel Co., Sorel Tracy, Quebec, Canada. Cylindrical specimens were machined with a diameter of 10mm and a height of 15mm. Hot compression tests were performed in Gleeble™ 3800 Thermomechanical Simulator at temperatures of 1150°C and 1200°C, strain rates of 0.25, 0.5, and 2s^{-1} and interpass times of 5,15 and 25s. The heating rate was 2°C/s till 1260°C where it was maintained for 300s so as to get homogenous temperature over the specimen. The specimen was then cooled to a respective deformation temperature at a cooling rate of 1°C/s. Water quench was done at the end of the deformation so as to freeze the microstructure.

Table 1: Composition of as-cast 42CrMo (Wt. %)

C	Mn	Si	Mo	Cr	Ni	Other
0.35	0.85	0.40	0.45	1.85	Added	Microalloying

3 Results and Discussions

3.1 Flow Stress-Strain Curves

For the present report, results from deformation temperature of 1200°C at strain rate of $0.25s^{-1}$ and interpass times of 5, 15 and 25s are considered. Stress-strain curve of double hit hot compression tests (Figure 1) reveals that the yield flow stress of the second hit is higher than that of the first hit deformation. It can be seen that in the second hit, the yield flow stress generally decreases with the increase in the interpass time. The results point out a fact that metadynamic softening increases with an increase in the interpass time. It is interesting to note that at interpass time of 25s, much of the metadynamic softening has taken place and the value of the yield flow stress of the second hit deformation is very close to the yield flow stress of the first hit deformation. This behavior can be explained by the fact that the dislocation density decreases with an increase in the interpass time. This results in low yield flow stress during second hit deformation [5].

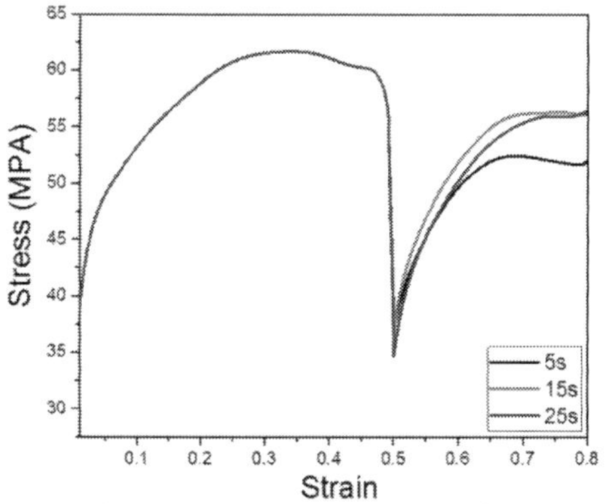

Figure 1. Flow stress strain curves in double hit hot compression at a temperature of 1200°C and strain rate of $0.25s^{-1}$.

3.2 Effect of Temperature on Metadynamic Softening

The metadynamic softening fraction X is determined by taking 0.2% offset-stress method. The value of X is calculated by:

$$X = \frac{\sigma_m - \sigma_2}{\sigma_m - \sigma_1} \qquad (1)$$

where σ_m (MPa) is the flow stress at the end point in the first hit deformation, σ_1(MPa) and σ_2 (MPa) are the offset yield flow stress for the first and second hit compression respectively.

The effect of temperature at the strain rate of $0.25s^{-1}$ is shown in Figure 2. It can be seen that with an increase in temperature, the metadynamic softening fraction increases. For different temperatures (1200°C and 1150°C) and interpass time of 15s, the metadynamic softening fraction values are 69.5% and 57.2%, respectively. This is due to the fact that, increase in the deformation temperature causes an increase in the diffusion of the atoms which make migration of the boundaries easier thereby benefitting the growth of the dynamically formed grains [10].

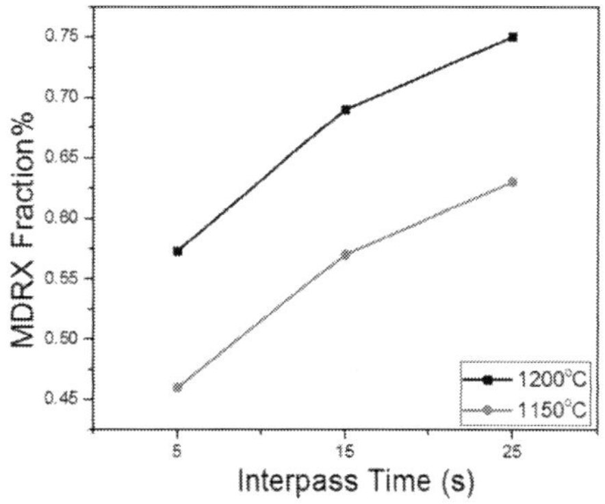

Figure 2. Effect of deformation temperature on metadynamic softening.

3.3 Kinetic Equation of Metadynamic Recrystallization

In order to characterize the metadynamic softening behavior of as-cast 42CrMo steels, the Avrami equation is applied. The equation is in the following form:

$$X_m = 1 - exp[-0.693\left(\frac{t}{t_{0.5}}\right)^n] \qquad (2)$$

Where X_m is the metadynamic softening fraction (%), t is the interpass time (s), n is the material dependent constant, $t_{0.5}$ is the time for metadynamic softening fraction of 50% (s) which is expressed as:

$$t_{0.5} = A\dot{\varepsilon}^p exp(\frac{Q_m}{RT}) \qquad (3)$$

where, Q_m is the activation energy for metadynamic recrystallization ($KJmol^{-1}$), T is the absolute deformation temperature (°K), and R is the universal gas constant ($Jmol^{-1o}K^{-1}$), p and A are the material dependent constants.

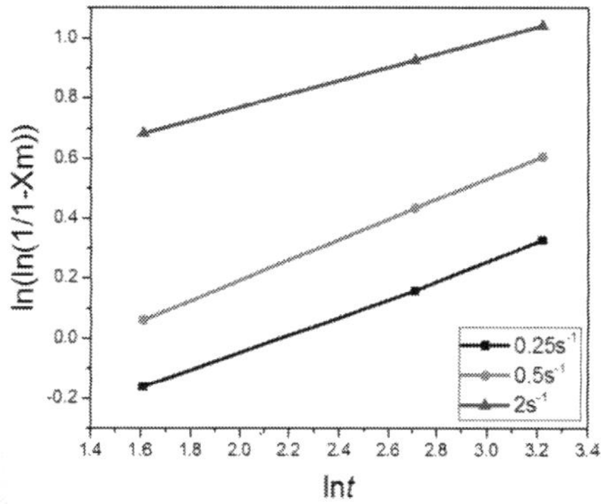

Figure 3. Relationship between **ln (ln (1/1-*Xm*))** and **ln*t*** at deformation temperature of 1200°C.

The values of $t_{0.5}$ and n in Eq. (2), A, Q_m and p in Eq. (3) were calculated by linear regression method presented in Figures 3 and 4 and as described in various published papers [5, 9, 10].

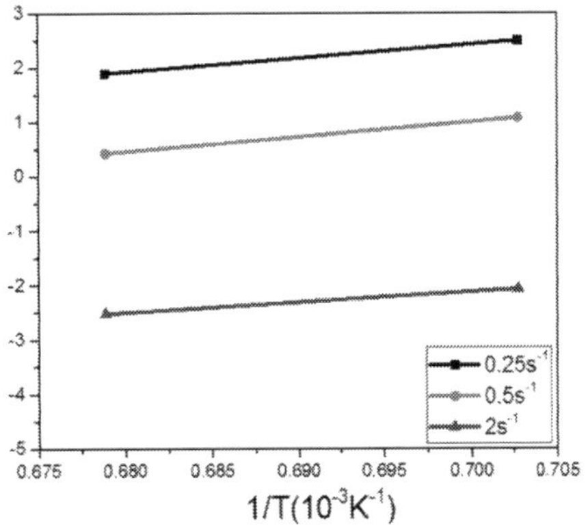

Figure 4. Relationship between **ln$t_{0.5}$** and **1/T**.

The values of the constants calculated using linear regression method are shown in Table 2.

Table 2: Values of constants derived from the regression analysis

n	Q_m	*p*	*LnA*
0.27	185853	-0.318	-2.019

The equations of metadynamic softening fraction and the time for 50% softening for as-cast 42CrMo steels are represented as follows:

$$X_m = 1 - exp[-0.693\left(\frac{t}{t_{0.5}}\right)^{0.27}] \quad (4)$$

$$t_{0.5} = 0.1328\dot{\varepsilon}^{-0.318}exp(\frac{185853}{RT}) \quad (5)$$

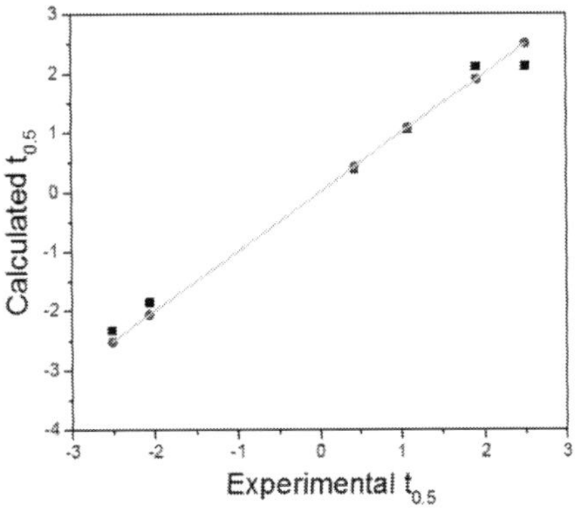

Figure 5. The comparison of the calculated and experimental $t_{0.5}$. Green line with brown symbols represent experimental values and black symbols represent calculated values.

The comparison of the experimental and calculated $t_{0.5}$ at various deformation parameters is shown in Figure 5. The calculated results are in good agreement with the experimental results, thereby confirming that the kinetic equation is able to predict the model for metadynamic behavior of as-cast 42CrMo steel.

3.4 FEM Simulation for Recrystallization Behavior of 42CrMo

In this study, the Forge NxT 1.1® software is used to carry out FEM simulation of hot compression of cylindrical specimen. An eighth of the 3D model is considered instead of a whole geometry to simulate the hot compression test because of the symmetrical features of the specimen, saving the computational time and getting the sufficient accuracy. The tetrahedral element is used to discrete the geometry model.

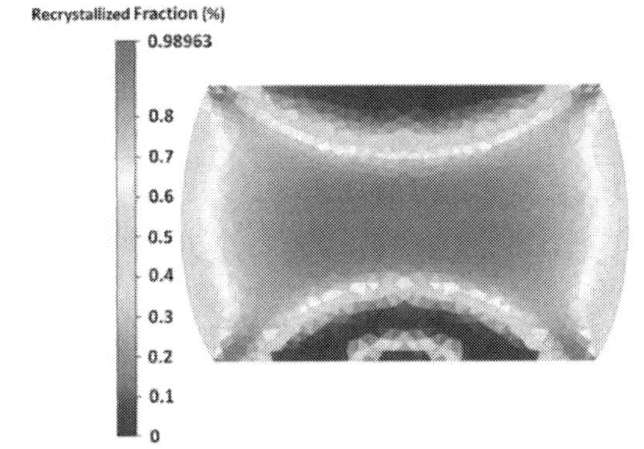

(a)

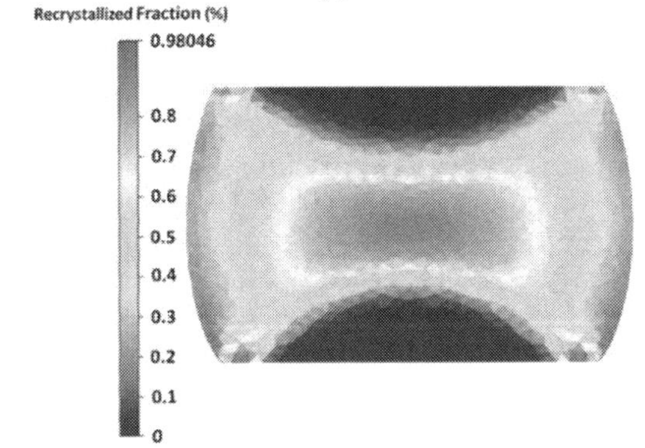

(b)

Figure 6. FEM simulation results of recrystallization fraction distribution under 1200 °C at the strain rate of (a) $0.25s^{-1}$ and (b) $2s^{-1}$.

The upper die moved along the central axis of model while both upper and lower dies are set as a rigid body. The friction coefficient between the dies and the sample is assumed as 0.4. Since the sample is very small and its heating during hot compression by Gleeble™ 3800 is in accordance with the accurate temperature control resistance heating method, an adiabatic heating condition is imposed in this simulation. Physical and thermo-physical properties are critical parameters for any hot deformation simulations. So, the material property data (as a function of temperature) are calculated by JMatPro® and then directly input into the FEM code. The implementation of the flow stress constitutive equations of as-cast 42CrMo steel according to Arrhenius constitutive model [11] and the recrystallization model in the FEM software are accomplished by user subroutine. Figure 6 (a & b) represents total recrystallization (RX) fraction under 1200°C at different strain rate. The recrystallization fractions in the center of the specimen under 1200°C at strain rate $2s^{-1}$ nearly achieve 90%. It can be seen that the recrystallized zone will enlarge as the strain rate reduces.

According to Figure 7, recrystallization region with strain rate $0.25s^{-1}$ at 1200°C in the center is almost complete (100%) and the size of this zone is approximately with a diameter 3 mm and a length of 1.5 mm.

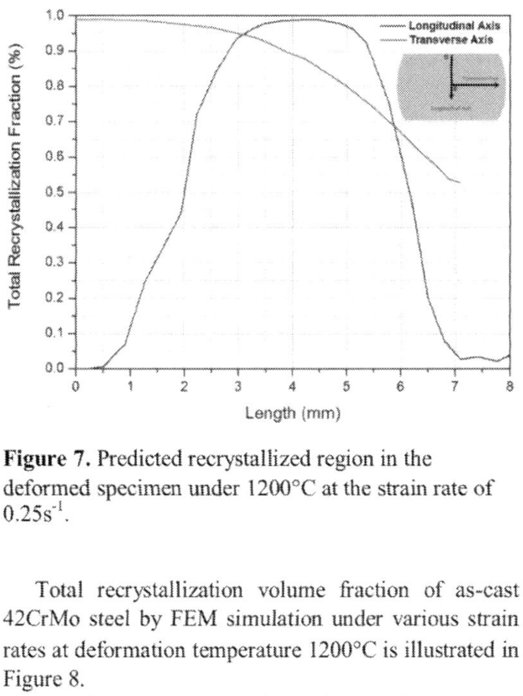

Figure 7. Predicted recrystallized region in the deformed specimen under 1200°C at the strain rate of $0.25s^{-1}$.

Total recrystallization volume fraction of as-cast 42CrMo steel by FEM simulation under various strain rates at deformation temperature 1200°C is illustrated in Figure 8.

It shows the total volume fraction of recrystallization increases with the increasing of the strain. It approves at lower strain rate total RX fraction tends to be complete and close to 100%. The main reason for this phenomenon is that higher strain rates lead to increase work hardening rate and decrease the time for recrystallization. So, the recrystallized nucleus at high strain rate has no enough time to fully grow up. Finally, there is a good agreement between the predicted value by FEM simulation results and the experimental data.

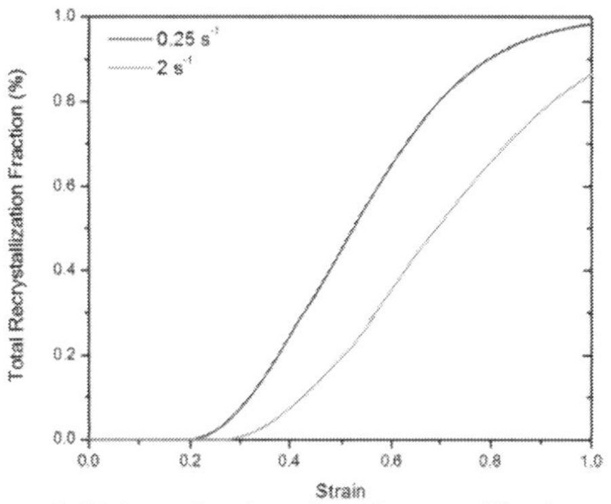

Figure 8. Volume fraction of total recrystallization fraction at the center of the deformed sample under 1200°C at the strain rate of $0.25s^{-1}$ and $2s^{-1}$.

4 Conclusions

Double hit hot compression tests of as-cast 42CrMo steel was conducted using Gleeble™ 3800 Thermomechanical Simulator. Deformation temperature of 1200°C and 1150°C, strain rates of 0.25,0.5 and $2s^{-1}$ and interpass times of 5,15 and 25s were selected so as to investigate the effect of deformation parameters on metadynamic recrystallization. Using experimental conditions, the Avrami equation was obtained to characterize the metadynamic behavior. The following are the main conclusions of the research:

a. With an increase in temperature, strain rate and interpass time, the metadynamic softening fraction increased.
b. Maximum metadynamic softening occurs at temperature of 1200°C at an interpass time of 25s.
c. Avrami equation developed is in good agreement with the experimental and calculated results.
d. Integrating with the material kinetic models in the Forge NxT 1.1® software is developed to analyze the microstructure evolution under various hot deformation conditions.

Acknowledgements

The authors are very much grateful to Finkl Steel Co. for providing the specimens for the present research. The authors are also grateful to NSERC for their support in the framework of a Collaborative Research and Development project (CRD), and Transvalor Americas Corp for providing Forge NxT 1.1® software.

References

1. Dieter, G.E., H.A. Kuhn, and S.L. Semiatin, *Handbook of Workability and Process Design*. 2003: Asm International.
2. McQueen, H.J. and J.J. Jonas, in *Treatise on Materials Science & Technology*, A. R.J, Editor. 1975, Elsevier. p. 393-493.

3. Lin, Y.C., M.-S. Chen, and J. Zhong, Computational Materials Science, 2008. **44**(2): p. 316-321.
4. Gao, F., et al., Materials Science and Engineering: A, 2000. **277**(1): p. 33-37.
5. Lin, Y.C., M.-S. Chen, and J. Zhong, Journal of Materials Processing Technology, 2009. **209**(5): p. 2477-2482.
6. Zhou, L. and T. Baker, Materials Science and Engineering: A, 1995. **196**(1): p. 89-95.
7. Beladi, H., P. Cizek, and P.D. Hodgson, Acta Materialia, 2011. **59**(4): p. 1482-1492.
8. Cho, S.-H. and Y.-C. Yoo, 2001. **36**(17): p. 4279-4284.
9. Elwazri, A., E. Essadiqi, and S. Yue, ISIJ international, 2004. **44**(4): p. 744-752.
10. Liu, J., et al., Materials Science and Engineering: A, 2013. **565**: p. 126-131.
11. K. Chadha, D.Shahriari, M. Jahazi, La Metallurgia Italiana - International Journal of the Italian Association for Metallurgy, 2016. **4**: p. 5-12.

Simulation of nonlinear benchmarks and sheet metal forming processes using linear and quadratic solid–shell elements combined with advanced anisotropic behavior models

Peng Wang[1,a], Hocine Chalal[1] and Farid Abed-Meraim[1]

[1] *LEM3, UMR CNRS 7239 - Arts et Métiers ParisTech, 4, rue Augustin Fresnel, 57078 Metz Cedex 03, France*

Abstract. A family of prismatic and hexahedral solid–shell (SHB) elements with their linear and quadratic versions is presented in this paper to model thin 3D structures. Based on reduced integration and special treatments to eliminate locking effects and to control spurious zero-energy modes, the SHB solid–shell elements are capable of modeling most thin 3D structural problems with only a single element layer, while describing accurately the various through-thickness phenomena. In this paper, the SHB elements are combined with fully 3D behavior models, including orthotropic elastic behavior for composite materials and anisotropic plastic behavior for metallic materials, which allows describing the strain/stress state in the thickness direction, in contrast to traditional shell elements. All SHB elements are implemented into ABAQUS using both standard/quasi-static and explicit/dynamic solvers. Several benchmark tests have been conducted, in order to first assess the performance of the SHB elements in quasi-static and dynamic analyses. Then, deep drawing of a hemispherical cup is performed to demonstrate the capabilities of the SHB elements in handling various types of nonlinearities (large displacements and rotations, anisotropic plasticity, and contact). Compared to classical ABAQUS solid and shell elements, the results given by the SHB elements show good agreement with the reference solutions.

1 Introduction

Nowadays, thin structures are increasingly used in engineering applications, and especially in automotive industries. For the design of new products while optimizing their forming processes with reasonable cost, the finite element simulation has become an indispensable practice. However, due to the large aspect ratio (length to thickness) of thin structures, the conventional solid and shell elements suffer from various locking phenomena both in linear and nonlinear analyses. In order to obtain accurate and reliable numerical results, much effort has been dedicated in recent decades to the development of efficient locking-free finite elements.

The recent concept of solid–shell elements attracted much attention due to their outstanding advantages compared to traditional solid and shell elements. They are based on a fully three-dimensional (3D) formulation with only displacements as degrees of freedom. Combined with the reduced-integration technique, various methods have been proposed in the literature to eliminate most locking phenomena [1–9], among which the assumed-strain method (ASM), the enhanced assumed strain (EAS) formulation, and the assumed natural strain (ANS) approach.

In this contribution, a family of assumed-strain based solid–shell elements (SHB elements) is briefly presented within a unified formulation. It consists of a linear eight-node hexahedral element (SHB8PS) and a linear six-node prismatic element (SHB6), and their quadratic counterparts (SHB20) and (SHB15), respectively. The SHB elements are combined with various types of constitutive equations, including classical isotropic elastic behavior, orthotropic elastic behavior for composite materials, and anisotropic plastic behavior for metallic materials. All SHB elements have been implemented into ABAQUS static/implicit and dynamic/explicit software packages in order to extend their applications to nonlinear quasi-static and dynamic analyses. A variety of nonlinear benchmark tests and a complex sheet metal forming process have been simulated to assess the performance of the proposed SHB elements.

2 SHB solid–shell elements

The proposed SHB elements are based on a fully three-dimensional formulation using an in-plane reduced-integration scheme and the assumed-strain method. This SHB family is composed of a linear six-node prismatic element (SHB6) and an eight-node hexahedral element (SHB8PS), along with their quadratic counterparts (SHB15) and (SHB20), respectively [3, 5, 10–12]. Figure 1 shows the reference geometry of the four SHB solid–shell elements as well as the location of their integration points. The latter are distributed along the local direction

[a] Corresponding author: peng.wang@ensam.eu

ζ, which is designated as the thickness direction (see Figure 1). Note that the number of through-thickness integration points is chosen freely, depending on the problem complexity, which avoids using several element layers for an accurate description of the various through-thickness phenomena.

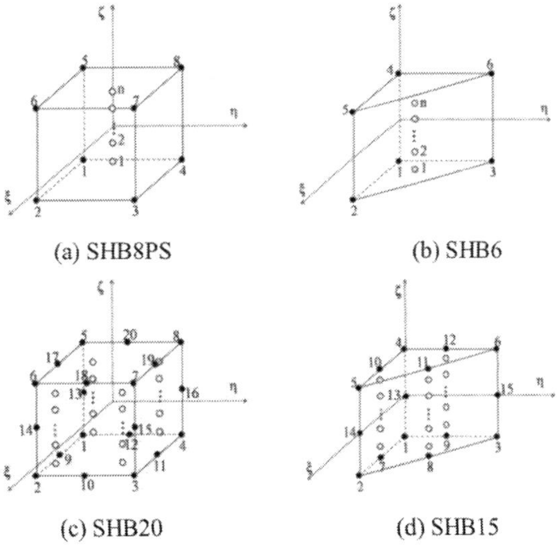

(a) SHB8PS (b) SHB6

(c) SHB20 (d) SHB15

Figure 1. Reference geometry and location of integration points for the SHB elements.

2.1 Finite element formulation

The assumed-strain formulation of the SHB solid–shell elements is derived from the simplified form of the Hu–Washizu variational principle [13]. In terms of assumed-strain rate $\dot{\bar{\boldsymbol{\varepsilon}}}$, interpolated stress $\boldsymbol{\sigma}$, nodal velocities $\dot{\mathbf{d}}$, and external nodal forces $\mathbf{f}^{ext}$, this principle writes

$$\pi(\dot{\bar{\boldsymbol{\varepsilon}}}) = \int_{\Omega_e} \delta\dot{\bar{\boldsymbol{\varepsilon}}}^T \cdot \boldsymbol{\sigma}\, d\Omega - \delta\dot{\mathbf{d}}^T \cdot \mathbf{f}^{ext} = 0 . \quad (1)$$

By introducing the discretized gradient operator $\mathbf{B}$, which contains the derivatives of the shape functions, the strain field $\boldsymbol{\varepsilon}$ can be related to the nodal displacement field $\mathbf{d}$ by the following expression:

$$\boldsymbol{\varepsilon} = \mathbf{B} \cdot \mathbf{d} . \quad (2)$$

In the case of linear SHB elements (i.e., SHB6 and SHB8PS), a $\overline{\mathbf{B}}$ matrix is introduced, which is obtained by projection of the original operator $\mathbf{B}$ onto an appropriate sub-space in order to eliminate most locking phenomena. Accordingly, the assumed-strain rate $\dot{\bar{\boldsymbol{\varepsilon}}}$ can be expressed in terms of the $\overline{\mathbf{B}}$ matrix as

$$\dot{\bar{\boldsymbol{\varepsilon}}} = \overline{\mathbf{B}} \cdot \dot{\mathbf{d}} . \quad (3)$$

Replacing the above assumed-strain rate in the expression of the variational principle (Eq. 1), the element stiffness matrix $\mathbf{K}_e$ and internal forces $\mathbf{f}^{\text{int}}$ are derived as

$$\mathbf{K}_e = \int_{\Omega_e} \overline{\mathbf{B}}^T \cdot \mathbf{C}^{ep} \cdot \overline{\mathbf{B}}\, d\Omega + \mathbf{K}_{GEOM} \quad (4)$$

$$\mathbf{f}^{\text{int}} = \int_{\Omega_e} \overline{\mathbf{B}}^T \cdot \boldsymbol{\sigma}\, d\Omega , \quad (5)$$

where $\mathbf{K}_{GEOM}$ is the geometric stiffness matrix, which originates from the nonlinear part of the strain tensor [5]. As to matrix $\mathbf{C}^{ep}$, it corresponds to the elastic–plastic tangent modulus.

In addition to the above formulation of SHB elements, a standard mass matrix is added to the dynamic version of each SHB element.

2.2 Constitutive modeling

The finite element implementation of the above SHB formulation requires introducing several local frames for the calculation of the stiffness matrix and internal forces associated with the constitutive law considered (see Figure 2). The "element frame" is used to define the thickness direction with respect to the global frame, while the "material frame" is specifically introduced to integrate the constitutive equations.

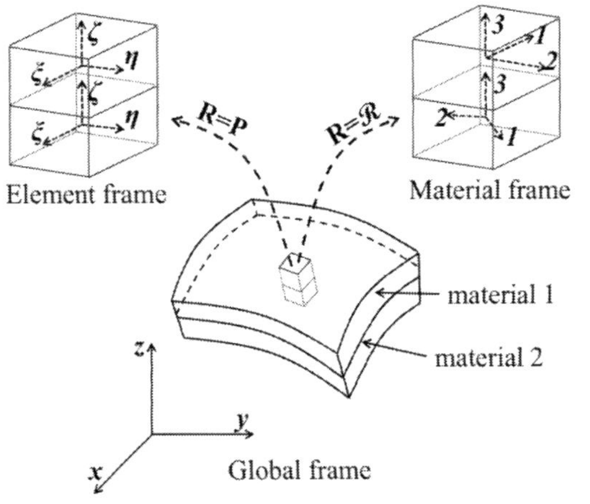

Figure 2. Illustration of the local frames used in the formulation of the SHB solid–shell elements.

2.2.1 Elastic material behavior

In this work, two typical 3D elastic behavior models are considered: an isotropic elastic behavior model for standard isotropic materials and an orthotropic elastic behavior model for laminated composites. The rate form of the stress–strain relationship can be expressed using the hypoelastic law defined in the local material frame by

$$\dot{\boldsymbol{\sigma}} = \mathbf{C} : \mathbf{D}^e , \quad (6)$$

where $\mathbf{D}^e$ is the elastic strain rate tensor. Matrix $\mathbf{C}$ is the fourth-order elasticity tensor, which is defined in the element local frame. In the case of isotropic elastic behavior, a degenerated elasticity tensor $\mathbf{C}$ is used in the SHB formulation in order to approach plane-stress-type behavior. Its expression is given by

$$\mathbf{C}=\begin{bmatrix} \bar{\lambda}+2\mu & \bar{\lambda} & 0 & 0 & 0 & 0 \\ \bar{\lambda} & \bar{\lambda}+2\mu & 0 & 0 & 0 & 0 \\ 0 & 0 & E & 0 & 0 & 0 \\ 0 & 0 & 0 & \mu & 0 & 0 \\ 0 & 0 & 0 & 0 & \mu & 0 \\ 0 & 0 & 0 & 0 & 0 & \mu \end{bmatrix},$$

$$\text{with } \mu=\frac{E}{2(1+\nu)},\quad \bar{\lambda}=\frac{E\nu}{1-\nu^2}, \tag{7}$$

where E is the Young modulus and ν is the Poisson ratio.

For laminated composite materials, a classical orthotropic elasticity law is adopted in the formulation of the SHB elements. Similar to the isotropic elasticity tensor, the orthotropic elasticity tensor $\overline{\mathbf{C}}$ is defined in the element local frame within the orthotropic axes as

$$\overline{\mathbf{C}}^{-1}=\begin{bmatrix} \frac{1}{E_1} & -\frac{\nu_{21}}{E_2} & -\frac{\nu_{31}}{E_3} & 0 & 0 & 0 \\ -\frac{\nu_{12}}{E_1} & \frac{1}{E_2} & -\frac{\nu_{32}}{E_3} & 0 & 0 & 0 \\ -\frac{\nu_{13}}{E_1} & -\frac{\nu_{23}}{E_2} & \frac{1}{E_3} & 0 & 0 & 0 \\ 0 & 0 & 0 & \frac{1}{G_{12}} & 0 & 0 \\ 0 & 0 & 0 & 0 & \frac{1}{G_{13}} & 0 \\ 0 & 0 & 0 & 0 & 0 & \frac{1}{G_{23}} \end{bmatrix},$$

$$\text{with } \nu_{ij}E_j=\nu_{ji}E_i \text{ for } i,j=1,2,3, \tag{8}$$

where E_i represents the Young modulus in the i th-direction, ν_{ij} is the Poisson ratio associated with the i th and j th-directions, and G_{12}, G_{13} and G_{23} are the shear moduli.

Considering an initial rotation angle θ of the orthotropic axes with respect to the global frame, the projected elasticity tensor can be obtained using the following expressions:

$$\hat{\mathbf{C}}=\mathbf{T}(\theta)\cdot\overline{\mathbf{C}}\cdot\mathbf{T}(\theta)^T, \tag{9}$$

where the rotation matrix $\mathbf{T}(\theta)$ has the following form:

$$\mathbf{T}(\theta)=\begin{bmatrix} c^2 & s^2 & 0 & 2cs & 0 & 0 \\ s^2 & c^2 & 0 & -2cs & 0 & 0 \\ 0 & 0 & 1 & 0 & 0 & 0 \\ -cs & cs & 0 & c^2-s^2 & 0 & 0 \\ 0 & 0 & 0 & 0 & c & -s \\ 0 & 0 & 0 & 0 & s & -s \end{bmatrix}, \tag{10}$$

with $c=\cos(\theta)$ and $s=\sin(\theta)$.

2.2.2 Anisotropic elastic–plastic behavior

When plastic behavior is considered for the material, the strain rate tensor $\mathbf{D}$ is partitioned into an elastic part $\mathbf{D}^e$ and a plastic part $\mathbf{D}^p$. The latter is defined using the classical plastic flow rule

$$\mathbf{D}^p=\dot{\lambda}\frac{\partial f}{\partial \boldsymbol{\sigma}}, \tag{11}$$

where f represents the plastic yield surface, and $\dot{\lambda}$ is the plastic multiplier whose expression is determined using the consistency condition.

In this work, anisotropic plastic behavior is taken into account using the Hill'48 quadratic yield surface. The yield condition is written in the following form:

$$f=\bar{\sigma}_{eq}-Y\le 0, \tag{12}$$

where $\bar{\sigma}_{eq}=\sqrt{(\boldsymbol{\sigma}'-\boldsymbol{\alpha}):\mathbf{H}:(\boldsymbol{\sigma}'-\boldsymbol{\alpha})}$ is the equivalent stress, in which $\boldsymbol{\sigma}'$ represents the deviatoric part of the Cauchy stress, and $\mathbf{H}$ is the fourth-order tensor that contains the six anisotropy coefficients of Hill'48 criterion. Isotropic and kinematic hardening is taken into account using the internal variables Y and $\boldsymbol{\alpha}$, respectively.

3 Numerical examples and results

In this section, several popular benchmark problems and a complex sheet metal forming process are simulated to evaluate the performance of the SHB elements in situations involving large rotations, large strain, material nonlinearity, and double-sided contact. In the case of elastic benchmark problems, only two integration points through the thickness are used in the simulations, while five integration points are considered in the case of elasto-plastic problems. The results obtained with the SHB elements are consistently compared with reference solutions as well as with the numerical results yielded by ABAQUS solid and shell elements.

In the following benchmark tests, the nomenclature used for the mesh is: $(N_1 \times N_2) \times N_3$ for hexahedral elements, where N_1 is the number of elements along the length, N_2 is the number of elements along the width and N_3 is the number of elements along the thickness direction. For the prismatic elements, however, the nomenclature used is $(N_1 \times N_2 \times 2) \times N_3$, due to the in-plane sub-division of a rectangular element into two triangles. Similarly, the nomenclature used for triangular ABAQUS shell elements is $N_1 \times N_2 \times 2$, while the nomenclature for quadrilateral ABAQUS shell elements is $N_1 \times N_2$.

3.1 Quasi-static benchmark problems

3.1.1 Simply supported elastic square plate

Figure 3 illustrates a simply supported isotropic elastic square plate subjected to a central concentrated force. The detailed information about the geometric dimensions and the elastic material parameters is given in Figure 3.

The maximum of the concentrated load F is equal to 300. Due to symmetry, only one quarter of the plate is analyzed.

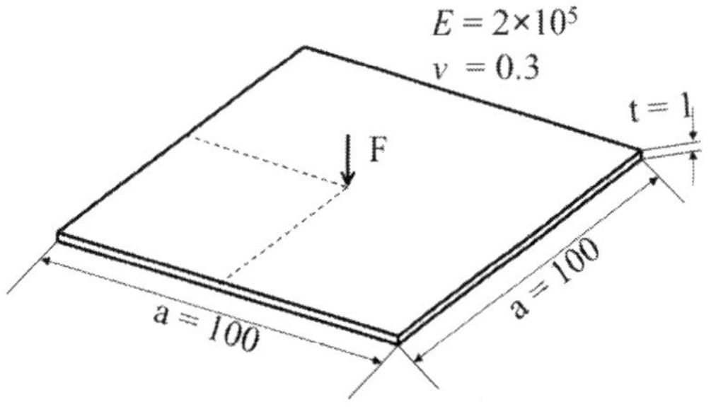

Figure 3. Geometry and elastic material parameters for the simply supported square plate subjected to a concentrated force.

The load–displacement results at the central point of the plate given by the SHB elements are compared in Figure 4 with the ABAQUS elements and also with the reference solutions taken from [14, 15].

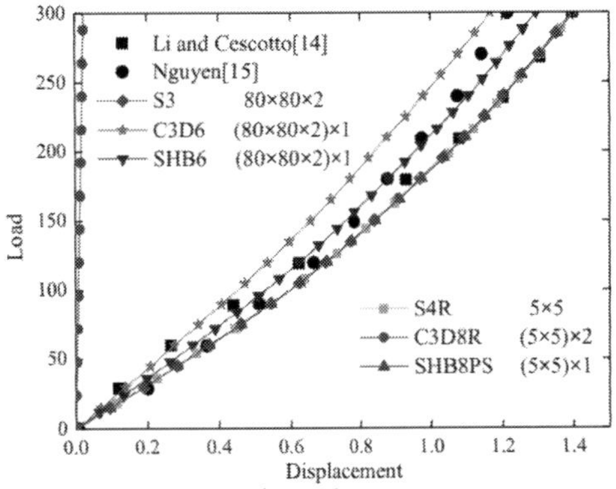

(a) Linear elements

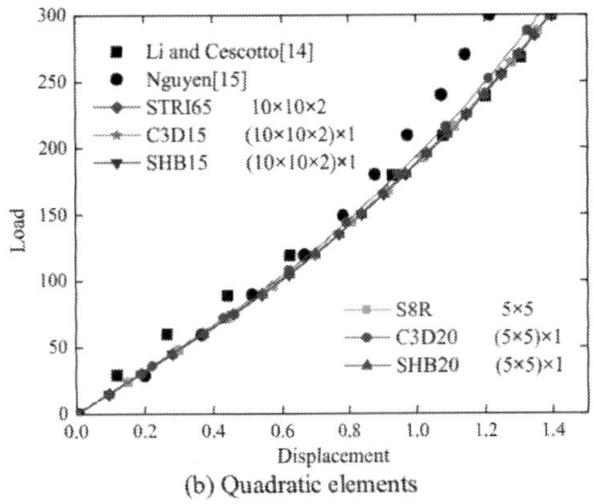

(b) Quadratic elements

Figure 4. Load–displacement curves for the simply supported square plate.

In the case of triangular-based linear elements (Figure 4a), the results given by the SHB6 solid–shell element are closer to the reference solutions than those given by the ABAQUS prismatic solid element (C3D6), while the results yielded by the ABAQUS triangular shell element (S3) are in agreement with one of the reference solutions (i.e., that taken from [14]). For the quadrangular-based linear elements, the load–displacement results given by the SHB8PS element are very close to those obtained with the ABAQUS quadrilateral shell element (S4R) as well as to the reference solution taken from [14]. Note that, using the same in-plane mesh as SHB8PS and S4R elements, the load–displacement curve given by the ABAQUS hexahedral solid element (C3D8R) lies far from the reference solution, which suggest resorting to much finer meshes for this element. In the case of quadratic elements (Figure 4b), the results given by the solid–shell elements SHB15 and SHB20 agree very well with the reference solution taken from [14], as well as with the results obtained with the ABAQUS quadratic elements (namely, the prismatic solid element C3D15, the triangular shell element STRI65, and the quadrilateral shell element S8R), while the hexahedral solid element C3D20 is slightly stiffer.

3.1.2 Pinched elastic–plastic hemispherical shell

A pinched hemispherical shell is considered here, which is loaded by alternating radial forces as shown in Figure 5. This test is very popular since it involves both geometric and material nonlinearities. Owing to the problem symmetry, only one quarter of the structure is modeled. In addition to the nomenclature defined previously, three partitions for the quarter of the hemispherical shell are created in order to achieve a relatively regular mesh (see Figure 5). This leads to the following nomenclature for this test: $(3\times(N_1\times N_2))\times N_3$ for hexahedral elements, $(3\times(N_1\times N_2\times 2))\times N_3$ for prismatic elements, $3\times(N_1\times N_2)$ for quadrilateral shell elements, and $3\times(N_1\times N_2\times 2)$ for triangular shell elements.

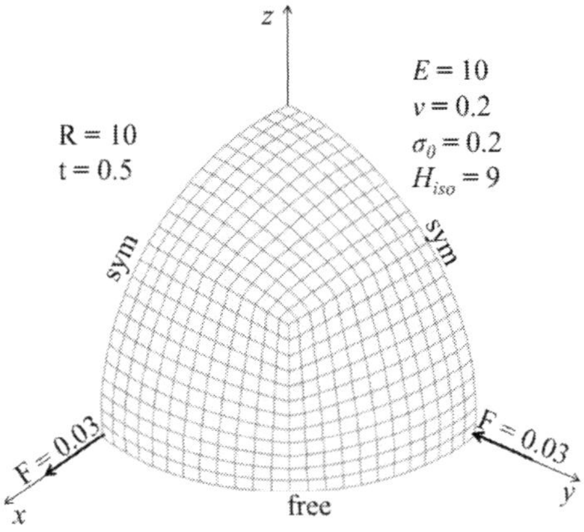

Figure 5. Geometry and material parameters for the pinched hemispherical shell.

In Figure 6, the displacements at the loading points in the *x* and *y* directions (namely, U and V, respectively) obtained with the SHB elements are compared with those given by ABAQUS elements as well as with the reference solution taken from [16]. Overall, all SHB and ABAQUS elements provide accurate results with respect to the reference solution, except for the SHB6 and C3D6 elements, which exhibit some locking and which require finer meshes for this test problem. Note also that the proposed solid–shell elements are more efficient than their counterparts Abaqus solid elements, since the latter require several layers of elements to provide comparable accuracy (see Figure 6).

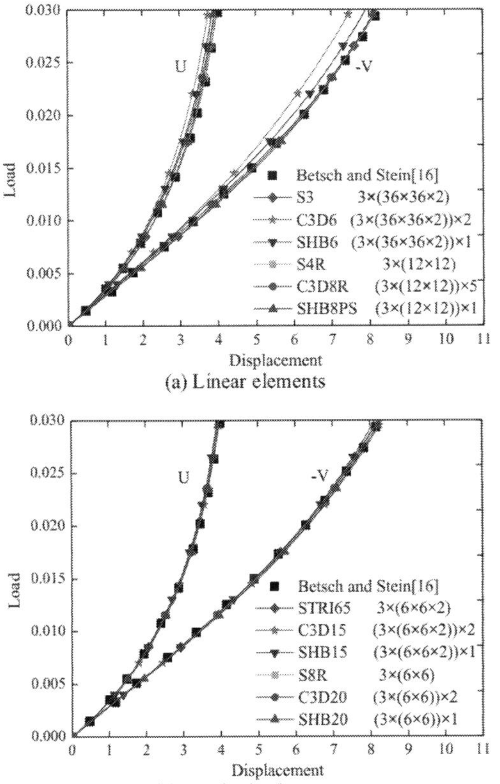

(a) Linear elements

(b) Quadratic elements

Figure 6. Load–displacement curves for the pinched hemispherical shell.

3.1.3 Clamped composite plate

A laminated composite plate with orthotropic elastic behavior (see Subsection 2.2.1) is considered here. The plate is fully clamped at one end and subjected to a bending force at the free end. All geometric dimensions as well as orthotropic elastic parameters are given in Figure 7. For comparison purposes, two configurations of laminated composites, which correspond to two different stacking sequences (0°/90°/0° and 90°/0°/90°), are considered. The load–deflection curves obtained with the SHB elements are compared in Figure 8 with those given by the ABAQUS elements as well as with the reference solutions taken from [17]. Note that the laminated composites are modeled with three layers in the case of prismatic and hexahedral elements due to the different stacking sequences through the thickness. The obtained results clearly show the excellent accuracy of the SHB elements when compared to ABAQUS elements and the reference solutions. This reveals the good performance of the proposed solid–shell elements when orthotropic behavior for laminated composites is considered.

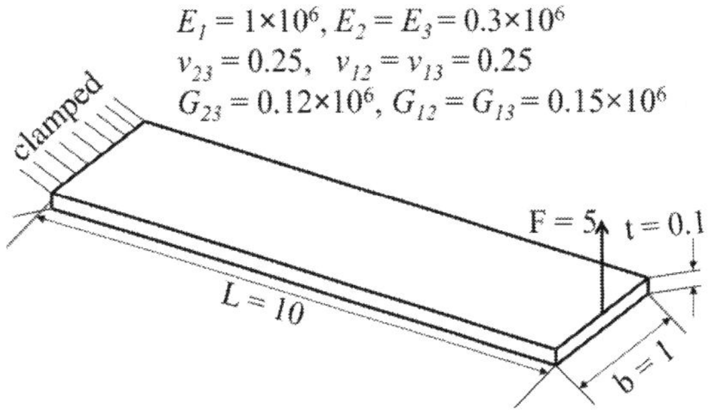

Figure 7. Orthotropic elastic parameters and geometric dimensions for the laminated composite cantilever plate.

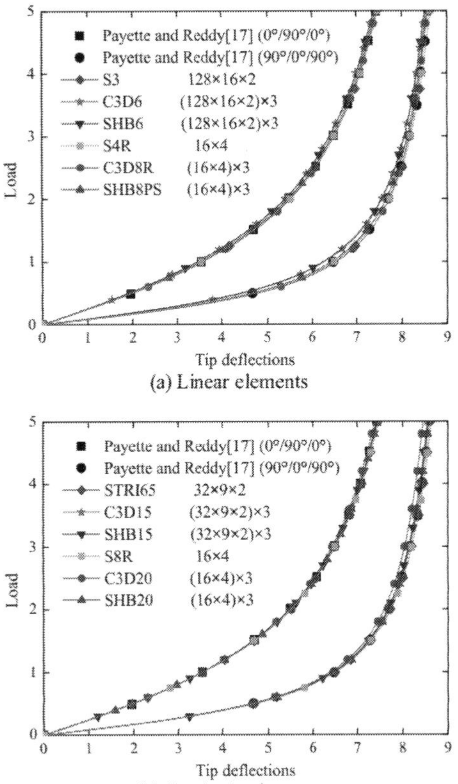

(a) Linear elements

(b) Quadratic elements

Figure 8. Load–displacement curves for the laminated composite cantilever plate.

3.2 Dynamic benchmark tests

3.2.1 Clamped elastic spherical cap

The performance of the SHB elements is evaluated here within the framework of explicit dynamic analysis. The first explicit/dynamic problem is illustrated in Figure 9, which consists of a fully clamped spherical cap loaded by a constant concentrated force at its apex. Owing to the symmetry, only one quarter of the cap is discretized. The dynamic response in terms of central vertical displacement is analyzed in Figure 10 using the dynamic versions of the proposed SHB elements as well as

ABAQUS explicit elements, which are compared to the reference solution taken from [18]. Note that, since no quadratic elements are available in ABAQUS/Explicit software package, the results given by the quadratic SHB elements are compared only with the reference solution (see Figure 10b). The results reveal that both the peak and the period of the response are well predicted using the proposed SHB elements as well as ABAQUS linear shell elements, while the solution yielded by the ABAQUS linear solid elements is found far from the reference solution during the second half-period.

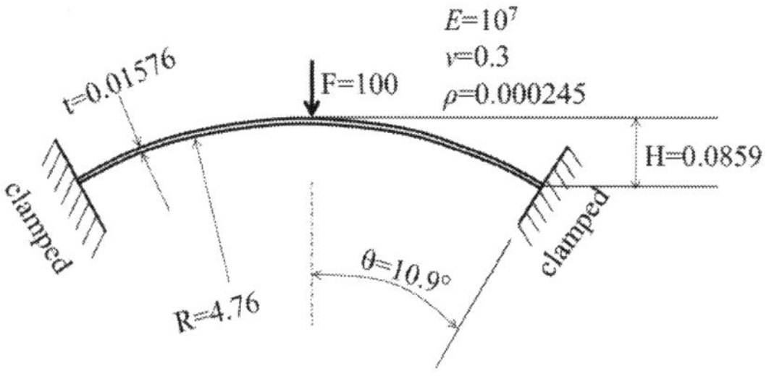

Figure 9. Elastic material parameters and geometric dimensions for the clamped spherical cap.

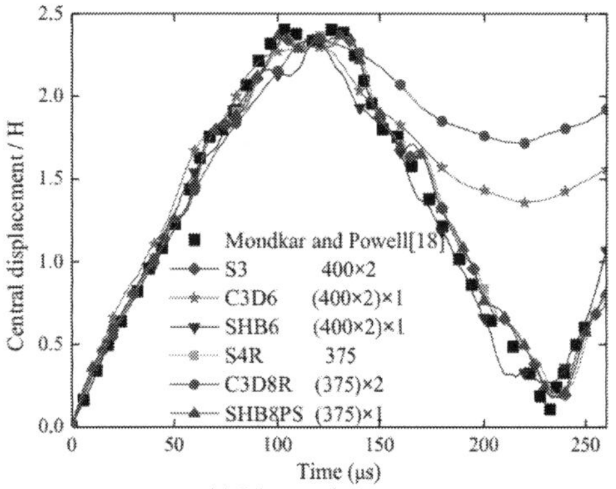

(a) Linear elements

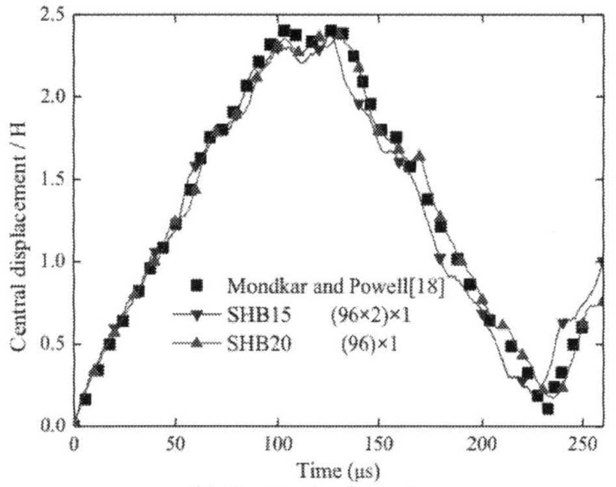

(b) Quadratic elements

Figure 10. Normalized displacement history for the point located at the apex of the clamped spherical cap.

3.2.2 Simply supported composite plate

The second dynamic benchmark test considered in this paper is a simply supported laminated plate, as shown in Figure 11, which is composed of two layers (0°/90°). A concentrated force is applied at the middle of the plate. Figure 12 presents the dynamic responses in terms of central deflection of the plate obtained using the SHB elements and ABAQUS linear elements. Once again, excellent agreement with the reference solution taken from [19] is observed with the dynamic versions of the proposed SHB elements. The ABAQUS linear elements show good performance as well, except for the prismatic element C3D6, which requires finer meshes to provide accurate solution.

$E_1 = 19.2\times10^6$, $E_2 = E_3 = 1.56\times10^6$
$\nu_{23} = 0.49$, $\nu_{12} = \nu_{13} = 0.24$
$G_{23} = 0.523\times10^6$, $G_{12} = G_{13} = 0.82\times10^6$
$\rho = 2.3763\times10^{-5}$

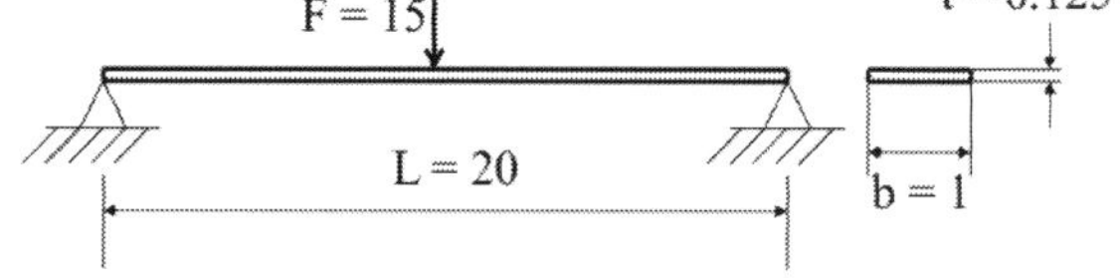

Figure 11. Geometry and material parameters for the simply supported laminated plate.

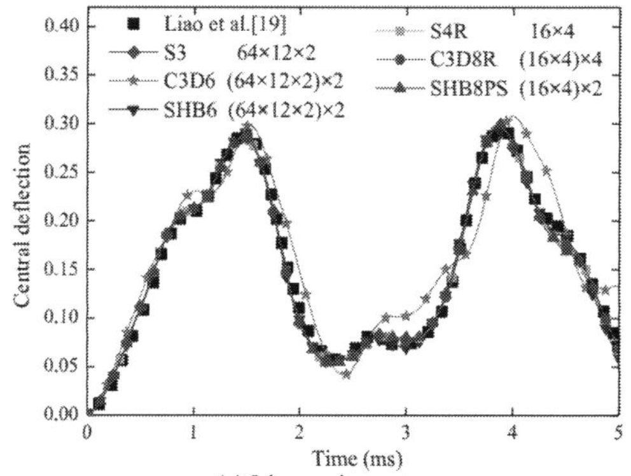

(a) Linear elements

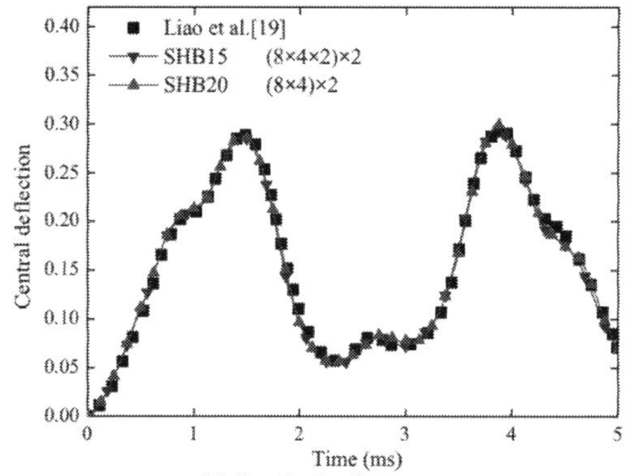

(b) Quadratic elements

Figure 12. Dynamic response curves for the simply supported laminated plate.

3.3 Sheet metal forming process

This subsection is devoted to the simulation of deep drawing of a hemispherical cup in order to evaluate the capabilities of the SHB elements in modeling sheet metal forming processes. The forming setup and the dimensions of the tools are presented in Figure 13. A circular steel sheet with initial blank radius R=126.4 mm and thickness t=0.7 mm is considered in the simulations. The elastic material properties of the sheet are given by Young's modulus E=210 GPa and Poisson's ratio ν=0.3. The stress–strain curve is described by the following Swift isotropic hardening law $\sigma = 545\left(0.003+\varepsilon^{p}\right)^{0.209}$, while kinematic hardening is not taken into account (see [20]). The Hill'48 plastic yield surface is considered here to model the plastic anisotropy of the studied material (see subsection 2.2.2). The associated coefficients are F=0.287, G=0.357, H=0.643, L=M=1.5, and N=1.306. Since no blank-holder is used in this test, the blank is clamped all around its circumferential surface during the forming process. The friction coefficient between the tools and the blank is set to 0.15. Owing to the axial symmetry, only one quarter of the circular sheet is modeled. The final deformed shape of the hemispherical cup using the SHB elements is shown in Figure 14.

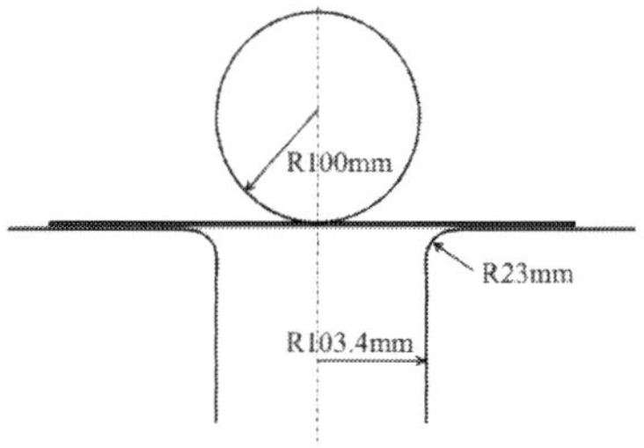

Figure 13. Hemispherical deep drawing setup.

(a) SHB6

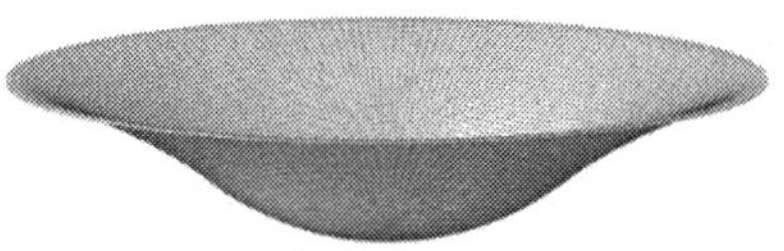

(b) SHB8PS

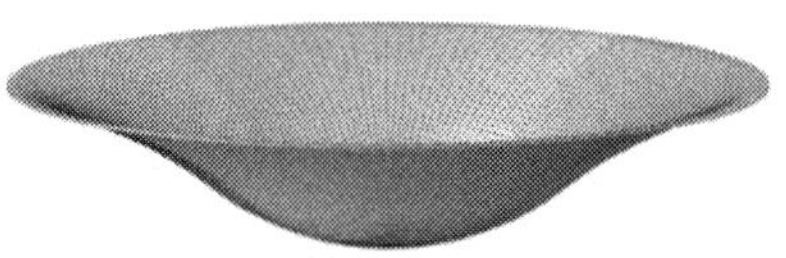

(c) SHB15

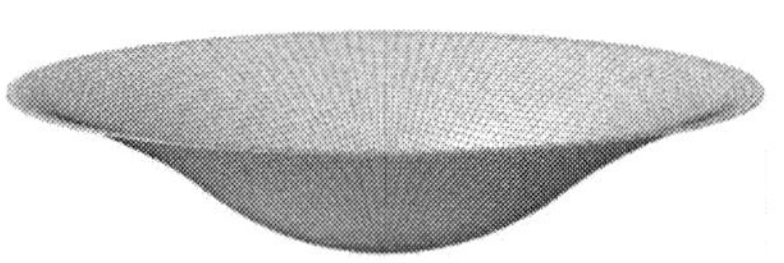

(d) SHB20

Figure 14. Final deformed shape for the hemispherical cup.

The profile of the thickness strain of the blank along the rolling direction and the reaction force of the punch are investigated in Figures 15 and 16, respectively, using the SHB solid–shell elements and only ABAQUS solid elements, and compared to the experimental results provided by Laurent [20]. Indeed, traditional shell elements do not allow outputting thickness strain variations due to their two-dimensional formulation.

From the predictions of the thickness strain profile in Figure 15, the SHB elements provide the closest results to the experimental one in terms of overall evolution and order of magnitude, while the results obtained with the ABAQUS solid elements are inaccurate, especially in the zone located about 40 mm from the centre of the blank. These contrasted results reveal that ABAQUS solid elements are not able to correctly handle the severe contact conditions between the blank and the punch in this zone, which lead to highly distorted meshes, as illustrated in Figure 16 with the ABAQUS quadratic element C3D20.

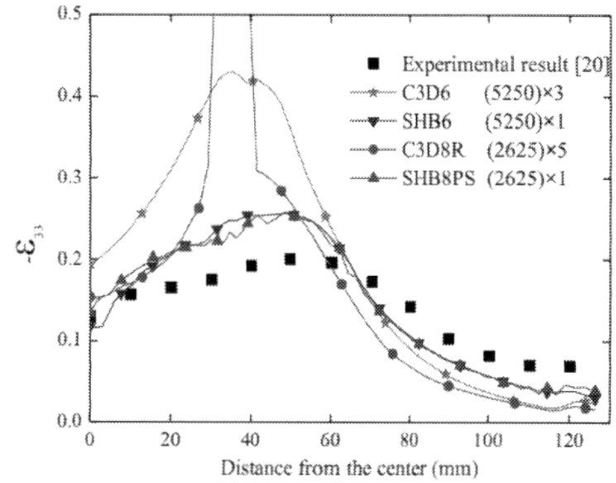

(a) Linear elements

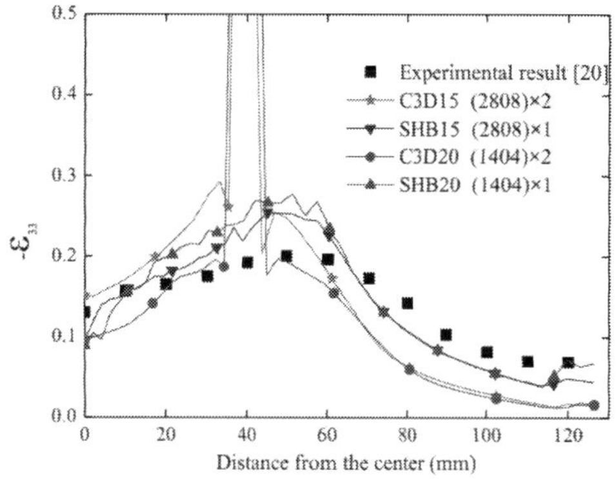

(b) Quadratic elements

Figure 15. Simulation results for the thickness strain.

Figure 17 shows the punch force, as predicted by all solid and solid–shell elements, together with the experimental results. Compared to ABAQUS elements, the results obtained with the SHB elements are in much better agreement with the experiments, which confirms once again the ability of the proposed solid–shell elements to model complex sheet metal forming processes using only a single layer of elements.

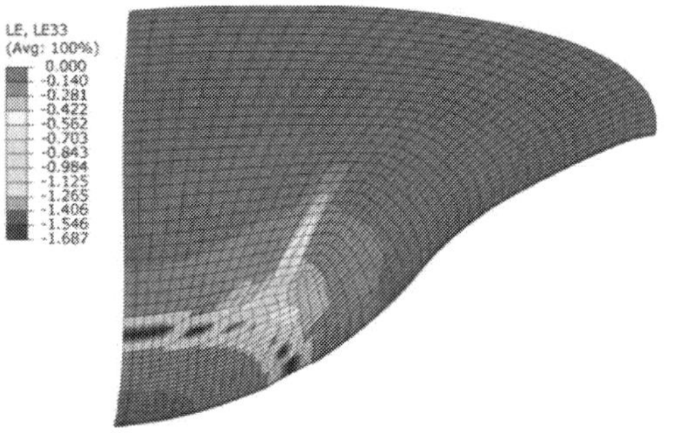

Figure 16. Final deformed cup using C3D20 element.

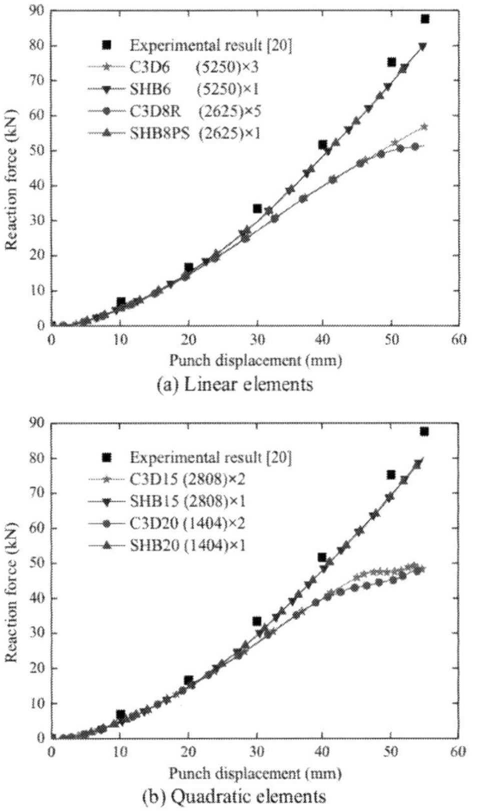

(a) Linear elements

(b) Quadratic elements

Figure 17. Punch force evolution during the deep drawing of the hemispherical cup.

4 Conclusions

In this paper, the formulation of a family of assumed-strain-based solid–shell elements has been described, which has been then evaluated through quasi-static and dynamic analyses as well as simulation of challenging sheet metal forming processes.

In the formulation of the SHB elements, several local frames have been defined and used in their numerical implementation. Taking advantage of such modularity, various constitutive models have been coupled with these SHB elements, including isotropic behavior, orthotropic elastic behavior for laminated composite materials, and anisotropic plastic behavior for metallic materials.

A series of selective and representative benchmark tests, involving different types of materials, loading and boundary conditions, has been conducted to assess the performance of the SHB elements in quasi-static and dynamics analyses. Also, complex deep drawing of a hemispherical cup has been considered to evaluate the capabilities of the proposed SHB elements in handling various types of nonlinearities (geometric, material, and due to contact). Adopting equivalent in-plane meshes and numbers of integration points in the thickness direction, the SHB elements performed much better than their ABAQUS counterparts, in both types of analyses (quasi-static and dynamic). Moreover, the variety of numerical results presented in this paper reveal the good capabilities of the SHB elements and their wide prospective applications in the simulation of thin 3D structures.

References

1. C. Cho, H.C. Park, S.W. Lee, Finite Elem. Anal. Des. **29**, 121–135 (1998)
2. R. Hauptmann, K. Schweizerhof, Int. J. Numer. Meth. Engng. **42**, 49–70 (1998)
3. F. Abed-Meraim, A. Combescure, Comput. & Struct. **80**, 791–803 (2002)
4. R.P.R Cardoso, J.W. Yoon, M. Mahardika, S. Choudhry, R.J. Alves de Sousa, R.A. Fontes Valente, Int. J. Numer. Meth. Engng. **75**, 156–187 (2008)
5. F. Abed-Meraim, A. Combescure, Int. J. Numer. Meth. Engng. **80**, 1640–1686 (2009)
6. M. Schwarze, I.N. Vladimirov, S. Reese, Comput. Methods Appl. Mech. Engng. **200**, 454–476 (2011)
7. F.G. Flores, Comput. Methods Appl. Mech. Engng. **266**, 81–97 (2013)
8. M. Pagani, S. Reese, U. Perego, Comput. Methods Appl. Mech. Engng. **268**, 141–159 (2014)
9. Q. Xie, K.Y. Sze, Y.X Zhou, Finite Elem. Anal. Des. **106**, 85–102 (2015)
10. V.D. Trinh, F. Abed-Meraim, A. Combescure, J. Mech. Sci. Technol. **25**, 2345–2364 (2011)
11. A. Salahouelhadj, F. Abed-Meraim, H. Chalal, T. Balan, Arch. Appl. Mech. **82**, 1269–1290 (2012)
12. F. Abed-Meraim, V.D. Trinh, A. Combescure, Computing **95**, 373–394 (2013)
13. J.C. Simo, T.J.R. Hughes. J. Appl. Mech. **53**, 51–54 (1986)
14. K.P. Li, J.L. Cescotto, Comput. Methods Appl. Mech. Engng. **141**, 157–204 (1997)
15. N.H. Nguyen, Ph.D. Thesis, Université de Liège, Belgium (2009)
16. P. Betsch, E. Stein, Comput. Mehods Appl. Mech. Engng. **179**, 215-245 (1999)
17. G.S. Payette, J.N. Reddy, Comput. Methods Appl. Mech. Engng. **278**, 664–704 (2014)
18. D.P. Mondkar, G.H. Powell, Int. J. Numer. Methods Engng. **11**, 499–520 (1977)
19. C.L. Liao, J.N. Reddy, S.P. Engelstad, Int. J. Numer. Methods Engng. **26**, 1843–1854 (1988)
20. H. Laurent. Ph.D. Thesis, Université du Maine, France (1996)

Automatic correction of the time step in implicit simulations of thermomechanical problems

J.M.P. Martins[1,a], D.M. Neto[1], J.L. Alves[2], M.C. Oliveira[1] and L.F. Menezes[1]

[1]*CEMUC, Department of Mechanical Engineering, University of Coimbra, Rua Luís Reis Santos, 3030-788 Coimbra, Portugal*
[2]*CMEMS, Center for MicroElectroMechanics Systems, University of Minho, Campus de Azurém, 4800-058 Guimarães, Portugal*

Abstract. The accurate numerical modelling of thermomechanical systems is important in several industrial applications. A new staggered coupling strategy is proposed to deal with thermo-elasto-plastic problems involving large deformations and rotations, which is combined with an automatic time-step control technique. The proposed strategy is based on the isothermal split approach. It differs from the classical scheme since, in each increment, there are two phases of interchange of information, instead of only one. In each increment, the solution procedure is divided in a prediction phase and a corrector phase, and the interchange of information is performed on both. The proposed strategy was implemented in the in-house *quasi*-static finite element code DD3IMP. Its performance is analysed and compared with the classical strategy and the iterative one, which are commonly employed for solving thermomechanical problems. The results indicate that the propose strategy contributes for improving the accuracy of the numerical solution, with a reasonable computational cost.

1 Introduction

The numerical modelling of thermomechanical systems is of high practical importance in many industrial applications. In fact, the prediction of temperature and stress fields (considering its reciprocal effects) is a key factor in various manufacturing processes, e.g. in the warm sheet metal forming. The interdependence between these two fields manifests itself through heat generation by plastic deformation and, consequently, by temperature increase and softening of the sheet material.

The finite element method has become a standard tool for the analysis of these problems. Nevertheless, the development and implementation of numerical models to deal with thermomechanical problems is a challenge, since the interdependence between thermal and mechanical fields requires an efficient coupling strategy. Intensive research has been conducted on this subject during the last years. Typically, the treatment of thermomechanical problems through the finite element method can be carried out using two different coupling strategies: (i) the monolithic strategy or (ii) the staggered strategy [1,2]. The monolithic strategy deals with both fields (temperature and stress) simultaneously within each increment, providing a single system of equations with all variables involved. The main advantage is the unconditional stability. Nevertheless, this strategy is rarely used because the resulting system of equations is very large and can be non-symmetric, depending on the formulation adopted. These characteristics result in a large computational cost [3–5]. Accordingly, the staggered strategy has been developed to reduce this cost. In this strategy, the two fields are treated separately, yielding two system of equations. This allows to take advantage of the usage of different time scales and different spatial discretization methods for each field [3]. Moreover, the implementation of this coupling strategy in the existing finite element codes is considerably easier [6].

The staggered strategy can be employed following two different approaches, which differ between them in the stability. The standard approach is called isothermal split, where the global thermomechanical problem is split into a mechanical problem, which is solved at a constant temperature, and a heat conduction problem, which is solved at a fixed configuration [7]. The drawback of this approach is the conditional stability, which arises when the thermoelastic effects play an important role. However, unstable approximations only occur for values of the thermal expansion coefficient considerably higher than the ones found in metals [3]. Therefore, several authors have adopted this approach to solve metal forming problems (e.g. [8–10]). In order to overcome the conditional stability of the isothermal split, Armero and Simo [11] proposed another approach, the so-called isentropic split (also referred as adiabatic split), which can retain the property of unconditional stability. In the isentropic split, the mechanical problem is solved at a constant entropy, followed by the heat conduction problem solved at a fixed configuration. This approach is usually adopted in works based on hyperelastic formulations and multiplicative decomposition of the deformation gradient (e.g. [12–14]).

The staggered strategy combined with the isothermal split approach is the one adopted in the present study. This strategy was implemented in the in-house *quasi*-static

[a] Corresponding author: joao.pmartins@dem.uc.pt

DOI: 10.1051/matecconf/20168007002

finite element code DD3IMP [15,16], which was specially developed to simulate the sheet metal forming process. The main characteristics of this finite element code are the adoption of a hypoelastic law for the material behaviour description, while the evolution of the deformation process is described by an update Lagrangian formulation. Regarding the algorithm used for solving the mechanical problem, an explicit approach is used to determine a trial solution for the nodal displacements. After that, in order to improve the convergence rate of the iterative Newton-Raphson algorithm, used to correct the trial solution, an automatic time-step control technique is employed to restrict the increment size, called r_{min} strategy [17,18].

This study focuses the thermo-elasto-plastic problem with large strains and rotations, neglecting the contact with friction conditions, although these are also important in the modelling of forming processes [19]. The outline of this paper is as follows. A brief introduction of the mechanical model formulation is presented in section 2. In section 3, the equation of heat conduction and the numerical methods used for its resolution are presented. The detailed description of the proposed thermomechanical coupling algorithm is presented in section 4. Section 5 contains two numerical examples, aiming to validate the numerical implementation and to assess the performance of the different coupling algorithms tested in terms of accuracy. The last section contains the concluding remarks.

2 Mechanical Model

2.1 Large deformation continuum mechanics

Let $\mathcal{B}$ be a continuum body, which consist of a set of particles defined by spatial points embedded on the domain $\Omega_0 \in \mathbb{R}^3$. At the time instant t_0, the continuum body assumes the configuration C_0, called reference configuration, where the position of a generic point is defined by the position vector $\mathbf{x}_0$. After the application of some external stimuli, the continuum body assumes another configuration C, called current configuration, at the time instant t, where the same point of the reference configuration can be defined by the position vector $\mathbf{X}$. Thus, the evolution of each point with time is defined by a deformation mapping, according to

$$\mathbf{x} = \mathbf{x}(\mathbf{x}_0, t). \tag{1}$$

The displacement of the spatial point, $\mathbf{x}_0$, between the reference and current configuration, can be defined by

$$\mathbf{u} = \mathbf{x}(\mathbf{x}_0, t) - \mathbf{x}_0, \tag{2}$$

and its velocity by

$$\mathbf{v} = \dot{\mathbf{x}} = \frac{\partial \mathbf{x}(\mathbf{x}_0, t)}{\partial t}. \tag{3}$$

The deformation gradient, $\mathbf{F}$, defined for the current configuration C, is

$$\mathbf{F} = \frac{\partial \mathbf{x}}{\partial \mathbf{x}_0} \quad \text{with } \det \mathbf{F} > 0, \tag{4}$$

and the corresponding velocity gradient is given by

$$\mathbf{L} = \frac{\partial \mathbf{v}}{\partial \mathbf{x}} = \dot{\mathbf{F}}\mathbf{F}^{-1}, \tag{5}$$

which can be decomposed in a symmetric and an antisymmetric part. The symmetric part is called the strain rate tensor, $\mathbf{D}$, and the antisymmetric part, is the spin rate tensor, $\mathbf{W}$. These can be defined as

$$\mathbf{D} = \frac{1}{2}\left(\mathbf{L} + \mathbf{L}^{\mathrm{T}}\right), \tag{6}$$

$$\mathbf{W} = \frac{1}{2}\left(\mathbf{L} - \mathbf{L}^{\mathrm{T}}\right). \tag{7}$$

2.2 Constitutive model

A hypoelastic stress-strain relation is adopted for the description of the material behaviour. In this case, the strain rate tensor can be addictively decomposed as [9,20,21]

$$\mathbf{D} = \mathbf{D}^{e} + \mathbf{D}^{th} + \mathbf{D}^{p}, \tag{8}$$

where $\mathbf{D}^{e}$, $\mathbf{D}^{th}$ and $\mathbf{D}^{p}$ are the elastic, the thermal and the inelastic (plastic) strain rate contributions, respectively. Thus, for describing a thermo-elasto-plastic material, the hypoelastic relation, which relates the strain rate with the stress rate, can be expressed as

$$\dot{\boldsymbol{\sigma}}^{J} = \mathbf{C}^{ep} : (\mathbf{D} - \mathbf{D}^{th} - \mathbf{D}^{p}), \tag{9}$$

where $\dot{\boldsymbol{\sigma}}^{J}$ is the Jaumann derivative of the Cauchy stress tensor $\boldsymbol{\sigma}$, which is defined by

$$\dot{\boldsymbol{\sigma}}^{J} = \dot{\boldsymbol{\sigma}} + \boldsymbol{\sigma}\mathbf{W} - \mathbf{W}\boldsymbol{\sigma}. \tag{10}$$

The $\mathbf{C}^{ep}$ in (9) is the elastoplastic module, which is a fourth-order tensor. The expression for this tensor depends of the algorithms adopted in the integration of the constitutive model and the type of relation considered between the states at the beginning and at the end of the loading increment. Thus, it is possible to consider the tangent elastoplastic modulus or the consistent elastoplastic modulus.

2.2.1 *Thermo-elasto-plastic state*

The associated flow rule is adopted, which implies that the plastic deformation rate tensor, $\mathbf{D}^{p}$, is given by the relation

$$\mathbf{D}^{p} = \dot{\lambda}\frac{\partial f(\bar{\sigma}, Y)}{\partial \boldsymbol{\sigma}}, \tag{11}$$

where $\dot{\lambda}$ is designated as the plastic multiplier, which can be determined by the so-called consistency condition i.e.

$\dot{f} = 0$. The scalar function f defines the yield surface and can be formulated as

$$f(\bar{\sigma}, Y) = \bar{\sigma} - Y = 0, \tag{12}$$

where Y denotes the flow stress in simple tension, while $\bar{\sigma}$ represents the equivalent stress, which for von Mises yield criterion is given by

$$\bar{\sigma} = \sqrt{\frac{3}{2}\boldsymbol{\sigma}' : \boldsymbol{\sigma}'}, \tag{13}$$

where $\boldsymbol{\sigma}'$ is the Cauchy deviatoric stress tensor.

The evolution of the flow stress is a combination of linear and saturation hardening [3], given by

$$Y(\bar{\varepsilon}^{\mathrm{p}}, T) = h\bar{\varepsilon}^{\mathrm{p}} + (Y_\infty - Y_0)\left(1 - \exp(-\delta\bar{\varepsilon}^{\mathrm{p}})\right), \tag{14}$$

where h, Y_0 and Y_∞ are the linear hardening modulus, initial yield stress and saturation stress, respectively. Besides, they depend on temperature by the following relations

$$h(T) = h_0\left[1 - \omega_h(T - T_0)\right], \tag{15}$$

$$Y_0(T) = Y_0\left[1 - \omega_0(T - T_0)\right], \tag{16}$$

$$Y_\infty(T) = Y_\infty\left[1 - \omega_h(T - T_0)\right]. \tag{17}$$

where T_0 is the reference temperature and ω_h and ω_0 are the parameters for the description of thermal softening. The description of the flow stress behaviour is completed with the saturation exponent δ.

2.2.2 Thermal strain rate

The thermal strain rate tensor involved in (8) can be written as

$$\mathbf{D}^{\mathrm{th}} = \alpha_{\mathrm{T}}^{*}\dot{T}\mathbf{I}, \tag{18}$$

$$\alpha_{\mathrm{T}}^{*} = \alpha_{\mathrm{T}} + \frac{\partial \alpha_{\mathrm{T}}}{\partial T}(T - T_0), \tag{19}$$

where α_{T} is the thermal expansion coefficient, T and $\dot{T}$ are the temperature and the temperature rate, respectively, while $\mathbf{I}$ is the second-order identity tensor.

2.3 Principle of virtual velocities

For the reference configuration, the region occupied by the deformable body is $\Omega_0 \in \mathbb{R}^3$ and its boundary is Σ_0. The boundary can be decomposed in two complementary parts: $\Sigma_{0|1}$, where the nominal stress vector is assigned, and $\Sigma_{0|2}$, where the velocity vector is prescribed. For this reference configuration, in an update Lagrangian scheme, the principle of virtual velocities is given by:

$$\int_{\Omega_0} \mathbf{S} : \delta\dot{\mathbf{F}} \mathrm{d}\Omega = \int_{\Sigma_{0|1}} \mathbf{s} \cdot \delta\mathbf{v} \mathrm{d}\Sigma, \tag{20}$$

which has to be valid for any virtual velocity field $\delta\mathbf{v}$ (with $\delta\mathbf{v} = 0$ on the boundary $\Sigma_{0|2}$). $\mathbf{S}$ is the first Piola-Kirchhoff stress tensor, $\dot{\mathbf{F}}$ is the time derivative of the deformation gradient and $\mathbf{s}$ corresponds to the external loading, i.e. the prescribed nominal stress vector.

3 Thermal model

The differential equation governing the heat conduction within a continuum body $\mathcal{B}$, defined by the open set $\Omega \in \mathbb{R}^3$ for the current configuration, can be derived from the first principle of thermodynamics, which combined with the Fourier's heat conduction law, assumes the following form:

$$\rho c \frac{\partial T}{\partial t} = \mathrm{div}\left(k \cdot \mathrm{grad}(T)\right) + \dot{W}^{\mathrm{pl}} + \dot{q}, \tag{21}$$

where ρ is the density, c is the specific heat, T is the temperature, k is the thermal conductivity, $\dot{W}^{\mathrm{pl}}$ is the internal heat generation term due to plastic deformation, and $\dot{q}$ corresponds to an heat source. This last term does not play any role in this work, thus it is neglected. Furthermore, the thermo-elastic coupling effect is not considered, since its effect is typically small, when compared with the one associated with heating induced by plastic deformation.

The heat generated by plastic deformation term, $\dot{W}^{\mathrm{pl}}$, usually involved in the modelling of thermomechanical problems, is the fraction of plastic power converted into heat. The remaining part of plastic power, which is not converted into heat power, is stored in the material [22]. This effect is often modelled using the assumption of Taylor and Quinney [23], which states that 5% to 15% of the plastic work is stored in the material. Thus, the heat generated by plastic deformation is expressed by

$$\dot{W}^{\mathrm{pl}} = \chi \boldsymbol{\sigma} : \mathbf{D}^{\mathrm{p}}, \tag{22}$$

where the parameter χ is called the Taylor-Quinney coefficient, which varies between 0.85 and 0.95.

The thermal boundary conditions, such as convection, radiation and heat conduction by thermal contact, are important aspects to be considered in numerical models of thermomechanical problems [24,25]. Nevertheless, they are not addressed in this study, since they can be neglected in the comparison of staggered algorithms performance.

Applying the principle of virtual temperature and proceeding to the decomposition of the domain, the initial boundary value problem defined in (21) can be given in the matrix form as [26]:

$$\mathbf{C}\dot{\mathbf{T}} + \mathbf{K}\mathbf{T} = \mathbf{f}, \tag{23}$$

where the matrixes and vector are given by

$$\mathbf{C} = \int_{\Omega} \rho c \mathbf{N}^{\mathrm{T}}\mathbf{N} \mathrm{d}\Omega, \tag{24}$$

$$\mathbf{K} = \int_{\Omega} k \mathbf{M}^{\mathrm{T}}\mathbf{M} \mathrm{d}\Omega, \tag{25}$$

$$\mathbf{f} = \int_{\Omega} \mathbf{N}^{\mathrm{T}}\dot{W}^{\mathrm{pl}} \mathrm{d}\Omega, \tag{26}$$

where $\mathbf{N}$ is the vector of shape functions and $\mathbf{M} = \text{grad}(\mathbf{N})$. The consistent form of the capacitance matrix $\mathbf{C}$ is replaced by the diagonal form, which avoids unstable results, incoherent with physical reality, and has the advantage of reducing the computational cost.

The numerical solution of the transient heat conduction equation (23) can be obtained throughout the adoption of the different numerical methods. Typically, the so-called α-method is adopted for time integration. This method is given by the following assumptions [26,27]:

$$\dot{\mathbf{T}}_{t+\alpha\Delta t} = \left(\mathbf{T}_{t+\Delta t} - \mathbf{T}_{t}\right)/\Delta t, \quad (27)$$

$$\mathbf{T}_{t+\alpha\Delta t} = \left(1-\alpha\right)\mathbf{T}_{t} + \alpha\mathbf{T}_{t+\Delta t}, \quad (28)$$

where the parameter α can vary between 0 and 1, which gives rises to different time integration methods, such as, explicit Euler forward ($\alpha = 0$), Crank-Nicolson ($\alpha = 1/2$), Galerkin ($\alpha = 3/2$) and implicit Euler backward ($\alpha = 1$).

4 Thermomechanical coupling

This section presents a description of the proposed thermomechanical coupling algorithm, as well as the technique used for the automatic time step control, called $r_{\min}$ strategy. The proposed algorithm is compared with the classical staggered algorithms, highlighting its advantages and drawbacks.

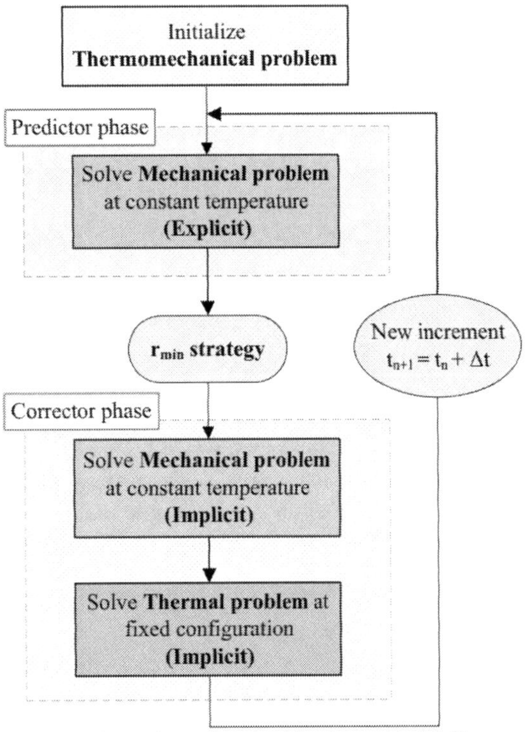

Figure 1. Classical coupling algorithm denoted by MMT.

The widely used classical staggered algorithm, comprises the interchange of information between the two fields only once, at each time increment. A diagram of this strategy, adapted to accommodate the $r_{\min}$ strategy, is presented in Figure 1. Since the mechanical problem in the corrector phase is solved using the temperature of the previous increment, there is always a delay between the two fields, which induces a loss of accuracy. On the other hand, the computational cost is relatively small due to the sequential solution of the mechanical and thermal problems. In order to improve the accuracy of the numerical solution, another coupling strategy has been developed, usually called iterative coupling algorithm [4]. In this case, the interchange of information between the two fields is performed through an iterative procedure, aiming to reach the fully converged solution. The mechanical and thermal problems are solved in the same sequence as presented in Figure 1, but the procedure is repeated, within each time increment, until attaining the convergence criterion. Consequently, higher computational cost is involved, due to the iterative cycle within each increment.

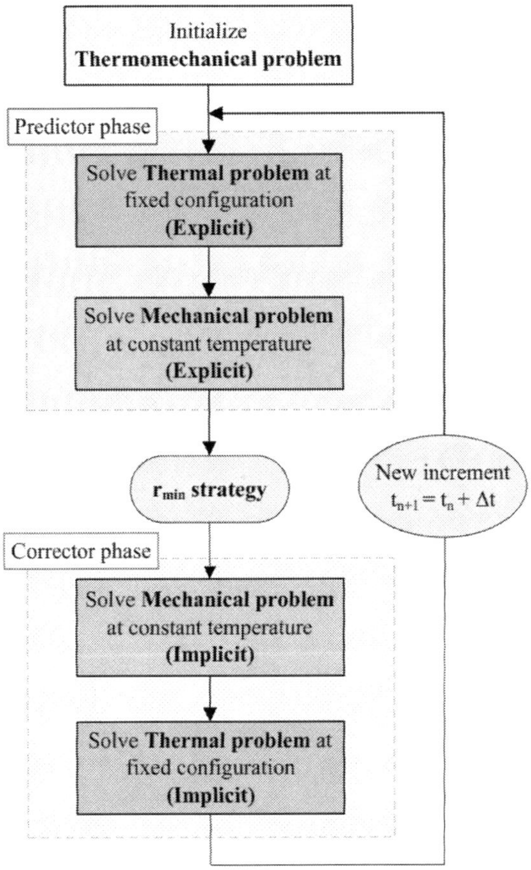

Figure 2. Proposed algorithm for the thermomechanical coupling, denoted by TMMT.

The proposed coupling algorithm was developed to reach a better compromise between computational cost and accuracy of results. The main difference to the classical staggered algorithms is the way the information between the two fields is interchanged. The proposed algorithm can be clearly divided in two phases, a prediction phase and a corrector phase, as shown in Figure 2. The first phase comprises solving the thermal problem for a user-defined time step, adopting an explicit or semi-implicit approach (depending on the value defined for the parameter α in (28)). In this first phase, the amount of heat generated by plastic deformation is assumed to be equal to the one of the last time increment. Thus, the result is a trial temperature field. After that, the mechanical problem is solved for this trial temperature field, adopting an explicit approach. At the end of this prediction phase, there is a trial displacement field and a trial temperature field. Since the integration methods adopted are explicit, the calculated variables are proportional to the time increment. Thus, in order to assure the convergence of the subsequent corrector phase, the r_{min} strategy is applied. This strategy adjusts the increment size according to some restrictions defined for the rotation, strain and stress increment. Based on that, it is possible to determine a weighting factor r, which is used to calculate a time increment that is expected to avoid convergence problems. The same weighting factor is used to calculate the new displacement field and temperature field, based on the trial solutions, since these are proportional to the time increment. More details about the application of this strategy to the mechanical problem can be found in Oliveira and Menezes [18]. Afterwards, the corrector phase is performed, where the mechanical problem is solved first, using the trial temperature field. The non-linear system of equations resultant from the large thermo-elasto-plastic deformation problem is solved with a Newton-Raphson iterative method. Thus, the previous displacement field, which corresponds to the initial solution for the Newton-Raphson method, is corrected until the equilibrium state is attained. Finally, the thermal problem is solved using an implicit approach (with the parameter $\alpha = 1$ in (28)), for the current equilibrium configuration.

5 Numerical examples

5.1 Expansion of a thick-walled cylinder

The purpose of this example is to validate the developed finite element code on the resolution of thermomechanical problems. This validation was carried out using the iterative coupling algorithm to assure the most accurate results. The example considered deals with the radial expansion of thick-walled infinite cylinder, which was proposed by Argyris and Doltsinis [28] and later adapted by Simo and Miehe [29].

A thick-walled cylinder with an inner radius of 100 mm and an outer radius of 200 mm is subjected to an internal pressure. In the presented model, the internal pressure is replaced by an imposed radial displacement of the inner surface, with a magnitude of 130 mm. The cylinder is assumed infinite long, along the axial direction, which allows the use of plain strain conditions in this direction. The thermal problem is solved assuming the external surface of the cylinder adiabatic and the only source of heat is the internal heat generated by plastic deformation. The cylinder as an initial temperature of 293 K.

The material parameters are the ones proposed by Simo and Miehe [29] and are listed in Table 1. Only one quarter of the cylinder was modelled, imposing symmetry conditions. The mesh was composed by linear hexahedral finite elements, using 10 elements through the cylinder thickness and 10 elements along the hoop direction. A selective reduced integration technique was adopted for the mechanical problem while for the thermal problem full integration was applied. The time step discretization was automatically adapted throughout the r_{min} strategy.

Table 1. Material properties of the cylinder.

Young' modulus	$E = 70000$ GPa
Poisson's ratio	$\nu = 0.3$
Initial yield stress	$Y_0 = 70$ MPa
Isotropic hardening modulus	$h = 210$ MPa
Thermal expansion coef.	$\alpha_T = 23.8 \times 10^{-6}$ K^{-1}
Conductivity	$k = 150$ Wm^{-1}K^{-1}
Density	$\rho = 2900$ kg m^{-3}
Heat capacity	$c = 900$ J kg^{-1} K^{-1}
Taylor-Quinney factor	$\chi = 0.9$
Thermal softening	$\omega_0 = 3.0 \times 10^{-4}$ K^{-1}
Hardening/softening	$\omega_h = 0$ K^{-1}
Reference temperature	$T_{ref} = 293$ K

Figure 3 shows the results obtained for the particular case of an adiabatic loading, i.e. since the internal conduction is neglected it corresponds to a loading duration $t \to 0$s. Both the internal pressure evolution and temperature evolution are in good agreement with the numerical results obtained by Simo and Miehe [29].

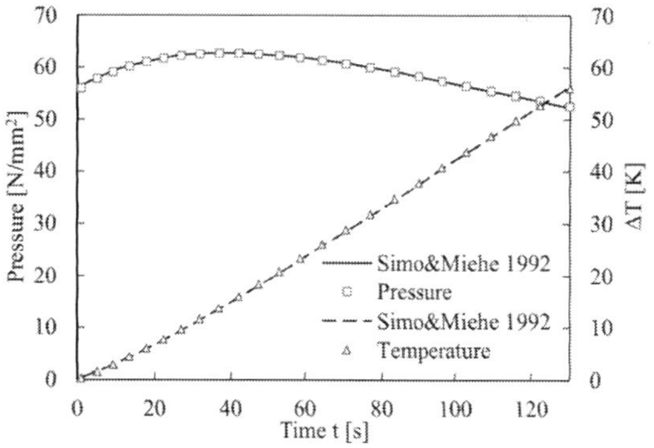

Figure 3. Evolution of the internal pressure and relative temperature ($\Delta T = T_I - T_{ref}$) at the inner surface of the cylinder for the adiabatic load case.

Figure 4 shows the temperature distribution along the radial position, considering different durations for the test, such as $t \to 0$, 1.3, 13 130 s and $\to \infty$. The increase in the total time enables a more homogeneous temperature distribution, resultant from the heat conduction within the material.

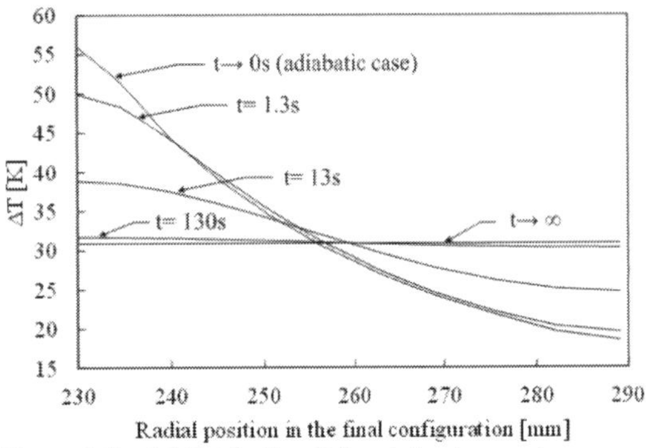

Figure 4. Spatial distribution of the relative temperature ($\Delta T = T_i - T_{ref}$) for the final configuration (radial displacement of 130 mm) for different values of total time.

5.2 Single element tensile test

The objective of this example is to assess the performance of the proposed staggered algorithm in terms of accuracy. Therefore, it is compared with the two classical algorithms described in section 4. In order to improve the results discussion, on a first, the test was performed with the $r_{\min}$ strategy deactivated. Afterwards, the tests were performed with this strategy activated. Thus, the impact of the $r_{\min}$ strategy can be evaluated separately. For the sake of clarity the following nomenclature is adopted: the proposed algorithm is denoted by TMMT, the algorithm presented in Figure 1 is denoted by MMT and the iterative algorithm is denoted by MMIT.

The example consists on a single finite element (linear hexahedral element) with dimensions $1 \times 1 \times 1\ \text{mm}^3$, which is submitted to a tensile stress state, by imposing a displacement $u = 2$ mm in one direction, at a constant velocity $v = 1\ \text{mm s}^{-1}$. The configuration of the single element is schematically represented in Figure 5. As for the previous example, selective reduced integration is considered for the mechanical problem and full integration for the thermal problem. The boundaries of the element are considered isolated from the environment and the only heat source is the heat generated by plastic deformation. The initial temperature was defined as 293 K. The material parameters used in this example are listed in Table 2.

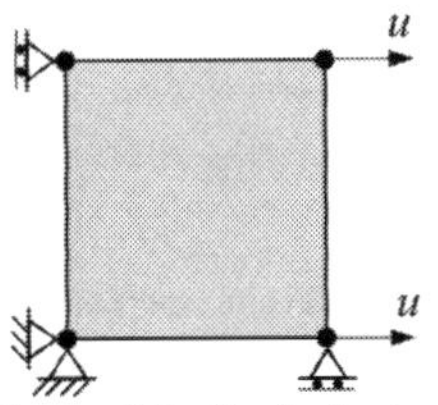

Figure 5. Configuration used for the single element test.

Table 2. Material properties used in the single element test.

Young' modulus	$E = 206900$ GPa
Poisson's ratio	$\nu = 0.29$
Initial yield stress	$Y_0 = 450$ MPa
Isotropic hardening modulus	$h = 129.24$ MPa
Thermal expansion coef.	$\alpha_T = 1.0\times10^{-5}\ \text{K}^{-1}$
Conductivity	$k = 45\ \text{Wm}^{-1}\text{K}^{-1}$
Density	$\rho = 7800\ \text{kg m}^{-3}$
Heat capacity	$c = 460\ \text{J kg}^{-1}\ \text{K}^{-1}$
Taylor-Quinney factor	$\chi = 0.9$
Thermal softening	$\omega_0 = 2.0\times10^{-3}\ \text{K}^{-1}$
Hardening/softening	$\omega_h = 2.0\times10^{-3}\ \text{K}^{-1}$
Reference temperature	$T_{\text{ref}} = 293$ K

As mentioned previously, initially the tests were performed considering a constant time step ($\Delta t = 0.2\text{s}$), corresponding to a total of 11 increments. The relative temperature evolution predicted by each coupling algorithm is presented in Figure 6, while the evolution of the equivalent (von Mises) stress is shown in Figure 7. The temperature increases during the loading due to the heat generated by plastic deformation. On the other hand, the von Mises stress decreases after the first two increments because the flow stress decreases with the temperature (softening). The norm of the relative error of the TMMT and MMT solutions is also presented in the same figures. The error was evaluated considering the iterative solution as reference (MMIT), since this is the most accurate.

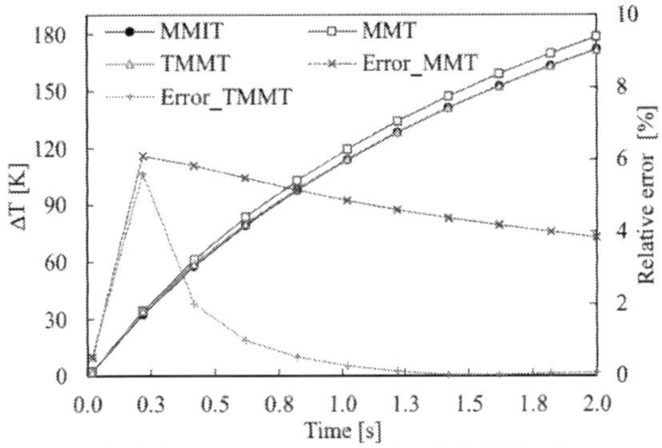

Figure 6. Relative temperature evolution (ΔT=Ti-Tref) obtained with the TMMT, MMT and MMIT algorithms, including the relative error calculated for the TMMT and MMT solutions (single element tensile test with constant time step).

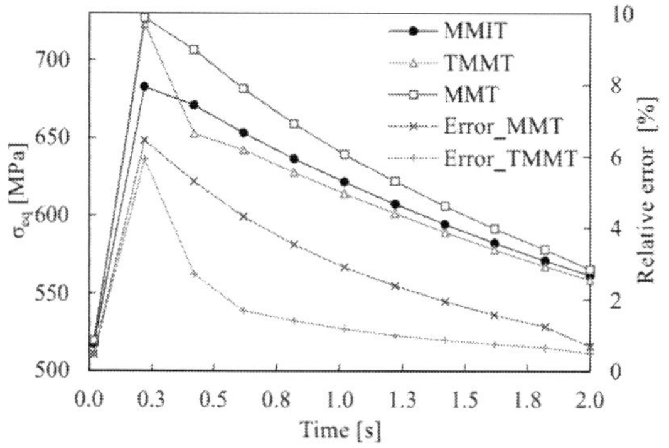

Figure 7. Equivalent (von Mises) stress evolution obtained with the TMMT, MMT and MMIT algorithms, including the relative error calculated for the TMMT and MMT solutions (single element tensile test with constant time step).

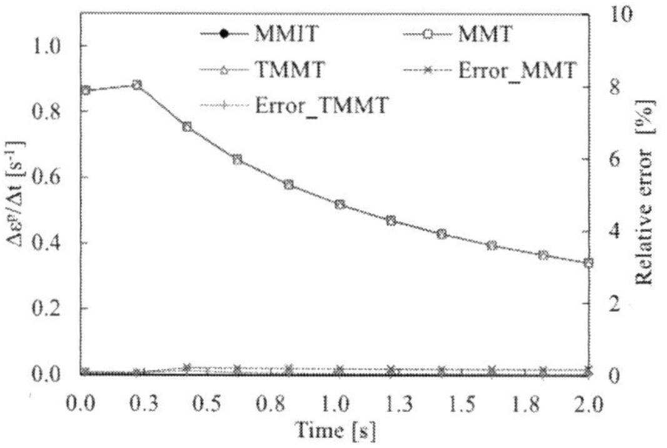

Figure 8. Plastic strain rate obtained with TMMT, MMT and MMIT algorithms, including the error calculated for the TMMT and MMT solutions (single element tensile test with constant time step).

For the first increment, the error of the algorithms TMMT and MMT is identical because there is no history of plastic deformation. Indeed, the TMMT algorithm uses the value of the plastic heat generated in the previous increment to predict the trial temperature field. The error of both algorithms attains the maximum value in the second increment, as shown in Figure 6 and Figure 7. This error is directly related with the temperature increment. Since the magnitude of the temperature increment is connected with the amount of heat generated by plastic deformation, the evolution of the plastic strain rate is presented in Figure 8, for both algorithms. For this variable the difference between coupling algorithms is negligible, because the increment of plastic strain is mainly dictated by the finite element deformed configuration, which is imposed by the prescribed displacement. Therefore, the maximum temperature increment occurs in the second increment, since both the equivalent stress (Figure 7) as the increment of plastic strain (Figure 8) attain their maximum values. However, this temperature increment is not taken into account by the MMT algorithm to induce the mechanical softening effect during this increment, leading to an overestimation of the von Mises stress. On the other hand, the prediction phase of the TMMT algorithm, which is used to generate a temperature field for the actual increment based on the last increment, provides a poor temperature estimative for the second increment. This results from the very small value of temperature increment in the first increment (a small amount of heat generated by plastic deformation). Nevertheless, after the second increment the error of the TMMT algorithm is significantly reduced, as shown in Figure 6 and Figure 7.

During the test, the equivalent stress and the rate of plastic strain show a decreasing trend. Thus, the temperature increment also decreases in each increment. Accordingly, the prediction phase of the TMMT algorithm overestimates the temperature field, which originates an underestimation of the equivalent stress, as shown in Figure 7. Since the MMT algorithm solves the mechanical field based on the last temperature field, equivalent stress is overestimated.

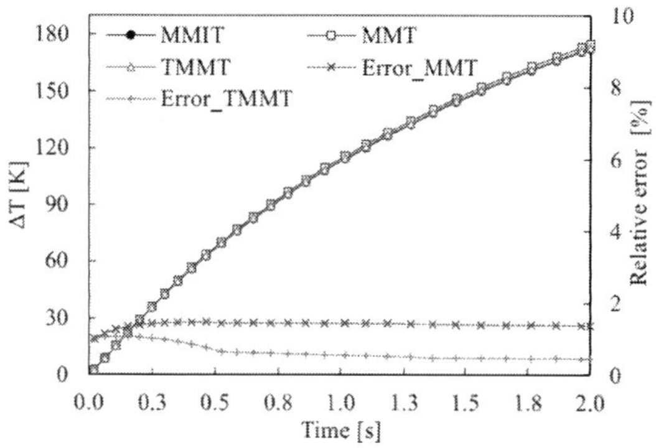

Figure 9. Relative temperature evolution (ΔT=Ti-Tref) obtained with the TMMT, MMT and MMIT algorithms and relative error calculated for the TMMT and MMT solutions (single element tensile test with the r_{min} strategy activated).

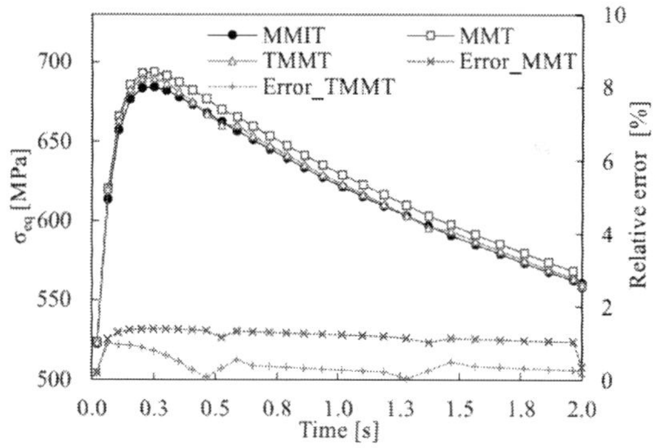

Figure 10 Equivalent von Mises stress evolution obtained with the TMMT, MMT and MMIT algorithms and relative error calculated for the TMMT and MMT solutions (single element tensile test, with the r_{min} strategy activated).

In order to assess the influence of the increment size in the accuracy of the coupling algorithms presented, the r_{min} strategy was activated. The obtained results for the evolutions of the relative temperature, the von Mises stress and the rate of plastic strain are shown in Figure 9, Figure 10 and Figure 11, respectively. The introduction of the r_{min} strategy to adjust the time step leads to a global decrease of the step size. Indeed, the number of increments increased from 11 to 29. Consequently, the accuracy of all coupling algorithms is improved, as shown in Figure 9 and

Figure 10. The relative error of the solution predicted by the TMMT algorithm is globally lower than half the one obtained with the MMT algorithm. Therefore, using the same increment size, the proposed coupling algorithm provides results more accurate than the ones obtained with the classical staggered strategy (MMT).

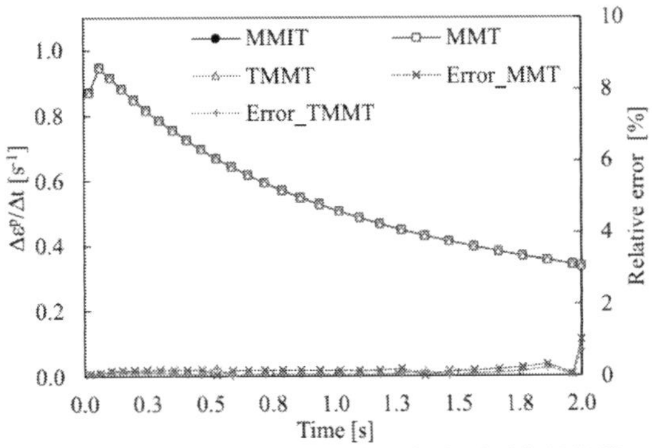

Figure 11. Rate of plastic deformation obtained with TMMT, MMT and MMIT algorithms, including the error calculated for the TMMT and MMT solutions (single element tensile test, with the $r_{\min}$ strategy activated).

6 Conclusion

The present study aims to improve the accuracy of thermomechanical simulations, within a reasonable computational cost. In this context, a new staggered coupling strategy was proposed, which involves a predictor and a correction phase, in which both the thermal and the mechanical problem are solved sequentially. In the predictor phase the thermal problem is solved previous to the mechanical, while in the correction phase the sequence is inverted. The finite element code developed was validated using a classical thermomechanical problem. The performance of the proposed strategy, in terms of accuracy, was discussed and compared with the one of classical strategies. The results achieved for a problem involving large thermo-elasto-plastic deformations and rotations show an improvement in the accuracy, when compared with the classical staggered algorithm. However, more complex examples need to be used to enable the assessment of the computational efficiency.

Acknowledgements

The authors gratefully acknowledge the financial support of the Portuguese Foundation for Science and Technology (FCT) under projects with reference UID/EMS/00285/2013, PTDC/EMS-TEC/0702/2014 (POCI-01-0145-FEDER-016779) and PTDC/EMS-TEC/6400/2014 (POCI-01-0145-FEDER-016876) by UE/FEDER through the program COMPETE2020. The second author is also grateful to the FCT for the Postdoctoral grant SFRH/BPD/101334/2014.

References

1. S. Rothe, P. Erbts, A. Düster, S. Hartmann, Comput. Methods Appl. Mech. Eng. 293 (2015) 375–410.
2. C.A. Felippa, K.C. Park, C. Farhat, Comput. Methods Appl. Mech. Eng. 190 (2001) 3247–3270.
3. F. Armero, J.C. Simo, Int. J. Plast. 9 (1993) 749–782.
4. M. Vaz, P.A. Muñoz-Rojas, M.R. Lange, Int. J. Mech. Sci. 53 (2011) 387–398.
5. T. Netz, S. Hartmann, Comput. Math. with Appl. 70 (2015) 1457–1480.
6. V. Dunić, N. Busarac, V. Slavković, B. Rosić, R. Niekamp, H. Matthies, R. Slavković, M. Živković, Contin. Mech. Thermodyn. (2015).
7. L. Adam, J. Ponthot, J. Eng. Mech. 128 (2002) 1222–1232.
8. M. Canadija, J. Brnic, Int. J. Plast. 20 (2004) 1851–1874.
9. H.L. Xing, a. Makinouchi, Eng. Comput. 19 (2002) 392–410.
10. L. Adam, J.P. Ponthot, Int. J. Solids Struct. 42 (2005) 5615–5655.
11. F. Armero, J. C. Simo, Int. J. Numer. Methods Eng. 35 (1992) 737–766.
12. C. Agelet De Saracibar, M. Cervera, M. Chiumenti, Int. J. Plast. 15 (1999) 1–34.
13. A. Ibrahimbegovic, L. Chorfi, Int. J. Solids Struct. 39 (2002) 499–528.
14. M.H. Ulz, Comput. Methods Appl. Mech. Eng. 198 (2009) 3262–3277.
15. L.F. Menezes, C. Teodosiu, J. Mater. Process. Technol. 97 (2000) 100–106.
16. M.C. Oliveira, J.L. Alves, B.M. Chaparro, L.F. Menezes, Int. J. Plast. 23 (2007) 516–543.
17. Y. Yamada, N. Yoshimura, T. Sakurai, Int. J. Mech. Sci. 10 (1968) 343–354.
18. M.C. Oliveira, L.F. Menezes, Finite Elem. Anal. Des. 40 (2004) 1995–2010.
19. D.M. Neto, M.C. Oliveira, L.F. Menezes, J.L. Alves, Comput. Methods Appl. Mech. Eng. 299 (2015) 283–315.
20. L. Adam, J.-P. Ponthot, J. Eng. Mech. 128 (2002) 1222–1232.
21. A. Andrade-Campos, F. da Silva, F. Teixeira-Dias, Int. J. Numer. Methods Eng. 70 (2007) 582–609.
22. J. Hodowany, G. Ravichandran, a Rosakis, P. Rosakis, Exp. Mech. 40 (2000) 113–123.
23. G.I. Taylor, H. Quinney, Proc. R. Soc. A Math. Phys. Eng. Sci. 143 (1934) 307–326.
24. J.M.P. Martins, J.L. Alves, D.M. Neto, M.C. Oliveira, L.F. Menezes, Int. J. Adv. Manuf. Technol. (2015).
25. D. Pantuso, K.-J. Bathe, P.A. Bouzinov, Comput. Struct. 75 (2000) 551–573.
26. K.J. Bathe, Finite Element Procedures, Prentice-Hall: Englewood Cliffs, 1996.
27. A. Ibrahimbegovic, (2009).
28. J.H. Argyris, J.S. Doltsinis, Comput. Methods Appl. Mech. Eng. 25 (1981) 195–253.
29. J.C. Simo, C. Miehe, Comput. Methods Appl. Mech. Eng. 98 (1992) 41–104.

DOI: 10.1051/matecconf/20168007003

Interaction of various functional elements in thin-walled cups formed by a sheet-bulk metal forming process

Robert Schulte[1,a], Thomas Schneider[1], Michael Lechner[1] and Marion Merklein[1]

[1] *Institute of Manufacturing Technology (LFT, Friedrich-Alexander Universität Erlangen-Nürnberg), Egerlandstrasse 13, 91058 Erlangen, Germany*

Abstract. Sheet-bulk metal forming allows the manufacturing of sheet metal parts with integrated functional elements. The investigated process combines deep drawing and upsetting. The occurring 2D and 3D stress and strain states lead to challenges regarding material flow control due to different geometries of functional elements. The numerical analysis shows the transferability for the results of the different functional elements for their combination and the interaction between those functional elements regarding the material flow. The results are validated by experimental tests.

1 Introduction

Legal frameworks and social change request low-emission products [1]. Especially the automotive industry is confronted with great challenges with regard to the application of resources both in the production and in the usage of modern vehicles [2]. For that reason, lightweight construction is increasingly important to enhance material efficiency to reduce fuel consumption [3]. However, a consistent implementation of lightweight construction is accompanied by increased functional integration of the parts manufactured especially in the powertrain [4]. This requires a higher load capacity. Therefore, the application of steel as workpiece material is favorable although current metal forming processes meet their limits with regard to cost and time expenditure due to insufficient process understanding [5]. The combination of conventional sheet and bulk metal forming processes to the class of sheet-bulk metal forming processes is a promising approach to realize lightweight parts with enhanced functional integration. A lack of process knowledge requires the investigation of sheet-bulk metal forming operations and process combinations as these have proofed their potential to expand the limits of conventional forming processes [6]. To shorten process chains and enhance the robustness of these processes fundamental aspects have to be investigated [7].

The investigated sheet-bulk metal forming process is a combined deep drawing and upsetting process which enables the single-step manufacturing of cups with various functional elements on the outer surface. These parts take up technical requirements of powertrain parts, such as synchronizer rings. The cups are manufactured with an external gearing, open carriers and a combination of both. For the forming of both types of functional elements the material flow control is the main challenge. The external gearing is a thick-walled element which requires a large thickening of material in the gear cavity [8] whereas for the open carrier, as a thin-walled element, excessive thinning of the material has to be prevented [9]. The combination of different functional elements now takes account of the increasing demand of functional integration and is presented in figure 1.

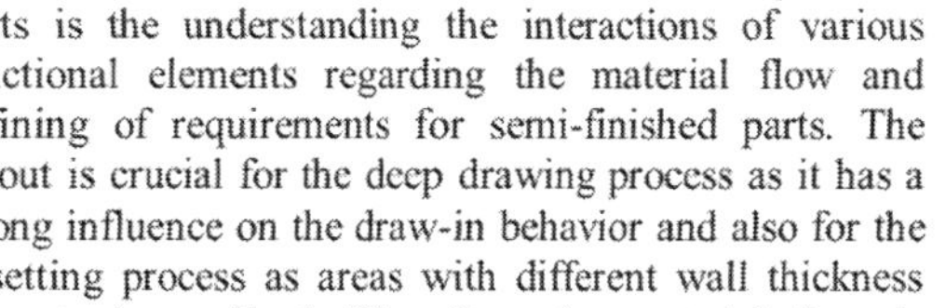

Figure 1: Combination of various functional elements in an integrated part geometry

Prerequisite for a successful manufacturing of the parts is the understanding the interactions of various functional elements regarding the material flow and defining of requirements for semi-finished parts. The layout is crucial for the deep drawing process as it has a strong influence on the draw-in behavior and also for the upsetting process as areas with different wall thickness have to be realized. Therefore, the material flow is investigated by numerical simulations and subsequently validated with experimental results. Relevant process limits for the forming of parts with an increased functional integration are shown and challenges are derived. An adaption of the blank layout is executed to enhance the material flow control and to meet the challenge.

[a] Corresponding author: robert.schulte@fau.de

2 Component characteristics and forming concept

This section gives an overview of part geometry, forming process, modelling and methodology to provide the necessary information for the investigations performed and discussed.

2.1. Part geometry

Figure 2 shows the investigated demo-part, which is designed according to a rising demand for parts with enhanced functional integration. The part investigated has three open carriers, arranged in an angle of 120°. Between those carriers the gearing is located. The number of teeth in a single section is 15. The various functional elements are arranged cyclically and symmetrically. The inner diameter of the cup with the external gearing amounts to $d_{dp} = 75.5$ mm whereas the open carrier has an inner diameter of $d_c = 85$ mm. The radius of the drawing punch is $r_{dp} = 1$ mm. For the cup wall and the open carrier the maximum thickness amounts to $t_{wmax} = 2.25$ mm which corresponds to the drawing gap, whereas the maximum thickness in the tooth tip is $t_{tmax} = 3.75$ mm. The outer transition radius from the functional elements to the round cup is R2 = 2 mm, the inner radius is R1 = 1 mm. The height of the cups depends on the geometry of the semi-finished product and on the forming force applied.

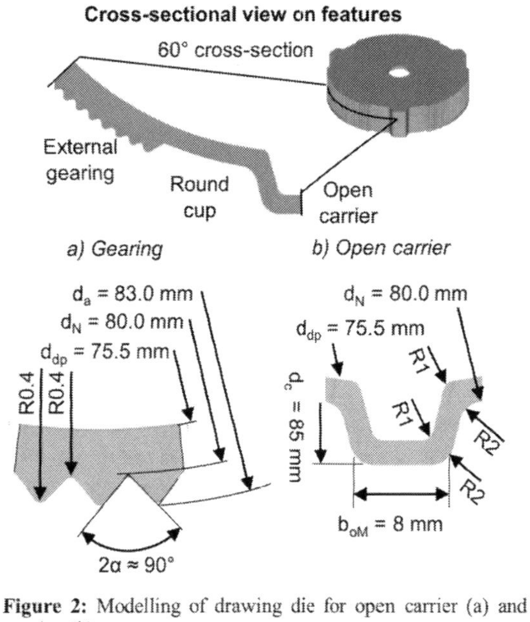

Figure 2: Modelling of drawing die for open carrier (a) and gearing (b)

2.2. Process set-up

To manufacture parts with an increased functional integration the tool concepts presented in [8] for the external gearing and [9] for open carriers are combined. The tool concept consists of an upper tool with drawing die and upsetting punch and a lower tool with upsetting plate and drawing punch. In addition, a hydraulic cylinder is connected to the upsetting punch. Within the manufacturing process the semi-finished product is positioned on the center of the drawing punch and then clamped by the upsetting punch in axial direction to avoid a lifting of the bottom of the cup from the drawing punch caused by bending stresses induced into the part at the beginning of the upsetting operation. The clamping force F_c is applied by the additional hydraulic cylinder. The tool setup is presented in figure 3. The cup is formed as the upper die cushion of the triple acting hydraulic press is continuously positively displaced under the drawing force F_d and the drawing die reaches a mechanical stop at the upsetting plate. Subsequently, the forming force is transmitted into the upsetting punch by a mechanical stop and the drawing punch is displaced by the upsetting force F_u. To perform the necessary kinematics a triple-acting hydraulic press Lasco TZP 400/3 is used. In the experimental set-up the deep drawing process proceeds stroke-controlled and the upsetting process proceeds force-controlled. The process set-up and tool components are shown in figure 3.

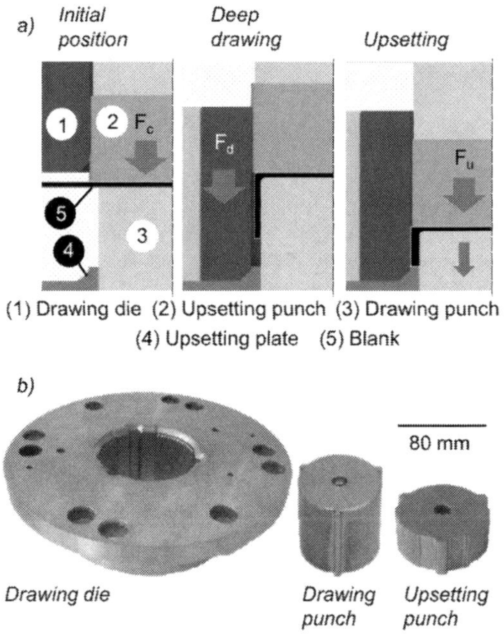

Figure 3: Cross-section of tool concept for deep drawing and upsetting (a) and tool components for experimental process set-up (b)

Both, the cup and the thin-walled element open carrier are formed in the deep drawing operation, whereas the gearing as a thick walled element is formed in the upsetting process. The open carrier is formed as workpiece material is drawn over the drawing punch. During upsetting the sheet thickness in round cup and open carrier exceeds the initial sheet thickness slightly as the thin-walled functional elements are formed during deep drawing and calibrated in the upsetting process. The gearing is formed under compression stress during upsetting, as the cup height is reduced with ongoing stroke the material is forced to flow radially into the gear cavity. With an increased contact zone between tool and workpiece the material flow into the bottom of the cup and burr formation increase as friction and yield stress are rising.

2.3. Modelling

For a fundamental investigation of the material flow the process set-up is transferred to a numerical model. For the numerical investigation the implicit FE-code simufact.forming 12 is used. The workpiece material applied is DC04 and the initial sheet thickness is $t_0 = 2$ mm. The flow curve is determined in a layer compression test and the flow criterion is modelled according to von Mises with isotropic hardening. The tools are modelled as rigid bodies. In the numerical simulation the forming process proceeds stroke-controlled, so that analyses at any force or stroke are possible.

Due to the cyclical and symmetrical arrangement of the functional elements a 60°-section of the workpiece geometry is used to model the process in the numerical simulation. The symmetry planes of the model are positioned in the middle of open carrier and gearing. To prevent contact problems at the interfaces and to enhance numerical stability a larger tool segment is used. The tool segment modelled is 90° wide. The numerical model allows a tool set-up without gaps so burr formation can be eliminated. As the material flow into the die cavity is investigated the material flow into the bottom of the cup is not considered. The inner diameter of the blanks applied is set to 70 mm and the inner section is filled with a dummy plate which is modelled as rigid body. This adaption allows a decreased number of elements and hereby a decreased calculating time.

The contact condition between modified part and dummy plate is defined as glue contact. In contrast to the experimental set-up the numerical process is divided into the two process steps deep drawing and upsetting to leave out passive components and decrease the calculation time. Furthermore, this enables a specific adaption of the meshing for sheet- and bulk metal forming operations.

The final conditions of the deep drawing process are the initial conditions for the upsetting operation. Tool position, workpiece geometry, mesh, forming history and residual stresses are transferred. For deep drawing the mesher Sheetmesh is applied with 6 elements in thickness-direction and an edge length of 0.35 mm. This mesher is chosen as the deep drawing process is a classical sheet metal forming process. In the upsetting operation a hexahedron mesh with an edge length of 0.3 mm is applied as upsetting belongs to the group of bulk metal forming processes. The meshing parameters are presented in Table 1.

Other than in the experimental set-up drawing die and upsetting plate are defined as active components.

Table 1: Parameters for the numerical model

	Deep drawing	Upsetting
Mesher	Sheetmesh	Overlay Hex
Edge length global	0.35 mm	0.30 mm
Edge length minimal	-	0 mm
Elements over thickness	6	-
Edge angle	30°	60°
Remesh	none	none
Friction model	Tresca	Tresca
Friction global	0.1	0.1
Workpiece material	DC04	DC04

2.4. Identification of the blank layout

The part geometry requires different material volumes in different areas to enable a complete die filling. The allocation of additional material for the functional elements shall enable an increased die filling and at the same time prevent tangential material flow and increased tool loads [9]. To adapt the blank layout the material necessary to enable a complete die filling is calculated. Figure 4 shows the material requirement of the different part areas.

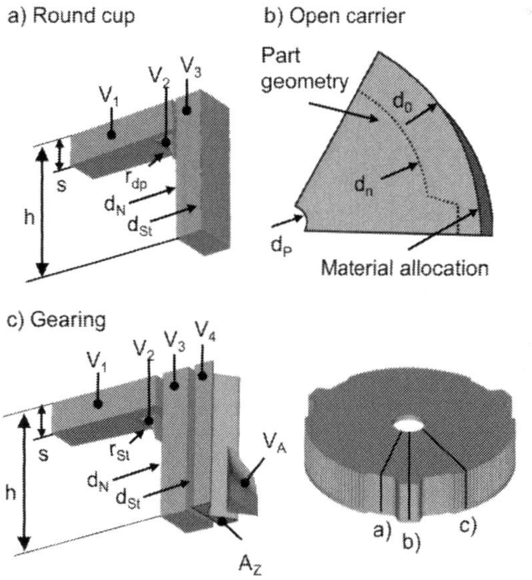

Figure 4: Determination of material requirement for round cup (a), open carrier (b) and gearing (c)

The adaption is realized by designing blanks in assumption of constant volume based on round semi-finished products with an initial diameter of $d_0 = 100$ mm. Therefore, the cup height h for a round cup without the arrangement of functional elements is determined and the required material volumes V_1, V_2 and V_3 are calculated. V_1 represents the material volume in the bottom of the cup, V_3 is the material required to fill the round cup. The volume V_2 is required to fill the gap at the drawing punch radius. Based on preliminary studies and the approach presented in [9] an adapted blank layout is identified to enhance the filling behavior for the open carrier and enable the manufacturing of a cup with homogeneous height. For the gearing the additional volumes V_4 and V_A are required. V_4 defines the material volume for the gear teeth and the volume V_A has to be considered due to the tool construction. The volumes are determined based on the cup height calculated for the round cup. Taking into account of these calculations additional material is provided by an adaption of the blank layout to allocate the material required. The material is intended to improve the die filling behavior of the open carrier and gearing during upsetting. Considering the calculations under assumption of constant volume the semi-finished product geometry shown in figure 5 is identified. The material provision for the gearing is realized by an expansion of the outer blank diameter from 100 mm to 110 mm. For the open carrier material is provided by increasing the initial blank diameter by an eccentric adaption with a maximum diameter of 106 mm. The angle of the transition between these zones is set to $\alpha = 40°$. The beginning of the transition set to the beginning of the gearing, proceeds over the first gear tooth and ends over the tooth tip of the second tooth. This ensures a complete allocation of material for the first gear tooth which lies in between the gearing and not at the transition to the round cup where less material volume is required.

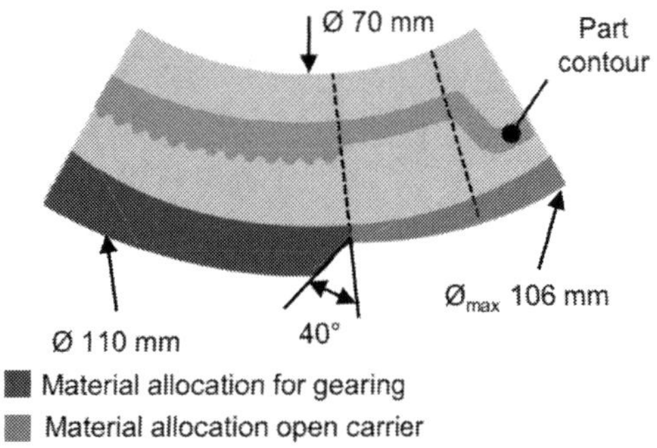

Figure 5: Initial semi-finished product geometry

An enhanced control of the induced material flow is the prerequisite for forming of near-net-shaped parts in the process combination of deep drawing and upsetting [8]. After defining the layout for the cup with increased functional integration the numerical investigation is performed to analyze the material flow in the forming process.

3 Numerical investigations of the material flow for manufacturing of cups with various functional elements

In the numerical investigation interactions regarding blank layout, draw-in behavior and resulting part properties after deep drawing and regarding die filling and part properties after upsetting are analyzed. To investigate the transferability of the results for cups with external gearing or open carrier and determine interactions between these functional elements an isolated consideration of the processes applied is necessary. The properties of the drawn cup have a strong influence on the material flow during upsetting and the resulting properties of the finished part. First of all, the force-displacement diagram is analyzed. Subsequently, the distribution of stress, tool contact and effective plastic strain are presented and discussed, to enhance the process knowledge. The numerical simulation thereby is an important tool as it enables an analysis during the forming process which is hardly possible in experimental tests.

3.1. Draw-in behaviour and material flow during deep drawing

The force-displacement curve for the deep drawing process is shown in figure 6. At the beginning of the deep drawing process the forming force increases as the material is drawn in over the outer radius of the drawing die r_o. As soon as the workpiece material gets in contact with the conical drawing profile with an angle of 45° the material is bent over the inner radius r_i. This results in a steeper increase of the force required due to a reduced length of the lever arm. A first maximum of the drawing force can be determined after most of the material is drawn in and afterwards the drawing force required decreases. Up to this point the curve shows the typical progression of a conventional deep drawing process. However, with ongoing process the forming of the open carrier results in a second increase of the drawing force as the material is uptight on both sides of the open carrier due to the draw-in behavior.

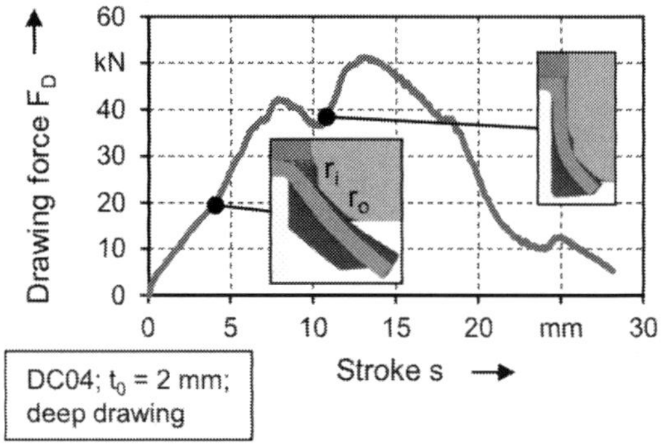

Figure 6: Force-displacement curve for deep drawing of the initial layout

The curve progression is caused by the specific stress and strain states in the functional elements and by interaction of these elements in the forming operation. In the beginning of the deep drawing process the area of the round cup is bulged between the areas with higher material volume for gearing and open carrier, as shown in figure 7. This leads to tangential compression on the outside of the cup wall and tangential tension on the inside. The forming of the open carrier leads to increased tangential tensile stresses during the ongoing drawing operation. The tangential material flow induced leads to increased material allocation in the outer area of the gearing in tangential direction. The angle of the transition zone amounts to 20° after deep drawing from initially 40° in the blank layout. The decreased transition angle leads to an increased material allocation for the first and last tooth in the gearing section, as the tangential material flow was not considered in the blank layout. The tool contact and the distribution of effective plastic strain have to be investigated during the process in the different sections, as the deep drawn cup defines the boundary conditions for the upsetting process.

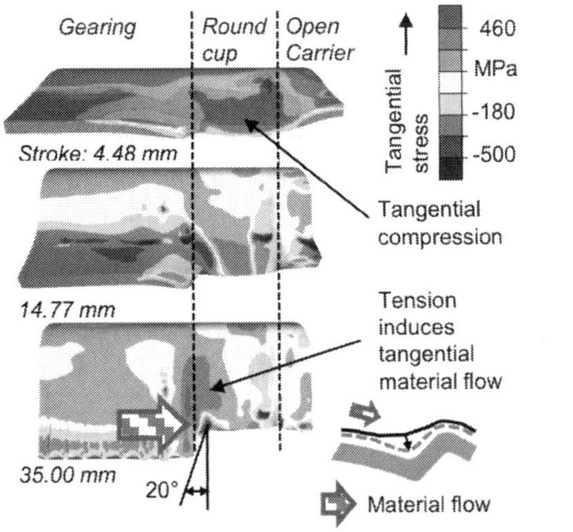

Figure 7: Draw in behavior and tangential stress states during deep drawing

The draw in behavior depends on the tool contact during the process. Table 2 shows the contact zones of tool and workpiece and the resulting development of the effective plastic strain in the part. It is shown that there is no circumferential contact between tool and workpiece. The areas of gearing and open carrier as well as the transition zones are in contact with the drawing die, whereas the area of the round cup is not due to bulging. At the beginning of the deep drawing process the highest effective plastic strain in the area of the drawing punch radius occurs at the open carrier as this zone faces radial and tangential stress and strain states. With ongoing process the drawing die moves downwards and forms the cup and open carrier. The effective plastic strain in the transition zones from round cup to the functional elements is significantly higher than in the other sections as small contact zones face high contact pressure. This leads to surface deformation due to high contact normal pressure. At the end of the deep drawing operation values of more than 1.5 are determined which are significantly higher than the values in the rest of the part. The effective plastic strain at the front of the open carrier locally surpasses 1 due to contact pressure caused by the forming of the carrier. In the outer area of the gearing in axial direction unintended material flow is induced by ironing of the cup wall. An increased wall thickness caused by tangential compression and the stiffness of the cup lead to a shaping of the gear contour into the workpiece material. Because of the tangential material flow and the smaller cavity of the round cup the values in the middle of the geared section of the part are lower than in the outer section. After deep drawing a significant difference in height between the gearing and the open carrier is detectable due to the blank geometry. The differences in height are required by the material volume necessary to fill the various functional elements in the upsetting process.

Table 2: Tool contact, distribution of effective plastic strain and stroke during deep drawing

Tool contact	Effective plastic strain	Stroke:
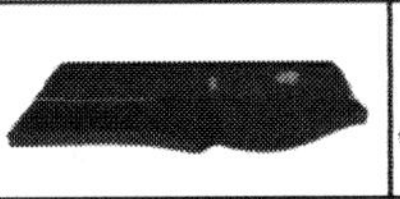		*5.95 mm*
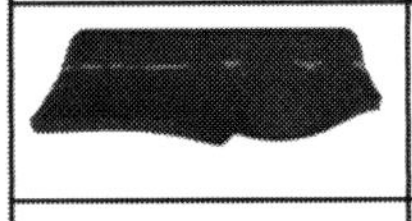		*8.89 mm*
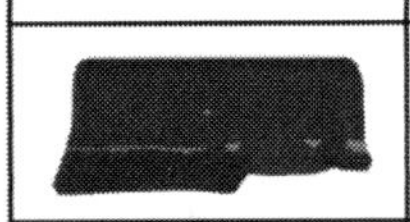	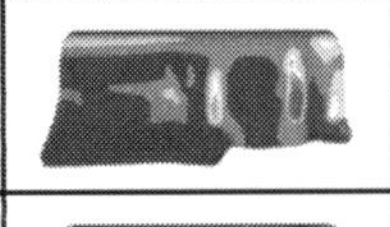	*16.24 mm*
	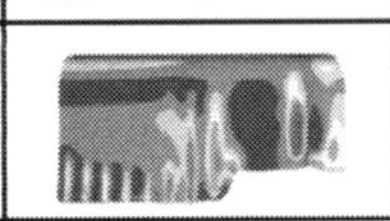	*28.00 mm*
No contact / Contact	Effective plastic strain → 0.0 0.5 - 1.5	*DC04; t_0 = 2 mm; deep drawing*

For a more detailed analysis of the part properties the distribution of effective plastic strain in the cross-section of the functional elements is investigated. The distribution of effective plastic strain in the cross section is shown in figure 8 for the different part. The cross-sections show the highest deformation in the open carrier and the increased thinning in the area of the drawing punch radius due to increased tangential tensile stress states described above. For round cup and gearing the values for effective plastic strain are increased only in the area of the drawing punch due to radial stress in the outer side and compressive stress in the inner side of the bending zone. For the round cup the effective plastic strain on the inside is higher than on the outside, whereas for the gearing higher effective plastic strain on the outside occurs. This is caused by the increased material volume in the drawing gap and the ironing of the cup wall. The material in the area of the drawing punch radius is supported on the inside surface as it is in contact with the punch and ironing causes additional tensile stress on the outside surface.

The results shown for cups with combined functional elements are transferable to the results for cups with similar functional elements presented in [8] and [9]. Increased thinning of the material under radial tensile stress in the area of the drawing punch radius and thickening in the wall due to tangential compression occurs. Also the intensified thinning at the top of the open carrier is determined. For the cup with various functional elements the minimum sheet thickness at the drawing punch radius is 0.9 mm and the maximum sheet thickness in the wall is 2.2 mm.

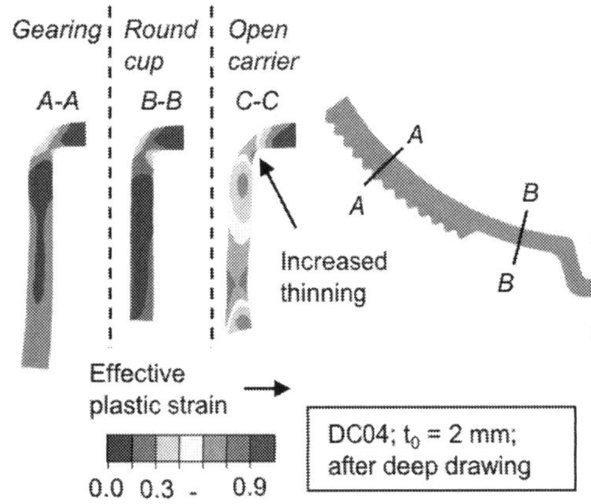

Figure 8: Effective plastic strain for the cross-sections of the functional elements after deep drawing for the initial layout

The geometry and the mechanical properties of the deep drawn cup are now transferred to the numerical model for the upsetting process. The influence of the part properties of the deep drawn cup are investigated as the cup is formed in the upsetting process.

3.2. Captions/numbering

To determine characteristic stages of the upsetting process the force-displacement graph is analyzed. The curve for upsetting is shown in figure 9 and rises after s = 2 mm when the cup gets in contact with the upsetting plate. In the beginning of the upsetting operation only the area of the external gearing is upset. An increase of the gradient is determined as an even cup height is reached and open carrier and cup wall are also upset.

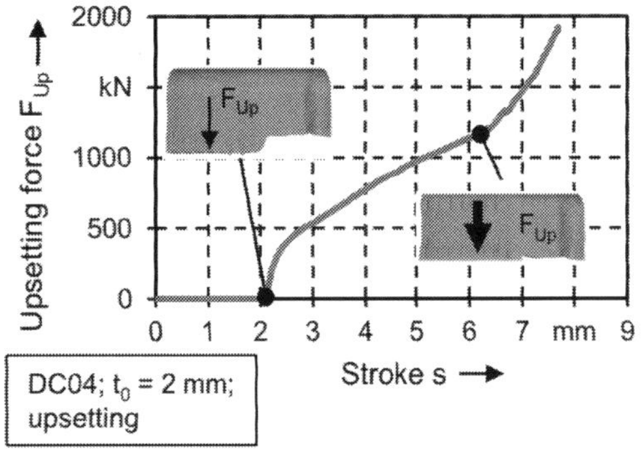

Figure 9: Force-displacement curve for upsetting of the initial layout

This is caused by the increased contact zone between tool and workpiece material which is accompanied by higher friction. Furthermore, the required upsetting force rises due to induced cold hardening in areas with high deformation. An increased die filling by applying higher upsetting forces is possible, yet not expedient as this causes high tool loads due to tangential material flow.

The distribution of elastic plastic strain is now investigated to analyze the material flow during the upsetting operation. The effective plastic strain distribution in the part as well as in the cross-sections of gearing, round cup and open carrier is analyzed. As the gearing is a thick-walled functional element it is formed by thickness increase. The material volume required is made available by radial material allocation. The resulting increased cup height after deep drawing faces axial compression and buckles. Due to the unfavorable upset ratio the cup wall buckles twice in the area of the external gearing and two grooves occur vertically to the gear teeth. This inhibits the material flow control due to inhomogeneous cold hardening and leads to worsened die filling behavior. Furthermore, the cross-sections of the part sections show high effective plastic strain at the drawing punch radius as presented in preliminary studies [8]. The decreased material thickness at the drawing punch and the high effective plastic strain are caused high deformation after deep drawing, as shown in [8]. The specific distribution of effective plastic strain for the various part sections is presented in figure 10.

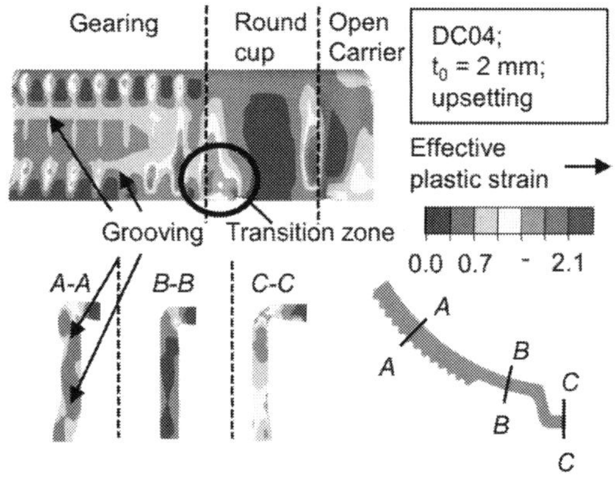

Figure 10: Effective plastic strain after upsetting of the initial layout

A zone with high effective plastic strain occurs at the transition zone from gearing to the round cup. This zone faces high tangential stress after deep drawing, as shown above. This increased tangential tensile stress and the induced lateral material flow lead to a decreased transition angle, as presented. For a closer investigation an analysis of the upsetting process at different strokes is necessary and is presented in figure 11. The transition angle amounts to $\alpha = 20°$ after deep drawing. In the upsetting operation not only the intended axial and radial material flow into the gear cavity occurs but also unintended tangential material flow in areas with minor flow resistance. As shown in the force-displacement diagram for the upsetting process only the additional material volume for the gearing is upset at the beginning of the process. The steep transition angle of $\alpha = 20°$ leads to a disadvantageous transmission of force into the area of the round cup and uncontrolled material flow. As upsetting induces lateral expansion of material the material allocated for the gearing becomes wider as the height is reduced. This is supported by the minor cup height in the area of the round cup as there is no material flow interference in tangential direction and causes folding of the material.

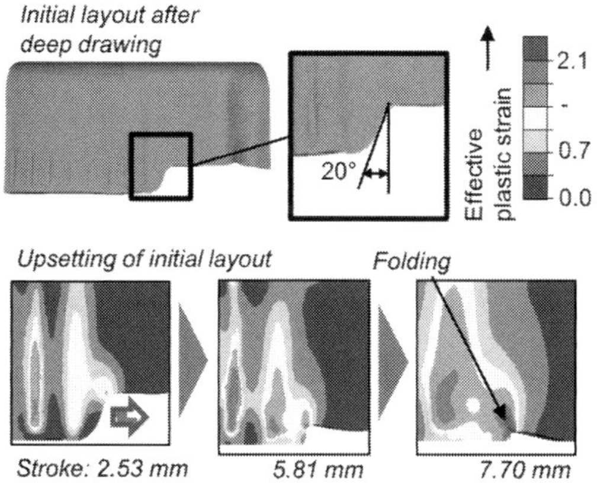

Figure 11: Folding of the initial layout in the upsetting process

Local inhomogeneous distribution of effective plastic strain in the transition zone causes problems with regard to material flow control. To prevent folding and achieve an increased homogeneity of the effective plastic strain distribution the transition angle of the blank layout is adapted.

3.3. Adaption of the transition angle to enhance the material flow control

The transition angle is reduced by drawing in the workpiece material in the deep drawing process and the lateral expansion during upsetting causes folding. This is considered as the layout of the semi-finished product is adapted. To prevent the folding in the transition zone the angle of the material provision for the gearing is flattened and set to $\alpha_a = 55°$, as shown in figure 12. The angle in the initial layout is 40° and was adopted according to the tangential material flow identified.

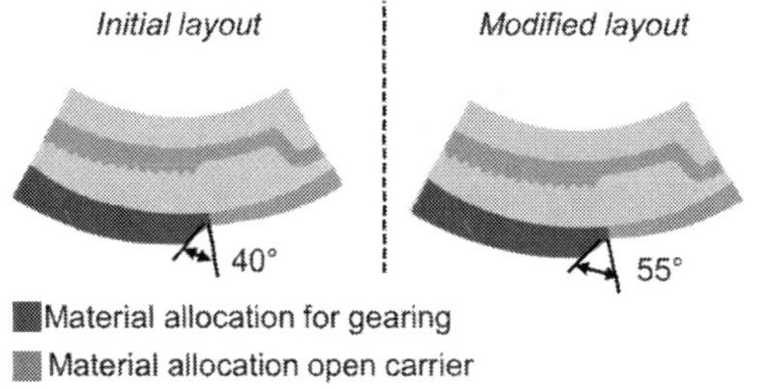

Figure 12: Adaption of blank layout

For the section between gearing and round cup a flatter transition is realized after deep drawing, as presented in figure 13. The transition angle is increased from 20° to 35°. The decrease of 20° for both cases can be explained by the tangential constant tensile stress induced by the open carrier.

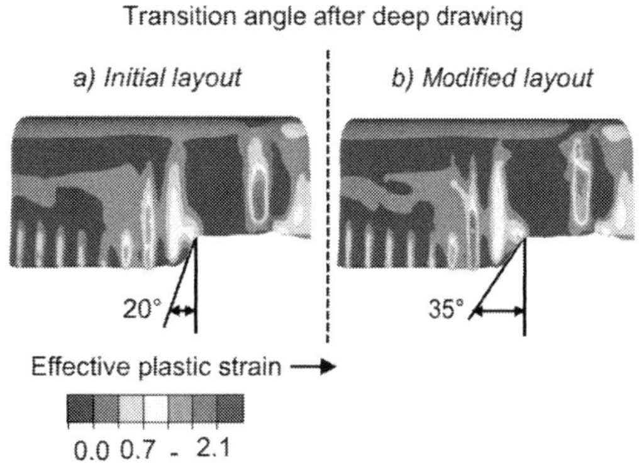

Figure 13: Transition angle for initial (a) and modified (b) layout after deep drawing

After deep drawing the adaption of the angle only minor deviations in the distribution of effective plastic strain occur due to the draw-in behavior. However, the adaption has a positive influence on the part properties after upsetting. Due to the adaption of the transition angle the lateral extension of the material allocation for the gearing does not cause folding, as shown in figure 14. The prevention of folding allows an increased radial material flow and thereby an improved die filling in the upsetting operation.

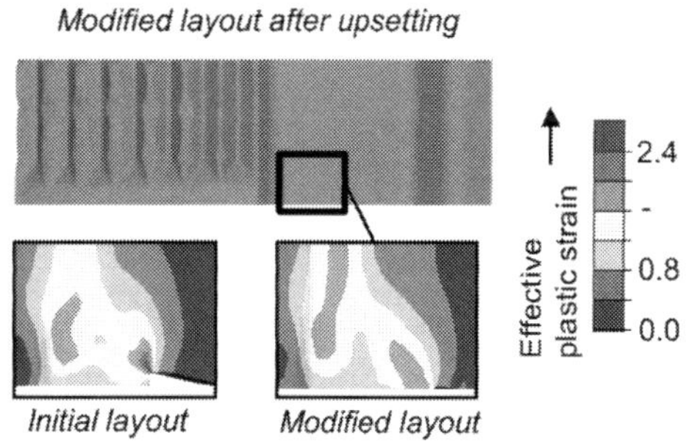

Figure 14: Effective plastic strain in the transition zone for initial (left) and modified layout (right)

4 Experimental verification of the numerical investigation

After numerical analysis an experimental investigation is performed to verify the numerical results as simplifications have to be made in the numerical investigation. Semi-finished products with modified layout are applied. The parts manufactured are digitalized with an optical scanner by strip projection. By measuring all 45 teeth the mean value and the standard deviation are calculated according to the arrangement of the teeth. Due to the cyclic symmetric arrangement half of the gearing is analyzed. As shown in the numerical investigation the die filling for the outer teeth of the geared section is significantly higher than for the rest of the gearing because of the interactions described above. The increased die filling of outer sides of the gearing is caused by tangential tensile stress which induces a tangential material flow and an increased material allocation at the outside of the gearing. The increase of the die filling in the middle of the gearing however is not significant. Figure 15 shows the mean values of the die filling for a manufactured part out of the deep drawing steel DC04 and confirms with the numerical results. The upsetting force applied is F_u = 400 kN.

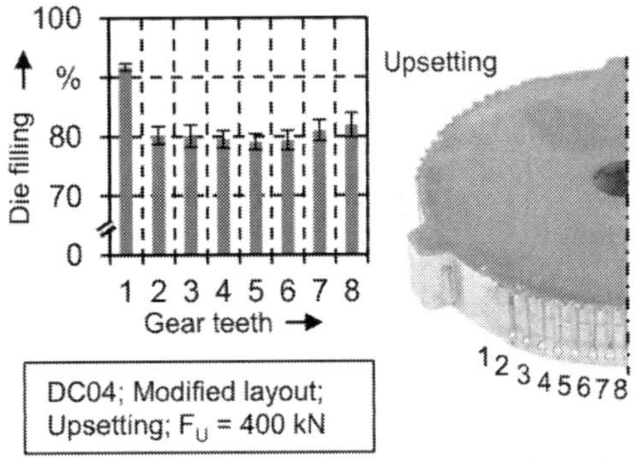

Figure 15: Die filling of gear teeth in the experimental test for DC04

The cross-sections of gearing and open carrier show the different material volume required to fill the die. The results confirm the numerical investigation as the geometrical part properties correspond to the results of the simulation. The outer contour of external gearing with circumferential grooving is shown in figure 16.

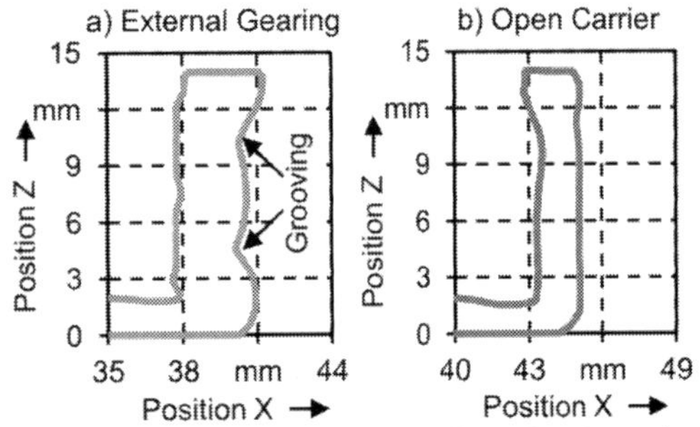

Figure 16: Cross-section of external gearing (a) and open carrier (b)

5 Conclusion

The functional integration of elements in a single part is shown in the numerical model as well as in the experimental test. A blank layout is constructed under assumption of constant volume. The investigation shows the interactions between different functional elements. The forming of the open carrier induces tangential stress states and tangential material flow which leads to folding in the transition zone from round cup and gearing and causes increased tool loads and insufficient die filling. An adaption of the transition angle prevents folding in this area of the part and enhances the material flow control. Also the grooving of the cup wall is analyzed. The adaption of the material allocation angle enables decreased folding in the upsetting operation. To prevent buckling of the cup wall the application of Tailored Blanks with process-adapted material thickness distribution is a promising approach. Pursuing this approach the application of Tailored Blanks has to be further investigated with regard to the material flow and the resulting part geometries.

Acknowledgement

This work was supported by the *German Research Foundation* (DFG) within the scope of the *Transregional Collaborative Research Centre* on Sheet-Bulk Metal Forming (*TCRC 73*, https://www.tr-73.de) in the subproject A1 (Deep Drawing).

References

1. M. Kleiner, M. Geiger, A. Klaus, CIRP Annals – Manufacturing Technology, Volume 52, Issue 2, pp. 521-542 (2007)
2. J. Rowe, *Advanced Materials in Automotive Engineering*, Woodhead Publishing, Cambridge, UK (2012)
3. J. M. Allwood, J. M. Cullen, *Sustainable Materials – With Both Eyes Open*, UIT Cambridge Ltd., Cambridge, UK (2011)
4. http://www.umformtechnikmagazin.de/umformtechnik-fachartikel/massivumformung-von-blech_19204_de/ (accessed January 26th 2016)
5. T. Takami, Procedia Engineering 81, pp. 5-17 (2014)
6. M. Merklein, J.M. Allwood, B.A. Behrens, A. Brosius, H. Hagenah, K. Kuzmann, K. Mori, A.E. Tekkaya, A. Weckenmann, CIRP Annals – Manufacturing Technology 61-2, pp. 725-745 (2012)
7. K. Mori, Steel Research International – Special Edition 14th International Conference Metal Forming, pp. 17-24 (2012)
8. T. Schneider, U. Vierzigmann, M. Merklein, AIP Conference Proceedings, Volume 1567, Issue 1, pp.930-933 (2013)
9. M. Merklein, M. Löffler, T. Schneider, CIRP Annals – Manufacturing Technology, 64-1, pp. 245–248 (2015)

A new shell element taking thickness-stretchability into account for mechanics-based springback compensation system

Hibiki Arashiyama[1], Tetsuo Oya[1] and Koichi Ito[2]

[1]*Graduate School of Science and Technology, Keio University, 14109 Kouhoku-ku, Yokohama-shi, Japan*
[2]*M&M Reserch Ltd, 13113 Shibuya-ku, Japan*

Abstract. To reduce springback, the coining process is often applied in sheet metal forming. However, the finite element method with a conventional shell element cannot formulate this process because of the plane stress condition. Therefore, in this study, a new shell element, which has extra nodes called pseudo-nodes, was developed to calculate the through-thickness stress. Using this element, a mechanics-based springback compensation system was constructed and tested.

1 Introduction

Springback is a serious problem in sheet metal forming. It has become increasingly important to solve this problem because of the use of AHSS (advanced high-strength steel), which exhibits a large amount of springback, particularly in the automobile industry. Springback is the elastic deformation driven by residual stress generated during the forming process. Thus, the bending moment before springback is the main factor causing springback. To solve the problem of springback, several tool compensation methods have been developed. Gan and Wagoner proposed the displacement adjustment (DA) method, which is the most effective means of compensating for the springback in tools [1], and Karafillis and Boyce developed the force descriptor method (FDM) [2][3]. Recently, various methods based on the DA method or FDM have been reported such as the comprehensive compensation method [4], the accelerated compensation method [5], and the smooth DA method [6], and so forth. However, the cause of springback, i.e., the bending moment before springback, is not considered in these methods. In the DA method, the compensated shape can be obtained using only geometric data. FDM has no ground for the variation of springforward obtained by reversing the nodal force at the end of the stamping process. Therefore, these methods are not based on springback mechanics. Thus, there is a high probability that the shape does not converge to the compensated shape. In addition, it takes considerable time to determine the compensated shape because their algorithms involve forming analysis. Moreover, in the conventional methods, although the shape can be obtained, other working parameters such as the sheet normal pressure and in-plane stretching force cannot be obtained. Thus, tool design cannot be conducted from the viewpoint of the designer.

Therefore, in our study, a new mechanics-based springback compensation system was developed. This system can take account of mechanical relationships account into springback compensation, by calculating the bending moment derived from not only the curvature but also the pressure and tension.

The finite element method with a conventional shell element cannot formulate the coining process because of the plane stress condition. By employing a solid element instead of a shell element, the normal stress can be calculated but it takes considerable time. Thus, we propose a new shell element, which has extra nodes called pseudo-nodes, to calculate the normal stress. In this study, using this element, a mechanics-based springback compensation system was constructed and tested.

2 Concept of springback compensation system

2.1 Springback mechanics [7]

Springback is an elastic deformation caused by residual stress generated at the end of the forming process. Thus, the bending moment before springback (at the end of forming analysis) is the main cause of springback. This means that the shape after springback is determined by the conditions at the end of forming analysis. Springback is calculated by the following mechanical equation:

$$\boldsymbol{D}\,(\,\boldsymbol{\kappa}_{\text{after}} - \boldsymbol{\kappa}_{\text{before}}\,) = -\,\boldsymbol{M}_{\text{before}}\,, \tag{1}$$

---

[a] Hibiki Arashiyama: arashiyamah@a6.keio.jp

where κ_{after}, κ_{before}, M_{before} and D are the curvature after springback, curvature before springback, bending moment before springback, and elastic bending rigidity, respectively. This equation is called the springback mechanical equation in this paper. To calculate κ_{before} and M_{before}, information such as the nodal coordinates and the stress tensor at the end of forming is essential.

2.2 Concept of springback compensation

Springback is expressed by eq.(1). By substituting $\kappa_{desired}$, the curvature of the desired shape, for κ_{after} in eq.(1), κ_{before} and M_{before} for the desired shape are obtained. This relationship is expressed as

$$D\ (\ \kappa_{desired} - \kappa_{before}\ ) = -\ M_{desired}\ (\kappa_{before}\ , P), \qquad (2)$$

where P and T is a force vectors describing the sheet normal pressure and the in-plane stretching force, respectively. Also, M_{before} is defined as a function of κ_{before}, P and T. To determine the conditions before springback, it is necessary that eq.(2) is satisfied. However, eq.(2) cannot be solved if a target has a complex structure. Therefore, M_{before} is determined by users by repeatedly adjusting the parameters κ_{before} and P and T. Figures 1 and 2 respectively show outlines of the conventional springback compensation method and the proposed method. Firstly, forming analysis and springback analysis are performed, then it is examined whether the shape is within the specified tolerance. If necessary, the bending moment before springback is modified by updating only the shape (curvature) in the conventional system. However in the proposed system, it is modified by updating the curvature distribution and the force vector in the subroutine.

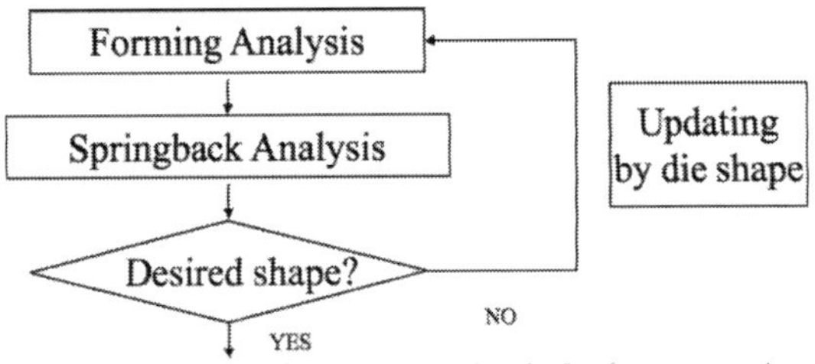

Figure 1. Outline of the conventional springback compensation system.

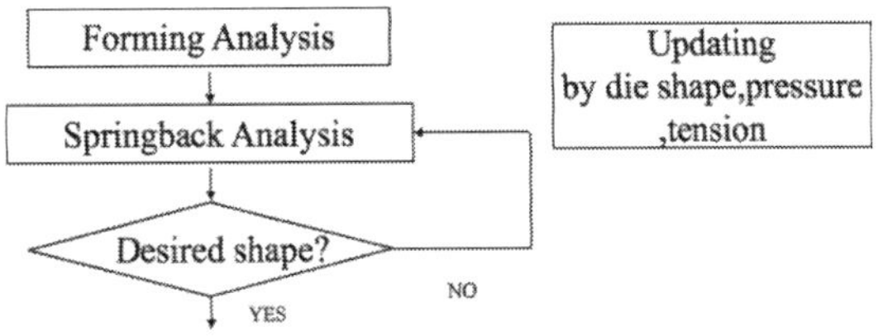

Figure 2. Outline of the proposed springback compensation system.

3 Formulation of the shell element

3.1 Outline

As mentioned above, it is necessary in springback compensation to predict the bending moment concisely in the subroutine of the proposed system. For this purpose, a new shell element is developed. The features of the proposed shell element are as follows.

a) The sheet-normal strain rate is introduced using an enhanced strain formulation.
b) A Fully three-dimensional constitutive relationship is used.
c) The formulation is completed by adding only the terms associated with the enhanced strain to the expressions for the conventional iso-parametric four-nodes shell element.

The enhanced strain formulation is a novel idea used to treat the normal stress for a shell element. When the shell element is applied to a sheet metal forming simulation, it is expected that the effect of the sheet normal pressure on the bending moment can be evaluated, furthermore, the constraint on both sides of the sheet (e.g. constraints by punch and the die) can be treated.

3.2 Formulation of the shell element

The sheet-normal strain rate, which is not considered in a conventional shell elements based on the Mindlin–Reissner's assumption, is introduced as an enhanced strain component in the iso-parametric four-nodes shell element. Assuming that the introduced strain rate varies linearly in the sheet thickness direction and is uniform over the sheet surface, the expression for the sheet-normal strain rate ε is

$$\varepsilon = \frac{2(\overline{v_U} + \overline{v_L})}{h}\xi_3 + \frac{\overline{v_U} - \overline{v_L}}{h}, \qquad (3)$$

where ξ_3, h, and $(\bar{v}_U\ , \bar{v}_L)$ are the natural coordinate in the sheet normal direction, the sheet thickness, and the relative velocities at strain evaluation points U and L from the midsurface, respectively (see Figure.3). The relative velocities $\bar{v}_U$ and $\bar{v}_L$ are defined as velocities consistent with the enhanced strain component and are independently of the nodal velocities for a conventional shell element.

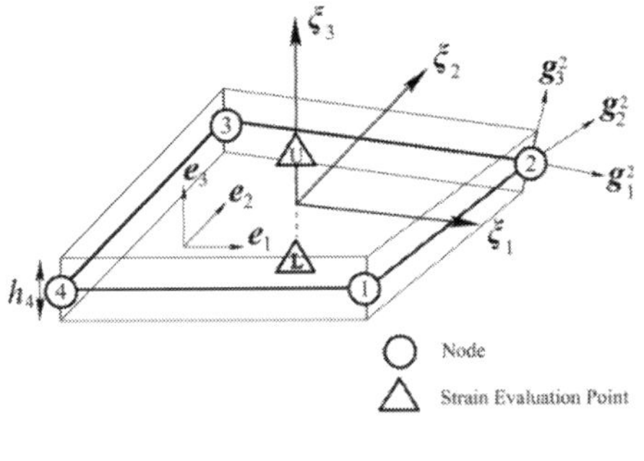

Figure 3. Configuration of the new element.

3.3 Velocity field consistent with the enhanced strain

Assuming $\varepsilon = \partial v/\partial(2h\,\xi_3)$ and integrating eq.(3) with respect to $h\,\xi_3$, we obtain the velocity field $v(\xi_3)$ consistent with the enhanced strain rate as

$$v(\xi_3)=\frac{\overline{v_U}+\overline{v_L}}{h}\xi_3^{\,2}+\frac{\overline{v_U}-\overline{v_L}}{h}\xi_3. \qquad (4)$$

Then, we assume that the sheet normal velocity at the midsurface v_M is defined by the mean value of the sheet-normal nodal velocity,

$$v_M=\frac{1}{4}\sum_{\alpha=1}^{4}v_3^{\alpha}, \qquad (5)$$

where $v^{\alpha}{}_3$, (α = 1, 2, 3, 4) are the sheet normal nodal velocity components. Denoting v_U, and v_L as the consistent velocities at the strain evaluation points and setting the strain evaluation points U and L at the center of the shell on the upper and lower surfaces, respectively, we obtain

$$\overline{v_U}=v_U-v_M,\overline{v_L}=v_L-v_M. \qquad (6)$$

Substituting eqs.(5) and (6) into eq.(4), we obtain the following expression for the consistent velocity field in the sheet thickness direction:

$$v(\xi_3)=-\frac{\xi_3^2}{4}\sum_{\alpha=1}^{4}v_3^{\alpha}+\frac{\xi_3(\xi_3-1)}{2}v_L+\frac{\xi_3(\xi_3+1)}{2}v_U. \qquad (7)$$

3.4 Velocity interpolation for the new shell element

The velocity field consistent with the enhanced strain $v(\xi_3)$ corresponds to the sheet-normal velocity field over the shell element. Therefore, we only need to add eq.(7) as the sheet normal velocity to the expression used for velocity interpolation for the conventional four-nodes iso-parametric shell element. Let indices 1, 2, and 3 correspond to in-plane variables and sheet normal variables. Then we can express the velocity interpolation in terms of the nodal velocities v_i^{α} and the consistent velocities at the enhanced strain evaluation points v_L and v_U as

$$v_i(\xi_1,\xi_2,\xi_3)=\sum_{\alpha=1}^{4}N^{\alpha}(\xi_1,\xi_2)[v_i^{\alpha}+\frac{1}{2}\xi_3h^{\alpha}(g_{1i}^{\alpha}\dot{\theta}_2^{\alpha}-g_{2i}^{\alpha}\dot{\theta}_1^{\alpha})]$$

$$\begin{aligned}v_3(\xi_1,\xi_2,\xi_3)=&\sum_{\alpha=1}^{4}[(N^{\alpha}(\xi_1,\xi_2)-\frac{\xi_3^2}{4})v_3^{\alpha}\\&+\frac{1}{2}N^{\alpha}(\xi_1,\xi_2)\xi_3h^{\alpha}(g_{13}^{\alpha}\dot{\theta}_2^{\alpha}-g_{23}^{\alpha}\dot{\theta}_1^{\alpha})]\\&+\frac{\xi_3(\xi_3-1)}{2}v_L+\frac{\xi_3(\xi_3+1)}{2}v_U\end{aligned} \qquad (8)$$

where $N^{\alpha}(\xi_1,\xi_2)$, g_{ij}^{α} and $\dot{\theta}_j^{\alpha}$ are the conventional shape function for a four-nodes iso-parametric shell element, the j-th component of the base vector $\boldsymbol{g}^{\alpha}{}_i$ of the nodal frame for node α (see. Figure 3), and the rotation angle around the $\boldsymbol{g}^{\alpha}{}_j$ axis, respectively. The new shell element has 22 degrees of freedom, which are composed of four nodal velocity vectors $\boldsymbol{v}^{\alpha}=(v_1^{\alpha}, v_2^{\alpha}, v_3^{\alpha}, \theta_1^{\alpha}, \theta_2^{\alpha})^T$ and the strain consistent velocity vector $\boldsymbol{v}^{c}=(v_L, v_U)^T$. The relationships between the velocity field over the shell element and these 22 element variables are completely expressed in eq.(8). On the basis of this expression, it is possible to use the same method as in the case of the conventional shell element to obtain the element stiffness equation. To describe the case of no transverse shear under a uniform bending deformation, it is recommended to adopt the assumed strain formulation for transverse shear strain components.

3.5 Contact and equivalent nodal forces

To treat the constraints on both sides of the sheet constraints, i.e., those on the upper surface and lower surface, only the strain evaluation points are subjected to the sheet normal constraints, in other words, the four nodes of the shell are free from sheet-normal contact conditions. We consider that the shell element is subjected to external forces at points U and L, as shown in Figure 4. The principle of the virtual work in relation to the external force and the equivalent nodal forces is expressed by

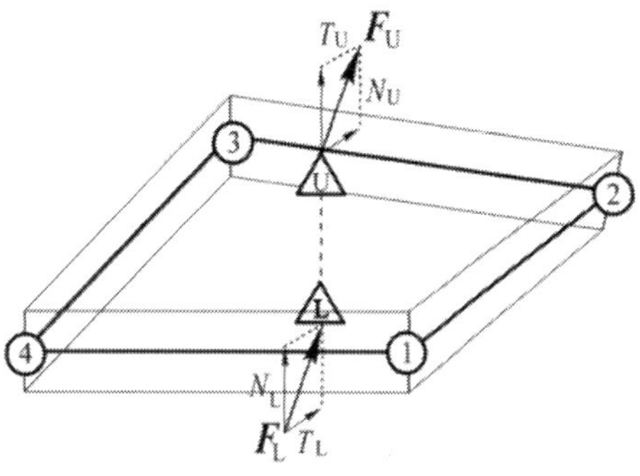

Figure 4. External forces applied to a shell element.

$$\sum_{\alpha=1}^{4}(\delta \boldsymbol{v})^T \cdot \boldsymbol{f} + \delta v_U f_U + \delta v_L f_L = \int \delta \boldsymbol{v}(\xi_1, \xi_2, 1) \cdot \boldsymbol{F}_U dS_U$$
$$+ \int \delta \boldsymbol{v}(\xi_1, \xi_2, -1) \cdot \boldsymbol{F}_L dS_L \qquad (9)$$

where $\boldsymbol{f}_\alpha$, $\boldsymbol{f}_U$ and $\boldsymbol{f}_L$ are the equivalent nodal forces, and the equivalent forces at strain evaluation points U and L, respectively. The first integration on the right-hand side of eq.(9) is performed over the upper surface and the second one is performed over the lower surface. Substituting eq.(8) into eq.(9) and assuming that $\boldsymbol{f}_U$ and $\boldsymbol{f}_U$ are the resultant forces in the case of a uniformly distributed surface load, some calculation leads to the following expressions.

$$f_i^\alpha = T_{Ui}\int N^\alpha(\xi_1,\xi_2)dS_U + T_{Li}\int N^\alpha(\xi_1,\xi_2)dS_L \qquad (i=1,2)$$
$$f_{\theta 1}^\alpha = \frac{1}{2}h^\alpha(g_{21}^\alpha + g_{22}^\alpha)[-T_{U1}\int N^\alpha(\xi_1,\xi_2)dS_U + T_{L1}\int N^\alpha(\xi_1,\xi_2)dS_L]$$
$$f_{\theta 2}^\alpha = \frac{1}{2}h^\alpha(g_{11}^\alpha + g_{12}^\alpha)[T_{U1}\int N^\alpha(\xi_1,\xi_2)dS_U - T_{L1}\int N^\alpha(\xi_1,\xi_2)dS_L]$$
$$f_U = N_U$$
$$f_L = N_L \qquad (10)$$

In the new shell element, the sheet normal force, for example, the hydrostatic pressure or contact force, applies only to the two strain evaluation points, while the four nodes, which are located on the midsurface are subjected to an in-plane force, such as a friction force.

4 Forming analysis with proposed element

To show that the proposed element behaves in the desired manner, the forming analysis of a beam with constant curvature is demonstrated. Setting the forced rotational displacement to the endpoint of the beam, a flat sheet is bent so that y, the rotational freedom of the endpoint θ, is $\pi/6$ as shown in Figure 5. Table 1 shows the material and geometrical properties of the sheet. Springback analysis was performed using the developed solver and the shape after springback was compared with the theoretical value. Ten shell elements were used in this analysis. Figure 6 shows the results for the curvature radius obtained by the analysis. The curvature radius after springback obtained by the developed solver is close to the theoretical value. Therefore, the proposed element behaves in the desired manner in forming analysis.

Table 1. Material and geometrical properties.

Young's modulus /GPa	Poisson's ratio	Length /mm	Width /mm	Thickness /mm
210	0.3	100	10	1

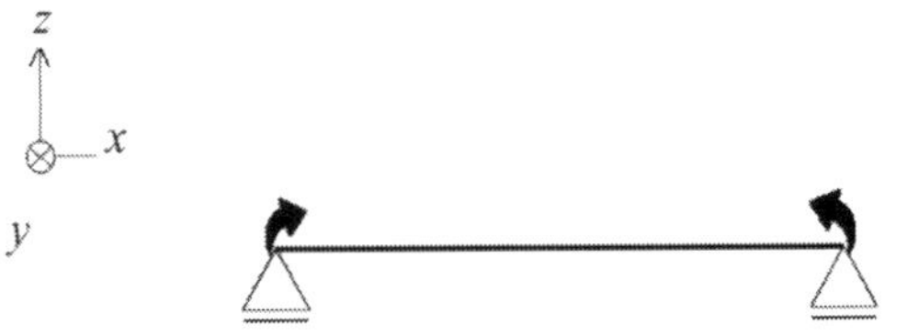

Figure 5. The schematic diagram of forming analysis.

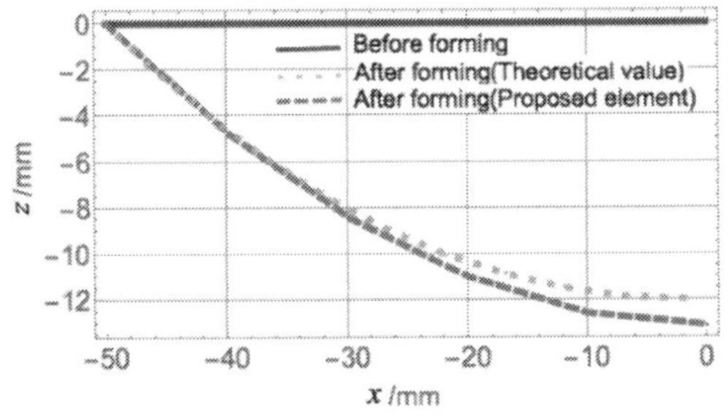

Figure 6. Results of forming analysis.

5. Coining simulation and comparison with solid element

To show that the proposed element behaves as desired in the coining process, a coining simulation is demonstrated. Figure 7 shows the FE model used in the coining simulation of bending. After carrying out forming analysis using LS-DYNA, coining analysis was carried out by applying pseudo-nodes to nodal forces. Table 2 shows the material properties of the sheet. Then, the coining simulation was performed using the solver and the radius of curvature after coining was compared with those when using a solid element and the conventional shell element. Ten shell elements were used. Figure 8 and Table 3 show the results for the radius of curvature obtained by the analysis. When springback is induced, the sheet unrolls from its original shape; therefore, the radius of curvature indicates the extent of springback. Although the result of analysis with the conventional shell shows that it cannot calculate the effect of coining that a bending moment decreases owing to pressure, the proposed element can express this effect similarly to a solid element. Therefore, this result shows that suggested shell element can treat normal stress similarly to a solid element, although the conventional shell element cannot calculate it.

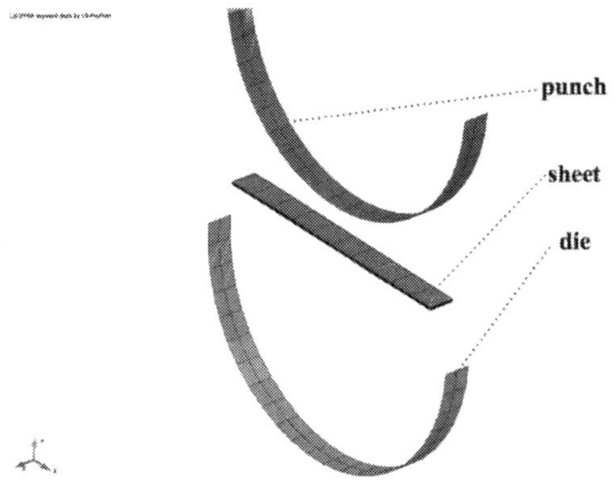

Figure 7. FE model used in coining simulation employing LS-DYNA.

Table 2. Material properties.

Young's modulus /GPa	Poisson's ratio	Plastcity coefficient /MPa	Work hardening coefficient
206	0.3	1300	0.45

Table 3. Results of coining simulation.

Punch displacement /mm	Radius of curvature /mm		
	Proposed shell	Conventional shell	Solid
92	83.05	83.05	83.33
92.25	80.91		80.90
95	80.69	84.54	80.65

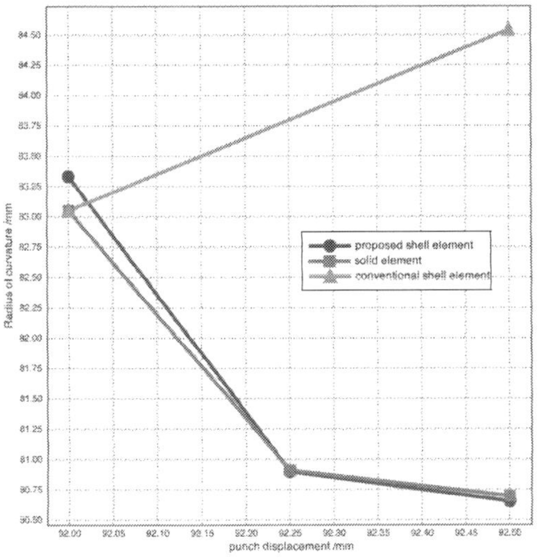

Figure 8. Results of coining simulation.

6. Conclusion

In this paper, new shell element with pseudo nodes for use in a mechanics-based springback compensation system was introduced. First, to test the basic function of the proposed element, the forming analysis of a beam with constant curvature was demonstrated. As a result, it was verified that this element behaves as desired by comparison with the theoretical value. Secondly, to show the effectiveness of the proposed element for calculating the normal stress in the coining process, a coining simulation of constant-curvature bending was demonstrated. It was verified that this element can calculate the effect by which the bending moment of a sheet decreases as it is subjected to pressure.

As future work, we intended to apply this element, to more complex examples such as a case including torsion. To more easily compensate for the springback using the proposed element, it may be necessary to integrate each computation.

References

1. W. Gan, R.H. Wagoner, *Die design method for sheet springback*, Int. J. Mech. Sci., **46**, (2004), 1097-1113.
2. A.P. Karafillis, M.C. Boyce, *Tooling Design in Sheet Metal Forming Using Springback Calculations*, Int. J. Mech. Sci., **34**, 2 (1992), 113-131.
3. A.P. Karafillis, M.C. Boyce, *Tooling and Binder Design for Sheet Metal Forming Compensating Springback Error*, Int. J. Mech. Tools Manu, **36**, 4 (1996), 503-526.
4. X.A. Yang, F. Ruan, *A die design method for springback compensation based on displacement adjustment*, Int. J. Mech. Sci., **53**, 5 (2011), 399-406.
5. H.S. Cheng , J. Cao, *An accelerated springback compensation method*, Int. J. Mech. Sci., **49**, 3 (2007), 267-279.
6. R. Lingbeek, J. Huetink, S. Ohnimus, M. Petzoldt, J.Weiher, *The development of a finite elements based on springback compensation tool for sheet metal product*, J. Mater. Process Technol, **169**, 1 (2005), 115 -125.
7. N. Doke, and T. Oya, *A Novel Springback Compensation Method Based on Finite Element Analysis and Springback Mechanics*, MetalForming, **14**,(2012), 335-338.

An Element Free Galerkin method for an elastoplastic coupled to damage analysis

Zohra Sendi[1, a], Hédi Belhadjsalah[1], Carl Labergere[2] and Khémais Saanouni[2]

[1] *University of Monastir, LGM/ENIM, Laboratory of Mechanical Engineering,National Engineers School of Monastir, Tunisia*
[2] *University of Technology of Troyes, ICD/LASMIS, Charles Delaunay Institute, France*

Abstract. In this work, a Meshless approach for nonlinear solid mechanics is developed based on the Element Free Galerkin method. Furthermore, Meshless is combined with an elastoplastic model coupled to ductile damage. The efficiency of the proposed methodology is evaluated through various numerical examples. Besides these, two-dimensional tensile tests under several boundary conditions were studied and solved by a Dynamic-Explicit resolution scheme. Finally, the results obtained from the numerical simulations are analyzed and critically compared with Finite Element Method results.

1 Introduction

The Finite Element Method (FEM) is the dominant technique for the simulation of forming processes in past decades. It has made significant contributions in computational mechanics. Nevertheless, this method is not well suited to problems including large deformation of materials and moving discontinuities such as crack propagation requiring considerable meshing and re-meshing in structural optimization problems.

Meshless methods have attracted much attention in recent decades. These methods aim at avoiding problems related to the use of a mesh eliminating the well-known drawbacks in the FEM.

While in the Finite Element Method, the shape functions are defined inside the elements, the Meshless methods are based on shape functions constructed the nodes positions.

The goal of this study is to develop a Meshless method known as Element Free Galerkin (EFG) based on Moving Least Square (MLS) approximation for solving structural problems involving large deformations in a more efficient manner than with finite elements.

A code incorporating an implementation of the MLS method has been developed and tested on several simulations.

The first step was to develop and describe the proposed method. Secondly, the used behaviour model was presented. Then, various numerical examples based on elastoplasticity coupled to damage were simulated using Meshless method.

To conclude, the results compare well to finite element simulations.

2 The Element Free Galerkin Method

The Element Free Galerkin (EFG) is different from the Finite Element method on how the shape functions are constructed. In this work, the shape functions in the EFG method are constructed using the Moving Least Square (MLS) approximation introduced by Lancaster and Salkauskas [1].

In what follows, the use of the symbols [ ], { } and < > means respectively the matrix, the column vector and the row vector.

2.1 Moving Least Square approximation

In this section, an MLS approximation, which was introduced by Lancanster and Salkauskaus [1] is described.

In the MLS approximation, the trial function around the point $\bar{X} \in \bar{\Omega}$ is:

$$\left\{U^h(X,\bar{X})\right\} = \underset{X}{L} \; U(X) = \langle p(X) \rangle \left\{\alpha(\bar{X})\right\} \tag{1}$$

Where $p_i(x)$, i=1, 2.....m are monomial basis functions, m is the number of terms in the basis and $\alpha_i(x)$ are the coefficient of the basis functions

In order to precisely obtain the local approximation of the function, the difference between the local approximation $\{U^h(X)\}$ and the function $\{U(X)\}$ must be minimized by a weighted least-squares method.

Define a functional

[a] Corresponding author: zohrasendii@yahoo.fr

$$J_X(\{\alpha\}) = \sum_{i=1}^{N} w(X - X_i)\left(\langle p(X_i)\rangle\left\{\alpha(\bar{X})\right\} - U_i\right)^2 \quad (2)$$

Where w(X-Xi) is a weight function defined on a domain Ω_X called domain of influence and N is the number of particles in Ω_X. From

$$\frac{\partial J_{\bar{X}}}{\partial\{a\}} = 0 \quad (3)$$

we obtain the system of equations of order m to solve:

$$\begin{cases} \frac{\partial J_{\bar{X}}}{\partial a_1} = 0 \Leftrightarrow \sum_{i=1}^{N} w(X - X_i)p_1(X_i)\left(\langle p(X_i)\rangle\left\{a(\bar{X})\right\} - U_i\right) = 0 \\ \frac{\partial J_{\bar{X}}}{\partial a_2} = 0 \Leftrightarrow \sum_{i=1}^{N} w(X - X_i)p_2(X_i)\left(\langle p(X_i)\rangle\left\{a(\bar{X})\right\} - U_i\right) = 0 \\ \vdots \\ \frac{\partial J_{\bar{X}}}{\partial a_m} = 0 \Leftrightarrow \sum_{i=1}^{N} w(X - X_i)p_m(X_i)\left(\langle p(X_i)\rangle\left\{a(\bar{X})\right\} - U_i\right) = 0 \end{cases} \quad (4)$$

The system can be written in vector form as follows:

$$\sum_{i=1}^{N} w(X - X_i)\{p(X_i)\}\langle p(X_i)\rangle\left\{a(\bar{X})\right\} = \sum_{i=1}^{N} w(X - X_i)\{p(X_i)\}U_i \quad (5)$$

If we replace $\alpha(\bar{x})$ in the local approximation, we can obtain:

$$\left\{U^h(X)\right\} = \langle p(X)\rangle[A(X)]^{-1}[B(X)]\{U\} \quad (6)$$

where

$$[A(X)] = \sum_{i=1}^{N} w(X - X_i)\{p(X_i)\}\langle p(X_i)\rangle \quad (7)$$

$$[B(X)] = \left[w(X - X_1)\{p(X_1)\} \ldots w(X - X_m)\{p(X_m)\}\right] \quad (8)$$

Then the expression of the local approximation $\{U^h(X)\}$ can be obtained as:

$$\left\{U^h(X)\right\} = \sum_{i=1}^{N} \varphi_i(X)u_i = [\Phi(X)]^T\{U\} \quad (9)$$

where $\phi(X)$ is called the shape function, i.e.

$$[\Phi(X)]^T = \langle p(X)\rangle[A(X)]^{-1}[B(X)] \quad (10)$$

The shape function of a particle ***i*** in a point X is then:

$$\varphi_i(X) = \langle p(X)\rangle[A(X)]^{-1} w(X - X_i)\{p(X_i)\} \quad (11)$$

2.2 Boundary conditions

In Meshless methods, imposing the essential (Dirichlet) boundary conditions is not evident as the Finite Element Method because MLS is non-interpolative. A number of techniques for the imposition of essential boundary conditions have been developed, including collocation methods [2, 3], Lagrange multiplier method[4], penalty method[4], Nitsche's method[4], coupled meshless-Finite Element Method[6, 14] and other methods [7].

The advantages and disadvantages of a variety for methods of imposing essential boundary conditions have been discussed briefly in [5].

2.3 Numerical integration

In Meshless methods, the numerical evaluation of integrals is more difficult than in the Finite Element Method (FEM). Indeed, the structure is only known by the description of its boundaries and a distributed set of nodes.

In contrast with the FEM, we do not have the concept of mesh used to build shape functions and which is suited as a basis for realising Gauss integration on each element.

In order to well understand and distinguish the most common integration methods, see [5, 14].

3 Elastoplasticity with damage

To further test our Meshless program, a plasticity model coupled to damage was implemented.

The used behavior model is decomposed into an elastic law and a plastic one. For the elastic law, we consider a linear isotropic behavior (Hooke taking into account damage) which is written as follows:

$$\underline{\sigma} = (1 - \theta D)\,\underline{\underline{\Lambda}}^e : \underline{\varepsilon}^e \quad (12)$$

where θ is a flag parameter, D is the variable describing the damage, Λ^e is the elasticity tensor and the symbol : means the double contracted product. f ($\underline{\sigma}$, R, D) is the charge function written as:

$$f(\underline{\sigma}, R, D) = \frac{\sigma_{eq}(\underline{\sigma})}{\sqrt{1 - \theta D}} - \frac{R(r)}{\sqrt{1 - \theta D^m}} - \sigma_Y \le 0 \quad (13)$$

where σ_y is the yield stress, R and σ_{eq} are respectively the hardening function and the equivalent stress. The equivalent stress is written in the case of von-Mises in the following way:

$$\sigma_{eq}(\underline{\sigma}) = \|\underline{S}\| = J_2(\underline{S}) = \sqrt{\tfrac{3}{2}\underline{S} : \underline{S}} \quad (14)$$

where $\underline{S}$ is the deviatoric part of $\underline{\sigma}$. We note n the derivative of σ_{eq} given as:

$$\underline{n} = \frac{3}{2}\frac{\underline{S}}{\|\underline{S}\|} \quad (15)$$

The hardening function is given as follows:

$$R(r) = \left(1 - \theta D^m\right)Qr \quad (16)$$

where r is the variable describing the isotropic hardening and Q is a constant of the material.

The evolution law of plastic deformation is written as follows:

$$\underline{\dot{\varepsilon}}^p = \frac{\dot{\lambda}}{\sqrt{(1-\theta D)}}\underline{n} \tag{17}$$

where λ is the plastic multiplier and the point refers to the time derivate.

The evolution law of the hardening variable is given by the following equation:

$$\dot{r} = \dot{\lambda}\left(\frac{1}{\sqrt{(1-\theta D^m)}} - br\right) \tag{18}$$

where b is a constant of the material. The evolution law of the damage variable is written as follows:

$$\dot{D} = \dot{\lambda}\left(\frac{Y-Y_0}{S_d}\right)^s \frac{1}{(1-\theta D)^\beta} \tag{19}$$

with S_d, s and β are constants of the material and Y is given by the following equation:

$$Y = \frac{\theta}{2}\underline{\varepsilon}^e : \Lambda^e : \underline{\varepsilon}^e + \frac{\theta}{2} m d^{m-1} Q r^2 \tag{20}$$

The problem is closed by the consistency conditions given by the following relations:

$$\dot{\lambda} \geq 0 \tag{21}$$

$$\dot{\lambda}\dot{f}(\underline{\sigma}, R, D) = \dot{\lambda} f(\underline{\sigma}, R, D) = 0 \tag{22}$$

4 Numerical examples and discussion

4.1 Equilibrium equation and boundary conditions

The basic equation of dynamic problems is

$$\{\nabla.\underline{\sigma}\} + \{f\} = \rho.\{\gamma\} \text{ on } \Omega. \tag{23}$$

where ∇. indicates the divergence, $\underline{\sigma}$ is the stress tensor, ρ is the density and $\{\gamma\}$ is the acceleration.

With the strain-displacement relationship:

$$\underline{\dot{\varepsilon}} = \frac{1}{2}\left(\nabla\dot{u} + (\nabla\dot{u})^T\right) \tag{24}$$

In the above, the operator ∇ is the gradient, T is the transpose, the point refers to the time derivate and $\underline{\dot{\varepsilon}}$ is the strain tensor, which is decomposed in elastic and plastic parts as follows:

$$\underline{\dot{\varepsilon}} = \underline{\dot{\varepsilon}}^e + \underline{\dot{\varepsilon}}^p \tag{25}$$

Noting that the large strain formulation is considered and the corotational (or Jaumann) frame well defined and developed in [9] is used.

4.2 Explicit scheme

The well-known dynamic explicit algorithm based on the finite difference discretization of the derivatives relative to the time variable is used in the resolution of the resultant system [8].

4.3 Implicit integration

An Euler implicit integration algorithm has been established for our behavior model. Therefore and according to [9, 10,12], knowing the different variables σ_n, R_n, ε_{pn} and D_n at time increment t_n, we propose to determine the parameters σ_{n+1}, R_{n+1}, εp_{n+1} and D_{n+1} at t_{n+1}.

4.4 MLS discretization

For a two-dimensional plane strain dynamic problem, the vectors of displacement, velocity and acceleration as developed in [11] are respectively,

$$\{U\} = [\Phi]\{U_n\} \tag{26}$$

$$\{\dot{U}\} = [\Phi]\{\dot{U}_n\} \tag{27}$$

and

$$\{\ddot{U}\} = [\Phi]\{\ddot{U}_n\} \tag{28}$$

where

$$[\Phi] = \begin{bmatrix} \phi_1 & 0 & \phi_2 & 0 & \phi_3 & 0 & \cdots & 0 & \phi_n & 0 \\ 0 & \phi_1 & 0 & \phi_2 & 0 & \phi_3 & 0 & \cdots & 0 & \phi_n \end{bmatrix} \tag{29}$$

Knowing that

$$\{\varepsilon\} = [D]\{U\} \tag{30}$$

where

$$[D] = \begin{bmatrix} \frac{\partial}{\partial x} & 0 \\ 0 & \frac{\partial}{\partial y} \\ \frac{\partial}{\partial y} & \frac{\partial}{\partial x} \end{bmatrix} \tag{31}$$

We can obtain

$$\{\varepsilon\} = [D][\Phi]\{U_n\} = [B]\{U_n\} \tag{32}$$

with

$$[B] = \begin{bmatrix} \frac{\partial\phi_1}{\partial x} & 0 & \frac{\partial\phi_2}{\partial x} & 0 & \frac{\partial\phi_3}{\partial x} & 0 & \cdots & \frac{\partial\phi_n}{\partial x} & 0 \\ 0 & \frac{\partial\phi_1}{\partial y} & 0 & \frac{\partial\phi_2}{\partial y} & 0 & \frac{\partial\phi_3}{\partial y} & \cdots & 0 & \frac{\partial\phi_n}{\partial y} \\ \frac{\partial\phi_1}{\partial y} & \frac{\partial\phi_1}{\partial x} & \frac{\partial\phi_2}{\partial y} & \frac{\partial\phi_2}{\partial x} & \frac{\partial\phi_3}{\partial y} & \frac{\partial\phi_3}{\partial x} & \cdots & \frac{\partial\phi_n}{\partial y} & \frac{\partial\phi_n}{\partial x} \end{bmatrix} \tag{33}$$

The principle of virtual power allows writing the following relation for any virtual velocity field $\{\dot{U}^*\}$ (zero on Γu):

$$\pi(U,\dot{U}^*) = -\int_{\Omega}\underline{\sigma}:\dot{\underline{\varepsilon}}^* d\Omega - \int_{\Omega}\rho\{\gamma\}\langle\dot{U}^*\rangle d\Omega + \int_{\Gamma_F}\{F\}.\langle\dot{U}^*\rangle d\Gamma = 0 \quad (34)$$

Substituting these expressions in the weak form in Eq. (34), the final system is written as follows:

$$[M].\{\ddot{U}_n\} + \{F_{\text{int}}\} - \{F_{ext}\} = 0 \quad (35)$$

Where

$$[M] = \int_{\Omega}\rho[\Phi]^T[\Phi]d\Omega \quad (36)$$

$$\{F_{\text{int}}\} = \{F_{\text{int}}(\{U_n\})\} = \int_{\Omega}[B]^T\underline{\sigma}d\Omega \quad (37)$$

$$\{F_{ext}\} = \int_{\Gamma_F}[\Phi]^T\{F\}d\Gamma \quad (38)$$

In this part, The Galerkin weak form was employed to obtain the discretized equations system. Since the shape functions arising from the MLS approximation do not possess the δ -property i.e., the basic unknown, quantities Ui (xi) contained in the array U in Eq. (9) are not real nodal variables.

Therefore, enforcement of essential boundary conditions needs to be dealt differently. A range of methods has been mentioned in the sub-section 2.2.

4.5 Examples and discussion

In this Section, the numerical solution strategy adopted in this work to perform the numerical simulation is summarized. Indeed, the results obtained by implementing numerical simulations of two specimens with the previously described constitutive formulations, will be presented and discussed. Firstly, the description of specimens with different geometries and boundary conditions is undertaken. Then, both qualitative and quantitative comparisons are proven in order to show the efficiency of our developed Meshless method. Finally, representative graphs are chosen to describe the evolution of internal variables.

Table1. Material properties for the Cu-DHP copper alloy

Description	Symbol	Value
Density	ρ	8.93 x 10^3 (kg/m^3)
Elastic Modulus	E	1.1 x 10^5 (MPa)
Poisson's ratio	ν	0.35
Initial yield stress	σ_Y	190 (MPa)
Damage data (exponent)	s	1.2
Damage data (denominator)	S_d	1.2 (MPa)
Isotropic Hardening parameter	Q	790 (MPa)
Isotropic Hardening parameter	b	2.5
Exponent parameter	m	3

First example considered is a plate with hole under concentrated load at the end (see Figure 1).
Second part will investigate the efficiency of the method for a butterfly specimen.

All the aforementioned material parameters are properly listed in table1.

In the case of the plate hole specimen, the geometry, the boundary conditions and the nodes distribution are presented respectively in Figure 1 and Figure 2. In order to capture the evolution of internal variables, a relatively condensate and irregular distribution is used in the region surrounding the critical zone of the specimen (see Figure 2).

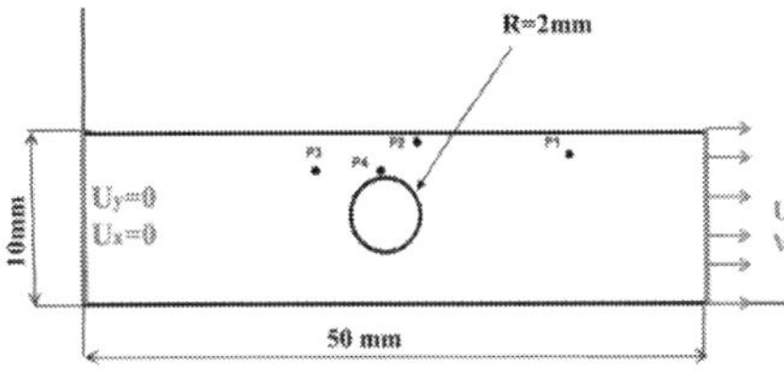

Figure 1. Geometry and boundary conditions of the plate hole specimen.

For the Meshless Method, we have used quadratic weight function with a circular support domain having as radius r=1.4*dx, see [5].

Concerning the integration, we have considered a quadratic cell with four Gauss point in each one.

For the Finite Element simulation, we have used Abaqus with a CPE4R element.

Also, it is very interesting to mention that the same number of nodes has been used for both methods.

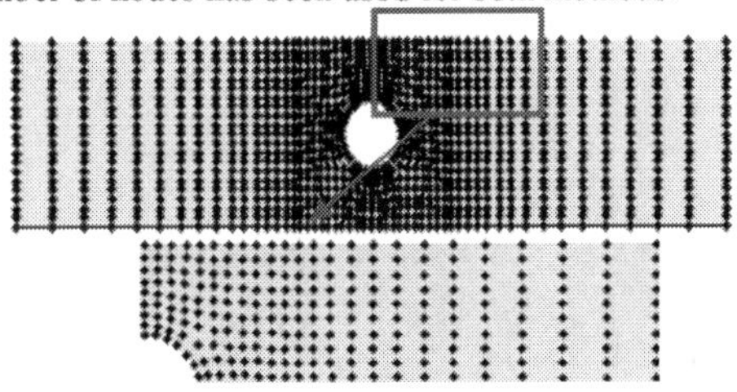

Figure 2. Node distribution in the plate hole specimen

The evolution of representative variables is analyzed. In particular and for different levels of displacement, the isovalues of the equivalent stress, the cumulative plasticity and the damage obtained by Meshless simulation are compared with those issued from the FEM especially in critical zones, knowing that we have used the same mesh for both methods.

Case 1: Imposed displacement U_f=0.025mm.

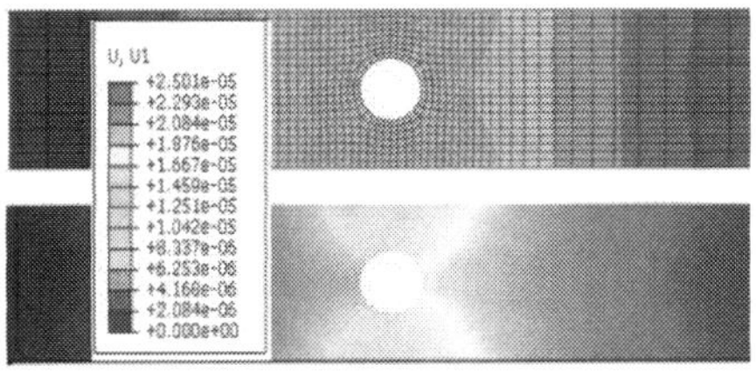

Figure 3. Displacement U_x isovalues by FEM and Meshless

 DOI: 10.1051/matecconf/20168007005

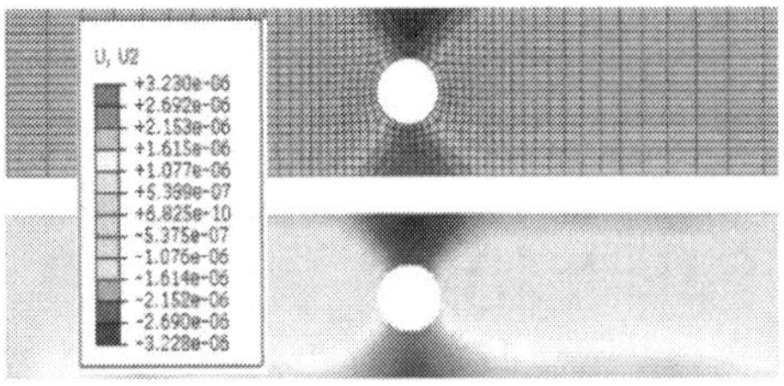

Figure 4. Displacement U_y isovalues by FEM and Meshless

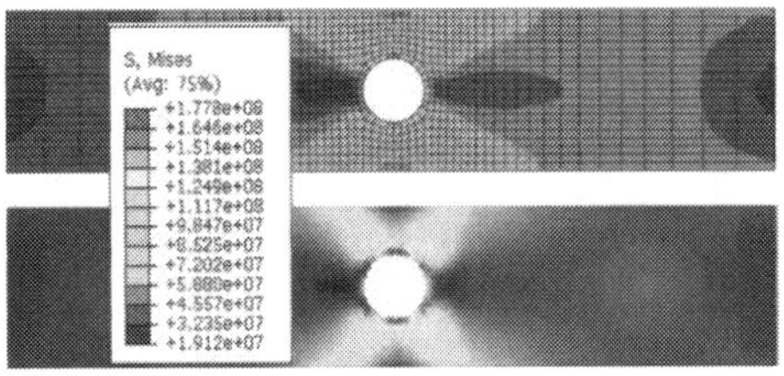

Figure 5. Equivalent stress isovalues by FEM and Meshless

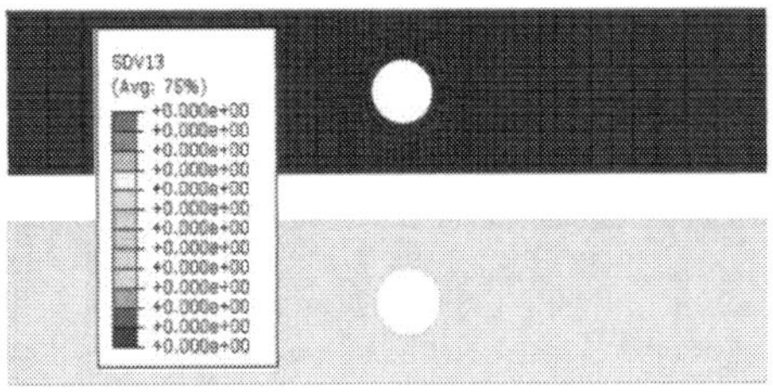

Figure 6. Cumulative plasticity isovalues by FEM and Meshless

As we consider the ductile damage, it follows the cumulative plasticity. Thus, it is equal to zero in the whole plate in this first case.

Case 2: Imposed displacement U_f=0.45mm.

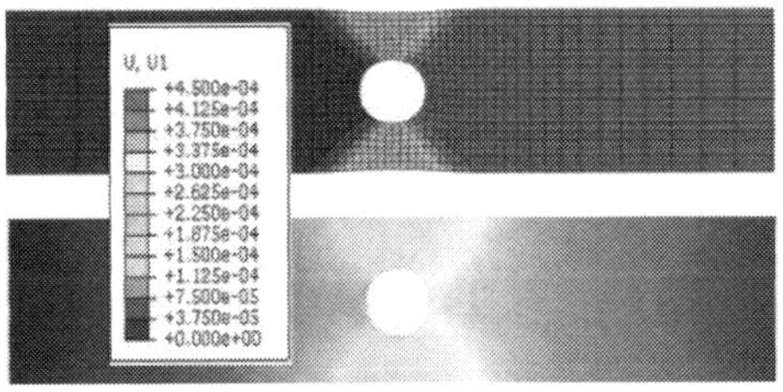

Figure 7. Displacement U_x isovalues by FEM and Meshless

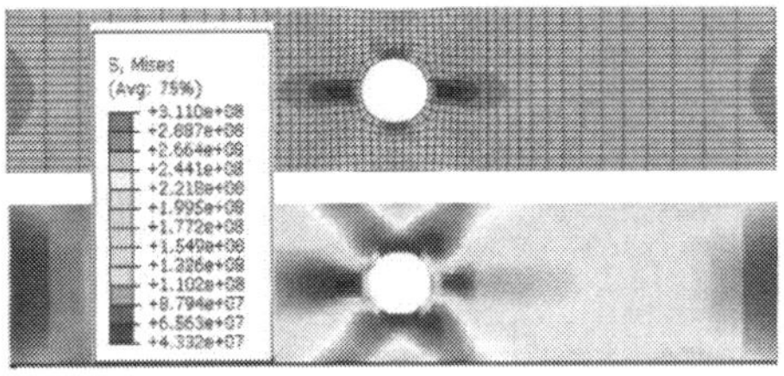

Figure 8. Equivalent stress isovalues by FEM and Meshless

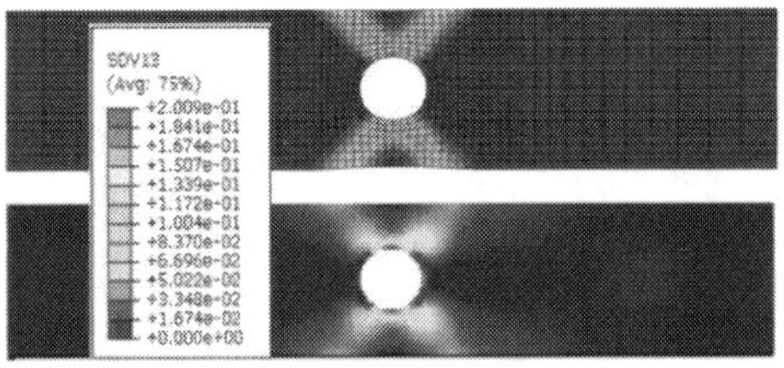

Figure 9. Cumulative plasticity isovalues by FEM and Meshless

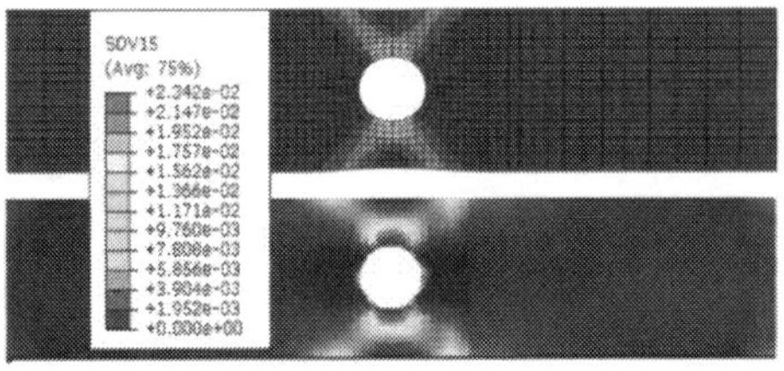

Figure 10. Damage isovalues by FEM and Meshless

Case 3: Imposed displacement U_f=2.5mm.

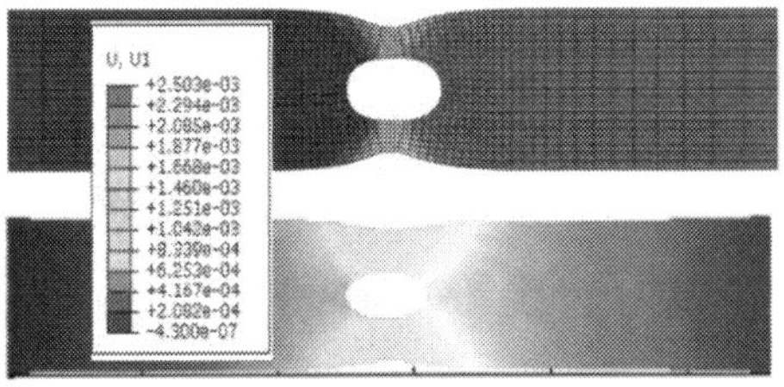

Figure 11. Displacement U_x isovalues by FEM and Meshless

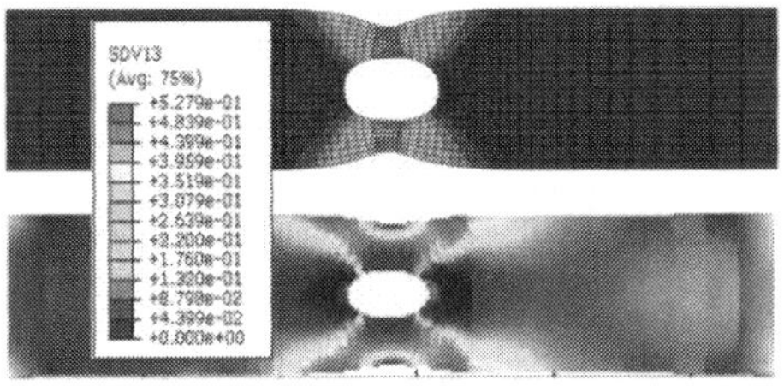

Figure 12. Cumulative plasticity isovalues by FEM and Meshless

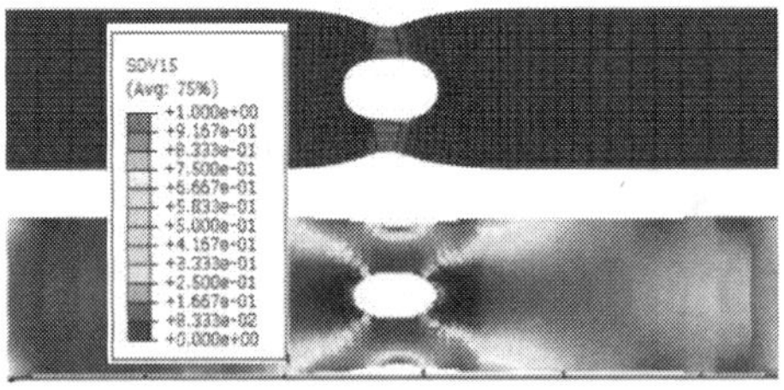

Figure 13. Damage isovalues by FEM and Meshless

In the three cases, it can be observed an agreement between results got from Meshless and those from FEM.

Certainly, the damage distribution for each method, when the critical damage is attained, can be seen on Figure 13.

In addition, the damage variable field obtained in the numerical simulation and illustrated by the contour plots, shown in Figures 10 and 13, improve clearly the ability of our developed and implemented method to predict damage growth under tensile loading conditions.

The curves depicted in Figures 14, 15, 16 and 17 describe respectively the evolution of the equivalent stress and the damage versus the cumulative plasticity. Then the damage and the cumulative plasticity according the time in different zones of the structure.

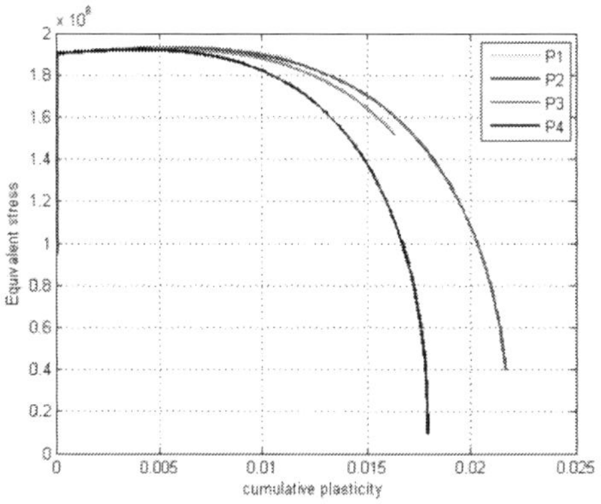

Figure 14. Equivalent stress vs. cumulative plasticity

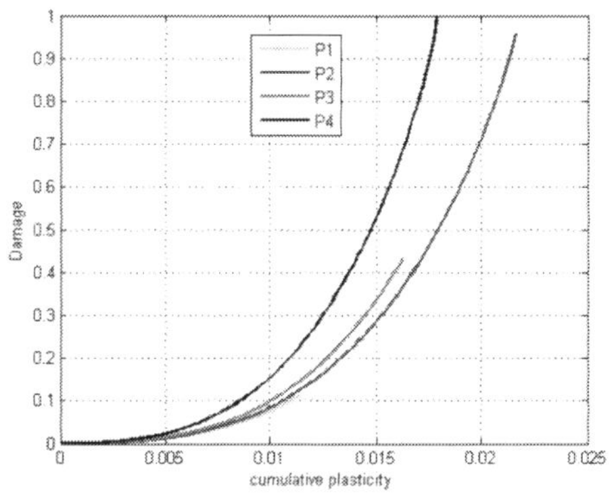

Figure 15. Damage vs. cumulative plasticity

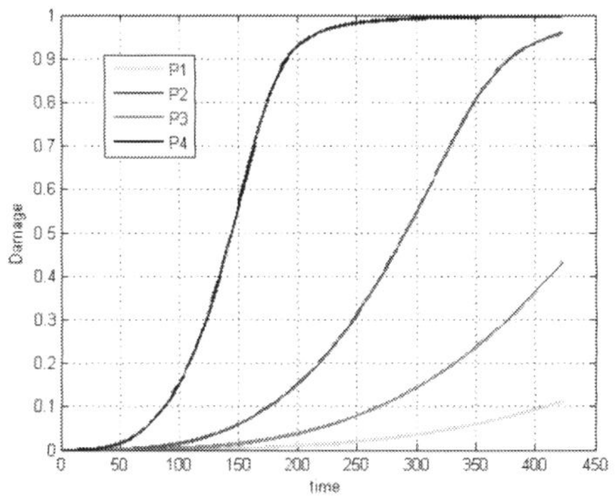

Figure 16. Damage evolution in different zones of the plate

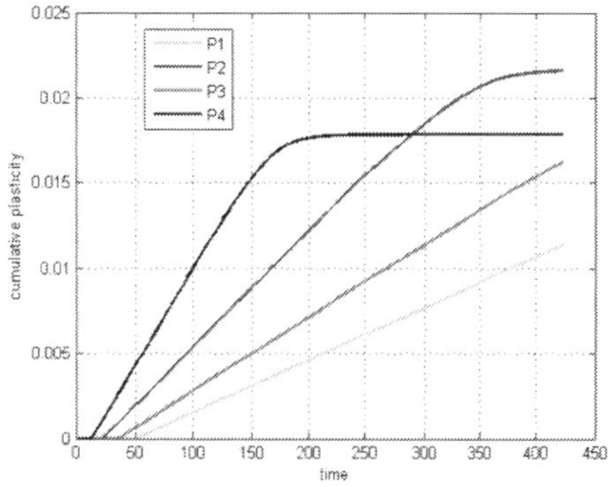

Figure 17. Cumulative plasticity evolution in different zones of the plate

As expected, the results obtained in the four points, p1, p2, p3 and p4 differently distributed in the plate (see Figure 1) demonstrates clearly the inhomogeneity of the quantities in the structure.

Certainly, the nearest point (p4) of the hole is the most quickly damaged.

In the next part, numerical results for the butterfly specimen, which is illustrated in Figure 18, are presented. The focus here is to study the behavior of different variables.

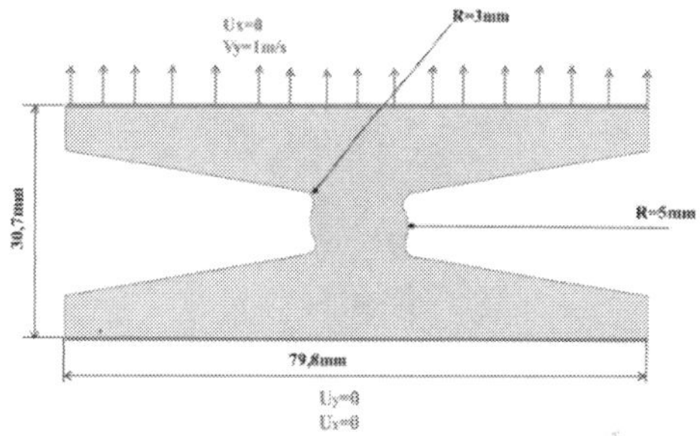

Figure 18. Geometry and boundary conditions of the butterfly specimen. The specimen was taken from [13]

With the aim to more push our study and prove that using our developed Meshless method, we can obtain the fracture location.

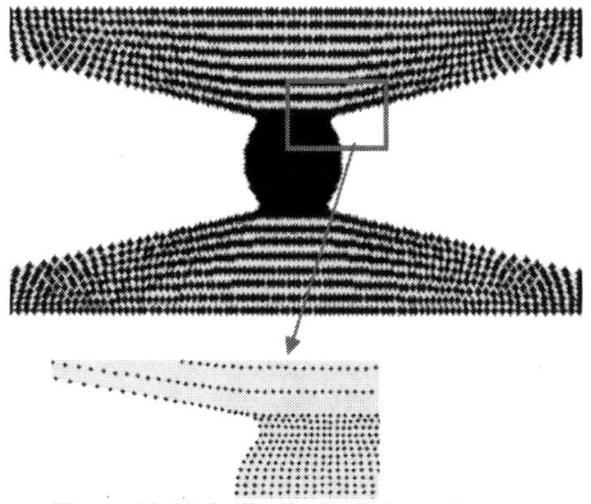

Figure 19. Node distribution in butterfly specimen

The butterfly specimen was simulated with both FEM and Meshless method. As expected, the results shown in Figures 20, 21 and 22 demonstrate a good agreement between the two methods. The imposed displacement is U=0.15mm.

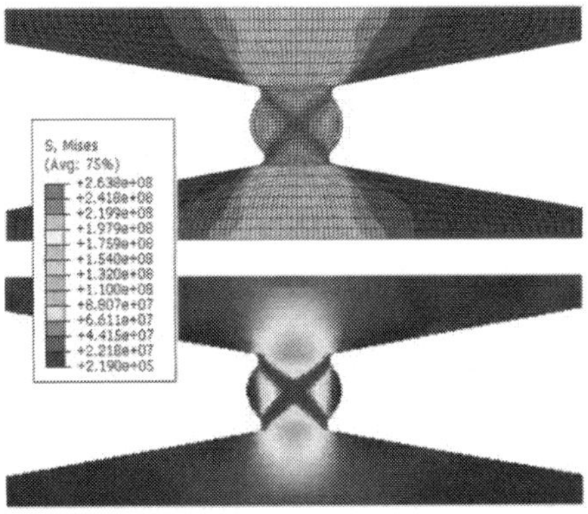

Figure 20. Equivalent stress contours by FEM and Meshless

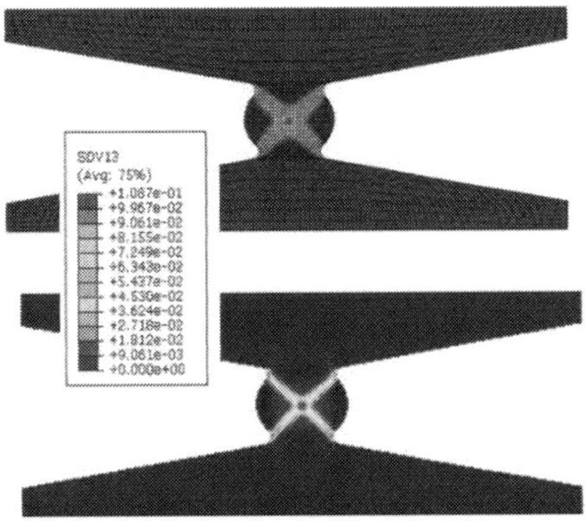

Figure 21. Cumulative plasticity contours by FEM and Meshless

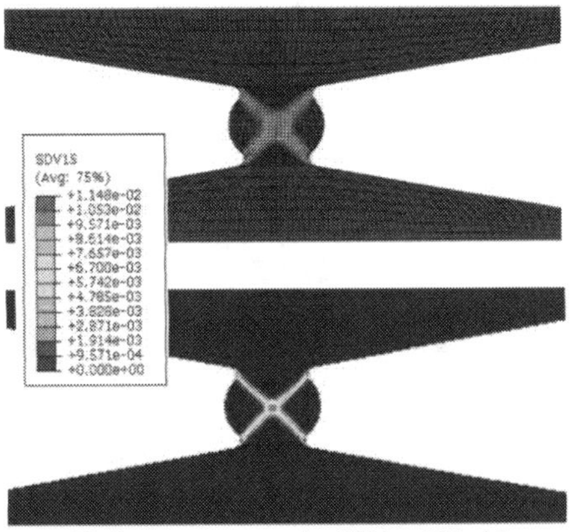

Figure 22. Damage contours by FEM and Meshless

When a crack propagates, a new frontier is forming. Consequently, the nodes of the both sides of the crack lips must be taken into account. This can explain the difference between FEM and Meshless method when damage reaches Dc (critical damage).

5 Conclusions

Based on the MLS approximation and the EFG method, the EFG method for two-dimensional elasto-plasticity problems is presented in this paper. The Galerkin weak form is employed to obtain the equations system, the penalty method is used to apply the essential boundary conditions. Some selected numerical examples are given to show that the EFG method has high computational precision and efficiency.

Further, this paper presents the application of Element Free Galerkin (EFG) method to evaluate the damage evolution. Evaluation is mainly on the efficiency of this method to capture the initiation of damage and its propagation for many steps of time.

By using the MLS approximation with consideration of the elastoplastic model coupled to damage, evolution of damage is obtained.

Validation of the method has been achieved through comparison of the simulated results with those of the Finite Element Method.

According to the comparison, both qualitative and qualitative agreement between the two methods can be found, indicating that the used Meshless method can reasonably predict the propagation of damage until the saturation.

References

1. P. Lancaster, K. Salkauskas, Math. Comput. , **37**,58-141 (1981).
2. J. Gregory, L. Kam, Int. J. Numer. Meth. Engng. **47**, 1367-1379 (2000)
3. C. Cheng-Kong, E. Plesha, Int. J. Numer. Meth. Engng, **53**, 499–514 (2002)
4. S. Fernandez-Mendez, A. Huerta, Comput. Methods Appl. Mech. Engrg. **193**, 1257–1275 (2004)
5. M. Duflot , thesis in applied science, University of Liège (2004)
6. T. Belytschko, D. Organ and Y., Comput. Mech., **17**, 186-195 (1995)
7. A. Huerta, T. Belytschko, S. Fernandez-Mendez, T. Rabczuk, Encyclopedia of Computational Mechanics. Edited by Erwin Stein, Rene de Borst and Thomas J.R. Hughes. (2004).
8. Z. Sendi, C. Labergere, H. Bel. Hadj.Salah, K. Saanouni, CMSM'2011, Sousse, Tunisia, (2011)
9. H. Badreddine., K. Saanouni, A. Dogui, Int.J. of Plasticity, 26, 1541-1575 (2010)
10. K. Saanouni. , N. Belamri., P. Autesserre, , Finite Elements in Analysis and Design, 46, 535-550 (2010)
11. Z. Sendi, C. Labergere, K. Saanouni, H. Bel. Hadj. Salah, Key Engineering Materials, 504-506, 413-418, (2012).
12. P. Lestriez, , thesis in mechanical science and material, University of technology of Troyes (2003)
13. L. Malcher, F.M. Andrade Pires, J.M.A. César de Sà, Int. J. of Plasticity, 30-31, 81-115 (2012)
14. Y. Ghozzi, C. Labergere, P. Proceedings of the 12th International Conference on Computational Plasticity - Fundamentals and Applications, COMPLAS 843–85 (2013).

Finite element modeling of deposition of ceramic material during SLM additive manufacturing

Qiang Chen[1,a], Gildas Guillemot[1], Charles-André Gandin[1] and Michel Bellet[1]

[1]*Cemef, Mines ParisTech, PSL Research University, UMR CNRS 7635, Sophia Antipolis, France*

Abstract. A three dimensional model for material deposition in Selective Laser Melting (SLM) with application to Al_2O_3-ZrO_2 eutectic ceramic is presented. As the material is transparent to laser, dopants are added to increase the heat absorption efficiency. Based on Beer-Lambert law, a volumetric heat source model taking into account the material absorption is derived. The Level Set method with multiphase homogenization is used to track the shape of deposed bead and the thermodynamic is coupled to calculate the melting-solidification path. The shrinkage during consolidation from powder to compact medium is modeled by a compressible Newtonian constitutive law. A semi-implicit formulation of surface tension is used, which permits a stable resolution to capture the gas-liquid interface. The formation of droplets is obtained and slight waves of melt pool are observed. The influence of different process parameters on temperature distribution, melt pool profiles and bead shapes is discussed.

1 Introduction

The Selective Laser Melting (SLM) process is a technique of additive manufacturing which heats and melts powder layer by layer with an energy concentrated laser beam. During each procedure, the feed piston moves upward a certain distance and the build piston moves downward a layer thickness (**Figure 1**). Then the roller deposes a new layer of powder and laser beam begins to scan the powder bed according to a trajectory predefined by a 3D model. The laser melts the powder which solidifies to form a new solid material layer constituted of multiple beads. The procedure is repeated to construct a 3D part layer by layer. The new molten and consolidated layer should be bound to the previous one to avoid mechanical defects. Besides the advantage of geometrical flexibility like any other additive manufacturing technologies, high density (up to 100%, Hagedorn *et al.* [1]) can be obtained with SLM, which means better mechanical properties. However, energy is so focused that process control becomes difficult. High thermal gradients usually exist and may result in defects such as cracking.

Ceramic material with eutectic composition of $Al_2O_3 - ZrO_2$ system (**Figure 4**) is studied. This material is potentially attractive for diverse applications (aeronautics, medical). It is light, refractory, and shows high resistance to creep. Regarding SLM processing, unlike metallic materials, which absorb most part of non-reflected laser energy on powder surface, it is transparent to Nd:YAG laser with wavelength 1064 nm. Dopants such as carbon particles are added in powder in order to increase and control absorption. Therefore, the heat source is on volume rather than on surface.

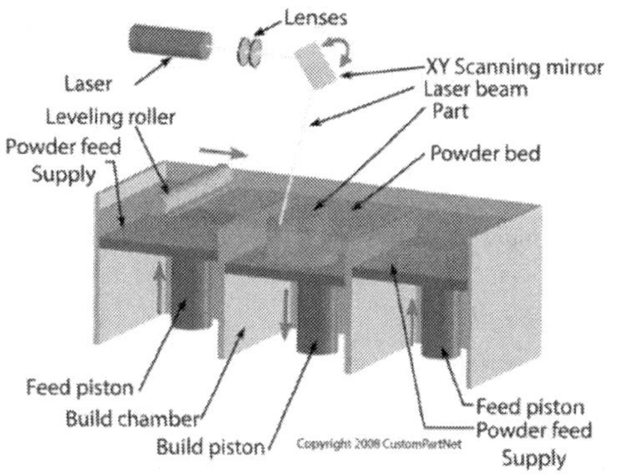

Figure 1. Schema of SLM (after CustomPartNet [2]).

Thermal phenomena in SLM are more complicated than in other processes. Unlike other traditional metallurgy processes, melting and consolidation take place locally and rapidly in SLM. A high cooling rate of 10^7 K/s was reported by Salehi and Dehghani [3] and refined phase size was obtained. Chou *et al.* [4] used pulsed laser beam rather than continuous one to increase the cooling rate. This resulted in higher undercooling and a significant microstructure deviation from equilibrium. The final part showed an increase in hardness compared with traditional cast alloy and other SLM studies.

Concerning the modeling of SLM processing, most researches are interested in heat transfer and prediction of bead shape. Hodge et al. [5] chose a complicated heat

[a] Corresponding author: qiang.chen@mines-paristech.fr

source model described by Gusarov et al. [6]. The temperature field was predicted, as well as the shape of melt pool. An interesting overhang case (melt pool over the powder rather than the constructed part or the substrate) was demonstrated, where two separate pools (in the overhung and non-overhung region) were obtained. They claimed that this phenomenon was caused by the insulating behavior of the unconsolidated powder under the overhung region. Li and Gu [7] predicted the temperature field in pure $Ti-6Al-4V$ powder bed, with a Gaussian heat source model and temperature dependent thermal properties. They found that most heat was eliminated through conduction in the cold substrate and that the material experienced a rapid quenching process. Roberts et al. [8] used the heat source model considering the absorption of laser energy described in [9] and employed the method of element birth and death to present the addition of multiple layers, also for $Ti-6Al-4V$ material. The same method is used by Marion et al. [10]. Quenching effect is also obtained in both [8] and [10].

In the present work, we aim to predict the temperature distribution, the bead formation and the dimension of melt pool. Their sensitivity to process parameter (scanning velocity) and material property (conductivity) is also studied.

2 Modeling method

In the present work the modeling of SLM processing is at the scale of the development of a single bead, for which a schematic drawing is illustrated in **Figure 2**. The scale is over the powder particle scale (several micrometers), so that continuous medium is considered for the material during the process. The modeling method should be able to track the bead shape during melting and solidification. The thermal resolution should take into account the phase transformation and use a dedicated heat source model, accounting for the interaction between laser radiation and the ceramic material in order to predict the temperature distribution.

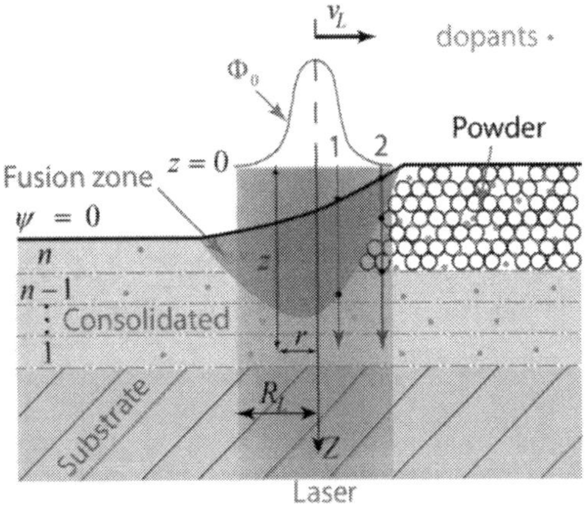

Figure 2. Schema of powder bed fusion and development of a solid bead.

2.1 Level Set method with multiphase modeling

2.1.1 Level Set method

Level Set (LS) method [11] is used to model the SLM processing. As shown in **Figure 3**, we consider the substrate (or, alternatively the previously consolidated solid material) and the powder bed as a material domain (D_0). It is surrounded by the gas domain (D_1). These two domains are separated by an interface $\psi = 0$, where $\psi = \psi(\mathbf{x}, t)$ is the signed distance (to the interface) function of position $\mathbf{x}$ and time t. Centered on this interface, we define a transition zone with half thickness ϵ, in which property (density, conductivity, etc.) of each domain is averaged by a Heaviside function $\mathcal{H}$ to get global property χ [12]:

$$\chi = \mathcal{H}\langle\chi\rangle^{D_1} + (1-\mathcal{H})\langle\chi\rangle^{D_0} \tag{1}$$

where $\langle\chi\rangle^{D_i}$ $(i = 0,1)$ is the multiphase average property in domain D_i and $\mathcal{H}$ is defined by:

$$\mathcal{H} = \begin{cases} 0 & if\ \psi < -\epsilon \\ \frac{1}{2}\left[1+\frac{\psi}{\epsilon}+\frac{1}{\pi}\sin\left(\frac{\pi\psi}{\epsilon}\right)\right] & if\ |\psi| \le \epsilon \\ 1 & if\ \psi > \epsilon \end{cases} \tag{2}$$

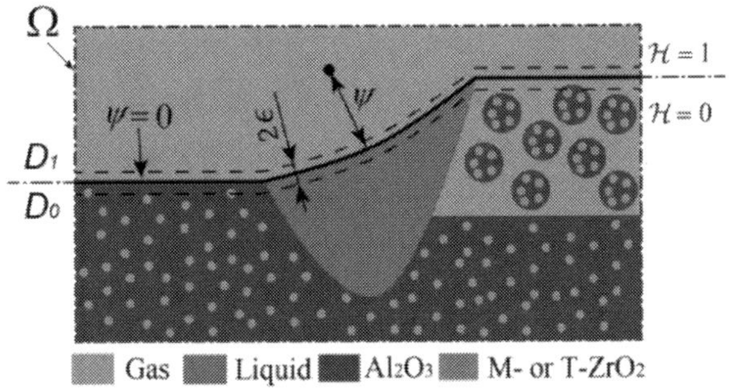

Figure 3. LS modeling of multiphase system.

The choice of ϵ is delicate because it should be small enough to capture precisely the interface and also big enough to have a smooth transition of property regarding the numerical resolution.

The distance function ψ should be adapted at each time step in order to track the interface. This step is just after the resolution of Navier-Stokes (NS) equation. It corresponds to the solution of the following equation:

$$\frac{d\psi}{dt} = \frac{\partial\psi}{\partial t} + \mathbf{u}^{LS} \cdot \nabla\psi = 0 \tag{3}$$

where $\mathbf{u}^{LS}$ denotes the velocity of the interface. However, this process does not guarantee the equality between the geometrical distance d from a point (except points at interface) to the interface and the value of $|\psi|$. Therefore, ψ is recalculated by a geometrical method [13] with respect to the new position of the interface resulting from Eq.(3), considering that the resolution of Eq.(3) is done only for the transportation of LS interface ($\psi = 0$).

2.1.2 Multiphase modeling

The material domain (D_0) is a multiphase system as several phases may coexist according to the phase diagram in **Figure 4**. **Figure 5** shows the melting and solidification steps in material domain. At initial state $T < T_f = 1860$ °C, there is only powder with 3 phases in D_0 (step 1). Once temperature attains the melting temperature T_f, powder begins to be melted and liquid fulfills gaps between particles. At this stage, T keeps constant as the system is in eutectic composition and latent heat is released (step 2). Once powder is totally melted (step 3) and begins to solidify, solid material with eutectic composition is obtained (step 4). During SLM, those four steps are chained extremely rapidly, with possible partial melting and solidification of powder.

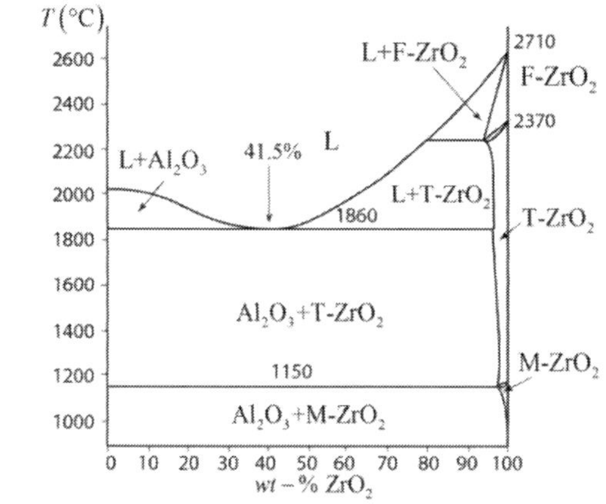

Figure 4. Phase diagram of $Al_2O_3 - ZrO_2$ system (after [14]). L, Liquid; F − ZrO_2, Fluorite-like (cubic) ZrO_2; T − ZrO_2, Tetragonal ZrO_2; M − ZrO_2, Monoclinic ZrO_2.

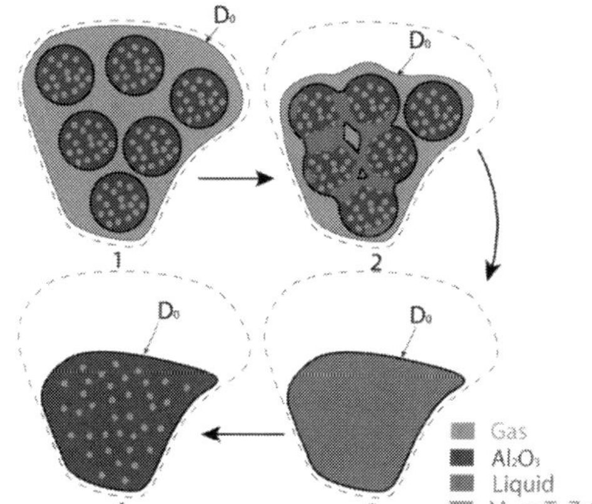

Figure 5. Scheme of phase transformation during melting and solidification of powder material.

Following the same procedure as in [15][16], we define zones, structures and phases in material domain as shown in **Table 1** and **Table 2**.

It should be pointed out that the transformation from Z_0 to Z_1 is non-reversible as the powder can never be recovered once melted. On the other hand, in Gas domain (D1), there is only one zone, one structure and one phase (g). They are not listed in **Table 1**.

Table 1. Definition of domains, zones, structures and phases

Domain	Material (D_0)							
Zone	Powder (Z_0)				Compact medium (Z_1)			
Structure	Solid (S_0)			Gas (S_1)	Solid (S_2)			Liquid (S_3)
Phase	m	t	a	g	m	t	a	l

Table 2. Abbreviation of phases

m	t	a	l	g
Monoclinic zirconia	Tetragonal zirconia	Alumina	Liquid	Gas

As we consider the material as a continuous medium, we use Representative Volume Element (RVE) [17] method to get the average property ($\langle\chi\rangle^{D_0}$) of a multiphase system which can be calculated by:

$$\langle\chi\rangle^{D_0} = \sum_i g_{D_0}^{\alpha_i} \chi^{\alpha_i} \tag{4}$$

where χ^{α_i} is the intrinsic property of phase α_i, $g_{D_0}^{\alpha_i}$ the volume fraction of phase α_i in domain D_0 and α_i could be any phase in **Table 2**. In the following, the notation g_A^B will refer to the volume fraction of B in A.

2.2 Governing equations

2.2.1 Mass conservation

Mass conservation should be respected in both the possibly multiphasic material and gas domains:

$$\begin{cases} \dfrac{\partial\langle\rho\rangle^{D_0}}{\partial t} + \nabla\cdot\langle\rho\mathbf{u}\rangle^{D_0} = 0 \\ \dfrac{\partial\langle\rho\rangle^{D_1}}{\partial t} + \nabla\cdot\langle\rho\mathbf{u}\rangle^{D_1} = 0 \end{cases} \tag{5}$$

where ρ is the local density, $\mathbf{u}$ the velocity deduced from NS equation. Multiplying the first equation by $1-\mathcal{H}$ and the second by $\mathcal{H}$ and using mass conservation along the interface [17], we get the global mass conservation:

$$\frac{\partial\rho}{\partial t} + \nabla\cdot(\rho\mathbf{u}) = 0 \tag{6}$$

In the classical velocity-pressure formulation, we express the divergence of velocity and couple it with the momentum equation. In fact, from Eq.(5), we can deduce:

$$\dot{\theta} = \nabla\cdot\mathbf{u} = \nabla\cdot(\mathcal{H}\langle\mathbf{u}\rangle^{D_1} + (1-\mathcal{H})\langle\mathbf{u}\rangle^{D_0}) = -(1-\mathcal{H})\frac{\frac{\partial\langle\rho\rangle^{D_0}}{\partial t}+\nabla\langle\rho\rangle^{D_0}\cdot\langle\mathbf{u}\rangle^{D_0}}{\langle\rho\rangle^{D_0}} - \delta(\langle\mathbf{u}\rangle^{D_1} - \langle\mathbf{u}\rangle^{D_0})\cdot\mathbf{n} \tag{7}$$

with $\delta = \frac{d\mathcal{H}}{d\psi}$ is the Dirac function and $\mathbf{n}$ is the unit vector normal to the interface. Here $\langle\rho\rangle^{D_1}$ is considered to be time and space independent. The velocity $\langle \mathbf{u} \rangle^{D_0}$ and $\langle \mathbf{u} \rangle^{D_1}$ may be assumed to be equal to $\mathbf{u}$ in the transition zone with $\psi \in [-\epsilon, \epsilon]$.

2.2.2 Energy conservation

The energy conservation in the whole system can be expressed by:

$$\frac{\partial(\rho h)}{\partial t} + \nabla \cdot (\rho h \mathbf{u}) - \nabla \cdot (\lambda \nabla T) = \dot{q} \qquad (8)$$

where T is the temperature, h the specific enthalpy (J/kg), λ the thermal conductivity and $\dot{q}$ the volume heat source. This is a non-linear equation as all properties including ρ, λ, h may depend on temperature. Therefore these parameters are tabulated in an input file of the solver [15].

As alumina and zirconia ceramics are transparent to the laser, dopants such as carbon particles are added in powder in order to augment heat absorption [18]. Consequently, the absorption varies in different areas. For example, the absorption of substrate is much lower than that of powder as the former is not doped. The absorption of consolidated material (melted from powder) depends on the atmosphere. It is low in air as dopants are combusted. The interaction between laser and particles (dopants and powder) includes conduction, convection and radiation, which is very complex to model. However, we aim to get a global heat source model taking into account the local absorption of material. A model is developed based on the Beer-Lambert law [18][19], assuming:

$$\Phi(x, y, z + \Delta z) = \Phi(x, y, z)e^{-\alpha \Delta z} \qquad (9)$$

where Φ is the flux $(\mathrm{W/m^2})$ which penetrates the material, z the penetration depth in the laser beam direction (**Figure 2**) and α the local absorption coefficient. On the other hand, we can get the volume heat source by calculating the attenuation of flux in the propagation direction:

$$\dot{q} = \frac{\Phi(x, y, z) - \Phi(x, y, z + \Delta z)}{\Delta z} = -\frac{d\Phi}{dz} \qquad (10)$$

Inserting Eq.(9) into Eq.(10) and considering the first order development of $e^{-\alpha \Delta z}$, we obtain a differential equation:

$$-\frac{d\Phi}{dz} = \alpha \Phi \qquad (11)$$

with boundary condition:

$$\Phi_0 = \Phi_{z=0} = (1 - R)\frac{2P_L}{\pi R_L^2} exp\left(-2\frac{r^2}{R_L^2}\right) \qquad (12)$$

$$(r^2 = x^2 + y^2 < R_L^2)$$

which is supposed to be a Gaussian distribution expressed in the local reference of the laser. Here R denotes the reflection coefficient at the powder bed surface, P_L the laser power and R_L the beam radius. Solving Eq.(12), the heat source model can be deduced:

$$\dot{q} = \Phi(x, y, 0)\alpha(x, y, z)e^{-\int_0^z \alpha(x,y,z)dz} \qquad (13)$$

Eq.(13) shows that this heat source model takes into account the attenuation of flux (1 and 2 in Figure 2) in the laser beam direction (uniaxial integration of α in exponent) and the local absorption of material (coefficient α).

2.2.3 Momentum conservation

In the considered SLM process, there exists 50% porosity in the powder bed at initial state. When the powder is heated and melted, the consolidation due to melting and the evacuation of gas will result in significant shrinkage and the displacement of gas-material interface represented by the LS with $\psi = 0$. Therefore, the momentum conservation should take into account compressibility:

$$\rho\left(\frac{\partial \mathbf{u}}{\partial t} + \mathbf{u} \cdot \nabla \mathbf{u}\right) - \nabla \cdot \underline{\underline{\sigma}} = \mathbf{f} \qquad (14)$$

where $\mathbf{u}$ is the velocity, $\mathbf{f}$ the volume force which includes gravity and surface tension, and $\underline{\underline{\sigma}}$ the stress tensor which respects the compressible Newtonian constitutive law:

$$\underline{\underline{\sigma}} = \underline{\underline{s}} - p\underline{\underline{I}} \qquad \underline{\underline{s}} = 2\eta\left(\underline{\underline{\dot{\epsilon}}} - \frac{1}{3}tr\left(\underline{\underline{\dot{\epsilon}}}\right)\underline{\underline{I}}\right) \qquad (15)$$

$$tr\left(\underline{\underline{\dot{\epsilon}}}\right) = \nabla \cdot \mathbf{u} = \dot{\theta} \qquad (16)$$

where p is the pressure, η the dynamic viscosity, $\underline{\underline{s}}$ the deviatoric stress tensor of $\underline{\underline{\sigma}}$, $\underline{\underline{I}}$ the identity tensor and $\underline{\underline{\dot{\epsilon}}}$ the strain rate tensor.

Another important driving force of the interface displacement results from surface tension $(\mathrm{N/m^2})$:

$$\mathbf{f}_s = \gamma \kappa \mathbf{n} \qquad (17)$$

where γ is the surface tension (N/m), $\kappa = -\nabla \cdot \mathbf{n}$ the average curvature and $\mathbf{n}$ the unit vector normal to the interface. In the present model, the tangential force resulting from the variation of γ along the interface (Marangoni effect) is not taken into account.

Numerical methods differ by the formulation of $\kappa\mathbf{n}$ [20]. The explicit method takes it at the previous time step while the implicit method takes it at the present time step, providing more stability. In fact, $\kappa\mathbf{n}$ can be related to the interface position $\mathbf{X}$ by a surface Nabla operator ∇_s:

$$\kappa \mathbf{n} = \nabla_s \cdot \underline{\underline{\nabla_s \mathbf{X}}} \qquad (18)$$

where $\underline{\underline{\nabla_s \mathbf{X}}} = \underline{\underline{\nabla \mathbf{X}}} - \left(\underline{\underline{\nabla \mathbf{X}}} \cdot \mathbf{n}\right) \otimes \mathbf{n}$. We can express $\mathbf{X}$ by the forward Euler method:

$$\mathbf{X} = \mathbf{X}^- + \mathbf{u}\Delta t \qquad (19)$$

where Δt is the time step. Replacing $\mathbf{X}$ by Eq.(19), Eq.(18) becomes:

$$\kappa\mathbf{n} = (\kappa\mathbf{n})^{-} + \Delta t \nabla_s \cdot \underline{\underline{\nabla_s \mathbf{u}}} \qquad (20)$$

It can be seen that compared with the explicit one, implicit method has an additional term related to the surface diffusion of velocity, which introduces additional stability of resolution. In the level set context, the surface force $\mathbf{f}_s$ must be multiplied by the discrete Dirac function to get a volume force contribution $\mathbf{f}$, according to the CSF method proposed by Brackbill *et al.* [21]. Following Hamide [22], such an implicit formulation was implemented by Khalloufi [23] in the numerical package CimLib used in the current study.

In the present work, we consider that surface tension appears when liquid and gas are present at the same time. This means that it is applied not only at the gas-liquid interface but also at the powder-liquid interface. The former is just one part of the interface $\psi = 0$, while the latter interface is immersed in domain D_0 so we will detect it by the contour $g^{l}_{D_0} = 0.5$ where gas is present. Note also that the evacuation of gas from the powder bed may affect the flow due to the shear stress (hydrodynamic drag) and possible additional normal load. These effects are not considered here, together with possible material vaporization.

3 One-pass simulation of SLM

3.1 Material properties

In order to calculate the average property $\langle\chi\rangle^{D_0}$ of domain D_0, we should know the intrinsic property χ^{α_i} of each phase α_i and its volume fraction $g^{\alpha_i}_{D_0}$ which may be a function of temperature and chemical composition.

3.1.1 Fraction evolution of zones, structures and phases

The melting temperature is constant (1860°C) for the eutectic composition of $Al_2O_3 - ZrO_2$ system. It results in a discrete change of enthalpy with respect to temperature, which should be avoided from the numerical point of view. In the present work, a large melting interval (1800 °C to 1900 °C) is taken as shown in **Figure 6**.

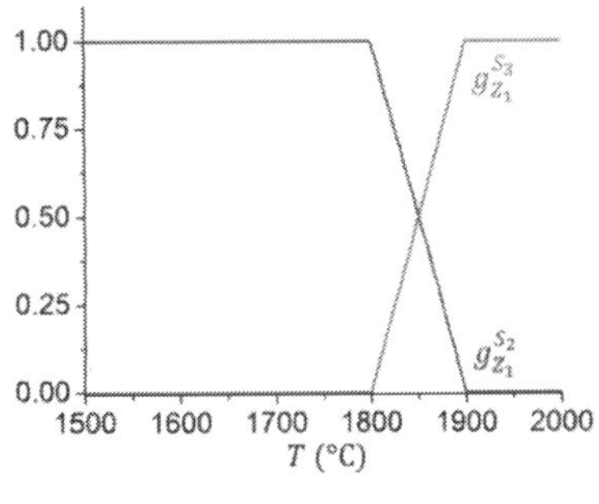

Figure 6. Fraction of structures in compact medium (Z_1).

Correspondingly, the transformation from powder (Z_0) to compact medium (Z_1) begins at the same temperature (1800 °C) as material is heated rapidly and the consolidation due to sintering is negligible. However, as powder has a low conductivity, the thermal gradient in powder around the laser is very significant. It means that a very fine mesh should be used in this area in order to well capture the shrinkage.

Another choice is to assume a greater temperature interval for Z_0 to Z_1 transformation (**Figure 7**), which is equivalent to suppose a sintering step before melting. This transformation is non-reversible, so arrows are used to indicate heating and cooling path.

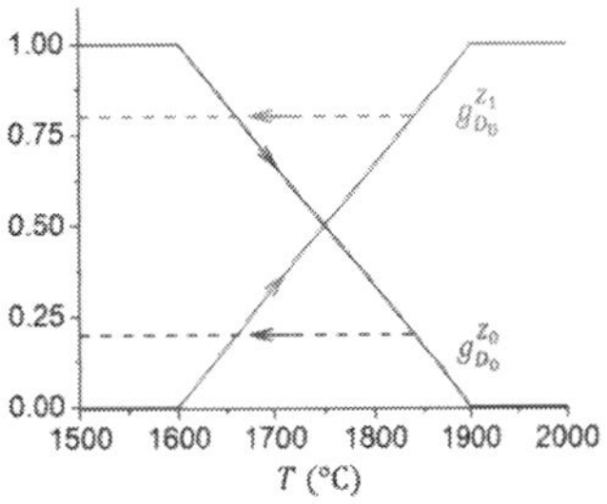

Figure 7. Fraction of zones in material (D_0). Cooling path is shown as dashed lines.

In powder, $g^{S_0}_{Z_0} = g^{S_1}_{Z_0} = 0.5$ as there is 50% porosity. The volume fraction of each phase in S_0 and S_2 can be easily deduced according to the eutectic composition and the density of each phase.

3.1.2 Properties

The specific enthalpy of Al_2O_3 and ZrO_2 phases as function of temperature is calculated by the model in [24]. Specific enthalpy out of the valuable interval is extrapolated. The specific enthalpy of liquid phase is averaged between liquid Al_2O_3 and liquid ZrO_2 by their mass fraction. The specific enthalpy evolution of material (**Figure 8**) is tabulated for the thermal resolution (Eq.(8)).

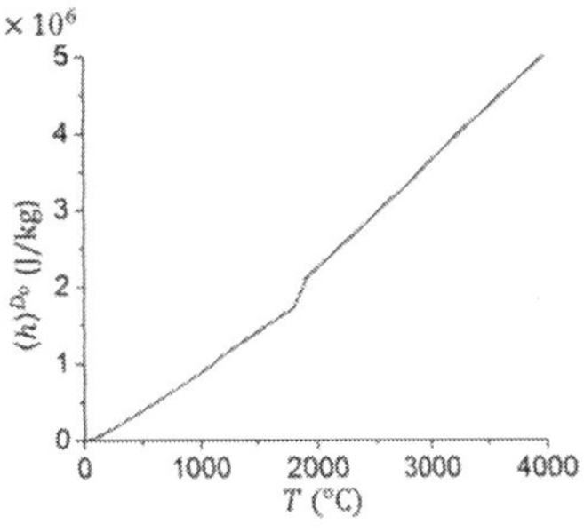

Figure 8. Specific enthalpy of material domain (D_0).

The absorption of material domain (D_0) is averaged between powder (Z_0) and compact medium (Z_1) by:

$$\langle\alpha\rangle^{D_0} = g^{Z_0}_{D_0}\langle\alpha\rangle^{Z_0} + g^{Z_1}_{D_0}\langle\alpha\rangle^{Z_1} \qquad (21)$$

where $\langle\alpha\rangle^{Z_0} = 70\ \mathrm{mm}^{-1}$ and $\langle\alpha\rangle^{Z_1} = 60\ \mathrm{mm}^{-1}$ are supposed [18]. The gas domain (D_1) is considered as transparent to laser. The coefficient of surface tension is $0.55\ \mathrm{N/m}$ for gas-liquid interface. The density of each phase is given in **Table 3**.

Table 3. Density of each phase [25]

	m	t	a	l	g
$\rho\ (\mathrm{kg/m^3})$	5680	5680	3800	4405	1.3

Thermal conductivity is a crucial parameter to the temperature distribution in SLM. The conductivity of powder is much lower than that of compact medium. As a consequence, most part of the heat is extracted by the substrate. Similar to the absorption $\langle\alpha\rangle^{D_0}$ in Eq.(21), we average the conductivity of powder (Z_0) and compact medium (Z_1) to get $\langle\lambda\rangle^{D_0}$ and $\langle\lambda\rangle^{Z_1}$:

$$\langle\lambda\rangle^{Z_1} = g_{Z_1}^{m}\lambda^m + g_{Z_1}^{t}\lambda^t + g_{Z_1}^{a}\lambda^a + g_{Z_1}^{l}\lambda^l \qquad (22)$$

The conductivity of Al_2O_3 could be found in [26] with expression $(T \in [25, 1300]\ °\mathrm{C})$:

$$\lambda^a = 5.5 + 34.5e^{-0.0033T} \qquad (23)$$

It is extrapolated out of the temperature interval. The conductivity of ZrO_2 (2 W/(m K) [27]) is considered constant. Several models exist for the calculation of powder conductivity and we have taken the Zehner-Schlunder model [28], which considers the conductivity of gas (S_1) to be dominant compared with that of solid (S_0). The conductivity evolution of powder is shown in **Figure 9**. **Figure 10** shows the conductivity of material domain (D_0), which follows the red line (same as $\langle\lambda\rangle^{Z_0}$ when there is no Z_1) when the material is heated. Once it is melted and then cooled, it follows the blue line ($\langle\lambda\rangle^{Z_1}$).

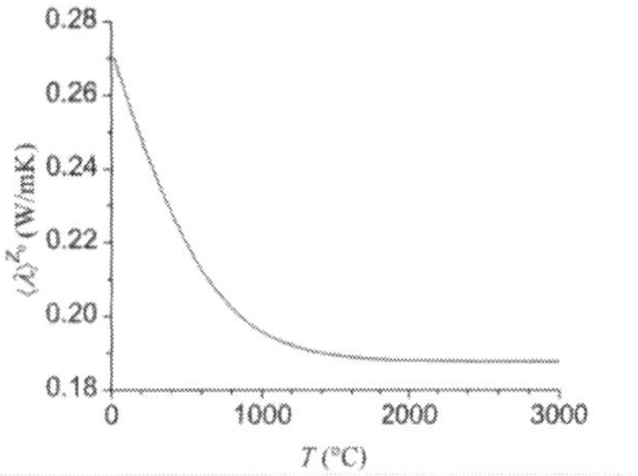

Figure 9. Conductivity of powder (Z_0).

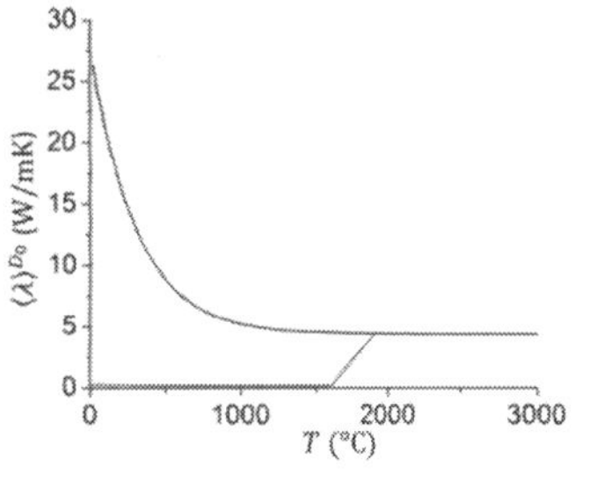

Figure 10. Conductivity of material (D_0).

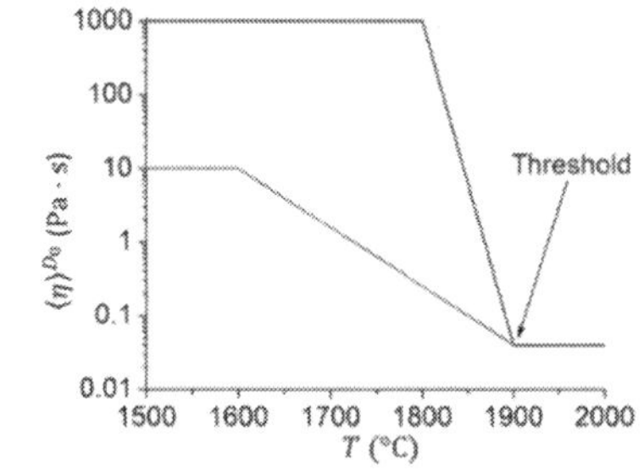

Figure 11. Dynamic viscosity of material (D_0).

The dynamic viscosity of liquid Al_2O_3 and ZrO_2 can be found in [29]. The dynamic viscosity of solid is taken as $1000\ \mathrm{Pa\cdot s}$. For compact medium (Z_1), $\langle\eta\rangle^{Z_1}$ is averaged by the geometric relation:

$$\langle\eta\rangle^{Z_1} = (\langle\eta\rangle^{S_2})^{g_{Z_1}^{S_2}} \cdot (\langle\eta\rangle^{S_3})^{g_{Z_1}^{S_3}} \qquad (24)$$

The dynamic viscosity of material (D_0, **Figure 11**) is then calculated by:

$$\langle\eta\rangle^{D_0} = \begin{cases} \langle\eta\rangle^{Z_0} & \text{if } g_{D_0}^{Z_0} > 0 \text{ (red line)} \\ \langle\eta\rangle^{Z_1} & \text{otherwise (blue line)} \end{cases} \qquad (25)$$

3.2 Simulation configuration

As illustrated in **Figure 12**, the whole system simulated is $0.5 \times 0.2 \times 0.3\ \mathrm{mm}^3$ and it is separated by the LS interface at z = 0.2 mm at initial state. Gas is above the interface and material is below. A substrate is located at the bottom ($z = 0 \sim 0.17$ mm) and a layer (30 µm) of powder is deposed above it. A laser beam with net power $P_L(1-R) = 15\ \mathrm{W}$ and radius $R_L = 35\ \mu\mathrm{m}$ heats the powder along a trajectory between locations $(0.1, 0.1, 0.2)$ and $(0.4, 0.1, 0.2)$ mm with velocity v_L.

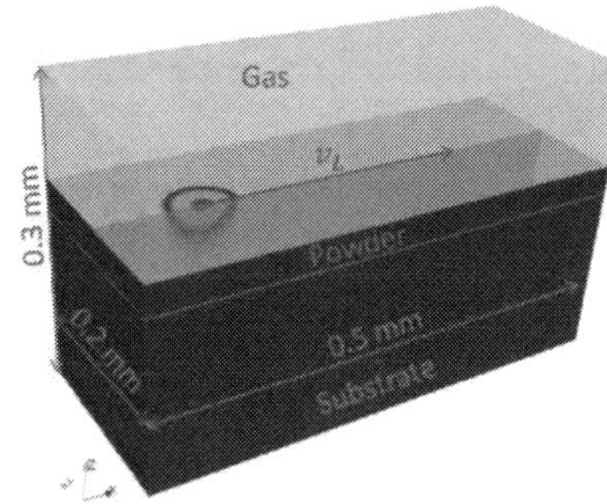

Figure 12. Dimension of the simulated system.

The initial temperature is 20 °C. Convection and radiation conditions are applied at the bottom of the substrate with heat transfer coefficient $h_T = 1000\ \mathrm{W/(m^2\ K)}$ and emissivity $\epsilon_T = 0.4$. Given the small thickness of the substrate in the system, an arbitrary augmented value for h_T is chosen, in order to simulate heat extraction from a real size substrate. Radiation of gas-material interface with $\epsilon_T = 0.6$ is also taken into account. By means of the CSF method, following Desmaison *et al.* [12], a contribution to the volumetric heat source term $\dot{q}$ in Eq.(8) can be expressed. Other surfaces are adiabatic. The bottom is fixed ($\mathbf{u} = 0$) and the top is a free surface through which gas can enter into the system. The normal velocity along the four lateral faces is supposed to be zero.

Table 4. Test parameters

	Test 1	Test 2	Test 3
v_L (m/s)	0.2	0.2	0.4
$\langle\lambda\rangle^{Z_1}$ (W/(m K))	Blue line **(Fig. 10)**	27	27

As the powder is almost adiabatic, the conductivity of compact medium ($\langle\lambda\rangle^{Z_1}$) is very important to temperature field. Other important parameters are the pair of process parameters (P_L, v_L). Therefore 3 tests are compared to investigate qualitatively the influence of v_L and $\langle\lambda\rangle^{Z_1}$. Parameters are listed in **Table 4**.

In order to simulate the effect of possible Marangoni flow on the homogenization of temperature in melt pool, $\langle\lambda\rangle^{Z_1}$ is multiplied by an arbitrary factor 5 when $T > 1900$ °C. This modification is directly inspired by the modeling of fusion zone in fusion welding [30]. The time step is 1 µs for all tests.

4 Results and discussion

At first view, the bead becomes narrower when $\langle\lambda\rangle^{Z_1}$ and v_L increases (**Figure 13**). Beside the bead, a tough edge is formed and it is still stuck to the powder bed. This is a zone partially melted and then solidified.

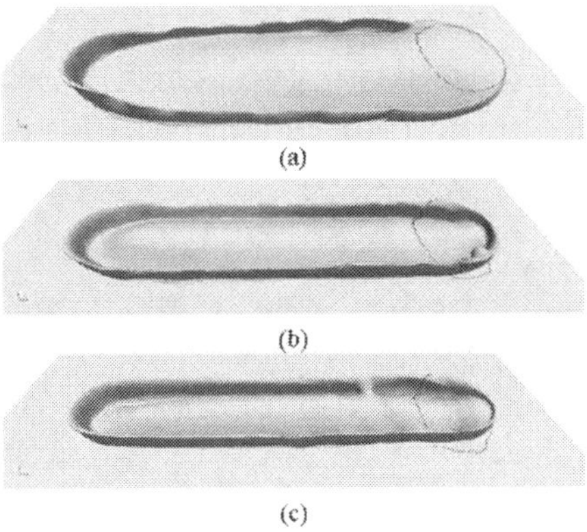

Figure 13. Calculated final bead shape for Tests 1~3 (a~c), as deposed on substrate (powder bed has been removed in this representation). Black line: zone affected by laser.

The dynamic of bead formation is different for the 3 tests. In Test 1, as the conductivity of compact medium Z_1 is very small, the melt pool (T > 1900 °C) expands laterally and is larger than the laser beam diameter. The powder in front and along lateral sides of melt pool is wetted by it. When the compact medium is more conductive, like in Test 2, the bead becomes narrower. The bead formation is then found periodical, as shown in Figure 14. At first (a), the powder is not wetted by the melt pool while it is affected by the laser beam. Latter (b) the affected powder is melted and droplets are formed and spheroidized by surface tension. Then (c), they fall down and join the melt pool. The local volume increase leads to move forward the melt pool, causing an augmented wetting of the powder. This phenomenon results in slight waves on the surface of the melt pool. At last (d), the melt pool comes back to the form in (a) due to surface tension and the waves disappear after solidification. These steps are continuously repeated while the laser moves forward. As seen from Figure 14, the hydrodynamics of the melt pool shows spatial variations with velocity varying between v_L and 2.5 v_L (when droplets joint into the melt pool).

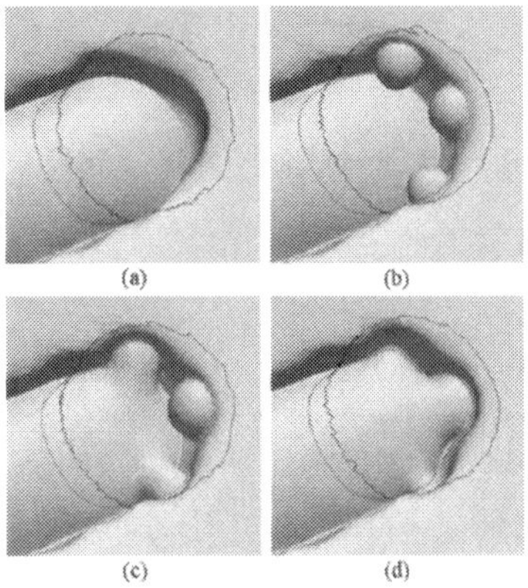

Figure 14. Periodic dynamics of bead formation in Test 2 at t = 730, 755, 800 and 810 µs (a~d). Black line (diameter $= 2R_L = 70$ µm): zone affected by laser. Red line: liquidus temperature (1900 °C).

In Test 3 (Figure 15), the bead is developed continuously and the melt pool appears to be more stable, being always in contact with the powder bed, and not showing a marked periodic regime like in Test 2. However, from time to time, it is disturbed by the formation of droplets. It may be thought that such perturbation is related to the asymmetry and the size of mesh, but this requires further investigation.

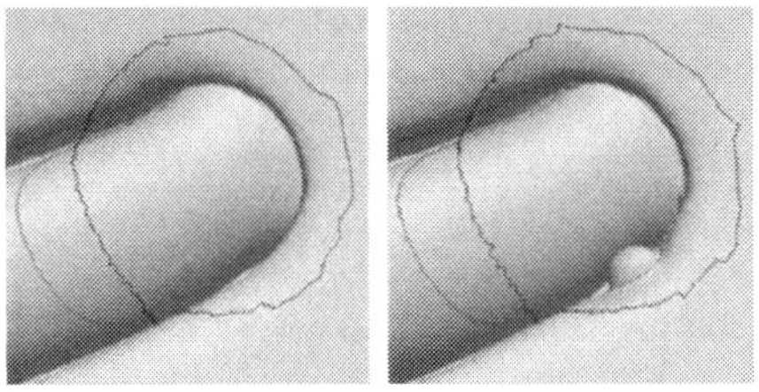

Figure 15. Dynamics of bead formation in Test 3 at t = 750 and 800 µs. Black line: zone affected by laser. Red line: liquidus.

The temperature distribution and the form of melt pool are compared in **Figure 16**. In Test 1, heat can't be extracted by the substrate and it is blocked in the melt pool due to the low conductivity of compact medium Z_1. The temperature in the melt pool exceeds 4000 °C. The melt pool is very long, large and deep. With the same scanning velocity, Test 2 shows a smaller melt pool with $T < 2200$ °C as the compact medium Z_1 is much more conductive. Comparing Test 2 and 3, we can see that the melt pool becomes longer but shallower and narrower when the scanning velocity increases. In all tests, the melt pool is lagged behind the laser beam. Thermal gradients are higher in powder than in compact medium due to different conductivities.

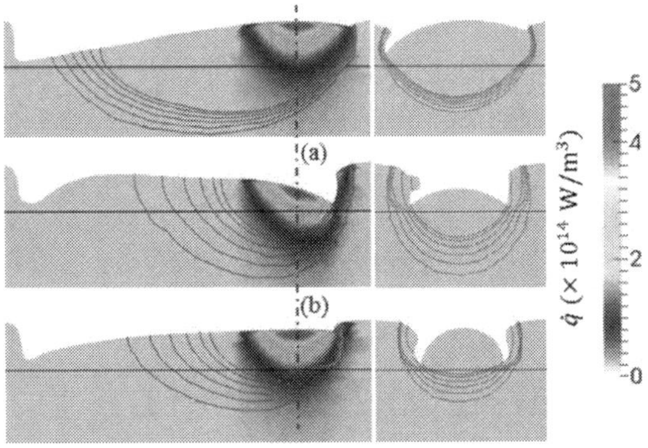

Figure 16. Temperature contours (1000, 1200, 1400, 1600, 1800, 1900 °C) and the position of heat source of Test 1~3 (a~c). The views in the longitudinal symmetry plane (left column), and transversal section after 0.15 mm beam displacement (right column) are shown. Black line: initial interface powder-substrate.

5 Conclusions

In the present work, the 3D finite element modeling of bead formation during SLM additive manufacturing has been presented. The Level Set method is used to track the gas-material interface and the multiphase homogenization takes into account the different phase transformations of the material. As dopants are added to monitor the laser absorption, the heat source is considered on volume rather than restricted to the material surface. Thereby, a heat source model based on Beer-Lambert law is deduced, which takes the absorption into consideration. The thermal resolution is coupled with thermodynamics.

The shrinkage during consolidation from powder to compact medium due to the evacuation of gas is modeled by means of a compressible Newtonian constitutive law. Semi-implicit formulation of surface tension is used, which permits a stable numerical resolution to capture the shape of melt pool.

Three tests with different conductivity and scanning velocity are compared to investigate qualitatively their influence on temperature distribution, dimension of melt pool and dynamics of bead formation. Although many parameters (thermal conductivity, absorption, reflection, etc.) need to be confirmed by experimental measurement or inverse method, our model shows an excellent sensitivity to different parameters. Future work will be focused on the calibration of parameters and their influence on multipass deposition.

Acknowledgement

This work has been conducted in the framework of the project "CEFALE", in the program "ACLAME" of Institute CARNOT MINES. The authors would like to thank Ch. Colin, J.-D. Bartout, and M.-H. Berger from Centre des Matériaux, Evry of Mines ParisTech for information regarding SLM processing and the studied ceramic material.

References

1. Y.-C. Hagedorn, J. Wilkes, W. Meiners, K. Wissenbach, R. Poprawe, Phy. Pro. 5, 587–594 (2010)
2. http://www.custompartnet.com/wu/selective-laser-sintering.
3. M. Salehi, K. Dehghani, J. All. Com. 457, 357–361 (2008)
4. R. Chou, J. Milligan, M. Paliwal, M. Brochu, JOM 67, 590–596 (2015)
5. N. E. Hodge, R. M. Ferencz, J. M. Solberg, Comp. Mech. 54, 33–51 (2014)
6. A. V. Gusarov, I. Yadroitsev, P. Bertrand, I. Smurov, J. H. Tran. 131, 72-101 (2009)
7. Y. Li, D. Gu, Add. Manu. 1–4, 99–109 (2014)
8. I. A. Roberts, C. J. Wang, R. Esterlein, M. Stanford, D. J. Mynors, Int. J. Mach. T. Manu. 49, 916–923 (2009)
9. J. P. Kruth, X. Wang, T. Laoui, L. Froyen, Assem. Auto. 23, 357–371 (2003)
10. [10]G. Marion, G. Cailletaud, C. Colin, A finite element model for the simulation of direct metal deposition (1801), ICALEO (2014).
11. S. Osher, J. A. Sethian, J. Comp. Phys. 79, 12–49 (1988).
12. O. Desmaison, M. Bellet, G. Guillemot, Comp. Mat. Sci. 91, 240–250 (2014).
13. M. Shakoor, B. Scholtes, P.-O. Bouchard, M. Bernacki, App. Math. Mod. 39, 7291–7302 (2015).
14. A. Z. Yttria, S. M. Lakiza, L. M. Lopato, J. Am. Ceram. Soc. 80, 893–902 (1997).
15. A. Saad, Ch.-A. Gandin, M. Bellet, Comp. Mat. Sci. 99, 221–231 (2015).
16. T. Carozzani, Ph.D. Thesis, Mines ParisTech, 2012.
17. M. Rappaz, M. Bellet, M. Deville, Numerical modeling in materials science and engineering. Springer, 2003.
18. G. Defillon, J. Gueusquin, N. Iooss, Modèle du bain de fusion de SLM sur céramique avec disparition de l'agent absorbant restreinte au bain, Internal report, Mines ParisTech, 2014.
19. J. F. Li, L. Li, F. H. Stott, Int. J. H. M. Tran. 47, 1159–1174 (2004).
20. S. Hysing, Int. J. Num. Meth. F. 51, 659–672 (2005).
21. J. U. Brackbill, D. B. Kothe, C. Zemach, J. Comp. Phys. 100, 335–354 (1992).
22. M. Hamide, Ph.D. Thesis, Mines ParisTech, 2008.
23. M. Khalloufi, Implementation of advanced numerical methods for multiphase flows, Internal report, Mines ParisTech, 2015.

24. M. W. Chase, Thermochemical tables, NIST-JANAF, 1998.
25. R. Morrell, Handbook of properties of Technical & Engeering Ceramics, 1987.
26. Y. S. Touloukian, R. K. Kirby, R. E. Taylor, T. T. R. Lee, Therm. prop. matt. 13, 176–177 (1984).
27. N. P. Bansal, D. Zhu, Thermal conductivity of Alumina-Toughened Zirconia Composites, NASA, 2003.
28. P. Zehner, E. U. Schlunder, Chem. Ing. Tech. 42, 933–941 (1970).
29. Y. Kawai, Y. Shiraishi, Hand. of Physico-chemical Properties at High Temperatures, Iron and Steel Institute of Japan, 1988.
30. O. Desmaison, Ph.D. Thesis, Mines ParisTech, 2013.

DOI: 10.1051/matecconf/20168008002

Application to the additive fabrication of Object Oriented Methodology

Jean-François Couturier[1] and Alexandre Schneider[1a]

[1] *Reims University FRANCE – CReSTIC - NUM3D AM platform*

Abstract. Based on some experimentation in NUM3D Platform, we integrate the possibilities of additive manufacturing processes and the characteristics of the materials. We extend to the specific property inherent in form and control to define a new tetraptych CPMF (Control/Process/Material/Form). We propose, from this tetraptych, an Oriented Object Modelization suited to Additive Manufacturing. In this paper, we set up a virtual representation specific to industrial manufacturing. The final objective, in fine, will be to use this representation in a decision-support tool.

1 Introduction

The objective of this paper is to produce an Object Oriented Methodology like structure suited to Additive Manufacturing (AM). Ultimately, we will use these attributes to implement a decision support tool. The result will be a numerical model integrating all the manufacturing constraints.

Our Object Oriented model integrate the manufacturing knowledge. In particular, the possibilities of manufacturing processes but also the characteristics of the materials and the ability to control the parts produced.

The proposed structure will permit to better take into account the specificity of additive manufacturing in numerical simulation algorithms (As meso-structures or bio-mimicry). That will also be able to validate and define an artifact part for validating a process or directly a machine.

The establishment of a detailed ontology of additive manufacturing was required to clearly explain the parameters who are, for time, implied or empirical. This ontology has two objectives, allow experienced users to better understand the field, and for novices, to get the ability to integrate this technology in their specialties (e.g. topology optimization).

2 State of art

2.1 Database

There are many specific database to materials as well as processes but few that combine the two. Marsden [10,11] highlights the role of metadata in manufacturing by integrating processes and material properties with the world of product development. Munguia [12], meanwhile, shows how to integrate a database into an artificial intelligence system dedicated to additive manufacturing.

Based on the work of Professor Ashby [13], the company GRANTA Design is the leading materials information management system. With their solution GRANTA MI *"you can apply this proven software to capture all of your vital data on Additive Manufacturing. Data is captured in one place, with full traceability"*.

The objective is to create an essential knowledge base to develop a better understanding of processes (especially additive manufacturing). This database must be expanding by integrating other essential criteria in order to serve as a tool for qualification and certification of manufactured parts.

Additive Manufacturing Template Database provides a data structure (Figure 1) developed by Granta based on extensive experience from AMAZE projects.

[a] Corresponding author: alexandre.schneider@univ-reims.fr

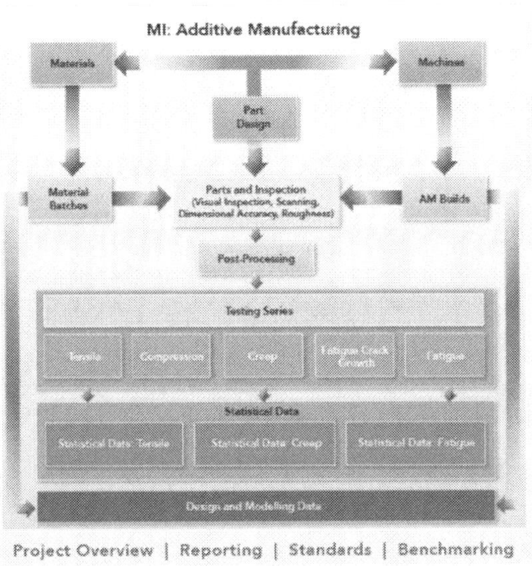

Figure 1. GRANTA AM Schema

It allows the integration of data from our own laboratory test or based on our experience, like:

- "Import 'logfiles' directly from Additive Manufacturing machines, store process parameters, extract logged data for specific builds
- Link build information directly to supplier data on the batches of material that were processed to make a part
- Capture and associate part finishing, inspection, and test data with individual parts, and feed these results into statistical analyses to determine mechanical properties and monitor reproducibility
- Export properties for use in simulation and design, capturing simulated outputs for use in optimizing part design and production."

2.2 Object Oriented Modelling

The object-orientation is a well know technique in informatics. This technique was introduce in 1966 by Ole-Johan Dahl and Kristen Nygaard [8], and was developed in the works of Alan Kay in 1968 [9]. For this work, they respectively win the Turing price in 2001 and 2003.
The main idea of object-orientation is "don't care how the others part of the system make for obtain the answer of your question, just have an answer". Object are see like some "black-box", only the public properties are in interest. Habitually, the language used for represent object is UML. The model abstracts the reality in some virtual object who interact each other. The real world is built from material or immaterial object, who are represent in the model.

The Object orientation is in fact a technique for broach a problem and cut it in smaller sub-problem.

The originality of our proposal is to use the OOM to virtualize the characteristics inherent in manufacturing production and in particular to additive manufacturing. A priori, we do not know any works developed in this sense.

2.3 Ontology in Additive Manufacturing

An Oriented Object model is a theoretical construction translating a practical case. The additive manufacturing is a new technology and, therefore, don't have the necessary objectivity for formalize this practice. We will base on the existent norm ISO CD/17296-2 that we reinforce in this paper with an AM dedicated ontology.

An ontology defines a common vocabulary for researchers who need to share information in a domain. [1] [2] [3–7]

- Explain what is considered implicit in a domain (e.g. AM)
- Distinguish field knowledge with respect to operational knowledge

3 Problem

Our goal is to link the operational properties and structural properties of classes modeling the constraints inherent in manufacturing production, particularly in the context of additive manufacturing.

All existing databases is based primarily on the characteristics of processes and materials.

However, and by feedback from multiple test on NUM3D Platform, we consider that we need to more characteristics.

Therefore, we propose two new characteristics (Control and Form) and define thus our "tetraptych CPMF" (detailed in Figure 0).

3.1 Generating a software

For a given part (forms / materials), we would like to develop a decision-making tool for the manufacturing process. As part of our work, we have a focus on additive manufacturing but the proposed methodology is applicable to other "traditional" processes.

Implementing a basic algorithm, that can adapt to the different elements of our "tetraptych CPMF" (Control / Process / Materials / Form) defining a manufactured object, requires a dynamic evolution of the program. Depending on the differences of each element of our tetraptych, the responses to the same need may require different methods. For example, for the realization of the part below (Figure 2) our need is to prevent the collapse of the cantilever. The provided method differs depending on the type of process, so the generation of a same support material is necessary in FDM, a filling with a support material in Material Jetting process but requires no action Binder Jetting SLS because the powder bed provides this function.

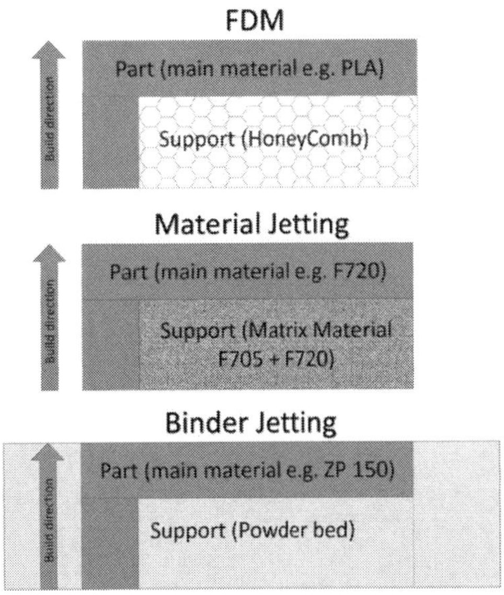

Figure 2. Part with cantilever

In principle, object programming is based on the existence of classes implementing different need for a more comprehensive algorithm. A class is defined virtually, via the methods as to answer a specific question, and this in "black box".

Due to the concept of heritage, it is possible to implement the same abstract class in several different ways, thus adapting dynamically to changing the algorithm based on internal or external parameters in the program.

It is for the reasons that the object-oriented methodology seemed appropriate to resolve our problem.

The same structure will also specify existing algorithms.

For example, topology optimization algorithms are not specifically adapted to additive manufacturing but are general. Our proposal will guide the seven standardized additive manufacturing processes and thus to best adapt the numerical models to mechanical stress in their environment. Work in support of the project OptiFabAdd allow to highlight a marked improvement forms topologically optimized by adding own specificities to AM [17,18].

Topological optimization algorithms are of course not the only ones to be adapted to additive manufacturing. Indeed, many studies concerned with meso-structures as well as mimicry [19], [20], [21], [22], [23], [24]. A development suited to additive manufacturing would specialize the algorithms and make them more efficient.

3.2 Consideration of dimensional control / geometric / topological

The vast majority of work who linked, by a general database system, the materials to processes do not take into account the ability of the process to realize dimensionally the parts (accuracy). Therefore, we propose the integration of a criteria "Control" able to take in to account the needed of accuracy and check the faculty of the device. There is no need to design a better part that the device can realize or that we can control!

The Faculty of additive manufacturing device make freeform forced us to integrate 3D scanning means [25,26] in addition to the "traditional" control means such as CMM [27].

The resulting program will allow users a decision support at various optimality parameters depending on our tetraptych. Thus a requirement of a given object, we can propose the better process (among the seven standardized additive manufacturing process) depending on the needs.

For example, a requirement of precision and a need for rapid manufacturing, the software will return us the fastest process capable of satisfying the required accuracy and associate the appropriate means of control. The result of software will give not only the method but also the associate devices.

3.3 Integration of knowledge in processes and specific materials for additive manufacturing

To validate your model, we based on the capitalization of knowledge realized on the NUM3D platform (CReSTIC laboratory from University of Reims Champagne Ardenne). We were able to realize different test allows us to determine the evolution in the mechanical characteristics in function of three criteria (materials, processes and orientation) on machines at our disposal [25]. The establishment DOE has given us the critical values on the limits of our machines. Thus, the minimum wall thickness or the minimum diameter of printable and cleaned without deterioration channels of the parts could be determined according to the materials and processes used in manufacturing [15].

4 Proposal for structuration

We wish to enrich and complete this type of bases by adding two new characteristics:

- The shape (this notion being just begun in some bases...): The geometric analysis of a 3D model allowing us, for example, fixed materials and processes, to predict differences in resistance of the part manufactured. It also allows, according to the methods, to define a better orientation of the workpiece manufactured in the device environment.
- Means of controls: we wish to integrate earlier in the manufacturing process, the possible constraints inherent to input controls. It is also important to align the precision of the manufacturing and control tools.

These two new characteristics (Shape and Control) added to conventional constraints (Process and Material) define what we call the "tetraptych CPMF" (Control / Process / Materials / Form).

4.1 Classes using tree

According to [6], we have detailed our proposition into 6 key steps:

- Definition of the Domain and its scope;
- Reuse of existing ontologies; (e.g. CES)
- Enumerate important terms in the ontology; (e.g. Brainstorming, FreeMind)
- Definition of classes and their prioritization;
- Properties of classes and attributes;
- Creating instances;

The manufacturing processes are all types (subtractive / additive / other). In our study, we have focused on the additive manufacturing processes.

4.2 Description of structuration

Our goal is to analyze a manufactured product split into four independent aspects. These aspects are (figure 3):

- The shape of the object and all its geometric characteristics modeled in 3D: class = "Form"
- The constituent materials with their mechanical characteristics: class = "Materials"
- The manufacturing processes have realized the piece manufactured with all these specific: class = "Process"
- A method (or more) to verify the compliance of the finished product: class = "Control"

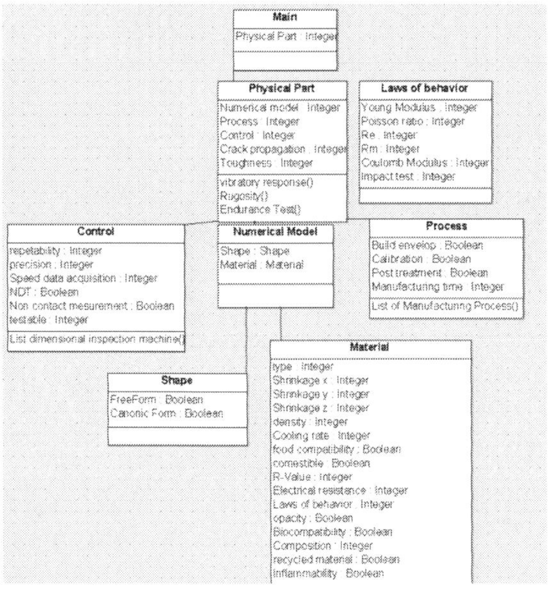

Figure 3. Object Tree

In the CAD modeling (CATIA, SW, PTC ...), we can numerically represent a product based on its shape and constituent material. Therefore, we call "numerical model" a class composed by the two classes "Shapes" and "materials" and which will represent the product DFN.

In fact, we imply that it is not usually to take into account the means of control or implementation when defining 3D modeling. Although many studies demonstrate the importance of this integration (Concept DFM) [28]

The finished manufactured product will be represented by the class "physical part" composed by classes "Process" – "Control" – "Numerical Model".

The "numerical model" based itself on classes "Shape" and "Material"; our physical part consists of classes that represented the four independent aspects of our structure.

The integration of our proposal at the heart of FEM software requires a lot of questioning of the laws of behavior. For easier using, we have chosen to group the "Laws of behavior" in a separate class who can easily be given as a parameter to specific programs.

4.3 Classes and heritage

One of the main objectives of our modeling is to allow dynamic adaptation of the calculation and optimization algorithms based on the consideration of criteria and parameters depending on the choice made on our tetraptych. The calculations of theoretical mechanical resistance, the forecast cost of a manufactured part or the production time, are intimately linked to the choices that have been made on the tetraptych (change materials, shape, manufacturing process or control).

The classes described in your model are abstract, and destined to be inherited according to the different opportunities available to each element of tetraptych. As in a conventional object modeling, the highest class in the interaction diagram are able to query their classes that are less based on attributes and predefined functions abstract classes. When a class is questioned, it behaves like a "black box" and a change in method or formula can be easily performed by inheriting inter-changing different classes of the same abstract class.

For example, we will focus on the abstract class "Process" of manufacturing. We know from the abstract class, a "process" must be able, among other things, an estimate of the volume of work available, or the time required to manufacture a part size and given shape. The manufacturing process should also be able to provide a list of its own characteristics to influence the strength of a part, or be able to validate the feasibility of a given digital model.

We consider as a first inheritance step does distinguish between different categories of manufacturing process, defining three new abstract classes inheriting the class manufacturing process:

- Subtractive manufacturing
- Additive manufacturing
- Other processes

Our study is focused on additive manufacturing (subtractive manufacturing and other process were not further).

As for additive manufacturing, we then describe concrete seven classes, based on the seven additive manufacturing processes as defined in the standard "Additive manufacturing General principles Part 2: Overview of the categories of processes and basic materials" [27].

Each of these classes must be able to answer any questions that may be asked in a manufacturing process. But each will have different function and method to answer these same questions, in fact inducing a dynamic modification effect of the global algorithm depending on the chosen manufacturing process, and therefore the instantiated class to represent it.

5 Opening

Our proposal aims to formalize the characteristics of our tetraptych bases in an abstract modeling to be completed and implemented in order to specialize in digital algorithms uses in engineering.

For example, and as an opening to our work, we search to adapt to the context of AM a topological optimization, bio-mimicry or meso-structures generation algorithms in the heart of our manufactured parts.

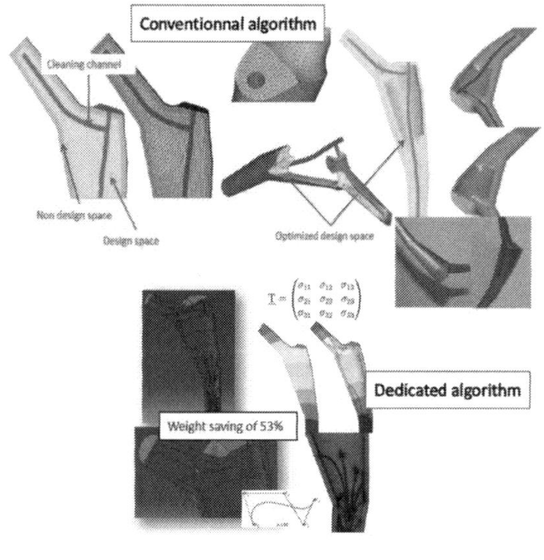

Figure 4 . Application OptiFabAdd© Project

Within the OptiFabAdd© industrial project, we were able to show that adaptations of generic algorithms (here topological optimization) by integrating the inherent parameters in the manufacturing used process (Figure 4), allow a significant gain in expectations (This dedicated optimization material allows savings of 53 per cent in terms of weight compared to a conventional topological optimization only 47 per cent).

Beyond this application, taking into account any changes in the tetraptych bases "CPMF" is likely to bring significantly improved responses in generic algorithms mentioned above.

This topological optimization application shows that taking into account inherent specificity of Additive Manufacturing generate convincing results. In OptiFabAdd©, this application was not automatized and could be easily due to our model.

6 Conclusion

This paper proposes an Object Oriented Methodology structure specifically adapted to additive manufacturing technologies. We propose a tree called "CPMF" (Control / Process / Material / Form) for, among other things, provide a dynamic response to various parameters of a tetraptych for conventional simulation algorithms (i.e. topology optimization) and dedicated to the AM.

The first results through OptiFabAdd© for topological optimization applied to the AM are promising and encourage us to expand our work for meso-structures and bio-mimicry like algorithms.

Acknowledgment

This project is partly based on the NUM3D platform (partners: UTT, ENSAM and CReSTIC for URCA). We also thank the Research Center DINCCS for OptifabAdd© Project. Advices on Topological Optimization from Nicolas Gardan CAD/CAE deserves our heartfelt thanks.

References

1. CW Holsapple, KD Joshi. Communications of the ACM, **45**:42–7 (2002).
2. D Wu, A Håkansson. Procedia Computer Science, **35**, 504–13 (2014).
3. R De Almeida Falbo, G Guizzardi, KC Duarte, Proceedings of the 14th international conference on Software engineering and knowledge engineering, ACM, **8**, p. 351 (2002).
4. R De Almeida Falbo, CS de Menezes, ARC da Rocha. Progress in Artificial Intelligence—IBERAMIA 98, Springer, **60** p. 349 (1998).
5. F Ameri, C Urbanovsky, C McArthur, Proc. 7th International Conference on FOIS 2012, Graz, Austria, Citeseer (2012).
6. NF Noy, DL McGuinness, and al., Stanford knowledge systems laboratory technical report KSL-01-05 and Stanford medical informatics technical report SMI-2001-0880 2001).
7. S Staab, R Studer, H-P Schnurr, Y Sure, IEEE Intelligent Systems **16** 26–34 (2001).

8. O-J Dahl, K Nygaard Communications of the ACM 9 , **8** 671 (1966).
9. A Kay Unpublished Master's Thesis, University of Utah, Salt Lake City, UT, United States (1968).
10. W Marsden, S Warde, E Cope, D Debon 56th AIAA/ASCE/AHS/ASC Structures, Structural Dynamics, and Materials Conference, American Institute of Aeronautics and Astronautics;
11. W Marsden, S Warde Granta Technical Paper (2010).
12. J Munguía, J Lloveras, S Llorens, T Laoui International Journal of Production Research **48** 2261–78 (2010).
13. GRANTA M. Granta Design Ltd. Cambridge, UK, www Grantadesign/products/mi
14. N Gardan International Journal of Manufacturing Engineering (2014).
15. N Gardan, A Schneider Journal of Manufacturing Systems.
16. CB Williams, JK Cochran, DW Rosen Int J Adv Manuf Technol **53** 231–9 (2010).
17. M Vesenjak Polymer Testing **29** 991–4 (2010).
18. W Miller, CW Smith, F Scarpa, KE Evans Composites Science and Technology **70** 1049–56 (2010).
19. Y Sugimura Mechanics of Materials **36** 715–21 (2004).
20. D Prall, RS Lakes. International Journal of Mechanical Sciences **39** 305–14 (1997).
21. H Abramovitch, M Burgard, L Edery-Azulay, KE Evans, M Hoffmeister, W Miller, et al. Composites Science and Technology **70** 1072–9 (2010).
22. BR Barbero, ES Ureta Computer-Aided Design **43** 188–206 (2011).
23. A Schneider, JM Coulon, F Boyer, CB Machado, N Gardan, N Dey, et al. IEEE Conference: Computing for Sustainable Global Development (2015).
24. J Kroll, S Botta, J Breuninger, A Verl **vol. 8650**, p. 11 (2013).
25. A Schneider, N Gardan, J Gardan 21ème CFM (2013).
26. K Swift, N Brown, Journal of Engineering Manufactured **vol.217** 827–33 (2003).
27. UNM920. Fabrication additive Principes généraux Partie 2 : vue d'ensemble des catégories de procédés et des matériaux de base (2015).

MATEC Web of Conferences **80**, 09001 (2016)
NUMIFORM 2016
DOI: 10.1051/matecconf/20168009001

Cross dies forging: A new method to reduce forging force & price up to 80% thanks to FEM method

Hamid Mansouri[1,a], Ahmad Nasri Mohajerani [2]

[1] *Iran University Science and Technology (www.iust.ac.ir)*
[2] *FIDEC company, Director Manager, No. 13, Erfan Alley, Mirzaye Shirazi Ave, Motahari St, Tehran, Iran*

Abstract. Purpose of this article was to introduce a new method of forging which is called "Cross Die Forging". In this method, the required force (load) is reduced to the greatest possible degree through elimination of flash channel; however, this would also decrease the positive effect of flash channel, namely filling the gaps and pores within the mold. Cross die forging procedure provides a way for providing a better preform design which ensures that the mold is filled without allowing the material to enter the flash channel. This method has been invented based on the need to decrease the production costs and to use lower tonnage pressing devices for production of heavy parts. This method is an economical method only for parts that: A) Has at least one plane of symmetry and the two ends that are perpendicular to the symmetry plane are flat; B) Has a weight that makes it impossible to be manufactured by rolling or roll forging processes. Examples of such parts are valve's body, T-junctions, etc.

1 Introduction

Forging is a manufacturing process in which the part undergoes plastic deformation under compressive load and temperature. This method not only deforms the part, but also improves the mechanical properties due to smaller grain size in this case.[4]

Generally, forging process is classified into closed and open die forging. Tools and shape of the die used in open forging process do not match the final shape of the product; while, in closed die forging, shape of the die is almost the same as the final shape of the part; and to give the part its final shape, various types of preform molds are used in a step by step approach.[4]

One of the conventional methods used to reduce cost of production is parting-off method in which the preform is turned into a profile whose shape is almost the same as final shape of the part using rolling process. Then, after the cutting process, the resulting part will be used as the preform for the closed die. But, producing the preform for heavy parts by parting-off method requires large force as well as rollers with large diameter which makes it practically impossible to use this method for production of the preform. [3]

In cross method, production of preform is performed by a shaped die during a step by step procedure with controlled loading which makes it possible to produce heavy parts. [5], [6]On the other hand, in the closed die forging method, a large force is needed to form (shape) heavy parts which requires application of high tonnage pressing device.[7]

Due to elimination of flash channel and providing appropriate preform design, as mentioned above, in cross die method, which is a combination of closed and open die approaches, a pressing device with lower tonnage is needed compared to closed die method, for the same weight. However, the part forged by cross method would be in semi-raw state with higher machining costs than closed die method; but, for small production volume, application of this method would be economical. [2]

2 Theoretical Frame Work

Current study aims to propose a new method called cross forging method. In this method, the parting line in molds is considered to be the plane of symmetry for the part and molds are assumed to be open-ended, in order to decrease pressure inside the mold. This greatly prevents the materials from entering the flash channel and causes the materials to flow in paths perpendicular to each other. However, this condition mitigates the positive effect of flash channels, i.e. filling the gaps and spaces in the molds. But, this method also greatly reduces the required deformation load. 0

In following, percent decrease in the required load is calculated using the available relations: [3]

$$F_{f_l} = 6.3 \times A^2{}_{f_l} \times \sqrt{\frac{1}{L_{f_l} \times V_{f_l}}} \tag{1}$$

$$F_{f_o} = 6.3 \times A^2{}_{f_o} \times (1+r) \times \sqrt{\frac{1}{L_{f_o} \times V_{f_o}}} \tag{2}$$

$$F_T = F_{f_l} + F_{f_o} \tag{3}$$

[a] Corresponding author: H.Mansouri@mail.com

$$S=\frac{(F_T-F_{f_l})}{F_T}\times 100=\frac{(F_{f_l}+F_{fo}-F_{f_l})}{(F_{fl}+F_{fo})}\times 100=\frac{F_{f_o}}{F_{f_l}+F_{fo}}\times 100$$

$$=\frac{1}{1+\frac{F_{fl}}{F_{fo}}}\times 100=\frac{1}{1+\underbrace{\left(\frac{A_{fl}}{A_{fo}}\right)^2\times\left(\frac{1}{1+r}\right)\times\sqrt{\frac{L_{fo}\times V_{fo}}{L_{fl}\times V_{fl}}}}_{K}}\times 100=\frac{1}{1+k}\times 100$$

$$S=\frac{1}{1+k}\times 100$$

$$r=\frac{w}{t}=3+1.2\times e^{-1.09W} \tag{4}$$

Neuberger &Mockel

$$t=0.89\times\sqrt{W}-0.017\times W+1.13 \tag{5}$$

Now, value of S is calculated for cylinders with diameter and length equal to D and L, respectively, positioned horizontally inside the closed and cross dies:

$$V_{fo}=\frac{\pi\times D^2}{4}\times L \tag{6}$$

$$V_{fl}=[(L+2\times w)\times w\times t]\times 2+2\times D\times w\times t \tag{7}$$

$$L_{fo}=L \qquad L_{fl}=L+2\times w \tag{8}$$

$$A_{fo}=D\times L \tag{9}$$

$$A_{fl}=[(L+2\times w)\times w]\times 2+2\times D\times w \tag{10}$$

$$V_{fo}=\frac{\pi\times D^2}{4}\times L \tag{11}$$

$$V_{fl}=[(L+2\times w)\times w\times t]\times 2+2\times D\times w\times t \tag{12}$$

$$L_{fo}=L \qquad L_{fl}=L+2\times w \tag{13}$$

$$A_{fo}=D\times L \tag{14}$$

$$A_{fl}=[(L+2\times w)\times w]\times 2+2\times D\times w \tag{15}$$

Table 1. Values of S, F0 and FT for Cylinders with Different Values of D and L (L=1.8×D)

Row	*D(mm)*	*L(mm)*	*W(kg)*	*t(mm)*	*w(mm)*	*k*	*S*	*F_T(N)*	*F_o(N)*	*F_l(N)*
1	*50*	*90*	*1.4*	*2.2*	*7.0*	*0.311*	*76.3*	*317508*	*136462*	*181046*
2	*70*	*126*	*3.8*	*2.8*	*8.5*	*0.270*	*78.7*	*525527*	*252052*	*273475*
3	*90*	*162*	*8.1*	*3.5*	*10.6*	*0.262*	*79.2*	*850049*	*414708*	*435340*
4	*110*	*198*	*14.8*	*4.3*	*12.9*	*0.262*	*79.2*	*1268492*	*619475*	*649017*
5	*130*	*234*	*24.4*	*5.1*	*15.3*	*0.264*	*79.1*	*1777163*	*865217*	*911946*
6	*150*	*270*	*37.4*	*5.9*	*17.8*	*0.266*	*79.0*	*2375592*	*1151916*	*1223676*
7	*170*	*306*	*54.5*	*6.8*	*20.3*	*0.267*	*78.9*	*3061889*	*1479572*	*1582317*
8	*190*	*342*	*76.1*	*7.6*	*22.8*	*0.268*	*78.8*	*3832745*	*1848185*	*1984560*
9	*210*	*378*	*102.7*	*8.4*	*25.2*	*0.269*	*78.8*	*4683498*	*2257755*	*2425743*
10	*230*	*414*	*135.0*	*9.2*	*27.5*	*0.268*	*78.9*	*5608187*	*2708282*	*2899905*

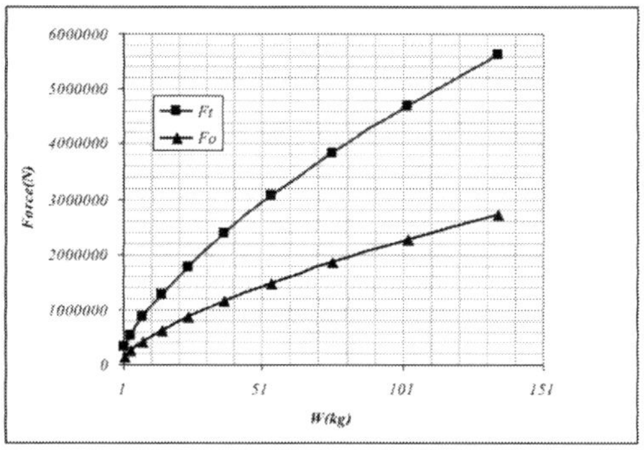

Figure 1. The required forging load for parts with different weights forged by closed dies (FT) and cross (F0)

As evident in Figure 1, when the pressure inside the mold decreases to the extent that material flows towards open ends of the mold instead of flowing through the flash channel, the required forging force (load) also decreases. The required forging load increases with increase in weight of the part. Thus, if all the gaps and pores within the mold are filled before material enters the flash channel, then there is no need to increase the pressure inside the mold through designing the flash channel. However, the main problem here is to fill the mold by the preform.

In cross forging method, preform is designed so that it completely fills the mold, and also it prevents materials from entering the flash channel. In this method, all the preforms are manufactures by profile method with two open molds using hammer or pressing device. Then, after being cooled, they are cut and used as the preform. The main goal is to produce a profile that can be used as the preform, when rotated 90 degrees around y axis.

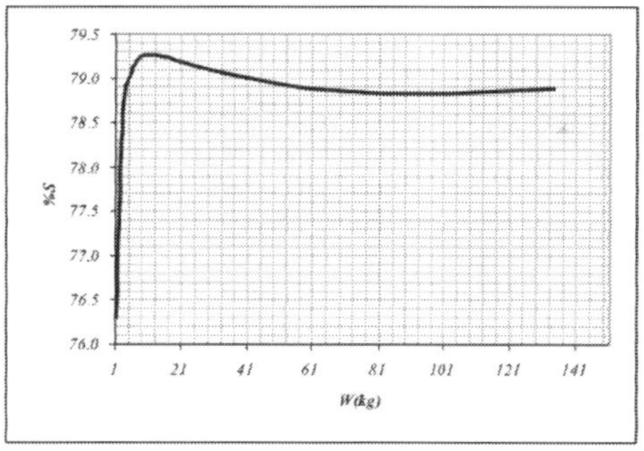

Figure 2. The percent decrease in the required forging load (%S) for parts with different weights forged by cross dies (F0)

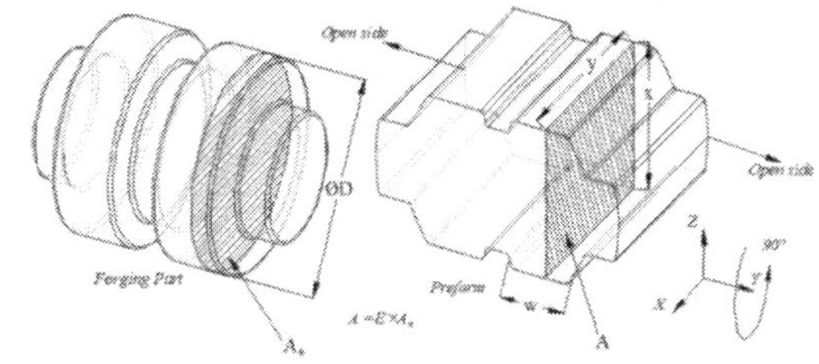

Figure 3. Each quadrilateral section in the profile must fill the circular sections in the mold without formation of any flash. in order to reduce the pressure inside the mold, both ends of the mold are assumed to be open

Now, since every each quadrilateral section in profile must fill the circular sections in the mold while producing least amount of flash, this question is raised: what kind of quadrilateral with what dimension is able to fill the molds with smallest possible force? To answer this question, shaped molds with internal diameter of 50-200 mm and anvil's width equal to 30-200 mm (Figure 4) were tested to be used as quadrilateral sections of different sizes according to Table II.

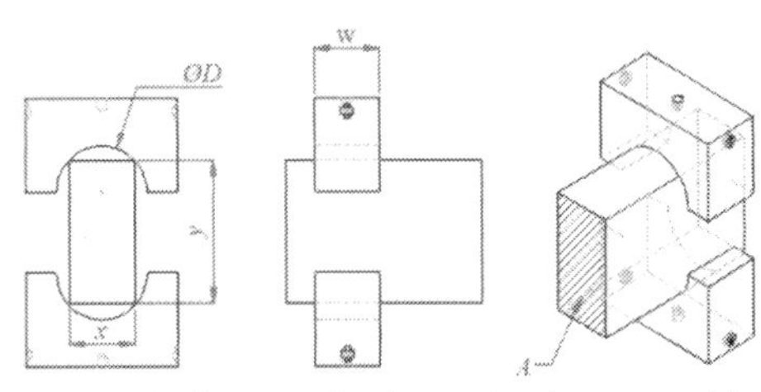

Figure 4. Schematic diagram related to testing the cross molds

Table 2. Dimensions of quadrilateral sections used to fill the shaped molds with internal diameter of 150 mm and width of 100 mm

A=E_m×A₀			*Row*	*X(mm)*	*Y(mm)*	*Accepted*	*Description*	*Force (N)*
1.4			*1*	*40*	*68.69*	*Yes*		
Y/X ≤ 2.8		*.A*	*2*	*35*	*78.5*	*Yes*		
X <D_n			*3*	*25*	*109.9*	*No*		
D_l	*50 mm*		*4*	*20*	*137.4*	*No*		
W_l=100 mm			*5*	*10*	*274.8*	*No*		
Material	*AISI 1018*		*6*	*0*	-	-		
Die Temperature	*300 c°*		*7*	-	-	-		
Work piece Temperature	*1100 c°*		*8*	-	-	-		
Press	***Hydrualic, 25mm/s***	*.B*	*9*	-	-	-		
Friction	*0.3*		*10*	-	-	-		
			11	-	-	-		

E_m: Escape confection D_n: Diameter of Round Bar (mm) A: Area of Square bar (mm^2) A_0: Area of Round Bar (mm^2)
Em=1.4. For each Em and D and different anvil width, W, quadrilaterals with different sizes were tested by considering the size limitations mentioned in left side of the table.

Table 3. Gide Table for Designing Prefome, By Knowing Values of W and D or A0, Dimensions of A Quadrilateral That Could Fully Fill the Shaped Open Die with Diameter of D (without Formation Of Flash) And Minimum Required Force Is Shown In The Below Table. (A=A0×E, K=Y/X, F=Fz(Forging Force))

Row	*W(m)*	*D(mm)*	*A_o(mm^2)*	*E*	*X*	*Y*	*K*	*F(N)*
1	30	50	1963	1.7	40	83	2.1	2.85E+05
2	30	60	2826	1.5	50	85	1.7	2.97E+05
3	30	70	3847	1.6	55	112	2	2.59E+05
4	30	80	5024	1.6	65	124	1.9	4.38E+05
5	30	90	6359	1.4	80	111	1.4	4.36E+05
6	30	100	7850	1.4	90	122	1.4	5.04E+05
7	30	150	17663	1.4	135	183	1.4	7.33E+05
8	30	200	31400	1.4	185	238	1.3	1.05E+06
9	60	50	1963	1.6	40	79	2	.30E+05
10	60	60	2826	1.6	50	90	1.8	6.40E+05
11	60	70	3847	1.6	60	103	1.7	7.46E+05
12	60	80	5024	1.5	70	108	1.5	8.02E+05
13	60	90	6359	1.5	80	119	1.5	9.04E+05
14	60	100	7850	1.5	90	131	1.5	1.01E+06
15	60	150	17663	1.5	135	196	1.5	1.50E+06
16	60	200	31400	1.4	190	231	1.2	1.99E+06
17	100	50	1963	1.5	40	74	1.8	9.12E+05
18	100	60	2826	1.5	50	85	1.7	1.06E+06
19	100	70	3847	1.6	55	112	2	1.23E+06
20	100	80	5024	1.5	65	116	1.8	1.34E+06
21	100	90	6359	1.6	75	136	1.8	1.56E+06
22	100	100	7850	1.6	85	148	1.7	1.75E+06
23	100	150	17663	1.6	135	209	1.6	2.62E+06
24	100	200	31400	1.7	180	297	1.65	2.83E+06
25	150	50	1963	1.5	40	74	1.8	1.50E+06

This table holds true for weights within range of 1.5 Kg ≤W≤100 Kg; All the measures obtained for the dimensions are related to hot dimensions (simulated with MSC. Superforge software)

1300 tests were performed on steel AIAI 1018, which is considered among the low-carbon steel, according to the following conditions:

Mold material was considered to be L6 and temperatures of die and work piece were set at 1100 and 300 degrees Celsius, respectively. Friction coefficient between die and work piece was considered to be 0.3. Also, a hydraulic pressing device has been used with ram speed of 25 mm/s.

Now, it is evident that increasing the anvil's width, considering that profile diameter is fixed, decreases E and also increases the required force. Moreover, during the tests, we found out that for each value of E, W and D, if a quadrilateral with dimensions of x1 and y1 fills a profile with diameter D, it would also fill a quadrilateral of dimensions x2 and y2, provided that y2≤y1.

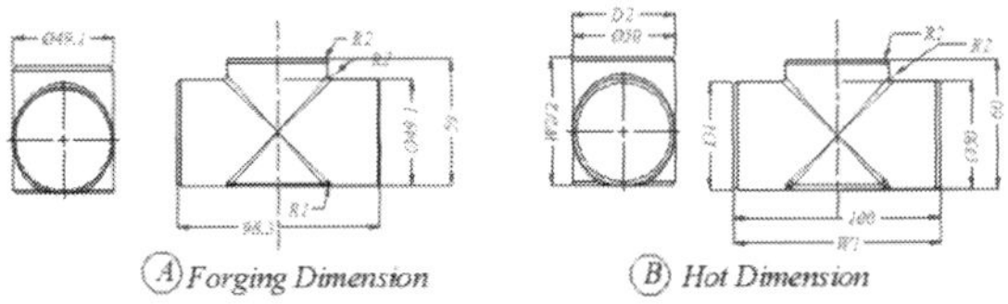

Figure 5. (A) Forging Dimensions, (B) Hot Dimensions

Assume that there is a forging part according to Figure 5 (A) with approximate weigh of 1.8 Kg and made of AISI 1018 for which we want to determine the preform to be used in cross forging method. First, hot dimensions of the part are calculated (Figure 5 (B)). Then, dimensions are calculated according to Table III, considering the fact that there must be quadrilateral with dimensions of X, Y for every circular section with cross section areas of A and width of W:

W1=100, D1=50 → X1=40, Y1=73.6→ F1=9.12×10^5 N (16)

W2/2=60→W2=120, D2=50 →W'2=100, D2=50→X'=40,Y'=73.6 F'=9.12×10^5 N (17)

W"2=150, D2=50→X"=40,Y"=73.6 →F"=15×10^5 N

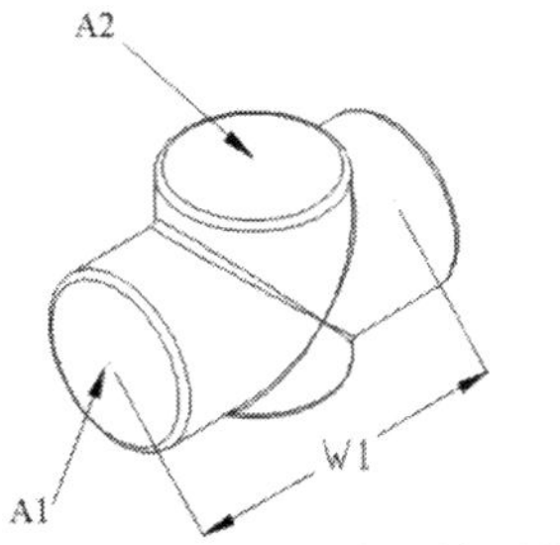

Figure 5.(C) Cross sections and width of Sample forging part

X2=40,Y2=73.6 F2= (F'+ F")/2=12.06×10^5N (18)

FT=F1+F2 =9.12×10^5 + 12.06×10^5=2.11×10^6 N (19)

$40\times73.6\times t_1 = 3.14\times50^2/4\times100+33382 \rightarrow t_1=78$ (20)

$40\times73.6\times(t_2-40) = 3.14\times50^2/4\times10 \rightarrow t_2 = 46$ (21)

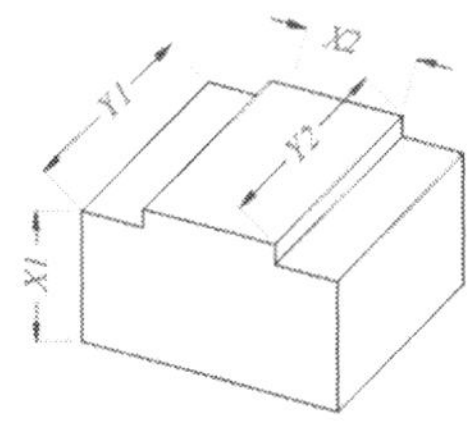

Figure 5.(D) Part In Its Raw Preform Shape

$40\times73.6\times t_1 = 3.14\times50^2/4\times100+33382 \rightarrow t_1=78$ (22)

$40\times73.6\times(t_2-40) = 3.14\times50^2/4\times10 \rightarrow t_2 = 46$ (23)

If y1≤y2, y1 is chosen as width of the profile and x1 is calculated according to E ratio; and dimensions of the profile is modified in this way. But if y2≤y1, y2 is considered as width of the profile, and x2 is calculated according to E ratio.

By comparing the force obtained from (19) and Figure 9 .it is evident that the required force for producing the part by cross forging method is approximately 70% less than other methods.

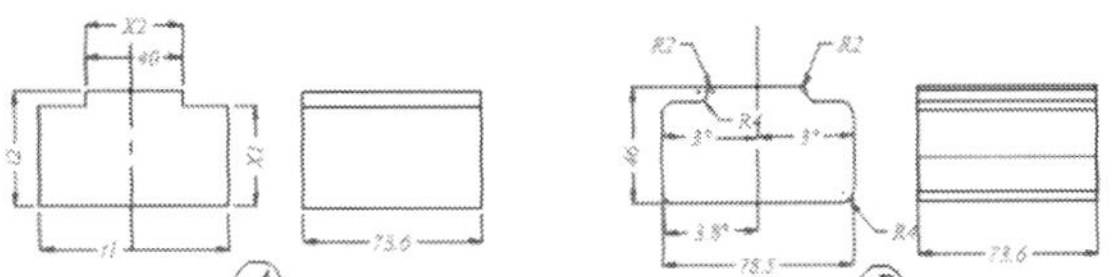

Figure 6.(a) Dimensions of raw preform and 6(b) final preform, considering slopes and R values for corners and edges

Now, according to standard DIN 7523, R related to edges and corners and values of slopes are calculated for the raw preform (Figure 6 (A)) to produce the part as shown in Figure 6 (B).

Now, since the preform must be manufactured from a section with a shaped open mold by a hammer or a pressing device, dimensions of the required quadrilateral can be calculated with respect to the cross section shown in Figure 6 (B) and anvil's width, according to Table III:

$A=A_0\times1.5 \rightarrow A_0=3475.96$ square millimeters →A=5213.94 square millimeters

X=70.5, Y=73 (24)

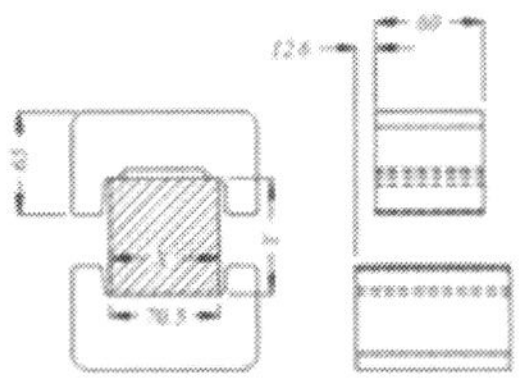

Figure 7. Method for producing the desired profile using shaped open die by a hydraulic pressing device with loading range of 20-30 mm

It must be noted that magnitude of the allowable loading at each pulse is set within 20-30 mm. Otherwise, problem of "overlapping" would occur on the produced profile which would appear as cracks or cavities on surface of the part after being cooled.

Now, the resulting preform is tested once on the cross die and once on the closed die as shown in the following Figure 10.

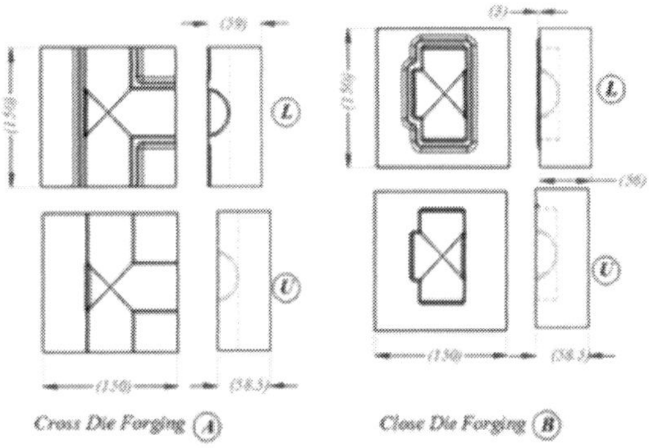

Figure 8. (A) schematics of the cross dies, right diagram,9(B) schematics of the closed die. as shown, the die is assumed to be open ended in this case. (l=lower die, u: upper die).

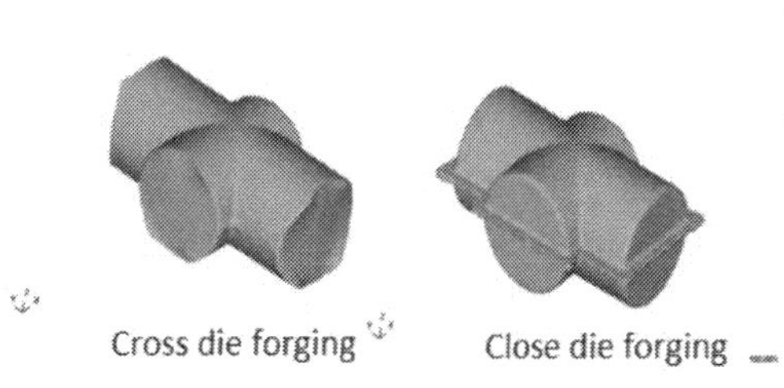

Figure 9. The part manufactured by cross die method

Following Figure 9 shows the parts manufactured by the close die and cross die. As expected, due to the decreased pressure inside the mold in the cross die method, material flows towards the open ends of the mold rather than flowing through the flash channel, thus material completely fills the pores and cavities in the mold and no flash is formed.

If we compare the force-time curves obtained for the two methods (Figure 9 (a)), we observe that the force required for manufacturing the part by cross die method is one tenth (1/10) of that required by the closed die method. The approximate force calculated by Table IV (Equations

1,2,4,5) is almost equal to the force provided by the software.

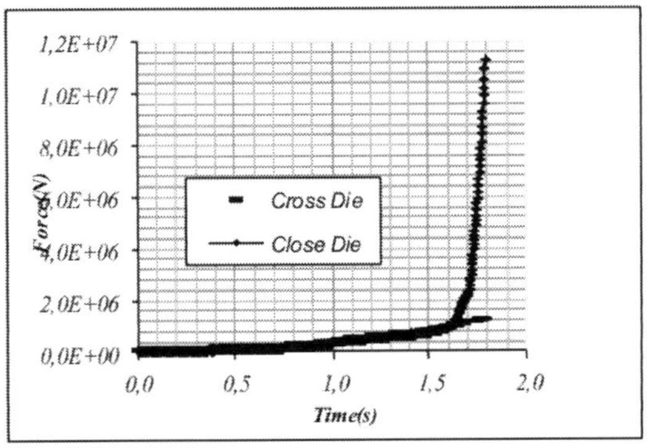

Figure 9. (a) Comparison of the force required for manufacturing the part (Figure 8 (A), 8 (B)) using closed die and cross die methods (simulated with MSC. Superforge software)

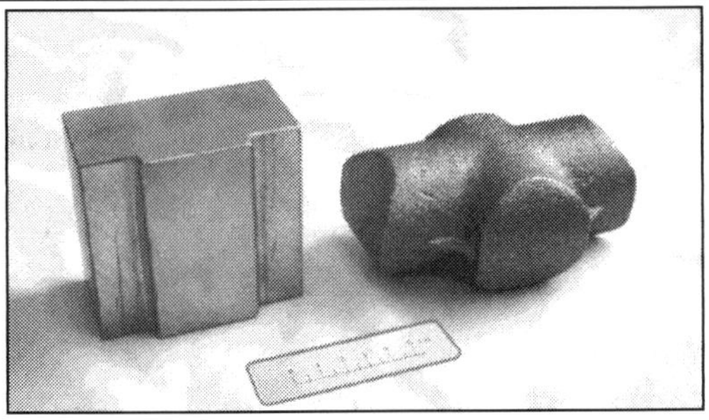

Figure 10. The produced part (valve sample) by mechanical press, 315 tons

$r=\frac{w}{t}=3+1.2\times e^{-1.09W}$ $t=0.89\times\sqrt{W}-0.017\times W+1.13$ Neuberger & Mockel

$$F_{f_l}=6.3\times A^2{}_{f_l}\times\sqrt{\frac{1}{L_{f_l}\times V_{f_l}}}\,, \tag{25}$$

$$F_{f_o}=6.3\times A^2{}_{f_o}\times(1+r)\times\sqrt{\frac{1}{L_{f_o}\times V_{f_o}}} \tag{26}$$

$$F_T=F_{f_l}+F_{f_o} \tag{27}$$

Table 4. Result Data for (Valve Sample)

W	t	w	V_{fo}	L_{fo}	V_{fl}	L_{fl}	A_{fo}	A_{fl}	k	$\%S$	$F_T(kg)$	$F_o(kg)$	$F_l(kg)$
1.8	2.3	7.3	230416.8	100	7522	115	5496.5	2711	0.302	44.8	373253.4	165297.7	207955.7

Figure 11. preform & final shape (cross die forging)

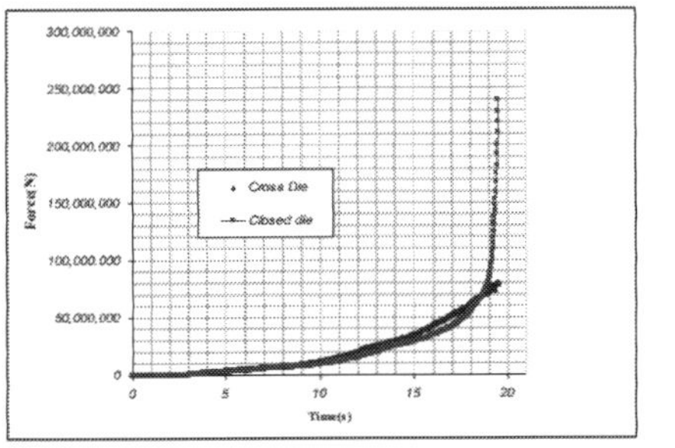

Figure 13. Comparison of the force required for manufacturing the part (Figure 10 (A), (B)) using closed die and cross die methods (Simulated with Msc.Superforge Software)

3 Practical Examples

Usually in the forging industry segments are designed to build them is not possible with existing presses. This repetition can be seen especially in the military industry, but here we are comparing two ways to produce Gate Valve 7 1/6 "3000 PSI economically:

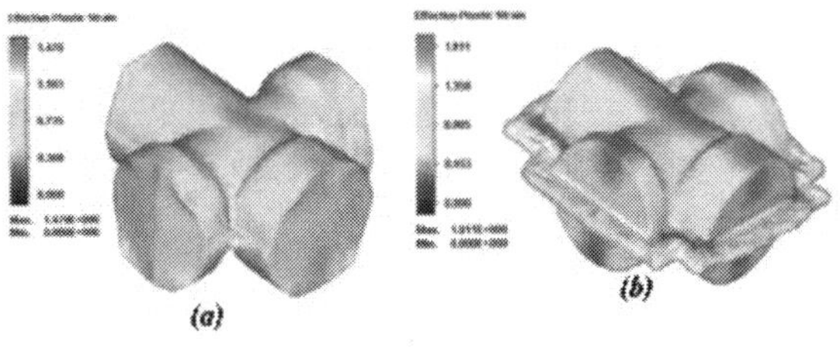

Figure 12 Cross (a) and closed (b) die forging

In order to compare costs associated with cross die method, with that of other conventional methods, a comparison was made on a profile related to a valve's body (7 1/6" 3000 Psi) between traditional and cross die methods.

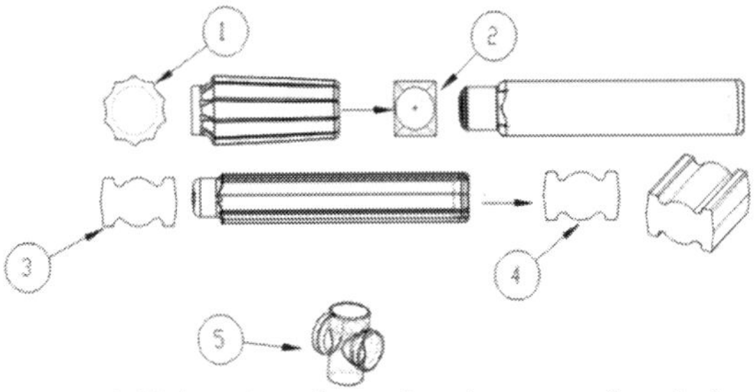

Figure 14. (a) Overview of manufacturing process for valve's body (7 1/16" 3000 Psi) by profile method

First, a 6.5 tons ingot was preheated and converted into a quadrilateral using open die forging. Then, the part was shaped into the desired profile using a shaped open mold and a hydraulic pressing device during several pulses. After being cooled down, the ingot was cut and the resulting profile was machined.

I. *Total Cost Related to Manufacturing the Valve by Conventional Method (A)*

Preparing the casted ingot; squeezing and converting the ingot into a quadrilateral; subjecting the ingot to tension to produce the profile; machining (rough-end mill); final machining.

Describing profile process Based on Figure14(a):

1- The average original cost for producing one 4130L dodecahedral ingot with net weight of 5320: 268925200 Rials
2- The cost of transforming a dodecahedral ingot to a quadrilateral ingot after cutting process: (actual price of the work 450,900+ production overhead costs 12,786,000+ service overhead costs 2,695,810)* effective hours of working with pressing device at each shift 5.5= 109,949,450 Rials
3- Four ingots can be pressed at each working shift: 109,949,450/4= 27,487,360 Rials
   Three valves can be produced from each ingot: Rials 296,412,560=27,487,360+268,925,200
   296,412,560/3=98,804,190 Rials
4- 4, 5 and 6- Machining and rough-end milling outside the machine manufacturing company (50000000)+CNC machining outside the machine manufacturing company (85 hours*1,200,000)=152,000,000 Rials
5- Original cost of producing a valve: 250,804,190=98,804,190+152,000,000 Rials~ **8360$**

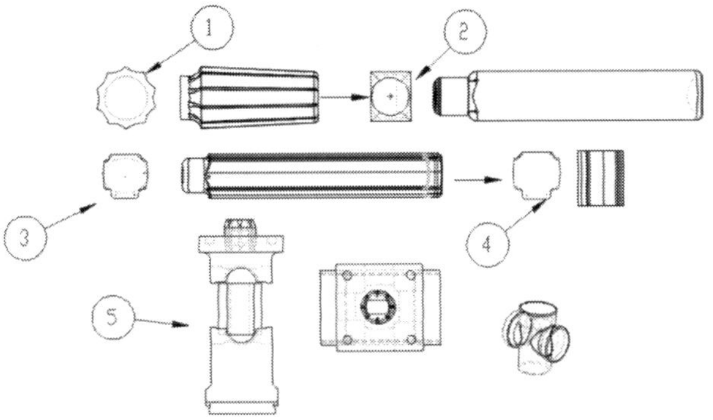

Figure 14. (b) Overview of process of producing a valve's body (7 1/16" 3000 Psi) using cross die method

Similar to previous method, in this method, the resulting profile, whose dimensions were calculated according to table (3), cools down; then, it is cut to a particular length to be used as a preform for cross dies.

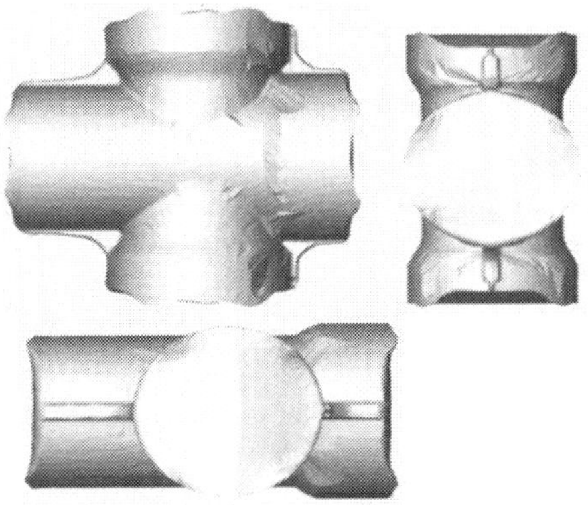

Figure 15. The part forged with cross die method- 7 1/16" 3000 Psi valve's body (Simulated with MSC.Superforge Software)

II. *Total Cost of Producing a Valve by Cross Die Method (B)*

Preparing the casted ingot; squeezing and converting the ingot into a quadrilateral; subjecting the ingot to tension to produce the profile; cutting and cross die forging; machining (rough-end mill); final machining.

Describing cross die forging process Based on Figure14(b):

1- The average original cost for producing one 4130L dodecahedral ingot with net weight of 5320: 268,925,200 Rials

2, 3 and 4- The cost of transforming a dodecahedral ingot to a quadrilateral ingot= 27,487,360Rials

In this method, 5 valves can be produced from one ingot: 296,412,560/5=59,282,510 Rials

5- There is 20 minutes overtime in this stage: 1/3(2,695,810+12,786,000+4,509,000)=6,663,600 Rials

Rials 65,946,110=59,282,510+6,663,600

6- Cost of machining for each valve: 50 hours*1,200,000=60,000,000

7- Original cost of producing a valve: 60,000,000+65,946,110=125,946,110Rials~ **4200 $**

III. *Benefits of Using Cross Die Methods for Production of Valve's Body (7 1/16" 3000Psi valve) in Comparison to Traditional Method:*

1- By using current conventional methods, three 7 1/16" 3K valve's bodies are manufactured from a dodecahedral ingot of 6.5 tons; while using the same ingot, 5 valves can be produced by cross die method.
2- Weight of the part to be forged decreases from 1247 kg to 900 kg, when using cross die method.
3- Time of machining the part decreases from 130 hours to 50 hours, in case of employing cross die method.
4- Better mechanical properties can be achieved by cross method due to higher squeezing ratio in this case. Also, it is possible to manufacture the product under PSL4 condition.

5- With regard to the oil & drilling industry's annual need to 1000 Qty. Gate Valves 7-1/16" 3000 & 5000 Psi, that much of it is imported. In case of using available presses in Iran, the amount will be about 124,858,080,000Rials (4,161,936 $) can be saved by cross die method.

4 Results

1- In this method, forging force can be reduce up to 80 percentage
2- This method can be used for some have forging parts which are not possible to produce theoretically
3- Cross die method is applicable only for parts that have at least one plane of symmetry.
4- Using cross die method is economical for forging parts with heavy weights and intermediate production volume.
5- In cross die method, the material flow through the flash channel is prevented by decreasing the pressure inside the mold and providing an appropriate preform design; thus, the required forging force in cross die method is 1/10 (one tenth of) closed dies.
6- In cross die method, the forged part must be in its semi-raw state and also requires machining after the final production.

Figure 16. Image of a cross die and the mechanism by which it is fixed on the work table of 315 tons mechanical press

Figure 17. Cross dies; the entire flash channel is considered to be on the lower die and other three ends of the upper and lower dies are assumed to be open

Figure 18. 315 Tons mechanical press used to test the part

Table 5. Abbreviations And Descriptions

Symbol	Description and Units
W	is the weight of the forging in Kg
W_f	is the width of the flash in mm
T_f	is the thickness of the flash in mm
T_g,T_f	are the thicknesses of the gutter and the flash, respectively in mm
W_g,W_f	are the widths of the gutter and flash,in mm
F_{f_l}	Forging force in N on forging part
F_{f_o}	Forging force in N on flash channel
F_T	Total Forging force in N on flash channel + Part
A_{f_o}	Cross section of forging part in mm^2
A_{f_l}	Cross section of flash channel in mm^2

References

[1] Taylan Altan, Gracious Ngaile, Gangshu Shen ,Cold and Hot Forging: Fundamentals and Applications, Volume 1 ,page 170
[2] Bauer, P., Korpert, K., Neuberger, M., &Raber, A. (1991). Risk Journal of the Acoustical Society of America, 90, 3086-3098.
[3] George E. Totten, Kiyoshi Funatani, Lin Xie, (1991) Handbook of Metallurgical Process Design 3,30,123.
[4] J. Beddoes, M. Bibb, Principles of Metal Manufacturing Processes, page 111
[5] N. Akgerman, J.R. Becker, T. Altan, Metalurgia and Metal Forming (1973) 135-138.
[6] G. Zhao, E. Wright, R.V. Grandhi, International Journal of Machine Tools & Manufacturing 36/7 (1996) 755-769.
[7] E. Doege, H. Naegele, Annals of the CIRP 43/1 (1994) 241-244
[8] N.C. Duffy, Technomorphology, Journal of Materials Processing Technologies (1983) 349-371.
[9] K. Hyunkee, T. Yagi, M. Yamanake, Journal of Materials Processing Technology 98 (2000) 143-149.

DOI: 10.1051/matecconf/20168010001

Research on optimization design of conformal cooling channels in hot stamping tool based on response surface methodology and multi-objective optimization

Bin He[1], Yanglei Si[1], Liang Ying[1] and Ping Hu[1,a]

[1]*School of Automotive Engineering, Faculty of Vehicle Engineering and Mechanics, State Key Laboratory of Structural Analysis for Industrial Equipment, Dalian University of Technology, Dalian 116024, P.R. China*

Abstract. In order to optimize the layout of the conformal cooling channels in hot stamping tools, a response surface methodology and multi-objective optimization technique are proposed. By means of an Optimal Latin Hypercube experimental design method, a design matrix with 17 factors and 50 levels is generated. Three kinds of design variables, the radius *Rad* of the cooling channel, the distance *H* from the channel center to tool work surface and the ratio *rat* of each channel center, are optimized to determine the layout of cooling channels. The average temperature and temperature deviation of work surface are used to evaluate the cooling performance of hot stamping tools. On the basis of the experimental design results, quadratic response surface models are established to describe the relationship between the design variables and the evaluation objectives. The error analysis is performed to ensure the accuracy of response surface models. Then the layout of the conformal cooling channels is optimized in accordance with a multi-objective optimization method to find the Pareto optimal frontier which consists of some optimal combinations of design variables that can lead to an acceptable cooling performance.

1 Introduction

Hot stamping technology which is widely applied in the automotive components manufacture provides the possibility to produce components with complex shape, high strength and less springback [1,2]. Cooling channels arranged inside hot stamping tools play an important role in the cooling efficiency and temperature uniformity of tools [3].

There have been a few researches about design and optimization of cooling channels. Straight hole design is the most common method which has been discussed by Hoffmann and Steinbeiss, and genetic algorithm was submitted to optimize the cooling system [4,5]. Liu et al. proposed a novel approach to design and manufacture the channels which were pre-embedded in a casting mould with pre-bending stainless pipes [6]. Sachs et al. first put forward the concept of conformal cooling channel design and produce the tool by 3D printing technology [7]. Xu et al. presented a systematic approach for conformal design which is evaluated by five criteria, including transient heat transfer condition, maximum distance from the mould surface to cooling channel, pressure drop, temperature drop and mould strength [8]. Park and Dang designed a cooling layout with an array of baffles to achieve the conformal characteristics [9]. Hu et al. raised four types of conformal designs and the longitudinal conformal cooling channel was proved as the optimal cooling design when inlet mass flow rate was large [10]. Wang et al. created a quadratic response surface model with sample data from Box-Behnken design and then optimized the regression model by a constrained particle swarm optimization method [11]. Jarrett et al. integrated GAMBIT and FLUENT by a MATLAB optimization program package to optimize the cooling layout in battery cells [12,13]. Although the conformal design was studied in many fields, the application for the optimal cooling layout of hot stamping tools has not yet been investigated.

In this paper, the radius of cooling channel, the distance from the channel center to tool work surface and the ratio of each channel center are considered as the design variables which can determine the layout of cooling channels. The average temperature and temperature deviation of work surface are chosen as evaluation indicators to assess the cooling performance. Based on the OLH experimental results, response surface models are established. Finally, the layout of cooling channels is optimized by multi-objective method and a Pareto frontier is obtained which reveals the optimal variables combination.

2 Conformal cooling channel design

During hot stamping process, the steel sheet is heated up to the austenite temperature range in a furnace until a

[a] Corresponding author: pinghu@dlut.edu.cn

DOI: 10.1051/matecconf/20168010001

homogeneous austenitic microstructure is obtained. Then the steel sheet is transferred to the press and subsequently formed and simultaneously quenched in the closed tools with a cooling system to produce the intended shape. The components finally exhibit a martensitic microstructure with 1500 MPa tensile strength [14]. Straight hole design is the traditional cooling configuration employed in hot stamping tools [15].

2.1 Comparative analysis of traditional and conformal design

2.1.1 Design structure

Take an electric vehicle B-pillar punch tool as an example, different cooling structures of straight hole design and conformal cooling channel design are illustrated in Figure 1. The cooling water first flows into the water manifold and then passes into the three inserts and finally flows out from the outlet on the manifold. Obviously, the main difference of cooling design occurs in the middle insert which has a curved shape. The straight hole design is realized only through drilling intersectant holes on two edge surfaces and it is a non-isometric design that will lead to inhomogeneous cooling performance. On the contrary, the conformal cooling channel design is an isometric design which can achieve better cooling performance.

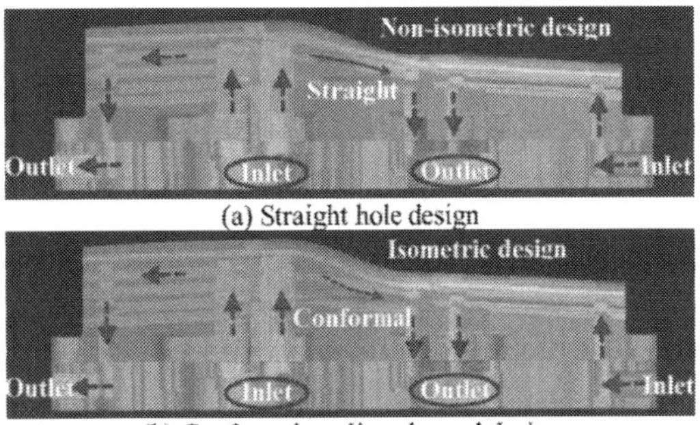

(a) Straight hole design

(b) Conformal cooling channel design

Figure 1. Structures of straight hole design and conformal cooling channel design.

2.1.2 Cooling performance

In order to prove the advantages of conformal cooling channel design, a heat transfer simulation is performed which can approximately describe the hot stamping process. Generally, the temperature distribution after achieving steady state is extracted as evaluation indicator and six stamping cycles are enough for hot stamping tool to reach a steady temperature fluctuation.

Table 1. Material property of tool.

Parameter	Value
Density / $kg{\cdot}m^{-3}$	7900
Thermal conductivity / $W{\cdot}m^{-1}{\cdot}K^{-1}$	$-3\times10^{-5}T^2+5.7\times10^{-3}T+42.624$
Special heat / $J{\cdot}kg^{-1}{\cdot}K^{-1}$	450
Initial tool temperature /°C	20
Mesh type	DC3D4: a 4-node linear heat transfer tetrahedron.

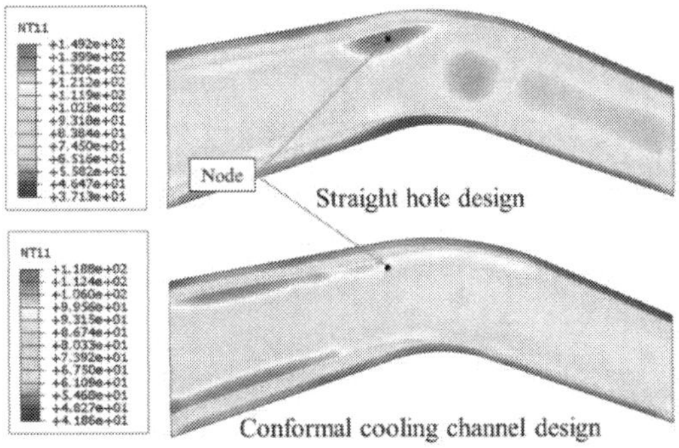

Figure 2. Temperature distributions of straight hole design and conformal cooling channel design.

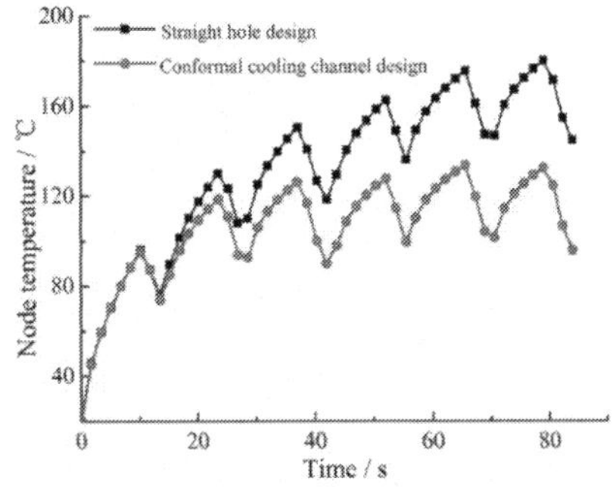

Figure 3. Temperature history of a node on the tool rounded corner.

The tool material is H13 steel. Its material property is listed in Table 1. Supposing that the total time of a stamping cycle is 14 s and among that the stamping and holding time, feeding time and reclaiming time are 10 s, 2 s, 2 s respectively. The time-averaged heat flux of 190,000 $W{\cdot}m^{-2}$ is loaded on the work surface to simulate the stamping and quenching stage, and it is regarded as 0 $W{\cdot}m^{-2}$ during the feeding and reclaiming stage. The smooth process by high order polynomial is implemented in the transition of heat flux load at different stages. According to the empirical equations and limit size of cooling channels, the convectional heat transfer coefficient between the channel wall and the cooling water is set as a constant of 5,417 $W{\cdot}m^{-2}{\cdot}K^{-1}$.

It is noted from Figure 2 that the same FEM model and boundary conditions are used, and for the straight hole design, the maximum work surface's temperature is 149.2 °C, but for the conformal cooling channel design, that is reduced to 118.8 °C. Besides, the temperature uniformity is obviously improved and especially shown in the middle insert. Due to the convex shape, the temperature is always concentrated on the rounded corner. In Figure 3, it is found after the first stamping cycle that the temperature at the specific node of straight hole design is always higher than that of conformal cooling channel design. Moreover, it is obvious that the tool using conformal cooling channel design takes less time to achieve temperature stability. The present study will focus on how to obtain the optimal layout of cooling channels with conformal design.

2.2 Experimental design by using Optimal Latin Hypercube method

2.2.1 Model verification

The accuracy of the FEM simulation generally increases with the increase of mesh number. But at the same time, the computing time is also greatly increased especially for optimization process which requires hundreds of computations. Therefore, before optimization, the FEM model should be verified with a mesh independence analysis to ascertain a mesh size that would achieve convincible results in an acceptable computing time.

Take the middle insert as an example, which is more sensitive to design structure, the same material property and boundary conditions are set up and different models with a mesh range of 2 mm to 25 mm are simulated. The relative error of work surface's temperature and the computing time for the FEM models with different mesh base sizes are shown in Figure 4.

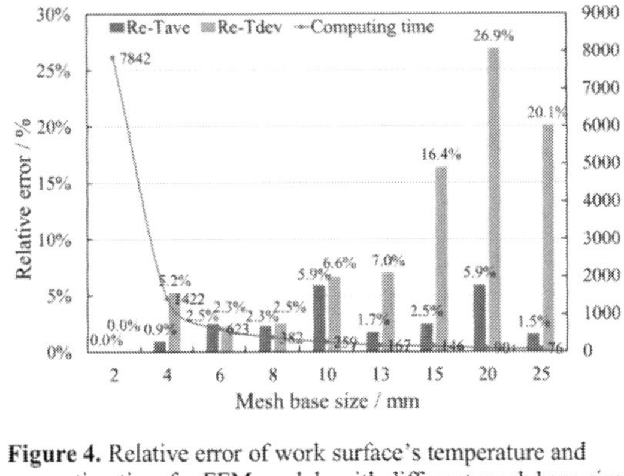

Figure 4. Relative error of work surface's temperature and computing time for FEM models with different mesh base sizes.

The Re-Tave and Re-Tdev in the Figure 4 stand for the relative error of average temperature and temperature deviation on the work surface comparing the model with 2 mm mesh base size. It is demonstrated that Re-Tdev increases with the increase of mesh base size but the Re-Tave presents random fluctuation. It is not surprising that the computing time reduces with the increase of mesh base size and the computing efficiency is improved sharply for the mesh size of 2 mm to 4 mm. Based on the independence analysis, it is easy to find that the optimal mesh base size is 8 mm for the present problem.

2.2.2 Range of design variables

The definition of design variables is depicted in Figure 5. It is noted that *Rad* symbol denotes the radius of the cooling channel, *H* symbol refers to the distance from the channel center to tool work surface and *rat* symbol stands for the ratio of each channel center.

Considering the design and assembly constraints, the design variables including the radius, distance and ratio must be restricted to a range to avoid the intersection and interference. Whether in the experimental design or in the optimization process, design variables always change within the range set out in Table 2.

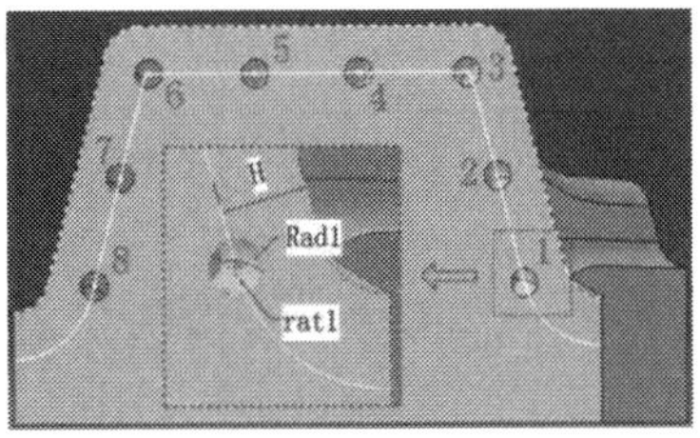

Figure 5. Schematic diagram of design variables.

Table 2. Range of design variables.

Design variables	Lower value	Upper value	Unit
H	13	17	mm
*Rad*1～*Rad*8	4	6	mm
*rat*1	0.05	0.127	-
*rat*2	0.203	0.237	-
*rat*3	0.313	0.347	-
*rat*4	0.423	0.457	-
*rat*5	0.533	0.567	-
*rat*6	0.643	0.677	-
*rat*7	0.753	0.787	-
*rat*8	0.863	0.95	-

2.2.3 Design matrix

Design of experiment (DOE) is an approach widely applied in industry that can reveal the relationship of design variables and response using a small amount of sample points. The Optimal Latin Hypercube is a method of DOE in which the number of levels of each factor is equal to the number of design points evenly spread within an n-dimensional space defined by the *n* factors through combination optimization. A design matrix with 17 factors i.e. design variables and 50 levels displayed in Table 3 is generated by OLH method. Due to space limitations, only some design points are listed in Table 3. The design variables adhere to the range placed in Table 2. The type of *H* and *Rad*1～*Rad*8 is integer, but *rat*1～*rat*8 is real.

Table 3. Design matrix generated by OLH method.

Design variables	Des.1	Des.2	Des.3～Des.48	Des.49	Des.50
H	14	14	…	15	14
*Rad*1	5	4	…	5	6
*Rad*2	4	5	…	5	5

 DOI: 10.1051/matecconf/20168010001

Rad3	5	5	...	5	6
Rad4	6	6	...	5	5
Rad5	5	4	...	6	5
Rad6	5	5	...	4	6
Rad7	5	5	...	4	6
Rad8	5	4	...	5	5
rat1	0.0516	0.1066	...	0.1019	0.1003
rat2	0.208	0.209	...	0.211	0.207
rat3	0.324	0.343	...	0.332	0.340
rat4	0.429	0.424	...	0.450	0.440
rat5	0.563	0.543	...	0.534	0.564
rat6	0.661	0.661	...	0.674	0.665
rat7	0.785	0.774	...	0.756	0.754
rat8	0.8648	0.9269	...	0.9447	0.934

2.3 Response surface model establishment

2.3.1 Samples from OLH

The sample points in the design matrix are executed automatically by an integration flow established by ISIGHT which can transfer the data information among the CATIA, ABAQUS and data process model. The workflow is shown in Figure 6.

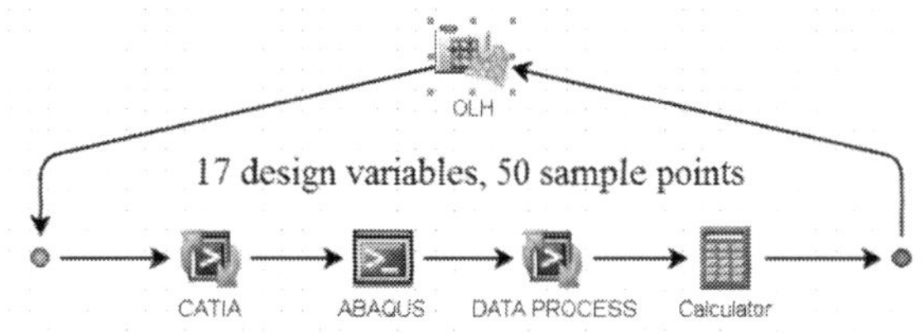

Figure 6. Workflow of OLH process.

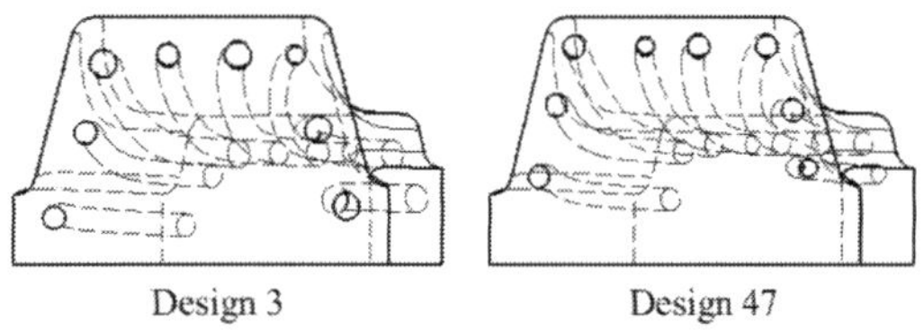

Figure 7. CAD models of Des. 3 and Des. 47.

The parameterized CAD model is created by CATIA, and then the CAD routine is mapped to ABAQUS. Next the temperature data simulated by ABAQUS are extracted to perform the data process. The response, i.e. evaluation indicators are solved in the Calculator. According to the design matrix including 17 design variables and 50 sample points, the workflow will auto run 50 times in total and the response values corresponding to the 50 sample points are listed in Table 4.

Tave and *Tdev* are the average temperature and temperature uniformity of work surface, respectively. It is noted that *Tave* and *Tdev* are different for different designs. In order to show more images to design differences, two different designs are illustrated in Figure 7.

Table 4. Response values corresponding to the 50 sample points.

Sample points	*Tave*	*Tdev*	Sample points	*Tave*	*Tdev*
Des.1	71.93	11.97	Des.26	75.94	14.78
Des.2	70.95	12.18	Des.27	78.09	14.90
Des.3	85.95	17.63	Des.28	86.38	17.32
Des.4	83.38	16.89	Des.29	75.98	14.59
Des.5	78.80	15.23	Des.30	66.74	10.61
Des.6	79.10	15.85	Des.31	78.56	14.56
Des.7	71.57	11.15	Des.32	65.91	10.48
Des.8	70.26	12.02	Des.33	66.74	11.50
Des.9	73.60	11.95	Des.34	63.15	10.08
Des.10	72.98	14.48	Des.35	75.41	14.78
Des.11	72.48	13.92	Des.36	84.59	15.26
Des.12	76.13	13.55	Des.37	85.82	17.53
Des.13	84.08	16.16	Des.38	70.63	13.38
Des.14	82.31	15.66	Des.39	74.66	13.95
Des.15	71.00	13.12	Des.40	81.59	14.81
Des.16	77.89	12.65	Des.41	68.84	10.60
Des.17	79.48	14.32	Des.42	78.89	15.62
Des.18	71.12	12.69	Des.43	68.64	10.39
Des.19	83.65	15.80	Des.44	77.12	14.05
Des.20	85.44	17.29	Des.45	73.15	12.57
Des.21	71.12	13.05	Des.46	70.35	13.00
Des.22	85.23	15.03	Des.47	65.70	9.87
Des.23	67.80	10.89	Des.48	70.21	13.51
Des.24	82.14	16.18	Des.49	75.63	15.12
Des.25	80.80	16.00	Des.50	70.43	11.09

2.3.2 Model fitting and error analysis

Response surface methodology (RSM) is a statistical and mathematical technique for relationship evaluation between design variables and response variables based on Taylor's expansion. The quadratic polynomial model can be expressed as,

$$y = \beta_0 + \sum \beta_i x_i + \sum \beta_{ii} x_i^2 + \sum^{i \neq j} \beta_{ij} x_i x_j + \varepsilon \quad (1)$$

where β_0 is constant term, β_i is linear coefficient, β_{ii} is quadratic coefficient, β_{ij} is interaction coefficient, i, j are the number of design variables, and ε is statistical error.

Depending on the design variables in Table 3 and the response variables in Table 4, the regression response surface models of *Tave* and *Tdev* are expressed as Eq. (2) and Eq. (3). Both of the two mathematical models described *Tave* and *Tdev* contain 25 terms. *Tave* model is composed by 1 constant term, 9 linear terms, 6 quadratic

DOI: 10.1051/matecconf/20168010001

terms and 9 interactive terms. *Tdev* model is composed by 1 constant term, 8 linear terms, 9 quadratic terms and 7 interactive terms.

$$Tave = 873.766 - 86.03H + 9.04Rad1 - 0.37Rad2 - 1.08Rad5 + 10.59Rad8 + 137.11rat1 + 6099.08rat4 - 5885.66rat5 + 41.51rat7 + 0.16H^2 - 0.68Rad8^2 - 1526.11rat2^2 - 6939.08rat4^2 + 4336.8rat5^2 - 662.83 rat6^2 - 0.62H*Rad1 - 0.11H*Rad3 - 0.03H*Rad4 - 0.06H*Rad7 - 0.32H*Rad8 - 10.18H*rat1 + 44.25H*rat2 + 77.99H*rat5 + 60.29H*rat6 \quad (2)$$

$$Tdev = -919.38 + 0.25Rad1 - 3.47Rad5 - 2.23Raad7 + 1563.37rat2 + 2451.49rat4 - 895.89rat5 - 2381.71 rat6 + 3201.53rat7 - 0.24Rad3^2 + 0.21Rad7^2 - 245.01rat1^2 - 3609.53rat2^2 - 2775.08rat4^2 + 675.75rat5^2 + 1825.76rat6^2 - 1978.71rat7^2 + 4.91rat8^2 + 0.11H*Rad3 + 0.22H*Rad5 - 0.04H*Rad6 + 2.67H*rat1 + 0.61H*rat3 + 10.16H*rat5 - 7.76H*rat7 \quad (3)$$

The cross validation method is used for error analysis (Figure 8) in this study. It is noted that match of the models are good and reliable, and the adjusted R square coefficient of *Tave* is 0.94 and *Tdev* is 0.95. It means that the two models present a good characterization of the relationship between design variables and response variables.

2.3.3 Pareto graph analysis

Pareto graph reflects the contribution degree of all items in the fitting model for each response. Figure 9 shows the Pareto graph analysis of *Tave* and *Tdev*. Red bar stands for negative effect, and blue bar represents positive effect. It is noted that whether for *Tave* or *Tdev*, the design variables of *rat* have the most significant effect. The effect of H is smaller than *rat* but bigger than all *Rad* variables. For *Tave*, *rat*4 and *rat*5 show significant impact and the effect degrees reach up to 35% and 39% respectively. However, effect degrees of other variables are all below 11%. For *Tdev*, effect degrees of *rat*7, *rat*4 and *rat*2 are 28%, 24% and 23% respectively. Similarly, effect degrees of other variables are all under 11%.

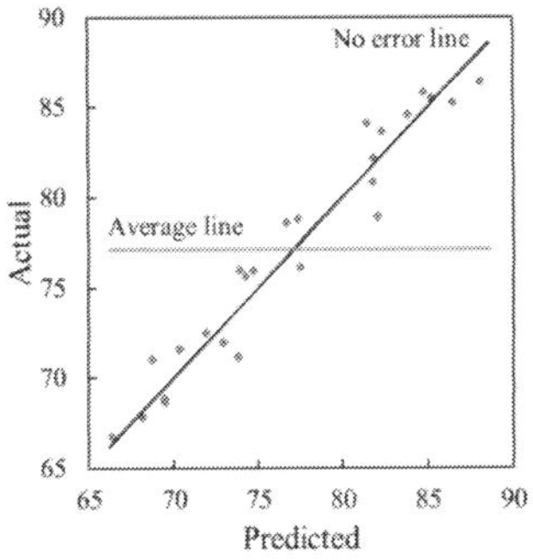

(a) Error analysis of *Tave*

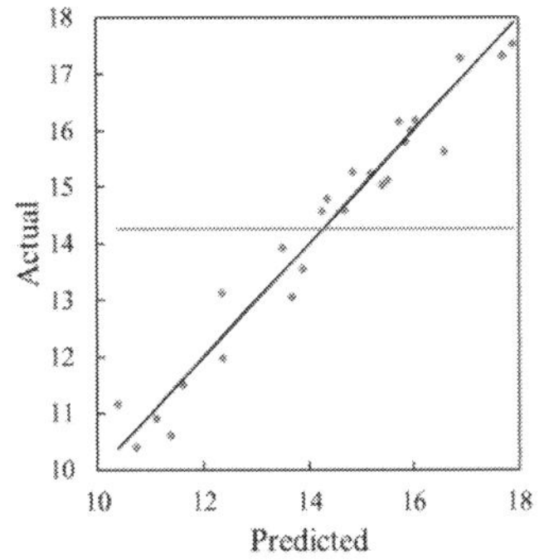

(b) Error analysis of *Tdev*

Figure 8. Error analysis of *Tave* and *Tdev*.

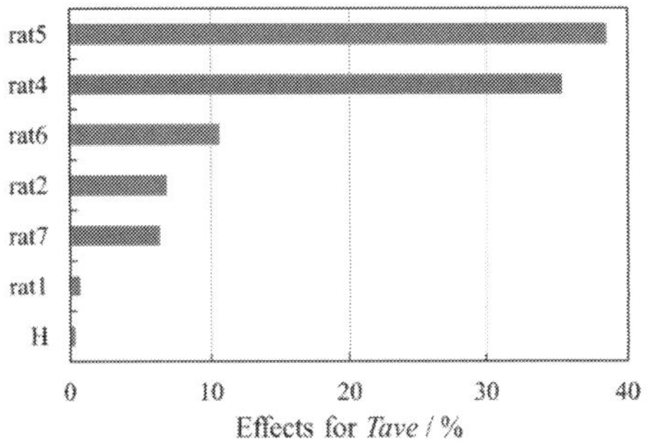

(a) Pareto graph of *Tave*

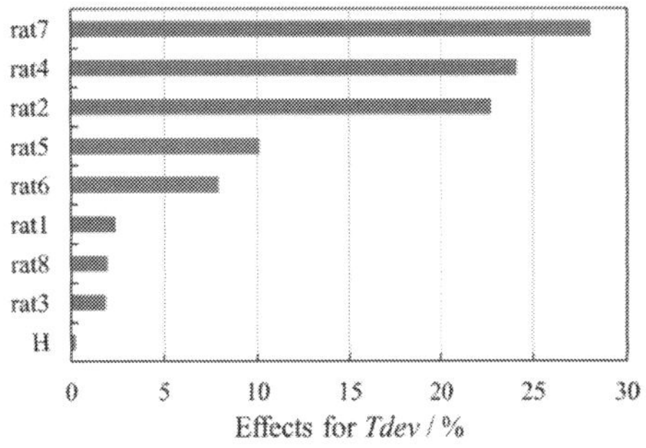

(b) Pareto graph of *Tdev*

Figure 9. Pareto graph analysis of *Tave* and *Tdev*.

3 Optimization

3.1. Model and process

Generally, *Tave* and *Tdev* are two conflicting response variables. When the average temperature attains the minimum value, the temperature uniformity is more obvious, and vice versa. Therefore, the multi-objective optimization method which will compromise with the two objectives is utilized to explore the optimal layout of cooling channels. The mathematical description of the optimization problem is given as,

Minimize *Tave* Eq.(2)
Minimize *Tdev* Eq.(3)
Variable range $\{H, Rad1 - Rad8, rat1 - rat8\}$ abide by Table 2

NSGA-II (Non-dominated Sorting Genetic Algorithm) which is well-suited for highly non-linear and discontinuous design spaces is used in the present study. By this method, each objective is treated separately. Standard genetic operation of mutation and crossover is performed on the designs. Selection process is based on two main mechanisms, "non-dominated sorting" and "crowding distance sorting". The population size which controls the number of solutions generated at each iteration is set to 400 and the number of generations which controls the number of iterations that the algorithm will execute before termination is 40. The total number of function evaluations exhausted by the algorithm equals the product of the population size and the number of generations. The process model created in ISIGHT is shown in Figure 10. The design variables of Des. 1 from OLH design matrix are used to initialize the optimization model.

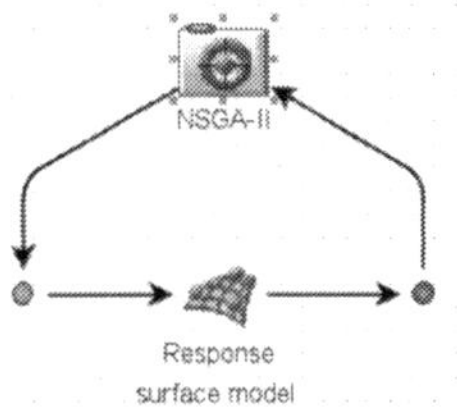

Figure 10. Optimization process.

3.2 Pareto frontier and feasibility analysis

A total of 16,000 evaluations are operated for objective functions. Due to the hot stamping simulation has been regressed as mathematical models by OLH and response surface method before optimization, the optimization efficiency is improved greatly and only 5 min is spent. A total of 329 Pareto-optimal solutions are obtained and the Pareto frontier results from Pareto-optimal solutions which present a convex shape is displayed in Figure 11. It is noted that after multi-objective optimization *Tave* will be controlled within the range of 54 °C to 58 °C, and *Tdev* is kept from 3.5 °C to 6.0 °C. The Pareto solutions are uniformly continuous and the trend line of these solutions is a quadratic polynomial. The point 1 marked as black in Figure 11 is the optimal solution revealed by ISIGHT. The *H*, *Rad*1 ~ *Rad*8 and *rat*1 ~ *rat*8 of point 1are, in order, 13, 4, 6, 6, 6, 6, 6, 6, 6, 0.127, 0.237, 0.313, 0.423, 0.562, 0.677, 0.753, 0.8631. According to the Pareto solutions, it can be found that design variables of *H* and *Rad*1 ~ *Rad*8 tend to unique peak values, but *rat*1 ~ *rat*8 have more options. It is a further validation that design variables of *rat*1 ~ *rat*8 have more effect on response variables.

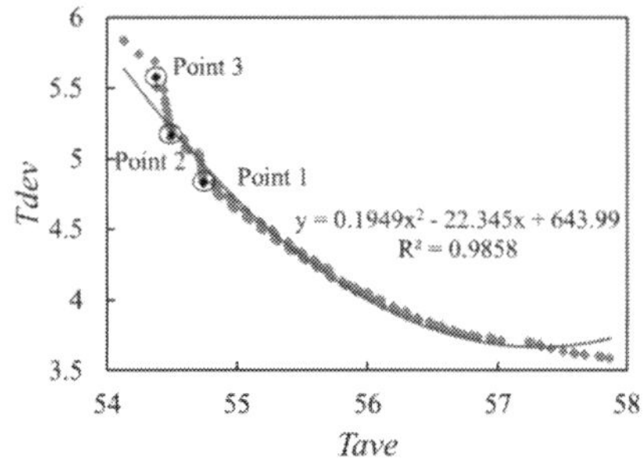

Figure 11. Pareto frontier graph.

In order to verify the accuracy of the optimization results, the design variables of three points including point 1, 2 and 3 are taken back to CAD routine and FEM model to carry out the hot stamping simulation separately. *Tave* and *Tdev* values getting from simulation and mathematical model are shown in Table 5. The results demonstrate that the optimization using regress models are not accurate enough comparing with direct simulation. But relatively speaking, *Tave* obtained by model optimization, which is always less than 15%, is more accurate than *Tdev*. The relative error of *Tdev* is below 45%, which means that though the *R* square of *Tdev* model reaches up to 0.95, the design space is still not entirely covered by the regress model. By increasing sample points or adjusting the experimental design process, the prediction precision and optimization accuracy will be improved. Although the prediction error is inevitable when using the response surface model to perform optimization, this method has a high cost-effective compared to the direct simulation optimization which represents a high computational cost.

Table 5. Feasibility analysis of the optimization results.

	Point 1		Point 2		Point 3	
	Tave	*Tdev*	*Tave*	*Tdev*	*Tave*	*Tdev*
Simulation	63.5	8.6	62.1	8.8	61.0	8.1
Mathematical model	54.7	4.8	54.5	5.2	54.4	5.6
Relative error	14%	44%	12%	41%	11%	31%

The temperature distributions before and after optimization are illustrated in Figure 12. Des. 1 listed in Table 3 is used as the initial design before optimization and Point 1 is the optimal solution recommended by Pareto solutions in ISIGHT. Obviously, the temperature distribution including average temperature and temperature uniformity has been improved after optimization. *Tave* and *Tdev* are separately reduced by 11.7% and 28.2%. Moreover, the design parameters of cooling channels after optimized become more uniform and helpful to manufacture the channels. The whole points on Pareto frontier are the optimal combinations of design variables, however, the design combinations which contribute to the convenience of processing and the low cost should be prioritized.

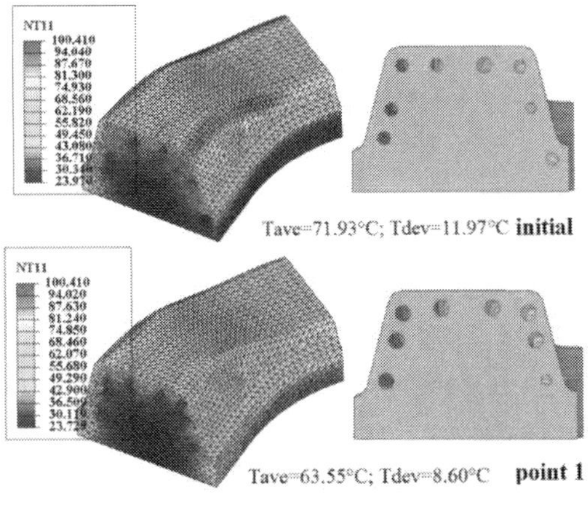

Figure 12. Temperature distribution before and after optimization.

4 Advanced manufacture

When adopting the conformal cooling channel design in hot stamping tools, how to realize the low-cost manufacture process is a tough issue. An advanced manufacture technology is proposed in this paper and its manufacturing process is displayed in Figure 13. The key step is to use the ceramic core model by extrusion to realize the channels' structure with conformal design. This advanced manufacture method is seen as a combination of ceramic core making and casting process.

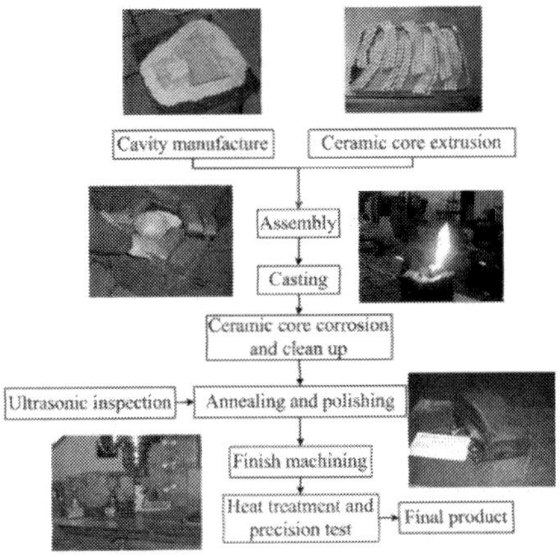

Figure 13. Manufacture process of hot stamping tools with conformal cooling channel design.

Look at Figure 13, first, the sand mould cavity and ceramic core with conformal shape are prepared. Second, the cavity, core, runner and riser are assembled into a gating system. Third, the casting process and the heat treatment after casted are performed. The survival state of ceramic core should be paid close attention in the casting process to prevent fracture. Fourth, after the cast mould cooling down, the ceramic core wrapped inside mould is cleaned by alkaline corrosion with a high temperature caldron. Fifth, ultrasonic inspection is used to find the severe defects, and if there are no serious defects, the casting mould will be annealed and polished subsequently, otherwise further evaluation of that mould will be carried out. Sixth, to operate the surface finishing and to quench hardened the casting mould. Finally, after a dimensional accuracy test, a finished hot stamping tool insert with conformal cooling channel design is produced successfully. By this method, the tool insert with any conformal structure is not difficult to be manufactured. Further verification tests of temperature distribution will be carried out in future studies.

Acknowledgements

This work was funded by the Key Project of the National Natural Science Foundation of China (No.11272075) and China Postdoctoral Science Foundation (2014M561223) and the Fundamental Research Funds for the Central Universities (DUT15TD31).

References

1. H. Karbasian, A.E. Tekkaya, J Mater Process Tech., **210** (2010) 2103-2118.
2. B. He, L. Ying, P. Hu, Y. Yu, X. Zhao, L.W. Zhang, Adv Mater Res., **1063** (2014) 186-189.
3. W. Lim, H. Choi, S. Ahn, B. Kim, Int J Adv Manuf Tech., **70** (2014) 1189-1203.
4. H. Hoffmann, H. So, H. Steinbeiss, CIRP Ann-Manuf Tech., **56** (2007) 269-272.
5. H. Steinbeiss, H. So, T. Michelitsch, H. Hoffmann, Production Engineering, **1** (2007) 149-155.
6. H. Liu, C. Lei, Z. Xing, Int J Adv Manuf Tech., **69** (2013) 211-223.
7. E. Sachs, E. Wylqnis, S. Allen, M. Cima, H. Guo, Polym Eng Sci., **40** (2000).
8. X. Xu, E. Sachs, S. Allen, Polym Eng Sci., **41** (2001).
9. H. Park, X. Dang, Int J Precis Eng Man., **11** (2010) 879-890.
10. P. Hu, B. He, L. Ying, Appl Therm Eng., **96** (2016) 338-351.
11. G. Wang, G. Zhao, H. Li, Y. Guan, Expert Syst Appl., **38** (2011) 6705-6719.
12. A. Jarrett, I.Y. Kim, J Power Sources, **196** (2011) 10359-10368.
13. A. Jarrett, I.Y. Kim, J Power Sources, **245** (2014) 644-655.
14. M. Merklein, J. Lechler, J Mater Process Tech., **177** (2006) 452-455.
15. F. Schieck, C. Hochmuth, S. Polster, A. Mosel, CIRP Journal of Manufacturing Science and Technology, **4** (2011) 189-199.

Semi-analytic parameter identification for complex yield functions

Niklas Küsters[1,a], Alexander Brosius[1]

[1]Institute of Manufacturing Technology, Chair of Forming and Machining Processes, 01062 Dresden, Germany

Abstract. This paper presents a new parameter identification scheme for complex yield criteria of sheet metals using experimentally determined load and strain distributions and combining analytical stress analysis on sectional strain data with common inverse analysis. The approach is suitable for specimen with a non-homogenous strain distribution in the specimen and reduces computing time compared to common methods like iteratively updated FEM considerably. It serves to identify a set of material parameters in a fast and precise manner, describing the material behaviour.

1 Introduction

For sheet metal forming industry, the numerical simulation of manufacturing processes is a competitive factor as reliable prediction allows the industry to shorten cycles of process development and to reduce time to production launch. Forming simulation supports engineers and toolmakers in the prediction of forming behaviour, formability and final part properties during the process design. In order to comply with the progressive demands in reducing the discrepancy between numerical prediction and experimental results, an improvement of the phenomenological modelling of forming behaviour combined with efficient experimental testing is of high relevance [1].

2 State of the art

In metal forming, the plastic material behaviour is usually modelled by a flow curve, describing the true stress response as a function of true strain for uniaxial loading. In combination with a yield function the material behaviour under multiaxial loading is described.

To overcome the drawbacks of isotropic yield functions as Tresca [2] or von Mises [3] complex yield criteria were developed in order to describe anisotropic behaviour more accurately. For many steel grades Hill's quadratic yield criterion [4] is sufficient. An advantage of this criterion is the possibility of direct parameter calculation based on experimental results from only three standard tensile tests. However, Hill's criterion is not appropriate for materials like high strength steels or aluminium that have a distinctive anisotropy. Therefore, more complex yield functions like Yld2000-2D [5], BBC2000 [6] and Vegter [7] were developed in order to describe the yield surface shape more accurately and improve the results of numerical simulations. As these yield functions need additional points in stress space, extensive efforts for experimental testing and parameter identification are required [1]. Additional testing data of biaxial, plain strain and shear specimen is commonly used to model an accurate material response with these complex yield criteria.

Compared to standard experiments like tensile tests, the challenge for most of the additional experiments is the appearance of non-homogenous strain and stress distributions, which bias the results of experimental analysis. For this reason, some experiments are specifically designed to improve the results. In shear test configurations [8], the geometric dimension of specimen is adapted, so that rotational effects are reduced and the quasi-homogenous centre of the shear zone is enlarged. Within these tests, the shear stress is often approximated analytically by using the testing force and assuming the stress distribution to be uniform over the shear zone. But even with special specimen design, the shear stress distribution is quite inhomogeneous and therefore leads to imprecise results for the stress states. Hence, a direct analytical calculation of material parameters based on global measurement of complex specimen is not appropriate for an accurate description of the yield surface.

To cope with the problems associated with non-homogenous strain distribution appearing in experimental testing, several approaches for parameter identification were proposed in the past. For instance, it is possible to transform the crystal anisotropy into a global anisotropy using crystal plasticity methods to obtain a phenomenological description of the yield surface [9]. Using this technique, inhomogeneity on grain scale is used to predict the macroscopic mechanical behaviour. A drawback of this method is, that only defects in the polycrystalline microstructure, but no further macroscopic defects of the sheet metal can be considered.

[a] Corresponding author: niklas.kuesters@tu-dresden.de

In contrast, identification methods based on kinematic full-field measurements and inverse analysis offer the possibility to model the material behaviour based on the observed macroscopic phenomena [10]. Within these approaches, a testing machine coupled to an optical strain measurement system records the experimental results as input for the identification scheme. Subsequently, possible strategies specially developed for parameter identification of plastic material behaviour are briefly described.

The most popular approach for parameter identification is finite element method updated (FEMU), using iterative finite element analysis within an inverse analysis [11-15]. The method uses the discrepancy between numerical results and experimentally determined loads and strain fields to define an objective function. Using an optimisation algorithm, numerical results with varying parameters are iteratively computed by finite element analysis until the objective function is minimised. The approach was utilised in diverse studies to determine material parameters for sheet metal forming processes. For instance, inverse analysis with iterative FEA was used by Güner et al. [11] to identify the yield function parameters of Yld2000-2d [5] using the FE-code Abaqus-Explicit and the Levenberg-Marquardt Algorithm. In a subsequent study of Aydın et al. [12], a parameter identification for the evolution of the yield surface Yld2000-2d [5] was performed using the optimisation tool Ls-Opt and the FE-code Ls-Dyna with experimental data of standard tensile tests and a plane strain tension test. Further studies calibrate the yield surface on experiments designed for the validation of specific forming conditions in order to get a process-oriented material response. Bäck et al. [13] presented a parameter identification strategy based on evolutionary algorithms and its application in calibration of diverse yield criteria with specially designed validation tools for stretch forming conditions. Using specifically designed experiments in combination with FEMU, it is also possible to identify material parameters for yield surface and flow curve simultaneously with one single test [14]. Furthermore, the FEMU approach can be used for the identification of material parameters for kinematic hardening models. Yin et al. [15] proposed an in-plane torsion test and used Abaqus-Explicit in combination with a trust region method to identify kinematic hardening parameters. The presented studies show, that FEMU can be used as a computationally expensive, but very powerful tool for parameter identification.

In order to increase the computational efficiency, the application of the virtual fields method (VFM) for parameter identification in elasto-plastic problems was proposed by Avril et al. [16]. The algorithm uses full-field surface measurements of the deformation to calculate a stress field with the predictor-corrector method of Sutton et al. [17]. Stress components are iteratively computed with varied and updated material parameters until the principle of virtual work used as optimisation criterion is satisfied. In the work of Avril et al. [16], four constitutive parameters for elastic and plastic material behaviour were identified on one single specimen assuming a von Mises yield criterion. According to the study, the approach was able to reduce the computing time by a factor of 100 compared to the FEMU approach. This promising approach was adjusted in a subsequent study for anisotropic yield criteria [18]. In the study, two specially designed tests were used to retrieve six constitutive parameters of the Hill yield criterion [4] and the Swift hardening law [19].

FEMU and VFM show good results for adjusting a complex yield surface to experimental results with arbitrary geometry. But as finite element simulations are costly, the widely-used iterative FEM shows only poor efficiency regarding the computation time. In contrast, the virtual field method presents an approach with considerably reduced computational efforts. Nevertheless, the potential in stress analysis and identification is not thoroughly used yet. The direct analytical calculation of stress states is still used in practice due to its simplicity and computational efficiency compared to the presented approaches. Hence, there is a need for improving the efficiency and accuracy of (semi-)analytical yield surface characterisation regarding computing time and reduced input of experimental results.

3 Objective and approach

3.1 Objective

To overcome the presented drawbacks of current parameter identification schemes, the development of an alternative identification strategy which significantly reduces computational costs is necessary. Within this strategy, field data, commonly recorded in standard setups of testing machines coupled with optical strain measurement, shall be used in an efficient way to identify material parameters. Unexploited potential in the combination of inverse analysis and direct analytical evaluation over cross-section area with simplified stress analysis shall be used in order to create a highly performant identification strategy. Subsequently, an alternative semi-analytical approach enabling fast parameter identification for complex yield functions is presented.

3.2 Approach

The proposed approach can be used for the determination of yield function parameters for experimental analysis of arbitrary sheet specimen containing non-homogenous distribution in the deformation zone. As no further finite element analysis is performed and a simplified computational procedure is used, a fast identification is possible. Prerequisites for the identification are the strain history distribution of one or more intersections through the plastic region, the initial sheet thickness, a flow curve approximation in uniaxial direction and a first estimation for the parameter set of the yield function. The approach can be subdivided into three parts, namely a simplified stress analysis, an internal load calculation and an inverse analysis.

As mentioned in section 2, a direct analytical stress calculation using global forces and the cross-section area is only possible if a uniform stress distribution can be assumed. In contrast, the calculation of resulting global forces by stress integration over the section area is straightforward for every kind of stress distribution.

Using this relation, we can analyse arbitrary sections with heterogeneous stress distribution on a specimen (see Fig. 1). For correct assumption of the stress distribution, the static equilibrium of forces has to be satisfied. Using stress analysis on surface measurements [17, 20] analogue to VFM, this relationship can be used in inverse analysis. Subsequently, the algorithm of the new approach is briefly described.

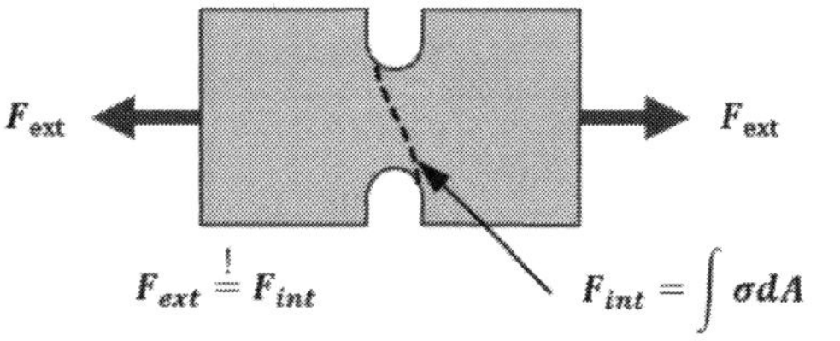

Figure 1. Equilibrium of internal and external forces

Before the identification, one or more cross-sections have to be defined as curves over the plastic region. Thereby, the time- and space-resolved strain field is reduced to time-resolved strain distributions over the predefined intersections. The strain distribution history subsequently is used as input data for the inverse analysis to identify yield function parameters.

Based on the strain history, the corresponding stress components are calculated. In contrast to VFM, that uses the elasto-plastic stress computation of Sutton et al. [17], a simplified direct stress analysis on incremental plastic strain is applied. Therefore, the idea of direct stress analysis on plastic strain measurements [20] is modified and formulated for general anisotropic plane stress yield functions. Within the framework, a stress response is computed for each point on the intersection curve using the incremental strain stress relationship and the given strain history. Furthermore, it is important to mention, that the stress analysis is performed in the coordinate system of the material orientation. The computational procedure presented in Fig. 2 is given below.

Assuming the total strain to be plastic, first the incremental plastic strain $\mathbf{d\varepsilon^p}$ is calculated based on the optically measured strain. Subsequently, the location on the yield surface is identified by finding the point on the yield surface with a normal vector **n** collinear to the incremental plastic strain $\mathbf{d\varepsilon^p}$ according to the associated flow rule. In a next step, the subsequent yield locus is computed for isotropic work hardening. Therefore, the yield surface is incrementally evolved until plastic work of the provided uniaxial flow curve equals the plastic work of the current stress point. Plastic work of this point is calculated using the measured strain and the stress corresponding to the current yield surface. The resulting accumulative strain $\varepsilon_{eq}{}^p$ is saved for consideration in the following time step as new initial yield surface. Furthermore, for the following procedure, the stress components are transformed to the global coordinate system, as the stress results are computed for the local material orientation system.

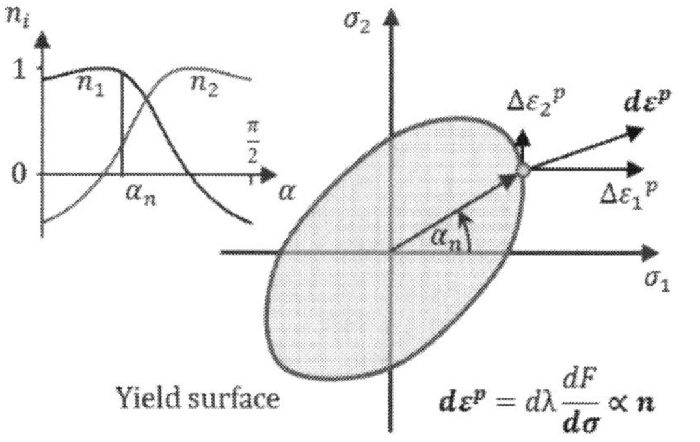

Figure 2. Stress computation based on incremental strain

Subsequently, internal loads are calculated in the global coordinate system by numerical integration of the identified stress components over the intersectional area (see Fig. 1). The intersectional area is defined piecewise by an interpolated curve and updated thickness.

The presented computational procedure of stress analysis and internal load calculation is performed for each time step and used within inverse analysis for the parameter identification. An optimisation algorithm [21] is performed with iteratively varying yield surface parameters until an objective function Φ is minimised:

$$\Phi = ||\boldsymbol{w} \cdot (F_{ext} - F_{int})||^2 \quad (1)$$

The objective function Φ satisfies the equilibrium of forces and is calculated as the weighted total mean square error of computed internal loads F_{int} compared to experimental determined external loads F_{ext}. Depending on the material model and the number of parameters to optimise, further constraining experimental data like the Lankford-coefficients has to be considered in the objective function. Result of the procedure is an unambiguous parameter set fitted to the material response of the examined intersections on the specimen.

4 Testing with virtual experiments

4.1 Virtual testing

For validation of the proposed parameter identification, the scheme is implemented in a software prototype and tested with reference data. Instead of using real experimental data, virtual experiments generated by FEA are used to develop the computational procedure. The quasi-ideal results of the fictitious experiment can be used to identify and redesign the effects modelled in the FE reference. Exemplary for the computational procedure, a Swift power law [19] for the flow curve σ_Y (2) and a Hill quadratic yield criterion F (3) for orthotropic materials [4] are used to model the hardening behaviour:

$$\sigma_Y = C_1 (C_2 + \varepsilon_{eq}{}^p)^n \quad (2)$$

$$F = [A_1\sigma_{xx}{}^2 - A_2\sigma_{xx}\sigma_{yy} + A_3\sigma_{yy}{}^2 + 3 \cdot A_4\tau_{xy}{}^2] - \sigma_Y{}^2 = 0 \quad (3)$$

Analogue to real experiments, strain data of an intersectional line in the plastic region computed for the virtual experiment is used as input for the identification scheme. Instead of using facets like in optical measurements the nodal FEA results are used for the following calculations. Compared to real experiments, the crucial advantage of using the FE reference is the possibility to observe internal variables as stress states and the certainty, that only numerical errors but no measurement mistakes will occur. This leads to easier testing and development of the procedure, as deviations between reference and new calculations can be measured.

Table 1. Parameters for assumed yield surface (for $r_0 = 1.6$; $r_{45} = 1.4$; $r_{90} = 2.0$)

A_1	A_2	A_3	A_4
1.0	1.23077	0.92308	0.87692

Table 2. Assumed flow curve parameters

C_1	C_2	n
527.0	0.0270703	0.22

4.2 Shear test

Subsequently, a shear test [22] is used for the validation of the new approach. Therefore a virtual specimen is generated by the commercial finite element code Marc 2013.1.0. using an Updated Lagrange procedure and implicit time integration. Fig. 3 shows the testing setup including geometry and mesh used for the validation. The specimen has a sheet thickness of 1.0 mm and a shearing zone with a length of 4.72 mm defined by the slots in the specimen. A bilinear thin shell formulation (type 139) is used. The mesh is refined in the shearing zone with an initial element size of 0.25 mm x 0.25 mm. The material orientation (R.D.) is collinear to the X-Axis.

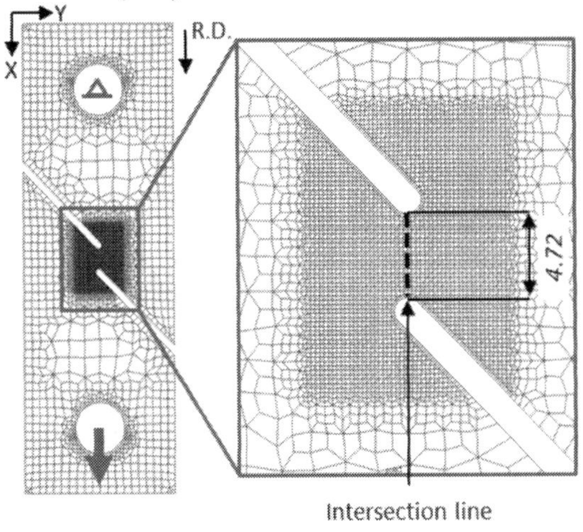

Figure 3. Virtual specimen – geometry and mesh

The material behaviour of the virtual specimen is modelled with the same Swift power law and an equivalent Hill yield criterion. The assumed material parameters for the virtual specimen are given in table 1 and table 2 for a mild steel (EN 1.0338 / DC04).

After FE simulation, the results are used to extract strain data over an intersectional curve in the shearing zone of the virtual specimen. Fig. 4 shows the deformed virtual specimen and the intersection through the shearing zone. Subsequently, the history of the corresponding non-homogenous strain distributions over the shearing zone (Fig. 5) is used as input for a validation of the computational procedure.

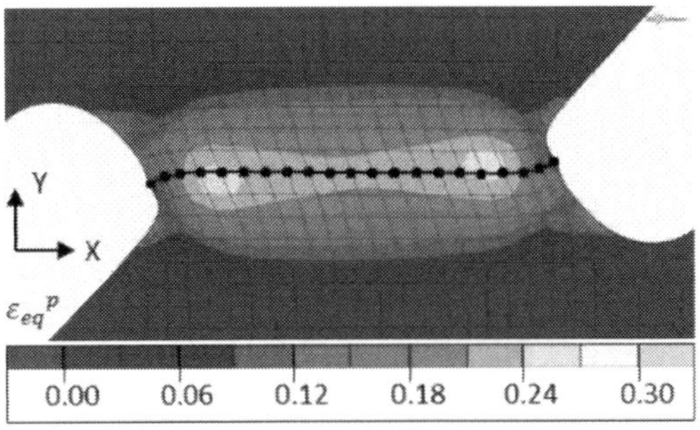

Figure 4. Deformed specimen: total equivalent plastic strain

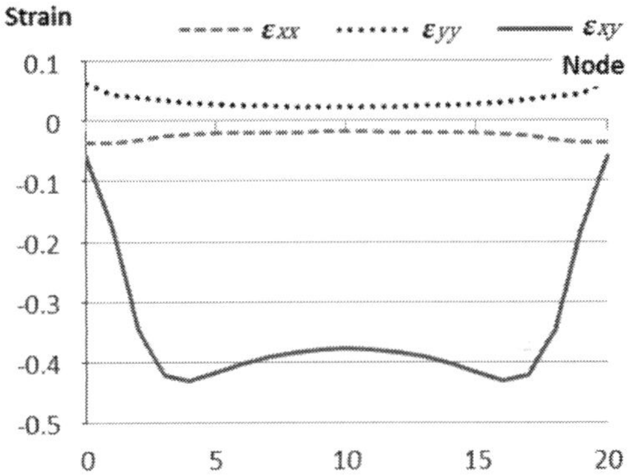

Figure 5. Non-homogenous strain distribution over intersection for the last time step

5 Evaluation of the new approach

Using this intersectional data of the virtual specimen as reference, the suitability of the new approach is evaluated in the following section. The evaluation focuses on stress analysis and internal load and briefly describes the inverse analysis for parameter identification.

5.1. Stress analysis

The simplified stress analysis proposed within the new approach is the key method which enables the computational procedure's performance as no further FEA is needed. As the computed stress is directly related to the internal force used as optimisation criterion, an accurate and reliable stress calculation is of major

 DOI: 10.1051/matecconf/20168010002

interest. Based on the strain distribution history of the virtual experiment, the stress components σ_{xx}, σ_{yy}, τ_{xy} are determined with the new approach and compared with the FE reference (Fig. 6). In accordance to the FE reference, stress estimations for actual A_4 parameters present good results for the inner part of the intersection. In the outer part, close to the edge, the deviations are up to 70 MPa higher compared to the FE reference (Fig. 6b).

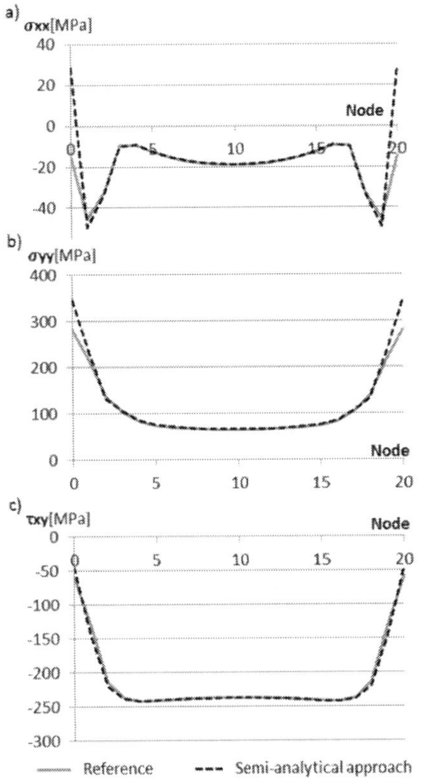

Figure 6. Stress distributions for actual parameter A_4

These deviations can be traced to simplifications of the nodal stress interpolation in the FE reference. Nodal results are usually interpolated from the integration points in direct neighbourhood. Therefore, the interpolated strain doesn't have to match with the interpolated stress regarding the nodal results. As the results for strain and rotation differ stronger among neighbouring integration points in the edge region, the calculation on nodal results in edge zones becomes less suitable than in the quasi-homogenous centre region.

To investigate the impact of this simplification, several elements were isolated (blue elements in Fig. 7). Then, the stress components on the integration points were estimated and compared with the FE reference (see Fig. 8). The accuracy of the stress analysis is significantly increased for a calculation on integration points. Mentionable deviations only occur on the first and last intersection point and are below 10 MPa compared to the FE reference. It is worth to mention, that the plot cannot be directly compared to the distribution on nodal results of Fig. 6, as it represents the results for an intersection translated to the integration points.

As the main deviation only appears in the relatively small edge regions and is homogenised by integration over the intersectional area later, the calculation on nodal results is sufficient for our further investigations, if a small total deviation for the internal load calculation is accepted. Furthermore, a comparatively poor stress calculation is observed for the first time steps, which can be traced to the simplification of only considering the plastic response. As elastic strains are evanescent small, this simplification is sufficient as long as the results are excluded from parameter identification for the first time steps with predominant elastic response.

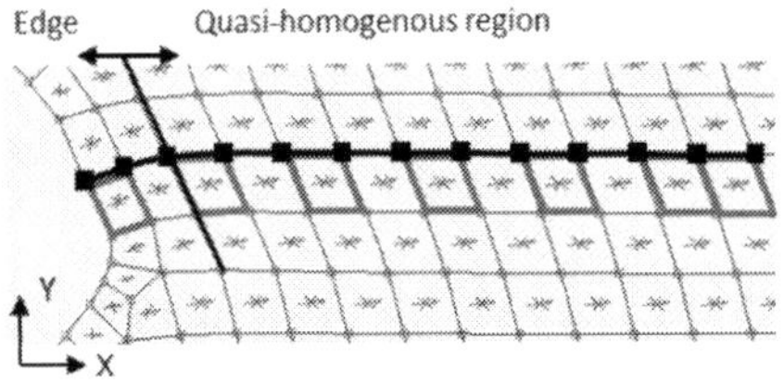

Figure 7. Material orientation during deformation

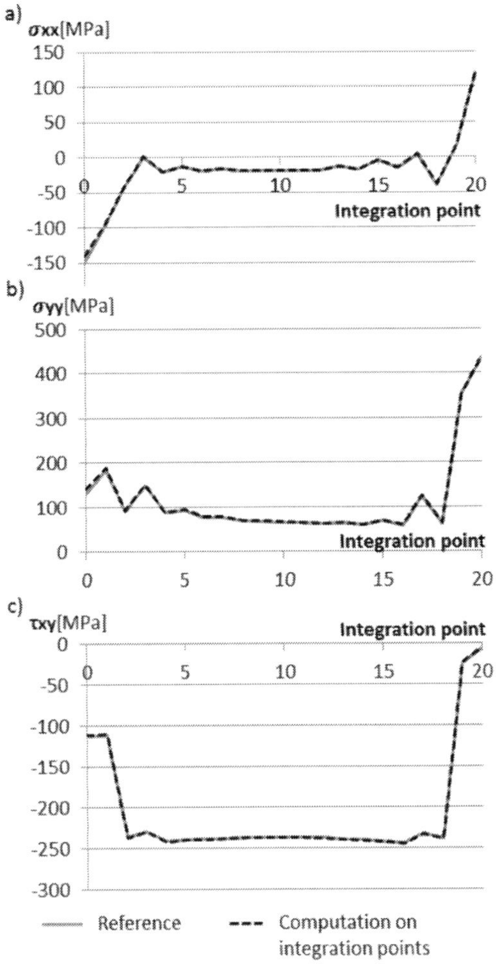

Figure 8. Stress results on integration points

5.2. Sensitivity of stress analysis

Subsequently, the variation of the Hill function parameters is examined in order to investigate their sensitivity on stress computation. Therefore, reference parameters of table 2 are varied by simply changing the A_4 parameter to a value of 1.15385. In the following sections, this new parameter set is called the initial estimation. By changing the A_4 parameter, the normal on the yield surface deviates drastically from its original direction and work hardening is estimated differently. In Fig. 9 the dashed curves shows the stress components calculated on a wrongly estimated initial parameter A_4. The wrong estimation leads to biased results for shear and normal stress. The strongly diverging value of the shear stress τ_{xy} is induced by changing the yield surface with an enlarged A_4 parameter. In contrast a smaller A_4 value leads to shear stress values higher than the actual reference.

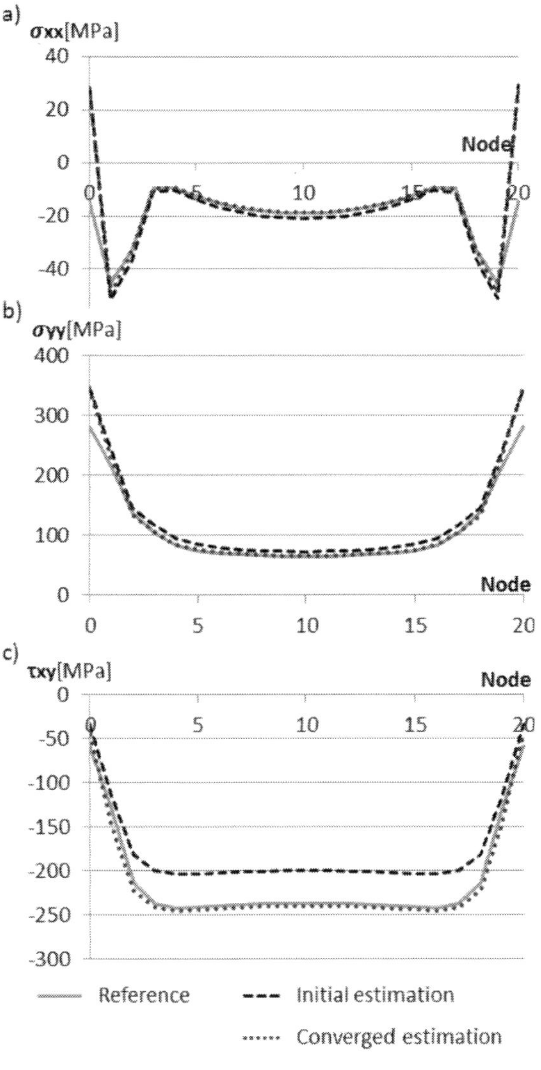

Figure 9. Stress distributions for initial and converged estimation of A_4

5.3 Internal load calculation

After stress analysis, internal loads can be calculated in the global coordinate system using the nodal coordinates. As stress analysis is performed with respect to the material orientation, every stress state has to be transformed from the system of material orientation to the global system before integration. In addition to the initial orientation, the rotation of the material during the deformation has to be considered. Reasons for this rotation are mainly related to rigid body rotation due to tilting of the specimen and to shear rotation induced by the rotation of local orthogonal element system of the used 3D shell during shear deformation. Fig. 7 shows the material orientation (red arrows) in comparison to the initially collinear global X-axis. Considering these rotations, an internal load in the global system was successfully computed for each time step of the virtual experiment. Fig. 10 shows the results for forces in X and Y direction based on the stress results calculated for the actual parameters in section 5.1. The comparison of internal and external loads shows only small deviations. For the estimated load in Y-direction a deviation of 2% can be observed (Fig. 10b). These deviations are mainly caused by inaccuracies of computed stress components. The second relevant aspect is the interpolation of the nodal results for rotation, analogue to section 5.1. But, as deviations mainly occur in small edge areas, their impact is not high for the internal load calculation, as they are homogenized after integration over the intersection area.

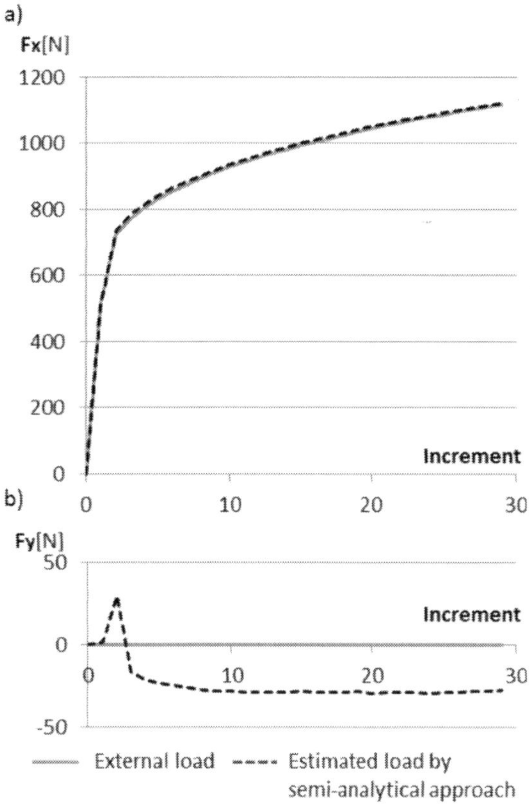

Figure 10. Load distribution over time

5.4 Inverse analysis

In order to investigate the suitability of the new approach, the implementation of the inverse analysis is tested using the virtual shear test and an initial estimation for the parameter A_4. The remaining parameters A_1, A_2 and A_3 are constrained to the values given in table 2.

Starting with a falsely estimated A_4 parameter as initial estimation, we can observe a wrong result for the internal load induced by incorrect stress estimation. For our assumed initial estimation of section 5.3, this leads to a computed force in X direction which is about 150 N smaller than the external load. In contrast, smaller A_4 values lead to force values higher than the external load.

During inverse analysis, the parameters are iteratively adapted until the internal load converges the external load. Fig. 11 shows the development of convergence for A_4 during the parameter identification.

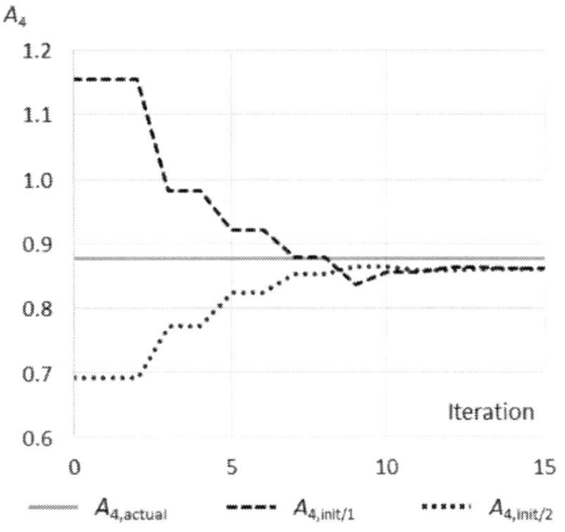

Figure 11. Convergence during identification procedure

After parameter optimisation, stress components (Fig. 9 dotted line) and internal loads (Fig.10) approximately result in a same response as the FE-reference. The dashed line in Fig. 11 shows the development of A_4 for an initial value of 1.15385. The final value of 0.86064 is reached at iteration step 30. An equivalent result for inverse analysis is computed for an initial value of 0.69231 (dotted line). The currently not runtime optimized implementation needs less than 5 minutes for this identification on a standard desktop computer. Compared to the FEMU approach, the computation time is reduced by a factor of more than 100.

As presented in Fig. 11, a small deviation of the final compared to the actual A_4 value is observed. As the optimisation considers forces in X and Y direction, biased force calculations for the Y direction like presented in Fig. 10b lead to this divergent results. By introducing weight factors and only considering the load in X direction, an A_4 value of 0.88668 with a smaller force deviation in X direction is calculated by the algorithm. This result shows, that accurate stress estimation, transformation and integration is essential for the parameter identification, as the computational procedure is very sensitive to small deviations of loads in the Y direction.

6 Summary and outlook

In this study, a new approach for efficient and performant parameter identification for complex yield functions is proposed. Compared to other approaches as finite element method updated, the numerical and experimental efforts for the evaluation of complex specimen with non-homogenous strain distribution can be reduced considerably.

Within the approach, a simplified stress analysis using the plastic strain stress relationship was used. To reduce computation time, the analysis is performed on the strain distribution of a user defined intersection in the plastic region. By simplified stress integration, a corresponding internal load is computed and used with measured testing forces in an inverse analysis to identify the yield function parameters for balanced forces. For testing, the computation procedure was developed and validated for a Hill yield criterion with fictitious virtual experiments generated by FEA. Using a parameter set describing the actual yield surface, computed stress components and corresponding internal loads only show little deviation compared to the reference data. An inverse analysis with one single varied parameter showed that the procedure is suitable for parameter identification.

With our approach, we introduce a performant tool for parameter identification for complex yield functions. Especially for yield functions needing a higher amount of sampling data to describe distinctive anisotropy, great potential is given. Further developments will focus on the implementation of complex yield functions like Yld2000-2D and on a transfer to experimental analysis of real specimen. Future studies should also consider phenomena like combined isotropic-kinematic hardening. Additionally, the scheme for refined stress computation based inside the approach could be used for other purposes using in-process analysis.

Acknowledgement

Parts of this research work were attended by the “European Research Association for Sheet Metal Working” (EFB) and supported by the “German Federation of Industrial Research Associations” (AiF) with the public funding of the “German Federal Ministry for Economic Affairs and Energy” (BMWi).

References

1. S. Bruschi, T. Altan, D. Banabic, P.F. Bariani, A. Brosius, J. Cao, A. Ghiotti, M. Khraisheh, M. Merklein, A.E. Tekkaya, Cirp Ann. Manuf. Techn. **23**, 727-749 (2014)
2. H. Tresca, Comptes Rendus Academie des Sciences **59,** 754 (1864)

3. R. von Mises, Göttinger Nachrichten, Math.-Phys. **4,** 582–592 (1913)
4. R. Hill, J. Mech. Phys. Solids **38.3**, 405-417 (1990)
5. F. Barlat, J.C. Brem, J.W. Yoon, K. Chung, R.E. Dick, D.J. Lege, F. Pourboghrat, S.-H. Choi, E. Chu, Int. J. Plasticity **19**, 1297-1319 (2003)
6. D. Banabic, T. Balan, D.S. Comsa, The 7th Conf. "TPR2000", Cluj Napoca, 217–224 (2000)
7. H. Vegter, A.H. van den Boogaard, Int. J. Plasticity **22**, 557–580 (2006)
8. Q. Yin, B. Zillmann, S. Suttner, G. Gerstein, M. Biasutti, A.E. Tekkaya, M. F.-X. Wagner, M. Merklein, M. Schaper, T. Halle, A. Brosius, Int. J. Solids Struct. **51**, 1066-1074 (2014)
9. M. Baiker, D. Helm, A. Butz, Steel Res. Int. **85** (2014)
10. S. Avril, M. Bonnet, A.-S. Bretelle, M. Grédiac, F. Hild, P. Ienny, F. Latourte, D. Lemosse, S. Pagano, E. Pagnacco, F. Pierron, Exp. Mech. **48**, 381–402 (2008)
11. A. Güner, Q. Yin, C., Soyarslan, A. Brosius, A.E. Tekkaya, Int. J. Mater. Form. **4**, 121-128 (2011)
12. M.-S. Aydin, J. Gerlach, L. Kessler, A.E. Tekkaya, J. Mater. Process. Technol. **211**, 1957–1964 (2011)
13. T. Bäck, L. Kessler, I. Heinle, Proc. GECCO '11, 1779–1786 (2011)
14. T. Pottier, P. Vacher, F. Toussaint, H. Louche, T. Coudert, Exp. Mech. **52**, 951–963 (2012)
15. Q. Yin, C. Soyarslan, A. Güner, A. Brosius, A.E. Tekkaya, Int. J. Mech. Sci. **59**, 31–43 (2012)
16. S. Avril, F. Pierron, Y. Pannier, R. Rotinat, Exp. Mech. **48**, 403-419 (2008)
17. M.A. Sutton, X. Deng, J. Liu, L. Yang, Exp. Mech. **36.2**, 99-112 (1996)
18. J.-H. Kim, F. Barlat, F Pierron, M.-G. Lee, Exp. Mech. **54**, 1189–1204 (2014)
19. H.W. Swift, J. Mech. Phys. Solids **1**, 1–18 (1952),
20. W. Voelkner, P. Feldmann, Conf. Proc. Sächsische Fachtagung Umformtechnik (1994)
21. D.W. Marquardt, SIAM J. Appl. Math. **11.2**, 431-441 (1963)
22. ASTM Standard B831-05, ASTM International, West Conshohocken, PA (2005)

Constrained design of sheet forming processes

Evripides G. Loukaides[1,a] and Julian M. Allwood[1]

[1]*Department of Engineering, University of Cambridge, CB2 1PZ, United Kingdom*

Abstract. New forming machines are typically incremental improvements on existing designs. Frequently tedious simulation and testing are required for insights on those possible improvements. A method for inventing features of flexible sheet forming processes is explored here. The inverse problem is solved by starting with a desired geometry, and "un-forming" the part while applying a selection of kinematic constraints and minimizing the plastic work on the workpiece. The resulting deformation reveals the required forces on the part and hence can inform machine design.

1 Introduction

Many modern forming technologies are inflexible which leads to significant material waste [1]. Reviews of existing flexible processes abound, especially with a focus in past decades on infusing modern technologies with the flexibility of traditional craftsmen [2-6]. However, progress remains slow and inconsistent. Sheet metal forming could greatly benefit from automated means of inventing new forming processes that can adhere to pre-determined constraints of deformation.

Such a method, based on the "un-forming" of target parts while minimizing the plastic work done on the workpiece was suggested in [7]. Here, the results of one iteration of the method are detailed and critiqued, with the intention of converging to a more robust method and eventually to a practically achievable process.

The idea of un-forming the part to reveal the underlying mechanics was inspired by similar work originating in the garment industry [8-9]. A simplified mass-and-spring model is used to describe a 3D surface and its elastic deformation, thus allowing evaluation of regions of high strain and subsequently of optimal cutting patterns. But the problem of parametrizing a 3D surface to a 2D plane is also an important field in applied mathematics, originating in cartography, but with modern implications for computer graphics. Again, many methods supply some simple material properties to the surface to define an objective function but do not strictly correspond to physical objects [10-12].

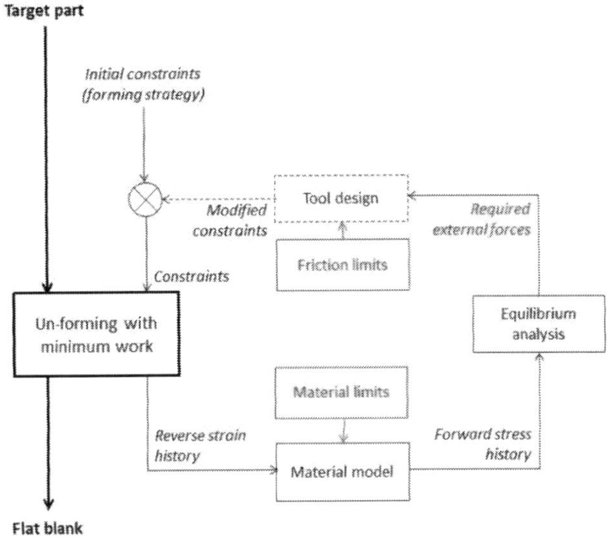

Figure 1. Block diagram of the "un-forming" method, adapted from [7]. The red text indicates the focus of this paper.

[a] Corresponding author: el249@cam.ac.uk

Inverse Finite Element Methods have been used in metal forming, but mostly for blank design. Minimization of plastic deformation is used to that end in [13]; a work guided by Hencky's deformation theory and Hill's anisotropic yield criterion. With the same objective Cai et al [14] used deformation non-uniformity as an objective function. The objective function in [7] draws from the work by Chung and Richmond [15, 16], which states that the optimal forming processes – under certain assumptions – are characterised by minimal deformation energy across the workpiece.

The assumptions and the implementation of the method in [7] are re-stated here for completeness, but the focus here is on the evaluation of the initial results and the potential for improvements.

2 Method for process design by un-forming

This paper examines the method shown in Figure 1 and originally suggested in [7]. An optimization strategy for the plastic work done is employed to "un-form" a selected part towards a flat blank. A combination of constraints are chosen to describe desired deformation characteristics and to direct the movement of the material towards a single plane.

The method in the diagram is iterative: After the un-forming completes, the un-forming is reversed to reveal what the corresponding forming process would look like. Then, the feasibility and requirements for such a process are evaluated and the constraints are adjusted to attempt a subsequent "un-forming." This is repeated until both the desired constraints are satisfied and a feasible process is described.

The algorithm is described in great detail in [7], but a brief overview of the formulation is presented here for clarity and completeness. The focus of this paper is on the evaluation of one such un-forming, and specifically the metrics shown in red on Figure 1. They will be described and calculated in later Sections.

2.1 Formulation for optimally un-forming subject to constraints

Although the method can be used on any initial geometry for a surface (i.e. for any formed part), the work here uses results from [7] and maintains the selection of the square cup shape. The particular geometry contains developable and non-developable regions and a flange at the perimeter, thus making it both a common and challenging part in the sheet forming industry.

The square cup shape along with the various dimensions is shown in Figure 2. In the same Figure, a triangular discretization is shown and is used throughout [7] and this work.

The material law does not allow for an elastic region and takes the form:

$$\sigma_f = \sigma_0(0.05 + \bar{\epsilon}_E)^n \tag{1}$$

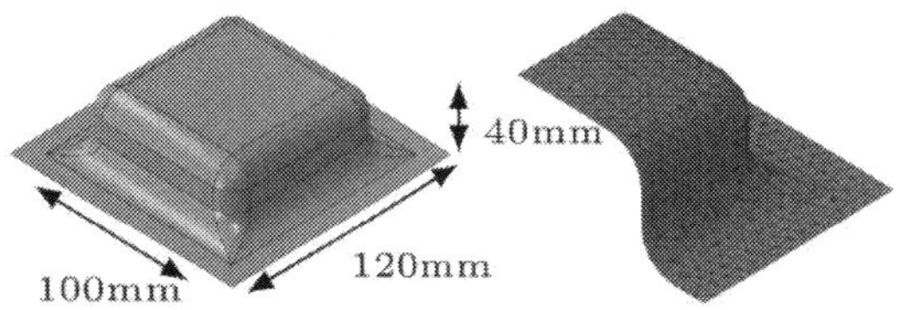

Figure 2. The square cup geometry used for all the simulations - modified from [7].

where $\bar{\epsilon}_E$ is the equivalent accumulated plastic strain. The parameters were modified from data in [17] and are based on the behaviour of aluminium 1100-O. These correspond to $\sigma_0 = 172\text{MPa}$ and $n = 0.25$. Only an isotropic material is considered.

Since the modelling beings with the formed part, a distribution of plastic strain has to be assumed. If correct, the ensuing deformation should result in a flat blank with no plastic strain. There is no direct way to obtain these strains and in [7] an iterative approach is used to estimate a reasonable starting strain field that results in a mean blank strain field close to zero. The first guess in this process is a uniform field of $\bar{\epsilon}_E = \bar{\epsilon}_{E0}^{(1)} = 0.5$. The algorithm then takes the final field, $\bar{\epsilon}_{Ef}$, from each iteration and subtracts is from the last starting field:

$$\bar{\epsilon}_{E0}^{(r)} = \bar{\epsilon}_{E0}^{(r-1)} - \bar{\epsilon}_{Ef}^{(r-1)} \tag{2}$$

with r being the current step in the iteration. The material law was adapted to prevent inadmissible negative strains from halting the calculation:

$$\sigma_f = \sigma_0(0.05 + \max\{\bar{\epsilon}_E, 0\})^n \tag{3}$$

The calculation of the plastic work for each element is given at each time step as:

$$dW = \sigma_f \, d\bar{\epsilon}_E \, dV \tag{4}$$

where dV is the element volume and $d\bar{\epsilon}_E$ is the equivalent plastic strain decrement. To calculate the equivalent strain for each element, the concept of the constant strain triangle is adopted. The strains for each edge of the triangular element can be calculated from the Cartesian coordinates $\boldsymbol{x}$, as the vector $\boldsymbol{\epsilon} = [\epsilon_1, \epsilon_2, \epsilon_3]$, such that

$$\epsilon_i = \ln\left(\left\|x_j^{t_n} - x_k^{t_n}\right\|_2 / \left\|x_j^{t_{n-1}} - x_k^{t_{n-1}}\right\|_2\right) \tag{5}$$

with x_j and x_k being node coordinate on edge i of the element, and the superscripts indicate the time step. The stretching effects are assumed to dominate the plastic work needed and no bending component is considered for the calculation.

The corresponding Cartesian strain field for the element can be calculated via the "rosette" transformation, T_e:

$$de = \begin{bmatrix} de_{xx} \\ de_{yy} \\ 2\, de_{xy} \end{bmatrix} = T_e \epsilon \quad (6)$$

where

$$T_e = \begin{bmatrix} c_1^2 & s_1^2 & s_1 c_1 \\ c_2^2 & s_2^2 & s_2 c_2 \\ c_3^2 & s_3^2 & s_3 c_3 \end{bmatrix} \quad (7)$$

and the subscripted c and s correspond to the cosines and sines of the element edges to a local Cartesian system of axes.

Once the Cartesian components of plastic strain are obtained, an equivalent plastic strain for each element is calculated as:

$$d\bar{\epsilon} = d\epsilon_{eq}^{p} = \sqrt{\frac{2}{3}\left[\left(de_{xx}^{p}\right)^2 + \left(de_{yy}^{p}\right)^2 + 2\left(de_{xy}^{p}\right)^2\right]} \quad (8)$$

Then, returning to Equation (4), the plastic work on the element can be calculated and summed to produce the total plastic strain across the workpiece. The minimization problem is defined as

$$\arg\min_{x \in \Omega} f = \sum_{i=1}^{n} dW_i \quad (9)$$

subject to

$$A_i^{t_k} = A_i^{t_0}, \forall i \in \Omega \quad (10)$$

$$x_p^{t_k} = C, \forall p \in \omega, \omega \in \Omega \quad (11)$$

The constraint in Equation (10) represents a constant area requirement for each element throughout the simulation, which in turn means the deformation is in pure shear and there is no change in thickness under the constant volume assumption from metal plasticity theory. Certain known processes, such as spinning [18], approach this behaviour. At the same time, Forming Limit Diagrams show no established failure boundary in the pure shear direction [19], thus this mode allows for maximum local deformation.

The second constraint was pasted here in a general form and represents the strategy that is followed in gradually un-forming the workpiece. This usually involved constraining a set of nodes, forming the node set ω, to a specific trajectory as to facilitate the un-forming, while allowing the rest of the nodes in the part, to move, as to minimize the objective function. Symmetry constraints are also applied to the xz and yz planes.

The solution to the optimization problem is attempted with the **fmincon** function in MATLAB which allows for a non-linear objective function as well as non-linear constraints. Within that, the 'interior-point' algorithm is used. The options give to **fmincon** are displayed in Table 1.

In practise, the optimization at each step is first attempted for a zero objective function as to obtain a feasible starting point for a subsequent attempt with the objective function in Equation (9).

Table 1. fmincon options for optimization

MaxFunEvals	1e8
TolCon	1e-3
TolFun	2e-1
TolX	1e-6
MaxIter	1e5
MaxFunEvals	1e8
TolCon	1e-3

The result of the optimization gives the updated geometry at each step and the equivalent decrement in equivalent plastic strain, which in turn allows for the calculation of the current plastic strain field:

$$\bar{\epsilon}^{t_k} = \bar{\epsilon}^{t_{k-1}} - d\bar{\epsilon}^{t_{k-1}} \quad (12)$$

2.2 Evaluation and refinement of constraints

After following the procedure in the previous Section, a complete history of displacement during un-forming is obtained. This can now be reversed and be considered the history of forming of the part from a flat blank. For each forward step in the forming, an incremental strain field can be calculated by comparing successive node coordinates, as defined in Equations (5-8).

A stress field at each step can then be produced at each step. The flow rule is defined as

$$d\epsilon_{ij}^{p} = \lambda \sigma_{ij}' \quad (13)$$

where σ_{ij}' represents the deviatoric stresses and λ is given by

$$\lambda = \frac{3 d\epsilon_{eq}^{p}}{2\ \sigma_f} = \frac{3 d\epsilon_{eq}^{p}}{2\ \sigma_0 \bar{\epsilon}^{n}} \quad (13)$$

Thus, the deviatoric stresses are given as

$$\sigma_{ij}' = \frac{2 d\epsilon_{ij}^{p}\, \sigma_0 \bar{\epsilon}^{n}}{3\ d\epsilon_{eq}^{p}} \quad (14)$$

However, to obtain a unique, complete stress field on the workpiece, an elastoplastic material model has to be considered. The calculation is straightforward in a displacement-controlled Finite Element model of the forming process. This not only produces the stress state of each element, but the required external nodal forces to achieve the suggested deformation.

The FE model was constructed in Abaqus as a single Dynamic Explicit step, with tabulated amplitude providing the location of each node at each time step. S3R elements were used. The material is also provided in tabulated form and follows the properties given for Al 1100 as above, but with a Young's Modulus of 71GPa

 DOI: 10.1051/matecconf/20168010003

and a Poisson's ratio of 0.33. The thickness of the sheet was set at 1mm.

3 Un-forming strategies

In order to un-form the part the model must be given some perturbation that encourages the motion of the nodes towards a common plane. Three strategies were tried in [7] and are all shown in Figure 3: (a) The flattening of nodes is attempted locally, starting from regions of high negative Gaussian curvature and proceeding accordingly; (b) Un-form the part by moving the peripheral nodes radially outwards from the centre.; (c) Similar to (b) but with a variation of the out-wards trajectory. Here, only results from (b) are considered for further analysis, hence the strategy is presented in greater detail.

3.1 Un-forming via radial boundary pull

Deep drawing is often used to form the square cup and some inspiration can be found in examining relevant results. The boundaries of the blank move towards the deeper locations on the die. Thus, in un-forming the same process would require the boundaries to move away from the centre of the workpiece.

One strategy in [7] is based on pulling the nodes on the periphery on a radial trajectory away from the centre of the cup. First, the radial path connecting each peripheral node to the centre is found and its length, u, is recorded. The end point for the trajectory is placed on the same xy-plane as the centre of the workpiece, such that it is radially at a distance u from the centre and on the same radial plane as the corresponding peripheral node. A linear path is defined between the two and the trajectory of the peripheral node is given at constant increments along this path. The same is done for all peripheral nodes and their motions are synchronised. The radial component of the peripheral motion ensures that gradually the depth of the part is reduced and it deforms towards a blank.

3.2 Results for "radial pulling" strategy

While the work in [7] demonstrated that the un-forming can be achieved with the method in Figure 1, the results of only the first iteration where shown. Those results are shown again here in Figure 4 for completeness. The diagrams (a-d) show steps 1, 29, 49 and 65 of the simulation (out of 65), starting in effect from the desired part geometry and ending with the blank. The remainder of this paper deals with the question "Which metrics are best to evaluate these results and inform the choice of constraints for the next iteration?"

4. Metrics for evaluation of optimization and current constraints

In Figure 1, the path to analysing the deformation history is shown on the right. In red lettering, specific metrics are noted that can inform us on the feasibility of the method and the practical implications of the results. The 'interior point' algorithm used for the optimization will not be evaluated here in depth, but the solver does return a feasible solution at each step, with the tolerances provided in Table 1, and a positive 'exitflag', indicating convergence or the step size reaching the 'TolX' tolerance.

The strains can be obtained directly, since the method produces a full deformation history. Then, the material model and equilibrium are employed to produce the stress field on the workpiece and the required external forces. In practise, an FEM model described in Section 2.2 is used to produce all output. Three potential ways of evaluating

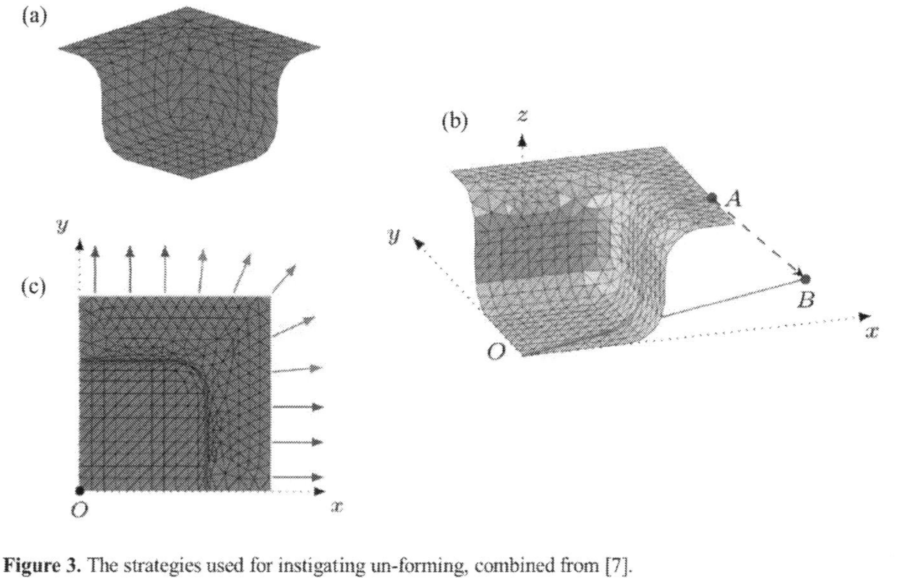

Figure 3. The strategies used for instigating un-forming, combined from [7].

DOI: 10.1051/matecconf/20168010003

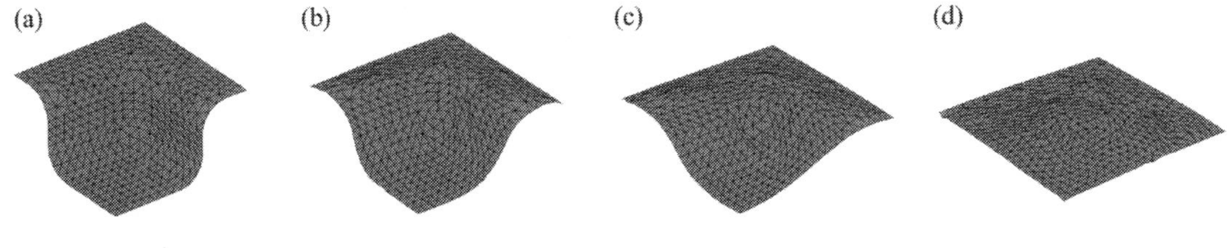

Figure 4. The deformation path for the "radial pull" strategy, modified from [7].

and improving the constraints on the process are listed below:

4.1 Plastic strain

There is little useful information in looking at the absolute values of strain, although empirically extreme values are troublesome. Instead, the standard technique for evaluating the material response and limitations is the Forming Limit Diagram (FLD). This examines the relative magnitudes of the major and minor strains for the metal. Much experimental work has been done to produce such plots for various materials. However, FLDs are usually based on the theoretical assumption that the strain path is linear or the corresponding experimental set up. According to Hosford and Caddel [17], changes in the strain path have a very large effect on the forming limits. Hence it is essential that the strain paths are also examined.

A plot of major and minor strains is used here to observe typical local deformation on the workpiece and evaluate the success of the optimization procedure. In Figure 5, the major and minor in-plane true plastic strains for each element and for all time steps (ϕ_1 and ϕ_2 respectively) are shown. Histograms were added on each axes to show the relative frequency of each value. In addition, the mean of all the strains was plotted as a continuous black line, while a dashed red line corresponding to $\phi_1 = -\phi_2$ is included.

The large number of data points, representing 780 elements at 65 time steps obfuscates any patterns, other than the fact that almost all the strain pairs are in the fourth quadrant, as expected for a "close to pure shear" deformation and close to the pure shear line. At the same time, it is apparent that some minor deviation from $\phi_1 = -\phi_2$ does exist at some locations. However, the average strain values across the workpiece show no noticeable deviation the pure shear line at any time step.

A closer look at this metric can be obtained by plotting a smaller subset of element strain paths. Representative elements are selected from the different geometrical regions on the part. Specifically, points of positive, negative and zero Gaussian curvature are selected. In Figure 6 the locations of the five points are shown, along with the corresponding strain paths. Similar to Figure 5, linear strain paths are observed and the behaviour appears independent of location. There is a noticeable jump in magnitude between the different elements. In particular, elements 3 and 4, with zero Gaussian curvature in the final part geometry seem to experience greater deformation than elements 1 and 2 with positive and negative GC respectively.

For the same set of points, the through thickness deformation is also evaluated in Figure 7, where a plot of the through-thickness plastic strain is shown. All the selected points show some deviation from the constant thickness constraint, but this is typically smaller than 0.5%. Most interestingly, elements 3 and 4 suffer the greatest through-thickness deformation – thickening and thinning respectively – but elements 1 and 2 are much closer to a constant thickness, throughout the process.

Figure 7 also allows a direct evaluation of the accuracy of the optimization algorithm. Given that the constraint tolerance was set at 0.001, it is clear that the optimizations succeeds in achieving the constant area constraint. Any discrepancies for ϕ_3, can perhaps be attributed to a cumulative increase in deviation from the constraint.

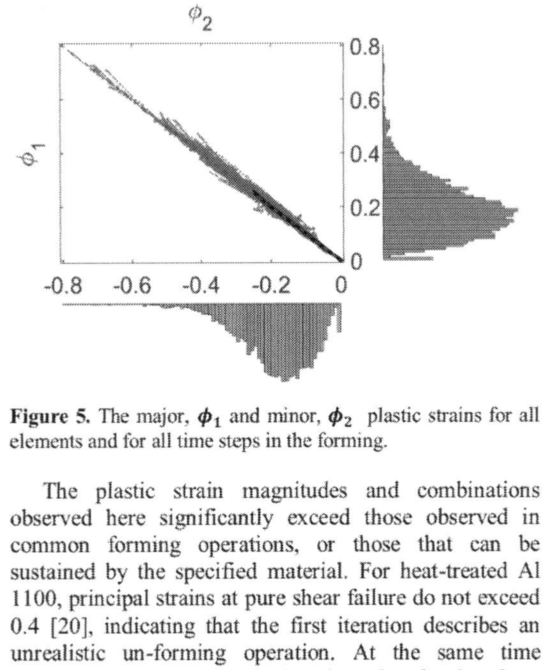

Figure 5. The major, ϕ_1 and minor, ϕ_2 plastic strains for all elements and for all time steps in the forming.

The plastic strain magnitudes and combinations observed here significantly exceed those observed in common forming operations, or those that can be sustained by the specified material. For heat-treated Al 1100, principal strains at pure shear failure do not exceed 0.4 [20], indicating that the first iteration describes an unrealistic un-forming operation. At the same time however, from Figure 5, the locations that develop these extreme values are very few and could perhaps be eliminated completely with improved un-forming strategies. The average strains across the workpiece are well-within tolerated values.

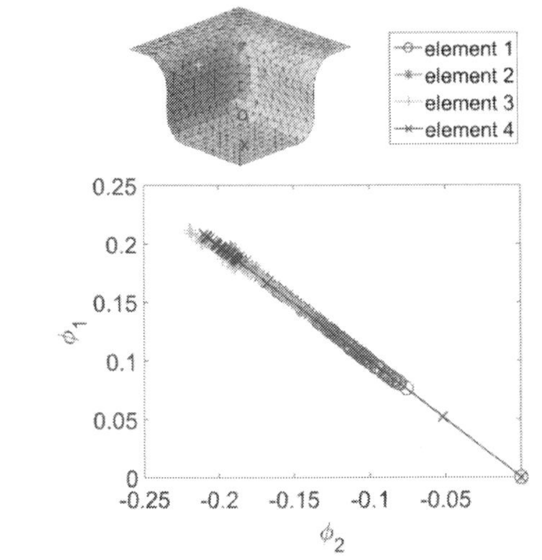

Figure 6. Strain paths of individual elements from representative geometrical locations.

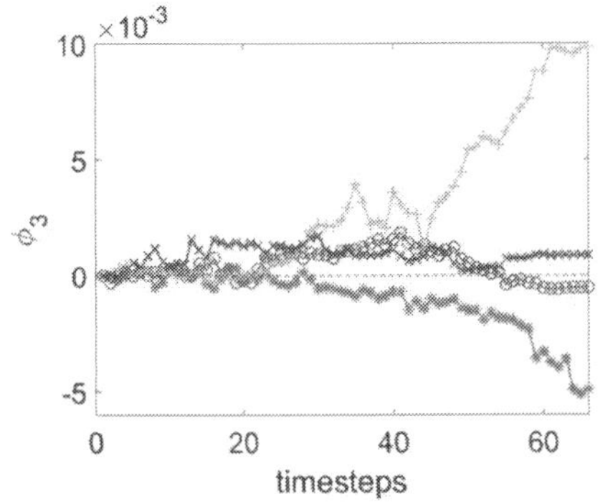

Figure 7. The through-thickness plastic strain for the same set of points as shown in Figure 6.

4.2 Required external forces

Beyond the material limits, for the process to be admissible, the required external forces need to be examined. In [7] a qualitative plot of the forces at one instant in the deformation where shown. A similar plot is shown in Figure 8(a).

The magnitudes of the forces have an apparent correlation with different geometrical features. For example, the highest force magnitudes are concentrated in the area of the brim, where the walls of the cup meet the flange. In contrast, and perhaps predictably, the bottom of the cup requires relatively low forces. This pattern seems to be consistent with knowledge from other forming processes, such as deep drawing.

A challenging result here however is that the direction of the forces with respect to the surface tends to alternate throughout the workpiece. Figure 8(c) displays a diagram of the nodal forces on a single element. Two of the nodal forces point to one side of the element, while the third points in the opposite side. This pattern is typical for the majority of the elements and perhaps a consequence of attempting to enforce pure shear everywhere. However, it makes the arrangement of conventional tooling impractical.

The directionality of the tangential component of the force is also an interesting result since there is no discernible pattern. In Figure 8(b) a plan view of the tangential forces during step 20 in the forming process confirms that even within specific regions, the range of directions needed for the suggested deformation does not point to any known tooling options. On the other hand, when the angle the tangential component of the force makes with the x-axes is plotted in Figure 9, an almost uniform distribution is observed.

The same Figure shows the corresponding quantitative distribution of magnitudes for the nodal forces – again for step 20. A sense of the magnitude of the force field throughout the forming process, can be obtained by looking at the average of the nodal forces. The median magnitude of nodal forces is 671N at the first forming step and 609N at the last step, and no great deviation beyond this range is observed at any other step.

Altan and Tekkaya [19] report punch forces for the deep drawing of a comparable part with similar material properties in the range of 50-200kN. Thus for a comparable mesh coarseness, an average nodal force in the deep drawing operation is in the 117-467N range. Hence, the forces suggested by the un-forming method are in proximity to those employed in existing processes, although their local application remains a challenge.

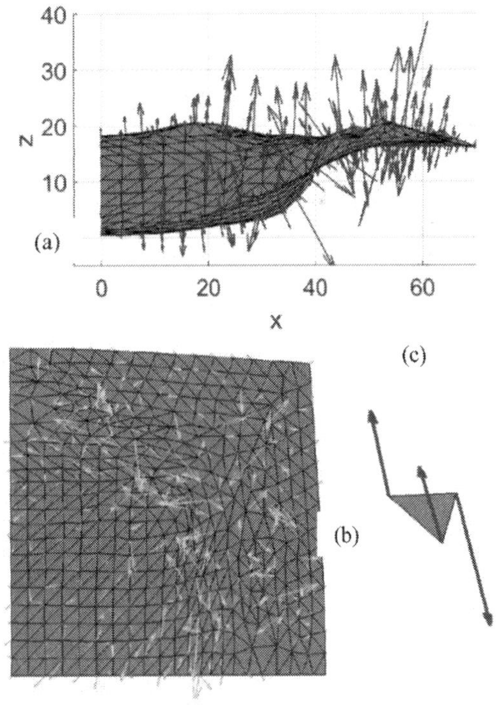

Figure 8. (a) Profile view of required nodal forces (step 20/65) and (b) plan view of tangential forces. Typical direction of forces on a single element is shown in (c).

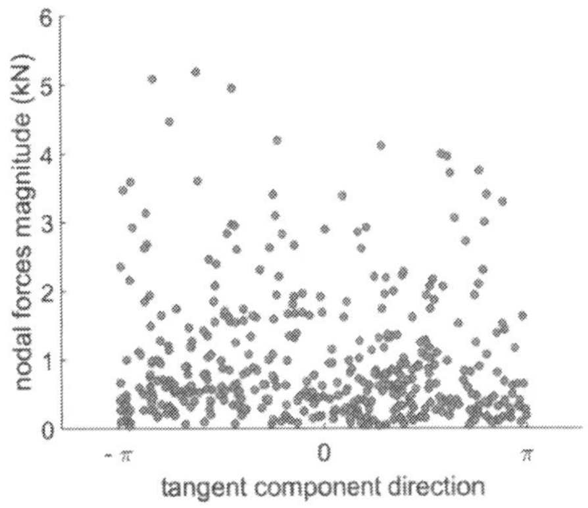

Figure 9. The magnitude of the nodal forces plotted against the tangent component's angle to the x-axes (step 20/65).

4.3 Required friction coefficient assuming single tool at each node

The process of coming up with practical tools necessitates an investigation of the friction required to achieve the external forces shown in the previous section. Such an investigation assumes that the forces will be provided at each node by a single tool. The tangential and normal components of nodal forces from the Abaqus simulation are first obtained and then divided to produce the required friction coefficient at each location on the surface.

Figure 10 is a "colormap" of the required friction coefficient at each node with the assumptions above. A large number of elements require infeasible values of friction. Thus, alternative tooling which allows the application of greater forces in the plane of the surface rather than out-of-plane need to be considered. In turn, this might mean the use of a holder (either locally or globally), as well as the necessity of fine local control of tool loads and directions.

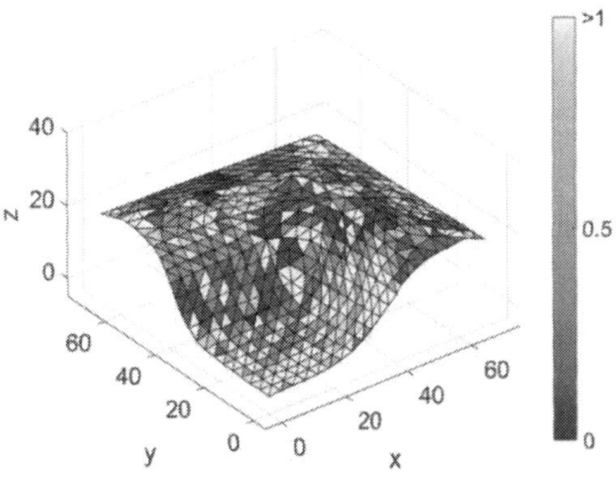

Figure 10. Colormap of required friction coefficients on surface of workpiece (time step 30 of 65)

5 Discussion

This paper extended work in [7], which proposed a new method for designing sheet forming processes by un-forming a desired part while minimizing the plastic work done and applying pre-selected kinematic constraints. Here, the results from [7] are evaluated with consideration for the practical implications.

This examination suggests that the current optimization method succeeds in satisfying the pure shear constraint across the workpiece. However, there is no direct way of measuring the proximity of the un-forming solution to the true minimum plastic work at each step.

Adherence to the constraints can potentially be improved further by a less naïve un-forming strategy. For example, instead of a hard constraint on the trajectory of peripheral nodes, their average distance from the centre might be gradually increased instead. But to guarantee the success of the optimization, a convex expression of the plastic work needs to be constructed and provided to a suitable solver that guarantees a global minimum. Similar efforts can be found in the literature [21].

For addressing the complexity of force directionality, an additional future constraint could prevent the deformation of certain regions, thus producing a more localized tool operation. Mathematically, this means allowing certain elements to only move without being distorted.

Finally, friction-based tools have been rejected as a practical approach to achieving such deformation. For pure shear deformation and the required in plane forces, gripping of the sheet might be necessary.

In summary, this paper analysed results from [7] and suggested possible steps in developing this method further and producing realistic processes in this automated manner.

Acknowledgements

This research was supported by the EPSRC grant EP/K018108/1.

References

1. R. L. Milford, J. M. Allwood, and J. M. Cullen, Resour. Conserv. Recycl., **55**, 12 (2011)
2. J. A. Polyblank and J. M. Allwood, CIRP Ann. - Manuf. Technol., **64**, 1 (2015).
3. J. M. Allwood, G. P. F. King, and J. R. Duflou, Proc. Inst. Mech. Eng. Part B J. Eng. Manuf., **219**, 2 (2005)
4. J. Jeswiet, M. Geiger, U. Engel, M. Kleiner, M. Schikorra, J. Duflou, R. Neugebauer, P. Bariani, and S. Bruschi, CIRP J. Manuf. Sci. Technol., **1**, 1 (2008)
5. J. M. Allwood and H. Utsunomiya, Int. J. Mach. Tools Manuf., **46**, 15 (2006).
6. P. Groche, D. Fritsche, E. a. Tekkaya, J. M. Allwood, G. Hirt, and R. Neugebauer, CIRP Ann. - Manuf. Technol., **56**, 2 (2007).
7. E. G. Loukaides, J. M. Allwood, Int. J. Mech. Sc. (submitted, 2016)
8. J. McCartney, B. K. Hinds, and B. L. Seow, Comput. Des., **31**, 4 (1999)
9. C. C. L. Wang, S. S. F. Smith, and M. M. F. Yuen, CAD Comput. Aided Des., **34**, 11 (2002)

10. B. K. Hinds, J. McCartney, and G. Woods, Comput. Des., **23**, 8 (1991)
11. P. Azariadis and N. Aspragathos, Computer-Aided Design, **29**, 10 (1997)
12. P. N. Azariadis, A. C. Nearchou, and N. A. Aspragathos, Comput. Ind., **47**, 3 (2002)
13. C. H. Lee and H. Huh, J. Mater. Process. Technol., **82**, 145–155 (1998)
14. Z.-Y. Cai, M.-Z. Li, and H.-M. Zhang, Finite Elem. Anal. Des., **43**, 4 (2007)
15. K. Chung and O. Richmond, Int. J. Mech. Sci., **34**, 7 (1992)
16. K. Chung and O. Richmond, Int. J. Mech. Sci., **34**, 8 (1992)
17. W. Hosford and R. Caddell, *Metal forming: mechanics and metallurgy* (2011)
18. K. Lange, *Handbook of metal forming* (McGraw-Hill, 1985)
19. T. Altan and A. E. Tekkaya, Eds., *Sheet Metal Forming: Fundamentals* (ASM International, 2012)
20. A. I., Durán, J. W. Signorelli, D. J. Celentano, M. A. Cruchaga and M. François, J. Mat. Eng. Perf. **24**, 10 (2015)
21. A. Raithatha and S. R. Duncan, Int. J. Numer. Methods Eng., **78**, 955–979 (2009).

Multi-objective optimization under uncertainty for sheet metal forming.

Pascal Lafon[1,a], Pierre Antoine Adragna[1], and Von Dim Nguyen[2]

[1] *Université de technoglogie de Troyes ICD-LASMIS UMR CNRS 6281*
[2] *Université de technologie de l'information et de la communication de Thai Nguyen*

Abstract. Aleatory uncertainties in material properties, blank thickness and friction condition are inherent and irreducible variabilities in sheet metal forming. Optimal design configurations, which are obtained by conventional design optimization methods, are not always able to meet the desired targets due to the effect of uncertainties. This paper proposes a multi-objective robust design optimization that aims to tackle this problem. Results obtained on a U shape draw bending benchmark show that spring-back effect can be controlled by optimizing process parameters.

1 Introduction

In Sheet metal forming processes, intrinsic variabilities in material properties, geometry of blank and process parameters are unavoidable. These variabilities will be considered here as aleatory uncertainties. As a consequence, these uncertainties may lead to instabilities in performance of manufactured parts such as shape variation due to spring-back, thinning, tearing and wrinkling. Here we are interested in instabilities produced by uncertainties on spring-back effects. We will consider the problem of tolerance specification in detailed design phase of stamped part. In this detailed design phase, to ensure the desired behaviour and functional requirements of a part in spite of uncertainties, the part features are assigned to a tolerance zone that must contain the part features. Therefore, tolerance design is a key element in industry for improving product quality and decreasing the manufacturing cost. Since the sheet metal part's performance variation induced by uncertainties of the inputs is quite large due to spring-back, geometric tolerance analysis for the sheet metal part is a problem to consider.

The goal of this paper is to propose a robust design optimization of forming processes. In this study we will take into account uncertainties of process parameter, material and geometry uncertainties of the blank. First we will introduce the two main approaches of dealing with uncertainties in an a optimization framework. Next, the general formulation of a robust optimization problem will be presented.

2 Uncertainties, robust or reliable optimization of processes

Dealing with uncertainties can be achieved by considering "Reliability Based Design Optimization (RBDO)" or "Robust Design Optimization (RDO)". Reliability analysis and optimization are two essential components of RBDO: (1) Reliability Analysis focuses on analyzing the probabilistic constraints to ensure the reliability levels are satisfied; (2) Optimization is seeking for the optimal performance subject to the probabilistic constraints. The RBDO approach handles noise variables in a probabilistic way. The RBDO was mainly used in optimization under uncertainty of sheet metal forming processes which the failure probability of wrinkling or excessive thinning was considered [1–4].Robust Design Optimization (RDO) aims at minimizing the variability of the part performance, while meeting the requirements of optimum performance and constraint conditions. Similar to the RBDO approach, uncertainties are handled in a probabilistic way. There is still a disagreement relating to the similarity between the RBDO and the RDO. According to Park et al. [5], the RBDO is similar to the RDO. Both approaches aim at incorporating uncertainty into the optimization study. Whereas Zang et al. [6] discussed that there is a conceptual difference between RDO and RBDO. The RDO rather aims at reducing the variability of performance caused by fluctuations in parameters than to avoid a catastrophe in an extreme event. In the case of RBDO, it can make a design that displays large variations as long as there are safety margins failure in the design, i.e. the variability is not minimized. Furthermore, the two approaches differ in some aspects [7]. The choice for using the RDO approach or the RBDO approach depends on the objective of the optimization study. In the RDO, insensitiveness of the objective function is emphasized. In the RBDO, reliability of constraints is important [5]. Thus, both approaches will obviously lead to different optimization outcomes.

[a]corresponding author :pascal.lafon@utt.fr

DOI: 10.1051/matecconf/20168010004

3 A general framework for robust optimization

We consider the robust optimization state as the following multi-objective optimization problem:

$$\begin{cases} \min\limits_{X} E\left[F(X,P)\right] \\ \text{and} \\ \min\limits_{X} \sigma\left[F(X,P)\right] \\ \text{Under the constraints} \\ \mathcal{P}\left[C(X,P) \leq 0\right] \geq \beta \end{cases}$$

Where X and P are vectors of random variables, which are modelled as normal probability distribution functions. Here X represents the design variables, which are controllable by the optimization process, and P represents the noise parameters which are not controllable. An expected value ($E[x_i]$, $E[p_i]$) and standard deviation $\sigma[x_i]$), $\sigma[p_i]$)) are associated to each component x_i and p_j of these vector of variable. Here the criteria to be optimized $F(X,P)$ and the constraints $C(X,P)$ are written as a vector of function. In this formulation, $\mathcal{P}[C(X,P) \leq 0]$ is the probability to respect the event "all constraints $C(X,P)$ are respected, i.e less than 0. In this kind of problem the evaluation of expected value $E[F(X,P)]$ and standard deviation $\sigma[F(X,P)]$ is known as "uncertainty propagation". This can be achieved by sampling methods (e.g. Monte-Carlo), approximation method like First or Second Order Method (FORM/SORM) or polynomial chaos. Practically, as soon as evaluation of function F and C requires finite elements calculation, surrogate model or meta model became essential. Thus, to obtain the solution for this kind of optimization problem, i.e the Pareto Front, we must go through the following steps:

- Setting up a "Design Of Experiment" on the domain of design variables X and noise parameters P, considering here that we deal with deterministic variables and parameters.
- Building a suitable meta-model for each single function of the vector of functions $F(X,P)$ and $C(X,P)$.
- Embedding uncertainty propagation into an optimization framework : calculation of expected value and standard deviation using meta-model within an stochastic optimization algorithm like NGSAII.

This procedure is now implemented in some commercial software as for example ModeFrontier^{TM}or Isight^{TM}. In this paper this procedure will be used on a NUMISHEET 2011 benchmark, and we will focus on spring-back effects for this draw bending process.

4 Case study : U shaped draw bending

The case study in this paper is a benchmark problem of NUMISHEET 2011 which the springback behavior of advanced high strength steels of DP780 steel in U-shaped draw bending is investigated [8]. The DP780 steel sheet with 1.4 mm thick, 360 mm long and 30 mm wide is considered. Experiments show that part profiles remain symmetric through the manufacturing process (see Fig. 1).

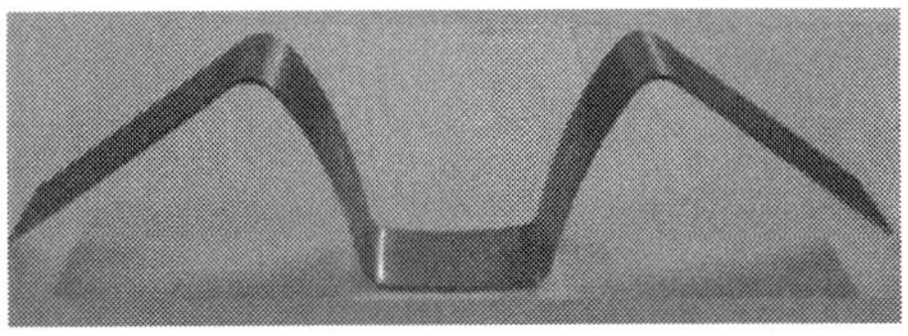

Figure 1. Effects of spring-back on U-shape drawing

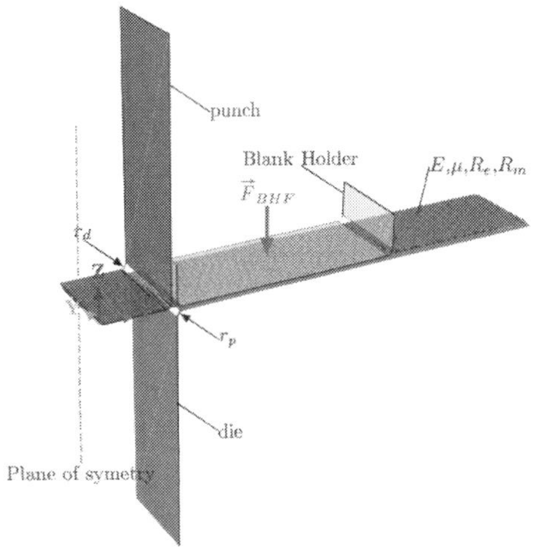

Figure 2. Numerical model of U-shape draw bending process

4.1 Numerical model of the draw bending process

The main tooling set-up and modelling for the draw bending process is shown in Fig. 2. In this study, modelling and numerical simulation of draw bending process are carried out by Finite Element Method using ABAQUS 6.12-2. The blank is modelled by 2709 S4R-shell elements with 7 integration points through thickness. The blank's strain-hardening behaviour is described by Swift law. Yield behaviour of the material is modelled by Hill'48 function. The nominal material properties of the blank are given as follows: Young's modulus $E = 198.8$ GPa, Yield strength $R_e = 550$ MPa, Tensile strength $R_m = 840$ MPa, Uniform elongation $\varepsilon_u = 13.1\,\%$, "R-value"= 0.781, Poisson's ratio $\nu = 0.3$. All tools are assumed to be rigid bodies. Boundary conditions applied on the tools are demonstrated in Fig.2. Contact interaction is modelled by penalty contact enforcement method in which type of contact of surface-to-surface is applied between blank surface and the tool's surfaces.

The process is executed through two different steps. The first step is to simulate forming operation via dynamic explicit. The punch moves downward a travel of 71.8 mm. Simulation of spring-back behaviour is performed in the second step through static implicit procedure. The final shape of the part is obtained after releasing the tools. An example of effects of spring-back is represent on Fig. 3. The U-shape after spring-back is defined by three parameters : the angles of θ_1, θ_2 and the side wall curl ρ (see Fig. 4.

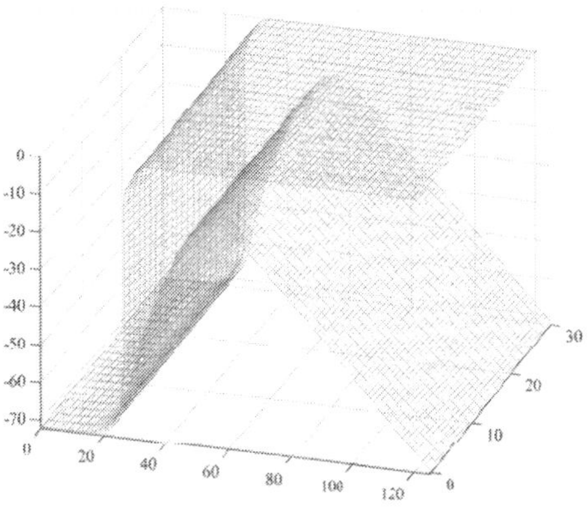

Figure 3. Example of the U-shape after spring-back

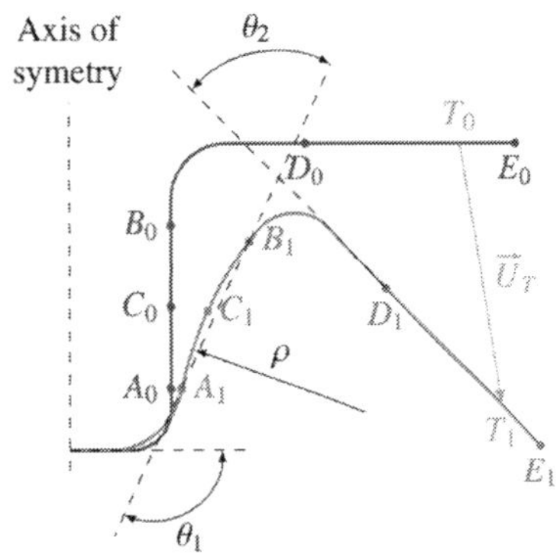

Figure 4. Parameters of the U-shape after spring-back

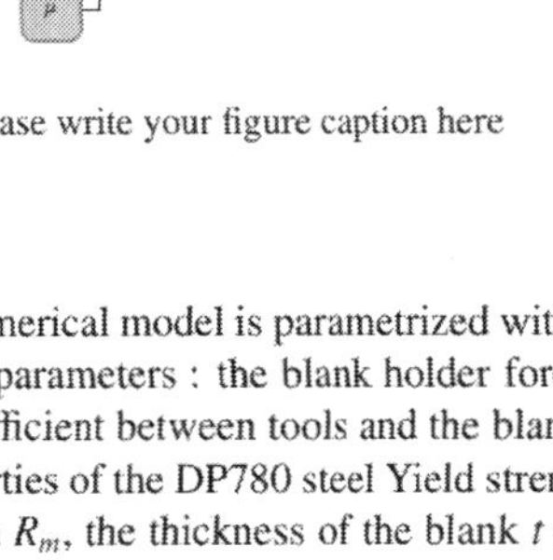

Figure 5. Please write your figure caption here

This numerical model is parametrized with the following 7 input parameters : the blank holder force F_{BHF}, the friction coefficient between tools and the blank μ, the material properties of the DP780 steel Yield strength R_e, Tensile strength R_m, the thickness of the blank t and the radii of die r_d and of punch r_p.

The output parameters under consideration include the angles of θ_1, θ_2, the side wall curl ρ and the displacement of a virtual hole U_T. These output parameters are computed with MATLAB™(Fig. 5).

4.2 Validation of the numerical model

In the numisheet 2011 benchmark [8] experimental measurements are reported and compared with the set of results from several simulation results. Table 2 shows the results of our model for values of parameters in Table 1. It can be observed, that there is rather small differences between our numerical results and experimental measurements from the numisheet 2011 benchmark.

It should be noticed that our model gives good results for the configuration of parameters given in Table 1. It is of course too expensive, to obtain experimental measurements for a wide range of paramter's values. So we must thrust in the numerical model to explore the physical response and specifically in terms of spring-back. In the robust optimization process developed further, it is assumed that the numerical model gives consistent response for spring-back effects.

Table 1. Input data for comparison with experimental results of Numisheet 2011.

Input parameters	
Blank thickness	1.4 mm
Yield strength	527 MPa
Tensile strength	831.5 MPa
Friction coef.	0.1
Blank Holder force	2940 N
Die radius	7 mm
Punch radius	5 mm

Table 2. Our numerical results and experimental measurements from Numisheet 2011.

Spring-back parameters	Numerical results	Experimental measurements [8]
θ_1	108.32°	110.62° to 118.3425°
θ_2	75.228°	77.61° to 80.00°
ρ	137.02 mm	118.36 mm to 120.00 mm

5 Robust optimization problem and meta-modelling

For this case study, we are interested in the effect of spring-back on the displacement of a virtual hole U_T on this U-shape part (Fig. 4). The goal is to evaluate the consequence of input parameters uncertainties on the hole displacement. All input parameters will be modelled as aleatory variables with normal distribution. Table 3 shows details of normal distributions used to model aleatory variables.

This problem will be formulated as a robust optimization problem, where the deviation of expected value of the displacement from a target is minimized, and where the standard deviation of the hole displacement will be also minimized. Among the 7 input parameters, F_{BHF}, r_d and r_p will be chosen as design variables X (i.e controllable parameters), and the other 4 parameter (μ, R_e, R_m, t) will be

Table 3. Probabilistic modelling and representation of uncertainties in the input parameters

Param.	Unit	Dist.	Std. dev. σ	Admissible Mean val.
F_{BHF}	kN	Normal	0.6667	2.94–50
r_d	mm	Normal	0.0167	2–10
r_p	mm	Normal	0.0167	2–10
R_e	MPa	Normal	16.667	500–600
R_m	MPa	Normal	20	780–900
μ		Normal	0.0033	0.04–0.16
t	mm	Normal	0.0167	1–2

considered as noise parameters P (i.e. uncontrollable parameters). This robust optimization problem can be stated as:

$$\begin{cases} \min_X |E[U_T(X,P)] - U_T^{\text{Cible}}| \\ \text{and} \\ \min_X \sigma[U_T(X,P)] \end{cases} \quad \text{with}: \begin{cases} X = \{F_{BHF}, r_d, r_p\} \\ P = \{R_e, R_m, t, \mu\} \end{cases}$$

The evaluation of the displacement of the virtual hole U_T for a (X, P) is based on the numerical model as presented previously and requires 2.5 h of calculation on a workstation running at 2.4 GHz.

In order to solve this optimization problem it is necessary to build a meta-model and set up an Design Of Experiment. We have defined a DOE based on a full factorial DOE with 3 levels for 7 parameters along a DOE of 2 supplementary levels for 2 parameters (F_{BHF}, μ). The overall DOE represent $3^7 + (3^5 \times 2^2) = 3159$ numerical simulations (see Table 4). The huge amount of simulations has been ran on a parallel computer with 1600 CPU's.

Table 4. Factors and levels for the DOE

Param.	levels
F_{BHF}	2.94, 14.705, 26.47, 38.325 , 50 kN
r_d	2, 6 , 10 mm
r_p	2, 6 , 10 mm
R_e	500, 550 , 600 MPa
R_m	780, 840 , 900 MPa
μ	0.04, 0.07, 0.1, 0.13 , 0.16
t	1, 1.5 , 2 mm

Table 5. Comparison of accuracy of meta-models for U_T

Meta-model	MAE	R^2
Kriging	1.0258×10^{-4}	0.9999
SVD2	36.6112	0.8268
SVD3	38.6366	0.9370
RBF	5.7494×10^{-10}	1
NN	16.9244	0.9988

Here several meta-models have been tested. The two metrics which are used to measure the meta-models' accuracy are the R-square (R^2) and the MAE.

$$\text{MAE} = \max |y_i - \hat{y}_i|, \quad i = 1 \ldots m$$

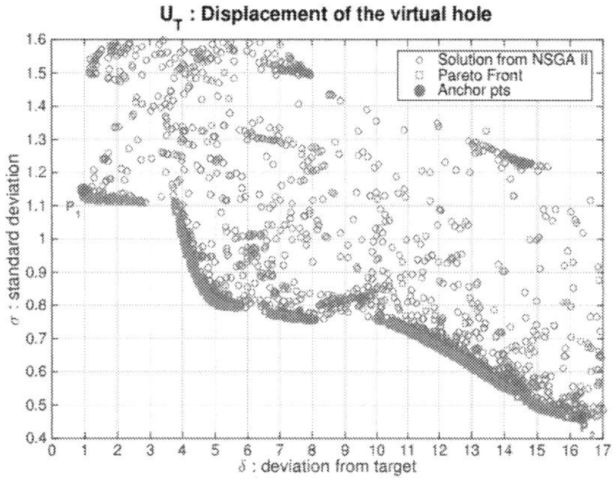

Figure 6. Please write your figure caption here

$$R^2 = 1 - \frac{\sum_{i=1}^{m} (y_i - \hat{y}_i)^2}{\sum_{i=1}^{m} (y_i - \bar{y})^2}$$

where m is the number of validation points; $\hat{y}_i$ is the predicted value for the observed value y_i. $\bar{y}$ is the mean of the observed values at the validation points. The accuracy of each kind of metamodels including Kriging, Singular Value Decomposition of degree 2 (SVD2), Singular Value Decomposition of degree 3 (SVD3), Radial Basis Functions (RBF) and Neural Networks (NN) is given in Table 5. It is clearly found that the accuracy of RBF is the best both in terms of R-square and in terms of MAE in this case study. Therefore, the meta-models built by using the RBF will be used.

6 Optimization Results

The optimization process uses the stochastic optimization algorithm NSGAII [9], in the ModeFrontier™software. The uncertainty propagation used here, is based on Latin Hypercube sampling with 1000 samples. Table 6 presents mean values and standard deviation of noise parameters used in the optimization problem modelling. The initial population comprises 125 individuals, and we ran 20 generations for the NSGAII algorithm.

The figure 6 shows the Pareto front obtained with the NSGAII algorithm. This Pareto front has 425 solutions. It can be seen that the two optimization criteria (deviation of expected value from target and standard deviation of hole displacement) are antagonist, thus this optimization problem is an actual robust optimization problem [10].

This optimization result has been obtained with the following noise parameters from Table 6.

Tables 7 and 8 give the solution of the two "anchor" points of the Pareto front, i.e. the minimum mean and the maximum mean values of U_T. For each of these optimal solutions, the value of design variables X is given. We can see that for the maximum mean value of U_T (Table 8) we have the minimum standard deviation for U_T, so under the uncertainty of noise parameters this optimum value is very stable. On the other hand, for the minimum value of

U_T (Table 7) we have the maximum standard deviation for U_T, so this optimum solution is not stable.

These results demonstrate that one can control the effect of uncertainty on input parameter. By controlling the blank holder force F_{BHF} we can reduce the effects of uncertainty of the hole displacement U_T.

Table 6. Mean and standard deviation values of noise parameters

Param.	Mean	Std. dev.
R_e [MPa]	550	16.667
R_m [MPa]	840	20
μ []	0.1	0.0033
t [mm]	2	0.0167

Table 7. Optimization result for anchor point P_1

Param.	Mean	Std. dev.	Min	Max
F_{BHF} [kN]	47.06	0.66	44.99	49.18
r_d [mm]	9.950	0.017	9.898	10.003
r_p [mm]	2.809	0.017	2.746	2.862
U_T [mm]	0.9541	1.1475	−2.3650	4.6207

Table 8. Optimization result for anchor point P_2

Param.	Mean	Std. dev.	Min	Max
F_{BHF} [kN]	19.23	0.66	17.16	21.36
r_d [mm]	4.042	0.017	3.990	4.095
r_p [mm]	4.914	0.017	4.851	4.966
U_T [mm]	16.3225	0.4614	14.4989	17.9437

7 Conclusions

In this paper we have proposed a robust optimization procedure of a draw bending process. In this work, uncertainties on the material properties, geometry of the blank and process's parameters have been modelled. Then among the set of parameters we have chosen design variables and noise parameters. This choice depend on the optimization context, here we have supposed that the geometry of the blank the material have been set previously in the design process of the part. Finally, assuming that the numerical model is consistent, the optimization results show that we can reduce effect of uncertainties by controlling the blank holder force.

References

1. M. Strano, Proceedings of the Institution of Mechanical Engineers, Part B: Journal of Engineering Manufacture **220**, 1305 (2006)
2. W. Zhang, R. Shivpuri, Reliability Engineering & System Safety **94**, 152 (2009)
3. G. Gantar, K. Kuzman, Journal of materials processing technology **167**, 237 (2005)
4. T. Buranathiti, J. Cao, Z.C. Xia, W. Chen, *Probabilistic design in a sheet metal stamping process under failure analysis*, in *NUMISHEET 2005: Proceedings of the 6 th International Conference and Workshop on Numerical Simulation of 3 D Sheet Metal Forming Processes; Part A* (2005), Vol. 778, pp. 867–872
5. G.J. Park, T.H. Lee, K.H. Lee, K.H. Hwang, AIAA journal **44**, 181 (2006)
6. C. Zang, M. Friswell, J. Mottershead, Computers & structures **83**, 315 (2005)
7. J.H. Wiebenga, Ph.D. thesis (2014)
8. H. H, C. K, H. SS, C. WJ, Proceedings of Numisheet 2011 (2011)
9. K. Deb, S. Agrawal, A. Pratap, T. Meyarivan, Lecture notes in computer science **1917**, 849 (2000)
10. V.D. Nguyen, Ph.D. thesis, Université de technologie de Troyes, ICD-LASMIS UMR 6281 (2015)

Sensitivity analysis of the Expansion Process for Alloy UNS N08028

Aitor Navarro[1,a], Mario Lechner[1], Unai Ruiz[1] and Alejandra Lopez[1]

[1]*Tubacex Group, Innovation Department, C/ Tres cruces 8, 01400 Llodio, Alava – Spain*

Abstract. Due to the good mechanical properties of forged parts, the forging process plays a decisive role in the manufacturing of seamless stainless steel pipes for oil country tubular goods (OCTG) lines. Tough competition between manufacturers gives them plenty of incentive to make their processes in raw material and energy usage more and more efficient. In this context the expansion process is one of the critical production steps in the manufacturing of seamless stainless steel pipes. This work presents a sensitivity analysis of a finite element method (FEM) for the simulation of the expansion of the alloy UNS N08028. The input parameters ram speed, tool angle, initial ID and final ID of the billet as well as temperature were used to describe responses like tool wear and material loss. With the aim to minimize the tool wear and to reduce the material waste, a study of influence of the input parameters on the mentioned responses were performed. This development is supported by experimental work in order to validate the simulation model. The sector demand for new materials with specific properties and the cost-intensive experimental trials justifies the use of such simulation tools and opens great opportunities for the industry.

1 Introduction

The extension of product portfolio with new grades and materials with specific properties demanded by the seamless stainless steel pipe sector significantly increases the number of tests in the production line of the companies. For this reason, a FEM model has been established as a highly valuable tool to detect process failures in early design stages and to reduce production standstill times.

Several techniques and processes exist in the field of metallurgy to produce high-alloyed seamless steel pipes, which were successfully developed in the 20th century. The extrusion process is one of these methods, which competes with different rolling processes, such as pierce rolling. The low billet weight and long dead cycle times are great disadvantages for the extrusion process but the low-cost of the tools gives an opportunity on small campaigns, for example, unusual materials or dimensions [2,4].

Round billets are usually used for the extrusion process. When high-alloyed steel pipes are manufactured, the billet is normally drilled, heated and the bore is expanded to a desired diameter [2]. The following picture shows an example of a typical manufacturing process chain for seamless pipe production.

The expansion process is a well-known method to produce expanded billet in order to reduce material waste during the pipe production. In this step, the billet is introduced in a container and compressed with an expansion tool – a so-called nose – characterised by its inclined contact area.

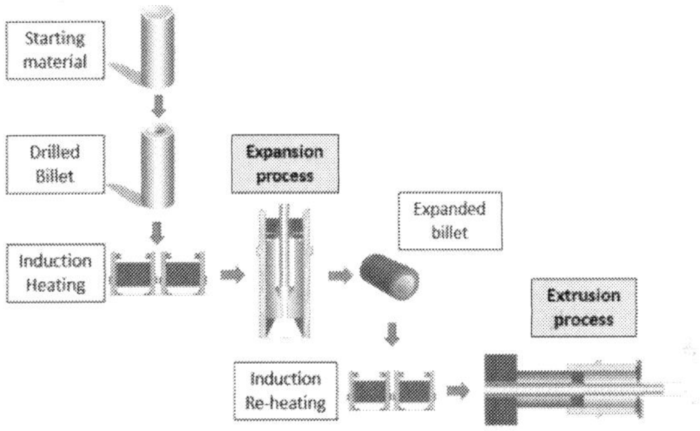

Figure 1. Sketch of seamless pipe manufacturing process chain.

As a result, the material flows along this tool expanding the inner diameter of the billet. The expansion process ends by cutting of a piece of material. This material loss - known as remainder - and its weight depends on the tool design, temperature of the billet, lubrication, tool velocity, material type and the expansion limit. The picture of Figure 2 shows a sketch of the expansion process [4].

Seamless stainless steel tubes are used in high resistance corrosion environments present in the oil and gas industry, the chemical sector, the mechanical engineering industry and the nuclear industry. One of these high corrosion resistance materials is the alloy UNS N08028 and therefore this grade has been chosen to carry out this study. The main application area of alloy UNS N08028 is the oil and gas industry for OCTG lines [1].

[a] Corresponding author: anavarro@tubacex.com

DOI: 10.1051/matecconf/20168010005

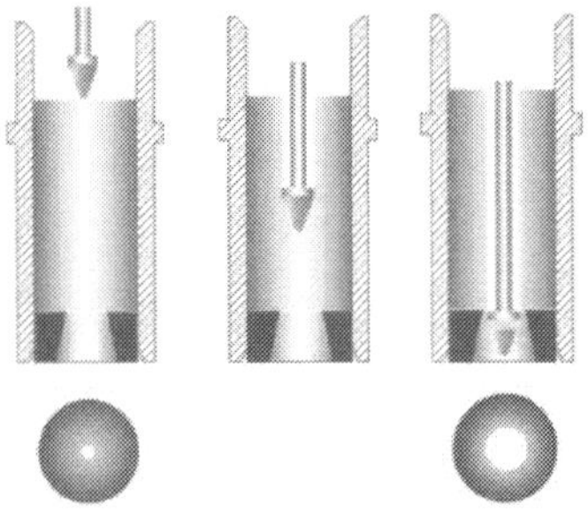

Figure 2. Sketch of expansion process steps.

For this reason, Tubacex developed a simulation model for the expansion process technology using alloy UNS N08028. This work focuses on increasing the tool life and reducing the material waste based on sensitivity analysis of the input parameters.

2 Model validation

2.1 Simulation model

In order to guarantee reliability of the results obtained from simulation, it is necessary to make a comparison with experimental results. Thus, a model validation was carried out first. Figure 3 shows the comparison between the experimental process and the simulation process.

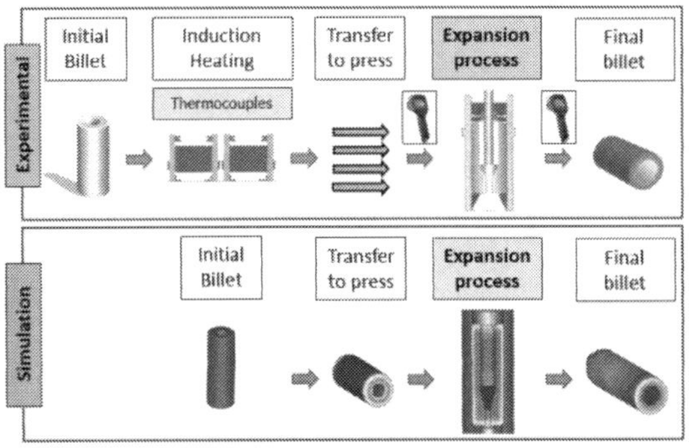

Figure 3. Simulation steps comparing with experimental steps.

Regarding the simulation, the commercial software FORGE NxT1.0 was chosen for this research project. High-performance workstation with Intel Xeon CPU E5-2643v2 processor was used. The uniaxial hot tensile test was selected to describe the alloy UNS N08028 deformation behaviour [3]. The test was carried out at temperatures between 900 - 1220°C and strain rates between 0.1 – 10s^{-1}. The temperature was obtained with thermocouples, and the strain and strain rate was measured with Gom Aramis method. The Hansel-Spittel approach was applied to extend the range of the flow stress curve. The material data was implemented in the database of the software with Point-by-point method. A 2D thermo-mechanical model was used in order to reduce the computation time. Glass was used as the lubrication medium and plays an important role for this type of processes. Its friction coefficient and heat exchange coefficient at a temperature of 1200°C are 0.03 - 0.13 and 1500 – 2200 W/m^{2}K, respectively [4,6,7]. Small variations in the friction and heat exchange coefficient give significant differences in expansion press load and billet temperature.

2.2 Experimental results

As for the experimental work, two trials were conducted for model validation. Furthermore, one billet was prepared to extract the temperature data from the induction heating system with five thermocouples placed inside. The experimental results reveal the homogeneity of the temperature achieved during the induction heating process. Even so few differences were found between bottom and top, and between inner and outer side of the billet.

After the induction heating process, the billets were transferred to the expansion press. With help of a thermographic camera, the thermal image was captured before the expansion of the billet and at the end of the expansion process. The following graphic shows the temperature comparison between experimental and numerical results before the billet is set into the container.

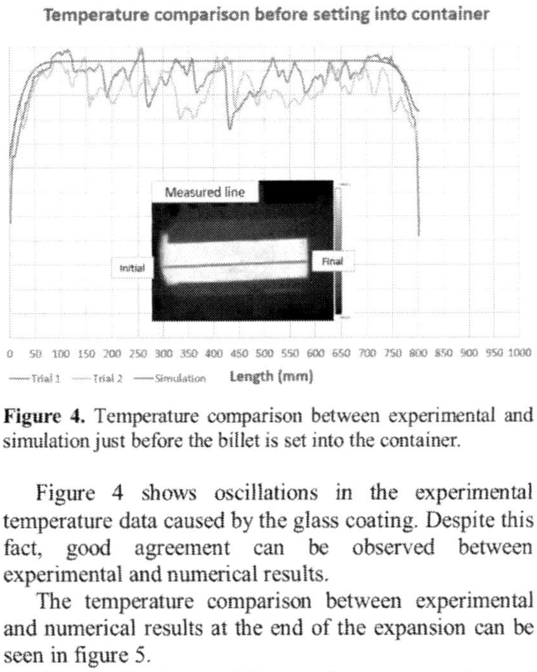

Figure 4. Temperature comparison between experimental and simulation just before the billet is set into the container.

Figure 4 shows oscillations in the experimental temperature data caused by the glass coating. Despite this fact, good agreement can be observed between experimental and numerical results.

The temperature comparison between experimental and numerical results at the end of the expansion can be seen in figure 5.

The results show a difference between experimental and numerical model of approximately 70°C at the left side of the billet. Extensive temperature testing was carried out to determine the veracity of the temperature curve obtained in the simulation. The collected results revealed that the glass coating had a great effect on the measurements obtained with the thermographic camera. Thus, care must be taken when using an infrared radiation system to measure the temperature on glass-lubricated billets.

 DOI: 10.1051/matecconf/20168010005

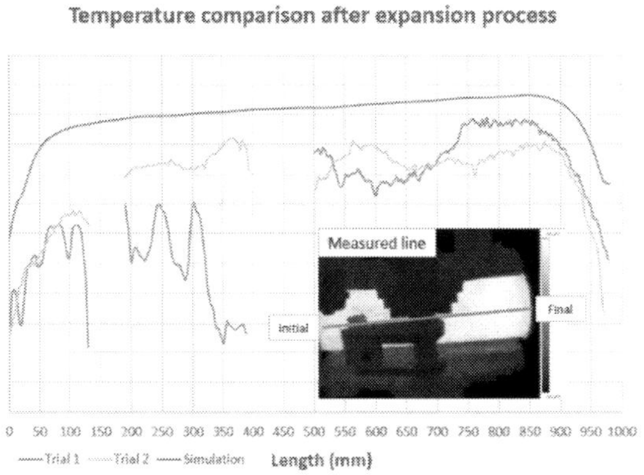

Figure 5. Temperature comparison between experimental and simulation after the expansion process.

The expansion process was carried out with a hydraulic press and with constant ram speed. The expansion press load was measured. The following diagram shows a comparison between experimental and numerical load.

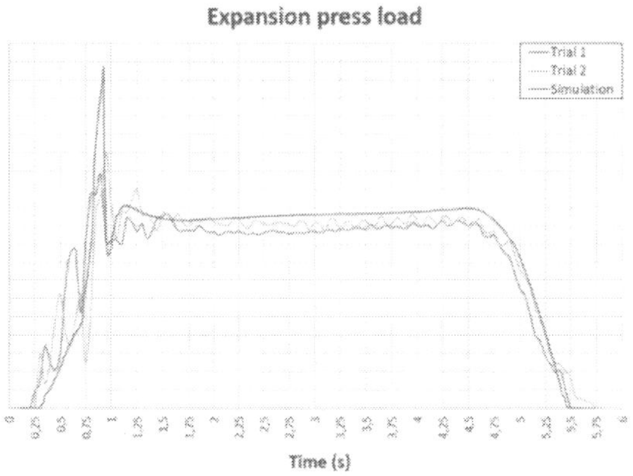

Figure 6. Load comparison between experimental and numerical results during the expansion process.

As can be seen, the result of the numerical model shows a good agreement with the experiment. The average value of the simulation load in the steady state is approx. 3-5% higher than the experimental load. Regarding the graphic analysis, two different forging behaviours can be clearly distinguished: At the beginning of the process, the press compresses the billet and material does not flow alongside the expansion nose; as a result, high peaks of load are observed. When the billet contacts with the cutting ring in the lowest part, the material starts to flow, the load drops and a steady state is attained. If the material does not flow before the bore hole of the billet is closed, the tool is trapped inside and it can cause a fracture of the mandrel. A rise on the heat exchange results in an increase on the gradient of the load, and an increase of friction reveals an increase on the global load.

2.3 Simulation results

During the expansion process, the expansion nose compresses the billet and the material flows along the tool surface. As a result, the temperature increases inside the billet due to the deformation heating. This leads to a rise in the temperature difference between the inner and outer surface. The following figure shows the temperature evolution during the expansion process.

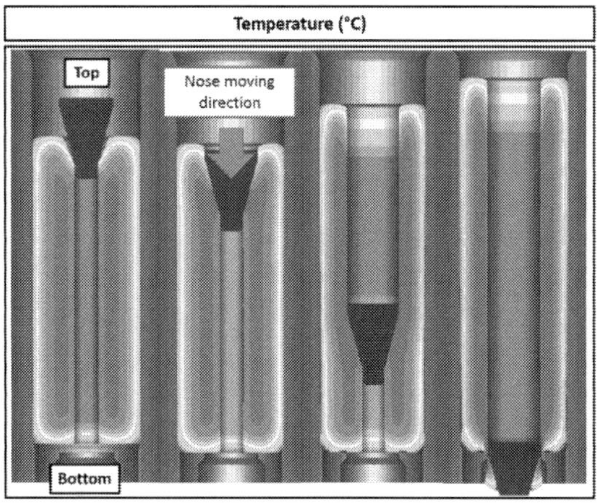

Figure 7. Temperature evolution during the expansion process.

As for deformation, high-accumulated strain values were obtained at the inside of the billet whereas it decreased towards the outside of the billet. Figure 8 shows the strain inside the billet obtained from simulation.

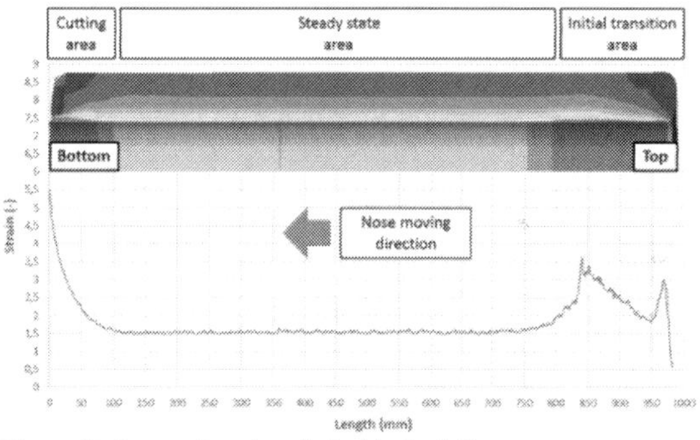

Figure 8. Accumulated strain inside the billet.

Regarding the strain in the inner surface, three areas can be clearly distinguished: an initial transition area with high level of strain due to the initial compression, a steady state with a constant strain of 1.5 and a cutting area with high strain values. This strain behaviour coincides with observations made in optical micrographs images.

The strain rate revealed the areas where significant deformation was located. The following picture shows the strain rate near the nose in the steady state.

According to the figure 9, the highest strain rate values are concentrated near the tool and two remarkable peaks can be detected. Hence, the deformation is located on the inner side of the billet.

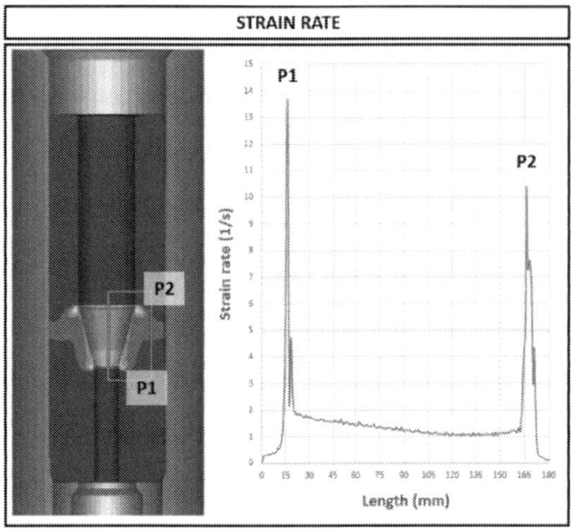

Figure 9. Strain rate near the nose in the steady state.

Regarding the defect predictions the Latham and Cockroft normalized criterion was applied. This model is used for cases with tensile failure.

$$C_{CR} = \int_0^{\overline{\varepsilon}_f} \left(\frac{\sigma_{max}}{\overline{\sigma}}\right) d\overline{\varepsilon} \qquad (1)$$

where σ_{max} is the maximum principal stress, $\overline{\varepsilon}_f$ is the strain at fracture, $\overline{\varepsilon}$ is the equivalent strain and $\overline{\sigma}$ is the equivalent stress. This criterion predicts defects when the material is under tensile stresses [3,5,8]. In accordance with this, the following picture shows the areas where the billet is prone to defects caused by the process.

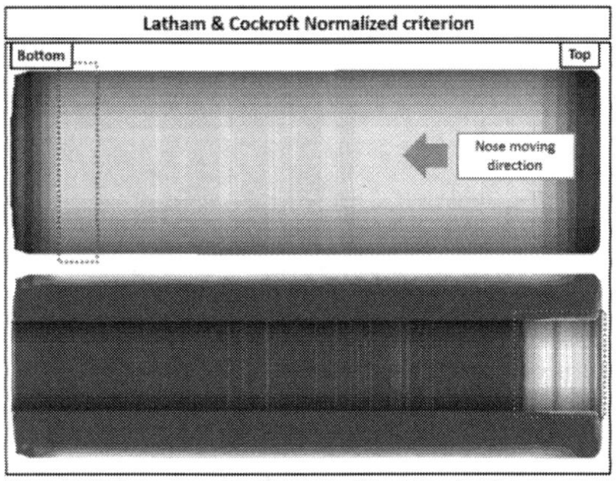

Figure 10. Damage critical areas predicted by simulation.

The simulation results shown in figure 10 reveal two critical areas; the initial contact area between the nose and the billet, and the external surface on the bottom side of the billet.

Figure 11 shows defects after pickling process. This figure reveals a surface defect in the bore hole at the top of the billet and scratches at the edge of the top, which agrees well with the simulation results. The experimental results carried out did not show defects at the external surface, although long-time experience specifies this as a critical area. High friction values or low billet temperature can cause cracks in this area.

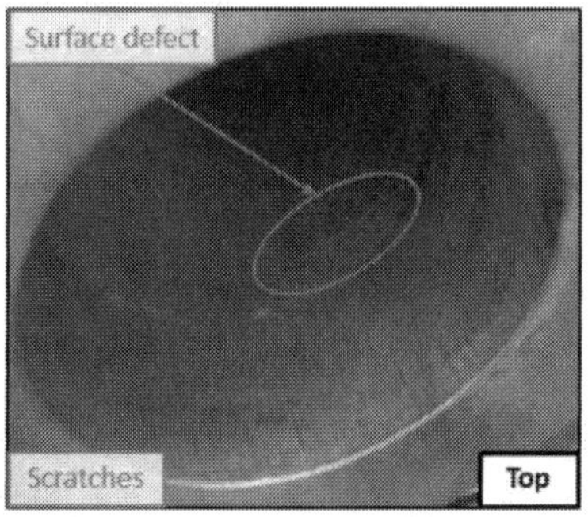

Figure 11. Defects in the top side after pickling process.

3 Sensitivity Analysis

With the help of validation work, a sensitivity analysis was carried out and critical parameters of the process were defined [4]. Some of these are the nose angle, the ram speed, the billet initial ID, the billet final ID and the temperature. To define the upper and lower limits of the input parameters some previous simulations were performed first.

Afterwards, a DOE for the simulation with the above five parameters was created using the software MINITAB, which was also used to analyse the sensitivity study [9]. Two main objectives were determined for the process optimisation: the reduction of the tool costs and the reduction of the raw material costs. Thus, the tool wear (TW) and the total material waste (TMW) were fixed as measurable output data.

Regarding the reduction of the tool costs, the maximum value of the TW was chosen for the sensitivity analysis. The results are shown in the following figure.

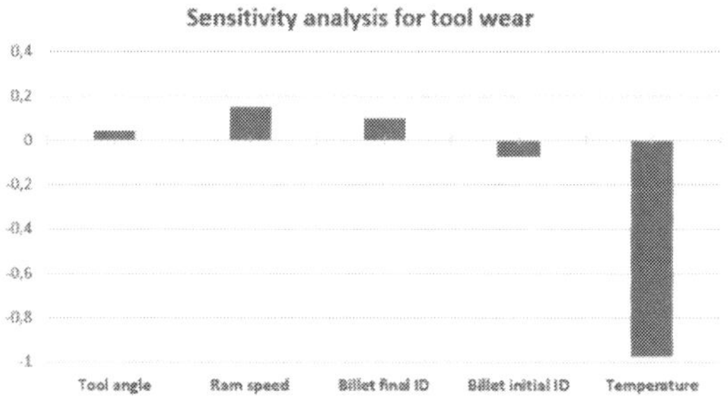

Figure 12. Sensitivity analysis for the TW.

The sensitivity analysis reveals the influence of the input parameters on the output. As can be clearly seen, the billet temperature has the strongest influence on the response of the TW. The ram speed and the billet final ID have a weak influence, whereas the nose angle and the billet initial ID do not show any relevant influence on TW.

According to the results aforementioned, a multiple linear model was found with the most important input parameters. These parameters were defined with help of the size of the R^2-value. The accuracy of the obtained model is approx. 97%. The following graphic shows the result.

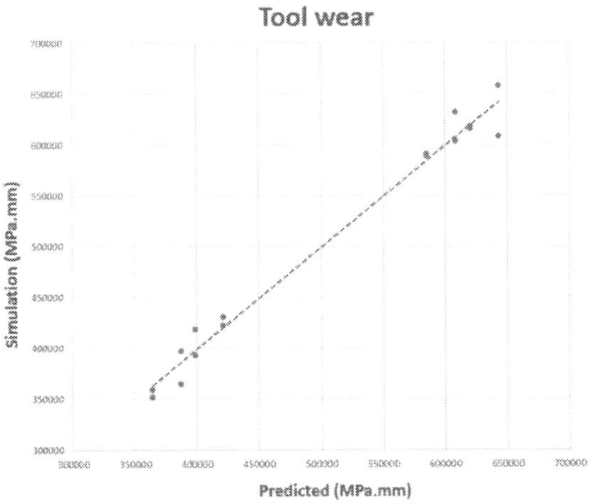

Figure 13. Predicted tool wear using multiple linear model compared to simulation results.

As for the reduction of the raw material costs, the maximum value of the TMW was chosen to carry out the sensitivity analysis using a multiple linear model. The TMW was defined as a combination of machining and remainder losses. The results are shown in the next figures.

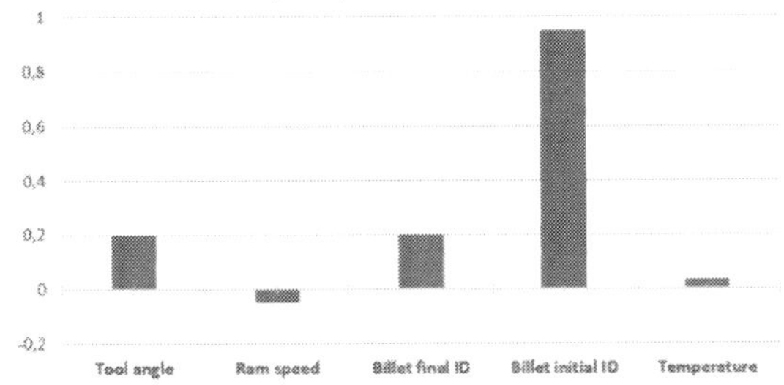

Figure 14. Sensitivity analysis for the total material waste.

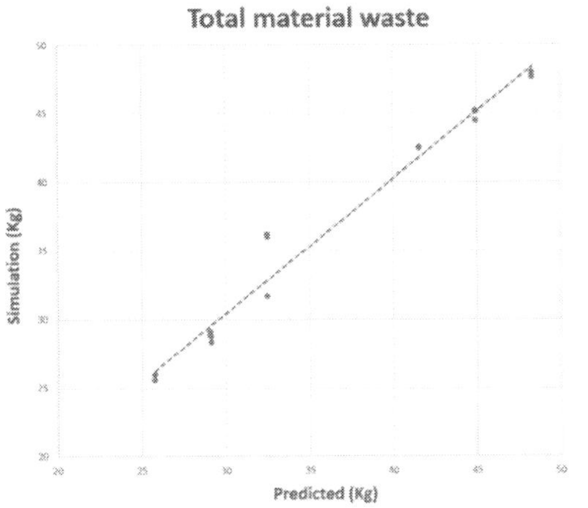

Figure 15. Predicted TMW using multiple linear model compared to simulation results.

As shown in figure 14, the billet initial ID has the strongest impact on the response of the TMW. The tool angle and the billet final ID have a weak influence, while the ram speed and the temperature do not show any relevant influence on TMW.

With the help of the relevant input parameters, a model with an accuracy of about 97% can be found. The result is shown in the following figure.

4 Conclusions

A simulation model for the expansion process was developed for alloy UNS N08028. The study showed good agreement between simulation and experimental results. Critical process parameters like friction, heat exchange, tool angle, ram speed, billet final ID, billet initial ID and temperature were identified. The model was able to predict with an acceptable accuracy defects, expansion press load and material wastes. High deformation areas were detected inside the billet with 150% of accumulated strain, whereas the accumulated strain at the outside was less than 30%. Two critical areas were detected where the billet is prone to defects.

Sensitivity analysis was carried out to reduce the tool costs and the raw material costs. In relation with the first target, the billet temperature was detected as the strongest influence parameter on the response of the TW, whereas the tool angle and the billet initial ID did not show any relevant influence. As for the second target, the billet initial ID showed the strongest influence on the response of the TMW, while the ram speed and the temperature did not show any relevant influence. For the two objectives, models were found with the most important input parameters and the error in both cases was less than 3%.

The study showed the capacity to predict defects using FEM and revealed the great possibilities that this type of method offers for the sensitivity analysis.

As for the future, the results of this study will be used to optimise the expansion process.

Acknowledgment

This work was supported by Tubacex Group. The author gratefully acknowledge the financial support from Tubacex Group.

References

1. https://www.tubacex.com
2. K. H. Brensing, B. Sommer, Res. Dpt., Salzgitter Mannesmann Röhrenwerke GmbH, Mülheim an der Ruhr, Germany, (2009)
3. S. Fanini, Ph.D. diss., Dpt. of Mech. Inn. and Mgmt., Univ. of Padova, Padova, Italy, (2008)
4. S. Hansson, Ph.D. diss., Dpt. of Appl. Ph. of Mech. Eng., Luleå Univ. of Tech., Luleå, Sweeden, (2009)

5. G. E. Dieter, H. A. Kuhn, S. L. Semiatin, *Handbook of workability and process Design*, ASM Inter., Mat. Park, Ohio, (2003)
6. J. Sejournet, J. Delcroix, *l.* Lubr. Eng. **11**, 389-396, (1995)
7. C. Liu, L. Zhang, F. Kang, Adv. M. Res. **668**, 856-860 (2013)
8. M. G. Cockroft, D. J. Latham, J. of the Inst. of Me. **96**, 33-39, (1996)
9. L. Eriksson, E. Johansson, N. Kettaneh-Wold, CA. Wikstrom, S.Wold, *Design of Experiments-Principles and Applications.* Umetrics AB, (2008)

Fast variable stiffness composite cylinder uncertainty analysis by using reanalysis assisted Copula function

Zeng Yang[1,3], Enying Li[2], Hu Wang[1,3,a]

[1] State Key Laboratory of Advanced Design and Manufacturing for Vehicle Body, Hunan University, Changsha 410082, PR China
[2] School of Logistics, Central South University of Forestry and Teleology, Changsha, 41004, P.R. China
[3] Joint Centre for Intelligent New Energy Vehicle, Shanghai, China

Abstract. There are lots of uncertainties in variable-stiffness composite materials such as material properties, fibre volume fraction, geometries at various scale and matrix porosity. Commonly, these uncertainties are not always mutually independent and there exist correlations among these random input variables. These correlations may affect the output of composite significantly. To address these correlations, a novel approach for uncertainty analysis based on copula function assisted by reanalysis method is suggested. The Copula function is utilized to address the correlations of random input variables. Monte Carlo simulation (MCS) is employed to obtain the uncertainty analysis. Therefore, a large number of samples should be generated and the expensive computational cost is not feasible when the popular finite element (FE) model is utilized. To save the computational cost and make the uncertainty analysis feasible in practice, an efficient fast computation method, reanalysis method is integrated in the frame. The numerical test demonstrates that the proposed approach is an efficient uncertainty analysis tool for the practical engineering problems.

1 Submitting the manuscript

Variable stiffness laminates can improve material performance considerably and simultaneously keep the geometry size unchanged. Recently, there are a series of studies to develop these structures [1, 2]. However, uncertainties such as material properties, fibre volume fraction, geometries at various scales, matrix porosity are still a challenge in real engineering [1]. Furthermore, these uncertainties are always not mutually independent and there exist correlations among the random input variables. These correlations may affect the output of the composite structure significantly. Fortunately, there exists a mathematical model—copula function [3, 4] which can address the correlations effectively. In this study, a D-vine copula [5] function is utilized to address the multivariate correlations among the random input variables.

In this work, Monte Carlo simulation (MCS) is utilized to propagate the uncertainty to the output of the structure. However, in this method, the accuracy of results mainly depends on the number of the samples. In other words, in order to obtain accurate results, it needs a large number of samples. These samples are commonly evaluated by FE simulations and the computational cost is too expensive to work. In order to address the time-consuming problem, a fast and accurate numerical tool, named reanalysis method [10] is applied. We hope to use reanalysis make the simulation-based uncertainty analysis feasible.

The paper is organized as follows: Section 2 presents optimized Variable-stiffness plate. In Section 3, the detail of copula function is presented. Section 4 presents the fundamental theory of reanalysis. The results are discussed in Section 5.

2 Optimized Variable stiffness plate

In this study, the investigation is based on an optimized composite square plate [6] as shown in Fig.1. Suppose that there are 4 fibre plies and each ply's thickness is 0.5. The material properties are referred from Ref.[7] and given in Table.1. The nodes on the left side are fixed and on the nodes on the right side are stretched. The forces are identical with the total value 200. Since the symmetrical (about x) structure and boundaries, the fibre paths are also symmetrical about axis x. Thus, relationship of path function of plies can be designed as: $g_2(x,y)=g_1(x,-y)$ and $g_4(x,y)=g_3(x,-y)$ (subs cripts represent the number of plies). The fibre paths are supposed to be a quadratic function of coordinate and given as

$$g(x,y,\mathbf{a})=x+a_1y+a_2xy+a_3x^2+a_4y^2 \quad (1)$$

According to Ref.[6], the fibre orientation angle at (x_0,y_0) can be given as

[a] Corresponding author: wanghu@hnu.edu.cn

$$\theta(x,y)=\begin{cases}\arctan\left(-\dfrac{1+a_2y+2a_3x}{a_1+a_2x+2a_4y}\right), a_1+a_2x+2a_4y\neq 0\\ \dfrac{\pi}{2}, a_1+a_2x+2a_4y\neq 0\end{cases} \quad (2)$$

In this study, the fibre orientation angle is supposed to be a constant in an element and calculated by the centroid of the element. The detailed process of optimization is referred to Ref.[6] and one of the Pareto solution of hole plate is presented in Table.2 and corresponding fibre path in each ply are illustrated in Fig.2.

However, in real engineering desgin, the ideal optimized fibre orientation angle is diffcult to be manufactured. There are always some deviations in the optimized fibre orientation angle. Since the manufacturing environment is the same, correlations will exist between the deviations in different ply in one plate. In this study, the deviations of fibre orientation angle in manufacturing are regarded as uncertainties. The details of the setting can be found in Section 3.

Table 1. Material parameters.

Material parameter	E_L	E_T	λ_{LT}	G_{LT}	G_{TN}	G_{LN}
Value (Mpa)	137.9	10.34	0.29	6.89	3.9	6.89

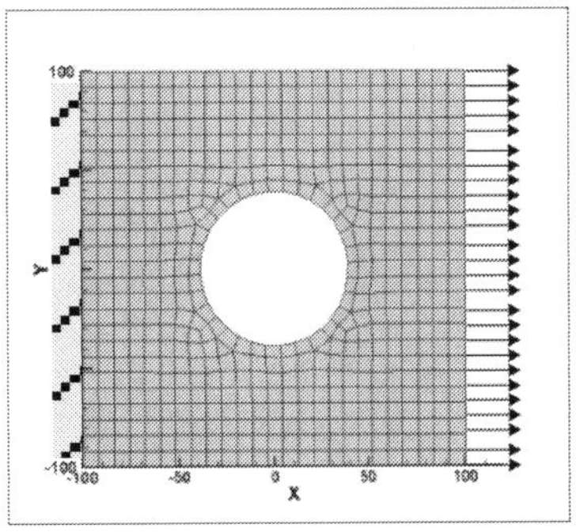

Figure 1. Square plate.

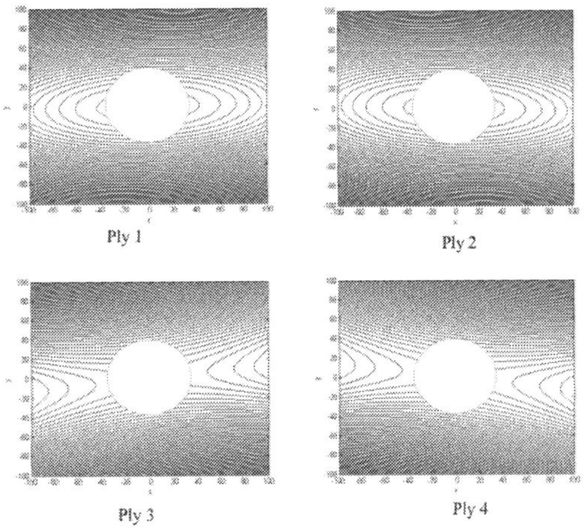

Figure 2. Fibre paths of given solution

Table 2. Optimal solution

Pareto solutions	Plies	a_1	a_2	a_3	a_4
solutions	Ply 1	-0.231	-0.424	0.846	6.613
	Ply 2	0.231	0.424	0.846	6.613
	Ply 3	2.803	-2.102	-0.558	7.363
	Ply 4	-2.803	2.102	-0.558	7.363

3 Copula function

Copula function is a mathematical model which can estimate the connection between the marginal CDFs and joint CDF of d-dimensional multivariate distribution. In Sklar's theorem [3], if the marginal CDFs are continuous, the multivariate CDF can be determined by the copula uniquely. The expression is

$$F(x_1,...,x_d)=C(F_1(x_1),...,F_d(x_d)) \quad (3)$$

where $F_i(x_i)(i=1,...d)$ is the ith continuous marginal CDF of the multivariate distribution, and $C(F_1(x_1),...,F_d(x_d))$ represents the copula function. According to the relationship between PDF and CDF, the joint density function of multivariate distribution can be given as

$$f(x_1,...,x_d)=c(F_1(x_1)...F_d(x_d))f_1(x_1)..f_d(x_d) \quad (4)$$

where $c(F_1(x_1)...F_d(x_d))$ is defined as the density function of $C(F_1(x_1),...,F_d(x_d))$ and given as

$$c(F_1(x_1)...F_d(x_d))=\frac{\partial C(F_1(x_1)...F_d(x_d))}{\partial F_1...\partial F_d} \quad (5)$$

Nevertheless, when d is large, an ideal $C(F_1(x_1),...,F_d(x_d))$ is difficult to be constructed directly. Commonly, most bivariate copula function can fit the bivariate data well. Therefore, the pair-copula constructions (PCC) [5] are considered in the application of multivariate copulas. In this study, a D-vine copula based on PCC is used. Suppose that the $U=(U_1,...,U_d)'$ is a random vector in which the $U_i, i=1,...,d$ represents marginal CDFs with uniformly $U(0,1)$. Suppose that p and q are integers which satisfy $(1\leq p\leq q\leq d)$ and the symbol $p:q$ represents all integers ranging from p to q, namely $p:q:=\{p,...,q\}$. Define that $U_{p:q}$ represents the variables $\{U_p,...,U_q\}$. According to a recursive decomposition, the d-dimensional density f can be written as

$$f(x_1,...,x_d)=f(x_1)\prod_{k=2}^{d}f(x_k\mid x_1,...,x_{k-1}) \quad (6)$$

As to the copula density c, we have

$$c(u_1,...,u_d)=1\prod_{k=2}^{d}c_{k|1:[k-1]}(u_k\mid u_{1:[k-1]}) \quad (7)$$

where **1** represents the marginal density of U_1.

According to Ref.[5], defining $i[i+1]\,|\,\phi := i[i+1]$, we have

$$
\begin{aligned}
c(u_1,\ldots,u_d) &= \prod_{j=1}^{d-1}\prod_{i=1}^{d-j} c_{i,i+j|[i+1]:[i+j-1]}(\mathbf{u},\mathbf{v}) \\
\mathbf{u} &= u_{i|[i+1]:[i+j-1]} \qquad (8) \\
\mathbf{v} &= u_{i+j|[i+1]:[i+j-1]}
\end{aligned}
$$

Equation (8) provides a D-vine copula which is factorized as the product of pair-copulas as shown in Fig.3.

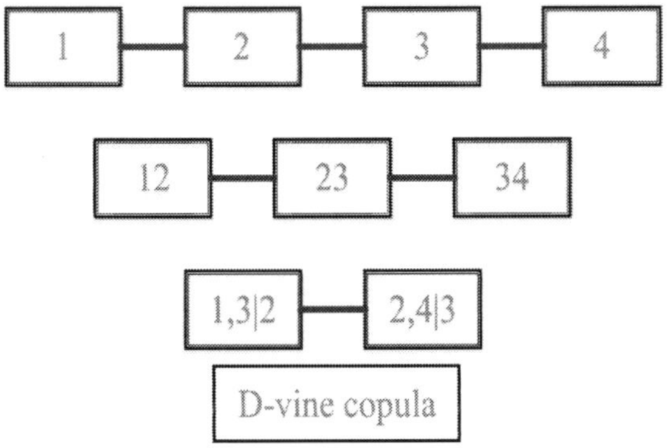

Figure 3. D-vine copula.

Commonly, the D-vine copula can be identified based on test samples via Akaike information criterion(AIC), goodness of fit (GOF) and Bayesian method [5,8]. In this study, in order to address the effect of correlations on the uncertainty analysis qualitatively, the correlations and distributions of the deviations are given. Suppose that the deviations of fibre orientation angle in a ply keep the same. Different plies subject to the same normal distribution with mean μ=0 and standard deviation $\sigma = 0.02$. This means that the deviations of fibre orientation is unbiased and only a relative small fluctuation. The correlations exist in different deviations from different plies. In order to generate the sample, suppose that all pair copulas are *Frank*. Since the aforementioned :

$$g_2(x,y) = g_1(x,-y) \quad \text{and} \quad g_4(x,y) = g_3(x,-y)$$

it's reasonable to assume that there is a negative correlation between ply 1 and ply 2 with a Kendall's tau $tau = -0.7$ as shown in Fig.4a. So is the correlation of ply 3 and ply 4. The Kendall's tau of the remaining correlations is supposed to be a small positive value with $tau = 0.3$ as shown in Fig.4b. MC simulation is used to uncertainty analysis based these assumption and D-vine copula function. Through calculation, it is proved that 20000 MC samples of these deviations which are drawn via Rosenblatt transformation [9] are enough to obtain the accurate results.

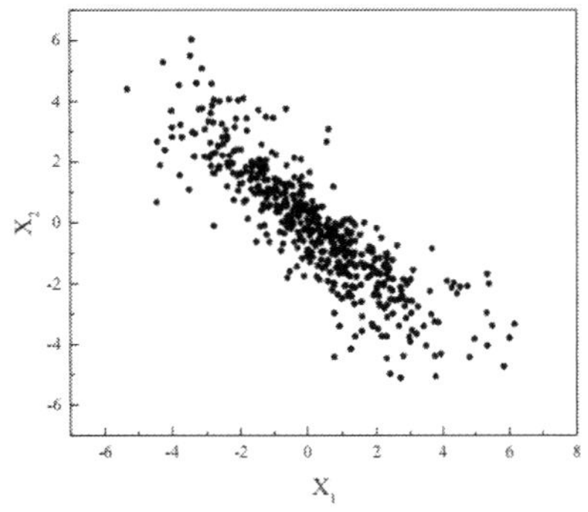

(a) $tau = -0.7$

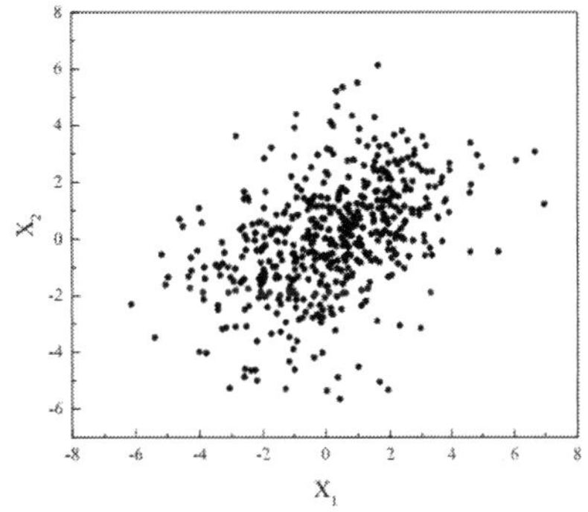

(b) $tau = 0.3$

Figure 4. Sample distributions with different *tau*

4 Reanalysis Method for uncertainty analysis

Some of uncertainties, such as the deviations of $\theta(x,y)$ in this study, are involved in the variation of the stiffness matrix. The reanalysis is a powerful method to address this variation of stiffness matrix. Among approximate reanalysis methods, the CA [10] can simultaneously satisfy the require of accuracy and efficiency. Therefore, the CA is introduced in this study.

Regarding the optimized path function in Section 2 as the initial fibre orientation, and the equilibrium equation is

$$\mathbf{K}_0\mathbf{r}_0 = \mathbf{R} \qquad (9)$$

where $\mathbf{K}_0 = \mathbf{K}(\theta(x,y))$ is the stiffness matrix, $\mathbf{R}$ is the load vector, and $\mathbf{r}_0$ is the initial displacement and can be solved as

$$\mathbf{r}_0 = \mathbf{K}_0^{-1}\mathbf{R} \qquad (10)$$

When there is uncertainty from the fibre orientation angle $\theta(x,y)$, the new equilibrium equation can be written as

$$\mathbf{Kr} = \mathbf{R} \tag{11}$$

Suppose

$$\Delta\mathbf{K} = \mathbf{K}\text{-}\mathbf{K}_0 \tag{12}$$

Equation(11) can be written as

$$(\mathbf{K}+\Delta\mathbf{K})\mathbf{r} = \mathbf{R} \tag{13}$$

Suppose that the displacement is a linear combination of a series of basis vectors as

$$\mathbf{r} = y_1\mathbf{v}_1 + y_2\mathbf{v}_2 + \cdots + y_s\mathbf{v}_s = \mathbf{r}_B\mathbf{y} \tag{14}$$

where $\mathbf{v}_i (i = 1,2,...s)$ is basis vector and $y_i (i = 1,2,...,s)$ is corresponding coefficients, s represents the number of basis vectors

$$\mathbf{r}_B = [\mathbf{v}_1\ \mathbf{v}_2\ \cdots\ \mathbf{v}_s] \\ \mathbf{y} = [y_1\ y_2\ \cdots\ y_s]^T \tag{15}$$

According to Ref.[10], we can construct the basis vectors as

$$\mathbf{v}_1 = \mathbf{r}_0 \\ \mathbf{v}_i = \text{-}\mathbf{B}\mathbf{v}_{i-1}, i = 2,3,...,s \\ \mathbf{B} = \mathbf{K}_0^{-1}\Delta\mathbf{K} \tag{16}$$

Substituting Eq.(14) into Eq.(11), then pre-multiplying both side with $\mathbf{r}_B^T$, we have

$$\mathbf{K}_r\mathbf{y} = \mathbf{R}_r \tag{17}$$

where

$$\mathbf{K}_r = \mathbf{r}_B^T\mathbf{K}\mathbf{r}_B \\ \mathbf{R}_r = \mathbf{r}_B^T\mathbf{R} \tag{18}$$

$\mathbf{y}$ can be obtained via solving Eq.(17). Thus, the approximation of $\mathbf{r}$ can be obtained via substituting $\mathbf{y}$ into Eq.(14)

In uncertainty analysis of variable-stiffness laminate, the optimized solution can be regard as the initial condition which can be solved by using full analysis according to Eq.(10). Then all samples of uncertainties can be added to the stiffness matrix and then solved by using the CA method. Since the scale of $\mathbf{K}$ is much larger than $\mathbf{K}_r$, the computational cost can be reduced significantly.

5 Results and discussions

The D-vine copula can address the correlation effectively. It does not need joint PDF of the random input variables. It only depends on the test samples. With the test samples, the pair-copula are quite easy to be constructed. Based on the identified vine copula, 20000 MC samples are generated to simulate the joint PDF in this study. Thus, the correlations exist in the samples and can propagate to results completely.

Commonly, some of the uncertainties are involved in the variation of part of stiffness matrix. Since the purpose of the reanalysis method is to analyze based on the changed part of stiffness matrix $\mathbf{K}_r$ which is much smaller than the $\mathbf{K}$, it presents a powerful performance in this kind of uncertainty analysis. With the full FE analysis, it takes 21 seconds for each evalutions. While with reanalysis method, it only takes less than 2 seconds. The reanalysis method greatly improves the efficiency of computation. 8 test samples are generated to test the accuracy of the reanalysis and the results are presented in Table.3. It is shown that the results of reanalysis are almost equal to the results of full analysis. The reanalysis is quite accurate.

Table. 3. The results of the test samples

	The maximum displacements (10^{-6})			
	Full analysis		Reanalysis	
Numbers	X	Y	X	Y
1	4.7070	1.2880	4.7069	1.2880
2	4.6967	1.2434	4.6967	1.2434
3	4.6296	1.1719	4.6296	1.1719
4	4.6102	1.2374	4.6102	1.2374
5	4.6173	1.2088	4.6173	1.2088
6	4.6197	1.1670	4.6197	1.1670
7	4.8146	1.3177	4.8146	1.3177
8	4.6577	1.1787	4.6576	1.1787

In the optimized variable-stiffness plate, the maximum displacement of X direction is 4.629×10^{-6} and the maximum displacement of Y direction is 1.209×10^{-6}. When the uncertainties are taken account, the displacements of the results are shown in Fig.5. The means of the corresponding maximum displacements are 4.634×10^{-6} and 1.211×10^{-6}, respectively. As for maximum displacement of X direction, the minimum value is 4.436×10^{-6} and the maximum value is 4.890×10^{-6}. This means that the uncertainty of maximum displacement of X direction may reach to 10% ($(4.890-4.436)/4.629=9.81\%$). For maximum displa-cement of Y direction, the minimu-m value is 1.041×10^{-6} and maximum value is 1.380×10^{-6}. The uncertainty of maximum displacement of Y direction is 28.04%. Either 9.81% or 28.04% is a quite large value of uncertainty in the real engineering problems since the deviations of the fibre orientation angle is quite small ($\sigma=0.02$). When in Variable-stiffness plate design, taking the uncertainties into account is quite important to design a reliable structure.

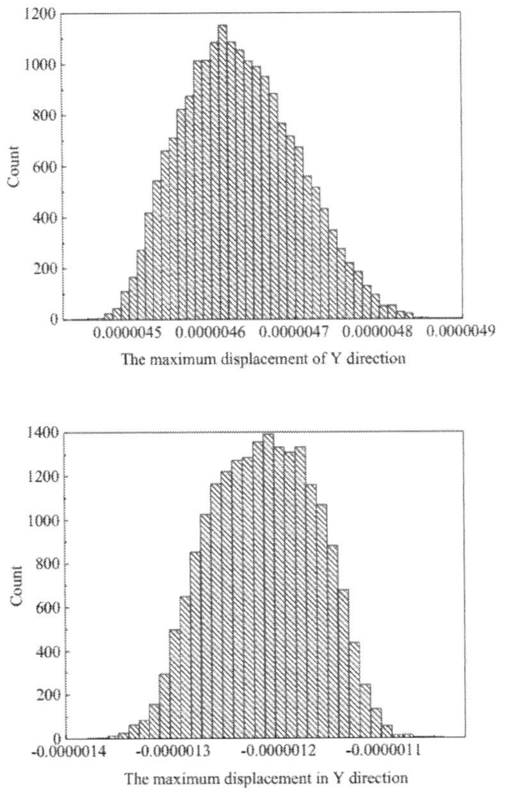

Figure 5. The maximum displacements with uncertainty

6 Remarks

- We present a novel approach for uncertainty analysis based on copula function assisted by reanalysis method.
- The correlations among random input variables are characterized as a D-vine copula function. In this way, the uncertainty analysis taking all kinds of factors of random input variables into account can be implemented conveniently.
- Reanalysis method is integrated in the frame to save the computational cost and make the uncertainty analysis feasible in practice.

Acknowledgements

This work has been supported by Project of the Program of National Natural Science Foundation of China under the Grant Numbers 11172097, 11302266 and 61232014.

References

1. Fayazbakhsh K, Nik M A, Pasini D, et al. Defect layer method to capture effect of gaps and overlaps in variable stiffness laminates made by automated fibre placement. Composite Structures, 2013, 97: 245-251.
2. Rouhi M, Ghayoor H, Hoa S V, et al. Effect of structural parameters on design of variable-stiffness composite cylinders made by fibre steering. Composite Structures, 2014, 118: 472-481.
3. Sklar, A. Fonctions de répartition à n dimensions et leurs marges. Publications de l'Institut de Statistique de l'Université de Paris 8, 1959: 229–231.
4. Nelsen, R B. An Introduction to Copulas, Springer,1999: New York.
5. Min A, Czado C. Bayesian model selection for D-vine pair-copula constructions. Canadian Journal of Statistics, 2011, 39(2): 239-258.
6. Guanxin H, H Wang, G.Y. Li. An Efficient Reanalysis for Variable-stiffness Composite Design by Using Path Function. Composite Strcutures. (Submitted).
7. Brampton CJ, Kim HA. Optimization of tow steered fibre orientation using the level set method. In: 10th World Congress on Structural and Multidisciplinary Optimization (WCSMO/10), Orlando, 19-24 May 2013.
8. Gruber L, Czado C. Sequential Bayesian model selection of regular vine copulas. Bayesian Analysis, 2015.
9. Rosenblatt, M. Remarks on a Multivariate Transformation. The Annals of Mathmatical Statistics, 1952; 23(3): 470–472.
10. Kirsch U, Moses F. An improved reanalysis method for grillage-type structures. Computers and Structures, 1998; 68:79-88.

DOI: 10.1051/matecconf/20168010007

Numerical investigation of manufacturing hollow preforms by combining the processes backward cup extrusion and piercing

Robinson Henry[1,a] and Mathias Liewald[1]

[1] *Institute for Metal Forming Technology, University Stuttgart, 70174 Stuttgart, Germany*

Abstract. Hollow or tubular preforms are used to meet lightweight requirements and reduce production costs of cold formed components. In order to increase efficiency of manufacturing, a combined forming process was developed, which will be presented in this paper: manufacturing of hollow preforms by combining processes backward cup extrusion and piercing. With regard to combined forming processes in the field of cold forging, disclosed work consists of numerical identification of significant process input variables, process outcomes, and the process limits.

1 Introduction

In general, cold forged components can contribute essentially to diverse lightweight strategies when it comes to increasing the efficiency of manufacturing processes with regard to the use of material. Furthermore, the use of material can be reduced using the occurring cold work hardening and the process-specific possibility of creating a continuous grain structure, which is designed for specific loads. Generally, a reduction of material use of cold forged components while maintaining the functional demands is realized by using hollow designs.

Drive shaft and main shaft are typical components for cold forged parts, which are applied in the power train and chassis systems of modern vehicles. These cold forged parts must withstand high specific operating loads and have to fulfill high demands regarding tolerances in form, dimension, and position.

A resource-efficient production is necessary in order to fulfill the high demands for environmental-friendly production. These days, when producing hollow cold or hollow hot forged parts, a resource-efficient and cost-saving process sequence inevitably starts using a hollow preforms or producing a hollow preforms in order to avoid subsequent machining operation after the process. Multi-stage forming processes are used frequently when producing preforms. For example, combining the processes upsetting, centering, backward cup extrusion, and piercing are used to produce a hollow preform [1].

Resulting from the functional separation of the inner and outer geometry of the relevant tool elements of sequential forming steps, the produced hollow preforms are not in the right position (e.g. figure 1a) This positional error (e.g. figure 1a) cannot be compensated for by further forming operations and therefore requires an additional machining process, which reduces the balance of the resource efficiency of the process and causes economical disadvantages. When producing hollow preforms the combination of two or more cold forming processes, regardless if they are the same processes or different ones (e.g. figure 1b), in one single tool has an advantage: during the same press stroke no functional separation of the inner and outer geometry of relevant tooling elements occurs resulting from multiple forming steps [2]. Additionally, the production costs can be decreased by reducing the number of forming steps and required forming tools in a physical meaning

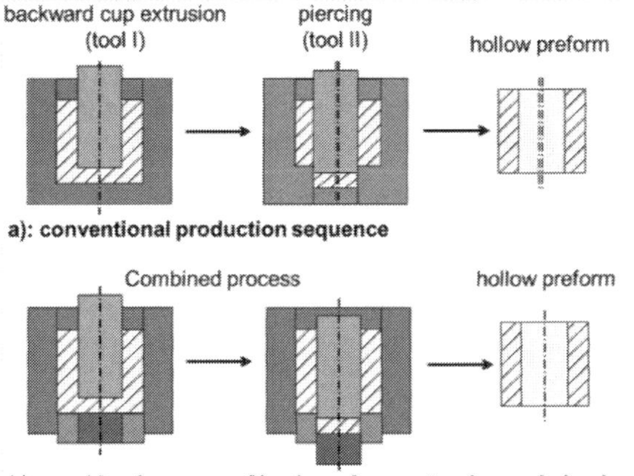

Figure 1. Conventional (a) and combined process (b) routes .

[a] Corresponding author: robinson.henry@ifu.uni-stuttgart.de

DOI: 10.1051/matecconf/20168010007

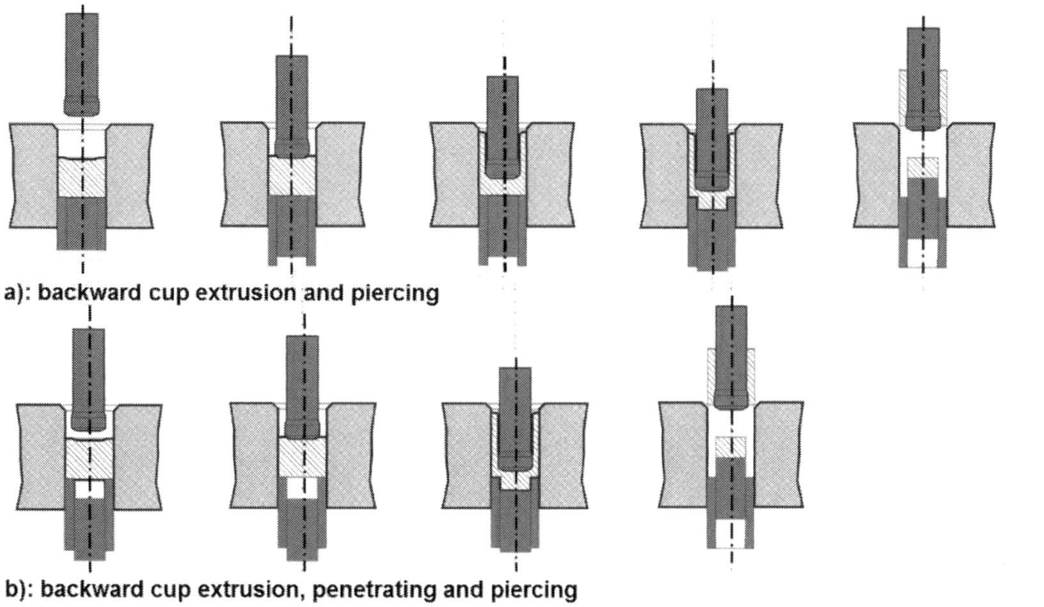

Figure 2. Combined process and tooling (punch)

In principle, two types of combined forming processes has been are investigated in this research project. The first variation of process is the combination of backward cup extrusion and piercing (e.g. figure 2a). A controlled movement of the counter punch can be performed to generate a hydrostatic stress state in the shearing zone during the piercing.

The second forming process is a combination of backward cup extrusion, penetrating, and piercing. A controlled movement of the counter punch is performed from the beginning of the backward cup extrusion process (e.g. figure 2b). The start of the piercing process is important with regard to the technical aspects. As part of the numerical investigation of the combination of backward cup extrusion and piercing, the influence of various parameters on the target variables (material flow, stroke force, process limits, temporal and local determination of crack initiation) during the piercing operation are investigated. The mechanical load of the experimental tooling is to be determined. Within this publication the combined process mentioned first will be discussed and first results will be presented.

2 State of the art

Currently, there only exists a low number of published studies available concerning the specific combination of backward cup extrusion and piercing, i.e. the performance of backward cup extrusion and piercing in one tool during single press stroke. Within investigations on combined extrusion processes of forward cup extrusion and backward cup extrusion it has been reported on certain geometric conditions of the piercing of the previously pressed pin in [3] and [4]. In [5] a forming of the shearing zone was observed for the investigated heights of the workpieces:

$$0{,}4 < \frac{h}{ØD} < 1 \quad (1)$$

and the choice of ratio of

$$\varepsilon_Z < \frac{ØD_{pin}^2}{ØD_i^2} < 1 \quad (2)$$

This procedure has been defined as a process limit of combined extrusion processes in all three publications and has therefore not been investigated any further. In [6] the topic combined extrusion using backward cup extrusion, forward cup extrusion and piercing, in which the remaining cup bottom is pierced by multiple-acting hydraulic press, is addressed. However, process-specific details of the combined processes of backward cup extrusion and piercing are not dealt with in [6]. In [7] a multi-stage production process of hollow preforms is described, in which a cylindrical workpiece is pierced in one forming stage. This operation leads to a material flow against the direction of the movement of the punch. However, no general information can be found regarding influencing factors and target parameters on the process. Another process description for combined extrusion and piercing can be found in [8].

In [9] hot-backward cup extrusion has been investigated with a controlled counterpunch for ferrous and non-ferrous metals. It is pointed out that the complexity of shape of component can be increased by a subsequent piercing of the cup. The focus of the investigations in [9] is on the reduction of the punch force by using a controlled counterpunch movement.

In [10, 11, 12, 13, 14] investigations on combined precision forging and piercing have been performed. Initially, basics of the process are examined using a forged disc having a central hole, which was produced in

DOI: 10.1051/matecconf/20168010007

a single forming operation. In [10, 11] these fundamental results are transferred to a suspension part.

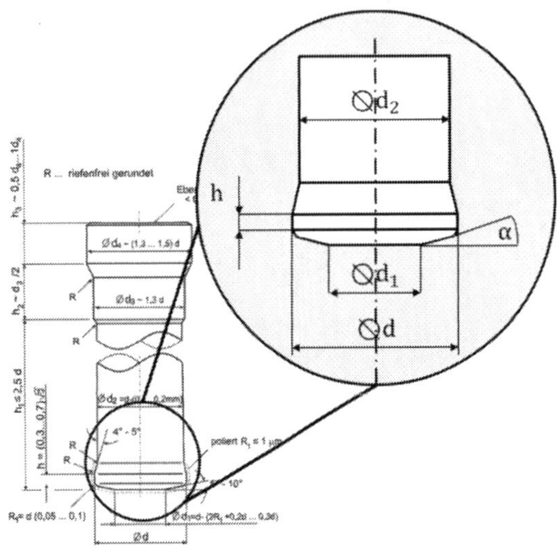

Figure 3. Geometrical properties of punch

Conducted investigation includes the influence of different punch geometries (e.g. figure 3), forming temperatures, tool and workpiece materials as well as the use of weight on the target parameter's hardness, microstructure, wear behavior, compliance with the shape and dimensional tolerances of the hole produced, and the surface quality.

[13] said that the hole produced by process combination is barrel-shaped. These observations are confirmed in [10, 11] and it was determined that an influence of the grain structure in interaction with the direction of the punch's movement exists. When arranging grain structure and punch movement direction parallel, barrel-shaped form deviations develop. However, grain structures of the forged parts which point vertically towards the moving direction of the punch lead to conical shape deviations with a larger diameter of hole in the area of the fractured surface. With regard to the shape tolerances of the hole produced it was described in [14] that the smaller the height of level in the area of the shearing zone, the greater the deviation in shape of the hole produced.

In [12, 10, 11, 14] neither influence of the punch's geometry nor an influence of residual mirror height on the surface quality of the hole was discovered.

The investigations of [12, 14, 13, 11, 10] with regard to combined precision forging and piercing show that little or only very few analogies exist between the punching of forged parts and the piercing of cold extruded parts. This becomes apparent for example in [10], where it is stated that the surface quality of the fracture surface of the hole is higher than the surface quality of the part produced when forming the holes. In investigations performed by the Institute for Metal Forming Technology differing results have been obtained for penetrating and piercing cold extruded parts.

3 Numerical investigations of new process arrangement

This chapter describes the development of simulation models and numerical development of the combination process. Numerical investigation of the combined processes is carried out to identify the significant process parameters. These results are used to define the design of the experimental tooling and the geometrical process parameters.

Within this publication the combination of backward cup extrusion and piercing will be presented as well as the corresponding numerical preliminary investigations such as design of the critical tools, force-displacement curve of the entire process, and a first look at the numerical investigation of the piercing process will be given.

In Figure 4 the concept of the combined process is shown, in particularly the forming process of backward cup extrusion and piercing. The experimental tooling assembly includes punch (1), die (2), counter punch (4), and counter sleeve (3). In this case the critical tooling part is the counter sleeve, which is analyzed in a numerical investigation using DEFORM 2D.

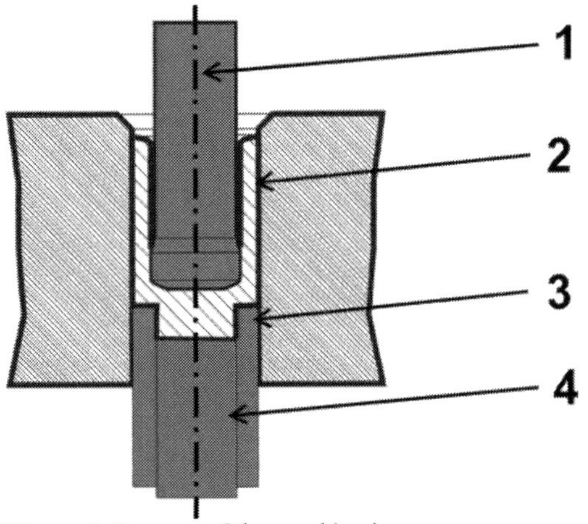

Figure 4. Concept of the combined process

The tools (punch, die and counter sleeve) were simulated as rigid objects in the numerical investigation. The workpiece was simulated as a plastic body having 3240 elements in the 2D simulation and the material 16MnCr5 was used. The diameter of the workpiece was 16mm and initial height was 18mm.

The velocity of the punch was chosen to 10mm/sec in simulation. The velocity of counterpunch was variated in the simulation to create hydrostatic stress in the shearing zone. The development of crack initiation can be controlled by the hydrostatic stress,

DOI: 10.1051/matecconf/20168010007

3.1 Force displacement curve of combined process sequence

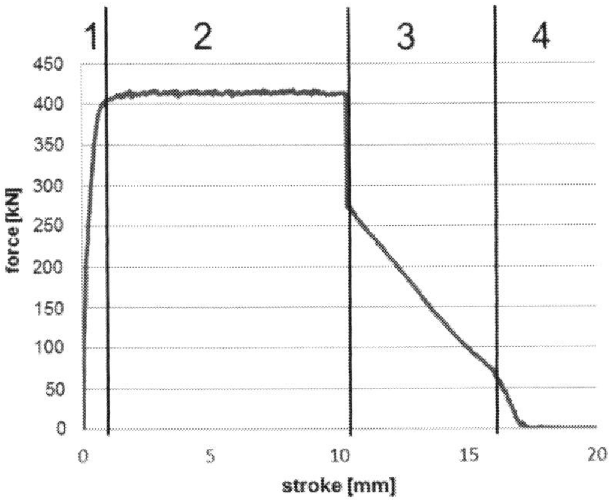

Figure 5. Force displacement curve of combined process sequence

In Figure 5 the force-displacement curve of the punch during the forming process is depicted in the numerical investigation. The force-displacement-curve can be divided into four sections. The first section is the instationary area which is characterized by a rapid increase in force to a maximum value of F = 417.6 kN. In the second section, the specimen is formed at constant punch force. The nearly vertical drop in force to ΔF=138.7 kN during the transition from the second to the third segment identifies the start of the counter punch movement at h1=10.3 mm. Thus, begins the piercing process and the force decreases linearly down to F=80.5 kN by the end of the third section. The actual piercing process begins as a result of the crack formation at the start of section four.

3.2 Stress distribution

In Fig. 6, the stress distribution during the forming process can be seen. Three different stress conditions are discovered in the workpiece. The first stress state develops during the backward cup extrusion. Furthermore, the highest stresses occur below the punch and are evenly distributed over the entire diameter. With increasing distance to the punch, stresses once more decrease. The high stresses around the punch mean that the material flow in this area is the largest and most of the material is pressed into the gap between the punch and the die by the punch itself.

As the counterpunch starts to move, the next part of the process begins. During this section different stress distributions are the result (see Fig.6). The highest stresses occur between the punch edge and the outer edge of the inside of the counter sleeve and reduce towards the center of the punch and towards the wall of the die. The stresses in the shearing zone increase the further the punch moves until the defined, critical value of the damage model is reached. In this numerical investigation the Cockcroft & Latham equation is used:

$$\int^{\bar{\varepsilon}} \frac{\bar{\sigma}}{\sigma^*} d\bar{\varepsilon} = 0{,}7 \qquad (3)$$

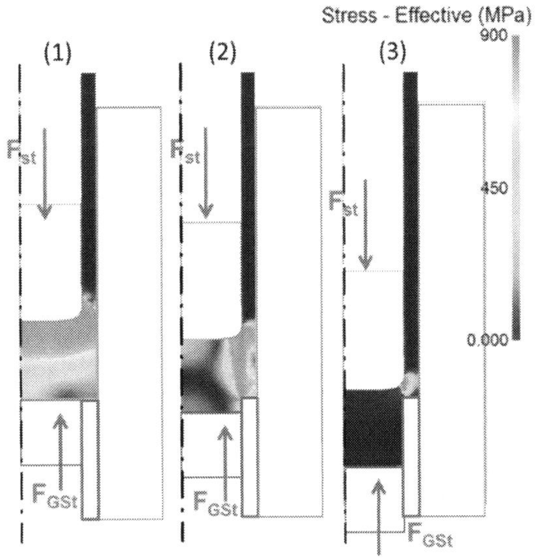

Figure 6. Stress distribution during piercing

After the initiation of the cracks, the stress distribution changes (Fig. 7 see.). While the cracks are spreading, the stresses relocate in areas, in which the workpiece is undamaged still. However, in these areas no further stresses can be generated.

As the forming process goes on, the stress changes from a hydrostatic stress state into a state of tension. The tensile stresses increase until reaching the point of separation resistance, which is why cracks occur and therefore, stresses and strains in the workpiece reduce. This sequence repeats itself throughout the piercing process and is responsible for the extension of the cracks.

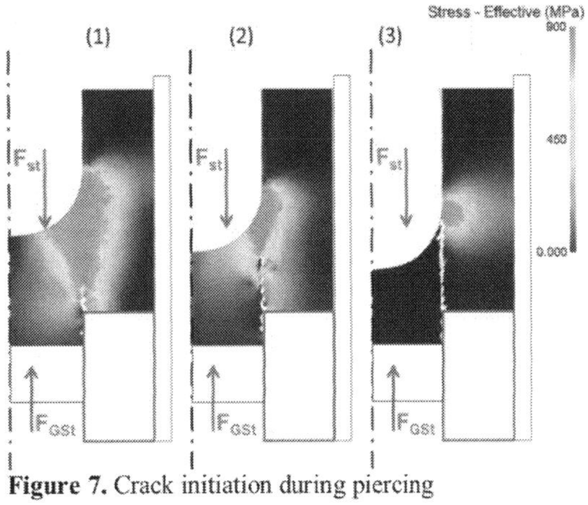

Figure 7. Crack initiation during piercing

3.3 Load simulation of the counter sleeve

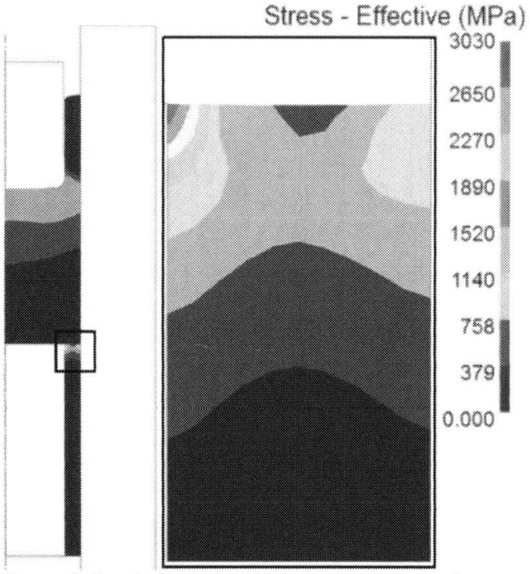

Figure 8. Load simulation of the counter sleeve for s=1,6mm

In Figure 8 the load of the counter sleeve is depicted. Therefore, different numerical investigations were performed prior to the process. The aim was reducing the wall thickness of the counter sleeve and reaching the minimum of the wall thickness. The numerical investigation has shown that a sufficient wall thickness of the counter sleeve is necessary.

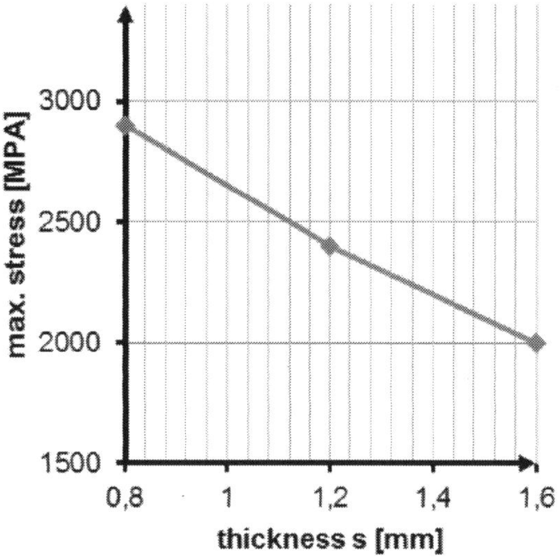

Figure 9. Load for different thickness of the counter sleeve

The maximum stress of the counter sleeve must be between 2000MPa and 2500MPa, in which case standard tool steel can be used. This means the wall thickness of the counter sleeve must be higher than 1,3mm (e.g. figure 9).

4 Outlook and further investigation

In future investigations numerical optimization of backward cup extrusion will be examined and numerical studies with regard to the influence of the controlled counterpunch will be carried out for the stress states in the shearing zone.

After the numerical investigation, the design and production of the experimental tooling (e.g. figure 10) will be performed. The process factors will be parametrized and integrated into the control of the tooling and the press control.

Furthermore, experimental investigation will be performed to identify significant process parameters and determine the process limits of combining processes and validate numerical investigations.

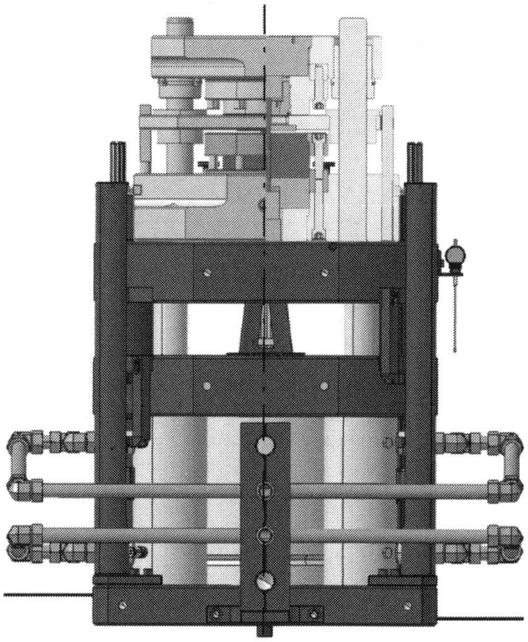

Figure 10. Newly developed tooling for combined extrusion and piercing

Evaluation of numerical and experimental results as well as the determination of technological and functional improvements of the hollow preforms with regard to the conventional process sequence will be carried out within the framework of this research project also.

Based on the results of the project, guidelines referring to dimensionless influencing and targeting parameters are to be developed, which makes designing those combined processes easier and makes transferring these results to industrial part geometries possible.

Acknowledgement

The IGF project 18431N of the "Forschungsvereinigung Stahlverformung e.V. (FSV)" was funded by the AIF within the framework of promoting industrial research and development (IGF) by the German Federal Ministry of Economics and Technology.

The authors would like to thank "Arbeitsgemeinschaft industrieller Forschungsvereinigungen (AiF)" for their financial support with regard to the research project „Erzeugung rohrförmiger Halbzeuge durch die Kombination der Umform-verfahren Napf-Rückwärtsfließpressen und Lochen".

References

1. M. Liewald, A. Felde, T. Schiemann, GCFG report (2009)
2. C. Mletzko et al., SchmiedeJOURNAL, H. 2 o. Jg. (2012)
3. J.-F. Adie, J.-M. Alexander, Proc. Instn. Mech. Engrs. **182**, 249-259 (1967-68)
4. W.-D. Graf, H. Schnorrbusch, Zeitschr. f. Techn. u. Organ. / Kammer d. Tech., Fachverband Maschinenbau, **23**, 535-539 (1973)
5. K. DaChang et al, J. of Materials Proc. Techn., **117**, 15-20 (2001)
6. H. Yoshimura, K. Tanaka, J. of Materials Proc. Techn., **98**, 196-204 (2000)
7. J. Klimczak, patent DE 10 2009 016 654 A1, (2010)
8. R. Mauermann, patent DD 231506 A1, (1986)
9. C.-W Jaschajaev, Kuznezno-stampowoznoje proizwodstwo, **9**, 4-6 (1966)
10. B.-A Behrens, J. Kerkling, Prod. Eng. Res. Devel., **5**, 201-207 (2011)
11. B.-A Behrens, R Nickel, J. Kerkling, Int. J Mater Form, **5**, 301-306 (2012)
12. K. Kerkling, K. Müller, B.-A. Behrens, Werkstattstechnik online, **100**, H. 6, 535-540 (2010)
13. B.-A Behrens, S. Reinsch, A Schott, E. Meyer, Prod. Eng., **13**, H.2, 87-90 (2006)
14. E. Meyer, A. Schott, Schmiedejournal, o. Jg., 16-18, (2005)

MATEC Web of Conferences **80**, 10008 (2016) DOI: 10.1051/matecconf/20168010008
NUMIFORM 2016

Multi-criteria optimization strategies for production chains

Jan Kusiak[1], Paweł Morkisz[2], Piotr Oprocha[2,a], Wojciech Pietrucha[3] and Łukasz Sztangret[1]

[1]*Faculty of Metals Engineering and Industrial Computer Science, AGH University of Science and Technology, al. A. Mickiewicza 30, 30-059 Kraków, Poland*
[2]*Faculty of Applied Mathematics, AGH University of Science and Technology, al. A. Mickiewicza 30, 30-059 Kraków, Poland*
[3]*Luxoft Poland Sp. z o.o., Krakowska 280, Z1, 32-080 Zabierzów, Poland*

Abstract. The aim of this paper is presentation of some optimization strategies applicable in the optimization of multi-stage and multi-threads chain structures (linear or tree-structured, acyclic graphs) with multiple intermediate goal functions. The inspirations for this type of analysis are production chains often seen in industrial plants. Production in these plants consists of sequences of multiple units connected linearly or in tree-structured graphs in which, at various production stages, intermediate quality criteria, related to the semi-products, may occur.

1 Introduction

Typical metallurgical industrial process can usually be divided into several smaller elements, corresponding to intermediate stages of production. They can arise in a natural way in correspondence to physical stand (e.g. furnace, hot rolling mill, laminar cooling, etc.) or representing smaller processes taking place in such stand (e.g. smelting and casting). In general, metallurgical production chains have a few elements in common:

- They do not have cycles. Delivered materials have to pass some production path without turning back. Branching is allowed, however, because some semi-products can be manufactured differently to reach the final form.
- Each stage requires some input parameters (materials, semi-products) with admissible range, demanded by technological restrictions for proper execution of reactions/processing of that stage.
- Each stage allows some controlling of performance by adjusting selected input parameters that can be controlled by an operator of the process at that stage.
- A number of quality output parameters are generated at the end of processing of each stage to judge performance or quality of the production.
- Output semi-products are described by some parameters characterizing their important technological features.

To illustrate the term of production chain better, an example of input-output parameters in two-stage model of production of blister copper (Cu_b) in Kennecott-Outokumpu process, considered in [1], is presented in Figure 1. In this particular case we have one global input parameter x (raw material – concentrate of Cu, Fe and S) that cannot be modified, two intermediate output-input parameters between two stages (Cu_2S, FeS) and global output parameter - blister copper Cu_b. Each stage has 3 control parameters (F – mass of fuel, O_2 – mass of oxygen in blast, N_2 – mass of nitrogen in blast) and two quality parameters (C_{Cu}, C_{FeO}; and C_{Fe3O4}, C_{Cu2O}) which are the concentrations of respective parameters (semi-products). Such a situation is typical in some sense. In general situation, each stage can be represented in the form of equations:

$$x_{i+1} = F_i(x_i, p_i), \quad (1)$$

$$q_i = Q_i(x_i, p_i), \quad (2)$$

where x_i, p_i, q_i are vectors representing input materials, control parameters and quality evaluation, respectively, and index i represents the i-th stage. While it is not written explicitly, each stage has attached a sequence of limiting equations on admissible values of input vector x_i as well as admissible range of control parameters p_i.

Clearly, when there are more than 2 stages then the production chain can have branching points, creating a tree-structured arrangement of production stages, as presented on Figure 2. Chosen strategies that can take advantage of the additional information about the structure of production chain will be presented in the paper.

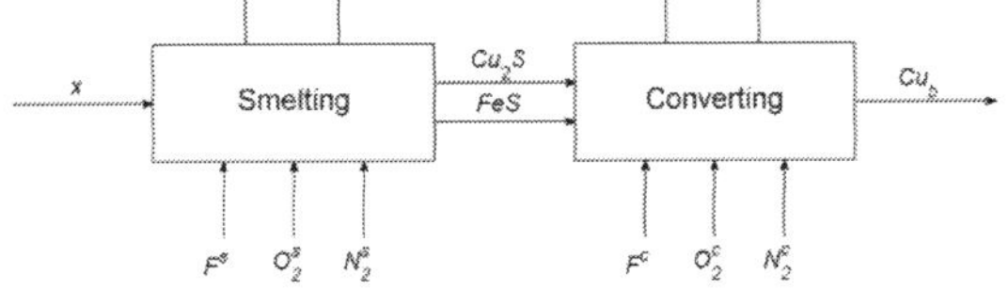

Figure 1. Example of two stage linear chain of industrial process (blister copper production [1,2]).

[a] Corresponding author: oprocha@agh.edu.pl

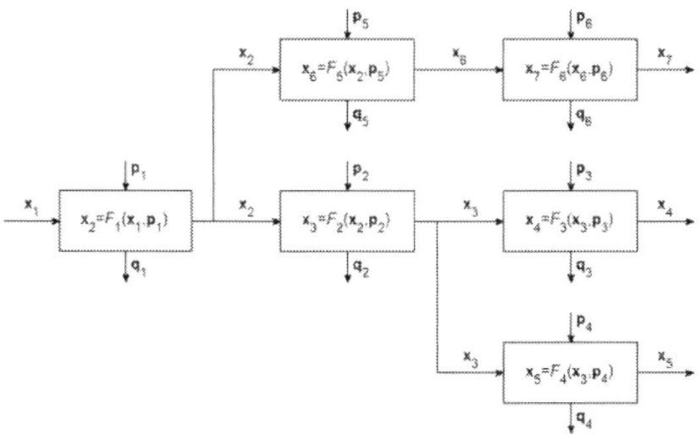

Figure 2. Example of hypothetical industrial process with a tree-structure.

It is also worth noting, that to calculate values q_i and x_i on deeper levels of the tree, it is necessary to set up an arrangement of stages. For example, to evaluate a value of q_4 (Figure 2) the following relationship has to be calculated:

$$q_4 = F_4(x_3, p_4) = F_4(F_2(F_1(x_1, p_1), p_2), p_4). \quad (3)$$

The indexing of nodes (production stages) of the tree will be performed using a standard Breadth-first Search (BFS) algorithm (see Figure 3).

2 Optimization Strategies

Since the knowledge about the structure of the production stages is known, such information can be used when choosing optimization strategy. When the structure is linear it is possible to look for optimal control parameters stage by stage. If there are branching points, the order in which the consecutive stages are dealt with, may be important for the final quality assessment. If the structure of relations between stages is not known (or is not at the point of interest), the **global approach** (GLO) can be applied. In this last approach, the global objective function composed by these stages is considered (see Section 2.2).

2.1. Standard searching in graphs

There are two standard graph algorithms (e.g. see [3]) allowing passing through all stages along the edges connecting them.

- Breadth-first search (BFS): in this approach all neighbouring (children) nodes are explored first, before going to next level of neighbours (Figure 3a).
- Depth-first search (DFS): in this approach tree is explored to the deepest possible level before backtracking (Figure 3b).

Note that while in DFS not all neighbouring nodes are checked simultaneously, their input-parameter bounds have to be checked. It can be simply explained. Let the output vector x_{i+1} is an input vector of all children nodes of i-th stage. If this output vector is not an admissible input for all children nodes, then it might happen that after backtracking the x_{i+1} cannot be used.

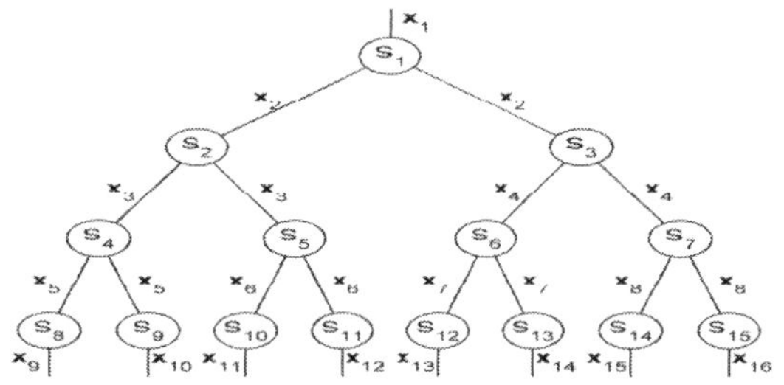

(a)

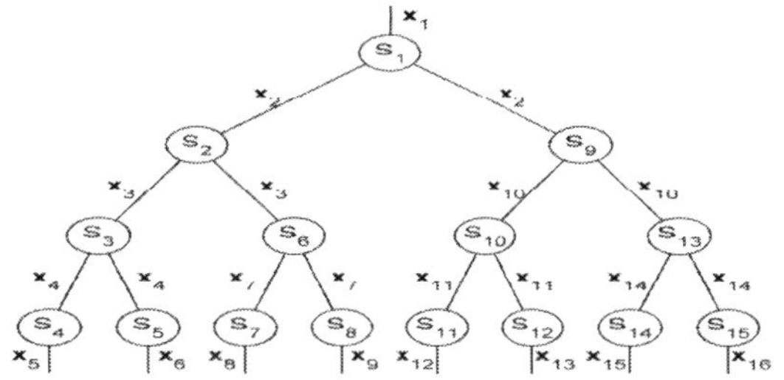

(b)

Figure 3. Indexing of nodes (stages) according to BFS (a) and DFS (b) algorithm.

It is worth noting, that we have to use BFS or DFS algorithm to order nodes, since we need some ordering when calculating values of functions F_i and Q_i. As we said, in our calculations we order nodes (order of propagation of x_i) by BFS algorithm, see Figure 3a.

2.2. Pareto-optimal solutions and order of children nodes

It is clear that in practice there is not a single (one) global process optimization criterion, since e.g. it is impossible to obtain best quality of final product, minimizing consumption of energy or fuel consumed in reactions, maximizing the speed of production and minimizing wastes, etc. Therefore, only the Pareto-optimal solution is the only aim of the research (the reader not familiar with this topic is referred to [4,5]).

A standard technique in this case is to transform multi-objective task into single objective task, assigning some weights w_j to each coordinate of each quality variable in vectors q_j. It results in the following objective function:

$$F = w_1 q_{1,1} + \cdots + w_k q_{1,k} + \cdots + w_s q_{n,r}, \quad (4)$$

where s is the total number of quality assessments in considered industrial process. Then it is reasonable to visit children nodes of i-th stage following their importance for the final value of function F. For this reason, for each child node its accumulated weight is calculated, summing up its weight w_j with weights of all of his children nodes on lower levels. These weights are used next in the DFS and BFS searches to order children nodes of stage i, that is children of node i are visited following the order induced by accumulated weights. Let the i-th stage has k children nodes and accumulative weights v of sub-trees induced by these children are:

$$v_1 \leq v_2 \leq \cdots \leq v_k. \quad (5)$$

Then the child with weight v_1 is visited first, next with v_2, and so on. Note that if weight w_i of each quality criterion q_i is the same (say $1/s$) then condition $v_i<v_j$ means that i-th subtree has less nodes than j-th. It means in our strategies, that sub-trees with the simplest structure are visited first, leaving more complex sub-trees for later.
While in standard DFS or BFS the order of children nodes is not that important, its importance will be much higher when transferring credits as described in the next section. Function F will be an objective function in GLO algorithm, hence GLO does not use additional knowledge about the structure of the industrial process and relations between stages (this knowledge is only used indirectly when calculating the value of F).

2.3. Credits

In our search for optimal solution, an upper limit N on the number of allowed evaluations of functions F_i and Q_i for i-th stage has to be set up. It may happen, however, that the optimal solution for this stage is found using n evaluations with $n<<N$. Saved credit $c_0=N-n$ of remaining evaluations can then be forwarded for the evaluation of children nodes.

2.3.1 DFS with transferable credits (DFS CRE)

This approach is almost the same as standard DFS with one simple modification. The limit of evaluations for the first child is set to $N+c_0$ and then is propagated onto consecutive levels, following recursion in DFS. In some situations, credit c may stack to quite large number compared to N, which in turn may help in finding solution when N evaluations are not sufficient. Now suppose that after visiting all descendants of the first child the search is finished with c_1 remaining evaluations. Then, this credit is forwarded to the next child, so its limit of executions is now $N+c_1$. Finally, suppose that there were k children and at the end there is c_k remaining evaluations. In other words, let it be assumed that the optimum was found for all l stages in sub-tree induced by i-th node in $r<lN$ iterations, that is below total lN limit. In that case the remaining number $c_k=lN-r+c_0$ is returned to the parent node and propagated to subsequent children nodes, following node of i-th stage in DFS recursion.

2.3.2 BFS with transferable credits (BFS CRE)

The case of BFS is a little more complex. Since all children nodes have to be visited simultaneously, it is necessary to split credit c_0 between these nodes. Let k children of i-th stage have accumulated weights given by (5). Then j-th child receives credit $c_j=c_0v_j$ and this credit is then propagated recursively into deeper levels using the same order of visit as in standard BFS algorithm.

2.4. Particle Swarm Optimization

In our tests the Particle Swarm Optimization (PSO), which is a global search method, was applied. PSO is a method representing probabilistic approach, performing global search for optimal solution. PSO belongs to a group of algorithms motivated by nature, and is widely used and well described in the literature, e.g. see [6,7,8,9]. Let us recall basic information about PSO algorithm. In PSO algorithm it is necessary to simultaneously execute optimization of k different instances, so-called particles, and then evolution of the algorithm is in some sense determined by interactions of these particles and their performance measured by quality expressed in terms of a objective function F whose minimum is searched for. The search begins with randomly chosen position vector z_0^k of each particle. Position vector z_i^k of k-th particle in i-th iteration is modified by the relation:

$$z_{i+1}^k = z_i^k + v_i^k, \quad (6)$$

where v^k_i is velocity vector updated in each iteration by the formula:

$$v_{i+1}^k = wv_i^k + c_1 r_{i,1}^k (p^g - z_i^k) + c_2 r_{i,2}^k (p^k - z_i^k), \quad (7)$$

where: p^g denotes the best position found so far by the whole swarm; vector p^k represents the best solution found so far by the k-th particle; w is defined as the inertia coefficient; c_1 and c_2 are acceleration coefficients (called also training coefficients); $r^k_{i,1}$and $r^k_{i,2}$ are numbers in [0;1] picked at random with the uniform distribution.
This approach is in some sense independent of the choice of functions F_i, Q_i and in practice works effectively for complex problems as well as for simple problems.

3 Hypothetical structures of process

3.1. Balanced and unbalanced binary trees

Our first testing ground will be balanced binary tree with four levels, that is: root stage has two children nodes, each of them has two children, etc., forming finally a tree with the number of nodes on each level 1-2-4-8 as presented on Figure 4.

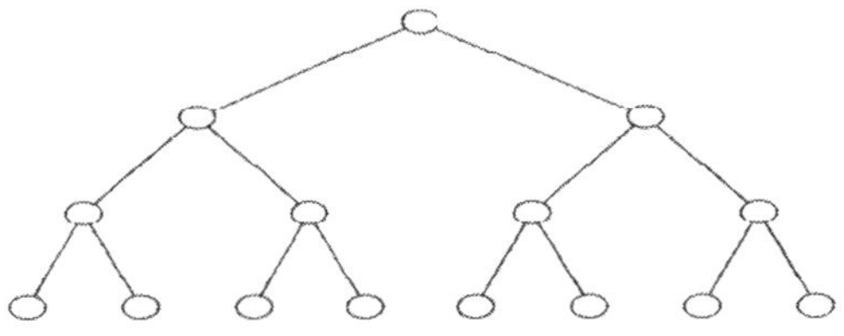

Figure 4. Process of balanced tree structure 1-2-4-8.

As a second example of hypothetical process tree with unbalanced structure will be examined. Such a situation seems more likely to happen in practice, because usually concrete industrial processes have different lengths measured by the number of intermediate stages. A tree with structure 1-2-4-4-4 as presented on Figure 5 will be considered. Note, that the number of nodes is exactly the same as in 1-2-4-8 tree.

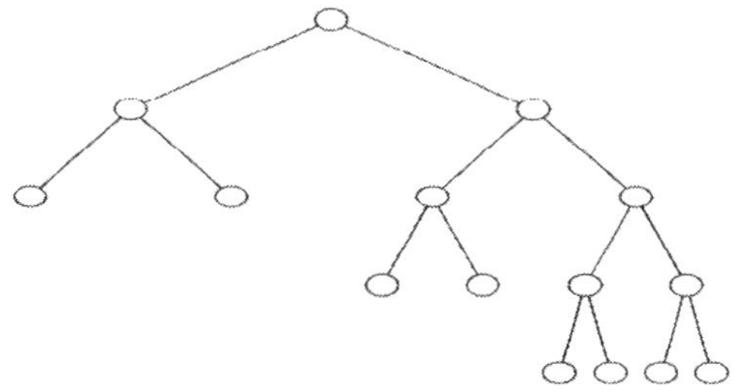

Figure 5. Process of unbalanced tree structure 1-2-4-4-4.

3.2. Functions Q_i and F_i

As it was said earlier, the aim of the paper is to investigate performance of the optimization strategies (and influence of credits in optimization). For this reason the simple, multi dimensional function was considered. This will minimize influence of randomness in our studies (there is zero probability of stuck of algorithm in local minimum) however there are still various complexity levels of the problem. Functions used in each node are the same, given by the formula:

$$q_i = Q_i(x_i, p_i) = \sum_{j=1}^{m} \left(x_{i,j} - p_{i,j}\right)^2 \qquad (8)$$

$$x_{i+1} = F_i(x_i, p_i) = p_i. \qquad (9)$$

Notice, that the control parameter p_i has a direct influence on the position of global minimum of function F_s for each children node s of i-th node, hence optimization in consecutive stages is not completely independent. The complexity of the process can be augmented by increasing number of dimensions m. When m is large, search space is also large, which makes that finding minimum by probabilistic algorithms is much more difficult.

4 Comparison of performance

Performance of the following five strategies was compared:

i. Global optimization (GLO - optimal control parameters of all stages set simultaneously),

ii. BFS without credits (BFS),

iii. DFS without credits (DFS),

iv. BFS with credits (BFS-CRE),

v. DFS with credits (DFS-CRE).

Let the number of evaluations of quality function Q_i for each stage be limited to N. In methods without credits this limit is strict. In methods with credits, remaining (not "consumed") evaluations can increase limits for consecutive nodes. In global approach we have global limit of $N{\cdot}s$ evaluations for the whole process (per particle), where s is the number of stages. The following 'global' limit of evaluations for the whole procedure is also introduced:

$$L_g = N \cdot s \cdot p \cdot t, \qquad (10)$$

where s is a number of stages, p is a number of particles, t is a limit for the 'time' of the optimization runs. When the algorithm calculates the value of any of the functions $Q_1, \dots, Q_s$, the index L_g becomes $L_g = L_g - 1$. In particular, to calculate value of F in (4) we need exactly s evaluations. It means that one step of GLO consumes s *'credits'* from L_g, and one step of each other method consumes 1 *'credit'*, however finally it is necessary to check all s stages (nodes of the tree). Then the limit in (10) allows at least t optimization attempts. However, if one run of the method is finished faster (or it is quickly decided that the whole optimization leads to a failure) the next optimization can be restarted faster. Thank that more runs than t can be performed. It is clear that the number of evaluations in one optimization run will be higher in methods with credits than in these without credits, and hence the average number of optimization attempts will be smaller. Since in every step of methods (i)-(v) it is necessary to preform evaluation of each particle of PSO method, the limit (10) may be simplified to the form:

$$L = N \cdot s \cdot t, \qquad (11)$$

having in mind that in fact 1 credit in L means p evaluations (one for each particle in PSO method), so p credits in terms of limit L_g.

5 Results

In all numerical experiments optimal solution using PSO algorithm with p=40 particles is searched for. Tests for different values of dimension parameter m in (8) were performed (m=1,5,7,10). Higher value of m is correlated with smaller chance of finding minimum by PSO.

5.1. Process of balanced tree structure

Every of the methods (i)-(v) was tested on the tree 1-2-4-8 (Figure 4) within the total limit of evaluations on functions F_i provided by (11) which in our case is L=1500000 (for N=1000 and t=100). As it was said before, each method can be executed at least 100 times within the limit L, however, this number can be changed if the method is terminated faster (due to success or failure). The number of full optimization runs of every method for different value of m is presented in Figure 6 and the number of attempts which were successful (method finished finding minimum) are presented on Figure 7.

DOI: 10.1051/matecconf/20168010008

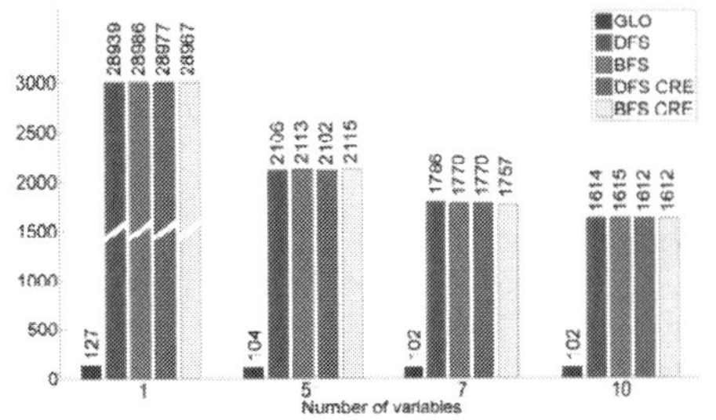

Figure 6. Average number of full optimization runs within limit L (tree 1-2-4-8).

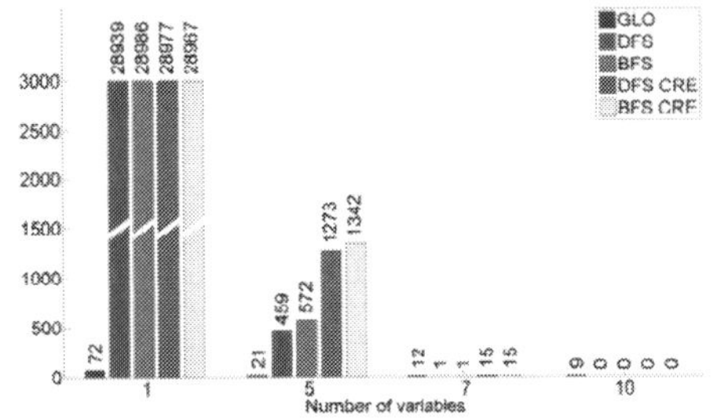

Figure 7. Average number of full optimization runs finished with success within limit L (tree 1-2-4-8).

Figure 8 presents relative number of successes of given method expressed by the ratio:

$$Sr = 100 \cdot {completed\_runs}/_{starts}\,[\%]. \quad (12)$$

In other words, Figure 8 shows relative percentage of chance that single instance of the method will finish with success.

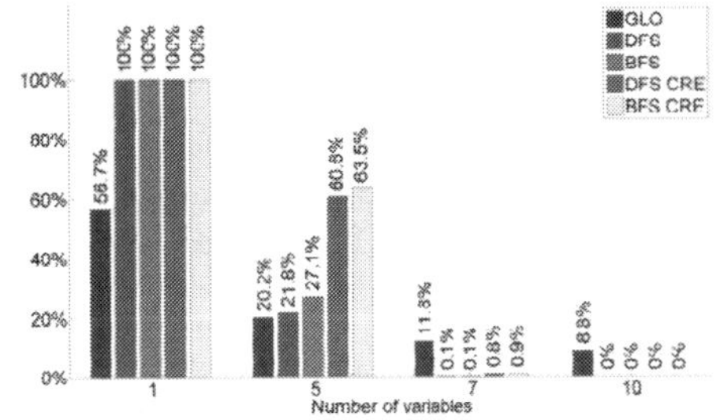

Figure 8. Relative success rate of every method (tree 1-2-4-8).

To decrease influence of randomness in our numerical experiment, every complete cycle of computations (sequence of optimizations within limit set *L*) was repeated 100-times for each method. Therefore, Figures 6-8 present average values calculated from 100 repetitions of every method.

5.2. Process of unbalanced tree structure

The same tests as in Section 4.1 were repeated for the process of the unbalanced tree structure 1-2-4-4-4 (Figure 5). Results of our numerical experiment are presented on Figures 9,10 and 11. Data presented in these figures follows exactly the same methodology as in section 4.1 in particular value of *L*, tested values of *m*, etc.

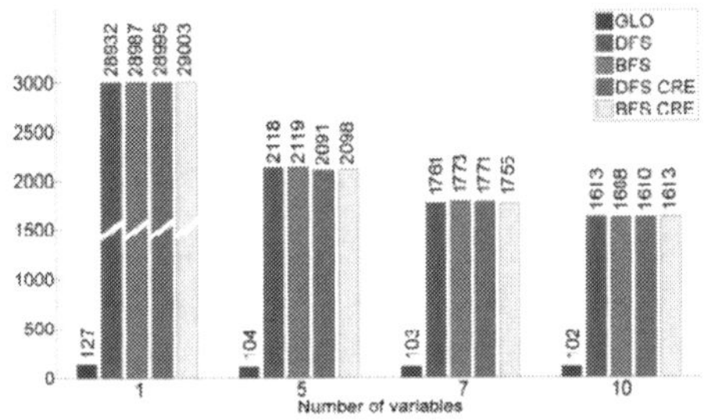

Figure 9. Average number of full optimization runs within limit L (tree 1-2-4-4-4).

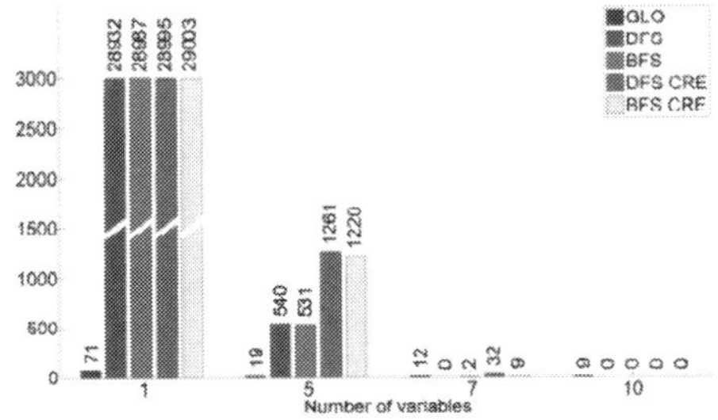

Figure 10. Average number of full optimization runs finished with success within limit L (tree 1-2-4-4-4).

The main difference now is that right side of the tree is much deeper, so in particular weights w_i defining division of credits between children nodes in BFS are now different. On the other hand, there is no visible change of the problem for GLO because the number of control parameters is the same (the same number of stages *s* as in tree 1-2-4-8). As before, each complete cycle of *L* evaluations was repeated 100 times.

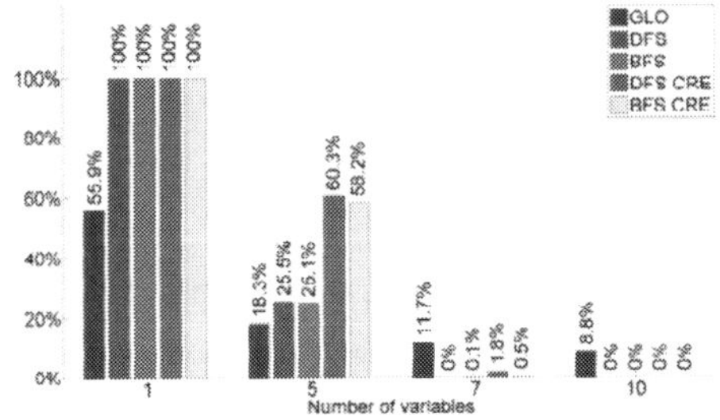

Figure 11. Relative success rate of every method (tree 1-2-4-4-4).

6 Conclusions

Five different optimization strategies of multi-stage industrial processes with two different tree structures were examined. According to the performed test, the following conclusions can be withdrawn.

- As it was mentioned earlier, GLO strategy does not take advantage of the knowledge about the structure of the process (relations between nodes/stages), therefore it should behave similarly on balanced or unbalanced tree structure, assuming that the number of stages and their dimensions are the same. It is confirmed by computational results presented in Figures 6-11. Average number of executed full optimization runs and the successful optimizations (and as a consequence, the success rate) were nearly the same in both cases. Comparison of the number of optimizations runs in sequential methods (ii)-(v) shows a huge difference between them and GLO. It is clear that these methods can terminate much faster, because in optimistic scenario they can decide that whole optimization is going to be a failure during first N evaluations for the first stage. If they do not succeed, they terminate the process, saving $N\cdot(s-1)$ evaluations. In the case of GLO it is necessary to wait until the real end to realize the failure, hence in such a case the whole $N\cdot s$ credits are spent. This situation is clearly visible in Figures 6 and 9, where number of started optimizations is close to minimum of t attempts (GLO can earn additional attempts only by finding optimal solution before spending the limit of $N\cdot s$ credits; then some credits within L are *``saved''*). Notice, that all the other (sequential) strategies are executed much more times than t.

- Comparison of strategies (ii), (iii) and (iv), (v). It was intuitively expected that (ii) and (iii) will be executed more often on average than their counterparts with credits, because they are able to continue computations longer. This statement is not supported by results of the experiment presented in Figure 6 and 9. On the other hand, success rate of DFS-CRE and BFS-CRE was much higher than for DFS and BFS. This suggests that while transferring *``credits''* between stages does not change expected execution time of the optimization much, it can highly decrease chances of failure. If the DFS and BFS (with of without *``credits''*) are compared, then it can be seen that visiting stages in optimization following order provided by BFS behaves slightly better when tree is balanced and DFS may be a better option in a case of unbalanced tree. The differences are not high however, so probably the above claim needs confirmation in a more extensive study with various different tree structures.
- Finally it can be seen that when complexity of the problem increases (the size of parameters space) GLO approach becomes more appropriate. It is most visible in the case of $m=10$ where it was the only strategy that was successful. This is situation in a sense similar to our previous observations from [10], where the influence of aggregation of stages on reliability of sequential optimization was considered.

Acknowledgements

The research leading to results presented in this paper was supported by National Science Centre in Poland (NCN), Grant no. DEC-2013/11/B/ST8/00352.

References

1. P. Jarosz, J. Kusiak, S. Małecki, P. Oprocha, Ł. Sztangret, M. Wilkus: Mathematical Problems in Engineering (2015) doi: 10.1155/2015/182679
2. P. Jarosz, J. Kusiak, S. Małecki, P. Morkisz, P. Oprocha, W. Pietrucha, Ł. Sztangret, *Metamodelling and optimization of copper blister two-stage production process*, JOMJ (2016), (*in print*)
3. T. H. Cormen, C. E. Leiserson, R. L. Rivest, C. Stein, *Introduction to algorithms*, 3rd ed. (MIT Press, Cambridge, 2009).
4. K. Deb, *Multi-objective optimization using evolutionary algorithms*. (John Wiley & Sons, Chichester, 2001).
5. S. Rao, *Engineering Optimization: Theory and Practice*. (John Wiley & Sons, New York,1996).
6. R.C. Eberhart, J. Kennedy, *Proceedings of the Sixth International Symposium on Micro Machine and Human Science* (IEEE Press,1995).
7. R.C. Eberhart, J. Kennedy, J., *Proceedings of IEEE International Conference on Neural Networks* (IEEE Press, 1995).
8. S. Helwig, R. Wanka, *Proceedings of the 2007 IEEE Swarm Intelligence Symposium* (IEEE Press, 2007).
9. I.C. Trelea, *Information Processing Letters*, **85** (2003).
10. J. Kusiak, P. Morkisz, P. Oprocha, W. Pietrucha, Ł. Sztangret, *On aggregation of stages in multi-criteria optimization of chain structured processes*, submitted to ICAISC 2016.

Rapid Tools Compensation in Sheet Metal Stamping Process

Lorenzo Iorio[1,a], Matteo Strano[2] and Michele Monno[2]

[1] *MUSP Lab, Piacenza, Italy*
[2] *Politecnico di Milano, Dipartimento di Meccanica, Milan, Italy*

Abstract. The sudden growth of additive manufacturing is generating a renovated interest towards the field of rapid tooling. We propose a geometrical compensation method for rapid tools made by thermoset polyurethane. The method is based on the explicit FEM simulation coupled to a geometrical optimization algorithm for designing the stamping tools. The compensation algorithm is enhanced by considering the deviations between the stamped and designed components. The FEM model validation has been performed by comparing the results of a DOE done at different values of press force.

1. Introduction

In the conventional deep drawing and stamping processes, a tooling setup made of a die, a punch and a blankholder is traditionally used. Over the years, many different types of flexible sheet forming processes have been developed in the industry, in order to improve the process, and especially in order to compress the tooling production times and costs. Single point and double point incremental forming processes 1 have been invented and are continuously being developed to this purpose. As another cost saving option, metal tools can be replaced with a rubber membrane. In the Guerin process, a movable thick rubber pad is pressed against a die 2 or a punch 3. The Marform variant of the Guerin process, an active blankholder 4 is used. In flexforming with a fluid cell, a rubber diaphragm is pressurized by a fluid or by a bladder 5. In deep drawing flexforming, a movable punch is used, too 6. In multi-point stretch forming 7, stretch forming clamps are used with the sheet bent over a flexible die, made by a raster of metallic movable pins.

In the last few years, the tremendously rapid growth of additive manufacturing is changing completely the way of thinking about and designing functional parts. At the same time, a renovated attention is being given to rapid tooling technologies 8, which offer cost-efficient and innovative solutions for improving the sheet metal forming processes.

The rapid tooling method proposed in the present work is to machine polymeric boards, made by thermoset polyurethane, which in a few minutes can be machined into a forming tooling setup (punches, dies and blank-holder). Very scarce scientific literature is available with this respect, to the authors' knowledge. One of the previous research works concerning all-polymeric forming tools with experimental and numerical analysis are by Park and Coulton of Georgia Tech 9, in the mid of years 2000. Sheet metal forming processes performed only with plastics tools are only used as prototyping methods, i.e. in prototyping job shops or for artistic uses by jewellers and metalsmiths 10.

The main advantage in using rapid polymeric tools is due to their low required machining energy and cost. The most expensive material used in this paper (commercial name Necuron 1300) has a comparable cost per unit volume, expressed in €/m3, to an AISI 1040 steel. However, the material removal rate in machining is more than 4 times faster, with a negligible tool wear in forming applications with thin sheet metals.

The main disadvantage of polymeric tools is that deflect elastically (or plastically, if they are not properly designed) under the forming forces. As a consequence, the final geometry of the formed part is difficult to predict, not only because of the usual springback, but also because of the deflection of the tools. For this reason, a numerical die compensation technique is required, able to suggest the correct shape for the tools, in order to keep the part within tolerances.

The most frequently used methods for die compensation are the displacement adjustment (DA), the surface controlled overbending (SCO) and the Force Descriptor Method (FDM). The displacement adjustment is very effective; the tool nodes are displaced in the opposite direction of the blank springback 11. The deviations calculation is done between the correspondent nodes of the simulated and the designed blank. This means that no remeshing is possible or, alternatively, any new mesh must be remapped with reference to the mesh of the designed part. For this reason, the DA method is

[a] Corresponding author: lorenzo.iorio@musp.it

frequently used for simple 2D forming cases 12, where a small number of nodes must be mapped or remeshing is not even required.

The surface controlled overbending algorithm performs the calculation directly on the tools CAD but, again, the calculation complexity makes the method inaccurate on a complex surface with high degree 13.

The Force Descriptor Method proposed by Karafillis and Boyce 14 is an iterative method based on the evaluation of the internal forces of the component, but the algorithm suffers from lack of convergence, especially in symmetric cases or limit values of springback.

The authors have proposed, at the ASME-MSEC 2015 conference, a compensation algorithm in combination with an FEM model 15, which allows to predict the deformation of the tools and to determine the required compensation in case the part falls outside the initial design tolerances. The proposed algorithm is inspired by the DA method, but improves the distance calculation by evaluating the normal distance from the tools node to the interpolated blank surfaces. Unlike the standard DA, there is no need to keep track of predetermined couples of nodes. The calculation of the normal distance solves the main disadvantage of the DA algorithm, allowing the applicability of the method also for 3D complex components. The results presented at the ASME paper were encouraging, but the method was tested with a relatively simple part. In the present work, a different test case, with greater geometrical complexity, has been used and the compensation algorithm has been significantly modified.

In the following Sections we will describe the experimental test case and the FEM model. Then, the development of the compensation algorithm will be described, in terms of mathematical formulation and algorithmic solutions. Finally, the validation and compensation results will be presented.

2. Description of the experimental test case

The test case (Figure 1) is a stamped component with a double symmetry plane and some geometrical features which are predictably hard to be obtained in a stamping process with flexible tools. The constant radii (R6 in Figure 1) are very small, i.e. difficult to be obtained in this kind of processes, because the deformable tools usually tend to slightly deform by compression in corner regions. The presence of the central, diamond shaped, depression adds complexity to the overall geometry of the process.

The material chosen for the test case is Al1050, annealed, with 1.5 mm wall thickness. The reference geometry of the part given in Figure 1, which will be called "designed geometry", is the starting point for determining the geometry of the deformable tools. An initial guess geometry of the deformable tools is built as an offset of the surfaces of the test case part; the iterative compensation algorithm will suggest a modification of this initial geometry. The deformable tooling setup (Figure 2) is made of a die, a punch (blue tools) and a blankholder (orange tool).

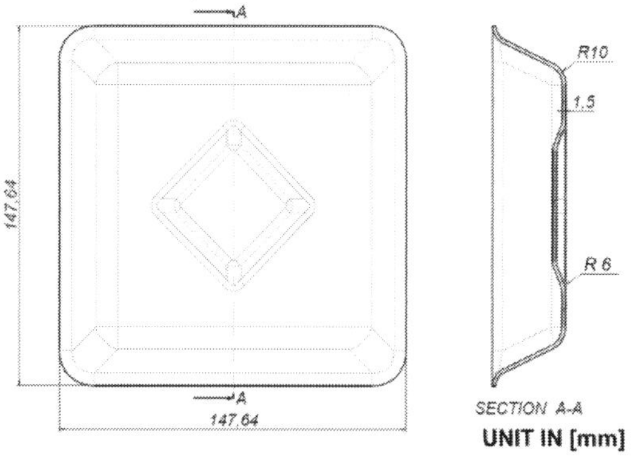

Figure 1: geometry of the test case used for developing the optimization algorithm.

The base of punch (orange) has been generated by using the material milled to the centre part of the blank holder, in order to reduce the cost of the required resin.

The polyurethane materials chosen for testing the deformable tools are: Necuron 1300 for the punch basement and blankholder; Miketool 1440 for the die and punch. The material properties are discussed in the following Section. The estimated cost of the polyurethane die is about € 990 (including machining costs). The same die, made by Zamak 2 (which is a low cost alloy typically used for prototyping tools) would cost about € 1860. In Figure 3 a quantitative comparison of the tools price in terms of total manufacturing and material costs has been presented, in order to show the economic advantage related to the use of polyurethane tools.

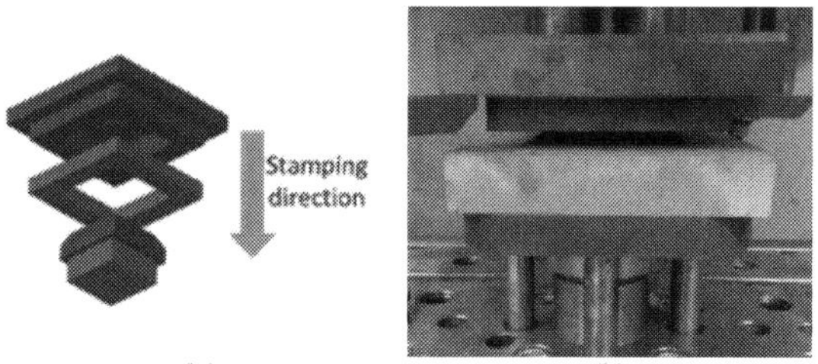

(a) (b)

Figure 2: (a) setup of the three deformable tools (punch, blankholder and die), along with the connecting plates (b).

Figure 2a shows the direction of the stamping process. The stamping process is made of three steps:

1 *Holding*: the blank is placed on the blankholder and then the die moves downward for holding it;
2 *Stamping*: the die pushes the blank against a fixed punch and against the blankholder with a maximum force of 980 kN; the maximum available reaction force of the blankholder is 392 kN.
3 *Springback*: the tools are released and both the tools and the blank recover the elastic deformation.

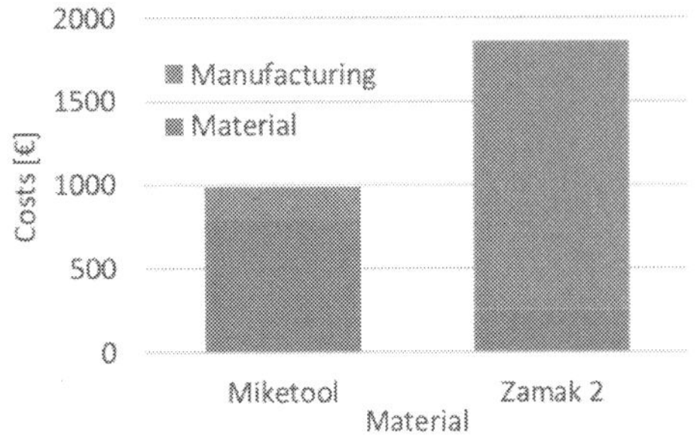

Figure 3: costs comparison between Miketool (polyurethane) and Zamak2 for making the die in the experimental case.

An experimental plan has been designed with variable levels of the blankholder force BHF, without replications (Table 1). The plan is aiming at determining which level of maximum BHF could determine failure of the parts (by either wrinkling or fracture) or failure of the tools (by fracture or plasticization). It is also aimed at determining the influence of BHF on the geometrical accuracy of the obtained parts.

Table 1: experimental plan

Experiment no.	**Maximum BHF** [kN]	**Maximum error measured on the A-A cross section** [mm]
1	4.9	-0.20
2	9.8	-0.21
3	14.7	+0.27
4	19.6	+0.14
5	23.5	-0.28
6	24.5	-0.26
7	25.5	Failure by fracture, part not measured

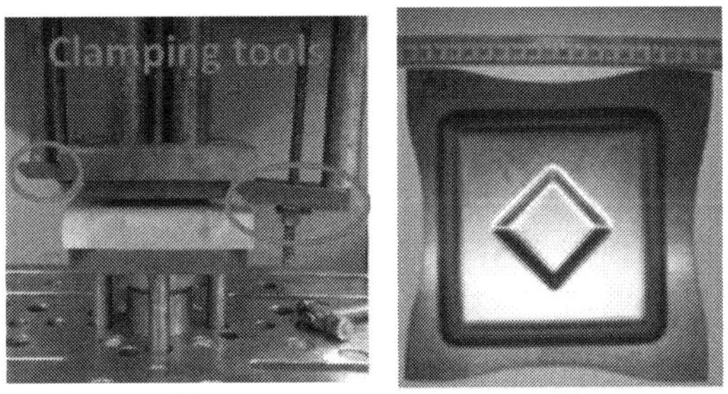

(a) (b)

Figure 4: (a) stamping tools mounted on the press; (b) the stamped component made with the initial geometry of the tools.

The experimental results showed that no macroscopic wrinkling occurred at any level of BHF_{max}.

The part at experimental condition no. 7 failed by fracture, hence the safe limit to fracture for BHF_{max} was assessed at 24.5 kN. All stamped parts have been measured with a CMM. All measured parts present some deviations (errors) from the designed part. The profiles measured at the symmetry plane (cross section A-A in Figure 1) are shown in Figure 5. Not all profiles are shown, in order to improve the clarity of the figure, not making too crowded with lines. The error profiles are not perfectly symmetrical around the centre of the chart because the clamping conditions of the blankholder were not perfectly symmetrical (Fig. 4a), hence different draw-in values have been allowed in different directions. Experiment number 6 (which has the largest admissible BHF value) presents the smallest possible error in the central region, but a relatively large error on the corners. The low error in the central part is due to increased stretching, while the larger error on the corners is due to compression of the plastic die corners. Experiment no. 4 seems to be the best compromise but it is not better than other profiles along the whole cross section. Indeed, this is the most interesting remark coming out the of the comparison of the error profiles: there is not a single BHF level able to generate an error which is consistently smaller throughout the whole cross section. The errors measured in some other regions of the part, away from the A-A cross section are even larger than the ones shown in Fig. 5. In order to reduce the amount of geometrical errors, a tool shape geometry compensation procedure is required, in combination with an FEM model of the process.

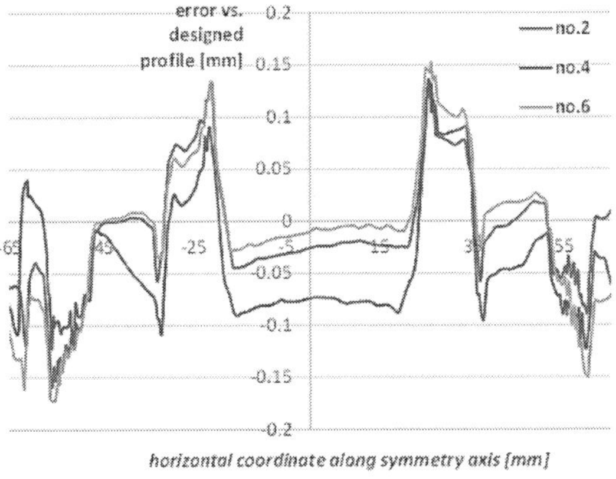

Figure 5: error on the nominal profile along cross section A-A

3. FEM model description

The FEM model has been implemented with the commercial code PAM-STAMP 2G V2015.1. In order to evaluate the deformation of the polyurethane tools, their mesh is made of four-node tetrahedral solid elements. The solid mesh of the plastic tools is created with VisualMesh 11 by using an automatic meshing module. The surface curvature criterion has been used to determine the element density (finer mesh is generated in areas of higher surface curvature). The number of elements and the number of nodes for each tool is shown in Table 2. A good discretization is important for the accuracy of the FEM simulation, but it is also very important for the application of the compensation algorithm described in the following Section.

The edge length of the squared blank is 200 mm and the surface area of each initial element is 400 mm^2 with a total number of 100 initial elements. Every refinement step splits 1 quadrangular element into 4 smaller quadrangular shells, by an automatic refinement algorithm. Each element can be split up to 5 times.

Table 2: number of elements and nodes used for meshing the tools.

Tool	N. elements	N. nodes
Die	304291	58952
Blankholder	19898	4797
Punch	387548	71129
Punch Base	12064	2641

Table 3: material properties of the polyurethane tools and aluminum blank.

Material	Young Modulus	Linear elastic limit	Yield Stress	Poisson coeff.	Density
	MPa	*MPa*	*MPa*		*g/cm³*
Al1050	72000		33	0.335	2.73
Necuron 1300	2317	35	68	0.340	1.15
Miketool 1440	2950	57	81	0.340	1.20

The tensile test data of the blank have been retrieved in the "CES EduPack" material database, since the Al 1050 is a very common material. Tensile data of the tool materials can be found in the literature too 16. Extensive quasi-static tensile and compressive tests have been performed on the polymeric tools in order to determine their elastic modulus. The resulting values for the elastic modulus and the Poisson coefficients are reported in

Table 3. The blank material is modelled as elastic-plastic. The coefficients of the Krupkowsky's strain hardening law used in the simulations:

$$\sigma = K \cdot (\varepsilon_0 + \varepsilon_p)^n \tag{1}$$

are K=126.4 MPa, n=0.193, ε_0= 0.00093.

The simulation starts at the end of the holding stage in which the blank is held between the die and the blankholder (Figure 6a).

The contact algorithm between the objects is automatically defined by the software taking into account a Coulomb friction coefficient of 0.2 9, in order to better simulate the interaction conditions between polyurethane and aluminium materials.

As shown in Figure 4a, the plastic tools are enclosed within metal clamping tools which connect and lock the tools to the press machine. The position of the locking tools along the diagonal of the die does not allow the application of the symmetry to the FEM model. In order to correctly simulate the effect of each clamp on the deformable tools shown in Figure 4a, proper boundary conditions have been applied only on specific nodes and surfaces.

For the die, an imposed displacement with constant velocity is applied on the upper and on some nodes corresponding to the clamping tools (Figure 4a). The BHF is applied on the lower surface, in order to simulate a uniform distribution of the pressure; finally, a fixed displacement is applied on the lower nodes of the punch base (Figure 6b). The nodes between the punch and the base have been fused in order to simulate the glued contact between the upper surface of the base (orange) and the lower surface of the punch (pink). The springback simulation is run by the FEM software with the "Advanced Implicit" algorithm, by locking three nodes on the stamped component.

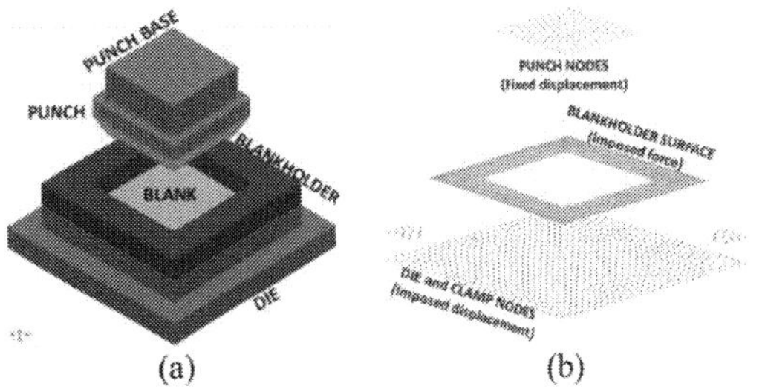

Figure 6: scheme of the simulation setup (a) and boundary conditions (b) applied to the FEM simulation.

4. Compensation algorithm

The main response variables in a stamping process with deformable tools are the deviations between the geometry of the real stamped component and the designed one. The proposed algorithm suggests the compensation to be applied on the plastic tools, in order to obtain the minimization of the deviations. The compensation algorithm iteratively runs the following two main steps:

STEP I. FEM simulation: the user runs a stamping simulation with springback, and a mesh refinement stage. The blank mesh is built with shell elements which represent the middle surface of the part, i.e. half of the thickness must be added in the normal direction in order to model the actual outer or inner surface of the part. As a first guess, in this initial run, the deformable tools are built by offsetting the designed part by half of the initial sheet thickness. During refinement, the meshes of the simulated component and the designed one are both regenerated with shell elements of approximately uniform side length. The final mesh refinement stage is very useful for performing simpler and more accurate computations at the next step II.

STEP II. Compensation: the proposed automatic algorithm has been coded within C++. The routine takes the following inputs: the refined triangular shell mesh of the simulated and the designed parts, the outer contour of the simulated part, the external surface shell mesh of the tools in ascii. The tool geometry is modified according to the following 4 sub-steps:

II.1 Blanks mesh import and surface fitting: the meshes of simulated and designed components are imported. The nodal cartesian coordinates of both

 DOI: 10.1051/matecconf/20168010009

meshes are converted into parametric coordinates, with the library implemented in 17, as:

$$\Sigma_s = \begin{cases} x_s = x_s(u,v) \\ y_s = y_s(u,v) \\ z_s = z_s(u,v) \end{cases} \quad \Sigma_d = \begin{cases} x_d = x_d(u,v) \\ y_d = y_d(u,v) \\ z_d = z_d(u,v) \end{cases} \tag{2}$$

The parametric surfaces are reconstructed with the Multilevel B-spline Approximation (MBA) 18, where an algorithm models the surface with a recursive refinement of the B-spline knots depending on the level of desired accuracy (k). This fitting allows handling the designed and simulated objects not as numerical meshes with their nodes, but as mathematical surfaces.

II.2 Tools and blank contour import: the external surface nodes of the tools and the outer contour of the simulated component are imported. The contour is useful to define the domain of the compensation.

II.3 Compensation. This step II.3 is repeated for both tools: the punch and the die. Figure 7 shows a 2D graphical representation of the computed quantities. The distances δ_d and δ_s between the generic tool (punch or die) node $\vec{T}_0(x_{t0}, y_{t0}, z_{t0})$, and the two fitted surfaces (simulated and designed) are computed through the conjugate gradient minimization algorithm 17. These distances are 3D vectors, with components in the cartesian coordinate system used for the simulation setup:

$$\vec{\delta}_s = [x_s, y_s, z_s] \tag{3}$$

$$\vec{\delta}_d = [x_d, y_d, z_d] \tag{4}$$

The deviation vector between the simulated and designed surfaces is defined as

$$\vec{\delta}_c = (\vec{\delta}_d - \vec{\delta}_s) \tag{5}$$

With the components:

$$\vec{\delta}_c = [x_{\delta c}, y_{\delta c}, z_{\delta c}] \tag{6}$$

The tool nodes are moved from the original position $\vec{T}_0$ by the quantity $\vec{\delta}_c$. The final position of the compensated tool node is defined as:

$$\vec{T}_c \begin{cases} x_{Tc} = x_{T0} + \lambda x_{\delta c} \\ y_{Tc} = y_{T0} + \lambda y_{\delta c} \\ z_{Tc} = z_{T0} + \lambda z_{\delta c} \end{cases} \tag{7}$$

II.4 Exporting: The meshes of both compensated tools are exported, the FEM simulation is run again and the deviations, after compensation, are calculated again.

The algorithm can be applied iteratively and the whole procedure is repeated until convergence, i.e. until the deviations between the simulated and designed parts are within the tolerance interval chosen by the designer.

5. Numerical Results

Before applying the compensation routine, the accountability of the FEM model must be verified. With respect to failure limits, experiment no. 7 is very useful, where the maximum force is applied on the blankholder. In the real tests, fracture occurred at this high level of draw-in restraint. The experiment and the simulation show the same location of the blank fracture localized at the deepest corners of the component (Figure 8). No other simulation nor experimental case had occurrence of fracture. Both in simulations and in real tests, no macroscopic wrinkling was detected. The errors across the cross section A-A between the experimental and numerical profiles are included in the range [-0.23; +0.24] for test no. 2 and [-0.20; +0.17] for test no. 3. These levels of geometrical accuracy are in the typical range of FEM models with shell elements; they are acceptable for general purposes but they seem to be too large for a reliable compensation routine. However, it is interesting to observe the comparison given in Figure 9, which is limited to test case no. 2 for brevity. In this figure, the error of the FEM profile with respect to the designed part is compared to the error (already given in Figure 2) between the actual CMM measured profile and the designed part.

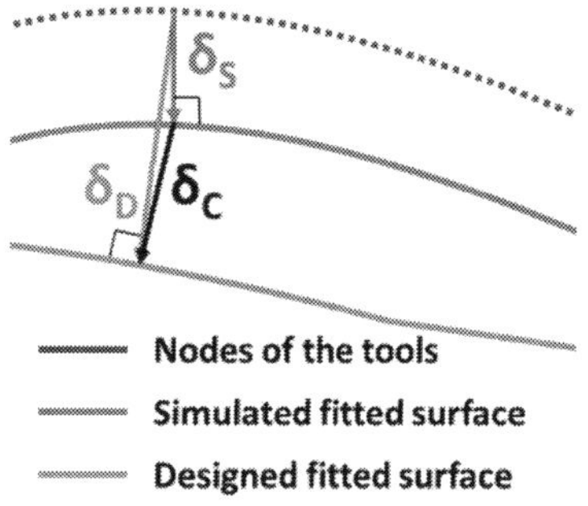

Figure 7: computation of the new tool nodes.

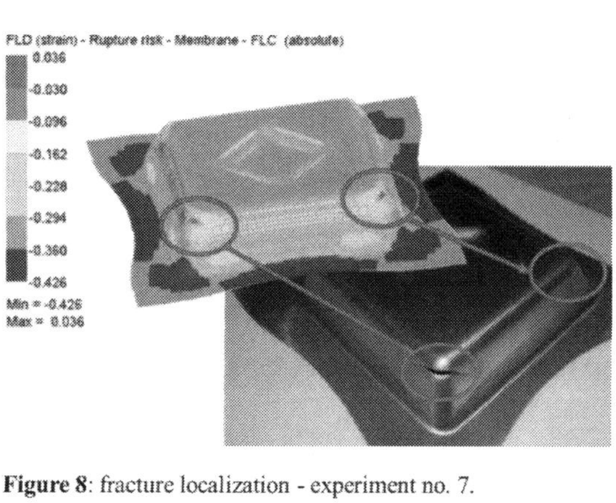

Figure 8: fracture localization - experiment no. 7.

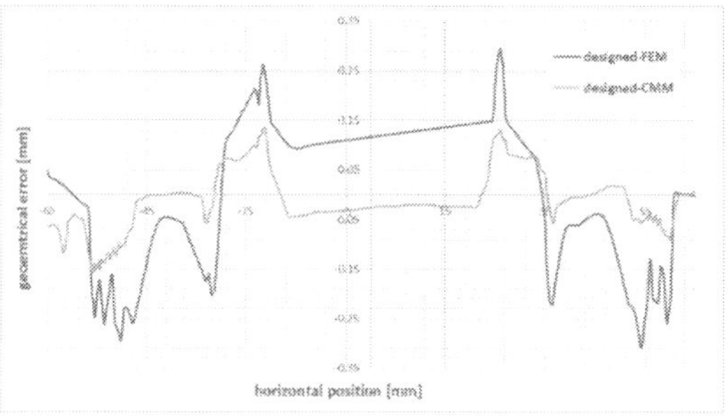

Figure 9: errors between FEM and experimental profile for experiment no.2.

Although the numbers are obviously different for the two profiles, (with a larger amplitude for the FEM error) the shape of the two profiles is surprisingly similar, with a clear indication of accountability of the FEM model.

In order to demonstrate the use of the proposed compensation technique, it has been applied to test cases no. 2 and no. 3.

TEST CASE no. 2. This test case is run with BHF=9.8 kN. As the first simulation run, the tools are simply built as an offset of the final designed part middle surface. The plot of the deviations between this simulated part and designed components is shown in Figure 10, for the whole part, excluding the flange. The simulated component deviates considerably from its designed geometry in a range between [-0.24; 0.44] mm and only 78.53% of the calculated deviations lays within the range of [-0.14; 0.15]. The maximum positive or negative values of deviation are located on both the convex and concave corners. This corresponds to a large deformation of both tools on the corner radii.

The algorithm performed an automatic compensation of the tools geometry, with λ=1. As a result of the first compensation, the punch and die corners radii have been modified by the algorithm in order to reduce the deviations shown in Figure 10.

After the compensation, the range of the deviations between the designed and simulated components is reduced. In Figure 11, the results of the first iteration are shown. The range of the deviations, calculated in the normal direction, lays within the range [-0.22; 0.15] mm.

This error range is not much different from the one in Figure 10 (obtained before compensation), but it must be noted that the 96.95% of the surface is within the range [-0.12; 0.10] mm. Furthermore, the increase of calculated effective stress on the tools is limited to only 1 MPa on the punch and 7 MPa on the die. This means the proposed algorithm is able not only to reduce the geometrical errors but also to save the impact on the expected tool life.

The time required for 1 simulation (run with 8 processors) and for 1 tool compensation (with 1 processor) is summarized in

Table 4.

The procedure described in Section 4 is iterative, it can be repeated with constant or variable values of the correction coefficient λ.

In Figure 12 the RMS value of the deviation vector is plotted vs. the number of iterations, with constant λ–value. The figure shows that the RMS values of the deviation vector decreases drastically after the first iteration and rapidly converges towards a stable solution. A very similar trend had been observed also in a previous work [15], and this indicates a typical behaviour of the proposed routine.

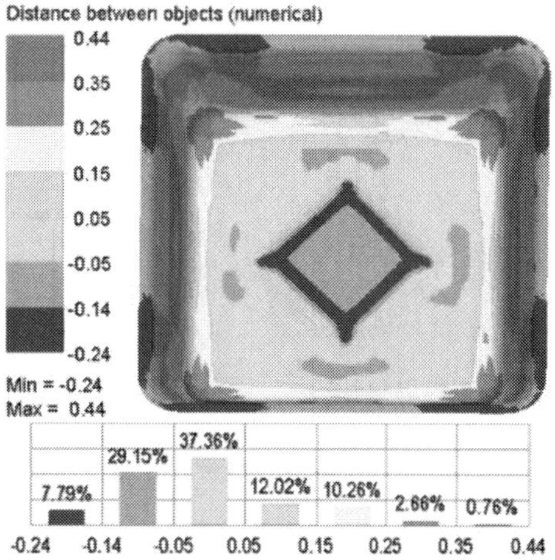

Figure 10: δ-plot before compensation; isometric and top views, units in [mm].

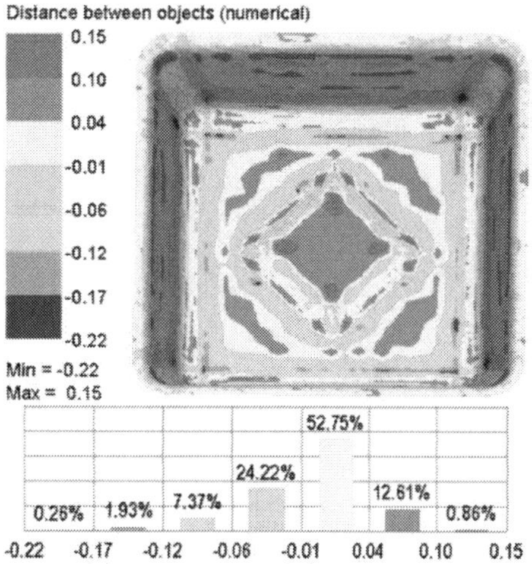

Figure 11: deviations between designed and optimized components after one iteration.

Table 4: time needed for 1 iteration in compensation of tests case no.1

Operation	Time needed
Simulation – 3 stages	4330 s
Reading blanks mesh	2 s
Parameterization	1 s
Compensation	60 s
Save tool mesh	0.1 s
Total	**11557 s (1.23 h)**

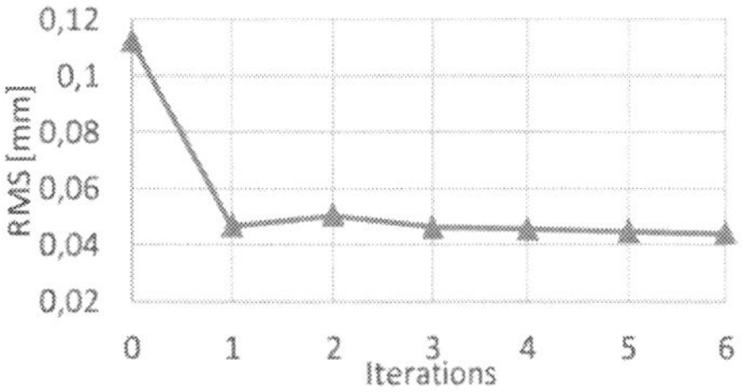

Figure 12: norm of deviation vector vs. iteration for test case no. 2

TEST CASE no. 3. This test case is run with BHF=14.7 kN. Again, in the first simulation run, the tools are built as an offset of the final designed part middle surface.

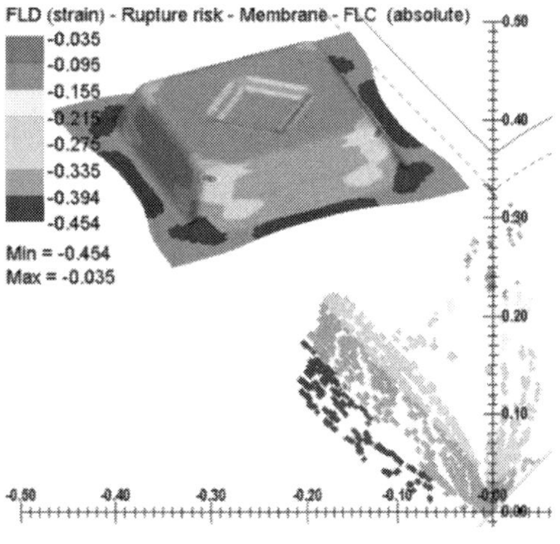

(a)

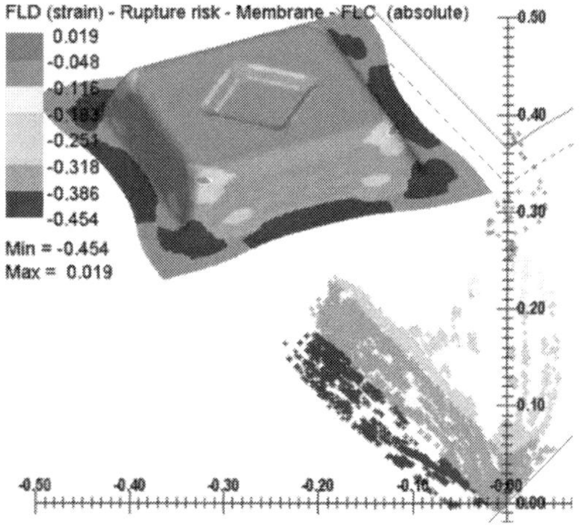

(b)

Figure 13: FLD map of test case no.3 before compensation (a) and after compensation (b).

The strain map, expressed in terms of distance from the FLD line, is shown in Fig. 13a. After the tool geometry compensation, the state of strain on the blank obviously changes, and the resulting FLD plot is shown in Fig. 13b, where a large risk of fracture is evident. In other words, in order to reduce the geometrical errors, the routine tends to increase the level of strain. This behaviour is typical since it has been observed also for test case no. 4 and for other tested geometries. As a result, the procedure described in Section 4 can be framed in a more general algorithm, which involves the selection of the correct level of BHF and the evaluation of the risk of fracture. This general algorithm is summarised in Fig. 14.

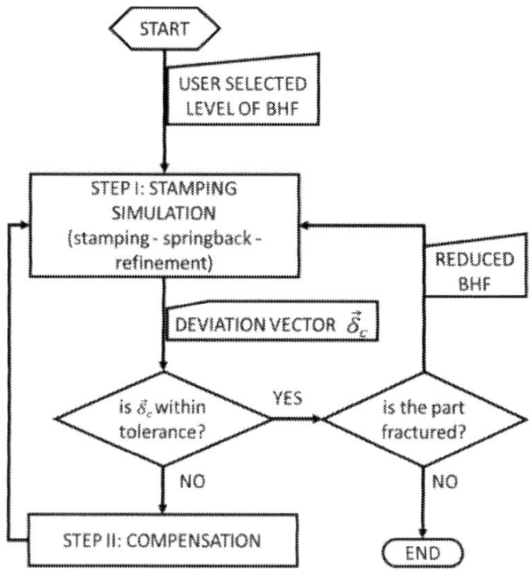

Figure 14: General compensation algorithm

6. Conclusions

In this paper, a compensation algorithm has been proposed, based on the displacement adjustment (DA) approach. The proposed method allows to evaluate and reduce the deviations between the simulated component and the designed one.

The algorithm, based on the reduction the normal deviation vector, can be applied iteratively. It generates an increase of the strain level in the final workpiece, as a consequence of the change in the shape of the tools. For this reason, the an iterative algorithm has been proposed, which takes the risk of fracture into account. When the results of the compensation algorithm are far from the risk of fracture, the algorithm rapidly yields a satisfactory solution already after the first iteration.

Acknowledgments

This project has been partly funded by the Italian National Cluster project "Fabbrica intelligente", CTN01_00163. Mr. Alessandro Donghi is gratefully acknowledged for his precious help in the development of the experimental tests case.

References

1. Malhotra R, Cao J, Ren F, Kiridena V, Cedric Xia ZZ, Reddy NV. ASME. J. Manuf. Sci. Eng. 2011;133(6):061019-061019-10. doi:10.1115/1.4005179.
2. Son C-Y, Jeon Y-P, Kim Y-T, Kang C-G., Journal of Engineering Manufacture. 2012 Jan 30;226(5):909–918.
3. Ramezani M, Ripin ZM, Ahmad R. CIRP Journal of Manufacturing Science and Technology. 2010 Jan;3(3):196–203.
4. Ramezani M, Ripin ZM. The International Journal of Advanced Manufacturing Technology. 2011; 59(5-8):491–505.
5. Strano M.: Journal of Engineering Manufacture. 2006; 220(8):1305–1315.
6. Vollertsen F, Breede R, Beckman M. Proceedings of the Institution of Mechanical Engineers, Part B: Journal of Engineering Manufacture. 2001 Jan 1;215(7):977–990.
7. Cai Z-Y, Wang S-H, Xu X-D, Li M-Z. Journal of Materials Processing Technology. 2009;209(1):396–407.
8. Chua CheeKai, Leong KahFai, Liu Zhong Hong, Handbook of Manufacturing Engineering and Technology, Spinger, 2014, p. 1-22.
9. Park, Y., Colton, J.S., 2005. J. Manuf. Sci. Eng. 127, 116.
10. Longhi B. H., Eid C., Brynmorgen Press, 2013.
11. YANG, Xiang An; RUAN, Feng. International Journal of Mechanical Sciences, 2011, 53.5: 399-406.
12. SHEN, Guozhe, et al. In: AIP Conference Proceedings. IOP INSTITUTE OF PHYSICS PUBLISHING LTD, 2005. p. 334.
13. W.H. Press, S.A. Teukolsky, W.T. Vetterling, B.P. Flannery, Numerical Recipes in C, Cambridge University Press, 1992.
14. Karafillis A.P., Boyce M.C., International Journal of Machine Tools & Manufacture 1996; 36 (4): 503–36.
15. Iorio, L., Strano, M., & Monno, M. (2015). In Proceedings of the ASME MSEC International Conference, Vol. 1: Processing (p. V001T02A089). Charlotte, NC (USA).
16. M. Amarandei, Antonio Virga, K-N. Berdich. Materiale Plastice, 2013; 50: 84-87.
17. SINTEF, Geometry/GoTools, https://github.com/SINTEF-Geometry/GoTools.
18. Lee, S., Wolberg, G., & Shin, S. Y. (1997). Visualization and Computer Graphics, IEEE Transactions on, 3(3), 228-244.

DOI: 10.1051/matecconf/20168010010

A new optimization procedure for the accurate characterization of thermal phase transformation curves based on controlled quenching experiments

Maurice Peterli[1,a], Minh-Trung Truong[1], Niko Manopulo[1] and Pavel Hora[1]

[1] *Institute of Virtual Manufacturing (IVP), ETH Zurich, Tannenstrasse 3, 8092, Zurich, Switzerland*

Abstract. Precise hardness and phase content prediction for quenched steel with the finite element method requires optimal material data, which is usually obtained from measured continuous cooling transformation (CCT) diagrams. However, most software packages that are able to predict final phase composition require time temperature transformation (TTT) diagrams. TTT diagrams can be calculated from the chemical composition of the material. With this methods the numerical prediction often result in deviations to reality. A newly developed optimization method can improve the accuracy of phase content and hardness prediction after quenching by optimizing the TTT diagram with measured data of controlled quenching experiments.

1 Introduction

Plenty of industrial components are subject to heat treatment in order to improve the parts mechanical properties. Prediction of the final phase content and hardness with finite element method can be difficult to achieve. Many software packages require a time temperature transformation diagram (TTT) which is difficult to obtain with standard measurement techniques. However it is more convenient to measure the continuous transformation diagram (CCT), which can be used to adjust the corresponding TTT diagram. Some deviations in the finite element results may come from this transformation, others can result from hard-won and imprecise phase content determination.

Realistic transformation kinetics are key to an accurate prediction of the current and final microstructure during quenching and can be represented with the TTT diagram. Although TTT diagrams contain only isothermal transformation data, they can be used to approximate the anisothermal quenching process by breaking it down into smaller isothermal steps. Together with the additivity principle, the theory "that of fictitious time" [1] and Sheil's [2] method, it is possible to access anisothermal transformation kinetics from isothermal data. Incubation and phase growth, which will only start when incubation is completed, are separated. The nucleation follows Sheil's summation and phase growth of diffusional transformations can be described with the Avrami equation [3, 4, 5], where y_n is the transformed volume fraction of bainite, ferrite or perlite, t is time , b and n are temperature dependent variables:

$$y_n = 1 - \exp(1 - bt^n) \qquad (1)$$

Diffusionless non-time dependent transformation from austenite to martensite can be expressed with the Koistinen and Marburger formula [6]:

$$y_{martensite} = y_{austenite}\{1 - \exp[-A(M_S - T)]\} \qquad (2)$$

$y_{austenite}$ represents the remaining austenite, T is the current temperature, A is a material dependent variable and M_S is the martensite start temperature. As soon as the temperature passes the M_S-temperature, the remaining austenite starts being transformed to martensite. There exist several tools to compute a TTT-diagram from the material's chemical composition and initial grain size [7, 8, 9]. The one used by the authors is presented in the following. The transformation kinetics can be predicted with the Kirkaldy method [10, 11], after the characteristic temperatures of phase change are estimated. The austenitization-end temperature (A_{C3}) is averaged over the estimation method from Andrews [12] and Kasatkin [13]. Both are based on the chemical composition. The austenitization start temperature (A_{C1}) is calculated with the Andrews model [12]. For the bainite-start-temperature B_S, the Kirkaldy model is used [10]. The martensite-start-temperature is averaged over Haynes [14] and Andrews [12].

The final hardness after quenching is calculated either by taking the final phase composition and match it with data from CCT-diagrams or by taking the transformation history into account. The latter can be done by determine the overall hardness through accumulation of the hardness values of each transformed phase component during each time step in the finite element simulation. The final hardness H can be formulated as [15]:

[a] Corresponding author: maurice.peterli@ivp.mavt.ethz.ch

$$H = \sum_k (\sum_i \Delta y_{ik} H_{ik}) \tag{3}$$

where y_{ik} is the phase component of phase i, that is formed during the time step k. H_{ik} is the hardness linked to phase i at the current temperature k.

This contribution aims to optimize the input data for finite element simulation of heat treatment processes for a better prediction of hardness and phase content in quenched parts.

2 Material Characterization

The material used in this work is a through-hardenable C60 steel. To characterize the thermo-mechanical properties, it is necessary to measure the standardized [16] CCT diagram. The experiments performed to obtain this data can later be used for the validation of the corresponding finite element simulation as later shown.

2.1 Measured CCT diagram

To characterize thermal characteristics a CCT-diagram was obtained with a DIL 805A/D quenching dilatometer from TA instruments (see Figure 1). Quenching experiments with the following cooling times (time from A_{C3}-temperature to 200°C) have been carried out; 10s, 15s, 20s, 40s, 60s, 100s, 200s, 500s and 1000s. By connecting the transformation points in the temperature-dilatation diagram, we get the phase regions and the martensite start temperature. With an additional slow heating experiment, the austenitization start and end temperatures are measured.

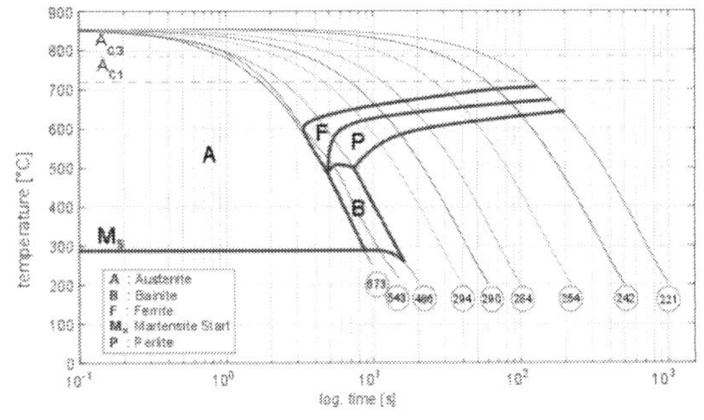

Figure 1. Measured-CCT diagram

The cylindrical specimen that were used for the quenching experiments were 4mm high and 4.5mm in diameter. After the experiments, final phase content and hardness (circled numbers at the end of each cooling curve) were measured.

2.2 Computed TTT-diagram

Additionally, the initial chemical composition of the material, the characteristic temperatures and grain size was determined as shown in Table 1.

Table 1. Chemical composition and characteristic temperatures

Fe	C	Cr	Mn	Si
98.209%	0.613%	0.27%	0.678%	0.204%
P	**Ni**	**Mo**	**S**	
0.017%	0.007%	0.001%	0.001%	
AC3	**AC1**	**BS**	**MS**	**Grain size**
785°C	720°C	572°C	261°C	7.5 μm

The following TTT diagram was calculated with the TTT-Generator from FORGE® using the material's chemical composition and grain size based on the method explained in the introduction. The computed TTT diagram is shown in Figure 2.

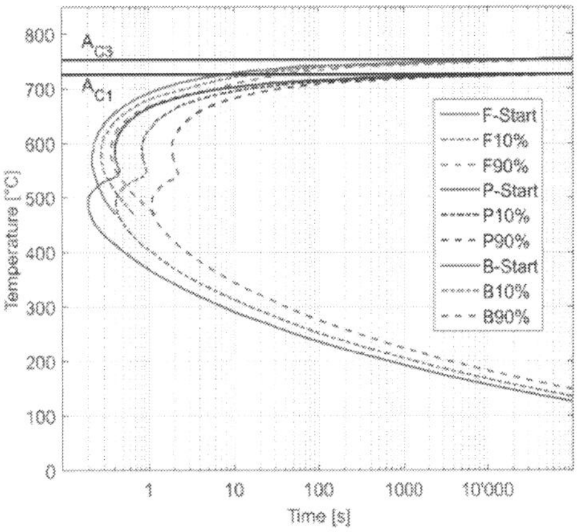

Figure 2. Computed TTT-diagram with the TTT Diagram Generator from FORGE®

This diagram is basis for the transformation kinetics during quenching. The solid lines represent the starting line for isothermal transformation for ferrite, perlite and bainite. The narrowly dotted line indicates the 10% and the coarse dotted line the 90% completion of transformation from austenite to the respective phase. For instance 10% bainite would have transformed from austenite after 15s, at 300°C under isothermal condition.

As mentioned in the introduction, hardness can be coupled with the phase transformation. The automatically generated temperature dependent hardness values for each phase is shown in Figure 3. For example the fraction of ferrite that is transformed from austenite to ferrite at 600°C has a hardness of 200HV. Every hardness of all fractions are summed-up for every time step according to equation (3) and result in the final hardness after quenching.

3 Quenching simulation

The finite element simulation was designed, so that the cylindrical specimens follow the measured temperature paths from the CTT diagram with FORGE® NxT 1.0.

After the simulations, final hardness and phase content was determined for all quenching times. The results of the experiments and the simulations are shown in Figure 4. Every dot represents a simulation or measurement for a certain quenching time.

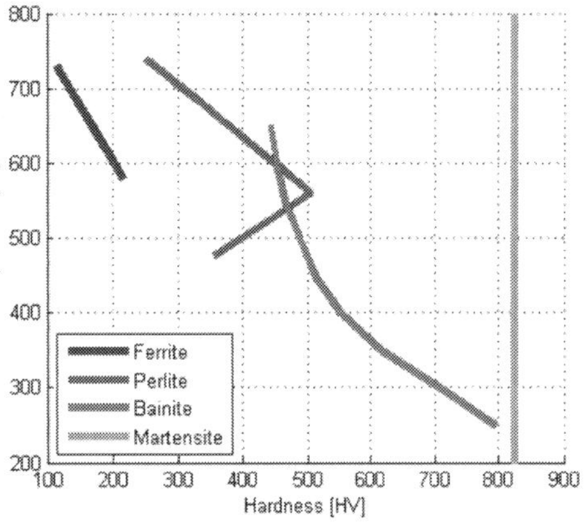

Figure 3. Calculated temperature dependent hardness

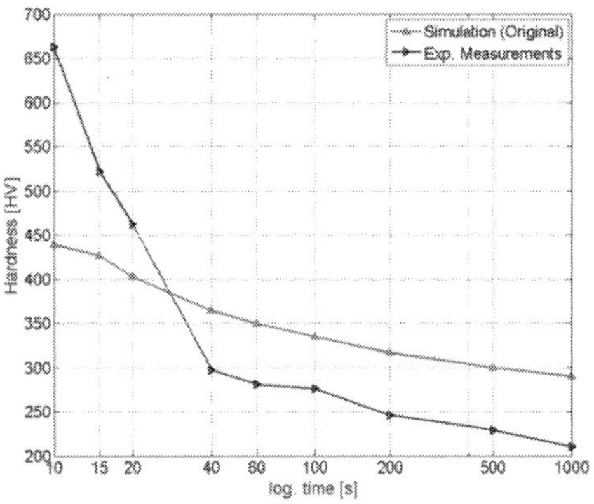

Figure 4. Measured versus simulated hardness

There is quite significant disagreement between measured and simulated hardness. The maximum hardness achieved in reality is over 200HV higher than in the simulation. This can be explained with the low martensite fraction of the simulation for fast cooling times. The micrographs show almost complete martensitic transformation for the quickest quenching specimen (10s). If we have look on the computed TTT-diagram in Figure 2 we can see that the ferrite, perlite and bainite 10% line is before the 1s mark, which means that it is impossible to get a full martensitic structure if the minimum quenching time is 10s. Those results show clearly that there is a potential for improvement.

4 Optimization process

For optimal prediction of phase content and hardness after quenching it is necessary to adjust the transformation curves in the TTT-diagram and change the hardness values, which are coupled with the phase evolution.

The determination of the phase composition by evaluating the corresponding micrographs is not trivial. Due to the light-dark contrast while using the etching method, the phase components and other metallic properties, such as grain boundaries, non-metallic inclusions or mixed phases etc., are visible. Furthermore, an accurate estimation of the percentage fractions of the phase composition is not possible with the naked eye. Although there exist picture analysis software to conquer this problem, the inaccuracies are still significant. A mathematical method can be used in order to identify the phase composition of the phase microstructure despite the uncertainties. A region-based optimization method can be used to identify the phase composition of the measured microstructure and optimizing the corresponding hardness values.

The uncertainties can be expressed with a minimum and a maximum possible phase fraction for each phase and quenching time.

4.1 Phase fraction and hardness optimization

The solution of the optimization problem is based on the assumption that a linear relationship between hardness values and the corresponding phase fractions exists. This is a simplification of reality and may cause some inaccuracies, but is necessary for the following optimization process. The linear dependency between phase fraction and hardness can be formulated as follows:

$$x_i h_F + x_{i+1} h_P + x_{i+2} h_B + x_{i+3} h_M = H_i \tag{4}$$

where h_F, h_P, h_B, h_M are the hardness values of the ferrite, pearlite, bainite and martensite phases. H_i is the macroscopic hardness for the corresponding quenching time i. $x_i, x_{i+1}, x_{i+2}, x_{i+3}$ represent the volume percentages of the phases, whereas their sum for each row results in 1:

$$x_i + x_{i+1} + x_{i+2} + x_{i+3} = 1 \tag{5}$$

Accordingly, a system of linear equations can be written as follows:

$$\begin{bmatrix} 0 & 0 & x_1 & x_2 \\ x_3 & 0 & x_4 & x_5 \\ x_6 & x_7 & x_8 & x_9 \\ x_{10} & x_{11} & 0 & 0 \\ x_{12} & x_{13} & 0 & 0 \\ x_{14} & x_{15} & 0 & 0 \\ x_{16} & x_{17} & 0 & 0 \\ x_{18} & x_{19} & 0 & 0 \\ x_{20} & x_{21} & 0 & 0 \end{bmatrix} * \begin{bmatrix} h_F \\ h_P \\ h_B \\ h_M \end{bmatrix} = \begin{bmatrix} 663 \\ 522 \\ 462 \\ 297 \\ 281 \\ 276 \\ 246 \\ 229 \\ 210 \end{bmatrix} \tag{6}$$

The left matrix contains the phase fractions. The first column is the ferrite, the second perlite, the third bainite and the last one the martensite fraction, that leads to the measured hardness values (vector on the right) at the end of the quenching experiments taken from the CCT-diagram (Figure 1). A zero means that no trace of this phase was found in the micrograph. Hardness measurements are, in contrast to the phase volume percentages, easily and reproducible to measure. The hardness values h_F, h_P, h_B, h_M are unknown and will be optimized together with the phase fractions. Because of the assumption in equation (5), the fractions of all phases in a specimen have to equal 1, we can rewrite this system of equations to:

$$\begin{bmatrix} 0 & 0 & x_1 & 1-x_1 \\ x_2 & 0 & x_3 & 1-x_2+x_3 \\ x_4 & x_5 & x_6 & 1-(x_4+x_5+x_6) \\ x_7 & 1-x_7 & 0 & 0 \\ x_8 & 1-x_8 & 0 & 0 \\ x_9 & 1-x_9 & 0 & 0 \\ x_{10} & 1-x_{10} & 0 & 0 \\ x_{11} & 1-x_{11} & 0 & 0 \\ x_{12} & 1-x_{12} & 0 & 0 \end{bmatrix} * \begin{bmatrix} h_F \\ h_P \\ h_B \\ h_M \end{bmatrix} = \begin{bmatrix} 663 \\ 522 \\ 462 \\ 297 \\ 281 \\ 276 \\ 246 \\ 229 \\ 210 \end{bmatrix} \tag{7}$$

Note that although (7) is written as a matrix equation, the problem is nonlinear in the unknowns. Furthermore it is underdetermined as we have 9 equations for totally 16 unknowns (12 phase fractions and 4 hardness values). In order to further restrict the problem, micrographs have been visually inspected to determine upper and lower bounds for the phase volume fractions (Table 2). The problem has been subsequently solved using the constrained optimization functionality in Matlab (function fmincon). As expected only three of the phase fractions (x_5, x_{10} and x_{11}) have been actively optimized in addition to the four hardness values leading to a fully determined system. The rest of the values have been simply set to the boundary values satisfying the constraints equations.

Table 2 Phase fraction boundaries

Quenching time	Ferrite	Perlite	Bainite	Mart.
10s	0%	0%	5-10%	90-95%
15s	0-5%	0%	40-50%	45-60%
20s	0-15%	0-20%	50-60%	5-50%
40s	20-60%	40-80%	0%	0%
60s	20-60%	40-80%	0%	0%
100s	20-60%	40-80%	0%	0%
200s	20-60%	40-80%	0%	0%
500s	20-60%	40-80%	0%	0%
1000s	20-60%	40-80%	0%	0%

After running the optimization, we get the hardness for each phase, as well as the volume fraction $x_1 - x_{12}$ for the different specimens see Table 3 and Table 4:

Table 3 Optimized hardness for each phase

Phase	Hardness [HV]
Ferrite	137.35
Perlite	322.24
Bainite	517.76
Martensite	653.74

Table 4 Optimized phase fractions

x_1	x_2	x_3	x_4	x_5	x_6
0.05	0.05	0.5	0.15.	0.19	0.6
x_7	x_8	x_9	x_{10}	x_{11}	x_{12}
0.2	0.2	0.2	0.37	0.52	0.6

Now that phase fractions and hardness values are known, the TTT-diagram hast to be changed in order to simulate the quenching experiments accurately. Possible modifications and their influence on hardness and phase transformation are discussed in the following.

4.2 TTT-diagram modifications

As discussed in the introduction, the TTT diagrams used in the Forge simulations are calculated solely based on the chemical composition of the material. Although this is a good starting point in absence of further information, it is unable of accurately mimicking the behavior observed in the CCT experiments. In order to make sure that the TTT delivers results within the measured observation limit, a local modification strategy based on translation and stretching of the evaluated curves has been implemented.

Increasing the grain size leads to a general increase in the hardness values, because the ferrite phase fraction is reduced and therefore the perlite and bainite phase content are increased (see Figure 5). Conversely, the hardness values for all quenching times will be reduced when the grain size is decreased. This particular behavior applies only in the simulation and doesn't correspond to the observable grain size – hardness relationship [17, 18].

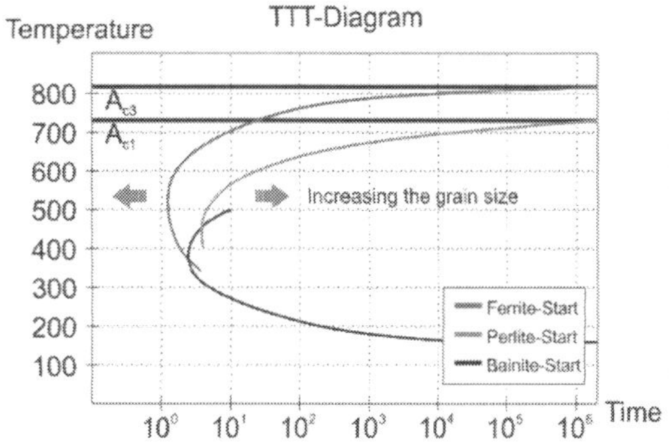

Figure 5. Influence of grain size

In order to influence the local hardness values, the TTT start curves must be modified (see Figure 6).

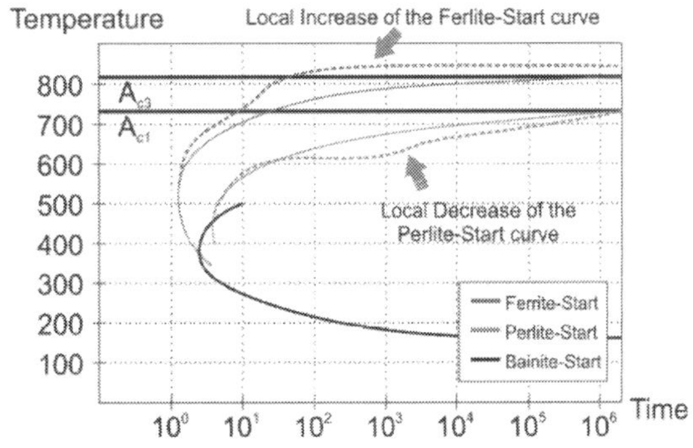

Figure 6. Modification of transformation curves

By lifting the ferrite-start curve, the hardness values for fast quenching (up to 20s) are reduced, lowering the curve leads to an increase of hardness. To change hardness for moderate cooling times (40s, 60s and 100s), the perlite start curve is most significant. The same rules as for the ferrite start curve applies. From micrographs it is known that bainite is only present for quenching times from 10s to 60s. Lifting the bainite curve results in a higher hardness after quenching and lowering the curve in a lesser hardness value.

The computed TTT-diagram (see Figure 2) was manually modified to fit the observations from the measured CCT-diagram (see Figure 1). Firstly the A_{C1}- and A_{C3}-temperature were corrected to fit the measurements. The transformation curves were generally shifted to the right, so that it is possible to transform martensite within 10s quenching time. The ferrite-start and bainite-start curve were lifted in order to have a lower hardness for moderate and higher hardness for faster quenching. The perlite transformation curves were not changed in shape. The final, optimized TTT-diagram is shown in Figure 7.

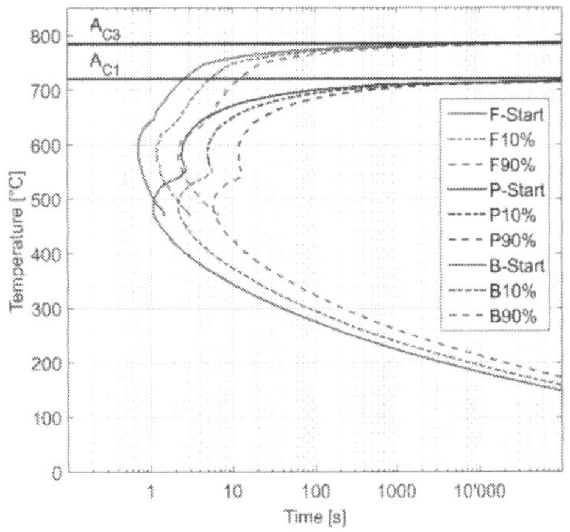

Figure 7. Optimized TTT-diagram

By combining the adjusted transformation kinetics and the simplified hardness values, an improved final hardness and phase prediction can be achieved as seen in Figure 8.

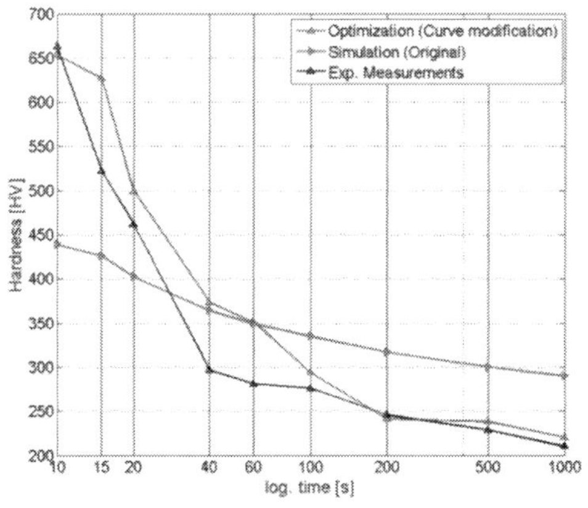

Figure 8. Optimized and original hardness prediction

There is a significant improvement for fast, mainly martensitic transformation (10s) and for slow ferrite/ perlite transformations (200-1000s). For quenching rates in between, further adjustments have to be done.

5 Conclusions

It was possible to show the flaws of using automatically generated TTT-diagrams based on the chemical composition for C60 steel. The impact of changes in the transformation curves on the final material hardness was also demonstrated. By optimizing the thermal material data with the presented optimization method, the prediction of final phase composition and hardness was more accurate for fast, pure martensitic and slow cooling rates.

With the proposed method it is no longer necessary to evaluate micrographs accurately, it is enough to define percentage boundaries for every phase. Additional effort has to be put in the optimization for more accurate results for all quenching rates.

References

1. W.I. Pumphrey, F.W. Jones, *ISIJ*, **189**, 137-144 (1948)
2. E. Scheil, *ZAAC*, **201(1)**, 259–264 (1931)
3. M. Avrami, *J. Chem. Phys.* **7**, 1103-1112 (1939)
4. M. Avrami, *J. Chem. Phys.* **8(2)**, 212 (1940)
5. M. Avrami, Granulation, *J. Chem. Phys.* **9(2)**, 177 (1941)
6. D.P. Koistinen, R.E. Marburger, *Acta Mater.*, **7(1)**, 59–60 (1959)
7. Z. Guo, J. Schillé, N. Saunders, et al, *2nd Int. Conf. on Heat Treat. and Surf. Eng. in Auto. Applic.* (2005)
8. N. Saunders, Z. Guo, AP Miodownik, J.P. Schillé, *J Mat Pro Software Literature* (2004)

9. P. Åkerström, M. Oldenburg, *J. Mater. Process. Technol.*, **174(1-3)**, 399–406 (2006)
10. J.S. Kirkaldy, D. Venugopalan, *Phase Transformations in ferrous alloys*, 125-148 (1984)
11. B. Buchmayr, J.S. Kirkaldy, *J. Heat Treating*, **8**, 127–136 (1990)
12. K. W. Andrews, J. *Iron Steel Inst*, **203(7)**, 721–727 (1965)
13. O.G. Kasatkin, B.B. Vinokur, V.L. Pilyushenko, *Met. Sci. Heat Treat.*, **26(1)**, 27–31 (1984)
14. W. Steven, A.G. Haynes, *J. Iron Steel Institute*, **183(8)**, 349–359 (1956)
15. *SEP 1680*, Aufstellung von Zeit-Temperatur-Umwandlungsschaubildern für Eisenlegierungen (1990)
16. M. Ehlers, Numerische und experimentelle Untersuchungen zur Eigenspannungs- und Verzugsausbildung beim martensitischen Härten von Stufenzylindern und Quadern in verdampfenden Flüssigkeiten, *Shaker* (2000)
17. N.J. Petch, *J. Iron Steel Inst.*, **174**, 25–28 (1953)
18. E.O. Hall, *Proc. Phys. Soc. Sect. B*, **64(9)**, 747 (1951)

Optimization of the single point incremental forming process for titanium sheets by using response surface

Badreddine Saidi [1,2*], Laurence Giraud-Moreau[1], Abel Cherouat [1]

[1]*Université de Technologie de Troyes France-GAMMA3, France*
[2]*Ecole Nationale d'Ingénieurs de Tunis-MAI Tunisia*

Abstract. The single point incremental forming process is well-known to be perfectly suited for prototyping and small series. One of its fields of applicability is the medicine area for the forming of titanium prostheses or titanium medical implants. However this process is not yet very industrialized, mainly due its geometrical inaccuracy, its not homogeneous thickness distribution... Moreover considerable forces can occur. They must be controlled in order to preserve the tooling. In this paper, a numerical approach is proposed in order to minimize the maximum force achieved during the incremental forming of titanium sheets and to maximize the minimal thickness. A surface response methodology is used to find the optimal values of two input parameters of the process, the punch diameter and the vertical step size of the tool path.

1 Introduction

The single point incremental forming process (ISF) is an emerging process with a high industrial interest. This process is based on the use of a spherical tool, which is moved along CNC controlled tool path (figure 1). During the incremental forming process, the sheet blank is fixed in sheet holder. The tool follows a certain tool path and progressively deforms the sheet. Because incremental forming is a dieless process, it is perfectly suited for prototyping and small volume production [1, 2].

Medical implants are one of the potential fields of application of the incremental forming process, due to the need of product customization to each patient. One of the major advantages of the ISF process is the ability to program several punch trajectories on the same machine in order to obtain different shapes perfectly suited to each patient. Titanium alloys are generally used by the biomedical industries because of their corrosive resistance and great biocompatibility with the human body. Titanium is recognized by its high strength-to-weight ratio. But its high cost and difficult formability compared with aluminum make that the incremental forming of titanium sheet is little studied. Before the incremental forming process of titanium sheets is really industrialized, studies are necessary. The ISF process indeed suffers from a big slowness, geometrical inaccuracy, a not homogeneous thickness distribution which reduces its industrial suitability [1, 3, 4, 5].

During this process, considerable forces can occur [5]. The control of the forming force is particularly important when titanium sheets are used to ensure the safe use of a CNC milling machine and preserve the tooling and machinery [5, 6].

The aim of this paper is to determine input parameters of the process in order to minimize the maximum force achieved during the incremental forming of titanium sheets and to optimize the thickness reduction. Several studies were developed to optimize the incremental forming of aluminum sheets [2, 8]. A little less was done for titanium sheets.

In this paper, a numerical procedure is developed in order to study the influence of two input parameters, the punch diameter and the vertical step on the forming force and the minimal thickness. To conduct this study, an experimental design combined to response surface methodology is used [2]. The experimental design is based on finite element analysis realized with ABAQUS/Explicit. Mathematical relations are defined from the response surfaces to predict the forming force and the thinning rate as function of the two input parameters and to define the input parameters allowing to maximize the forming force and maximize the minimal thickness.

2 Incremental forming of titanium sheets

The single point incremental sheet forming (SPIF) process is illustrated on the figure 1. Different trajectories along which the tool moves have been studied in the literature. The tool path most frequently used and that the

[a] Corresponding author: laurence.moreau@utt.fr

authors of the paper have already studied during experimental works [3] is composed of rotational movements in the horizontal plane combined with small steps in the vertical direction (see figure 2). This tool path is used in this study.

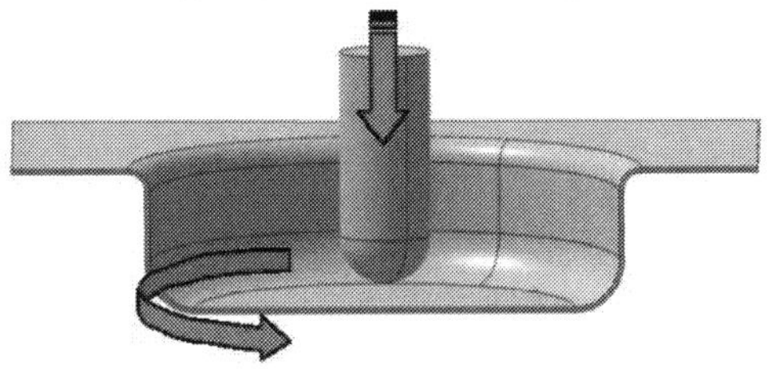

Figure 1. Incremental forming process.

This paper concerns the incremental forming of titanium ASTM grade 2 sheets with a thickness of 0.5 mm. A truncated cone with a wall angle of 60°, a maximum diameter of 100 mm and a depth of 20 mm was chosen for this preliminary study about the optimization of the incremental forming process with titanium sheets. The sheet has a size of 300 mm x 300 mm.

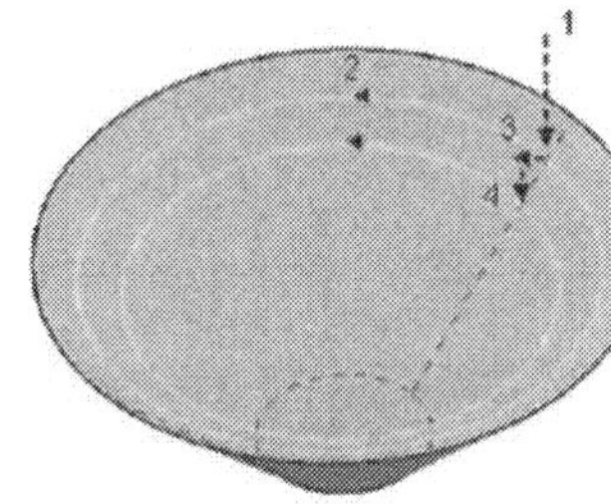

Figure 2. Tool path.

3 Numerical simulation

During the incremental forming process, the sheet is held firmly by the clamping system so that there is no slipping. A full model of blank holder modeled by rigid surfaces has been chosen to represent the experimental clamping system. A vertical pressure of 4.3MPA [9] has been applied. Figure 3 presents the numerical model at the beginning of the process. The initial and final mesh of the sheet for a punch diameter of 10mm and a step size of 1 mm is shown on figure 4.

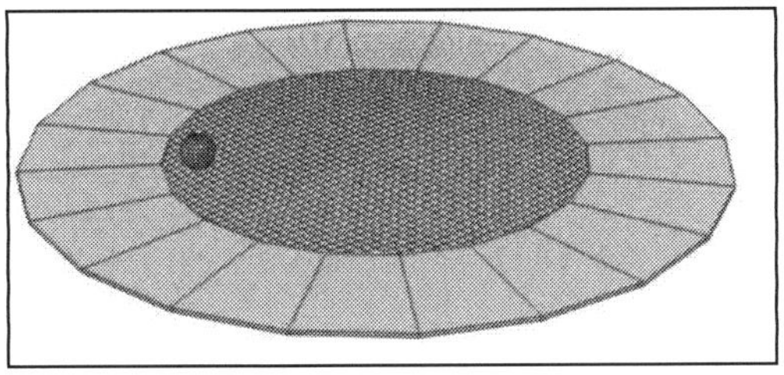

Figure 3. Numerical model.

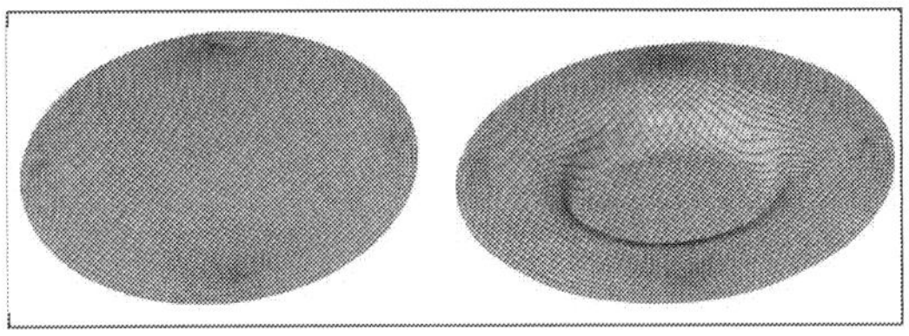

Figure 4. Initial and final mesh of the sheet.

In what concerns material modeling, an elasto-plastic model was chosen for the simulation. The titanium behavior was modeled by means of a Swift type hardening law :

$$\sigma = k\,(\varepsilon_0 + \varepsilon_p)^n \quad (1)$$

A tensile test allowed generating the strain-stress curve for the titanium sheet. Figure 5 presents this stress-strain curve.

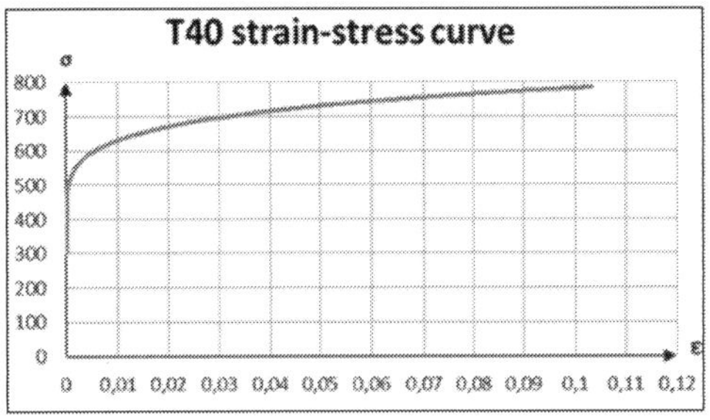

Figure 5. Stress-strain curve.

An inverse analysis using the data from this tensile test has been used to identify the mechanical parameters. Table 1 summarizes the mechanical properties of the material obtained thanks to the inverse method.

Table 1. Mechanical properties of the Titanium grade 2.

Modulus of elasticity : ***E***	100 Gpa
Yield strength: σ	210 Mpa
Hardening coefficient: ***n***	0.084
Coefficient of proportionality: ***K***	697,7 MPa
Strain : ε_0	0.00157

4 Experimental design

The goal of this present study is to determine the values of the punch diameter (dp), and the vertical step size (Δz) which allow to minimize the forming force and maximize the minimal thickness. As the number of input variables is low, a full factorial design at three levels has been chosen to realize the experimental design. Table 2 presents the forming factors and their respected levels.

The numerical simulation of the incremental forming process has been performed with the ABAQUS explicit solver. The finite elements chosen to mesh the sheet are shell elements S4R with 4 nodes and reduced integration. The spherical tool is modeled by a rigid surface.

Table 2. Forming factors and levels.

Factors	Factor notation	Level 1	Level 2	Level 3
dp : Tool diameter (mm)	A	10	12.5	15
Δz : Step size (mm)	B	0.5	0.75	1

Nine numerical simulations of the forming process were carried out. The responses values are the minimal thinning and maximal forming force. Table 3 presents the numerical results performed according to the full factorial design.

Table 3. Numerical results.

n⁰	Dp (mm)	Δz (mm)	Minimal thickness (mm)	Maximal forming force (N)
1	10	0.5	0,255909	870
2	10	0.75	0,222006	950
3	10	1	0,172874	1030
4	12.5	0.5	0,258584	920
5	12.5	0.75	0,106811	980
6	12.5	1	0,197069	1110
7	15	0.5	0,205474	1060
8	15	0.75	0,205505	1120
9	15	1	0,274236	1180

5 Optimization results

A quadratic polynomial equation has been considered in order to fit the numerical results and get a relationship between the response values and the input parameters.

The two polynomial models (2) and (3) of second degree considered for the forming force and minimal thickness are the following [10]:

$$F_{max} = a_0 + a_1 d_p + a_2 \Delta z + a_3 d_p^2 + a_4 d_p \Delta z + a_5 \Delta z^2 \quad (2)$$

$$STH = b_0 + b_1 d_p + b_2 \Delta z + b_3 d_p^2 + b_4 d_p \Delta z + b_5 \Delta z^2 \quad (3)$$

Table 4 presents the values of the coefficients, which allow to predict the forming force and the thinning rate according to the two input parameters. These numerical results have been obtained with ANOVA.

Table 4. Coefficients of the fit functions.

a_0	a_1	a_2	a_3	a_4	a_5
1082,2	- 80,7	233,3	5,1	- 16,0	186,7
b_0	b_1	b_2	b_3	b_4	b_5
1,48	- 0,12	- 1,40	0,0037	0,03	0,6

The 3D response surfaces showing the variations of the forming force and the minimal thickness with respect to the punch diameter and the step size are given in figure 6 and 7. These surfaces are useful to understand the interaction properties between both the punch diameter and step size and the output parameters. We can see on these figures that the increase of the tool diameter and the step size induces an increase of force forming. As regards the minimal thickness, a nonlinear response is observed. The maximal thickness for the considered input parameters is obtained for the minimal tool diameter and minimal step size.

These response surfaces have been used to determine the optimal values of the tool diameter and the vertical step size which allow to minimize the forming force and maximize the minimal thickness. The results are presented in the table 5.

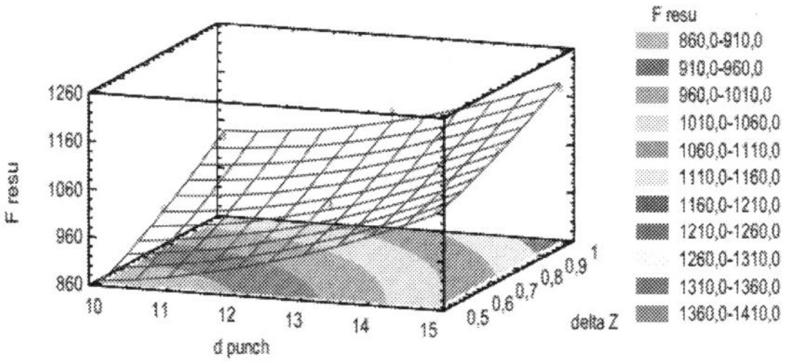

Figure 6. Response surface associated to the forming force.

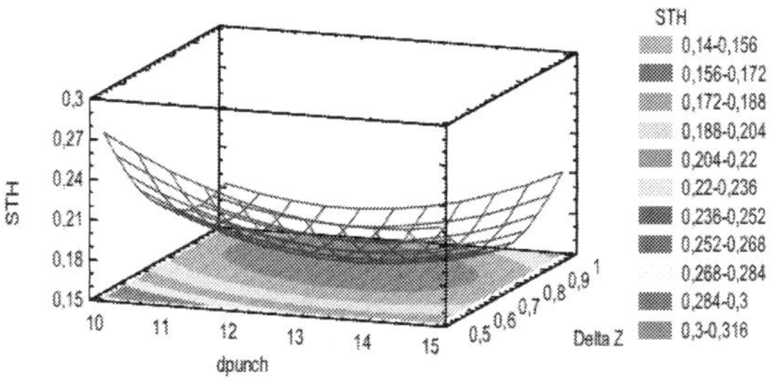

Figure 7. Response surface associated to the minimal thickness.

Table 5. Coefficients of the fit functions.

Factor	*Low*	*High*	*Optimum*
dp (mm)	10,0	15,0	10,0
Δz (mm)	0,5	1,0	0,5
		STH $_{max}$ (mm)	0,273923
		Fres $_{min}$ (N)	865,556

An analysis of variance was conducted with ANOVA in order to analyze the effects of the input parameters on the value of the forming force and minimal thickness. They are presented in figures 8 and 9.

A quasi linear effect of the punch diameter and the step size on the forming force is observed. Figure 9 shows a quadratic effect of the input parameters on the thickness response. Figures 10 and 11 presenting the standardized PARETO chart show that the punch diameter and the vertical step size have a similar effect on the forming force but the vertical step size has a more important effect than the punch diameter on the thickness.

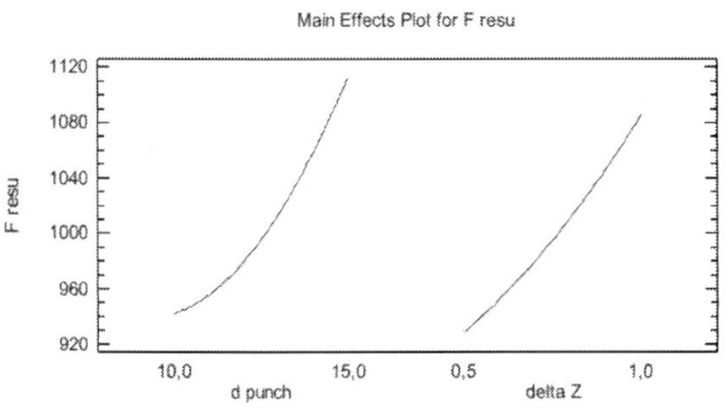

Figure 8. Effects of the input parameters on the value of the forming force.

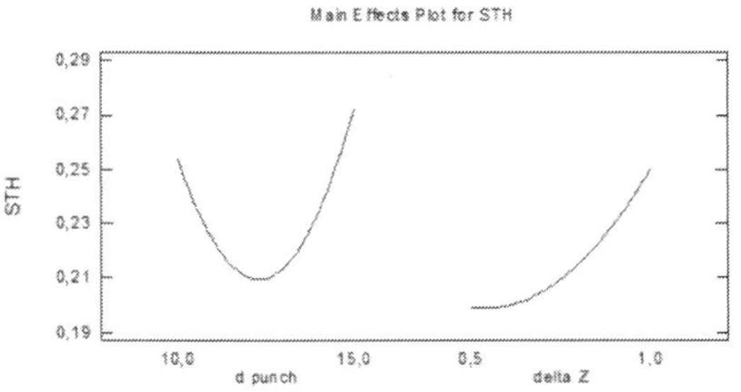

Figure 9. Effects of the input parameters on the value of minimal thickness.

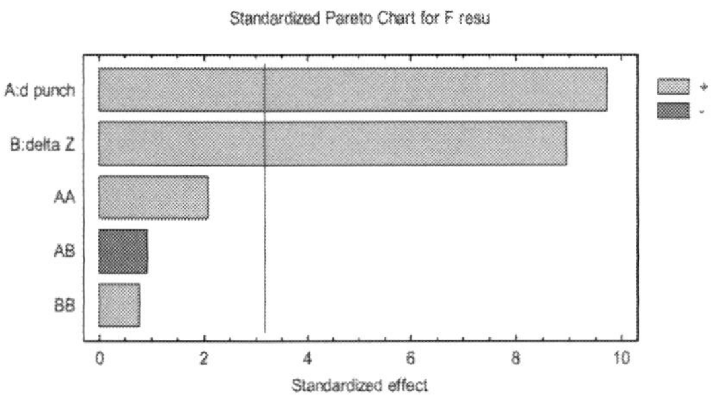

Figure 10. Standardized PARETO chart for force forming.

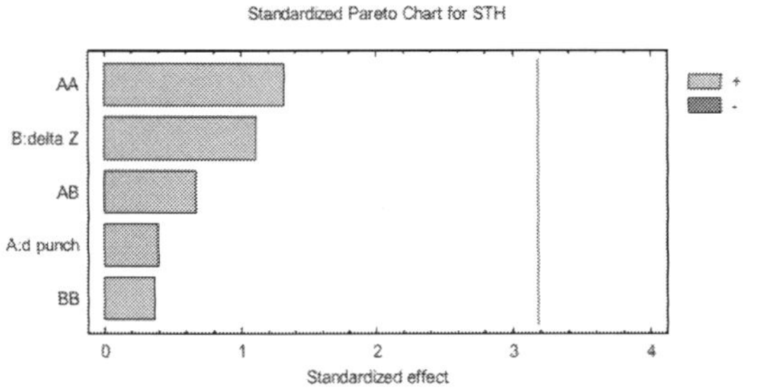

Figure 11. Standardized PARETO chart for the thickness.

Figures 12 and 13 present the normal probability plot. The process parameters are considered significance to the experiment as there is a large angled slope of the straight line. We can see that all the points are relatively close to the straight line and thus that the data follow a normal distribution [8].

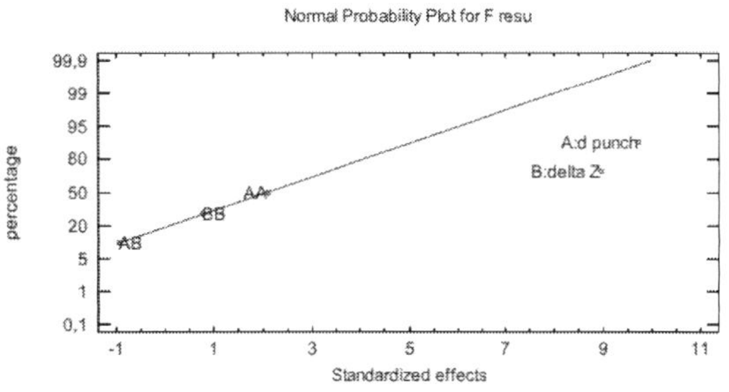

Figure 12. Normal probability distribution for force forming.

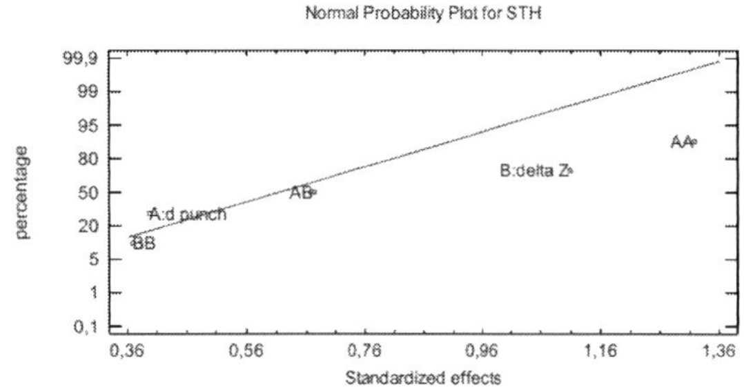

Figure 13. Normal probability distribution for thickness.

6 Conclusion

In this paper, a numerical approach based on finite element analysis and response surface methodology has been applied to optimize the incremental forming process of titanium sheets. A second-order quadratic function has been defined in order to predict the maximal forming force and minimal thickness as function of the punch diameter and the vertical step size.

The impact of the punch diameter and the vertical step size on the maximal forming force and minimal thickness has been studied.

The optimal values of the input parameters allowing minimizing the forming force and maximizing the minimal thickness have been defined from the response surfaces.

This study about the optimization of the incremental forming process with titanium sheets is a preliminary study. Only two input parameters were considered and the material low used in the finite element model doesn't take into account the general anisotropy. This study will be pursued in future works.

References

1. G.Ambrogio, I. Corstantino, L. De Napoli, L. Filice, L. Fratini, M. Muzzupappa, J. of Materials Proc. Technol., **1153–154**, p. 501–507, (2004)

2. V. Gulati, A. Aryal, P. Katyal, A. Goswami, J.l of The Institution of Engineers (India): Series C, (2015)
3. L. Giraud-Moreau, A. Cherouat, A. Merat *IDDRG 2014*, (2014)
4. J. Duflou, Y. Tunckol,A. Szekeres, P. Vanherck, J. of Mat. Proc. Technol. , **1189**, p. 65–72, (2007)
5. R. Aerens, P. Eyckens, A. Van Bael, J. R. Duflou, Int. J. Adv. Manuf. Technol., (2009)
6. L. Giraud-Moreau, A. Cherouat, J. Zhang, H. Borouchaki, Key Engin. Mater., **554-557**, pp 1375-1381, (2013)
7. B. Saidi, A. Boulila, A. Mahfoubh, N. Rachid, Mechanics & Industry, **16**, n° 1410, (2015)
8. S. Kurra, N. H. Rahman, S. P. Regalla, A. K. Gupta, Comput. Science and Software Engin., (2015)
9. J. Belchior, Thèse de l'INSA Rennes, France, (2013)
10. C. Raju, S. Narayanan, Measurement, **78**, 296–308, (2016)

DOI: 10.1051/matecconf/20168010012

Identification of material parameters using indentation test —study of the intrinsic dimensionality of P-h curves and residual imprints

Liang Meng[1,a], Piotr Breitkopf[1] and Guenhael Le Quilliec[2]

[1] *Sorbonne Universités, Université de Technologie de Compiègne*
UMR UTC-CNRS 7337, Compiègne, France
[2] *University of Tours, E.A. 2640 Laboratoire de Mécanique et Rhéologie, Tour Cedex, France*

Abstract. Instrumented indentation test has been used to determine material parameters with two different sources of information. Based on the recent shape-manifold identification approach, we formally demonstrate, in the scope of manifold, the non-uniqueness of the solution to the inverse problem based on load-displacement curves. The identifiability is also compared for P-h curve and imprint mapping, highlighting the manifold's ability to estimate the maximum number of independent material parameters that may be determined with a given experimental setup.

1 Introduction

Instrumented indentation is a popular technique used for the assessment of material properties. Fruitful research results have been reported [1–5] for a variety of materials, such as metals, alloys, ceramic, concrete and bone, among others. In a typical identification procedure, it is expected that there exists a one-to-one correspondence between the input pattern (usually the indentation curve, also called a P-h curve) and the desired material parameters defining a postulated constitutive law. Unfortunately, the indentation problem is frequently ill-posed, and the existence of different "sibling" materials with distinct material properties but almost identical indentation behaviors has been reported more than once [6–8].

The often used inverse method, employing Finite Element (FE) simulation and optimization algorithms, tries to estimate the material parameters by minimizing the discrepancy between measured and simulated quantities. However, the uniqueness of the solution still remains a matter of discussion since different material parameter sets may give almost identical indentation curves [7, 8]. Therefore, it is of high interest to improve the reliability of the inverse analysis by involving more information for identification. Accompanied by the development of high-resolution Scanning probe microscopy (e.g., Atomic Force Microscopy (AFM)), [2, 9, 10] reported several methods for assessing material properties, taking into consideration not only the indentation curve, but also the residual imprint measured with high precision and accuracy, such that material's unknown properties are evaluated in a much more elegant manner in that more parameters are identifiable. Other solutions, like adopting indenters with complex profiles (i.e., a dual sharp indenter), were also proposed to alleviate the non-uniqueness property of the identification problem[11–13].

More recently, due to the significant development of manifold learning algorithms and its applications in the field of computational and applied mechanics, biomechanics, and especially in material science[14–18], our most recent work [19] recommended that the identification procedure be carried out within a α-space constructed by using *only* the residual imprint of an indentation test. The core idea is based on the fundamental hypothesis that, there exists a low-dimensional space in which the original higher-dimensional data could be embedded and the identification is implemented over a manifold defined within the constructed shape space. The efficacity of this approach has been demonstrated by successfully evaluating two hardening properties of a piece of C100 steel and an AU4G alloy.

Regarding the remarkable advancements in material characterization, to the knowledge of the authors, the number of parameters that can be probed by an instrumented indentation test has never been systematically investigated, despite the awareness of the presence of uniqueness. In this work, we extend the aforementioned imprint-based manifold identification approach to the measured indentation curve to investigate the non-uniqueness phenomenon. In addition, the manifold is also studied in order to obtain a cursory estimation of the number of parameters that can be identified.

2 Material characterization using manifold learning

In our former work [19], we proposed an original approach for evaluating diverse material properties *solely* based on the residual imprint after withdrawing the indenter. This information, usually high-dimensional, was not employed

[a] Corresponding author: liang.meng@utc.fr

in calculating the point-wise distance so as to characterize the discrepancy between simulated and experimental measured quantities. Instead, we first seek a low-dimensional embedded domain in which it can be described in a more compact pattern. There is a common sense notion in the community of signal processing that *"Real life data does not uniformly fill the space in which it is represented"*. Instead, their distributions are observed to concentrate to a manifold (linear/non-linear) of lower intrinsic dimensionality.

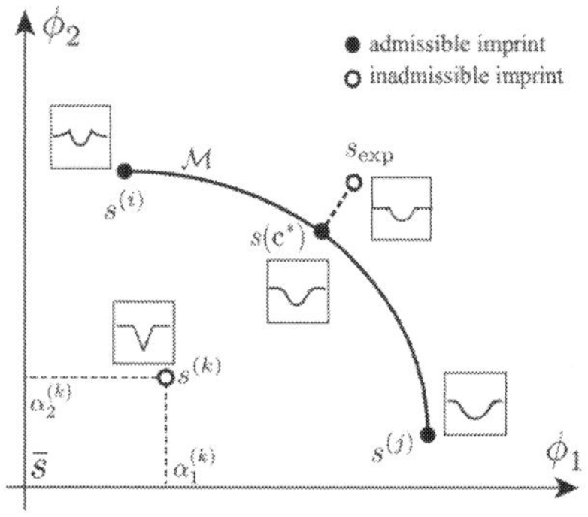

Figure 1. Illustration of admissible/inadmissible imprints in a low dimensional embedded space.

The manifold is generally problem-dependent. For an indentation test, which is our case, the manifold connects all of these simulated imprints together. As a consequence, we are able to distinguish between admissible/inadmissible shapes by referring them to the manifold. Generally, we consider an imprint as admissible under the condition that it is attainable by manipulating the parameter set **c** in a postulated constitutive law. Two admissible imprints $\mathbf{s}^{(i)}$ and $\mathbf{s}^{(j)}$, as noted by solid circles in Fig.1, lie exactly on the the *manifold* $\mathcal{M}$. On the other hand, all of those imprints outside the manifold may then be inadmissible, e.g., $\mathbf{s}^{(k)}$, marked by a round circle. It is worth noting that the experimental imprint $\mathbf{s}_{exp}$ is *inadmissible* in consideration of measurement noise and model error.

In practice, find the proper embedded system can be done in various ways [18, 20, 21], and we adopted the Principal Component Analysis (PCA), which is routinely used in image analysis [22]. Note here, the embedded space is a POD coefficient space, the origin of which is the mean imprint $\bar{\mathbf{s}}$ of all these in the database, and its principle axis are orthogonal basis derived from PCA. Thus the embedded system is also called an α-space and the hyper-surface an α-manifold. Their constructions were detailed in [19] thus omitted here.

The identification procedure is quite straightforward in the constructed α-space: find the parameter set that gives an admissible shape $\mathbf{s}(\mathbf{c}^*)$ on the manifold $\mathcal{M}$ which is closest (minimum distance) to the experimental shape

$$\mathbf{c}^* = Argmin\|(\alpha_{\rm exp} - \alpha(\mathbf{c}))\|, \quad \alpha(\mathbf{c}) \in \mathcal{M}, \qquad (1)$$

here $\alpha_{\rm exp}$ is the coordinate of the experimental imprint in this shape space; the constraint $\alpha(\mathbf{c}) \in \mathcal{M}$ confines automatically that the imprint shape should be an admissible one. We summarize the complete manifold-based identification processus as follows:

1. Create a database of imprints by a proper Design Of Experiment (DOE) varying material parameters in a constitutive law;
2. Find the embedded lower-dimensional space by PCA and project the imprint database into the new system;
3. Interpolate a manifold in α-space and built a one-to-one relationship between the design space and the point on manifold;
4. Project experimental imprint on the manifold and identify material parameters referring to Eq.(1).

It is worth mentioning that the pair-wise distance between two different imprint shapes is transformed into a single Euclidean distance in α-space.

3 Description of measured quantities by indentation

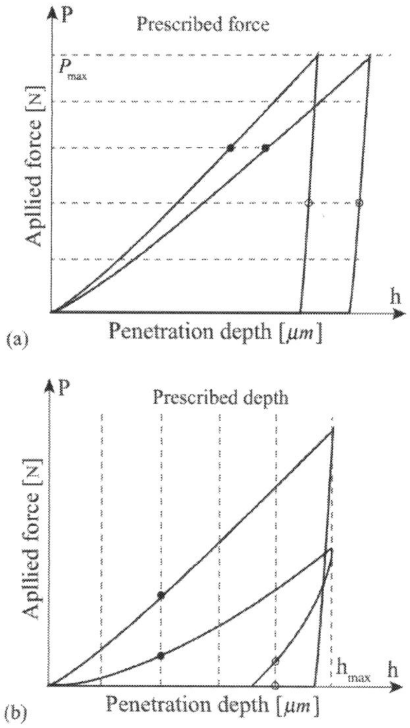

Figure 2. Characterization of different indentation curves.

The manifold-based identification approach does not restrict itself to imprint mapping only. We can also consider the obtained P-h curve as a special 'imprint' since the ultimate goal is to distinguish these curves/shapes one from another.

Unlike imprint mapping, P-h curve is a multivalued function and it makes no sense to compare the loading phase of one curve with the unloading phase of another in identification procedure, thus these two phases should be properly separated. Take a typical indentation curve obtained by a prescribed force as an example, the two parts are characterized by the penetration depths at a series of

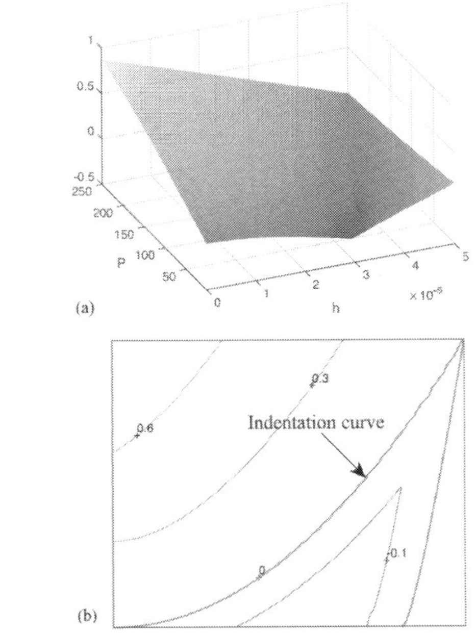

Figure 3. A typical indentation curve described by the zero iso-contour of a Level-set function.

given forces, noted by $\mathbf{s}_l$ and $\mathbf{s}_u$, respectively. Then the whole curve is described by a vector which concatenates the two, giving $\mathbf{s}_c = [\mathbf{s}_l, \mathbf{s}_u]$, guaranteeing the comparability of different curves. For the case where a maximum penetration depth is predefined, the same routine should be followed except that we compare forces for different penetration depths. For the latter, it should be noticed also that the unloading part includes also the segment that overlies on the h-axis. Fig.2 illustrates the characterization of the P-h curves obtained from two experimental set ups: (a) prescribed force; (b) prescribed penetration depth. By consequence, our manifold protocol can be extended to indentation curve also, allowing for a comparison between the identifiabilities of two source of information.

This concatenating description, while feasible, is setup-dependent. Thus it is of great interest to describe the P-h curve with an Eulerian representation, examples of which may include pixel/voxel maps, and the more popular Level-set approach which is first proposed in [23]. The basis idea is the use of a background grid points and define the Level-set function simply as the signed minimum distance of these points to the curve. By convention, the distance is positive if the point lies outside the curve while negative for inside ones. As a result, the zero iso-contour describes the boundary of the curve. A typical indentation curve which is characterized by such method can be found in Fig.3.

4 Numerical examples

4.1 Forward analysis using FE

First, by adopting the same FE model used in our last work [19], two databases, each containing $21 \times 21 = 441$ samples, were built up for indentation curves and imprint mappings by simulations with various hardening parameter combinations. The simulation parameters are equal-spaced sampled from the design domain ($n \in [0.2, 0.3]$, $\sigma_y \in [200, 240]$) allowing a straightforward demonstration of connectivity in embedded α-space.

The material hardening law is characterized by two parameters: elastic limit σ_y and hardening coefficient n, having the form

$$\sigma = \sigma_y \left(\frac{E}{\sigma_y}\right)^n \varepsilon^n. \tag{2}$$

4.2 Identifiability analysis using manifold

In this section, the two database were employed separately so as to detect the identifiability of P-h curve and imprint mapping. It should be noticed that, the bellowing results are obtained from purely numerical data, and the experimental imprint is chosen from one of the 441 simulation results and no noise is introduced.

Errors of identification criterions employing different measured quantities are first included in Fig.4. Clearly, for conventional point-wise distance error, the parameter sets (σ_y, n) having almost the same minimal error are found in a thin-long valley. This is highly consistent with the one observed by [24]. Similar result is also obtained for the imprint mapping.

On the other hand, by adopting the innovate identification method referring to the manifold, the Euclidean distance between the measured quantities and the corresponding experimental one is calculated. Its distribution is found in the second row of Fig.4. To our delight, from the isoline of the error, we probably yield a convergent iterative value. However, in this situation, there is no way to guarantee that the solution is a global minimal rather than a local one. Similar result is obtained if we employ the same manifold approach for P-h curves. Thus we would like to conclude that, the manifold based method takes more advantage of the origin measured data, while no clear difference is observed for P-h curve and imprint mapping.

Moreover, since the manifold method interprets well the measured quantities compared with conventional method, it is of great interest to compare the two manifolds constructed from P-h curve and imprint mapping respectively. It is observed from Fig.5.(a) and Fig.6.(a) that both the manifolds are thin and long, illustrating that only one of the sensitivities with respect to the two parameters has a significant value and the other is almost insensible. This explain quite clear the reason why it is difficulty to identify the two parameters simultaneously.

That said, slight difference are also observed for the two manifolds. We noticed that the α-manifold of the imprint shapes seems to expand along two directions despite its long-thin property. Motivated by this, we consider to normalize the α-coordinates of each embedded point in shape space to bring the error sensitivity with respect to the two parameters to a comparable level. The scaled α-manifolds are given in Fig.5.(b) and Fig.6.(b).

Clearly, the scaled manifold proved that the imprint interprets the material properties in the most elegant man-

 DOI: 10.1051/matecconf/20168010012

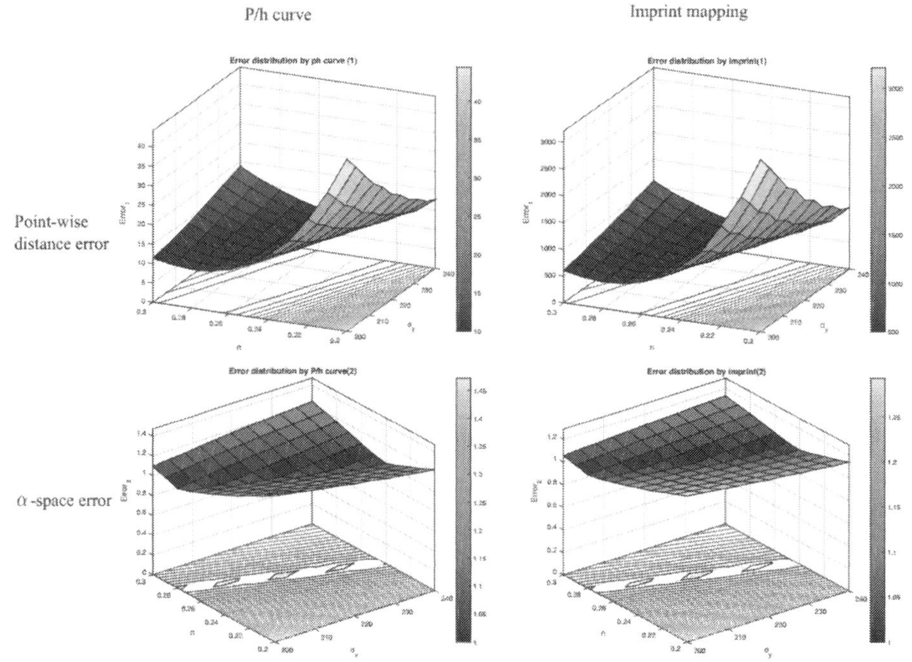

Figure 4. Comparisons of error distribution within different spaces employing different information sources.

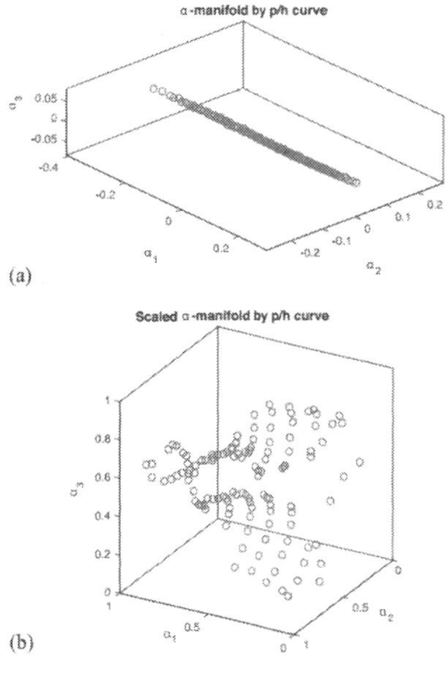

Figure 5. (a)Original manifold profile; (b) Manifold in normalized α-coordinate system obtained from indentation P-h curve.

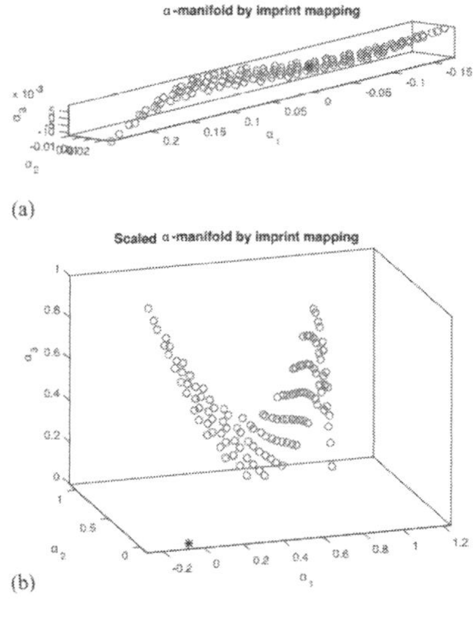

Figure 6. (a)Original manifold profile; (b) Manifold in normalized α-coordinate system obtained from imprint mapping.

ner since the manifold has a dimensionality of 2, which is right the number of parameters for the construction of database. Also, similar with their distributions in parametric space, different imprints are embedded regularly in the shape space (Fig.7 left), allowing the identification of the parameter set. Contrarily, the embedding of P-h curve is kind of disordered make it inappropriate for material charcterization.

5 Conclusion

In this work, we demonstrated the identifiability of different measured indentation quantities using a shape manifold and observed that the residual imprint of an indentation test outperforms the P-h curve during identification. In addition, the identification procedure based on the manifold has the additional ability to estimate the maximum number of identifiable parameters d_{in}, especially for those with a dimensionality d_{in} <= 3. This outstanding property of the

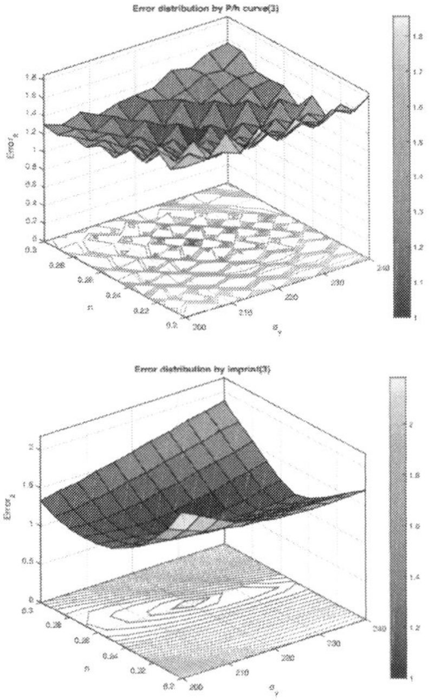

Figure 7. Error distributions computed from scaled α-manifold.

manifold could provide valuable guidance for designing the indentation test and choosing the quantities to be measured.

That said, for complex material law which involves more parameters, the manifold can no longer be visualized. Thus the quantitive estimation of intrinsic dimensionality is of great necessity.

References

1. C.K. Moy, M. Bocciarelli, S.P. Ringer, G. Ranzi, Materials Science and Engineering: A **529**, 119 (2011)
2. M. Bocciarelli, G. Bolzon, Materials Science and Engineering: A **448**, 303 (2007)
3. S. Kucharski, Z. Mroz, Journal of engineering materials and technology **123**, 245 (2001)
4. D. Carnelli, R. Lucchini, M. Ponzoni, R. Contro, P. Vena, Journal of biomechanics **44**, 1852 (2011)
5. M. Rodríguez, J.M. Molina-Aldareguía, C. González, J. LLorca, Acta Materialia **60**, 3953 (2012)
6. Y.T. Cheng, C.M. Cheng, Journal of Materials Research **14**, 3493 (1999)
7. K. Tho, S. Swaddiwudhipong, Z. Liu, K. Zeng, J. Hua, Journal of materials research **19**, 2498 (2004)
8. J. Alkorta, J. Martinez-Esnaola, J.G. Sevillano, Journal of materials research **20**, 432 (2005)
9. G. Bolzon, G. Maier, M. Panico, International Journal of Solids and Structures **41**, 2957 (2004)
10. M. Bocciarelli, G. Bolzon, G. Maier, Mechanics of Materials **37**, 855 (2005)
11. M. Futakawa, T. Wakui, Y. Tanabe, I. Ioka, Journal of Materials Research **16**, 2283 (2001)
12. J.L. Bucaille, S. Stauss, E. Felder, J. Michler, Acta materialia **51**, 1663 (2003)
13. Y. Cao, N. Huber, Journal of materials research **21**, 1810 (2006)
14. B. Raghavan, P. Breitkopf, Y. Tourbier, P. Villon, Structural & Multidisciplinary Optimization **48**, 987 (2013)
15. B. Raghavan, L. Xia, P. Breitkopf, A. Rassineux, P. Villon, Computer Methods in Applied Mechanics and Engineering **265**, 174 (2013)
16. B. Raghavan, G. Le Quilliec, P. Breitkopf, A. Rassineux, J.M. Roelandt, P. Villon, International Journal of Material Forming **7**, 487 (2014)
17. G. Le Quilliec, B. Raghavan, P. Breitkopf, Computer Methods in Applied Mechanics and Engineering **285**, 621 (2015)
18. D. González, E. Cueto, F. Chinesta, Annals of biomedical engineering pp. 1–11 (2015)
19. L. Meng, P. Breitkopf, B. Raghavan, G. Mauvoisin, O. Bartier, X. Hernot, Computer Methods in Applied Mechanics and Engineering (2015)
20. L.K. Saul, S.T. Roweis, unpublished. Available at: http://www. cs. toronto. edu/~ roweis/lle/publications. html (2000)
21. L.K. Saul, K.Q. Weinberger, J.H. Ham, F. Sha, D.D. Lee, Semisupervised learning pp. 293–308 (2006)
22. I. Jolliffe, *Principal component analysis* (Wiley Online Library, 2002)
23. S. Osher, J.A. Sethian, Journal of computational physics **79**, 12 (1988)
24. C. Moussa, X. Hernot, O. Bartier, G. Delattre, G. Mauvoisin, Materials Science and Engineering: A **606**, 409 (2014)

Elastic-plastic transition: A universal law

Zhong Chen[1], Hyuk Jong Bong[2], Dayong Li[3] and R.H. Wagoner[2,a]

[1]*Research and Innovation Center, Ford Motor Company, Michigan, 48131, United States*
[2]*Department of Materials Science and Engineering, The Ohio State University, Columbus, 43210, United States*
[3]*School of Mechanical Engineering, Shanghai Jiao Tong University, 200240, China*

Abstract.
Although the initial stress-strain behavior in a tensile test is often characterized as linear elastic up to a yield stress and nonlinear plastic thereafter, the pre-yield transition region is known to exhibit significant curvature and hysteresis. Hundreds of high-precision loading-unloading-loading tensile tests were performed using 26 commercial sheet alloys exhibiting a wide range of strength, ductility and crystal structure. Analysis of the results reveals the following:

1. There is no significant linear elastic region; the proportional limit is ~0 MPa when measured with sufficient sensitivity.
2. Each of the hundreds of measured transitional stress-strain curves can be characterized by a single parameter, here called the "modulus reduction rate." The corresponding equation captures ~80% of the observed variation, a factor of 3 to 6 better than a one-parameter linear approximation.
3. Most interestingly, the transitional behavior for all alloys follows a "Universal Law" requiring no fit parameters. The law depends only upon the strength of the material and its Young's modulus, both of which are can be measured by independent tests or adopted from handbooks. The Universal Law captures ~90% of the variation represented by the one-parameter representation and eliminates the need for mechanical testing to implement and apply.

The practical and theoretical implications of these results are discussed. The results provide a simple path to significantly improving applied constitutive models in the transitional regime. The consistency of the effect for such a wide range of metals and suggests that the origin of the behavior lies in the pile-up and relaxation of dislocation arrays.

1 Introduction

While it has long been known that mechanical testing produces stress-strain slopes lower than physically-measured Young's modulus, the phenomenon was trivialized as a testing problem, or an anelastic rate effect. It was only of passing academic interest until the 1990's when Morestin and Boivin [1] showed that springback after sheet forming occurred according to the mechanically measured, reduced modulus, not the handbook / physically measured value.

Following that discovery, in the 2000's many researchers interested in springback following sheet forming advocated characterizing the behavior as a reduced Young's modulus (10-30% reductions were typically reported) in the pre-yield region, and presented results showing that improved simulation accuracy using this approach. [2,3]

In the 2010's, Sun and Wagoner [4] measured the transitional behavior in detail, not only upon loading or unloading, but also in unloading/loading loops, with partial and full cycles, singly and in multiple sequences, and at various rates (none of which were significant to first order). They showed that the deformation was not classically elastic or plastic, nor was it was time-dependent anelasticity. Instead, the pre-yield deformation was elastic in the sense that the original shape (no strain) was recovered after a full loop. However, the pre-yield deformation was plastic in the sense that energy was dissipated hysteretically. However, it had a remarkable and mysteriously simple feature: it was quantitatively identical in two distinct alloys having the same strength and Young's modulus even though the strength was attained in divergent ways (e.g. by alloy content / precipitates vs. strain hardening).

Sun and Wagoner proposed a general, 3-D mechanics formulation called QPE (*Quasi-Plastic Elastic*) that reproduced the transition region according to the observations noted above. It added a third regime of behavior between a purely elastic region and an elastic-plastic one. The shape of pre-yield stress-strain function was controlled by up to 4 parameters including the initial slope E_0 (if not the Young's modulus) and the strain extent of the linear region (strain proportional limit), ε_c.

Extending Morestin and Boivin's conclusion, Sun and Wagoner also showed that incorporating the nonlinear behavior in FE simulations was essential to predicting springback behavior accurately, i.e. that using a chord modulus, no matter how attractive and convenient, was insufficient. In fact, using the QPE formulation reduced springback error by factors of 2-6 compared with chord modulus and handbook modulus formulations.

The most obvious (but incorrect) candidate mechanism for the *modulus effect* is related to nonlinear

[a] Corresponding author: wagoner.2@osu.edu

DOI: 10.1051/matecconf/20168011001

elasticity. An equation proposed for the tangent modulus E_T [5] based on atomic binding energy curves [6] takes the following form:

$$E_T(\varepsilon) = E_0(1-\alpha\varepsilon)\exp(-\alpha\varepsilon) \quad (1)$$

where E_0 is the initial slope (Young's Modulus normally) and α is a material constant. However, atomic bond stretching is purely elastic, i.e. reversible and energy-preserving, with no hysteresis. Furthermore, in metals, which yield at small fractions of their theoretical strength, elastic behavior is essentially linear. Therefore, bond-stretching can be ruled out as the source of pre-yield nonlinearity.

A simple modification of Eq. (1) to allow for hysteresis within a nonlinear elastic framework was proposed by Zhu et al. [7]. At the same time, the (1-α term) in Eq. (1), which is inconsequential in metals, was removed as a simplification.

Most recently, Chen et al. [8] performed hundreds of tests using 12 diverse steels to identify the range of pre-yield behavior. They reported the following simplifications that are used in the current work:

- The initial tangential modulus is Young's modulus.
- Loading and unloading nonlinearity are the same, although unloading is preferred for testing because of reduced scatter.
- The only factor altering the pre-yield behavior of 12 steels quantitatively is the strength of the material.
- The transitional behavior is essentially isotropic.
- The transitional behavior is nearly independent of prestrain (larger than 2%) except for the effect strain hardening on the strength of the material.

The current work was motivated by questions raised primarily by the Sun and Wagoner [4] and Chen et al. [8] results. The main questions in this study are as follows:

- What is the nature of the elastic-plastic transition? What is the key mechanism induces the nonlinearity?
- Is it possible to represent the nonlinear transition behavior by a simple and single equation?

Based on the foregoing, the extensive testing was streamlined by assuming that the the transition behavior is independent of strain rate, pre-strain (except for pre-strain hardening), multiple cycles, direction of testing, loading vs. unloading. In particular, only unloading legs at one prestrain were used.

2 Theoretical Background

2.1 QPE representation

Sun and Wagoner [4] proposed, developed, and implemented a three-regime general constitutive model. For uniaxial tension, the scalar total strain increment can be decomposed into three parts.

$$d\varepsilon = d\varepsilon_e + d\varepsilon_p + d\varepsilon_{QPE} \quad (2)$$

where $d\varepsilon_e$ is the elastic strain, $d\varepsilon_p$ is the plastic strain and $d\varepsilon_{QPE}$ is the so-called QPE strain. Figure 1 shows a typical unloading-loading hysteresis loop originated by the continuous changes of tangential modulus during unloading followed by reloading. The elastic strain is readily calculated from the initial slope (which in the case shown was sufficiently represented by Young's modulus) and the plastic irreversible strain component using Eq. (2).

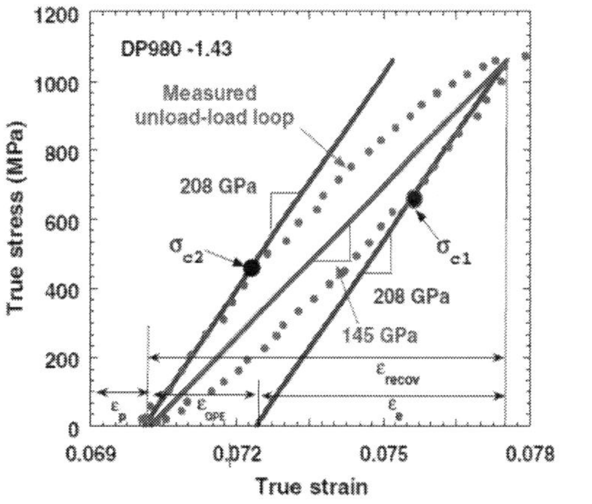

Figure 1. Expanded view of the unload-reload cycle for DP980 steel showing conceptual components of strain. After [4]

In the QPE constitutive model, the slope of the transitional stress-strain curve in uniaxial tension is represented as follows:

$$E(\Delta\varepsilon) = \begin{cases} E_0 & for\,\Delta\varepsilon \le \varepsilon_c \\ E_0 - A_1\left[1-\exp\{-A_2(\Delta\varepsilon-\varepsilon_c)\}\right] & for\,\Delta\varepsilon \ge \varepsilon_c \end{cases} \quad (3)$$

where E_0 is the physical Young's modulus (or can be taken as an undetermined parameter), ε_c is a critical transition strain separating the linear and nonlinear regions, $\Delta\varepsilon$ is the absolute value of the increment of strain from the strain rate reversal, i.e., the unloading strain, and A_1, A_2 are material parameters. The equation has up to 4 undetermined parameters, i.e., E_0, A1, A_2 and ε_c.

Integration of Eq. (3) leads to the equivalent stress form as follows:

$$\Delta\sigma(\Delta\varepsilon) = \begin{cases} E_0\Delta\varepsilon & for\,\Delta\varepsilon \le \varepsilon_c \\ (E_0 - A_1)(\Delta\varepsilon-\varepsilon_c) & \\ +\dfrac{A_1}{A_2}\{1-\exp[-A_2(\Delta\varepsilon-\varepsilon_c)]\} + E_0\varepsilon_c & for\,\Delta\varepsilon \le \varepsilon_c \end{cases} \quad (4)$$

where $\Delta\sigma$ is the absolute value of the stress drop during the unloading.

2.2 Simple 1-Parameter representation

In order to compare transition behavior in a wider range of alloys, it is highly advantageous to reduce the characterization of the nonlinearity to a single parameter. To do so, we make use of the simplifications noted above. In the current work, Eq. (1) is adopted with the small-

DOI: 10.1051/matecconf/20168011001

strain / metal simplification, which we nickname the 1-P form (for one parameter):

$$E_T(\varepsilon) = E_0 \exp(-A\varepsilon) \tag{5}$$

where E_0 is taken here to be the independently-measured physical Young's modulus, and we call A the modulus reduction rate. A is a material constant representing the entirety of the transitional behavior for a single unloading leg. Note that this representation is a significant simplification of the more general QPE equation, Eq. 3. Although there is not space here to discuss it further, Chen et al. [9] recently showed that Eq. (5) adequately represents the material-to-material, approximately 80% of the variation represented by the 4-parameter Eq. 4. Therefore, it is useful for looking at broad comparisons.

Integration of Eq. (5) produces the equivalent stress form as follows:

$$\Delta\sigma = \frac{E_0}{A}\left[1-\exp(-A\,\Delta\varepsilon)\right] \tag{6}$$

3 Experimental Procedures

3.1 Materials

Twenty-six typical but diverse sheet forming alloys were tested. Tensile tests shown in Figure 2 illustrate the diverse properties in terms of tensile strength (213-1199 MPa) and total elongation (5-49%). Moreover, the assembled material set contains metals with a variety crystal structure, BCC, FCC and HCP, and also a variety of hardening mechanisms and deformation mechanisms including slip, twinning (Mg, TWIP), and phase transformation (TRIP).

The "handbook" values of metals were adopted for following analyses as follows: 208 GPa for steels, 70 GPa for aluminium alloys, 110 GPa for brass and titanium alloy, and 46 GPa for magnesium alloys.

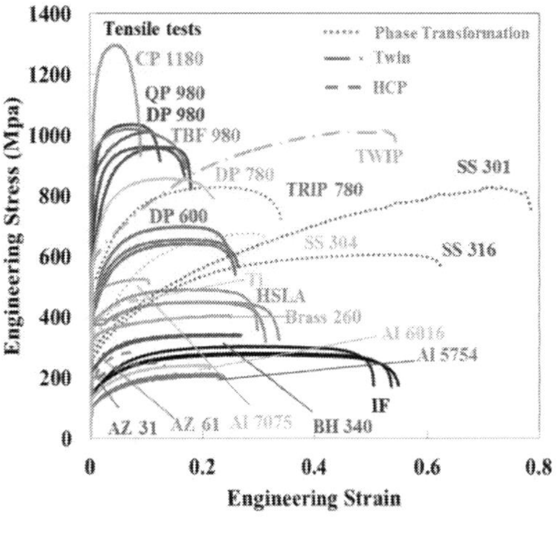

Figure 2. Tensile test results for 26 sheet alloys tested.

3.2 Mechanical Test

Note: For more information on testing and analysis procedures, refer to [9], upon which the current paper is largely based.

Standard ASTM E-8 tensile tests (RD) and unloading tests were carried out with an MTS 810 tensile frame. Strain was measured using a Class A Epsilon extensometer (3452-0200-030-ST), 2" gage, +30%/-10% extension range. This extensometer is a commercial extensometer with the highest available resolution and accuracy. Data was collected at a sampling frequency of 10 Hz up to 0.005 strain and at 1 Hz for subsequent loading.

Tests were conducted at constant grip velocity of 0.08 mm/s, which is equivalent to nominal strain rate of $\sim 10^{-3}$/s and at room temperature. Two or three duplicated tests were conducted for core 12 steels[a] to measure the test-to-test scatter.

Similar tests aimed to characterize the transition behaviors were conducted with 5 LUL cycles at 2, 4, 6, 8 and 10% in engineering strain. Among them, only the unloading-loading cycle at 8% prestrain was used in the current work. Each instantaneous tangential slope point $E_T(=d\Delta\sigma/d\Delta\varepsilon)$ was taken as the mean value of the 6 adjacent ($\Delta\varepsilon$, $\Delta\sigma$) data points. Other alternatives were considered to convert stress-strain data to slope-strain, but the procedure above led to reasonable combination of resolution and scatter for most experiments.

4 Results and Analysis

4.1 Individual Fits of Unloading Legs

In order to probe the extent of the linear region (i.e. the proportional limit), two versions of the QPE form, Eqs. (3) and (4), were utilized with the data for the 12 core steels. "QPE-4" refers to the 4 parameters being optimized. "QPE-3" eliminates ε_c by setting it equal identically to zero, thereby eliminating any linear region from the stress-strain representation (or, equivalently, any region of constant E_T).

The commercially software, Origin 8 SR3, was used to obtain optimized parameters by minimizing the least-squares error relative to unloading stress-strain data ($\Delta\varepsilon$, $\Delta\sigma$). All data was used in the optimization although a subset of all data points is shown in the following figures, for clarity of presentation. The optimized equation was then differentiated to obtain a corresponding tangential modulus, E_T, as a function of $\Delta\varepsilon$. Note that for this procedure the E_T representation are not precisely optimal, but they are very similar.

The process and results are illustrated in Figure 3 along with the standard error of fit, $\langle\Delta\sigma\rangle$. Figure 3 (a) compares QPE-3 and QPE-4 fits with experimental unloading data for one core steel, DP980-1. A remarkable improvement, by a factor of 10 to 30, is obtained by QPE

[a] 12 core steels represents 4 DP980 steels, 4 DP600 steels, 4 HSLA steels, and 4 IF steels.

fits as compared to the handbook physical or chord moduli. Figure 3 (b) shows E_T vs. $\Delta\varepsilon$ plot using the same fit equation as used in Figure 3 (a). The results reveal that the QPE fits (to stress-strain) also reproduce the E_T data better than the scatter of repeated tests with improvement by factors of ~8 and ~4, compared to handbook physical and chord moduli, respectively.

Most significantly, as can be seen in Figure 3 (a) and (b), the QPE-3 and 4 fits gave identical results! (The two lines overlay one another, so that only one can be seen.) That is, the optimal value of ε_c in the QPE-4 fits was found to be 0, such there is no distinction between the QPE-3 and QPE-4 forms. Therefore, no significant linear elastic region can be discerned for DP980-1. To make this more precise, the QPE-4 fits for 8 of the 12 core steels produced ε_c=0. For the other 4 alloys, very small values of ε_c were determined, but none produced significant improvement of $\langle\Delta\sigma\rangle$. Therefore, it was concluded that the high-precision data available in this study suggests that the proportional limit is 0 MPa. That is, nonlinearity starts at the smallest stress excursions. This upsets nearly all standard teaching and learning on this subject.

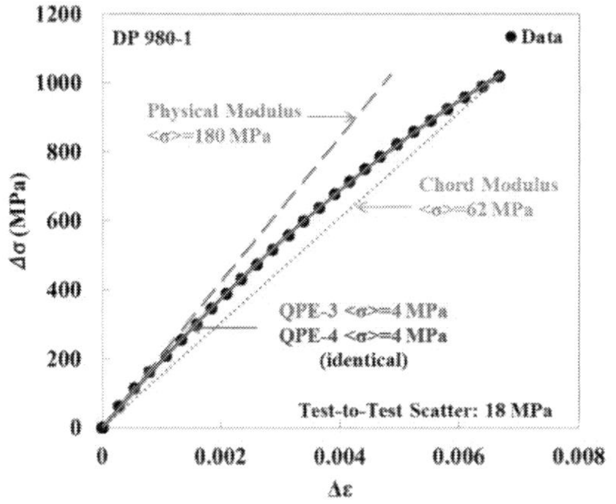

(a)

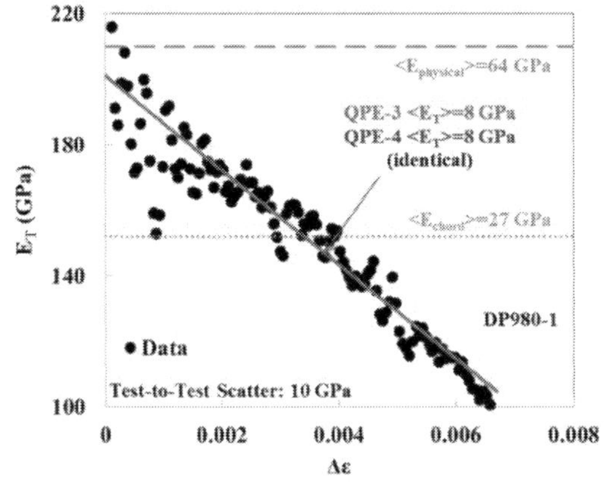

(b)

Figure 3. Comparison of QPE-3 and 4 for DP 980-1, (a) $\Delta\varepsilon$ vs. $\Delta\sigma$ and (b) $\Delta\varepsilon$ vs. E_T

As noted above, for use in comparing transitional behavior among various alloys the 1-P form is preferable. In order to test the ability of this simpler form to capture the essential transitional behavior, optimized 1-P fits were compared with the optimized QPE-3/4 fits, which are now known to reproduce the experimental precisely.

Figure 4 shows that for the same data as in Figure 3 (b), the 1-P fits introduce an insignificantly higher standard error of fit. It is insignificant because it remains lower than the test-to-test scatter for the tests upon which the analysis is based.

The fitting result for DP980-1 by 1-P in E_T vs. $\Delta\varepsilon$ base is shown in Figure 4 together with results by QPE-3 and 4 fits. The 1-P fit leads to slightly higher standard deviation than QPE-3 and 4, but remains less than the test-to test scatter (9 GPa vs. 10 GPa). This is particularly convincing in view of the fact that the 1-P fit is derived from stress-strain data, it was not optimized to the tangential modulus – strain result.

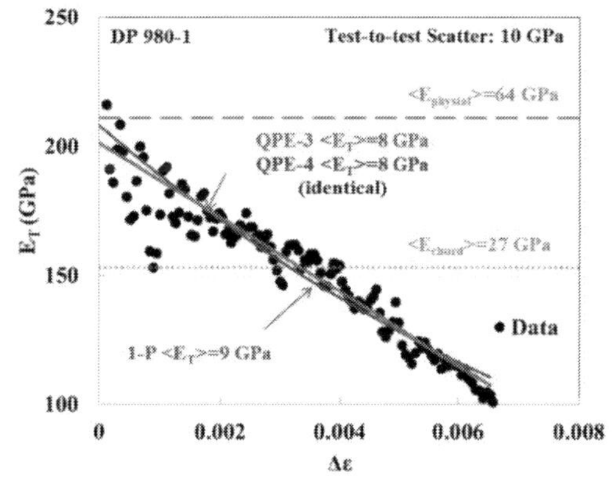

Figure 4. Comparison of fit equations for DP980-1

The corresponding analyses and comparison for the 26 materials are shown in Table 1. The results clearly reveal the tremendous improvements by any of the nonlinear equations compared with linear representations, i.e., either standard handbook or chord moduli. As expected, the 1-P fit leads to slightly higher standard error, but again remain less than the test-to-test scatter of the data itself. Based on the results, the use of 1-P form was deemed justified for further work.

Table 1 Summary of optimization using 1-P equation for 26 metals

	$\langle\sigma\rangle$ (MPa)	$\langle E_T\rangle$ (GPa)
$E_{physicl}$	72	41
E_{chord}	25	19
1-P	5	8
QPE-3P	2	6
QPE-4P	2	6
Test-to-test scatter[b]	10	11

[b] The test-to-test scatter is calculated from duplicated data for 12 core steels only.

4.2 A Universal Law

The simple 1-P representation, in view of the Sun and Wagoner and Chen observation that the transitional behavior in steels depends only on strength, offers the intriguing possibility of representing the entire phenomenon with zero fit parameters. If justified, this would render application of this significant improvement without any testing at all. (Recall, that the 1-P form improved the accuracy by factors of 4-8 as compared with the standard choices: zero-parameter Young's modulus or the 1-parameter chord modulus.)

In order to probe this idea, first a broader range of 21 steels was considered, including all of those shown in Table 1. These steels show the full range of hardening and deformation mechanisms and two of the three typical metal crystal structures, fcc and bcc. For this work we also included results for all pre-strains that were available.

Figure 5 (a) plots the modulus reduction rate A in the individual 1-P fits as a function of material strength[c] obtained from hundreds of loading-unloading-loading tests. The data shows a clear trend of a first-order dependency of modulus reduction rate on the material strength. The modulus reduction rate was found to fit the data by a curve as follows:

$$A = \frac{15000\ MPa^{3/4}}{\sigma_u^{3/4}\ MPa^{3/4}} \tag{7}$$

The variation in A is captured within $\langle A \rangle = 39$ (as compared with a total variation of 350) by using only the independently determined "strength," measuring anything with respect to the stress-strain transition.

The foregoing analysis was extended to all 26 tested materials with other base elements and crystal structures. Closer examination revealed that the A values for the other 5 metals exhibit same trend, i.e., a similar dependence on the unloading stress. However, the comparison also showed changes of A proportional to the physical Young's modulus for each metal. For instance, the A value for Brass 260 is approximately 100 whereas the A value for a steel of the same strength is approximately 200, for a ratio is 1:2, approximately the same as the ratio of the physically measured Young's moduli for brass and steel: 110 GPa/208 GPa.

Using this second observed dependence of the modulus reduction rate on Young's modulus, the following "Universal Law" was derived taking into account both independently measured quantities, strength and Young's modulus:

$$\Delta\sigma = \frac{208(GPa)\ \left[\sigma_u (MPa)\right]^{3/4}}{15000\ (MPa)^{3/4}} \left[1 - \exp\left(\frac{-15000\ (MPa)^{3/4}}{\left[\sigma_u\ (MPa)\right]^{3/4}} \frac{E_0\ GPa}{208\ GPa} \Delta\varepsilon\right)\right] \tag{8}$$

[c] Throughout the current work, the material strength is measured by the flow stress just prior to unloading, σ_u.

$$E = E_0\ (GPa) \exp\left(-\frac{15000\ (MPa)^{3/4}}{\left[\sigma_u\ (MPa)\right]^{3/4}} \frac{E_0\ \text{GPa}}{208\ GPa} \Delta\varepsilon\right) \tag{9}$$

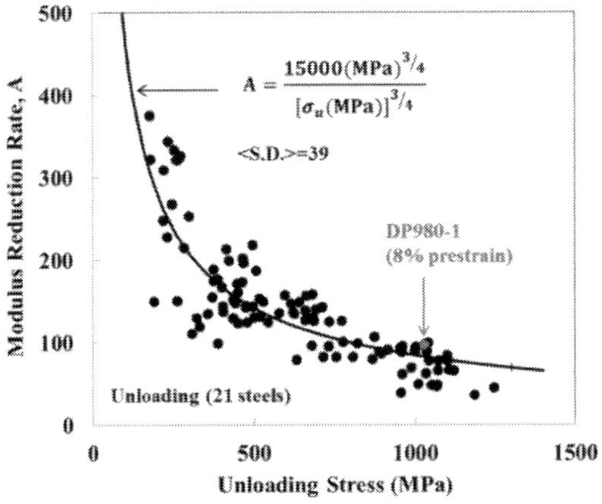

(a)

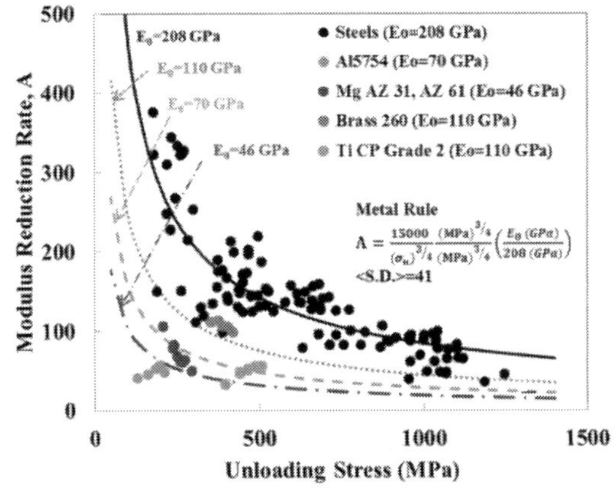

(b)

Figure 5. The variation of modulus reduction rate A vs. Unloading stress for all 26 metals (a) measured modulus reduction rates, A (b) modulus reduction rates, A, compared with a metal rule

The Universal Law is identical to the rules found for 12 steels or 21 steels, all of which had handbook Young's moduli of approximately 208 GPa.

To illustrate the use of the new relationship, the case of DP 980 with a pre-strain of 0.077 was again considered. Figure 6 compares the zero-parameter Universal Law with the custom 1-P fit for that data alone. The results show that there is only a slight loss of accuracy by using Eq. (9) compared with the 1-P fit (from $\langle E_T \rangle = 7$ to $\langle E_T \rangle = 12$ GPa). Moreover, the zero-parameter Universal Law reduces the standard error relative to the 1-parameter chord modulus by a factor of greater than 2, i.e., $\langle E_T \rangle = 12$ to 27 GPa.

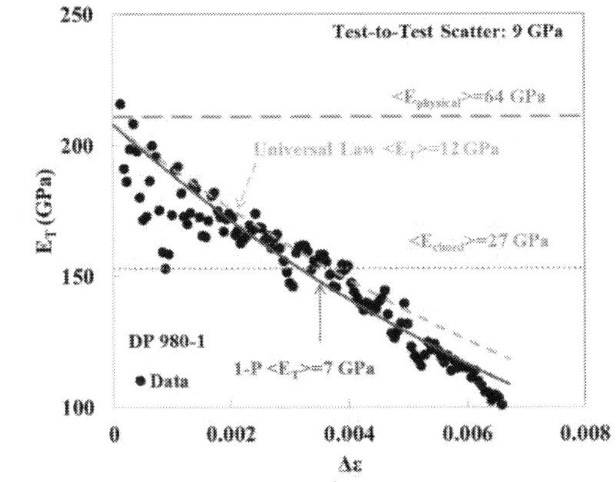

Figure 6. A comparison of 1-P and the Universal Law for DP980-1

The accuracy of Eqs. (8) and (9) as compared with the 1-P custom fit for all 26 metals is compared in terms of average standard errors of fit in Table 2. The results reveal that the Universal Law can be used profitably in applications. It exhibits little or no loss of accuracy compared with test-to-test scatter. Moreover, it greatly improves on the commonly-used linear representations, i.e., physical Young' modulus or chord modulus. Furthermore, it does not require any special test to implement profitably.

Table 2. Comparison of Universal Law and custom 1-P fit for 26 metals

	<σ> (MPa)	<E_T> (GPa)
$E_{physicl}$	72	41
E_{chord}	25	19
Universal Law	*13*	11
1-P	*5*	8
Test-to-test scatter	10	11

Using the Eqs. (8) and (9), the hysteresis loop width, δ, can be derived as follows:

$$\delta = \frac{2}{A}\ln\left[1-\frac{A}{E_0}\sigma_u+\frac{A}{E_0}\left(\sigma_u-\frac{E_0}{A}-\frac{\sigma_u}{\ln\left(1-\frac{A\sigma_u}{E_0}\right)}\right)\right]$$
$$-\frac{2}{A}\left[\frac{\sigma_u-\left(\sigma_u-\frac{E_0}{A}-\frac{\sigma_u}{\ln\left(1-\frac{A\sigma_u}{E_0}\right)}\right)}{\sigma_u}\right]\ln\left(1-\frac{A}{E_0}\sigma_u\right) \quad (10)$$

A broad comparison of measured and Universal-Law-predicted loop widths is shown in Figure 7 for all 27 materials tested. Figure 7 (a) shows the predicted loop width from the Universal Law together with measured loop width as a function of the unloading stress. The results are in good agreement except for two steels which exhibit exceptionally small hysteresis relative to the trends of the other alloy. In order to assess realistically how good the agreement is, please note that either of the current standard approaches, Young's modulus or chord modulus, exhibit NO loop width. That is, plotting those data would put them all on the X-axis of Figure 7 (a).

Figure 7 (b) directly compares measured and predicted loop width. Although there is some systematic deviation, the Universal Law represents the highly-sensitive loop-width data within the scatter of the data itself, with the possible exception of two "special" steels, i.e., CP 1180 and TBF 980. These metals showed (almost) linear-like unloading behavior, with less reduction of the modulus than for the other alloys. These two alloys are known to have extremely fine microstructures, which suggest a possible mechanism for this behavior and their divergence from the norms. This will be discussed in Section 5.

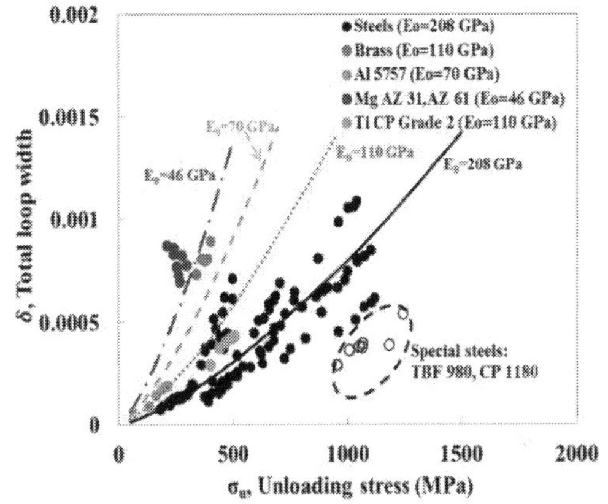

(a)

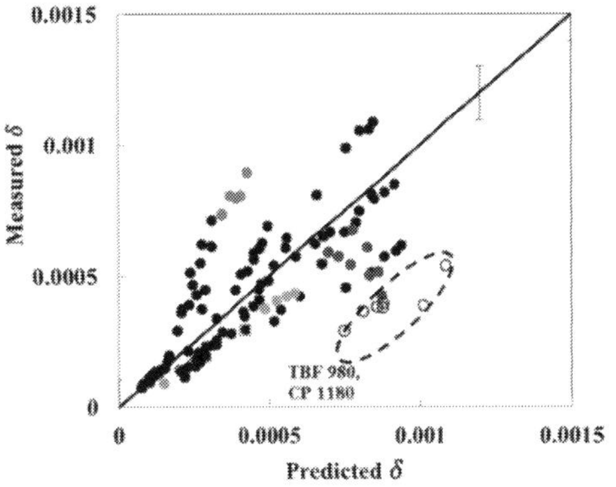

(b)

Figure 7 Summary of loop width (δ) measurements at various pre-strains compared with Universal Law predictions: (a) as a function of unloading stress and Young's modulus, and (b) compared with Universal Law predicted values.

5 Application

The Universal Law was applied to Toyota pillar compression tests, as are presented in detail elsewhere [8]. 600 mm long structural pillars was formed in two pieces from SPC 400 steel sheet (0.8 mm thick steel cold-rolled with an ultimate tensile strength of 440 MPa) and spot welded. The welded pillars were then compressed in a universal testing machine with behavior as shown in Figure 8.

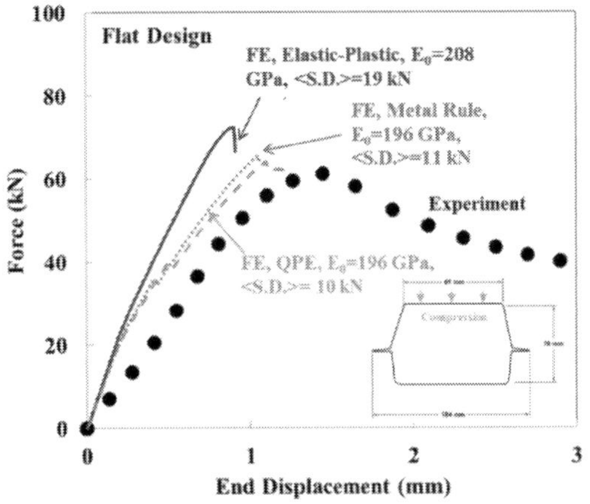

Figure 8. Toyota compression test data and FE prediction by QPE equation and Universal Law

The pillar compression was also simulated using Abaqus/Standard incorporating either the QPE-4 form or the Universal Law. For QPE-4, a custom value of E_0 obtained by ultrasonic testing was adopted. In keeping with the simplest application, the Universal Law was applied using the handbook value of Young's modulus.

The predicted force-displacement curves are compared in Figure 8 with the experimental data. The advanced models, i.e., QPE-4 and Universal Law, improve the prediction accuracy compared with the standard FE result as a factor of ~2. Moreover, the Universal Law applied with no adjustable parameters produces almost identical fidelity as compared with 4-paramter QPE-4. This application confirms the usefulness of the Universal Law for engineers who need to accurately predict springback or automotive structural crush without incurring a corresponding time and effort for additional experiments.

6 Discussion

Many mechanisms have been proposed as accounting for the modulus effect in metals, whether it is considered a linear departure from the physical Young's modulus, or a nonlinearity. The current work suggests an answer to the question of "why?"

The simplest proposal is second-order nonlinear elasticity related to bond stretching [10]. As already discussed, the well-established energy absorption and nonlinearity as vanishingly small strains as confirmed here rules out this mechanism.

A twinning and detwinning [11–13] mechanism cannot explain experimentally observed nonlinearity in non-HCP structured metals.

Textural changes during the plastic pre-straining [13,14] can be also considered. However, these changes occur too slowly with strain to account for the observed changes. Furthermore, textural changes would not be expected to introduce nonlinearities, and would be expected to lead to higher modulus changes at higher stain in many cases, which is never observed.

Material damage by porosity or void evolution has been also proposed [15,16]. However, the reduction of density by the damage mechanism was found as insufficient to explain experimentally observed modulus degradation [17].

Heterogeneity arising from internal and residual stresses [13,14,18] established by microstructural inhomogeneities during pre-straining can cause additional micro-strain or stress in the subsequent unloading. However, these effects would not be expected to reverse during an unloading leg with no plastic deformation.

The mechanism that seems the most plausible on the surface is a generalized dislocation pile-up and relaxation mechanism [2,19,20]. Dislocations are known to interact elastically with one another but also dissipate energy whenever they move [20]. Thus, the main features are satisfied. In this concept, mobile dislocations move on slip planes until stopped by obstacles such as grain boundary or inclusions. When the external force is removed, the dislocations repel each other and provide additional micro-strain concurrent with elastic unloading strain. Such a mechanism reproduces all of the qualitative features observed. For example, all moving dislocations dissipate work by exciting lattice phonons [21].

The two "special" steels, CP 1180 and TBF 980, having abnormally small loop widths and are known to have very fine microstructures as is beneficial for increasing flow strength. However, extremely fine microstructures also introduce the constraint of extremely limited pile-up lengths, which can potentially limit the storage and release of strain.

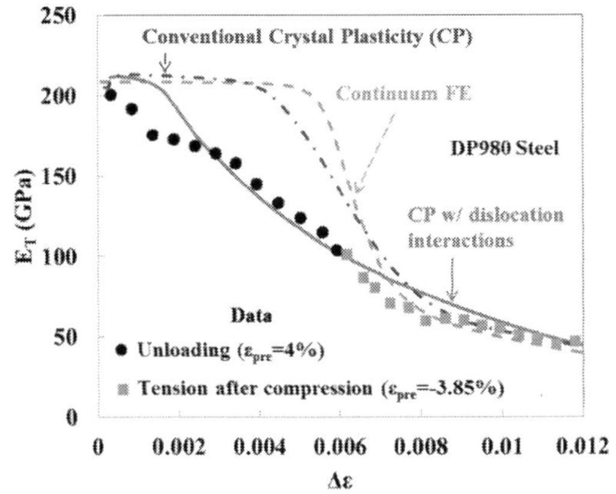

Figure 9. Comparison of the measured tangential modulus with FE simulations incorporating alternate constitutive models.

As part of other, so far unpublished work, the authors simulated unloading and reverse loading (i.e. unloading and then compression after a tensile prestrain) using a carefully-constructed RVE based on a DP980 steel microstructure. Care was taken to use the proper microstructural properties, sizes and proportions in the RVE. Figure 9 shows the compares the measured tangential slope change to simulations using three constitutive models: a standard Continuum elastic-plastic model, a standard Crystal Plasticity (CP) model, and CP model with dislocation elastic interactions [22–24]. The results suggest that the dislocation pile-up and relaxation mechanism dominates the nonlinear transition behavior.

7 Conclusions

The elastic-plastic transition behavior of 26 metals was analysed using advanced nonlinear models and hundreds of special tensile test data using a Class A extensometer. Based on the analyses and observations, following conclusions were reached:

- A zero-parameter Universal Law predicting the transitional stress-strain response in all metals with no undetermined parameters was introduced.
- Implementing the Universal Law requires no testing but was shown to improve the accuracy of simulating automotive crush, by a factor of two.
- The proportional limit of metals is 0 MPa. That is, nonlinearity starts for vanishingly small strains.
- A simple 1-parameter equation describes the transition response well for all transition regions, without significant loss of accuracy compared with multi-parameter models appearing in the literature.
- The dominant mechanism of transitional behavior appears to be a dislocation pile-up and release.

Acknowledgements

This material is based upon work supported by the U.S. Department of Energy, Office of Science, Office of Basic Energy Sciences, Materials Science and Engineering Division, and particularly Program Director John S. Vetrano, for their support of the current work under Award Number DE-SC0012483 and DE-SC0012587. The 12 core steels were provided by ArcelorMittal, Nucor, ThyssenKrupp, and U. S. Steel, as arranged by the Steel Marketing and Development Institute. Six other steels were provided by Posco.

References

1. F. Morestin, M. Boivin, Nucl. Eng. Des. **162,** 107–16 (1996)
2. R. Cleveland, A. Ghosh, Int. J. Plast. **18,** 769-85 (2002)
3. F. Yoshida, T. Uemori, K. Fujiwara, Int. J. Plast. **18,** 633-59 (2002)
4. L. Sun, R.H. Wagoner, Int. J. Plast. **27,** 1126-44 (2011)
5. A.S. Argon, *Physical Metallurgy*. (Elsevier; 1996)
6. J. H. Rose, J.R. Smith, J. Ferrante, Phys. Rev. B. **28,** 1835-45 (1983)
7. H. Zhu, S. Sriram, B. Yan, P. Duroux, SAE Int. J. Mater. Manuf. **3,** 691-701 (2010)
8. Z. Chen, U. Gandhi, J. Lee, R.H. Wagoner, J. Mater. Process. Technol. **227,** 227-43 (2016)
9. Z. Chen, H.J. Bong, D. Li, R.H. Wagoner, Int. J. Plast. (in final review)
10. B.E. Powell, J. Appl. Phys. **53**, 764 (1982)
11. C. Cáceres, T. Sumitomo, M. Veidt, Acta. Mater. **51**6, 211-8 (2003)
12. T. Hama, N. Kitamura, H. Takuda, Mater. Sci. Eng. A. **583,** 232-41 (2013)
13. H. Wang, P.D. Wu, J. Wang, Int. J. Plast. **47**, 49-64 (2013)
14. S.R. Agnew, J.R. Weertman, Mater. Sci. Eng. A. **242,** 174-80 (1998)
15. S. Murakami, *Continuum damage mechanics* (Springer, 2012)
16. K. Yamaguchi, H. Adachi, N. Takakura, Met. Mater. Int. **4**, 420-5 (1998)
17. M. Yang, Y. Akiyama, T. Sasaki, J. Mater. Process. Technol. **151,** 232-6 (2004)
18. A. Govik, R. Rentmeester, L. Nilsson, Mater. Sci. Eng. A. **602,** 119-26 (2014)
19. J.H. Kim, D. Kim, F. Barlat, M.-G. Lee, Mater. Sci. Eng. A. **539**, 259-70 (2012)
20. L. Luo, A.K. Ghosh, J. Eng. Mater. Technol. **125,** 237 (2003)
21. J.P. Hirth, J. Lothe. *Theory of Dislocations* (Krieger Publishing Company, 1982)
22. M.-G. Lee MG, H. Lim, B.L. Adams, J.P. Hirth, R.H. Wagoner, Int. J. Plast. **20**, 925-38, (2010)
23. H. Lim, M.-G. Lee, J.H. Kim, B.L. Adams, R.H. Wagoner, Int. J. Plast. **27**, 1328-53 (2011)
24. H. Lim, S. Subedi, D.T. Fullwood, B.L. Adams, R.H. Wagoner, Mater Trans, **55**, 35-8, (2014)

Principles of polymer processing modelling

Jean-François Agassant[1a] and Malcolm. R. Mackley[2]

[1] *MINES ParisTech, PSL Research University, CEMEF, UMR CNRS 7635, 06560 Valbonne ,France*
[2] *Department of Chemical Engineering and Biotechnology, University of Cambridge, UK*

Abstract. Polymer processing involves three thermo-mechanical stages: Plastication of solid polymer granules or powder to an homogeneous fluid which is shaped under pressure in moulds or dies and finally cooled and eventually drawn to obtain the final plastic part. Physical properties of polymers (high viscosity, non-linear rheology, low thermal diffusivity) as well as the complex shape of most plastic parts make modelling a challenge. Several examples (film blowing extrusion dies, injection moulding, blow moulding) are presented and discussed.

1 Introduction

Numerical modeling of polymer processing has now become an important tool in relation to the cost effective design and operation of polymer processes. In the past, the empirical development of new polymer grades has been motivated largely by searching for new or improved properties and it was assumed that the forming processes used to produce end products could be adapted through laboratory based adjustments made by simple trial and error. This worked effectively for a number of early cases; however sometimes it was discovered that new polymers with predicted "exceptional" properties were very difficult to process, either because the required pressure or torque could not be obtained with existing machines, or because the process exhibited flow or drawing instabilities at production rates inconsistent with their economic viability.

More recently, polymer producers have integrated the forming process into their development of new materials by using at an early stage of the development mini-processing machines (extruders or injection molding) that could discern potential processing problems using just a few hundred grams of polymer. Sometimes, the extrapolation to industrial size machines has proved uncertain, particularly because "scaling" does not obey the same scaling rules for mechanical and thermal behavior.

Numerical modeling, used initially in the field of injection mold design, has now become an important design aid tool in different sectors of the plastics industry, but it only makes sense to do this if the modeling is based on sound physical principles and the applied numerical techniques can capture the relevant process physics.

2 Fundamentals relevant to polymer Processing modelling

Shaping thermoplastic polymers is a thermo-mechanical process which requires generally three successive steps:

- Melting, that is to say the passage of the solid state, powder or granules, to a molten state in order to give a sufficiently fluid and homogeneous melt.
- Flow of the molten polymer under pressure through a die or into a mold, shaping cooling, possibly coupled with drawing operations, biaxial stretching, blowing.
- Solidification into the final component shape and form.

The properties of the produced part will depend, of course, on the chosen polymer, but also on the "thermo-mechanical path" that was followed to shape the object. In the light of figure 1, it can be imagined that polymer processing is a problem of fluid mechanics coupled to heat transfer. In reality it is more complicated because rheology and heat transfer of thermoplastic polymers produce both experimental and modeling challenges.

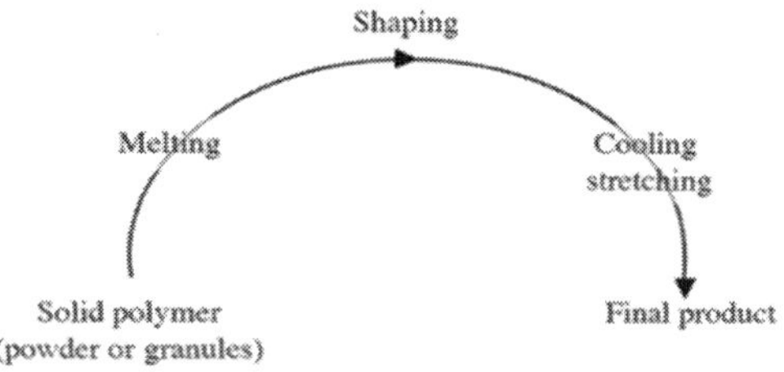

Figure 1. General procedure for polymer processing

- At processing temperatures, molten polymers generally have a viscosity as high as 10^3 Pas, which is a million times larger than water's viscosity and this property has a number of significant consequences: the Reynolds number is low and generally below unity. As a consequence a molten polymer flow will never be turbulent, and often the inertial terms of the force balance equations can be neglected. The gravity force is negligible in most confined flows, however it may

[a] Corresponding author: jean-francois.agassant@mines-paristech.fr

become important in free surface processes as in film blowing, fiber spinning and extrusion blow molding for large hollow bodies.

Heating of the polymer by viscous flow dissipation can be very significant and this can be used with benefit for example in plasticization units: In many common extrusion operating conditions, energy introduced into the polymer by heat conduction and by viscous dissipation are of the same order of magnitude and therefore it is essential when modelling these situations to solve both the thermal and force balance coupled equations.

- Molten polymer present shear thinning as well as viscoelastic effects which lead to complex non-linear constitutive equations. The most sophisticated constitutive equation is not necessarily the most relevant or useful for modeling engineering processes. The choice depends on the process to be solved, on the type of flow, on the available computer facilities and, of course, on the available rheological measurements. If the objective of the calculation is to know the stress distribution, the orientation of macromolecules or of short glass reinforcing fibers, it is necessary to introduce a viscoelastic behavior law. If, however, the aim is only to establish pressure flow-rate relationships in a confined shear flow or shear dominant flow it is necessary to take into account accurately the shear rate and temperature dependence of the viscosity. In that kind of geometry, viscoelasticity modifies only marginally the velocity field and the normal stress at the wall. In an extensional flow with a free surface, it is usually essential to use a viscoelastic law because only this type of behavior is able to account for the often experimentally observed increase in extensional viscosity with the elongation rate.
- The low thermal conductivity of polymers as well as the temperature dependence of the viscosity introduce specific problems which may need to be addressed when solving polymer processing problems. Heating by viscous dissipation is localized where both rate of strain and stress are important and the low thermal conductivity of polymer will consequently promote the existence of high thermal gradients. Given the risk of polymer degradation due to these localized overheating phenomena, this limits the amount of energy that can be provided by viscous dissipation and the flow rates that can be obtained.

3 Base equations and modelling strategies

3.1 A set of ten base equations…

… has to be solved in a 3D geometry: a mass balance (generally, but not always, incompressible), three force balance equations where mass and inertia forces are neglected, constitutive equations for the rheology (5 equations), a thermal balance equation.

3.2 Boundary conditions

In confined flows (extrusion dies, injection molds), it is generally assumed that the polymer adheres to the metal tool wall. Slip may occur when processing reinforced polymers, rubbers or PVC compounds. The problem is then to define a realistic sliding law. This gives rise, even today, to scientific debate. For standard thermoplastic polymers slip can also occur at high throughput; it is then usually associated with the existence of extrusion defects.

In free surface flows (fiber spinning, blown film, for example), a velocity tangent at the free surface and a zero normal stress to this free surface are imposed.

In unsteady free surface flows (mold filling, but also blow molding), the stress component normal at the surface is again zero, but now the free surface propagates along the normal to this free surface.

Thermal boundary conditions can also present difficulties as the geometry of polymer processing equipment can be complex and wherever there is a free surface involved appropriate heat transfer coefficients must be chosen involving free or forced convection, radiation.

3.3 Model simplification

Solving the above full set of equations is still a very challenging numerical problem and so in the past it has been traditional to make wherever possible a number of approximations [1,2].

Geometric approximations consist for example in unrolling the flow geometry. This is carried out to analyze the flow between the screw and the barrel of a single screw extruder (Tadmor and Klein [3]). They may also consist of decomposing a complex flow geometry into several more simple flow geometries where local mass balance and pressure equilibrium equations can be written (this will be described for simplifying a film blowing die).

Hydrodynamic lubrication or Hele-Shaw Approximations [4]: In many polymer processing geometries, the transverse dimension of the flow is small compared to the other dimensions and slightly varies in the flow direction. Lubrication approximation helps to reduce the problem of solving 2D or 3D Stokes equations to a single pressure dependent equation.

Membrane approximation: When considering, for example, the blowing of a membrane, it is assumed that the velocity components are uniform in the thickness.

Approximations can also be made to the **thermal equation**. For thin layer geometries an average temperature balance equation may be written on a material volume. Dimensionless numbers (Cameron [5], Brinkman [6]) allow obtaining rapidly relevant order of magnitude values of the temperature field in most polymer processes.

3.4 Numerical schemes

In any three-dimensional flow there are ten equations that provide access to the ten unknowns that have to be solved. Several numerical methods are available.

- In "direct" methods, a CAD definition of the die or mold geometry allows meshing the 3D flow volume and then solving the set of equations using generally finite elements. Meshing is a key point for obtaining an accurate result while controlling the computation time. Figure 2 presents the mesh of a complex cavity mold where capturing the flow in narrow flow regions, as the one around the injection gate, requires a very refined mesh. Extending this refined mesh to the whole geometry would lead to several millions of nodes and very large computation times and facilities. The development of anisotropic meshes (refined in the thickness and coarse in the flow direction) and of local mesh refinement governed by the local flow metric (Coupez [7]) allows reducing significantly computation time and improving accuracy. One difficult point is to choose appropriate finite elements and the related approximations spaces in order to respect the compatibility conditions for velocity, pressure (and stress components when using viscoelastic constitutive equations) (Arnold et al. [8]). Another difficulty is to deal with the convective terms, especially for viscoelastic constitutive equations (Brooks and Hughes [9]; Silva et al. [10]).

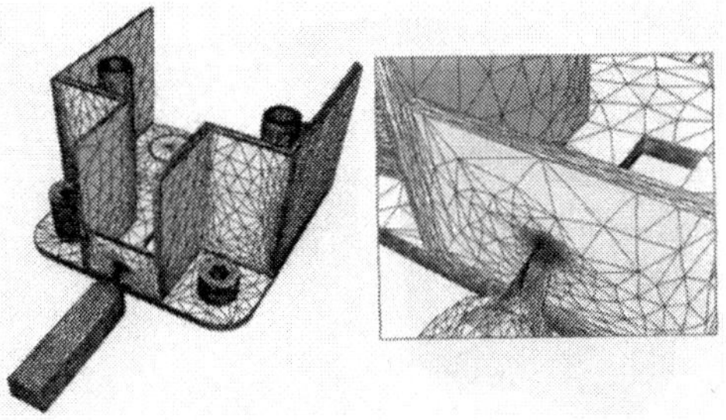

Figure 2. Example of a mold cavity meshing with sharp mesh refinement in the vicinity of the injection gate (Silva et al.[11])

- In many cases, however, the tooling geometries are such that geometric and kinematic approximations may be used. Moreover, the different thermal dimensionless numbers listed above allow simplifying the temperature balance equation and provide a good order of magnitude of thermal phenomena.
- The choice between direct numerical methods and approximation methods is not obvious: Direct methods are difficult to implement and costly in computation time and memory space, which makes them tedious for parametric studies that require forming tool optimization. Moreover, the accuracy of the result depends on the accuracy of the mesh and, in the case of complex three-dimensional flows, mesh refinement until a stabilized numerical solution is not always possible. Approximation methods are obviously much faster, but the results depend on the validity of the approximations made. Nevertheless, these quick calculations allow easier parametric optimization.

4. Some examples of industrial flow geometries

Three examples are selected in order to demonstrate different features of simplifications and complexities that can arise when attempting to model polymer melt processing.

4.1. Optimization of a film-blowing die

A film-blowing die must provide a uniform flow and temperature distribution around the circumference of the die and any thickness and temperature heterogeneity would be enhanced in the subsequent film blowing process leading to unacceptable film properties. The geometry of a film-blowing die is therefore complex as shown schematically in figure3. Molten polymer delivered by the extruder is distributed into different radial channels that open into helical channels machined into the central mandrel. The clearance between the mandrel and the axisymmetric die shaft is gradually increased in the vertical direction thus ensuring a homogeneous distribution of the polymer at the die outlet, independent on the position of the radial channels.

The use of a direct numerical method would require a refined mesh, especially between the die shaft and the top restrictions between two successive helical channels and this will result in significant meshing sizing. Moreover, the number of geometrical parameters that can be varied will induce a very expensive parametric study. Much simpler modeling can be achieved by unrolling the flow between the central mandrel and the axisymmetric die shaft and by schematizing the flow in helical channels by tube Poiseuille flows and the flow above the restrictions between two successive channels by plane Poiseuille flows, and then writing local flow and pressure balances on local finite volumes.

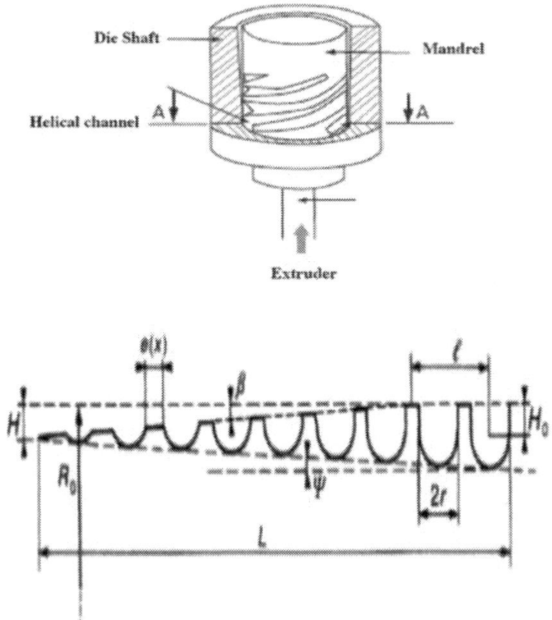

Figure 3. Geometry of a film-blowing die: (top) overall view; (bottom) cut in the vertical plane

By using this method, Figure 4 shows for example the influence of a modification of the initial channel depth H_0 on the final flow rate distribution output.

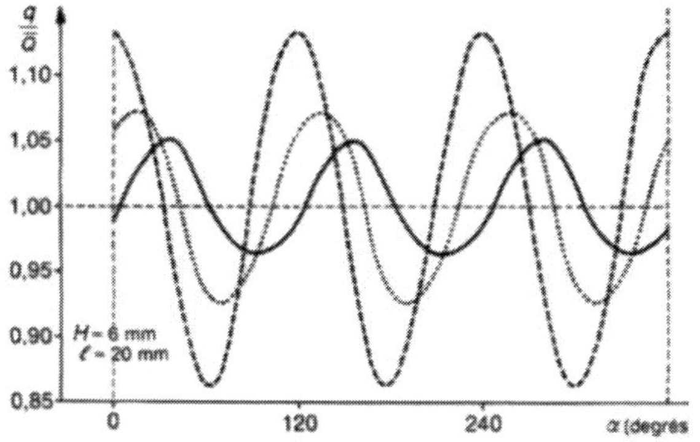

Figure 4. Influence of the initial depth of the channel H_0 on the polymer flow distribution at film-blowing die exit.. q is the local flow along the circumference of the die , $\bar{Q}$ is the average flow rate (Vergnes et al.[12];Agassant and Vergnes [13]).

Thermal phenomena can significantly influence the polymer flow distribution. An average temperature calculation is presented here assuming adiabatic conditions along the mandrel and isothermal conditions along the die shaft. The coupling between mechanical and thermal resolution is carried out iteratively on the whole flow domain. Figure 5 shows that flow homogeneity is improved by increasing this die temperature. However, a too high die exit temperature will make the subsequent polymer blowing step more difficult

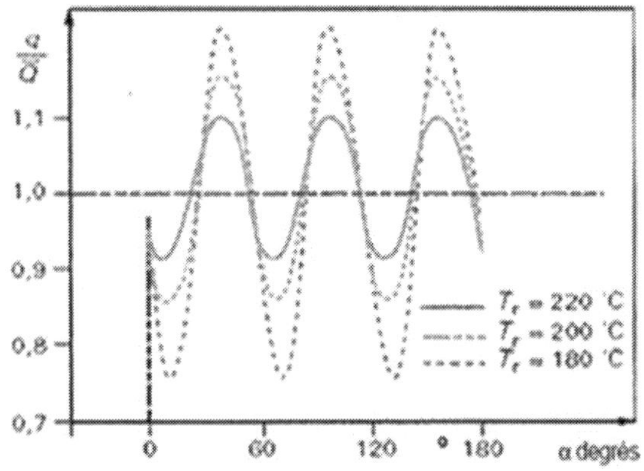

Figure 5. Influence of the thermal regulation of the die shaft on the polymer flow distribution at film-blowing die exit.. q is the local flow along the circumference of the die , $\bar{Q}$ is the average flow rate (Vergnes et al.[12];Agassant and Vergnes [13])

4.2 Injection moulding

It is in this area that the first polymer processing numerical simulation software in the mid-1980s was developed. The first objective is to fill the cavity as uniformly as possible (that is to say such as the ends of the mold cavity are reached at equivalent filling times), at a temperature as constant as possible (i.e. in conditions where the viscous energy dissipation is roughly balanced by heat conduction to the walls). A key problem is to predict the location of weld lines in such a way that they are not visible from the outside and submitted to low stresses during the life time of the plastic part. This cavity filling optimization in highly non-isothermal conditions is now properly resolved by many commercial software using purely viscous temperature dependent behavior.

Often commercial software assumes the mold to be a "thin layer" and then use the Hele -Shaw approximations. This type of approach, commonly named 2.5D, was introduced by Kamal et al.[14], Hieber and Shen[15], Wang et al. [16]. The average temperature approximation is not necessarily consistent given the important temperature gradient between the molten polymer and the mold walls. This requires coupling the mechanical and thermal equations everywhere in the cavity. Moreover this "thin layer" approximation is not able to capture the "fountain flow" effect at flow front (Castro and Macosko [17]; Mavridis et al. [18]). In order to capture such phenomena and to account for complex mold geometry as the one presented figure 2, 3D viscous compressible modeling is required (Haagh et al.[19]; Silva et al.[20]). Figure 6 shows successive steps of the filling of the cavity where the meshing has been shown in figure 2.

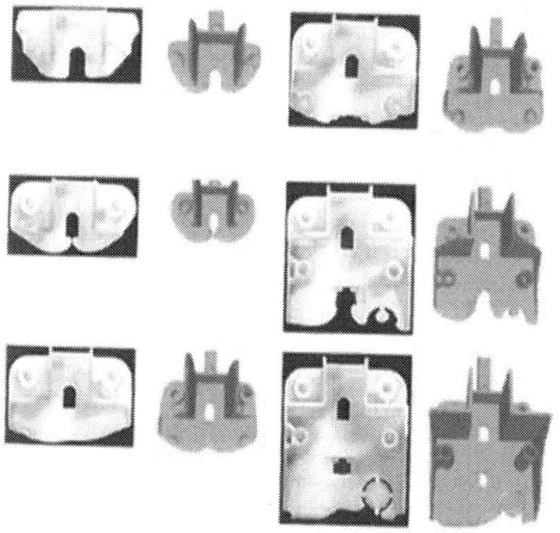

Figure 6. Computed (on the right in each column) and experimental (on the left in each column) short shots for the filling of the cavity presented Figure 2 (Silva et al.[11])

Nowadays the question in injection molding is not only to fill the mold in appropriate conditions, but also to predict the shape and the properties of the produced plastic part that is to say to master:

- The packing stage, which consists in feeding the cavity already filled with additional molten polymer to compensate for the variation in specific volume due to temperature and pressure decrease. This requires introducing relevant data for the dependence of the density as a function of temperature and pressure. Measuring and modeling rheology around the Glass transition temperature for amorphous polymer and in the crystallization range for semi-crystalline polymers remains a challenge and this explains using a mysterious "no flow temperature" in commercial software which is just a fitting parameter. The early models have been applied to simple plaque or disk geometries (Kamal and Kenig [21]; Huilier et al.[22]; Titomanlio et al.[23]) then generalized to Hele-Shaw models (Chiang et al.[24]) and to 3D models (Silva et al.[11]).
- The solidification stage, which starts during the filling phase and continues during the packing phase for semi-crystalline polymers, requires measuring the crystallization temperature as a function of the cooling rate (which is obviously very high near the mold wall) and the flow conditions. During this phase, residual stresses develop and lead to the deformation of the plastic part at the mold's opening and thus result in final dimensions being different from those of the mold cavity. Purely elastic models have been first developed (Titomanlio et al.[25]; Denizart et al.[26]; Boitout et al.[27]; Farhoudi and Kamal [28]) followed by more sophisticated solid viscoelastic models (Douven et al.[29]; Kabanemi et al.[30]; Kamal et al.[31]). The accurate prediction of residual stresses and strains remains a challenge.

Figure 7 compares the calculated and the measured pressure in the mold during the whole injection cycle, including filling, packing and cooling.

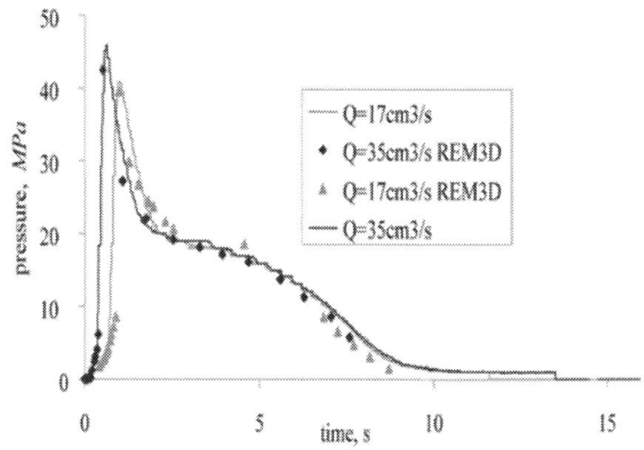

Figure 7. Pressure as a function of time in the cavity during the entire injection cycle for two different flow rates, the polymer is a polystyrene (Silva et al.[11]).

4.3 Blow moulding

The blow-molding process consists of extruding a vertical tube, the parison, pinching it at its base within a cooled mold and then blowing air inside to stretch the hot tube along the cold mold cavity where it is solidified, then the mold opens, the part is extracted and a new cycle can begin. When manufacturing small hollow bodies, these operations are performed at high speed in a multi-station mold rotating below the tube die. For bigger parts (automotive tanks, for example), the extruder continuously fills up a pot from which the molten polymer is discharged at regular intervals through a tube die, and the blowing operation is then performed more slowly and in a single mold . The challenge is to master the thickness distribution of the final hollow body and so to adjust the initial thickness of the tube by "parison programming" (Langkamp and Michaeli [32]) Most mathematical models use a membrane approach and hyperelastic constitutive equations (De Lorenzi and Nied [33]) which are suitable for the stretch blow-molding of PET bottles in the rubbery plateau temperature. For the blow molding of Polyethylene which occurs in the molten state, a viscoelastic Maxwell constitutive equation has been used (Rodriguez-Villa et al.[33]).

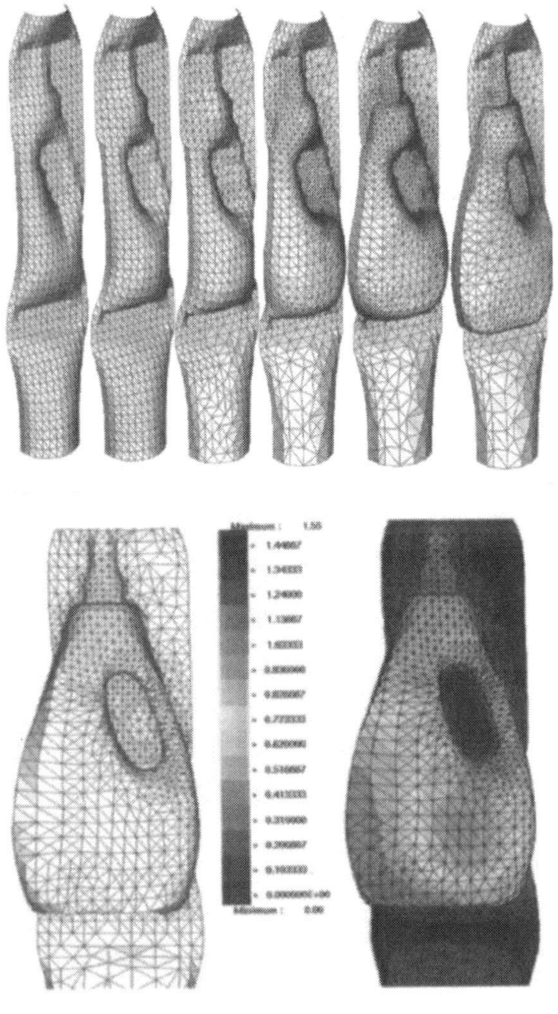

Figure 8. (top) Shape and meshing of the membrane at different steps of simulation of the extrusion; (bottom) Final mesh and thickness distribution in the bottle (Bellet et al.[34])

Successive simulated deformation of the parison during the blowing process are shown in figure 8a as well as the corresponding refinement and coarsening of the mesh surface. This allows mastering the computation time while having a good accuracy in areas that are heavily deformed and/or in the vicinity of the contact with the mold.

Figure 8b presents the final thickness distribution of the bottle: significant over-thicknesses are observed at the inner portion of the handle that will be removed later, but also at the middle of the hollow body, which has no justification in terms of mechanical strength. Instead the lower left portion of the hollow body has an insufficient thickness that can cause the bottle breakage in case of falling. The modelling therefore allows optimization of the parison shape and, to a lesser extent, the blowing parameters in order to obtain a more appropriate bottle thickness distribution.

5 Conclusions

Currently it is now possible to use numerical software to address many polymer processes and there is an increasing trend to apply a "black box" approach to using this software. This may be appropriate for some situations, however accurately modelling commercial polymer processes remains both challenging and still often requires making "intelligent" simplifying assumptions and approximations. In the future computing power and advanced numerical techniques will help reduce computation times and add to the precision of numerical solutions, however there is still value in making geometric simplifications, realistic rheology assessments and developing a good physical understanding of a process before and during embarking on a "black box" simulation.

At present there is no universally accepted constitutive equation that describes all polymer single phase melt behaviour and increasingly more complex polymer compositions are being used with ever more complex constitutive behaviour. Numerical simulations are in general still unable to predict the onset or form of most extrusion instabilities. There therefore remains much to be done both in terms of constitutive equation development and modelling rheologically difficult, but important polymeric materials.

When a non isothermal phase transition occurs from solid to melt or melt to solid, the problem becomes more complex. In particular the transition from solid state polymer pellet or powder to a full melt still offers many challenges in relation to accurate modelling as too does the transition from a melt to a final solid product. If the polymer crystallises during solidification there are even greater difficulties and of course flow can also influence both kinetics and the way the polymer crystallises with quite dramatic effects on final product properties (see for example Roozemond and Peters [35]).

The numerical simulation of polymer processing has now reached the stage where it is used successfully as a genuine design tool for the optimisation of existing processes and in the future it will be able to help develop new processes that will evolve or be invented. The future challenge is to model the macromolecule orientation (amorphous polymer) the structure development (semi-crystalline polymers, size and shape of spherulites, shish-kebab....) and so to predict the mechanical properties of the produced part (elastic modulus but also yield stress, fatigue behavior...).

References

1. J.F.Agassant, P. Avenas., P.J. Carreau, J.P. Sergent, Polymer Processing, Principles and Modeling, Hanser, Munich (1991).
2. J.F.Agassant, M.R.Mackley, Intern. Polym. Proc., 30:121-140 (2015).
3. Z. Tadmor, C.G. Gogos, Principles of Polymer Processing, 2nd edition, Wiley, New-York (2006).
4. 4. H.S. Hele-Shaw, Notes on Proceedings of the Royal Institution of Great Britain, 16: 49-64 (1899).
5. 5. A. Cameron , Principles of lubrication, Longmans Green, London (1966).
6. 6. H.C. Brinkmann , Appl. Sci. Research, 2: 120-124 (1951).
7. 7. T. Coupez, J.Comp.Phys., 230: 2391-2405 (2011).
8. 8. D. Arnold, F. Brezzi F.,M. Fortin, Calcolo, 23: 337-344 (1984).
9. 9. A.N. Brooks, T.J.R. Hughes., Computer Methods in Applied Mechanics and Engineering, 32: 199-259 (1982)
10. 10. L. Silva, R. Valette, P. Laure, T. Coupez, International journal of material forming, 5: 55-72 (2012).
11. 11. L. Silva, J.F. Agassant, T. Coupez, Three dimensional injection molding simulation, In: Injection molding. Technology and fundamentals, M.R. Kamal, A.Isayev, S.J. Liu editors, Hanser, Munich, 599-651 (2009).
12. 12. B. Vergnes, P. Saillard, J.F. Agassant., Polym. Eng. Sci., 24: 980-987 (1984).
13. 13. B. Vergnes, J.F. Agassant, Modélisation des écoulements dans les filières d'extrusion, Techniques de l'Ingénieur, AM 3 655, (2008).
14. 14. M.R. Kamal, Y. Kuo, P.H. Doan, Polym. Eng. Sci., 15: 863–868 (1975).
15. 15. C.A. Hieber, S.F. Shen, J. Non-Newt. Fluid Mech., 7: 1-32 (1980).
16. 16. V.W. Wang, C.A. Hieber, K.K. Wang, J. Polym. Eng., 7: 21-45 (1986).
17. 17. J.W. Castro, C.W. Macosko, AICHE J., 28: 250-260 (1982).
18. 18. H. Mavridis, A.N. Hrymak, J. Vlachopoulos, Polym. Eng. Sci., 26, 449-454 (1986).
19. 19. G.A.A.V. Haagh, G.W.M. Peters, F.N. Van De Vosse, H.E.H. Meijer, Polym. Eng. Sci., 41: 449- 461 (2001).
20. 20. L. Silva, C. Gruau, J.F. Agassant, T. Coupez, J. Mauffrey, Intern.Polym. Proc., 20: 265-273 (2005).
21. 21. M.R. Kamal, S. Kenig, Polym. Eng. Sci., 12: 294-301 (1972).
22. 22. D. Huilier, J. Terrisse, M.E. de la Lande, A. Latrobe, Intern. Polym. Proc., 4: 184-190 (1988).
23. 23. G. Titomanlio, S. Piccarolo, G.J. Levati, Appl. Polym. Sci., 35:1483-1495 (1988).
24. 24. H.H. Chiang, C.A. Hieber, K.K. Wang, Polym. Eng. Sci., 31: 116-124 (1991).
25. 25. G. Titomanlio, V. Brucato, M.R. Kamal, Intern. Polym. Proc., 2: 55-59 (1987).
26. 26. O. Denizart, M. Vincent, J.F. Agassant, J.Mat.Sci., 30: 552-560 (1995).
27. 27. F. Boitout, J.F. Agassant, M. Vincent, Intern. Polym. Proc., 3: 237-242 (1995).
28. 28. Y. Farhoudi, M.R. Kamal, Int. J. Forming Proc., 2: 277-306 (1999).
29. 29. L.F.A. Douven,F.T.P. Baaijen, H.E.H. Meijer, Prog. Polym. Sci., 20 : 403-457 (1995).
30. 30. K.K. Kabanemi, H.Vaillancourt, H. Wang, G. Salloum, Polym. Eng. Sci., 38: 21-37 (1998).
31. 31. M.R. Kamal, R.A. Lai-Fook, J.R. Hernandez-Aguilar, Polym. Eng. Sci., 42: 1098-1114 (2002).
32. 32. U. Langkamp, W. Michaeli, Plast. Eng., 52: 29-31 (1996).
33. 33. H.G. De Lorenzi, H.F. Nied, Comp. Struct., 26: 197-206 (1987).
34. 34. A. Rodriguez-Villa, J.F. Agassant, M. Bellet, Finite element simulation of the extrusion blow-molding process, in : Simulation of material processing, theory, methods and applications, S.D. Shen and P. Dawson editors, Balkerma, Amsterdam, 1053-1058 (1995).
35. 35. M. Bellet, B. Monasse, J.F. Agassant, Simulation numérique des procédés de soufflage, Techniques de l'Ingénieur, AM 3705 (2000).
36. 36. P.C. Roozemond, , G.W.M. Peters., J. Rheology, 57:1633-1652 (2013)

MATEC Web of Conferences **80**, 11003 (2016) DOI: 10.1051/matecconf/20168011003
NUMIFORM 2016

Significance of the local sheet curvature in the prediction of sheet metal forming limits by necking instabilities and cracks

Pavel Hora[1,a], Longchang Tong [1], Maysam Gorji, Niko Manopulo[1] and Bekim Berisha[1]

[1] *ETH Zurich, Institute of virtual Manufacturing, Tannenstrasse 3, 8092, Zurich, Switzerland*

Abstract. The industrial based prediction in sheet metal forming bases still on the Forming Limit Diagrams (FLD) as formally proposed by Keeler 1. The FLD are commonly specified by the Nakajima tests and evaluated with the so called cross section method. Although widely used, the FLC concept has numerous serious limitations. In the paper the influences of bending on the FLC as well as the later crack limits will be discussed. Both criteria will be combined to an extended FLC concept (X-FLC). The new concept demonstrates that the Nakajima tests are not only appropriate for the evaluation of the necking instability but for the detection of the real crack strains too. For the evaluation of the crack strains a new local thinning method is proposed and tested for special 6xxx Al-alloys.

1 Introduction

Nowadays forming limits in sheet metal forming are mostly predicted based on the necking initiation. This limit will be usually evaluated with the Nakajima test. The FLC's are in this way only valid for linear strain path and for negligible curvature radii.

It is well known, that under conditions like hemming or in cases of deep drawing (DD) with small radii fracture limits arise, which needs different experimental as well as theoretical approaches.

Well known are the investigations of Hasek 3 who demonstrated the influence of different punch geometries on the position of the FLC, 0.

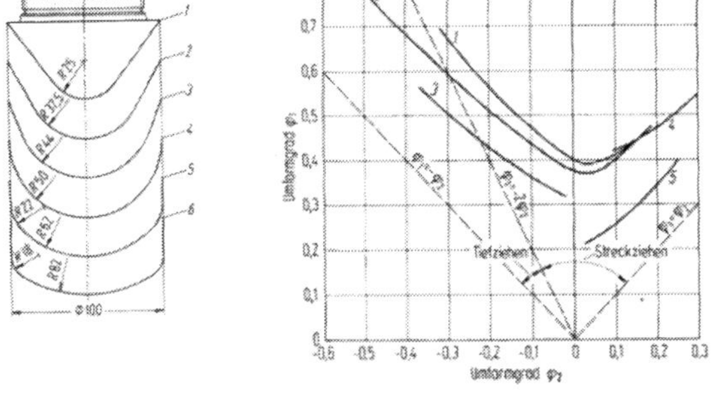

Figure 1 Influence of the curvature on the position of the FLC, left: used punch geometries. Hasek 3

Theoretical models like the M-K Model 2 do not consider the bending influence at all.

2 Processes with bending impact on formability

That the influence of the curvature is of industrial relevance can be demonstrated on different examples.

2.1 Influence of bending effects on deep drawing behaviour

0shows a deep drawing example with a die radius of r=3.0 mm for the material AA6016. The DD process was done with different square blanks and different blank holder forces.

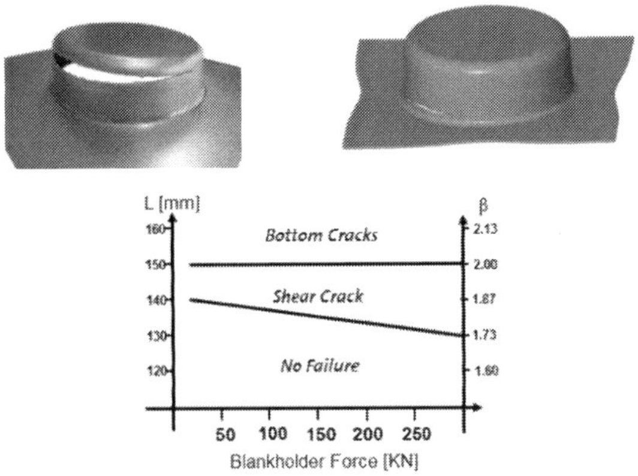

Figure 2 Formability diagram of AA6016 sheet sample, Gorji 17

[a] Corresponding author: phora@ivp.mavt.ethz.ch

Dependent on the selected parameter combination the localized necking occurs as classical bottom neck or as a upper radius crack. By increasing the die radius to r=5.0mm the upper neck disappears.

Gorji 17 demonstrated that a FEM simulation based on shell elements and classical FLC is not able to predict this influence correctly.

2.2 Prediction of DD behaviour of multilayer materials

A more complex behaviour is given when multi-layer materials like the multilayer composite FUSION™ [Novelis] is applied, 0.

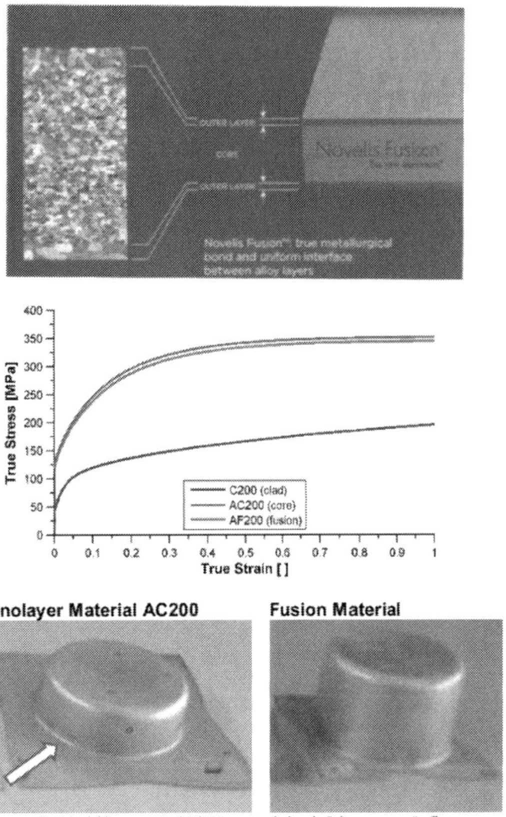

Figure 3 Multilayer FUSION material, yield curves. Influence on the forming behaviour. Gorji 17

The material is composed of a soft AA5005 alloy outside (clad) and a hard AA6016 alloy inside (core). 0bottom demonstrates again clearly how significantly the DD behaviour changes. Again a correct prediction cannot be done on the base of the classical FLC-method.

As reasons for the increase of the strain limits the change of the strain distribution in the sheet plane as well as strain gradients over the thickness can be seen as the influencing effects.

The thickness strain gradients depend on the ratio χ=t/r. In the current DD examples with the die radius of 3.0 mm the critical ratio was about 0.3.

2.3 Prediction of crack limits for hemming operations

From many former investigations it is well known, that in pure bending – corresponding to hemming operations - the fracture strains lies even above the crack limits for flat sheets.

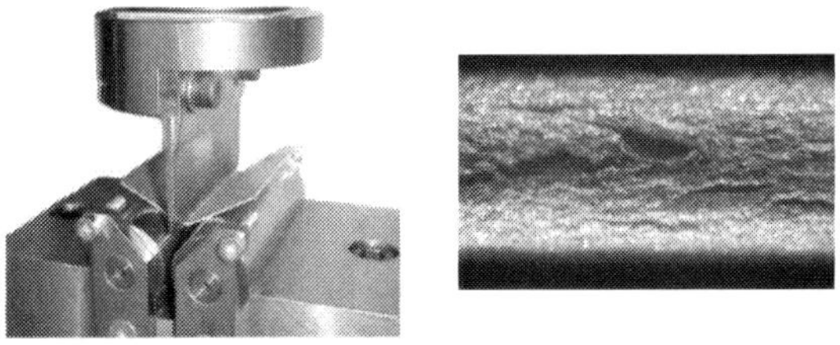

Figure 4 Evaluation of the fracture strain by bending tests. Occurrence of edge cracks (right). Gorji 17

0 demonstrates the bending results for the alloy AA6016 under different pre-strain conditions. It is worth to mention that the critical strain of the pure bending test surmount even the crack strains of the quasi flat Nakajima tests with $\chi = 1/50 = 0.02$ by nearly 20%.

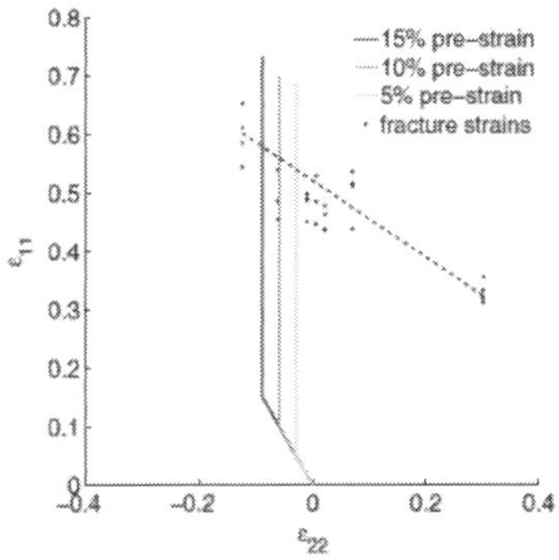

Figure 5 Fracture points determined by the Nakajima tests compared with the strains in bending at the critical state for AA6016. Gorji 16

3 Prediction of FLC's under bending influences

The already mentioned investigations by Hasek, 0predicts higher strains with increasing curvatures of the punches. This effect can be caused by stabilizing effects in the global strain distribution as well as by strain gradients in the thickness direction.

3.1 MMFC criterion under consideration of strain rate and bending effects

A basic description of the bending effect was formulated in the framework of the MMFC criterions by Hora et al.,5 to 12 and 18.

The criterion can be expressed mathematically as

$$\frac{\partial\sigma_{11}}{\partial\varepsilon_{11}}+\frac{\partial\sigma_{11}}{\partial\beta}\cdot\frac{\partial\beta}{\partial\varepsilon_{11}}+\frac{\partial\sigma_{11}}{\partial\dot{\varepsilon}}\cdot\frac{\partial\dot{\varepsilon}}{\partial\varepsilon_{11}}\leq\sigma_{11}\ , \tag{1}$$

where $\beta=\Delta\varepsilon_{22}/\Delta\varepsilon_{11}$ denotes the strain increment ratio. Similarly, the stress ratio α is defined as $\alpha=\sigma_{22}/\sigma_{11}$. Despite of different descriptions of yield loci , the following relations can be generally established with functions:

$$\sigma_{11}=f(\alpha)\bar{\sigma}\ \text{, and} \qquad \Delta\bar{\varepsilon}=g(\beta)\Delta\varepsilon_{11}. \tag{2}$$

where $\bar{\sigma}$ and $\Delta\bar{\varepsilon}$ are the equivalent stress and the equivalent strain increment respectively.

With the definitions above, the description of the modified maximum force criterion takes the general form as

$$H'\leq\frac{f(\alpha)+\dfrac{f'(\alpha)g(\beta)\beta}{\beta'(\alpha)\bar{\varepsilon}}}{f(\alpha)g(\beta)}H \tag{3}$$

where $H=f(\varepsilon_v)$ and $H'=d\bar{\sigma}/d\bar{\varepsilon}$ are the functions for the description of hardening behaviours of the sheet materials.

3.1.1 Influence of curvature

If significant bending effects occurs the strain distribution is no more constant over the thickness. This effect was included into MMFC in following simplified way 4.

Setting the bottom layer as reference the strain distribution over the thickness can be expressed with a linear function:

$$\Delta\varepsilon_{11}{}^{\eta}=\Delta\varepsilon_{11}{}^{0}(1+\frac{\eta}{\rho_0}) \tag{4}$$

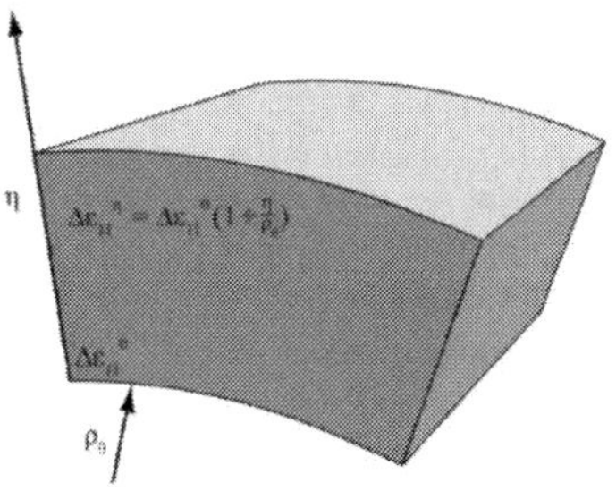

Figure 6 Strain distribution of a shell under bending conditiones

Substitute the relation into MMFC, the stress increment is expressed as:

$$d\sigma_{11}=\frac{\partial\sigma_{11}{}^{\eta}}{\partial\varepsilon_{11}{}^{\eta}}d\varepsilon_{11}{}^{\eta}+\frac{\partial\sigma_{11}}{\partial\beta}d\beta+\frac{\partial\sigma_{11}}{\partial\dot{\varepsilon}}d\dot{\varepsilon} \tag{5}$$

Substitute (4) into (5) and make the approximation that

$$\frac{\partial\sigma_{11}{}^{\eta}}{\partial\varepsilon_{11}{}^{\eta}}\approx\frac{\partial\sigma_{11}}{\partial\varepsilon_{11}}$$

(5) can be transformed into:

$$d\sigma_{11}=\frac{\partial\sigma_{11}}{\partial\varepsilon_{11}}d\varepsilon_{11}(1+\frac{\eta}{\rho_0})+\frac{\partial\sigma_{11}}{\partial\beta}d\beta+\frac{\partial\sigma_{11}}{\partial\dot{\varepsilon}}d\dot{\varepsilon} \tag{6}$$

Rewrite the original maximum force equation $d\sigma_{11}A+\sigma_{11}dA=0$ into the integral form

$$\frac{\partial\sigma_{11}}{\partial\varepsilon_{11}}d\varepsilon_{11}\int_0^t(1+\frac{\eta}{\rho_0})Bd\eta+(\frac{\partial\sigma_{11}}{\partial\beta}d\beta+\frac{\partial\sigma_{11}}{\partial\dot{\varepsilon}}d\dot{\varepsilon})A+\sigma_{11}dA=0 \tag{7}$$

where $A=Bt$ is the current section area of the sheet. The final form of MMFC with the consideration of curvature and strain rate effects is obtained as:

$$\frac{\partial\sigma_{11}}{\partial\varepsilon_{11}}(1+\frac{t}{2\rho_0})+\frac{\partial\sigma_{11}}{\partial\beta}\cdot\frac{\partial\beta}{\partial\varepsilon_{11}}+\frac{\partial\sigma_{11}}{\partial\dot{\varepsilon}}\cdot\frac{\partial\dot{\varepsilon}}{\partial\varepsilon_{11}}\leq\sigma_{11}\ . \tag{8}$$

The influence of the curvature is shown later in 0.

3.1.2 Influence of strain rate

It is well-known that steel materials in contrast to Al-alloys have a positive strain rate dependency. The positive strain rate effects significantly slows down the necking and leads consequently to higher FLC values.

The description of strain rate effect in the framework of the analytical MMFC criterion – which bases on material point evaluations – requires a functional description of the strain rate increase in function of the equivalent strain, eq. (11).

A preliminary approach of the strain rate function can be defined by the 1D tensile conditions

$$\begin{aligned}\varepsilon_{11}\leq n:\ \ \dot{\varepsilon}&=\dot{\varepsilon}_{hom}\approx v/L\\ \varepsilon_{11}=2n:\ \ \dot{\varepsilon}&=\dot{\varepsilon}_{loc}\approx v/2t\end{aligned} \tag{9}$$

A more general material specific description of the strain rate evolution by the terms of eq. (11) can be achieved by mapping the localization strain rate of the Nakajima tests. 0shows the history of the strain rate for different Nakajima specimens for the material DC05.

 DOI: 10.1051/matecconf/20168011003

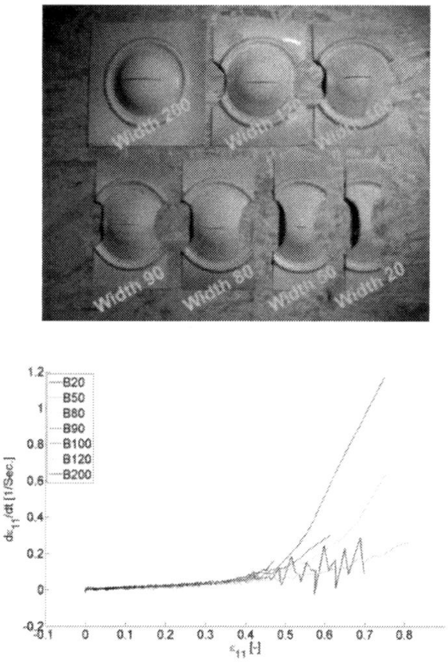

Figure 7 Increase of the strain rate for different shapes of Nakajima specimens. Material DC05

In a very approximative way the strain rate behaviour can mapped with the function

$$\dot{\varepsilon}_{11} = \dot{\varepsilon}_{11}^{hom} + A(\beta)*[(\varepsilon_{11} - \varepsilon_{11}^{uni}) / \varepsilon_{11}^{uni}]^{p} \quad (10)$$

Due to the fact, that the strain rate evolution varies very strongly for the different sizes of the Nakajima specimens A(β) was fitted based on the measured strain rates of 0and is given in 0. Parameter p was assumed as p=2.

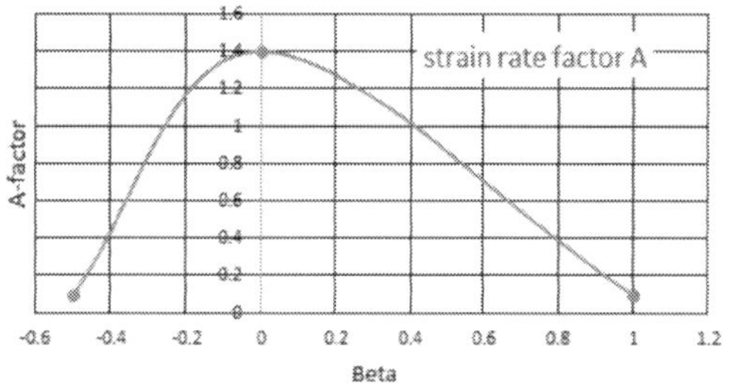

Figure 8 Distribution of the strain rate factor A(β) in eq. (10) for a DC05 material.

In further investigations it shall be verified, if the same A(β)-Function can be used for specific materials groups.

3.1.3 Validation for DC05

The numerical evaluation of the MMFC, eq. (8), bases on an incremental calculation. The strain rate influence can then be evaluated directly based on the both dependencies

$$\dot{\varepsilon}_{eq} = g(\varepsilon_{eq}) \quad (11)$$

$$\sigma_y = f(\varepsilon_{eq}, \dot{\varepsilon}_{eq}) \quad (12)$$

Material DC05 14:

$$\sigma_y = \left[670.4*(0.01+\varphi)^{0.174} - 146\right](\dot{\varepsilon} + 0.001)^{0.05} \quad (13)$$

R-values:

	R_0	R_{45}	R_{90}	R_b
DC05	2.1	1.5	2.5	0.85

Yield stresses:

	Y_0	Y_{45}	Y_{90}	Y_b
DC05	154.4	158.9	156	192

The FLCs evaluated for the material DC05 are shown in 0.

0top shows the experimental values of DCO5. Under the consideration of the strain rate influence as well as without it are visualized. 0bottom demonstrates the influence of the curvature as predicted with the MMFC approach. The different curves shows that an increase of FLC due to curvature becomes significant only for relatively large curvatures. For $\chi \leq 0.1$ the effect can be neglected

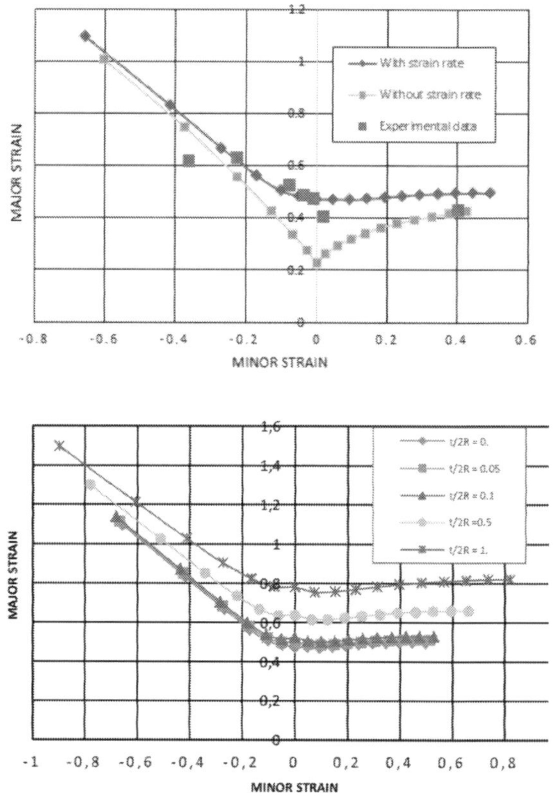

Figure 9 MMFC prediction of FLC's for a DC05 material. Top: FLC prediction with and without strain rate influence. Bottom: Influence of curvatures

4. Prediction of forming limits by cracks with and without bending effects

4.1 Experimental detection of crack limits based on Nakajima tests

As demonstrated with the bending test and also related to other DD test with overlaid bending effects, not only the onset of necking instability but also the final crack have to be detected.

For the experimental detection of crack strains for sheet recent investigations by Wierzbicki, Bao, Bai and others 15 found a large interest. The authors proposed up to 15 different specimens to evaluate the so called triaxiality diagrams. As shown in 0.the proposed specimens have a very complex geometry which are difficult to be realized with sheet materials.

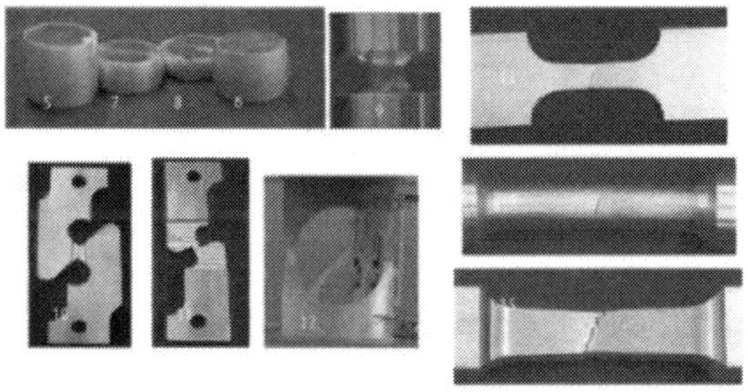

Figure 10 Proposed specimens for the evaluation of the triaxiality diagrams, 15

An alternative procedure was recently proposed by Gorji et al. 19. He uses a so called thinning method, where he evaluates with micrographs the crack thinning strain directly on broken Nakajima specimens, 0

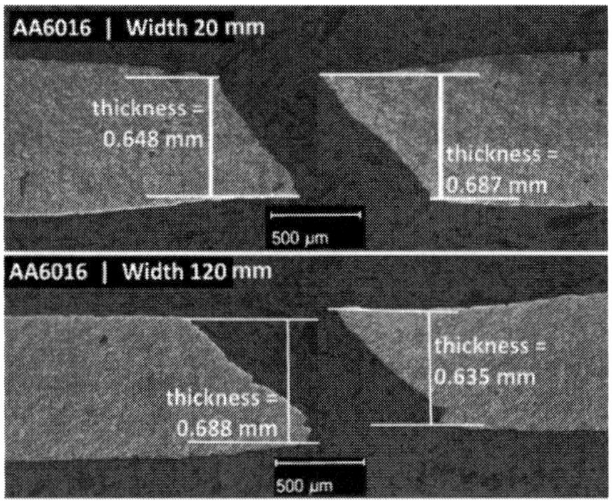

Figure 11 Evaluation of the fracture strain by the local detection of fracture thinning on the Nakajima specimens. Gorji 16,17

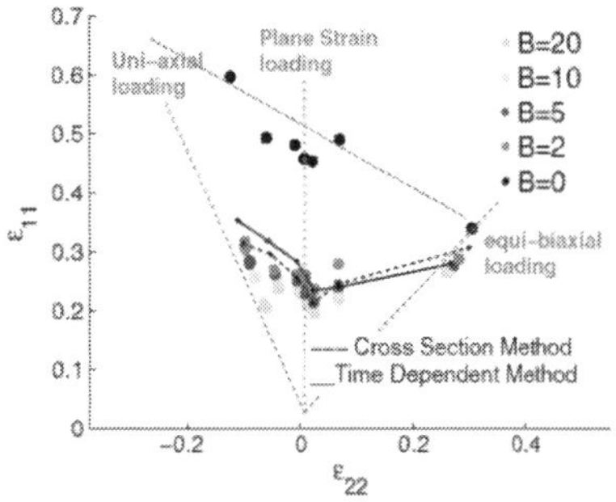

Figure 12 Evaluation of the fracture strain by the local detection of fracture thinning on the Nakajima specimens and comparison with the FLC. Gorji 16,17

4.2 Experimental detection of crack limits based on Deep Drawing tests

The Nakajima based test are restricted to the stress range

$$0 \le \alpha = \frac{\sigma_2}{\sigma_1} \le 1.0$$

In many deep drawing applications the largest strains occur on the left side of the FLC. For those reasons a special DD test with a quadratic blank and a relatively small die curvature of r=3.0 mm was applied to get an additional "point" specifically in the deep drawing (compression-tension combination) range. The experiment with the corresponding strain distribution at the crack time step shows 0

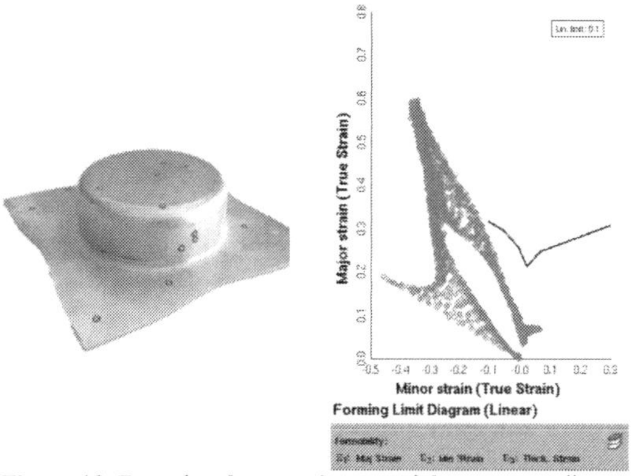

Figure 13 Deep drawing experiment and the corresponding strain distribution at the crack time step. Gorji 17

The combination of the Nakajima fracture strains with the additional DD fracture strains can be used as data base for the determination of a generalized fracture line.

4.3 Comparison with different failure criteria

The so experimentally predefined strain have been compared with the theoretical limits of different failure criteria. 0shows the fits of the theoretical fitted Failure curves with the experimental points. Remarkable is the fact, that all criterions fits the position of the measures points quite well, but that the criterions deviate significantly in the left "deep drawing" range. Especially this range influences strongly the virtual results for deep drawing operations.

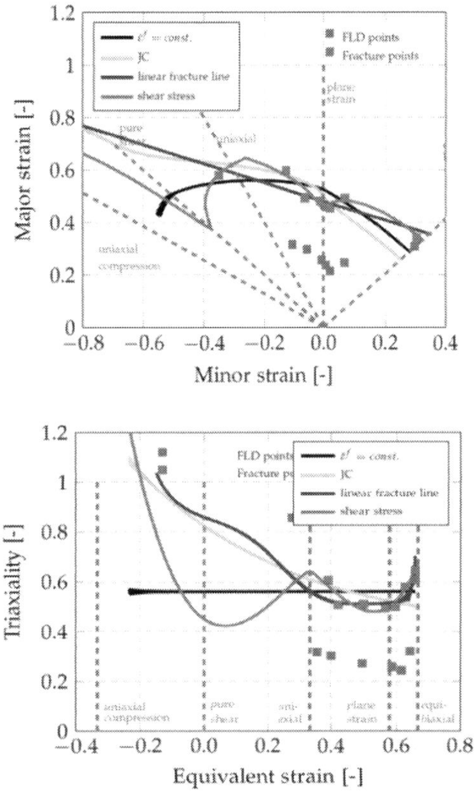

Figure 14 Comparison of different fracture criteria in principal strain space and in triaxiality-equivalent strain space, 17

Gorji et al. 17 demonstrated on comparisons with real parts that the Linear Fracture Line describes the behaviour in the most accurate way.

5. Multilayer based consideration of the failure prediction

5.1 Layer based failure prediction with shell elements

The classical failure predictions bases on a mono-layer FLC prediction. Typically the range above the FLC will be – from the real material perspective incorrectly – considered as crack region.

If the fracture is initiated by a surface crack, as it is by small t/r ratios, an extended X-FLC method has to be applied. The crack develops and initiates only when the outer layer reached the crack limit

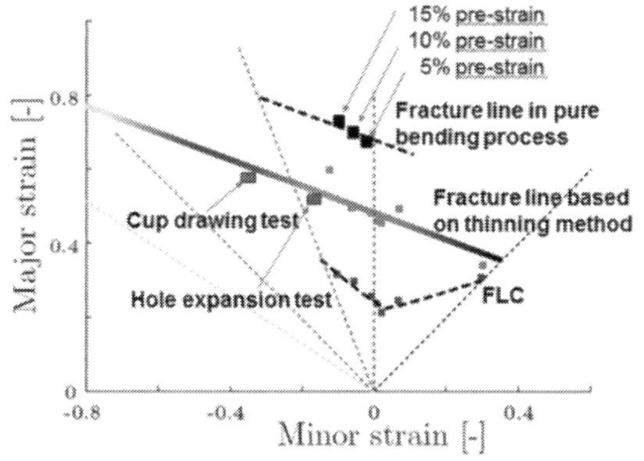

Figure 15 Extended X-FLC description considering the fracture ("crack") limits by the bending strains

5.2 Prediction of crack limits for parts with small die radii and a monolayer structure

For the validation of the above implementation a so called "triangle" test part was used.

Fig. 16 shows the crack position on the real part. The critical depth was at H ≈ 43 mm

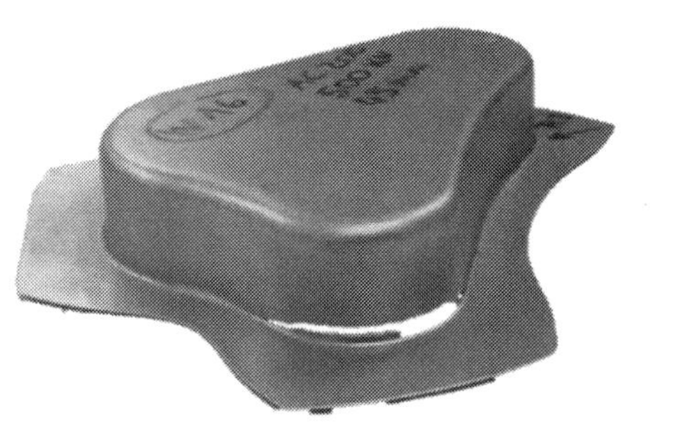

Figure 16 Triangle part. Monolayer material AA6016. Critical depth H = 43 mm

In the case of a simple "monolayer" FLC evaluation the predicted forming depth lies at H= 29 mm. It is obvious that not only the predicted part depth of 29 mm instead of 43 mm but also the position of the crack are not correctly predicted.

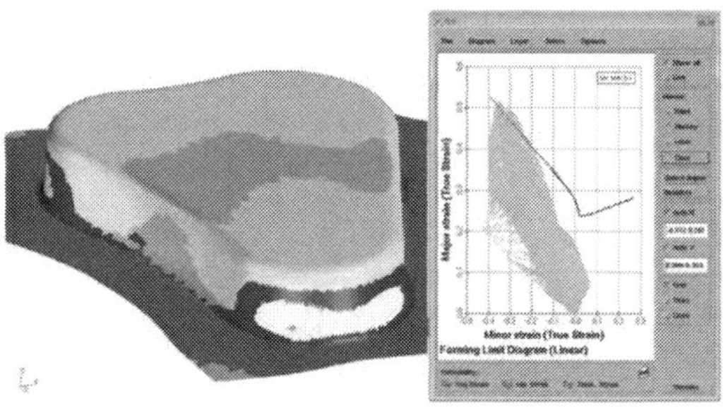

Figure 17 Triangle part. Monolayer material AA6016. Prediction of critical depth based on a monolayer FLC: H = 29 mm

With the alternative method - applying the extended X-FLC, 0- the failure can be identified correctly, 0

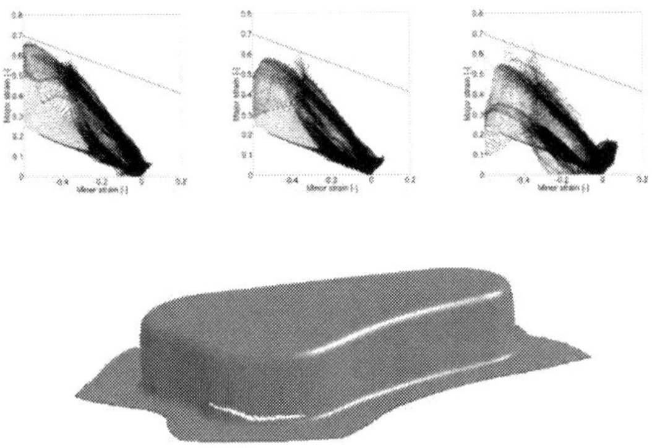

Figure 18 Material AA6016. FE Simulation of triangular experiment with Linear fracture criterion (strain distribution of lower, middle and upper-layer)

5.3 Prediction of crack limits for multilayer FUSION materials

If the monolayer material is replaced by the FUSION multilayer material as introduced in 0, the failure can be completely avoided.

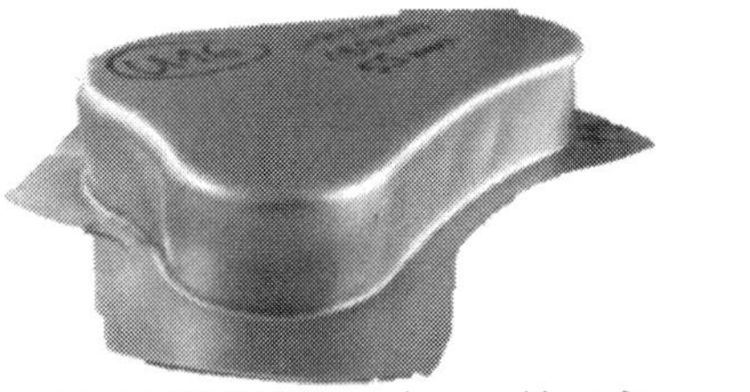

Figure 19 Material FUSION. Experiment with no fracture

For the simulation of the FUSION behaviour the X-FLC diagrams have to be specified separately for the core and clad material. Interesting in this case is that the FLC's are the same for the soft (clad) and the hard (core) materials but the fracture lines reach strongly different levels, 0.

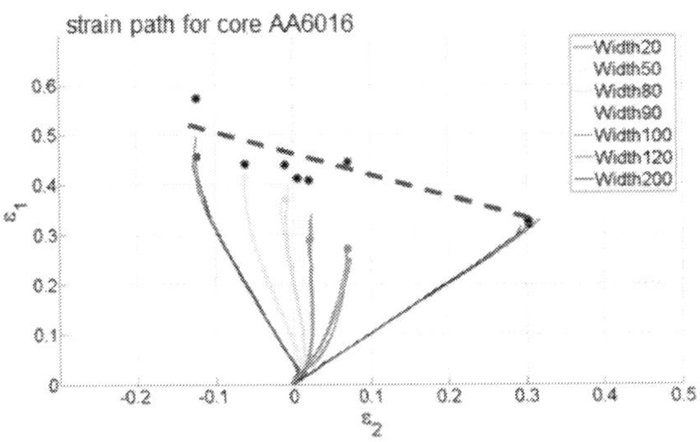

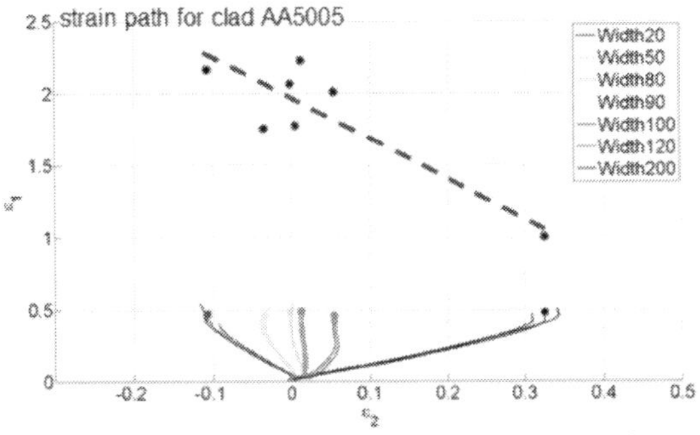

Figure 20 Comparing of strain history and fracture line of core and clad material.

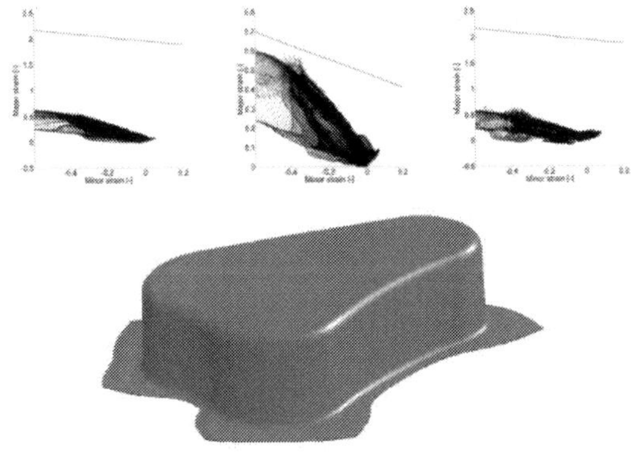

Figure 21 Forming behaviour of FUSION material: FE simulation of triangular experiment with linear fracture criterion specifically defined for the layers (strain distribution of lower, middle and upper-layer), Gorji 16.

5.4 Prediction of crack limits for hemming operations

The experimental detected crack limits based on the Nakajima test corresponds to the curvature of $\chi = t / r = 1.0 / 50 = 0.02$. In this case the curvature can be practically neglected. In cases like the pure bending with $\chi = t / r = 1.0 / 1.0 = 1$. the conditions are strongly different.

5.4.1 Dependency of crack limits on curvature

As demonstrated in 0experimentally evaluated strains lie above the fracture strain detected by the nearly flat Nakajima (χ=0.02) test.

Simulation of bending limits based on the previously determined X-FLC (Nakajima-cup limits) leads to a weak underestimate of the critical closing angle (92° instead of 74°)

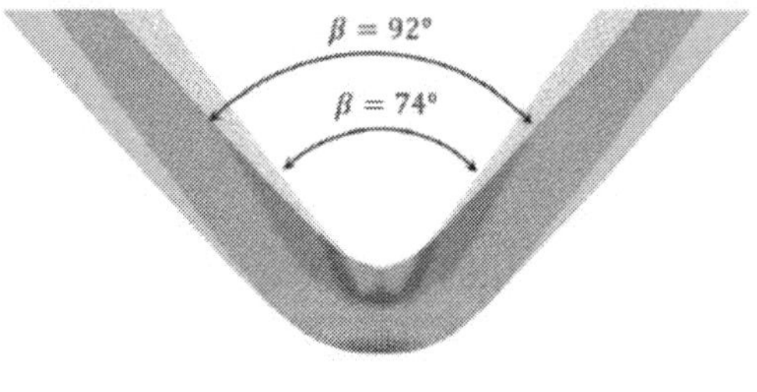

Figure 22 Evaluation of the fracture strain by the local detection of fracture thinning on the Nakajima specimens, Gorji 16.

For a better approach Hora and Gorji proposed a following correction. Correlation between membrane and bending fracture can be modelled by following relationship:

$$\bar{\varepsilon}^{f^{Bending}} = \bar{\varepsilon}^{f^{Membrane}} + c\chi$$

where $\chi = t / \rho$. Parameter c can be defined based on the linear graph in 0

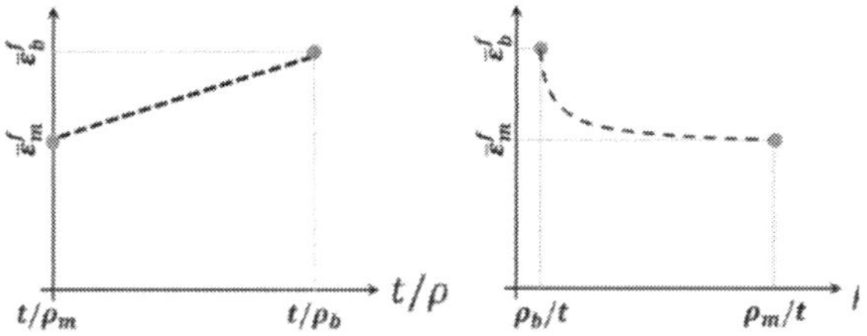

Figure 23 Correlation between membrane and bending fracture; ρ_m and ρ_b are punch radii in the Nakajima and 3-point bending experiment, respectively, Gorji 16.

Conclusions

The contribution demonstrates on selected examples that the classical FLC prediction is not applicable if the parts have either small die radii or are composed by layers with different properties.

In this case beside the FLC a crack limit curve has to be specified too. For the detection of such critical strain the recently presented thinning method, evaluating the fracture strains on Nakajima specimens, have been used.

The multi-layer failure identification was implemented in to the explicit FEM code LS-Dyna.

Acknowledgements

The authors are very grateful to the CTI (The Swiss Innovation Promotion Agency) for the financial support of part of this work within the project 13082.1 PFIW-IW. Also thanks to J. Timm and E. Combaz (Novelis Switzerland SA) for providing the material and performing a part of the experiments. K. Wiegand (Daimler AG), M. Selig (AutoForm Development GmbH and E. Müller (GOM International AG) are also gratefully acknowledged for their contribution on the project results.

References

1. S. P. Keeler, Soc. of Automotive Engineering, Nr. 650 535, pp. 1-9, 1965.
2. Z. Marciniak and K. Kuczynski, Int. J. Mech. Sci., 9, pp. 609-620, 1967.
3. V.Hasek: Buch Lange Umfortmtech. Bd. 3, p.51-57
4. P. Hora et al: IFU-Krupp, Teilbericht II, 1993.
5. P. Hora et al:, IDDRG, 1994.
6. P. Hora, et al.:, NUMISHEET, Dearborn, MI, USA, pp. 252-256, 1996.
7. Tong, et al.:, NUMISHEET, Jeju, Korea, pp. 623-627, 2002.
8. P. Hora et al.: Plasticity, Quebec, Canada, pp. 364-366, 2003.
9. P.Hora, M. Merklein, L. Tong, J. Lechler:. Proc of IDDRG 2007
10. J. Krauer et al.: NumiForm Porto, Portugal, pp. 1263-1268, 2007.
11. P. Hora, T. Longchang:. Numisheet 2008
12. Numisheet Benchmark 1: Numisheet, Interlaken, Switzerland, 2008.
13. S.P. Keeler: pp. 4. 2.12 ,1989.
14. P.Peters: Diss. ETH 22707 (2015).
15. Y.Bai, T.Wierzbicki:. Int J Fract (2010) 161:1–20
16. M.Gorji, B. Berisha, P.Hora, F. Barlat: Int.J. of Mater. Form., DOI 10.1007/s12289-015-1242-y, 2015.
17. M.Gorji:. Diss ETH 23066, 2015
18. N. Manopulo, P. Hora, P. Peters, M. Gorji, F.Barlat: Int. J. Plasticity,75, 189-203, 2015
19. M. Gorji, B. Berisha, N. Manopulo, P. Hora. J Materials Processing Technology, 232: 19-33, 2016

Support Tool for Anchoring System Optimization of Titanium Craniofacial Prostheses

Claudio Ciancio[1], Maria Vittoria Caruso[1,2,a], Gionata Fragomeni[2] and Giuseppina Ambrogio[1]

[1]*Department of Mechanical, Energetic and Management Engineering (DIMEG), University of Calabria, 87036 Rende, Italy*
[2]*Department of Medical and Surgical Science, University of Catanzaro, 88100 Catanzaro, Italy*

Abstract. Titanium prostheses are artificial components used during cranioplasty, which can be made with the Incremental Sheet Forming (ISF). To repair the craniofacial defects, the implant must be fixed to the skull, so an adequate anchoring system is necessary. Because there are no specific guidelines regarding this system, the aim of this study was to present a methodology for the identification of the optimal screw shank diameter and the overlap length, considering different damage areas and accidental external loads. In this method, a numerical model for structural analysis, a statistical approach and a metamodeling technique were coupled. The results demonstrate that this methodology can be used as a decision support tool in the titanium prostheses design, ensuring the best anchorage.

1 Introduction

Cranioplasty is a surgical procedure aimed at repairing craniofacial defects, which are changes in the anatomy and morphology of skull and facial bones due to different causes, such as traumas, congenital malformations or diseases that require neurosurgical interventions [1].

Initially, during this surgery, bone grafts were employed and the first one was implanted in the 1668, even if the use of autografts for cranioplasty became popular in the 20th century [2]. This type of intervention requires a second surgery and, so, a good amount of substitutive bone [3], which can be both considered as disadvantages.

So, in order to replace the human bone, artificial prostheses may be used to correct the defect and to fill the hole. These implants are made with substitutive materials, which must be inert, thermally non-conductive, non-corrosive, stable, resistant to infection, with good biomechanical properties malleable and easily contoured, inexpensive, readily available, and can be sterilized [4]. The most adopted material is the titanium and its alloys, in particular the bio-alloy Ti6Al4V, characterized by very good properties [5]. Furthermore, this material favors the osseointegration with the bone that indicates the direct contact at microscopic level between living bone and implant, creating a permanent anchorage that improves the long-term function of the prosthesis [6].

Furthermore, the prosthesis is anchored to the craniofacial bone using screws, which can be of different types, such as self-tapping or self-drilling (for example: Stryker Global Headquarters, Kalamazoo, Michigan, USA and Synthes, Inc., West Chester, Pennsylvania, USA).

Because the face is the most important aesthetic part of the human body [4], craniofacial prostheses have both functional and aesthetic aspects, which must not be compromised. For this reason, the anchoring system can be considered as one of the major issues for the surgical treatment success. Indeed, this system allows the fixation of prosthesis to the bone and, consequently, the correct positioning between them, and ensures the protection of intracranial structures, as the brain, from mechanical impacts and a normal and appreciable appearance [1].

Thanks to the computational advantages in medical image acquisition and to new design methodologies, recently custom-made prostheses can be made starting from the reconstruction of patient-specific 3D models obtained processing the computer tomography (CT) and the magnetic resonance imaging (MRI) scans [7].

Moreover, titanium custom-made craniofacial prostheses can be made using the Incremental Sheet Forming (ISF), a manufacturing process with significant vantages [8-11]. Generally, with this manufacture, the prosthesis has a greater area and, so, a larger perimeter in respect to the defect [11], but no information regarding the anchoring system is available, in terms of overlap length and diameter of screw shank.

Since cranioplasty is a research branch where surgical expertise has to meet engineering knowledge for a good success of craniofacial reconstruction, the aim of this study is to present a methodology for the optimal design of the anchoring system in case of patient-specific titanium prostheses made using ISF. In details, different defect areas have been analyzed in a numerical

[a] Corresponding author: mariavittoria.caruso@unical.it

environment in order to identify the better overlap length and the diameter of screw shank. So, a "ready to use" decision support (DS) tool was developed in order to put into closer contact the players involved in the clinical process for obtaining custom and faster implants. The idea is to make easier the integration between bioengineers and surgeons during the preliminary phase by adopting the DS tool, which merges the expert knowledge and helps to converge toward the optimal solution.

2 Methodology

To pursue the aim previously reported, a hybrid approach has been developed, based on the coupled use of numerical model for structural analysis, statistical approach and metamodeling technique. Actually, this choice results necessary for enriching the robustness and wideness of methodology.

Hybrid approaches are usually implemented in engineering problems [12]. Focusing to the interested application field, a dataset derived by the implementation of a robust finite element model is typically analyzed by ANOVA or Response Surfaces [13] as far as used for training Neural Network or optimization tools [14]. Their use is aimed at describing and predicting the behavior of such response variable.

A similar methodology is proposed in this study, by following the sequential steps described in Figure 1.

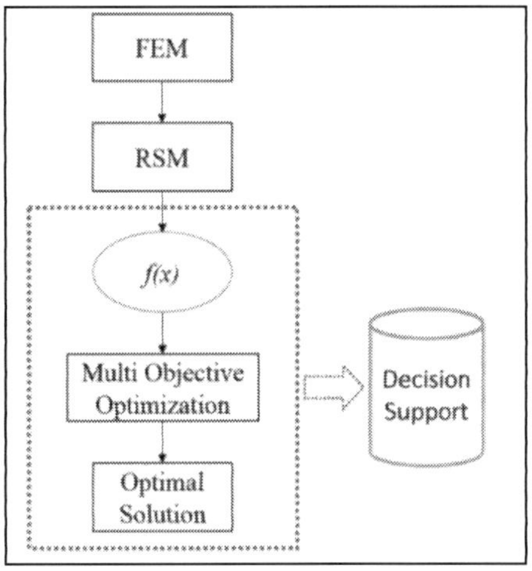

Figure 1. Adopted methodology.

First of all, a finite element model (FEM) of the anchoring system has been implemented in order to analyze the stress and displacements distributions and to generate a base of knowledge. The outcomes of the FEM analysis represent the incomes for the Response Statistical Methodology (RSM), used both for understating of the factors significance on the investigated output and for defining a prediction function that analytically describes the input-output relationship.

After that, a multi-objective model has been set and developed for selecting the best value of different features of the cranial prosthesis for each damage area value.

All details are provided in the next sections.

3 Design of experiment

The first step is to define the boundaries of the analysis fixing a lower-upper value of each investigated feature. So, the range of each input was chosen as reported in Table 1.

Subsequently, it was necessary to define the sampling strategy with which a set of data was collected in order to test and train the model. For the specific case, a full factorial Design of Experiment (DOE) was created, considering three levels for all the variables, except for the damage area, in which only two possible values have been tested (Table 1).

Table 1. Design factors information of FEM simulations.

Factor	Unit	Values
Screw Shank	mm	1.5 - 2.0 - 2.5
Overlap Length	mm	7 - 10 - 13
Load Tilt Angle	Degree	0 - 45° - 90°
Load Intensity	N	100 - 300 - 500
Damage Area	mm^2	2100 - 2600

The number of experiments necessary to carry out this design is $N = L^k$, where L represents the number of levels for the investigation and k represents the number of variables, or factors (five in our case). The number of simulations for the specific case study is therefore equal to 162 ($3^4 \cdot 2^1$).

4 Numerical model and structural analysis

In order to design the optimal anchoring system for a specific defect, 162 different numerical simulations were carried out using the finite element method (FEM) and five variation factors (Table 1). All structural simulations were performed by means of COMSOL 5.0 (COMSOL Inc, Stockholm, Sweden), a finite-element-based commercial software package.

4.1. Geometrical modeling

An assembly, including the skull bone, a prosthesis and a screw, was used as geometry in each structural analysis.

An easy way to obtain a volumetric model of head and of defects is to use the segmentation process of medical images (or slices), available in DICOM format, generated by CT or MRI. Using the segmentation process, a 3D virtual model of the patient skull can be obtained by means of different software [15]. Within this work, Invesalius, an open source software [16], was chosen for the 3D/volumetric reconstruction of a skull bone (Figure 2). After that, the reverse engineering process was applied in order to obtain a 3D solid model.

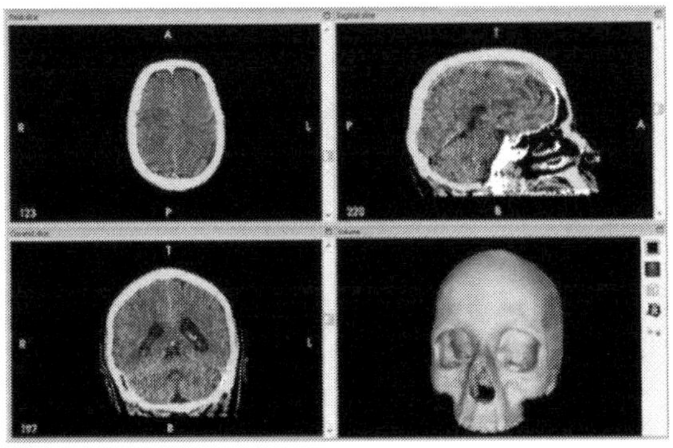

Figure 2. Segmentation of a skull bone by means of Invesalius.

Since the reconstructed head has no defects, a cylindrical hole was created in the upper part (frontal-parietal bone), considering two damage areas, of 2100 and of 2600 mm^2, respectively. Moreover, the "removed bone" was used to model the prosthesis (Figure 3A), that was obtained considering a thickness of 1.5 mm [8] and different overlap lengths (7, 10 and 13 mm). Furthermore, only a little part of the skull bone was considered (Figure 3B). Because the defect and the prosthesis were axisymmetric, a simplified model of 20 degrees was considered (Figure 3C). These two assumptions were necessary in order to reduce the computational cost obtaining realistic solutions anyway.

Regarding the screw, it was modeled considering a length of 4.5 mm, three different screw shanks (1.5, 2.0 and 2.5 mm) and ignoring the thread. Moreover, it was placed in the center of overlap length (Figure 3C).

The final considered model of the assembly is reported in Figure 3C.

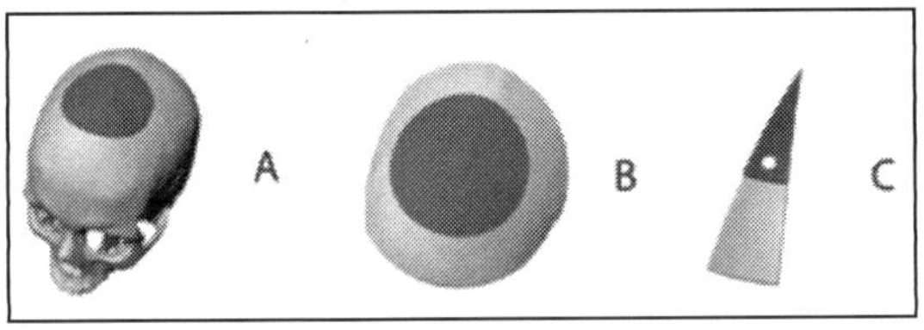

Figure 3. Geometry assembly for FEM analysis: real (A), total simplified (B) and 20° simplified (C).

4.2 Mechanical properties

Skull bone was approximated as a cortical one [17], with an ultimate compression stress of 145 MPa [17], whereas prosthesis and screw was set as Ti6Al4V [18]. Both materials were assumed to be homogeneous and isotropic [17], defined by linear elastic laws.

4.3 Boundary conditions

Because the exact fit between bone and implant is very important and is correlated with the osseointegration and, consequently, with the long-term function of the prosthesis [6], the screw-bone interface was modeled considering a perfect osseointegration. To numerically reproduce this condition, the identity pair was applied between bone and implant and bone and screw. Moreover, as the osseointegration was numerically reproduced, a rigid connection was assumed between prosthesis and screw.

Titanium and bone have different elastic modules (reported in [18] and [17] respectively), so, the anchoring system must be correctly designed in order to avoid the skull bone rupture in case of mechanical impacts. For this reason, a distributed compressive load was applied to the top surface of prosthesis (Figure 4) in order to simulate an accidental load of 100, 300 and 500 N, considering also three tilt angles (0°, 45° and 90°).

Moreover, skull bone was fixed in the bottom part in order to simulate the connection with the spine.

Finally, as reported before, a simplified assembly of 20° was modeled (Figure 3C), so symmetry boundary conditions were applied in the lateral faces of skull and prosthesis.

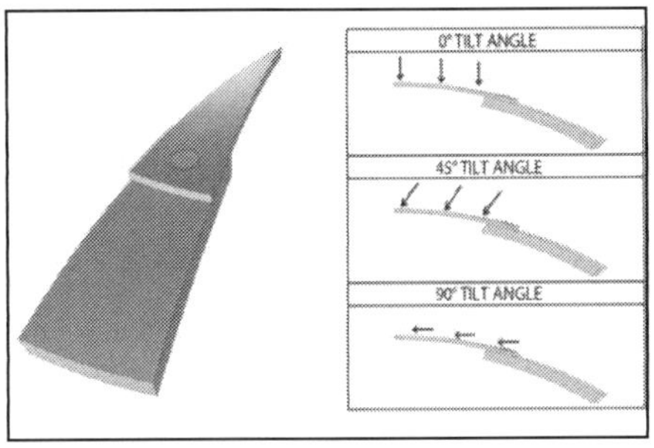

Figure 4. Load boundary condition, with the three considered tilt angles.

4.4 Simulations details

In all cases, the grid was created using fine tetrahedral mesh.

Since the anchoring system must be design to resist to accidental loads, the maximum stress on skull bone due to loaded prosthesis was evaluated by means of the Von Mises criteria [19]. An example of stress distribution is reported in Figure 5.

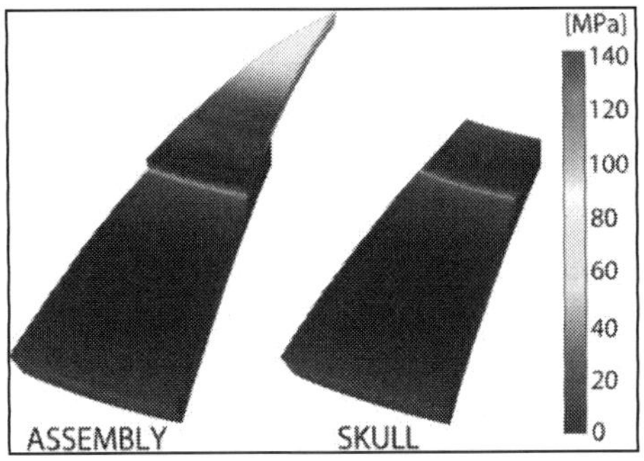

Figure 5. Von Mises stress distribution, considering a defect of 2100 mm^2, a load of 500N and 0°, an overlap length of 10 mm and a screw shank of 2 mm.

DOI: 10.1051/matecconf/20168011004

5 Statistical and mathematical approaches

5.1 Response Surface Methodology

The numerical data provided by FEM analyses were used to build a model using the response surface methodology (RSM) [20]. It consists of a group of mathematical and statistical techniques adopted in the development of an adequate functional relationship between a response of interest, y, (in our case the stress of skull) and a number of associated control (or input) variables denoted by x_1, x_2, ... , x_k.

In general, such a relationship is unknown but can be approximated by a low-degree polynomial model of the form:

$$y = f'(x)\beta + \varepsilon \quad (1)$$

where $x = (x_1, x_2, \dots, x_k)$, $f(x)$ is a vector function that consists of powers and cross-products of powers of x_1, x_2, ... , x_k up to a certain degree, β is a vector of unknown constant coefficients, and ε is a random experimental error assumed to have zero mean. In this case, the quantity $f'(x)\beta$ represents the expected value of y.

The purposes of building a model such as (1) are the following:

- to establish a relationship, albeit approximate, between y and x that can be used to predict response values for given settings of the control variables.
- to determine the relevance of the factors analyzed.
- to determine the optimum settings of x.

5.2 Optimization model

The input-output relationship defined through the RSM can be used to select the best features of the craniofacial prosthesis to optimize one or a set of criteria. In details, the prosthesis will not reach high stress if subjected to an accidental load, considering the maximum allowed stress equal to the ultimate compression stress of bone [17]. Furthermore, screws will select from an available set. So, the considered optimization problem has two objective functions. To find the efficient frontier, a multi-objective algorithm is necessary. Moreover, due to the non-linear input-output relationship, the problem is a multi-objective problem with non-linear constraints.

One of the most popular algorithms to solve this kind of problems is the NSGAII. The first version of this algorithm was proposed by [21]. It is a very effective algorithm, but has been generally criticized for its computational complexity and lack of elitism. A new version of this algorithm was implemented later with a minor complexity of $O(MN^2)$, where M is the number of objective functions and N the size of the population [22]. This new algorithm incorporates also elitism ensuring the survival of the best solutions. Furthermore, the algorithm can be briefly described as follow. First of all, a random population is generated, and then the population is sorted according to non-domination in several fronts. The first front is totally non-dominated in the current population, while the second front is dominated only by the first front. Each individual in each front has a value of rank (fitness) based on the front to which it belongs. In addition to the fitness value is considered a function called crowding distance [23]. The crowding distance is a measure of the distance of an individual from its two closest neighbors. Large average crowding distance creates a population with a greater diversification.

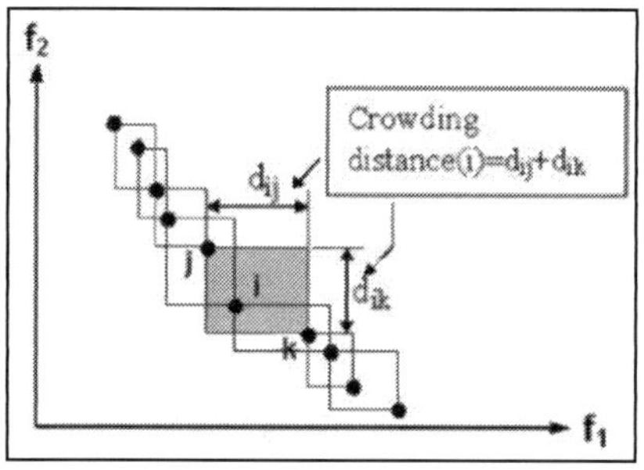

Figure 6. Crowding distance.

In the version of the NSGAII adopted in this study, the simulated binary crossovers were chosen for the crossover phase and the polynomial mutation for the mutation phase. One of the most important aspects of this technique is given by the choice of two parameters: the size of the population and the number of the generations. A too high value of these two parameters leads to an increase of the computational time. However, decreasing the number of generations the quality of the solutions found deteriorates while decreasing the size of the population the number of optimal solutions found is lower.

6 Discussion of the results

The mathematical approach described in section 5 was implemented to build a guided user interface, which can be considered as a useful decision making support tool.

The data collected through a full factorial DOE, obtained by means of a set of FEM simulations of structural analysis, were used to build a polynomial model.

The design and statistical analysis of experiments were done through the Design Expert (State Ease, Inc, trial version) software.

For the investigated factors, a reduced third order polynomial (regression) model was chosen to describe the relationship for the response and tested factors. A p_{value} threshold of 0.05 was adopted to select the statistically relevant parameters interactions to build the model. The selected model is characterized by a high level of accuracy (Table 2). The overall adequacy of the model in describing the numerical data may be evaluated by the coefficient of multiple determination, R^2. It is a measure of the proportion of the overall variability in the response- bone stress in this study -, that is accounted for by the regressor variables and interactions that have been selected. However, R^2 will always increase when a variable is added to the model, whether or not the variable is significant, and thus models with large values of R^2 may yield poor predictions of future observations. A more meaningful measure is the adj R^2 (the adjusted value of R^2), which will only increase when a new variable is added if there is enough reduction in the residual sum of squares (SSE) to compensate for the loss of a degree of freedom.

In our case, the "Predicted R-Squared" of 99.8% is in reasonable agreement with the "Adjusted R-Squared" of 99.6% (Table 2). As a result, this model can be chosen to navigate the design space.

Table 2. Polynomial model accuracy

Parameter	Simplify Third order Polynomial model
RMSE	3.81
F_{model}	1915.81
p_{model}	<0.001
R^2	0.998
Adj R^2	0.996

Analyzing the measured response, the fit summary of the output indicates that the reduced cubic model is statistically significant. Furthermore, the normal plot of the residuals was employed to verify the normal distribution of data, that was confirmed (Figure 7A). Also Gaussian probability density function was applied to the residuals to check the normality and detect outliers, which were confirmed since all data were in the interval of $[-3\sigma, 3\sigma]$ (Figure 7B). Finally, to investigate the reliability of the prediction, the predicted versus actual diagnostic plot was analyzed (Figure 7C). All data were properly predicted, also around the ultimate compressive stress value (145 MPa).

The response surface of bone stress is reported in Figure 8, considering a damage area of 2500 mm^2 and an accidental load of 350 N and 0°. The response decreases if the overlap length increases, regardless of the diameter of screw shank.

To select a suitable value of screw diameter and overlap length, the NSGAII algorithm was applied to the cubic model. The mathematical model is the following:

$$\begin{gathered} \min S \min O \\ Stress < 145 \\ I=I_0,\ L=L_0,\ D=D_0 \\ s \in \underline{S} \end{gathered} \tag{2}$$

where I_0, L_0 and D_0 are the input values of tilt angle, load and damage area and S is the set of available types of screws. In the present case study, the preliminary hypothesis is that three types of screws are available, with a diameter equal to 1.5, 2 and 2.5 mm, chosen since they are the most used in neurosurgery.

The problem was to select the best screw shank diameter and the overlap length for a given damage area, ensuring that the stress due to an external load is less than the ultimate compression stress of bone.

In a first scenario, a tilt angle equal to 0° and a damage area of 2600 mm^2 were considered. The NSGAII algorithm was applied to define the efficient frontier for all the possible screw types. The results are reported in Figure 9. It is not possible to find an overlap length which allows to obtain a stress less than 145 MPa for all the analyzed screw diameters. Moreover, the three curves are really close, showing that the diameter value is not a critical variable (at least in the analyzed range of 1.5-2.5 mm).

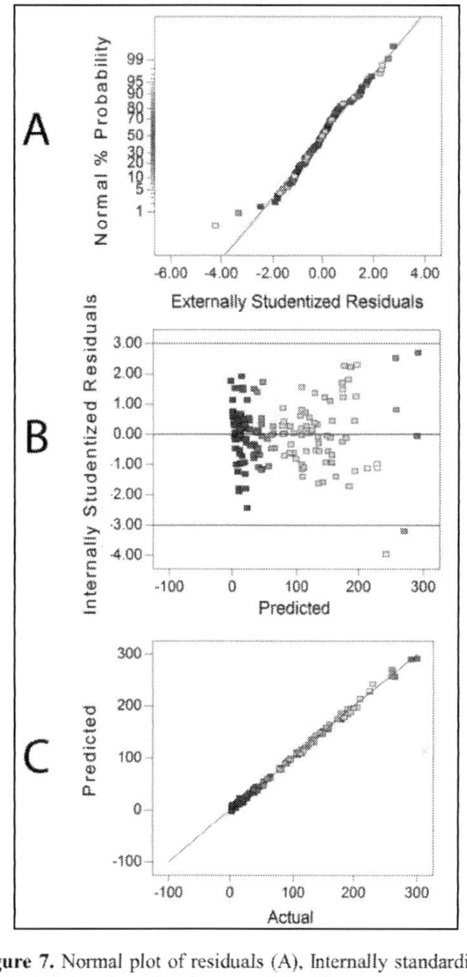

Figure 7. Normal plot of residuals (A), Internally standardized residuals (B) and predicted vs actual plot (C).

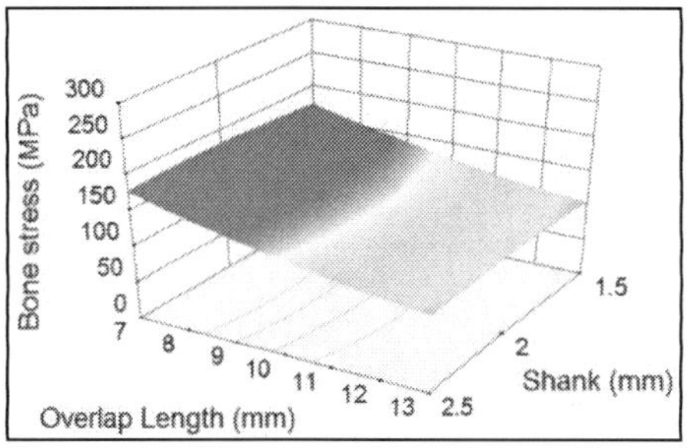

Figure 8. Surface response for bone stress, considering a damage area of 2500 mm^2 and an accidental load of 350 N and 0°.

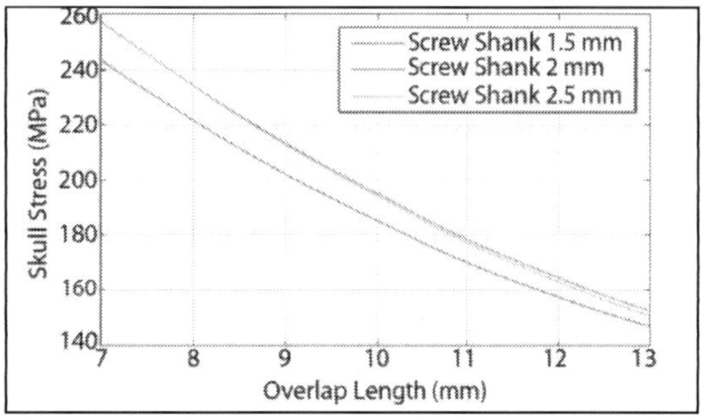

Figure 9. Efficient frontier (Tilt angle 0°, Damage 2600 mm^2)

In a second test, an inclination of 40° and a damage of 2400 mm^2 were considered. The efficient frontier of this scenario is reported in Figure 10. Also in this case the three curves are almost identical, confirming that the stress is mainly due to the level of overlap length between prosthesis and cranium. However, in this case, it is possible to respect the constraint on the maximum stress by using an overlap value greater than 10 mm.

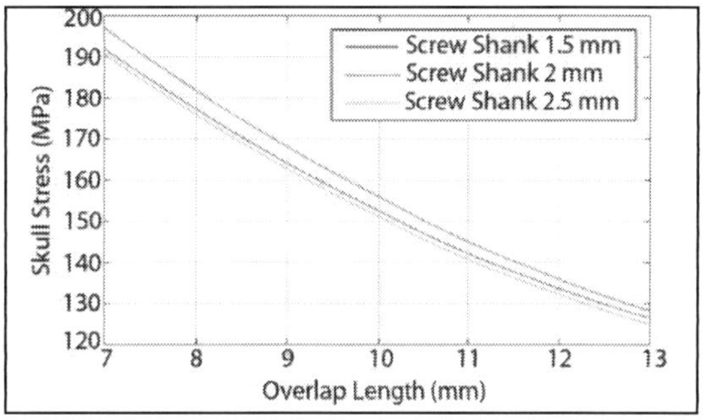

Figure 10. Efficient frontier (Tilt angle 40°, Damage 2400 mm^2)

7 Conclusions

During cranioplasty, patient-specific titanium prostheses, made using ISF, can be adopted to reconstruct the skull or facial defects. The presented methodology, that coupled a numerical model for structural analysis, a statistical approach and a metamodeling technique, allows to identify the best anchoring system, in terms of overlap length and diameter of screw shank, so it can be considered as a useful decision support tool for the design of titanium prostheses.

Acknowledgments

This research is funded by the Italian Ministry of University and Research within the project PRIN "Bioforming".

References

1. P. Solaro, E. Pierangeli, C. Pizzoni, P. Boffi, P., G. Scalese, Journal of neurosurgical sciences, **52(4)**, 113 (2008)
2. A. Sanan, S.J. Haines, Neurosurgery, **40(3)**, 588-603 (1997)
3. A.L. Fernandes da Silva, A. M. Borba, N. R. Simão, F.L.M. Pedro, A.H. Borges, M. Miloro, Case reports in surgery, 2014
4. M.C. Goiato, R.B. Anchieta, M.S., Pita, D.M. dos Santos, Journal of Craniofacial Surgery, **20(5)**, 1512-1518 (2009)
5. M.A.H Gepreel, M. Niinomi, Journal of the mechanical behavior of biomedical materials, **20**, 407-415 (2013)
6. T. Albrektsson, P. I. Brånemark, H. A. Hansson, J. Lindström, Acta Orthopaedica, **52(2)**, 155-170 (1981)
7. J.F.I. Saldarriaga, S.C. Velez, M.D. Posada, American Journal of Engineering and Applied Sciences, **4(1)** (2011)
8. G. Ambrogio, R. Conte, L. De Napoli, G. Fragomeni, F. Gagliardi, Key Engineering Materials, **651-653**, 925-931 (2015)
9. J. Castelan, L. Schaeffer, A. Daleffe, D. Fritzen, V. Salvaro, F.P.D. Silva, Revista Brasileira de Engenharia Biomédica, **30(3)**, 265-273 (2014).
10. B. Lu, H. Ou, S.Q, Shi, H. Long, J. Chen, International Journal of Material Forming, 1-10 (2014).
11. B. Lu, D. K. Xu, R.Z. Liu, H.A. Ou, H. Long, J. Chen, Key Engineering Materials, **639**, 535-542 (2015)
12. D. Umbrello, G. Ambrogio, L. Filice, R. Shivpuri, Materials and Design, **29(4)**, 873-883 (2008)
13. I. Asiltürk, S. Neseli, M.A. Ince, Measurements, **78**, 120-128 (2016)
14. S.Chanda, S.Gupta, D.K. Pratihar, Applied soft computing, **38**, 296-307 (2016)
15. G. L. Presti, M. Carbone, D. Ciriaci, D. Aramini, M. Ferrari, V. Ferrari, Journal of digital imaging, 1-10 (2015).
16. T. F., Moraes, P. H. J., Amorim, F. S., Azevedo, and J. V. L. Silva, Computational Vision and edical Image Processing, 405–408 (2011).
17. J. S. Raul, D. Baumgartner, R. Willinger, B. Ludes, International Journal of Legal Medicine, **120(4)**, 212-218 (2006)
18. M. Vosough, F. Schultheiss, M. Agmell, J. E Ståhl, The International Journal of Advanced Manufacturing Technology, **67(1-4)**, 339-348 (2013)
19. L. Baggi, I. Cappelloni, M. Di Girolamo, F. Maceri, G.Vairo, The Journal of prosthetic dentistry, **100(6)**, 422-431 (2008)
20. R. Myers, D.C. Montgomery, C.M. Anderson-Cook, 705, John Wiley & Sons, 2009
21. N. Srinivas, D. Kalyanmoy, Evolutionary computation, **2(3)**, 221-248 (1994)
22. D. Kalyanmoy, S. Agrawal, A. Pratap, T. Meyarivan, Lecture notes in computer science, 849-858 (2000)
23. C. R. Raquel, P. C. Naval Jr., Proceedings of the 7th Annual conference on Genetic and Evolutionary Computation, ACM (2005)

Dynamic DOE for porthole die extrusion optimisation

Claudio Ciancio[1], Francesco Gagliardi[1], Giuseppina Ambrogio[1] and Luigi Filice[1]

[1]*Department of Mechanical, Energetic and Management Engineering (DIMEG), University of Calabria, 87036 Rende, Italy*

Abstract. Extruded profiles are widely used but their design is still difficult. Several parameters have to be considered. This is particularly true when complex geometries, i.e. non-symmetric or hollow profiles have to be obtained. A novel metamodel technique was proposed to define the relationship between input and output process parameters in porthole die extrusion. An iterative dynamic DOE was finally used to improve the accuracy of the model around the most promising area of the process domain.

1 Introduction

Nowadays, complex profiles made of lightweight materials, such as aluminium or magnesium alloys, are produced largely by extrusion processes [1]. In particular, the extrusion with a porthole die is the most widespread manufacturing process as it allows to manufacture hollow profiles with complex shapes [2,3]. In porthole die extrusion, the extruded billet is split by some bridges which are designed properly to link the mandrel to the die walls. The material streams, crossing the porthole, get inside the so called welding chamber where they join if suitable pressure and temperature conditions are guaranteed [4]. A good weld quality has to be ensured to avoid failed extruded with a careful die design [5]. After the streams welding, the material flows through the bearing orifice, obtaining its final shape. In order to have undistorted profiles, the local lengths of the bearing can be variable along the exit perimeter [6]. Moreover, the bearing length has to be set taking into account grain growth [7] and surface defects [8,9].

Sometimes, it is also helpful to introduce an additional chamber, called pocket, between the welding chamber and the bearing, which section plays an important role in the flow optimization [10,11]. On the other side, the pocket introduction brings to higher extrusion loads [12]. Therefore, a right combination of the bearing land and pocket design has to be defined properly.

The optimization complexity increases if sections with non-symmetric shape have to be produced. According to that, the use of numerical tools is always more widespread and commercial finite element software have been developed to investigate the forming phases [13,14]. However, the problem complexity is so high that the exclusive use of finite element model (FEM) can result poorly effective being the numerical approach "case dependent"; on the contrary, the combined use of numerical tools and mathematical regression models can allow defining wider results.

According to that, this paper aims at investigating the influence of geometrical parameters (i.e. bearing length and pocket geometry) to produce undistorted profiles. To make the example more complex and interesting from an industrial point of view, dies with unsymmetrical flow have been considered. More in detail, different die geometries characterised by changing velocity flow at the exit of the die have been tested; both the flow velocity along the profile and the steady-state ram load are measured as response factors. As concerns the former one, it is required that velocity gradient along the exit profile converges to zero for ensuring an homogeneous material flow and preventing the profile distortion; at the same time, the steady-state ram load has to be minimized for limiting the die wear and the required press capacity. The analysis has been based on numerical simulations carried out by the commercial code, Altair HyperXtrude (V-13), to generate the base of knowledge; subsequently, dynamic sampling techniques have been used to build a metamodel starting by a previous model designed for a similar problem as a reference point. All the details are reported in the following sections.

2 Methodology

Statistical methods have been used widely in manufacturing to ensure quality and robust process design [15]. Identification of input-output relationships of manufacturing processes is usually a complex task, which involves integration of physical-driven (or even simulation-driven) experiments and data mining algorithms. This task is costly and time consuming. Many approaches have been proposed to apply mathematical methods, techniques and paradigms to the solution of

[a] Corresponding author: claudio.ciancio@unical.it

manufacturing problems. Mathematical model can in fact be used to understand how to improve efficiently a given system configuration analyzing and identifying the factors that affect majorly the investigated responses [16]. In this chapter a new methodology has been presented to perform a faster identification of these relationships by integrating response surface models and similarity functions. The proposed approach allows generating intelligent decision support systems that can be used to study and manage complex manufacturing processes. The system learns basing on accumulated data and adapts itself during the analysis. Therefore, the proposed methodology can be a helpful tool to reduce the cost (time and experiments) and to perform the design of a reengineered process. The aim of the proposed methodology is to identify the relationship that links a set of controlled process parameters and a set of output responses; this one reflects the critical aspects of the manufacturing system. More formally, a transfer function is defined for each output $y_j \in Y$, $y_j = f_j^k(x_1, x_2, .., x_n)$, where n is the number of controlled parameters, $X = \{x_1, x_2, ... x_n\}$. The transfer function y_j approximates the process behavior under a certain process condition k. The proposed technique is used to estimate a new model $f_j^{k'}$ if a new process condition k' has to be investigated. A set of data is usually collected and used to train and validate a machine learning model. However, a different approach has been used to reduce the total number of experiments adding the necessary feedbacks with a model (previously defined and used to map the process before the process reengineering) and a sequential sampling procedure (Figure 1). The sequential approach, moreover, offers the possibility to achieve different "milestone" during the learning phase identifying the magnitude of the effect of the analyzed parameters with a small dataset. Other data can be collected subsequently to achieve the required accuracy.

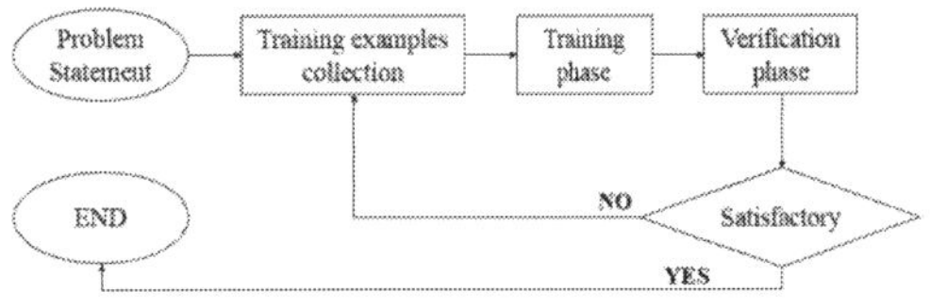

Figure 1. Mathematical approach to define input-output relationships.

The proposed approach allows obtaining the following benefits compared to benchmark methods: 1) ability to map complex shapes using simple transformations (overfitting risk reduction); 2) use of historical data on similar problems; 3) greater flexibility given by the integration between the learning algorithm and the experimental design. The proposed methodology has been performed through the following steps:

- sampling strategy (Phase 1)
- surrogate model preliminary analysis
- surrogate model
- sampling strategy (Phase 2)
- stopping criteria

2.1 Sampling strategy (Phase 1)

The sampling strategy used to train the surrogate model is composed by two phases. In the first phase, a small set of data is collected using a benchmark approach to capture the first indications about the influence of each controlled parameters on each output response. In this study, the data are selected using a latin hypercube design [17].

2.2 Surrogate model preliminary analysis

A preliminary analysis has been used to identify the complexity of the function to approximate. This phase has been performed to introduce a regularization parameter (or a penalty) to avoid over-fitting. Given a model f_j^k so that one output component y_j of a problem k can be predicted through the equation $y_j = f_j^k(x_1, x_2, .., x_n)$, the optimal value of a vector has been identified:

$$\beta = \{\beta_0, \beta_1, ..., \beta_{n+1}\} \quad (1)$$

that allow minimizing the error obtained on a second problem, k' using a model $f_j^{k'}$ defined as follows:

$$f_j^{k'} = \beta_0 f_1 (x_1 + \beta_1, x_2 + \beta_2, .., x_n + \beta_n) + \beta_{n+1} (2)$$

The error E_0 obtained on the training data by this model will be used as an input parameter to define a risk coefficient.

2.3 Surrogate model

In this phase, an alternative and general theory of learning with similarity functions has been developed. The idea of the proposed algorithm is based on the computation of a linear regression function in a high dimensional feature space where the input data are mapped via a nonlinear function. In the nonlinear setting, therefore, the optimization problem corresponds to finding the flattest function in the feature space. The methodology developed in this section can be considered as a generalization of the kernel method [18]. The theory behind kernel functions is based on the fact that many standard algorithms, such as SVR and Perceptron algorithm [19], can be written in such a way that the only way they interact with their data is via computing dot-products on pairs of examples. Kernel functions are usually used to fit a set of data (training data) trying to build a model that can be generalized for a second set of data (validation data). However, the problem addressed in this phase is different. Given two problems k and k' for which two sets of data are available, a surrogate model $f_j^{k'}$ has been defined through a transformation (denoted henceforth as similarity transformation) on a second model f_j^k previously defined. It has been denoted as X_t^k the t-th datum available for the process condition k and by $x_{i,t}^k$ the value of the i-th input of X_t^k. The proposed approach has been used as the one which allows obtaining the following benefits compared to classic kernel methods:

- ability to map complex shapes using simple transformations (overfitting risk reduction);
- use of historical data on similar problems.

A similarity function can be considered as a function that is estimated from the data, or that is chosen to fit the data. However, the similarity function should be as simple as possible to reduce the overfitting risk. Simple similarity transformations can be built considering a weighted Euclidean distance:

$$d_w(X_t^k, X_p^{k'}) = \sqrt{\sum_{i=1}^{n} w_i \left(x_{i,t}^k - x_{i,p}^{k'}\right)^2} \quad (3)$$

where w_i is a weight assigned to the i-th process parameter. The similarity function s_w is expected to increase decreasing the value d_w reaching the value 1 for $d_w = 0$. So typical similarity function can be:

- Exponential function: $s_w = e^{-d_w + d_0}$ (4)
- Linear function: $s_w = \frac{1}{1+d_w+d_0}$ (5)
- Logarithmic function: $s_w = \frac{1}{\log(1+d_w+d_0)+1}$ (6)
- Logarithmic function: $s_w = cos\,(d_w + d_0)$ (7)

where d_0 is used to map translational changes of the response surface. The output of a new point $X_p^{k'}$ is estimated trough the following equation:

$$y_{j,p}^{k'} = \sum_{t=1}^{T} \alpha_t\, s_w\left(X_t^k, X_p^{k'}\right) y_{j,t}^k \quad (8)$$

where the optimal value of the vector $\alpha = \{\alpha_1, \alpha_2, .., \alpha_T\}$ is determined through a mathematical model. This one is solved by means of genetic algorithms where it is minimized:

$$\min \lambda\, E + \sum_{t=1}^{T} \alpha_t^2. \quad (9)$$

The parameter λ is inversely proportional to the risk coefficient E_0 determined in section 2.2. Moreover E_0 is used to tune a parameter α_{max} that determines the maximum number of variables which can assume any value in addition to 0.

2.4 Sampling strategy (Phase 2)

The similarity function technique has been used to update the response surface of the problem defining a new fitting model. However, the prediction abilities of the new tool could be not yet satisfactory due to the lack of data in some area of the process domain. A new set of experiments has been defined to reduce the inaccuracy of the tool. Each new experiment has been designed using the results of the current metamodel until the accuracy of this tool can be considered statistically satisfactory. More in particular, at each iteration, a set of new experiments is performed considering new samples localized in the so called ϵ-tube. The ϵ-tube is composed by all the possible solution

$$X_\epsilon^{k'} \in \Omega : \left|P\left(x \in \Omega^{k'}\right) - P\left(x \notin \Omega^{k'}\right)\right| \leq 2\epsilon \quad (10)$$

where $\Omega^{k'}$ is a promising area of the process domain that satisfies some quality constraints.

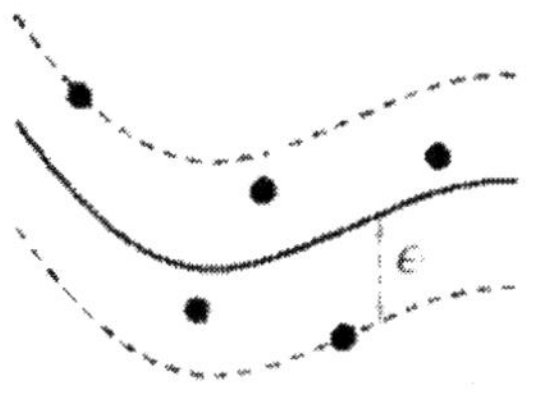

Figure 2. ϵ-tube schematization.

The new set of points to analyze has been selected trying to maximize the distance between the new point and the previous training data. The ϵ value has been reduced gradually during the iterations. The motivation of this choice is related to the fact that the accuracy of the prediction increases gradually so the subsequent iterations lead to small changes. The proposed approach owns two degrees of freedom that allows reducing the number of required experiments compared to benchmark approaches:

- *Parameters weighting*: the interaction with the surrogate model allows to identify the most relevant process parameters;
- *Data concentration*: more experiments are performed in the most promising area of the process domain.

2.5 Stopping criteria

According to the preliminary testing, the use of two criteria to stop each iteration of the algorithm has been found profitable, namely:

- the number of experiments performed exceeds a maximum threshold;
- the accuracy of the predictive tool is greater than a given threshold ε_{min} .

3 The case study

Feasibility of asymmetric profile by porthole die extrusion is guaranteed homogenising the material flow. This aim can be pursued by playing with two variables: the bearing length and the pocket geometry. The pocket is a die extension below the welding chamber that is sometime used to get material flow homogenisation. The pocket has to be designed properly taking into account the porthole die geometry in all its parts. According to that, the present research has been focalised on the manufacturing of complex L profile by using a multi-gates die. This choice has been performed to take also into account the asymmetry between the axis of the extruded billet and the die exit as further variable to consider in the flow homogenization. In detail, the process has been designed for the simultaneous extrusion of two L profiles whose dimensions are 20 mm and 47 mm for the short and long length respectively, and 5 mm for the part thickness (Figure 3.a). The starting billet

DOI: 10.1051/matecconf/20168011005

diameter is equal to 146 mm and the extrusion ratio is therefore of about 27; finally, the billet length is of 150 mm.

Two die configurations that correspond to different material flows into the pocket have been investigated; the distance between billet axis and die gate is the geometric variable (Figure 3.b) which allows the configuration changing.

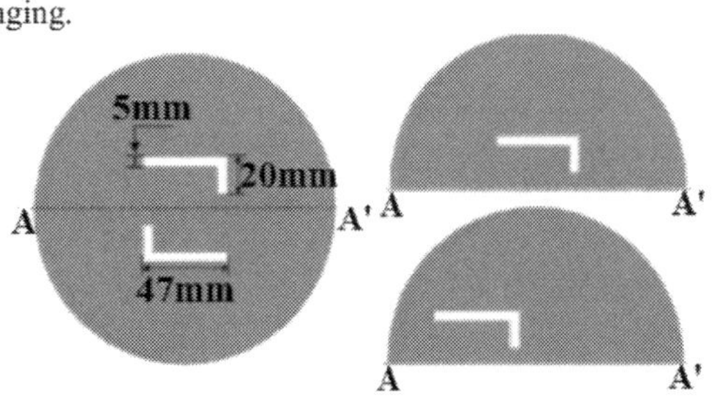

Figure 3. 2D Sketch a) of the investigated geometry and b) of the die configurations.

The parameterized dimensions of the bearing lengths and of the main pocket sizes have been planned and multiplicative constants have been set for changing logically the extrusion die. More in detail, the parameters taken into account are: the local bearing length, $H_{bearing}$, the pocket height, H_{pocket}, and the pocket cross section D_{pocket}. Concerning the last investigated factor, this one has been set by starting from the L shape going to widen the section where lower material flow velocity is expected (Figure 4).

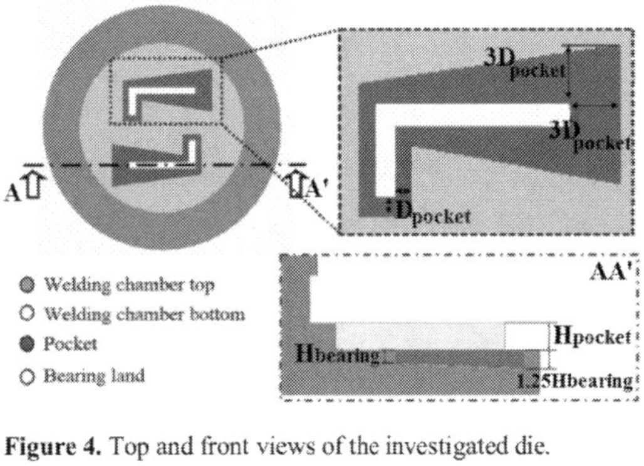

Figure 4. Top and front views of the investigated die.

The proportionality between the lengths of the bearing and the dimensions of the pocket has been established to avoid an unbalanced flow. The multiplicative constants used for varying the investigated variables are summarized in Table 1.

Table 1. Multiplicative constants.

$H_{bearing}$ [mm]	H_{pocket} [mm]	D_{pocket} [mm]
3	2.5	3
6	5	7
9	7.5	1

4 Numerical model

A reduced number of simulations have been performed for all the process configurations in order to find a general guideline, which can allow the homogeneous distribution of the material flow into the pocket, by properly choosing the pocket geometry and bearing height.

The numerical investigation has been carried out by means of the software Altair HyperXtrude V-13. HyperXtrude is based on an Arbitrary Lagrandian-Eulerian (ALE) approach which is used to get locally material information inside the extrusion die and to predict also the flow of the extruded profile in the bearing region and at its exit.

The numerical model has been designed by considering six elements: the billet, the porthole, the welding chamber, the pocket, the bearing and the finale profile (Figure 5).

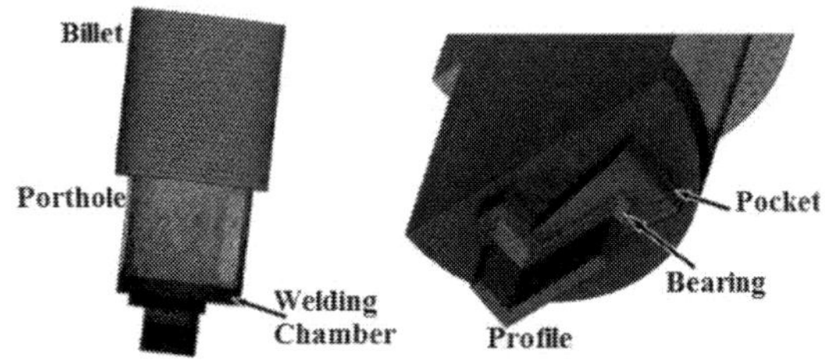

Figure 5. Meshed model.

Half geometry has been simulated by exploiting the symmetry of full model; this choice has been performed to reduce the computational time. For the same reason, the meshed model is characterized by a gradually coarser refinement from the extruded profile to the billet. By doing that, 78000 prismatic elements have been used between bearing and profile; tetra elements have been instead adopted to mesh the remaining zones (about 440000 elements).

The constitutive equation used to describe the aluminium profile behaviour during the extrusion process is the sine hyperbolic law. Regarding the boundary conditions, a ram speed of 1 mm/s was utilized; taking into account the extrusion ratio value, the average value expected as profile exit velocity is 27 mm/s. The initial billet temperature has been set to 500°C, while the temperatures of the container and the other regions have been set to 450°C and 420°C, respectively. Sticking conditions have been considered at the interface material-dies except of the bearing surface where viscoplastic conditions have been set with a friction coefficient of 0.3, to consider the material sliding in the orifice [20].

The distribution of the profile velocity at the exit of the bearing zone and the steady state ram load necessary for the extrusion have been the analysed outcomes.

As concerns the profile distortion, this one has been investigated by focusing the attention on the velocity discrepancy between two points on the extruded cross section. The points have been chosen to maximize the difference (ΔV) in the velocity distribution (Figure 6).

DOI: 10.1051/matecconf/20168011005

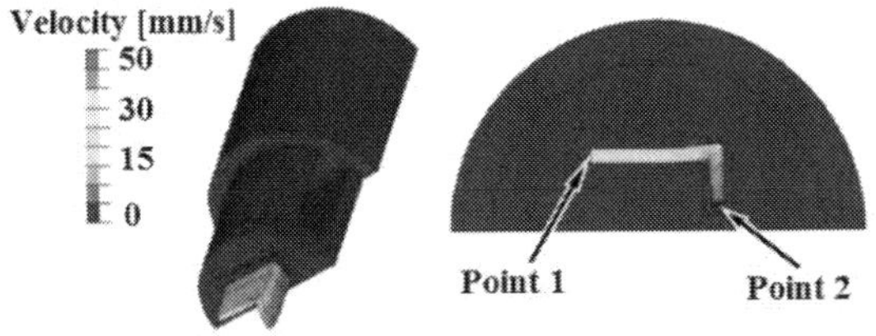

Figure 6. Velocity distribution and detail of the points used for ΔV measure.

5 Discussion of the result

The optimization procedure described in section 2 has been used to define the most suitable input parameters of a porthole extrusion process. In particular, a polynomial model has been defined to set the relationship between the local bearing length, $H_{bearing}$, the pocket height, H_{pocket}, the pocket cross section D_{pocket} and the velocity discrepancy ΔV and the ram load F. The proposed approach based on response surface projection has been then used to redefine the model for a different position of the L section. The two objective functions (minimization of the velocity discrepancy and ram load) have been weighted differently and joined into a single objective function:

$$z = \lambda_1 \Delta V' + \lambda_2 F' \tag{11}$$

where $\Delta V'$ and F' are the normalized value of ΔV and F in the range 0-1. A value equal to 0.8 is assigned to λ_1 and a value equal to 0.2 to λ_2. The objective function z is therefore also defined in the range 0-1. The first section has been investigated in a previous study [21] using a full factorial DoE and a second order model for both responses. The input parameters have been investigated inside the range reported in Table 2

Table 2. Design factors information of FEM simulations.

Factor	Unit	Values
$H_{bearing}$	mm	1.5 - 3.0 - 4.5
H_{pocket}	mm	3 - 7 – 11
D_{pocket}	mm	2.5 - 5 – 7.5

The accuracy of the model has been measured through the R^2. A value equal to 0.953 has been reached for the ΔV and a value equal to 0.988 for the F. The dynamic sampling and the metamodel based on similarity functions has been then used to analyse the geometry in Figure 3.b.

5.1 Sampling strategy (Phase 1)

To start the algorithm it is necessary to select a first set of data to train the model. 9 process configuration have been analyzed using a latin hypercube design sampling. In this stage the selected experiments are not influenced by the model developed for the first geometry.

5.2 Surrogate Model Preliminary analysis

The surrogate model for the first section has been used as reference point to develop a new model for the second one not yet investigated using a combination of simple transformations. The only changes allowed in this phase are translations along the x and y axes or using a different magnitude level.

Table 3. Surrogate model (Preliminary analysis)

Parameter	R^2
ΔV	0.769
Ram Load	0.710

5.3 Surrogate model

The results achieved using the preliminary analysis has been used to define the λ showed in section 2.3. Different similarity transformations have been tested on this problem trying to map the relationship between the problems analyzed maximizing the value of the available information. The results of this phase are summarized in Table 4:

Table 4. Surrogate model (Preliminary analysis)

Parameter	R^2 Phase 1	Similarity Function
ΔV	0.792	Exponential
Ram Load	0.851	Exponential

As it is possible to see, the best results have been achieved using the exponential function for all the responces. This model has been then used to develop the new DoE strategy.

5.4 Sampling strategy (Phase 2)

The results obtained have been utilized to define a second set of experiments. The experiments have been selected iteratively around a ϵ-tube (ϵ =0.05) until the accuracy has been greater than a predetermined threshold. In this case-study, ϵ-tube has been defined as the portion of the process domain with a predicted value of z less than 0.1. The ϵ value has been iteratively modified at each iteration t according to the residual number of maximum experiments available:

$$\epsilon_t = \epsilon_0 - \epsilon_0 \frac{N_t}{N_{Max}} \tag{12}$$

where N_t and N_{Max} are respectively the current and the maximum number of performed experiments.

5.5 Stopping criteria

Two stopping criteria have been used in this analysis. The accuracy of the tool has been measured using the R^2 with a leave-one-out cross validation. The threshold value of this parameter has been fixed equal to 0.97 equal to the average R^2on the two responses obtained with the second order model and a full factorial DoE for the first Stack-up. The maximum number of experiments has been set equal to 27 (dataset size of the full factorial DoE used in the previous analysis).

It is possible to see that the proposed algorithm has reached a superior level of accuracy using the same dataset size. The algorithm has been arrested after 18 experiments with a reduction of 33% of the dataset size compared to the one used for the first geometry. The value of ΔV ranges between 0.19-15.75 while the value of F between 7.134-7.822. The model has been finally used to select an optimal configuration which minimize both objectives.

Table 5. Model accuracy

	Geometry 1	Geometry 2
DoE	Full Factorial	Adaptive sampling
Methodology	Polynomial Model	Similarity Function
N° Data	27	18
Accuracy ΔV (R^2)	0.953	0.968
Accuracy F (R^2)	0.988	0.991

Table 6. Optimal Configuration geometry 2 (input values)

Factor	Value
$H_{bearing}$	1.50mm
H_{pocket}	6.72mm
D_{pocket}	6.93mm

Table 7. Optimal Configuration geometry 2 (output values)

Parameter	Predicted	Real value
ΔV	0.431	0.700
Ram Force	7.142	7.134
z	0,011	0.016

6 Conclusions

Porthole die extrusion has been utilized widely for producing lightweight material profiles characterized by complex shapes. Different variables have to be constantly monitored for guaranteeing good product quality; for maintaining these output values inside desirable ranges, several geometric and process parameters (input values) have to be set properly for each investigated geometry. This aspect drastically increases the complexity of the design phase as far as postpones the lead time before a new production rump up.

In the proposed work, an innovative procedure based on the recursive and intelligent use of already known knowledge in generating new one for new process design has been applied to solve the above specified problem.

A new algorithm able to determine automatically input-output relationships in manufacturing applications has been applied for mapping the porthole die extrusion process of complex hollow profiles. In detail, three input and two output variables have been taken into account.

The discussed results demonstrate that the proposed method based on an adaptive sampling DoE and a prediction tool based on shape similarities between a set of surfaces leads to better prediction compared to benchmark approaches used for this kind of problems.

The proposed methodology could be used to solve problems on different domains. However, this technique is particularly effective if at least one of this conditions is satisfied:

- restricted interest in a particular subarea of the solutions space;
- some interactions between input parameters are negligible.

All these conditions are exploited in order to obtain the desired information reducing the number of required data which is directly related to the cost and time necessary to perform the parameter selection for the process start up.

References

1. G. Liu, J. Zhou, J. Duszczyk, J. of Mat. Proc. Tech. **200/1-3**, 185-198 (2008);
2. H. Valberg, Int. J. of Mat. & Prod. Techn. **17**, 497–556 (2002);
3. L. Li, H. Zhang, J. Zhou, J. Duszczyk, G.Y. Li, Z.H. Zhong, Mater. & Des, **29/6**, 1190-1198 (2008);
4. K.J. Kim, C.H. Lee, D.Y. Yang, Jour. of Mater. Proc. Techn., **130-131**, 426-431 (2002);
5. J. Liu, G. Lin, D. Feng, Y. Zou, L. Sun, Central South University Technology, **17**, 688-696 (2010);
6. R. Mayavaram, U. Sajja, C. Secli, S. Niranjan, Proc. CIRP, **12**, 276-281 (2013).
7. F. Gagliardi, T. Citrea, G. Ambrogio and L. Filice, Mater. & Des. **60**, 274-281 (2014);
8. G. Fang, J. Zhou, J. Duszczyk, Jour. of Mater. Proc. Techn., **209**/6, 3050–3059 (2009);
9. M. Bauser, K. Siegert, Extrusion: Second Edition, ASM International 242-243 (2006);
10. Y. He, S. Xie, L. Cheng, G. Huang, Y. Fu, Trans of Nonf. Met. Soc. of China, **20**/6, 1067–1071 (2010);
11. J. Lof, Developments in finite element simulations of aluminum extrusion, thesis (2000);
12. Q. Li, C.J. Smith, C. Harris, M.R. Jolly, Jour. of Mater. Proc. Techn., **135**/2–3, 189–196 (2003);
13. J. Zhou, L. Li, J. Duszczyk, Jour. of Mater. Proc. Techn., **134**, 383–397 (2003);
14. A.J. Williams, A.K. Slone, T.N. Croft, M. Cross, Comp. Meth. in Appl. Mech. and Eng, **199/33–36**, 2123–2134 (2010);
15. M. J. Anderson, "Response Surface Methods (RSM) Achieve Design for Six Sigma (DFSS) Goals for Medical Device Manufacturing;
16. L. Monostori, Eng. Appl. Artif. Intel. **16.4** (2003): 277-291;
17. M.D. Shields, J. Zhang, Rel. Eng. & Syst. Saf. **148**, 96-108 (2016);
18. D. Basak, S. Pal, D.C. Patranabis, Neur. Inf. Processing-Letters and Reviews, **11.10** 203-224 (2007);
19. M. Collins, in Proc. ACL-02 Conf. Emp. Met. in Nat. Lan. Proc. 10, (2002);
20. X. Ma, M.B. de Rooij, D.J. Shipper, Wear, **278–279**, 1–8, (2012).
21. T. Citrea, F. Gagliardi, G. Ambrogio, L. Filice, in Proc. XII A.I.Te.M Conf. (2015).

Computer system for identification of tool wear model in hot forging

Marek Wilkus[1], Łukasz Rauch[1], Zbigniew Gronostajski[2], Sławomir Polak[2], Maciej Pietrzyk[1]

[1]*AGH University of Science and Technology, al. Mickiewicza 30, 30-059 Kraków, Poland*
[2]*Wrocław University of Technology, ul. Łukasiewicza 5-7, 50-370 Wrocław, Poland*

Abstract. The aim of the research was to create a methodology that will enable effective and reliable prediction of the tool wear. The idea of the hybrid model, which accounts for various mechanisms of tool material deterioration, is proposed in the paper. The mechanisms, which were considered, include abrasive wear, adhesive wear, thermal fatigue, mechanical fatigue, oxidation and plastic deformation. Individual models of various complexity were used for separate phenomena and strategy of combination of these models in one hybrid system was developed to account for the synergy of various mechanisms. The complex hybrid model was built on the basis of these individual models for various wear mechanisms. The individual models expanded from phenomenological ones for abrasive wear to multi-scale methods for modelling micro cracks initiation and propagation utilizing virtual representations of granular microstructures. The latter have been intensively developed recently and they form potentially a powerful tool that allows modelling of thermal and mechanical fatigue, accounting explicitly for the tool material microstructure.

1 Introduction

In industrial hot forging processes, prediction of die wear is critical for optimization of costs and materials usage. Proper die wear model allows to design production process with minimal amount of material waste and minimal time losses due to technical stops [1]. Die wear is an effect of many physical phenomena, such as friction (abrasive wear), material sticking (adhesive wear), thermal and mechanical fatigue, oxidation of working layers of material, etc. [3,2,4]. These mechanisms depend on many factors, both related to process and material. Majority of wear models used today in engineering applications is focused on one mechanism, while effects of other are ignored. This approach makes the wear prediction less useful, especially if wear in large product series (industrial application) is calculated. In processes with different geometries and parameters, wear depends on different mechanisms to various degrees. Even in a single process, in different parts of die different wear mechanisms may be of primary effect. Dies for hot forging usually have surfaces enhanced some way (thermal processing, nitriding, etc.), so after enriched layer wears off, other wear mechanisms occur and different die material and surface parameters have to be used. In modelling of tool wear, it is not always easy to identify all coefficients in models. Beyond identification of a particular model, an importance of different wear mechanisms has to be analysed in various parts of the die.

The objectives of the paper have been formulated having above remarks in mind. Creating a methodology that will enable effective and reliable prediction of the tool wear was the main objective.

2 Wear models

Die wear is a phenomenon which depends on many factors such as friction, thermal conditions (temperature, heat transfer coefficients), contact time, number of forging cycles in the same die, chemical parameters of die or die geometry. Various wear mechanisms are usually affecting properties of the die. Accounting for the synergy of these mechanisms (a single wear mechanism changes effects of other mechanisms) is the main challenge in modelling. Archard model of abrasive wear [5] is the most commonly used tool wear model in engineering applications. Similar model was proposed by Shaw [6]. This model covers only wear related to friction. The wear depth W is described in as:

$$W = \int C \frac{\mu p v}{HV} dt \tag{1}$$

where C is a non-dimensional coefficient, μ is friction coefficient, p is normal pressure, v is sliding velocity and HV is a hardness of a die material. In Authors previous work [7] it was shown that using this model sometimes requires changing of parameters after some wear threshold, which may be caused by changes of the die surface properties after enhanced layer (e.g. nitrided) wears off. To solve problem of large number of forgings, the model should be modified by introducing hardness dependent on various factors such as number of forgings, temperature, residual wear etc. [8].

Formula (1) can be also adapted to work with adhesive wear [9]. It was done by introducing hardness of the workpiece and adding exponential coefficients for friction, velocity and hardness parameters.

Simple models such as Paris equation [10] or more complex crack growth equations may be used for modelling of thermomechanical fatigue crack initiation and propagation. There is a variety of models available, from statistical or phenomenological ones, through formulas including deformation thresholds [11], to advanced formulas including deformation zones and crystallographic parameters changes [12]. Some of these formulas use parameters, which are difficult to obtain even from computer simulation, e.g. microstructural parameters of analysed part.

Another approach to modelling thermomechanical fatigue cracks is multi-scale analysis. In this approach statistically similar representative volume element (SSRVE) [13] of the die is analysed using Finite Element (FE) method (or e.g. cellular automata (CA) based on microstructure, meshless methods, etc.) or on a lower level (e.g. microstructure) returning material or crack properties to higher-level analysis. This approach may give more precise results, but it will require more computations (different parts of the die have different wear mechanisms) and must include changes of material properties (layer wear-off). Additionally, during subsequent forging cycles plastic deformation of crack as well as crack closure may occur, which strongly depends on materials ductility. Crack closing happens until some crack width is reached. After this, plastic deformation of workpiece occurs and fosters crack propagation. Thus, plastic deformation is another important wear mechanism of die itself. Plastic deformation increases by a small increment at each cycle, therefore, it is difficult to modelling. In many applications this mechanism has been simulated using model to suspected cumulative load [14], cycle simulation or steady-state simulation if possible [15]. Simulation of industrial cycle of forging is time and resource consuming. Thus, simple extrapolation-based models with geometry analysis have been used in the present work.

3 Computer system idea

To compute wear values, a hybrid computer system has been proposed, which combines numerical simulation of forging process with wear models. The schematic idea of this system is shown in figure 1. Parts of this system were created with user procedures implemented in commercial software, while other parts are standalone implementations. This approach allows further extension and quick testing of modifications. Execution of the system starts by specifying analysed problem (*Problem* component in figure 1). The problem consists of the following two parts:

- The numerical simulation of analysed process, prepared by using commercial tools.
- Information needed for wear modelling, e.g. initial die hardness used by abrasive wear module or chemical parameters for oxidation wear module.

The simulation is passed to the *Compiler* module which prepares the simulation and customizes it to compute the wear.

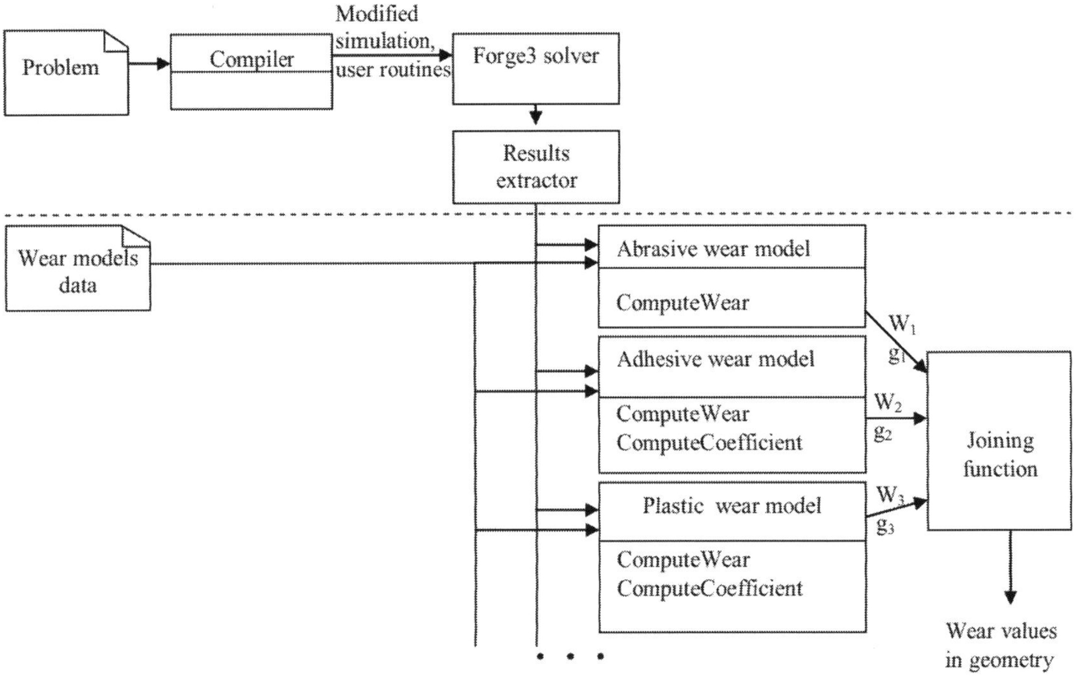

Figure 1. Schematic diagram of the computer system simulating hybrid wear model.

The most important parts of these customizations are:

- Introduction of specific output variables definitions from solver for further computations (e.g. sliding distance).
- Copying modified solver's user routines if crack initiation or propagation is computed.
- Verification of a die mesh regarding surface elements and problems resulting from improper surface mesh; Optional surface remeshing to improve accuracy.
- Compiling simulation to a commercial software (*Forge 3* component in figure 1); Optionally, generation of machine-specific files and start-up configurations.
- Execution of simulations.

After simulation finishes, post-processing routines are executed. Not only typical wear is computed, but also significance of different wear mechanisms must be resolved. Each post-processing module consists of procedure computing wear value and weight coefficient. Such models are getting information from data acquisition module. Separating solver from model using data acquisition and starting interface-like modules allows to develop the model with deeply modified solvers or different ones, as well as test multi-solver approach in future research. This architecture is commonly used in industry. The last stage of wear computation is applying joining function on wear models' results. This allows to obtain the final wear value.

4 Numerical simulation

Numerical simulation with Finite Element (FE) method was used to verify and evaluate wear models. There are many commercial FE packages for simulation of metal forming processes. All simulations in this work have been made using Forge 2011 package, which is widely used in engineering applications. This software uses Norton-Hoff constitutive law and has a large base of rheological models. It is also possible to add user routines, mainly for computation of scalar variables. To calculate wear, post-processing calculation has been used. The script-driven software for simulation automation has been implemented to form system based on a scheme shown in Figure 1. It allowed to prepare simulations with specially defined parameters as well as to perform multi-stage simulations with different variants of these parameters. Finally, the results are exported to compute wear using own models. Post-processing computations allow to find simple abrasive or other formula-defined wear equations, while e.g. special crack propagation models will have to be started with simulation. Evaluation of wear mechanisms begins with execution of a single simulation results of which are analysed to find importance of different wear mechanisms. Geometry properties are taken into account during analysis. For example, abrasive wear (computed from Archard model) depends on sliding distance, velocity and force. Contrary, the wear caused by plastic deformation of the die is in many cases caused by forces acting on specific parts of the die, prone to plastic deformation. The plastic effect of these forces is cumulative and it becomes visible when many forgings are produced. Plastic deformations are very small in a single forging simulation. Thus, the data need to be filtered to extract only those forces which cause plastic deformation. Such situations is recognized by analysing minimal displacements and material which is under deformation.

Tangent forces and forces parallel to workpiece flow direction are the most important factors in crack closing. This direction is obtained by analysing die and workpiece mesh from the two simulation steps. The distance difference for a forging node, which is the closest to analysed die node or surface, is considered. Then direction of material flow is the direction in which forces are analysed.

5 Hybrid model

The main idea of a hybrid model is usage of different dedicated models to simulate different wear mechanisms and combining their results in a single wear value. This approach requires not only computation of wear using different models, but also identification of coefficients of merging function, which in this research was a simple weighted sum function defined as:

$$W_{total} = g_{abrasive}W_{abrasive}+g_{adhesive}W_{adhesive}+g_{plastic}W_{plastic}... \quad (2)$$

where g_i are weight coefficients and W_i are results from different wear models. This merging function is simple, but it requires identification of weights. It is possible to use prior knowledge or analyse the case itself, e.g. by using an expert system. Results of this system are combined with an analysis of the die geometry. A simple processing routine has been developed to identify die areas, in which plastic deformation is likely to occur. Impact of this deformation is determined by analysing stresses acting on the die and small deformations which are repeated in subsequent forgings. Small deformation is used as a vector, which forms a line in die cross-cut. The line goes through a die volume until the surface (usually outside face of the die, if not, sum of other vectors is used) is inversely proportional to possible plastic deformation. The idea is based on the fact that if less material forms a protruding part exposed to high stresses, the more likely it will be affected by plastic deformation. The resulting lengths are normalized to form a coefficient for impact of the plastic deformation. Examples of measurements in two points in which impact of plastic deformation is different are shown in figure 2.

The system allows to use a case-based expert system to recognize wear mechanisms. In this type of system, die geometry is segmented into rules in which different wear mechanisms apply. It requires prior knowledge and exact identification of parameters in large number of rules, what is usually obtained from empirical data. The lesser data is in the system rules, the lower the quality of mechanisms identification will be. To combine rules, probability-based fuzzy approach was used in [16]. Such an attempt of classification by dividing typical geometry elements requires empirical data, which is not always accessible. The flowchart of hybrid model concept is shown in figure 3.

DOI: 10.1051/matecconf/20168011006

Figure 2. Measuring of lengths corresponding to amount of material prone to plastic deformation.

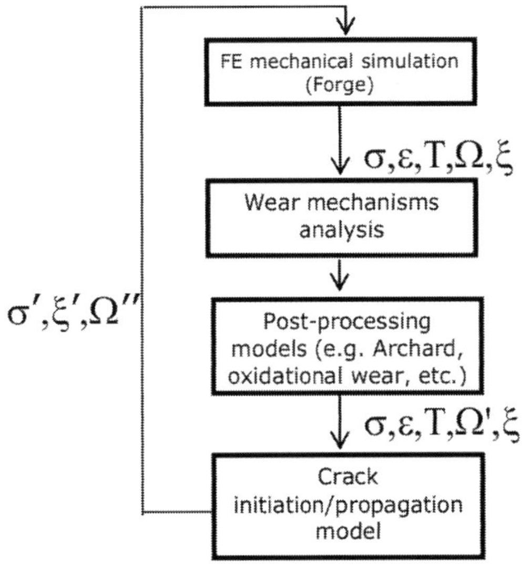

Figure 3. Data flow in hybrid wear model based on numerical simulation.

Work of the system begins with the FE simulation of a single process. Then, results are analysed and post-processing models such as Archard model are applied. Geometry is analysed to determine importance of wear mechanisms. Some models, such as multi-scale approach to prediction of propagation and initiation of cracks, have to be started in the next stage, when other mechanisms of wear are applied on geometry. Then, wear is computed and the procedure is optionally restarted. It should be emphasized, however, that restarting the procedure should be avoided and prediction based on a single computations should be used to minimize the computing costs.

6 Verification and identification

To verify hybrid model concept, simulations of the industrial forging were performed. Lower die in forging of a front wheel of the clutch described in [17] was considered. Nine measurement points, in which different wear mechanisms occur, were selected in this die, see. figure 4. Measurements of the die wear in this process were available.

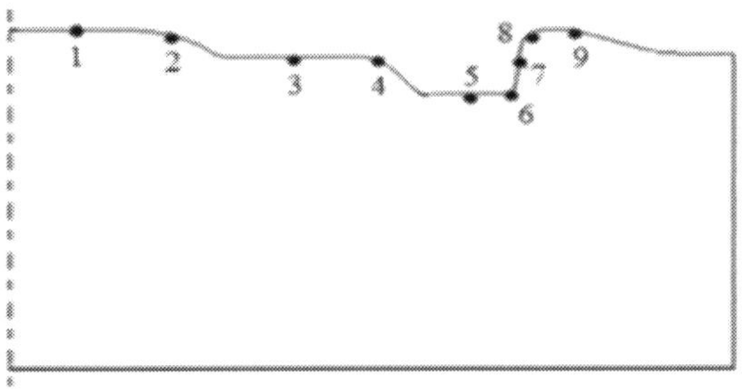

Figure 4. Analysed lower die with measurement points.

Simulation was a 3D FE and it included the die, which allowed to perform geometry analysis. The die material is X40Cr13 steel with hardness of 52 HRC after thermal processing. In the industrial forging cycle the wear was measured after 500, 6900 and 9000 forgings. Numerical computations of the abrasive wear have been performed. Following this parameters of the abrasive wear model have been identified as shown in [7]. Afterwards, plastic deformation model was computed and its impact was added giving a simple hybrid approach. Coefficients for equation (2) for two wear mechanisms have been identified by using mean square root error and geometry analysis. Values of plastic deformations at the die cross section are shown in figure 5.

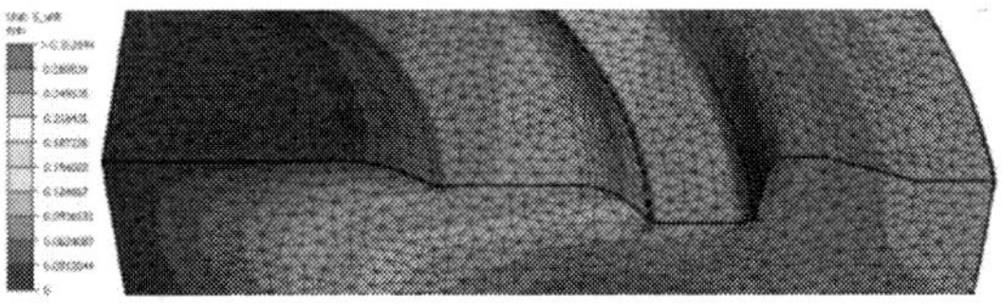

Figure 5. Visualization of die deformation during forging in a cross-cut plane.

Result of identification of equation (2) parameters is: $g_{abrasive}$ = 0.999535, $g_{plastic}$ = 0.000465. The results of comparison of measured and calculated abrasive wear for 9000 forgings are shown in figure 6. In points 1-4 the difference between hybrid and abrasive wear is relatively small, since plastic deformation is small. Calculated wear in points 5 and 6 is closer to measurements, but the plastic deformation in these points is probably caused mostly by force near point 8, giving a "lever effect" on the die. It should be noticed that wear in point 8 is closer to measured values. Then, the geometry-based coefficients adaptation has been performed using method described in Chapter 4. The wear obtained from this adaptation is also shown in figure 6 and it is different in each point. It should be noticed that these results are closer to measured values even in points 4, 5 and 6.

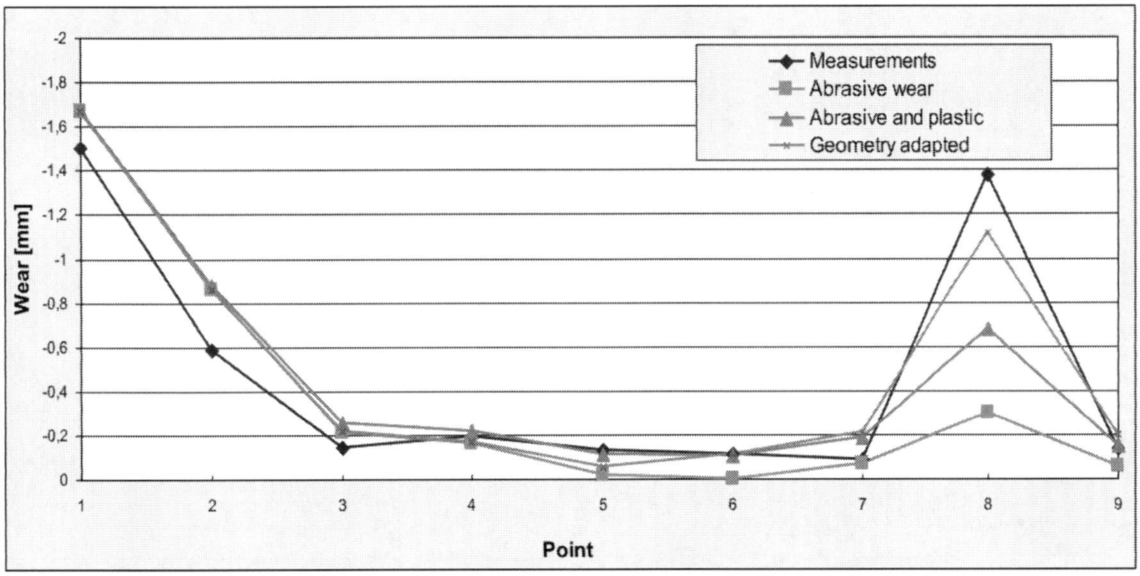

Figure 6. Results of abrasive and plastic wear computations with coefficient adaptation based on error and geometry

The values of computed coefficients are shown in Figure 7. It can be seen that in points 7, 8 and 9 plastic deformation have the biggest impact on wear, so the coefficient will be much higher in these points. In point 9, plastic deformation is mostly caused by material flash flowing outside a die, when deformed in the last stage of forging.

It is seen in figure 6 that application of the adaptation based on the error and the geometry resulted in a noticeable improvement of the accuracy of the model prediction in the point 8 comparing to the work [7]. The difference between measurements and predictions in other points than 8 are small and they may be a result of assumption that plastic deformation occurs uniformly in each of 9000 forgings.

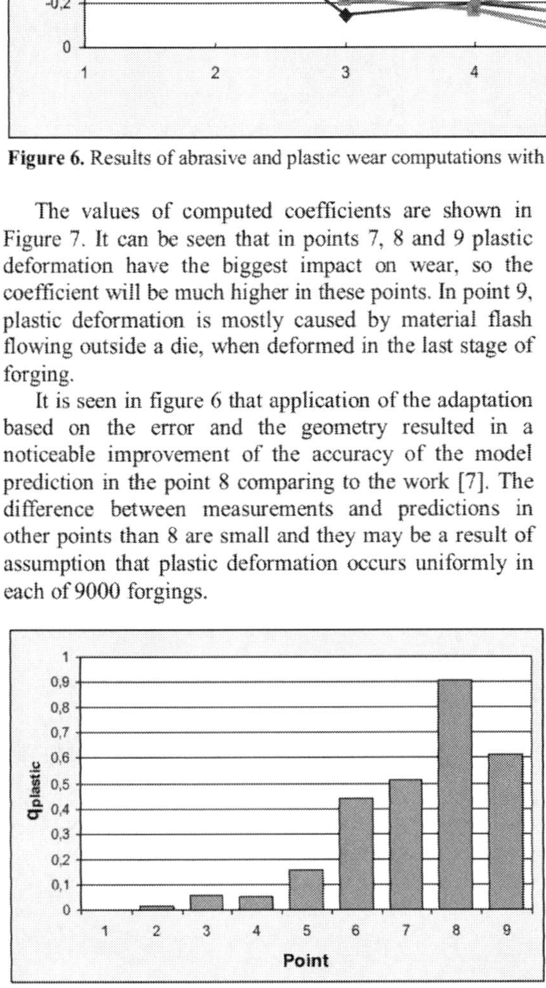

Figure 7. Values of plastic displacement wear coefficients at selected points.

General assumption, that plastic deformation occurs in a similar way in each cycle, is questionable. Exact computing of such deformations accounting for the die material rheology would require many cycles of simulations or metamodel trained from data acquired from real or simulated forging cycle. Part of the die with point 8 was indicated by geometry analysis as prone to plastic deformation (protruding part on workpiece expansion direction). The weight coefficient is relatively high, but computed values are still lower than measured.

7 Conclusions

It was shown in the paper that it is possible to combine different wear models to form a complex hybrid model, which covers different mechanisms of die destruction. In such approach even simple joining functions like weighted sum of components are sufficient. It is important to properly identify both parameters for different models of wear mechanisms, as well as coefficients for joining function. Analysis of results of simulations allowed evaluation of importance of different wear mechanisms in specified parts of the die. Wear mechanisms could be distinguished. These coefficients can also be computed using an expert system, which detects similarities in simulation results or parameters, but it requires empirical knowledge to be used as a case base.

It was shown in the paper that hybrid model approach is an efficient solution to combined wear mechanisms modelling. Further research would require an extension of the hybrid by including more single mechanisms models. More measurements of forging tools, including oxidation layers, crack parameters and surface properties, would be needed to reach this goal. Factors responsible for the impact of other wear mechanisms in parts of the die would have to be identified.

Acknowledgements

Financial support of the Polish National Science Foundation NCN, project, no. 2014/15/B/ST8/00187, is acknowledged.

References

1 B.-A. Behrens, F. Schäfer, F., Steel Res. Int. **80** (2009) 887-891.
2. S. Abachi, M. Akkok, M.I. Gokler, Tribology International **43** (2010) 467–473.
3. Y.-J. Kim, C.-H. Choi, Int. J. Precision Eng. Manuf., **10** (2009) 105-113..
4. Z. Gronostajski, M. Kaszuba, M. Hawryluk, M. Zwierzchowski, Arch. Civil. Mech., Eng. **14** (2014) 528-539.
5. J.F. Archard, J. Appl. Phys. **24** (1953) 981-988.
6. M.C. Shaw, Wear **43** (1977) 263-266.
7. M. Wilkus, S. Polak, Z. Gronostajski, M. Kaszuba, Ł. Rauch, M. Pietrzyk, Comp. Meth. Mater. Sci. **15** (2015) 311-321.
8. J.H. Kang, I.W Park, J.S. Jae, S.S. Kang, J. Mat. Proc. Techn. **94** 1999 183-188.
9. C. Dahl, V.H. Vazquez, T. Altan, Proc. XX Conf. SENAFOR, Porto Allegre, 2000.
10. P.C. Paris M.P. Gomez, Anderson W.E., The Trends in Engineering, 13, 1961 9-14.
11. B. Tomkins, Philosophical Magazine **18** (1968) 1041-1066
12. M.M. Mirsayar, Materials and Design **86** (2015) 941-947.
13. J. Schroeder, D. Balzani, D. Brands, Arch. Appl. Mech. **81** (2011) 975-997.
14. J.L. Chaboche, European J. of Mechanics **13** (1994) 501-518.
15. C. Choi, A. Groseclose, T. Altan, J. Mat. Proc. Techn. **212**, 2012, 1742-1752.
16. Z. Gronostajski, M. Hawryluk, M. Kaszuba, M. Marciniak, A. Niechajowicz, S. Polak, M. Zwierzchwoski, A. Adrian, B. Mrzygłód, J. Durak, Int. J. Adv. Manuf. Techn. **82** (2016) 1973–1991.
17. Z. Gronostajski, T. Będza, M. Kaszuba, M. Marciniak, S. Polak, Obróbka Plastyczna **25** (2014) 301-315.

Case study on the influence of kinematic hardening within a parameter-free and non-invasive form finding approach

Philipp Landkammer[1,a] and Paul Steinmann[1]

[1]*Chair of Applied Mechanics, Friedrich-Alexander University Erlangen-Nuremberg, 91058 Erlangen, Germany*

Abstract. Inverse form finding – as a type of shape optimization – aims in determining the optimal preform design of a workpiece for a specific forming process, whereby the desired target geometry is known. Recently, a novel parameter-free and heuristic approach was developed to tackle this nonlinear optimization problem. Benchmark tests already delivered promising results. As a particular note-worthy feature of the approach, a coupling to an arbitrary finite element software is feasible in a non-invasive fashion. The aim of this contribution is to investigate the effect of kinematic hardening and cyclic loading on the convergence behavior of the algorithm.

1 Introduction and related works

Within the metal forming industry, trial-and-error methods of earlier days have since long been replaced by simulative predictions and simulation based optimization strategies. This leads not only to a cost reduction and an acceleration in the development process of new products, but is nowadays also a cornerstone for the manufacturing of highest quality and near-net-shaped functional components. Optimization in order to improve product properties is thereby often the main purpose of numerical forming simulations and consequently a key point in the field of future research, see Andrietti et al. [1]. Thereby, the main issue of optimization, besides the inverse problem of parameter identification, is the determination of an optimal tool and workpiece design, see Chenot et al. [2].

The latter constitutes the investigated type of optimization in this contribution and is also frequently denoted as inverse form finding. It aims in determining the optimal preform design of a workpiece for a specific forming process, whereby the desired target shape is known a priori. Especially in the context of sheet and bulk metal forming, or forging respectively, several concepts have been developed since decades to solve this non-linear optimization problem. Since path-dependency is occurring in plasticity, a similar iterative procedure of all shape optimization approaches can be recognized. It consists of updating the geometry of the material space, before re-starting a new simulation and comparing again the forming results of the spatial space to the desired target shape.

Form finding strategies are commonly distinguished between parameter-free and parameter-based approaches. In the former, a computer aided design (CAD) geometry is given and its control points serve as design variables, as initially proposed by Braibant and Fleury [3]. For example, Fourment and Chenot [4] already proposed a parameter-based optimization approach for determining an optimal workpiece and die design in non-steady-state forming processes. Within parameter-free approaches, as considered in this contribution, the geometry is given in a discretized setting as a finite element (FE) model and its nodal positions serve as design variables. This leads to a larger number of design parameters. Furthermore, nodal-based mesh moving strategies often suffer from distorted elements when updating the material configuration so that regularization and filter techniques have to be deployed, see Bletzinger et al. [5] and Scherer et al. [6].

As a further distinction, different strategies can be found for sheet and for bulk metal forming. Within sheet metal forming, the sought optimal material configuration renders commonly a planar sheet pre-cut and the aim is to determine an optimal cut-out contour, e.g. described by spline functions as in Kim et al. [7]. Before re-starting the forming simulation in the iterative optimization procedure, the surface inside the cut-out contour is automatically re-meshed in most cases, which is a feasible task for two-dimensional structural elements like shells. A different strategy which repositions the nodal coordinates of a discretized planar sheet pre-form without remeshing is presented by the inverse approach of Gou and Batoz [8]. In contrast, within bulk metal forming simulations three dimensional geometries can only be remeshed automatically with great difficulties, so parameter-free mesh moving strategies are predominated in this context.

It should be emphasized that several types of sensitivity analyses (see Haftka and Adelmann [9] for an overview) exist within both, parameter-based and parameter-free approaches, and have already been successfully adopted to the context of inverse form

[a] Corresponding author: philipp.landkammer@fau.de

finding. Michaleris and Tortorelli [10] for example provided a general framework for the calculation of elasto-plastic tangent operators or Archajee and Zabaras [11] already presented a sensitivity approach for three-dimensional contact problems. However, in most cases it is cumbersome to compute the required sensitivity derivations of the system responses.

A further interesting and promising approach is proposed by Germain et al. [12] for elasto-plastic problems. Here an inverse mechanical formulation, parametrized with spatial coordinates, is solved with respect to the material coordinates, as introduced by Govindjee et al. [13] for elasticity. For the extension to plasticity, the internal plastic variables are recursively transferred to circumvent the path-dependency.

Recently, a novel heuristic and parameter-free approach was developed for inverse form finding problems, see Landkammer and Steinmann [14]. The updating operation is thereby performed by a reverse tangent mapping after analyzing the nodal distances to the target shape in the final deformation stage. As a benefit of the algorithm, a coupling by subroutines to any desired external FE code is feasible. Benchmark tests already demonstrated an excellent numerical efficiency. Also the applicability to metal forming processes have been already successfully shown [15]. The main goal of this contribution is to evaluate, whether a sophisticated hardening model or cyclic loading with reverse strain path changes affect the convergence behavior. This is realized by a case study with computational examples.

The paper is structured as follows: **Sec. 2** explains the functional principle of the non-invasive form finding algorithm. Afterwards, **Sec. 3** reiterates some basics from plasticity regarding isotropic and kinematic hardening, before the benchmark model for the computational examples is presented. A subsequent parameter study regarding the constitutive behavior and the loading path is conducted to demonstrate the applicability to combined hardening and cyclic loading. Finally, **Sec. 4** draws a conclusion to summarize the findings.

2 Functional principle of the iterative update procedure

In this section, the basic framework of the update operation is explained, while a more detailed description is presented in Landkammer and Steinmann [14]. The iterative optimization algorithm is purely based on the nodal positions of the material and the spatial configurations. Almost every FE software provides options to export these data by subroutines. After the determination of improved material nodal positions within the update procedure, they are imported again and the next re-computation of the forming simulation starts automatically until suited spatial nodal positions are computed. Due to the ability of coupling the optimization program as a black box to the user's habitual forming software without interfering with the FE code of the simulation itself, the approach is denoted as non-invasive.

Figure 1 additionally illustrates the nodal based inverse form finding strategy.

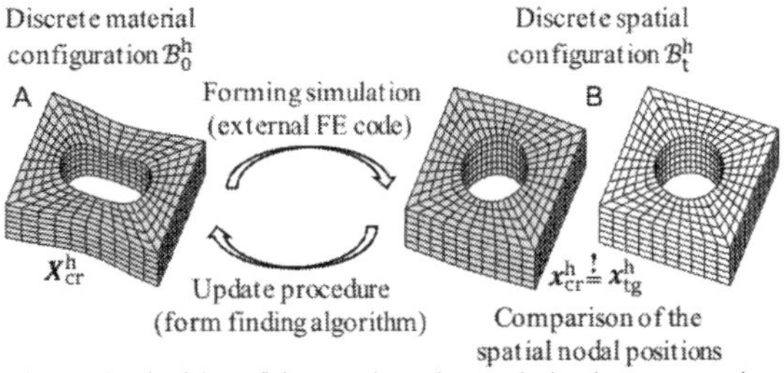

Figure 1. The idea of the non-invasive optimization strategy is to update the mesh of the material configuration (A) until the nodal positions computed by a forming simulation are close enough to the nodal positions of a target mesh in the discretized spatial configuration (B)

Mathematically, the forming process of a body renders a mapping from the material to the spatial configuration. With respect to a FE simulation, this can be formulated by a map $\boldsymbol{\varphi}$ of the coordinates $\boldsymbol{X}^h$ of a discretized material configuration $\mathcal{B}_0^h$ to the coordinates $\boldsymbol{x}^h$ of the spatial configuration $\mathcal{B}_t^h$:

$$\boldsymbol{x}^h = \boldsymbol{\varphi}(\boldsymbol{X}^h) : \mathcal{B}_0^h \rightarrow \mathcal{B}_t^h \tag{1}$$

As an objective for the optimization, a discretized target configuration with spatial coordinates $\boldsymbol{x}_{tg}^h$ has to be defined a priori. The optimal material coordinates are then determined, if the nodal distances between both spatial configurations are close enough for all element nodes ($I = 1, \cdots, n_{nodes}$). This is evaluated by the objective function δ_{rms}, which renders the root mean square over all spatial nodal distances, whereby $\boldsymbol{d}^I$ refers to the nodal difference vectors in the spatial setting.

$$\delta_{rms}(\mathbf{x}_{tg}^h, \mathbf{x}_{cr}^h) = \sqrt{\frac{1}{n_{nodes}} \sum_{I=1}^{n_{nodes}} |\boldsymbol{d}^I|^2} < tol. \tag{2}$$

$$\text{with } \boldsymbol{d}^I = \boldsymbol{x}_{tg}^I - \boldsymbol{x}_{cr}^I(\boldsymbol{X}_{cr}^h) \tag{3}$$

For a suited repositioning of the material coordinates, the final deformation state after the forming simulation has to be analyzed. The idea is to exploit the linear tangent map $\boldsymbol{F}^h$, which arises from the derivative of the deformation map $\boldsymbol{\varphi}$ with respect to the design parameters $\boldsymbol{X}^h$, for a node-wise fixed-point iteration to minimize the objective function $\delta_{rms}(\boldsymbol{x}_{tg}^h, \boldsymbol{x}_{cr}^h)$.

$$\boldsymbol{F}^h = \nabla_{\boldsymbol{X}^h} \boldsymbol{\varphi}(\boldsymbol{X}^h) : T\mathcal{B}_0^h \rightarrow T\mathcal{B}_t^h \tag{4}$$

However, the tangent map cannot be computed directly at the nodal positions $\boldsymbol{X}^I$, but the deformation gradient $\boldsymbol{F}^j$ is only given at the Gauss points. The computation of the deformation gradient is feasible with the Jacobian matrices of the material and the spatial configurations, whereby $N^a(\boldsymbol{\xi}^j)$ refers to isoparametric shape functions evaluated at the j-th Gauss point:

$$\boldsymbol{F}^j = J_x^j \cdot [J_X^j]^{-1} \tag{5}$$

with $J_X^j = \sum_{a=1}^{n_{en}} X^a \otimes \nabla_\xi N^a(\xi^j)$

and $J_x^j = \sum_{a=1}^{n_{en}} x^a \otimes \nabla_\xi N^a(\xi^j)$

To achieve an approximation $\tilde{F}$ of the deformation gradient at the nodal positions, a least squares smoothing is performed, as described by Hinton and Campbell [16]. This smoothing technique is also commonly used in FE codes for mapping the entries of a stress tensor from the Gauss points to the element nodes. With the inverse of the smoothed nodal deformation gradient, a difference vector d^I can be mapped from the spatial configuration $\mathcal{B}_t^h$ to the discretized material configuration $\mathcal{B}_0^h$.

$$D^I = \left[\tilde{F}\right]^{-1} \cdot d^I \tag{6}$$

The update of the material configuration is then performed by adding all mapped difference vectors D^I to their material nodal positions:

$$X^I \leftarrow X^I + \alpha D^I \tag{7}$$

In order to control the update procedure in the case of mesh distortions, a global relaxation parameter α is introduced in Eq. (7). Before importing the updated material nodal positions and restarting the FE simulation, the mesh quality is analyzed by an elementwise evaluation of discrete Jacobian matrices as proposed in Knupp et al. [17]. If needed, the relaxation parameter is then automatically reduced, which helps to render the algorithm more robust.

3 Example: Tension and compression of a cuboid with a cylindric hole

The investigated model is a cuboid with a cylindric hole in its center and has already been used for benchmark tests within inverse form finding problems in [14] and [18].

The aim in this contribution is to analyze the effect of a material law with kinematic hardening to the convergence behavior of the algorithm. The dual phase steel DP600, which is a typical material in automotive industries or for Sheet-Bulk metal formed components [20] respectively, is used for the simulative investigations.

3.1. The material model of DP600 steel with isotropic and kinematic hardening

Some basics from plasticity theory are firstly reiterated. Within an elasto-plastic analysis, a yield function

$$\Phi(\bar{\sigma}, \sigma_y) = \bar{\sigma}(\boldsymbol{\sigma}, \boldsymbol{B}) - \sigma_y(\epsilon^p) \tag{8}$$

has to be evaluated for the current stress state to prove if plastic flow occurs. Herein, the equivalent stress $\bar{\sigma}$ is compared to the current yield stress σ_y and the yield condition $\Phi = 0$ has to be satisfied at any time in case of plasticity. The equivalent stress renders e. g. for the commonly used orthotropic Hill yield criterion

$$\bar{\sigma}(\boldsymbol{\sigma}, \boldsymbol{B}) = \sqrt{\frac{3}{2}}\sqrt{[\boldsymbol{\sigma} - \boldsymbol{B}]^{\mathrm{T}} : \mathbb{H} : [\boldsymbol{\sigma} - \boldsymbol{B}]} \tag{9}$$

with the Cauchy stress tensor $\boldsymbol{\sigma}$, the back stress tensor $\boldsymbol{B}$ and the fourth order Hill tensor $\mathbb{H}$ (respectively $\mathbb{H} = \mathbb{I}_{\mathrm{dev}}^{\mathrm{sym}}$ for the isotropic v.Mises yield criterion with the symmetric deviatoric fourth order unit tensor $\mathbb{I}_{\mathrm{dev}}^{\mathrm{sym}}$). For the sake of an accurate modelling of the physical behavior of elasto-plastic materials, two main types of hardening are usually considered in the context of metal forming.

Firstly, isotropic hardening is used to model an expansion of the yield surface. Thereby, the hardening function $\sigma_y(\epsilon^p)$ of the current yield stress with respect to the current equivalent plastic strain ϵ^p is evaluated. A frequently used model for isotropic hardening is for example the Voce type saturation law:

$$\sigma_y(\epsilon^p) = \sigma_0 + \sigma_{sat}[1 - \exp(-C\epsilon^p)] \tag{10}$$

Secondly, kinematic hardening is used to model a translation of the yield surface in the stress space. Here, the back stress tensor has to be determined simultaneously by an additional evolution equation:

$$\partial_t \boldsymbol{B} = B\left[\frac{\sigma_B}{\bar{\sigma}}\,\mathrm{dev}(\boldsymbol{\sigma} - \boldsymbol{B}) - \boldsymbol{B}\right]\partial_t \epsilon^p \tag{11}$$

Kinematic hardening has to be considered to capture the so-called Bauschinger effect. This refers to a differing yield stress in case of a reverse strain path change when loading is applied in the opposite direction than before, as illustrated in **Figure 2**.

The parameters for the DP600 dual phase steel, see **Table 1**, are determined by Bouvier et al. [19] by weighting the results of several mechanical tests. Oliveira et al. [21] further utilized the parameter set for analyzing the springback prediction of sheets, in which the modelling of the Bauschinger effect has a major impact. As a feature, the parameter sets are thereby separately identified in [19] on the one hand for a pure isotropic hardening ($\boldsymbol{B} = \boldsymbol{0}$) with the Voce law, see Eq. (10) and on the other hand for a combined hardening model. In the latter, the Voce law is again used for the isotropic hardening and the Frederick-Armstrong law, see Eq. (11), is additionally applied for modelling the kinematic hardening effect.

Table 1. Material parameters of DP600 high strength steel determined by experimental test in Bouvier et al. [19]

Material data	Isotropic	Combined
Voce, see Eq. (1)	$\sigma_0 = 330.3$ MPa, $C = 16.3$, $\sigma_{sat} = 516.4$ MPa	$\sigma_0 = 308.3$ MPa, $C = 6.75$, $\sigma_{sat} = 365.6$ MPa
Frederick Armstrong, see Eq. (2)	-	$\sigma_B = 225.3$ MPa, $B = 73.7$

The Bauschinger effect is additionally demonstrated for the material parameters of the isotropic and the combined material law by a cyclic loading (tension / compression / tension) applied to a single element within a displacement controlled computation. In the case of a pure isotropic material model, the absolute value of the yield stress stays constant (*blue curve,* 790.4 MPa) when applying the displacement load in the opposite direction. In contrast, in case of a combined hardening model a lower absolute value of the yield stress (*green curve,* 453.5 MPa instead of 678.7 MPa) is observed upon reverse loading due to the translation of the yield surface **by the back stress tensor (*red curve,* 225.2 MPa), see Figure 2.**

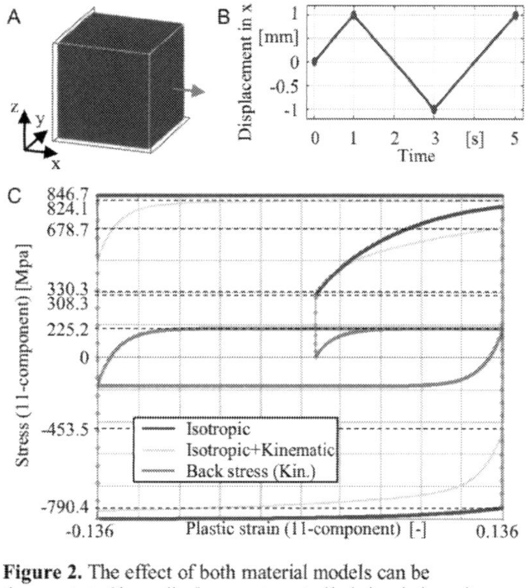

Figure 2. The effect of both material models can be demonstrated by a displacement controlled simulation of a single element (A) with a cyclic loading (B). Only the combined hardening law is able to capture the Bauschinger effect (C)

3.2 Simulation

For evaluating the influence of an isotropic and a combined material modelling on the convergence behavior of the form finding algorithm, the benchmark model of a cuboid with a cylindric hole is analyzed. **Figure 3** depicts the dimension of the workpiece and illustrates the symmetry planes, which corresponds to the boundary conditions. The geometry is discretized in the FE model with 150 solid elements (264 nodes). In the subsequent optimization the material start configuration equals the desired spatial target configuration.

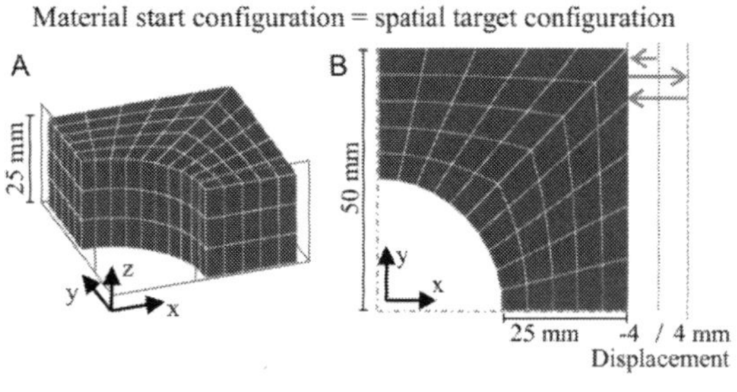

Figure 3. The material start configuration (A) equals the desired spatial target configuration (B). The dimensions of the FE model are stated and the symmetry planes are depicted

The material model of the DP600 steel is adopted according to **Table 1**. The computation is carried out with multiplicative large strain plasticity with the commercial FE software *MSC.MarcMentat.*

For a comparison of the convergence behavior, the simulations are performed four times combining both hardening models with two different loadings, as listed in **Table 2**. Firstly, a monotonic loading of a pure compression (displacement of -4 mm in x-direction, 0-1 s) is applied for the isotropic and the combined material model. Secondly, a cyclic loading by a changing compression and tension (displacements of -4/ 4/ -4 mm in x-direction, 0-5 s) is applied for the hardening models. The according load path of the displacement controlled forming simulation is depicted in **Figure 4**.

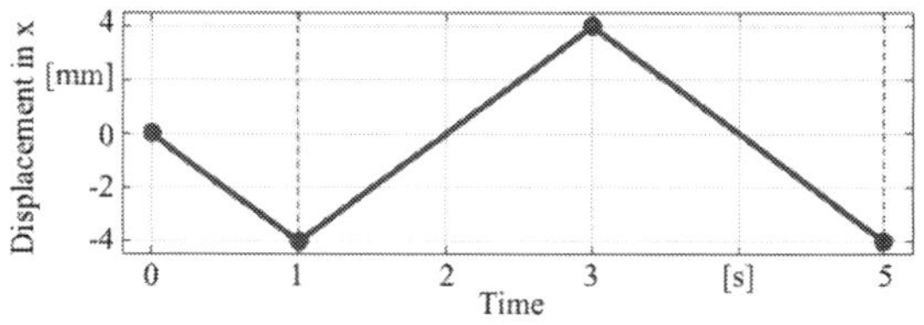

Figure 4. The path for the monotonic (0-1 s) and cyclic loading (0-5 s) of the displacement controlled forming simulation

Table 2. Case study: Combinations of an isotropic and a combined hardening law, respectively a monotonic and a cyclic loading are investigated in order to evaluate the effects on the convergence rate of the form finding algorithm

Case	Hardening law	Loading path (time)
1	Isotropic	Monotonic (0-1 s)
2	Combined	Monotonic (0-1 s)
3	Isotropic	Cyclic (0-5 s)
4	Combined	Cyclic (0-5 s)

3.3 Optimization

The optimization algorithm is implemented in *Matlab*. For the update of the material configuration of the workpiece, a constant relaxation parameter ($\alpha = 1$) is used in all iterations and for all four modelled variants.

Figure 5 shows the determined optimal material configuration for the pure compression (monotonic loading, 4 mm in x-direction) with the underlying isotropic (A) and the combined hardening model (B). Inputting the determined optimal material configurations, subsequent forming simulations compute the desired spatial nodal positions of the given target configuration (C, D).

For comparison, **Figure 6** shows the determined optimal material configuration for the cyclic loading with reverse strain path changes (-4/ 4/ -4 mm in x-direction) with the underlying isotropic (A) and the combined hardening model (B). Subsequent forming simulations after inputting the determined optimal material configurations compute again the desired spatial nodal positions of the given target configuration (C, D). The equivalent plastic strains ϵ^{p}, the equivalent stresses $\bar{\sigma}$ and the equivalent back stresses $\bar{B}$ are uniformly scaled and depicted.

DOI: 10.1051/matecconf/20168011007

The computed maximal equivalent plastic strains vary between 0.265 and 0.287, whereas the maximal equivalent stresses range from 772 MPa to 848 MPa.

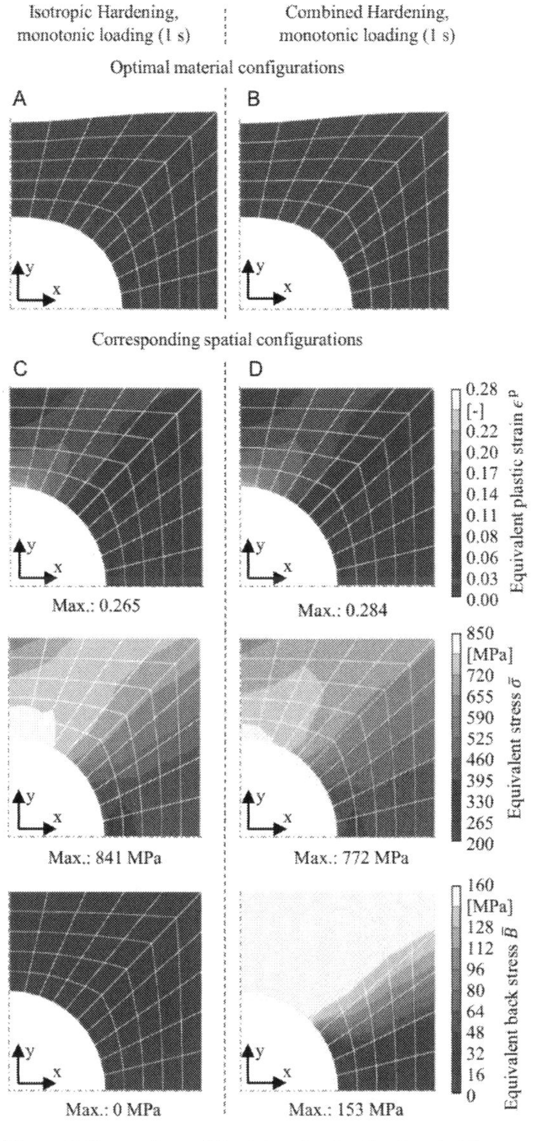

Figure 5. Comparison of the determined optimal material configurations (A, B) with the isotropic and the combined hardening model and their corresponding spatial configurations (C, D) with equivalent strains and stresses after applying a monotonic loading (see **Figure 4**, 0-1 s)

In all cases, the algorithm continuously decreases the objective functions δ_{rms}, which can also be interpreted as the nodal standard deviation between the computed and the target configuration.

Furthermore, nearly linear convergence rates can be observed, see **Figure 7**. The investigation demonstrates that applying a combined hardening model and a cyclic loading does not poses problems to the optimization approach.

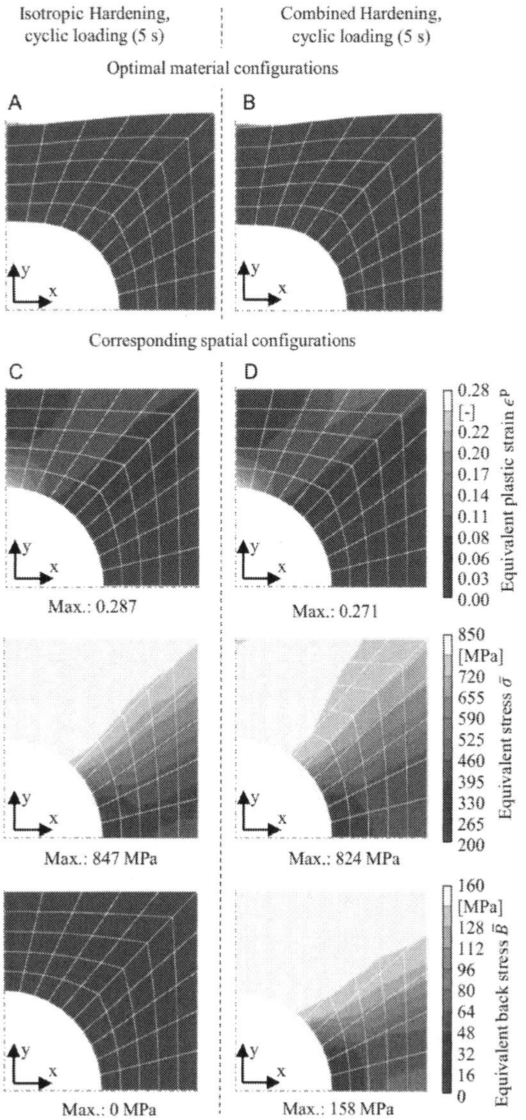

Figure 6. Comparison of the determined optimal material configurations (A, B) with the isotropic and the combined hardening model and their corresponding spatial configurations (C, D) with equivalent strains and stresses after applying a cyclic loading (see **Figure 4**, 0-5 s)

A minimal error value δ_{rms} close to 10^{-15} is achieved after 25 (case 1: isotropic hardening / monotonic loading) to 30 iterations (case 3: isotropic hardening / cyclic loading). In the case of employing the combined hardening model no tendency regarding the efficiency is apparent (slightly more iterations for monotonic loading, but less iterations for cyclic loading).

In contrast, a slight difference can be observed, when applying a cyclic loading (more iterations are needed) in comparison to the case of a monotonic loading. A clear relation between the amount of the equivalent stresses, or equivalent plastic strains respectively, and the convergence rates is not recognizable.

Table 3 lists additionally the error values of the objective function and shows again clearly the excellent efficiency of the non-invasive approach.

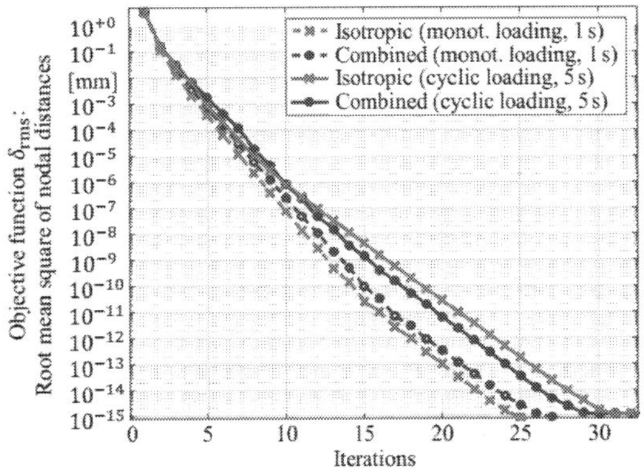

Figure 7. For all cases the objective function δ_{rms} of the root mean square of nodal distances decreases continuously within a nearly linear convergence rate during the optimization procedure

Table 3. The error values of the objective function $\boldsymbol{\delta_{rms}}$ show a good agreement of the computed and the target spatial configuration within only a few optimization steps

Iter.	Objective function δ_{rms}: Root mean square error in mm			
	Isotropic / Monotonic	Combined / Monotonic	Isotropic / Cyclic	Combined / Cyclic
1	3.34	3.38	3.46	3.53
2	$1.1\cdot 10^{-1}$	$1.4\cdot 10^{-1}$	$1.6\cdot 10^{-1}$	$1.6\cdot 10^{-1}$
3	$1.2\cdot 10^{-2}$	$1.9\cdot 10^{-2}$	$2.7\cdot 10^{-2}$	$3.1\cdot 10^{-2}$
4	$2.0\cdot 10^{-3}$	$3.6\cdot 10^{-3}$	$6.4\cdot 10^{-3}$	$6.9\cdot 10^{-3}$
5	$3.6\cdot 10^{-4}$	$7.3\cdot 10^{-4}$	$8.8\cdot 10^{-4}$	$1.8\cdot 10^{-3}$
6	$6.4\cdot 10^{-5}$	$1.4\cdot 10^{-4}$	$2.1\cdot 10^{-4}$	$4.2\cdot 10^{-4}$
⋮	⋮	⋮	⋮	⋮
30	$5.0\cdot 10^{-16}$	$6.0\cdot 10^{-16}$	$1.6\cdot 10^{-15}$	$1.1\cdot 10^{-15}$

4 Conclusion

Kinematic hardening is of crucial importance when the forming of a workpiece includes strain-path changes, as typically occurring e. g. in bulk forming of a sheet-metal. Therefore, the convergence behavior of a non-invasive iterative approach for inverse form finding, whose operating principle is briefly summarized, is investigated by a case study with a pure isotropic and a combined (isotropic and kinematic) hardening law. It is demonstrated that different hardening models does not affect the performance of the algorithm. Furthermore, only slightly more iterations are needed when applying a cyclic instead of a monotonic loading.

Acknowledgement

The work was supported by the *German Research Foundation* (DFG) within the scope of the Transregional Collaborative Research Center (SFB/TR 73) *Manufacturing of complex functional components with variants by using a new metal forming process - Sheet-Bulk Metal Forming* (http://www.tr-73.de).

References

1. S. Andrietti, J.L. Chenot, M. Bernacki, P.O. Bouchard, L. Fourment, E. Hachem, E. Perchat, Comp. Meth. Mat. Sc. **15-2**, pp. 265-293 (2015)
2. J.L. Chenot, E. Massoni, L. Fourment, Engrg. Comp. **13 2-4**, pp. 190-225 (1996)
3. V. Braibant, C. Fleury, Comp. Meth. Appl. Mech. Engrg. **44-3**, pp. 247-267 (1984)
4. L. Fourment, J.L. Chenot, Int. J. Num. Meth. Engrg. **39**, pp. 33-50 (1990)
5. K.U. Bletzinger, R. Wuechner, F. Daoud, N. Camprubi, Comp. Meth. Appl. Mech. Engrg. **194**, pp 3438-3452 (2005)
6. M. Scherer, R. Denzer, P. Steinmann, Num. Meth. Engrg. **82-3**, pp. 269-302 (2009)
7. J.Y. Kim, N. Kim, M.S. Huh, J. Mat. Proc. Tech. **101**, pp. 31-41 (2000)
8. Y.Q. Guo, J.L. Batoz, J.M. Detraux, P. Duroux, Int. J. Num. Mech. Engrg. **30**, pp. 1385-1401 (1990)
9. R.T. Haftka, H.M. Adelmann, Struct. Opt. **1**, pp. 137-151 (1989)
10. P. Michaleris, D. Tortorelli, C. Vidal, Int. J. Num. Meth. Engrg. **37-14**, pp. 2471-2499 (1994)
11. S. Acharjee, N. Zabaras, Comp. Meth. Appl. Mech. Engrg. **195**, pp. 6822-6842 (2006)
12. S. Germain, P. Landkammer, P. Steinmann, Adv. Mod. Sim. Engrg. **10-1** (2013)
13. S. Govindjee, P.-A. Mihalic, Comp. Meth. Appl. Mech. Engrg. **136-1/2**, pp. 47-57 (1996)
14. P. Landkammer, P. Steinmann, Comp.Mech., **57-2**, pp. 169-191 (2015)
15. P. Landkammer, T. Schneider, R. Schulte, P. Steinmann, M. Merklein, Prod. Eng. Res. Dev., **10-1**, pp. 93-102 (2016)
16. E. Hinton, J.-S. Campbell, Int. J. Num. Meth. Engrg. **8-3**, pp. 461-480 (1974)
17. P. Knupp, Int. J. Num. Meth. Engrg. **48**, pp. 1165-1185 (2000)
18. S. Germain, Ph.D. thesis, Schriftenreihe Technische Mechanik **14**, (2015)
19. S. Bouvier, C. Teodosiu, C. Maier, M. Banu, V. Tabacaru, WP3, Progress Report of 3DS (1999)
20. M. Merklein, J.M. Allwood, B.A. Behrens, A. Brosius, H. Hagenah, K. Kuzmann, K. Mori, A.E. Tekkaya, A. Weckenmann, CIRP Annals – Manuf. Techn. **61-2**, pp. 725-745 (2012)
21. M.C. Oliveira, J.L. Alves, B.M. Chaparro, L.F. Menezes, Int. J. Plast. **23**, pp. 516-543 (2007)

Simulation of Friction Stir Processing in 304L Stainless Steel

M.P. Miles[1a], T.W. Nelson[2], F.C. Liu[2], C. Gunter[1], and L. Fourment[3]

[1]*Manufacturing Engineering Technology, Brigham Young University, Provo, UT 84602, USA*
[2]*Mechanical Engineering, Brigham Young University, Provo, UT 84602, USA*
[3]*Mines ParisTech, Centre de Mise en Forme des Materiaux (CEMEF), 06904 Sophia-Antipolis Cedex, France*

Abstract. A major dilemma facing the nuclear industry is repair or replacement of stainless steel reactor components that have been exposed to neutron irradiation. When conventional fusion welding is used for weld repair, the high temperatures and thermal stresses inherent in the process enhance the growth of helium bubbles, causing intergranular cracking in the heat-affected zone (HAZ). Friction stir processing (FSP) has potential as a weld repair technique for irradiated stainless steel, because it operates at much lower temperatures than fusion welding, and is therefore less likely to cause cracking in the HAZ. Numerical simulation of the FSP process in 304L stainless steel was performed using an Eulerian finite element approach. Model input required flow stresses for the large range of strain rates and temperatures inherent in the FSP process. Temperature predictions in three locations adjacent to the stir zone were accurate to within 4% of experimentally measure values. Prediction of recrystallized grain size at a location about 6mm behind the tool center was less accurate, because the empirical model employed for the prediction did not account for grain growth that occurred after deformation in the experiment was halted.

1 Introduction

The nuclear industry is facing challenges in repair or replacement of stainless steel reactor components, which have been exposed to neutron irradiation. Irradiated stainless steel contains helium, which forms primarily by n/α reactions with alloying elements nickel and boron [1-6]. Helium gas is insoluble in metals, so it precipitates and forms bubbles (at the nanoscale), nucleating at dislocations, point defects, and grain boundaries [4]. When conventional fusion welding is used for weld repair, the high temperatures and thermal stresses inherent in the process enhance the growth of helium bubbles, causing intergranular cracking in the heat-affected zone (HAZ) [6-8].

Repair of nuclear components encompasses both the replacement of a failed component, in which case a new component must be joined to existing structures, and the in situ repair of arc welds that have developed stress corrosion cracks in service [2, 9, 10]. In both cases the issue of helium embrittlement, accelerated by the temperatures and stresses of fusion welding, presents a serious difficulty that must be overcome. While low heat input laser welding can be a partial solution to the problem [7], reactors with 30-40 years of service may have irradiated components which cannot be repaired by any of the conventional methods [11]. It is anticipated that many nuclear reactors will function for up to 80 years [12], so the repairability of irradiated stainless steel is of great importance to this industry.

Friction stir processing (FSP) is similar to friction stir welding (FSW), except that its purpose is not to join parts together, but to modify material microstructures and properties. The repair of a stress corrosion crack in an existing weld could be done by passing a tool along an existing weld or HAZ and "healing" the crack. The process of FSP is shown in Figure 1.

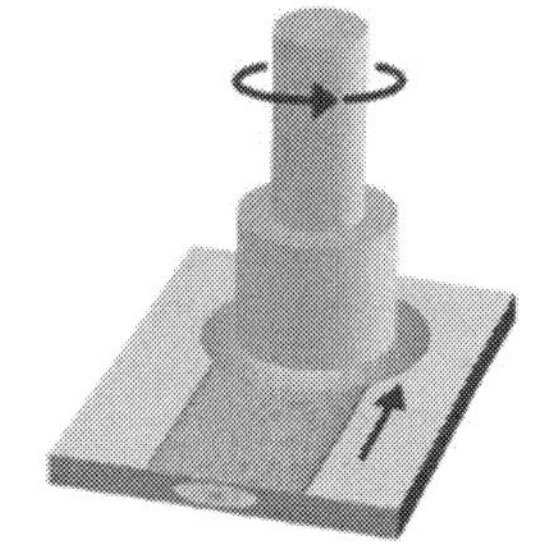

Figure 1. FSP could be used to repair a crack by processing an existing arc weld.

Much of the prior work done in FSW can be applied to FSP. Both FSW and FSP have been studied extensively for their ability to create favorable properties in metal alloys [13-19], most of which have been aluminum alloys. While we have not found research literature on the use of FSP to "heal", or repair, stress corrosion cracks in existing welds, one study by Sterling demonstrated that FSP can be used to process over existing arc welds in austenitic stainless steel, to enhance the mechanical properties of the weld [20]. For stress corrosion crack

[a] Corresponding author: mmiles@byu.edu

repair of irradiated stainless steel, lower peak temperatures would favor the use of FSP over fusion welding processes like GTAW, because the helium bubbles present in the material would have a lower driving force to diffuse to grain boundaries in the HAZ. But if FSP is a potential solution for weld repair of irradiated stainless steel, there remains the potential challenge of sigma phase and chromium carbide formation, which have been reported in friction stir welded 304L stainless steel along the advancing side of the stir zone [21]. Although others have reported preferential etching along this same region of the stir zone, they also report large variability, depending on the welding conditions [20, 22]. Sigma and chromium carbide phases are well known to reduce the corrosion resistance of austenitic stainless steel [21-23], and need to be suppressed if FSP is to be employed for weld repair. Based on these prior studies it is thought that dynamic recrystallization during processing may be responsible for the precipitation of these phases in the stir zone.

Most modeling work of the FSW/FSP process has been done under the assumption of steady state conditions, using an Eulerian approach. Some representative prior work is found in [18, 24-27]. Other investigators have used an arbitrary Lagrangian-Eulerian (ALE) formulation, where the velocity of the mesh and the velocity of the material are coupled at the free boundary, allowing the position of the free surfaces to evolve, but decoupled inside the free boundary, where the mesh velocity and material velocities can be different. This allows for adjustment of the mesh in order to avoid excessive distortions [28-30]. In all of these cases the weld interface, or faying surface, is not included explicitly in the model. Rather, the plates to be joined are modeled as one solid part, referred to as a bead on plate approach. This reduces the numerical complexity of managing contact problem at an interface, while still allowing for the computation of temperatures and material flow patterns in the stir zone. In the current work, an Eulerian approach was used to simulate FSP of 304L stainless steel using a bead on plate model. This is computationally more efficient than an ALE approach, while effectively predicting temperatures in the stir zone as long as friction is properly modeled between the workpiece and the tool [29, 31, 32]. Predicted temperatures and strain rates were used to estimate recrystallized grain size at three different locations in the stir zone, and these estimates were compared to an FSP experiment on 304L plate.

2 Experimental Methods

FSP experiments were performed on 12mm thick 304L stainless steel using a tool speed of 250 rpm and a feed rate of 100 mm/min. The tool had a pin length of 8 mm and a convex shoulder with a diameter of 25 mm, as shown in Figure 2.

Figure 2. Friction stir processing tool, with convex shoulder and conical pin.

Initial average grain size for the as-received plate was 40 μm as measured by electron backscatter diffraction (EBSD). The composition of the plates that were processed is shown in Table 1.

Table 1. Composition of 304L stainless steel (wt %).

C	Mn	P	S	Si	Cr	Ni	N	Fe
0.08	2.0	0.05	0.03	0.75	18	10	0.1	Balance

Holes were drilled in one side of the plate so that thermocouples could be embedded for temperature measurements. In addition to measuring temperatures in the plate, a thermocouple was also embedded in the FSP tool, near the outside diameter a few millimeters from the tool surface. The temperature in the tool was used to develop operating parameters for processing, prior to the experiment where plate temperatures were measured.

At the end of the FSP experiment, using a tool speed of 250 rpm and a feed rate of 100 mm/min, the tool was retracted and ice water was immediately applied to the stir zone where the pin was removed. This was done in order to prevent grain growth, so that recrystallized grain size in different areas of the weld could be measured. Recrystallized grain sizes from the experiment were compared to predictions from the model.

3 Model Description

The modeling of FSP was done using a research version [28] of the Forge® software package. An Eulerian approach was used to simulate temperatures and material flow for 12mm thick FSP of 304L stainless steel. This approach assumes that contact conditions between the workpiece and the tool are known in advance, which is reasonable for friction stir processing, as long as the feeds (translation rate) and speeds (rotation rate) of the tool have been studied experimentally first and that the quality of the stir zone was verified (i.e. no voids were present).

The material flow stress was modeled as a function of strain, strain rate, and temperature using JMatPro [33] data for 304L stainless steel. The flow stresses were modeled for a temperature range of 25-1300°C, for strain rates from 10^{-3} to 10^{3} s^{-1}, and for true strains of 0 to 4.

DOI: 10.1051/matecconf/20168012001

Friction conditions at the sliding interface between the workpiece and the tool were modeled by a viscoplastic Norton law:

$$\tau_f = -\alpha_f K(T, \bar{\varepsilon}) \|\Delta v_s\|^{q-1} \Delta v_s \quad (1)$$

where α_f is friction coefficient, q is sensitivity to the sliding velocity at the tool/workpiece interface, Δv_s is the relatively sliding velocity between the tool and the workpiece, and K is a hardening parameter, function of temperature and effective strain, with the properties of the workpiece alloy.

Material flow prediction in the workpiece is based on a finite element discretization using an enhanced (P1+/P1) 4-node tetrahedron element, where equilibrium equations are solved iteratively until steady-state is reached in the velocity field. The velocity field and the temperature calculations are coupled, where flow stresses are updated after each iteration using the latest temperature estimates, until convergence. The final velocity field satisfies equilibrium conditions, and at the same time corresponds to the correct temperatures. The unilateral contact condition is applied to the workpiece surfaces by means of a nodal penalty formulation, where the FSP tool and the backing plate are considered to be rigid.

The heat generated by simultaneous rotation and translation of the tool in the 304L workpiece causes temperatures to rise to approximately 1000°C at steady state. For the model, heat caused by plastic deformation is calculated as follows:

$$\dot{q}_p = f\bar{\sigma}\dot{\bar{\varepsilon}} \quad (2)$$

where f is the fraction of work converted into heat, taken as 0.9, $\bar{\sigma}$ is effective stress, and $\dot{\bar{\varepsilon}}$ is effective strain rate. Heat caused by the frictional sliding at the tool/workpiece interface is given by:

$$\dot{q}_f = \tau \cdot \Delta v_s \quad (3)$$

where τ is the shear stress given by equation (1). Frictional heat is shared between the tool and the plate as a function of the effusivities of each, where the material with higher effusivity receives a greater portion of the frictional heat. The thermal conductivity, heat capacity, and density of both the tool and the 304L plate were modeled as a function of temperature, over the range of temperatures from 25°C to 1300°C.

The recrystallized grain size can also be estimated from the strains, strain rates, and temperatures that emerge from the steady-state FSP simulation. We follow the approach of Venugopal [34], who calculated average recrystallized grain size, using the following equation adapted to 304L stainless steel:

$$d = 20{,}560\dot{\varepsilon}^{-0.3} e^{-0.25\left(\frac{Q}{RT}\right)} \quad (4)$$

where d is the recrystallized grain diameter (μm), $\dot{\varepsilon}$ is strain rate, Q=310 kJ/mol, and R=8.314×10^{-3} kJ/mol-K.

4 Results and Discussion

The model consisted of a 304L plate with 28,401 elements, a carbon steel backing plate with 23,270 elements, and a PCBN/W-Re tool with 23,270 elements. The heat transfer coefficient between the tool and the plate was 50,000 W/m^2-°C, and between the 304L and the carbon steel backing plate was 5,000 W/m^2-°C. A tool holder imposed a boundary condition of 19°C on the shank of the tool, which simulated the cooled tool holder on the friction stir processing machine. Figure 3 shows the initial condition of the model.

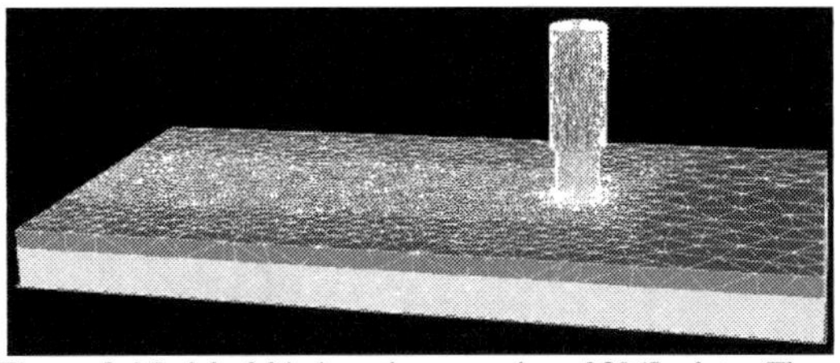

Figure 3. Model of friction stir processing of 304L plate. The tool (green) is embedded in the 304L plate (red). A constant temperature boundary condition of 19°C is imposed by the tool holder (blue) on the shank of the tool. The backing plate (yellow) is supported by a rigid support with a constant temperature boundary condition of 50°C.

The tool speed was 250 rpm while the feed rate was 100 mm/min. Friction coefficients were varied in order to accurately predict the steady-state temperatures of 3 thermocouples that were embedded in the 304L plate. The model of the tool was not threaded, even though the physical tool had threads. However, the velocity boundary condition on the pin of the tool imposed a vertical component that essentially simulates action of the mechanical threads on the physical tool [28].

Thermocouples embedded in the 304L plate were 7mm, 9mm, and 11mm from the centerline of the stir zone. Steady state temperatures were reached in the experiment after a translation of approximately 200 mm. The comparison of experiment and simulation for the steady-state condition is shown in Figure 4.

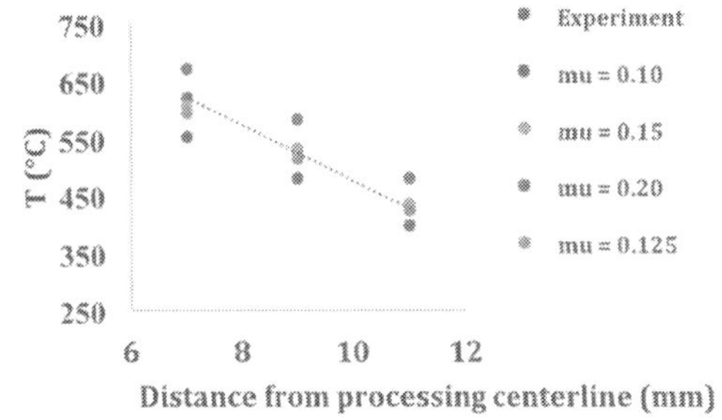

Figure 4. Comparison of predicted and experimentally measured temperatures at steady state. A friction coefficient of 0.125 provided the best prediction.

A point by point comparison of temperatures in the 304L plate and in the simulation is shown in Table 2.

Table 2. Comparison of predicted and experimentally measured temperatures at steady state, for a friction coefficient of 0.125.

Distance from center of stir zone	Experiment	Simulation
7mm	620°C	595°C
9mm	529°C	514°C
11mm	425°C	423°C

Heat generated by friction at steady state was about 4.3 kW, versus about 0.2 kW for the heat dissipated by plastic deformation. Therefore, about 96% of the heat generated by FSP was created by friction between the tool and the workpiece. This is consistent with prior simulation work in 304L. The steady state temperature profile for a friction coefficient of 0.125 is shown in figure 5.

Figure 5. Temperature profile in 304L plate at steady state (isovalues on right are in degrees Celsius). The maximum predicted temperature in the stir zone was 928°C.

A specimen was taken from behind the keyhole left in the plate after the tool was retracted. After retracting the tool, ice water was poured over the keyhole in order to arrest grain growth so that recrystallized grain size could be measured. At a location about 6mm behind the center of the tool the average grain size was about 7.4 μm. An EBSD image showing the grains at this location is shown in figure 6.

Figure 6. EBSD grain map from 6mm behind the pin in the location where the tool was retracted, at the end of the experiment. Average recrystallized grain size at this location was 7 mm, compared to an average grain size of 25 mm in the base material.

An estimate of the recrystallized grain size from the simulation was calculated from equation (4). The strain rate and temperature 6mm behind tool center, at steady state, were 0.06 s^{-1} and 894°C, respectively. Using these values the predicted grain size is 1.4 μm, which is only 20% of the measured grain size. However, the time that it took to cool the keyhole area at the end of the experiment was approximately 10 seconds, which is enough time for significant grain growth to occur, after deformation was halted. In order to more accurately predict final grain size the static grain growth that occurs when stirring is completed, and before cooling is applied, must also be modeled.

5 Conclusions

A steady state simulation of friction stir processing was carried out for 12mm thick 304L stainless steel. For a tool speed of 250 rpm and a feed rate of 100 mm/min the peak steady-state experimental temperatures at 7, 9, and 11mm from the centreline of the stir zone were 620, 529, and 425°C respectively. Simulated temperatures in the same locations were 595, 514, and 423°C. Thus, the greatest difference between simulation and experiment was 4%, which represents a good level of accuracy. Temperatures and strain rates from the model were used to predict recrystallized grain size at a location about 6mm behind the tool, at the end of the experiment when the plate was cooled rapidly with ice water. The predicted value was only about 20% of the measured grain size. The reason for the discrepancy is the grain growth that occurred during about 10 seconds between the end of the experiment and the application of the ice water. More accurate grain size predictions will require modelling of static grain growth that occurs after deformation has stopped.

Acknowledgement

This work was supported by National Science Foundation grant CMMI-1405508.

References

1. K. Asano, S. Nishimura, Y. Saito, H. Sakamoto, Y. Yamada, T. Kato, T. Hashimoto, J Nucl Mater, 264 (1999) 1-9.
2. W.R. Kanne, M.R. Louthan, D.T. Rankin, M.H. Tosten, Mater Charact, 43 (1999) 203-214.
3. K. Tsuchiya, H. Kawamura, G. Kalinin, J Nucl Mater, 283 (2000) 1210-1214.
4. S. Li, M.L. Grossbeck, Z. Zhang, W. Shen, B.A. Chin, Weld J, 90 (2011) 19S-26S.
5. M.H. Tosten, S.L. West, W.R. Kanne, B.J. Cross, Weld J, 86 (2007) 245S-252S.
6. Z. Feng, G. Wilkowski, Repair Welding of Irradiated Materials: Modeling of Helium Bubble Distributions for Determining Crack-Free Welding Procedures, 2002.
7. N. Yurioka, Y. Horii, Sci Technol Weld Joi, 11 (2006) 255-264.
8. W.R. Kanne, G.T. Chandler, D.Z. Nelson, E.A. Francoferreira, J Nucl Mater, 225 (1995) 69-75.

9. C.A. Wang, M.L. Grossbeck, H. Aglan, B.A. Chin, J Nucl Mater, 239 (1996) 85-89.
10. C.A. Wang, M.L. Grossbeck, N.B. Potluri, B.A. Chin, J Nucl Mater, 233 (1996) 213-217.
11. E. Willis, personal communication, 2013.
12. S. Li, M.L. Grossbeck, Z. Zhang, W. Shen, B.A. Chin, Weld J, 90 (2011) 7.
13. Q. Yang, B.L. Xiao, Z.Y. Ma, Metall Mater Trans A, 43A (2012) 2094-2109.
14. F.Y. Tsai, P.W. Kao, Mater Lett, 80 (2012) 40-42.
15. N. Sun, D. Apelian, Jom-Us, 63 (2011) 44-50.
16. X.L. Feng, H.J. Liu, S.S. Babu, Scripta Mater, 65 (2011) 1057-1060.
17. F.C. Liu, Z.Y. Ma, Scripta Mater, 62 (2010) 125-128.
18. B.C. Liechty, B.W. Webb, International Journal of Machine Tools & Manufacture, 48 (2008) 1474-1485.
19. R.S. Mishra, Z.Y. Ma, Materials Science and Engineering: R: Reports, 50 (2005) 1-78.
20. C.J. Sterling, T.W. Nelson, C.D. Sorensen, M. Posada, Research Report, Office of Naval Research, 2004, pp. 7.
21. S.H.C. Park, Y.S. Sato, H. Kokawa, K. Okamoto, S. Hirano, M. Inagaki, Scripta Mater, 49 (2003) 1175-1180.
22. C.D. Sorensen, T.W. Nelson, Proceedings of the 7th International Conference on Trends in Welding Research, ASM, Pine Mountain, GA, 2005, pp. 441-446.
23. S.H.C. Park, Y.S. Sato, H. Kokawa, K. Okamoto, S. Hirano, M. Inagaki, Scripta Mater, 51 (2004) 101-105.
24. S. Perivilli, J. Peddieson, J. Cui, Journal of Manufacturing Science and Engineering-Transactions of the Asme, 131 (2008) 1-8.
25. J.H. Kim, F. Barlat, C. Kim, K. Chung, Metals and Materials International, 15 (2009) 125-132.
26. C. Cox, D. Lammlein, A. Strauss, G. Cook, Materials and Manufacturing Processes, 25 (2010) 1278-1282.
27. H. Atharifar, D.C. Lin, R. Kovacevic, Journal of Materials Engineering and Performance, 18 (2009) 339-350.
28. S. Guerdoux, L. Fourment, Modelling and Simulation in Materials Science and Engineering, 17 (2009) 143-155.
29. M. Assidi, L. Fourment, S. Guerdoux, T. Nelson, International Journal of Machine Tools & Manufacture, 50 (2010) 143-155.
30. L. Fratini, G. Buffa, Journal of Engineering Materials and Technology-Transactions of the ASME, 130 (2008)
31. S. Guerdoux, L. Fourment, M. Miles, T. Nelson, Proceedings of the Annual TMS Conference, Orlando, FL, 2007, pp. 83-88.
32. M. Miles, T. Nelson, L. Fourment, S. Guerdoux, Materials Science & Technology 2008 - Joining of Advances and Specialty Metals, Pittsburgh, PA, 2008, pp. 2319-2330.
33. Sente Software Ltd., Surrey Technology Centre, 40 Occam Road GU2 7YG, United Kingdom, 2015.
34. S. Venugopal, P.V. Sivaprasad, Journal of Materials Engineering and Performance, 12 (2003) 674-686.

Numerical Simulation of Temperature Distribution and Material Flow During Friction Stir Welding 2017A Aluminum Alloys

Oussama Mimouni[1,a], *Riad Badji*[2], *Mohamed Hadji*[1], *Afia Kouadri-David*[3], *Hamel Rachid*[4], *Nabil Chekroun*[5]

[1]*Laboratory of Aircrafts, University of Saad Dahlab, Blida 1, Blida, Algeria.*
[2]*Center of Industrial Technologies CRTI, P.O Box 64, Cheraga, Algeria*
[3]*Laboratory of Civil and Mechanical Engineering INSA Rennes, France.*
[4]*Polytechnic military school Bordj El Bahri Algeria- Algeria.*
[5]*Research Laboratory of Manufacturing Technology Mechanical / ENPO- Algeria.*

Abstract. This study describes the use of fluid dynamic code, FLUENT to model the flow of metal in the AA2017A case around the welding tool pin (FSW). A standard threaded tool profile is used for the analysis of phenomena during welding such as heat generation and flow of the material are included. The main objective is to gain a better understanding of the flow of material around a tool. The model showed a large number of phenomena similar to those of the real process. The model has also generated a sufficient amount of heat, which leads to a good estimate of the junction temperature. These results were obtained using a viscosity which is near the solidus softening.

1 Intoduction

Friction stir welding (FSW) is recent method of welding in solid state, created and patented by The Welding Institute (TWI) in 1991.

In FSW a cylindrical, shouldered tool with a profiled probe, also called pin, is rotated and slowly plunged into the joint line between two pieces of sheet or plate material, which are butted together. The parts have to be clamped onto a backing bar in a manner that prevents the abutting joint faces from being forced apart. Once the probe has been completely inserted, it is moved with a small tilt angle in the welding direction. The shoulder applies a pressure on the material to constrain the plasticised material around the probe tool. Due to the advancing and rotating effect of the probe and shoulder of the tool along the seam, an advancing side and a retreating side are formed and the softened and heated material flows around the probe to its backside where the material is consolidated to create a high-quality solid-state weld.

2 fluid model

2.1 The Arbitrary Lagrangian–Eulerian (ALE) for fluids problems

Because of its larger capacity, compared to the Eulerian description, of dealing with interfaces between materials and the moving boundaries, the description of the Arbitrary Lagrangian–Eulerian (ALE) is widely used for spatial discretization problems in fluid dynamics and structural. In fact, the method is frequently used in "hydrocodes" to simulate the response of large distortion / deformation of the materials.

The most obvious return of an Arbitrary Lagrangian–Eulerian (ALE) formulation in flow problems is that the convective term must take into account the mesh movement. Also, the grid movement may increase or decrease the effects of convection.

It is important to note that the stress tensor depends on the pressure or on the velocity field according to the viscous or non-viscous flow. This is not the case in solid mechanics. Thus the determination of the stress is not a major concern in the dynamics of Arbitrary Lagrangian–Eulerian (ALE) fluids.

[a] Corresponding author: oussama.mimouni@hotmail.fr

2.2 Use of the dynamic mesh method with Fluent

In many works, the simulation and modelling of the FSW welding process are limited through the use of a fixed tool (pin) where the material is modelled as a high viscosity flow of a fluid around the pin. In reality, it's the tool which moves along the welding line. It's this difference which is considered in our simulation (2D/3D). To deal with this problem of the displacement of the pin in the material, the use of re-meshing method is needed, which allows to gather and re-mesh locally the cells. If the new created cells meet the distortion criterion, the meshing is updated locally with new cells. In this work two methods are used : " local re-meshing " and "local face re-meshing" for 2D /3D simulation.

2.3 Governing equations

In order to simplify the numerical simulation process, some supposed conditions are fulfilled:

- The rotational tool is perpendicular to the surface of a weldment;
- The material is considered to be incompressible fluid;

In the FSW process the mass of material is supposed to be a constant , so the equation of mass conservation is as follows:

$$\frac{\partial u}{\partial x}+\frac{\partial v}{\partial y}+\frac{\partial w}{\partial z}=0 \qquad (1)$$

where u, v and w mean the velocity of x direction, y direction and z direction, respectively.

The momentum conservation equation can be described with the Navier-Stokes equation. The expression of the equation is as follows:

$$\begin{cases} \frac{\partial u}{\partial t}+u\frac{\partial u}{\partial x}v+\frac{\partial u}{\partial y}+w\frac{\partial u}{\partial z}=F_x-\frac{1}{\rho}\frac{\partial p}{\partial x}+\mu\left(\frac{\partial^2 u}{\partial x^2}+\frac{\partial^2 u}{\partial y^2}+\frac{\partial^2 u}{\partial z^2}\right) \\ \frac{\partial v}{\partial t}+u\frac{\partial v}{\partial x}v+\frac{\partial v}{\partial y}+w\frac{\partial v}{\partial z}=F_y-\frac{1}{\rho}\frac{\partial p}{\partial y}+\mu\left(\frac{\partial^2 v}{\partial x^2}+\frac{\partial^2 v}{\partial y^2}+\frac{\partial^2 v}{\partial z^2}\right) \\ \frac{\partial w}{\partial t}+u\frac{\partial w}{\partial x}v+\frac{\partial w}{\partial y}+w\frac{\partial w}{\partial z}=F_x-\frac{1}{\rho}\frac{\partial p}{\partial z}+\mu\left(\frac{\partial^2 w}{\partial x^2}+\frac{\partial^2 w}{\partial y^2}+\frac{\partial^2 w}{\partial z^2}\right) \end{cases} \qquad (2)$$

where F_x, F_y and F_z are the volume force of x direction, y direction and z direction, respectively, p is the static pressure of the flow field, μ is the viscosity of fluid, ρ means the mass of material.

The energy conservation equation can be expressed as follows:

$$\rho c\left(\frac{\partial T}{\partial t}+u\frac{\partial T}{\partial x}v+\frac{\partial T}{\partial y}+w\frac{\partial T}{\partial z}\right)=\lambda\left(\frac{\partial^2 T}{\partial x^2}+\frac{\partial^2 T}{\partial y^2}+\frac{\partial^2 T}{\partial z^2}\right) \quad (3)$$

where c is the specific heat of material, λ is the thermal conductivity of material and T is the temperature of fluid.

In this work, a 2D/3D model of welding FSW of aluminum alloy AA2017A was developed and validated by a transient simulation with fully coupled materials flow and taking into account the heat flow generated by friction. The material is assumed to be viscoelastic, of which the viscosity was obtained from tests on a dynamic mechanical analyzer (DMA). By combining the two formulations Euler and Lagrange, a dynamic mesh method (see figures 1 and 2) was used to model the flow of material induced by the movement of the threaded pin. The use of transient model allows not only displaying the complex movement of threaded pin, but also the simulation of heat and material flow during welding. This approach aims getting better prediction and representation of thermomechanical history of the material, which is necessary in the microstructure and final texture analysis. The equations governing the 2D/3D transient heat transfer and material flows are discretized using the finite volume method and solved using fluid dynamics solver « ANSYS / FLUENT ».

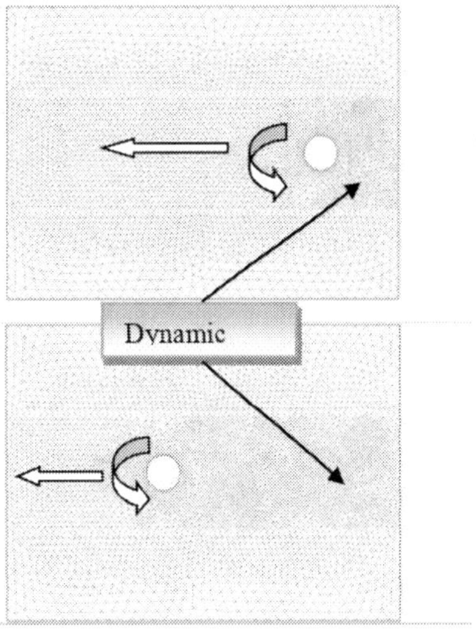

Figure 1. Evolution of 2D dynamic mesh.

DOI: 10.1051/matecconf/20168012002

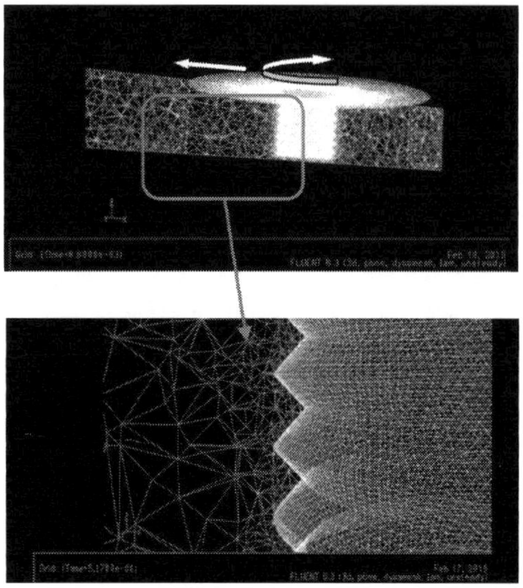

Figure 2. Evolution of 3D dynamic mesh.

2.4 Heat generation model

In the present model, a heat flux is applied on the outer surfaces of the pin and the shoulder which are in contact with the material.

In the FSW process, the heat is generated near the contact surfaces, which may have complex geometries depending on the tool shape. However, for the used model, the surface of the tool are assumed to be cylindrical with a horizontal shoulder. The conical shoulder surface is characterized by the angle α, which is set to zero in the case of a flat shoulder.

The simplified tool is shown in Figure 3, where $R_{shoulder}$ is the radius of the shoulder, and R_{PIN} H_{PIN} and are respectively the radius and height of the pin. In this way total heat output can be calculated as follows :

$$Q_{total}=2/\pi\omega\tau_{contact}(R^{3}_{shoulder}+ 3R^{3}_{probe} H_{probe}) \qquad (4)$$

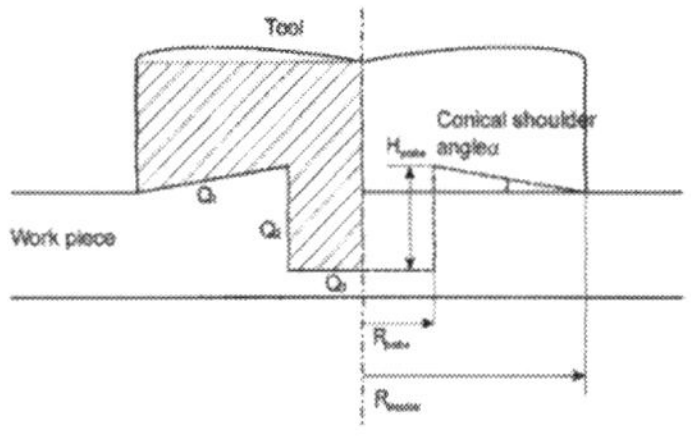

Figure 3. Heat flow represented in a simplified FSW tool.[3]

The numerical modeling of FSW flow can be based on models and technics used for other processes such as friction welding, extrusion, machining, forging, rolling and ballistic impact [2]. As for the analysis of heat flow, numerical flow models can use for meshing either the Eulerian formulation, Lagrangian formulation, or a combination of both (hybrid solution Lagrange-Eulerian).

Through the use of fluid dynamic analysis models (CFD) instead of solid mechanical models, some mechanical effects are excluded from the analysis, such the forging effect. In addition, these models cannot predict the needed force because the elasticity is neglected.

One of the difficulties of the numerical analysis is the strong gradient of the flow velocity near the tool. To solve this problem, the mesh areas is divided in several regions, as shown in Figure 4. The flow around the tool is a predominantly rotational, the meshing is adapted to the tool displacement. This rotation region is made large enough to contain the entire deformation zone where the mesh size is much finer [4].

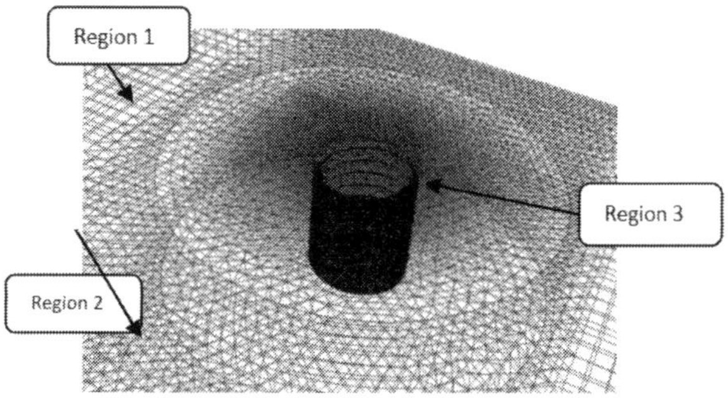

Figure 4. Discretization of the plate into differents zones.

A 3D analysis using the finite volume method based on an Arbitrary Lagrangian–Eulerian (ALE) formulation, provided results with an interesting physical insight. Also, the 3-D analysis is able to handle some of the process complexities such as a concave shoulder, tool tilt, and different threaded pin profiles.

A numerical simulation of the 3-D FSW method, presented by Riahi and Nazari [5], relates to studying the impact of tool speed, the heat flux and the residual stress distribution. Another interesting study shows a 3-D thermomechanical model FSW based on CFD analysis [6]. This model describes the material flow around the tool during the welding operation. An alloy of AA2024 aluminum plate of 3.2 mm thick is used. The minimum and maximum values of the workpiece temperature (near the tool shoulder) are shown in Figure 5,

DOI: 10.1051/matecconf/20168012002

where one can see that the maximum temperature value decreases as the welding speed increases. On the other hand, the maximum value of the temperature increases when the tool rotation speed increases.

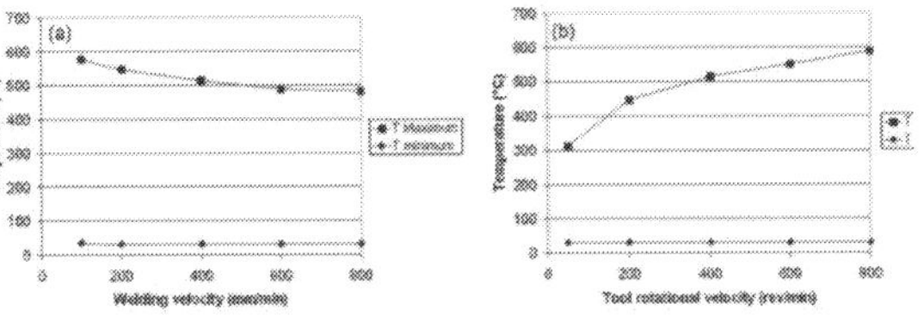

Figure 5. Extreme temperatures in the welds: (a) as a function of the welding velocity of the tool for a tool rotational velocity equal to 400 rpm; (b) as a function of the tool rotational velocity for a welding velocity equal to 400 mm/min. [6]

The modeling of the FSW welding process allows to visualize the fundamental behavior of the welded materials, and to study the influence of different welding parameters, including the design of the tool and boundary conditions, without performing costly experiments. The FSW modeling is a difficult task because of its multiphysics features. The process combines heat flow, high temperature, plastic deformation and microstructure evolution. Nowadays, the numerical simulation of the FSW process may not only be used to optimize the process. Increased knowledge produced based on the FSW process simulation can lead in the near future, to replace experimental tests. This will help promoting and expanding the FSW process to a wider range of different applications.

3 Geometry and boundary conditions

The purpose is to perform the simulation using 2D and 3D modeling with FLUENT software, and to present the different results obtained.

The sheet is modeled as a rectangle of length L = 70 mm and b = 50 width and thickness t = 6mm. A ring of outer diameter D = 22mm is drawn on the upper surface of the sheet, which corresponds to the shoulder. The figure 6 shows the surfaces designation that will be used thereafter. The pin is modeled as a cylinder of diameter d = 6 mm and height h = 5.7mm.

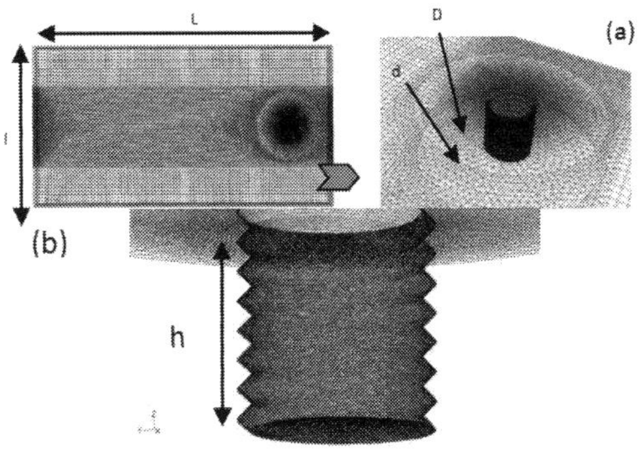

Figure. 6 Perspective of model mesh (a) and details of pin meshed zones (b) Thread tool.

In FLUENT 3D model, the origin is in the center of the sheet and at mid-thickness. Thus, knowing the rotation direction of the tool and its advancing speed, the AS-side and the RS-side can be differentiated by positive or negative Y values.

The meshing of reference used in the calculations is shown in figure 7. It consists of 394427 cells (tetrahedral / square) with minimum edge size of 0.1mm.

A dynamic mesh was used to capture the two movements of the axis of the tool rotation and translation. Thus, the mesh is modified continuously over the time as shown in Figure 7. The dynamic mesh is composed of two regions inner and outer regions. The inner region contains the contour of the tool pin (engaged with tetrahedral elements) and the outer region contains the tool shoulder. The use of such a dynamic mesh combines the advantages of the two formulations of Lagrange and Euler. In other words, the movement of the tool is correctly taken into account with the movement of Lagrangian elements, while at the same time the formulation of Euler allows the material to flow through the mesh, by maintaining good mesh quality.

The transient model was used to simulate the movement of the tool. Heat flow, defined in equation (4) was applied to the tool surfaces. For the upper and side surfaces of the plate, the thermal boundary condition was fixed by convection between the air and the plate, with a heat transfer coefficient of 30 W / m^2 °C.

For the lower surface of the plate, the heat transfer coefficient by convection between the plate and the anvil plate was set at 150 W / m^2 °C. The value of the heat transfer coefficient is found to be similar to that reported in the literature for modeling FSW aluminum alloy 2017A. The plate was initially at room temperature of 27 ° C.

In the simulation process using the FLUENT software, the approach is to consider the metal material as a fluid, which is under the effect of the rotation and advance of the tool. The fluid is stationary and the tool has an advancing speed of 0.6mm / s and a rotational speed of 1250tr / min.

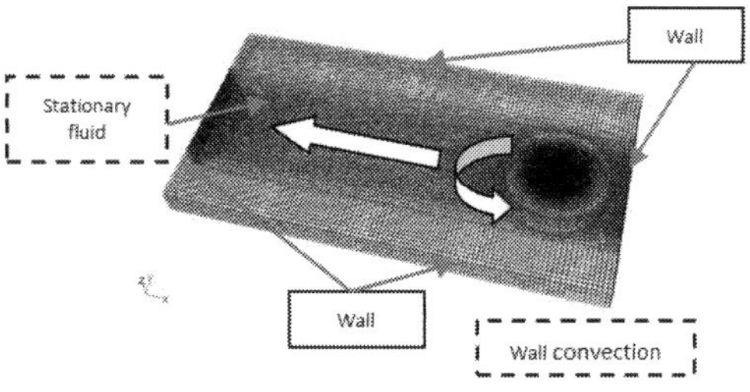

Figure7. Mesh generation of model used in simulation.

4 Material parameters

For the model based on the finite volume Euler formulation, the main parameters are the specific heat material, the thermal conductivity and the viscosity coefficient as a function of the temperature. The latter is derived from the experiment performed on the DMA machine (see figure 8), that allows identifying and quantifying the dynamic viscosity of the material depending on the temperature, which is used directly into the numerical model.

4.1Dynamic mechanical analysis

The DMA is used to determine the evolution of the viscoelastic properties as a function of temperature, frequency or time, by applying cyclic deformation [1]. The viscoelastic behaviour is generally described as a complex modulus $E^* = E' + iE''$. The real part E′ called storage modulus, reflects the conservative elastic behaviour of the material. As for the imaginary part E′′ called loss modulus, characterizes the viscous and dissipative behaviour. The quotient $E''/E' = \tan\delta$ is called loss factor.

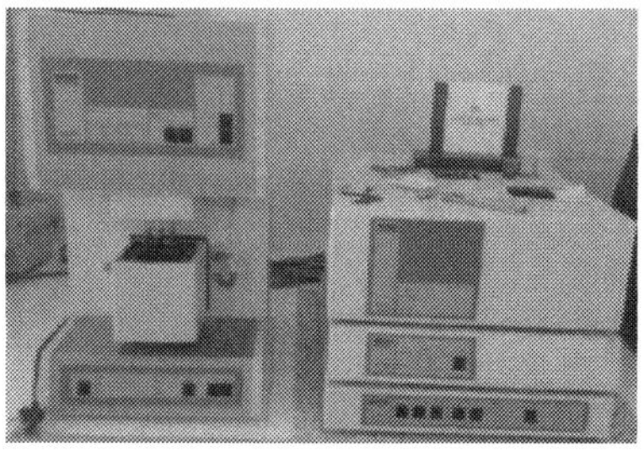

Figure 8. Dynamic Mechanical Analyzer (DMA).

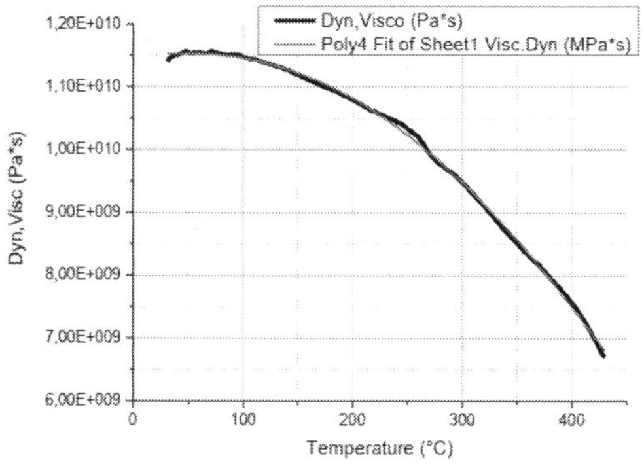

Figure 9. Dynamic viscosity if AA2017A obtained from DMA test.

The viscosity in term of the temperature can be expressed in polynomial form as :

$$\mu\,(Pa.s) = 1{,}14632\ 10^{+10} + 2{,}488410^{+6}T - 22117{,}33216T^2 - 43{,}29134x\ T^3 + 0{,}05236T^4 \quad (5)$$

The density is 2800 kg / m^3. The specific heat (C_p) and the thermal conductivity (λ) at different temperatures are calculated using equations. (6) and (7) [7].

$$\lambda = 103.264 + 0.241T \quad (6)$$

$$C_p = 754.08 + 0.3729T + 0.0012T^2 \quad (7)$$

5 Results and discussion

As shown in figure 10, the solution reaches the stationary state when the mechanical equilibrium is attained after a number of additional iterations. It also shows the evolution of minimum, maximum and average temperatures calculated on the entire field of study.

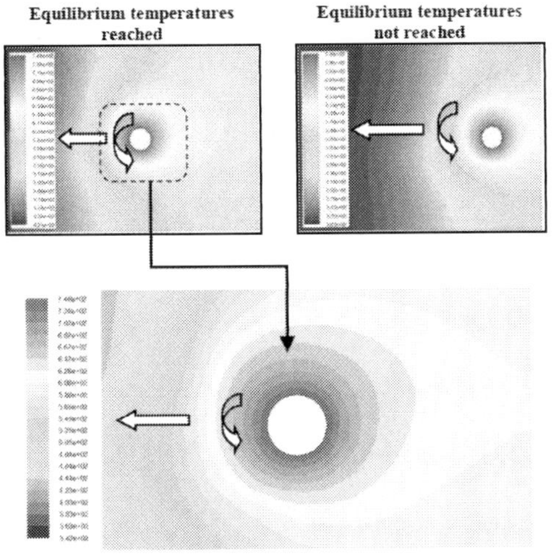

Figure 10. Temperature distribution during FSW welding.

The comparison between the predicted temperatures with the experimental data is presented in Figure 12. The temperature values are measured on the upper surface at the points T1 and T2 of the plate, as represented in Figure 11. One can observe a reasonable agreement between the temperature values obtained with the model with those measured in experiment. The heat generated during FSW must be sufficiently enough to enable softening, stirring and mixing the material.

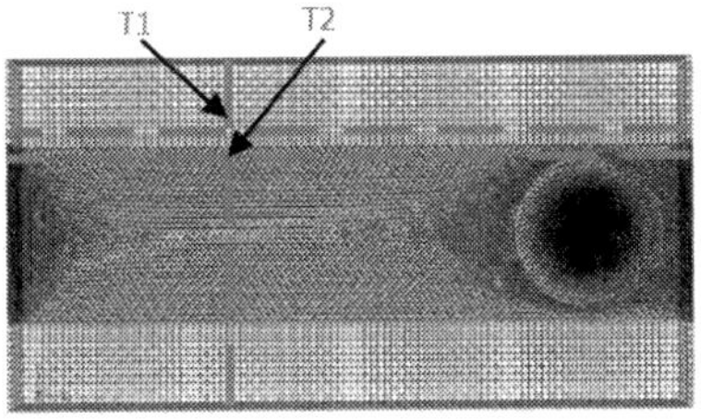

Figure 11. Thermocouple's positions along plate.

The calculated temperature profiles along the upper surface sections of the cross-piece through the tool axis are shown in Figure 12.

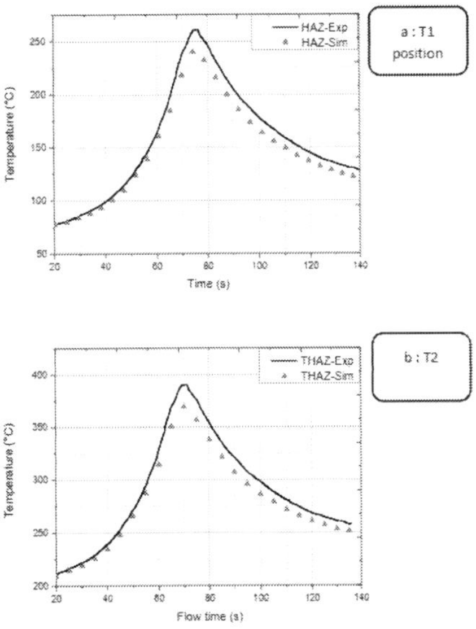

Figure 12. Comparison between the measured and predicted temperatures: (a) HAZ zone, (b) THAZ zone.

The maximum temperature is located in the area where the deformation is greater. A temperature peak of 475 ° C is recorded in the area close to the tool (shoulder, pin). The temperature decreases when going far from the mixing zone. The temperature distribution is asymmetric, which is a characteristic of the welding process.

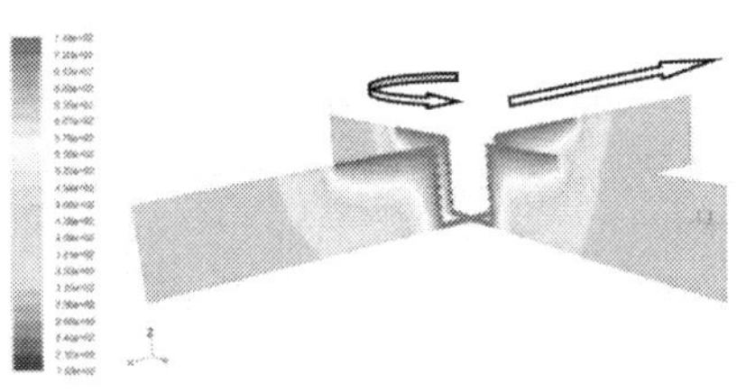

Figure 13. Cross-section along the axis (X) and (Y).

The contours of the velocity vectors through a cross-sectional below the tool shoulder are shown in Figures 14 and 15 respectively for 2D / 3D cases. This analysis is only intended to be illustrative. The slipping was simulated by prescribing the speed to the shoulder and to the tool pin. These figures illustrate several flow characteristics. First, the greater speed of rotation is always in the region of the greater deformation, namely under the shoulder and around the pin. Also, one can distinguish a phenomenon rarely shown in the literature, namely the creation of vortices between each pin thread pitch, as seen in Figure 16.

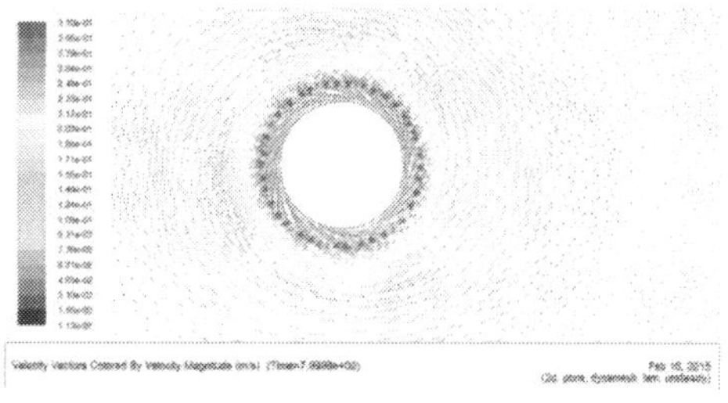

Figure 14. Velocity vectors for the 2D case.

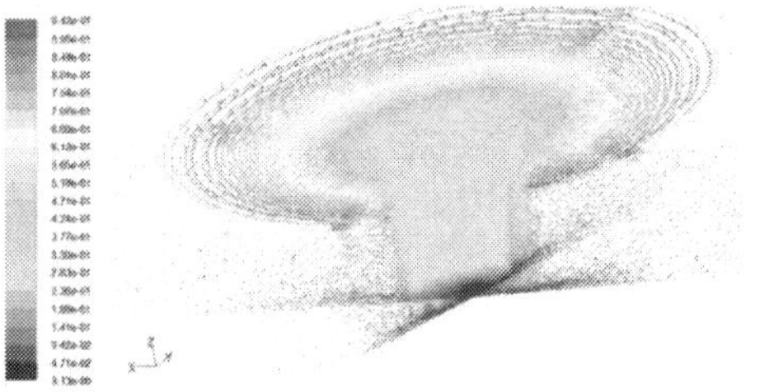

Figure 15. Velocity vectors for the 3D case.

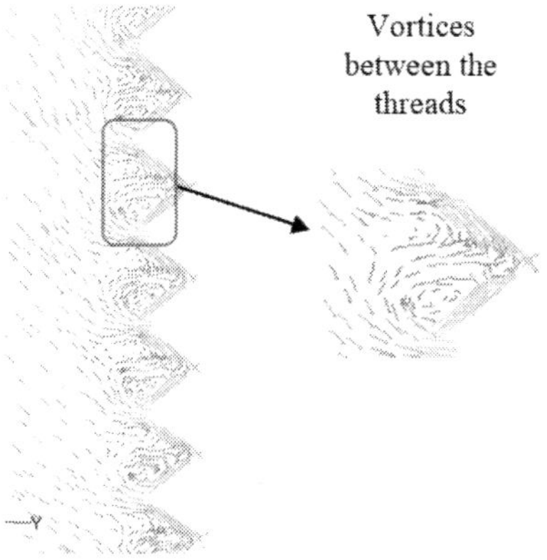

Figure 16. Vortices apparition between threads.

The figure 17 shows an instantaneous material flow caused by the displacement of FSW welding tool. The 3D visualization shows precisely the material flow around the tool and through the pin threads.

This figure also shows the streamlines around the tool, which illustrate the path of the material particles passing through the plate stirring region. The upward effect of the material flow produced by the pin thread is well illustrated too. We can observe that once the flow reaches the upper surface in contact with the tool shoulder, it takes a descending path and turns several times before continuing its course. Accordingly, the material remains near the surface of the tool shoulder. This kind of flow patterns was observed by Guerra et al [8]. It has been demonstrated the creation of a rotational area around the shoulder, where the material undergoes several turns before deposited in the pin wake. In addition, one can observe how the flow in the lower half of the plate is directed through the interaction with the pin threads, while in the upper part is driven by the interaction with the tool shoulder. Thus, it is possible to visualize the flow patterns without using tracing particles.

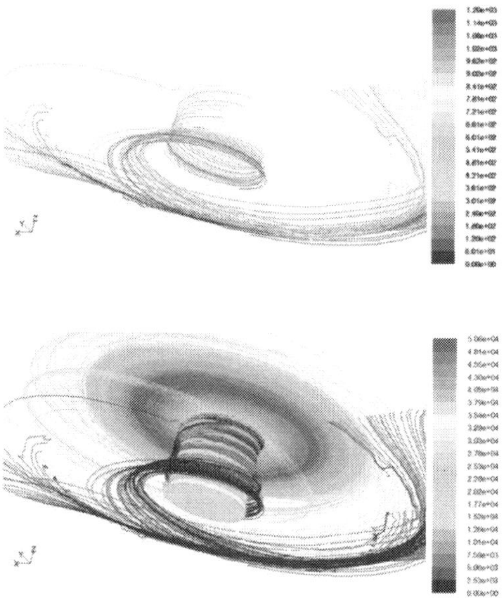

Figure 17. Streamlines of the material flow.

6 Conclusion

A 2D/3D simulation for FSW process model based on fluid mechanics has been presented in this paper. A friction stir tool with threaded tapered pin, similar to the real friction stir tool, was used in the mathematical modeling to explore the general material flow rule. The method of dividing the welding zone in several regions combining the velocity field and viscosity field is used in this paper.

A numerical procedure for the visualization of material flow during the FSW process is presented. The technique based on particle streamline tracing is applied to track particle trajectories, at the post process level once the velocity field is obtained.

The proposed method is validated through comparing the results of a 2D FSW modeling to experimental data. The 2D simulation shows that the material flow is not symmetric about the welding line. The flow patterns in the advancing and retreating sides are different. The welding join is created straightly at the center line and no material mixing is observed. These results are found to be in agreement with experimental data.

The particle streamline tracing method is also applied to a 3D simulation including the complete tool geometry and threaded pin. This simulation has shown some distinctive 3D features of the flow. The material flow around the pin has a centrifugal shape across the thickness. The material stirred on the advancing side performs more than one revolution around the tool. The up and down of the material movement in the thickness direction is also observed.

In both 2D and 3D simulations, agreement with experimental evidence is remarkably good, showing that particle tracing is a useful tool giving good insight of the FSW process.

References

1. K.P. Menard: CRC PRESS Boca,Library of Congress Cataloging-in-Publication Data, USA, (1999)

2. Mishra RS, Ma ZY (2005) Friction stir welding and processing. Mater Sci Eng R 50:1–78. doi: 10.1016/j.mser.2005.07.001

3. Assidi M, Fourment L (2009) Accurate 3D friction stir welding simulation tool based on friction model calibration. Int J Mater Form 2:327–330. doi: 10.1007/s12289-009-0541-6

4. Colegrove PA, Shercliff HR (2005) 3-Dimensional CFD modelling of flow round a threaded friction stir welding tool profile. J Mater Proces s Tech 169(2):320–327. doi: 10.1016/j.jmatprotec.2005.03.015

5. Riahi M, Nazari H (2011) Analysis of transient temperature and residual thermal stresses in friction stir welding of aluminum alloy 6061-T6 via numerical simulation. Int J Adv Manuf Technol 55:143–152. doi: 10.1007/s00170-010-3038-z

6. Jacquin D, de Meester B, Simar A, Deloison D, Montheillet F, Desrayaud C (2011) A simple Eulerian thermomechanical modeling of friction stir welding. J Mater Process Tech 211(1):57–65. doi: 10.1016/j.jmatprotec.2010.08.016

7. S.d.ji,and Al , "numerical simulation of material flow behaviour of friction stir welding influenced by rotational tool geometry", computational materials science 63 218-226, 2012

8. Guerra, M. et al, Flow patterns during friction stir welding, Materials Characterizations, v.49, p.95-101, 2003.

Sensitivity analysis of size effect on the performance of hydrostatic bearing

Chen Dongju[1a], Dong Lihua[1], Zhou Shuai[1] and Fan Jinwei[1]

[1] *College of Mechanical Engineering and Applied Electrics Technology, Beijing University of Technology, Beijing 100124, China*

Abstract. For size effect on solid-liquid interface of hydrostatic bearing oil film gap flow in two-dimension, fluid dynamic method is applied to investigate the influence of size effect on bearing capacity, dynamic stiffness and other performances. With the consideration of size effect, Reynolds equation is modified by adopting velocity slip boundary condition into Reynolds equation. The sensitivity factors are used to make a quantitative and qualitative analysis. Numerical simulation results show that size effect will affect bearing performances to a certain degree and the effect curve of size effect on bearing performances are given. The four maximum oil film pressures reduce with the increase of slip length. The maximum sensitivity of bearing capacity is 81.94%.

1 Introduction

Hydrostatic bearings are widely used in precision machines. Oil film of hydrostatic bearing plays an important role in balancing external load, reducing friction, eliminating wear and so on. With the consideration of the scale division of conventional liquid flow, the research of oil film flow of hydrostatic bearing has entered into micro-scale. However, the performance of liquid flow in micro-scale will be different with the one in macro-scale. Size effect on the one hand behaves of the ratio increase of fluid surface area to volume ratio, when the scale is transferred from macro-scale (cm ~ m) to micro-scale (um ~ mm). Compared to the volume, the ratio of surface area increases from 102 m-1to 106m-1, which will accelerate the process of heat transfer on the surface, thereby the temperature rise of the film will be reduced; on the other hand shows a velocity gradient effect, i.e. the shear stress, which along the surface normal is increased since size shrink. At the same time, shear force improves 103 times in micro-scale. If the shear stress in solid-liquid interface exceeds the limit value, the velocity slip occurs inevitably.

Theoretically, oil film in micro-scale is more likely to slip in solid-liquid interface than in macro conditions, and slip is validated by experiments [1, 2]. Velocity slip will affect the shape and pressure distribution of oil film [3], thereby affecting oil film flow, bearing capacity, bearing dynamic stiffness and other performances parameters of hydrostatic bearing, and more seriously, oil film is prone to collapse and failure [4-6]. Janez Lupse et al [7] extended boundary element method for the discretisation of the incompressible Navier-Stokes equations in the velocity-vorticity form. Paper [8] also points out that because of the size effect of oil film thickness, macroscopic tribology law in a way is no longer applicable. In specific friction conditions, material conventional properties cannot be used as the research basis in the analysis of tribology performance, therefore, the influence of size effect on the material properties should be fully considered [9]. At present, the research of the size impact on friction performance and temperature rise of hydrostatic bearings has not been conducted extensively. Thus, it is very necessary to put slip theory to the performance calculation of hydrostatic bearing.

In this paper, the influence of size effect on the performance of hydrostatic bearing is researched. First, the Navier velocity-slip boundary conditions are applied to fluid continuous flow equation, thus the Reynolds equation is modified in micro-scale. Next, according to the principle of oil film flow conservation and modified Reynold, bearing performances are obtained. Then, comparing the result with non-slip model, the influences of size effect on hydrostatic bearing performances are given. This provides a new method for further research on liquid flow in micro-scale.

2 Analysis of oil film flow characteristics in micro-scale

2.1 Model the oil film flow

Hydrostatic bearing can be divided into two basic categories: recessed hydrostatic bearing and unrecessed hydrostatic bearing. This article puts four pockets recessed liquid radial hydrostatic bearing as the investigated object and its structure is shown in Figure 1. Lubricating oil is pressured into the bearing pocket

[a] Corresponding author: djchen@bjut.edu.cn

DOI: 10.1051/matecconf/20168012003

through the restrict orifice (as shown in Figure 2 (a)), until the oil pocket is full, and then discharges from circumferential, axial sealing surface and is returned to channel. The lubricating oil flow in bearing clearance is just like the flow in two plate gap shown in Figure 2 (b), and the coordinate system is established as shown in Fig 2 (b).1, 2 represent journal and bush surface and V represents line speed of rotating spindle. X, Y, and Z respectively represent circumferential, radial and axial direction of bearing oil film.

Figure 1 The structure of hydrostatic bearing

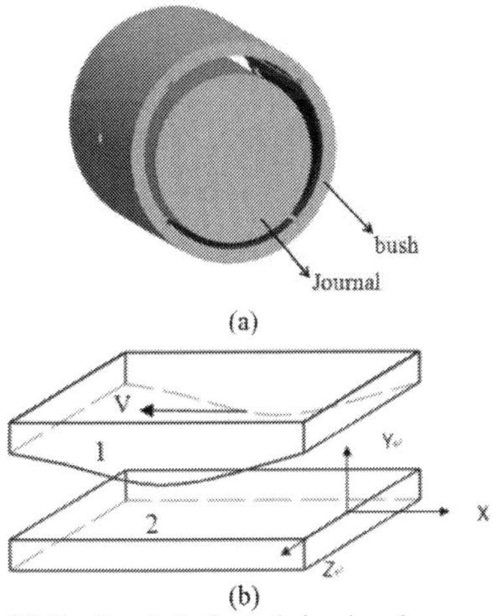

Figure 2 Oil film flow in hydrostatic bearing clearance and coordinates establishment

Assuming lubricating oil is Newtonian fluid, lies in laminar flow state, and the pressure along film thickness direction remains constant. So the Navier–Stokes equation can be simplified as Eq. (1) and the isoviscous 2-D Reynolds equation is derived from it.

$$\begin{cases} \dfrac{\partial p}{\partial x} = \dfrac{\partial \tau_{yx}}{\partial y} = \dfrac{\partial}{\partial y}\left(\eta \dfrac{\partial u}{\partial y}\right) = \eta \dfrac{\partial^2 u}{\partial y^2} \\ \dfrac{\partial p}{\partial y} = 0 \\ \dfrac{\partial p}{\partial z} = \dfrac{\partial \tau_{yz}}{\partial y} = \dfrac{\partial}{\partial y}\left(\eta \dfrac{\partial w}{\partial y}\right) = \eta \dfrac{\partial^2 w}{\partial y^2} \end{cases} \quad (1)$$

where p is oil film pressure, τ_{yx} and τ_{yz} is oil film shear stress along X direction and Y direction, u and w is oil film speed along X direction and Z direction, η is fluid viscosity.

It requires fixed boundary conditions to solve Eq. (1), namely reflecting the effect of micro-scale velocity slip boundary conditions.

2.2 Velocity slip boundary condition in two directions

With the improvement of testing techniques in micro-nano level, it was found that the size effect of micro-scale makes it easier to the occurrence of velocity slip in the solid-liquid interface.

Currently, a lot of researches on the model of velocity slip have been made, they are: Navier slip model [10], nonlinear slip model [11], ultimate shear force model [4] and so on. Navier slip model is suitable for the occasion of lower shearing rate and the most widely used at present. In this paper, Navier slip model is adopted to analyze oil film flow in hydrostatic bearing.

In Navier slip model, it assumes that fluid velocity is proportional to the shear stress , that is

$$v_s = b \left.\frac{\partial v_x}{\partial y}\right|_{wall} \quad (2)$$

where v_x is fluid velocity along X direction, Y is the normal direction of the interface, b is called slip length, which is the distance between actual surface and assumed surface, shown in Figure 3. The size of slip length is determined by some attribute parameters such as material properties, shear rate and gas layer existing in solid-liquid interface, and these factors cannot be simply coupled [12], so slip length is very difficult to determined, generally cannot be obtained by numberical calculation but experiments. Li Xinming used a method of theory combined with practice to study the method of slip length measurement and established a model [13]. According to [14] the slip length in this paper ranges from 0~7μm.

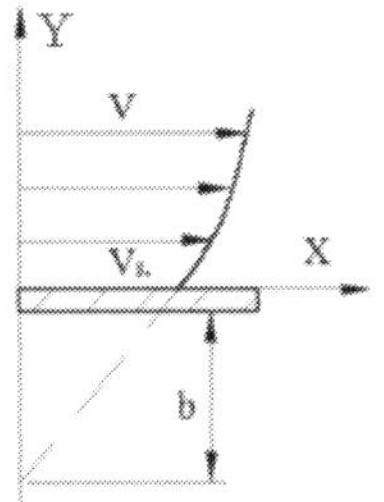

Figure 3 Navier slip mode

According to Navier slip velocity boundary condition, velocity boundary condition of oil flow film on the sealing surface can be written as:

$$\begin{aligned} &\text{at } y=0,\; u=u_s=-b_{1x}\frac{\partial u}{\partial y},\; w=w_s=-b_{1z}\frac{\partial w}{\partial y} \\ &\text{at } y=h,\; u=u_2-b_{2x}\frac{\partial u}{\partial y},\; w=w_s=-b_{2z}\frac{\partial w}{\partial y} \end{aligned} \quad (3)$$

where u_2 is line-speed of rotating spindle; b_{1x}, b_{1z} are the slip length in X and Z directions of bush; b_{2x}, b_{2z} are slip length in X and Z directions of journal.

2.3 Modified Reynolds equation by velocity slip boundary condition

Taken slip boundary condition into consideration, oil film velocity equations in the circumferential and axial direction are gained by the integration of Eq. (1).

$$\begin{aligned} u &= \frac{1}{2\eta}\frac{\partial p}{\partial x}y^2 - \frac{h^* b_{1x}}{\eta}\frac{\partial p}{\partial x} + \\ &\frac{y}{h}\left(u_2 - \frac{b_{2x}(h-h^*)}{\eta}\frac{\partial p}{\partial x} - \frac{h^2}{2\eta}\frac{\partial p}{\partial x} + \frac{h^* b_{1x}}{\eta}\frac{\partial p}{\partial x}\right) \\ w &= \frac{1}{2\eta}\frac{\partial p}{\partial z}\left(y^2 - yh - hb_{1z}\right) \end{aligned} \tag{4}$$

where h^* is oil film thickness in axial direction when the shear stress is zero.

The volume of fluid flow through unit width is gained by integrating Eq. (4) along the way of film thickness,

$$\begin{aligned} q_x &= \int_0^h u dy = \\ &\left[-\frac{h^3}{12\eta} - \frac{hb_{2x}}{2\eta}\left(h-h^*\right) - \frac{hh^* b_{1x}}{2\eta}\right]\frac{\partial p}{\partial x} + \frac{hu_2}{2} \\ q_z &= \int_0^h w dy = \left(-\frac{h^3}{12\eta} - \frac{h^2 b_{1z}}{2\eta}\right)\frac{\partial p}{\partial z} \end{aligned} \tag{5}$$

The modified Reynolds is from mass conservation equations [15]:

$$\frac{\partial}{\partial x}(q_x) + \frac{\partial}{\partial z}(q_z) = -\frac{\partial h}{\partial t} \tag{6}$$

Assuming that spindle rotated in steady state, modified Reynolds equation is gained by bringing the equation (5) into Eq. (6):

$$\begin{aligned} &\frac{\partial}{\partial x}\left\{\left[-\frac{h^3}{12\eta} - \frac{hb_{2x}}{2\eta}\left(h-h^*\right) - \frac{hh^* b_{1x}}{2\eta}\right]\frac{\partial p}{\partial x} + \frac{hu_2}{2}\right\} \\ &+\frac{\partial}{\partial z}\left[\left(-\frac{h^3}{12\eta} - \frac{h^2 b_{1z}}{2\eta}\right)\frac{\partial p}{\partial z}\right] = 0 \end{aligned} \tag{7}$$

It can be seen from the modified Reynolds equation that slip in the Z direction of journal is not introduced into the Eq.(7), that is, the slip in the Z direction of journal does not affect oil film flow. However, some assumptions must be made about the equation for the further study. Assumptions should be based on current study and specific working conditions, such as material conditions and fluid properties alike. Slip length is assumed to be equal in isotropic material [15, 16]. This paper assumes $b = b_{1x} = b_{1z} = b_{2x} = b_{2z}$.

In order to carry out calculations, Eq.(7) should be dimensionless combined with the hypothesis of slip length. Characteristic pressure is p_0, characteristic slip length and characteristic oil film thickness are bearing radius clearance c, characteristic axial length of oil film is half the length of bearing $L/2$, characteristic axial length of oil film is radius R of bearing. Suppose $p = p_0 P$, $p_0 = \frac{6w\eta R^2}{h_0^2}$, $b = b_0 c$, $h = Hc$, $z = \lambda\frac{L}{2}(-1 \le \lambda \le 1)$, $x = R\varphi(0 \le \varphi \le 2\pi)$, φ is position angle of oil film, the dimensionless Reynolds equation is as equation (8):

$$\left(1+6b_0\right)\frac{\partial}{\partial\varphi}\left(H^3\frac{\partial P}{\partial\varphi}\right) + \left(1+6b_0\right)\left(\frac{2R}{L}\right)^2\frac{\partial}{\partial\lambda}\left(H^3\frac{\partial P}{\partial\lambda}\right) = \frac{\partial H}{\partial\varphi} \tag{8}$$

3 Calculation of bearing performance

3.1 The calculation of oil film flow

On the calculation of flow of hydrostatic bearing, the existing studies do not take the impact of size effect into account, which have a great difference with actual working conditions. To solve the problem, this paper takes orifice compensated hydrostatic bearing with four pockets as research object, and the main structural parameters is shown in Figure 4. Under the external load the spindle finally stays in the equilibrium position O_1, the eccentricity of the spindle is e, deviation angle is θ, θ_1, θ_2 is oil chamber half-width and oil pad half-width, R is radius of spindle, P_s is oil supply pressure.

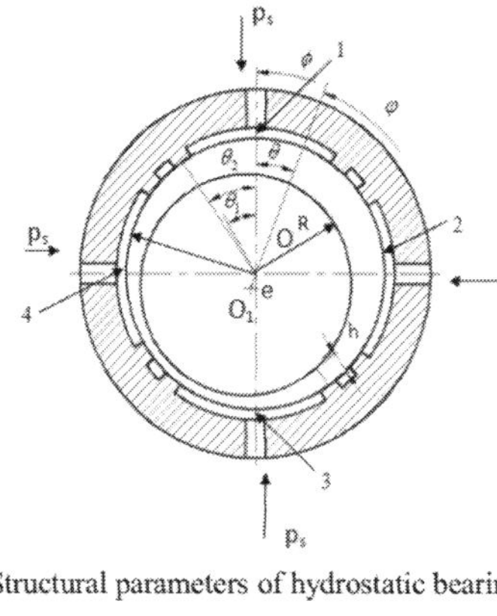

Figure 4 Structural parameters of hydrostatic bearing

When the spindle occurs eccentric, oil film thickness is not constant, and it will change as angular position changes. Oil film thickness equation is given as follow:

$$h_1 = h_0 + e\cos(\varphi) = h_0\left(1 + \varepsilon\cos(\phi - \theta)\right) \tag{9}$$

where h_0 is the initial film thickness, φ is the position angle, ε is eccentricity ratio, $\varepsilon = e/h_0$.

Combining with the Figure 7, mass flow of each oil pocket can be calculated:

$$Q_{iao}=2\int_{-\frac{\theta_1+\theta_2}{2}+\theta}^{\frac{\theta_1+\theta_2}{2}+\theta}\int_0^h Rwdyd\varphi =$$

$$2\int_{-\frac{\theta_1+\theta_2}{2}+\theta}^{\frac{\theta_1+\theta_2}{2}+\theta}\int_0^h\left(\frac{1}{2\eta}\left(y^2-yh-bh\right)\frac{\partial p}{\partial z}\right)dyd\varphi$$

$$Q_{iro}=\mathrm{L}\left(-\frac{h_3}{12\eta}-\frac{h^2b}{2\eta}\right)\frac{\partial p}{\partial x}\bigg|_{h=h_l}+ \tag{10}$$

$$\mathrm{L}\left(-\frac{h_3}{12\eta}-\frac{h^2b}{2\eta}\right)\frac{\partial p}{\partial x}\bigg|_{h=h_r}+\frac{hu_2}{2}\bigg|_{h=h_r}-\frac{hu_2}{2}\bigg|_{h=h_l}$$

$$Q_{io}=Q_{iao}+Q_{iro}$$

where Q_{iao} is volume of lubricating oil flow along axial sealing side, Q_{iro} is the volume of lubricating oil flow along circumferential sealing side, L is the length of bearing, h_r is upstream midpoint clearance of oil sealing side of a certain chamber, h_l is downstream midpoint clearance of oil sealing side of oil sealing side of a certain chamber.

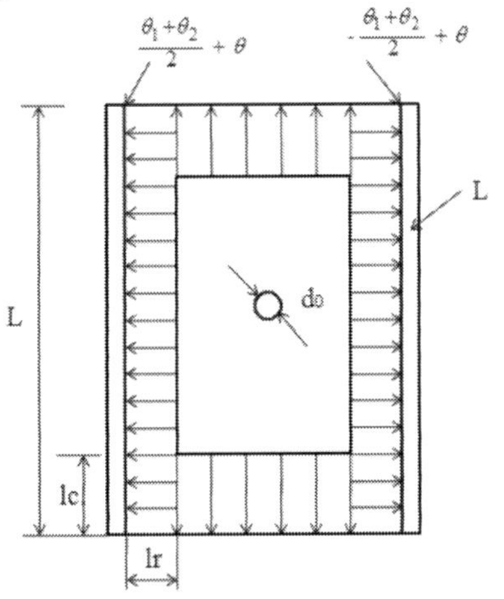

Figure 5 Oil flow out of the hydrostatic bearing chamber

This paper uses orifice restrictor. According to hydrodynamic equations, the flow through orifice restrictor is [17]:

$$Q_{ii}=\alpha\frac{\pi d_0^2}{4}\sqrt{\frac{2(p_s-p_{ri})}{\rho}} \tag{11}$$

where Q_{ii} is the flow that flows into a certain chamber through orifice restrictor, α is discharge coefficient (0.6~0.7), ρ is the density of lubricating oil, d_0 is the diameter of orifice restrictor.

Assuming the oil cannot be compressed. The flow flowing into the bearing oil pocket through restrictor is equal to the outflow through sealing surface, thus the flow conservation equation is below:

$$Q_{ii}=Q_{io}=Q_{iro}+Q_{iao} \tag{12}$$

3.2 Calculation of bearing capacity

Bearing capacity means when oil film formats pressure, the synthetic hydraulic pressure floats the bearing and balances external load acting on the spindle, the synthetic liquid pressure is called load capacity.

The equation of load capacity is as follows:

$$W_x=\int_0^L\int_0^{2\pi}RP\cos(\varphi)d\phi dz$$

$$W_y=\int_0^L\int_0^{2\pi}RP\sin(\varphi)d\phi dz \tag{13}$$

$$W=\sqrt{W_x^2+W_y^2}$$

where oil film pressure P is interval function of staging areas, when the integration areas are the chambers($z=[0,l]$, $\phi=[-\theta_1,\theta_1]$, $\left[\frac{\pi}{2}-\theta_1,\frac{\pi}{2}+\theta_1\right]$, $[\pi-\theta_1,\pi+\theta_1]$, $\left[\frac{3\pi}{2}-\theta_1,\frac{3\pi}{2}+\theta_1\right]$), the film pressure P is P_{ri} (i = 1,2,3,4), When the integration areas are return chutes, the oil film pressure P is zero, when the integration areas are sealing surfaces, the oil film pressure P values as the linear distribution.

According to above analysis, it is easy to determine bearing capacity. This paper provides a convenient method for designing the validation of hydrostatic bearing.

3.3 Calculation of bearing dynamic stiffness

Dynamic stiffness of hydrostatic bearing is the ability to resist displacement of the spindle. Bearing oil film stiffness is defined as: when the axis has a tiny displacement, the ratio of the increase of oil film force to the amount of displacement. The displacement and is decomposed into two components Δx (along X direction) and Δy (along Y direction), and the corresponding increase of oil film strength can be decomposed into two components ΔW_x and ΔW_y . Four oil film stiffness coefficients are as follows:

$$K_{xx}=\frac{\partial W_x}{\partial x},K_{xy}=\frac{\partial W_x}{\partial y},$$

$$K_{yx}=\frac{\partial W_y}{\partial x},K_{yy}=\frac{\partial W_y}{\partial y} \tag{14}$$

4 Simulation of oil film performance

4.1 Simulation of oil film pressure

The sensitivity factor of oil film flow Δ_1 is defined as following:

$$\Delta_1 = \frac{\left|(P_{max})_{slip} - (P_{max})_{no}\right|}{(P_{max})_{no}} \quad (15)$$

where, $(P_{max})_{slip}$ is the maximum pressure of oil film when the slip occurs, $(P_{max})_{no}$ is the maximum pressure of oil film when the slip does not happen.

Figure 6 shows the variation of the dimensionless maximum pressure of four pockets P_1, P_2, P_3, P_4 with the eccentricity in different slip level. It can be seen from Figure 6 that, with the eccentricity increasing, the maximum pressures of the oil pockets P_1, P_2, P_4 decrease, and the maximum pressure P_3 increases. When the eccentricity is constant, the four maximum oil film pressures reduces with the increase of slip length, but the reduction is constantly changing, the sensitivity of maximum oil film pressure can be used to describe the reduced degree of pressure and quantitatively analyzes the influence degree of slip on the maximum pressure. Figure 7 shows the variation of the pressure sensitivities of four pockets Δ_{P1}, Δ_{P2}, Δ_{P3}, Δ_{P4} with the eccentricity in different slip level. The maximum pressure sensitivities are 59.31%, 21.14%, 7.54%, 22.31% . Compared Figure 6 with Figure 7, it can be concluded that with the increase of slip length, the four maximum oil pressure sensitivities will increase accordingly.

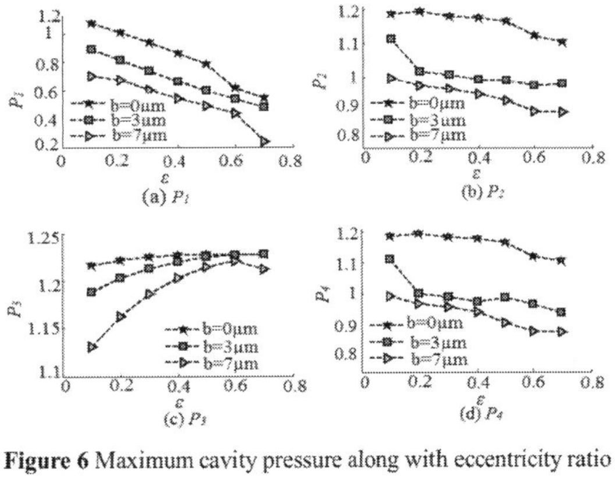

Figure 6 Maximum cavity pressure along with eccentricity ratio

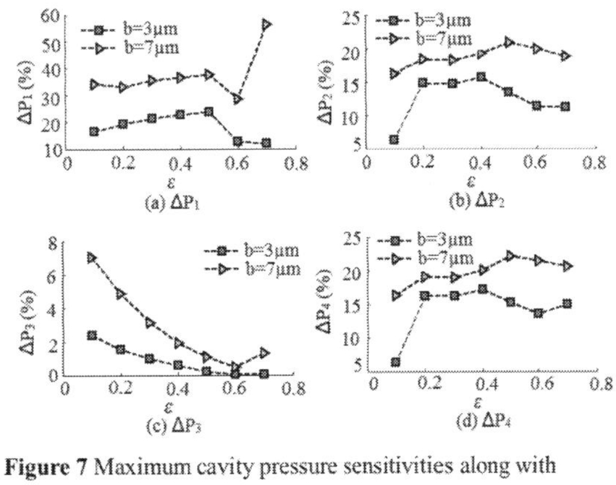

Figure 7 Maximum cavity pressure sensitivities along with eccentricity ratio

4.2 Simulation of oil film flow

$$\Delta_2 = \frac{\left|(Q)_{slip} - (Q)_{no}\right|}{(Q)_{no}} \quad (16)$$

where $(Q)_{slip}$ is the oil film flow when the slip occurs, $(Q)_{no}$ is oil film flow when the slip does not happen. Q_1, Q_2, Q_3, Q_4 are the dimensionless oil film flow of four pockets of the bearing. Figure 8 shows the variation of dimensionless oil film flow Q_1, Q_2, Q_3, Q_4 with the eccentricity of the journal in different slip level. It can be seen from Figure 8 that, with the eccentricity increasing, the flow Q_1, Q_2, Q_4 increase, and the flow Q_3 decreases. When the eccentricity is constant, the four flows increase with the increase of slip length, but the increase rate is constantly changing, the sensitivity factor of flow can be used to describe the increase degree of oil film flow.

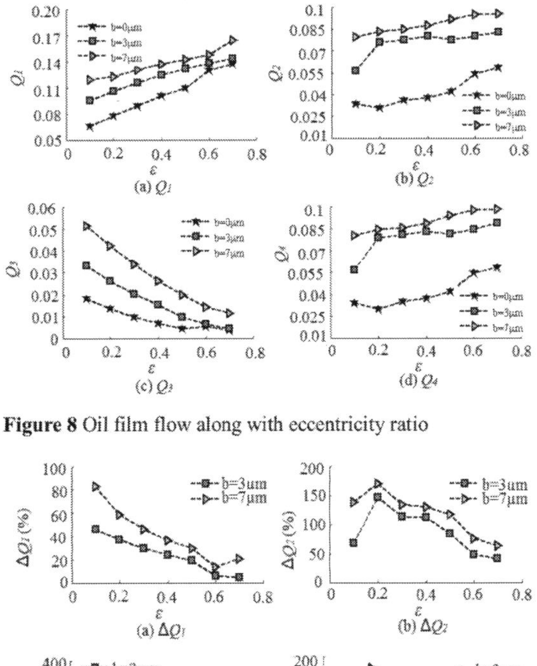

Figure 8 Oil film flow along with eccentricity ratio

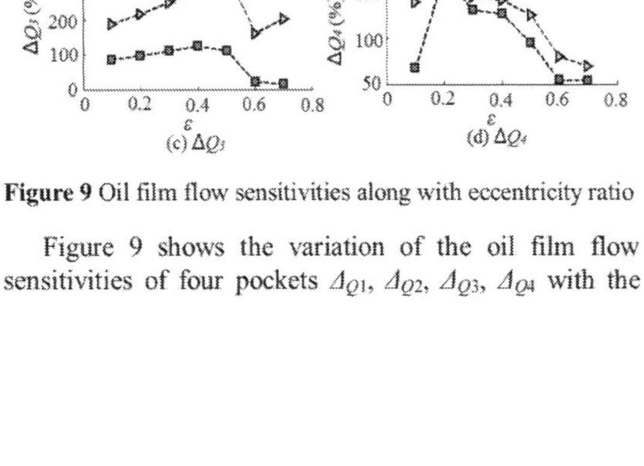

Figure 9 Oil film flow sensitivities along with eccentricity ratio

Figure 9 shows the variation of the oil film flow sensitivities of four pockets Δ_{Q1}, Δ_{Q2}, Δ_{Q3}, Δ_{Q4} with the eccentricity in different slip level. The maximum flow sensitivities are 84.14%, 183.33%, 337.24%, 183.33% and the minimum flow sensitivities are 4.67%, 40.81%, 16.09%, 53.31%. By comparing Figure 8 with Figure 9, it can be concluded that with the increase of slip length, the four oil film flow sensitivities will increase accordingly.

Figure 10 shows the relationship between the total flow Q of the hydrostatic bearing and the rate of eccentricity in different slip case. Under different cases of slip, the total flow increases with increasing eccentricity. The largest relative increase rate of total flow is 57.78% when the slip does not occur and the slip length is 7um, largest relative increase rate is 22.22%, although the slip does not change the relationship between eccentricity and total flow, the happen of slip weakens the degree of increase of total flow. When the spindle eccentricity is constant, the flow increases with increasing slip length, maximum total flow sensitivity was 100%. This means that the occurrence of slip accelerate flow in bearing clearance, which is conducive to high temperature lubricant rapid outflow of bearings, there is more likely to have a dynamic pressure effect on seal and improve the bearing capacity.

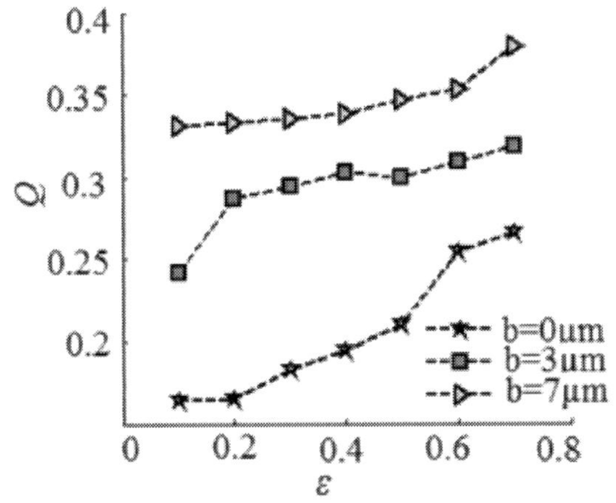

Figure 10 Total flow along with eccentricity ratio

4.3 Simulation of bearing capacity

The sensitivity factor of oil film flow Δ_3 is defined as following:

$$\Delta_3 = \frac{\left|(W)_{slip} - (W)_{no}\right|}{(W)_{no}} \quad (17)$$

where $(W)_{slip}$ is bearing capacity when the slip occurs, $(W)_{no}$ is bearing capacity when the slip does not happen.

Figure 11 shows the relationship between the dimensionless bearing capacity and the eccentricity. It can be seen from Figure 9, when no slip occurs, bearing capacity increases with increase of eccentricity, and the relationship cannot be changed by velocity slip occurred in solid - liquid interface. But when the eccentricity is constant, bearing capacity increases with the increase of slip length, the maximum sensitivity of bearing capacity is 81.94%. Therefore, considering the slip in the design of hydrostatic bearing is different from does not consider, and this shows it is necessary to research the mechanism and influencing factors of slip in oil film clearance flow.

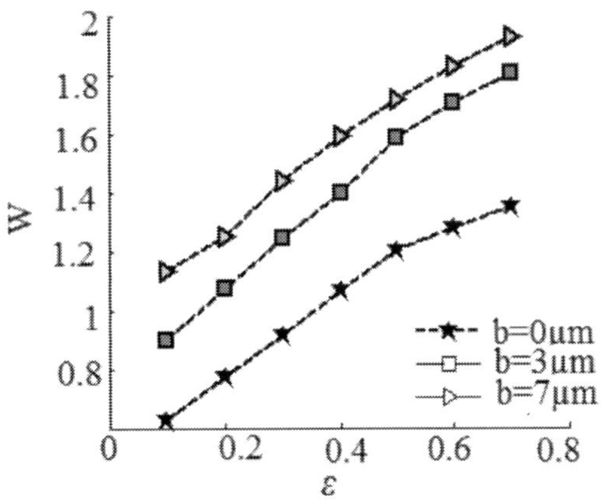

Figure 11 Dimensionless bearing capacity along with eccentricity ratio

4.4 Simulation of bearing dynamic stiffness

Dynamic stiffness of hydrostatic bearing is the ability to resist displacement of the flexible shaft moving. In this paper, an incremental method is used to solve the four bearing dynamic stiffness K_{xx}, K_{xy}, K_{yx}, K_{yy}. Figure 12 shows the relationship between bearing dynamic stiffness and eccentricity when the slip does not occur. It can be seen from Figure 12, the four bearing dynamic stiffness has the same relationship with eccentricity; with the eccentricity increase, the four dynamic stiffness decreases and proves the cross stiffness K_{xy} and K_{yx} are equal. When the eccentricity is constant, K_{xx} is the largest, followed by the cross stiffness K_{xy} and K_{yx}, K_{yy} is the minimum.

The sensitivity factor of oil film flow Δ_4 is defined as following:

$$\Delta_4 = \frac{\left|(D)_{slip} - (D)_{no}\right|}{(D)_{no}} \quad (18)$$

where $(D)_{slip}$ is bearing dynamic stiffness when the slip occurs, $(D)_{no}$ is bearing dynamic stiffness when the slip does not happen.

Figure 13 shows the relationship between four dynamic stiffness K_{xx}, K_{xy}, K_{yx}, K_{yy} and slip length when the eccentricity is 0.3. Four stiffness increase linearly with slip length decreases, the maximum sensitivities are 3.02%, 2.14%, 3.30%, 3.26%, through quantitative analysis shows that the velocity slip has a certain influence on dynamic stiffness. Therefore, slip should be taken into consideration in the design of

DOI: 10.1051/matecconf/20168012003

hydrostatic bearing, especially in the special working conditions the dynamic stiffness is very important.

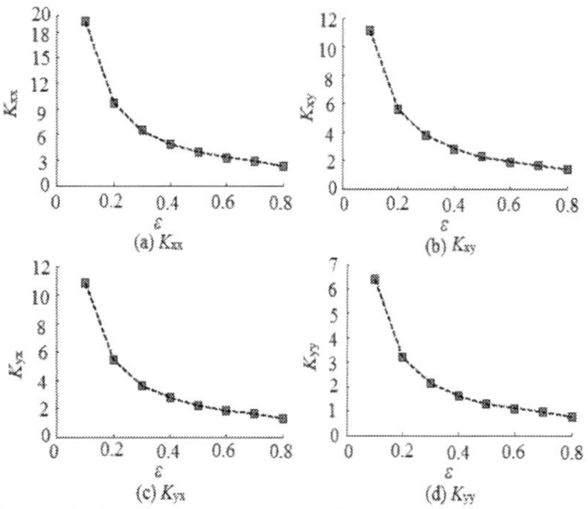

Figure 12 Dynamic stiffness of no slip along with eccentricity ratio

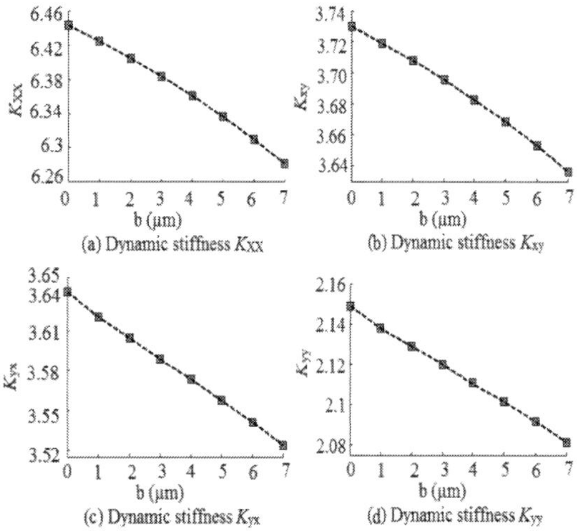

Figure 13 Dynamic stiffness along with slip length for ε=0.3

5 Conclusions

This paper analyzes the properties of two-dimensional oil film flow of hydrostatic bearing in micro-scale, and successfully gives flow regularities using Navier slip model. The Reynolds equation is modified by definite boundary conditions in the consideration of velocity slip. Finite difference method is used to solve Reynolds equation, and oil film pressure distribution, bearing capacity, oil film flow, bearing dynamic stiffness are gained. Sensitivity factors of oil film pressure, load capacity, flow and dynamic stiffness are deduced and the variation of them with the eccentricity in different slip length is showed, and the result of the velocity slip reduces bearing dynamic characteristics and stability to some extent is obtained. This proposed the slip should be taken into consideration in the design of hydrostatic bearing, especially in the special working conditions the dynamic stiffness is very important.

Acknowledgements

This research was funded by the National Natural Science Foundation of China Grant No.51475010, the China Nova program of Beijing and Natural Science Foundation of Beijing Municipality Grant No.3142005.

References

1. OI Vinogradova, International Journal of Mineral Processing, **56**,31-60 (1999).
2. G Zhang, P Huang, etc., Tribology, **18**,243-247 (1998).
3. G Zhang Gu, HR Meng etc., Lubrication. Lubrication Engineering, **4**,48-51 (1999).
4. C Wu, Journal of Dalian University of Technology, **33**,172-178 (1993).
5. P Huang,J Luo,S Wen, Acta Mechanica Sinica, **31**,745-752 (1999).
6. Z Liu, Journal of Mechanical Engineering,**47**, 104-110(2011).
7. J Lupse, L Skerget, J Ravnik, Journal of Mechanical Engineering, **61**,254-264 (2015).
8. Q Xue, J Zhang, Tribology, **14**,360-369 (1994).
9. Z Zhou, Chinese Journal of Mechanical Engineering, **39**,22-26 (2003).
10. C Navier, C Navier, Mem.acad.r.sci, **6**,389-440(1823).
11. P.A Thompson, S.M Troian,etc., Nature, **389**,360-362(1997).
12. C. Neto, DR Evans., E Bonaccurso, etc., Reports on Progress in Physics, **68**,2859-2897 (2005).
13. LI Xin-.Ming, F.Guo, W Pat-Lam, Mocaxue Xuebao/tribology, **32**,34-39 (2012).
14. L Ehrenworth, Journal of Fluid Mechanics. **489**,55-77(2003).
15. CY Chen, QD Chen, WL Li, Tribology International, **61**,144-155 (2013).
16. WL Li, HM Chu, MD Chen, Tribology International, **39**,1428-1435, (2006).

FEM Analysis of the Multi-Wedge Helical Rolling Process for a Workholding Bolt

Zbigniew Pater[1,a]

[1] *Lublin University of Technology, Faculty of Mechanical Engineering, 36 Nadbystrzycka Str, Lublin, Poland*

Abstract. This paper presents the results of a numerical analysis of a helical rolling process for producing a 25 mm diameter workholding bolt using multi-wedge helical tools. The numerical modeling was performed by the finite element method (FEM) using the commercial simulation software Simufact.Forming v. 12. The modeling was performed in a three-dimensional state of strain and included a full thermal analysis. As a result, the changes in workpiece shape as well as the distributions of strains, temperature, damage, loads and torque were determined. The numerical results confirm that workholding bolts can be produced by multi-wedge helical rolling.

1 Introduction

Metal forming methods are widely employed for producing axi-symmetric parts that are used in the automotive, aircraft, machine building and mining industries as well as in other branches of economy. Axi-symmetric parts can be produced by methods such as die forging, rotary forging, machine forging, extrusion, forge rolling and others [1, 2]. Despite the availability of the numerous methods for producing axi-symmetric parts, research has been continued to develop new methods for forming these parts. The focus is here on developing methods which will help reduce material and energy consumption of the manufacturing process, as well as ensuring at the same time increased efficiency. One of these new methods is helical wedge rolling (HWR).

The assumptions of the HWR process were developed at the Lublin University of Technology as part of the research and development work on new highly efficient methods for producing balls for ball mills [3-4]. In the research, it was assumed that balls can be formed by wedges that are helically wound (in a single- or multi-coil manner) on the roll face. The numerical results and experimental tests confirmed the correctness of this solution [5-8]. The development of the HWR method involved the design of single-coil tools for producing such parts as stepped shafts [9, 10], bodies of rotary cutters [11] and workholding bolts [12]. The developed solutions were verified by numerical simulation.

This paper describes a new way of forming parts such as a workholding bolt (Fig. 1). The method is based on the helical wedge rolling (HWR) process which is performed using rolls with three helically wound forming wedges. Named as multi-wedge helical rolling (MWHR), this process is characterized by very high efficiency and high values of force parameters. The proposed solution was verified in a numerical simulation performed using the finite element method-based simulation program Simufact.Forming v.12.

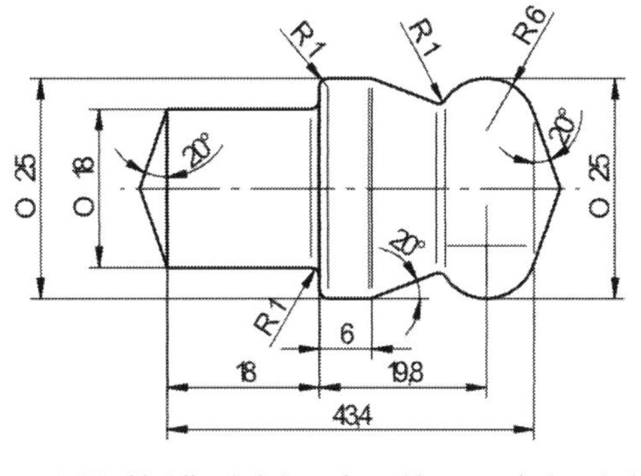

Figure 1. Workholding bolt investigated by numerical analysis

2 Application of FEM for analysis of helical rolling processes

The MWHR process is a variant of helical rolling. It is performed using rolls with helically wound wedges on their perimeter. The first studies on numerical modeling of helical rolling were conducted in the 1990s; they concerned a piercing process for thick-walled tubes and were based on two-dimensional FEM models [13]. However, given the nature of metal flow in these processes, they should be simulated in a three-dimensional (3D) state of strain. The first attempts at 3D modeling of piercing processes were made by Pietisch and Thieren [14], Ceretti et al. [15], and Komori [16]. The numerical models developed by the above authors included numerous simplifications, for instance they did not take into account thermal phenomena in the material.

[a] Corresponding author: z.pater@pollub.pl

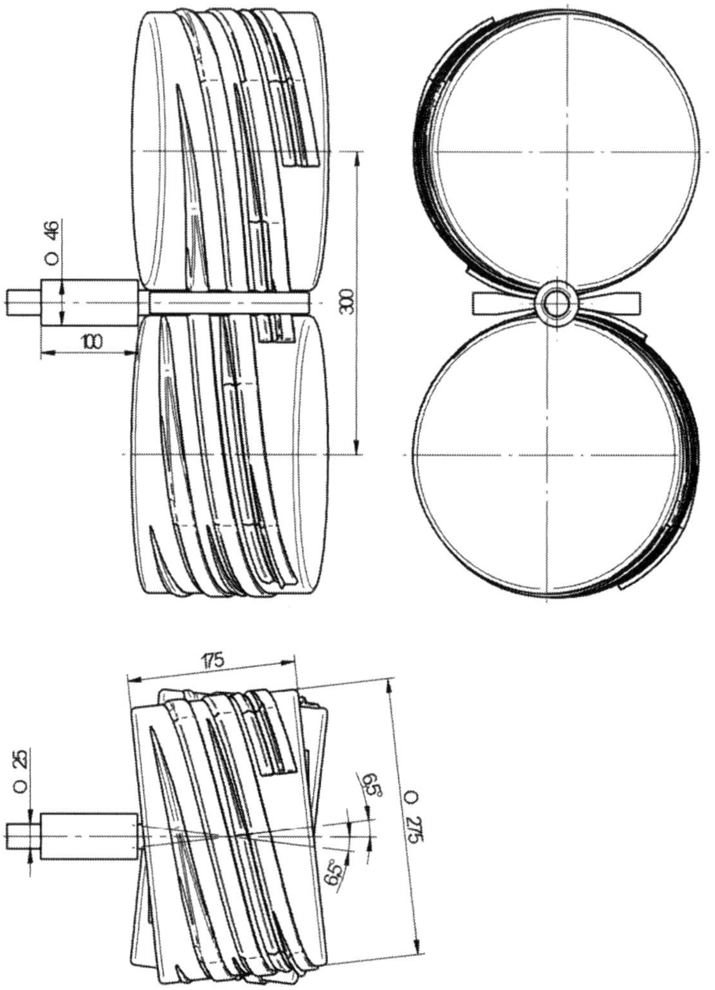

Figure 2. Schematic design of the multi-wedge helical rolling (MWHR) process for producing a workholding bolt (with selected dimensions shown in the figure)

In addition, the computations were limited either to the steady-state stage of the process or to the beginning of the forming process. Complete numerical models of the piercing process for thick-walled tubes taking into consideration thermal phenomena were constructed by several authors only in the last ten years [17-25]. The computations were performed with commercial simulation programs such as MSC. SuperForm [15, 21-23], Forge [18-20], Deform-3D [25].

Other studies were devoted to numerical analysis of elongation processes in helical rolling mills. One can mention here publications on elongation of such parts as thick-walled tubes [26-30], round ingots [31, 32] and round bimetallic bars [33, 34]. The fact that there is a scarce number of publications on elongation processes can be attributed to long computational time, which imposes limits on the number of forming cases for analysis.

It can be observed that the smallest number of publications concerns the modeling of helical rolling process with helical rolls. Aside from the above mentioned publications by the present author [5-12], there are only few other publications available on producing shafts [35-37] and balls produced by cold rolling [38-40].

3 Design of the MWHR process for workholding bolts

Fig. 2 shows a schematic design of the MWHR process for producing workholding bolts using three-coil wedge rolls. Aside from the rolls, the model also consists of two guiding blades for maintaining the billet in the workspace of the rolls and a sleeve for feeding the billet into the workspace of the rolls. To make the process automatic, the rolls are twisted relative to the axis of the billet by an angle θ set to 6.5°.

The forming wedges are wound on the roll face with a variable pitch applied such that the material adheres to the surface of the working (lateral) wedges during the entire forming process. The spreading angle and the angle of inclination were selected intuitively based on the

DOI: 10.1051/matecconf/20168013001

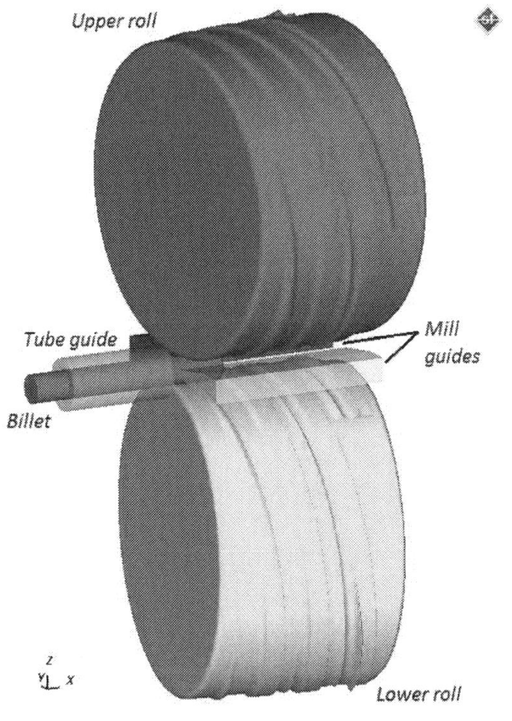

Figure 3. Geometrical model of the multi-wedge helical rolling process for producing workholding bolts

author's considerable experience in modeling of typical cross wedge rolling (CWR) processes. The wedges are located every 120° on the perimeter of the rolls.

The finite element method analysis verifying the correctness of the developed solution was performed on a geometrical model of the MWHR process for workholding bolts shown in Fig. 3. The billet was assumed to be made of a corrosion-resistant steel grade X12CrNi17-12 and its material model (yield stress σ_p versus temperature T, effective strain ε and strain rate $\dot{\varepsilon}$) was obtained from the material database of the applied software, i.e. Simufact.Forming v. 12. As for other parameters applied in the simulation, the billet temperature was set to 1100 °C, the temperature of the rolls was 50 °C, the temperature of other tools was 250 °C, the friction factor was set to 0.9 for the rolls and to 0.4 for other tools, the material-tools heat transfer coefficient was 10 kW/m^{2}K, and the rotational speed of the rolls was 15 rev/min.

The billet (with a diameter of 25 mm and a length of 185 mm) was modeled using eight-node cubic elements, with the mesh refined in the forming zone. Due to the change of material shape during rolling, the model had to be remeshed, and the number of elements was gradually increased from 12414 at the beginning to 47659 at the end (for time t=6 s), as shown in Fig. 4.

4 Numerical results

The application of FEM enabled numerical verification of the developed MWHR process for producing a workholding bolt. Fig. 4 shows the changes in workpiece shape in the advanced (steady state) stage of forming after producing the first part. The developed process is highly efficient (three bolts are produced during one revolution of the rolls) and characterized by reduced material waste. One billet generates two end scraps, which means that increased workpiece length could lead to higher material efficiency. After forming, the parts are separated from one another by a helical cutter that is mounted at the end of the wedge. In the solution described in this paper, the cutter's height is decreased on purpose; in this way, the cutters only make a necking (thereby indicating where the cutting is to be made) and they do not separate the material (this would stop the numerical simulation). As a result, it was unequivocally found that the MWHR process can be used to produce parts such as workholding bolts.

Fig. 5 shows the distributions of effective strain in the axial section and on the surface of the produced part. The data given in this figure demonstrate that the strains in the cross section are not uniformly distributed. The highest strains are located at the part's surface and they gradually decrease the closer it is to the axis of the workpiece. Such a distribution of strains results from a rapid flow of the material in the tangential direction caused by the friction forces. This tangential flow is characteristic of cross and helical rolling processes. In addition, it was observed that cross sectional reduction leads to a higher effective strain

Figure 4. Workpiece after 1.5 revolution of the rolls, with division into finite elements

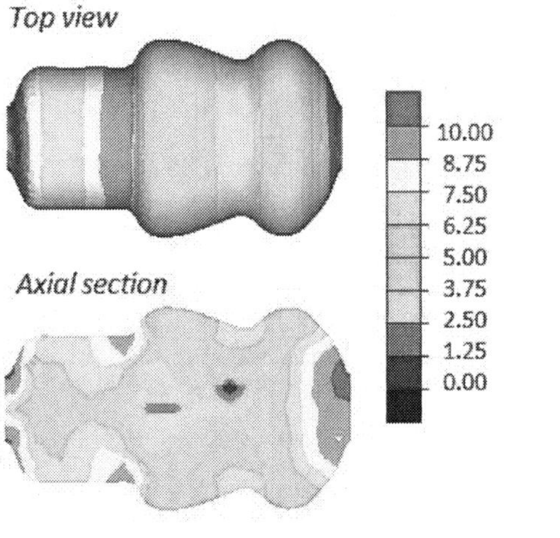

Figure 5. Distribution of effective strain in a part produced by MWHR

in the semi-finished part.

In the analyzed process, a single part is formed during 1.5 revolution of the rolls, which is equal to 6 s. In this period of time, the material has contact with cold tools which absorb heat. At the same time, the exchange of friction work into strain work leads to generation of considerable amounts of heat, which partly compensates for heat losses resulting from the heat transfer to the tools (Fig. 6). As a consequence, the temperature of the workpiece decreases in the surface regions (up to 150 °C) while it increases in the region of bridges that separate individual bolts (up to 75 °C). It is important here that the temperature of the material is maintained during the whole forming process in a range suitable for hot metal forming processes (Fig. 7).

Interesting observations can be made with respect to

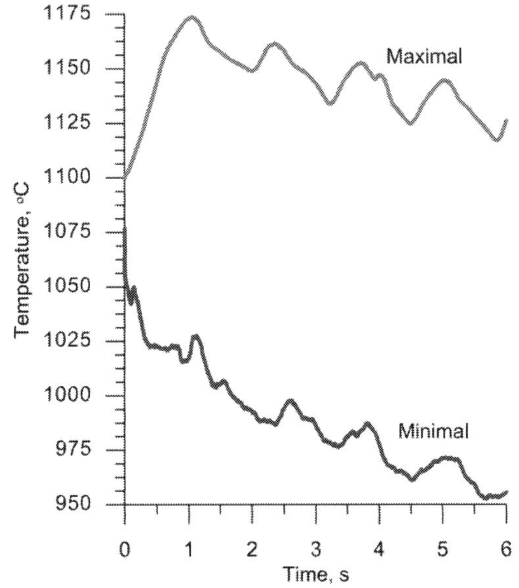

Figure 7. Maximum and minimum temperatures of the workpiece during MWHR

the distribution of the damage function (calculated in compliance with the Cockcroft-Latham criterion), shown in Fig. 8. The highest values of this function occur at the ends of the part, i.e. in the places affected by the cutters. It is in these places that material cohesion can be broken, so the bolt is separated from the rest of the material.

The application of FEM also allows us to determine distributions of force parameters during MWHR. They are illustrated in Fig. 9 (radial and tangential loads) and in Fig. 10 (torque on the roll). These distributions have a characteristic oscillatory character, which reflects the

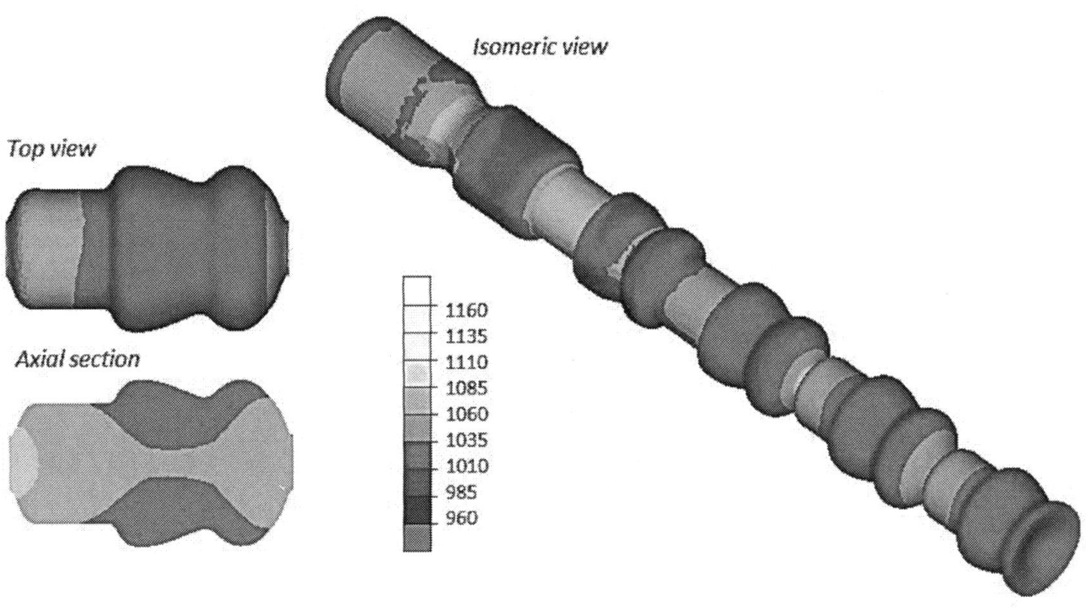

Figure 6. Distribution of temperature (in °C) in a single part and in the entire semi-finished product after 1.5 revolution of the rolls

DOI: 10.1051/matecconf/20168013001

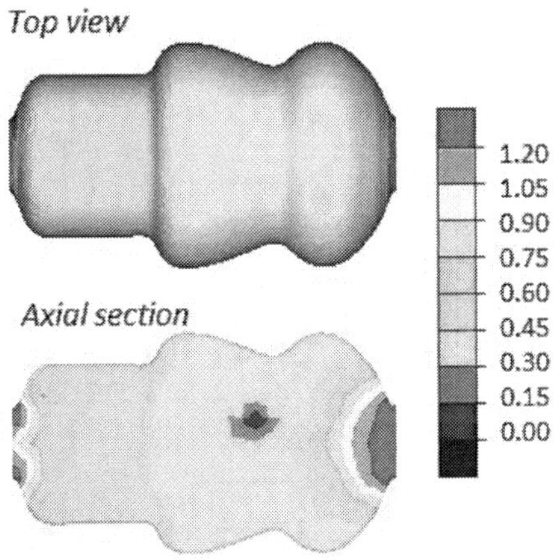

Figure 8. Distribution of the Cockcroft-Latham damage function in a part produced by MWHR

cyclical nature of the contact between the material being formed and the helical wedges. Owing to this kind of distribution of force parameters in the MWHR process, it is suggested that the process should be run using roll stands equipped with flywheels accumulating energy in the rotational motion.

4 Conclusions

The paper investigated the possibility of producing workholding bolts by a new process of multi-wedge helical rolling (MWHR) developed at the Lublin University of Technology. The results led to the following conclusions:

- parts such as workholding bolts can be produced by three-coil wedge rolls;

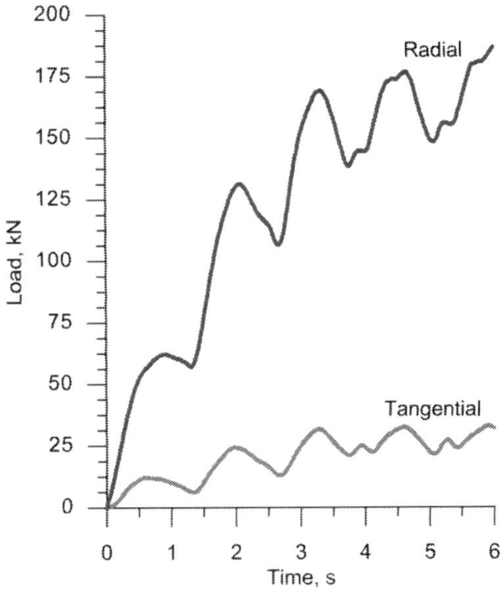

Figure 9. Distribution of loads acting on the tool in the investigated MWHR process

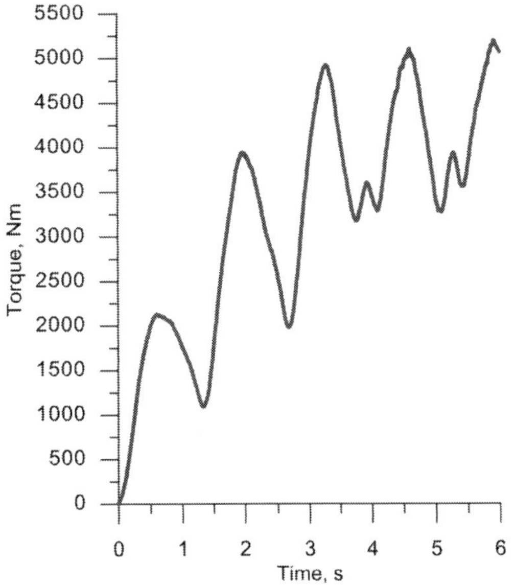

Figure 10. Distribution of torque in the investigated MWHR process

- strains in workholding bolts produced by MWHR are distributed in a way that is typical of cross and helical rolling processes;
- the temperature of the investigated workholding bolt material is maintained in a range that is suitable for hot metal forming;
- workholding bolts produced by MWHR are free from internal cracking, which is often the case with parts produced by helical rolling (the so-called Mannesmann effect);
- force parameters in MWHR are distributed in a characteristic oscillatory manner, which reflects the nature of this forming process in which the wedges cyclically cut in the material being formed;
- it is necessary to construct a laboratory stand that will enable experimental verification of the described MWHR process.

Acknowledgements

This research was made under agreement no. INNOTECH-K3/IN3/12/226286/NCBR/14 on conducting and financing projects under the "INNOTECH" program, IN-TECH program path.

References

1. K. Lange, *Handbook of Metal Forming* (Ed. McGraw-Hill Book Company, 1985)
2. Z. Pater, G. Samołyk, *Podstawy obróbki plastycznej metali* (Ed. Lublin Univ. of Technology, Lublin, 2013)
3. Z. Pater, J. Tomczak, *Walcowanie poprzeczno-klinowe kul* (Ed. Lublin Univ. of Technology, Lublin, 2012)

4. Z. Pater, J. Tomczak, *Walcowanie śrubowe kul do młynów kulowych* (Ed. Lublin Univ. of Technology, Lublin, 2012)
5. Z. Pater Z., J. Tomczak, J. Kazanecki, J. Iron Steel Res. Int. (2012) Spec. Ed. Metal Forming 2012, 79-82
6. Z. Pater, Int. J. Mater. Prod. Tec. **47**, 1/2/3/4 (2013)
7. Z. Pater, J. Tomczak, J. Bartnicki, M.R. Lovell, P.L. Menezes, Int. J. Mach. Tool. Manu. **67** (2013)
8. J. Tomczak, Z. Pater, J. Bartnicki, Arch. Metall. Mater. **58**, 4 (2013)
9. Z. Pater, J. Tomczak, T. Bulzak, Mechanik **86**, 11 (2013)
10. Z. Pater, Key Eng. Mat. **622-623** (2014)
11. Z. Pater, A. Tofil, Hutnik – Wiadomości Hutnicze **80**, 10 (2013)
12. Z. Pater, Metalurgija **53**, 4 (2014)
13. S. Urbański, J. Kazanecki, J. Mater. Process. Tech. **45** (1994)
14. J. Pietsh, P. Thievien, MPT International **2** (2003)
15. E. Ceretti, C. Giardini, A. Attanasio, F. Brisotto, G. Capoferri, TPT, March/April (2004)
16. K. Komori, Int. J. Mech. Sci. **47** (2005)
17. Z. Pater, J. Kazanecki, J. Bartnicki, J. Mater. Process. Tech. **177** (2006)
18. R. Pschera, J. Klarner, C. Sommitsch, BHM **152**, 7 (2007)
19. Y. Chastel, A. Diop, S. Fanini, P.P. Bouchard, K. Mocellin, Int. J. Mater. Form, Suppl. 1 (2008)
20. A. Ghiotti , S. Fanini, S. Bruschi, P.F. Bariani, CIRP Ann-Manuf. Techn. **58** (2009)
21. L. Lu, F. Wang, Z. Wang, G. Zhu, X. Zhang, J. Iron Steel Res. Int. **18**, 7 (2011)
22. L. Lu, Z. Wang, F. Wang, G. Zhu, X. Zhang, J. Shanghai Jiaotong Univ. (Sci.) **16**, 3 (2011)
23. Z. Pater, J. Bartnicki, J. Kazanecki, Metalurgija **51**, 4 (2012)
24. Z. Pater, J. Kazanecki, Arch. Metall. Mater. **58**, 3 (2013)
25. F. Wang, Y. Shuang, J. Hu, Q. Wang, J. Sun, J. Mater. Process. Tech. **214** (2014)
26. Z. Malinowski., J. Kazanecki, S. Urbański, J. Mater. Process. Tech. **60** (1996)
27. R.J. Montecinos, E.S. Arauco, J. Mater. Process. Tech. **95** (1999)
28. Z. Pater, J. Kazanecki, Hutnik – Wiadomości Hutnicze **74**, 9 (2007)
29. B. Li, S.H. Zhang, G.L. Zhang, H.Q. Zhang, J. Mater. Process. Tech. **205** (2008)
30. F. Wang, Y. Shuang, J. Hu, Q. Wang, J. Sun, J. Mater. Process. Tech. **214** (2014)
31. E.I. Panov, Metallurgist **48**, 1-2 (2004)
32. E.I. Panov, Metallurgist **48**, 3-4 (2004)
33. S. Sawicki, P. Szota, H. Dyja, AMSE **32**, 1 (2008)
34. S. Sawicki, P. Szota, S. Mróz, A. Stefanik, H. Dyja, Obróbka Plastyczna Metali **20**, 1 (2009)
35. Y. Kang, Z. Hu, J. Liu, C. Yang, H. Du, J. Mater. Sci. Technol. **19**, Suppl. 1 (2003)
36. G. Fang, P. Zeng, J. Mater. Sci. Technol. **19**, Suppl. 1 (2003)
37. H. Yang, L. Zhang, Z. Hu, Adv. Mat. Res. **538-541** (2012)
38. X. Shi, B. Wang, Mater. Sci. Forum **704-705** (2012)
39. X. Ren, Y. Liu, AMM **117-119** (2012)
40. Q. Cao, L. Hua, D. Qian, J. Cent. South Univ. **22** (2015)

DOI: 10.1051/matecconf/20168013002

Theoretical study on Cold Open Die Forging Process Optimization for Multipass Workability

Ajitkumar Gaikwad[1-a], Shreyas Kirwai[1], Provat Koley[2], Dr. G. Balachandran[3] and Dr. Rajkumar Singh[1]

[1]*Kalyani Centre for technology and innovation, Bharat Forge Ltd., Pune, India*
[2]*Heavy Forge Division, Bharat Forge Ltd., Pune, India*
[3]*Kalyni Carpenter Special Steels ltd., Pune, India*

Abstract. Cold Workability limits strength enhancement of austenitic materials through cold deformation. The intrinsic workability is the material characteristic whereas state-of-stress workability is governed by nature of applied stress, strain rate and geometry of deformation zone. For Cold Open Die Forging (CODF), multipass workability is essential. In this work, FEM tool FORGE-3 is used to optimize CODF on hydraulic press by analysis of stress-strain profiles and use of Latham-Cockroft damage criterion. Study recommends optimized process parameters, die combinations and pass-schedules.

1 Introduction

The material properties of Austenitic Stainless Steels (ASS) are sensitive to plastic deformation. A significant improvement in the yield strength of ASS can be achieved by its cold deformation as a result of work/strain hardening. Austenitic stainless steels have a very high strain hardening coefficient and hence, can be effectively strain hardened to a very high strength.

This concept is used in some industrial applications where long bars of ASS are required to be cold forged to achieve certain strength levels. Having very high Length to Diameter (L/D) ratio, these are required to be processed through incremental or Open Die Forging route. Though, specialized machines/presses are developed for these kinds of processes, a thought was given to use general purpose open die forging hydraulic press with proper process design.

The amount of plastic deformation, and so the corresponding strength enhancement, without defect formation in deforming body, is limited by workability. In addition to intrinsic workability of the material, workability is also the characteristic of the metal working process followed.

In the present study, open die forging process with different process parameters and die combinations is analysed using FORGE-3. The objective of the study is to recommend, based on Finite Element Method (FEM) studies, an optimized process in order to get the desired degree of deformation without centre burst. Exact degree of deformation to achieve the desired enhancement in strength level can be decided only after actual shop floor trials. With this in mind, here, different process routes are analysed and compared on the basis of stress-strain profiles and Latham-Cockroft damage criterion.

Workability is the extent to which material can be shaped by plastic flow without onset of fracture. It is well understood that cold workability is not only influenced by microstructure of the material (ductility of the material), strain and the strain rate, but also by the stress state in the deformation zone [1]. Hence it is always convenient to consider workability to consist of two independent parts i.e. State of stress (SOS) workability and intrinsic workability. Intrinsic workability is governed by material properties like chemistry, microstructure, prior processing history and its response to the strain and strain rate at room temperature. The SOS workability depends on geometry of deformation zone in which workpiece is subjected to a three dimensional state of stress. The state of stress is represented by stress tensor with shear stress components and hydrostatic components. The shear stress components contribute to the plastic flow of the material while hydrostatic components decide the workability. The Hydrostatic components need to be essentially compressive as tensile components will open up the weak interfaces within deforming material causing the internal fracture. Hence, SOS workability is controlled by nature of applied stress and geometry of deformation zone and is specific to particular mechanical working process. One can play with this SOS workability by playing with die design and other process parameters in order to achieve the desired degree of plastic deformation.

In a typical open die forging, workpiece is hold at one end in manipulator, which is capable of rotating the job around its axis and also provides linear motion in axial direction. In this way manipulator positions the job between the dies and the press. The process is carried out

[a] Corresponding author: ajitkumar.gaikwad@bharatforge.com

incrementally, with only a part of workpiece being deformed at a time. The tool i.e. die sizes are quite small than the overall size of the forging. Dies used are of simple shape like flat die, 'V' dies or simple contoured dies. As the only part of the workpiece is being deformed at a time, the requisite forging loads are less than those needed to deform entire workpiece at a time. The reduced forging load means lower capacity machine can be used to produce given part.

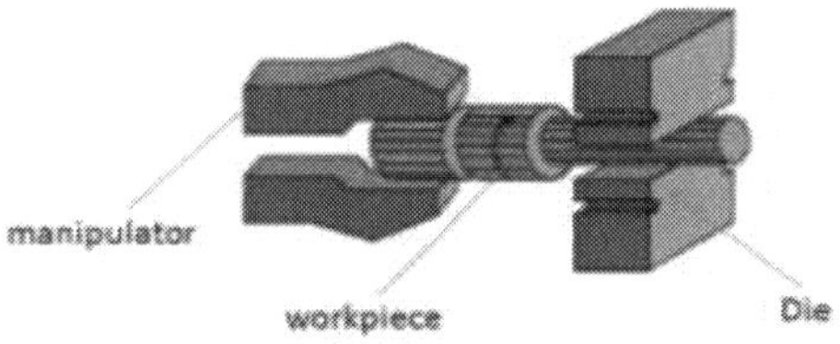

Fig. 1. Open die forging system

In open die forging, in addition to SOS workability, the ability of the metal to withstand the fracture or defect development in the face of progressive straining in multiple drawing/cogging passes is important criterion. This should be achieved without the benefit of intermediate annealing as annealing will again bring down the enhanced strength. Generally, this Multipass workability is tied to the resistance to ductile fracture. It is important to know that in some materials 'Multipass' workability may be so limited that fracture occurs at strains or reductions below those commonly undertaken in a single drawing pass (up to about 40%). During such Multipass cogging operations, this ductile fracture generally occurs at centre of deforming body called as centerburst [2]. Figure 2 shows typical centerburst.

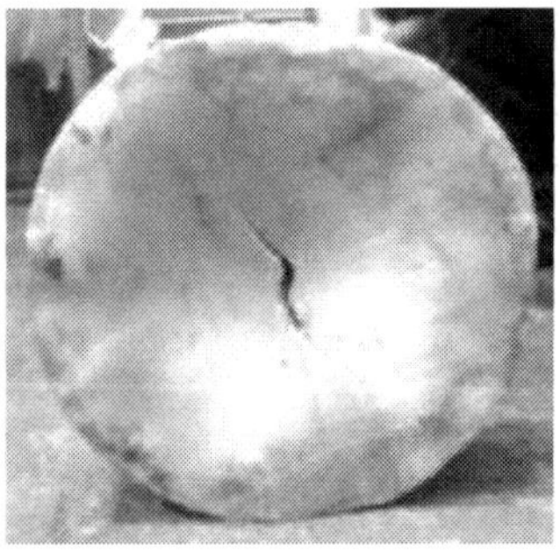

Figure 2. Typical centerburst

Centerbursts in their repulsive/coarse form are chevron like shear fractures that form in the centre of workpiece. In finer form, they may be little more than aggregates of pores along the centreline of the product. They may develop spontaneously or may slowly build up with progressive working. The basic cause of centerbusrting in nonuniform metal flow through dies and associated development of tensile stresses in the center of workpiece. Fracture criterions can be used to predict these Centerbursts during any metal forming operation [2].

1.1 Fracture Criterion:

Different fracture criteria are developed for evaluating potential of fracture during particular process. These criterions are generally developed through some workability tests. Fracture criterion extends the workability established in standard conditions to the workability in the deformation process of interest.

Latham-Cockroft Damage Criterion:

It is well reported in a literature that cracking in metal-working processes is associated with tensile stresses; even processes like forging are essentially compressive [2]. However, as both stress and strain can decide the onset of fracture, that criterion should be based on some combination of stress and strain, rather than on either of these quantities [1]. Based on this, there are indications reported in literature that total plastic work done till the fracture is an important factor. If we apply this to tensile test, total plastic work done per unit volume at the fracture point is,

$$\int_0^{\varepsilon_f} \sigma d\varepsilon$$

Where, σ is current yield stress and ε_f is fracture strain.

This criterion takes into consideration the yield stress at the onset of fracture and does not takes into consideration the peak stress. Unlike the peak stress, the current yield stress is not influenced by shape of the necked region. The pick stress need to be taken into consideration as it is affected by neck shape. To overcome this limitations, Latham and cockroft has proposed fracture criterion for ductile material. They proposed that fracture occurs in ductile material when the quantity

$$\int_0^{\varepsilon_f} \bar{\sigma} \left(\frac{\sigma^*}{\bar{\sigma}}\right) d\bar{\varepsilon}$$

reaches a constant value (c) for a given temperature and strain rate, where,

$\bar{\sigma}$ = equivalent stress,

ε_f= fracture strain,

$\bar{\varepsilon}$= equivalent strain and

$\left(\frac{\sigma^*}{\bar{\sigma}}\right)$ is a non-dimensional stress-concentration factor representing the effect of highest tensile stress- σ^*. The reduced form, given below, is used for calculations, whose expression has the dimensions of work done-

$$\int_0^{\varepsilon_f} \sigma^* d\varepsilon = C \ldots\ldots\ldots\ldots\ldots\ldots (1)$$

This criterion implies that ductile fracture is dependent on both on shear stresses (giving rise to plastic deformation and work hardening) and tensile stresses. Fracture also depends both on the stresses imposed and on the strains developed-neither stress or strain alone[2].

1.2 FEM application in Metal Forming Simulation

Recent developments in application of Finite Element Analysis (FEA) to metal forming processes through simulation software's is helping a lot in new product development process. Numerical simulation yields a deeper insight into the mechanics of the forming process than that provided by classic calculations and testing techniques. To development of the product, physical tests on a press are necessary; however, with the help of simulation one is in a position to develop the product more quickly and systematically [3].

Transvalor has developed a commercial metal forming simulation software-FORGE-3. The said software is capable doing simulation of hot as well as cold forging process. Dedicated template for Open Die Forging simulation is provided. The said software enables user to carry out stress, strain, temperature analysis. It is also incorporated with built in Latham-Cockroft damage criterion. This FORGE-3 by Transvalor is used for analysis in the present study [4].

2 Simulations - Design of Experiments (DOE)

Table 1 gives the details of simulation DOE. As a case for typical open die forging process, process variables under consideration are die combinations, depth of stroke, rotation angle after each pass and forging sequence. The starting billet dimensions are Ø335 (mm) × 2000 (mm). The starting billet temperature is 30°C. The input Ø335 cross-section is reduced to Ø315 in each case. This gives total reduction ratio of 11.58%.

In case no. 1.1 and 1.2, top and bottom dies are flat dies with the die width of 300 mm and corner radius of 30mm. After first pass 10mm throughout length, except the part hold on manipulator, the job is rotated by 90°. Then for the third pass, job will rotated through 45° and again through 90° to make an octagon. The octagon is then converted into round by pressing across the edges. The same process is followed twice for case no. 1.1 i.e. with depth of stroke 10mm, so as to get final Ø315mm.

In case no. 2.1 and 2.2, with the same die combination, instead of making it square-octagon-round, the desired reduction is achieved through 22.5° rotations per stroke and making the job round locally. Then job is given an axial increment to forge the next part. The same process sequence is followed 4 times for case no. 2.1 i.e. with 5 mm depth of stroke and 2 times for case no. 2.2 i.e. 10 mm depth of stroke.

For the case no. 3.1 and 3.2, the similar process sequence is followed with the top V-die instead of flat die.

For the case no. 4.1 and 4.2, the similar process sequence is used with clapper dies in top and bottom. Clapper dies are semi-circular dies with diameter close to operating diameter.

3 Deformation Model

Initially, 3D model of the input billet and the dies were prepared using Solid Edge CAD modeling software. Using FORGE software, mesh generation and setting up of input parameters was carried out in a pre-processor stage. The metal flow was evaluated in computation phase. The flow stress equations for deformation, used in FORGE software is shown in Table 2. The flow stress for a chosen material i.e. SS304 as a function of strain, strain rate and temperature was evaluated using Hansel–Spittel equation. The average mesh size used is 15 with remeshing option ON. The FEM mesh of the billet is consisted of 15,811 nodes and 90145 elements. For the individual cases shown in table 2.1, depending upon the no. of passes involved, simulation was completed in 10 to 40 hours.

The friction between the dies and the billet surface depends upon on the amount of reduction, temperature, surface roughness of the dies and die-billet contact time. For present study, is considered coulomb's friction law with friction coefficient of 0.05. For dissipation of heat generated as a result of cold deformation, heat transfer model considered for present study is with overall heat-transfer coefficient of 10000w/mm^{2}K.

Table 1. Simulation DOE

Sr. No.	Iteration	Die Combination		Depth of Stroke	Rotation Angle	Final Dimensions
		Top Die	Bottom Die			
1	1.1	Flat	Flat	10	90-45-22.5	315
	1.2	Flat	Flat	20	90-45-22.5	315
2	2.1	Flat	Flat	5	22.5	315
	2.2	Flat	Flat	10	22.5	315
3	3.1	Flat	V	5	22.5	315
	3.2	Flat	V	10	22.5	315
4	4.1	Clapper	Clapper	5	22.5	315
	4.2	Clapper	Clapper	10	22.5	315

DOI: 10.1051/matecconf/20168013002

Table 2. Rheology and flow stress model used for forging simulation

Flow Stress Model	$\sigma_f = A_1 e^{m1T} T^{m9} \varepsilon^{m2} e^{m4/\varepsilon} (1+\varepsilon)^{m5T} e^{m7\varepsilon} \dot{\varepsilon}^{m3} \dot{\varepsilon}^{m8T}$	σ_f - Flow stress ε - Equivalent strain $\dot{\varepsilon}$ - Strain rate, A, m1 to m9 are constants

Note: Material database in FORGE software has inbuilt values for the constants

4 Results and Discussion

In the present study, process parameters affecting the multipass workability are analysed. This includes the analysis of effect of process design on tensile stresses i.e. first Principle stress, hydrostatic pressure, strain penetration, and Latham-Cockroft damage criterion. Since the objective of the exercise is to predict the failure, the values of worst scenario i.e. maximum value of Latham-Cockroft damage criterion is taken and corresponding stress and strain values are taken from same location.

4.1. Latham-Cockroft Damage criterion

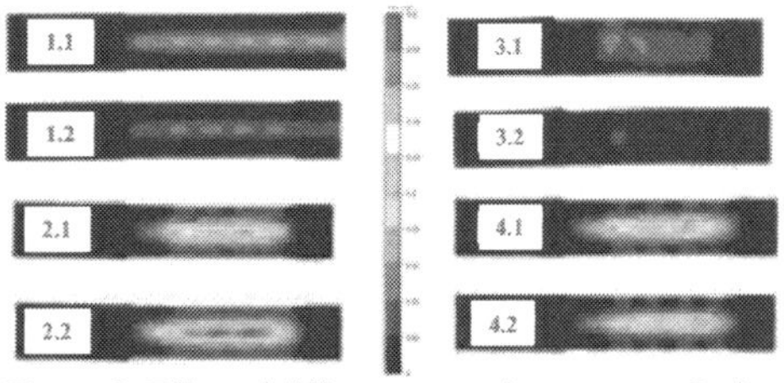

Figure 3. Effect of different processing routes on Latham-Cockroft Damage criterion

Figure 3 gives the Latham-Cockroft damage criterion profiles for the last pass of all the 4 cases. These values are taken at interior of the workpiece using a cutting plane. The scale with contours blue to red shows the magnitude from 0 to 0.4. The magnitude goes on increasing from Blue to red colour. In all the cases, material is not likely to have surface damage but has potential for center burst. Since the threshold value for centre burst for a particular material and processing route can be defined only after some shop floor trials, only relative measures can be analysed. Within the analysed options, Flat-V die combination (Case 3.1 and 3.2) has resulted in least damage. It is then followed by flat-flat die combination with 90°-45°-22.5° rotation i.e. Square-octagon-round forging sequence (case no. 1.1 and 1.2). Rest of the two cases has very high chances of center burst.

4.2. Strain analysis

Figure 4 gives the strain profiles for the last pass of all the 4 cases. These values are taken at interior of the workpiece using a cutting plane. The scale with contours blue to red shows the magnitude from 0 to 1. The In all the cases, the strain maps show non-zero values; this means that material is plastically deformed till core. In case of flat-flat die combination with 22.5° rotations i.e. round-round forging sequence (case no.2.1 and 2.2), the highest strain penetration is observed.

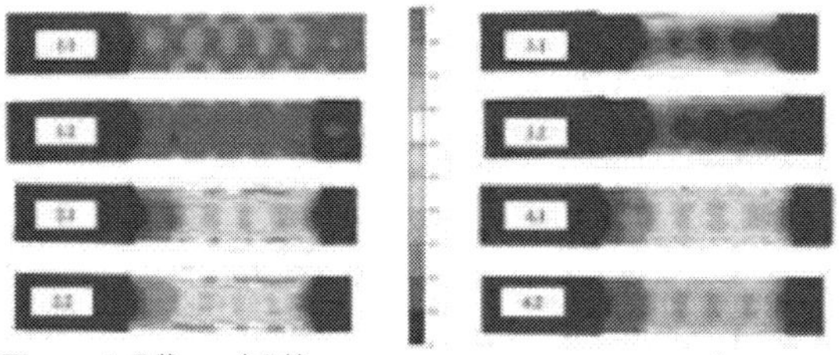

Figure 4. Effect of different process parameters on strain profiles

Similarly, clapper die combination (case 4.1 and 4.2) is also resulted in higher strain penetration. In case of Flat-Flat Die combination with 90°-45°-22.5° rotation i.e. Square-octagon-round forging sequence (case no. 1.1 and 1.2), minimal strain penetration is observed. In case of Flat-V die combination (case no. 3.1 and 3.2), Strain is concentrated in the periphery and less strain is penetrated in core. This is useful scenario, where product is subsequently subjected to drilling at the core. Since the magnitude of the strain is going to decide the amount of work-hardening and hence the resulting strength enhancement, shop floor trials at different strain levels need to be carried out in order to decide the required strain values. As minimal strain at the core is beneficial in order to control the center burst, process sequence in case 1 and 3 is recommended. It is also observed that for case no. 1, 3, 4, finishing the operation in minimal no. of passes i.e. higher depth of stroke, is turning out beneficial

4.3. Stress Analysis

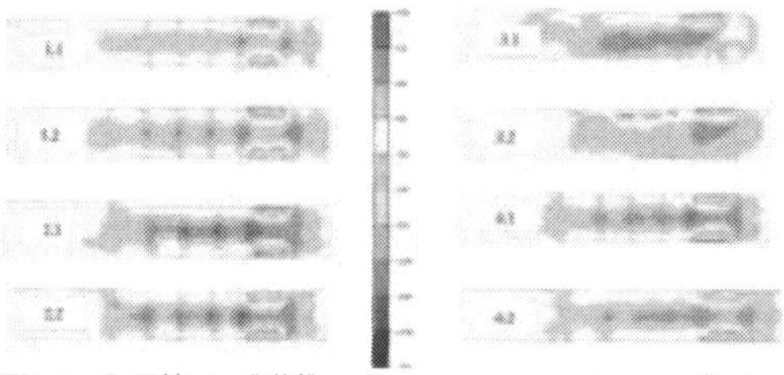

Figure 5. Effect of different process parameters on first principal stress profiles

Figure 5 gives the first principal stress profiles for the last pass of all the 4 cases. These values are taken at interior of the workpiece using a cutting plane. The scale with contours blue to red shows the magnitude from -2000Mpa to1000Mpa. It is clear from the above analysis that, Case no. 1 and 3 are producing lower first principal stress. Being a cold deformation process, the stresses induced in the material keeps on accumulating. To nullify these stresses intermediate annealing is recommended for heavy reductions. However, in the present study, since purpose of the deformation is to enhance the strength using work hardening, intermediate annealing is not recommended.

4.4 Compiled results and discussion

Table 3 compiles all the results measured during last pass of each process-as per DOE. These values are taken where the Latham-cockroft criterion is showing maximum value, at a cutting plane at the centre of the width of the billet. Figure 6 to 8 gives graphical representation of effect of process parameters- selected as per DOE-on parameters selected for analysis.

From the analysis, it is clear that the high tensile stress and high strain penetration results in higher Latham-cockroft damage criterions value. This is analogues with the logic and equation of the Latham-cockroft damage criterion.

Table 3. Compiled Results

Sr. No.	Iteration	Equivalent Strain at Centre	Latham Cockroft damage Criterion	First Principal Stress (σ_1)	Second Principal Stress(σ_2)	Third Principal Stress(σ_3)	Hydrostatic Stress ($(\sigma_1+\sigma_2+\sigma_3)/3$)
1	1.1	0.22	0.13	460	110	-200	123.33
	1.2	0.15	0.09	490	130	-320	100
2	2.1	0.37	0.31	950	340	-40	416.67
	2.2	0.42	0.39	850	230	-450	210
3	3.1	0.09	0.1	495	250	-10	245
	3.2	0.05	0.04	580	250	-40	263.33
4	4.1	0.4	0.32	820	230	-180	290
	4.2	0.35	0.24	690	180	-90	260

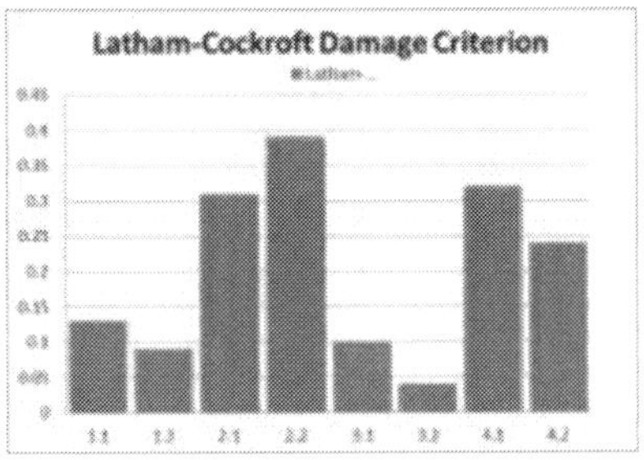

Figure 6. Comparison of Latham-cockroft Damage criteria

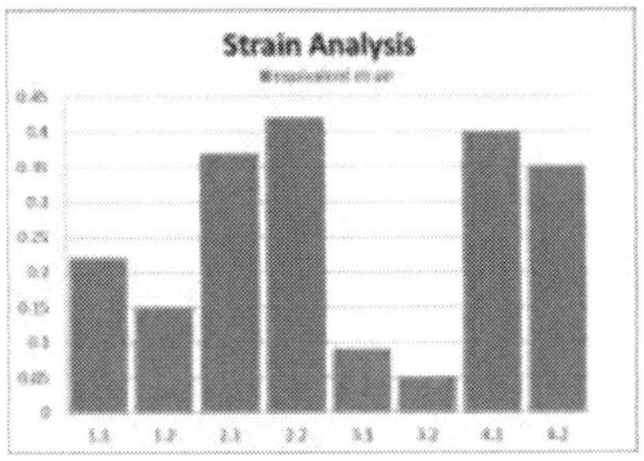

Figure 7. Comparison of strain magnitudes

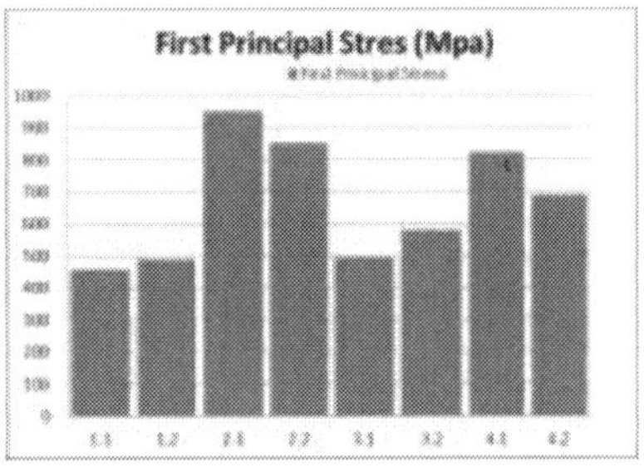

Figure 8. Comparison of first principal stresses

It is clear from the above exercise that for selected application, to avoid centerburst, strain need to be accumulated in the periphery. Also, the process should generate minimum possible tensile stresses at the core. Process which satisfies above conditions is recommended for this application. The threshold value for magnitude of the strain level can be finalised only after shop-floor trials.

The probable reason why case 2 is worse than case 1 may be the strain accumulation due to repetitive strokes at same location. This also results in higher stresses at core, because being a cold deformation; stresses keep on accumulating at the core. Same has happened with in case of clapper dies i.e. case no 4. Due to repetitive strokes,

 DOI: 10.1051/matecconf/20168013002

accumulated strain and stress levels are high which results in higher damage possibility at centre. In case 3 with Flat-V die combination, geometry of deformation zone helps in getting lesser stresses and strain at core. Also, from analysis of hydrostatic stresses (table 3), it is clear that magnitude of highest tensile stresses i.e. first principal stresses is most influencing factor on damage.

With the above analysis, it is clear that FEM tool gives better insights of the mechanics of the deformation process. It is concluded that, among the analysed options, process with Flat and V die combination i.e. case no. 3 is producing ideal results. It is then followed with case no. 1. Since, in case no. 1.2 and 3.2, with higher depth of stroke, possibility of damage is low compared to 1.1 and 3.1 respectively, these processes are recommended.

References

1. M.G. Cockroft, D. J. Latham, J. Inst. Met., **96**, 33–39 (1968).
2. Dieter: Handbook of workability and process design (2003).
3. Ajit Gaikwad: M.Tech. Thesis: material and process optimization, by simulation, for forging spool body (2011)
4. Web.: www.transvalor.com

The process parameters effect of ovality in cross wedge rolling for hollow valve without mandril

Hongchao Ji[1], Jinping Liu[1a], Baoyu Wang[1], Jianguo Lin[2], Xuefeng Tang[1]

[1] *School of Mechanical Engineering, University of Science and Technology Beijing, Beijing 100083, China*
[2] *Department of Mechanical Engineering, Imperial College London, London SW7 2AZ, UK*

Abstract. This paper presents the experimental and numerical results of the effect process parameters on ovality in cross wedge rolling (CWR) for hollow engine valve without mandrel. Numerical simulation model for ovality was established by means of the rigid-plastic finite element modeling (FEM) method for hollow engine valve. The experiments and numerical analyses suggest that the following parameters represent the best conditions for CWR of hollow engine valve: 30°-34° for the forming angle(α), 5°-7° for the stretching angle(β), 0.2-0.3mm for the mold void width(L), and 65%-70% for the area reduction(Ψ).

Key words: Hollow engine valve; Ovality; Cross wedge rolling; Experiment; Finite element modeling

1 Introduction

With the development of automotive technology, more and more hollow engine valves are needed. The CWR process is a new metal forming technique for producing hollow engine valve[1]. Many studies have been carried out on hollow parts by experimental, analytical and simulation methods. Z.Pater [2-4] *et al* studied the three-roll CWR without mandrel forming hollow shafts, and the parameters on CWR process were analyzed using numerical simulation. In addition, the plate wedge rolling of hollow shafts were comparatively analyzed. M.Lovell [5-6] *et al* studied the CWR theory and technology on the hollow shaft parts. Experiments and FE simulation was carried out to study the instable conditions on CWR process.

Ovality defects significantly affect the mechanical properties of hollow shaft parts and yield additional stresses under the action of internal pressure, in which the former causes uneven and hollow shafts. Compressive strength decreases under external pressure, thereby reduces the stability of the hollow engine valve. A degree of ovality is observed in hollow shaft parts. Installation of mechanical vibrations is caused by equipment or components, which seriously affects the use of shaft parts. In this paper, we used CWR technology to produce hollow engine valve. FEM of the CWR process was performed using the software Deform-3D. Experimental and FEM were used to study the ovality of hollow engine valve in different parameters.

2 Experimental details

Experimental tests on the CWR process were conducted using the H500 mill available at the University of Science and Technology Beijing in Beijing, China. The simplified model of cross wedge rolling is shown in Figure. 1a. A picture of the H500 mill is shown in Figure.1b. During the research, the workpieces of 4Cr9Si2 were heated to 1150°C and formed. The experimental process was follows:

(1) A GB4025B horizontal metal band saw and a hollow steel rod were used. The rod is 25 mm in diameter and was cut to 100mm in length.

(2) Workpieces were heated in the tube furnace for 30min at a suitable temperature to conserve electricity and evenly distribute the heat within the pieces. The procedure was based on previous cross-valve wedge rolling experiments.

(3) The workpieces were uniformly heated and processed in H500 mill. Subsequently, the workpieces were cooled after rolling to ordinary temperature.

(4) The formation and quality of rolling after cooling were observed and recorded.

[a] Corresponding author: Jinping Liu E-mail: liujp@ustb.edu.cn

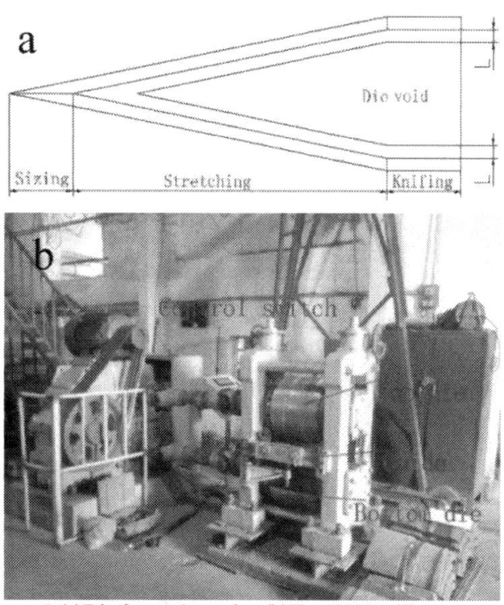

Figure.1 (a)Die for engine valve (b)The H500 mill

3 Finite element modeling

FEM of CWR was performed using the deform. Figure.2 shows the geometrical model of the analyzed process. The material accuracy of constitutive models directly determined the accuracy of the simulation results. No flow stress of 4Cr9Si2 alloys existed in the library of deform software. Hence, Gleeble-1500 thermal simulations for the alloys were characterized under different deformation conditions to analyze the behavior of thermal deformation. Actual stress–strain curves for a temperature range from 950°C to 1180 °C and strain rate values of 0.1, 1, 5, and 10 s^{-1} were obtained. (Figure.3). Constitutive relationships were imported into the Deform-3D software to investigate the cooling process. Figure.4 is the forming process of the rolled piece.

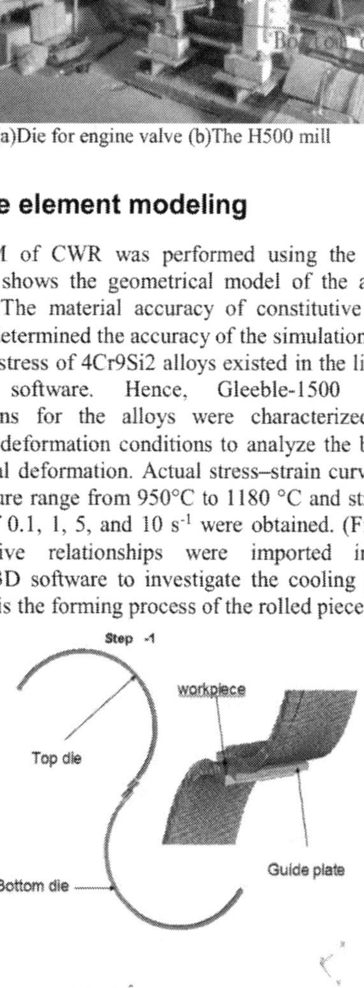

Figure.2 FEM model of cross wedge rolling (Deform-3D)

Friction coefficient m on the material–mold interface is constant. The friction force in the constant shear model is defined as follows:

$$f_s = m*k \qquad (1)$$

where f_s is the frictional stress and k is the shear yield stress. Eq. (1) indicates that friction is a function of k in the deforming body.

Table.1 illustrates the main rolling conditions and information of the workpieces.

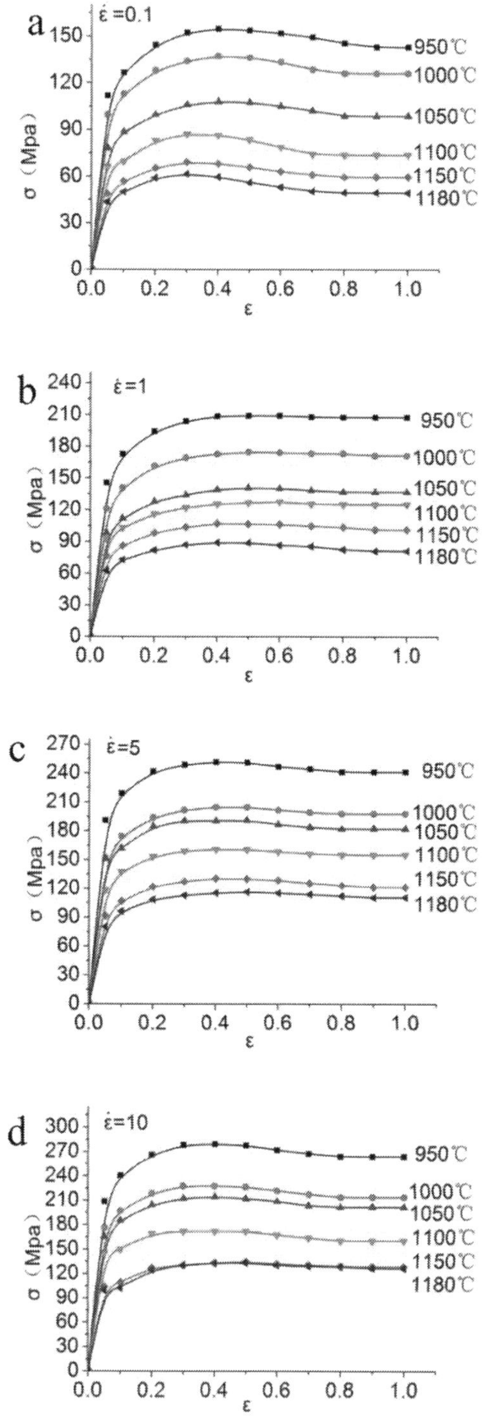

Figure.3 Flow curves of 4Cr9Si2 at strain rates of (a) $0.1s^{-1}$ (b) $1s^{-1}$(c)$5s^{-1}$and (d) $10s^{-1}$.

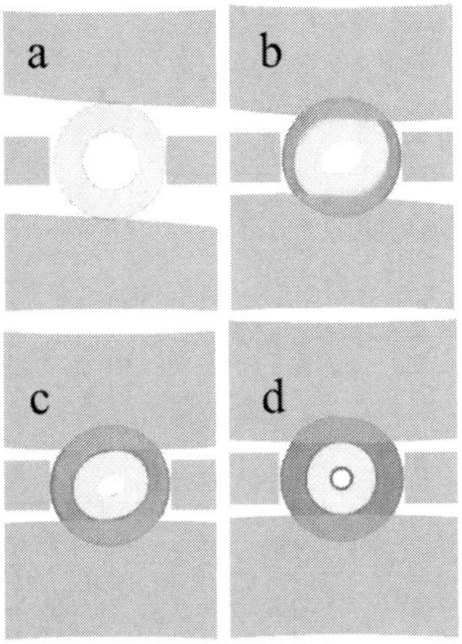

Figure.4 Forming process of the rolled piece

Table.1 Parameter of cross wedge rolling simulation

Parameter	Value
Speed of roll (rpm)	10
Workpiece temperature (°C)	1,150
Molds temperature (°C)	20
Environment temperature (°C)	20
Heat transfer coefficient (N/s/mm/°C)	40
Convection coefficient (N/s/mm/°C)	0.02
Friction factor(m)	0.9
Mesh number for billet	90000
Mold material	H3
Billet material	4Cr9Si2

The simulation is divided into four groups and the calculation parameters are shown in Table.2.

Table 2 main process parameters for simulation

Item	Forming angle α(°)	Stretching angle β(°)	Die void width *L(mm)*	Area reduction Ψ(%)
1	26,28,30,32, 34	6.5	0.25	67.89
2	30	4,6.5,7.5,8.5, 9.5	0.25	67.89
3	30	6.5	0.20,0.25,0. 30	67.89
4	30	6.5	0.25	30.56,40. 22, 51,59.89, 67.89

4 Results and discussion

Experimental results reveal that the hollow engine valve formed by CWR can be accurately pierced and rolled. Figure.5 illustrates the hollow engine valve produced by CWR.

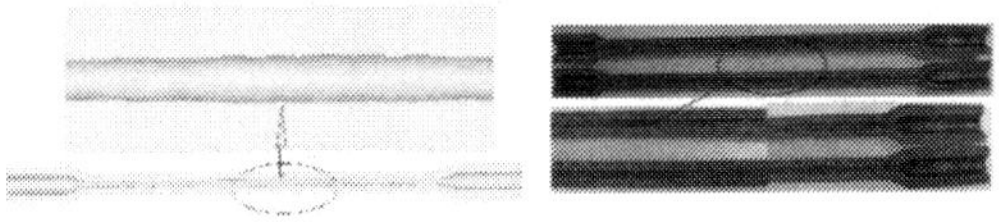

Figure.5 FEM and experimental test samples

Given the maximum and minimum inner diameter D_{max} and D_{min}, the bore ovality e is expressed as follows:

$$e = \frac{2(D_{\max} - D_{\min})}{D_{\max} + D_{\min}} \times 100\% \qquad (2)$$

From FEM of hollow engine valve, e was measured in the symmetrically center section of the roll. The ovality of hollow engine valve in different parameters was analyzed.

4.1 Effects of forming angle

The workpiece is bitten and rotated in the knifing zone. The model gradually shifts from the radial state to rolling state. The radial force is large, the radial flow of the metal rolling near the symmetry plane is relatively intense, and the axial flows relatively flat. Given the condition of axial flow, metal flows in the radial direction and results in a large radial force. Moreover, the axial force is relatively small and can easily be formed in the ovality-induced section of the wedge hole. Figure.6 shows the effect of α on e.

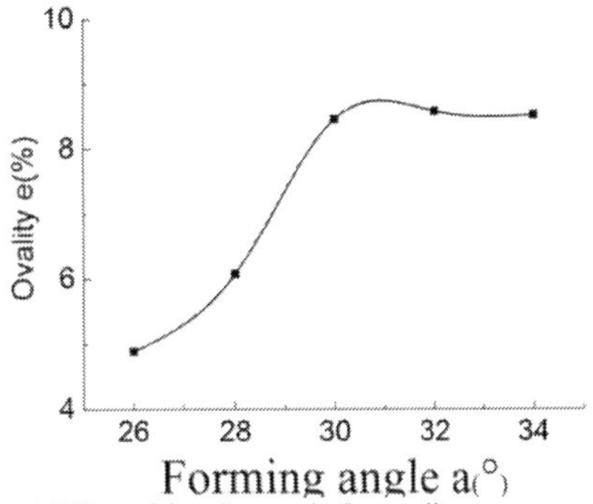

Figure.6 Effect of forming angle for ovality
β=6.5°,L=0.25mm,Ψ=67.89%,T=1150°C

The turning point for the curve occurs at α =32°. The ovality reaches a maximum, but the curve is symmetrical oval near the surface of the workpiece position in the hole with increasing α.

Collected, comparison, and in-depth analysis of the radial and axial forces of rolling wedge segments were performed on deform software under different α of the ovality to determine the effect of the law. Figure.7 indicate the temporal changes in the radial and axial forces of rolling wedge segments.

Figure.7 reveals that the rolling radial force increases with time in the wedge segments. On the other hand, the axial force initially increases with time and slowly reduces thereafter. Meanwhile, rolling subjected to axial force increases with α. The rolling radial force suffered before and after 0.35s increase and decrease with increasing α, respectively. For rolling under different α in

DOI: 10.1051/matecconf/20168013003

the inner diameter of the wedge segments before 0.35 s, the rolling force remains constant with time. However, the radial rolling force is generated with increasing α because of rapid rolling by wedging effect. Hence, the angle formed by rolling at a large radial force is greater than that at a small radial force. The rolling force acts on the metal roll at an exactly symmetrical position near the surface, In addition, the increase in radial force increases e of the rolling hole within specific α values. After 0.35s, the radial force decreases with α and decreases e. The wedging segment bore the narrow ovality, but the rolling before 0.35s exhibits a specific e. Radial forces in different angles after 0.35s reduces e in the inner hole. However, the effect on the nearby rolling plane of symmetry of the oval hole is limited.

The increase in axial force reduces e of the workpiece. In wedge segments, the axial force is significantly less than the radial force. Moreover, the magnitude of the increase in radial force is greater than that of the axial force. By contrast, the axial force in the local time decreases with time; hence, the axial flow of the metal workpiece is relatively flat. Therefore, increased α increases the rolling plane of symmetry near the location of hole ovality in wedging segments. The radial force in the rolling region is relatively stable.

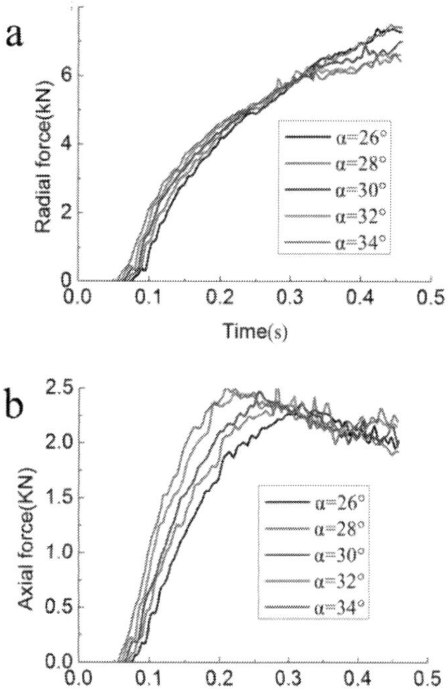

Figure.7 The influence of forming angle for force at knifing (a) radial force (b) axial force

4.2 Effect of stretching angle

The e curve (Figure.8) reveals that the hole ovality near the position of the rolling symmetry gradually increases with β.

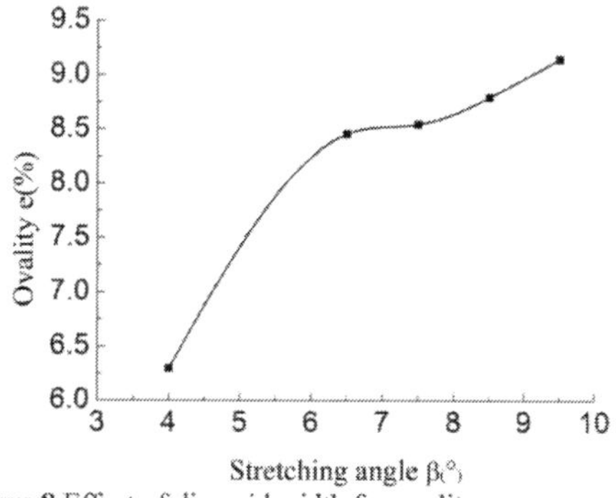

Figure.8 Effect of die void width for ovality
α=30°, L=0.25mm, Ψ=67.89%, T=1150°C

Figure.9 show the rolling radial and axial forces in the wedging segment with time. The radial force in the wedge segment increases with time and β. Meanwhile, the axial force in the wedging segment initially increases with time and slightly elevates thereafter. For the increasing stage, the axial force increases with β; for gentle stages, the force decreases with increasing β.

Increased β increases the magnitude of metal flow of metal in the rolling deformation zone. Moreover, the magnitude of axial compression and the rolling resistance to the extension increase. The oppositely directed friction to metal flow increases; however, the axial metal flow weakens and the axial force gradually reduces in gentle stages. Rolling metal flow and the contact area of the deformed shaft radially increases. The resultant radial force increases, which emphasizes the behavior of e. The radial force is relatively stable and the roundness of the formation zone in the rolling forming region.

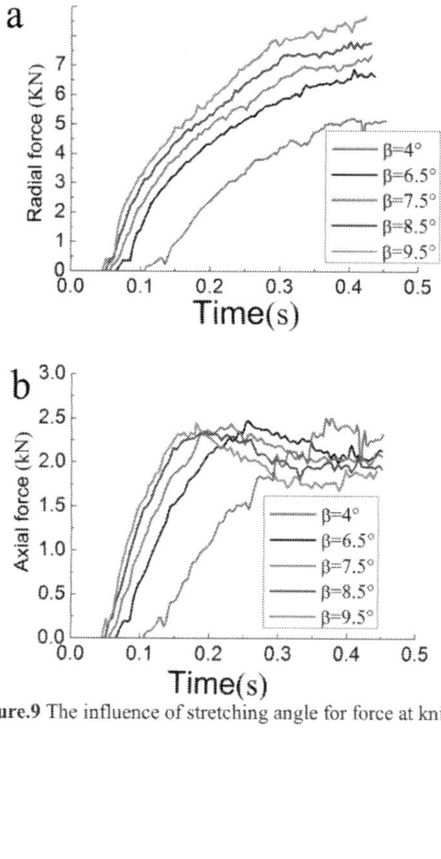

Figure.9 The influence of stretching angle for force at knifing

(a) radial force (b) axial force

4.3 Effect of die void width

Figure.10 shows that e near the rolling location slowly increases with L. Figure.11 represent the temporal variations in radial and axial forces in the rolling wedging segment. The radial force increases with time; the axial force initially increases with time and decreases thereafter. For various L, rolling radial and axial forces in the wedge segment remain constant with time. Therefore, in the wedge segments, force parameters under various L do not primarily affect e near the location of symmetry after rolling.

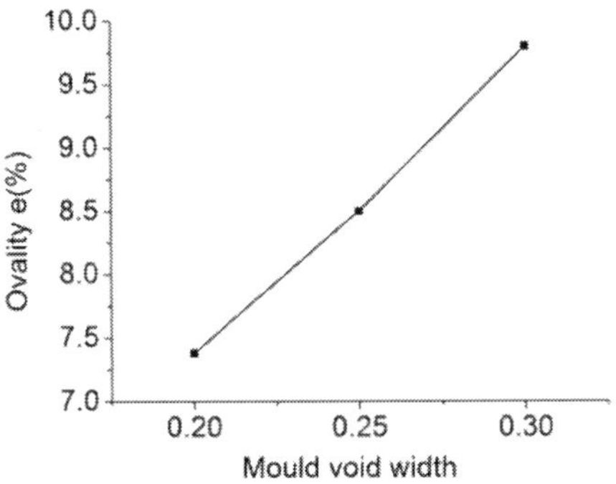

Figure.10 Effect of die void width for ovality
α=30°, β=6.5, Ψ=67.89%, T=1150°C

During wedge deformation, L does not affect e near the symmetry location; however, e in this part influences the mold wedge segment. The rolling environment is highly stable after stretching deformation. e in the hole near the position of rolling symmetry plane is maintained. However, e on the position of the rolling plane of symmetry is affected because of changes in L. When the metal enters the void position of the stretching segment, it flows from the pressure of the large area to that of the small area. The void range increases with decreasing L, which is unfavorable to axial efflux of the metal. A portion of the metal has refluxed to the rolling plane of symmetry. After rolling to the location near the symmetry plane, a portion of the metal refluxes to the small radial flow rather than the lateral flow. Hence, the hole is gradually rounded and e decreases at the original location near the rolling plane of symmetry. While refluxing occurs in some metals, e slightly varies with L. The result explains the increase in e near the location of the rolling plane of symmetry with increasing L. Stretched formation region relatively stabilizes the radial force and increases the probability of bore round holes in this region.

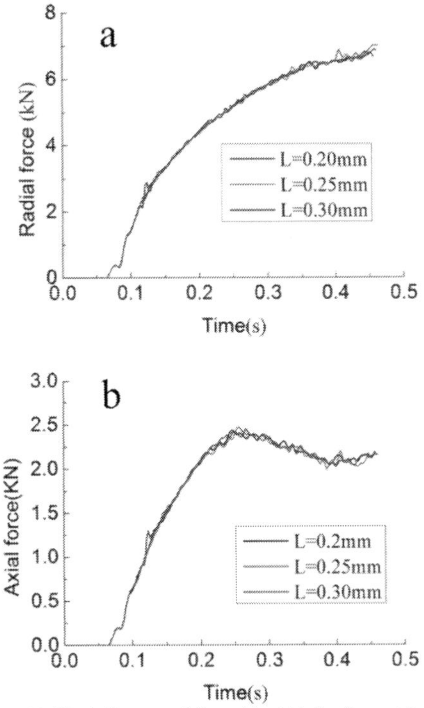

Figure.11 The influence of die void width for force at knifing
(a) radial force (b) axial force

4.4 Effect of area reduction

e near the position of the symmetry plane initially increases then decreases thereafter with increasing section shrinkage (Figure.12). The reduction rate is 41.22%, which is the turning point.

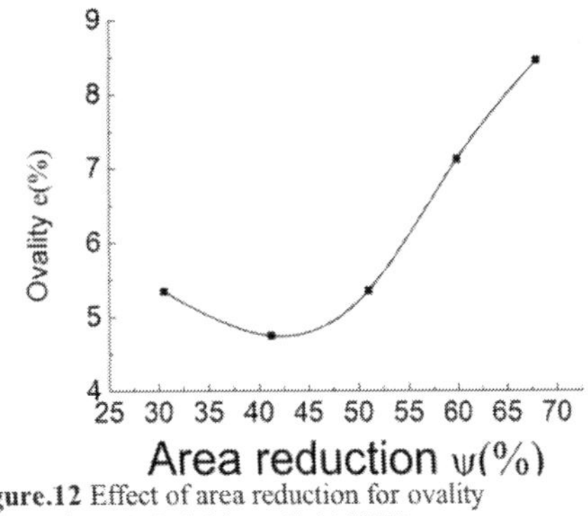

Figure.12 Effect of area reduction for ovality
α=30°, β=6.5°, L=0.25mm, T=1150°C

Figure.13 show that rolling radial force in the wedge segment increases with time; the axial force initially increases and then slowly decreases with time. The high extent of reduction and small rolling radial force increase the rolling axial force.

After mold wedge rolling with increasing mold amount relative to the axial exclusion, rolling increases, axial force increases, radial flow decreases, and radial force reduces.

Increased axial force and reduced radial force

decreases *e*. However, *e* constantly decreases with increasing cross-sectional shrinkage rate between 30.56% and 41.22%. Given the increase in section shrinkage rate between 41.22% and 67.89%, the thinning rate of the rolling wall thinning increases, the mold roll surface is close to the inner surface of the rolling, and the radial mobility of the metal to the mold increases. Radial shrinkage forces affect the results of the accuracy of the hole shape in the workpiece are highly significant. Although the radial force is reduced or slightly decreased, variation in *e* are large and the degree of the ovality increases. The radial force being stretched in rolling forming region is relatively stable, and roundness in the formation zone increases.

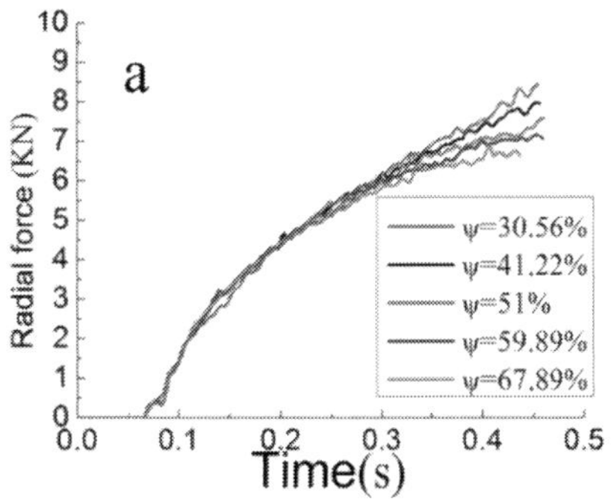

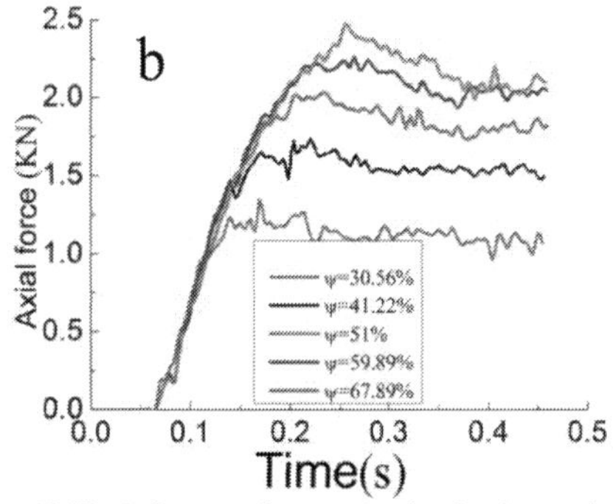

Figure.13 The influence of area reduction for force at knifing (a) radial force (b) axial force

5. Summary and conclusions

The following conclusions were drawn from conducted experiments and numerical analyses:

(1) *e* in the cross-section near the symmetry location increases with *α* of the mold from 26° to 34° after hollow rolling mill.

(2) Given the increase in *β* of the mold from 4° to 9.5°, *e* in the cross-section near the symmetry location increases after hollow rolling mill.

(3) Given the increase in *L* from 20mm to 30mm, *e* in the cross-section near the symmetry location around the inner rolling surface slightly increases after hollow rolling mill.

(4) Increased area reduction rate from 30.56% to 67.89% initially increases *e* in the cross-section near the position of the rolling symmetry plane and then decreases. The turning point occurs at an area reduction rate of 41.22%.

(5) We determine the following rolling process parameters given their effects on *e* as follows: *α* = 30°–34°; *β*=5 –7°; *L* = 20mm–30mm; area reduction 65%–70%. *e* significantly improves in line with target product requirements. Hence, CWR of hollow engine valve is feasible.

Acknowledgement

This work is supported by YangFan Innovative & Entepreneurial Research Team Project (No.201312G02).

References

1.H. Ji, J. Liu, B. Wang, Z. Zheng, J. Huang, Z. Hu, Cross-wedge rolling of a 4Cr9Si2 hollow valve: explorative experiment and finite element simulation, Int. J. Adv. Manuf. Technol. **77** (2015) 15-26.

2.J. Bartnicki, Z. Pater, The influence of tool geometry on the CWR process of hollowed shafts, STEEL-GRIPS J. Steel. Relat. Mater. (2004) **2** 103–107, Suppl. metal forming.

3.Z. Pater, A. Gontarz, J. Tomczak, T. Bulzak, Producing hollow drive shafts by rotary compression, Arch. Civ. Mech. Eng. (2015). DOI: 10.1016/j.acme.2014.10.002

4.J. Tomczak, T. Bulzak, Z. Pater, The Effect of Billet Wall Thickness on the Rotary Compression Process for Hollow Parts, Strojniški vestnik – J. Mech. Eng. **61** (2015) 149-156.

5.S. Urankar, M. Lovell, C. Morrow, Q. Li, K. Kawada, Establishment of failure conditions for the cross-wedge rolling of hollow shafts, J. Mater. Process. Technol. **177** (2006) 545-549.

6. Q. Li, M. Lovell, The establishment of a failure criterion in cross wedge rolling. Int. J. Adv. Manuf. Technol. (2004) **24**:180–189.

Numerical analysis of bevel gear forming

Jaroslaw Bartnicki[1]

[1]*Lublin University of Technology 38D Nadbystrzycka, 20-618 Lublin, Poland*

Abstract. In this paper the numerical analysis of bevel gear forming is presented. Simulations are made by means of FEM using Deform 3D software. Distributions of stress, strain, temperature and Cockroft - Latham damage criterion are presented at various process parameters. Obtained shape of final part and calculated loads during the process course can be useful for the further industrial applications.

1 Introduction

The problematic of different mechanical elements with teeth forming is really complex. Among several manufacturing technologies there is a choice between machining or metal forming methods. Rarely the casting technologies are applied, but their role begins to be marginal.

The choice between different machining technologies is connected with time of manufacturing, which is really the most important factor in cost calculation. Due to this reason, in wider scope of applications (especially in mass production) the metal forming technologies find their place.

The other requirements are also connected with different fields of applications as aviation or automotive industry with special norms concerning manufacturing methods. Presently, thinking about durability and elevated mechanical properties of different parts with teeth the forging or rolling processes are widely applied. In the case of forging technology many parts are formed as preform by rolling or forging. Finally these parts are forged with flash. In minority of forging plants forging with flash is replaced by process without flash. This tendency permits to reduce wastes and time loss due to additional flash trimming operation. The advantage of this process is also in continuity of material into the forging.

2 Bevel gear forming

2.1 Traditional forging

Presently in one of the Polish industry plant bevel gear part, presented in Figure 1, is traditionally forged in two operations. This automotive element is firstly prepared as cylindrical bar, which is heated and upset in the first operation. In the second operation it is possible to obtain final part with flash which is trimmed in additional stand. Normally, these kinds of parts are made in series between 10000 – 20000 pieces on one command.

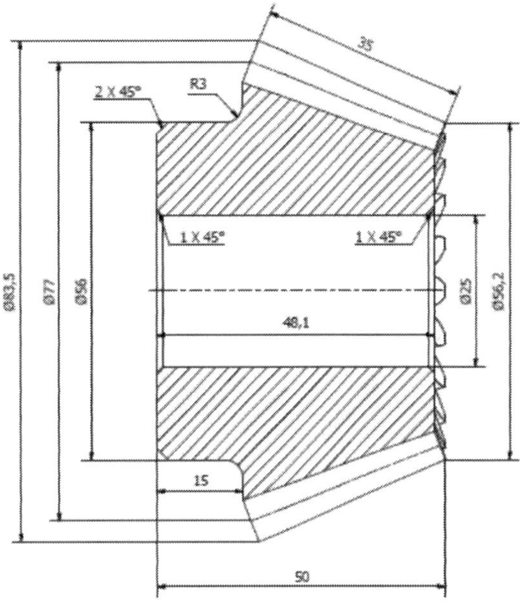

Figure 1. Schema of bevel gear.

This "traditional" forging technology has many advantages. Among them it is worth to notice that it is possible to apply bars or preforms with some height and diameter tolerances. The process can be realized using different machines – presses and hammers and obtained final parts have relatively good repeatability. This very important factor is connected with possible to obtain close manufacturing tolerances, which permit to decrease final machining of obtained bevel gears. Normally in the presented case the material allowances can be limited to 0.3 – 0.4 mm. During machining ,due to non continuous

[1] Corresponding author: j.bartnicki@pollub.pl

contact between cutting blades of tools, in this case these allowances are bigger (usually about 0.8 mm) This allows obtaining better machining conditions, limiting vibrations and increasing lifetime of machining tools [1-5].

2.2 Proposed solution

One of the biggest problems in industry is the proper choice of manufacturing technological parameters. The optimal range of forming forces, speeds and temperatures decide about costs and time of the proposed process. Parallely, the same rule have additional operations, which are also very important.

The modern forging technologies are focused on reduction of wastes, operations and operating time. One of the solutions for obtaining these gains concern close die forging technologies without flash, the second are connected with near net shape. The terms informs about very close to the final (net) shape of parts. It reduces ,of course ,the need for surface finishing as machining or grinding. The proposition for bevel gear parts is the numerical calculations presented in this paper. Analysis concerns stage of preform manufacturing from upset cylindrical bar – shown in Figure 2 and final bevel gear forging –shown in Figure 3. The dies for final forging stage are presented in Figure 4.

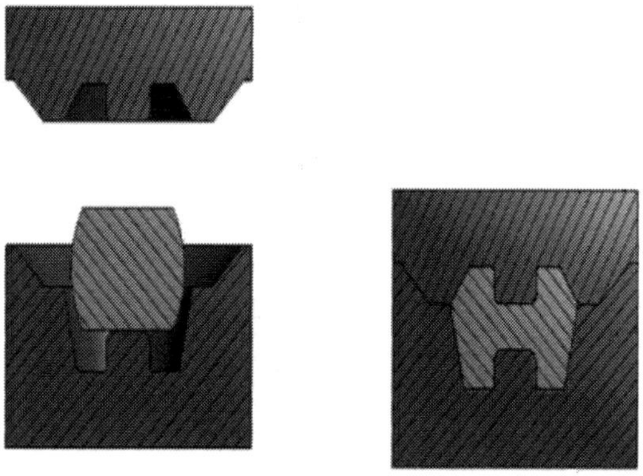

Figure 2. Preform forging from upset cylindrical bar (initial position and final stage)

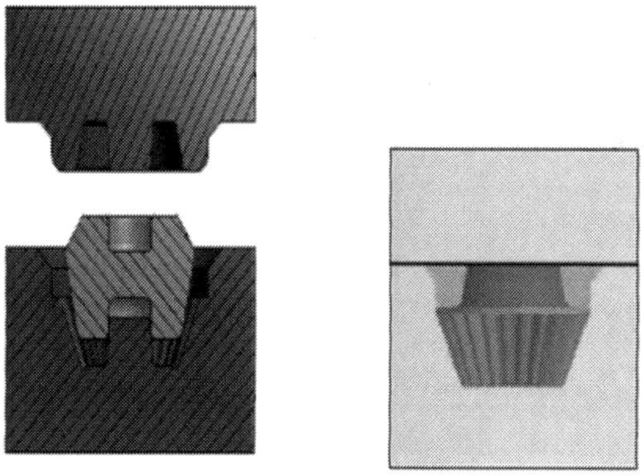

Figure 3. Final bevel gear forging from preform (initial position and final stage)

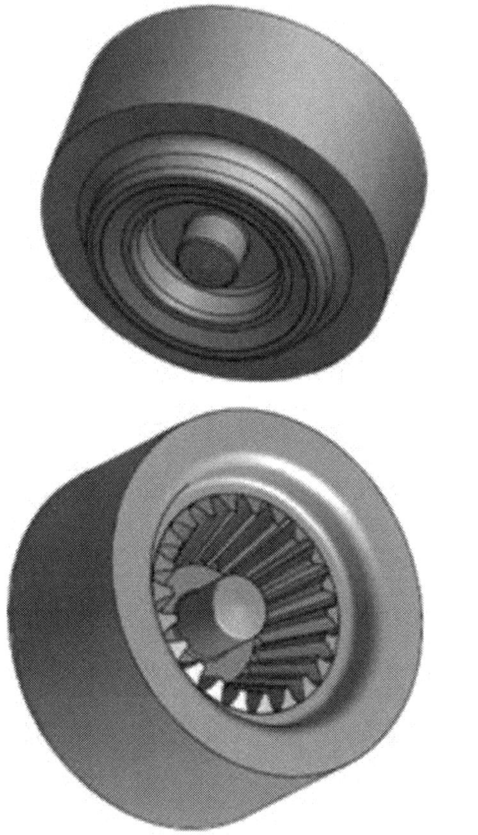

Figure 4. Final bevel gear dies: bottom die (left) and upper die (right)

3 Numerical simulation

3.1 Software, process conditions and numerical model

Numerical calculations of the close die forging without flash process of bevel gear realized with the application of software Deform3D, according to the model of final part given in Figure 5, were discussed. In order to achieve the result the most similar to the real one, it was decided that the process would take place in conditions of heat exchange with the environment. Material model of steel 20MnCr5 (Table 1) was taken from software library, modified and finally described as:

$$\sigma = 3599.85 e^{-0.003 \cdot T} \varphi^{-4.7e-05 \cdot T - 0.043} \cdot e^{\left(\frac{-3.4e-05 \cdot T - 0.028}{\varphi}\right)} \dot{\varphi}^{(5.3e-05 \cdot T + 0.032)} \quad (1)$$

where: σ – flow stress; φ - strain; $\dot{\varphi}$ – strain rate and T – temperature.

The initial temperature of the billet was form the range between (1000÷1150)°C with steps by 50°C. The forming dies (modelled as rigid bodies) temperature was assumed 150°C. The heat exchange coefficient between tools and the billet was assumed equal 4000 W/m^{2}K and between the billet and the air it was 200 W/m^{2}K. Shear friction model with friction factor m = 0.3 – 0.5 was taken into account during calculations [6, 7].

Table 1. Chemical composition of 20MnCr5 steel

C	Si	Mn	Cr	Ni	Cu	P
0.17	0.15	1.10	1.00	Max	Max	Max
0.22	0.40	1.40	1.30	0.30	0.30	0.035

Numerical model of the close die forging without flash was build into the Deform3D software. The workpiece was divided into 210000 tetrahedral elements. The automatically remeshing option was switched on with criterion of maximal effective strains. Due to this fact, very often remeshing calculations needs more time. Application of volume control option during forming and remeshing permits to preserve stable material volume. Without this option, during remeshing the proposed software systematically decreases material volume. In the case of close die forging without flash it would provoke infilling of dies, especially in the zones of created teeth. The initial position for final forging of bevel gear without flash is presented in Figure 6.

3.2 Distribution of effective stress and strains

Figure 7 presents the forming process with distributions of effective stress with process advancement given in percents. The maximum values of effective stress during this process reach the values about (820 – 1350) MPa at the area of contact tools – workpiece. After finishing of teeth forming, effective stresses values increase instantanely due to closed material volume between the dies. Considering the really high values of effective stresses inside the formed workpiece, it is possible to predict very high load on the dies surfaces. Especially, it can be dangerous near the spliting line between dies at the end of teeth forming areas. This fact can be unfavourable for the die durability due to risk of cracking initiation or faster wear usage with elevated temperature. Because of that, it is higly recommended in die manufacturing to make special pads with impressions segments, which allow to change these damaged elements very fast. The main problem in designing of this king of forging processes can be practical prevision of durability of impressions connected with machine characteristics, workpiece temperature and distributions of loads during forming [8].

Apart from the effective stress distribution into the workpiece and on the dies surfaces, distribution of strain in the final product is also the important factor . The fields of effective strains inform about material flow into the workpiece and show, for example possiblity of overlaping creation during the forging process. In the Figure 8 the distribution of effective strain in this process is shown. The highest values of effective strain are observed in the area of teeth forming and at the punches acting areas. The core of the formed part stays without strain, yet, it gives part rigidity. In the same cases of forming from tube shaped billets material flowed also into the billet axis, which disturbed final shape of obtained parts. Finally, all calculation cases were realized using full billets.

First simulations were made in the most favourable conditions recommended to these types of forging process designing. Friction factor m=0.3 applied in this case was chosen basing on laboratory experiments and observations of industrial processes with different lubrications of dies surfaces.

Figure 5. Bevel gear – model of final part.

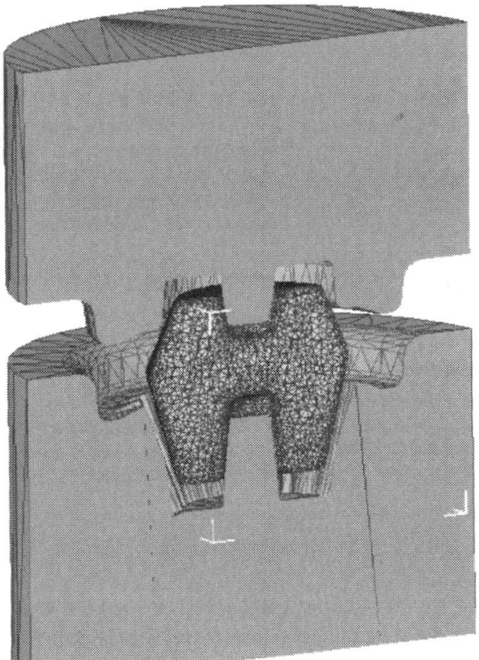

Figure 6. Numerical model of the bevel gear forging without flash and workpiece with initial mesh

DOI: 10.1051/matecconf/20168013004

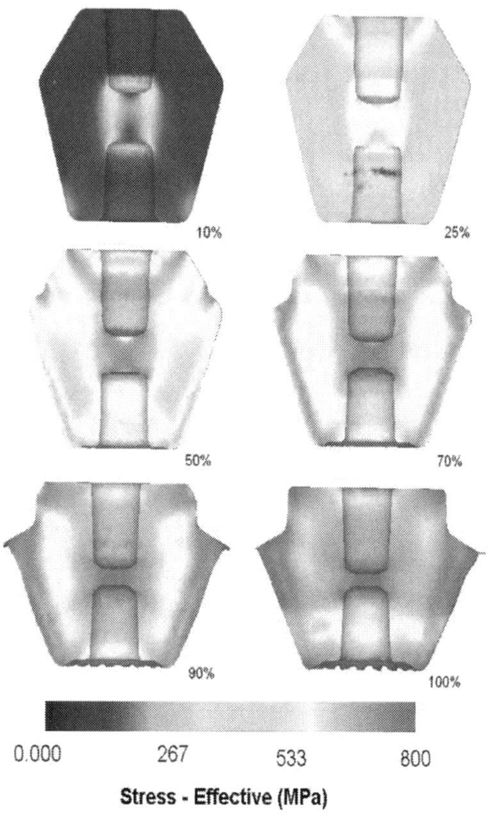

Figure 7. Effective stress distribution during forging with initial temperature 1150°C and friction factor m = 0.3

3.3 Influence of temperature range and friction factor values

Presented in the Figure 7 and 8 distributions of effective stress and strains were obtained in the most favourable process conditions taken into consideration in assumptions. Beginning the process with recommended temperature of material 1150°C it was possible to finish forging at 940°C. All forming cycle took 4,5 seconds, which was in accordance with industrial possibilities of the press proposed for this process. Changing the conditions as lowering the temperature of workpiece or taking into calculations dies with initial environment temperature 20°C can be dangerous for virtually designed impressions. For the initial workpiece temperature 1000°C the end of the forming process was calculated with temperature of the final part about 830°C. At the same time, the effective stress into the final part reached 1350 MPa and observed loads can be dangerous for dies impressions. In the Figure 9 and 10 distributions of effective stresses in workpiece during forming and loads on tools with different initial temperatures are shown. In all presented in this diagram effective stress distributions initial temperature of dies was 150°C. The biggest values of effective stresses were noticed at the punch front area and near the biggest part's diameter. Analysing the process beginning with "cold" dies at 20°C, the final temperature of the obtained bevel gear part will be relatively lower than in stable manufacturing stage. With the part temperature about 720°C (workpiece initial temperature in this case was 1000°C) it is almost certain to damage fixed in the press tools. Applying during the forming process different friction factors allowed to verify the influence of lubrication on dies impressions filled by metal. Proposed to the simulations friction factor value m = 0.3 in all different temperatures variants was satisfactory and provided to obtain final part with correct shapes. Almost the same results were observed with the higher friction factor value m = 0.4. Comparing the same simulations steps, it was possible to observe a little slower filling of the dies impressions. Greater change was noticed with application of friction factor m = 0.5. In this case it was really problematic to finish the process before final dies closing. Actually, it cannot be observed in practise because of constant material volume in all the analysed cases. However, in worked out simulations this observation informs about possible difficulties in real process without proper lubrication [9-10].

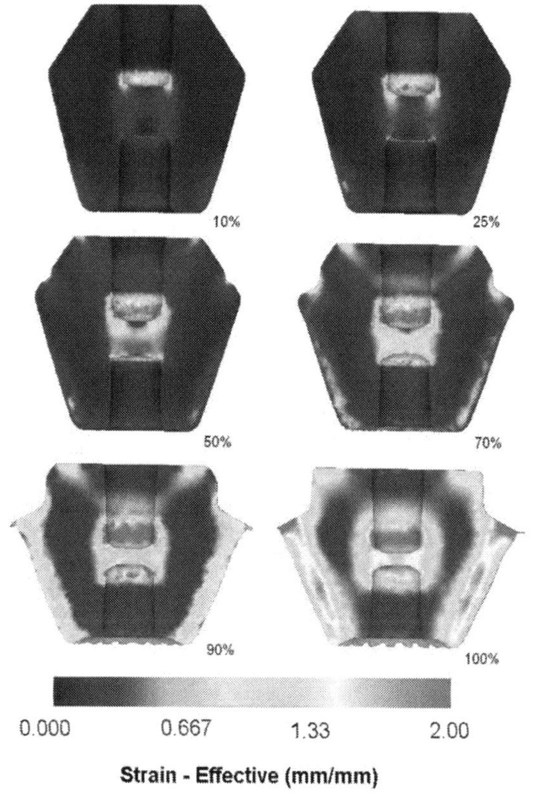

Figure 8. Effective strain distribution during forging with initial temperature 1150°C and friction factor m = 0.3

3.4 Cockroft – Latham criterion

During the calculation also the damage criterion was verified. In the Deform3D software there are several damage criteria but the most popular is Cockroft Latham integral, which calculated value informs about the risk of crack presence inside the formed part.

Basing on experience and literature research it can be stated that in the case of hot metal forming conditions for steels during forging process accepted values of this integral are about 0.7 – 0.8. In the Figure 11, the results of numerical calculations using Cockroft–Latham integral at final simulations steps, are presented.

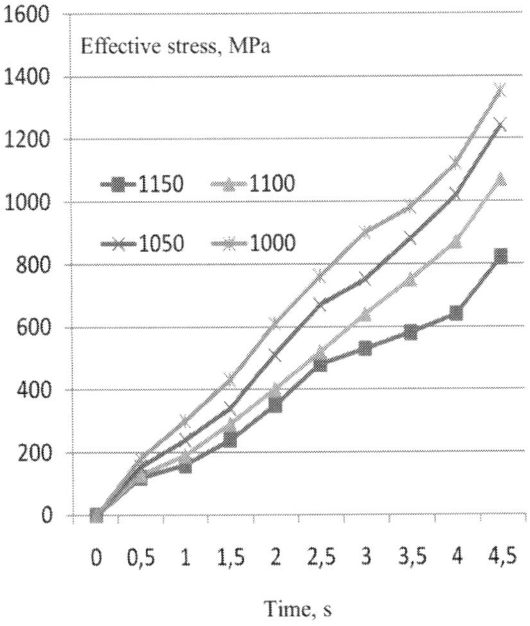

Figure 9. Effective stress distribution during forging with marked initial workpiece temperatures

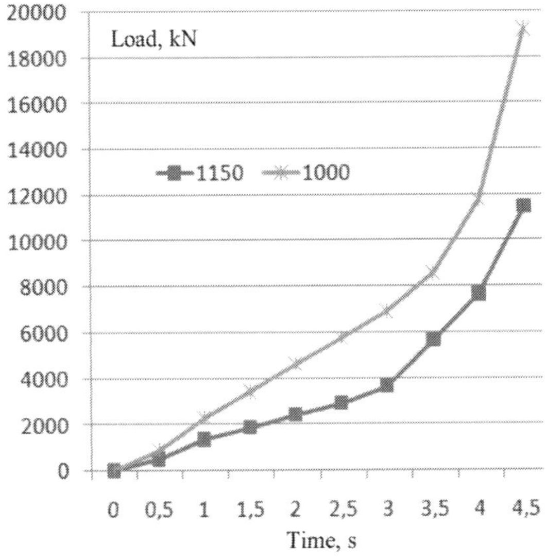

Figure 10. Loads on dies distribution during forging with marked initial workpiece temperatures

Observed distribution of Cockroft – Latham damage criterion allows for crack initiation predictions. In the most favourable composition of the highest initial temperature of workpiece and the lowest friction factor it can be possible to obtain part with relatively reduced risk of internal faults presence. But all changes of the mentioned above process parameters can disturb the process by increasing probability of internal cracks propagation. Due to this fact, all parts applied in automotive and aviation industry must be checked by non destructive tests. Presented simulations results showed only that the risk of part failure during manufacturing process is real and must be verified in laboratory and further industrial tests before application in manufacturing program of forging plant.

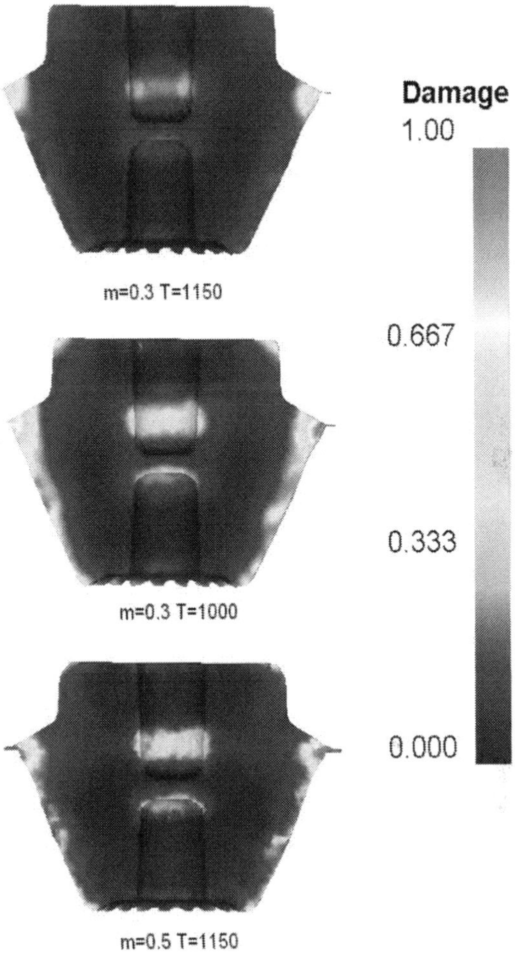

Figure 11. Cockroft – Latham integral values calculated in simulations realised with marked friction factors and initial workpiece temperatures

4. Summary

The conducted numerical calculations show the possibility of forming of bevel gear by means of close die forging without flash on parts from steel 20MnCr5. Presented in Figure 12 final shape of bevel gear part modelled as forging without flash show very close dimensional accuracy to the proposed model of the final part. Presented in this Figure teeth have a correct contour, which reflect die impressions. In many proposed solution the last stage of this kind of forming is additional sizing process. This operation permits to obtain straight surfaces and smaller material allowances. In practice ,before this operation additional heating is needed and the cost of this added energy is elevated in comparison with total cost of manufacturing process.

On the basis of the obtained numerical results, it was planned to make experiments in the hydraulic press, allowing for practical realization of laboratory tests for this kind of parts. The further works on determining the dependency between the process parameters and the achieved dimensional deviations of the formed products should lead to obtain products of shape close to the final outline.

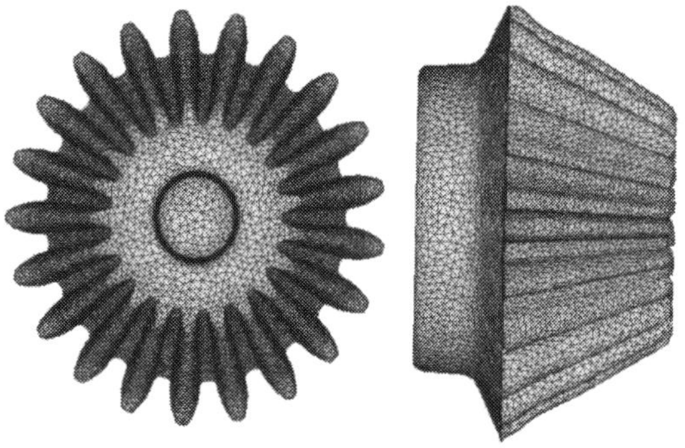

Figure 12. View of final shape of bevel gear part modelled as forging without flash

References

1. R. Glass, F. Hahn, M. Kolbe, L.W. Meyer, , Journal of Materials Processing Technology, **80-81**, 174 (1998).
2. D. P. Townsend, Dudley's Gear Handbook—the Design, Manufacture, and Application of Gears, 2nd Ed., McGraw-Hill, New York, 1991
3. J. Bartnicki, J. Tomczak, Archives of Metallurgy and Materials, **60**, 2739 (2015)
4. J. Bartnicki, J. Tomczak, Z. Pater Key Engineering Materials – **622-623**,1215 (2014)
5. Z. Pater , J. Bartnicki, Metalurgija, **51,** 501 (2012)
6. A. Stefanik, H. Dyja, P. Szota, S. Mróz, , Archives of Metallurgy and Materials, **56**, 543 (2011).
7. Z. Pater, J. Tomczak, T. Bulzak, Metalurgija, **54**, 419 (2015)
8. Jun-song Jin, Ju-chen Xia, Xin-yun Wang, Guo-an Hu, Hua Liu, Journal of Central South University of Technology, **16**, 546 (2009)
9. Z. Pater, A. Gontarz, A. Tofil, Journal of Shanghai Jiaotong University (Science), **16**, 162 (2011)
10. A. Gontarz, A. Dziubińska, Ł Okoń, Archives of Metallurgy and Materials, **56**, 379 (2011)

Numerical investigations of multicomponent process lightweight design for joining gearwheels by lateral extrusion

Robert Meissner[1,a] and Mathias Liewald[1]

[1] *Institute for Metal Forming Technology, University of Stuttgart, Germany*

Abstract. Multicomponent processes for production of high performance gearwheels combines both high power density and high lightweight potential. In this paper, the focus is on manufacturing an assembled gearwheel using cold forging and simultaneously joining the gearwheel body and a gear ring. In this study, FEA results regarding the joining process of two different materials for assembled gearwheels are presented. The aim of the numerical investigation is to achieve a robust forging process and define appropriate tool parameters.

1 Introduction

Funded by the Federal Ministry of Economic Affairs and Energy BMWi via the German Industrial Federation of Industrial Research Associations AiF, the joint research project "massiver Leichtbau" started in May 2015. In the context of this project, lightweight potentials of forged components are investigated in five work packages using different methods. In order to identify and to make use of lightweight potential of forged parts, networking of several stages of production becomes essential. The work package „Intelligent lightweight design through multi-component processes", run by the Gear Research Centre (FZG) of the Technical University of Munich, investigates the potential of using assembled gearwheels. A common gearwheel is divided into three parts – gear ring, gearwheel body, and shaft – and the substitution of the gearwheel body will be investigated using different manufacturing processes. In the investigations, special heat treated gear rings provided by the Institute of Material Science in Bremen (IWT) will be used. The Institute of Metal Forming and Casting (Utg Munich) of the Technical University of Munich deals with a gearwheel body made of joint sheet metal and deep drawn gearwheel body. The assembled gearwheel will be manufactured at the Institute for Metal Forming Technology (IFU Stuttgart) of the University of Stuttgart by forging the gearwheel body into the gear ring.

The forging process of an assembled gearwheel will be performed by laterally extruding a blank, placed between shaft and gear ring. Therefore, numerical simulations are necessary to define the process limits and design the forging tools. Joining by forming can be performed in various ways. Former investigations focused on forming a shaft into a hub, for example by internal high pressure forming or by using a heated hub and a cold shaft. [1,2] Dörr and Funk examined joining by forming processes by laterally extruding a shaft into a hub. [3] Funk used several inner hub geometries such as polygons (P3G) as well as simple and complex cycloids. Regarding the punch force, the roundness deviation of the hub and the torque transmission a profile with six extensions leads to the best result. [4,5] According to Funk, epitrochoids are most promising concerning process requirements during the forging process and while used. [6] Hence, in order to achieve a force-fit and form-fit joint of the gear ring and the gearwheel body, several inner gear ring geometries must be investigated.

The design of the inner geometry of the gear ring will adapt to these results. Due to a higher axial work load, in case of the built gearwheel coming from a helical gearing, an additional axial form-fit will be generated by creating a cavity on the inside of the gear ring. Next, the investigations in this paper will show how the inner geometry affects material flow and work load during the forging process to predict a promising geometry for experimental tests. In order to design the tools for experimental investigations, the work load was determined considering maximum lightweight potential of the gearwheel body. Further investigations were conducted regarding tribology behaviour and elastic behaviour of the gear ring during the process and after the process respectively.

2 Numerical model setup

The fundamental tool concept is shown in Figure 1. Experiments will be performed on a one-stage hydraulic press. In order to initiate lateral extrusion material flow, a closing device will be used. The synchronization of the upper and lower closing device is achieved by adjusting the closing force of both acting against each other. In [7], it is shown that the synchronization has significant

[a] Corresponding author: robert.meissner@ifu.uni-stuttgart.de

influence on part quality regarding axial symmetry of forged parts produced by lateral extrusion. Therefore, a hydraulic closing device must adjust the upper and lower closing forces precisely. Using hydraulic chambers, nitrogen storages are mounted on each hydraulic chamber and separated by a membrane. The nitrogen storages themselves cause an increase of pressure in the hydraulic chamber. Therefore, the synchronization is performed by adjusting the upper and lower pressure in the nitrogen storages.

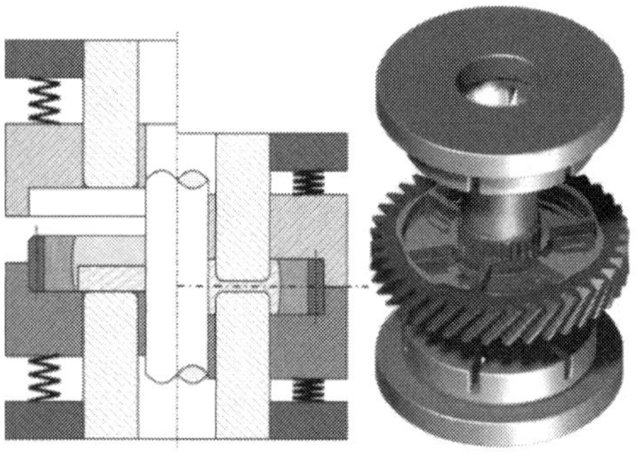

Figure 1: Tool concept for lateral extrusion (left) and built gearwheels (right)

In order to determine process properties and design the tool concept, only tools directly touching the part during the forging process were considered. The investigation was divided into two parts. First, the effect of geometrical modifications on material flow and work load was examined. In the next stage, elastic behaviour of the gear ring during and after the process was investigated. With regard to all process simulation runs, the parameters in table 1 were considered and kept constant.

Table 1. Geometrical conditions

Part	Dimension
Diameter of inner gear ring D_{Zi}	77 mm
Diameter of outer gear ring D_{Zo}	94.12 mm
Diameter of shaft D_W	30 mm

During the forging process, a constant punch velocity of 100 mm/s was used. In order to save computing time, the smallest geometry model was used considering axial and rotational symmetries.

MATERIAL DATA

The investigations were conducted using steel C15. In order to determine the lightweight potential, steel was replaced with material Al 1070A and was used for the comparison. The material properties are shown in table 2. Flow curves were obtained up to a deformation degree of 0.8 by conducting compression tests. For calculating higher deformation degrees, flow stress was extrapolated linearly.

Table 2. Material parameters for simulation

Properties	Unit	C15	Al 1070A
E-Module	[N/mm²]	210,000	68,900
Tensile Strength	[N/mm²]	360	57
Poisson Ratio	[-]	0.3	0.33
Density	[g/cm³]	7.87	2.70

2.1 Investigations on material flow and tool load

The joining of a gearwheel body by lateral extrusion is an axial symmetric process. Therefore, only the upper tools were investigated regarding material flow and tool work load. Punch, shaft, and gear ring were considered rigid, while the workpiece was modelled as a plastic model with a mesh of 80.000 elements. DEFORM 3D™ was used for the simulation.

2.1.1 Standard geometry

In order to investigate material flow and punch work load in particular, the geometrical parameters "outer diameter of blank to blank height ratio", "extensions of inner gear ring profile", "punch segmentation", "wall thickness", "eccentricity", "cavity", "ratio profile extensions to punch segmentation", "positioning of profile extensions to punch segmentation" and "positioning of upper punch to lower punch" were varied.

At the beginning, a basic value was determined. In order to investigate the effects of each parameter, only one parameter was varied regarding size. In table 3, the basic parameters are listed with the corresponding standard values.

Table 3. Standard parameter values

Parameter	dimension
Outer diameter of blank to blank height ratio	8.1
Friction	0.12
Extensions in inner gear ring profile	6
Punch segmentation	6
Wall thickness	3 mm
Eccentricity	1.4 mm
Cavity depth	0.7 mm
Ratio profile extensions to punch segmentation	1
Positioning of profile extensions to punch segmentation	punch spacing between two extensions
Positioning of upper punch to lower punch	Parallel

 DOI: 10.1051/matecconf/20168013005

2.1.2 Layout of blank geometry

The inner diameter of the blanks was kept constant at 30 mm. In order to direct material flow on the inner (shaft side) and on the outer side (gear ring side), outer diameter to blank height ratios were investigated to achieve a constant part height after forging several blanks. The maximum outer diameter of the blank was 74.5 mm being the smallest considered inner diameter of the gear ring. Seven different outer diameters were investigated starting at 74.5 mm and decreased about 1 mm.

2.1.3 Design of punch segmentation

The punch segmentation causes different shapes of gearwheel bodies, and therefore, a different number of bridges affecting the mechanical behaviour are produced. A punch having four to nine segments was investigated, shown in figure 2. A conventional flow shoulder, as used in forward extrusion processes, was not considered for usage since the effective contacting length of the punch and the blank was 4 mm at a maximum. The distance between two segments was kept constant at each point of the segments.

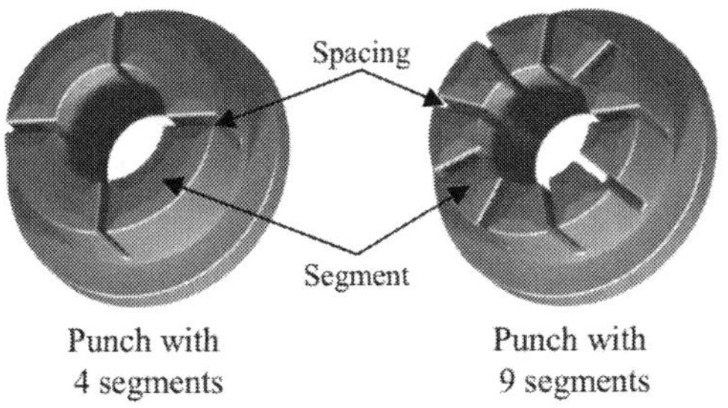

Figure 2: Design of punch segments

The variation of the wall thickness of the gearwheel body was performed by adjusting the distance between two segments of a punch as well as the distance between punch and gear ring and between punch and shaft. The thickness of the gear wheel bottom was adjusted to the other thicknesses. Wall thicknesses of 2 mm, 3 mm, and 4 mm were investigated. The wall thickness was kept constant at each part of the gearwheel body.

2.1.4 Profile of inner gear ring

Dörr conducted numerous investigations on the geometrical design of inner hub profiles by joining a shaft with a hub using lateral extrusion. [4,5] He showed that an epitrochoid profile with six extensions results in high form filling and sufficient work load of the hub. Based on these investigations, epitrochoid profiles were used for the inner gear ring geometry in the presented investigations. The profile contour is expressed by equation (1).

$$x = (R + r) \cdot cos(\alpha) - a \cdot cos\left(\left(1 + \frac{R}{r}\right) \cdot \alpha\right) \quad (1)$$
$$y = (R + r) \cdot sin(\alpha) - a \cdot sin\left(\left(1 + \frac{R}{r}\right) \cdot \alpha\right)$$

An epitrochoid is created by tracing a point P attached to a circle of a radius r rolling around the outside of a fixed circle of radius R. The measure of this point is indicated as a multiple of the rolling circle radius. [8] Following the definitions were made: parameter a has to be radian measure and was set to 0.7 mm. The inner circle radius R is fixed to a radius of 37.1 mm. The diameter of the epitrochoid outer envelope was kept constant at 77 mm in order to maintain a gear ring thickness comparable to the other investigations in this project. The profiles in figure 3 have as many profile extensions as punch segmentations (4-9S), while the inner and outer envelope remain constant.

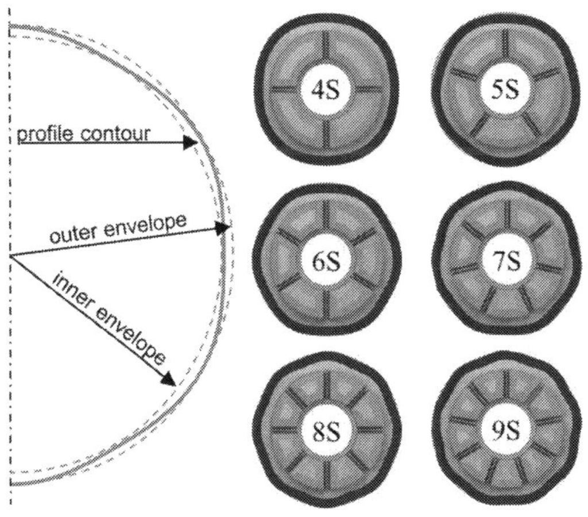

Figure 3: Various punch segmentations having four (4S) to nine (9S) segments with corresponding epitrochoid profiles

In order to increase a force fit during usage and to determine the effect on punch force and material flow, the eccentricity of the inner and outer envelopes, shown in figure 3, was varied by altering parameter R and a in the epitrochoid equation (1). These alterations change the distance between inner and outer envelope. The eccentricities 0.7 mm, 1.4 mm, and 2.0 mm were investigated.

In order to investigate the cavity at the inner gear ring, a fixed inner diameter had to be determined first. Secondly, the cavity cannot exceed a diameter of 77.0 mm, since the stability of the gear ring would be affected significantly. Therefore, the cavity was referenced to a fixed diameter of 74.2 mm and enters the gear ring up to 1.4 mm as shown in figure 4. Five cavities were investigated having a maximum distance to inner diameter of 0.0 mm, 0.3 mm, 0.7 mm and 1.4 mm.

DOI: 10.1051/matecconf/20168013005

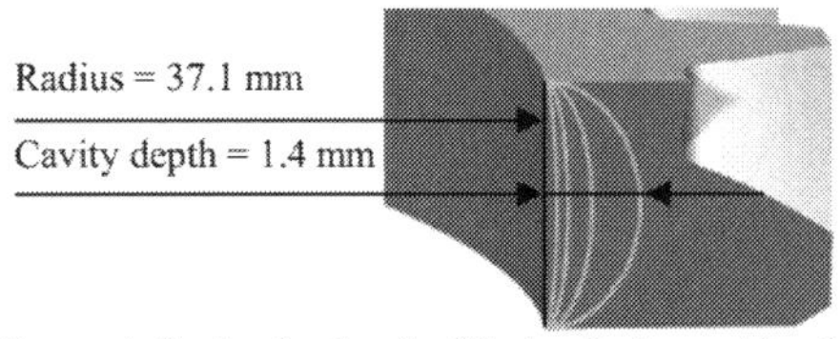

Figure 4: Cavity depth to be filled at the inner side of the gear ring

2.1.5 Positioning of upper punch to lower punch and punch to gear ring

In three different positions, the influence of punch positions to each other as well as the positioning of punch segments to the gear ring was investigated with regard to material flow, see figure 5. Firstly, both upper punch and lower punch were positioned parallel and the extensions of the gear ring were placed in the middle of a punch segment. Afterwards, the lower punch was turned by half the angle of a segment. Finally, the gear ring was turned by half the angle of a segment: more explicitly, the profiles' extensions were placed between two punch segments.

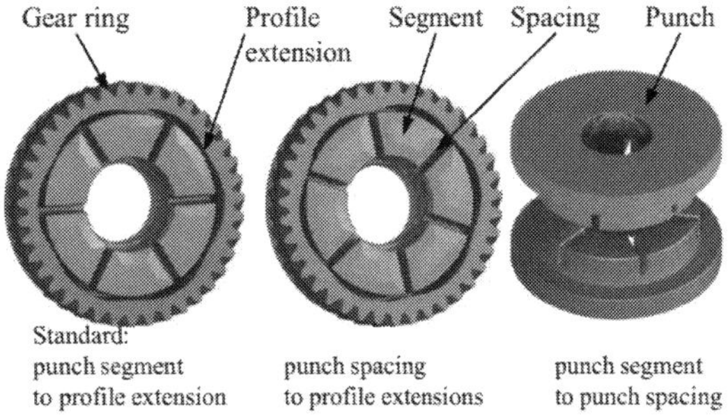

Figure 5: Tool and gear ring alignments to be investigated

2.1.6 Filling of shaft gearing

In order to manufacture a completely assembled gearwheel, a final investigation of joining by forming of a gearwheel body with a gear ring and a shaft was conducted. Therefore, geometrical data of the shaft having a toothing was provided by FZG. The material flow was investigated by altering the inner punch diameter regarding a wall thickness of 3 mm. Next, the related diameter of the shaft was referenced to the tooth root diameter, the pitch circle, and tip diameter.

2.2 Elastic behaviour and stress distribution

During the forging process, radial compression stresses are applied on the gear ring, resulting in tangential tensile stresses on the outside. In order to prevent cracks, elastic numerical simulations were carried out. These investigations were performed to predict both radial deformations and stresses, and to make a point for a tool concept to implement a system for application of a radial preload. Furthermore, the elastic behaviour after ejection of the part was investigated to analyse the remaining pressure between the gearwheel body and the gear ring. Two setups were prepared for the investigation. First, the gear ring was fixed by its symmetric boundary conditions, but was free to expand to the outside. For the second setup, a rigidly modelled die was placed on the outside of the gear ring to simulate a fix bearing during the forging process.

3 Results and discussion

3.1 Material flow and tool load

3.1.1 Influence of material on punch force

In figure 6, the punch force versus punch stroke diagram is shown for two materials regarding the basic parameter setup. Up to a punch stroke of 0.8 mm, a rather linear punch force increase can be detected. At this point the blank touches the gear ring, which results in a higher punch force, and therefore, increases to a punch stroke of 1.2 mm. Next, the punch force increases slightly to a punch stroke of 2.6 mm. The slight increase of punch force is caused by a growing friction surface on the punch, shaft, and gear ring. Using aluminium Al 1070A to forge the gearwheel body, it was determined that the maximum punch force was 82 % lower in comparison to steel C15. In order to evaluate the influence of friction on punch forces, three cold forging friction values were investigated regarding material C15. At the beginning of the process, the influence of friction cannot be detected. The influence of friction increases at a punch stroke of approx. 1.3 mm when the contacting surface of the inner gear ring increases. Thus, friction forces increase as well, but only have a minor effect on the maximum punch force.

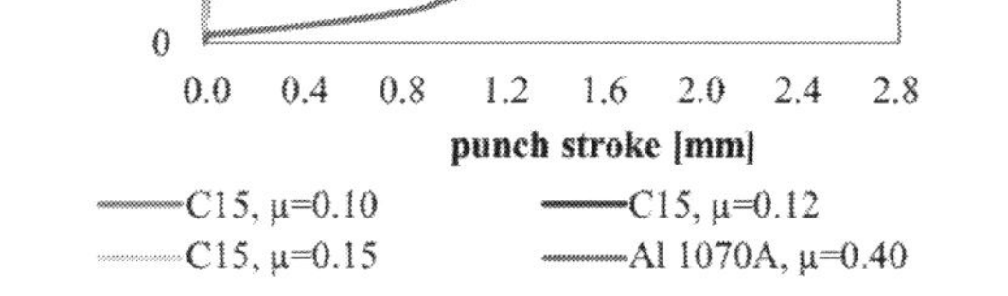

Figure 6: Punch force to punch stroke curve for steel C15, aluminium Al 1070A and different friction values

3.1.2 Influence of blank geometry layout on part height

The influence of the layout of blank geometry is shown in figure 7. Four measurements were taken on the outer edges of the investigated symmetric part and one to

DOI: 10.1051/matecconf/20168013005

determine the bridge height in order to calculate the average height of a part.

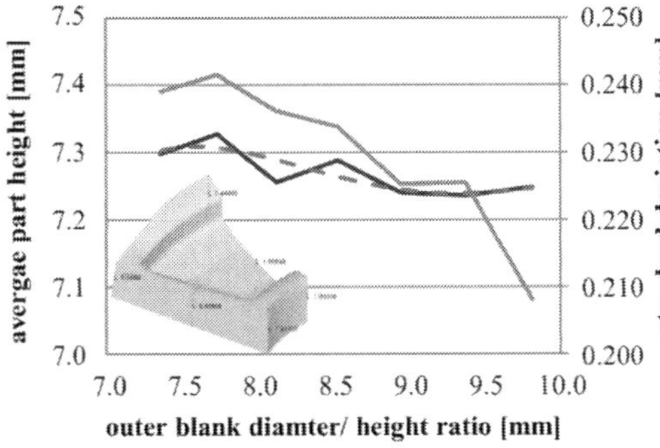

Figure 7: Average part height and standard deviation for different initial outer blank diameter to height ratios

The average value, as well as the standard deviation, is depicted on the ordinate for an increasing outer blank diameter to height ratio on the abscissa. Therefore, it can be concluded that a higher outer blank diameter results in a more balanced part height. A trendline representing a polynom grade four shows a minimum at a ratio of 9.5 in figure 7. The outer blank diameter to height ratio is limited since the maximum outer blank diameter has to be smaller than 75 mm in order to be able to place the sample in the gear ring before starting the forging process.

3.1.3 Design of punch segmentation

In order to investigate the influence of various punch segmentations (figure 3) on punch forces, both punch segment and gear ring profiles were varied to have the same segments and extensions respectively. In figure 7, punch force to punch stroke diagram is depicted for segments varying from four (4S) to nine (9S).

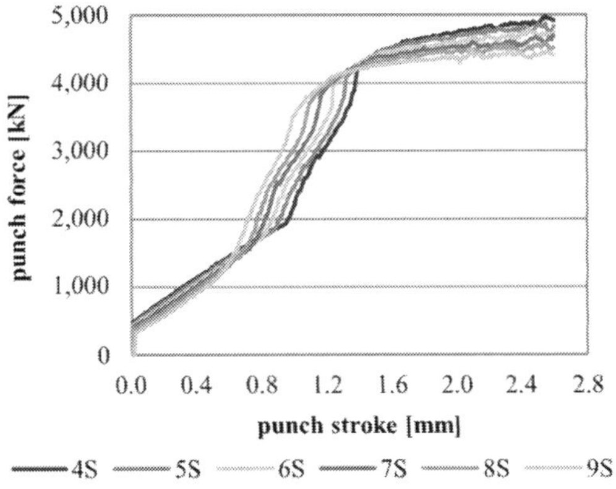

Figure 8: Punch forces for different punch segmentations

It can be concluded that fewer segments cause higher maximum punch forces, but on the other hand, the punch force increases at a later point during the process. In order to evaluate the occurring loads, the punch compression stresses have to be taken into account. In doing so, a small incline of the punch stress for numerical setup with more punch segments was detected.

In figure 9 a bar chart displays maximum compression stresses of different punch designs having four (4S) to nine (9S) segments. The punch stresses increase with a higher number of segments. However, this increase remains below 1 % when comparing two punches with four and nine segments respectively. As opposed to this, weight reduction potential decreases from 54 % to 49 % in reference to a conventional gearwheel body.

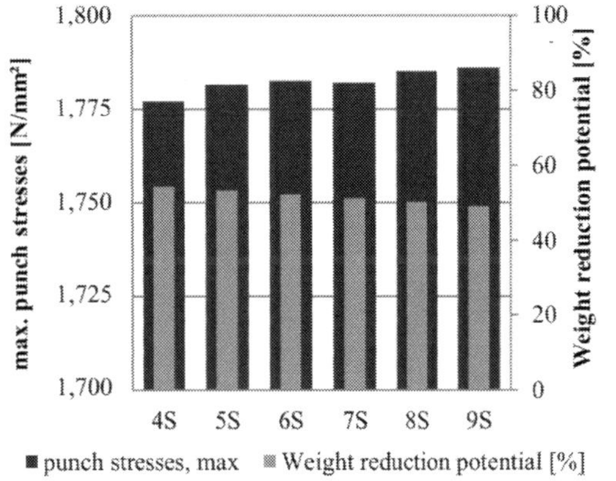

Figure 9: Maximum punch stresses and gearwheel body lightweight potentials related to full gearwheel bodies

Varying the wall thickness (distances between punch to gear ring, punch segment to punch segment and punch to shaft) causes a significant change of maximum punch force as can be seen in figure 10. Decreasing the wall thickness to 2 mm results in an increase of maximum punch force of 30,4 % (+1.421 kN) and an increase of punch stress of 15.0 % (+313.9 N/mm²). However, lightweight potential of the gearwheel body could reach 66 %.

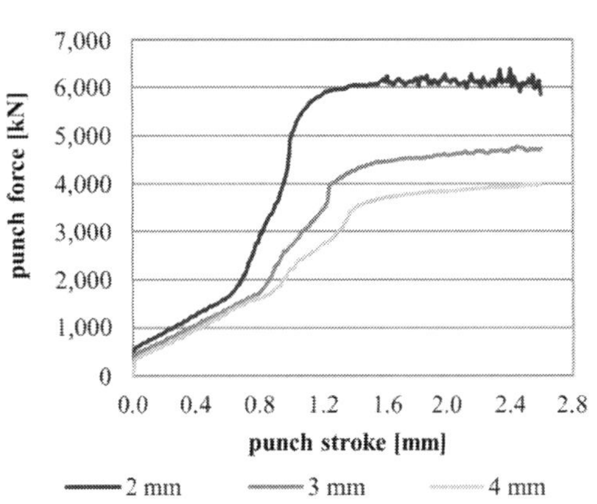

Figure 10: Calculated punch forces for different wall thicknesses of the gearwheel body

On the contrary, the lightweight potential of the gearwheel body decreases to 40 % when increasing the wall thickness to 4 mm. Therefore, the maximum punch force decreases by 16.2 % (-757 kN) and the maximum punch compression stresses decrease by 0.1 % (-0.9 N/mm²). This small decrease results from a smaller cross sectional area of the punch.

3.1.4 Alteration of profile of inner gear ring

In order to investigate the influence of the inner gear ring profile on punch forces, the profile's eccentricity and cavity were varied. In figure 11, the punch force to punch stroke course is displayed for different eccentricities (parameter e) as well as different cavity depths (parameter c), representing the inner gear ring profile.

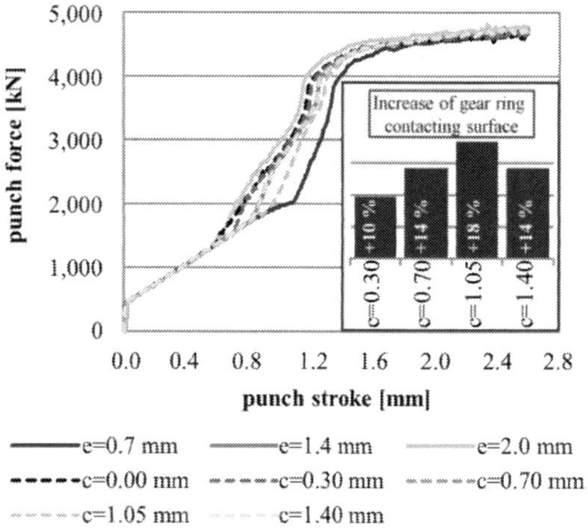

Figure 11: Calculated punch force for different eccentricities and cavity depths of the inner gear ring profile

By increasing eccentricity, the material touches the gear ring at an early stage of the process, which results in a higher punch force at a smaller punch stroke. On the contrary, the maximum punch force level is equal for all eccentricities. Hence, eccentricity has minor effects on punch compression stresses. In comparison to eccentricity, a cavity volume in the inner gear ring affects the punch force even less. However, the bar chart depicted in figure 10 shows the potential benefits of adding a cavity volume to the inner profile of the gear ring. A cavity depth of c = 0.3 mm results in an increase of 10 % when contacting the surface of the gearwheel body at the inner gear ring. For a cavity depth of c =1.05 mm, a maximum increase of 18 % was determined. A higher shape filling does not automatically result in higher performance of a joint shaft hub assembly [5], but regarding a higher axial form-fit a higher axial force transmission is expected.

3.2.5 Positioning of upper punch to lower punch and punch to the profile of the gear ring

Turning the upper punch to the lower punch by half the angle of a punch segment, the spacing of each punch are now directed to a punch segment causing a lower force of the punch segment to extrude the material into the opponent spacing. Hence, the displaced punch segment to punch spacing positioning results in lower punch forces and punch compression stress, as depicted in figure 12.

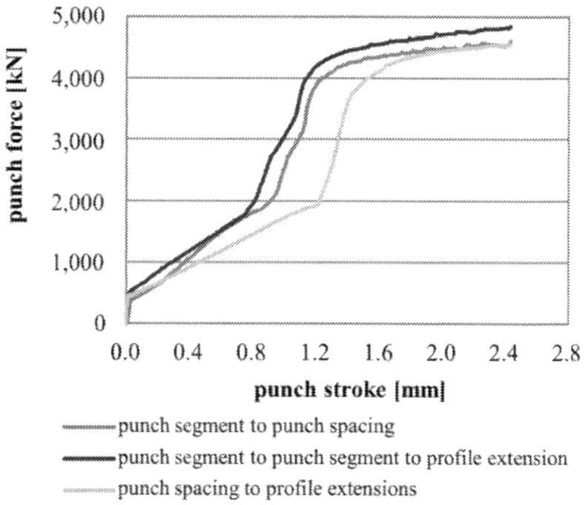

Figure 12: Punch force versus punch stroke curve for different positionings of upper to lower punch and punch to gear ring

Considering a parallel positioning of upper to lower punch, a higher punch force is required, but a symmetric punch assembly also results in a more evenly distributed part height of the gearwheel body. A second investigated positioning evaluated the orientation of gear ring profile extensions to punch segment spacings (gearwheel body wall). Placing the profile extensions to a spacing of punch segment significantly affects both material flow and punch force. The gear ring is touched at a later stage of the process since the distance between punch and gear ring profile extension increased. Despite this, the maximum punch force remains at the same level. However, manufactured gearwheel bodies show better results regarding average part height for a positioning of the punch segment to profile extension of the gear ring.

3.1.6 Filling of shaft gearing

Altering the inner punch diameter has minor effects on filling the shaft gearing and on radial stress in the gearwheel body volume, shown in figure 13. A lower distance of punch to shaft seems to result in a straighter axial material flow along the shaft. However, the smallest investigated inner diameter of the punch does not have a higher contacting surface area responsible for transmitting torsion on both the shaft and the gearwheel body.

 DOI: 10.1051/matecconf/20168013005

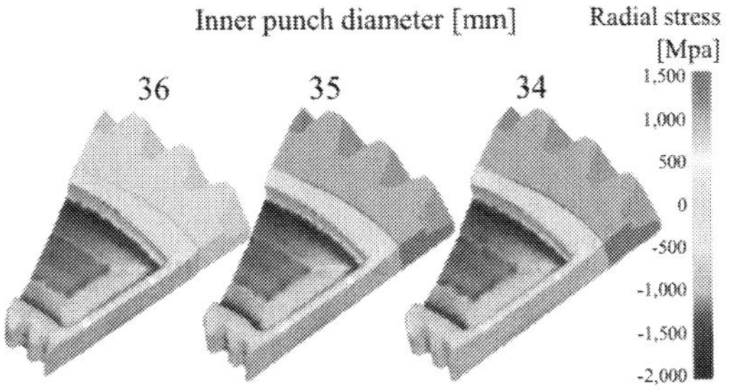

Figure 13: Distribution of radial stress for different inner punch diameters at the end of forging process

3.2 Investigations on elastic gear ring behaviour

In figure 14, the elastic behaviour of the gear ring after the forging process and the release of work load are depicted. The gear ring was modelled elastic-plastically to have 80.000 mesh elements. As can be seen, radial compression stress decreases significantly after work load release. Tangential tensile stress on the outside of the gear ring must be evaluated since cracks will occur if the material property limits are exceeded. Despite the release of work load, tangential tensile stress remains at the same level, which is likely to result in cracking.

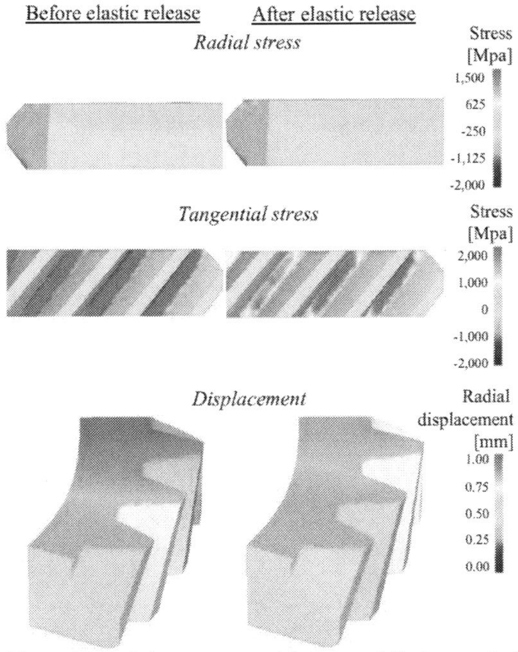

Figure 14: Radial stress, tangential stress and displacement of gear ring after forging process and after work load release for a non-fixed gear ring on the outside

When evaluating gear ring displacement, both inner gear ring profile and gear ring gearing need to be considered in order to result in a non-linear radial deformation. The maximum deformation during the process is up to 0.9 mm. After elastic spring back occurs, a maximum deformation of approx. 0.8 mm remains in some areas of the gear ring. The contact surface increases by roughly 1 % after work load release.

Assuming an ideal rigid bearing at the outside of the gear ring, the elastic behaviour changes significantly before and after the release of the work load. In figure 15, radial stress, tangential stress, and radial displacement of the gear ring after the forging process and after the release of work load for a fixed gear ring on the outside are depicted. Before the release of work load, radial stress reaches up to higher values. The distribution seems to be quite even on the complete inside, and a higher contact surface can be detected in comparison to a non-fixed gear ring (figure 14).

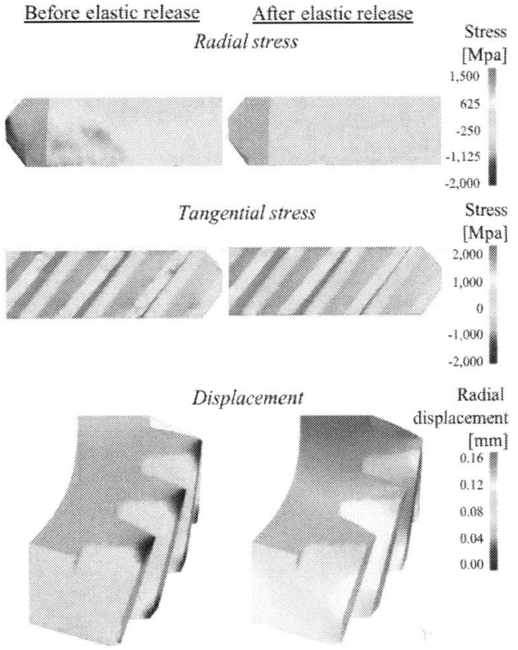

Figure 15: Radial stress, tangential stress and displacement of gear ring after forging process and after work load release for a fixed gear ring on the outside

Furthermore, tangential stress on the outside of the gear ring could be decreased significantly. On the contrary, tangential stress increases after the release of work load, but remains below the values of tangential stress of a non-fixed gear ring. In this case, displacement behaviour is similar to tangential stress. Before releasing the work load, the radial displacement is significantly lower, since the gear ring is fixed at the outside. However, the radial displacement increases after the release of work load. Furthermore, the area of the inner gear ring profile peaks undergoes a larger displacement.

4 Conclusion

In this paper, fundamental investigations on material flow, emerging punch forces, and elastic behaviour of a gear ring for joining a built gearwheel body in a multi-component process were presented. Regarding material flow, the blank geometry and the punch positioning showed clear effects on constant part height at the inside and outside. While friction has no considerable influence on the punch force, the wall thickness mostly affects punch forces as well as lightweight potential. The investigations on the inner gear ring profile regarding different eccentricities and cavity depths outline a trend, especially of the performance of built component during usage. The comparison of non-fixed and fixed gear ring bearings show significant differences regarding radial and tangential stresses.

Using a fixed bearing, a bigger decrease of radial stress in the inner gear ring was determined. In contrast, tangential stress and radial displacement were decreased distinctly.

Acknowledgement

Based on a decision of the German Bundestag, the IGF project 18229 N of the research associations FOSTA, AWT, FSV, and FVA is supported via the AiF within the program of promoting joint industrial research projects by the Federal Ministry of Economy and Energy.

Supported by:

Federal Ministry
for Economic Affairs
and Energy

on the basis of a decision
by the German Bundestag

References

1. M. Hagedorn; K. Weinert, Journal of Materials Processing Technology 153–154 (2004) p. 323–329
2. R. Matsumoto; S. Hanami; A. Ogura; H. Yoshimura; K. Osakada, *CIRP Annals - Manufacturing Technology* **57**, p. 279–282 (2008)
3. F. Dörr, M. Funk, M. Liewald, H. Binz, *11th International Conference on Numerical Methods in Industrial Forming Processes (NUMIFORM 2013)*, July 6-10, , Shenyang, Liaoning, China (2013)
4. F. Dörr; M. Funk; M. Liewald , H. Binz, In: Neugebauer, R.; Drossel, W.-G. (Hrsg.): *4th International Conference on Accuracy in Forming Technology.Chemnitz*, p. 437-450 (2012)
5. F. Dörr; M. Funk; M. Liewald M., H. Binz, In, *11th International Conference on Technology of Plasticity*, ICTP 2014, *Procedia Engineering* **81**; Nagoya, Japan, p. 1988-1993 (2014)
6. M. Funk; F. Dörr; H. Binz; M. Liewald, NAFEMS World Congress 2015, *2nd International SPDM Conference*, San Diego (2015)
7. J. Wälder; A. Felde; M. Liewald: Hollow lateral extrusion of tubular billets – Further development of the cold forging process *Applied Mechanics and Materials*, **794**, pp 160-165 (2015)
8. J. D. Lawrence: *A Catalog of Special Plane Curves*. New York: Dover, pp. 168-170 (1972)

Interpolation of final geometry and result fields in process parameter space

Grzegorz Misiun[1,a], Chao Wang[1], Hubert Geijselaers [1] and Ton van den Boogaard [1]

[1]*University of Twente, Section of Applied Mechanics, Enschede, Netherlands*

Abstract. Different routes to produce a product in a bulk forming process can be described by a limited set of process parameters. The parameters determine the final geometry as well as the distribution of state variables in the final shape. Ring rolling has been simulated using different parameter settings. The final state variable fields have been subjected to Principal Component Analysis. Interpolation of amplitudes in process parameter space gives a good representation of the fields and is useful as metamodel of the process.

1 Introduction

In engineering process design it is often needed to find an optimal solution for a given design parameters space. This type of research requires providing information about influence of different process parameters on final part results. Usually to avoid time and cost consuming experimental approach, numerical modelling is applied. Performing simulations significantly reduces costs and allows to give better insight into the process. Unfortunately in case of material nonlinearity which is characteristic for large variety of metal production processes numerical calculation may take even much more time than production itself. If there is a need for obtaining information from this process with large number of different parameter values it may be beneficial to construct its approximation response model known as surrogate model or metamodel.

Different mathematical techniques may be used for meta-modelling purposes. In [1] four different variants are compared for a simple FEM simulation: Response Surface Methodology, Proper Orthogonal Decomposition combined with Radial Basis Function, Moving Least Squares and Neighborhood Approximation. POD-RBF gives the best results when the number of sampling points is relatively small. This method will be used in this research as well.

In literature many examples regarding application of POD-RBF technique for metamodelling purposes may be found. Mostly they may be divided into two main categories. The most popular is inverse analysis where based on the system response compared to its metamodel some of the systems parameters such as material properties may be identified [2, 3, 4]. The second group consists of direct problems where based on POD-RBF methodology calculation time is improved by coupling it with FEM [5, 6] or where it is used to support the design process [7, 8].

In the present work the possibility of application of the POD-RBF technique is investigated to describe distributed FEM results for plastic metal forming process where large mesh distortions occur. Due to significant mesh deformations it is necessary to apply a mesh refinement algorithm during the simulations. This results in a newly created mesh with independent element distribution compared to its previous configuration. It means that simulations for different parameters will have different mesh number and configuration thus they cannot be directly used for metamodel creation. The possibility of describing part shape results independently from mesh distribution is investigated in this work for large deformation nonlinear problems.

2 POD-RBF metamodeling methodology

Three different methods are distinguished that can be used to perform POD: Karhunen-Loève Decomposition (KLD), Principal Component Analysis (PCA) and Singular Value Decomposition (SVD). The similarities and differences between them have been exhaustively described in [9]. In the present work decomposition is performed with use of the PCA methodology.

Apart from the possibility of applying different numerical approaches the effect of POD is always the same. It results in a subspace $\mathbf{\Phi}$ described by orthonormal basis vectors $\boldsymbol{\varphi}_i$ with corresponding amplitudes $\mathbf{y}_i$, for k orthogonal vectors:

$$\mathbf{X} = \sum_{i=1}^{k} \boldsymbol{\phi}_i \mathbf{y}_i = \mathbf{\Phi}\mathbf{Y} \quad (1)$$

The snapshot matrix $\mathbf{X}$ is created by collecting and concatenating data from simulations with different process parameters called snapshot vectors $\mathbf{x}_i$. Before

[a] Corresponding author: g.s.misiun@utwente.nl

decomposition data should be preprocessed to have zero mean.

In case of PCA based decomposition a subspace $\mathbf{\Phi}$ is found by performing an eigendecomposition of a matrix created based on **X**. Assuming that the snapshot matrix is a rectangular $m \times n$ matrix, two different matrices may be used for decomposition depending on the size of m and n:

- when $m > n$, the correlation matrix $\mathbf{D} = \mathbf{X}^{\mathrm{T}}\mathbf{X}$
- when $m < n$, the covariance matrix $\mathbf{C} = \mathbf{X}\mathbf{X}^{\mathrm{T}}$

should be used. In both cases an eigendecomposition is performed. In this process it is stated that for any $p \times p$ square matrix **A** eigenvectors $p \times p$ **V** can be found such that:

$$\mathbf{AV} = \mathbf{\Lambda V} \tag{2}$$

$\mathbf{\Lambda}$ is a diagonal matrix storing λ_i - the eigenvalues of matrix **V.** Depending on m and n different matrices for **A** are used and their eigenvalues and eigenvector are obtained. Then for creation of a subspace $\mathbf{\Phi}$ simply eigenvectors **V** are used.

Next step of decomposition is to obtain the amplitudes **Y** corresponding to the new basis:

$$\mathbf{Y} = \mathbf{\Phi}^{\mathrm{T}}\mathbf{X} \tag{3}$$

After the decomposition, model reduction is performed. Based on eigenvalues λ_i the first l basis vectors and amplitudes are chosen to retain in a metamodel. The eigenvalues are sorted in descending order. Each eigenvalue corresponds to the energy accumulated within given eigenvector. Usually most of the energy is held in the few first vectors so that the rest can be considered as redundant. In order to reduce the amount of unnecessary data the cumulative energy factor is used. The cumulative energy e of the l first eigenvectors from a total of n, can be described as sum:

$$e = \frac{\sum_{i=1}^{l} \lambda_i}{\sum_{i=1}^{n} \lambda_i} \tag{4}$$

According to this factor some threshold of cumulative energy is set above which reduction is being performed, for example 90%. The reduced set of vectors can be then used in (3), with assurance of the most optimal way of the data reduction.

Last step of metamodeling is interpolation of reduced amplitudes. Many different methods may be applied but the most common for POD metamodeling approach and used also in this research is Radial Basis Functions. With this method an approximation function is build based on the distance from known centers – x_i, and corresponding to them weights – w_i. The general approximation function for Nc centers takes form:

$$f(x) = \sum_{i=1}^{Nc} w_i \psi(\|x - x_i\|) \tag{5}$$

The radial basis function ψ can be any function that values depends on distance. In presented case the cubic function is being used:

$$\psi(x, x_i) = \|x - x_i\|^3 \tag{6}$$

The full method of obtaining weight values based on approximation function, data point locations and their values has been described in [10].

The metamodel created in that way may be used to predict values for new data points. Based on the interpolation functions, amplitudes values for a new data point may be obtained. When the amplitude values are known the whole process may be easily reversed with use of (1).

3 Process description

The process of preparing a preform for ring rolling can be divided into 4 stages – heat treatment, upsetting, forming and piercing (Fig. 1).

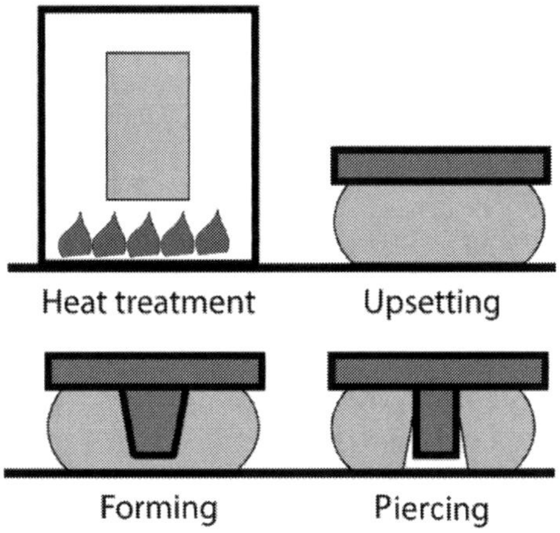

Figure 1. Stages of ring rolling operations: heat treatment, upsetting, forming and punching.

During these steps axisymmetrical, cylindrical initial billet shape changes. First, the part gets upset on the chosen height between two flat dies. Then the forming operation, takes place. The punch is pressed in the middle of a billet, displacing the material radially. The operation is carried out until the cavity bottom is 4 mm thick. Next the part is flipped over and piercing is performed. The thin 4 mm bottom is removed by the punch due to shearing and the hole in its middle is formed. Besides removing the bottom, piercing has one more important role. The flat tool surface to which punch is connected, is used for element flattening, because during forming and piercing the part shape may change. The punch moving distance is prescribed such, that its flat surface final height is 20% lower than the final upsetting height. It assures that element compression will take place. The part in its shape is then ready for the ring rolling operation [11, 12].

During the ring rolling process the initial preform is deformed simultaneously in radial and axial direction (Fig. 2). The main roll applies rotational movement on the outside diameter of the ring. The feed of the mandrel in main roll direction is responsible for radial deformation. The effect of radial deformation is reduced

DOI: 10.1051/matecconf/20168013006

part thickness and increased diameter but also increased height. In order to control ring height two axial rolls, upper and lower are applied. Additional guide rolls assure process stability.

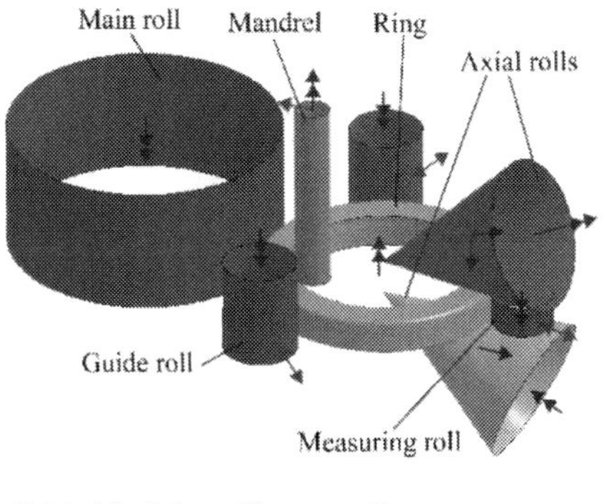

Figure 2. Model of ring rolling operation.

4 Metamodeling of numerical analysis

Numerical modeling is performed in LS-Dyna. The FEM simulations of preform forming are carried out with use of a 2D axisymmetrical model. The results of the final preform shape, after piercing are then mapped from 2D axisymmetrical state into 3D and then are used as an initial state for a ring rolling process simulation [11]. During the ring rolling process a large number of radial and axial deformations is applied until the final part geometry is reached.

In the presented work two different metamodels have been created. The first one provides information about shape and distributed part results for the preform stage. There are two parameter settings varied in this model: the part initial height and the height after upsetting. Due to constant volume constraint the part initial diameter varies with initial height. The height after upsetting defines its preform final height which is 80% of it.

The second metamodel gives information about the product after the rolling stage. Due to the fact that final part shape is always the same but only achieved with different tool paths the metamodel will focus only on the part distributed results. Beside initial part and upsetting height two additional parameters are implemented. The ring grow rate and the rolling path. The ring grow rate defines the rate of ring outer diameter growth during ring rolling process as a function of part shape [12]. The rolling path is a number that defines curve of a ring rolling deformation. It may vary from -1 to 1. When the number is equal -1 the ring is first fully formed in axial direction then radially, 0 means proportional height and thickness reduction rates and 1 stands for first radial and then axial reduction.

4.1. Preform metamodel

As mentioned for the preform metamodel two variables are used: the initial billet height and the height after upsetting. In this parameter space 18 data points are defined for metamodeling purposes. Their spread is presented in Fig. 3.

Beside the data points used for creation of the metamodel, also the location of a test point which will be used for metamodel validation has been marked.

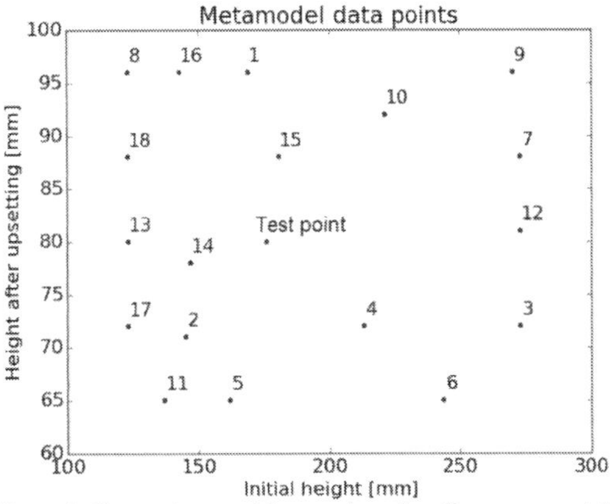

Figure 3. Data points used for creation of preform metamodel.

As mentioned in the presented metamodel there is a need of providing quantitative information about the part cross section shape and distributed results that is independent from mesh configuration. For this, a method is proposed which connects the part shape with its geometrical center coordinate, characteristic angles and lengths of lines from the center point to the part boundary (Fig. 4).

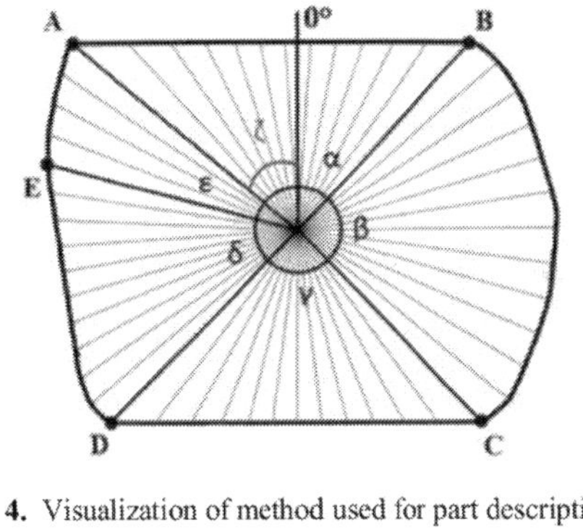

Figure 4. Visualization of method used for part description.

The part description begins with identification of the geometrical center. Next 5 characteristic points responsible for describing shape are identified. Four of them, A-D are corner nodes. E is a characteristic node located at the sharp corner, which is created during the piercing operation. The corner nodes assure that location of sampling points will change with the change of part dimensions. On the other hand a necessity of the point at the sharp corner E is dictated by inconsistency of shape correlation between different upsetting heights resulting from same shape of the forming stamp used for second, forming operation (length EA is always the same). The role of E is to prevent unnecessary data loss.

Based on this set of points the cross section is divided into 5 slices and the information about their size is stored as polar coordinates. Five angles describing size of every slice are stored. Additionally angle ζ is introduced. It describes angle between point A and the vertical. It locates shape position in space. Last stage of part description consists of dividing each of slice by equally distributed lines from the center to part boundaries. The length of each line is stored which combined with other information is sufficient for part shape description.

Next data sampling point locations are determined. To assure sampling point independence from the element mesh grid and consistency between different shapes, the previously defined lines are used. On each line 10 points are equally distributed. At these points the distributed data results are sampled. Due to the fact that the distribution is based on polar coordinates, the density of points will increase with decreasing distance to cross section geometric center. That is why it has been decided to use every n-th data point within the data point row while the row distance decreases. According to this rule it has been decided for the farthest four rows to use 50 points, next four 25 points, then 10 points and finally 5 points (Fig. 5). The effect of reduction is 316 sampling points, while initial number of points was 501.

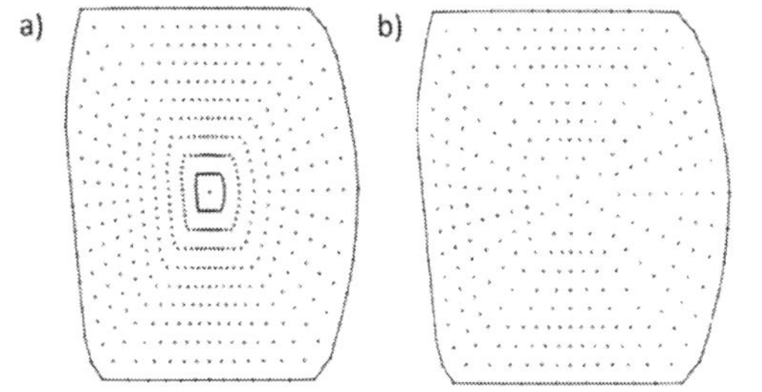

Figure 5. Distribution of sampling points a) before and b) after reduction.

Based on the presented method different distributed results may be sampled and used for metamodeling. It has been decided to create the metamodel in Python due to its large library of built-in functions. The procedure of metamodel construction is illustrated in Fig. 6.

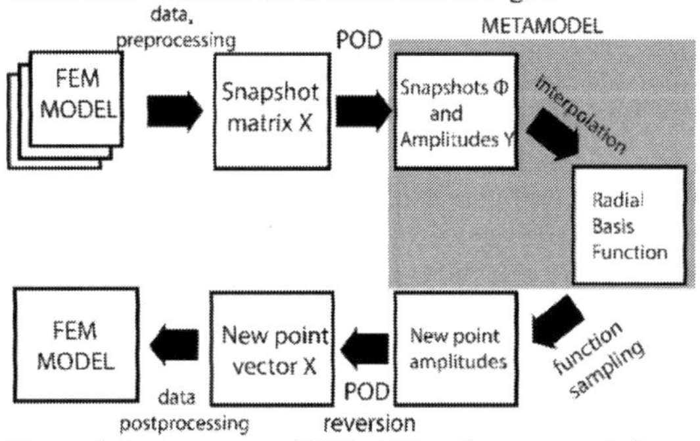

Figure 6. Process stages of POD-RBF preform metamodeling.

First the result from the FEM model are read into the program. The information is interpolated to allow its sampling according to sampling point locations. For this purposes quadric Radial Base Functions are used with centers corresponding to integration point locations. From the result fields obtained in this way, data is gathered and collected in separate vectors called snapshots. Separate snapshots are stored for different type of results. Next same snapshots from different simulation parameters are gathered and concatenated into one snapshot matrix **X**. The data in this form is reduced by POD and then chosen amplitudes are interpolated with use of RBF. From interpolation function amplitude values for new data point may be obtained and the process is reversed according to (1). The results stored in the form of parameters describing the new shape and sampling point values are then interpolated with RBF and results obtained this way may be used for new FEM model creation.

In case of preform metamodeling investigation of 9 distributed results is performed: 4 stress components, plastic strain, 3 damage indicators and temperature. Separate POD decomposition has been performed for each of them and for the vector describing shape. Eigenvalues of each mode (Fig. 7) and cumulative eigenvalues energy (Fig. 8) for each of the decompositions is presented.

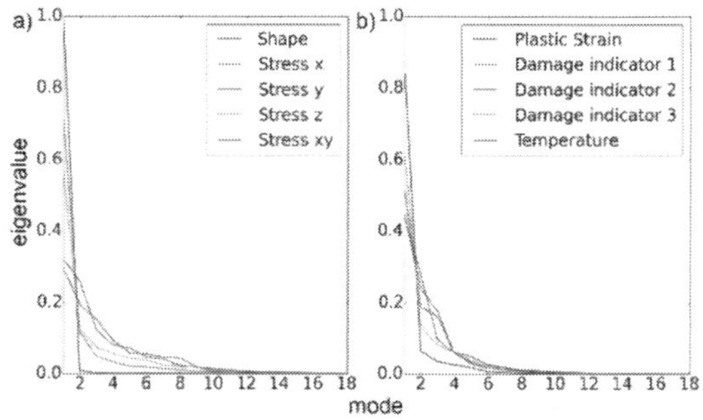

Figure 7. Eigenvalue of each mode for a) shape and stress results, b) plastic strain, damage indicators and temperature.

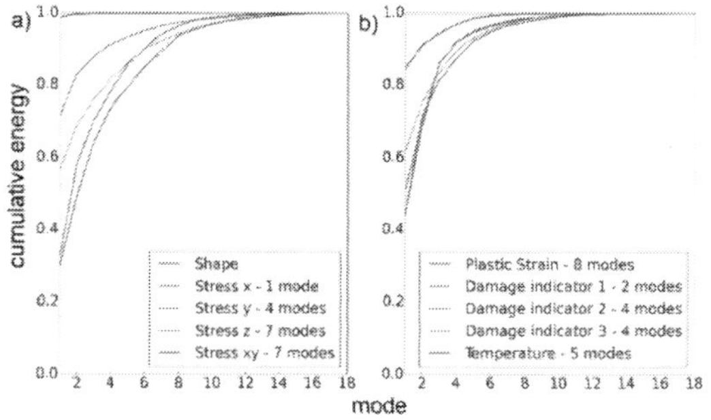

Figure 8. Cumulative eigenvalues energy of each mode for decomposition of a) shape and stress results b) plastic strain, damage indicators and temperature.

It can be noticed that in every case only first few modes contain most of the data information. According to cumulative energy a threshold e from equation (4) equal to 90% of total decomposition energy has been set and applied for data reduction. Exception has been made for shape decomposition and plastic strain. In their cases first mode and first two modes over 90% of total energy. It has been established that this number vectors is to less to provide correct information about results and minimum value of required modes has been set to 3. Number of first modes which contains 90% of total energy for different results type is listed in legend of figure 8.

Creation of a metamodel requires amplitudes interpolation in reduced subspace. For this purpose the RBF are used. To check correctness of results, smoothness of interpolation functions has been checked for every type of results. As an example, the interpolation functions for the shape discretization are presented at figure 9.

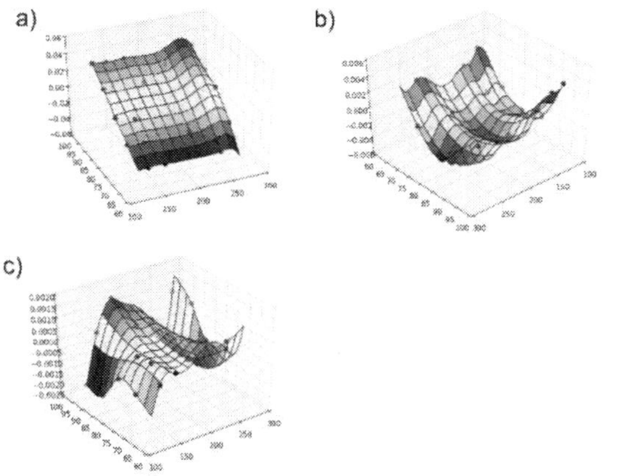

Figure 9. Comparison of RBF smoothness for shape function decomposition of a) first, b) second and c) third mode.

It can be noticed that for every type of results the smoothness of interpolation function decreases with increasing mode number. This seems correct and can be explained by the decreasing covariance of the data.

For results verification FEM calculation for the test data point (Fig. 2) has been performed. Results for the same parameters have been obtained from metamodel. Both result fields are compared. As the results before interpolation are preprocessed the error comes not only from data reduction but also from the method applied for sampling points location. In presented research the focus is only on the error coming from model reduction, assuming that different points numbers and distributions for better part description may be achieved. The part shape and results for stress in x direction are presented at figure 10.

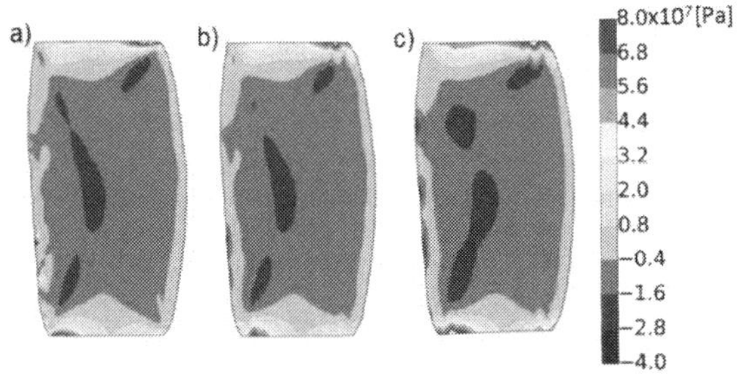

Figure 10. Comparison of stress in x direction for results obtained from a) numerical modeling, b) preprocessed data and c) metamodeling.

Visual inspection indicates high accuracy of metamodel results compared to standard FEM calculation. In order to verify results values at sampling point locations from numerical model has been obtained and compared to metamodel. A global error measure is proposed:

$$E_g = \sqrt[2]{\frac{\sum_{i=1}^{m}(x_i-\bar{x}_i)^2}{\sum_{i=1}^{m}(x_i-\mu_x)^2}} \quad (7)$$

where:

m – number of sampling points
x_i – result values obtained from numerical model
$\bar{x}_i$- result values obtained from metamodel
$\mu_x - x_i$ mean value

A summary of cumulative error different result types is presented in table 1:

Table 1. Relative cumulative error of preform metamodel

Result type	E_g[%]	Result type	E_g[%]
Stress x	36.22	Damage ind. 1	34.21
Stress y	89.15	Damage ind. 2	48.41
Stress z	37.47	Damage ind. 3	8.84
Stress xy	109.94	Temperature	32.41
Plastic strain	16.88		

Beside the cumulative error, a local relative error distribution is calculated:

$$e = \frac{x_i-\bar{x}_i}{\sqrt{\frac{1}{m}\sum_{i=1}^{m}(x_i-\bar{x}_i)^2}} \quad (8)$$

It has been decided that for an error distribution positive and negative values should be introduced for its better assessment. Error distribution for chosen results has been shown (Fig. 11).

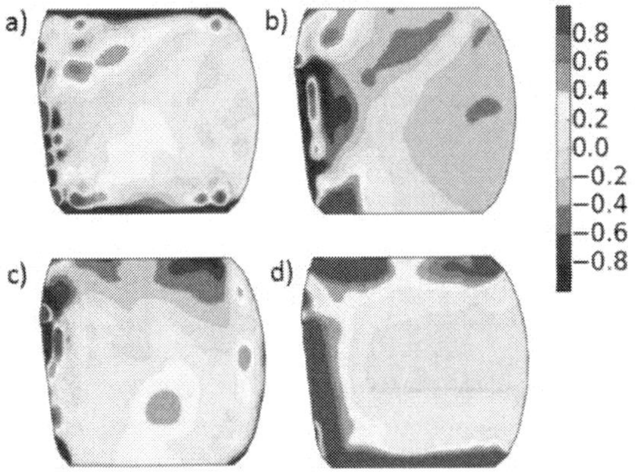

Figure 11. Error distribution for a) stress in x direction, b) plastic strain, c) damage indicator 1 and d) temperature.

Beside metamodeling verification an algorithm has been developed which based on metamodel results creates input file for LS-Dyna. It uses commercial mesh algorithm and includes all of obtained results into file. Comparison of one of the results obtained for test point from numerical modeling and metamodel in LS-Dyna environment is presented in Fig. 12.

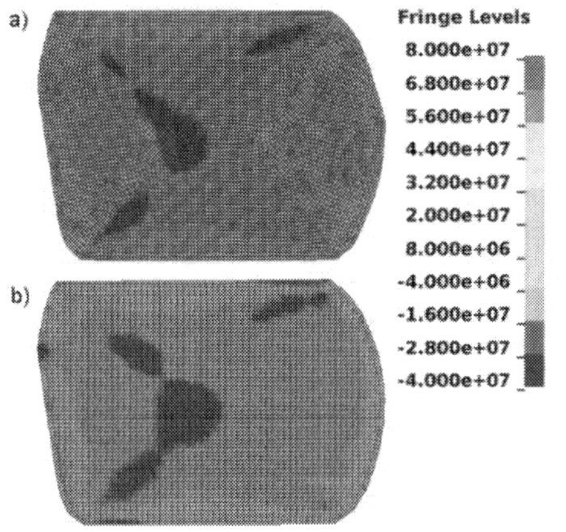

Figure 12. Comparison of results obtained for a) numerical modeling, b) metamodeling.

4.2. Final results metamodel

The metamodel for final results is created in 4 dimensional parameter space. In this case 17 data points have been used. Beside initial billet height and height after upsetting the ring rolling growth and rolling path are used. Due to thesmall number of data points their location has been carefully chosen to assure their optimal distribution.

The purpose of performing numerical calculations of ring rolling is to provide information about final distributed results according to different initial conditions and deformation histories. The final part shape is assumed to be constant. For creation of metamodel 5 of the most significant results has been identified – plastic strain, three damage indicators and the temperature field.

The results obtained from FEM ring rolling operation are in 3 dimensions. Despite spatial deformation history, overall part results are axisymmetrical. That's why it has been decided to perform metamodeling for a cross section instead of full three dimensional model.

Despite identical final tools positions for different types of parameters slight differences in the part shape may occur. It has been decided to compensate it by interpolation of the distributed results and describing the cross section shape by rectangle defined by maximum and minimum given cross section nodal points positions. Within this rectangle a uniform distribution of sampling points is generated (Fig. 13).

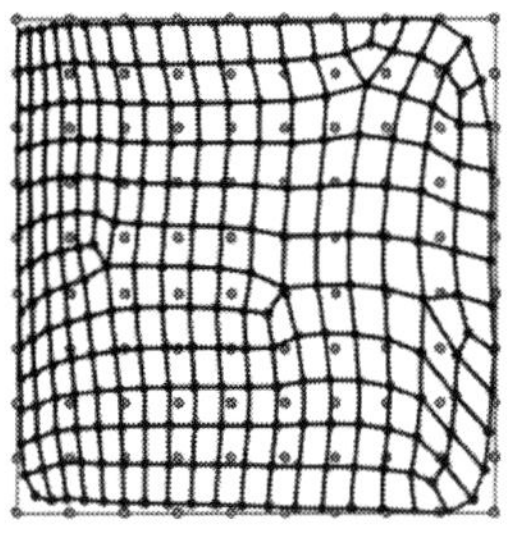

Figure 13. Exemplary cross section obtained from FEM model with sampling point distribution.

It can be noticed that some of the sampling points are located outside of the cross section geometry. It means that in their case sampled data will be obtained from extrapolated results and their values may under or overestimate real data. In case of metamodeling and from engineering points of view it does not matter. After ring rolling operation about 10% of the outside part cross section area is being removed by machining to assure good part quality. The points on the outside are used only to provide information about metamodeling and overall process quality but their values will not have influence on the final product.

Proper orthogonal decomposition of 17 data points for 5 distributed results has been performed. Based on their cumulative fluctuation energy (Fig. 14) number of reduced modes has been determined. Again the lower bound of 3 modes has been applied.

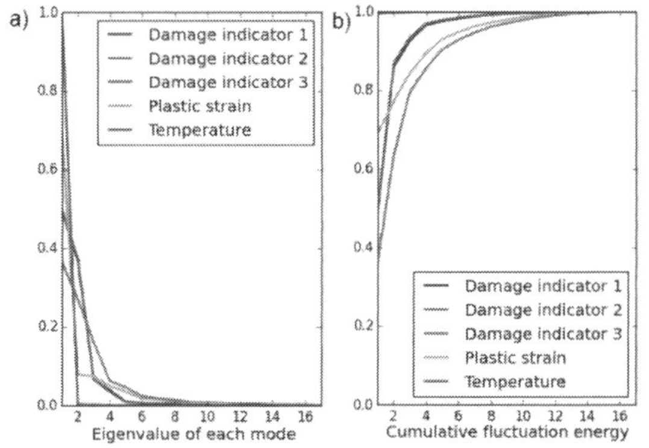

Figure 14. Second metamodel a) eigenvalues of each mode and b) cumulative eigenvalues energy.

For results verification numerical calculations for a new point has been performed. Point location has been carefully chosen to lay in a middle of the parameter range. Result of a numerical model has been compared with metamodel output for the same point. Field distribution for of the results has been plotted and compared.

Two of damage indicators - second and third and plastic strain has been chosen for more detailed check (Fig. 15). Both of them described damage as a contribution of hydrostatic tension within given plastic strain increment. The only difference between them is that the second assumes possibility of material healing.

For every decomposition global (7), and local (8) errors has been calculated. The summary of cumulative errors is presented in table 2.

Table 2. Errors of final part metamodel

Result type	E_g [%]
Plastic strain	34.88
Damage ind. 1	43.83
Damage ind. 2	43.21
Damage ind. 3	29.58
Temperature	105.83

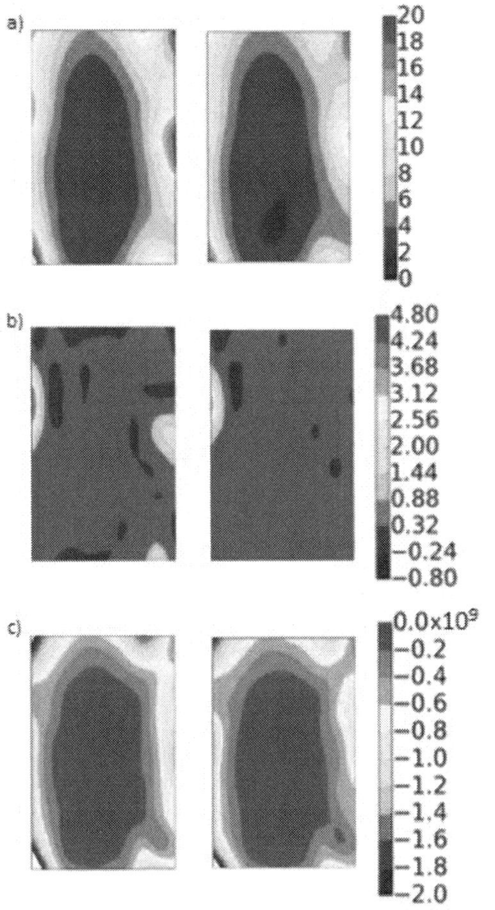

Figure 15. Result obtained from numerical model(left) and metamodel(right) for a) plastic strain, b) second damage indicator c) third damage indicator.

Moreover local error distribution for investigated results has been plotted and compared (Fig. 16).

5 Results discussion

Despite the fact that presented metamodels shows big similarity during their visual comparison their global error values are surprisingly big for both metamodels. It is probably caused by the sampling points inconsistency between different data points. Theirs localization does not fully correspond between different shapes and it results in preserving overall field trends but with slightly different local values.

Based on the result type different accuracy was achieved, what is indicated by different global error values. For preform metamodel the highest inaccuracy occurs when two stress fields are obtained. Stress in y and xy direction. The probable reason for this is a large variance of these stress fields compared to other type of results. The stress results values usually show higher sensitivity for process parameters and they are affected by recent strain elastic deformation while cumulative results

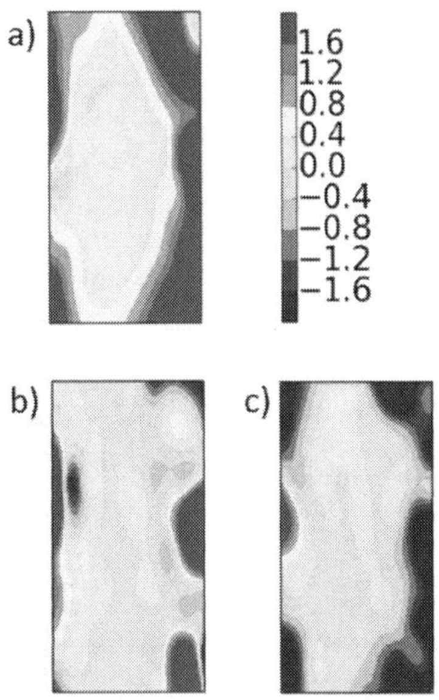

Figure 16. Local error distribution for a) plastic strain and b) second and c) third damage indicator

like plastic strain, damage indicators or temperature depend on whole deformation history. Moreover due to high field variance they may be more influenced by shape and sampling point location inconsistency. Visual comparison of fields distribution connected with local error distribution confirms it. Locally errors may be high but overall trends are preserved.

The results with smaller field variations like plastic strain, damage indicators and temperature shows higher accuracy. It may be noticeable that there is a difference in error level between first and second damage indicator compared to third. This is probably caused by their definition. The latter include material healing in them which mean that whole deformation history consist on it. The formers does not, which means that chosen deformation steps contributes to it which leads to the field with better outlined field. After inspection of local error distribution (Fig. 11c) it can be noted that biggest error occurs in upper part of element where shape inconsistency occurs which confirms this theory.

The enlargement trend of error values for the same results in the second metamodel may be noticed. It probably results from smaller number of data points compared to metamodel dimensions. Reduced number of data point per dimension may lead to poor interpolation function which will in consequence provide wrong amplitudes estimations. Visual comparison of different results again indicates big similarity of results despite the global errors. For plastic strain most of the field trends is preserved and overall results are good (Fig. 15a). Same tendency as for preform metamodel may be noticed in case of damage indicator. Material healing again leads to bigger nonlinearity of damage fields. Effect of it is that most of the result area oscillates around zero values which is harder to metamodel (Fig. 15b). The third

DOI: 10.1051/matecconf/20168013006

damage indicator only contributes to damage which in consequence gives easier to describe trends (Fig. 15c).

By inspecting the local error distributions it may be noticed that in both metamodels higher error deviation occurs in case of local maxima and minima. The dominant behavior is described well but the error increases when the field departs from overall trend. It may be effect of metamodeling inaccuracy. Preservation of bigger number of modes connected with increasing number of data points should prevent that.

6 Conclusion

Presented results show that application of proper orthogonal decomposition connected with radial basis function may serve as an efficient tool for metamodeling of some distributed nonlinear FEM results. The overall field trends have been preserved in every case but global error values are high. Probable reason for that is inconsistency of sampling point locations between different data points. It is expected to increase model accuracy with different, deformation dependent point localization methodology.

Surrogate models for two stages of ring rolling forming – preform and final results, has been built and investigated. For highly nonlinear results the values of calculated errors are big. The reason for higher error values besides sampling points inconsistency lays probably within relatively small number of data points that do not describes nonlinear behavior fully. High local fields nonlinearity that with model subspace reduction leads to higher error values. It can be especially noted in second metamodel. Additionally first of investigated operations – forming of preform, shows shape inconsistency that influences some of results(damage indicators with healing).

Despite relatively high error values, visual inspection of results shows high correlation between data. Overall trends has been preserved and the fields extremum values lay within reasonable ranges. Created metamodels after some corrections, may be used as fast and computationally cheap method of result assessment.

Acknowledgments

This research was carried out under the project number M41.1.11418 in the framework of the Research Program of the Materials innovation institute M2i (www.m2i.nl). The industrial partner, SKF Group is gratefully acknowledged for the contributions to this research.

References

1. K. Khaledi, T. Schanz, S. Miro, J. Theor. Appl. Mech. **44**, 45-54 (2014)
2. V. Buljak, G. Maier, Eng. Struct. **33**, 492-501 (2011)
3. G. Bolzon, V. Buljak, Comput. Mech. **48**, 675–687 (2011)
4. Z. Ostrowski, R.A. Bialecki, A.J. Kassab, Inverse. Probl. Sci. En. 16, 39-54 (2008)
5. A. Radermacher, S. Reese, Comput. Mech. 54, 677–687 (2014)
6. R. A. Białecki, A.J. Kassab, A. Fic, Int. J. Numer. Meth. Eng. 62, 774–797 (2005)
7. M.J. Mifsud, S.T. Shaw, D.G. MacManus, Int. J. Numer. Meth. Fluids 63, 468–494 (2010)
8. E. Iuliano, D. Quagliarella, Comput. Fluids 84, 327-350 (2013)
9. Y.C. Liang, H.P. Lee, S.P. Lim, W.Z. Lin, K.H. Lee C.G. Wu, J. Sound. Vib. 252, 527-544 (2002)
10. V. Buljak, Inverse Analyses with Model Reduction: Proper Orthogonal Decomposition in Structural Mechanics (Springer Verlag, Berlin, 2012)
11. C. Wang, H.J.M. Geijselaers, A.H. Van Den Boogaard, NUMIFORM 2013, 1532, 1014–1019 (2013)
12. C. Wang, H.J.M. Geijselaers, E. Omerspahic, V. Recina, A.H. Van Den Boogard, J. Mater. Process. Tech. 227, 268-280 (2016)

Implementing Digital Image Correlation for Determining the Tensile Characteristics of Post-Processed Thin Sheet Metal

Andrew D. Evans[1,a], Amit Das[1] and Nick Croft[1]
[1]College of Engineering, Bay Campus, Swansea University, Fabian Way, Swansea, United Kingdom, SA1 8EN

Abstract. Acquiring material properties for finite element analysis is a necessity for producing accurate response outputs. In the sheet metal forming industry, there are many challenges in gaining material property definition from post-processed and/or extremely thin sheet metal. The difficulty in obtaining this information is the founding reason for applying digital image correlation in the characterisation process of post-processed beverage cans. These beverage cans have been cupped, drawn, re-drawn, ironed and stoved, so these can be described as "pre-necked" beverage cans. Due to the residual stress/strain worked into the material, it means that the specimens curl, hence, there are difficulties in attaching contact extensometers. This experiment has shown that the preparation and testing of test specimens is critical in obtaining accurate and robust results. The robustness has been fulfilled by ensuring the data matches contact extensometers, and the number of test specimens used is sufficient. This experiment has proven that the application of digital image correlation has produced reliable and useful tensile properties for the finite element model, whilst demonstrating the the thermal effects imposed during the stoving process, which in-turn has raised questions over the impact of thermal effects in the processing of cans.

1 Introduction

To manufacture a beverage can, there is a number of metal processing techniques required to manipulate the sheet metal into the cylindrical cup shape. These forming techniques include cupping, drawing, re-drawing, ironing, and necking.

In the "die necking" process used in the beverage can industry, the body is diametrically reduced to accommodate the joining of the beverage can end - which is financially beneficial to the process. This process is completed over a number of stages to reduce the likely hood of wrinkling.

The general physics of the "Die necking" process comprises of compression in the axial, circumferential and radial directions. In essence, this processing is exploiting the natural compressive instability of sheet metal.

Wrinkling is a phenomenon that is caused by compressive instability. This instability causes the metal to fold and collapse in a manner similar to pleating. This failure can cause the can to be rejected from the line, cause blocking of the equipment, increased tool wear and financial implications on the process.

There is huge research investment into reducing wrinkling effects in the "die necking" process, utilising numerical tools such as finite element analysis (FEA).

FEA is a tool that requires the user to accurately input the physics, geometry and material characteristics of the problem to gain desired outputs, such as stress, strain, forces etc. The inputted values are important due to the fact that the reliability and accuracy of the output is heavily reliant on the input parameters i.e. garbage-in, garbage-out (GIGO).

Due to the number of stages used prior to the "die necking" phase, re-characterisation of the material properties is required to ensure accuracy and reliability of the model outputs. It is generally well known that FEA has a corresponding error, this will be magnified over a number of simulations. The magnitude of this error will naturally limit the number of simulations that can take place.

Re-characterising pre-processed material poses huge obstacles caused by the thin nature of the metal (0.155 mm), the curved nature of the cup and the residual stress in the metal.

2 Process definition

The process of necking is relatively straight forward to understand. It is important to fully comprehend the geometric setup and material characteristics of the can body - in addition to the dynamics of the process.

It is common to expect the necking process to produce approximately 3.5 cans every second. At this rate, the loads on the beverage can are highly significant to the structural integrity of the beverage can bodies and the robustness/reliability of the process.

[a] Corresponding author: 487056@swansea.ac.uk

2.1 Body wall geometry

As shown in Figure 1, the cross-section of the body wall will show that the wall has no uniform thickness distribution. The thickness distribution in the axial direction will show that the top wall thickness, mid wall thickness and bottom wall thickness all differ. This difference has been applied for a purpose.

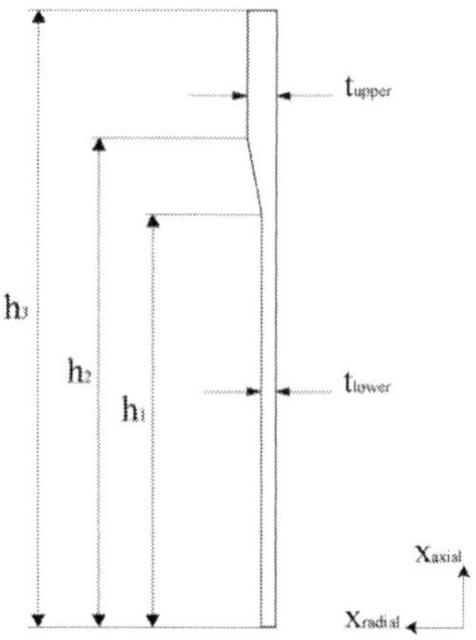

Figure 1. Demonstrating the thickness distribution of the beverage can in the axial direction.

If a cross-section of the body wall is taken on the radial – circumferential plane, there will be a non-uniform distribution in the thickness. This distribution is much harder to measure as the magnitude is approximately 10^{-3} mm; however, the existence of this thickness distribution can be confirmed using the poisons ratio relations when measuring the wall height of the can as shown in Figure 2. During the drawing/ironing processes, the observing party will notice a considerable change in the axial wall height. This correlates directly the the anisotropic nature of the aluminium material [1].

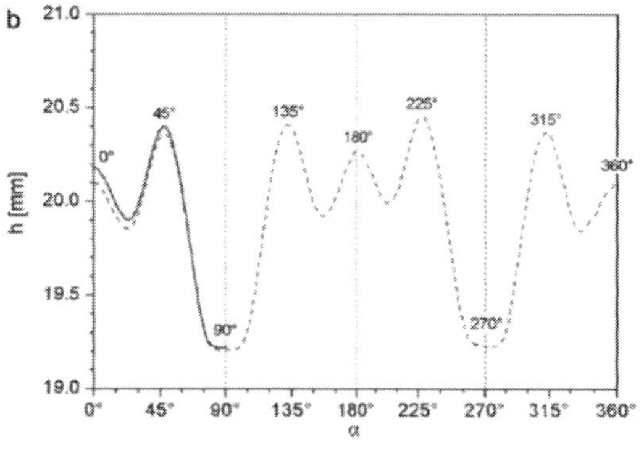

Figure 2. Showing the trend of the earing profile of the beverage height caused by anisotropy [1].

2.2 The "Die Necking" process

The die necking process usually comprises of multiple manufacturing stages. Each stage will gradually reduce the diameter of the body ready for the attachment of the lid. Generally, the process is identified as a "bottle neck" in the process, in-terms of speed, cost and reliability. For these reason, there is increased research investment in the optimisation of the process.

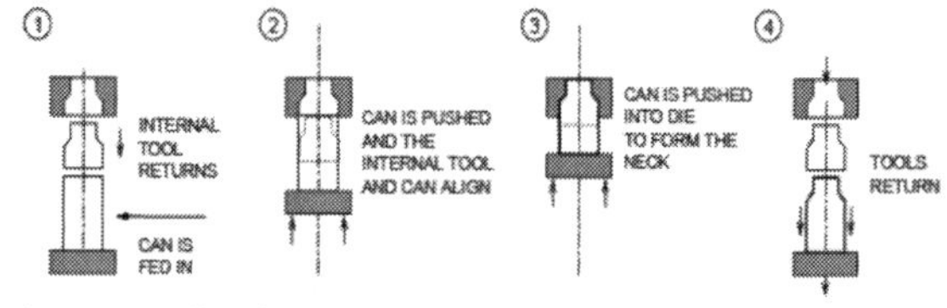

Figure 3. Showing the general forming operation of the beverage can in the "die necking" stage [2].

The tooling for the process requires a die, knockout and pusher/feeder. These tools work in tandem to conduct the forming operation required. The die is fully constrained, and the feeder and the knockout share the same constraints where UR1 = UR2 = UR3 = U1 = U2 = 0. U3 is free to move and represents the axial direction.

As shown in Figure 3, the process operates in four main stages, where:

- Stage 1 – in this part, the can body is displaced towards the die by the pusher/feeder working in the U3 (axial) direction.
- Stage 2 – in this part the body and the die make contact, forcing the can body to be relatively concentric. Once the seal is airtight between the top edge of the can body and the wall of the die, internal air pressure is increased.
- Stage 3 – the body wall makes contact with the die, and subsequently follows the profile of the tooling. During this forming stage, the can wall deflects off the knockout during contact.
- Stage 4 – The internal air pressure is used to force the can off the die, once forming is complete, and the can body is moved to the next staged.

2.3 Pre-processed material data

Characterising the material is vital for fully understanding and defining the mechanical limits for manufacture. With this in mind, a fully explored materials characterisation program must be introduced. The pre-processed material properties have been defined using standard contact tensile testing for the materials characterisation, and provides a benchmark for further investigations. This is shown in Figure 4, where the engineering stress and strain is plotted.

The pre-processed material tensile data shown in Figure 4, must be compared with the post-processed material data to fully understand the changes of the material between the pre-processed and pre-necking stages.

The pre-processed material data demonstrates that there is an anisotropic relationship present within the material.

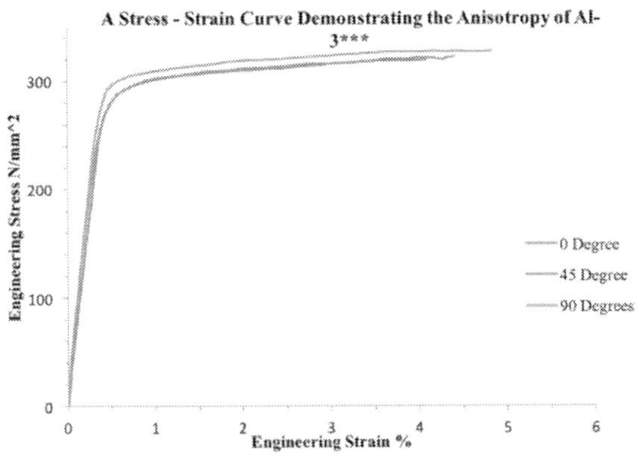

Figure 4. Showing the trend of the engineering stress-strain curve for pre-processed aluminium 3*** series.

3 Characterisation requirements

In order to produce a material input card that can provide an accurate representation of the state of the material, certain test parameters need to be chosen to define the experiment. These parameters/experiments are:

- Top wall characterisation must be used for the material input card. The top wall is where the material will be necked, therefore a greater load and deformation will occur in this region – making it critical for robustness and accuracy.
- Cauchy Stress/Strain data including the plastic and elastic regions of the curve, UTS (ultimate tensile strength), yield stress and maximum strain - This is the basic material data that is required for any material characterisation.
- Comparison of multiple specimens and mean data profile – a comparison of multiple specimens will produce a quantifiable representation of the extreme results. The mean data will be required for the majority of simulations.
- Comparison between pre-processed, and pre-necked material – it will be interesting to show the change in profile of the material once it has been processed.
- Comparison between stoved and non-stoved material to quantify the difference (if any), that may exist – stoved can body stock may have had some release in residual stress/strain, which will incur some changes to the material properties.
- Anisotropic material data – it is vital to build a data bank that will include anisotropic data. Anisotropy is an inherent property in sheet metals, so it is important to capture this behaviour.

4 Characterisation options

Characterising thin sheet metal is very problematic, especially when mechanic data such as stress-strain data is required. Largely, the main challenges faced with thin sheet metal is witnessed in the use of contact extensometers.

Before a relevant method for characterisation can be selected for the experiment, there must be a sufficient exploration into the viable options available for the experiment.

4.1 Extensometers

There are two main types of extensometers under consideration, which include contact, and non-contact extensometers.

4.1.1 Contact Extensometers

These are a more traditional type of extensometer which is designed for most routine applications where surface normal forces may not be an issue, high energy events are not used and damage from the extensometer being contact with the test specimen is not an issue. Contact extensometers are available in generally two variations, clip-on and sensor arms.

The clip-on variation is an individual unit that can be fixed onto the surface of the specimen. This is suitable for most applications; however, when applied to very thin samples below 1 mm in this instance, it is expected that the extensometer will produce a surface normal force, and may cause inaccuracies in the result. During initial testing, the clip-on extensometers proved to be inaccurate due to slippage of the contact feet. This caused the data to be unusable for any practical uses.

The sensor arm type of extensometer consists of two prongs that are placed at the top and bottom of the gauge of the specimen. The prongs are fixed to the edge of the sample rather than the surface to reduce any influence on the data. In some instances, slippage of the arms can be witnessed due to natural bends in the test specimens caused by residual stresses.

4.1.2 Non-Contact Extensometers

Non-contact extensometers are usually applied to scenarios where explosive failure or high strain-rates where specimens maybe influenced by surface forces or non-standard geometries may exist. In-terms of accuracy, non-contact extensometers can be better in high strain-rates. Non-contact extensometers are usually more expensive than contact extensometers.

In this project, due to availability, the non-contact extensometers under consideration are laser and video-based extensometers such as digital image correlation.

Laser extensometers can provide a good alternative to standard contact methods. They have the capability of measuring strain to a similar standard, but with the advantage of being able to operate in high temperatures. These operate by projecting a grid pattern onto the surface of the specimen and tracking the deformation. More modern systems can be integrated into the tensometer rigs and provide very accurate results in the micro and macro scale.

Video extensometers are available in different forms, but the type of most interest to the project is called Digital Image Correlation (DIC). DIC uses two cameras to record images at high frame rates of 5,400 at a 1-million-pixel density. The specimen must be coated in a matt based paint with maximum contrast. In other words, the specimen must have a white speckled pattern on a black background or vice versa. The DIC software is programmed to track unique patterns, so there is no need for a repetitive pattern being applied. DIC is a great alternative to laser extensometers, and can provide more data through its software. It is suitable for strains greater than 0.01%. DIC is expensive to buy with a cost of £150,000 per unit; however, the equipment is available for free rental through the EPSRC Loan pool [3].

4.2 Extensometers conclusion

Due to the restrictions imposed by specimen curvature caused by residual stress, and extremely thin aluminium specimens, it seems only reasonable to implement a non-contact extensometer. With this in mind, the use of a video based extensometer or laser may prove to be beneficial. Due to issues obtaining laser extensometers, the only option available is to use digital image correlation equipment provided by the EPSRC loan pool.

5 Experimental Apparatus, Setup and Sample Preparation

Using DIC requires an investment in time to setup the equipment correctly. The majority of the setup is simple to conduct, and is a case of connecting wires. There are some portions of the setup that are very important, which may produce errors if completed poorly. Usually the ideal setup cannot be arranged due to space and movement limitations, and this setup is no different. As shown in Figure 5, the setup of the cameras is positioned either side of the pillars.

Figure 5. showing the general setup of the cameras. :- note the placement of the pillar and cameras.

5.1 Apparatus

The DIC equipment is supplied through the EPSRC loan pool at no cost to the project. The equipment supplied is:

- 2 x Photron SA1 Camera Processors – This is the two high speed cameras shooting at 1MP and up to 5,400 frames per second (fps).
- Adaptor rings for lenses – These are required for different mounting rings for different brands of lenses.
- Cable remote shutter cable – This is used to start/stop the recording of the camera.
- Tokina 100mm f\2.8 macro lens – There is a choice of different lenses, but due to the size of the specimens, a larger focal length with a smaller focus distance is required.
- Heavy duty tripod and camera mounts – The tripod must be adequate to withstand the mass of two cameras. The cameras must be mounted at the same height, so the large camera mount is used.
- Continuous lighting with barn door attachments – The lighting is used in poorer lighting conditions where the light may not be stable or intense enough. Higher f-stops may be required, so more light will be required to illuminate the subject.
- Calibration grids – Calibration grids are required for the setup of the software.
- VIC-3D Software – This conducts the analysis, and is used to extract the strain data from the experiment.

5.2 Apparatus Setup

The initial setup of the DIC equipment is important to ensure that there is minimal opportunity for discrepancies to influence the result.

5.2.1 Camera Position

Firstly, the DIC equipment needs to be setup correctly. As shown in Figure 6, the angle between the two cameras and the specimen must be between 15° and 60° ensuring that the cameras are symmetric relative to the specimen.

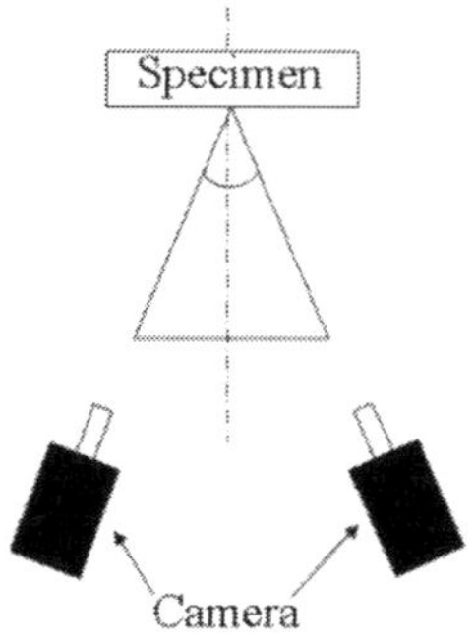

Figure 6. showing the general setup of the two cameras and the specimen, with particular focus of the angle and symmetry between the two cameras.

5.2.2 Lens Selection

In this case, there is a choice between a 100mm macro and a 24-85mm lens. The choice of lens is dependant on the size of the specimen, and the distance between the camera and the specimen. Due to the large distance required between the specimen and the camera, caused by the pillar of the tensometer obscuring the line of sight, the larger 100mm lens is required.

5.3 Camera settings

At this point, the calibration grid can be placed into the region where the tensile specimen will be loaded for the correct exposure. The exposure can be controlled with a number of different settings on the camera or by using controlled lighting. This is similar to standard photography, where there is usually a balance between a number of factors in-order to gain the correct exposure. In this particular scenario, the exposure was controlled by frame rates, aperture and continuous lighting. camera and the specimen.

To achieve the correct output data, the user must prioritise the exposure factors:

1. Frame-rates – this is usually measure in frames per second (fps), and is the primary factor for the data output. Due to the tensometer having an output of 50 data points per second, the cameras must be set to this number, or a divisional of the tensometer data output. A scenario where the fps must be set as a divisional value is if the aperture of the lighting reaches its upper limit. In the experiment, the frame-rate is set to 250 fps, with the intention of saving 1 in 5 frames as an output.
2. Aperture – The aperture of the camera is determined by the amount of light the lens can allow to enter the camera sensor. A lower f number (i.e. f2.8) will let in more light than a higher f number (i.e. f22). Also, the lower f number will have a much greater depth of field than that the higher value, which means that more of the background will be softer/out of focus in comparison to the high f number. In this experiment, the tensile specimen must be in focus, and due to a relatively close focus distance, the depth of field is exaggerated, so, f16 is used. This is also the highest number that is run on the experiment, even though the maximum is f22. This is due to lens imperfections such as dust or scratches, become visible over f16, and may skew the results.
3. Continuous lighting – fortunately, the testing laboratory chosen for the tensile testing has low ambient light, which means that the light output can be controlled with a much greater accuracy. The lighting is controlled using a swivel switch, but there is no measurement output of the light being emitted. Therefore, this is controlled by the user. The light output is strong, and even on a matt white surface, some reflections occur. Therefore, it is recommended to take the light output to the highest possible light output with no reflections visible on screen.

5.4 Sample preparation

The accuracy of the output data produced by the DIC system is partly reliant on the preparation of the tensile specimen. The specimen dimensions are shown in Figure 7.

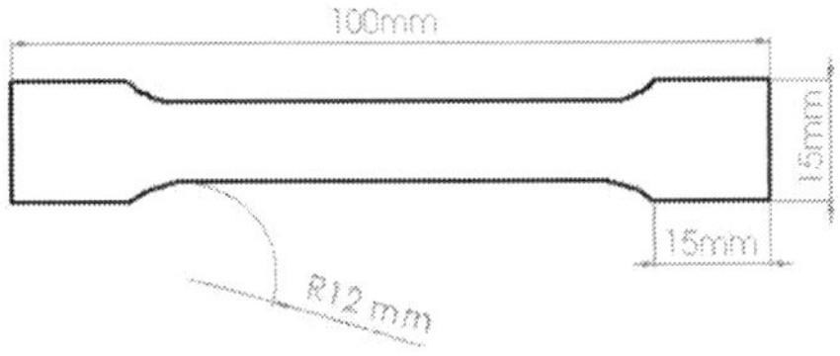

Figure 7. Showing the schematic of the tensile specimen used in the DIC testing.

The specimen chosen in Figure 7 was selected due to the size being perfectly suited to the dimensions of the can body. The can body was required to be altered slightly to accommodate any specimen. This alteration extends the top-wall thickness down the entire wall of the can. The top wall thickness is required for testing because it is the region that is necked during the process. The extension of the top-wall thickness is completed using protrusion dies, which will produce the same thickness, and the (approximate) same strain as the top-wall. This profile is demonstrated in Figure 8.

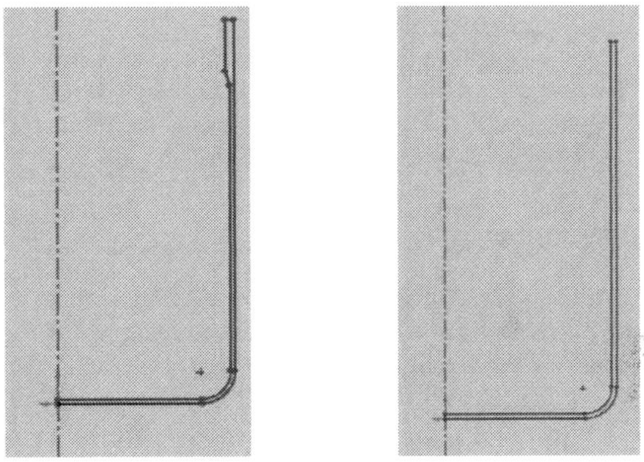

Figure 8. (Left) showing the Original can cross-sectional thickness profile and (Right) showing the can protrusion of the cross-sectional thickness.

Once the test specimens have been punched and the thickness measurements have been recorded, the paint must be applied. This step is crucial for producing accurate results, and the requirements include having a high contrast speckled pattern applied to the specimen [4].

For maximum contrast, it is recommended that the tester uses either white speckles on a black undercoat, or visa versa. The speckles must not be applied using a specific uniform pattern, they must be random. This random pattern ensures that the software can identify a unique area of the specimen and track it for the duration of the trial. The paint should be applied with reference to Figure 9 [4].

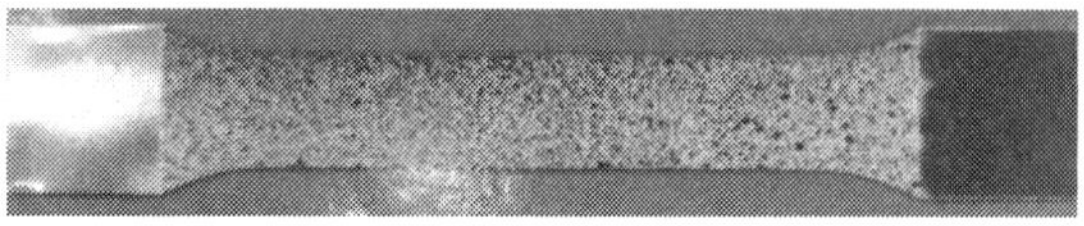

Figure 9. Illustrating the desired speckle pattern.

5.5 Experiment sample size

To ensure that the experiment is conducted to the required standard, a variety of different experiments will take place. The main focus of this program is to produce a material input card, with a secondary focus on the thermal effects that the stoving process may have on the metal. With this in mind, three tests were constructed:

1. Benchmarking – the benchmark test is used to compare standard pre-processed material using a standard arm extensometer for direct comparison with the digital image correlation strain data. This will ensure that the test data is reliable. In this test, 10 specimens will be used for each extensometer.
2. Heat treatment comparison – This will compare the data produced from stoved can bodies with non-stoved bodies. This will highlight whether there is any release in residual stress/strain.
3. Anisotropy – There will be a comparison between the axial and circumferential specimens to characterise the anisotropic nature of the material.
4. For the remaining experiments, 40 test specimens will be used for stoved and unstoved with a total of 80 for both scenarios. These will have an equal proportion of 20 axial and circumferential specimens for both.

6 Results

As discussed in the previous Section 5.3, there will be a total of 3 experiments to ensure that the test has been conducted thoroughly.

6.1 Benchmarking - DIC comparison with contact extensometer

As shown in Figure 10, both curves are closely matched in the plastic region of the graph with some difference in the elastic region. At the start of the DIC curve at approximately 10 N, there us a slight change in gradient before the curve becomes linear until Yield. There is very little difference between the gradient between the two curves otherwise.

The yield point for both methods is at 260 MPa and the Ultimate Tensile Strength (UTS) is measure at 216 MPa for both. The maximum Engineering Strain is measured at 6.2% and 6.03% for DIC and Contact Extensometer respectively.

There are important observations to understand when using DIC, which may have a noticeable impact on the results. In these circumstances, this data can be disregarded. Such discrepancies include the lighting conditions, paint quality and camera knocks.

Overall, the reliability of DIC when compared to standard contact extensometers has proven to shown negligible discrepancies when observing the results shown in Figure 10.

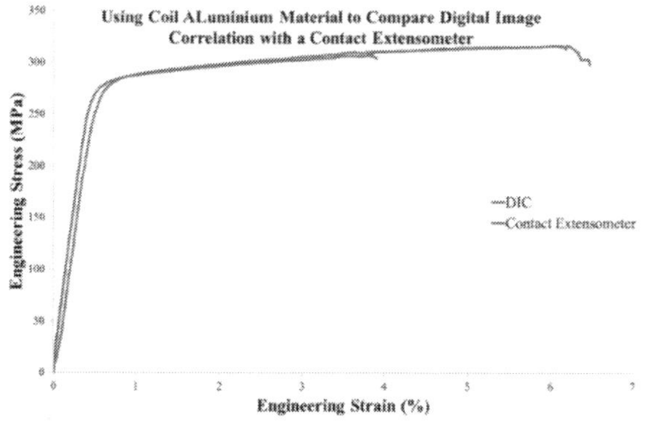

Figure 10. Comparing DIC with the Zwick contact extensometer.

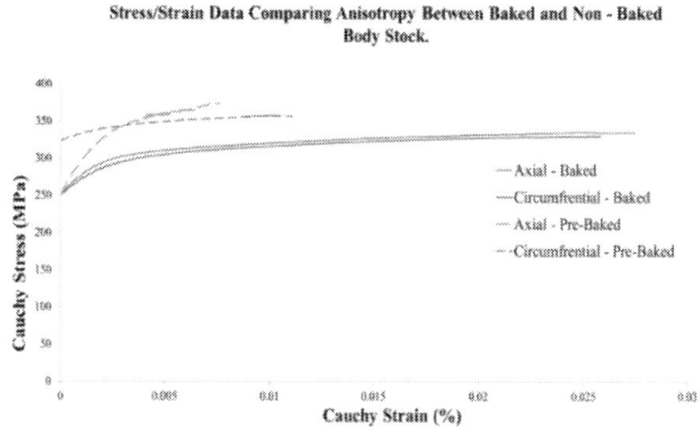

Figure 11. Showing the input Cauchy stress/strain data past the yield point for pre-processed (dashed lines) and post-processes (solid lines) material data for the circumferential and axial orientations. This is the format that Abaqus requires for its material input card definition.

6.2 Anisotropy – Axial vs Circumferential

As shown in Figure 11, the pre-baked material data (dashed lines) shows a distinct anisotropic trend after the yield point. The material yield occurs at 320 MPa for the circumferential and 250 MPa for the axial orientations.

The pre-baked material data for the axial specimens continues on a steeper gradient – when compared to the circumferential specimen - before failing at 0.065 strain and 375 MPa stress.

The pre-baked material data for the circumferential specimens has a shallower gradient, when compared to the axial specimen. The failure of the circumferential specimen occurs at 0.012 strain at 350 MPa.

The post-baked material data for the circumferential and axial specimens follows similar trends, where the yield point is witnessed at 250 MPa, and the maximum stress is 320 MPa. The maximum plastic Cauchy strain is shown to be 0.025 for the circumferential and 0.027 for the axial specimens.

6.3 Comparing heat treatment

As shown in Figure 11, there is a significant difference between the heat treated and non-heat treated samples. There is a minimum gain in strain of 0.015 and a maximum of 0.0175 when comparing axial samples and circumferential. There is also a significant decrease in stress of 50 MPa for the circumferential samples and 65 MPa difference for the axial samples.

7 Conclusions and perspectives

The digital image correlation experiment has provided new means for testing materials where test specimens have a thin profile.

There are difficulties associated with the use of digital image correlation that need to be noted, which include, paint quality, lighting and compiling the strain data with the stress data. In other words, there is no real integrated approach for data aggregation.

The reliability and comparability of the DIC data with a proven contact extensometer has proven to be consistent. Over a test batch of 10 specimens per extensometer, there has been very good consistency, which reduces/eliminates elements of doubt.

The anisotropic tendencies found in the sheet metal post-baking and pre-baking, has shown that there is a significant change in the characteristics of the material to significant proportions. This change needs to be verified by observing the microstructure of the material; however, it is unlikely any phase transformations have occurred, and it is more likely that the changes have been merely releases of residuals.

Overall, the use of DIC has given an incite into the manufacturing process that has never been looked into in the past; which has subsequently opened new questions to be answered in-terms of the application of DIC in other processes and research topics into other stages. DIC has been important in providing an FEA model up-to-date and reliable data to conduct simulations, which will inevitably reduce the tendencies for errors to occur. This paper has highlighted the importance of materials characterisation for FEA and its unexpected influences on other projects.

This project has covered many aspects of the materials characterisation of thin sheet metals, and the challenges that are part of it; However, there hare some questions that have become visible due to the results of the experiment.

An investigation into the cause of the changes seen before and after stoving. This can be completed by viewing the microstructure, and searching for phase transformations.

It would be useful to conduct experiments using high strain rates. DIC will be able to be implemented for this once a suitable high strain rate tensometer can be located.

References

1. O. Engler, N. Mertens, P. van Dam, J. of Mat. Proc. Tech, **211** (2011), pp. 1278 – 1284.
2. J.A. Jordan-Cordera, *8th International LS-DYNA Users Conference*, pp. 37-46 (2004).
3. S. Amini, R. Kumar, Mat. Sci. & Eng, *Structural Materials: Properties, Microstructures, and Processing*, **594** (2014), pp. 394-403
4. V.P. Rajan, M.N. Rossol, F.W. Zok, Exp. Mech., **52** (2011), pp. 1355-1373

Application of Multivariate Adaptive Regression Splines to Sheet Metal Bending Process for Springback Compensation

Rasim Aşkın Dilan[1,a], Tuna Balkan[2] and Bülent E. Platin[2]

[1]*Servo and Stabilization Technologies Design Department, ASELSAN A.S., 06370, Ankara, Turkey*
[2]*Mechanical Engineering Dept., Middle East Technical University, 06530, Ankara, Turkey*

Abstract. An intelligent regression technique is applied for sheet metal bending processes to improve bending performance. This study is a part of another extensive study, automated sheet bending assistance for press brakes. Data related to material properties of sheet metal is collected in an online manner and fed to an intelligent system for determining the most accurate punch displacement without any offline iteration or calibration. The overall system aims to reduce the production time while increasing the performance of press brakes.

1 Introduction

Mechanical systems such as press brakes and punches are widely used production machines in the industry. Especially, press brakes performing sheet metal forming are one of the major production machines providing a variety of basic parts to several industries. Sheet metal forming is a very effective and highly developed process for high output production situations. Though, the process itself has some impediments and these impediments would increase the production time, which in return would increase the cost of the process in today's competitive world of production.

Press brakes, like the several other mechanical systems, are composed of mechanical systems, hydraulic systems and electrical control systems. Furthermore, these subsystems are also composed of primitive components. For proper functioning, every component should be configured and the harmony of these individual components in overall system should be maintained as well. One of the crucial components of these systems is the material that is being processed on them. For press brakes, processed material is the sheet metal being bent. Sheet metal processed on press brakes actually completes the overall mechanical model of the press brake and affects the performance and functioning of the press brake in every manner. Correct evaluation of material behavior is the utmost important aspect to consider for obtaining suitable metal-forming analysis, simulations and results [1]. Moreover, behavior of the material affects the quality and pace of the production drastically. It is due to the fact that, during the production process some iterations or calibrations would be required to be performed for different types of sheet metals (even the different batches of the same material) in order to get the correct bending result and this would increase the production time. In other words, physical properties of even the different batches of the same material would deviate from the nominal values, and for every different batch of material, a production calibration becomes necessary. Therefore, the cost of the production would increase as well as the time spent on the production.

To sum up, the physical properties of the raw material used in press brakes affects the overall system significantly and the automatisation of the whole bending process with compensating the effects of different material properties of the raw material would be the key purpose for increasing the efficiency and speed of bending process.

2 Background

Press brake system can be described as a servo, electro, hydro, mechanical system. The desired target is the final bending angle of a sheet metal processed by the press brake. Therefore, the input of the system is the desired angle. This desired bending reference is taken by the press brake system and a displacement command is generated by the system and this command is executed on sheet metal. The result is the bended material. However, in this system there is no feedback that would be taken and fed to the system. The overall described system is depicted in Figure 1.

Figure 1. Overview of Sheet Metal Bending in Press Brake

[a] Corresponding author: radilan@aselsan.com.tr

The adverse effects that would deteriorate the quality and accuracy of bending should be predicted beforehand and a feedforward mechanism or module (see Figure 2) should be added to the system. The adverse effects would stem from the change of material properties, elastic and plastic behaviors of the materials as well as the tooling and the body. Moreover, the changes in environmental conditions would also affect the performance of the press brake. In other words, quality analytical or black box models should be generated for each adverse effect or for their overall effect and the bending reference should be regulated according to these models before feeding to the press brake system.

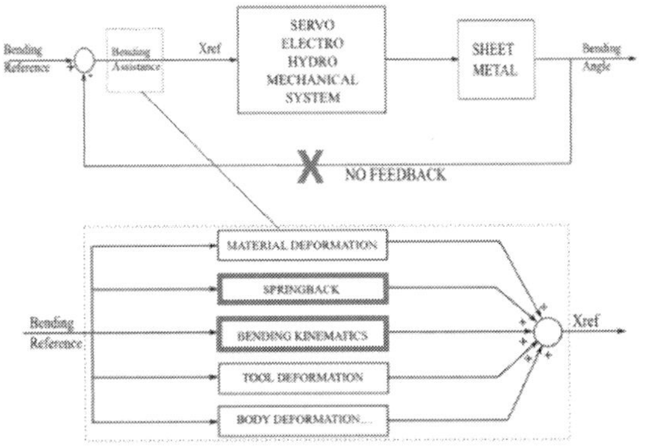

Figure 2. Sheet Metal Bending Assistance

2.1 Adverse Effects on Accurate Bending

There are several factors affecting the bending process. As a result of these effects, the shape of bended sheet metal deviates from the desired final shape. Understanding all of the factors truly has a great importance for this study, since the aim of this study involves reducing these effects so far as possible to enhance the bending performance. Predicting or evaluating the adverse effects, which deteriorates the quality and accuracy of bending, can create an opportunity for us to add a feedforward mechanism (see Figure 2) or module to the system enhancing the performance. The adverse effects can be collected under three titles, namely material properties, springback effect and other adverse effects interfering an accurate bending.

Material properties are intensely interfering the quality and accuracy of the bending. The input parameters for bending has to be based on values provided by the steel manufacturers, or values from standards or material certificates. However, these values are never exact, but reflect ranges that describe between which boundaries a certain parameter can vary. This undoubtedly presents a large uncertainty in the input of the bending model, which also has a profound influence on both parameters like the calculated punch displacement and the calculated sheet length correction [6].

One of the physical properties affecting bending performance is modulus of elasticity. In the initial stage of the bending process, the forces imposed by the punch are too small to cause plastic deformation in the sheet. Consequently, no strain results in the sheet if the punch force is removed. The extent to which the sheet deforms under influence of the small punch force is related to the modulus of elasticity of the sheet material [6]. Obviously, the modulus of elasticity influences the required punch displacement, as the ratio between elastic and plastic deformation is an important indication for this. Evidently this has large influence on the springback of a sheet.
Another important physical property is yield stress. The transition point between elastic and plastic deformation is related to the yield stress of the material. With increasing deformation, certain regions in the sheet located at the surface of the sheet are subject to stresses that exceed the yield stress. In these regions, the plastic deformation occurs first, and the applicable material model changes from elastic to plastic behavior. The value of yield stress is crucial for bending process.

In a piece of material, a number of material properties can alter in value, depending on the direction in which this property is determined. This phenomenon, known as anisotropy, can only be manifested by more-dimensional properties. A one-dimensional property (e.g. the density of a material), being a scalar value, merely assigns a value to a certain point in the material [8]. This does not imply the property to be a constant within the specimen, but only that the property is direction-independent. During the sheet forming operation, the metal crystals are enforced a certain prestraining. Due to this, the material shows an orientation of the crystals called rolling texture. This crystal orientation causes the varying results in the tensile test. Therefore, it is clear that the effects of anisotropy are important for sheet metal bending.

The Ludwik–Nadai relation is used to describe the plastic material behavior and it is also a crucial parameter for bending process. The advantage of this model is the non-linear description of work hardening. Plastic material behavior is important for bending.

Bauschinger effect is a physical property of material that also affects bending process. Not only the bending direction in a sheet is important in tensile experiments, the direction of the force that is applied to the test specimen plays an important role as well. This effect signifies that the behavior of a test specimen under pressure differs from a specimen in tension. The Bauschinger effect is an anisotropic material property and it is affecting the quality of bending [11].

Friction is also an important aspect of the bending process, which cannot be related to sheer material behavior is the friction. During the process, friction forces arise at two locations: the contact between sheet and punch and between sheet and die. From a kinematic point of view, the first friction force is very difficult to prove. The friction between the sheet and the die is often modeled according to the friction force by Coulomb friction and it is affecting the bending process.

Another vital phenomenon affecting the quality of a bending process is springback effect. For predicting the final geometry of produced sheet metal part as well as the determination of value for springback is an important feature. A simple description of springback is that after the punch removal, generated dimensional change is because of material elastic recovery [15].

The amount of springback is related to the material thickness, material type, bending angle, bending direction, etc. as presented in lots of the studies [15]-[21]. Moreover there are several quality springback models for compensation the springback. Some of these are empirical models and some of them are parametric models.

In this study, the compensation for the springback effect would be studied.

2.2 Geometric Interpretation of Three Point Bending

Bending kinematics of air bending (three point bending / V-bending) will be discussed with the assumption of no slippage of the sheet on the press brake V-die. Bending kinematics is a geometric calculation that provides information for the preliminary value of punch displacement regarding the material and punch are rigid and it is independent of the type and physical properties of the material used in the bending process.

First of all, in Figure 3, a sheet metal while being bended by press brake punch is depicted with important features presented. Nomenclature for this figure is listed in Table 1.

Table 1. Nomenclature for Bending Kinematics.

Symbol	Description
V	width of V-opening of the die
R_p	tip radius of the punch
R_d	radius of the die sides
t	thickness of the sheet metal
β	angle that supplements the bending angle
z	needed punch penetration to bend the sheet metal to (180-ß) degrees

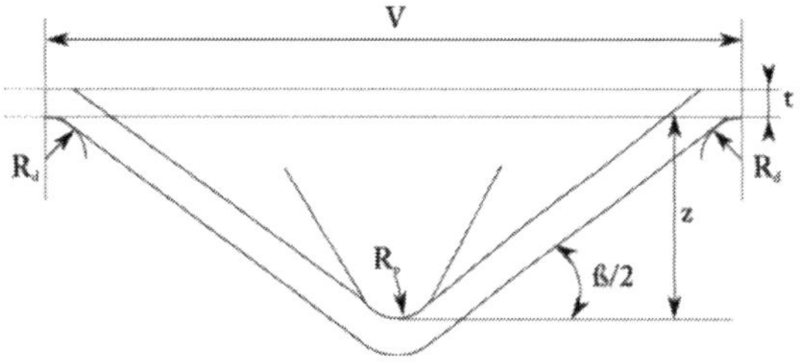

Figure 3. Bending Operation on a Punch-Die Configuration.

In several studies, bending kinematics calculation of V-bending is presented [22]-[25]. In literature, there are several versions for punch displacement equation that generates the same results. The geometrical result is presented in equations (1)-(3). The formulation takes die and punch types, bending angle and sheet metal thickness as inputs and returns the necessary punch displacement without taking material type of the sheet metal into consideration.

$$z = \frac{V}{2}\tan\left(\frac{\beta}{2}\right) + \left[1 - \cos\left(\frac{\beta}{2}\right) - \tan\left(\frac{\beta}{2}\right)\sin\left(\frac{\beta}{2}\right)\right](R_d + R_p + t) \quad (1)$$

$$z = \frac{V}{2}\tan\left(\frac{\beta}{2}\right) + \left[\frac{\cos\left(\frac{\beta}{2}\right) - \cos^2\left(\frac{\beta}{2}\right) - \sin^2\left(\frac{\beta}{2}\right)}{\cos\left(\frac{\beta}{2}\right)}\right](R_d + R_p + t) \quad (2)$$

$$z = \frac{V}{2}\tan\left(\frac{\beta}{2}\right) + \left[1 - \sec\left(\frac{\beta}{2}\right)\right](R_d + R_p + t) \quad (3)$$

This formulation is a starting point for searching the correct and accurate punch penetration value. One should take this result as a starting point or initial value for searching the correct punch displacement. Therefore any automation system exploring the true punch displacement for a given bending angle should take this formulation as an input. However, it should be noted that, this formulation does not involve any information on material properties of bended sheet metal.

3 Data Collection and Evaluation

Improvement of bending process on press brake requires a thorough observation of the process on the system, extensive collection of data during the process and good evaluation of the collected data. During the studies on bending process, a hydraulic press brake was used and for several different configurations, bending data had been collected for good understanding of the bending practice. In this section, data collected on the press brake is presented as well as the evaluation and preprocessing of data before feeding the intelligent system.

3.1 Data Collected on Press Brake

During the studies on press brake system, for several configurations, bending data had been collected. This data is categorized in Table 2. Data is collected for 90 different bending angles at 4 different die configurations and for 3 different material thicknesses. This adds up to 90 × 4 × 3 = 1080 data points. For these 1080 different data points, punch displacement after material touch with 1 µm resolution, cylinder chamber pressures of the press brake and the final bent angle of the materials with 0.1 degrees resolution are collected.

Table 2. Data Collected on Press Brake

Bending Angle	Die Config. (Die Width/Die Radius [mm/mm]	Material Thickness [mm]	Collected Data
1 to 90 degrees (with 1 degree res.)	16/2	1	Punch Displacement [mm]
	22/2.5	1.5	
	35/3	3	Cylinder Pressures [bar]
	50/4	-	Bent Angles [deg]

 DOI: 10.1051/matecconf/20168014002

Punch Displacements and the bent angles are used for the construction of a truth table for the intelligent system. The use of this truth table is described in the following sections. Cylinder pressure of the press brake are used for calculating the force transmitted to the material during bending operation. It is easier and cheaper to place a pressure transducer to cylinders when compared to the use of a sensor or strain gauge to be placed within die or press brake body. Therefore the usability of the noisy data of cylinder pressures is tested in this study. A sample data on force transmitted to material during bending phase is depicted on Figure 4. As it can be seen from the figure, force versus displacement data is similar to strain-stress curve which contains lots of information about physical properties of materials. This data is also used for feeding the intelligent system as well as the other available data.

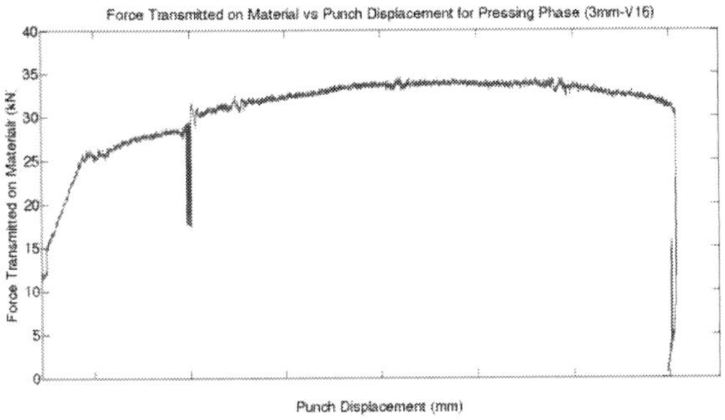

Figure 4. A Sample Force Transmitted on Material vs Punch Displacement Data extracted from Cylinder Pressure Data Collected on Press Brake

3.2 Evaluation of Collected Data

Using the bent angle and punch displacement data collected, a truth table for punch displacement for the correct bending angle had been constructed for 1080 data points. Throughout this paper, mentioned bent angle versus punch displacement table is used as the truth table.

For understanding the behavior of the material during bending, the geometrical interpretation of the punch displacement mentioned in section 2.2 is compared with this truth table. This comparison gave an insight for the rest of the study presented in this paper. Geometrical relation mentioned before lacks material information. However, on the other side, truth table constructed using the data collected has the material information that the former one lacks. The difference between these two relations constructs the punch displacement error results in the deviation between the bending angle and bent angle, in other words, the springback effect. The error between these two relations are given in Figures 5 to 7 for different material thicknesses, for different die configurations and depicted as a function of bending angle.

As the given figures are examined thoroughly, it gives us several good features to be used. These features are given below.

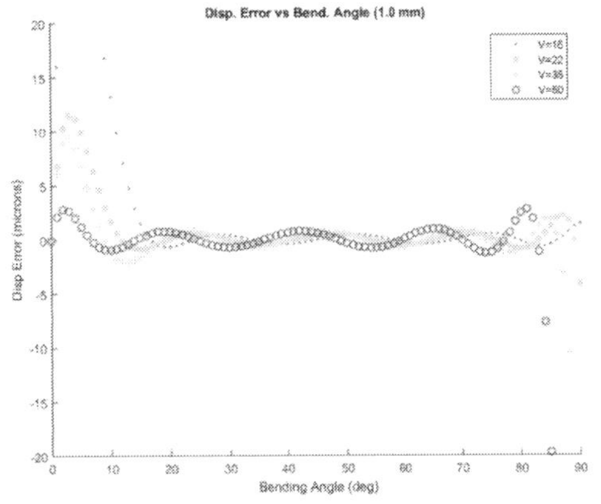

Figure 5. Punch displacement error for material with 1 mm thickness

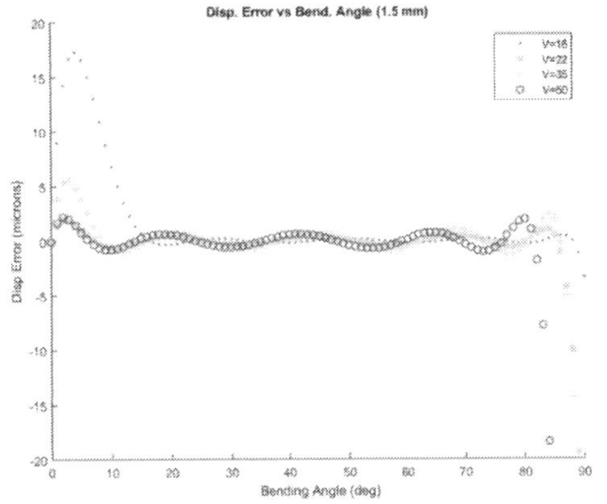

Figure 6. Punch displacement error for material with 1.5 mm thickness

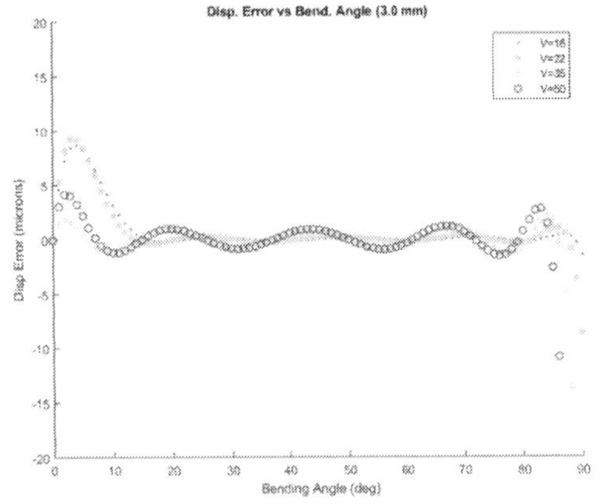

Figure 7. Punch displacement error for material with 3 mm thickness

Displacement error between the geometric relation and the truth table constructed

- increases as the material gets thinner
- increases as the die width gets smaller
- is large for small bending angles as well as large bending angles.

Moreover, as the figures are examined it can be seen that the displacement error resembles a *half sinc* function for both ends of the figures. Using all of this information, it is conceivable that a good function fitting (avoiding overfitting) would be a solution possibility for compensation of springback effect This brings the study to the research of intelligent fitting techniques, and the technique selected is presented at Chapter 4.

3.3 Preprocessing of Collected Data for Feeding the Intelligent System

The data collected on the press brake is preprocessed and constructed with a specific structure before being fed to the intelligent system. Preprocessed input data is composed of 7 different variables constructed from 9 different variables using some normalizations. Preprocessed data is listed in Table 3.

Table 3. Preprocessing of Collected Data

Collected Data	Unit	Normalization →	Processed Input	Unit
θ	deg	θ/90(pi/180)	θn	-
t	mm	t/t_{max}	tn	-
TPD	mm	TPD/TPD_{max}	TPDn	-
V	mm	$V/R/D_{max}$	D	-
R	mm			
F_1	N	$(F_2-F_1)/F_1$	MP_1	-
F_2	N	$(F_3-F_2)/F_2$	MP_2	-
F_3	N	$(F_4-F_3)/F_3$	MP_3[a]	-
F_4	N			

The unprocessed variables are bending angle, material thickness, theoretical punch displacement value, die width, die radius, and four different force values collected on specific four different punch displacement points. All of the inputs are converted into unitless forms in a suitable meaningful way in order to improve convergence ability of the data. The bending angle data is converted into radians. Die widths and die radii are specific to each other for a specific type of die, therefore their ratio will be fed to the system. Material thicknesses normalized to 5mm. For this study, materials with thicker than 5 mm is left beyond the scope. The theoretical punch displacement value is normalized by dividing the value maximum possible theoretical punch displacement value.

For adapting material information force reactions on material is retrieved for 4 points during pressing. These points [P1, P2, P3, P4] are decided after the examination of force versus punch displacement data collected on press brake. Most of the bending results show that the transition between elastic to plastic region of material is reached around 10 degrees of bending. In other words, if data until the end of the elastic region would be covered, this data would represent the elastic features of the material. For that reason, for 4 points until this transition point force data will be collected and will be fed to system. Last of these points are selected as the theoretical punch displacement value for 10 degrees of bending and this punch displacement point is marked as P4, the other 3 points are calculated such that the elastic region is divided into 3 parts equally. Force transmitted to material at these points are taken as input to the intelligent system using the normalization given in the table.

As mentioned before, it all of the preprocessed data has no units and normalized. This feature will allow the model to converge easier to a specific solution since the order of magnitude between variables are removed.

4 Proposed Method

Several neural network and regression techniques are examined for the compensation of the springback effect using priori information. The comparison of these techniques are subject of another paper to be presented. Among several techniques, one of the most successful results acquired by the use of Multivariate Adaptive Regression Splines technique. This technique is suitable for the presented problem, since, as mentioned before, the error of the output of the geometric relation and the truth values are in shape of a common function and easily can be fitted. In this chapter, a brief introduction to the technique and the application of the technique to bending problem is presented.

4.1 Introduction of Multivariate Adaptive Regression Splines

Multivariate Adaptive Regression Splines is a method for nonparametric approximating and inferring general functions of a high-dimensional argument noisy data. The basic assumption of the method is that the function to be estimated is locally smooth [28-31]. Multivariate Adaptive Regression Splines is a faster method in terms of training times compared to that of Feedforward Neural Networks.

Multivariate Adaptive Regression Splines algorithm uses two sided truncated functions of predictors (x) as basis functions and the models built out of these basis functions. Representation of the basis function is given in equation (4). The generalized expression of the method using the mentioned basis function is given in equation (5).

[a] MP : Material Property

$$(x-t)_+ = \begin{cases} x-t & x>t \\ 0 & otherwise \end{cases} \quad (4)$$

$$y = f(x) = \beta_0 + \sum_{m=1}^{M} \beta_m H_{km}(x_{v(k,m)}) \quad (5)$$

In Equation (5) it can be seen that the summation is over the M terms (number of basis functions) in the model, and β_o, β_m and t (knot values in equation (4)) are all parameters of the model. These parameters are estimated using the training data. Function H is represented as in Equation (6).

$$H_{km}(x_{v(k,m)}) = \prod_{k=1}^{K} h_{km} \quad (6)$$

In Equation (6), $x_v(k,m)$ is the predictor in the kth of the mth product. If K is selected as 1, the model becomes additive, if K is selected as 2, the model becomes pairwise interative.

There are two main phases during the application of the methods. These are forward stepwise procedure and backward pruning procedure. During forward stepwise procedure, several basis functions are added to the model according to a pre-determined number. The forward procedure is a brute one and at the end of the procedure, the model would overfit to data. To avoid this, backward pruning procedure is executed.

A backward pruning procedure is applied in which the model is pruned by removing those basis functions that are associated with the smallest increase in the goodness-of-fit. For measuring the goodness of fit, relation given in the equation (7), namely Generalized Cross Validation, is used. The given relation measures the goodness of fit as well as the complexity of the model.

$$GCV = \frac{\sum_{i=1}^{N}(y_i - f(x_i))}{\left(1-\frac{C}{N}\right)^2}, \quad C = 1 + cd \quad (7)$$

N is the number of cases in the data set (number of training samples) and d is the number of the number of independent basis functions, namely effective degrees of freedom. c is the penalty for adding a basis function. Experiments have shown that the best value for C can be found somewhere in the range 2 < d < 3.

4.2 Application of Multivariate Adaptive Regression Splines to Bending

Using the above basic equations, and "*earth-py*" [32] python library Multivariate Adaptive Regression Splines Method is applied to bending problem.

Regression is performed on data input presented in Table 3 and data output corresponding truth value constructed. Total number of data points available is 1080 as mentioned before. Several training data set sizes and validation data set sizes are used for the application of the method. The number of data points in the different training data sets can be seen in Table 4. Validation data set sizes for these training sets are 80 for the first set and 100 for the rest of the sets.

To validate the performance of the model, for each training session, training set and validation set are selected randomly among 1080 data points for 1000 times, model is trained 1000 times for these selected data set couple and validated. The results are presented in the next chapter.

5 Results

Multivariate Adaptive Regression Splines Techniques has been applied for 12 different training sessions. 12 different training and validation data set sizes are tried during these sessions. For each session, training and data evaluation sets are selected randomly for 1000 times and training is repeated for 1000 times for each session. A sample training validation result for 200 training data points and 100 validation data points is presented in Figure 8. In this figure, 1000 different colors represent 1000 different training model and as it can be seen from the figure, the maximum displacement error of the trained model is around 70 nanometers.

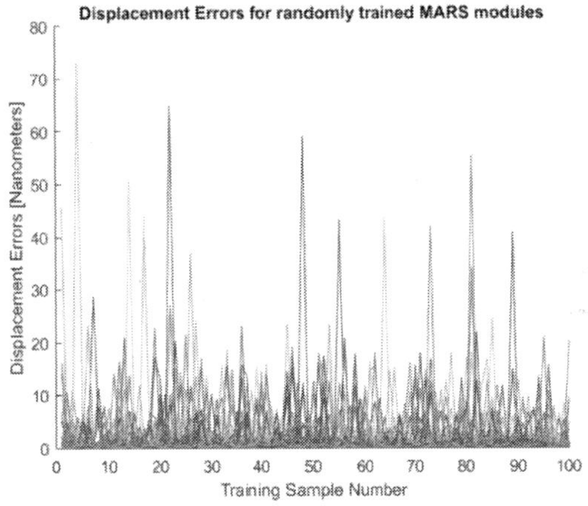

Figure 8. Displacement Errors for Randomly Trained MARS Modules (200 training samples)

For the comparison of the training performances, for each training sessions, mean square error (MSE) and maximum error of the trained models have been calculated. Among 1000 trained models, the maximum value of these two parameters are selected as the performance index of that training session.

As the performance index table is examined, it can be seen that the mean squared error of the models trained decreased gradually as the number of training data size increases. It should also be noted that, maximum error decreases for training sets with size greater than 500 data points; however it seems stable for training sets with size smaller than 500 data points.

It should also be noted that, the punch displacement errors depicted in results table in units of nanometers. Most of the press brakes can provide position control accuracy around 1 micrometer minimum, therefore the output of all of the trained models are acceptable.

It can be concluded that the selected identifying data inputs are selected so that it well covers the data space and using Multivariate Adaptive Regression Splines technique the bending procedure can be defined with success.

Table 4. Mean Squared Error and Maximum Error values for different Training Sample (TS) Numbers.

# of TS	MSE (nm^2)	Max Error (nm)	# of TS	MSE (nm^2)	Max Error (nm)
1000	10.58	27.11	180	112.70	73.70
800	26.85	43.22	160	93.28	74.01
500	60.99	73.04	140	94.90	73.84
300	77.35	72.76	120	87.77	73.88
280	99.49	73.43	100	91.59	74.87
260	96.89	73.14	50	118.43	74.04
240	85.57	73.23	30	164.10	74.63
220	78.01	72.99	20	183.04	74.23
200	85.13	73.35	10	251.55	74.28

6 Conclusion

In this study, using the data collected on a real press brake and by the help of an intelligent regression technique, namely Multivariate Adaptive Regression Splines, a bending assistance model for one type of material of different thicknesses has been generated. Model is generated using different sizes of data and for each case, it has presented satisfactory results.

7 Future Work

In this study, improvement of bending procedure for compensation of springback is achieved for one type of material with different thicknesses using an intelligent regression technique. In the future, the model developed in this study is planned to be extended into a model working for more than one type of material.

Moreover, several studies has also been carried out for compensation of adverse effects like springback in sheet metal bending. In another study, an extrapolating model is being developed, in which the material information is extracted at the early phase of bending and estimates the required punch displacement without a priori information.

References

1. L. Antonelli, P. Salvini, F. Vivio, and V. Vullo, *Journal of Materials Processing Technology*, **183-1**, 127 (2007).
2. H. Wagener, *Journal of Materials Processing Technology*, **72-3**, 342 (1997)
3. Y. Huang, D. Leu, *International Journal of Mechanical Sciences*, **40-7**, 631 (1998)
4. K. Sato, T. Inazumi, A. Yoshitake, S. Liu, *International Journal of Impact Engineering*, **54**, 1 (2013)
5. G. D'Urso, G. Pellegrini, G. Maccarini, *Key Engineering Materials*, **344**, 277 (2007)
6. P. Koc, B. Štok, *Computational Materials Science*, **31**, 155 (2004)
7. B.K. Chun, J.T. Jinn, J.K. Lee, *International Journal of Plasticity*, **18**, 571 (2002)
8. Zhao, K. M.,Lee, J. K., *Numerical Methods in Engineering*, **20**, 105 (2004)
9. P. Becker, H.J. Jeon, C.C. Chang, A.N. Bramley, *CIRP Annals - Manufacturing Technology*, **52**, 209 (2003)
10. A.H. Streppel, D. Lutters, E. Brinke, H.H. Pijlman, H.J.J. Kals, *Journal of Materials Processing Technology*, **115 (1)**, 76 (2001)
11. A. Anderla, D. Stefanović, M. Mirković, *Machine Design*, **3(3)**, 229 (2011)
12. H. Baseri, B. Bakhshi-Jooybari, B. Rahmani, *Expert Systems with Applications*, **38(7)**, 8894 (2011)
13. M. Tanbun, "Springback Prediction of Mild Steel on V-Bending", *PhD Thesis, Faculty of Mechanical Engineering, Universiti Malaysia Pahang*, (2000)
14. J. Liao, C. Zhou, F. Ruan, Y. Zhu, *Advanced Materials Research*, **97-101**, 130 (2010)
15. S.K. Panthi, N. Ramakrishnan, M. Ahmed, S.S. Singh, M.D. Goel, *Materials & Design*, **31(2)**, 657 (2010)
16. K. Yilamu, R. Hino, H. Hamasaki, F. Yoshida, *Journal of Materials Processing Technology*, **210(2)**, 272 (2010)
17. L.C. Zhang, Z. Lin, *Journal of Materials Processing Technology*, **63(1-3)** 49 (1997)
18. M Samuel, *Journal of Materials Processing Technology*, **122(1)**, 94 (2002)
19. M.V. Inamdar, P.P. Date, S.V. Sabnis, *Journal of Materials Processing Technology*, **123(3)** 459 (2002)
20. L. Courvoisier, M. Martiny, G. Ferron, *Journal of Materials Processing Technology*, **133(3)**, 359 (2003)
21. U.P Singh, S.K Maiti, P.P Date, K Narasimhan, *Journal of Materials Processing Technology*, **145(3)**, 269 (2004)
22. L.J. de Vin, A.H. Streppel, U.P. Singh, H.J.J. Kals, *Journal of Materials Processing Technology*,**57(1-2)**, 48 (1996)
23. K. A. Stelson, D. C. Gossard, *Journal of Engineering for Industry*, **104(4)**, 389 (1982)
24. N. Asnafi, *Materials and Design*, **21(3)**, 217 (2000)
25. L. J. De Vin, *Journal of Materials Processing Technology*, **117(1)**, 244 (2001)
26. R.J Mentink, D Lutters, A.H Streppel, H.J.J Kals, *Journal of Materials Processing Technology*, **141(1)**, 143 (2003)
27. Z. Boger, H. Guterman, *IEEE Systems, Man, and Cybernetics Conference*, Orlando, FL, USA (1997)
28. J. Friedman, "Fast MARS", *Dept of Statistics, Standford Univ, Technical Report*, No.110, (1993)
29. J. Friedman, "Estimating Functions of Mixed Ordinal and Categorical Variables Using Adaptive Splines",

Dept of Statistics, Standford Univ, Technical Report, No.108 (1991)
30. L Francis, "Martian Chronicles: Is MARS better than Neural Networks?", *Dept of Statistics, Standford Univ, Technical Report*, No.108 (1991)
31. D. Steinberg, *PC AI*, **1**, 38 (2001)
32. D. Steinberg, "An Introduction to MARS", *Salford Systems* (1999)
33. https://github.com/jcrudy/py-earth, last visited on 14.02.2016.

MATEC Web of Conferences **80**, 14003 (2016) DOI: 10.1051/matecconf/20168014003
NUMIFORM 2016

Finite element analysis of non-isothermal warm deep drawing of dual phase steel

T. Pepelnjak[1a], B. Kaftanoglu[2]

[1]*Forming Laboratory, Faculty of Mechanical Engineering, University of Ljubljana, Ljubljana, Slovenia*
[2]*Dept. of Manufacturing Engineering and Metal Forming Center of Excellence, ATILIM University, Incek, Ankara, Turkey*

Abstract. Improving the formability of the material is an important issue in the deep drawing process. Heating the material above its recrystallization temperature drastically increases formability but in the case of dual phase (DP) steels it results in the loss of their mechanical properties. To improve the drawing ratio, only the heating of the flange region in the warm temperature range up to 300°C was studied on DP600 sheet steel by numerical simulation. Thermo-elastic-plastic FEM analysis of deep drawing at several drawing ratios was performed and compared with experimental results.

1 Introduction

Modern industrial practice tends towards the minimisation of material use with concurrent improvement of product properties what is in particular emphasised in the automotive industry. In several cases the most important properties are strength, crashworthiness, dynamic properties, bending resistance. These properties are to achieve with proper part design and/or forming technology. However, in the last decade new material development has drastically speeded-up the increase of sheet metal strength and formability of produced parts [1, 2]. This rapid development of new forming technologies, design concepts and new materials with high hardness or/and good formability was stipulated mainly in the automotive sector where added value of sheet metal parts allow high degree of implemented innovations. Demands on improved strength-to-weight ratio led to the development of various new steel types with improved strengths. Among those steels nowadays mostly so called Advanced High Strength steels (AHSS) are used [2]. They can be classified according to their crystallographic structure into dual phase (DP), complex phase (CP), transformation induced plasticity (TRIP), twinning induced plasticity (TWIP) and martensitic (MART) steels as shown in Figure 1. Additional to here mentioned steels already being used in the industrial practice the third generation AHSS steels [3] with improved strength and ductility are developed which should even exceed the tensile strength of martensitic steels. Despite the intensive development of various steel grades and their processing techniques the predictions of use of various AHSS materials in the automotive industry favorizes the DP steels as the most used materials until year 2020 [4].

The modern production of stamped parts is composed from several forming processes ranging from cutting, bending, deep drawing etc. where various loading conditions of the material appear simultaneously. In the presented paper, the deep drawing process with a tension-compression stress state will be analysed more in detail.

Deep drawing process is one of the most widely used sheet metal forming operations enabling the producers manufacturing of complex hollow shapes. The process is mostly performed at room temperature. However, in some cases the material formability needs to be increased to achieve the demanded part geometry. The increase of forming temperature above the recrystallization temperature offers one solution to raise the formability limits of the material. This concept is mainly used when materials with extremely low ductility are applied as it is the case of implementation of martensitic steels – Figure 1.

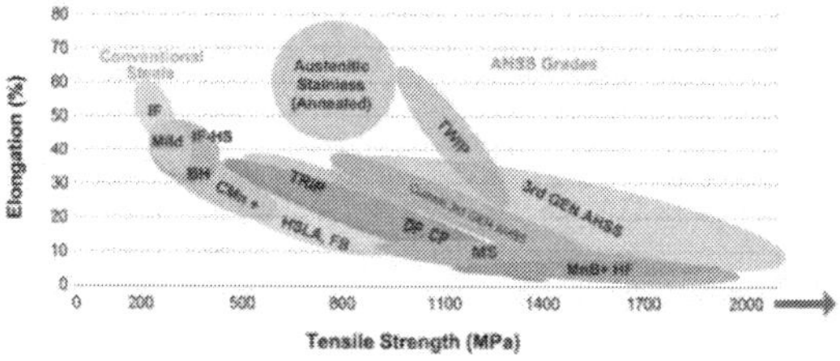

Figure 1. Global formability diagram with position of advanced high strength steels [1].

These steels are formed mainly above the recrystallization temperature with so-called hot stamping

[a] Corresponding author: tomaz.pepelnjak@fs.uni-lj.si

[4]. On the other side classic dual phase steels could not be formed in hot condition since their characteristic properties results from the combination of the hard martensitic microstructure in the soft ferritic base [2] and this structure is destroyed with material heated above the recrystallization temperature.

The process of deep drawing in one or several operations can be analytically described for simple round geometries. However, the forming of geometrically complex shapes can be predicted in advance only by the numerical simulations [5, 6] using the Finite Element Method (FEM) as a proven tool for evaluations of forming processes [7]. The FEM used for analyses of various metal forming applications [6] enables to cope complex material behaviour at different stress-strain states, material formability, crystallographic behaviour etc. up to the influence of strain rate and temperature of the formed part when forming at elevated temperatures is analysed [7, 8]. The coupling of different multi-physics phenomena like in the case of hot stamping simulations represents one of the most demanding topics of digital evaluation of forming processes [9]. Here also the appropriate techniques and hardware resources are necessary.

To raise the low formability limits of AHSS steels several concepts and technological improvements are presented in the literature. Kim et. al. worked on lubricants used in stamping process of AHSS [10]. In their work, they analysed the deep drawing and ironing process. Further on, the influence of the tool was analysed by the Mackensen et. al. at cutting of AHSS steels [11]. Similarly, Kim [12] and Cora [13] worked on the effects of different tool materials and various coatings as well as lubricants used in forming of AHSS materials.

An alternative approach for the increase of the formability of AHSS steels and other materials with low formability in cold state is to raise the temperature in forming process [1]. The hot stamping process is the most common solution for the increase of the formability of martensitic steels which express the lowest attainable elongation at room temperature [14, 15]. As already mentioned is this concept unfortunately not applicable to other AHSS steels where the material properties are achieved by combination of two or more crystallographic phases in the same material and heating above the recrystallization temperature would diminish their characteristics. Therefore, if the DP, CP or TRIP materials should be heated only to the level of warm working is a reasonable solution to improve the material formability, decreasing the level of flow curve and also lowering the forming forces. At the same time when using only the temperatures of warm forming, the material structure remains intact. Furthermore, the partial heating of blank only results in targeted local changing of material properties and softening the material in the area of large deformations. Local heating can be achieved by the laser beam heating [16] or with inductive preferential heating [17]. The later was selected as a heating solution in the research of deep drawing of DP600 steel at elevated temperatures. This is in particular important at deep drawing process where with the local heating of the flange area the formability is increased while at the same time unheated area below the punch could remain colder and material there has higher strength.

Application of flange heating is used in forming of several materials (which are brittle in cold forming conditions) using deep drawing as presented by Lehtinen et. al.[16].

In the presented study, the main focus is the FEM simulation of flange heating in deep drawing process in comparison with the experimental results presented by Kayhan [17].

Since one of the indicators to determine the material's formability of sheet metals is the deep drawing test performed as a round cup-drawing with a cylindrical punch - Figure 2. The diameter of the workpiece for the formed specimen is increased up to the fracture limit of drawn cup on its bottom. The obtained limit is called limit drawing ratio which is calculated as the ratio:

$$\beta_{max} = \frac{D_0}{D_p} \tag{1}$$

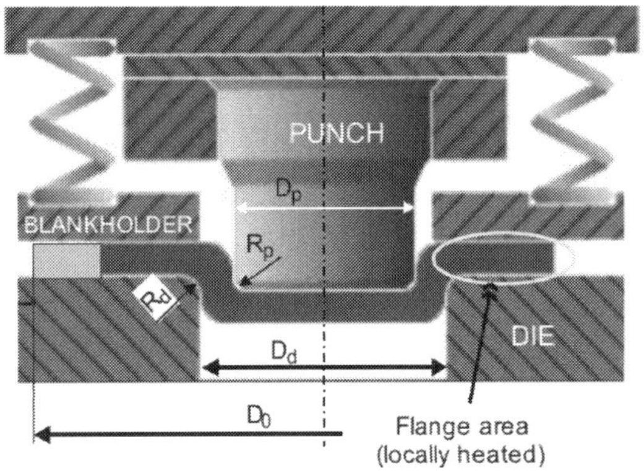

Figure 2. Schematic deep drawing of a round cup according to [18].

2 Material

The material considered in the experimental research and numerical simulations is sheet metal dual phase steel DP600 with thickness of 1.6 mm mostly used for the automotive applications at room temperature. The material is positioned in the softer third of the DP steels - Figure 1. With its formability it delivers reasonable formability for forming analyses in cold and express necessary improvements of formability during the warm forming with locally heated flange area.

The forming properties of the used steel were analysed at room temperature and at elevated temperatures up to 300°C. It can be expected that heating above the level of 300°C can start to change the crystallographic structure of the DP steel. Due to this reason, the 300°C temperature was selected as the upper working level of warm preferential heating during the deep drawing process. It needs to be considered that the limit of 300°C is suitable for inductive heating and plastic deformation simultaneously in the warm range.

In order to obtain the flow curves for the DP 600 steel at room temperature and at elevated temperatures the uniaxial tensile tests were used. The tensile test at room temperature was performed on Zwick Z300 universal testing machine having 300 kN nominal force.

The flow curves at elevated temperatures were performed parallel at Gleeble 1500 and Gleeble 3800 machines – Figure 3. The Gleeble 3800 machine has better performances while at Gleeble 1500 a self-designed extensometer placed direct on the specimen could be used to measure the elongation of the specimens very precisely. The flow curves were measured at room temperature, at 150°C, 200°C, 250°C and 300°C respectively. Additionally, the flow curves were measured at various strain rates ranging between 0.01 and 2.5 with following values: $\dot{\varepsilon}$ = 0.01, 0.1, 1.0 and 2.5 respectively.

Figure 3. Gleeble 3800 testing machine.

Different temperature and strain rate combinations were verified in order to assure sufficient input data for the numerical simulations as well as to analyse the influence of the forming conditions on the level of the flow curve. It was found out that the flow curve at 150°C and $\dot{\varepsilon}$ = 0.01 differs only slightly from the flow curve at room temperature. With the temperature increase at small strain rates, the true stress of the flow curve decrease but the drop of the flow curve level is more evident at higher strain rates where it reaches down to 80% of the level at room temperature. At strain rates of $\dot{\varepsilon}$ = 2.5, also the attainable maximal elongation drops with the temperature – Figure 4 – in comparison to the room temperature flow curve performed at $\dot{\varepsilon}$ = 0.01. It was found out that at higher strain rates, the flow curves in the range between 200°C and 300°C are very similar. At lower strain rates this phenomena was not so drastically emphasised. Furthermore, the data from the tensile test are only the base data for the material characterisation and further analysed need to be performed for comprehensive material behaviour at elevated temperatures. From Figure 4, it is also evident that at elevated temperatures with high strain rates the attainable elongation is also decreased in comparison to the flow curve at room temperature obtained with significant lower strain rates. Last but not least, the anisotropic material behaviour was not yet implemented at this research level since the rotational-symmetric case was analysed by FE method.

3 FE model

The flow curves obtained experimentally were implemented into the FEM program ABAQUS in order to simulate the preferential heating of the flange during the deep drawing of a round cup.

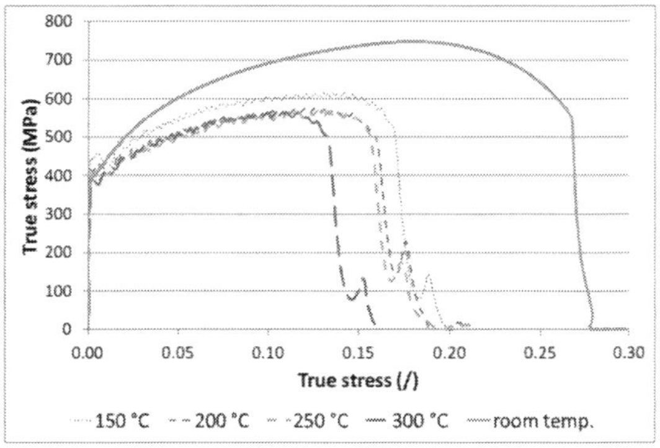

Figure 4. Tensile tests at room temperature ($\dot{\varepsilon}$ = 0.01) and at elevated temperatures ($\dot{\varepsilon}$ = 2.5).

Numerical simulations were comparatively evaluated with the experimental work in order to obtain a reliable tool for further predicting of locally heated forming behaviour of DP steels and other interesting materials at elevated temperatures formed below the temperature of recrystallization.

The FE data of round cup were set-up as rotational-symmetric simulation since the anisotropic material behaviour was neglected at this research level. The FE model consists of:

- Geometrical data
- Mechanical properties
- Temperature and strain rate dependant formability properties (flow curves)
- Thermal properties (conductivity, heat transfer, etc.)
- Contact conditions between tool and the workpiece
- Boundary conditions (initial temperatures, temperatures during the forming, blankholder force, ram speeds during deep drawing).

The finite element (FE) model consists of heated die, heated blankholder, local heated blank flange, cooled punch, and cooled central area of the blank – Figure 5 (top). In order to shorten the calculation times, the thickness of the die and the blankholder were decreased in comparison to the real tool geometry. Experimentally the temperature was measured at the die radii of 56 mm and 21 mm respectively as presented on Figure 5 (bottom). Between the heated die and blankholder the temperatures of 260°C was measured while at the drawing radius R_d only 180°C was measured. For the simulation, the heated zones between above determined radii were divided into ten sections on the die and the blankholder with sequential increasing of the

DOI: 10.1051/matecconf/20168014003

temperature. The punch was considered as semi-cold with the contact temperature of 80°C during the heating of the flange area. The bottom of the inner part of the specimen is cooled with the water spray – Figure 5 - during the drawing process. Therefore, in this area the surface temperature was assumed to be 40°C during the deep drawing.

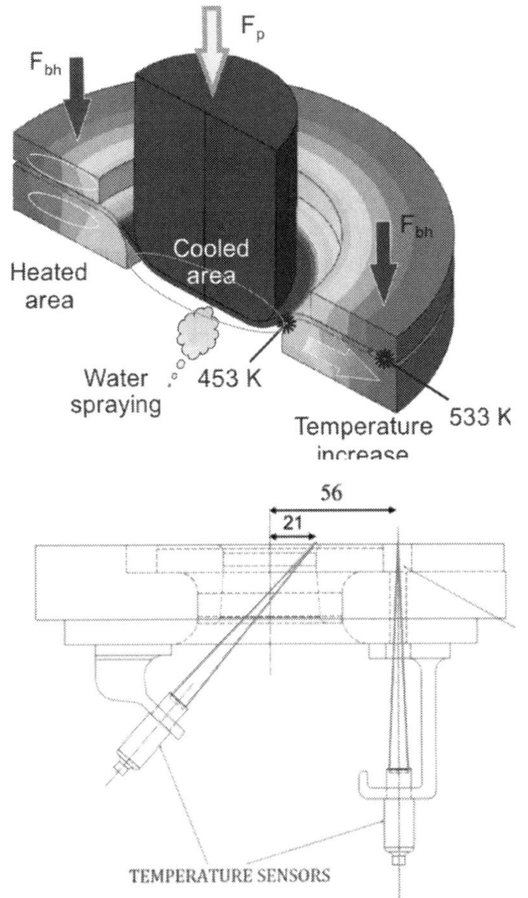

Figure 5. Concept of the sequential heating of the die and the blankholder (top) and the location of temperature sensors at experimental set-up (bottom).

The FEM model consists of fully thermo-elasto-plastic simulation where also the tool parts are considered as thermo-elastic. The sheet thickness of 1.6 mm was divided into 5 layers of rotational-symmetric elements. The model of the tool parts and the blank is meshed with CAX4RT elements. The mesh of the FEM model of a blank having Ø96 mm in diameter and the drawn cup are presented on Figure 6 prior (top) and after the deep drawing (bottom).

Numerical simulation was performed in three steps: heating of the entire system by the temperature increase at the blankholder and the die, deep drawing and springback analysis. Despite the springback in the flange is minimised through the heating it still needs to be analysed in order to verify its influence on residual stresses.

Numerical simulation of the deep drawing process with localised heating shows that the sequential heating through the entire die and blankholder thickness delivers good heat distribution prior to the drawing process observed by temperature distribution of individual nodes as shown on Figure 7. All temperatures in the FE model are considered and visualised in degrees of Kelvin.

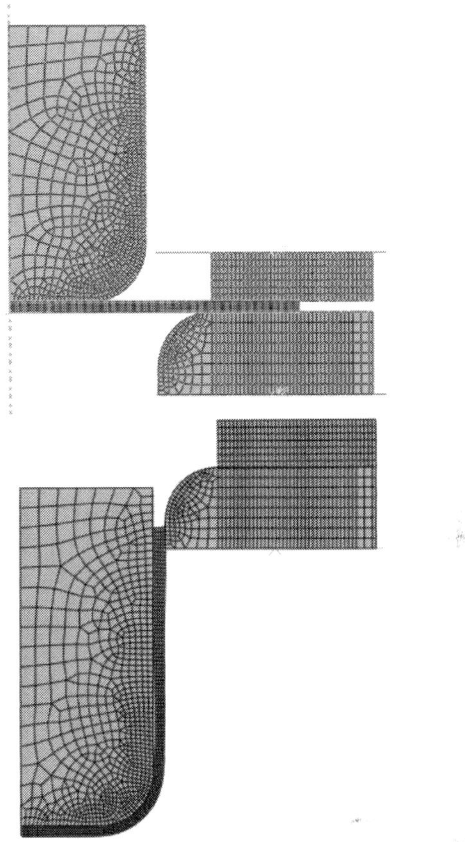

Figure 6. FEM mesh of the deep drawn cup: initial state (top) and after the drawing process (bottom).

The explicit simulation of the deep drawing process was performed without any time acceleration or mass scaling using the true ram speed calculated according to the data of the applied mechanical (eccentric) press.

In order to determine the maximal attainable blank diameter which could be successfully drawn with the heating of the flange area following blank diameters were analysed: 88 mm, 90 mm, 92 mm, 94 mm, 96 mm, 98 mm, 100 mm, 101 mm, and 102 mm respectively. The heated flange area of the specimens was lubricated with graphite and/or politetrafluoretilen (PTFE) foil. The friction coefficient for this lubricants was experimentally determined and varies between μ=0.035 and μ=0,055. For the PTFE foil also friction coefficients at elevated temperatures were determined. The values are ranging between μ=0.02 and μ=0.035 and are vary with the temperature. The lowest friction coefficient express PTFE at 200°C while with further increase of the temperature the friction coefficient increase as well. This can be explained with extensive softening of PTFE foil having its melting point at 326.8°C. When using this lubricant for warm deep drawing the melting point of PTFE represent additional temperature limitation beside the crystallographic structure of DP600 steel.

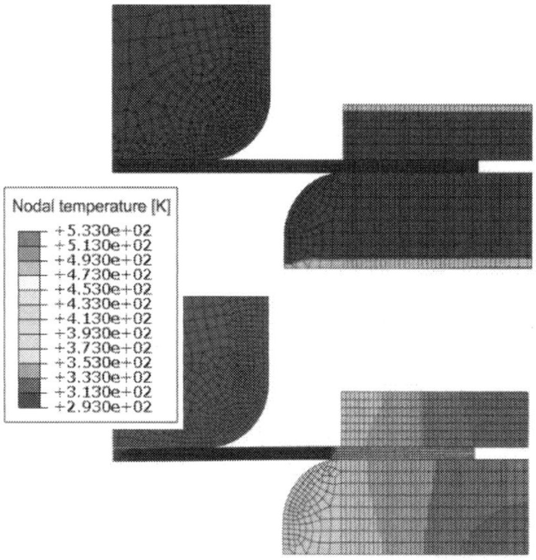

Figure 7. Nodal temperature distribution at the differential heating of the flange – initial stage (top) and at the end of the heating phase (bottom).

The unheated specimen area below the punch was not lubricated. Therefore, the friction coefficient of μ =0.15 was selected here.

The locally heated flange expresses lower stresses and through this improved formability in comparison to the forming at room temperature. This results in higher attainable drawing ratios in comparison to the room temperature conditions. Additional to the inductive heating prior to the forming operation the material heats due to the plastic strains. Therefore, the temperature in the preliminary heated area increases during the deep drawing process. Figure 8 shows the temperature distribution during the deep drawing for the specimen with the diameter of Ø96 mm.

The comparative analyses with different blank diameters have shown that with the temperature variable friction coefficient in the flange area the largest successfully drawn diameter is Ø100 mm while the cups with larger diameters of 101 mm and 102 mm were torn at the cup bottom – Figure 9.

The evaluation of the springback shows, that the springback has geometrically the highest influence on the cup diameter at its top edge which change with the relaxation outside of the tool. The relaxation of the specimen with a blank diameter of Ø98 mm and the diagram with values for various drawing ratios are presented on Figure 10. It is obvious that the springback effect has only a minor role from the geometrical point of view and could be neglected in further analyses.

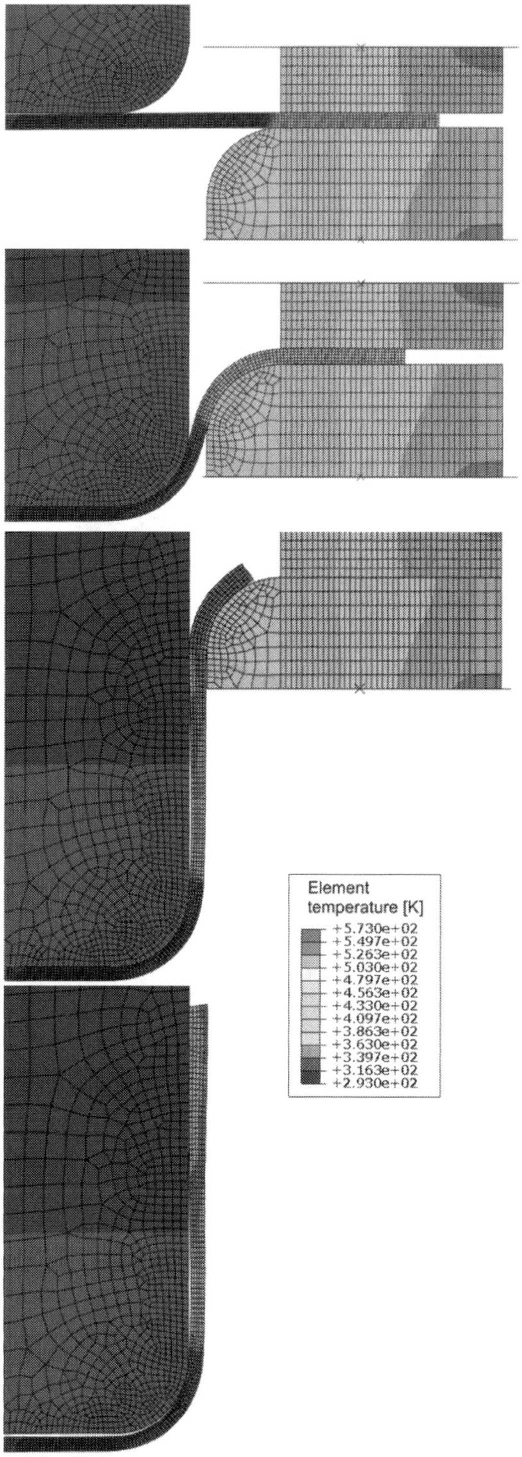

Figure 8. Temperature distribution during the forming process.

Obtained results have been compared with experimental data of deep drawn cups. In particular the maximal drawing ratio was evaluated which is mainly influenced by the proper determination of the flow curves at elevated temperatures and friction coefficient in the flange area of the drawn cup.

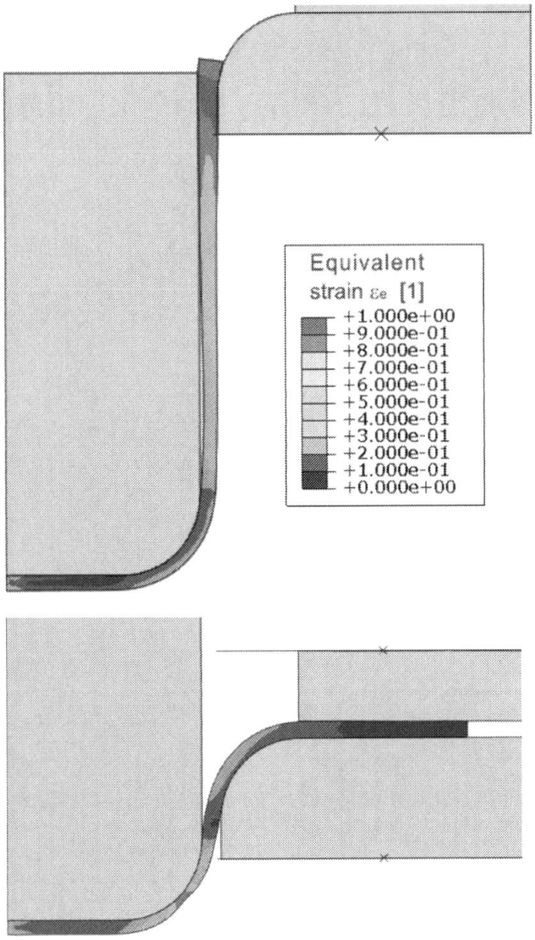

Figure 9. Strain distribution at deep drawing: blank diameter Ø100 mm (top) and Ø102 mm (bottom).

4 Experimental verification

The experimental verification of the performed FE simulations was performed on special adopted heated device positioned on the eccentric press [17] – Figure 11. The heating was provided by the inductance coil inserted in the die while the cooling of the central area of the blank was assured by spraying the water from the bottom side. The data of drawn specimens at various conditions are presented in Table 1.

Experimentally obtained succesful drawn parts are presented in Figure 12 together with the drawing data obtained by the numerical simulation. It is evident that the limiting drawing ratio obtained with experiments is comparable with numerical results for the analyses at room temperature. At elevated temperatures, the experimentally obtained limit drawing ratio of DP600 steel is higher than the limit obtained by the numerical simulations.

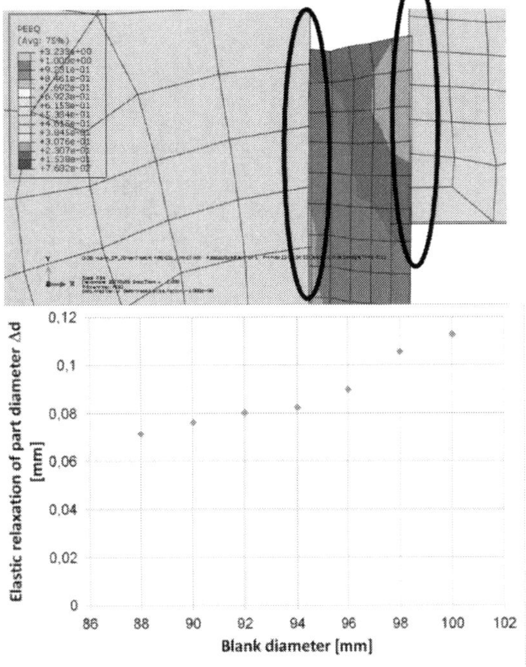

Figure 10. Elastic distortion for the cup drawn from the blank of Ø98 mm (top) and correlation of springback versus drawing ratio (bottom).

Table 1: Experimental data of heated deep drawing [17].

Blank diameter (mm)	Lubricant / Temperature	β (/)	Condition	Max. Force (kN)
84	Graphite / Room	2.05	OK Room	130.3
86	Graphite / Room	2.10	OK Room	136.4
88	Graphite / Room	2.15	Room limit	132.8
90	Graphite / High	2.20	OK High Temp	120.2
94	Graphite / High	2.30	OK High Temp	131.5
96	Graphite / High	2.35	OK High Temp	134.1
98	Graphite + PTFE / High	2.40	OK High Temp	135.2
102	Graphite + PTFE / High	2.49	OK High Temp	132.9
104	Graphite + PTFE / High	2.54	OK High Temp	135.1
106	Graphite + PTFE / High Graph+Tefl	2.59	High Temp Fail	--

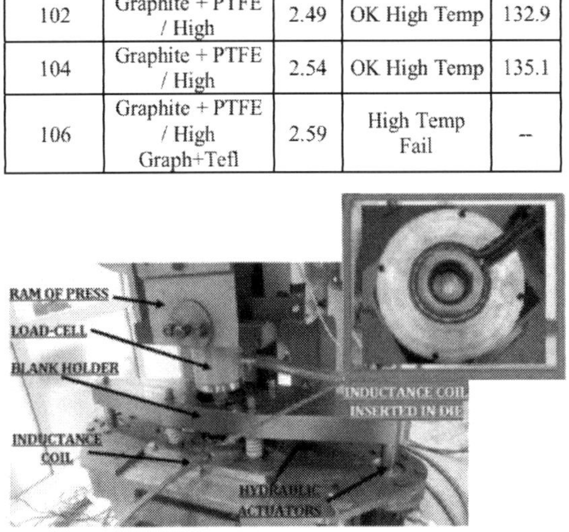

Figure 11. General construction showing the inductance heating device in the die [17].

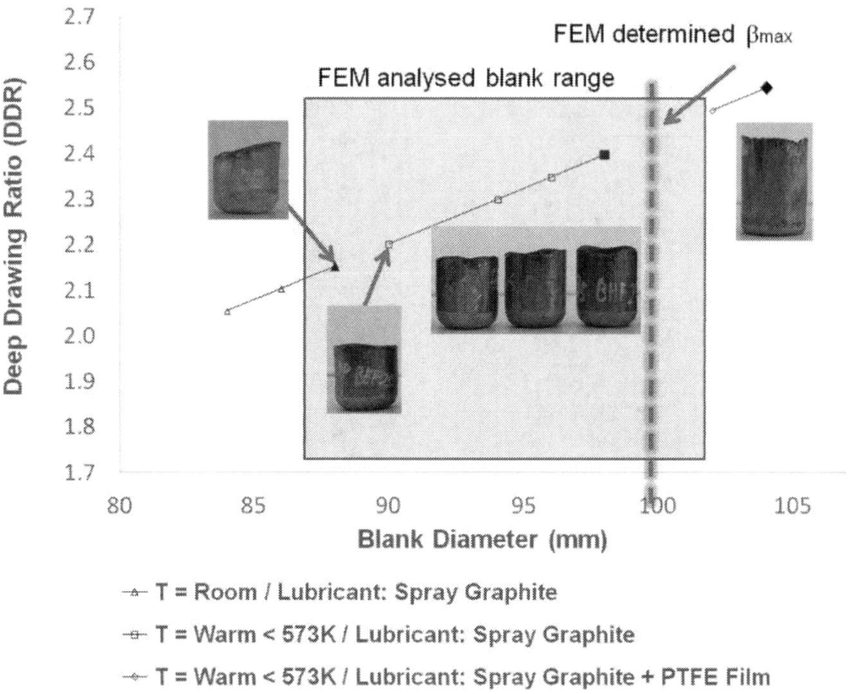

Figure 12. Experimentally obtained maximal drawing ratios with graphite lubrication.

5 Conclusion

In presented work, simulations of preferential flange heating at deep drawing of round cup are performed and compared with experimentally obtained results. The input parameters for the simulation were determined experimentally; in particular the material behaviour at room temperature and at elevated temperatures up to 300°C. Additionally, friction coefficient and its temperature dependence for PTFE foil was determined in a range of μ=0.02 to 0.035. By applying the PTFE foil the experiments are limited to 300°C since the solid film lubricant has its melting point at 326.8°C.

Flow curves for DP600 were analysed for various temperatures and strain rates. The increase in temperature up to 300°C decrease the level of flow curves what is in particular emphasised when the high strain rates are considered. The highest decrease of flow curve was observed $\dot{\varepsilon}$=2.5. However, performed simulations have shown that the strain rates in the flange region rarely exceed the value of $\dot{\varepsilon}=1$. Therefore, extensive measurements of flow curves for strain rates below $\dot{\varepsilon}=1$ needs to be performed.

Numerical simulations were performed with explicit FEM solver using fully thermo-elasto-plastic model of the tool and the formed specimen. Simulations were carried out in three steps included thermal heating, deep drawing process and springback analysis. The flange region was heated to 260°C in its outer region and to 180°C on the die radius with graduate temperature decrease between both limiting values.

Numerical simulations have shown increase of maximal drawing ration from 2.15 obtained at room temperature up to 2.45 obtained with the heated flange area.

Experimental analyses with heated tool have shown even higher limits of the drawing ratio reaching the value of β_{max} = 2.54 for the observed DP600 material with sheet thickness of 1.6 mm. This represents an increase of 18.18% of the maximal drawing ratio in comparison to the forming at room temperature.

In order to improve the fit of numerical and experimental results, additional evaluation of the flow curves as well as the behaviour of the solid lubricants at elevated temperatures needs to be performed.

Here presented research work has shown promising influence on increase of material draw ability with a local heating of the flange area in the temperature range between 200°C and 300°C. Further research will be dedicated to the analyses of non-rotational parts demanding also additional extensive computer resources for thermo-elasto-plastic analyses.

Acknowledgements

The authors would like to acknowledge the financial support received from The Scientific and Technological Research Council of Turkey (TUBITAK) and Slovenia(ARRS) under Project No: 111M448 and from ATILIM University under the project No: ATÜ-BAP-1011-13. The use of facilities in the Metal Forming Center of Excellence at ATILIM University as well as

Faculty of Natural Sciences and Engineering, University of Ljubljana is also acknowledged and thanks are extended to the personnel for their contributions on the material tests and the use of HPC.

References

1. S. Sadagopan, D. Urban, American Iron and Steel Institute/Department of Energy Technology Roadmap Project TRP#0012 (2003)
2. N.N.: *Advanced High-Strength Steels application guidelines Version 5.0*, WorldAutoSteel (2014)
3. H. Aydin, E. Essadiqi, I.-H. Jung, S. Yue, Mat. Sci. & Eng., A **564,** 501–508 (2013)
4. R. Krupitzer, AHSS Technology Workshop - Southfield MI, Steel Market Development Institute (2012)
5. K. Bandyopadhyay, S.K. Panda, P. Saha, G. Padmanabham, J. Mat. Proc. Technol., **217**, 48–64 (2015)
6. G. Ingarao, R. Di Lorenzo, F. Micari, Mat. & Design, **30**, 4421–4433 (2009)
7. A.E. Tekkaya et. all, CIRP Annals , **64**, 629–653 (2015)
8. C. Lei, J. Cui, Z. Xing, H. Fu, H. Zhao, , Phys.Proc. **25** 118 – 124 (2012)
9. R.K.Verma, P.Biswas, T. Kuwabara, K. Chung, Mat. Sci. & Eng. A, **604** 98–102 (2014)
10. H. Kim,T. Altan, Q. Yan, J. Mat. Proc. Technol., **209,** 4122–4133 (2009)
11. Mackensen, M. Golle, R. Golle, H. Hoffmann, CIRP Annals **59,** 283–286 (2010)
12. Kim, H., Han., S., Yan, Q., Altan, T., CIRP Annals, **57**, 299-304 (2008)
13. Ö.N. Cora, A. Ağcayazı, K. Namiki, H. Sofuoğlu, M. Koc, Tribol. Int. **52,** 50–60 (2012)
14. A. Abdollahpoor, X. Chen, M. P. Pereira, N.Xiao, B.F. Rolfe, J. Mat. Proc. Technol., **228,** 125-136 (2016)
15. M. Merklein, M.Wielanda, M. Lechnera, S. Bruschib,A. Ghiotti, J. Mat. Proc. Technol., **228,** 11–24 (2016)
16. P. Lehtinen, T. Väisänen, M. Salmi, Phys. Proc., **78,** 312-319 (2015)
17. E. Kayhan, *The development of a method to improve the limit drawing ratio of blanks using preferential heating*, Ph.D. Thesis, Atılım University, Turkey (2015)

Method for Deep Drawing Process Control Using Segmented-Multiple Active Drawbeads

Catalina Maier[1,a], Viorel Paunoiu[1], Vasile Marinescu[1] and Alexandru Epureanu[1]

[1]*Dunarea de Jos University of Galati, Manufacturing Engineering Department, 111 Domneasca Street 800201 Galati, Romania*

Abstract. This paper presents a new method for deep drawing process control. The new method consists in: *i)* using of a segmented-multiple moving drawbeads; *ii)* imposing an optimal reference trajectory for drawbeads position; *iii)* in-process control where controlled variable is chosen the amplitude of deviation δ of the actual instant workpiece surface geometry with respect to the surface geometry during simulation and as manipulated variable is chosen the deviation Δh_{DB} of the actual drawbeads position with respect to the programmed one; *iv)* periodical validation, calibration and updatation both of the control model and reference trajectory of drawbeads position; *v)* using a reduced order model of the complex system design. An experimental device conception and testing is presented. The results obtained using this device confirms the validity of the new proposed method for deep drawing process control.

1 Introduction

Sheet metal forming is one of the primary manufacturing processes which are continuously under new developments. The deep drawn process produces high-strength, lightweight parts more cost-effectively than other methods. Among the advantages offered by deep drawing are: rapid press cycle times, fewer operations required to finish a part, the ability to create complex geometries unattainable through other processes.

The performance in manufacturing of complex parts by thin sheet metal forming is crucial affected by the risk that the material strain-stress state, during deformation process, to exceed the allowable limits and consequently to appear the material fracture or wrinkling.

The sheet metal deep drawing technology is one of the most challenging processes in manufacturing. The size, shape, thickness and deep drawn metal used to produces sheet metal deep drawn part cover a diverse range of variable. Individual variables should be evaluated carefully in order to determine the optimum manufacturing method. Criteria of this optimum take account three quality considerations [1]: *i)* formability (wrinkling due to excessive compression and tearing because of high local tensile stress that cause thinning and failure); *ii)* dimensional accuracy (springback caused by elastic recovery); *iii)* consistency, that represent, minimizing dimensional variations due to the variation of different parameters (lubrication, material properties, thickness).

The new challenges in this field are: *i)* the use of new materials, and *ii)* a new concept of metal forming like an intelligent next generation technology considering user needs and future technology [2].

The new tendencies in design, optimization and control of the metal forming process determined the need of using new modelling techniques [3-8]. The main characteristic of the new models is their very short response time, this being the reason why they can be used for real-time control of the process. By analogy with the finite element method the new developed modelling techniques are called metamodelling [6], reduce order modelling [8] or macromodelling of the deep drawing process [9]. The idea of reduce order modelling is to use two models for optimization: a fine model (finite element model) of the process which has high accuracy with computationally expensive to solve and a reduced order model associated with the adaptive space mapping, fast to solve but less accurate.

This all considerations determine the motivation for researchers to continue those studies.

In the context of the work conducted in this field by different researchers [1, 10], the present paper propose a new method to control deep drawing process by using segmented-multiple active drawbeads and using a combination of different process control procedures.

The study consider a new device for a reduce order model of the deep drawing process and experimental and numerical results are presented.

2 Description of the method

The new method tries to improve significantly deep drawing process control considering following

[a] Corresponding author: catalina.maier@ugal.ro

requirements: *i)* selection of the correct process variables in order to monitor real state of the material during the process and command the correction; *ii)* reduce time consumption for deep drawing process and control system design by reduced order modelling; *iii)* generation of accurate reference trajectories for the process variables control using FEA and experiments; *iv)* combine different process control procedures and use those principal capabilities.

In this context, the method's steps are:

- **first step** - we establish the process variables for complex deep-drawing process;
- **second step** consist in the design of the reduce order model of the deep drawing process;
- **third step** represents the identification of the reference trajectories and control models, using finite element simulation;
- the **last step** in this algorithm (figure 1), consisting in reference trajectories and control models validation it is optional and is applied only for one deep drawn piece belongs to a dimensional group; the decision point selects, function on the answer YES/NO to the question "is it the first piece for the dimensional group?", if we proceed or no the experimental validation.

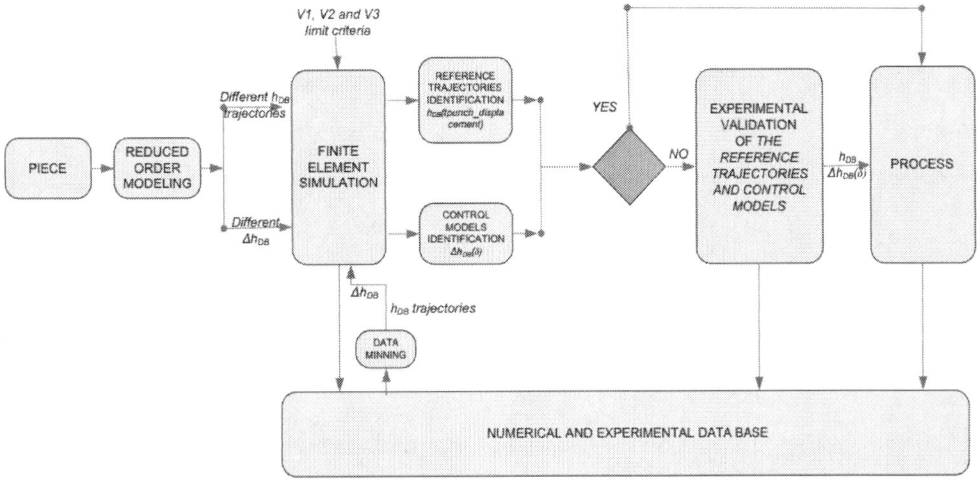

Figure 1. Algorithm of the new control process method

2.1 Process variables

The ideal in-process control of complex deep drawn would be the stress field control throughout the sheet metal based on cases presented in figure 2, where σ_{11} and σ_{22} represent Cauchy stress tensor components. The limit values of σ_{22}/σ_{11} stress ratio, respectively V1, V2 and V3, depend on the material properties and represent the criteria to define the following stress state types:

- stress state 1 – tension+compression with compression dominance determining wrinkles during the process;
- stress state 2 – tension+compression with tension dominance, determining wrinkles after the process, due to the springback;
- stress state 3 – biaxial tension when tearing appears.

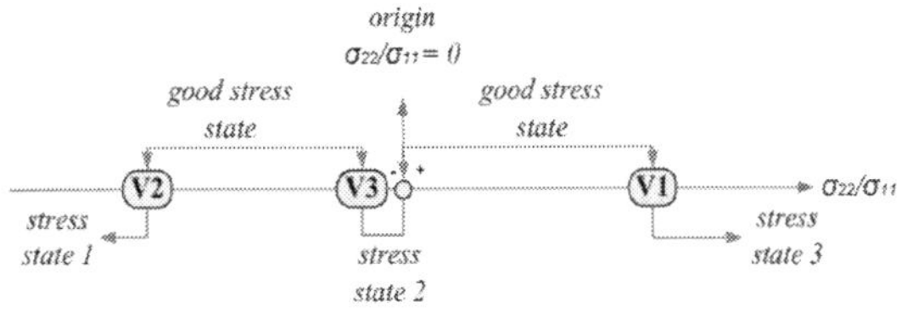

Figure 2. Characteristic stress states during complex deep drawing process

The good stress states (figure 2) represent the objective of the new method in order to respect quality considerations – formability, dimensional accuracy and consistency – presented at the beginning of this paper.

2.2 Reduced order modelling

In order to reduce time consumption for deep drawing process and control system design a reduce order model of the deep drawing process is applied. The main idea is to decompose complex deep drawing process in macroelements - elementary deep drawing processes - and in their identification. Each one of these macroelements represents a particular case of the generalized elementary deep drawing process. The steps of real deep drawing process decomposition consists in: *i)* the *delimitation* of each elementary process and *ii)* his *identification* starting to a generalized elementary deep drawing process in order to establish its characteristics. The resulted process macromodel is less time consuming in comparison with finite element model, assures a very good precision and he is easier to use for the real-time control of deep drawing process [9].

The macromodelling of deep drawing process of any part provides new macroelements for enrich the numerical and experimental database. The decomposition of every real deep drawing process can use this database in order to build its macromodel. Each macromodel is characterized by: dimensions, boundaries conditions, reference trajectories, control model. For a large number of macroelements the finite element simulation is performed and the information obtained is also stored in the database.

2.3 Reference trajectories identification

Therefore, for each of the macroelement in the deep drawing process of the considered piece, from database is selected a situation that is the closest, so far, to the macroelement dimensions, new possible reference trajectories for process parameters are added and, finally FEM simulation is performed.

Analyzing the results obtained on the evolution of stress ratio and considering the limit criteria V1, V2 and V3, the optimal reference trajectory is determined. For example, FEM analysis of macroelement with a nonlinear contour was performing in terms of the following various trajectories (figure 3) of the punch and drawbeads:

- successive action of drawbeads (penetration up to a predefined value and maintain this position until the end of the process) and punch (whose action starts when reaching preset value the depth of drawbeads penetration), considered classical evolution used in deep drawing process;
- simultaneous action of drawbeads and punch with consideration of the various trajectories of drawbeads position.

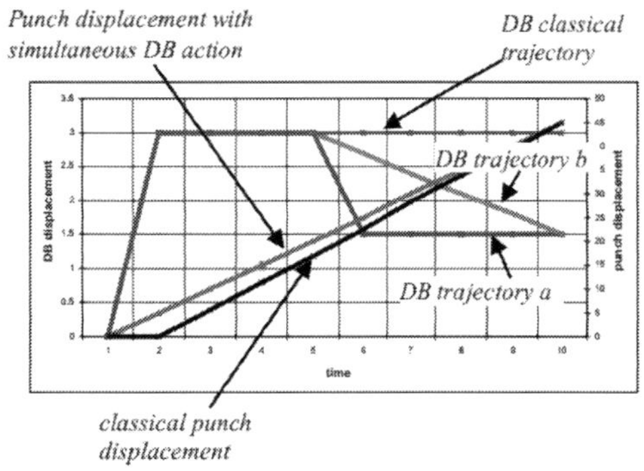

Figure 3. Different trajectories of drawbeads and punch

Numerical simulation results show a differential evolution of the ratio σ_{22}/σ_{11} during the forming process, depending on the type of action (simultaneous or successive) of the drawbeads and punch and its reference trajectory (figure 4).

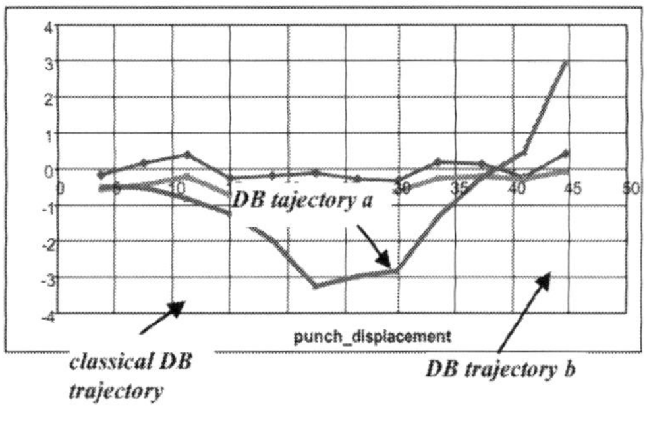

Figure 4. Stress state evolution function drawbead trajectory

From this analysis resulted that in classic case, used so far in deep-drawing processes while the drawbead enters into the material before the punch displacement starts and remains in this position until the end of the process, there is a substantial critical values exceeding 1.5 of the stress ratio that vary in this case between 2.94742 and -3.26625.

In contrast, when simultaneous action of the punch and drawbead and gradual withdrawal after 50% of the active trajectory of the punch, the stress ratio is between -0.23 and 0.39 in the case of *DB trajectory a* and between -0.22 and -0.72, if the *DB trajectory b* case is considered. Because the *DB trajectory a* determinate the stress state 2 that can induce wrinkles formation after metal forming process (caused by springback), it follows that the drawbead optimal trajectory is *b* which meet the limit criteria V1, V2 and V3 and also offers a reserve to the stress ratio variation before the loss of material stability.

We demonstrate that among various possible trajectories of drawbead position, we can determine one to meet the limit criteria V1, V2 and V3 and is considered the optimal reference trajectory.

2.4 Control model determination

As previously shown, process variable which will be controlled and particularly will have zero value is δ. This variable is the deviation of real surface of the part from the surface obtained through numerical simulation. If the stress state, characterized by the ratio σ_{22}/σ_{11}, is becoming favourable to wrinkles formation, then variable δ will be nonzero and equal with deviation of the current surface of the part from the reference surface generated through numerical simulation. The correction of this deviation is done by a supplementary displacement of the drawbeads, Δh_{DB}, added to the programmed trajectory of its position. If a new deviation δ is recorded during deformation process, a new supplementary displacement will be given. The control model of the deformation process will be given by the function $\Delta h_{DB}(\delta)$. This relation is determined through numerical simulation with supplementary displacements at different steps along draw beads trajectory.

Finally, this information is transferred to the forming process carrying out an off-line control through programming of the reference trajectories and an on-line control through control models involvement.

A numerical and experimental database is created using the results obtained by numerical simulation, experimental research and from measurements and observations during real processes. It is constantly updated by data obtained from the execution of the actual metal forming process of complex parts.

Exploiting numerical and experimental database is performed by knowledge extraction techniques with aim of updating continuously both the reference trajectories and the process control model.

2.5 Reference trajectories and control model validation

We chose between macroelements one that characterizes the most complete the process and the complex stress and strain evolution in the sheet. The macroelement who respect these conditions correspond to the nonlinear contour shape and include drawbeads. In our conception this macroelement represent the reduced order model of the real complex deep drawing process.

It is used to: *i)* validate the proposed method, *ii)* design and validation of metal forming process control model; *iii)* updatation of the database. Using the reduced order model of the complex deep-drawing process lead to reduced time required to its design and control as well as reduced costs for die try out procedure.

The conception of the device used, in this paper, to validate the new method is based on the: *i)* reduced order model of the complex deep drawing process; *ii)* different solutions used by researchers [11 - 15]; *iii)* functions to be assessed – control and monitor of the process variables h_{DB} and δ.

The components of the new proposed device are: *i)* controllable metal forming device (figure 5), *ii)* gear mechanism with ball screws for the control of drawbeds; *iii)* monitoring and controling system of the manipulated process variables – programmed value h_{DB} and its deviation Δh_{DB} (monitoring and control system of metal forming process using a programable logic controller); *iv)* optical system measurement with machine vision algorithms.

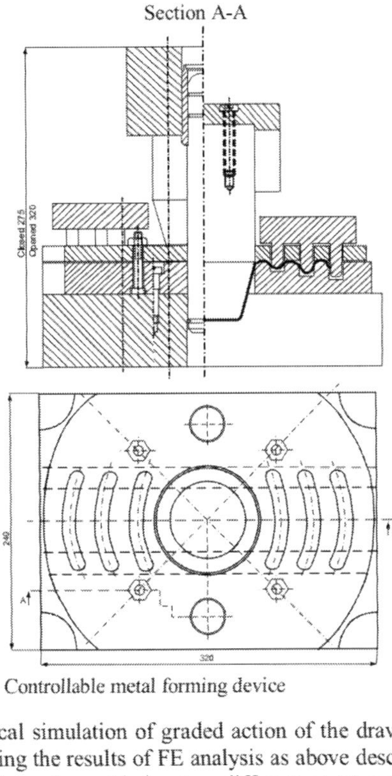

Figure 5. Controllable metal forming device

Physical simulation of graded action of the drawbeds, considering the results of FE analysis as above described, was performed considering two different construction of the drawbeds: multiple and single drawbeds. The use of multiple drawbeds in the methodology proposed in this paper is different from its use proposed by Papaioanu A, Schleich [12, 13], D. Vlahovic [16, 17] and Liewald M. R. [18] in that it is *active* (acting simultaneously with the punch and its position can be adjusted depending on the evolution of monitored parameter δ) and its application is *generalized* for complex shape of parts.

3 Results and conclusions

Experimental researches are performed using the device previously presented, considering different thickness of the blank (0.4 ; 0.6 ; 0.8 ; 1 mm) and conditions of the forming process development (drwbeads maximal penetration, drawbeads and punch reference trajectories identified after the study presented in the chapter 2.3). In figure 6 we present the pieces obtained following these 3 conditions:

a). classical trajectory of drawbeads - penetration up to a predefined value and maintain this position until the end of the process – without in-process control (figure 6.a)

b). reference drawbeads trajectory b, considered optimal after FE analysis, without in-process control (figure 6.b);

c). reference drawbeads trajectory b, considered optimal after FE analysis, with in-process control (figure 6.c);

Maximal programed value of drawbeads penetration is 2 mm for a blank thickness of 0.4 mm. It results the following observations :

- piece ***a*** presents wrinkles in the uncontrollable deformed zone during the process and large springback, after the finish of the forming process, in flange zone ;
- piece ***b*** presents a small level of the wrinkles in the uncontrollable deformed zone during the process and accepted springback of the flange after the forming process finish ;
- piece ***c*** results without wrinkles during the process and springback of the flange after the finish of the forming process.

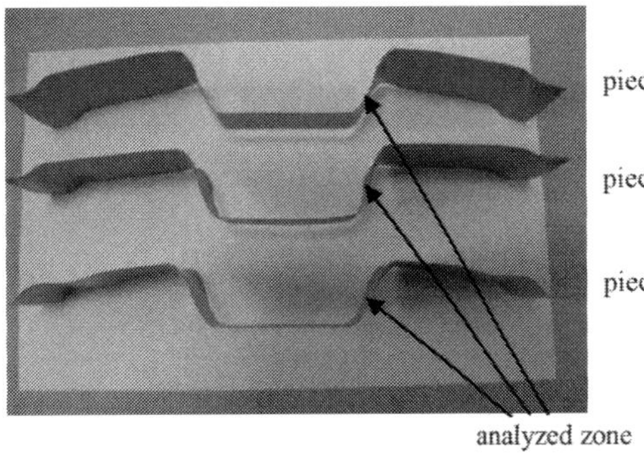

Figure 6. Experimental results

We analyze the deformed shape in the central zone of the free deformable part of the specimen, where the instability of the material is highest, for the considered conditions, as above. This instability is due to the un-contact conditions between the specimen and the tools. In figure below is presented the deformed shape at the end of the process for the studied cases.

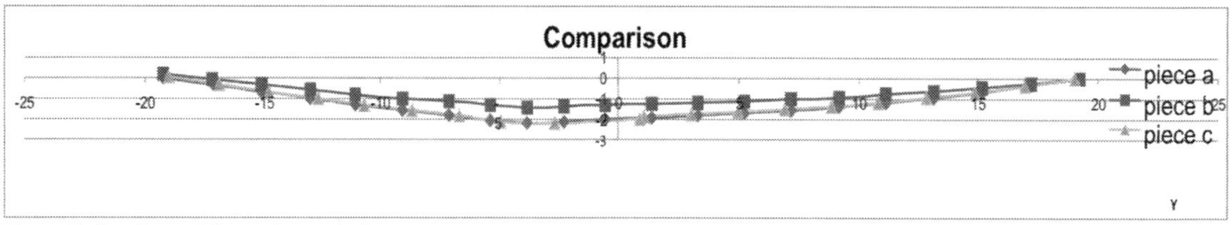

Figure 7. Specimen deformation analysis

We can observe that the minimum amplitude of deformed shape deviation is obtained in the case of the piece ***b*** (figure 7).

The results obtained with the device previously presented confirm the validity of the proposed new method.

The new methodology proposed has the following characteristics:

- high degree of generality by allowing only the adaptation of the macroelements dimensions used for the reduced order model of the forming process of complex parts;
- combining multiple process control techniques;
- use multiple active drawbeds differentiated in different segments of the part contour;
- optimal reference identification for programmed parameter, h_{DB}, of the process considering its effect on the stress rate σ_{22}/σ_{11}, the ideal feedback information for in-process control;
- continuous updating of the reference trajectory and control model using the database.

These characteristics provide the following advantages of the proposed method:

- reducing the time required for the process and control system design in case of complex deep-drawn parts;
- increased precision of metal forming process control by using the ratio σ_{22}/σ_{11} as criterion for obtaining the reference trajectory parameter monitored h_{DB};
- reducing costs represented by die try out stage using reduced order model of the process.

References

1. Y. Lim, R. Venugopal, G.A. Ulsoy, *Proc. of the 17th Int. Federation of Automatic Control World Congress*, 1875 (2008).
2. H. Hayashi, H. Nishimura, The Annals of Dunarea de Jos University of Galati, Fasc. V, 3 (2009).
3. C.W. Hsu, A.G. Ulsoy, M.Y. Demeri, ASME J. Manuf. Sci. Eng. **122**, 717 (2000).
4. C.W. Hsu, A.G. Ulsoy, M.Y. Demeri, J. of Mater. Proc. Tech. **127**, 361 (2002).
5. V. Viswanathan, B. Kinsey, J. Cao, J Eng Mater-T ASME, 141 (2003).
6. M.H.A. Bonte, A.H. van den Boogaard, Int. J. Forming Proc., 1 (2005).
7. C. Maier, Al. Epureanu, V. Marinescu, V. Paunoiu, M. Afteni, F.B. Marin. *Proc. of ModTech Int. Conf.*, 363 (2010).
8. C. Maier, Al. Epureanu, V. Marinescu, F.B. Marin, *6th Int. Conf. on Advanced in Dynamical Systems and Control*, 133 (2010).
9. C. Maier, Al. Epureanu, F.B. Marin, J. of Engineering and Architecture **1**, 1 (2013).
10. C. Maier, V. Paunoiu, V. Marinescu, The Annals of Dunarea de Jos University of Galati, Fasc. V, 25 (2013).
11. Michler J.R., Weinmann K.J., Kashani A.R., Majlessi S.A., J. of Mater. Proc. **43**, 177 (1994).
12. A. Papaioanu, R. Schleich, M. Liewald, *7th European LS-DYNA Conference* (2009).
13. A. Papaioanu, M. Liewald, J. Mater. Form. **3**, 155 (2010).
14. M. Samuel, J. of Mater. Proc. Tech. **122**, 94 (2002).
15. L.R. Sanchez, Int. J. Mech. Sciences **42**, 705 (2000).
16. D. Vlahovic, M. Liewald, *7th International Stuttgart Symposium "Automotive and Engine Technology"*, (2007).
17. D. Vlahovic, M. Liewald, *Proc. of 11th ESAFORM International Conference*, MS03, 175 (2008).
18. M. Liewald, D. Vlahovic, *Patent No. DE 10 2006 047 484 A1* (2006).

Influence of Surface-profile and Movement-path of Roller on Thickness Thinning during Multi-pass Deep Drawing Spinning

Qinxiang Xia[1,a], Junhao Zhang[1], Chenglong Huang[1] and Xiuquan Cheng[2]

[1]*School of Mechanical and Automotive Engineering, South China University of Technology, China*
[2]*Aircraft Maintenance Engineering College, Guangzhou Civil Aviation College, China*

Abstract. Over thinning is a serious defect influencing the forming quality of spun workpiece during multi-pass deep drawing spinning. Surface-profile and movement-path of roller are the key factors influencing the thinning ratio of wall thickness of spun workpiece. The influence of surface-profile and movement-path of roller on thickness thinning were studied based on numerical simulation and experimental research, four groups of forming experiments were carried out under the combination of the different surface-profile of roller (R12 and R25-12) and movement-path of roller (spinning from the bottom of the blank and spinning from the middle of the blank). The results show that both the surface-profile and movement-path of roller have great influence on wall thickness thinning during multi-pass deep drawing spinning; and compared with the movement-path of roller, the influence of surface-profile of roller is more significant. The experimental results conform well to the simulation ones. It indicates that the FEA model established is reasonable and reliable.

1 Introduction

Deep drawing spinning is a kind of forming process that the diameter of the blank is reduced mainly by the radial drawing, which is the most widely used conventional spin-forming process[1-2]. Shielding case of high voltage switch is a kind of thin-walled rotational part with deep cavity, which is usually manufactured by welding after stamping. Compared with the welding after stamping, multi-pass deep drawing spinning has the advantages of low production cost and high material utilization[3]. However, due to the characteristics of deep cavity and thin wall of shielding case, over thinning defect occurs easily during multi-pass deep drawing spinning.

For solving the over thinning problem during deep drawing spinning, Wang and Long[4] studied the influence of the roller paths on tool forces, part wall thickness and stress variations in conventional metal spinning numerically. The results show that using the concave roller path tends to cause higher reductions of wall thickness of the spun part and using the convex roller path helps to maintain the original wall thickness unchanged. A greater curvature of the concave path would result in more thinning in wall thickness of the spun part. Zeng et al.[5] researched the influence of roller paths and processing parameters on multi-pass drawing spinning of 6061 and SPCC plates experimentally. The results show that the distribution of wall thickness of the spun part is the most uniform when spinning with the movement-path of convex and concave curves under the forward and backward alternately spinning method. Lin et al.[6] researched the influence of the movement-path of roller on multi-process drawing spinning of Q235 and 1060 plates experimentally. The results show that spinning from the middle of the blank is beneficial the material flow in the subsequent passes. Ma et al.[7] analyzed the influence of the feed rate, fillet radius and entry tangential angle of roller on the forming quality during conventional spinning numerically. The results show that the wall thickness thinning decreases with the increasing of the fillet radius of roller. In order to improve the limit drawing ratio of one-pass deep drawing spinning, Xia et al.[8] designed a kind of roller with composite surface-profile, which was composed of two different fillet radius, and pointed out that the large fillet radius was used to press down the blank previously, which was beneficial the material flow, therefore, the deformation turned to be more easier under the large fillet radius. On the other hand, the small fillet radius was mainly used for the part forming.

Due to the limitations of the researches, it is not effective to solve the problem of over thinning of the thin-walled shielding case with deep cavity in practical applications. In this paper, the feasibility and influence of surface-profile and movement-path of roller on wall thickness thinning during multi-pass deep drawing spinning of thin-walled shielding case with deep cavity are studied.

2 Feasibility analyses of surface-profile and movement-path of roller

[a] Corresponding author: meqxxia@scut.edu.cn

2.1 Analysis of surface-profile of roller

The fillet radius of roller has a great influence on the forming quality of parts during multi-pass deep drawing spinning[1]. The composite surface-profile roller composes of two different fillet radius, R_1 and R_2 (as shown in Fig.1), as reported by Xia et al.[8]. The main function of R_1 is to press down the blank previously and that of the R_2 is to form the blank. For the aluminum blank with diameter between 150~300 mm, the range of the fillet radius R_2 is 12~15 mm[2]. Therefore, the value of R_2 was selected as 12 mm, and the values of R_1 were selected as 25 mm, 30 mm and 45 mm respectively (as shown in Fig.2).

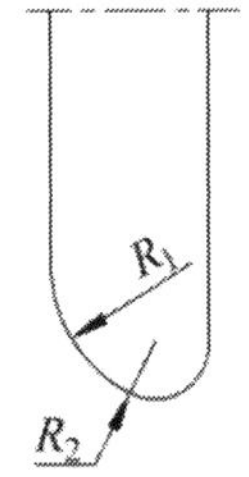

Figure 1. Composite surface-profile roller

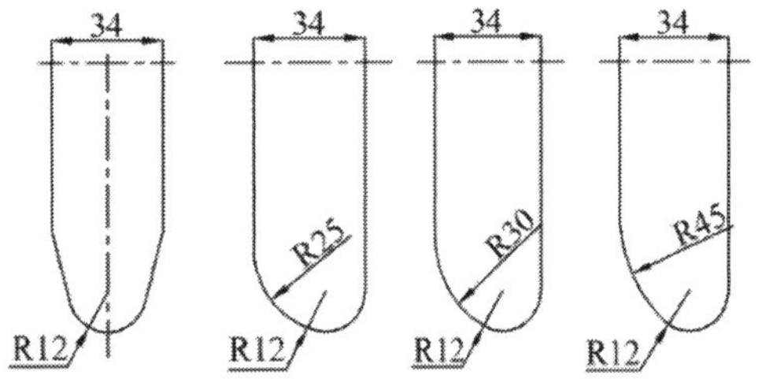

Figure 2. Rollers with different surface-profile

2.2 Analysis of the movement-path of roller

Fig.3 shows the principle of traditional multi-pass deep drawing spinning process. It shows that in the initial stage of the forming process, the shape of blank is still closed to the plane (the diameter of blank is still large), which increases the forming resistance of the material on the blank edge. As a result, local forming and over thinning occur easily[9].

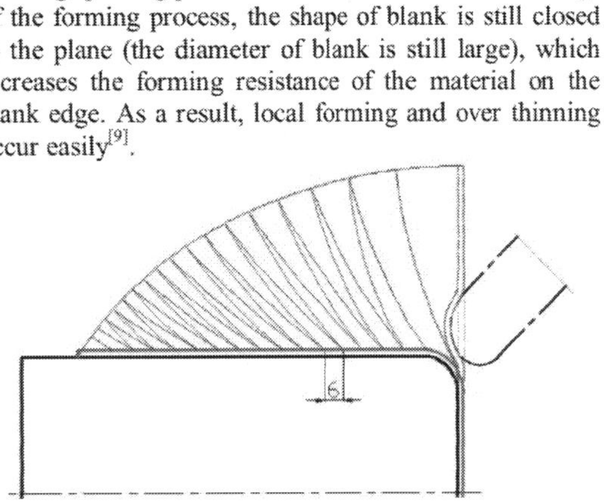

Figure 3. Principle of traditional multi-pass deep drawing spinning process

In order to reduce the negative influence of local forming, the initial spinning point O_1 starts from the middle of the blank was carried out during the multi-pass deep drawing spinning[6] (as shown in Fig.4), hence to increase the drawing coefficient m of the first pass (The drawing coefficient changes from d_0/D_0 to d_s/D_0). In order to reduce the influence of local forming, according to the range of complete drawing[9], d_s/D_0 was selected as 0.7. In this case, the distance between the starting point O_1 of spinning and the edge of the blank A is 30 mm.

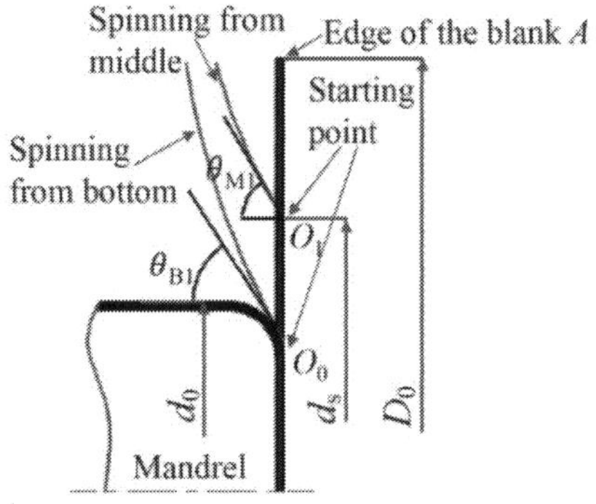

Figure 4. Diagram of movement-path of roller

The second pass of spinning starts from the bottom of the blank (as shown in Fig.5). The deformation of the second pass is smaller than that of the first pass spinning from the bottom of the blank, therefore, the negative influence of local forming is decreased.

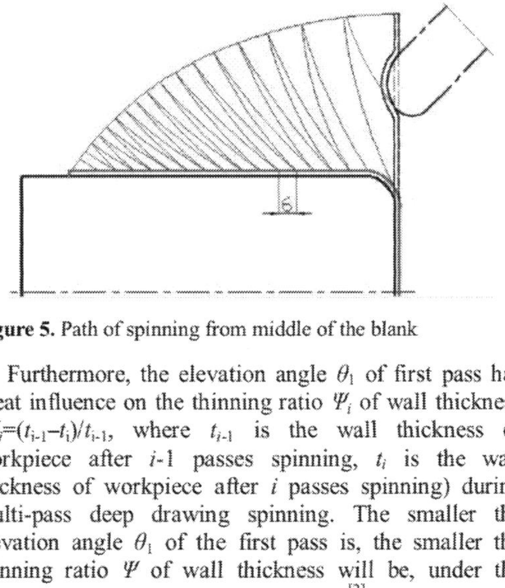

Figure 5. Path of spinning from middle of the blank

Furthermore, the elevation angle θ_1 of first pass has great influence on the thinning ratio Ψ_i of wall thickness ($\Psi_i=(t_{i-1}-t_i)/t_{i-1}$, where t_{i-1} is the wall thickness of workpiece after i-1 passes spinning, t_i is the wall thickness of workpiece after i passes spinning) during multi-pass deep drawing spinning. The smaller the elevation angle θ_1 of the first pass is, the smaller the thinning ratio Ψ of wall thickness will be, under the condition of no wrinkling occurring[2]. The range of elevation angle θ_1 of the first pass for aluminum alloy is usually from 50° to 60°[2]. Wrinkling may occur if θ_1 is less than 50°, while fracture may occur if θ_1 is larger than 60°[2]. Therefore, for the simulation and experimental, the elevation angles θ_M of the first and the second pass were selected as θ_{M1}=55°, θ_{M2}=60° respectively during spinning from the middle of the blank; the elevation angle θ_{B1} of the first pass was selected as θ_{B1}=60° during the simulation of the spinning from the bottom of the blank.

DOI: 10.1051/matecconf/20168014005

3 Finite element simulation results

3.1 Establishment of finite element model

A three dimensional finite element model of multi-pass deep drawing spinning was established based on software-MSC.MARC (as shown in Fig.6), which compose of mandrel, roller, tail stock and blank. According to the deformation characteristics of the deep drawing spinning, the contact pairs are established between the mandrel, roller, tail stock and roller. In the FEA model, the blank is fixed between mandrel and tail stock, and its deformation is restricted by mandrel and tail stock. The part of blank which does not contact with the mandrel and tail stock is free, and plastic deformation occurs under the action of the roller. The mandrel, roller and tail stock are set as rigid body due to their negligible deformation during spinning and without discretization [10]. In the FEA model, the modified Coulomb friction model was adopted. The Coulomb friction coefficient (μ) between the roller and the blank is taken as 0.1[11]. The tail stock and the blank are tied together by using the glue function of the Marc i. e, a large separating force is defined to exist between the tail stock and blank, the action force between the tail stock and blank might not surpass it, so that the tail stock and blank glued firmly together. During the spinning, the deformation of the materials clamped between the mandrel and tail stock is too small and can be ignored, therefore this part is not discretized, which can reduce the number of grid and improve the computational efficiency. The blank is discretized into 22320 8-node structural 3-D solid shell elements with 45360 nodes. The parameters of blank (1060-O aluminum plate) used for the finite element model are shown in Table 1.

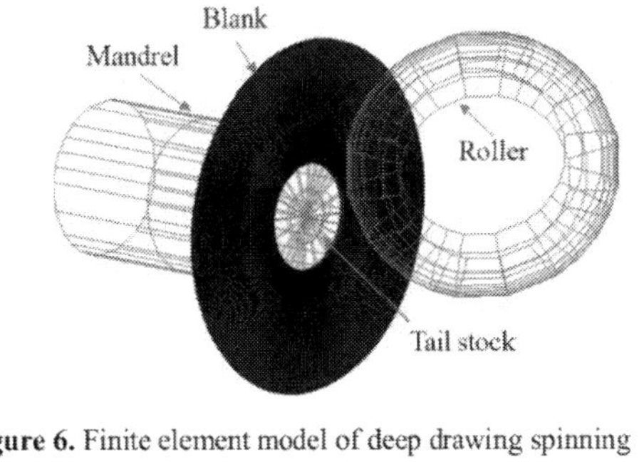

Figure 6. Finite element model of deep drawing spinning

Table 1. Parameters of the 1060-O aluminum blank

Diameter of blank /D_0/mm	Thickness of blank /t/mm	Yield strength /σ_s/MPa	Tensile strength /σ_b/MPa	Elongation /δ/%	Young's modulus /E/GPa	Poisson's ratio / μ	Hardening exponent /n	Enhancement coefficient /K
200	1.8	38	107	46	69	0.3	0.28	120.3

3.2 Analysis of simulation results

3.2.1 Influence of surface-profile of roller on thinning ratio of wall thickness of workpiece

Table 2 shows the simulation results of wall thickness of workpiece after the first and second passes spinning under the different surface-profile of roller when spinning from the bottom of the blank (as shown in Fig.3), and the main processing parameters adopted in the simulation are as follows: the elevation angle θ_{B1}=60°, feed rate of roller f=2.5 mm/r, the rotational speed of mandrel n=500 rpm and the diameter of roller D_R=160 mm, the distance between every two passes p=6 mm.

As shown in Table 2, after the first pass spinning, the thinning ratio Ψ of wall thickness is the smallest when using the roller with composite surface-profile of R30-12, and is the secondary smallest when using the roller with composite surface-profile of R25-12. This is because that the blank is closed to the plane shape during the first pass spinning, and all the contact positions between the blank and rollers locate in the range of R_1 (as shown in Fig.7), which is equivalent that the spin-forming occurs under the roller with single surface-profile R_1. The bigger the R_1 is, the smaller the wall thickness reduction would be. Therefore, the thinning ratio Ψ of wall thickness under the roller with composite surface-profile of R30-12 is smaller than that of R25-12. With the increasing of R_1, the contact area between R_2 and blank increases, and between R_1 and blank decreases. The R_1 doesn't contact the blank anymore when R_1 is over large, and the spin-forming hereby occurs only under the roller with the single surface-profile R_2, such as the roller R45-12. Therefore, the thinning ratio Ψ under the roller R45-12 and R12 is almost the same.

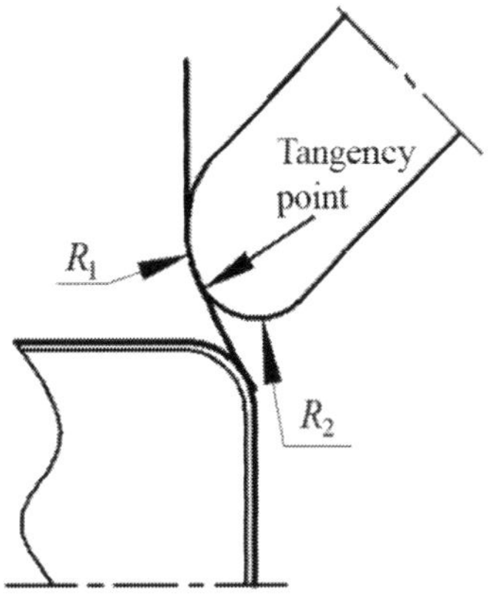

Figure 7. Contact situation between the roller and blank during the first pass spinning

Table 2. Simulation results of wall thickness after the first and second passes spinning under different surface-profile of roller

	The first pass				The second pass			
	Minimum wall thickness t_{min} /mm	Maximum thinning ratio Ψ_{max}/%	Wall thickness deviation Δt/mm	Mean wall thickness /mm	Minimum wall thickness t_{min} / mm	Maximum thinning ratio Ψ_{max}/ %	Wall thickness deviation Δt/ mm	Mean wall thickness /mm
R12	1.652	8.22	0.154	1.737	1.500	16.67	0.335	1.661
R25-12	1.679	6.72	0.129	1.749	1.590	11.67	0.251	1.693
R30-12	1.689	6.17	0.119	1.753	1.559	13.33	0.282	1.689
R45-12	1.654	8.11	0.152	1.741	1.52	15.56	0.32	1.675

Table 2 shows also that after the second pass spinning, the thinning ratio Ψ of wall thickness of workpiece is the smallest when using the roller with composite surface-profile of R25-12. With the proceeding of passes, the thinning ratio Ψ under roller R25-12 would be ultimately smaller than that of the roller R30-12. This is because that the contact area between R_2 and blank increases gradually with the blank pressing down (the blank is no longer a flat shape), and the R_1 and R_2 contact with the blank simultaneously (as shown in Fig.8). The main function of R_1 is to press down the blank previously during the first spinning pass, the contact area between the blank and roller during the following passes decreases with the decreasing of R_1, hence to decrease the flow resistance of the material, therefore, the thinning ratio Ψ of wall thickness decreases. The maximum thinning ratio Ψ_{max} under the roller with composite surface-profile of R25-12 after two-pass spinning is 11.67%, which is reduced by 30% compared with spinning under the roller with the single surface-profile of R12 only.

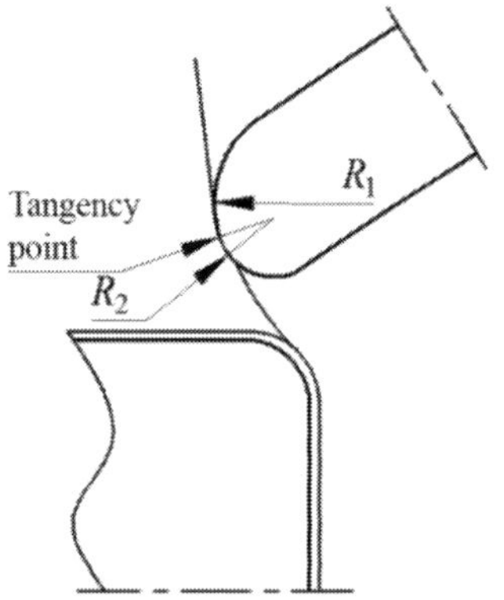

Figure 8. Contact situation between the roller and blank during the second pass spinning

3.2.2 Influence of the movement-path of roller on the thinning ratio of wall thickness of workpiece

Table 3 shows the simulation results of wall thickness of workpiece after the first and second passes spinning under the two different movement-path of roller shown in Fig.4. The constant processing parameters adopted in the simulation are as follows: the rotational speed of mandrel n=500 rpm, the diameter of roller D_R=160 mm, the fillet radius of rollers r_ρ=12 mm, the distance between every two passes p=6 mm; and the variable processing parameters adopted in the simulation are listed in Table 4.

Table 3. Simulation results of wall thickness

Movement-path of roller	Number of the pass	Minimum wall thickness /t_{max}/mm	Maximum thickness thinning ratio /Ψ_{max}/%	Deviation of wall thickness /Δt/mm
Spinning from middle	First pass	1.764	2	0.046
	Second pass	1.736	3.56	0.076
Spinning from bottom	First pass	1.652	8.11	0.153

Table 4. Variable processing parameters used for simulation

	Spinning from bottom	Spinning from middle	
		The first pass	The second pass
Elevation angle/θ/°	60	55	60
Feed ration /f/(mm/r)	2.5	3	2.5

Table 3 shows that the thinning ratio Ψ of wall thickness when spinning from the middle of blank in the first pass is very small. The maximum thinning ratio Ψ_{max} of the first pass is only 2%, and is increased to 5.56% (the total maximum thinning ratio Ψ of the first and two passes) after the second pass (the minimum of the wall thickness t_{min} is 1.736 mm), while the maximum thinning ratio Ψ_{max} of first pass spinning from the bottom of blank was 8.11% (the minimum of the wall thickness t_{min} is 1.652 mm). It indicates that the thinning ratio Ψ of wall thickness can be decreased effectively by spinning from the middle of the blank.

4 Experimental research

4.1 Experimental Conditions

Fig.9 shows the diagram of the shielding case of the high voltage switch. The dimensions of the shielding case are as follows: the inner diameter d=88 mm, the height H=96.8 mm, the inner fillet radius r=10 mm, the wall thickness t=1.8 mm.

The experimental was carried out on a two-wheel numerical controlled spinning machine and the processing parameters used in the experimental are the same as that of the simulation.

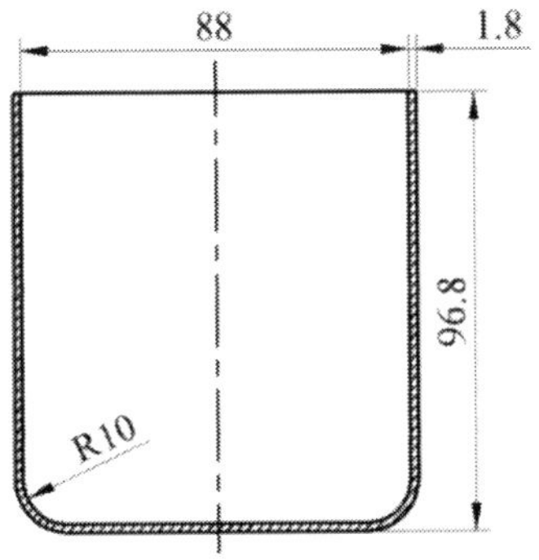

Figure 9. Diagram of shielding case of high voltage switch

4.2 Comparison of experimental and simulation results

Fig.10 shows the comparison of wall thickness between the experimental and simulation results. It shows that the distributions of wall thickness between the experimental and the simulation results conform well to each other, the maximum error is only 0.02 mm (the relative error is 1.1%). It indicates that the finite element model established is reasonable and reliable.

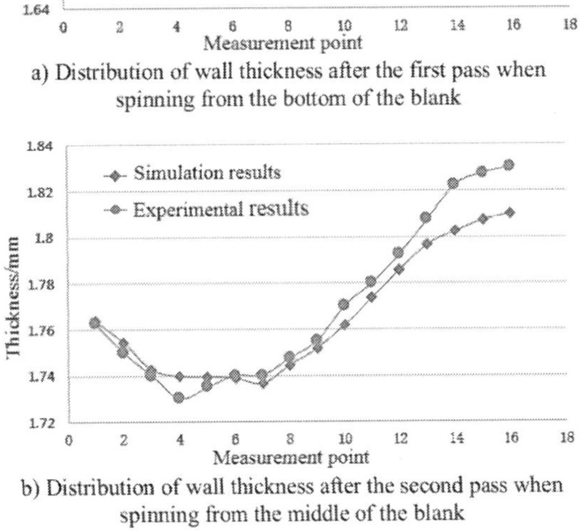

Figure 10. Comparison of the distribution of wall thickness of workpiece between experimental and simulation results

4.3 Trial production of shielding case of the high voltage switch

Four groups of forming experiments were carried out under the combination of the different surface-profile of roller (R12 and R25-12) and movement-path of roller (spinning from the bottom of the blank and spinning from the middle of the blank), as listed in Table 5. The total number of the passes is 31 when spinning from the bottom of the blank, while that of the number is 32 passes when spinning from the middle of the blank.

Table 5. Experimental scheme

No.	Surface-profile of roller	Movement-path of roller
1	R12	Spinning from bottom
2	R12	Spinning from middle
3	R25-12	Spinning from bottom
4	R25-12	Spinning from middle

Fig.11 shows the distribution of the wall thickness of spun workpiece under four combinations of the different surface-profile and movement-path of roller, and Table 6 shows the totally maximum thickness thinning ratio Ψ_{max} and the relative maximum thickness thinning ratio $\Delta\Psi_i$ ($\Delta\Psi_i = \Psi_{maxi}/\min(\Psi_{maxi})$, where Ψ_{maxi} is the maximum thickness thinning ratio of the experiment i). As shown in Fig.11 and Table 6, both of the surface-profile and movement-path of roller have great influence on wall thickness thinning during multi-pass deep drawing spinning; and compared with the movement-path of roller, the influence of surface-profile of roller is more significant.

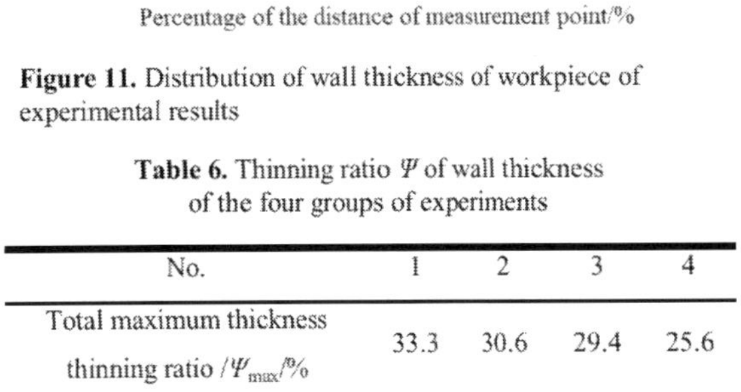

Figure 11. Distribution of wall thickness of workpiece of experimental results

Table 6. Thinning ratio Ψ of wall thickness of the four groups of experiments

No.	1	2	3	4
Total maximum thickness thinning ratio /Ψ_{max}/%	33.3	30.6	29.4	25.6
Relative maximum thickness thinning ratio/$\Delta\Psi_i$	1.30	1.20	1.15	1

DOI: 10.1051/matecconf/20168014005

Fig.12 shows the workpiece obtained by the multi deep-drawing spinning under the four combinations of the different surface-profile and movement-path of roller listed in Table 5. As shown in Fig.12, the height of the spun workpiece obtained by roller R12 and spinning from the bottom of the blank is obviously higher than that of the three other combinations. It indicates the wall thickness thinning under the R12 and spinning from the bottom of the blank is the most serious.

(a) (b) (c) (d)

Figure 12. Spun workpiece: (a) R12, Spinning from bottom (b) R12, Spinning from middle (c) R25-12, Spinning from bottom (d) R25-12, Spinning from middle

The experimental results show that the thinning ratio Ψ of wall thickness can be effectively reduced by spinning with composite surface-profile roller and spinning from the middle of the blank, but the influence decreases with the increasing of the spinning pass (as shown in Fig.11 and Table 2). The total thinning ratio Ψ of wall thickness increases rapidly with the increasing of the spinning pass, which leads to the decline of the influence of surface-profile and movement-path of roller on the thinning ratio Ψ of wall thickness of workpiece. Therefore, the number of the spinning pass should be controlled strictly to avoid the over thinning defect.

5 Conclusions

The influence of surface-profile and movement-path of roller on thinning ratio Ψ of wall thickness of workpiece during multi-pass deep drawing spinning is researched, the conclusions are as follows:

(1) The surface-profile and movement-path of roller have great influence on wall thickness thinning during multi-pass deep drawing spinning, and compared with the movement-path of roller, the influence of surface-profile of roller is more significant.

(2) For the 1060-O pure aluminum, the thinning ratio Ψ of wall thickness is the smallest when spinning with the composite surface-profile roller of R25-12 and spinning from the middle of the blank.

(3) The influence of the surface-profile and movement-path of roller on the thinning ratio of wall thickness decreases with the increasing of the spinning pass, the number of the spinning pass should be controlled strictly to avoid the over thinning defect.

Acknowledgments

This project was financially supported by National Natural Science Foundation of China (51075153) and Open Foundation of National Engineering Research Centre of Near-net shape Forming for Metallic Materials, South China University of Technology(2012001).

References

1. Q.X. Xia, G.F. Xiao, H. Long, X.Q. Cheng, X.F. Sheng. Int J Mach Tools Manuf, **85**(2014): 100-121.
2. O. Music, J.M. Allwood, K. Kawai, A review of the mechanics of metal spinning, J Mater Process Technol, **210** (1) (2009) : 3-23.
3. C.C. Wong, T.A. Dean, J. Lin. Int. J. Mach Tools Manuf, **43**(2003): 1419-1435.
4. L. Wang, H. Long, J. Mater. Process. Technol., **211**(2011): 2140-2151.
5. C. Zeng, S.J. Zhang, Q.X. Xia, H.X. Xie. Forging & Stamping technology, 2014, **39**(1): 58-63 (In Chinese).
6. B. Lin, S.R. Zhou, J.J. Liu, J.M. Gu, D.H. Wang, W. Cai. Forging & Stamping technology, 2010, **35(4)**: 56-61 (In Chinese).
7. F. Ma, H.Yang, M. Zhan. Mechanical Science and Technology for Aerospace Engineering, 2007, **26(3)**: 309-313 (In Chinese).
8. Q.X. Xia, S. Shima, H. Kotera, D. Yasuhuku. J. Mater. Process. Technol., **159** (2005): 397–400.
9. X.P. Wang. *Stamping handbook*. Beijing: China Machine Press, 2004 (In Chinese).
10. K. Essa, P. Hartley. Int. J. Mater. Form., **2**(2009): 271-281.
11. Q.X. Xia, X.Q. Cheng, H. Long, F. Ruan. Int J Adv Manuf Technol, **59**(1-4) : 263~272.

MATEC Web of Conferences **80**, 14006 (2016) DOI: 10.1051/matecconf/20168014006
NUMIFORM 2016

Finite Element Analysis of hot Single Point Incremental forming of hip prostheses

Manel Sbayti[1,a], Andrea Ghiotti[2], Riadh Bahloul[1], Hedi Belhadjsalah[1] and Stefania Bruschi[2]

[1]*University of Monastir, National Engineering School, Mechanical Engineering Laboratory, Avenue Ibn El Jazzar, 5000, Monastir, Tunisia*
[2]*University of Padua, Department of Industrial Engineering, via Venezia 1, 35131, Padova, Italy*

Abstract. The titanium alloy Ti-6Al-4V is one of the most frequently used materials for biomedical applications due to its biocompatibility and excellent mechanical properties. However, wide usage of Ti-6Al-4V sheet is limited by its poor room-temperature formability. Therefore, hot Single Point Incremental Forming (SPIF) has been used to improve its formability. This paper aims at proving the feasibility of the hot incremental forming of an acetabular component of hip prosthesis by Finite Element simulations. The effect of process parameters, namely the forming temperature and the punch diameter on the geometric accuracy and the failure of the final product are investigated.

Keywords: Single Point Incremental Forming / High Temperature/ Formability / Ti-6Al-4V

1 Introduction

Single Point Incremental Forming (SPIF) is a variant of incremental forming suitable for prototyping and small batch production due to its flexibility and low tool effort. The flat sheets metals are progressively deformed using a hemispherical end forming tool whose path is generated in a CNC machining center or robot [1].

The selection of suitable technique for manufacturing of a specific product represents a complex issue. It is a multicriteria decision that depends on several factors regarding part design and complexity, material processing ability, desired accuracy, physical and mechanical properties of part, production quantity and cost, etc.

The technology of Incremental Sheet Forming (ISF) was primarily applied in the automobile and aerospace's industry [2, 3]. However, there are other branches with an important potential for the technology, such as the biomedical field. When addressing its use in medicine, it is possible to produce thin walled custom-made metal components. Araújo et al. [4] evaluated the SPIF process for producing titanium customized maxillofacial implants and demonstrated the role of this technology in manufacturing customized medical parts. Castelan et al. [5] presented the manufacturing by ISF of the custom-made cranial implants from a sheet of pure grade 2 titanium from Digital Image and Communications in Medicine (DICOM) images using 3D printing, CAD/CAM technology and incremental sheet forming. From Computerized Tomography (CT) images of a fractured skull, a CAD model of the skull BioModeling and a restorative implant were constructed digitally. Duflou et al. [6] used three different materials: AA1050, AISI 304 and medical grade titanium for accurate manufacture of cranio-facial implant using single point incremental forming process.

Fiorentino et al. [7] manufactured a plate prosthesis using a titanium alloy and Polycaprolactone (PCL). Although their results were promising, an optimization of the process parameters should be done in order to increase the accuracy of the part. The work of Milutinović et al. [8] has been focused on the manufacturing of a denture base (framework) of a complete denture by SPIF technology. The research included two different materials: carbon and stainless steel. They concluded that the new procedure enabled an additional mass reduction of the dental base (the final sheet thickness is 0.5mm instead 0.8mm of the original thickness) which is highly desirable from viewpoint of patient and comfort of wear.

Among biocompatible metallic materials, titanium and its alloys such as Ti-6Al-4V are increasingly used in biomedical applications due to their excellent properties such as corrosion resistance, strength-to-weight ratio and high toughness. However, Ti-6Al-4V possess low formability at room temperature. In order to improve the forming behavior of Ti6Al4V sheets, deformation under elevated temperatures is needed. Therefore, multiple concepts for warm ISF processes were recently developed and introduced in the technical literature [9, 10]. It was shown that, by using hot ISF process, the formability of lightweight alloys among them Ti-6Al-4V can be significantly improved [11, 12, 13, 14].

[a] Manel Sbayti: manelasbayti@gmail.com

DOI: 10.1051/matecconf/20168014006

The present paper aims to provide a preliminary numerical study of hot single point incremental forming technology using the commercial finite element computer program ABAQUS. The main objective of these research works is to evaluate the feasibility of producing customized acetabular component of hip prosthesis with an acceptable geometrical accuracy with this process and based on the use of biocompatible materials such as Ti-6Al-4V titanium alloy.

Figure 1 represents the component of hip prosthesis to be studied in the present work.

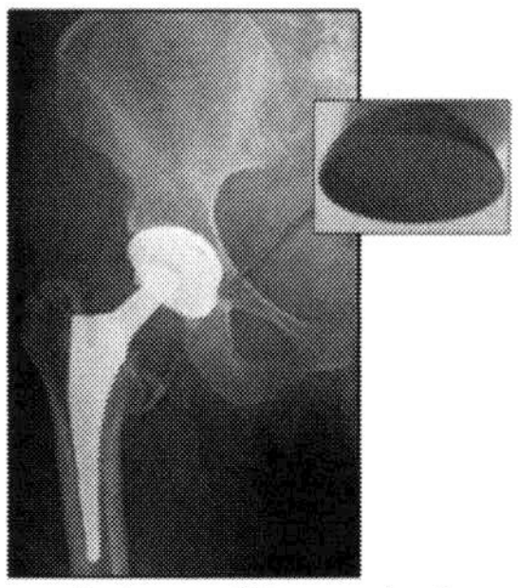

Figure 1. The studied part of hip prosthesis.

2 Finite element aspects and material modeling

The accuracy of Finite Element Method (FEM) in the simulation of metal forming processes depends both on the constitutive laws used and their material parameters identification. In order to simulate the SPIF process, temperature dependence is taken into consideration in the current simulation.

2.1 Material

2.1.1 Punch material

The material used for the hemispherical tool was the SM30 tungsten carbide. Table 1 shows the thermos-physical properties of the punch material.

Table 1. Thermo-physical material parameters [15].

Material parameters of the tool	**Values**	
Density ρ [Kg/m³]	12800	
Poisson's ratio υ	0.22	
Young's modulus E [MPa]	630000	
Heat conductivity λ [W/(m·°C)]	T (°C)	Value
	23	44.6
	600	59
	800	51.2
Specific heat C_p [J/(kg·°C)]	T (°C)	Value
	20	226
	320	296
	520	326
	820	342

2.1.2 Sheet material

The sheet material used in the numerical simulation is the Ti6Al4V alloy. The mechanical and thermo-physical properties of the Ti-6Al-4V alloy are defined as temperature (T) dependent. The thermo-physical material parameters used for the work piece are compiled in Table 2.

Table 2. Thermo-physical material parameters [16].

Material parameters of Ti-6Al-4V sheet	**Values**
Density ρ [Kg/m³]	4430
Poisson's ratio υ	0.22
Young's modulus E [MPa]	E(T)= 0.7412T+113375
Thermal expansion α [1/°K]	8.9E-6
Heat conductivity λ [W/(m·°C)]	$\lambda(T)= 7.039\ e^{0.0011T}$
Specific heat C_p [N/(mm²·°C)]	$C_p(T)= 2.24\ e^{0.0007T}$

2.1.3 Material constitutive model

The parameters for the Johnson-Cook (J-C) constitutive model of a biomedical grade Ti6Al4V alloy were determined for the numerical modelling of its deformation in the warm SPIF process. So, the J-C flow stress is proposed as the applied material model for the workpiece in this study. It considers simple forms of empirical relations of stress with strain, strain rate and temperature as follows:

$$\sigma_f = \left(A + B \cdot \bar{\varepsilon}_{pl}^{\,n}\right) \cdot \left[1 + C \cdot \ln\left(\frac{\dot{\bar{\varepsilon}}}{\dot{\varepsilon}_0}\right)\right] \cdot \left[1 - \left(\frac{T - T_{room}}{T_m - T_{room}}\right)^m\right] \quad (1)$$

where, A, B, n, C and m are the constants of material, $\dot{\bar{\varepsilon}}$ is the strain rate which is normalized with a reference strain rate $\dot{\varepsilon}_0$ ($1s^{-1}$). T_{room} represents the room temperature and T_m is the melting temperature of the material equal to 1604°C. The parameter *n* takes into account the strain hardening effect and the two coefficients *m* and *C* model respectively the thermal softening effect and the strain rate sensitivity. The identified Johnson-Cook constitutive model parameters relative to the used material are illustrated in Table 3.

Table 3. Constants of the Johnson-Cook constitutive model for the Ti-6Al-4V alloy.

A (MPa)	B (MPa)	C	n	m
460	1450	0.08	1.31	0.85

DOI: 10.1051/matecconf/20168014006

2.2 Description of the numerical model

As the SPIF process involves large material deformation without the presence of the dies, suitable algorithm should be employed in the FEA. In this investigation, a three-dimensional thermo-viscoplastic FE model is set up for the simulation of the SPIF process to obtain thermos-mechanical responses of the process. The numerical computations were performed with the commercial code ABAQUS/EXPLICIT software.

2.2.1 Tool trajectory

The tool path design is fundamental in the design of the SPIF processes, since it impacts on the dimensional accuracy, surface finish, formability, thickness variation and processing time. The punch movement must be gradual in order to progressively deform the sheet. The tool path is prescribed by the Numerical Control (NC) data that is generated from a CAD model of the component to be formed. The forming strategy consists of a single forming stage where the tool traces along a sequence of contour lines with a vertical increment step size ($\Delta z = 0.2$ mm) in between. In all the simulations, the diameter of the tool was 10 mm. The shape of the titanium alloy sheet Ti-6Al-4V has been considered square for different temperature values with a size of 90 mm× 90 mm × 1mm.

The part is formed according to the tool trajectory presented in Figure 2. The considered tool path is made up of a series of contours generated transverse to the long axis of the real part. The forming tool follows to the predetermined tool path and gradually forms the sheet metal in a series of incremental steps until the final depth is reached.

Figure 2. Schematic description of the tooling path generated by CATIA and integrated into ABAQUS.

2.2.2 Boundary conditions

Modeling the interaction between the tool and the sheet is one of the most important considerations necessary to simulate the incremental forming process correctly. The interaction properties between the sheet and the contact area with the punch are defined by the Coulomb's friction model expressed as follows :

$$\tau_{fric} = \mu P \tag{2}$$

where τ_{fric} is the friction shear stress, μ the friction coefficient and P the normal contact pressure. Hence, a fraction of dissipated energy converted into heat is tacked into account. the heat generation from the friction is described as follows: So, the rate of frictional energy dissipation is described as follows:

$$q_{fric} = \eta \dot{\gamma}\, \tau_{fric} = \eta \dot{\gamma} \mu P \tag{3}$$

where $\dot{\gamma}$ is the slipping rate and η the fraction of dissipation energy converted into heat.

The sheet blank is experimentally supported on a four-sided fixture and is clamped rigidly to this fixture with a backing plate to support the punch displacement. From numerical point of view, the sides of the initial blank are fixed in all directions and for each simulation, a uniform temperature has been applied to the entire sheet (Figure 3).

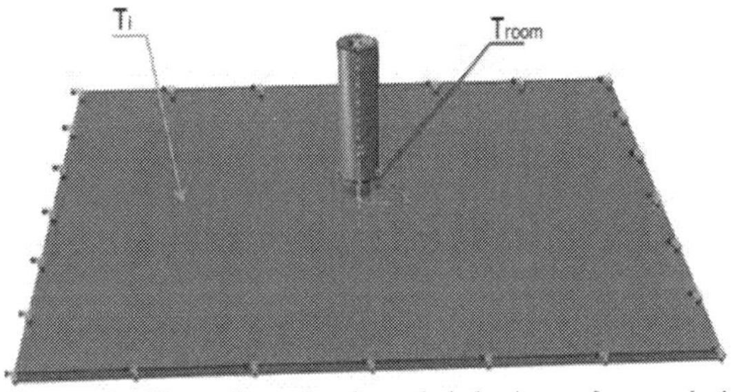

Figure 3. Three-dimensional exploded view of numerical model used in SPIF at high temperatures.

2.2.3 Finite element mesh

The finite element meshing subdivision of the initial sheet is depicted in Figure 4. In the FE model, eight-node thermally coupled brick, with trilinear displacement and temperature, reduced integration and hourglass control (C3D8RT) were applied in the fully coupled thermal analysis to model the sheet blank. In the numerical calculations, the sheet was meshed in three layers in the thickness direction to take into account of bending stiffness and the global size of elements is 0.8 mm×0.8 mm

Also, in this study, the forming tool has been meshed using 4-node thermally coupled tetrahedron elements (C3D4T), linear in displacement and temperature.

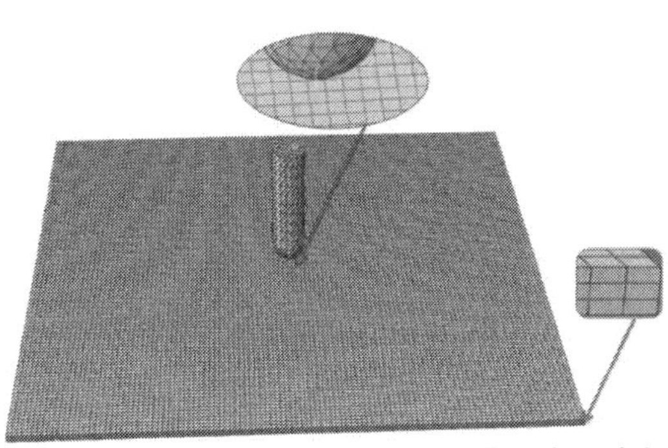

Figure 4. The finite element meshing configuration of the initial blank.

3 Results and discussions

3.1 Forming temperature influence

In Figure 5 are pictured the 3D iso-values of the final reached temperature distributions for various initial forming temperatures. Figures 6a$_1$, a$_2$, a$_3$ and a$_4$ report the simulated incremental forming sequences, and highlight the calculated temperature histories of the final part respectively for the forming temperatures T of 20°C, 400°C, 500°C and 600°C.

As it can be concluded from figures 6ai, the calculated final temperatures at the end of the process simulations increase due to initial sheet heating and to heat generation during the process. It can also be observed from these figures that the temperature distributions agree with the position of the tool on the sheet.

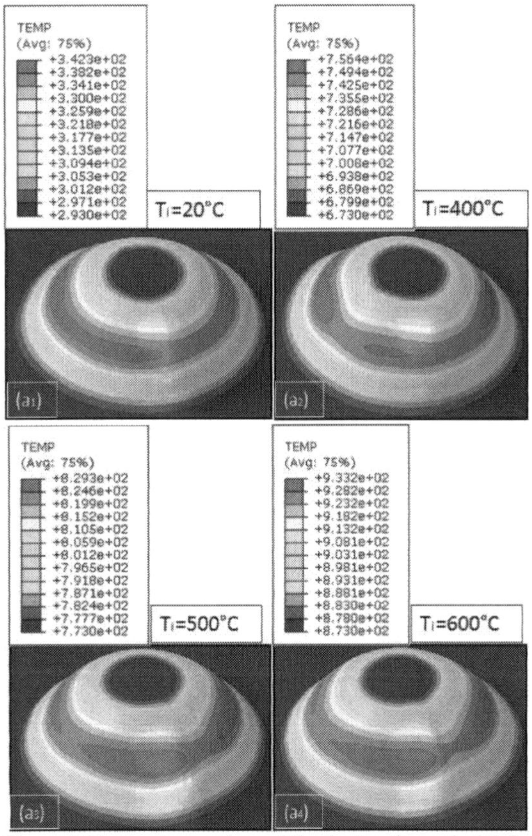

Figure 5. Temperature contours at the sheet surface calculated by FE modeling for various temperatures.

A coordinate system is assigned to the simulated model to measure set of points on profile of upper mesh layer (the model is meshed in three layers). Results are taken along a radial path after the removal of the backing plate, which is shown in Figure 6.

Figure 6. Final shape of parts with a global heating: radial path definition monitored during the simulation for measuring the depth profiles.

Figure 7 illustrates the thickness distribution versus radial distance (Figure 6) formed at various temperatures: Ti = 20°C, 400°C and 600°C. There is a rigorous reduction in thickness in the stressed areas and a fixing thickness to its initial value at the bottom of the sheet. At high temperature the distribution of the thickness is more homogeneous.

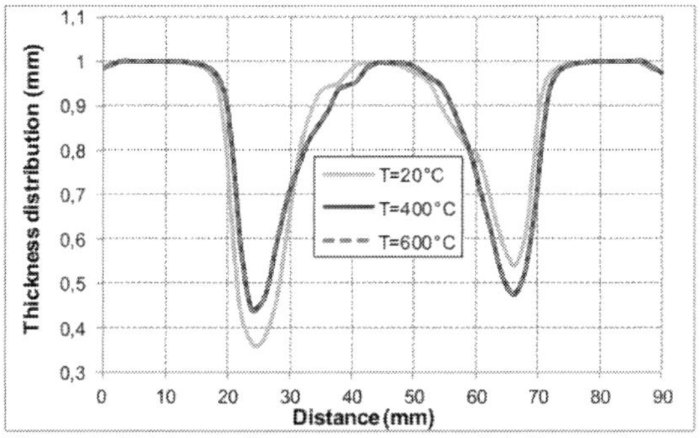

Figure 7. Thickness distribution versus radial distance formed at various temperatures: Ti = 20°C, 400°C and 600°C.

Figure 8 represents the final profiles of the part according to the radial distance from the clamped corner described in Figure 6. Compared to the conventional process, this strategy can dramatically improve geometric accuracy, since the springback effects decrease as the temperature increases.

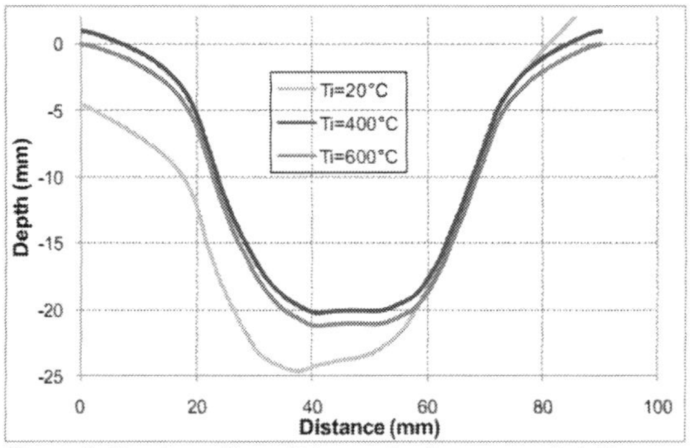

Figure 8. Depth profiles of hip prosthesis versus radial distance formed at various temperatures: T_i = 20°C, 400°C and 600°C.

3.2 Punch diameter influence

Figure 9 shows the evolution of the temperature along a radial path at T = 400°C with the variation in punch diameter of 6, 8 and 10 mm values. Thus, the dependence of the punch position on distribution of the final state of a temperature against the sheet and after the accumulation of numerous incremental deformation passes throughout the process proves to be logical. It can be noted that the smaller the tool size the higher the temperature calculated by the FE model.

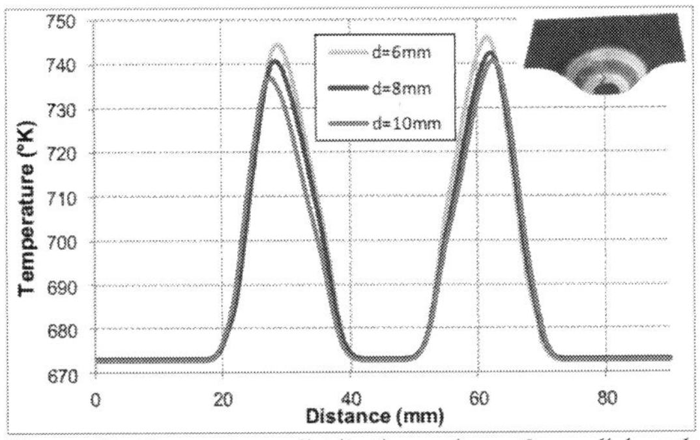

Figure 9. Temperature distributions along the radial path predicted by FEM for different punch diameter variations : a_1) d = 6mm, a_2) d = 8mm, a_3) d = 10mm at T = 400°C.

3.3. Thermo-viscoplastic material behaviour coupled with damage

Damage model proposed by Johnson-Cook is used in conjunction with J-C yield model. According to classical damage law, the theoretical expression of this response in a finite element representative of fracture or macro-crack initiation resulting from the material damage of workpiece is defined by:

$$D = \sum \frac{\Delta\bar{\varepsilon}}{\bar{\varepsilon}_f} \quad (4)$$

where $\Delta\bar{\varepsilon}$ is the increment of equivalent plastic strain during an integration step, and $\bar{\varepsilon}_f$ is the equivalent strain value at fracture under current conditions. The initiation of cracking and damage of product is then allowed to occur when this damage variable reached its critical value (D = 1) in any element and the concerned one is removed from computation. According to Johnson-Cook damage law, the general formulation of the fracture strain at critical damage can be written as:

$$\bar{\varepsilon}_f = \left[D_1 + D_2.\exp(D_3\eta)\right] \times \left[1 + D_4.\ln\left(\frac{\dot{\bar{\varepsilon}}}{\dot{\bar{\varepsilon}}_0}\right)\right] \times \left[1 + D_5.\left(\frac{T - T_{room}}{T_m - T_{room}}\right)\right] \quad (5)$$

where the Johnson-Cook damage parameters are represented in the following forms:

D_1: Initial failure strain,
D_2: Exponential factor,
D_3: Triaxiality factor,
D_4: Strain rate factor,
D_5: Temperature factor,
η: The ratio of the hydrostatic pressure to the effective stress

The Titanium 6Al-4V failure strain parameters are summarized in Table 4.

Table 4. Johnson-Cook failure parameters for Ti-6Al-4V.

D_1	D_2	D_3	D_4	D_5
-0.09	0.25	-0.5	0.014	3.87

3.3.1 Influence of heating temperature on damage field distribution

Figure 10 shows the iso-values of the scalar damage parameter modeled by the Stiffness Degradation SDEG during the forming process at four different temperatures. It should be noted that SDEG = 1.0 means that the corresponding elements failed and a crack occurred. According to these results, the critical damage appears at room temperature since the SDEG value is very close to 1 in the contact region between punch and sheet. As expected in these conditions, the damage variable increased which confirm the previous result that the forming of the part doesn't occur at Ti = 20°C.

These results are presented at the same time when the maximum degradation appeared in the part at room temperature. Hence, the damage variable SDEG is investigated at the same frame for the different temperature values (approximately the middle of the step). As expected, at T_i = 20°C the part is completely damaged in a while frame, because as we have cited that the Ti-6Al-4V titanium alloy has a very poor formability at room temperature. However, at T ≥ 400°C and at the same time of the step calculation, the part is not damaged since the damage parameter SDEG is much lower to 1. This conclusion is proved by the experiences of several authors among them we can cite G. Fan et al. [11].

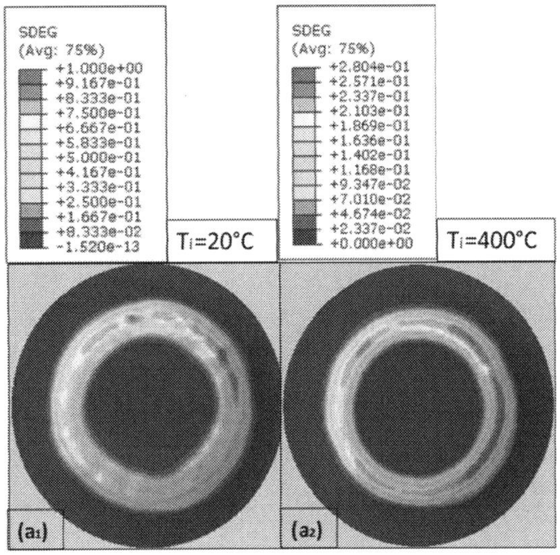

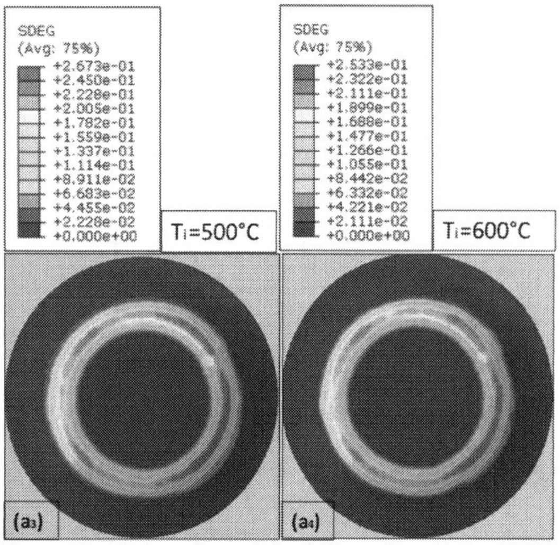

Figure 10. 3D iso-values of the Stiffness Degradation SDEG obtained by FEM calculation for various temperatures: a_1) T_i = 20°C, a_2) T_i= 400°C, a_3) T_i = 500°C, a_4) T_i = 600°C at the middle of the step.

According to the same path previously indicated in Figure 6, the final geometry was predicted at three temperatures after removal of the backing plate, as shown in Figure 11. Such a thermos-mechanical problem results in appearance of springback effect after the stress relaxation during the forming of thin Ti-6Al-4V sheets. Figure 11 reports the numerical results in terms of the final depth reached along the meridian direction in parametric form.

So, taking into account the damage effect, it was observed that the maximum final depth cannot exceed 8.77 mm at room temperature (Ti =20°C). The plasticity of the Ti-6Al-4V alloy at 500°C is better than that at room temperature, therefore the ductility is greatly enhanced, and the yield stress is reduced in half [17]. In the investigated range, a temperature from 400 to 600°C should be selected to achieve a better formability in terms of forming penetration and geometrical accuracy.

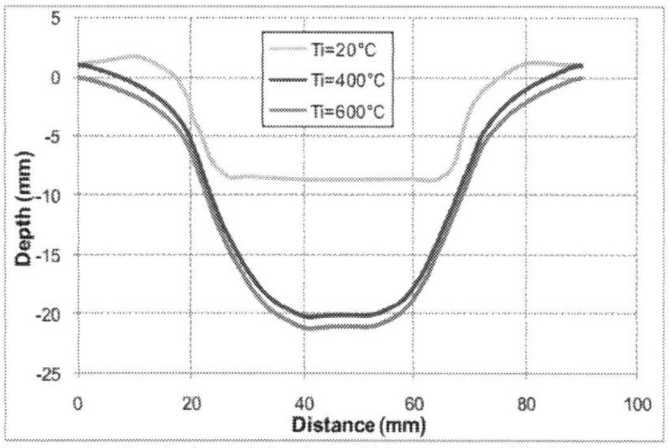

Figure 11. Evolution of depth profiles versus radial distance.

3.3.2 Influence of punch diameter on damage field distribution at T = 400°C

Figures 12a1, a2 et a3 show the evolution of the accumulated damage defined by stiffness degradation SDEG for various tool diameters. As illustrated in Figure 12, there are no damaged elements at T = 400°C with the three tool diameters (6, 8 and 10 mm). At the end of sheet forming operation, the SDEG values are close and doesn't exceed 0.547, proving that the parameter punch size has a small effect on the damage generation. Finally, it can be concluded that the effect of forming temperature is more important on the geometric accuracy of the part.

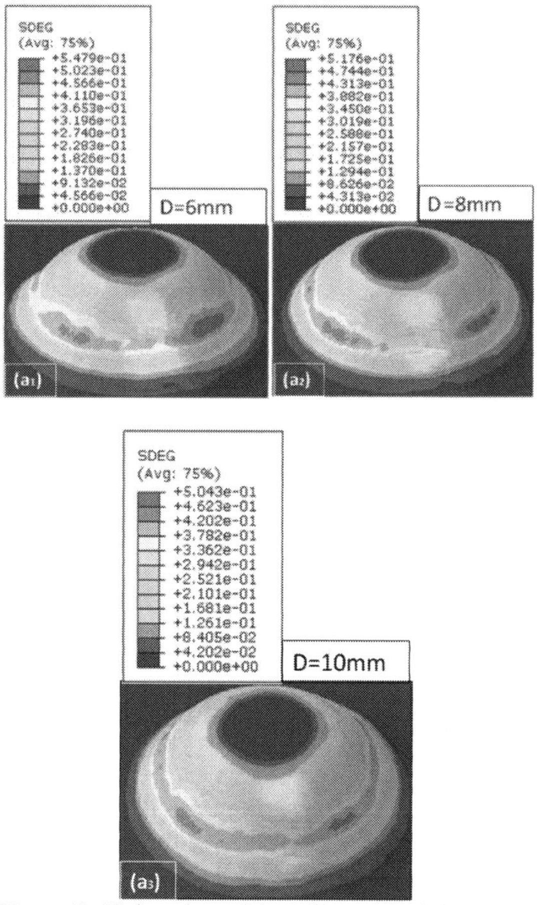

Figure 12. 3D iso-values of the Stiffness degradation SDEG obtained by FEM calculation for various punch diameters: a_1) d = 6 mm, a_2) d = 8 mm, a_3) d =10 mm at the middle of the step.

4 Conclusions

In this work, by evaluating FE simulation of hot SPIF process for an acetabular component, the feasibility of manufacturing customized titanium hip by using the ISF approach is studied and the particular issues and possible solutions were discussed. The conclusions of this work may be summarized as follows:

1) ISF of acetabular by using Grade 5 titanium sheets is a feasible solution and it shows clearly the potential for real medical application.
2) The geometry accuracy is improved at T ≥ 400°C
3) First results show promise and the process seems to be satisfactory for the application of hip prosthesis.
4) Further development is still needed to evaluate the effect of the majority of process parameters on the formability of the part to optimize these parameters for the experimental study.

In the paper presented, possibilities of applying SPIF technology to produce a metal acetabular of hip prosthesis have been numerically investigated.

References

1. M. Callegari, D. Amodio, E. Ceretti, C. Giardini, In: Low, K. (Ed.). Pro Literatur Verlag, Germany, pp. 493–514 (2006)
2. W.B. Lievers, A.K. Pilkey, D.J. Lloyd, Acta Mater **52**(10):3001–3007 (2004)
3. G. Hussain , H. R. Khan , L. Gao, N. Hayat. Mater Manuf Process **28**(3):324–329 (2012)
4. R. Araújo, P. Teixeira, M.B. Silva, A. Reis, , P.A.F. Martins, Key Eng. Mat., 544–557: 1388–1393 (2013).
5. J. Castelan, L. Schaeffer, A. Daleffe, D. Fritzen, V. Salvaro, FP. Silva. Revista Brasileira de Engenharia Biomédica. **30**(3):265-73 (2014)
6. Duflou JR, Behera AK, Vanhove H, Bertol LS. Key Engineering Materials. 549:223-30 (2013)
7. A. Fiorentino, G.P. Marenda, R. Marzi, E. Ceretti. In: Proceedings of the 5th International Conference on Advanced Research in Virtual and Rapid Prototyping. p. 589-594 (2011)
8. M. Milutinović , R. Lenđel, M. Potranb , D. Vilotić, P. Skakuna, M. Plančak. Journal for Technology of Plasticity, Vol. **39** Number 2, (2014)
9. G. Fan, L. Gao, G. Hussain, Zhaoli Wu. International Journal of Machine Tools & Manufacture **48** 1688–1692 (2008)
10. A. Göttmann, J. Diettrich, G. Bergweiler, M. Bambach, G. Hirt, P. Loosen, R. Poprawe. Production Engineering - Research and Development, **5** (3), 263-271.(2011)
11. G. Fan, F. Sun, X. Meng, L. Gao, G. Tong. Int J Adv Manuf Technol **49**:941–947 (2010)
12. P. Homola, L. Novakova, V. Kafka, Mariluz P. Oscoz, Int J of Chemical, Molecular, Nuclear, Materials and Metallurgical Engineering Vol:**8**, No:1, 2014
13. M. Honarpisheh, M. J. Abdolhoseini, S. Amini. Int J Adv Manuf Technol (2015)
14. R. Liu, B. Lu, D. Xu, J. Chen, F. Chen, H. Ou, H.Long. Int J Adv Manuf Technol (2015)
15. F. Zemzemi. PhD thesis: Central School of Lyon, pp. 92-93, (2007)
16. T. Ozel, M. Sima, A. Srivastava, Finite Element Simulation of High Speed Machining Ti-6Al-4V Alloy using Modified Material Models, Transactions of the NAMRI/SME, Vol.**38**, pp. 49-56, (2010)
17. R.S. Lee, H.C. Lin. Journal of Materials Processing Technology **79** 224–235 (1998)

Thinning behavior of laminated sheets metal in warm deep-drawing process under various grain sizes

Mehran Kadkhodayan[a] and Ehsan Afshin

Department of Mechanical Engineering, The Ferdowsi University of Mashhad, Mashhad, Iran

Abstract. The purpose of present research is to investigate the thickness distribution on the warm deep-drawing process of laminated sheets consisting of aluminum alloy series 1050, 5052 and stainless steel 304 (SUS), experimentally. Individually for each layer, the influences of blank temperature and grain size on thinning behavior are clearly demonstrated. In order to survey the thinning behavior in laminate sheet behavior during warm deep-drawing process; three blank temperatures namely, 25° C, 100° C and 160° C are examined. Moreover, to obtain different grain sizes, the aluminum sheets are annealed at 350° C, 400° C and 450° C for 1 hour. Results indicate that increasing temperature and grain size lead to maximum thinning in all layers in Al 1050/SUS and Al 5052/SUS specimens increase. In addition, the most susceptible zone to fracture in aluminum sheets (Al 1050 and Al 5052) is punch profile radius region; nevertheless, for stainless steel sheets this zone switch to central zone of formed cup. These can be attributed to the fact that the adhesive layer play a crucial role in thickness distribution of steel 304 layer, therefore the distribution of thickness strain for adhesive layer is also investigated.

1 Introduction

Nowadays, the warm deep-drawing process is used extensively because of their application advantages such as manufacturing of complex shape, improved formability, reduced production time and controlled plastic flow. The temperature has significant effect on the material behavior during process and accuracy of finished parts. Moreover, due to wide range of application and unique characteristics of laminated sheets such as; high strength, low density, damping covering structures and corrosion resistibility researchers have been paying more attention to laminated sheets. Consequently, they are used in various industrial fields such as aerospace, automobile, chemical and electrical industries. The laminated sheets formed by warm deep-drawing process can be utilized in manufacturing of parts with different inner and outer conditions, i.e. corrosion, wear resistance and thermal and electrical conductivities [1, 2]. In the recent years, numerous researchers have paid more attention to deep-drawing process of laminated sheets [3-5]. Nevertheless, their surveys were confined to processes conducted in room temperature.

It is known that aluminum alloys are favorable for aerospace and automotive industry because of their, (i) excellent high-strength to weight ratio, (ii) superior corrosion resistance, (iii) high weldability, and (iv) recyclability. Furthermore, stainless steel sheets are extensively used for manufacturing various parts in the kitchen, vessels, transformation, etc., because of their high corrosion-resistivity and proper appearance. Warm deep-drawing for aluminum alloy has already been dealt with by many scholars [6-8]. Ethiraj and Kumar surveyed the St 304 deep-drawing process through finite element simulation [9] and empirical observation [10].

The most commonly observed defects in deep-drawing process are thinning, fracture and wrinkling of the material. Thinning and fracture are directly related to part performance and safety. Additionally, annealing and grain size are significant factors in deep-drawing to improve the quality of product, formability, wrinkle pattern, forming load, limit drawing ratio and forming limit diagram (FLD). The grain size is strongly reliant on annealing and re-crystallization. Furthermore, the effect of cold rolling is modified by heat treatment; therefore, properties of metal are changed. Sherbiny et al. [11] clarified the influences of punch and die radii on thinning and residual stresses of sheet metal in the deep-drawing process experimentally. Thinning, thickness distribution and spring back of sheet metal on deep-drawing process successfully were predicted by Zein et al. [12]. Zoesch et al. [13] detected thinning and crack of material during deep-drawing process.

The objective of present work is to investigate the thickness distribution of laminated sheets in warm deep-drawing process. In addition, effect of temperature, grain size (GS) and Blank Holder Force (BHF) on load-displacement curves were studied in the authors previous work [14, 15]. The thickness distribution and fracture zone for each individual layer is also studied in

[a] Mehran Kadkhodayan: kadkhoda@um.ac.ir

the current study. In summary, the objectives are: (1) performance a heat treatment to obtain various grain sizes on aluminum alloys 1050 and 5052, and revelation their microstructures (2) conduction of several tests to observe the effect of grain size on some parameters such as stress and elongation, (3) determination of effect of increasing temperature and grain size on thickness distribution.

2 Experimental procedures

2.1 Warm deep-drawing equipment description

The scheme of experimental equipment utilized for warm deep-drawing is illustrated in Figure 1. Figure 2 shows the test apparatus which consists of a draw die flange with circular shape, a blank-holder, 8 springs which supply blank-holder force (BHF), punch and heat equipment. Some features of the test rig are shown in Table 1. MO40 steel and SPK steel with appropriate mechanical properties and hardening treatments are used for the materials of the punch and die, respectively. Two annular electrical heaters with a power of 1400 W are utilized to raise the temperature and the required time to reach the temperature of 160° C is about 30 min. According to the tool design, the annular electrical heaters are placed on outer circumference of drawing die and blank-holder. The heaters warm the drawing die and blank-holder; therefore, the blank is heated, even though it has no direct contact with the heaters. A thermocouple is placed in a die, and a digital thermostat with the precision of 1° C controls the temperature. A hydraulic press with a maximum load capacity of 60 tons and a punch speed of 540 mm/min is used for this process.

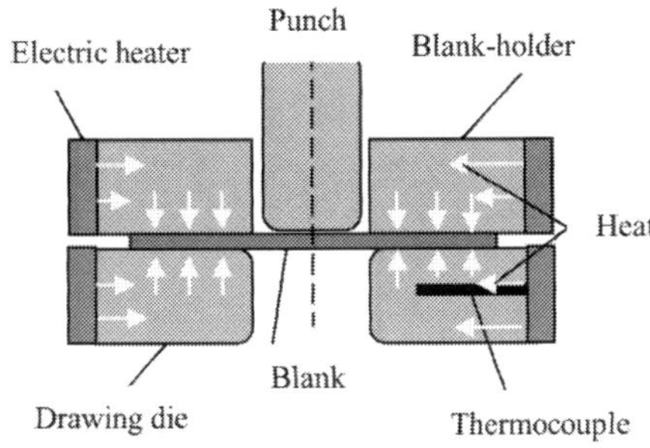

Figure 1. Schematic representation of the equipment for the deep-drawing test.

2.2 Material and specimen preparation

The materials of layers used in this process are stainless steel 304 (SUS), aluminum alloys 1050 and 5052 with the chemical compositions as given in Table 2. Two grades of aluminum alloys of 1050 and 5052 are combined with stainless steel 304 to make two various types of laminated sheets (Al 1050/SUS and Al 5052/SUS). Taking forming load, die diameter and drawing depth into account, the optimal diameter of the samples is selected as 132 mm, and thicknesses for aluminum alloys 1050 and 5052 and stainless steel 304 are as 0.8, 1 and 0.4 mm, respectively. Loctite 5368 adhesive is utilized to stick the sheets together due to some advantages such as (i) high flexibility, (ii) medium to high initial strength (iii) and high heat resistant up to 300° C. Some of the mechanical properties of Loctite 5368 adhesive based on the technical data sheet are presented in Table 3 [16]. In addition, the adhesive layer is uniformly distributed between sheets layer by rig press. Ultimate thickness for Al 5052/SUS and Al 1050/SUS specimen become 2.2 and 2 mm, respectively. Figure 3 exhibits specimen after forming. In all the samples, aluminum sheets are set as upper layer to be in contact with punch and stainless steel sheets are set as lower layer in contact with drawing die. Moreover, process performed in dry condition.

Figure 2. Experimental test equipment.

Table 1. Dimensions of various parts of the deep-drawing die set.

Dimension	Value (mm)
Punch diameter	65.5
Die inner diameter	70
Die outer diameter	153.5
Punch and die profile	5
Blank-holder inner	67
Blank-holder outer	164
Clearance	2.25

2.3 Annealing

One of the noticeable factors in forming process is grain size (GS). Influence of grain size on parameters such as flow stress, machining, friction coefficient and electrical properties are undeniable. Hence, specimens are annealed at 350° C, 400° C and 450° C for 1 hour, and then cooling process is performed within the furnace for 24 hours. In order to measure the grain size, samples are etched by (50 ml Poulton reagent, 25 ml HNO_3, 40 ml of solution of 3gr chromic acid per 10 ml of H_2O) etchant. The excellent resolution of individual grains and microstructure of Al 5052 and Al 1050 are revealed by Polarized Light Microscope (PLM), see Figures 4 and 5. The PLM microstructure results indicated that the crystallization has taken place for all samples.

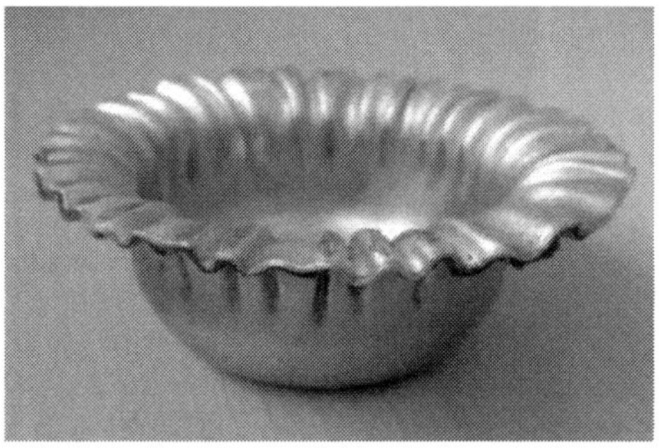

Figure 3. Sample of laminated sheet after forming.

Table 2. Chemical composition (wt. %) of aluminum alloys and stainless steel.

Element	Al 1050	Al 5052	St 304
Si	0.12	0.25	0.44
Fe	0.28	0.40	70.55
Cu	0.02	0.10	0.14
Mn	0.02	0.10	1.38
Mg	0.02	2.82	-
Zn	0.011	0.10	-
Ti	-	0.15	-
Cr	-	0.10	19.97
Al	99.49	95.98	-
V	-	-	0.12
Ni	-	-	7.00

2.4 Tensile test and flow stress

To determine the engineering stress-strain behavior of aluminum alloys and St 304 tensile tests are carried out on a Zwick-Z250 uniaxial tensile test machine at room temperature. Tensile samples are prepared to have the dimension with gauge length of 50.0 mm, width of 12.5 mm based on the ASTM-0557M-02 standard and are cut in the direction of 0° which conforms directly to the rolling direction. The heat treatment process on aluminums sheets provides various grain sizes for samples which give the possibility to study the influence of grain size on the flow stress. The testing speed has remarkable effect on engineering stress-strain behavior as reported by Rezaei Ashtiani et al. [17]. Their results indicated that magnitude of ultimate stress (S_{uts}) and yield stress (S_y), when the grain size increase, depends on testing speed. Based on the punch speed, the tensile tests are carried out with testing speed of 0.002 $(1/S)$. The engineering stress-strain curves for Al 1050, Al 5052 and St 304 are shown in Figure 6.

Figure 4. PLM microstructure of Al 5052 specimen with three annealing temperatures of (a) 350°C, (b) 400°C and (c) 450°C.

Figure 5. PLM microstructure of Al 1050 specimen with three annealing temperatures of (a) 350°C, (b) 400°C and (c) 450°C.

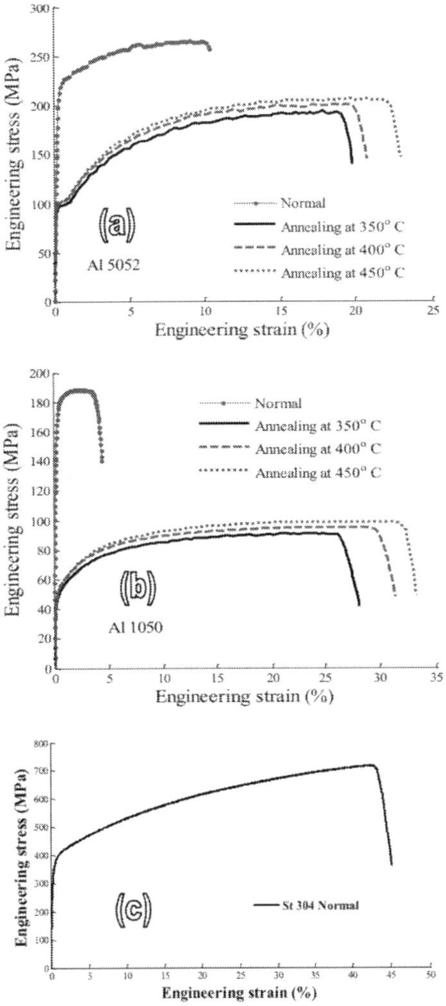

Figure 6. Engineering stress-strain behavior for (a) Al 5052, (b) Al 1050 and, (c) St 304.

3 Results and discussion

3.1 Effect of annealing on grain size and mechanical properties

According to the PLM microstructure for Al 5052 and Al 1050, it can be concluded that re-crystallization of aluminum alloys 5052 and 1050 occurred in the 350° C, for all samples. The grain sizes are measured based on the ASTM-E122-12 standard for different annealing temperatures and are reported in Table 4. Moreover, in Figures 4 and 5 effects of annealing on microstructure and grain size are illustrated. Results indicate that the grain size in Al 1050 and 5052 rise when annealing temperatures increase. Furthermore, the results of engineering stress-strain curves demonstrate that elongation, yield stress and ultimate stress rise with increasing the grain size, Table 5.

Table 3. Mechanical properties for Loctite 5368 adhesive.

Temperature	Tensile Strength (MPa)	Elongation (%)
25°C	2.2	435
100°C	0.54	450
150°C	0.32	580

Table 4. Grain size for aluminum alloys.

Annealing Temperature	GS for Al 5052 (μm)	GS for Al 1050 (μm)
350°C	10	37
400°C	12	40
450°C	14	44

Table 5. Mechanical properties for (a) Al 5052 and, (b) Al 1050.

(a) Al 5052

Type	S_y (MPa)	S_{uts} (MPa)	Elongation (%)
Normal	223.8	266.5	10.37
Annealing at 350°C	98.2	193.9	19.8
Annealing at 400°C	101.2	200.8	20.7
Annealing at 450°C	102.1	206.8	23

(b) Al 1050

Type	S_y (MPa)	S_{uts} (MPa)	Elongation (%)
Normal	173.1	189.2	4.5
Annealing at 350°C	46.4	91.2	28.1
Annealing at 400°C	48.3	95.1	31.2
Annealing at 450°C	49.1	98.5	33.3

3.2 Thinning

3.2.1 Effect of blank temperature on thinning

One of the most significant factors in warm deep-drawing process is effect of blank temperature on thickness distribution in final product. Hence, for Al 5052/SUS and Al 1050/SUS samples, thickness strain variations of both layers are determined. Because the thickness variation in circumferential direction is almost symmetrical, it is investigated only diametrically. For this aim, the drawn component is sectioned diametrically by a water jet machine. The thickness of aluminum and stainless steel layer are measured at various positions from the centerline of the component by a micrometer, Figure 7. Some experiments with different blank temperatures namely, 25° C, 100° C and 160° C on samples contain Al 1050 and stainless steel 304 are carried out to study the effect of temperature on thinning. The variations of thickness strain for each layer is also determined in the radial direction and are presented in Figures 8 and 9. Maximum thinning for Al 1050 is taken place at radial distance of about 30 mm in the region of punch profile radius in all samples. Moreover, the increase in temperature coincides with rising the maximum thinning.

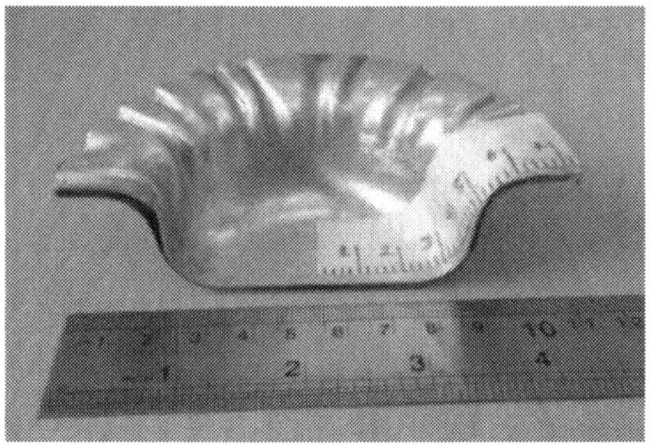

Figure 7. The section formed cup sectioned for evaluation of thickness strain.

Variations of thickness strain for St 304 at Al 1050/SUS are illustrated in Figure 9. It is seen that the maximum thinning for St 304 increases with raising the temperature. Maximum thinning for St 304 takes place in central region of cup. To show that tearing occurs in the bottom of cup, where significant thickness strain happens; some tests with higher BHF are conducted. One may see an unusual behavior here that tearing occurs in the central area of cup for St 304, see Figure 10 (a). These can be attributed to the function of adhesive layer in thickness distribution of St 304 layer. In fact, radial strain increases when blank temperature rises in the punch profile radius. Hence, the adhesive in this area moves and accumulates in the center area of cup which is especially true in for higher temperatures, and this adhesive movement causes more pressure in the center part. Additionally, the punch acts as a constraint on upper layer (Al 1050) which is not the case in the lower one (St 304). Therefore, rising pressure in the central area of cup has significant effect on thinning of lower layer, and maximum thinning takes place in center area for lower layer (St 304) of cup. To clarify this phenomenon, thickness strain of adhesive layer on sample Al 1050/SUS as an example is illustrated in Figure 11. It is seen that the maximum thickening for adhesive layer occurs at the center of punch. Hence, tearing for St 304 layer takes place at the center of cup. Figure 12 is provided for comparison between thickness strain in all layers in Al 1050/SUS specimen and final cup. It can be concluded that the maximum thickening among all layers corresponds with punch center in adhesive layer, and maximum thinning occurs in the punch radius in adhesive layer.

In continuing, the experiments are carried out on specimen of Al 5052/SUS. Generally, thickness distribution of Al 5052/SUS in all layers closely resembles to that of Al 1050/SUS. Figure 13 shows the distribution of thickness strain for Al 5052 layer under varying temperatures. The obtained results show that the maximum thinning takes place in the region of punch profile radius and by increasing the temperature the tendency of thinning in this region raises. On the other hand, the variation of thickness strain in the bottom of cup is negligible.

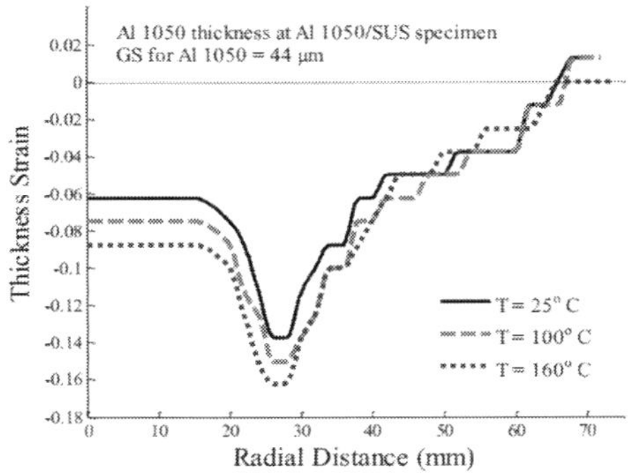

Figure 8. Effect of various temperatures on distribution of thickness strain for Al 1050 layer at Al 1050/SUS specimen (BHF = 15 kN).

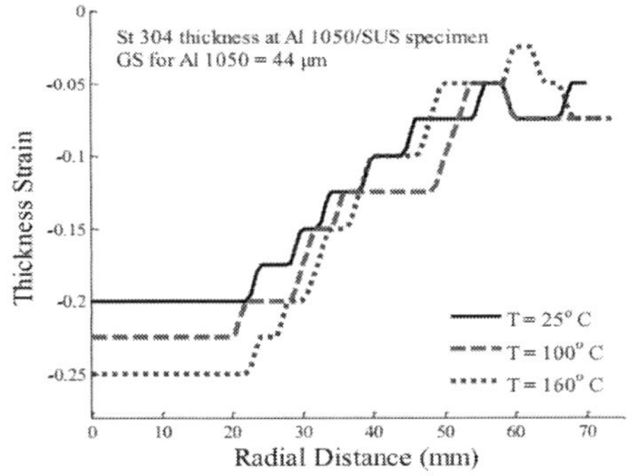

Figure 9. Effect of various temperatures on distribution of thickness strain for St 304 layer at Al 1050/SUS specimen (BHF = 15 kN).

Figure 10. Tearing in the central area of cup for St 304 layer in (a) Al 1050/SUS sample (BHF = 17 kN, T = 100°C), and (b) Al 5052/SUS sample (BHF = 19 kN, T = 100°C).

Variation of thickness strain for St 304 layer in Al 5052/SUS is illustrated in Figure 14. The thickness distribution of St 304 layer in Al 5052/SUS samples resembles the St 304 layer in Al 1050/SUS specimen in that the maximum thinning occurs in the central area of cup; moreover, raising temperature increase it. In St 304 layer, tearing begins in the central area of cup and as aforementioned it can be attributed to the accumulation of adhesive in central region, Figure 10 (b). Figure 15 shows the distribution of adhesive layer in Al 5052/SUS. Behavior of thickness strain in adhesive layer indicates that the center of cup has maximum thickening which decreases throughout the punch up to punch profile radius where the maximum thinning of adhesive layer occurs. Figure 16 clearly illustrates the thickness strains for all layers. It is seen that their lowest and highest values correspond to the punch radius and center zone in adhesive layer, respectively.

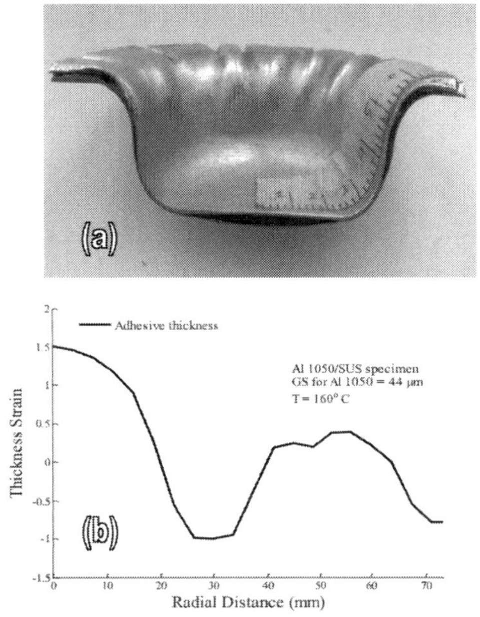

Figure 11. Distribution of thickness strain for adhesive layer at Al 1050/SUS specimen (GS for Al 1050 = 44 μm, T= 160°C, BHF = 15 kN), (a) real sample, (b) curve of adhesive thickness strain.

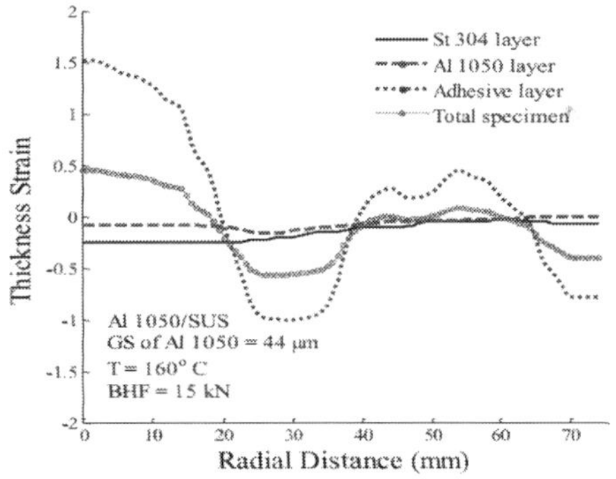

Figure 12. Distribution of thickness strain for all layers and specimen Al 1050/SUS specimen.

3.2.2 Effect of grain size on thinning

Effect of various grain sizes on thickness strain of Al 1050/SUS specimens are investigated here. The results indicate that various grain sizes and temperatures have similar effect on strain thickness. Figure 17 represents the influence of different grain sizes of Al 1050 sheet on thickness strain for specimen of Al 1050/SUS. As expected, the punch profile radius zone is susceptible to tearing; because the maximum thinning happens there, and this tendency increases with rising grain size of Al 1050 layer. Moreover, when a material with more elongation is used, the maximum thinning increases (Figure 17). Distribution of thickness strain for St 304 layer in Al 5052/SUS layer shows that the maximum thinning takes place in the central region of cup and increase of thickness strain corresponds to rise of grain size for all specimens, Figure 18. As aforementioned, the distribution of adhesive layer may have the key role here, Figures 11 and 12.

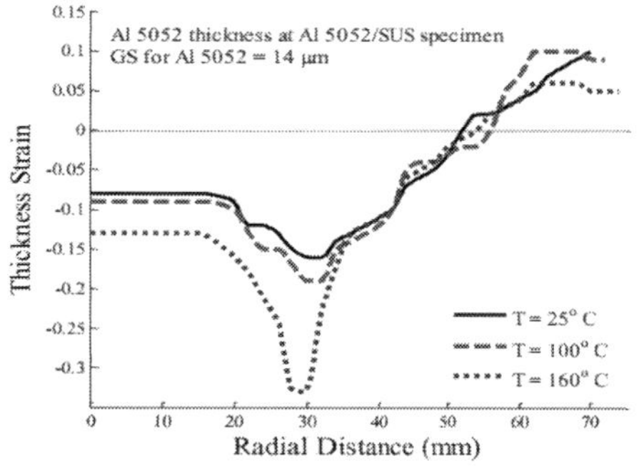

Figure 13. Effect of various temperatures on distribution of thickness strain for Al 5052 layer at Al 5052/SUS specimen (BHF = 18 kN).

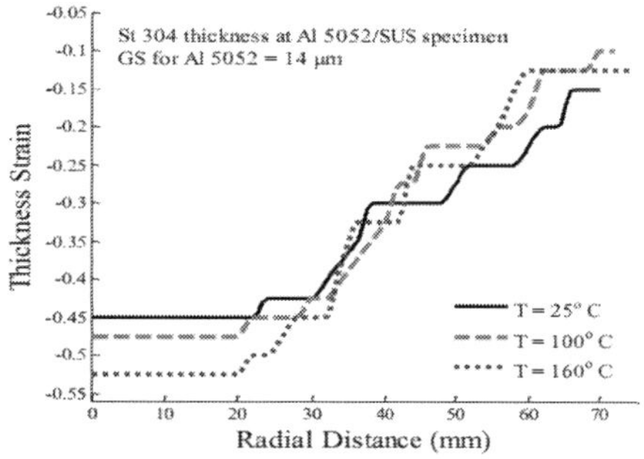

Figure 14. Effect of various temperatures on distribution of thickness strain for St 304 layer at Al 5052/SUS specimen (BHF = 18 kN).

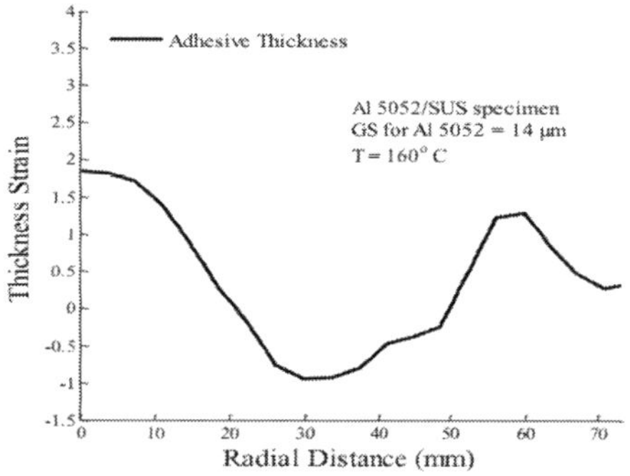

Figure 15. Distributions of the thickness strain for adhesive layer at Al 5052/SUS specimen (GS for Al 5052 = 14 μm, T= 160°C, BHF = 18 kN).

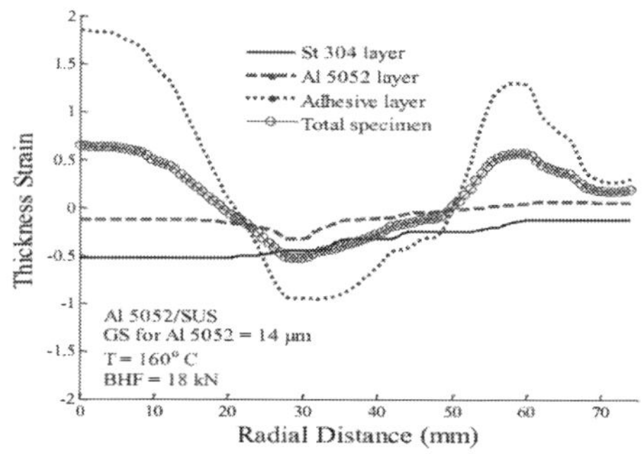

Figure 16. Distributions of the thickness strain for all layers and final specimen Al 5052/SUS.

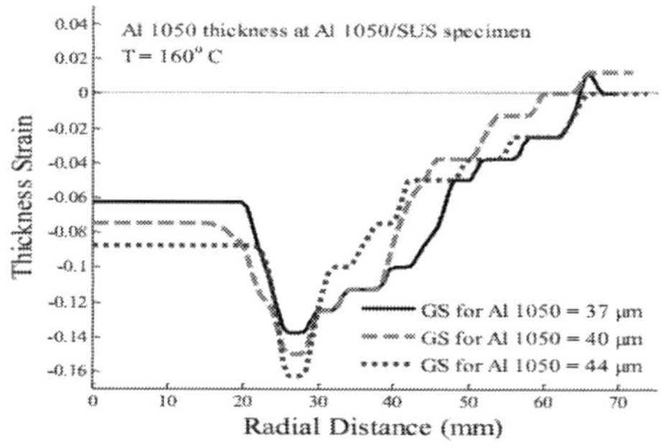

Figure 17. Effect of various grain sizes on distributions of the thickness strain for Al 1050 layer at Al 1050/SUS specimen (BHF = 15 kN).

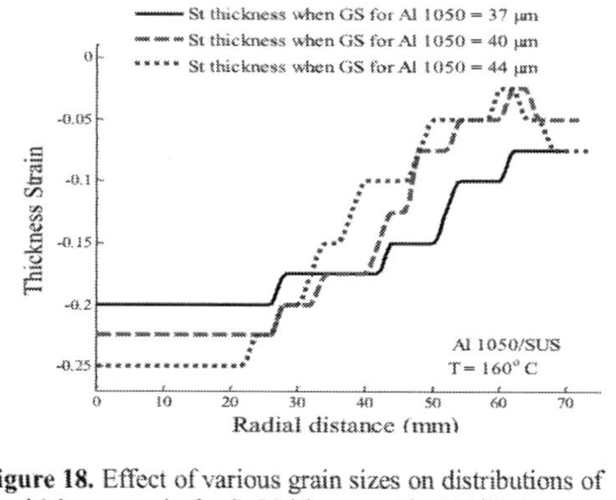

Figure 18. Effect of various grain sizes on distributions of the thickness strain for St 304 layer at Al 1050/SUS.

Finally, the influence of grain size on thickness strain for Al 5052/SUS sample is also investigated. Comparisons of thickness distribution in various grain sizes of Al 5052 and St 304 layer in Al 5052/SUS samples are shown in Figures 19 and 20, respectively. Thickness strain of St 304 layer varies from Al 5052 layer in that the maximum thinning in the former occurs in the central area of cup while in the latter it takes place on the shoulder of cup, especially for the larger grain size. However, increasing of maximum thinning with growth of grain size is clearly visible in both layers.

4 Conclusions

To provide guidelines and to extend fundamental understanding of warm deep-drawing process on laminated sheets, some experimental tests were conducted. The effects of temperature and grain size as main factors on thinning for each layer were investigated under various conditions. In addition, tearing and adhesive behavior were surveyed. The behaviors of Al 1050/SUS and Al 5052/SUS specimen under various conditions were almost similar. In summary, the most remarkable obtained results are as followings.

- Recrystallization occurs at the lowest studied temperature of 350° C for aluminum sheets. Moreover, increasing annealing temperature causes higher elongation, ultimate stress, yield stress and grain size.
- Distribution of thickness strain indicates that increasing temperature and grain size lead to maximum thinning rises in all layers in Al 1050/SUS and Al 5052/SUS specimens.
- The most susceptible areas to fracture in aluminum and stainless steel sheets are at the punch profile radius and central area of cup, respectively.
- Adhesive layer has notable effect on rupture in St 304 layers. Furthermore, the highest thinning and thickening takes place on punch radius and central area of cup in adhesive layer.

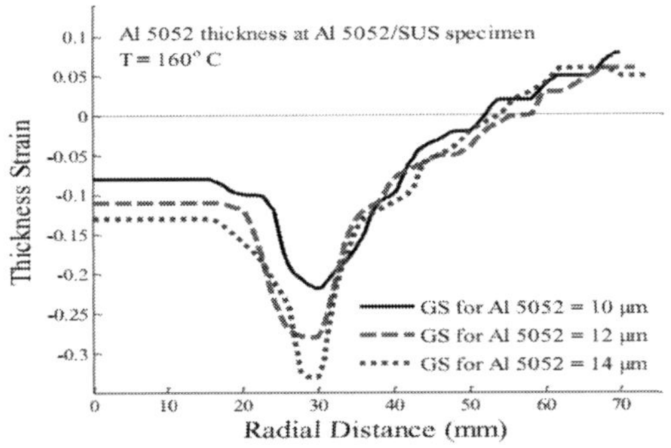

Figure 19 Effect of various grain sizes on distributions of the thickness strain for Al 5052 layer at Al 5052/SUS specimen (BHF = 18 kN).

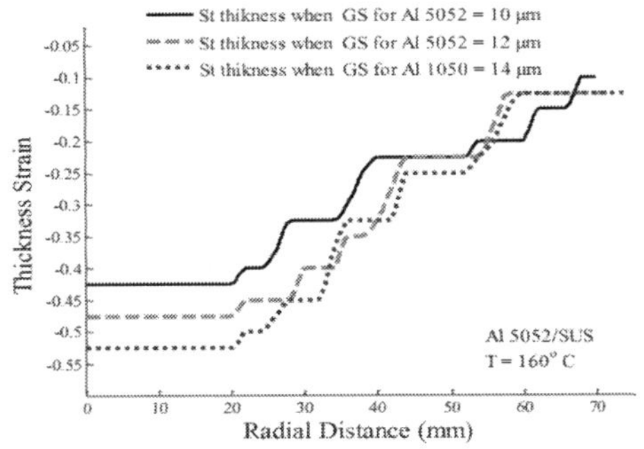

Figure 20. Effect of various grain sizes on distributions of the thickness strain for St 304 layer at Al 5052/SUS specimen (BHF = 18 kN).

References

1. S. Bagherzadeh, B. Mollaei-Darianib, K. Malekzadehc, J. Mater. Process. Technol **212,** 1840 (2012)
2. F. Stachowicz, T. Trzepieciński, T. Pieja, Arch. Civ. Mech. Eng **10**, 85 (2010)
3. M.H. Parsa, K. Yamaguchi, N. Takakura, Int. J. Mech. Sci **43**, 2331 (2001)
4. A. Atrian, F. Fereshteh-Saniee, Compos. Part B **47**, 75 (2013)
5. A. Rajabi, M. Kadkhodayan, M. Manoochehri, R. Farjadfar, J. Mater. Process. Technol **215**, 159 (2015)
6. P. Bolt, N. Lamboo, P. Rozier, J. Mater. Process. Technol **115**, 118 (2001)
7. H.S. Kim, M. Koç, J. Ni, J. Mater. Process. Technol **197**, 393 (2008)
8. G. Palumbo, L. Tricarico, J. Mater. Process. Technol **184**, 115 (2007)
9. N. Ethiraj, V. Kumar, Procedia. Eng **38**, 1836 (2012)
10. N. Ethiraj, V. Kumar, Appl. Mech. Mater **26**, 436 (2010)
11. M. El Sherbiny, H. Zein, M. Abd-Rabou, Mater. Des **55**, 869 (2014)
12. H. Zein, M. El Sherbiny, M. Abd-Rabou, Mater. Des **53**, 797 (2014)
13. A. Zoesch, T. Wiener, M. Kuhl, Procedia. CIRP **33**, 179 (2015)
14. M. Kadkhodayan, E. Afshin, *F T F 2014*, 8 (2014)
15. E. Afshin, M. Kadkhodayan, Mater. Des **87**, 25 (2015)
16. http://www.loctite.co.uk/homepage.htm, data sheet, (2015)
17. H. Rezaei Ashtiani, M. Parsa, H. Bisadi, Mater. Des **42**, 478 (2012)

Suitability of the electromagnetic ring expansion test to characterize materials under high strain rate deformation

Kang Yang[1], Geoffrey Taber[2], Thaneshan Sapanathan[1, a], Anupam Vivek[2], Glenn S Daehn[2], Rija Nirina Raoelison[1, 3], Nicolas Buiron[1] and Mohamed Rachik[1]

[1]*Sorbonne universités, Université de Technologie de Compiègne, Laboratoire Roberval, CNRS UMR 7337, Centre de recherche Royallieu, CS 60319, 60203 Compiègne cedex, France*
[2]*Department of Materials Science and Engineering, The Ohio State University, 2041 College Road, Columbus, OH 43210, USA*
[3]*Université de Bourgogne Franche Comté, IRTES-LERMPS, Université de Technologie de Belfort Montbéliard, 90100 Belfort, France*

Abstract. Characterisation of materials under high strain rate is always challenging, and requires sophisticated apparatus. Magnetic pulse forming techniques can involve high strain rate deformation and the electromagnetic ring expansion test (ERET) is considered here to determine its suitability to characterise materials at high strain rate. The multi-physics and high speed nature of the process requires advanced numerical modelling techniques to gain insight and understanding of stress development when used as a material characterisation technique. In order to evaluate the suitability of the ERET to fully characterise the material behaviour at high strain rate, stress states were investigated using 3D coupled mechanical-electromagnetic simulations on a validated numerical model. The numerical simulations were performed in LS-Dyna® with a modified Johnson-Cook (MJC) material model in this study. Albeit the preliminary results show the development of biaxial stress during the test, a shielding mechanism is proposed to eliminate the axial component of the stress on the rings. This study indicates that the ERET has a potential to be used for characterisation of the material parameters under the influence of high strain rate.

1 Introduction

In general, manufacturing processes such as forging, forming, stamping and cutting are typically involved with strain rate of 10^2 – 10^4 s^{-1} [1]. Further increase in strain rate is experienced by materials under high speed manufacturing processes such as electromagnetic pulse forming [2-4] and electromagnetic pulse welding [5-14], vaporising foil actuator welding [15-17], friction stir processing [18-20], explosive welding [21], and other high speed impact conditions like explosion or cold spray processing [22, 23]. Material characterisation at high strain rate also requires precise measurement techniques to capture the behaviour within the short duration of a mechanical test [1].

The present study proposes and evaluates the suitability of an electromagnetic ring expansion test (ERET) to be used in the characterisation of material under high strain rate conditions including the damage behaviour. The ERET is based on a ring expansion which is facilitated by the Lorentz force due to the change in magnetic field. Furthermore, this technique could especially be suitable for the characterisation of tube materials that contain residual stresses created during the fabrication of tubes from sheet metals.

In order to fully characterise a material, one should consider capturing the elasto-plastic and damage behaviours of the material under the exact nature of the process condition. A material characteristic could be simply altered by the working conditions of temperature and strain rate. The stress state, that can be interpreted based on Lode angle parameter and triaxiality, play a major role in damage behaviours of a material [24-26]. In order to investigate the suitability of the ERET for material characterisation, the stress state has been investigated in this study for various scenarios of the test. Triaxiality and Lode angle parameters are important, and require being consistent during the damage to determine the fracture strain of a material to treat it as a one dimensional tensile test.

A validated coupled mechanical-electromagnetic 3D model has been used under various conditions to evaluate the stress cases during the ERET. The validation step of the model is also described and the experimental result shows reasonable agreement with the numerical simulation. The coupled simulations performed with LS-DYNA®, which uses both finite element method and boundary element to solve such problems. The preliminary predictions indicate that the ring experiences bi-axial stresses resulted from radial and axial components of the Lorentz forces. However, minimising the height of the ring reduces the influence of biaxiality of the stress state that cannot eliminate the longitudinal component of the compressive stress on the rings.

[a] Corresponding author: thaneshan.sapanathan@utc.fr

Therefore, this study has been extended by using a shielding mechanism to control the axial component of stress experienced by the ring, which is promising for the characterisation of materials under high strain rate condition in a uniaxial fashion.

2 Materials and method

2.1 Experiment design

Applied source current in a helix coil creates the required magnetic field and Lorentz force during electromagnetic forming processes. In this case, the coil was designed to be fixed and made of copper alloy Cu–Zr–Cr 18150 with a cross section of 6.35 mm × 6.35 mm. This helix coil has 5 turns with a helix pitch of 8 mm. It has an inner diameter of 25.45 mm. A schematic illustration of the coil used in this study is shown in Figure 1.

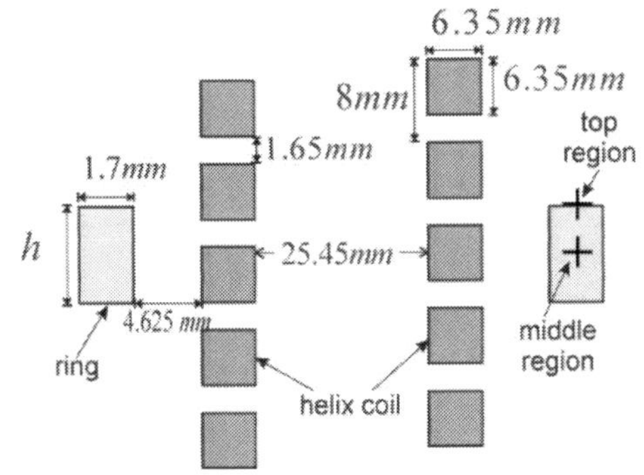

Figure 1. Schematic illustration showing the cross-section of the electromagnetic ring expansion setup.

The expansion ring used in the simulation was made of aluminum alloy Al 6061 - T6 with an inner diameter of 47.4 mm and thickness of 1.7 mm. In order to analyse the influence of height on the stress state, 3 different heights (h in Figure 1) of the ring were chosen with 3.2 mm, 6.4 mm and 9.6 mm.

An experimental case with 6.4 mm height of the ring was used to validate the model. The experiment was performed at Ohio State University with the aluminum rings made from Al 6061-T6 tube stock. Figure 2 shows the design of the helix coil with leads, which is potted with epoxy in a fiberglass tube. Leads of the same material were brazed to the helix. The velocity was measured using photonic Doppler velocimetry (PDV) system [27]. The experimental current flow through the cross section of the helix was measured using Rogowski coil and used in the numerical simulations. The time versus current evolution is shown in Figure 3.

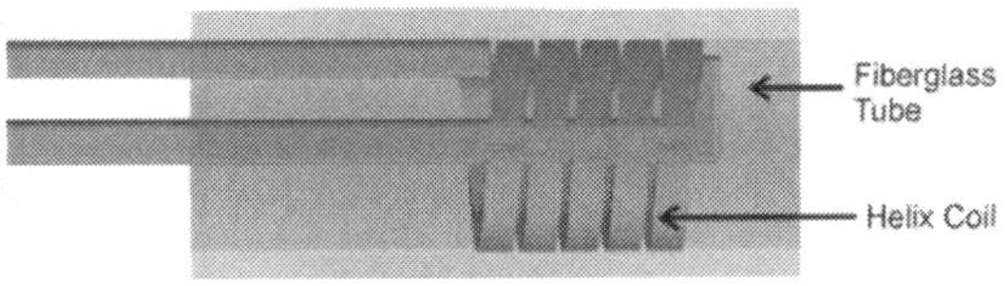

Figure 2: An illustration showing the helix coil inside a fiberglass tube.

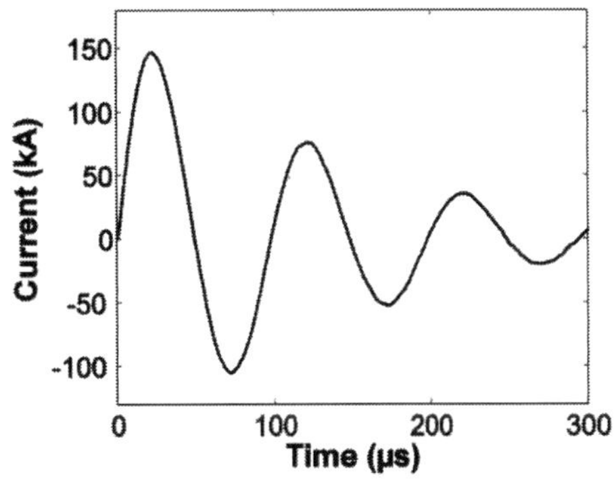

Figure 3: Current flow in the expansion coil.

From preliminary simulations, it was identified that the stress states are not the same at various locations of the ring due to the uneven distribution of the Lorentz force on the ring. The stress states of top and middle regions of the ring are used to explain the development of stresses during a ring expansion test and the suitability to use the test for high strain rate material characterisation.

Later models are used with a controlling mechanism using shield blocks to eliminate the development of compressive stress during the ERET. The controlling shield blocks setup is shown in Figure 4. The shield used in this study consists of two thick ring shapes made of copper and fixed in place during the simulations (i.e. the deformation on these shield blocks are neglected and considered to be used here to block the vertical force components acting on the ring). The main function of the shield is to eliminate magnetic fields in the radial direction, thus the compressive Lorentz force on the ring is minimized. More explanation on this topic is included in Section 3.3.

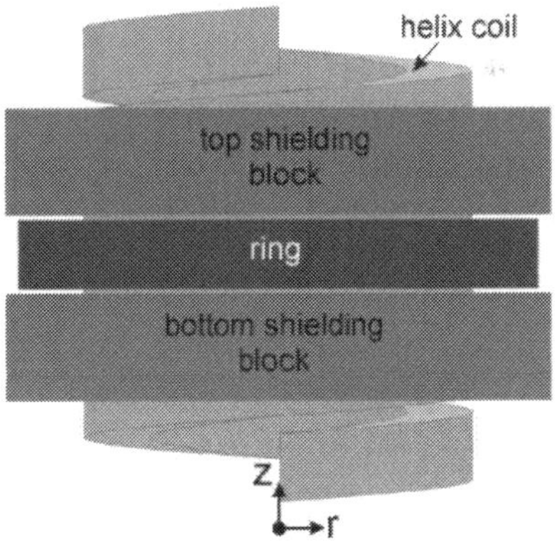

Figure 4. Numerical simulation interface showing the shield set up used in the ring expansion test.

2.2 Simulation method

A coupled mechanical-electromagnetic numerical simulation was carried out to investigate the development of stresses during the ring expansion test. This simulation used LS-DYNA numerical package with integrated solver version R8 [28].

Based on Maxwell's equations and Biot–Savart's law, a combined finite element method (FEM) with boundary element method (BEM) is used during the calculation of magnetic field and Lorentz forces in LS-DYNA [29]. BEM is used to compute the surface current and electromagnetic field thus the magnetic field in the air is not required in LS Dyna® simulations. FEM is used during the computation of eddy current and Lorentz force in the ring. At each electromagnetic time step, the electromagnetic and mechanical computations are coupled.

In order to implement the constitutive behavior of the material in the model a modified Johnson-Cook model [28] was chosen as expressed in Equation 1.

$$\bar{\sigma} = (A + B\bar{\varepsilon}^{n})\left[1 + \left(\frac{\dot{\bar{\varepsilon}}}{\dot{\bar{\varepsilon}}_0}\right)\right]^{C}(1 - T^{*m}) \quad (1)$$

where, $\bar{\sigma}$ and $\bar{\varepsilon}$ are the von Mises equivalent stress and strain respectively, $\dot{\bar{\varepsilon}}$ is the strain rate. Here $\dot{\bar{\varepsilon}}_0$ is equal to *1/s* and T^* is a dimensionless temperature parameter. *A*, *B*, *C*, *n*, and *m* are constants, obtained from literature, given in Table 1 [30]. The coil was fixed and set as rigid in the model.

Table 1. Johnson-Cook parameters of Al 6061-T6

Johnson-Cook parameters	*A* (MPa)	*B* (MPa)	*C*	*n*	m
Numerical values	324	114	0.002	0.42	1

The influences of both temperature increase and high strain rate are considered in this model. However, the electrical conductivity change due to the temperature increase is neglected in this present work. Longitudinal and transverse views of the meshed model used in the simulation are respectively shown in Figure 5a and 5b. The model used solid 8 nodes brick elements for both coil and the expansion ring. The element size is 0.8 mm in height and 0.57 mm in thickness, which is smaller than skin depth 0.7 mm in this case. Details of the skin depth can be found elsewhere [31]. It is also ensured to capture the electromagnetic skin depth effect during the simulation.

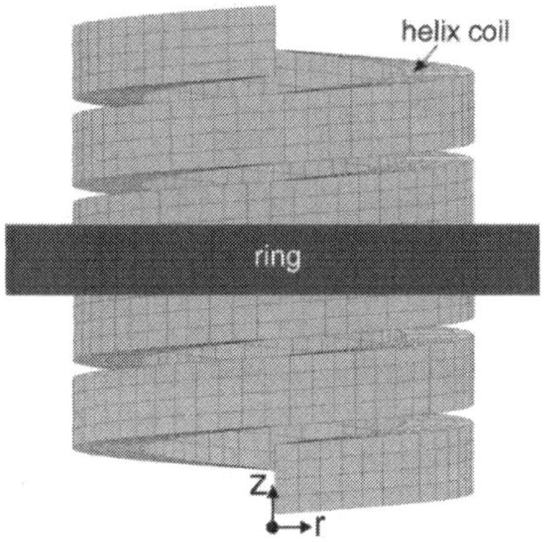

(a)

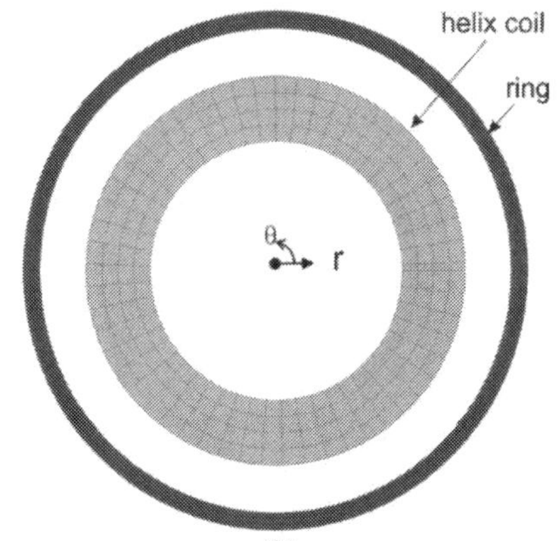

(b)

Figure 5. Meshed model with the 6.4 mm height ring used in preliminary investigation without shielding blocks, views along a longitudinal (a) and a transverse (b) planes.

2.3 Material

The electromagnetic ring expansion test has a multi physics nature which requires effective coupling between mechanical, electrical, magnetic and thermal components.

Physical quantities of the Al 6061–T6, the density, Young's modulus, Poisson's ratio are respectively set as 2700 kg/m³, 68.9 GPa and 0.33. Electromagnetic and thermal properties of both the ring and the coil are given in Table 2.

Table 2. Electromagnetic and thermal properties of the components

Part name	Electrical conductivity (S/m)	Thermal conductivity (W/m·K)	Specific heat Capacity (J/kg·K)	Melting temperature (K)
Coil and shield	5.62×10^7	377	390	1356
Ring	2.51×10^7	230	880	933

3 Results and Discussion

3.1 Validation of numerical simulation

Figure 6 shows the comparison between predicted and measured velocities for the case of 6.4 mm height. Although fracture occurs at the time of ~100 μs during the experiment while the damage parameters are not yet included in the present study, this model still can be used to investigate the stress state during the ring expansion process. Thus, the simulation was carried out for the first 100 μs.

The simulation result shows a good agreement with the experiment. Nevertheless, the comparison shows a

DOI: 10.1051/matecconf/20168015002

deviation of 10% in the maximum velocity, the tendency is well captured during the simulation. This could be due to high strain rate effect, while the aforementioned Johnson-Cook parameters may fail to represent the accurate behavior of the material. This is also in agreement with another study, where lower energy cases (i.e. lower strain rate case) agree well compared to higher energy cases for same material parameter [32]. This situation also highlights the definite requirement of the material calibration for respective strain rates. More information on the material calibration using ring expansion test could be found elsewhere [32-35].

The discrepancy could also arise from the numerical conditions such as mesh convergence and the time step used in this study. In this present study a single point PDV was used, the results could also be improved using multipoint PDV system for this measurement to improve the accuracy.

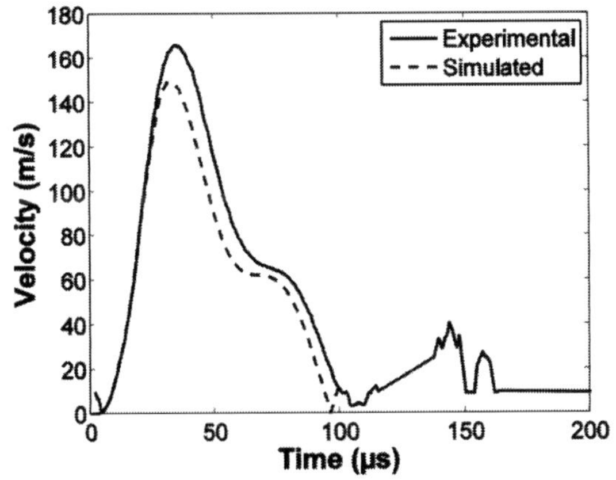

Figure 6. Comparison of experimental and simulated radial velocity

3.2 Stress state analysis of the ring during preliminary simulation

During the ring expansion, the radial component of the force drives the deformation and causes angular stress component in the ring. In ERET, there exists compression force along the longitudinal direction which makes the case as biaxial. Both the radial and longitudinal components of Lorentz forces are proportional to the square values of the input current, but due to the longitudinal symmetry of the force, this test allows to have only radial motion of the ring. Besides, the longitudinal compression force is a body force so that has varying influences on the top (or bottom) and middle region of the ring. Therefore, in order to study the stress state, the middle and top regions of ring are considered separately.

In order to study the damage behavior of materials, one could consider investigating Lode parameter and triaxiality [28]. Therefore, these parameters are carefully investigated for the test to determine the suitability of the ERET for material characterization. These parameters are computed from Cauchy stress tensor $\boldsymbol{\sigma}$ and the stress deviator $\mathbf{S}$ using the Equations 2-4.

$$I_1 = \mathrm{tr}(\boldsymbol{\sigma}) \tag{2}$$

$$J_2 = \frac{1}{2}\left[\mathrm{tr}(\boldsymbol{\sigma}^2) - \frac{1}{3}\mathrm{tr}(\boldsymbol{\sigma})^2\right] = \frac{1}{2}\|\mathbf{S}\|^2 \tag{3}$$

$$J_3 = \det(\mathbf{S}) \tag{4}$$

The Lode parameter ξ is defined by Equation 5.

$$\xi = \frac{\sqrt{27}}{2}\frac{J_3}{J_2^{\frac{3}{2}}} \tag{5}$$

The Lode parameter shows the relationship between shear stress and compression or tensile stress, when $\xi = 1$, $\xi = 0$ and $\xi = -1$ respectively indicates an expansion case, a pure shear case and a compression case.

The triaxiality Γ is defined by Equation 6. The triaxiality is used to determine the complexity of stress cases, where $\Gamma = 0.33$ indicate a uniaxial tensile test case. Figure 7 shows the map of triaxiality and lode parameter variance for various tests and stress cases [36].

$$\Gamma = \frac{I_1}{3\sqrt{3J_2}} \tag{6}$$

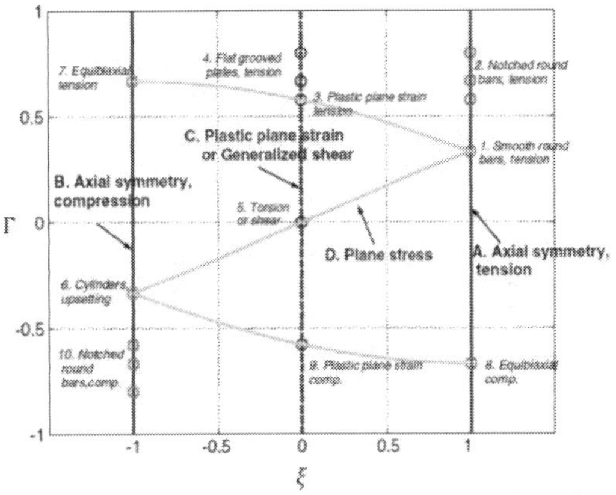

Figure 7. Triaxiality and Lode parameter variance for various test cases [36].

The Lode parameter versus plastic strain for the middle and top regions of the ring are respectively shown in Figure 8a and 8b. During the simulation, the ring with 9.6 mm height only reach ~15% plastic strain because it requires more energy for its expansion than that of other rings with shorter height.

 DOI: 10.1051/matecconf/20168015002

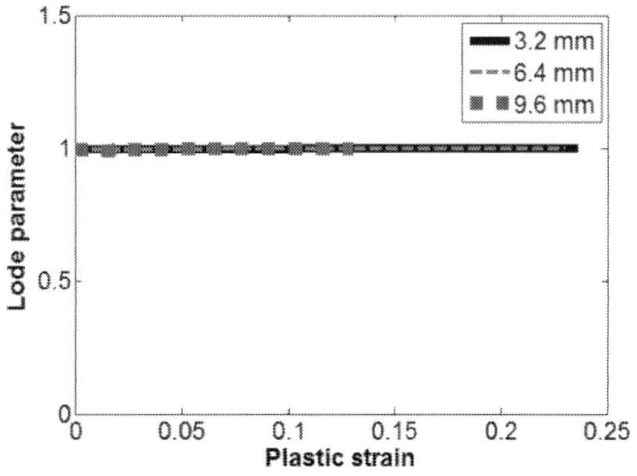

(a)

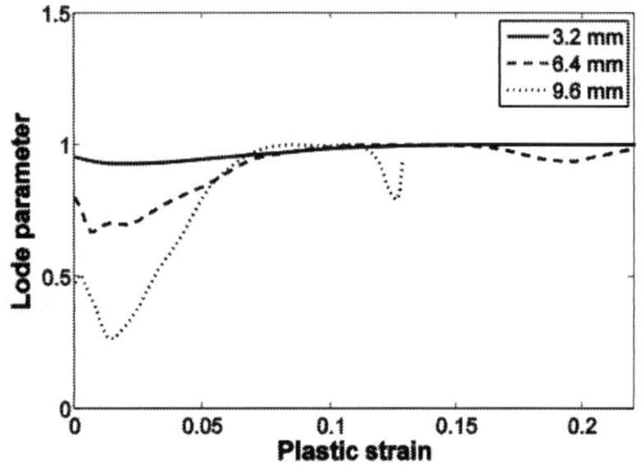

(b)

Figure 8. Lode parameter against equivalent plastic strain for each case on the top region (a) and the middle region (b) of the ring.

Figure 8a shows, the Lode parameter is always equal to 1 at the top region of the ring, which signifies that the stress development in the top region of the ring is similar to an axial symmetric tensile mode (Figure 7). The bottom side of the ring also has the similar results due to the symmetry.

In contrast, Lode parameter varies in the middle region of the ring (Figure 8b), while 9.6 mm height ring shows the maximum variation (~ 0.3 to 1) and 3.2 mm height ring has the minimum variation in comparison among these 3 cases. It is clear that due to the magnetic field decrease from the middle towards both top and bottom ends of the ring that significantly influences the stress development at various locations of the ring during the ERET. This difference also increases with the increase in ring's height. Consequently, the ERET with smaller height of the ring better represents a pure (radial) compression or expansion stress case.

Figure 9a shows triaxiality parameter of the top region that varies between 0.23 and 0.34, which covers close to the value 0.33. Combine with the Lode parameter property, the stress on the top and bottom of the ring could be considered as similar to a case of uniaxial tension on smooth round bars (Figure 7). The triaxiality of 9.6 mm ring shows more closeness to 0.33 among these three heights because of the role of compressive Lorentz force caused by the radial component of magnetic field. This radial magnetic field on the top surface of 9.6 mm ring is smaller than that of other cases that leads to have minimum compressive force among these 3 cases.

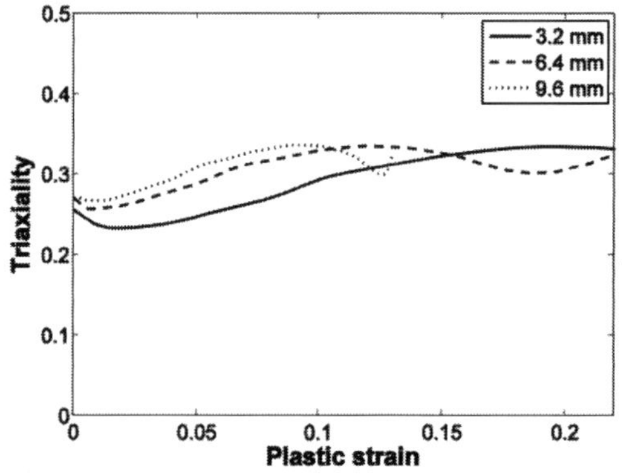

(a)

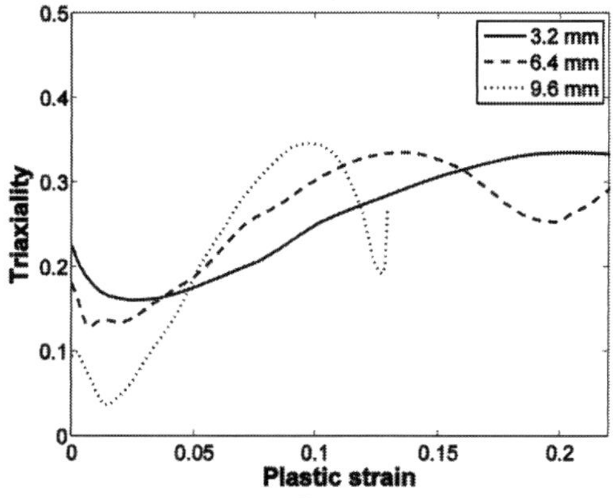

(b)

Figure 9. Triaxiality in each case on the top region (a) and middle region (b) of the ring.

Triaxiality in the middle of the ring (Figure 9b) varies between 0.03 and 0.35 in different cases. The compressive Lorentz force is also a body force. As a result, the longitudinal compressive stress on the top or bottom of the ring is negligible while it is comparable to the angular expansion stress in the middle region of the ring. This force also significantly influences the rings with large height, which bears more force and experiences a very unstable triaxiality.

According to the aforementioned results, the stress state of top and bottom regions of the ring shows a close connection with a uniaxial tensile test case while the stress state of middle region indicates a complex scenario; 3.2 mm height indicates closeness with tensile-compressive stress case and 9.6 mm height shows tensile-compressive-shear stresses.

3.3 Improvement of stress state towards uniaxial using shield blocks

A control mechanism using shield blocks is proposed and investigated to improve the stress state by eliminating the

 DOI: 10.1051/matecconf/20168015002

axial compressive Lorentz force components on the expansion ring. A 6.4 mm height ring sample is investigated in this simulation using the model described in Figure 4 to evaluate the proposed shielding mechanism.

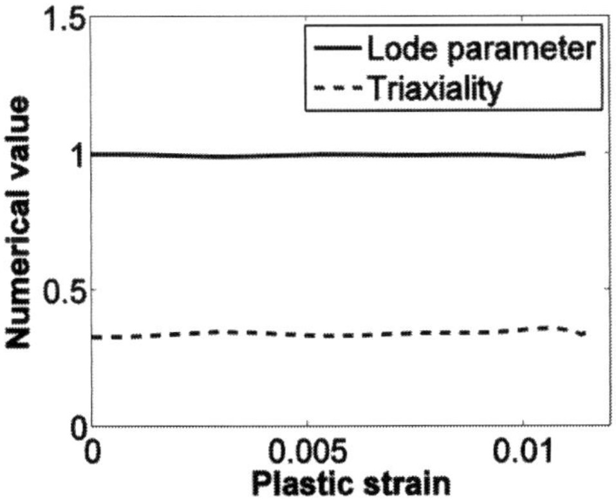

Figure 10. Triaxiality and Lode parameter in the middle region of 6.4 mm height ring with shields with the input current shown in Figure 3.

Figure 10 shows that the triaxiality is more stable and close to 0.33 (~ 0.32-0.35) and Lode Parameter is always equal to one for the entire simulation with the shield blocks. The results indicate that the shielding mechanism eliminates the radial component of the magnetic field which creates the compressive Lorentz force in previous cases. Therefore, stress state obtained in this case of electromagnetic ring expansion test agrees well with a uniaxial tensile test case. Therefore, the ERET with the proposed shielding mechanism could be considered to characterise materials in a similar manner to a tensile test with the implemented role of high strain rate behaviour.

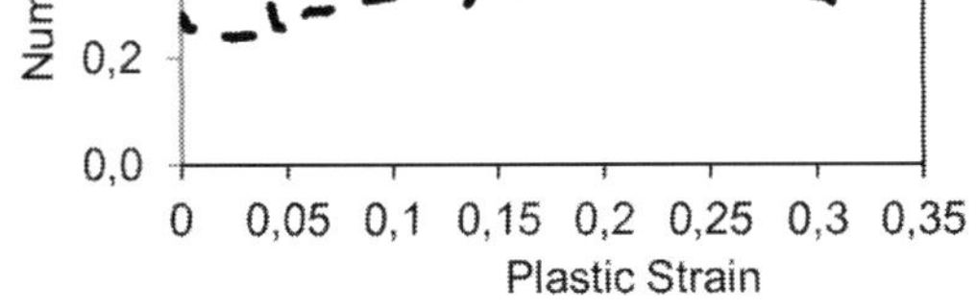

Figure 11. Triaxiality and Lode parameter in the middle region of 6.4 mm height ring with shields with an increase in input current.

Although the maximum plastic strain is significantly reduced because of the presence of shield blocks, several solutions could be considered to increase the plastic strain by increasing the input current or creating a notch in the ring or using a smaller height rings than considered in this study. In order to verify and achieve large plastic stain, the same model was investigated with an increased input current. The ordinate values (y values) of the electric current data shown in Figure 3 are scaled by 2.5 times without altering the abscissa values (i.e without changing the frequency of the current), to investigate the influence of the electric current. The triaxiality and Lode parameter results obtained from the new model are shown in Figure 11.The results in Figure 11 show the increase in plastic strain with a stable triaxiality (~ 0.33) and load parameter (~ 1.0) that obtained with an increased input current. Although, the triaxiality and Lode parameter fluctuate at low plastic stain due to the biaxial stress, one could consider other options of reducing the height of the ring or using a notch, to closely resemble the required uniaxial stress state to cause failure for a test criterion.

4 Conclusions

Suitability of the electromagnetic ring expansion test to fully characterise material at high strain rate is investigated using coupled 3D numerical modelling. Firstly, the validation of the numerical method was made with the experimental results. After that, stress states were investigated to determine the stress development during the test using the validated model that indicates the development of biaxial stress components and which significantly differ from a uniaxial tensile test cases. It was also identified that the compressive Lorentz force on the ring that influences the stress state and leads to the biaxial nature. Moreover, the biaxiality is significantly apparent in the middle region of the ring as compared to top and bottom edges. Furthermore, the complexity of the stress state increased with the increase in height of the ring. Finally, a shielding mechanism was proposed to avoid the compressive Lorenz force and that enables the development of a stress state closely resembling that during a uniaxial tensile test case. This setup could be used experimentally to characterise materials under high strain rate conditions.

References

1. L. Meyer, 1st International Conference on High Speed Forming, Institut für Umformtechnik - Technische Universität Dortmund, Dortmund, 2004, pp. 45-56.
2. V. Psyk, D. Risch, B.L. Kinsey, A.E. Tekkaya, M. Kleiner, Journal of Materials Processing Technology, 211 (2011) 787-829.
3. Y.U. Haiping, L.I. Chunfeng, D.E.N.G. Jianghua, Journal of Materials Processing Technology, 209 (2009) 707-713.
4. H. Yu, C. Li, Z. Zhao, Z. Li, Journal of Materials Processing Technology, 168 (2005) 245-249.
5. A. Kapil, A. Sharma, Journal of Cleaner Production, 100 (2015) 35-58.
6. R. Raoelison, M. Rachik, N. Buiron, D. Haye, M. Morel, B. Dos Sanstos, D. Jouaffre, G. Frantz, 5th International Conference on High Speed Forming, Dortmund, 2012, pp. 207-216.

7. M. Marya, S. Marya, D. Priem, Welding in the World, 49 (2005) 74-84.
8. R. Raoelison, N. Buiron, M. Rachik, D. Haye, G. Franz, M. Habak, Journal of Materials Processing Technology, 213 (2013) 1348-1354.
9. M. Geyer, S. Böhm, IIW Denver Annual Assembly, (2012).
10. R. Raoelison, D. Racine, Z. Zhang, N. Buiron, D. Marceau, M. Rachik, Journal of Manufacturing Processes, 16 (2014) 427-434.
11. C. Pabst, P. Groche, 6th International Conference on High Speed Forming, Institut für Umformtechnik - Technische Universität Dortmund, South Korea, 2014, pp. 77-88.
12. R. Raoelison, N. Buiron, M. Rachik, D. Haye, G. Franz, Journal of Manufacturing Processes, 14 (2012) 372-377.
13. M. Marya, S. Marya, Science and Technology of Welding and Joining, 9 (2004) 541-547.
14. R.N. Raoelison, T. Sapanathan, N. Buiron, M. Rachik, Journal of Manufacturing Processes, 20, Part 1 (2015) 112-127.
15. A. Vivek, S. Hansen, B. Liu, G.S. Daehn, Journal of Materials Processing Technology, 213 (2013) 2304-2311.
16. A. Vivek, B. Liu, S. Hansen, G. Daehn, Journal of Materials Processing Technology, 214 (2014) 1583-1589.
17. A. Nassiri, G. Chini, A. Vivek, G. Daehn, B. Kinsey, Materials & Design, 88 (2015) 345-358.
18. Z. Ma, Metallurgical and Materials Transactions A, 39 (2008) 642-658.
19. R.S. Mishra, P.S. De, N. Kumar, Friction stir processing, Springer, 2014.
20. Y. Ahuja, R. Ibrahim, A. Paradowska, D. Riley, Journal of Materials Processing Technology, 217 (2015) 222-231.
21. F. Findik, Materials & Design, 32 (2011) 1081-1093.
22. P.C. King, S.H. Zahiri, M. Jahedi, Acta Materialia, 56 (2008) 5617-5626.
23. P.C. King, G. Bae, S.H. Zahiri, M. Jahedi, C. Lee, Journal of Thermal Spray Technology, 19 (2010) 620-634.
24. Y. Bao, T. Wierzbicki, International Journal of Mechanical Sciences, 46 (2004) 81-98.
25. J.R. Rice, D.M. Tracey, Journal of the Mechanics and Physics of Solids, 17 (1969) 201-217.
26. T. Wierzbicki, Y. Bao, Y.-W. Lee, Y. Bai, International Journal of Mechanical Sciences, 47 (2005) 719-743.
27. J.R. Johnson, G. Taber, A. Vivek, Y. Zhang, S. Golowin, K. Banik, G.K. Fenton, G.S. Daehn, steel research international, 80 (2009) 359-365.
28. LS-DYNA Keyword User's Manual R8.0, Livermore Software Technology Corporation (LSTC), Livermore, CA, USA, 2015.
29. I. Çaldichoury, P. L'Eplattenier, Electromagnetism and Linear Algebra in LS-DYNA, Livemore Software Technology Corporation, 2012.
30. C. Anderson Jr, I.S. Chocron, A. Nicholls, Journal de Physique IV (Proceedings), EDP sciences, 2006, pp. 331-337.
31. W.H. Hayt, J.A. Buck, Engineering electromagnetics, McGraw-Hill New York, 2001.
32. K. Yang, G. Taber, T. Sapanathan, A. Vivek, G. Daehn, R. Raoelison, N. Buiron, M. Rachik, 7th International Conference on High Speed Forming, Dortmund, 2016, pp. 223-234.
33. A.-C. Jeanson, G. Avrillaud, G. Mazars, G. TABER, G.S. Daehn, F. Bay, N. Jacques, M. Arrigoni, 6th International Conference on High Speed Forming, Institut für Umformtechnik - Technische Universität Dortmund, 2014, pp. 31-36.
34. W.H. Gourdin, Journal of Applied Physics, 65 (1989) 411-422.
35. W. Gourdin, S. Weinland, R. Boling, Review of Scientific Instruments, 60 (1989) 427-432.
36. F.J.P. Reis, A.F.X. C., 11th German LS-DYNA Forum, Ulm, Germany, 2012.

Material flow analysis in dissimilar friction stir welding of AA2024 and Ti6Al4V butt joints

Gianluca Buffa[1a], Michele De Lisi[1], Antonio Barcellona[1] and Livan Fratini[1]

[1] *Department of Chemical, Management, Computer Science and Mechanical Engineering, University of Palermo, Viale delle Scienze, 90128 Palermo, Italy*

Abstract. The complex material flow occurring during the weld of dissimilar AA2024 to Ti6Al4V butt and lap joints was highlighted through a dedicated numerical model able to take into account the effects of the different materials as well as the phase transformation of the used titanium alloy.

Keywords: Friction Stir Welding, titanium alloys, aluminum alloys, FEM

1. Introduction

Nowadays aluminum and titanium alloys are extensively used in industrial processes. These alloys offer good mechanical properties and considerable mechanical resistance to weight ratios, which are key parameters in order to improve and optimize the design of vehicles in the aerospace, aeronautical and transportation fields. Moreover, since environment pollution is becoming a significant issue, the reduction of weight, where it is possible through proper materials, is an efficient countermeasure against fuel consumption and, thus, a good strategy to adopt for protecting the environment.

AA2024 is a high-resistant lightweight aluminum alloy, with copper as the primary alloying element, characterized by good atmospheric corrosion resistance and good mechanical properties. It is applied into many fields, typically in aircraft structures, rivets, connectors and miscellaneous structural parts. It is a difficult-to-weld alloy, especially with fusion welding techniques, like ones of AA7xxx series, because of formation and precipitation of hard oxides that brittle the welding and weaken the joining. Porosity can be detected too.

Ti-6Al-4V is an alpha+beta titanium alloy with excellent mechanical properties and corrosion resistance. Due to its phase evolution consisting of three different phases, microstructure and morphology of the alloy are very sensitive to heat treatments changing its mechanical and metallurgical properties. Ti-6Al-4V oxides, whose precipitation is favourite at high temperature, offer strength to the alloy at the expense of its workability.

Traditional fusion welding techniques can be used, but problems as weld cracks, porosities and contaminations appear. FSW is again a suitable solution. Ti-6Al-4V is used mostly in aerospace and aeronautical fields, for building aircraft structures with high added value and compressor blades, but it also used in the biomedical field as main material of prosthesis.

Unfortunately, as known, high-resistant aluminum alloys, such as 2xxx and 7xxx ones, and titanium alloys are difficult to weld. In fact, these alloys were considered almost "unweldable" until 1991, when The Welding Institute (TWI) patented the friction stir welding (FSW) process [1]. FSW is a solid-state joining process in which melting temperature is not reached, avoiding the formation of oxides, brittle cast structures, metallurgy porosity and cracking, "hot cracking", large distortion and residual stresses. Because of this, FSW is preferable over traditional fusion welding processes to join aluminum and titanium alloys. After its invention, FSW was applied to several materials and alloys, e.g. aluminum alloys, magnesium alloys, titanium alloys, nickel-copper super-alloys, with applications in the aerospace field, shipbuilding and automotive. Researchers and industries have shown interest in studying the process, making it engineered and foretelling eventual defects caused by inappropriate heat input or incorrect plunging settings [2, 3]. Different kinds of geometric set-up were explored changing process parameters such as feed speed, tool rotation rate, tilt angle, plunging depth and relative position between tool and materials and results of various configurations have been scanned. The relevance of tool geometry was investigated and it was found out that it was crucial for the correct material flow and for increasing tensile strength [4, 5]. In these cases, the tool design influence and pin offsets were analyzed and highlighted too [6, 7].

Finally, FSW of dissimilar joints has been studied and demonstrated feasible with the aim to put right materials

[a] Corresponding author: gianluca.buffa@unipa.it

in right places giving more resistance where it is needed and reducing weight where it is possible.

Many studies focused on FSW between different lightweight alloys as magnesium [8] and aluminum ones [9], analyzing different material arrangements and the effects of varying welding parameters on hardness distribution, tensile resistance and joint microstructure. The sheets mutual position, i.e. the position of one sheet in the advancing or in the retreating side, and tool offsets on the tensile strength was investigated for dissimilar joints in which magnesium alloys were welded with aluminum ones [10]. It was obtained that material position deeply influences the final microstructure of the weld and its mechanical and metallurgical properties. The right positioning is determinant in order to avoid inappropriate heat inputs and, hence, unwanted liquid formation and intermetallic compounds [11]. Additionally, choosing the most suitable offset is fundamental for obtaining higher tensile strength. Studies were conducted about the relevance of the cooling conditions in dissimilar Mg-Al joints and the different effects of underwater cooling rather than air cooling on hardness (which resulted lower in the first case), tensile strength (which decreased in the second case) and formation of intermetallic compounds (which was reduced in the first case) [12].

A few researches focused on the applicability of friction stir welding to titanium alloys and on its characterization [13, 14], trying to find out engineered process parameters. Mostly because of their thermal properties, it was deduced that low tool rotation rates and low feed rates are eligible for titanium alloys [15]. Hence, it was widely demonstrated the feasibility of FSW for Ti6Al4V, which is largely used overall in aerospace field. In addition to this, friction stir welded joints with super alloys have been realized with successful results [16].

Only a very limited number of papers is found on the FSW of dissimilar aluminum to titanium joints. Li et al. [17] studied the influence of process variables on the weld interfaces and tensile properties of Ti–6Al–4V alloy to Al–6Mg alloy. It was found that higher joint mechanical tensile strength can be reached wen the Ti-Al diffusion bonding interlayer is obtained. Wu et al. [18] examined the influence of welding parameters on the interface and the properties of a Ti-6Al-4V alloy to AA6061 alloy joint investigating the macrostructure of the joint, the fracture surfaces and the reaction layer. Continuous $TiAl_3$ intermetallic compounds at the interface of the two materials were found with the welding parameters used. The interface thickness, key factor for obtaining good tensile strength and appropriate fracture locations, resulted associated to the rotating rate. The joints bonded at the rotating rate of 750 rpm failed in the TMAZ/HAZ and had the highest tensile strength, accompanied with the perfect bonded and thin interface. However, the complex material flow occurring the two largely different materials is still unclear. A properly designed numerical model of the process can help in the understanding of the occurring process mechanics enabling the use of these joints for industrial applications.

In the last years, different research groups have been working on the development of a dedicated numerical model for FSW. Thermo-mechanically coupled FEM models have been developed [19, 20]. Good agreement and fine matching have been achieved between numerical and experimental results in terms of predicting the effect of process parameters on process thermo-mechanics and temperature distribution. Similar and dissimilar joints have been analyzed with finite element methods, in order to understand material flow, strain, strain rate and temperature distributions, verified by experimental validation [21, 22]. FEM modelling have been successfully adopted also for investigating the feasibility of the friction stir welding of tailor-welded blanks [23], finding that joint morphology, heat input and forces are strictly dependent from the increase of the thickness ratio. However, FEA of FSW is still not enough mature in terms of analyzing and predicting material flow, temperature distribution and effects of geometrical and material arrangements of dissimilar butt and lap joints of the most used alloys such as aluminum, magnesium and titanium ones. As far as the authors know, there are no papers on the FEM models of FSW between Ti-6Al-4V and an aluminum high-resistant alloy highlighting the occurring material flow.

In this paper, a dedicated numerical model capable to take into account the complex material flow occurring in dissimilar FSW of titanium and aluminum alloys, in different configurations, was developed. The model is able to predict material flow, temperature, strain and strain rate distributions.

2. Numerical model set up

A Lagrangian, thermo-mechanically coupled 3D FEM model with a rigid visco-plastic material model was developed. The analysis was conducted using the 3D version of the implicit lagrangian code DEFORM™. The implemented FEM formulation is based on the variational approach where the actual velocity (the solution) among all admissible velocities that satisfy the compatibility conditions, incompressibility and the boundary condition cause a functional to assume a stationary value. Workpiece and tool geometries were imported into DEFORM™ from a commercial CAD software. FSW of AA2024 and Ti-6Al-4V dissimilar joint, 1.6 mm in thickness, in butt configuration was modeled. Temperature, strain, and strain rate material flow were considered for the two materials. AA2024 material data were taken from the material library of the software, Ti-6Al-4V material data were taken from previous studies [24, 25]. Fig.1 shows the utilized curves, as a function of temperature. It is worth noticing material flow stress was input in the model in the form of tabular data.

Tool and workpiece were meshed with tetrahedral elements. The workpiece model consisted of about 45,000 elements, whereas the tool, characterized by shoulder of 13 mm, conical pin with base diameter of 5 mm and tip diameter of 3.5 mm, was modelled as a rigid object (in tungsten carbide) and meshed with about 10,000 elements for the thermal analysis only. Both butt and lap configurations were analyzed. Tool was tilted of 2° for all the simulations. In order to simulate a dissimilar friction stir welded joint, a fictitious phase transformation had to be set. In fact, the workpiece is formed by a single material object in both simulations. The material model of the

DOI: 10.1051/matecconf/20168015003

workpiece was set as a bi-phasic material in which the two phases are Ti-6Al-4V and AA2024. The single-block workpiece was made of one of the two phases, i.e. AA2024. Then a fictitious transformation is induced in such a way that the part corresponding to the titanium sheet changes its phase into titanium one. After it was completed, this transformation was disabled, obtaining a starting workpiece composed of the two alloys considered. A similar approach was used by some of the authors to simulate dissimilar FSW of Ti-6Al-4V and AISI 304 [21]. Further details on the numerical approach used for the phase change can be found in the above mentioned paper.

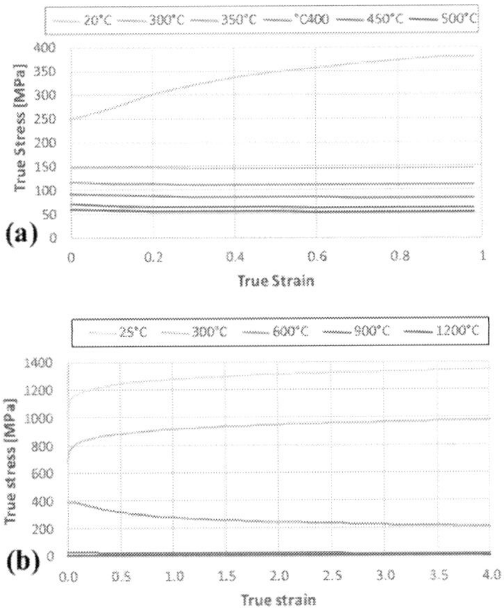

Figure 1. flow stress curves for fixed strain rate of 0.3/s for (a) AA2024 and (b)Ti6Al4V

Furthermore, pressurized argon impinging jet shielding has been implemented with a toroidal environment window. The heat exchange coefficient used was 1.9 $N \cdot s^{-1} \cdot mm^{-1} \cdot °C^{-1}$. Heat transfer with the environment at the room temperature of 20°C was enabled. The heat exchange coefficient used was 0.02 $N \cdot s^{-1} \cdot mm^{-1} \cdot °C^{-1}$.

The initial configuration of the butt joint model is displayed in Fig.2 indicating also materials position.

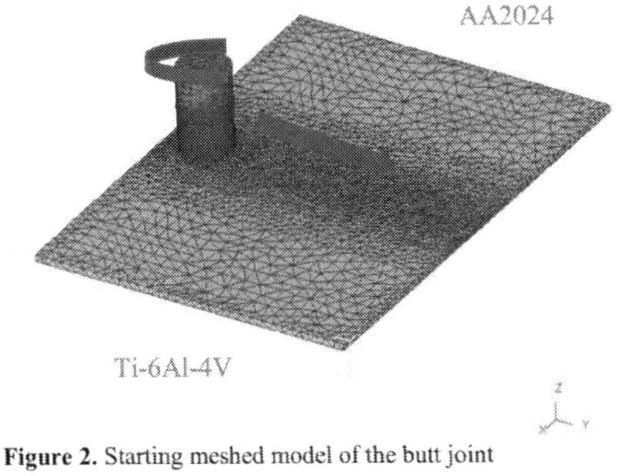

Figure 2. Starting meshed model of the butt joint

On the left side of the tool, which is the retreating one, AA2024 was placed. Ti-6Al-4V was in the advancing side (AS). This choice was due based results of previous studies demonstrating that the harder material must be placed in the AS in order to maximize joint performances [11, 17, 18]. Tool rotation rate was adjusted at 900 rpm and feed rate at 3.33 mm/s. Finer mesh windows were implemented in order to have a better resolution of the results along the weld seam. In addition to this, an adaptive remesh was enabled and deformation and step triggers were considered. A large number of remeshing step occurs during the process due to the large deformation induced by the tool action. The automatic remeshing algorithm of the utilized softhearted was used and remeshing was triggered by assigning maximum element distortion of 50%.

Regarding inter-objects relations, the shear model, according to which the shear stress is equal to a certain percentage of the material shear flow stress, was utilized. In particular, a constant factor of 0.45 was used to model the contact between the tool and the aluminum alloy sheet, whereas a constant factor of 0.35 was applied at the contact between the titanium alloy and the tool. These values are taken from previous studies of the authors on similar welds for the considered materials [19-25]. Furthermore, a constant interface heat exchange coefficient of 11 $N \cdot s^{-1} \cdot mm^{-1} \cdot °C^{-1}$ was used for the parts of the workpiece that were in contact with the tool. As the CPU time ids considered, only a few minutes were needed in order to simulate the phase change while the welding process, including the tool plunge and the tool movement along the weld seam, took about 20 hours. Two different configurations have been examined. First, zero tool offset was considered, i.e. the tool axis corresponds to the separation line between the two materials. Then, an offset equal to 1.75 mm was given to the tool shifting it towards the aluminum side. Fig.3 shows a sketch of the considered configurations. The position of the two different materials with respect to the advancing and retreating side of the joint is also highlighted.

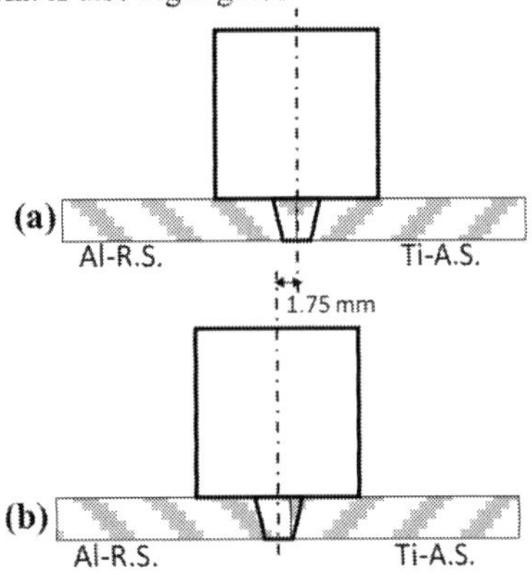

Figure 3. Sketch of the considered configurations: (a) zero-offset and (b) offset

The different material flow and main field variables distributions were used to highlight the changes in the process mechanics due to the offset.

3 Results

First, temperature dsitributions have been studied. Fig.4 shows the temperature distribution in a top view of the joint for the two configurations. The initial separation line between the two materials is highlighted. A few considerations can be made on the figure. As expected, larger temperatures are found in the titanium side, with peak values of about 990°C. However, although similar peak values are found for both configurations, the extension of the area involved by large temperature is smaller when an offset is given to the tool. This is due to the reduced contact surface between the tool and the titanium sheet. In particular, in the zero-offset configuration the red area, corresponding to the highest temperatures, is found also in the side of the joint where aluminum should be present. This would result in local melting of the AA2024 alloy, with detrimental effects on the joint nugget integrity. Nevertheless, as it will be better explained in the following paragraph, titanium is ofund also in the RS of the joint due to the occurring material flow. Additionally, it arises that large temperatures are found close to the tool pin also for the offset configuration (orange areas). Finally, larger areas characterized by temperature higher that room temperature, are found in the aluminum side due to the larger thermal conductivity of this alloy with respect to the titanium alloy.

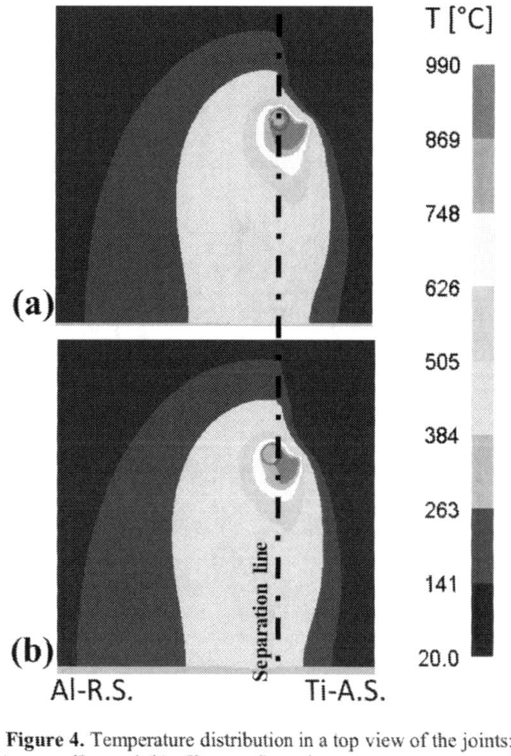

Figure 4. Temperature distribution in a top view of the joints: (a) zero-offset and (b) offset configuration

Temperature profiles in a cross section of the joint have been calculated. It is worth noticing that the cross section was taken after the tool advanced 40 mm in order to be sure that the process reached the steady state. Fig.5 shows the obtained curves as well as a sketch of the abscissa used for data extraction, which was at mid height of the cross section, i.e. 0.8mm from the top surface of the joint. The analysis of the curves confirms that similar maximum temperature is reached in the titanium side, while larger temperature is reached in the aluminum side when the zero-offset configuration is used. Additionally, temperature below 600°C are reached, in the aluminum side, at a distance from the initial separation line of about 4.5 mm and 5.5 mm for the zero-offset and offset configurations, respectively.

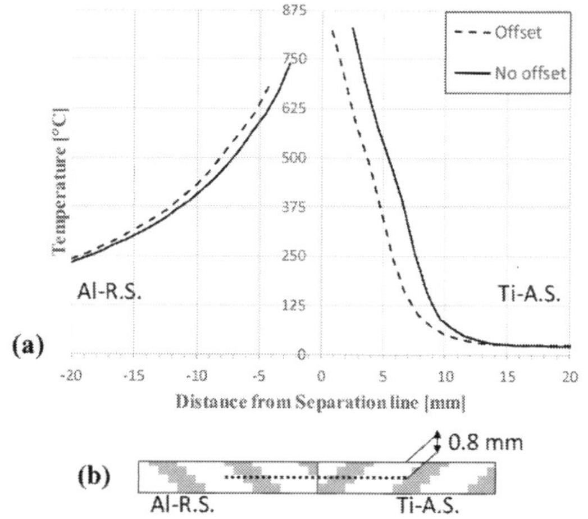

Figure 5. (a) temperature profiles in a cross section for the two configurations and (b) sketch of the cross section highlighting the position of the abscissa used for data extraction.

In order to explain this behavior and assess whether local melting occurs, the material flow was investigated. Fig.6 shows the strain distribution in a cross section, taken after 40 mm of weld length, for the two considered case studies. The position of the tool and the initial separation line between the two sheets are highlighted.

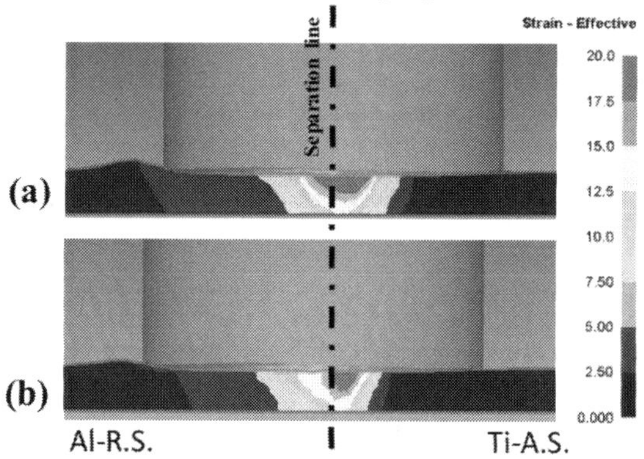

Figure 6. Strain distribution in a cross section of the joint for the (a) zero-offset and (b) offset configurations.

Similarly to what observed for the temperature, maximum values, consistent with what found in literature [21], are found in the titanium side. Again, the area involved by the highest values of accumulated strain is larger for the zero-offset case study, affecting part of the retreating side of the joint, i.e. where aluminum was initially placed. In a similar way, also the area of the retreating side affected by significant accumulated strain values, i.e. in excess of 2.5, is larger for the zero-offset case study.

As described in the previous paragraph, the presence of the two different materials in the "single block" model was taken into account by creating a new biphasic material in which the two phases correspond to the AA2024 and Ti6Al4V. Hence, the position of the two materials can be analyzed by calculating the volume fraction of one of the two phases. Fig. 7 shows the distribution of the volume fraction of the AA2024 in a top view of the joint, while Fig. 8 shows the nodal velocity vectors during the process, for both the configurations.

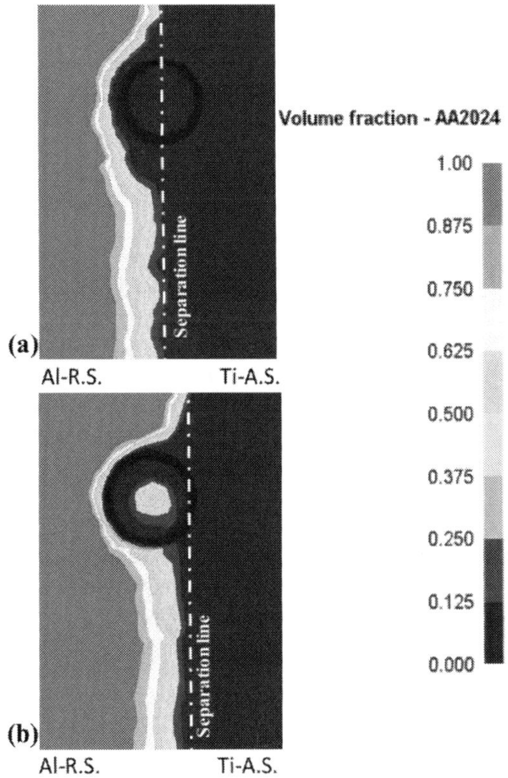

Figure 7. AA2024 volume fraction during the weld for the (a) zero-offset and (b) offset case studies

A significantly different material flow is observed. As the zero-offset case study is taken into account, it can be observed that in the retreating side, i.e. where aluminum was initially placed, titanium is found. The material flow can be described as follows: titanium rotates together with the tool from the advancing side to the retreating side pushing away from the tool axis, the aluminum. Then, titanium closes right back the tool pin and final bond with aluminum, represented by the area in which the two materials are mixed and AA2024 volume fraction is between 0.25 and 0.875, is obtained in the advancing side "far" from the tool pin, i.e. at a distance of about 4 mm in the trailing edge of the joint. On the other hand, when an offset is given to the tool, AA2024 is not pushed towards the periphery of the retreating side but bonds with the titanium coming from the advancing side in front of the tool. This is visible observing the red area in front of the tool pin also in the advancing side. Then, the two materials rotate with the tool and are left right behind it forming a mixed materials band centered with the tool axis. The lack of material mixing and material flow observed for the zero-offset configuration behind the tool in the retreating side, is highlighted by the light grey area in Fig.8a, as compared to the same area shown Fig. 8b for the offset configuration. The combination of larger velocity and material rotation close to the tool pin indicates that proper conditions for solid bonding are achieved for the offset case study.

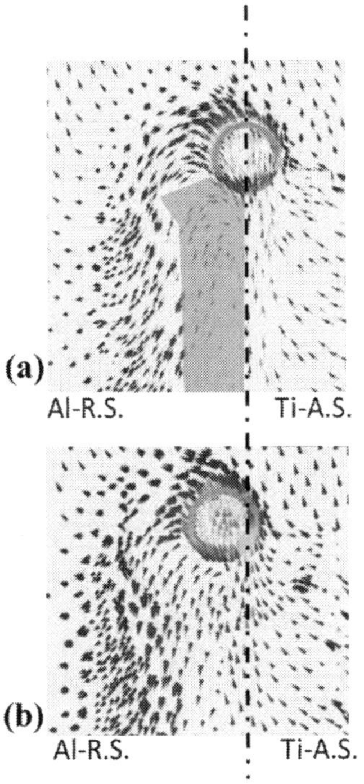

Figure 8. Nodal velocity vectors during the weld for the (a) zero-offset and (b) offset case studies

Finally, it should be observed that, for the offset case study, due to the occurring process mechanics, "pure" AA2024 is found at a distance of about 6 mm from the separation line. This value is the sum of the pin major diameter and the mixed material band rotating with the tool in the cross section containing the pin. In this way, it can be stated that local melting, detrimental to the joint integrity because of the deformation of brittle intermetallics, does not take place in this process conditions.

4 Conclusions

A numerical analysis on the process mechanics in FSW of dissimilar AA2024 and Ti6Al4V butt joints is presented.

From the obtained results it can be concluded that, both the zero-offset and the offset condition result in proper local values of temperature and accumulated strain in order to obtain sound welds. However, shifting the tool towards the low melting material results in a more effective material flow behind the tool and better materials mixing highlighted by a mixed materials band formed close to the tool.

References

1. W.M. Thomas, E.D. Nicholas, J.C. Needham, M.G. Murch, P. Temple Smith, C.J. Dawes, *G.B. Patent Application No. 9125978.8*, (1991)
2. R. Nandan, T. DebRoy, H.K.D.H. Bhadeshia, Prog Mater Sci, *Recent advances in friction-stir welding – Process, weldment structure and properties*, **53** (6), pp. 980–1023, (2008), doi:10.1016/j.pmatsci.2008.05.001
3. Y.G. Kima, H. Fujii, T. Tsumura, T. Komazaki, K. Nakata, Mat Sci Eng A-Struct, *Three defect types in friction stir welding of aluminum die casting alloy*, **415** (1–2), pp. 250–254, (2006), doi:10.1016/j.msea.2005.09.072
4. R.S. Mishra, Z.Y. Mab, Mat Sci Eng R, *Friction stir welding and processing*, **50** (1–2), pp. 1–78, (2005), doi:10.1016/j.mser.2005.07.001
5. S. Tiwari, H. Chelladurai, A. K. Shukla, 5th International & 26th All India Manufacturing Technology, Design and Research Conference (AIMTDR 2014), *Parametric Analysis of Friction Stir Welding*, (IIT Guwahati, Assam, India, December 12th – 14th, 2014), http://www.iitg.ernet.in/aimtdr2014/PROCEEDINGS/papers/528.pdf
6. K. P. Mehta, V. J. Badheka, Int J Adv Manuf Tech, *Influence of tool design and process parameters on dissimilar friction stir welding of copper to AA6061-T651 joints*, **80** (9), pp. 2073–2082, (2015), doi:10.1007/s00170-015-7176-1
7. M. Aonumaa, K. Nakata, Mat Sci Eng B-Solid, *Dissimilar metal joining of ZK60 magnesium alloy and titanium by friction stir welding*, **177** (7), pp. 543–54, (2012), doi:10.1016/j.mseb.2011.12.031
8. D. Liu, R. Xin, X. Zheng, Z. Zhou, Q. Liu, Mat Sci Eng A-Struct, *Microstructure and mechanical properties of friction stir welded dissimilar Mg alloys of ZK60 – AZ31*, **561**, pp. 419–426, (2013), doi:10.1016/j.msea.2012.10.052
9. S. A. Khodir, T. Shibayanagi, Mat Sci Eng B-Solid, *Friction stir welding of dissimilar AA2024 and AA7075 aluminum alloys*, **148** (1–3), pp. 82–87, (2008), doi:10.1016/j.mseb.2007.09.024
10. G. Buffa, D. Baffari, A. Di Caro, L. Fratini, Sci Technol Weld Joi, *Friction stir welding of dissimilar aluminum–magnesium joints: sheet mutual position effects*, **20** (4), pp. 271–279, (2015), doi:10.1179/1362171815Y.0000000016
11. V. Firouzdor, S. Kou, Metall Mater Trans A, *Al-to-Mg Friction Stir Welding: Effect of Material Position, Travel Speed, and Rotation Speed*, **41** (11), pp. 2914–2935, (2010), doi:10.1007/s11661-010-0340-1
12. Y. Zhao, S. Jiang, S. Yang, Z. Lu1, K. Yan, Int J Adv Manuf Tech, *Influence of cooling conditions on joint properties and microstructures of aluminum and magnesium dissimilar alloys by friction stir welding*, pp. 1–7, (2015), doi:10.1007/s00170-015-7624-y
13. M. Ramulu, P.D. Edwards, D.G. Sanders, A.P. Reynolds, T. Trapp, Mater Design, *Tensile properties of friction stir welded and friction stir welded-superplastically formed Ti–6Al–4V butt joints*, **31** (6), pp. 3056–3061, (2010), doi:10.1016/j.matdes.2010.01.023
14. S, Nimera, J. Wolkb, M. Zupan, Acta Mater, *Local property characterization of friction stir welded Ti-5111: Transverse orientation measurements*, **61** (8), pp. 3050–3059, (2013), doi:10.1016/j.actamat.2013.01.065
15. L. Zhou, H.J. Liu, Q.W. Liu, Mater Design, *Effect of rotation speed on microstructure and mechanical properties of Ti–6Al–4V friction stir welded joints*, **31** (5), pp. 2631–2636, (2010), doi:10.1016/j.matdes.2009.12.014
16. M. M. Z. Ahmed, B. P. Wynne, J. P. Martin, Sci Technol Weld Joi, *Effect of friction stir welding speed on mechanical properties and microstructure of nickel based super alloy Inconel 718*, **18** (8), pp. 680–687, (2013), doi: 10.1179/1362171813Y.0000000156
17. B. Li, Z. Zhang, Y. Shen, W. Hub, L. Luo, Mater Design, *Dissimilar friction stir welding of Ti–6Al–4V alloy and aluminum alloy employing a modified butt joint configuration: Influences of process variables on the weld interfaces and tensile properties*, **53**, pp. 838–848, (2014), doi:10.1016/j.matdes.2013.07.019
18. A. Wua, Z. Song, K. Nakata, J. Liao, L. Zhou, Mater Design, *Interface and properties of the friction stir welded joints of titanium alloy Ti6Al4V with aluminum alloy 6061*, **71**, pp. 85–92, (2015), doi:10.1016/j.matdes.2014.12.015
19. G. Buffa, J. Hua, R. Shivpuri, L. Fratini, Mat Sci Eng A-Struct, *A continuum based fem model for friction stir welding—model development*, **419** (1–2), pp. 389–396, (2006), doi:10.1016/j.msea.2005.09.040
20. K. Gök, M. Aydin, Int J Adv Manuf Tech, *Investigations of friction stir welding process using finite element method*, **68**, pp. 775–780, (2013), doi:10.1007/s00170-013-4798-z
21. G. Buffa, Int J Mater Form, *Joining Ti6Al4V and AISI 304 through friction stir welding of lap joints: experimental and numerical analysis*, **9** (1), pp. 59–70, (2014), doi:10.1007/s12289-014-1206-7
22. M. Abbasi, B. Bagheri, R. Keivani, J Mech Sci Tech, *Thermal analysis of friction stir welding process and investigation into affective parameters using simulation*, **29** (2), pp. 861–866, (2015), doi:10.1007/s12206-015-0149-3
23. G. Buffa, L. Fratini, R. Shivpuri, Comput Struct, *Finite element studies on friction stir welding processes of tailored blanks*, **86** (1–2), pp. 181–189, (2008), doi:10.1016/j.compstruc.2007.04.007

24. G. Buffa, A. Ducato, L. Fratini, Mat Sci Eng A-Struct, *FEM based prediction of phase transformations during Friction Stir Welding of Ti6Al4V titanium alloy*, **581**, pp. 56–65, (2013), doi:10.1016/j.msea.2013.06.009
25. G. Buffa, A. Ducato, L. Fratini, AIP Conference Proceedings, *Dissimilar material lap joints by friction stir welding of steel and titanium sheets: Process modelling*, **1532**, pp. 491–498, (2013), doi:10.1063/1.4806866

Finite element modeling of tube deformation during cold pilgering

Yağız Azizoğlu[1,3,a], *Mattias* Gärdsback[2], *Bengt* Sjöberg[2] and *Lars-Erik* Lindgren[3]

[1] *Dalarna University, Sandbacka Park, SE - 811 32 Sandviken, Sweden*
[2] *Sandvik Materials Technology, R & D, SE - 811 81 Sandviken, Sweden*
[3] *Luleå University of Technology, Department of Engineering Sciences and Mathematics, SE - 971 87 Luleå, Sweden*

Abstract. A three-dimensional finite element model of cold pilgering of stainless steel tubes is developed in this paper. The objective is to use the model to increase the understanding of forces and deformations in the process. The focus is on the influence of vertical displacements of the roll stand and axial displacements of the mandrel and tube. Therefore, the rigid tools and the tube are supported with elastic springs. Additionally, the influences of friction coefficients in the tube/mandrel and tube/roll interfaces are examined. A sensitivity study is performed to investigate the influences of these parameters on the strain path and the roll separation force. The results show the importance of accounting for the displacements of the tube and rigid tools on the roll separation force and the accumulative plastic strain.

1 Introduction

Cold pilgering is one of two major cold forming methods for producing seamless tubes. These tubes are used in a wide range of industries where high precision is needed, e.g.: nuclear, aerospace, automotive, oil and gas. This method enables cost efficient production of tubes with tight geometric tolerances and high surface quality. Most metals are suitable for this method, even though they have different ductility, which limits the maximum possible cross-sectional area reduction. Typical area reductions for stainless steels are up to 80 percent [1].

The ingoing material is a tube preform obtained from extrusion. The preform is subjected to repeated compressions using a mandrel on the inside and roll dies pressing from the outside. In a stroke, the diameter and wall thickness of the tube are reduced according to the profiles of the mandrel and the roll dies. After the stroke, the tube is fed forward and turned. A full stroke consists of a forward and backward stroke. The possibilities for feeding and turning in the two ends of a stroke vary between different mills. These small forming steps deform the tube in many small strain increments in various directions to ensure virtually homogenous wall thicknesses and material properties.

Understanding the relation between forces and deformations during pilgering of tubes is important in order to design a process with increased productivity and quality. However, the deformation process is complicated because the material is subjected to large plastic deformation with frequent changes of strain direction. The elastic deflections of the roll stand and the mandrel, together with the response of the formed tube, are very important parameters as they influence the gap between the mandrel and the groove of the roll dies. This gap is critical as it determines the amount of deformation at each axial position of the formed tube.

In the literature, several papers have been published where the roll separation force and deformation history during pilgering have been studied using experimental and/or computational methods [2-14]. However, there are only a few studies where the influence of roll stand deflection and/or the displacement of the mandrel and the tube are included [4, 5, 6, 10]. Yet, none of these studies have investigated the influence of displacements of the tube and the rigid tools on the roll separation force and the accumulative plastic strain by use of a three-dimensional (3D) finite element (FE) model.

In the present work, a 3D mechanical FE model of cold pilgering of stainless steel tubes is used to investigate the influence of vertical displacements of the roll stand and axial displacements of the mandrel and tube. The displacements are achieved by supporting them with elastic springs in the model. Furthermore, the influence of friction in the tube/mandrel and tube/roll interfaces is investigated. A sensitivity study is performed to investigate the influence on the strain path and the roll separation force.

1.1 Background

The first numerical model of cold pilgering was developed by Siebel and Neumann in 1954 by using an analytical axisymmetric 2D approach [2]. The vertical rolling force and contact lengths were estimated. A strip rolling analogy was used without considering the oval shape of the roll grooves.

[a] Corresponding author: yaz@du.se, yagiz.azizoglu@ltu.se

In 1975, Yoshida *et al.* [3] published a comprehensive experimental study in which they measured the vertical and axial rolling forces, the contact pressure and the contact length by using pressure sensors buried in the groove surface. They also measured the strains from a deforming grid printed on the preform surface and calculated the stress tensor.

Later on, in 1984, Furugen and Hayashi developed an analytical 3D model for the cold pilgering process [4]. The tube deformation was modeled by tracing the axial and tangential movements of a metal piece during the process. To include the ovalisation, the cross-section of the tube was modeled as two parts: the groove and the flange. Analytical equations were used to calculate stresses from estimated strain values. The roll separating force was also calculated and compared with experiments. This model was the first attempt to account for the elastic deformation of the roll stand as they estimated the gap between the groove and the mandrel.

Aubin *et al.* [5] made an important study where a number of experiments and numerical analyses were performed. They measured the roll separation force, the axial forces applied to the mandrel and the preform and the axial displacements of the preform, tube end and mandrel end. By comparing the forces and the displacements, the stiffnesses of the mandrel-mandrel rod system and the preform-carriage system were estimated and integrated in the numerical model giving axial displacements of the mandrel and tube. They stated, but did not show, that these motions are very important for the state of stress in the tube. Moreover, the roll gap during the pilgering was estimated by the deformation of an indium wire positioned between the lateral treads of the two rolls. The measured gap was related to the measured roll separation force so that the roll stand stiffness could be estimated. Additionally, the roll flattening due to the elastic deformation of the rolls and the wave in front of the compression area were included by modifying the contact length formula of Siebel and Neumann [2], using the Hitchcock formula for the flattened roll radius and a wave factor for the increased contact length. They used their results to update the analytical model by Furugen and Hayashi [4].

Harada *et al.* [6] employed a 2D generalised plane strain model for the simulation of cold pilgering of a Zircaloy tube. They traced the tube-fed portion during the pilgering. They included the deflection of the roll stand in the estimation of the distance between the mandrel and the rolls. They also included the flattened roll radius and the wave factor as in [5]. With this model, they investigated the influence of tool design, the flattened roll radius and the wave factor on the roll separation force. They concluded that the wave factor has a large influence but the flattened roll radius has a fairly low influence, which is not in agreement with Aubin *et al.* [5]. They also examined the effect of tube spring-back on the roll separation force and showed that the model is in better agreement with experiments once the spring-back is considered.

In 1996, Mulot *et al.* [7] presented the first 3D finite element simulation of cold pilgering. They simulated one forward stroke. The main focus of their work was to evaluate if FE analysis (FEA) is a feasible method for the modeling of the cold pilgering process. They showed that FEA can predict the stress and strain states in the flange and bottom of the groove, which other models cannot. They also showed the influence of friction, at the inner and outer tube surfaces, on the horizontal forces acting on the mandrel and the preform.

Montmitonnet *et al.* [8] further developed the 3D FE model in [7] and simulated three strokes. They concluded that stresses and rolling forces were almost identical in all strokes for an ideal material without plastic hardening. Work hardening was included in the model by Lodej *et al.* [9]. A post-processing procedure was created to decrease the calculation cost of their model. This procedure was employed to estimate the strain and stress trajectories of a material point for multiple strokes from the FE results of a single stroke.

Nakanishi *et al.* [10] developed a 3D FE model of the pilgering process where the rigid rollers were supported by springs in the vertical direction. Thus they included the roll stand stiffness. Moreover, the rigid mandrel and the tube were supported by springs in the axial direction to capture axial motions during pilgering. The tube elongation was measured by laser equipment in an industrial pilger mill. The elongation of the tube in the numerical model was fit to the measurements by changing the friction coefficients between the tube and the mandrel and the tube and the roll dies. The displacements of the mandrel, tube inlet, tube outlet and the elongation were calculated using the model. The elongation was in a good agreement with measured values but the tube and mandrel displacements were uncertain

Pociecha *et al.* [11] used a measurement system based on the principle of photogrammetry to measure the deformation on the surface of an aluminum tube by printed grids and non-destructive testing using a calibrated ultrasonic gauge to measure thicknesses. They also developed a 3D FE model of the pilgering process. Strain distributions along the tube in a single cycle were computed and compared with measurements. Recently, the present authors developed a 3D thermo-mechanical FE model of the cold pilgering process to investigate the influence of temperature and strain rate on the roll separating force [12].

2 Model

2.1 Material

The material of the tube is subjected to multi-axial, and quasi-random non-proportional loadings during the pilgering operation. The flow stress depends on the deformation path, strain rate and temperature [12-14]. There exists a wide variety of material models used for simulating pilgering. However, there is still no material model applicable for macroscopic simulations that can describe various load paths and evolution of texture.

In the present work, the material behavior is simplified and only total accumulated plastic strain is accounted for, i.e. the classical isotropic von Mises

 DOI: 10.1051/matecconf/20168015004

plasticity model is used. A high-alloy austenitic stainless steel, Alloy 28, is considered. The flow stress has been evaluated from uniaxial compression tests at room temperature and strain rate of 0.01 s^{-1}, see Fig. 1. Young's Modulus is 195 GPa and Poisson's ratio is 0.29.

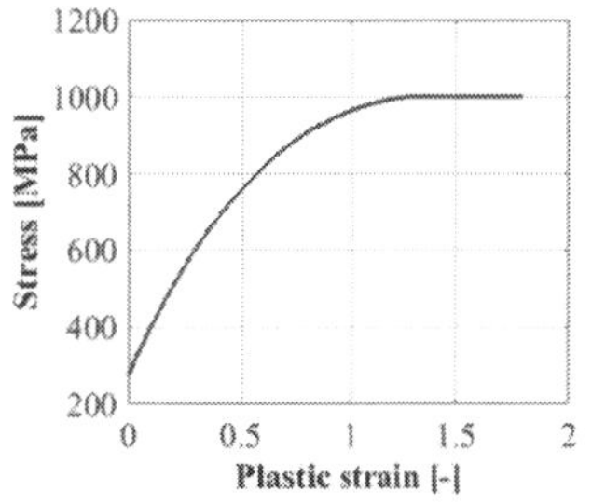

Figure 1. Flow stress curve.

2.2 Geometry

The geometry of the 3D mechanical FE model is shown in Figures 2 and 3.

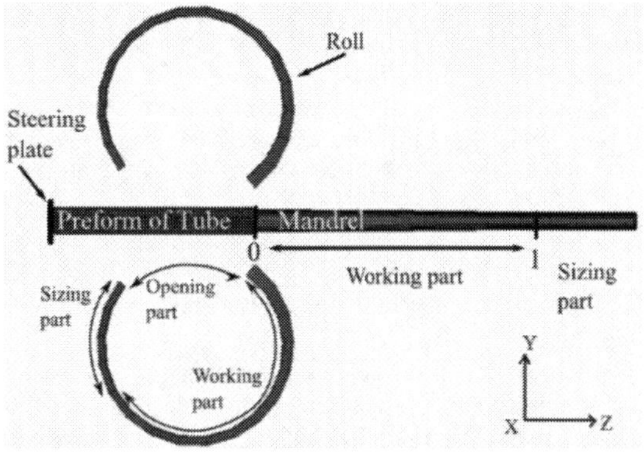

Figure 2. Geometry used in the FE model.

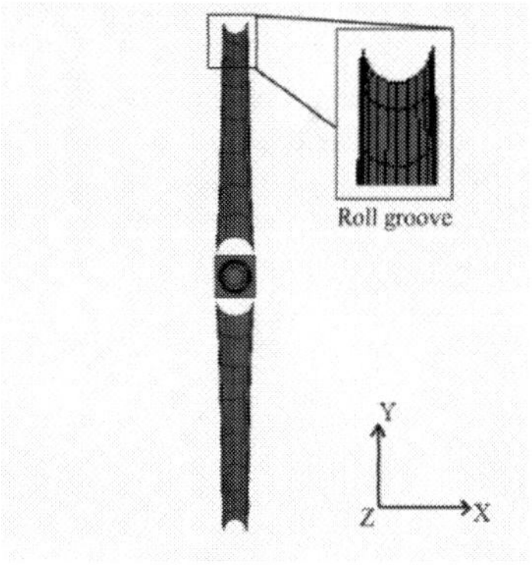

Figure 3. Details of roll groove in working part of roll.

This model consists of two rolls (upper and lower), the mandrel, the steering plate and the tube preform. The rolls consist of three parts: the working, sizing and opening part. In the working part, the distance between the roll grooves and the mandrel is decreasing. In the sizing part, this distance is constant. Similarly, the mandrel has a conical shape in the working part and a constant radius in the sizing part. The length of working part is normalized to 1. In the opening part, there is no contact with the tube; instead feeding and rotation of the preform is possible. The profiles of the mandrel and roll dies are the most important design parameters to determine the quality of the final product. The virtual steering plate is only used in the model to feed forward and rotate the preform when the rolls are in the beginning and end of the stroke.

2.3 Boundary conditions

Figure 4 illustrates the components and the operation positions in the pilgering process. The motion of the rolls is obtained by prescribed displacements and rotations. The rolls are assumed to be rigid and supported with a spring in the vertical direction to account for the deflection of the roll stand during the operation.

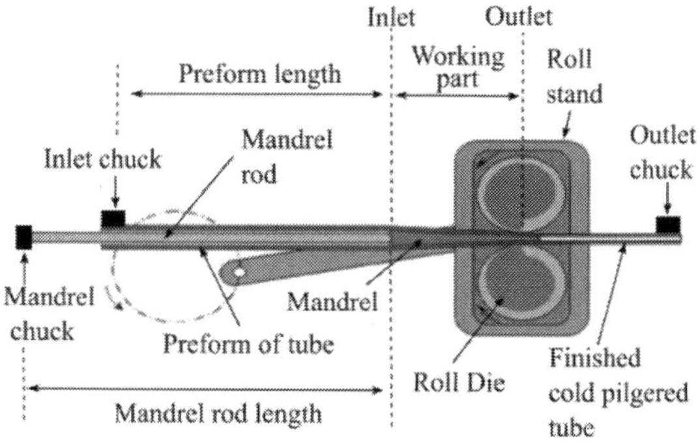

Figure 4. Configuration of pilgering process.

The mandrel is also assumed to be rigid and supported with a spring in the horizontal direction to include the elastic deflection of the mandrel rod. A third spring is inserted between the preform and the steering plate accounting for the elastic deflection of the preform. The locations of the springs are indicated in Figure 5.

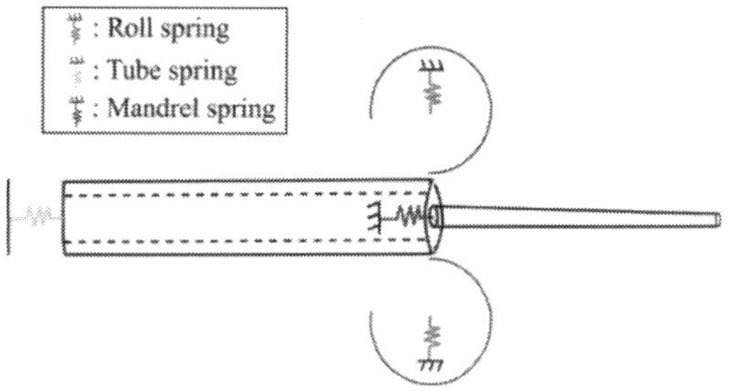

Figure 5. Inserted springs to mimic deflections of rigid tools.

The spring stiffnesses of the mandrel rod and preform are calculated by use of Equation 1, where E is the elastic modulus, A is the cross-sectional area and l is the length of the rod or preform.

$$k = EA/l \qquad (1)$$

The distance between the mandrel chuck and the inlet position is constant during the process. Thus, only one stiffness value is calculated, 9 MN/m, for the mandrel rod. Contrary, the inlet chuck of the preform moves forward while feeding the preform. Therefore, two limiting cases are considered for the preform: the maximum and minimum lengths, where the stiffnesses are 13 and 58 MN/m, respectively. The stiffness of the roll stand, 6000 MN/m, is estimated by using measurements of the roll deflection for various roll separation forces.

The friction depends on the contact surfaces, the lubrication and the contact conditions [15]. A simple Coulomb friction model has been used in the present work. The friction coefficients 0.075 and 0.030 are used between both tube and rolls and tube and mandrel surfaces.

2.4 Process parameters

The process parameters used in the simulations are shown in Table 1.

Table 1. Process parameters.

Feed step (mm/stroke)	9
Stroke rate (rpm)	200
Total number of strokes	32
Dimensions of preform tube diameter x thickness (mm)	42 x 5
Finishing dimensions of tube (mm)	25.4 x 2.4

The feed step is the total feed of the tube, distributed as 60% before the forward stroke and 40% before the backward stroke.

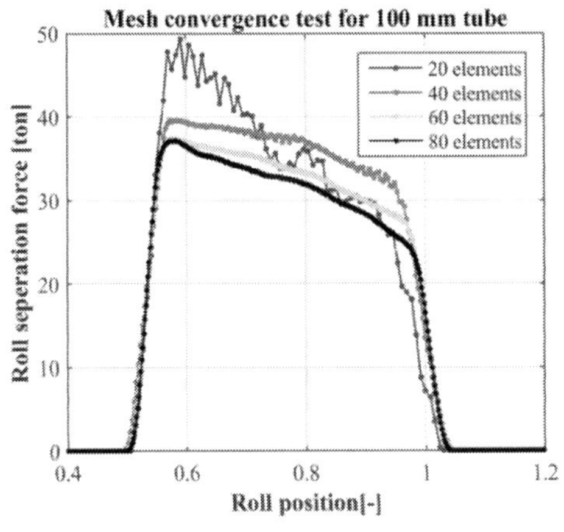

Figure 6. Computed roll separation forces used to choose mesh in the axial direction.

2.5 Mesh and computation

The preform was meshed with eight-node solid elements. Mesh convergence tests were performed for a smaller preform (100 mm). First, tests were performed with 20, 40, 60 and 80 elements in the rolling direction, whereas three elements in the radial direction and 24 elements in the tangential direction were used. The calculated roll separation forces are shown in Figure 6. Numerical oscillations in the roll separation forces were smoothed to simplify the comparison between the different cases for the mesh convergence test and the final model. The number of elements in the rolling direction is important since the contact zone affects the rolling force. Furthermore, the deformation causes the elements to become three times longer at the end of deformation. To keep the CPU time at a reasonable level, the mesh size of the 60 elements case was used in the axial direction of the final model.

Mesh sensitivity studies were also performed in the radial and tangential directions, with 60 elements in the rolling direction. Figure 7 shows the calculated roll separation forces where three, four and five elements were used in the radial direction. Three elements were used in the radial direction of the final model.

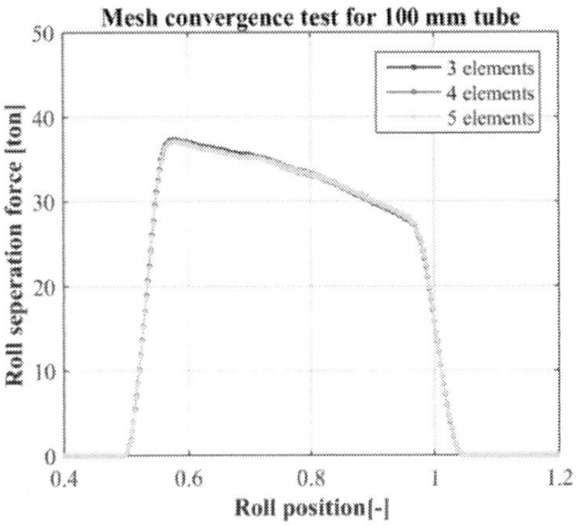

Figure 7. Computed roll separation forces used to choose mesh in the radial direction.

In Figure 8, roll separation forces are given for three different numbers of elements in the tangential direction: 24, 36 and 48. No significant difference was obtained in the tangential direction. Thus, 24 elements was used in the tangential direction. Similar convergence trends were observed for the axial forces acting on the mandrel and the steering plate which determine the axial displacements.

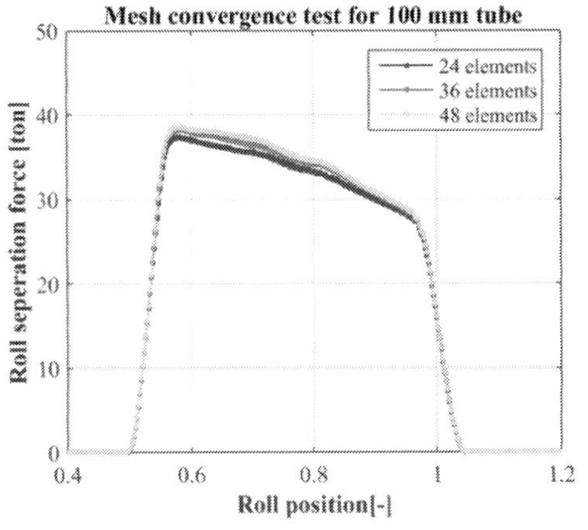

Figure 8. Computed roll separation forces used to choose mesh in the tangential direction.

The model was solved using the finite element code MSC.Marc, version 2013.r1, with implicit time stepping. The nonlinear system of equations was solved by in an incremental, iterative procedure using the full Newton-Raphson method. The model consists of 12960 elements eight-node brick elements for the 300 mm long preform. Adaptive time-stepping was used, where 0.145 ms was assigned for almost all time steps. The cpu time for 32 strokes was 58 h using a computer with an Intel Xeon E5-2680 v3 processor (24 cores, 3.07 GHz, 128 GB RAM).

3 Results and discussion

The results are presented in two groups. The influence of the spring stiffnesses on the tool displacement, the accumulative strain and the roll separation force are examined in the first group, Figures 9-10, 12-14. Four cases were simulated for this study. Two different stiffnesses, 9.0 and 4.5 MN/m, were used for the mandrel rod (ms). For the stiffness of the preform (ts), the values 13.6 and 58.0 MN/m were used to represent the maximum and minimum preform length, respectively. The stiffness of the roll (rs) was 6000 MN/m. The case with a rigid roll stand (rigid rs) was also investigated. For comparison the model without springs (rigid) is presented in Figures 12-14. A feed step of 9 mm per stroke was used in all cases. The case where ts = 13.6 MN/m, ms = 9.0 MN/m, rs = 6000 MN/m is used as the reference case.

The influence of the friction coefficient μ is investigated in the second group, Figures 15-18. In the first case, μ=0.075 was assigned for both contacts between roll/tube (μ_{rt}) and mandrel/tube (μ_{mt}). In the second case, 0.030 and 0.075 were assigned for μ_{rt} and μ_{mt}, respectively. For the last case, 0.075 and 0.030 were used for μ_{rt} and μ_{mt}, respectively.

Figures 9 and 10 show axial displacements of the tube inlet, relative to the ground and relative to the mandrel inlet, in the last stroke for different spring stiffnesses. In these figures, solid and dashed lines represent the forward and backward part of the last stroke, respectively. During the forward stroke, the displacement of the tube is first positive and then negative, Figure 9. The direction of the friction force depends on the direction of the velocity of the roll surface. The velocity of the roll surface depends on the distance from the groove to the roll center. This distance changes around the circumference of the roll die, Figure 11. At one point, the neutral point P, the sum of the friction forces is zero. In area A, the sum of the friction forces is in the same direction as the stroke. Contrary, in area B, the sum is in the opposite direction. During the forward stroke the tube is allowed to elongate in the same direction as the tube is pushed. Therefore, the direction of the friction force mainly determines the direction of the axial force acting on the tube.

During the backward stroke, the displacement of the tube is always negative. This is because the friction force is not the only force acting on the tube from the roll. The roll also pushes material in front of it which causes a large compressive force on the constrained tube inlet. This affects the axial force acting on the tube.

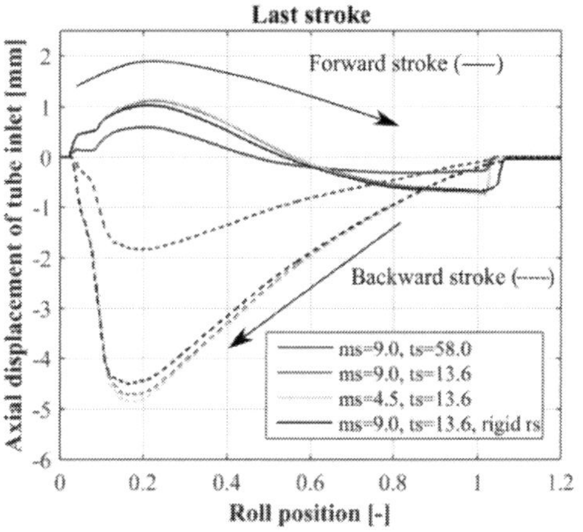

Figure 9. Axial displacement of tube at the inlet during the forward and backward stroke.

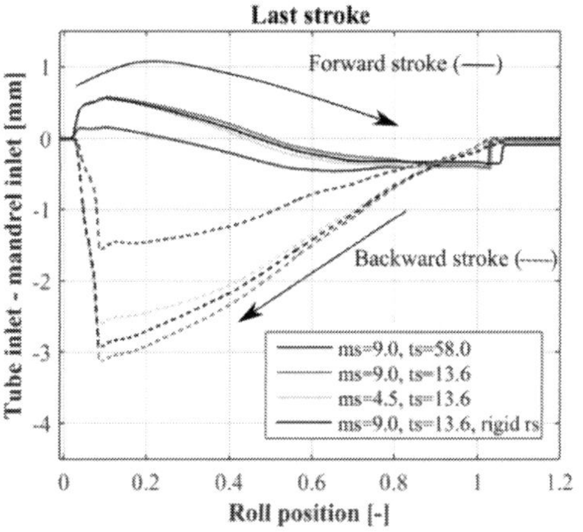

Figure 10. Axial displacement of tube with respect to mandrel movement during the forward and backward stroke.

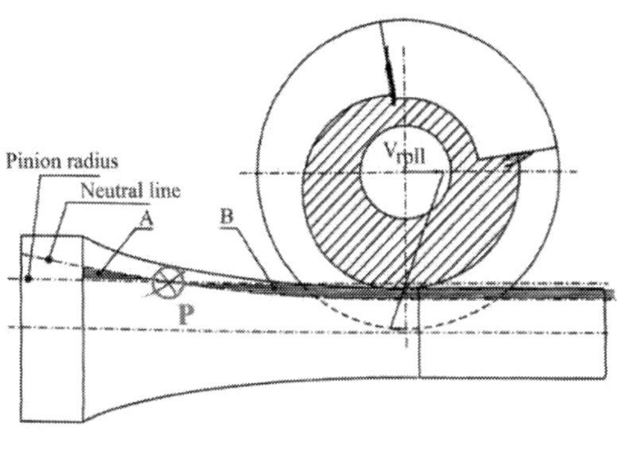

Figure 11. Illustration of the neutral point P.

Yoshida *et al.* [3] measured the reaction force acting on the tube inlet during the forward and backward strokes. The displacement curves in Figure 9 are in good agreement with their measured results. The stiffness of the preform (ts) has great influence on the displacements in both Figures 9 and 10. The mandrel stiffness (ms) influences only the tube movement relative to the mandrel movement during the backward stroke, Figure 10. The stiffness of the roll stand has limited influence on the displacements, since the results for the reference case and the case with rigid roll stand are fairly close.

Investigation of the strain history in the pilgering process is of interest for several reasons. Greater accumulative strain yields greater redundant work, which may affect the temperature in the material. Another reason is that the work hardening depend on the strain path, strain rate and temperature [13,14]. Knowledge of the plastic strain makes it possible to estimate damage and failure during forming [5]. Figure 12 shows the accumulated equivalent plastic strain of a material point on the tube surface that is traced from the beginning to the end of the process. The results show that the use of rigid tools overestimates the plastic deformation. There is no significant difference between the four other cases.

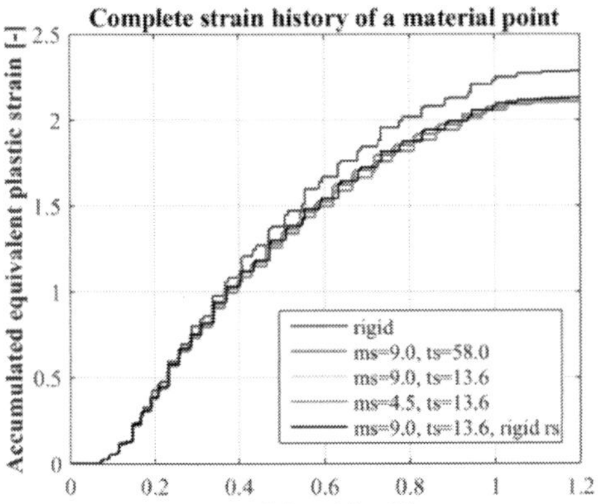

Figure 12. Accumulated strain of a material point during the cold pilgering process.

Figures 13 and 14 show the roll separation force during the forward and backward part of the last stroke. The curves, both for the forward and backward stroke, are consistent with [3, 5, 7]. In the forward stroke, there is no significant deviation between the cases.

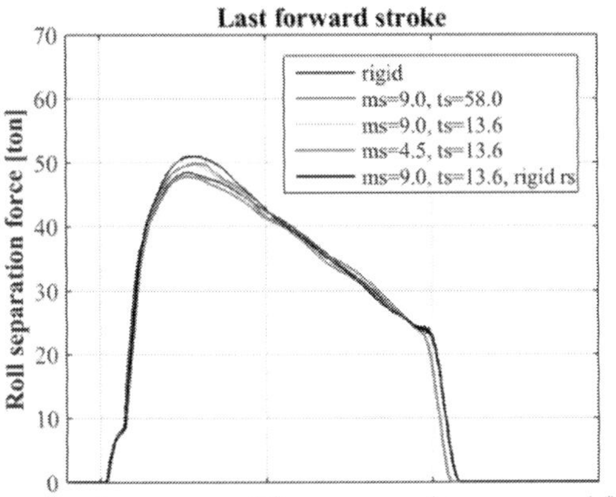

Figure 13. Roll separation force vs roll position during the forward stroke.

Figure 14 shows that the influences of spring stiffnesses on the roll separation force are more significant during the backward stroke. This is due to the larger displacements in this stroke (see Figures 9 and 10). The spring stiffnesses also affect the axial position of the maximum force because of the tube displacement relative to the mandrel (see Figure 10). This underlines the importance of mandrel and tool design.

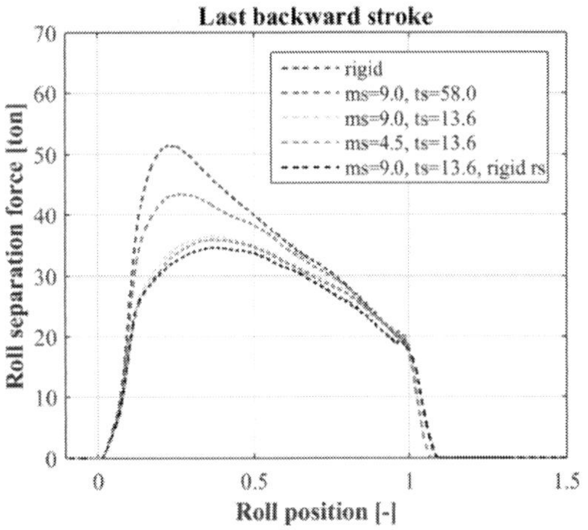

Figure 14. Roll separation force vs roll position during the backward stroke.

Figures 15 and 16 show relative displacements of the tube inlet, for varying friction constants, on the outer (μ_{rt}) and inner (μ_{mt}) surfaces of the tube.

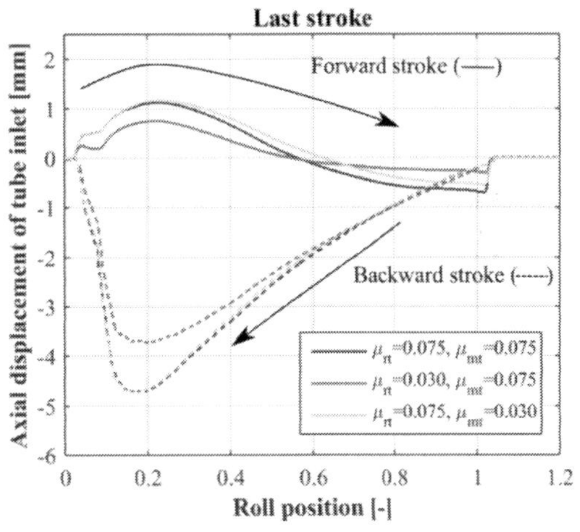

Figure 15. Axial displacement of tube during the forward and backward stroke.

It is evident in these figures that μ_{mt} has limited influence on the tube displacement relative to the inlet. However, it has considerably large influence on the tube displacement relative to the displacement of the mandrel inlet. Contrary, μ_{rt} has an influence on both of the displacement curves.

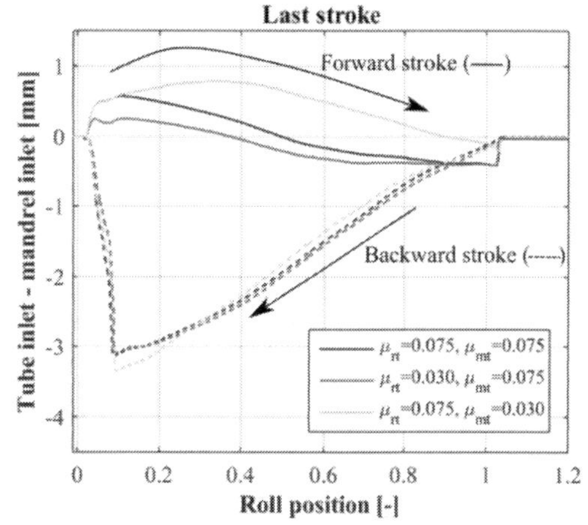

Figure 16. Axial displacement of tube with respect to mandrel movement during the forward and backward stroke.

In Figure 16, the direction of displacement during the forward stroke almost never changes for the case when μ_{rt} is greater than μ_{ms}, which does not agree with the expected position of the neutral point. The experiments in [5] suggested that μ_{mt} is greater than μ_{rt} in the pilgering process. It can also be seen in this figure that the friction coefficient affects the axial position, where the direction of displacement changes, in the forward stroke.

Figure 17 shows accumulative plastic strains for the various friction cases. There are only small differences, even though the case with high friction on the both surfaces has slightly greater accumulative plastic strain.

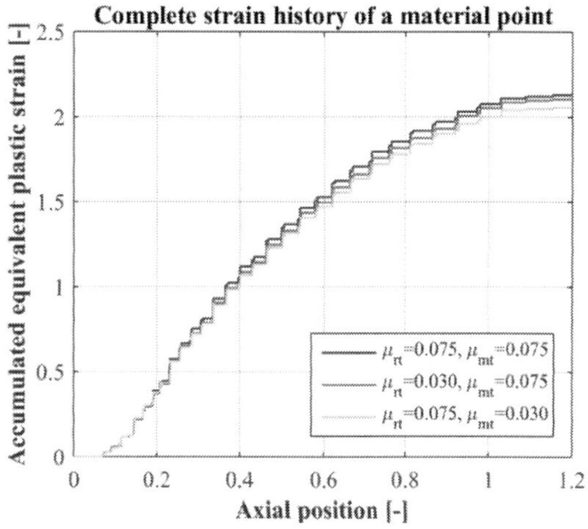

Figure 17. Accumulative strain of a material point during the cold pilgering process.

Figure 18 shows roll separation forces for the various friction cases during the forward and backward strokes. The difference is small between the three cases in the forward stroke, which is consistent with [5], and relatively larger in the backward stroke. The reason is that the material flow is more complex and the friction causes even more inhomogeneous deformation in the backward stroke. The friction coefficient also influences the axial position of the maximum force during the backward stroke.

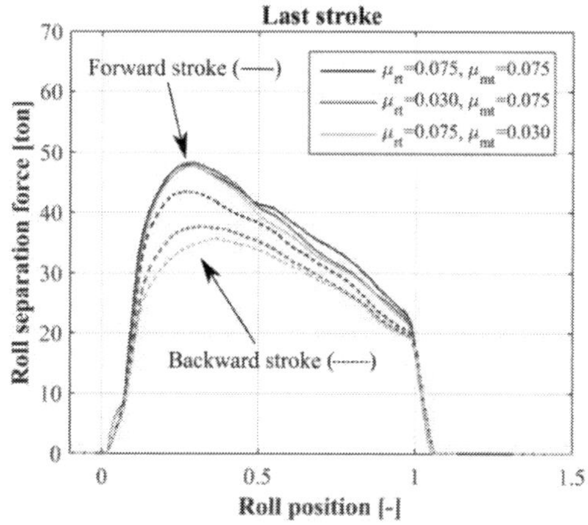

Figure 18. Roll separation force vs roll position during the forward and backward stroke.

4 Conclusions

The influence of tube and mandrel displacements and roll stand deflection on the roll separation force and strain history were investigated by use of a 3D FE model. The influences of friction in the tube/mandrel and tube/roll interfaces were also investigated.

The displacements and roll separation forces showed good agreement with measurements in the literature [3, 5, 7]. The stiffness of the tube preform had a significant impact on the roll separation force in the backward stroke. The effect of the supporting spring stiffnesses in the tools (the rolls and the mandrel) seemed to be of minor importance. However, the cases with springs had considerably different roll separation forces and accumulative plastic strains compared to the rigid case. Therefore, it is recommended that the displacements of the tools and the tube should be included in FE models for more accurate prediction of the roll separation force and the strain history.

Acknowledgement

The financial supports from Sandvik Materials Technology, Högskolan Dalarna, Jernkontoret, Region Gävleborg, Sandvikens Kommun, Länsstyrelsen Gävleborg, Region Dalarna and Stiftelsen Axel Hultgrens Fond are gratefully acknowledged.

References

1. O. Strehlau, *"Introducing cold pilger mill technology"*, TPJ (July 11, 2006)
2. E. Siebel, F.W. Neumann, *"Das Kaltpilgern von Rohren - versuchergebniße und Untersuchungen über dem Walzvorgang"*, Stahl und Eisen, **74**, 3, 139-145 (1954)

3. H. Yoshida, T. Matsui, T. Otani, K. Mandai, CIRP, **24**, 197 (1975)
4. M. Furugen, C. Hayashi, J. Mech. Work. Tech., **10**, 3, 273-286 (1984)
5. J. L. Aubin, E. Girard, P. Montmitonnet, A.M. Garde & E.M. Bradley (Eds), ASTM STP 1245, 245-263 (1994)
6. M. Harada, A. Honda, S. Toyoshima, J. of ASTM International, **2**, 3, 233-247 (2005)
7. S. Mulot, A. Hacquin, P. Montmitonnet, J. L. Aubin, J. Mater. Process. Technol., **60**, 505-512 (1996)
8. P. Montmitonnet, R. Logé, M. Hamery, Y. Chastel, J. L. Doudoux, J. L. Aubin, J. Mater. Process. Technol., **125-126**, 814-820 (2002)
9. B. Lodej, K. Niang, P. Montmitonnet, J. L. Aubin, J. Mater. Process. Technol., **177**, 1-3, 188-191 (2006)
10. H. Nakanishi, S. Toyoshima, M. Harada, A. Honda, COMPLAS X, CIMNE (2009)
11. D. Pociecha, B. Boryczko, J. Osiko, M. Mroczkowski, Arch. Civ. Mech. Eng, **14**, 3, 376-382 (2014)
12. Y. Azizoğlu, M. Gärdsback, B. Sjöberg and L.-E. Lindgren, COMPLAS XIII, CIMNE, 716-726 (2015)
13. P. Huml, M. Lindegren, CIRP Ann-Manuf. Techn., **41**, 1, 267-270 (1992)
14. E. Vanegas-Márquez, K. Mocellin, L. Toualbi, Y. de Carlan, R.E. Logé, J. Nucl. Mater., **420**, 1-3, 479-490 (2012)
15. H. Abe, T. Nomura, Y. Kubota, J. Mater. Process. Technol., **214**, 8, 1627-1637 (2014)

Formability Effects of Variable Blank Holder Force on Deep Drawing of Stainless Steel

Pramote Koowattanasuchat[1], Numpon Mahayotsanun[1], Sukunthakan Ngernbamrung[2], and Sasawat Mahabunphachai[2,a]

[1]*Department of Mechanical Engineering, Faculty of Engineering, Khon Kaen University, 123 Moo 16 Mittraphap Road, Nai-Muang, Muang District, Khon Kaen 40002 Thailand*
[2]*National Metal and Materials Technology Center (MTEC), 114 Thailand Science Park, Phahonyothin Road, Klong 1, Klong Luang, Phathumthani 12120 Thailand*

Abstract. This paper investigates the formability effects of variable blank holder force on deep drawing of AISI 304 rectangular cup. Various sets of blank holder forces were set to observe the formability, which was indicated by the percentage of sheet thinning in this study. The results showed that the blank holder forces at the locations of the sheet edges surrounding the cavity areas were considered dominant to reduce sheet thinning and enhance the formability.

1 Introduction

A recent trend in deep drawing of stainless steel sheets is to acquire deep drawn depths, sharp corners, and highly sophisticated shapes in order to add value to the products. To successfully form such parts, understanding and controlling the flow of the sheet material are critical.

Many techniques have been utilized in deep drawing to enhance material flow, for instances, forming at elevated temperatures, annealing between stamping steps, electric shimming, applying ultrasonic vibration, etc. Using variable blank holder force (BHF) is also a common technique to effectively control the material flow and enhance the formability [1-10]. This paper investigates the formability effects under different blank holder forces on deep drawing of AISI 304 rectangular cup. Various sets of blank holder forces were applied to observe the effect of sheet formability, which was indicated by the sheet thinning percentage. The results and analysis of this study could be used to acquire optimized sets of blank holder forces that would yield good formability.

2 Deep drawing experiment

The experimental setup of the deep drawing experiment using variable blank holder force control carried out in this study is illustrated in Figure 1. There were three sets (A, B, and C) of variable blank holder force located at set positions as shown in that figure. These variable blank holder forces were pressure-controlled by nitrogen-gas springs. In this study, nine variable blank holder force conditions were considered as shown in Figure 2. The values of the set pressure were ranged from low (L, 50 Bar) to medium (M, 100 Bar) to high (H, 150 Bar).

In this study, the sheet dimensions were 337.5 mm x 237.0 mm x 0.4 mm with the longitudinal direction parallel to the rolling direction. Since AISI 304 sheet under the annealed condition is quite isotropic [11], the effect of anisotropy on the formability of the sheet is neglected in this study. The punch dimensions were 99.6 mm x 112.1 mm x 90.9 mm. The die dimensions were 347.5 mm x 440.0 mm x 20 mm. The blank holder dimensions were 347.5 mm x 440.0 mm x 30 mm. The tool (punch, die and blank holder) material was SKD11 and the sheet material was AISI 304. The mechanical properties of the considered tool and sheet materials are presented in Table 1. Prior to each deep drawing test, a lubricant was applied on both sides of a sheet. The properties of the lubricant used in this study are presented in Table 2. During the test, a lubricated sheet was placed in between the blank holder and the die. The die was then moved vertically downward for 30 mm to form the sheet into the rectangular cup shape.

In order to investigate the formability under different variable blank holder forces, the percentage of thinning of sheet thickness after being deep drawn was used as an indicator. After each test, the sheet thickness at three measuring lines (0°, 45° and 90°) was measured at 10 locations as illustrated in Figure 3. The percentage of thinning can be calculated by Equation 1.

[a] Corresponding author: sasawatm@mtec.or.th

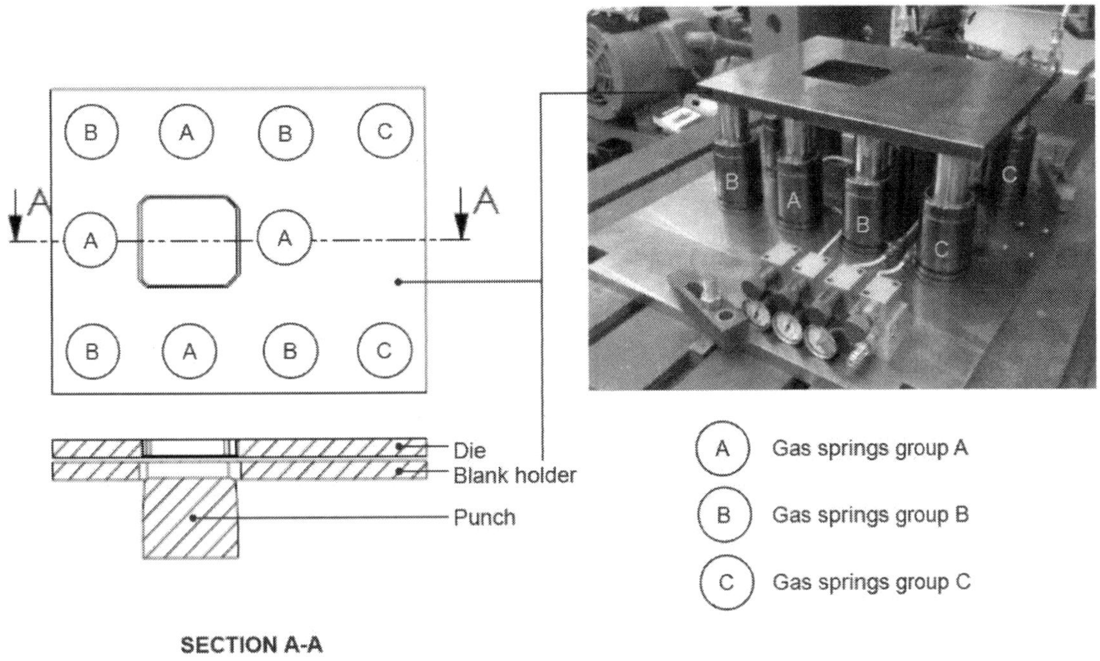

Figure 1. Deep drawing setup using variable blank holder force control.

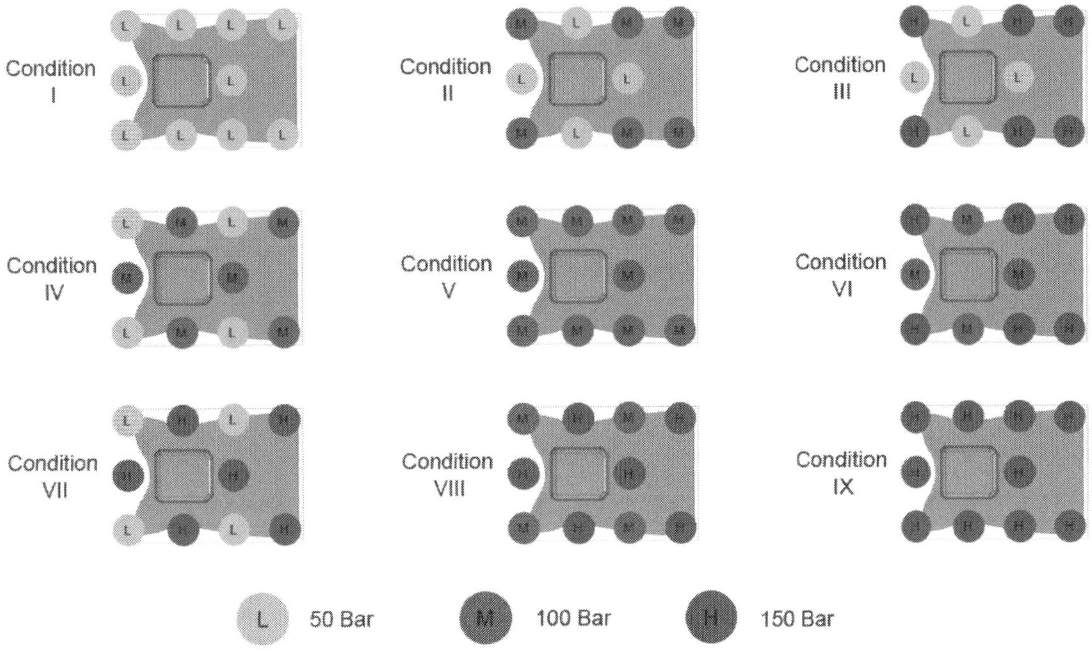

Figure 2. Considered variable blank holder force conditions.

Table 1. Mechanical properties of the considered tool and sheet materials.

Part	Material	Tensile strength (N/mm^2)	Yield strength (N/mm^2)	Elongation (%)	Hardness (HRC)
Tool	SKD11	-	-	-	58-62
Sheet	AISI 304	734	305	50	-

Table 2. Properties of the lubricant used in this study.

Active elements	Base oil group	Appearance	Specific gravity (g/cm^3, 15°C)	Viscosity (mm^2/s, 40°C)	Copper corrosion test (100 °C ×1hr)	Application
Chlorine (<20 wt%), Sulfur (<10 wt%), Fatty oil (<20 wt%)	Group II	Dark red transparent	1.0	135	2 (Inactive)	Forming oil

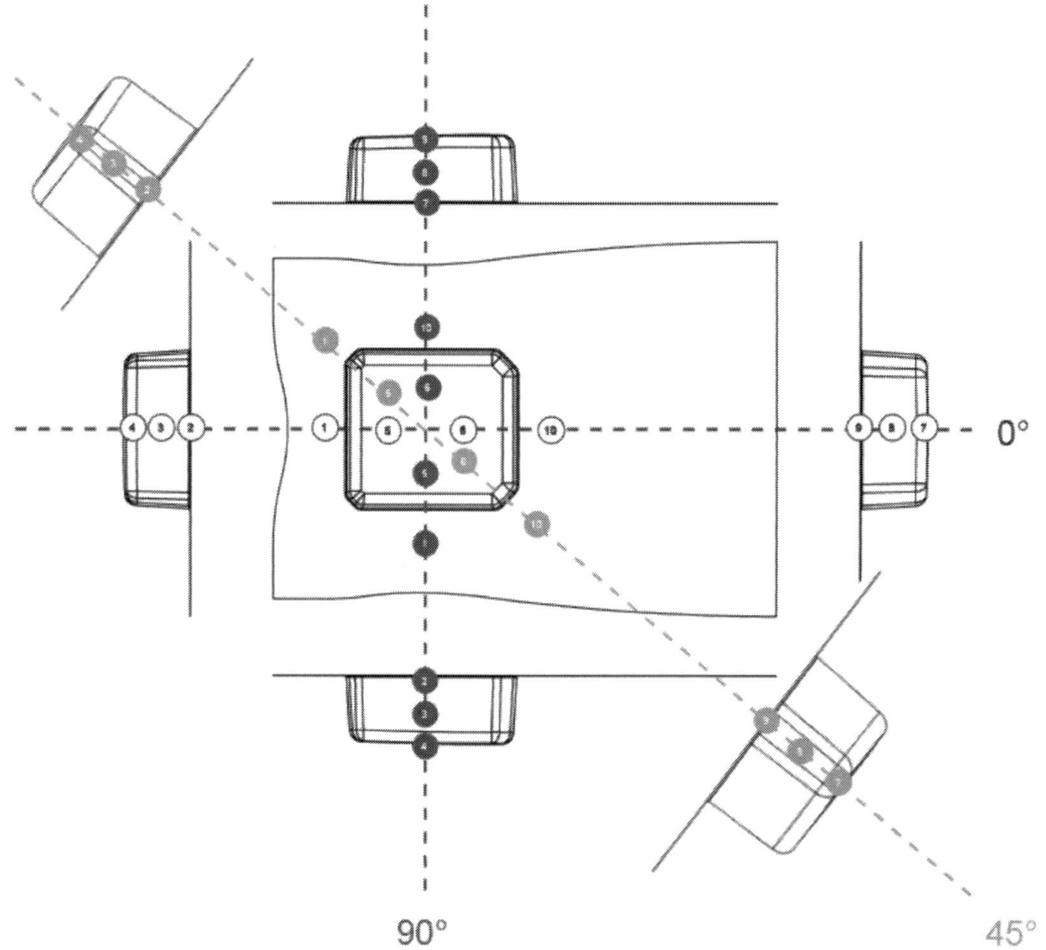

Figure 3. Sheet thickness measurement locations after deep drawing tests.

$$T_{\%} = (T_{\text{initial}} - T_{\text{final}}) \times 100\% \,/\, T_{\text{initial}} \quad (1)$$

where ($T_{\%}$) is the percentage of thinning, (T_{initial}) is the initial sheet thickness, and (T_{final}) is the final sheet thickness.

3 Results and discussions

The results of percentage of thinning of all the conditions are shown in Figure 4. In each graph, the calculated percentage of thinning values was plotted against measured points (10 locations) at three measured lines (0°, 45°, 90°). Although none of the conditions provided any defects (tearing or wrinkling) after the deep drawing tests, the locations where the values of percentage of thinning were high (reduced sheet thickness) might lead to tearing if higher drawn depths were required. According to the results, it could be observed that the locations number 8 were prone to thinning because the nearby areas (locations number 7 and locations number 9) were the cup corners, restricting the material flow and causing the sheet to elongate with reduced thickness. Location numbers 1 and 10 provided the lowest percentage of thinning values due to the fact that these areas were tightly compressed between the die and blank holder, prohibiting the sheet to move and deform.

In order to observe the effect of blank holder force, particularly at the locations prone to thinning (numbers 7, 8 and 9), nine various sets of conditions were compared as follows:

- *Conditions I, II and III.* The main difference among these three conditions is the increased blank holder forces in Groups B and C from Condition I, to Condition III, respectively while blank holder forces in Group A were kept at 50 Bar (Low). It could be observed that the percentage of thinning at locations number 7, 8 and 9 tended to reduce if the variable blank holder forces increased from 50 Bar (Condition I) to 100 Bar (Condition II), indicating that increasing the blank holder force in the sheet edges allowed the sheet to flow well if the blank holder forces at the cavity edges were kept low. However, increasing the variable blank holder forces from 100 Bar (Condition II) to 150 Bar (Condition III) did not further decrease the percentage of thinning.
- *Conditions I, IV and VII.* The main difference among these three conditions is the increased blank holder forces in Groups A and C from Conditions I, IV and VII, respectively while blank holder forces in Group B were kept at 50 Bar (low). It could be observed the no significant effect could be observed through the variations of blank holder forces in the areas close to the cavity edges.
- *Conditions IV, V and VI.* The main difference among these three conditions is the increased blank holder forces in Groups B and C from Conditions I, to Condition III, respectively while blank holder forces in Group A were kept at medium (100 Bar). It could be observed that the percentage of thinning at locations number 7, 8 and 9 tended to reduce if the variable blank holder forces increased from 50 Bar (Condition IV) to 100 Bar (Condition V), indicating that increasing variable blank holder force in the sheet edges allowed the sheet to flow well if the blank holder forces at the cavity edges were kept medium. However, increasing the variable blank holder forces from 100 Bar (Condition V) to 150 Bar (Condition VI) did not further decrease the percentage of thinning.
- *Conditions VII, VIII and IX.* The main difference among these three conditions is the increased blank holder forces in Groups B and C from Conditions VII, to Condition IX, respectively while blank holder forces in Group A were kept a 150 Bar (High). It could be observed that the percentage of thinning at locations number 7, 8 and 9 tended to reduce if the variable blank holder forces increased from 50 Bar (Condition VII) to 100 Bar (Condition VIII), indicating that increasing variable blank holder force near the sheet edges allowed the sheet to flow easier into the die cavity when the blank holder forces at the cavity edges were kept medium. However, increasing the blank holder forces from 100 Bar (Condition VIII) to 150 Bar (Condition IX) was increase the percentage of thinning.
- *Conditions III, VI and IX.* The main differences among these three conditions were the increased blank holder forces in Groups A from Condition III, VI and IX, respectively while blank holder forces in Groups B and C were kept at 150 Bar (High). No significant effect is observed through the variations of blank holder forces in the areas close to the cavity edges if the blank holder forces at the sheet edges were kept high.
- *Conditions I, V and IX.* The main difference among these three conditions is the overall increased blank holder forces in Groups A, B and C from Conditions I, V and IX, respectively. No significant effect is observed through the increased of the overall blank holder forces in all the areas.

According to the comparisons of the aforementioned conditions, it is observed that increasing the blank holder force at the sheet edges from low (50 Bar) to medium (100 Bar) tended to reduce the percentage of thinning. However, increasing the blank holder force further from medium (100 Bar) to high (150 Bar) did not decrease the percentage of thinning further. In addition, increasing the blank holder force at the cavity edges did not significantly affect the percentage of thinning.

In order to observe the effect of variable blank holder force to all the locations (numbers 1 to 10), the percentage of thinning of locations number 1 to 10 were averaged and plotted against all the conditions as shown in Figure 5. The graphs of the left to right columns show the ranges of the percentage of thinning at 0°, 45°, 90° and the average (mean of 0°, 45°, and 90°), respectively. It could be clearly seen that the average values of percentage of thinning measured at 90° is the lowest compared to those of 0° and 45°. This could be due to the fact that the rectangular cup examined in this study provided more areas for the sheet material to flow in the 90° direction than those of 0° and 45° directions. By observing the values of the percentage of thinning of average (mean of 0°, 45°, and 90°), it could be noticed that Conditions II, V and VIII provided low percentage of thinning values. These are the conditions where the blank holder forces of Group B are kept at medium value (100 Bar). Group B controlled the blank holder locations of the sheet edges surrounding the cavity areas, which was considered the dominant set of locations to control the variable blank holder forces.

In order to obtain the optimum amounts and locations of blank holder forces, a finite element simulation and theoretical modelling are currently being carried out by the authors. However, the analysis of the simulation results and theoretical models will be published in a future work. Nevertheless, the experimental results obtained from this study can still be used as a guideline to design and select the appropriate distribution blank holder forces that can reduce sheet thinning and enhance the formability.

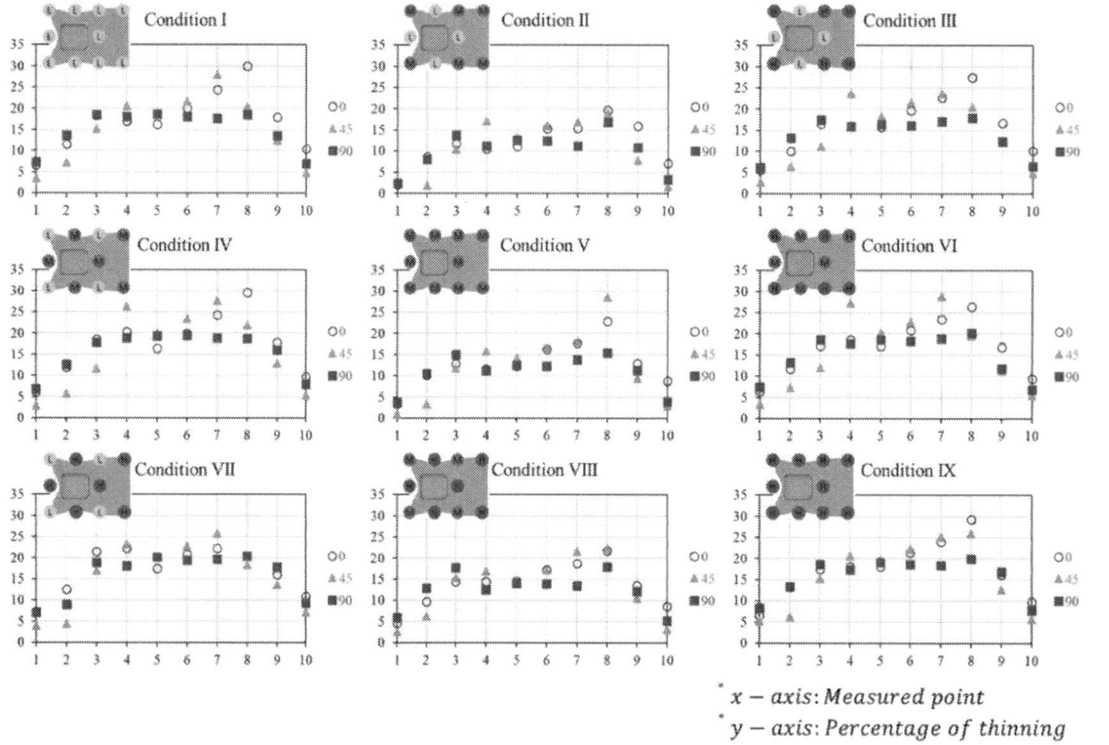

Figure 4. Percentage of thinning vs. measured points of all the considered conditions.

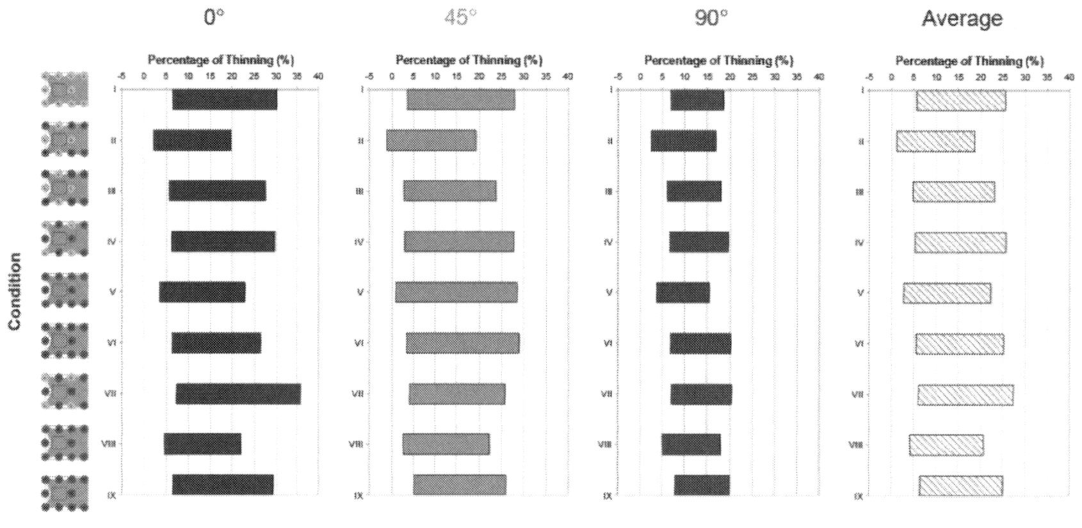

Figure 5. Average percentage of thinning of all the considered conditions.

4 Conclusions

This study experimentally investigated the effect of the distribution of the blank holder forces on the formability of AISI 304 in the deep drawing under lubricated conditions. The ranges of blank holder forces ranged from 50 Bar to 150 Bar and nine conditions of blank holder forces were configured under three groups of pressure-controlled positions. The sheet thickness values after the deep drawing tests were measured at three set lines (0°, 45° and 90°) each having 10 different points. The formability of each condition was observed through the percentage of sheet thinning. The results showed that the blank holder forces at the locations of the sheet edges surrounding the box corner areas are found dominant to reduce the percentage of thinning and thus enhance the formability.

[a] Corresponding author: sasawatm@mtec.or.th

Acknowledgement

The study was financially supported by the NSTDA University Industry Research Collaboration (NUI-RC). The authors would also like to acknowledge the supports of Diamond Brand Co., Ltd., National Metal and Materials Technology Center (MTEC), and Department of Mechanical Engineering, Faculty of Engineering, Khon Kaen University.

References

1. W.-r. Wang, G.-l. Chen and Z.-q. Lin, Trans. Nonferrous Met. Soc. China **20**, 471 (2010).
2. S. Qin, L. Yang and B. Yang, ICA. **2**, 450-455 (2011).
3. H. Koyama, R. H. Wagoner and K.-i. Manabe, J. Mater. Process. Technol. **152**, 190 (2004).
4. S.-x. Tian, Y.-x. Feng and Y.-c. Gao, Procedia CIRP **27**, 309 (2015).
5. L. Zhong-qin, W. Wu-rong and C. Guan-long, J. Mater. Process. Technol. **183**, 339 (2007).
6. B. Endelt, S. Tommerup and J. Danckert, J. Mater. Process. Technol. **213**, 36 (2013).
7. L. B. Shulkin, R. A. Posteraro and M. A. Ahmetoglu, J. Mater. Process. Technol. 98, 7 (2000).
8. T. Intarakumthornchai, S. Jirathearanat, S. Thongprasert and P. Dechaumphai, *APIEMS Conference* (2006).
9. H. Koyama, R. H. Wagoner and K.-i. Manabe, J. Mater. Process. Technol. **152**, 190 (2004).
10. E.J. Obermeyer and S. A. Majlessi, J. Mater. Process. Technol. **75**, 222 (1998).
11. H. G. Steyn, J. Mater. Eng. Perform. **21**, 1491 (2012)

DOI: 10.1051/matecconf/20168015006

FEM Analysis of Punching-Process in Consideration of Micro Die Wear

Takashi Ueda[1,a], Takashi Iizuka[1] and Shinichi Enoki[2]

[1] *Department of Mechanical and System Engineering, Kyoto Institute of Technology, Matsugasaki Goshokaido-cho, Sakyo-ku, 606-8585 Kyoto, Japan*
[2] *National Institute of Technology, Nara college, Yata-cho, Yamatokooriyama-shi, 639-1080 Nara, Japan*

Abstract. Detailed punching-process analysis is difficult because the punch diameter, sheet thickness, clearance, and die wear of the punching-process are approximately 10 mm, 1 mm, 100 μm, and 10 μm, respectively. Furthermore, such analyses have rarely considered die wear. Therefore, in the present study, we set the mesh size in the neighborhood of the die corner radius and the punch to 1/10 μm. Moreover, we analyze the punching process considering die wear. We investigated the stress, incremental strain, and equivalent strain for different levels of die wear.

1 Introduction

Burr formation in the punching process has adverse effects on post-processing, product accuracy, and worker safety. Thus, a burr-free punching method is desired. Therefore, burr-free blanking using servo press machine on a few steps punching method was reported[1, 2]. However, preventing burr formation during the one step punching process is difficult due to the nature of the formation mechanism of the punching process. Reducing burrs by a simpler method may be possible if the mechanism of burr formation can be clarified. A number of experimental and analytical investigations dealing with burr shape formation have been conducted [3]. The results of these studies revealed that the clearance [4] and die wear [5] primarily affected burr shape formation. However, the formation mechanism of burr shape in the punching process has not yet been clarified. Therefore, Iizuka [6, 7] and Sakamoto [8] focused on die wear that affects the formation of burrs and investigated the burr formation mechanism in the punching process. The burr formation mechanism in the punching process has been investigated experimentally while focusing on die wear. However, in order to support the experimental results, a detailed numerical analysis of the burr formation mechanism in the punching process is needed.

Finite element method (FEM) analysis has been reported as being effective in plastic formation. A number of FEM analyses have been conducted in order to clarify the formation mechanism [9, 10] and crack growth [11, 12] of the punching process. However, detailed analysis of the punching process is more difficult than the analysis of other formation mechanisms. Since orders of magnitude on the punch diameter, the sheet thickness, the clearance, and the die wear of the punching-process are approximately 10 mm, 1 mm, 100 μm, and 10 μm, respectively, they are very difference order of magnitude. So, there are reports that the mesh size in the neighborhood of the die corner radius and punch was set 10 μm [4, 10]. But, Conventional analyses of the punching process rarely consider micro die wear. Therefore, we set the mesh size in the neighborhood of the die corner radius and punch to 1/10 μm. Moreover, we analyze the punching process while considering die wear. We varied the die wear amount as 10 μm, 20 μm, 30 μm, and 40 μm. We investigated the hydrostatic pressure distribution, the equivalent plastic strain distribution, and the plastic incremental strain distribution for different levels of die wear in the neighborhood of the die corner radius.

2 Analysis method

The punching process analysis of the present study was conducted using Simufact.forming12.0.3 (Simufact Engineering System). In the present study, an axisymmetric two-dimensional model was used as an analysis model, as shown in Figure 1. At this time, R_d, R_p and R_h are die corner radius, punch corner radius and die holder shoulder round. We simulated the case in which a blank was punched by a metal punch with a diameter of 4 mm. Cold-rolled steel sheet (SPCC) with a thickness of 1.6 mm was used as blank. The punch and die holder are rigid bodies. The coefficient of friction between the punch and die and the blank was 0.2. The blank was subjected to a forced displacement of 1.4 mm with a processing velocity of 500 mm/s. The base mesh size of the blank in this analysis was set to 37.5 μm. The mesh size in the neighborhood of the die corner radius and the

[a] Corresponding author: m5623104@edu.kit.ac.jp

punch was set to 4.7 μm. The clearance was 4.69%. In the analysis, the die corner radius R_d was varied as 10 μm, 20 μm, 30 μm, and 40 μm, and the punch corner radius R_p and die holder shoulder round R_h were set to 20 μm.

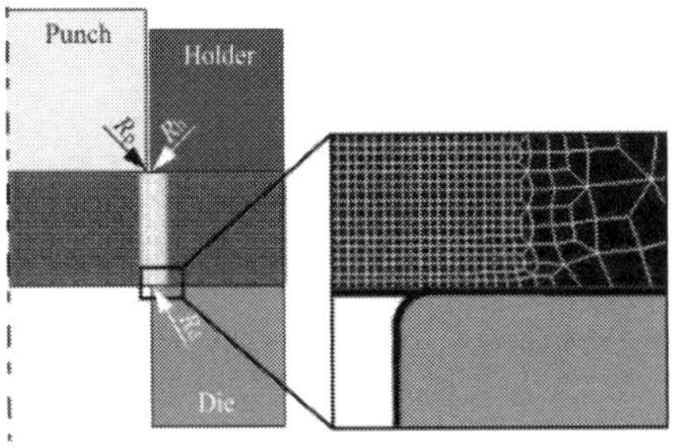

Figure 1. Analysis model.

3 Analysis result

3.1 Plastic incremental strain

For a stroke of 1 mm, the plastic incremental strain distribution is shown in Figure 2 for die corner radius radii of 10 μm, 20 μm, 30 μm, and 40 μm.

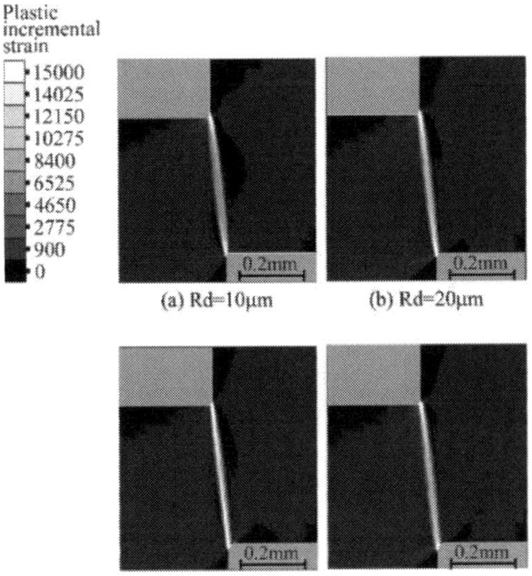

Figure 2. Plastic incremental strain distribution with a stroke of 1mm.

As shown in Figure 2, thin layer shear is thought to occur in the domain linking the punch corner radius and the die corner radius because high plastic incremental strain is confirmed in this region. Next, we confirm plastic incremental strain in the neighborhood of the die corner radius for stroke of 1 mm, 1.2 mm, or 1.28 mm. The plastic incremental strain distributions are shown for die corner radius radii of 20 μm and 40 μm in Figure 3.

As shown in Figure 3, we confirm that the plastic incremental strain that occurs in the neighborhood of the die corner radius is distributed on the die corner radius and the die side in the case of R_d = 20 μm. However, the plastic incremental strain is distributed on the die corner radius only in case of R_d = 40 μm. Next, we confirm the plastic incremental strain distribution for a stroke of 1.2 mm. In this case, a gap between the die

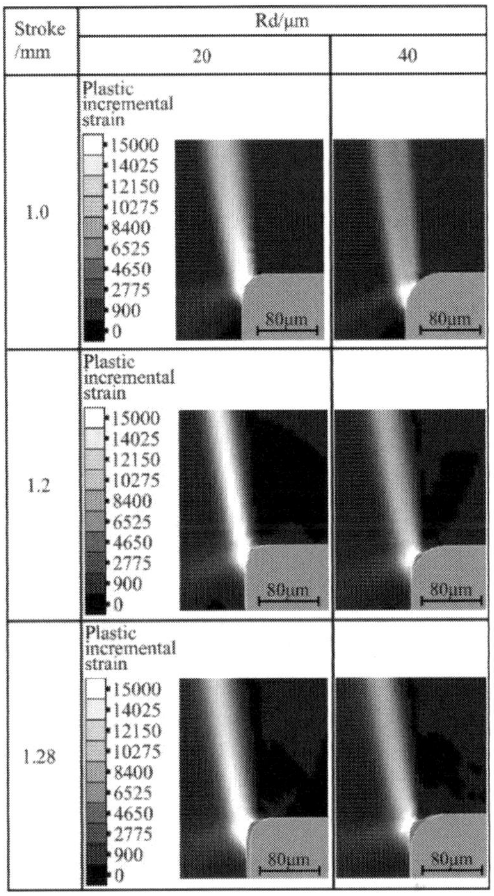

Figure 3. Plastic incremental strain distribution on neighborhood of die edge with every stroke.

and the blank occurs at the die corner radius in the cases of R_d = 20 μm and 40 μm. The gap is bigger in the case of R_d = 40 μm than in case of R_d = 20 μm. For a stroke of 1.28 mm, we can confirm that the gap is distributed on the die corner radius and the die side in the case of R_d = 20 μm. However, for the case in which R_d = 40 μm, the gap is larger than that in the case of a stroke of 1.2 mm. We cannot confirm the existence of a gap on the die side. Next, the plastic incremental strain distributions for a stroke of 1.2 mm in the cases of R_d = 10 μm, 20 μm, 30 μm, and 40 μm are shown in Figure 4.

As shown in Figures 4(a) and 4(b), in the cases of R_d = 10 μm and 20 μm, the gap is distributed on the die corner radius and the die side. In case of R_d = 10 μm, the gap is focused on the die side. However, in the case of R_d = 20 μm, the gap is focused on the die corner radius. As shown in Figures 4(c) and 4(d), in the cases of R_d = 30 μm and 40 μm, the gap is distributed on only on the die corner radius. We cannot confirm the existence of a gap on the die side. Therefore, when the die corner radius R_d increases, the domain that the gap is focused moves from the die corner radius to the die side. The gap is thought to satisfy the fracture condition, because the contact pressure is not focused in the gap. As shown in Figure 3, fracture is thought to occur from the end part of the gap for all die conditions, because, for a stroke of 1.28 mm, high plastic incremental strain occurs at the end of the gap.

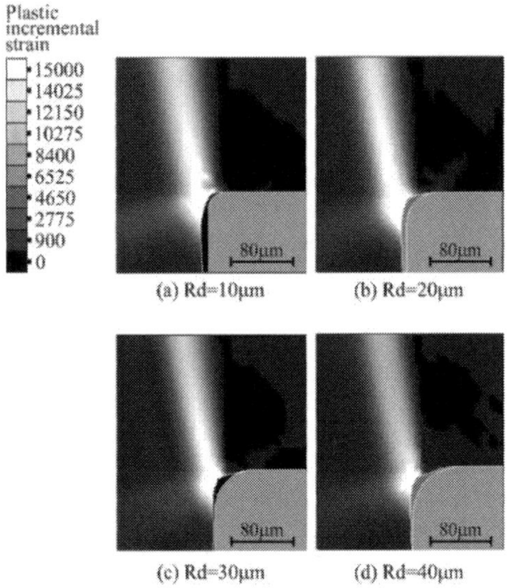

Figure 4. Plastic incremental strain distribution with a stroke of 1.28mm on ever die abrasion status.

3.2 Hydrostatic pressure

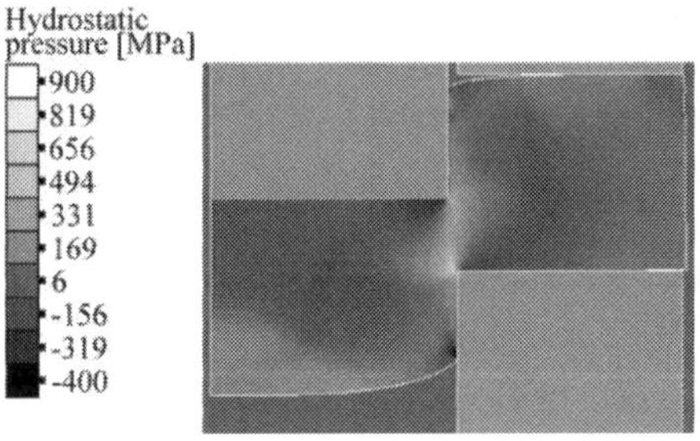

Figure 5. Hydrostatic pressure distribution with a stroke of 1mm in case of $R_d = 20\,\mu$ m

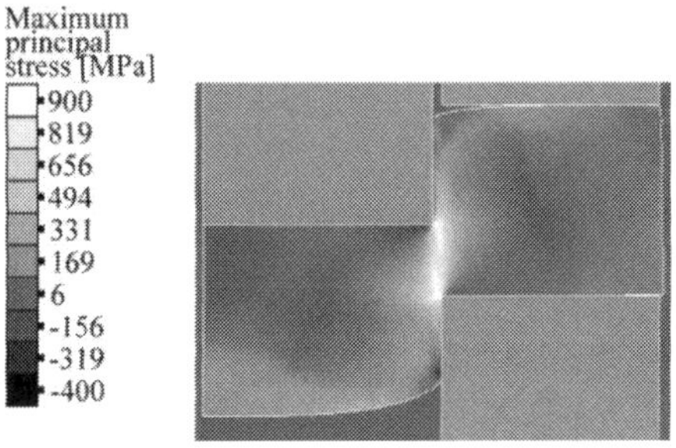

Figure 6. Maximum principal stress distribution with a stroke of 1mm in case of $R_d = 20\,\mu$ m

The hydrostatic pressure distribution, the maximum principal stress distribution, and the minimum principal stress distribution with a stroke of 1 mm in the case of R_d = 20 μm are shown in Figure 5, Figure 6 and Figure 7.

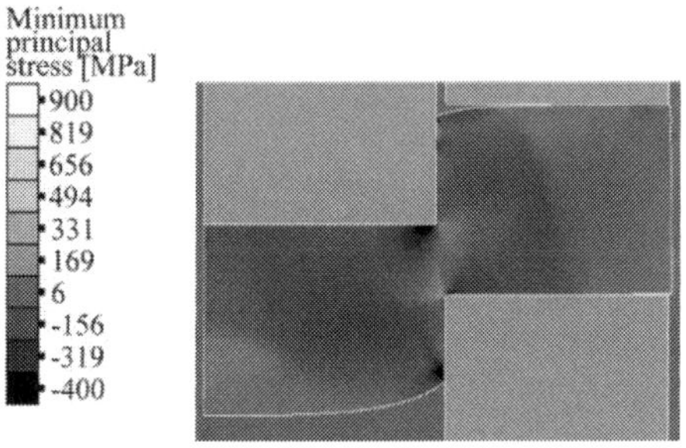

Figure 7. Minimum principal stress distribution with a stroke of 1mm in case of $R_d = 20\,\mu$ m

Stroke /mm	Rd/μm 20
1.0 (Step 1)	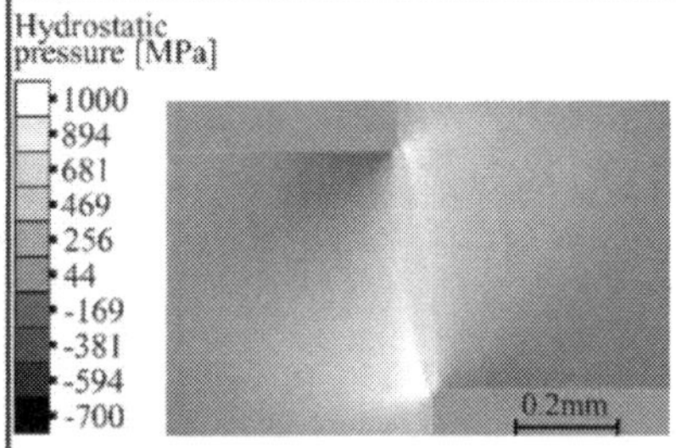
1.2 (Step 2)	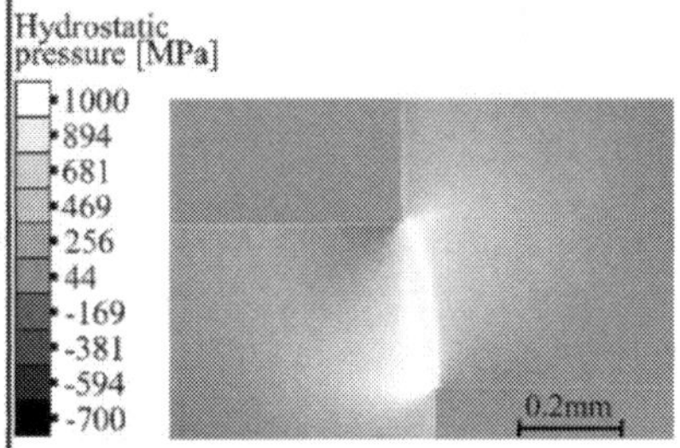
1.28 (Step 3)	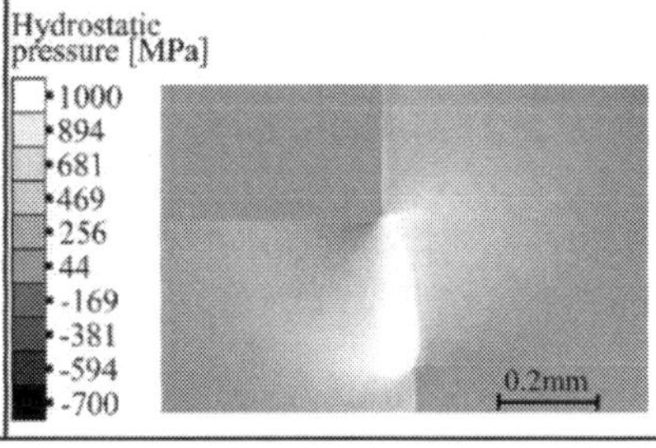

Figure 8. Hydrostatic pressure distribution with every stroke in case of $R_d = 20\,\mu$ m

In the present study, the punching process exhibits a circumferential plane strain field because, as shown in Figure 5, Figure 6 and Figure 7, the mean of the maximum principal stress distribution and the minimum principal stress distribution is the same as the hydrostatic pressure distribution. Therefore, the mechanism of the punching process can be explained by the hydrostatic pressure. As shown in Figure 5, tensile stress occurs in the neighborhood of the die corner radius and the punch corner radius. The tensile stress in the neighborhood of the die edge is shown as a tensile stress field in Figure 6 and Figure 7. Compression stress occurred at the in bottom face of the punch, at the top face of the die, at the die corner radius is shown as a tensile stress field the punch side, and at the die side. Therefore, the blank on the die top face is expected to flow toward the punch side. As for the blank on punch top face is toward die side. Next, the hydrostatic pressure distributions for strokes of 1 mm, 1.2 mm, and 1.28 mm in the case of R_d = 20 μm are shown in Figure 8.

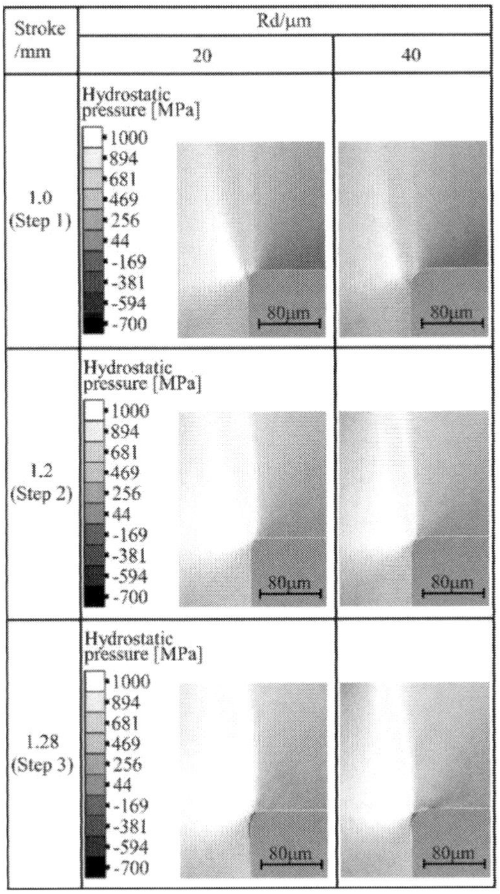

Figure 9. Hydrostatic pressure distribution on neighborhood of die edge.

As shown in Figure 8, we confirmed that the stress distribution for the punching process is divided into three steps. In Step 1, the hydrostatic pressure in the domain linking the punch corner radius and the die corner radius is approximately 0 MPa. This domain is thought to be a pure shear domain. Next, the hydrostatic pressure distributions in the neighborhood of the die corner radius for strokes of 1 mm, 1.2 mm, and 1.28 mm in the case of R_d = 20 μm and 40 μm are shown in Figure 9.

The tensile stress in the neighborhood of the domain is shown. The tensile stress in the neighborhood of the die corner radius and the punch corner radius is higher than elsewhere. In Step 2, the tensile stress appears in the pure shear domain in Step 1. In Step 3, the tensile stress in this domain is approximately the same as that in the neighborhood of the die corner radius. Next, the hydrostatic pressure distributions in the neighborhood of the die corner radius for R_d = 20 μm and 40 μm are shown in Figure 9. As shown in Figure 9, in Step 1, the tensile stress is distributed only on the end part of the die corner radius. Compression stress occurs at locations other than the die corner radius. The tensile stress in Step 2 is higher than that in Step 1. Moreover, a gap appears between the die corner radius and the blank. The high tensile stress domain in Step 3 is larger than that in Step 2. The tensile stress domain for R_d = 40 μm is larger than that for R_d = 20 μm. Therefore, it's expected that gap form varies depending on the level of die corner radius.

3.2 Equivalent plastic strain

The equivalent plastic strain distributions for a stroke of 1 mm in the cases of R_d = 10 μm, 20 μm, 30 μm, and 40 μm are shown in Figure 10. As shown in Figure 10, high equivalent plastic strain is distributed over the range of the die side from the die corner radius. Therefore, the blank over the range from the die corner radius to the die side is thought to be brittle. We confirmed that die wear has little effect on equivalent plastic strain.

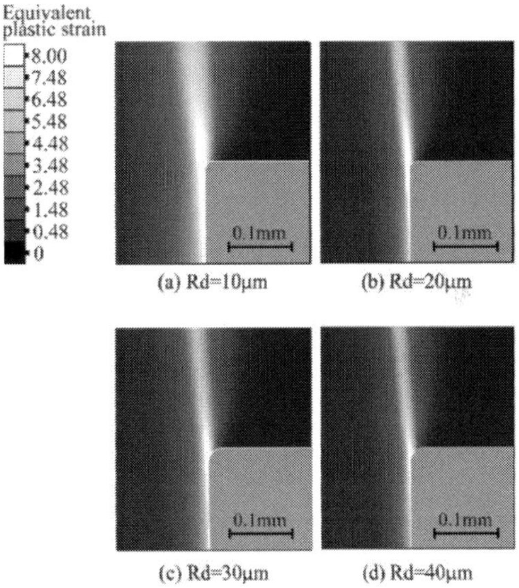

Figure 10. Equivalent plastic strain distribution with a stroke of 1mm.

4 Discussion

The plastic incremental strain, hydrostatic pressure, and equivalent plastic strain are high for a stroke of 1 mm, as shown in Figure 11. As shown in Figure 11, the high plastic incremental strain occurred at the die corner radius. And, the high equivalent plastic strain is distributed over the range of the die side from the die corner radius. However, high hydrostatic pressure occurred at the neighborhood of the die corner radius. The domain that high equivalent plastic strain was shown is thought to be brittle. The plastic incremental strain shows the deformation volume. Therefore, the location at which the equivalent plastic strain overlaps the plastic incremental strain is likely to exhibit crack formation. Cracks are thought to progresses as a result of the hydrostatic pressure of the tensile field. Therefore, in case of Rd = 20 μm, crack is thought to occurred over the range of the die side from the die corner radius. In case of Rd = 40 μm, crack is thought to occurred on die corner radius. The origin of crack formation differs according to the level of die wear.

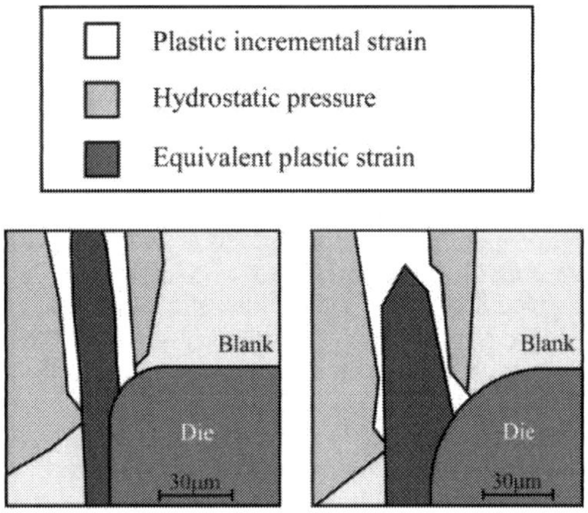

(a) Rd=20μm (b) Rd=40μm

Figure 11. High value of plastic incremental strain, hydrostatic pressure and equivalent plastic strain with a stroke of 1mm.

5 Conclusion

We analyze the punching process while considering die wear. We varied the die wear amount as 10 μm, 20 μm, 30 μm, and 40 μm. We investigated the hydrostatic pressure distribution, the equivalent plastic strain distribution, and the plastic incremental strain distribution for different levels of die wear in the neighborhood of the die corner radius. The results obtained are summarized as follows.

(1) In the plastic incremental strain distribution, the thin layer shear is thought to occur in the domain linking the punch corner radius and the die corner radius.

(2) The gap between the die and the blank occurs at the die corner radius for a stroke of 1.2 mm. Moreover, the gap differs depending of the level of die corner radius.

(3) In this punching process, the stress distribution of the thin layer shear is divided into three steps.

(4) We confirmed that the level of die wear has little effect on the equivalent plastic strain.

(5) The origin of crack formation differs according to the level of die wear.

References

1. K. Junlapen, M. Kubo, K. Mizuochi, N. Koga, P. Keawtatip, ICTP, 158-163, (2008)
2. S. Thipprakmas, M. Jin, M. Murakawa, ICTP, Oct. 28-31, 889-894, (2002)
3. T. Maeda, I. Aoki, J. Fac. Eng. Univ. Tokyo (B), 34-3,537-577 (1978)
4. N. Hatanaka, K. Yamaguchi, N. Takakura, T. Iizuka, J. Mat. Pro. Tec., 140, 628-634(2003)
5. H. Jaafar, K. Mori, Y. Abe, K. Nakanishi, Journal of Materials Processing Technology, 227, 190-199, (2016)
6. T. Iizuka, N. Hatanaka, T. Ishihara, N. Takakura, ICTP, 4-1, 2330-2335, (2008)
7. T. Iizuka, Wei. Yu, N. Hatanaka, N. Takakura, J. Jpn. Soc. Technol., *(2nd report)*, 337, (2009)
8. D. Sakamoto, T. Iizuka, N. Hatanaka, N. Takakura, Metal Forming, 537, (2012)
9. S. H. Jeong, J. J. Kang, S. I. Oh, J. ICTP, 631-634, (1996)
10. N. Hatanaka, K. Yamaguchi, N. Takakura, Journal of Materials Technology, 139, 64-70, (2003)
11. Y. Yoshida, S. Sumikawa, N. Yukawa, T. Ishikawa, Ad. Tec. Pla., 2 883-888, (2002)
12. A. Ahmed, H. Muhamad, IIUM, Vol. 13, 153-159, (2012)

Biomedical Titanium alloy prostheses manufacturing by means of Superplastic and Incremental Forming processes

Antonio Piccininni[1], Francesco Gagliardi[2], Pasquale Guglielmi[1], Luigi De Napoli[2], Giuseppina Ambrogio[2], Donato Sorgente[1], Gianfranco Palumbo[1,a]

[1]*Politecnico di Bari, Department of Mechanics, Mathematics and Management, viale Japigia 182, 70126, Bari, Italy.*
[2]*Università della Calabria, Dipartimento di Meccanica Energia e Gestionale, via Pietro Bucci, 87036, Rende (CS), Italy.*

Abstract. The present work collects some results of the three-years Research Program "BioForming", funded by the Italian Ministry of Education (MIUR) and aimed to investigate the possibility of using flexible sheet forming processes, i.e. Super Plastic Forming (SPF) and Single Point Incremental Forming (SPIF), for the manufacturing of patient-oriented titanium prostheses. The prosthetic implants used as case studies were from the skull; in particular, two different Ti alloys and geometries were considered: one to be produced in Ti-Gr23 by SPF and one to be produced in Ti-Gr2 by SPIF. Numerical simulations implementing material behaviours evaluated by characterization tests were conducted in order to design both the manufacturing processes. Subsequently, experimental tests were carried out implementing numerical results in terms of: (i) gas pressure profile able to determine a constant (and optimal) strain rate during the SPF process; (ii) tool path able to avoid rupture during the SPIF process. Post forming characteristics of the prostheses in terms of thickness distributions were measured and compared to data from simulations for validation purposes. A good correlation between numerical and experimental thickness distributions has been obtained; in addition, the possibility of successfully adopting both the SPF and the SPIF processes for the manufacturing of prostheses has been demonstrated.

1 Introduction

The process of replacing parts of the skeletal system with the adoption of prostheses is not completely optimized, especially in the case of complex bone geometries such as those belonging to the maxillofacial or mandibular regions. One of the most adopted approach is currently based on the reconstruction of a 3D model starting from CT scanning. Through the mirroring technique, it is possible to virtually build the missing part and, thus, to design a prosthetic implant fully customized on the single patient [1]. Despite such process chain is almost completely arranged, the choice of the most suitable material remains still an open question: several examples are well reported in literature as being suitable for biomedical implants. Magnesium and its alloys have been studied over the last century for prosthetic systems thanks to their higher corrosion resistance and biocompatibility [2]. Austenitic stainless steel has been adopted due to its high structural properties and pitting corrosion resistance [3]. Beside metallic materials, bio-ceramics and ceramic coatings have been extensively adopted in the last years [4]: hydroxyapatite is one of the most diffused due to its properties which are very similar to bone's ones, thus promoting the body integration of the implant [5]. Actually, a promising alternative to the above mentioned materials is represented by the adoption of polymers for prosthetic implants: methyl-methacrylate, being cost-effective and reducing the occurrence of intoxication, is widely adopted in cranioplasty [6]. On the other hand, the proper selection of the manufacturing process is a key point: stereolitography, as well as the CAD/CAM approach, is widely adopted since it facilitates, speeds up and improves the quality of surgical procedures [7–10]. Additive manufacturing (3D printing) represents one of the newest process, widely adopted for ABS plastic implants [10,11]. Despite the above described state-of-art, research in such a field is still looking for the best combination of material and process. As concerns the former, lack of structural properties, risks of intoxications or difficulties to obtain complete customized implants make necessary to move the attention towards other materials: Titanium (Ti) alloys are actually considered the best solution, able to match both the aesthetic and the functional requirements to be guaranteed by the implant;

[a] Corresponding author: gianfranco.palumbo@poliba.it

in particular, Ti alloys exhibit low values of the Young's modulus, very similar to that of the cortical bone, thus ensuring a homogeneous stress transfer between the implant and the bone [12,13]; Ti alloys also offer a wide span of properties according to the adopted grades, showing higher ductility and formability from room temperature (commercially pure Ti) up to higher temperature (Ti6Al4V, especially if it is characterised by a controlled microstructure and quantity of alloying elements). In addition, Ti alloys are versatile since they can be easily processed via different manufacturing technologies: (i) powder metallurgy [14], (ii) forming processes [15], (iii) additive techniques [16]. The present work presents some results of the three-years Research Program "BioForming", funded by the Italian Ministry of Education (MIUR) and aimed to investigate the possibility of using flexible sheet metal forming processes, like Super Plastic Forming (SPF) and Single Point Incremental Forming (SPIF), for the manufacturing of Ti alloy prostheses. The prosthetic implants used as case studies in the present work were from the skull; in particular, two different Ti alloys and geometries have been considered: one to be produced by SPF in Ti-Gr23 (or Ti6Al4V-ELI, i.e. the Extra Low Interstitial version of the Ti6Al4V) and one to be produced by SPIF in Ti-Gr2 (i.e. the commercially pure Ti). Numerical simulations, implementing the material behaviours evaluated in the plastic and superplastic field, were preliminary conducted in order to design both the manufacturing processes. Subsequently, experimental tests were carried out implementing numerical results in terms of: (i) gas pressure profile able to determine a constant (and optimal) strain rate during the SPF process; (ii) tool path able to avoid rupture during the SPIF process. Post forming characteristics (thickness distributions) of the prostheses were measured and compared to the correspondent values predicted by simulations for validation purposes.

2 Material and Methods

2.1 Material

In the present work the attention was focused on materials largely preferred as implant materials because of their elastic modulus (similar to the bone's one), low density, non-magnetic properties, low thermal conductivity, high corrosion resistance and good biocompatibility. Ti-Gr2 is considered the most adopted implant material when high mechanical strength is not a priority, since its cost is lower than the Ti6Al4V one [17]. Ti-Gr2 and Ti-Gr23 have been used in this work for manufacturing the prostheses chosen as case studies. The chemical composition of Ti-Gr2 is presented in Table 1.

Table 1. Chemical composition of Ti-Gr2

Fe [%]	C [%]	N [%]	H [%]	O [%]	Ti [%]
0.05	0.00	0.00	0.001	0.13	Bal.

Ti-Gr2 is characterised by good formability at room temperature; for this reason is was chosen for SPIF investigations. SPIF experiments were carried out on 1.5 mm thick square blanks (240 mm x 240 mm) extracted from a commercial sheet purchased in the annealed condition.

SPF experiments were carried out on rectangular blanks (225 mm x 190 mm) extracted from a 1 mm thick Ti-Gr23 sheet. The chemical composition of the sheet, purchased in the annealed condition (790°C for 68 min and then air cooled) is presented in Table 2.

Table 2. Chemical composition of the adopted Ti-Gr23

Al [%]	V [%]	Fe [%]	C [%]	N [%]	H [%]	O [%]	Ti [%]
5.88	3.87	0.14	0.22	0.006	0.002	0.112	Bal.

The material behaviour in the superplastic field was preliminary evaluated by means of an extensive experimental campaign based on free inflation tests: circular blanks extracted from the above mentioned sheet were heated up to 850°C according to literature [18]; free inflation tests were thus conducted setting either a constant gas pressure or several pressure jumps during the same test. Using data from free inflation tests (dome height as a function of the time), material constants (C and m) for defining the *Backofen* power law ($\sigma = C\dot{\varepsilon}^m$) could be evaluated in a strain condition very similar to the one in the real forming process. In particular, material constants to be implemented in the FE model were evaluated both analytically, using the approach proposed by Enikeev and Kruglov [19], and through an inverse analysis based on a genetic algorithm coupled to dome height data from jump pressure tests [20].

2.2 Case studies

One of the main objectives of the present research activity concerns the set-up of a standard procedure that, starting from the damage reconstruction, allows to define the prosthesis profile, also taking into account information about the manufacturing process.

The damage reconstruction was obtained starting from the DICOM image of the skull and implementing a mirroring technique for the skull rebuilding [21]. Great attention was paid to this step of the procedure in order to have the geometry of the prosthesis very similar to the one of the missing anatomical part.

As concerns the implant material, Ti-Gr2 (characterized by high formability at room temperature) was used for manufacturing a prosthesis at room temperature by SPIF, while Ti-Gr23 was used for manufacturing a prosthesis by SPF. Two damage areas were thus considered due to the limitations imposed by the manufacturing process: a frontal prosthesis, to be produced by SPF, and a lateral one, to be produced by SPIF. In Figure 1 the damage areas considered for the prostheses reconstruction have been highlighted.

DOI: 10.1051/matecconf/20168015007

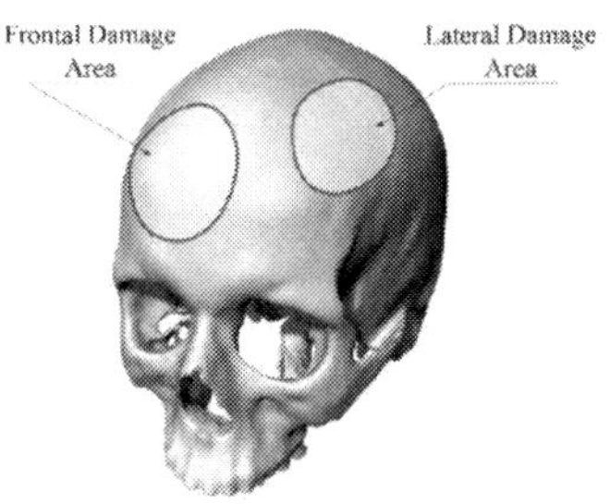

Figure 1. Damage areas.

The lateral prosthesis is shown in Figure 2.a: it is characterized by quite a flat profile and is suitable to be produced at room temperature using the SPIF process. On the contrary, the frontal prosthesis is shown in Figure 2.b: it is characterized by more severe curvature and it can be produced using the SPF process.

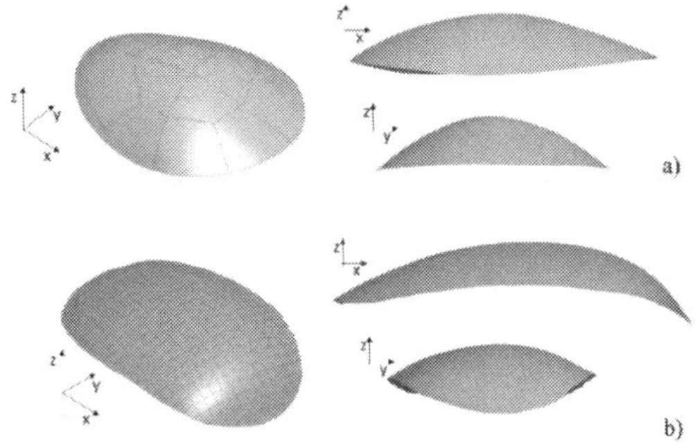

Figure 2. a) Lateral and b) Frontal prostheses.

As concerns the rebuilding of the anatomical part, the steps executed to complete the prosthesis design for both the forming processes have been resumed in Figure 3.

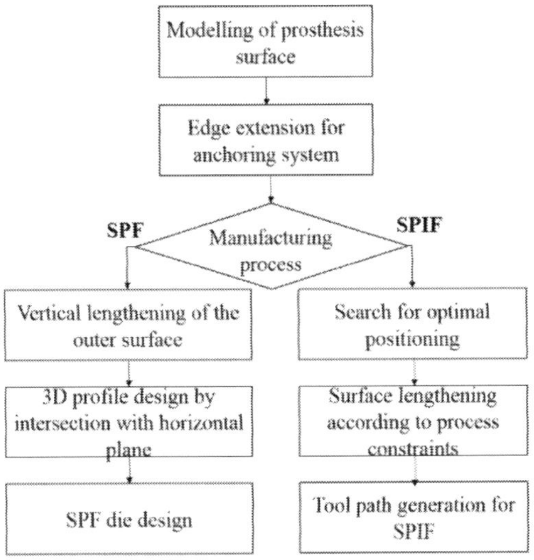

Figure 3. Workflow of adopted procedure

First, the isolated surface was processed to obtain a suitable analytical model. After that, the built surface was extended in order to ensure adequate anchoring of the prosthesis. As a consequence the original periphery of the prosthesis was lengthened in the tangential direction to guarantee the smoothness of the final part. The output of such phase was a larger part suitable to be applied by overlapping on the patient defect and anchored by screws to the skull. In this research an offset of about 13mm was used for both the case studies.

Subsequently the procedure divides into two branches, due to the peculiarity of each forming process: for SPF process the extended periphery was further lengthened in the vertical direction up to the horizontal plane on which the original blank is clamped by the Blank Holder. The sharp angularities at the intersection were smoothed by means of proper fillet radii. The final geometry of the prosthesis for the SPF process is reported in Figure 4.

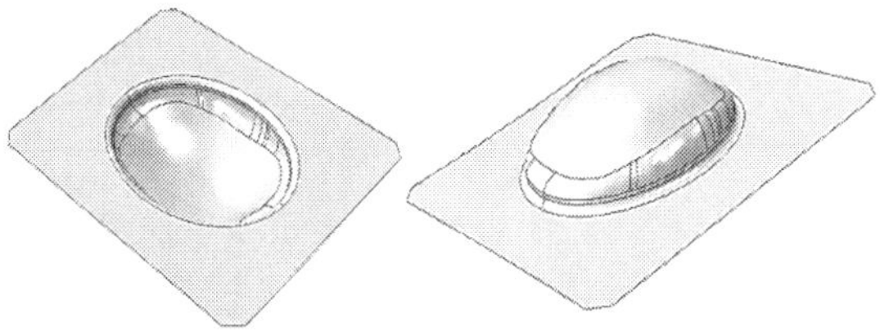

Figure 4. Prosthesis reconstruction for SPF process.

As concerns the SPIF process, the right shape positioning on the sheet plane was performed considering the technological constraints of the SPIF process [22]. More deeply, an optimization loop was carried out to maintain the orientation of the geometry to be SPIFed, with respect to the clamping plane, below the formability limits of the investigated material. Once the clamping plane was set, the analytical surface of the prosthesis was extended up to it, tangentially from the extended periphery. The final geometry of the prosthesis to be produced by the SPIF process is shown in Figure 5.

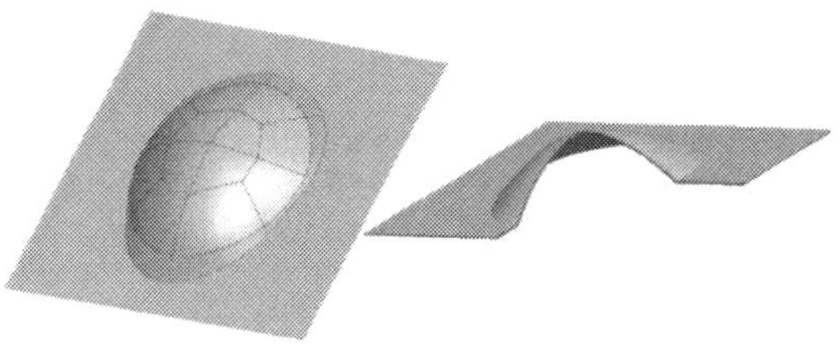

Figure 5. Prosthesis reconstruction for SPIF process.

Finally, the obtained geometries of the prostheses were used to define the die geometry (in the case of the SPF process) and the CAM part program (in the case of the SPIF process).

2.3 Experimental equipment

2.3.1 SPF equipment

SPF tests were carried out using a 2500 kN prototype hybrid electro-hydraulic press machine. The closing force is ensured by four electric motors supported, when necessary, by two additional hydraulic cylinders. SPF experimental tests were carried out at 850°C adopting heated tools (maximum temperature: 1000°C). As shown in Figure 6, the upper tool was equipped with a 310S die insert in which the geometry of the prosthesis was created by machining.

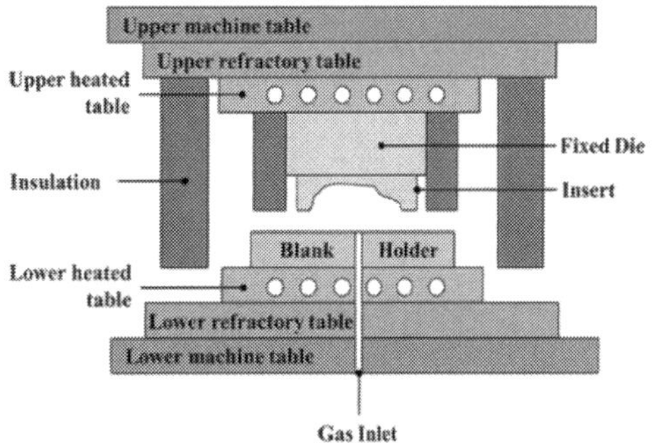

Figure 6. Main component of the press machine for SPF tests.

Tools were heated by means of six electric cartridges (13.2 kW per single tool) managed by a Programmable Logic Control (PLC). Temperature monitoring was provided by K-type thermocouples placed close to both the die and the blankholder surfaces: when a uniform temperature distribution was reached (5°C between the upper and the lower tool) the heating was stopped. Once reached the working temperature, the blank was thus introduced between the tools and then clamped (480 kN); experimental test were carried out setting the same temperature used for characterization purposes (850°C), being this value reported in literature as favourable to emphasize the superplastic properties of the investigated alloy [18]. Before the test, boron nitride was applied on both the surfaces of each blank, in order to make easier the extraction of the formed part. SPF test were carried out implementing optimized pressure laws directly coming from the numerical simulation.

2.3.2 SPIF equipment

Due to its cheapness, the use of the SPIF technology can be strategic in fields like the medical one, since natural differences in terms of anthropometrical items of each individual is a fundamental production constraint. Moreover, SPIF process seems the natural candidate for skull manufacturing since the use of customised dies has to be absolutely rejected [23].

As introduced in the previous section, a part program was generated using a CAD/CAM system. More in detail, the manufacturing module of Pro-Engineering was adopted for generating the tool path. The sheet was positioned on the table of a 3-axes CNC work centre, where both the SPIF step and the trimming one could be executed. An image of the experimental set up is shown in Figure 7.

In addition, a "general purpose" backing plate was positioned under the sheet to support the same during the process. A simple HSS tool having an hemispherical profile with 12mm of diameter was used. After the forming step, the subsequent cutting operation was executed to remove the sheet periphery; in particular, a small milling tool, having a diameter of 5 mm, was adopted.

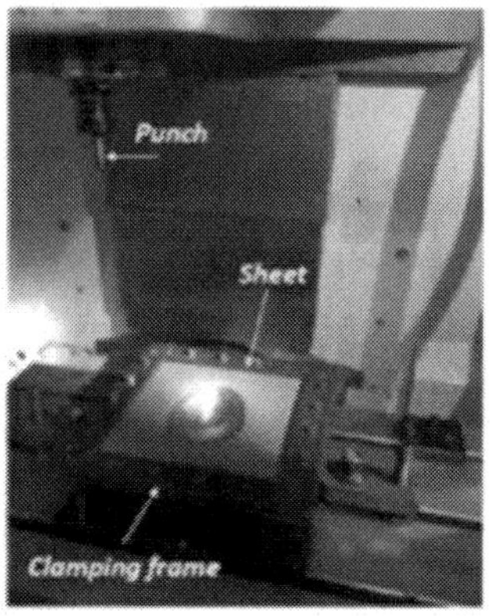

Figure 7. SPIF equipment.

2.4 Numerical model

2.4.1 SPF model

A 3D numerical model shown in Figure 8 was created using the FE commercial code Abaqus.

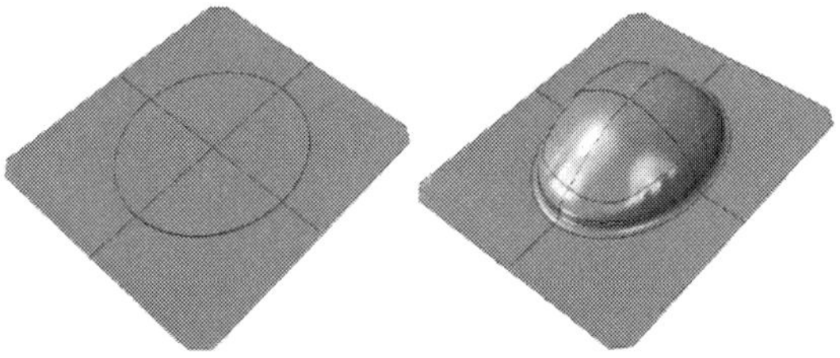

Figure 8. 3D modelling of the die cavity and the Ti blank

Die geometry was modelled through discrete rigid shell elements while the blank as an elliptical deformable body (orange region in Figure 8) meshed using about 8000 linear quadrilateral shell elements (S4R type) with 5 through-thickness integration points, as a good compromise between the accuracy of numerical results and simulation time. In order to correctly simulate the experimental clamping action by the blank holder, the periphery of the blank was pinned. Material superplastic behaviour was modelled according to the *Backofen* power law: in particular, material constants obtained analysing free-inflation tests data using both the analytical approach [19] and the inverse analysis [20] were implemented. Coulomb's friction law was implemented in all simulations, setting the Coefficient of Friction to the value of 0.1.

2.4.2 SPIF model

SPIF numerical simulations were set out by *Hypermesh*, creating the FE model (LS-Dyna) shown in Figure 9.

DOI: 10.1051/matecconf/20168015007

The sheet (only the region deformed by the punch) was uniformly meshed by quad elements having a side length equal to 4mm; 50 elements were utilized for each sheet side in its initial configuration. In order to take into account the clamping action exerted by the blank holder in the SPIF experiments, the periphery of the modelled blank was pinned.

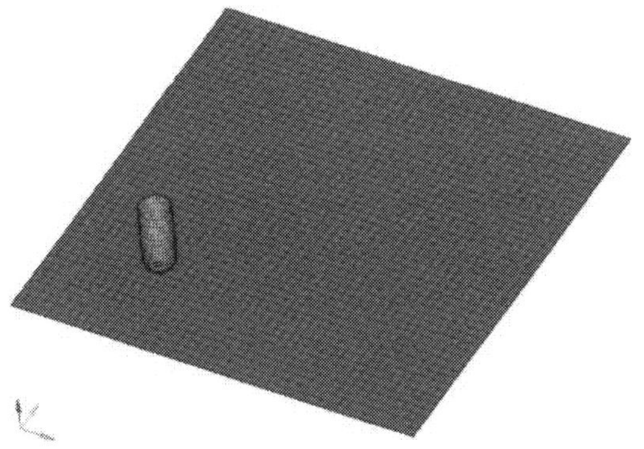

Figure 9. FE model in LS-Dyna.

Blank elements with the Tsay–Hill formulation and five integration points along the thickness were used. An adaptive mesh refinement with four maximum refinements was implemented to adapt the solution accuracy superimposing finer sub-grids in the most deformed regions. The strain rate dependent isotropic plasticity law (MAT-019) was used to consider the plastic deformation of the sheet. The punch, modelled as rigid, was moved according to prescribed boundary curves along the three Cartesian axes. Finally, the contact interface between the punch and the sheet was automatically defined using contact surfaces (Forming-Surface-To-Surface). Coulomb's friction model with a constant value (0.1) was implemented in SPIF simulations. The carried out numerical analysis was based on the explicit formulation: both the time scaling and the mass scaling approaches were utilized, verifying the energy condition of the model during the simulation. In particular, considering the adopted explicit formulation, a check on the total kinetic energy was carried out, verifying that the quota added virtually was lower than the 5 per cent of the internal energy [24]. Finally, with regard to time scaling, the simulation time was about 30 minutes with a time step of 2.5 E-5 s (a dual processor Xeon 3.2GHz with 16 GB RAM was used).

3 Results

3.1 Numerical results

3.1.1 SPF numerical results

Since the investigated alloy exhibited a highly pronounced superplastic behaviour when subjected to a strain rate of 2E-4 s^{-1} at the working temperature of 850°C [18], a pressure profile able to keep such value was numerically determined using the Abaqus built in routine. Gas pressure was varied ranging from a minimum value of 0.01 MPa up to the maximum pressure level of the press machine (3 MPa).

Numerical simulations were run implementing the two sets of material constants: one coming from the analytical model [19] and the other coming from the inverse analysis [20]. In Figure 10 numerical results have been compared in terms of thickness distributions along the longitudinal path of the formed component, being the equivalent time (t*) equal to 0.4 (i.e. 40% of the total cycle time).

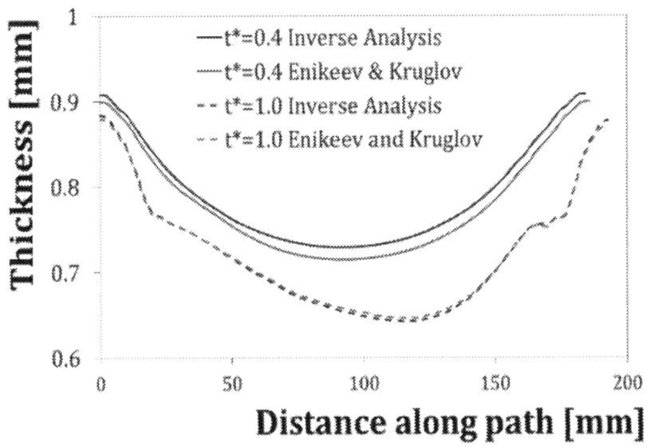

Figure 10. Numerical thickness distributions

It is worthy of notice that the difference between the two distributions is evident. In particular, the simulation implementing the material constants obtained using the analytical approach predicts higher thinning of the part. On the contrary, when t* is equal to 1.0 (i.e. at the end of the SPF process) the two distributions are almost overlapped. In Figure 11 results concerning the same conditions (t* equal to both 0.4 and 1) but in terms of vertical displacements have been compared (the curves have been arranged so that the maximum displacement corresponds to an abscissa equal to 0).

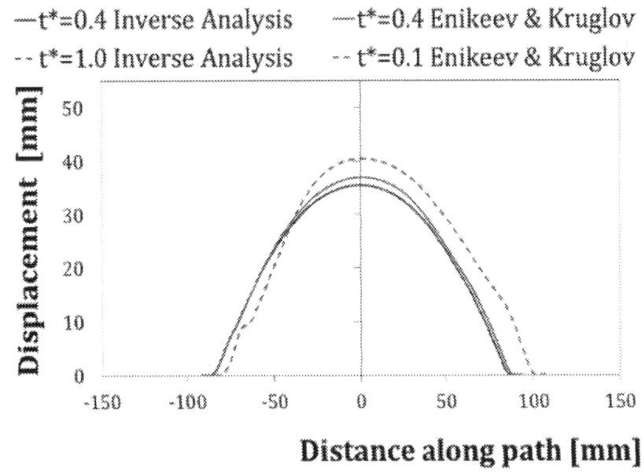

Figure 11. Comparison in terms of vertical displacements

It is worthy of notice that, when the blank (which is deforming under the gas pressure action) touches the die cavity, its profile has to copy the one of the die cavity. For this reason, when t* is equal to 1, the profiles predicted by both models are overlapped.

DOI: 10.1051/matecconf/20168015007

3.1.2 SPIF numerical results

Each SPIF numerical simulation required about 30min if using a CPU of 3.2GHz and a RAM memory of 16 GB. The final thickness, plotted on the formed sheet, has been reported in Figure 12.

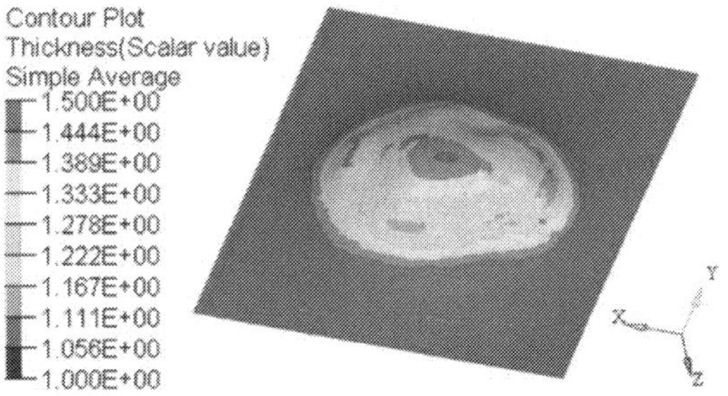

Figure 12. Thickness distribution at the end of SPIF process (mm).

The thickness trend along the symmetry section (x direction) of the prosthesis is furthermore reported in Figure 13.

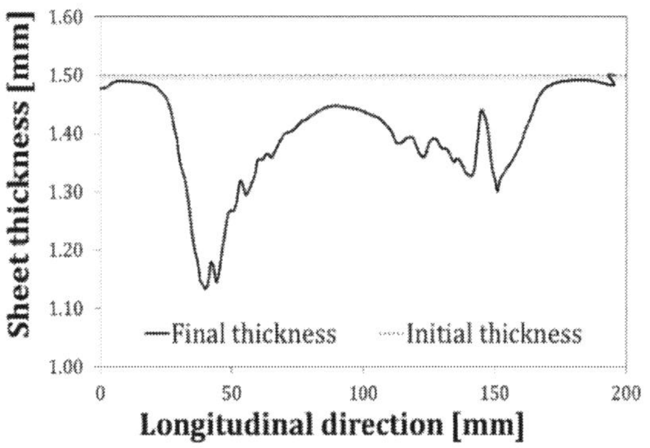

Figure 13. Numerical thickness distribution at the end of the SPIF process

Looking at the results, a minimum thickness of about 1.17mm was reached with a maximum sheet thinning of 22%, which was observed in a small area of the most curved side of the prosthesis.

3.2 Experimental results

3.2.1 SPF experimental results

The final shape of the SPF-ed component was acquired by means of the Digital Image Correlation (DIC) system Aramis. The formed components were painted to create a stochastic pattern able to be recognized by the DIC cameras as a virtual grid (Figure 14. a) DIC cameras, b) acquired shape of the formed component
a).
In such a way it was possible to acquire the final shape of the component (Figure 14. a) DIC cameras, b) acquired shape of the formed component
b) and compare the vertical displacement along both the longitudinal and the transversal direction with the correspondent numerical ones.
Formed components were also cut into two halves: in such a way it was possible to measure the thickness distribution along the longitudinal and the transversal direction by means of a calliper.

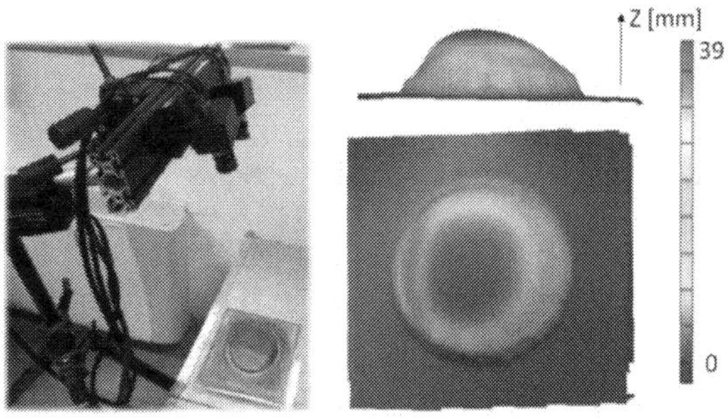

a) b)
Figure 14. a) DIC cameras, b) acquired shape of the formed component

3.2.2 SPIF experimental results

The Ti-Gr2 blank was deformed according to the designed tool trajectory and adopting the following process parameters: a constant tool depth step equal to 0.1 mm and a tool feed rate of about 2000 mm/min. More in detail, the tangential movement of the tool was fully described by the product profile up to obtain the 3D part. The flange region of the SPIFed component was removed and the resulting part has been shown in Figure 15.

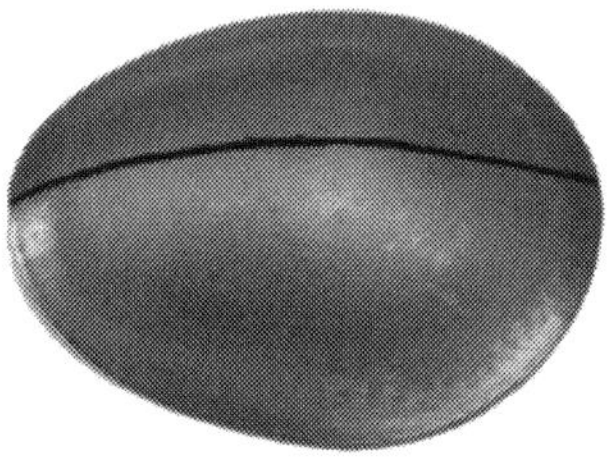

Figure 15. SPIF formed component

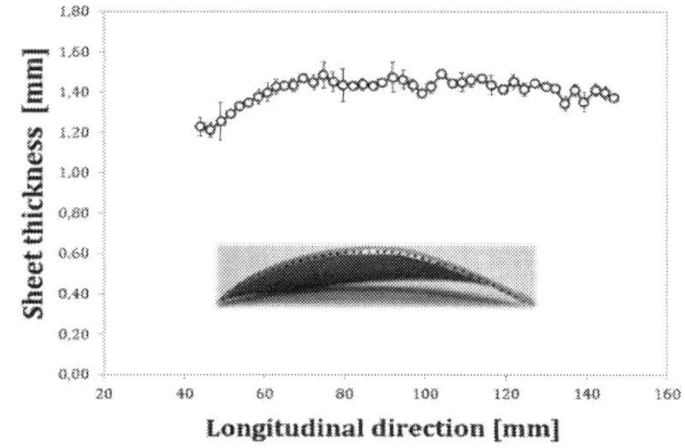

Figure 16. Experimental thickness distribution at the end of the SPIF process

Also in this case the prosthesis was cut into two halves in order to evaluate the thickness distribution: equally spaced points were measured by calliper along the longitudinal direction of the prosthesis. The experimentally measured thickness evolution along the longitudinal direction is reported in Figure 16.

As can be observed, even if a maximum thinning of 0.29 mm was measured at the periphery of the prosthesis (where the effect of the cutting phase is more important) the thickness reduction is less than 1% for the investigated profile.

4 Discussion

4.1 Comparison between experimental and numerical results

Numerical results were compared to the experimental measurements collected in terms of thickness distributions for both the prostheses.

4.1.1 SPF numerical vs experimental

In Figure 17 the numerical thickness distribution has been plotted together with the correspondent experimental one.

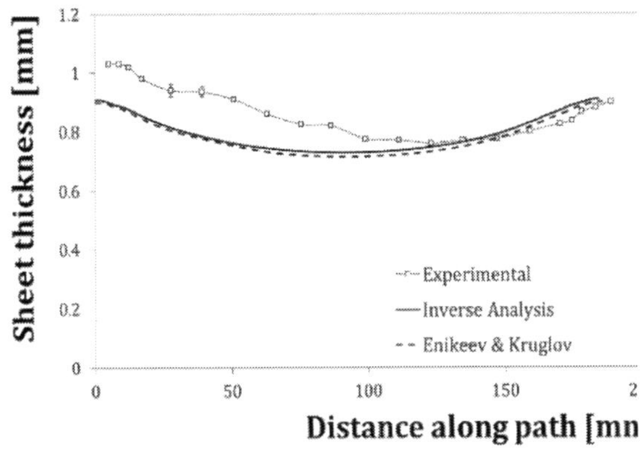

Figure 17. Numerical vs. experimental thickness distribution (t^*=0.4)

It is important to put in evidence that a better agreement was found when implementing the material constants coming from the inverse analysis (with respect to the one implementing the ones obtained using the analytical approach by *Enikeev and Kruglov*). This is particularly evident when the parameter t^* is set to 0.4.

At the end of the simulation, the numerical thickness distributions according to the two above mentioned approaches gave the same results, as also confirmed in the bar chart proposed in Figure 18.

In Figure 19 simulation results and experimental data (at the end of the SPF process) in terms of vertical displacements have been compared (like in Figure 11 the curves have been arranged so that the maximum displacement corresponds to an abscissa equal to 0).

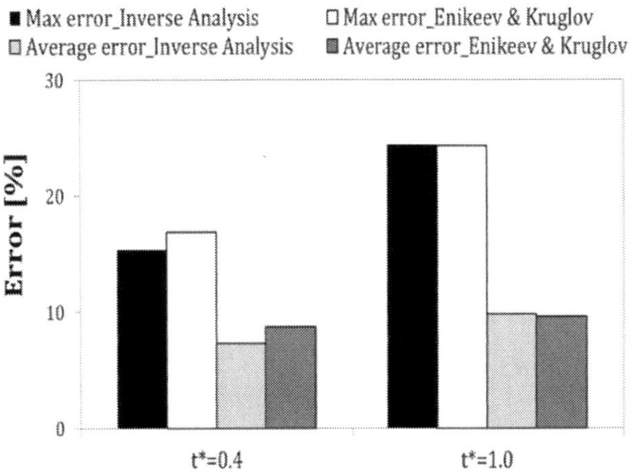

Figure 18. Resume of the comparison between numerical and experimental results.

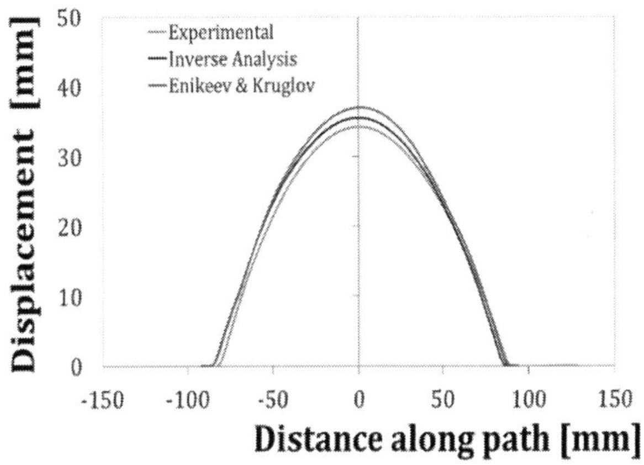

Figure 19. Shape of the formed component: numerical vs. experimental results at t^*=0.4

Such a comparison once again proves the accuracy of the numerical model and the robustness of the approach adopted for the evaluation of the material constants. In fact, as for the thickness distribution, the numerically evaluated shape of the SPF-ed component is closer to the experimental one when implementing in the FE model the material constants obtained from the inverse analysis.

4.1.2 SPIF numerical vs experimental

In Figure 20 the thickness distribution measured on the SPIF-ed component along the longitudinal direction has been compared with the correspondent numerical one.

It is worthy of notice that a suitable overlapping between the numerical and experimental measurements was obtained in the region of the prostheses (external parts where measurements are not present concern the portion of the blank which was trimmed).

Maximum and average prediction errors of about 8% and 4% respectively could be obtained.

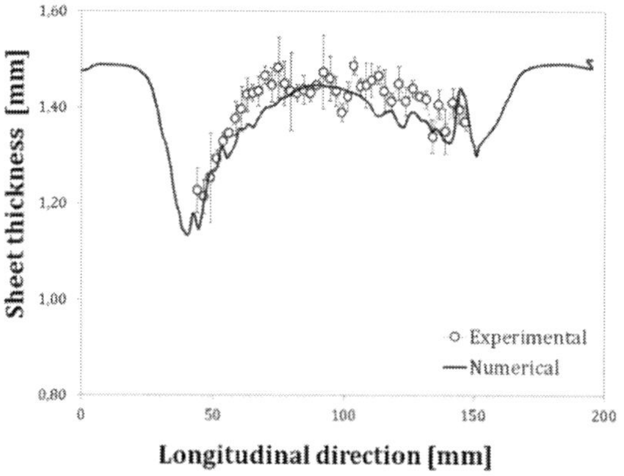

Figure 20. Numerical vs. experimental thickness distribution for SPIF process.

5 Conclusions

Both the materials investigated in the present work, largely preferred as implant materials because of their properties (elastic modulus very similar to the bone's one, low density, non-magnetic properties, low thermal conductivity, high corrosion resistance and good biocompatibility) have confirmed to be suitable for producing prostheses either by Super Plastic Forming (Ti-Gr23) and by Single Point Incremental Forming (Ti-Gr2).

The procedure created in the present research program for producing the prosthesis revealed to be valid: both the die for Super Plastic Forming process and the tool path for Single Point Incremental Forming were obtained.

Finite Element models created for simulating both the forming processes revealed to be robust and accurate, since a good correlation between numerical and experimental thickness distributions could be obtained.

Finally, the possibility of successfully adopting both the Super Plastic Forming and the Single Point Incremental Forming for the manufacturing of Titanium medical prostheses has been demonstrated.

Acknowledgements

The activities in the present work were funded by the Italian Ministry of Education, Universities and Research through the PRIN Project 2012 (project acronym: BIOFORMING).

References

1. S. Katsumura, K. Sato, T. Ikawa, K. Yamamura, E. Ando, Y. Shigeta, et al., Leg. Med. 18 (2016) 37–43.
2. M.P. Staiger, A.M. Pietak, J. Huadmai, G. Dias, Biomaterials. 27 (2006) 1728–1734.
3. S.V. Muley, A.N. Vidvans, G.P. Chaudhari, S. Udainiya, An assessment of ultra fine grained 316L stainless steel for implant applications, Acta Biomater. 30 (2015) 408–19.
4. B.J. McEntire, B.S. Bal, M.N. Rahaman, J. Chevalier, J. Eur. Ceram. Soc. 35 (2015) 4327–4369.
5. G. Staffa, a. Nataloni, C. Compagnone, F. Servadei, 149 (2007) 161–170.
6. J. Jaberi, K. Gambrell, P. Tiwana, C. Madden, R. Finn, J. Oral Maxillofac. Surg. 71 (2013) e81–e88.
7. F.P.W. Melchels, J. Feijen, D.W. Grijpma, Biomaterials. 31 (2010) 6121–30.
8. P. Kasprzak, G. Tomaszewski, G. Wróbel-Wiśniewska, M. Zawirski, Clin. Neurol. Neurosurg. 113 (2011) 311–315.
9. W.-M. Xi, A.-M. Wang, Q. Wu, C.-H. Liu, J.-F. Zhu, F.-F. Xia, Med. Eng. Phys. 37 (2015) 911–915.
10. H.H. Malik, A.R.J. Darwood, S. Shaunak, P. Kulatilake, A.A. El-Hilly, O. Mulki, et al., J. Surg. Res. 199 (2015) 512–522.
11. J. Cohen, S. A. Reyes, Am. J. Otolaryngol. 36 (2015) 1–6.
12. H.J. Rack, J.I. Qazi, Mater. Sci. Eng. C. 26 (2006) 1269–1277.
13. M. Niinomi, J. Mech. Behav. Biomed. Mater. 1 (2008) 30–42.
14. L. Bolzoni, E.M. Ruiz-Navas, E. Gordo, Mater. Sci. Eng. C. 49 (2015) 400–407.
15. F.X. Tan, M.Z. Li, Z.Y. Cai, J. Mater. Process. Technol. 187-188 (2007) 453–457.
16. A.L. Jardini, M.A. Larosa, R. Maciel Filho, C.A. de C. Zavaglia, L.F. Bernardes, C.S. Lambert, et al., J. Craniomaxillofac. Surg. 42 (2014) 1877–84.
17. D. Teker, F. Muhaffel, M. Menekse, N.G. Karaguler, M. Baydogan, H. Cimenoglu, Mater. Sci. Eng. C. 48 (2015) 579–585.
18. H.S. Lee, J.H. Yoon, C.H. Park, Y.G. Ko, D.H. Shin, C.S. Lee, J. Mater. Process. Technol. 187–188 (2007), 526–529.
19. F.U. Enikeev, A.A. Kruglov, Int. J. Mech. Sci. 37 (1995) 473–483.
20. D. Sorgente, A. Piccininni, V. Piglionico, P. Guglielmi, D. Grossi, G. Palumbo, et al., Mater Sci. Forum 839 (2016) 177–182.
21. F. Chen, G.M. Brown, M. Song, Opt. Eng. 39 (2000) 10–22.
22. G. Ambrogio, L. De Napoli, L. Filice, F. Gagliardi, M. Muzzupappa, J. Mater. Process. Technol. 162-163 (2005) 156–162.
23. G. Ambrogio, R. Conte, L. De Napoli, G. Fragomeni, F. Gagliardi, , Key Eng. Mater. 651-653 (2015) 925–931.
24. J. Hallquist, LS-Dyna theory manual, 2001 (STC, Livermore, CA).

DOI: 10.1051/matecconf/20168015008

Finite Element Modelling of Combined Process of Plate Rolling and Stamping

Alexander Pesin[1,a], Ernst Drigunt[2], Denis Pustovoytov[1] and Ilya Pesin[1]

[1]*Nosov Magnitogorsk State Technical University, 38, Lenin pr., Magnitogorsk, 455000, Russia*
[2]*IIC "ChermetInformSystemy", 11-54, Metallurgov pr., Magnitogorsk, 455000, Russia*

Abstract. This article presents the results of finite element modelling of a fundamentally new combined process of plate rolling and stamping. This process was first used to produce large-sized bodies of revolution, such as the bottom of a vacuum vessel, bottom of a ladle furnace, and others. Traditionally, such products (with the wall thickness exceeding 40 mm and the diameter width of up to 4000 mm) are produced in machine-building enterprises using stamping processes. Various schemes of implementing this combined process are investigated in the article.

1 Introduction

In such energy-intensive industries as metallurgy and heavy engineering, technologies of manufacturing and processing thick-plated metal products with wall thickness exceeding 40 mm and diameter up to 4000 mm are especially important. Such products are large-sized bodies of revolution (e.g. bottom of a vacuum vessel) formed by hot stamping. They are used in metallurgical, petrochemical, atomic, and gas industries in construction of storage tanks, vessels, high-pressure machinery, and other industrial machines. Current methods of manufacturing these metal products have several technological flaws, such as: 1) high energy intensity of manufacturing processes; 2) low efficiency; 3) operating conditions of large-sized products demand a high degree of resistance to stress, yet thick-plate metal parts are susceptible to stress and have a high degree of variation in their width, which increases the probability of eventual cracks, folds and other defects, lowering the reliability of the resulting metal structures. All these issues decrease the economic efficiency of manufacturing high-duty large-sized metal parts.

Plate stamping process is described in the works [1-2]. The basic principles of combined process of asymmetric rolling and plastic bending are studied and presented extensively in academic literature [2-11]. This article is directed towards developing the theory and technology of new, efficient and productive, deformation methods that are based on combining the processes of thick-plate rolling and stamping [12-13]. These new methods will allow us to produce large-sized metal parts with curvilinear surfaces (with wall thickness lying between 40 and 120 mm and diameter up to 4000 mm), from metallic materials of enhanced physical properties, under the conditions of a rolling mill without the use of pressure equipment. The article cites the FE modelling results of a drastically new combined process of rolling and stamping. Various methods of implementing this combined process are investigated.

2 Numerical study

FE modelling of the combined rolling-and-stamping process was given with the help of DEFORM 3D software. The boundary conditions were as follows : 1) boundary surfaces of operating tools (lower die, upper die, rolls) ; 2) the surfaces of the deformed material that interact with the environment ; 3) the law of friction on contact with an operating tool ; 4) rheological model of the deformed material.

The following parameters were given as initial conditions of the model: 1) diameters and rotating speed of working rolls; 2) flow curve of the material from which the workpiece is made; 3) heating temperature of the workpiece in the oven; 4) initial dimensions of the workpiece; 5) friction coefficient m on contact with an operating tool; 6) pause time; 7) temperature of the environment ; 8) coefficients of thermal conductivity, heating capaciy and ferrocity of the workpiece's material ; 9) coefficients of thermal conductivity, heating capacity and ferrocity of the operating tool's material ; 10) convection coefficient during thermal exchange with the environment ; 11) coefficient of deformational heating ; 12) dimensions and the amount of finite elements.

The following assumptions were made in 3-D modelling of the process : 1) the deformed material is viscoplastic ; 2) the upper die, the lower die and the rolls are rigid ; 3) the stresses of contact friction are proportional to the yield strength of the shear.

[a]Corresponding author: pesin@bk.ru

As a result of solving the three-dimensional task, the following values are determined : 1) temperature field; 2) stress field; 3) strain field.

The flow curve (deformation resistance) of 09G2S steels is described by the equation (1), and the equation (2) describes the flow curve of 10HSND steels.

$$\sigma_s = k_0\,\varepsilon^{k_1}\dot{\varepsilon}^{k_2} exp(-k_3 T), \tag{1}$$

where σ_s is the resistance to deformation, in MPa; ε is the true strain; $\dot{\varepsilon}$ – is the strain rate, s-1; T is the metal temperature, in °C; k0, k1, k2, k3 are the coefficients (see table 1).

$$\sigma_s = k_0 (10\varepsilon)^{k_1}\dot{\varepsilon}^{k_2}(0{,}001T)^{k_3}, \tag{2}$$

where k0, k1, k2, k3 are the coefficients (see table 2).

Table 1. Coefficients in the equation (1)

Coefficient				Range of deformation parameters change		
k0	k1	k2	k3	ε	$\dot{\varepsilon}$, s-1	T,°C
3727	0,1229	0,0718	0,00357	0,03 to 0,8	0,2…25	500 to 800
1265	0,1592	0,1048	0,00223	0,03 to 0,8	0,2…25	800 to 1200

Table 2. Coefficients in the equation (2)

Coefficient				Range of deformation parameters change		
k0	k1	k2	k3	ε	$\dot{\varepsilon}$, c-1	T,°C
82,01	0,226	0,122	-2,900	0,03…0,8	1…55	700…1200

2.1 Mathematical modelling of metal stress-strain state in the manufacturing of curvilinear bottoms

2.1.1 Modelling of thick-plate stamping process on a press

The diameter of the initial workpiece was 2660 mm, its thickness was 40 mm, it was made of 09G2C type steel. The height of the resulting metal part was 200 mm. The heating temperature of the workpiece was 900 °C. The deformation speed was 300 mm/sec, the friction coefficient was 0,7. Transporting the workpiece from the oven to the stamp took 60 sec. Deformation conditions were non-isothermal. Modelling conditions were symmetrical in one plane.

Fig. 1 - 2 show the changing stress field in the volume of the workpiece under deformation. Hot metal stamping is distinguished by significant variation of temperature across the volume of the workpiece. Therefore, considering that mechanic and plastic characteristics of a metal significantly depend on its temperature and deformation speed, the changing temperature of the workpiece during stamping, the non-uniformity of the temperature field, the intensity of heat emission, and the friction conditions should all be taken into account when analysis the processes of hot stamping.

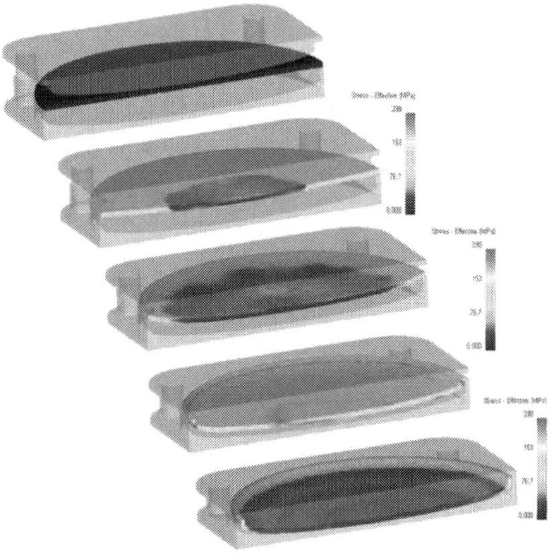

Figure 1. Stress-strain state of metal during thick-plate stamping of a curviliear bottom

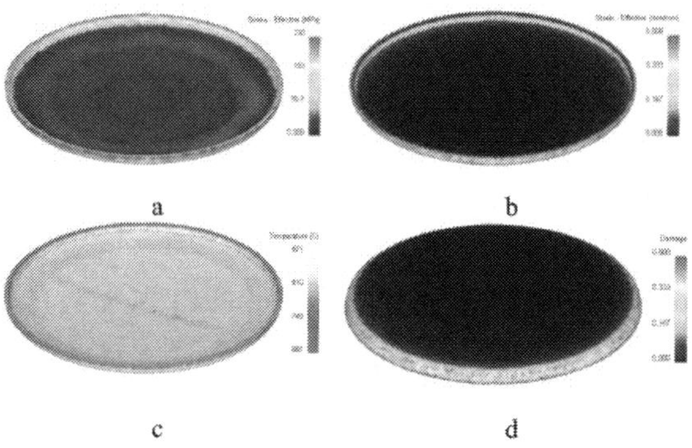

Figure 2. Curvilinear bottom: its stress field (a), deformation field (b), temperature field (c), damage probability field (d)

The deformation force was 26,1 MN (figure 3). Therefore, the mathematical model allows us to optimize the existing technologies of large-size metal part production, and to develop new ones, taking into account the metal stress-strain state.

2.1.2 Combined rolling-stamping process modelling

Two fundamentally new methods of thick plate rolling-stamping combined process for manufacturing large-sized bodies of revolution without using pressure equipment were developed. The main principle of this process is rolling of a pack consisting of two foundations - the upper die and the lower die - and the thick-plated workpiece between them. The first method of the combined process consists of picking up and gradual rolling of the pack from the leading edge to the end section with the roll gap of pressure screws remaining in the same position. The second method includes the reverse rolling of the pack in working drive rolls, with the pack being placed in the beforehand separated rolls so that the center of the pack would be on the line that connects the rolls' axes. The rolling in this case is conducted with the changing position of pressure screws (the roll gap changes during the rolling process).

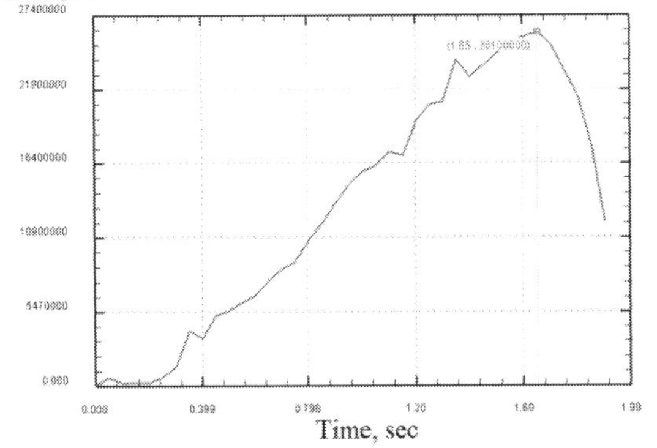

Figure 3. Change of deformation effort, in Newtons, relative to time, in seconds, in thick-plated curvilinear bottom stamping

The following parameter values were used in modelling:

The diameter of the initial workpiece was 2660 mm, its thickness was 40 mm, it was made of 09G2S steel. The height of the resulting metal part was 200 mm. The heating temperature of the workpiece was 900 ^{0}C. The strain rate was 300 mm/sec, the friction coefficient was 0,7. Transporting the workpiece from the oven to the stamp took 60 sec. Deformation conditions were non-isothermal. Modelling conditions were symmetrical in one plane.

The shapes of upper and lower dies and the FE model of the pack is shown in figures 4-6.

The modelling results have shown that the main issue that arises is the possible loss of pack fixity during rolling, and the consequent formation of defects in the form of folds. Accordingly, increasing the quality of manufactured parts is a primary problem during the process. In order to solve it, compliance with the requirements of material durability, form accuracy and surface quality. A necessary condition for this compliance is the absence of technological failures, e.g. flaw signs in the form of deformation localization with the proceeding breakdown; breakdowns die to the metal of the workpiece lacking plasticity; formation of folds, etc.

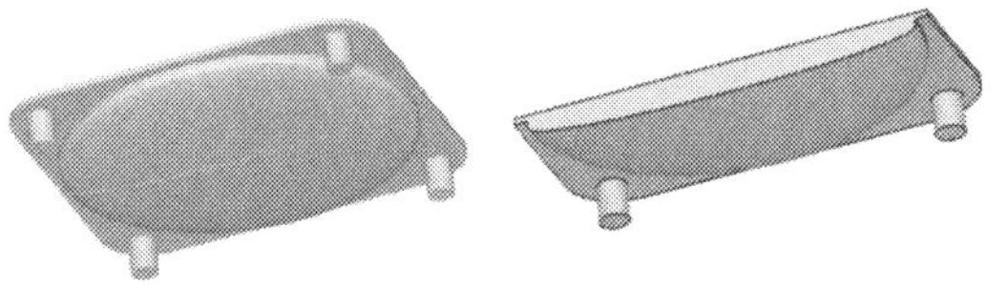

Figure 4. Upper die

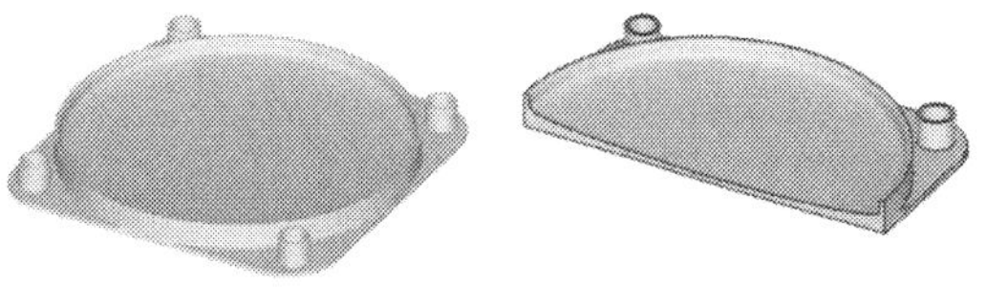

Figure 5. Lower die

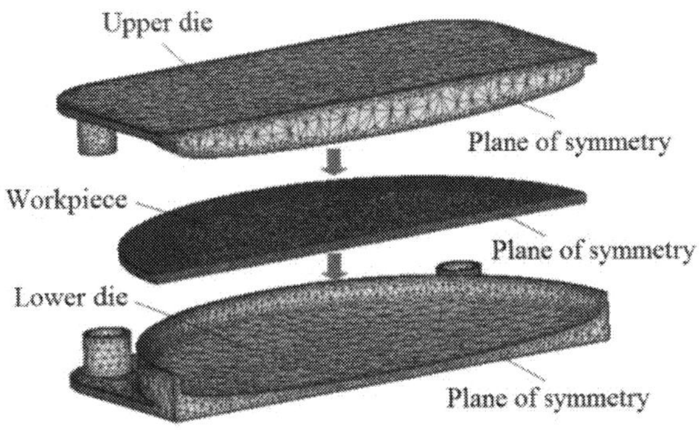

Figure 6. FE model of the packet

Figures 7 to 9 show cases of fixity loss. The deformation of the metal workpiece and the subsequent appearance of folds under the influence of tangential compressive stresses, eventually lead to spoilage of the workpiece. Folds, for instance, are formed in the workpiece's flange during the flanging of the convex curvilinear profile of the part. This flaw is theoretically repairable, however the technological process should be conducted as to prevent the conditions under which the compression becomes unfixed. Prevention of undesirable folds is a main task that arises during both planning and adjusting the combined rolling-stamping process.

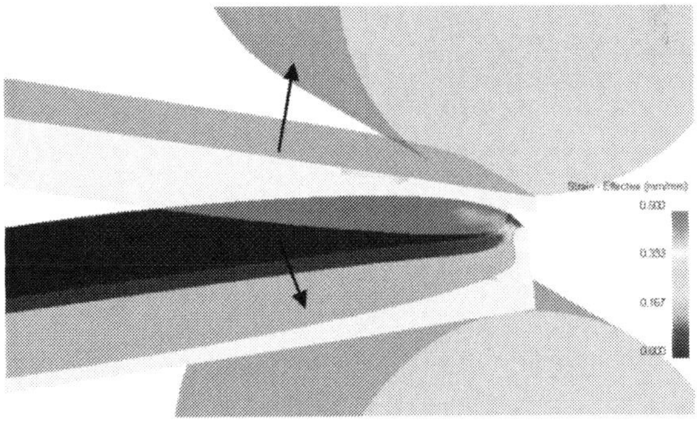

Figure 7. Loss of pack fixity during rolling

DOI: 10.1051/matecconf/20168015008

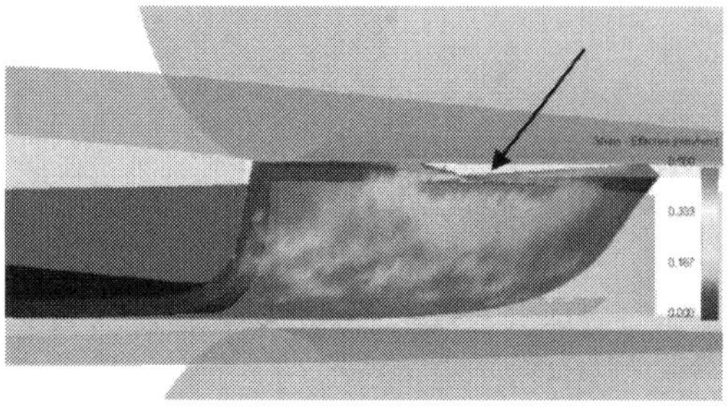

Figure 8. Detail deformation, appearance of folds

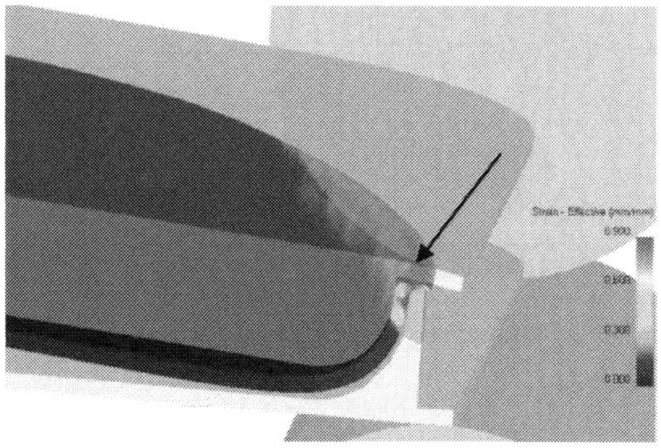

Figure 9. Detail deformation with the formation of a fold

Our investigation of stress and strain metal state was conducted for two deformation methods: direct and reversed. The direct method consists of picking up and gradual rolling of the pack from the front edge to the end section with the roll gap of pressure screws remaining in the same position. The 3-D modelling of the combined process was conducted with the assumption of the process being symmetrical about the vertical plane, in order to speed up the calculations (figure 10).

The model of the process is shown in figure 11. It includes assembling a pack from the upper die, the lower die and the sheet workpiece between them, and rolling the pack in the working rolls of the plate mill.

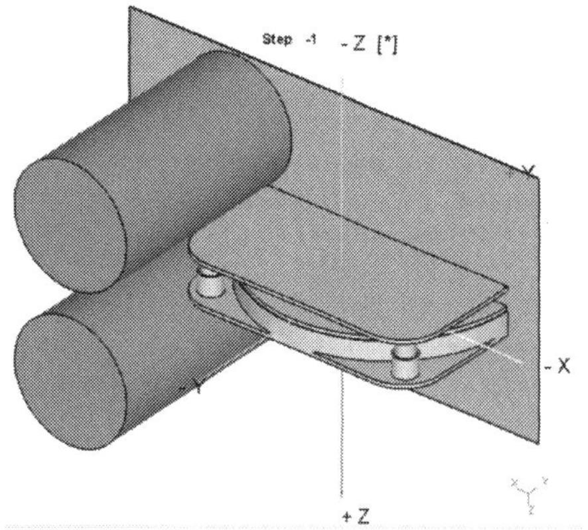

Figure 10. The symmetry plane of the combined rolling-stamping process

A vital part of the process being realized successfully is fulfilling the conditions of picking up the pack in its hot rolling (figure 12):

$$\Delta H_{pack} \le 2R(1-\cos\alpha), \quad (3)$$

$$\alpha = arctg(m), \quad (4)$$

where ΔH_{pack} is the absolute compression of a pack, in mm ; m is the friction coefficient, and R is the radius of a roll, in mm.

$$H_{pack} = h_{top} + h_{blank} + h_{lower}, \quad (5)$$

where h_{top} is the height of the upper die, in mm; h_{blank} is the thickness of the workpiece, in mm; h_{lower} – is the height of the lower die, in mm.

The formula (3) can be rewritten as thus:

$$\Delta H_{pack} = h_{top}, \quad (6)$$

$$h_{top} \le 2R(1-\cos\alpha), \quad (7)$$

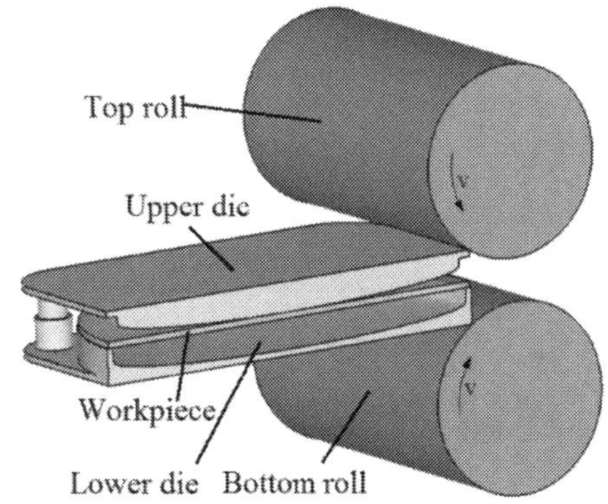

Figure 11. Forward method of thick plate rolling-and-stamping combined process

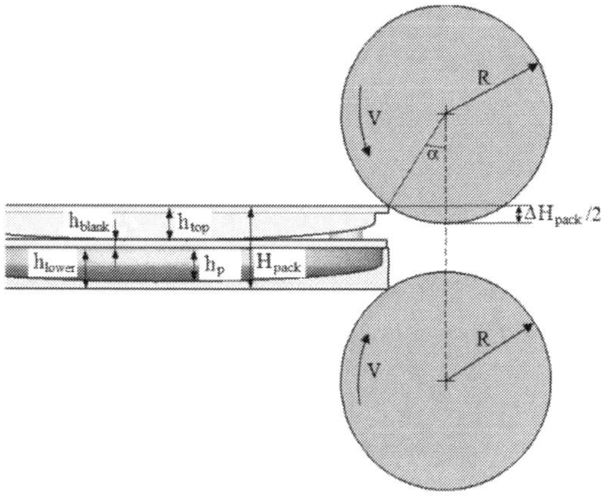

Figure 12. A scheme describing the conditions of the successful process

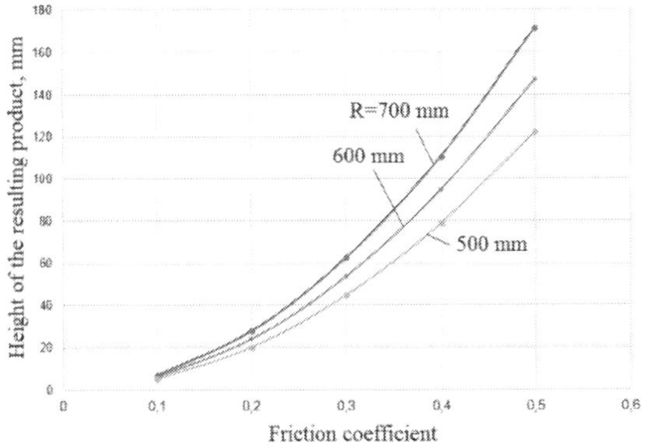

Figure 13. The influence of friction coefficient and roll radius on the maximally possible height of the resulting product

The height of the resulting product is determined by the height of the upper die, and thus also determined by the conditions in which the pack is packed up at the initial state of the combined thick-plated rolling-and-stamping process.

The interrelation between the maximum possible height of the resulting product, the radius of working rolls and the friction coefficient is shown in Figure 13.

An analysis of the temperature field has shown that during transportation time, the temperature of edge areas of a workpiece decreases significantly: from 900 ^{0}C to 764 ^{0}C. This can lead to higher stress in these areas during metal deformation, since the most intensive plastic strain takes place exactly there.

This testifies to the non-uniformity of plastic strain in the workpiece's volume about its central symmetry axis. The high intensity of stresses on the foremost area of the workpiece leads to a higher probability of cracks, folds there. On the foremost area of the workpiece the strain happens in a very short period of time, right during the pack is picked up by the rolls. As a result of this, the force rapidly increases, which can negatively affect the process stability and even lead to equipment damage.

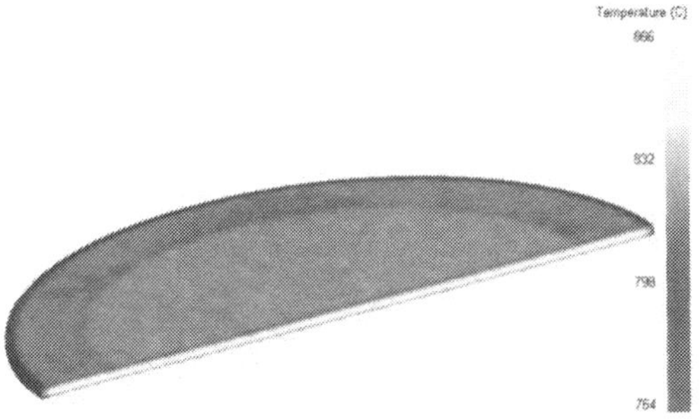

Figure 14. Temperature field of the initial sheet workpiece before deformation

An analysis of the stress and strain metal state in different stage of the combined thick plate rolling-and-stamping process according to the forward method has shown that stresses localize in the area where the elliptic bottom undergoes a plastic bend, and that this localization is consecutive in time. Thus, in the first strain stage, during the pick-up, the highest stress intensity in the foremost area of the workpiece, and then it radially moves in the direction of the rear area of the workpiece (as illustrated by the arrows in Figure 15).

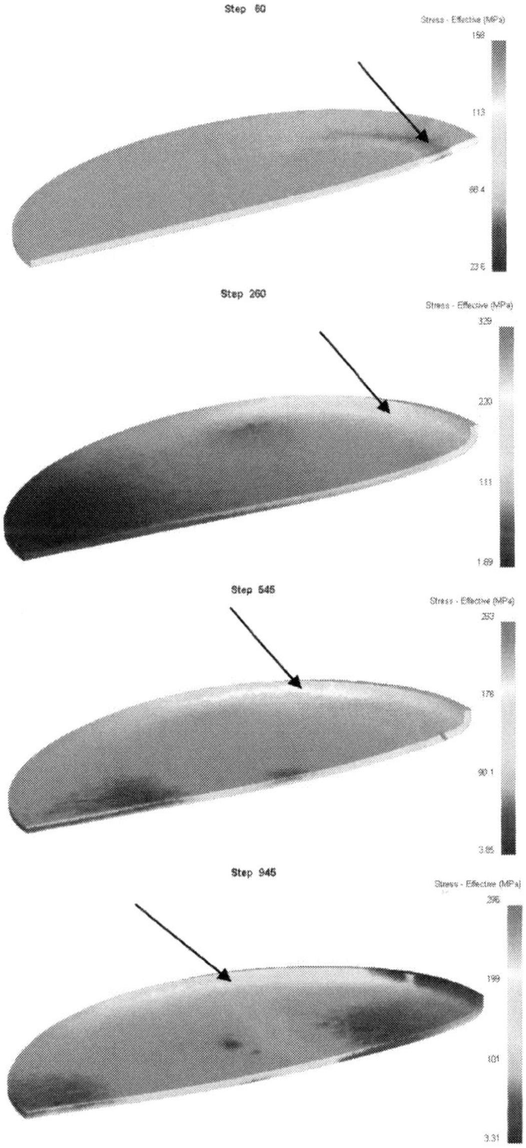

Figure 15. Change of stressed workpiece state in different calculation steps

The strain at the level 0,4 is localized at the external surface of a workpiece (Figure 16). The high strain here, combined with workpiece thickness of 40 mm can be critical in relation to the possibility of metal discontinuity flaw in this area during the time the workpiece is picked up by the rolls.

In manufacturing elliptical bottoms by the combined rolling and stamping method, the influence of the workpiece's relative thickness is especially important. If during the thin plate deformation the relative thickness

DOI: 10.1051/matecconf/20168015008

influences the flange fixity and leads to a certain change in the limiting drawing coefficient, during the deformation of a thick plate it influences the area of contact friction, the degree of wall thinning, and the circumstances of contact between the workpiece and the instrument.

Eventually, this leads to an increased danger of production defects due to excessive thinning of the wall and the formation of folds. Extensive mathematical modelling of the combined rolling-and-stamping process with different process parameters (changing the workpiece thickness from 40 to 60 mm, the strain rate, the temperature and friction conditions) has shown that in all cases the largest deformation localizes in the external radius of the bottom.

A particularity of the reverse rolling-and-stamping method (Figure 17) is that the process of tension and deformation distribution in the workpiece volume occurs differently. In particular, our analysis has shown that during the vertical-plane convergence of the rolls (due to pressuring mechanisms) the deformations are localized in the central part of the workpiece (Figure 18). During the further stages, tensions localize on the flanges of the elliptical bottom (Figures 19-20).

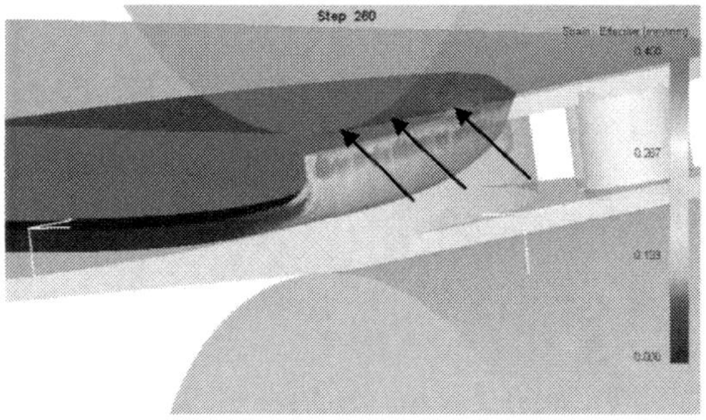

Figure 16. Intensity field of deformation during the moment the pack is picked up by the rolls

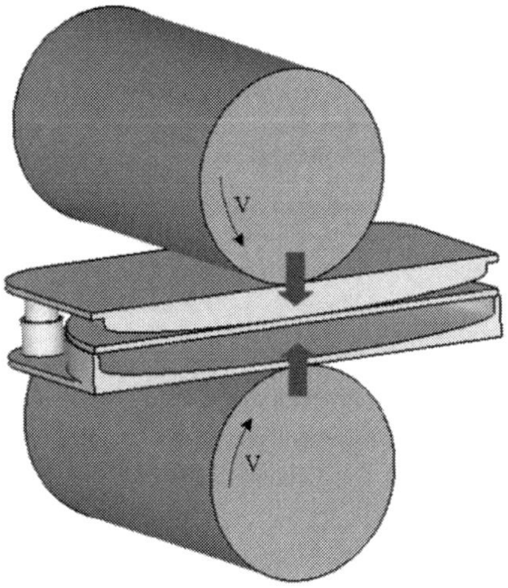

Figure 17. Reverse method of the combined thick plate rolling and stamping

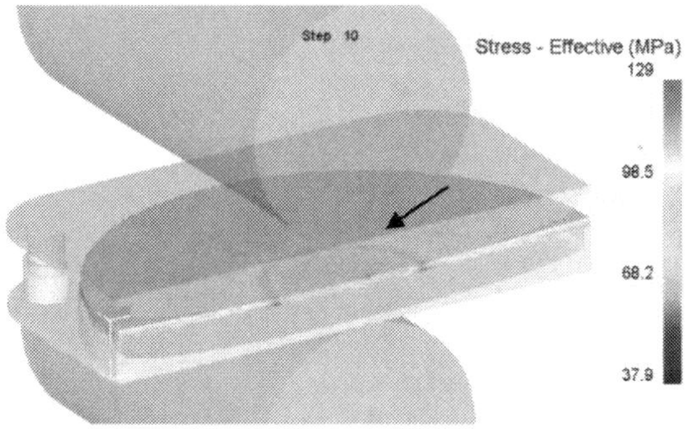

Figure 18. Stress intensity field of the workpiece's volume during the first deformation stage

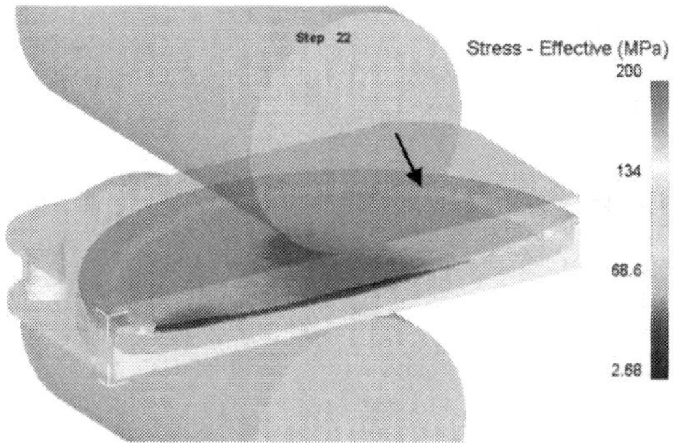

Figure 19. Intensity field of tensions in the volume of the workpiece during the second stage of deformation

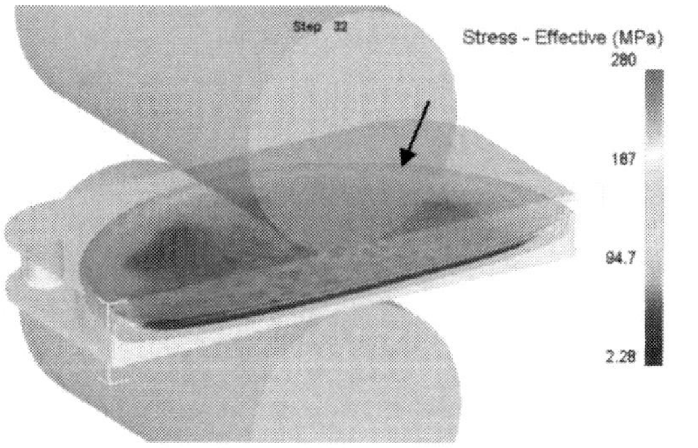

Figure 20. Intensity field of tensions in the volume of the workpiece during the second stage of deformation

The geometry of the resulting solutions is in good agreement (deviation is not higher than 6%) with the experimental results.

3 Conclusion

Two fundamentally new methods of combined thick plate rolling-and-stamping for manufacturing bodies of rotation (without using pressure equipment) were developed. The principle of rolling a pack of upper die, lower die, and the thick plate workpiece between them is the foundation of the whole process. The first method of the combined process involves the pick-up and successive rolling of the pack from the foremost edge to the end area with the roll gap (pressure screw position) being kept the same all the way. The second method involves the reverse rolling of the pack in drive working rolls, with the pack being placed in the separated rolls so that its center is on the line connecting the rolls' axes, and the rolling being conducted with the changing positions of pressure devices (meaning, a changing roll gap).

The results of FE modelling of the combined rolling-and-stamping process with the help of DEFORM 3D software have shown that the key problems of the process is the loss of pack fixity during the rolling, and the corresponding defects in the form of folds, increasing the consumption index and increasing the production cost of the resulting detail.

The investigation and 3D modelling of the two methods of the combined process of thick plate rolling and stamping process were conducted. A 3D analysis of the metal stressed-strained state during the combined rolling and stamping process is worked. It was shown that on the 0,4 level the strain localizes at the external surface of the workpiece. A high strain value combined with workpiece thickness of 40 mm or more can be critical in relation to the possibilities of discontinuity flaws appearing in this area.

It was established that during the forward-method rolling the maximal pulling stresses arise in the forefront are of the pack, in particular, on the forefront areas of the top and the lower dies. The stress levels there can reach 1000 MPa. Such high values of tensile stresses are in large part explained by the fact that the workpiece is hot, and the instruments (the upper and lower dies, the rolls) are cold, having the same temperature as the environment.

For the reverse deformation method, the patterns of stressed-strained metal (workpiece and both dies) state change, depending on the process parameters (temperature of workpiece heating, its thickness, the speed of deformation, friction conditions) were established.

The modelling results allowed us to optimize the shapes of the upper and lower dies, and also to optimize the construction of the pack, taking the stability of the rolling-and-stamping process into account. The benefits and the flaws of both combined process methods were established. The reverse deformation method was determined to be preferable for practical realization, from the point of view of process stability and energy efficiency.

A complex technical solutions that allow the use of rolling mill for manufacturing large-size bodies of revolution was developed. Two combined processes - rolling and stamping, and also asymmetric rolling and bending - were proposed for this.

Acknowledgements

The reported study was supported by grant (program START 2014), research project No 190GS1/8745.

References

1. V.A. Demin, Development of a method of the plate stamping processes designing on the basis of forecasting of technological failure: D. Sc. Thesis. Moscow (2003) 342.
2. H. Karbasian, A.E. Tekkaya, A review on hot stamping, Journal of Materials Processing Technology. 210 (2010) 2103-2118.
3. A. Pesin, V. Salganik, E. Trahtengertz, M. Cherniahovsky, V. Rudakov, Mathematical modeling of the stress-strain state in asymmetric flattening of metal band, Journal of Materials Processing Technology. 125-126 (2002) 689-694.
4. A. Pesin, Practical results of modeling asymmetric rolling, Steel in Translation. 33 (2003) 46-49.
5. A. Pesin, New solutions on basis of non-symmetric rolling model, Stal. (2003) 66-68.
6. A. Pesin, V. Salganik, E. Drigun, D. Chikishev, Device for asymmetrical rolling metal plate, RU Patent 38646 (2004).
7. A. Pesin, V. Salganik, E. Drigun, D. Chikishev, Device for asymmetrical rolling metal plate, RU Patent 2254943 (2005).
8. A. Pesin, V. Salganik, D. Chikishev, Device for asymmetrical rolling metal plate, RU Patent 54831 (2006).
9. A. Pesin, D. Chikishev, S. Blinov, Device for measuring a radius of curvature of the cylindrical surface of large parts, RU Patent 56592 (2006).
10. A. Pesin, D. Chikishev, S. Blinov, D. Pustovoytov, Device for asymmetrical rolling metal plate, RU Patent 87649 (2009).
11. M Sverdlik, A. Pesin , D. Pustovoytov. Advanced Materials Research (2012)
12. A. Pesin, E. Drigun, D. Pustovoytov, I. Pesin. MATEC Web of Conferences. 26 03007 (2015)
13. A. Pesin, E. Drigun, D. Pustovoytov, I. Pesin. Key Engineering Materials. Vol. 685, (2016) 375-379

MATEC Web of Conferences **80**, 15009 (2016) DOI: 10.1051/matecconf/20168015009
NUMIFORM 2016

The Role of Evolutive Elastic Properties in the Performance of a Sheet Formed Spring Applied in Multimedia Car Industry

Joel Silva[2,a], *João* Faria[1], *Rita* Ferreira[1], *Pedro* Bernardo[2] and *J. L* Alves[1]

[1]CMEMS, University of Minho Campus of Azurém, 4800-058 Guimarães, Portugal
[2]Bosch Car Multimedia Portugal S.A., Apartado 2458, 4705-820 Braga, Portugal

Abstract. The manufacturing process and the behavior of a sheet formed spring manufactured from an aluminum sheet is described and investigated in this work considering the specifications for the in-service conditions. The sheet formed spring is intended to be applied in car multimedia industry to replace bolted connections. Among others, are investigated the roles of the constitutive parameters and the hypothesis of evolutive elastic properties with the plastic work in the multi-step forming process and in working conditions.

1 Introduction

The competitiveness of the automotive industry has led to a simultaneous increase of innovation and a more efficient costs control, which are the key to achieve better products and consequently meet strategic business goals. The tendency to use lightweight materials and innovative design concepts are usually focused on car body panels and chassis. However, this design concept should be extended to all metallic car components, even in auxiliary parts like the ones in multimedia chassis applications.

The metallic chassis of multimedia systems are usually sheet metal parts and the assembling of its components is mostly made with bolts. Bolted connections are responsible for joining the parts together, besides they must provide ground contact and ensure an electromagnetic shielding of the assembled system.

Bolted connections have several drawbacks since each bolt must be carefully and individually bolted at the right position and with a pre-determined tightening torque. Having in consideration that each chassis may contain several bolts, the screwing duration is a significant amount of time consumed during the assembling process. Because the used bolts are the self-tapping type, during the screwing process occurs the production of metallic shavings, which once deposited in the printed circuit boards or other electronic components may provoke short circuit and consequently malfunctions.

The bolt nonconformity is also a problem because it makes the process repeatability very hard to ensure.

The present study aims to investigate an innovative solution for fixing a printed circuit board under a metallic chassis without bolted connections. This geometry must be capable of submitting the printed circuit board under a compressive load, to ensure the ground contact between components.

The compressive load has a key role in this geometry: it must be capable to replace the bolt's compressive load and ensure a good performance in several tests specified in automotive standards.

The main methodologies applied to evaluate and predict the material and geometry's performance are experimental tests and numerical simulations. With the rapid development of computational capacity, the finite element analysis has been widely used by scientific and industrial communities in product design and to predict material's behaviour in manufacturing process. Finite element analysis reduces the number of trials, reducing the cost of product development [1,2].

The problem investigated in this paper is to evaluate, by means of numerical simulation, the usage of a new sheet formed spring to replace bolted connections, and predict its compressive load.

The research objectives are to evaluate the effects of the manufacturing process, the evolution of the elastic properties due to the plastic work, and the hardening models, in the prediction of the proposed geometry behaviour. To evaluate these effects several numerical simulations were performed applying different parameters, and finally the main results are compared and discussed.

2 The new sheet formed spring: geometry description

The proposed geometry was designed having in mind the industrial problem previously addressed. This geometry is located at the cover corners of a metallic chassis, and it consists in a recess with a trapezium format where the contact between the printed circuit board and chassis will occur (see more details on Figure 1).

[a] Corresponding author: vieiradasilvajf@gmail.com

The innovation presented in this geometry is focused on the geometry around the recess. This geometry consists of material removal in the shape of tabs and it was designed to have a similar behaviour to a spring. The geometry shown in (Figure 1) was subject of previous numerical simulations aiming to find the best geometry's shape of the tabs. The geometry selected has shown the better ratio force/displacement.

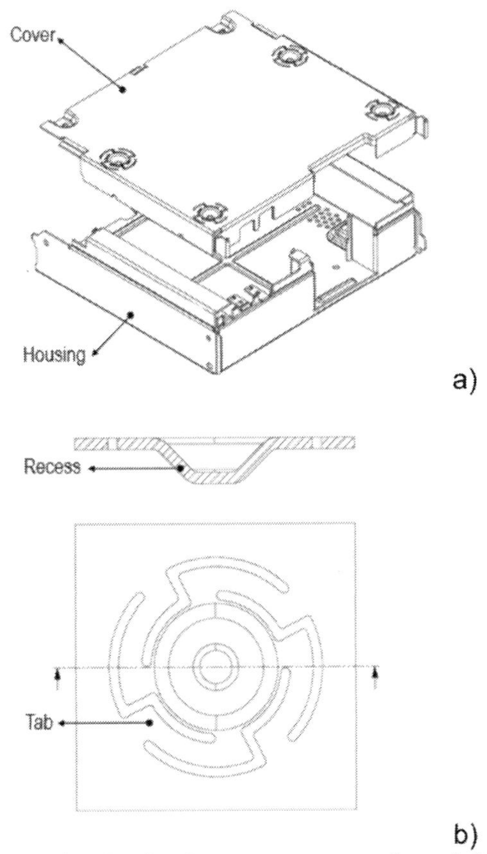

Figure 1. Developed geometry to replace bolts: a) its placement in the chassis top cover, and b) detailed geometry of the sheet formed spring, comprising a recess and four tabs.

3 Numerical modelling

Through finite element analysis the numerical studies are an important tool allowing the shortening of several resources, such as time, money and useless prototypes.

3.1 DD3IMP

Deep Drawing 3D IMPlicit Code (DD3IMP) is a finite element code which is the result from continuous development and optimizations aimed for the simulation of sheet metal forming processes [3,4]. This FE solver considers the evolution of deformation process through an updated Lagrangian scheme. This 3-D elastoplastic finite element code allows large elastoplastic strains and rotations, following a fully implicit time integration scheme.

The configuration of the sheet, tools and the state variables are updated at the end of each time increment Δt, and the new configuration is taken as reference for the subsequent time step. To calculate an approximate first solution for incremental displacements, stresses and frictional contact forces is used an explicit approach. Because this first solution satisfies neither the variational principle for the coherence condition, due to the strong non-linearities of the problem, a fully- implicit algorithm of Newton-Raphson type is used for the correction of the first approximation at instant $t+\Delta t$.

DD3IPM models the geometry of the forming tools by parametric Bézier surfaces or Nagata patches [5]. Contact is established between the deformable body (blank sheet) and the tools, which are assumed to behave rigidly. The frictional contact between the two bodies is interpreted with a classical Coulomb's law with an augmented Lagrangian approach. The numerical schemes follow a frictional contact algorithm that operates directly on the parametric Bézier surfaces or the Nagata patches [6].

3.2 Constitutive Models

Several efforts have been done by scientific community to improve and create new constitutive models in finite element codes. These efforts are now providing a more efficient description of the mechanical behaviour of sheet metals [7]. Developments are achieved through the implementation of new and more accurate yield criteria associated to the isotropic and kinematic hardening behaviour. The orthotropic yield criterion is currently used to model the planar anisotropy of the blank. In turn, the isotropic and kinematic hardening describes the evolution of the flow stress and the yield surface with the plastic work.

The study of the kinematic behaviour is essential in case of Bauschinger effect. This effect represents the material's answer to the multiple tension-compression cycles to which material points are submitted during the forming process.

In the present work it is used a Voce law to describe the isotropic work hardening since it is the law that better describes the mechanical behaviour of the aluminum alloy used in the proposed geometry. The equation of the mentioned law is given by:

$$Y = Y_0 + R_{sat}[1 - \exp(-C_Y \bar{\varepsilon}^p)] \qquad (1)$$

where R_{sat} and C_Y are material parameters, and $\bar{\varepsilon}^p$ is the equivalent plastic strain. Y_0 is the initial yield stress and Y is the yield stress.

Kinematic hardening is computed by Lemaître & Chaboche evolution law, which is defined by:

$$\dot{\mathbf{X}} = C_X\left[\frac{X_{sat}}{\bar{\sigma}}(\boldsymbol{\sigma}' - \mathbf{X}) - \mathbf{X}\right]\dot{\bar{\varepsilon}}^p, \mathbf{X}(0) = 0 \qquad (2)$$

being C_X and X_{sat} the material parameters. $\mathbf{X}$ is the back stress tensor, $\boldsymbol{\sigma}'$ the deviatoric Cauchy stress tensor and

$\bar{\sigma}$ the equivalent tensile stress as computed from the isotropic von Mises yield criterion.

For this case study, it is intended to manufacture the multimedia system chassis in aluminum alloy, in the grade AA5182-O [7]. For this specific material, the constitutive parameters to be used in equations 1 and 2 are depicted on Table 1.

Table 1. Material parameters of Voce with Kinematic Hardening law for AA5181-O.

	Y_0	C_R	R_{sat}	C_X	X_{sat}
AA5182-O (Hill'48)	148.5	9.7	192.4	152.7	26.0

3.3 Elastic properties evolution

The majority of studies of sheet metal forming considers the elastic behavior as isotropic and constant, assuming that plastic work has no effect in the elastic properties. In this connection, it is considered that when a permanent plastic deformation takes place, upon unloading, the elastic strain will be recovered. Thereby the unloading line is parallel to the initial elastic loading line. Other studies claim that the elastic properties can evolve with the plastic work, assuming that the elastic behavior remains isotropic but not constant with the plastic deformation [8-11].

Currently there are no consensus in which models have higher influence in the prediction of materials' behavior. The accuracy of each study is defined through the mathematical models implemented and depends on the studied material. Alves *et. al* [10] point out that for some materials, taking into account the evolutionary isotropic elasticity can have a larger effect on springback prediction than taking into account the kinematic hardening. This results were plainly observed for the A6016-T4 aluminum alloy. Moreover, in case of DP600 steel, the influence is almost similar. Zang *et. al* [8] arrived to similar conclusions, refereeing that for DP780 steel, the unloading modulus is more important than Bauschinger effect in springback prediction.

In finite element codes, mechanical behavior is described in a macro approach based on phenomenological laws. The numerical tools consider different behavior laws of the elastic properties and their variation with the equivalent plastic strain.

Several studies referred an inelastic recovery behavior after plastic deformation. Such behavior was already proved for several metals, such as copper, some aluminum alloys and some ferrous alloys. With this study exists the possibility of being demonstrated the evolutionary isotropic elasticity of the materials. This results will allow a more accurate performance prediction of the developed component.

Yu [12] referred the elastic modulus variation with plastic strain as a macroscopic phenomenon. However, to this author, from the microscopic view this condition could be inducted by microstructural evolution, textural change and internal stress during plastic deformation. Other authors, as Yang *et al.* [8,9] also adopted microscopic and macroscopic measurements and concluded that movement of mobile dislocation and dislocation pileup are the main sources for the decrease of the elastic modulus.

In the present study, only macroscopic evaluations will be considered because the goal is to realize how this variation affects the component performance and not why it does it.

The evolution of Young modulus as a function of the plastic strain is given in this work throughout the law proposal by Alves *et. al.* [10] This law is based on the experimental work of Yang et Al [8,9], who proposed an evolution law based on experimental measurements given by a polynomial equation, however its interval of validity is restricted to less than 25% of plastic strain. In sheet metal forming the maximum equivalent plastic strain is usually higher than 25%, and because of that Alves *et. al.* introduced an exponential equation with saturation which describes better the evolution of the elastic properties in a wider range of plastic strains. The evolution law was implemented in DD3IMP FE code and it is given in a generic form by:

$$E = E_0 + k_E . E_0[1 - \exp(-C_E \bar{\varepsilon}^p)] \qquad (3)$$

where E_0 is the value of the initial Young modulus, k_E and C_E are the parameters of the evolution law and $\bar{\varepsilon}^p$ is the equivalent plastic strain. $k_E \in [0,1]$ defines the saturation value of the Young modulus, $\lim_{\bar{\varepsilon}^p \to +\infty} \left(\frac{E_0 - E(\bar{\varepsilon}^p)}{E_0}\right) = k_E$, and C_E measures the rate of approximation to the saturation value. Figure 2 shows a variation of Young modulus as a function of the equivalent plastic strain given by the exponentially-based evolution law introduced by Alves *et. al.* [10] in equation 3. For this case, k_E=0.25 and C_E=12.5.

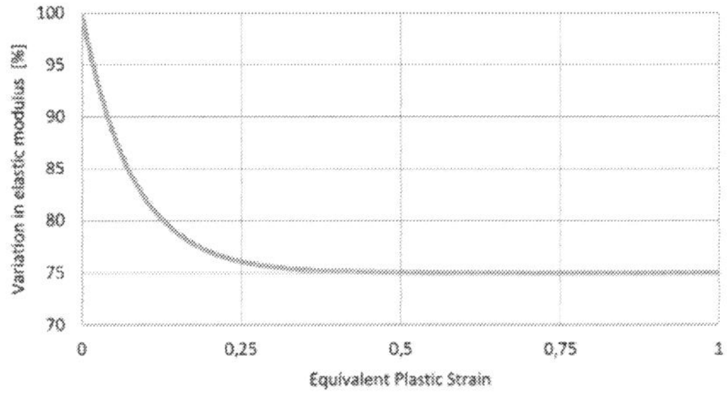

Figure 2. A variation of Young modulus as a function of the equivalent plastic strain given by the exponentially-based evolution law introduced by Alves *et. al* and implemented in DD3IMP FE solver.

4 Numerical results

4.1 Simulation Setup

The numerical study was separated in two main stages. In the first stage the stamping process was numerically simulated. This process is responsible for the major geometrical changes and plastic work field in the raw sheet metal. A generic scheme of the stamping tooling as well the general positioning of the sheet and tools in the first step is shown in Figure 3.

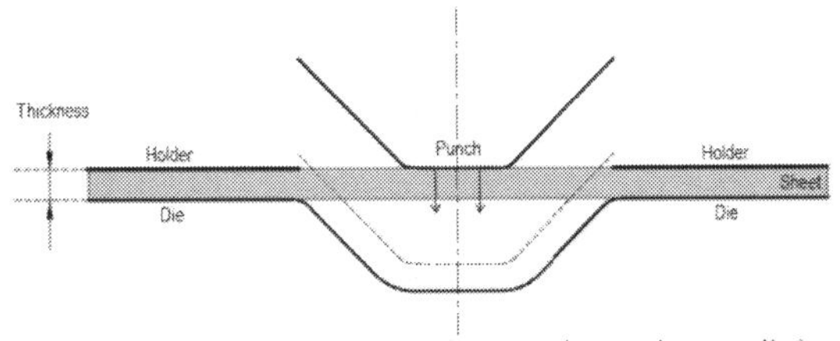

Figure 3. Schematic diagram of the stamping tools as well the general positioning of the sheet and tools at the begging of the forming step.

In the first step a blank sheet with a rectangular shape of 30 mm x 30 mm x 1.2 mm is positioned between the holder and the die. The punch moves vertically giving shape to the sheet metal, and when the punch meets to the final position it moves back in the opposite direction, allowing the springback of formed part.

The final geometry obtained in this simulation, as well of the state variables (stress and backstress tensors, equivalent plastic strain, ...) will be used as an input data to create the FE mesh for the second stage simulation. The equivalent stress and strain generated during the manufacturing process are remapped into this new FE mesh, allowing the simulation of the operating conditions taking into account, or not, the full history of the manufacturing process. In more detail, a new FE mesh is generated considering both the final geometry of the forming stage, i.e. the first stage, and the trimming geometry to obtain the new FE mesh corresponding to the geometry of the sheet formed spring as shown in Figure 1b). After, the values of the state variables are remapped from the original FE mesh to the new one. It is worth noting that the local effects of the trimming operation are not taken into account.

The second stage simulates the operating conditions of the proposed geometry, in order to investigate the role of the elastic properties evolution on the mechanical behavior of the sheet formed spring. A scheme with the general positioning of the sheet formed spring and the so-called tools used in the simulation of operating conditions is shown in Figure 4. In this stage, the sheet formed spring is loaded by a tool representing the printed circuit board. As boundary conditions, a displacement of 1 mm is imposed to the tool, and the reaction force is evaluated.

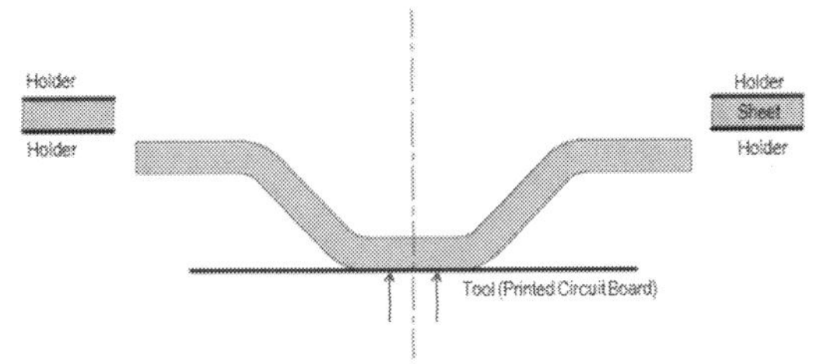

Figure 4. Schematic diagram of the sheet formed spring and the tool, i.e. the printed circuit board, as well the general positioning of the sheet formed spring and tool in the so-called working conditions (sheet formed spring being loaded).

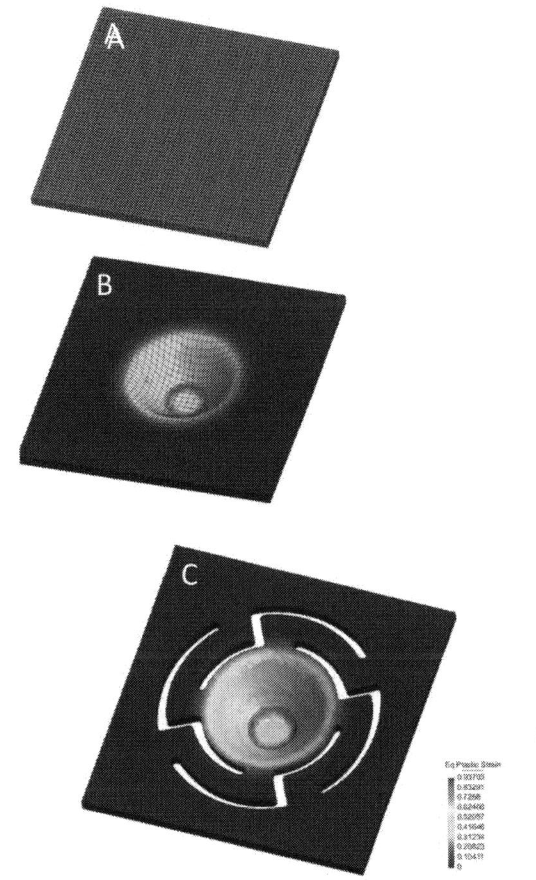

Figure 5. Mesh geometry for the two-stage simulation: a) Initial FE mesh; b) FE mesh after the stamping process and recess formation (displayed isocontour corresponds to the equivalent plastic strain); and c) FE mesh for the second stage, after trimming and remapping of the full-history state variables.

In order to precisely compare numerical results, and the role of either constitutive models or evolutive elasticity, several simulations were carried out using the same boundary conditions and FE meshes. Table 2 shows the difference between the performed simulations. There are two main types of numerical simulations: the first group only considers the assembly conditions (AC label) during the assembly of the cover in the chassis, which corresponds to pressing the sheet formed spring; the second group consider all the forming history since manufacturing process (MP label) until assembly conditions (MP.AC). For each of these groups, different material laws are used.

Table 2 shows the possible combination between first (AC) and second (MP.AC) groups with both evolutive elastic properties (EPE) and/or kinematic hardening (KH). For example, "MP.AC.EPE.KH" is a simulation in which all condition are activated.

Table 2. Performed numerical simulations and the difference between them in terms of material laws and pre strain information.

Assembly Conditions (AC)		Manufacturing Process + Assembly Conditions (MP.AC)	
Elastic Properties Evolution (EPE)	Kinematic Hardening (KH)	Elastic Properties Evolution (EPE)	Kinematic Hardening (KH)
-	-	-	-
√	-	√	-
-	√	-	√
√	√	√	√

4.2 Results and discussion

The results of each simulation are compared based on the reaction force of the tool named printed circuit board versus the displacement of that same tool, as shown in Figure 6. To evaluate the influence of elastic properties evolution and kinematic hardening in numerical simulation of assembly conditions (AC), Figure 6 shows the force reaction along a 1 mm displacement in four different cases. The four simulations were conducted taking into account the following phenomena:

- with only elastic properties evolution (AC.EPE),
- with only kinematic hardening (AC.KH),
- with both phenomena (AC.EPE.KH) and
- without any of them (AC).

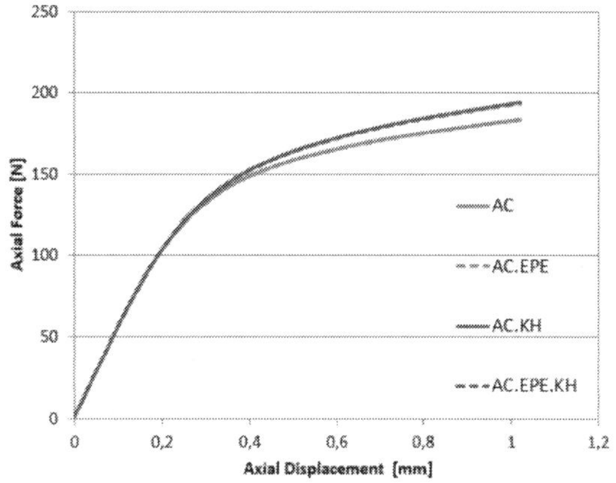

Figure 6. Curve force versus displacement of the AC, AC.EPE, AC.KH and AC.EPE.KH simulations during the assembling process.

In the present case, the MP condition is not activated, i.e. the forming history is not taken into account given that the state variables were not remapped from the first to the second FE mesh. In this case, the material of the sheet formed spring is considered as new material without any plastic work history.

The AC and AC.EPE curves show a very similar behavior. In the previous figure both curves have the same evolution as they are almost perfectly superposed, as well as the AC.KH and AC.EPE.KH curves which are also almost indistinguible. Therefore, this figure allows to conclude that elastic properties evolution does not influence the spring behavior. On the other hand, the results indicate that the simulations where kinematic hardening was considered displays a lower maximum force values for the same displacement of the tool (or the printed circuit board). The difference between this curves starts after around 0.2 mm of displacement. The maximum force is obtained in AC case, and was of approximately 183 N, while for AC.KH curve was of 194 N. This difference only represents about 6%, and for the application of this sheet formed spring the value is not relevant. However, and taking into account the slope of both curves, for displacements higher than 1mm this difference increase.

The evolution of the elastic properties, i.e. of the value of the Young modulus with the plastic work has almost no influence. This behavior is due to the low levels of plastic strain observed in the case where only the assembly conditions are simulated, i.e. the history is not taken into account. Table 3 shows the maximum plastic strain for the four cases. The table shows that the maximum plastic strain is consistently low. There is no substantial difference in the maximum plastic strain amongst the four different studied cases. The maximum equivalent plastic strain is of about 4.7%.

Table 3. Maximum equivalent plastic strain at the end of the assembly process, for the AC, AC.EPE, AC.KH and AC.EPE.KH cases.

Studied Cases	Maximum plastic strain [%]
AC	4.7
AC.EPE	4.4
AC.KH	4.4
AC.EPE.KH	4.6

To assess the difference between the previous cases, a new set of simulations were carried out taking into consideration the stress and strain history from the manufacturing process (label MP). Figure 7 shows the force versus displacement curves of the simulations where the manufacturing process was considered.

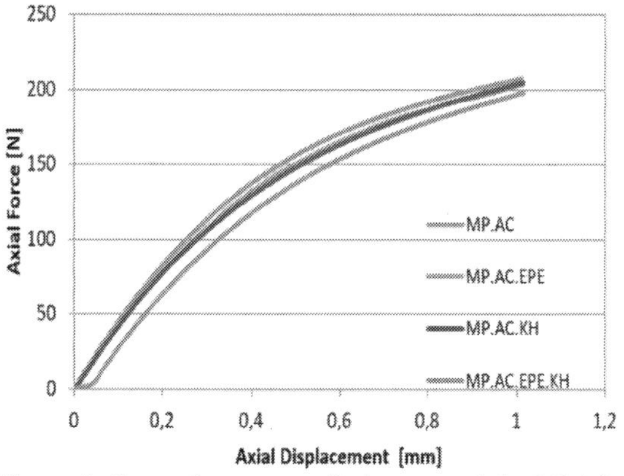

Figure 7. Curves force versus displacement of the MP.AC, MP.AC.EPE, MP.AC.KH and MP.AC.EPE.KH simulations during the assembling process.

The results show that there is a difference between the considered cases. In the case where the material does not considers Bauschinger effect (i.e. the kinematic hardening is not taken into account) and the elastic properties evolution, the force required to perform the same displacement is about 5% higher (208 N/198 N). When the evolution of Young modulus and the Bauschinger effect are considered, the maximum force is lower, showing that these phenomena have influence in the force/displacement curve when the full loading history is considered. The higher levels of platic work developed during the manufacturing process leads to a strong variation of the elastic properties of the material.

To evaluate the influence of the manufacturing process history in the working conditions of the sheet formed spring, two simulations are compared: in the first one the deformation history from the manufacturing process was considered (MP.AC.EPE.KH) and the second one only the operating conditions was considered (AC.EPE.KH), i.e. no history. Figure 8 shows a direct comparison for these two cases, considering the obtained force versus displacement curves.

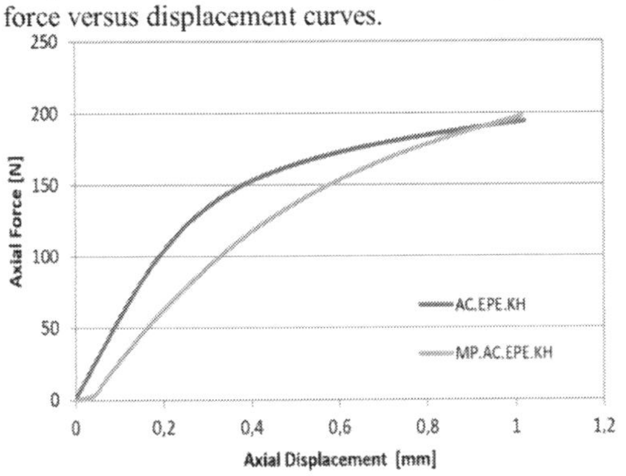

Figure 8. Curve force versus displacement of the AC.EPE.KH and MP. AC.EPE.KH simulations during assembly process.

Figure 8 indicates that there is a non-negligible difference between these two cases. For about 0.4 mm displacement the difference between the curves is of 22% (121N /155 N), but AC.EPE.KH curve starts to saturate earlier and, at the end of 1 mm displacement, the MP.AC.EPE.KH displays a higher maximum force. The difference between the maximum force values at the end of displacement is of 3% (198N / 194N). Although the difference, at the end of displacement is small, it could be higher if the displacement was larger, once the curves' slopes are totally different and they start to diverge at the final moments of displacement.

Figure 9 depicts the evolution of the equivalent plastic strain throughout the different manufacturing process stages and in the assembly process, with two different ranges, 0 to 0.92 and 0 to 0.044, allowing to emphasize the deformation evolution on the tabs.

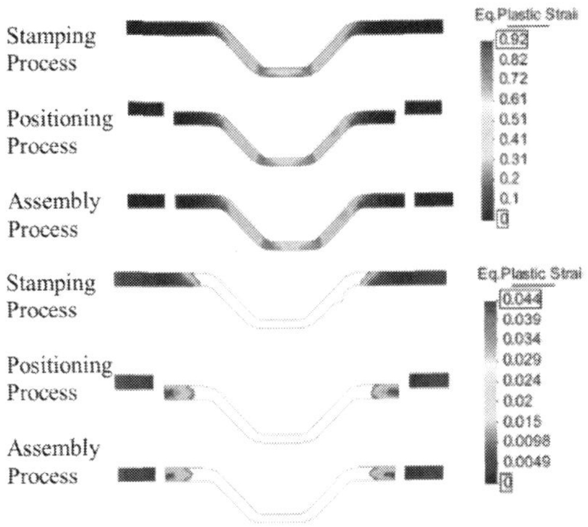

Figure 9. Evolution of the equivalent plastic strain in the investigated geometry, with two different ranges, respectively 0-0.92 and 0-0.044.

The maximum plastic strain takes place during the manufacturing process stage as expected. The plastic work caused by the remaining steps is too low when compared with the first stage. In the remaining steps, deformation is concentrated in the extremities of the tabs, and mostly in the elastic regime.

Figure 10 shows a comparison between the equivalent plastic strain for the cases presented in Figure 8.

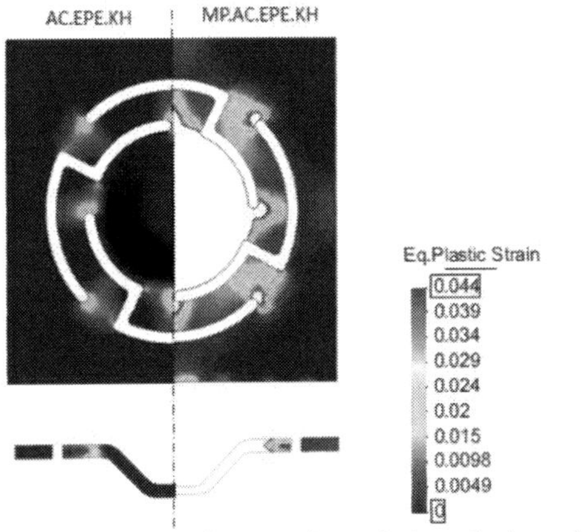

Figure 10. Comparison between the equivalent plastic strain state for the AC.EPE.KH and MP.AC.EPE.KH simulated cases. Range between 0 and 0.044.

The region where occurs the highest plastic work is at the middle of the recess in the MP simulation, and due to the stamping process. Although the tab extremities still exhibit high levels of plastic strain. The maximum equivalent plastic strain in the tabs is between 0.30 and 0.40, this value coincides with the point where the evolution curve do describe the evolution of the elastic properties with the plastic work starts to saturate (see Figure 1). The value of the equivalent plastic strain in the tabs is high enough to, according to the implemented law, saturate to the maximum decrease of the Young modulus.

5 Conclusion

The study of the mechanical behaviour and manufacturing process of a new sheet formed spring to be applied in car multimedia industry was presented. The role of both constitutive modelling and evolutive elastic properties was analysed in order to determine their in-service conditions and specifications.

The new sheet formed spring is intended to replace several bolted connections, and thus the spring forces plays a paramount role. The analysis of the numerical results presented in the previous figures allows to conclude:

- For the investigated cases where the equivalent plastic strain is low, the evolution of the elastic properties, due to the plastic deformation, has almost no influence in the behavior of the sheet formed spring;
- The study of loading history due to the manufacturing process is important to obtain reliable results during assembling and working conditions. The effect of pre-deformation on the geometry behavior occurs, likely, due to the change of the elastic behavior of the material. A study without taking into account the loading history of the material generates different results that can lead to wrong predictions concerning the spring's behavior;
- It has been shown previously by scientific community the importance of kinematic hardening and elastic evolution properties in springback prediction of sheet metal parts. This study shows that, for the specific application studied, the role of the loading history and evolutive elastic properties is more important than the role of the kinematic hardening;
- Finally, this work demonstrates the enormous importance of a correct description of features of the mechanical behavior to accurately predict the performance of a sheet metal formed part.

Acknowledgements

This research was sponsored by:
a) Portugal Incentive System for Research and Technological Development. Project in co-promotion nº 36265/2013 (Project HMIExcel - 2013-2015), and
b) FCT with the reference project UID/EEA/04436/2013, by FEDER funds through the COMPETE 2020 – Programa Operacional Competitividade e Internacionalização (POCI) with the reference project POCI-01-0145-FEDER-006941.

References

1. D.M Neto, M.C Oliveira, J.L Alves, L.F Menezes, *Influence of the plastic anisotropy modelling in the reverse deep drawing process simulation,* MD, **60**,368–379, (2014).
2. H Laurent, R Grèze, M.C Oliveira, L.F Menezes, P.Y Manach, J.L Alves, *Numerical study of springback using the split-ring test for an AA5754 aluminium alloy*, FEAD **46** 51–759, (2010).
3. M.C Oliveira, J.L Alves, L.F Menezes, *Algorithms and strategies for treatment of large deformation frictional contact in the numerical simulation of deep drawing process*, ACME, **15**, 113-162, (2008)
4. L.F Menezes, C. Teodosiu, *Three dimensional numerical simulation of the deep drawing process using solid finite elements*, JMPT, **97**, 100-106, (2000).
5. D.M Neto, M.C Oliveira, L.F Menezes, J.L Alves, *Nagata patch interpolation using surface normal vectors evaluated from the IGES file*, FEAD, **72**, 35-46, (2013).
6. D.M Neto, M.C Oliveira, L.F Menezes, J.L Alves, *Applying Nagata patches to smooth discretized surfaces used in 3D frictional contact problems*, CMAME, **271**, 296-320, (2014).
7. L.P.M.T.M, *Selection and identification of elastoplastic models for the materials used in the benchmarks*, 3DS IMS, (2001).
8. M. Yang, Y. Akiama, T. Sasaki, *Evaluation of change in material properties due to plastic deformation*. JMPT, **151**, 232-236, (2004).
9. M. Yang, Y. Akiama, T. Sasaki, *Microscopic Evaluation of Change in Springback Characteristics Due to Plastic deformation*, NUMIFORM, Columbus, EUA, (2004).
10. J.L Alves, M.C Oliveira, L.F Menezes, S. Bouvier, *Numerical evaluation of the influence of the elastic properties evolution in springback prediction*, IDDRG, 161-168, (2006).
11. R Padmanabhan, M.C Oliveira, A. Ramalho, J.L Alves, L.F Menezes, *Influence of plastic work on the elastic properties of steel*, AMPT, p. 22, (2009).

MATEC Web of Conferences **80**, 15010 (2016) DOI: 10.1051/matecconf/20168015010
NUMIFORM 2016

SOLUTIONS FOR SAFE HOT COIL EVACUATION AND COIL HANDLING IN CASE OF THICK AND HIGH STRENGTH STEEL

Stefan Sieberer[1], Lukas Pichler[1a] and Manfred Hackl[1]

[1] *Primetals Technologies Austria GmbH, Turmstraße 44, 4031 Linz, Austria*

Abstract. Currently hot rolling plants are entering the market segment for thick gauges and high strength steel grades where the elastic bending property of the strip leads to internal forces in the coil during coiling operation. The strip tends to open. Primetals is investigating several possibilities to facilitate safe coil evacuation and coil handling under spring-back conditions. The contribution includes finite element models of such mechanical solutions. Results of parameter variation and stability limits of case studies are presented in the paper.

1 Introduction

At present, there is an increasing market demand for solutions for safe coiling and coil evacuation for hot rolling mills. This is due to the fact that currently hot rolling plants are entering the market segment for thick gauges and high strength material. In case of thick-gauge strip, the elastic bending property of the strip leads to internal forces in the coil during coiling operation. The strip tends to open. This is called 'spring-back'. Primetals is investigating several possibilities to facilitate coil evacuation and coil handling under spring-back conditions. This contribution includes finite element models to develop new mechanical solutions as well as results of parameter studies and stability investigations.

For providing a mechanism of safe evacuation and transport of spring-back coils from the downcoiler to the binding station, the so-called omega car was established. In the omega car, a circumferentially tensioned strip secures the coil during transport. In the development process of the omega car, various investigations including finite element simulations were used for development of an appropriate design concept as well as for design optimization.

Another approach to reduce the spring-back tendency is the pre-bending of the strip. Pre-bending can be achieved with the pinch roll unit in front of the downcoiler as well as with a 3-roll bending unit. Simulations were carried out with an explicit dynamic finite element solver using a 2D-model of a strip section entering the downcoiler. The results give an indication for obtainable plastic deformations with such installations.

The application of the presented numerical methods and accompanying test installations yield solutions with high potential to reduce the tendency for coils to spring-back, facilitating safe coil evacuation and coil handling operation.

2 Mechanical issues and main influence parameters

The main influence parameters on the coil stability are (i) the elastic bending moment resp. the stored elastic energy of the strip, (ii) the stabilizing counterforces (weight, pre-bending status of the strip) and the friction between the strip layers. The portion of elastic bending moment is a function of the elastic material properties, i.e. Young's modulus and moment of inertia. In the case of strip rolling this means that the thickness of the final product has the highest impact on its tendency to coil spring-back.

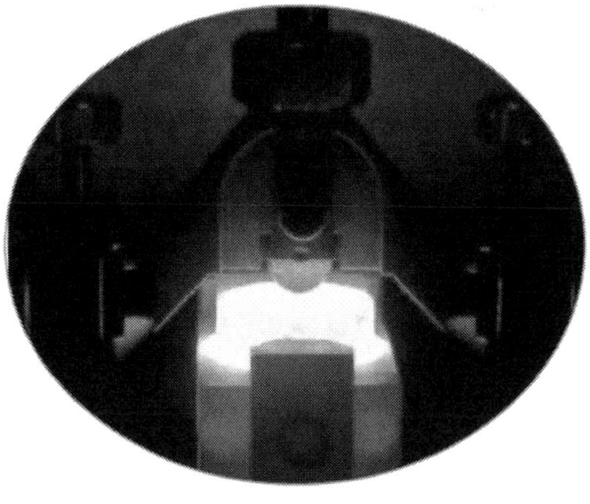

Figure 1. Tribometer – friction between cylinder and plane [1].

The stabilizing effect of the coil self-weight is limited by the maximum admissible slab length for furnace

[a] Corresponding author: Lukas.Pichler@primetals.com

operation as well as the maximum admissible force acting on the coil transportation system after coiling. Thirdly, the amount of friction stabilizes the strip windings inside the coil against loosening. An experimental investigation was carried out in order to evaluate exact values for the friction coefficients (sticking and sliding friction) for the strip, depending on the current temperature range. Friction is an important parameter in hot rolling, where the coiling temperature can vary in a range of several hundred °C. Thus, the according friction curves were evaluated for the following temperatures: 800°C, 600°C, 400°C, 200°C and room temperature. The tribometer used for the experimental investigation is shown in Figure 1.

For the finite element simulation, a proper setting of friction values depending on the contact forces are obtained from the experiments (see Section 4). The contact pressure and forces were evaluated based on the experimental conditions based on the Hertz formula.

(1)

$$\frac{1}{R} = \frac{1}{R_a} + \frac{1}{R_b}$$

$$E' = \frac{1}{\left(\frac{1-v_a^2}{E_a} + \frac{1-v_b^2}{E_b}\right)}$$

$$w = 4\left(\frac{F_n' R}{\pi E'}\right)^{\frac{1}{2}}$$

$$p_{max} = \left(\frac{F_n' E'}{\pi R}\right)^{\frac{1}{2}}$$

$$\delta = \frac{2F_n'}{\pi}\left(\frac{1-v_a^2}{E_a}\left(\ln\frac{8R_a}{w} + \frac{1}{2}\right) + \frac{1-v_b^2}{E_b}\left(\ln\frac{8R_b}{w} + \frac{1}{2}\right)\right)$$

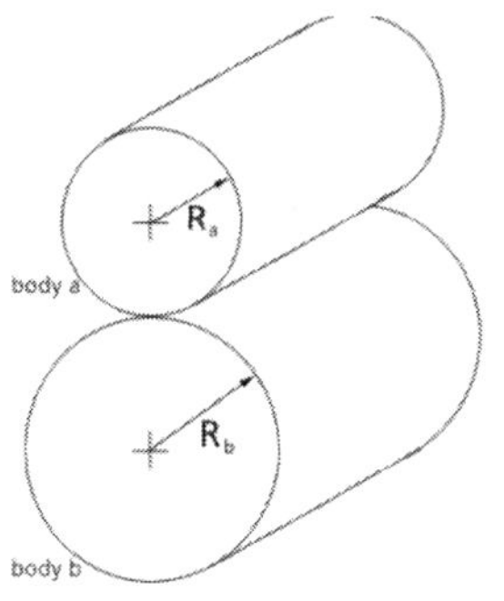

where

R_a ... radius of cylinder, $R_b \rightarrow \infty$ (plane), F_N ... normal force, E, v ... Young's modulus, A = l δ ... contact area, p = F_N/A ... contact pressure, τ = F_R/A = contact shear force.

3 Pre-bending and 3-roll unit

Figure 2 shows the model of the downcoiler assembly. The main components are (1) mandrel with core-coil, (2) pinch roll unit, (3) wrapper rolls mounted on pivot arms, (4) cradle rolls and (5) roller table. In state-of-the-art downcoilers, strip bending is performed by winding the strip around the mandrel to form the coil. The strip is tensioned by the driven mandrel and the pinch roll.

To achieve the necessary plastic strip deformation to reduce spring-back of thick-gauge steel coils, two different approaches using rollers in the downcoiler area were investigated: pre-bending of the strip (i) by lowering the roller table, and (ii) with a dedicated 3-roll unit.
For the first investigation, the roller table can be displaced vertically and a hold-down roll is placed on the strip (Figure 3). Thus, the roller table can be lowered together with the hold-down roll towards the end of the strip to create a pre-bending of the strip in the pinch-roll area. The effect on plastic strip deformation is investigated for the roller table as a standalone unit and for the assembly in the downcoiler.

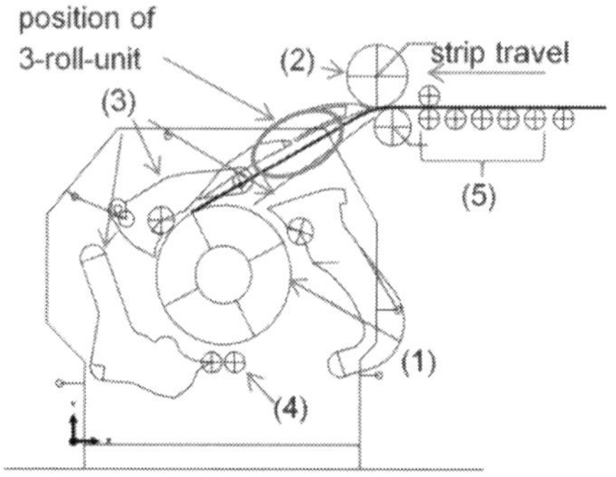

Figure 2. Model of the downcoiler assembly

For the second investigation, a 3-roll unit is modelled first to investigate the potential of the unit for induction of plastic bending of the strip (Figure 4). A 3-roll unit induces plastic deformation dependent on the relative position of the three rolls. In a second step, such a 3-roll unit is positioned in the downcoiler model and the obtained bending performance is evaluated.

3.1 Modelling of the bending units and the downcoiler assembly

Initially, the influence of the roller table and the three-roll unit were evaluated using 2D-Abaqus models where only the respective area was modelled. Figure 3 shows the model of pinch roll and roller table, and Figure 4 shows the model of the 3-roll unit. In these models, the strip is modelled using deformable bilinear quadrilateral elements with six elements over the strip thickness.

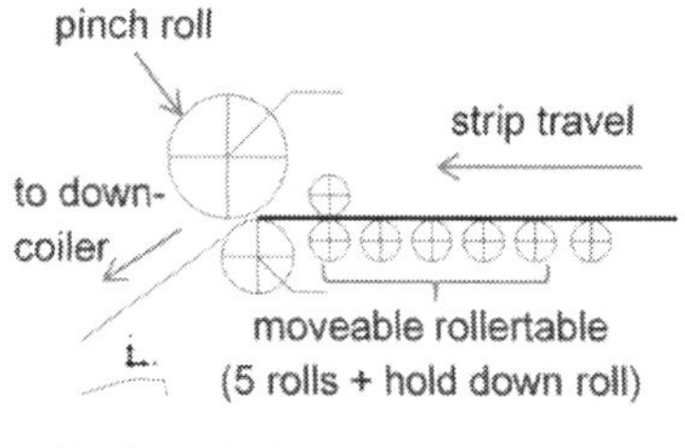

Figure 3. Model of the pinch roll unit with moveable roller table.

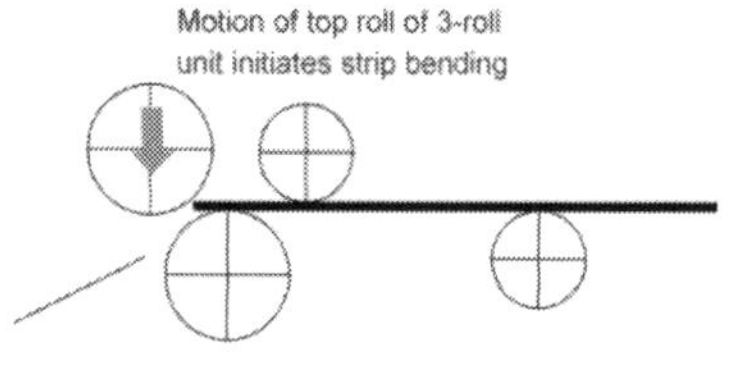

Figure 4. Model of the pinch roll unit with a moveable top roll to form a 3-roll unit.

For simulations with the complete downcoiler, an extended 2D-Abaqus model was set-up. This model is shown in Figure 2. For performance reasons and due to the fact that only the deformation of the tail end of the strip is of interest, only the last two windings are coiled in the simulation. The mandrel with the inner windings of the coil is modelled as a solid body having equivalent mass and rotary inertia applied to it, and can be driven as in real coiling operation. Surrounding the coil are one or more wrapper rolls depending on the simulation. These wrapper rolls are standard equipment in downcoilers and facilitate the coiling of the first and last couple of windings of a coil by adjusting and retracting during the process.

The strip is fed to the coiler by means of the pinch roll unit. The pinch roll unit comprises two rollers, one top and one bottom roller of different diameter, which can be utilised to tension the strip and induce some bending of the strip. For additional guidance of the strip to the coil, two chute parts are implemented in the model.

During coiling, the coil-core is fixed in x- and y-direction, and can then be released to assess the spring-back. At release, the coil is lowered onto the cradle rolls, and the wrapper rolls are retracted. Care is taken to apply smooth steps where possible as to minimise the potential for spring-back of the coiled strip induced by excessive force variations during solving with Abaqus/Explicit.

All modelled parts of the downcoiler are modelled using wire-features, with only the strip being modelled as a deformable 2D-structure using bilinear quadrilateral elements. To reduce hour-glassing-tendency, four elements were used over the thickness of the strip [2].

3.2 Simulation steps

In the initial simulations, the strip is moving through the units in one step, and the achieved bending is evaluated at a suitable point towards the end of the step.

For the extended model, the simulations are performed in a number of subsequent steps. Following the coiling and release steps is a settling step to allow the coil to fully perform eventual spring-back. The evaluation of spring back is done at the end of the settling step. To evaluate the efficiency of the different approaches to reduce spring-back, the elastic and plastic energy in the strip can be used alternatively to assessing the spring-back at the end of the simulation.

In all simulations, a X100 pipe-grade steel strip with thickness 25.4mm, yield strength σ_y=700MPa and friction μ=0.15 (steel against steel) is used.

3.3 Simulation results

3.3.1 Simulation results for roller table variation

For the simulations including the roller table variation, the strip radius reduction is achieved by including roller table motion at the end of the strip run. Figure 5 shows plots of a simulation run. A variation of the final position of the roller table is performed, and for comparison, the radius achieved with pinch roll bending only is evaluated.

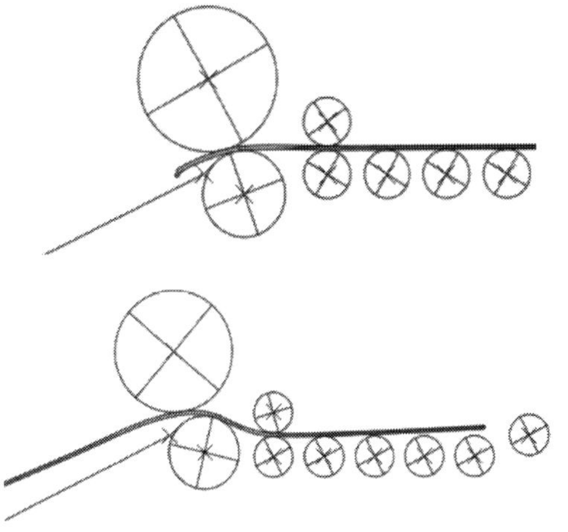

Figure 5. Simulation plots at the strip head (top) and strip tail (bottom) with the roller table lowered towards the end of the strip.

Results for the achieved bending radii with the lowered roller table are shown in Table 1. The large effect of some variation of the roller table position is not increased when the lowering is excessive, leading to sticking of the strip in the pinch roll under the assumed operational tension applied to the strip in coiling.

Table 1. Results for the roller table simulations.

simulation	1	2	3	4
roller table lowered (mm)	0	100	150	200
bending radius (m)	4.82	1.92	1.81	sticking

3.3.2 Simulation results for the 3-roll unit

Figure 6 shows the 3-roll unit, which was modelled in the place of the top pinch-roll and the strip motion was reduced to an initial velocity in the 3-roll unit simulations. The achieved radii for a variation of the upper roll motion are shown in Figure 7. For the considered range of values, a linear relationship of roll motion to bending radius can be found.

The results show the effectivity of the 3-roll unit in strip bending. However, the simulations with the 3-roll unit do not include contact with the chute as no such contact would occur when placed in the chute area close to the coil. To evaluate the performance of both, the lowered roller table and the 3-roll unit in a downcoiler assembly, simulations were performed which are shown in the next subsection.

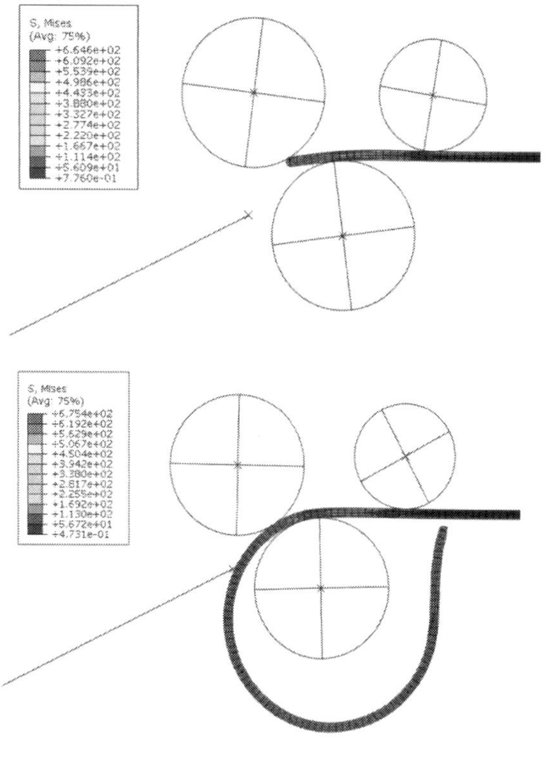

Figure 6. Simulation plots of the pinch roll used as 3-roll unit for pre-bending. Strip entering the unit (top) and strip with plastic bending (bottom).

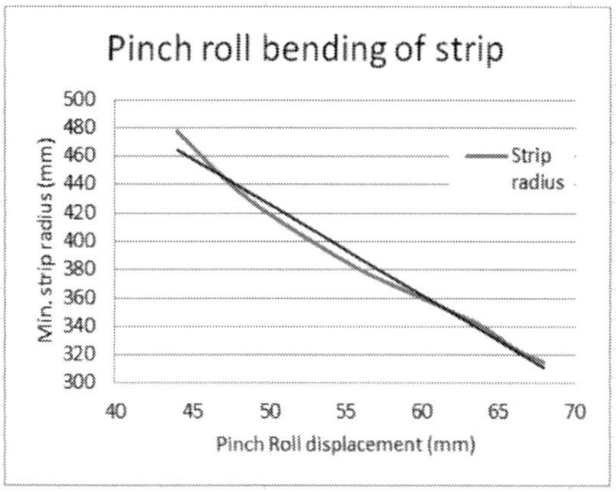

Figure 7. Resulting obtained minimum strip radius as a function of the motion of the moveable roll of the 3-roll unit (blue). The linear fit curve is shown in black.

3.3.3 Simulations of the downcoiler assembly

For simulations with the downcoiler assembly, (i) the movable roller table was included or (ii) a 3-roll unit was placed in the chute area (see highlighted region in Figure 2). Simulations of both arrangements were performed and the resulting outer coil diameters evaluated. The coil shape and outer diameter is an easy measure of spring-back.

Also, the amount of elastic energy stored in the system and the plastic energy applied to the strip is evaluated at the end of the coiling. However, it is difficult to yield a reliable spring-back measure from an elastic energy measure for reasons of complicated friction-related interaction of the strip windings. Also, the amount of plastic deformation gives a measure of the overall plastic deformation applied to the strip, including bending against the direction of coiling (e.g. in the 3-roll unit).

The result of a simulation with no pre-bending is shown in Figures 8 and 9. The resulting elastic and plastic energies the strip and the coil diameter at the end of the simulations are summarised in Table 2 as normalised values.

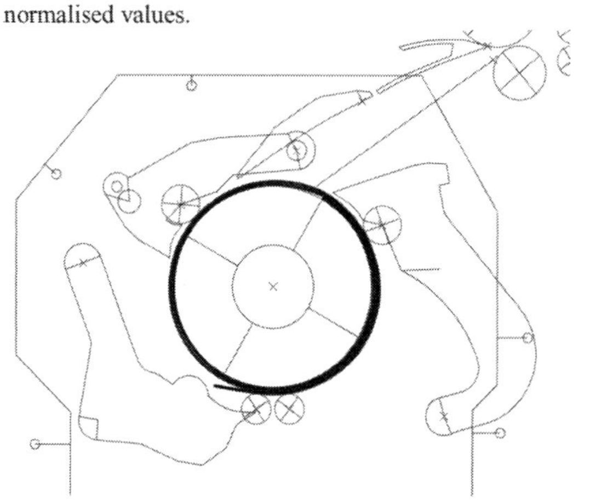

Figure 8. Simulation result of the coil with no pre-bending. Coiling is finished and the wrapper rolls are still adjusted.

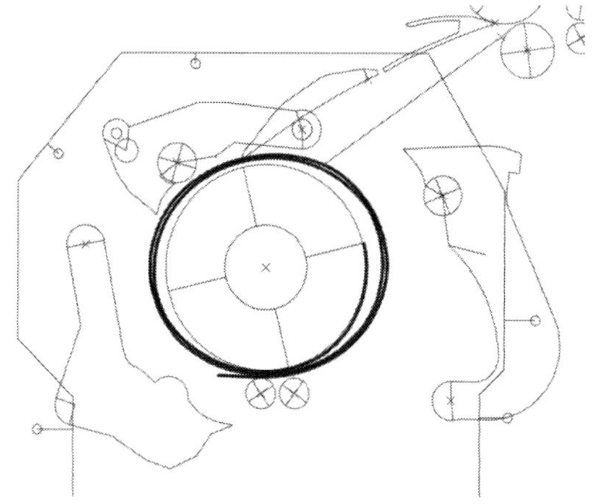

Figure 9. Simulation result of the coil with no pre-bending. The pronounced spring-back is visible at the end of the coiling with the wrapper rolls retracted.

The relative spring-back, which is defined as the maximum coil diameter compared to the diameter obtained with no spring-back, has been reduced by either of the two pre-bending measures by 34%. The strain energy in the coil compared to the simulation with no pre-bending is slightly less for the version with lowered roller table, whereas more strain energy is retained in the case of the 3-roll unit.

Table 2. Results of the process simulations.

	rel. spring-back	rel. strain energy	rel. plastic def. energy
No pre-bending	1	1	1
lower roller table	0.66	0.97	1.91
3-roll unit	0.66	1.11	7.60

The relative plastic deformation is significantly higher in both cases, as the lowered roller table as well as the 3-roll unit induce plastic deformation in the strip.

4 Coiling simulation considering friction according to experimental results

Two different friction models are used in order to simulate the coiling of hot strip, with the goal to reproduce the real hot rolling conditions as close as possible: (i) consideration of static friction only between the windings - μ=0.15 for steel against steel and (ii) an advanced friction model considering the friction coefficient, the slip rate and the contact pressure according to experimental data. The numerical investigation is carried out for two different pipe grade materials. The first material corresponds to X70 (σ_y=490 MPa), the second to X100 (σ_y=700 MPa). The strip thickness is assumed to be 12 mm, the total coil weight to be 35 tons.

A comparison is carried out first for material 1 (steel grade X70). The used factor μ=0.15 steel-steel for contact provides conservative results (see Figure 10). The coil is not stable after retraction of the wrapper rolls from the coil and springs open. When the advanced friction model is applied and compared to industrial data; this results in stable conditions which are in good agreement with the observed real life situation (see Figure 11). Thus, the advanced friction model is used for assessment of coil stability. The according material and friction data is shown in Tables 3 and 4.

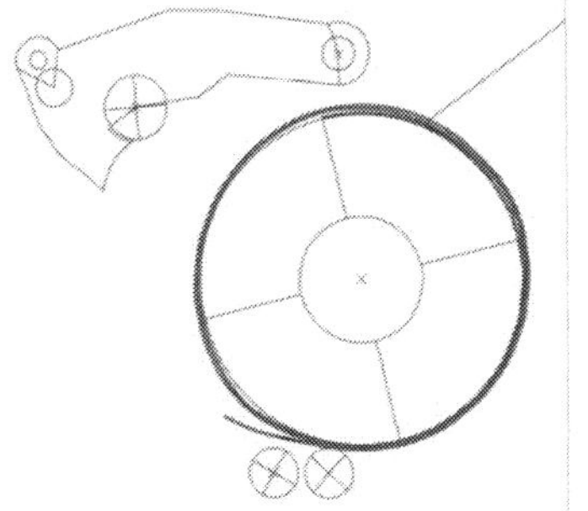

Figure 10. Friction model (i) sticking friction μ=0.15; the coil falls off the support rolls.

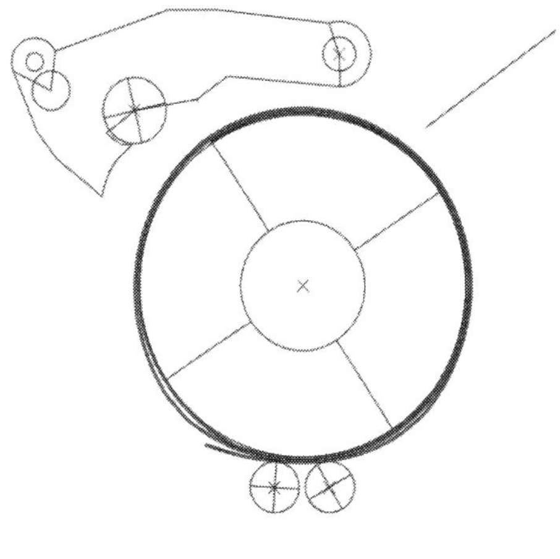

Figure 11. Advanced friction model (ii) combined friction model (sticking and gliding); the coil stays on the support rolls.

Table 3. Sticking friction

	Normal force 30kN	Normal force 80kN
Material 1 (X70)	μ=3	μ=1.5
Material 2 (X100)	μ=2	μ=1.2

A contact pressure of 68MPa is obtained from a normal force of 30kN and a contact pressure of 110MPa from 80kN according to evaluation of the measurement data. The normal force acting on the coil windings increases with the coil diameter and the associated increasing coil weight during the coiling process.

Table 4. Friction model.

	Friction coefficient	Slip rate	Contact pressure (MPa)
Material 1 (X70)	3	0	68
	0.15	1	68
	1.5	0	110
	0.15	1	110

The conclusion of the numerical investigation is that consideration of the detailed information from the experimental friction tests leads to results which are close to reality. Consideration of sticking friction of μ=0.15 only leads to a coil which drops from the support roll unit. Consideration of the experimental friction data delivers a stable coil which is in correspondence to the situations on-site a hot strip mill. Coils of material X70 and a thickness of 12mm do not fall off the cradle rolls in daily operation.

5 Concept for safe coil evacuation and safe coil handling

A mechanical equipment was recently developed, which provides a viable approach in order to assure safe evacuation of the coil as well as safe transport to the binding machine. The so-called "omega-car" (see also Ref. [3] for the according patent application) as shown in Figure 12 is an appropriate device to solve the abovementioned requirements. A securing strap is guided inside a U-profile. The role of the securing strap is to tighten and thus secure the coil. The entire omega shaped frame is mounted on guiding rails and is moved into the downcoiler by means of a winch drive.

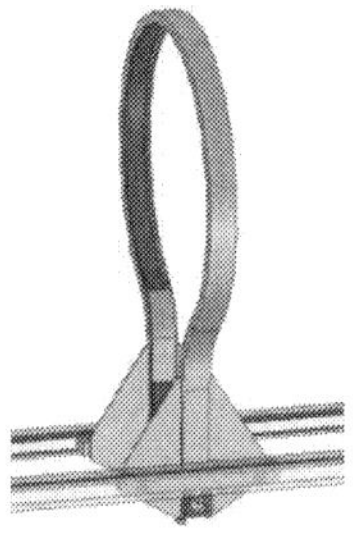

Figure 12. Omega car - frame

Figure 13 shows the omega car, which has already entered the downcoiler and is ready to tighten the finally coiled hot strip.

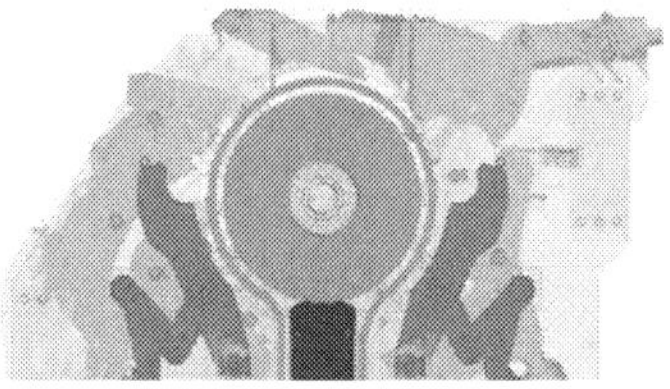

Figure 13. Omega car inside downcoiler – front view

The secured coil is transported to the binding station (see Figure 14). Inside the downcoiler, the omega frame is docked to the coil stripper car (shown in red), which acts from this step on as the master equipment to further transport the coil. Clock-springing is avoided due to the fact that a strap secures the outer windings of the coil.

Figure 14. Omega car at binding station – 3d view

As a first step to realise this concept for practical use, the basic design for a test rig installation has been carried out (see Figure 15).

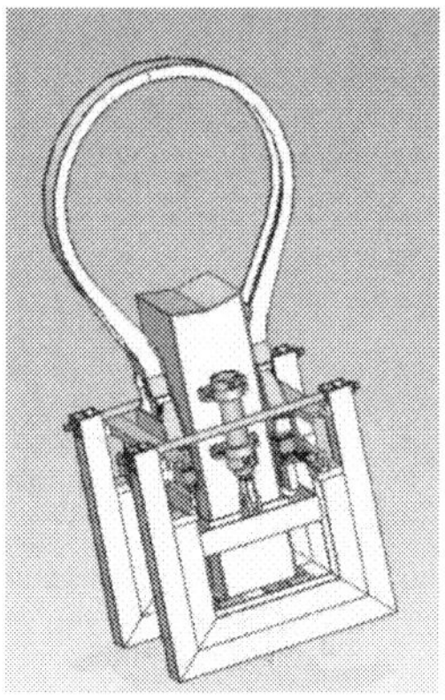

Figure 15. Basic design of test rig installation

Checks on structural integrity of the frame as well as the stability of banding strip were carried out. Also, the deformations of the rollers which connect the frame with the rail system were considered for the analysis. The forces acting on the frame and the support structure as well are shown in Figure 16.

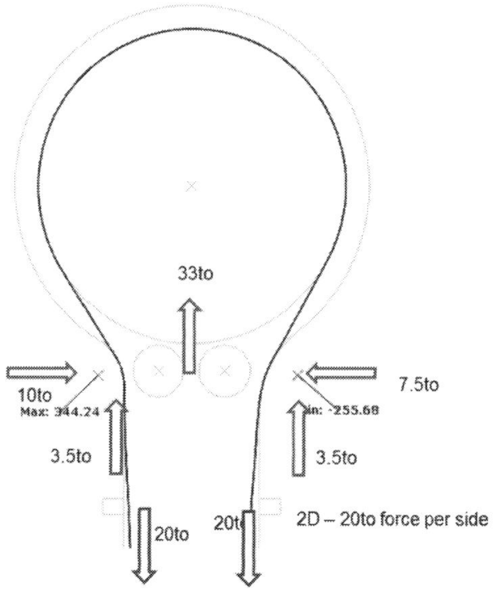

Figure 16. Loads and reaction forces

The appropriate design of the base structure was verified by simulations, and the stability of the securing strip was evaluated by finite element simulations in Abaqus/Standard (see Figures 17 and 18). Table 5 shows the admissible radius R to avoid plastic deformation of strip,

$$R=(E\ t)/(2\ \sigma_y) \qquad (2)$$

where R denots the admissible radius, E the Young's modulus, t the strip thickness and σ_y the yield strength of the material.

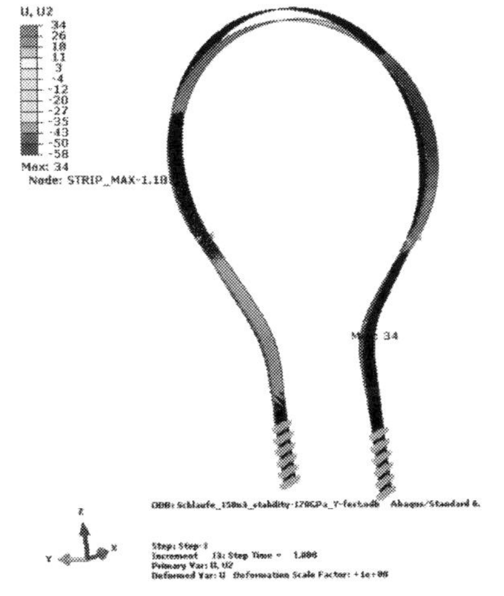

Figure 17. Deformation of the strap

As shown in Figure 17, sufficient stability is given when using a strap width of 150mm and thickness of 3mm. The Young's modulus of hot steel with a temperature around 400°C is E=170GPa.

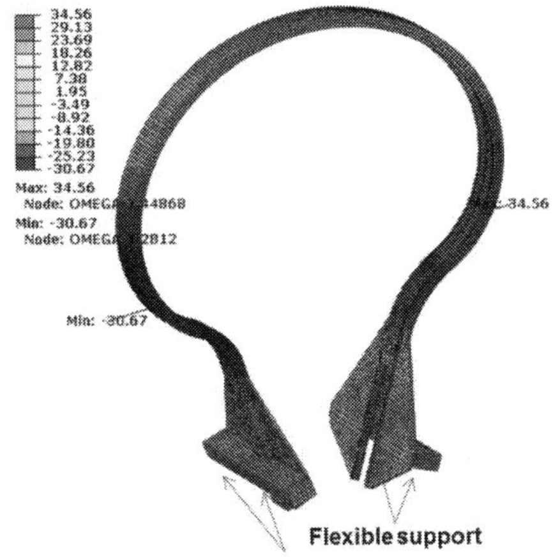

Figure 18. Deformation of the frame

Table 5. Possible dimensions for the binding strap – admissible radius R to avoid plastic deformation of strip

Yield strength σ_y [MPa]	t=2mm	t=3mm	t=4mm
400	525	788	1050
600	350	525	700
800	263	394	525

6 Summary and Outlook

For safe coil evacuation, pre-bending of the strip using a lowered roller table and a 3-roll unit can yield improvements in terms of reduced coil diameter. However, the simulations so far have not achieved total elimination of spring-back. This could be because of simulation-related issues as: (i) the hourglass-control utilised in the simulations which could yield increased stiff elements, (ii) the initial conditions with a core-coil and a rigid connection of strip to core-coil, and (ii) instabilities during the settling steps

In addition, the absolute position of the 3-roll unit and the relative position of the rolls are not optimised. There is further potential to reduce spring-back by using both measures, the lowered roller table and the 3-roll unit, in simulations.

Finally, there is scope to reduce spring-back by including the friction model described in Section 4 in the Abaqus/Explicit model for more realistic friction behaviour of the coil. Such simulations are scheduled for the near future.

The concept of the omega car provides an alternative for safe coil evacuation and transport of the coil. The requirements to this equipments would, however, be reduced when pre-bending of the strip can be achieved by the above mentioned installations in the downcoiler area.

References

1. A. Pauschitz, F. Franek, Austrian competence center for tribology AC2T; www.ac2t.at/ac2t/
2. ABAQUS/CAE V6.13-5, User's manual. Online documentation help: Dassault Systèmes (2013)
3. C. Salzmann, L. Pichler et. al., World patent, Holding together and transporting a metal band rolled up into a coil, WO2015172965A1 (2015)

Investigation of a composite ring rolling process by FEM and experiment

Joachim Seitz[1,a], Gideon Schwich[1], Stefan Guenther[1] and Gerhard Hirt[1]

[1]*IBF-Institute of Metal Forming, RWTH Aachen University, Intzestraße 10, 52056 Aachen, Germany*

Abstract. Roll bonding is a well-known process to produce composite sheet metals. Transferring this principle to ring rolling would allow to produce seamless radial composite rings, which combine the advantages of different material properties. This process is studied, both experimentally and using FEM-simulation. For the FEM an explicit model is applied including the kinematic control algorithms of the radial-axial ring rolling machine. In the used model the occurrence of unexpected asymmetrical joints can be reproduced, which are observed during the rolling experiments. By varying the influencing parameters on the asymmetric joint in the FEM a plausible reason was found.

1 Introduction

Ring rolling is an incremental bulk metal forming process for manufacturing seamless rings of different diameters and cross-sections. Seamless rings are typically used in bearing, aircraft, energy, and automotive industry [1]. The conventional radial-axial ring rolling process holds eight degrees of freedom and reduces the cross-section of a ring within two roll gaps. The radial roll gap consists of a rotatory driven main roll (1) and a non-driven mandrel (2), which moves in the direction of the main roll to reduce the wall thickness. Two rotationally driven conical axial rolls (3, 4) constitute the axial roll gap, which is located opposite to the radial roll gap. The upper axial roll moves vertically downwards to the lower one (5) reducing the ring height. The axial rolls are fixed in the axial frame (6), which can move in radial direction with growing ring diameter. Two guide rolls control the position of the ring (7, 8) and ensure its circularity during the process. The guide rolls are situated at the side of the ring [1]. Figure 1 provides a schematic overview of the radial-axial ring rolling process.

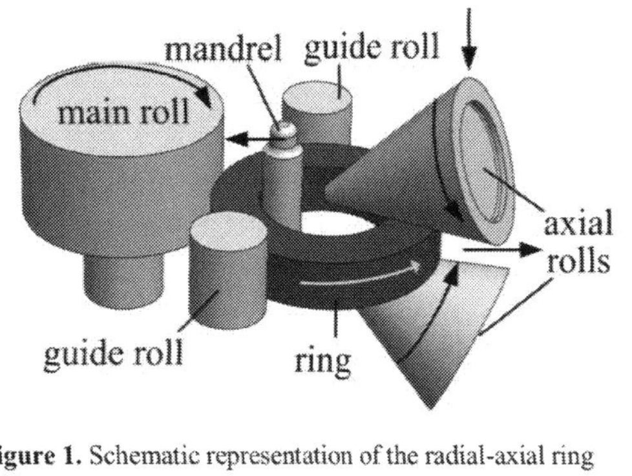

Figure 1. Schematic representation of the radial-axial ring rolling process.

Composite materials provide the opportunity to satisfy locally variable material demands by combining the respective advantages of different materials in one workpiece. This combination of materials enables customized material properties. Through the combination of high quality materials with low-cost materials the component costs can be reduced. Within this investigation metallic seamless composite rings are considered, where the bonding mechanism is characterized by a material bond. While today roll-bonding is a basically established technology for producing planar semi-finished products [2], the manufacturing of composite rings by ring rolling is only poorly studied[3, 4]. In order to enable the manufacture of composite rings by ring rolling, the relevant correlations between process parameters and material properties have to be identified and described by models.

Within this paper, first findings of the material flow are presented based on investigations by ring rolling experiments as well as further investigations using FEM-simulations. Using FEM-models will enable to derive suitable rolling strategies and to identify valid process windows for a reliable and reproducible formation of a material bond by ring rolling.

2 State of the Art

2.1 Metallic composites joined by a material bond

Apart from all kinds of surface coating, the basic concept of metallic composite materials by material bonding can amongst others be realized by roll bonding [5], composite casting, composite forging [6], cladding [7], powder metallurgy [3, 8] and modern additive manufacturing. For thin material bonded coatings, various thermal spraying processes are industrial state of the art. Thermal spraying

[a] Corresponding author: seitz@ibf.rwth-aachen.de

can be divided according to the type of spray additive material, to the production process and according to the type of energy source (e.g. plasma spraying, flame spraying, laser spraying). However, the achievable layer thicknesses are limited to a few millimeters [9]. In contrast, composite materials with relatively thick layers can be achieved by cladding, for which various fusion welding methods are available [7]. In particular for high alloyed high-temperature materials problems like hot cracks, shrinkage cracks and diffusion-related alloy changes are observed. In general, the thermal coating of layers is limited with respect to the coating rate and therefore more suitable for smaller volumes.

Rotationally symmetrical seamless hollow composites can be produced by centrifugal casting processes [10]. However, for this process an accumulation of nonmetallic inclusions and less dense components such as oxides, slag and gases on the inside can occur due to the centrifugal force and the solidification direction. These nonmetallic inclusions generate a harmful segregation layer, which is difficult to remove for the composite casting because the second (inner) layer should be casted immediately after the solidification of the first to achieve a good metallic bond between the layers [11]. In summary, all melt metallurgical processes cause "cast structures" in the broader sense with the associated microstructures and the adverse mechanical properties compared to forming processes.

Various forming or thermal joining processes for rotationally symmetrical parts exist, in which the composite is mainly realized by force fit or material bonding. For example, seamless composite pipes are produced by hydroforming (wall thickness about 10 mm, outer diameter about 150 mm), where the resultant composite is realized by force fit. This process is amongst others used in the chemical industry [12].

The generation of material bonded composites for seamless tubular semi-finished products by forming processes can be carried out using more recent approaches like the simultaneous extrusion of different metals [13] or the joining by metal spinning (so-called "spin-bonding") [14]. Seamless composite rings can also be produced by a process-integrated powder coating during the radial-axial ring rolling process. In this case, the powder substrate is filled into a previously welded capsule space. Here, the process steps compacting, forming and heat treatment are carried out simultaneously during the ring rolling. Depending on the composition of the coating powder, a wear and / or corrosion-resistant layer can be obtained [3]. For this process, the advantage of reducing the necessary production steps is accompanied by the high procedural effort for attaching the necessary capsule space. In addition, for a large ring growth cracks can occur in the coating during the forming process, which significantly limit the process window compared to conventional ring rolling. Despite the compression by rolling, residual porosity is not completely avoidable and the achievable coating thickness is limited.

Roll bonding is probably the most industrially established and quantitatively widespread method for producing material bonded composites [2] and it has many analogies to the composite ring rolling. Roll bonding is a pressure welding process, which requires a close contact and a high pressure of the materials to be joined. Inserting forming energy into the contact area initializes a material bond of the composite layers at the atomic level [5, 15]. Roll bonding is used in cold and hot rolling, wherein the clad can by realized both by reversible and continuous rolling. As described by Bay in [15], roll bonding enables the realization of metal combinations, which cannot be bonded by melting processes due to the formation of intermetallic phases. The application areas of hot roll bonded composites are mainly the chemical industry, where corrosion resistant materials are combined with low-cost construction materials. Thin layered composites are mainly used in the automotive industry, electrical industry and heat technology. Here, material properties such as electrical and thermal conductivity or corrosion resistance of non-ferrous metals are combined with the strength of steel or other alloys [16].

Transferring the concepts of roll bonding to radial-axial ring rolling could basically allow producing seamless composite rings with outer diameters up to 9,500 mm and ring heights up to 4,000 mm with individual weights up to 150 tons [1]. The composite ring rolling has hardly been studied, although all other joining methods are not suitable for these target geometries by forming processes. Therefore, the process is not applied in industry yet.

The composite ring rolling of concentrically arranged rings has been studied as an aspect of a research project in the early 1990s [4]. Figure 2 provides a schematic representation of the composite ring rolling process. As a result of the investigations, it was postulated that the inner ring has to grow faster than the outer ring. This requirement has to be satisfied for the creation and maintenance of a material bond. The flow resistance is the parameter to evaluate when plastic deformation and therefore ring growth occurs. The flow resistance in the circumferential direction depends on the flow stress of the material σ_y as well as on the radial shape factor c_r (ratio of contact length l_D to the wall thickness s of the ring, cf. chapter 4). Consequently, for the use of materials with different strengths, the higher strength material should be located outside.

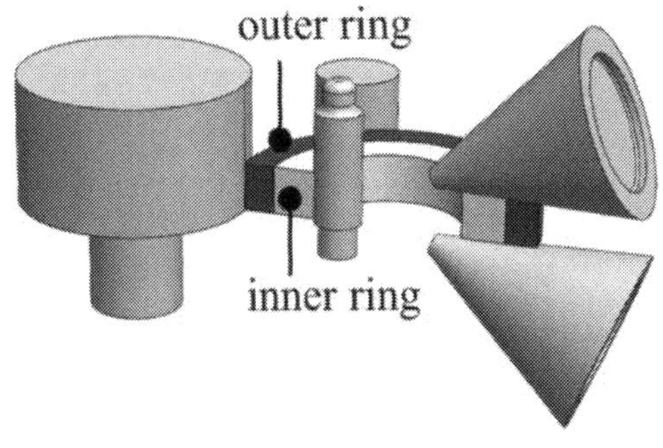

Figure 2. Schematic representation of the composite ring rolling process of concentrically arranged rings.

2.2 FEM-simulation of the ring rolling process

The analysis of local stress and strain states and the thermal conditions is required in order to investigate the elaboration of a material bond in composite ring rolling. Due to the asymmetrical and transient process conditions this can only be achieved by the finite element method.

As a result of the increased computing power and more efficient contact algorithms, today ring rolling can be simulated using FEM with both implicit [17, 18] as well as with explicit solution methods [19]. The FEM-simulation of the radial-axial ring rolling process uses increasingly integrated closed-loop control algorithms for most or all of the at least eight degrees of freedom. The control algorithms of the simulation are based on the process control of real ring rolling machines using virtual sensors to measure current process information. Based on this approach, no additional information from previously conducted rolling experiments are required anymore. In some publications the controlled kinematics of the guide rolls are described in the model [19–22]. Also some approaches for a force control of the guide rolls were realized, using hydraulic cylinders as drives for the centering arms [21]. Successively all eight degrees of freedom have been automated through the implementation of virtual sensors and control algorithms [23]. Since the control algorithms of the ring rolling mills belong to the protected know-how of the supplier, in past investigations of the IBF [24, 25] the control algorithms were implemented in a first step as a black box provided by the machine supplier. In a further step, own fully documented control algorithms were integrated into the FEM-simulation. The existing FEM-simulation model allows ring rolling processes without prior knowledge of the expected kinematics as well as the development of new rolling strategies [24, 25].

3 Methods and procedure

3.1 Experimental equipment

The rolling experiments were conducted on a radial axial ring rolling mill Banning H100/V80 with a maximal radial force of 1000 kN and a maximal axial force of 800 kN. For the conducted experiments the materials 1.7131 (inner ring) and 1.4401 (outer ring) were used. The rings were heated up to 1200°C and held at 1200°C for 90 min. The transport time was approx. 60 s and the process time approx. 15 s.

3.2 Description of the FEM-model

The flow curves of the materials are determined using frictionless isothermal single step compression tests at different strain rates and temperatures for the above mentioned two materials. The determined flow curve data are stored in tabular form depending on the strain rate, the temperature and the equivalent strain. Therefore isotropic hardening as a function of strain, strain rate and temperature is considered. The dependency on strain rate and temperature is covered by linear interpolation between experimental values measured at $d\varepsilon/dt$ = 0,01/s, 0,1/s, 1/s, up to 10/s and temperatures between ϑ = 950°C and 1200°C in steps of 50°C. The additional material data, such as density, Young's modulus, specific heat capacity and thermal conductivity were taken from literature [26]. The simulation uses a thermo-mechanical coupled model with an explicit solver in ABAQUS. Between the tools and the rings, as well as between the rings among each other a "surface to surface" contact condition is defined using a "hard contact" condition in normal direction and a Coulombian friction coefficient of μ = 0.3 in tangential direction. The ring temperature at the beginning of the process is $\vartheta_{0,ring}$ = 1200 °C and the tool temperatures are ϑ_{tool} = 100°C throughout the process. The two rings are meshed each with approx. 7500 hexahedral reduced elements (C3D8RT) using an ALE formulation. The simulation considers gravity and the existence of a rolling table. The conducted simulations use a fully closed-loop controlled model, which is described in detail in [24, 25]. The control loop for the eight degrees of freedom is realized in FORTRAN via an ABAQUS user subroutine called VUAMP. Figure 3 shows the basic structure of the control loop and the FEM-simulation model containing all the used virtual sensors and actuators.

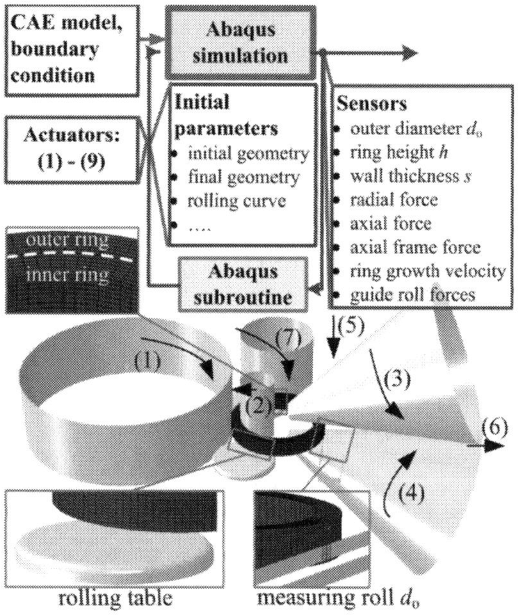

Figure 3. FEM-model of the composite ring rolling process with integrated closed-loop control.

4 Results and Discussion

In order to enable the formation of a permanent material bond by ring rolling the inner ring has to grow faster than the outer ring.

Considering the material, the following condition must be satisfied: The forming resistance of the inner ring close to the mandrel is less than that of the outer ring next to the main roll. To satisfy this condition, a material combination of a case hardening steel 1.7131 (inner ring) and an austenitic stainless steel 1.4401 (outer ring) was examined. The difference of the two flow stresses increases with decreasing temperature and with increasing equivalent strain. Therefore the difference increases during the ring rolling process. Figure 4 provides an exemplary comparison of the flow curves of the inner ring and the outer ring at two different temperatures, which point out the increasing difference with decreasing temperature.

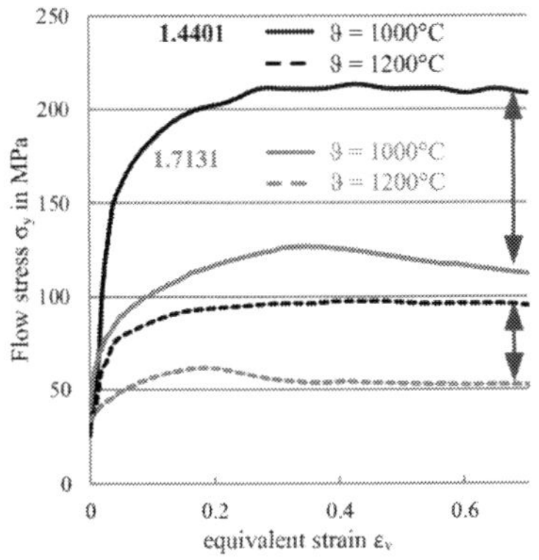

Figure 4. Exemplary comparison of flow curves of the inner ring (1.7131) and the outer ring (1.4401) for a strain rate of $d\varepsilon/dt$ = 1/s at the temperatures ϑ = 1000°C and ϑ = 1200°C.

In addition, the choice of the mandrel diameter and the inner diameter of the preform is restricted based on geometrical considerations. The basic requirement is that the contact length l_M of the main roll has to be at least as large as the contact length of the mandrel l_D, since an increase of the contact length effects a higher ring growth rate. The concept of contact length is transferred from conventional flat rolling, where l_c is a function of the height reduction $\Delta h = (h_1 - h_0)$ and the roll radius r. Figure 5 provides the concept of contact length l_c and shape factor c_s for conventional flat rolling.

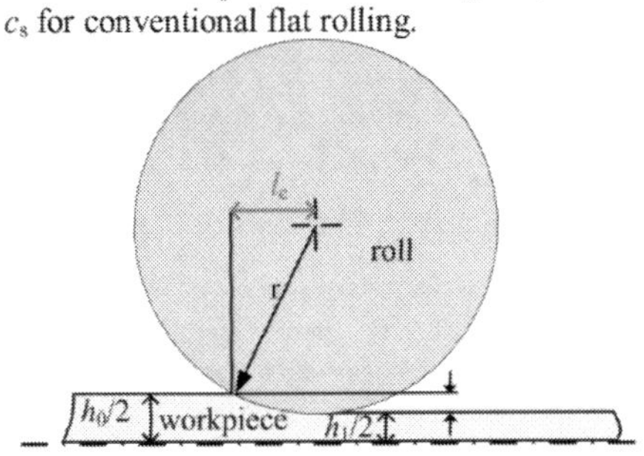

Figure 5. Concept of contact length l_c and shape factor $c_s = l_c/[(h_1+h_0)/2]$ for conventional flat rolling.

In ring rolling, the geometric relationships of radial shape factor and contact length are more complex due to the curvature of the ring, the different diameters of main roll and mandrel as well as the diameter increase of the ring. Therefore, a notional diameter of the main roll d'_M and the mandrel d'_D, which depends on the current ring diameter and the tool geometry can be calculated according to [27] through the entire process.

Figure 6 provides the evolution of the measured outer diameter d_o, the inner diameter d_i and the calculated evolution of the notional diameters of the main roll d'_M and the mandrel d'_D for the performed rolling operations (used mandrel roll diameter d_D = 75 mm, main roll diameter d_M = 748 mm). Since the notional diameter of the mandrel d'_D is reduced with increasing ring outer diameter d_o whereas the notional diameter of the main roll d'_M is increased, it is possible to determine a valid preform based on fixed tool dimensions, which satisfy the geometrical requirement for the entire process.

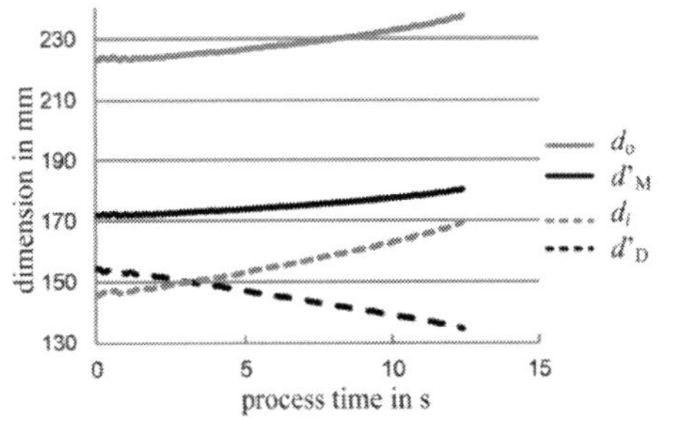

Figure 6. Evolution of the measured diameters d_o and d_i and the notional diameters d'_D and d'_M.

Figure 7 provides valid preform geometries for two different mandrel diameters according to the described geometrical considerations. The area above the dashed line of one color, which corresponds to one mandrel diameter, can be chosen for the inner diameter of the preform. The outer diameter can be determined by adding twice the total wall thickness to the inner diameter. In this Figure, the wall thicknesses of the inner and outer ring are equal.

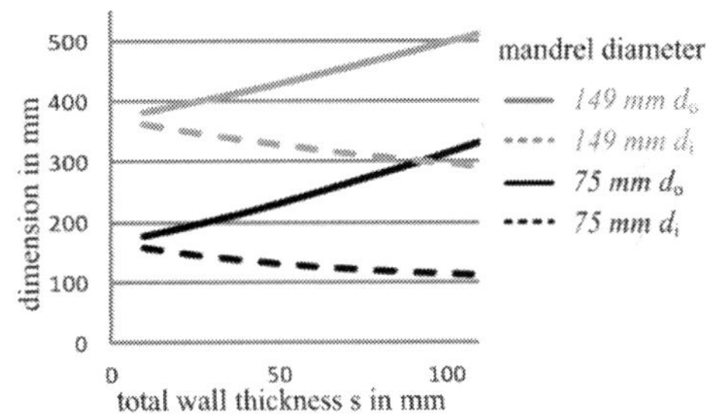

Figure 7. Process window for the choice of the preform geometry using different mandrel diameters.

Based on the above mentioned considerations, the following preform geometries were selected, in order to induce a higher ring growth of the inner ring compared to the outer ring.

Table 1. Chosen preform and final geometries

d_D = 75mm	preform outside	preform inside	final geometry
d_o in mm	217	177	242
d_i in mm	177	136	174
h in mm	39	39	39

DOI: 10.1051/matecconf/20168015011

4.1 Results of rolling experiments and FEM-simulations

The results of the first rolling experiments and FEM-simulations using a mandrel diameter of $d_D = 75$ mm show that an excessive ring growth of the inner ring can be harmful. In this case, the inner ring grows too fast compared to the outer ring, so the inner ring outgrows the outer one. In consequence, the two rings are separated. Figure 8 shows the final geometries of the first rolling experiments in reality and FEM-simulation. This first result already shows that a reproducible successful control of composite ring rolling requires additional process boundaries to the above mentioned considerations, which will be examined within the running research.

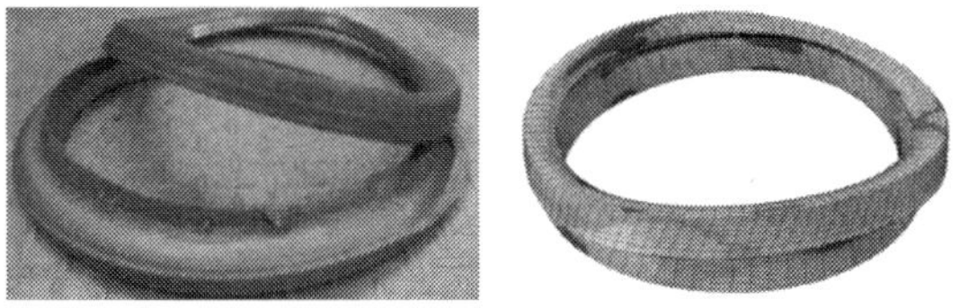

Figure 8. Final geometries in case, the inner ring grows too fast and outgrows the outer ring. left: rolling experiment right: FEM-simulation.

In the subsequent rolling experiments the inner diameter of the inner ring has been reduced and a larger mandrel diameter of d_D = 120 mm has been selected, which increases the notional diameter of the mandrel d'_D. This leads to a smaller ring growth of the inner ring.

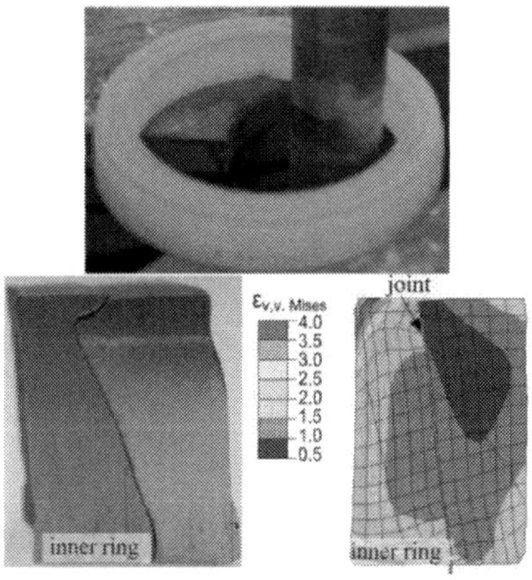

Figure 9. Composite ring after rolling. left: cross section of the rolled ring, right: Cross-section of the rolled ring in FEM.

According to a first examination of the rolled ring, which is provided in Figure 9, it can be assumed that the flow behavior satisfies all the requirements for an adequate material bond. However, a radial cut of the rings shows that the material flow of the inner ring and the outer ring is not uniform in axial direction. This phenomenon can also be reproduced in the FEM-simulation (see Figure 9). In addition, the resulting composite has no sufficient bonding and can be separated manually. In order to understand the reason for asymmetrical joints and the influence of process parameters, FEM-simulations have been conducted.

4.2 Influence of the ratio between wall thickness and ring height

It is known from flat rolling that the tendency of spreading increases with increasing shape factor c_s (ratio of contact length to height of the workpiece) and with decreasing width-to-height ratio b_m / h_m. In order to investigate the influence of the width-to-height ratio b_m / h_m in the radial roll gap, the height of the rings was varied using otherwise unchanged rolling conditions. Transferring the conditions of flat rolling to the radial roll gap of the ring rolling process, the ring height corresponds to the width of the workpiece b_m (the dimension perpendicular to the roll feed). In all ring rolling processes the ring height was kept constant by the axial rolls during the process. Figure 10 shows the final cross-sectional geometry for three different initial and final ring heights. The lower the ratio of height to wall thickness of the ring, the more asymmetric the joint becomes.

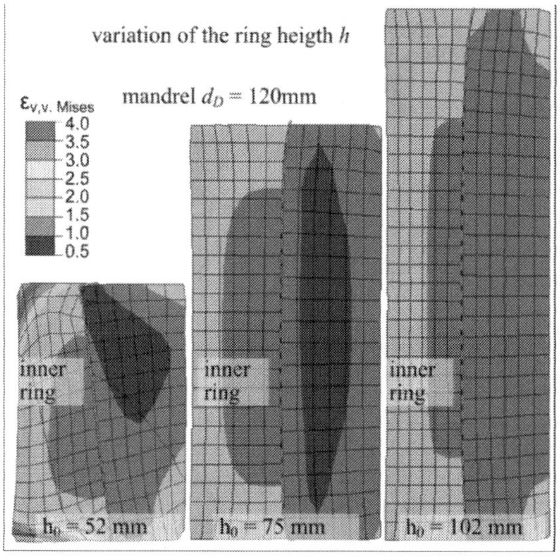

Figure 10. Final cross-sections for different ring heights h.

4.3 Influence of the flow stress ratio between inner and outer ring and the mandrel diameter

In order to investigate the influence of the flow stress ratio between inner and outer ring on the asymmetric joint, notional flow curves have been used for the inner ring. These flow curves were generated using a scaled flow curve of the outer ring (1.4401) by factor 100%, 90%, 80% and 70%. This flow curve variation was performed for two different mandrel diameters of d_D = 75 mm and d_D = 120 mm. Figure 11 provides the cross-sections at the end of the rolling process. The smaller the notional diameter of the mandrel d'_D and the lower the flow stress, the earlier the inner ring starts growing plastically compared to the outer ring and the more asymmetric is the formed joint.

 DOI: 10.1051/matecconf/20168015011

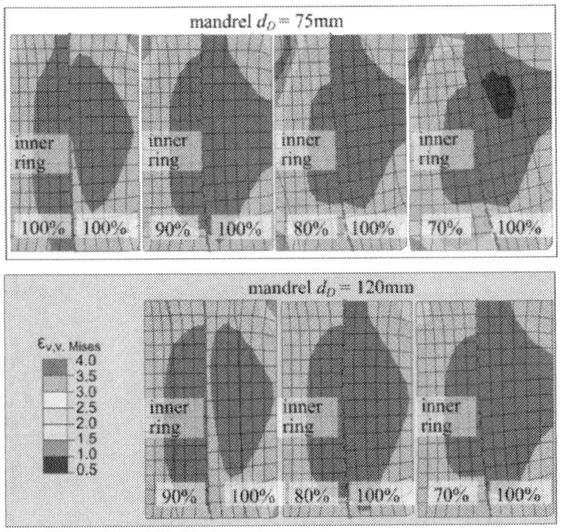

Figure 11. Comparison of the final cross-sections for the variation of the flow stress of the inner ring relative to the flow stress of the outer ring (precentage values) at the example of two mandrel roll diameters.

4.4 Influence of the rolling table and the gravity

After having determined how the occurring asymmetrical joints may be affected, the reason for its occurrence is still unknown. In order to find the reason, successively the conditions, which act in different ways up and downwards, were excluded keeping all other conditions constant. Random process disturbances or different tool stiffnesses of the axial rolls by differences in the guiding of the axial slide can be excluded, since the tools are modelled in FEM as rigid bodies and the asymmetrical joint can be reproduced by the FEM-model. The two remaining asymmetric boundary conditions are gravity and the rolling table. In Figure 12, the final cross-sections of three FEM-simulations are shown, in which the above mentioned influencing parameters were investigated. Accordingly, the rolling table is the reason for the asymmetrical joint.

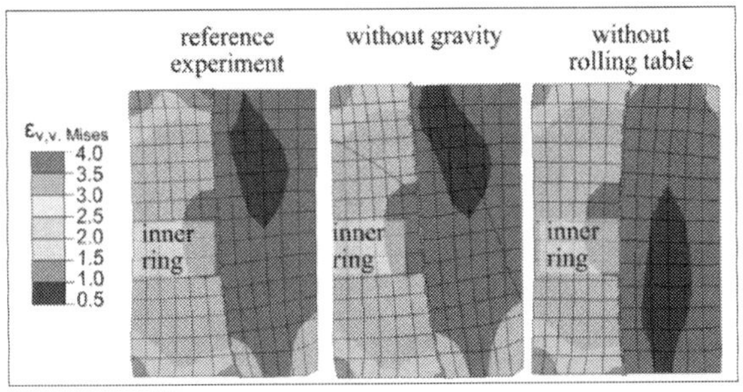

Figure 12. left: final cross-section of the reference experiment, taking into account gravity and the rolling table, center: without considering gravity with rolling table, right: without rolling table with gravity.

5 Conclusion and Outlook

Composite materials provide the opportunity to realize customized material properties by combining the advantages of different materials in one workpiece. The producible layer thicknesses in composite ring rolling are not limited unlike the industrially established processes for seamless composite rings. The composite ring rolling process allows joining and forming in only one process.

In this paper a FEM-model for the composite ring rolling using simple contact algorithms was developed. This model provides important information on the stress and strain state, i.e. the radial and tangential material flow. The results of the investigations show that simple analytical considerations are not sufficient for the process layout and a reproducible realization of composite ring rolling. Even composite rolling experiments, which seem to be successful according to a first consideration, not necessarily meet the required demands on a uniform material distribution or a specific joint strength. Notably, the following effects are observed:

- The inner ring may outgrow from the outer ring
- No sufficient material bond
- Occuring of an asymmetric joint, which can be influenced by:
  - The chosen material (flow stress) combination of the rings
  - The ratio of wall thickness to ring height
  - The existence of a rolling table

The realized investigations generate first relationships between preform, tool geometries and the expected ring growths. Nevertheless, no criteria are known yet for describing these process boundaries qualitatively or even quantitatively. For a reproducible material bond with thick wall thicknesses and large diameters by composite ring rolling a detailed understanding of all influencing parameters of the material flow and the bond formation is required.

The focus of future investigations will therefore concentrate on modelling the resulting joint strength in the FEM. For this purpose it will be necessary to extend the ring rolling model to a contact formulation that considers the material bond creation.

Acknowledgement

The authors would like to thank the "Deutsche Forschungsgemeinschaft" (DFG) for the financial support of these works within the project HI 790/54-1 "Verbundringwalzen - Methode zur Herstellung nahtloser Verbundringe auf einem Radial-Axial Ringwalzwerk".

References

1. U. Koppers, D. Michl, in Handbuch Umformen, ed. by H. Hoffmann, R. Neugebauer, G. Spur (Hanser München, 2012), p. 187
2. Roll-bonded clad plates: our cost-efficient answer to corrosion (Linz) (https://www.voestalpine.com/division_stahl/conten

t/download/4248/34321/file/Folder_Roll-bonded-clad-plates_E_291014.pdf)
3. R. Kebriaei, J. Frischkorn, S. Reese, T. Husmann, H. Meier, H. Moll, W. Theisen, J. Mater. Process. Technol. 213, 2015 (2013)
4. A. Kluge, H. Wiegels, R. Kopp, Stahl und Eisen, 37 (1995)
5. R. Jamaati, M.R. Toroghinejad, Mater. Sci. Technol. 27, 1101 (2011)
6. B.-A. Behrens, K.-G. Kosch, Mat.-wiss. u. Werkstofftech. 42, 973 (2011)
7. H.J. Fahrenwaldt, H. Schuler, J. Twrdek, Praxiswissen Schweißtechnik - Werkstoffe, Prozesse, Fertigung (Springer Vieweg, Wiesbaden, 2014)
8. H.V. Atkinson, S. Davies, Metall. and Mat. Trans. A 31, 2981 (2000)
9. L. Pawlowski, The science and engineering of thermal spray coatings (Wiley, Chichester, England, Hoboken, NJ, 2008)
10. M. Eitschberger, C. Körner, R.F. Singer, in Verbundwerkstoffe und Werkstoffverbunde, ed. by B. Wielage, G. Leonhardt (Deutsche Gesellschaft für Matrialkunde e.V. Weinheim, 2001), p. 114
11. P. Kapranos, C. Carney, A. Pola, M. Jolly, in Comprehensive Materials Processing (Elsevier, 2014), p. 39
12. X. Wang, P. Li, R. Wang, Int. J. Mach. Tool Manu. 45, 373 (2005)
13. L. Madej, H. Paul, L. Trebacz, W. Wajda, M. Pietrzyk, CIRP Annals - Manufacturing Technology 61, 235 (2012)
14. M.S. Mohebbi, A. Akbarzadeh, J. Mater. Process. Technol. 210, 510 (2010)
15. N. Bay, H. Bjerregaard, S.B. Petersen, C. dos Santos, J. Mater. Process. Technol. 45, 1 (1994)
16. Plattiertes Stahlblech: Merkblatt 383
17. M. Ramadan, L. Fourment, H. Digonnet, Int J Mater Form 2, 581 (2009)
18. H. Schafstall, C. Barth, in Proceedings of the 13th International Conference on Metal Forming, ed. by Steel research international (Stahleisen GmbH Düsseldorf, 2010), p. 202
19. M. Wang, H. Yang, Z. Sun, L. Guo, X. Ou, Trans. Nonferrous Met. Soc. China 16, 1274 (2006)
20. Z.W. Wang, S.Q. Zeng, X.H. Yang, C. Cheng, J. Mater. Process. Technol. 182, 374 (2007)
21. L. Li, H. Yang, L. Guo, Z. Sun, J. Mater. Process. Technol. 1-3, 205 (2008)
22. L. Guo, H. Yang, M. Zhan, H. Li, L. Li, Materials Science Forum, 471-764 (2004), p. 760-764
23. Z. Wang, J. Fan, D. Hu, C. Tang, C. Tsui, Int. J. Mech. Sci., 52, (2010), p. 1325-1333
24. V. Jenkouk, G. Hirt, M. Franzke, T. Zhang, CIRP Annals - Manufacturing Technology 61, 1 (2012)
25. V. Jenkouk, G. Hirt, J. Seitz, AIP Conference Proceedings 1532, 695 (2013)
26. M. Spittel, T. Spittel., Materials, Landolt-Boernstein - Group VIII Advanced Materials and Technologies Volume 2, 98/162 (2009)
27. U. Koppers, Geometrie, Kinematik und Statik beim Walzen von Ringen mit Rechteckquerschnitten. Dissertation (Aachen, RWTH) (1987)

A numerical study of multi-pass design based on Bezier curve in conventional spinning of spherical components

Tian Gan[1], Qingshuai Kong[1], Zhongqi Yu[1], Yixi Zhao[1] and Xinmin Lai[1]

[1]*Shanghai Jiao Tong University, China*

Abstract. Bezier curve has been widely used as the roller path in conventional spinning. In this paper, Finite Element model with parameterized 2-pass conventional spinning tool paths based on quadratic Bezier curve is developed to explore the influence on the wall thickness reduction of spherical components. The results show that the first path has significant effect on wall thickness distribution and the selection of parameter values in tool path is crucial for thickness distribution.

1 Introduction

Conventional spinning is defined as the process with a constant thickness but a changeable diameter. Its inherent advantages, such as high quality surface finish, low forming forces and process flexibility [1], make it widely used in automotive and aerospace industries. Single one-pass conventional spinning, limited by forming limits, tends to wrinkles formation and fracture occurrence. In the view of Runge [2] and Lange [3] works, multi-pass should be performed in conventional spinning in order to form the blank without defects. In multi-pass conventional spinning, the shape of tool path in each pass plays an important role in achieving the targeted geometries.

Many studies have been carried out on roller tool path design. Various tool paths such as linear, concave and convex have been analysed. Hayama et al. [4-6] concluded that concave tool paths give the best results and involute paths were recommended. Kang et al. [7] experimentally analysed the above three types of tool paths and stated that the deformation in the first pass has a decisive effect on the wall thickness distribution of the product when the original blank is a plate. Wang and Long [8] numerically analysed the effects of four different roller paths on tool forces, wall thickness and stress distribution in conventional spinning. They have concluded that concave roller path resulted in higher reduction of wall thickness. Li et al. [9] developed parameterized roller paths with three times Bezier curve during the first pass of die-less spinning, and observed that with the increase of curvature, the location of severe thinning area moves backwards from the middle part of the blank to the end part. James et al. [10] performed a set of experiments to investigate how tool force and part geometry evolved with the change of parameters of quadratic Bezier curve during the first pass.

The published studies give an insight into tool path design in conventional spinning. However, in the previous works, roller paths design were almost focused on one-pass conventional spinning, and limited investigations were carried out on multi-pass processes. Previous works have already demonstrated that concave tool paths gave better results. In this study, Bezier curve is simply applied during the deformation of the flange. Therefore, thickness distributions of spherical components are investigated based on two-pass conventional spinning simulations with concave quadratic Bezier curve in forward pass.

2 Methodology

2.1 Roller paths parameterization

Control points of the roller paths are defined at the center of roller nose. The tool path is parameterized referring to the study by James et al. [10]. As Figure 2 shows, each forward pass consists of two parts: one is circular arc, which keeps constant clearance between mandrel and roller, the other is quadratic Bezier curve. The definition of the quadratic Bezier curve is:

$$P(t)=\sum_{i=1}^{2}P_iB_{i,2}(t)=(1-t)^2P_0+2t(1-t)P_1+t^2P_2 \quad t\in[0,1] \quad (1)$$

where P_i are control points. To facilitate the generation of Bezier curve in Matlab, the matrix form of Bezier curve is used:

$$P(t)=[t^2 \quad t \quad 1]\begin{bmatrix}1 & -2 & 1\\ -2 & 2 & 0\\ 1 & 0 & 0\end{bmatrix}\begin{bmatrix}P_0\\ P_1\\ P_2\end{bmatrix} \qquad t\in[0,1] \quad (2)$$

[a] Corresponding author: gantian426833@sjtu.edu.cn

The quadratic Bezier curve (Figure 1) is determined by three control points, which constitute the characteristic polygon.

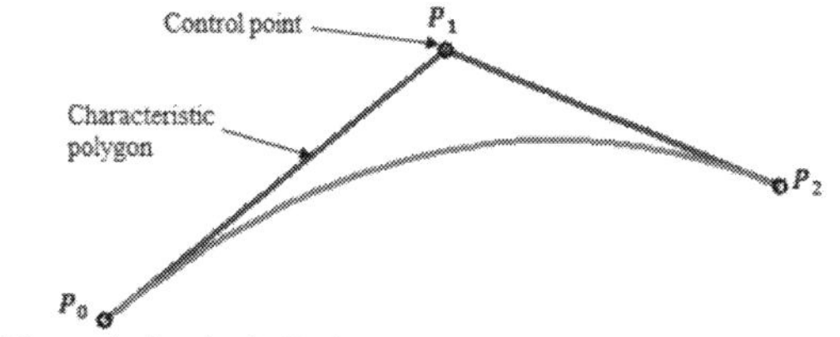

Figure 1. Quadratic Bezier curve.

The Bezier curves used in this study are parameterized in Figure 2. The starting point P_0 is defined by the end of the constant clearance part of the previous pass and the angle θ_0, which is the deformation of adjacent paths. The ending point P_2 is determined by the axial coordinate Z_2 measured from the ending point of the previous pass to the end of this pass. Control point P_1 is defined relative to the mid-point of line P_0P_2, the mid-point moves with U_d in the direction of $\mathbf{P_0P_2}$, then moves with U_n in the direction perpendicular to $\mathbf{P_0P_2}$. The tool path between path 1 and path 2 consists of two straight lines and an arc. The roller does not contact with the workpiece during this process, therefore the control points can be set at any appropriate places.

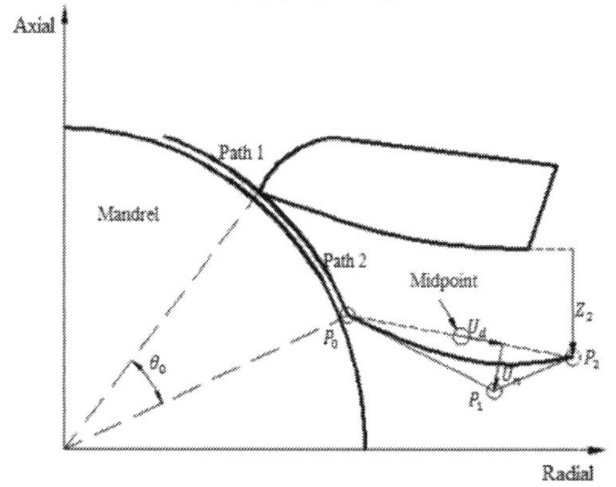

Figure 2. Toopath parameterization.

2.2 Setting of basic parameters

The material used in this study is 2024 aluminium alloy, the dimension of the blank is a diameter of 200 mm with constant thickness of 1.8 mm. A spinning speed of 200r/min and a feed rate of 200mm/min are used in both experiment and simulation. The angle between the axis of the mandrel and the axis of the roller is 40 degree. The experiments are all conducted on the horizontal spinning machine. A schematic diagram of spinning experiments is shown in Figure 3.

2.3 Finite element analysis

Finite Element models are developed by using commercial FE software Abaqus/Explicit. Due to the large strains and complex contact conditions during this process, the dynamic explicit solver is chosen to analyse the forming process. The roller, back plate and mandrel are modelled as 3D analytical rigid bodies. The blank, the only deformable component, is meshed with 8-node 3D reduced integration continuum shell element. The material properties of the blank are calculated by the mass density as 2780 kg/m^3, elastic modulus as 71000 MPa, and the Poisson's ratio as 0.33. A friction coefficient of 0.5 is set up between the blank and the back plate, 0.02 between the blank and the roller, and 0.2 between the blank and the mandrel. To assure the blank is clamped between the mandrel and the back plate a compressive force of 60 KN is set above the back plate. A mass scaling factor of 25 is used in all models to speed up the Finite Element analysis. The Finite Element model is shown in Figure 4.

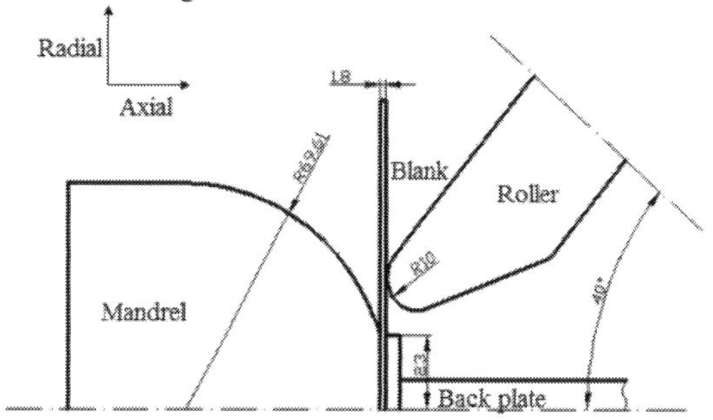

Figure 3. Schematic diagram of the spinning experiment (unit: mm)

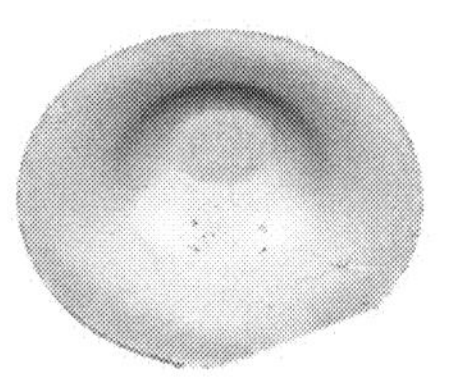

Figure 4. Finite Element model. **Figure 5.** Data points of experiment

3 Results and discussion

3.1. Verification of the Finite Element model

The simulation models are verified by comparing with experimental data. A conventional spinning experiment with single pass to 45 degree is applied. A laser scanning 3D measurement system FARO is used to obtain the coordinates of points at the inner and outer surface of the experiment component, then with the aid of the software Geomagic Qualify the wall thickness distribution is analysed. The data points scanned by FARO are shown in Figure 5. The wall thickness distributions of three Finite Element models with different grid densities of the blank are studied as shown in Figures 6a-6c. The grid densities of model (a), (b) and (c) are 1936, 4956 and 9000 elements. At the same time, the tool paths used in Finite Element models are consistent with the experiments. Figure 6 shows the wall thickness distribution of FE analysis.

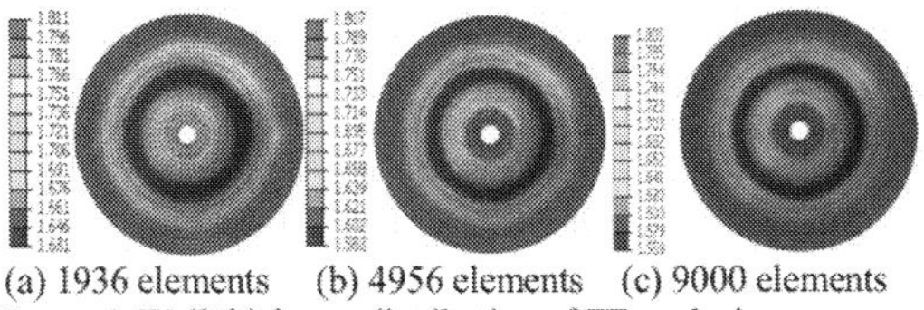

(a) 1936 elements (b) 4956 elements (c) 9000 elements
Figure 6. Wall thickness distribution of FE analysis.

The wall thickness reductions obtained from the FE analysis models are compared with the result of the experimental sample, as shown in Figure 7. The deviation of the former region is smaller than that of the latter region, and as the grid density increases, the deviation becomes smaller. In order to keep the experiment component fixed during the measurment process, the bottom of the component need to be glued to the holder, thus the wall thickness deviation increases obviously during the radial distance of 0mm to 20mm. The maximum error of the wall thickness reduction between the experimental and simulated values is about 4%. Deviations within 5% are considered acceptable. Table 1 compares the maximum wall thickness reduction of the experimental part with the simulation results. The result of model (c) shows a good agreement with experimental data.

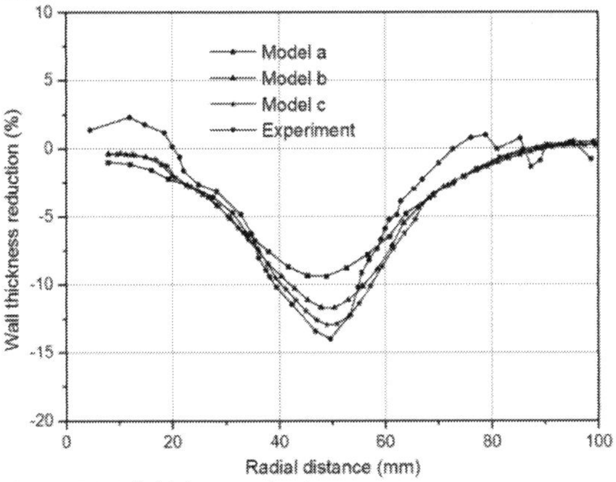

Figure 7. Wall thickness reduction

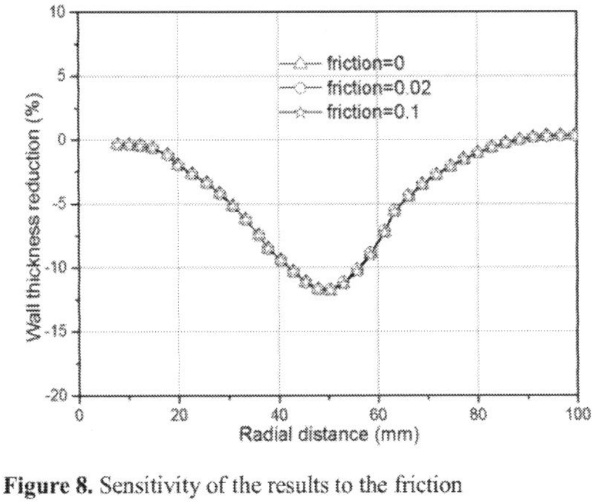

Figure 8. Sensitivity of the results to the friction

In the works of Xia [12], which have investigated the forming forces of tubes spinning using the material of 6061 aluminium alloy, the friction coefficient between the roller and the workpiece is set as 0.1. The sensitivity of the wall thickness reduction to the friction between the roller and the workpiece is necessary to be analysed. Three FE models with different friction coefficients of 0, 0.02 and 0.1 are established. As revealed in Figure 8, wall thickness reduction curves with various friction coefficients are almost completely coincide. The friction coefficients of 0.02 used in this study is appropriate.

Table 1. Comparison of FE and experimental results

	Model a	Model b	Model b	Experiment
Max thinning	9.37%	11.69%	12.97%	13.97%

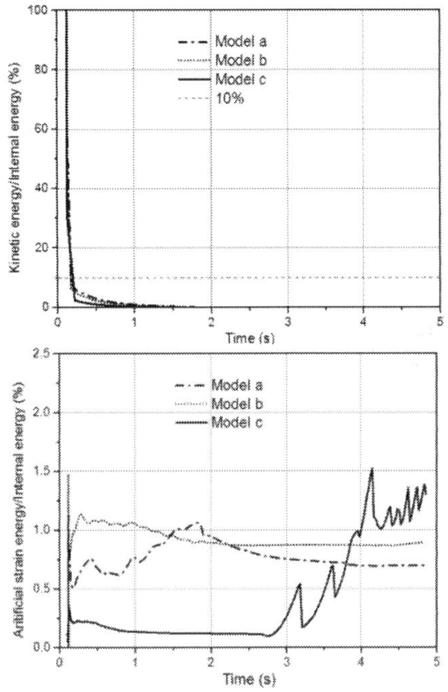

Figure 9. Evaluation of energy histories

The energy histories of the FE models are studied to evaluate the simulation results. To ensure that the simulation results are not significantly affected by mass scaling, the ratio of kinetic energy to the internal energy of the blank should not be greater than 10% during most time of the process. Moreover, the artificial strain energy should also be less than 5% of the internal energy to avoid the hourglass problem, which causes the shear locking phenomenon in the reduced integration linear elements and reduces simulation accuracy. As shown in Figure 9, at the beginning of the process, because of the rotational acceleration of the blank at the beginning of the process, the ratio of the kinematic energy to internal energy is much larger than 10% at that time. However, during most of the process this ratio is below 10%. The ratio of the artificial strain energy to the internal energy is below 5% throughout the process. Therefore, the dynamic problem caused by mass scaling and the hourglass problem are well controlled. Considering the accuracy of the simulation, model (c) is used for the following Finite Element analysis.

3.2 Analysis of the first pass

The deformation in the first pass has a significant effect on the wall thickness distribution of the product when the original blank is a plate [7], meanwhile a wide ring-shaped contact region between blank and mandrel can avoid failures such as wrinkling or fracture. Experiments carried out by Quigley and Monaghan [11] indicated that there was some degree of shear forming involved in the first roller pass of conventional spinning. Therefore, it is certain that clearance exists between mandrel and the inner face of the component. In order to reduce the wall thickness variation of the first pass, three different shapes of the first pass with the original thickness of 1.8mm are analysed: (i) conventional spinning to 35 degree with the clearance of 1.8 mm; (ii) conventional spinning to 35 degree with the clearance of 1.65 mm; (iii) shear spinning to 35 degree. Figure 10 presents the thickness reduction of the models, and Figure 11 shows the comparison of clearance. Table 2 is the maximum thickness reduction of the three different first passes. Concluding from the figures that conventional spinning with the clearance of 1.65 mm has the smallest clearance and the thickness reduction is less than shear spinning. The conventional spinning to 35 degree with the clearance of 1.65 mm is selected as the first pass.

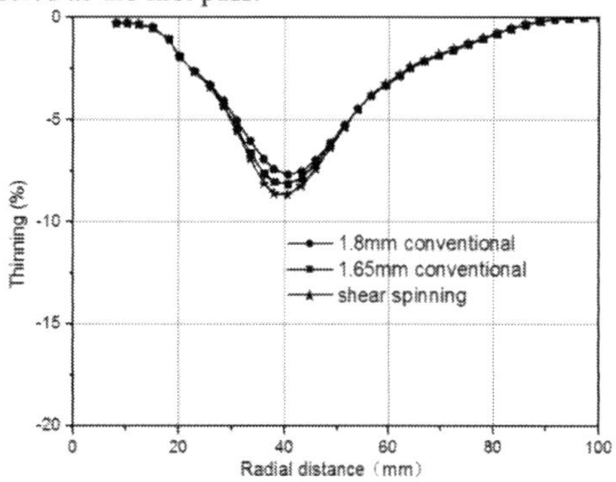

Figure 10. Wall thickness reduction of first pass

Table 2. Maximum thinning comparison

	1.8 conventional	1.65 conventional	shear spinning
Max thinning	7.72%	8.20%	8.72%

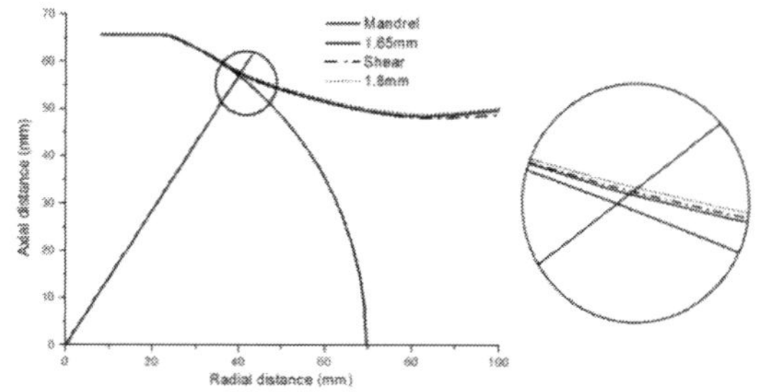

Figure 11. Clearance comparison of first pass

3.3 Analysis of the second pass

The thickness reduction varies with different parameters. Figure 12 shows the thinning along the blank of various passes. The result demonstrates that for the second pass, increasing parameters (θ_0, Z_2, U_n) leads to increased thinning, particularly in the middle part of the blank. Li Y [9] and James[10] gave the similar conclusion about the thinning phenomenon in die-less spinning. Parameter θ_0 is the deformation of adjacent paths, Figure 13 (a) shows the different roller paths with the change of θ_0. The radial distance of starting points of Bezier curve, as shown in Figure 13 (a), is between 55 and 60. Figure 13 (a) and Figure 12 (a) illustrate that θ_0 affects not only the thickness of the circular arc area but also the thickness of flange, and the thinning is highly sensitive to parameter θ_0.

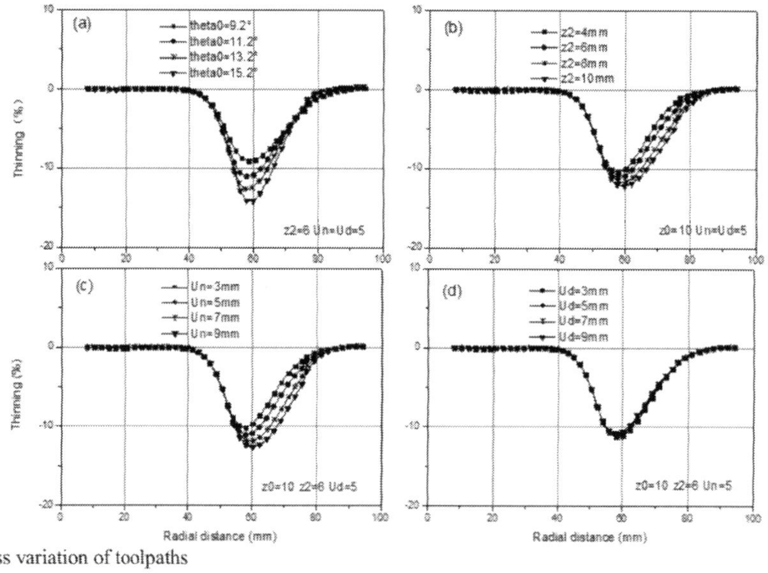

Figure 12. Thickness variation of toolpaths

DOI: 10.1051/matecconf/20168015012

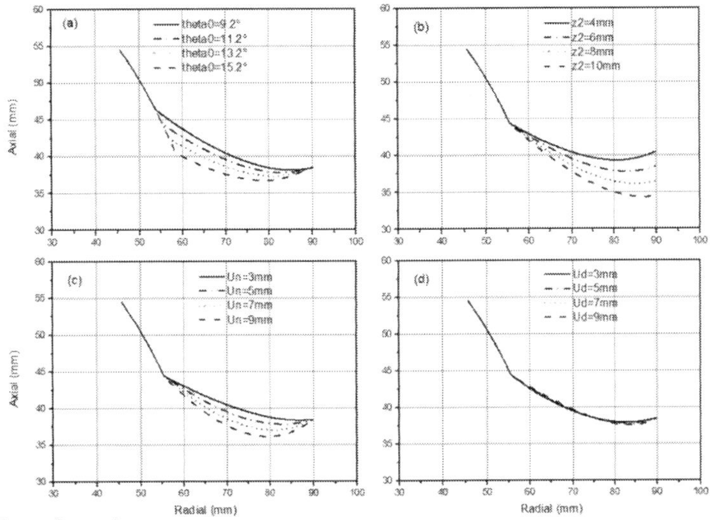

Figure 13. Shape of different forward passes

The sloping degree of the flange is defined by Z_2, while the increasing of U_n leads to a more concave toolpath. All the starting points of the Bezier curves shown in Figure 13 (b) and (c) have the same radial coordinate of 55.4. As shown in Figure 12 (b), the increasing of Z_2 only affects the thickness of the flange, and U_n has the similar effects on thinning. However, the wall thickness reduction is insensitive to U_d. As revealed in Figure 13 (d), with the change of U_d, the shape of toolpaths have very little variations. We can conclude that the shape of tool path, which results in thinning, is insensitive to parameter U_d.

Considering the wall thickness variation, this study tries to find a more satisfying path, which has a well-proportioned distribution of thickness, to conduct the following analysis. The expected path is chosen from the sample points selected by the design of experiments method. This paper presents the Latin hypercube as a sampling strategy. Latin hypercube sampling (LHS) [13] is a stratified random procedure that provides an efficient way of sampling variables from their multivariate distributions. It provides a full coverage of the range of each variable by maximally stratifying the marginal distribution. In this section, the abovementioned 4 parameters are chosen as variables, the range of each variable is equally divided into 15 intervals, then 15 sample points are selected. The sample points' distribution is shown in Figure 14, and the *X* and *Y* axes represent Z_2 and U_d respectively.

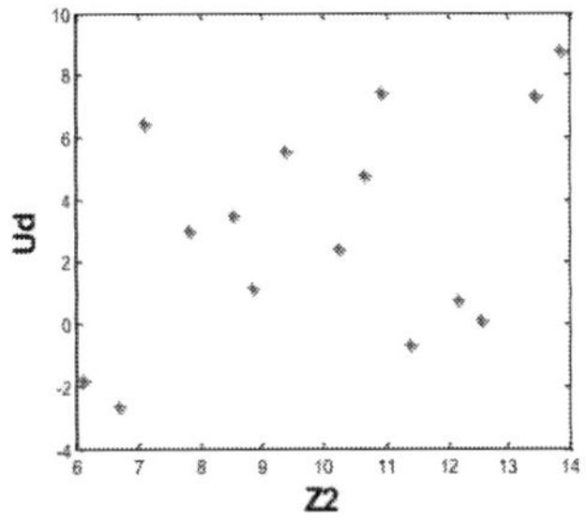

Figure 14. Sample points' distribution

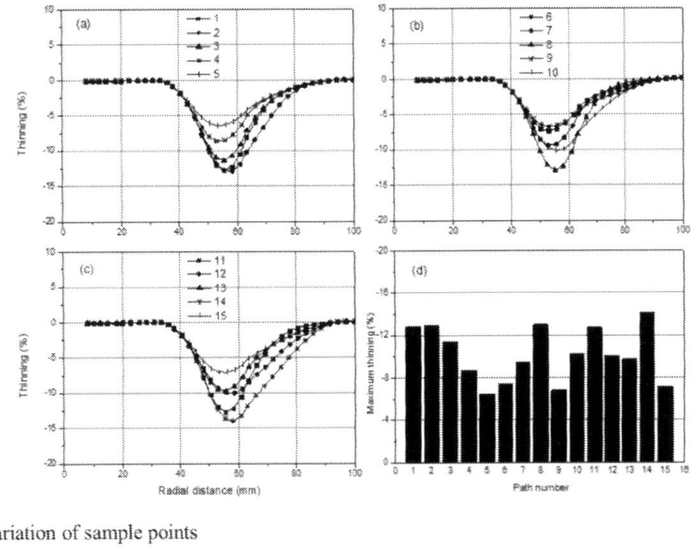

Figure 15. Thickness variation of sample points

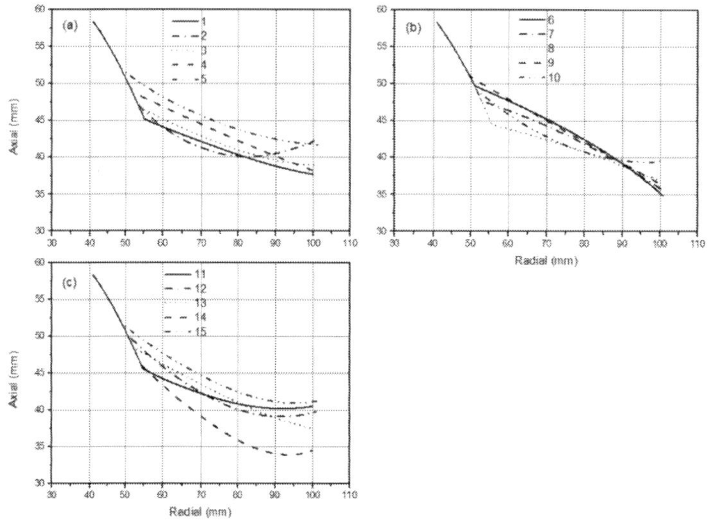

Figure 16. Roller paths of different sample points

As can be seen in Figure 15 (d), in terms of the maximum wall thickness reduction path 5 has the best result, but path 14 has the worst result. Analyzing Figure 15 we summarize that path 5, 9 and 15 get the thinning of below 8%, while path 1, 2, 8, 11 and 14 get the thinning of above 12%. Comparing different shape of roller paths in Figure 16 and Table 3, we find that path 5, 9 and 15 all have small values of θ_0, in addition, path 1, 8, 11 and 14 have large values of θ_0. For path 2, a large value of U_n induces the serious thinning of wall thickness, although the value of θ_0 is relatively small. Path 5 gets the best result due to the smallest value of θ_0 and a small value of U_n. Moreover, a large value of θ_0 and a very large value of U_n result in the worst wall thickness reduction of path 14. The above results of tool paths indicate that the deformation of adjacent paths (θ_0) has a crucial influence on thinning, and another important factor is U_n, the increasing of which will lead to a more aggressive toolpath.

Table 3. Sample points of LHS (unit: mm)

	θ_0	Z_2	U_n	U_d
1	15.61°	10.66	1.39	4.77
2	13.56°	6.12	8.72	-1.83
3	13.93°	9.41	2.71	5.53
4	11.66°	10.26	0.79	2.41
5	8.08°	6.69	4.33	-2.64
6	10.67°	13.46	-2.67	7.31
7	12.96°	12.57	-1.98	0.09
8	16.25°	11.40	-0.88	-0.68
9	9.30°	12.17	0.19	0.75
10	11.25°	8.86	5.65	1.15
11	15.17°	7.82	3.79	2.99
12	9.98°	8.55	8.18	3.51
13	12.00°	10.92	2.16	7.42
14	14.61°	13.85	7.18	8.74
15	8.36°	7.11	6.04	6.42

4 Conclusions

Based on the Finite Element model with parameterized 2-pass conventional spinning process, the effects of tool path with quadratic Bezier curve on the thickness variation have been studied thoroughly. The following conclusions can be drawn from the process analysed above:

1. When the original blank is a plate, the first path in conventional spinning has significant effect on wall thickness distribution, and the toolpath with a clearance that is less than the original thickness of the blank, can obtain a good thickness result to avoid failures such as wrinkling or fracture.

2. For forward pass, the larger value of the parameters of θ_0 , Z_2 and U_n , the more severe the thinning phenomenon is. It is interpreted like this: the increase of the deformation between adjacent passes, the increase of the sloping degree of the flange and the more aggressive of the Bezier curve all result in a more serious thinning.

These conclusions could be used to conduct processes with multi-pass conventional spinning to produce better dimensional precision of spherical components.

Acknowledgement

The research was supported by National Basic Research Program of China (973 Program, No.2014CB046601).

References

1. C.C. Wong, T.A. Dean, J. Lin. International Journal of Machine Tools &Manufacture, **43** (2004): 1419-1435.
2. M. Runge. *Spinning and flow forming.* Werkzeugmaschinenbau/Verlag Moderne Industrie AG, **D-86895** (1994).

3. K. Lange, *Handbook of metal forming*, McGraw-Hill Book Company, **1216** (1985).
4. M. Hayama, Hi. Kudo, T. Shinokura. Bulletin of JSME, **13**(1970) : 1358-1366.
5. M. Hayama, M. Nakamura, T. Watanabe, H. Hamano. JSTP, **27**(1986): 1053-1059.
6. M. Hayama. JSTP, **30**:1403-1410
7. D.C. Kang, X.C. Gao, X.F. Meng, et al. Journal of Materials Processing Technology. **91**(1999): 226-230.
8. L. Wang, H. Long. Journal of Materials Processing Technology, **211**(2011) : 2140-2151.
9. Y. Li, J. Wang, G. Lu, et al. Journal of Zhejiang University SCIENCE A, **15**(2014): 432–446.
10. J.A. Polyblank, J.M. Allwood. CIRP Annals-Manufacturing Technology, **64**(2015) : 301-304.
11. E. Quigleg, J. Monaghan. Journal of Materials Processing Technology, **103**(2000) : 114-119.
12. Q.X. Xia, X.Q. Cheng, Y. Hu, F. Ruan. International Journal of Mechanical Sciences, **48**(2006) : 726-735.
13. M.D. McKay, R.J. Beckman, W.J. Conover. Technometrics, **21**(1979) : 239-245.

DOI: 10.1051/matecconf/20168015013

Advanced constitutive modeling and application to industrial forming processes

Frédéric Barlat[1,a] and Dong-Jin Kim[2]

[1]*Pohang University of Science and Technology, Pohang, Republic of Korea*
[2]*POSCO Materials Forming Research Group, Incheon, Republic of Korea*

Abstract. Continuum constitutive descriptions of plasticity suitable for finite element simulations of sheet forming processes are succinctly discussed in this presentation. Although multi-scale approaches allow for a more explicit representation of the physical deformation mechanisms occurring at microscopic scales, they are usually not suitable for industrial applications because of the quick turnaround time needed for process design simulations. Therefore, advances in classical concepts such as plastic anisotropy and strain hardening are still very much in demand. Actual forming simulation examples conducted for the steel industry are presented for illustration purposes.

1 Introduction

Numerical simulations for industrial forming processes have been very useful in the last decades, since the start of the NUMIFORM conference series. They have strongly contributed to the decrease of time for process optimization. Progress in deformation mechanics, heat transfer and microstructure evolution and their time and space discretization have allowed a number of commercial codes to provide robust numerical tools for this purpose [1, 2]. There are still a number of challenging aspects regarding the code themselves not only on the mechanics of finite elements but also on boundary and interface conditions and on material constitutive descriptions. This articles only deals with the latter aspect but if the goal of numerical simulations is to achieve accurate predictions of processes above a certain scale, much effort is still needed in other areas as well.

The constitutive description must not only account for the physical aspects to describe the proper mechanisms but also for the input data needed and the identification of the material coefficients. It must translate in mathematical terms the relationship between a number of variables with their evolutions and the stress-strain response within a framework set by mechanics and thermodynamic principles. A pure numerical description, which specifies the state of a material with a set of data is more empirical. It leads to poor extrapolation of states, which are not included in the database. However, sometimes, it is not practically possible to exclude this type of input as, for instance, for the case of hot press forming (HPF), such as schematically shown in Fig. 1. HPF simulations using thermo-mechanical-metallurgical numerical models require input information such as temperature and strain rate-dependent stress-strain relationship, which are not the only material characteristics needed. Thermal properties such as conductivity, latent heat, phase or temperature-transformation-cooling diagram, and metallurgical states such as composition, grain size, phase or atomic diffusivity. It is also worth mentioning the coupling, sometime strong, between all these variables as well as with temperature and strain rate. Therefore, for such a problem, it is not wise to reject a priori any form of empirical input.

Another consideration necessary to choose a constitutive model is the scale of the material features needed, the size of the problem in terms of node or element number and the associated computation time. The scale is dictated by the microstructural information needed to characterize a product as, for instance, the grain size, which requires simulations at the meso-scale. It is possible to envision cases were details at a finer scale are necessary, for instance for the manufacturing of micro-devices. However, in general, the description of lower scales conflicts with computation time. In sheet metal forming processes for automotive applications, the purpose of the simulations is to optimize the process geometry, boundary and interface conditions, load and, possibly, the material itself. Therefore, FE simulations must be run a number of times before an acceptable optimized solution is achieved. This prohibits the consideration of any material microstructural features explicitly. Thus, even if the idea of defining the constitutive description from the atomic scale is scientifically and philosophically very attractive, it is usually not practical.

The size of the simulation and the product itself also influence the nature of the constitutive description for the

[a] Corresponding author: f.barlat@postech.ac.kr

 DOI: 10.1051/matecconf/20168015013

same reason, that is, computation time. An automotive panel of the size of a car door might require a large number of elements if the goal of the simulation is to detect plastic localization or fracture zones. In contrast, the stretch-forming of a panel for the fuselage of an airplane, several meters in length, may not be such a large size simulation because plastic deformations are usually small. Thus, the size of the FE element may be relatively large. Nevertheless, large product sizes imply generally that the number of elements or nodes is large, that is, the simulation of only one process condition long.

Other issues for practical industrial applications are the ability of a constitutive model to be well understood by metal forming analysts, to be identified at the lowest possible cost and to require a minimum number of experiments. Ideally, the best constitutive model for industry is a description that provides good accuracy with low computation time and necessitates only a few experiments for the coefficient identification.

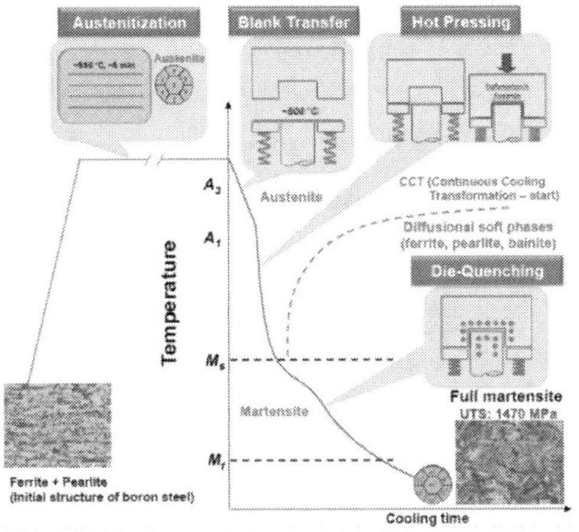

Figure 1. Hot press forming process schematics for boron steel (adapted from [3]).

2 Constitutive description example

In this example, some of the authors' own models are presented but there are many ways to describe a flow theory of plasticity. More emphasis is put on the reasons for the authors' assumptions since the detailed formulation can be found in the relevant publications [4-7]. Room temperature deformation for rate-independent plasticity is considered in this section. Since the addition of a viscous effect can be done relatively simply, rate independent plasticity only is discussed. The goal is to define a yield condition, a flow rule and a hardening model.

The von Mises yield condition is the most popular for isotropy because it is one of the simplest. It contains some physics because it describes yielding as the limit of the elastic shear distortion energy, is in reasonable agreement with many experimental results and is independent of the hydrostatic stress as observed in metals. In this sense, it is as physical as the Schmid law, except that it is not expressed at the same scale. Similarly, Tresca expresses yielding based on a critical macroscopic shear stress, which, as a consequence, is independent of the hydrostatic stress. However, von Mises is more popular in numerical simulations because it leads to a smooth yield surface. Another isotropic yield condition is that of Hershey [8]

$$\bar{\xi}(\boldsymbol{\sigma})=\left\{\tfrac{1}{2}\left(\left|\sigma_1'-\sigma_2'\right|^a+\left|\sigma_2'-\sigma_3'\right|^a+\left|\sigma_3'-\sigma_1'\right|^a\right)\right\}^{1/a}=\sigma_R(\bar{\varphi}) \quad (1)$$

where $\boldsymbol{\sigma}$ is the Cauchy stress tensor and σ_k' the principal values of the deviator. This yield condition is similar to that of von Mises but non-quadratic and is again based on an average of the principal shear stresses. Again, it contains as much physics as the Schmid law but no microstructure information. In addition, the yield surface shapes predicted with this criterion are in excellent agreement with those calculated with crystal plasticity model when the exponent a is set appropriately, usually significantly larger than 2 [9]. The point of this paragraph is to say that von Mises, Tresca, Hershey and other suitable yield criteria are equally valid to represent isotropy and contain some physics but no microstructural information.

In order to introduce anisotropic formulations, a simple method is to use linear transformations of the type

$$\tilde{\boldsymbol{\sigma}}'=\mathbf{C}:\boldsymbol{\sigma}'=\mathbf{L}:\boldsymbol{\sigma} \quad (2)$$

and to substitute the principal deviatoric stresses in Eq. (1) by the principal values of the linearly transformed deviator [4, 5]. The advantages of this method are numerous. The yield condition reduces to the isotropic formulation if the 4th order tensor $\mathbf{C}$ (or $\mathbf{L}$) reduces to the corresponding identity tensor. Moreover, the yield surface remains convex if the anisotropic yield surface associated with the isotropic yield function is convex. Usually, the convexity condition is more easily verified with principal stresses. $\mathbf{C}$ and $\mathbf{L}$ have a structure, which is consistent with the macroscopic symmetry of the material. Finally, when yield is defined conventionally with an offset strain of a fraction of a percent, it leads to yield surface shapes in agreement with experiments [10, 11].

Under the isotropic hardening assumption, the yield surface only expands because it is controlled by one parameter, the reference stress σ_R or, equivalently, the effective strain, $\bar{\varphi}$. The former is defined by the plastic work equivalence, i.e.,

$$\bar{\xi}\, d\bar{\varphi}=\sigma_{pq}\, d\varepsilon_{pq} \quad (3)$$

Note that when $\bar{\xi}$ is von Mises, it is possible to retrieve the corresponding effective strain from Eq. (3). However, if the effective stress is defined by Eq. (1), Hershey's approach, or any isotropic or anisotropic yield function, the von Mises effective strain is not proper. The effective (or accumulated) strain always increases, unless specific recovery mechanisms are active, because plastic strains accumulate irrespective of the current strain increment. This corresponds to the physical observation that the density of dislocations, which produces slip before being trapped, increases. The reference curve $\sigma_R(\bar{\varphi})$ is mathematically expressed by an equation, which can take numerous forms and may or may not contain microstructural information. However, very often the stress is represented as a monotonic function of the effective strain because the material hardens as the accumulated strain increases. Thus, the effective stress and strain are physically motivated although they do not necessarily contain microstructural information.

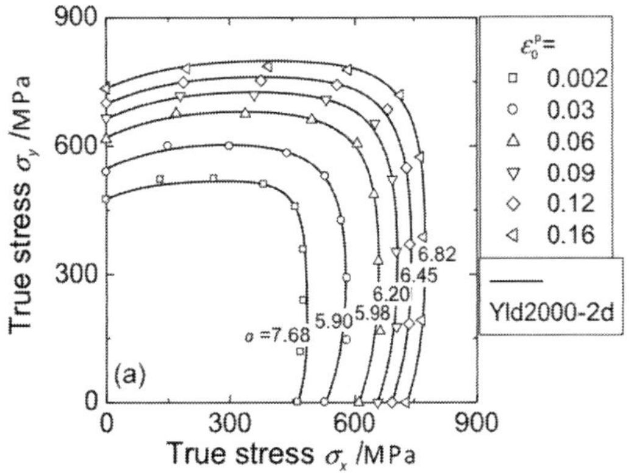

Figure 2. Experimental of iso-plastic work contours for DP590 steel approximated by Yld2000-2d (reprinted from [11] with permission from Springer).

Isotropic hardening is particularly convenient for industrial applications because it requires only a limited number of tests to identify the constitutive coefficients. In addition, the use of shell elements for sheet forming problem reduces the computation time drastically. As a result, a state of plane stress is usually assumed. Thus, a form of Eq. (1) modified with two linear transformations to describe plastic anisotropy, the so-called plane stress Yld2000-2d yield function, is more convenient. It contains eight anisotropy coefficients, which require a minimum of four mechanical tests to identify. Three uniaxial tension tests in different directions provide six data since each test allows the determination of a flow stress and r-value (width-to-thickness strain ratio in uniaxial tension). Since the bulge test has recently become a standard, it can be used in addition to provide input data for balanced biaxial tension. However, other tests could be used as well. In addition, new techniques, such as the virtual fields method, are now emerging to identify a full set of constitutive coefficients [12]. These methods are based on the measurement of full inhomogeneous displacement fields with digital image correlation. Researches are conducted to extract all the anisotropy coefficients from one single test, which would reduce considerably the number of experiments to conduct. However, this approach, although well established for elasticity, requires much more research for plasticity.

Finally, the question of the flow rule has been debated for several decades and a resurgence of the discussion has occurred recently [13, 14]. The associated flow rule (AFR) assumes that the strain increment is normal to the yield surface while the non-associated flow rule (NAFR) assumes it is not. The constraint of the AFR is so high that it does not appear to be reasonable. However, general considerations of crystal plasticity with only the Schmid law for the crystal behavior as the main assumption indicates that metals, which deforms by microscopic shearing, should deform according to the AFR. Thus, even if the AFR is not strictly respected, the physics of plastic deformation indicates that the strain increment should be reasonably close to the normal of the yield surface. Because it is difficult to precisely define AFR and NAFR, the low scale physics for metals tends to justify the use of the AFR. This is why in this work, the AFR is considered more suitable, that is

$$d\boldsymbol{\varepsilon} = d\bar{\varphi}\frac{\partial\bar{\xi}}{\partial\boldsymbol{\sigma}} \tag{4}$$

The isotropic hardening assumption leads to reasonable simulation results when loading is roughly proportional. However, when sharp changes in strain path occur, as often the case in forming, the accuracy of the predictions is usually not as good. This is particularly true when load reversals occur such as bending and unbending around a die radius, leading to the so-called Baushinger effect. The springback, that is, the elastic shape distortion of a part due to the removal of forming loads, strongly depends on the Bauschinger effect and the stress-strain behavior after reversal. Since springback is a critical problem for high strength sheet materials such as modern steels, constitutive models that account for the Bauschinger and associated effects are important for industrial applications. This effect can be captured by non-linear kinematic hardening rules [15, 16], in which the yield surface translates in stress space. Kinematic hardening has been used very efficiently during a half of a century to account for anisotropic hardening. However, a different approach was recently proposed in an effort to make industrial applications easier. More than the model itself, the arguments to justify the use of this model are developed below.

The so-called homogeneous anisotropic hardening (HAH) approach makes use of the isotropic hardening concept discussed above. The yield condition is described by

$$\bar{\sigma}(\boldsymbol{\sigma}') = \left\{ \bar{\xi}^{\,q}(\boldsymbol{\sigma}') + f_1^q \left| \hat{\mathbf{h}} : \boldsymbol{\sigma}' - \left| \hat{\mathbf{h}} : \boldsymbol{\sigma}' \right| \right|^q + f_2^q \left| \hat{\mathbf{h}} : \boldsymbol{\sigma}' + \left| \hat{\mathbf{h}} : \boldsymbol{\sigma}' \right| \right|^q \right\}^{\frac{1}{q}} = \sigma_R(\bar{\varepsilon}) \tag{5}$$

where f_1, f_2 are two scalar state variables that evolve rapidly as a function of strain and distort the yield surface. When $f_1 = f_2 = 0$, the standard isotropic yield

DOI: 10.1051/matecconf/20168015013

condition evolving under the isotropic hardening assumption as seen above is recovered, that is, $\bar{\xi}(\boldsymbol{\sigma}')=\sigma_R(\bar{\varepsilon})$. $\hat{\mathbf{h}}$, a tensorial state variable called microstructure deviator, controls the location of the distortion on the yield surface. This variable evolves slowly as a function of the strain because it is assumed to represent the dislocation structure that develops in a material. Although this model is macroscopic, it is consistent with a number of observations made about dislocation patterning under monotonic and non-monotonic loading conditions, which indicate that dislocation structures tend to evolve slowly towards the structure corresponding to a particular loading direction. Thus, the material is physical, although it does not explicitly include microstructural features. The material behavior predicted with this approach when the strain path is linear is strictly the same as that obtained using the anisotropic yield function considered under isotropic hardening. Although, there is no back-stress or kinematic hardening component, this approach successfully describes the Bauschinger effect as illustrated in Fig. 3 for a balanced biaxial stress state followed by load reversal.

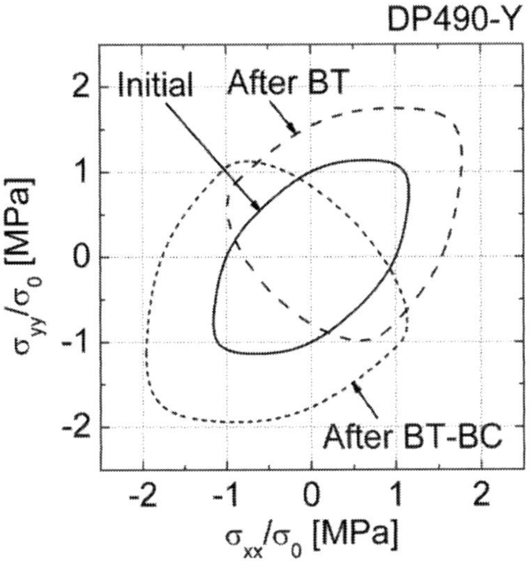

Figure 3. Yield surface evolution for DP490 steel calculated for as received material, after balanced biaxial tension and after subsequent balanced biaxial compression (reprinted from [17] with permission from Elsevier).

3 Applications

A few examples are given in this section that underline the progress and current limits of macroscopic material models.

3.1 Indentation

The first example is an application to indentation [17]. The process is schematically explained in Fig. 4a. The process consists of two steps: 1) A flat sheet is first pre-stretched in balanced biaxial tension with a large punch to a given strain and subsequently deformed by an indenter operating in a quasi-static mode. Preliminary simulations indicated that elastic deformation is preponderant in the whole specimen since the maximum plastic deformation near the indented area is less than 10%, loading is highly non-linear with biaxial reversals and the elasto-plastic behavior of the material near balanced biaxial tension requires special attention.

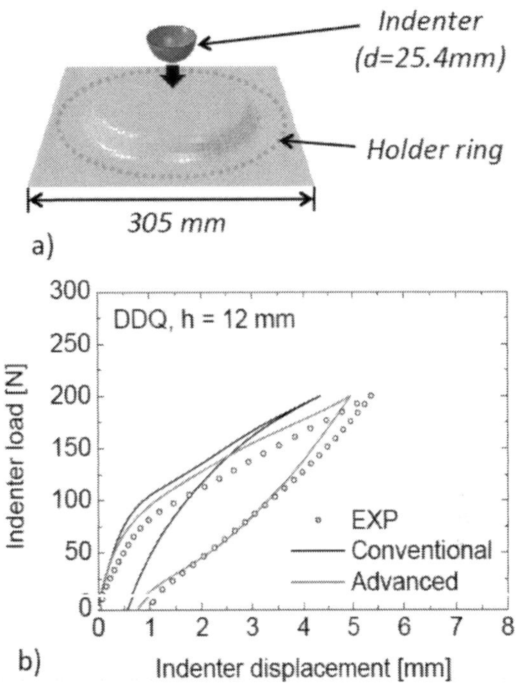

Figure 4. Indentation of panel for drawing quality steel (reprinted from [17] with permission from Elsevier); a) Testing schematics and; b) Load-displacement curve.

Thus, in order to accurately describe the behavior of the material, the biaxial elastic modulus as well as its evolution as a function of the accumulated strain has been accounted for. Plastic anisotropy, strain hardening and its transient stage during reversal require a good description based on biaxial measurements, as well as simple shear testing with load reversal. The former allows the characterization of the biaxial elasto-plasticity in proportional loading and the latter, the transient hardening effects. Additional uniaxial tension tests are necessary to characterize plastic anisotropy. In order to compare the effect of the constitutive description, two models were employed in the simulations. The first, referred to as conventional, consisted of a constant uniaxial elastic modulus, von Mises yield condition with the associated flow rule and Swift isotropic hardening description. The other, referred to as advanced constitutive model, consisted of the biaxial elastic modulus and its decrease as a function of the accumulated plastic strain, the Yld2000-2d non quadratic anisotropic yield function, and the HAH distortional framework to account for the Bauschinger effect and other transient hardening phenomena. The simulations were conducted with Abaqus/Standard (implicit).

Figure 4b compares the experimental indentation load for a DDQ steel as a function of the indenter displacement with the predicted results computed with the conventional and advanced models. This figure clearly indicates the superiority of the advanced model. Of course, the better material description leads to the better results. This implies that a careful material characterization is necessary and requires tests, which are not necessarily standard. This implies an additional significant effort to gain accuracy. For practical purpose, the permanent indent depth is about 30% higher with the advanced model compared with the standard. However, the computation time is significantly higher with the advanced model. If this is not an issue for this example, this could be prohibitively different for much larger problems. As a whole, accuracy comes with a cost that industry may or may not be willing to assume. The important question is what maximum error/budget ratio is allowed for a particular process and product.

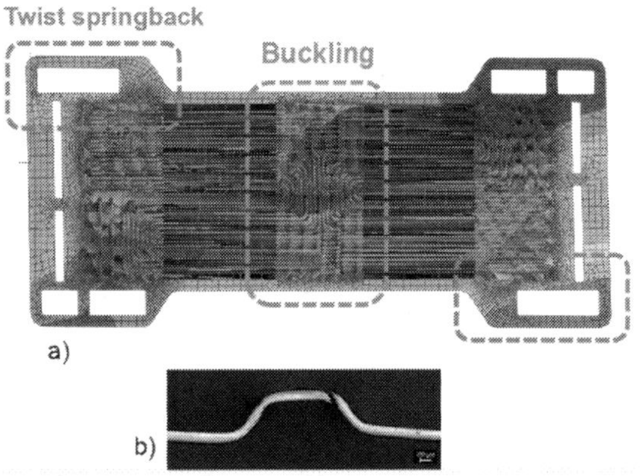

Figure 5. PEMFC fuel cell cathode made of ferritic stainless steel (scale marker, 10 microns); a) FE mesh; b) Channel cross-section [18].

3.2 Fuel cell separator

The next example is about forming of ultra-thin sheets (less than 0.15 mm thick) for PEM fuel cells (PEMFC) [18]. Fig. 5 shows the geometry of the mesh of the cathode, which has a size of 400 by 220 mm. The blank, initially flat, becomes corrugated with micro-channels after forming. A channel is less than 1 mm wide, approximately 0.5 mm deep and features corner radii of 0.3 mm. The forming consists of three steps, namely, forming, trimming and springback. Experimentally, the PEMFC cathode was formed on a HSK direct-drive digital servo-press, which, because of its high accuracy compared to traditional presses, is necessary to successfully make product from ferritic stainless steels.

This problem is very challenging from many viewpoints. The material itself is so thin that it is very difficult to measure the mechanical properties. In fact, the standards for sheet metal characterization are not valid for thickness values lower than 0.5 mm. Tests for extra-thin sheet materials require extreme care and are sometimes impossible to conduct as, for instance, reversals for the characterization of the Bauschinger effect. In addition, since the grain size is not negligible compared to the thickness, this feature might be needed in the characterization input data, as well as a model to transfer this information to the continuum scale. In passing, another issue with many thin materials is that they are often heavily deformed and do not provide sufficient elongation in uniaxial tension for characterization. Thus, in spite of all the advances made with material property measurements, it is expected that additional efforts in this area will be necessary in the future.

For the simulations, even though the size of the product is modest, this problem is already very large. The simulation were run with about 1.5 millions shell elements although many more are required because the mesh size of 0.25 mm is barely smaller than the radius of the channel corners. In spite of that, the computation is prohibitively too large for process optimization in industry. As an aside, sheet forming process optimization is almost always conducted with shell elements and explicit solvers in industry. One of the reasons is computation time while the other is numerical convergence. Due to the many tool/workpiece contacts that occur in forming, it is very difficult to solve a complete problem, even a small one, using an implicit solver. The use of solid elements is generally not considered because of computation time.

In this example, in order to evaluate buckling and twist springback, and to optimize drawbead positions, a simple constitutive description only, similar to the conventional model of the previous section (von Mises, etc.), was considered in order to keep the effort at a reasonable level. Nevertheless, a realistic solution of the drawbead positions was obtained. The plastic flow localization, which occurs during forming (Fig. 5b), was analyzed using a plane strain analysis throughout the channel section. The search of an optimum process condition was reached for this particular example but real products are expected to be commercialized only after down-gauging. Therefore, a number of serious challenges still remain in numerical and experimental methods in industrial forming processes for such products.

3.3 Hot press forming

The hot press technology (HPF) is very important for the automotive industry because it allows the manufacturing of products such a B-pillar from boron-added steel, which leads to very high strengths because of the martensitic structure. However, the material must be formed at high temperature in the austenite domain and die-quenched at a high cooling rate to avoid diffusive and softer phases to transform. Also of importance is to tailor the properties of the product by producing zones with full martensitic microstructure and others interspersed with softer phases. The use of tailored-welded blank is another option [19,

DOI: 10.1051/matecconf/20168015013

20]. For instance, for the B-pillar example, the upper end should be very strong for, during a crash, resistance to intrusion in the passenger cabin and the bottom part more ductile for energy absorption. The partial cooling technology is among a number of technical solutions to this end. In this process, forming takes place approximately from about 800 °C down to 600 °C. Then, the critical die cooling stage occurs from the temperature at the end of forming stage down to about 200 °C. Segments of the die are cooled using water circulating in channels at various flow rates while other are not (Fig. 6a). This produces cooling at different rates for different locations in the blank.

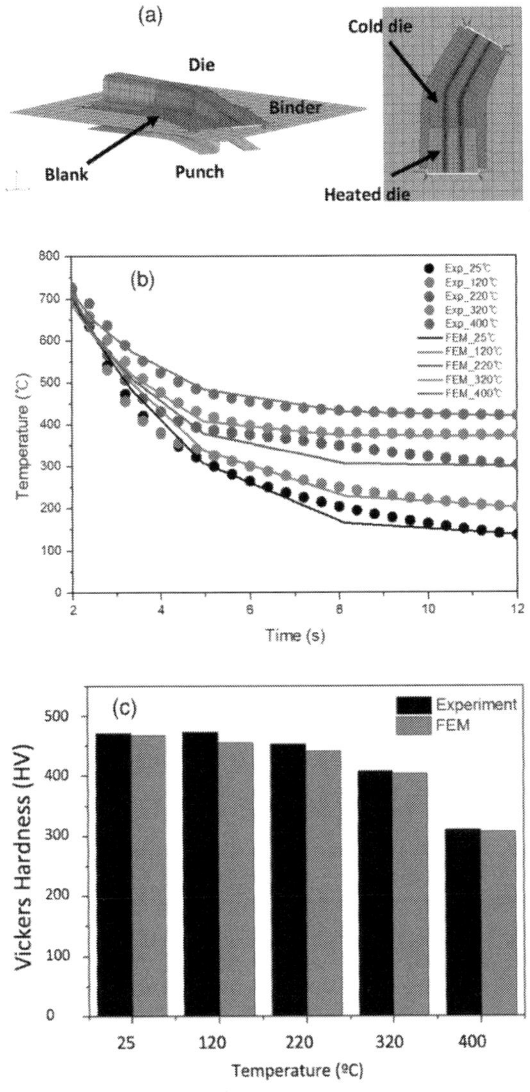

Figure 6. Vickers hardness as function of die temperature [24].

This process is difficult to model because it is time and temperature dependent. This requires a fully coupled thermo-mechanical-metallurgical solution [21, 23]. The plastic behavior itself, stress-strain and localization, must be temperature and strain rate dependent. Then, phase transformation kinetics must be known accurately, which demand the knowledge of a heat transfer description. All these parameters are coupled since heat transfer coefficients, for instance, are at least pressure and temperature dependent. In addition, the material itself is subjected to drastic microstructural changes associated with the transformation kinetics. Of course, all this requires a very large quantity of input data, some of it very empirical, based on composition, or extracted from thermodynamics databases and software.

An example of such simulation is given in Fig. 6a for a non-straight channel with a U-shape cross-section. The die was segmented in order to produce different cooling rates. The cold die zone leads to full martensite in the blank, while the heated zone can produce softer phases, depending on the die segment temperature. Fig. 6b compares the cooling curves from the start of forming to a time after which, the microstructure remains stable. The predicted cooling curves are in good agreement with the experiments but the preliminary results were not. As a result, it was necessary to measure the heat transfer coefficients between a particular tool steel and the blank as a function of the temperature and the contact pressure.

The Vickers hardness can be calculated, yet from an empirical formula based on phase volume fraction and composition. It leads to predicted values that are in good agreement with experiments (Fig. 6c) but at the expense of additional costs and delays. This example indicates that empirical methods are necessary to completely solve such a problem. A more scientific approach to this problem with FE and more physically-based models, namely, constitutive, heat transfer, metallurgical states and reaction kinetics, is not readily available yet. This example suggests that there are still huge efforts to conduct just from the material behavior viewpoint. In spite of this, the results of thermo-mechanical-metallurgical finite element simulations can be a tremendous help for process designers and practitioners.

4 Discussion

Advances in continuum constitutive modeling have improved the accuracy of numerical simulation methods in industrial forming processes. In the sheet metal forming area, a number of obstacles still remain in order to produce reasonably accurate results in a time and cost efficient manner, even with continuum constitutive models. The requirement of high turnaround time and low cost requires the use of compromises that prevent the achievement of higher accuracy. Lower-scales approaches (Fig. 7) are not suitable yet for direct applications in sheet metal forming codes because huge amounts of microscopic information should be pre-processed first, one way or another. Nevertheless, meso- and micro-scopic material models can be used to guide and refine the development of higher scale constitutive descriptions (Fig. 8).

In addition, new experimental techniques and procedures are still required in order to provide sufficiently accurate input data to the numerical forming process codes. Again, limitations arise because of the complexity of the database to prepare, such as for HPF, or because of the nature of the product itself, such as extra thin sheet materials, which are hurdles to overcome. At the same time, all these obstacles constitute new challenges for science and engineering. Alternate routes for the description of the macroscopic behavior are showing up such as full field measurements and virtual field methods [12], which provide a wealth of information about the material behavior. However, this information has to be processed first before it can be used optimally for forming application. Finally, other forms of input data can be generated from virtual experiments conducted with lower scale microstructure-based models.

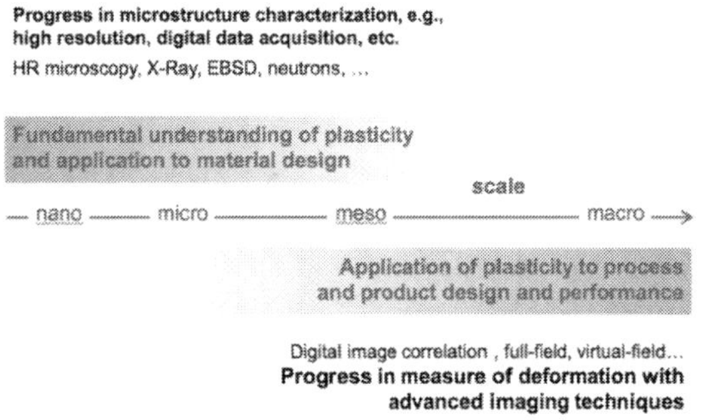

Figure 7. Multi-scale description showing the huge amount of input data to encapsulate in simple concepts for application to manufacturing.

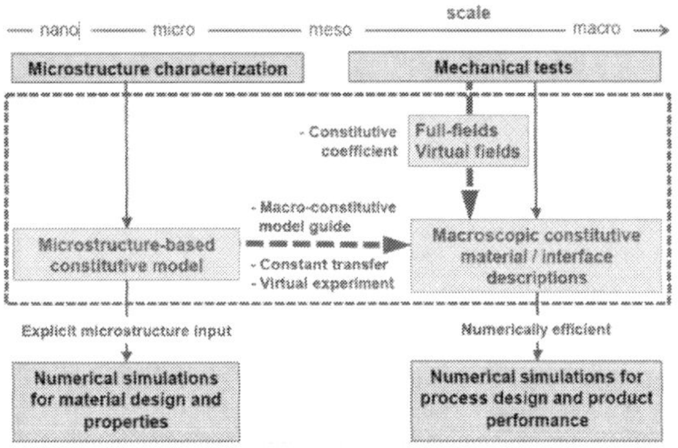

Figure 8. Application of multi-scale approaches through transfer towards continuum constitutive plasticity models.

In any case, even though the field of simulation for metal forming and advanced constitutive modeling, particularly in industrial forming processes, is not an emerging technology, this field is still vibrant and full of innovations yet to discover. This is also "fuelled" by the imperative necessity of sustainable development, which drives the emergence of lightweight materials with enhanced attributes. This exciting time is coming with challenges for the development of advanced constitutive descriptions as well.

Acknowledgments

The authors are grateful to POSCO for providing material and sharing experimental and numerical results.

References

1. J.-L. Chenot, M. Bernacki, L. Fourment, R. Ducloux, in F. Barlat et al. (Eds.), Proc. NUMIFORM, AIP 1252, 2010, pp. 27-38.
2. E. Oñate, in F. Barlat et al. (Eds.), Proc. NUMIFORM, AIP 1252, 2010, pp. 11-17.
3. H. Bok, PhD Thesis, Pohang University of Science and Technology, 2014.
4. F. Barlat, J.C. Brem, J.Y. Yoon, K. Chung, R.E. Dick, D.J. Lege, F. Pourboghrat, S.-H. Choi, S.-H., E. Chu, Int. J. Plasticity, **19**, 1297-1319 (2003).
5. F. Barlat, H. Aretz, J.W. Yoon, M.E. Karabin, J.C. Brem, R.E. Dick, Int. J. Plasticity, **21**, 1009-1039 (2005).
6. F. Barlat, J.J. Gracio, M.G. Lee, E.F. Rauch, G. Vincze, Int. J. Plasticity, **27**, 1309-1327 (2011).
7. F. Barlat, G. Vincze, J.J. Grácio, M.G. Lee, E.F. Rauch, C. Tomé, Int. J. Plasticity, **58**, 201-218 (2014).
8. A.V. Hershey, J. Appl. Mech., **21**, 241-249 (1954).
9. R.W. Logan, W.F. Hosford, Int. J. Mech. Sci., **22**, 419-430 (1980).
10. N. Tozawa, in: D.P. Koistinen, N.M. Wang, (Eds.), *Mechanics of Sheet Metal Forming*, Plenum Press, 1978, pp. 81-110.
11. T. Hakoyama, T. Kuwabara, in: H. Altenbach, T. Matsuda, D. Okumura, (Eds.), *From Creep Damage Mechanics to Homogenization Methods*, Springer, 2015, pp. 67-95.
12. M. Grediac, F. Pierron, *The VFM Method*, Springer, 2012.
13. T.B. Stoughton, Int. J. Plasticity, **18**, 687-714 (2002).
14. T.B. Stoughton, J.W. Yoon, Int. J. Plasticity, **25**, 1777–1817 (2009).
15. J.L. Chaboche, Int. J. Plasticity, **2**, 149-188 (1986).
16. F. Yoshida, T. Uemori, K. Fujiwara, Int. J. Plasticity, **18**, 633-659 (2002).
17. J.Y. Lee, M.G. Lee, F. Barlat, K.H., Chung, D.J. Kim, Int. J. Solids Structures, **87**, 254–266 (2016).
18. H.J. Bong, F. Barlat, J. Lee, M.G. Lee, J.H. Kim, Metals Mater. International, **22**, 276-287 (2016).
19. J.W. Choi, M.G. Lee, F. Barlat, H.S. Son, J.B. Nam, ISIJ International, **52**, 2059-2068 (2012).
20. J. Choi, H. Bok, F. Barlat, H. Son, D.J. Kim, M.G. Lee, ISIJ International, **53**, 1047–1056 (2013).
21. H. Bok, J.W. Choi, F. Barlat, D.W. Suh, M.G. Lee, Int. J. Plasticity, **58**, 154-183 (2014).
22. H. Bok, S.K. Kim, D.W. Suh, F. Barlat, M.G. Lee, Mat. Sci. Engineering, **A626**, 67-73 (2015).
23. H. Bok, J.W. Choi, F. Barlat, M.G. Lee, Int. J. Plasticity, **73**, 142-170 (2015).
24. H. Kim, Unpublished work, GIFT, Pohang University of Science and Technology, 2016.

DOI: 10.1051/matecconf/20168016001

Finite element simulation of stretch forming of aluminium-polymer laminate foils used for pharmaceutical packaging

Simon Müller[1,a] and Sabine M. Weygand[2,b]

[1] *Karlsruhe University of Applied Sciences, Institute of Applied Research, Moltkestr. 30, 76133 Karlsruhe, Germany*
[2] *Karlsruhe University of Applied Sciences, Faculty of Mechanical Engineering and Mechatronics, Moltkestr. 30, 76133 Karlsruhe, Germany*

Abstract. Pharmaceutical high barrier blister packages are manufactured from aluminium-polymer laminate foils (e.g. consisting of PA-Al-PVC layers). By a cold stretch forming process cavities are formed. The aim of this work is to determine a homogenized elastic-plastic description of the laminate by micromechanics. Therefore, a microstructural model is developed where the layers are mapped in a representative volume element. The obtained homogenized material model is applied to simulate the stretch forming to gain more insight into the forming process.

1 Introduction

Blister packs are often used for the packaging of pharmaceutical tablets. Currently, there are two main types of pharmaceutical blister packages on the market. According to the process of manufacturing they can be divided into thermoformed and cold-formed packages. The thermoformed blister packs are produced by thermoforming of a thermoplastic polymer foil into cavity moulds. The other type of package is manufactured by cold stretching of aluminium-polymer laminate foils. The aluminium in this composite material serves as barrier layer. Consequently, the packed tablet is protected against moisture, light, oxygen and other gases. However, the composite foil possesses less formability which leads to bigger cavities than needed for the product being packed. Figure 1 shows the difference in size of thermoformed and cold-formed blisters.

The companies developing forming machines for cold forming blister packs often lack knowledge about the exact specification of foils their costumers will use. Since they have limited information about the deformation behavior of aluminium-polymer laminate foils, they have to design the punches with high safety factors. Because of the endless variety of tablet forms the definition of the optimal punch geometry is an ongoing problem and one of the critical development stages for the developer of cold-forming blister packaging machines.

A direct approach for evaluating the suitability of a punch geometry for forming aluminium-polymer laminate foil is to test the punch on the packaging machine. Actually, this is the current procedure. However, this approach is only possible when the material the customer will use is

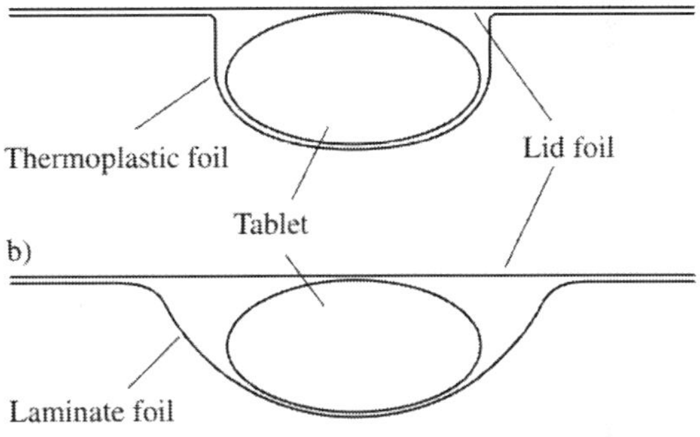

Figure 1. A thermoformed blister is shown in a) and a cold stretch formed blister is displayed in b). The difference of the dimensions of the cavities is clearly visible.

known and available. In addition, this approach requires a lot of time.

A numerical approach through the finite element (FE) method is a possible alternative way to optimize the punch geometry. Such a method allows to design the punches in an early stage of development and no machine time, for setting up the tools and testing, is wasted. One challenge in such a FE-modeling is the description of the mechanical behavior of aluminium-polymer laminate foils, since due to their layered structure their forming behavior and failure behavior is not yet well understood [1]. Furthermore, representing all different layers and their interfaces between them in the FE model leads to high computational time when simulating the stretch forming process. Homogenization techniques that help in reducing the computer time would be useful [2].

This work is a first attempt to describe the cold stretch forming behavior of aluminium-polymer laminate

[a] e-mail: simon.mueller@hs-karlsruhe.de
[b] e-mail: sabine.weygand@hs-karlsruhe.de

foil through the finite element method. As reference material an aluminium-polymer laminate foil, consisting of an outer layer of orientated polyamide (PA), a core layer of aluminium (Al), an inner layer of polyvinyl chloride (PVC) and adhesive and primer layers between is used. This combination of materials is common and finds great use in the pharmaceutical packaging industry. For the numerical investigations a micromechanical model is developed where the layers are mapped with simple material properties. Based on calculations with this model two homogenized material models are presented for the foil. The stretch forming process is simulated with a geometry close to an actually used one. There, three cases are studied. In the first one the foil is subdivided into a layer-by-layer structure. Each layer is assigned with simple material properties like in the micromechanical model. In the second and the third case the material behavior of the foil is described with homogenized behavior. The three cases with different material descriptions are compared by hypotheses regarding the reasons for failure.

2 Material models for the individual layers

In our numerical study each layer is described as a linear elastic, rate independent plastic material with isotropic hardening. To identify the material parameters of each layer uniaxial tensile tests were carried out on the individual layers. Thereby, nominal stress-strain curves were measured for PA, Al and PVC and converted into true stress-strain curves. Because this translation is only possible up to the tensile strength the curves are tangentially linearly extended above this point up to a true strain $\varepsilon_{tr} = 1$. The obtained flow curves are shown in Figure 2 and they provide the starting point for the FE simulations. It has to be noted that PVC shows strain softening right after yield initiation. That means that after the elastic deformation the nominal stress-strain curve includes a stress overshoot [3]. In this work, softening is neglected. This means that the stress overshoot of the measured nominal stress-strain curve is cut off. Because of that, we will denote the material model for the modified PVC with an asterisk. Additionally, the values for the Young's modulus E for each layer were determined from the tensile tests. For the Poisson's ratio ν of aluminium the well-known value is used [4]. For the Poisson's ratio of PVC and PA assumptions are made. Table 1 summarizes the applied elastic constants of the three materials.

Table 1. Linear elastic properties of the three material models

	Al	PA	PVC*
E in MPa	70000	2000	1800
ν	0.33	0.34	0.35

Details regarding the tensile experiments have not yet been published. However, the results show significant similarities with experiments carried out in other studies on equivalent materials [5].

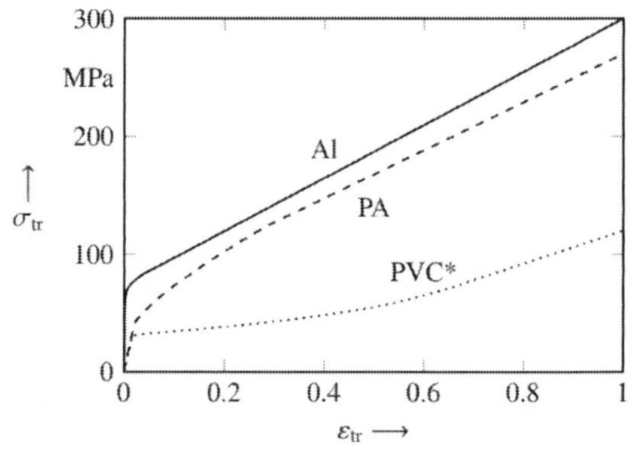

Figure 2. True stress-strain curves of the layer materials of the aluminium-polymer laminate. These three curves were derived from nominal stress-strain curves and represent the base for the numerical calculations. Because of simplifications regarding the mechanical behavior of PVC it is denoted with an asterisk.

3 Numerical modeling

To obtain the overall behavior of the aluminium-polymer laminate foil a micromechanical model has been developed. In this so called representative volume element (RVE) all three single layers are mapped individually, and the previously described material models are assigned to them. This model allows to study the influence of the layered structure on the elastic and plastic behavior of the foil. Furthermore, it enables to apply load cases, which are not accessible in experiments, due to the fact that the composite material is a thin foil.

In addition, a FE stretch forming model close to real process was designed. In the first simulation (called reference case) the single layers are again represented with their individual material behavior. Then the simulations are repeated with homogenized material models obtained from the studies on the representative volume element. For the modeling and the following simulations the commercial FE software Abaqus is used.

3.1 Representative volume element

The developed microstructural model is based on the representative volume element technique. The RVE employed in this work is shown in Figure 3. It was build up as a cube according to the geometry given in the figure. As can be seen, the model represents the laminar structure with its different layers (PA-Al-PVC). The material models presented in the previous section were assigned according to the labels. It is assumed that the individual layers are perfectly bounded and that the adhesives and primer layer thicknesses are negligible. Therefore, they are not included into the model.

The foil can be described as a periodic arrangement of a set of RVEs in the (x, y)-plane. To ensure this behavior under load, periodic boundary conditions have to be applied via the following constraints. The difference of the displacement vectors in two corresponding points situated at opposite volume element boundaries is a constant displacement vector corresponding pairs of points. If one looks at

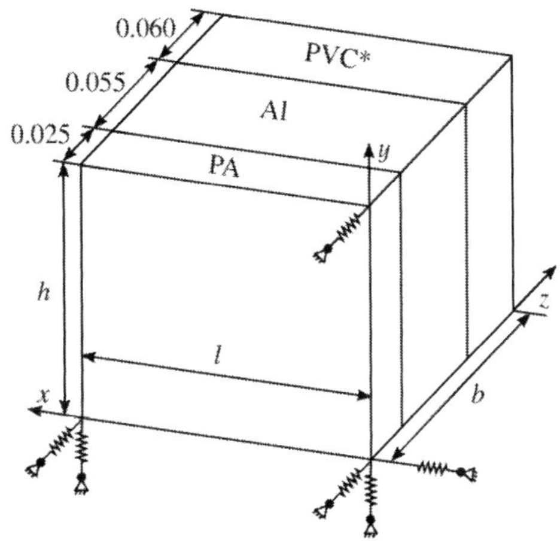

Figure 3. Model of the representative volume element (RVE). Depicted are the three layers of the aluminium-polymer laminate with their dimensions given in mm. The spring elements hinder the rigid body motion.

Figure 3, these constraints can be expressed by the relations

$$\begin{aligned} \mathbf{u}(l, y, z) - \mathbf{u}(0, y, z) &= \mathbf{k}_1, \\ \mathbf{u}(x, h, z) - \mathbf{u}(x, 0, y) &= \mathbf{k}_2, \\ \mathbf{u}(x, y, b) - \mathbf{u}(x, y, 0) &= \mathbf{k}_3. \end{aligned} \tag{1}$$

$\mathbf{u}(x, y, z)$ is the displacement vector of a point with the coordinates x, y and z. The vectors $\mathbf{k}_1$, $\mathbf{k}_2$ and $\mathbf{k}_3$ are constant displacement vectors which may be predefined or follow from the global equilibrium conditions of the RVE.

Instead of using displacement boundary conditions for hindering the rigid body motion, spring elements with a very high stiffness are employed. Here, the benefit lies in a more homogeneous stress distribution in the model than using displacement boundary conditions. The RVE was meshed with first-order, full-integrated, hexahedral elements. The resulting model has a mesh consisting of 21952 elements with the same dimensions.

3.2 Stretch forming model

The stretch forming model is designed according to a tool for industrial use. It consists of one punch, a die plate and the aluminium-polymer laminate foil. The foil has a length of 32 mm, a width of 26 mm and again a thickness of 0.14 mm. The dimensions of the single layers are the same and will be included into the model, too. The adhesive and primer layers are neglected. The dimensions of the other employed parts are shown in Figure 4. Due to geometrical symmetry only a quarter of the tool and the foil have to be realized. Furthermore, the punch and the die plate are assumed to be rigid and are meshed with three-dimensional bilinear rigid quadrilateral elements to reduce the computation time.

The foil is meshed with first-order, full-integrated, hexahedral elements. Thereby, it is ensured that over the thickness of each layer five elements are generated. The foil is placed with the PA layer facing the die plate as can be seen in Figure 4. To establish the contact between the foil and the other parts, a hard surface to surface contact is used. Frictionless behavior between the different parts is assumed.

The applied boundary conditions are such that all translational and rotational degrees of freedom of the die plate are blocked. The punch is only allowed to do translational movements in z-direction. Thus, the loading is applied there. Because only a quarter model of the tool was created, symmetric boundary conditions have to be defined on the foil. In addition, the edges of the laminate are fixed in all directions.

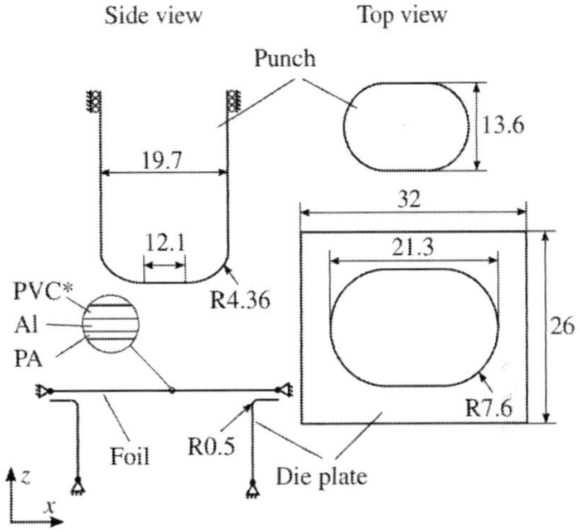

Figure 4. Side view and top view of the stretch forming model. The die plate, the punch and the laminate foil are shown. Their boundary conditions are indicated in the side view. The dimensions of the die plate and the punch are given in mm.

4 Finite element analyses and results

To obtain the overall mechanical behavior of the aluminium-polymer laminate foil and to gain insight into the internal stress variation, the RVE was applied to simulate various different load cases. This was done by defining force vectors $\mathbf{F}_i$ on reference nodes, which are coupled to the displacement vectors $\mathbf{k}_i$.

In the first set of calculations the RVE was applied to simulate uniaxial loading in three different directions, namely in x-, y- and z-direction. As can be seen in Figure 3, the x and y-directions lie in the foil plane while the z-direction corresponds to the transverse direction. The resulting three nominal stress-strain curves in tension are shown in Figure 5. It can be clearly seen, that loads in x- and y-directions lead to identical curves. In contrast, a load in z-direction shows a slightly different behavior. Here, the low value for the Young's modulus is noticeable. Another point is the elastic limit, which is approximately 21 MPa when loaded in x- and y-directions and 38 MPa when loaded in z-direction. After a strain of 0.02, the nominal stress-strain behavior is almost similar. The hardening only differs in a few units between the load cases.

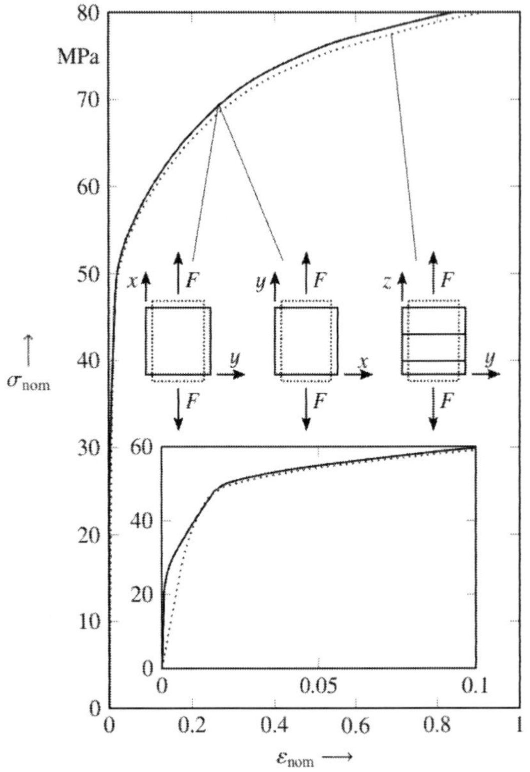

Figure 5. Nominal stress-strain curves of tensile loads on the RVE. The diagram shows that the mechanical behavior under load in x- and y- direction shows similar behavior. The small diagram shows the stress-strain behavior at small strains.

In the second set of calculations the RVE was applied to simple shear. The results are shown in Figure 6. Shearing on the x-plane in z-direction and shearing on the y-plane in z-direction shows an identical mechanical behavior. In contrast, shear load on the x-plane in y-direction shows a different trend. In this load case the shear modulus is higher than in the other two cases. The work hardening behavior differs, too.

The computed stress-strain curves together with the geometrical changes transverse to the loading direction are used to determine the elastic moduli of the composite foil. The whole 12 material constants are summarized in Table 2. E_i stands for the Young's modulus along axis i. G_{ij} is the shear modulus in direction j on the plane whose normal is in direction i. ν_{ij} is the Poisson's ratio that corresponds to a contraction in direction j when an extension is applied in direction i. It is noticeable that a lot of these constants are similar. As the x- and y-directions lie in the foil plane, their corresponding Young's moduli E_1 and E_2 are identical. Its value roughly corresponds to the arithmetical average of the individual elastic moduli of the layers (upper bound). Is the foil loaded transverse to its plane (i.e. in z-direction), the Young's modulus E_3 is smaller as the layers are loaded in row (lower bound).

As our investigated foil is symmetric with respect to a rotation about the z-axis, its overall elastic behavior is transversely isotropic. Therefore, the number of independent constants in the elasticity tensor is reduced to 5. These will be used in the following stretch forming simulations.

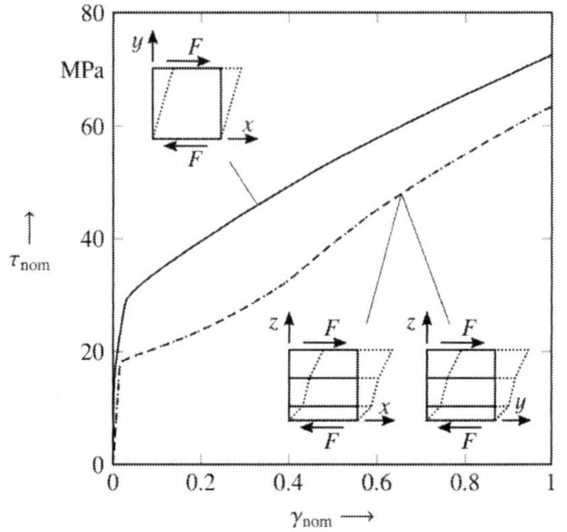

Figure 6. Nominal shear stress-strain of shear loading on the RVE. The diagram shows that shearing on the x-plane in z-direction and shearing on the y-plane in z-direction leads to the same mechanical behavior.

The simulations of the stretch forming process have been grouped in three different cases. These cases differ in the material description of the foil. In the first case the laminar structure of the foil was maintained. Each layer was assigned to its individual material properties given in Figure 2 as done in the calculations with the RVE. As in this case the layered structure is represented, it is closest to reality. That's why it is considered as reference case. Two more cases are investigated, where the laminar structure is not taken into consideration. Here the material behavior of the foil is described with homogenized elastic-plastic material models.

Table 2. Computed elastic moduli of the foil. The indexes 1, 2 and 3 correspond to the x-, y-, z-direction defined in Figure 3.

E_1 in MPa	E_2 in MPa	E_3 in MPa
28625	28625	4490
ν_{12}	ν_{13}	ν_{23}
0.3306	0.3455	0.3455
ν_{21}	ν_{31}	ν_{32}
0.3306	0.0542	0.0542
G_{12} in MPa	G_{13} in MPa	G_{23} in MPa
10757	1114	1114

In the second case the material behavior of the foil is described with linear elasticity with isotropic hardening, rate independent Mises plasticity. To derived the true stress-strain behavior the nominal stress-strain curve from the uniaxial loading of the RVE in x-direction is taken and transferred to the true stress-plastic strain curve (hardening curve) shown in Figure 7. The plastic strain ε_{pl} was calculated subtracting the elastic strain by using the Young's

DOI: 10.1051/matecconf/20168016001

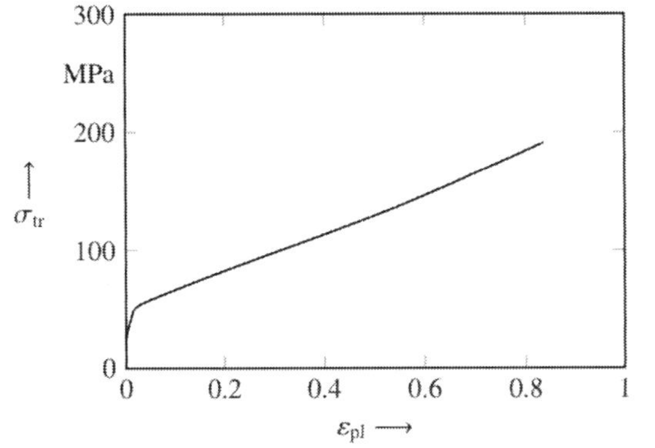

Figure 7. Hardening curve determined from the nominal stress-strain curve which was calculated by tensile loading of the RVE in x-direction. For translation the Young's modulus E_1 was used. This curve describes the plastic material behavior in the second and the third case of the stretch forming simulations.

modulus E_1. To describe the transverse contraction the Poisson's ratio ν_{12} is used.

In the third case the elastic material behavior of the foil is defined by the 5 parameters of the transversely isotropic elasticity, which had been determined from the calculations on the RVE. The plastic material behavior is specified by the true stress-plastic strain curve in Figure 7.

All stretch forming simulations started with the punch touching the foil. Then the load was applied through a translational punch displacement of 7.14 mm in negative z-direction. The detailed analysis of the stress and strain distribution is performed at maximum loading (i.e. at punch displacement of 7.14 mm). At the moment, the crucial failure mechanism in stretch forming aluminium-polymer laminate foils is not yet known. To obtain a better understanding of possibly involved mechanisms some hypotheses regarding the reasons for failure are made. These hypotheses will be evaluated with the results from experimental investigations in a future work.

The first hypothesis is that a crack occurs at the location of the maximal principal stress. The interest does not only lie in the maximum value, but also in its location and whether the more simple material descriptions in case 2 and case 3 predict the same location. The outcomes of the simulations are shown in Figure 8. The two cases with the homogenized material behavior show nearly the same maximal principal stress at the same location. This was expected since in both cases the hardening is described by the same curve. A closer examination shows that the maximal value occurs in the area where the punch has an edge. In the second case the maximal principal stress is approximately 241 MPa. In the case with the transversely isotropic elasticity the value is greater by 270 MPa. Both stresses appear on the surface of the foil. In the first case, where the foil is described by a multilayer model, the maximal principal stress occurred with 325 MPa in the area where the punch has an edge. However, a detailed investigation into the laminate structure shows that this stress doesn't occur on the surface of the foil. Here the maximum stress value appears in the aluminium core layer.

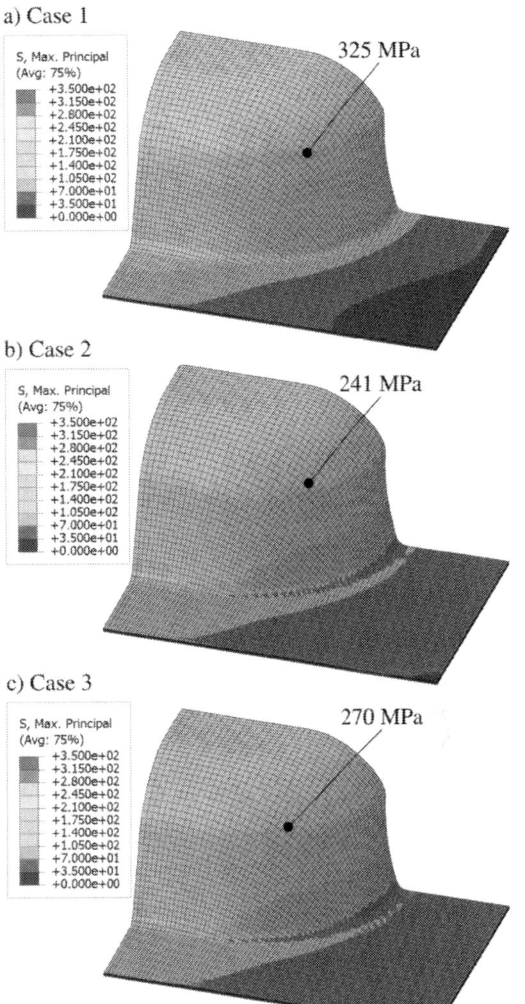

Figure 8. Computed maximal principal stress in the simulation of the stretch forming process at highest load. The first case, where the foil is described by a laminate structure, is shown in a). The second case, where the foil was defined with linear elasticity and rate independent plastic material behavior with isotropic hardening, is visible in b). The third case is shown in c). Here, the linear elasticity was replaced with transversely isotropic elasticity. The marks show the location of the maximal principal stresses.

The second hypothesis is the appearance of failure due to a critical thickness. Is the Al layer too thin, permeation can occur which is equal to failure. Therefore, the strain in thickness direction is analyzed. The results of the simulations are visible in Figure 9. The given strain component in the legend refers to the material orientation. It was chosen such that the shown component is equal to the strain in thickness direction. Visible is the fact that in all three cases high negative strains occur. Furthermore, they appear at nearly the same location. In the first case, which is the model with the laminate structure, a minimal strain value of -0.911 is reached. The location of appearance is at the tran-

 DOI: 10.1051/matecconf/20168016001

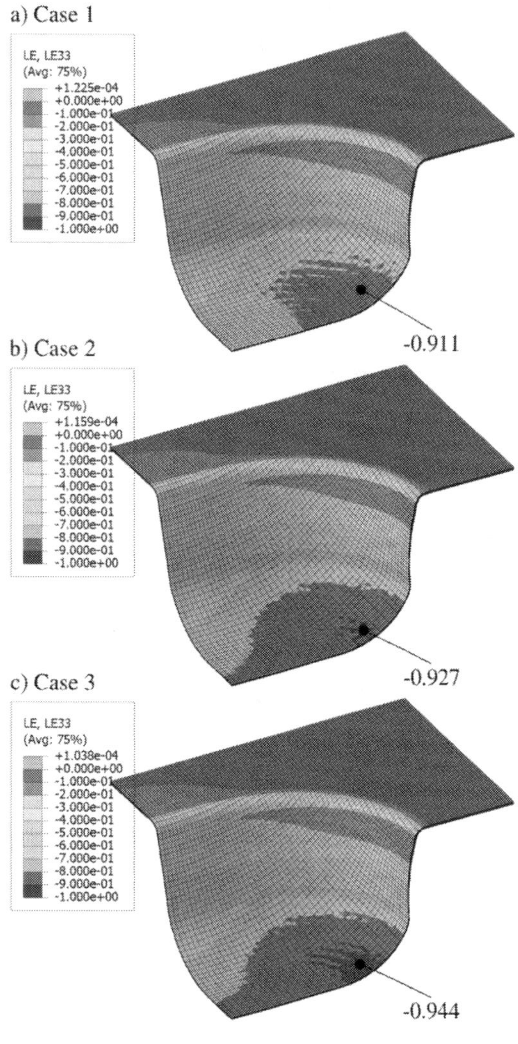

Figure 9. Computed minimal strain in thickness direction in the stretch forming simulations. The first case is shown in a), the second in b) and the third case in c). The marks show the location of the minimal strain in thickness direction.

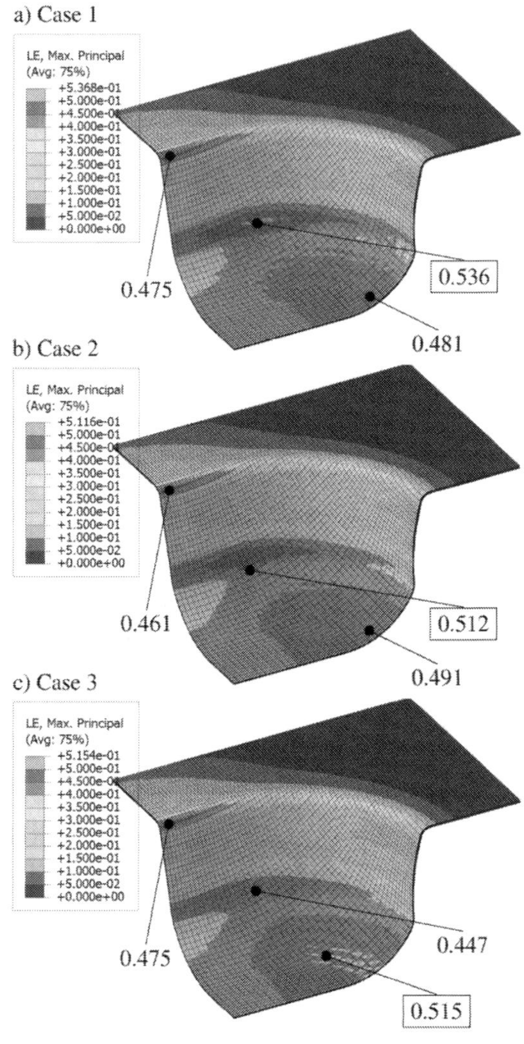

Figure 10. Computed maximal principal strain in the simulation of the stretch forming process at highest load. The first case is shown in a), the second in b) and the third case in c). The numbers with the boxes around indicate the highest values.

sition from the flat area at the bottom of the cavity to the curvature. A detailed analysis of the different layers in this area shows, that nearly the same strain occurs in every material layer. This was expected since the foil has to deform with the displacement movement of the punch. By contrast, the stresses are different in every layer. In the second case, where linear elasticity and rate independent plastic behavior with isotropic hardening was defined, a minimal strain of -0.927 is reached. In the case with the transversely isotropic elasticity the lowest strains in thickness direction occur with a value of -0.944.

The third hypothesis refers to the forming limit of the foil. Therefore, the maximal principal strains are evaluated. The outcome of the simulation is shown in Figure 10. Here, the location of the highest value is not that clearly as for the other hypotheses. In the first case the maximal value of strain reaches 0.536 and is located at the area, where the punch has an edge. But on the bottom of the cavity and near the edge of the die plate high strains occur, too. The second case has a maximal principal strain of 0.512. The location is the same as in the first case. Again high strains on the bottom of the cavity and near the edge of the die plate occur. In the third case the maximal principal strain of 0.515 appears on the bottom of the cavity near the radius. At the edge of the punch and at the edge of the die plate are other high values, too.

5 Conclusion

For numerical investigations of the deformation behavior of aluminium-polymer laminate foils a micromechanical FE model was developed. This model is based on the repre-

DOI: 10.1051/matecconf/20168016001

sentative volume element technique. It consists of a layered structure with periodic boundary conditions. Different linear elastic, rate independent plastic material behaviors with isotropic hardening have been assigned to the single layers of the RVE.

By simulations with the RVE nominal stress-strain curves and shear stress-shear strain curves were gained. These curves give an insight into the anisotropy that results only through the laminate structure. With the help of these curves the 12 parameters of the orthotropic linear elasticity have been determined. The values of these parameters show that the elastic behavior of the RVE can be described with the special case of transversely isotropic elasticity. Furthermore, the RVE was used to develop two simple homogenized material models on the basis of the composite behavior under tension load and the anisotropic linear elasticity.

In addition a stretch forming model close to the real cold stretch forming process of blister packaging was designed. That model offers the possibility to define a foil with single layers with different material models. Thereby, three simulations of the forming process were done. In the first case the three layer of the foil were assigned with the same material properties used in the simulations with the RVE. For the second case a homogenized material behavior was used which corresponds to a linear elastic, rate independent plastic material behavior with isotropic hardening. In the third case the linear elasticity of the homogenized material model was changed to transversely isotropic elasticity with determined parameters from the calculations with the RVE.

Three hypothesis regarding the reason for failure while stretch forming the foil are made. The first hypothesis is that a crack may nucleate at the location of the maximal principal stress. All simulations showed that the location of the maximal principal stress is where the punch has an edge. In the second case and in third case these stresses appear at the surface of the foil. In the first case it occurs in the aluminium core layer of the laminate foil. The stress values differ in every case. The second hypothesis is that a critical thickness reduction is responsible for failure. Regarding this the strain component in thickness direction has been evaluated. The results of the simulations show that high negative strains occur between the bottom of the cavity and the curvature. It is worth noting that in the first case in every point of the foil the different layers underlie the same strains. The third hypothesis refers to failure of the foil at the area of the maximal principal strain. The outcome of the simulations shows that the maximal values are similar. But the locations differ between the cases. To mention is that in every calculation three areas with high maximal principal strains occur. One is the transition from the bottom of the cavity to the curvature. The other are at the edge of the punch and at the edge of the die plate.

References

[1] P.N. Giannakakos, K.L. Yam, Packaging Technology and Science **5**, 31 (1992)

[2] M. Geiger, M. Kleiner, R. Eckstein, N. Tiesler, U. Engel, CIRP Annals-Manufacturing Technology **50**, 445 (2001)

[3] K. Chen, K.S. Schweizer, Macromolecules **44**, 3988 (2011)

[4] J. Gere, B. Goodno, *Mechanics of Materials* (Cengage Learning, 2012)

[5] F.Ó. Dubhlaing, D.J. Browne, R. Rennicks, C. Rennicks, Light Metals 2013 **1**, 405 (2013)

Numerical investigation of blanking for metal polymer sandwich sheets

Florian Gutknecht[1,a], David Übelacker[2], Till Clausmeyer[1] and A. Erman Tekkaya[1]

[1] *Institute of Forming Technology and Lightweight Construction (IUL), TU Dortmund, Germany*
[2] *Institut für Produktionstechnik und Umformmaschinen (PtU), TU Darmstadt, Germany*

Abstract. Metal polymer sandwich sheets consist of materials with drastically different mechanical properties. Due to this fact and because of high local gradients in the cutting zone during the blanking process, traditional process strategies and empirical knowledge are difficult to apply. A finite-element simulation of the shear cutting process is used to predict the necessary force and the geometry of the cutting surface. A fully-coupled ductile damage model is used for the description of the material behaviour. This model considers the influence of shear and compression-dominated stress states on the initiation of damage. Experimental tensile and compression test data is used for the identification of material parameters. The results of the blanking simulation are compared with experimental data. Furthermore, the evolution of the stress state is analysed to gain understanding of the underlying physics. Finally this model enables the prediction of core compression and other quantities such as the acting stresses and corresponding triaxilities, which provide valuable information for the development of analytical models.

1 Introduction

Sandwich sheets are excellently suited to fulfil conflicting requirements for weight reduction and increasing bearing capacity, and as a consequence replace monolithic sheets in some applications. Usually their structure consists of two thin metal sheets on the outside and a light and inexpensive core of polymer. For the sandwich investigated in this paper, the thickness and compliance is several times larger than the outer sheets. Due to the combination of materials with drastically different properties and high local gradients in the cutting zone during blanking process, traditional process strategies and empirical knowledge are no longer applicable. The selection of different materials with varying properties and thicknesses for the core, as well as the surface layers leads to a vast spectrum of possible effective behaviour, and thus offers a wide range of technological applications. But to establish this new class of material the users of this technology need to know how to process these panels. As single or multiple blanking occur in the process chain of all sheet metal parts, a physical understanding of the blanking process has high technological relevance. In particular, the knowledge of the stress state is important because it provides information on the load of tools, but is also assumed to be closely connected to the attributes of the finished product, such as the cut surface quality [2].

Even though several authors investigated the blanking of sandwich panels experimentally [1, 2, 7], the number of studies on the simulation of the blanking process of sandwich sheets is limited. However, Chen et al. [3] have presented a numerical investigation of half-sandwich-panels with only one skin sheet.

Complementary numerical investigation is advised, as those studies can support existing theories and might give new insights, because experimental investigation cannot give any statement about stress states. Moreover, information about strains is difficult to obtain due to the small deformation zone and high local gradients. Besides experimental and numerical investigations analytical methods may be used for analysis; yet up to now they are not able to provide an adequate insight in surface quality, or force requirements. However, a first experimental observation to determine the maximum force during blanking of sandwich panels is presented by Übelacker et al. [1].

Due to the complexity of both, material and process the single layers of the panel are modelled instead of an average effective composite behaviour. Thus, this paper starts with characterization of the single layers. Afterwards the setup for shear cutting tests is presented.

In section three the numerical process model built is presented. Moreover, details on the choice of constitutive models are given.

Section four presents the surface quality and force displacement curve of different setups in experimental shear cutting. The results of a single setup are compared to the corresponding simulation results. An analysis of the stress state evolution reveals insights into the physics of shear cutting sandwich panels.

2 Experimental investigations

An aluminium-polymer sandwich panel, commercially available as Alucobond from 3A Composites, is examined.

[a] Corresponding author: Florian.Gutknecht@iul.tu-dortmund.de

As depicted in Figure 1, the sandwich panel has a symmetrical structure.

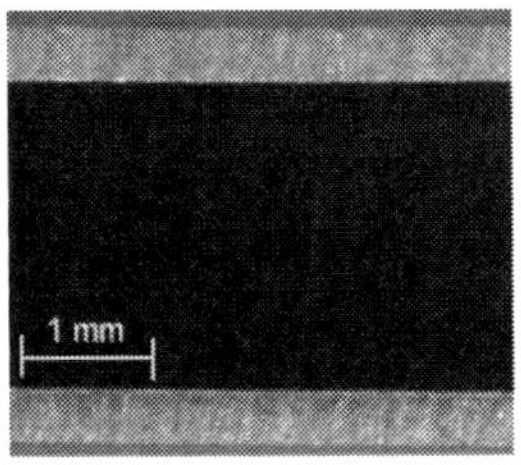

Figure 1. Microsection of the used aluminium-polymer sandwich panel.

The skin of the sandwich is manufactured from 0.5 mm sheets of Aluminium EN AW 5005A H22/H42 and the core is taken from a 3.3 mm thick panel of low density Polyethylene (LDPE). During the manufacturing process the single layers are glued together. To this end, two thin adhesive foils are placed between the skin sheets and the core material. The bonding process is initialized by high temperatures and a constant pressure. Thus the thickness of the composite is reduced from 4.3 mm to a final thickness of 4 mm.

2.1 Material characterisation

For characterisation of the sandwich behaviour single unprocessed layers of core and skin material are delivered by the manufacturer. These layers are tested in different setups, given below. Some deviations in the mechanical behaviour of the single layers compared to the complete sandwich are expected, because the layers in the sandwich panel have already undergone some forming steps. For investigation of the elastic-plastic behaviour of the aluminium uniaxial tensile tests according to DIN 50125 are performed. Thus the specimens have been loaded at a constant strain-rate of 0.0025 1/s. Modelling of damage requires the determination of fracture strains for different triaxialities [8], e.g. ratio of hydrostatic stress versus von Mises stress. Due to physical limitations of testing the 0.5 mm thick sheets in shear and compressive loading, fracture strains from Sun et al. [5] for a similar material are used as fracture locus.

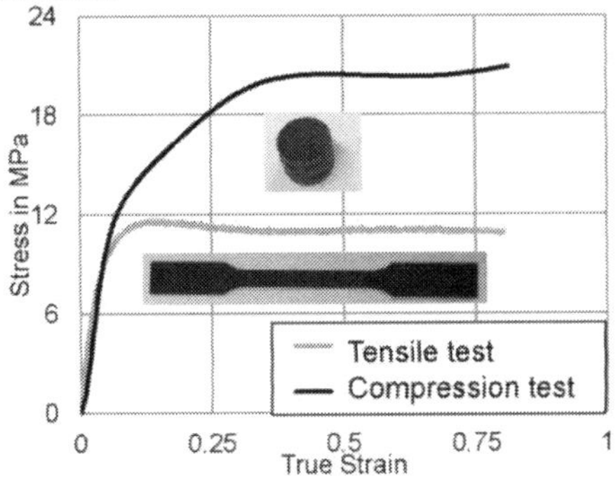

Figure 2. Polymer stress strain curve for tensile and compressive loading. [3]

For identical LDPE sheets uniaxial tensile and compressive test have been performed by Chen et al. [3] according to DIN EN ISO 527 and DIN EN ISO 604 respectively. Results are shown in Figure 2. The initial yield stress is similar in tension and compression. For larger levels of strain a substantial difference is observed.

2.2 Blanking tests

A conventional shear cutting tool is used for the experimental investigation. A detailed view a) and an illustration/photography b) of the tool are depicted in Figure 3. The cylindrical punch (B) is guided by a standardized guidance (C). During the downward stroke the sandwich sheet is cut along the die (F). The sheets are positioned in a material inlet (E). During the backward stroke the sheet and the punch are separated by a stripper (D). To measure the cutting forces a Kistler 9051A force measuring ring is used (A). The displacement curve is measured by an inductive displacement transducer HBM WA/100 (G). A HBM Quantum is used as measuring amplifier.

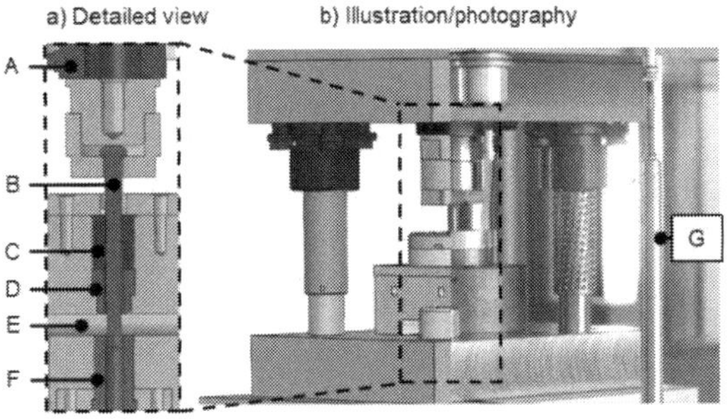

Figure 3. Experimental setup of the shear cutting tool [1].

All experiments are performed on a Synchropress SWP 2500 direct-driven servo press. The experimental parameters are listed in Table 1. As listed, a sine curve with amplitude of 25 mm and an average cutting speed of 50 mm/s is used. Within the parameter variation, two different punch diameters and five clearances are investigated. A lubricant free shear cutting process is used to prevent an absorption of the lubricant by the composite.

Table 1. Experimental parameters.

Experimental parameters	
Punch diameter [mm]	8 and 32
Punch and die material	1.3343 (64 ± 2HRC)
Cutting edge radius [µm]	15
Clearance u [µm]	25/ 50/100/200/400
Avg. Cutting Speed [mm/s]	50
Stroke curve type	sine
Amplitude [mm]	25

3 Simulation model

Previous investigations [4] have shown that even conventional shear cutting of monolithic sheets poses some challenges. On the one hand the constitutive model has to account for different aspects of the material behaviour (e.g. plasticity, damage depending on the stress state). On the other hand the numerical method has to be capable to deal with severe deformation and high gradients in space and time (fast increase of plastic strain and damage in a narrow zone).

3.1 FE-Setup

Due to the characteristics of the process, namely strong nonlinearity of the constitutive behaviour, an explicit FEM approach is chosen. The framework presented in [3] is briefly reviewed. Assuming an additive split of the symmetric part $\mathbf{d}=\text{sym}(\mathbf{l})$ of the velocity gradient $\mathbf{l}=\dot{\mathbf{F}}\cdot\mathbf{F}^{-1}$ into an elastic part ($\mathbf{d}^e$) and a plastic part ($\mathbf{d}^p$) $\mathbf{d}=\mathbf{d}^e+\mathbf{d}^p$, one obtains the Jaumann-type stress update

$$\overset{J}{T} = (1-D)\left(\frac{E\nu}{(1+\nu)(1-2\nu)}\text{tr}\left(\mathbf{d}^e\right)\mathbf{1}+\frac{E\nu}{(1+\nu)}\mathbf{d}^e\right) \quad (1)$$

for the general case of fully-coupled elasto-plasticity, with $\mathbf{F}$ the deformation gradient, D the damage variable as well as E the Young's modulus and ν the Poisson's ratio. $\text{tr}(\bullet)$ denotes the trace operator. The general form of the yield condition for coupled damage and plasticity with isotropic hardening is

$$\Phi = \frac{1}{(1-D)}\sigma_{eqv} - \sigma_y(\alpha) \le 0 \quad (2)$$

where σ_y and α represent the deformation-dependent yield stress and the equivalent plastic strain, respectively.

All simulations are performed with the software ABAQUS/explicit. Assuming isotropic material behaviour, an axisymmetric setup is chosen. The minimum element-size is set to 0.02 mm in a structured mesh (Figure 4).

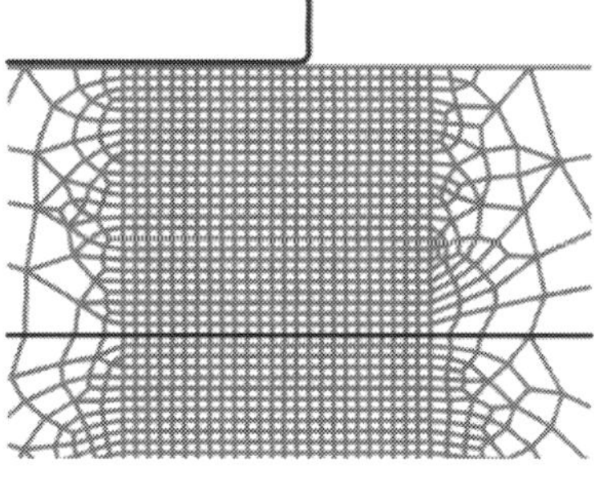

Figure 4. Detail of spatial discretization

Compared to the shear cutting of monolithic steel the usage of an Arbitrary-Lagrange-Euler (ALE) step for adjusting mesh topology is not necessarily required, as the mesh distortion is smaller compared to blanking simulation of monolithic sheet and thus omitted. Contact with the tools is considered via a kinematic formulation and a friction coefficient of 0.1 in the framework of Coulomb's model. The tools are modelled as discrete rigid bodies with an edge-radius of 0.03 mm.

3.2 Material model

The heterogeneous material properties of the sandwich-panel in combination with strong local effects motivate the modelling of the individual sandwich-layers. Any softening in the individual layers is captured with the damage routines provided by ABAQUS. Initially a damage initiation criteria has to be defined. When the criteria is fulfilled, an evolution of damage variable D, coupled to plasticity according to equation 2, may be defined. As soon as the damage reaches a value of 1 the stiffness of the corresponding element is set to a small value and does no longer contribute to the solution. To counter the natural mesh-dependence of local damage-models, either an energy or displacement at fracture may be defined.

3.2.1 Aluminium

For the aluminium the equivalent stress in equation 2 is chosen as the von Mises stress, thus $\sigma_{eqv} = \sigma_{eqv}^{vM} := \sqrt{3/2}\,\|\text{dev}(\mathbf{T})\|$. Furthermore the hardening is assumed to be isotropic. From the tensile tests the elastic parameter Young's Modulus E and poison ratio ν, as well as the ultimate tensile strength R_m and the corresponding hardening exponent n are determined. The latter two values can be used to determine the coefficients for the applied hardening law $\sigma_y = C\alpha^n$. The values are given in Table 2.

Table 2. Elastic and plastic material parameters for aluminium.

Material parameter	Value
Young's Modulus E	75 000 MPa
Poison ratio ν	0.3
Max. uniform strain n	0.037
Extrapolation coefficient C	176 MPa

The equivalent plastic strain at the onset of damage $\bar{\varepsilon}_D^{pl}$, according to [6] is assumed to be a function of stress triaxiality $\eta = \sigma_{hyd}/\sigma_{eq}$. With σ_{hyd} and σ_{eq} being the hydrostatic and von Mises stress respectively. At each increment during the analysis the incremental increase $\Delta\omega_D$ is computed.

$$\Delta\omega_D = \frac{\Delta\bar{\varepsilon}^{pl}}{\bar{\varepsilon}_D^{pl}(\eta)} \ge 0 \quad (3)$$

The values of $\bar{\varepsilon}_D^{pl}(\eta)$ used in this analysis, can be found in [5]. For the evolution of damage a fracture energy of 10 kJ has been found iteratively.

3.2.2 Low density Polyethylene (LDPE)

Preliminary simulations [4] revealed that during the blanking process of sandwich sheets the stress state is predominantly compressive, i.e. negative values of the hydrostatic stress are observed in the polymer. Therefore it is assumed that accurate modelling of the behaviour during compression is more important, than the prediction of the behaviour during tensile stress states, i.e. for positive values of normal stress. Consequently, a von Mises plasticity model with isotropic hardening, which is determined from compression test, is chosen for the polymer (Figure 2).

From tensile tests a failure strain of 0.4 could be determined. This value was extrapolated for other triaxialities, to obtain a function $\bar{\varepsilon}_D^{pl}(\eta)$ for damage initiation. The corresponding values are given in Table 3. For the evolution of damage, a displacement at fracture of 0.05 has been determined iteratively.

Table 3. Failure strain of LDPE.

Triaxiality	Failure strain
-3.0	1.0
-1.0	0.5
0.0	0.4
0.33	0.4
1	0.1

3.2.3 Interface

Experiments of Chen et al. [3] show that the interface under mode I loading will fail at a maximum stress of 24 MPa. Recalling that the maximum flow stress of LDPE is below 12 MPa under tensile loading, the strength of the interface is assumed to be superior to the core material. Early simulations have shown dominance of pressure stress state in the entire process-cycle. Both arguments support the assumption that the behaviour of the interface has a small influence on the cutting force and the cutting surface. As the interface is supposed to have negligible effect on the overall performance of the shear cutting process, the constitutive behaviour of the interface is modelled by an ideal strong bonding. Therefore explicit prediction of delamination is not possible. However, the focus is on prediction of blanking force and shape deviations.

4 Results and Discussion

4.1 Blanking tests

The experimental data used for the validation of the numerical model is part of a greater experimental study presented by Stein et. al [2]. In Figure 5, microsections of the cutting surfaces are shown. As depicted, the cutting surface is strongly influenced by the cutting parameters. In general, the face bending increases with larger punch diameters. This can be explained by the acting tangential stresses in the core. In case of a smaller punch diameter the tangential stresses are higher compared to the larger punch. As a consequence, the resistance of the core material towards a deformation rises. This fact leads to a better support (i.e. transfer of axial forces) of the upper skin sheet and an earlier crack initiation.

Besides the punch diameter, the clearance between punch and die has a significant influence on the face bending. Because of the higher bending component, larger clearances lead to a higher face bending. Smaller clearances lead to parallel cutting surfaces.

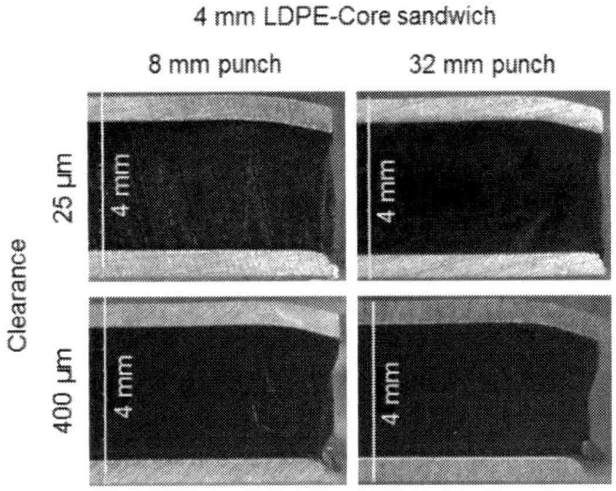

Figure 5. Cutting surfaces of LDPE-Core Sandwich [2]

In addition to the analysis of the cutting surfaces the cutting force is measured. Figure 6 shows typical force displacement curves for metal polymer sandwich sheets. The force displacement curve is separated in four main sections. The first section is dominated by a compression of the polymer core. In the second section the shearing process of the skin sheets occurs. When the skin sheets are cut, the third section begins and the core material is sheared. In the final section the force is dominated by the friction between the different layers and the shell surface of the punch. However, the clearance influences the blanked surface as discussed before. In general, a high level of parallelism caused by small clearances leads to more friction between the punch and the surface of the hole. As depicted in the force displacement curve this effect is dominant at the end of the process.

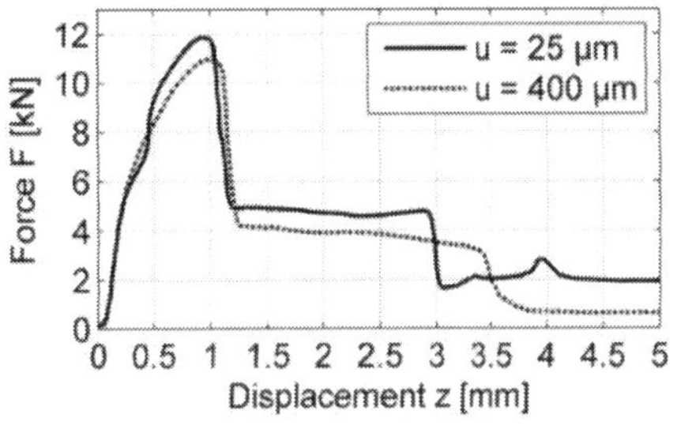

Figure 6. Force displacement curve for different clearences, D=32 mm [2]

In Figure 7, a specific force displacement curve for different punch diameters is shown. As depicted a delayed rupture of the face sheets can be seen when cutting with smaller punch diameters. This can be explained by the influence of the axial stresses. In case of a smaller punch diameter the ratio of the hole area to the cutting length decreases and the axial forces increase. Therefore, the core material will be more compressed before the rupture of the lower face sheet occurs.

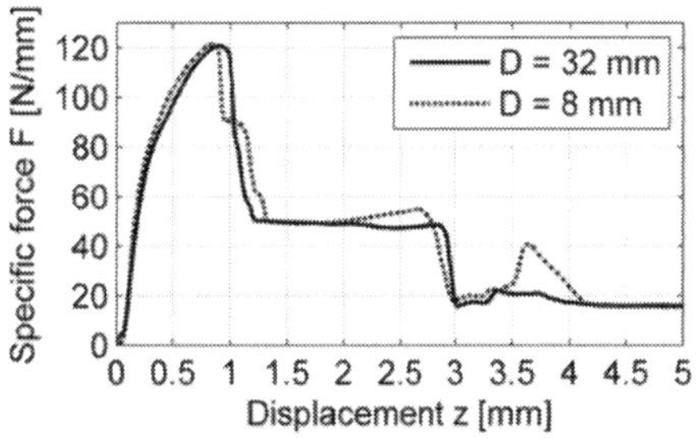

Figure 7. Specific force displacement curve for different punch diameters, u=50 μm [2]

4.2 Blanking simulation

4.2.1 Validation

With the setup given in previous sections, the results of the simulation are compared to the experimental ones.

For punch diameter D=8 mm and clearance u=100 μm a face bending of 290 μm could be determined (Figure 8 left). The corresponding simulation predicts a face bending of 340 μm (Figure 8 right). The relative error of 17% might be improved using a finer discretization in the cutting zone, as the precise identification on severely deformed elements is difficult. Other measures for cutting surface quality, as used for monolithic sheets, such as height of burnish or fracture are not examined, as they are no longer applicable.

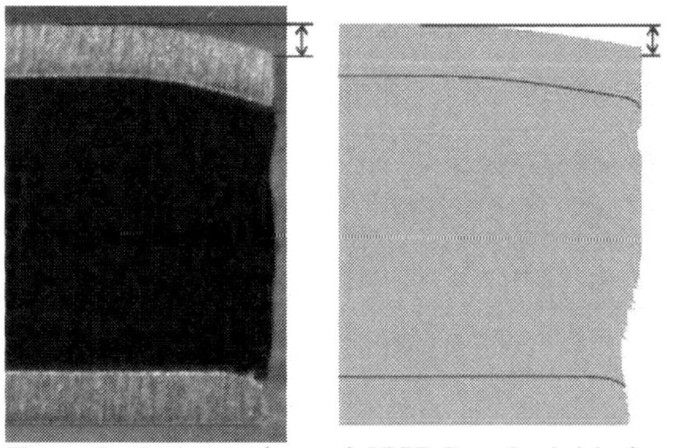

Figure 8. Cutting surfaces of LDPE-Core Sandwich from experiment (left) and simulation (right) for D=8mm and u=100μm

The force displacement curve is shown in Figure 9. Three distinct decreases in the force-displacement curve are identified. These are characteristic for shear cutting of sandwich panels. The first decrease indicates the rupture of the upper skin sheet, the second the separation of the lower skin sheet and the last descent occurs, when the polymer and thus the entire sandwich-sheet is separated.

While the characteristic shape of the curve for the punch force in the simulation meets the experimental very well, the maximum force has a relative error of about 20%. This discrepancy is expected to improve, with a more precise parameter identification for the fracture curve of the aluminium. A tougher fracture curve would result in later separation of the skin sheets and thus improve both maximum force and punch displacement of decrease towards the values observed in experiments.

The last decrease in the force-displacement curve is determined by the constitutive behaviour of the polymer. The maximum force is reached quite satisfactorily, but decreases quasi instantly. This is a result of the pragmatic approach for the extrapolation of the fracture curve and failure energy. Both fracture strain and energy were determined with rather small values, as a much more ductile behaviour would result in numerical difficulties that are hard to taken into account, even with an ALE approach. Beside that use of ALE is questionable when multiple materials are included. Alternative approaches like remeshing result in both very small elements and wide spread of these elements due to the enormous gradients in the cutting zone. Thus the computational cost would increase enormously.

The qualitative behaviour of the blanking of the sandwich sheet is well predicted. However, the model can be improved by using more sophisticated material models for damage and failure. Due to the good qualitative agreement, further analysis of simulation results with respect to the qualitative behaviour and general physics is applicable.

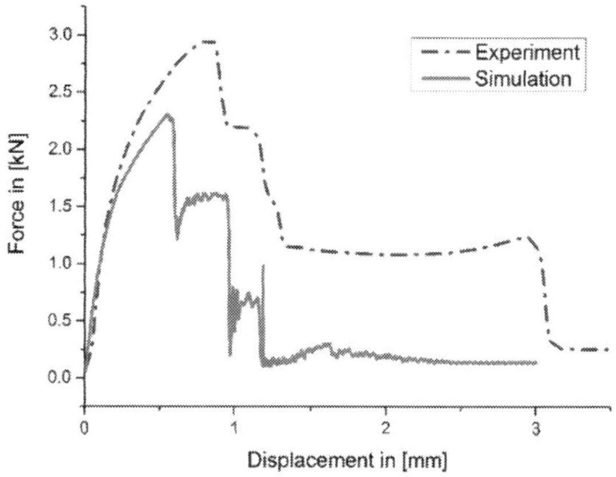

Figure 9. Force displacement curve for D=8mm and u=100μm

4.2.2 Analysis

A detailed analysis of the evolution of stress state, reveals that the physics of shear cutting sandwich-sheets differs significantly from the stress state of shear cutting monolithic sheets. The kinematics of the cutting process are described by the relative punch displacement $x = x_P/t$. x_P is the absolute displacement of the punch and t the sheet thickness. Furthermore, it is known that stress

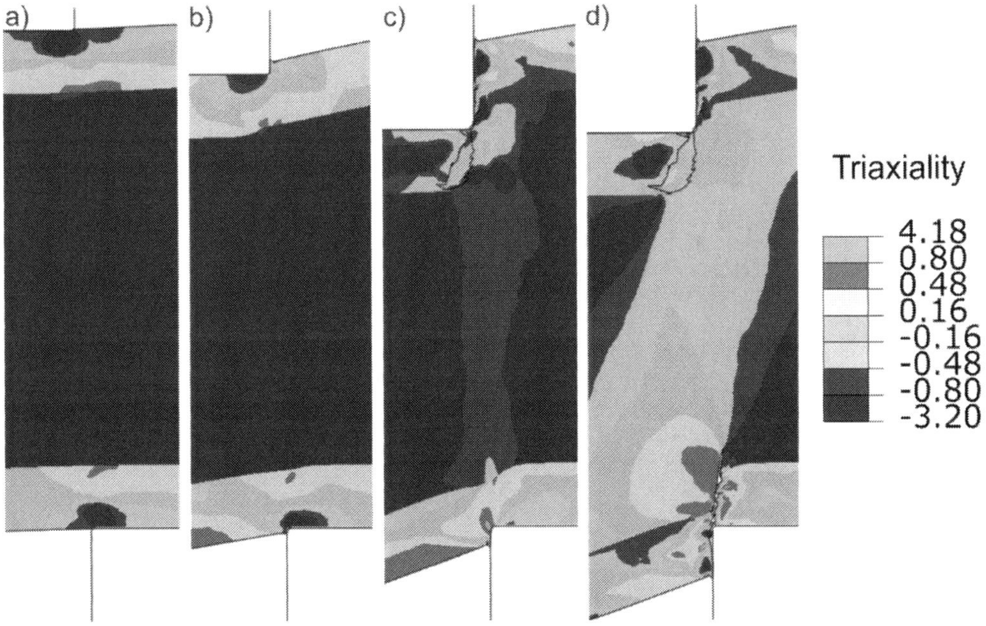

Figure 10. Distribution of triaxiality in the sandwich for relative punch-displacement a) 4%, b) 14%, c) 24% and d) 25%

triaxiality η is characteristic for certain states of stress. Thus $\eta = 0$ corresponds to shear, $\eta = \pm 1/3$ is equivalent to uniaxial tension/ compression and an equi-biaxial stress state in tension/ compression corresponds to $\eta = \pm 2/3$. For small punch penetrations ($x \approx 4$ %) the triaxiality indicates a bending of the skin sheets, while the core is compressed approximately under a low triaxiality (Figure 10 a)). This state is predominately retained up to the point before rupture of the upper skin-sheet ($x \approx 14$ %), with slight reduction of compressive states (Figure 10 b)). After the upper skin-sheet has been separated ($x \approx 24$ %), the triaxiality in the core changes to only biaxial compression and the lower skin sheet is loaded with pure tensile stresses (Figure 10 c)). When both skin-sheets are separated ($x \approx 25$ %) the triaxiality in the core adopts the pattern typical for shear-cutting of monolithic sheets (Figure 10 d)). This pattern has been described in detail by Gutknecht et al. in [4]. Compared to the processing of monolithic sheets the following differences are found for the skin sheets:

- The loading state is comparable to a bending case.
- Significantly reduced compressive and shear stress states, thus there is a vanishing influence of those stress states in damage evolution.
- Increased occurrence of uniaxial tensile stresses.
- Damage initiation in biaxial tensile stress states.

Knowing that the failure strain for high triaxiality may be determined at moderate cost via tensile and notched tensile specimens, while failure strains for low triaxialities in thin sheets are much harder to determine, these findings are pleasant. As a consequence characterisation tests for failure strain of the skin sheets might be easier as it is sufficient to consider tests with positive triaxiality.

For investigation of axial stresses transferred by the core the stress component in axial direction is depicted in Figure 11. The analysis reveals that the zone with a minimum transferred compressive stress of 20 MPa (maximum flow stress of LDPE) has a width of approximately 3 mm, which is identical to the core's undeformed thickness. This zone is not symmetrical with respect to the clearance. In the skeleton this zone extends for 1 mm and is almost constant over the height of the core. On the other hand in the coin the stress distribution depends significantly on the height. Below the punch this zone ranges 2 mm into the coin and decreases to 0.3 mm on the opposite.

After all one should keep in mind that the constitutive behaviour of the layers, that have been characterized and used for simulation may differ from the final behaviour in the sandwich due to thermal processing and the applied plastic strain by compression during sandwich manufacturing.

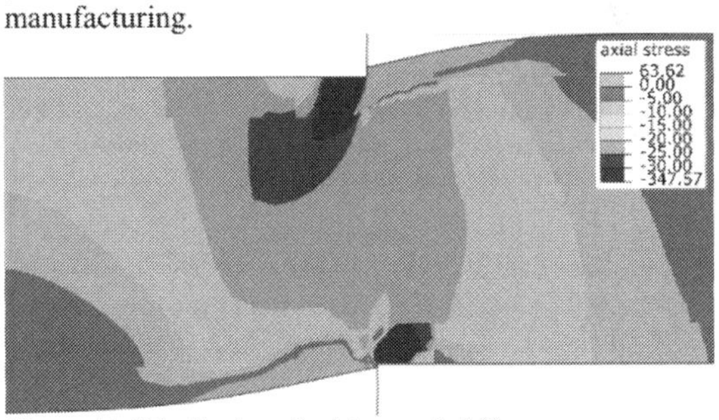

Figure 11. Distribution of axial stress in MPa.

5 Conclusion and outlook

An approach on modelling sandwich panels for blanking in the framework of the Finite-Element Method is presented. Experimental data for the characterisation of raw layers are either gained from uniaxial tensile tests or taken from literature. Based on this test data, choices for the material models are made. The polymer is assumed to be linear elastic and the plasticity is accounted for with von

Mises plasticity criterion. Furthermore, arguments are given to assume ideal bonding between each layer. An axisymmetric process model of shear cutting is built in the software ABAQUS/explicit. The process parameters are chosen to match those of physical process. Results of physical tests imply a significant effect of core compressibility on the cut-surface quality and the shape of force-displacement-curve.

Analysis of force-displacement curve and cutting-surface shows good qualitative compliance of simulation results with physical tests. The analysis of stress state in the sandwich reveals that the physics are different from shear cutting of monolithic sheets. The stresses in the skin sheet correspond to those for bending. As a consequence the relevance of shear stresses for deformation and failure decreases, while tensile stresses become more important. Axial stresses in the core are found to be asymmetrical. Significant stresses (in the range of the yield stress of LDPE) are only transferred in a zone, which is approximately as wide, as the core's thickness.

In further investigations different punch diameters and clearances, as well as material thickness will be examined to confirm the presented findings.

As the constitutive behaviour of the core is found to be crucial for the characteristics of the shear cutting results, special emphasis should be paid on the characterization of LDPE.

Acknowledgements

The authors are grateful for the support of the German Research Foundation (DFG) for funding the projects DFG Te 508/37-1 and DFG Gr 1818/44-1. Both projects are embedded in the Research-Cluster PAK 678/0 "Dry Shear Cutting of Metal Laminated Composite Material" supported by DFG and Arbeitsgemeinschaft industrieller Forschungsvereinigungen "Otto von Guericke" e.V. (AiF).

Moreover, the authors would like to thank 3A Composites for their support with Alucobond as well as all other contributors of the joint research project. Frank Steinbach (IUL, TU Dortmund) is thanked for providing the tensile data for aluminium.

References

1. D. Übelacker, J. Hohmann, P. Groche, Adv. Mat. Res. **1018**, 137-144 (2014)
2. P. Stein, D. Übelacker, D. Holke, P. Groche, Mater. Sci. Forum **825-826**, 433-440 (2015)
3. L. Chen, T. Clausmeyer, A.E. Tekkaya, Comput. Meth. Mater. Sci. **15**, 162-168 (2015)
4. F. Gutknecht, F. Steinbach, T. Clausmeyer, A.E. Tekkaya, *Proceedings of COMPLAS*, **13**, 170-181 (2015)
5. D.-Z. Sun, A. Ockewitz, *Proceedings of the Euro LightMAT*, 6ff (2013)
6. Dassault Systems, ABAQUS 6.14 Analysis User's Guide (2014)
7. M. Liewald, C. Bolay, S. Thullner, J. Manuf. Proc. **15**, 364-373 (2013)
8. Y. Bao, T. Wierzbicki, Int. J. Mech. Sci. **46**, 81-98 (2004)

Numerical investigations for simultaneously processing metal and plastic using impact extrusion

Jonas Wälder[1,a], Jochen Wellekötter[2], Alexander Felde[1], Mathias Liewald[1], and Christian Bonten[2]

[1]*Institute for Metal Forming Technology (IFU), University of Stuttgart, Germany*
[2]*Insitut für Kunststofftechnik (IKT), University of Stuttgart, Germany*

Abstract. Combining different materials for producing high performance parts opens a broad range of high functionality and applicability. Scope of this project is the investigation of a new method to process metal and plastic within one single production step using impact extrusion. In this study, first FEM results of the combined impact extrusion of metal and plastic will be presented. The aim of the numerical investigation is to achieve a plastic flow of the polymer material during the metal forming process.

1 Introduction

Hybrid components are gaining more and more importance in the field of mechanical engineering due to the fact that weight has to be reduced, demands for special component properties are growing, and the pressure to lower manufacturing costs is rising. Due to their complementary material properties the main focus is on hybrid components made from plastic and metal. Usually, metals are less resistant to aggressive media, but can transform vibrations very well, and are good thermal and electrical conductors. In contrast, plastics normally have inferior mechanical properties, but combine better thermal insulation, media resistance and good damping properties [1 - 3].

Down to the present day, the production of hybrid components has been characterized by serial production. Basically, there are two different ways of production [2]. In both cases, the forming process of the metal component takes place separately from the production of the plastic component. Figure 1a) shows a component produced in a form fit joining/bonding process. This process consists of at least three steps: the production of the metal component, the separate production of the plastic component, and the following joining process. A connection of the complementary properties is only partially possible by simultaneous processing.

a)  b) 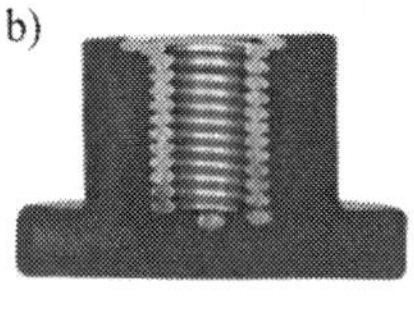

Figure 1. Manufacturing of hybrid components by a) joining and b) injection molding around metal inserts

Figure 1b) shows an alternative, but very common option for producing hybrid components. As described in the production process before, the metal component is produced before joining. The finished metal component is afterwards back injection molded and thereby encapsulated with molten plastic. Therefore, classic joining operations as shown in in figure 1a) have become redundant. It is, however, necessary to have at least two process steps (forming and back injection molding). Furthermore, the geometric requirements of the metal component have to be taken into account; e. g. a way of demolding must be implemented and the flow path of the plastic has to be optimized in order to achieve acceptable mechanical properties. This results in a more complex tool structure [2 - 5].

1.1 Process description

A new manufacturing process for hybrid components, developed at the Institute for Metal Forming Technology (IFU) and the Institut für Kunststofftechnik (IKT) of the University of Stuttgart, Germany, provides the possibility of manufacturing both material components within one single processing step. This new method allows to precisely include plastic granulates into the forming process in order to simultaneously form the metal blank and the plastic (cf. figure 2). Fundamentally, this approach to production opens a new spectrum of light weight constructional components, isolation- / damping components, and high performance components.

This new process also allows to reduce the wall thickness of metal components in the field of lightweight construction without losing rigidity. Components in less stress loaded areas can contain a light plastic phase. Due to the low density of plastics (e. g. 0.917 g/cm³ for polyethylene [6] and 1.13 g/cm³ for polyamide [7]), the

[a] Corresponding author: jonas.waelder@ifu.uni-stuttgart.de

weight can be reduced as well while the component properties remain the same.

Another new field of application is the manufacturing of NVH-components (noise, vibration, harshness) which offer vibration damping properties. The components are made of an outer metal shape and an inner plastic core. This kind of components allow to transfer specific properties of plastic on the whole component. Similar concepts have already been implemented successfully in the field of sheet metal forming (damping plate) [6].

Further scopes of application for metal-plastic-components are high performance components subjected to special requirements. For example, it is possible to produce pipes with a protective metal coating to increase properties regarding mechanical wear. Another possibility of use is the production of plastic components with a metal and thus electrical conductive outer layer. The processing of plastic and metal in one processing step has further advantages such as shorter cycle times, better utilization of machines, and increased production efficiency.

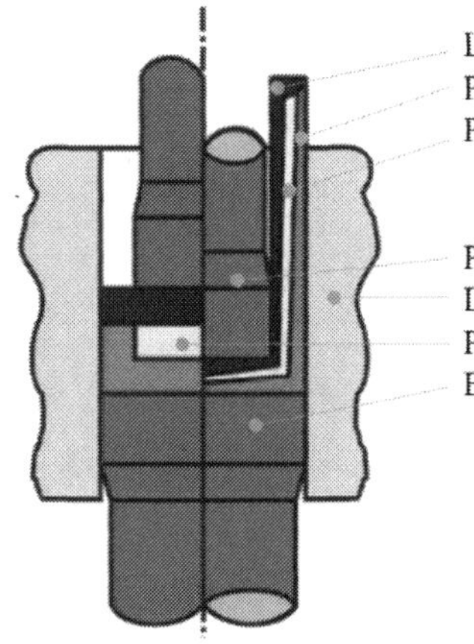

Figure 2. Scheme of the cup-backward-extrusion; left: initial state; right: process sequence during forming process

As described, the new approach developed by IFU and IKT deals with the combined processing of plastic granulates and bulk metal forming by impact extrusion, which is well-established in bulk metal forming technology. The forming process can be seen in Figure 2. The figure shows the initial situation of the process on the left and the process sequence during the forming process on the right. For the manufacturing of hybrid components, two forming processes can be considered: full-forward extrusion and cup-backward-extrusion, which will be described in more detail. At the beginning of the forming process the prepared raw part is inserted in the forming tool. This raw part consists of three components: a metallic lid (dark blue), a metallic pre-cup (light blue), and plastic which is inserted as granulate (yellow) without being compacted before.

At the beginning of the forming process, the raw part is upset and the granulate is compacted before the material begins to yield. After the compaction is complete (bulk density ≈ 1), the basic forming process starts and the cup begins to flow (figure 3). The temperature developed by the forming process and the initiated shear lead to plasticization of the granulate.

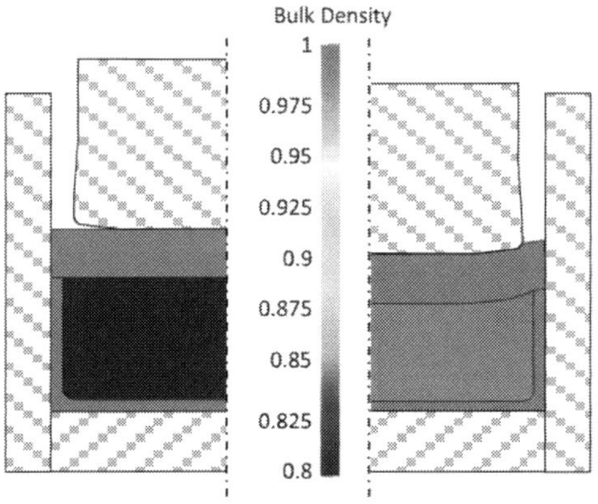

Figure 3. Compression of the bulk good

As the degree of plasticization depends on the resulting heat, it is attempted to achieve the highest temperature possibly attained in order to ensure a complete in-situ processing of the polymers.
The principle of full forward extrusion is identical to backward extrusion. First, the plastic granulate is compressed and plasticizes due to high pressure and temperature development. After processing, cooling of the plastic and metal phase take place.
The materials used are aluminum alloy EN AW 6082 and steel alloy 16MnCr5. The lid and the pre-cup always consist of the same metallic material. The two used thermoplastics are polyethylene and polyamide, which are often used standard and technical plastics, respectively [7, 8, 9]. The plastic is always added to the process in granular shape. Following, the material characterization of metal and plastic will be described and the forming simulations will be analyzed.

2 Material characterization

A simplified simulation of the forming process is to be executed using the software STFC Deform. Therefore, all material groups used (metal and plastic) have to be characterized in terms of modelling its material behaviour during the forming process. In order to display the metal materials, the established upsetting test was used. The plastics can be displayed with a rheometer shear test, which is well-established in plastic processing. The observed amplitude sweep will be converted into a stress-strain curve.

2.1 Recording of Flow Curves

The flow curves were determined with simple cylinder upsetting tests in order to characterize the aluminum alloy AlMgSi1 (EN AW 6082) and the steel 16MnCr5. The flow curves were recorded using the Institute's thermo-mechanical testing system (Gleeble 3800). The forming velocities for both materials were chosen to 1 s^{-1}. The temperature ranged from room temperature to 400 °C.

DOI: 10.1051/matecconf/20168016003

The upsetting samples were heated up straight before the forming process. For the regulation of the sample temperature thermocouples were welded on the test specimen's lateral surface. These thermocouples allow recording the temperature of the test specimen´s shell during the forming process. In order to minimize the friction at the sample's front surface, graphite foil was placed between the front surface and the carbide crush areas. The heating rate was given to 4 Ks^{-1}. The maximum temperature was afterwards held was for 60 seconds in order to achieve a homogeneous temperature distribution in each sample. The elastic strain of the machine can be corrected depending on force. Every flow curve was recorded three times.

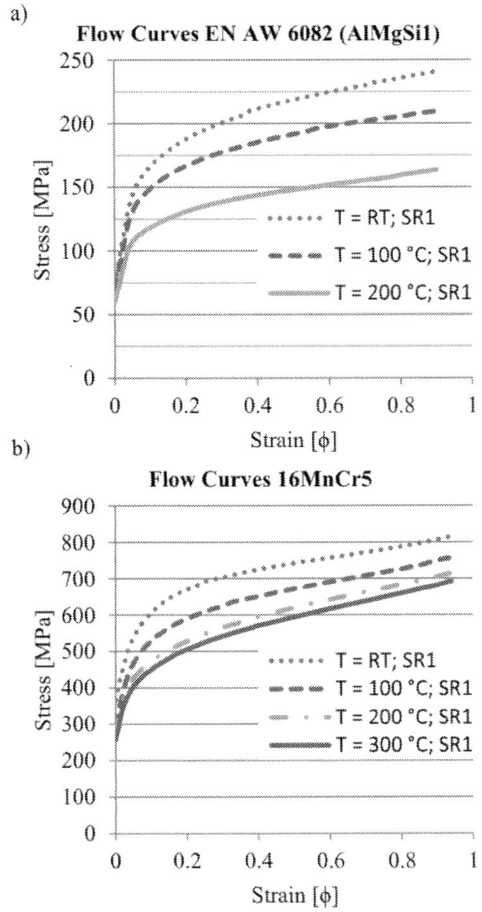

Figure 4. Determined flow curves of a) EN AW 6082 and b) 16MnCr5 at different temperature levels and constant strain rates of 1 s^{-1}

Figure 4 shows the determined flow curves of the aluminium alloy above and steel below. The starting yield point of the aluminium alloy ranges near 90 MPa at room temperature. As the forming process goes on the yield stress increases to 235 MPa ($\varphi = 0.8$) at room temperature. Steel has a significantly higher yield stress of 400 MPa right at the beginning, while the yield stress is rising up to 800 MPa ($\varphi = 0.8$) during the further course of forming. Due to the different starting stress curves, the pressure for compressing the plastic granulate can be varied. A high initial yield stress generates high pressure (and a high temperature) on the plastic granulate. A smaller initial yield stress creates a smaller pressure (smaller temperature development) before the forming process starts. Therefore, the material of the lid and the pre-cup influence the plasticizing of the plastic granulates during forming.

2.2 Characterization of the plastic phase

The behaviour of plastics fundamentally differs from that of metals. Characteristic temperatures of plastics such as melting temperature usually are significantly lower than their equivalent temperatures of metal. Obviously, this is both disadvantageous and advantageous for the plastics industry: disadvantageous due to the lack of high temperature applications and advantageous because of lower energy consumption (and therefore costs) during processing [9].

In order to manufacture a plastic part, a plastic granulate usually is heated to temperatures higher than its melting temperature; the resulting fluid is then deformed under pressure in order to achieve the shape of the final part and cooled afterwards. For the simultaneous impact extrusion of metal and plastic it is essential to determine whether the plastic will melt during the process. Melting temperatures of plastics may vary significantly due to the huge variety of different kinds of thermoplastics. Potential plastics for impact extrusion like polyethylene and polyamide have melting temperatures of 108 °C and 220 °C, respectively [7, 8].

The temperature dependent heat uptake of a plastic can be determined using differential scanning calorimetry (DSC) as shown in figure 5 for low density polyethylene (PE-LD; Lupolen 1800 S, LyondellBasell Industries, Rotterdam, NL).

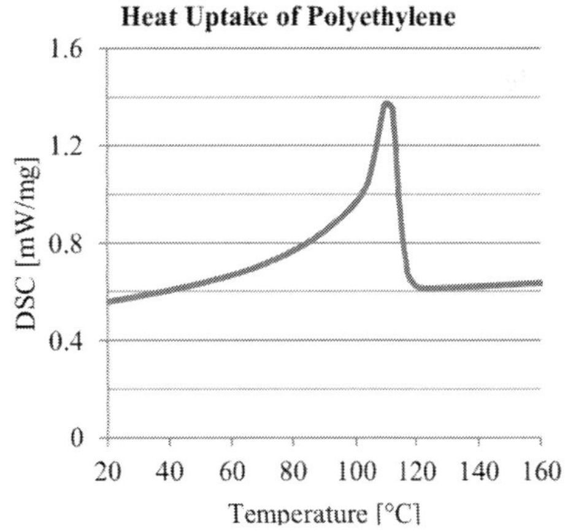

Figure 5. Heat flow versus temperature of PE-LD at a heat rate of 10 K/min

The peak near 120 °C is due to the melting of the crystalline phase of the plastic, which consumes huge amounts of energy [9]. The DSC data can be implemented into the simulation to set a specific heat uptake for the plastic at each temperature.

If temperatures higher than the melting temperature of the plastic are achieved, fluid state can be expected.

Assuming fluid state during processing, the plastic shows shear thinning behaviour. This means that the force required for deforming the fluid depends on the shear rate, and therefore directly on deformation speed. This behaviour is equivalent to the behaviour of solid state metals. For simulation purposes an equivalent to the flow curve of a metal needs to be discovered. Therefore, viscosity curves using specific displacements were measured at the IKT at different shear rates, as shown in figure 6 for low density polyethylene.

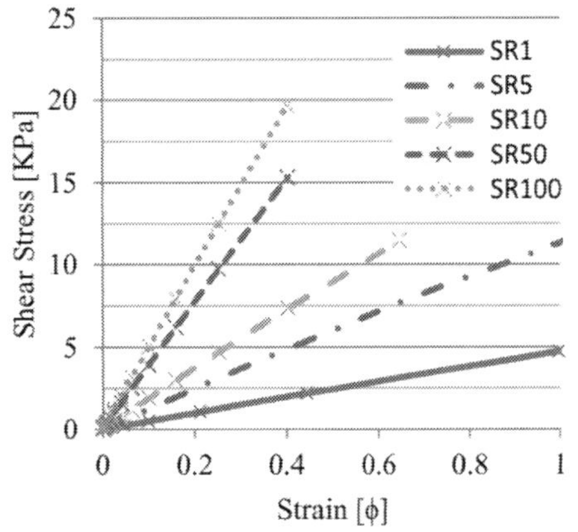

Figure 6. Shear stress versus displacement for PE-LD (T = 190 °C), measured for strain rates of 1–100 s^{-1}

The tests were conducted using a parallel plate rheometer SR 500 (Rheometrics Inc., Piscataway, NY, USA) with set deformations up to 100 % and deformation rates of 1 s^{-1} to 100 s^{-1}. The corresponding momentum was measured, and the resulting stress was calculated. Using such obtained data, the dependency of the shear stress versus different displacements can be determined and transformed into a flow curve using equation 1 [10]:

$$\bar{\sigma} = \sqrt{3} * \tau \qquad \varepsilon = \frac{\gamma}{\sqrt{3}} \tag{1}$$

Where $\bar{\sigma}$ is the resulting strain and ε the corresponding elongation of the flow curve, τ is the shear stress and γ the displacement. However, it has to be mentioned that the data is acquired at a set temperature of 190 °C (for polyethylene). Since the temperature is not constant during impact extrusion the approach is more difficult to handle. Hence, in future works temperature dependent flow curves will be implemented.

3 Design of the Forming Process

For the simulation of the forming process, the commercial program STFC DEFORM was used. Due to the component's rotational symmetry, it is possible to reduce the numeric model down to two dimensions. As material properties, the determined characteristic values (see chapter 2) are used. The metallic components were modulated as all-plastic and the polymer was modulated as porous (density 0.8) because of its insertion as loose material. Furthermore, the friction model to shear was used with a friction coefficient of 0.3. First of all, screening simulations were carried out to study the functionality of different raw part geometries. Based on these findings, a detailed raw part design was carried out by means of DOE in the framework of this paper.

3.1 Screening

Using model properties as shown above, numerical tests were carried out to find an optimal geometry of the raw part. Therefore, four different raw part variations were simulated and evaluated at the beginning. The aim was to choose the initial geometry in a way in which the polymer is completely encapsulated between the raw part cover and pre-cup as well as to reach a satisfying form closure. The important parameters for the plasticizing will not be discussed at this point.

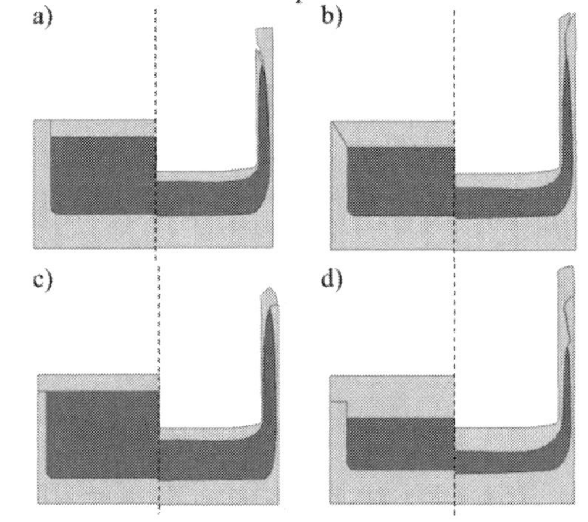

Figure 7. Depiction of the raw parts concerning the resulting form closure.

The figures a-d depicted in figure 7 show the initial situation of the processes on the left side and process sequence during forming on the right side. The preliminary tests revealed the following behavior:

a) The lid was designed as large as the inner diameter of the pre-cup. Hereby, the lid was pressed inwards at the beginning of the process. During the following forming process, a gap occurs. This gap prohibits the form closure to arise and the polymer may leak out.

b) The lid was designed as large as the outer diameter of the pre-cup. Additionally, it was provided with a bevel. This supports an intrusion of the lid into the polymer, preventing the occurrence of a satisfying form closure. However, the form closure tends to be more suitable at wide angles.

c) The lid was designed as large as the outer diameter of the pre-cup. This prevents the penetration of the lid into the pre-cup, causing a perfect form closure between the components.

d) The lid was also designed as large as the outer diameter of the pre-cup. However, a step has been added in order to get an optimized form closure. As in the previous variation c), no gap occurs permitting an escape of the plastic phase.

As shown, set up c) and d) are suited best for the manufacturing of hybrid components by impact extrusion. As a result of the raw part's geometry in these two variations, a complete inclusion of the polymer as well as a good form fit was achieved. Due to the similar behavior in the inpatient sector of the forming process, only variation c) is optimised via DOE. Therefore, the results achieved can be transferred directly on the raw part geometry of variation d).

3.2 Design of Experiments

In order to ensure a complete plasticization, it is important to achieve the highest possible temperature during the forming process (melting point of polyethylene: 108 °C: melting point of polyamide: 220 °C). Therefore, the temperature for the selected raw part by means of DOE was optimized.

Figure 8 shows the raw part to be investigated. Gap width (S), wall thickness (W), base thickness (B) as well as lid thickness (D) are varied as parameters. Corresponding variation ranges of the parameters are shown in table 1.

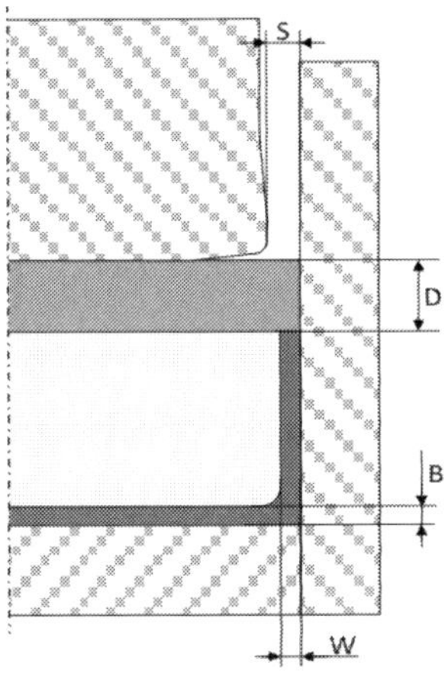

Figure 8. Definition of the used parameters for backward extrusion

Table 1. Parameter range corresponding to DOE calculation

Parameter	Min. [mm]	Max. [mm]
S (gap width)	1	3
D (lid thickness)	2	4
B (base thickness)	1	3
W (wall thickness)	1	3

The objective was to allow the inclusion of high amounts of plastic granulate without causing a decrease of the support effect, which is dominated by the metallic components of the raw part assembly. Furthermore, considerations concerning production were investigated. The influence of the parameter on the resulting temperature was determined by full factorial design.

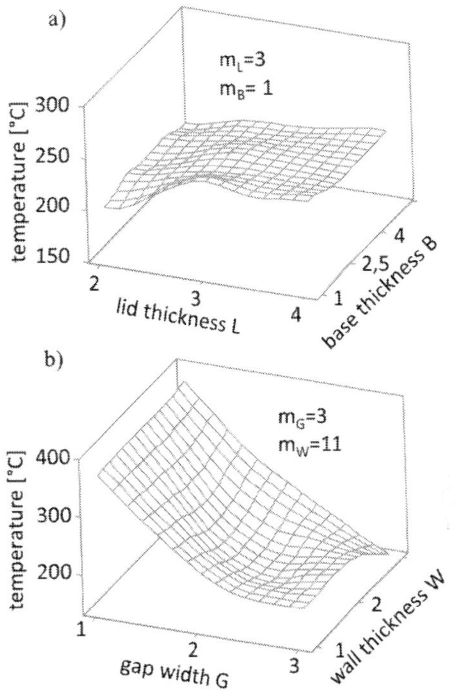

Figure 9. Response surface of the DOE

Figure 9 illustrates the response surface of the DOE as well as the strengths of influence. The resulting temperature is displayed on the applicate. The parameters lid thickness (figure 9a) and gap width (figure 9b) become apparent on the abscissae. The parameter's base thickness (figure 9a) and the wall thickness (figure 9b) are shown on the ordinate. Both unembodied parameters are kept constant (wall thickness = 1 mm, base thickness = 5 mm or gap width = 2 mm, lid thickness = 4 mm) in order to prevent a distortion by cross influences at the response field.

As shown in the figure, the resulting temperature ranges from 157 °C to 391 °C. Therefore, a processing of polyamide is not possible with regard to all variations (melting point 220 °C). As becomes evident in figure 9a, a rise of the temperature at increasing values of lid thickness and base thickness can be observed. The figure on the right side also shows a steady rise of the temperature for increasing values of the wall thickness or for decreasing values of the gap width. Based on the ratio (figure 9, m_L, m_B, m_G, m_W) of the average temperature increase and the parameter changes, the respective variable's influence can be determined: the higher the slope, the greater the influence of the value. This shows

that the gap width has the greatest influence on the temperature of the forming zone. Furthermore, it becomes apparent that the values wall thickness and lid thickness have a considerable influence on temperature, even though they are weaker than the gap width. The parameter base thickness has a negligible effect on the resulting temperature. Therefore, the parameters S, D, and W are set with regard to achieving a high resulting process temperature. The base thickness is defined in a way of reaching the greatest possible contribution of the plastic granulates due to its minor effect on temperature.
For the processing of polyethylene, all parameter variations are permitted due to the low melting point of 108 °C. For the production of components made of polyamide (melting point 220 °C) a restriction of the parameter values must be conducted.

Possible resulting raw part geometries for polyamide are shown in table 2. The table contains values for the defined parameters as well as the reachable temperature in the forming zone. The maximum achievable temperature referring to test number 8 is 391 °C. Based on the results of the DOE, experimental investigations with these defined parameters will be carried out in the next stage of experiment.

Table 2. Test parameters for a resulting temperature higher than 220 °C

Test number	B [mm]	D [mm]	W [mm]	S [mm]	Temp. [°C]
1	1	2	3	2	224
2	1	3	1	2	242
3	1	4	1	2	238
4	1	4	3	2	236
5	1	4	1	2	226
6	1	4	1	1	367
7	1	4	2	1	387
8	1	4	3	1	391

4 Conclusion and outlook

This paper deals with the manufacturing of hybrid components in one single production step via impact extrusion. It became obvious that the forming temperature has a large influence on plasticisation of the used plastics. With the assistance of the parameter variation, the raw part geometry which allows reaching a maximum of process temperature was defined. By optimising this measure, it is attempted to achieve the best possible plasticization of the polymers during impact extrusion.

Future research work will focus on the experimental validation of the numerical investigations as well as on analysis of the hybrid components. For example, the adhesive strength and mechanism between metal and plastic have to be investigated.

The authors like to thank the Deutsche Forschungsgemeinschaft (DFG) for their financial support with regard to the research project entitled 'Metall-Kunststoff-Fließpressen'.

References

1. W. Michaeli, O. Grönlund, A. Neuss, M. Gründler, J. Wunderle. Neuer Prozess für Kunststoff-Metall-Hybride, *Kunststoffe* **9**, Page 136–140 (2010),
2. U. Endemann, S. Glaser, M. Völker, Kunststoff und Metall im festen Verbund, *Kunststoffe* **11**, (2012), Pages 110–113
3. C. Bonten: Kunststofftechnik – Einführung und Grundlagen, Carl Hanser Verlag (2014)
4. M. Stumpf, Maßgenaues umspritzen metallischer Gewindeeinsätze, *Maschinenmarkt*, (2012)
5. W. Michaeli, W.M. Hoffmann.: Hybride Verbindungen, *Kunststoffe* **6**, Page 50–53 (2009)
6. M.M. Matsco: Injection Molding and Metal Fabrication Combined, *Design News* **1**, Page 96 (1996),
7. H. Friedrich : Leichtbau in der Fahrzeugtechnik, Springer Verlag, (2013)
8. Material Data Sheet: Lupolen 1800 S, LyondellBasell Industries, (2013)
9. Material Data Sheet: Ultramid B3S, BASF SE, Ludwigshafen, (2014)
10. J.D. Bressan, U. Kirchhof, Construction and validation tests of a torsion test machine, *Journal of Materials Processing Technology* **179**, Page 23–29, (2006)

Effect of normal stresses on the results of thermoplastic mold filling simulation

Alexander Bakharev, David Astbury, Shishir Ray, Franco S. Costa and Russell Speight

[1]*Autodesk, 259-261 Colchester Rd., Kilsyth, Victoria 3137, Australia*

Abstract. The paper deals with the effect of the normal stresses on the predicted flow front during the filling stage of thermoplastic injection molding. The normal stresses are predicted using the non-linear Criminale-Ericksen-Filbey model (a variant of the second-order fluid rheological model with viscosity, first and second normal stress coefficients dependent upon magnitude of shear rate) incorporated into a comprehensive 3D simulation software for mold-filling analysis. The additional stress term allows the prediction of the so called ear-flow effect (melt racing on the edges of the cavity).
Keywords: Finite Elements, Injection Molding, Simulation, Rheology, Normal stress difference, Ear-flow.

1. Introduction

Thermoplastic injection molding is the most common manufacturing process for producing plastic parts. Material is fed into a heated barrel, mixed, and forced into a mold cavity where it cools and hardens to the configuration of the cavity. Significant progress has been achieved in three dimensional finite element simulation of plastic filling the mold (mold filling analysis) [1]. Typically in commercial simulations polymer melt is considered as generalized Newtonian fluid, there the deviatoric stress tensor is proportional to the deviatoric deformation rate, the scalar coefficient connecting the shear rate and shear stress, known as viscosity, is dependent upon temperature, shear rate invariants, pressure and other factors [2]. However, the generalized Newtonian fluid model does not predict any normal stress differences during simple shear flow whereas real polymers usually exhibit significant normal stress differences.

In this current work, we develop a finite element mold-filling program that allows incorporation of the Criminale-Ericksen-Filbey viscoelastic model that can accurately predict normal stress differences in a wide range of temperatures and shear rates [3]. One of the motivators for this development was potentially improving the prediction of ear-flow, a little understood phenomenon of the more rapid advance of the flow front on the edge of a mold cavity than in the center of the cavity [4].

2. Ear-Flow Phenomenon

Numerous cases have been observed in industrial injection molding practice of amorphous materials which exhibit a race-track or ear-flow effect. This is the more rapid advance of the flow front at the edges of the molding cavity than in the center of the cavity, usually observed at elevated injection speed. This flow-leading effect at the edge cannot be explained by differences in cavity wall section thickness. In the worst cases, this race-tracking leads to air-traps and visual defects in the molded part. A typical flow front propagation demonstrating the ear-flow in a polystyrene material is shown on Figure **1**.

As was originally suggested by experimental observations of Murata *et al.* [5], and confirmed by mold-filling simulations of Costa *et al.* [4], in many cases the ear flow is caused by the higher polymer temperatures in the edge region. The temperature rise is caused by shear heating in the high-shear region (close to the boundary) of the runners and gates. This temperature rise is then convected into the cavity, favoring the cavity edge due to the distribution pattern in the gate. The effect of the preferential convection of the temperature rise from the shear regions of the runners is also similar to the development of flow imbalance in geometrically balanced feed systems explained by Cook *et al.* [6].

However in many practical injection molding cases ear flow phenomenon is observed for conditions where very little shear heating is predicted, thus, rising suspicions that there is also another mechanism responsible for the ear flow effect. Since normal stress differences caused by in-plane shear near the cavity edges may push polymer perpendicular to the flow they are a

[a] Corresponding author: Alex.Bakharev@autodesk.com

candidate for such a mechanism. Our simulation was used to test this possibility.

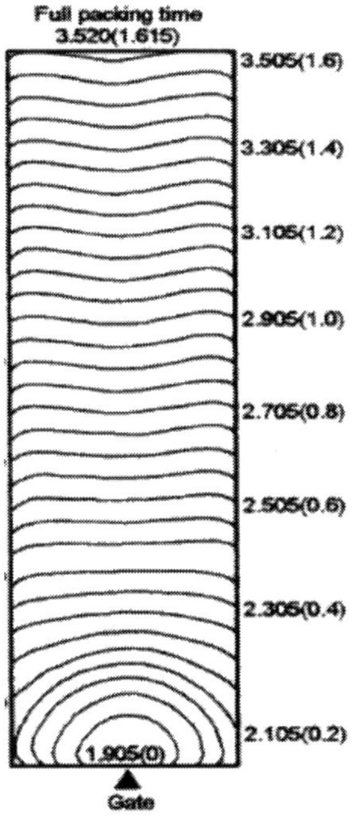

Figure 1 Flow front shapes obtained by Murata *et al.* **[5]** for a Polystyrene material in a glass insert mold.

3. Mathematical model

Filling stages of injection molding process are described by the combination of the momentum equation 1:

$$\frac{\partial}{\partial t}(\rho\vec{v}) = \rho\vec{g} + (\nabla\cdot\underline{\sigma}) + \nabla p - (\nabla\cdot(\rho\vec{v}\vec{v})) \qquad (1)$$

mass continuity (equation 2):

$$\frac{\partial\rho}{\partial t} + \nabla\cdot(\rho\vec{v}) = 0 \qquad (2)$$

and the energy equation (3):

$$\begin{aligned} &\rho c_p\left(\frac{\partial T}{\partial t} + \vec{v}\cdot\nabla T\right) = \\ &-\frac{1}{\rho}\left(\frac{\partial\rho}{\partial T}\right)_p T(\frac{\partial p}{\partial t} + \vec{v}\cdot\nabla p) \\ &+ p\nabla\cdot\vec{v} \\ &+ (\underline{\sigma}:\nabla\vec{v}) \\ &+ \nabla\cdot(k\nabla T) \end{aligned} \qquad (3)$$

together with appropriate material equations and boundary conditions described in [1] and [2]. In the simplest mold filling case the material equations include generalized Newtonian rheological equation:

$$\underline{\sigma} = 2\eta(T,\dot{\gamma},p)\underline{D} \qquad (4)$$

and density or PVT equation:

$$\rho = \rho(p,T) \qquad (5)$$

Boundary conditions are set on the mold-plastic interface, on the melt flow front and on the injection surfaces.

On the mold-plastic interface the boundary conditions are set as:

$$T = T_{mold} \qquad (6)$$

$$\vec{v} = \vec{0} \qquad (7)$$

On the melt flow front (F) the boundary conditions are set as

$$\nabla T\bullet\vec{n}_F = 0 \qquad (8)$$

$$\underline{p + \sigma}\bullet\vec{n}_F = 0 \qquad (9)$$

where $\vec{n}_F$ is the normal to the flow front F.

On the injection surface (I) the boundary conditions are set as

$$T = T_{melt} \qquad (10)$$

$$\underline{p + \sigma}\bullet\vec{n}_I = const \qquad (11)$$

$$\iint_I \vec{v}\bullet\vec{n}_I dI = Q(t) \qquad (12)$$

where $\vec{n}_I$ is the normal to the injection surface I and $Q(t)$ is the injection flow rate.

The system of equations (1)-(12) is solved using a specialized finite element method customized for the typical conditions of the injection molding process described in [1].

In order to incorporate the effect of normal stresses we implemented material rheological function connecting stresses (σ) with flow conditions that follows Criminale-Ericksen-Filbey model [3]:

$$\begin{aligned} \underline{\sigma} &= 2\eta(T,\dot{\gamma},p)\underline{D} \\ &-\psi_1(T,\dot{\gamma},p)\overset{\nabla}{\underline{D}} + 4\psi_2(T,\dot{\gamma},p)\underline{D}\bullet\underline{D} \end{aligned} \qquad (13)$$

In the equation (4) $\underline{\sigma}$ is the deviatoric stress tensor; $\underline{D}$ is the deformation rate tensor, $\eta(T,\dot{\gamma},p)$ is viscosity dependent upon temperature T, scalar shear rate $\dot{\gamma} = \|\underline{D}\|$ and pressure p; $\overset{\nabla}{\underline{D}}$ denotes the upper convective derivative of the deformation rate tensor:

$$\overset{\nabla}{\underline{D}} = \frac{\partial}{\partial t}\underline{D} + \vec{v}\bullet\nabla\underline{D} - (\nabla\vec{v})^T\bullet\underline{D} - \underline{D}\bullet\nabla\vec{v} \qquad (14)$$

In equation (14), $\vec{v}$ is velocity and functions $\psi_1(T,\dot{\gamma},p)$ and $\psi_2(T,\dot{\gamma},p)$ are correspondingly the first and the second normal stress difference functions. Equation (13) is an extension of the generalized Newtonian equation (4).

Following established practice for mold-filling simulation we use Cross-WLF model [2] for the viscosity model:

$$\eta(T,\dot{\gamma},p)=\frac{\eta_0(T,p)}{1+\left(\dfrac{\eta_0(T,p)\dot{\gamma}}{\tau^*}\right)^{1-n}} \quad (15)$$

Where

$$\eta_0(T,p)=D_1\exp\left[-\frac{A_1(T-T^*)}{A_2+(T-T^*)}\right],$$

T^* is the glass transition temperature,

$A_2=A_3+D_3p$,

D_1, A_1, A_3, D_3 and n are empirical coefficients. Autodesk Moldflow material library stores quite an extensive data of the Cross-WLF parameters for thousands of material grades.

To evaluate functions $\psi_1(T,\dot{\gamma},p)$ and $\psi_2(T,\dot{\gamma},p)$ when viscosity is known we followed the so called Cox-Mertz Abnormal Rule as described by V. Sharma and G.H. McKinley [7].

$$\psi_1(T,\dot{\gamma},p)\approx 2\dot{\gamma}^{-1}\eta\left(1-\frac{\eta_c^2}{\eta^2}\right)^{0.5}\left(\frac{\eta_c^2}{\eta^2}\right)^{0.7} \quad (16)$$

In the equation 7, $\eta=\eta(T,\dot{\gamma},p)$ is the steady state viscosity and $\eta_c=\frac{\partial}{\partial\dot{\gamma}}\left(\eta(T,\dot{\gamma},p)\dot{\gamma}\right)$ is the tangential or consistency viscosity. As shown in [7] equation 7 allows relatively accurate estimation of the first normal stress difference function if Cross-WLF parameters are known.

Finally, to estimate the second normal stress difference function $\psi_2(T,\dot{\gamma},p)$ we assume that it is proportional to the first normal stress difference function:

$$\psi_2(T,\dot{\gamma},p)=K_\psi\psi_1(T,\dot{\gamma},p) \quad (17)$$

where the coefficient K_ψ takes values between -0.5 and 0 [3].

The set of equations 1-17 was integrated into a special build of 3D flow solver of Autodesk Moldflow Insight 2017 software and the resulting program was used for simulation of injection molding processes. The addition included calculation of the additional normal stress tensor field:

$$\underline{\sigma_n}=-\psi_1(T,\dot{\gamma},p)\overset{\nabla}{\underline{D}}+4\psi_2(T,\dot{\gamma},p)\underline{D}\bullet\underline{D} \quad (18)$$

using equations (15)-(18). Then we apply additional nodal forces $\vec{N}_i$ to each of the filled nodes i calculated as:

$$\vec{N}_i=\oiint_{S_i}\underline{\sigma_n}\bullet\vec{n}_{S_i}dS_i \quad (19)$$

where S_i is the surface of the control volume of the node i. At the each time step the algorithm iterated the velocity-pressure solver together with the calculations of the normal stress by equations (18) and (19) until they converge. The rest of the mold filling simulation algorithm described in [1] was left intact.

4. Filling of a thin rectangular cavity

4.1 Case study

In order to estimate the effect of normal stresses on flow front propagation we use a filling simulation of a simple thin rectangular plaque of 100mm in length, 20mm in width and 2mm in thickness shown on Figure 2. The plague being filled by the generic polymethyl-methacrylate (PMMA) material. The filling time is 1 second. The cavity is meshed by 4 node tetrahedral elements with at least 10 layers of elements through the thickness. Melt inlet boundary conditions are applied along one of the short edges of the plaques in the style of a film or fan gate.

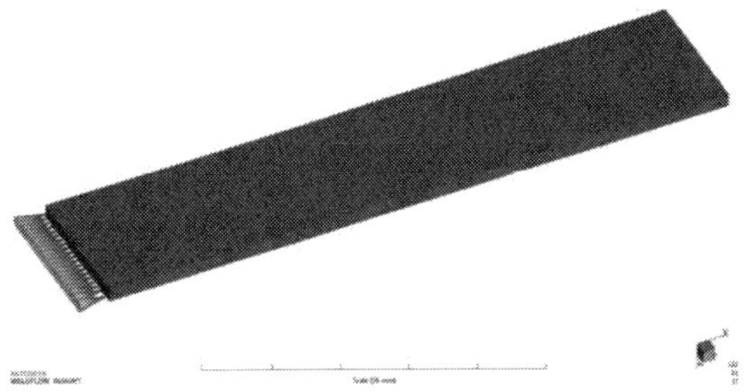

Figure 2 Illustrative molding case

Three rheological models were considered:

- No normal stresses
- First normal stress difference estimated from the Cox-Mertz Abnormal Rule but no Second normal stress differences: coefficient K_ψ =0
- First normal stress difference estimated from the Cox-Mertz Abnormal Rule and large Second normal stress differences K_ψ =-0.5.

All other rheological, thermal and mechanical material parameters were taken from the standard material library.

All three cases simulation do not show any significant shear heating in the edge area (see Figure 3) as was expected because the runners and gates were not modeled.

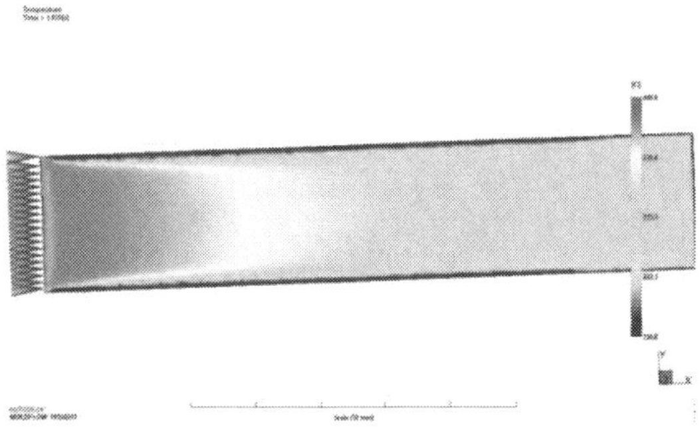

Figure 3 Temperature distribution at the end of fill. No normal stresses case. The cutting plane is in the middle-thickness plane of the cavity.

4.2 No normal stresses case

The simulation without normal stresses does not predict any ear flow phenomenon as shown in Figure 4.

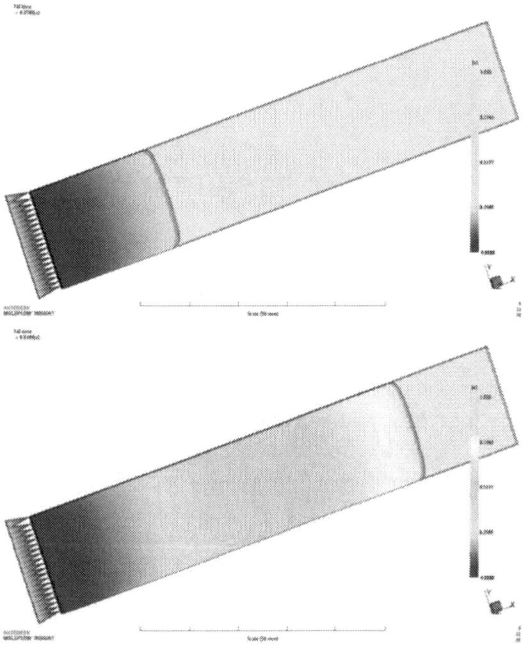

Figure 4 Flow front positions for the case with no normal stress.

4.3 First normal stress differences only

When only the first normal stress difference is included, the flow front propagation pattern is very similar to the case of no normal stress. As shown in Figure 5 there is no ear flow phenomenon in this case either.

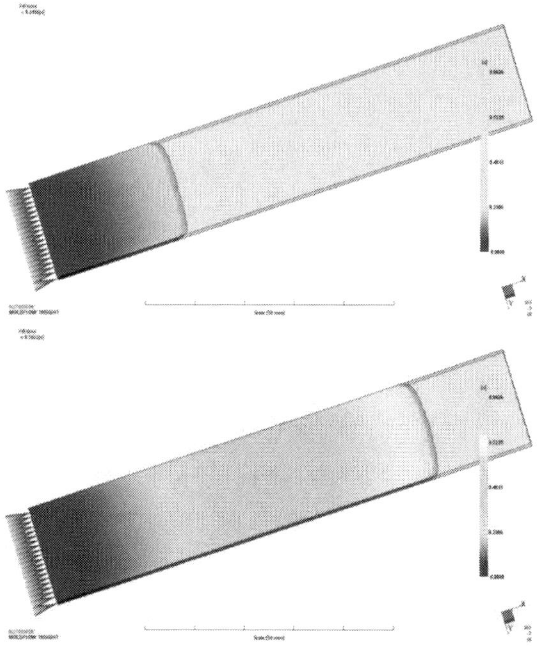

Figure 5 Flow front positions for the case with the first normal stress differences only.

4.4 First normal differences and maximal second normal differences

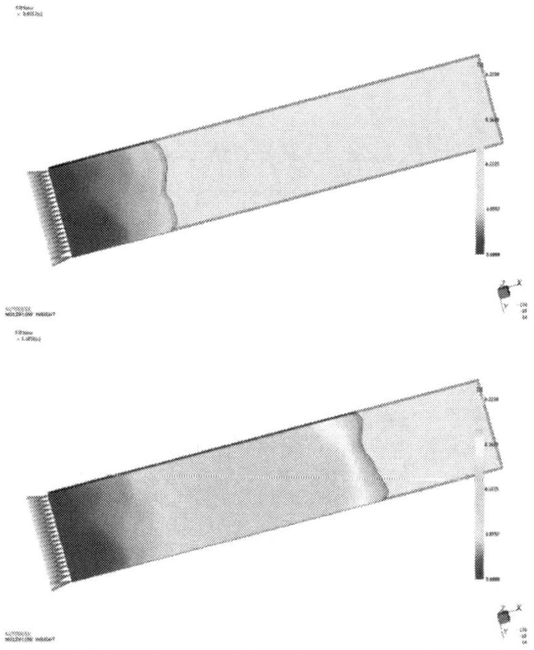

Figure 6 Flow front positions for the case with the first and second normal stress differences.

The flow front distribution from the simulation using first and second normal stress differences is shown in Figure 6. A significant ear flow effect can be seen. The effect is quite prominent despite a moderate magnitude of normal stresses up to ~30kPa (see Figure 7).

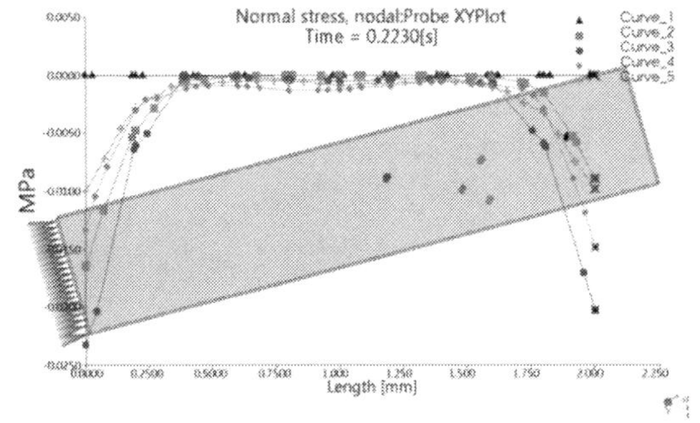

Figure 7 Normal stress difference $\sigma_{yy} - \sigma_{zz}$ component, thickness probe.

The results of these simulations show that the second normal stress difference can be an important contributing factor to the ear flow phenomenon.

5. Conclusions

An integrated system of 3D mold filling simulation that takes into account nonlinear rheological properties of normal stress differences is presented.

The second normal stress difference appears to be an important contributing factor to the ear flow phenomenon while the first normal stress difference does not affect much the flow front propagation.

References

1. Talwar, K., Costa, F., Rajupalem, V., Antanovski, L and Friedl, C., *Three Dimensional Simulation of Plastic Injection Molding.* SPE ANTEC Proceedings (1998)

2. Kennedy, P. and Zheng, R., *Flow Analysis of Injection Molds 2d edition.* Munich : Carl Hanser Verlag (2013)

3. Macosko, Christopher W. *Rheology: principles, measurements and applications.* New York : Willey-VCH, (1994)

4. Costa, F.S., Yokoi, H., Murata, Y. and Kennedy, P.K., *Numerical Simulation of Ear-flow: The Faster Advance of the Flow Front at the Edge of a Cavity.* Japan : Proc.22nd Annual Conference of the Polymer Processing Society (2006)

5. Murata, Y., Abe, S. and Yokoi, H., *Experimental Analysis of Flow Front Advancing Phenomena.* Seikei-Kakou **15(10)**, pp.706-713 (2003)

6. Cook, P. S., Yu, H. Kietzmann, C. V., Costa, F. S., *Prediction of Flow Imbalance in Geometrically Balanced Feed Systems.* Boston : SPE ANTEC Proceedings, 2005.

7. Sharma, V. and McKinley, G.H., *An Intriguing Empirical Rule for Computing the First Normal Stress Difference from Steady Shear Viscosity Data for Polymer Solutions and Melts.* Rheol. Acta (2012).

Prediction of acoustic foam properties by numerical simulation of polyurethane foaming process

Hichem Abdessalam[1], Boussad Abbès[1,a], Yuming Li[1], Ying-Qiao Guo[1], Elvis Kwassi[2] and Jean-Luc Romain[2]

[1]*Université de Reims Champagne-Ardenne, GRESPI, MPSE, Campus Moulin de la Housse, BP 1039, 51687 Reims Cedex 2, France*
[2]*Faurecia, Route de Villemontry, 08210 Mouzon, France*

Abstract. This work aims to model and to simulate the polyurethane foaming process. Models taking into account the two main chemical reactions of the formation of polyurethane, the exothermic effect of these reactions as well as the thermo-rheo-kinetic coupling characterizing this process are proposed and implemented in the software NOGRID-points based on a meshless method (Finite Pointset Method). A prediction of some acoustic foam characteristics is also proposed based on the results of the numerical simulation of the foaming process and semi-phenomenological models.

1 Introduction

The automotive suppliers produce large amounts of soundproofing parts in flexible polyurethane foam whose shapes are more and more complex in order to satisfy customer demands. The complexity of the parts may cause several types of product defects such as air entrapment, weld lines, unfilled regions in the mould ...etc. affecting the quality of the final part. Thus, the numerical simulation of the foaming process is important for the overall design process to define suitable injection points and air vents to avoid these defects. Different authors [1-5] addressed these issues.

On the other hand, the foamed parts are used in the automotive industry to absorb the maximum amount of sound waves to improve the comfort inside the car. Acoustical performance of the porous materials can be characterized by sound absorption and sound transmission loss. The measurement of these parameters takes lot of time and money. So it is necessary to predict acoustic properties of porous materials by simulation. To predict acoustic performance, five intrinsic parameters i.e., porosity, airflow resistivity, tortuosity, viscous characteristic length and thermal characteristic length are required [6-7]. However, these parameters are highly dependent on the material nature and the foaming process conditions. There is a need to connect the process conditions to the aforementioned parameters to improve the acoustic efficiency of the parts during the part and process design.

This paper is a first tentative to estimate the acoustic properties of the foam using the results of the numerical simulation of foaming process. To do so, a thermo-rheo-kinetic model was developed to simulate the polyurethane foaming process by the Finite Poinset Method (FPM) and the Johnson-Champoux-Allard semi-phenomenological model was used to predict the acoustic foam properties.

2 Simulation of mould filling with polyurethane foams

Describing polyurethane (PU) foams as a pseudo-homogeneous fluid with continuously changing material properties is a common approach when modelling the mould filling processes [1-4].

The PU foam expansion is governed by the evolution of its density obtained by an empirical equation [8] or by considering the contribution of the chemical reactions, involving the temperature and the viscosity evolution [9-11].

2.1 Polyurethane thermo-rheo-kinetic models

The reactions taking place in PU foams are complex, but can be represented by only two global exothermic reactions: the first gelling reaction, where polyol and isocyanate react to form highly cross-linked polyurethane; and blowing reaction where water and isocyanate react to form urea and carbon dioxide.

Both global reactions are governed by chemical kinetics and are supposed in this work to follow a Piloyan law [12]:

$$\frac{dr}{dt} = \frac{1}{\tau_r} r^{m_r} (1-r)^{n_r} \qquad (1)$$

where r is the chemical conversion rate of CO_2 creation ($r=\alpha$) or of the gelling reaction ($r=\beta$), τ_r is the

[a] Corresponding author: boussad.abbes@univ-reims.fr

characteristic time of the reaction, m_r and n_r are the exponents of the reaction rate.

The foam is considered in this work as a generalized Newtonian fluid with its viscosity depending on the temperature (T), the chemical conversion rate of the gelling reaction (β) and the porosity of the foam (ϕ):

$$\eta(T,\phi,\beta)=\eta_0\exp\left[\frac{E_\eta}{R}\left(\frac{1}{T}-\frac{1}{T_0}\right)\right]\times \exp\left(\frac{k(T)}{\phi}\right)\left(\frac{\beta_{gel}}{\beta_{gel}-\beta}\right)^{a(T)\beta+b(T)} \quad (2)$$

where η_0 is the viscosity at a reference temperature (T_0), R is ideal gas constant, E_η is the activation energy, β_g is the conversion at gel point, $k(T)$, $a(T)$ and $b(T)$ are parameters of the model which generally depend on the temperature.

2.2 Governing equations

The general governing equations for compressible Newtonian fluid based on the Stokes' hypothesis include mass conservation, momentum conservation, and energy conservation equations respectively:

$$\nabla\cdot\mathbf{v}=\phi\left(\frac{1}{\alpha}\frac{d\alpha}{dt}+\frac{1}{T}\frac{dT}{dt}\right) \quad (3)$$

$$\frac{d(\rho\mathbf{v})}{dt}=-\nabla p+\nabla\cdot\mathbf{s}+\rho\mathbf{g} \quad (4)$$

$$\frac{d(\rho C_p T)}{dt}=\nabla\cdot(\lambda\nabla T)-(\mathbf{s}:\nabla\mathbf{v})+Q \quad (5)$$

with:

$$\mathbf{s}=\eta(T,\phi,\beta)\left[(\nabla\mathbf{v})+(\nabla\mathbf{v})^T\right] \quad (6)$$

$$Q=\Delta H_\beta\frac{d\beta}{dt}+\Delta H_\alpha\frac{d\alpha}{dt} \quad (7)$$

where $\mathbf{v}$ is the velocity vector, $\mathbf{g}$ is the gravity acceleration vector, $\mathbf{s}$ is the deviator part of the stress tensor, $\rho=(1-\phi)\rho_{PU}+\phi\,\rho_{CO_2}$ is the density of the foam, $C_p=(1-\phi)C_{PU}+\phi C_{CO_2}$ is the thermal capacity of the foam, $\lambda=(1-\phi)\lambda_{PU}+\phi\lambda_{CO_2}$ is the thermal conductivity of the foam, ΔH_α is the enthalpy of the reaction of isocyanate with water and ΔH_β is the enthalpy of the gelling reaction.

2.3 Numerical implementation

Since the 1970s, the meshless formulation was developed from with the emergence of SPH (smoothed particle hydrodynamics) as an alternative to grid-based numerical methods [13]. Later, a set of methods based on this approach have been developed [14]. To avoid the interpolation consistency at the boundary conditions, several solutions have been proposed such as the finite point method (FPM) [15-16]. The fluid domain is discretized by finite number of Lagrangian particles which move with fluid velocity and carry along with them all fluid properties such as density, viscosity, velocity and temperature. The list of neighbour points is determined for each point at each time step through the definition of a zone of influence. Indeed, a smoothing length h is attached to each particle and used to create an interpolation function using a Moving Least Square approximation.

The set of differential equations defined by conservation equations (3)-(5) with appropriate boundary conditions are solved in the NoGrid-points software, where the chemical kinetics, the foam viscosity and the source terms defined by equations (1), (2) and (7) respectively were implemented. A splitting technique is used to solve this highly coupled and non-linear system.

3 Acoustic properties of PU foam

The homogeneous foam layer is rather described by the following two intrinsic acoustical properties: the wave number $k(\omega)=\omega(\rho(\omega)/K(\omega))^{1/2}$ and the characteristic impedance $Z_c(\omega)=(\rho(\omega)K(\omega))^{1/2}$. The well-known Johnson-Champoux-Allard semi-phenomenological model [6-7] is generally used to predict the frequency behaviour of ρ and K as:

$$\rho(\omega)=\rho_0\left\{\alpha_\infty+\frac{\phi\sigma}{j\omega\rho_0}\left[1+\frac{j\omega}{\nu}\left(\frac{2\alpha_\infty\eta}{\Lambda^2\phi\sigma}\right)\right]^{1/2}\right\} \quad (8)$$

$$K(\omega)=P_0\left\{1-\frac{\gamma-1}{\gamma}\left(\frac{8\nu'}{j\omega\Lambda'^2}\left[1+\frac{j\omega}{\nu'}\left(\frac{\Lambda'}{4}\right)^2\right]^{1/2}+1\right)^{-1}\right\}^{-1} \quad (9)$$

where ω is the angular frequency, ρ_0 is the fluid density, η is the fluid viscosity, P_0 is the atmospheric pressure, γ is the specific heat ratio and $\nu=\eta/\rho_0=P_r\nu'$, P_r being the Prandtl number.

The macroscopic non-acoustical parameters involved in this model (i.e., static airflow resistivity σ, tortuosity α_∞, viscous Λ, and thermal Λ' characteristic lengths) were estimated by the semi-phenomenological model of Dourtes et al. [17] where the foam cell is considered as a tetrakaidecahedron made of 14-sided polyhedron, with six squares faces and eight hexagonal faces (Fig. 1).

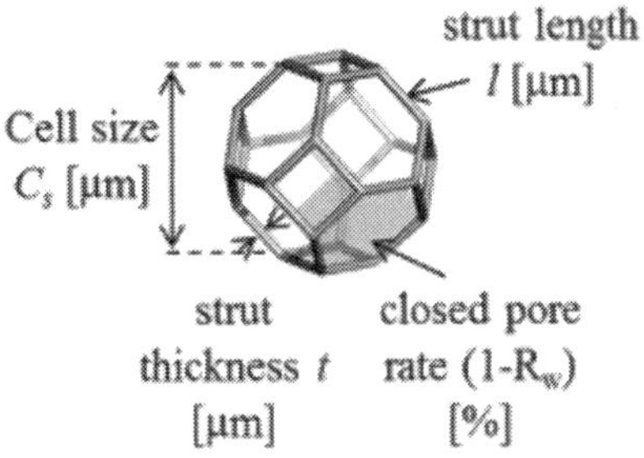

Figure 1. Shape of the idealized unit- [16].

The airflow resistivity is given by:

$$\sigma = 128\eta\left(C_r^{\rho}\frac{A\sqrt{2}}{B}\right)^2\left(\frac{1}{C_s}\right)^2\left(\frac{1}{R_W}\right)^{1.1166} \qquad (10)$$

where C_s is the cell size, R_W is the reticulation rate, $A = C_s/\sqrt{2}l$ and $B = l/t$ are two geometric ratios, η is the air viscosity, and C_s^{ρ} is associated to strut with circular cross-section shape $\left(C_s^{\rho} = 3\pi/8\sqrt{2}\right)$.

The tortuosity increase with the close pore content $(1-R_W)$ and is given empirically as:

$$\alpha_\infty = 1.05\left(\frac{1}{R_W}\right)^{0.3802} \qquad (11)$$

The thermal characteristic length Λ' (also referred to as the hydraulic radius) can be determined from simple geometrical calculations since it is defined as twice the average ratio of the cell volume to their wet surface. It can be written in terms of cell size, reticulation rate and the two geometric ratios (*A* and *B*) as:

$$\Lambda' = C_s\frac{8\left[1-\left(2\sqrt{3}-\pi\right)/B^2\sqrt{2}\right]/3A}{\left(1+2\sqrt{3}\right)-R_W\left(1+2\sqrt{3}-4\pi/B\sqrt{3}\right)} \qquad (12)$$

The viscous characteristic length Λ is given by the following empirical expression in terms of the closed pore content:

$$\Lambda = \frac{\Lambda'}{1.55\left(\frac{1}{R_W}\right)^{0.6763}} \qquad (13)$$

In this work, the foam density ρ and the foam porosity ϕ are directly obtained from the simulation of the foaming process.

The Young's modulus of the foam was estimated by the model of Gibson and Ashby [18] where the foam cell is idealized by a cubic shape:

$$\frac{E^*}{E_s} = C_1\left(\frac{\rho}{\rho_s}\right)^2 \qquad (14)$$

where E^* is the Young's modulus of the foam, E_s is the Young's modulus of the solid polyurethane, ρ is the foam density, ρ_s is the density of the solid polyurethane, C_1 is the model parameter.

The normal-incidence sound absorption coefficient of the foam is given by:

$$\alpha = 1 - \left|\frac{Z_s - \rho_0 c_0}{Z_s + \rho_0 c_0}\right|^2 \qquad (15)$$

where c_0 is the speed of sound in the air and Z_s is the normal-incidence surface impedance. For a layer of fluid of thickness h and backed by an impervious rigid wall, this impedance is given by [19]:

$$Z_S = -jZ_C \cot(kh)/\phi \qquad (16)$$

4 Results and discussion

4.1 Mould filling results

The square panel mould shown in Fig. 2 is used to validate the implemented models. By fixing the mass flow to 200 g/s, short shot foams are obtained by using different injection times and by varying the mould cavity thickness from 8 mm to 20 mm.

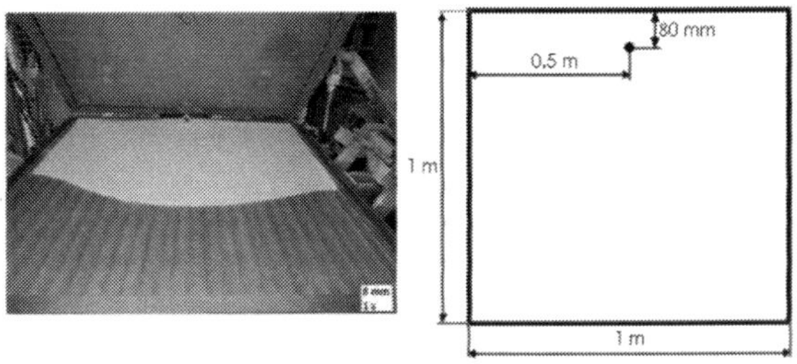

Figure 2. Square panel mould.

Thanks to the symmetry of the problem only a half of the mould has been modelled. Initially, the fluid domain is free from particles and a cylindrical injector is filled with a set of particles defining the free surface at the injection point. A constant volumetric flow rate is imposed at the injector inlet. A set of fixed particles are automatically affected to the walls in NoGrid-points. When the mould is completely filled the number of particles reaches 305000.

An inverse identification procedure for the model parameters was developed in a previous work [18] and herein used for the simulation of the PU foaming process.

The experimental and numerical results obtained for the mould cavity thickness of 12 mm and injections times of 1s and 2 s are shown in Fig. 3. One can notice that

 DOI: 10.1051/matecconf/20168016005

FPM simulations capture accurately the front position showing a good agreement with the corresponding experimental results.

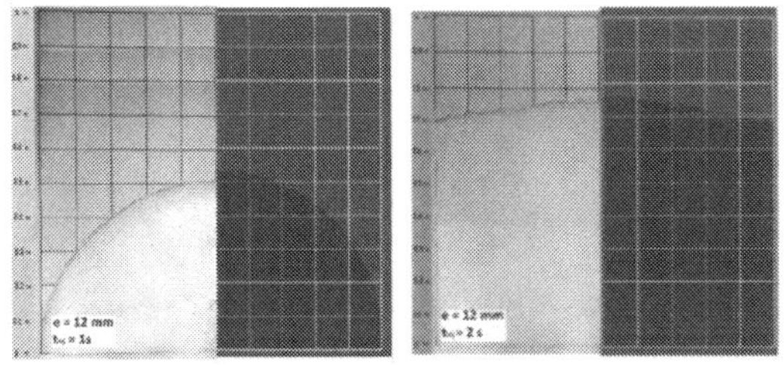

Figure 3. Comparison between experimental and numerical results for short shot foams of the mould cavity.

Fig. 4 shows the temperature, viscosity and density profiles of through the thickness of the foam. The temperature increases from the fixed mould temperature (60°C) to its maximum value at the middle of the thickness because of the exothermic effects of the two global reactions. Since the viscosity depends on the temperature, there is also a viscosity gradient in the thickness. The profile of the foam density is inverted with respect to that of the temperature, since the density is inversely proportional to the porosity which is proportional to the temperature. It is then found that the density is minimum at middle of the thickness. This result was confirmed by using an optical microscope observation of the foam structure across the section of the short shot (Fig. 5).

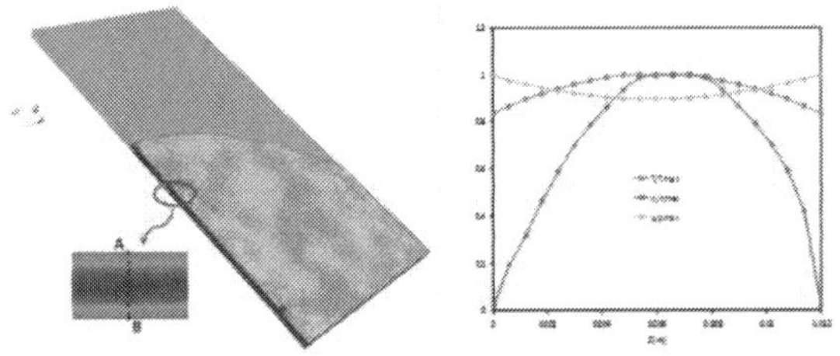

Figure 4. Temperature, viscosity and density profiles through foam thickness.

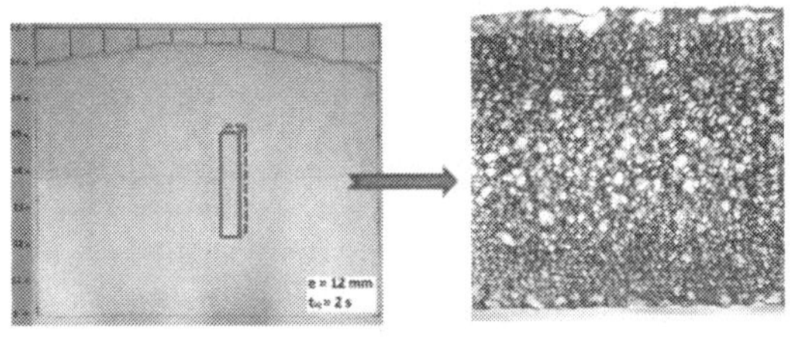

Figure 5. Optical microscope observation of the foam structure across section of the short shot.

4.2 Acoustic results

The model parameters values of the acoustic semi-phenomenological model are listed in Table 1. Fig. 6 shows the iso-values of the porosity and the Young's modulus obtained at the end of the short shot simulation showing in particular the presence of a gradient across the thickness.

Table 1. Semi-phenomenological model parameters.

Parameter	Value
E_s	0.03 GPa
ρ_s	1200 kg/m^3
η	1.81 10^{-5} Pa.s
C_s	900 μm
C_1	0.63
R_W	1
A	2.33
B	2.13

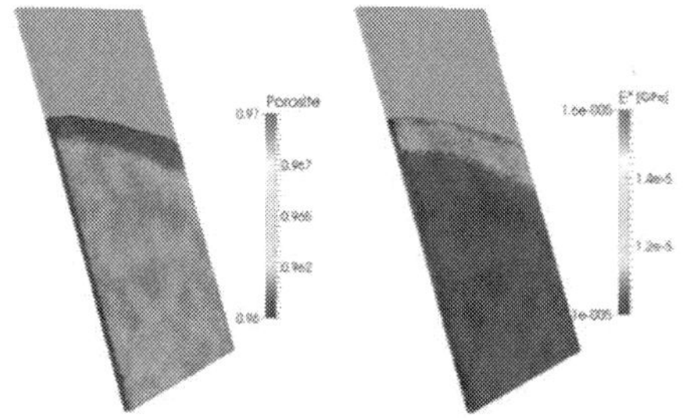

Figure 6. Iso-values of porosity and Young's modulus.

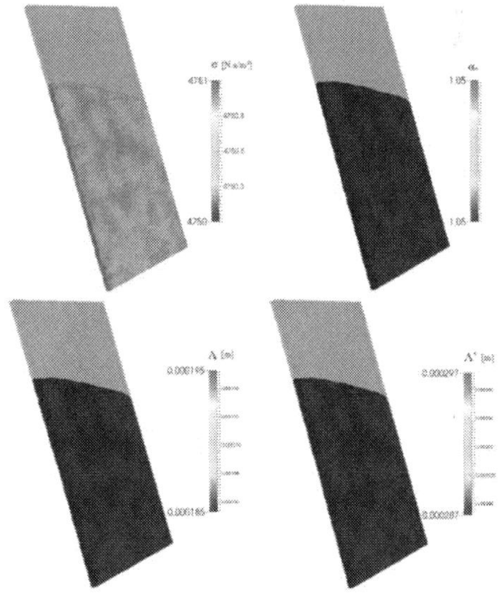

Figure 7. Iso-values of static airflow resistivity, tortuosity, viscous and thermal characteristic lengths.

Constant iso-values of the macroscopic non-acoustical parameters (i.e., static airflow resistivity σ, tortuosity α_∞, viscous Λ, and thermal Λ' characteristic lengths) are shown in Fig. 7.

Fig. 8 presents the simulation of the sound absorption coefficient of the 12 mm thick foam derived from the Johnson-Champoux-Allard model using the computed non-acoustic properties.

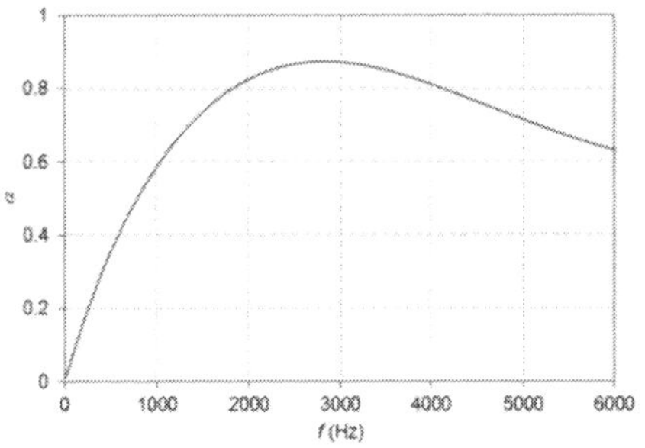

Figure 8. Sound absorption coefficient of PU foam.

5 Conclusions

The foaming process of polyurethane was physically modelled by considering the expansion of a compressible quasi-homogeneous continuous mixture. The model was implemented in the FPM framework. The comparison between the simulation and the experimental results was achieved through the foaming process of set of short shots using a panel mould cavity. The validation consisted in comparing the flow front positions obtained experimentally with the numerical ones. The results show a good agreement between the experimental and the numerical results. A prediction of some acoustic foam characteristics based on the results of the numerical simulation of the foaming process by using a semi-phenomenological model were also proposed. The proposed models and methodology enable to largely reduce the development delay and costs by quickly determining the foaming process parameters and by using less mould prototypes.

References

1. Y.B. Kim, K.D. Kim, S.E. Hong, J.G. Kim, M.H. Park, J.H. Kim, J.K. Kweon, J. Cell. Plast., **41**, 251 (2005).
2. S. Seo, J.A. Youn, Polymer, **46**, 6482 (2005).
3. J. Bikard, J. Bruchon, T. Coupez, L. Silva, Colloids Surf. A, **49**, 309 (2007).
4. S. Geier, C. Winkler, M. Piesche, Chem. Eng. Technol., **32**, 1438 (2009).
5. H. Abdessalam, B. Abbès, Y.M. Li, Y.Q. Guo, E. Kwassi, J.L. Romain, Int. J. Mater. Form., **9**(1), 85 (2016).
6. D.L. Johnson, J. Koplik, R. Dashen, J. Fluid Mech., **176**, 379 (1987).
7. Y. Champoux, J.-F. Allard, J. Appl. Phys., **70**(4), 1975 (1991).
8. W.H. Lee, S.W. Lee, T.J. Kang, K. Chung, J.R. Youn, Fibers Polym., **3**, 159 (2002).
9. (9)S.A. Baser, D.V. Khakhar, Polym. Eng. Sci., **34**, 642 (1994).
10. J. Bikard, J. Bruchon, T. Coupez, B. Vergnes, J. Mater. Sci., **40**, 5875 (2005).c
11. R. Bouayad, J. Bikard, J. F. Agassant, Int. J. Mater. Form., **2**, 243 (2009).
12. G.O. Piloyan, I.D. Ryabchikov, O.S. Novikora, Nature, **212**, 1229 (1966).
13. L.B. Lucy, Astron. J., **83**, 1013 (1977).
14. S. Li and W.K. Liu, *Meshfree Particle Methods* (Springer, Berlin, 2007).
15. E. Oñate, S. Idelsohn, O.C. Zienkiewicz OC, L.R. Taylor and C. Sacco, Comput. Meth. Appl. Mech. Eng., **139**, 315 (1996).
16. J. Kuhnert, *General smoothed particle hydrodynamics*, PhD Thesis, (University of Kaiserslautern, Germany, 1999).
17. O. Doutres, N. Atalla, K. Dong, J. Appl. Phys., **113**, 054901 (2013).
18. L.J. Gibson, M.F. Ashby, *Cellular solids: Structure and Properties*, (Cambridge University Press, Cambridge, 1988).
19. J.F. Allard, N. Atalla, *Propagation of Sound in Porous Media: Modeling Sound Absorbing Materials*, 2nd ed. (Willey, New York, 2009).

DOI: 10.1051/matecconf/20168016006

Development and identification of rheological model adapted for Powder Injection Moulding

CLAUDEL Dimitri, SAHLI Mohamed, BARRIERE Thierry and GELIN Jean-Claude

FEMTO-ST Institute / Applied Mechanics Department, 24 rue de l'Epitaphe, 25000 Besançon, France

Abstract. An Inconel based feedstock for manufacturing super alloy components by powder injection moulding is introduced. The relationship between the rheological properties and different parameters of the feedstock is investigated, based on a prescribed model. The method to determine the parameters of the rheological model for powder injection moulding is presented, including the important thermal effect.

1 Introduction

The Inconel based super alloys represents the important value in aeronautic and other industries. Powder injection moulding (PIM) is one of the most advantageous techniques for manufacturing super alloy components. The preparation of feedstock of Inconel based super alloys for injection moulding is the most crucial stage.

Researches on PIM process have had important developments in the last 20 years [1,2,3]. The feedstock plays a primary and decisive role on dimensions and mechanical properties of the PIM products. It dominates the process parameters for different stages of the PIM process, including injection moulding, debinding and sintering.

Numerical simulation of injection moulding is a very important way for designing the process and products. Accurate rheological model of the feedstock is crucial for correct simulations. The feedstock is injected with large shear rates up to 10^5 s^{-1}, and the melting temperature without degradation of the polymer binder.

The research on rheological model for PIM feedstock had been affected in the research team. Hidalgo Garcia proposed the suitable model, with the proofs of its application [4]. The model is shown in equation (1).

$$\eta = \frac{C_u}{d_{50}} . exp\left(\frac{E_a}{R.T}\right) . \left(\frac{\phi_{crit}}{\phi_{crit}-\phi}\right)^m . \eta_{binder} \qquad (1)$$

Where η is viscosity, E_a is the Arrhenius activation energy ($kJ.mol^{-1}$), R represents perfect gas constant ($J\ K^{-1}\ mol^{-1}$), T is the temperature (K), C_u is the ratio d_{60}/d_{10}, d_{10}, d_{50} and d_{60} are the particle sizes respectively at 10%, 50% and 60%, η_{binder} is the viscosity of the binder formulation, ϕ_{crit} is critical load rate and ϕ is load rate.

2 Materials and methods

2.1 Description of powders, binders and formulations

This powder is a nickel-chromium-based superalloy, typically used in high-temperature and high-performance applications, particularly in the aeronautic industry. However, this nickel alloy is rather expensive; it is why it is very important to reduce the amount of wasted material [5]. In this context, the powder injection moulding is suitable.

A multi-component binders system was used to elaborate the feedstock. The binder composition is given table 1 and the relative volumetric fractions of powder are about 70%. The surfactant acts as lubricant that enhances the dispersion of powder in the binder during mixing and also enhances the powder loading and green strength without sacrificing the flow properties of the mixtures

Table 1. Binder composition.

Component	Volumetric fraction
Polypropylene (PP)	40
Polyethylene glycol (PEG)	55
Stearic acid (SA)	5

Fig. 1 shows scanning electron microscopy (SEM) photography of Inconel powder without binder. The shape of powder particles is quasi-spherical. This shape is ideal for powder injection moulding.

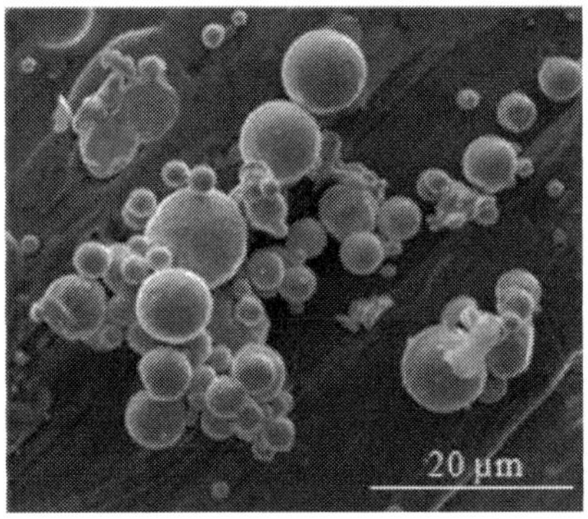

Figure 1. SEM analysis of Inconel powders.

2.1 Processing methods

2.1.1 Description of mixing test

Experiments related to the mixing of binders and feedstocks were carried out using a twin-screw

[a] Corresponding author: thierry.barriere@univ-fcomte.fr

 DOI: 10.1051/matecconf/20168016006

Brabender® Plastograph EC mixer with a pair of roller blades. The maximum capacity of the mixing chamber was 55 cm^3. The mixer was connected with software to measure the mixing torque. The same processing conditions were used for each mixture, including a mixing temperature of 175 °C, a mixing time equal to 30 min and a mixing rotation speed of 30 rpm.

The powders were loaded when the temperature had stabilised at the required mixing temperature to facilitate homogeneous mixing of the powder and binder components. During a mixing test, the mixture homogeneity can be evaluated through the mixing torque curves. Uniform mixing is achieved when the torque reaches a stable, steady state value. After mixing, the feedstock was granulated in pellet form in order to be used during rheological characterization and to be fed easily in the injection process.

Figure 2 shows SEM photography of Inconel feedstock. The binder includes powder particles.

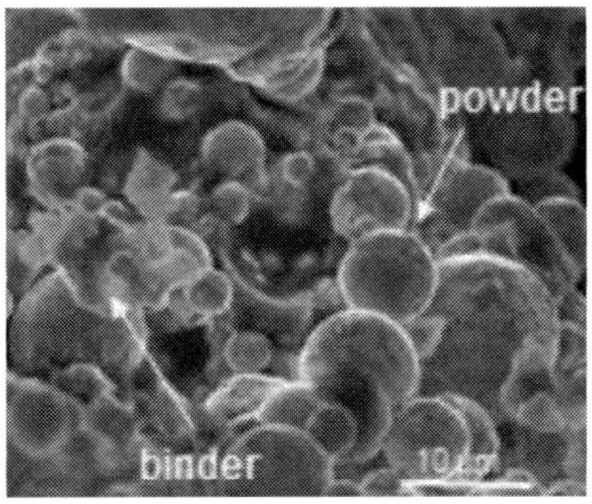

Figure 2. SEM photography of the Inconel feedstock

2.1.2 Thermal characterisation

The thermal characterization of each component of the binder and mixed feedstock is made by differential scanning calorimeter (DSC) and thermogravimetric analysis (TGA). The temperatures of glass transition, melting and degradation of each components of the binder were determined by DSC.

The tests were performed from room temperature to 500°C with heating and cooling rates 10°C/min.

Then, TGA was conducted for identical objects under inert atmospheric conditions. The samples were heated over a range of temperature from 25°C to 500°C with three different heating rates 2, 5 and 10°C/min. These TGA scans allow inspection of the homogeneity for the feedstock after mixing.

2.1.3 Rheology and particle size characterization

The rheological properties of the resulting binder formulations and feedstocks were characterized using a capillary rheometer. First, all the binder granules were filled into rheometer barrel heated to 170, 180 or 190°C, beyond the melting temperature of the polymer and a slight pressure was applied using a piston and allowed for 5-10 min to attain equilibrium temperature throughout the materials. Then the binders were extruded through several tungsten carbide dies with 0.5, 1 and 2 mm diameters and measuring respectively 8, 16 and 32 mm in length. The viscosity of binder was measured over a shear rate varying from 10^3 to 10^5 s^{-1}, using the dies from 0,5 to 2 mm corresponding to 16 to 32 mm giving a ratio (L/D) equal to 16.

To get viscosity curves with the largest possible shear rate range, a Mars rotational rheometer is employs. This kind of rheometer is suitable for low shear rate measures, opposite of the capillary rheometer. Thus, shear rates varying from 1 to $10^3 s^{-1}$ were applied.

The particle size distribution was measured with a laser scattering particle analyser and the specific area was determined using ASAP2020 accelerated surface area and porosimetry system.

3 Results and discussions

Figure 3 illustrates the particle size distribution for an Inconel super alloy powder. The particle size distribution is shown in terms of the number of particles. As the graph shows, the d10, d50 and d90 diameters that equivalent at 3,53 µm, 6,24 µm and 10,97 µm, respectively. The standard deviation in this case, was 0,86 µm. The measure of SSA indicates a specific area of 0,095 m^2/g.

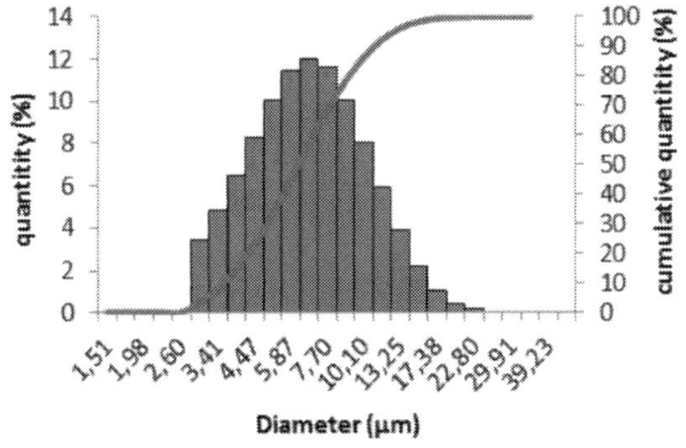

Figure 3. Particle size distribution for Inconel super alloy powders

3.1. Viscosity of binder

First, the viscosities of each component in the binder were measured separately by a rotational rheometer. The tests were realised for shear rates from 10^{-1} to 10^3 /sec at a temperature of 180°C. Then, identification of the parameters for Carreau-Yasuda model, adapted for pure polymer [7], was realized (see equation. 2). The identified models, compared to the experimental values, are shown in Figures 4 and 5.

$$\eta = \eta_0 (1 + (\lambda\dot{\gamma})^a)^{(n-1)/a} \qquad (2)$$

η_0 shear apparent viscosity at zero shear rate (Pa.s), λ is (Cross) time constant, $\dot{\gamma}$ is shear rate (/s), a is a constant.

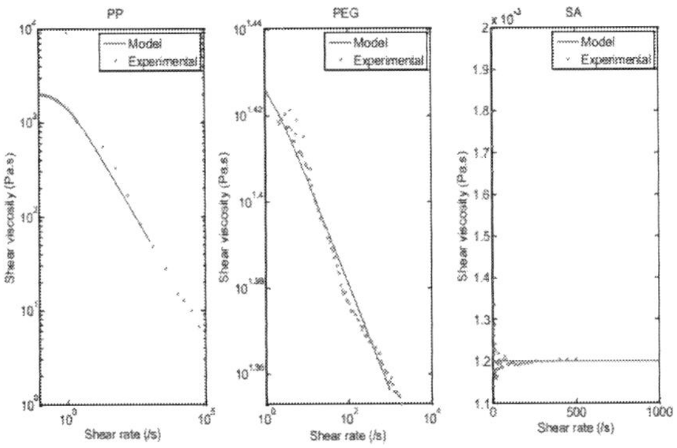

Figure 4. Identifications of Carreau-Yasuda law PP, PEG and SA

Then, three binders were elaborated with the components of different quantities. The binder compositions are represented in Table 2.

Table 2. Binder composition.

	PP	PEG	SA
Mixture 1	40	40	20
Mixture 2	60	15	25
Mixture 3	70	20	10

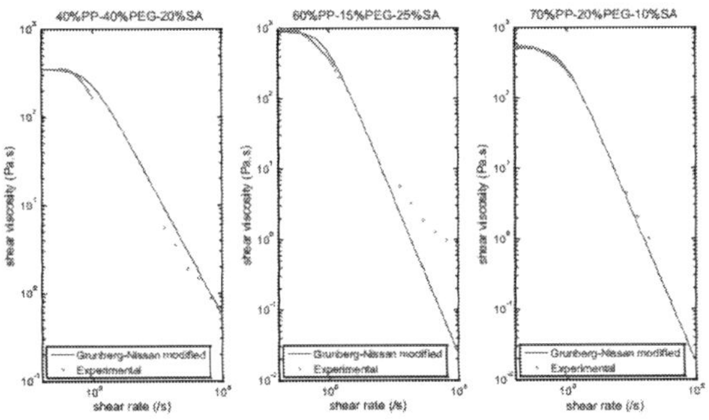

Figure 5. Identified Grunberg-Nissan models for 3 different binders.

The viscosities of different binders were measured. The Grunberg-Nissan modified law was identified for each of the binder. This law was used by Kukla [8], expressed by Eq. 3. The identified models for each binder are shown in Fig. 4.

$$\log(\eta) = x_1.\log(\eta_1) + x_2.\log(\eta_2) + x_3.\log(\eta_3) + x_1x_2d_{12} + x_1x_3d_{13} + x_2x_3d_{23} + x_1x_2x_3d_{123} \quad (3)$$

Where η is the viscosity of binder, $\eta_{1,2,3}$ are viscosities of components, $x_{1,2,3}$ are mole fractions of components and $d_{12,13,23,123}$ are curve fitted parameters.

These identifications allowed obtaining significant correlations among experimental and numerical behaviours for binders with different proportions. So, the Grunberg-Nissan modified law is suitable modelling viscosities of binder system.

3.3 Influence of the powder loading rate on the viscosity

To observe the influence of powder loading rate on viscosity, three feedstocks with different powder loading rates respectively of 68, 70 and 72% were use. Feedstock viscosities are measured using capillary rheometer with a 1 mm diameter die and shear rate including between 100 and 10 000/s.

The shear viscosity of feedstock versus the power loading and shear rate is shown in Figure 6. The viscosity increases more significantly when the powder loading becomes more than 70%. At 72%, the powder charge approaches to the maximal value [6].

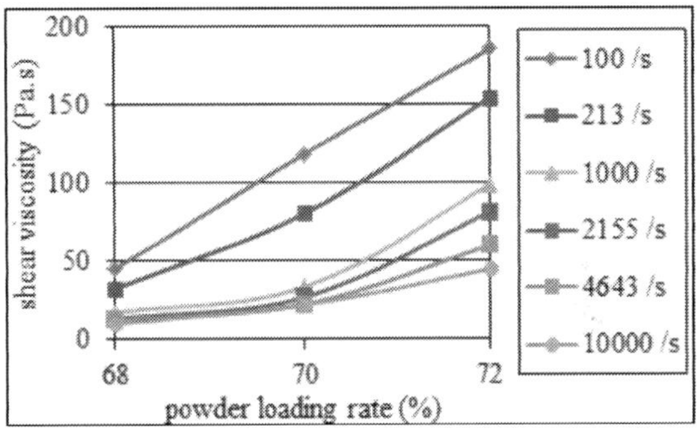

Figure 6. Evolution of shear viscosity of feedstock versus powder loading rate and shear rate for Inconel feedstock

3.4 Apparent viscosity and identification of the rheological parameters

The results of the viscosity measurements of the Inconel elaborated feedstocks are shown in fig. 6. It exhibited pseudo-plastic flow behaviour and the viscosity decreased as the shear rate increased at all working temperatures. This viscosity is plotted for 170, 180 and 190°C. Shear viscosity decreases with shear rate. This behaviour matches with pseudo-elastic behaviour.

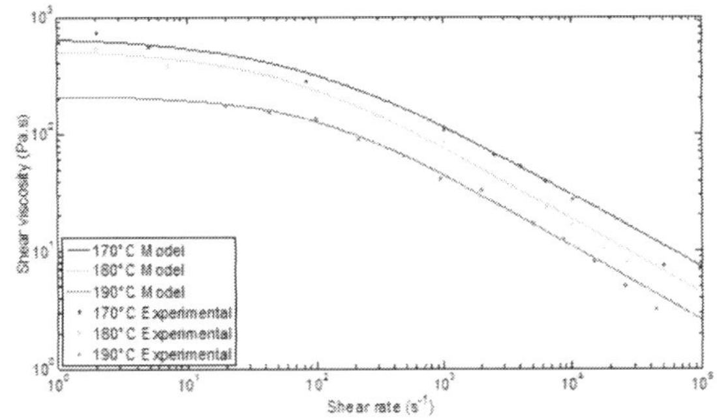

Figure 7. Comparison of rheological constitutive modeling after parameter identification of shear viscosity for Inconel feedstock

The first identification was carried out with the power-law model. This identification allows defining the flow behaviour index equivalent to a value of 0.36. For this model, only the experimental measuring with shear rate higher as 10^3 s^{-1} were considerate.

The second identification was perform with the generalized constitutive law model and is given on the figure 7. With this model, the experimental measuring and analytical modeling are almost. The parameters allowing this proper correlation between analytical and experimental curves are given in the table 3.

Table 3. Identified parameters.

T(°C)	a	λ	η_0 (Pa.s)
170	0.6148	0.0119	207.5
180	0.7344	0.01755	158.9
190	0.8591	0.00995	64.47
n	**Ea (kJ/mol)**	**m**	**R^2**
0.36	20	0.009	0.998
0.36	20	0.009	0.984
0.36	20	0.009	0.988

4 Conclusions

The determination of rheological parameters using data obtained through standard viscosimetrics flows opens the possibility of building reliable analytical models, which can be used in the injection simulations. The present work aims at the identification of rheological parameters associated to a nonlinear constitutive equation. Firstly, a robust analysis has been used, in order to determine the constitutive coefficient values, starting from experimental data obtained by rheological tests. Then, an extensive sensitivity analysis has been performed.

The aim of this work is to examine a proper constitutive model to describe more accurately the influence of the shear rates and temperature on the feedstock behaviour in large shear rates ranges.

To improve the rheological sensitivity on the injection simulations, in a future work, we propose to analyze the influence of a more complex law. It is then necessary to take into account the solids loading, binder ingredient and particle size of powders and then interaction between powders and binders.

References

1. Ö. Özgün, H. Özkan Gülsoy, R. Yilmaz, and F. Findik, J. Alloy Compd. 546 (2013) 192–207.
2. C. Quinard, J. Song, T. Barriere, and J. C. Gelin, Powder Technol. 208 (2011) 383–389.
3. T. Shivashankar, S. Ravi, K. Enneti, S-J. Park, R.M. German and S.V. Atre, Powder Technol. 243 (2013) 79–84.
4. J. Hidalgo, A. Jiménez-Morales, T. Barriere, J.C. Gelin, J.M. Torralba, Leganés, November 2013, 111-113.
5. J. Lambarri, J. Leunda, V. García Navas, C. Soriano and C. Sanz, Opt Laser Eng. 51 (2013) 813–821.
6. X. Kong T. Barriere and J. C. Gelin, J. Mater Process Tech. 212 (2012) 2173-2182.
7. K.Y. Yasuda, R.C Armstrong and R.E. Cohen, Rheol. Acta.20 (1981) 163-178.
8. C. Kukla, I. Duretek and C. Holzer, EPMA. (2013) 305-310.

Determination of the activation energy of silicone rubbers using different kinetic analysis methods

OU Huibin SAHLI Mohamed BARRIERE Thierry[a] and GELIN Jean-Claude

FEMTO-ST Institute / Applied Mechanics Department, 24 rue de l'Epitaphe, 25000 Besançon, France

Abstract. The solidification of silicone rubbers is a crosslinking reaction called vulcanization. It follows that the physical and mechanical properties of silicone rubbers increase significantly due to the growth of molecular weight and were affected strongly by the vulcanization process, such as the temperatures and times for curing. In this study, the kinetics of the crosslinking reaction of silicone rubbers was investigated by two current techniques: differential scanning calorimeter (DSC) and oscillating disk rheometry (ODR). From the kinetic experimental data, different kinetic analysis methods were used to determine the activation energy of silicone rubbers, such as the Kissinger, the Ozawa, the Friedman, the Flynn-Wall-Ozawa, the Kissinger-Akahira-Sunose and the integral methods. The results show the influence of the chemical composition of silicone rubbers, the current technique used and the applied kinetic analysis methods on the activation energy and the evolution of activation energy during the crosslinking reaction.

1 Introduction

Silicone rubbers is a specific category of rubbery material, which have many interesting physical and chemical properties, such as good thermal, chemical and mechanical stabilities, good biocompatibility and high elasticity [1]. These unique properties have resulted silicone rubbers to be used in a wide variety of field of application in the industries: aeronautic and aerospace, automotive and transport, electrical and electronic, health and medical, building and construction [2].
Silicone rubber is a thermoset elastomer, mainly based on polydimethylsiloxane (PDMS) and solidified exothermally by a crosslinking reaction called vulcanization. The vulcanization process has a crucial influence on the performance and quality of the final product of silicone rubber. A suitable vulcanization process can improve the properties of silicone rubbers, such as the tear strength, the elongation at break, the hardness and the compression modulus [3].

To optimize and improve the properties of silicone rubbers, the choice of vulcanization process parameters is particularly important, such as the vulcanization temperature and time. In this investigation, two current techniques DSC and ODR were used to characterize the evolution of the crosslinking reaction of silicone rubbers in different heating cycle. In DSC measurement, the vulcanization degree is related to the heat released during the crosslinking reaction. Otherwise, in ODR measurement, the vulcanization degree is attached to the evolution of the elastic modulus. From the experimental data, several kinetic methods were used to determine the activation energy of silicone rubbers following the evolution of crosslinking reaction during the vulcanization process.

2 Theoretical

During the crosslinking reaction of silicone rubbers, the reaction rate $d\alpha/dt$ is assumed to be a linear function of the rate constant $k(T)$ and a function of the vulcanization degree $f(\alpha)$ [4]:

$$\frac{d\alpha}{dt} = k(T)f(\alpha) = k_0 \exp\left(-\frac{E_a}{RT}\right)f(\alpha) \quad (1)$$

where k_0 is the pre-exponential factor, E_a is the activation energy, R is the gas constant and T is the absolute temperature. In DSC measurements, the vulcanization degree can be related to the heat released during the reaction and calculated by [5]:

$$\alpha = \frac{Q(t)}{Q_T} \quad (2)$$

where $Q(t)$ is the heat released up to time t and Q_T is the total heat released during the crosslinking reaction. In ODR measurements, the vulcanization degree can be defined as [6]:

$$\alpha = \frac{G'(t) - G'_{min}}{G'_{max} - G'_{min}} \quad (3)$$

where $G'(t)$, G'_{max}, G'_{min} are respectively the elastic modulus at time t, the maximum elastic modulus and the minimum elastic modulus during the reaction.

3. Kinetic analysis methods

In anisothermal mode, the Kissinger method assumes that the reaction rate reached the maximum at the peak temperature and determines the activation energy simply without precise knowledge of the reaction mechanism involved:

$$\ln\left(\frac{\beta}{T_P^2}\right) = \ln\left(\frac{k_0 R}{E_a}\right) - \frac{E_a}{RT_P} \quad (4)$$

where β is the heating rate and T_P is the peak temperature. The activation energy can be determined from the slope of the plot of $\ln(\beta/T_P^2)$ against $1/T_P$ [7]. The Ozawa method assumes that the reaction order is a constant value independent of the heating rate at the peak temperature, and is represented by the following equation:

[a] Corresponding author: thierry.barriere@univ-fcomte.fr

$$\ln(\beta) = const - \frac{1.052E_a}{RT_P} \quad (5)$$

By the plot of ln(β) against $1/T_P$, the activation energy can be calculated from the slope [8]. The Friedman method is probably the most general of the differential techniques and uses the following natural logarithmic equation:

$$\ln\left(\frac{d\alpha}{dt}\right) = \ln(k_0 f(\alpha)) - \frac{E_a}{RT} \quad (6)$$

By plotting ln(dα/dt) against 1/T for a constant vulcanization degree α value, the activation energy can be obtained from the slope [9]. The Flynn-Wall-Ozawa method can determine the activation energy without knowledge of reaction order:

$$\ln(\beta) \cong \ln\left(\frac{k_0 E_a}{R}\right) - \ln\left(\int_0^\alpha \frac{d\alpha}{f(\alpha)}\right) - 5.3305 - \frac{1.052E_a}{RT} \quad (7)$$

For a given value α, the activation energy can be calculated by the plot of ln(β) against 1/T [10]. The Kissinger-Akahira-Sunose method is the extension of the Kissinger method, and is represented by the following equation:

$$\ln\left(\frac{\beta}{T^2}\right) \cong \ln\left(\frac{k_0 R}{E_a}\right) - \ln\left(\int_0^\alpha \frac{d\alpha}{f(\alpha)}\right) - \frac{E_a}{RT} \quad (8)$$

It determine the activation energy by the plot of $\ln(\beta/T^2)$ against 1/T for a constant value α [11]. In isothermal mode, the integral method results in:

$$\ln(t_{\alpha,i}) = \ln\left(\int_0^\alpha \frac{d\alpha}{f(\alpha)}\right) - \ln(k_0) + \frac{E_a}{RT_i} \quad (9)$$

where $t_{\alpha,i}$ is the time corresponding to a degree α for an isothermal reaction at the temperature T_i. The activation energy can be obtained by plotting $\ln(t_{\alpha,i})$ against $1/T_i$ for a given constant α [6].

4. Experimental

4.1. Materials

The silicone rubbers studied in this work are originally liquid and two-part, which include three types of Liquid Silicone Rubber (LSR) provided by Bluestar Silicones and three types of Room Temperature Vulcanization (RTV) silicone rubbers provided by Wacker Chemie.

Table 1. Main properties of silicone rubbers studied

Materials		Den sity	Hard ness Shore A	Tensile stress [MPa]	Elong ation at break [%]	Mix ratio A:B
L	LSR4350	1.12	50	8.4	570	1:1
S	LSR4370	1.14	68	9.0	450	1:1
R	TCS7550	1.25	52	5.5	200	1:1
R	M4370	1.43	55	3.0	130	9:1
T	M4641	1.07	43	4.5	300	10:1
V	M4670	1.28	55	5.5	250	10:1

The main properties of these rubber materials are summarized in Table 1.

4.2. DSC tests

Differential scanning calorimetry tests were performed on Setaram DSC 92. The mixture of silicone rubbers was heated from 25 to 200 °C at various heating rates from 1 to 20 °C/min. The samples were about 46 mg. The dynamic heat flow signals were recorded during the heating cycle.

4.3. ODR tests

The vulcanization measurement of silicone rubbers was carried out in dynamic mode using a rotational rheometer HAAKE MARS III with a plate-and-plate geometry of a diameter 20 mm. The crosslinking reaction of silicone rubbers were characterized by mean of elastic modulus curves obtained under isothermal conditions. Test temperatures used were from 100 to 140 °C for LSR and from 70 to 100 °C for RTV. The distance between the two plates is 0.5 mm and the frequency used in the measurement is 1 Hz with a deformation of 1 %.

5. Results and discussion

5.1. Kinetic curves

The kinetics curves of reaction rate versus temperature at different heating rates obtained by DSC measurements for silicone rubber LSR4350 are shown in Fig. 1. It is clear that the peak temperature increased with increasing heating rate.

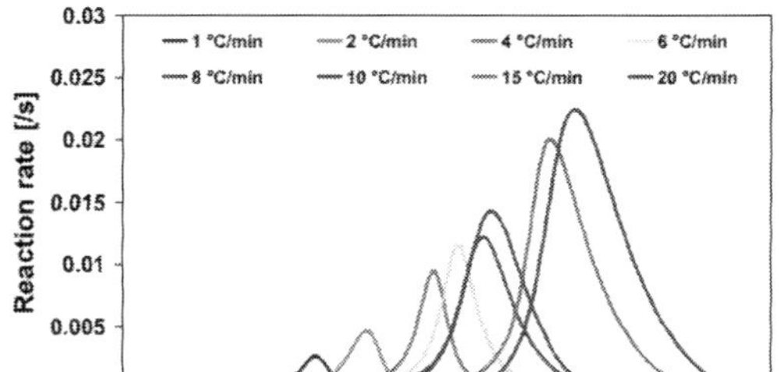

Figure 1. Kinetic curves at different heating rates for silicone rubber LSR4350 obtained by DSC measurements

The evolution of reaction degree versus time at different temperatures obtained by ODR measurements for silicone rubber LSR4350 are illustrated in Fig. 2. It appears that the variation of vulcanization time records an exponential decrease with the increasing temperature for the rubber materials.

DOI: 10.1051/matecconf/20168016007

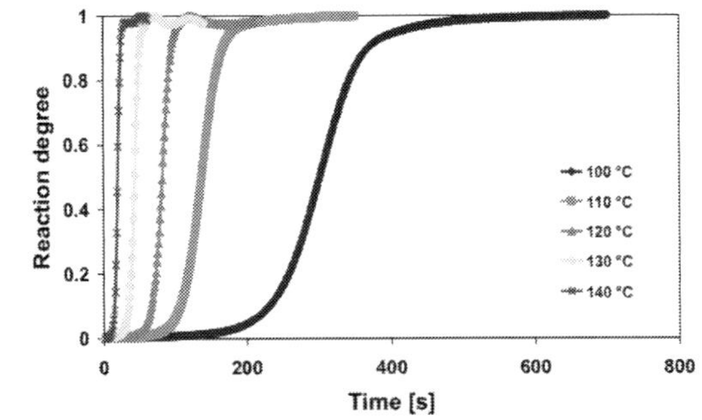

Figure 2. Kinetic curves at different temperatures for silicone rubber LSR4350 obtained by ODR measurements

5.2. Activation energy

From DSC experimental data, Kissinger and Ozawa methods were employed to calculate the activation energies of silicone rubbers at the peak temperature, which are shown in Figs. 3 and 4. It appears that Kissinger and Ozawa methods fit well the results a correlation coefficient at least 0.985. The values of activation energy were obtained from the slope and summarized in Table 2. We can note that the activation energy of LSR are larger that those of RTV.

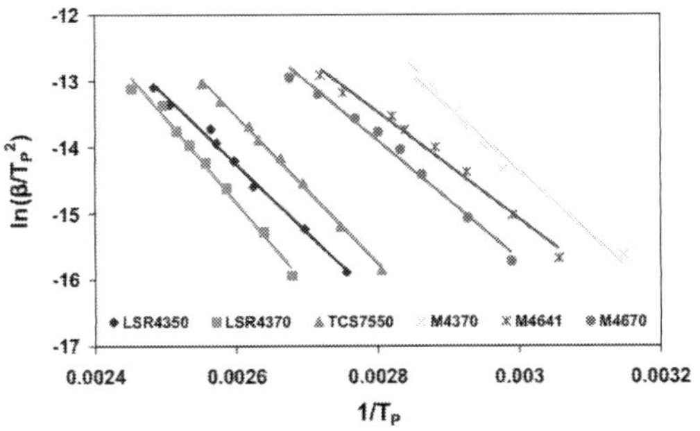

Figure 3. Kissinger plot for silicone rubbers ($1/T_p$: K^{-1})

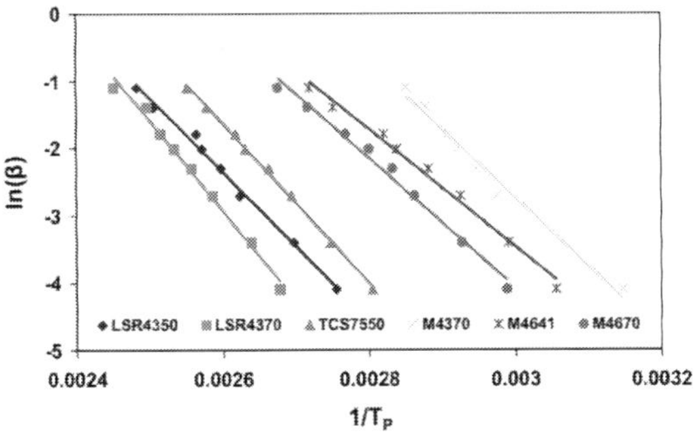

Figure 4. Ozawa plot for silicone rubbers ($1/T_p$: K^{-1})

Table 2. The values of activation energy of silicone rubbers obtained by the Kissinger and the Ozawa methods

	Rubber materials and values of estimated activation energy [kJ/mol]					
	LSR 4350	LSR 4370	TCS 7550	M 4370	M 4641	M 4670
1	85.5	105.4	92.6	79.2	67.1	74.3
2	87.3	78.5	66.8	68.6	58.9	64.9

(Method: 1 Kissinger and 2 Ozawa)

The Friedman method, the Flynn-Wall-Ozawa method and the Kissinger-Akahira-Sunose method were also performed to calculate the activation energy of silicone rubber LSR4350 with the evolution of reaction degree, which are shown in Figs 5, 6 and 7. The values of activation energy were obtained from the slope and summarized in Fig 8. One can remark that the values obtained by the Friedman method decrease sharply from 100 to 40 kJ/mol with the increasing of reaction degree. While the values obtained by the Flynn-Wall-Ozawa method and the Kissinger-Akahira-Sunose method are around 80 kJ/mol for the reaction degree varying from 0.1 to 0.9.

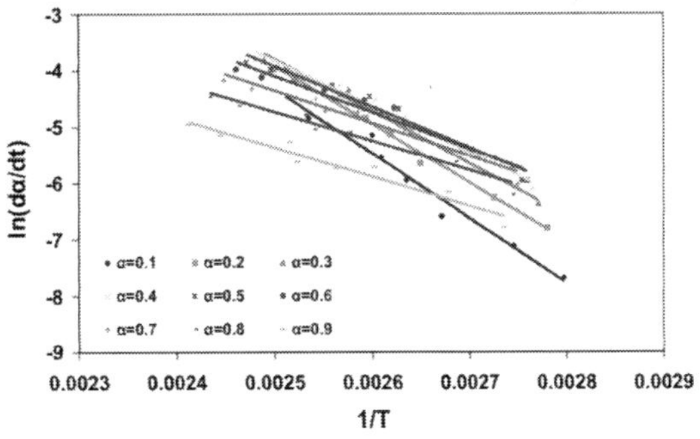

Figure 5. Friedman plot for silicone rubber LSR4350 in different reaction degrees (1/T: K^{-1})

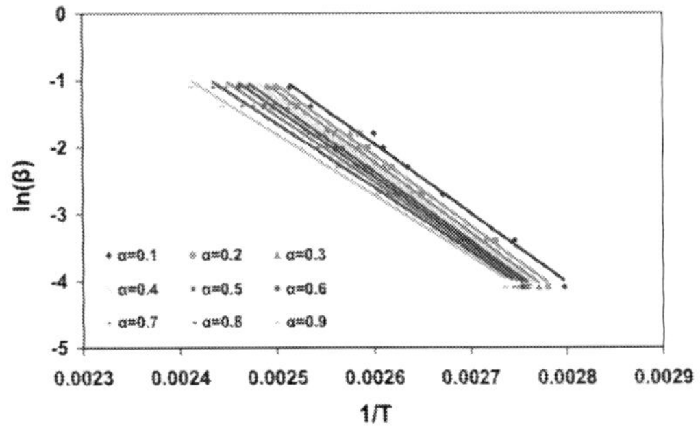

Figure 6. Flynn-Wall-Ozawa plot for silicone rubber LSR4350 in different reaction degrees (1/T: K^{-1})

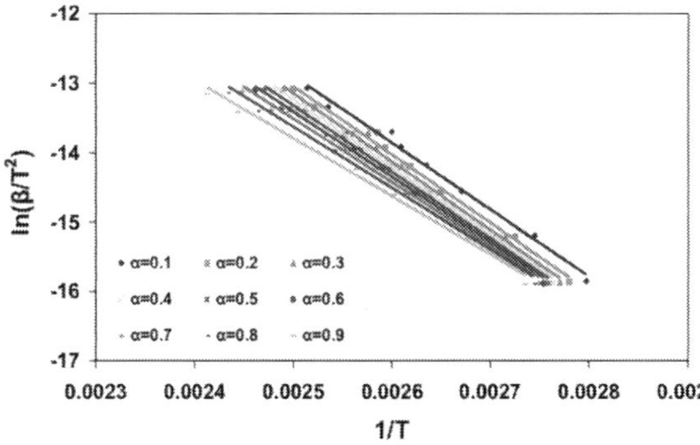

Figure 7. Kissinger-Akahira-Sunose plot for silicone rubber LSR4350 in different reaction degrees (1/T: K^{-1})

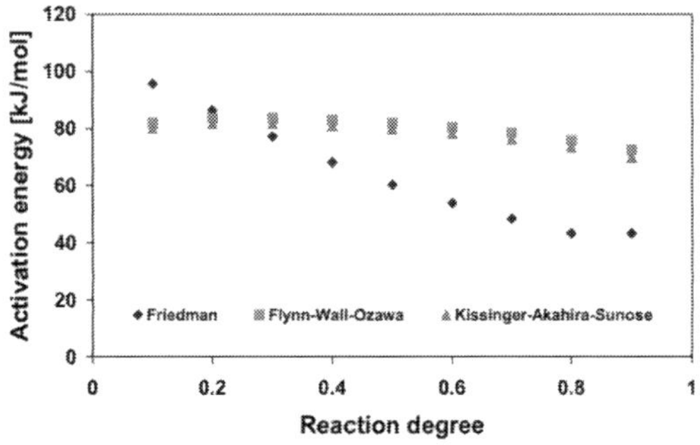

Figure 8. Evolution of the values of activation energy as a function of reaction degree for silicone rubber LSR4350 obtained by the Friedman, the Flynn-Wall-Ozawa and the Kissinger-Akahira-Sunose methods, from the kinetic curves obtained by DSC measurements

From the ODR experiment results in isothermal conditions, the integral method was applied to determine the activation energy of silicone rubber by plotting $\ln(t_{\alpha,i})$ against $1/T_i$ for reaction degree varying from 0.1 to 0.9, Fig. 9. It shows that the straight lines are parallel for the different reaction degree and have a correlation coefficient over 0.99. In Table 3, the average values of activation energy were calculated and compared for the six types of silicone rubbers by the Friedman, the Flynn-Wall-Ozawa, the Kissinger-Akahira-Sunose and the isothermal integral methods. We can find that the values of activation energy of LSR silicone rubbers are always larger than those of RTV silicone rubbers. In additions, the average values obtained by ODR measurements in isothermal conditions are almost identical to those obtained by DSC measurements in anisothermal conditions.

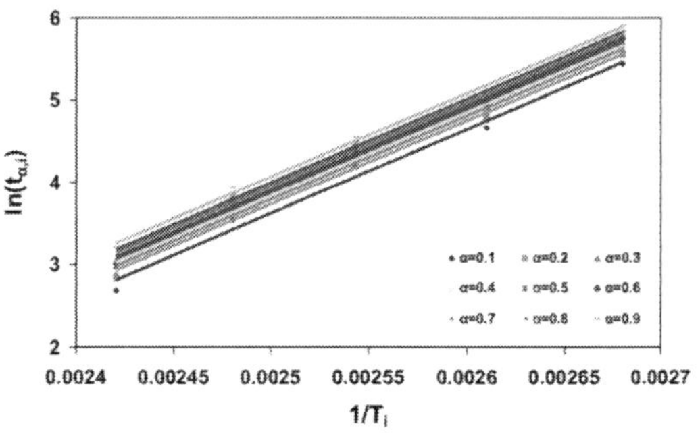

Figure 9. Isothermal integral plot for silicone rubber LSR4350 in different reaction degrees ($1/T_i$: K^{-1})

Table 3. The values of activation energy of silicone rubbers obtained by the different kinetic analysis methods

	Rubber materials and values of estimated activation energy [kJ/mol]					
	LSR4 350	**LSR4 370**	**TCS7 550**	**M 4370**	**M 4641**	**M 4670**
1	70.1	85.9	80.3	61.8	50.4	54.9
2	80.2	98.1	87.0	73.9	64.5	67.0
3	77.9	96.7	85.3	72.2	62.1	64.4
4	84.6	89.2	91.2	67.9	63.9	73.8

(Method: 1 Friedman, 2 Flynn-Wall-Ozawa, 3 Kissinger-Akahira-Sunose and 4 Isothermal Integral)

6. Conclusions

The curing kinetic study was carried out by DSC and ODR measurements in isothermal and anisothermal conditions in order to investigate the crosslinking reaction of silicone rubbers. From the kinetic curves obtained, the kinetic parameters of activation energy were determined by six methods. The Kissinger and the Ozawa methods show that the maximum activation energies were reached at the peak temperature in anisothermal mode. In additions, the isothermal integral method almost gives the same values of activation energies as the Flynn-Wall-Ozawa and the Kissinger-Akahira-Sunose methods. At the end, the values of activation energy of LSR silicone rubbers were larger than those of RTV silicone rubbers studied in this investigation.

References

1. L. Wang, Q. Liu, D. Jing, S. Zhou, L. Shao, Journal of Dentistry, Progress in Organic Coatings, **42**, (2014), 475-483.
2. W. Brachaczek, **77**, (2014), 859-867.
3. X. Ye, H. Liu, Y. Ding, H. Li, B. Lu, Microelectronic Engineering, **86**, (2009), 310-313,.
4. I.K. Hong, S. Lee, vol. 19, pp. 42-47, 2013.
5. J.P. Hernández-Ortiz, T.A. Osswald, Journal of Applied Polymer Science, **119**, (2011), 1864-1871.
6. E. Leroy, A. Souid, A. Sarda, R. Deterre, Polymer Testing, **32**, 2013, 9-14.

7. H.E. Kissinger, Reaction kinetics in differential thermal analysis, Analytical Chemistry, **29**, (1957), 1702-1706.
8. T. Ozawa, Journal of Thermal Analysis, **2**, (1970), 301-324,
9. H.L. Friedman, Journal of Polymer Science Part C, **6**, (1964), 183-195.
10. J.H. Flynn, Journal of Thermal Analysis, **27**, (1983), 92-102.
11. T. Akahira, T. Sunose, Research report, Chiba Institute of Technology, Journal of Science Education and Technology, **16**, (1971), 22-31.

Mechanical behavior of fiber/matrix interfaces in CFRP sheets subjected to plastic deformation

Ryuta Kamiya[1,a] and Tetsuo Oya[1]

[1]*Keio University, 3-14-1 Hiyoshi, Kohoku-ku, Yokohama, Kanagawa, Japan*

Abstract. The use of Carbon Fiber Reinforced Plastic (CFRP) is increasing markedly, partially in the aviation industry, but it has been considered that CFRP sheets cannot be formed by press-forming techniques owing to the low ductility of CFRP. Since the mechanical characteristics of CFRP are dominated by the microscale structure, it is possible to improve its formability by optimizing the material structure. Therefore, to improve the formability, the interaction between the carbon fibers and the matrix must be clarified. In this study, microscale analyses were conducted by a finite-element model with cohesive zone elements.

1 Introduction

The demand for lower manufacturing costs, increased safety, improved fuel consumption and a shorter delivery time is increasing for transport equipment. In recent years, carbon fiber reinforced plastic (CFRP) has been focused on as one of the most promising lightweight materials that can meet these requirements. Although it has higher specific strength and higher specific stiffness than metals, CFRP has low ductility. Because of its low ductility, a forming method that can process a large number of CFRP sheets in a short time such as press forming has not been established. If a plastic working method for CFRP (particularly thermosetting CFRP) were constructed, CFRP sheets could be applied to various products. Since the macroscopic mechanical characteristics of CFRP sheets are dominated by the microscopic structure, such as the material properties of the fiber/matrix interface, the internal fiber orientations and the laminated form, it maybe possible to improve the formability of CFRP sheets by modifying the internal mechanical characteristics without changing the macroscopic shape. To improve the formability, the interaction mechanism between the fibers and the matrix must be clarified.

It has already been confirmed that the interfacial properties play a crucial role in the load transfer between the fibers and the matrix and affect the global performance of a composite structure. The effect of the interfacial properties on both the microscale and macroscale composite behavior and their relationships has been studied by various researchers. Pisanova et al.[1] determined the relationships between the micromechanical interfacial shear strength and macroscopic performance of composites by performing transverse tensile and compression shear tests on glass fiber reinforced plastics. The degree of fiber matrix adhesion has been measured using micromechanical experimental methods including single-fiber pull out, fiber push out, fiber fragmentation and microdroplet tests [2]. Nishikawa et al. [3] attempted to quantify the fracture properties (strength and toughness) of the fiber/matrix interface in composites, by considering the fragmentation process and debonding growth for two types of composite. Their method was based on numerical modeling of the microscopic damage and fragmentation process in fragmentation tests using a cohesive zone model, which is often employed to express the interfacial debonding growth in fracture mechanics [4]. However, their study discussed the interactions between the fibers and the matrix (i.e., fiber breaks, matrix plasticity and cracking, and interfacial debonding) in a single fiber composite, not a CFRP sheet, during plastic deformation. Our final goal is to elucidate mechanisms of the plastic deformation of CFRP sheets using cohesive zone elements. Therefore, the present study was designed to assess the effects of the interfacial properties on the behavior of the entire material by finite-element analyses using a microscale CFRP model to express the interfacial behavior, which is considered as one of the factors that can be modified to improve the formability of CFRP sheets.

2 Experimental procedures

2.1 Uniaxial tensile test

First, a uniaxial tensile test was performed on a unidirectional CFRP specimen to determine the properties of the fibers and the matrix for use in microstructure analyses. Carbon-epoxy prepreg layers comprising continuous carbon fibers (T700SC) and epoxy resin (#2592) were supplied by Toray Industries

[a] Ryuta Kamiya: seiko48057@a5.keio.jp

Inc. The thickness of each prepreg layer was 0.1 mm and the diameter of the carbon fibers was 5 μ m. The weight fraction of carbon fibers in the CFRP sheet was 67 wt%. The ply laminates of the CFRP sheet were unidirectional and parallel to the tensile direction. The CFRP sheet was composed of six prepreg layers with the same thickness, and the CFRP layers were laminated by vacuum bag molding. The dimensions of the CFRP specimen are shown in Fig.1.

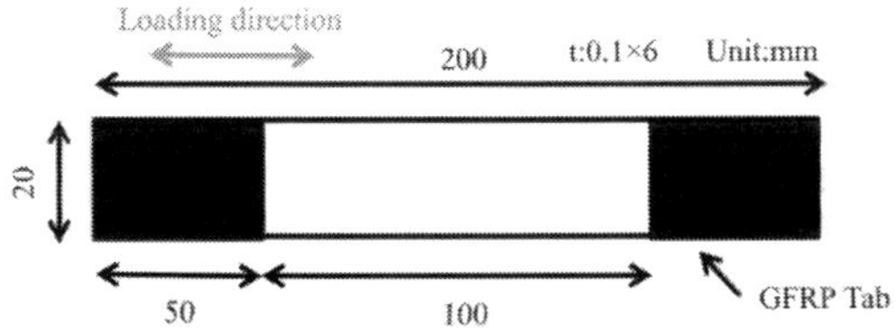

Figure 1. Dimensions of the CFRP specimen.

The tensile test was performed at room temperature using servo-hydraulic tensile testing machine with 200 kN capacity. The loading speed in the tensile test was fixed at 1 mm/min up to the fracture point.

2.2 Experimental results

Fig.2 and Table 1 respectively show the nominal stress-strain curve and material properties obtained from the experiment. The Young's modulus was measured in accordance with the JIS K 7161 standard.

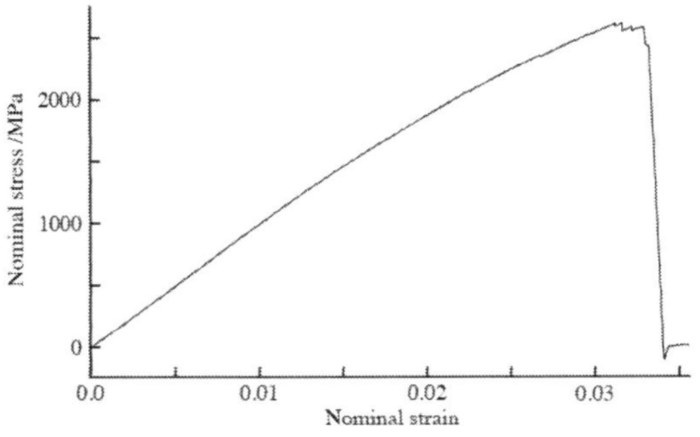

Figure 2. Nominal stress-strain curve of the specimen.

Table 1. Material properties of the specimen.

Property	CFRP specimen
Young's modulus /GPa	99.0
Fracture stress /MPa	2438
Fracture strain	0.033

3 Microscale analyses of the tensile test

3.1 Determination of material properties for use in the microscale analysis

Fibers and resin should be modeled separately to express the interaction between the fibers and the matrix, but this is almost impossible from the view of the computational cost. Thus, a microscale unit cell model was constructed. A microscale unit cell model, which has a periodicity in three dimensions, is often used to determine the quantitative mechanical properties of CFRP subjected to elastic deformation. J. M. Guedes et al. [5] discussed the homogenization method to determine the effective average elastic constants of linear elasticity of the composite material, which is considered as the spatial repetition of a microscale unit cell model. In the homogenization method, periodic boundary conditions are often set to propagate the mechanical behavior of unit cell model. However, this study aims to observe the vicinity of a broken fiber in the entire material. Therefore, we set the boundary condition so that the opposite boundary planes are free to move but remain planer in a parallel way for assuming generality with adjacent cells.

The material properties of the fiber and the matrix were estimated by fitting them to the experimental data by the rule of mixtures. This estimation assumes that the fiber/matrix interfaces are completely fixed and that the Young's moduli of the fibers and resin follow the rule of mixture. Properties that could not be estimated from the experiment were not used in the microscale analysis. The estimated material properties are shown in Table 2.

Table 2. Material properties estimated by the rule of mixtures.

Property	Fiber	Resin
Young's modulus /GPa	155.6	2.65
Poisson's ratio	0.34	0.36
Mass density /g/cm^3	1.80	1.24
Yield stress /MPa	-	18
Tangent modulus /MPa	-	2300

Next, a microscale analysis without considering interfacial behavior was performed to confirm that the estimated parameters were accurate by comparison with the elastic deformation in the experiment. Fig.3 shows the microscale model used in this section. This model was created and analyzed using the finite element code LS-DYNA.

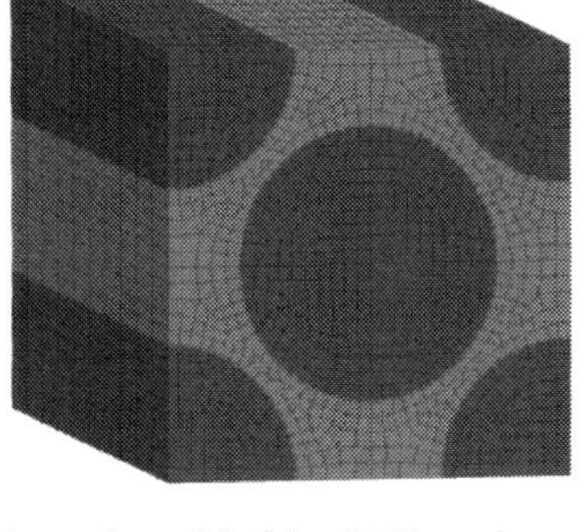

Figure 3. Microscale model of the CFRP specimen without considering interfacial behavior.

The whole model was meshed with eight-node hexahedral solid elements. The interface between the fibers and the matrix was assumed to be well bonded. The circular parts in Fig.3 are fibers. Although fibers are normally orthotropic elastic materials, they are modeled as an isotropic elastic material in this study. The resin was modeled as an isotropic elastic-plastic material. The load in this simulation was applied by controlling the displacement of the surface perpendicular to the fiber direction and fixing the opposite surface until the nominal strain of the model reached 0.015. Fig.4 shows a comparison of the results of the experiment and the microscale analysis without considering interfacial behavior.

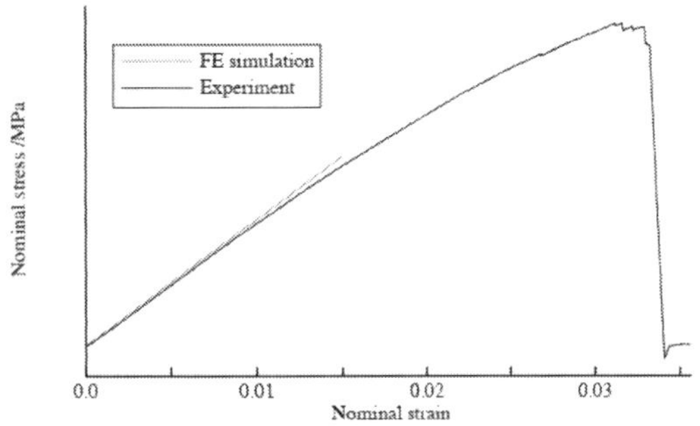

Figure 4. Comparison of the results of the experiment and microscale analysis without considering interfacial behavior.

It was found that the stress-strain curve obtained in the simulation was almost identical to that obtained in the experiment under elastic deformation. Therefore, the material properties shown in Table 2 were used in the analyses considering interfacial behavior.

3.2 Microscale analyses by modeling with cohesive zone elements

3.2.1 Cohesive zone elements

To express the fiber/matrix interfacial debonding, the application of cohesive zone elements (CZEs) to the interface was proposed. The analysis of fracture development has changed considerably in the last few decades. Instead of the traditional approaches of fracture mechanics, various predictive schemes such as CZEs are now increasingly being used, particularly in numerical simulations. The use of CZEs is a relatively simple method, making it easier to implement in finite-element codes and, as a result, CZEs are now available in various commercial FE packages such as LS-DYNA used in this simulation.

One of the first studies of cohesive zones was performed by Dugdale [6], who established the fundamental idea of CZEs. In his study, a plastic region was set around a crack tip and the relationship between the extent of plastic yielding and the applied external load was obtained. Needleman [7] was one of the first to use polynomial and exponential traction–separation relations to simulate the particle debonding in metal matrices. An exponential fit was used for a normal traction while a trigonometric fit was used for a shear traction. In short, CZEs have been found to be very useful for approaching problems involving nonlinear interfacial behavior.

This study includes the implementation of CZEs based on a bilinear traction–separation law. The mechanical response of the interface is described by constitutive equations, which relate the stresses t and s, normal and tangential to the interface, respectively, to the relative displacements in the same directions at the interface δ by a bilinear relationship. Fig.5 shows the traction–separation law defined in this simulation. The cohesive zone model used in this study defines the same traction–separation law in the normal and tangential directions.

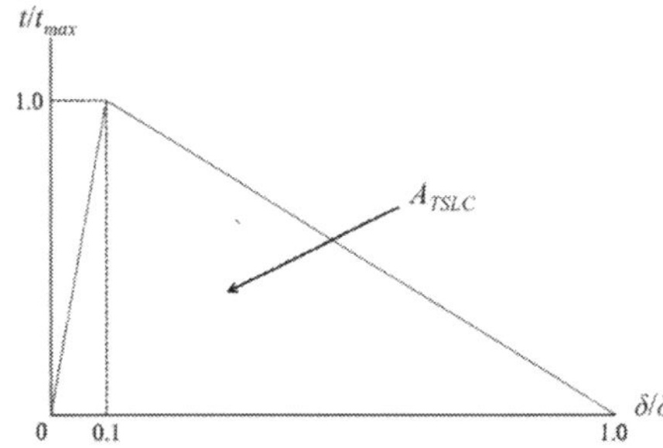

Figure 5. Normalized traction–separation law of CZEs.

This figure indicates that interfacial debonding starts at t_{max} and ends at δ_{max}. The area between the curve and the horizontal axis represents A_{TSLC}, which is the normalized energy release rate. Interfacial properties are determined by the energy release rate G_{I}^{c} (mode Ⅰ), the maximum normal stress T in the normal direction (mode Ⅰ), the energy release rate G_{II}^{c} (mode Ⅱ) and the maximum normal stress S in the tangential direction (mode Ⅱ) as follows.

$$\delta_{\mathrm{I}}^{F} = \frac{G_{\mathrm{I}}^{c}}{A_{TSLC}T}, \quad \delta_{\mathrm{II}}^{F} = \frac{G_{\mathrm{II}}^{c}}{A_{TSLC}S} \quad (1)$$

3.2.2 Conditions of microscale analyses with CZEs

In this section, analyses of the tensile test with CZEs are discussed, where it was assumed that the central fiber in the microscale model ruptured at a strain of 0.015. The effect of the properties of the cohesive zone on the fiber/matrix interfacial behavior after the breakage of a fiber was examined. Fig.6 shows the model constructed for this simulation.

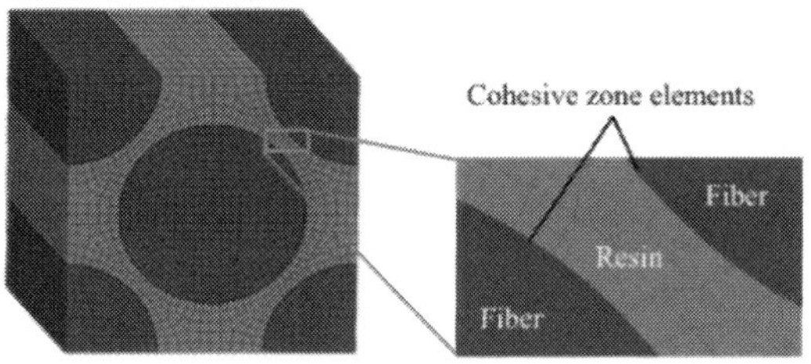

Figure 6. Microscale model with CZEs.

CZEs were inserted between the fibers and the resin. The other geometric conditions were the same as those in the model in section 3.1. To simulate the process of fiber breakage in the composite, a load was applied as follows. First, a uniform tensile displacement in the fiber direction was imposed on the entire top surface of the model until the nominal strain reached 0.015 as in section 3.1. After obtaining the applied composite stress, the nodes on the surface of the center fiber was released and become traction-free, thus simulating a fiber break. A tensile load was applied until the fracture strain observed in the experiment was reached. This simulation was performed for the four sets of cohesive zone properties shown in Table 3.

Table 3. Material properties of CZEs.

Set	G_{I}^{c} /N/mm	*T*/MPa	G_{II}^{c} /N/mm	*S*/MPa
(1)	0.645	86	0.51	34
(2)	1.29	86	0.51	34
(3)	0.645	86	0.765	34
(4)	0.645	86	0.51	68

One parameter was varied from the basic model (1) in each set of parameters with G_{I}^{c} , G_{II}^{c} and *S* changed in sets (2) to (4), respectively. Actually, interfacial properties should have been determined by characteristic tests. However, we used estimated values instead because it is extremely difficult to measure cohesive zone properties in experiments. Therefore, we qualitatively discuss the effect of changing the interfacial parameters.

3.3 *Analytical results of interfacial debonding*

3.3.1 Effect of the energy release rate normal to the fibers

The results for sets (1) and (2) were compared to observe the effect of G_{I}^{c} . Fig.7 shows a comparison of the normal-direction stress-strain curve of the CZEs at specified distances from the fiber break point. Points with zero stress in the figure indicate complete debonding at corresponding distance.

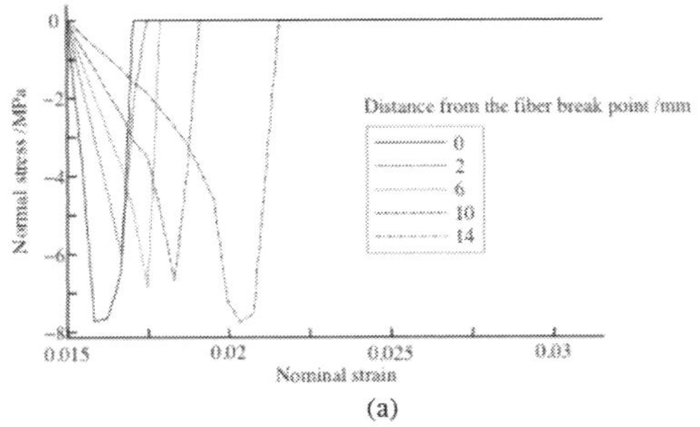

(a)

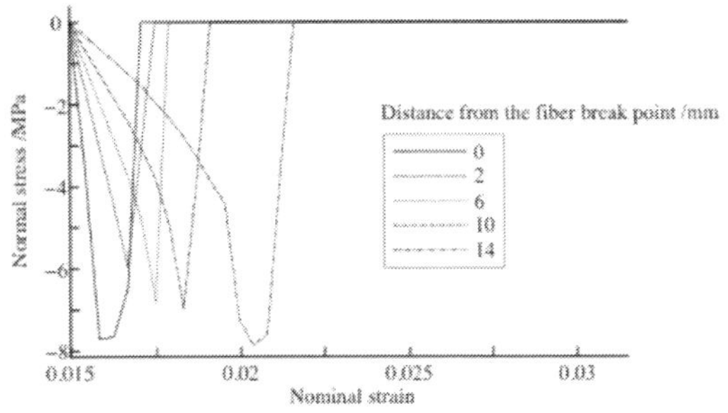

(b)

Figure 7. Normal-direction stress-strain curves of CZEs at specified distances from the fiber break point: (a) set (1) and (b) set (2).

By considering the complete debonding points, it was found that no major differences occurred upon changing the value of G_{I}^{c} . This result clearly shows that the result of a tensile test on unidirectional CFRP sheets does not depend on the fiber/matrix interfacial characteristics normal to the fibers.

3.3.2 Effect of the energy release rate tangential to the fibers

The results for sets (1) and (3) were compared to observe the effect of G_{II}^{c} . Fig.8 shows a comparison of the tangential-direction stress-strain curve of the CZEs at specified distances from the fiber break point.

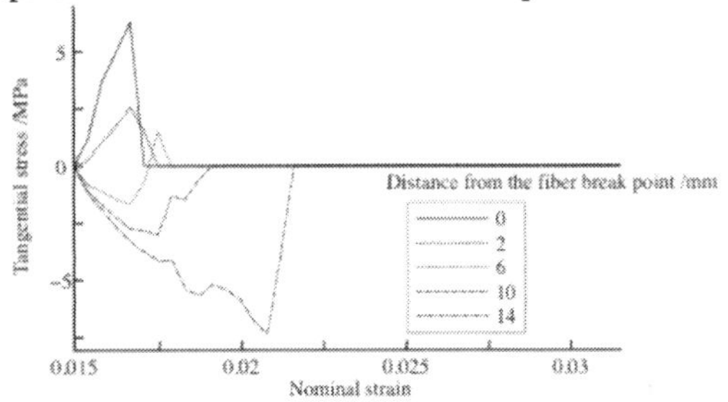

(a)

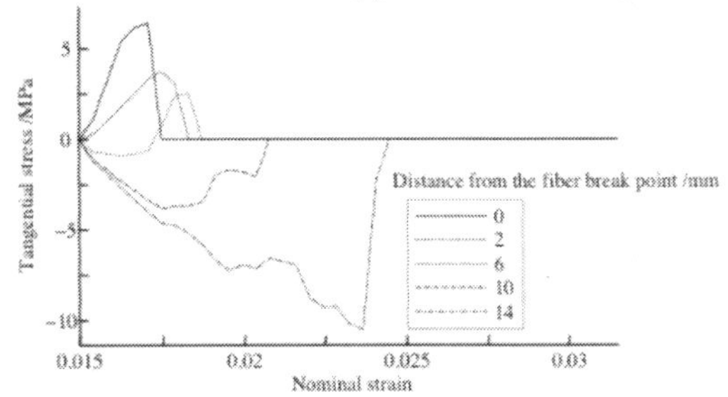

(b)

Figure 8. Tangential-direction stress-strain curves of CZEs at specified distances from the fiber break point: (a) set (1) and (b) set (3).

Fig.8 indicates that the complete debonding strain of set (3) is larger than that of set (1) at all distances. It can be concluded that the tangential-direction energy release rate G_{II}^{c} substantially dominates the degree of resistance to interfacial debonding.

 DOI: 10.1051/matecconf/20168016008

3.3.3 Effect of the maximum stress tangential to the fibers

The results for sets (1) and (4) were compared to observe the effect of the maximum shear stress *S*. Fig.9 shows a comparison of the tangential-direction stress-strain curve of the CZEs at specified distances from the fiber break point.

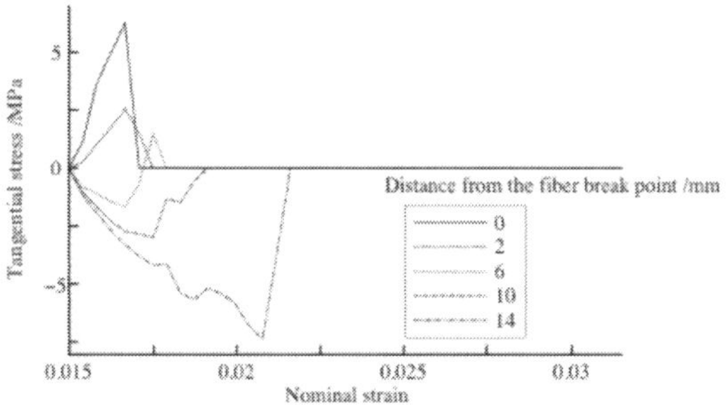

(a)

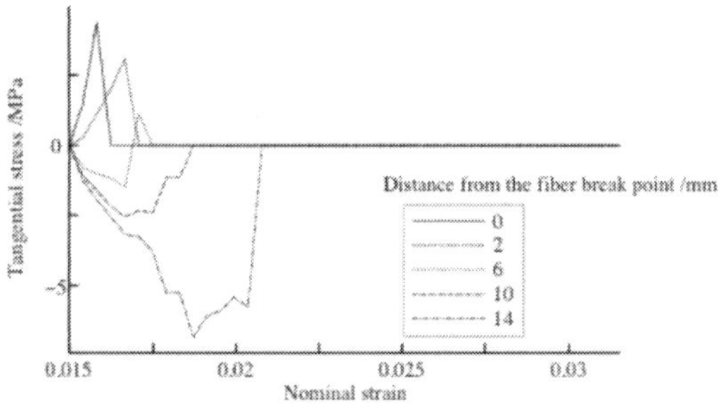

(b)

Figure 9. Tangential-direction stress-strain curves of CZEs at specified distances from the fiber break point: (a) set (1) and (b) set (4).

Fig.9 shows that the interfacial debonding of set (4) ends at a lower strain. This indicates that an increase in the tangential maximum stress makes it easy to debond the fiber/matrix interface.

3.4 *Analytical results of the stress distributions around the fibers*

In general, a CFRP breaks as a result of local stress concentrations caused by a complicated mixture of internal fiber breakage, interfacial debonding and matrix cracking. Thus, understanding the stress distribution around a fiber is also important for investigating the mechanism of plastic deformation for CFRP sheets. Fig.10 shows the stress distributions of the bottom surface of the resin for each set after the simulation.

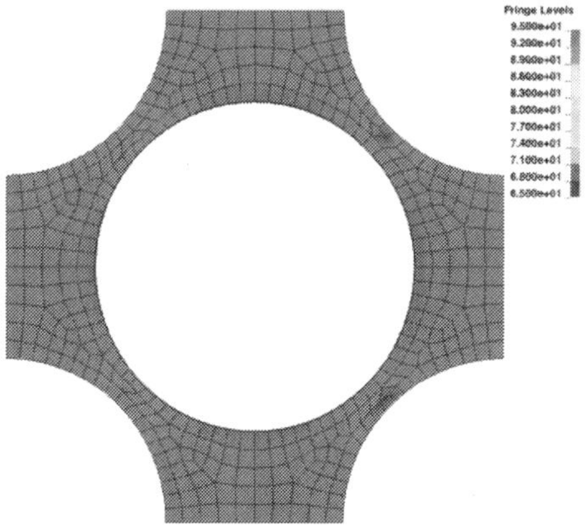

(a)

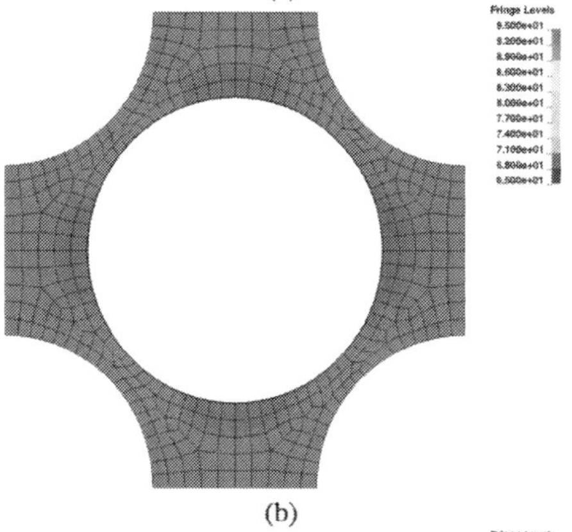

(b)

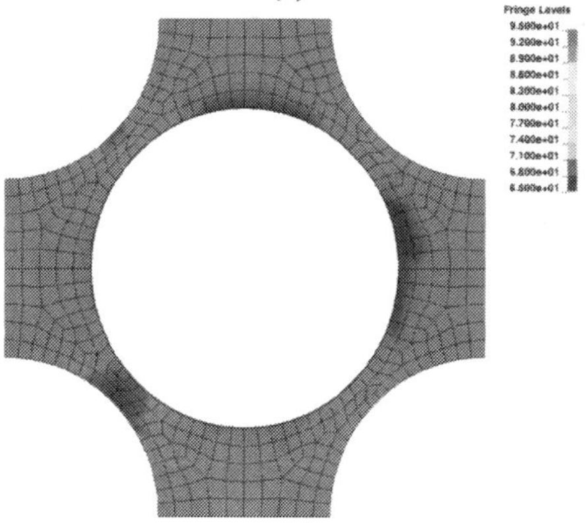

(c)

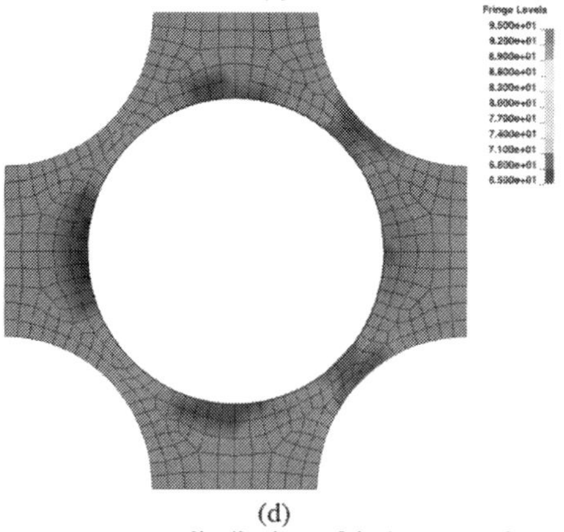

(d)

Figure 10. Stress distributions of the bottom surface of the resin: (a) set (1), (b) set (2), (c) set (3) and (d) set (4).

Fig.10 (d), which was obtained at a large maximum tangential stress, shows the most scattering of the stress distribution among the sets. In other words, the cohesive zone property of set (4) tends to concentrate the local stress. This is consistent with the experimental result that composites have a stronger fiber/matrix interface, making matrix cracking around the fiber break point more likely to occur [3]. Therefore, the CZEs in this model qualitatively well express the fiber/matrix interfacial behavior.

4 Conclusion

A uniaxial tensile test on a CFRP specimen and microscale analyses were performed. Fibers and matrix properties were estimated by the rule of mixtures for composites using the experimental results.

CZEs were applied to the fiber/matrix interfaces in the microscale unit cell model. Analytical results indicated that the properties of cohesive zone elements qualitatively well represented the actual fiber/matrix interfacial behavior. This result will be helpful in understanding and modeling the deformation processes in CFRP sheets.

As a future task, a means of identifying quantitative interfacial properties under various thermal environments and considering other interactive factors, such as the friction after debonding or the viscosity of the resin, should be established using CZEs.

References

1. E. Pisanova, S. Zhandarov, E. Mader, I. Ahmad, R. J. Young, Composites Part A **32**, 435 (2001).
2. M. J. Pitkethly, J. P. Favre, U. Gaur, J. Jakubowski, S. F. Mudrich, D. L. Caldwell, L. T. Drzal, M. Nardin, H. D. Wagner, L. Di Landro, A. Hampe, J. P. Armistead, M. Desaeger, I. Verpoest, Compos. Sci. Technolo. **48**, 205 (1993).
3. M. Nishikawa, T. Okabe, N. Takeda, Mater. Sci. Eng. A **480**, 549 (2008).
4. B. W. Kim, J. A. Nairn, J. Compos. Mater. **36**, 1825 (2002).
5. J. M. Guedes, N. Kikuchi, Comput. Methods in Appl. Mech. Eng. **83**, 143 (1990).
6. D. S. Dugdale, J. Mech. Phys. Solids **8**, 100 (1960).
7. A. Needleman, J. Mech. Phys. Solids **38**, 289 (1990).

 DOI: 10.1051/matecconf/20168016009

Homogenization on Multi-Materials' Elements: Application to Printed Circuit Boards and Warpage Analysis

Manuel Araújo[1,a], J. L. Alves[1], Paulo Silva[1] and Pedro Delgado[2]

[1]*MEMS, University of Minho, Dep. of Mechanical Engineering, Campus de Azurém, 4800-058 Guimarães, Portugal*
[2]*Bosch Car Multimedia Portugal S.A., Apartado 2458, 4705-820 Braga, Portugal*

Abstract. Multi-material domains are often found in industrial applications. Modelling them can be computationally very expensive due to meshing requirements. The finite element properties comprising different materials are hardly accurate. In this work, a new homogenization method that simplifies the computation of the homogenized Young modulus, Poisson ratio and thermal expansion coefficient is proposed, and applied to composite-like material on a printed circuit board. The results show a good properties correspondence between the homogenized domain and the real geometry simulation.

1 Introduction

One of the aims of the microelectronics field is the performance and reliability improvement [1]. In industry, weight reduction also plays an important role in a product's development cycle. Owed to the attempt of manufacturing lighter and thinner components the warpage phenomena became more noticeable.

PCBs (Printed Circuit Boards) comprise conductive and dielectric layers, consisting of copper and epoxy resin and an epoxy-woven glass fibres composite (mostly FR4), respectively (Figure 1). During the reflow soldering process, PCBs are exposed to temperatures ranging from 230°C to 250°C [2].

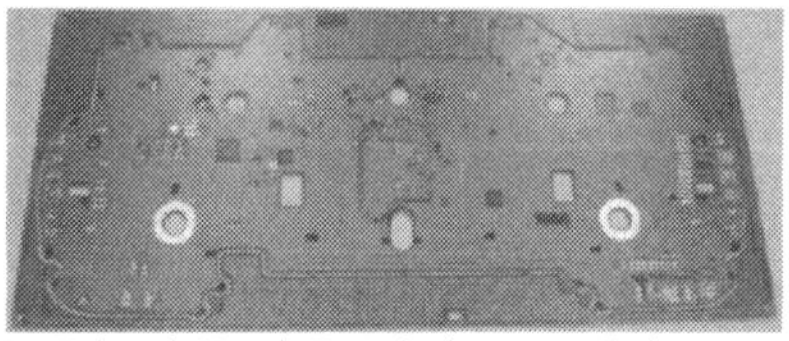

Figure 1 – Printed Circuit Board of a car cockpit.

When subjected to temperature changes, the mismatch in the thermomechanical properties of the constituent materials of the PCBs leads to the appearance of significant thermal stresses [3, 4]. Warpage is responsible for misalignments during the package assembly stage, smaller tolerances [4], imperfect soldering and consequent detachment of components [4]. In addition, it may induce cracks and the separation of the layers [5].

1.1 Warpage problem

The main materials of PCBs are fiberglass-reinforced (known as FR4), copper and resin. Copper and resin are isotropic materials. They have the same elastic property values in every direction [6]. On the other hand, FR4 is an orthotropic material, i.e. it has three mutually perpendicular planes of elastic symmetry [7]. The key thermomechanical properties required to estimate the warpage of a PCB are the Young modulus (E), shear modulus (G), bulk modulus (K), Poisson ratio (v) and the coefficient of thermal expansion (α). These properties fully define the materials' thermoelastic behaviour.

A PCB has always more than one conductive layer and it can contain both copper and resin. The layers have not the same geometry or copper distribution (Figure 2). For this reason there is unbalanced forces between layers.

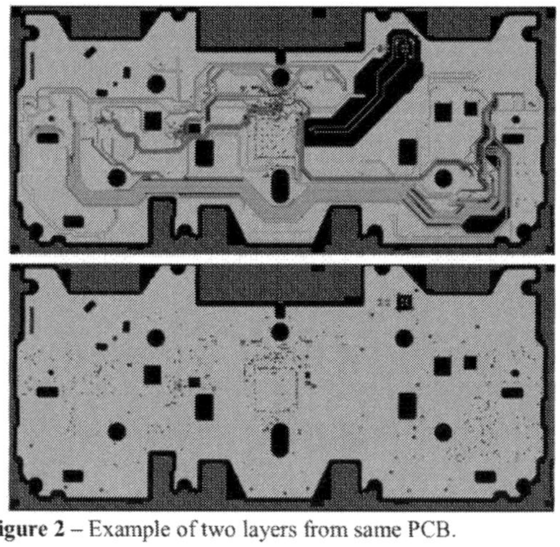

Figure 2 – Example of two layers from same PCB.

Therefore, the selection of the thermoelastic properties of the materials at the reflow soldering temperature is of great importance.

[a] Corresponding author: m.araujo@dem.uminho.pt

1.2 Warpage simulation

For the purpose of evaluating the warpage of a PCB, the layers of the board are discretized using a finite element mesh, and then a FE analysis is performed. The fact that the conductive layers can contain both copper and resin or only one of these makes it more complex to characterize. On the other hand, the dielectric layers always have only FR4 (mainly).

The conductive layers can have copper artworks of just 80 μm wide. The simulation of those geometries is very time consuming due to the complexity of the copper distribution. Furthermore, in the most cases, the computer memory space is not enough. The finite element size has to be much larger than the copper tracks wide. Thus, the calculation of homogenized thermomechanical properties is required. The methods usually applied for modelling warpage do not consider the orientation of the copper traces to calculate the PCB's homogenized properties.

Indeed, present warpage prediction techniques use volume averaging to estimate these properties [3]. Nevertheless, studies by Hutapea and Grenestedt [8] show that the copper trace orientation has a noteworthy influence on the PCBs' warpage even when the board is completely copper balanced.

1.3 The aim of the work

This study aims to analyse different methods for the homogenization of PCBs' conductive layers. The PCBs' constituent materials are described, as well as their most relevant thermomechanical properties.

From all the considered information, values for the different thermomechanical properties at the reflow soldering temperature (approximately 250°C) were selected (Table 1). All the FEA present in this work is performed using these values. The homogenization analysis is applied to the copper layers only, and it does not take into account the FR4 layers.

First, this work takes 4 examples of simple elements of a conductive layer. The simple rule of mixtures is used and some results shown.

Then, it is presented the homogenization method, which is developed in order to become able to homogenize more properly the properties of more complex copper distribution elements. This homogenization method assumes some of the principles of the rule of mixtures.

Finally, the global results of 28 copper distribution elements are analysed. It includes some considerations about the elastic moduli, the time spent in the computation and a statistical analysis of the performance of this method.

1.4 FEM calculation

All models were composed by 25x25x1 8-node hexahedron finite elements. Models were created according to the original and "pixelated" Gerber file (see examples in Figure 3 and Figure 7). Its global dimensions are 1 mm x 1 mm x 0.04 mm.

Table 1 – Thermomechanical properties of copper and resin.

Property	Copper	Epoxy Resin
E [GPa]	117	0.070
ν [-]	0.30	0.45
α [10^{-6}/°C]	21	197

To simulate the mechanical behaviour of the model in tension, a displacement of 4 μm on both in-plane directions was carried out. This value was established so the copper remained on the elastic domain. The total force was set to be null in the perpendicular directions to the imposed displacement, keeping the surfaces parallel to each other during the deformation process as a real homogeneous material would behave.

The relevant properties of the materials are the Young modulus and the Poisson ratio and the values considered for these properties are depicted in Table 1. The global Young modulus and Poisson ratio are obtained through eq. (1) and (2), respectively.

$$E = \frac{\sigma}{\varepsilon} \quad (1)$$

$$\nu = -\frac{\varepsilon_T}{\varepsilon_L} \quad (2)$$

2 Homogenization: the rule of mixtures

Different element configurations were imported from a Gerber file (original file describing a PCB). The simpler evaluated configurations are presented in Figure 3. Those four elements, referenced as e426, e610, e614 and e625, have different distributions and area fractions of copper: 19.8%, 24%, 13.9% and 5.9%.

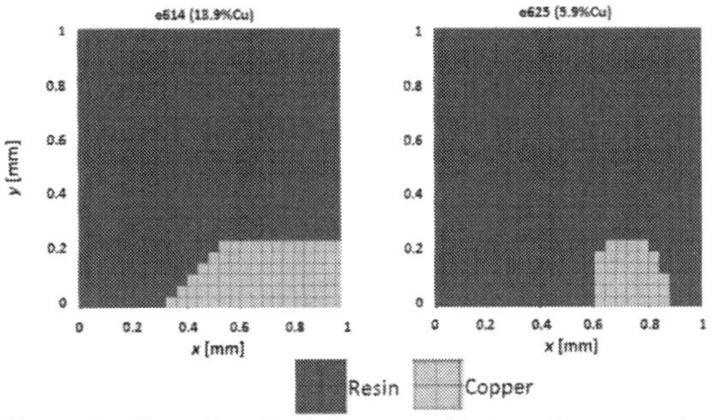

Figure 3 – Examples of real elements with the real copper and resin distribution. The area fractions of copper are of 19.8%, 24%, 13.9% and 5.9%, respectively.

2.1 Equivalence to a rectangle of copper

The calculation of the equivalent rectangle is based on the moment of inertia tensor. The coordinates of copper's centre of mass are previously determined. The components of the inertia tensor are given by

$$\mathbf{I} = \begin{bmatrix} I_{xx} & I_{xy} \\ I_{yx} & I_{yy} \end{bmatrix}, \quad \begin{cases} I_{xx} = \sum_{i=1}^{n} {y_i}^2 a_i \\ I_{yy} = \sum_{i=1}^{n} {x_i}^2 a_i \\ I_{xy} = -\sum_{i=1}^{n} x_i\, y_i a_i \\ I_{yx} = I_{xy} \end{cases} \tag{3}$$

I_{xx} and I_{yy} are the moments of inertia about the x- and y-axis, respectively. I_{xy} and I_{xy} have the same value and represent products of inertia. The distances from the copper's centre of mass to the centre of each point in the x and y axes are denoted as x_i and y_i, respectively. The eigenvalues (λ_1 and λ_2) are used to calculate the dimensions of a rectangle with an equivalent area of copper (see Figure 5) and the eigenvectors are used to determine its spatial orientation. For simplicity of calculation, the orientation of the equivalent rectangle is assumed to be 0 or 90 degrees according to the eigenvectors.

The area of the equivalent rectangle of copper is $b \times h$, with b and h given by,

$$b = \sqrt[8]{144\frac{{\lambda_1}^3}{\lambda_2}}, \quad h = \sqrt[8]{144\frac{{\lambda_2}^3}{\lambda_1}} \tag{4}$$

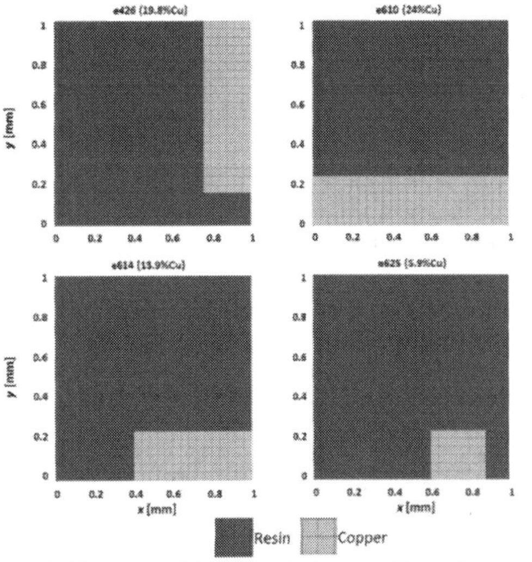

Figure 4 –Examples of the elements shown in Figure 3, in which the real copper distribution is replaced by an equivalent rectangle of copper determined to display the same inertia tensor and the same area fraction of copper. The area fractions of copper are of 19.8%, 24%, 13.9% and 5.9%, respectively.

The centre of mass and the copper area fraction, f_{cu}, are the same as the original copper distribution. The sum of copper fraction and resin fraction, f_{re}, is equal to 1. The dimensions b and h are updated if the copper fraction of the equivalent rectangle if different from the copper fraction of the original configuration. The equivalent rectangles of the real element configurations (see Figure 3) are shown in Figure 4.

2.2 Mathematical model

The mathematical model for the determination of the homogenized Young modulus was based on the rule of mixtures. The rule of mixtures is based on the assumption of the equivalent stiffness, R_{eq}, which depends on the position of the bodies, i.e., if they are in parallel [eq. (5)] or in series [eq. (6)].

$$R_{eq} = R_1 + R_2 \tag{5}$$

$$R_{eq} = \left(\frac{1}{R_1} + \frac{1}{R_2}\right)^{-1} \tag{6}$$

The relation between the stiffness, R, and the Young modulus, E, is expressed in eq. (7) and it depends on the area of the body's section, A, and the longitudinal length, L.

$$R = \frac{EA}{L} \tag{7}$$

For each direction, the element can be divided in 3 parts: one containing resin and copper (α-zone) and two others with resin only (see Figure 5). When tensioned, it is assumed that both materials of the α-zone display the same displacement as they are tensioned in parallel.

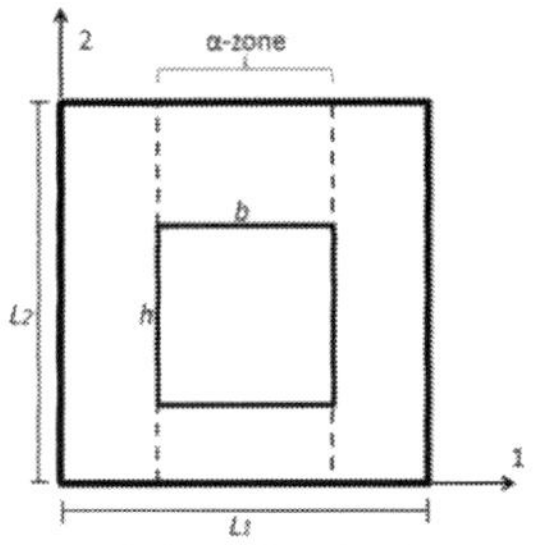

Figure 5 – Schema of the rule of mixtures for tension in direction 1.

The total force is affected by the force applied to each material. For direction 1, the Young modulus of α-zone ($E_{1\alpha}$) can be calculated with a simple relationship between the areas [see eq. (8)]. Along direction 1, the portions of resin and the α-zone are placed in series and they are submitted to the same load. The element's total Young modulus along direction 1 (E_1) is given by eq. (9).

$$E_{1\alpha} = \frac{L_2 - h}{L_2} E_{res} + \frac{h}{L_2} E_{Cu} \tag{8}$$

$$E_1 = \left(\frac{b}{E_{1\alpha}L_1} + \frac{(L_1 - b)}{E_{res}L_1}\right)^{-1} \quad (9)$$

2.3 Comparing results

The values of the Young modulus obtained with FEM and with the rule of mixtures are exposed in Figure 6. For each configuration, the Young modulus along x and y direction is analysed.

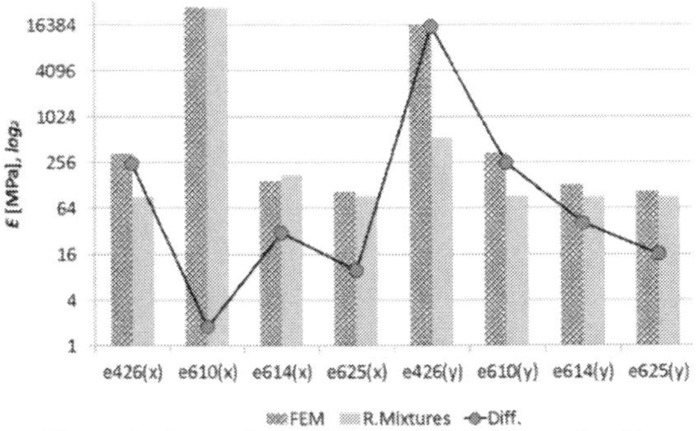

Figure 6 – Comparison of Young modulus determined by i) finite element computation and by using the ii) homogenization technique based on the rule of mixtures.

The best result obtained is the Young modulus for element e610 in x direction (Figure 6). On the other hand, the worst results were obtained for element e426 in x direction and for element e426 and e610 in y direction. This is an indicator that the rule of mixtures provides good predictions when the distribution of copper is aligned with the tension test.

However, the copper can have infinite number of different distributions. And the homogenization method should provide better predictions in a much larger number of cases. Due to this reason, more complex copper distribution elements are used to improve a homogenization method in the next section.

3 Homogenization procedure

This homogenization method takes into account not only the calculation of equivalent properties but the characteristics of the elastic moduli matrix as well. In this section it is explained the homogenization method used to predict the Young modulus, Poisson ratio and thermal expansion properties of several multi-material elements of the copper layer of an arbitrary PCB.

This method has in consideration if the equivalent rectangle reaches opposite sides of the element for each direction. For each element it is calculated the dimensions of the equivalent rectangle by the same procedure as in the rule of mixtures. The complex copper distribution elements are shown in Figure 7 and its equivalent rectangles in Figure 8. The labels of these elements are e416, e424, e430 and e438.

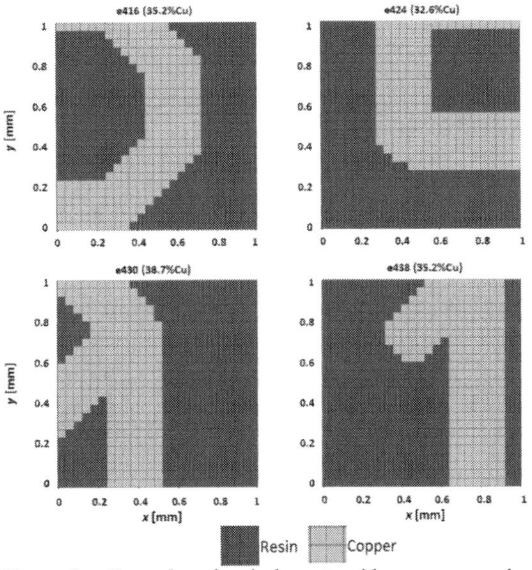

Figure 7 – Examples of real elements with a more complex geometrical distribution of copper and resin.

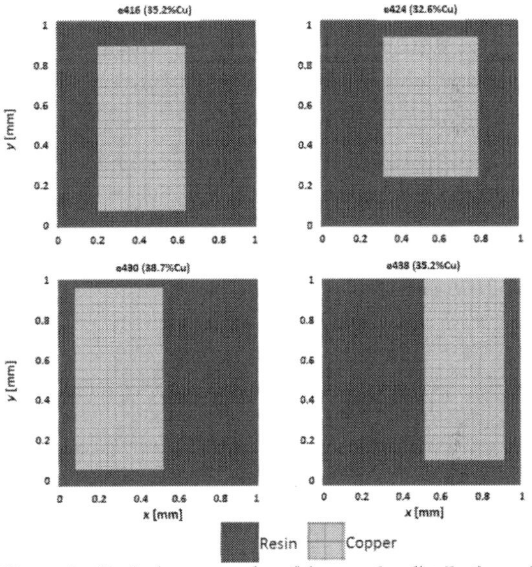

Figure 8 – Equivalent rectangles of the complex distributions of copper depicted in Figure 7.

In this part of the work it is studied the equivalent properties prediction of Young modulus, Poisson ratio and thermal expansion coefficient. However, it is explained only the procedure when the equivalent rectangle and the original copper distribution do not cross the element through opposite sides in any direction.

3.1 Equivalent Young modulus

The equivalent Young modulus prediction starts with the equivalent stiffness calculation through eq. (5) and (6). This previous equivalent stiffness, R_{eq}, is obtained for direction 1 and 2, R_{eq1}, and R_{eq2}, respectively. For the direction 1, the procedure follows to eq. (10):

$$E_1 = R_{M1}\frac{L_1}{L_2L_3} \tag{10}$$

where:

$$R_{M1} = R_{eq1}\left(1 + \frac{(E_{cu})^{K_0} - (E_{re})^{K_0}}{(E_{re})^{K_0}}\right)C_1 \tag{11}$$

K_m is obtained by eq. (12) and C_1 by eq. (13):

$$K_m = \left(\frac{E_{re}}{1000}\right)^{k_1}\left(\frac{E_{cu}}{E_{re}}\right)^{k_2}\left(\frac{v_{cu}}{v_{re}}\right)^{k_3} \tag{12}$$

$$\begin{cases} C_1 = \frac{h_r}{15}(1 + b_r - b_r^3)(3 - b_r)\left(4 + \frac{1}{h_r}\right) \\ b_r = b/L_1 \\ h_r = h/L_2 \end{cases} \tag{13}$$

The non-dimensional constants k_1, k_2, k_3 are set to 0.35, 0.505 and 3.8, respectively, in order to get a better fitting to the FEM results. The constants were optimized according to an objective function that corresponds to the minimization of the differences between FEM and homogenization results. This homogenization method also checks if the original copper distribution crosses the element through opposite sides. This is done due to the good performance of the rule of mixtures in those cases, as it is possible to check the Young modulus (x direction) prediction in Figure 6. Due to this reason, the Young modulus in z direction is calculated by the rule of mixtures as well.

3.1.1 Comparing methods: simple elements

The homogenized values of the Young modulus were obtained for each one of the simple elements (see Figure 3) and compared with the values from rule of mixtures, and summarized in Figure 9, with the values of the Young modulus in x and y directions are displayed as bars and the differences between the methods on the lines. The line of "ΔRM" means the difference or error between FEM and rule of mixtures results. The line of "ΔHomog." indicates the error between FEM and homogenization method.

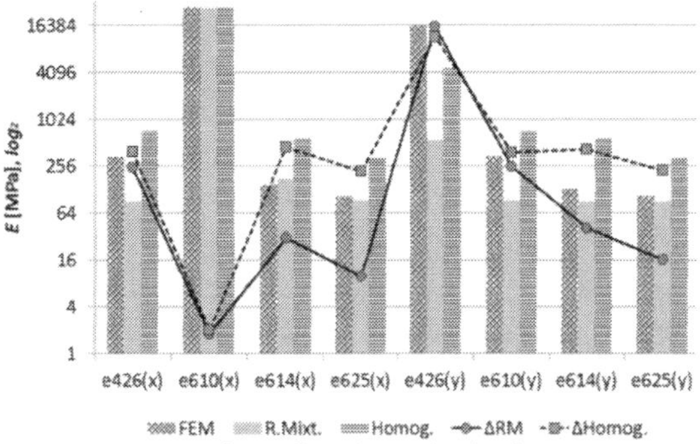

Figure 9 – Comparison of results in simple elements between the rule of mixtures and homogenization.

Most of the values of the rule of mixtures have lower differences than the homogenized ones. However, the rule of mixtures has the higher error: the E_y of e426. The FEM value is 16206 [MPa], while the homogenization prediction is 4632, but the rule of mixtures estimates only 553 [MPa]. This element displays the largest error of the new homogenization method.

3.1.2 Comparing methods: complex elements

For each one of the complex elements (Figure 7), the homogenized values of the Young modulus are compared with the values from the rule of mixtures (Figure 6).

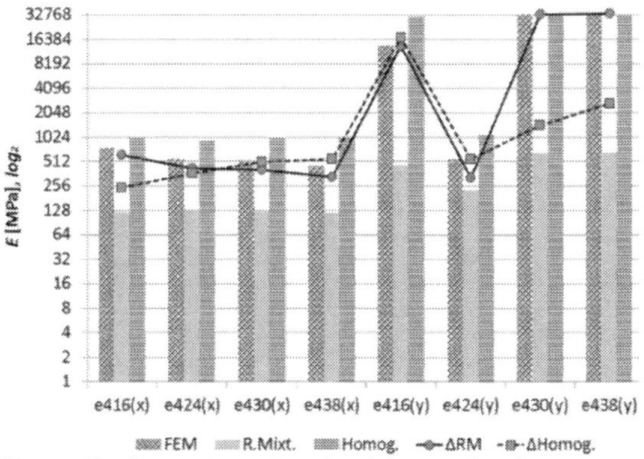

Figure 10 – Comparison of results of the complex elements between the rule of mixtures and the new homogenization technique.

Figure 10 indicates that the homogenization has a better performance than the rule of mixtures. The values of the rule of mixtures are always under 1 GPa while some of the FEM values are over 10 GPa.

3.2 Equivalent Poisson ratio

In this section it is explained the homogenized Poisson ratio when a tensile deformation is applied along direction 1. Two values of Poisson ratio are calculated: v_{12} and v_{13} (eq. (14) and (15), respectively).

$$v_{12} = (v_{cu} + C_2)[1 - k_4(h_r - k_5)^2] \tag{14}$$

$$v_{13} = b_r v_{cu}\left(1 + f_{re}\frac{E_{re}}{E_{cu}}\right)\frac{E_1}{E_3} \tag{15}$$

where C_2 is obtained by the eq. (16),

$$C_2 = f_{re}\left(\frac{v_{re} - v_{cu}}{v_{cu}}\right)^{k_6}\left(\frac{E_{re} - E_{cu}}{E_{cu}}\right)^{k_7}\sqrt{\frac{1 - b_r}{K_m}} \tag{16}$$

The non-dimensional constants k_4, k_5, k_6 and k_7 are set to 1.2, 0.1, 1.15 and 4, respectively. The constants were optimized for the minimization of the differences between FEM and homogenization results. This method was developed taking into consideration that the range of resin's Young modulus is lower than the range of copper's Young modulus. It was also assumed that the range of resin's Poisson ratio is higher than the range of copper's Poisson ratio.

Poisson ratio has a crucial impact on the mathematical calculation of FEM. The elastic moduli matrices have to be positive definite. That is taken into account in the equivalent Poisson ratio method: the error of Poisson ratio prediction by homogenization forces the elastic moduli matrix to be positive definite (check "4.1 Elastic moduli matrices"). Otherwise, no FEM simulation can be performed using homogenized properties.

The results for Poisson ratios ν_{xy} and ν_{yx} are shown in Figure 11 and Figure 12. The first four elements are the elements of Figure 3 and the last ones are the elements of Figure 7.

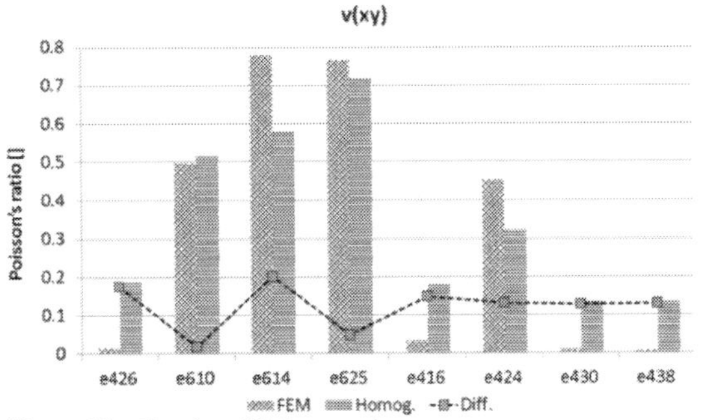

Figure 11 – Results of Poisson ratio ν_{xy}.

These results show that the differences ("Diff." series) between homogenization and FEM method are mostly higher when Poisson ratio from FEM is lower than 0.2 and higher than 0.5. It means that the homogenization method tends to provide a lower range of Poisson ratios. This is established in order to try to ensure that the elastic moduli are positive definite matrices.

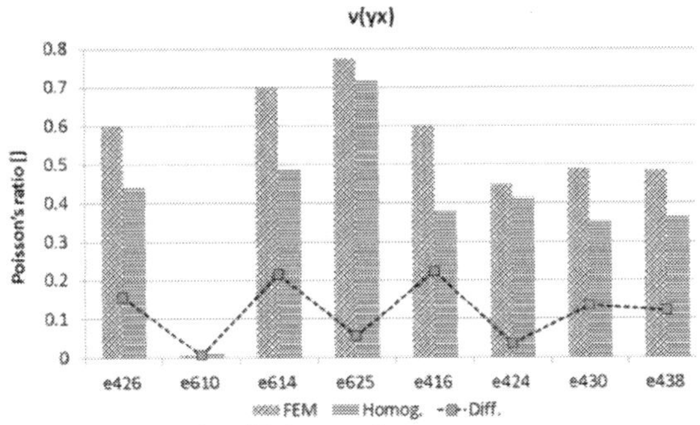

Figure 12 – Results of Poisson ratio ν_{yx}.

The results for Poisson ratios ν_{xz} and ν_{yz} are shown in Figure 13 and Figure 14, respectively. These values have special concerns because the copper crosses the element through opposite sides in z direction.

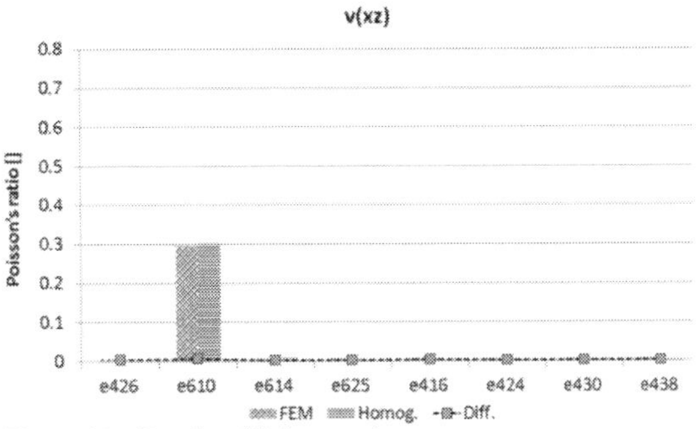

Figure 13 – Results of Poisson ratio ν_{xz}.

When the extension is applied (in x or y direction), the contraction of the element is defined by the copper. The stiffness of copper is much higher than the resin stiffness. For this reason most of the Poisson ratios of Figure 13 and Figure 14 are nearly zero.

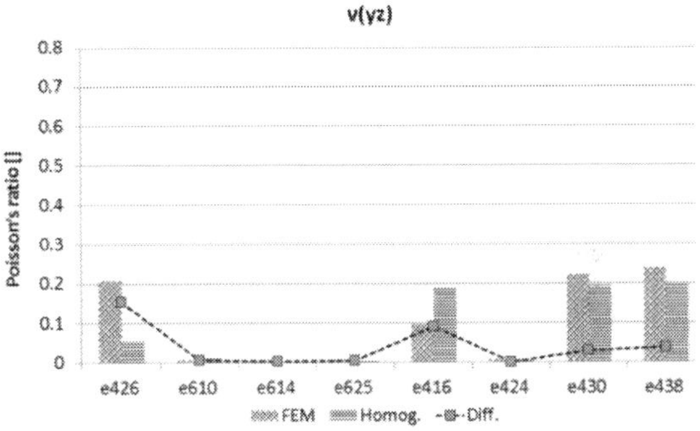

Figure 14 – Results of Poisson ratio ν_{yz}.

However, the error of ν_{yz} in element e426 can be improved because the copper distribution crosses the element in y direction. And that is not taken into account in this homogenization process.

3.3 Equivalent thermal expansion

The homogenized thermal expansion plays a paramount role on the analysis of warpage in PCBs. The calculation of the thermal expansion in direction 1, α_1, is based on eq. (17):

$$\alpha_1 = \frac{C_3(\alpha_{re} + 1) + b(\alpha_{cu} + 1) - L_1}{L_1} \quad (17)$$

where C_3 is obtained by the eq. (18):

$$C_3 = L_1 - b + k_8\sqrt{L_1^2 - b^2} + k_9(L_1 - b) \quad (18)$$

The dimensions of variables α_1, α_{re} and α_{cu}, are [m.m^{-1}.K^{-1}]. The non-dimensional constants k_8 and k_9 are set to 0.020 and 0.005, respectively, in order to get a better matching between FEM results and the homogenization method. The error function corresponds to the minimization of the differences between FEM and homogenization results.

DOI: 10.1051/matecconf/20168016009

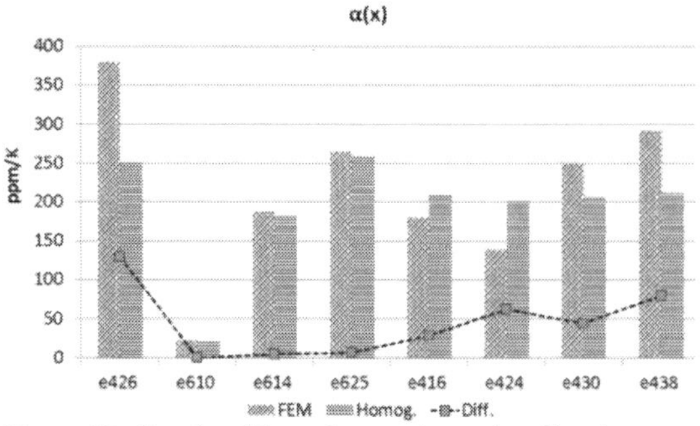

Figure 15 – Results of thermal expansion, α, in x direction.

The results of the thermal expansion in x and y direction is shown in Figure 15 and Figure 16, respectively.

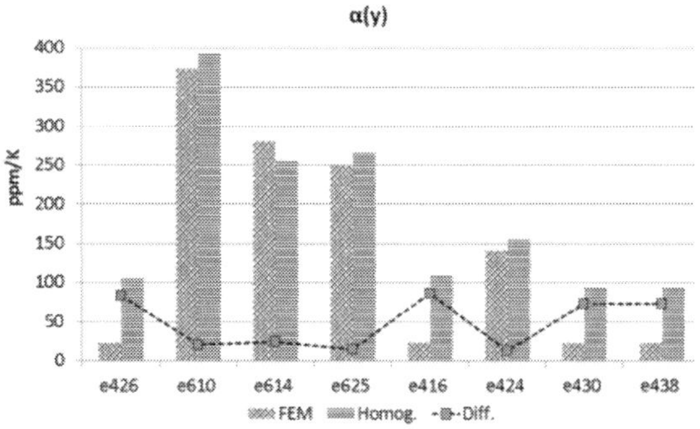

Figure 16 – Results of thermal expansion, α, in y direction.

The error of homogenization results is considerably high in some cases, e.g. see elements e426 and e438. A reason for that is the nonattendance to the original copper distribution. Once again, and similarly to the analysis of Figure 13 and Figure 14, the homogenization of the thermal expansion should take into consideration if the original copper distribution crosses the element at any direction.

4 Global results analysis

The homogenized values of the Young modulus, Poisson ratio and thermal expansion were obtained for each one of the 28 studied elements. In this section it is analysed the requirements of the elastic moduli, the time spent for each method calculation and the statistical analysis of all set of elements.

4.1 Elastic moduli matrices

The elastic moduli require special attention in a homogenization process. These matrices have to be positive definite. This aspect is hard to ensure during the homogenization process. When the elastic moduli is not a positive definite matrix, the global stiffness matrix of the FE problem becomes singular.

The worst results of homogenization are obtained with element e416 (see Figure 7). The first nine terms non-zero of the elastic moduli (D) obtained by FEM and by homogenization are shown in eq. (19) and (20).

$$D_{FEM} = \begin{bmatrix} 763 & 500 & 444 \\ 504 & 14571 & 4513 \\ 445 & 4511 & 44049 \end{bmatrix} \tag{19}$$

$$D_{hom} = \begin{bmatrix} 1073 & 450 & 252 \\ 6362 & 35234 & 9818 \\ 1763 & 9061 & 45273 \end{bmatrix} \tag{20}$$

The elastic moduli obtained by homogenization has all the diagonal terms higher than the moduli obtained by FEM. And some of the other terms are lower (see eq. (20)). However, matrix (20) is not a positive definite matrix given that the value of $D_{hom}(2,1)$ is too high. That is shown in eq. (21):

$$\left(D_{\text{hom}(2,1)}\right)^2 > \left(D_{\text{hom}(1,1)}\right)\left(D_{\text{hom}(2,2)}\right) \tag{21}$$

Nevertheless, the elastic moduli has also to be a symmetric matrix. The one obtained by FEM is nearly symmetric. That is an indicator that the properties obtained by FEM may be more reliable. The elastic modulus D_{hom} is forced to become symmetric, and thus symmetric terms are averaged. So, one obtains

$$D_{hom} = \begin{bmatrix} 1073 & 3406 & 1008 \\ 3406 & 35234 & 9439 \\ 1008 & 9439 & 45273 \end{bmatrix} \tag{22}$$

After this process, the elastic moduli of element e416 is a positive definite matrix. It is interesting to take a look to the issue expressed in eq. (21). Due to inaccuracy of homogenization method, the value of $D_{hom}(2,1)$ was too high. However, the same accuracy of the method forced the values of $D_{hom}(1,1)$ and $D_{hom}(2,2)$ to be higher and the value of $D_{hom}(2,1)$ to be lower. So, after symmetrisation, the elastic moduli became a positive definite matrix.

4.2 Time spent for each method

The CPU time spent to calculate the elastic moduli is much lower through homogenization method than by FEM. The time spent by FEM is very dependent of the number of elements. In this analysis, the number of elements was 625 (25x25x1) and the homogenization process was nearly 90 times faster than the FEM analysis.

4.3 Statistical analysis

The statistical analysis of the data obtained from the homogenization model and FEM can be performed using the coefficient of determination, R^2, correlation and the analysis of the mean values and variance, as summarized in Table 2.

The requirements for R^2 analysis are assumed. The Spearman's correlation indicates that the homogenization is a consistent model. However, the variances of the thermal expansion models are very different, i.e. higher than 20%.

Table 2 – Statistical data of the results

		α(x)	α(y)	E(x)	E(y)
R^2		76.5%	92.9%	99.6%	94.6%
Spearman's correlation		0.896	0.894	0.981	0.984
Mean	FEM	181	194	5.12	11.73
	Homo.	179	217	5.39	12.08
Variance	FEM	9794	20284	98.58	323.33
	Homo.	6799	13998	97.50	322.20

When the differences between models of thermal expansion (x direction) and Young modulus (y direction), several outliers are found in the respective boxplot (Figure 17). This is an indicator that the homogenization method still needs to be improved because it clearly fails in some very particular cases: when the elastic property is under 1GPa and when the copper distribution crosses the element in x or y direction.

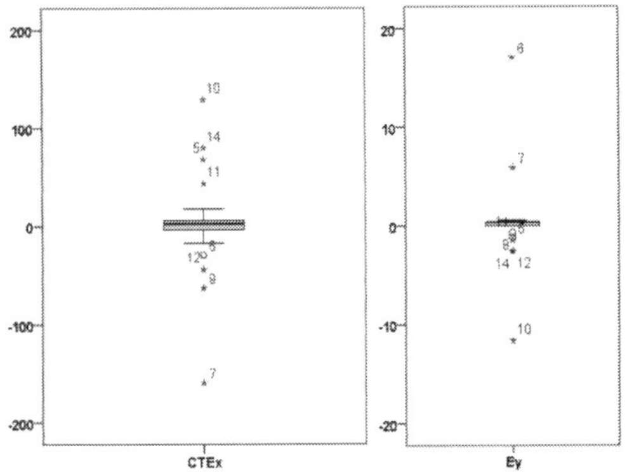

Figure 17 – Boxplot of the differences between FEM and homogenization model for α_x (left) and E_y (right).

The number of outliers in the boxplots of Figure 17 also indicates that the sample should be bigger and the elements should represent a larger range of copper distributions.

5 Conclusions

The homogenization method is a reliable alternative to the FEM simulation for equivalent properties calculation. However, there are opportunities to develop the method regarding less accurate aspects.

Concerning Young modulus calculation, there is a lack of accuracy in the homogenised properties when the elastic property is under 1GPa. Maybe k_1 optimization should be improved in eq. (12). A similar observation can be made for Poisson ratio. When the Poisson ratio from FEM is lower than 0.2 and higher than 0.5, the homogenised property loses accuracy.

The error of Poisson ratio and thermal expansion can be reduced because in some elements (e.g. element e416 and e430) the copper distribution crosses the element in y direction. The homogenization process does not have this aspect in consideration. By this way, the accuracy of the model will be improved for some copper distributions (e.g. for element e426).

The homogenization method provides an effective way of properties prediction. In general, the values obtained through this method are acceptable for warpage prediction because R^2 and Spearman's correlation are higher than 0.75 and 0.8, respectively. Moreover, the homogenization method is much faster (90 times faster) than FEM method and provides positive definite matrices of the elastic moduli.

Finally, other copper distributions should be used to test the homogenization model in order to get a better analysis of the range and the potential of this method.

Acknowledgements

This research is sponsored:
- by the Portugal Incentive System for Research and Technological Development. Project in co-promotion nº 36265/2013 (Project HMIExcel - 2013-2015), and
- by FCT with the reference project UID/EEA/04436/2013, by FEDER funds through the COMPETE 2020 – Programa Operacional Competitividade e Internacionalização (POCI) with the reference project POCI-01-0145-FEDER-006941.

References

1. S.T. Pantelides, et al., Performance, reliability, radiation effects, and aging issues in microelectronics – From atomic-scale physics to engineering-level modeling **54**, 9 (2010)
2. F. Clyde and J. Coombs, *Printed Circuits Handbook.* (2008): McGraw-Hill.
3. P. Hutapea and J. Grenestedt, Effect of temperature on elastic properties of woven-glass epoxy composites for printed circuit board applications **32**, (2003)
4. L. McCaslin and K. Sitaraman. *Methodology for modeling substrate warpage using copper trace pattern implementation.* in *Electronic Components and Technology Conference*. (2008).
5. S.-W. Kim, et al., Simulation of warpage during fabrication of printed circuit boards **1**, (2011)
6. S. Timoshenko, *Theory of Elasticity.* (1951): McGraw-Hill.
7. A.K. Kaw, *Mechanics of Composite Materials.* (2006): CRC Press.
8. P. Hutapea and J. Grenestedt, Tuning of electric artworks of printed circuit boards to reduce warpage (2004)

 DOI: 10.1051/matecconf/20168016010

Integrated thermomechanical model for forming of glass containers

Bruno Martins[1, a], Ana Reis[1], Pedro Teixeira[1], Margarida Machado[1], Jaime Rodrigues[2] and J. Cesar de Sá[2]

[1]*INEGI, Institute of Science and Innovation in Mechanical and Industrial Engineering, 4200-465 Porto, Portugal*
[2]*University of Porto, Mechanical Department, 4200-465 Porto, Portugal*

Abstract. In this work an integrated model for all the process of glass forming of glass containers is presented. The model addresses press-and-blow and blow-and-blow procedures from the gob forming to the final product. Glass forming involves coupled thermal-mechanical physical phenomena, in which the material viscous flow is highly dependent on heat transfer which, in turn, is dependent changes of shape due to material flow. Numerical aspects associated with the model implementation and its validation with real formed products are discussed.

1 Introduction

The growth of competitiveness on the industries that use glass containers puts demands on the glass industry to develop new designs. In this context, simulations models are essential as virtual prototyping tools, reducing time and costs. To be effective, the model must give some information for the optimisation of the process, namely thickness evolution, residual stresses and mould temperatures distributions. The forming of hollow-ware glass articles is a complex coupled thermal/mechanical problem with interaction between the heat transfer analysis and the viscous flow of molten glass. The glass viscosity, which governs the glass flow during the forming process, is strongly dependent on the temperature [1-2] and the contact with the moulds along with the constant geometry changes alters significantly the heat transfer process. The final product thickness distribution is one of the aspects which is more affected by changing process conditions [2, 4-7].

Simulating glass forming with FEM is particularly difficult task since the material behaviour lies somewhere between a fluid and a solid. One of the first models was developed by Cesar de Sa [2]. In his approach, the thermal problem is coupled to the mechanical one at every time step, in order to handle the strong coupled phenomena. The model has provided good predictions of glass flow, resulting from gravity and blow pressure. The model was the basis for the development of a commercial 2D code [9]. But, nowadays, manufacturers are facing an evolving and increasing competition in the development of new complex 3D products. Tools are available which are capable of dealing with the most important aspects of these forming processes [10-11]. Nevertheless, some particular aspects requires a dedicated software or some user intervention in order to adapt commercial codes to this particular field (which may not be an easy task). Besides the referred commercial code, a new approach has been recently released, [12], based on meshless methods, with 3D modelling. Nevertheless, in those software some important features in terms of process knowledge still lack or are simplified like, for example, the plunger movement defined by a prescribed velocity, which does not represent the real case, where a pressure must be set. A software suite for glass forming should include the calculation of gob[1] initial geometry and temperature distribution, as well as moulds heat transfer with the glass and with external ambient temperature, in order to optimize the final geometry and process efficiency.

This last feature is not available in the mentioned software. Besides the paramount importance in the final container thickness distribution, it can be also extremely useful for machine developers since it allows to design dedicated cooling systems that could push the process further to its limits. Therefore the main objective is to build a robust and reliable suite of numerical modelling programs designed to provide glassmakers with decision support system tools to aid in the design, production and quality control of all types of glass containers. The programs should combine the latest numerical technology to provide practical, representative simulation tools that glassmakers can use and benefit from. They should be rooted in sound theoretical formulations to deal with the complex coupled thermal and mechanical phenomena, as glass viscosity is dramatically influenced by temperature distribution that itself is highly influenced by shape modification and contact with the tools and moulds. The suite of programs should also comprise pre and post processors and analysis modules for the complete definition of the problem, analysis and viewing of the results. It should be used to simulate the forming process

[a] Corresponding author: bmartins@inegi.up.pt

of round and non-round glass containers on an IS (individual section) forming machine from gob delivery through parison formation, invert and final blow until mould open.

The types of forming process modelled should include BB (Blow and Blow), PB (Wide Mouth Press and Blow) and NNPB (Narrow Neck Press and Blow) processes. As final result it should be possible to: predict the glass thickness distribution in the final product under a variety of conditions; analyse the thermal behaviour of the blank and blow moulds; investigate the influence of process conditions like glass composition, IS Machine timing, gob temperature, mould temperatures upon final container wall thickness results; evaluate the effects of mould cooling on the forming process and container wall thickness results. The final program should be not only a design tool for improvement and optimisation of existing products but also a means of understanding the main process parameters role to provide the assessment of new ideas for new products and innovation.

2 Formulation problem

2.1 Constitutive model

For the whole range of the forming process a non-linear viscoelastic model, of a Maxwell type, is assumed for the constitutive behaviour of the glass, [13] and [14]. Therefore, assuming an additive decomposition of the strain rate tensor D in the elastic and the viscous strain rate tensors, $\mathbf{D}_e$ and $\mathbf{D}_v$ respectively, [15], and assuming that the viscous deformation is only deviatoric then:

$$\mathbf{D} = \mathbf{D}_e + \mathbf{D}_v \tag{1}$$

$$\mathbf{D}_e = \frac{1}{3K}\left[\frac{(\dot{\sigma}:\mathbf{I})}{3} + \alpha\dot{T}\right]\mathbf{I} + \frac{1}{2\mu}\dot{\boldsymbol{\sigma}}' \tag{2}$$

$$\mathbf{D}_v = \frac{1}{2\eta(T)}\boldsymbol{\sigma}' \tag{3}$$

where $\dot{\boldsymbol{\sigma}}$ and $\dot{\boldsymbol{\sigma}}'$are the time derivative of the stress tensor and of the stress deviator tensor, K and μ are the bulk and the shear elastic modulus, α is the coefficient of thermal expansion and $\eta(T)$ is the viscosity, dependent on the temperature T.

The dependence of glass viscosity on the temperature is assumed to follow the relationship proposed by Fulcher, [16]:

$$\log_{10}\eta(T) = A + \frac{B}{T - T_0} \tag{4}$$

where A, B and T_0 are empirical constants.

In the forming range of temperatures the elastic part of the deformation is almost negligible and therefore the constitutive behaviour is similar to the one of an incompressible Newtonian fluid. In the cooling stages after the containers are formed it is the viscous part of the deformation that is negligible and the constitutive behaviour is then mainly thermoelastic.

2.2 Equilibrium equations

The finite element model is established from the equilibrium equations for the thermo-viscoelastic problem. Neglecting inertia forces the mechanical equilibrium equations are:

$$\mathrm{div}(\boldsymbol{\sigma}) + \mathbf{b} = 0 \tag{5}$$

where **b** are body forces per unit volume. In the forming range of temperatures, as referred, the deformation is almost incompressible, and, therefore, some care must be taken in the finite element model, when evaluating the volumetric terms, in order to avoid locking, [17]. The thermal equilibrium may be stated, for a Lagrangian set of coordinates, by the Fourier law, assuming that there is no heat source in the material, as:

$$k(T)\nabla^2 T = \rho(T)c(T)\dot{T} + 3K\alpha T(\mathbf{D}:\mathbf{I}) + r(\dot{\boldsymbol{\sigma}}':\mathbf{D}) \tag{6}$$

k is the thermal conductivity, c is the specific heat and ρ is the density. The last term of the right hand side of equation [6] represents the fraction r of viscous work that is converted into heat.

In the finite element solution the mesh representing the glass, is updated for each time step and therefore in the heat transfer equation [6] the convective terms were disregarded, i.e., the heat is convected each time step to the new mesh position, from where a diffusion process takes place. In the moulds the last two terms of equation [6] are not considered. The thermal natural boundary conditions may include a flux due to convection of the contact fluid, air in this case:

$$q_c = -\alpha_c(T - T_a) \tag{7}$$

and a flux due to radiation:

$$q_r = -\beta\varepsilon(T^4 - T_\infty^4) \tag{8}$$

In the above T_a is the ambient temperature, α_c is the convection heat transfer coefficient, β is the Stefan-Boltzmann constant, ε the emissivity, and T_∞ the absorbing surface temperature. Thus, the two boundary conditions combine to give:

$$q_{rc} = q_r + q_c \tag{9}$$

The essential thermal boundary conditions are prescribed temperatures in the contact zones between the glass and the mould and are imposed in the finite element model by a penalty function technique, as explained in the next section.

2.3 Boundary conditions

As the simulation of the forming process takes place the boundary conditions must be changed whenever the material touches the mould. Initially, a contact algorithm deals with this problem by redefining the time step, so that the nodes of the mesh representing the glass do not penetrate into the mesh that represents the mould, [17]. A node achieving contact with the mould is considered to be fixed, which seems to be a reasonable assumption due

DOI: 10.1051/matecconf/20168016010

to the sudden drop of temperature and the consequent local large increase on viscosity.

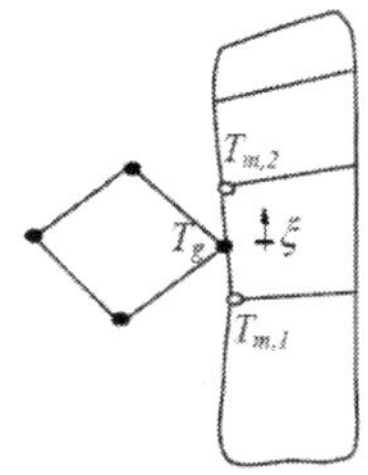

Figure 1. Thermal contact

The thermal boundary conditions must also be changed and the node temperature is constrained to be the same as the mould temperature in the point of contact. This is achieved by imposing, with a penalty function technique, the condition:

$$T_g = N_{m,1}(\zeta)T_{m,1} + N_{m,2}(\zeta)T_{m,2} \quad (10)$$

where T_g is the temperature at the contacting node of the glass mesh, $T_{m,}1$ and $T_{m,2}$ are the node temperatures on the side of the mould element where contact is made, and $N_{m,1}$ and $N_{m,2}$ are the respective shape functions, Figure 1

3 Numerical Modelling

In the production of glass containers there are two ways to perform a final product, the press and blow or blow and blow operations. Glass containers are manufactured by melting raw materials and recycled glass in tank furnaces which melt the glass at an elevated temperatures. Then, the molten glass is cut into gobs of the same shape and size and sent to a high speed machines. These gobs fall into a first mould, the blank mould, and are pressed (press and blow) or blown (blow and blow) against the mould wall forming the parison. The parison in then transferred to a second and final mould where, by gravity first and then pressure (or sometimes suction), the final product is finished.

A software to simulate the production of glass containers was developed. The software is able to reproduce the blow and blow and press and blow processes allowing to simulate all the stages, since gob loading to final blow.

Initial gob shape was assumed a "spherocylinder" with an initial velocity of 6m/s. In the blow and blow process a gob of a flint glass with 0.400Kg of mass was considered. The initial gob temperature was 1135°C and assumed as uniform. Tools temperatures considered were: for plunger and neckring a temperature of 550°C; for the blank mould and baffle a temperature of 450°C and finally for blow mould a temperature of 480°C. The air pressure used was of 2.5.

In the press and blow the same material was used. The initial temperature were taken as 1170°C for the gob, 550°C for the plunger, 420°C for the neckring, 450°C for the blank mould and baffle, 480°C for the blow mould. To the plunger a force of 400N was applied.

A remeshing procedure is necessary when the elements become too distorted and consequently a capability to interpolate state variables (thermal and mechanical) from the old to the new mesh is necessary, allowing for a continuous process modelling.

4 Results

The first example refers to all stages of the blow and blow process. The figures show results for important thermomechanical stages namely shape and temperature distribution.

Figures 2 and 3 represent gob loading and settle blow. In Figure 3 can be seen the cooling effect after filling completely the neckring.

After the settle blow the plunger is removed to allow the corkage reheat in plunger zone, i.e. glass nearby the plunger will reheat and soften for better counter blow pressure effect. Figure 4 shows the end of counter blow stage. The following stages are the reheat and invert. After the counter blow the blank mould is opened and the parison is allowed to reheat before inverting into the blow mould. During this time, the parison continues its reheat or temperature equalization. Then, the parison is transferred to the blow mould.

At the rundown gravity is the only force acting in this stage, Figure 5. A cooling effect will appear when the glass contacts the baffle and is followed by the final blow.

In final blow stage inflated pressurize air forces the parison to finalize the container shape. Additionally, vacuum is also applied from the blow mould. Figure 6 represents the temperature evolution in final blow stage. It can be noted that the neckring region is cooler than the other zones. Results present uniform thickness along the container profile, with more material in the bottom and thinner thickness in the shoulder, as expected.

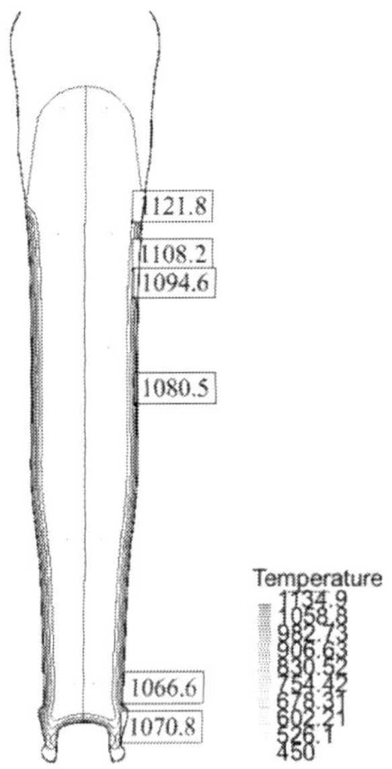

Figure 2. Gob loading in blow and blow process

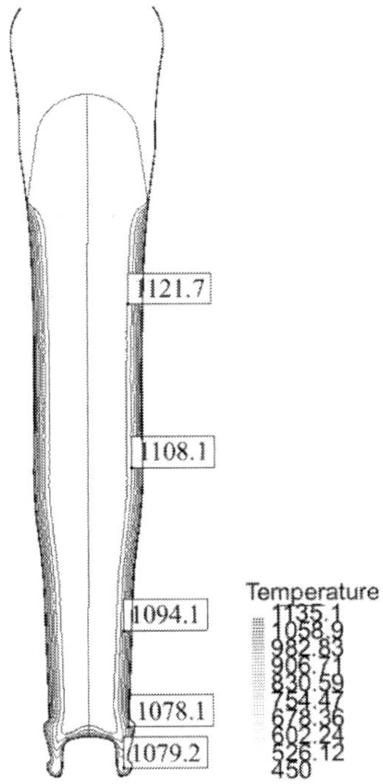

Figure 3. Settle blow in blow and blow process.

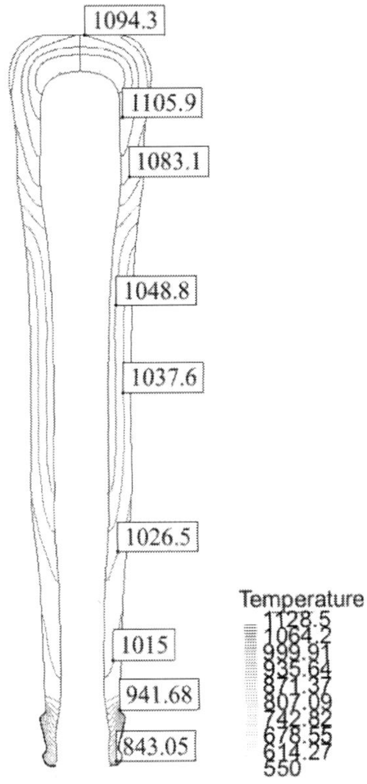

Figure 4. Counter blow in blow and blow process

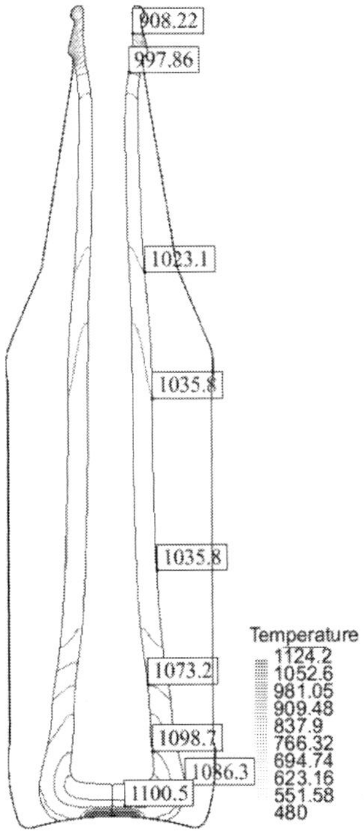

Figure 5. Rundown in blow and blow process

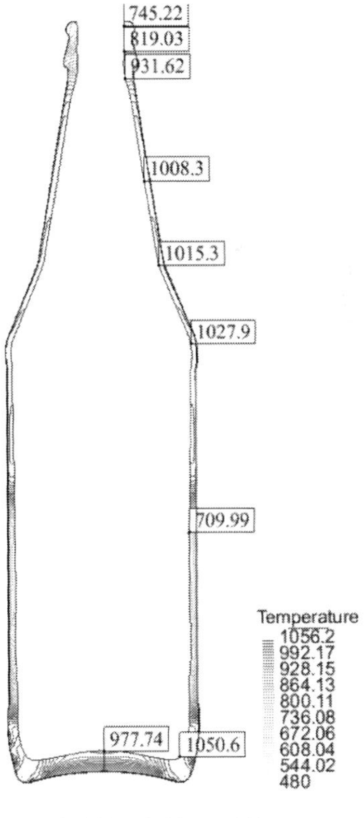

Figure 6. Final blow in blow and blow process

The first example refers to all stages of the press and blow process. Figure 7 represents the gob loading stage, glass is dropped directly against the plunger and starts to cool. Then, the baffle is positioned on top of the blank mould and the plunger starts move towards the baffle with a prescribed force. In the last instants the glass fills the neckring zone. Figure 8 shows the final step of plunger up stage. Comparing to the similar blow and blow process, Figure 4, the glass presents a different temperature evolution in consequence to contact to the plunger. It can be noted a more 'horizontal' isothermal evolution, presenting a cooling effect from the top to the bottom while in counter blow of the blow and blow process present a 'vertical' isothermal evolution, showing the cooling effect from the mould to inside of parison, is more clear.

After plunger up stage follows the reheat and invert stages which are critical to perform a successful rundown stage. Too much cooling can result in a deficient rundown. Not enough cooling, in other hand, would have the opposite effect, i.e., excessive stretching in the wall resulting in a very thinner wall glass during the rundown. Thus, the time of invert should be in according to the desire container final shape. Figure 8 represents the rundown. As the cooling advances the bottom of the container presents a higher temperature distribution than the top, which is a result of a bigger thickness.

In Figure 10 the final blow stage is presented. Note that the neckring and the shoulder are cooler than the other zones, as expected. Also, thickness is well distributed showing a larger quantity of material at the bottom and being uniform along the wall.

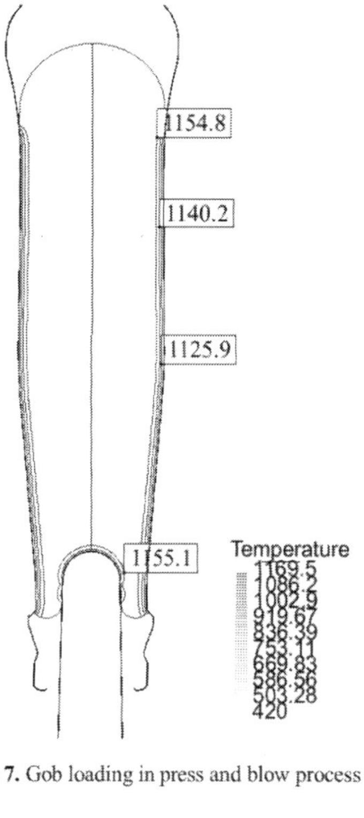

Figure 7. Gob loading in press and blow process

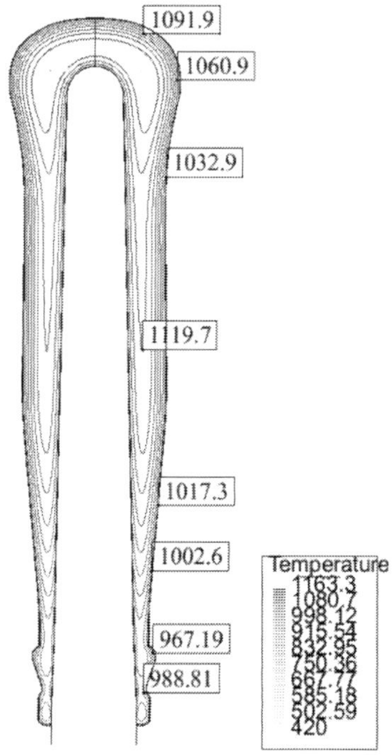

Figure 8. Plunger up in press and blow process.

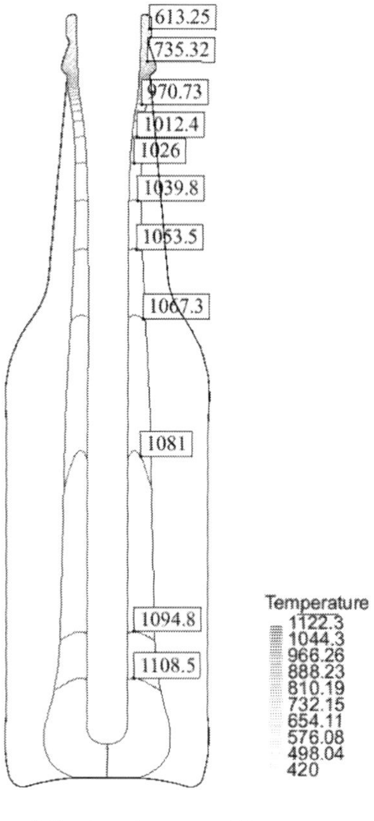

Figure 9. Rundown in press and blow process.

DOI: 10.1051/matecconf/20168016010

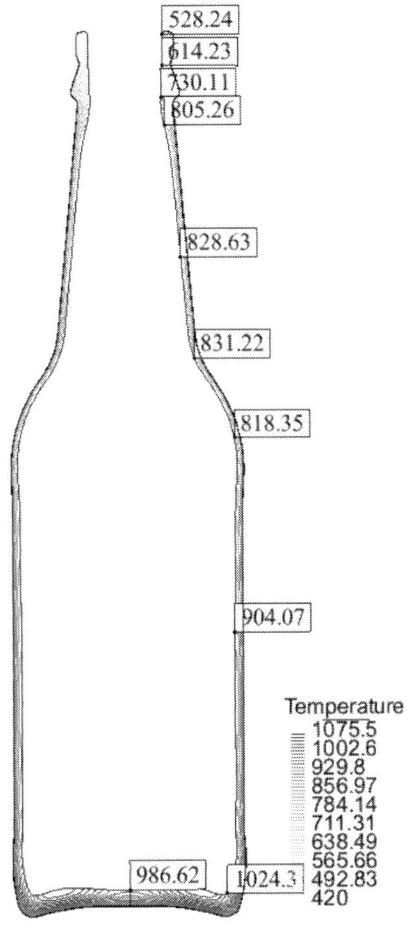

Figure 10. Final blow in press and blow process.

5 Conclusion

A development of numerical software for forming of glass containers has been performed. Two different glass containers have been simulated for two typical blow and blow and press and blow processes.

The main goal is to perform automatically all the stages without user handling. The software is able to efficiently choose the timestep, boundary conditions, loads, remeshing parameters and to transfer all information to a new mesh as well as all calculation to perform the simulation successfully.

The results show reliability and robustness of the developed computational tool

Acknowledgements

Authors gratefully acknowledge the funding of Project NORTE-01-0145-FEDER-000022 - SciTech - Science and Technology for Competitive and Sustainable Industries, cofinanced by Programa Operacional Regional do Norte (NORTE2020), through Fundo Europeu de Desenvolvimento Regional.

References

1. JH. Williams, DRJ. Owen, JMA. Cesar de Sa, Proceedings of the XIVth Inter. Congress on Glass, (1986).
2. JMA. César de Sá,, Engineering Computations; **3(6)**: 266-275, (1986)
3. K. Storck, Studies in Science and Technology. Dissertation nº 542, (1998).
4. D. Lochegnies, C. Maréchal, In: Proceedings of the 2nd international colloquium Modelling of Glass Forming and Tempering. Presses, (2002).
5. JMA. Cesar de Sa, RM. Natal-Jorge, RPR. Cardoso, In: Proceedings of the Euromech 388 Colloqium on Modelling of Glass Forming Processes. Lez Valenciennes nº **27**, (1999).
6. JMA. Cesar de Sa, RM. Natal-Jorge, CMC. Silva, RPR. Cardoso, Proceedings of ECCM'99- European Conference on Computational Mechanics, (1999).
7. D . Lochegnies, P. Moreau, R. Guilbaut, Glass Technology, **46(2)**: 116-120, (2005).
8. C. Maréchal, P. Moreau, D. Lochegnies, Engineering with Computers, **19**: 233-240, (2004).
9. Elfen GD, Glass Design Modelling Software, Rockfield Software Ltd
10. ABAQUS, Finite Element Analysis, Dassault Systèmes
11. Ansys Fluent, CFD software, ANSYS, Inc.
12. NoGRID, Flow simulation for industrial processes and operations, NOGRID GmbH
13 Babcock, C.L., *Silicate glass technology methods*, (John Wiley & Sons, New York, 1977)
14 Fung, Y.C., *Foundations of Solid Mechanics*, (Prentice-Hall, New Jersey, USA, 1965)
15 Owen, D.R.J., Hinton, E., *Finite Elements in Plasticity: Theory and Practice*, (1980).
16 Fulcher, G.S.., J. Amer. Ceram. Soc., Vol 8/6, p. 339-355, (1925).

AUTHOR INDEX

AUTHOR INDEX

AUTHOR INDEX

AUTHOR INDEX